Arithmetic Operations:

$$ab + ac = a(b + c)$$

$$\frac{a}{b} + \frac{c}{d} = \frac{ad + bc}{bd}$$

$$\frac{a + b}{c} = \frac{a}{c} + \frac{b}{c}$$

$$\frac{\left(\frac{a}{b}\right)}{\left(\frac{c}{d}\right)} = \frac{ad}{bc}$$

$$a\left(\frac{b}{c}\right) = \frac{ab}{c}$$

$$\frac{a - b}{c - d} = \frac{b - a}{d - c}$$

$$\frac{ab + ac}{a} = b + c, \ a \neq 0$$

$$\frac{\left(\frac{a}{b}\right)}{c} = \frac{a}{bc}$$

$$\frac{a}{\left(\frac{b}{c}\right)} = \frac{ac}{b}$$

Exponents and Radicals:

$$a^0 = 1, \ a \neq 0$$

$$\frac{a^x}{a^y} = a^{x-y}$$

$$\left(\frac{a}{b}\right)^x = \frac{a^x}{b^x}$$

$$\sqrt[n]{a^m} = a^{m/n} = \left(\sqrt[n]{a}\right)^m$$

$$a^{-x} = \frac{1}{a^x}$$

$$(a^x)^y = a^{xy}$$

$$\sqrt{a} = a^{1/2}$$

$$\sqrt[n]{ab} = \sqrt[n]{a}\sqrt[n]{b}$$

$$a^x a^y = a^{x+y}$$

$$(ab)^x = a^x b^x$$

$$\sqrt[n]{a} = a^{1/n}$$

$$\sqrt[n]{\left(\frac{a}{b}\right)} = \frac{\sqrt[n]{a}}{\sqrt[n]{b}}$$

Algebraic Errors to Avoid:

$\frac{a}{x + b} \neq \frac{a}{x} + \frac{a}{b}$ (To see this error, let $a = b = x = 1$.)

$\sqrt{x^2 + a^2} \neq x + a$ (To see this error, let $x = 3$ and $a = 4$.)

$a - b(x - 1) \neq a - bx - b$ [Remember to distribute negative signs. The equation should be $a - b(x - 1) = a - bx + b$.]

$\frac{\left(\frac{x}{a}\right)}{b} \neq \frac{bx}{a}$ [To divide fractions, invert and multiply. The equation should be

$$\frac{\left(\frac{x}{a}\right)}{b} = \frac{\left(\frac{x}{a}\right)}{\left(\frac{b}{1}\right)} = \left(\frac{x}{a}\right)\left(\frac{1}{b}\right) = \frac{x}{ab}.]$$

$\sqrt{-x^2 + a^2} \neq -\sqrt{x^2 - a^2}$ (The negative sign cannot be factored out of the square root.)

$\frac{\not{a} + bx}{\not{a}} \neq 1 + bx$ (This is one of many examples of incorrect cancellation. The equation should be $\frac{a + bx}{a} = \frac{a}{a} + \frac{bx}{a} = 1 + \frac{bx}{a}$.)

$\frac{1}{x^{1/2} - x^{1/3}} \neq x^{-1/2} - x^{-1/3}$ (This error is a more complex version of the first error.)

$(x^2)^3 \neq x^5$ [This equation should be $(x^2)^3 = x^2x^2x^2 = x^6$.]

Conversion Table:

1 centimeter ≈ 0.394 inch	1 joule ≈ 0.738 foot-pound	1 mile ≈ 1.609 kilometers
1 meter ≈ 39.370 inches	1 gram ≈ 0.035 ounce	1 gallon ≈ 3.785 liters
≈ 3.281 feet	1 kilogram ≈ 2.205 pounds	1 pound ≈ 4.448 newtons
1 kilometer ≈ 0.621 mile	1 inch ≈ 2.540 centimeters	1 foot-lb ≈ 1.356 joules
1 liter ≈ 0.264 gallon	1 foot ≈ 30.480 centimeters	1 ounce ≈ 28.350 grams
1 newton ≈ 0.225 pound	≈ 0.305 meter	1 pound ≈ 0.454 kilogram

Instructor's Annotated Edition

Precalculus

Fifth Edition

► **Ron Larson**

► **Robert P. Hostetler**

The Pennsylvania State University
The Behrend College

► **With the assistance of David C. Falvo**

The Pennsylvania State University
The Behrend College

Houghton Mifflin Company Boston New York

Sponsoring Editor: Jack Shira
Managing Editor: Cathy Cantin
Senior Associate Editor: Maureen Ross
Associate Editor: Laura Wheel
Assistant Editor: Carolyn Johnson
Supervising Editor: Karen Carter
Project Editor: Patty Bergin
Editorial Assistant: Kate Hartke
Art Supervisor: Gary Crespo
Marketing Manager: Michael Busnach
Senior Manufacturing Coordinator: Sally Culler
Composition and Art: Meridian Creative Group

Cover designer: Gary Crespo
Cover image: Meridian Creative Group

Printed in the U.S.A.

Library of Congress Catalog Card Number: 00-103027

ISBN: 0-618-06646-2

456789–DOW–04 03 02

Contents

A Word from the Authors

Welcome to *Precalculus*, Fifth Edition. In this revision we focus on student success, accessibility, and flexibility.

Student Success: During the past 30 years of teaching and writing, we have learned many things about the teaching and learning of mathematics. We have found that students are most successful when they know what they are expected to learn and why it is important to learn. With that in mind, we have restructured the Fifth Edition to include a thematic study thread in every chapter.

Each chapter begins with a study guide called *How to Study This Chapter*, which includes a comprehensive overview of the chapter concepts (*The Big Picture*), a list of *Important Vocabulary* that is integral to learning *The Big Picture* concepts, a list of study resources, and a general study tip. The study guide allows students to get organized and prepare for the chapter.

An old pedagogical recipe goes something like this: "First I'm going to tell you what I'm going to teach you, then I will teach it to you, and finally I will go over what I taught you." Following this recipe, we have also included a set of learning objectives in every section that outlines what students are expected to learn, followed by an interesting real-life application that illustrates why it is important to learn the concepts in that section. Finally, the chapter summary (*What did you learn?*), which reinforces the section objectives, and the chapter *Review Exercises*, which are correlated to the chapter summary, provide additional study support at the conclusion of each chapter.

Our new *Student Success Organizer* supplement takes this study thread one step further, providing a content-based study aid.

Accessibility: Over the years we have taken care to write our texts for the student. We have paid careful attention to the presentation, using precise mathematical language and clear writing, to create an effective learning tool. We believe that every student can learn mathematics and we are committed to providing a text that makes the mathematics within it accessible to all students. In the Fifth Edition, we have revised and improved many text features designed for this purpose. The *Technology*, *Exploration*, and *Study Tip* features have been expanded. *Chapter Tests*, which give students an opportunity for self-assessment, now follow every chapter in the Fifth Edition. The exercise sets now include both *Synthesis* exercises, which check students' conceptual understanding, and *Review* exercises, which reinforce skills learned in previous sections and chapters. Also, students have access to several media resources that accompany this text—videotapes, *Interactive Precalculus* CD-ROM, and a *Precalculus* website—that provide additional text-specific support.

Flexibility: From the time we first began writing in the early 1970s, we have always viewed part of our authoring role as that of providing instructors with flexible teaching programs. The optional features within the text allow instructors with different pedagogical approaches to design their courses to meet both their instructional needs and the needs of their students. Instructors who stress applications and problem solving, or exploration and technology, or more traditional methods, will be able to use this text successfully. In addition, we provide several print and media resources to support instructors, including a new *Instructor Success Organizer.*

We hope you enjoy the Fifth Edition.

Ron Larson

Robert P. Hostetler

Acknowledgments

We would like to thank the many people who have helped us at various stages of this project to prepare the text and supplements package. Their encouragement, criticisms, and suggestions have been invaluable to us.

Fifth Edition Reviewers

James Alsobrook, Southern Union State Community College; Sherry Biggers, Clemson University; Charles Biles, Humboldt State University; Randall Boan, Aims Community College; Jeremy Carr, Pensacola Junior College; D. J. Clark, Portland Community College; Donald Clayton, Madisonville Community College; Linda Crabtree, Metropolitan Community College; David DeLatte, University of North Texas; Gregory Dlabach, Northeastern Oklahoma A & M College; Joseph Lloyd Harris, Gulf Coast Community College; Jeff Heiking, St. Petersburg Junior College; Celeste Hernandez, Richland College; Heidi Howard, Florida Community College at Jacksonville; Wanda Long, St. Charles County Community College; Wayne F. Mackey, University of Arkansas; Rhonda MacLeod, Florida State University; M. Maheswaran, University of Wisconsin–Marathon County; Valerie Miller, Georgia State University; Katharine Muller, Cisco Junior College; Bonnie Oppenheimer, Mississippi University for Women; James Pohl, Florida Atlantic University; Hari Pulapaka, Valdosta State University; Michael Russo, Suffolk County Community College; Cynthia Floyd Sikes, Georgia Southern University; Susan Schindler, Baruch College–CUNY; Stanley Smith, Black Hills State University. In addition, we would like to thank all the college algebra instructors who took the time to respond to our survey.

We would like to extend a special thanks to Hari Pulapaka for his contributions to this revision.

We would like to thank the staff of Larson Texts, Inc. and the staff of Meridian Creative Group, who assisted in proofreading the manuscript, preparing and proofreading the art package, and typesetting the supplements.

On a personal level, we are grateful to our wives, Deanna Gilbert Larson and Eloise Hostetler, for their love, patience, and support. Also, a special thanks goes to R. Scott O'Neil.

If you have suggestions for improving this text, please feel free to write to us. Over the past two decades we have received many useful comments from both instructors and students, and we value these comments very much.

Ron Larson
Robert P. Hostetler

Features Highlights

Student Success Tools

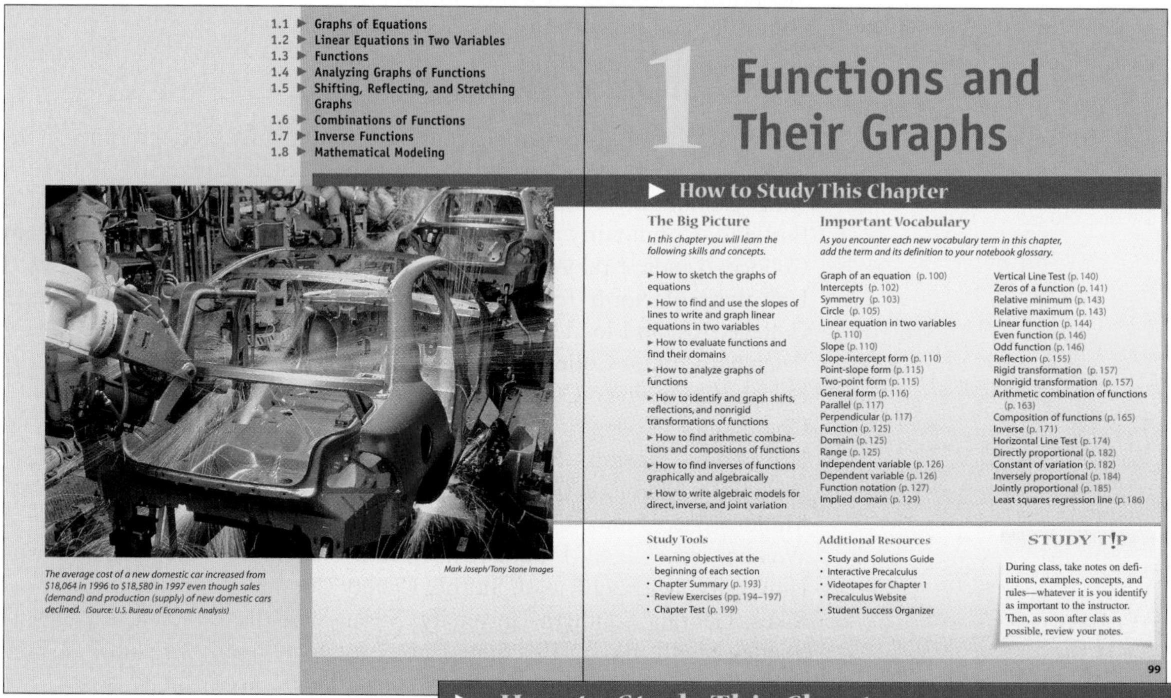
1.1 ▶ Graphs of Equations
1.2 ▶ Linear Equations in Two Variables
1.3 ▶ Functions
1.4 ▶ Analyzing Graphs of Functions
1.5 ▶ Shifting, Reflecting, and Stretching Graphs
1.6 ▶ Combinations of Functions
1.7 ▶ Inverse Functions
1.8 ▶ Mathematical Modeling
1 Functions and Their Graphs
The average cost of a new domestic car increased from $18,064 in 1996 to $18,580 in 1997 even though sales (demand) and production (supply) of new domestic cars declined. (Source: U.S. Bureau of Economic Analysis)
Mark Joseph/Tony Stone Images
99

▶ "How to Study This Chapter"

The new chapter-opening study guide includes:

- *The Big Picture*—an objective-based overview of the main concepts of the chapter
- *Important Vocabulary*—mathematical terms integral to learning *The Big Picture* concepts
- *Study Tools*
- *Additional Resources*
- *Study Tip*

▶ How to Study This Chapter

The Big Picture

In this chapter you will learn the following skills and concepts.

- ▶ How to sketch the graphs of equations
- ▶ How to find and use the slopes of lines to write and graph linear equations in two variables
- ▶ How to evaluate functions and find their domains
- ▶ How to analyze graphs of functions
- ▶ How to identify and graph shifts, reflections, and nonrigid transformations of functions
- ▶ How to find arithmetic combinations and compositions of functions
- ▶ How to find inverses of functions graphically and algebraically
- ▶ How to write algebraic models for direct, inverse, and joint variation

Important Vocabulary

As you encounter each new vocabulary term in this chapter, add the term and its definition to your notebook glossary.

Graph of an equation (p. 100)
Intercepts (p. 102)
Symmetry (p. 103)
Circle (p. 105)
Linear equation in two variables (p. 110)
Slope (p. 110)
Slope-intercept form (p. 110)
Point-slope form (p. 115)
Two-point form (p. 115)
General form (p. 116)
Parallel (p. 117)
Perpendicular (p. 117)
Function (p. 125)
Domain (p. 125)
Range (p. 125)
Independent variable (p. 126)
Dependent variable (p. 126)
Function notation (p. 127)
Implied domain (p. 129)
Vertical Line Test (p. 140)
Zeros of a function (p. 141)
Relative minimum (p. 143)
Relative maximum (p. 143)
Linear function (p. 144)
Even function (p. 146)
Odd function (p. 146)
Reflection (p. 155)
Rigid transformation (p. 157)
Nonrigid transformation (p. 157)
Arithmetic combination of functions (p. 163)
Composition of functions (p. 165)
Inverse (p. 171)
Horizontal Line Test (p. 174)
Directly proportional (p. 182)
Constant of variation (p. 182)
Inversely proportional (p. 184)
Jointly proportional (p. 185)
Least squares regression line (p. 186)

Study Tools

- Learning objectives at the beginning of each section
- Chapter Summary (p. 193)
- Review Exercises (pp. 194–197)
- Chapter Test (p. 199)

Additional Resources

- Study and Solutions Guide
- Interactive Precalculus
- Videotapes for Chapter 1
- Precalculus Website
- Student Success Organizer

STUDY TIP

During class, take notes on definitions, examples, concepts, and rules—whatever it is you identify as important to the instructor. Then, as soon as after class as possible, review your notes.

New Section Openers include:

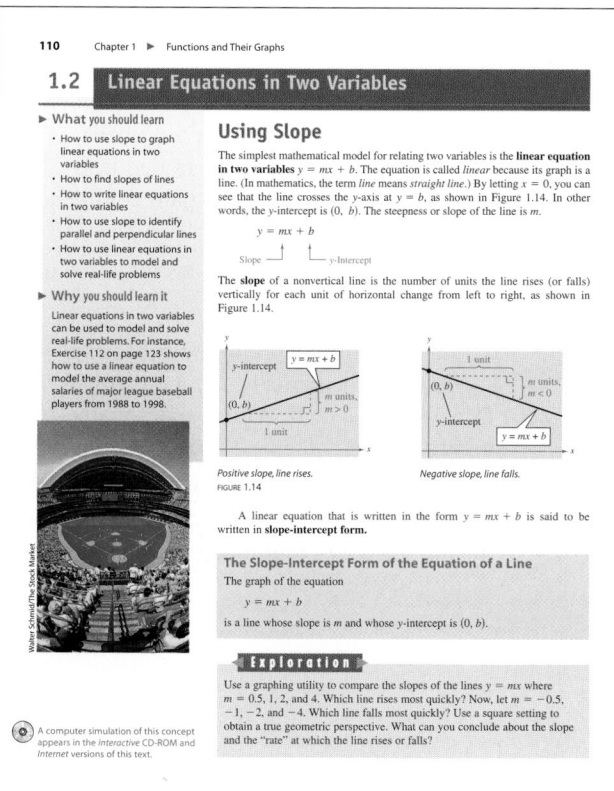

1.2 Linear Equations in Two Variables

► **What you should learn**

- How to use slope to graph linear equations in two variables
- How to find slopes of lines
- How to write linear equations in two variables
- How to use slope to identify parallel and perpendicular lines
- How to use linear equations in two variables to model and solve real-life problems

► **Why you should learn it**

Linear equations in two variables can be used to model and solve real-life problems. For instance, Exercise 112 on page 123 shows how to use a linear equation to model the average annual salaries of major league baseball players from 1988 to 1998.

Walter Schmid/The Stock Market

A computer simulation of this concept appears in the *Interactive* CD-ROM and *Internet* versions of this text.

Using Slope

The simplest mathematical model for relating two variables is the **linear equation in two variables** $y = mx + b$. The equation is called *linear* because its graph is a line. (In mathematics, the term *line* means *straight line*.) By letting $x = 0$, you can see that the line crosses the y-axis at $y = b$, as shown in Figure 1.14. In other words, the y-intercept is $(0, b)$. The steepness or slope of the line is m.

$$y = mx + b$$

Slope — y-Intercept

The **slope** of a nonvertical line is the number of units the line rises (or falls) vertically for each unit of horizontal change from left to right, as shown in Figure 1.14.

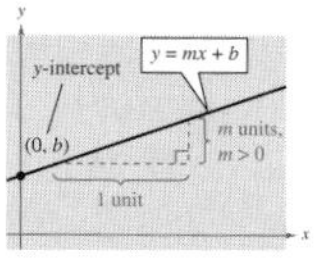

Positive slope, line rises.

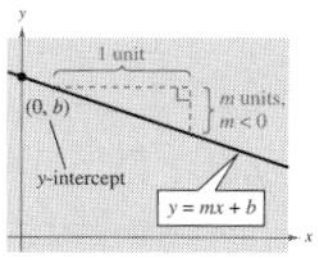

Negative slope, line falls.

FIGURE 1.14

A linear equation that is written in the form $y = mx + b$ is said to be written in **slope-intercept form.**

The Slope-Intercept Form of the Equation of a Line

The graph of the equation

$$y = mx + b$$

is a line whose slope is m and whose y-intercept is $(0, b)$.

Exploration

Use a graphing utility to compare the slopes of the lines $y = mx$ where $m = 0.5$, 1, 2, and 4. Which line rises most quickly? Now, let $m = -0.5$, -1, -2, and -4. Which line falls most quickly? Use a square setting to obtain a true geometric perspective. What can you conclude about the slope and the "rate" at which the line rises or falls?

► "What you should learn"

Objectives outline the main concepts and help keep students focused on *The Big Picture*.

► "Why you should learn it"

A real-life application or a reference to other branches of mathematics illustrates the relevance of the section's content.

► "What did you learn?" Summary

The chapter summary provides a concise, section-by-section review of the section objectives. These objectives are correlated to the chapter Review Exercises.

Chapter Summary

What did you learn?

	Review Exercises
Section 1.1	
☐ How to sketch graphs of equations	1–12
☐ How to use intercepts and symmetry to sketch graphs of equations	5–12
☐ How to find equations and sketch graphs of circles	13–18
☐ How to use graphs of equations in real-life problems	19, 20
Section 1.2	
☐ How to find slopes and use slope to graph linear equations	21–29
☐ How to write linear equations and identify parallel and perpendicular lines	30–39
☐ How to use linear equations to model and solve real-life problems	40, 41
Section 1.3	
☐ How to decide whether relations between two variables are functions	42–45
☐ How to use function notation and evaluate functions	46, 47
☐ How to find the domains of functions	48–51
☐ How to use functions to model and solve real-life problems	52, 53
Section 1.4	
☐ How to use the Vertical Line Test and find the zeros of functions	54–61
☐ How to determine intervals on which functions are increasing or decreasing	62–65
☐ How to identify and graph linear and piecewise-defined functions	66–71
☐ How to identify even and odd functions	72–75
Section 1.5	
☐ How to recognize graphs of common functions	76, 77
☐ How to use transformations to sketch graphs of functions	78–83
Section 1.6	
☐ How to add, subtract, multiply, and divide functions	84, 85
☐ How to find compositions and combinations of functions	86–89
Section 1.7	
☐ How to find inverse functions informally	90–93
☐ How to use graphs of functions to decide whether functions have inverses	94–97
☐ How to find inverse functions algebraically	98–103
Section 1.8	
☐ How to use mathematical models to approximate sets of data points	104
☐ How to write mathematical models for direct variation, inverse variation, and joint variation	105–109
☐ How to use the least squares regression feature of a graphing utility to find mathematical models	110

FEATURES

Revised Exercises and Applications

42. $h(x) = \begin{cases} 9 - x^2, & x < 3 \\ x - 3, & x \geq 3 \end{cases}$

x	1	2	3	4	5
$h(x)$					

In Exercises 43–50, find all real values of x such that $f(x) = 0$.

43. $f(x) = 15 - 3x$
44. $f(x) = 5x + 1$
45. $f(x) = \dfrac{3x - 4}{5}$
46. $f(x) = \dfrac{12 - x^2}{5}$
47. $f(x) = x^2 - 9$
48. $f(x) = x^2 - 8x + 15$
49. $f(x) = x^3 - x$
50. $f(x) = x^3 - x^2 - 4x + 4$

In Exercises 51–54, find the value(s) of x for which $f(x) = g(x)$.

51. $f(x) = x^2, \quad g(x) = x + 2$
52. $f(x) = x^2 + 2x + 1, \quad g(x) = 3x + 3$
53. $f(x) = \sqrt{3x} + 1, \quad g(x) = x + 1$
54. $f(x) = x^4 - 2x^2, \quad g(x) = 2x^2$

In Exercises 55–68, find the domain of the function.

55. $f(x) = 5x^2 + 2x - 1$
56. $g(x) = 1 - 2x^2$
57. $h(t) = \dfrac{4}{t}$
58. $s(y) = \dfrac{3y}{y + 5}$
59. $g(y) = \sqrt{y - 10}$
60. $f(t) = \sqrt[3]{t + 4}$
61. $f(x) = \sqrt[4]{1 - x^2}$
62. $f(x) = \sqrt[4]{x^2 + 3x}$
63. $g(x) = \dfrac{1}{x} - \dfrac{3}{x + 2}$
64. $h(x) = \dfrac{10}{x^2 - 2x}$
65. $f(s) = \dfrac{\sqrt{s - 1}}{s - 4}$
66. $f(x) = \dfrac{\sqrt{x + 6}}{6 + x}$
67. $f(x) = \dfrac{\sqrt[3]{x - 4}}{x}$
68. $f(x) = \dfrac{x - 5}{x^2 - 9}$

In Exercises 69–72, assume that the domain of f is the set $A = \{-2, -1, 0, 1, 2\}$. Determine the set of ordered pairs that represents the function f.

69. $f(x) = x^2$
70. $f(x) = \dfrac{2x}{x^2 + 1}$
71. $f(x) = \sqrt{x + 2}$
72. $f(x) = |x + 1|$

Exploration **In Exercises 73–76, determine the function from**

$f(x) = cx$, $g(x) = cx^2$, $h(x) = c\sqrt{|x|}$, and $r(x) = \dfrac{c}{x}$

and the value of the constant c that will make the function fit the data in the table.

73.

x	−4	−1	0	1	4
y	−32	−2	0	−2	−32

74.

x	−4	−1	0	1	4
y	−1	$-\frac{1}{4}$	0	$\frac{1}{4}$	1

75.

x	−4	−1	0	1	4
y	−8	−32	Undef.	32	8

76.

x	−4	−1	0	1	4
y	6	3	0	3	6

In Exercises 77–84, find the difference quotient and simplify your answer.

77. $f(x) = x^2 - x + 1, \quad \dfrac{f(2 + h) - f(2)}{h}, h \neq 0$
78. $f(x) = 5x - x^2, \quad \dfrac{f(5 + h) - f(5)}{h}, h \neq 0$
79. $f(x) = x^3, \quad \dfrac{f(x + c) - f(x)}{c}, c \neq 0$
80. $f(x) = 2x, \quad \dfrac{f(x + c) - f(x)}{c}, c \neq 0$
81. $g(x) = 3x - 1, \quad \dfrac{g(x) - g(3)}{x - 3}, x \neq 3$
82. $f(t) = \dfrac{1}{t}, \quad \dfrac{f(t) - f(1)}{t - 1}, t \neq 1$
83. $f(x) = \sqrt{5x}, \quad \dfrac{f(x) - f(5)}{x - 5}, x \neq 5$
84. $f(x) = x^{2/3} + 1, \quad \dfrac{f(x) - f(8)}{x - 8}, x \neq 8$
85. *Geometry* Express the area A of a square as a function of its perimeter P.
86. *Geometry* Express the area A of a circle as a function of its circumference C.

The symbol indicates an example or exercise that highlights algebraic techniques specifically used in calculus.

▶ Exercises

- Each exercise set contains a variety of computational, conceptual, and applied problems.
- Each exercise set is carefully graded in difficulty to allow students to gain confidence as they progress.
- Each exercise set now concludes with two new types of exercises:
  - *Synthesis* exercises promote further exploration of mathematical concepts, critical thinking skills, and writing about mathematics. These exercises require students to synthesize the main concepts presented in the section and chapter.
  - *Review* exercises reinforce previously learned skills and concepts.

Synthesis

True or False? **In Exercises 93 and 94, determine whether the statement is true or false. Justify your answer.**

93. A fifth-degree polynomial can have five turning points in its graph.
94. It is possible for a sixth-degree polynomial to have only one solution.
95. *Graphical Analysis* Describe a polynomial function that could represent the graph. (Indicate the degree of the function and the sign of its leading coefficient.)

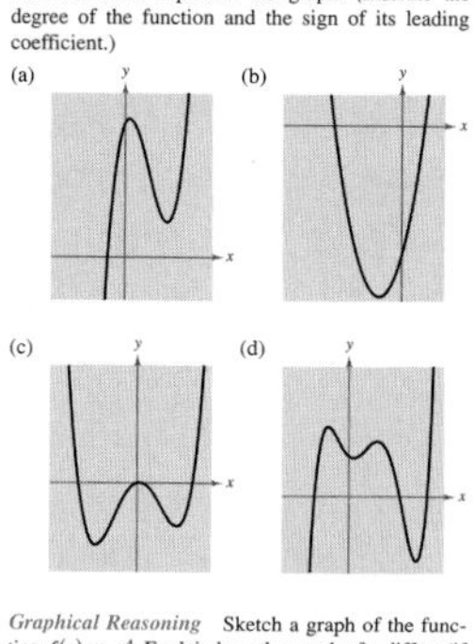

96. *Graphical Reasoning* Sketch a graph of the function $f(x) = x^4$. Explain how the graph of g differs (if it does) from the graph of f. Determine whether g is odd, even, or neither.
(a) $g(x) = f(x) + 2$
(b) $g(x) = f(x + 2)$
(c) $g(x) = f(-x)$
(d) $g(x) = -f(x)$
(e) $g(x) = f\left(\frac{1}{2}x\right)$
(f) $g(x) = \frac{1}{2}f(x)$
(g) $g(x) = f\left(x^{3/4}\right)$
(h) $g(x) = (f \circ f)(x)$

97. *Exploration* Explore the transformations of the form $g(x) = a(x - h)^5 + k$.
(a) Use a graphing utility to graph the functions
$$y_1 = -\frac{1}{3}(x - 2)^5 + 1$$
and
$$y_2 = \frac{3}{5}(x + 2)^5 - 3.$$
Determine whether the graphs are increasing or decreasing. Explain.
(b) Will the graph of g always be increasing or decreasing? If so, is this behavior determined by a, h, or k? Explain.
(c) Use a graphing utility to graph the function $H(x) = x^5 - 3x^3 + 2x + 1$.
Use the graph and the result of part (b) to determine whether H can be written in the form $H(x) = a(x - h)^5 + k$. Explain.

Review

In Exercises 98–101, solve the equation by factoring.

98. $2x^2 - x - 28 = 0$
99. $3x^2 - 22x - 16 = 0$
100. $12x^2 + 11x - 5 = 0$
101. $x^2 + 24x + 144 = 0$

In Exercises 102–105, solve the equation by completing the square.

102. $x^2 - 2x - 21 = 0$
103. $x^2 - 8x + 2 = 0$
104. $2x^2 + 5x - 20 = 0$
105. $3x^2 + 4x - 9 = 0$

In Exercises 106–109, factor the expression completely.

106. $5x^2 + 7x - 24$
107. $6x^3 - 61x^2 + 10x$
108. $4x^4 - 7x^3 - 15x^2$
109. $y^3 + 216$

110. Use the graphs of the functions f and g to answer each question.

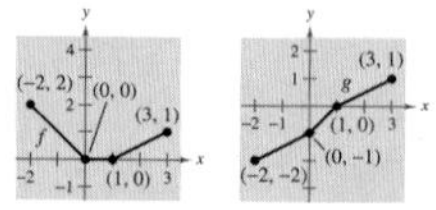

(a) Explain why f does not have an inverse.
(b) Find $g^{-1}(1)$.

▶ Real-Life Applications

- A wide variety of real-life applications, many using current, real data, are integrated throughout examples and exercises.
- The icon indicates an example that involves a real-life application.

In Exercises 69–74, use the functions $f(x) = \frac{1}{8}x - 3$ and $g(x) = x^3$ to find the indicated value or function.

69. $(f^{-1} \circ g^{-1})(1)$ **70.** $(g^{-1} \circ f^{-1})(-3)$

71. $(f^{-1} \circ f^{-1})(6)$ **72.** $(g^{-1} \circ g^{-1})(-4)$

73. $(f \circ g)^{-1}$ **74.** $g^{-1} \circ f^{-1}$

In Exercises 75–78, use the functions $f(x) = x + 4$ and $g(x) = 2x - 5$ to find the specified function.

75. $g^{-1} \circ f^{-1}$ **76.** $f^{-1} \circ g^{-1}$

77. $(f \circ g)^{-1}$ **78.** $(g \circ f)^{-1}$

79. *Hourly Wage* Your wage is $8.00 per hour plus $0.75 for each unit produced per hour. So, your hourly wage y in terms of the number of units produced is

$y = 8 + 0.75x.$

(a) Find the inverse of the function.

(b) What does each variable represent in the inverse function?

(c) Determine the number of units produced when your hourly wage is $22.25.

80. *Cost* Suppose you need a total of 50 pounds of two commodities costing $1.25 and $1.60 per pound, respectively.

(a) Verify that the total cost is

$y = 1.25x + 1.60(50 - x)$

where x is the number of pounds of the less expensive commodity.

(b) Find the inverse of the cost function. What does each variable represent in the inverse function?

(c) Use the context of the problem to determine the domain of the inverse function.

(d) Determine the number of pounds of the less expensive commodity purchased if the total cost is $73.

81. *Diesel Mechanics* The function

$y = 0.03x^2 + 245.50, \quad 0 < x < 100$

approximates the exhaust temperature y in degrees Fahrenheit where x is the percent load for a diesel engine.

(a) Find the inverse of the function. What does each variable represent in the inverse function?

(b) Use a graphing utility to graph the inverse function.

(c) Determine the percent load interval if the exhaust temperature of the engine must not exceed 500 degrees Fahrenheit.

82. *New Car Sales* The total value of new car sales f (in billions of dollars) in the United States from 1992 through 1997 is shown in the table. The time (in years) is given by t, with $t = 2$ corresponding to 1992. (Source: National Automobile Dealers Association)

t	2	3	4	5	6	7
$f(t)$	333.8	377.3	430.6	456.2	490.0	507.5

(a) Does f^{-1} exist?

(b) If f^{-1} exists, what does it mean in the context of the problem?

(c) If f^{-1} exists, find $f^{-1}(456.2)$.

83. If the table in Exercise 82 were extended to 1998 and if the total value of new car sales for that year was $430.6 billion, would f^{-1} exist? Explain.

84. *Cellular Phones* The average local bill (in dollars) for cellular phones in the United States from 1990 to 1997 is shown in the table. The time (in years) is given by t, with $t = 0$ corresponding to 1990. (Source: Cellular Telecommunications Industry Association)

t	0	1	2	3
$f(t)$	80.90	72.74	68.68	61.48

t	4	5	6	7
$f(t)$	56.21	51.00	47.70	42.78

(a) Find $f^{-1}(51)$.

(b) What does f^{-1} mean in the context of the problem?

(c) Use the regression feature of a graphing utility to find a linear model for the data, $y = mx + b$. Round m and b to two decimal places.

(d) Algebraically find the inverse of the linear model in part (c).

(e) Use the inverse of the linear model you found in part (d) to approximate $f^{-1}(11)$.

▶ Algebra of Calculus

- Special emphasis is given to the algebraic techniques used in calculus.
- Algebra of Calculus examples and exercises are integrated throughout the text.
- The symbol indicates an example or exercise in which the Algebra of Calculus is featured.

Example 8 ▶ Direct Mail Advertising

The money C (in billions of dollars) spent for direct mail advertising in the United States increased in a linear pattern from 1990 to 1992, as shown in Figure 1.30. Then, in 1993, the money spent took a jump and, until 1996, increased in a *different* linear pattern. These two patterns can be approximated by the function

$$C(t) = \begin{cases} 23.40 + 1.01t, & 0 \le t \le 2 \\ 19.85 + 2.50t, & 3 \le t \le 6 \end{cases}$$

where $t = 0$ represents 1990. Use this function to approximate the total amount spent for direct mail advertising between 1990 and 1996. (Source: McCann-Erickson)

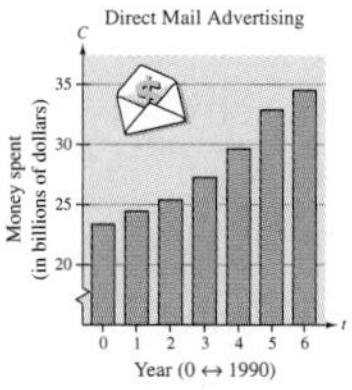

FIGURE 1.30

Solution

From 1990 to 1992, use the formula $C(t) = 23.40 + 1.01t$.

$\underbrace{\$23.40}_{1990}, \underbrace{\$24.41}_{1991}, \underbrace{\$25.42}_{1992}$

From 1993 to 1996, use the formula $C(t) = 19.85 + 2.50t$.

$\underbrace{\$27.35}_{1993}, \underbrace{\$29.85}_{1994}, \underbrace{\$32.35}_{1995}, \underbrace{\$34.85}_{1996}$

The total of these seven amounts is $197.63, which implies that the total amount spent was approximately $197,630,000,000.

One of the basic definitions in calculus employs the ratio

$$\frac{f(x + h) - f(x)}{h}, \quad h \neq 0.$$

This ratio is called a **difference quotient,** as illustrated in Example 9.

Example 9 ▶ Evaluating a Difference Quotient

For $f(x) = x^2 - 4x + 7$, find $\dfrac{f(x + h) - f(x)}{h}$.

Solution

$$\begin{aligned} \frac{f(x+h) - f(x)}{h} &= \frac{[(x+h)^2 - 4(x+h) + 7] - (x^2 - 4x + 7)}{h} \\ &= \frac{x^2 + 2xh + h^2 - 4x - 4h + 7 - x^2 + 4x - 7}{h} \\ &= \frac{2xh + h^2 - 4h}{h} \\ &= \frac{h(2x + h - 4)}{h} \\ &= 2x + h - 4, \quad h \neq 0 \end{aligned}$$

The symbol indicates an example or exercise that highlights algebraic techniques specifically used in calculus.

FEATURES

Flexibility and Accessibility

Section 3.5 ▶ Exponential and Logarithmic Models **329**

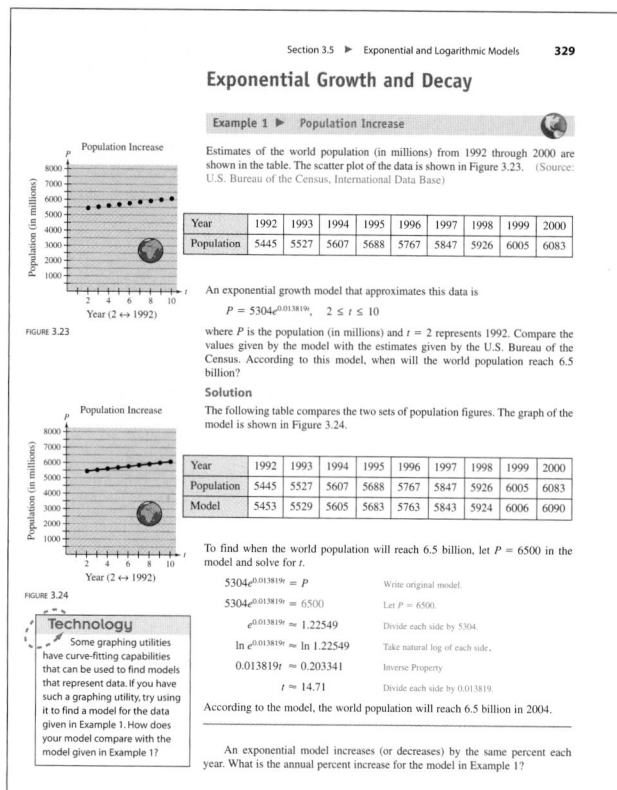

Exponential Growth and Decay

Example 1 ▶ Population Increase

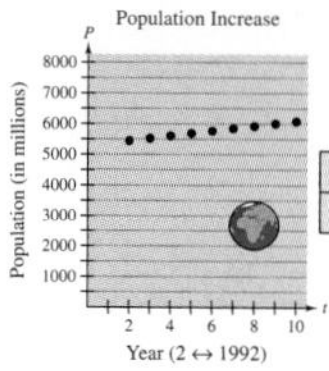

FIGURE 3.23

Estimates of the world population (in millions) from 1992 through 2000 are shown in the table. The scatter plot of the data is shown in Figure 3.23. (Source: U.S. Bureau of the Census, International Data Base)

Year	1992	1993	1994	1995	1996	1997	1998	1999	2000
Population	5445	5527	5607	5688	5767	5847	5926	6005	6083

An exponential growth model that approximates this data is

$$P = 5304e^{0.013819t}, \quad 2 \le t \le 10$$

where P is the population (in millions) and $t = 2$ represents 1992. Compare the values given by the model with the estimates given by the U.S. Bureau of the Census. According to this model, when will the world population reach 6.5 billion?

Solution

The following table compares the two sets of population figures. The graph of the model is shown in Figure 3.24.

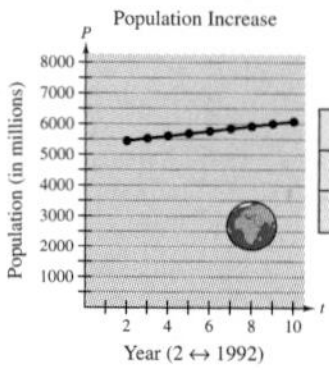

FIGURE 3.24

Year	1992	1993	1994	1995	1996	1997	1998	1999	2000
Population	5445	5527	5607	5688	5767	5847	5926	6005	6083
Model	5453	5529	5605	5683	5763	5843	5924	6006	6090

To find when the world population will reach 6.5 billion, let $P = 6500$ in the model and solve for t.

$5304e^{0.013819t} = P$ Write original model.

$5304e^{0.013819t} = 6500$ Let $P = 6500$.

$e^{0.013819t} \approx 1.22549$ Divide each side by 5304.

$\ln e^{0.013819t} \approx \ln 1.22549$ Take natural log of each side.

$0.013819t \approx 0.203341$ Inverse Property

$t \approx 14.71$ Divide each side by 0.013819.

According to the model, the world population will reach 6.5 billion in 2004.

Technology

Some graphing utilities have curve-fitting capabilities that can be used to find models that represent data. If you have such a graphing utility, try using it to find a model for the data given in Example 1. How does your model compare with the model given in Example 1?

An exponential model increases (or decreases) by the same percent each year. What is the annual percent increase for the model in Example 1?

Exploration

- Before introduction of selected topics, *Exploration* engages students in active discovery of mathematical concepts and relationships, often through the power of technology.
- *Exploration* strengthens students' critical thinking skills and helps them develop an intuitive understanding of theoretical concepts.
- *Exploration* is an optional feature and can be omitted without loss of continuity in coverage.

▶ Additional Features

Carefully crafted learning tools designed to create a rich learning environment can be found throughout the text. These learning tools include Study Tips, Historical Notes, Writing About Mathematics, Chapter Projects, Chapter Review Exercises, Chapter Tests, Cumulative Tests, and an extensive art program.

Technology

- Point-of-use instructions for using graphing utilities appear in the margins, encouraging the use of graphing technology as a tool for visualization of mathematical concepts, for verification of other solution methods, and for facilitation of computations.
- The use of technology is optional in this text. This feature and related exercises can easily be omitted without loss of continuity in coverage. Exercises that require the use of a graphing utility are identified by the symbol ▦.

▶ Examples

- Each example was carefully chosen to illustrate a particular mathematical concept or problem-solving skill.
- All examples contain step-by-step solutions, most with side-by-side explanations that lead students through the solution process.

146 Chapter 1 ▶ Functions and Their Graphs

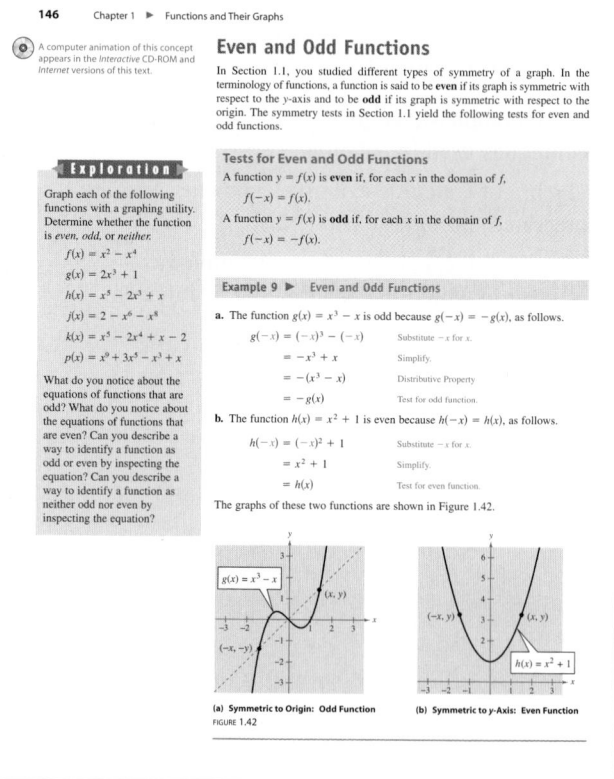

A computer animation of this concept appears in the *Interactive* CD-ROM and *Internet* versions of this text.

Even and Odd Functions

In Section 1.1, you studied different types of symmetry of a graph. In the terminology of functions, a function is said to be **even** if its graph is symmetric with respect to the y-axis and to be **odd** if its graph is symmetric with respect to the origin. The symmetry tests in Section 1.1 yield the following tests for even and odd functions.

Tests for Even and Odd Functions

A function $y = f(x)$ is **even** if, for each x in the domain of f,

$f(-x) = f(x)$.

A function $y = f(x)$ is **odd** if, for each x in the domain of f,

$f(-x) = -f(x)$.

Exploration

Graph each of the following functions with a graphing utility. Determine whether the function is *even*, *odd*, or *neither*.

$f(x) = x^2 - x^4$

$g(x) = 2x^3 + 1$

$h(x) = x^5 - 2x^3 + x$

$j(x) = 2 - x^6 - x^8$

$k(x) = x^5 - 2x^4 + x - 2$

$p(x) = x^9 + 3x^5 - x^3 + x$

What do you notice about the equations of functions that are odd? What do you notice about the equations of functions that are even? Can you describe a way to identify a function as odd or even by inspecting the equation? Can you describe a way to identify a function as neither odd nor even by inspecting the equation?

Example 9 ▶ Even and Odd Functions

a. The function $g(x) = x^3 - x$ is odd because $g(-x) = -g(x)$, as follows.

$g(-x) = (-x)^3 - (-x)$ Substitute $-x$ for x.

$= -x^3 + x$ Simplify.

$= -(x^3 - x)$ Distributive Property

$= -g(x)$ Test for odd function.

b. The function $h(x) = x^2 + 1$ is even because $h(-x) = h(x)$, as follows.

$h(-x) = (-x)^2 + 1$ Substitute $-x$ for x.

$= x^2 + 1$ Simplify.

$= h(x)$ Test for even function.

The graphs of these two functions are shown in Figure 1.42.

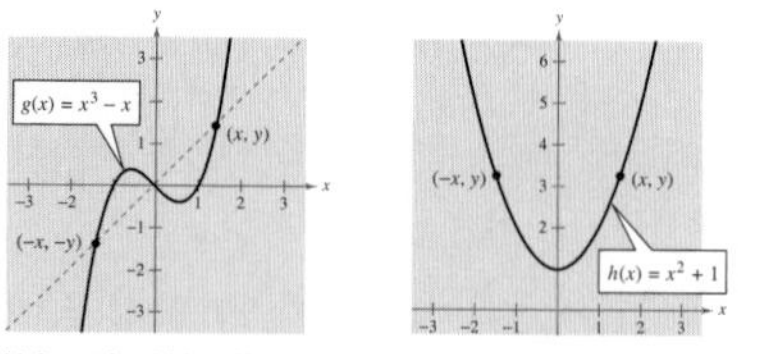

(a) Symmetric to Origin: Odd Function (b) Symmetric to y-Axis: Even Function

FIGURE 1.42

Supplements

Resources

Website (*college.hmco.com*)
Many additional text-specific study and interactive features for students and instructors can be found at the Houghton Mifflin website.

For the Student

Student Success Organizer

Study and Solutions Guide by Dianna L. Zook (Indiana University/Purdue University–Fort Wayne)

Graphing Technology Guide by Benjamin N. Levy and Laurel Technical Services

Instructional Videotapes by Dana Mosely

Instructional Videotapes for Graphing Calculators by Dana Mosely

For the Instructor

Instructor's Annotated Edition

Instructor Success Organizer

Complete Solutions Guide by Dianna L. Zook (Indiana University/Purdue University–Fort Wayne), Laurel Technical Services, and Mike Jones

Test Item File

Problem Solving, Modeling, and Data Analysis Labs by Wendy Metzger (Palomar College)

Computerized Testing (Windows, Macintosh)

HMClassPrep Instructor's CD-ROM

An Introduction to Graphing Utilities

Graphing utilities such as graphing calculators and computers with graphing software are very valuable tools for visualizing mathematical principles, verifying solutions to equations, exploring mathematical ideas, and developing mathematical models. Although graphing utilities are extremely helpful in learning mathematics, their use does not mean that learning algebra is any less important. In fact, the combination of knowledge of mathematics and the use of graphing utilities allows you to explore mathematics more easily and to a greater depth. If you are using a graphing utility in this course, it is up to you to learn its capabilities and to practice using this tool to enhance your mathematical learning.

In this text there are many opportunities to use a graphing utility, some of which are described below.

Some Uses of a Graphing Utility

A graphing utility can be used to

- check or validate answers to problems obtained using algebraic methods.
- discover and explore algebraic properties, rules, and concepts.
- graph functions, and approximate solutions to equations involving functions.
- efficiently perform complicated mathematical procedures such as those found in many real-life applications.
- find mathematical models for sets of data.

In this introduction, the features of graphing utilities are discussed from a generic perspective. To learn how to use the features of a specific graphing utility, consult your user's manual or the website for this text found at *college.hmco.com.* Additionally, keystroke guides are available for most graphing utilities, and your college library may have a videotape on how to use your graphing utility.

The Equation Editor

Many graphing utilities are designed to act as "function graphers." In this course, you will study functions and their graphs in detail. You may recall from previous courses that a function can be thought of as a rule that describes the relationship between two variables. These rules are frequently written in terms of x and y. For example, the equation $y = 3x + 5$ represents y as a function of x.

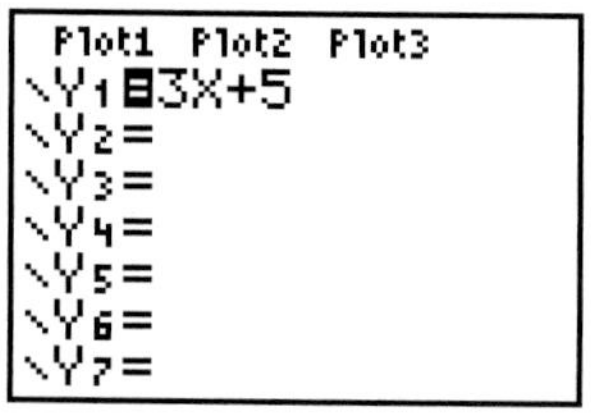

FIGURE 1

Many graphing utilities have an equation editor that requires an equation to be written in "$y =$" form in order to be entered, as shown in Figure 1. (You should note that your equation editor screen may not look like the screen shown in Figure 1.) To determine exactly how to enter an equation into your graphing utility, consult your user's manual.

The Table Feature

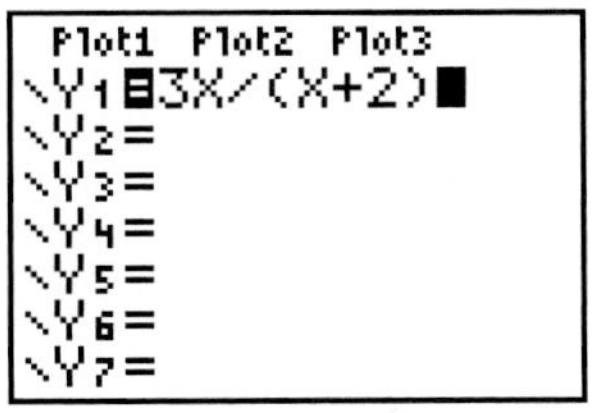

FIGURE 2

Most graphing utilities are capable of displaying a table of values with x-values and one or more corresponding y-values. These tables can be used to check solutions of an equation and to generate ordered pairs to assist in graphing an equation.

To use the *table* feature, enter an equation into the equation editor in "$y =$" form. The table may have a setup screen, which allows you to select the starting x-value and the table step or x-increment. You may then have the option of automatically generating values for x and y or building your own table using the *ask* mode. In the *ask* mode, you enter a value for x and the graphing utility displays the y-value.

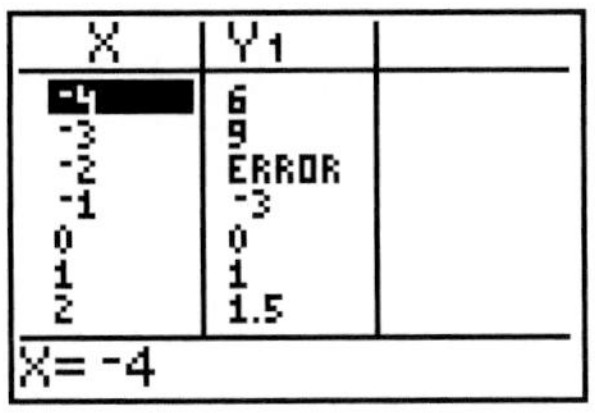

FIGURE 3

For example, enter the equation

$$y = \frac{3x}{x + 2}$$

into the equation editor, as shown in Figure 2. In the table setup screen, set the table to start at $x = -4$ and set the table step to 1. When you view the table, notice that the first x-value is -4 and each value after it increases by 1. Also notice that the Y_1 column gives the resulting y-value for each x-value, as shown in Figure 3. The table shows that the y-value when $x = -2$ is ERROR. This means that the variable x may not take on the value -2 in this equation.

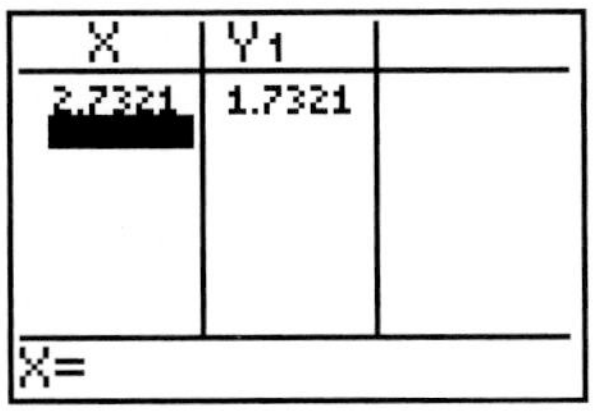

FIGURE 4

With the same equation in the equation editor, set the table to *ask* mode. In this mode you do not need to set the starting x-value or the table step, because you are entering any value you choose for x. You may enter any real value for x—an integer, fraction, decimal, irrational number, and so forth. If you enter $x = 1 + \sqrt{3}$, the graphing utility may rewrite the number as a decimal approximation, as shown in Figure 4. You can continue to build your own table by entering additional x-values in order to generate y-values.

If you have several equations in the equation editor, the table may generate y-values for each equation.

Creating a Viewing Window

A **viewing window** for a graph is a rectangular portion of the coordinate plane. A viewing window is determined by the following six values.

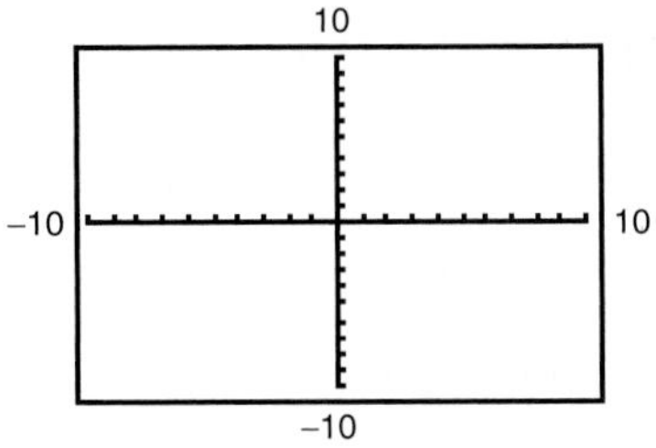

FIGURE 5

Xmin = the smallest value of x
Xmax = the largest value of x
Xscl = the number of units per tick mark on the x-axis
Ymin = the smallest value of y
Ymax = the largest value of y
Yscl = the number of units per tick mark on the y-axis

When you enter these six values into a graphing utility, you are setting the viewing window. Some graphing utilities have a standard viewing window, as shown in Figure 5.

By choosing different viewing windows for a graph, it is possible to obtain very different impressions of the graph's shape. For instance, Figure 6 shows four different viewing windows for the graph of

$$y = 0.1x^4 - x^3 + 2x^2.$$

Of these, the view shown in part (a) is the most complete.

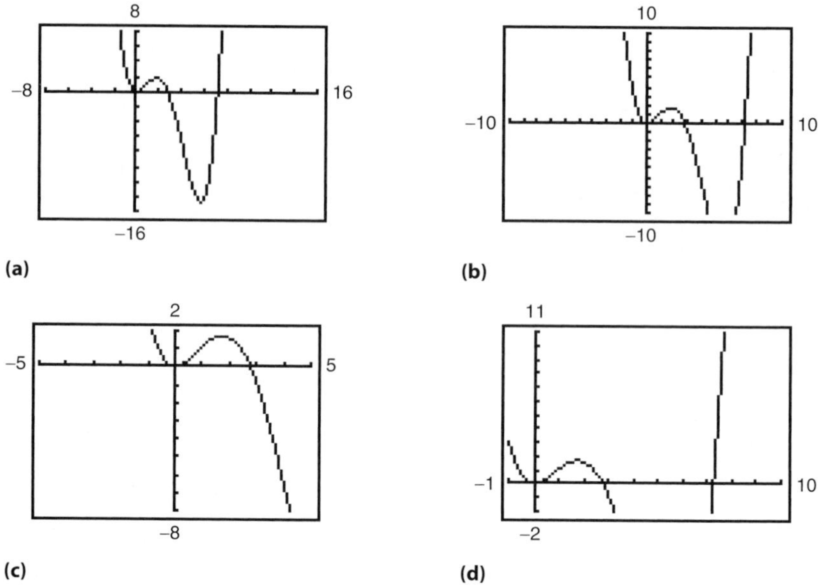

FIGURE 6

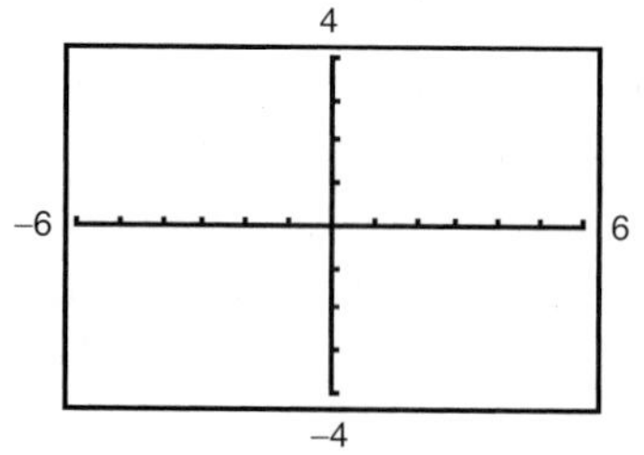

FIGURE 7

On most graphing utilities, the display screen is two-thirds as high as it is wide. On such screens, you can obtain a graph with a true geometric perspective by using a **square setting**—one in which

$$\frac{\text{Ymax} - \text{Ymin}}{\text{Xmax} - \text{Xmin}} = \frac{2}{3}.$$

One such setting is shown in Figure 7. Notice that the x and y tick marks are equally spaced on a square setting, but not on a standard setting.

To see how the viewing window affects the geometric perspective, graph the semicircles $y_1 = \sqrt{9 - x^2}$ and $y_2 = -\sqrt{9 - x^2}$ in a standard viewing window. Then graph y_1 and y_2 in a square window. Note the difference in the shapes of the circles.

Zoom and Trace Features

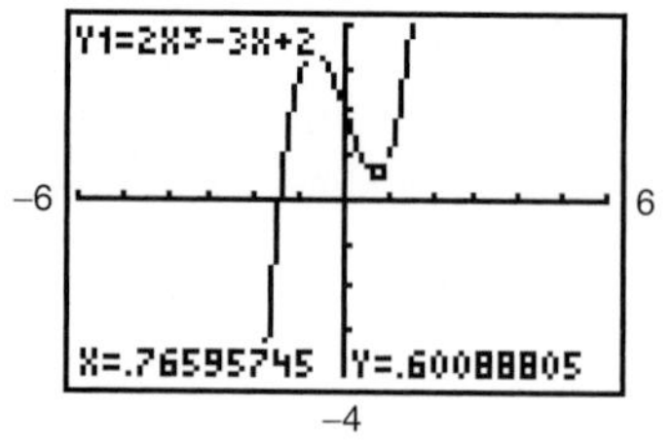

FIGURE 8

When you graph an equation, you can move from point to point along its graph using the *trace* feature. As you trace the graph, the coordinates of each point are displayed, as shown in Figure 8. The *trace* feature combined with the *zoom* feature allows you to obtain better and better approximations of desired points on a graph. For instance, you can use the *zoom* feature of a graphing utility to approximate the x-intercept(s) of a graph [the point(s) where the graph crosses the x-axis]. Suppose you want to approximate the x-intercept(s) of the graph of $y = 2x^3 - 3x + 2$.

Begin by graphing the equation, as shown in Figure 9(a). From the viewing window shown, the graph appears to have only one x-intercept. This intercept lies between -2 and -1. By zooming in on the intercept, you can improve the approximation, as shown in Figure 9(b). To three decimal places, the solution is $x \approx -1.476$.

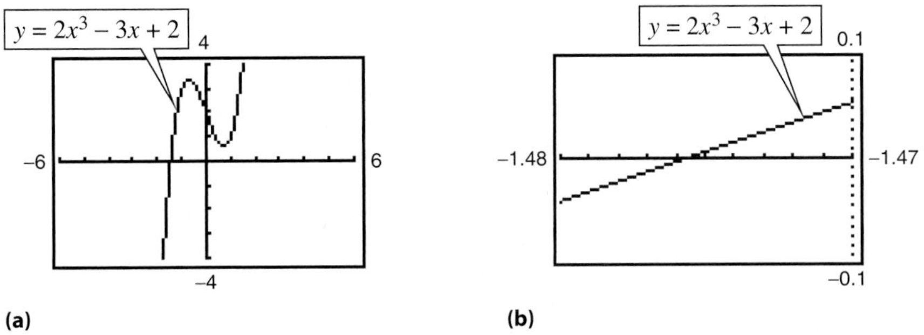

FIGURE 9

Here are some suggestions for using the *zoom* feature.

1. With each successive zoom-in, adjust the x-scale so that the viewing window shows at least one tick mark on each side of the x-intercept.
2. The error in your approximation will be less than the distance between two scale marks.
3. The *trace* feature can usually be used to add one more decimal place of accuracy without changing the viewing window.

Figure 10(a) shows the graph of $y = x^2 - 5x + 3$. Figures 10(b) and 10(c) show "zoom-in views" of the two x-intercepts. From these views, you can approximate the x-intercepts to be $x \approx 0.697$ and $x \approx 4.303$.

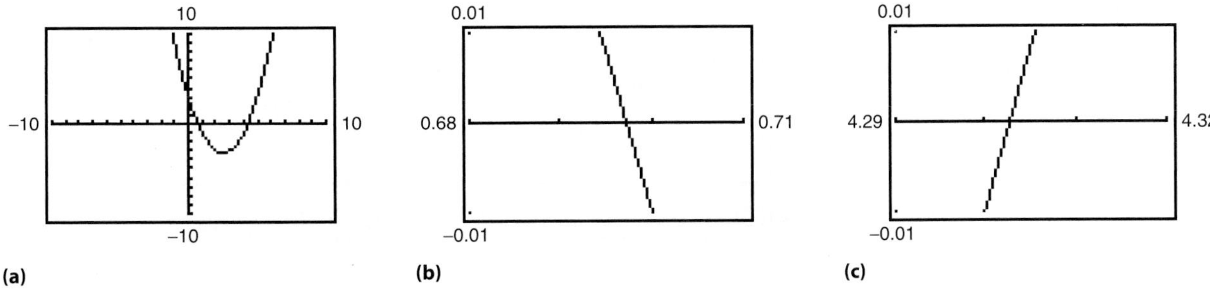

FIGURE 10

Zero or Root Feature

Using the *zero* or *root* feature, you can find the real zeros of functions of the various types studied in this text—polynomial, exponential, logarithmic, and trigonometric functions. To find the zeros of a function such as $f(x) = \frac{3}{4}x - 2$, first enter the function as $y_1 = \frac{3}{4}x - 2$. Then use the *zero* or *root* feature, which may require entering lower and upper bound estimates of the root, as shown in Figure 11.

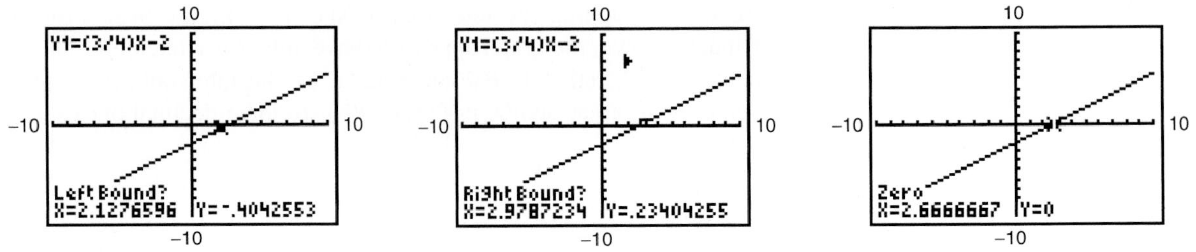

(a) (b) (c)

FIGURE 11

In Figure 11(c), you can see that the zero is $x = 2.6666667 \approx 2\frac{2}{3}$.

Intersect Feature

To find the points of intersection of two graphs, you can use the *intersect* feature. For instance, to find the points of intersection of the graphs of $y_1 = -x + 2$ and $y_2 = x + 4$, enter these two functions and use the *intersect* feature, as shown in Figure 12.

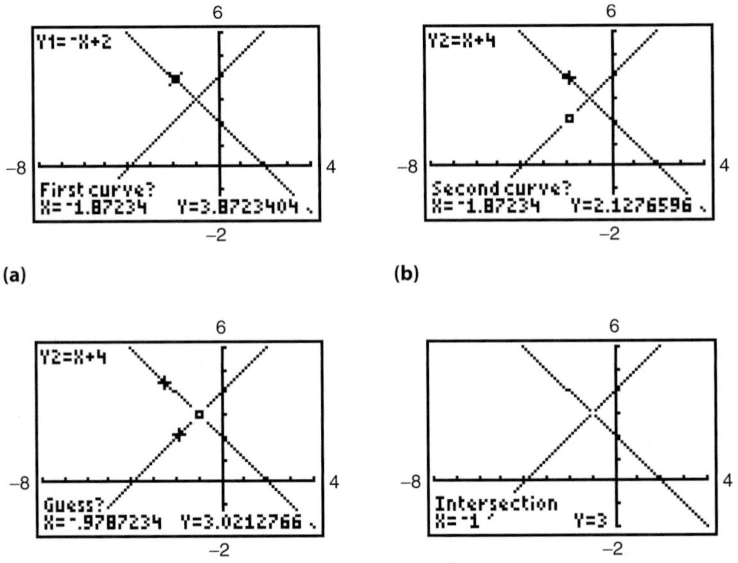

(a) (b) (c) (d)

FIGURE 12

From Figure 12(d), you can see that the point of intersection is $(-1, 3)$.

Regression Capabilities

Throughout the text, you will be asked to use the regression capabilities of a graphing utility to find models for sets of data. Most graphing utilities have built-in regression programs for the following.

Regression	*Form of Model*
Linear	$y = ax + b$
Quadratic	$y = ax^2 + bx + c$
Cubic	$y = ax^3 + bx^2 + cx + d$
Quartic	$y = ax^4 + bx^3 + cx^2 + dx + e$
Logarithmic	$y = a + b\ln(x)$
Exponential	$y = ab^x$
Power	$y = ax^b$
Logistic	$y = \dfrac{c}{1 + ae^{-bx}}$
Sine	$y = a\sin(bx + c) + d$

For instance, you can find the linear regression model for the average hourly wages y (in dollars per hour) of production workers in manufacturing industries from 1987 through 1997 shown in the table. (Source: U.S. Bureau of Labor Statistics)

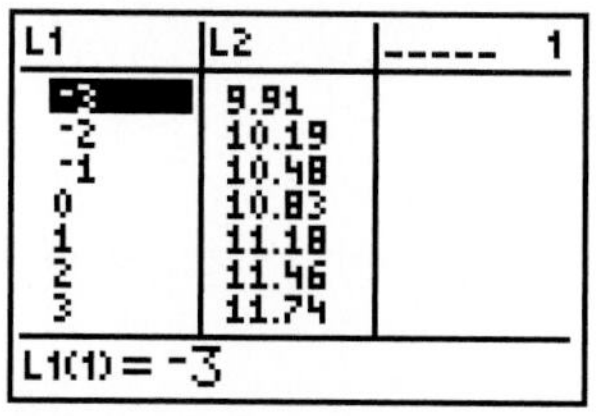

FIGURE 13

Year	1987	1988	1989	1990	1991	1992
y	9.91	10.19	10.48	10.83	11.18	11.46

Year	1993	1994	1995	1996	1997
y	11.74	12.06	12.37	12.78	13.17

First, let $x = 0$ correspond to 1990 and enter the data into the list editor, as shown in Figure 13. Note that the list in the first column contains the years and the list in the second column contains the hourly wages that correspond to the years. Run your graphing utility's built-in linear regression program to obtain the coefficients a and b for the model $y = ax + b$, as shown in Figure 14. So, a linear model for the data is

$$y \approx 0.321x + 10.83.$$

When you run some regression programs, you may obtain an "r-value," which gives a measure of how well the model fits the data. The closer the value of $|r|$ is to 1, the better the fit. For the data in the table above, $r \approx 0.999$, which implies that the model is a very good fit.

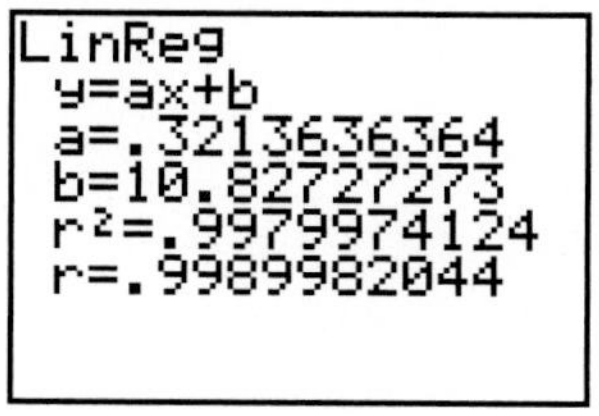

FIGURE 14

Andy Sachs/Tony Stone Images

In 1997, over $24 billion worth of corrugated and solid fiber boxes were produced to create shipping containers for products such as globes as well as containers for food and beverage products. *(Source: U.S. Bureau of the Census)*

P Prerequisites

▶ How to Study This Chapter

The Big Picture

In this chapter you will learn the following skills and concepts.

▶ How to order real numbers, use inequalities, and evaluate algebraic expressions

▶ How to add, subtract, multiply, and factor polynomials

▶ How to determine the domains of algebraic expressions and simplify rational expressions

▶ How to solve linear, quadratic, polynomial, radical, and absolute value equations

▶ How to solve linear, polynomial, rational, and absolute value inequalities

▶ How to use algebraic techniques common in calculus

▶ How to plot points in the coordinate plane and find the distance between two points

Important Vocabulary

As you encounter each new vocabulary term in this chapter, add the term and its definition to your notebook glossary.

Real numbers (p. 2)
Real number line (p. 2)
Order (p. 3)
Inequality (p. 3)
Infinity (p. 4)
Absolute value (p. 5)
Variables (p. 6)
Algebraic expressions (p. 6)
Coefficient (p. 6)
Evaluate (p. 6)
Factors (p. 8)
Exponential form (p. 12)
Exponent (p. 12)
Base (p. 12)
Scientific notation (p. 14)
Square root (p. 15)
Cube root (p. 15)
Principal *n*th root (p. 15)
Index (p. 15)
Radicand (p. 15)
Conjugate (p. 18)
Rational exponent (p. 19)
Polynomial (p. 25)
Degree (p. 25)
Domain (p. 38)
Rational expression (p. 38)
Complex fraction (p. 42)
Equation (p. 49)
Identity (p. 49)
Conditional equation (p. 49)
Linear equation (p. 49)
Equivalent equations (p. 50)
Extraneous solution (p. 51)
Quadratic equation (p. 52)
Linear inequality (p. 64)
Critical numbers (p. 67)
Rectangular coordinate system (p. 81)
Ordered pair (p. 81)
Distance Formula (p. 83)
Midpoint Formula (p. 85)

Study Tools

- Learning objectives at the beginning of each section
- Chapter Summary (p. 91)
- Review Exercises (pp. 92–95)
- Chapter Test (p. 97)

Additional Resources

- Study and Solutions Guide
- Interactive Precalculus
- Videotapes for Chapter P
- Precalculus Website
- Student Success Organizer

STUDY T!P

A good rule of thumb to use is to study 2 to 4 hours for every hour in class. Increase your study time if you do not make the grade you want on your first major test.

P.1 Real Numbers

▶ **What you should learn**

- How to represent and classify real numbers
- How to order real numbers and use inequalities
- How to find the absolute values of real numbers and find the distance between two real numbers
- How to evaluate algebraic expressions
- How to use the basic rules and properties of algebra

▶ **Why you should learn it**

Real numbers are used to represent many real-life quantities, such as the variance of a budget (see Exercises 83–86 on page 10).

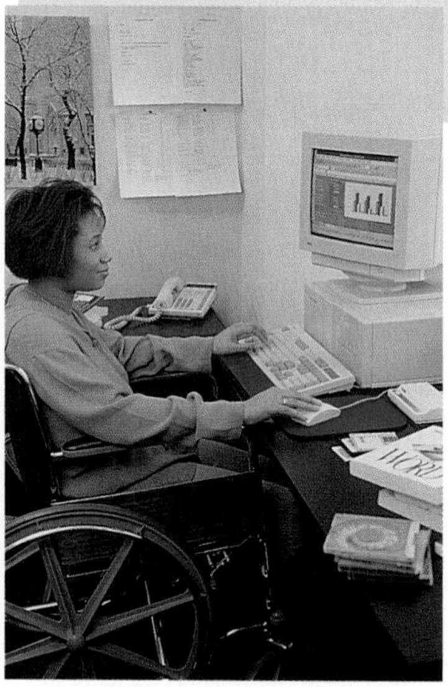

Don Smetzer/Tony Stone Images

Real Numbers

Real numbers are used in everyday life to describe quantities such as age, miles per gallon, container size, and population. Real numbers are represented by symbols such as

$$-5, 9, 0, \frac{4}{3}, 0.666\ .\ .\ ., 28.21, \sqrt{2}, \pi, \text{ and } \sqrt[3]{-32}.$$

Here are some important subsets of the real numbers.

$$\{1, 2, 3, 4, \ldots\} \qquad \text{Set of natural numbers}$$

$$\{0, 1, 2, 3, 4, \ldots\} \qquad \text{Set of whole numbers}$$

$$\{\ldots, -3, -2, -1, 0, 1, 2, 3, \ldots\} \qquad \text{Set of integers}$$

A real number is **rational** if it can be written as the ratio p/q of two integers, where $q \neq 0$. For instance, the numbers

$$\frac{1}{3} = 0.3333\ .\ .\ . = 0.\overline{3}, \frac{1}{8} = 0.125, \text{ and } \frac{125}{111} = 1.126126\ .\ .\ . = 1.\overline{126}$$

are rational. The decimal representation of a rational number either repeats (as in $\frac{173}{55} = 3.1\overline{45}$) or terminates (as in $\frac{1}{2} = 0.5$). A real number that cannot be written as the ratio of two integers is called **irrational.** Irrational numbers have infinite nonrepeating decimal representations. For instance, the numbers

$$\sqrt{2} \approx 1.4142136 \quad \text{and} \quad \pi \approx 3.1415927$$

are irrational. (The symbol $\approx$ means "is approximately equal to.")

Real numbers are represented graphically by a **real number line.** The point 0 on the real number line is the **origin.** Numbers to the right of 0 are positive, and numbers to the left of 0 are negative, as shown in Figure P.1. The term **nonnegative** describes a number that is either positive or zero.

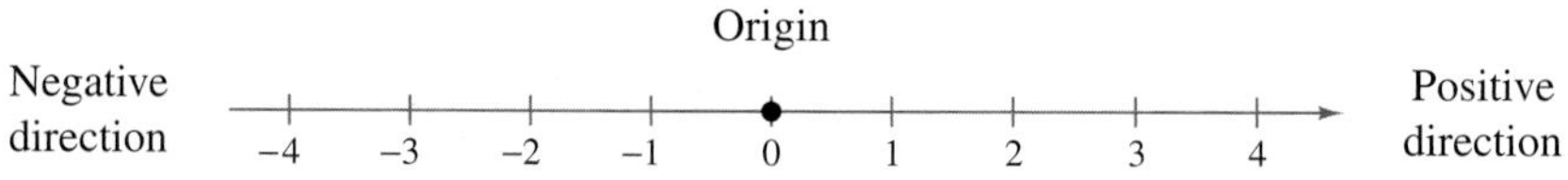

FIGURE P.1 *The Real Number Line*

As illustrated in Figure P.2, there is a *one-to-one correspondence* between real numbers and points on the real number line.

Every real number corresponds to exactly one point on the real number line.

Every point on the real number line corresponds to exactly one real number.

FIGURE P.2 *One-to-One Correspondence*

Ordering Real Numbers

One important property of real numbers is that they are *ordered.*

Definition of Order on the Real Number Line

If a and b are real numbers, a is *less than* b if $b - a$ is positive. The **order** of a and b is denoted by the **inequality**

$$a < b.$$

This relationship can also be described by saying that b is *greater than* a and writing $b > a$. The inequality $a \le b$ means that a is *less than or equal to* b, and the inequality $b \ge a$ means that b is *greater than or equal to* a. The symbols $<$, $>$, $\le$, and $\ge$ are *inequality symbols.*

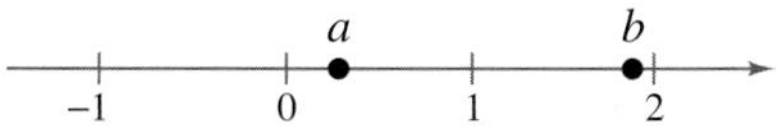

FIGURE P.3 *$a < b$ if and only if a lies to the left of b.*

Geometrically, this definition implies that $a < b$ if and only if a lies to the *left* of b on the real number line, as shown in Figure P.3.

Example 1 ▶ Interpreting Inequalities

Describe the subset of real numbers represented by each inequality.

a. $x \le 2$ **b.** $-2 \le x < 3$

Solution

a. The inequality $x \le 2$ denotes all real numbers less than or equal to 2, as shown in Figure P.4(a).

b. The inequality $-2 \le x < 3$ means that $x \ge -2$ and $x < 3$. This "double inequality" denotes all real numbers between -2 and 3, including -2 but not including 3, as shown in Figure P.4(b).

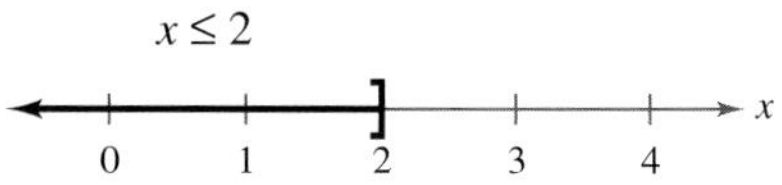

(a)

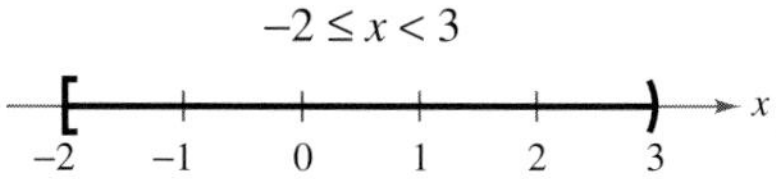

(b)

FIGURE P.4

A computer animation of this example appears in the *Interactive* CD-ROM and *Internet* versions of this text.

Inequalities can be used to describe subsets of real numbers called **intervals.** In the bounded intervals below, the real numbers a and b are the **endpoints** of each interval.

Bounded Intervals on the Real Number Line

Notation	*Interval Type*	*Inequality*	*Graph*
$[a, b]$	Closed	$a \le x \le b$	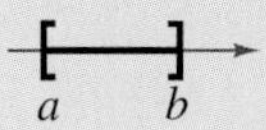
(a, b)	Open	$a < x < b$	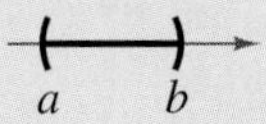
$[a, b)$		$a \le x < b$	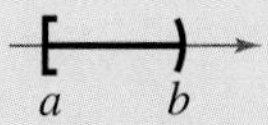
$(a, b]$		$a < x \le b$	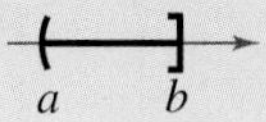

The symbols ∞, **positive infinity,** and $-\infty$, **negative infinity,** do not represent real numbers. They are simply convenient symbols used to describe the unboundedness of an interval such as $(1, \infty)$ or $(-\infty, 3]$.

Unbounded Intervals on the Real Number Line

Notation	*Interval Type*	*Inequality*	*Graph*
$[a, \infty)$		$x \geq a$	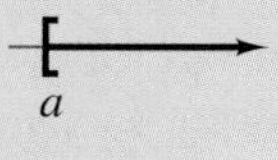
(a, ∞)	Open	$x > a$	
$(-\infty, b]$		$x \leq b$	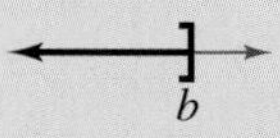
$(-\infty, b)$	Open	$x < b$	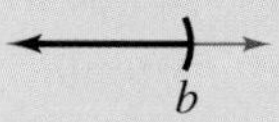
$(-\infty, \infty)$	Entire real line		

Example 2 ▶ Using Inequalities to Represent Intervals

Additional Examples

Use inequality notation to describe each of the following.

a. All x in the interval $(-2, 2]$

b. w is at least 1 and at most 5

c. All q in the interval $(-12, 0)$

Solution

a. $-2 < x \leq 2$

b. $1 \leq w \leq 5$

c. $-12 < q < 0$

Use inequality notation to describe each of the following.

a. c is at most 2.

b. m is at least -3.

c. All x in the interval $(-3, 5]$

Solution

a. The statement "c is at most 2" can be represented by $c \leq 2$.

b. The statement "m is at least -3" can be represented by $m \geq -3$.

c. "All x in the interval $(-3, 5]$" can be represented by $-3 < x \leq 5$.

Example 3 ▶ Interpreting Intervals

Give a verbal description of each interval.

a. $(-1, 0)$ **b.** $[2, \infty)$ **c.** $(-\infty, 0)$

Solution

a. This interval consists of all real numbers that are greater than -1 and less than 0.

b. This interval consists of all real numbers that are greater than or equal to 2.

c. This interval consists of all real numbers that are less than zero (the negative real numbers).

The **Law of Trichotomy** states that for any two real numbers a and b, *precisely* one of three relationships is possible:

$$a = b, \quad a < b, \quad \text{or} \quad a > b. \qquad \text{Law of Trichotomy}$$

Exploration

Absolute value expressions can be evaluated on a graphing utility. When an expression such as $|3 - 8|$ is evaluated, parentheses should surround the expression, as in abs(3 − 8). Evaluate each expression below. What can you conclude?

a. $|6|$ **b.** $|-1|$

c. $|5 - 2|$ **d.** $|2 - 5|$

Absolute Value and Distance

The **absolute value** of a real number is its *magnitude*, or the distance between the origin and the point representing the real number on the real number line.

Definition of Absolute Value

If a is a real number, then the absolute value of a is

$$|a| = \begin{cases} a, & \text{if } a \geq 0 \\ -a, & \text{if } a < 0. \end{cases}$$

Notice in this definition that the absolute value of a real number is never negative. For instance, if $a = -5$, then $|-5| = -(-5) = 5$. The absolute value of a real number is either positive or zero. Moreover, 0 is the only real number whose absolute value is 0. So, $|0| = 0$.

Example 4 ▶ Evaluating the Absolute Value of a Number

Evaluate $\dfrac{|x|}{x}$ for (a) $x > 0$ and (b) $x < 0$.

Solution

a. If $x > 0$, then $|x| = x$ and $\dfrac{|x|}{x} = \dfrac{x}{x} = 1$.

b. If $x < 0$, then $|x| = -x$ and $\dfrac{|x|}{x} = \dfrac{-x}{x} = -1$.

Properties of Absolute Values

1. $|a| \geq 0$ **2.** $|-a| = |a|$

3. $|ab| = |a||b|$ **4.** $\left|\dfrac{a}{b}\right| = \dfrac{|a|}{|b|}, \quad b \neq 0$

Distance Between Two Points on the Real Line

Let a and b be real numbers. The **distance between a and b** is

$$d(a, b) = |b - a| = |a - b|.$$

Absolute value can be used to define the distance between two points on the real number line. For instance, the distance between -3 and 4 is

$$|-3 - 4| = |-7| = 7,$$

as shown in Figure P.5.

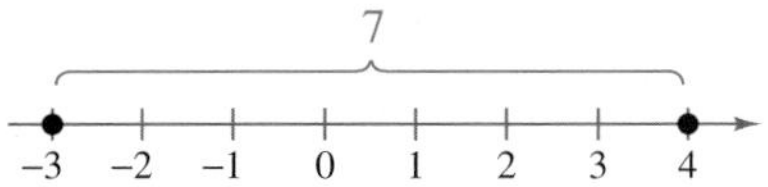

FIGURE P.5 *The distance between −3 and 4 is 7.*

Common Error
A common error is to use the wrong order of operations when evaluating expressions. Remind students that the order of operations is as follows.

1. First do operations that occur within symbols of grouping.
2. Then evaluate powers.
3. Then do multiplications and divisions from left to right.
4. Then do additions and subtractions from left to right.

A computer animation of this concept appears in the *Interactive* CD-ROM and *Internet* versions of this text.

STUDY T!P

Try to formulate a verbal description of each property. For instance, the first property states that *the order in which two real numbers are added does not affect their sum.*

Algebraic Expressions

One characteristic of algebra is the use of letters to represent numbers. The letters are **variables,** and combinations of letters and numbers are **algebraic expressions.** Here are a few examples of algebraic expressions.

$$5x, \qquad 2x - 3, \qquad \frac{4}{x^2 + 2}, \qquad 7x + y$$

Definition of an Algebraic Expression

An **algebraic expression** is a collection of letters (**variables**) and real numbers (**constants**) combined using the operations of addition, subtraction, multiplication, division, and exponentiation.

The **terms** of an algebraic expression are those parts that are separated by *addition.* For example,

$$x^2 - 5x + 8 = x^2 + (-5x) + 8$$

has three terms: x^2 and $-5x$ are the **variable terms** and 8 is the **constant term.** The numerical factor of a variable term is the **coefficient** of the variable term. For instance, the coefficient of $-5x$ is -5, and the coefficient of x^2 is 1.

To **evaluate** an algebraic expression, substitute numerical values for each of the variables in the expression. Here are two examples.

Expression	*Value of Variable*	*Substitute*	*Value of Expression*
$-3x + 5$	$x = 3$	$-3(3) + 5$	$-9 + 5 = -4$
$3x^2 + 2x - 1$	$x = -1$	$3(-1)^2 + 2(-1) - 1$	$3 - 2 - 1 = 0$

When an algebraic expression is evaluated, the **Substitution Principle** is used. It states that "If $a = b$, then a can be replaced by b in any expression involving a." In the first evaluation shown above, for instance, 3 is *substituted* for x in the expression $-3x + 5$.

Basic Rules of Algebra

There are four arithmetic operations with real numbers: *addition, multiplication, subtraction,* and *division,* denoted by the symbols $+$, $\times$ or $\cdot$, $-$, and $\div$. Of these, addition and multiplication are the two primary operations. Subtraction and division are the inverse operations of addition and multiplication, respectively.

Subtraction: Add the opposite.

$$a - b = a + (-b)$$

Division: Multiply by the reciprocal.

If $b \neq 0$, then $a/b = a\left(\frac{1}{b}\right) = \frac{a}{b}$.

In these definitions, $-b$ is the **additive inverse** (or opposite) of b, and $1/b$ is the **multiplicative inverse** (or reciprocal) of b. In the fractional form a/b, a is the **numerator** of the fraction and b is the **denominator.**

Because the properties of real numbers on page 7 are true for variables and algebraic expressions as well as for real numbers, they are often called the **Basic Rules of Algebra.**

Basic Rules of Algebra

Let *a*, *b*, and *c* be real numbers, variables, or algebraic expressions.

Property		*Example*
Commutative Property of Addition:	$a + b = b + a$	$4x + x^2 = x^2 + 4x$
Commutative Property of Multiplication:	$ab = ba$	$(4 - x)x^2 = x^2(4 - x)$
Associative Property of Addition:	$(a + b) + c = a + (b + c)$	$(x + 5) + x^2 = x + (5 + x^2)$
Associative Property of Multiplication:	$(ab)c = a(bc)$	$(2x \cdot 3y)(8) = (2x)(3y \cdot 8)$
Distributive Properties:	$a(b + c) = ab + ac$	$3x(5 + 2x) = 3x \cdot 5 + 3x \cdot 2x$
	$(a + b)c = ac + bc$	$(y + 8)y = y \cdot y + 8 \cdot y$
Additive Identity Property:	$a + 0 = a$	$5y^2 + 0 = 5y^2$
Multiplicative Identity Property:	$a \cdot 1 = a$	$(4x^2)(1) = 4x^2$
Additive Inverse Property:	$a + (-a) = 0$	$5x^3 + (-5x^3) = 0$
Multiplicative Inverse Property:	$a \cdot \frac{1}{a} = 1, \quad a \neq 0$	$(x^2 + 4)\left(\frac{1}{x^2 + 4}\right) = 1$

Because subtraction is defined as "adding the opposite," the Distributive Properties are also true for subtraction. For instance, the "subtraction form" of $a(b + c) = ab + ac$ is

$$a(b - c) = ab - ac.$$

Properties of Negation

Let *a* and *b* be real numbers, variables, or algebraic expressions.

Property	*Example*
1. $(-1)a = -a$	$(-1)7 = -7$
2. $-(-a) = a$	$-(-6) = 6$
3. $(-a)b = -(ab) = a(-b)$	$(-5)3 = -(5 \cdot 3) = 5(-3)$
4. $(-a)(-b) = ab$	$(-2)(-x) = 2x$
5. $-(a + b) = (-a) + (-b)$	$-(x + 8) = (-x) + (-8)$
	$= -x - 8$

STUDY TIP

Be sure you see the difference between the *opposite of a number* and a *negative number.* If a is already negative, then its opposite, $-a$, is positive. For instance, if $a = -5$, then

$$-a = -(-5) = 5.$$

Properties of Equality

Let *a*, *b*, and *c* be real numbers, variables, or algebraic expressions.

1. If $a = b$, then $a + c = b + c$. Add c to each side.

2. If $a = b$, then $ac = bc$. Multiply each side by c.

3. If $a + c = b + c$, then $a = b$. Subtract c from each side.

4. If $ac = bc$ and $c \neq 0$, then $a = b$. Divide each side by c.

STUDY TIP

The "or" in the Zero-Factor Property includes the possibility that either or both factors may be zero. This is an **inclusive or,** and it is the way the word "or" is generally used in mathematics.

Activities

1. Evaluate: $4 - |-3|$.
   Answer: 1
2. Find the distance between -41 and 16.
   Answer: 57
3. Use inequality notation to describe the set of nonnegative numbers.
   Answer: $x \geq 0$
4. Use interval notation to describe the inequality $-6 < x \leq 13$.
   Answer: $(-6, 13]$

STUDY TIP

In Property 1, the phrase "if and only if" implies two statements. One statement is: If $a/b = c/d$, then $ad = bc$. The other statement is: If $ad = bc$, where $b \neq 0$ and $d \neq 0$, then $a/b = c/d$.

Point out to students that to add or subtract fractions with unlike denominators, they can use Property 5 of fractions as in Example 5(c), or they can rewrite the fractions with like denominators using the least common denominator (LCD) of the fractions.

Properties of Zero

Let a and b be real numbers, variables, or algebraic expressions.

1. $a + 0 = a$ and $a - 0 = a$ **2.** $a \cdot 0 = 0$

3. $\dfrac{0}{a} = 0, \quad a \neq 0$ **4.** $\dfrac{a}{0}$ is undefined.

5. Zero-Factor Property: If $ab = 0$, then $a = 0$ or $b = 0$.

Properties and Operations of Fractions

Let a, b, c, and d be real numbers, variables, or algebraic expressions such that $b \neq 0$ and $d \neq 0$.

1. Equivalent Fractions: $\dfrac{a}{b} = \dfrac{c}{d}$ if and only if $ad = bc$.

2. Rules of Signs: $-\dfrac{a}{b} = \dfrac{-a}{b} = \dfrac{a}{-b}$ and $\dfrac{-a}{-b} = \dfrac{a}{b}$

3. Generate Equivalent Fractions: $\dfrac{a}{b} = \dfrac{ac}{bc}, \quad c \neq 0$

4. Add or Subtract with Like Denominators: $\dfrac{a}{b} \pm \dfrac{c}{b} = \dfrac{a \pm c}{b}$

5. Add or Subtract with Unlike Denominators: $\dfrac{a}{b} \pm \dfrac{c}{d} = \dfrac{ad \pm bc}{bd}$

6. Multiply Fractions: $\dfrac{a}{b} \cdot \dfrac{c}{d} = \dfrac{ac}{bd}$

7. Divide Fractions: $\dfrac{a}{b} \div \dfrac{c}{d} = \dfrac{a}{b} \cdot \dfrac{d}{c} = \dfrac{ad}{bc}, \quad c \neq 0$

Example 5 ▶ Properties and Operations of Fractions

a. Equivalent fractions: $\dfrac{x}{5} = \dfrac{3 \cdot x}{3 \cdot 5} = \dfrac{3x}{15}$ **b.** Divide fractions: $\dfrac{7}{x} \div \dfrac{3}{2} = \dfrac{7}{x} \cdot \dfrac{2}{3} = \dfrac{14}{3x}$

c. Add fractions with unlike denominators: $\dfrac{x}{3} + \dfrac{2x}{5} = \dfrac{5 \cdot x + 3 \cdot 2x}{15} = \dfrac{11x}{15}$

If a, b, and c are integers such that $ab = c$, then a and b are **factors** or **divisors** of c. A **prime number** is an integer that has exactly two positive factors: itself and 1, such as 2, 3, 5, 7, and 11. The numbers 4, 6, 8, 9, and 10 are **composite** because they can be written as the product of two or more prime numbers. The number 1 is neither prime nor composite. The **Fundamental Theorem of Arithmetic** states that every positive integer greater than 1 can be written as the product of prime numbers in precisely one way (disregarding order). For instance, the *prime factorization* of 24 is $24 = 2 \cdot 2 \cdot 2 \cdot 3$.

P.1 Exercises

The *Interactive* CD-ROM and *Internet* versions of this text contain step-by-step solutions to all odd-numbered Section and Review Exercises. They also provide Tutorial Exercises that link to Guided Examples for additional help.

In Exercises 1–6, determine which numbers are (a) natural numbers, (b) integers, (c) rational numbers, and (d) irrational numbers.

1. $-9, -\frac{7}{2}, 5, \frac{2}{3}, \sqrt{2}, 0, 1, -4, 2, -11$

2. $\sqrt{5}, -7, -\frac{7}{3}, 0, 3.12, \frac{5}{4}, -3, 12, 5$

3. $2.01, 0.666\ .\ .\ .\ , -13, 0.010110111\ .\ .\ .\ , 1, -6$

4. $2.3030030003\ .\ .\ .\ , 0.7575, -4.63, \sqrt{10}, -75, 15, 31$

5. $-\pi, -\frac{1}{3}, \frac{6}{3}, \frac{1}{2}\sqrt{2}, -7.5, -1, 8, -22$

6. $25, -17, -\frac{12}{5}, \sqrt{9}, 3.12, \frac{1}{2}\pi, 7, -11.1, 13$

In Exercises 7–10, use a calculator to find the decimal form of the rational number. If it is a nonterminating decimal, write the repeating pattern.

7. $\frac{5}{8}$ **8.** $\frac{1}{3}$

9. $\frac{41}{333}$ **10.** $\frac{6}{11}$

In Exercises 11–16, use a calculator with a fraction feature to write the rational number as the ratio of two integers.

11. 4.1 **12.** 8.5

13. $10.\overline{2}$ **14.** $5.\overline{45}$

15. $-2.01\overline{2}$ **16.** $-1.6\overline{5}$

In Exercises 17 and 18, approximate the numbers and place the correct symbol (< or >) between them.

17.

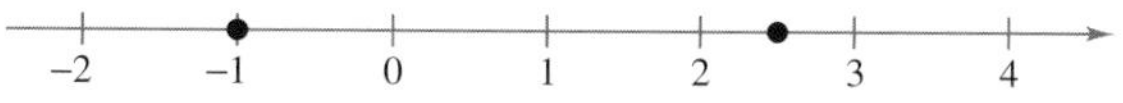

18.

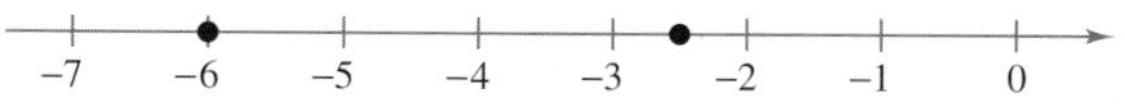

In Exercises 19–24, plot the two real numbers on the real number line. Then place the appropriate inequality symbol (< or >) between them.

19. $-4, -8$ **20.** $-3.5, 1$

21. $\frac{3}{2}, 7$ **22.** $1, \frac{16}{3}$

23. $\frac{5}{6}, \frac{2}{3}$ **24.** $-\frac{8}{7}, -\frac{3}{7}$

In Exercises 25–34, verbally describe the subset of real numbers represented by the inequality. Then sketch the subset on the real number line. State whether the interval is bounded or unbounded.

25. $x \le 5$ **26.** $x \ge -2$

27. $x < 0$ **28.** $x > 3$

29. $x \ge 4$ **30.** $x < 2$

31. $-2 < x < 2$ **32.** $0 \le x \le 5$

33. $-1 \le x < 0$ **34.** $0 < x \le 6$

In Exercises 35 and 36, use a calculator to order the numbers from smallest to largest.

35. $\frac{7071}{5000}, \frac{584}{413}, \sqrt{2}, \frac{47}{33}, \frac{127}{90}$

36. $\frac{26}{15}, \sqrt{3}, 1.7320, \frac{381}{220}, \frac{2103}{1214}$

In Exercises 37–44, use inequality notation to describe the set.

37. All x in the interval $(-2, 4]$

38. All y in the interval $[-6, 0)$

39. y is nonnegative.

40. y is no more than 25.

41. t is at least 10 and at most 22.

42. k is less than 5 but no less than -3.

43. The dog's weight W is more than 65 pounds.

44. The annual rate of inflation r is expected to be at least 2.5% but no more than 5%.

In Exercises 45–48, give a verbal description of the interval.

45. $[0, 8)$ **46.** $[-5, 7]$

47. $(-6, \infty)$ **48.** $(-\infty, 4]$

In Exercises 49–58, evaluate the expression.

49. $|-10|$ **50.** $|0|$

51. $|3 - x|$ **52.** $|4 - x|$

53. $|-1| - |-2|$ **54.** $-3 - |-3|$

55. $\dfrac{-5}{|-5|}$ **56.** $-3|-3|$

57. $\dfrac{|x+2|}{x+2}, \quad x < -2$ **58.** $\dfrac{|x-1|}{x-1}, \quad x > 1$

In Exercises 59–64, place the correct symbol (<, >, or =) between the pair of real numbers.

59. $|-3| \quad \square \quad -|-3|$ **60.** $|-4| \quad \square \quad |4|$

61. $-5 \quad \square \quad -|5|$ **62.** $-|-6| \quad \square \quad |-6|$

63. $-|-2| \quad \square \quad -|2|$ **64.** $-(-2) \quad \square \quad -2$

In Exercises 65–72, find the distance between *a* and *b*.

65. 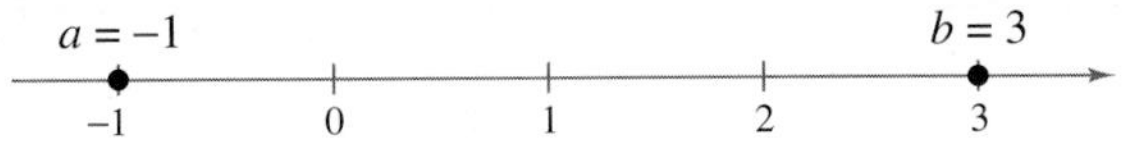

66. 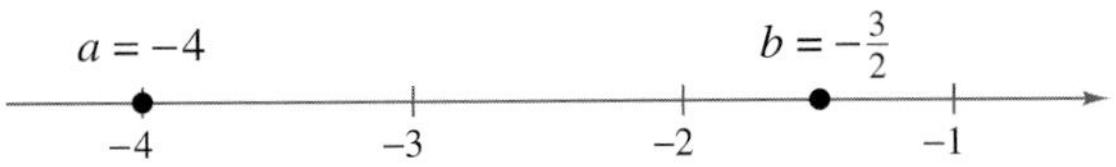

67. $a = 126, b = 75$

68. $a = -126, b = -75$

69. $a = -\frac{5}{2}, b = 0$

70. $a = \frac{1}{4}, b = \frac{11}{4}$

71. $a = \frac{16}{5}, b = \frac{112}{75}$

72. $a = 9.34, b = -5.65$

In Exercises 73 and 74, use the real numbers *A*, *B*, and *C* shown on the number line. Determine the sign of each expression.

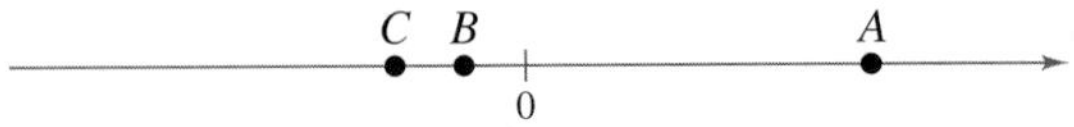

73. (a) $-A$
(b) $B - A$

74. (a) $-C$
(b) $A - C$

In Exercises 75–82, use absolute value notation to describe the situation.

75. While traveling, you pass milepost 7, then milepost 18. How far do you travel during that time period?

76. While traveling, you pass milepost 103, then milepost 86. How far do you travel during that time period?

77. The temperature was 60° at noon, then 23° at midnight. What was the change in temperature over the 12-hour period?

78. The temperature was 48° last night at midnight, then 82° at noon today. What was the change in temperature over the 12-hour period?

79. The distance between x and 5 is no more than 3.

80. The distance between x and -10 is at least 6.

81. y is at least 6 units from 0.

82. y is at most 2 units from a.

Budget Variance **In Exercises 83–86, the accounting department of a company is checking to see whether the actual expenses of a department differ from the budgeted expenses by more than \$500 or by more than 5%. Fill in the missing parts of the table, and determine whether the actual expense passes the "budget variance test."**

	Budgeted Expense, b	*Actual Expense, a*	$\lvert a - b \rvert$	$0.05b$
83. Wages	\$112,700	\$113,356		
84. Utilities	\$9,400	\$9,772		
85. Taxes	\$37,640	\$37,335		
86. Insurance	\$2,575	\$2,613		

Federal Deficit **In Exercises 87–90, use the bar graph, which shows the receipts of the federal government (in billions of dollars) for selected years from 1960 through 1998. In each exercise you are given the outlay of the federal government. Find the magnitude of the surplus or deficit for the year.** (Source: U.S. Treasury Department)

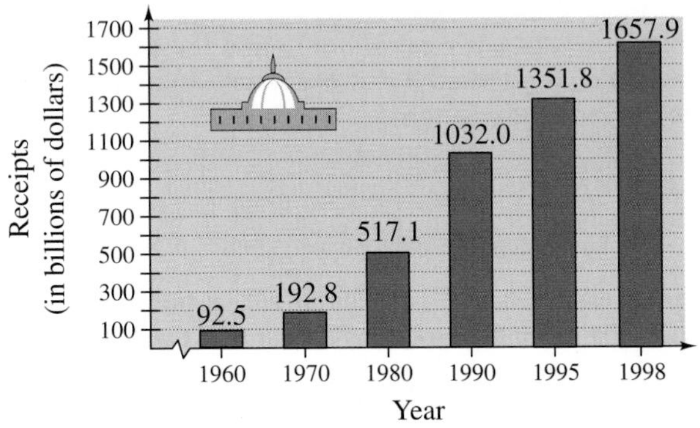

	Receipts	*Outlay*	$\lvert Receipts - Outlay \rvert$
87. 1960		\$92.2 billion	
88. 1980		\$590.9 billion	
89. 1990		\$1253.2 billion	
90. 1998		\$1667.8 billion	

In Exercises 91–96, identify the terms. Then identify the coefficients of the variable terms of the expression.

91. $7x + 4$

92. $6x^3 - 5x$

93. $\sqrt{3}x^2 - 8x - 11$

94. $3\sqrt{3}x^2 + 1$

95. $4x^3 + \frac{x}{2} - 5$

96. $3x^4 - \frac{x^2}{4}$

In Exercises 97–102, evaluate the expression for each value of *x*. (If not possible, state the reason.)

Expression	*Values*	
97. $4x - 6$	(a) $x = -1$	(b) $x = 0$
98. $9 - 7x$	(a) $x = -3$	(b) $x = 3$
99. $x^2 - 3x + 4$	(a) $x = -2$	(b) $x = 2$

	Expression	*Values*	
100.	$-x^2 + 5x - 4$	(a) $x = -1$	(b) $x = 1$
101.	$\dfrac{x+1}{x-1}$	(a) $x = 1$	(b) $x = -1$
102.	$\dfrac{x}{x+2}$	(a) $x = 2$	(b) $x = -2$

In Exercises 103–112, identify the rule(s) of algebra illustrated by the equation.

103. $x + 9 = 9 + x$

104. $2\left(\frac{1}{2}\right) = 1$

105. $\dfrac{1}{h+6}(h + 6) = 1, \quad h \neq -6$

106. $(x + 3) - (x + 3) = 0$

107. $2(x + 3) = 2x + 6$

108. $(z - 2) + 0 = z - 2$

109. $1 \cdot (1 + x) = 1 + x$

110. $x + (y + 10) = (x + y) + 10$

111. $x(3y) = (x \cdot 3)y = (3x)y$

112. $\frac{1}{7}(7 \cdot 12) = \left(\frac{1}{7} \cdot 7\right)12 = 1 \cdot 12 = 12$

In Exercises 113–120, perform the operation(s). (Write fractional answers in simplest form.)

113. $\frac{3}{16} + \frac{5}{16}$

114. $\frac{6}{7} - \frac{4}{7}$

115. $\frac{5}{8} - \frac{5}{12} + \frac{1}{6}$

116. $\frac{10}{11} + \frac{6}{33} - \frac{13}{66}$

117. $12 \div \frac{1}{4}$

118. $-\left(6 \cdot \frac{4}{8}\right)$

119. $\dfrac{2x}{3} - \dfrac{x}{4}$

120. $\dfrac{5x}{6} \cdot \dfrac{2}{9}$

In Exercises 121–124, use a calculator to evaluate the expression. (Round your answer to two decimal places.)

121. $-3 + \frac{3}{7}$

122. $3\left(-\frac{5}{12} + \frac{3}{8}\right)$

123. $\dfrac{11.46 - 5.37}{3.91}$

124. $\dfrac{\frac{1}{5}(-8 - 9)}{-\frac{1}{3}}$

125. (a) Use a calculator to complete the table.

n	1	0.5	0.01	0.0001	0.000001
$5/n$					

(b) Use the result from part (a) to make a conjecture about the value of $5/n$ as n approaches 0.

126. (a) Use a calculator to complete the table.

n	1	10	100	10,000	100,000
$5/n$					

(b) Use the result from part (a) to make a conjecture about the value of $5/n$ as n increases without bound.

Synthesis

127. *Exploration* Consider $|u + v|$ and $|u| + |v|$.

(a) Are the values of the expressions always equal? If not, under what conditions are they unequal?

(b) If the two expressions are not equal for certain values of u and v, is one of the expressions always greater than the other? Explain.

128. *Think About It* Is there a difference between saying that a real number is positive and saying that a real number is nonnegative? Explain.

129. *Think About It* Because every even number is divisible by 2, is it possible that there exist any even prime numbers? Explain.

True or False? **In Exercises 130 and 131, determine whether the statement is true or false. Justify your answer.**

130. If $a < b$, then $\dfrac{1}{a} < \dfrac{1}{b}$, where $a \neq b \neq 0$.

131. Because $\dfrac{a+b}{c} = \dfrac{a}{c} + \dfrac{b}{c}$, then $\dfrac{c}{a+b} = \dfrac{c}{a} + \dfrac{c}{b}$.

132. *Writing* Describe the differences among the sets of natural numbers, integers, rational numbers, and irrational numbers.

133. *Writing* You may hear it said that to take the absolute value of a real number you simply remove any negative sign and make the number positive. Can it ever be true that $|a| = -a$ for a real number a? Explain.

P.2 Exponents and Radicals

▶ What you should learn

- How to use properties of exponents
- How to use scientific notation to represent real numbers
- How to use properties of radicals
- How to simplify and combine radicals
- How to rationalize denominators and numerators
- How to use properties of rational exponents

▶ Why you should learn it

Real numbers and algebraic expressions are often written with exponents and radicals. For instance, in Exercise 122 on page 23, you will use an expression involving a radical to find the size of a particle that can be carried by a stream moving at a certain velocity.

Marc Muench/Tony Stone Images

Exponents

Repeated *multiplication* can be written in **exponential form.**

Repeated Multiplication	*Exponential Form*
$a \cdot a \cdot a \cdot a \cdot a$	a^5
$(-4)(-4)(-4)$	$(-4)^3$
$(2x)(2x)(2x)(2x)$	$(2x)^4$

In general, if a is a real number and n is a positive integer, then

$$a^n = \underbrace{a \cdot a \cdot a \cdots a}_{n \text{ factors}}$$

where n is the **exponent** and a is the **base.** The expression a^n is read "a to the nth **power.**" In Property 3 below, be sure you see how to use a negative exponent.

Properties of Exponents

Let a and b be real numbers, variables, or algebraic expressions, and let m and n be integers. (All denominators and bases are nonzero.)

Property	*Example*
1. $a^m a^n = a^{m+n}$	$3^2 \cdot 3^4 = 3^{2+4} = 3^6 = 729$
2. $\dfrac{a^m}{a^n} = a^{m-n}$	$\dfrac{x^7}{x^4} = x^{7-4} = x^3$
3. $a^{-n} = \dfrac{1}{a^n} = \left(\dfrac{1}{a}\right)^n$	$y^{-4} = \dfrac{1}{y^4} = \left(\dfrac{1}{y}\right)^4$
4. $a^0 = 1, \quad a \neq 0$	$(x^2 + 1)^0 = 1$
5. $(ab)^m = a^m b^m$	$(5x)^3 = 5^3x^3 = 125x^3$
6. $(a^m)^n = a^{mn}$	$(y^3)^{-4} = y^{3(-4)} = y^{-12} = \dfrac{1}{y^{12}}$
7. $\left(\dfrac{a}{b}\right)^m = \dfrac{a^m}{b^m}$	$\left(\dfrac{2}{x}\right)^3 = \dfrac{2^3}{x^3} = \dfrac{8}{x^3}$
8. $\lvert a^2 \rvert = \lvert a \rvert^2 = a^2$	$\lvert(-2)^2\rvert = \lvert -2 \rvert^2 = (-2)^2 = 4$

It is important to recognize the difference between expressions such as $(-2)^4$ and -2^4. In $(-2)^4$, the parentheses indicate that the exponent applies to the negative sign as well as to the 2, but in $-2^4 = -(2^4)$, the exponent applies only to the 2. So,

$$(-2)^4 = 16 \quad \text{and} \quad -2^4 = -16.$$

STUDY TIP

Rarely in algebra is there only one way to solve a problem. Don't be concerned if the steps you use to solve a problem are not exactly the same as the steps presented in this text. The important thing is to use steps that you understand and, of course, steps that are justified by the rules of algebra. For instance, you might prefer the following steps for Example 2(d).

$$\left(\frac{3x^2}{y}\right)^{-2} = \left(\frac{y}{3x^2}\right)^2 = \frac{y^2}{9x^4}$$

Note how Property 3 is used in the first step of this solution. The fractional form of this property is

$$\left(\frac{a}{b}\right)^{-m} = \left(\frac{b}{a}\right)^m.$$

Additional Examples

a. $4x^{-1} = \frac{4}{x}$

b. $\frac{3}{2x^{-2}} = \frac{3x^2}{2}$

c. $\left(\frac{2}{5}\right)^{-1} = \frac{5}{2}$

Technology

You can use a calculator to evaluate expressions with exponents. For instance, evaluate -3^{-2} as follows.

Scientific:

3 [+/−] [y^x] 2 [+/−] [=]

Graphing:

[(−)] 3 [^] [(−)] 2

The display will be as follows.

−.1111111111

The properties of exponents listed on the preceding page apply to *all* integers m and n, not just to positive integers. For instance, by Property 2, you can write

$$\frac{3^4}{3^{-5}} = 3^{4-(-5)} = 3^{4+5} = 3^9.$$

Example 1 ▶ Using Properties of Exponents

Use the properties of exponents to simplify each expression.

a. $(-3ab^4)(4ab^{-3})$ **b.** $(2xy^2)^3$ **c.** $3a(-4a^2)^0$ **d.** $\left(\frac{5x^3}{y}\right)^2$

Solution

a. $(-3ab^4)(4ab^{-3}) = -12(a)(a)(b^4)(b^{-3}) = -12a^2b$

b. $(2xy^2)^3 = 2^3(x)^3(y^2)^3 = 8x^3y^6$

c. $3a(-4a^2)^0 = 3a(1) = 3a, \quad a \neq 0$

d. $\left(\frac{5x^3}{y}\right)^2 = \frac{5^2(x^3)^2}{y^2} = \frac{25x^6}{y^2}$

Example 2 ▶ Rewriting with Positive Exponents

Rewrite each expression with positive exponents.

a. x^{-1} **b.** $\frac{1}{3x^{-2}}$ **c.** $\frac{12a^3b^{-4}}{4a^{-2}b}$ **d.** $\left(\frac{3x^2}{y}\right)^{-2}$

Solution

a. $x^{-1} = \frac{1}{x}$ — Property 3

b. $\frac{1}{3x^{-2}} = \frac{1(x^2)}{3} = \frac{x^2}{3}$ — The exponent −2 does not apply to 3.

c. $\frac{12a^3b^{-4}}{4a^{-2}b} = \frac{12a^3 \cdot a^2}{4b \cdot b^4}$ — Property 3

$= \frac{3a^5}{b^5}$ — Property 1

d. $\left(\frac{3x^2}{y}\right)^{-2} = \frac{3^{-2}(x^2)^{-2}}{y^{-2}}$ — Properties 5 and 7

$= \frac{3^{-2}x^{-4}}{y^{-2}}$ — Property 6

$= \frac{y^2}{3^2x^4}$ — Property 3

$= \frac{y^2}{9x^4}$ — Simplify.

Historical Note

The French mathematician Nicolas Chuquet (ca. 1500) wrote *Triparty en la science des nombres*, in which a form of exponent notation was used. Our expressions $6x^3$ and $10x^2$ were written as $.6.^3$ and $.10.^2$. Zero and negative exponents were also represented, so x^0 would be written as $.1.^0$ and $3x^{-2}$ as $.3.^{2m}$. Chuquet wrote that $.72.^1$ divided by $.8.^3$ is $.9.^{2m}$. That is, $72x \div 8x^3 = 9x^{-2}$.

Activities

1. Rewrite using the Distributive Property: $3(x - y)$.
   Answer: $3x - 3y$
2. Simplify: $\left(\frac{6x^{-1}y}{y^3}\right)^{-2}$
   Answer: $\frac{x^2y^4}{36}$
3. Write in scientific notation: 39,000,000.
   Answer: 3.9×10^7

The *Interactive* CD-ROM and *Internet* versions of this text offer a built-in graphing calculator, which can be used in the Examples, Explorations, Technology notes, and Exercises.

Scientific Notation

Exponents provide an efficient way of writing and computing with very large (or very small) numbers. For instance, there are about 326 billion billion gallons of water on earth—that is, 326 followed by 18 zeros.

$$326{,}000{,}000{,}000{,}000{,}000{,}000$$

It is convenient to write such numbers in **scientific notation.** This notation has the form $\pm c \times 10^n$, where $1 \le c < 10$ and n is an integer. So, the number of gallons of water on earth can be written in scientific notation as

$$3.26 \times 100{,}000{,}000{,}000{,}000{,}000{,}000 = 3.26 \times 10^{20}.$$

The *positive* exponent 20 indicates that the number is *large* (10 or more) and that the decimal point has been moved 20 places. A *negative* exponent indicates that the number is *small* (less than 1). For instance, the mass (in grams) of one electron is approximately

$$9.0 \times 10^{-28} = 0.0000000000000000000000000009.$$

28 decimal places

Example 3 ▶ Scientific Notation

a. $1.345 \times 10^2 = 134.5$

b. $0.0000782 = 7.82 \times 10^{-5}$

c. $9.36 \times 10^{-6} = 0.00000936$

d. $836{,}100{,}000 = 8.361 \times 10^8$

Most calculators switch to scientific notation when they are showing large (or small) numbers that exceed the display range. Try evaluating $86{,}500{,}000 \times 6000$. If your calculator follows standard conventions, its display should be

[5.19 11] or [5.19 E 11]

which is 5.19×10^{11}.

Example 4 ▶ Using Scientific Notation with a Calculator

Use a calculator to evaluate $65{,}000 \times 3{,}400{,}000{,}000$.

Solution

Because $65{,}000 = 6.5 \times 10^4$ and $3{,}400{,}000{,}000 = 3.4 \times 10^9$, you can multiply the two numbers using the following graphing calculator steps.

6.5 [EE] 4 [×] 3.4 [EE] 9 [ENTER]

After entering these keystrokes, the calculator display should read [2.21 E 14]. Therefore, the product of the two numbers is

$$(6.5 \times 10^4)(3.4 \times 10^9) = 2.21 \times 10^{14}$$
$$= 221{,}000{,}000{,}000{,}000.$$

Use pattern recognition to help students identify perfect squares, cubes, etc., of both positive and negative integers when simplifying radicals. Have students construct a table of powers for several integers. For example:

n	n^2	n^3	n^4
-3	9	-27	81
-2	4	-8	16
-1	1	-1	1
0	0	0	0
1	1	1	1
2	4	8	16
3	9	27	81

Radicals and Their Properties

A **square root** of a number is one of its two equal factors. For example, 5 is a square root of 25 because 5 is one of the two equal factors of 25. In a similar way, a **cube root** of a number is one of its three equal factors, as in $125 = 5^3$.

Definition of *n*th Root of a Number

Let a and b be real numbers and let $n \geq 2$ be a positive integer. If

$$a = b^n$$

then b is an ***n*th root of *a*.** If $n = 2$, the root is a **square root.** If $n = 3$, the root is a **cube root.**

Some numbers have more than one nth root. For example, both 5 and -5 are square roots of 25. The *principal square root* of 25, written as $\sqrt{25}$, is the positive root, 5. The **principal *n*th root** of a number is defined as follows.

Principal *n*th Root of a Number

Let a be a real number that has at least one nth root. The **principal *n*th root of *a*** is the nth root that has the same sign as a. It is denoted by a **radical symbol**

$$\sqrt[n]{a}. \qquad \text{Principal } n\text{th root}$$

The positive integer n is the **index** of the radical, and the number a is the **radicand.** If $n = 2$, we omit the index and write $\sqrt{a}$ rather than $\sqrt[2]{a}$. (The plural of index is *indices.*)

A common misunderstanding is that the square root sign implies both negative and positive roots. This is not correct. The square root sign implies only a positive root. When a negative root is needed, you must use the negative sign with the square root sign.

Incorrect: ~~$\sqrt{4} = \pm 2$~~ *Correct:* $-\sqrt{4} = -2$ *and* $\sqrt{4} = 2$

Example 5 ▶ Evaluating Expressions Involving Radicals

a. $\sqrt{36} = 6$ because $6^2 = 36$.

b. $-\sqrt{36} = -6$ because $6^2 = 36$.

c. $\sqrt[3]{\dfrac{125}{64}} = \dfrac{5}{4}$ because $\left(\dfrac{5}{4}\right)^3 = \dfrac{5^3}{4^3} = \dfrac{125}{64}$.

d. $\sqrt[5]{-32} = -2$ because $(-2)^5 = -32$.

e. $\sqrt[4]{-81}$ is not a real number because there is no real number that can be raised to the fourth power to produce -81.

Here are some generalizations about the nth roots of a real number.

Generalizations About nth Roots of Real Numbers

Real number a	Integer n	Root(s) of a	Example
$a > 0$	$n > 0$, n is even.	$\sqrt[n]{a},\ -\sqrt[n]{a}$	$\sqrt[4]{81} = 3,\ -\sqrt[4]{81} = -3$
$a > 0$ or $a < 0$	n is odd.	$\sqrt[n]{a}$	$\sqrt[3]{-8} = -2$
$a < 0$	n is even.	No real roots	$\sqrt{-4}$ is not real.
$a = 0$	n is even or odd.	$\sqrt[n]{0} = 0$	$\sqrt[5]{0} = 0$

Additional Examples

a. $\sqrt{50} = \sqrt{25 \cdot 2} = 5\sqrt{2}$

b. $\sqrt{\dfrac{3}{4}} = \dfrac{\sqrt{3}}{\sqrt{4}} = \dfrac{\sqrt{3}}{2}$

c. $(\sqrt{10})^2 = 10$

Integers such as 1, 4, 9, 16, 25, and 36 are called **perfect squares** because they have integer square roots. Similarly, integers such as 1, 8, 27, 64, and 125 are called **perfect cubes** because they have integer cube roots.

Properties of Radicals

Let a and b be real numbers, variables, or algebraic expressions such that the indicated roots are real numbers, and let m and n be positive integers.

Property		*Example*
1. $\sqrt[n]{a^m} = (\sqrt[n]{a})^m$		$\sqrt[3]{8^2} = (\sqrt[3]{8})^2 = (2)^2 = 4$
2. $\sqrt[n]{a} \cdot \sqrt[n]{b} = \sqrt[n]{ab}$		$\sqrt{5} \cdot \sqrt{7} = \sqrt{5 \cdot 7} = \sqrt{35}$
3. $\dfrac{\sqrt[n]{a}}{\sqrt[n]{b}} = \sqrt[n]{\dfrac{a}{b}}$,	$b \neq 0$	$\dfrac{\sqrt[4]{27}}{\sqrt[4]{9}} = \sqrt[4]{\dfrac{27}{9}} = \sqrt[4]{3}$
4. $\sqrt[m]{\sqrt[n]{a}} = \sqrt[mn]{a}$		$\sqrt[3]{\sqrt{10}} = \sqrt[6]{10}$
5. $(\sqrt[n]{a})^n = a$		$(\sqrt{3})^2 = 3$
6. For n even, $\sqrt[n]{a^n} = \lvert a \rvert$.		$\sqrt{(-12)^2} = \lvert -12 \rvert = 12$
For n odd, $\sqrt[n]{a^n} = a$.		$\sqrt[3]{(-12)^3} = -12$

A common special case of Property 6 is $\sqrt{a^2} = \lvert a \rvert$.

Example 6 ▶ Using Properties of Radicals

Use the properties of radicals to simplify each expression.

a. $\sqrt{8} \cdot \sqrt{2}$ **b.** $(\sqrt[3]{5})^3$ **c.** $\sqrt[3]{x^3}$ **d.** $\sqrt[6]{y^6}$

Solution

a. $\sqrt{8} \cdot \sqrt{2} = \sqrt{8 \cdot 2} = \sqrt{16} = 4$

b. $(\sqrt[3]{5})^3 = 5$

c. $\sqrt[3]{x^3} = x$

d. $\sqrt[6]{y^6} = \lvert y \rvert$

Simplifying Radicals

An expression involving radicals is in **simplest form** when the following conditions are satisfied.

1. All possible factors have been removed from the radical.
2. All fractions have radical-free denominators (accomplished by a process called *rationalizing the denominator*).
3. The index of the radical is reduced.

To simplify a radical, factor the radicand into factors whose exponents are multiples of the index. The roots of these factors are written outside the radical, and the "leftover" factors make up the new radicand.

STUDY TIP

When you simplify a radical, it is important that both expressions are defined for the same values of the variable. For instance, in Example 7(b), $\sqrt{75x^3}$ and $5x\sqrt{3x}$ are both defined only for nonnegative values of x. Similarly, in Example 7(c), $\sqrt[4]{(5x)^4}$ and $5|x|$ are both defined for all real values of x.

Example 7 ▶ Simplifying Even Roots

Perfect 4th power ↓ Leftover factor ↓

a. $\sqrt[4]{48} = \sqrt[4]{16 \cdot 3} = \sqrt[4]{2^4 \cdot 3} = 2\sqrt[4]{3}$

Perfect square ↓ Leftover factor ↓

b. $\sqrt{75x^3} = \sqrt{25x^2 \cdot 3x}$ Find largest square factor.

$= \sqrt{(5x)^2 \cdot 3x}$

$= 5x\sqrt{3x}$ Find root of perfect square.

c. $\sqrt[4]{(5x)^4} = |5x| = 5|x|$

In Example 7(b), the expression $\sqrt{75x^3}$ makes sense only for nonnegative values of x.

Example 8 ▶ Simplifying Odd Roots

Perfect cube ↓ Leftover factor ↓

a. $\sqrt[3]{24} = \sqrt[3]{8 \cdot 3} = \sqrt[3]{2^3 \cdot 3} = 2\sqrt[3]{3}$

Perfect cube ↓ Leftover factor ↓

b. $\sqrt[3]{24a^4} = \sqrt[3]{8a^3 \cdot 3a}$ Find largest cube factor.

$= \sqrt[3]{(2a)^3 \cdot 3a}$

$= 2a\sqrt[3]{3a}$ Find root of perfect cube.

c. $\sqrt[3]{-40x^6} = \sqrt[3]{(-8x^6) \cdot 5}$ Find largest cube factor.

$= \sqrt[3]{(-2x^2)^3 \cdot 5}$

$= -2x^2\sqrt[3]{5}$ Find root of perfect cube.

Radical expressions can be combined (added or subtracted) if they are **like radicals**—that is, if they have the same index and radicand. For instance, $\sqrt{2}$, $3\sqrt{2}$, and $\frac{1}{2}\sqrt{2}$ are like radicals, but $\sqrt{3}$ and $\sqrt{2}$ are unlike radicals. To determine whether two radicals can be combined, you should first simplify each radical.

Example 9 ▶ Combining Radicals

a. $2\sqrt{48} - 3\sqrt{27} = 2\sqrt{16 \cdot 3} - 3\sqrt{9 \cdot 3}$ Find square factors.

$= 8\sqrt{3} - 9\sqrt{3}$ Find square roots.

$= (8 - 9)\sqrt{3}$ Combine like terms.

$= -\sqrt{3}$

b. $\sqrt[3]{16x} - \sqrt[3]{54x^4} = \sqrt[3]{8 \cdot 2x} - \sqrt[3]{27 \cdot x^3 \cdot 2x}$ Find cube factors.

$= 2\sqrt[3]{2x} - 3x\sqrt[3]{2x}$ Find cube roots.

$= (2 - 3x)\sqrt[3]{2x}$ Combine like terms.

Rationalizing Denominators and Numerators

To rationalize a denominator or numerator of the form $a - b\sqrt{m}$ or $a + b\sqrt{m}$, multiply both numerator and denominator by a **conjugate:** $a + b\sqrt{m}$ and $a - b\sqrt{m}$ are conjugates of each other. If $a = 0$, then the rationalizing factor for $\sqrt{m}$ is itself, $\sqrt{m}$. For cube roots, choose a rationalizing factor that generates a perfect cube.

Example 10 ▶ Rationalizing Single-Term Denominators

Rationalize the denominator of each expression.

a. $\dfrac{5}{2\sqrt{3}}$ **b.** $\dfrac{2}{\sqrt[3]{5}}$

Solution

a. $\dfrac{5}{2\sqrt{3}} = \dfrac{5}{2\sqrt{3}} \cdot \dfrac{\sqrt{3}}{\sqrt{3}}$ $\sqrt{3}$ is rationalizing factor.

$= \dfrac{5\sqrt{3}}{2(3)}$

$= \dfrac{5\sqrt{3}}{6}$

b. $\dfrac{2}{\sqrt[3]{5}} = \dfrac{2}{\sqrt[3]{5}} \cdot \dfrac{\sqrt[3]{5^2}}{\sqrt[3]{5^2}}$ $\sqrt[3]{5^2}$ is rationalizing factor.

$= \dfrac{2\sqrt[3]{5^2}}{\sqrt[3]{5^3}}$

$= \dfrac{2\sqrt[3]{25}}{5}$

Additional Examples

a. $\dfrac{3}{\sqrt{7}} = \dfrac{3}{\sqrt{7}} \cdot \dfrac{\sqrt{7}}{\sqrt{7}} = \dfrac{3\sqrt{7}}{\sqrt{7}}$

b. $\dfrac{2}{\sqrt[3]{4}} = \dfrac{2}{\sqrt[3]{4}} \cdot \dfrac{\sqrt[3]{2}}{\sqrt[3]{2}}$

$= \dfrac{2\sqrt[3]{2}}{2}$

$= \sqrt[3]{2}$

c. $\dfrac{6}{\sqrt{2} + \sqrt{3}}$

$= \dfrac{6}{\sqrt{2} + \sqrt{3}} \cdot \dfrac{\sqrt{2} - \sqrt{3}}{\sqrt{2} - \sqrt{3}}$

$= \dfrac{6(\sqrt{2} - \sqrt{3})}{-1}$

$= -6\sqrt{2} + 6\sqrt{3}$

Example 11 ▶ Rationalizing a Denominator with Two Terms

$$\frac{2}{3 + \sqrt{7}} = \frac{2}{3 + \sqrt{7}} \cdot \frac{3 - \sqrt{7}}{3 - \sqrt{7}}$$ Multiply numerator and denominator by conjugate of denominator.

$$= \frac{2(3 - \sqrt{7})}{3(3) + 3(-\sqrt{7}) + \sqrt{7}(3) - (\sqrt{7})(\sqrt{7})}$$ Use Distributive Property.

$$= \frac{2(3 - \sqrt{7})}{(3)^2 - (\sqrt{7})^2}$$ Simplify.

$$= \frac{2(3 - \sqrt{7})}{9 - 7}$$ Square terms of denominator.

$$= \frac{2(3 - \sqrt{7})}{2} = 3 - \sqrt{7}$$ Simplify.

Sometimes it is necessary to rationalize the numerator of an expression. For instance, in Section P.4 you will use the technique shown in the next example to rationalize the numerator of an expression from calculus.

STUDY TIP

Do not confuse the expression $\sqrt{5} + \sqrt{7}$ with the expression $\sqrt{5 + 7}$. In general, $\sqrt{x + y}$ does not equal $\sqrt{x} + \sqrt{y}$. Similarly, $\sqrt{x^2 + y^2}$ does not equal $x + y$.

Example 12 ▶ Rationalizing a Numerator

$$\frac{\sqrt{5} - \sqrt{7}}{2} = \frac{\sqrt{5} - \sqrt{7}}{2} \cdot \frac{\sqrt{5} + \sqrt{7}}{\sqrt{5} + \sqrt{7}}$$ Multiply numerator and denominator by conjugate of numerator.

$$= \frac{(\sqrt{5})^2 - (\sqrt{7})^2}{2(\sqrt{5} + \sqrt{7})}$$ Simplify.

$$= \frac{5 - 7}{2(\sqrt{5} + \sqrt{7})}$$ Square terms of numerator.

$$= \frac{-2}{2(\sqrt{5} + \sqrt{7})} = \frac{-1}{\sqrt{5} + \sqrt{7}}$$ Simplify.

Rationalizing the numerator is especially useful when finding limits in calculus.

Rational Exponents

Definition of Rational Exponents

If a is a real number and n is a positive integer such that the principal nth root of a exists, then $a^{1/n}$ is defined as

$a^{1/n} = \sqrt[n]{a}$, where $1/n$ is the **rational exponent** of a.

Moreover, if m is a positive integer that has no common factor with n, then

$$a^{m/n} = (a^{1/n})^m = (\sqrt[n]{a})^m \quad \text{and} \quad a^{m/n} = (a^m)^{1/n} = \sqrt[n]{a^m}.$$

The symbol indicates an example or exercise that highlights algebraic techniques specifically used in calculus.

STUDY T!P

Rational exponents can be tricky, and you must remember that the expression $b^{m/n}$ is not defined unless $\sqrt[n]{b}$ is a real number. This restriction produces some unusual-looking results. For instance, the number $(-8)^{1/3}$ is defined because $\sqrt[3]{-8} = -2$ but the number $(-8)^{2/6}$ is undefined because $\sqrt[6]{-8}$ is not a real number.

The *Interactive* CD-ROM and *Internet* versions of this text show every example with its solution; clicking on the *Try It!* button brings up similar problems. Guided Examples and Integrated Examples show step-by-step solutions to additional examples. Integrated Examples are related to several concepts in the section.

Technology

There are four methods of evaluating radicals on most graphing calculators. For square roots, you can use the square root key $\boxed{\sqrt{\ }}$. For cube roots, you can use the cube root key $\boxed{\sqrt[3]{\ }}$. For other roots, you can first convert the radical to exponential form and then use the exponential key $\boxed{\wedge}$, or you can use the *n*th root key $\boxed{\sqrt[x]{\ }}$.

Additional Examples

a. $\dfrac{9^{2/3}}{9^{1/6}} = 9^{(2/3)-(1/6)}$
$= 9^{1/2}$
$= 3$

b. $\sqrt[3]{6} + \sqrt[3]{48} = \sqrt[3]{6} + \sqrt[3]{8 \cdot 6}$
$= \sqrt[3]{6} + 2\sqrt[3]{6}$
$= 3\sqrt[3]{6}$

The numerator of a rational exponent denotes the *power* to which the base is raised, and the denominator denotes the *index* or the *root* to be taken.

Power (numerator m), Index (denominator n):

$$b^{m/n} = \left(\sqrt[n]{b}\right)^m = \sqrt[n]{b^m}$$

When you are working with rational exponents, the properties of integer exponents still apply. For instance,

$$2^{1/2}2^{1/3} = 2^{(1/2)+(1/3)} = 2^{5/6}.$$

Example 13 ▶ Changing from Radical to Exponential Form

a. $\sqrt{3} = 3^{1/2}$ **b.** $\sqrt{(3xy)^5} = \sqrt[2]{(3xy)^5} = (3xy)^{(5/2)}$

c. $2x\sqrt[4]{x^3} = (2x)(x^{3/4}) = 2x^{1+(3/4)} = 2x^{7/4}$

Example 14 ▶ Changing from Exponential to Radical Form

a. $(x^2 + y^2)^{3/2} = \left(\sqrt{x^2 + y^2}\right)^3 = \sqrt{(x^2 + y^2)^3}$

b. $2y^{3/4}z^{1/4} = 2(y^3z)^{1/4} = 2\sqrt[4]{y^3z}$

c. $a^{-3/2} = \dfrac{1}{a^{3/2}} = \dfrac{1}{\sqrt{a^3}}$ **d.** $x^{0.2} = x^{1/5} = \sqrt[5]{x}$

Rational exponents are useful for evaluating roots of numbers on a calculator, for reducing the index of a radical, and for simplifying calculus expressions.

Example 15 ▶ Simplifying with Rational Exponents

a. $(-32)^{-4/5} = \left(\sqrt[5]{-32}\right)^{-4} = (-2)^{-4} = \dfrac{1}{(-2)^4} = \dfrac{1}{16}$

b. $(-5x^{5/3})(3x^{-3/4}) = -15x^{(5/3)-(3/4)} = -15x^{11/12}, \quad x \neq 0$

c. $\sqrt[9]{a^3} = a^{3/9} = a^{1/3} = \sqrt[3]{a}$ Reduce index.

d. $\sqrt[3]{\sqrt{125}} = \sqrt[6]{125} = \sqrt[6]{(5)^3} = 5^{3/6} = 5^{1/2} = \sqrt{5}$

e. $(2x - 1)^{4/3}(2x - 1)^{-1/3} = (2x - 1)^{(4/3)-(1/3)}$

$= 2x - 1, \quad x \neq \dfrac{1}{2}$

f. $\dfrac{x - 1}{(x - 1)^{-1/2}} = \dfrac{x - 1}{(x - 1)^{-1/2}} \cdot \dfrac{(x - 1)^{1/2}}{(x - 1)^{1/2}}$

$= \dfrac{(x - 1)^{3/2}}{(x - 1)^0}$

$= (x - 1)^{3/2}, \quad x \neq 1$

P.2 Exercises

In Exercises 1–4, write the expression as a repeated multiplication problem.

1. 8^5
2. $(-2)^7$
3. -0.4^6
4. 11.3^4

In Exercises 5–8, write the expression using exponential notation.

5. $(4.9)(4.9)(4.9)(4.9)(4.9)(4.9)$
6. $(2\sqrt{5})(2\sqrt{5})(2\sqrt{5})(2\sqrt{5})$
7. $(-10)(-10)(-10)(-10)(-10)$
8. $-\left(\frac{3}{2} \times \frac{3}{2} \times \frac{3}{2} \times \frac{3}{2}\right)$

In Exercises 9–16, evaluate each expression.

9. (a) $3^2 \cdot 3$ (b) $3 \cdot 3^3$
10. (a) $\dfrac{5^5}{5^2}$ (b) $\dfrac{3^2}{3^4}$
11. (a) $(3^3)^2$ (b) -3^2
12. (a) $(2^3 \cdot 3^2)^2$ (b) $\left(-\frac{3}{5}\right)^3\left(\frac{5}{3}\right)^2$
13. (a) $\dfrac{3 \cdot 4^{-4}}{3^{-4} \cdot 4^{-1}}$ (b) $32(-2)^{-5}$
14. (a) $\dfrac{4 \cdot 3^{-2}}{2^{-2} \cdot 3^{-1}}$ (b) $(-2)^0$
15. (a) $2^{-1} + 3^{-1}$ (b) $(2^{-1})^{-2}$
16. (a) $3^{-1} + 2^{-2}$ (b) $(3^{-2})^2$

In Exercises 17–20, use a calculator to evaluate the expression. (If necessary, round your answer to three decimal places.)

17. $(-4)^3(5^2)$
18. $(8^{-4})(10^3)$
19. $\dfrac{3^6}{7^3}$
20. $\dfrac{4^3}{3^{-4}}$

In Exercises 21–28, evaluate the expression for the value of x.

	Expression	*Value*
21.	$-3x^3$	2
22.	$7x^{-2}$	4
23.	$6x^0$	10
24.	$5(-x)^3$	3
25.	$2x^3$	-3
26.	$-3x^4$	-2
27.	$4x^2$	$-\frac{1}{2}$
28.	$5(-x)^3$	$\frac{1}{3}$

In Exercises 29–34, simplify each expression.

29. (a) $(-5z)^3$ (b) $5x^4(x^2)$
30. (a) $(3x)^2$ (b) $(4x^3)^2$
31. (a) $6y^2(2y^4)^2$ (b) $\dfrac{3x^5}{x^3}$
32. (a) $(-z)^3(3z^4)$ (b) $\dfrac{25y^8}{10y^4}$
33. (a) $\dfrac{7x^2}{x^3}$ (b) $\dfrac{12(x+y)^3}{9(x+y)}$
34. (a) $\dfrac{r^4}{r^6}$ (b) $\left(\dfrac{4}{y}\right)^3\left(\dfrac{3}{y}\right)^4$

In Exercises 35–42, rewrite each expression with positive exponents and simplify.

35. (a) $(x+5)^0, \quad x \neq -5$ (b) $(2x^2)^{-2}$
36. (a) $(2x^5)^0, \quad x \neq 0$ (b) $(z+2)^{-3}(z+2)^{-1}$
37. (a) $(-2x^2)^3(4x^3)^{-1}$ (b) $\left(\dfrac{x}{10}\right)^{-1}$
38. (a) $(4y^{-2})(8y^4)$ (b) $\left(\dfrac{x^{-3}y^4}{5}\right)^{-3}$
39. (a) $(4a^{-2}b^3)^{-3}$ (b) $\left(\dfrac{5x^2}{y^{-2}}\right)^{-4}$
40. (a) $[(x^2y^{-2})^{-1}]^{-1}$ (b) $(5x^2z^6)^3(5x^2z^6)^{-3}$
41. (a) $3^n \cdot 3^{2n}$ (b) $\left(\dfrac{a^{-2}}{b^{-2}}\right)\left(\dfrac{b}{a}\right)^3$
42. (a) $\dfrac{x^2 \cdot x^n}{x^3 \cdot x^n}$ (b) $\left(\dfrac{a^{-3}}{b^{-3}}\right)\left(\dfrac{a}{b}\right)^3$

In Exercises 43–54, fill in the missing form of the equation.

	Radical Form	*Rational Exponent Form*
43.	$\sqrt{9} = 3$	
44.	$\sqrt[3]{64} = 4$	
45.		$32^{1/5} = 2$
46.		$-(144^{1/2}) = -12$

	Radical Form	*Rational Exponent Form*
47.		$196^{1/2} = 14$
48.	$\sqrt[3]{614.125} = 8.5$	
49.	$\sqrt[3]{-216} = -6$	
50.		$(-243)^{1/5} = -3$
51.		$27^{2/3} = 9$
52.	$(\sqrt[4]{81})^3 = 27$	
53.	$\sqrt[4]{81^3} = 27$	
54.		$16^{5/4} = 32$

In Exercises 55–64, evaluate each expression without using a calculator.

55. (a) $\sqrt{9}$ (b) $\sqrt[3]{8}$

56. (a) $\sqrt{49}$ (b) $\sqrt[3]{\frac{27}{8}}$

57. (a) $-\sqrt[3]{-27}$ (b) $\dfrac{4}{\sqrt{64}}$

58. (a) $\sqrt[3]{0}$ (b) $\dfrac{\sqrt[4]{81}}{3}$

59. (a) $(\sqrt[3]{-125})^3$ (b) $27^{1/3}$

60. (a) $\sqrt[4]{562^4}$ (b) $36^{3/2}$

61. (a) $32^{-3/5}$ (b) $\left(\frac{16}{81}\right)^{-3/4}$

62. (a) $100^{-3/2}$ (b) $\left(\frac{9}{4}\right)^{-1/2}$

63. (a) $\left(-\dfrac{1}{64}\right)^{-1/3}$ (b) $\left(\dfrac{1}{\sqrt{32}}\right)^{-2/5}$

64. (a) $\left(-\dfrac{125}{27}\right)^{-1/3}$ (b) $-\left(\dfrac{1}{125}\right)^{-4/3}$

In Exercises 65–70, use a calculator to approximate the number. (Round your answer to three decimal places.)

65. (a) $\sqrt{57}$ (b) $\sqrt[5]{-27^3}$

66. (a) $\sqrt[3]{45^2}$ (b) $\sqrt[6]{125}$

67. (a) $(1.2^{-2})\sqrt{75} + 3\sqrt{8}$ (b) $\dfrac{-3 + \sqrt{21}}{3}$

68. (a) $(15.25)^{-1.4}$ (b) $(3.4)^{2.5}$

69. (a) $(-12.4)^{-1.8}$ (b) $(5\sqrt{3})^{-2.5}$

70. (a) $\dfrac{7 - (4.1)^{-3.2}}{2}$ (b) $\left(\dfrac{13}{3}\right)^{-3/2} - \left(-\dfrac{3}{2}\right)^{13/3}$

In Exercises 71–76, simplify by removing all possible factors from each radical.

71. (a) $\sqrt{8}$ (b) $\sqrt[3]{24}$

72. (a) $\sqrt[3]{\frac{16}{27}}$ (b) $\sqrt{\frac{75}{4}}$

73. (a) $\sqrt{72x^3}$ (b) $\sqrt{\dfrac{18^2}{z^3}}$

74. (a) $\sqrt{54xy^4}$ (b) $\sqrt{\dfrac{32a^4}{b^2}}$

75. (a) $\sqrt[3]{16x^5}$ (b) $\sqrt{75x^2y^{-4}}$

76. (a) $\sqrt[4]{(3x^2)^4}$ (b) $\sqrt[5]{96x^5}$

In Exercises 77–82, perform the operations and simplify.

77. $5^{4/3} \cdot 5^{8/3}$ **78.** $\dfrac{8^{12/5}}{8^{2/5}}$

79. $\dfrac{(2x^2)^{3/2}}{2^{1/2}x^4}$ **80.** $\dfrac{x^{4/3}y^{2/3}}{(xy)^{1/3}}$

81. $\dfrac{x^{-3} \cdot x^{1/2}}{x^{3/2} \cdot x^{-1}}$ **82.** $\dfrac{5^{-1/2} \cdot 5x^{5/2}}{(5x)^{3/2}}$

In Exercises 83–86, rationalize the denominator of each expression. Then simplify your answer.

83. (a) $\dfrac{1}{\sqrt{3}}$ (b) $\dfrac{8}{\sqrt[3]{2}}$

84. (a) $\dfrac{5}{\sqrt{10}}$ (b) $\dfrac{5}{\sqrt[3]{(5x)^2}}$

85. (a) $\dfrac{2x}{5 - \sqrt{3}}$ (b) $\dfrac{3}{\sqrt{5} + \sqrt{6}}$

86. (a) $\dfrac{5}{\sqrt{14} - 2}$ (b) $\dfrac{5}{\sqrt{10} - 5}$

In Exercises 87–90, rationalize the numerator of each expression. Then simplify your answer.

87. (a) $\dfrac{\sqrt{8}}{2}$ (b) $\sqrt[3]{\dfrac{9}{25}}$

88. (a) $\dfrac{\sqrt{2}}{3}$ (b) $\sqrt[4]{\dfrac{5}{4}}$

89. (a) $\dfrac{\sqrt{5} + \sqrt{3}}{3}$ (b) $\dfrac{\sqrt{7} - 3}{4}$

90. (a) $\dfrac{\sqrt{3} - \sqrt{2}}{2}$ (b) $\dfrac{2\sqrt{3} + \sqrt{3}}{3}$

In Exercises 91 and 92, reduce the index of each radical.

91. (a) $\sqrt[4]{3^2}$ (b) $\sqrt[6]{(x + 1)^4}$

92. (a) $\sqrt[6]{x^3}$ (b) $\sqrt[4]{(3x^2)^4}$

The symbol indicates an example or exercise that highlights algebraic techniques specifically used in calculus.

In Exercises 93 and 94, write each expression as a single radical. Then simplify your answer.

93. (a) $\sqrt{\sqrt{32}}$ (b) $\sqrt{\sqrt[4]{2x}}$

94. (a) $\sqrt{\sqrt{243(x+1)}}$ (b) $\sqrt{\sqrt[3]{10a^7b}}$

In Exercises 95–100, simplify each expression.

95. (a) $2\sqrt{50} + 12\sqrt{8}$ (b) $10\sqrt{32} - 6\sqrt{18}$

96. (a) $4\sqrt{27} - \sqrt{75}$ (b) $\sqrt[3]{16} + 3\sqrt[3]{54}$

97. (a) $5\sqrt{x} - 3\sqrt{x}$ (b) $-2\sqrt{9y} + 10\sqrt{y}$

98. (a) $8\sqrt{49x} - 14\sqrt{100x}$
(b) $-3\sqrt{48x^2} + 7\sqrt{75x^2}$

99. (a) $3\sqrt{x+1} + 10\sqrt{x+1}$
(b) $7\sqrt{80x} - 2\sqrt{125x}$

100. (a) $-\sqrt{x^3-7} + 5\sqrt{x^3-7}$
(b) $11\sqrt{245x^3} - 9\sqrt{45x^3}$

In Exercises 101–104, complete the statement with <, =, or >.

101. $\sqrt{5} + \sqrt{3}$ ☐ $\sqrt{5+3}$

102. $\sqrt{\frac{3}{11}}$ ☐ $\frac{\sqrt{3}}{\sqrt{11}}$

103. 5 ☐ $\sqrt{3^2 + 2^2}$

104. 5 ☐ $\sqrt{3^2 + 4^2}$

In Exercises 105–108, write the number in scientific notation.

105. Land area of the earth: 57,300,000 square miles

106. Light year: 9,460,000,000,000,000 kilometers

107. Relative density of hydrogen: 0.0000899 gram per cubic centimeter

108. One micron (millionth of a meter): 0.00003937 inch

In Exercises 109–112, write the number in decimal form.

109. U.S. daily Coca-Cola consumption: 6.048×10^8 servings (Source: The World of Coca-Cola Pavilion)

110. Interior temperature of the sun: 1.5×10^7 degrees Celsius

111. Charge of an electron: 1.602×10^{-19} coulomb

112. Width of a human hair: 9.0×10^{-5} meter

In Exercises 113 and 114, evaluate each expression without using a calculator.

113. (a) $\sqrt{25 \times 10^8}$ (b) $\sqrt[3]{8 \times 10^{15}}$

114. (a) $(1.2 \times 10^7)(5 \times 10^{-3})$ (b) $\dfrac{(6.0 \times 10^8)}{(3.0 \times 10^{-3})}$

In Exercises 115–118, use a calculator to evaluate each expression. (Round your answer to three decimal places.)

115. (a) $750\left(1 + \dfrac{0.11}{365}\right)^{800}$

(b) $\dfrac{67{,}000{,}000 + 93{,}000{,}000}{0.0052}$

116. (a) $(9.3 \times 10^6)^3(6.1 \times 10^{-4})$

(b) $\dfrac{(2.414 \times 10^4)^6}{(1.68 \times 10^5)^5}$

117. (a) $\sqrt{4.5 \times 10^9}$ (b) $\sqrt[3]{6.3 \times 10^4}$

118. (a) $(2.65 \times 10^{-4})^{1/3}$ (b) $\sqrt{9 \times 10^{-4}}$

119. *Exploration* List all possible digits that occur in the units place of the square of a positive integer. Use that list to determine whether $\sqrt{5233}$ is an integer.

120. *Think About It* Square the real number $2/\sqrt{5}$ and note that the radical is eliminated from the denominator. Is this equivalent to rationalizing the denominator? Why or why not?

121. *Period of a Pendulum* The period T (in seconds) of a pendulum is

$$T = 2\pi\sqrt{\frac{L}{32}}$$

where L is the length of the pendulum (in feet). Find the period of a pendulum whose length is 2 feet.

122. *Erosion* A stream of water moving at the rate of v feet per second can carry particles of size $0.03\sqrt{v}$ inches. Find the size of the largest particles that can be carried by a stream flowing at the rate of $\frac{3}{4}$ foot per second.

123. *Mathematical Modeling* A funnel is filled with water to a height of h centimeters. The time t (in seconds) for the funnel to empty is

$$t = 0.03[12^{5/2} - (12 - h)^{5/2}], \quad 0 \le h \le 12.$$

Find t for $h = 7$ centimeters.

124. *Speed of Light* The speed of light is 11,160,000 miles per minute. The distance from the sun to the earth is 93,000,000 miles. Find the time for light to travel from the sun to the earth.

125. *Depreciation* Find the annual depreciation rate r from the bar graph below. To find r by the declining balances method, use the formula

$$r = 1 - \left(\frac{S}{C}\right)^{1/n}$$

where n is the useful life of the item (in years), S is the salvage value (in dollars), and C is the original cost (in dollars).

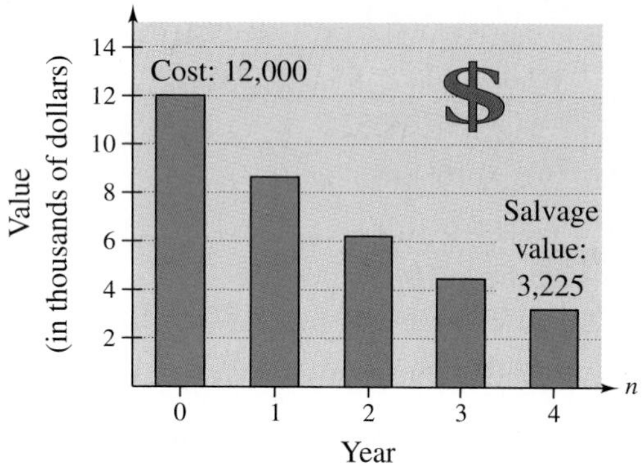

126. *Ecology* There were 2.097×10^8 tons of municipal waste generated in the United States in 1996. Find the number of tons for each of the categories in the figure. (Source: U.S. Environmental Protection Agency)

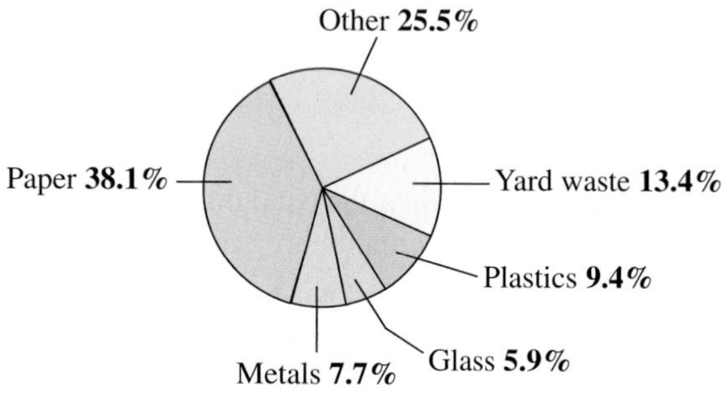

Synthesis

True or False? **In Exercises 127 and 128, determine whether the statement is true or false. Justify your answer.**

127. $\dfrac{x^{k+1}}{x} = x^k$ **128.** $(a^n)^k = a^{n^k}$

129. Verify that $a^0 = 1$, $a \neq 0$. (*Hint*: Use the property of exponents $\dfrac{a^m}{a^n} = a^{m-n}$.)

130. Explain why each of the following pairs is not equal.

(a) $(3x)^{-1} \neq \dfrac{3}{x}$ (b) $y^3 \cdot y^2 \neq y^6$

(c) $(a^2b^3)^4 \neq a^6b^7$ (d) $(a + b)^2 \neq a^2 + b^2$

(e) $\sqrt{4x^2} \neq 2x$ (f) $\sqrt{2} + \sqrt{3} \neq \sqrt{5}$

131. Is the real number 52.7×10^5 written in scientific notation? Explain.

132. *Writing* Johannes Kepler (1571–1630), a well-known German astronomer, discovered a relationship between the average distance of a planet from the sun and the time (or period) it takes the planet to orbit the sun. People then knew that planets that are closer to the sun take less time to complete an orbit than planets that are farther from the sun. Kepler discovered that the distance and period are related by an exact mathematical formula.

The table shows the average distance x (in astronomical units) and period y (in years) for the five planets that are closest to the sun. By completing the table, can you rediscover Kepler's relationship? Write a paragraph that summarizes your conclusions.

Planet	Mercury	Venus	Earth	Mars	Jupiter
x	0.387	0.723	1.0	1.523	5.203
$\sqrt{x}$					
y	0.241	0.615	1.0	1.881	11.861
$\sqrt[3]{y}$					

P.3 Polynomials and Factoring

▶ **What you should learn**

- How to write polynomials in standard form
- How to add, subtract, and multiply polynomials
- How to use special products to multiply polynomials
- How to remove common factors from polynomials
- How to factor special polynomial forms
- How to factor trinomials as the product of two binomials
- How to factor by grouping

▶ **Why you should learn it**

Polynomials can be used to model and solve real-life problems. For instance, in Exercise 178 on page 36, a polynomial is used to model the stopping distance of an automobile.

Nicholas DeVore/Tony Stone Images

Polynomials

The most common type of algebraic expression is the **polynomial.** Some examples of polynomials are

$$2x + 5, \quad 3x^4 - 7x^2 + 2x + 4, \quad \text{and} \quad 5x^2y^2 - xy + 3.$$

The first two are *polynomials in x* and the third is a *polynomial in x and y.* The terms of a polynomial in x have the form ax^k, where a is the **coefficient** and k is the **degree** of the term. For instance, the polynomial

$$2x^3 - 5x^2 + 1 = 2x^3 + (-5)x^2 + (0)x + 1$$

has coefficients 2, -5, 0, and 1.

Definition of a Polynomial in *x*

Let $a_0, a_1, a_2, \ldots, a_n$ be real numbers and let n be a nonnegative integer. A polynomial in x is an expression of the form

$$a_nx^n + a_{n-1}x^{n-1} + \cdot \cdot \cdot + a_1x + a_0$$

where $a_n \neq 0$. The polynomial is of **degree** n, a_n is the **leading coefficient,** and a_0 is the **constant term.**

Polynomials with one, two, and three terms are called **monomials, binomials,** and **trinomials,** respectively. In **standard form,** a polynomial is written with descending powers of x.

Example 1 ▶ Writing Polynomials in Standard Form

	Polynomial	*Standard Form*	*Degree*
a.	$4x^2 - 5x^7 - 2 + 3x$	$-5x^7 + 4x^2 + 3x - 2$	7
b.	$4 - 9x^2$	$-9x^2 + 4$	2
c.	8	$8 \ (8 = 8x^0)$	0

A polynomial that has all zero coefficients is called the zero polynomial, denoted by 0. No degree is assigned to this particular polynomial. For polynomials in more than one variable, the degree of a *term* is the sum of the exponents of the variables in the term. The degree of the *polynomial* is the highest degree of its terms. The leading coefficient of the polynomial is the coefficient of the highest-degree term. Expressions such as the following are not polynomials.

$x^3 - \sqrt{3x} = x^3 - (3x)^{1/2}$ — Exponent in $\sqrt{3x}$ is not an integer.

$x^2 + 5x^{-1}$ — Exponent in $5x^{-1}$ is not a nonnegative integer.

Operations with Polynomials

You can add and subtract polynomials in much the same way you add and subtract real numbers. Simply add or subtract the *like terms* (terms having the same variables to the same powers) by adding their coefficients. For instance, $-3xy^2$ and $5xy^2$ are like terms and their sum is

$$-3xy^2 + 5xy^2 = (-3 + 5)xy^2 = 2xy^2.$$

STUDY TIP

A common mistake is to fail to change the sign of each term inside parentheses preceded by a negative sign. For instance, note that

$$-(x^2 - x + 3) = -x^2 + x - 3$$

and

$$-(x^2 - x + 3) \neq -x^2 - x + 3.$$

Example 2 ▶ Sums and Differences of Polynomials

Perform the operation on the polynomials.

a. $(5x^3 - 7x^2 - 3) + (x^3 + 2x^2 - x + 8)$

b. $(7x^4 - x^2 - 4x + 2) - (3x^4 - 4x^2 + 3x)$

Solution

a. $(5x^3 - 7x^2 - 3) + (x^3 + 2x^2 - x + 8)$

$= (5x^3 + x^3) + (2x^2 - 7x^2) - x + (8 - 3)$ Group like terms.

$= 6x^3 - 5x^2 - x + 5$ Combine like terms.

b. $(7x^4 - x^2 - 4x + 2) - (3x^4 - 4x^2 + 3x)$

$= 7x^4 - x^2 - 4x + 2 - 3x^4 + 4x^2 - 3x$ Distributive Property

$= (7x^4 - 3x^4) + (4x^2 - x^2) + (-3x - 4x) + 2$ Group like terms.

$= 4x^4 + 3x^2 - 7x + 2$ Combine like terms.

To find the **product** of two polynomials, use the left and right Distributive Properties.

When using the FOIL Method, the following scheme may be helpful.

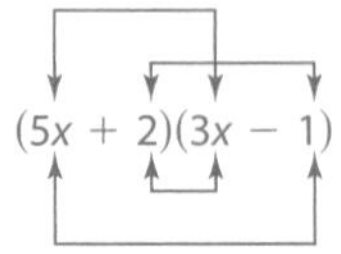

Example 3 ▶ Multiplying Polynomials: The FOIL Method

Multiply $(3x - 2)$ by $(5x + 7)$.

Solution

$$(3x - 2)(5x + 7) = 3x(5x + 7) - 2(5x + 7)$$

$$= (3x)(5x) + (3x)(7) - (2)(5x) - (2)(7)$$

$$= 15x^2 + 21x - 10x - 14$$

Product of First terms | Product of Outer terms | Product of Inner terms | Product of Last terms

$$= 15x^2 + 11x - 14$$

Note in this **FOIL Method** that for binomials, the outer (O) and inner (I) terms are like terms and can be combined.

A computer simulation to accompany this example appears in the *Interactive* CD-ROM and *Internet* versions of this text.

To understand the individual patterns of special products, have students "derive" each product. Then explain how special products save time and that the pattern of the product must be recognized to factor expressions.

Special Products

Some binomial products have special forms that occur frequently in algebra.

Special Products

Let u and v be real numbers, variables, or algebraic expressions.

Special Product	*Example*
Sum and Difference of Same Terms	
$(u + v)(u - v) = u^2 - v^2$	$(x + 4)(x - 4) = x^2 - 4^2$ $= x^2 - 16$
Square of a Binomial	
$(u + v)^2 = u^2 + 2uv + v^2$	$(x + 3)^2 = x^2 + 2(x)(3) + 3^2$ $= x^2 + 6x + 9$
$(u - v)^2 = u^2 - 2uv + v^2$	$(3x - 2)^2 = (3x)^2 - 2(3x)(2) + 2^2$ $= 9x^2 - 12x + 4$
Cube of a Binomial	
$(u + v)^3 = u^3 + 3u^2v + 3uv^2 + v^3$	$(x + 2)^3 = x^3 + 3x^2(2) + 3x(2^2) + 2^3$ $= x^3 + 6x^2 + 12x + 8$
$(u - v)^3 = u^3 - 3u^2v + 3uv^2 - v^3$	$(x - 1)^3 = x^3 - 3x^2(1) + 3x(1^2) - 1^3$ $= x^3 - 3x^2 + 3x - 1$

Example 4 ▶ Sum and Difference of Same Terms

Find the product of $(5x + 9)$ and $(5x - 9)$.

Solution

The product of a sum and a difference of the *same* two terms has no middle term and takes the form $(u + v)(u - v) = u^2 - v^2$.

$$\begin{aligned}(5x + 9)(5x - 9) &= (5x)^2 - 9^2\\ &= 25x^2 - 81\end{aligned}$$

Example 5 ▶ The Product of Two Trinomials

Find the product of $(x + y - 2)$ and $(x + y + 2)$.

Solution

By grouping $x + y$ in parentheses, you can write the product of the trinomials as a special product.

$$\begin{aligned}(x + y - 2)(x + y + 2) &= [(x + y) - 2][(x + y) + 2]\\ &= (x + y)^2 - 2^2\\ &= x^2 + 2xy + y^2 - 4\end{aligned}$$

Activities

1. Explain what happens when the parentheses are removed from the algebraic statement: $-(3x^2 - 4x + 2)$.
2. Multiply using special products: $(\sqrt{2} + \sqrt{3})^2$.
   Answer: $5 + 2\sqrt{6}$
3. Multiply: $(2x + 1)(x - 5)$.
   Answer: $2x^2 - 9x - 5$

Factoring

The process of writing a polynomial as a product is called **factoring.** It is an important tool for solving equations and for simplifying rational expressions.

Unless noted otherwise, when you are asked to factor a polynomial, you can assume that you are hunting for factors with integer coefficients. If a polynomial cannot be factored using integer coefficients, then it is **prime** or **irreducible over the integers.** For instance, the polynomial $x^2 - 3$ is irreducible over the integers. Over the *real numbers,* this polynomial can be factored as

$$x^2 - 3 = (x + \sqrt{3})(x - \sqrt{3}).$$

Example 6 ▶ Recognizing Completely Factored Polynomials

Determine whether the polynomial is completely factored.

a. $x^3 - x^2 + 4x - 4 = (x - 1)(x^2 + 4)$

b. $x^3 - x^2 - 4x + 4 = (x - 1)(x^2 - 4)$

Solution

a. $x^3 - x^2 + 4x - 4 = (x - 1)(x^2 + 4)$ is completely factored.

b. $x^3 - x^2 - 4x + 4 = (x - 1)(x^2 - 4)$ is not completely factored. Its complete factorization would be

$$x^3 - x^2 - 4x + 4 = (x - 1)(x + 2)(x - 2).$$

The simplest type of factoring involves a polynomial that can be written as the product of a monomial and another polynomial. The technique used here is the Distributive Property, $a(b + c) = ab + ac$, in the *reverse* direction.

$$ab + ac = a(b + c)$$ a is a common factor.

Removing (factoring out) a common factor is the first step in completely factoring a polynomial.

Example 7 ▶ Removing Common Factors

Factor each expression.

a. $6x^3 - 4x$ **b.** $-4x^2 + 12x - 16$ **c.** $(x - 2)(2x) + (x - 2)(3)$

Solution

a. $6x^3 - 4x = 2x(3x^2) - 2x(2)$ $2x$ is a common factor.

$$= 2x(3x^2 - 2)$$

b. $-4x^2 + 12x - 16 = -4(x^2) + (-4)(-3x) + (-4)4$ -4 is a common factor.

$$= -4(x^2 - 3x + 4)$$

c. $(x - 2)(2x) + (x - 2)(3) = (x - 2)(2x + 3)$ $x - 2$ is a common factor.

Factoring Special Polynomial Forms

Some polynomials have special forms that you should learn to recognize so that you can factor the polynomials easily.

Factoring Special Polynomial Forms

Factored Form	*Example*
Difference of Two Squares	
$u^2 - v^2 = (u + v)(u - v)$	$9x^2 - 4 = (3x)^2 - 2^2 = (3x + 2)(3x - 2)$
Perfect Square Trinomial	
$u^2 + 2uv + v^2 = (u + v)^2$	$x^2 + 6x + 9 = x^2 + 2(x)(3) + 3^2 = (x + 3)^2$
$u^2 - 2uv + v^2 = (u - v)^2$	$x^2 - 6x + 9 = x^2 - 2(x)(3) + 3^2 = (x - 3)^2$
Sum or Difference of Two Cubes	
$u^3 + v^3 = (u + v)(u^2 - uv + v^2)$	$x^3 + 8 = x^3 + 2^3 = (x + 2)(x^2 - 2x + 4)$
$u^3 - v^3 = (u - v)(u^2 + uv + v^2)$	$27x^3 - 1 = (3x)^3 - 1^3 = (3x - 1)(9x^2 + 3x + 1)$

One of the easiest special polynomial forms to factor is the difference of two squares. Think of this form as follows.

$$u^2 - v^2 = (u + v)(u - v)$$

Difference (the minus sign in $u^2 - v^2$) Opposite signs (the signs in $(u + v)(u - v)$)

To recognize perfect square terms, look for coefficients that are squares of integers and variables raised to *even powers.*

STUDY TIP

In Example 8, note that the first step in factoring a polynomial is to check for a common factor. Once the common factor is removed, it is often possible to recognize patterns that were not immediately obvious.

Example 8 ▶ Removing a Common Factor First

Factor $3 - 12x^2$.

Solution

$$\begin{aligned} 3 - 12x^2 &= 3(1 - 4x^2) && \text{3 is a common factor.} \\ &= 3[1^2 - (2x)^2] \\ &= 3(1 + 2x)(1 - 2x) && \text{Difference of two squares} \end{aligned}$$

Example 9 ▶ Factoring the Difference of Two Squares

a. $$\begin{aligned} (x + 2)^2 - y^2 &= [(x + 2) + y][(x + 2) - y] \\ &= (x + 2 + y)(x + 2 - y) \end{aligned}$$

b. $$\begin{aligned} 16x^4 - 81 &= (4x^2)^2 - 9^2 \\ &= (4x^2 + 9)(4x^2 - 9) && \text{Difference of two squares} \\ &= (4x^2 + 9)[(2x)^2 - 3^2] \\ &= (4x^2 + 9)(2x + 3)(2x - 3) && \text{Difference of two squares} \end{aligned}$$

A perfect square trinomial is the square of a binomial, and it has the following form.

$$u^2 + 2uv + v^2 = (u + v)^2 \quad \text{or} \quad u^2 - 2uv + v^2 = (u - v)^2$$

Like signs   Like signs

Note that the first and last terms are squares and the middle term is twice the product of u and v.

Example 10 ▶ Factoring Perfect Square Trinomials

Factor each trinomial.

a. $x^2 - 10x + 25$ **b.** $16x^2 + 8x + 1$

Solution

a. $x^2 - 10x + 25 = x^2 - 2(x)(5) + 5^2$

$= (x - 5)^2$

b. $16x^2 + 8x + 1 = (4x)^2 + 2(4x)(1) + 1^2$

$= (4x + 1)^2$

Exploration

Rewrite $u^6 - v^6$ as the difference of two squares. Then find a formula for completely factoring $u^6 - v^6$. Use your formula to factor completely $x^6 - 1$ and $x^6 - 64$.

The next two formulas show the sums and differences of cubes. Pay special attention to the signs of the terms.

Like signs   Like signs

$$u^3 + v^3 = (u + v)(u^2 - uv + v^2) \qquad u^3 - v^3 = (u - v)(u^2 + uv + v^2)$$

Unlike signs   Unlike signs

Example 11 ▶ Factoring the Difference of Cubes

Factor $x^3 - 27$.

Solution

$x^3 - 27 = x^3 - 3^3$ Rewrite 27 as 3^3.

$= (x - 3)(x^2 + 3x + 9)$ Factor.

Example 12 ▶ Factoring the Sum of Cubes

a. $y^3 + 8 = y^3 + 2^3$ Rewrite 8 as 2^3.

$= (y + 2)(y^2 - 2y + 4)$ Factor.

b. $3(x^3 + 64) = 3(x^3 + 4^3)$ Rewrite 64 as 4^3.

$= 3(x + 4)(x^2 - 4x + 16)$ Factor.

You can draw arrows to find the correct middle term. (Encourage students to find the middle term mentally.)

$(2x - 1)(x + 15)$

$-1x$

$+30x$

$+29x$ Middle term

$(2x - 3)(x + 5)$

$-3x$

$+10x$

$+7x$ Middle term

Trinomials with Binomial Factors

To factor a trinomial of the form $ax^2 + bx + c$, use the following pattern.

Factors of a

$$ax^2 + bx + c = (\square x + \square)(\square x + \square)$$

Factors of c

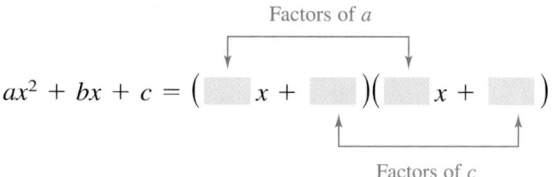

The goal is to find a combination of factors of a and c such that the outer and inner products add up to the middle term bx. For instance, in the trinomial $6x^2 + 17x + 5$, you can write

F O I L

$$(2x + 5)(3x + 1) = 6x^2 + 2x + 15x + 5$$

O + I

$$= 6x^2 + 17x + 5.$$

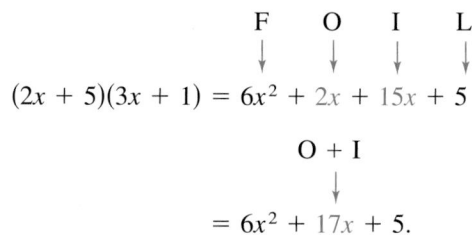

Note that the outer (O) and inner (I) products add up to $17x$.

Example 13 ▶ Factoring a Trinomial: Leading Coefficient Is 1

Factor $x^2 - 7x + 12$.

Solution

The possible factorizations are

$$(x - 2)(x - 6), \quad (x - 1)(x - 12), \quad \text{and} \quad (x - 3)(x - 4).$$

Testing the middle term, you will find the correct factorization to be

$$x^2 - 7x + 12 = (x - 3)(x - 4).$$

A computer animation of this example appears in the *Interactive* CD-ROM and *Internet* versions of this text.

Example 14 ▶ Factoring a Trinomial: Leading Coefficient Is Not 1

Factor $2x^2 + x - 15$.

Solution

The eight possible factorizations are as follows.

$(2x - 1)(x + 15)$ $\quad$ $(2x + 1)(x - 15)$

$(2x - 3)(x + 5)$ $\quad$ $(2x + 3)(x - 5)$

$(2x - 5)(x + 3)$ $\quad$ $(2x + 5)(x - 3)$

$(2x - 15)(x + 1)$ $\quad$ $(2x + 15)(x - 1)$

Testing the middle term, you will find the correct factorization to be

$$2x^2 + x - 15 = (2x - 5)(x + 3). \qquad O + I = 6x - 5x = x$$

Factoring by Grouping

Sometimes polynomials with more than three terms can be factored by a method called **factoring by grouping.** It is not always obvious which terms to group, and sometimes several different groupings will work.

Example 15 ▶ Factoring by Grouping

Use factoring by grouping to factor

$$x^3 - 2x^2 - 3x + 6.$$

Solution

$x^3 - 2x^2 - 3x + 6$	$= (x^3 - 2x^2) - (3x - 6)$	Group terms.
	$= x^2(x - 2) - 3(x - 2)$	Factor groups.
	$= (x - 2)(x^2 - 3)$	Distributive Property

Factoring a trinomial can involve quite a bit of trial and error. Some of this trial and error can be lessened by using factoring by grouping.

Example 16 ▶ Factoring a Trinomial by Grouping

Use factoring by grouping to factor $2x^2 + 5x - 3$.

Solution

In the trinomial $2x^2 + 5x - 3$, $a = 2$ and $c = -3$, which implies that the product ac is -6. Now, -6 factors as $(6)(-1)$ and $6 - 1 = 5 = b$. So, you can rewrite the middle term as $5x = 6x - x$. This produces the following.

$2x^2 + 5x - 3$	$= 2x^2 + 6x - x - 3$	Rewrite middle term.
	$= (2x^2 + 6x) - (x + 3)$	Group terms.
	$= 2x(x + 3) - (x + 3)$	Factor groups.
	$= (x + 3)(2x - 1)$	Distributive Property

Therefore, the trinomial factors as

$$2x^2 + 5x - 3 = (x + 3)(2x - 1).$$

Activities

1. Completely factor: $4x^2 - 4x + 1$.
   Answer: $(2x - 1)^2$
2. Completely factor: $9x^4 - 144$.
   Answer: $9(x^2 + 4)(x - 2)(x + 2)$
3. What order should you follow when completely factoring polynomials?

Guidelines for Factoring Polynomials

1. Factor out any common factors using the Distributive Property.
2. Factor according to one of the special polynomial forms.
3. Factor as $ax^2 + bx + c = (mx + r)(nx + s)$.
4. Factor by grouping.

P.3 Exercises

In Exercises 1–6, match the polynomial with its description. [The polynomials are labeled (a), (b), (c), (d), (e), and (f).]

(a) $3x^2$ (b) $1 - 2x^3$
(c) $x^3 + 3x^2 + 3x + 1$ (d) 12
(e) $-3x^5 + 2x^3 + x$ (f) $\frac{2}{3}x^4 + x^2 + 10$

1. A polynomial of degree zero
2. A trinomial of degree five
3. A binomial with leading coefficient -2
4. A monomial of positive degree
5. A trinomial with leading coefficient $\frac{2}{3}$
6. A third-degree polynomial with leading coefficient 1

In Exercises 7–10, write a polynomial for the description. (There are many correct answers.)

7. A third-degree polynomial with leading coefficient -2
8. A fifth-degree polynomial with leading coefficient 6
9. A fourth-degree binomial with a negative leading coefficient
10. A third-degree binomial with an even leading coefficient

In Exercises 11–16, find the degree and leading coefficient of the polynomial.

11. $3 + 2x$
12. $-3x^4 + 2x^2 - 5$
13. $1 - x + 6x^4 - 4x^5$
14. 3
15. $4x^3y - 3xy^2 + x^2y^3$
16. $-x^5y + 2x^2y^2 + xy^4$

In Exercises 17–22, is the expression a polynomial? If so, write the polynomial in standard form.

17. $2x - 3x^3 + 8$
18. $2x^3 + x - 3x^{-1}$
19. $\dfrac{3x + 4}{x}$
20. $\dfrac{x^2 + 2x - 3}{2}$
21. $y^2 - y^4 + y^3$
22. $\sqrt{y^2 - y^4}$

In Exercises 23–38, perform the operations and write the result in standard form.

23. $(2x^2 + 1) - (x^2 - 2x + 1)$
24. $-(5x^2 - 1) - (-3x^2 + 5)$
25. $(15x^2 - 6) - (-8.3x^3 - 14.7x^2 - 17)$
26. $(15.2x^4 - 18x - 19.1) - (13.9x^4 - 9.6x + 15)$
27. $5z - [3z - (10z + 8)]$
28. $(y^3 + 1) - [(y^2 + 1) + (3y - 7)]$
29. $3x(x^2 - 2x + 1)$
30. $y^2(4y^2 + 2y - 3)$
31. $-5z(3z - 1)$
32. $(-3x)(5x + 2)$
33. $(1 - x^3)(4x)$
34. $-4x(3 - x^3)$
35. $(2.5x^2 + 3)(3x)$
36. $(2 - 3.5y)(2y^3)$
37. $-4x\left(\frac{1}{8}x + 3\right)$
38. $2y\left(4 - \frac{7}{8}y\right)$

In Exercises 39–72, multiply or find the special product.

39. $(x + 3)(x + 4)$
40. $(x - 5)(x + 10)$
41. $(3x - 5)(2x + 1)$
42. $(7x - 2)(4x - 3)$
43. $(2x + 3)^2$
44. $(4x + 5)^2$
45. $(2x - 5y)^2$
46. $(5 - 8x)^2$
47. $(x + 10)(x - 10)$
48. $(2x + 3)(2x - 3)$
49. $(x + 2y)(x - 2y)$
50. $(2x + 3y)(2x - 3y)$
51. $[(m - 3) + n][(m - 3) - n]$
52. $[(x + y) + 1][(x + y) - 1]$
53. $[(x - 3) + y]^2$
54. $[(x + 1) - y]^2$
55. $(2r^2 - 5)(2r^2 + 5)$
56. $(3a^3 - 4b^2)(3a^3 + 4b^2)$
57. $(x + 1)^3$
58. $(x - 2)^3$
59. $(2x - y)^3$
60. $(4x^3 - 3)^2$
61. $\left(\frac{1}{2}x - 3\right)^2$
62. $\left(\frac{2}{3}t + 5\right)^2$
63. $\left(\frac{1}{3}x - 2\right)\left(\frac{1}{3}x + 2\right)$
64. $\left(2x + \frac{1}{5}\right)\left(2x - \frac{1}{5}\right)$
65. $(1.2x + 3)^2$
66. $(1.5y - 3)^2$
67. $(1.5x - 4)(1.5x + 4)$
68. $(2.5y + 3)(2.5y - 3)$
69. $5x(x + 1) - 3x(x + 1)$
70. $(2x - 1)(x + 3) + 3(x + 3)$

71. $(u + 2)(u - 2)(u^2 + 4)$

72. $(x + y)(x - y)(x^2 + y^2)$

In Exercises 73–76, find the product. The expressions are not polynomials, but the formulas can still be used.

73. $(\sqrt{x} + \sqrt{y})(\sqrt{x} - \sqrt{y})$

74. $(5 + \sqrt{x})(5 - \sqrt{x})$

75. $(x - \sqrt{5})^2$

76. $(x + \sqrt{3})^2$

In Exercises 77–80, determine whether the polynomial is completely factored. If not, give the complete factorization.

77. $x^3 + 2x^2 + x + 2 = (x + 2)(x^2 + 1)$

78. $x^3 + 3x^2 - 9x - 27 = (x + 3)(x^2 - 9)$

79. $x^3 + x^2 - 7x - 7 = (x^2 - 7)(x + 1)$

80. $4x^4 + 12x^3 - x^2 - 3x = (x^2 + 3x)(4x^2 - 1)$

In Exercises 81–88, factor out the common factor.

81. $3x + 6$

82. $5y - 30$

83. $2x^3 - 6x$

84. $4x^3 - 6x^2 + 12x$

85. $x(x - 1) + 6(x - 1)$

86. $3x(x + 2) - 4(x + 2)$

87. $(x + 3)^2 - 4(x + 3)$

88. $(3x - 1)^2 + (3x - 1)$

In Exercises 89–92, find the greatest common factor such that the remaining factors have only integer coefficients.

89. $\frac{1}{2}x^3 + 2x^2 - 5x$

90. $\frac{1}{3}y^4 - 5y^2 + 2y$

91. $\frac{2}{3}x(x - 3) - 4(x - 3)$

92. $\frac{4}{5}y(y + 1) - 2(y + 1)$

In Exercises 93–100, factor the difference of two squares.

93. $16y^2 - 9$

94. $49 - 9y^2$

95. $16x^2 - \frac{1}{9}$

96. $\frac{4}{25}y^2 - 64$

97. $(x - 1)^2 - 4$

98. $25 - (z + 5)^2$

99. $9u^2 - 4v^2$

100. $25x^2 - 16y^2$

In Exercises 101–108, factor the perfect square trinomial.

101. $x^2 - 4x + 4$

102. $x^2 + 10x + 25$

103. $36y^2 - 108y + 81$

104. $9x^2 - 12x + 4$

105. $9u^2 + 24uv + 16v^2$

106. $4x^2 - 4xy + y^2$

107. $x^2 - \frac{4}{3}x + \frac{4}{9}$

108. $z^2 + z + \frac{1}{4}$

In Exercises 109–120, factor the trinomial.

109. $x^2 + x - 2$

110. $x^2 + 5x + 6$

111. $s^2 - 5s + 6$

112. $t^2 - t - 6$

113. $20 - y - y^2$

114. $24 + 5z - z^2$

115. $3x^2 - 5x + 2$

116. $2x^2 - x - 1$

117. $5x^2 + 26x + 5$

118. $12x^2 + 7x + 1$

119. $-9z^2 + 3z + 2$

120. $-5u^2 - 13u + 6$

In Exercises 121–128, factor the sum or difference of cubes.

121. $x^3 - 8$

122. $x^3 - 27$

123. $y^3 + 64$

124. $z^3 + 125$

125. $8t^3 - 1$

126. $27x^3 + 8$

127. $u^3 + 27v^3$

128. $64x^3 - y^3$

In Exercises 129–134, factor by grouping.

129. $x^3 - x^2 + 2x - 2$

130. $x^3 + 5x^2 - 5x - 25$

131. $2x^3 - x^2 - 6x + 3$

132. $6 + 2x - 3x^3 - x^4$

133. $6x^3 - 2x + 3x^2 - 1$

134. $8x^5 - 6x^2 + 12x^3 - 9$

In Exercises 135–138, factor the trinomial by grouping.

135. $3x^2 + 10x + 8$

136. $2x^2 + 9x + 9$

137. $15x^2 - 11x + 2$

138. $12x^2 - 13x + 1$

In Exercises 139–160, completely factor the expression.

139. $6x^2 - 54$

140. $12x^2 - 48$

141. $x^3 - 4x^2$

142. $x^3 - 9x$

143. $2x^2 + 4x - 2x^3$

144. $2y^3 - 7y^2 - 15y$

145. $3x^3 + x^2 + 15x + 5$

146. $13x + 6 + 5x^2$

147. $\frac{1}{81}x^2 + \frac{2}{9}x - 8$

148. $\frac{1}{8}x^2 - \frac{1}{96}x - \frac{1}{16}$

149. $x^4 - 4x^3 + x^2 - 4x$

150. $3u - 2u^2 + 6 - u^3$

151. $(x^2 + 1)^2 - 4x^2$

152. $(x^2 + 8)^2 - 36x^2$

153. $2t^3 - 16$

154. $5x^3 + 40$

155. $4x(2x - 1) + (2x - 1)^2$

156. $5(3 - 4x)^2 - 8(3 - 4x)(5x - 1)$

157. $7(3x + 2)^2(1 - x)^2 + (3x + 2)(1 - x)^3$

158. $7x(2)(x^2 + 1)(2x) - (x^2 + 1)^2(7)$

159. $3(x - 2)^2(x + 1)^4 + (x - 2)^3(4)(x + 1)^3$

160. $5(x^6 + 1)^4(6x^5)(3x + 2)^3 + 3(3x + 2)^2(3)(x^6 + 1)^5$

In Exercises 161–164, find all values of b for which the trinomial can be factored.

161. $x^2 + bx - 15$

162. $x^2 + bx + 50$

163. $x^2 + bx - 12$

164. $x^2 + bx + 24$

In Exercises 165–168, find two integer values of c such that the trinomial can be factored. (There are many correct answers.)

165. $2x^2 + 5x + c$

166. $3x^2 - 10x + c$

167. $3x^2 - x + c$

168. $2x^2 + 9x + c$

169. *Business* A manufacturer can produce and sell x radios per week. The total cost (in dollars) for producing x radios is

$C = 73x + 25{,}000$

and the total revenue (in dollars) is

$R = 95x.$

Find the profit P obtained by selling 5000 radios per week.

170. *Business* An artist can produce and sell x craft items per month. The total cost (in dollars) for producing x craft items is

$C = 460 + 12x$

and the total revenue (in dollars) is

$R = 36x.$

Find the profit P obtained by selling 42 craft items per month.

171. *Finance* After 2 years, an investment of \$500 compounded annually at an interest rate r will yield an amount of

$500(1 + r)^2.$

(a) Write this polynomial in standard form.

(b) Use a calculator to evaluate the polynomial for the values of r in the table.

r	$2\frac{1}{2}\%$	3%	4%	$4\frac{1}{2}\%$	5%
$500(1 + r)^2$					

(c) What conclusion can you make from the table?

172. *Finance* After 3 years, an investment of \$1200 compounded annually at an interest rate r will yield an amount of

$1200(1 + r)^3.$

(a) Write this polynomial in standard form.

(b) Use a calculator to evaluate the polynomial for the values of r in the table.

r	2%	3%	$3\frac{1}{2}\%$	4%	$4\frac{1}{2}\%$
$1200(1 + r)^3$					

(c) What conclusion can you make from the table?

173. *Volume of a Box* An open box is made by cutting squares from the corners of a piece of metal that is 18 centimeters by 26 centimeters (see figure). If the edge of each cut-out square is x centimeters, find the volume when $x = 1$, $x = 2$, and $x = 3$.

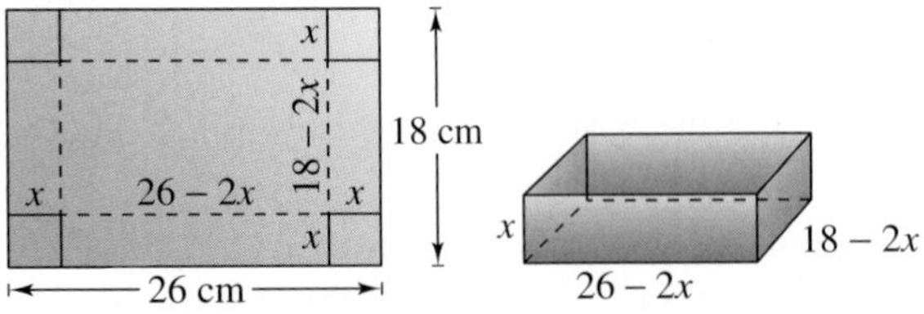

174. *Geometry* Find the area of the shaded region in each figure. Write your result as a polynomial in standard form.

(a)

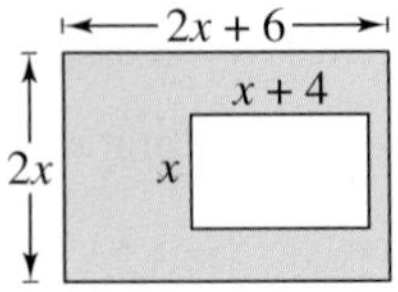

(b)

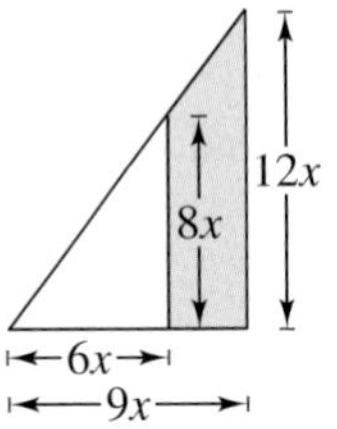

175. *Geometry* Find the area of the shaded region in each figure. Write your result as a polynomial in standard form.

(a)

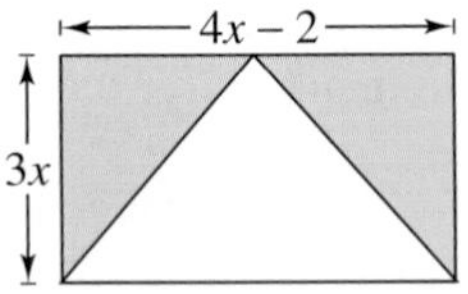

(b)

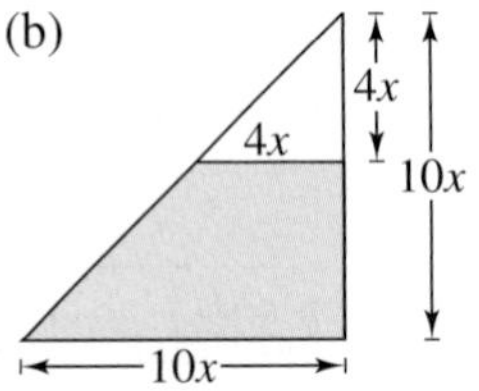

Geometry **In Exercises 176 and 177, find a polynomial that represents the total number of square feet for the floor plan shown in the figure.**

176.

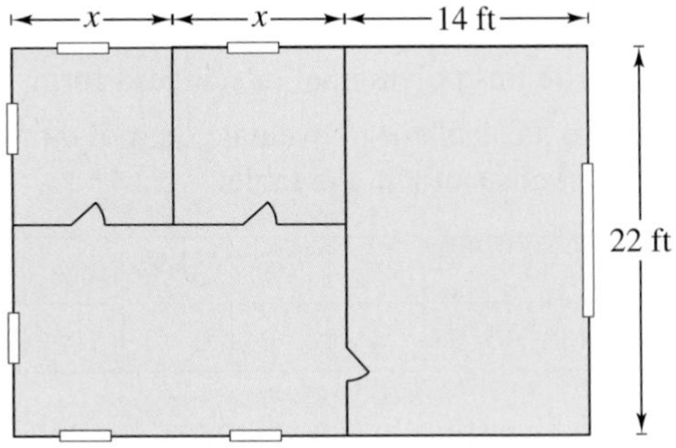

177.

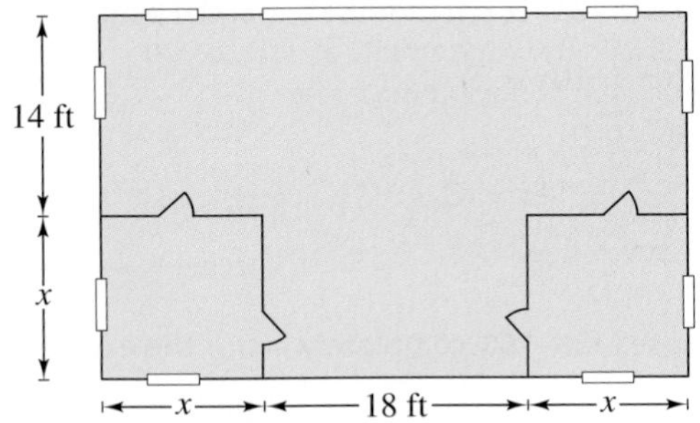

178. *Stopping Distance* The stopping distance of an automobile is the distance traveled during the driver's reaction time plus the distance traveled after the brakes are applied. In an experiment, these distances were measured (in feet) when the automobile was traveling at a speed of x miles per hour, as shown in the bar graph. The distance traveled during the reaction time was $R = 1.1x$, and the braking distance was

$B = 0.14x^2 - 4.43x + 58.40$.

(a) Determine the polynomial that represents the total stopping distance T.

(b) Use the result of part (a) to estimate the total stopping distance when $x = 30$, $x = 40$, and $x = 55$ miles per hour.

(c) Use the bar graph to make a statement about the total stopping distance required for increasing speeds.

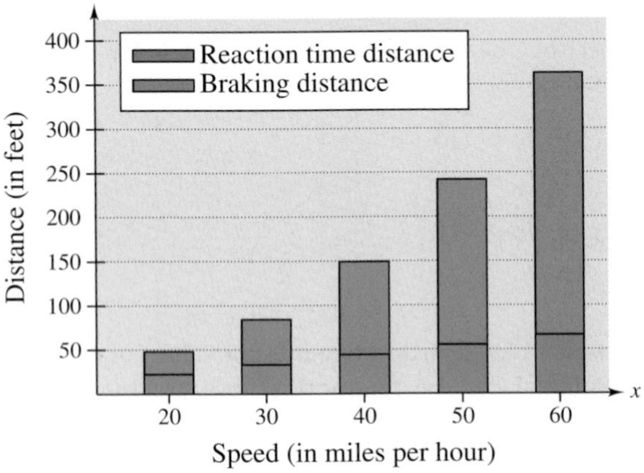

Geometric Modeling **In Exercises 179 and 180, draw a "geometric factoring model" to represent the factorization. For instance, a factoring model for**

$$2x^2 + 3x + 1 = (2x + 1)(x + 1)$$

is shown in the figure.

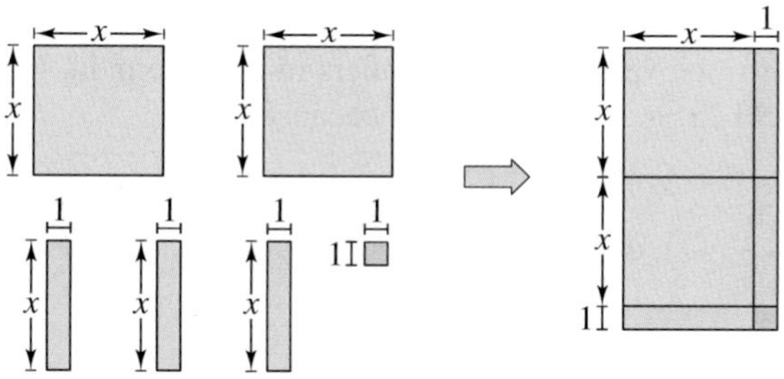

179. $3x^2 + 7x + 2 = (3x + 1)(x + 2)$

180. $2x^2 + 7x + 3 = (2x + 1)(x + 3)$

Geometry **In Exercises 181–184, write an expression in factored form for the shaded portion of the figure.**

181.

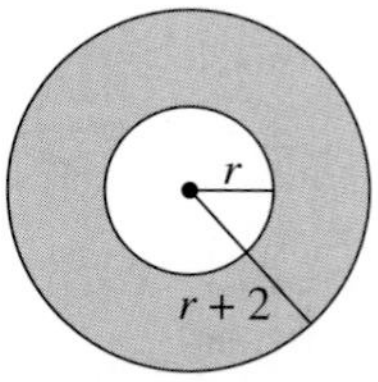

182.

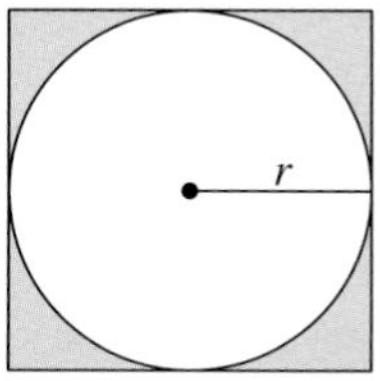

183.

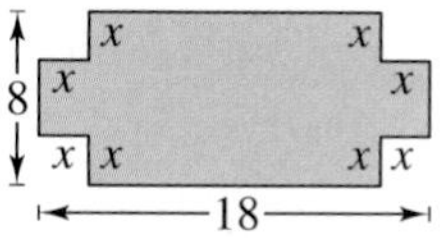

184.

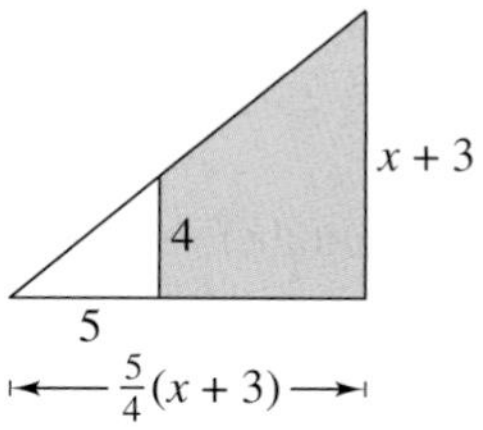

185. *Geometry* The volume of concrete used to make the cylindrical concrete storage tank shown in the figure is

$$V = \pi R^2h - \pi r^2h.$$

(a) Factor the expression for the volume.

(b) From the result of part (a), show that the volume of concrete is

$$2\pi \text{ (average radius)(thickness of the tank)}h.$$

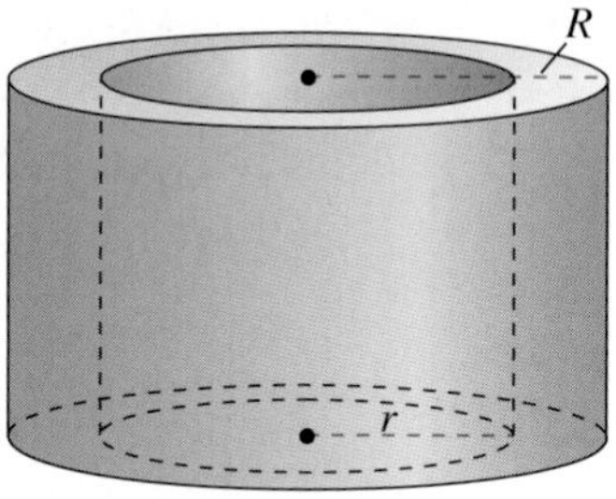

FIGURE FOR 185

186. *Chemistry* The rate of change of an autocatalytic chemical reaction is $kQx - kx^2$, where Q is the amount of the original substance, x is the amount of substance formed, and k is a constant of proportionality. Factor the expression.

Synthesis

True or False? **In Exercises 187–190, determine whether the statement is true or false. Justify your answer.**

187. The product of two binomials is always a second-degree polynomial.

188. The sum of two binomials is always a binomial.

189. The difference of two perfect squares can be factored as the product of conjugate pairs.

190. The sum of two perfect squares can be factored as the binomial sum squared.

191. Find the degree of the product of two polynomials of degrees m and n.

192. Find the degree of the sum of two polynomials of degrees m and n if $m < n$.

193. *Logical Reasoning* Verify that $(x + y)^2$ is not equal to $x^2 + y^2$ by letting $x = 3$ and $y = 4$ and evaluating both expressions. Are there any values of x or y for which $(x + y)^2 = x^2 + y^2$? Explain.

194. *Pattern Recognition* Perform the multiplications.

(a) $(x - 1)(x + 1)$

(b) $(x - 1)(x^2 + x + 1)$

(c) $(x - 1)(x^3 + x^2 + x + 1)$

From the pattern formed by these products, can you predict the result of $(x - 1)(x^4 + x^3 + x^2 + x + 1)$?

195. Factor $x^{2n} - y^{2n}$ completely.

196. Factor $x^{3n} + y^{3n}$ completely.

197. Factor $x^{3n} - y^{2n}$ completely.

P.4 Rational Expressions

▶ **What you should learn**

- How to find domains of algebraic expressions
- How to simplify rational expressions
- How to add, subtract, multiply, and divide rational expressions
- How to simplify complex fractions

▶ **Why you should learn it**

Rational expressions can be used to solve real-life problems. For instance, in Exercise 102 on page 48, a rational expression is used to model the cost per ounce of precious metals from 1992 through 1997.

Patrick Cocklin/Tony Stone Images

Alert students to the importance of the domain of an expression in graphing functions later in this course and in calculus.

Domain of an Algebraic Expression

The set of real numbers for which an algebraic expression is defined is the **domain** of the expression. Two algebraic expressions are **equivalent** if they have the same domain and yield the same values for all numbers in their domain. For instance, $(x + 1) + (x + 2)$ and $2x + 3$ are equivalent because

$$\begin{aligned}(x + 1) + (x + 2) &= x + 1 + x + 2\\ &= x + x + 1 + 2\\ &= 2x + 3.\end{aligned}$$

Example 1 ▶ Finding the Domain of an Algebraic Expression

a. The domain of the polynomial

$$2x^3 + 3x + 4$$

is the set of all real numbers. In fact, the domain of any polynomial is the set of all real numbers, unless the domain is specifically restricted.

b. The domain of the radical expression

$$\sqrt{x - 2}$$

is the set of real numbers greater than or equal to 2, because the square root of a negative number is not a real number.

c. The domain of the expression

$$\frac{x + 2}{x - 3}$$

is the set of all real numbers except $x = 3$, which would produce an undefined division by zero.

The quotient of two algebraic expressions is a *fractional expression.* Moreover, the quotient of two *polynomials* such as

$$\frac{1}{x}, \qquad \frac{2x - 1}{x + 1}, \qquad \text{or} \qquad \frac{x^2 - 1}{x^2 + 1}$$

is a **rational expression.** Recall that a fraction is in simplest form if its numerator and denominator have no factors in common aside from ± 1. To write a fraction in simplest form, divide out common factors.

$$\frac{a \cdot \cancel{c}}{b \cdot \cancel{c}} = \frac{a}{b}, \quad c \neq 0$$

The key to success in simplifying rational expressions lies in your ability to *factor* polynomials.

Simplifying Rational Expressions

When simplifying rational expressions, be sure to factor each polynomial completely before concluding that the numerator and denominator have no factors in common.

Example 2 ▶ Simplifying a Rational Expression

Write $\frac{x^2 + 4x - 12}{3x - 6}$ in simplest form.

Solution

$$\frac{x^2 + 4x - 12}{3x - 6} = \frac{(x + 6)\cancel{(x - 2)}}{3\cancel{(x - 2)}} \qquad \text{Factor completely.}$$

$$= \frac{x + 6}{3}, \quad x \neq 2 \qquad \text{Divide out common factors.}$$

Note that the original expression is undefined when $x = 2$ (because division by zero is undefined). To make sure that the simplified expression is *equivalent* to the original expression, you must restrict the domain of the simplified expression by excluding the value $x = 2$.

STUDY TIP

In Example 2, do not make the mistake of trying to simplify further by dividing out terms.

$$\frac{x + 6}{3} \neq \frac{x + \cancel{6}^{\,2}}{\cancel{3}} = x + 2$$

Remember that to simplify fractions, divide out common *factors*, not terms.

Sometimes it may be necessary to change the sign of a factor to simplify a rational expression, as shown in Example 3(b).

Example 3 ▶ Simplifying Rational Expressions

Write each expression in simplest form.

a. $\frac{x^3 - 4x}{x^2 + x - 2}$ **b.** $\frac{12 + x - x^2}{2x^2 - 9x + 4}$

Solution

a. $$\frac{x^3 - 4x}{x^2 + x - 2} = \frac{x(x^2 - 4)}{(x + 2)(x - 1)}$$

$$= \frac{x\cancel{(x + 2)}(x - 2)}{\cancel{(x + 2)}(x - 1)} \qquad \text{Factor completely.}$$

$$= \frac{x(x - 2)}{(x - 1)}, \quad x \neq -2 \qquad \text{Divide out common factors.}$$

b. $$\frac{12 + x - x^2}{2x^2 - 9x + 4} = \frac{(4 - x)(3 + x)}{(2x - 1)(x - 4)} \qquad \text{Factor completely.}$$

$$= \frac{-\cancel{(x - 4)}(3 + x)}{(2x - 1)\cancel{(x - 4)}} \qquad (4 - x) = -(x - 4)$$

$$= -\frac{3 + x}{2x - 1}, \quad x \neq 4 \qquad \text{Divide out common factors.}$$

The factoring technique used in equating $(4 - x)$ and $-(x - 4)$ is not always obvious to students. Identify this technique in one or two examples.

Operations with Rational Expressions

To multiply or divide rational expressions, use the properties of fractions discussed in Section P.1. Recall that to divide fractions, you invert the divisor and multiply.

Example 4 ▶ Multiplying Rational Expressions

$$\frac{2x^2+x-6}{x^2+4x-5}\cdot\frac{x^3-3x^2+2x}{4x^2-6x}=\frac{\cancel{(2x-3)}(x+2)}{(x+5)\cancel{(x-1)}}\cdot\frac{\cancel{x}(x-2)\cancel{(x-1)}}{2\cancel{x}\cancel{(2x-3)}}$$

$$=\frac{(x+2)(x-2)}{2(x+5)},\qquad x\neq 0,\ x\neq 1,\ x\neq \tfrac{3}{2}$$

Common Error
A common error is to divide out common terms rather than common factors when simplifying rational expressions after multiplying or dividing. Remind students that only factors that occur in both the numerator and denominator can be divided out.

In Example 4, note that when multiplying rational expressions you must follow the convention of listing *by the product* all values of x that must be specifically excluded from the domain in order to make the domain of the product agree with the domains of the factors.

Example 5 ▶ Dividing Rational Expressions

$$\frac{x^3-8}{x^2-4}\div\frac{x^2+2x+4}{x^3+8}=\frac{x^3-8}{x^2-4}\cdot\frac{x^3+8}{x^2+2x+4}\qquad \text{Invert and multiply.}$$

$$=\frac{\cancel{(x-2)}\cancel{(x^2+2x+4)}}{\cancel{(x+2)}\cancel{(x-2)}}\cdot\frac{\cancel{(x+2)}(x^2-2x+4)}{\cancel{x^2+2x+4}}$$

$$=x^2-2x+4,\qquad x\neq\pm 2$$

To add or subtract rational expressions, you can use the LCD (least common denominator) method or the basic definition

$$\frac{a}{b}\pm\frac{c}{d}=\frac{ad\pm bc}{bd},\qquad b\neq 0 \text{ and } d\neq 0.\qquad \text{Basic definition}$$

This definition provides an efficient way of adding or subtracting *two* fractions that have no common factors in their denominators.

Example 6 ▶ Subtracting Rational Expressions

$$\frac{x}{x-3}-\frac{2}{3x+4}=\frac{x(3x+4)-2(x-3)}{(x-3)(3x+4)}\qquad \text{Basic definition}$$

$$=\frac{3x^2+4x-2x+6}{(x-3)(3x+4)}\qquad \text{Distributive Property}$$

$$=\frac{3x^2+2x+6}{(x-3)(3x+4)}\qquad \text{Combine like terms.}$$

For three or more fractions, or for fractions with a repeated factor in the denominators, the LCD method works well. Recall that the least common denominator of several fractions consists of the product of all prime factors in the denominators, with each factor given the highest power of its occurrence in any denominator. Here is a numerical example.

$$\frac{1}{6}+\frac{3}{4}-\frac{2}{3}=\frac{1\cdot 2}{6\cdot 2}+\frac{3\cdot 3}{4\cdot 3}-\frac{2\cdot 4}{3\cdot 4} \qquad \text{The LCD is 12.}$$

$$=\frac{2}{12}+\frac{9}{12}-\frac{8}{12}$$

$$=\frac{3}{12}$$

$$=\frac{1}{4}$$

Sometimes the numerator of the answer has a factor in common with the denominator. In such cases the answer should be simplified. For instance, in the example above, $\frac{3}{12}$ was simplified to $\frac{1}{4}$.

Example 7 ▶ Combining Rational Expressions: The LCD Method

Perform the operations and simplify.

$$\frac{3}{x-1}-\frac{2}{x}+\frac{x+3}{x^2-1}$$

Solution

Using the factored denominators $(x-1)$, x, and $(x+1)(x-1)$, you can see that the LCD is $x(x+1)(x-1)$.

$$\frac{3}{x-1}-\frac{2}{x}+\frac{x+3}{(x+1)(x-1)}$$

$$=\frac{3(x)(x+1)}{x(x+1)(x-1)}-\frac{2(x+1)(x-1)}{x(x+1)(x-1)}+\frac{(x+3)(x)}{x(x+1)(x-1)}$$

$$=\frac{3(x)(x+1)-2(x+1)(x-1)+(x+3)(x)}{x(x+1)(x-1)}$$

$$=\frac{3x^2+3x-2x^2+2+x^2+3x}{x(x+1)(x-1)} \qquad \text{Distributive Property}$$

$$=\frac{3x^2-2x^2+x^2+3x+3x+2}{x(x+1)(x-1)} \qquad \text{Group like terms.}$$

$$=\frac{2x^2+6x+2}{x(x+1)(x-1)} \qquad \text{Combine like terms.}$$

$$=\frac{2(x^2+3x+1)}{x(x+1)(x-1)} \qquad \text{Factor.}$$

Complex Fractions

Fractional expressions with separate fractions in the numerator, denominator, or both are called **complex fractions.** Here are two examples.

$$\frac{\left(\frac{1}{x}\right)}{x^2+1} \quad \text{and} \quad \frac{\left(\frac{1}{x}\right)}{\left(\frac{1}{x^2+1}\right)}$$

A complex fraction can be simplified by combining its numerator and denominator into single fractions, then inverting the denominator and multiplying.

Additional Example

$$\frac{\left(\frac{x}{4}+\frac{3}{2}\right)}{\left(2-\frac{3}{x}\right)} = \frac{\left(\frac{x+6}{4}\right)}{\left(\frac{2x-3}{x}\right)}$$

$$= \left(\frac{x+6}{4}\right)\cdot\left(\frac{x}{2x-3}\right)$$

$$= \frac{x(x+6)}{4(2x-3)}$$

Example 8 ▶ Simplifying a Complex Fraction

$$\frac{\left(\frac{2}{x}-3\right)}{\left(1-\frac{1}{x-1}\right)} = \frac{\left[\frac{2-3(x)}{x}\right]}{\left[\frac{1(x-1)-1}{x-1}\right]} \qquad \text{Combine fractions.}$$

$$= \frac{\left(\frac{2-3x}{x}\right)}{\left(\frac{x-2}{x-1}\right)} \qquad \text{Simplify.}$$

$$= \frac{2-3x}{x}\cdot\frac{x-1}{x-2} \qquad \text{Invert and multiply.}$$

$$= \frac{(2-3x)(x-1)}{x(x-2)}, \quad x \neq 1$$

Another way to simplify a complex fraction is to multiply the numerator and denominator by the LCD of all fractions in its numerator and denominator. This method is applied to the fraction in Example 8 as follows.

$$\frac{\left(\frac{2}{x}-3\right)}{\left(1-\frac{1}{x-1}\right)} = \frac{\left(\frac{2}{x}-3\right)}{\left(1-\frac{1}{x-1}\right)}\cdot\frac{x(x-1)}{x(x-1)} \qquad \text{LCD is } x(x-1).$$

$$= \frac{\left(\frac{2-3x}{\cancel{x}}\right)\cdot \cancel{x}(x-1)}{\left(\frac{x-2}{\cancel{x-1}}\right)\cdot x(\cancel{x-1})}$$

$$= \frac{(2-3x)(x-1)}{x(x-2)}, \quad x \neq 1$$

The next three examples illustrate some methods for simplifying rational expressions involving radicals and negative exponents. These types of expressions occur frequently in calculus.

To simplify an expression with negative exponents, one method is to begin by factoring out the common factor with the smaller exponent. Remember that when factoring, you subtract exponents. For instance, in $3x^{-5/2} + 2x^{-3/2}$ the smaller exponent is $-\frac{5}{2}$ and the common factor is $x^{-5/2}$.

$$\begin{aligned} 3x^{-5/2} + 2x^{-3/2} &= x^{-5/2}[3(1) + 2x^{-3/2-(-5/2)}] \\ &= x^{-5/2}(3 + 2x^1) \\ &= \frac{3 + 2x}{x^{5/2}} \end{aligned}$$

Example 9 ▶ Simplifying an Expression

Simplify the following expression containing negative exponents.

$$x(1 - 2x)^{-3/2} + (1 - 2x)^{-1/2}$$

Solution

Begin by factoring out the common factor with the *smaller exponent*.

$$\begin{aligned} x(1 - 2x)^{-3/2} + (1 - 2x)^{-1/2} &= (1 - 2x)^{-3/2}[x + (1 - 2x)^{(-1/2)-(-3/2)}] \\ &= (1 - 2x)^{-3/2}[x + (1 - 2x)^1] \\ &= \frac{1 - x}{(1 - 2x)^{3/2}} \end{aligned}$$

A second method for simplifying an expression with negative exponents is shown in the next example.

Example 10 ▶ Simplifying a Complex Fraction

$$\frac{(4 - x^2)^{1/2} + x^2(4 - x^2)^{-1/2}}{4 - x^2}$$

$$\begin{aligned} &= \frac{(4 - x^2)^{1/2} + x^2(4 - x^2)^{-1/2}}{4 - x^2} \cdot \frac{(4 - x^2)^{1/2}}{(4 - x^2)^{1/2}} \\ &= \frac{(4 - x^2)^1 + x^2(4 - x^2)^0}{(4 - x^2)^{3/2}} \\ &= \frac{4 - x^2 + x^2}{(4 - x^2)^{3/2}} \\ &= \frac{4}{(4 - x^2)^{3/2}} \end{aligned}$$

Activities

1. What is the domain of $\sqrt{3 - x^2}$?
   Answer: The set of all reals less than or equal to $\sqrt{3}$.
2. The implied domain excludes what values of x from the domain of $\dfrac{x^2 + 6x + 9}{x^2 - 9}$?
   Answer: $x = 3, x = -3$
3. Simplify: $\dfrac{\frac{1}{x} - \frac{1}{x + 1}}{x + 1}$
   Answer: $\dfrac{1}{x(x + 1)^2}$

Example 11 ▶ Rewriting a Difference Quotient

The expression from calculus

$$\frac{\sqrt{x+h}-\sqrt{x}}{h}$$

is an example of a *difference quotient.* Rewrite this expression by rationalizing its numerator.

Solution

$$\frac{\sqrt{x+h}-\sqrt{x}}{h} = \frac{\sqrt{x+h}-\sqrt{x}}{h} \cdot \frac{\sqrt{x+h}+\sqrt{x}}{\sqrt{x+h}+\sqrt{x}}$$

$$= \frac{\left(\sqrt{x+h}\right)^2-\left(\sqrt{x}\right)^2}{h\left(\sqrt{x+h}+\sqrt{x}\right)}$$

$$= \frac{h}{h\left(\sqrt{x+h}+\sqrt{x}\right)}$$

$$= \frac{1}{\sqrt{x+h}+\sqrt{x}}, \quad h \neq 0$$

Notice that the original expression is undefined when $h = 0$. So, you must exclude $h = 0$ from the domain of the simplified expression so that the expressions are equivalent.

Difference quotients, such as that in Example 11, occur frequently in calculus. Often, they need to be rewritten in an equivalent form that can be evaluated when $h = 0$. Note that the equivalent form is not simpler than the original form, but it has the advantage that it is defined when $h = 0$.

Writing ABOUT MATHEMATICS

Comparing Domains of Two Expressions Complete the table by evaluating the expressions

$$\frac{x^2-3x+2}{x-2} \quad \text{and} \quad x-1$$

for the values of x. If you have a graphing utility with a table feature, use it to help create the table. Write a short paragraph describing the equivalence or nonequivalence of the two expressions.

x	-3	-2	-1	0	1	2	3
$\frac{x^2-3x+2}{x-2}$							
$x-1$							

P.4 Exercises

In Exercises 1–8, find the domain of the expression.

1. $3x^2 - 4x + 7$
2. $2x^2 + 5x - 2$
3. $4x^3 + 3, \quad x \geq 0$
4. $6x^2 - 9, \quad x > 0$
5. $\dfrac{1}{x - 2}$
6. $\dfrac{x + 1}{2x + 1}$
7. $\sqrt{x + 1}$
8. $\sqrt{6 - x}$

In Exercises 9–14, find the missing factor in the numerator such that the two fractions are equivalent.

9. $\dfrac{5}{2x} = \dfrac{5(\quad)}{6x^2}$
10. $\dfrac{3}{4} = \dfrac{3(\quad)}{4(x + 1)}$
11. $\dfrac{x + 1}{x} = \dfrac{(x + 1)(\quad)}{x(x - 2)}$
12. $\dfrac{3y - 4}{y + 1} = \dfrac{(3y - 4)(\quad)}{y^2 - 1}$
13. $\dfrac{3x}{x - 3} = \dfrac{3x(\quad)}{x^2 - 3x}$
14. $\dfrac{1 - z}{z^2} = \dfrac{(1 - z)(\quad)}{z^3 + z^2}$

In Exercises 15–32, write the rational expression in simplest form.

15. $\dfrac{15x^2}{10x}$
16. $\dfrac{18y^2}{60y^5}$
17. $\dfrac{3xy}{xy + x}$
18. $\dfrac{2x^2y}{xy - y}$
19. $\dfrac{4y - 8y^2}{10y - 5}$
20. $\dfrac{9x^2 + 9x}{2x + 2}$
21. $\dfrac{x - 5}{10 - 2x}$
22. $\dfrac{12 - 4x}{x - 3}$
23. $\dfrac{y^2 - 16}{y + 4}$
24. $\dfrac{x^2 - 25}{5 - x}$
25. $\dfrac{x^3 + 5x^2 + 6x}{x^2 - 4}$
26. $\dfrac{x^2 + 8x - 20}{x^2 + 11x + 10}$
27. $\dfrac{y^2 - 7y + 12}{y^2 + 3y - 18}$
28. $\dfrac{x^2 - 7x + 6}{x^2 + 11x + 10}$
29. $\dfrac{2 - x + 2x^2 - x^3}{x^2 - 4}$
30. $\dfrac{x^2 - 9}{x^3 + x^2 - 9x - 9}$
31. $\dfrac{z^3 - 8}{z^2 + 2z + 4}$
32. $\dfrac{y^3 - 2y^2 - 3y}{y^3 + 1}$

In Exercises 33 and 34, complete the table. What can you conclude?

33.

x	0	1	2	3	4	5	6
$\dfrac{x^2 - 2x - 3}{x - 3}$							
$x + 1$							

34.

x	0	1	2	3	4	5	6
$\dfrac{x - 3}{x^2 - x - 6}$							
$\dfrac{1}{x + 2}$							

35. *Error Analysis* Describe the error.

$$\frac{5x^3}{2x^3 + 4} = \frac{5\cancel{x^3}}{2\cancel{x^3} + 4} = \frac{5}{2 + 4} = \frac{5}{6} \quad \times$$

36. *Error Analysis* Describe the error.

$$\frac{x^3 + 25x}{x^2 - 2x - 15} = \frac{x(x^2 + 25)}{(x - 5)(x + 3)} \quad \times$$

$$= \frac{x\cancel{(x - 5)}(x + 5)}{\cancel{(x - 5)}(x + 3)}$$

$$= \frac{x(x + 5)}{x + 3}$$

Geometry **In Exercises 37 and 38, find the ratio of the area of the shaded portion of the figure to the total area of the figure.**

37.

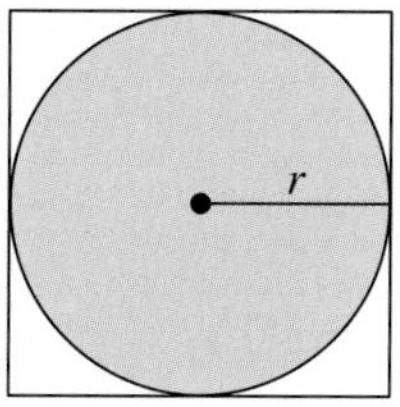

38. 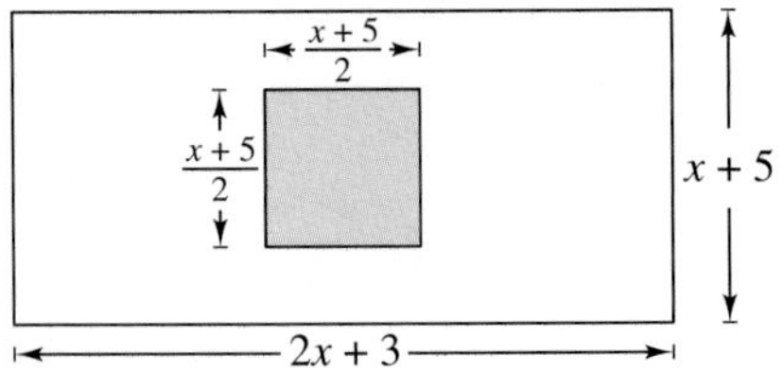

In Exercises 39–54, perform the multiplication or division and simplify.

39. $\frac{5}{x-1} \cdot \frac{x-1}{25(x-2)}$

40. $\frac{x+13}{x^3(3-x)} \cdot \frac{x(x-3)}{5}$

41. $\frac{(x+5)(x-3)}{x+2} \cdot \frac{1}{(x+5)(x+2)}$

42. $\frac{(x-9)(x+7)}{x+1} \cdot \frac{x}{9-x}$

43. $\frac{r}{r-1} \cdot \frac{r^2-1}{r^2}$

44. $\frac{4y-16}{5y+15} \cdot \frac{2y+6}{4-y}$

45. $\frac{t^2-t-6}{t^2+6t+9} \cdot \frac{t+3}{t^2-4}$

46. $\frac{y^3-8}{2y^3} \cdot \frac{4y}{y^2-5y+6}$

47. $\frac{x^2+xy-2y^2}{x^3+x^2y} \cdot \frac{x}{x^2+3xy+2y^2}$

48. $\frac{x^3-1}{x+1} \cdot \frac{x^2+1}{x^2-1}$

49. $\frac{3(x+y)}{4} \div \frac{x+y}{2}$

50. $\frac{x+2}{5(x-3)} \div \frac{x-2}{5(x-3)}$

51. $\frac{x^2+16}{5x+20} \div \frac{x+4}{5x^2-20}$

52. $\frac{3x+18}{x^4} \div \frac{x+6}{x^2}$

53. $\frac{x^2-36}{x} \div \frac{x^3-6x^2}{x^2+x}$

54. $\frac{x^2-14x+49}{x^2-49} \div \frac{3x-21}{x+7}$

In Exercises 55–66, perform the addition or subtraction and simplify.

55. $\frac{5}{x-1} + \frac{x}{x-1}$

56. $\frac{2x-1}{x+3} + \frac{1-x}{x+3}$

57. $6 - \frac{5}{x+3}$

58. $\frac{3}{x-1} - 5$

59. $\frac{3}{x-2} + \frac{5}{2-x}$

60. $\frac{2x}{x-5} - \frac{5}{5-x}$

61. $\frac{2}{x^2-4} - \frac{1}{x^2-3x+2}$

62. $\frac{x}{x^2+x-2} - \frac{1}{x+2}$

63. $\frac{1}{x^2-x-2} - \frac{x}{x^2-5x+6}$

64. $\frac{2}{x^2-x-2} + \frac{10}{x^2+2x-8}$

65. $-\frac{1}{x} + \frac{2}{x^2+1} + \frac{1}{x^3+x}$

66. $\frac{2}{x+1} + \frac{2}{x-1} + \frac{1}{x^2-1}$

In Exercises 67–74, factor the expression by removing the common factor with the smaller exponent.

67. $x^5 - 2x^{-2}$

68. $x^5 - 5x^{-3}$

69. $3x^{3/2} - 2x^{-1/2}$

70. $5x^5 - 3x^{-3/2}$

71. $x^2(x^2+1)^{-5} - (x^2+1)^{-4}$

72. $2x(x-5)^{-3} - 4x^2(x-5)^{-4}$

73. $2x^2(x-1)^{1/2} - 5(x-1)^{-1/2}$

74. $4x^3(2x-1)^{3/2} - 2x(2x-1)^{-1/2}$

Error Analysis **In Exercises 75 and 76, describe the error.**

75. $$\frac{x+4}{x+2} - \frac{3x-8}{x+2} = \frac{x+4-3x-8}{x+2} = \frac{-2x-4}{x+2} = \frac{-2(x+2)}{x+2} = -2$$ ✗

76. $$\frac{6-x}{x(x+2)} + \frac{x+2}{x^2} + \frac{8}{x^2(x+2)} = \frac{x(6-x)+(x+2)^2+8}{x^2(x+2)} = \frac{6x-x^2+x^2+4+8}{x^2(x+2)} = \frac{6(x+2)}{x^2(x+2)} = \frac{6}{x^2}$$ ✗

In Exercises 77–92, simplify the complex fraction.

77. $\frac{\left(\frac{x}{2}-1\right)}{(x-2)}$

78. $\frac{(x-4)}{\left(\frac{x}{4}-\frac{4}{x}\right)}$

79. $\dfrac{\left[\dfrac{x^2}{(x+1)^2}\right]}{\left[\dfrac{x}{(x+1)^3}\right]}$

80. $\dfrac{\left(\dfrac{x^2-1}{x}\right)}{\left[\dfrac{(x-1)^2}{x}\right]}$

81. $\dfrac{\left(\dfrac{1}{x}-\dfrac{1}{x+1}\right)}{\left(\dfrac{1}{x+1}\right)}$

82. $\dfrac{\left(\dfrac{5}{y}-\dfrac{6}{2y+1}\right)}{\left(\dfrac{5}{y}+4\right)}$

83. $\dfrac{\left(\dfrac{x+3}{x-3}\right)^2}{\left(\dfrac{1}{x+3}+\dfrac{1}{x-3}\right)}$

84. $\dfrac{\left(\dfrac{x+4}{x+5}-\dfrac{x}{x+1}\right)}{4}$

85. $\dfrac{\left[\dfrac{1}{(x+h)^2}-\dfrac{1}{x^2}\right]}{h}$

86. $\dfrac{\left(\dfrac{x+h}{x+h+1}-\dfrac{x}{x+1}\right)}{h}$

87. $\dfrac{\left(\sqrt{x}-\dfrac{1}{2\sqrt{x}}\right)}{\sqrt{x}}$

88. $\dfrac{\left(\dfrac{t^2}{\sqrt{t^2+1}}-\sqrt{t^2+1}\right)}{t^2}$

89. $\dfrac{3x^{1/3}-x^{-2/3}}{3x^{-2/3}}$

90. $\dfrac{-x^3(1-x^2)^{-1/2}-2x(1-x^2)^{1/2}}{x^4}$

91. $\dfrac{x(x+1)^{-3/4}-(x+1)^{1/4}}{x^2}$

92. $\dfrac{(2x+1)^{1/3}-\dfrac{4x}{3(2x+1)^{2/3}}}{(2x+1)^{2/3}}$

In Exercises 93 and 94, rationalize the numerator of the expression.

93. $\dfrac{\sqrt{x+2}-\sqrt{x}}{2}$

94. $\dfrac{\sqrt{z-3}-\sqrt{z}}{3}$

95. *Rate* A photocopier copies at a rate of 16 pages per minute.

(a) Find the time required to copy one page.

(b) Find the time required to copy x pages.

(c) Find the time required to copy 60 pages.

96. *Rate* After working together for t hours on a common task, two workers have done fractional parts of the job equal to $t/3$ and $t/5$, respectively. What fractional part of the task has been completed?

97. *Average* Determine the average of the two real numbers $x/3$ and $2x/5$.

98. *Partition into Equal Parts* Find three real numbers that divide the real number line between $x/3$ and $3x/4$ into four equal parts.

Finance **In Exercises 99 and 100, use the formula that gives the approximate annual interest rate r of a monthly installment loan**

$$r = \frac{\left[\dfrac{24(NM-P)}{N}\right]}{\left(P+\dfrac{NM}{12}\right)}$$

where N is the total number of payments, M is the monthly payment, and P is the amount financed.

99. (a) Approximate the annual interest rate for a 4-year car loan of \$16,000 that has monthly payments of \$400.

(b) Simplify the expression for the annual interest rate r, and then rework part (a).

100. (a) Approximate the annual interest rate for a 5-year car loan of \$20,000 that has monthly payments of \$400.

(b) Simplify the expression for the annual interest rate r, and then rework part (a).

101. *Refrigeration* When food (at room temperature) is placed in a refrigerator, the time required for the food to cool depends on the amount of food, the air circulation in the refrigerator, the original temperature of the food, and the temperature of the refrigerator. Consider the model that gives the temperature of food that has an original temperature of 75°F and is placed in a 40°F refrigerator

$$T = 10\left(\frac{4t^2+16t+75}{t^2+4t+10}\right)$$

where T is the temperature (in degrees Fahrenheit) and t is the time (in hours).

(a) Complete the table.

t	0	2	4	6	8	10
T						

t	12	14	16	18	20	22
T						

(b) What value of T does the mathematical model appear to be approaching?

102. *Precious Metals* The costs per fine ounce of gold and silver for the years 1992 through 1997 are given in the table. (Source: U.S. Bureau of Mines, U.S. Geological Survey)

Year	1992	1993	1994	1995	1996	1997
Gold	\$345	\$361	\$385	\$386	\$389	\$333
Silver	\$3.94	\$4.30	\$5.29	\$5.15	\$5.19	\$4.90

Mathematical models for these data are

$$\text{Cost of gold} = \frac{-38.5t + 310.1}{0.007t^2 - 0.176t + 1}$$

and

$$\text{Cost of silver} = \frac{0.242t^2 - 1.86t + 4.02}{0.056t^2 - 0.45t + 1}$$

where $t = 2$ corresponds to the year 1992.

(a) Create a table using the models to estimate the prices of the two metals for the given years. Compare the estimates given by the models with the actual prices.

(b) Determine a model for the ratio of the price of gold to the price of silver. Use the model to find this ratio over the given years. Over this period of time, did the price of gold increase or decrease relative to the price of silver?

Probability **In Exercises 103–106, consider an experiment in which a marble is tossed into a box whose base is shown in the figure. The probability that the marble will come to rest in the shaded portion of the box is equal to the ratio of the shaded area to the total area of the figure. Find the probability.**

103.

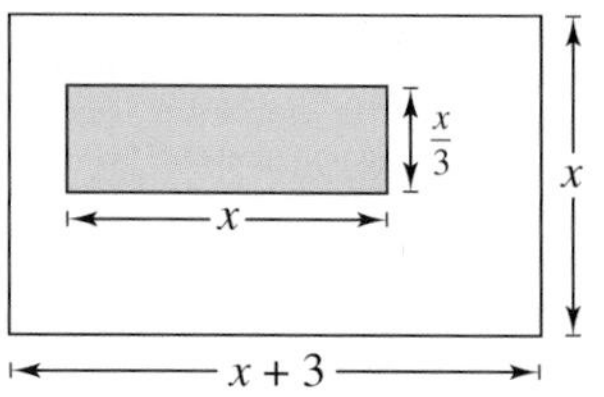

104.

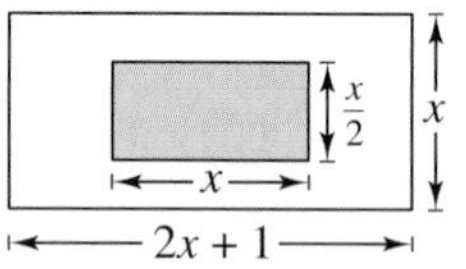

105.

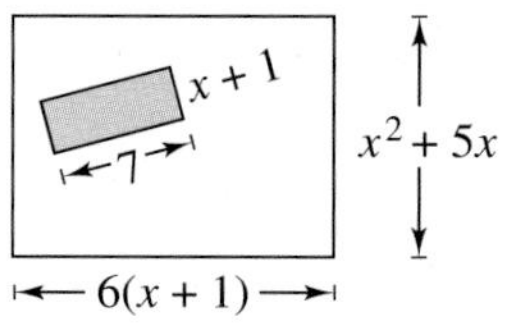

106.

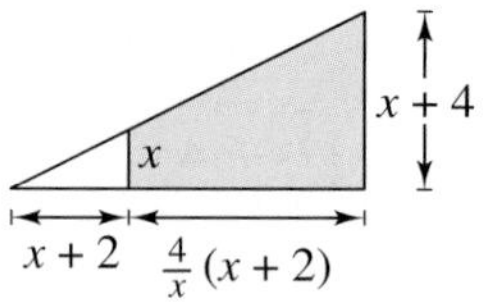

Synthesis

True or False? **In Exercises 107–109, determine whether the statement is true or false. Justify your answer.**

107. $\dfrac{x^{2n} - 1^{2n}}{x^n - 1^n} = x^n + 1^n$

108. $\dfrac{x^2 - 3x + 2}{x - 1} = x - 2$ for all values of x.

109. The least common denominator of two or more fractions is always the product of all the denominators of the fractions.

110. How do you determine whether a rational expression is in simplest form?

111. *Think About It* Is the following statement true for all nonzero real numbers a and b? Explain.

$$\frac{ax - b}{b - ax} = -1$$

P.5 Solving Equations

▶ **What you should learn**

- How to identify different types of equations
- How to solve linear equations in one variable
- How to solve quadratic equations by factoring, extracting square roots, and using the Quadratic Formula
- How to solve polynomial equations of degree three or greater
- How to solve equations involving radicals
- How to solve equations involving absolute values

▶ **Why you should learn it**

Linear equations are used in many real-life applications. For example, Exercises 185 and 186 on page 60 show how linear equations can model the relationship between the length of a thigh bone and the height of a person, helping researchers learn about ancient cultures.

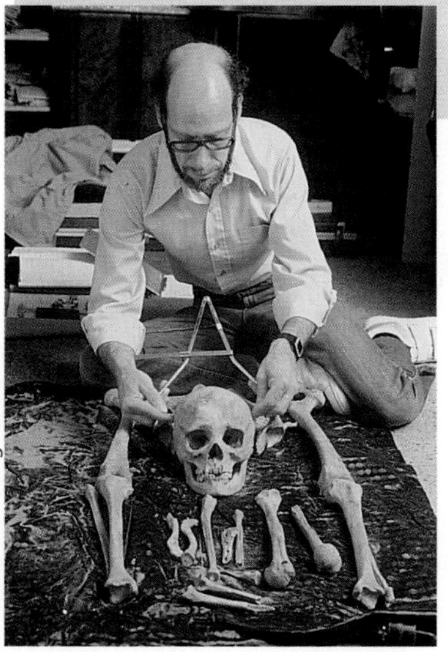

M. Greenlar/The Image Works

Equations and Solutions of Equations

An **equation** in x is a statement that two algebraic expressions are equal. For example,

$$3x - 5 = 7,\ x^2 - x - 6 = 0,\ \text{and}\ \sqrt{2x} = 4$$

are equations. To **solve** an equation in x means to find all values of x for which the equation is true. Such values are **solutions.** For instance, $x = 4$ is a solution of the equation

$$3x - 5 = 7,$$

because $3(4) - 5 = 7$ is a true statement.

The solutions of an equation depend on the kinds of numbers being considered. For instance, in the set of rational numbers, $x^2 = 10$ has no solution because there is no rational number whose square is 10. However, in the set of real numbers, the equation has the two solutions $\sqrt{10}$ and $-\sqrt{10}$.

An equation that is true for *every* real number in the domain of the variable is called an **identity.** For example,

$$x^2 - 9 = (x + 3)(x - 3) \qquad \text{Identity}$$

is an identity because it is a true statement for any real value of x, and

$$\frac{x}{3x^2} = \frac{1}{3x} \qquad \text{Identity}$$

where $x \neq 0$, is an identity because it is true for any nonzero real value of x.

An equation that is true for just *some* (or even none) of the real numbers in the domain of the variable is called a **conditional equation.** For example, the equation

$$x^2 - 9 = 0 \qquad \text{Conditional equation}$$

is conditional because $x = 3$ and $x = -3$ are the only values in the domain that satisfy the equation. The equation $2x - 4 = 2x + 1$ is conditional because there are no real values of x for which the equation is true. Learning to solve conditional equations is the primary focus of this section.

Linear Equations

Definition of Linear Equation

A **linear equation** in one variable x is an equation that can be written in the standard form

$$ax + b = 0$$

where a and b are real numbers with $a \neq 0$.

A linear equation has exactly one solution. To see this, consider the following steps. (Remember that $a \neq 0$.)

$ax + b = 0$ Write original equation.

$ax = -b$ Subtract b from each side.

$x = -\frac{b}{a}$ Divide each side by a.

To solve a conditional equation in x, isolate x on one side of the equation by a sequence of **equivalent** (and usually simpler) **equations,** each having the same solution(s) as the original equation. The operations that yield equivalent equations come from the Substitution Principle and simplification techniques.

Generating Equivalent Equations

An equation can be transformed into an *equivalent equation* by one or more of the following steps.

	Given Equation	*Equivalent Equation*
1. Remove symbols of grouping, combine like terms, or simplify fractions on one or both sides of the equation.	$2x - x = 4$	$x = 4$
2. Add (or subtract) the same quantity to (from) *each* side of the equation.	$x + 1 = 6$	$x = 5$
3. Multiply (or divide) *each* side of the equation by the same *nonzero* quantity.	$2x = 6$	$x = 3$
4. Interchange the two sides of the equation.	$2 = x$	$x = 2$

Example 1 ▶ Solving a Linear Equation

$3x - 6 = 0$ Write original equation.

$3x = 6$ Add 6 to each side.

$x = 2$ Divide each side by 3.

Check

After solving an equation, you should check each solution in the original equation.

$3x - 6 = 0$ Write original equation.

$3(2) - 6 \stackrel{?}{=} 0$ Substitute 2 for x.

$0 = 0$ Solution checks. ✓

So, the solution is 2.

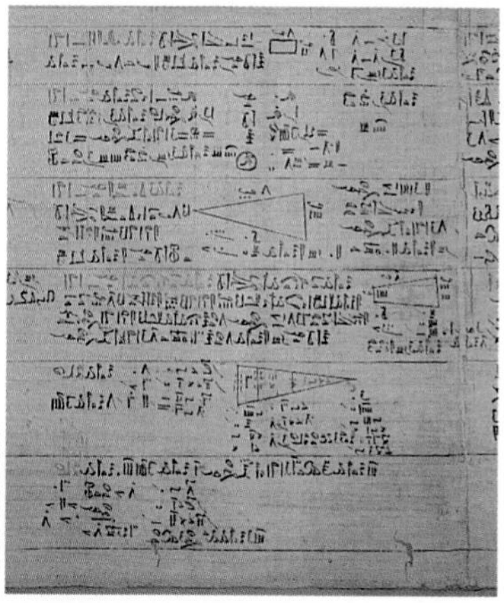

British Museum

Historical Note

This ancient Egyptian papyrus, discovered in 1858, contains one of the earliest examples of mathematical writing in existence. The papyrus itself dates back to around 1650 B.C., but it is actually a copy of writings from two centuries earlier. The algebraic equations on the papyrus were written in words. Diophantus, a Greek who lived around A.D. 250, is often called the Father of Algebra. He was the first to use abbreviated word forms in equations.

To solve an equation involving fractional expressions, find the least common denominator of all terms and multiply every term by this LCD.

Example 2 ▶ An Equation Involving Fractional Expressions

Solve $\frac{x}{3} + \frac{3x}{4} = 2$.

Solution

$$\frac{x}{3} + \frac{3x}{4} = 2 \qquad \text{Write original equation.}$$

$$(12)\frac{x}{3} + (12)\frac{3x}{4} = (12)2 \qquad \text{Multiply each term by the LCD of 12.}$$

$$4x + 9x = 24 \qquad \text{Divide out and multiply.}$$

$$13x = 24 \qquad \text{Combine like terms.}$$

$$x = \frac{24}{13} \qquad \text{Divide each side by 13.}$$

The solution is $\frac{24}{13}$. Check this in the original equation.

When multiplying or dividing an equation by a *variable* quantity, it is possible to introduce an extraneous solution. An **extraneous solution** is one that does not satisfy the original equation.

Example 3 ▶ An Equation with an Extraneous Solution

Solve $\frac{1}{x-2} = \frac{3}{x+2} - \frac{6x}{x^2-4}$.

Solution

The LCD is $x^2 - 4$, or $(x + 2)(x - 2)$. Multiply each term by this LCD.

$$\frac{1}{x-2}(x+2)(x-2) = \frac{3}{x+2}(x+2)(x-2) - \frac{6x}{x^2-4}(x+2)(x-2)$$

$$x + 2 = 3(x - 2) - 6x, \qquad x \neq \pm 2$$

$$x + 2 = 3x - 6 - 6x$$

$$x + 2 = -3x - 6$$

$$4x = -8$$

$$x = -2$$

In the original equation, $x = -2$ yields a denominator of zero. Therefore, $x = -2$ is an extraneous solution, and the original equation has *no solution.*

Activities

1. Solve and check:
   $4(x - 2) - 13 = 5(6x + 1)$.
   Answer: $x = -1$
2. Have students explain the differences between solving an equation and simplifying an expression.
3. Have students solve the following equation and then write one or two sentences explaining their answers.
   $\frac{3}{x(x+1)} + \frac{7}{x+1} = \frac{3}{x}$
   Answer: The solution yields $x = 0$, which is extraneous. Ask students to explain why this solution is extraneous.

Quadratic Equations

A **quadratic equation** in x is an equation that can be written in the general form

$$ax^2 + bx + c = 0$$

where a, b, and c are real numbers, with $a \neq 0$. A quadratic equation in x is also known as a **second-degree polynomial equation in x.**

You should be familiar with the following four methods for solving quadratic equations.

Solving a Quadratic Equation

Factoring: If $ab = 0$, then $a = 0$ or $b = 0$.

Example:

$$x^2 - x - 6 = 0$$
$$(x - 3)(x + 2) = 0$$
$$x - 3 = 0 \quad \Rightarrow \quad x = 3$$
$$x + 2 = 0 \quad \Rightarrow \quad x = -2$$

Square Root Principle: If $u^2 = c$, where $c > 0$, then $u = \pm\sqrt{c}$.

Example:

$$(x + 3)^2 = 16$$
$$x + 3 = \pm 4$$
$$x = -3 \pm 4$$
$$x = 1 \quad \text{or} \quad x = -7$$

Completing the Square: If $x^2 + bx = c$, then

$$x^2 + bx + \left(\frac{b}{2}\right)^2 = c + \left(\frac{b}{2}\right)^2$$
$$\left(x + \frac{b}{2}\right)^2 = c + \frac{b^2}{4}.$$

Example:

$$x^2 + 6x = 5$$
$$x^2 + 6x + 3^2 = 5 + 3^2$$
$$(x + 3)^2 = 14$$
$$x + 3 = \pm\sqrt{14}$$
$$x = -3 \pm \sqrt{14}$$

Quadratic Formula: If $ax^2 + bx + c = 0$, then $x = \dfrac{-b \pm \sqrt{b^2 - 4ac}}{2a}$.

Example:

$$2x^2 + 3x - 1 = 0$$
$$x = \frac{-3 \pm \sqrt{3^2 - 4(2)(-1)}}{2(2)}$$
$$= \frac{-3 \pm \sqrt{17}}{4}$$

Additional Example: Completing the Square Within an Algebraic Expression

Rewrite the denominator as the sum or difference of two squares.

$$\frac{1}{x^2 - 2x - 3}$$

Solution

Complete the square of the denominator.

$$x^2 - 2x - 3 = x^2 - 2x + 1^2 - 3 - 1^2$$
$$= (x^2 - 2x + 1) - 4$$
$$= (x - 1)^2 - 2^2$$

The original expression can be written as

$$\frac{1}{(x - 1)^2 - 2^2}.$$

A common error occurs when there are only two terms in an equation—as in Example 4(b).

Sample Error:

$6x^2 - 3x = 0$

$6x^2 = 3x$

$2x = 1$

$x = \frac{1}{2}$

The root $x = 0$ is lost when $6x^2 = 3x$ is divided by $3x$.

Example 4 ▶ Solving a Quadratic Equation by Factoring

a. $2x^2 + 9x + 7 = 3$ Write original equation.

$2x^2 + 9x + 4 = 0$ Write in general form.

$(2x + 1)(x + 4) = 0$ Factor.

$2x + 1 = 0 \Rightarrow x = -\dfrac{1}{2}$ Set 1st factor equal to 0.

$x + 4 = 0 \Rightarrow x = -4$ Set 2nd factor equal to 0.

The solutions are $-\frac{1}{2}$ and -4. Check these in the original equation.

b. $6x^2 - 3x = 0$ Write original equation.

$3x(2x - 1) = 0$ Factor.

$3x = 0 \Rightarrow x = 0$ Set 1st factor equal to 0.

$2x - 1 = 0 \Rightarrow x = \dfrac{1}{2}$ Set 2nd factor equal to 0.

The solutions are 0 and $\frac{1}{2}$. Check these in the original equation.

Note that the method of solution in Example 4 is based on the Zero-Factor Property given in Section P.1. Be sure you see that this property works *only* for equations written in general form (in which the right side of the equation is zero). Therefore, all terms must be collected on one side *before* factoring. For instance, in the equation

$$(x - 5)(x + 2) = 8$$

it is *incorrect* to set each factor equal to 8. Can you solve this equation correctly?

Example 5 ▶ Extracting Square Roots

Solve each equation by extracting square roots.

a. $4x^2 = 12$ **b.** $(x - 3)^2 = 7$

Solution

a. $4x^2 = 12$ Write original equation.

$x^2 = 3$ Divide each side by 4.

$x = \pm\sqrt{3}$ Extract square roots.

The solutions are $\sqrt{3}$ and $-\sqrt{3}$. Check these in the original equation.

b. $(x - 3)^2 = 7$ Write original equation.

$x - 3 = \pm\sqrt{7}$ Extract square roots.

$x = 3 \pm \sqrt{7}$ Add 3 to each side.

The solutions are $3 \pm \sqrt{7}$. Check these in the original equation.

A computer animation of this example appears in the *Interactive* CD-ROM and *Internet* versions of this text.

Example 6 ▶ The Quadratic Formula: Two Distinct Solutions

Use the Quadratic Formula to solve

$$x^2 + 3x = 9.$$

Solution

$x^2 + 3x = 9$ — Write original equation.

$x^2 + 3x - 9 = 0$ — Write in general form.

$x = \dfrac{-b \pm \sqrt{b^2 - 4ac}}{2a}$ — Quadratic Formula

$x = \dfrac{-3 \pm \sqrt{(3)^2 - 4(1)(-9)}}{2(1)}$ — Substitute 1 for a, 3 for b, and -9 for c.

$x = \dfrac{-3 \pm \sqrt{45}}{2}$ — Simplify.

$x = \dfrac{-3 \pm 3\sqrt{5}}{2}$ — Simplify.

The equation has two solutions:

$$x = \frac{-3 + 3\sqrt{5}}{2} \quad \text{and} \quad x = \frac{-3 - 3\sqrt{5}}{2}.$$

Check these in the original equation.

Activities

1. Solve $3x^2 = 5x$.

   Answer: $x = 0, \frac{5}{3}$

2. Solve $(x + 3)^2 = 12$.

   Answer: $x = -3 \pm 2\sqrt{3}$

3. Complete the square for the quadratic portion of $\dfrac{1}{\sqrt{x^2 - 6x + 13}}$.

   Answer: $\dfrac{1}{\sqrt{(x - 3)^2 + 4}}$

4. Use the Quadratic Formula to solve $3x^2 = 3x + 1$.

   Answer: $x = \dfrac{3 \pm \sqrt{21}}{6}$

5. If the perimeter of a right triangle is 42 units and the length of the hypotenuse is 20 units, find the lengths of the other two sides of the triangle.

   Answer: 19.888; 2.112

Example 7 ▶ The Quadratic Formula: One Solution

Use the Quadratic Formula to solve

$$8x^2 - 24x + 18 = 0.$$

Solution

$8x^2 - 24x + 18 = 0$ — Write original equation.

$4x^2 - 12x + 9 = 0$ — Divide out common factor of 2.

$x = \dfrac{-b \pm \sqrt{b^2 - 4ac}}{2a}$ — Quadratic Formula

$x = \dfrac{-(-12) \pm \sqrt{(-12)^2 - 4(4)(9)}}{2(4)}$ — Substitute.

$x = \dfrac{12 \pm \sqrt{0}}{8}$ — Simplify.

$x = \dfrac{3}{2}$ — Simplify.

This quadratic equation has only one solution: $\frac{3}{2}$. Check this in the original equation.

STUDY TIP

A common mistake that is made in solving an equation such as that in Example 8 is to divide both sides of the equation by the variable factor x^2. This loses the solution $x = 0$. When solving an equation, be sure to write the equation in general form, then factor the equation and set each factor equal to zero. Don't divide both sides of an equation by a variable factor in an attempt to simplify the equation.

Polynomial Equations of Higher Degree

The methods used to solve quadratic equations can sometimes be extended to polynomials of higher degree.

Example 8 ▶ Solving a Polynomial Equation by Factoring

Solve $3x^4 = 48x^2$.

Solution

First write the polynomial equation in general form with zero on one side, factor the other side, and then set each factor equal to zero.

$$3x^4 = 48x^2 \quad \text{Write original equation.}$$
$$3x^4 - 48x^2 = 0 \quad \text{Write in general form.}$$
$$3x^2(x^2 - 16) = 0 \quad \text{Factor.}$$
$$3x^2(x + 4)(x - 4) = 0 \quad \text{Write in factored form.}$$
$$3x^2 = 0 \Rightarrow x = 0 \quad \text{Set 1st factor equal to 0.}$$
$$x + 4 = 0 \Rightarrow x = -4 \quad \text{Set 2nd factor equal to 0.}$$
$$x - 4 = 0 \Rightarrow x = 4 \quad \text{Set 3rd factor equal to 0.}$$

You can check these solutions by substituting in the original equation, as follows.

Check

$$3x^4 = 48x^2 \quad \text{Write original equation.}$$
$$3(0)^4 = 48(0)^2 \quad \text{0 checks. ✓}$$
$$3(-4)^4 = 48(-4)^2 \quad \text{−4 checks. ✓}$$
$$3(4)^4 = 48(4)^2 \quad \text{4 checks. ✓}$$

After checking, you can conclude that the solutions are 0, −4, and 4.

Example 9 ▶ Solving a Polynomial Equation by Factoring

Solve $x^3 - 3x^2 - 3x + 9 = 0$.

Solution

$$x^3 - 3x^2 - 3x + 9 = 0 \quad \text{Write original equation.}$$
$$x^2(x - 3) - 3(x - 3) = 0 \quad \text{Factor by grouping.}$$
$$(x - 3)(x^2 - 3) = 0 \quad \text{Distributive Property}$$
$$x - 3 = 0 \Rightarrow x = 3 \quad \text{Set 1st factor equal to 0.}$$
$$x^2 - 3 = 0 \Rightarrow x = \pm\sqrt{3} \quad \text{Set 2nd factor equal to 0.}$$

The solutions are 3, $\sqrt{3}$, and $-\sqrt{3}$.

Radical Equations

The steps involved in solving the remaining equations in this section will often introduce *extraneous solutions*. Extraneous solutions occur during operations such as squaring each side of an equation, raising each side of an equation to a rational power, and multiplying each side by a variable quantity. So, when you use any of these operations, checking is crucial.

Example 10 ▶ Solving an Equation Involving a Rational Exponent

$4x^{3/2} - 8 = 0$	Write original equation.
$4x^{3/2} = 8$	Add 8 to each side.
$x^{3/2} = 2$	Divide each side by 4.
$x = 2^{2/3}$	Raise each side to the $\frac{2}{3}$ power.
$x \approx 1.587$	Round to three decimal places.

The solution is $2^{2/3}$. Check this in the original equation.

STUDY TIP

The essential operations in Example 10 are isolating the factor with the rational exponent and raising each side to the reciprocal power. In Example 11, this is equivalent to isolating the square root and squaring each side.

Example 11 ▶ Solving Equations Involving Radicals

a.

$\sqrt{2x+7} - x = 2$	Write original equation.
$\sqrt{2x+7} = x + 2$	Add x to each side.
$2x + 7 = x^2 + 4x + 4$	Square each side.
$0 = x^2 + 2x - 3$	Write in general form.
$0 = (x+3)(x-1)$	Factor.
$x + 3 = 0 \Rightarrow x = -3$	Set 1st factor equal to 0.
$x - 1 = 0 \Rightarrow x = 1$	Set 2nd factor equal to 0.

By checking these values, you can determine that the only solution is 1.

b.

$\sqrt{2x-5} - \sqrt{x-3} = 1$	Write original equation.
$\sqrt{2x-5} = \sqrt{x-3} + 1$	Add $\sqrt{x-3}$ to each side.
$2x - 5 = x - 3 + 2\sqrt{x-3} + 1$	Square each side.
$2x - 5 = x - 2 + 2\sqrt{x-3}$	Combine like terms.
$x - 3 = 2\sqrt{x-3}$	Add $-x + 2$ to each side.
$x^2 - 6x + 9 = 4(x-3)$	Square each side.
$x^2 - 10x + 21 = 0$	Write in general form.
$(x-3)(x-7) = 0$	Factor.
$x - 3 = 0 \Rightarrow x = 3$	Set 1st factor equal to 0.
$x - 7 = 0 \Rightarrow x = 7$	Set 2nd factor equal to 0.

The solutions are 3 and 7. Check these in the original equation.

To show why the radical should be isolated, have students square both sides of the equation before isolating the radical. Compare problem-solving strategies to help convince students of the need to isolate the radical.

Additional Examples

a. $\frac{7}{x} - \frac{1}{3x} = \frac{8}{3}$

$3x\left(\frac{7}{x}\right) - (3x)\left(\frac{1}{3x}\right) = (3x)\frac{8}{3}$

$21 - 1 = 8x$

$\frac{20}{8} = x$

$\frac{5}{2} = x$

b. $\frac{3x}{x+1} = \frac{12}{x^2-1} + 2$

$(x^2-1)\frac{3x}{x+1} = (x^2-1)\frac{12}{x^2-1} + (x^2-1)2$

$3x(x-1) = 12 + (x^2-1)2$

$3x^2 - 3x = 12 + 2x^2 - 2$

$x^2 - 3x - 10 = 0$

$(x+2)(x-5) = 0$

$x + 2 = 0 \Rightarrow x = -2$

$x - 5 = 0 \Rightarrow x = 5$

Activities

1. Solve for x: $9x^4 - 9x^2 + 2 = 0$.

   Answer: $x = \pm\frac{\sqrt{3}}{3}, \pm\frac{\sqrt{6}}{3}$

2. Solve for x: $\sqrt{3x+10} - x = 4$.

   Answer: $x = -2, -3$

3. Solve for x: $|5x + 2| = 22$.

   Answer: $x = -\frac{24}{5}, 4$

Absolute Value Equations

To solve an equation involving an absolute value, remember that the expression inside the absolute value signs can be positive or negative. This results in *two* separate equations, each of which must be solved. For instance, the equation

$$|x - 2| = 3$$

results in the two equations $x - 2 = 3$ and $-(x - 2) = 3$ which implies that the equation has two solutions: 5 and -1.

Example 12 ▶ Solving an Equation Involving Absolute Value

Solve $|x^2 - 3x| = -4x + 6$.

Solution

Because the variable expression inside the absolute value signs can be positive or negative, you must solve the following two equations.

First Equation

$x^2 - 3x = -4x + 6$		Use positive expression.
$x^2 + x - 6 = 0$		Write in general form.
$(x + 3)(x - 2) = 0$		Factor.
$x + 3 = 0$ ⇨	$x = -3$	Set 1st factor equal to 0.
$x - 2 = 0$ ⇨	$x = 2$	Set 2nd factor equal to 0.

Second Equation

$-(x^2 - 3x) = -4x + 6$		Use negative expression.
$x^2 - 7x + 6 = 0$		Write in general form.
$(x - 1)(x - 6) = 0$		Factor.
$x - 1 = 0$ ⇨	$x = 1$	Set 1st factor equal to 0.
$x - 6 = 0$ ⇨	$x = 6$	Set 2nd factor equal to 0.

Check

$\|(-3)^2 - 3(-3)\| \stackrel{?}{=} -4(-3) + 6$	Substitute -3 for x.
$18 = 18$	-3 checks. ✓
$\|(2)^2 - 3(2)\| \stackrel{?}{=} -4(2) + 6$	Substitute 2 for x.
$2 \neq -2$	2 does not check.
$\|(1)^2 - 3(1)\| \stackrel{?}{=} -4(1) + 6$	Substitute 1 for x.
$2 = 2$	1 checks. ✓
$\|(6)^2 - 3(6)\| \stackrel{?}{=} -4(6) + 6$	Substitute 6 for x.
$18 \neq -18$	6 does not check.

The solutions are -3 and 1.

P.5 Exercises

In Exercises 1–10, determine whether the equation is an identity or a conditional equation.

1. $2(x - 1) = 2x - 2$
2. $3(x + 2) = 5x + 4$
3. $-6(x - 3) + 5 = -2x + 10$
4. $3(x + 2) - 5 = 3x + 1$
5. $4(x + 1) - 2x = 2(x + 2)$
6. $-7(x - 3) + 4x = 3(7 - x)$
7. $x^2 - 8x - 5 = (x - 4)^2 - 11$
8. $x^2 + 2(3x - 2) = x^2 + 6x - 4$
9. $3 + \dfrac{1}{x + 1} = \dfrac{4x}{x + 1}$
10. $\dfrac{5}{x} + \dfrac{3}{x} = 24$

In Exercises 11–26, solve the equation and check your solution.

11. $x + 11 = 15$
12. $7 - x = 19$
13. $7 - 2x = 25$
14. $7x + 2 = 23$
15. $8x - 5 = 3x + 20$
16. $7x + 3 = 3x - 17$
17. $2(x + 5) - 7 = 3(x - 2)$
18. $3(x + 3) = 5(1 - x) - 1$
19. $x - 3(2x + 3) = 8 - 5x$
20. $9x - 10 = 5x + 2(2x - 5)$
21. $\dfrac{5x}{4} + \dfrac{1}{2} = x - \dfrac{1}{2}$
22. $\dfrac{x}{5} - \dfrac{x}{2} = 3 + \dfrac{3x}{10}$
23. $\frac{3}{2}(z + 5) - \frac{1}{4}(z + 24) = 0$
24. $\dfrac{3x}{2} + \dfrac{1}{4}(x - 2) = 10$
25. $0.25x + 0.75(10 - x) = 3$
26. $0.60x + 0.40(100 - x) = 50$

In Exercises 27–48, solve the equation and check your solution. (If not possible, explain why.)

27. $x + 8 = 2(x - 2) - x$
28. $8(x + 2) - 3(2x + 1) = 2(x + 5)$
29. $\dfrac{100 - 4x}{3} = \dfrac{5x + 6}{4} + 6$
30. $\dfrac{17 + y}{y} + \dfrac{32 + y}{y} = 100$
31. $\dfrac{5x - 4}{5x + 4} = \dfrac{2}{3}$
32. $\dfrac{10x + 3}{5x + 6} = \dfrac{1}{2}$
33. $10 - \dfrac{13}{x} = 4 + \dfrac{5}{x}$
34. $\dfrac{15}{x} - 4 = \dfrac{6}{x} + 3$
35. $\dfrac{x}{x + 4} + \dfrac{4}{x + 4} + 2 = 0$
36. $3 = 2 + \dfrac{2}{z + 2}$
37. $\dfrac{1}{x} + \dfrac{2}{x - 5} = 0$
38. $\dfrac{7}{2x + 1} - \dfrac{8x}{2x - 1} = -4$
39. $\dfrac{2}{(x - 4)(x - 2)} = \dfrac{1}{x - 4} + \dfrac{2}{x - 2}$
40. $\dfrac{4}{x - 1} + \dfrac{6}{3x + 1} = \dfrac{15}{3x + 1}$
41. $\dfrac{1}{x - 3} + \dfrac{1}{x + 3} = \dfrac{10}{x^2 - 9}$
42. $\dfrac{1}{x - 2} + \dfrac{3}{x + 3} = \dfrac{4}{x^2 + x - 6}$
43. $\dfrac{3}{x^2 - 3x} + \dfrac{4}{x} = \dfrac{1}{x - 3}$
44. $\dfrac{6}{x} - \dfrac{2}{x + 3} = \dfrac{3(x + 5)}{x^2 + 3x}$
45. $(x + 2)^2 + 5 = (x + 3)^2$
46. $(x + 1)^2 + 2(x - 2) = (x + 1)(x - 2)$
47. $(x + 2)^2 - x^2 = 4(x + 1)$
48. $(2x + 1)^2 = 4(x^2 + x + 1)$

In Exercises 49–54, write the quadratic equation in general form.

49. $2x^2 = 3 - 8x$
50. $x^2 = 16x$
51. $(x - 3)^2 = 3$
52. $13 - 3(x + 7)^2 = 0$
53. $\frac{1}{5}(3x^2 - 10) = 18x$
54. $x(x + 2) = 5x^2 + 1$

In Exercises 55–68, solve the quadratic equation for x by factoring.

55. $6x^2 + 3x = 0$
56. $9x^2 - 1 = 0$
57. $x^2 - 2x - 8 = 0$
58. $x^2 - 10x + 9 = 0$
59. $x^2 + 10x + 25 = 0$
60. $4x^2 + 12x + 9 = 0$
61. $3 + 5x - 2x^2 = 0$
62. $2x^2 = 19x + 33$
63. $x^2 + 4x = 12$
64. $-x^2 + 8x = 12$

65. $\frac{3}{4}x^2 + 8x + 20 = 0$ **66.** $\frac{1}{8}x^2 - x - 16 = 0$

67. $x^2 + 2ax + a^2 = 0$ **68.** $(x + a)^2 - b^2 = 0$

In Exercises 69–82, solve the equation by extracting square roots. List both the exact solution and the decimal solution rounded to two decimal places.

69. $x^2 = 49$ **70.** $x^2 = 169$

71. $x^2 = 11$ **72.** $x^2 = 32$

73. $3x^2 = 81$ **74.** $9x^2 = 36$

75. $(x - 12)^2 = 16$ **76.** $(x + 13)^2 = 25$

77. $(x + 2)^2 = 14$ **78.** $(x - 5)^2 = 30$

79. $(2x - 1)^2 = 18$ **80.** $(4x + 7)^2 = 44$

81. $(x - 7)^2 = (x + 3)^2$ **82.** $(x + 5)^2 = (x + 4)^2$

In Exercises 83–92, solve the quadratic equation by completing the square.

83. $x^2 - 2x = 0$ **84.** $x^2 + 4x = 0$

85. $x^2 + 4x - 32 = 0$ **86.** $x^2 - 2x - 3 = 0$

87. $x^2 + 6x + 2 = 0$ **88.** $x^2 + 8x + 14 = 0$

89. $9x^2 - 18x = -3$ **90.** $9x^2 - 12x = 14$

91. $8 + 4x - x^2 = 0$ **92.** $4x^2 - 4x - 99 = 0$

In Exercises 93–116, use the Quadratic Formula to solve the equation.

93. $2x^2 + x - 1 = 0$ **94.** $2x^2 - x - 1 = 0$

95. $16x^2 + 8x - 3 = 0$ **96.** $25x^2 - 20x + 3 = 0$

97. $2 + 2x - x^2 = 0$ **98.** $x^2 - 10x + 22 = 0$

99. $x^2 + 14x + 44 = 0$ **100.** $6x = 4 - x^2$

101. $x^2 + 8x - 4 = 0$ **102.** $4x^2 - 4x - 4 = 0$

103. $12x - 9x^2 = -3$ **104.** $16x^2 + 22 = 40x$

105. $9x^2 + 24x + 16 = 0$

106. $36x^2 + 24x - 7 = 0$

107. $4x^2 + 4x = 7$ **108.** $16x^2 - 40x + 5 = 0$

109. $28x - 49x^2 = 4$ **110.** $3x + x^2 - 1 = 0$

111. $8t = 5 + 2t^2$ **112.** $25h^2 + 80h + 61 = 0$

113. $(y - 5)^2 = 2y$ **114.** $(z + 6)^2 = -2z$

115. $\frac{1}{2}x^2 + \frac{3}{8}x = 2$ **116.** $(\frac{5}{7}x - 14)^2 = 8x$

In Exercises 117–124, use the Quadratic Formula to solve the equation. (Round your answers to three decimal places.)

117. $5.1x^2 - 1.7x - 3.2 = 0$

118. $2x^2 - 2.50x - 0.42 = 0$

119. $-0.067x^2 - 0.852x + 1.277 = 0$

120. $-0.005x^2 + 0.101x - 0.193 = 0$

121. $422x^2 - 506x - 347 = 0$

122. $1100x^2 + 326x - 715 = 0$

123. $12.67x^2 + 31.55x + 8.09 = 0$

124. $-3.22x^2 - 0.08x + 28.651 = 0$

In Exercises 125–134, solve the equation for *x* by any convenient method.

125. $x^2 - 2x - 1 = 0$ **126.** $11x^2 + 33x = 0$

127. $(x + 3)^2 = 81$ **128.** $x^2 - 14x + 49 = 0$

129. $x^2 - x - \frac{11}{4} = 0$ **130.** $x^2 + 3x - \frac{3}{4} = 0$

131. $(x + 1)^2 = x^2$ **132.** $a^2x^2 - b^2 = 0$

133. $3x + 4 = 2x^2 - 7$

134. $4x^2 + 2x + 4 = 2x + 8$

In Exercises 135–152, find all real solutions of the equation. Check your solutions in the original equation.

135. $4x^4 - 18x^2 = 0$ **136.** $20x^3 - 125x = 0$

137. $x^4 - 81 = 0$ **138.** $x^6 - 64 = 0$

139. $x^3 + 216 = 0$ **140.** $27x^3 - 512 = 0$

141. $5x^3 + 30x^2 + 45x = 0$

142. $9x^4 - 24x^3 + 16x^2 = 0$

143. $x^3 - 3x^2 - x + 3 = 0$

144. $x^3 + 2x^2 + 3x + 6 = 0$

145. $x^4 - x^3 + x - 1 = 0$

146. $x^4 + 2x^3 - 8x - 16 = 0$

147. $x^4 - 4x^2 + 3 = 0$

148. $x^4 + 5x^2 - 36 = 0$

149. $4x^4 - 65x^2 + 16 = 0$

150. $36t^4 + 29t^2 - 7 = 0$

151. $x^6 + 7x^3 - 8 = 0$

152. $x^6 + 3x^3 + 2 = 0$

In Exercises 153–170, find all solutions of the equation. Check your solutions in the original equation.

153. $\sqrt{2x} - 10 = 0$ **154.** $4\sqrt{x} - 3 = 0$

155. $\sqrt{x - 10} - 4 = 0$ **156.** $\sqrt{5 - x} - 3 = 0$

157. $\sqrt[3]{2x + 5} + 3 = 0$ **158.** $\sqrt[3]{3x + 1} - 5 = 0$

159. $-\sqrt{26 - 11x} + 4 = x$

160. $x + \sqrt{31 - 9x} = 5$

161. $\sqrt{x + 1} = \sqrt{3x + 1}$

162. $\sqrt{x + 5} = \sqrt{x - 5}$

163. $(x-5)^{3/2}=8$

164. $(x+3)^{3/2}=8$

165. $(x+3)^{2/3}=8$

166. $(x+2)^{2/3}=9$

167. $(x^2-5)^{3/2}=27$

168. $(x^2-x-22)^{3/2}=27$

169. $3x(x-1)^{1/2}+2(x-1)^{3/2}=0$

170. $4x^2(x-1)^{1/3}+6x(x-1)^{4/3}=0$

In Exercises 171–184, find all solutions of the equation. Check your solutions in the original equation.

171. $\dfrac{20-x}{x}=x$

172. $\dfrac{4}{x}-\dfrac{5}{3}=\dfrac{x}{6}$

173. $\dfrac{1}{x}-\dfrac{1}{x+1}=3$

174. $\dfrac{x}{x^2-4}+\dfrac{1}{x+2}=3$

175. $x=\dfrac{3}{x}+\dfrac{1}{2}$

176. $4x+1=\dfrac{3}{x}$

177. $\dfrac{4}{x+1}-\dfrac{3}{x+2}=1$

178. $\dfrac{x+1}{3}-\dfrac{x+1}{x+2}=0$

179. $|2x-1|=5$

180. $|3x+2|=7$

181. $|x|=x^2+x-3$

182. $|x^2+6x|=3x+18$

183. $|x+1|=x^2-5$

184. $|x-10|=x^2-10x$

Anthropology **In Exercises 185 and 186, use the following information. The relationship between the length of an adult's thigh bone and the height of the adult can be approximated by the linear equations**

$y=0.432x-10.44$ **Female**

$y=0.449x-12.15$ **Male**

where y is the length of the femur (thigh bone) in inches and x is the height in inches (see figure).

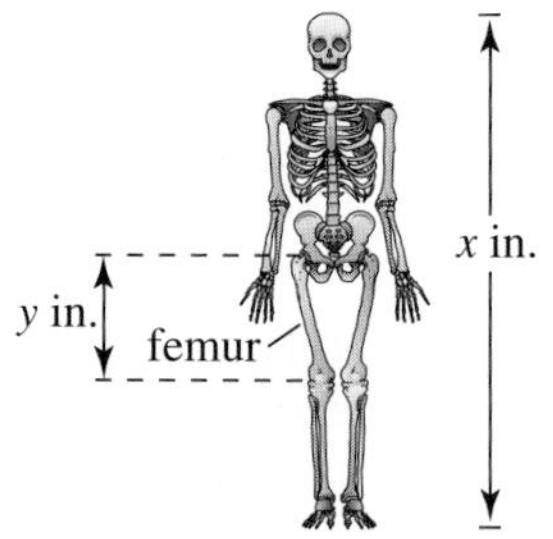

185. An anthropologist discovers a thigh bone belonging to an adult human female. The bone is 16 inches long. Estimate the height of the female.

186. From the foot bones of an adult human male, an anthropologist estimates that the person's height was 69 inches. A few feet away from the site where the foot bones were discovered, the anthropologist discovered a male adult thigh bone that was 19 inches long. Is it likely that both the foot bones and the thigh bone came from the same person?

187. *Operating Cost* A delivery company has a fleet of vans. The annual operating cost per van is

$C=0.32m+2500$

where m is the number of miles traveled by a van in a year. What number of miles will yield an annual operating cost that is equal to \$10,000?

188. *Flood Control* Suppose a river has risen 8 feet above its flood stage. The water begins to recede at a rate of 3 inches per hour. Write a mathematical model that shows the number of feet above flood stage after t hours. If the water continually recedes at this rate, when will the river be 1 foot above its flood stage?

189. *Floor Space* The floor of a one-story building is 14 feet longer than it is wide. The building has 1632 square feet of floor space.

(a) Draw a diagram that gives a visual representation of the floor space. Represent the width as w and show the length in terms of w.

(b) Write a quadratic equation in terms of w.

(c) Find the length and width of the building floor.

190. *Packaging* An open box with a square base (see figure) is to be constructed from 84 square inches of material. What should be the dimensions of the base if the height of the box is to be 2 inches? (*Hint:* The surface area is $S=x^2+4xh$.)

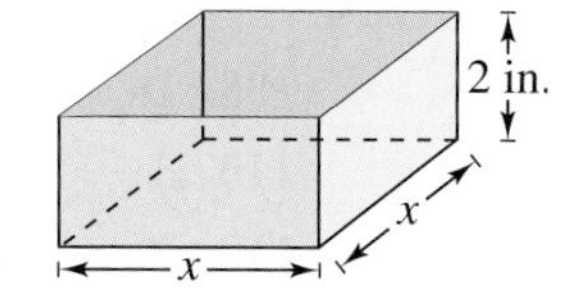

191. *Geometry* The hypotenuse of an isosceles right triangle is 5 centimeters long. How long are its sides?

192. *Geometry* An equilateral triangle has a height of 10 inches. How long are its sides? (*Hint:* Use the height of the triangle to partition the triangle into two congruent right triangles.)

193. *Flying Speed* Two planes leave simultaneously from the same airport, one flying due north and the other due east (see figure). The northbound plane is flying 50 miles per hour faster than the eastbound plane. After 3 hours the planes are 2440 miles apart. Find the speed of each plane.

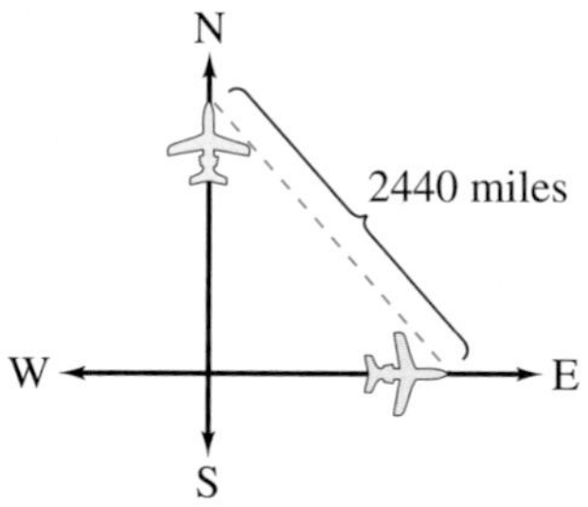

194. *Airline Passengers* An airline offers daily flights between Chicago and Denver. The total monthly cost of these flights is

$$C = \sqrt{0.2x + 1}$$

where C is the cost (in millions of dollars) and x is the number of passengers (in thousands). The total cost of the flights for a certain month is 2.5 million dollars. How many passengers flew that month?

195. *Economics* The demand equation for a certain product is $p = 20 - 0.0002x$, where p is the price per unit and x is the number of units sold. The total revenue for selling x units is

$$\text{Revenue} = xp = x(20 - 0.0002x).$$

How many units must be sold to produce a revenue of \$500,000?

196. *Economics* The demand equation for a certain product is modeled by

$$p = 40 - \sqrt{0.01x + 1}$$

where x is the number of units demanded per day and p is the price per unit. Approximate the demand if the price is \$37.55.

197. *Saturated Steam* The temperature T (in degrees Fahrenheit) of saturated steam increases as pressure increases. This relationship is approximated by

$$T = 75.82 - 2.11x + 43.51\sqrt{x}, \quad 5 \le x \le 40$$

where x is the absolute pressure (in pounds per square inch). Approximate the pressure if the temperature of the steam is 240°F.

Synthesis

True or False? **In Exercises 198–200, determine whether the statement is true or false. Justify your answer.**

198. The equation $x(3 - x) = 10$ is a linear equation.

199. If $(2x - 3)(x + 5) = 8$, then $2x - 3 = 8$ or $x + 5 = 8$.

200. When solving an absolute value equation, you will always have to check more than one solution.

201. To solve the equation

$$2x^2 + 3x = 15x$$

a student divides both sides by x and solves the equation $2x + 3 = 15$. The resulting solution $(x = 6)$ satisfies the given equation. Is there an error? Explain.

202. *Think About It* What is meant by "equivalent equations"? Give an example of two equivalent equations.

203. *Writing* In your own words, describe the steps used to transform an equation into an equivalent equation.

204. *Exploration* Solve $3(x + 4)^2 + (x + 4) - 2 = 0$ in two ways.

(a) Let $u = x + 4$, and solve the resulting equation for u. Then solve the u-solution for x.

(b) Expand and collect like terms in the equation, and solve the resulting equation for x.

(c) Which method is easier? Explain.

205. *Exploration* Solve the equations, given that a and b are not zero.

(a) $ax^2 + bx = 0$

(b) $ax^2 - ax = 0$

In Exercises 206 and 207, consider an equation of the form $x + |x - a| = b$, where a and b are constants.

206. *Exploration* Find a and b if the solution to the equation is $x = 9$. (There are many correct answers.)

207. *Writing* Write a short paragraph listing the steps required to solve this equation involving absolute value.

P.6 Solving Inequalities

▶ **What you should learn**

- How to recognize solutions of linear inequalities
- How to use properties of inequalities to solve linear inequalities
- How to solve inequalities involving absolute values
- How to solve polynomial and rational inequalities

▶ **Why you should learn it**

Inequalities can be used to model and solve real-life problems. For instance, Exercise 144 on page 72 shows how to use a linear inequality to analyze data about the maximum weight a weightlifter can bench press.

Brian Smith/Stock Boston

Introduction

Simple inequalities were reviewed in Section P.1. There, you used the inequality symbols $<$, $\leq$, $>$, and $\geq$ to compare two numbers and to denote subsets of real numbers. For instance, the simple inequality

$$x \geq 3$$

denotes all real numbers x that are greater than or equal to 3.

In this section you will expand your work with inequalities to include more involved statements such as

$$5x - 7 < 3x + 9$$

and

$$-3 \leq 6x - 1 < 3.$$

As with an equation, you **solve an inequality** in the variable x by finding all values of x for which the inequality is true. Such values are **solutions** and are said to **satisfy** the inequality. The set of all real numbers that are solutions of an inequality is the **solution set** of the inequality. For instance, the solution set of

$$x + 1 < 4$$

is all real numbers that are less than 3.

The set of all points on the real number line that represent the solution set is the **graph of the inequality.** Graphs of many types of inequalities consist of intervals on the real number line. You can review the nine basic types of intervals on the real number line by turning to pages 3 and 4 in Section P.1. On those pages, note that each type of interval can be classified as *bounded* or *unbounded.*

Example 1 ▶ Intervals and Inequalities

Write an inequality to represent each interval and state whether the interval is bounded or unbounded.

a. $(-3, 5]$

b. $(-3, \infty)$

c. $[0, 2]$

d. $(-\infty, \infty)$

Solution

a. $(-3, 5]$ corresponds to $-3 < x \leq 5$. Bounded

b. $(-3, \infty)$ corresponds to $-3 < x$. Unbounded

c. $[0, 2]$ corresponds to $0 \leq x \leq 2$. Bounded

d. $(-\infty, \infty)$ corresponds to $-\infty < x < \infty$. Unbounded

Properties of Inequalities

The procedures for solving linear inequalities in one variable are much like those for solving linear equations. To isolate the variable, you can make use of the **properties of inequalities.** These properties are similar to the properties of equality, but there are two important exceptions. When each side of an inequality is multiplied or divided by a negative number, the direction of the inequality symbol must be reversed. Here is an example.

$$-2 < 5 \qquad \text{Write original inequality.}$$

$$(-3)(-2) > (-3)(5) \qquad \text{Multiply each side by } -3 \text{ and reverse inequality.}$$

$$6 > -15 \qquad \text{Simplify.}$$

Two inequalities that have the same solution set are **equivalent.** For instance, the inequalities

$$x + 2 < 5$$

and

$$x < 3$$

are equivalent. To obtain the second inequality from the first, you can subtract 2 from each side of the inequality. The following list describes the operations that can be used to create equivalent inequalities.

Additional Examples
To emphasize Property 4, show several examples.

a. $-4 > -16$
$(-4)(-3) < (-16)(-3)$
$12 < 48$

b. $4 - 3x \le 13$
$-3x \le 9$
$x \ge -3$

c. $5 + 2x > 1$
$2x > -4$
$x > -2$

Properties of Inequalities

Let a, b, c, and d be real numbers.

1. Transitive Property

$a < b$ and $b < c \implies a < c$

2. Addition of Inequalities

$a < b$ and $c < d \implies a + c < b + d$

3. Addition of a Constant

$a < b \implies a + c < b + c$

4. Multiplication by a Constant

For $c > 0$, $a < b \implies ac < bc$

For $c < 0$, $a < b \implies ac > bc$

Each of the properties above is true if the symbol $<$ is replaced by $\le$ and $>$ is replaced by $\ge$. For instance, another form of the multiplication property would be as follows.

For $c > 0$, $a \le b \implies ac \le bc$

For $c < 0$, $a \le b \implies ac \ge bc$

Linear Inequalities

The simplest type of inequality is a **linear inequality** in a single variable. For instance, $2x + 3 > 4$ is a linear inequality in x.

In the following examples, pay special attention to the steps in which the inequality symbol is reversed. Remember that when you multiply or divide by a negative number, you must reverse the inequality symbol.

STUDY TIP

Checking the solution set of an inequality is not as simple as checking the solutions of an equation. You can, however, get an indication of the validity of a solution set by substituting a few convenient values of x.

Example 2 ▶ Solving Linear Inequalities

Solve each inequality.

a. $5x - 7 > 3x + 9$ **b.** $1 - \dfrac{3x}{2} \geq x - 4$

Solution

a.

$5x - 7 > 3x + 9$	Write original inequality.
$5x > 3x + 16$	Add 7 to each side.
$5x - 3x > 16$	Subtract $3x$ from each side.
$2x > 16$	Combine like terms.
$x > 8$	Divide each side by 2.

The solution set is all real numbers that are greater than 8, which is denoted by $(8, \infty)$. The graph of this solution set is shown in Figure P.6.

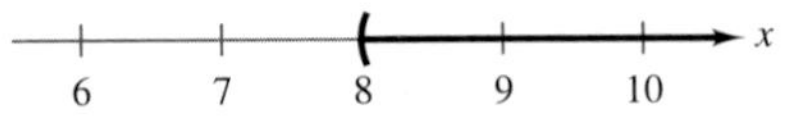

Solution interval: $(8, \infty)$

FIGURE P.6

b.

$1 - \dfrac{3x}{2} \geq x - 4$	Write original inequality.
$2 - 3x \geq 2x - 8$	Multiply each side by 2.
$-3x \geq 2x - 10$	Subtract 2 from each side.
$-5x \geq -10$	Subtract $2x$ from each side.
$x \leq 2$	Divide each side by -5 and reverse the inequality.

The solution set is all real numbers that are less than or equal to 2, which is denoted by $(-\infty, 2]$. The graph of this solution set is shown in Figure P.7.

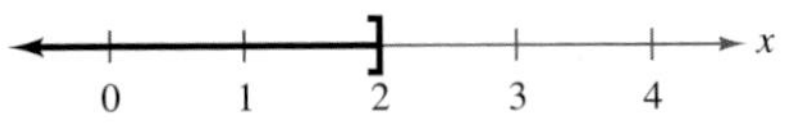

Solution interval: $(-\infty, 2]$

FIGURE P.7

Sometimes it is possible to write two inequalities as a **double inequality.** For instance, you can write the two inequalities $-4 \le 5x - 2$ and $5x - 2 < 7$ more simply as

$$-4 \le 5x - 2 < 7.$$

This form allows you to solve the two inequalities together, as demonstrated in Example 3.

Additional Examples

a. $-2 < -2 - x < 1$

$0 < -x < 3$

$0 > x > -3$

b. $17 < 5 - 3x < 29$

$12 < -3x < 24$

$-4 > x > -8$

Example 3 ▶ Solving a Double Inequality

To solve a double inequality, you can isolate x as the middle term.

$$-3 \le 6x - 1 < 3 \qquad \text{Write original inequality.}$$

$$-3 + 1 \le 6x - 1 + 1 < 3 + 1 \qquad \text{Add 1 to each part.}$$

$$-2 \le 6x < 4 \qquad \text{Simplify.}$$

$$\frac{-2}{6} \le \frac{6x}{6} < \frac{4}{6} \qquad \text{Divide each part by 6.}$$

$$-\frac{1}{3} \le x < \frac{2}{3} \qquad \text{Simplify.}$$

The solution set is all real numbers that are greater than or equal to $-\frac{1}{3}$ and less than $\frac{2}{3}$, which is denoted by $\left[-\frac{1}{3}, \frac{2}{3}\right)$. The graph of this solution set is shown in Figure P.8.

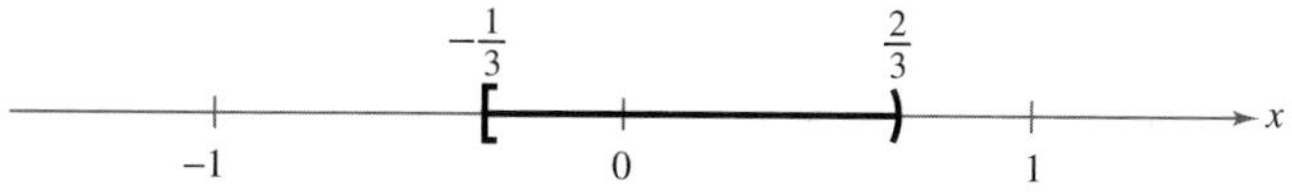

Solution interval: $\left[-\frac{1}{3}, \frac{2}{3}\right)$

FIGURE P.8

The double inequality in Example 3 could have been solved in two parts as follows.

$$-3 \le 6x - 1 \quad \text{and} \quad 6x - 1 < 3$$

$$-2 \le 6x \qquad\qquad 6x < 4$$

$$-\frac{1}{3} \le x \qquad\qquad x < \frac{2}{3}$$

The solution set consists of all real numbers that satisfy *both* inequalities. In other words, the solution set is the set of all values of x for which

$$-\frac{1}{3} \le x < \frac{2}{3}.$$

When combining two inequalities to form a double inequality, be sure that the inequalities satisfy the Transitive Property. For instance, it is *incorrect* to combine the inequalities $3 < x$ and $x \le -1$ as $3 < x \le -1$. This "inequality" is obviously wrong because 3 is not less than -1.

Absolute Value Inequalities

Solving an Absolute Value Inequality

Let x be a variable or an algebraic expression and let a be a real number such that $a \geq 0$.

1. The solutions of $|x| < a$ are all values of x that lie between $-a$ and a.

$|x| < a$ if and only if $-a < x < a$.

2. The solutions of $|x| > a$ are all values of x that are less than $-a$ or greater than a.

$|x| > a$ if and only if $x < -a$ or $x > a$.

These rules are also valid if $<$ is replaced by $\leq$ and $>$ is replaced by $\geq$.

Example 4 ▶ Solving an Absolute Value Inequality

Solve each inequality.

a. $|x - 5| < 2$ **b.** $|x + 3| \geq 7$

Solution

a. $|x - 5| < 2$ Write original inequality.

$-2 < x - 5 < 2$ Write equivalent inequalities.

$-2 + 5 < x - 5 + 5 < 2 + 5$ Add 5 to each part.

$3 < x < 7$ Simplify.

The solution set is all real numbers that are greater than 3 and less than 7, which is denoted by $(3, 7)$. The graph of this solution set is shown in Figure P.9.

b. $|x + 3| \geq 7$ Write original inequality.

$x + 3 \leq -7$ or $x + 3 \geq 7$ Write equivalent inequalities.

$x + 3 - 3 \leq -7 - 3$ $x + 3 - 3 \geq 7 - 3$ Subtract 3 from each side.

$x \leq -10$ $x \geq 4$ Simplify.

The solution set is all real numbers that are less than or equal to -10 *or* greater than or equal to 4. The interval notation for this solution set is $(-\infty, -10] \cup [4, \infty)$. The symbol $\cup$ is called a *union* symbol and is used to denote the combining of two sets. The graph of this solution set is shown in Figure P.10.

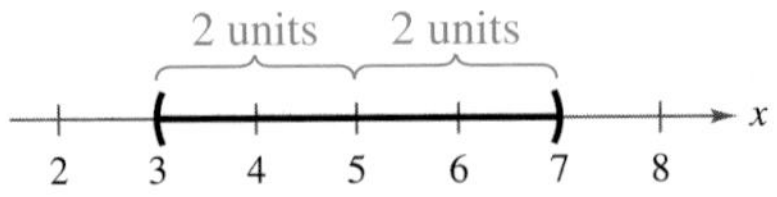

$|x - 5| < 2$: *Solutions lie inside* $(3, 7)$

FIGURE P.9

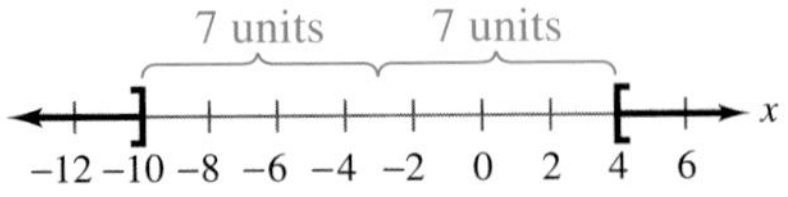

$|x + 3| \geq 7$: *Solutions lie outside* $(-10, 4)$

FIGURE P.10

Technology

A graphing utility can be used to identify the solution set of an inequality. For instance, to find the solution set of $|x - 5| < 2$ (see Example 4a), enter

Y1 = abs (X − 5) < 2

and press the graph key. The graph should look like the one shown below.

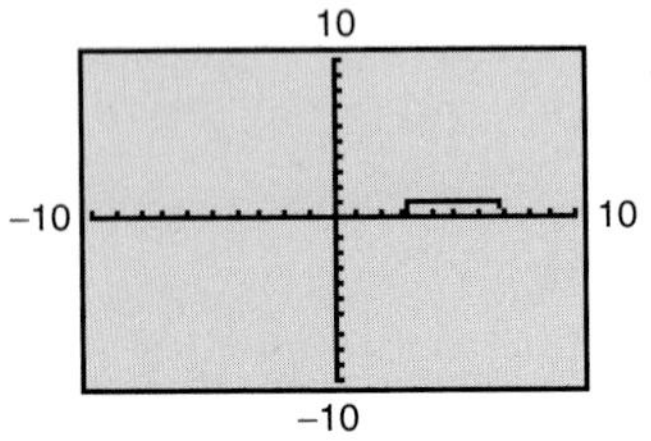

The solution set is indicated by the line segment above the x-axis.

Common Error

A common error is to write $|x + 3| \geq 7$ as $-7 \geq x + 3 \geq 7$. Point out that this is incorrect because the "inequality" does not satisfy the Transitive Property—i. e., -7 is not greater than or equal to 7.

STUDY TIP

Note that the graph of the inequality $|x - 5| < 2$ can be described as all real numbers within 2 units of 5, as shown in Figure P.9.

Remind students that critical numbers are numbers at which the polynomial can change signs and are not necessarily parts of the solutions of the inequality.

Other Types of Inequalities

To solve a polynomial inequality, you can use the fact that a polynomial can change signs only at its zeros (the x-values that make the polynomial equal to zero). Between two consecutive zeros a polynomial must be entirely positive or entirely negative. This means that when the real zeros of a polynomial are put in order, they divide the real number line into intervals in which the polynomial has no sign changes. These zeros are the **critical numbers** of the inequality, and the resulting intervals are the **test intervals** for the inequality.

STUDY TIP

As with linear inequalities, you can check the reasonableness of a solution by substituting x-values into the original inequality. For instance, to check the solution found in Example 5, try substituting several x-values from the interval $(-2, 3)$ into the inequality

$$x^2 - x - 6 < 0.$$

Regardless of which x-values you choose, the inequality should be satisfied.

Example 5 ▶ Solving a Polynomial Inequality

Solve $x^2 - x - 6 < 0$.

Solution

By factoring the quadratic as

$$x^2 - x - 6 = (x + 2)(x - 3)$$

you can see that the critical numbers are

$$x = -2 \quad \text{and} \quad x = 3.$$

So, the polynomial's test intervals are

$$(-\infty, -2), \quad (-2, 3), \quad \text{and} \quad (3, \infty). \qquad \text{Test intervals}$$

In each test interval, choose a representative x-value and evaluate the polynomial.

Interval	*x-Value*	*Polynomial Value*	*Conclusion*
$(-\infty, -2)$	$x = -3$	$(-3)^2 - (-3) - 6 = 6$	Positive
$(-2, 3)$	$x = 0$	$(0)^2 - (0) - 6 = -6$	Negative
$(3, \infty)$	$x = 4$	$(4)^2 - (4) - 6 = 6$	Positive

From this you can conclude that the polynomial is positive for all x-values in $(-\infty, -2)$ and $(3, \infty)$ and is negative only for all x-values in $(-2, 3)$. This implies that the solution of the inequality $x^2 - x - 6 < 0$ is the interval $(-2, 3)$, as shown in Figure P.11.

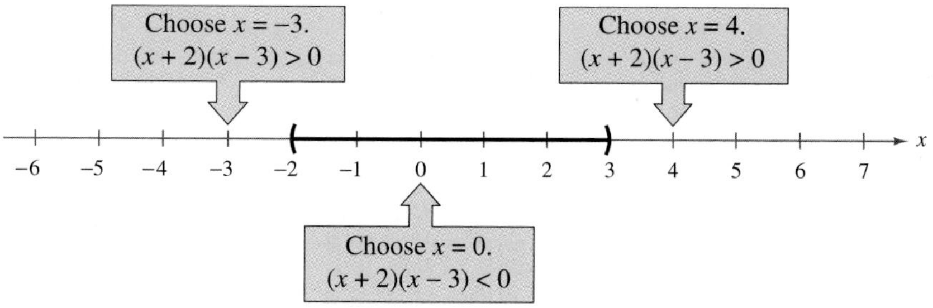

FIGURE P.11

The concepts of critical numbers and test intervals can be extended to rational inequalities. To do this, use the fact that the value of a rational expression can change sign only at its *zeros* (the x-values for which its numerator is zero) and its *undefined values* (the x-values for which its denominator is zero). These two types of numbers make up the *critical numbers* of a rational inequality.

Example 6 ▶ Solving a Rational Inequality

Solve $\dfrac{2x-7}{x-5} \le 3$.

Solution

Begin by writing the rational inequality in general form.

$$\frac{2x-7}{x-5} \le 3 \qquad \text{Write original inequality.}$$

$$\frac{2x-7}{x-5} - 3 \le 0 \qquad \text{Write in general form.}$$

$$\frac{2x-7-3x+15}{x-5} \le 0 \qquad \text{Add fractions.}$$

$$\frac{-x+8}{x-5} \le 0 \qquad \text{Simplify.}$$

Critical numbers: $x = 5, x = 8$ Zeros and undefined values of rational expression

Test intervals: $(-\infty, 5), (5, 8), (8, \infty)$

Test: Is $\dfrac{-x+8}{x-5} \le 0$?

After testing these intervals, as shown in Figure P.12, you can see that the rational expression $(-x+8)/(x-5)$ is negative in the open intervals $(-\infty, 5)$ and $(8, \infty)$. Moreover, because $(-x+8)/(x-5) = 0$ when $x = 8$, you can conclude that the solution set consists of all real numbers in the intervals $(-\infty, 5) \cup [8, \infty)$. (Be sure to use a closed interval to indicate that x can equal 8.)

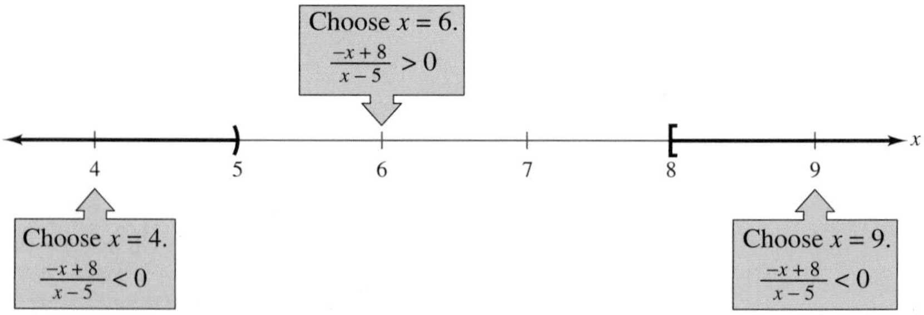

FIGURE P.12

Activities

1. Solve the inequality, and give the answer in interval notation.
   $(x-3)(x+1)(x-5) > 0$
   Answer: $(-1, 3) \cup (5, \infty)$
2. Find the domain.
   $\sqrt{x^2 - x - 2}$
   Answer: $(-\infty, -1] \cup [2, \infty)$
3. Solve the inequality, and give the answer in interval notation.
   $\dfrac{1}{x-3} \le \dfrac{3}{2x+1}$
   Answer: $\left(-\frac{1}{2}, 3\right) \cup [10, \infty)$
4. A projectile is fired straight upward from ground level with an initial velocity of 384 feet per second. During what time period will its height exceed 2000 feet?
   Solution
   $s = -16t^2 + v_0t + s_0$
   $s_0 = 0$
   $v_0 = 384$
   So, solve
   $-16t^2 + 384t > 2000$
   $t^2 - 24t < -125$
   $t^2 - 24t + 125 < 0.$
   Using the Quadratic Formula, determine that the critical numbers are $t = 12 \pm \sqrt{19} \approx 7.64$ or 16.36.
   A test will verify that the height of the projectile will exceed 2000 feet during the time interval 7.64 seconds $< t <$ 16.36 seconds.

Additional Examples

a. Find the domain of $\sqrt[4]{2x^2 - 18}$.

Solution

The domain is given by

$$2x^2 - 18 \ge 0$$
$$2(x^2 - 9) \ge 0$$
$$2(x - 3)(x + 3) \ge 0.$$

The inequality has two critical numbers: -3 and 3. When you test the inequality, you can see that $2x^2 - 18$ is greater than or equal to 0 in the intervals $(-\infty, -3]$ and $[3, \infty)$. So, the domain of $\sqrt[4]{2x^2 - 18}$ is $(-\infty, -3]$ and $[3, \infty)$.

b. Find the domain of $\dfrac{2}{\sqrt{27 - 3x^2}}$.

Solution

The domain is given by

$$\sqrt{27 - 3x^2} > 0$$
$$27 - 3x^2 > 0$$
$$3(9 - x^2) > 0$$
$$3(3 - x)(3 + x) > 0.$$

The inequality has two critical numbers : -3 and 3. When you test the inequality, you can see that $27 - 3x^2$ is greater than 0 in the open interval $(-3, 3)$. So, the domain of $\dfrac{2}{\sqrt{27 - 3x^2}}$ is $(-3, 3)$.

A common application of inequalities is finding the domain of an expression that involves a square root, as shown in Example 7.

Example 7 ▶ Finding the Domain of an Expression

Find the domain of $\sqrt{64 - 4x^2}$.

Solution

Remember that the domain of an expression is the set of all x-values for which the expression is defined. Because $\sqrt{64 - 4x^2}$ is defined (has real values) only if $64 - 4x^2$ is nonnegative, the domain is given by $64 - 4x^2 \ge 0$.

$64 - 4x^2 \ge 0$ — Write in general form.

$16 - x^2 \ge 0$ — Divide each side by 4.

$(4 - x)(4 + x) \ge 0$ — Write in factored form.

So, the inequality has two critical numbers: -4 and 4. You can use these two numbers to test the inequality as follows.

Critical numbers: $x = -4, x = 4$

Test intervals: $(-\infty, -4), (-4, 4), (4, \infty)$

Test: Is $(4 - x)(4 + x) \ge 0$?

A test shows that $64 - 4x^2$ is greater than or equal to 0 in the *closed interval* $[-4, 4]$. So, the domain of the expression $\sqrt{64 - 4x^2}$ is the interval $[-4, 4]$, as shown in Figure P.13.

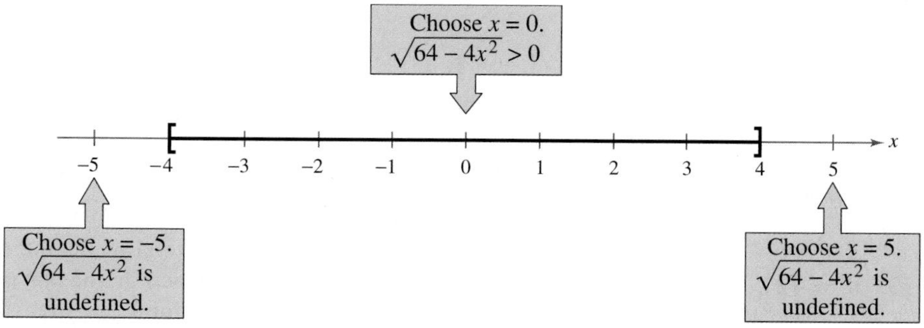

FIGURE P.13

Writing ABOUT MATHEMATICS

Communicating Mathematically Four different properties of inequalities are listed on page 63. For each property, (a) translate the mathematical statement into a *verbal statement*, (b) compile a list of several numerical examples that demonstrate the property, and (c) construct a number line or series of number lines that graphically illustrates the property.

P.6 Exercises

In Exercises 1–6, write an inequality to represent the interval, and state whether the interval is bounded or unbounded.

1. $[-1, 5]$
2. $(2, 10]$
3. $(11, \infty)$
4. $[-5, \infty)$
5. $(-\infty, -2)$
6. $(-\infty, 7]$

In Exercises 7–12, match the inequality with its graph. [The graphs are labeled (a), (b), (c), (d), (e), and (f).]

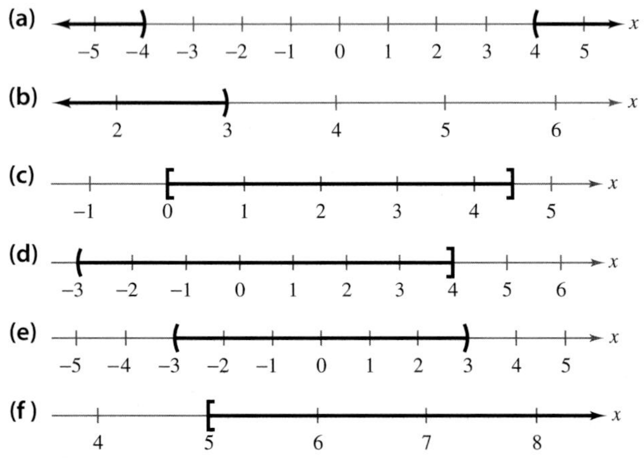

7. $x < 3$
8. $x \geq 5$
9. $-3 < x \leq 4$
10. $0 \leq x \leq \frac{9}{2}$
11. $|x| < 3$
12. $|x| > 4$

In Exercises 13–18, determine whether each value of *x* is a solution of the inequality.

Inequality	*Values*	
13. $5x - 12 > 0$	(a) $x = 3$	(b) $x = -3$
	(c) $x = \frac{5}{2}$	(d) $x = \frac{3}{2}$
14. $2x + 1 < -3$	(a) $x = 0$	(b) $x = -\frac{1}{4}$
	(c) $x = -4$	(d) $x = -\frac{3}{2}$
15. $0 < \frac{x-2}{4} < 2$	(a) $x = 4$	(b) $x = 10$
	(c) $x = 0$	(d) $x = \frac{7}{2}$
16. $-1 < \frac{3-x}{2} \leq 1$	(a) $x = 0$	(b) $x = -5$
	(c) $x = 1$	(d) $x = 5$
17. $\|x - 10\| \geq 3$	(a) $x = 13$	(b) $x = -1$
	(c) $x = 14$	(d) $x = 9$
18. $\|2x - 3\| < 15$	(a) $x = -6$	(b) $x = 0$
	(c) $x = 12$	(d) $x = 7$

In Exercises 19–44, solve the inequality and sketch the solution on the real number line.

19. $4x < 12$
20. $-10x < 40$
21. $2x > 3$
22. $-6x > 15$
23. $x - 5 \geq 7$
24. $x + 7 \leq 12$
25. $2x + 7 < 3 + 4x$
26. $3x + 1 \geq 2 + x$
27. $2x - 1 \geq 1 - 5x$
28. $6x - 4 \leq 2 + 8x$
29. $4 - 2x < 3(3 - x)$
30. $4(x + 1) < 2x + 3$
31. $\frac{3}{4}x - 6 \leq x - 7$
32. $3 + \frac{2}{7}x > x - 2$
33. $\frac{1}{2}(8x + 1) \geq 3x + \frac{5}{2}$
34. $9x - 1 < \frac{3}{4}(16x - 2)$
35. $3.6x + 11 \geq -3.4$
36. $15.6 - 1.3x < -5.2$
37. $1 < 2x + 3 < 9$
38. $-8 \leq -(3x + 5) < 13$
39. $-4 < \frac{2x - 3}{3} < 4$
40. $0 \leq \frac{x + 3}{2} < 5$
41. $\frac{3}{4} > x + 1 > \frac{1}{4}$
42. $-1 < 2 - \frac{x}{3} < 1$
43. $3.2 \leq 0.4x - 1 \leq 4.4$
44. $4.5 > \frac{1.5x + 6}{2} > 10.5$

In Exercises 45–60, solve the inequality and sketch the solution on the real number line. (Some equations have no solution.)

45. $|x| < 6$
46. $|x| > 4$
47. $\left|\frac{x}{2}\right| > 5$
48. $\left|\frac{x}{5}\right| > 3$
49. $|x - 5| < -1$
50. $|x - 5| \geq 0$
51. $|x - 20| \leq 6$
52. $|x - 7| < -5$
53. $|3 - 4x| \geq 9$
54. $|1 - 2x| < 5$
55. $\left|\frac{x - 3}{2}\right| \geq 5$
56. $\left|1 - \frac{2x}{3}\right| < 1$
57. $|9 - 2x| - 2 < -1$
58. $|x + 14| + 3 > 17$
59. $2|x + 10| \geq 9$
60. $3|4 - 5x| \leq 9$

Graphical Analysis **In Exercises 61–68, use a graphing utility to graph the inequality and identify the solution set.**

61. $6x > 12$

62. $3x - 1 \le 5$

63. $5 - 2x \ge 1$

64. $3(x + 1) < x + 7$

65. $|x - 8| \le 14$

66. $|2x + 9| > 13$

67. $2|x + 7| \ge 13$

68. $\frac{1}{2}|x + 1| \le 3$

In Exercises 69–74, find the interval(s) on the real number line for which the radicand is nonnegative (greater than or equal to zero).

69. $\sqrt{x - 5}$

70. $\sqrt{x - 10}$

71. $\sqrt{x + 3}$

72. $\sqrt{3 - x}$

73. $\sqrt[4]{7 - 2x}$

74. $\sqrt[4]{6x + 15}$

75. *Think About It* The graph of $|x - 5| < 3$ can be described as all real numbers within 3 units of 5. Give a similar description of $|x - 10| < 8$.

76. *Think About It* The graph of $|x - 2| > 5$ can be described as all real numbers more than 5 units from 2. Give a similar description of $|x - 8| > 4$.

In Exercises 77–84, use absolute value notation to define the interval (or pair of intervals) on the real number line.

77. 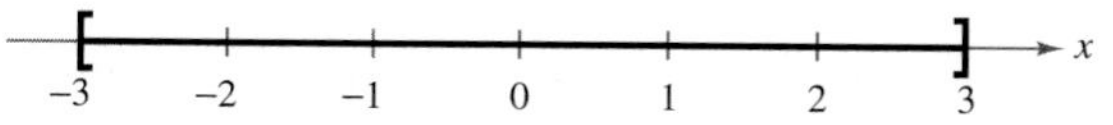

78. 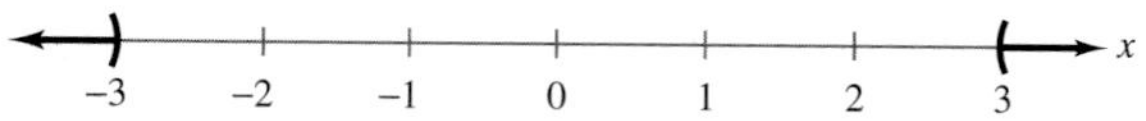

79. 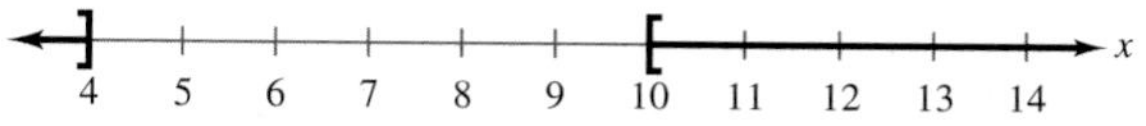

80. 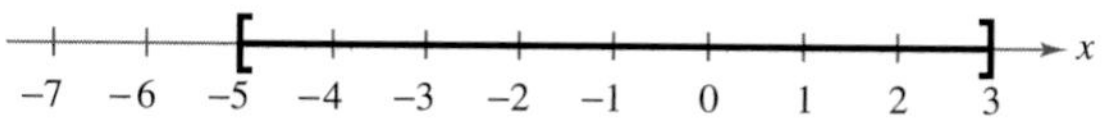

81. All real numbers within 10 units of 12

82. All real numbers at least 5 units from 8

83. All real numbers more than 5 units from -3

84. All real numbers no more than 7 units from -6

In Exercises 85–88, determine whether each value of *x* is a solution of the inequality.

Inequality	*Values*	
85. $x^2 - 3 < 0$	(a) $x = 3$	(b) $x = 0$
	(c) $x = \frac{3}{2}$	(d) $x = -5$
86. $x^2 - x - 12 \ge 0$	(a) $x = 5$	(b) $x = 0$
	(c) $x = -4$	(d) $x = -3$
87. $\dfrac{x + 2}{x - 4} \ge 3$	(a) $x = 5$	(b) $x = 4$
	(c) $x = -\frac{9}{2}$	(d) $x = \frac{9}{2}$
88. $\dfrac{3x^2}{x^2 + 4} < 1$	(a) $x = -2$	(b) $x = -1$
	(c) $x = 0$	(d) $x = 3$

In Exercises 89–92, find the critical numbers.

89. $2x^2 - x - 6$

90. $9x^3 - 25x^2$

91. $2 + \dfrac{3}{x - 5}$

92. $\dfrac{x}{x + 2} - \dfrac{2}{x - 1}$

In Exercises 93–108, solve the inequality and graph the solution on the real number line.

93. $x^2 \le 9$

94. $x^2 < 5$

95. $(x + 2)^2 < 25$

96. $(x - 3)^2 \ge 1$

97. $x^2 + 4x + 4 \ge 9$

98. $x^2 - 6x + 9 < 16$

99. $x^2 + x < 6$

100. $x^2 + 2x > 3$

101. $x^2 + 2x - 3 < 0$

102. $x^2 - 4x - 1 > 0$

103. $x^2 + 8x - 5 \ge 0$

104. $-2x^2 + 6x + 15 \le 0$

105. $x^3 - 3x^2 - x + 3 > 0$

106. $x^3 + 2x^2 - 4x - 8 \le 0$

107. $x^3 - 2x^2 - 9x - 2 \ge -20$

108. $2x^3 + 13x^2 - 8x - 46 \ge 6$

In Exercises 109–114, solve the inequality and write the solution set in interval notation.

109. $4x^3 - 6x^2 < 0$

110. $4x^3 - 12x^2 > 0$

111. $x^3 - 4x \ge 0$

112. $2x^3 - x^4 \le 0$

113. $(x - 1)^2(x + 2)^3 \ge 0$

114. $x^4(x - 3) \le 0$

The symbol indicates an exercise or parts of an exercise in which you are instructed to use a graphing utility.

In Exercises 115–128, solve the inequality and graph the solution on the real number line.

115. $\frac{1}{x} - x > 0$

116. $\frac{1}{x} - 4 < 0$

117. $\frac{x+6}{x+1} - 2 < 0$

118. $\frac{x+12}{x+2} - 3 \geq 0$

119. $\frac{3x-5}{x-5} > 4$

120. $\frac{5+7x}{1+2x} < 4$

121. $\frac{4}{x+5} > \frac{1}{2x+3}$

122. $\frac{5}{x-6} > \frac{3}{x+2}$

123. $\frac{1}{x-3} \leq \frac{9}{4x+3}$

124. $\frac{1}{x} \geq \frac{1}{x+3}$

125. $\frac{x^2+2x}{x^2-9} \leq 0$

126. $\frac{x^2+x-6}{x} \geq 0$

127. $\frac{5}{x-1} - \frac{2x}{x+1} < 1$

128. $\frac{3x}{x-1} \leq \frac{x}{x+4} + 3$

In Exercises 129–134, find the domain of x in the expression.

129. $\sqrt{4-x^2}$

130. $\sqrt{x^2-4}$

131. $\sqrt{x^2-7x+12}$

132. $\sqrt{144-9x^2}$

133. $\sqrt{\frac{x}{x^2-2x-35}}$

134. $\sqrt{\frac{x}{x^2-9}}$

In Exercises 135–140, solve the inequality. (Round your answers to two decimal places.)

135. $0.4x^2 + 5.26 < 10.2$

136. $-1.3x^2 + 3.78 > 2.12$

137. $-0.5x^2 + 12.5x + 1.6 > 0$

138. $1.2x^2 + 4.8x + 3.1 < 5.3$

139. $\frac{1}{2.3x - 5.2} > 3.4$

140. $\frac{2}{3.1x - 3.7} > 5.8$

141. *Car Rental* You can rent a midsize car from Company A for \$250 per week with unlimited mileage. A similar car can be rented from Company B for \$150 per week plus 25 cents for each mile driven. How many miles must you drive in a week in order for the rental fee for Company B to be greater than that for Company A?

142. *Copying Costs* Your department sends its copying to the photocopy center of your company. The center bills your department \$0.10 per page. You have investigated the possibility of buying a departmental copier for \$3000. With your own copier, the cost per page would be \$0.03. The expected life of the copier is 4 years. How many copies must you make in the 4-year period to justify buying the copier?

143. *Investment* In order for an investment of \$1000 to grow to more than \$1062.50 in 2 years, what must the annual interest rate be? $[A = P(1 + rt)]$

144. *Weightlifting* For 60 men enrolled in a weightlifting class, the relationship between body weight x (in pounds) and maximum bench-press weight y (in pounds) can be modeled by the equation $y = 1.266x - 35.766$. Use this model to estimate the range of body weights of the men in this group that can bench press more than 200 pounds.

145. *Height* The heights h of two-thirds of the members of a certain population satisfy the inequality

$$\left|\frac{h - 68.5}{2.7}\right| \leq 1$$

where h is measured in inches. Determine the interval on the real number line in which these heights lie.

146. *Meteorology* A certain electronic device is to be operated in an environment with relative humidity h in the interval defined by

$$|h - 50| \leq 30.$$

What are the minimum and maximum relative humidities for the operation of this device?

147. *Geometry* A rectangular playing field with a perimeter of 100 meters is to have an area of at least 500 square meters. Within what bounds must the length of the rectangle lie?

148. *Geometry* A rectangular parking lot with a perimeter of 440 feet is to have an area of at least 8000 square feet. Within what bounds must the length of the rectangle lie?

149. *Investment* P dollars, invested at interest rate r compounded annually, increases to an amount

$$A = P(1 + r)^2$$

in 2 years. If an investment of \$1000 is to increase to an amount greater than \$1100 in 2 years, then the interest rate must be greater than what percent?

150. *Economics* The revenue and cost equations for a product are

$$R = x(50 - 0.0002x)$$

$$C = 12x + 150{,}000$$

where R and C are measured in dollars and x represents the number of units sold. How many units must be sold to obtain a profit of at least \$1,650,000?

151. *Resistors* When two resistors of resistances R_1 and R_2 are connected in parallel (see figure), the total resistance R satisfies the equation

$$\frac{1}{R} = \frac{1}{R_1} + \frac{1}{R_2}.$$

Find R_1 for a parallel circuit in which $R_2 = 2$ ohms and R must be at least 1 ohm.

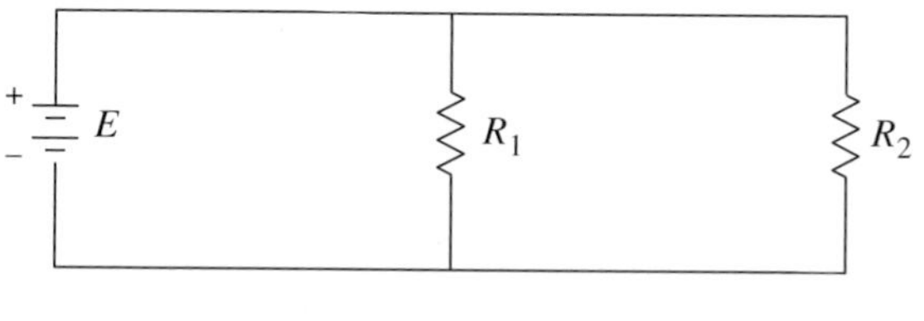

152. *Safe Load* The maximum safe load uniformly distributed over a 1-foot section of a 2-inch-wide wooden beam is approximated by the model

$$\text{Load} = 168.5d^2 - 472.1$$

where d is the depth of the beam.

(a) Evaluate the model for $d = 4, d = 6, d = 8, d = 10$, and $d = 12$. Use the results to create a bar graph.

(b) Determine the minimum depth of the beam that will safely support a load of 2000 pounds.

Synthesis

True or False? **In Exercises 153–156, determine whether the statement is true or false. Justify your answer.**

153. If a, b, and c are real numbers, and $a \le b$, then $ac \le bc$.

154. If $-10 \le x \le 8$, then $-10 \ge -x$ and $-x \ge -8$.

155. The zeros of the polynomial $x^3 - 2x^2 - 11x + 12 \ge 0$ divide the real number line into four test intervals.

156. The solution set of the inequality $\frac{3}{2}x^2 + 3x + 6 \ge 0$ is the set of real numbers.

157. Identify the graph of the inequality $|x - a| \ge 2$.

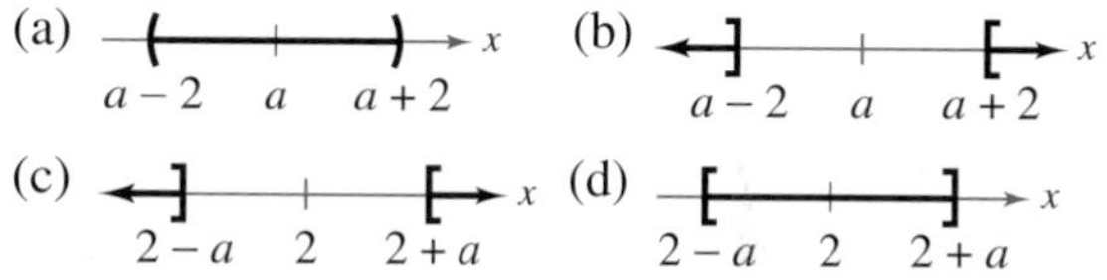

158. *Exploration* Find sets of values for a, b, and c such that $0 \le x \le 10$ is a solution of the inequality $|ax - b| \le c$.

Exploration **In Exercises 159–162, find the interval for b such that the equation has at least one real solution.**

159. $x^2 + bx + 4 = 0$

160. $x^2 + bx - 4 = 0$

161. $3x^2 + bx + 10 = 0$

162. $2x^2 + bx + 5 = 0$

163. *Conjecture* Write a conjecture about the interval for b in Exercises 159–162. Explain your reasoning.

164. *Think About It* What is the center of the interval for b in Exercises 159–162?

165. Consider the polynomial $(x - a)(x - b)$ and the real number line shown below.

(a) Identify the points on the line at which the polynomial is zero.

(b) In each of the three subintervals of the line, write the sign of each factor and the sign of the product.

(c) For what x-values does the polynomial change signs?

P.7 Errors and the Algebra of Calculus

▶ **What you should learn**

- How to avoid common algebraic errors
- How to recognize and use algebraic techniques that are common in calculus

▶ **Why you should learn it**

An efficient command of algebra is critical in the study of calculus.

Algebraic Errors to Avoid

This section contains five lists of common algebraic errors: errors involving parentheses, errors involving fractions, errors involving exponents, errors involving radicals, and errors involving dividing out. Many of these errors are made because they seem to be the *easiest* things to do.

Errors Involving Parentheses

Potential Error	*Correct Form*	*Comment*
$a-(x-b) \neq a-x-b$	$a-(x-b)=a-x+b$	Change all signs when distributing minus sign.
$(a+b)^2 \neq a^2+b^2$	$(a+b)^2=a^2+2ab+b^2$	Remember the middle term when squaring binomials.
$\left(\frac{1}{2}a\right)\left(\frac{1}{2}b\right) \neq \frac{1}{2}(ab)$	$\left(\frac{1}{2}a\right)\left(\frac{1}{2}b\right)=\frac{1}{4}(ab)=\frac{ab}{4}$	$\frac{1}{2}$ occurs twice as a factor.
$(3x+6)^2 \neq 3(x+2)^2$	$(3x+6)^2=[3(x+2)]^2$ $=3^2(x+2)^2$	When factoring, apply exponents to all factors.

Errors Involving Fractions

Potential Error	*Correct Form*	*Comment*
$\frac{a}{x+b} \neq \frac{a}{x}+\frac{a}{b}$	Leave as $\frac{a}{x+b}$.	Do not add denominators when adding fractions.
$\frac{\left(\frac{x}{a}\right)}{b} \neq \frac{bx}{a}$	$\frac{\left(\frac{x}{a}\right)}{b}=\left(\frac{x}{a}\right)\left(\frac{1}{b}\right)=\frac{x}{ab}$	Multiply by the reciprocal when dividing fractions.
$\frac{1}{a}+\frac{1}{b} \neq \frac{1}{a+b}$	$\frac{1}{a}+\frac{1}{b}=\frac{b+a}{ab}$	Use the property for adding fractions.
$\frac{1}{3x} \neq \frac{1}{3}x$	$\frac{1}{3x}=\frac{1}{3}\cdot\frac{1}{x}$	Use the property for multiplying fractions.
$(1/3)x \neq \frac{1}{3x}$	$(1/3)x=\frac{1}{3}\cdot x=\frac{x}{3}$	Be careful when using a slash to denote division.
$(1/x)+2 \neq \frac{1}{x+2}$	$(1/x)+2=\frac{1}{x}+2=\frac{1+2x}{x}$	Be careful when using a slash to denote division.

Errors Involving Exponents

Potential Error	*Correct Form*	*Comment*
$(x^2)^3 \neq x^5$	$(x^2)^3 = x^{2 \cdot 3} = x^6$	Multiply exponents when raising a power to a power.
$x^2 \cdot x^3 \neq x^6$	$x^2 \cdot x^3 = x^{2+3} = x^5$	Add exponents when multiplying powers with like bases.
$2x^3 \neq (2x)^3$	$2x^3 = 2(x^3)$	Exponents have priority over coefficients.
$\dfrac{1}{x^2 - x^3} \neq x^{-2} - x^{-3}$	Leave as $\dfrac{1}{x^2 - x^3}$.	Do not move term-by-term from denominator to numerator.

Errors Involving Radicals

Potential Error	*Correct Form*	*Comment*
$\sqrt{5x} \neq 5\sqrt{x}$	$\sqrt{5x} = \sqrt{5}\sqrt{x}$	Radicals apply to every factor inside the radical.
$\sqrt{x^2 + a^2} \neq x + a$	Leave as $\sqrt{x^2 + a^2}$.	Do not apply radicals term-by-term.
$\sqrt{-x + a} \neq -\sqrt{x - a}$	Leave as $\sqrt{-x + a}$.	Do not factor minus signs out of square roots.

Errors Involving Dividing Out

Potential Error	*Correct Form*	*Comment*
$\dfrac{a + bx}{a} \neq 1 + bx$	$\dfrac{a + bx}{a} = \dfrac{a}{a} + \dfrac{bx}{a} = 1 + \dfrac{b}{a}x$	Divide out common factors, not common terms.
$\dfrac{a + ax}{a} \neq a + x$	$\dfrac{a + ax}{a} = \dfrac{a(1 + x)}{a} = 1 + x$	Factor before dividing out.
$1 + \dfrac{x}{2x} \neq 1 + \dfrac{1}{x}$	$1 + \dfrac{x}{2x} = 1 + \dfrac{1}{2} = \dfrac{3}{2}$	Divide out common factors.

For many people, a good way to avoid errors is to *work slowly, write neatly*, and *talk to yourself.* Each time you write a step, ask yourself why the step is algebraically legitimate. For instance, you can justify the step written below because *dividing the numerator and denominator by the same nonzero number produces an equivalent fraction.*

$$\frac{2x}{6} = \frac{\cancel{2} \cdot x}{\cancel{2} \cdot 3} = \frac{x}{3}$$

Example 1 ▶ Using the Property for Adding Fractions

Describe and correct the error. $\dfrac{1}{2x} + \dfrac{1}{3x} = \dfrac{1}{5x}$ ✕

Solution

When adding fractions, use the property for adding fractions: $\dfrac{1}{a} + \dfrac{1}{b} = \dfrac{b + a}{ab}$.

$$\frac{1}{2x} + \frac{1}{3x} = \frac{3x + 2x}{6x^2} = \frac{5x}{6x^2} = \frac{5}{6x}$$

Some Algebra of Calculus

In calculus it is often necessary to take a simplified algebraic expression and "unsimplify" it. See the following lists, taken from a standard calculus text.

Unusual Factoring

Expression	*Useful Calculus Form*	*Comment*
$\dfrac{5x^4}{8}$	$\dfrac{5}{8}x^4$	Write with fractional coefficient.
$\dfrac{x^2 + 3x}{-6}$	$-\dfrac{1}{6}(x^2 + 3x)$	Write with fractional coefficient.
$2x^2 - x - 3$	$2\left(x^2 - \dfrac{x}{2} - \dfrac{3}{2}\right)$	Factor out the leading coefficient.
$\dfrac{x}{2}(x + 1)^{-1/2} + (x + 1)^{1/2}$	$\dfrac{(x + 1)^{-1/2}}{2}[x + 2(x + 1)]$	Factor out factor with least power.

Writing with Negative Exponents

Expression	*Useful Calculus Form*	*Comment*
$\dfrac{9}{5x^3}$	$\dfrac{9}{5}x^{-3}$	Move the factor to the numerator and change the sign of the exponent.
$\dfrac{7}{\sqrt{2x - 3}}$	$7(2x - 3)^{-1/2}$	Move the factor to the numerator and change the sign of the exponent.

Writing a Fraction as a Sum

Expression	*Useful Calculus Form*	*Comment*
$\dfrac{x + 2x^2 + 1}{\sqrt{x}}$	$x^{1/2} + 2x^{3/2} + x^{-1/2}$	Divide each term by $x^{1/2}$.
$\dfrac{1 + x}{x^2 + 1}$	$\dfrac{1}{x^2 + 1} + \dfrac{x}{x^2 + 1}$	Rewrite the fraction as the sum of fractions.
$\dfrac{2x}{x^2 + 2x + 1}$	$\dfrac{2x + 2 - 2}{x^2 + 2x + 1}$	Add and subtract the same term.
	$= \dfrac{2x + 2}{x^2 + 2x + 1} - \dfrac{2}{(x + 1)^2}$	Rewrite the fraction as the difference of fractions.
$\dfrac{x^2 - 2}{x + 1}$	$x - 1 - \dfrac{1}{x + 1}$	Use long division. (See Section 2.3.)
$\dfrac{x + 7}{x^2 - x - 6}$	$\dfrac{2}{x - 3} - \dfrac{1}{x + 2}$	Use the method of partial fractions. (See Section 2.7.)

Inserting Factors and Terms

Expression	*Useful Calculus Form*	*Comment*
$(2x-1)^3$	$\frac{1}{2}(2x-1)^3(2)$	Multiply and divide by 2.
$7x^2(4x^3-5)^{1/2}$	$\frac{7}{12}(4x^3-5)^{1/2}(12x^2)$	Multiply and divide by 12.
$\frac{4x^2}{9}-4y^2=1$	$\frac{x^2}{9/4}-\frac{y^2}{1/4}=1$	Write with fractional denominators.
$\frac{x}{x+1}$	$\frac{x+1-1}{x+1}=1-\frac{1}{x+1}$	Add and subtract the same term.

The next five examples demonstrate many of the steps in the preceding lists.

Example 2 ▶ Factors Involving Negative Exponents

Factor $x(x+1)^{-1/2}+(x+1)^{1/2}$.

Solution

When multiplying factors with like bases, you add exponents. When factoring, you are undoing multiplication, and so you *subtract* exponents.

$$\begin{aligned} x(x+1)^{-1/2}+(x+1)^{1/2} &= (x+1)^{-1/2}[x(x+1)^0+(x+1)^1] \\ &= (x+1)^{-1/2}[x+(x+1)] \\ &= (x+1)^{-1/2}(2x+1) \end{aligned}$$

Here is another way to simplify the expression in Example 2.

$$\begin{aligned} x(x+1)^{-1/2}+(x+1)^{1/2} &= x(x+1)^{-1/2}+(x+1)^{1/2}\cdot\frac{(x+1)^{1/2}}{(x+1)^{1/2}} \\ &= \frac{x(x+1)^0+(x+1)^1}{(x+1)^{1/2}} \\ &= \frac{2x+1}{\sqrt{x+1}} \end{aligned}$$

Example 3 ▶ Inserting Factors in an Expression

Insert the required factor: $\frac{x+2}{(x^2+4x-3)^2}=(\quad)\frac{1}{(x^2+4x-3)^2}(2x+4)$.

Solution

The expression on the right side of the equation is twice the expression on the left side. To make both sides equal, insert a factor of $\frac{1}{2}$.

$$\frac{x+2}{(x^2+4x-3)^2}=\left(\frac{1}{2}\right)\frac{1}{(x^2+4x-3)^2}(2x+4)$$

Right side is multiplied and divided by 2.

Example 4 ▶ Rewriting Fractions

Explain the following.

$$\frac{4x^2}{9} - 4y^2 = \frac{x^2}{9/4} - \frac{y^2}{1/4}$$

Solution

To write the expression on the left side of the equation in the form given on the right side, multiply the numerators and denominators of both terms by $\frac{1}{4}$.

$$\frac{4x^2}{9} - 4y^2 = \frac{4x^2}{9}\left(\frac{1/4}{1/4}\right) - 4y^2\left(\frac{1/4}{1/4}\right)$$

$$= \frac{x^2}{9/4} - \frac{y^2}{1/4}$$

Example 5 ▶ Rewriting with Negative Exponents

Rewrite each expression using negative exponents.

a. $\dfrac{-4x}{(1 - 2x^2)^2}$ **b.** $\dfrac{2}{5x^3} - \dfrac{1}{\sqrt{x}} + \dfrac{3}{5(4x)^2}$

Solution

a. $\dfrac{-4x}{(1 - 2x^2)^2} = -4x(1 - 2x^2)^{-2}$

b. Begin by writing the second term in exponential form.

$$\frac{2}{5x^3} - \frac{1}{\sqrt{x}} + \frac{3}{5(4x)^2} = \frac{2}{5x^3} - \frac{1}{x^{1/2}} + \frac{3}{5(4x)^2}$$

$$= \frac{2}{5}x^{-3} - x^{-1/2} + \frac{3}{5}(4x)^{-2}$$

Activities

1. Simplify: $(2x - 1)^{1/2} + \frac{2}{5}(2x - 1)^{-1/2}$.

   Answer: $\dfrac{10x - 3}{5(2x - 1)^{1/2}}$

2. Find the error(s):

   $3(x - 1)^2 = (3x - 3)^2$
   $= 9x^2 - 18x + 9$

   Answer:
   $3(x - 1)^2 \neq (3x - 3)^2$
   $3(x - 1)^2 = 3(x^2 - 2x + 1)$
   $= 3x^2 - 6x + 3$

Example 6 ▶ Writing a Fraction as a Sum of Terms

Rewrite each fraction as the sum of three terms.

a. $\dfrac{x^2 - 4x + 8}{2x}$ **b.** $\dfrac{x + 2x^2 + 1}{\sqrt{x}}$

Solution

a. $$\frac{x^2 - 4x + 8}{2x} = \frac{x^2}{2x} - \frac{4x}{2x} + \frac{8}{2x}$$

$$= \frac{x}{2} - 2 + \frac{4}{x}$$

b. $$\frac{x + 2x^2 + 1}{\sqrt{x}} = \frac{x}{x^{1/2}} + \frac{2x^2}{x^{1/2}} + \frac{1}{x^{1/2}}$$

$$= x^{1/2} + 2x^{3/2} + x^{-1/2}$$

P.7 Exercises

In Exercises 1–22, describe and correct the error.

1. $2x - (3y + 4) = 2x - 3y + 4$

2. $5z + 3(x - 2) = 5z + 3x - 2$

3. $\dfrac{4}{16x - (2x + 1)} = \dfrac{4}{14x + 1}$

4. $\dfrac{1 - x}{(5 - x)(-x)} = \dfrac{x - 1}{x(x - 5)}$

5. $(5z)(6z) = 30z$

6. $x(yz) = (xy)(xz)$

7. $a\left(\dfrac{x}{y}\right) = \dfrac{ax}{ay}$

8. $(4x)^2 = 4x^2$

9. $\left(\dfrac{x}{y}\right)^3 = \dfrac{x^3}{y}$

10. $\sqrt{25 - x^2} = 5 - x$

11. $\sqrt{x + 9} = \sqrt{x} + 3$

12. $\dfrac{2x^2 + 1}{5x} = \dfrac{2x + 1}{5}$

13. $\dfrac{6x + y}{6x - y} = \dfrac{x + y}{x - y}$

14. $\dfrac{1}{a^{-1} + b^{-1}} = \left(\dfrac{1}{a + b}\right)^{-1}$

15. $\dfrac{1}{x + y^{-1}} = \dfrac{y}{x + 1}$

16. $(x^2 + 5x)^{1/2} = x(x + 5)^{1/2}$

17. $x(2x - 1)^2 = (2x^2 - x)^2$

18. $(3x^2 - 6x)^3 = 3x(x - 2)^3$

19. $\sqrt[3]{x^3 + 7x^2} = x^2\sqrt[3]{x + 7}$

20. $\dfrac{7 + 5(x + 3)}{x + 3} = 12$

21. $\dfrac{3}{x} + \dfrac{4}{y} = \dfrac{7}{x + y}$

22. $\dfrac{1}{2y} = (1/2)y$

In Exercises 23–44, insert the required factor in the parentheses.

23. $\dfrac{3x + 2}{5} = \dfrac{1}{5}(\quad)$

24. $\dfrac{7x^2}{10} = \dfrac{7}{10}(\quad)$

25. $\frac{2}{3}x^2 + \frac{1}{3}x + 5 = \frac{1}{3}(\quad)$

26. $\frac{3}{4}x + \frac{1}{2} = \frac{1}{4}(\quad)$

27. $\frac{5}{2}z^2 - \frac{1}{4}z + 2 = (\quad)(10z^2 - z + 8)$

28. $x^2(x^3 - 1)^4 = (\quad)(x^3 - 1)^4(3x^2)$

29. $x(1 - 2x^2)^3 = (\quad)(1 - 2x^2)^3(-4x)$

30. $\dfrac{4x + 6}{(x^2 + 3x + 7)^3} = (\quad)\dfrac{1}{(x^2 + 3x + 7)^3}(2x + 3)$

31. $\dfrac{x + 1}{(x^2 + 2x - 3)^2} = (\quad)\dfrac{1}{(x^2 + 2x - 3)^2}(2x + 2)$

32. $\dfrac{1}{(x - 1)\sqrt{(x - 1)^4 - 4}} = \dfrac{(\quad)}{(x - 1)^2\sqrt{(x - 1)^4 - 4}}$

33. $\dfrac{3}{x} + \dfrac{5}{2x^2} - \dfrac{3}{2}x = (\quad)(6x + 5 - 3x^3)$

34. $\dfrac{(x - 1)^2}{169} + (y + 5)^2 = \dfrac{(x - 1)^3}{169(\quad)} + (y + 5)^2$

35. $\dfrac{9x^2}{25} + \dfrac{16y^2}{49} = \dfrac{x^2}{(\quad)} + \dfrac{y^2}{(\quad)}$

36. $\dfrac{3x^2}{4} - \dfrac{9y^2}{16} = \dfrac{x^2}{(\quad)} - \dfrac{y^2}{(\quad)}$

37. $\dfrac{x^2}{1/12} - \dfrac{y^2}{2/3} = \dfrac{12x^2}{(\quad)} - \dfrac{3y^2}{(\quad)}$

38. $\dfrac{x^2}{4/9} + \dfrac{y^2}{7/8} = \dfrac{9x^2}{(\quad)} + \dfrac{8y^2}{(\quad)}$

39. $x^{1/3} - 5x^{4/3} = x^{1/3}(\quad)$

40. $3(2x + 1)x^{1/2} + 4x^{3/2} = x^{1/2}(\quad)$

41. $(1 - 3x)^{4/3} - 4x(1 - 3x)^{1/3} = (1 - 3x)^{1/3}(\quad)$

42. $\dfrac{1}{2\sqrt{x}} + 5x^{3/2} - 10x^{5/2} = \dfrac{1}{2\sqrt{x}}(\quad)$

43. $\dfrac{1}{10}(2x + 1)^{5/2} - \dfrac{1}{6}(2x + 1)^{3/2} = \dfrac{(2x + 1)^{3/2}}{15}(\quad)$

44. $\dfrac{3}{7}(t + 1)^{7/3} - \dfrac{3}{4}(t + 1)^{4/3} = \dfrac{3(t + 1)^{4/3}}{28}(\quad)$

In Exercises 45–50, write the fraction as a sum of two or more terms.

45. $\dfrac{16 - 5x - x^2}{x}$

46. $\dfrac{x^3 - 5x^2 + 4}{x^2}$

47. $\dfrac{4x^3 - 7x^2 + 1}{x^{1/3}}$

48. $\dfrac{2x^5 - 3x^3 + 5x - 1}{x^{3/2}}$

49. $\dfrac{3 - 5x^2 - x^4}{\sqrt{x}}$

50. $\dfrac{x^3 - 5x^4}{3x^2}$

In Exercises 51–62, simplify the expression.

51. $\dfrac{-2(x^2-3)^{-3}(2x)(x+1)^3 - 3(x+1)^2(x^2-3)^{-2}}{[(x+1)^3]^2}$

52. $\dfrac{x^5(-3)(x^2+1)^{-4}(2x) - (x^2+1)^{-3}(5)x^4}{(x^5)^2}$

53. $\dfrac{(6x+1)^3(27x^2+2) - (9x^3+2x)(3)(6x+1)^2(6)}{[(6x+1)^3]^2}$

54. $\dfrac{(4x^2+9)^{1/2}(2) - (2x+3)\left(\frac{1}{2}\right)(4x^2+9)^{-1/2}(8x)}{[(4x^2+9)^{1/2}]^2}$

55. $\dfrac{(x+2)^{3/4}(x+3)^{-2/3} - (x+3)^{1/3}(x+2)^{-1/4}}{[(x+2)^{3/4}]^2}$

56. $(2x-1)^{1/2} - (x+2)(2x-1)^{-1/2}$

57. $\dfrac{2(3x-1)^{1/3} - (2x+1)\left(\frac{1}{3}\right)(3x-1)^{-2/3}(3)}{(3x-1)^{2/3}}$

58. $\dfrac{(x+1)\left(\frac{1}{2}\right)(2x-3x^2)^{-1/2}(2-6x) - (2x-3x^2)^{1/2}}{(x+1)^2}$

59. $\dfrac{1}{(x^2+4)^{1/2}} \cdot \dfrac{1}{2}(x^2+4)^{-1/2}(2x)$

60. $\dfrac{1}{x^2-6}(2x) + \dfrac{1}{2x+5}(2)$

61. $(x^2+5)^{1/2}\left(\frac{3}{2}\right)(3x-2)^{1/2}(3) +$
$(3x-2)^{3/2}\left(\frac{1}{2}\right)(x^2+5)^{-1/2}(2x)$

62. $(3x+2)^{-1/2}(3)(x-6)^{1/2}(1) +$
$(x-6)^3\left(-\frac{1}{2}\right)(3x+2)^{-3/2}(3)$

63. (a) Verify that $y_1 = y_2$ analytically.

$$y_1 = x^2\left(\tfrac{1}{3}\right)(x^2+1)^{-2/3}(2x) + (x^2+1)^{1/3}(2x)$$

$$y_2 = \frac{2x(4x^2+3)}{3(x^2+1)^{2/3}}$$

(b) Complete the table and demonstrate the equality in part (a) numerically.

x	-2	-1	$-\frac{1}{2}$	0	1	2	$\frac{5}{2}$
y_1							
y_2							

64. (a) Verify that $y_1 = y_2$ analytically.

$$y_1 = -\frac{\sqrt{9-x^2}}{x^2} - \frac{1}{\sqrt{9-x^2}}$$

$$y_2 = \frac{-9}{x^2\sqrt{9-x^2}}$$

(b) Complete the table and demonstrate the equality in part (a) numerically.

x	-2	-1	$-\frac{1}{2}$	$\frac{1}{4}$	1	2	$\frac{5}{2}$
y_1							
y_2							

65. *Logical Reasoning* Verify that $y_1 \neq y_2$ by letting $x = 0$ and evaluating y_1 and y_2.

$$y_1 = 2x\sqrt{1-x^2} - \frac{x^3}{\sqrt{1-x^2}}$$

$$y_2 = \frac{2-3x^2}{\sqrt{1-x^2}}$$

Change y_2 so that $y_1 = y_2$.

Synthesis

True or False? **In Exercises 66–69, determine whether the statement is true or false. Justify your answer.**

66. $x^{-1} + y^{-2} = \dfrac{y^2 + x}{xy^2}$

67. $\dfrac{1}{x^{-2} + y^{-1}} = x^2 + y$

68. $\dfrac{1}{\sqrt{x}+4} = \dfrac{\sqrt{x}-4}{x-16}$

69. $\dfrac{x^2-9}{\sqrt{x}-3} = \sqrt{x}+3$

In Exercises 70–73, find and correct any errors.

70. $x^n \cdot x^{3n} = x^{3n^2}$

71. $(x^n)^{2n} + (x^{2n})^n = 2x^{2n^2}$

72. $x^{2n} + y^{2n} = (x^n + y^n)^2$

73. $\dfrac{x^{2n} \cdot x^{3n}}{x^{3n} + x^2} = \dfrac{x^{5n}}{x^{3n} + x^2}$

74. *Think About It* Suppose you are taking a course in calculus, and for one of the homework problems you obtain the following answer.

$\dfrac{1}{10}(2x-1)^{5/2} + \dfrac{1}{6}(2x-1)^{3/2}$

The answer in the back of the book is

$\dfrac{1}{15}(2x-1)^{3/2}(3x+1).$

Are these two answers equivalent? If so, show how the second answer can be obtained from the first.

P.8 Graphical Representation of Data

▶ What you should learn

- How to plot points in the Cartesian plane
- How to use the Distance Formula to find the distance between two points
- How to use the Midpoint Formula to find the midpoint of a line segment
- How to use a coordinate plane to model and solve real-life problems

▶ Why you should learn it

The Cartesian plane can be used to represent relationships between two variables. For instance, Exercise 68 on page 90 shows how to represent graphically the number of recording artists elected to the Rock and Roll Hall of Fame from 1986 to 1999.

Alex Bartel/Tony Stone Images

The Cartesian Plane

Just as you can represent real numbers by points on a real number line, you can represent ordered pairs of real numbers by points in a plane called the **rectangular coordinate system,** or the **Cartesian plane,** named after the French mathematician René Descartes (1596–1650).

The Cartesian plane is formed by using two real number lines intersecting at right angles, as shown in Figure P.14. The horizontal real number line is usually called the ***x*-axis,** and the vertical real number line is usually called the ***y*-axis.** The point of intersection of these two axes is the **origin,** and the two axes divide the plane into four parts called **quadrants.**

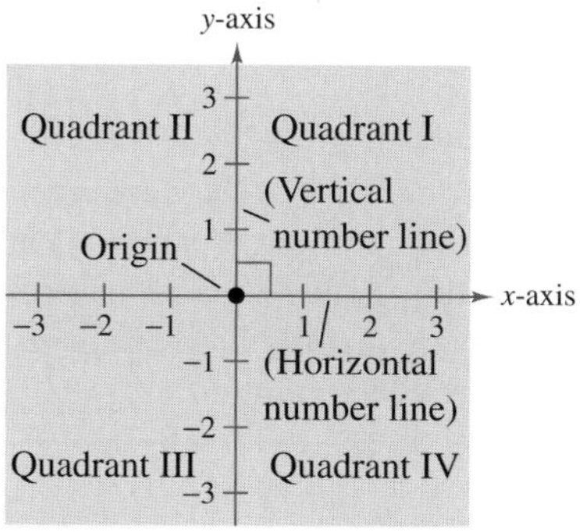

FIGURE P.14

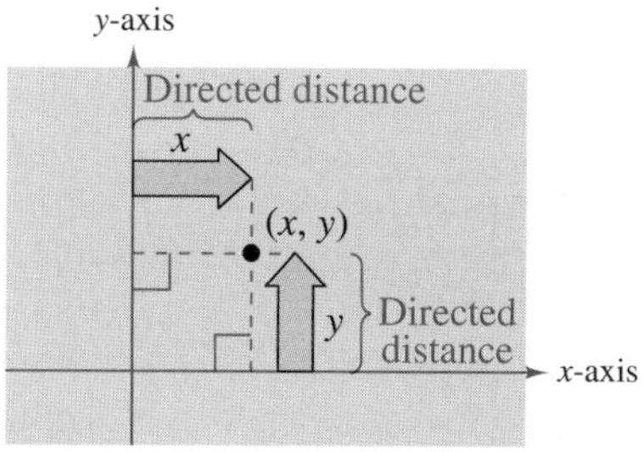

FIGURE P.15

Each point in the plane corresponds to an **ordered pair** (x, y) of real numbers x and y, called **coordinates** of the point. The ***x*-coordinate** represents the directed distance from the y-axis to the point, and the ***y*-coordinate** represents the directed distance from the x-axis to the point, as shown in Figure P.15.

Directed distance from y-axis ↗ (x, y) ↖ Directed distance from x-axis

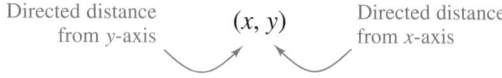

The notation (x, y) denotes both a point in the plane and an open interval on the real number line. The context will tell you which meaning is intended.

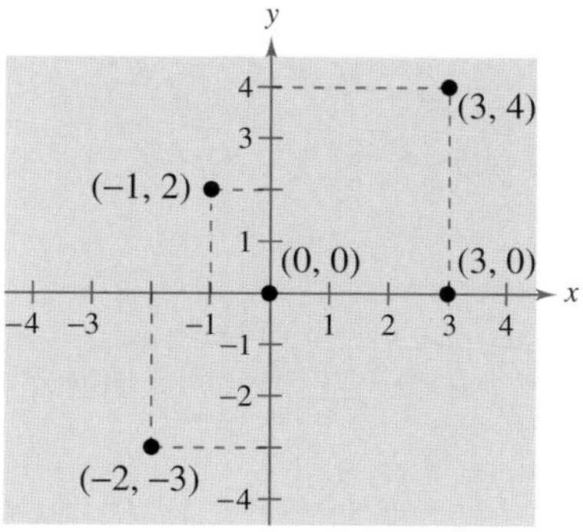

FIGURE P.16

Example 1 ▶ Plotting Points in the Cartesian Plane

Plot the points $(-1, 2)$, $(3, 4)$, $(0, 0)$, $(3, 0)$, and $(-2, -3)$.

Solution

To plot the point $(-1, 2)$, imagine a vertical line through -1 on the x-axis and a horizontal line through 2 on the y-axis. The intersection of these two lines is the point $(-1, 2)$. The other four points can be plotted in a similar way, as shown in Figure P.16.

The beauty of a rectangular coordinate system is that it allows you to see relationships between two variables. It would be difficult to overestimate the importance of Descartes's introduction of coordinates to the plane. Today, his ideas are in common use in virtually every scientific and business-related field.

A computer animation of this example appears in the *Interactive* CD-ROM and *Internet* versions of this text.

Example 2 ▶ Sketching a Scatter Plot

From 1988 through 1997, the amount A (in millions of dollars) spent on archery equipment in the United States is given in the table, where t represents the year. Sketch a scatter plot of the data. (Source: National Sporting Goods Association)

t	1988	1989	1990	1991	1992	1993	1994	1995	1996	1997
A	235	261	265	270	334	285	306	287	272	273

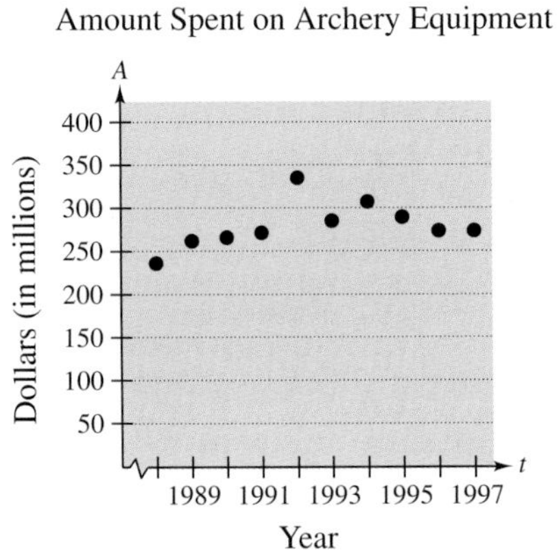

FIGURE P.17

Solution

To sketch a *scatter plot* of the data given in the table, you simply represent each pair of values by an ordered pair (t, A) and plot the resulting points, as shown in Figure P.17. For instance, the first pair of values is represented by the ordered pair (1988, 235). Note that the break in the t-axis indicates that the numbers between 0 and 1988 have been omitted.

In Example 2, you could have let $t = 1$ represent the year 1988. In that case, the horizontal axis would not have been broken, and the tick marks would have been labeled 1 through 10 (instead of 1988 through 1997).

Technology

The scatter plot in Example 2 is only one way to represent the data graphically. Two other techniques are shown at the right. The first is a bar graph and the second is a line graph. All three graphical representations were created with a computer. If you have access to a graphing utility, try using it to represent graphically the data given in Example 2.

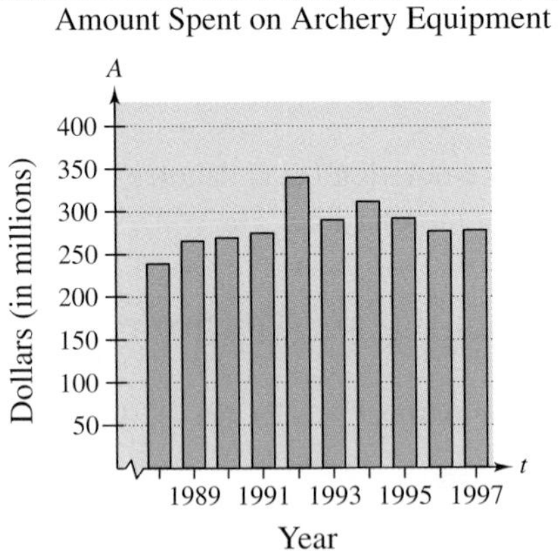

Bar graph

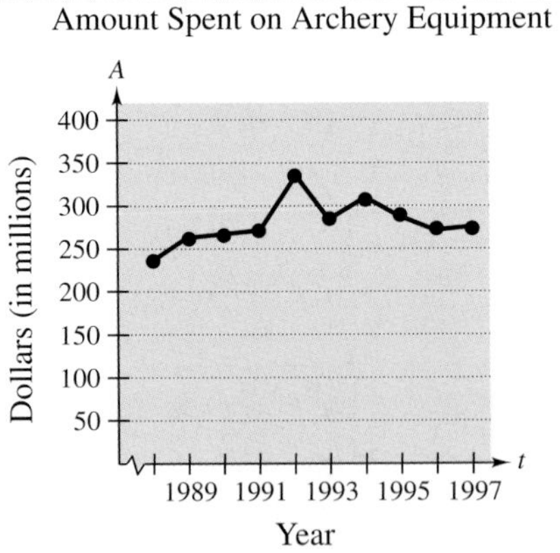

Line graph

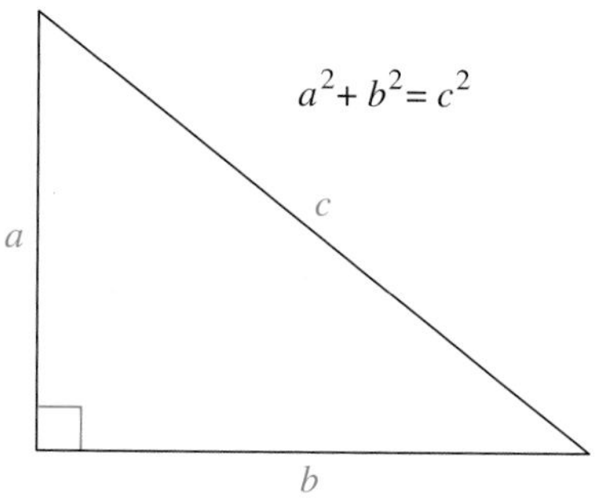

FIGURE P.18

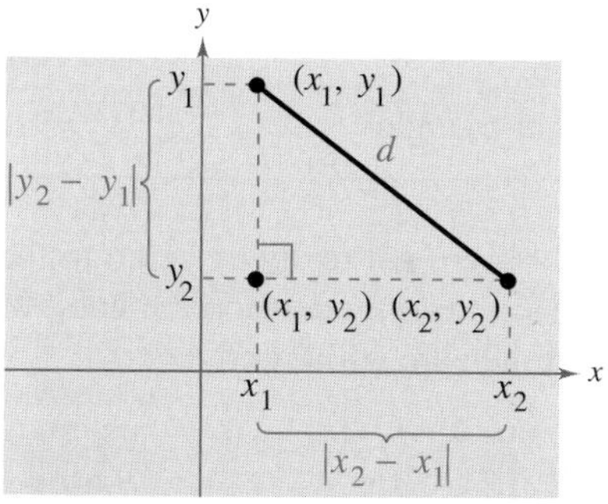

FIGURE P.19

The Distance Formula

Recall from the Pythagorean Theorem that, for a right triangle with hypotenuse of length c and sides of lengths a and b, you have

$$a^2 + b^2 = c^2 \qquad \text{Pythagorean Theorem}$$

as shown in Figure P.18. (The converse is also true. That is, if $a^2 + b^2 = c^2$, then the triangle is a right triangle.)

Suppose you want to determine the distance d between two points (x_1, y_1) and (x_2, y_2) in the plane. With these two points, a right triangle can be formed, as shown in Figure P.19. The length of the vertical side of the triangle is $|y_2 - y_1|$, and the length of the horizontal side is $|x_2 - x_1|$. By the Pythagorean Theorem, you can write

$$d^2 = |x_2 - x_1|^2 + |y_2 - y_1|^2$$
$$d = \sqrt{|x_2 - x_1|^2 + |y_2 - y_1|^2}$$
$$d = \sqrt{(x_2 - x_1)^2 + (y_2 - y_1)^2}.$$

This result is the **Distance Formula.**

The Distance Formula

The distance d between the points (x_1, y_1) and (x_2, y_2) in the plane is

$$d = \sqrt{(x_2 - x_1)^2 + (y_2 - y_1)^2}.$$

Example 3 ▶ Finding a Distance

Find the distance between the points $(-2, 1)$ and $(3, 4)$.

Solution

Let $(x_1, y_1) = (-2, 1)$ and $(x_2, y_2) = (3, 4)$. Then apply the Distance Formula.

$$d = \sqrt{(x_2 - x_1)^2 + (y_2 - y_1)^2} \qquad \text{Distance Formula}$$
$$= \sqrt{[3 - (-2)]^2 + (4 - 1)^2} \qquad \text{Substitute for } x_1, y_1, x_2, \text{ and } y_2.$$
$$= \sqrt{(5)^2 + (3)^2} \qquad \text{Simplify.}$$
$$= \sqrt{34}$$
$$\approx 5.83 \qquad \text{Use a calculator.}$$

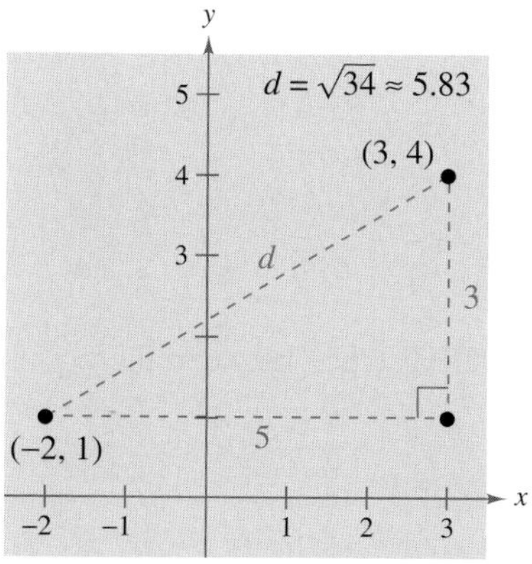

FIGURE P.20

Note in Figure P.20 that a distance of 5.83 looks about right.

You can use the Pythagorean Theorem to check that the distance is correct.

$$d^2 \stackrel{?}{=} 3^2 + 5^2 \qquad \text{Pythagorean Theorem}$$
$$(\sqrt{34})^2 \stackrel{?}{=} 3^2 + 5^2 \qquad \text{Substitute for } d.$$
$$34 = 34 \qquad \text{Distance checks. ✓}$$

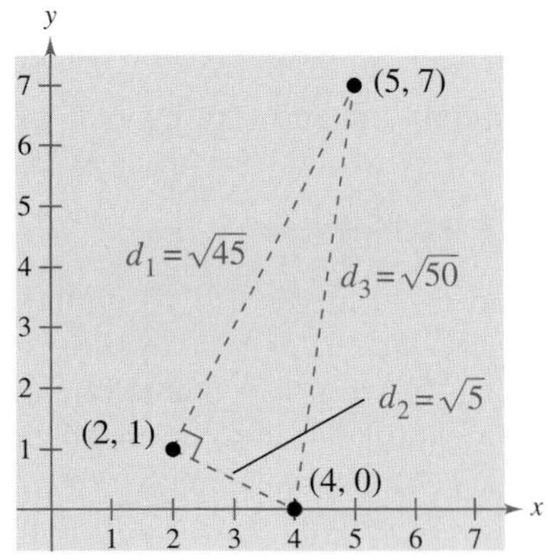

FIGURE P.21

Example 4 ▶ Verifying a Right Triangle

Show that the points (2, 1), (4, 0), and (5, 7) are vertices of a right triangle.

Solution

The three points are plotted in Figure P.21. Using the Distance Formula, you can find the lengths of the three sides as follows.

$$d_1 = \sqrt{(5-2)^2 + (7-1)^2} = \sqrt{9+36} = \sqrt{45}$$
$$d_2 = \sqrt{(4-2)^2 + (0-1)^2} = \sqrt{4+1} = \sqrt{5}$$
$$d_3 = \sqrt{(5-4)^2 + (7-0)^2} = \sqrt{1+49} = \sqrt{50}$$

Because

$$d_1^2 + d_2^2 = 45 + 5 = 50 = d_3^2,$$

you can conclude that the triangle must be a right triangle.

An overhead projector is useful for showing how to plot points and equations. Try projecting a grid onto the chalkboard and then plotting points on the chalkboard, or try using overhead markers and graph directly on the transparency. A viewscreen, a device used with an overhead projector to project a graphing calculator's screen image, is also useful.

The figures provided with Examples 3 and 4 were not really essential to the solution. Nevertheless, it is strongly recommended that you develop the habit of including sketches with your solutions—even if they are not required.

Example 5 ▶ Finding the Length of a Pass

A football quarterback throws a pass from the 5-yard line, 20 yards from the sideline. The pass is caught by a wide receiver on the 45-yard line, 50 yards from the same sideline, as shown in Figure P.22. How long is the pass?

STUDY TIP

In Example 5, the scale along the goal line does not normally appear on a football field. However, when you use coordinate geometry to solve real-life problems, you are free to place the coordinate system in any way that is convenient to the solution of the problem.

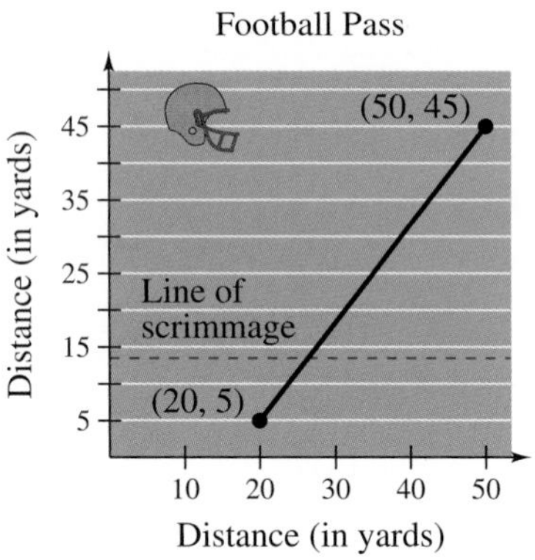

FIGURE P.22

Solution

You can find the length of the pass by finding the distance between the points (20, 5) and (50, 45).

$$d = \sqrt{(50-20)^2 + (45-5)^2} \qquad \text{Distance Formula}$$
$$= \sqrt{900 + 1600}$$
$$= 50 \qquad \text{Simplify.}$$

So, the pass is 50 yards long.

The Midpoint Formula

To find the **midpoint** of the line segment that joins two points in a coordinate plane, you can simply find the average values of the respective coordinates of the two endpoints using the **Midpoint Formula.** (See Appendix A for a proof of the Midpoint Formula.)

Exercises 61–64 on page 89 help develop a general understanding of the Midpoint Formula.

The Midpoint Formula

The midpoint of the segment joining the points (x_1, y_1) and (x_2, y_2) is given by the Midpoint Formula

$$\text{Midpoint} = \left(\frac{x_1 + x_2}{2}, \frac{y_1 + y_2}{2}\right).$$

Example 6 ▶ Finding a Line Segment's Midpoint

Find the midpoint of the line segment joining the points $(-5, -3)$ and $(9, 3)$, as shown in Figure P.23.

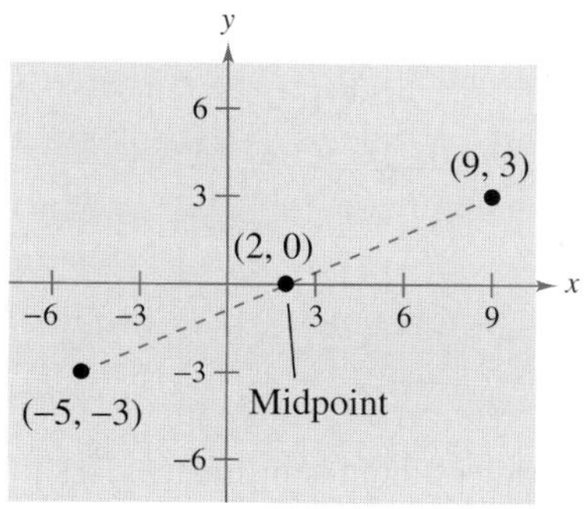

FIGURE P.23

Solution

Let $(x_1, y_1) = (-5, -3)$ and $(x_2, y_2) = (9, 3)$.

$$\text{Midpoint} = \left(\frac{x_1 + x_2}{2}, \frac{y_1 + y_2}{2}\right) \quad \text{Midpoint Formula}$$

$$= \left(\frac{-5 + 9}{2}, \frac{-3 + 3}{2}\right) \quad \text{Substitute for } x_1, y_1, x_2, \text{ and } y_2.$$

$$= (2, 0) \quad \text{Simplify.}$$

Example 7 ▶ Estimating Annual Sales

Winn-Dixie Stores had annual sales of \$1.30 billion in 1996 and \$1.36 billion in 1998. Without knowing any additional information, what would you estimate the 1997 sales to have been? (Source: Winn-Dixie Stores, Inc.)

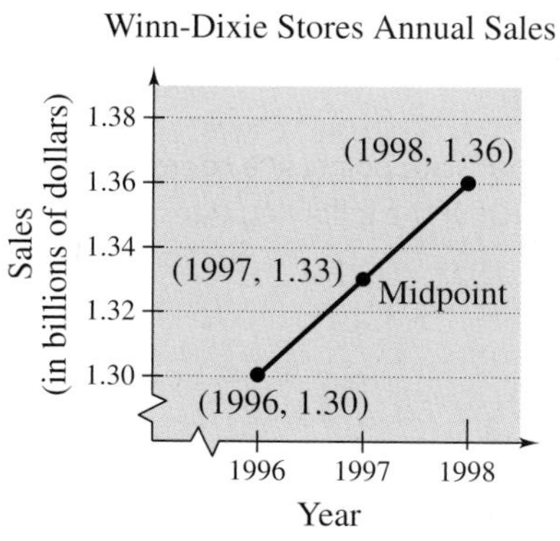

FIGURE P.24

Solution

One solution to the problem is to assume that sales followed a linear pattern. With this assumption, you can estimate the 1997 sales by finding the midpoint of the segment connecting the points (1996, 1.30) and (1998, 1.36).

$$\text{Midpoint} = \left(\frac{1996 + 1998}{2}, \frac{1.30 + 1.36}{2}\right)$$

$$= (1997, 1.33)$$

So, you would estimate the 1997 sales to have been about \$1.33 billion, as shown in Figure P.24. (The actual 1997 sales were \$1.32 billion.)

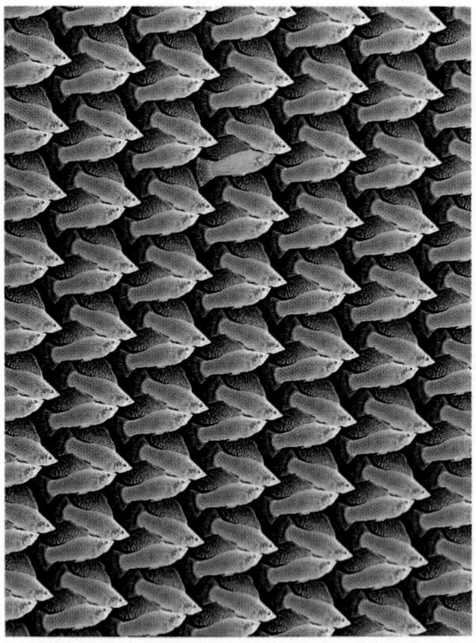

Paul Morrell

Much of computer graphics, including this computer-generated goldfish tesselation, consists of transformations of points in a coordinate plane. One type of transformation, a translation, is illustrated in Example 8. Other types include reflections, rotations, and stretches.

Activities

1. Set up a Cartesian plane and locate the points $(3, 0)$ and $(-4, 1)$.
2. Find x such that the distance between $(x, 5)$ and $(-1, 2)$ is 5.
   Answer: $x = 3, -5$
3. Find the midpoint of the line segment joining the points $(-1, -4)$ and $(3, -2)$.
   Answer: $(1, -3)$

Alternate Writing About Mathematics
Use your school's library or some other reference source to locate a set of real data that can be considered as ordered pairs. Decide which variables should be the x- and y-variables, and explain why you made this decision. List each ordered pair. Sketch a scatter plot of the data, choosing appropriate axes and scale. Be sure to label each axis and give the units for each variable if necessary. Describe any apparent trends in the data. What may be learned from your graph?

Application

A computer animation of this example appears in the *Interactive* CD-ROM and *Internet* versions of this text.

Example 8 ▶ Translating Points in the Plane

The triangle in Figure P.25(a) has vertices at the points $(-1, 2)$, $(1, -4)$, and $(2, 3)$. Shift the triangle 3 units to the right and 2 units up and find the vertices of the shifted triangle, as shown in Figure P.25(b).

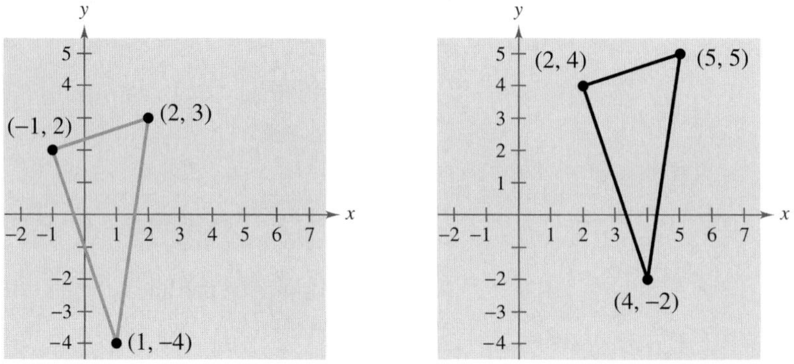

(a) **(b)**

FIGURE P.25

Solution

To shift the vertices 3 units to the right, add 3 to each of the x-coordinates. To shift the vertices 2 units up, add 2 to each of the y-coordinates.

Original Point	*Translated Point*
$(-1, 2)$	$(-1 + 3, 2 + 2) = (2, 4)$
$(1, -4)$	$(1 + 3, -4 + 2) = (4, -2)$
$(2, 3)$	$(2 + 3, 3 + 2) = (5, 5)$

Writing ABOUT MATHEMATICS

Extending the Example Example 8 shows how to translate points in a coordinate plane. Write a short paragraph describing how each of the following transformed points is related to the original point.

Original Point	*Transformed Point*
(x, y)	$(-x, y)$
(x, y)	$(x, -y)$
(x, y)	$(-x, -y)$

P.8 Exercises

In Exercises 1–4, sketch the polygon with the indicated vertices.

1. Triangle: $(-1, 1), (2, -1), (3, 4)$
2. Triangle: $(0, 3), (-1, -2), (4, 8)$
3. Square: $(2, 4), (5, 1), (2, -2), (-1, 1)$
4. Parallelogram: $(5, 2), (7, 0), (1, -2), (-1, 0)$

In Exercises 5 and 6, approximate the coordinates of the points.

5.

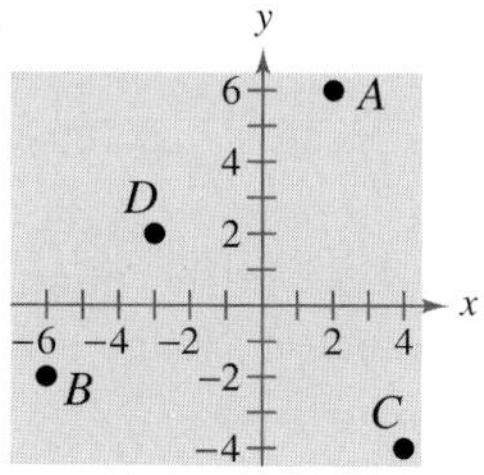

6.

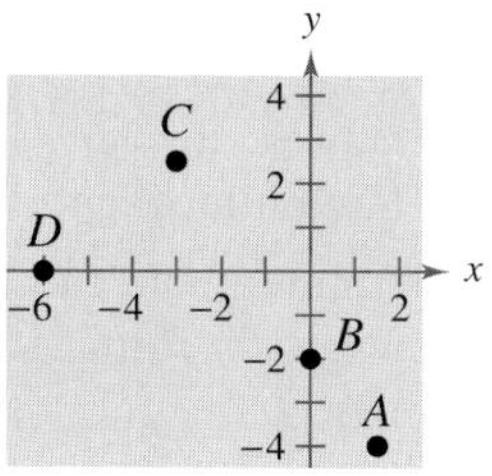

In Exercises 7–10, find the coordinates of the point.

7. The point is located 3 units to the left of the y-axis and 4 units above the x-axis.
8. The point is located 8 units below the x-axis and 4 units to the right of the y-axis.
9. The point is located 5 units below the x-axis and the coordinates of the point are equal.
10. The point is on the x-axis and 12 units to the left of the y-axis.

In Exercises 11–20, determine the quadrant(s) in which (x, y) is located so that the condition(s) is (are) satisfied.

11. $x > 0$ and $y < 0$
12. $x < 0$ and $y < 0$
13. $x = -4$ and $y > 0$
14. $x > 2$ and $y = 3$
15. $y < -5$
16. $x > 4$
17. $(x, -y)$ is in the second quadrant.
18. $(-x, y)$ is in the fourth quadrant.
19. $xy > 0$
20. $xy < 0$

In Exercises 21–24, the polygon is shifted to a new position in the plane. Find the coordinates of the vertices of the polygon in its *new* position.

21.

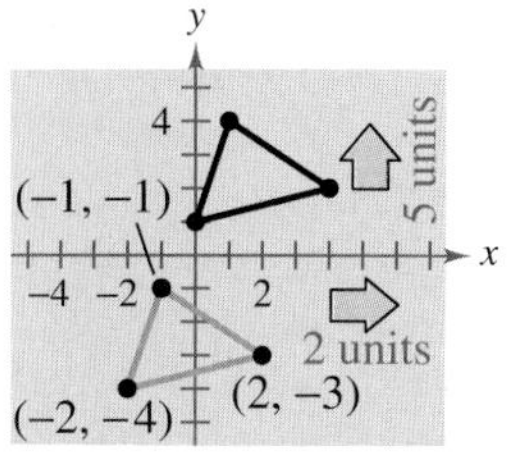

22.

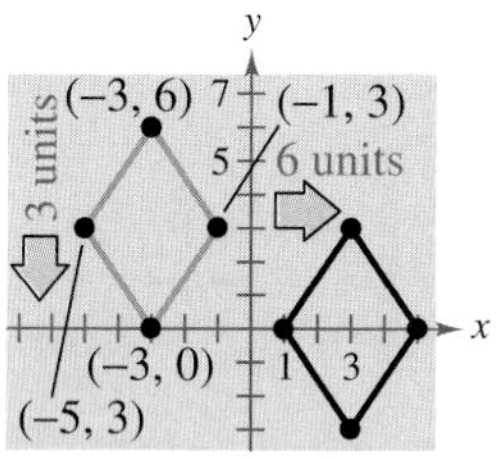

23. Original coordinates of vertices:

    $(-7, -2), (-2, 2), (-2, -4), (-7, -4)$

    Shift: 8 units up, 4 units to the right

24. Original coordinates of vertices:

    $(5, 8), (3, 6), (7, 6), (5, 2)$

    Shift: 6 units down, 10 units to the left

In Exercises 25 and 26, sketch a scatter plot of the data given in the table.

25. *Meteorology* The table shows the lowest temperature of record y (in degrees Fahrenheit) in Duluth, Minnesota, for each month x, where $x = 1$ represents January. (Source: NOAA)

x	1	2	3	4	5	6
y	−39	−33	−29	−5	17	27

x	7	8	9	10	11	12
y	35	32	22	8	−23	−34

26. *Business* The table shows the number y of Wal-Mart stores for each year x from 1992 through 1999. (Source: Wal-Mart Stores, Inc.)

x	1992	1993	1994	1995
y	2136	2440	2759	2943

x	1996	1997	1998	1999
y	3054	3406	3630	3815

Milk Prices **In Exercises 27 and 28, use the graph below, which shows the average retail price of one-half gallon of milk from 1992 to 1997.** (Source: U.S. Bureau of Labor Statistics)

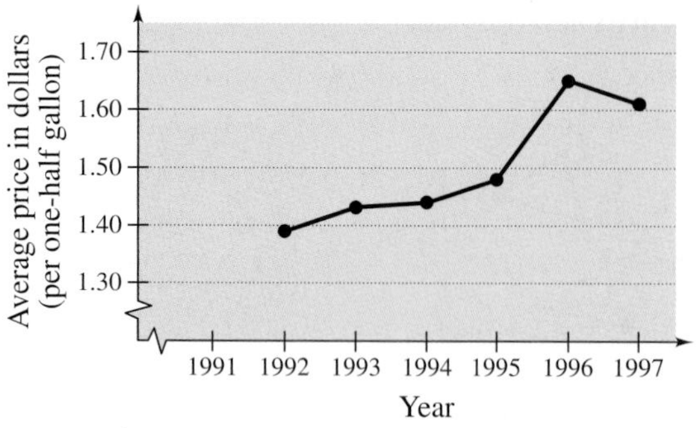

27. Approximate the highest price of one-half gallon of milk shown in the graph. When did this occur?

28. Approximate the percent change in the price of milk from the price in 1992 to the highest price shown in the graph.

Advertising **In Exercises 29 and 30, use the graph below, which shows the cost of a 30-second television spot (in thousands of dollars) during the Super Bowl from 1987 to 1999.** (Source: USA Today Research)

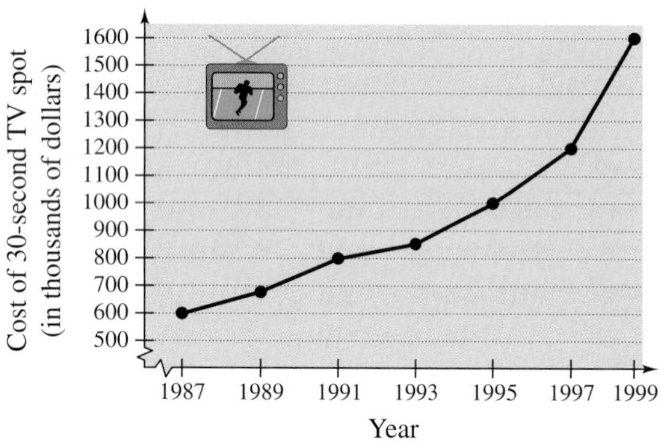

29. Approximate the percent increase in the cost of a 30-second spot from Super Bowl XXI in 1987 to Super Bowl XXXIII in 1999.

30. Estimate the increase in the cost of a 30-second spot (a) from Super Bowl XXI to Super Bowl XXVII and (b) from Super Bowl XXVII to Super Bowl XXXIII.

Labor Force **In Exercises 31 and 32, use the graph below, which shows the minimum wage in the United States (in dollars) from 1950 to 1999.** (Source: U.S. Employment Standards Administration)

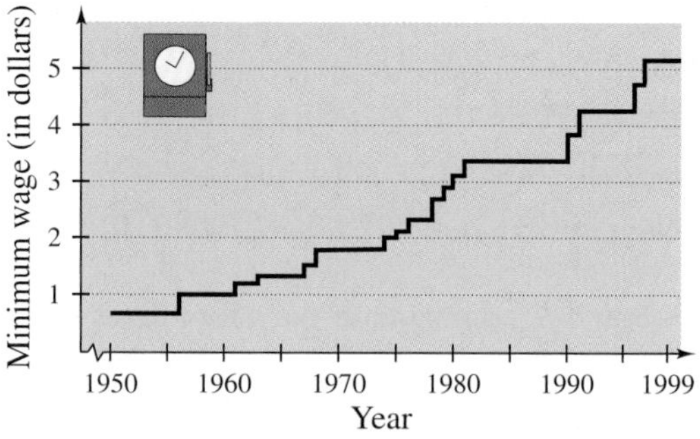

31. Which decade shows the greatest increase in minimum wage?

32. Approximate the percent increase in the minimum wage (a) from 1990 to 1995 and (b) from 1955 to 1995.

Data Analysis **In Exercises 33 and 34, use the graph below, which shows the mathematics entrance test scores *x*, and the final examination scores *y*, in an algebra course for a sample of 10 students.**

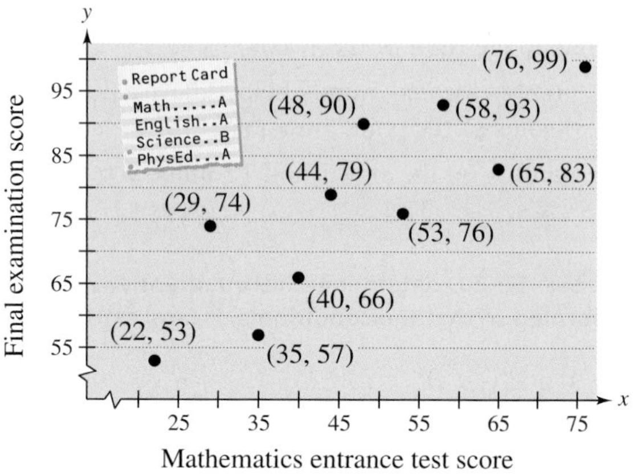

33. Find the entrance exam score of any student with a final exam score in the 80s.

34. Does a higher entrance exam score imply a higher final exam score? Explain.

In Exercises 35–38, find the distance between the points. (*Note:* In each case the two points lie on the same horizontal or vertical line.)

35. $(6, -3), (6, 5)$

36. $(1, 4), (8, 4)$

37. $(-3, -1), (2, -1)$

38. $(-3, -4), (-3, 6)$

In Exercises 39–42, (a) find the length of each side of the right triangle, and (b) show that these lengths satisfy the Pythagorean Theorem.

39.

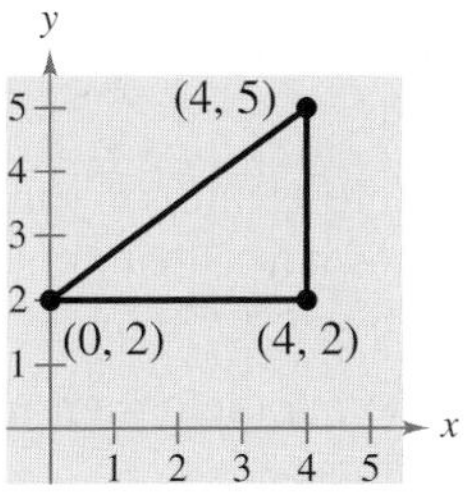

40.

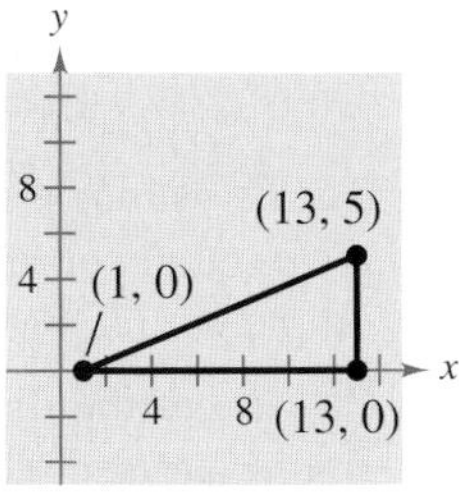

41.

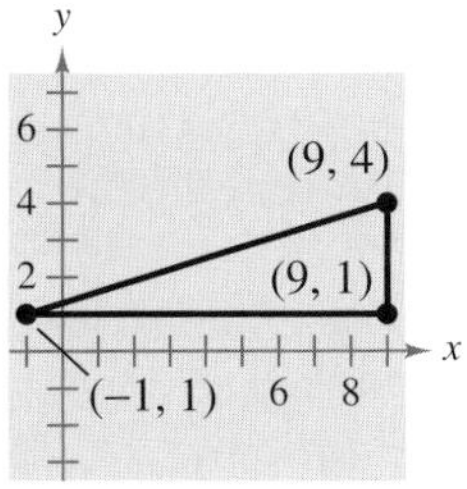

42.

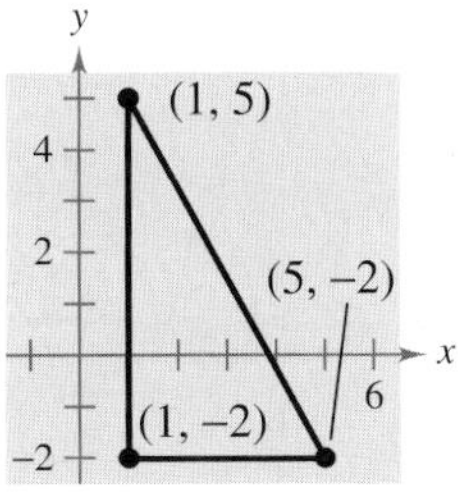

In Exercises 43–54, (a) plot the points, (b) find the distance between the points, and (c) find the midpoint of the line segment joining the points.

43. $(1, 1), (9, 7)$

44. $(1, 12), (6, 0)$

45. $(-4, 10), (4, -5)$

46. $(-7, -4), (2, 8)$

47. $(-1, 2), (5, 4)$

48. $(2, 10), (10, 2)$

49. $\left(\frac{1}{2}, 1\right), \left(-\frac{5}{2}, \frac{4}{3}\right)$

50. $\left(-\frac{1}{3}, -\frac{1}{3}\right), \left(-\frac{1}{6}, -\frac{1}{2}\right)$

51. $(6.2, 5.4), (-3.7, 1.8)$

52. $(-16.8, 12.3), (5.6, 4.9)$

53. $(-36, -18), (48, -72)$

54. $(1.451, 3.051), (5.906, 11.360)$

Business **In Exercises 55 and 56, use the Midpoint Formula to estimate the sales of a company in 1998, given the sales in 1996 and 2000. Assume that the sales followed a linear pattern.**

55.

Year	1996	2000
Sales	\$520,000	\$740,000

56.

Year	1996	2000
Sales	\$4,200,000	\$5,650,000

In Exercises 57–60, show that the points form the vertices of the polygon.

57. Right triangle: $(4, 0), (2, 1), (-1, -5)$

58. Isosceles triangle: $(1, -3), (3, 2), (-2, 4)$

59. Parallelogram: $(2, 5), (0, 9), (-2, 0), (0, -4)$

60. Parallelogram: $(0, 1), (3, 7), (4, 4), (1, -2)$

61. A line segment has (x_1, y_1) as one endpoint and (x_m, y_m) as its midpoint. Find the other endpoint (x_2, y_2) of the line segment in terms of x_1, y_1, x_m, and y_m.

62. Use the result of Exercise 61 to find the coordinates of the endpoint of a line segment if the coordinates of the other endpoint and midpoint are, respectively,

(a) $(1, -2), (4, -1)$ and (b) $(-5, 11), (2, 4)$.

63. Use the Midpoint Formula three times to find the three points that divide the line segment joining (x_1, y_1) and (x_2, y_2) into four parts.

64. Use the result of Exercise 63 to find the points that divide the line segment joining the given points into four equal parts.

(a) $(1, -2), (4, -1)$ (b) $(-2, -3), (0, 0)$

65. *Sports* In a football game, a quarterback throws a pass from the 15-yard line, 10 yards from the sideline, as shown in the figure. The pass is caught on the 40-yard line, 45 yards from the same sideline. How long is the pass?

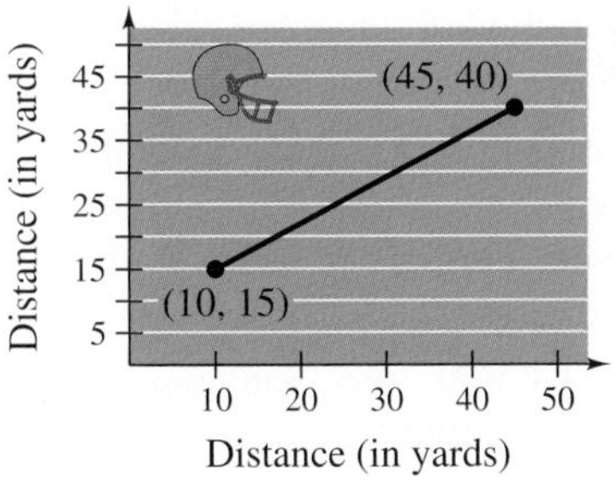

66. *Flying Distance* A plane flies in a straight line to a city that is 100 kilometers east and 150 kilometers north of the point of departure. How far does it fly?

67. *Make a Conjecture* Plot the points $(2, 1)$, $(-3, 5)$, and $(7, -3)$ on a rectangular coordinate system. Then change the sign of the x-coordinate of each point and plot the three new points on the same rectangular coordinate system. Make a conjecture about the location of a point when each of the following occurs.

(a) The sign of the x-coordinate is changed.

(b) The sign of the y-coordinate is changed.

(c) The signs of both the x- and y-coordinates are changed.

68. *Rock and Roll Hall of Fame* The graph below shows the numbers of recording artists who were elected to the Rock and Roll Hall of Fame in the years from 1986 to 1999.

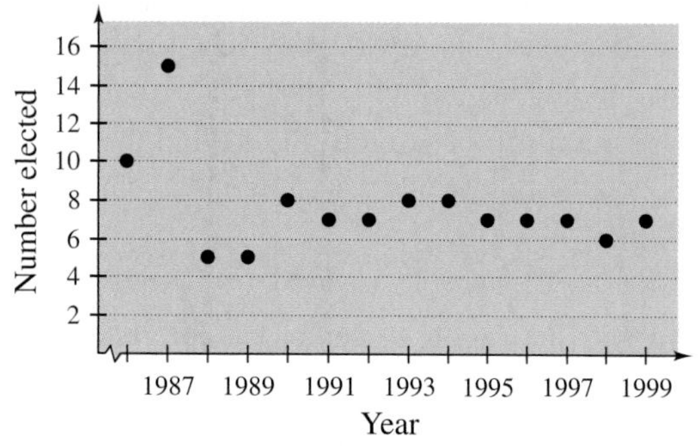

(a) Describe any trends in the data. From these trends, predict the number of artists elected in 2001.

(b) Why do you think the numbers elected in 1986 and 1987 were greater than in other years?

69. *Business* Starbucks Corp. had annual sales of \$696.5 million in 1996 and \$1308.7 million in 1998. Use the Midpoint Formula to estimate the sales in 1997. (Source: Starbucks Corp.)

70. *Business* Lands' End, Inc. had annual sales of \$1118.7 million in 1996 and \$1371.4 million in 1998. Use the Midpoint Formula to estimate the sales in 1997. (Source: Lands' End, Inc.)

Synthesis

True or False? **In Exercises 71 and 72, determine whether the statement is true or false. Justify your answer.**

71. In order to divide a line segment into 16 equal parts, you would have to use the Midpoint Formula 16 times.

72. The points $(-8, 4)$, $(2, 11)$, and $(-5, 1)$ represent the vertices of an isosceles triangle.

73. *Think About It* What is the y-coordinate of any point on the x-axis? What is the x-coordinate of any point on the y-axis?

74. *Think About It* When plotting points on the rectangular coordinate system, is it true that the scales on the x- and y-axes must be the same? Explain.

75. Prove that the diagonals of the parallelogram in the figure intersect at their midpoints.

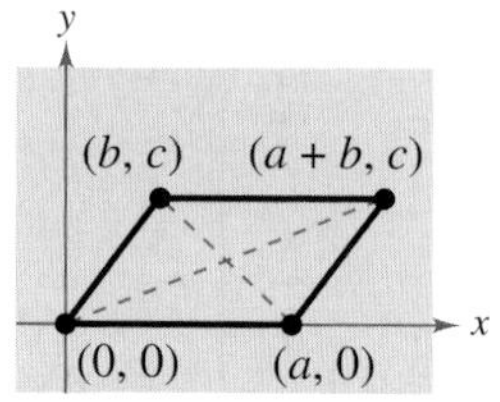

In Exercises 76–79, use the plot of the point (x_0, y_0) in the figure. Match the transformation of the point with the correct plot. [The plots are labeled (a), (b), (c), and (d).]

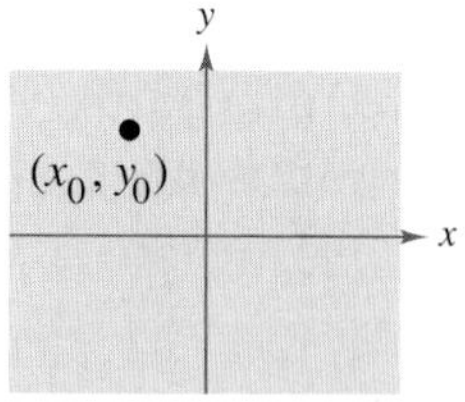

(a)

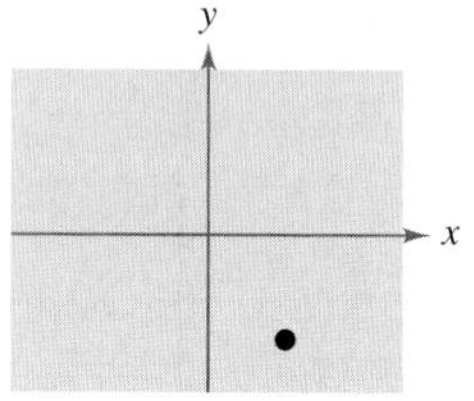

(b)

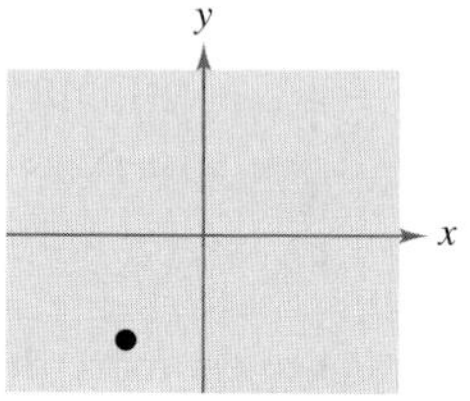

(c)

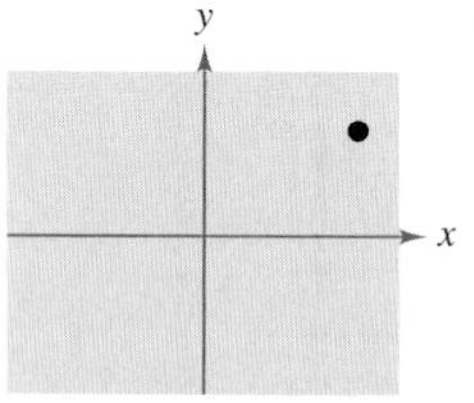

(d)

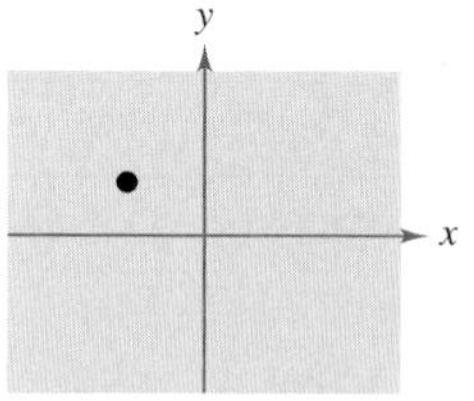

76. $(x_0, -y_0)$

77. $(-2x_0, y_0)$

78. $\left(x_0, \frac{1}{2}y_0\right)$

79. $(-x_0, -y_0)$

Chapter Summary

What did you learn?

Section P.1	Review Exercises
☐ How to represent, classify, and order real numbers and use inequalities	1–6
☐ How to find the absolute values of real numbers and find the distance between two real numbers	7–13
☐ How to evaluate algebraic expressions	14, 15
☐ How to use the basic rules and properties of algebra	16–21
Section P.2	
☐ How to use properties of exponents	22, 23
☐ How to use scientific notation to represent real numbers	24, 25
☐ How to use properties of radicals to simplify and combine radicals	26–31
☐ How to rationalize denominators and numerators	32, 33
☐ How to use properties of rational exponents	34–39
Section P.3	
☐ How to write polynomials in standard form	40–43
☐ How to add, subtract, multiply, and factor polynomials	44–62
Section P.4	
☐ How to find domains of algebraic expressions	63, 64
☐ How to simplify, add, subtract, multiply, and divide rational expressions	65–72
☐ How to simplify complex fractions	73, 74
Section P.5	
☐ How to identify different types of equations	75–78
☐ How to solve a linear equation in one variable	79–83
☐ How to solve polynomial equations	84–97
☐ How to solve equations involving radicals and absolute values	98–110
Section P.6	
☐ How to recognize solutions of linear inequalities	111, 112
☐ How to use properties of linear inequalities to solve linear inequalities	113–118, 123
☐ How to solve inequalities involving absolute values	119–122
☐ How to solve polynomial and rational inequalities	124–134
Section P.7	
☐ How to avoid common algebraic errors	135–142
☐ How to recognize and use algebraic techniques that are common in calculus	143–152
Section P.8	
☐ How to plot points in the Cartesian plane	153–166
☐ How to use the Distance Formula and the Midpoint Formula	159–166
☐ How to use a coordinate plane to model and solve real-life problems	167

Review Exercises

P.1 In Exercises 1 and 2, determine which numbers in the set are (a) natural numbers, (b) integers, (c) rational numbers, and (d) irrational numbers.

1. $\left\{11, -14, -\frac{8}{9}, \frac{5}{2}, \sqrt{6}, 0.4\right\}$

2. $\left\{\sqrt{15}, -22, -\frac{10}{3}, 0, 5.2, \frac{3}{7}\right\}$

In Exercises 3 and 4, use a calculator to find the decimal form of each rational number. If it is a nonterminating decimal, write the repeating pattern. Then plot the numbers on the real number line and place the appropriate inequality sign (< or >) between them.

3. (a) $\frac{5}{6}$ (b) $\frac{7}{8}$

4. (a) $\frac{9}{25}$ (b) $\frac{5}{7}$

In Exercises 5 and 6, give a verbal description of the subset of real numbers represented by the inequality, and sketch the subset on the real number line.

5. $x \le 7$

6. $x > 1$

In Exercises 7 and 8, find the distance between *a* and *b*.

7. $a = -92, b = 63$

8. $a = -12.4, b = -27.13$

9. At 5:00 P.M. the temperature is 23.7°F, and it drops to −0.9°F by 5:00 A.M. What is the change in temperature over the 12-hour period?

In Exercises 10–13, evaluate the expression.

10. $|-37.234|$

11. $|0.017|$

12. $|34 + (-6.8)|$

13. $-5|14 - 21|$

In Exercises 14 and 15, evaluate the expression for each value of *x*. (If not possible, state the reason.)

	Expression	*Values*	
14.	$-x^2 + x - 1$	(a) $x = 1$	(b) $x = -1$
15.	$\dfrac{x}{x-3}$	(a) $x = -3$	(b) $x = 3$

In Exercises 16–21, perform the operations without using a calculator.

16. $|-3| + 4(-2) - 6$

17. $\dfrac{|-10|}{-10}$

18. $\frac{5}{18} \div \frac{10}{3}$

19. $(16 - 8) \div 4$

20. $6[4 - 2(6 + 8)]$

21. $-4[16 - 3(7 - 10)]$

P.2 In Exercises 22 and 23, simplify each expression.

22. (a) $\dfrac{6^2u^3v^{-3}}{12u^{-2}v}$ (b) $\dfrac{3^{-4}m^{-1}n^{-3}}{9^{-2}mn^{-3}}$

23. (a) $(x + y^{-1})^{-1}$ (b) $\left(\dfrac{x^{-3}}{y}\right)\left(\dfrac{x}{y}\right)^{-1}$

In Exercises 24 and 25, write the number in scientific notation.

24. *Sales of K-Mart Corporation in 1998:* \$33,674,000 (Source: K-Mart Corporation)

25. *Number of meters in 1 foot:* 0.3048

In Exercises 26–29, use the properties of radicals to simplify the expression.

26. $\left(\sqrt[3]{216}\right)^3$

27. $\sqrt[4]{32^4}$

28. $\sqrt{18u}\sqrt{2u}$

29. $\dfrac{\sqrt[3]{24u^4}}{\sqrt[3]{3u}}$

In Exercises 30 and 31, simplify each expression.

30. (a) $\sqrt{50} - \sqrt{18}$ (b) $2\sqrt{32} + 3\sqrt{72}$

31. (a) $\sqrt{8x^3} + \sqrt{2x}$ (b) $\sqrt{18x^5} - \sqrt{8x^3}$

In Exercises 32 and 33, rewrite the expression by rationalizing the denominator or numerator. Simplify your answer.

32. $\dfrac{1}{2 - \sqrt{3}}$

33. $\dfrac{\sqrt{x} - 1}{2}$

In Exercises 34–37, simplify the expression.

34. $(16)^{3/2}$

35. $(64)^{-2/3}$

36. $(3x^{2/5})(2x^{1/2})$

37. $(x - 1)^{1/3}(x - 1)^{-1/4}$

In Exercises 38 and 39, fill in the missing form of the equation.

	Radical Form	*Rational Exponent Form*
38.	$\sqrt{16} = 4$	▒ $= 4$
39.	▒ $= 2$	$16^{1/4} = 2$

P.3 In Exercises 40-43, write the polynomial in standard form.

40. $3 - 11x^2$

41. $3x^3 - 5x^5 + x - 4$

42. $-4 - 12x^2$

43. $12x - 7x^2 + 6$

In Exercises 44–47, perform the operations and write the result in standard form.

44. $-(3x^2 + 2x) + (1 - 5x)$

45. $8y - [2y^2 - (3y - 8)]$

46. $(3x - 6)(5x + 1)$

47. $\left(x - \dfrac{1}{x}\right)(x + 2)$

In Exercises 48–51, find the product.

48. $(2x - 3)^2$

49. $(6x + 5)(6x - 5)$

50. $(3\sqrt{5} + 2)(3\sqrt{5} - 2)$

51. $(x - 4)^3$

52. *Business* The revenue from selling x units of a product at a price of p dollars per unit is $R = xp$. For a particular product the revenue is

$R = 1600x - 0.50x^2.$

(a) Find the revenue when the number of units sold is 3000.

(b) Find the price when the number of units sold is 2000.

In Exercises 53–62, factor completely.

53. $x^3 - x$

54. $x(x - 3) + 4(x - 3)$

55. $25x^2 - 49$

56. $x^2 - 12x + 36$

57. $x^3 - 64$

58. $8x^3 + 27$

59. $2x^2 + 21x + 10$

60. $3x^2 + 14x + 8$

61. $x^3 - x^2 + 2x - 2$

62. $x^3 - 4x^2 + 2x - 8$

P.4 In Exercises 63 and 64, find the domain of the expression.

63. $\dfrac{1}{x + 6}$

64. $\sqrt{x + 4}$

In Exercises 65 and 66, write the rational expression in simplest form.

65. $\dfrac{x^2 - 64}{5(3x + 24)}$

66. $\dfrac{x^3 + 27}{x^2 + x - 6}$

In Exercises 67–72, perform the operations and simplify.

67. $\dfrac{x^2 - 4}{x^4 - 2x^2 - 8} \cdot \dfrac{x^2 + 2}{x^2}$

68. $\dfrac{4x - 6}{(x - 1)^2} \div \dfrac{2x^2 - 3x}{x^2 + 2x - 3}$

69. $2x + \dfrac{3}{2(x - 4)}$

70. $\dfrac{1}{x} - \dfrac{x - 1}{x^2 + 1}$

71. $\dfrac{1}{x - 1} + \dfrac{1 - x}{x^2 + x + 1}$

72. $\dfrac{3x}{x + 2} - \dfrac{4x^2 - 5}{2x^2 + 3x - 2}$

In Exercises 73 and 74, simplify the complex fraction.

73. $\dfrac{\left[\dfrac{3a}{(a^2/x) - 1}\right]}{\left(\dfrac{a}{x} - 1\right)}$

74. $\dfrac{\left(\dfrac{1}{2x - 3} - \dfrac{1}{2x + 3}\right)}{\left(\dfrac{1}{2x} - \dfrac{1}{2x + 3}\right)}$

P.5 In Exercises 75–78, determine whether the equation is an identity or a conditional equation.

75. $6 - (x - 2)^2 = 2 + 4x - x^2$

76. $3(x - 2) + 2x = 2(x + 3)$

77. $-x^3 + x(7 - x) + 3 = x(-x^2 - x) + 7(x + 1) - 4$

78. $3(x^2 - 4x + 8) = -10(x + 2) - 3x^2 + 6$

In Exercises 79–82, solve the equation (if possible) and check your solution.

79. $3x - 2(x + 5) = 10$

80. $4x + 2(7 - x) = 5$

81. $4(x + 3) - 3 = 2(4 - 3x) - 4$

82. $\frac{1}{2}(x - 3) - 2(x + 1) = 5$

83. *Mixture Problem* A car radiator contains 10 liters of a 30% antifreeze solution. How many liters will have to be replaced with pure antifreeze if the resulting solution is to be 50% antifreeze?

In Exercises 84–93, use any method to solve the equation.

84. $15 + x - 2x^2 = 0$

85. $2x^2 - x - 28 = 0$

86. $6 = 3x^2$

87. $16x^2 = 25$

88. $(x + 4)^2 = 18$

89. $(x - 8)^2 = 15$

90. $x^2 - 12x + 30 = 0$

91. $x^2 + 6x - 3 = 0$

92. $-2x^2 - 5x + 27 = 0$

93. $-20 - 3x + 3x^2 = 0$

In Exercises 94–109, find all solutions of the equation. Check your solutions in the original equation.

94. $5x^4 - 12x^3 = 0$

95. $4x^3 - 6x^2 = 0$

96. $x^4 - 5x^2 + 6 = 0$

97. $9x^4 + 27x^3 - 4x^2 - 12x = 0$

98. $\sqrt{x + 4} = 3$

99. $\sqrt{x - 2} - 8 = 0$

100. $2\sqrt{x} - 5 = x$

101. $\sqrt{3x - 2} = 4 - x$

102. $(x - 1)^{2/3} - 25 = 0$

103. $(x + 2)^{3/4} = 27$

104. $(x + 4)^{1/2} + 5x(x + 4)^{3/2} = 0$

105. $8x^2(x^2 - 4)^{1/3} + (x^2 - 4)^{4/3} = 0$

106. $|x - 5| = 10$ **107.** $|2x + 3| = 7$

108. $|x^2 - 3| = 2x$ **109.** $|x^2 - 6| = x$

110. *Economics* The demand equation for a product is

$$p = 42 - \sqrt{0.001x + 2}$$

where x is the number of units demanded per day and p is the price per unit. Find the demand if the price is set at \$29.95.

P.6 In Exercises 111 and 112, determine whether each value of x is a solution of the inequality.

111. $6x - 17 > 0$ (a) $x = 3$ (b) $x = -4$ (c) $x = 2.9$ (d) $x = 1$

112. $-3 \le \frac{x - 3}{5} < 2$ (a) $x = 3$ (b) $x = -12$ (c) $x = 13$ (d) $x = -18$

In Exercises 113–122, solve the inequality.

113. $9x - 8 \le 7x + 16$ **114.** $\frac{15}{2}x + 4 > 3x - 5$

115. $4(5 - 2x) \le \frac{1}{2}(8 - x)$

116. $\frac{1}{2}(3 - x) > \frac{1}{3}(2 - 3x)$

117. $-19 < 3x - 17 \le 34$ **118.** $-3 \le \frac{2x - 5}{3} < 5$

119. $|x| \le 4$ **120.** $|x - 2| < 1$

121. $|x - 3| > 4$ **122.** $\left|x - \frac{3}{2}\right| \ge \frac{3}{2}$

123. *Business* The revenue for selling x units of a product is $R = 125.33x$. The cost of producing x units is $C = 92x + 1200$. To obtain a profit, the revenue must be greater than the cost. Determine the smallest value of x for which this product returns a profit.

124. *Geometry* The side of a square is measured as 19.3 centimeters with a possible error of 0.5 centimeter. Using these measurements, determine the interval containing the area of the square.

In Exercises 125–132, solve the inequality.

125. $x^2 - 6x - 27 < 0$ **126.** $x^2 - 2x \ge 3$

127. $6x^2 + 5x < 4$ **128.** $2x^2 + x \ge 15$

129. $\frac{2}{x + 1} \le \frac{3}{x - 1}$ **130.** $\frac{x - 5}{3 - x} < 0$

131. $\frac{x^2 + 7x + 12}{x} \ge 0$ **132.** $\frac{1}{x - 2} > \frac{1}{x}$

133. *Investment* P dollars invested at interest rate r compounded annually increases to an amount

$$A = P(1 + r)^2$$

in 2 years. If an investment of \$5000 is to increase to an amount greater than \$5500 in 2 years, then the interest rate must be greater than what percent?

134. *Biology* A biologist introduces 200 ladybugs into a crop field. The population P of the ladybugs is approximated by the model

$$P = \frac{1000(1 + 3t)}{5 + t}$$

where t is the time in days. Find the time required for the population to increase to at least 2000 ladybugs.

P.7 In Exercises 135–142, describe and correct the error.

135. $10(4 \cdot 7) = 40 \cdot 70$ ✗ **136.** $\left(\frac{1}{3}x\right)\left(\frac{1}{3}y\right) = \frac{1}{3}xy$ ✗

137. $(2x)^4 = 2x^4$ ✗ **138.** $(-x)^6 = -x^6$ ✗

139. $(3^4)^4 = 3^8$ ✗

140. $\sqrt{3^2 + 4^2} = 3 + 4$ ✗

141. $(5 + 8)^2 = 5^2 + 8^2$ ✗

142. $(9x + 12)^2 = 3(3x + 4)^2$ ✗

In Exercises 143–146, insert the missing factor.

143. $\frac{2}{3}x^4 - \frac{3}{8}x^3 + \frac{5}{6}x^2 = \frac{1}{24}x^2(\quad)$

144. $\frac{t}{\sqrt{t + 1}} - \sqrt{t + 1} = \frac{1}{\sqrt{t + 1}}(\quad)$

145. $2x(x^2 - 3)^{1/3} - 5(x^2 - 3)^{4/3} = (x^2 - 3)^{1/3}(\quad)$

146. $y(y - 1)^{5/4} - y^2(y - 1)^{1/4} = y(y - 1)^{1/4}(\quad)$

In Exercises 147 and 148, factor the expression.

147. $x(x + 2)^{-1/2} + (x + 2)^{1/2}$

148. $\frac{2}{3}x(4 + x)^{-1/2} - \frac{2}{15}(4 + x)^{3/2}$

In Exercises 149–152, write the fraction as a sum of terms.

149. $\frac{x^2 - 4x + 2}{x}$ **150.** $\frac{3x^3 + 4x^2 + x - 5}{x^2}$

151. $\frac{2x^3 - 4x^2 + 3}{\sqrt{x}}$ **152.** $\frac{8x^4 + 5x^3 - 10x + 5}{5x^2}$

P.8 *Geometry* **In Exercises 153 and 154, plot the points and verify that the points form the polygon.**

153. Right triangle: (2, 3), (13, 11), (5, 22)

154. Parallelogram: (1, 2), (8, 3), (9, 6), (2, 5)

In Exercises 155–158, determine the quadrant(s) in which (x, y) is located so that the condition(s) is (are) satisfied.

155. $x > 0$ and $y = -2$ **156.** $y > 0$

157. $(-x, y)$ is in the third quadrant.

158. $xy = 4$

In Exercises 159–162, (a) plot the points and (b) find the distance between the points.

159. $(-3, 8), (1, 5)$ **160.** $(14, -3), (-9, 7)$

161. $(5.6, 0), (0, 8.2)$

162. $(-2.3, 4.8), (6.1, -5.2)$

In Exercises 163–166, (a) plot the points and (b) find the midpoint of the line segment joining the points.

163. $(-2, 6), (4, -3)$ **164.** $(12, 2), (2, 8)$

165. $(0, -1.2), (-3.6, 0)$

166. $(-3.2, 4), (-4.5, -6.8)$

167. *Meteorology* The apparent temperature is a measure of relative discomfort to a person from heat and high humidity. The scatter plot shows the apparent temperatures (in degrees Fahrenheit) for a relative humidity of 75%.

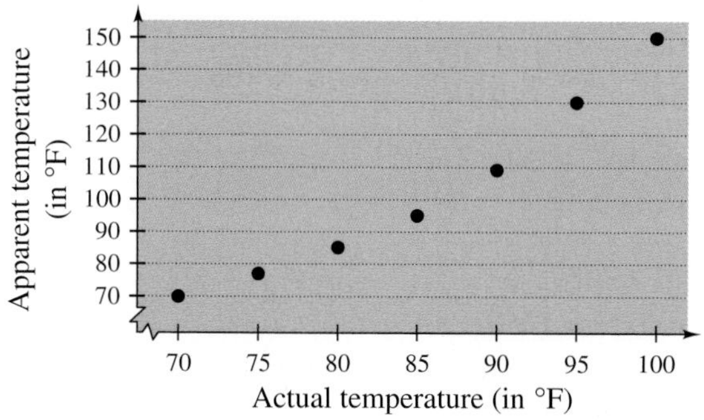

Find the change in the apparent temperature when the actual temperature changes from 70°F to 100°F.

Synthesis

True or False? **In Exercises 168–172, determine whether the statement is true or false. Justify your answer.**

168. $\dfrac{x^3 - 1}{x - 1} = x^2 + x + 1$ for all values of x.

169. A binomial sum squared is equal to the sum of the terms squared.

170. $x^n - y^n$ factors as conjugates for all values of n.

171. The Quadratic Formula can be used to solve any quadratic equation, but not all quadratic equations can be factored.

172. The equation $325x^2 - 717x + 398 = 0$ has no solution.

173. The graphs show the solutions of equations plotted on the real number line. In each case, determine whether the solution(s) is (are) for a linear equation, a quadratic equation, both, or neither. Explain.

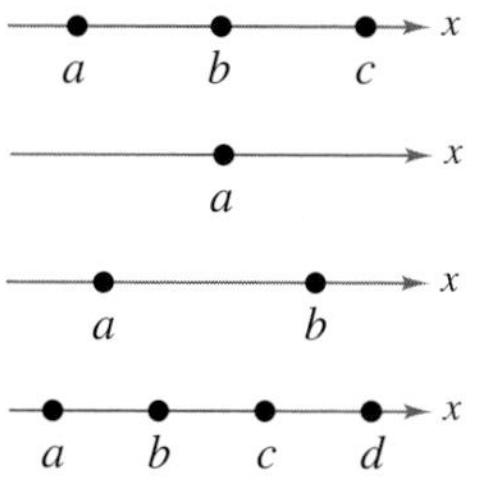

174. *Exploration* The surface area of a right circular cylinder is $S = 2\pi r^2 + 2\pi rh$.

(a) Draw a right circular cylinder of radius r and height h. Use the figure to explain how the surface area formula was obtained.

(b) Find the surface area when the radius is 6 inches and the height is 8 inches.

175. *Writing* Explain why $\sqrt{5u} + \sqrt{3u} \neq 2\sqrt{2u}$.

176. Explain why it is important to check your solutions to certain types of equations.

177. Create a solution that solves a real-life problem using a linear model.

178. *Error Analysis* What is wrong with the following solution?

$$|11x - 4| = -26 \quad \times$$

$$11x - 4 = -26 \quad \text{or} \quad 11x - 4 = 26$$

$$11x = -22 \qquad\qquad 11x = 30$$

$$x = -2 \qquad\qquad x = \frac{30}{11}$$

Chapter Project ▶ Numerically Finding a Maximum Volume

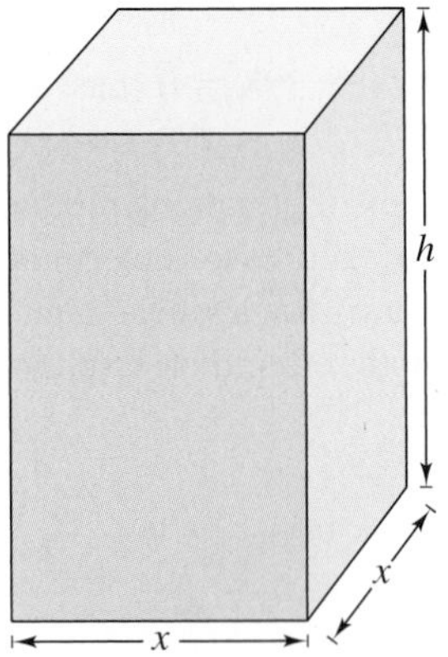

A computer simulation to accompany this project appears in the *Interactive* CD-ROM and *Internet* versions of this text.

Base, x	*Height*	*Surface Area*	*Volume*
1.0	53.5	216.0	53.5
1.5	35.3	216.0	79.3
2.0	26.0	216.0	104.0
2.5	20.4	216.0	127.2
3.0	16.5	216.0	148.5
3.5	13.7	216.0	167.6
4.0	11.5	216.0	184.0
4.5	9.8	216.0	197.4
5.0	8.3	216.0	207.5
5.5	7.1	216.0	213.8
6.0	6.0	216.0	216.0
6.5	5.1	216.0	213.7
7.0	4.2	216.0	206.5
7.5	3.5	216.0	194.1
8.0	2.8	216.0	176.0
8.5	2.1	216.0	151.9
9.0	1.5	216.0	121.5
9.5	0.9	216.0	84.3
10.0	0.4	216.0	40.0

The table shows the surface areas and volumes of boxes with different bases and heights. From the table, it appears that the cube (the box whose dimensions are 6 inches by 6 inches by 6 inches) has the greatest volume.

Many mathematical results are discovered by calculating examples and looking for patterns. Prior to the 1950s, this mode of discovery was very time-consuming because the calculations had to be done by hand. The introduction of computer and calculator technology has removed much of the drudgery of calculation.

Using technology to conduct mathematical experiments usually involves creation of an algebraic **model** to represent the quantity under question.

Example ▶ Modeling the Volume of a Box

Consider a rectangular box with a square base and a surface area of 216 square inches. Let x represent the length (in inches) of each side of the base. Use the variable x to write a model for the volume of the box.

Solution

Begin by expressing the areas of the base, top, and sides in terms of x and h. The expression for the surface area of the box in terms of x and h is $S = x^2 + x^2 + 4xh$. Express the variable h in terms of x.

$$216 = 2x^2 + 4xh \qquad \text{Substitute 216 for } S.$$

$$\frac{54}{x} - \frac{x}{2} = h$$

Find an expression for the volume of the box in the figure above in terms of x.

$$V = x \cdot x \cdot \left(\frac{54}{x} - \frac{x}{2}\right) = 54x - \frac{1}{2}x^3, \quad 0 \le x \le \sqrt{108}$$

So, a model for the volume is $V = 54x - \frac{1}{2}x^3$.

Chapter Project Investigations

1. In the example, what happens to the height of the boxes as x gets closer and closer to 0? Of all boxes with square bases and surface areas of 216 square inches, is there a tallest? Explain your reasoning.
2. In the example, what happens to the height of the boxes as x gets closer and closer to $\sqrt{108}$? Is there a shortest box that has a square base and a surface area of 216 square inches? Explain your reasoning.
3. Complete the table. Why does it support the conclusion in the margin?

x	5.9	5.99	5.999	6.001	6.01	6.1
V	?	?	?	?	?	?

4. Of all rectangular boxes with surface areas of 216 square inches and bases of x inches by $2x$ inches, which has the maximum volume? Explain.

Chapter Test

The *Interactive* CD-ROM and *Internet* versions of this text provide answers to the Chapter Tests and Cumulative Tests. They also offer Chapter Pre-Tests (which test key skills and concepts covered in previous chapters) and Chapter Post-Tests, both of which have randomly generated exercises with diagnostic capabilities.

Take this test as you would take a test in class. After you are done, check your work against the answers given in the back of the book.

1. Place < or > between the real numbers $-\frac{10}{3}$ and $-|-4|$.
2. Find the distance between the real numbers -5.4 and $3\frac{3}{4}$.

In Exercises 3–6, evaluate each expression without using a calculator.

3. (a) $27\left(-\frac{2}{3}\right)$ (b) $\frac{5}{18} \div \frac{15}{8}$
4. (a) $\left(-\frac{3}{5}\right)^3$ (b) $\left(\frac{3^2}{2}\right)^{-3}$
5. (a) $\sqrt{5} \cdot \sqrt{125}$ (b) $\frac{\sqrt{72}}{\sqrt{2}}$
6. (a) $\frac{5.4 \times 10^8}{3 \times 10^3}$ (b) $(3 \times 10^4)^3$

In Exercises 7 and 8, simplify each expression.

7. (a) $3z^2(2z^3)^2$ (b) $(u-2)^{-4}(u-2)^{-3}$ (c) $\left(\frac{x^{-2}y^2}{3}\right)^{-1}$
8. (a) $9z\sqrt{8z} - 3\sqrt{2z^3}$ (b) $-5\sqrt{16y} + 10\sqrt{y}$ (c) $\sqrt[3]{\frac{16}{v^5}}$

In Exercises 9 and 10, perform the operations and simplify.

9. $(x^2 + 3) - [3x + (8 - x^2)]$
10. $(x + \sqrt{5})(x - \sqrt{5})$
11. Factor (a) $2x^4 - 3x^3 - 2x^2$ and (b) $x^3 + 2x^2 - 4x - 8$ completely.
12. Rationalize each denominator. (a) $\frac{16}{\sqrt[3]{16}}$ (b) $\frac{6}{1 - \sqrt{3}}$

In Exercises 13–18, solve the equation (if possible).

13. $\frac{2}{3}(x - 1) + \frac{1}{4}x = 10$
14. $(x - 3)(x + 2) = 14$
15. $\frac{x - 2}{x + 2} + \frac{4}{x + 2} + 4 = 0$
16. $x^4 + x^2 - 6 = 0$
17. $2\sqrt{x} - \sqrt{2x + 1} = 1$
18. $|3x - 1| = 7$

In Exercises 19–22, solve the inequality. Sketch the solution on the real number line.

19. $-3 \le 2(x + 4) < 14$
20. $\frac{2}{x} > \frac{5}{x + 6}$
21. $2x^2 + 5x > 12$
22. $|x - 15| \ge 5$
23. Simplify the expression $\frac{3x^2 + 5x - 2}{2x^2 + x - 6}$.
24. After working for x hours together on a carpentry job, two workers have done fractional parts of the job equal to $x/4$ and $2x/7$, respectively. What fractional part of the job has been completed? What fractional part of the work that has been completed was done by the first worker?
25. The positions of two boats on a lake are given by the coordinates $(-2, 5)$ and $(6, 0)$, where x and y are measured in miles. Find the distance between the boats, and the coordinates of the point that is halfway between them.

Mark Joseph/Tony Stone Images

The average cost of a new domestic car increased from \$18,064 in 1996 to \$18,580 in 1997 even though sales (demand) and production (supply) of new domestic cars declined. *(Source: U.S. Bureau of Economic Analysis)*

1 Functions and Their Graphs

▶ How to Study This Chapter

The Big Picture

In this chapter you will learn the following skills and concepts.

- ▶ How to sketch the graphs of equations
- ▶ How to find and use the slopes of lines to write and graph linear equations in two variables
- ▶ How to evaluate functions and find their domains
- ▶ How to analyze graphs of functions
- ▶ How to identify and graph shifts, reflections, and nonrigid transformations of functions
- ▶ How to find arithmetic combinations and compositions of functions
- ▶ How to find inverses of functions graphically and algebraically
- ▶ How to write algebraic models for direct, inverse, and joint variation

Important Vocabulary

As you encounter each new vocabulary term in this chapter, add the term and its definition to your notebook glossary.

Study Tools

Additional Resources

- Study and Solutions Guide
- Interactive Precalculus
- Videotapes for Chapter 1
- Precalculus Website
- Student Success Organizer

STUDY T!P

During class, take notes on definitions, examples, concepts, and rules—whatever it is you identify as important to the instructor. Then, as soon after class as possible, review your notes.

1.1 Graphs of Equations

▶ **What you should learn**

- How to sketch graphs of equations
- How to find *x*- and *y*-intercepts of graphs
- How to use symmetry to sketch graphs of equations
- How to find equations and sketch graphs of circles
- How to use graphs of equations in real-life problems

▶ **Why you should learn it**

The graph of an equation can help you see relationships between real-life quantities. For example, Exercise 74 on page 109 shows how a graph can be used to estimate the life expectancies of children who are born in the years 2002 and 2004.

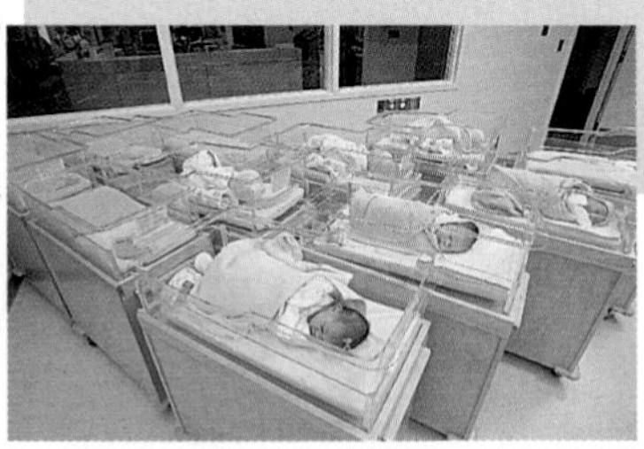

Bruce Forster/Tony Stone Images

The Graph of an Equation

In Section P.8, you used a coordinate system to represent graphically the relationship between two quantities. There, the graphical picture consisted of a collection of points in a coordinate plane.

Frequently, a relationship between two quantities is expressed as an **equation in two variables.** For instance, $y = 7 - 3x$ is an equation in x and y. An ordered pair (a, b) is a **solution** or **solution point** of an equation in x and y if the equation is true when a is substituted for x and b is substituted for y. For instance, $(1, 4)$ is a solution of $y = 7 - 3x$ because $4 = 7 - 3(1)$ is a true statement.

In this section, you will review some basic procedures for sketching the graph of an equation in two variables. The **graph of an equation** is the set of all points that are solutions of the equation.

Example 1 ▶ Sketching the Graph of an Equation

Sketch the graph of $y = 7 - 3x$.

Solution

The simplest way to sketch the graph of an equation is the *point-plotting method.* With this method, you construct a table of values that consists of several solution points of the equation. For instance, when $x = 0$,

$$y = 7 - 3(0) = 7,$$

which implies that $(0, 7)$ is a solution point of the graph.

x	0	1	2	3	4
$y = 7 - 3x$	7	4	1	-2	-5

From the table, it follows that $(0, 7)$, $(1, 4)$, $(2, 1)$, $(3, -2)$, and $(4, -5)$ are solution points of the equation. After plotting these points, you can see that they appear to lie on a line, as shown in Figure 1.1. The graph of the equation is the line that passes through the five plotted points.

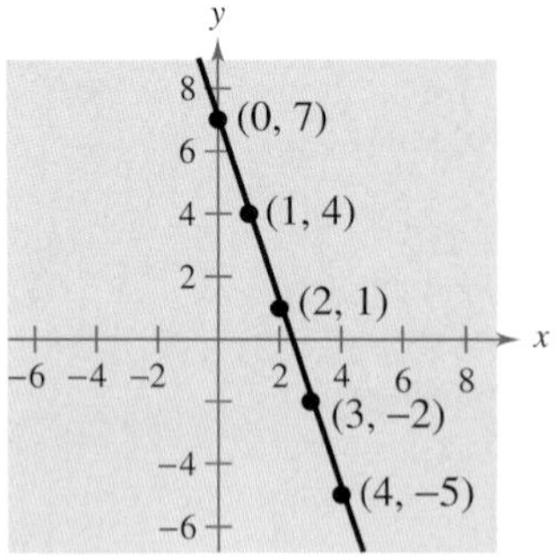

FIGURE 1.1

The *Interactive* CD-ROM and *Internet* versions of this text offer a built-in graphing calculator, which can be used in the Examples, Explorations, Technology notes, and Exercises.

STUDY TIP

One of your goals in this course is to learn to classify the basic shape of a graph from its equation. For instance, you will learn that the *linear equation* in Example 1 has the form

$$y = mx + b$$

and its graph is a straight line. Similarly, the *quadratic equation* in Example 2 has the form

$$y = ax^2 + bx + c$$

and its graph is a parabola.

Example 2 ▶ Sketching the Graph of an Equation

Sketch the graph of $y = x^2 - 2$.

Solution

Begin by constructing a table of values.

x	-2	-1	0	1	2	3
$y = x^2 - 2$	2	-1	-2	-1	2	7

Next, plot the points given in the table, as shown in Figure 1.2(a). Finally, connect the points with a smooth curve, as shown in Figure 1.2(b).

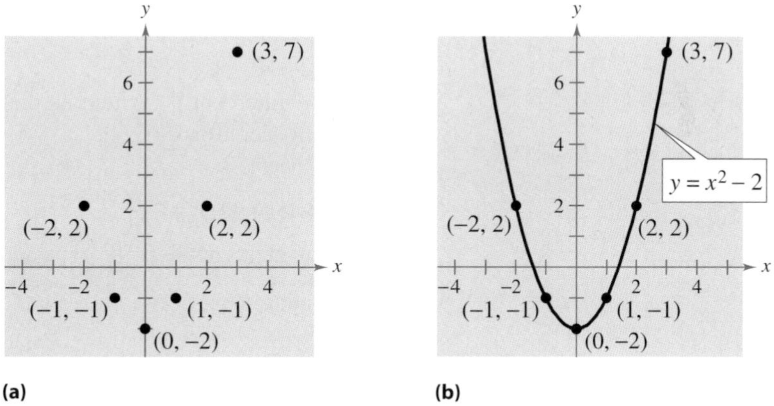

FIGURE 1.2

A computer animation of this example appears in the *Interactive* CD-ROM and *Internet* versions of this text.

The point-plotting technique demonstrated in Examples 1 and 2 is easy to use, but it has some shortcomings. With too few solution points, you can badly misrepresent the graph of an equation. For instance, using only the four points

$$(-2, 2), (-1, -1), (1, -1), \text{ and } (2, 2)$$

in Figure 1.2, any one of the three graphs in Figure 1.3 would be reasonable.

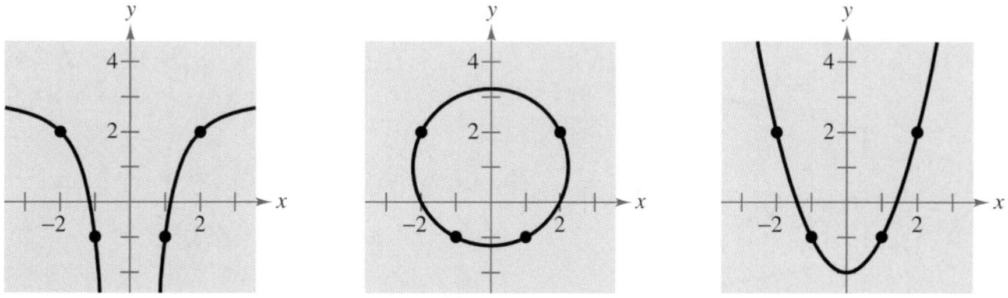

FIGURE 1.3

Intercepts of a Graph

It is often easy to determine the solution points that have zero as either the x-coordinate or the y-coordinate. These points are called **intercepts** because they are the points at which the graph intersects the x- or y-axis. It is possible for a graph to have no intercepts or several intercepts, as shown in Figure 1.4.

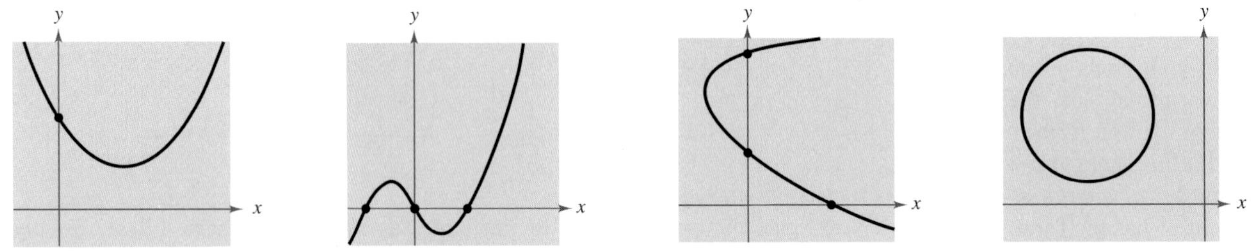

No x-intercept
One y-intercept

Three x-intercepts
One y-intercept

One x-intercept
Two y-intercepts

No intercepts

FIGURE 1.4

Note that an x-intercept is written as the ordered pair $(x, 0)$ and a y-intercept is written as the ordered pair $(0, y)$.

Finding Intercepts

1. To find x-intercepts, let y be zero and solve the equation for x.
2. To find y-intercepts, let x be zero and solve the equation for y.

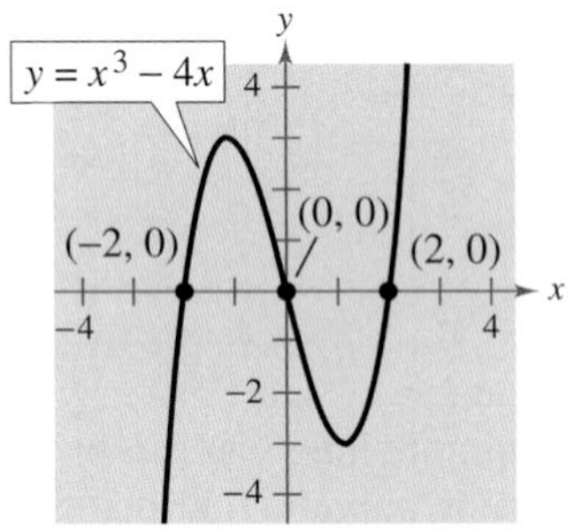

FIGURE 1.5

Example 3 ▶ Finding x- and y-Intercepts

Find the x- and y-intercepts of the graph of each equation.

a. $y = x^3 - 4x$ **b.** $y^2 = x + 4$

Solution

a. Let $y = 0$. Then

$$0 = x^3 - 4x = x(x^2 - 4)$$

has solutions $x = 0$ and $x = \pm 2$.

x-intercepts: $(0, 0), (2, 0), (-2, 0)$

Let $x = 0$. Then $y = (0)^3 - 4(0) = 0$.

y-intercept: $(0, 0)$ (See Figure 1.5.)

b. Let $y = 0$. Then $(0)^2 = x + 4$, and $-4 = x$.

x-intercept: $(-4, 0)$

Let $x = 0$. Then $y^2 = 0 + 4 = 4$ has solutions $y = \pm 2$.

y-intercepts: $(0, 2), (0, -2)$ (See Figure 1.6.)

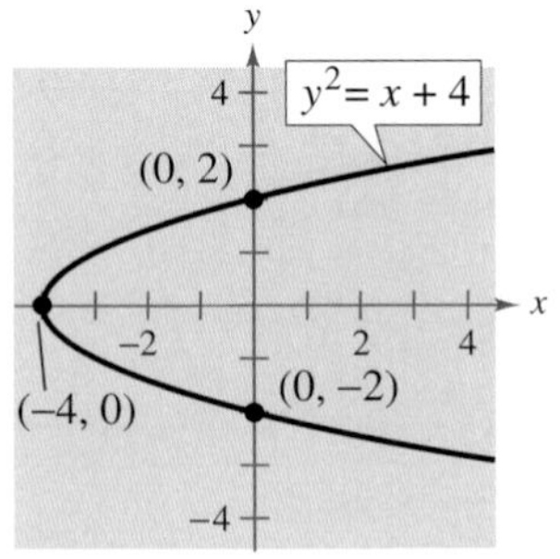

FIGURE 1.6

Symmetry

The graphs shown in Figures 1.2(b), 1.5, and 1.6 each have **symmetry** with respect to one of the coordinate axes or with respect to the origin.

Figure 1.2(b)	$y = x^2 - 2$	y-Axis symmetry
Figure 1.5	$y = x^3 - 4x$	Origin symmetry
Figure 1.6	$y^2 = x + 4$	x-Axis symmetry

Symmetry with respect to the x-axis means that if the Cartesian plane were folded along the x-axis, the portion of the graph above the x-axis would coincide with the portion below the x-axis. Symmetry with respect to the y-axis or the origin can be described in a similar manner, as shown in Figure 1.7.

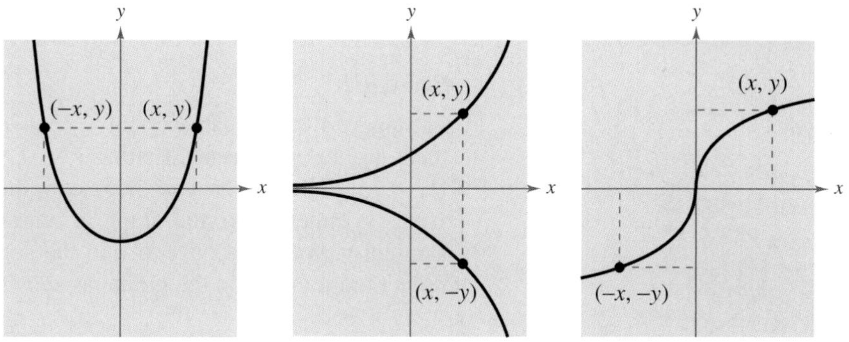

y-Axis symmetry *x-Axis symmetry* *Origin symmetry*

FIGURE 1.7

A computer animation of this concept appears in the *Interactive* CD-ROM and *Internet* versions of this text.

Knowing the symmetry of a graph *before* attempting to sketch it is helpful, because then you need only half as many solution points to sketch the graph. There are three basic types of symmetry. A graph is *symmetric with respect to the y-axis* if, whenever (x, y) is on the graph, $(-x, y)$ is also on the graph. A graph is *symmetric with respect to the x-axis* if, whenever (x, y) is on the graph, $(x, -y)$ is also on the graph. A graph is *symmetric with respect to the origin* if, whenever (x, y) is on the graph, $(-x, -y)$ is also on the graph.

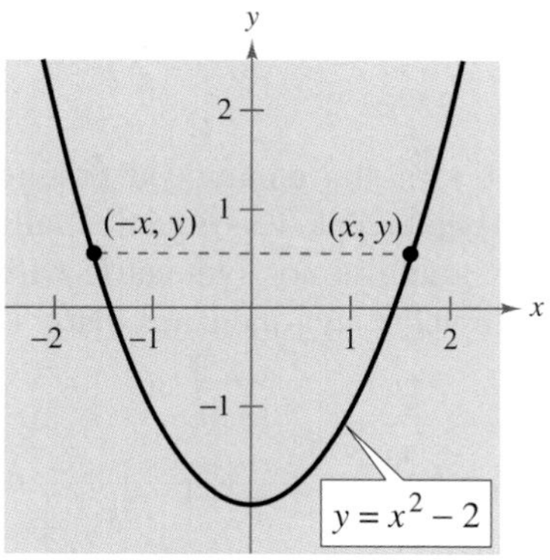

y-Axis symmetry

FIGURE 1.8

Example 4 ▶ Testing for Symmetry

The graph of $y = x^2 - 2$ is symmetric with respect to the y-axis because the point $(-x, y)$ satisfies the equation.

$y = x^2 - 2$	Write original equation.
$y = (-x)^2 - 2$	Substitute $(-x, y)$ for (x, y).
$y = x^2 - 2$	Replacement yields equivalent equation.

See Figure 1.8.

Tests for Symmetry

1. The graph of an equation is symmetric with respect to the *y-axis* if replacing x with $-x$ yields an equivalent equation.
2. The graph of an equation is symmetric with respect to the *x-axis* if replacing y with $-y$ yields an equivalent equation.
3. The graph of an equation is symmetric with respect to the *origin* if replacing x with $-x$ *and* y with $-y$ yields an equivalent equation.

A computer animation of this example appears in the *Interactive* CD-ROM and *Internet* versions of this text.

Example 5 ▶ Using Intercepts and Symmetry as Sketching Aids

Use intercepts and symmetry to sketch the graph of

$$x - y^2 = 1.$$

Solution

Letting $x = 0$, you can see that $-y^2 = 1$ or $y^2 = -1$ has no real solutions. So, there are no y-intercepts. Letting $y = 0$, you obtain $x = 1$. So, the x-intercept is $(1, 0)$. Of the three tests for symmetry, the only one that is satisfied is the test for x-axis symmetry. So, the graph is symmetric with respect to the x-axis. Using symmetry, you need only to find the solution points above the x-axis and then reflect them to obtain the graph, as shown in Figure 1.9.

y	0	1	2
$x = y^2 + 1$	1	2	5

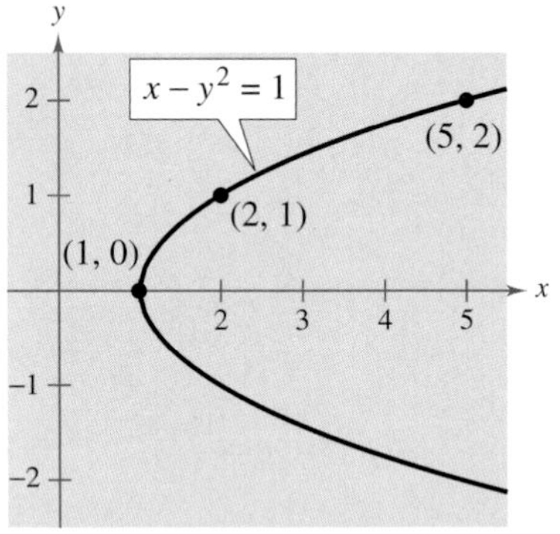

FIGURE 1.9

Example 6 ▶ Sketching the Graph of an Equation

Sketch the graph of

$$y = |x - 1|.$$

Solution

Letting $x = 0$ yields $y = 1$, which means that $(0, 1)$ is the y-intercept. Letting $y = 0$ yields $x = 1$, which means that $(1, 0)$ is the x-intercept. This equation fails all three tests for symmetry and consequently its graph is not symmetric with respect to either axis or to the origin. The absolute value sign indicates that y is always nonnegative.

x	-2	-1	0	1	2	3	4
$y = \|x - 1\|$	3	2	1	0	1	2	3

The graph is shown in Figure 1.10.

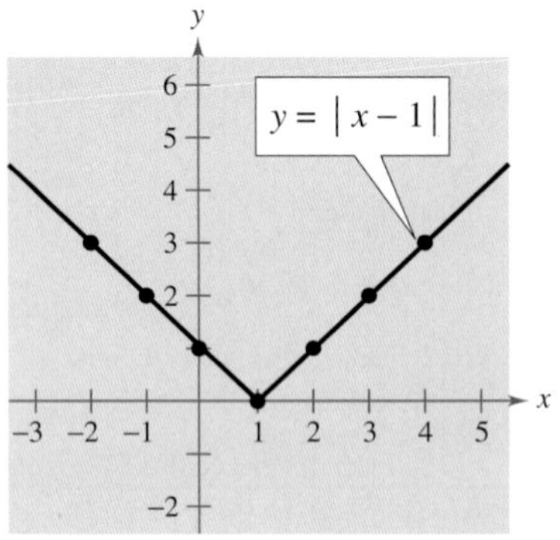

FIGURE 1.10

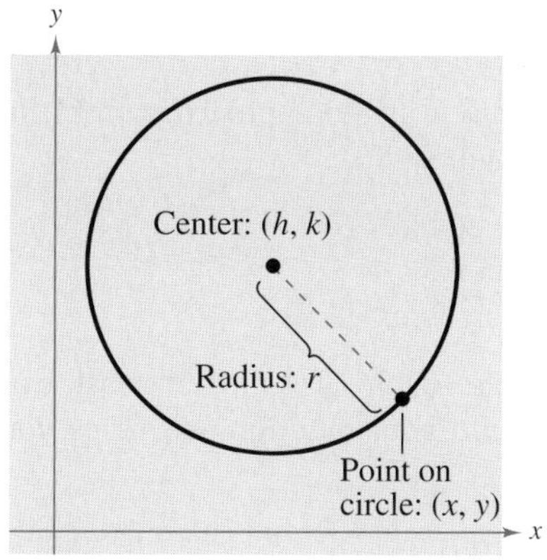

FIGURE 1.11

Throughout this course, you will learn to recognize several types of graphs from their equations. For instance, you will learn to recognize that the graph of a second-degree equation of the form

$$y = ax^2 + bx + c$$

is a parabola (see Example 2). Another easily recognized graph is that of a **circle.**

Circles

Consider the circle shown in Figure 1.11. A point (x, y) is on the circle if and only if its distance from the center (h, k) is r. By the Distance Formula,

$$\sqrt{(x - h)^2 + (y - k)^2} = r.$$

By squaring both sides of this equation, you obtain the **standard form of the equation of a circle.**

Standard Form of the Equation of a Circle

The point (x, y) lies on the circle of radius r and center (h, k) if and only if

$$(x - h)^2 + (y - k)^2 = r^2.$$

STUDY TIP

To find the correct h and k, it may be helpful to rewrite the quantities $(x + 1)^2$ and $(y - 2)^2$.

$$(x + 1)^2 = [x - (-1)]^2,$$
$$h = -1$$
$$(y - 2)^2 = [y - (2)]^2,$$
$$k = 2$$

From this result, you can see that the standard form of the equation of a circle *with its center at the origin*, $(h, k) = (0, 0)$, is simply

$$x^2 + y^2 = r^2.$$ Circle with center at origin

Example 7 ▶ Finding the Equation of a Circle

The point $(3, 4)$ lies on a circle whose center is at $(-1, 2)$, as shown in Figure 1.12. Find the standard form of the equation of this circle.

Solution

The radius of the circle is the distance between $(-1, 2)$ and $(3, 4)$.

$$r = \sqrt{(x - h)^2 + (y - k)^2}$$ Distance Formula

$$r = \sqrt{[3 - (-1)]^2 + (4 - 2)^2}$$ Substitute for x, y, h, and k.

$$= \sqrt{4^2 + 2^2}$$ Simplify.

$$= \sqrt{16 + 4}$$ Simplify.

$$= \sqrt{20}$$ Radius

Using $(h, k) = (-1, 2)$ and $r = \sqrt{20}$, the equation of the circle is

$$(x - h)^2 + (y - k)^2 = r^2$$

$$[x - (-1)]^2 + (y - 2)^2 = \left(\sqrt{20}\right)^2$$ Substitute for h, k, and r.

$$(x + 1)^2 + (y - 2)^2 = 20.$$ Standard form

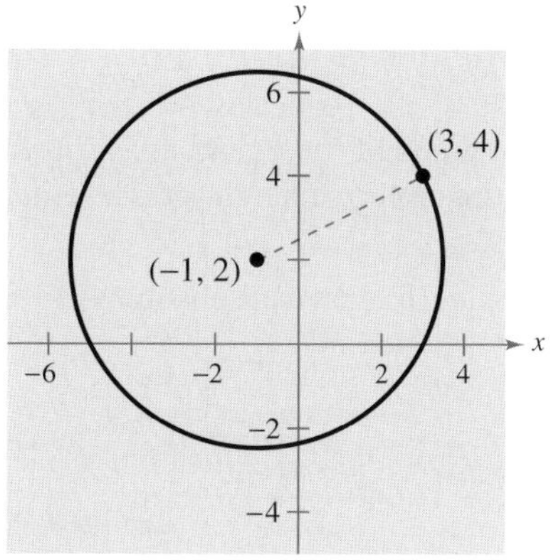

FIGURE 1.12

Activities

1. Write the equation of the circle with center $(-1, 2)$ and radius 3 in general form.

   Answer: $x^2 + y^2 + 2x - 4y - 4 = 0$

2. Find the intercepts: $y = 2x^2 - 3x - 2$.

   Answer: $(0, -2), (2, 0), (-0.5, 0)$

3. Identify any symmetry: $y = \sqrt{4 - x^2}$.

   Answer: y-axis symmetry

Application

In this course, you will learn that there are many ways to approach a problem. Three common approaches are illustrated in Example 8.

A Numerical Approach: Construct and use a table.
A Graphical Approach: Draw and use a graph.
An Analytical Approach: Use the rules of algebra.

We strongly recommend that you develop the habit of using at least two approaches with every problem. This helps build your intuition and helps you check that your answer is reasonable.

Example 8 ▶ Recommended Weight

The median recommended weight y (in pounds) for men of medium frame who are 25 to 59 years old can be approximated by the mathematical model

$$y = 0.073x^2 - 6.986x + 288.985, \quad 62 \le x \le 76$$

where x is the man's height in inches. (Source: Metropolitan Life Insurance Company)

a. Construct a table of values that shows the median recommended weights for men with heights of 62, 64, 66, 68, 70, 72, 74, and 76 inches.

b. Use the table of values to sketch a graph of the model. Then use the graph to estimate *graphically* the median recommended weight for a man whose height is 71 inches.

c. Use the model to confirm *analytically* the estimate you found in part (b).

Solution

a. You can use a calculator to complete the table, as shown below.

x	62	64	66	68	70	72	74	76
y	136.5	140.9	145.9	151.5	157.7	164.4	171.8	179.7

b. The table of values can be used to sketch the graph of the function, as shown in Figure 1.13. From the graph, you can estimate that a height of 71 inches corresponds to a weight of about 160 pounds.

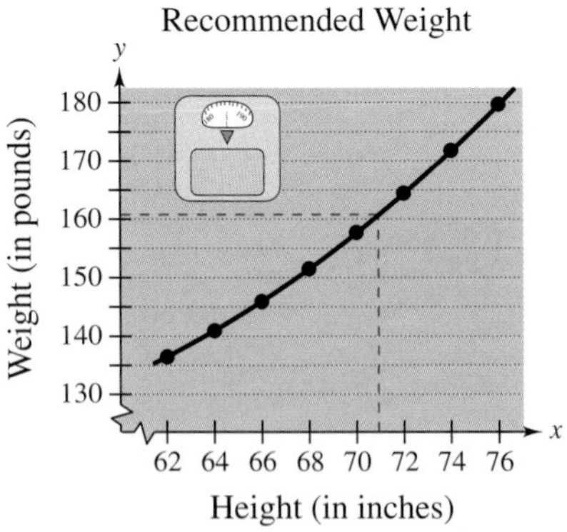

FIGURE 1.13

c. To confirm analytically the estimate found in part (b), you can substitute 71 for x in the model.

$$\begin{aligned} y &= 0.073x^2 - 6.986x + 288.985 && \text{Write original model.} \\ &= 0.073(71)^2 - 6.986(71) + 288.985 && \text{Substitute 71 for } x. \\ &\approx 160.97 && \text{Use a calculator.} \end{aligned}$$

So, the graphical estimate of 160 pounds is fairly good.

1.1 Exercises

The *Interactive* CD-ROM and *Internet* versions of this text contain step-by-step solutions to all odd-numbered Section and Review Exercises. They also provide Tutorial Exercises that link to Guided Examples for additional help.

In Exercises 1–4, determine whether each point lies on the graph of the equation.

Equation	*Points*	
1. $y = \sqrt{x + 4}$	(a) $(0, 2)$	(b) $(5, 3)$
2. $y = x^2 - 3x + 2$	(a) $(2, 0)$	(b) $(-2, 8)$
3. $y = 4 - \lvert x - 2 \rvert$	(a) $(1, 5)$	(b) $(6, 0)$
4. $y = \frac{1}{3}x^3 - 2x^2$	(a) $\left(2, -\frac{16}{3}\right)$	(b) $(-3, 9)$

In Exercises 5–8, complete the table. Use the resulting solution points to sketch the graph of the equation.

5. $y = -2x + 5$

x	-1	0	1	2	$\frac{5}{2}$
y					

6. $y = \frac{3}{4}x - 1$

x	-2	0	1	$\frac{4}{3}$	2
y					

7. $y = x^2 - 3x$

x	-1	0	1	2	3
y					

8. $y = 5 - x^2$

x	-2	-1	0	1	2
y					

In Exercises 9–16, check for symmetry with respect to both axes and the origin.

9. $x^2 - y = 0$

10. $x - y^2 = 0$

11. $y = x^3$

12. $y = x^4 - x^2 + 3$

13. $y = \dfrac{x}{x^2 + 1}$

14. $y = \sqrt{9 - x^2}$

15. $xy^2 + 10 = 0$

16. $xy = 4$

In Exercises 17–20, assume that the graph has the indicated type of symmetry. Sketch the complete graph of the equation.

17.

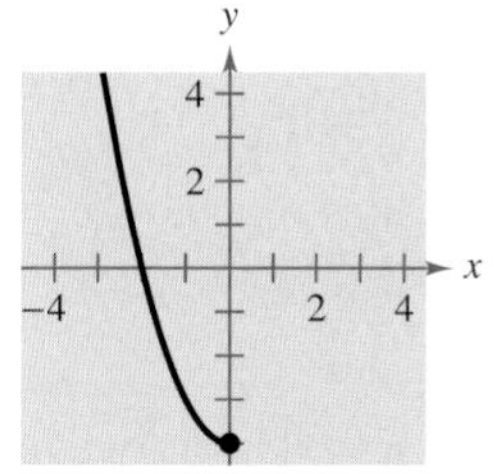

y-Axis symmetry

18.

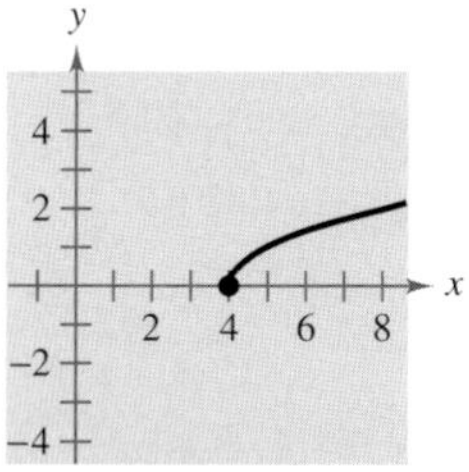

x-Axis symmetry

19.

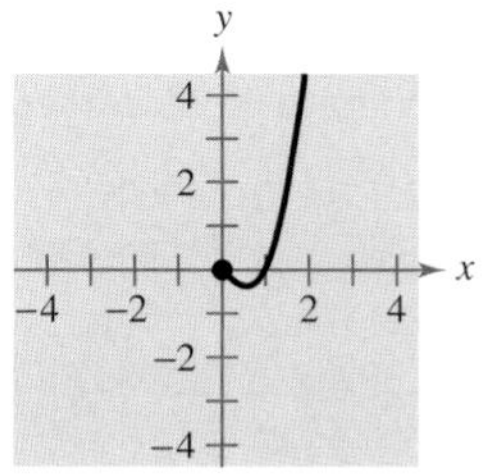

Origin symmetry

20.

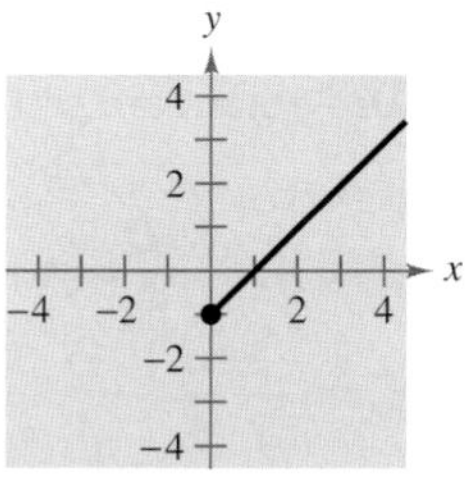

y-Axis symmetry

In Exercises 21–24, match the equation with its graph. [The graphs are labeled (a), (b), (c), and (d).]

(a)

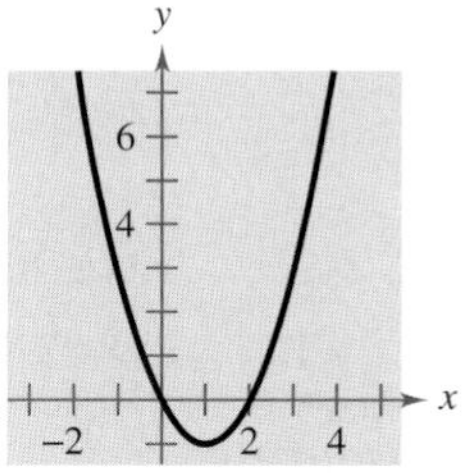

(b)

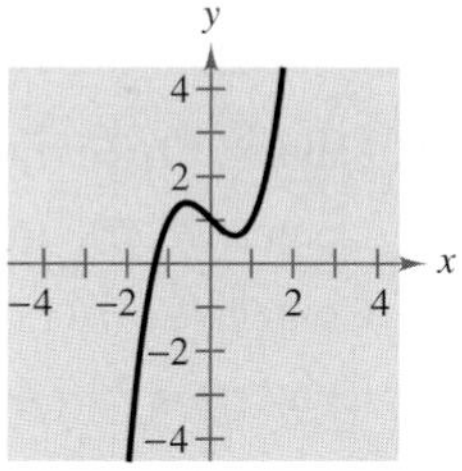

(c)

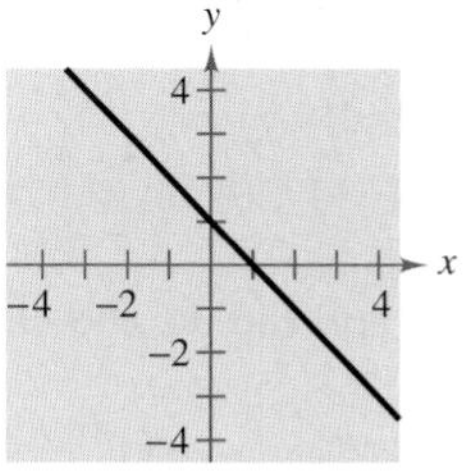

(d)

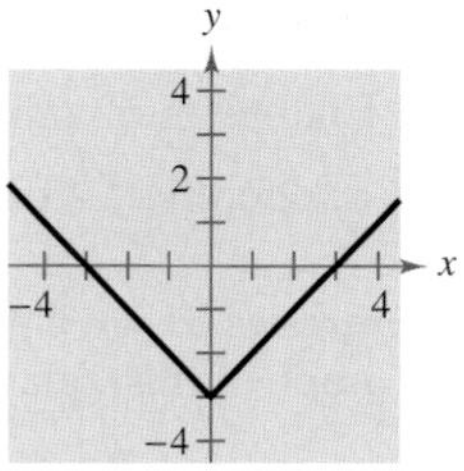

21. $y = 1 - x$

22. $y = x^2 - 2x$

23. $y = x^3 - x + 1$

24. $y = \lvert x \rvert - 3$

In Exercises 25–28, find the x- and y-intercepts of the graph of the equation.

25. $y = 16 - 4x^2$ **26.** $y = (x + 2)^2$

27. $y = 2x^3 - 5x^2$ **28.** $y^2 = x + 1$

In Exercises 29–40, use intercepts and symmetry to sketch the graph of the equation.

29. $y = -3x + 1$ **30.** $y = 2x - 3$

31. $y = x^2 - 2x$ **32.** $y = -x^2 - 2x$

33. $y = x^3 + 3$ **34.** $y = x^3 - 1$

35. $y = \sqrt{x - 3}$ **36.** $y = \sqrt{1 - x}$

37. $y = |x - 6|$ **38.** $y = 1 - |x|$

39. $x = y^2 - 1$ **40.** $x = y^2 - 5$

In Exercises 41–52, use a graphing utility to graph the equation. Use a standard setting. Approximate any intercepts.

41. $y = 3 - \frac{1}{2}x$ **42.** $y = \frac{2}{3}x - 1$

43. $y = x^2 - 4x + 3$ **44.** $y = x^2 + x - 2$

45. $y = \dfrac{2x}{x - 1}$ **46.** $y = \dfrac{4}{x^2 + 1}$

47. $y = \sqrt[3]{x}$ **48.** $y = \sqrt[3]{x + 1}$

49. $y = x\sqrt{x + 6}$ **50.** $y = (6 - x)\sqrt{x}$

51. $y = |x + 3|$ **52.** $y = 2 - |x|$

In Exercises 53–60, find the standard form of the equation of the specified circle.

53. Center: $(0, 0)$; radius: 4

54. Center: $(0, 0)$; radius: 5

55. Center: $(2, -1)$; radius: 4

56. Center: $(-7, -4)$; radius: 7

57. Center: $(-1, 2)$; solution point: $(0, 0)$

58. Center: $(3, -2)$; solution point: $(-1, 1)$

59. Endpoints of a diameter: $(0, 0)$, $(6, 8)$

60. Endpoints of a diameter: $(-4, -1)$, $(4, 1)$

In Exercises 61–66, find the center and radius, and sketch the graph of the equation.

61. $x^2 + y^2 = 25$ **62.** $x^2 + y^2 = 16$

63. $(x - 1)^2 + (y + 3)^2 = 9$

64. $x^2 + (y - 1)^2 = 1$

65. $\left(x - \frac{1}{2}\right)^2 + \left(y - \frac{1}{2}\right)^2 = \frac{9}{4}$

66. $(x - 2)^2 + (y + 1)^2 = 3$

In Exercises 67 and 68, use a graphing utility to graph y_1 and y_2. Use a square setting. Identify the graph.

67. $y_1 = 4 + \sqrt{25 - x^2}$

$y_2 = 4 - \sqrt{25 - x^2}$

68. $y_1 = 2 + \sqrt{16 - (x - 1)^2}$

$y_2 = 2 - \sqrt{16 - (x - 1)^2}$

69. ***Business*** A manufacturing plant purchases a new molding machine for \$225,000. The depreciated value y after t years is

$$y = 225{,}000 - 20{,}000t, \qquad 0 \le t \le 8.$$

Sketch the graph of the equation.

70. ***Consumerism*** You purchase a jet ski for \$8100. The depreciated value y after t years is

$$y = 8100 - 929t, \qquad 0 \le t \le 6.$$

Sketch the graph of the equation.

71. ***Geometry*** A rectangle of length x and width w has a perimeter of 12 meters.

(a) Draw a rectangle that gives a visual representation of the problem. Use the specified variables to label the sides of the rectangle.

(b) Show that the width of the rectangle is $w = 6 - x$ and its area is $A = x(6 - x)$.

(c) Use a graphing utility to graph the area equation.

(d) From the graph in part (c), estimate the dimensions of the rectangle that yields a maximum area.

72. ***Geometry*** A rectangle of length x and width w has a perimeter of 22 yards.

(a) Draw a rectangle that gives a visual representation of the problem. Use the specified variables to label the sides of the rectangle.

(b) Show that the width of the rectangle is $w = 11 - x$ and its area is $A = x(11 - x)$.

(c) Use a graphing utility to graph the area equation. Be sure to adjust your window settings.

(d) From the graph in part (c), estimate the dimensions of the rectangle that yields a maximum area.

The symbol indicates an exercise or parts of an exercise in which you are instructed to use a graphing utility.

Data Analysis **In Exercises 73 and 74, (a) sketch a scatter plot of the data, (b) graph the model for the data and compare the scatter plot and the graph, and (c) use the model to estimate the values of y for the years 2002 and 2004.**

73. *Federal Debt* The table shows the per capita federal debt of the United States for several years. (Source: U.S. Treasury Department, U.S. Bureau of the Census)

Year	1950	1960	1970	1980
Per capita debt	\$1688	\$1572	\$1807	\$3981

Year	1990	1994	1997	1998
Per capita debt	\$12,848	\$15,750	\$20,063	\$20,513

A model for the per capita debt during this period is

$$y = 0.223t^3 - 0.733t^2 - 78.255t + 1837.433$$

where y represents the per capita debt and t is the time in years, with $t = 0$ corresponding to 1950.

74. *Population Statistics* The table shows the life expectancy of a child (at birth) in the United States for selected years from 1920 to 2000. (Source: U.S. National Center for Health Statistics, U.S. Bureau of the Census)

Year	1920	1930	1940	1950
Life expectancy	54.1	59.7	62.9	68.2

Year	1960	1970	1980	1990	2000
Life expectancy	69.7	70.8	73.7	75.4	76.4

A model for the life expectancy during this period is

$$y = \frac{66.93 + t}{1 + 0.01t}$$

where y represents the life expectancy and t is the time in years, with $t = 0$ corresponding to 1950.

75. *Electronics* The resistance y in ohms of 1000 feet of solid copper wire at 77 degrees Fahrenheit can be approximated by the model

$$y = \frac{10{,}770}{x^2} - 0.37, \qquad 5 \le x \le 100$$

where x is the diameter of the wire in mils (0.001 in.). Use the model to estimate the resistance when $x = 50$. (Source: American Wire Gage)

Synthesis

True or False? **In Exercises 76 and 77, determine whether the statement is true or false. Justify your answer.**

76. In order to find the y-intercepts of the graph of an equation, let $y = 0$ and solve the equation for x.

77. The graph of a linear equation of the form $y = mx + b$ has one y-intercept.

78. *Think About It* Suppose you correctly enter an expression for the variable y on a graphing utility. However, no graph appears on the display when you graph the equation. Give a possible explanation and the steps you could take to remedy the problem. Illustrate your explanation with an example.

79. *Think About It* Find a and b if the graph of $y = ax^2 + bx^3$ is symmetric with respect to (a) the y-axis and (b) the origin. (There are many correct answers.)

80. In your own words, explain how the display of a graphing utility changes if the maximum setting for x is changed from 10 to 20.

Review

True or False? **In Exercises 81 and 82, determine whether the statement is true or false. Justify your answer.**

81. $\dfrac{1}{3 \cdot 4^{-1}} = 3 \cdot 4$

82. $(3 + 4)^2 = 3^2 + 4^2$

83. Identify the terms: $9x^5 + 4x^3 - 7$.

84. Write the expression using exponential notation: $-(7 \times 7 \times 7 \times 7)$

In Exercises 85–90, simplify the expression.

85. $\sqrt{18x} - \sqrt{2x}$

86. $\sqrt[4]{x^5}$

87. $\dfrac{70}{\sqrt{7x}}$

88. $\dfrac{55}{\sqrt{20} - 3}$

89. $\sqrt[6]{t^2}$

90. $\sqrt[3]{\sqrt{y}}$

1.2 Linear Equations in Two Variables

▶ What you should learn

- How to use slope to graph linear equations in two variables
- How to find slopes of lines
- How to write linear equations in two variables
- How to use slope to identify parallel and perpendicular lines
- How to use linear equations in two variables to model and solve real-life problems

▶ Why you should learn it

Linear equations in two variables can be used to model and solve real-life problems. For instance, Exercise 112 on page 123 shows how to use a linear equation to model the average annual salaries of major league baseball players from 1988 to 1998.

Walter Schmid/The Stock Market

A computer simulation of this concept appears in the *Interactive* CD-ROM and *Internet* versions of this text.

Using Slope

The simplest mathematical model for relating two variables is the **linear equation in two variables** $y = mx + b$. The equation is called *linear* because its graph is a line. (In mathematics, the term *line* means *straight line*.) By letting $x = 0$, you can see that the line crosses the y-axis at $y = b$, as shown in Figure 1.14. In other words, the y-intercept is $(0, b)$. The steepness or slope of the line is m.

$$y = mx + b$$

Slope ↑ (under m) ↑ y-Intercept (under b)

The **slope** of a nonvertical line is the number of units the line rises (or falls) vertically for each unit of horizontal change from left to right, as shown in Figure 1.14.

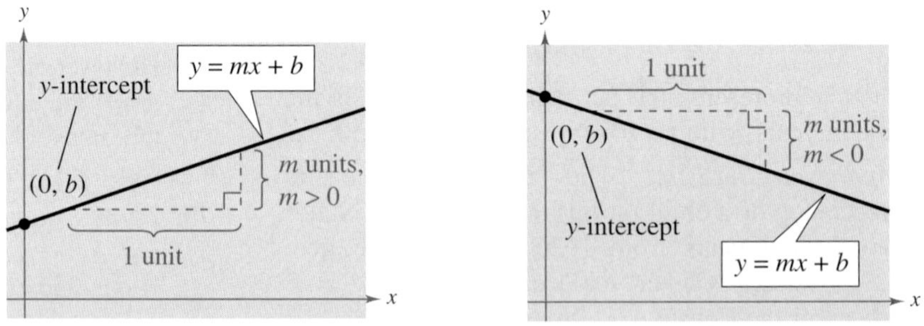

Positive slope, line rises. *Negative slope, line falls.*

FIGURE 1.14

A linear equation that is written in the form $y = mx + b$ is said to be written in **slope-intercept form.**

The Slope-Intercept Form of the Equation of a Line

The graph of the equation

$$y = mx + b$$

is a line whose slope is m and whose y-intercept is $(0, b)$.

Exploration

Use a graphing utility to compare the slopes of the lines $y = mx$ where $m = 0.5$, 1, 2, and 4. Which line rises most quickly? Now, let $m = -0.5$, -1, -2, and -4. Which line falls most quickly? Use a square setting to obtain a true geometric perspective. What can you conclude about the slope and the "rate" at which the line rises or falls?

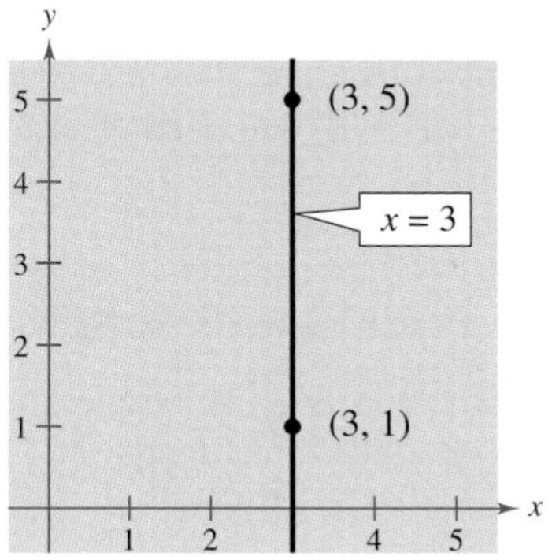

FIGURE 1.15 *Slope is undefined.*

Common Error

Many students confuse the line $x = a$ with the point $x = a$ on the real number line, *or* the line $y = b$ with the point $y = b$. Point out to students that they need to be aware of the context in which $x = a$ or $y = b$ is presented to know whether it refers to the line in the plane or the point on the number line.

Once you have determined the slope and the y-intercept of a line, it is a relatively simple matter to sketch its graph. In the following example, note that none of the lines is vertical. A vertical line has an equation of the form

$$x = a. \qquad \text{Vertical line}$$

The equation of a vertical line cannot be written in the form $y = mx + b$ because the slope of a vertical line is undefined, as indicated in Figure 1.15.

Example 1 ▶ Graphing a Linear Equation

Sketch the graph of each linear equation.

a. $y = 2x + 1$

b. $y = 2$

c. $x + y = 2$

Solution

a. Because $b = 1$, the y-intercept is $(0, 1)$. Moreover, because the slope is $m = 2$, the line *rises* 2 units for each unit the line moves to the right, as shown in Figure 1.16(a).

b. By writing this equation in the form $y = (0)x + 2$, you can see that the y-intercept is $(0, 2)$ and the slope is zero. A zero slope implies that the line is horizontal—that is, it doesn't rise *or* fall, as shown in Figure 1.16(b).

c. By writing this equation in slope-intercept form

$$x + y = 2 \qquad \text{Write original equation.}$$

$$y = -x + 2 \qquad \text{Subtract } x \text{ from each side.}$$

$$y = (-1)x + 2 \qquad \text{Write in slope-intercept form.}$$

you can see that the y-intercept is $(0, 2)$. Moreover, because the slope is $m = -1$, the line *falls* 1 unit for each unit the line moves to the right, as shown in Figure 1.16(c).

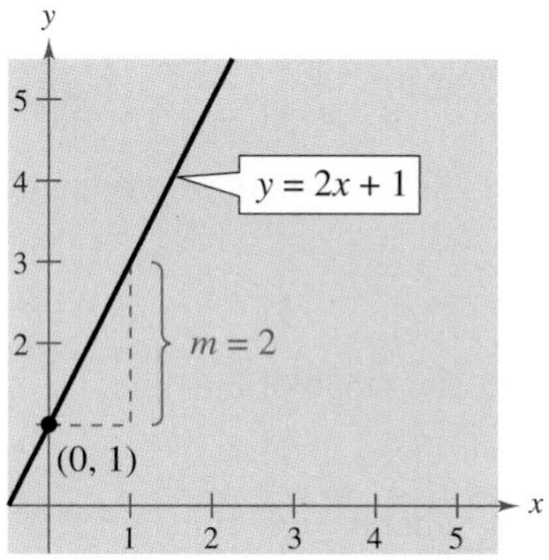

(a) When *m* is positive, the line rises.

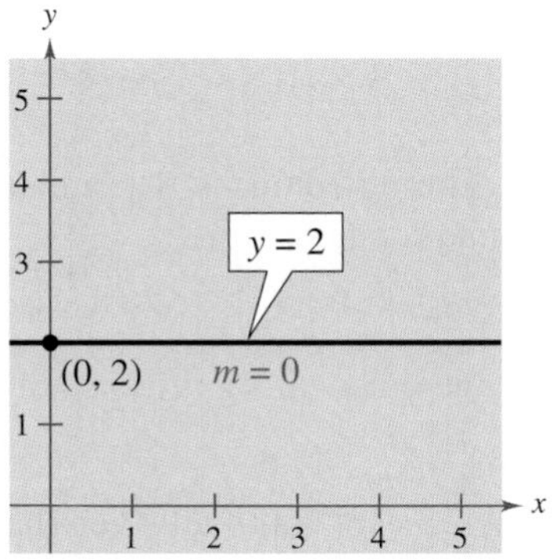

(b) When *m* is 0, the line is horizontal.

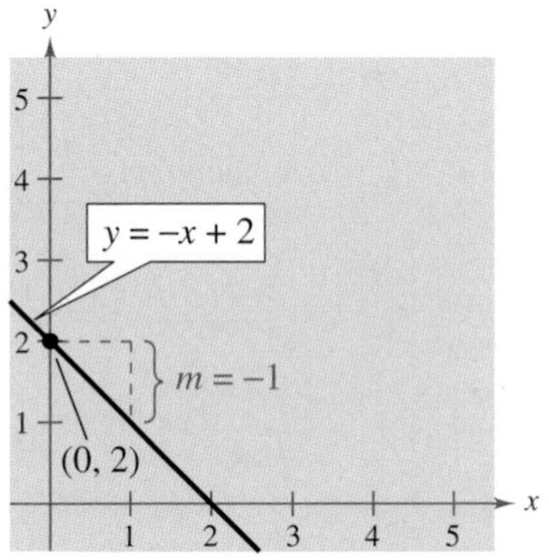

(c) When *m* is negative, the line falls.

FIGURE 1.16

In real-life problems, the slope of a line can be interpreted as either a *ratio* or a *rate*. If the x-axis and y-axis have the same unit of measure, then the slope has no units and is a **ratio.** If the x-axis and y-axis have different units of measure, then the slope is a **rate** or **rate of change.**

Example 2 ▶ Using Slope as a Ratio

The maximum recommended slope of a wheelchair ramp is $\frac{1}{12}$. A business is installing a wheelchair ramp that rises 22 inches over a horizontal length of 24 feet. Is the ramp steeper than recommended? (Source: Americans with Disabilities Act Handbook)

Solution

The horizontal length of the ramp is 24 feet or $12(24) = 288$ inches, as shown in Figure 1.17. So, the slope of the ramp is

$$\text{Slope} = \frac{\text{vertical change}}{\text{horizontal change}} = \frac{22 \text{ in.}}{288 \text{ in.}} \approx 0.076.$$

Because $\frac{1}{12} \approx 0.083$, the slope of the ramp is not steeper than recommended.

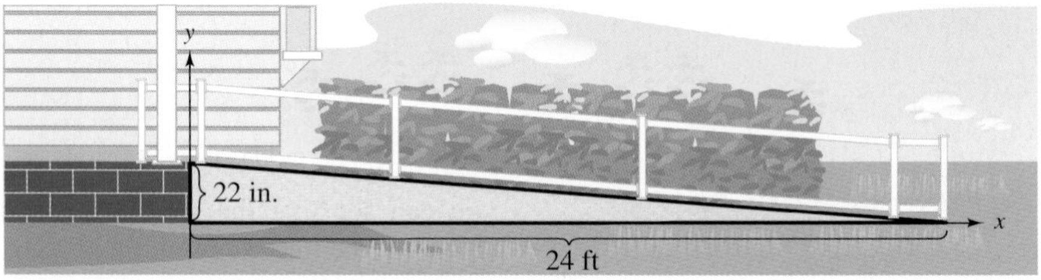

FIGURE 1.17

Example 3 ▶ Using Slope as a Rate of Change

A manufacturing company determines that the total cost in dollars of producing x units of a product is

$$C = 25x + 3500. \qquad \text{Cost equation}$$

Describe the practical significance of the y-intercept and slope of this line.

Solution

The y-intercept $(0, 3500)$ tells you that the cost of producing zero units is \$3500. This is the *fixed cost* of production—it includes costs that must be paid regardless of the number of units produced. The slope of $m = 25$ tells you that the cost of producing each unit is \$25, as shown in Figure 1.18. Economists call the cost per unit the *marginal cost*. If the production increases by 1 unit, then the "margin," or extra amount of cost, is \$25. So, the cost increases at a rate of \$25 per unit.

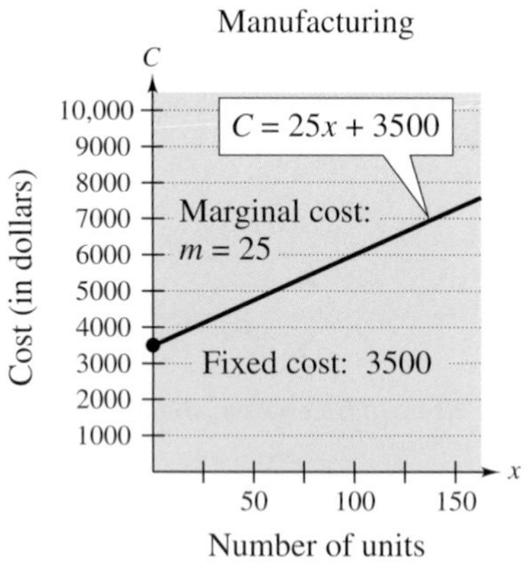

FIGURE 1.18 *Production Cost*

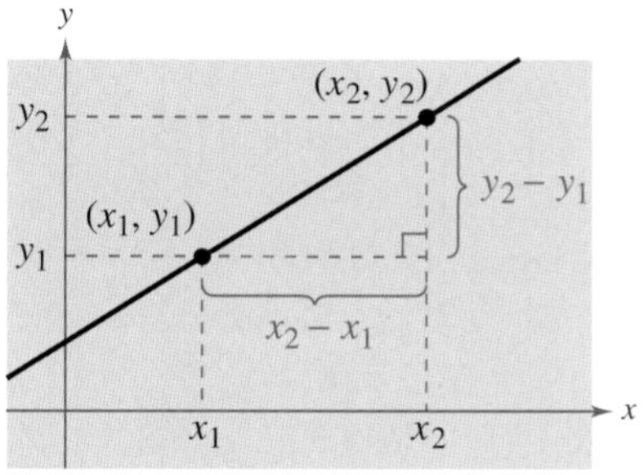

FIGURE 1.19

Finding the Slope of a Line

Given an equation of a line, you can find its slope by writing the equation in slope-intercept form. If you are not given an equation, you can still find the slope of a line. For instance, suppose you want to find the slope of the line passing through the points (x_1, y_1) and (x_2, y_2), as shown in Figure 1.19. As you move from left to right along this line, a change of $(y_2 - y_1)$ units in the vertical direction corresponds to a change of $(x_2 - x_1)$ units in the horizontal direction.

$$y_2 - y_1 = \text{the change in } y = \text{rise}$$

and

$$x_2 - x_1 = \text{the change in } x = \text{run}$$

The ratio of $(y_2 - y_1)$ to $(x_2 - x_1)$ represents the slope of the line that passes through the points (x_1, y_1) and (x_2, y_2).

$$\begin{aligned} \text{Slope} &= \frac{\text{change in } y}{\text{change in } x} \\ &= \frac{\text{rise}}{\text{run}} \\ &= \frac{y_2 - y_1}{x_2 - x_1} \end{aligned}$$

Corbis-Bettmann

Historical Note

René Descartes (1596–1650) was largely responsible for the development of analytic geometry. His use of the symbol m for slope came from the French verb *monter*, meaning to mount, to climb, or to rise.

The Slope of a Line Passing Through Two Points

The **slope** m of the nonvertical line through (x_1, y_1) and (x_2, y_2) is

$$m = \frac{y_2 - y_1}{x_2 - x_1}$$

where $x_1 \neq x_2$.

When this formula is used for slope, the *order of subtraction* is important. Given two points on a line, you are free to label either one of them as (x_1, y_1) and the other as (x_2, y_2). However, once you have done this, you must form the numerator and denominator using the same order of subtraction.

$$\underbrace{m = \frac{y_2 - y_1}{x_2 - x_1}}_{\text{Correct}} \qquad \underbrace{m = \frac{y_1 - y_2}{x_1 - x_2}}_{\text{Correct}} \qquad \underbrace{\cancel{m = \frac{y_2 - y_1}{x_1 - x_2}}}_{\text{Incorrect}}$$

For instance, the slope of the line passing through the points (3, 4) and (5, 7) can be calculated as

$$m = \frac{7 - 4}{5 - 3} = \frac{3}{2}$$

or

$$m = \frac{4 - 7}{3 - 5} = \frac{-3}{-2} = \frac{3}{2}.$$

The *Interactive* CD-ROM and *Internet* versions of this text show every example with its solution; clicking on the *Try It!* button brings up similar problems. Guided Examples and Integrated Examples show step-by-step solutions to additional examples. Integrated Examples are related to several concepts in the section.

Example 4 ▶ Finding the Slope of a Line Through Two Points

Find the slope of the line passing through each pair of points. (See Figure1.20.)

a. $(-2, 0)$ and $(3, 1)$ **b.** $(-1, 2)$ and $(2, 2)$

c. $(0, 4)$ and $(1, -1)$ **d.** $(3, 4)$ and $(3, 1)$

Solution

a. Letting $(x_1, y_1) = (-2, 0)$ and $(x_2, y_2) = (3, 1)$, you obtain a slope of

$$m = \frac{y_2 - y_1}{x_2 - x_1} = \frac{1 - 0}{3 - (-2)} = \frac{1}{5}.$$

b. The slope of the line passing through $(-1, 2)$ and $(2, 2)$ is

$$m = \frac{2 - 2}{2 - (-1)} = \frac{0}{3} = 0.$$

c. The slope of the line passing through $(0, 4)$ and $(1, -1)$ is

$$m = \frac{-1 - 4}{1 - 0} = \frac{-5}{1} = -5.$$

d. The slope of the vertical line passing through $(3, 4)$ and $(3, 1)$ is

$$m = \frac{1 - 4}{3 - 3} = \frac{-3}{0}.$$

Because division by 0 is undefined, the slope is undefined.

Common Error

A common error when finding the slope of a line is combining x and y coordinates in either the numerator or denominator, as in

$$m = \frac{y_2 - x_1}{x_2 - y_1}.$$

STUDY T!P

In Figure 1.20, note the relationships between slope and the description of the line.

a. Positive slope; line rises from left to right

b. Zero slope; line is horizontal

c. Negative slope; line falls from left to right

d. Undefined slope; line is vertical

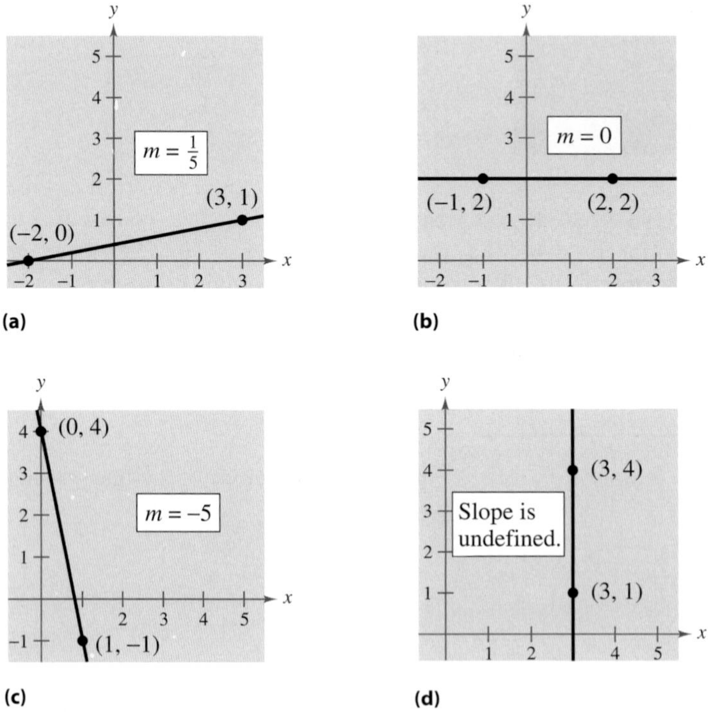

FIGURE 1.20

Writing Linear Equations in Two Variables

If (x_1, y_1) is a point on a line of slope m and (x, y) is *any other* point on the line, then

$$\frac{y - y_1}{x - x_1} = m.$$

This equation, involving the variables x and y, can be rewritten in the form

$$y - y_1 = m(x - x_1),$$

which is the **point-slope form** of the equation of a line.

Point-Slope Form of the Equation of a Line

The equation of the line with slope m passing through the point (x_1, y_1) is

$$y - y_1 = m(x - x_1).$$

The point-slope form is most useful for *finding* the equation of a line. You should remember this formula.

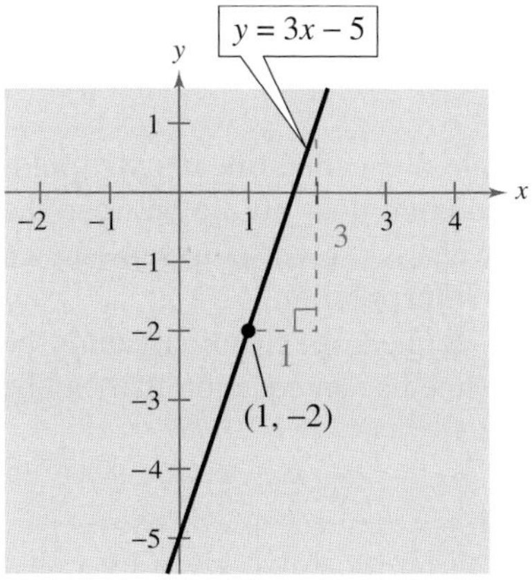

FIGURE 1.21

Example 5 ▶ Using the Point-Slope Form

Find the slope-intercept form of the equation of the line that has a slope of 3 and passes through the point $(1, -2)$, as shown in Figure 1.21.

Solution

Use the point-slope form with $m = 3$ and $(x_1, y_1) = (1, -2)$.

$y - y_1 = m(x - x_1)$	Point-slope form
$y - (-2) = 3(x - 1)$	Substitute for m, x_1, and y_1.
$y + 2 = 3x - 3$	Simplify.
$y = 3x - 5$	Write in slope-intercept form.

The point-slope form can be used to find an equation of the line passing through points (x_1, y_1) and (x_2, y_2). To do this, first find the slope of the line

$$m = \frac{y_2 - y_1}{x_2 - x_1}, \qquad x_1 \neq x_2$$

and then use the point-slope form to obtain the equation

$$y - y_1 = \frac{y_2 - y_1}{x_2 - x_1}(x - x_1). \qquad \text{Two-point form}$$

This is sometimes called the **two-point form** of the equation of a line.

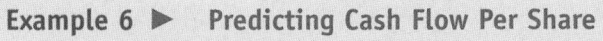

Example 6 ▶ Predicting Cash Flow Per Share

The cash flow per share for Yahoo! Inc. was \$0.03 in 1997 and \$0.30 in 1998. Using only this information, write a linear equation that gives the cash flow per share in terms of the year. Then predict the cash flow for 1999. (Source: Yahoo! Inc.)

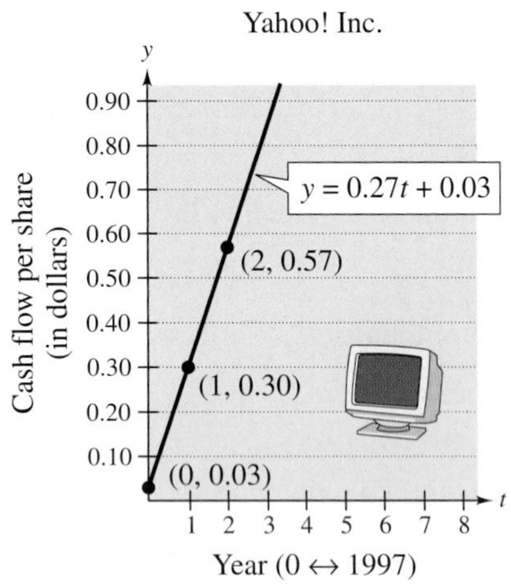

FIGURE 1.22

Solution

Let $t = 0$ represent 1997. Then the two given values are represented by the data points $(0, 0.03)$ and $(1, 0.30)$. The slope of the line through these points is

$$m = \frac{0.30 - 0.03}{1 - 0}$$

$$= 0.27.$$

Using the point-slope form, you can find the equation that relates the cash flow y and the year t to be

$$y = 0.27t + 0.03.$$

According to this equation, the cash flow in 1999 was \$0.57, as shown in Figure 1.22. (In this case, the prediction is quite good—the actual cash flow in 1999 was \$0.55.)

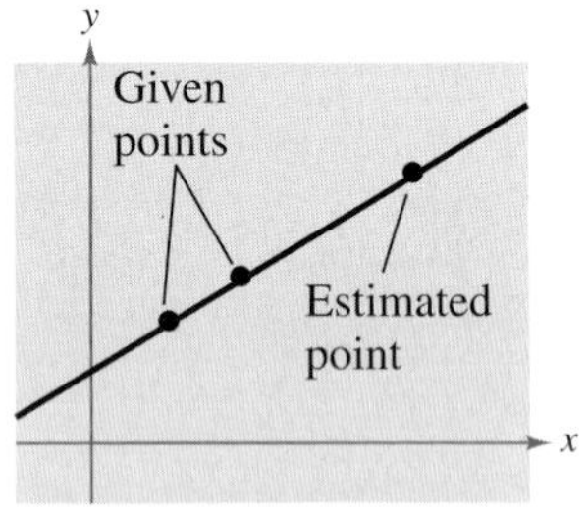

(a) Linear extrapolation

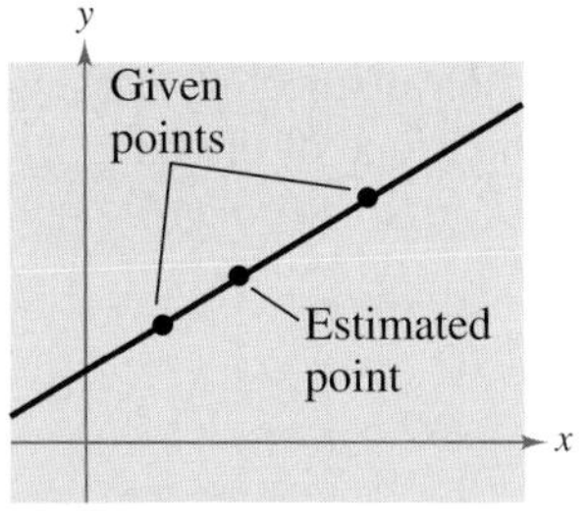

(b) Linear interpolation

FIGURE 1.23

The prediction method illustrated in Example 6 is called **linear extrapolation.** Note in Figure 1.23(a) that an extrapolated point does not lie between the given points. When the estimated point lies between two given points, as shown in Figure 1.23(b), the procedure is called **linear interpolation.**

Because the slope of a vertical line is not defined, its equation cannot be written in slope-intercept form. However, every line has an equation that can be written in the **general form**

$$Ax + By + C = 0 \qquad \text{General form}$$

where A and B are not both zero. For instance, the vertical line given by $x = a$ can be represented by the general form

$$x - a = 0.$$

Equations of Lines

1. General form: $Ax + By + C = 0$
2. Vertical line: $x = a$
3. Horizontal line: $y = b$
4. Slope-intercept form: $y = mx + b$
5. Point-slope form: $y - y_1 = m(x - x_1)$
6. Two-point form: $y - y_1 = \dfrac{y_2 - y_1}{x_2 - x_1}(x - x_1)$

> **Technology**
>
> On a graphing utility, lines will not appear to have the correct slope unless you use a viewing window that has a square setting. For instance, try graphing the lines in Example 7 using the standard setting $-10 \le x \le 10$ and $-10 \le y \le 10$. Then reset the viewing window with the square setting $-9 \le x \le 9$ and $-6 \le y \le 6$. On which setting do the lines $y = \frac{2}{3}x - \frac{5}{3}$ and $y = -\frac{3}{2}x + 2$ appear perpendicular?

Parallel and Perpendicular Lines

Slope can be used to decide whether two nonvertical lines in a plane are parallel, perpendicular, or neither.

> **Parallel and Perpendicular Lines**
>
> 1. Two distinct nonvertical lines are **parallel** if and only if their slopes are equal. That is, $m_1 = m_2$.
> 2. Two nonvertical lines are **perpendicular** if and only if their slopes are negative reciprocals of each other. That is, $m_1 = -1/m_2$.

Example 7 ▶ Finding Parallel and Perpendicular Lines

Find the slope-intercept form of the equation of the line that passes through the point $(2, -1)$ and is (a) parallel to and (b) perpendicular to the line $2x - 3y = 5$.

Solution

By writing the equation of the given line in slope-intercept form

$2x - 3y = 5$ Write original equation.

$-3y = -2x + 5$ Subtract $2x$ from each side.

$y = \frac{2}{3}x - \frac{5}{3}$ Write in slope-intercept form.

you can see that it has a slope of $m = \frac{2}{3}$, as shown in Figure 1.24.

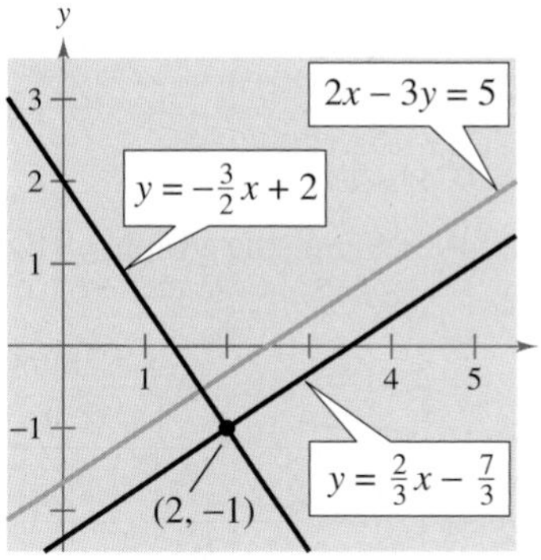

FIGURE 1.24

a. Any line parallel to the given line must also have a slope of $\frac{2}{3}$. So, the line through $(2, -1)$ that is parallel to the given line has the following equation.

$y - (-1) = \frac{2}{3}(x - 2)$ Write in point-slope form.

$3(y + 1) = 2(x - 2)$ Multiply each side by 3.

$3y + 3 = 2x - 4$ Distributive Property

$2x - 3y - 7 = 0$ Write in general form.

$y = \frac{2}{3}x - \frac{7}{3}$ Write in slope-intercept form.

b. Any line perpendicular to the given line must have a slope of $-1/(2/3)$ or $-3/2$. So, the line through $(2, -1)$ that is perpendicular to the given line has the following equation.

$y - (-1) = -\frac{3}{2}(x - 2)$ Write in point-slope form.

$2(y + 1) = -3(x - 2)$ Multiply each side by 2.

$2y + 2 = -3x + 6$ Distributive Property

$3x + 2y - 4 = 0$ Write in general form.

$y = -\frac{3}{2}x + 2$ Write in slope-intercept form.

Activities

1. Write an equation of the line that passes through the points $(-2, 1)$ and $(3, 2)$.
   Answer: $x = 5y + 7 = 0$
2. Find the slope of the line that is perpendicular to the line $4x - 7y = 12$.
   Answer: $m = -\frac{7}{4}$
3. Write the equation of the vertical line that passes through the point $(3, 2)$.
   Answer: $x = 3$

Application

Most business expenses can be deducted in the same year they occur. One exception is the cost of property that has a useful life of more than 1 year. Such costs must be *depreciated* over the useful life of the property. If the *same amount* is depreciated each year, the procedure is called *linear* or *straight-line depreciation.* The *book value* is the difference between the original value and the total amount of depreciation accumulated to date.

Example 8 ▶ Straight-Line Depreciation

Your company has purchased a \$12,000 machine that has a useful life of 8 years. The salvage value at the end of 8 years is \$2000. Write a linear equation that describes the book value of the machine each year.

Solution

Let V represent the value of the machine at the end of year t. You can represent the initial value of the machine by the data point (0, 12,000) and the salvage value of the machine by the data point (8, 2000). The slope of the line is

$$m = \frac{2000 - 12{,}000}{8 - 0} = -\$1250$$

which represents the annual depreciation in *dollars per year.* Using the point-slope form, you can write the equation of the line as follows.

$$V - 12{,}000 = -1250(t - 0) \quad \text{Write in point-slope form.}$$

$$V = -1250t + 12{,}000 \quad \text{Write in slope-intercept form.}$$

The table shows the book value at the end of each year, and the graph of the equation is shown in Figure 1.25.

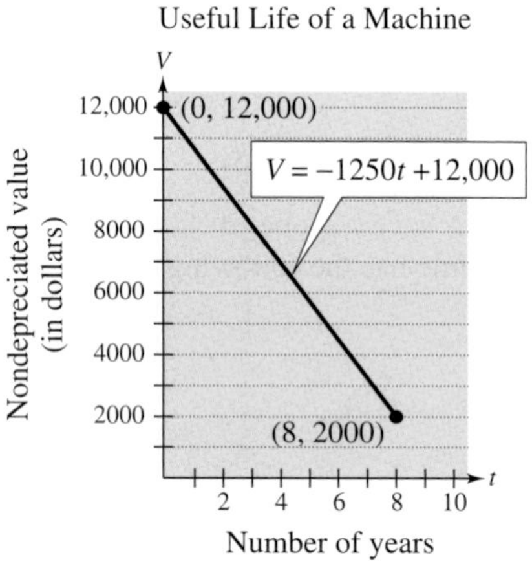

FIGURE 1.25 *Straight-Line Depreciation*

t	0	1	2	3	4	5	6	7	8
V	12,000	10,750	9500	8250	7000	5750	4500	3250	2000

Writing ABOUT MATHEMATICS

x	y
0	80.90
1	72.74
2	68.68
3	61.48

x	y
4	56.21
5	51.00
6	47.70
7	42.78

Modeling Linear Data The table at the left shows the average monthly cellular phone bill y (in dollars) for subscribers in the United States from 1990 through 1997, where x is the time in years, with $x = 0$ corresponding to 1990. (Source: Cellular Telecommunications Industry Association)

Sketch a scatter plot of the data, and use a straightedge to sketch the best-fitting line through the points. Find the equation of the line. Explain the procedure you used. Write a short paragraph explaining the meaning of the slope and the y-intercept of the line in terms of the data. Compare the values obtained using your model with the actual values in the table. Use your model to estimate the average monthly bill in the year 2000.

1.2 Exercises

In Exercises 1 and 2, identify the line that has each slope.

1. (a) $m = \frac{2}{3}$
(b) m is undefined.
(c) $m = -2$

2. (a) $m = 0$
(b) $m = -\frac{3}{4}$
(c) $m = 1$

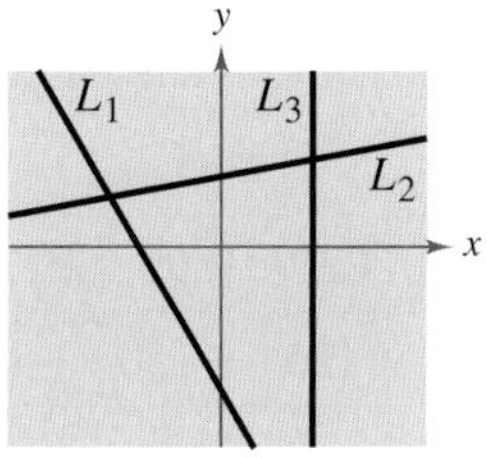

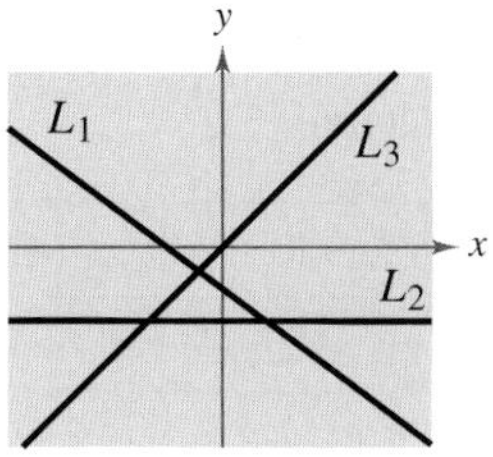

In Exercises 3 and 4, sketch the graph of the line through the point with each indicated slope on the same set of coordinate axes.

	Point	*Slopes*			
3.	(2, 3)	(a) 0	(b) 1	(c) 2	(d) −3
4.	(−4, 1)	(a) 3	(b) −3	(c) $\frac{1}{2}$	(d) Undefined

In Exercises 5–10, estimate the slope of the line.

5.

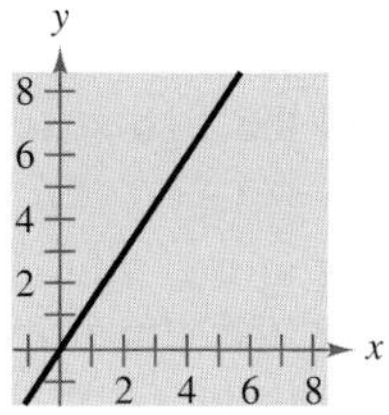

6.

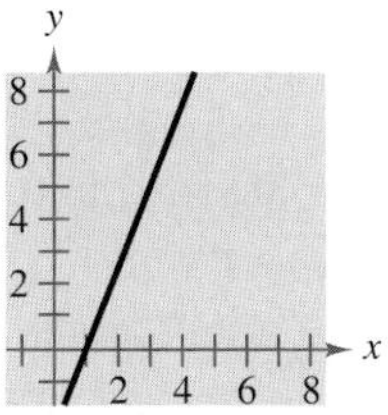

7.

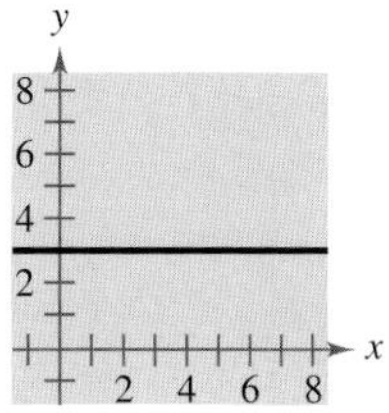

8.

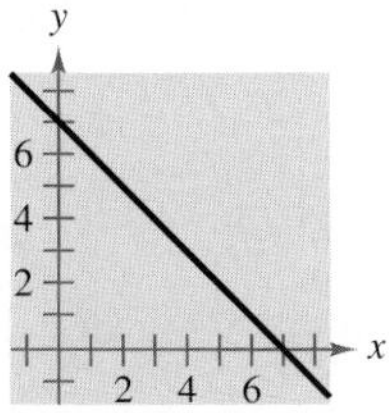

9.

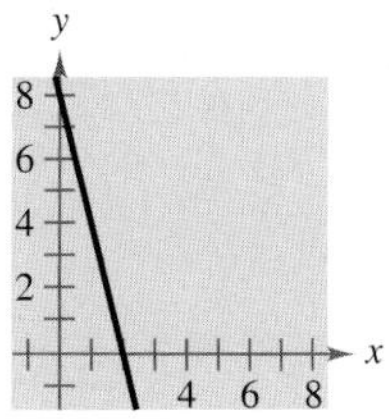

10.

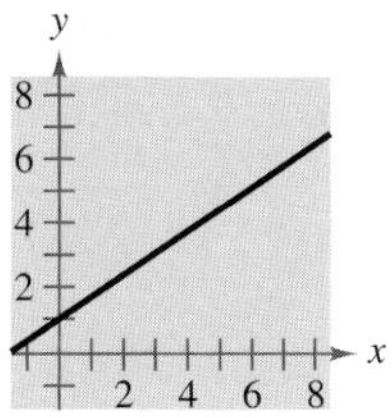

In Exercises 11–18, plot the points and find the slope of the line passing through the pair of points.

11. $(-3, -2), (1, 6)$

12. $(2, 4), (4, -4)$

13. $(-6, -1), (-6, 4)$

14. $(0, -10), (-4, 0)$

15. $\left(\frac{11}{2}, -\frac{4}{3}\right), \left(-\frac{3}{2}, -\frac{1}{3}\right)$

16. $\left(\frac{7}{8}, \frac{3}{4}\right), \left(\frac{5}{4}, -\frac{1}{4}\right)$

17. $(4.8, 3.1), (-5.2, 1.6)$

18. $(-1.75, -8.3), (2.25, -2.6)$

In Exercises 19–28, use the point on the line and the slope of the line to find three additional points through which the line passes. (There are many correct answers.)

	Point	*Slope*
19.	(2, 1)	$m = 0$
20.	(−4, 1)	m is undefined.
21.	(5, −6)	$m = 1$
22.	(10, −6)	$m = -1$
23.	(−8, 1)	m is undefined.
24.	(−3, −1)	$m = 0$
25.	(−5, 4)	$m = 2$
26.	(0, −9)	$m = -2$
27.	(7, −2)	$m = \frac{1}{2}$
28.	(−1, −6)	$m = -\frac{1}{2}$

In Exercises 29–32, determine whether the lines L_1 and L_2 passing through the pairs of points are parallel, perpendicular, or neither.

29. L_1: $(0, -1), (5, 9)$
L_2: $(0, 3), (4, 1)$

30. L_1: $(-2, -1), (1, 5)$
L_2: $(1, 3), (5, -5)$

31. L_1: $(3, 6), (-6, 0)$
L_2: $(0, -1), \left(5, \frac{7}{3}\right)$

32. L_1: $(4, 8), (-4, 2)$
L_2: $(3, -5), \left(-1, \frac{1}{3}\right)$

33. *Business* The following are the slopes of lines representing annual sales y in terms of time x in years. Use the slopes to interpret any change in annual sales for a 1-year increase in time.

(a) The line has a slope of $m = 135$.

(b) The line has a slope of $m = 0$.

(c) The line has a slope of $m = -40$.

34. ***Business*** The following are the slopes of lines representing daily revenues y in terms of time x in days. Use the slopes to interpret any change in daily revenues for a 1-day increase in time.

(a) The line has a slope of $m = 400$.

(b) The line has a slope of $m = 100$.

(c) The line has a slope of $m = 0$.

35. ***Business*** The graph shows the earnings per share of stock for the Kellogg Company for the years 1988 through 1998. (Source: Kellogg Company)

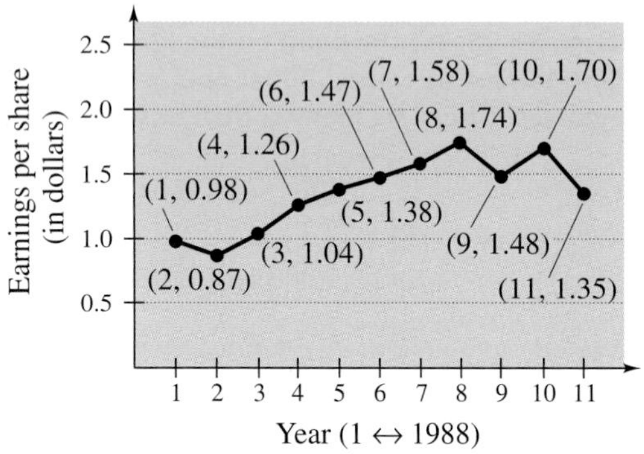

(a) Use the slopes to determine the years when the earnings per share showed the greatest increase and decrease.

(b) Find the slope of the line segment connecting the years 1988 and 1998.

(c) Interpret the meaning of the slope in part (b) in the context of the problem.

36. ***Business*** The graph shows the dividends declared per share of stock for the Colgate-Palmolive Company for the years 1988 through 1998. (Source: Colgate-Palmolive Company)

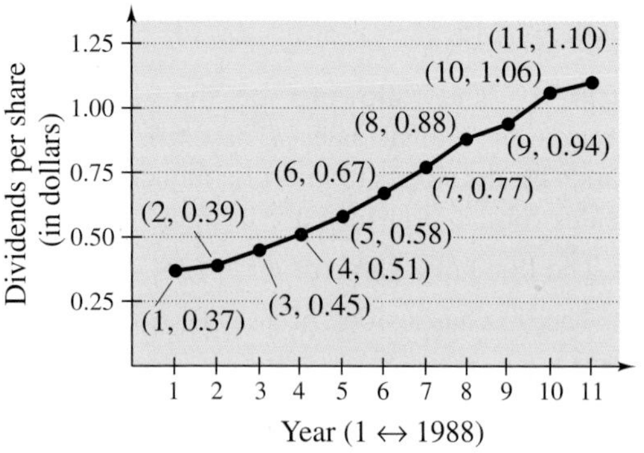

(a) Use the slopes to determine the years when the dividends declared per share showed the greatest increase and the smallest increase.

(b) Find the slope of the line segment connecting the years 1988 and 1998.

(c) Interpret the meaning of the slope in part (b) in the context of the problem.

37. ***Road Grade*** From the top of a mountain road, a surveyor takes several horizontal measurements x and several vertical measurements y, as shown in the table.

x	300	600	900	1200	1500	1800	2100
y	-25	-50	-75	-100	-125	-150	-175

(a) Sketch a scatter plot of the data.

(b) Use a straightedge to sketch the best-fitting line through the points.

(c) Find an equation for the line you sketched in part (b).

(d) Interpret the meaning of the slope of the line in part (c) in the context of the problem.

(e) The surveyor needs to put up a road sign that indicates the steepness of the road. For instance, a surveyor would put a sign that states "8% grade" on a road with a downhill grade that has a slope of $-\frac{8}{100}$. What should the sign state for the road in this problem?

38. ***Road Grade*** You are driving on a road that has a 6% uphill grade. This means that the slope of the road is $\frac{6}{100}$. Approximate the amount of vertical change in your position if you drive 200 feet.

39. *Height of an Attic* The "rise to run" in determining the steepness of the roof on a house is 3 to 4 (see figure). Determine the maximum height in the attic if the house is 32 feet wide.

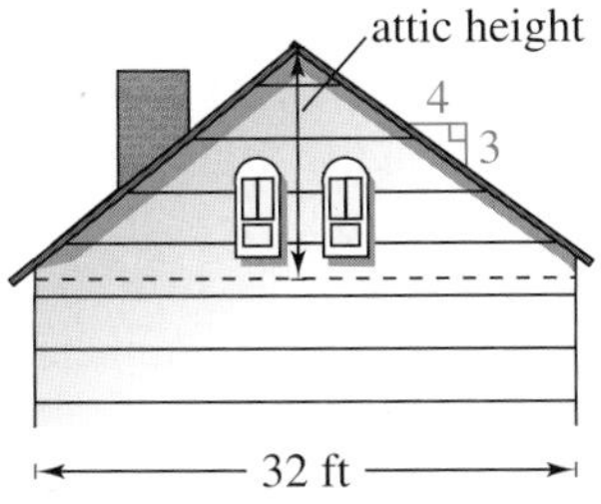

In Exercises 40–45, find the slope and *y*-intercept (if possible) of the equation of the line. Sketch a graph of the line.

40. $y = 5x + 3$ **41.** $y = x - 10$

42. $5x - 2 = 0$ **43.** $3y + 5 = 0$

44. $7x + 6y = 30$ **45.** $2x + 3y = 9$

In Exercises 46–57, find the slope-intercept form of the equation of the line that passes through the given point and has the indicated slope. Sketch a graph of the line.

	Point	*Slope*
46.	$(0, -2)$	$m = 3$
47.	$(0, 10)$	$m = -1$
48.	$(-3, 6)$	$m = -2$
49.	$(0, 0)$	$m = 4$
50.	$(4, 0)$	$m = -\frac{1}{3}$
51.	$(-2, -5)$	$m = \frac{3}{4}$
52.	$(6, -1)$	m is undefined.
53.	$(-10, 4)$	$m = 0$
54.	$\left(4, \frac{5}{2}\right)$	$m = \frac{4}{3}$
55.	$\left(-\frac{1}{2}, \frac{3}{2}\right)$	$m = -3$
56.	$(-5.1, 1.8)$	$m = 5$
57.	$(2.3, -8.5)$	$m = -\frac{5}{2}$

In Exercises 58–67, find the slope-intercept form of the equation of the line passing through the points. Sketch a graph of the line.

58. $(5, -1), (-5, 5)$ **59.** $(4, 3), (-4, -4)$

60. $(-8, 1), (-8, 7)$ **61.** $(-1, 4), (6, 4)$

62. $\left(2, \frac{1}{2}\right), \left(\frac{1}{2}, \frac{5}{4}\right)$ **63.** $(1, 1), \left(6, -\frac{2}{3}\right)$

64. $\left(-\frac{1}{10}, -\frac{3}{5}\right), \left(\frac{9}{10}, -\frac{9}{5}\right)$ **65.** $\left(\frac{3}{4}, \frac{3}{2}\right), \left(-\frac{4}{3}, \frac{7}{4}\right)$

66. $(1, 0.6), (-2, -0.6)$ **67.** $(-8, 0.6), (2, -2.4)$

In Exercises 68–73, use the *intercept form* to find the equation of the line with the given intercepts. The intercept form of the equation of a line with intercepts $(a, 0)$ and $(0, b)$ is

$$\frac{x}{a} + \frac{y}{b} = 1, \quad a \neq 0, \quad b \neq 0.$$

68. x-intercept: $(2, 0)$
y-intercept: $(0, 3)$

69. x-intercept: $(-3, 0)$
y-intercept: $(0, 4)$

70. x-intercept: $\left(-\frac{1}{6}, 0\right)$
y-intercept: $\left(0, -\frac{2}{3}\right)$

71. x-intercept: $\left(\frac{2}{3}, 0\right)$
y-intercept: $(0, -2)$

72. Point on line: $(1, 2)$
x-intercept: $(c, 0)$
y-intercept: $(0, c), \quad c \neq 0$

73. Point on line: $(-3, 4)$
x-intercept: $(d, 0)$
y-intercept: $(0, d), \quad d \neq 0$

In Exercises 74–81, write the slope-intercept forms of the equations of the lines through the given point (a) parallel to the given line and (b) perpendicular to the given line.

	Point	*Line*
74.	$(2, 1)$	$4x - 2y = 3$
75.	$(-3, 2)$	$x + y = 7$
76.	$\left(-\frac{2}{3}, \frac{7}{8}\right)$	$3x + 4y = 7$
77.	$\left(\frac{7}{8}, \frac{3}{4}\right)$	$5x + 3y = 0$
78.	$(-1, 0)$	$y = -3$
79.	$(2, 5)$	$x = 4$
80.	$(2.5, 6.8)$	$x - y = 4$
81.	$(-3.9, -1.4)$	$6x + 2y = 9$

Graphical Interpretation **In Exercises 82–85, identify any relationships that exist among the lines, and then use a graphing utility to graph the three equations in the same viewing window. Adjust the viewing window so that the slope appears visually correct.**

82. (a) $y = 2x$ (b) $y = -2x$ (c) $y = \frac{1}{2}x$

83. (a) $y = \frac{2}{3}x$ (b) $y = -\frac{3}{2}x$ (c) $y = \frac{2}{3}x + 2$

84. (a) $y = -\frac{1}{2}x$ (b) $y = -\frac{1}{2}x + 3$ (c) $y = 2x - 4$

85. (a) $y = x - 8$ (b) $y = x + 1$ (c) $y = -x + 3$

Rate of Change **In Exercises 86 and 87, you are given the dollar value of a product in 2001 and the rate at which the value of the product is expected to change during the next 5 years. Use this information to write a linear equation that gives the dollar value V of the product in terms of the year t. (Let $t = 1$ represent 2001.)**

	2001 Value	*Rate*
86.	\$2540	\$125 increase per year
87.	\$156	\$4.50 increase per year

Graphical Interpretation **In Exercises 88–91, match the description of the situation with its graph. Also determine the slope of each graph and interpret the slope in the context of the situation. [The graphs are labeled (a), (b), (c), and (d).]**

(a)

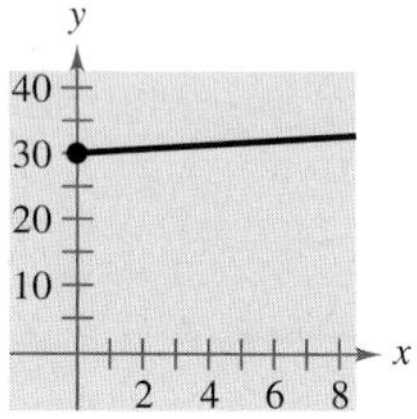

(b)

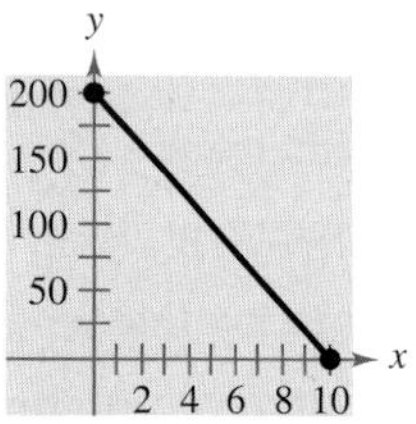

(c)

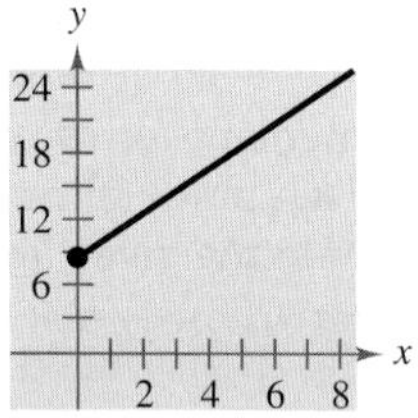

(d)

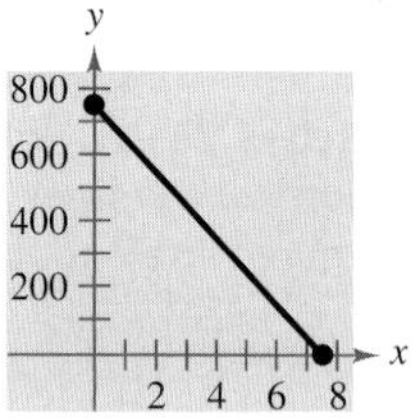

88. A person is paying \$20 per week to a friend to repay a \$200 loan.

89. An employee is paid \$8.50 per hour plus \$2 for each unit produced per hour.

90. A sales representative receives \$30 per day for food plus \$0.32 for each mile traveled.

91. A word processor that was purchased for \$750 depreciates \$100 per year.

In Exercises 92–95, find a relationship between x and y such that (x, y) is equidistant from the two points.

92. $(4, -1), (-2, 3)$

93. $(6, 5), (1, -8)$

94. $\left(3, \frac{5}{2}\right), (-7, 1)$

95. $\left(-\frac{1}{2}, -4\right), \left(\frac{7}{2}, \frac{5}{4}\right)$

96. *Cash Flow per Share* The cash flow per share for America Online, Inc. was \$0.26 in 1998 and \$0.70 in 1999. Write a linear equation that gives the cash flow per share in terms of the year. Let $t = 0$ represent 1998. Then predict the cash flows for the years 2000 and 2001. (Source: America Online, Inc.)

97. *Number of Stores* In 1996 there were 3927 J.C. Penney stores and in 1997 there were 3981 stores. Write a linear equation that gives the number of stores in terms of the year. Let $t = 0$ represent 1996. Then predict the numbers of stores for the years 1999 and 2000. (Source: J.C. Penney Co.)

98. *Temperature* Find the equation of the line that shows the relationship between the temperature in degrees Celsius C and degrees Fahrenheit F. Remember that water freezes at 0° Celsius (32° Fahrenheit) and boils at 100° Celsius (212° Fahrenheit).

99. *Temperature* Use the result of Exercise 98 to complete the table.

C		−10°	10°			177°
F	0°			68°	90°	

100. *Annual Salary* Your salary was \$28,500 in 1998 and \$32,900 in 2000. If your salary follows a linear growth pattern, what will your salary be in 2003?

101. *College Enrollment* A small college had 2546 students in 1998 and 2702 students in 2000. If the enrollment follows a linear growth pattern, how many students will the college have in 2004?

102. *Business* A business purchases a piece of equipment for \$875. After 5 years the equipment will be outdated and have no value. Write a linear equation giving the value V of the equipment during the 5 years it will be used.

103. *Business* A business purchases a piece of equipment for \$25,000. After 10 years the equipment will have to be replaced. Its value at that time is expected to be \$2000. Write a linear equation giving the value V of the equipment during the 10 years it will be used.

104. *Sales* A store is offering a 15% discount on all items. Write a linear equation giving the sale price S for an item with a list price L.

105. *Hourly Wage* A manufacturer pays its assembly line workers \$11.50 per hour. In addition, workers receive a piecework rate of \$0.75 per unit produced. Write a linear equation for the hourly wage W in terms of the number of units x produced per hour.

106. *Business* A contractor purchases a piece of equipment for \$36,500. The equipment requires an average expenditure of \$5.25 per hour for fuel and maintenance, and the operator is paid \$11.50 per hour.

(a) Write a linear equation giving the total cost C of operating this equipment for t hours. (Include the purchase cost of the equipment.)

(b) Assuming that customers are charged \$27 per hour of machine use, write an equation for the revenue R derived from t hours of use.

(c) Use the formula for profit $(P = R - C)$ to write an equation for the profit derived from t hours of use.

(d) Use the result of part (c) to find the break-even point—that is, the number of hours this equipment must be used to yield a profit of 0 dollars.

107. *Rental Demand* A real estate office handles an apartment complex with 50 units. When the rent per unit is \$580 per month, all 50 units are occupied. However, when the rent is \$625 per month, the average number of occupied units drops to 47. Assume that the relationship between the monthly rent p and the demand x is linear.

(a) Write the equation of the line giving the demand x in terms of the rent p.

(b) Use this equation to predict the number of units occupied if the rent is \$655.

(c) Predict the number of units occupied if the rent is \$595.

108. *Geometry* The length and width of a rectangular garden are 15 meters and 10 meters, respectively. A walkway of width x surrounds the garden.

(a) Draw a diagram that gives a visual representation of the problem.

(b) Write the equation for the perimeter y of the walkway in terms of x.

(c) Use a graphing utility to graph the equation for the perimeter.

(d) Determine the slope of the graph in part (c). For each additional 1-meter increase in the width of the walkway, determine the increase in its perimeter.

109. *Monthly Salary* A salesperson receives a monthly salary of \$2500 plus a commission of 7% of sales. Write a linear equation for the salesperson's monthly wage W in terms of monthly sales S.

110. *Business Costs* A sales representative of a company using a personal car receives \$120 per day for lodging and meals plus \$0.31 per mile driven. Write a linear equation giving the daily cost C to the company in terms of x, the number of miles driven.

111. *Investment* An inheritance of \$12,000 is invested in two different mutual funds. A less risky fund pays $2\frac{1}{2}\%$ simple interest and a more risky fund pays 4% simple interest.

(a) If x dollars is invested in the fund paying $2\frac{1}{2}\%$, how much money is invested in the fund paying 4%?

(b) Write the total annual interest y in terms of x.

(c) Use a graphing utility to graph the function in part (b) over the interval $0 \leq x \leq 12{,}000$.

(d) Explain why the slope of the line in part (c) is negative.

112. *Sports* The average annual salaries of major league baseball players (in thousands of dollars) from 1988 to 1998 are shown in the scatter plot. Find the equation of the line that you think best fits these data. (Let y represent the average salary and let t represent the year, with $t = 0$ corresponding to 1988.) (Source: Major League Baseball Player Relations Committee)

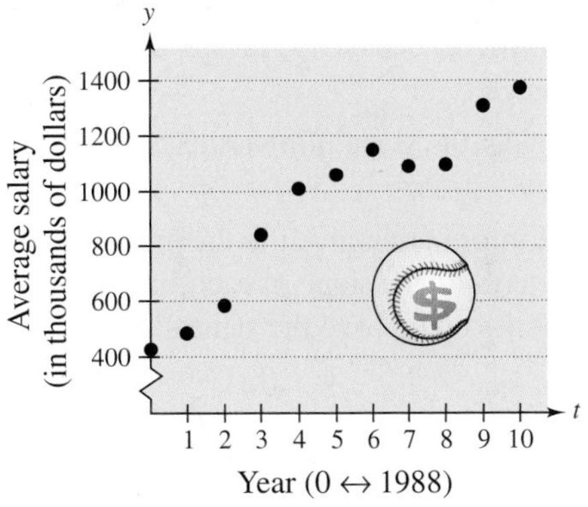

113. *Data Analysis* An instructor gives regular 20-point quizzes and 100-point exams in a mathematics course. Average scores for six students, given as data points (x, y) where x is the average quiz score and y is the average test score, are $(18, 87)$, $(10, 55)$, $(19, 96)$, $(16, 79)$, $(13, 76)$, and $(15, 82)$. [*Note*: There are many correct answers for parts (b)–(d).]

(a) Sketch a scatter plot of the data.

(b) Use a straightedge to sketch the best-fitting line through the points.

(c) Find an equation for the line sketched in part (b).

(d) Use the equation in part (c) to estimate the average test score for a person with an average quiz score of 17.

(e) If the instructor adds 4 points to the average test score of everyone in the class, describe the changes in the positions of the plotted points and the change in the equation of the line.

Synthesis

True or False? **In Exercises 114 and 115, determine whether the statement is true or false. Justify your answer.**

114. A line with a slope of $-\frac{5}{7}$ is steeper than a line with a slope of $-\frac{6}{7}$.

115. The line through $(-8, 2)$ and $(-1, 4)$ and the line through $(0, -4)$ and $(-7, 7)$ are parallel.

116. Explain how you could show that the points $A(2, 3)$, $B(2, 9)$, and $C(7, 2)$ are the vertices of a right triangle.

117. Explain why the slope of a vertical line is said to be undefined.

118. With the information given in the graphs, is it possible to determine the slope of each line? Is it possible that the lines could have the same slope? Explain.

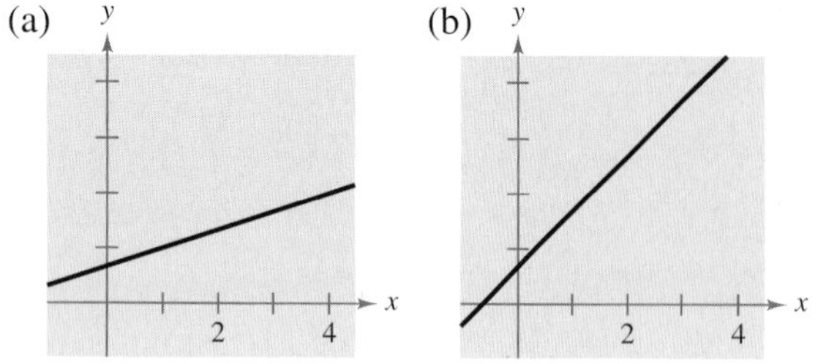

119. The slopes of two lines are -4 and $\frac{5}{2}$. Which is steeper? Explain.

120. The value V of a machine t years after it is purchased is

$$V = -4000t + 58{,}500, \quad 0 \le t \le 5.$$

Explain what the V-intercept and slope measure.

121. *Writing* Write a brief paragraph explaining whether or not any pair of points on a line can be used to calculate the slope of the line.

122. *Think About It* Is it possible for two lines with positive slopes to be perpendicular? Explain.

Review

In Exercises 123–126, match the equation with its graph. [The graphs are labeled (a), (b), (c), and (d).]

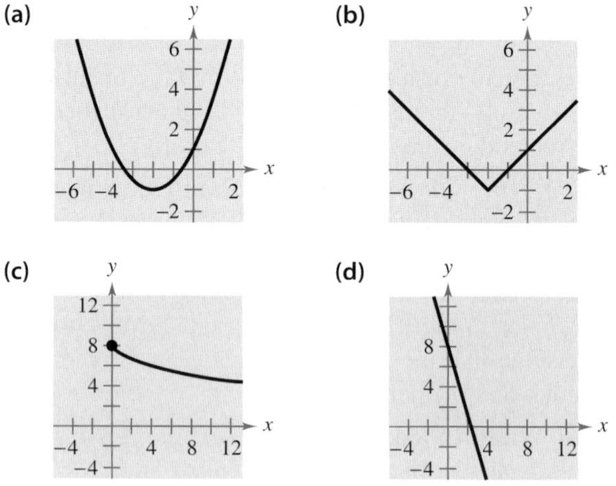

123. $y = 8 - 3x$

124. $y = 8 - \sqrt{x}$

125. $y = \frac{1}{2}x^2 + 2x + 1$

126. $y = |x + 2| - 1$

In Exercises 127–132, find all the solutions of the equation. Check your solution(s) in the original equation.

127. $-7(3 - x) = 14(x - 1)$

128. $\dfrac{8}{2x - 7} = \dfrac{4}{9 - 4x}$

129. $2x^2 - 21x + 49 = 0$

130. $x^2 - 8x + 3 = 0$

131. $\sqrt{x - 9} + 15 = 0$

132. $3x - 16\sqrt{x} + 5 = 0$

1.3 Functions

▶ **What you should learn**

- How to decide whether relations between two variables are functions
- How to use function notation and evaluate functions
- How to find the domains of functions
- How to use functions to model and solve real-life problems

▶ **Why you should learn it**

Functions can be used to model and solve real-life problems. For instance, Exercise 98 on page 137 shows how to use a function to find the force of water against the face of a dam.

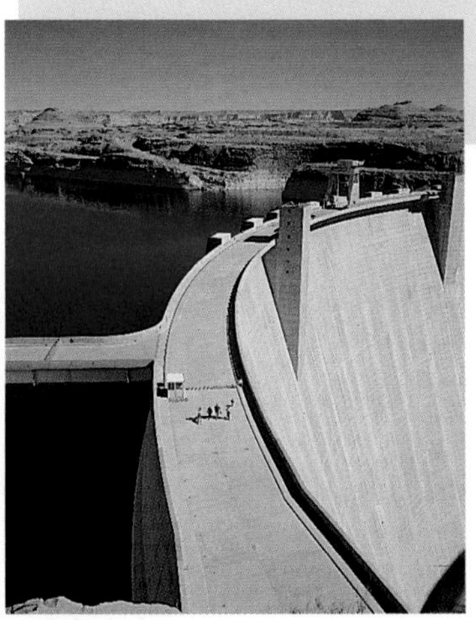

Kunio Owaki/The Stock Market

Introduction to Functions

Many everyday phenomena involve two quantities that are related to each other by some rule of correspondence. The mathematical term for such a rule of correspondence is a **relation.** Here are two examples.

1. The simple interest I earned on \$1000 for 1 year is related to the annual interest rate r by the formula $I = 1000r$.
2. The area A of a circle is related to its radius r by the formula $A = \pi r^2$.

Not all relations have simple mathematical formulas. For instance, people commonly match up NFL starting quarterbacks with touchdown passes and hours of the day with temperature. In cases 1 and 2 above, however, there is some relation that matches each item from one set with exactly one item from a different set. Such a relation is called a **function.**

Definition of a Function

A **function** f from a set A to a set B is a relation that assigns to each element x in the set A exactly one element y in the set B. The set A is the **domain** (or set of inputs) of the function f, and the set B contains the **range** (or set of outputs).

To help understand this definition, look at the function in Figure 1.26.

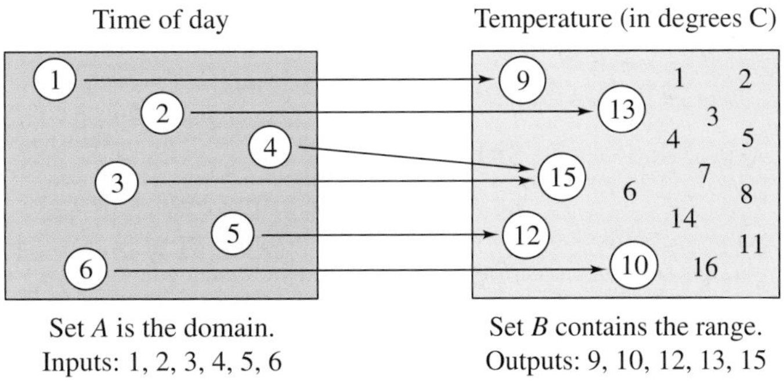

FIGURE 1.26

This function can be represented by the following ordered pairs.

$$\{(1, 9°), (2, 13°), (3, 15°), (4, 15°), (5, 12°), (6, 10°)\}$$

In each ordered pair, the first coordinate is the input and the second coordinate is the output. In this example, note the following characteristics of a function.

1. Each element in A must be matched with an element of B.
2. Some elements in B may not be matched with any element in A.
3. Two or more elements of A may be matched with the same element of B.

The converse of the third statement is not true. That is, an element of A (the domain) cannot be matched with two different elements of B.

Have your students pay special attention to the concepts of function, domain, and range, because they will be used throughout this text and in calculus.

Functions are commonly represented in four ways.

1. *Verbally* by a sentence that describes how the input variable is related to the output variable
2. *Numerically* by a table or a list of ordered pairs that matches input values with output values
3. *Graphically* by points on a graph in a coordinate plane in which the input values are represented by the horizontal axis and the output values are represented by the vertical axis
4. *Algebraically* by an equation in two variables

In the following example, you are asked to decide whether or not the given relation is a function. To do this, you must decide whether each input value is matched with exactly one output value. If any input value is matched with two or more output values, the relation is not a function.

Example 1 ▶ Testing for Functions

Decide whether the description represents y as a function of x.

a. The input value x is the number of representatives from a state, and the output value y is the number of senators.

b.

Input x	2	2	3	4	5
Output y	11	10	8	5	1

c.

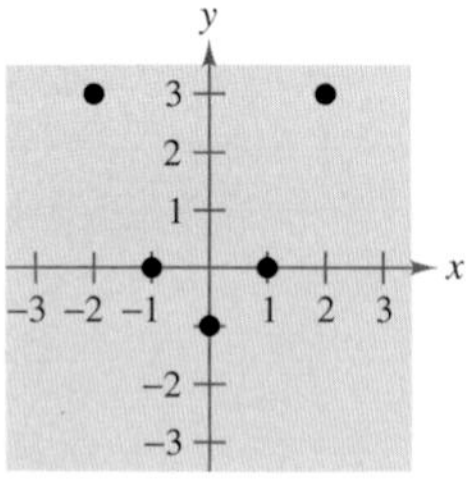

FIGURE 1.27

Solution

a. This verbal description *does* describe y as a function of x. Regardless of the value of x, the value of y is always 2. Such functions are called *constant functions*.

b. This table *does not* describe y as a function of x. The input value 2 is matched with two different y-values.

c. The graph in Figure 1.27 *does* describe y as a function of x. No input value is matched with two output values.

Representing functions by sets of ordered pairs is common in *discrete mathematics*. In algebra, however, it is more common to represent functions by equations or formulas involving two variables. For instance, the equation

$$y = x^2 \qquad y \text{ is a function of } x.$$

represents the variable y as a function of the variable x. In this equation, x is the **independent variable** and y is the **dependent variable.** The domain of the function is the set of all values taken on by the independent variable x, and the range of the function is the set of all values taken on by the dependent variable y.

The Granger Collection

Historical Note
Leonhard Euler (1707–1783), a Swiss mathematician, is considered to have been the most prolific and productive mathematician in history. One of his greatest influences on mathematics was his use of symbols, or notation. The function notation $y = f(x)$ was introduced by Euler.

Example 2 ▶ Testing for Functions Represented Algebraically

Which of the equations represent(s) y as a function of x?

a. $x^2 + y = 1$

b. $-x + y^2 = 1$

Solution

To determine whether y is a function of x, try to solve for y in terms of x.

a. Solving for y yields

$x^2 + y = 1$ Write original equation.

$y = 1 - x^2.$ Solve for y.

To each value of x there corresponds exactly one value of y. So, y is a function of x.

b. Solving for y yields

$-x + y^2 = 1$ Write original equation.

$y^2 = 1 + x$ Add x to each side.

$y = \pm\sqrt{1 + x}.$ Solve for y.

The $\pm$ indicates that to a given value of x there correspond two values of y. So, y is not a function of x.

Function Notation

When an equation is used to represent a function, it is convenient to name the function so that it can be referenced easily. For example, you know that the equation $y = 1 - x^2$ describes y as a function of x. Suppose you give this function the name "f." Then you can use the following **function notation.**

Input	*Output*	*Equation*
x	$f(x)$	$f(x) = 1 - x^2$

Understanding the concept of functions is essential. Be sure students understand function notation. Frequently $f(x)$ is misinterpreted as "f times x" rather than "f of x."

The symbol $f(x)$ is read as *the value of f at x* or simply *f of x*. The symbol $f(x)$ corresponds to the y-value for a given x. So, you can write $y = f(x)$. Keep in mind that f is the *name* of the function, whereas $f(x)$ is the *value* of the function at x. For instance, the function

$$f(x) = 3 - 2x$$

has *function values* denoted by $f(-1)$, $f(0)$, $f(2)$, and so on. To find these values, substitute the specified input values into the given equation.

For $x = -1$, $f(-1) = 3 - 2(-1) = 3 + 2 = 5.$
For $x = 0$, $f(0) = 3 - 2(0) = 3 - 0 = 3.$
For $x = 2$, $f(2) = 3 - 2(2) = 3 - 4 = -1.$

Although f is often used as a convenient function name and x is often used as the independent variable, you can use other letters. For instance,

$$f(x) = x^2 - 4x + 7, \quad f(t) = t^2 - 4t + 7, \quad \text{and} \quad g(s) = s^2 - 4s + 7$$

all define the same function. In fact, the role of the independent variable is that of a "placeholder." Consequently, the function could be described by

$$f(\quad) = (\quad)^2 - 4(\quad) + 7.$$

Example 3 ▶ Evaluating a Function

Let $g(x) = -x^2 + 4x + 1$ and find

a. $g(2)$ **b.** $g(t)$ **c.** $g(x + 2)$.

Solution

a. Replacing x with 2 in $g(x) = -x^2 + 4x + 1$ yields the following.

$$g(2) = -(2)^2 + 4(2) + 1 = -4 + 8 + 1 = 5$$

b. Replacing x with t yields the following.

$$g(t) = -(t)^2 + 4(t) + 1 = -t^2 + 4t + 1$$

c. Replacing x with $x + 2$ yields the following.

$$\begin{aligned} g(x+2) &= -(x+2)^2 + 4(x+2) + 1 \\ &= -(x^2 + 4x + 4) + 4x + 8 + 1 \\ &= -x^2 - 4x - 4 + 4x + 8 + 1 \\ &= -x^2 + 5 \end{aligned}$$

STUDY TIP

In Example 3, note that $g(x + 2)$ is not equal to $g(x) + g(2)$. In general, $g(u + v) \neq g(u) + g(v)$.

A function defined by two or more equations over a specified domain is called a **piecewise-defined function.**

Example 4 ▶ A Piecewise-Defined Function

Evaluate the function when $x = -1, 0,$ and 1.

$$f(x) = \begin{cases} x^2 + 1, & x < 0 \\ x - 1, & x \geq 0 \end{cases}$$

Solution

Because $x = -1$ is less than 0, use $f(x) = x^2 + 1$ to obtain

$$f(-1) = (-1)^2 + 1 = 2.$$

For $x = 0$, use $f(x) = x - 1$ to obtain

$$f(0) = (0) - 1 = -1.$$

For $x = 1$, use $f(x) = x - 1$ to obtain

$$f(1) = (1) - 1 = 0.$$

Additional Examples

a. Evaluate at $x = 0, 1, 3$

$$f(x) = \begin{cases} \dfrac{x}{2} + 1, & x \leq 1 \\ 3x + 2, & x > 1 \end{cases}$$

Solution

Because $x = 0$ is less than or equal to 1, use $f(x) = (x/2) + 1$ to obtain

$$f(0) = \frac{0}{2} + 1 = 1.$$

For $x = 1$, use $f(x) = (x/2) + 1$ to obtain

$$f(1) = \frac{1}{2} + 1 = 1\frac{1}{2}.$$

For $x = 3$, use $f(x) = 3x + 2$ to obtain

$$f(3) = 3(3) + 2 = 11.$$

b. Evaluate at $x = 0, 3, 5$.

$$f(x) = \begin{cases} x^2 + 3, & x < 2 \\ 7, & 2 \leq x \leq 4 \\ 2x - 1, & x > 4 \end{cases}$$

Solution

Because $x = 0$ is less than 2, use $f(x) = x^2 + 3$ to obtain

$$f(0) = 0^2 + 3 = 3.$$

For $x = 3$, use $f(x) = 7$ to obtain

$$f(3) = 7.$$

For $x = 5$, use $f(x) = 2x - 1$ to obtain

$$f(5) = 2(5) - 1 = 9.$$

> **Technology**
>
> Use a graphing utility to graph $y = \sqrt{4 - x^2}$. What is the domain of this function? Then graph $y = \sqrt{x^2 - 4}$. What is the domain of this function? Do the domains of these two functions overlap? If so, for what values?

The Domain of a Function

The domain of a function can be described explicitly or it can be *implied* by the expression used to define the function. The **implied domain** is the set of all real numbers for which the expression is defined. For instance, the function

$$f(x) = \frac{1}{x^2 - 4}$$ Domain excludes x-values that result in division by zero.

has an implied domain that consists of all real x other than $x = \pm 2$. These two values are excluded from the domain because division by zero is undefined. Another common type of implied domain is that used to avoid even roots of negative numbers. For example, the function

$$f(x) = \sqrt{x}$$ Domain excludes x-values that result in even roots of negative numbers.

is defined only for $x \geq 0$. So, its implied domain is the interval $[0, \infty)$. In general, the domain of a function *excludes* values that would cause division by zero *or* that would result in the even root of a negative number.

Example 5 ▶ Finding the Domain of a Function

Find the domain of each function.

a. f: $\{(-3, 0), (-1, 4), (0, 2), (2, 2), (4, -1)\}$ **b.** $g(x) = \dfrac{1}{x + 5}$

c. Volume of a sphere: $V = \frac{4}{3}\pi r^3$ **d.** $h(x) = \sqrt{4 - x^2}$

Solution

a. The domain of f consists of all first coordinates in the set of ordered pairs.

$$\text{Domain} = \{-3, -1, 0, 2, 4\}$$

b. Excluding x-values that yield zero in the denominator, the domain of g is the set of all real numbers $x \neq -5$.

c. Because this function represents the volume of a sphere, the values of the radius r must be positive. So, the domain is the set of all real numbers r such that $r > 0$.

d. This function is defined only for x-values for which

$$4 - x^2 \geq 0.$$

Using the methods described in Section P.6, you can conclude that $-2 \leq x \leq 2$. So, the domain is the interval $[-2, 2]$.

In Example 5(c), note that the domain of a function may be implied by the physical context. For instance, from the equation $V = \frac{4}{3}\pi r^3$, you would have no reason to restrict r to positive values, but the physical context implies that a sphere cannot have a negative radius.

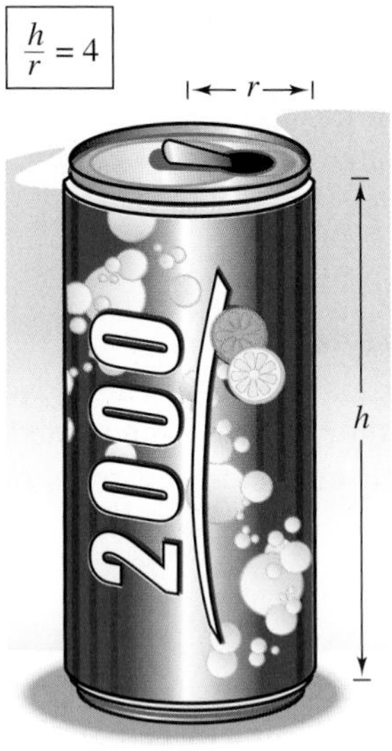

FIGURE 1.28

Applications

Example 6 ▶ The Dimensions of a Container

You work in the marketing department of a soft-drink company and are experimenting with a new soft-drink can that is slightly narrower and taller than a standard can. For your experimental can, the ratio of the height to the radius is 4, as shown in Figure 1.28.

a. Express the volume of the can as a function of the radius r.

b. Express the volume of the can as a function of the height h.

Solution

a. $V(r) = \pi r^2 h = \pi r^2(4r) = 4\pi r^3$ Write V as a function of r.

b. $V(h) = \pi\left(\dfrac{h}{4}\right)^2 h = \dfrac{\pi h^3}{16}$ Write V as a function of h.

Example 7 ▶ The Path of a Baseball

A baseball is hit at a point 3 feet above ground at a velocity of 100 feet per second and an angle of 45°. The path of the baseball is given by the function

$$f(x) = -0.0032x^2 + x + 3$$

where y and x are measured in feet, as shown in Figure 1.29. Will the baseball clear a 10-foot fence located 300 feet from home plate?

Solution

When $x = 300$, the height of the baseball is

$$\begin{aligned} f(300) &= -0.0032(300)^2 + 300 + 3 \\ &= 15 \text{ feet.} \end{aligned}$$

So, the ball will clear the fence.

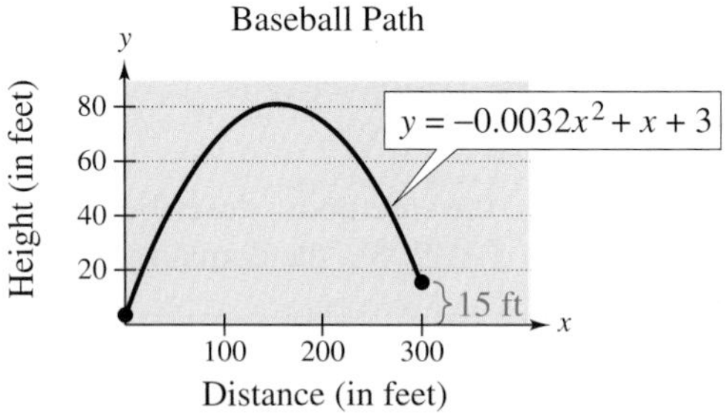

FIGURE 1.29

In the equation in Example 7, the height of the baseball is a function of the distance from home plate.

Example 8 ▶ Direct Mail Advertising

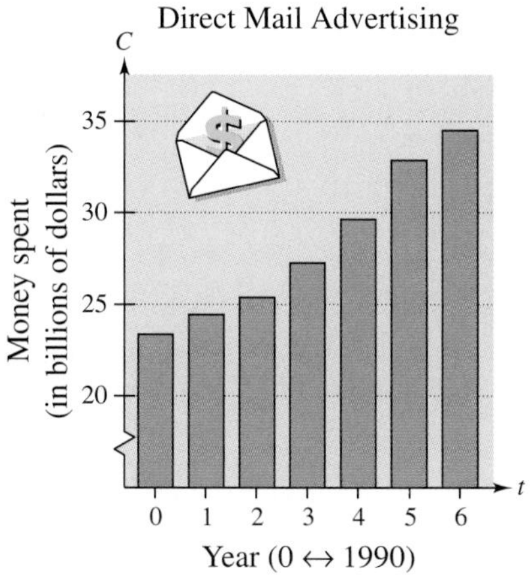

FIGURE 1.30

The money C (in billions of dollars) spent for direct mail advertising in the United States increased in a linear pattern from 1990 to 1992, as shown in Figure 1.30. Then, in 1993, the money spent took a jump and, until 1996, increased in a *different* linear pattern. These two patterns can be approximated by the function

$$C(t) = \begin{cases} 23.40 + 1.01t, & 0 \le t \le 2 \\ 19.85 + 2.50t, & 3 \le t \le 6 \end{cases}$$

where $t = 0$ represents 1990. Use this function to approximate the total amount spent for direct mail advertising between 1990 and 1996. (Source: McCann-Erickson)

Solution

From 1990 to 1992, use the formula $C(t) = 23.40 + 1.01t$.

$$\underbrace{\$23.40,}_{1990}\ \underbrace{\$24.41,}_{1991}\ \underbrace{\$25.42}_{1992}$$

From 1993 to 1996, use the formula $C(t) = 19.85 + 2.50t$.

$$\underbrace{\$27.35,}_{1993}\ \underbrace{\$29.85,}_{1994}\ \underbrace{\$32.35,}_{1995}\ \underbrace{\$34.85}_{1996}$$

The total of these seven amounts is \$197.63, which implies that the total amount spent was approximately \$197,630,000,000.

One of the basic definitions in calculus employs the ratio

$$\frac{f(x+h) - f(x)}{h}, \qquad h \neq 0.$$

This ratio is called a **difference quotient,** as illustrated in Example 9.

Example 9 ▶ Evaluating a Difference Quotient

For $f(x) = x^2 - 4x + 7$, find $\dfrac{f(x+h) - f(x)}{h}$.

Solution

$$\begin{aligned} \frac{f(x+h) - f(x)}{h} &= \frac{[(x+h)^2 - 4(x+h) + 7] - (x^2 - 4x + 7)}{h} \\ &= \frac{x^2 + 2xh + h^2 - 4x - 4h + 7 - x^2 + 4x - 7}{h} \\ &= \frac{2xh + h^2 - 4h}{h} \\ &= \frac{h(2x + h - 4)}{h} \\ &= 2x + h - 4, \qquad h \neq 0 \end{aligned}$$

The symbol [calculus icon] indicates an example or exercise that highlights algebraic techniques specifically used in calculus.

Activities

1. Evaluate $f(x) = 2 + 3x - x^2$ for
   a. $f(-3)$
   b. $f(x + 1)$
   c. $f(x + \Delta x) - f(x)$.

   Answers: a. -16
   b. $-x^2 + x + 4$
   c. $3\Delta x - 2x\Delta x - (\Delta x)^2$

2. Determine whether y is a function of x.

   $2x^3 + 3x^2y^2 + 1 = 0$

   Answer: No

3. Find the domain: $f(x) = \dfrac{3}{x + 1}$.

   Answer: $(-\infty, -1), (-1, \infty)$

Summary of Function Terminology

Function: A **function** is a relationship between two variables such that to each value of the independent variable there corresponds exactly one value of the dependent variable.

Function Notation: $y = f(x)$

f is the *name* of the function.

y is the **dependent variable.**

x is the **independent variable.**

$f(x)$ is the *value of the function at x.*

Domain: The **domain** of a function is the set of all values (inputs) of the independent variable for which the function is defined. If x is in the domain of f, f is said to be *defined* at x. If x is not in the domain of f, f is said to be *undefined* at x.

Range: The **range** of a function is the set of all values (outputs) assumed by the dependent variable (that is, the set of all function values).

Implied Domain: If f is defined by an algebraic expression and the domain is not specified, the **implied domain** consists of all real numbers for which the expression is defined.

Writing ABOUT MATHEMATICS

Modeling with Piecewise-Defined Functions The table below shows the monthly revenue y (in thousands of dollars) for one year of a landscaping business, with $x = 1$ representing January.

x	1	2	3	4	5	6
y	5.2	5.6	6.6	8.3	11.5	15.8

x	7	8	9	10	11	12
y	12.8	10.1	8.6	6.9	4.5	2.7

A mathematical model that represents these data is

$$f(x) = \begin{cases} -1.97x + 26.33 \\ 0.5x^2 - 1.47x + 6.3 \end{cases}.$$

What is the domain of each part of the piecewise-defined function? How can you tell? Explain your reasoning.

Find $f(5)$ and $f(11)$, and interpret your results in the context of the problem. How do these model values compare with the actual data values?

1.3 Exercises

In Exercises 1–4, is the relationship a function?

1. *Domain* *Range*

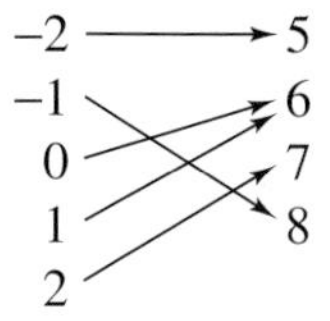

2. *Domain* *Range*

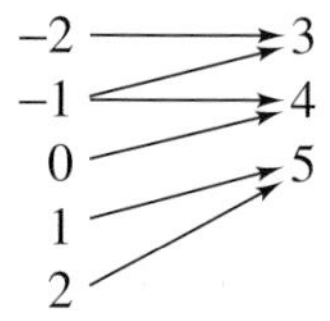

3. *Domain* *Range*

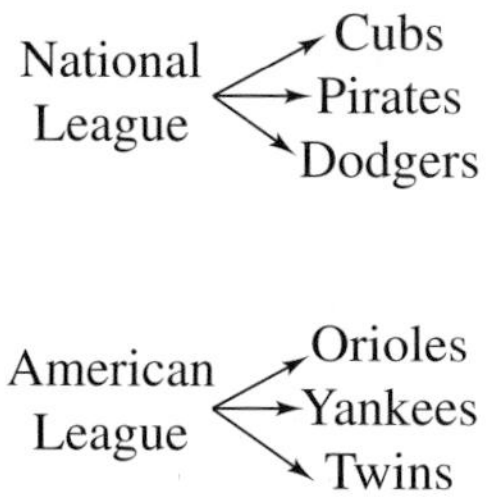

4. *Domain* *Range*

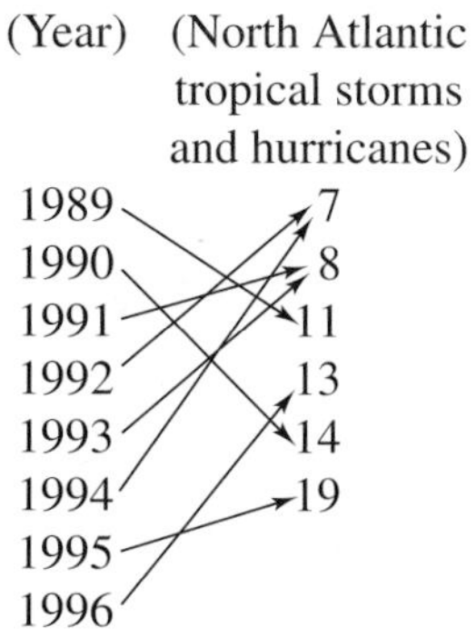

In Exercises 5–8, does the table describe a function? Explain your reasoning.

5.

Input value	−2	−1	0	1	2
Output value	−8	−1	0	1	8

6.

Input value	0	1	2	1	0
Output value	−4	−2	0	2	4

7.

Input value	10	7	4	7	10
Output value	3	6	9	12	15

8.

Input value	0	3	9	12	15
Output value	3	3	3	3	3

In Exercises 9 and 10, which sets of ordered pairs represent functions from *A* to *B*? Explain.

9. $A = \{0, 1, 2, 3\}$ and $B = \{-2, -1, 0, 1, 2\}$

(a) $\{(0, 1), (1, -2), (2, 0), (3, 2)\}$

(b) $\{(0, -1), (2, 2), (1, -2), (3, 0), (1, 1)\}$

(c) $\{(0, 0), (1, 0), (2, 0), (3, 0)\}$

(d) $\{(0, 2), (3, 0), (1, 1)\}$

10. $A = \{a, b, c\}$ and $B = \{0, 1, 2, 3\}$

(a) $\{(a, 1), (c, 2), (c, 3), (b, 3)\}$

(b) $\{(a, 1), (b, 2), (c, 3)\}$

(c) $\{(1, a), (0, a), (2, c), (3, b)\}$

(d) $\{(c, 0), (b, 0), (a, 3)\}$

Circulation of Newspapers **In Exercises 11 and 12, use the graph, which shows the circulation (in millions) of daily newspapers in the United States.** (Source: Editor & Publisher Company)

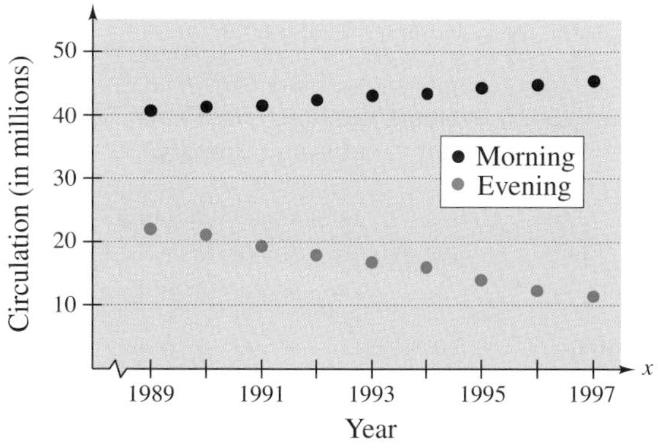

11. Is the circulation of morning newspapers a function of the year? Is the circulation of evening newspapers a function of the year? Explain.

12. Let $f(x)$ represent the circulation of evening newspapers in year x. Find $f(1994)$.

In Exercises 13–22, determine whether the equation represents *y* as a function of *x*.

13. $x^2 + y^2 = 4$

14. $x = y^2$

15. $x^2 + y = 4$

16. $x + y^2 = 4$

17. $2x + 3y = 4$ **18.** $(x - 2)^2 + y^2 = 4$

19. $y^2 = x^2 - 1$ **20.** $y = \sqrt{x + 5}$

21. $y = |4 - x|$ **22.** $|y| = 4 - x$

In Exercises 23 and 24, fill in the blanks using the specified function and the given values of the independent variable.

23. $f(s) = \dfrac{1}{s + 1}$

(a) $f(4) = \dfrac{1}{(\quad) + 1}$

(b) $f(0) = \dfrac{1}{(\quad) + 1}$

(c) $f(4x) = \dfrac{1}{(\quad) + 1}$

(d) $f(x + c) = \dfrac{1}{(\quad) + 1}$

24. $g(x) = x^2 - 2x$

(a) $g(2) = (\quad)^2 - 2(\quad)$

(b) $g(-3) = (\quad)^2 - 2(\quad)$

(c) $g(t + 1) = (\quad)^2 - 2(\quad)$

(d) $g(x + c) = (\quad)^2 - 2(\quad)$

In Exercises 25–36, evaluate the function at each specified value of the independent variable and simplify.

25. $f(x) = 2x - 3$

(a) $f(1)$ (b) $f(-3)$ (c) $f(x - 1)$

26. $g(y) = 7 - 3y$

(a) $g(0)$ (b) $g\left(\frac{7}{3}\right)$ (c) $g(s + 2)$

27. $V(r) = \frac{4}{3}\pi r^3$

(a) $V(3)$ (b) $V\left(\frac{3}{2}\right)$ (c) $V(2r)$

28. $h(t) = t^2 - 2t$

(a) $h(2)$ (b) $h(1.5)$ (c) $h(x + 2)$

29. $f(y) = 3 - \sqrt{y}$

(a) $f(4)$ (b) $f(0.25)$ (c) $f(4x^2)$

30. $f(x) = \sqrt{x + 8} + 2$

(a) $f(-8)$ (b) $f(1)$ (c) $f(x - 8)$

31. $q(x) = \dfrac{1}{x^2 - 9}$

(a) $q(0)$ (b) $q(3)$ (c) $q(y + 3)$

32. $q(t) = \dfrac{2t^2 + 3}{t^2}$

(a) $q(2)$ (b) $q(0)$ (c) $q(-x)$

33. $f(x) = \dfrac{|x|}{x}$

(a) $f(2)$ (b) $f(-2)$ (c) $f(x - 1)$

34. $f(x) = |x| + 4$

(a) $f(2)$ (b) $f(-2)$ (c) $f(x^2)$

35. $f(x) = \begin{cases} 2x + 1, & x < 0 \\ 2x + 2, & x \ge 0 \end{cases}$

(a) $f(-1)$ (b) $f(0)$ (c) $f(2)$

36. $f(x) = \begin{cases} x^2 + 2, & x \le 1 \\ 2x^2 + 2, & x > 1 \end{cases}$

(a) $f(-2)$ (b) $f(1)$ (c) $f(2)$

In Exercises 37–42, complete the table.

37. $f(x) = x^2 - 3$

x	-2	-1	0	1	2
$f(x)$					

38. $g(x) = \sqrt{x - 3}$

x	3	4	5	6	7
$g(x)$					

39. $h(t) = \frac{1}{2}|t + 3|$

t	-5	-4	-3	-2	-1
$h(t)$					

40. $f(s) = \dfrac{|s - 2|}{s - 2}$

s	0	1	$\frac{3}{2}$	$\frac{5}{2}$	4
$f(s)$					

41. $f(x) = \begin{cases} -\frac{1}{2}x + 4, & x \le 0 \\ (x - 2)^2, & x > 0 \end{cases}$

x	-2	-1	0	1	2
$f(x)$					

42. $h(x) = \begin{cases} 9 - x^2, & x < 3 \\ x - 3, & x \geq 3 \end{cases}$

x	1	2	3	4	5
$h(x)$					

In Exercises 43–50, find all real values of x such that $f(x) = 0$.

43. $f(x) = 15 - 3x$

44. $f(x) = 5x + 1$

45. $f(x) = \dfrac{3x - 4}{5}$

46. $f(x) = \dfrac{12 - x^2}{5}$

47. $f(x) = x^2 - 9$

48. $f(x) = x^2 - 8x + 15$

49. $f(x) = x^3 - x$

50. $f(x) = x^3 - x^2 - 4x + 4$

In Exercises 51–54, find the value(s) of x for which $f(x) = g(x)$.

51. $f(x) = x^2, \quad g(x) = x + 2$

52. $f(x) = x^2 + 2x + 1, \quad g(x) = 3x + 3$

53. $f(x) = \sqrt{3x} + 1, \quad g(x) = x + 1$

54. $f(x) = x^4 - 2x^2, \quad g(x) = 2x^2$

In Exercises 55–68, find the domain of the function.

55. $f(x) = 5x^2 + 2x - 1$

56. $g(x) = 1 - 2x^2$

57. $h(t) = \dfrac{4}{t}$

58. $s(y) = \dfrac{3y}{y + 5}$

59. $g(y) = \sqrt{y - 10}$

60. $f(t) = \sqrt[3]{t + 4}$

61. $f(x) = \sqrt[4]{1 - x^2}$

62. $f(x) = \sqrt[4]{x^2 + 3x}$

63. $g(x) = \dfrac{1}{x} - \dfrac{3}{x + 2}$

64. $h(x) = \dfrac{10}{x^2 - 2x}$

65. $f(s) = \dfrac{\sqrt{s - 1}}{s - 4}$

66. $f(x) = \dfrac{\sqrt{x + 6}}{6 + x}$

67. $f(x) = \dfrac{\sqrt[3]{x - 4}}{x}$

68. $f(x) = \dfrac{x - 5}{x^2 - 9}$

In Exercises 69–72, assume that the domain of f is the set $A = \{-2, -1, 0, 1, 2\}$. Determine the set of ordered pairs that represents the function f.

69. $f(x) = x^2$

70. $f(x) = \dfrac{2x}{x^2 + 1}$

71. $f(x) = \sqrt{x + 2}$

72. $f(x) = |x + 1|$

Exploration **In Exercises 73–76, determine the function from**

$$f(x) = cx,\ g(x) = cx^2,\ h(x) = c\sqrt{|x|},\ \text{and } r(x) = \frac{c}{x}$$

and the value of the constant c that will make the function fit the data in the table.

73.

x	-4	-1	0	1	4
y	-32	-2	0	-2	-32

74.

x	-4	-1	0	1	4
y	-1	$-\frac{1}{4}$	0	$\frac{1}{4}$	1

75.

x	-4	-1	0	1	4
y	-8	-32	Undef.	32	8

76.

x	-4	-1	0	1	4
y	6	3	0	3	6

In Exercises 77–84, find the difference quotient and simplify your answer.

77. $f(x) = x^2 - x + 1, \quad \dfrac{f(2 + h) - f(2)}{h}, h \neq 0$

78. $f(x) = 5x - x^2, \quad \dfrac{f(5 + h) - f(5)}{h}, h \neq 0$

79. $f(x) = x^3, \quad \dfrac{f(x + c) - f(x)}{c}, c \neq 0$

80. $f(x) = 2x, \quad \dfrac{f(x + c) - f(x)}{c}, c \neq 0$

81. $g(x) = 3x - 1, \quad \dfrac{g(x) - g(3)}{x - 3}, x \neq 3$

82. $f(t) = \dfrac{1}{t}, \quad \dfrac{f(t) - f(1)}{t - 1}, t \neq 1$

83. $f(x) = \sqrt{5x}, \quad \dfrac{f(x) - f(5)}{x - 5}, x \neq 5$

84. $f(x) = x^{2/3} + 1, \quad \dfrac{f(x) - f(8)}{x - 8}, x \neq 8$

85. *Geometry* Express the area A of a square as a function of its perimeter P.

86. *Geometry* Express the area A of a circle as a function of its circumference C.

The symbol indicates an example or exercise that highlights algebraic techniques specifically used in calculus.

87. *Geometry* Express the area A of an isosceles triangle with a height of 8 inches and a base of b inches as a function of the length s of one of its two equal sides.

88. *Geometry* Express the area A of an equilateral triangle as a function of the length s of its sides.

89. *Maximum Volume* An open box of maximum volume is to be made from a square piece of material 24 centimeters on a side by cutting equal squares from the corners and turning up the sides (see figure).

(a) Complete six rows of a table. (The first two rows are shown.) Use the result to estimate the maximum volume.

Height x	Width	Volume V
1	$24 - 2(1)$	$1[24 - 2(1)]^2 = 484$
2	$24 - 2(2)$	$2[24 - 2(2)]^2 = 800$

(b) Plot the points (x, V). Is V a function of x?

(c) If V is a function of x, write the function and determine its domain.

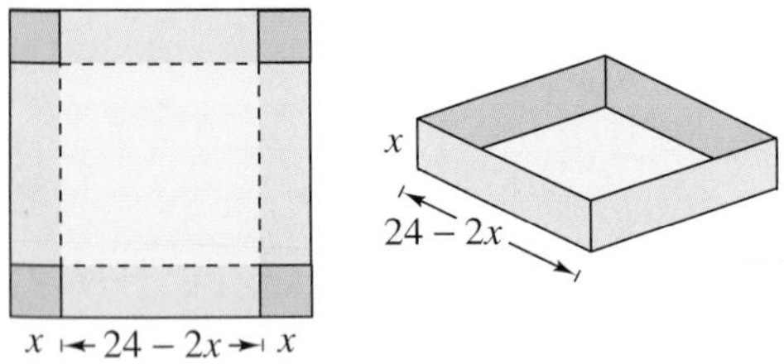

90. *Maximum Profit* The cost per unit in the production of a certain radio model is \$60. The manufacturer charges \$90 per unit for orders of 100 or less. To encourage large orders, the manufacturer reduces the charge by \$0.15 per radio for each unit ordered in excess of 100 (for example, there would be a charge of \$87 per radio for an order size of 120).

(a) Complete six rows of the table. Use the result to estimate the maximum profit.

Units x	Price p	Profit P
110	$90 - 10(0.15)$	$xp - 110(60)$
120	$90 - 20(0.15)$	$xp - 120(60)$

(b) Plot the points (x, P). Is P a function of x?

(c) If P is a function of x, write the function and determine its domain.

91. *Geometry* A right triangle is formed in the first quadrant by the x- and y-axes and a line through the point $(2, 1)$ (see figure). Write the area A of the triangle as a function of x, and determine the domain of the function.

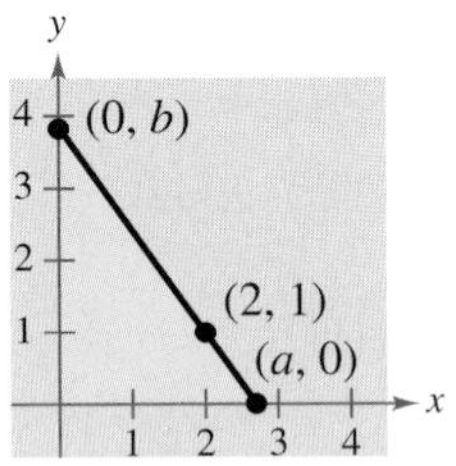

92. *Geometry* A rectangle is bounded by the x-axis and the semicircle $y = \sqrt{36 - x^2}$ (see figure). Write the area A of the rectangle as a function of x, and determine the domain of the function.

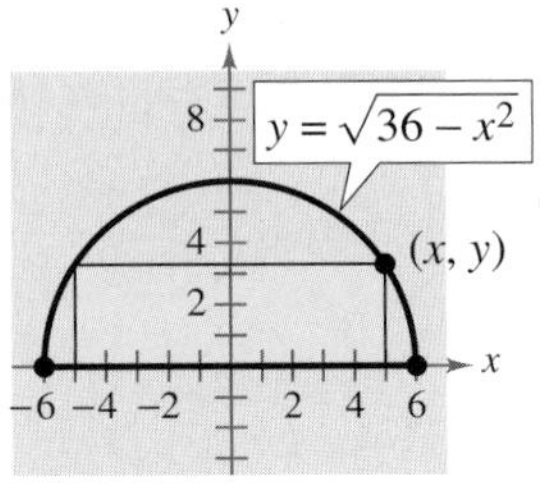

93. *Average Price* The average price p (in thousands of dollars) of a new mobile home in the United States from 1974 to 1997 (see figure) can be approximated by the model

$$p(t) = \begin{cases} 17.27 + 1.036t, & -6 \le t \le 11 \\ -4.807 + 2.882t - 0.011t^2, & 12 \le t \le 17 \end{cases}$$

where $t = 0$ represents 1980. Use this model to find the average prices of a mobile home in 1978, 1988, 1993, and 1997. (Source: U.S. Bureau of Census, Construction Reports)

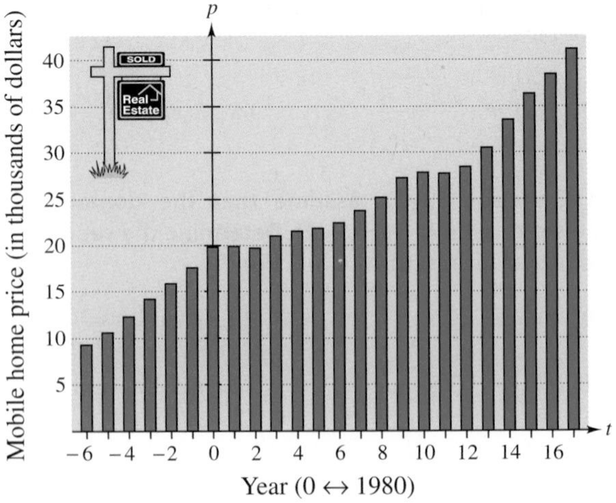

94. *Postal Regulations* A rectangular package to be sent by the U.S. Postal Service can have a maximum combined length and girth (perimeter of a cross section) of 108 inches (see figure).

(a) Write the volume V of the package as a function of x.

(b) What is the domain of the function?

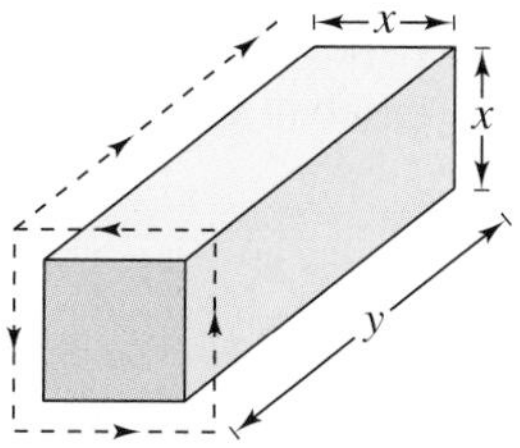

 (c) Use a graphing utility to graph your function. Be sure to use the appropriate window setting.

(d) What dimensions will maximize the volume of the package? Explain your answer.

95. *Business* A company produces a product for which the variable cost is \$12.30 per unit and the fixed costs are \$98,000. The product sells for \$17.98. Let x be the number of units produced and sold.

(a) The total cost for a business is the sum of the variable cost and the fixed costs. Write the total cost C as a function of the number of units produced.

(b) Write the revenue R as a function of the number of units sold.

(c) Write the profit P as a function of the number of units sold. (*Note*: $P = R - C$.)

96. *Total Average Cost* The inventor of a new game believes that the variable cost for producing the game is \$0.95 per unit and the fixed costs are \$6000. The inventor sells each game for \$1.69. Let x be the number of games sold.

(a) The total cost for a business is the sum of the variable cost and the fixed costs. Write the total cost C as a function of the number of games sold.

(b) Write the average cost per unit $\overline{C} = C/x$ as a function of x.

97. *Transportation* For groups of 80 or more people, a charter bus company determines the rate per person according to the formula

$$\text{Rate} = 8 - 0.05(n - 80), \qquad n \geq 80$$

where the rate is given in dollars and n is the number of people.

(a) Express the revenue R for the bus company as a function of n.

(b) Use the function in part (a) to complete the table. What can you conclude?

n	90	100	110	120	130	140	150
$R(n)$							

98. *Physics* The force F (in tons) of water against the face of a dam is estimated by the function

$$F(y) = 149.76\sqrt{10}y^{5/2}$$

where y is the depth of the water in feet.

(a) Complete the table. What can you conclude from the table?

y	5	10	20	30	40
$F(y)$					

(b) Use the table to approximate the depth at which the force against the dam is 1,000,000 tons. How could you find a better estimate?

99. *Height of a Balloon* A balloon carrying a transmitter ascends vertically from a point 3000 feet from the receiving station.

(a) Draw a diagram that gives a visual representation of the problem. Let h represent the height of the balloon and let d represent the distance between the balloon and the receiving station.

(b) Express the height of the balloon as a function of d. What is the domain of the function?

100. *Lynx Population* A study was done on the lynx population of the Yukon Territory in Canada. The graph shows the lynx population from 1988 through 1995 in a 350-square-kilometer region of the Yukon Territory. Let $f(t)$ represent the number of lynx in year t. (Source: Kluane Boreal Forest Ecosystem Project)

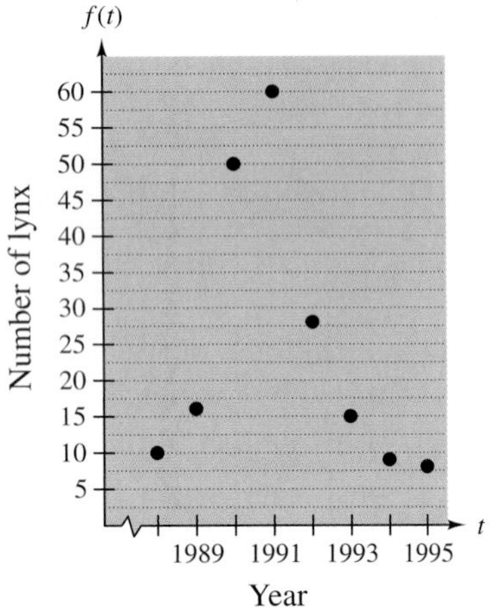

(a) Find

$$\frac{f(1994) - f(1991)}{1994 - 1991}$$

and interpret the result in the context of the problem.

(b) An approximate formula for the function $f(t)$ is

$$N(x) = \frac{434x + 4387}{45x^2 - 55x + 100}$$

where N is the number of lynx and x is time in years, with $x = 0$ corresponding to 1990. Complete the table and compare the result with the data.

x	-2	-1	0	1
N				

x	2	3	4	5
N				

101. *Path of a Ball* The height y (in feet) of a baseball thrown by a child is

$$y = -\frac{1}{10}x^2 + 3x + 6,$$

where x is the horizontal distance (in feet) from where the ball was thrown. Will the ball fly over the head of another child 30 feet away trying to catch the ball? (Assume that the child who is trying to catch the ball holds a baseball glove at a height of 5 feet.)

Synthesis

True or False? **In Exercises 102 and 103, determine whether the statement is true or false. Justify your answer.**

102. The domain of the function $f(x) = x^4 - 1$ is $(-\infty, \infty)$, and the range of $f(x)$ is $(0, \infty)$.

103. The set of ordered pairs $\{(-8, -2), (-6, 0), (-4, 0), (-2, 2), (0, 4), (2, -2)\}$ represents a function.

104. Does the relationship shown in the figure represent a function from set A to set B? Explain.

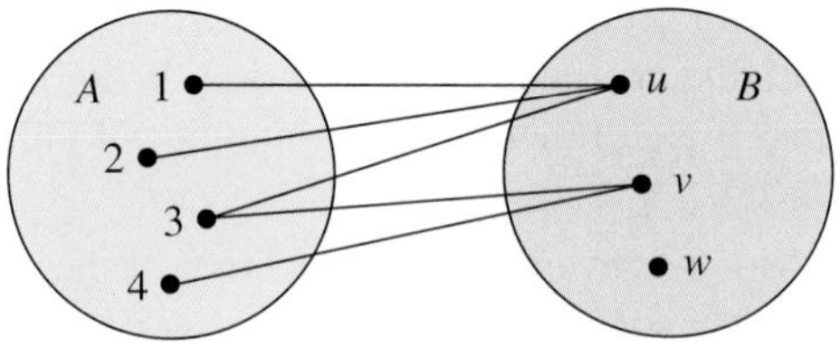

105. *Think About It* In your own words, explain the meanings of *domain* and *range*.

106. *Think About It* Describe an advantage of function notation.

Review

In Exercises 107–110, solve the equation.

107. $\frac{t}{3} + \frac{t}{5} = 1$

108. $\frac{3}{t} + \frac{5}{t} = 1$

109. $\frac{3}{x(x + 1)} - \frac{4}{x} = \frac{1}{x + 1}$

110. $\frac{12}{x} - 3 = \frac{4}{x} + 9$

In Exercises 111–114, find the equation of the line passing through the given pair of points.

111. $(-2, -5), (4, -1)$

112. $(10, 0), (1, 9)$

113. $(-6, 5), (3, -5)$

114. $\left(-\frac{1}{2}, 3\right), \left(\frac{11}{2}, -\frac{1}{3}\right)$

1.4 Analyzing Graphs of Functions

▶ **What you should learn**

- How to use the Vertical Line Test for functions
- How to find the zeros of functions
- How to determine intervals on which functions are increasing or decreasing
- How to identify and graph linear functions
- How to identify and graph step functions and other piecewise-defined functions
- How to identify even and odd functions

▶ **Why you should learn it**

Graphs of functions can help you visualize relationships between variables in real life. For instance, Exercise 103 on page 149 shows how the graph of a step function visually represents the cost of a telephone call.

The Graph of a Function

In Section 1.3, you studied functions from an algebraic point of view. In this section, you will study functions from a graphical perspective.

The **graph of a function** f is the collection of ordered pairs $(x, f(x))$ such that x is in the domain of f. As you study this section, remember that

$$x = \text{the directed distance from the } y\text{-axis}$$

$$f(x) = \text{the directed distance from the } x\text{-axis}$$

as shown in Figure 1.31.

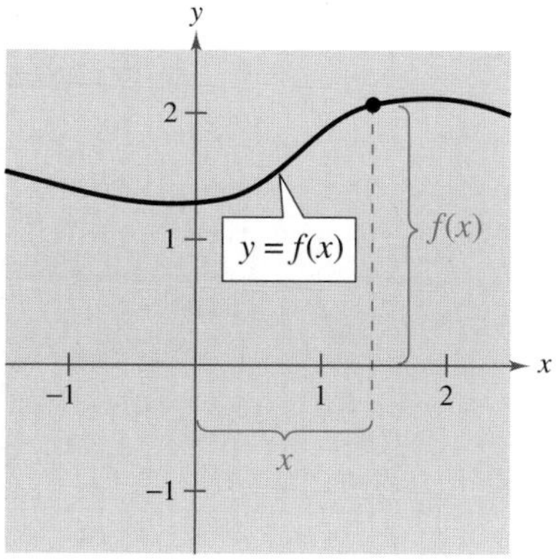

FIGURE 1.31

Example 1 ▶ Finding the Domain and Range of a Function

Use the graph of the function f, shown in Figure 1.32, to find (a) the domain of f, (b) the function values $f(-1)$ and $f(2)$, and (c) the range of f.

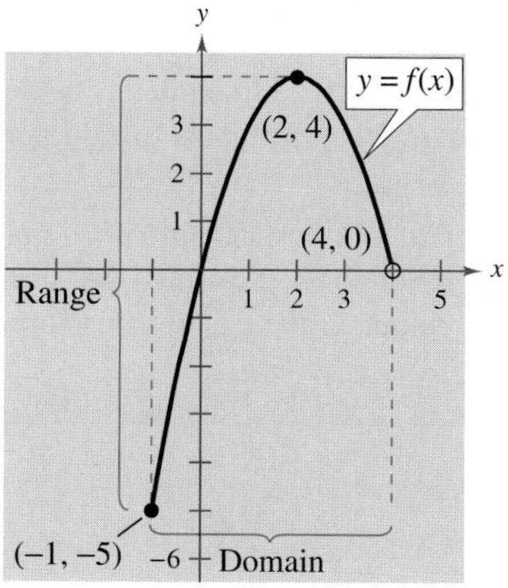

FIGURE 1.32

Solution

a. The closed dot at $(-1, -5)$ indicates that $x = -1$ is in the domain of f, whereas the open dot at $(4, 0)$ indicates $x = 4$ is not in the domain. So, the domain of f is all x in the interval $[-1, 4)$.

b. Because $(-1, -5)$ is a point on the graph of f, it follows that $f(-1) = -5$. Similarly, because $(2, 4)$ is a point on the graph of f, it follows that $f(2) = 4$.

c. Because the graph does not extend below $f(-1) = -5$ or above $f(2) = 4$, the range of f is the interval $[-5, 4]$.

The use of dots (open or closed) at the extreme left and right points of a graph indicates that the graph does not extend beyond these points. If no such dots are shown, assume that the graph extends beyond these points.

By the definition of a function, at most one y-value corresponds to a given x-value. This means that the graph of a function cannot have two or more different points with the same x-coordinate, and no two points on the graph of a function can be vertically above and below each other. It follows, then, that a vertical line can intersect the graph of a function at most once. This observation provides a convenient visual test called the **Vertical Line Test** for functions.

Vertical Line Test for Functions

A set of points in a coordinate plane is the graph of y as a function of x if and only if no vertical line intersects the graph at more than one point.

Example 2 ▶ Vertical Line Test for Functions

Use the Vertical Line Test to decide whether the graphs in Figure 1.33 represent y as a function of x.

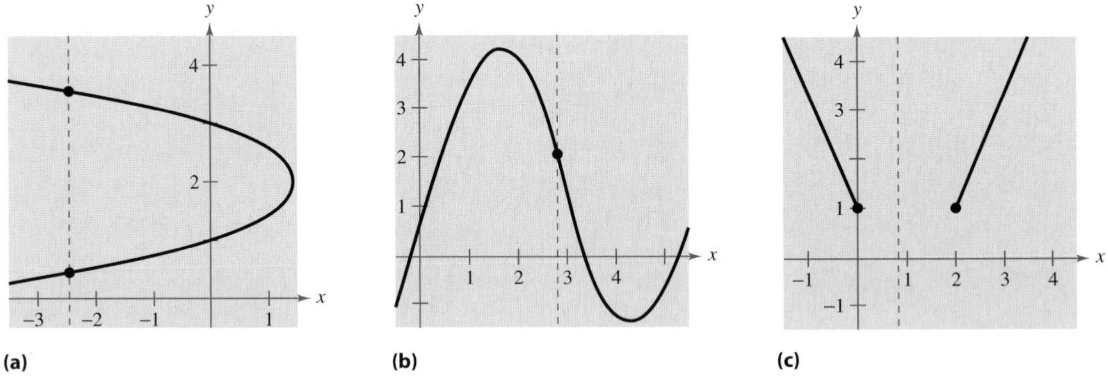

FIGURE 1.33

Solution

a. This *is not* a graph of y as a function of x because you can find a vertical line that intersects the graph twice. That is, for a particular input x, there is more than one output y.

b. This *is* a graph of y as a function of x because every vertical line intersects the graph at most once. That is, for a particular input x, there is at most one output y.

c. This *is* a graph of y as a function of x. (Note that if a vertical line does not intersect the graph, it simply means that the function is undefined for that particular value of x.) That is, for a particular input x, there is at most one output y.

Zeros of a Function

If the graph of a function of x has an x-intercept at $(a, 0)$, then a is a **zero** of the function.

Zeros of a Function

The **zeros of a function** f of x are the x-values for which $f(x) = 0$.

Example 3 ▶ Finding Zeros of a Function

Find the zeros of each function.

a. $f(x) = 3x^2 + x - 10$ **b.** $g(x) = \sqrt{10 - x^2}$ **c.** $h(t) = \dfrac{2t - 3}{t + 5}$

Solution

To find the zeros of a function, set the function equal to zero and solve for the independent variable.

a.

$3x^2 + x - 10 = 0$		Set $f(x)$ equal to 0.
$(3x - 5)(x + 2) = 0$		Factor.
$3x - 5 = 0$	⟹ $x = \frac{5}{3}$	Set 1st factor equal to 0.
$x + 2 = 0$	⟹ $x = -2$	Set 2nd factor equal to 0.

The zeros of f are $\frac{5}{3}$ and -2. In Figure 1.34(a), note that the graph of f has $\left(\frac{5}{3}, 0\right)$ and $(-2, 0)$ as its x-intercepts.

b.

$\sqrt{10 - x^2} = 0$	Set $g(x)$ equal to 0.
$10 - x^2 = 0$	Square each side.
$10 = x^2$	Add x^2 to each side.
$\pm\sqrt{10} = x$	Extract square root.

The zeros of g are $-\sqrt{10}$ and $\sqrt{10}$. In Figure 1.34(b), note that the graph of g has $\left(-\sqrt{10}, 0\right)$ and $\left(\sqrt{10}, 0\right)$ as its x-intercepts.

c.

$\dfrac{2t - 3}{t + 5} = 0$	Set $h(t)$ equal to 0.
$2t - 3 = 0$	Set numerator equal to 0.
$2t = 3$	Add 3 to each side.
$t = \dfrac{3}{2}$	Divide each side by 2.

The zero of h is $\frac{3}{2}$. In Figure 1.34(c), note that the graph of h has $\left(\frac{3}{2}, 0\right)$ as its x-intercept.

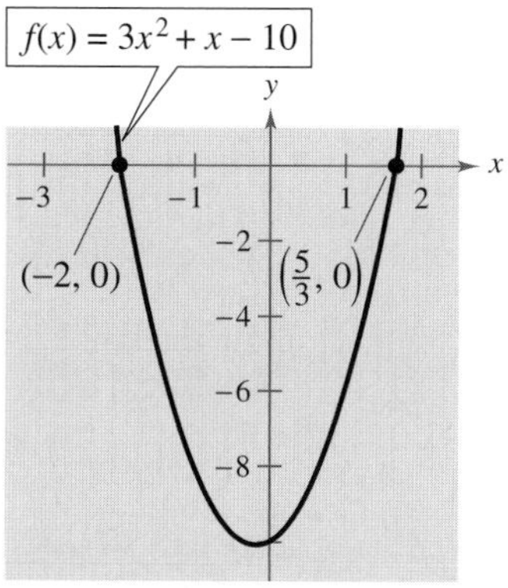

(a) Zeros of f: $x = -2, x = \frac{5}{3}$

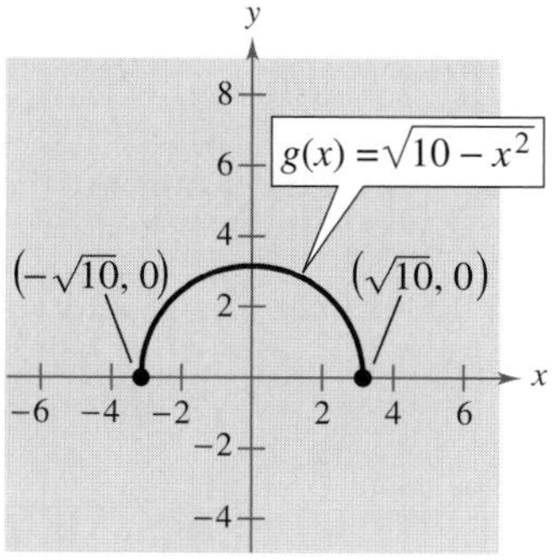

(b) Zeros of g: $x = \pm\sqrt{10}$

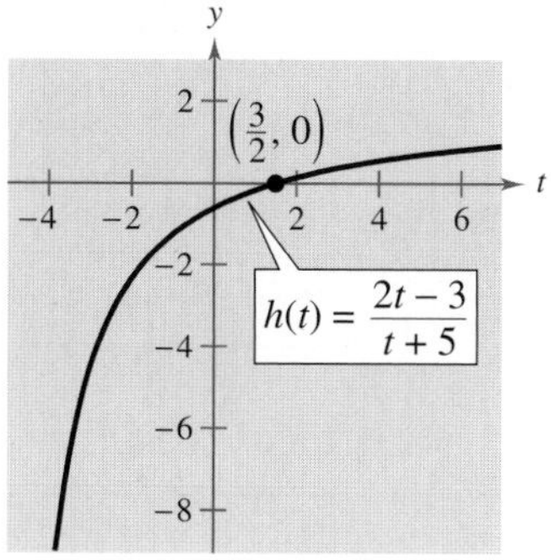

(c) Zero of h: $t = \frac{3}{2}$

FIGURE 1.34

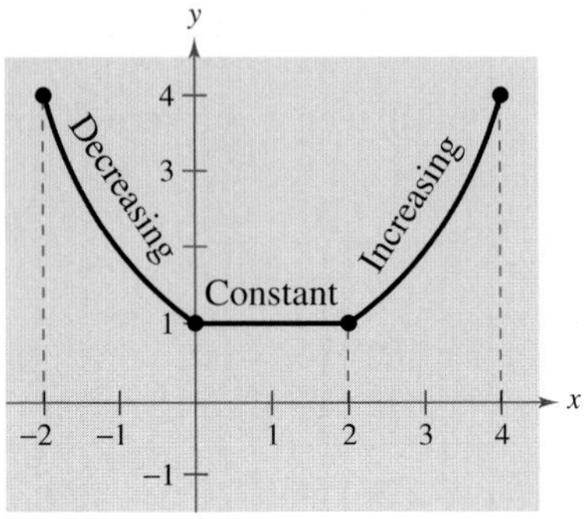

FIGURE 1.35

Increasing and Decreasing Functions

The more you know about the graph of a function, the more you know about the function itself. Consider the graph shown in Figure1.35. As you move from *left to right,* this graph decreases, then is constant, and then increases.

Increasing, Decreasing, and Constant Functions

A function f is **increasing** on an interval if, for any x_1 and x_2 in the interval, $x_1 < x_2$ implies $f(x_1) < f(x_2)$.

A function f is **decreasing** on an interval if, for any x_1 and x_2 in the interval, $x_1 < x_2$ implies $f(x_1) > f(x_2)$.

A function f is **constant** on an interval if, for any x_1 and x_2 in the interval, $f(x_1) = f(x_2)$.

Example 4 ▶ Increasing and Decreasing Functions

In Figure 1.36, use the graphs to describe the increasing or decreasing behavior of each function.

Solution

a. This function is increasing over the entire real line.

b. This function is increasing on the interval $(-\infty, -1)$, decreasing on the interval $(-1, 1)$, and increasing on the interval $(1, \infty)$.

c. This function is increasing on the interval $(-\infty, 0)$, constant on the interval $(0, 2)$, and decreasing on the interval $(2, \infty)$.

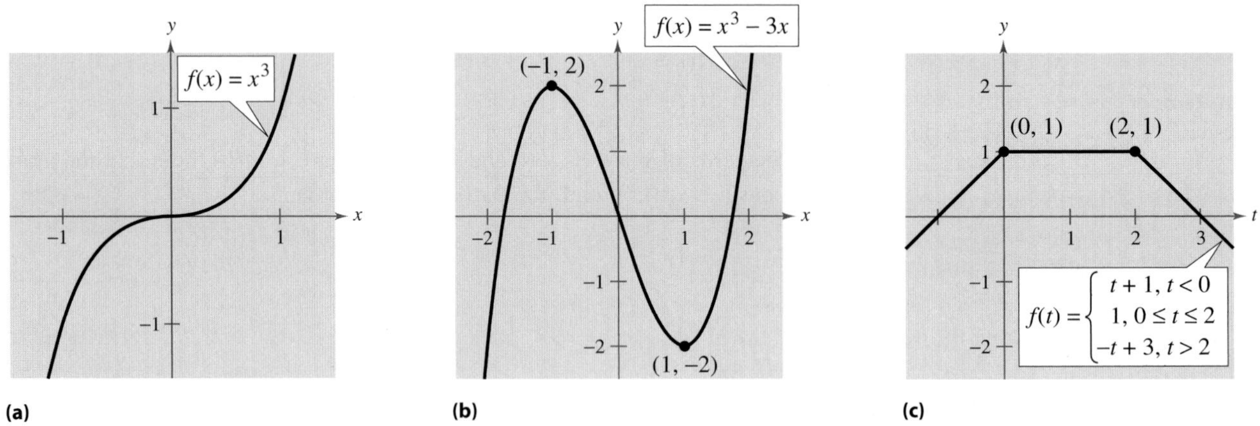

FIGURE 1.36

To decide whether a function is increasing, decreasing, or constant, you can evaluate the function for several values of x. For instance, the table below verifies that the function in Example 4(a) is increasing over the entire real line.

x	-100	-10	-1	0	1	10	100
$f(x) = x^3$	$-1{,}000{,}000$	-1000	-1	0	1	1000	1,000,000

The points at which a function changes its increasing, decreasing, or constant behavior are helpful in determining the **relative maximum** or **relative minimum** values of the function.

Definition of Relative Minimum and Relative Maximum

A function value $f(a)$ is called a **relative minimum** of f if there exists an interval (x_1, x_2) that contains a such that

$$x_1 < x < x_2 \quad \text{implies} \quad f(a) \le f(x).$$

A function value $f(a)$ is called a **relative maximum** of f if there exists an interval (x_1, x_2) that contains a such that

$$x_1 < x < x_2 \quad \text{implies} \quad f(a) \ge f(x).$$

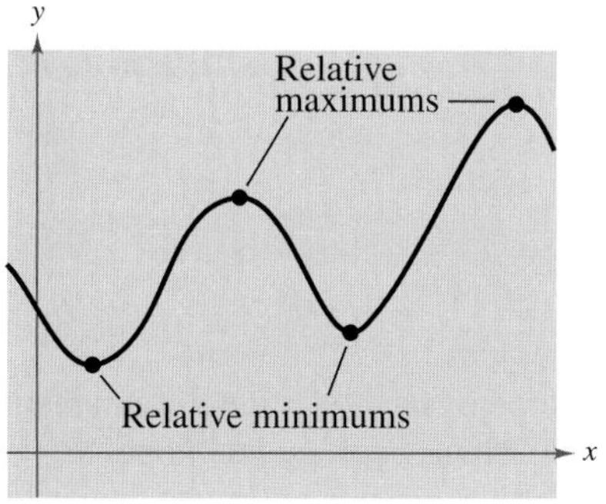

FIGURE 1.37

Figure 1.37 shows several different examples of relative minimums and relative maximums. In Section 2.1, you will study a technique for finding the *exact point* at which a second-degree polynomial function has a relative minimum or relative maximum. For the time being, however, you can use a graphing utility to find reasonable approximations of these points.

Example 5 ▶ Approximating a Relative Minimum

Use a graphing utility to approximate the relative minimum of the function $f(x) = 3x^2 - 4x - 2$.

Solution

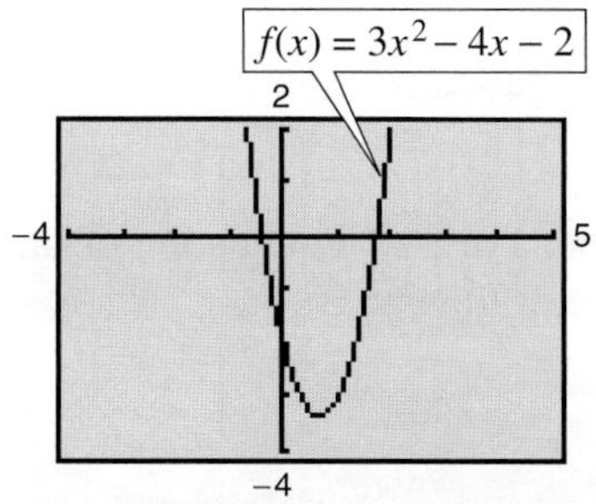

FIGURE 1.38

The graph of f is shown in Figure 1.38. By using the zoom and trace features of a graphing utility, you can estimate that the function has a relative minimum at the point

$(0.67, -3.33)$. Relative minimum

Later, in Section 2.1, you will be able to determine that the exact point at which the relative minimum occurs is $\left(\frac{2}{3}, -\frac{10}{3}\right)$.

Note in Example 5 that you can also use the table feature of a graphing utility to approximate numerically the relative minimum of the function. Using a table that begins at 0.6 and increments the value of x by 0.01, you can approximate the minimum of $f(x) = 3x^2 - 4x - 2$ to be $(0.67, -3.33)$.

Technology

If you use a graphing utility to estimate the x- and y-values of a relative minimum or relative maximum, the automatic zoom feature will often produce graphs that are nearly flat. To overcome this problem, you can manually change the vertical setting of the viewing window. The graph will stretch vertically if the values of Ymin and Ymax are closer together.

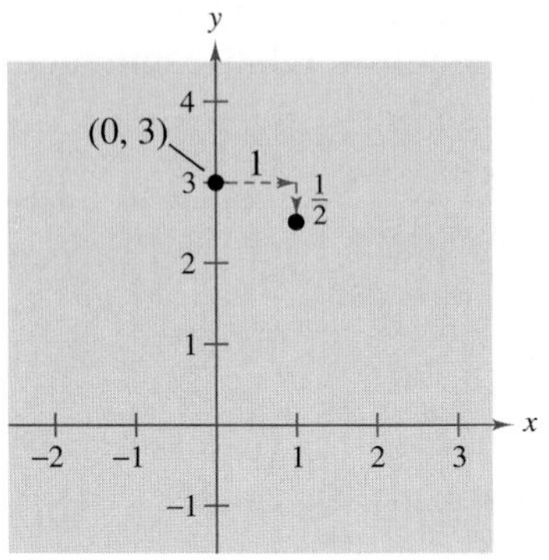

(a)

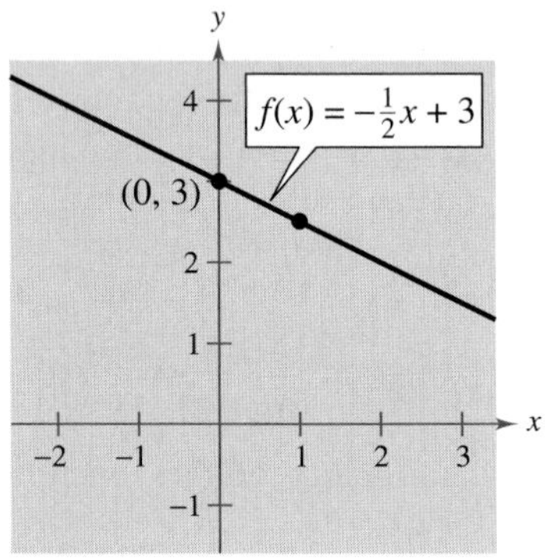

(b)

FIGURE 1.39

Additional Example

Write the linear function f for which $f(0) = -1$ and $f(5) = 8$.

Solution

$$m = \frac{y_2 - y_1}{x_2 - x_1} = \frac{8 - (-1)}{5 - 0} = \frac{9}{5}$$

$$y - y_1 = m(x - x_1)$$

$$y - (-1) = \frac{9}{5}(x - 0)$$

$$y + 1 = \frac{9}{5}x$$

$$y = \frac{9}{5}x - 1$$

Linear Functions

A **linear function** of x is a function of the form

$$f(x) = mx + b. \qquad \text{Linear function}$$

In Section 1.2, you learned that the graph of such a function is a line that has a slope of m and a y-intercept of $(0, b)$.

Example 6 ▶ Graphing a Linear Function

Sketch the graph of the linear function

$$f(x) = -\frac{1}{2}x + 3.$$

Solution

The graph of this function is a line that has a slope of $m = -\frac{1}{2}$ and a y-intercept of $(0, 3)$. To sketch the line, plot the y-intercept. Then, because the slope is $-\frac{1}{2}$, move 1 unit to the right and $\frac{1}{2}$ unit *down* and plot a second point, as shown in Figure 1.39(a). Finally, draw the line that passes through these two points, as shown in Figure 1.39(b).

Example 7 ▶ Writing a Linear Function

Write the linear function f for which $f(1) = 3$ and $f(4) = 0$.

Solution

The graph of a linear function is a line. So, you need to find an equation of the line that passes through $(1, 3)$ and $(4, 0)$. The slope of the line is

$$m = \frac{y_2 - y_1}{x_2 - x_1}$$

$$= \frac{0 - 3}{4 - 1}$$

$$= -1.$$

Using the point-slope form of the equation of a line, you can write

$$y - y_1 = m(x - x_1) \qquad \text{Point-slope form}$$

$$y - 3 = -1(x - 1) \qquad \text{Substitute for } y_1, m, \text{ and } x_1.$$

$$y = -x + 4. \qquad \text{Simplify.}$$

So, the linear function is $f(x) = -x + 4$. You can check this result as follows.

$$f(1) = -(1) + 4 = 3 \ \checkmark$$

$$f(4) = -(4) + 4 = 0 \ \checkmark$$

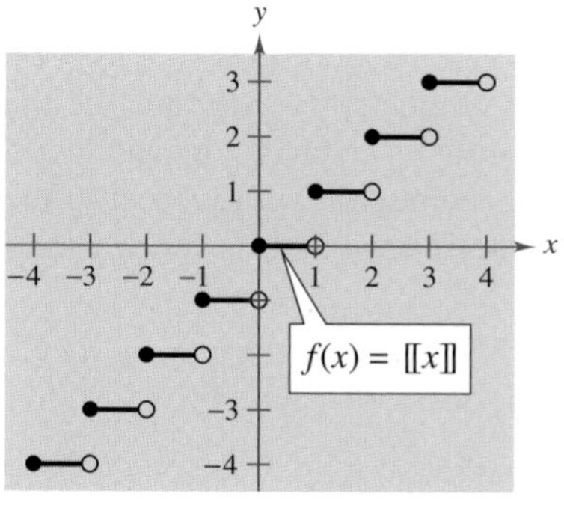

FIGURE 1.40

Step Functions and Piecewise-Defined Functions

The **greatest integer function** is denoted by $[\![x]\!]$ and is defined as follows.

$$[\![x]\!] = \text{the greatest integer less than or equal to } x$$

The graph of this function is shown in Figure 1.40. Note that the graph of the greatest integer function jumps vertically 1 unit at each integer and is constant (a horizontal line segment) between each pair of consecutive integers. The greatest integer function is an example of a category of functions called **step functions** whose graphs resemble a set of stair steps. Some values of the greatest integer function are as follows.

$[\![-1]\!] = -1$ $\quad$ $[\![-0.5]\!] = -1$ $\quad$ $[\![0]\!] = 0$

$[\![0.5]\!] = 0$ $\quad$ $[\![1]\!] = 1$ $\quad$ $[\![1.5]\!] = 1$

The range of the greatest integer function is the set of all integers. (If you use a graphing utility to sketch a step function, you should set the utility to *dot* mode rather than *connected* mode.)

In Section 1.3, you learned that a piecewise-defined function is a function that is defined by two or more equations over a specified domain. To sketch the graph of a piecewise-defined function, you need to sketch the graph of each equation on the appropriate portion of the domain.

Technology

When graphing a step function, you should set your graphing utility to dot mode.

Example 8 ▶ Graphing a Piecewise-Defined Function

Sketch the graph of $f(x) = \begin{cases} 2x + 3, & x \le 1 \\ -x + 4, & x > 1 \end{cases}$.

Solution

This piecewise-defined function is composed of two linear functions. At $x = 1$ and to the left of $x = 1$ the graph is the line $y = 2x + 3$, and to the right of $x = 1$ the graph is the line $y = -x + 4$, as shown in Figure 1.41.

Demonstrate the real-life nature of step functions by discussing Exercises 103 and 104 on page 149. If writing is a part of your course, this section provides a good opportunity for students to find other examples of step functions and write brief essays on their applications.

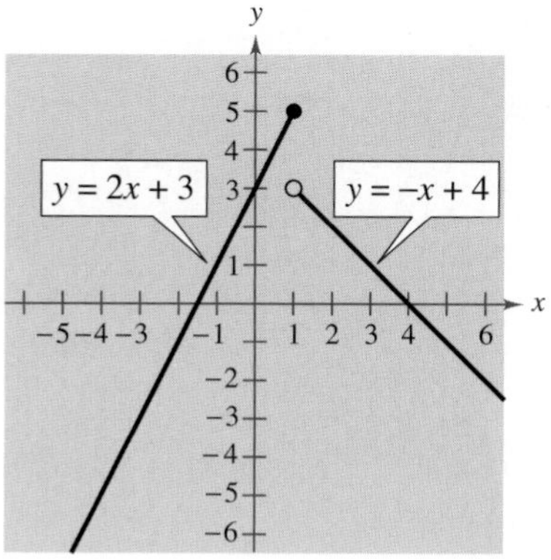

FIGURE 1.41

A computer animation of this concept appears in the *Interactive* CD-ROM and *Internet* versions of this text.

Even and Odd Functions

In Section 1.1, you studied different types of symmetry of a graph. In the terminology of functions, a function is said to be **even** if its graph is symmetric with respect to the y-axis and to be **odd** if its graph is symmetric with respect to the origin. The symmetry tests in Section 1.1 yield the following tests for even and odd functions.

Tests for Even and Odd Functions

A function $y = f(x)$ is **even** if, for each x in the domain of f,

$$f(-x) = f(x).$$

A function $y = f(x)$ is **odd** if, for each x in the domain of f,

$$f(-x) = -f(x).$$

Exploration

Graph each of the following functions with a graphing utility. Determine whether the function is *even, odd,* or *neither.*

$f(x) = x^2 - x^4$

$g(x) = 2x^3 + 1$

$h(x) = x^5 - 2x^3 + x$

$j(x) = 2 - x^6 - x^8$

$k(x) = x^5 - 2x^4 + x - 2$

$p(x) = x^9 + 3x^5 - x^3 + x$

What do you notice about the equations of functions that are odd? What do you notice about the equations of functions that are even? Can you describe a way to identify a function as odd or even by inspecting the equation? Can you describe a way to identify a function as neither odd nor even by inspecting the equation?

Example 9 ▶ Even and Odd Functions

a. The function $g(x) = x^3 - x$ is odd because $g(-x) = -g(x)$, as follows.

$$g(-x) = (-x)^3 - (-x) \qquad \text{Substitute } -x \text{ for } x.$$
$$= -x^3 + x \qquad \text{Simplify.}$$
$$= -(x^3 - x) \qquad \text{Distributive Property}$$
$$= -g(x) \qquad \text{Test for odd function.}$$

b. The function $h(x) = x^2 + 1$ is even because $h(-x) = h(x)$, as follows.

$$h(-x) = (-x)^2 + 1 \qquad \text{Substitute } -x \text{ for } x.$$
$$= x^2 + 1 \qquad \text{Simplify.}$$
$$= h(x) \qquad \text{Test for even function.}$$

The graphs of these two functions are shown in Figure 1.42.

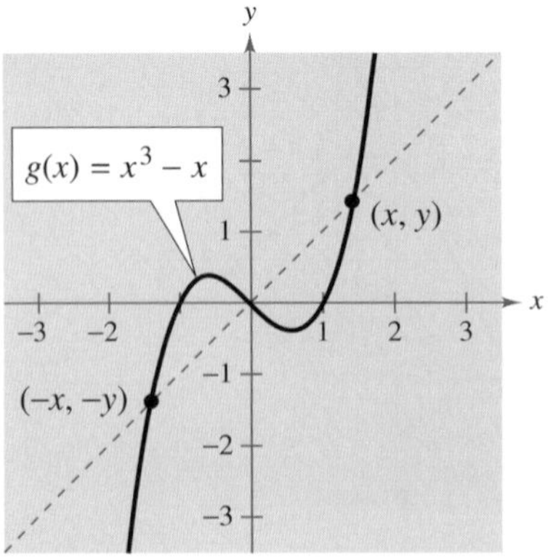

(a) Symmetric to Origin: Odd Function

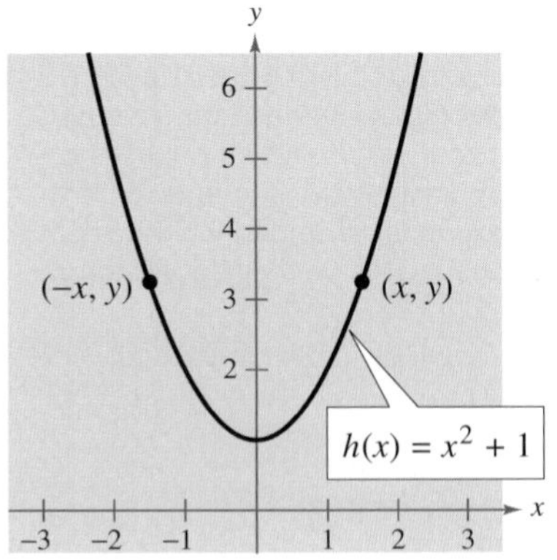

(b) Symmetric to y-Axis: Even Function

FIGURE 1.42

Additional Example

Is the function $f(x) = x^3 - 1$ even, odd, or neither?

Solution

Substituting $-x$ for x,

$f(-x) = (-x)^3 - 1 = -x^3 - 1.$

Because $f(x) = x^3 - 1$ and $f(-x) = -x^3 - 1$, $f(-x) \neq f(x)$ and $f(-x) \neq -f(x)$. So, the function is neither even nor odd.

1.4 Exercises

In Exercises 1–8, find the domain and range of the function.

1. $f(x) = \frac{2}{3}x - 4$

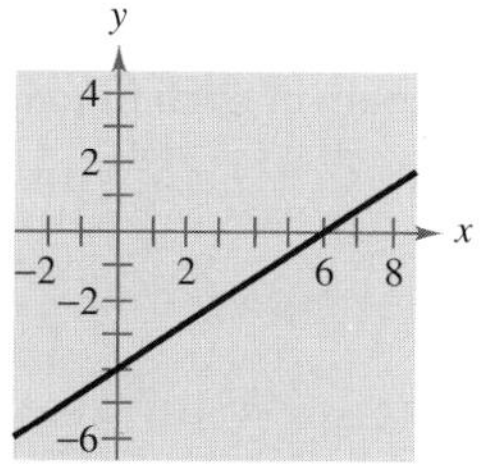

2. $f(x) = x^3 - 3x + 2$

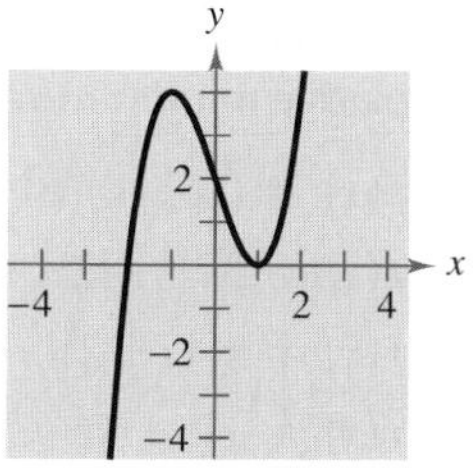

3. $f(x) = 1 - x^2$

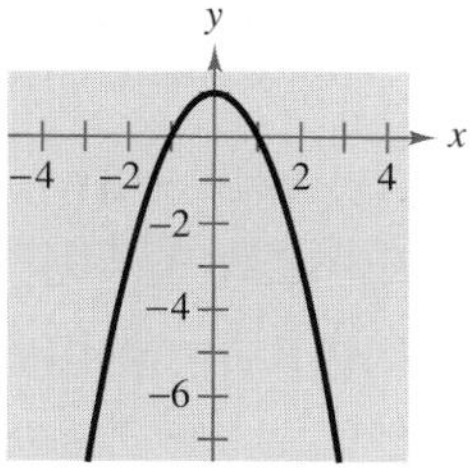

4. $f(x) = \sqrt{x - 1}$

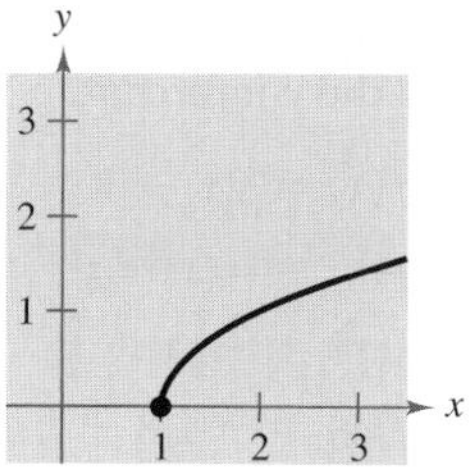

5. $f(x) = \sqrt{x^2 - 1}$

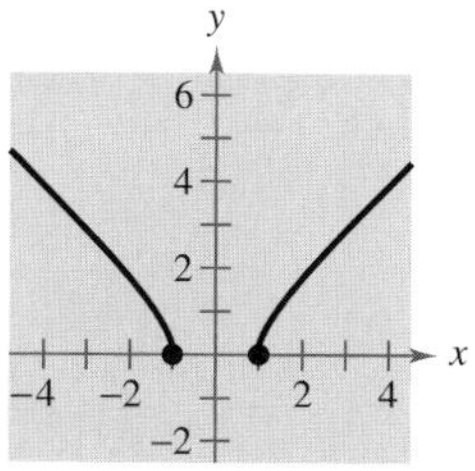

6. $f(x) = \frac{1}{2}|x - 2|$

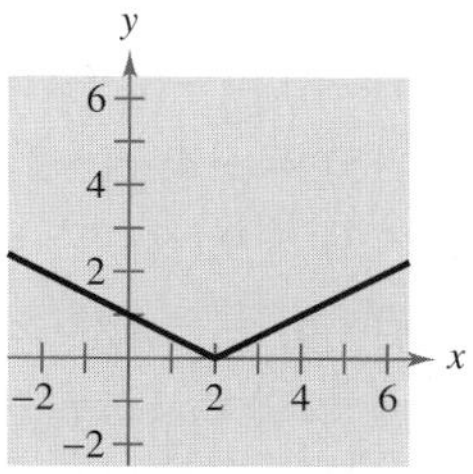

7. $h(x) = \sqrt{16 - x^2}$

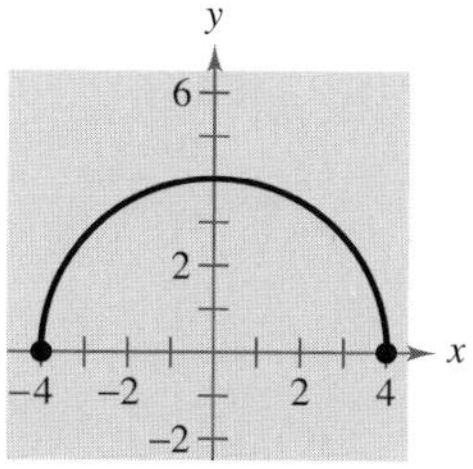

8. $g(x) = \dfrac{|x - 1|}{x - 1}$

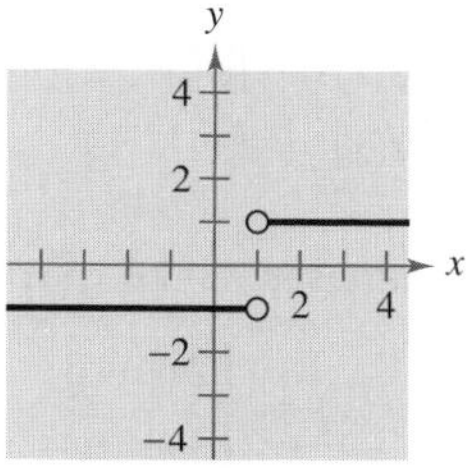

In Exercises 9–14, use the Vertical Line Test to determine whether *y* is a function of *x*.

9. $y = \frac{1}{2}x^2$

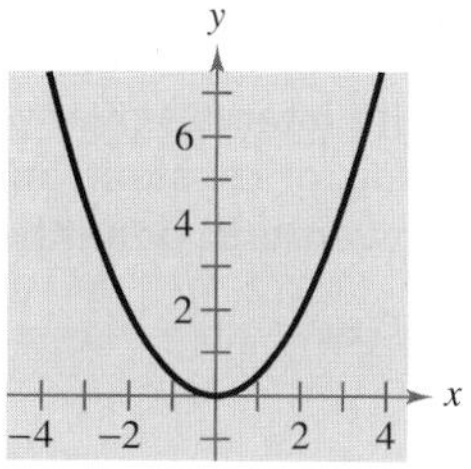

10. $y = \frac{1}{4}x^3$

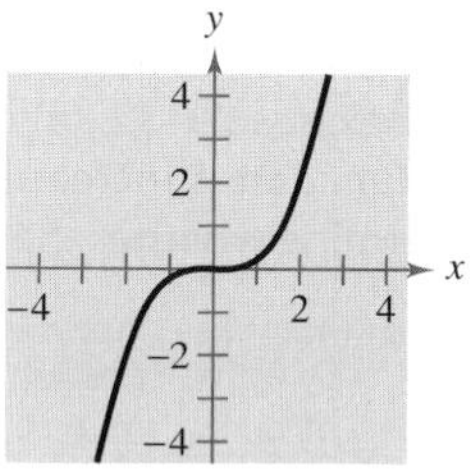

11. $x - y^2 = 1$

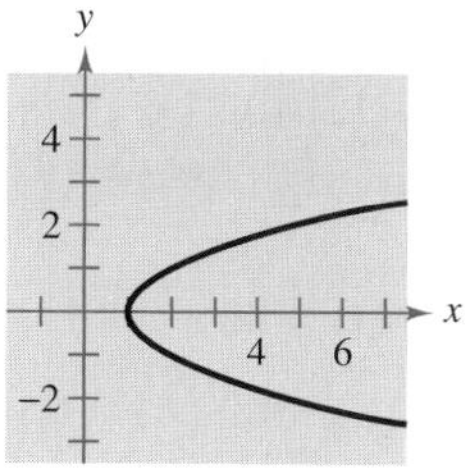

12. $x^2 + y^2 = 25$

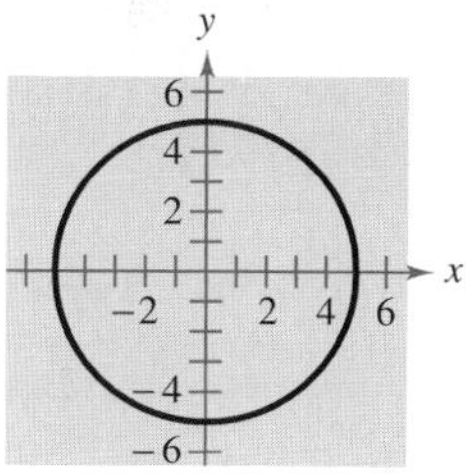

13. $x^2 = 2xy - 1$

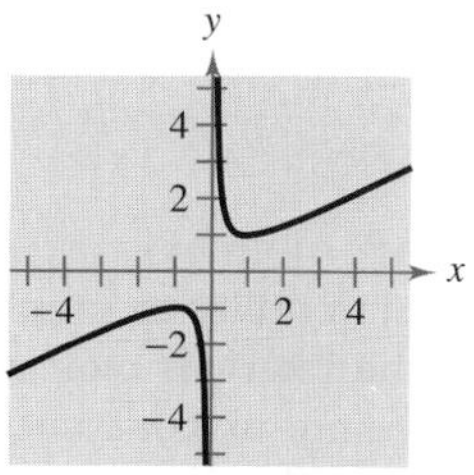

14. $x = |y + 2|$

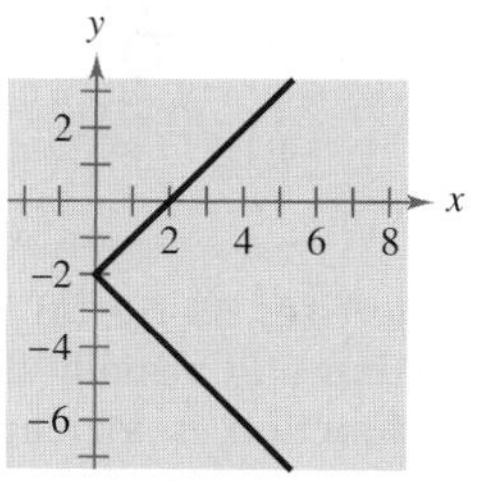

In Exercises 15–22, find the zeros of the function by factoring.

15. $f(x) = 2x^2 - 7x - 30$

16. $f(x) = 3x^2 + 22x - 16$

17. $f(x) = \dfrac{x}{9x^2 - 4}$

18. $f(x) = \dfrac{x^2 - 9x + 14}{4x}$

19. $f(x) = \frac{1}{2}x^3 - x$

20. $f(x) = x^3 - 4x^2 - 9x + 36$

21. $f(x) = 4x^3 - 24x^2 - x + 6$

22. $f(x) = 9x^4 - 25x^2$

In Exercises 23–28, find the zeros of the function algebraically. Verify your results graphically.

23. $f(x) = 3 + \dfrac{5}{x}$

24. $f(x) = x(x - 7)$

25. $f(x) = \sqrt{2x + 11}$

26. $f(x) = \sqrt{3x - 14} - 8$

27. $f(x) = \dfrac{3x - 1}{x - 6}$

28. $f(x) = \dfrac{2x^2 - 9}{3 - x}$

In Exercises 29–32, (a) determine the intervals over which the function is increasing, decreasing, or constant, and (b) determine whether the function is even, odd, or neither.

29. $f(x) = \frac{3}{2}x$

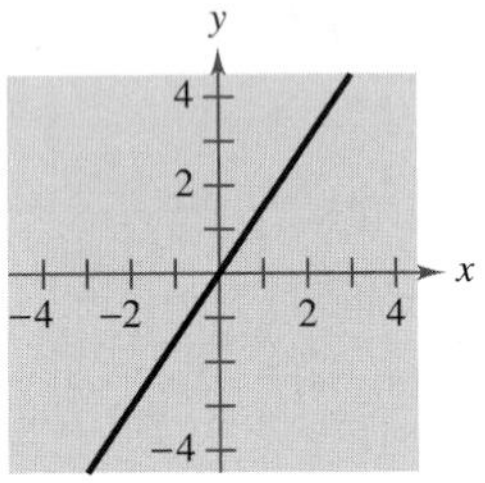

30. $f(x) = x^2 - 4x$

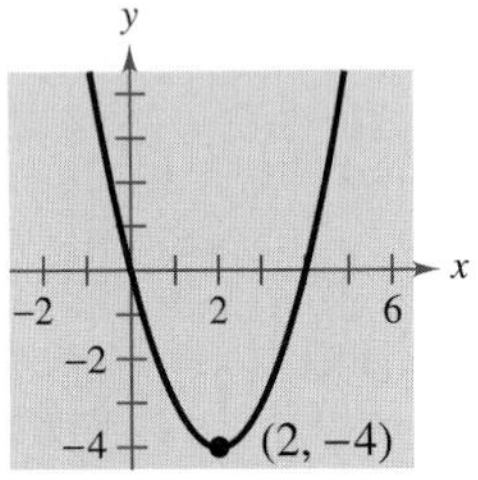

31. $f(x) = x^3 - 3x^2 + 2$

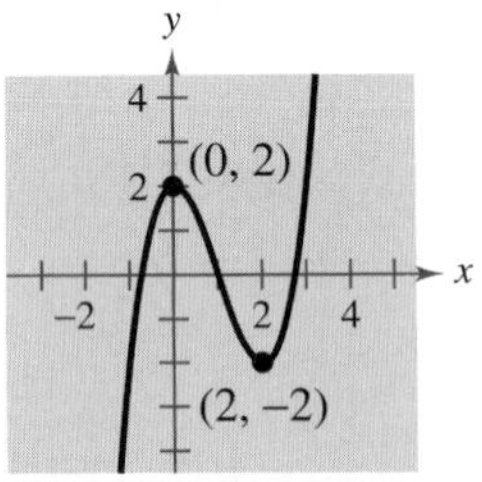

32. $f(x) = \sqrt{x^2 - 1}$

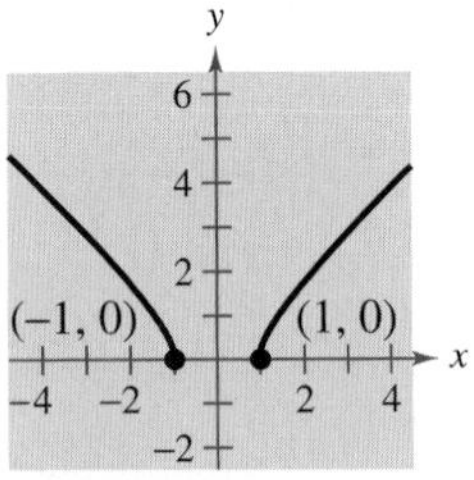

In Exercises 33–50, (a) use a graphing utility to graph the function and visually determine the intervals over which the function is increasing, decreasing, or constant, and (b) make a table of values to verify whether the function is increasing, decreasing, or constant.

33. $f(x) = 3$

34. $g(x) = x$

35. $f(x) = 5 - 3x$

36. $f(x) = -x - 10$

37. $g(s) = \dfrac{s^2}{4}$

38. $h(x) = x^2 - 4$

39. $f(t) = -t^4$

40. $f(x) = 3x^4 - 6x^2$

41. $f(x) = \sqrt{1 - x}$

42. $f(x) = x\sqrt{x + 3}$

43. $f(x) = x^{3/2}$

44. $f(x) = x^{2/3}$

45. $g(t) = \sqrt[3]{t - 1}$

46. $f(x) = \sqrt[4]{x + 5}$

47. $f(x) = |x + 2|$

48. $f(x) = |x + 1| + |x - 1|$

49. $f(x) = \begin{cases} x + 3, & x \le 0 \\ 3, & 0 < x \le 2 \\ 2x - 1, & x > 2 \end{cases}$

50. $f(x) = \begin{cases} 2x + 1, & x \le -1 \\ x^2 - 2, & x > -1 \end{cases}$

In Exercises 51–56, use a graphing utility to approximate the relative minimum/relative maximum of each function.

51. $f(x) = (x - 4)(x + 2)$

52. $f(x) = 3x^2 - 2x - 5$

53. $f(x) = x(x - 2)(x + 3)$

54. $f(x) = x^3 - 3x^2 - x + 1$

55. $f(x) = 2x^3 - 5x^2 - 4x - 1$

56. $f(x) = 8x^4 - 3x - 1$

In Exercises 57–64, sketch the graph of the linear function. Label the *y*-intercept.

57. $f(x) = 2x - 1$

58. $f(x) = 11 - 3x$

59. $f(x) = -x - \frac{3}{4}$

60. $f(x) = 3x - \frac{5}{2}$

61. $f(x) = -\frac{1}{6}x - \frac{5}{2}$

62. $f(x) = \frac{5}{6} - \frac{2}{3}x$

63. $f(x) = -1.8 + 2.5x$

64. $f(x) = 10.2 + 3.1x$

In Exercises 65–72, write the linear function that has the indicated function values.

65. $f(1) = 4, f(0) = 6$

66. $f(-3) = -8, f(1) = 2$

67. $f(5) = -4, f(-2) = 17$

68. $f(3) = 9, f(-1) = -11$

69. $f(-5) = -5, f(5) = -1$

70. $f(-10) = 12, f(16) = -1$

71. $f\left(\frac{1}{2}\right) = -6, f(4) = -3$

72. $f\left(\frac{2}{3}\right) = -\frac{15}{2}, f(-4) = -11$

In Exercises 73 and 74, sketch the graph of the function. Describe how it differs from the graph of $f(x) = [\![x]\!]$.

73. $f(x) = [\![x]\!] - 2$

74. $f(x) = [\![x - 1]\!]$

In Exercises 75–78, graph the function.

75. $f(x) = \begin{cases} 2x + 3, & x < 0 \\ 3 - x, & x \ge 0 \end{cases}$

76. $f(x) = \begin{cases} \sqrt{4 + x}, & x < 0 \\ \sqrt{4 - x}, & x \ge 0 \end{cases}$

77. $f(x) = \begin{cases} x^2 + 5, & x \le 1 \\ -x^2 + 4x + 3, & x > 1 \end{cases}$

78. $f(x) = \begin{cases} 1 - (x - 1)^2, & x \le 2 \\ \sqrt{x - 2}, & x > 2 \end{cases}$

In Exercises 79–90, graph the function and determine the interval(s) for which $f(x) \ge 0$.

79. $f(x) = 4 - x$

80. $f(x) = 4x + 2$

81. $f(x) = x^2 - 9$

82. $f(x) = x^2 - 4x$

83. $f(x) = 1 - x^4$

84. $f(x) = \sqrt{x + 2}$

85. $f(x) = x^2 + 1$

86. $f(x) = -(1 + |x|)$

87. $f(x) = -5$

88. $f(x) = \frac{1}{2}(2 + |x|)$

89. $f(x) = \begin{cases} 1 - 2x^2, & x \le -2 \\ -x + 8, & x > -2 \end{cases}$

90. $f(x) = \begin{cases} \sqrt{x - 5}, & x > 5 \\ x^2 + x - 1, & x \le 5 \end{cases}$

In Exercises 91 and 92, use a graphing utility to graph the function. State the domain and range of the function. Describe the pattern of the graph.

91. $s(x) = 2\left(\frac{1}{4}x - [\![\frac{1}{4}x]\!]\right)$

92. $g(x) = 2\left(\frac{1}{4}x - [\![\frac{1}{4}x]\!]\right)^2$

In Exercises 93–98, determine whether the function is even, odd, or neither.

93. $f(x) = x^6 - 2x^2 + 3$

94. $h(x) = x^3 - 5$

95. $g(x) = x^3 - 5x$

96. $f(x) = x\sqrt{1 - x^2}$

97. $f(t) = t^2 + 2t - 3$

98. $g(s) = 4s^{2/3}$

Think About It **In Exercises 99–102, find the coordinates of a second point on the graph of a function f if the given point is on the graph and the function is (a) even and (b) odd.**

99. $\left(-\frac{3}{2}, 4\right)$

100. $\left(-\frac{5}{3}, -7\right)$

101. $(4, 9)$

102. $(5, -1)$

103. *Communications* The cost of using a telephone calling card is \$1.05 for the first minute and \$0.38 for each additional minute or portion of a minute.

(a) A customer needs a model for the cost C of using a calling card for a call lasting t minutes. Which of the following is the appropriate model? Explain.

$$C_1(t) = 1.05 + 0.38[\![t - 1]\!]$$

$$C_2(t) = 1.05 - 0.38[\![-(t - 1)]\!]$$

(b) Graph the appropriate model. Determine the cost of a call lasting 18 minutes and 45 seconds.

104. *Delivery Charges* Suppose that the cost of sending an overnight package from New York to Atlanta is \$9.80 for a package weighing up to but not including 1 pound and \$2.50 for each additional pound or portion of a pound. Use the greatest integer function to create a model for the cost C of overnight delivery of a package weighing x pounds, $x > 0$. Sketch the graph of the function.

105. *Electronics* The number of lumens (time rate of flow of light) L from a fluorescent lamp can be approximated by the model

$$L = -0.294x^2 + 97.744x - 664.875, \quad 20 \le x \le 90$$

where x is the wattage of the lamp. Use a graphing utility to graph the function and estimate the wattage necessary to obtain 2000 lumens.

In Exercises 106–109, write the height h of the rectangle as a function of x.

106.

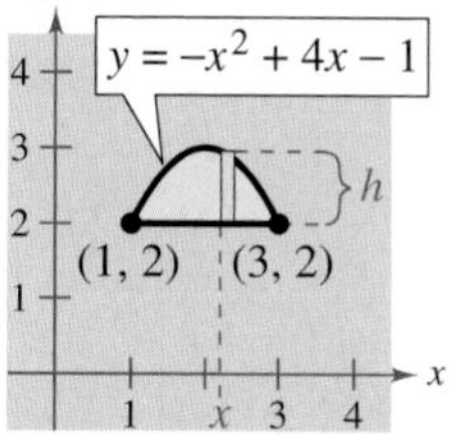

107.

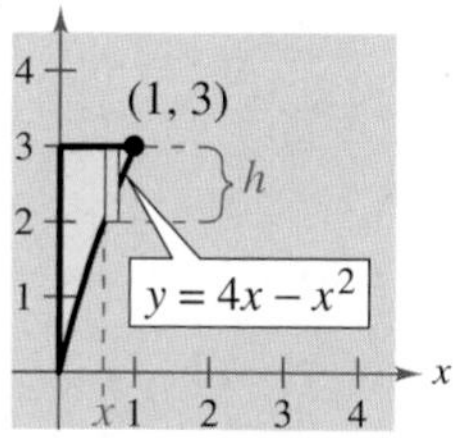

108.

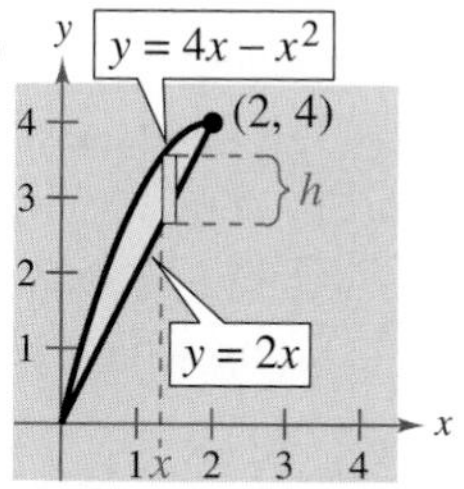

109.

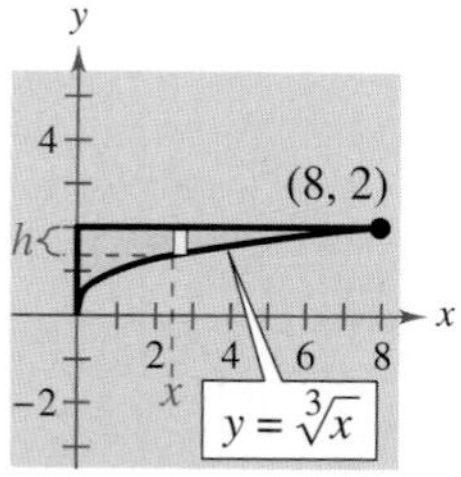

In Exercises 110–113, write the length L of the rectangle as a function of y.

110.

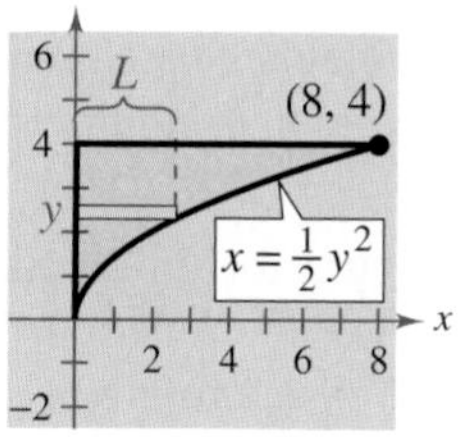

111.

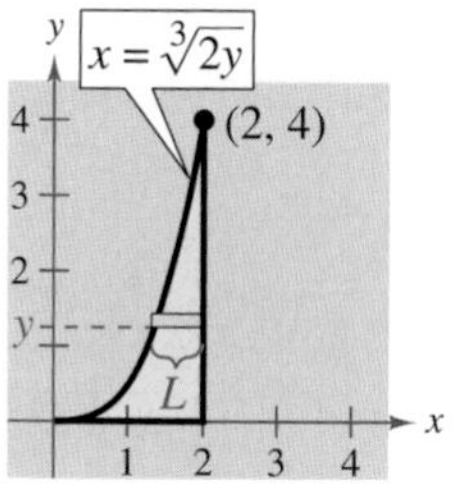

112.

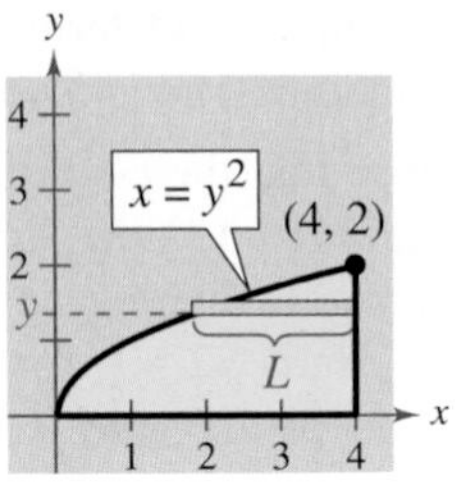

113.

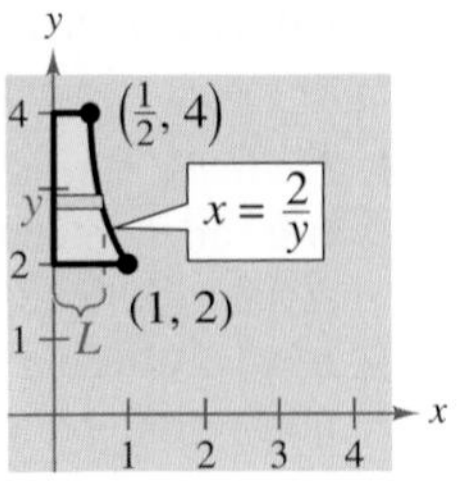

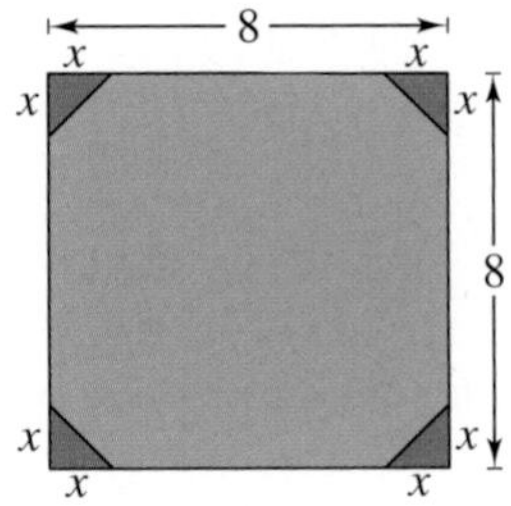

FIGURE FOR 115

114. *Data Analysis* The table shows the amount y (in billions of dollars) of the merchandise trade balance of the United States for the years 1990 through 1997. The merchandise trade balance is the difference between the values of exports and imports. A negative merchandise trade balance indicates that imports exceeded exports. (Source: U.S. International Trade Administration)

Year	1990	1991	1992	1993
y	−101.7	−66.7	−84.5	−115.6

Year	1994	1995	1996	1997
y	−150.6	−158.7	−170.2	−181.5

(a) Use the regression feature of a graphing utility to find a cubic model (a model of the form $y = ax^3 + bx^2 + cx + d$) for the data. Let x be the time (in years), with $x = 0$ corresponding to 1990.

(b) What is the domain of the model?

(c) Use a graphing utility to graph the data and the model in the same viewing window.

(d) For which year does the model most accurately estimate the actual data? During which year is it least accurate?

115. *Geometry* Corners of equal size are cut from a square with sides of length 8 meters (see figure).

(a) Write the area A of the resulting figure as a function of x. Determine the domain of the function.

(b) Use a graphing utility to graph the area function over its domain. Use the graph to find the range of the function.

(c) Identify the resulting figure for the maximum value of x in the domain of the function. What is the length of each side of the figure?

116. *Coordinate Axis Scale* Each function models the specified data for the years 1993 through 2000, with $t = 3$ corresponding to 1993. Estimate a reasonable scale for the vertical axis (e.g., hundreds, thousands, millions, etc.) of the graph and justify your answer. (There are many correct answers.)

(a) $f(t)$ represents the average salary of college professors.

(b) $f(t)$ represents the U.S. population.

(c) $f(t)$ represents the percent of the civilian work force that is unemployed.

117. *Fluid Flow* The intake pipe of a 100-gallon tank has a flow rate of 10 gallons per minute, and two drainpipes have flow rates of 5 gallons per minute each. The figure shows the volume V of fluid in the tank as a function of time t. Determine the combination of the input pipe and drain pipes in which the fluid is flowing in specific subintervals of the 1 hour of time shown on the graph. (There are many correct answers.)

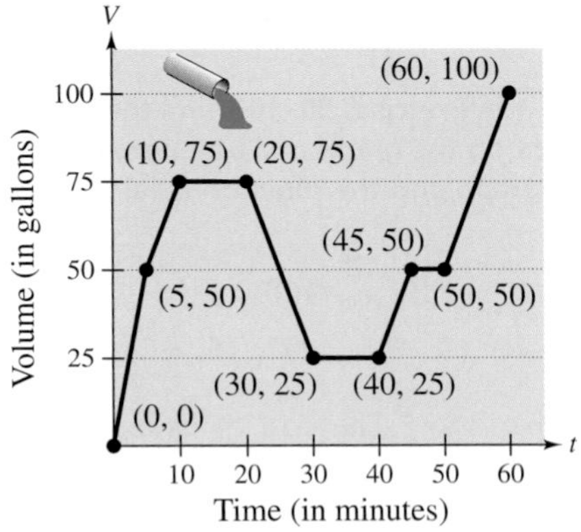

Synthesis

True or False? **In Exercises 118 and 119, determine whether the statement is true or false. Justify your answer.**

118. A function with a square root cannot have a domain that is the set of real numbers.

119. A piecewise-defined function will always have at least one x-intercept or at least one y-intercept.

120. Prove that a function of the following form is odd.

$$y = a_{2n+1}x^{2n+1} + a_{2n-1}x^{2n-1} + \cdots + a_3x^3 + a_1x$$

121. Prove that a function of the following form is even.

$$y = a_{2n}x^{2n} + a_{2n-2}x^{2n-2} + \cdots + a_2x^2 + a_0$$

122. If f is an even function, determine whether g is even, odd, or neither. Explain.

(a) $g(x) = -f(x)$ (b) $g(x) = f(-x)$

(c) $g(x) = f(x) - 2$ (d) $g(x) = f(x - 2)$

123. *Think About It* Does the graph in Exercise 11 represent x as a function of y? Explain.

124. *Think About It* Does the graph in Exercise 12 represent x as a function of y? Explain.

125. *Writing* Use a graphing utility to graph each function. Write a paragraph describing any similarities and differences you observe among the graphs.

(a) $y = x$ (b) $y = x^2$

(c) $y = x^3$ (d) $y = x^4$

(e) $y = x^5$ (f) $y = x^6$

126. *Conjecture* Use the results of Exercise 125 to make a conjecture about the graphs of $y = x^7$ and $y = x^8$. Use a graphing utility to graph the functions and compare the results with your conjecture.

Review

In Exercises 127–130, solve the equation.

127. $x^2 - 10x = 0$

128. $100 - (x - 5)^2 = 0$

129. $x^3 - x = 0$

130. $16x^2 - 40x + 25 = 0$

In Exercises 131–134, evaluate the function at each specified value of the independent variable and simplify.

131. $f(x) = 5x - 8$

(a) $f(9)$ (b) $f(-4)$ (c) $f(x - 7)$

132. $f(x) = x^2 - 10x$

(a) $f(4)$ (b) $f(-8)$ (c) $f(x - 4)$

133. $f(x) = \sqrt{x - 12} - 9$

(a) $f(12)$ (b) $f(40)$ (c) $f(-\sqrt{36})$

134. $f(x) = x^4 - x - 5$

(a) $f(-1)$ (b) $f(\frac{1}{2})$ (c) $f(2\sqrt{3})$

In Exercises 135 and 136, find the difference quotient and simplify your answer.

135. $f(x) = x^2 - 2x + 9, \quad \dfrac{f(3 + h) - f(3)}{h}, \quad h \neq 0$

136. $f(x) = 5 + 6x - x^2, \quad \dfrac{f(6 + h) - f(6)}{h}, \quad h \neq 0$

1.5 Shifting, Reflecting, and Stretching Graphs

▶ What you should learn

- How to recognize graphs of common functions
- How to use vertical and horizontal shifts to sketch graphs of functions
- How to use reflections to sketch graphs of functions
- How to use nonrigid transformations to sketch graphs of functions

▶ Why you should learn it

Knowing the graphs of common functions and knowing how to shift, reflect, and stretch graphs of functions can help you sketch a wide variety of simple functions by hand. This skill is useful in sketching graphs of functions that model real-life data, such as in Exercise 59 on page 161, where you are asked to sketch a function that models the amount of fuel used by trucks from 1980 through 1996.

Chuck Keeler/The Stock Market

Summary of Graphs of Common Functions

One of the goals of this text is to enable you to recognize the basic shapes of the graphs of different types of functions. For instance, from your study of lines in Section 1.2, you can determine the basic shape of the graph of the linear function $f(x) = mx + b$. Specifically, you know that the graph of this function is a line whose slope is m and whose y-intercept is b.

The six graphs shown in Figure 1.43 represent the most commonly used functions in algebra. Familiarity with the basic characteristics of these simple graphs will help you analyze the shapes of more complicated graphs.

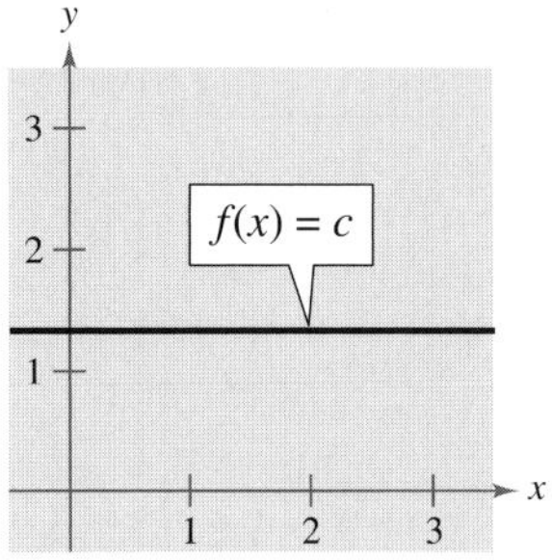

(a) Constant Function

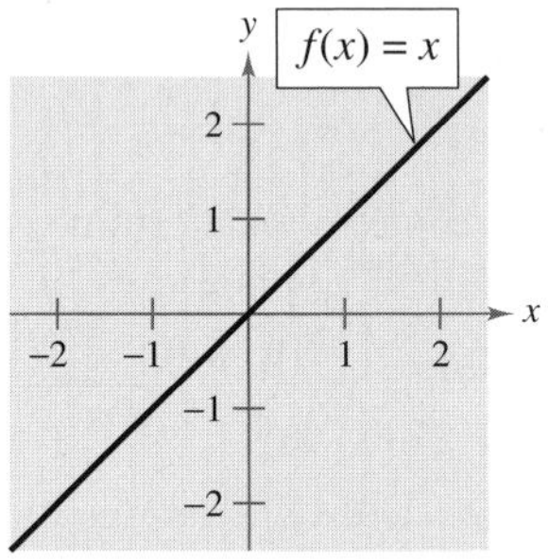

(b) Identity Function

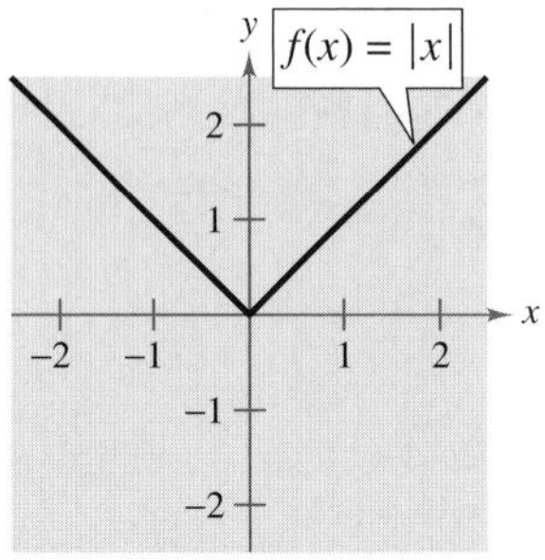

(c) Absolute Value Function

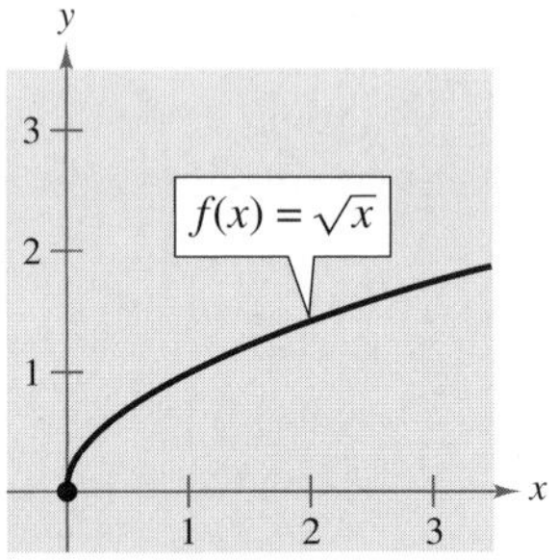

(d) Square Root Function

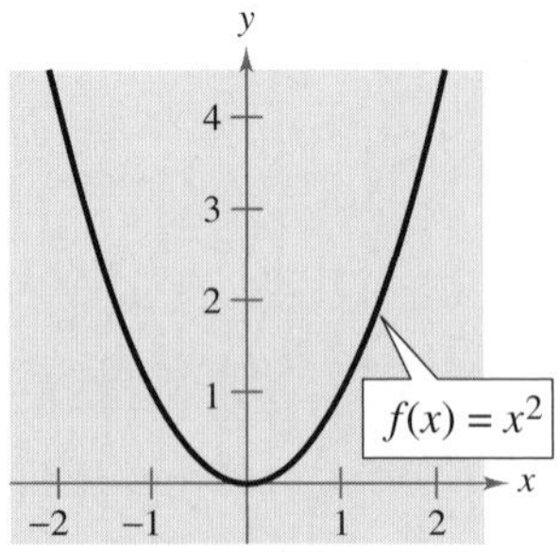

(e) Quadratic Function

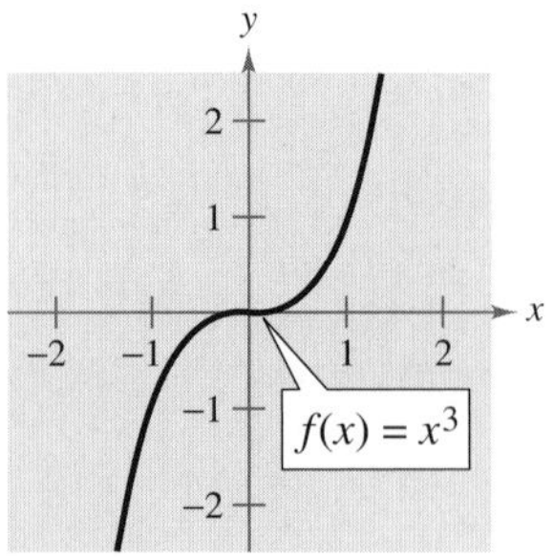

(f) Cubic Function

FIGURE 1.43

A computer animation of this concept appears in the *Interactive* CD-ROM and *Internet* versions of this text.

Shifting Graphs

Many functions have graphs that are simple transformations of the common graphs summarized on page 152. For example, you can obtain the graph of

$$h(x) = x^2 + 2$$

by shifting the graph of $f(x) = x^2$ *up* 2 units, as shown in Figure 1.44. In function notation, h and f are related as follows.

$$h(x) = x^2 + 2$$

$$= f(x) + 2 \qquad \text{Upward shift of 2}$$

Similarly, you can obtain the graph of

$$g(x) = (x - 2)^2$$

by shifting the graph of $f(x) = x^2$ to the *right* 2 units, as shown in Figure 1.45. In this case, the functions g and f have the following relationship.

$$g(x) = (x - 2)^2$$

$$= f(x - 2) \qquad \text{Right shift of 2}$$

You might also wish to illustrate simple transformations of functions numerically using tables to emphasize what happens to individual ordered pairs. For instance, if you have $f(x) = x^2$, $h(x) = x^2 + 2 = f(x) + 2$, and $g(x) = (x - 2)^2 = f(x - 2)$, you can illustrate these transformations with the following tables.

x	$f(x)$	$h(x) = f(x) + 2$
-2	4	$4 + 2 = 6$
-1	1	$1 + 2 = 3$
0	0	$0 + 2 = 2$
1	1	$1 + 2 = 3$
2	4	$4 + 2 = 6$

x	$x - 2$	$g(x) = f(x - 2)$
0	$0 - 2 = -2$	4
1	$1 - 2 = -1$	1
2	$2 - 2 = 0$	0
3	$3 - 2 = 1$	1
4	$4 - 2 = 2$	4

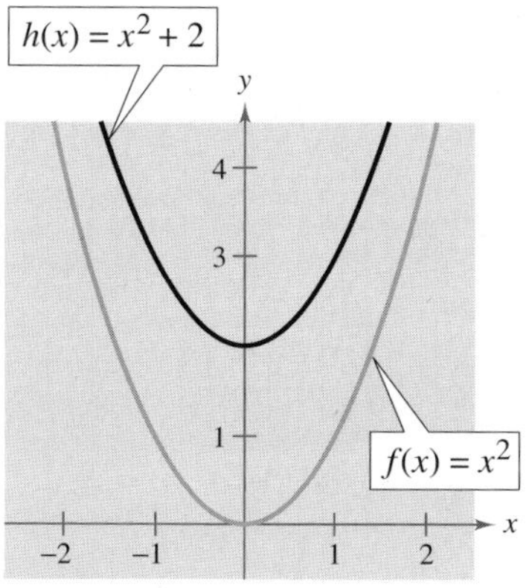

FIGURE 1.44

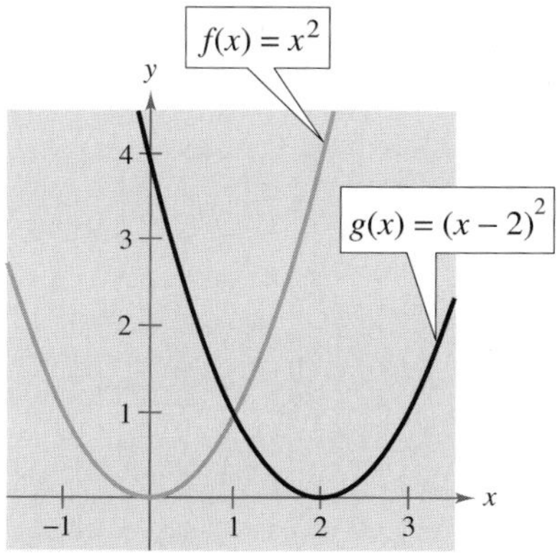

FIGURE 1.45

The following list summarizes this discussion about horizontal and vertical shifts.

Vertical and Horizontal Shifts

Let c be a positive real number. **Vertical and horizontal shifts** in the graph of $y = f(x)$ are represented as follows.

1. Vertical shift c units *upward:* $\quad h(x) = f(x) + c$
2. Vertical shift c units *downward:* $\quad h(x) = f(x) - c$
3. Horizontal shift c units to the *right:* $\quad h(x) = f(x - c)$
4. Horizontal shift c units to the *left:* $\quad h(x) = f(x + c)$

In items 3 and 4, be sure you see that $h(x) = f(x - c)$ corresponds to a *right* shift and $h(x) = f(x + c)$ corresponds to a *left* shift for $c > 0$.

Some graphs can be obtained from a combination of vertical and horizontal shifts, as demonstrated in Example 1(b). Vertical and horizontal shifts generate a *family of functions*, each with the same shape but at different locations in the plane.

Example 1 ▶ Shifts in the Graph of a Function

Use the graph of $f(x) = x^3$ to sketch the graph of each function.

a. $g(x) = x^3 - 1$

b. $h(x) = (x + 2)^3 + 1$

Solution

Relative to the graph of $f(x) = x^3$, the graph of $g(x) = x^3 - 1$ is a downward shift of 1 unit. The graph of $h(x) = (x + 2)^3 + 1$ involves a left shift of 2 units *and* an upward shift of 1 unit. The graphs of both functions are compared with the graph of $f(x) = x^3$ in Figure 1.46.

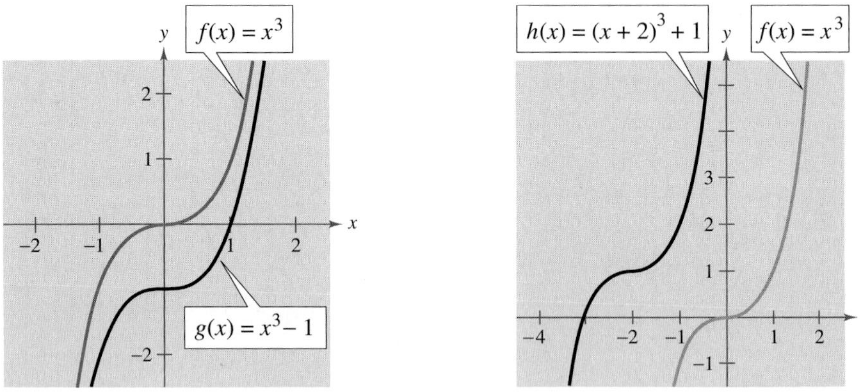

(a) (b)

FIGURE 1.46

In part (b) of Figure 1.46, notice that the same result is obtained if the vertical shift precedes the horizontal shift *or* if the horizontal shift precedes the vertical shift.

Exploration

Graphing utilities are ideal tools for exploring translations of functions. Graph *f*, *g*, and *h* on the same screen. Before looking at the graphs, try to predict how the graphs of *g* and *h* relate to the graph of *f*.

a. $f(x) = x^2$, $g(x) = (x - 4)^2$, $h(x) = (x - 4)^2 + 3$

b. $f(x) = x^2$, $g(x) = (x + 1)^2$, $h(x) = (x + 1)^2 - 2$

c. $f(x) = x^2$, $g(x) = (x + 4)^2$, $h(x) = (x + 4)^2 + 2$

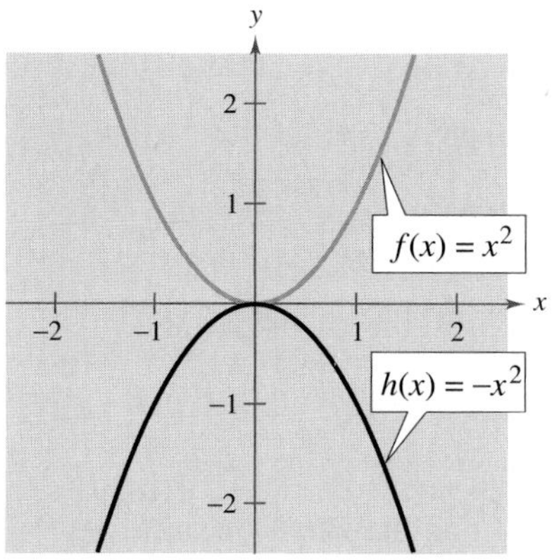

FIGURE 1.47

A computer animation of this concept appears in the *Interactive* CD-ROM and *Internet* versions of this text.

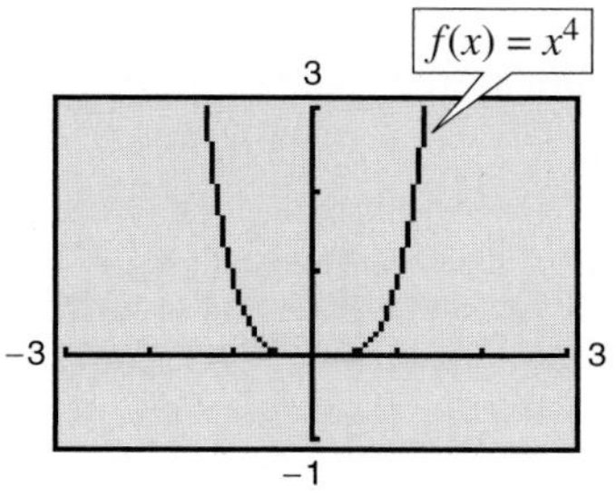

FIGURE 1.48

Reflecting Graphs

The second common type of transformation is a **reflection.** For instance, if you consider the x-axis to be a mirror, the graph of

$$h(x) = -x^2$$

is the mirror image (or reflection) of the graph of $f(x) = x^2$, as shown in Figure 1.47.

Reflections in the Coordinate Axes

Reflections in the coordinate axes of the graph of $y = f(x)$ are represented as follows.

1. Reflection in the x-axis: $h(x) = -f(x)$

2. Reflection in the y-axis: $h(x) = f(-x)$

Example 2 ▶ Finding Equations from Graphs

The graph of the function

$$f(x) = x^4$$

is shown in Figure 1.48. Each of the graphs in Figure 1.49 is a transformation of the graph of f. Find an equation for each of these functions.

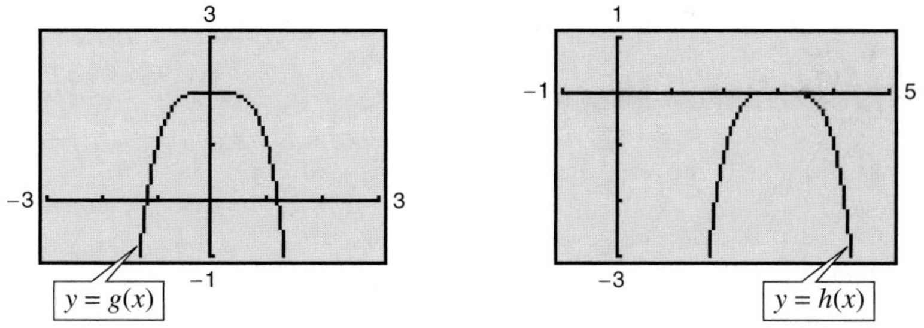

(a) **(b)**

FIGURE 1.49

Solution

a. The graph of g is a reflection in the x-axis *followed by* an upward shift of 2 units of the graph of $f(x) = x^4$. So, the equation for g is

$$g(x) = -x^4 + 2.$$

b. The graph of h is a horizontal shift of 3 units to the right *followed by* a reflection in the x-axis of the graph of $f(x) = x^4$. So, the equation for h is

$$h(x) = -(x - 3)^4.$$

Activities

1. How are the graphs of $f(x)$ and $g = -f(x)$ related?

   Answer: They are reflections in the x-axis of each other.

2. Compare the graph of $f(x) = |x|$ to the graph of $g(x) = |x - 9|$.

   Answer: $g(x)$ is $f(x)$ shifted to the right 9 units.

3. Does the graph of $f(x) = -(x + 1)^3 + 4$ represent a horizontal shift of 1 unit to the left, followed by a vertical shift of 4 units up, followed by a reflection in the x-axis?

   Answer: No, it represents a horizontal shift of 1 unit to the left, followed by a reflection in the axis, followed by a vertical shift of 4 units up.

Example 3 ▶ Reflections and Shifts

Compare the graph of each function with the graph of $f(x) = \sqrt{x}$.

a. $g(x) = -\sqrt{x}$ **b.** $h(x) = \sqrt{-x}$ **c.** $k(x) = -\sqrt{x+2}$

Solution

a. The graph of g is a reflection of the graph of f in the x-axis because

$$\begin{aligned} g(x) &= -\sqrt{x} \\ &= -f(x). \end{aligned}$$

b. The graph of h is a reflection of the graph of f in the y-axis because

$$\begin{aligned} h(x) &= \sqrt{-x} \\ &= f(-x). \end{aligned}$$

c. The graph of k is a left shift of 2 units, followed by a reflection in the x-axis because

$$\begin{aligned} k(x) &= -\sqrt{x+2} \\ &= -f(x+2). \end{aligned}$$

The graphs of all three functions are shown in Figure 1.50.

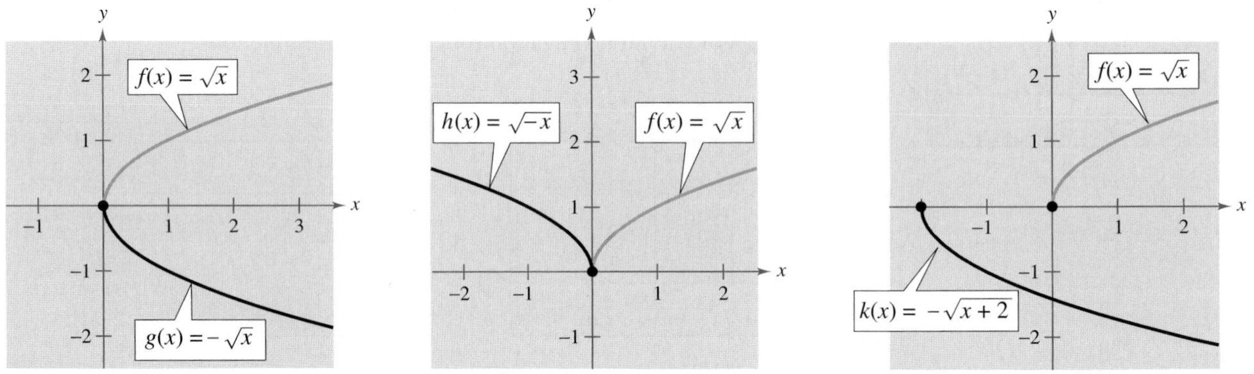

(a) (b) (c)

FIGURE 1.50

When sketching the graphs of functions involving square roots, remember that the domain must be restricted to exclude negative numbers inside the radical. For instance, here are the domains of the functions in Example 3.

Domain of $g(x) = -\sqrt{x}$:	$x \geq 0$
Domain of $h(x) = \sqrt{-x}$:	$x \leq 0$
Domain of $k(x) = -\sqrt{x+2}$:	$x \geq -2$

A computer animation of this concept appears in the *Interactive* CD-ROM and *Internet* versions of this text.

Nonrigid Transformations

Horizontal shifts, vertical shifts, and reflections are **rigid transformations** because the basic shape of the graph is unchanged. These transformations change only the *position* of the graph in the xy-plane. **Nonrigid transformations** are those that cause a *distortion*—a change in the shape of the original graph. For instance, a nonrigid transformation of the graph of $y = f(x)$ is represented by $g(x) = cf(x)$, where the transformation is a **vertical stretch** if $c > 1$ and a **vertical shrink** if $0 < c < 1$.

Exploration

Sketch the graphs of $f(x) = 2x^2$ and $h(x) = x^2$ on the same set of axes. Describe the effect of multiplying x^2 by a number greater than 1. Then sketch the graphs of $g(x) = \frac{1}{2}x^2$ and $h(x) = x^2$ on the same set of axes. Describe the effect of multiplying x^2 by a number less than 1. Can you think of an easy way to remember this generalization?

Example 4 ▶ Nonrigid Transformations

Compare the graph of each function with the graph of $f(x) = |x|$.

a. $h(x) = 3|x|$

b. $g(x) = \dfrac{1}{3}|x|$

Solution

a. Relative to the graph of $f(x) = |x|$, the graph of

$$h(x) = 3|x| = 3f(x)$$

is a vertical stretch (each y-value is multiplied by 3) of the graph of f.

b. Similarly, the function

$$g(x) = \frac{1}{3}|x| = \frac{1}{3}f(x)$$

indicates that the graph of g is a vertical shrink $\left(\text{each } y\text{-value is multiplied by } \frac{1}{3}\right)$ of the graph of f.

The graphs of both functions are shown in Figure 1.51.

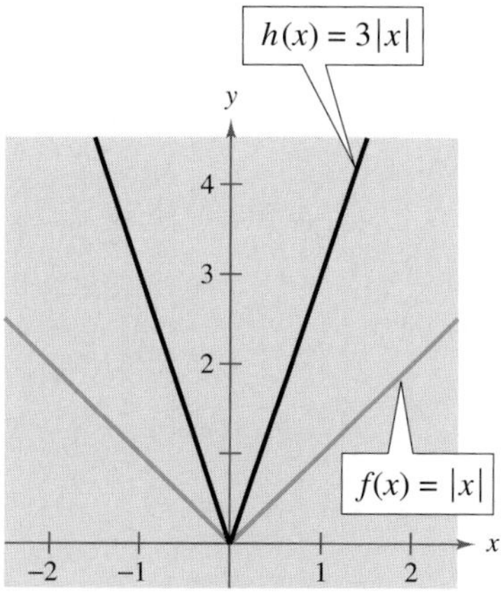

(a)

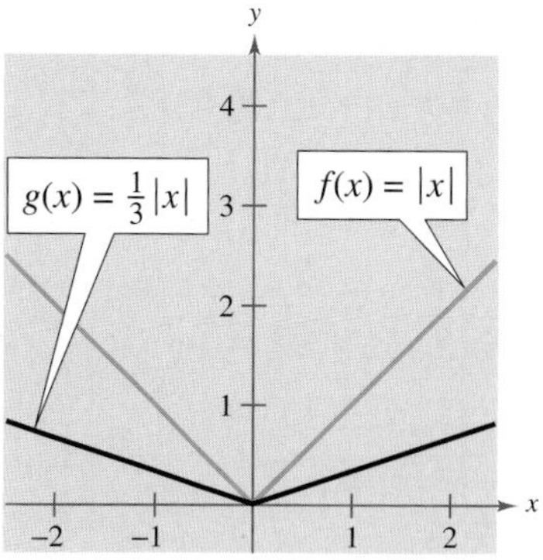

(b)

FIGURE 1.51

1.5 Exercises

1. Sketch (on the same set of coordinate axes) a graph of each function for $c = -2, 0,$ and 2.

(a) $f(x) = x^3 + c$

(b) $f(x) = (x - c)^3$

2. Sketch (on the same set of coordinate axes) a graph of each function for $c = -1, 1,$ and 3.

(a) $f(x) = |x| + c$

(b) $f(x) = |x - c|$

(c) $f(x) = |x + 4| + c$

3. Sketch (on the same set of coordinate axes) a graph of each function for $c = -3, -1, 1,$ and 3.

(a) $f(x) = \sqrt{x} + c$

(b) $f(x) = \sqrt{x - c}$

(c) $f(x) = \sqrt{x - 3} + c$

4. Sketch (on the same set of coordinate axes) a graph of each function for $c = -3, -1, 1,$ and 3.

(a) $f(x) = \begin{cases} x^2 + c, & x < 0 \\ -x^2 + c, & x \geq 0 \end{cases}$

(b) $f(x) = \begin{cases} (x + c)^2, & x < 0 \\ -(x + c)^2, & x \geq 0 \end{cases}$

5. Use the graph of f to sketch each graph.

(a) $y = f(x) + 2$

(b) $y = f(x - 2)$

(c) $y = 2f(x)$

(d) $y = -f(x)$

(e) $y = f(x + 3)$

(f) $y = f(-x)$

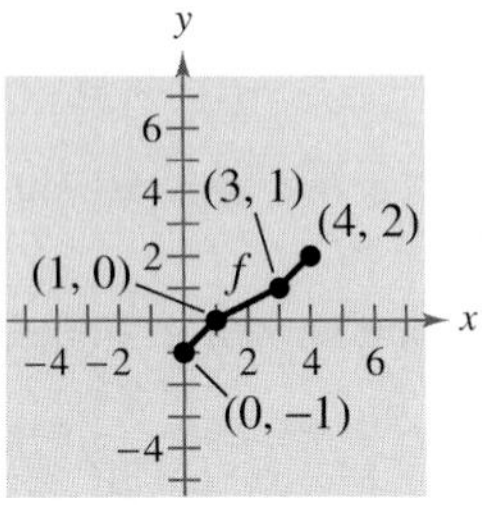

6. Use the graph of f to sketch each graph.

(a) $y = f(-x)$

(b) $y = f(x) + 4$

(c) $y = 2f(x)$

(d) $y = -f(x - 4)$

(e) $y = f(x) - 3$

(f) $y = -f(x) - 1$

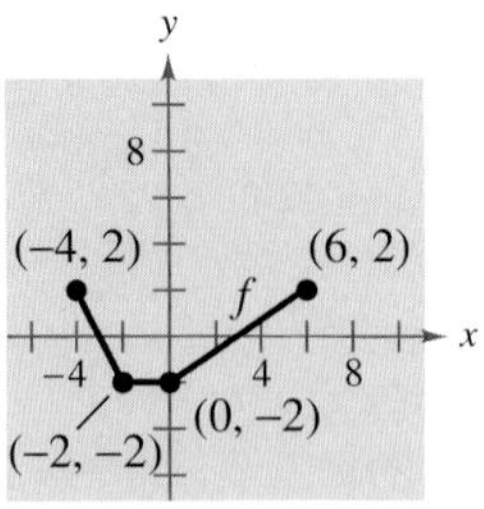

7. Use the graph of f to sketch each graph.

(a) $y = f(x) - 1$

(b) $y = f(x - 1)$

(c) $y = f(-x)$

(d) $y = f(x + 1)$

(e) $y = -f(x - 2)$

(f) $y = \frac{1}{2}f(x)$

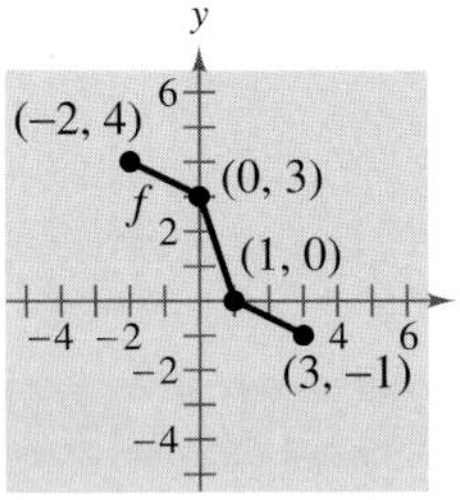

8. Use the graph of f to sketch each graph.

(a) $y = f(x - 5)$

(b) $y = -f(x) + 3$

(c) $y = \frac{1}{3}f(x)$

(d) $y = -f(x + 1)$

(e) $y = f(-x)$

(f) $y = f(x) - 10$

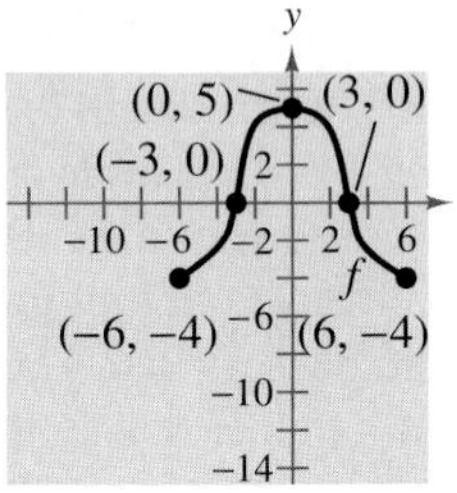

9. Use the graph of $f(x) = x^2$ to write an equation for each function whose graph is shown.

(a)

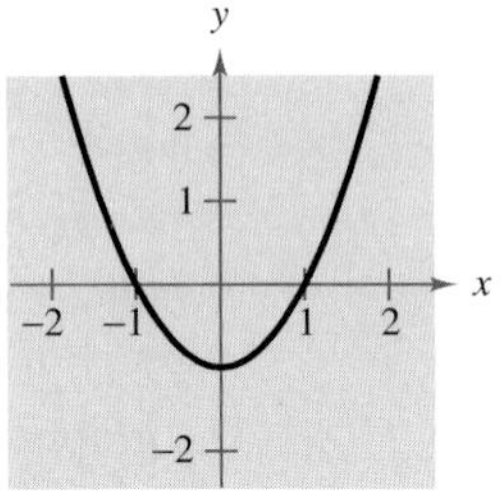

(b)

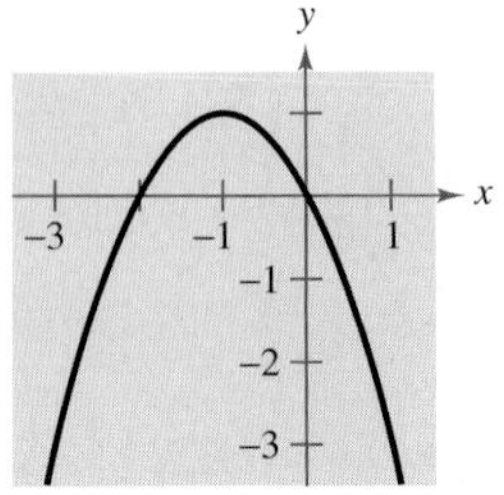

(c)

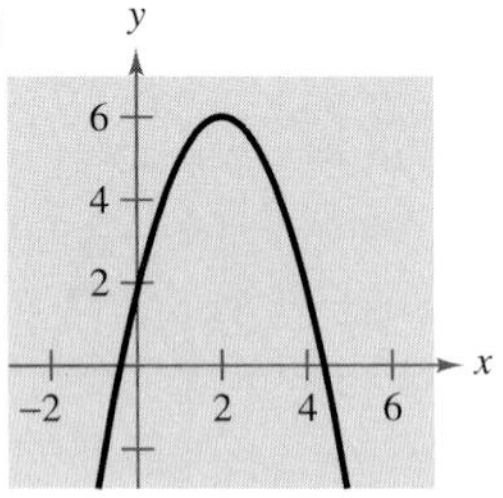

(d)

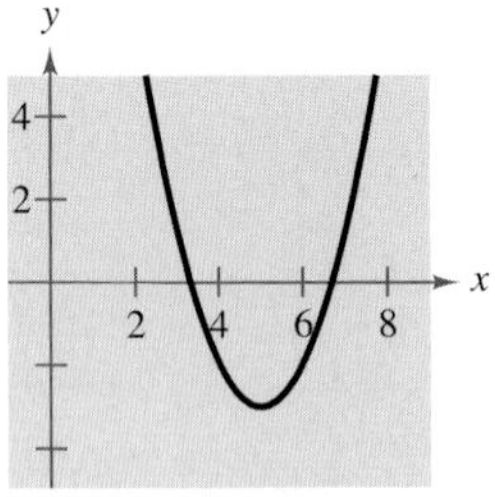

10. Use the graph of $f(x) = x^3$ to write an equation for each function whose graph is shown.

(a)
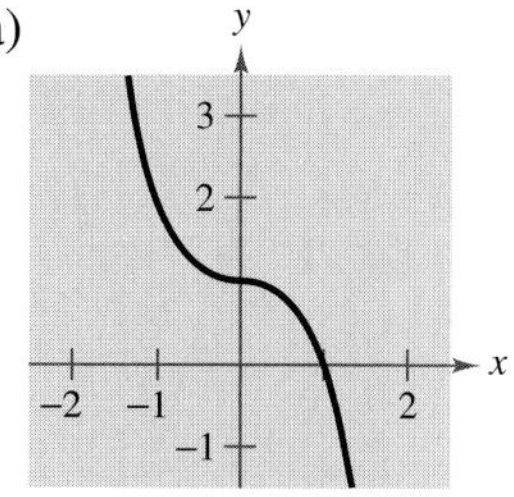

(b)
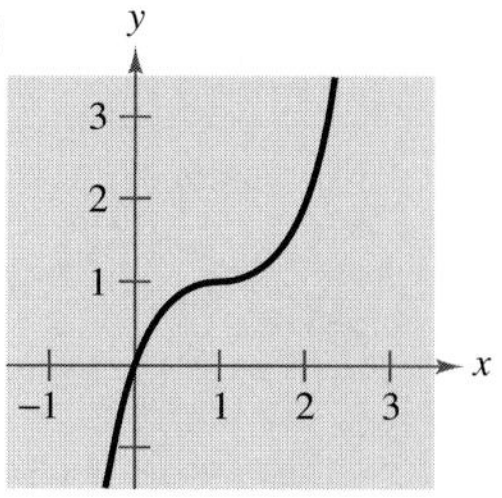

(c)
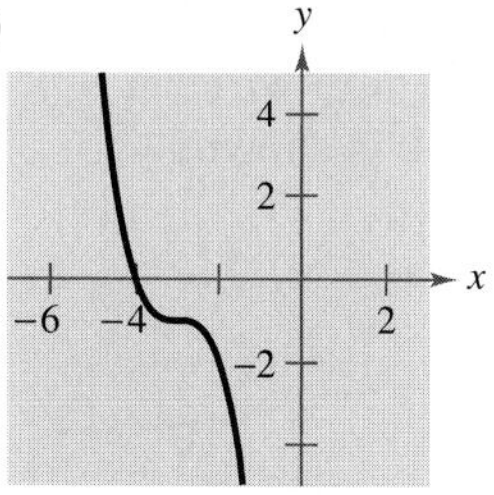

(d)
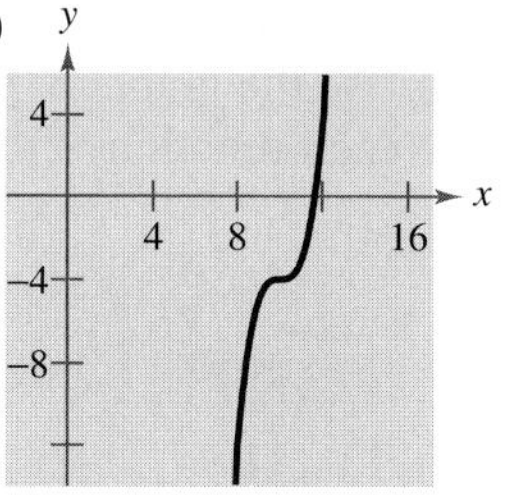

11. Use the graph of $f(x) = |x|$ to write an equation for each function whose graph is shown.

(a)
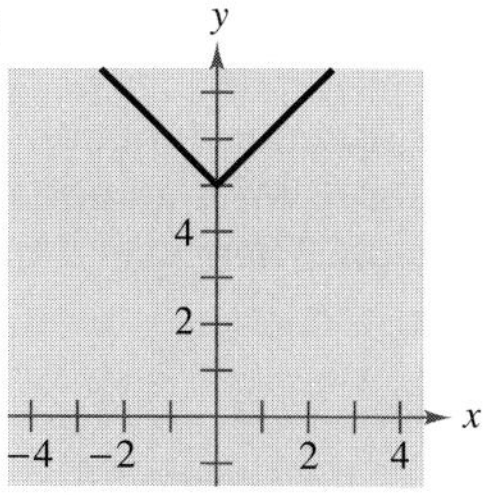

(b)
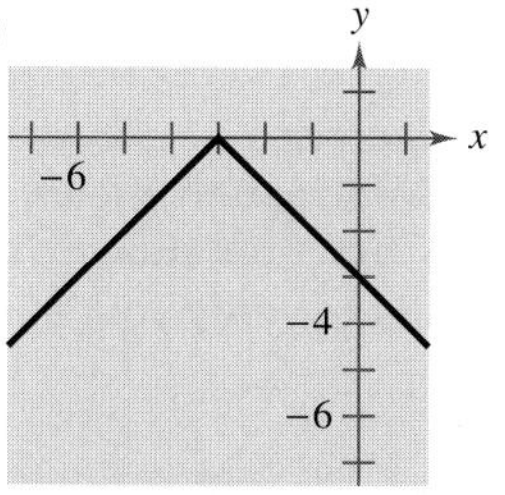

(c)
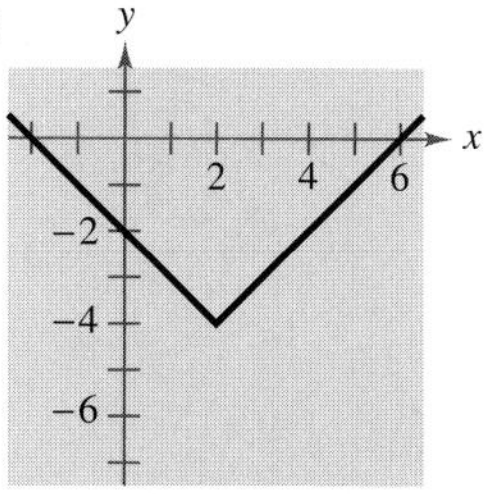

(d)
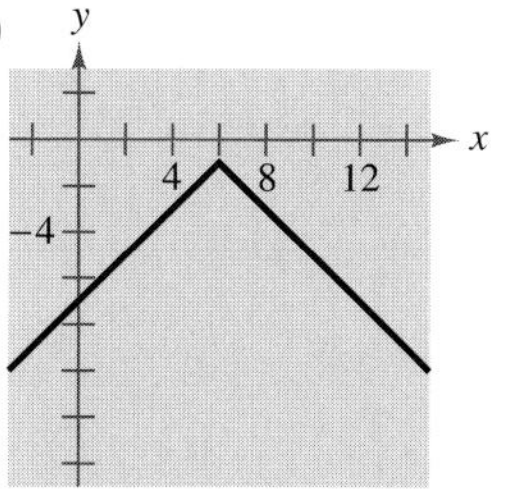

12. Use the graph of $f(x) = \sqrt{x}$ to write an equation for each function whose graph is shown.

(a)
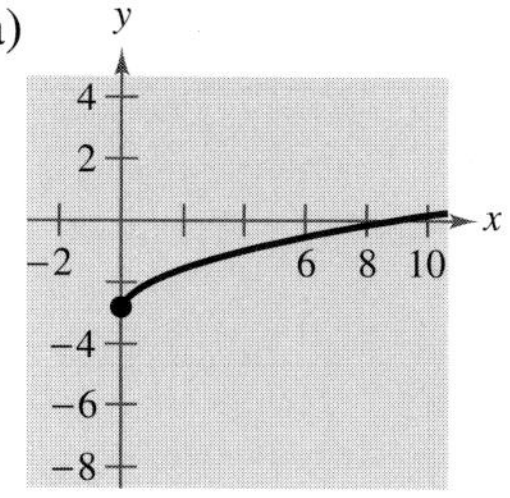

(b)
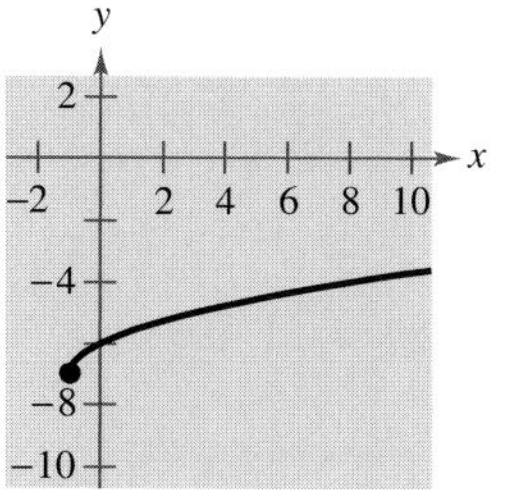

(c)
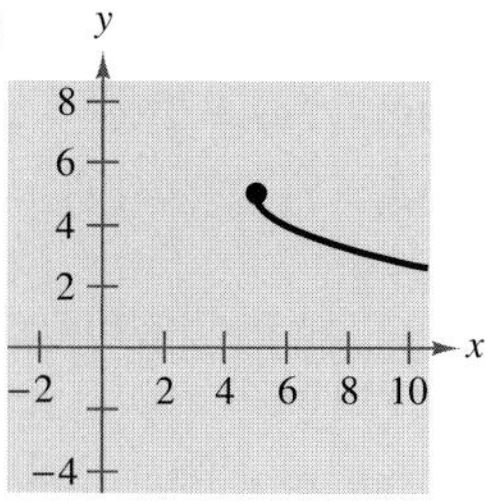

(d)
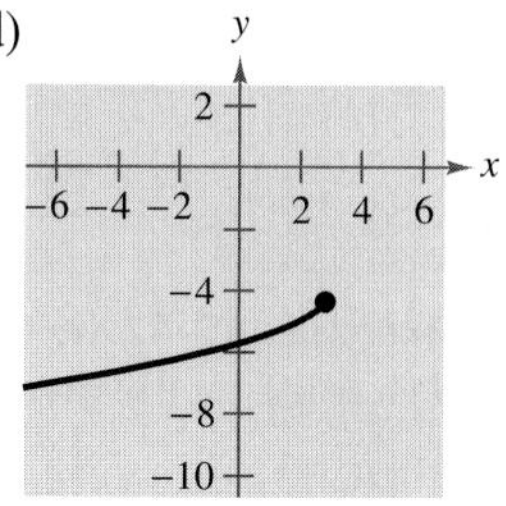

In Exercises 13–18, identify the common function and the transformation shown in the graph. Write an equation for the function shown in the graph.

13.
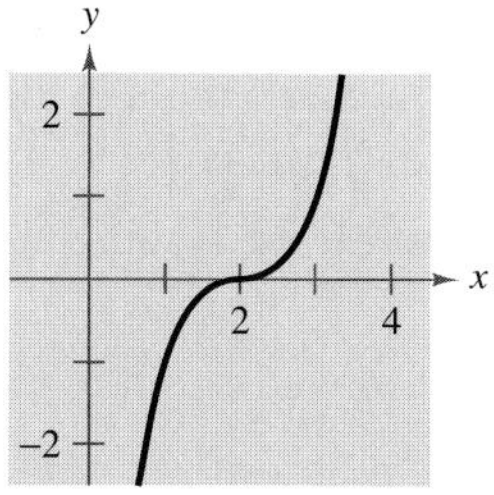

14.
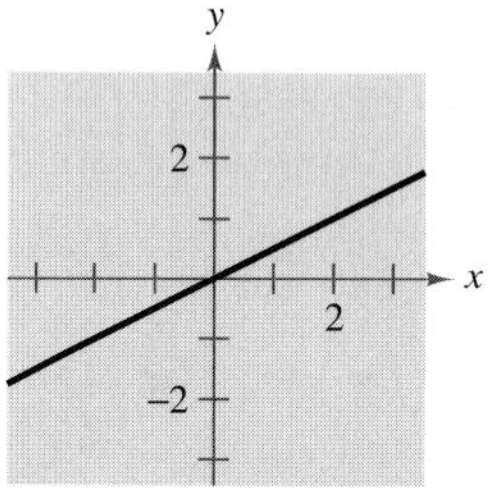

15.
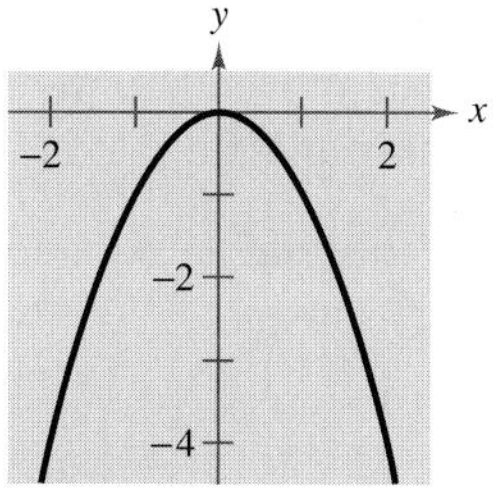

16.
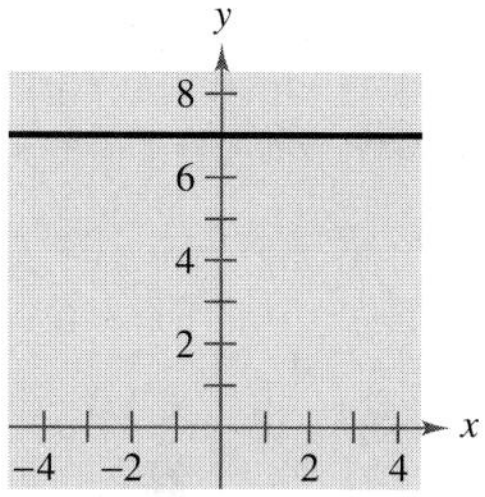

17.
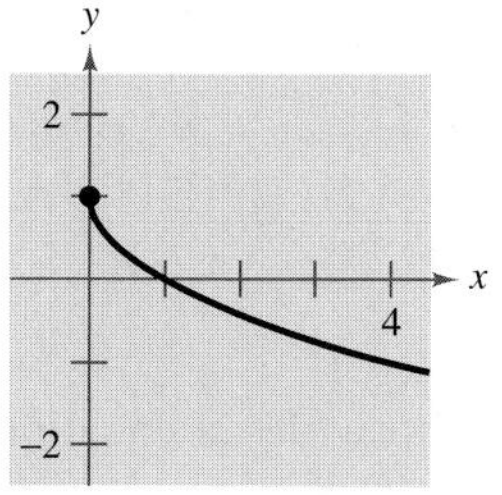

18.
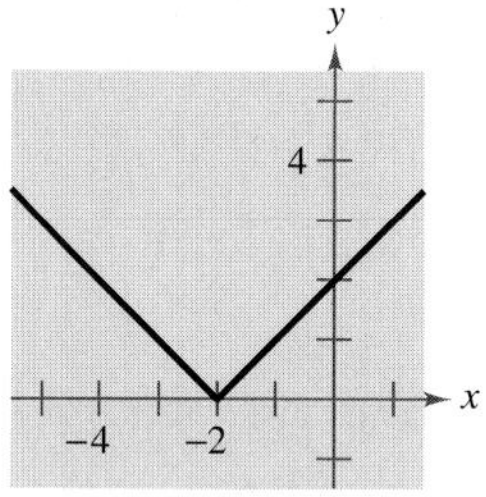

In Exercises 19–34, describe the transformation that occurs in the function. Then sketch its graph.

19. $f(x) = 12 - x^2$

20. $f(x) = (x - 8)^2$

21. $f(x) = x^3 + 7$

22. $f(x) = -x^3 - 1$

23. $f(x) = 2 - (x + 5)^2$

24. $f(x) = -(x + 10)^2 + 5$

25. $f(x) = (x - 1)^3 + 2$

26. $f(x) = (x + 3)^3 - 10$

27. $f(x) = -|x| - 2$

28. $f(x) = 6 - |x + 5|$

29. $f(x) = -|x + 4| + 8$ **30.** $f(x) = |-x + 3| + 9$

31. $f(x) = \sqrt{x - 9}$ **32.** $f(x) = \sqrt{x + 4} + 8$

33. $f(x) = \sqrt{7 - x} - 2$ **34.** $f(x) = -\sqrt{x + 1} - 6$

In Exercises 35–42, write an equation for the function that is described by the given characteristics.

35. The shape of $f(x) = x^2$, but moved 2 units to the right and 8 units down

36. The shape of $f(x) = x^2$, but moved 3 units to the left, 7 units up, and reflected in the x-axis

37. The shape of $f(x) = x^3$, but moved 13 units to the right

38. The shape of $f(x) = x^3$, but moved 6 units to the left, 6 units down, and reflected in the y-axis

39. The shape of $f(x) = |x|$, but moved 10 units up and reflected in the x-axis

40. The shape of $f(x) = |x|$, but moved 1 unit to the left and 7 units down

41. The shape of $f(x) = \sqrt{x}$, but moved 6 units to the left and reflected in both the x-axis and the y-axis

42. The shape of $f(x) = \sqrt{x}$, but moved 9 units down and reflected in both the x-axis and the y-axis

43. Use the graph of $f(x) = x^2$ to write an equation for each function whose graph is shown.

(a)

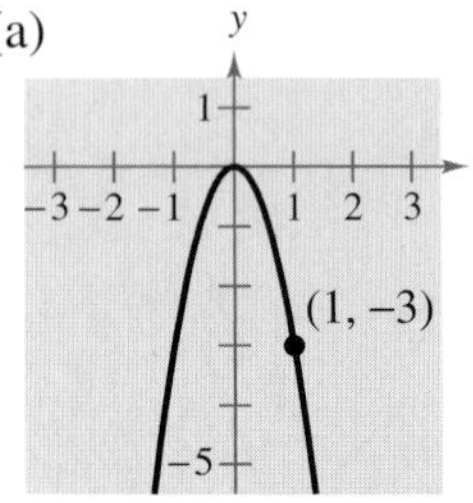

(b)

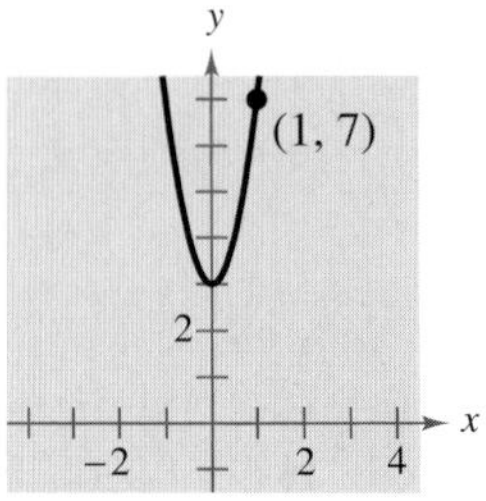

44. Use the graph of $f(x) = x^3$ to write an equation for each function whose graph is shown.

(a)

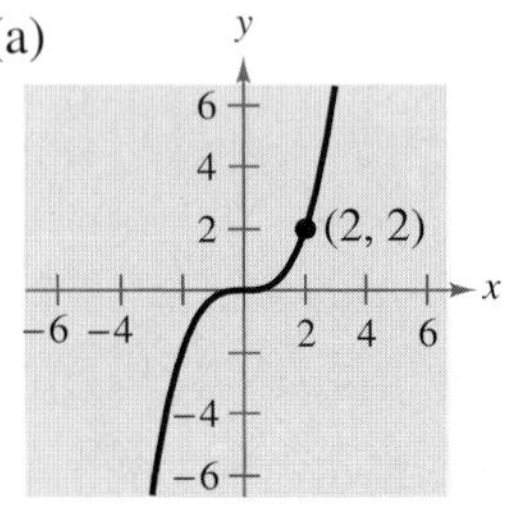

(b)

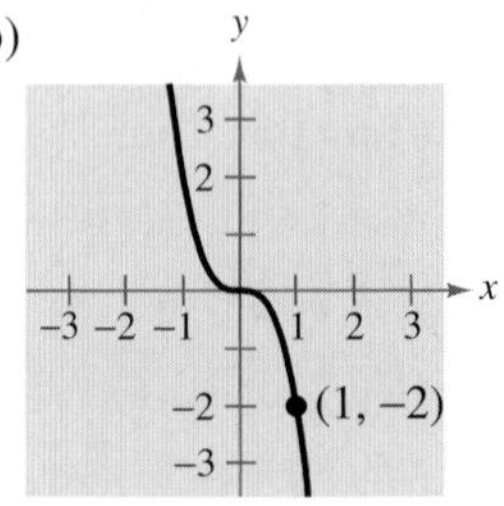

45. Use the graph of $f(x) = |x|$ to write an equation for each function whose graph is shown.

(a)

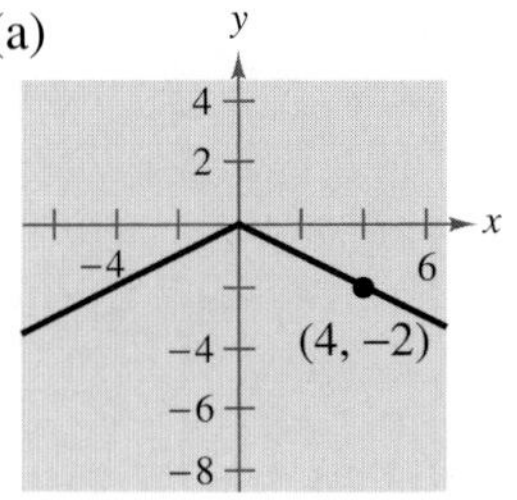

(b)

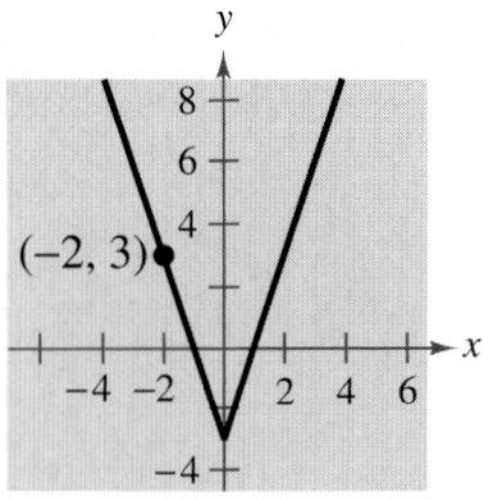

46. Use the graph of $f(x) = \sqrt{x}$ to write an equation for each function whose graph is shown.

(a)

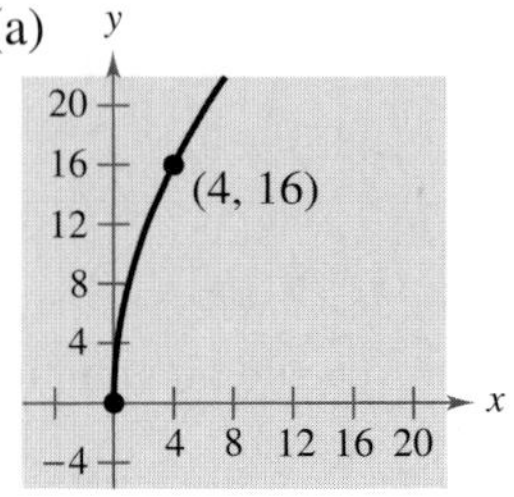

(b)

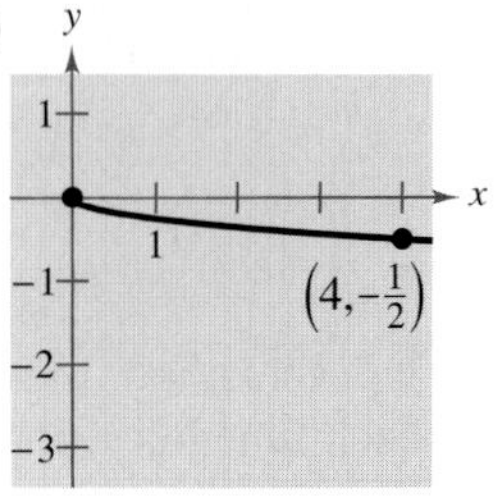

In Exercises 47–52, identify the common function and the transformation shown in the graph. Write an equation for the function shown in the graph. Then use a graphing utility to verify your answer.

47.

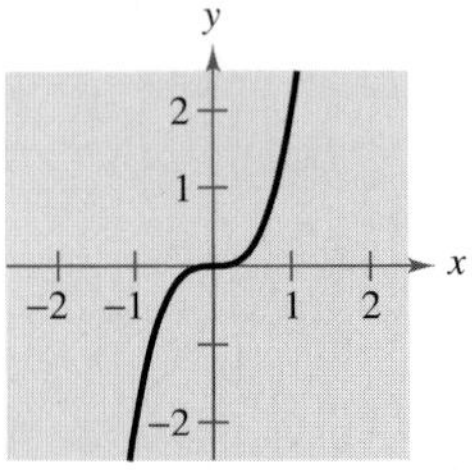

48.

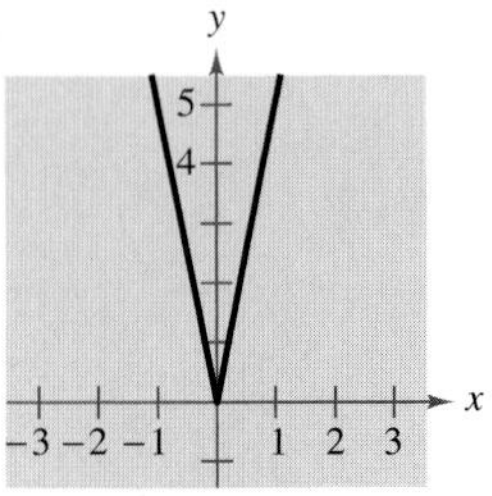

49.

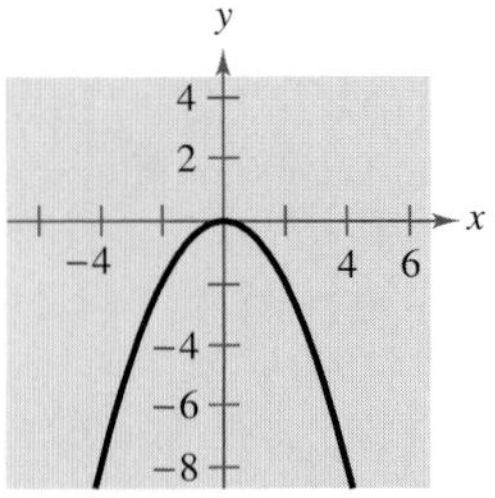

50.

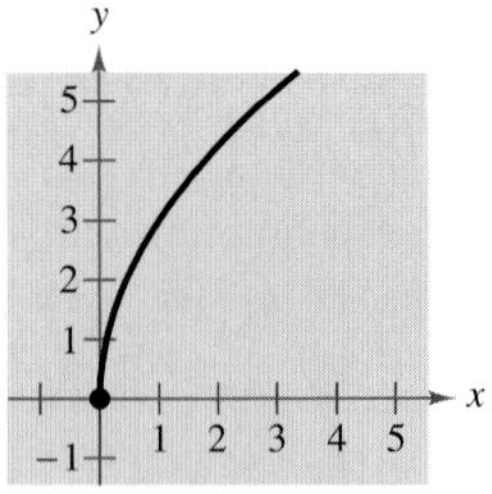

51.

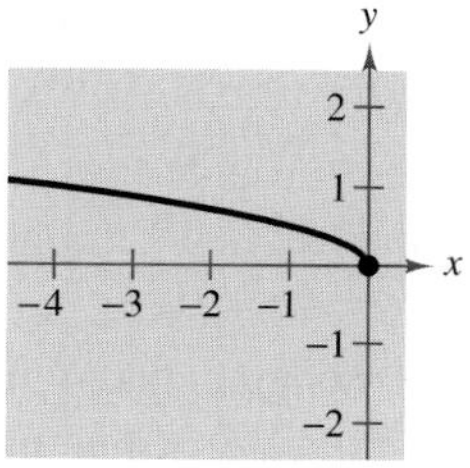

52.

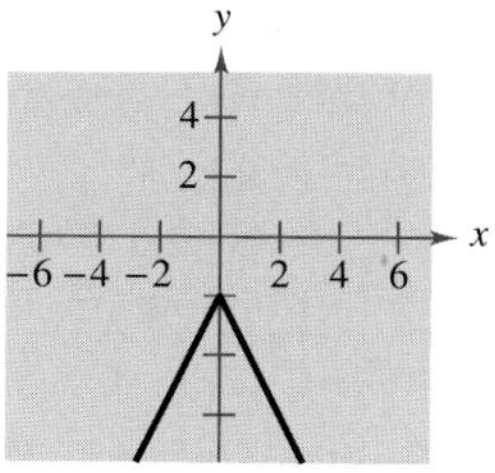

Graphical Analysis **In Exercises 53–56, use the viewing window shown to write a possible equation for the transformation of the common function.**

53.

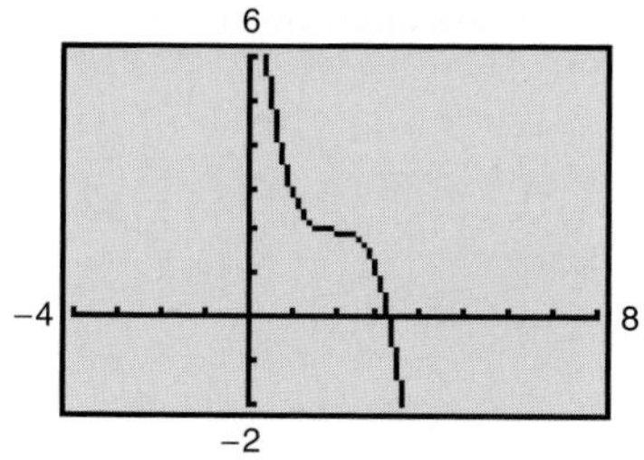

54.

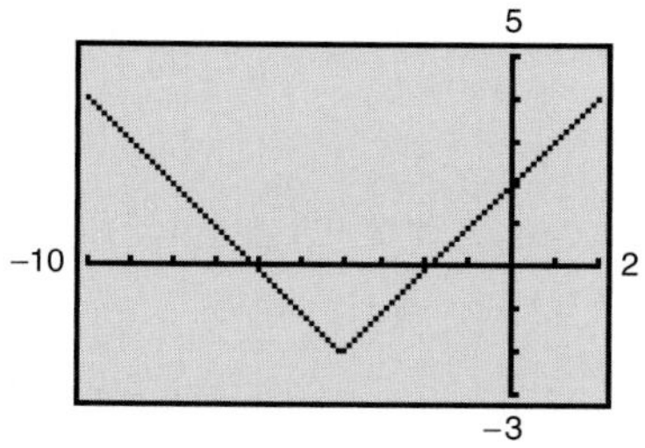

55.

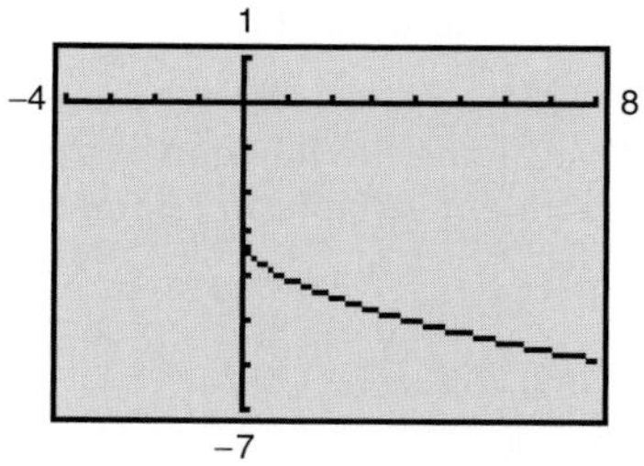

56.

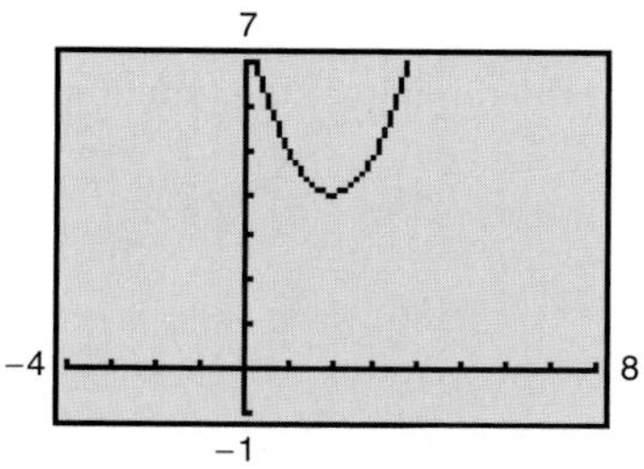

Graphical Reasoning **In Exercises 57 and 58, use the graph of *f* to sketch the graph of *g*.**

57.

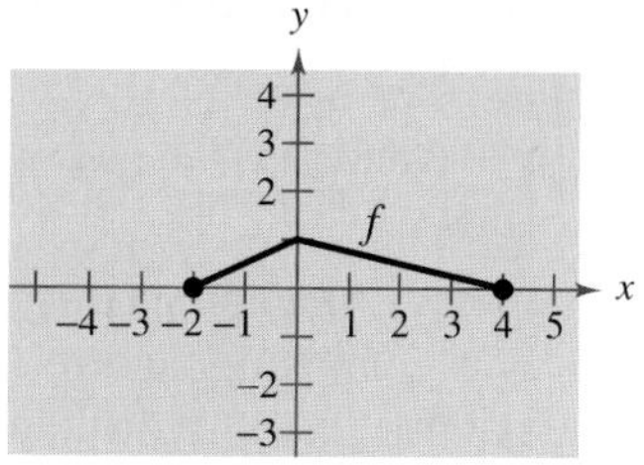

(a) $g(x) = f(x) + 2$ (b) $g(x) = f(x) - 1$

(c) $g(x) = f(-x)$ (d) $g(x) = -2f(x)$

58.

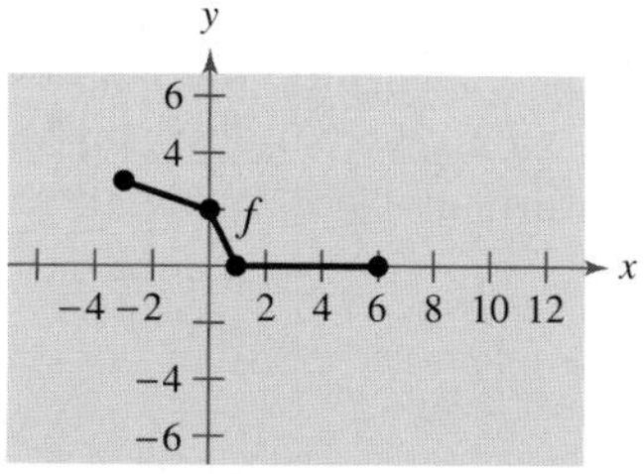

(a) $g(x) = f(x) - 5$ (b) $g(x) = f(x) + \frac{1}{2}$

(c) $g(x) = f(-x)$ (d) $g(x) = -4f(x)$

59. *Fuel Use* The amount of fuel F (in billions of gallons) used by trucks from 1980 through 1996 can be approximated by the function

$$F = f(t) = 20.46 + 0.04t^2$$

where $t = 0$ represents 1980. (Source: U.S. Federal Highway Administration)

(a) Describe the transformation of the common function $f(x) = x^2$. Then sketch the graph over the interval $0 \le t \le 16$.

(b) Rewrite the function so that $t = 0$ represents 1990. Explain how you got your answer.

60. *Finance* The amount (in trillions of dollars) of mortgage debt M outstanding in the United States from 1985 through 1997 can be approximated by the function

$$M = f(t) = 1.5\sqrt{t} - 1.25$$

where $t = 5$ represents 1985. (Source: Board of Governors of the Federal Reserve System)

(a) Describe the transformation of the common function $f(x) = \sqrt{x}$. Then sketch the graph over the interval $5 \le t \le 17$.

(b) Rewrite the function so that $t = 5$ represents 1995. Explain how you got your answer.

Synthesis

In Exercises 61 and 62, determine whether the statement is true or false. Justify your answer.

61. The graphs of $f(x) = |x| + 6$ and $f(x) = |-x| + 6$ are identical.

62. If the graph of the common function $f(x) = x^2$ is moved 6 units to the right, 3 units up, and reflected in the x-axis, then the point $(-2, 19)$ will lie on the graph of the transformation.

63. *Describing Profits* Management originally predicted that the profits from the sales of a new product would be approximated by the graph of the function f shown. The actual profits are shown by the function g along with a verbal description. Use the concepts of transformations of graphs to write g in terms of f.

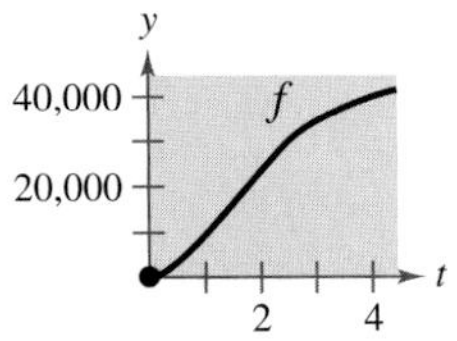

(a) The profits were only three-fourths as large as expected.

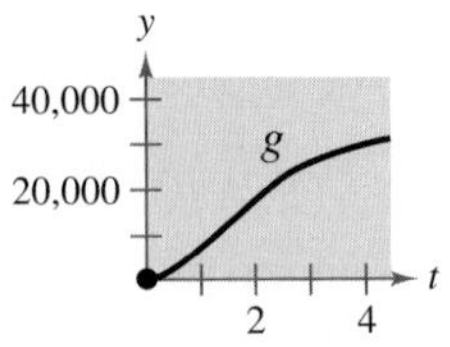

(b) The profits were consistently \$10,000 greater than predicted.

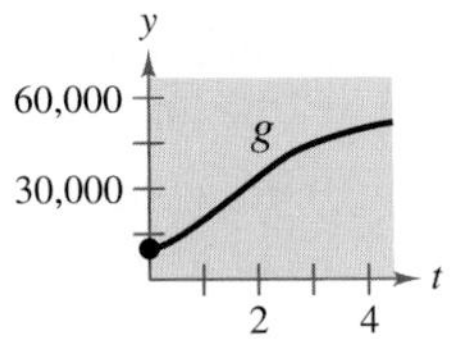

(c) There was a 2-year delay in the introduction of the product. After sales began, profits grew as expected.

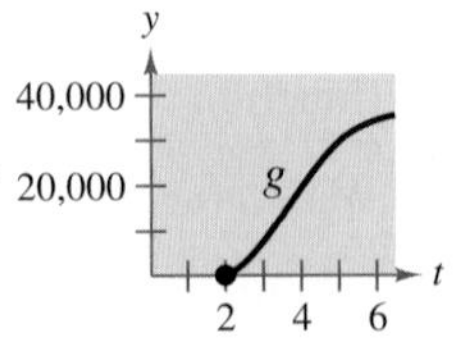

64. Explain why the graph of $y = -f(x)$ is a reflection of the graph of $y = f(x)$ about the x-axis.

65. The graph of $y = f(x)$ passes through the points $(0, 1)$, $(1, 2)$, and $(2, 3)$. Find the corresponding points on the graph of $y = f(x + 2) - 1$.

66. *Think About It* You can use two methods to graph a function: plotting points, or translating a common function as shown in this section. Which method of graphing do you prefer to use? Explain.

Review

In Exercises 67–74, perform the operations and simplify.

67. $\dfrac{4}{x} + \dfrac{4}{1 - x}$

68. $\dfrac{2}{x + 5} - \dfrac{2}{x - 5}$

69. $\dfrac{3}{x - 1} - \dfrac{2}{x(x - 1)}$

70. $\dfrac{x}{x - 5} + \dfrac{1}{2}$

71. $(x - 4)\left(\dfrac{1}{\sqrt{x^2 - 4}}\right)$

72. $\left(\dfrac{x}{x^2 - 4}\right)\left(\dfrac{x^2 - x - 2}{x^2}\right)$

73. $(x^2 - 9) \div \left(\dfrac{x + 3}{5}\right)$

74. $\left(\dfrac{x}{x^2 - 3x - 28}\right) \div \left(\dfrac{x^2 + 3x}{x^2 + 5x + 4}\right)$

In Exercises 75 and 76, evaluate the function at the specified values of the independent variable and simplify.

75. $f(x) = x^2 - 6x + 11$

(a) $f(-3)$ (b) $f\left(-\frac{1}{2}\right)$ (c) $f(x - 3)$

76. $f(x) = \sqrt{x + 10} - 3$

(a) $f(-10)$ (b) $f(26)$ (c) $f(x - 10)$

In Exercises 77–80, find the domain of the function.

77. $f(x) = \dfrac{2}{11 - x}$

78. $f(x) = \dfrac{\sqrt{x - 3}}{x - 8}$

79. $f(x) = \sqrt{81 - x^2}$

80. $f(x) = \sqrt[3]{4 - x^2}$

1.6 Combinations of Functions

▶ **What you should learn**

- How to add, subtract, multiply, and divide functions
- How to find compositions of one function with another function
- How to use combinations of functions to model and solve real-life problems

▶ **Why you should learn it**

Combinations of functions can be used to model and solve real-life problems. For instance, Exercises 33 and 34 on page 169 show how to use combinations of functions to analyze U.S. health expenditures.

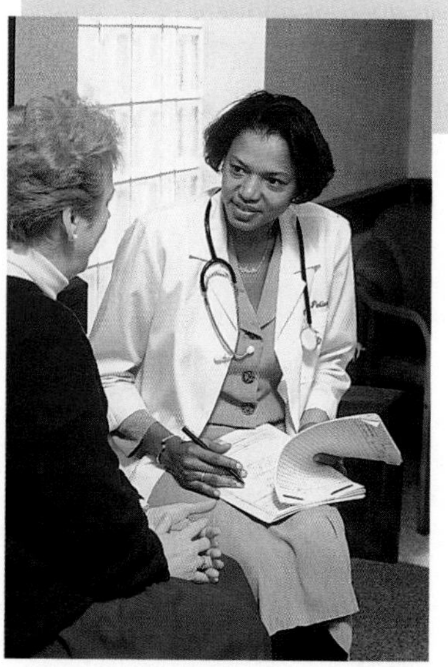

Charles Gupton/Tony Stone Images

A computer animation of this concept appears in the *Interactive* CD-ROM and *Internet* versions of this text.

Arithmetic Combinations of Functions

Just as two real numbers can be combined by the operations of addition, subtraction, multiplication, and division to form other real numbers, two *functions* can be combined to create new functions. For example, the functions $f(x) = 2x - 3$ and $g(x) = x^2 - 1$ can be combined to form the sum, difference, product, and quotient of f and g.

$$f(x) + g(x) = (2x - 3) + (x^2 - 1)$$
$$= x^2 + 2x - 4 \qquad \text{Sum}$$
$$f(x) - g(x) = (2x - 3) - (x^2 - 1)$$
$$= -x^2 + 2x - 2 \qquad \text{Difference}$$
$$f(x)g(x) = (2x - 3)(x^2 - 1)$$
$$= 2x^3 - 3x^2 - 2x + 3 \qquad \text{Product}$$
$$\frac{f(x)}{g(x)} = \frac{2x - 3}{x^2 - 1}, \quad x \neq \pm 1 \qquad \text{Quotient}$$

The domain of an **arithmetic combination** of functions f and g consists of all real numbers that are common to the domains of f and g. In the case of the quotient $f(x)/g(x)$, there is the further restriction that $g(x) \neq 0$.

Sum, Difference, Product, and Quotient of Functions

Let f and g be two functions with overlapping domains. Then, for all x common to both domains, the *sum, difference, product,* and *quotient* of f and g are defined as follows.

1. *Sum:* $(f + g)(x) = f(x) + g(x)$
2. *Difference:* $(f - g)(x) = f(x) - g(x)$
3. *Product:* $(fg)(x) = f(x) \cdot g(x)$
4. *Quotient:* $\left(\frac{f}{g}\right)(x) = \frac{f(x)}{g(x)}, \quad g(x) \neq 0$

Example 1 ▶ Finding the Sum of Two Functions

Given $f(x) = 2x + 1$ and $g(x) = x^2 + 2x - 1$, find $(f + g)(x)$. Then evaluate the sum when $x = -1$.

Solution

$$(f + g)(x) = f(x) + g(x)$$
$$= (2x + 1) + (x^2 + 2x - 1)$$
$$= x^2 + 4x$$

When $x = -1$, the value of the sum is $(f + g)(-1) = (-1)^2 + 4(-1) = -3$.

Example 2 ▶ Finding the Difference of Two Functions

Given $f(x) = 2x + 1$ and $g(x) = x^2 + 2x - 1$, find $(f - g)(x)$. Then evaluate the difference when $x = 2$.

Solution

The difference of f and g is

$$\begin{aligned}(f - g)(x) &= f(x) - g(x)\\ &= (2x + 1) - (x^2 + 2x - 1)\\ &= -x^2 + 2.\end{aligned}$$

When $x = 2$, the value of this difference is

$$\begin{aligned}(f - g)(2) &= -(2)^2 + 2\\ &= -2.\end{aligned}$$

In Examples 1 and 2, both f and g have domains that consist of all real numbers. So, the domains of $(f + g)$ and $(f - g)$ are also the set of all real numbers. Remember that any restrictions on the domains of f and g must be considered when forming the sum, difference, product, or quotient of f and g.

Additional Examples

a. Find $fg(x)$ given that $f(x) = x + 5$ and $g(x) = 3x$.

Solution

$$\begin{aligned}fg(x) &= f(x) \cdot g(x)\\ &= (x + 5)(3x)\\ &= 3x^2 + 15x\end{aligned}$$

b. Find $gf(x)$ given that $f(x) = \frac{1}{x}$ and $g(x) = \frac{x}{x + 1}$.

Solution

$$\begin{aligned}gf(x) &= g(x) \cdot f(x)\\ &= \left(\frac{x}{x + 1}\right)\left(\frac{1}{x}\right)\\ &= \frac{1}{x + 1}, \quad x \neq 0\end{aligned}$$

Example 3 ▶ Finding the Domains of Quotients of Functions

Find the domains of $\left(\frac{f}{g}\right)(x)$ and $\left(\frac{g}{f}\right)(x)$ for the functions

$$f(x) = \sqrt{x} \quad \text{and} \quad g(x) = \sqrt{4 - x^2}.$$

Solution

The quotient of f and g is

$$\left(\frac{f}{g}\right)(x) = \frac{f(x)}{g(x)} = \frac{\sqrt{x}}{\sqrt{4 - x^2}}$$

and the quotient of g and f is

$$\left(\frac{g}{f}\right)(x) = \frac{g(x)}{f(x)} = \frac{\sqrt{4 - x^2}}{\sqrt{x}}.$$

The domain of f is $[0, \infty)$ and the domain of g is $[-2, 2]$. The intersection of these domains is $[0, 2]$. So, the domains of $\left(\frac{f}{g}\right)$ and $\left(\frac{g}{f}\right)$ are as follows.

$$\text{Domain of } \left(\frac{f}{g}\right): [0, 2) \qquad \text{Domain of } \left(\frac{g}{f}\right): (0, 2]$$

Can you see why these two domains differ slightly?

A computer animation of this concept appears in the *Interactive* CD-ROM and *Internet* versions of this text.

Composition of Functions

Another way of combining two functions is to form the **composition** of one with the other. For instance, if $f(x) = x^2$ and $g(x) = x + 1$, the composition of f with g is

$$f(g(x)) = f(x + 1)$$
$$= (x + 1)^2.$$

This composition is denoted as $(f \circ g)$.

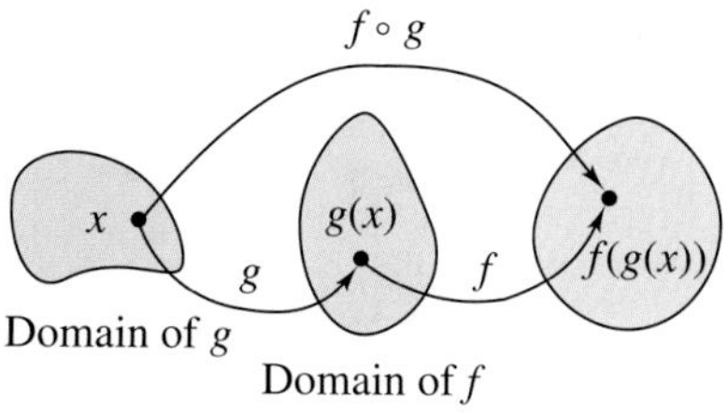

FIGURE 1.52

Definition of Composition of Two Functions

The **composition** of the function f with the function g is

$$(f \circ g)(x) = f(g(x)).$$

The domain of $(f \circ g)$ is the set of all x in the domain of g such that $g(x)$ is in the domain of f. (See Figure 1.52.)

Study Tip

The following tables of values help illustrate the composition of the functions f and g given in Example 4.

x	0	1	2	3
$g(x)$	4	3	0	-5

$g(x)$	4	3	0	-5
$f(g(x))$	6	5	2	-3

x	0	1	2	3
$f(g(x))$	6	5	2	-3

Note that the first two tables can be combined (or "composed") to produce the values given in the third table.

Example 4 ▶ Composition of Functions

Given $f(x) = x + 2$ and $g(x) = 4 - x^2$, find the following.

a. $(f \circ g)(x)$ **b.** $(g \circ f)(x)$ **c.** $(g \circ f)(-2)$

Solution

a. The composition of f with g is as follows.

$(f \circ g)(x) = f(g(x))$ Definition of $f \circ g$

$= f(4 - x^2)$ Definition of $g(x)$

$= (4 - x^2) + 2$ Definition of $f(x)$

$= -x^2 + 6$ Simplify.

b. The composition of g with f is as follows.

$(g \circ f)(x) = g(f(x))$ Definition of $g \circ f$

$= g(x + 2)$ Definition of $f(x)$

$= 4 - (x + 2)^2$ Definition of $g(x)$

$= 4 - (x^2 + 4x + 4)$ Expand.

$= -x^2 - 4x$ Simplify.

Note that, in this case, $(f \circ g)(x) \neq (g \circ f)(x)$.

c. Using the result of part (b), you can write the following.

$(g \circ f)(-2) = -(-2)^2 - 4(-2)$ Substitute.

$= -4 + 8$ Simplify.

$= 4$ Simplify.

Additional Example

Given the functions

$f(x) = \dfrac{1}{x - 2}$

and

$g(x) = \sqrt{x}$

find the composition of f with g. Then find the domain of the composition.

Solution

The composition of f with g is

$$\begin{aligned}(f \circ g)(x) &= f(g(x)) \\ &= f(\sqrt{x}) \\ &= \frac{1}{\sqrt{x} - 2}.\end{aligned}$$

The domain of $(f \circ g)$ is the set of all nonnegative real numbers except $x = 4$.

Example 5 ▶ Finding the Domain of a Composite Function

Find the composition $(f \circ g)(x)$ for the functions

$$f(x) = x^2 - 9 \quad \text{and} \quad g(x) = \sqrt{9 - x^2}.$$

Then find the domain of $(f \circ g)$.

Solution

$$\begin{aligned}(f \circ g)(x) &= f(g(x)) \\ &= f\left(\sqrt{9 - x^2}\right) \\ &= \left(\sqrt{9 - x^2}\right)^2 - 9 \\ &= 9 - x^2 - 9 \\ &= -x^2\end{aligned}$$

From this, it might appear that the domain of the composition is the set of all real numbers. This, however, is not true because the domain of g is $-3 \le x \le 3$.

In Examples 4 and 5 you formed the composition of two given functions. In calculus, it is also important to be able to identify two functions that make up a given composite function. For instance, the function h given by

$$h(x) = (3x - 5)^3$$

is the composition of f with g, where $f(x) = x^3$ and $g(x) = 3x - 5$. That is,

$$h(x) = (3x - 5)^3 = [g(x)]^3 = f(g(x)).$$

Basically, to "decompose" a composite function, look for an "inner" and an "outer" function. In the function h above, $g(x) = 3x - 5$ is the inner function and $f(x) = x^3$ is the outer function.

Activities

1. Find $(f + g)(-1)$ and $\left(\dfrac{f}{g}\right)(2)$ for $f(x) = 3x^2 + 2$ and $g(x) = 2x$.

   Answer: $3; \dfrac{7}{2}$

2. Given $f(x) = 3x^2 + 2$ and $g(x) = 2x$, find $f \circ g$.

   Answer: $(f \circ g)(x) = 12x^2 + 2$

3. Find two functions f and g such that $(f \circ g)(x) = h(x)$. (There are many correct answers.)

   a. $h(x) = \dfrac{1}{\sqrt{3x + 1}}$

      Answer: $f(x) = \dfrac{1}{\sqrt{x}}$ and $g(x) = 3x + 1$

   b. $h(x) = (2x + 3)^4$

      Answer: $f(x) = x^4$ and $g(x) = 2x + 3$

Example 6 ▶ Identifying a Composite Function

Express the function

$$h(x) = \frac{1}{(x - 2)^2}$$

as a composition of two functions.

Solution

One way to write h as a composition of two functions is to take the inner function to be $g(x) = x - 2$ and the outer function to be

$$f(x) = \frac{1}{x^2} = x^{-2}.$$

Then you can write

$$h(x) = \frac{1}{(x - 2)^2} = (x - 2)^{-2} = f(x - 2) = f(g(x)).$$

Exploration

You are buying an automobile whose price is \$18,500. Which of the following options would you choose? Explain.

a. You are given a factory rebate of \$2000, followed by a dealer discount of 10%.

b. You are given a dealer discount of 10%, followed by a factory rebate of \$2000.

Let $f(x) = x - 2000$ and let $g(x) = 0.9x$. Which option is represented by the composite $f(g(x))$? Which is represented by the composite $g(f(x))$?

Writing About Mathematics

To expand upon this activity, you might consider asking your students to use the tables they created in parts (a) and (b), along with a table of values for x and $f(x)$, to demonstrate and explain how the tables can be manipulated to yield tables of values for $h(x)$ and $g(x)$.

Application

Example 7 ▶ Bacteria Count

The number N of bacteria in a refrigerated food is

$$N(T) = 20T^2 - 80T + 500, \qquad 2 \le T \le 14$$

where T is the temperature of the food. When the food is removed from refrigeration, the temperature is

$$T(t) = 4t + 2, \qquad 0 \le t \le 3$$

where t is the time in hours. (a) Find the composite $N(T(t))$ and interpret its meaning in context. (b) Find the time when the bacterial count reaches 2000.

Solution

a.
$$\begin{aligned} N(T(t)) &= 20(4t + 2)^2 - 80(4t + 2) + 500 \\ &= 20(16t^2 + 16t + 4) - 320t - 160 + 500 \\ &= 320t^2 + 320t + 80 - 320t - 160 + 500 \\ &= 320t^2 + 420 \end{aligned}$$

The composite function $N(T(t))$ represents the number of bacteria in the food as a function of time.

b. The bacterial count will reach 2000 when $320t^2 + 420 = 2000$. Solve this equation to find that the count will reach 2000 when $t \approx 2.2$ hours. When you solve this equation, note that the negative value is rejected because it is not in the domain of the composite function.

Writing ABOUT MATHEMATICS

Analyzing Arithmetic Combinations of Functions

a. Use the graphs of f and $(f + g)$ in Figure 1.53 to make a table showing the values of $g(x)$ when $x = 1, 2, 3, 4, 5,$ and 6. Explain your reasoning.

b. Use the graphs of f and $(f - h)$ in Figure 1.53 to make a table showing the values of $h(x)$ when $x = 1, 2, 3, 4, 5,$ and 6. Explain your reasoning.

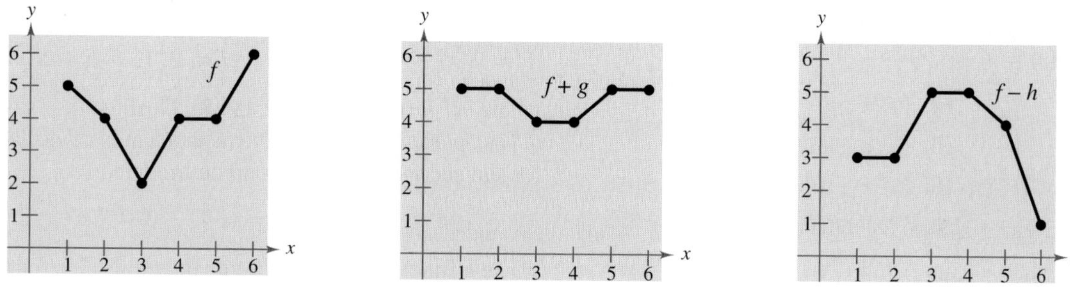

FIGURE 1.53

1.6 Exercises

In Exercises 1–4, use the graphs of f and g to graph $h(x) = (f + g)(x)$.

1.

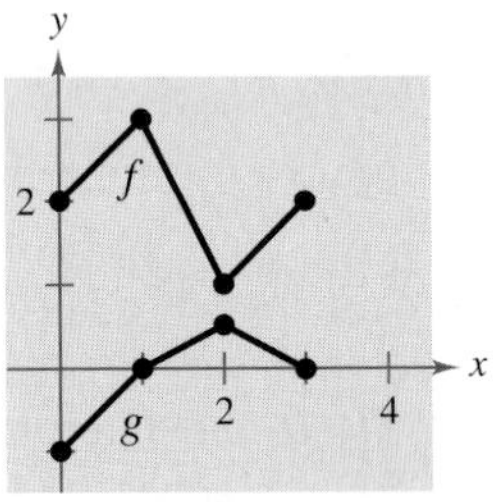

2.

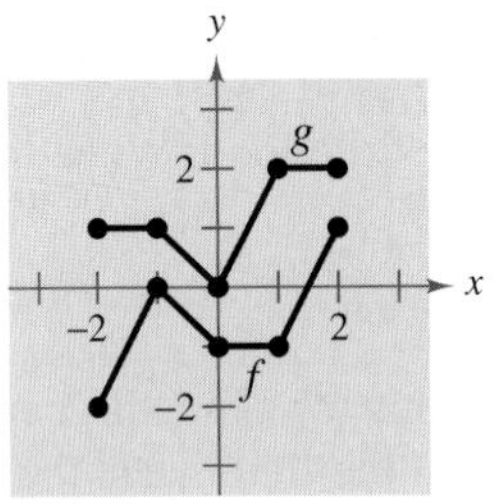

3.

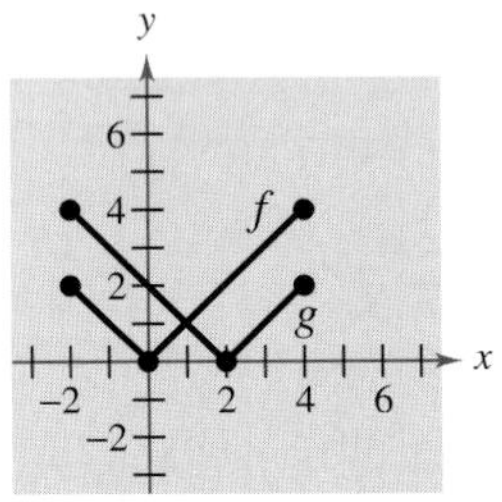

4.

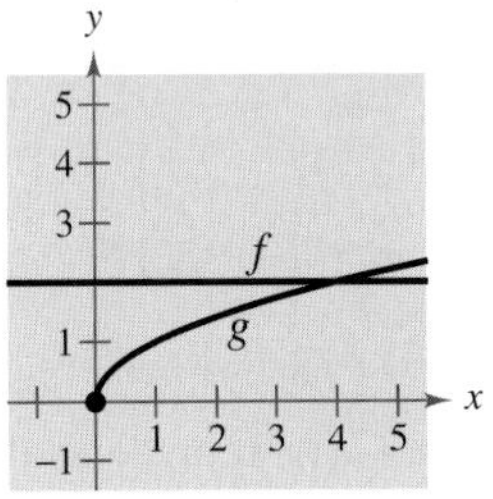

In Exercises 5–12, find (a) $(f + g)(x)$, (b) $(f - g)(x)$, (c) $(fg)(x)$, and (d) $(f/g)(x)$. What is the domain of f/g?

5. $f(x) = x + 2, \quad g(x) = x - 2$

6. $f(x) = 2x - 5, \quad g(x) = 2 - x$

7. $f(x) = x^2, \quad g(x) = 2 - x$

8. $f(x) = 2x - 5, \quad g(x) = 4$

9. $f(x) = x^2 + 6, \quad g(x) = \sqrt{1 - x}$

10. $f(x) = \sqrt{x^2 - 4}, \quad g(x) = \dfrac{x^2}{x^2 + 1}$

11. $f(x) = \dfrac{1}{x}, \quad g(x) = \dfrac{1}{x^2}$

12. $f(x) = \dfrac{x}{x + 1}, \quad g(x) = x^3$

In Exercises 13–24, evaluate the indicated function for $f(x) = x^2 + 1$ and $g(x) = x - 4$.

13. $(f + g)(2)$ **14.** $(f - g)(-1)$

15. $(f - g)(0)$ **16.** $(f + g)(1)$

17. $(f - g)(3t)$ **18.** $(f + g)(t - 2)$

19. $(fg)(6)$ **20.** $(fg)(-6)$

21. $\left(\dfrac{f}{g}\right)(5)$ **22.** $\left(\dfrac{f}{g}\right)(0)$

23. $\left(\dfrac{f}{g}\right)(-1) - g(3)$ **24.** $(2f)(5)$

In Exercises 25–28, graph the functions f, g, and $f + g$ on the same set of coordinate axes.

25. $f(x) = \frac{1}{2}x, \quad g(x) = x - 1$

26. $f(x) = \frac{1}{3}x, \quad g(x) = -x + 4$

27. $f(x) = x^2, \quad g(x) = -2x$

28. $f(x) = 4 - x^2, \quad g(x) = x$

Graphical Reasoning **In Exercises 29 and 30, use a graphing utility to sketch the graphs of f, g, and $f + g$ in the same viewing window. Which function contributes most to the magnitude of the sum when $0 \le x \le 2$? Which function contributes most to the magnitude of the sum when $x > 6$?**

29. $f(x) = 3x, \quad g(x) = -\dfrac{x^3}{10}$

30. $f(x) = \dfrac{x}{2}, \quad g(x) = \sqrt{x}$

31. *Stopping Distance* While traveling in a car at x miles per hour, you are required to stop quickly to avoid an accident. The distance (in feet) the car travels during your reaction time is given by $R(x) = \frac{3}{4}x$. The distance (in feet) traveled while you are braking is

$$B(x) = \frac{1}{15}x^2.$$

Find the function that represents the total stopping distance (in feet) T. Graph the functions R, B, and T on the same set of coordinate axes for $0 \le x \le 60$.

32. *Business* You own two restaurants. From 1995 to 2000, the sales R_1 (in thousands of dollars) for one restaurant can be modeled by

$$R_1 = 480 - 8t - 0.8t^2, \qquad t = 0, 1, 2, 3, 4, 5$$

where $t = 0$ represents 1995. During the same 6-year period, the sales R_2 (in thousands of dollars) for the second restaurant can be modeled by

$$R_2 = 254 + 0.78t, \qquad t = 0, 1, 2, 3, 4, 5.$$

Write a function that represents the total sales for the two restaurants. Use a graphing utility to graph the total sales function.

Health Care Costs **In Exercises 33 and 34, use the table, which gives the total amount spent (in billions of dollars) on health services and supplies in the United States (including Puerto Rico) for the years 1990 through 1996. The variables y_1, y_2, and y_3 represent out-of-pocket payments, insurance premiums, and other types of payments, respectively.** (Source: U.S. Health Care Financing Administration)

Year	1990	1991	1992	1993	1994	1995	1996
y_1	144.4	151.6	159.5	163.6	164.8	166.7	171.2
y_2	238.6	259.4	282.5	303.3	315.6	326.9	337.3
y_3	21.8	24.0	25.1	27.3	29.6	31.7	32.4

33. Use a graphing utility to find a mathematical model for each of the variables. Let $t = 0$ represent 1990. Find a quadratic model for y_1 and linear models for y_2 and y_3.

34. Use a graphing utility to graph y_1, y_2, y_3, and $y_1 + y_2 + y_3$ in the same viewing window. Use the model to estimate the total amount spent on health services and supplies in the year 2000.

35. *Graphical Reasoning* An electronically controlled thermostat in a home is programmed to lower the temperature automatically during the night. The temperature in the house T (in degrees Fahrenheit) is given in terms of t, the time in hours on a 24-hour clock (see figure).

(a) Explain why T is a function of t.

(b) Approximate $T(4)$ and $T(15)$.

(c) Suppose the thermostat were reprogrammed to produce a temperature H where $H(t) = T(t - 1)$. How would this change the temperature?

(d) Suppose the thermostat were reprogrammed to produce a temperature H where $H(t) = T(t) - 1$. How would this change the temperature?

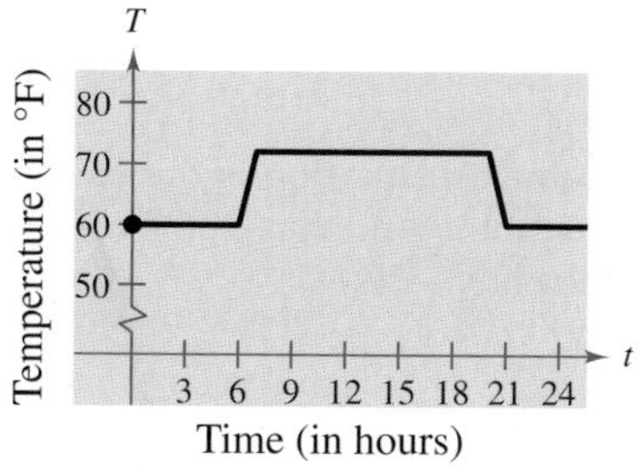

36. *Think About It* Write a piecewise-defined function that represents the graph in Exercise 35.

In Exercises 37–40, find (a) $f \circ g$, (b) $g \circ f$, and (c) $f \circ f$.

37. $f(x) = x^2$, $\quad g(x) = x - 1$

38. $f(x) = \sqrt[3]{x - 1}$, $\quad g(x) = x^3 + 1$

39. $f(x) = 3x + 5$, $\quad g(x) = 5 - x$

40. $f(x) = x^3$, $\quad g(x) = \dfrac{1}{x}$

In Exercises 41–50, find (a) $f \circ g$ and (b) $g \circ f$. Find the domain of each function and each composite function.

41. $f(x) = \sqrt{x + 4}$, $\quad g(x) = x^2$

42. $f(x) = \sqrt[3]{x - 5}$, $\quad g(x) = x^3 + 1$

43. $f(x) = \frac{1}{3}x - 3$, $\quad g(x) = 3x + 1$

44. $f(x) = x^2 + 1$, $\quad g(x) = \sqrt{x}$

45. $f(x) = x^4$, $\quad g(x) = x^4$

46. $f(x) = \sqrt{x}$, $\quad g(x) = 2x - 3$

47. $f(x) = |x|$, $\quad g(x) = x + 6$

48. $f(x) = x^{2/3}$, $\quad g(x) = x^6$

49. $f(x) = \dfrac{1}{x}$, $\quad g(x) = x + 3$

50. $f(x) = \dfrac{3}{x^2 - 1}$, $\quad g(x) = x + 1$

In Exercises 51–54, use the graphs of f and g to evaluate the functions.

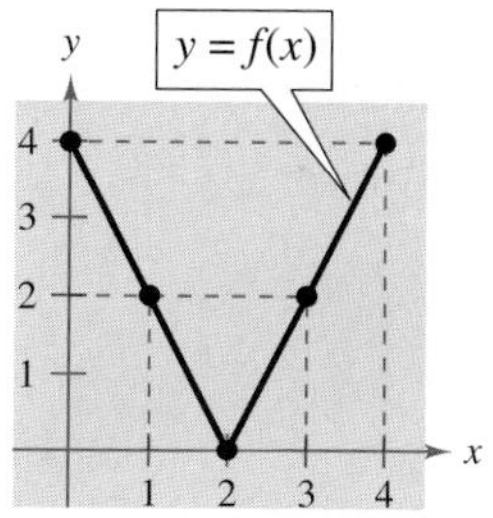

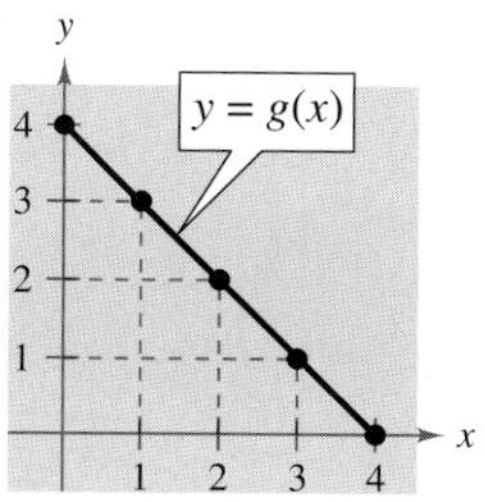

51. (a) $(f + g)(3)$ $\quad$ (b) $\left(\dfrac{f}{g}\right)(2)$

52. (a) $(f - g)(1)$ $\quad$ (b) $(fg)(4)$

53. (a) $(f \circ g)(2)$ $\quad$ (b) $(g \circ f)(2)$

54. (a) $(f \circ g)(1)$ $\quad$ (b) $(g \circ f)(3)$

In Exercises 55–62, find two functions f and g such that $(f \circ g)(x) = h(x)$. (There is more than one correct answer.)

55. $h(x) = (2x + 1)^2$

56. $h(x) = (1 - x)^3$

57. $h(x) = \sqrt[3]{x^2 - 4}$

58. $h(x) = \sqrt{9 - x}$

59. $h(x) = \dfrac{1}{x + 2}$

60. $h(x) = \dfrac{4}{(5x + 2)^2}$

61. $h(x) = \dfrac{-x^2 + 3}{4 - x^2}$

62. $h(x) = \dfrac{27x^3 + 6x}{10 - 27x^3}$

63. *Geometry* A square concrete foundation was prepared as a base for a cylindrical tank (see figure).

(a) Express the radius r of the tank as a function of the length x of the sides of the square.

(b) Express the area A of the circular base of the tank as a function of the radius r.

(c) Find and interpret $(A \circ r)(x)$.

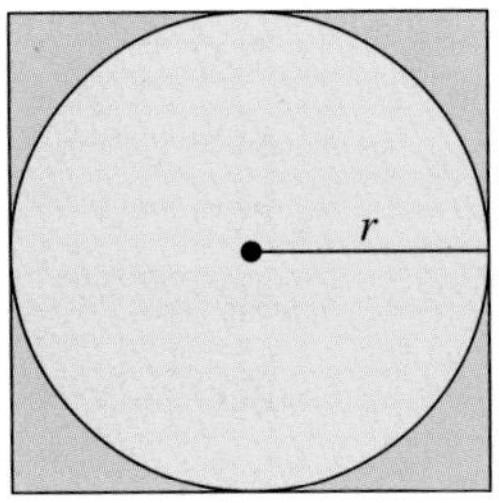

x

64. *Physics* A pebble is dropped into a calm pond, causing ripples in the form of concentric circles. The radius (in feet) of the outer ripple is $r(t) = 0.6t$, where t is the time in seconds after the pebble strikes the water. The area of the circle is given by the function $A(r) = \pi r^2$. Find and interpret $(A \circ r)(t)$.

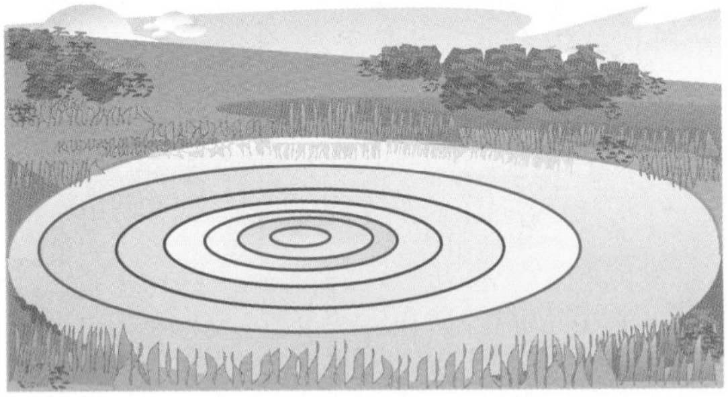

65. *Economics* The weekly cost of producing x units in a manufacturing process is given by the function

$C(x) = 60x + 750.$

The number of units produced in t hours is $x(t) = 50t$. Find and interpret $(C \circ x)(t)$.

Synthesis

True or False? **In Exercises 66 and 67, determine whether the statement is true or false. Justify your answer.**

66. If $f(x) = x + 1$ and $g(x) = 6x$, then $(f \circ g)(x) = (g \circ f)(x)$.

67. If you are given two functions $f(x)$ and $g(x)$, you can calculate $(f \circ g)(x)$ if and only if the range of g is a subset of the domain of f.

68. *Think About It* You are a sales representative for an automobile manufacturer. You are paid an annual salary, plus a bonus of 3% of your sales over \$500,000. Consider the two functions

$f(x) = x - 500{,}000$ and $g(x) = 0.03x$.

If x is greater than \$500,000, which of the following represents your bonus? Explain your reasoning.

(a) $f(g(x))$ (b) $g(f(x))$

69. Prove that the product of two odd functions is an even function, and that the product of two even functions is an even function.

70. *Conjecture* Use examples to hypothesize whether the product of an odd function and an even function is even or odd. Then prove your hypothesis.

Review

Average Rate of Change **In Exercises 71–74, find the difference quotient**

$$\frac{f(x + h) - f(x)}{h}$$

and simplify your answer.

71. $f(x) = 3x - 4$

72. $f(x) = 1 - x^2$

73. $f(x) = \dfrac{4}{x}$

74. $f(x) = \sqrt{2x + 1}$

In Exercises 75–78, find an equation of the line that passes through the given point and has the indicated slope. Sketch a graph of the line.

75. $(2, -4), m = 3$

76. $(-6, 3), m = -1$

77. $(8, -1), m = -\frac{3}{2}$

78. $(7, 0), m = \frac{5}{7}$

1.7 Inverse Functions

▶ **What you should learn**

- How to find inverse functions informally and verify that two functions are inverses of each other
- How to use graphs of functions to decide whether functions have inverses
- How to find inverse functions algebraically

▶ **Why you should learn it**

Inverse functions can be used to model and solve real-life problems. For instance, Exercise 81 on page 179 shows how the inverse of a function can be used to determine when damage to a diesel engine may occur.

David J. Sams/Stock Boston

The Inverse of a Function

Recall from Section 1.3 that a function can be represented by a set of ordered pairs. For instance, the function $f(x) = x + 4$ from the set $A = \{1, 2, 3, 4\}$ to the set $B = \{5, 6, 7, 8\}$ can be written as follows.

$$f(x) = x + 4: \ \{(1, 5), (2, 6), (3, 7), (4, 8)\}$$

In this case, by interchanging the first and second coordinates of each of these ordered pairs, you can form the **inverse function** of f, which is denoted by f^{-1}. It is a function from the set B to the set A, and can be written as follows.

$$f^{-1}(x) = x - 4: \ \{(5, 1), (6, 2), (7, 3), (8, 4)\}$$

Note that the domain of f is equal to the range of f^{-1}, and vice versa, as shown in Figure 1.54. Also note that the functions f and f^{-1} have the effect of "undoing" each other. In other words, when you form the composition of f with f^{-1} or the composition of f^{-1} with f, you obtain the identity function.

$$f(f^{-1}(x)) = f(x - 4) = (x - 4) + 4 = x$$

$$f^{-1}(f(x)) = f^{-1}(x + 4) = (x + 4) - 4 = x$$

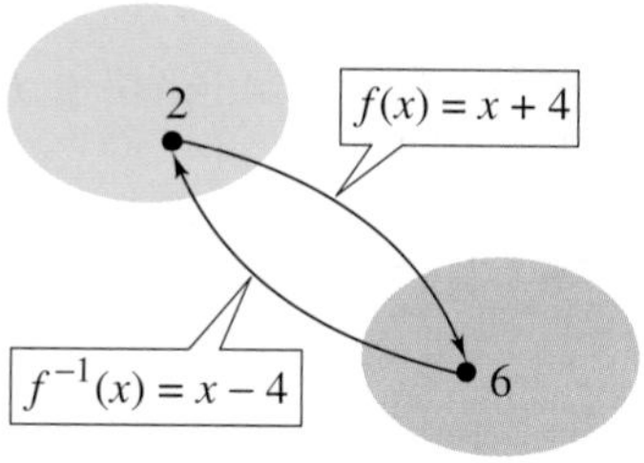

FIGURE 1.54

Example 1 ▶ Finding Inverse Functions Informally

Find the inverse of $f(x) = 4x$. Then verify that both $f(f^{-1}(x))$ and $f^{-1}(f(x))$ are equal to the identity function.

Solution

The function f *multiplies* each input by 4. To "undo" this function, you need to *divide* each input by 4. So, the inverse function of $f(x) = 4x$ is

$$f^{-1}(x) = \frac{x}{4}.$$

You can verify that both $f(f^{-1}(x))$ and $f^{-1}(f(x))$ are equal to the identity function as follows.

$$f(f^{-1}(x)) = f\left(\frac{x}{4}\right) = 4\left(\frac{x}{4}\right) = x \qquad f^{-1}(f(x)) = f^{-1}(4x) = \frac{4x}{4} = x$$

Exploration

Consider the functions

$$f(x) = x + 2$$

and

$$f^{-1}(x) = x - 2.$$

Evaluate $f(f^{-1}(x))$ and $f^{-1}(f(x))$ for the indicated values of x. What can you conclude about the functions?

x	-10	0	7	45
$f(f^{-1}(x))$				
$f^{-1}(f(x))$				

Definition of the Inverse of a Function

Let f and g be two functions such that

$$f(g(x)) = x \quad \text{for every } x \text{ in the domain of } g$$

and

$$g(f(x)) = x \quad \text{for every } x \text{ in the domain of } f.$$

Under these conditions, the function g is the **inverse** of the function f. The function g is denoted by f^{-1} (read "f-inverse"). So,

$$f(f^{-1}(x)) = x \quad \text{and} \quad f^{-1}(f(x)) = x.$$

The domain of f must be equal to the range of f^{-1}, and the range of f must be equal to the domain of f^{-1}.

Don't be confused by the use of -1 to denote the inverse function f^{-1}. In this text, whenever f^{-1} is written, it *always* refers to the inverse of the function f and *not* to the reciprocal of $f(x)$.

If the function g is the inverse of the function f, it must also be true that the function f is the inverse of the function g. For this reason, you can say that the functions f and g are *inverses of each other.*

Example 2 ▶ Verifying Inverse Functions

Which of the functions is the inverse of $f(x) = \dfrac{5}{x-2}$?

$$g(x) = \frac{x-2}{5} \qquad h(x) = \frac{5}{x} + 2$$

Solution

By forming the composition of f with g, you have

$$f(g(x)) = f\left(\frac{x-2}{5}\right)$$

$$= \frac{5}{\left(\dfrac{x-2}{5}\right) - 2} \qquad \text{Substitute } \frac{x-2}{5} \text{ for } x.$$

$$= \frac{25}{x-12} \qquad \text{Simplify.}$$

$$\neq x.$$

Because this composition is not equal to the identity function x, it follows that g *is not* the inverse of f. By forming the composition of f with h, you have

$$f(h(x)) = f\left(\frac{5}{x} + 2\right) = \frac{5}{\left(\dfrac{5}{x} + 2\right) - 2} = \frac{5}{\left(\dfrac{5}{x}\right)} = x.$$

So, it appears that h *is* the inverse of f. You can confirm this by showing that the composition of h with f is also equal to the identity function. (Try doing this.)

Additional Examples

Determine whether the two functions are inverses of each other.

a. $f(x) = 3x - 2,\ g(x) = \frac{1}{3}x + 2$

b. $f(x) = \frac{1}{4}x - 3,\ g(x) = 12 + 4x$

c. $f(x) = 2x + 4,\ g(x) = \frac{1}{2}x - 2$

Solution

The functions in parts b and c are inverses of each other because $f(g(x)) = g(f(x)) = x$.

The functions in part a are not inverses of each other.

A computer animation of this concept appears in the *Interactive* CD-ROM and *Internet* versions of this text.

The Graph of the Inverse of a Function

The graphs of a function f and its inverse f^{-1} are related to each other in the following way. If the point (a, b) lies on the graph of f, then the point (b, a) must lie on the graph of f^{-1}, and vice versa. This means that the graph of f^{-1} is a *reflection* of the graph of f in the line $y = x$, as shown in Figure 1.55.

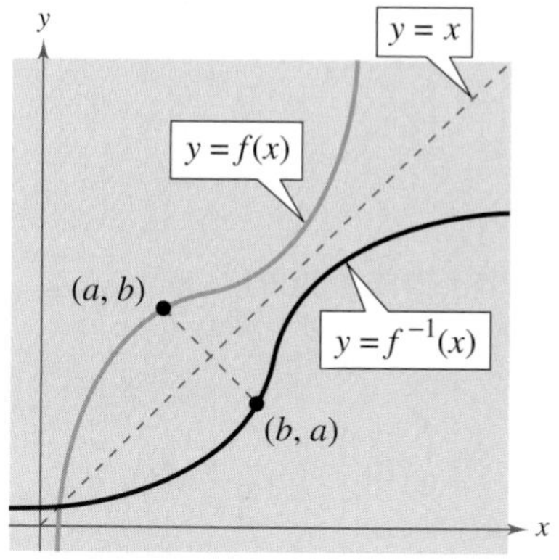

FIGURE 1.55

Example 3 ▶ The Graphs of f and f^{-1}

Sketch the graphs of the inverse functions $f(x) = 2x - 3$ and $f^{-1}(x) = \frac{1}{2}(x + 3)$ on the same rectangular coordinate system and show that the graphs are reflections of each other in the line $y = x$.

Solution

The graphs of f and f^{-1} are shown in Figure 1.56. It appears that the graphs are reflections of each other in the line $y = x$. You can further verify this reflective property by testing a few points on each graph. Note in the following list that if the point (a, b) is on the graph of f, the point (b, a) is on the graph of f^{-1}.

Graph of $f(x) = 2x - 3$	*Graph of* $f^{-1}(x) = \frac{1}{2}(x + 3)$
$(-1, -5)$	$(-5, -1)$
$(0, -3)$	$(-3, 0)$
$(1, -1)$	$(-1, 1)$
$(2, 1)$	$(1, 2)$
$(3, 3)$	$(3, 3)$

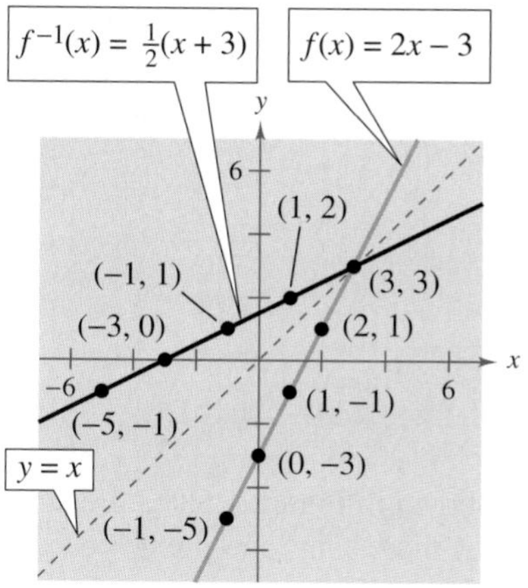

FIGURE 1.56

Example 4 ▶ Finding Inverse Functions Graphically

Sketch the graphs of the inverse functions $f(x) = x^2, x \geq 0$ and $f^{-1}(x) = \sqrt{x}$ on the same rectangular coordinate system and show that the graphs are reflections of each other in the line $y = x$.

Solution

The graphs of f and f^{-1} are shown in Figure 1.57. It appears that the graphs are reflections of each other in the line $y = x$. You can further verify this reflective property by testing a few points on each graph. Note in the following list that if the point (a, b) is on the graph of f, the point (b, a) is on the graph of f^{-1}.

Graph of $f(x) = x^2, \quad x \geq 0$	*Graph of* $f^{-1}(x) = \sqrt{x}$
$(0, 0)$	$(0, 0)$
$(1, 1)$	$(1, 1)$
$(2, 4)$	$(4, 2)$
$(3, 9)$	$(9, 3)$

Try showing that $f(f^{-1}(x)) = x$ and $f^{-1}(f(x)) = x$.

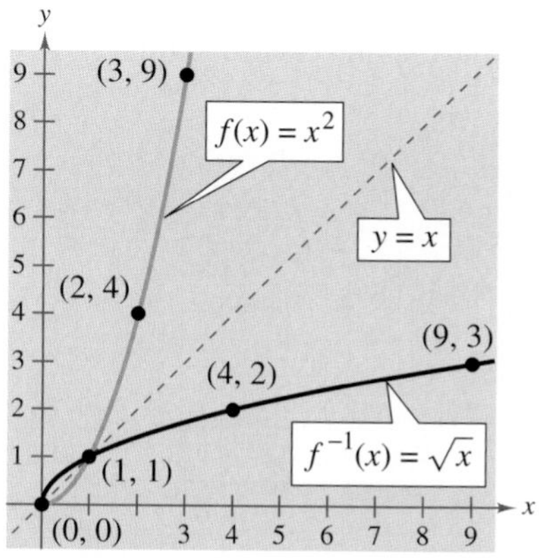

FIGURE 1.57

The reflective property of the graphs of inverse functions gives you a nice *geometric* test for determining whether a function has an inverse. This test is called the **Horizontal Line Test** for inverse functions.

Horizontal Line Test for Inverse Functions

A function f has an inverse function if and only if no *horizontal* line intersects the graph of f at more than one point.

STUDY TIP

Not every function has an inverse function. Consider the following table of values for the function $f(x) = x^2$.

x	-2	-1	0	1	2	3
$f(x)$	4	1	0	1	4	9

The table of values made up by interchanging the rows of the table does not represent a function because the input $x = 4$ is matched with two different outputs: $y = -2$ and $y = 2$.

x	4	1	0	1	4	9
y	-2	-1	0	1	2	3

So, $f(x) = x^2$ does not have an inverse function.

Example 5 ▶ Applying the Horizontal Line Test

a. The graph of the function $f(x) = x^3 - 1$ is shown in Figure 1.58(a). Because no horizontal line intersects the graph of f at more than one point, you can conclude that f *does* have an inverse function.

b. The graph of the function $f(x) = x^2 - 1$ is shown in Figure 1.58(b). Because it is possible to find a horizontal line that intersects the graph of f at more than one point, you can conclude that f *does not* have an inverse function.

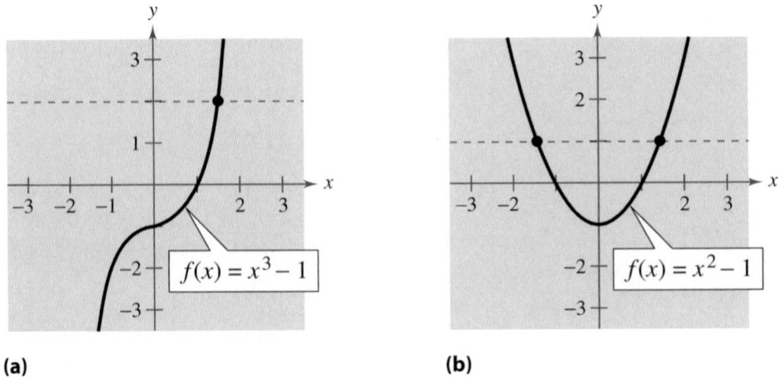

FIGURE 1.58

Activities

1. Given $f(x) = 5x - 7$, find $f^{-1}(x)$.

   Answer: $f^{-1}(x) = \dfrac{x+7}{5}$

2. Show that f and g are inverse functions by showing that $f(g(x)) = x$ and $g(f(x)) = x$.

   $f(x) = 3x^3 + 1$

   $g(x) = \sqrt[3]{\dfrac{x-1}{3}}$

3. Describe the graphs of functions that have inverses and show how the graph of a function and its inverse are related.

Finding the Inverse of a Function Algebraically

For simple functions (such as the one in Example 1), you can find inverse functions by inspection. For more complicated functions, however, it is best to use the following guidelines. The key step in these guidelines is Step 3—interchanging the roles of x and y. This step corresponds to the fact that inverse functions have ordered pairs with the coordinates reversed.

Finding the Inverse of a Function

1. Use the Horizontal Line Test to decide whether f has an inverse.
2. In the equation for $f(x)$, replace $f(x)$ by y.
3. Interchange the roles of x and y, and solve for y.
4. Replace y by $f^{-1}(x)$ in the new equation.
5. Verify that f and f^{-1} are inverses of each other by showing that the domain of f is equal to the range of f^{-1}, the range of f is equal to the domain of f^{-1}, and $f(f^{-1}(x)) = x = f^{-1}(f(x))$.

Example 6 ▶ Finding the Inverse of a Function Algebraically

Find the inverse of $f(x) = \dfrac{5-3x}{2}$.

Solution

The graph of f is a line, as shown in Figure 1.59. This graph passes the Horizontal Line Test. So, you know that f has an inverse.

$f(x) = \dfrac{5-3x}{2}$ — Write original function.

$y = \dfrac{5-3x}{2}$ — Replace $f(x)$ by y.

$x = \dfrac{5-3y}{2}$ — Interchange x and y.

$2x = 5 - 3y$ — Multiply each side by 2.

$3y = 5 - 2x$ — Isolate the y-term.

$y = \dfrac{5-2x}{3}$ — Solve for y.

$f^{-1}(x) = \dfrac{5-2x}{3}$ — Replace y by $f^{-1}(x)$.

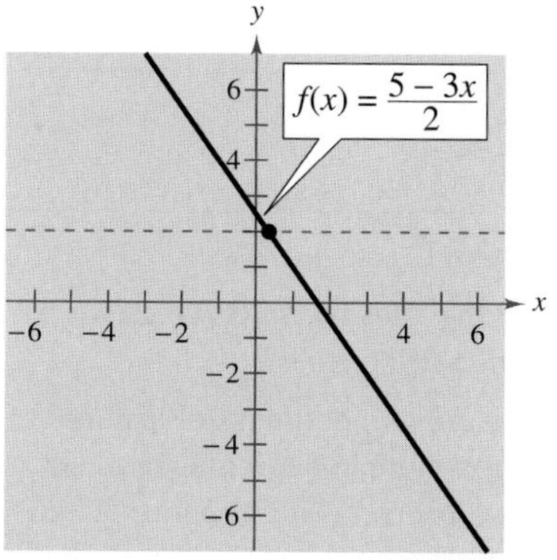

FIGURE 1.59

Note that both f and f^{-1} have domains and ranges that consist of the entire set of real numbers. Check that $f(f^{-1}(x)) = x$ and $f^{-1}(f(x)) = x$.

Example 7 ▶ Finding the Inverse of a Function

Find the inverse of $f(x) = \sqrt[3]{x + 1}$.

Solution

The graph of f is a curve, as shown in Figure 1.60. Because this graph passes the Horizontal Line Test, you know that f has an inverse function.

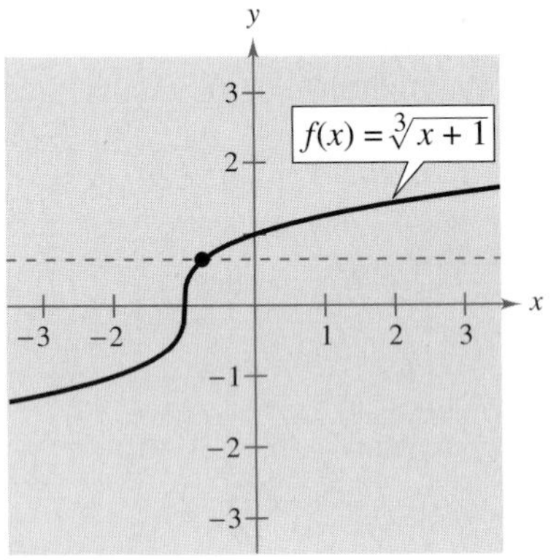

FIGURE 1.60

$$f(x) = \sqrt[3]{x + 1}$$ Write original function.

$$y = \sqrt[3]{x + 1}$$ Replace $f(x)$ by y.

$$x = \sqrt[3]{y + 1}$$ Interchange x and y.

$$x^3 = y + 1$$ Cube each side.

$$x^3 - 1 = y$$ Solve for y.

$$x^3 - 1 = f^{-1}(x)$$ Replace y by $f^{-1}(x)$.

Both f and f^{-1} have domains and ranges that consist of the entire set of real numbers.

Check

$$\begin{aligned} f(f^{-1}(x)) &= f(x^3 - 1) \\ &= \sqrt[3]{(x^3 - 1) + 1} \\ &= \sqrt[3]{x^3} \\ &= x \checkmark \end{aligned}$$

$$\begin{aligned} f^{-1}(f(x)) &= f^{-1}\left(\sqrt[3]{x + 1}\right) \\ &= \left(\sqrt[3]{x + 1}\right)^3 - 1 \\ &= x + 1 - 1 \\ &= x \checkmark \end{aligned}$$

You can verify the result of Example 7 numerically as shown in the tables below.

x	-28	-9	-2	-1	0	7	26
$f(x)$	-3	-2	-1	0	1	2	3

x	-3	-2	-1	0	1	2	3
$f^{-1}(x)$	-28	-9	-2	-1	0	7	26

Writing ABOUT MATHEMATICS

The Existence of an Inverse Function Write a short paragraph describing why the following functions do or do not have inverse functions.

a. Let x represent the retail price of an item (in dollars), and let $f(x)$ represent the sales tax on the item. Assume that the sales tax is 6% of the retail price *and* that the sales tax is rounded to the nearest cent. Does this function have an inverse? (*Hint:* Can you undo this function? For instance, if you know that the sales tax is \$0.12, can you determine *exactly* what the retail price is?)

b. Let x represent the temperature in degrees Celsius, and let $f(x)$ represent the temperature in degrees Fahrenheit. Does this function have an inverse? (*Hint:* The formula for converting from degrees Celsius to degrees Fahrenheit is $F = \frac{9}{5}C + 32$.)

1.7 Exercises

In Exercises 1–4, match the graph of the function with the graph of its inverse. [The graphs of the inverse functions are labeled (a), (b), (c), and (d).]

(a)

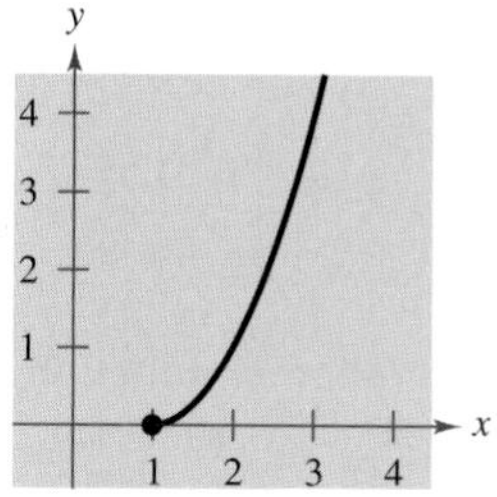

(b)

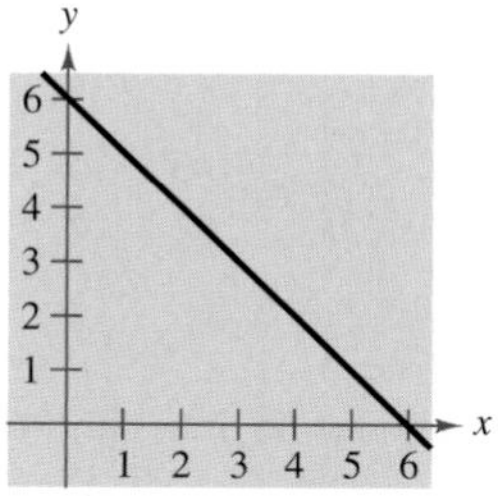

(c)

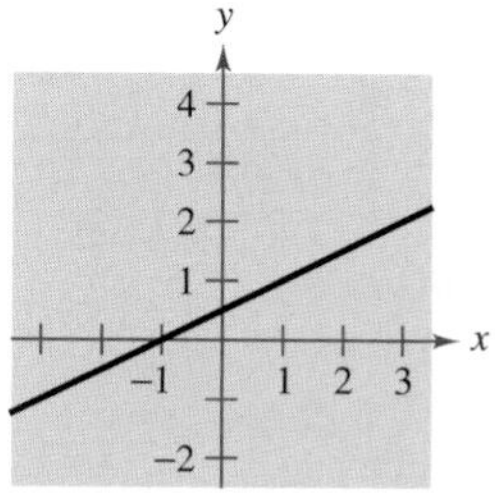

(d)

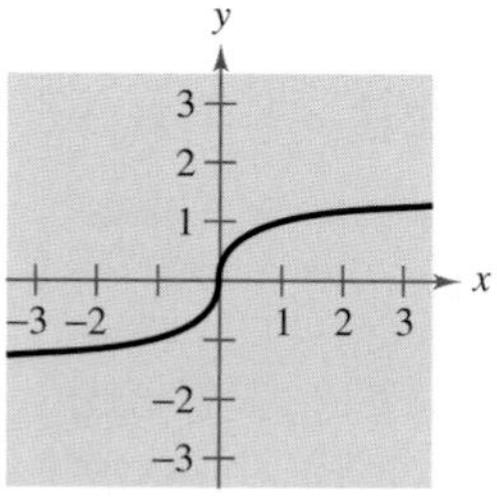

1.

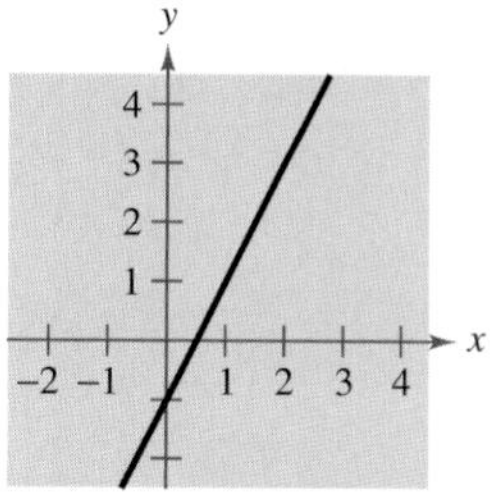

2.

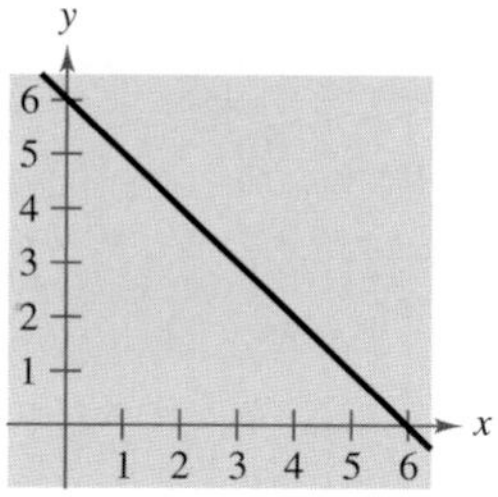

3.

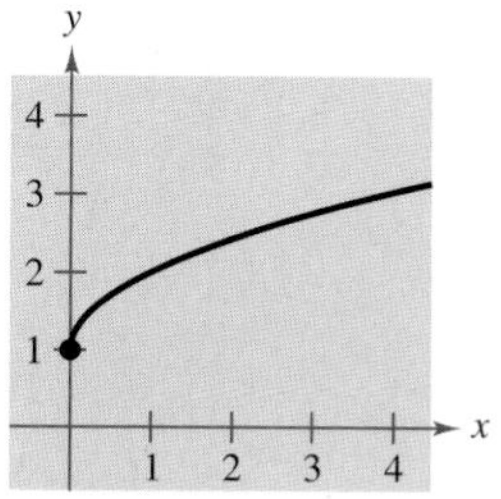

4.

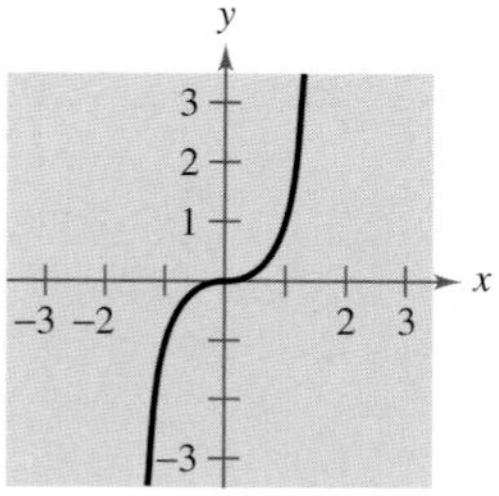

In Exercises 5–12, find the inverse of f informally. Verify that $f(f^{-1}(x)) = x$ and $f^{-1}(f(x)) = x$.

5. $f(x) = 6x$

6. $f(x) = \frac{1}{3}x$

7. $f(x) = x + 9$

8. $f(x) = x - 4$

9. $f(x) = 3x + 1$

10. $f(x) = \dfrac{x - 1}{5}$

11. $f(x) = \sqrt[3]{x}$

12. $f(x) = x^5$

In Exercises 13–24, show that f and g are inverse functions (a) algebraically and (b) graphically.

13. $f(x) = 2x,\quad g(x) = \dfrac{x}{2}$

14. $f(x) = x - 5,\quad g(x) = x + 5$

15. $f(x) = 5x + 1,\quad g(x) = \dfrac{x - 1}{5}$

16. $f(x) = 3 - 4x,\quad g(x) = \dfrac{3 - x}{4}$

17. $f(x) = x^3,\quad g(x) = \sqrt[3]{x}$

18. $f(x) = \dfrac{1}{x},\quad g(x) = \dfrac{1}{x}$

19. $f(x) = \sqrt{x - 4},\quad g(x) = x^2 + 4,\quad x \ge 0$

20. $f(x) = 1 - x^3,\quad g(x) = \sqrt[3]{1 - x}$

21. $f(x) = 9 - x^2,\quad x \ge 0,\quad g(x) = \sqrt{9 - x},\quad x \le 9$

22. $f(x) = \dfrac{1}{1 + x},\quad x \ge 0$

$g(x) = \dfrac{1 - x}{x},\quad 0 < x \le 1$

23. $f(x) = \dfrac{x - 1}{x + 5},\quad g(x) = -\dfrac{5x + 1}{x - 1}$

24. $f(x) = \dfrac{x + 3}{x - 2},\quad g(x) = \dfrac{2x + 3}{x - 1}$

In Exercises 25 and 26, does the function have an inverse?

25.

x	−1	0	1	2	3	4
$f(x)$	−2	1	2	1	−2	−6

26.

x	−3	−2	−1	0	2	3
$f(x)$	10	6	4	1	−3	−10

In Exercises 27 and 28, use the table of values for $y = f(x)$ to complete a table for $y = f^{-1}(x)$.

27.

x	-2	-1	0	1	2	3
$f(x)$	-2	0	2	4	6	8

28.

x	-3	-2	-1	0	1	2
$f(x)$	-10	-7	-4	-1	2	5

In Exercises 29–32, does the function have an inverse?

29.

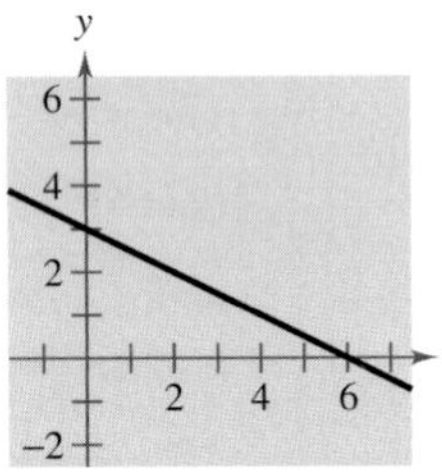

30.

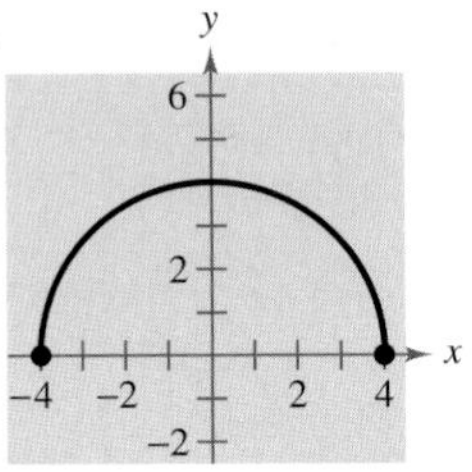

31.

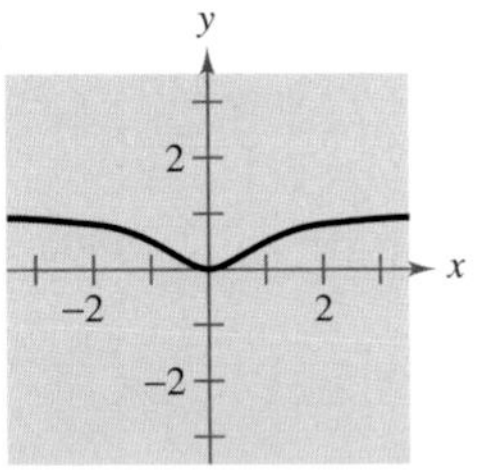

32.

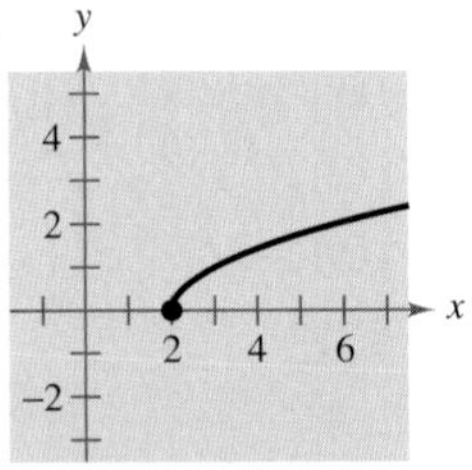

In Exercises 33–38, use a graphing utility to graph the function and use the Horizontal Line Test to determine whether the function has an inverse.

33. $g(x) = \dfrac{4 - x}{6}$

34. $f(x) = 10$

35. $h(x) = |x + 4| - |x - 4|$

36. $g(x) = (x + 5)^3$

37. $f(x) = -2x\sqrt{16 - x^2}$

38. $f(x) = \frac{1}{8}(x + 2)^2 - 1$

In Exercises 39–54, find the inverse of the function *f*. Then graph both *f* and f^{-1} on the same set of coordinate axes.

39. $f(x) = 2x - 3$

40. $f(x) = 3x + 1$

41. $f(x) = x^5 - 2$

42. $f(x) = x^3 + 1$

43. $f(x) = \sqrt{x}$

44. $f(x) = x^2, \quad x \ge 0$

45. $f(x) = \sqrt{4 - x^2}, \quad 0 \le x \le 2$

46. $f(x) = x^2 - 2, \quad x \le 0$

47. $f(x) = \dfrac{4}{x}$

48. $f(x) = -\dfrac{2}{x}$

49. $f(x) = \dfrac{x + 1}{x - 2}$

50. $f(x) = \dfrac{x - 3}{x + 2}$

51. $f(x) = \sqrt[3]{x - 1}$

52. $f(x) = x^{3/5}$

53. $f(x) = \dfrac{6x + 4}{4x + 5}$

54. $f(x) = \dfrac{8x - 4}{2x + 6}$

In Exercises 55–68, determine whether the function has an inverse. If it does, find the inverse.

55. $f(x) = x^4$

56. $f(x) = \dfrac{1}{x^2}$

57. $g(x) = \dfrac{x}{8}$

58. $f(x) = 3x + 5$

59. $p(x) = -4$

60. $f(x) = \dfrac{3x + 4}{5}$

61. $f(x) = (x + 3)^2, \quad x \ge -3$

62. $q(x) = (x - 5)^2$

63. $f(x) = \begin{cases} x + 3, & x < 0 \\ 6 - x, & x \ge 0 \end{cases}$

64. $f(x) = \begin{cases} -x, & x \le 0 \\ x^2 - 3x, & x > 0 \end{cases}$

65. $h(x) = \dfrac{1}{x}$

66. $f(x) = |x - 2|, \quad x \le 2$

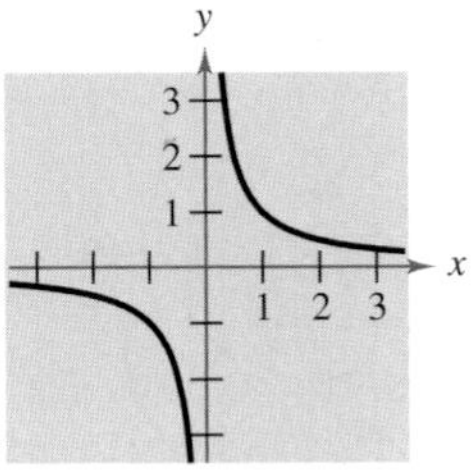

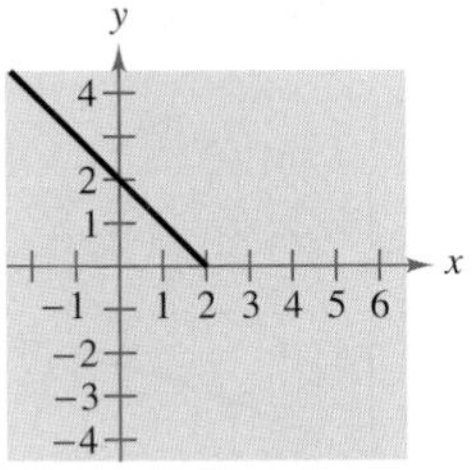

67. $f(x) = \sqrt{2x + 3}$

68. $f(x) = \sqrt{x - 2}$

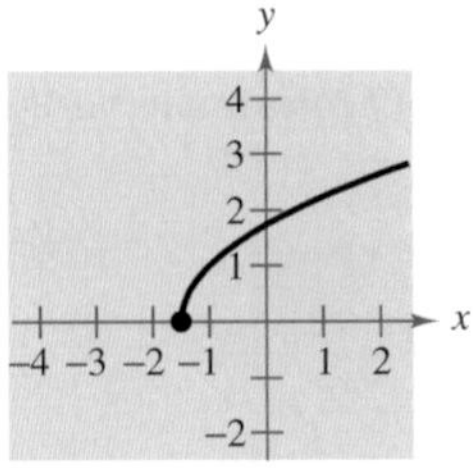

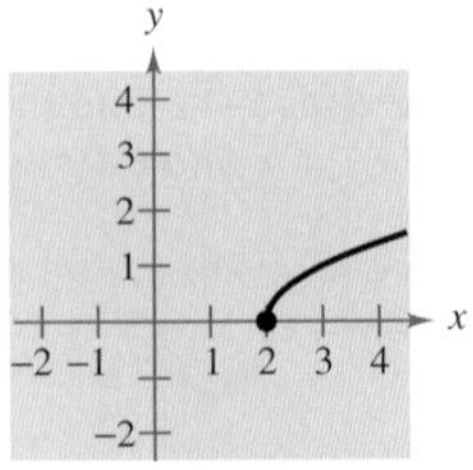

In Exercises 69–74, use the functions $f(x) = \frac{1}{8}x - 3$ and $g(x) = x^3$ to find the indicated value or function.

69. $(f^{-1} \circ g^{-1})(1)$

70. $(g^{-1} \circ f^{-1})(-3)$

71. $(f^{-1} \circ f^{-1})(6)$

72. $(g^{-1} \circ g^{-1})(-4)$

73. $(f \circ g)^{-1}$

74. $g^{-1} \circ f^{-1}$

In Exercises 75–78, use the functions $f(x) = x + 4$ and $g(x) = 2x - 5$ to find the specified function.

75. $g^{-1} \circ f^{-1}$

76. $f^{-1} \circ g^{-1}$

77. $(f \circ g)^{-1}$

78. $(g \circ f)^{-1}$

79. *Hourly Wage* Your wage is \$8.00 per hour plus \$0.75 for each unit produced per hour. So, your hourly wage y in terms of the number of units produced is

$$y = 8 + 0.75x.$$

(a) Find the inverse of the function.

(b) What does each variable represent in the inverse function?

(c) Determine the number of units produced when your hourly wage is \$22.25.

80. *Cost* Suppose you need a total of 50 pounds of two commodities costing \$1.25 and \$1.60 per pound, respectively.

(a) Verify that the total cost is

$$y = 1.25x + 1.60(50 - x)$$

where x is the number of pounds of the less expensive commodity.

(b) Find the inverse of the cost function. What does each variable represent in the inverse function?

(c) Use the context of the problem to determine the domain of the inverse function.

(d) Determine the number of pounds of the less expensive commodity purchased if the total cost is \$73.

81. *Diesel Mechanics* The function

$$y = 0.03x^2 + 245.50, \qquad 0 < x < 100$$

approximates the exhaust temperature y in degrees Fahrenheit where x is the percent load for a diesel engine.

(a) Find the inverse of the function. What does each variable represent in the inverse function?

(b) Use a graphing utility to graph the inverse function.

(c) Determine the percent load interval if the exhaust temperature of the engine must not exceed 500 degrees Fahrenheit.

82. *New Car Sales* The total value of new car sales f (in billions of dollars) in the United States from 1992 through 1997 is shown in the table. The time (in years) is given by t, with $t = 2$ corresponding to 1992. (Source: National Automobile Dealers Association)

t	2	3	4	5	6	7
$f(t)$	333.8	377.3	430.6	456.2	490.0	507.5

(a) Does f^{-1} exist?

(b) If f^{-1} exists, what does it mean in the context of the problem?

(c) If f^{-1} exists, find $f^{-1}(456.2)$.

83. If the table in Exercise 82 were extended to 1998 and if the total value of new car sales for that year was \$430.6 billion, would f^{-1} exist? Explain.

84. *Cellular Phones* The average local bill (in dollars) for cellular phones in the United States from 1990 to 1997 is shown in the table. The time (in years) is given by t, with $t = 0$ corresponding to 1990. (Source: Cellular Telecommunications Industry Association)

t	0	1	2	3
$f(t)$	80.90	72.74	68.68	61.48

t	4	5	6	7
$f(t)$	56.21	51.00	47.70	42.78

(a) Find $f^{-1}(51)$.

(b) What does f^{-1} mean in the context of the problem?

(c) Use the regression feature of a graphing utility to find a linear model for the data, $y = mx + b$. Round m and b to two decimal places.

(d) Algebraically find the inverse of the linear model in part (c).

(e) Use the inverse of the linear model you found in part (d) to approximate $f^{-1}(11)$.

85. *Soft Drink Consumption* The per capita consumption of regular soft drinks f (in gallons) in the United States from 1991 through 1996 is shown in the table. The time (in years) is given by t, with $t = 1$ corresponding to 1991. (Source: U.S. Department of Agriculture)

t	1	2	3	4	5	6
$f(t)$	36.3	36.9	38.4	39.5	39.8	40.2

(a) Does f^{-1} exist? If so, what does it represent in the context of the problem?

(b) If f^{-1} exists, what is $f^{-1}(39.8)$?

Synthesis

True or False? **In Exercises 86–89, determine whether the statement is true or false. Justify your answer.**

86. If f is an even function, f^{-1} exists.

87. If the inverse of f exists and the y-intercept of the graph of f exists, the y-intercept of f is an x-intercept of f^{-1}.

88. If $f(x) = x^n$ where n is odd, f^{-1} exists.

89. There exists no function f such that $f = f^{-1}$.

In Exercises 90–93, use the graph of the function f to create a table of values for the given points. Then create a second table that can be used to find f^{-1} and sketch the graph of f^{-1} if possible.

90.

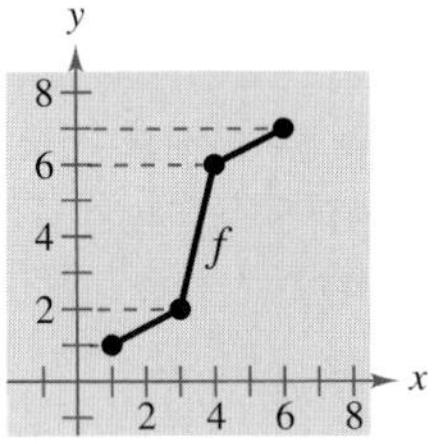

91.

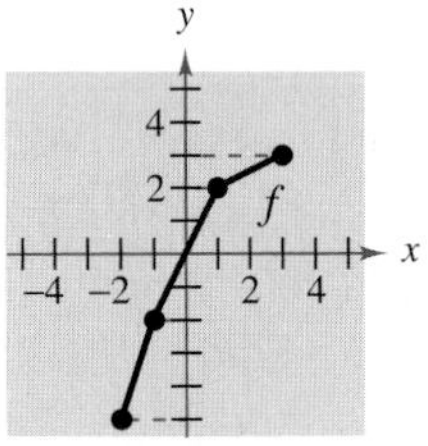

92.

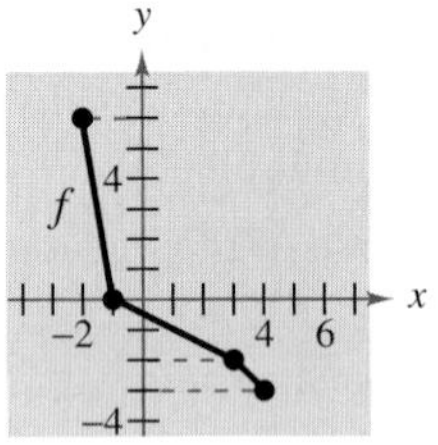

93.

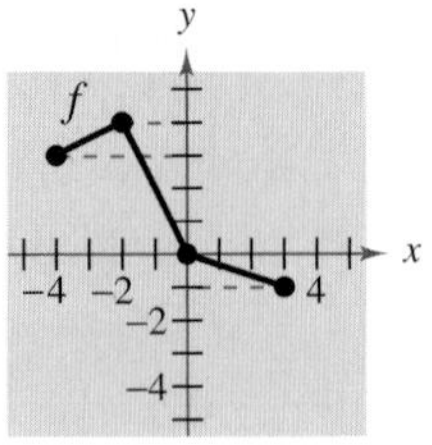

94. *Think About It* The function

$$f(x) = k(2 - x - x^3)$$

has an inverse, and $f^{-1}(3) = -2$. Find k.

Review

In Exercises 95–102, solve the equation by any convenient method.

95. $x^2 = 64$

96. $(x - 5)^2 = 8$

97. $4x^2 - 12x + 9 = 0$

98. $9x^2 + 12x + 3 = 0$

99. $x^2 - 6x + 4 = 0$

100. $2x^2 - 4x - 6 = 0$

101. $50 + 5x = 3x^2$

102. $2x^2 + 4x - 9 = 2(x - 1)^2$

In Exercises 103–106, find the domain of the function.

103. $f(x) = \sqrt[3]{x + 4}$

104. $f(x) = \sqrt{x + 6}$

105. $g(x) = \dfrac{2}{x^2 - 4x}$

106. $h(x) = \dfrac{x}{5x + 7}$

107. Find two consecutive positive even integers whose product is 288.

108. *Landscaping* Two people must mow a rectangular lawn measuring 100 feet by 200 feet. The second person agrees to mow three-fourths of the lawn and starts by mowing around the outside. How wide a strip must the person mow on each of the four sides? If the mower has a 24-inch cut, approximate the required number of trips around the lawn.

109. *Geometry* A triangular sign has a height that is equal to its base. The area of the sign is 10 square feet. Find the base and height of the sign.

110. *Geometry* A triangular sign has a height that is twice its base. The area of the sign is 10 square feet. Find the base and height of the sign.

1.8 Mathematical Modeling

▶ What you should learn

- How to use mathematical models to approximate sets of data points
- How to write mathematical models for direct variation
- How to write mathematical models for direct variation as an *n*th power
- How to write mathematical models for inverse variation
- How to write mathematical models for joint variation
- How to use the least squares regression feature of a graphing utility to find mathematical models

▶ Why you should learn it

You can use functions as models to represent a wide variety of real-life data sets. For instance, Exercise 69 on page 191 shows how a linear function can be used to model the salaries of professional hockey players from 1990 through 1996.

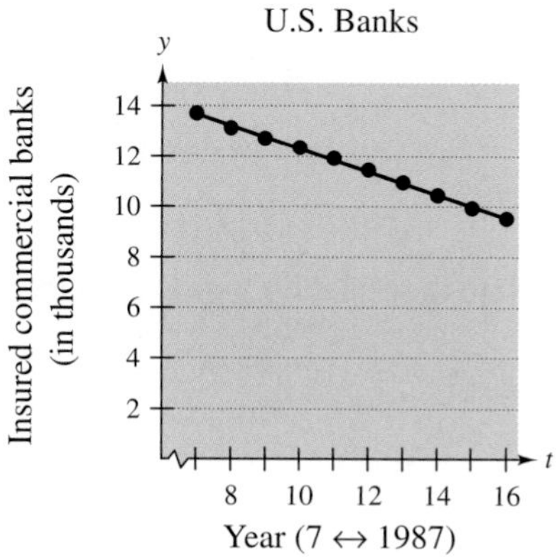

FIGURE 1.61

Introduction

You have already studied some techniques for fitting models to data. For instance, in Section 1.2, you learned how to find the equation of a line that passes through two points. In this section, you will study other techniques for fitting models to data: *direct and inverse variation* and *least squares regression.* The resulting models are either polynomial functions or rational functions. (Rational functions will be studied in Chapter 2.)

Example 1 ▶ A Mathematical Model

The numbers of insured commercial banks y (in thousands) in the United States for the years 1987 to 1996 are shown in the table. (Source: Federal Deposit Insurance Corporation)

Year	1987	1988	1989	1990	1991	1992	1993	1994	1995	1996
y	13.70	13.12	12.71	12.34	11.92	11.46	10.96	10.45	9.94	9.53

A linear model that approximates this data is

$$y = -0.459t + 16.89, \qquad 7 \le t \le 16$$

where $t = 7$ corresponds to 1987. Plot the actual data *and* the model on the same graph. How closely does the model represent the data?

Solution

The actual data is plotted in Figure 1.61, along with the graph of the linear model. From the graph, it appears that the model is a "good fit" for the actual data. You can see how well the model fits by comparing the actual values of y with the values of y given by the model. The values given by the model are labeled y^* in the table below.

t	7	8	9	10	11	12	13	14	15	16
y	13.70	13.12	12.71	12.34	11.92	11.46	10.96	10.45	9.94	9.53
y^*	13.68	13.22	12.76	12.30	11.84	11.38	10.92	10.46	10.01	9.55

Note in Example 1 that you could have chosen any two points to find a line that fits the data. However, the linear model above was found using the regression feature of a graphing utility and is the line that *best* fits the data. This concept of a "best-fitting" line is discussed later in this section.

Activity

In physics, Hooke's Law for springs states that the distance a spring is stretched or compressed from its natural, or equilibrium, length varies directly as the force on the spring. Distance is measured in inches (or meters) and force is measured in pounds (or newtons). One newton is equivalent to 0.225 pound.

a. Apply the concept of direct variation to find an equation relating the distance stretched (or compressed) to the force applied.

Answer: $d = kF$, where d is the distance stretched or compressed, F is the force, and k is the "spring constant"

b. If a force of 100 newtons stretches a spring 0.25 meter, how far will a force of 80 newtons stretch the spring?

Answer: 0.20 meter

Direct Variation

There are two basic types of linear models. The more general model has a y-intercept that is nonzero.

$$y = mx + b, \quad b \neq 0$$

The simpler model

$$y = kx$$

has a y-intercept that is zero. In the simpler model, y is said to **vary directly** as x, or to be **directly proportional** to x.

Direct Variation

The following statements are equivalent.

1. y **varies directly** as x.
2. y is **directly proportional** to x.
3. $y = kx$ for some nonzero constant k.

k is the **constant of variation** or the **constant of proportionality.**

Example 2 ▶ Direct Variation

In Pennsylvania, the state income tax is directly proportional to *gross income*. Suppose you were working in Pennsylvania and your state income tax deduction was \$42 for a gross monthly income of \$1500. Find a mathematical model that gives the Pennsylvania state income tax in terms of gross income.

Solution

Verbal Model: State income tax $=$ k $\cdot$ Gross income

Labels:
State income tax $= y$ (dollars)
Gross income $= x$ (dollars)
Income tax rate $= k$ (percent in decimal form)

Equation: $y = kx$

To solve for k, substitute the given information into the equation $y = kx$, and then solve for k.

$$y = kx \qquad \text{Write direct variation model.}$$

$$42 = k(1500) \qquad \text{Substitute } y = 42 \text{ and } x = 1500.$$

$$0.028 = k \qquad \text{Simplify.}$$

So, the equation (or model) for state income tax in Pennsylvania is

$$y = 0.028x.$$

In other words, Pennsylvania has a state income tax rate of 2.8% of gross income. The graph of this equation is shown in Figure 1.62.

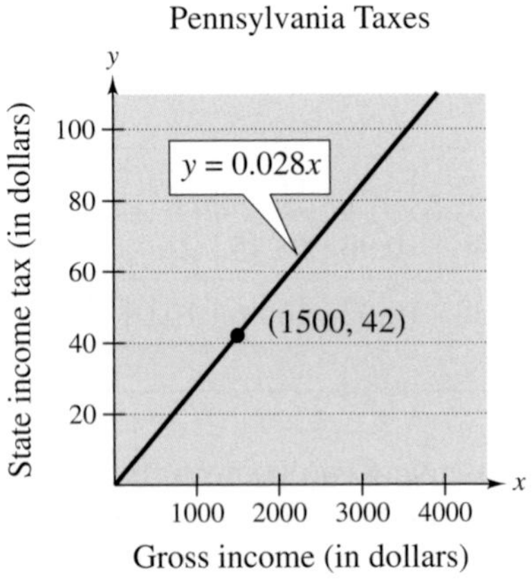

FIGURE 1.62

Direct Variation as *n*th Power

Another type of direct variation relates one variable to a *power* of another variable. For example, in the formula for the area of a circle

$$A = \pi r^2$$

the area A is directly proportional to the square of the radius r. Note that for this formula, π is the constant of proportionality.

STUDY TIP

Note that the direct variation model $y = kx$ is a special case of $y = kx^n$ with $n = 1$.

Direct Variation as *n*th Power

The following statements are equivalent.

1. y **varies directly as the** n**th power** of x.
2. y is **directly proportional to the** n**th power** of x.
3. $y = kx^n$ for some constant k.

Example 3 ▶ Direct Variation as *n*th Power

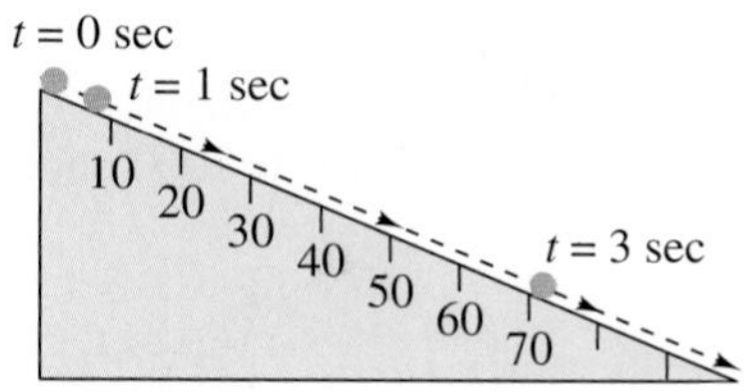

FIGURE 1.63

The distance a ball rolls down an inclined plane is directly proportional to the square of the time it rolls. During the first second the ball rolls 8 feet. (See Figure 1.63.)

a. Write an equation relating the distance traveled to the time.

b. How far will the ball roll during the first 3 seconds?

Solution

a. Letting d be the distance (in feet) the ball rolls and letting t be the time (in seconds), you have

$$d = kt^2.$$

Now, because $d = 8$ when $t = 1$, you can see that $k = 8$, as follows.

$$d = kt^2$$
$$8 = k(1)^2$$
$$8 = k$$

So, the equation relating distance to time is

$$d = 8t^2.$$

b. When $t = 3$, the distance traveled is $d = 8(3^2) = 8(9) = 72$ feet.

In Examples 2 and 3, the direct variations are such that an *increase* in one variable corresponds to an *increase* in the other variable. This is also true in the model $d = \frac{1}{5}F$, $F > 0$, where an increase in F results in an increase in d. You should not, however, assume that this always occurs with direct variation. For example, in the model $y = -3x$, an increase in x results in a *decrease* in y, and yet y is said to vary directly as x.

Inverse Variation

Inverse Variation

The following statements are equivalent.

1. y **varies inversely** as x.
2. y is **inversely proportional** to x.
3. $y = \frac{k}{x}$ for some constant k.

If x and y are related by an equation of the form $y = k/x^n$, then y varies inversely as the nth power of x (or y is inversely proportional to the nth power of x).

Example 4 ▶ Inverse Variation

A gas law states that the volume of an enclosed gas varies directly as the temperature *and* inversely as the pressure, as shown in Figure 1.64. The pressure of a gas is 0.75 kilogram per square centimeter when the temperature is 294 K and the volume is 8000 cubic centimeters.

a. Write an equation relating pressure, temperature, and volume.

b. Find the pressure when the temperature is 300 K and the volume is 7000 cubic centimeters.

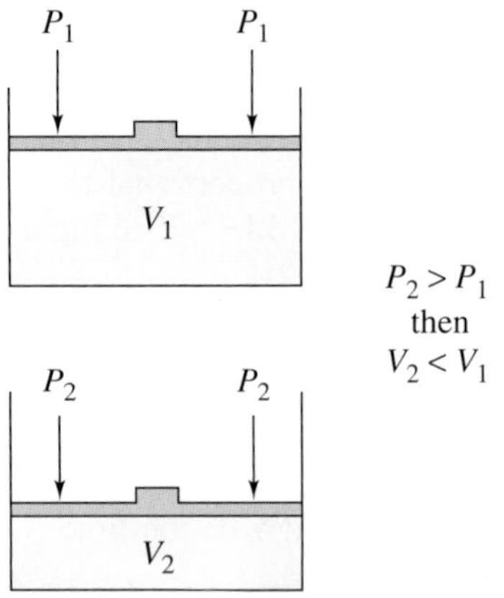

FIGURE 1.64 *If the temperature is held constant and pressure increases, volume decreases.*

Solution

a. Let V be volume (in cubic centimeters), let P be pressure (in kilograms per square centimeter), and let T be temperature (in Kelvin). Because V varies directly as T and inversely as P,

$$V = \frac{kT}{P}.$$

Now, because $P = 0.75$ when $T = 294$ and $V = 8000$,

$$8000 = \frac{k(294)}{0.75}$$

$$\frac{8000(0.75)}{294} = k$$

$$k = \frac{6000}{294} = \frac{1000}{49}.$$

So, the equation relating pressure, temperature, and volume is

$$V = \frac{1000}{49}\left(\frac{T}{P}\right).$$

b. When $T = 300$ and $V = 7000$, the pressure is

$$P = \frac{1000}{49}\left(\frac{300}{7000}\right) = \frac{300}{343} \approx 0.87 \text{ kilogram per square centimeter.}$$

Joint Variation

In Example 4, note that when a direct variation and an inverse variation occur in the same statement, they are coupled with the word "and." To describe two different *direct* variations in the same statement, the word **jointly** is used.

Joint Variation

The following statements are equivalent.

1. z **varies jointly** as x and y.
2. z is **jointly proportional** to x and y.
3. $z = kxy$ for some constant k.

If x, y, and z are related by an equation of the form

$$z = kx^n y^m$$

then z varies jointly as the nth power of x and the mth power of y.

Example 5 ▶ Joint Variation

The *simple* interest for a certain savings account is jointly proportional to the time and the principal. After one quarter (3 months), the interest on a principal of \$5000 is \$43.75.

a. Write an equation relating the interest, principal, and time.

b. Find the interest after three quarters.

Solution

a. Let I = interest (in dollars), P = principal (in dollars), and t = time (in years). Because I is jointly proportional to P and t,

$$I = kPt.$$

For $I = 43.75$, $P = 5000$, and $t = \frac{1}{4}$,

$$43.75 = k(5000)\left(\frac{1}{4}\right)$$

which implies that $k = 4(43.75)/5000 = 0.035$. So, the equation relating interest, principal, and time is

$$I = 0.035Pt$$

which is the familiar equation for simple interest where the constant of proportionality, 0.035, represents an annual interest rate of 3.5%.

b. When $P = \$5000$ and $t = \frac{3}{4}$, the interest is

$$I = (0.035)(5000)\left(\frac{3}{4}\right) = \$131.25.$$

Activities

1. Find a mathematical model for "a is jointly proportional to y and z."
   Answer: $a = kyz$
2. Determine the constant of proportionality if x is inversely proportional to the square of y, when $y = 7$ and $x = 3$.
   Answer: 147
3. Write a sentence using variation terminology to describe the formula $F = k\dfrac{m_1 m_2}{t^2}$.
   Answer: F is jointly proportional to m_1 and m_2 and inversely proportional to the square of t.

Technology

Most graphing utilities have a built-in linear regression program. When you run such a program, the "r-value" or **correlation coefficient,** gives a measure of how well the model fits the data. The closer the value of $|r|$ is to 1, the better the fit. For the data in Example 6, $r \approx 0.955$, which implies that the model is a good fit. This is confirmed in the table at the right.

Least Squares Regression

So far in this text, you have worked with many different types of mathematical models that approximate real-life data. For instance, in Example 1 on page 181 you analyzed a model for data on the number of insured commercial banks in the United States.

To find such a model, statisticians use a measure called the **sum of square differences,** which is the sum of the squares of the differences between actual data values and model values. The "best-fitting" linear model is the one with the least sum of square differences. This best-fitting linear model is called the **least squares regression line.** You can approximate this line visually by plotting the data points and drawing the line that appears to fit best—or you can enter the data points into a calculator or computer and use the calculator's or computer's linear regression program.

Example 6 ▶ Finding a Least Squares Regression Line

A computer simulation of this example appears in the *Interactive* CD-ROM and *Internet* versions of this text.

The amounts of total annual prize money p (in millions of dollars) awarded at the Indianapolis 500 race from 1989 to 1997 are shown in the table. Construct a scatter plot that represents the data and find a linear model that approximates the data. (Source: Indianapolis Motor Speedway Hall of Fame)

Year	1989	1990	1991	1992	1993	1994	1995	1996	1997
p	5.71	6.33	7.01	7.53	7.68	7.86	8.06	8.11	8.61

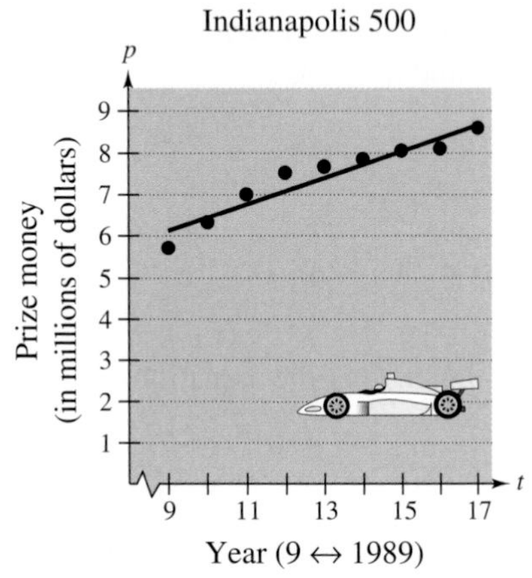

FIGURE 1.65

Solution

Let $t = 9$ represent 1989. The scatter plot for the points is shown in Figure 1.65. Using the least squares regression feature of a graphing utility, you can determine that the equation of the line is

$$p = 0.323t + 3.24.$$

To check this model, compare the actual p-values with the p-values given by the model, which are labeled $p*$ in the table below.

t	9	10	11	12	13	14	15	16	17
p	5.71	6.33	7.01	7.53	7.68	7.86	8.06	8.11	8.61
$p*$	6.15	6.47	6.79	7.12	7.44	7.76	8.09	8.41	8.73

1.8 Exercises

1. *Employment* The total numbers of employees (in thousands) in the United States from 1990 to 1997 are given by the following ordered pairs.

(1990, 125,840) (1994, 131,056)
(1991, 126,346) (1995, 132,304)
(1992, 128,105) (1996, 133,943)
(1993, 129,200) (1997, 136,297)

A linear model that approximates this data is

$y = 125{,}151.5 + 1495.68t, \quad 0 \le t \le 7$

where y represents the number of employees (in thousands) and $t = 0$ represents 1990. Plot the actual data and the model on the same graph. How closely does the model represent the data? (Source: U.S. Bureau of Labor Statistics)

2. *Sports* The winning times (in minutes) in the women's 400-meter freestyle swimming event in the Olympics from 1936 to 1996 are given by the following ordered pairs.

(1936, 5.44) (1972, 4.32)
(1948, 5.30) (1976, 4.16)
(1952, 5.20) (1980, 4.15)
(1956, 4.91) (1984, 4.12)
(1960, 4.84) (1988, 4.06)
(1964, 4.72) (1992, 4.12)
(1968, 4.53) (1996, 4.12)

A linear model that approximates this data is

$y = 5.35 - 0.027t, \quad -4 \le t \le 56$

where y represents the winning time in minutes and $t = 0$ represents 1940. Plot the actual data and the model on the same graph. How closely does the model represent the data? (Source: ESPN)

Think About It **In Exercises 3 and 4, use the graph to determine whether y varies directly as some power of x or inversely as some power of x. Explain.**

3.

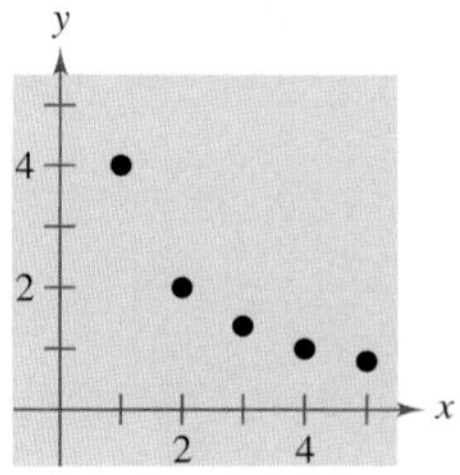

4.

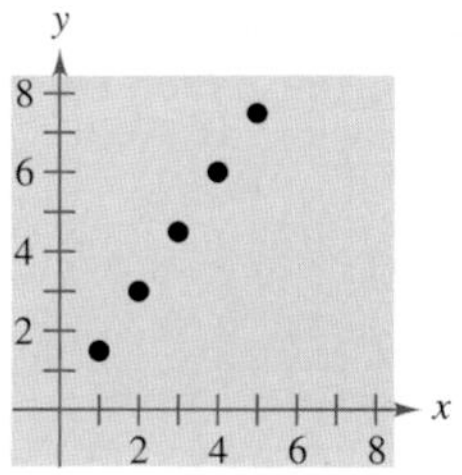

In Exercises 5–8, use the given value of k to complete the table for the direct variation model $y = kx^2$. Plot the points on a rectangular coordinate system.

x	2	4	6	8	10
$y = kx^2$					

5. $k = 1$ **6.** $k = 2$

7. $k = \frac{1}{2}$ **8.** $k = \frac{1}{4}$

In Exercises 9–12, use the given value of k to complete the table for the inverse variation model

$$y = \frac{k}{x^2}.$$

Plot the points on a rectangular coordinate system.

x	2	4	6	8	10
$y = \frac{k}{x^2}$					

9. $k = 2$ **10.** $k = 5$

11. $k = 10$ **12.** $k = 20$

In Exercises 13–16, determine whether the variation model is of the form

$$y = kx \quad \text{or} \quad y = \frac{k}{x}$$

and find k.

13.

x	5	10	15	20	25
y	1	$\frac{1}{2}$	$\frac{1}{3}$	$\frac{1}{4}$	$\frac{1}{5}$

14.

x	5	10	15	20	25
y	2	4	6	8	10

15.

x	5	10	15	20	25
y	−3.5	−7	−10.5	−14	−17.5

16.

x	5	10	15	20	25
y	24	12	8	6	$\frac{24}{5}$

Direct Variation **In Exercises 17–20, assume that y is directly proportional to x. Use the given x-value and y-value to find a linear model that relates y and x.**

	x-Value	*y-Value*		*x-Value*	*y-Value*
17.	$x = 5$	$y = 12$	**18.**	$x = 2$	$y = 14$
19.	$x = 10$	$y = 2050$	**20.**	$x = 6$	$y = 580$

21. *Simple Interest* The simple interest on an investment is directly proportional to the amount of the investment. By investing $2500 in a certain bond issue, you obtained an interest payment of $87.50 after 1 year. Find a mathematical model that gives the interest I for this bond issue after 1 year in terms of the amount invested P.

22. *Simple Interest* The simple interest on an investment is directly proportional to the amount of the investment. By investing $5000 in a municipal bond, you obtained an interest payment of $187.50 after 1 year. Find a mathematical model that gives the interest I for this municipal bond after 1 year in terms of the amount invested P.

23. *Measurement* On a yardstick with scales in inches and centimeters, you notice that 13 inches is approximately the same length as 33 centimeters. Use this information to find a mathematical model that relates centimeters to inches. Then use the model to complete the table.

Inches	5	10	20	25	30
Centimeters					

24. *Measurement* When buying gasoline, you notice that 14 gallons of gasoline is approximately the same amount of gasoline as 53 liters. Use this information to find a linear model that relates gallons to liters. Use the model to complete the table.

Gallons	5	10	20	25	30
Liters					

25. *Taxes* Property tax is based on the assessed value of the property. A house that has an assessed value of $150,000 has a property tax of $5520. Find a mathematical model that gives the amount of property tax y in terms of the assessed value x of the property. Use the model to find the property tax on a house that has an assessed value of $200,000.

26. *Taxes* State sales tax is based on retail price. An item that sells for $145.99 has a sales tax of $10.22. Find a mathematical model that gives the amount of sales tax y in terms of the retail price x. Use the model to find the sales tax on a $540.50 purchase.

Physics **In Exercises 27–30, use Hooke's Law for springs, which states that the distance a spring is stretched (or compressed) varies directly as the force on the spring.**

27. A force of 265 newtons stretches a spring 0.15 meter (see figure).

(a) How far will a force of 90 newtons stretch the spring?

(b) What force is required to stretch the spring 0.1 meter?

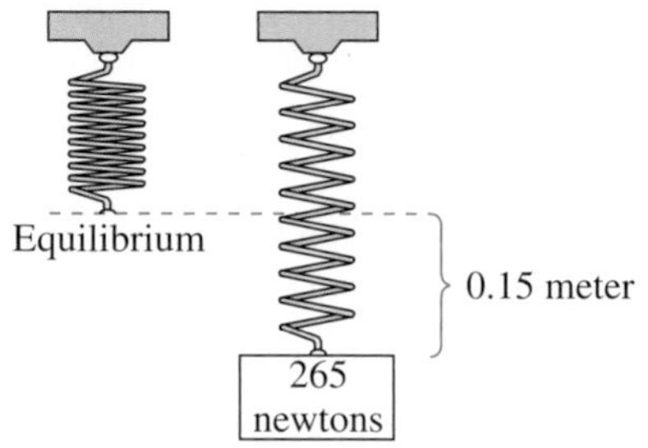

28. A force of 220 newtons stretches a spring 0.12 meter. What force is required to stretch the spring 0.16 meter?

29. The coiled spring of a toy supports the weight of a child. The spring is compressed a distance of 1.9 inches by the weight of a 25-pound child. The toy will not work properly if its spring is compressed more than 3 inches. What is the weight of the heaviest child who should be allowed to use the toy?

30. An overhead garage door has two springs, one on each side of the door (see figure). A force of 15 pounds is required to stretch each spring 1 foot. Because of a pulley system, the springs stretch only one-half the distance the door travels. The door moves a total of 8 feet, and the springs are at their natural length when the door is open. Find the combined lifting force applied to the door by the springs when the door is closed.

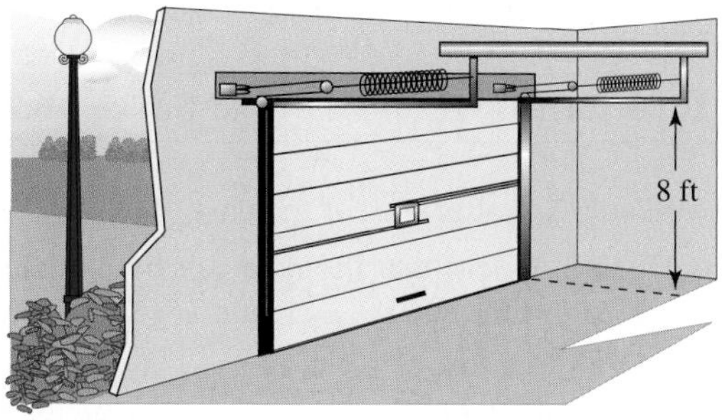

FIGURE FOR 30

In Exercises 31–40, find a mathematical model for the verbal statement.

31. A varies directly as the square of r.

32. V varies directly as the cube of e.

33. y varies inversely as the square of x.

34. h varies inversely as the square root of s.

35. F varies directly as g and inversely as r^2.

36. z is jointly proportional to the square of x and y^3.

37. Boyle's Law: For a constant temperature, the pressure P of a gas is inversely proportional to the volume V of the gas.

38. Newton's Law of Cooling: The rate of change R of the temperature of an object is proportional to the difference between the temperature T of the object and the temperature T_e of the environment in which the object is placed.

39. Newton's Law of Universal Gravitation: The gravitational attraction F between two objects of masses m_1 and m_2 is proportional to the product of the masses and inversely proportional to the square of the distance r between the objects.

40. Logistic growth: The rate of growth R of a population is jointly proportional to the size S of the population and the difference between S and the maximum population size L that the environment can support.

In Exercises 41–46, write a sentence using the variation terminology of this section to describe the formula.

41. Area of a triangle: $A = \frac{1}{2}bh$

42. Surface area of a sphere: $S = 4\pi r^2$

43. Volume of a sphere: $V = \frac{4}{3}\pi r^3$

44. Volume of a right circular cylinder: $V = \pi r^2 h$

45. Average speed: $r = \dfrac{d}{t}$

46. Free vibrations: $\omega = \sqrt{\dfrac{kg}{W}}$

In Exercises 47–54, find a mathematical model representing the statement. (In each case, determine the constant of proportionality.)

47. A varies directly as r^2. ($A = 9\pi$ when $r = 3$.)

48. y varies inversely as x. ($y = 3$ when $x = 25$.)

49. y is inversely proportional to x. ($y = 7$ when $x = 4$.)

50. z varies jointly as x and y. ($z = 64$ when $x = 4$ and $y = 8$.)

51. F is jointly proportional to r and the third power of s. ($F = 4158$ when $r = 11$ and $s = 3$.)

52. P varies directly as x and inversely as the square of y. $\left(P = \frac{28}{3}\right.$ when $x = 42$ and $\left.y = 9.\right)$

53. z varies directly as the square of x and inversely as y. ($z = 6$ when $x = 6$ and $y = 4$.)

54. v varies jointly as p and q and inversely as the square of s. ($v = 1.5$ when $p = 4.1$, $q = 6.3$, and $s = 1.2$.)

Ecology **In Exercises 55 and 56, use the fact that the diameter of the largest particle that can be moved by a stream varies approximately directly as the square of the velocity of the stream.**

55. A stream with a velocity of $\frac{1}{4}$ mile per hour can move coarse sand particles about 0.02 inch in diameter. Approximate the velocity required to carry particles 0.12 inch in diameter.

56. A stream of velocity v can move particles of diameter d or less. By what factor does d increase when the velocity is doubled?

Resistance **In Exercises 57 and 58, use the fact that the resistance of a wire carrying an electrical current is directly proportional to its length and inversely proportional to its cross-sectional area.**

57. If #28 copper wire (which has a diameter of 0.0126 inch) has a resistance of 66.17 ohms per thousand feet, what length of #28 copper wire will produce a resistance of 33.5 ohms?

58. A 14-foot piece of copper wire produces a resistance of 0.05 ohm. Use the constant of proportionality from Exercise 57 to find the diameter of the wire.

59. *Free Fall* Neglecting air resistance, the distance s an object falls varies directly as the square of the duration t of the fall. An object falls a distance of 144 feet in 3 seconds. How far will it fall in 5 seconds?

60. *Stopping Distance* The stopping distance d of an automobile is directly proportional to the square of its speed s. A car required 75 feet to stop when its speed was 30 miles per hour. Estimate the stopping distance if the brakes are applied when the car is traveling at 50 miles per hour.

61. *Spending* The prices of three sizes of pizza at a pizza shop are as follows.

9-inch: \$8.78
12-inch: \$11.78
15-inch: \$14.18

You would expect that the price of a certain size of pizza would be directly proportional to its surface area. Is that the case for this pizza shop? If not, which size of pizza is the best buy?

62. *Economics* A company has found that the demand for its product varies inversely as the price of the product. When the price is \$3.75, the demand is 500 units. Approximate the demand when the price is \$4.25.

63. *Fluid Flow* The velocity v of a fluid flowing in a conduit is inversely proportional to the cross-sectional area of the conduit. (Assume that the volume of the flow per unit of time is held constant.)

(a) Determine the change in the velocity of water flowing from a hose when a person places a finger over the end of the hose to decrease its cross-sectional area by 25%.

(b) Use the fluid velocity model in part (a) to determine the effect on the velocity of a stream when it is dredged to increase its cross-sectional area by one-third.

64. *Beam Load* The maximum load that can be safely supported by a horizontal beam varies jointly as the width of the beam and the square of its depth, and inversely as the length of the beam. Determine the change in the maximum safe load under the following conditions.

(a) The width and length of the beam are doubled.

(b) The width and depth of the beam are doubled.

(c) All three of the dimensions are doubled.

(d) The depth of the beam is halved.

65. *Data Analysis* An experiment in a physics lab requires a student to measure the compressed length x (in centimeters) of a spring when a force of F pounds is applied. The data is shown in the table.

F	0	2	4	6	8	10	12
x	0	1.15	2.3	3.45	4.6	5.75	6.9

(a) Sketch a scatter plot of the data.

(b) Does it appear that the data can be modeled by Hooke's Law? If so, estimate k. (See Exercises 27–30.)

(c) Use the model in part (b) to approximate the force required to compress the spring 9 centimeters.

66. *Data Analysis* An oceanographer took readings of the water temperature C (in degrees Celsius) at depth d (in meters). The data collected is shown in the table.

d	1000	2000	3000	4000	5000
C	4.2°	1.9°	1.4°	1.2°	0.9°

(a) Sketch a scatter plot of the data.

(b) Does it appear that the data can be modeled by the inverse proportion model $C = k/d$? If so, estimate k.

(c) Use a graphing utility to plot the data points and the inverse model in part (b).

(d) Use the model to approximate the depth at which the water temperature is 3°C.

67. *Data Analysis* A light probe is located x centimeters from a light source, and the intensity y (in microwatts per square centimeter) of the light is measured. The results are shown in the table.

x	30	34	38	42	46	50
y	0.1881	0.1543	0.1172	0.0998	0.0775	0.0645

A model for the data is $y = 262.76/x^{2.12}$.

(a) Use a graphing utility to plot the data points and the model in the same viewing window.

(b) Use the model to approximate the light intensity 25 centimeters from the light source.

68. *Illumination* The illumination from a light source varies inversely as the square of the distance from the light source. When the distance from a light source is doubled, how does the illumination change? Discuss this model in terms of the data given in Exercise 67. Give a possible explanation of the difference.

69. *Hockey Salaries* The average annual salaries of professional hockey players (in thousands of dollars) from 1990 to 1996 are shown in the table. (Source: The News and Observer Publishing Company)

Year	1990	1991	1992	1993	1994	1995	1996
Salary	253	351	434	560	733	892	982

(a) Use the regression feature of a graphing utility to find the least squares regression line that fits this data. [Let y represent the average salary (in thousands of dollars) and let $t = 0$ represent 1990.]

(b) Sketch a scatter plot of the data and graph the linear model you found in part (a) on the same set of axes.

(c) Use the model to estimate the average salaries in 1997, 1998, and 1999.

(d) Use your school's library or some other reference source to analyze the accuracy of the salary estimates in part (c).

70. *Sports* The lengths (in feet) of the winning men's discus throws in the Olympics from 1904 to 1996 are listed below. (Source: ESPN)

1904	128.9	1936	165.6	1972	211.3
1908	134.2	1948	173.2	1976	221.4
1912	148.3	1952	180.5	1980	218.7
1920	146.6	1956	184.9	1984	218.5
1924	151.3	1960	194.2	1988	225.8
1928	155.3	1964	200.1	1992	213.7
1932	162.3	1968	212.5	1996	227.7

(a) Use the regression feature of a graphing utility to find the least squares regression line that fits this data. [Let y represent the length of the winning discus throw (in feet) and let $t = 4$ represent 1904.]

(b) Sketch a scatter plot of the data and graph the linear model you found in part (a) on the same set of axes.

(c) Use the model to estimate the winning men's discus throw in the year 2000.

(d) Use your school's library or some other reference source to analyze the accuracy of the estimate in part (c).

71. *Business* The total assets (in millions of dollars) for First Virginia Banks, Inc. from 1990 to 1998 are listed below. (Source: First Virginia Banks, Inc.)

1990	5384.2	1993	7036.9	1996	8236.1
1991	6119.3	1994	7865.4	1997	9011.6
1992	6840.6	1995	8221.5	1998	9564.7

(a) Use the regression feature of a graphing utility to find the least squares regression line that fits this data. [Let y represent the total assets (in millions of dollars) and let $t = 0$ represent 1990.]

(b) Use a graphing utility to sketch a scatter plot of the data and the graph of the model in the same viewing window.

(c) Use the model to estimate the assets of First Virginia Banks, Inc. in 1999.

(d) Use your school's library or some other reference source to analyze the accuracy of the estimate in part (c).

72. *Energy* The table gives the oil production x (in thousands of barrels per day) in Canada and the oil production y (in thousands of barrels per day) in the United States for the years 1991 through 1996. (Source: U.S. Energy Information Administration)

x	1548	1605	1679	1746	1805	1837
y	7417	7171	6847	6662	6560	6465

(a) Use the regression feature of a graphing utility to find the least squares regression line that fits this data.

(b) Sketch a scatter plot of the data and graph the linear model on the same set of axes.

(c) Use the model to estimate oil production in the United States if oil production in Canada is 2000 thousand barrels per day.

(d) Interpret the meaning of the slope of the linear model in the context of the problem.

73. *Sales* The table gives the amounts x (in millions of dollars) of home computer sales by factories and the amounts y (in millions of dollars) of personal word processor sales by factories for the years 1991 through 1996 in the United States. (Source: Electronic Industries Association)

x	4287	6825	8190	10,088	12,600	15,040
y	600	555	558	504	451	404

(a) Use the regression feature of a graphing utility to find the least squares regression line that fits this data.

(b) Sketch a scatter plot of the data and graph the linear model on the same set of axes.

(c) Use the model to estimate the amount of personal word processor sales if the amount of home computer sales is $18,000 million.

Synthesis

True or False? **In Exercises 74 and 75, decide whether the statement is true or false. Justify your answer.**

74. If y varies directly as x, then if x increases, y will increase as well.

75. In the equation for kinetic energy, $E = \frac{1}{2}mv^2$, the amount of kinetic energy E is directly proportional to the mass m of an object and the square of its velocity v.

In Exercises 76–79, discuss how well the data shown in the scatter plot can be approximated by a linear model.

76.

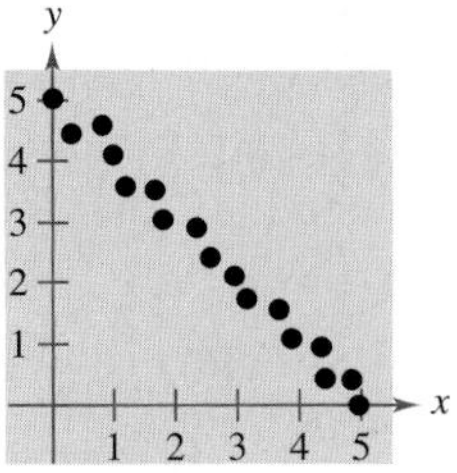

77.

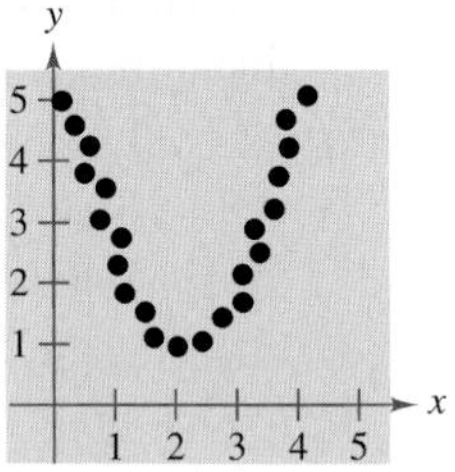

78.

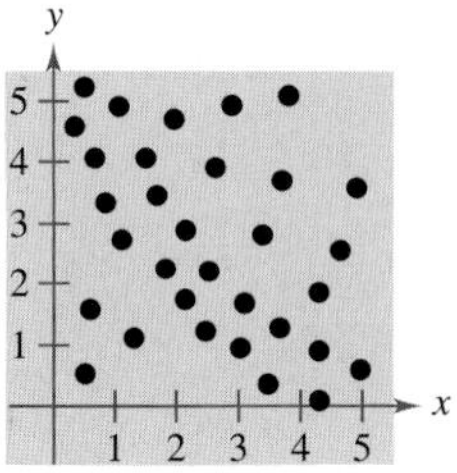

79.

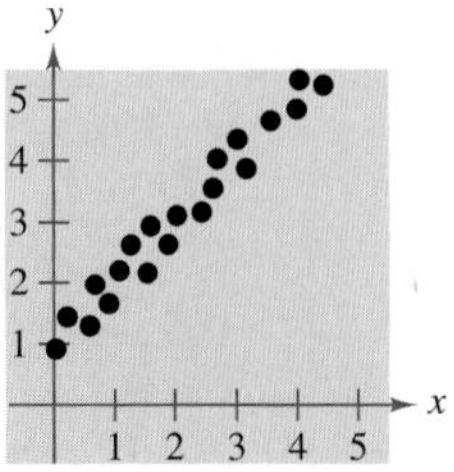

In Exercises 80–83, sketch the line that you think best approximates the data in the scatter plot. Then find an equation of the line.

80.

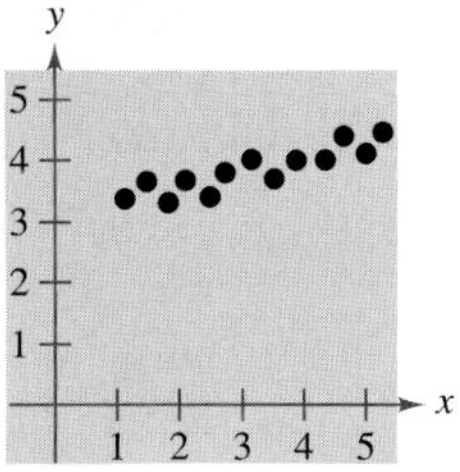

81.

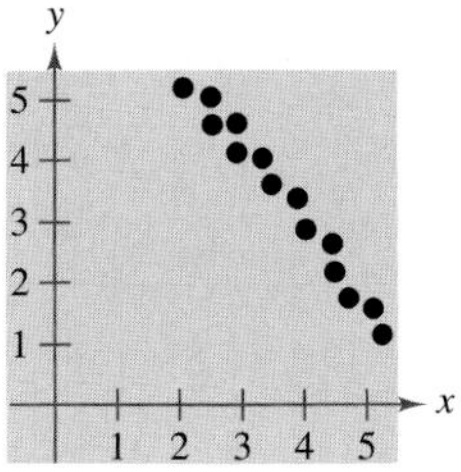

82.

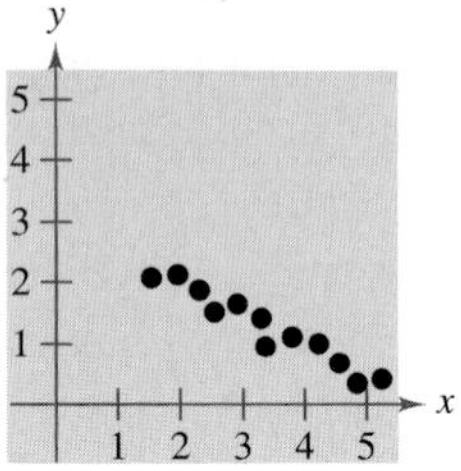

83.

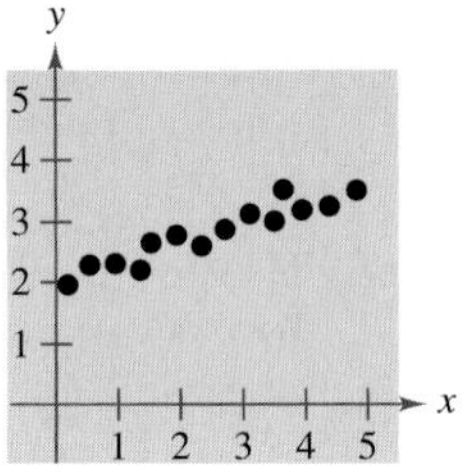

84. *Writing* A linear mathematical model for predicting prize winnings at a race is based on data for 3 years. Write a paragraph discussing the potential accuracy or inaccuracy of such a model.

Review

In Exercises 85–88, solve the inequality and graph the solution on the real number line.

85. $(x - 5)^2 \geq 1$

86. $3(x + 1)(x - 3) < 0$

87. $6x^3 - 30x^2 > 0$

88. $x^4(x - 8) \geq 0$

In Exercises 89 and 90, evaluate the function at each value of the independent variable and simplify.

89. $f(x) = \dfrac{x^2 + 5}{x - 3}$

(a) $f(0)$ (b) $f(-3)$ (c) $f(4)$

90. $f(x) = \begin{cases} -x^2 + 10, & x \geq -2 \\ 6x^2 - 1, & x < -2 \end{cases}$

(a) $f(-2)$ (b) $f(1)$ (c) $f(-8)$

In Exercises 91–94, find the domain of the function.

91. $f(x) = -10x^2 - x - 1$

92. $f(x) = \sqrt[3]{x - 2}$

93. $f(x) = \dfrac{x - 1}{x + 7}$

94. $f(x) = \dfrac{\sqrt{x - 3}}{x - 6}$

Chapter Summary

What did you learn?

Section 1.1	Review Exercises
☐ How to sketch graphs of equations	1–12
☐ How to use intercepts and symmetry to sketch graphs of equations	5–12
☐ How to find equations and sketch graphs of circles	13–18
☐ How to use graphs of equations in real-life problems	19, 20
Section 1.2	
☐ How to find slopes and use slope to graph linear equations	21–29
☐ How to write linear equations and identify parallel and perpendicular lines	30–39
☐ How to use linear equations to model and solve real-life problems	40, 41
Section 1.3	
☐ How to decide whether relations between two variables are functions	42–45
☐ How to use function notation and evaluate functions	46, 47
☐ How to find the domains of functions	48–51
☐ How to use functions to model and solve real-life problems	52, 53
Section 1.4	
☐ How to use the Vertical Line Test and find the zeros of functions	54–61
☐ How to determine intervals on which functions are increasing or decreasing	62–65
☐ How to identify and graph linear and piecewise-defined functions	66–71
☐ How to identify even and odd functions	72–75
Section 1.5	
☐ How to recognize graphs of common functions	76, 77
☐ How to use transformations to sketch graphs of functions	78–83
Section 1.6	
☐ How to add, subtract, multiply, and divide functions	84, 85
☐ How to find compositions and combinations of functions	86–89
Section 1.7	
☐ How to find inverse functions informally	90–93
☐ How to use graphs of functions to decide whether functions have inverses	94–97
☐ How to find inverse functions algebraically	98–103
Section 1.8	
☐ How to use mathematical models to approximate sets of data points	104
☐ How to write mathematical models for direct variation, inverse variation, and joint variation	105–109
☐ How to use the least squares regression feature of a graphing utility to find mathematical models	110

Review Exercises

1.1 In Exercises 1–4, complete a table of values. Use the solution points to sketch the graph of the equation.

1. $y = 3x - 5$
2. $y = -\frac{1}{2}x + 2$
3. $y = x^2 - 3x$
4. $y = 2x^2 - x - 9$

In Exercises 5–8, find the x- and y-intercepts of the graph of the equation.

5. $y = 2x - 9$
6. $y = |x - 4| - 4$
7. $y = (x + 1)^2$
8. $y = x\sqrt{9 - x^2}$

In Exercises 9–12, use intercepts and symmetry to sketch the graph of the equation.

9. $y = 5 - x^2$
10. $y = x^3 + 3$
11. $y = \sqrt{x + 5}$
12. $y = 1 - |x|$

In Exercises 13–16, find the center and radius of the circle and sketch its graph.

13. $x^2 + y^2 = 25$
14. $x^2 + y^2 = 4$
15. $(x + 2)^2 + y^2 = 16$
16. $x^2 + (y - 8)^2 = 81$

17. Find the standard form of the equation of the circle for which the endpoints of a diameter are $(0, 0)$ and $(4, -6)$.
18. Find the standard form of the equation of the circle for which the endpoints of a diameter are $(-2, -3)$ and $(4, -10)$.
19. *Business* The dividends declared per share of the Clorox Company from 1990 to 1998 can be approximated by the model

$$y = 0.073t + 0.644$$

where y is the dividend (in dollars) and t is the time (in years), with $t = 0$ corresponding to 1990. Sketch a graph of this equation. Use the graph to estimate the year in which the dividend per share will be \$1.50. (Source: Clorox Company)

20. *Fence* You have 100 feet of fencing to use for three sides of a rectangular fence, with your house enclosing the fourth side. The area of the enclosure is given by $A = -2x^2 + 100x$. Graph the equation to find the maximum area possible, and how long each side needs to be to obtain that area.

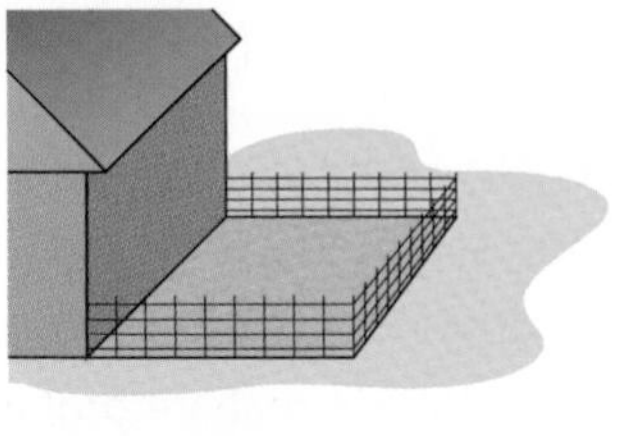

FIGURE FOR 20

1.2 In Exercises 21–24, give the slope and y-intercept of the line given by the equation. Graph the line.

21. $y = 6$
22. $x = -3$
23. $y = 3x + 13$
24. $y = -10x + 9$

25. Match each value of slope m with the corresponding line in the figure.

(a) $m = \frac{3}{2}$ (b) $m = 0$

(c) $m = -3$ (d) $m = -\frac{1}{5}$

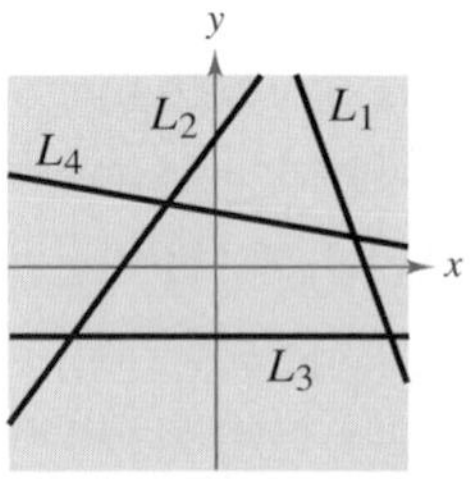

In Exercises 26–29, plot the points and find the slope of the line passing through the pair of points.

26. $(3, -4), (-7, 1)$
27. $(-1, 8), (6, 5)$
28. $(-4.5, 6), (2.1, 3)$
29. $(-3, 2), (8, 2)$

In Exercises 30–33, find an equation of the line that passes through the points.

30. $(0, 0), (0, 10)$
31. $(2, 5), (-2, -1)$
32. $(-1, 4), (2, 0)$
33. $(11, -2), (6, -1)$

In Exercises 34–37, find an equation of the line that passes through the given point and has the specified slope. Sketch the graph of the line.

	Point	*Slope*		*Point*	*Slope*
34.	$(0, -5)$	$m = \frac{3}{2}$	35.	$(-2, 6)$	$m = 0$
36.	$(10, -3)$	$m = -\frac{1}{2}$	37.	$(-8, 5)$	Undefined

In Exercises 38 and 39, write an equation of the line through the point (a) parallel to the given line and (b) perpendicular to the given line.

	Point	*Line*
38.	$(3, -2)$	$5x - 4y = 8$
39.	$(-8, 3)$	$2x + 3y = 5$

Rate of Change **In Exercises 40 and 41, you are given the dollar value of a product in the year 2000 *and* the rate at which the value of the item is expected to change during the next 5 years. Write a linear equation that gives the dollar value V of the product in terms of the year t. (Let $t = 0$ represent 2000.)**

	2000 Value	*Rate*
40.	\$12,500	\$850 increase per year
41.	\$72.95	\$5.15 increase per year

1.3 In Exercises 42–45, determine whether the equation represents y as a function of x.

42. $16x - y^4 = 0$ **43.** $2x - y - 3 = 0$

44. $y = \sqrt{1 - x}$ **45.** $|y| = x + 2$

In Exercises 46 and 47, evaluate the function as indicated. Simplify your answers.

46. $f(x) = x^2 + 1$

(a) $f(2)$ (b) $f(-4)$

(c) $f(t^2)$ (d) $-f(x)$

47. $g(x) = x^{4/3}$

(a) $g(8)$ (b) $g(t + 1)$

(c) $\dfrac{g(8) - g(1)}{8 - 1}$ (d) $g(-x)$

In Exercises 48–51, determine the domain of the function. Verify your result with a graph.

48. $f(x) = \sqrt{25 - x^2}$ **49.** $f(x) = 3x + 4$

50. $h(x) = \dfrac{x}{x^2 - x - 6}$ **51.** $h(t) = |t + 1|$

52. *Physics* The velocity of a ball thrown vertically upward from ground level is $v(t) = -32t + 48$, where t is the time in seconds and v is the velocity in feet per second.

(a) Find the velocity when $t = 1$.

(b) Find the time when the ball reaches its maximum height. [*Hint:* Find the time when $v(t) = 0$.]

(c) Find the velocity when $t = 2$.

53. *Mixture Problem* From a full 50-liter container of a 40% concentration of acid, x liters is removed and replaced with 100% acid.

(a) Write the amount of acid in the final mixture as a function of x.

(b) Determine the domain and range of the function.

(c) Determine x if the final mixture is 50% acid.

1.4 In Exercises 54–57, use the Vertical Line Test to determine whether y is a function of x.

54. $y = (x - 3)^2$

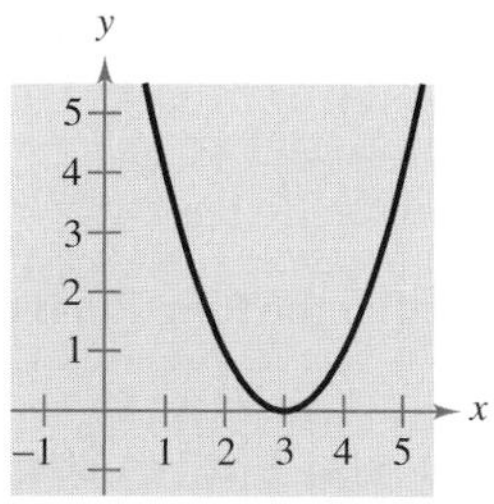

55. $y = -\frac{3}{5}x^3 - 2x + 1$

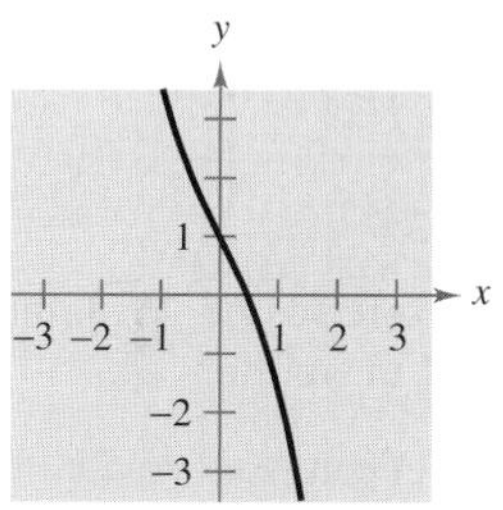

56. $x - 4 = y^2$

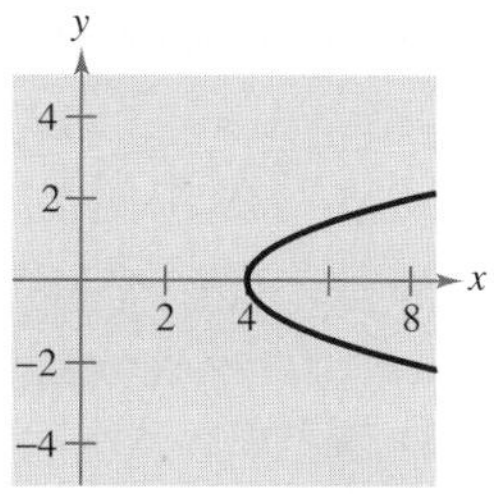

57. $x = -|4 - y|$

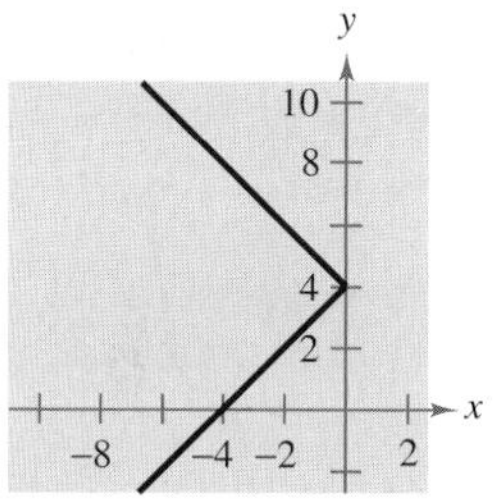

In Exercises 58–61, find the zeros of the function.

58. $f(x) = 3x^2 - 16x + 21$

59. $f(x) = 5x^2 + 4x - 1$

60. $f(x) = \dfrac{8x + 3}{11 - x}$

61. $f(x) = x^3 - x^2 - 25x + 25$

Graphical Analysis **In Exercises 62–65, use a graphing utility to graph the function and approximate the intervals in which the function is increasing, decreasing, or constant.**

62. $g(x) = |x + 2| - |x - 2|$

63. $f(x) = (x^2 - 4)^2$

64. $h(x) = 4x^3 - x^4$

65. $g(x) = \sqrt[3]{x(x + 3)^2}$

In Exercises 66–69, write the linear function f such that the following are true. Then use a graphing utility to graph the function.

66. $f(2) = -6,\ f(-1) = 3$

67. $f(0) = -5,\ f(4) = -8$

68. $f(-\frac{4}{5}) = 2,\ f(\frac{11}{5}) = 7$

69. $f(3.3) = 5.6,\ f(-4.7) = -1.4$

In Exercises 70 and 71, graph the function.

70. $f(x) = \begin{cases} 5x - 3, & x \ge -1 \\ -4x + 5, & x < -1 \end{cases}$

71. $f(x) = \begin{cases} x^2 - 2, & x < -2 \\ 5, & -2 \le x \le 0 \\ 8x - 5, & x > 0 \end{cases}$

In Exercises 72–75, determine whether the function is even, odd, or neither.

72. $f(x) = x^5 + 4x - 7$

73. $f(x) = x^4 - 20x^2$

74. $f(x) = 2x\sqrt{x^2 + 3}$

75. $f(x) = \sqrt[5]{6x^2}$

1.5 In Exercises 76 and 77, identify the common function and describe the transformation shown in the graph.

76.

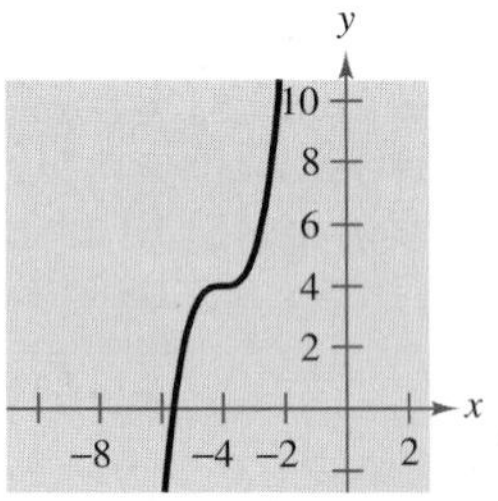

77.

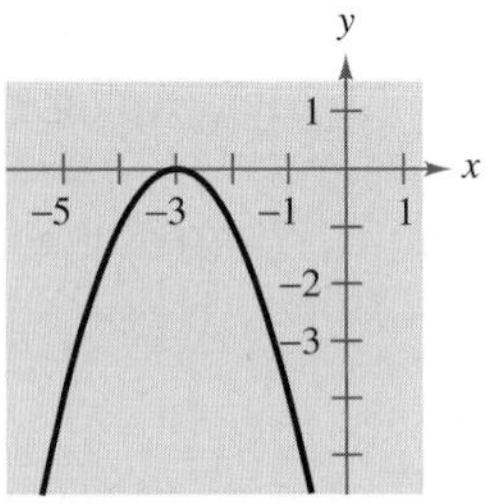

In Exercises 78–83, identify the transformation of the graph of f and sketch the graph of h.

78. $f(x) = x^2,\quad h(x) = x^2 - 9$

79. $f(x) = \sqrt{x},\quad h(x) = \sqrt{x - 7}$

80. $f(x) = |x|,\quad h(x) = |x + 3| - 5$

81. $f(x) = x^2,\quad h(x) = -(x + 3)^2 + 1$

82. $f(x) = \sqrt{x},\quad h(x) = -\sqrt{x + 1} + 9$

83. $f(x) = x^3,\quad h(x) = -\frac{1}{3}x^3$

1.6 In Exercises 84–87, let $f(x) = 3 - 2x$, $g(x) = \sqrt{x}$, and $h(x) = 3x^2 + 2$. Find the indicated value.

84. $(f - g)(4)$

85. $\left(\dfrac{f}{h}\right)(0)$

86. $(h \circ g)(7)$

87. $(g \circ f)(-2)$

Data Analysis **In Exercises 88 and 89, use the table, which shows the total values (in billions of dollars) of U.S. imports from Mexico and Canada for the years 1992 through 1997. The variables y_1 and y_2 represent the total values of imports from Mexico and Canada, respectively.** (Source: U.S. Bureau of the Census)

Year	1992	1993	1994	1995	1996	1997
y_1	35.2	39.9	49.5	62.1	74.3	85.9
y_2	98.6	111.2	128.4	144.4	155.9	168.2

88. Use a graphing utility to find quadratic models for each of the variables. Let $t = 2$ represent 1992.

89. Use a graphing utility to graph y_1, y_2, and $y_1 + y_2$ in the same viewing window. Use the model to estimate the total value of U.S. imports from Canada and Mexico in 2002.

1.7 In Exercises 90–93, find the inverse of f informally. Verify that $f(f^{-1}(x)) = x = f^{-1}(f(x))$.

90. $f(x) = 6x$

91. $f(x) = \frac{1}{12}x$

92. $f(x) = x - 7$

93. $f(x) = x + 5$

In Exercises 94–97, use a graphing utility to graph each function and determine whether the function has an inverse.

94. $f(x) = 3x^3 - 5$

95. $f(x) = -\frac{1}{4}x^2 - 3$

96. $f(x) = -\sqrt{4 - x}$

97. $f(x) = -|x + 2| + |7 - x|$

In Exercises 98–101, (a) find f^{-1}, (b) sketch the graphs of f and f^{-1} on the same coordinate system, and (c) verify that $f^{-1}(f(x)) = x = f(f^{-1}(x))$.

98. $f(x) = \frac{1}{2}x - 3$

99. $f(x) = 5x - 7$

100. $f(x) = \sqrt{x + 1}$

101. $f(x) = x^3 + 2$

In Exercises 102 and 103, restrict the domain of the function f to an interval over which the function is increasing and determine f^{-1} over that interval.

102. $f(x) = 2(x - 4)^2$

103. $f(x) = |x - 2|$

1.8 104. *Data Analysis* The sales S (in billions of dollars) of recreational vehicles in the United States for the years 1988 through 1997 are shown in the table. (Source: National Sporting Goods Association)

Year	8	9	10	11	12
S	4.8	4.5	4.1	3.6	4.4

Year	13	14	15	16	17
S	4.8	5.7	5.9	6.3	6.5

A model for this data is

$$S = 6.7 + 2.60t - 0.742t^2 + 0.0611t^3 - 0.00156t^4$$

where t is the time in years, with $t = 8$ corresponding to 1988.

(a) Use a graphing utility to sketch a scatter plot of the data and the model in the same viewing window. How do they compare?

(b) The table shows that sales were down from 1989 through 1991. Give a possible explanation. Does the model show the downturn in sales?

(c) Use a graphing utility to approximate the magnitude of the decrease in sales during the slump described in part (b). Was the actual decrease more or less than indicated by the model?

(d) Use the model to estimate sales in 2001. Is this model accurate in predicting future sales? Explain.

105. *Measurement* You notice a billboard indicating that it is 2.5 miles or 4 kilometers to the next restaurant of a national fast-food chain. Use this information to find a linear model that relates miles to kilometers. Use the model to complete the table.

Miles	2	5	10	12
Kilometers				

106. *Energy* The power P produced by a wind turbine is proportional to the cube of the wind speed S. A wind speed of 27 miles per hour produces a power output of 750 kilowatts. Find the output for a wind speed of 40 miles per hour.

107. *Frictional Force* The frictional force F between the tires and the road required to keep a car on a curved section of a highway is directly proportional to the square of the speed s of the car. If the speed of the car is doubled, the force will change by what factor?

In Exercises 108 and 109, find a mathematical model representing the statement. (In each case, determine the constant of proportionality.)

108. y is inversely proportional to x. ($y = 9$ when $x = 5.5$.)

109. F is jointly proportional to x and to the square root of y. ($F = 6$ when $x = 9$ and $y = 4$.)

110. *Employment* The table shows the average hourly wages (y_1) for workers in the mining industry and the average hourly wages (y_2) for workers in the construction industry in the United States for the years 1994 through 1997, where t is the time in years, with $t = 4$ corresponding to 1994. (Source: U.S. Bureau of Labor Statistics)

t	4	5	6	7
y_1	\$14.89	\$15.30	\$15.60	\$16.17
y_2	\$14.69	\$15.08	\$15.43	\$16.03

(a) Use the regression feature of a graphing utility to find the least squares regression lines for mining wages versus time and for construction wages versus time.

(b) Use a graphing utility to sketch a scatter plot of the data. Graph the linear models you found in part (a) on the same set of axes.

(c) Interpret the slope of each model in the context of the problem.

(d) Use the models to estimate the wages in each industry for the year 2002.

Synthesis

True or False? **In Exercises 111 and 112, determine whether the statement is true or false. Justify your answer.**

111. Relative to the graph of $f(x) = \sqrt{x}$, the function $h(x) = -\sqrt{x + 9} - 13$ is shifted 9 units to the left and 13 units down, then reflected in the x-axis.

112. If f and g are two inverse functions, then the domain of g is equal to the range of f.

113. Explain how to tell whether a relation between two variables is a function.

114. Explain the difference between the Vertical Line Test and the Horizontal Line Test.

115. If y is directly proportional to x for a particular linear model, what is the y-intercept of the graph of the model?

Chapter Project ▶ A Graphical Approach to Maximization

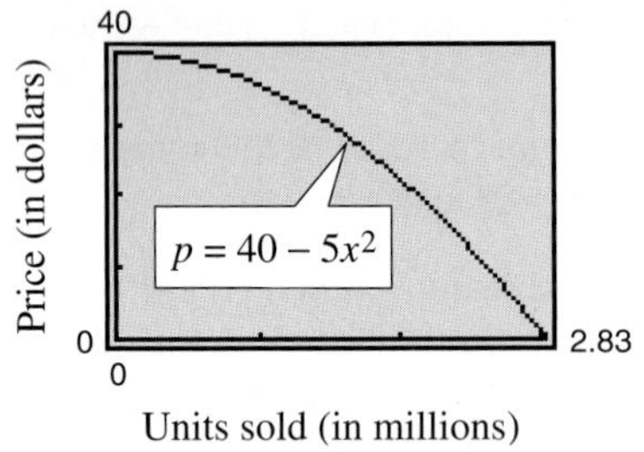

In business, a **demand function** gives the price per unit p in terms of the number of units sold x. The demand function whose graph is shown at the left is

$$p = 40 - 5x^2, \qquad 0 \le x \le \sqrt{8} \qquad \text{Demand function}$$

where x is measured in millions of units. Note that as the price decreases, the number of units sold increases. The **revenue** R (in millions of dollars) is determined by multiplying the number of units sold by the price per unit. So,

$$R = xp = x(40 - 5x^2), \qquad 0 \le x \le \sqrt{8}. \qquad \text{Revenue function}$$

Example ▶ Finding the Maximum Revenue

Use a graphing utility to sketch the graph of the revenue function $R = 40x - 5x^3$ for $0 \le x \le \sqrt{8}$. How many units should be sold to obtain maximum revenue? What price per unit should be charged to obtain maximum revenue?

Solution

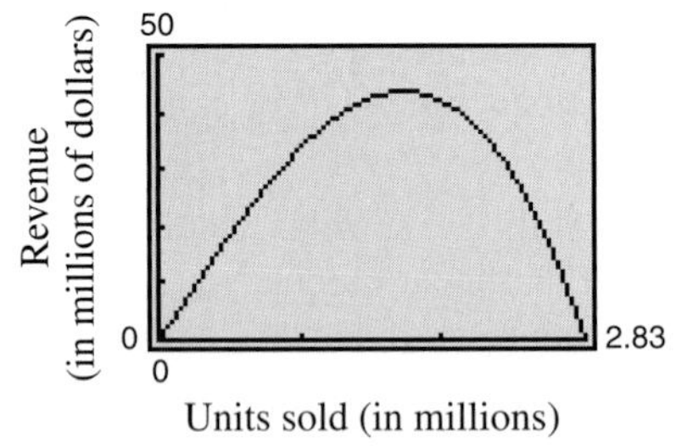

To begin, you need to determine a viewing window that will display the part of the graph that is important to this problem. The domain is given, so you can set the x-boundaries of the graph between 0 and $\sqrt{8}$. To determine the y-boundaries, however, you need to experiment a little. After calculating several values of R, you could decide to use y-boundaries between 0 and 50, as shown in the graph at the left. Next, you can use the trace key to find that the maximum revenue of about \$43.5 million occurs when x is approximately 1.64 million units. To find the price per unit that corresponds to this maximum revenue, you can substitute $x = 1.64$ into the demand function to obtain

$$p = 40 - 5(1.64)^2 \approx \$26.55.$$

Chapter Project Investigations

1. For the demand function $p = 40 - 5x^2$, match each of the points $(0, 40)$ and $(\sqrt{8}, 0)$ with statement (a) or (b). Explain your reasoning.
   (a) No one will buy the product at this price.
   (b) You can't give more than this number away.
2. Use a graphing utility to zoom in on the maximum point of the revenue function in the example. (Use a setting of $1.62 \le x \le 1.65$ and $43.5 \le y \le 43.6$.) Use the trace feature to improve the accuracy of the approximation obtained in the example. Do you think this improved accuracy is appropriate in the context of this particular problem? Does it change the price?
3. For the revenue function discussed in the example, the cost of producing each unit is \$15, so the total cost of producing x million units is $C = 15x$. Use a graphing utility to graph the profit function $P = R - C$ to determine how many units should be sold to obtain maximum profit. What price per unit should be charged to obtain maximum profit?

Chapter Test

The *Interactive* CD-ROM and *Internet* versions of this text provide answers to the Chapter Tests and Cumulative Tests. They also offer Chapter Pre-Tests (which test key skills and concepts covered in previous chapters) and Chapter Post-Tests, both of which have randomly generated exercises with diagnostic capabilities.

Take this test as you would take a test in class. After you are done, check your work against the answers given in the back of the book.

In Exercises 1–3, use intercepts and symmetry to sketch the graph of the equation.

1. $y = 4 - \frac{3}{4}x$ **2.** $y = 4 - \frac{3}{4}|x|$ **3.** $y = 4 - (x - 2)^2$

In Exercises 4 and 5, find an equation of the line passing through the given points.

4. $(2, -3), (-4, 9)$ **5.** $(3, 0.8), (7, -6)$

6. Find an equation of the line that passes through the point $(3, 8)$ and is (a) parallel to and (b) perpendicular to the line $-4x + 7y = -5$.

7. Evaluate $f(x) = \dfrac{\sqrt{x + 9}}{x^2 - 81}$ at each value: (a) $f(7)$ (b) $f(-5)$ (c) $f(x - 9)$

In Exercises 8 and 9, determine the domain of the function.

8. $f(x) = \sqrt{100 - x^2}$ **9.** $f(x) = |-x + 6| + 2$

In Exercises 10–12, (a) find the zeros of the function, (b) use a graphing utility to graph the function, (c) approximate the intervals over which the function is increasing, decreasing, or constant, and (d) determine whether the function is even, odd, or neither.

10. $f(x) = 2x^6 + 5x^4 - x^2$ **11.** $f(x) = 4x\sqrt{3 - x}$ **12.** $f(x) = |x + 5|$

13. Sketch the graph of $f(x) = \begin{cases} 3x + 7, & x \le -3 \\ 4x^2 - 1, & x > -3 \end{cases}$.

In Exercises 14–16, sketch a graph of the function.

14. $h(x) = -x^3 - 7$ **15.** $h(x) = -\sqrt{x + 5} + 8$ **16.** $h(x) = \frac{1}{4}|x + 1| - 3$

In Exercises 17–20, let $f(x) = 3x^2 - 7$ and $g(x) = -x^2 - 4x + 5$. Find the indicated value.

17. $(f + g)(2)$ **18.** $(f - g)(-3)$ **19.** $(fg)(0)$ **20.** $(g \circ f)(-1)$

In Exercises 21–23, find the inverse function, if possible.

21. $f(x) = x^3 + 8$ **22.** $f(x) = |x^2 - 3| + 6$ **23.** $f(x) = \dfrac{3x\sqrt{x}}{8}$

In Exercises 24–26, find a mathematical model representing the statement. (In each case, determine the constant of proportionality.)

24. v varies directly as the square root of s. ($v = 24$ when $s = 16$.)

25. A varies jointly as x and y. ($A = 500$ when $x = 15$ and $y = 8$.)

26. b varies inversely as a. ($b = 32$ when $a = 1.5$.)

27. It costs a company \$58 to produce 6 units of a product and \$78 to produce 10 units. How much does it cost to produce 25 units, assuming that the cost function is linear?

Greg Probst/Tony Stone Images

Production of wheat in the United States increased from 2183 million bushels in 1995 to 2527 million bushels in 1997. During that time the price per bushel dropped from \$4.55 to \$3.45. *(Source: U.S. Department of Agriculture)*

2 Polynomial and Rational Functions

▶ How to Study This Chapter

The Big Picture

In this chapter you will learn the following skills and concepts.

- ▶ How to sketch and analyze graphs of functions
- ▶ How to use long division and synthetic division to divide polynomials by other polynomials
- ▶ How to perform operations with complex numbers
- ▶ How to determine the numbers of rational and real zeros of polynomial functions, and find the zeros
- ▶ How to determine the domains of rational functions and find asymptotes of rational functions
- ▶ How to sketch the graphs of rational functions
- ▶ How to recognize and find partial fraction decompositions of rational expressions

Important Vocabulary

As you encounter each new vocabulary term in this chapter, add the term and its definition to your notebook glossary.

Polynomial function (p. 202)
Constant function (p. 202)
Linear function (p. 202)
Quadratic function (p. 202)
Parabola (p. 202)
Axis (p. 203)
Vertex (p. 203)
Standard form of a quadratic function (p. 205)
Continuous (p. 213)
Leading Coefficient Test (p. 215)
Repeated zero (p. 217)
Intermediate Value Theorem (p. 220)
Division algorithm (p. 227)
Improper rational expression (p. 227)
Proper rational expression (p. 227)
Synthetic division (p. 229)
Remainder Theorem (p. 230)
Factor Theorem (p. 230)
Complex number (p. 236)
Imaginary number (p. 236)
Complex conjugates (p. 239)
Fundamental Theorem of Algebra (p. 243)
Linear Factorization Theorem (p. 243)
Rational Zero Test (p. 244)
Irreducible over the reals (p. 248)
Descartes's Rule of Signs (p. 250)
Variation in sign (p. 250)
Upper bound (p. 251)
Lower bound (p. 251)
Rational function (p. 258)
Vertical asymptote (p. 259)
Horizontal asymptote (p. 259)
Slant (or oblique) asymptote (p. 264)
Partial fraction (p. 271)
Basic equation (p. 272)

Study Tools

- Learning objectives at the beginning of each section
- Chapter Summary (p. 279)
- Review Exercises (pp. 280–283)
- Chapter Test (p. 285)

Additional Resources

- Study and Solutions Guide
- Interactive Precalculus
- Videotapes for Chapter 2
- Precalculus Website
- Student Success Organizer

STUDY T!P

The best time to do homework is right after class, when concepts are still fresh in your mind. This increases your chances of retaining the information in long-term memory.

2.1 Quadratic Functions

▶ **What you should learn**

- How to analyze graphs of quadratic functions
- How to write quadratic functions in standard form and use the results to sketch graphs of functions
- How to use quadratic functions to model and solve real-life problems

▶ **Why you should learn it**

Quadratic functions can be used to model data to analyze consumer behavior. For instance, Exercise 91 on page 212 shows how a quadratic function can model VCR usage in the United States.

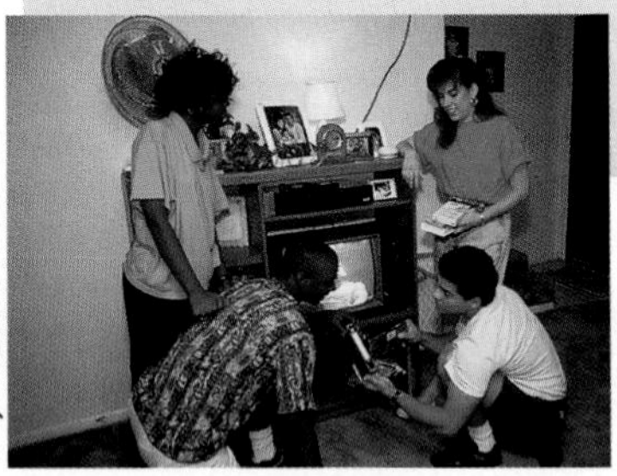

Tony Freeman/PhotoEdit

The Graph of a Quadratic Function

In this and the next section, you will study the graphs of polynomial functions.

Definition of Polynomial Function

Let n be a nonnegative integer and let $a_n, a_{n-1}, \ldots, a_2, a_1, a_0$ be real numbers with $a_n \neq 0$. The function

$$f(x) = a_n x^n + a_{n-1}x^{n-1} + \cdots + a_2x^2 + a_1x + a_0$$

is called a **polynomial function of x with degree n.**

Polynomial functions are classified by degree. For instance, the polynomial function

$$f(x) = a, \qquad a \neq 0 \qquad \text{Constant function}$$

has degree 0 and is called a **constant function.** In Chapter 1, you learned that the graph of this type of function is a horizontal line. The polynomial function

$$f(x) = ax + b, \qquad a \neq 0 \qquad \text{Linear function}$$

has degree 1 and is called a **linear function.** In Chapter 1, you learned that the graph of the linear function $f(x) = ax + b$ is a line whose slope is a and whose y-intercept is $(0, b)$. In this section you will study second-degree polynomial functions, which are called **quadratic functions.**

For instance, each of the following functions is a quadratic function.

$$f(x) = x^2 + 6x + 2$$

$$g(x) = 2(x + 1)^2 - 3$$

$$h(x) = 9 + \tfrac{1}{4}x^2$$

$$k(x) = -3x^2 + 4$$

$$m(x) = (x - 2)(x + 1)$$

Definition of Quadratic Function

Let a, b, and c be real numbers with $a \neq 0$. The function

$$f(x) = ax^2 + bx + c \qquad \text{Quadratic function}$$

is called a **quadratic function.**

The graph of a quadratic function is a special type of "U"-shaped curve that is called a **parabola.** Parabolas occur in many real-life applications—especially those involving reflective properties of satellite dishes and flashlight reflectors. You will study these properties in Section 10.2.

All parabolas are symmetric with respect to a line called the **axis of symmetry,** or simply the **axis** of the parabola. The point where the axis intersects the parabola is the **vertex** of the parabola, as shown in Figure 2.1. If the leading coefficient is positive, the graph of $f(x) = ax^2 + bx + c$ is a parabola that opens upward. If the leading coefficient is negative, the graph of $f(x) = ax^2 + bx + c$ is a parabola that opens downward.

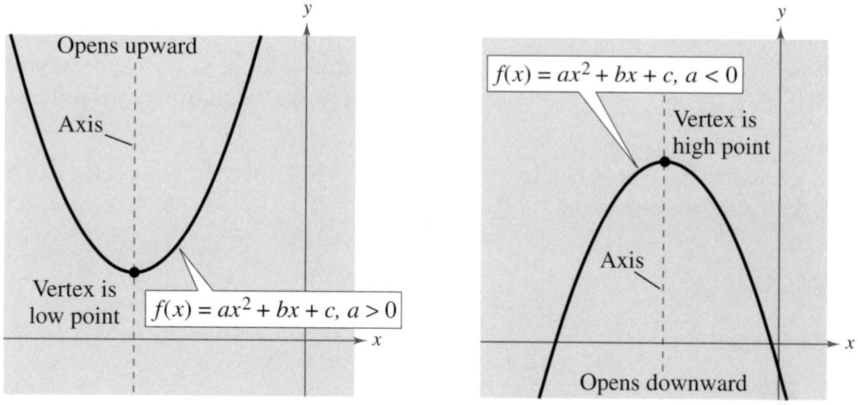

Leading coefficient is positive. *Leading coefficient is negative.*

FIGURE 2.1

The simplest type of quadratic function is

$$f(x) = ax^2.$$

Its graph is a parabola whose vertex is (0, 0). If $a > 0$, the vertex is the point with the *minimum* y-value on the graph, and if $a < 0$, the vertex is the point with the *maximum* y-value on the graph, as shown in Figure 2.2.

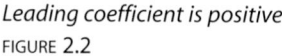

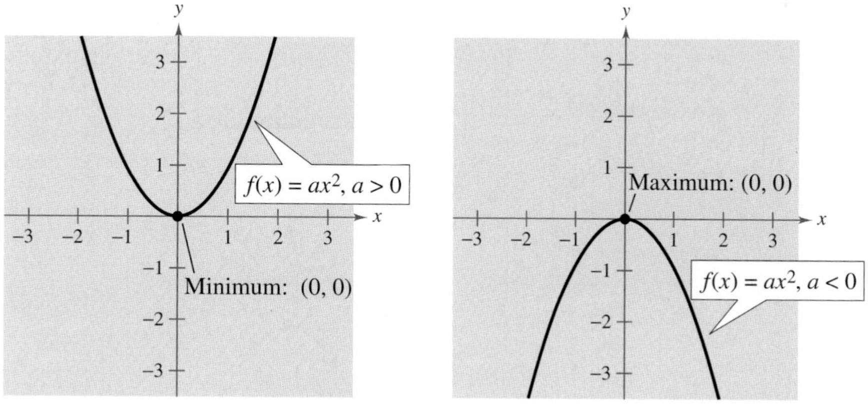

Leading coefficient is positive. *Leading coefficient is negative.*

FIGURE 2.2

When sketching the graph of $f(x) = ax^2$, it is helpful to use the graph of $y = x^2$ as a reference, as discussed in Section 1.5.

Exploration

Graph $y = ax^2$ for $a = -2, -1, -0.5, 0.5, 1$, and 2. How does changing the value of a affect the graph?

Graph $y = (x - h)^2$ for $h = -4, -2, 2$, and 4. How does changing the value of h affect the graph?

Graph $y = x^2 + k$ for $k = -4, -2, 2$, and 4. How does changing the value of k affect the graph?

Example 1 ▶ Sketching Graphs of Quadratic Functions

a. Compare the graphs of $y = x^2$ and $f(x) = \frac{1}{3}x^2$.

b. Compare the graphs of $y = x^2$ and $g(x) = 2x^2$.

Solution

a. Compared with $y = x^2$, each output of $f(x) = \frac{1}{3}x^2$ "shrinks" by a factor of $\frac{1}{3}$, creating the broader parabola shown in Figure 2.3(a).

b. Compared with $y = x^2$, each output of $g(x) = 2x^2$ "stretches" by a factor of 2, creating the narrower parabola shown in Figure 2.3(b).

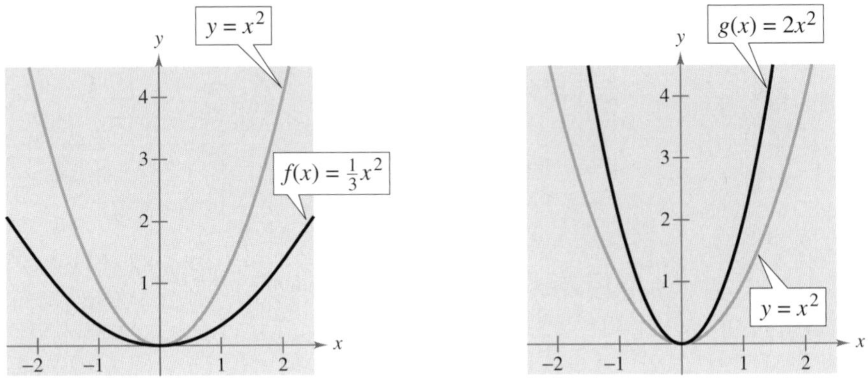

(a) (b)

FIGURE 2.3

In Example 1, note that the coefficient a determines how widely the parabola given by $f(x) = ax^2$ opens. If $|a|$ is small, the parabola opens more widely than if $|a|$ is large.

Recall from Section 1.5 that the graphs of $y = f(x \pm c)$, $y = f(x) \pm c$, $y = f(-x)$, and $y = -f(x)$ are rigid transformations of the graph of $y = f(x)$. For instance, in Figure 2.4, notice how the graph of $y = x^2$ can be transformed to produce the graphs of $f(x) = -x^2 + 1$ and $g(x) = (x + 2)^2 - 3$.

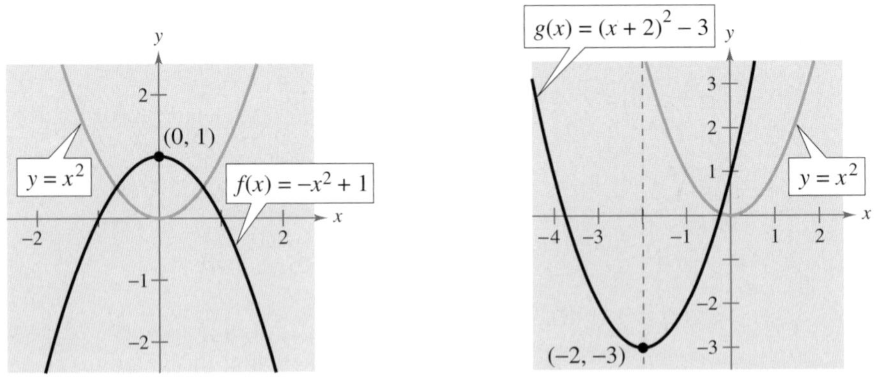

FIGURE 2.4

STUDY TIP

The standard form of a quadratic function identifies three basic transformations of the graph of $y = x^2$.

1. The factor a produces a vertical stretch or shrink. If $a < 0$, the graph is reflected in the x-axis.
2. The factor $(x - h)^2$ represents a horizontal shift of h units.
3. The term k represents a vertical shift of k units.

The Standard Form of a Quadratic Function

The **standard form** of a quadratic function is

$$f(x) = a(x - h)^2 + k.$$

This form is especially convenient for sketching a parabola because it identifies the vertex of the parabola.

Standard Form of a Quadratic Function

The quadratic function

$$f(x) = a(x - h)^2 + k, \qquad a \neq 0$$

is in **standard form.** The graph of f is a parabola whose axis is the vertical line $x = h$ and whose vertex is the point (h, k). If $a > 0$, the parabola opens upward, and if $a < 0$, the parabola opens downward.

To write a quadratic function in standard form, you can use the process of *completing the square,* as illustrated in Example 2.

To prepare for rewriting a function $f(x)$ in standard form, review the process of completing the square for an algebraic expression, paying special attention to problems in which $a \neq 1$.

Example 2 ▶ Graphing a Parabola in Standard Form

Sketch the graph of

$$f(x) = 2x^2 + 8x + 7$$

and identify the vertex and the axis of the parabola.

Solution

Begin by writing the quadratic function in standard form. Notice that the first step in completing the square is to factor out any coefficient of x^2 that is not 1.

$$\begin{aligned} f(x) &= 2x^2 + 8x + 7 && \text{Write original function.}\\ &= 2(x^2 + 4x) + 7 && \text{Factor 2 out of } x\text{-terms.}\\ &= 2(x^2 + 4x + 4 - 4) + 7 && \text{Add and subtract 4 within parentheses.}\\ &\quad (2^2) && \\ &= 2(x^2 + 4x + 4) - 2(4) + 7 && \text{Regroup terms.}\\ &= 2(x^2 + 4x + 4) - 8 + 7 && \text{Simplify.}\\ &= 2(x + 2)^2 - 1 && \text{Write in standard form.} \end{aligned}$$

From this form, you can see that the graph of f is a parabola that opens upward and has its vertex at $(-2, -1)$. This corresponds to a left shift of 2 units and a downward shift of 1 unit relative to the graph of $y = 2x^2$, as shown in Figure 2.5. In the figure, you can see that the axis of the parabola is the vertical line through the vertex, $x = -2$.

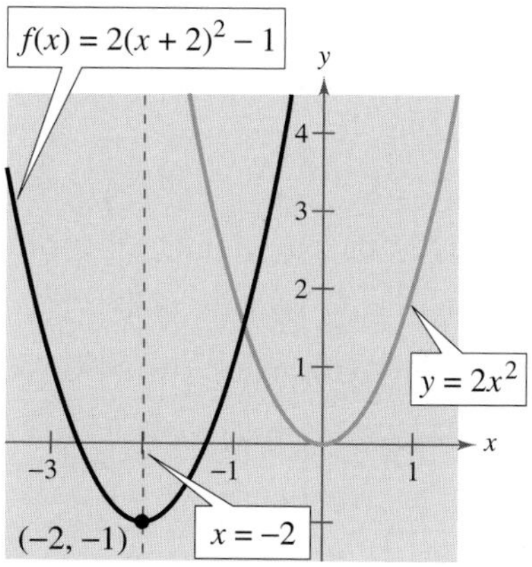

FIGURE 2.5

To find the x-intercepts of the graph of $f(x) = ax^2 + bx + c$, you must solve the equation $ax^2 + bx + c = 0$. If $ax^2 + bx + c$ does not factor, you can use the Quadratic Formula to find the x-intercepts. Remember, however, that a parabola may have no x-intercepts.

Example 3 ▶ Finding the Vertex and x-Intercepts of a Parabola

Sketch the graph of $f(x) = -x^2 + 6x - 8$ and identify the vertex and x-intercepts.

Solution

As in Example 2, begin by writing the quadratic function in standard form.

$$f(x) = -x^2 + 6x - 8 \qquad \text{Write original function.}$$
$$= -(x^2 - 6x) - 8 \qquad \text{Factor } -1 \text{ out of } x\text{-terms.}$$
$$= -(x^2 - 6x + 9 - 9) - 8 \qquad \text{Add and subtract 9 within parentheses.}$$
$$(-3)^2$$
$$= -(x^2 - 6x + 9) - (-9) - 8 \qquad \text{Regroup terms.}$$
$$= -(x - 3)^2 + 1 \qquad \text{Write in standard form.}$$

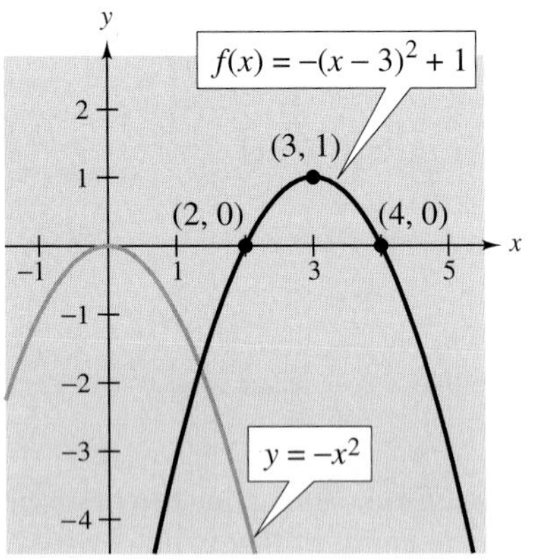

FIGURE 2.6

From this form, you can see that the vertex is $(3, 1)$. To find the x-intercepts of the graph, solve the equation $-x^2 + 6x - 8 = 0$.

$$-x^2 + 6x - 8 = 0 \qquad \text{Write original equation.}$$
$$-(x^2 - 6x + 8) = 0 \qquad \text{Factor out } -1.$$
$$-(x - 2)(x - 4) = 0 \qquad \text{Factor.}$$
$$x - 2 = 0 \quad \Rightarrow \quad x = 2 \qquad \text{Set 1st factor equal to 0.}$$
$$x - 4 = 0 \quad \Rightarrow \quad x = 4 \qquad \text{Set 2nd factor equal to 0.}$$

The x-intercepts are $(2, 0)$ and $(4, 0)$. So, the graph of f is a parabola that opens downward, as shown in Figure 2.6.

Example 4 ▶ Finding the Equation of a Parabola

Find the standard form of the equation of the parabola whose vertex is $(1, 2)$ and that passes through the point $(0, 0)$, as shown in Figure 2.7.

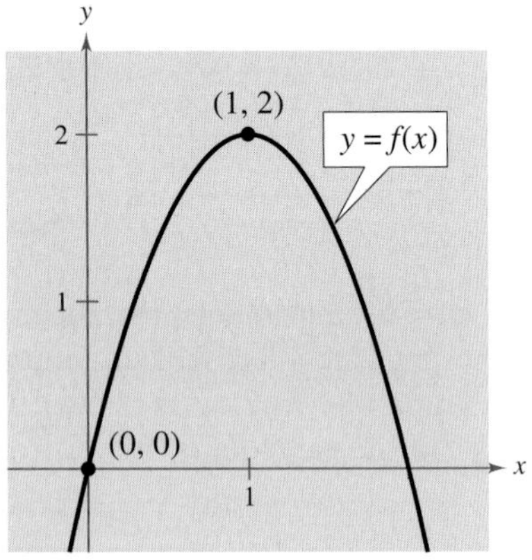

FIGURE 2.7

Solution

Because the vertex of the parabola is at $(h, k) = (1, 2)$, the equation has the form

$$f(x) = a(x - 1)^2 + 2. \qquad \text{Substitute for } h \text{ and } k \text{ in standard form.}$$

Because the parabola passes through the point $(0, 0)$, it follows that $f(0) = 0$. So,

$$0 = a(0 - 1)^2 + 2 \quad \Rightarrow \quad a = -2 \qquad \text{Substitute 0 for } x\text{; solve for } a.$$

which implies that the equation is

$$f(x) = -2(x - 1)^2 + 2. \qquad \text{Substitute for } a \text{ in standard form.}$$

So, the equation of this parabola is $y = -2(x - 1)^2 + 2$.

You may wish to mention that much of beginning calculus deals with finding maxima and minima.

Applications

Many applications involve finding the maximum or minimum value of a quadratic function. Some quadratic functions are not easily written in standard form. For such functions, it is useful to have an alternative method for finding the vertex. For a quadratic function in the form $f(x) = ax^2 + bx + c$, the vertex occurs when $x = -b/2a$.

Vertex of a Parabola

The vertex of the graph of $f(x) = ax^2 + bx + c$ is $\left(-\dfrac{b}{2a}, f\left(-\dfrac{b}{2a}\right)\right)$.

A computer simulation of this example appears in the *Interactive* CD-ROM and *Internet* versions of this text.

Example 5 ▶ The Maximum Height of a Baseball

A baseball is hit at a point 3 feet above the ground at a velocity of 100 feet per second and at an angle of 45° with respect to the ground. The path of the baseball is given by the function

$$f(x) = -0.0032x^2 + x + 3$$

where $f(x)$ is the height of the baseball (in feet) and x is the horizontal distance from home plate (in feet). What is the maximum height reached by the baseball?

Solution

For this quadratic function, you have

$$\begin{aligned} f(x) &= ax^2 + bx + c \\ &= -0.0032x^2 + x + 3. \end{aligned}$$

So, $a = -0.0032$ and $b = 1$. Because the function has a maximum at $x = -b/2a$, you can conclude that the baseball reaches its maximum height when it is x feet from home plate, where x is

$$\begin{aligned} x &= -\frac{b}{2a} \\ &= -\frac{1}{2(-0.0032)} \qquad \text{Substitute for } a \text{ and } b. \\ &= 156.25 \text{ feet.} \end{aligned}$$

To find the maximum height, you must determine the value of the function when $x = 156.25$.

$$\begin{aligned} f(156.25) &= -0.0032(156.25)^2 + 156.25 + 3 \\ &= 81.125 \text{ feet.} \end{aligned}$$

The path of the baseball is shown in Figure 2.8. You can estimate from the graph in Figure 2.8 that the ball hits the ground at a distance of about 320 feet from home plate. The actual distance is the x-intercept of the graph of f, which you can find by solving the equation $-0.0032x^2 + x + 3 = 0$ and taking the positive solution, $x \approx 315.5$.

Activities

1. Describe the effect of a on the graph of $f(x) = ax^2 + bx + c$.
2. Rewrite the quadratic function $f(x) = -x^2 + 4x - 1$ in standard form.
   Answer: $f(x) = -(x - 2)^2 + 3$
3. Find the quadratic function whose graph has vertex $(3, -1)$ and passes through the point $(0, 2)$.
   Answer: $f(x) = \frac{1}{3}x^2 - 2x + 2$
4. Fit a quadratic model to the points $(0, -2)$, $(2, 0)$, and $(-0.5, 0)$.
   Answer: $y = 2x^2 - 3x - 2$

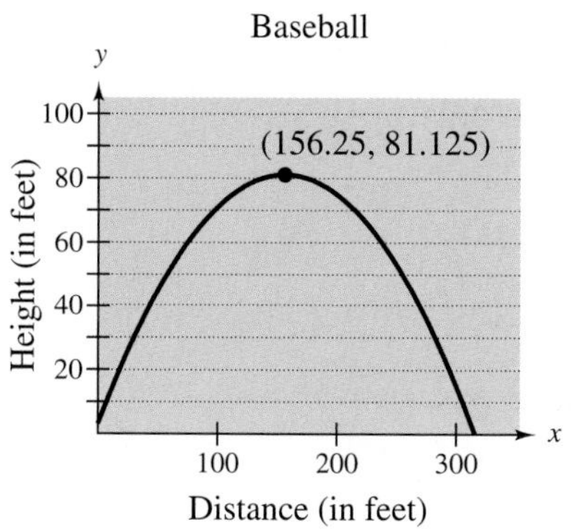

FIGURE 2.8

2.1 Exercises

The *Interactive* CD-ROM and *Internet* versions of this text contain step-by-step solutions to all odd-numbered Section and Review Exercises. They also provide Tutorial Exercises that link to Guided Examples for additional help.

In Exercises 1–8, match the quadratic function with its graph. [The graphs are labeled (a), (b), (c), (d), (e), (f), (g), and (h).]

(a)

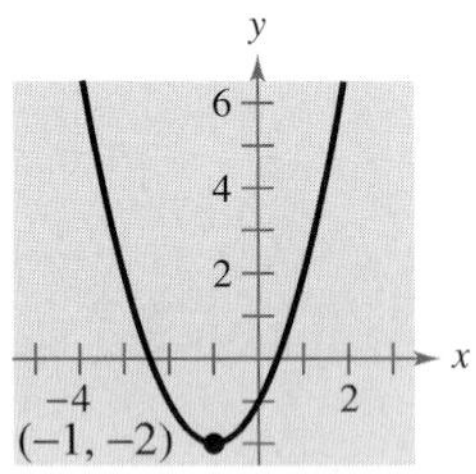

(b)

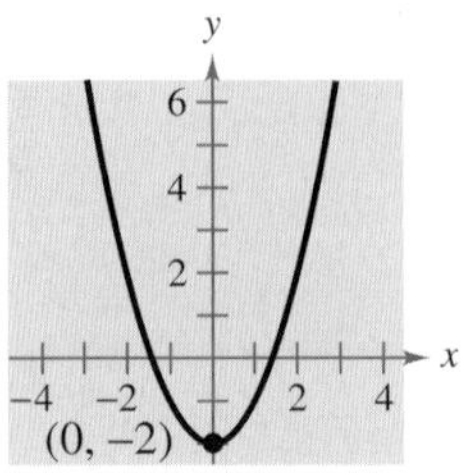

(c)

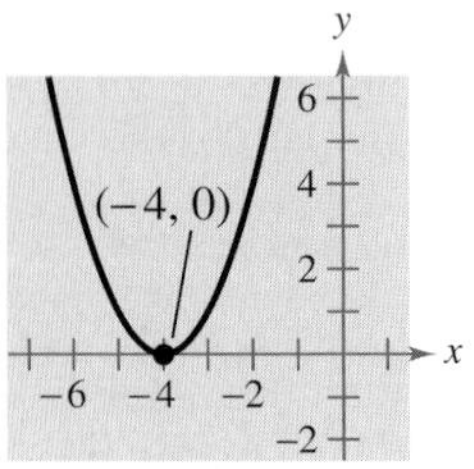

(d)

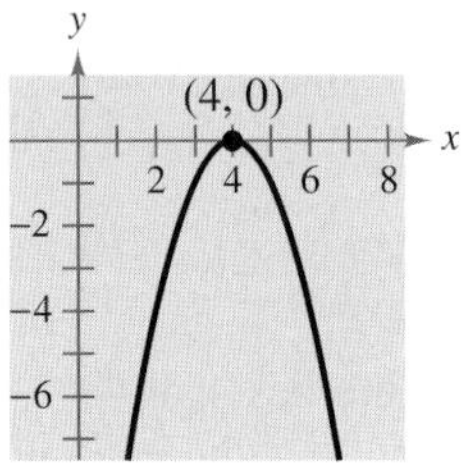

(e)

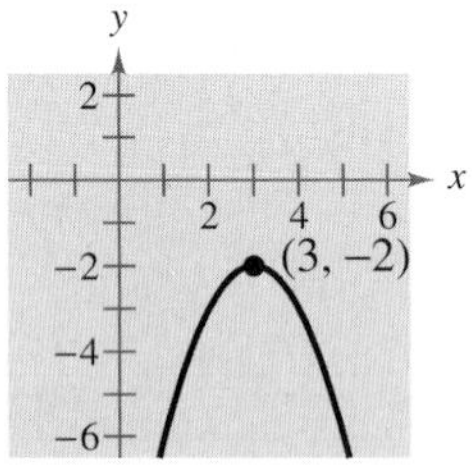

(f)

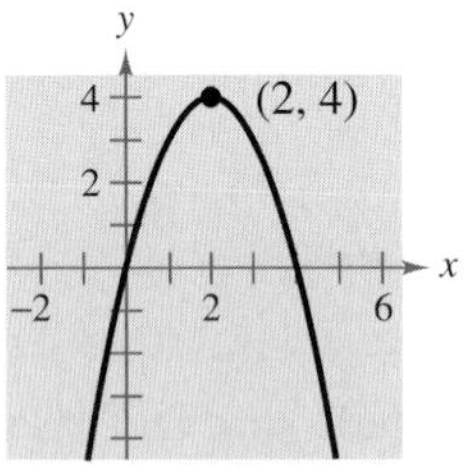

(g)

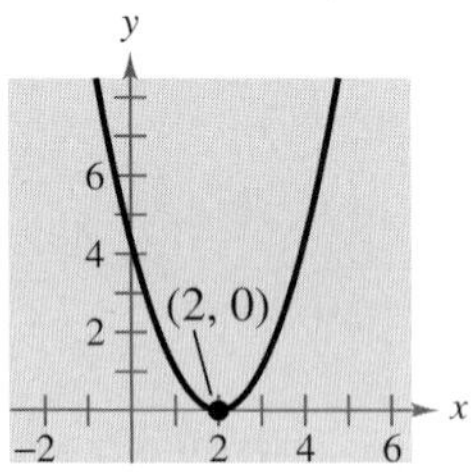

(h)

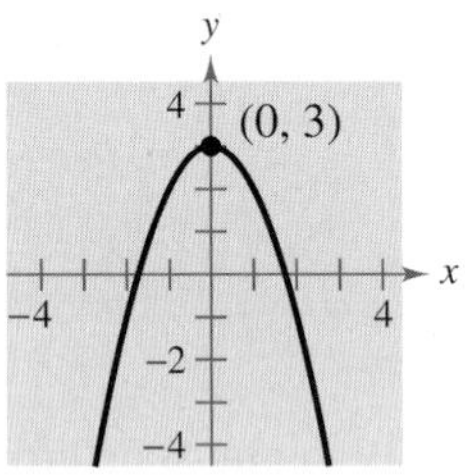

1. $f(x) = (x - 2)^2$ **2.** $f(x) = (x + 4)^2$

3. $f(x) = x^2 - 2$ **4.** $f(x) = 3 - x^2$

5. $f(x) = 4 - (x - 2)^2$ **6.** $f(x) = (x + 1)^2 - 2$

7. $f(x) = -(x - 3)^2 - 2$ **8.** $f(x) = -(x - 4)^2$

Exploration **In Exercises 9–12, graph each equation. Compare the graph of each function with the graph of $y = x^2$.**

9. (a) $f(x) = \frac{1}{2}x^2$ (b) $g(x) = -\frac{1}{8}x^2$
(c) $h(x) = \frac{3}{2}x^2$ (d) $k(x) = -3x^2$

10. (a) $f(x) = x^2 + 1$ (b) $g(x) = x^2 - 1$
(c) $h(x) = x^2 + 3$ (d) $k(x) = x^2 - 3$

11. (a) $f(x) = (x - 1)^2$ (b) $g(x) = (x + 1)^2$
(c) $h(x) = (x - 3)^2$ (d) $k(x) = (x + 3)^2$

12. (a) $f(x) = -\frac{1}{2}(x - 2)^2 + 1$
(b) $g(x) = \frac{1}{2}(x - 2)^2 + 1$
(c) $h(x) = -\frac{1}{2}(x + 2)^2 - 1$
(d) $k(x) = \frac{1}{2}(x + 2)^2 - 1$

In Exercises 13–28, sketch the graph of the quadratic function without using a graphing utility. Identify the vertex and x-intercepts.

13. $f(x) = x^2 - 5$ **14.** $h(x) = 25 - x^2$

15. $f(x) = \frac{1}{2}x^2 - 4$ **16.** $f(x) = 16 - \frac{1}{4}x^2$

17. $f(x) = (x + 5)^2 - 6$ **18.** $f(x) = (x - 6)^2 + 3$

19. $h(x) = x^2 - 8x + 16$ **20.** $g(x) = x^2 + 2x + 1$

21. $f(x) = x^2 - x + \frac{5}{4}$ **22.** $f(x) = x^2 + 3x + \frac{1}{4}$

23. $f(x) = -x^2 + 2x + 5$ **24.** $f(x) = -x^2 - 4x + 1$

25. $h(x) = 4x^2 - 4x + 21$ **26.** $f(x) = 2x^2 - x + 1$

27. $f(x) = \frac{1}{4}x^2 - 2x - 12$

28. $f(x) = -\frac{1}{3}x^2 + 3x - 6$

In Exercises 29–36, use a graphing utility to graph the quadratic function. Identify the vertex and x-intercepts. Then check your results algebraically by completing the square.

29. $f(x) = -(x^2 + 2x - 3)$

30. $f(x) = -(x^2 + x - 30)$

31. $g(x) = x^2 + 8x + 11$

32. $f(x) = x^2 + 10x + 14$

33. $f(x) = 2x^2 - 16x + 31$

34. $f(x) = -4x^2 + 24x - 41$

35. $g(x) = \frac{1}{2}(x^2 + 4x - 2)$

36. $f(x) = \frac{3}{5}(x^2 + 6x - 5)$

In Exercises 37–42, find the standard form of the equation of the parabola.

37.
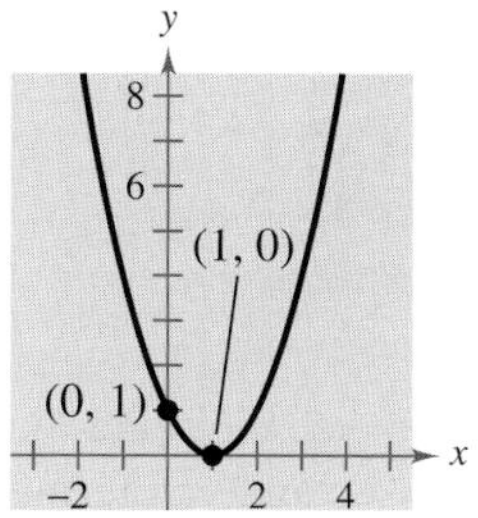

38.
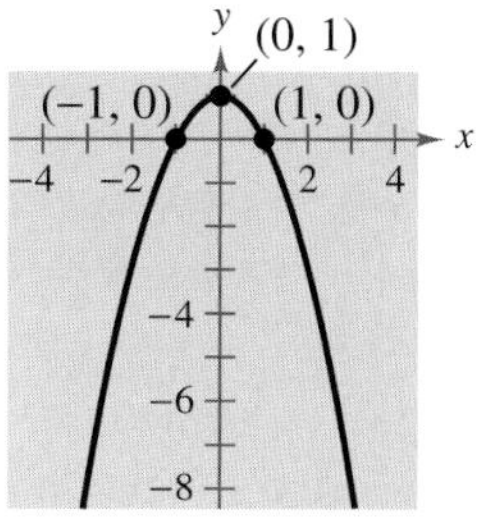

39.
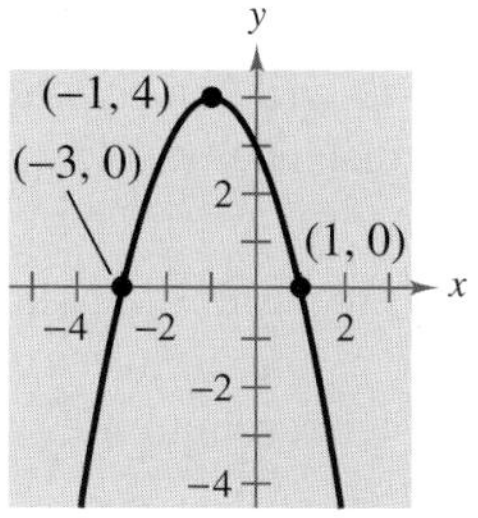

40.
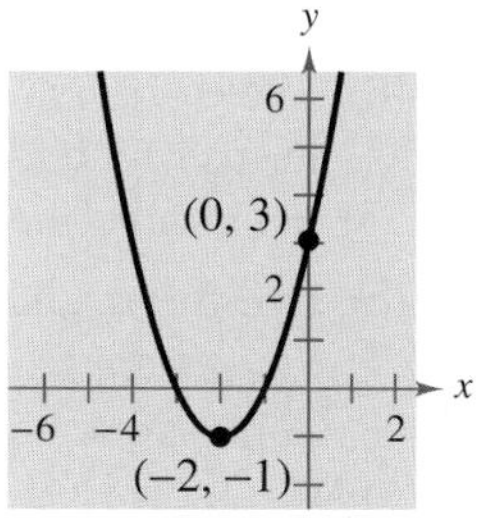

41.
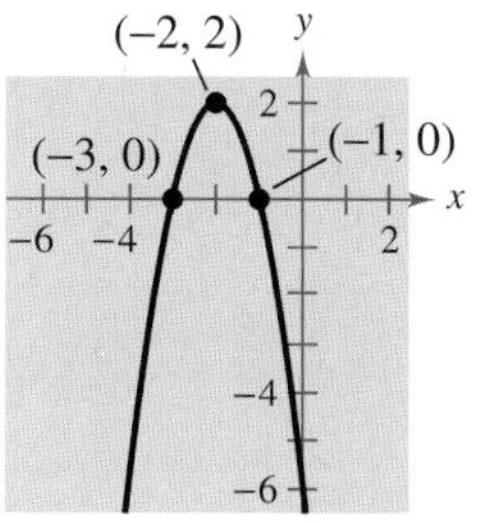

42.
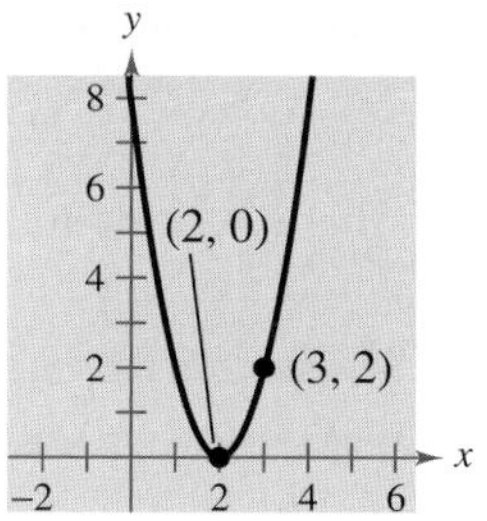

In Exercises 43–52, find the quadratic function that has the indicated vertex and whose graph passes through the given point.

43. Vertex: $(-2, 5)$; Point: $(0, 9)$

44. Vertex: $(4, -1)$; Point: $(2, 3)$

45. Vertex: $(3, 4)$; Point: $(1, 2)$

46. Vertex: $(2, 3)$; Point: $(0, 2)$

47. Vertex: $(5, 12)$; Point: $(7, 15)$

48. Vertex: $(-2, -2)$; Point: $(-1, 0)$

49. Vertex: $\left(-\frac{1}{4}, \frac{3}{2}\right)$; Point: $(-2, 0)$

50. Vertex: $\left(\frac{5}{2}, -\frac{3}{4}\right)$; Point: $(-2, 4)$

51. Vertex: $\left(-\frac{5}{2}, 0\right)$; Point: $\left(-\frac{7}{2}, -\frac{16}{3}\right)$

52. Vertex: $(6, 6)$; Point: $\left(\frac{61}{10}, \frac{3}{2}\right)$

Graphical Reasoning **In Exercises 53–56, determine the x-intercepts of the graph visually. Explain how the x-intercepts relate to the solutions of the quadratic equation when $y = 0$. Then find the x-intercepts algebraically to confirm your results.**

53. $y = x^2 - 16$

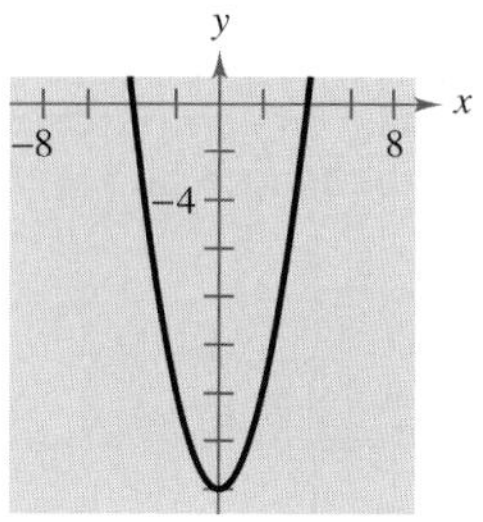

54. $y = x^2 - 6x + 9$

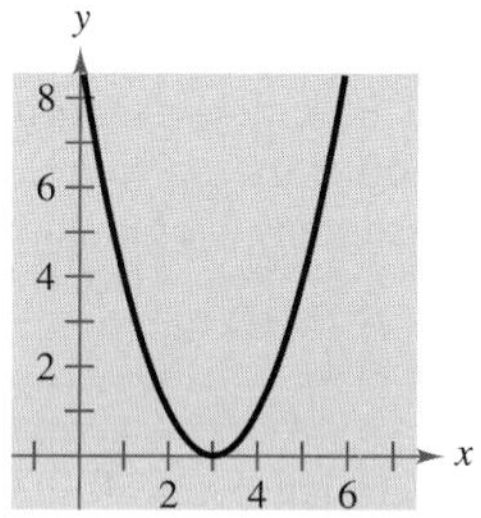

55. $y = x^2 - 4x - 5$

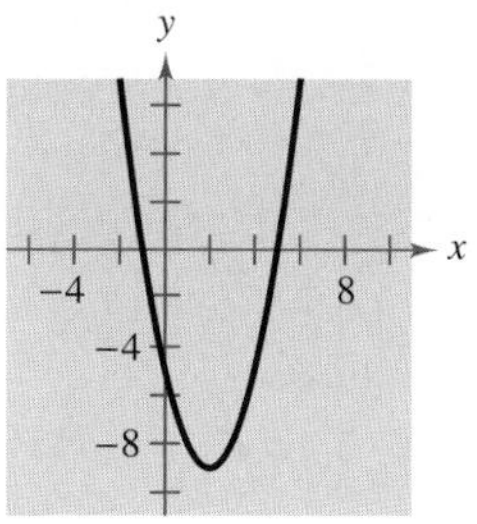

56. $y = 2x^2 + 5x - 3$

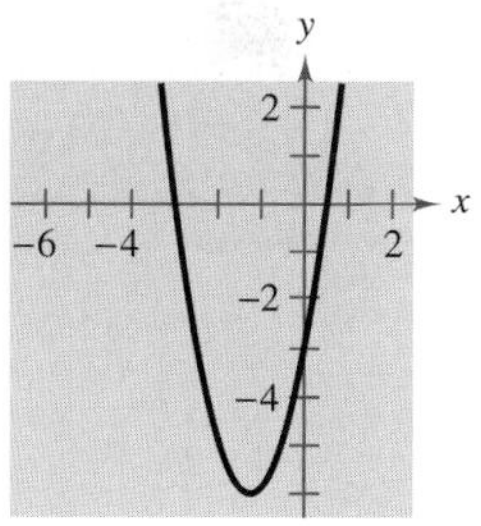

In Exercises 57–64, use a graphing utility to graph the quadratic function. Find the x-intercepts of the graph and compare them with the solutions of the corresponding quadratic equation when $y = 0$.

57. $f(x) = x^2 - 4x$

58. $f(x) = -2x^2 + 10x$

59. $f(x) = x^2 - 9x + 18$

60. $f(x) = x^2 - 8x - 20$

61. $f(x) = 2x^2 - 7x - 30$

62. $f(x) = 4x^2 + 25x - 21$

63. $f(x) = -\frac{1}{2}(x^2 - 6x - 7)$

64. $f(x) = \frac{7}{10}(x^2 + 12x - 45)$

In Exercises 65–70, find two quadratic functions, one that opens upward and one that opens downward, whose graphs have the given x-intercepts. (There are many correct answers.)

65. $(-1, 0), (3, 0)$

66. $(-5, 0), (5, 0)$

67. $(0, 0), (10, 0)$

68. $(4, 0), (8, 0)$

69. $(-3, 0), \left(-\frac{1}{2}, 0\right)$

70. $\left(-\frac{5}{2}, 0\right), (2, 0)$

In Exercises 71–74, find two positive real numbers whose product is a maximum.

71. The sum is 110.

72. The sum is S.

73. The sum of the first and twice the second is 24.

74. The sum of the first and three times the second is 42.

Geometry **In Exercises 75 and 76, consider a rectangle of length x and perimeter P. (a) Express the area A as a function of x and determine the domain of the function. (b) Graph the area function. (c) Find the length and width of the rectangle of maximum area.**

75. $P = 100$ feet

76. $P = 36$ meters

77. *Numerical, Graphical, and Analytical Analysis* A rancher has 200 feet of fencing to enclose two adjacent rectangular corrals (see figure).

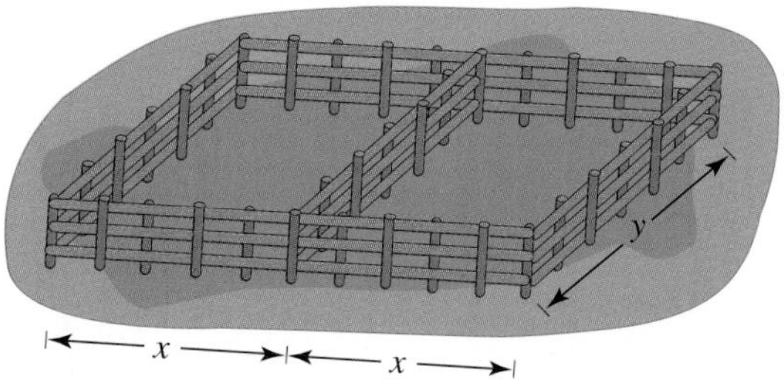

(a) Complete six rows of a table such as the one below, showing possible values for x, y, and the area of the corral.

x	y	Area
2	$\frac{1}{3}[200 - 4(2)]$	$2xy = 256$
4	$\frac{1}{3}[200 - 4(4)]$	$2xy \approx 491$

(b) Use a graphing utility to generate additional rows of the table. Use the table to estimate the dimensions that will enclose the maximum area.

(c) Write the area A as a function of x.

(d) Use a graphing utility to graph the area function. Use the graph to approximate the dimensions that will produce the maximum enclosed area.

(e) Write the area function in standard form to find analytically the dimensions that will produce the maximum area.

78. *Geometry* An indoor physical fitness room consists of a rectangular region with a semicircle on each end (see figure). The perimeter of the room is to be a 200-meter single-lane running track.

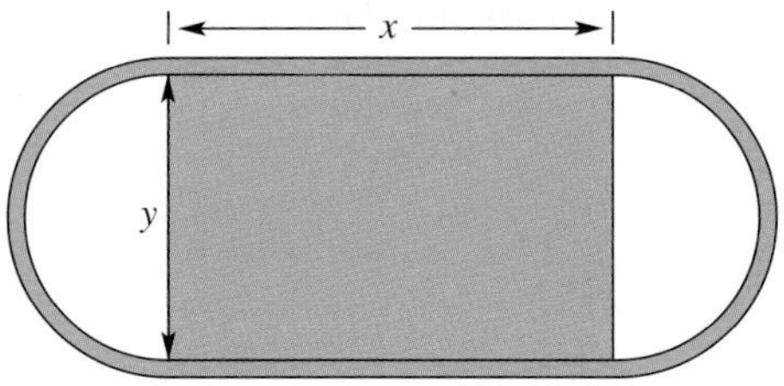

(a) Determine the radius of the semicircular ends of the room. Determine the distance, in terms of y, around the inside edge of the two semicircular parts of the track.

(b) Use the result of part (a) to write an equation, in terms of x and y, for the distance traveled in one lap around the track. Solve for y.

(c) Use the result of part (b) to write the area A of the rectangular region as a function of x. What dimensions will produce a maximum area of the rectangle?

79. *Maximum Revenue* Find the number of units sold that produces a maximum revenue

$$R = 900x - 0.1x^2$$

where R is the total revenue (in dollars) and x is the number of units sold.

80. *Maximum Revenue* Find the number of units sold that produces a maximum revenue

$$R = 100x - 0.0002x^2$$

where R is the total revenue (in dollars) and x is the number of units sold.

81. *Minimum Cost* A manufacturer of lighting fixtures has daily production costs of

$$C = 800 - 10x + 0.25x^2$$

where C is the total cost (in dollars) and x is the number of units produced. How many fixtures should be produced each day to yield a minimum cost?

82. *Minimum Cost* A textile manufacturer has daily production costs of

$$C = 100{,}000 - 110x + 0.045x^2$$

where C is the total cost (in dollars) and x is the number of units produced. How many units should be produced each day to yield a minimum cost?

83. *Maximum Profit* The profit for a company is

$$P = -0.0002x^2 + 140x - 250{,}000$$

where x is the number of units sold. What sales level will yield a maximum profit?

84. *Maximum Profit* The profit P (in hundreds of dollars) that a company makes depends on the amount x (in hundreds of dollars) the company spends on advertising according to the model

$$P = 230 + 20x - 0.5x^2.$$

What expenditure for advertising will yield a maximum profit?

85. *Physics* The height y (in feet) of a ball thrown by a child is

$$y = -\frac{1}{12}x^2 + 2x + 4$$

where x is the horizontal distance (in feet) from the point at which the ball is thrown.

(a) How high is the ball when it leaves the child's hand? (*Hint:* Find y when $x = 0$.)

(b) What is the maximum height of the ball?

(c) How far from the child does the ball strike the ground?

86. *Forestry* The number of board feet in a 16-foot log is approximated by the model

$$V = 0.77x^2 - 1.32x - 9.31, \qquad 5 \le x \le 40$$

where V is the number of board feet and x is the diameter (in inches) of the log at the small end. (One board foot is a measure of volume equivalent to a board that is 12 inches wide, 12 inches long, and 1 inch thick.)

(a) Sketch a graph of the function.

(b) Estimate the number of board feet in a 16-foot log with a diameter of 16 inches.

(c) Estimate the diameter of a 16-foot log that produced 500 board feet.

87. *Physics* The path of a diver is

$$y = -\frac{4}{9}x^2 + \frac{24}{9}x + 12$$

where y is the height (in feet) and x is the horizontal distance from the end of the diving board (in feet). What is the maximum height of the diver?

88. *Physics* The number of horsepower y required to overcome wind drag on a certain automobile is approximated by

$$y = 0.002s^2 + 0.005s - 0.029, \qquad 0 \le s \le 100$$

where s is the speed of the car (in miles per hour).

(a) Use a graphing utility to graph the function.

(b) Graphically estimate the maximum speed of the car if the power required to overcome wind drag is not to exceed 10 horsepower. Verify analytically.

89. *Graphical Analysis* From 1950 to 1990, the average annual consumption C of cigarettes by Americans (18 and older) for selected years can be modeled by

$$C = 3248.89 + 108.64t - 2.97t^2, \qquad 0 \le t \le 40$$

where t is the year, with $t = 0$ corresponding to 1950. (Source: U.S. Department of Agriculture)

(a) Use a graphing utility to graph the model.

(b) Use the graph of the model to approximate the maximum average annual consumption. Beginning in 1966, all cigarette packages were required by law to carry a health warning. Do you think the warning had any effect? Explain.

(c) In 1960, the U.S. population (18 and over) was 116,530,000. Of those, about 48,500,000 were smokers. What was the average annual cigarette consumption *per smoker* in 1960? What was the average daily cigarette consumption *per smoker*?

90. *Maximum Fuel Economy* A study was done to compare the speed x (in miles per hour) with the mileage y (in miles per gallon) of an automobile. The results are shown in the table. (Source: Federal Highway Administration)

Speed x	15	20	25	30	35	40	45
Mileage y	22.3	25.5	27.5	29.0	28.8	30.0	29.9

Speed x	50	55	60	65	70	75
Mileage y	30.2	30.4	28.8	27.4	25.3	23.3

(a) Use a graphing utility to plot the data and find the quadratic model that best fits the data. Graph the model in the same viewing window as the data.

(b) Estimate the speed for which the miles per gallon is greatest.

91. *Data Analysis* The numbers y (in millions) of VCRs in use in the United States for the years 1987 through 1996 are shown in the table. The variable t represents time (in years), with $t = 7$ corresponding to 1987. (Source: Television Bureau of Advertising, Inc.)

t	7	8	9	10	11	12	13	14	15	16
y	43	51	58	63	67	69	72	74	77	79

(a) Use a graphing utility to sketch a scatter plot of the data.

(b) Use the regression feature of a graphing utility to find a quadratic model for the data.

(c) Use a graphing utility to graph the model in the same viewing window as the scatter plot.

(d) Do you think the model can be used to predict VCR use in 2005? Explain.

Synthesis

True or False? **In Exercises 92 and 93, determine whether the statement is true or false. Justify your answer.**

92. The function $f(x) = -12x^2 - 1$ has no x-intercepts.

93. The graphs of $f(x) = -4x^2 - 10x + 7$ and $g(x) = 12x^2 + 30x + 1$ have the same axis of symmetry.

94. Write the quadratic equation $f(x) = ax^2 + bx + c$ in standard form to verify that the vertex occurs at

$$\left(-\frac{b}{2a}, f\left(-\frac{b}{2a}\right)\right).$$

95. *Business* The profit P (in millions of dollars) for a company is modeled by a quadratic function of the form

$$P = at^2 + bt + c$$

where t represents the year. If you were president of the company, which of the models below would you prefer? Explain your reasoning.

(a) a is positive and $-b/(2a) \le t$.

(b) a is positive and $t \le -b/(2a)$.

(c) a is negative and $-b/(2a) \le t$.

(d) a is negative and $t \le -b/(2a)$.

96. Is it possible for a quadratic equation to have only one x-intercept? Explain.

97. Assume that the function $f(x) = ax^2 + bx + c$ $(a \ne 0)$ has two real zeros. Show that the x-coordinate of the vertex of the graph is the average of the zeros of f. (*Hint:* Use the Quadratic Formula.)

98. Use a graphing utility to demonstrate the result of Exercise 97 for each of the following functions.

(a) $f(x) = \frac{1}{2}(x - 3)^2 - 2$

(b) $f(x) = 6 - \frac{2}{3}(x + 1)^2$

Review

In Exercises 99–102, find the equation of the line in slope-intercept form that has the given characteristics.

99. Contains the points $(-4, 3)$ and $(2, 1)$

100. Contains the point $\left(\frac{7}{2}, 2\right)$ and has a slope of $\frac{3}{2}$

101. Contains the point $(0, 3)$ and is perpendicular to the line $4x + 5y = 10$

102. Contains the point $(-8, 4)$ and is parallel to the line $y = -3x + 2$

In Exercises 103–108, let $f(x) = 14x - 3$ and let $g(x) = 8x^2$. Find the indicated value.

103. $(f + g)(-3)$

104. $(g - f)(2)$

105. $(fg)\left(-\frac{4}{7}\right)$

106. $\left(\frac{f}{g}\right)(-1.5)$

107. $(f \circ g)(-1)$

108. $(g \circ f)(0)$

2.2 Polynomial Functions of Higher Degree

▶ **What you should learn**

- How to use transformations to sketch graphs of polynomial functions
- How to use the Leading Coefficient Test to determine the end behavior of graphs of polynomial functions
- How to use zeros of polynomial functions as sketching aids
- How to use the Intermediate Value Theorem to help locate zeros of polynomial functions

▶ **Why you should learn it**

You can use polynomial functions to model real-life processes, such as the growth of a red oak tree, as discussed in Exercise 92 on page 224.

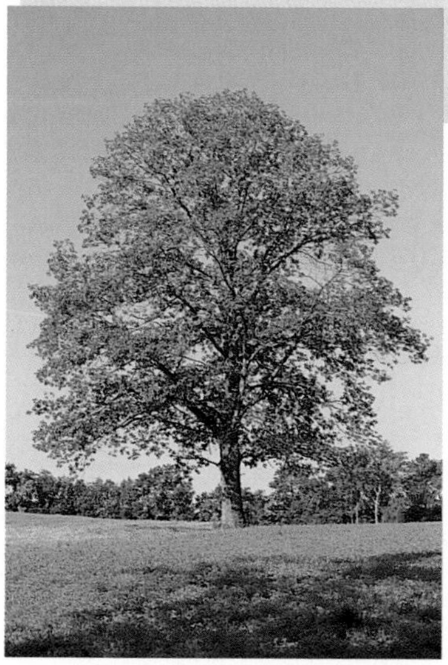

Leonard Lee Rue III/Earth Scenes

Point out to students that although all polynomial functions are continuous and have rounded turns, not all graphs that are continuous and have rounded turns are polynomials.

Graphs of Polynomial Functions

In this section, you will study basic features of the graphs of polynomial functions. The first feature is that the graph of a polynomial function is **continuous.** Essentially, this means that the graph of a polynomial function has no breaks, holes, or gaps, as shown in Figure 2.9(a).

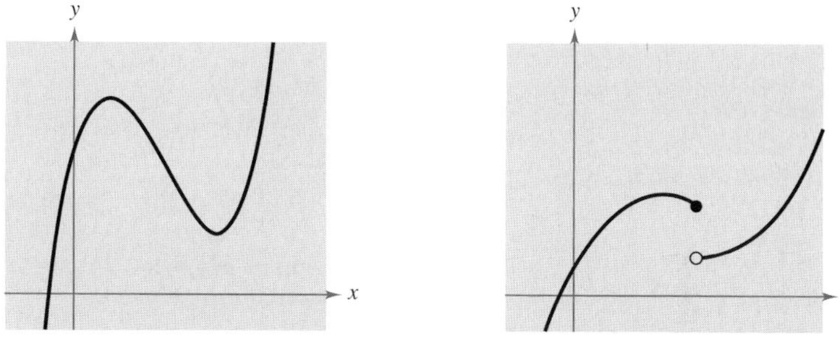

(a) Polynomial functions have continuous graphs.

(b) Functions with graphs that are not continuous are not polynomial functions.

FIGURE 2.9

The second feature is that the graph of a polynomial function has only smooth, rounded turns, as shown in Figure 2.10(a). A polynomial function cannot have a sharp turn. For instance, the function $f(x) = |x|$, which has a sharp turn at the point $(0, 0)$, as shown in Figure 2.10(b), is not a polynomial function.

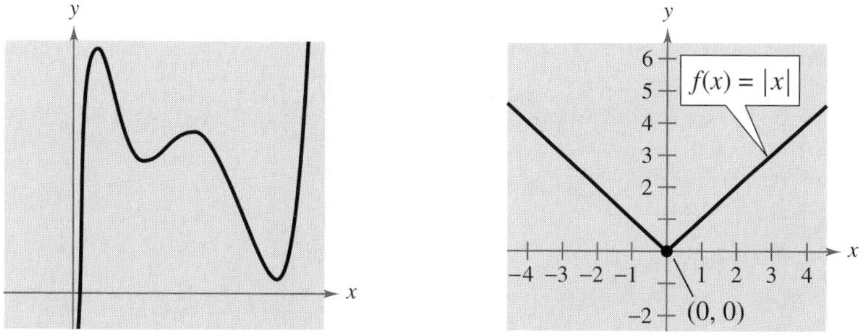

(a) Polynomial functions have graphs with rounded turns.

(b) Functions whose graphs have sharp turns are not polynomial functions.

FIGURE 2.10

The graphs of polynomial functions of degree greater than 2 are more difficult to analyze than the graphs of polynomials of degree 0, 1, or 2. However, using the features presented in this section, together with point plotting, intercepts, and symmetry, you should be able to make reasonably accurate sketches *by hand*.

The polynomial functions that have the simplest graphs are monomials of the form $f(x) = x^n$, where n is an integer greater than zero. From Figure 2.11, you can see that when n is *even* the graph is similar to the graph of $f(x) = x^2$ and when n is *odd* the graph is similar to the graph of $f(x) = x^3$. Moreover, the greater the value of n, the flatter the graph near the origin.

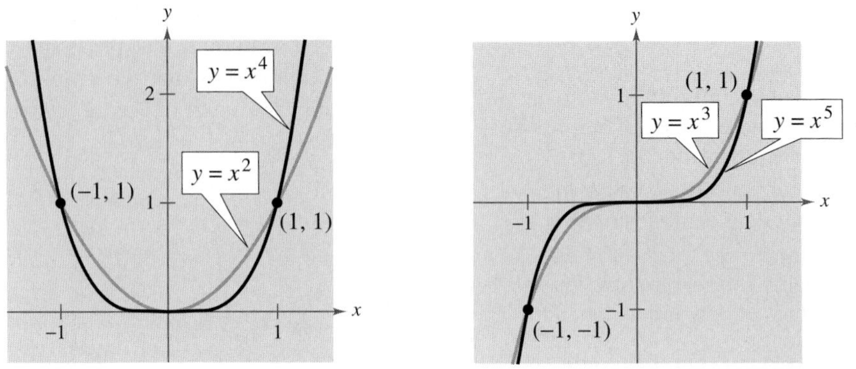

(a) If n is even, the graph of $y = x^n$ touches the axis at the x-intercept.

(b) If n is odd, the graph of $y = x^n$ crosses the axis at the x-intercept.

FIGURE 2.11

The *Interactive* CD-ROM and *Internet* versions of this text show every example with its solution; clicking on the *Try It!* button brings up similar problems. Guided Examples and Integrated Examples show step-by-step solutions to additional examples. Integrated Examples are related to several concepts in the section.

Example 1 ▶ Sketching Transformations of Monomial Functions

Sketch the graph of each function.

a. $f(x) = -x^5$ **b.** $h(x) = (x + 1)^4$

Solution

a. Because the degree of $f(x) = -x^5$ is odd, its graph is similar to the graph of $y = x^3$. In Figure 2.12(a), note that the negative coefficient has the effect of reflecting the graph in the x-axis.

b. The graph of $h(x) = (x + 1)^4$ as shown in Figure 2.12(b), is a left shift by 1 unit of the graph of $y = x^4$.

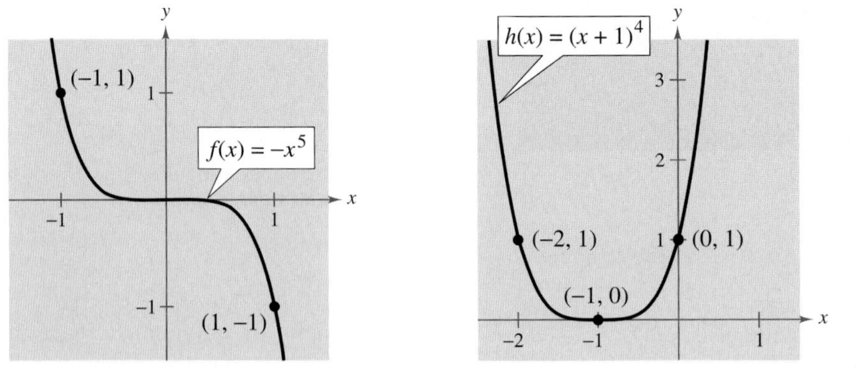

(a) **(b)**

FIGURE 2.12

Exploration

For each function, identify the degree of the function and whether the degree of the function is even or odd. Identify the leading coefficient and whether the leading coefficient is greater than 0 or less than 0. Use a graphing utility to graph each function. Describe the relationship between the degree and leading coefficient of the function and the right- and left-hand behavior of the graph of the function.

a. $f(x) = x^3 - 2x^2 - x + 1$
b. $f(x) = 2x^5 + 2x^2 - 5x + 1$
c. $f(x) = -2x^5 - x^2 + 5x + 3$
d. $f(x) = -x^3 + 5x - 2$
e. $f(x) = 2x^2 + 3x - 4$
f. $f(x) = x^4 - 3x^2 + 2x - 1$
g. $f(x) = x^2 + 3x + 2$

The Leading Coefficient Test

In Example 1, note that both graphs eventually rise or fall without bound as x moves to the right. Whether the graph of a polynomial eventually rises or falls can be determined by the function's degree (even or odd) and by its leading coefficient, as indicated in the **Leading Coefficient Test.**

Leading Coefficient Test

As x moves without bound to the left or to the right, the graph of the polynomial function $f(x) = a_n x^n + \cdot \cdot \cdot + a_1 x + a_0$ eventually rises or falls in the following manner.

1. When n is *odd:*

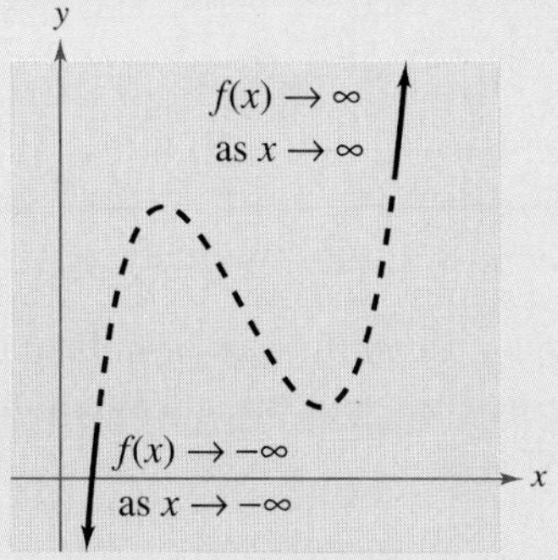

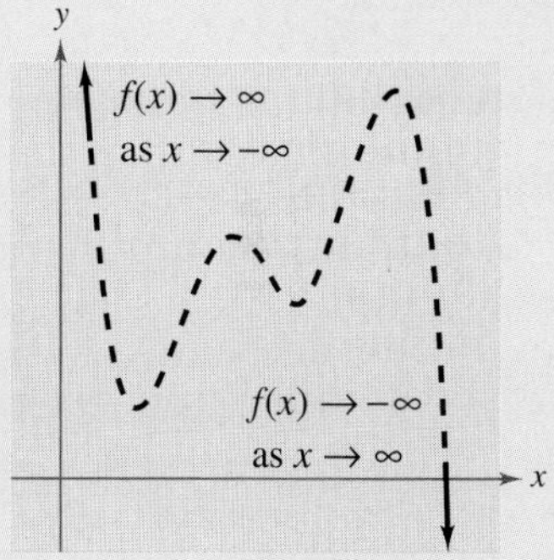

If the leading coefficient is positive ($a_n > 0$), the graph falls to the left and rises to the right.

If the leading coefficient is negative ($a_n < 0$), the graph rises to the left and falls to the right.

2. When n is *even:*

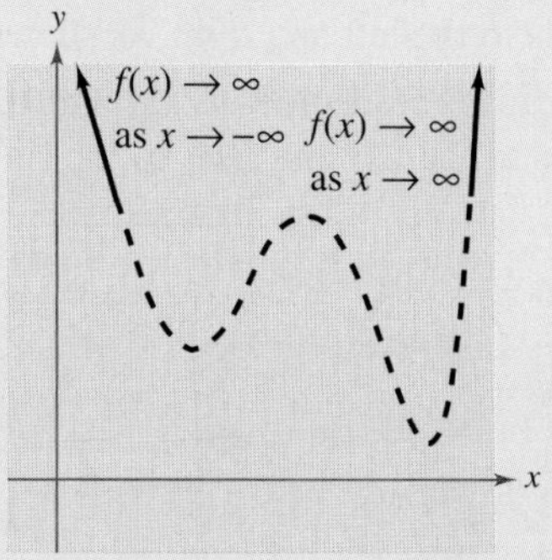

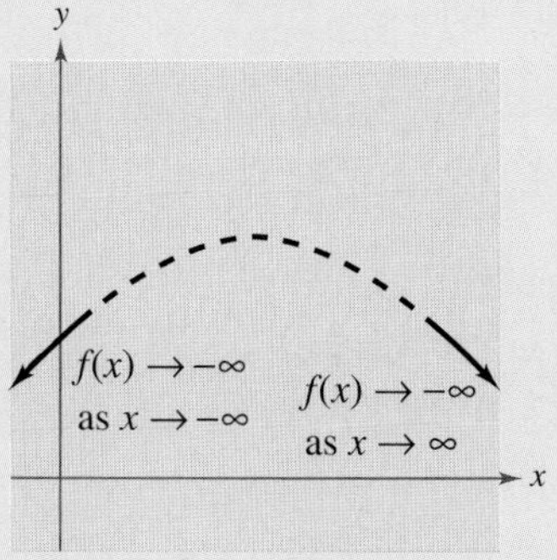

If the leading coefficient is positive ($a_n > 0$), the graph rises to the left and right.

If the leading coefficient is negative ($a_n < 0$), the graph falls to the left and right.

The dashed portions of the graphs indicate that the test determines *only* the right-hand and left-hand behavior of the graph.

STUDY TIP

The notation "$f(x) \to -\infty$ as $x \to -\infty$" indicates that the graph falls to the left. The notation "$f(x) \to \infty$ as $x \to \infty$" indicates that the graph rises to the right.

A review of the shapes of the graphs of polynomial functions of degrees 0, 1, and 2 may be used to illustrate the Leading Coefficient Test.

Example 2 ▶ Applying the Leading Coefficient Test

Describe the right-hand and left-hand behavior of the graph of $f(x) = -x^3 + 4x$.

Solution

Because the degree is odd and the leading coefficient is negative, the graph rises to the left and falls to the right, as shown in Figure 2.13.

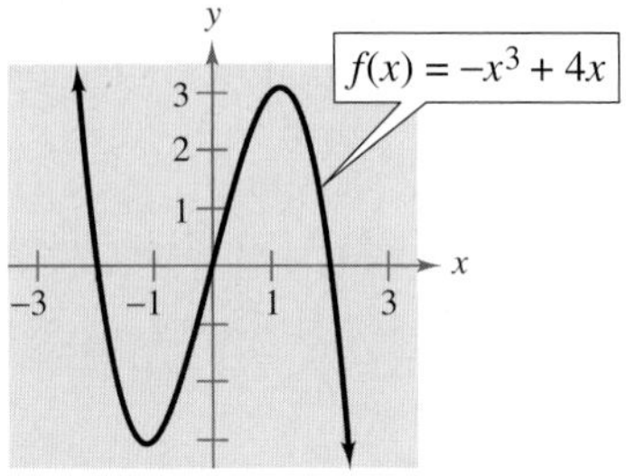

FIGURE 2.13

Additional Examples

Describe the right- and left-hand behavior of the graph of each function.

a. $f(x) = x^4 + 2x^2 - 3x$

b. $f(x) = -x^5 + 3x^4 - x$

c. $f(x) = 2x^3 - 3x^2 + 5$

d. $f(x) = -x^6 - x^4 + 2$

Solution

a. The graph rises to the left and to the right.

b. The graph rises to the left and falls to the right.

c. The graph falls to the left and rises to the right.

d. The graph falls to the left and to the right.

In Example 2, note that the Leading Coefficient Test only tells you whether the graph *eventually* rises or falls to the right or left. Other characteristics of the graph, such as intercepts and minimum and maximum points, must be determined by other tests.

Example 3 ▶ Applying the Leading Coefficient Test

Describe the right-hand and left-hand behavior of the graph of each function.

a. $f(x) = x^4 - 5x^2 + 4$ **b.** $f(x) = x^5 - x$

Solution

a. Because the degree is even and the leading coefficient is positive, the graph rises to the left and right, as shown in Figure 2.14(a).

b. Because the degree is odd and the leading coefficient is positive, the graph falls to the left and rises to the right, as shown in Figure 2.14(b).

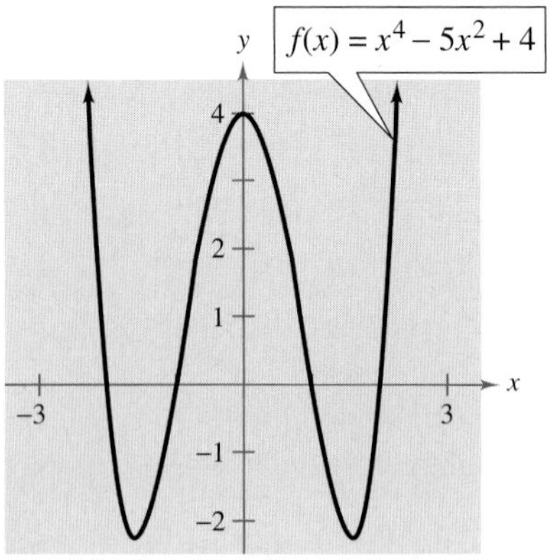

(a)

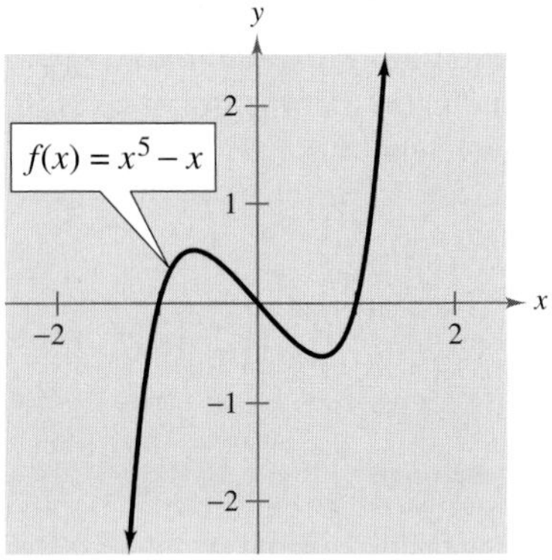

(b)

FIGURE 2.14

Zeros of Polynomial Functions

It can be shown that for a polynomial function f of degree n, the following statements are true. (Remember that the *zeros* of a function of x are the x-values for which the function is zero.)

1. The graph of f has, at most, $n - 1$ turning points. (Turning points are points at which the graph changes from increasing to decreasing or vice versa.)
2. The function f has, at most, n real zeros. (You will study this result in detail in Section 2.5 on the Fundamental Theorem of Algebra.)

Finding the zeros of polynomial functions is one of the most important problems in algebra. There is a strong interplay between graphical and algebraic approaches to this problem. Sometimes you can use information about the graph of a function to help find its zeros, and in other cases you can use information about the zeros of a function to help sketch its graph.

Real Zeros of Polynomial Functions

If f is a polynomial function and a is a real number, the following statements are equivalent.

1. $x = a$ is a *zero* of the function f.
2. $x = a$ is a *solution* of the polynomial equation $f(x) = 0$.
3. $(x - a)$ is a *factor* of the polynomial $f(x)$.
4. $(a, 0)$ is an *x-intercept* of the graph of f.

Example 4 ▶ Finding the Zeros of a Polynomial Function

Find all real zeros of $f(x) = -2x^4 + 2x^2$. Use the graph in Figure 2.15 to determine the number of turning points of the graph of the function.

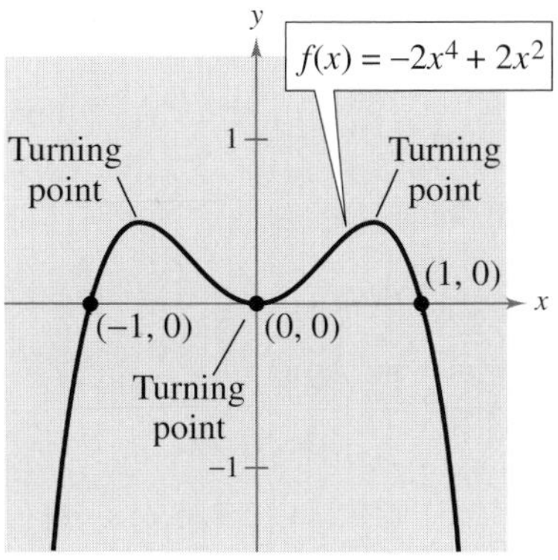

FIGURE 2.15

Solution

In this case, the polynomial factors as follows.

$$f(x) = -2x^4 + 2x^2 \qquad \text{Write original function.}$$

$$= -2x^2(x^2 - 1) \qquad \text{Remove common monomial factor.}$$

$$= -2x^2(x - 1)(x + 1) \qquad \text{Factor completely.}$$

So, the real zeros are $x = 0$, $x = 1$, and $x = -1$, and the corresponding x-intercepts are $(0, 0)$, $(1, 0)$, and $(-1, 0)$, as shown in Figure 2.15. Note in the figure that the graph has three turning points. This is consistent with the fact that a fourth-degree polynomial can have *at most* three turning points.

In Example 4, the real zero arising from $-2x^2 = 0$ is called a **repeated zero.** In general, a factor $(x - a)^k$, $k > 1$, yields a repeated zero $x = a$ of **multiplicity** k. If k is odd, the graph *crosses* the x-axis at $x = a$. If k is even, the graph *touches* the x-axis (but does not cross the x-axis) at $x = a$. In Example 4, the factor $-2x^2$ yields the repeated zero $x = 0$. Because k is even, the graph touches the x-axis at $x = 0$, as shown in Figure 2.15.

Technology

Example 5 uses an "algebraic approach" to describe the graph of the function. A graphing utility is a complement to this approach. Remember that an important aspect of using a graphing utility is to find a viewing window that shows all significant features of the graph. For instance, which of the graphs below shows all of the significant features of the function in Example 5?

a.

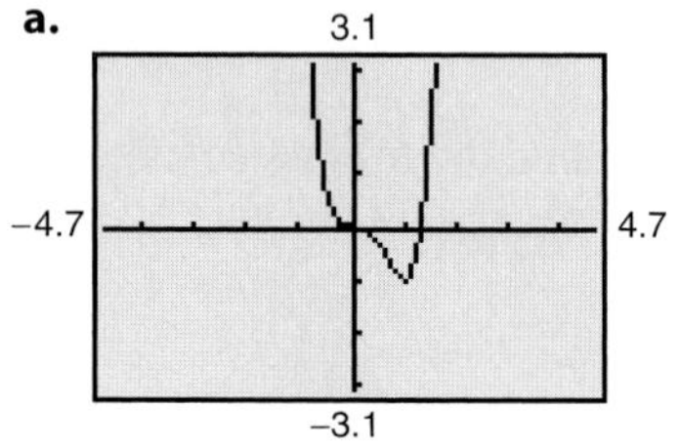

b.

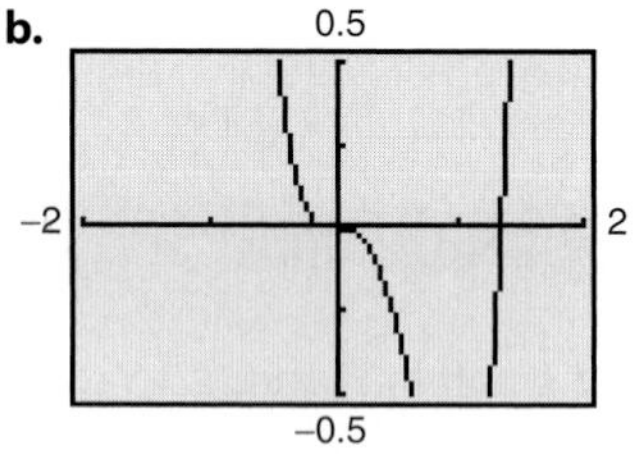

Example 5 ▶ Sketching the Graph of a Polynomial Function

Sketch the graph of $f(x) = 3x^4 - 4x^3$.

Solution

1. *Apply the Leading Coefficient Test.* Because the leading coefficient is positive and the degree is even, you know that the graph eventually rises to the left and to the right (see Figure 2.16a).
2. *Find the Zeros of the Polynomial.* By factoring

$$f(x) = 3x^4 - 4x^3 = x^3(3x - 4) \qquad \text{Remove common factor.}$$

you can see that the zeros of f are $x = 0$ and $x = \frac{4}{3}$ (both of odd multiplicity). So, the x-intercepts occur at $(0, 0)$ and $\left(\frac{4}{3}, 0\right)$. Add these points to your graph, as shown in Figure 2.16(a).

3. *Plot a Few Additional Points.* To sketch the graph by hand, find a few additional points, as shown in the table. Then plot the points (see Figure 2.16b).

x	-1	0.5	1	1.5
$f(x)$	7	-0.3125	-1	1.6875

4. *Draw the Graph.* Draw a continuous curve through the points, as shown in Figure 2.16(b). Because both zeros are of odd multiplicity, you know that the graph should cross the x-axis at $x = 0$ and $x = \frac{4}{3}$. If you are unsure of the shape of that portion of the graph, plot some additional points.

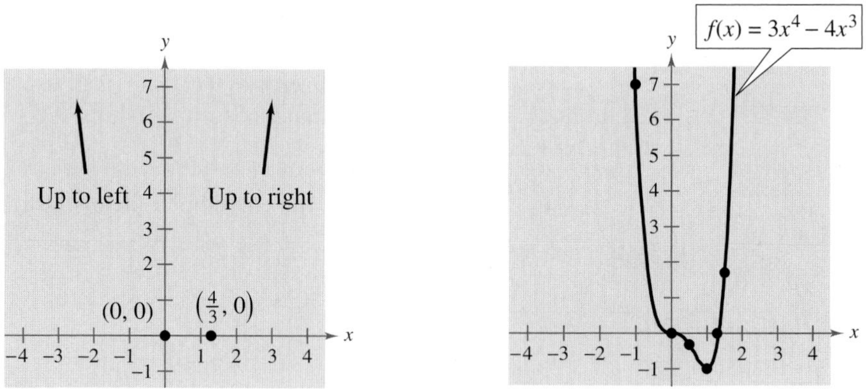

(a) **(b)**

FIGURE 2.16

A polynomial function is written in **standard form** if its terms are written in descending order of exponents from left to right. Before applying the Leading Coefficient Test to a polynomial function, it is a good idea to check that the polynomial function is written in standard form. For instance, if the function in Example 5 had been given as $f(x) = -4x^3 + 3x^4$, it might have appeared that the leading coefficient was negative.

The *Interactive* CD-ROM and *Internet* versions of this text offer a built-in graphing calculator, which can be used in the Examples, Explorations, Technology notes, and Exercises.

In Exercises 27–42, find all the real zeros of the polynomial function.

27. $f(x) = x^2 - 25$
28. $f(x) = 49 - x^2$
29. $h(t) = t^2 - 6t + 9$
30. $f(x) = x^2 + 10x + 25$
31. $f(x) = \frac{1}{3}x^2 + \frac{1}{3}x - \frac{2}{3}$
32. $f(x) = \frac{1}{2}x^2 + \frac{5}{2}x - \frac{3}{2}$
33. $f(x) = 3x^2 - 12x + 3$
34. $g(x) = 5(x^2 - 2x - 1)$
35. $f(t) = t^3 - 4t^2 + 4t$
36. $f(x) = x^4 - x^3 - 20x^2$
37. $g(t) = \frac{1}{2}t^4 - \frac{1}{2}$
38. $f(x) = x^5 + x^3 - 6x$
39. $g(t) = t^5 - 6t^3 + 9t$
40. $f(x) = 2x^4 - 2x^2 - 40$
41. $f(x) = 5x^4 + 15x^2 + 10$
42. $f(x) = x^3 - 4x^2 - 25x + 100$

Graphical Analysis **In Exercises 43–46, use a graphing utility to graph the function. Use the graph to approximate any *x*-intercepts of the graph. Set $y = 0$ and solve the resulting equation. Compare the result with any *x*-intercepts of the graph.**

43. $y = 4x^3 - 20x^2 + 25x$
44. $y = 4x^3 + 4x^2 - 7x + 2$
45. $y = x^5 - 5x^3 + 4x$
46. $y = \frac{1}{4}x^3(x^2 - 9)$

In Exercises 47–56, find a polynomial function that has the given zeros. (There are many correct answers.)

47. 0, 10
48. 0, −3
49. 2, −6
50. −4, 5
51. 0, −2, −3
52. 0, 2, 5
53. 4, −3, 3, 0
54. −2, −1, 0, 1, 2
55. $1 + \sqrt{3}, 1 - \sqrt{3}$
56. $2, 4 + \sqrt{5}, 4 - \sqrt{5}$

In Exercises 57–66, find a polynomial of degree *n* that has the given zeros. (There are many correct answers.)

57. Zero: $x = -2$	Degree: $n = 2$
58. Zeros: $x = -8, -4$	Degree: $n = 2$
59. Zeros: $x = -3, 0, 1$	Degree: $n = 3$
60. Zeros: $x = -2, 4, 7$	Degree: $n = 3$
61. Zeros: $x = 0, \sqrt{3}, -\sqrt{3}$	Degree: $n = 3$
62. Zero: $x = 9$	Degree: $n = 3$
63. Zeros: $x = -5, 1, 2$	Degree: $n = 4$
64. Zeros: $x = -4, -1, 3, 6$	Degree: $n = 4$
65. Zeros: $x = 0, -4$	Degree: $n = 5$
66. Zeros: $x = -3, 1, 5, 6$	Degree: $n = 5$

In Exercises 67–80, sketch the graph of the function by (a) applying the Leading Coefficient Test, (b) finding the zeros of the polynomial, (c) plotting sufficient solution points, and (d) drawing a continuous curve through the points.

67. $f(x) = x^3 - 9x$
68. $g(x) = x^4 - 4x^2$
69. $f(t) = \frac{1}{4}(t^2 - 2t + 15)$
70. $g(x) = -x^2 + 10x - 16$
71. $f(x) = x^3 - 3x^2$
72. $f(x) = 1 - x^3$
73. $f(x) = 3x^3 - 15x^2 + 18x$
74. $f(x) = -4x^3 + 4x^2 + 15x$
75. $f(x) = -5x^2 - x^3$
76. $f(x) = -48x^2 + 3x^4$
77. $f(x) = x^2(x - 4)$
78. $h(x) = \frac{1}{3}x^3(x - 4)^2$
79. $g(t) = -\frac{1}{4}(t - 2)^2(t + 2)^2$
80. $g(x) = \frac{1}{10}(x + 1)^2(x - 3)^3$

In Exercises 81–84, use a graphing utility to graph the function. Use the *zero* or *root* feature to approximate zeros of the function. Determine the multiplicity of each zero.

81. $f(x) = x^3 - 4x$
82. $f(x) = \frac{1}{4}x^4 - 2x^2$
83. $g(x) = \frac{1}{5}(x + 1)^2(x - 3)(2x - 9)$
84. $h(x) = \frac{1}{5}(x + 2)^2(3x - 5)^2$

In Exercises 85–88, use the Intermediate Value Theorem and a graphing utility to find intervals 1 unit in length in which the polynomial function is guaranteed to have a zero. (See Example 7.)

85. $f(x) = x^3 - 3x^2 + 3$
86. $f(x) = 0.11x^3 - 2.07x^2 + 9.81x - 6.88$
87. $g(x) = 3x^4 + 4x^3 - 3$
88. $h(x) = x^4 - 10x^2 + 3$

89. *Numerical and Graphical Analysis* An open box is to be made from a square piece of material, 36 inches on a side, by cutting equal squares of length x from the corners and turning up the sides (see figure).

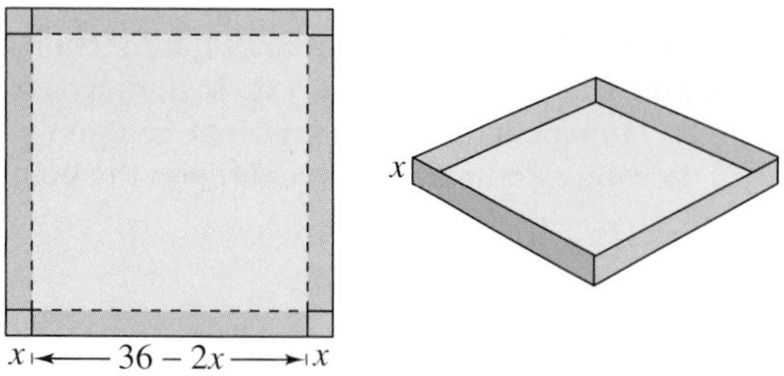

(a) Complete four rows of a table such as the one below.

Box height	Box width	Box volume
1	$36 - 2(1)$	$1[36 - 2(1)]^2 = 1156$
2	$36 - 2(2)$	$2[36 - 2(2)]^2 = 2048$

(b) Use a graphing utility to generate additional rows of the table. Use the table to estimate the dimensions that will produce a maximum volume.

(c) Verify that the volume of the box is given by the function

$$V(x) = x(36 - 2x)^2.$$

Determine the domain of the function.

(d) Use a graphing utility to graph V and use the graph to estimate the value of x for which $V(x)$ is maximum. Compare your result with that of part (b).

90. *Maximum Volume* An open box with locking tabs is to be made from a square piece of material 24 inches on a side. This is to be done by cutting equal squares from the corners and folding along the dashed lines shown in the figure.

(a) Verify that the volume of the box is given by the function

$$V(x) = 8x(6 - x)(12 - x).$$

(b) Determine the domain of the function V.

(c) Sketch the graph of the function and estimate the value of x for which $V(x)$ is maximum.

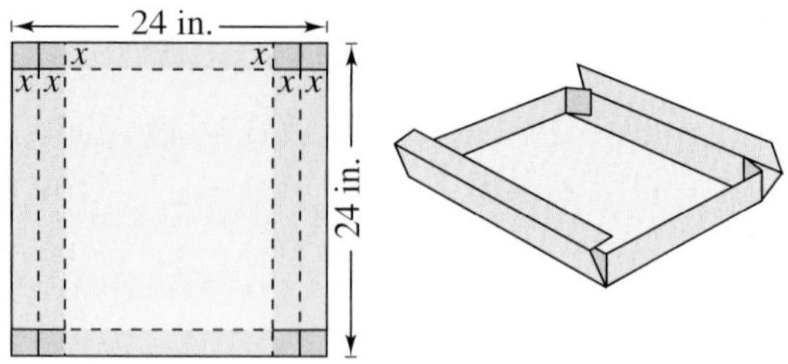

FIGURE FOR 90

91. *Business* The total revenue R (in millions of dollars) for a company is related to its advertising expense by the function

$$R = \frac{1}{100{,}000}(-x^3 + 600x^2), \qquad 0 \le x \le 400$$

where x is the amount spent on advertising (in tens of thousands of dollars). Use the graph of this function, shown in the figure, to estimate the point on the graph at which the function is increasing most rapidly. This point is called the *point of diminishing returns* because any expense above this amount will yield less return per dollar invested in advertising.

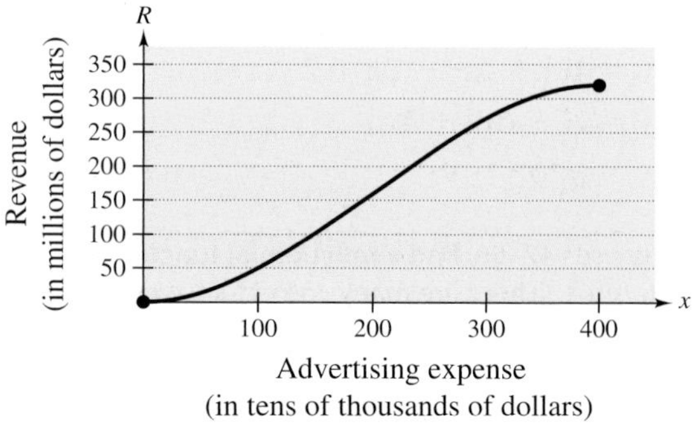

92. *Environment* The growth of a red oak tree is approximated by the function

$$G = -0.003t^3 + 0.137t^2 + 0.458t - 0.839$$

where G is the height of the tree (in feet) and t ($2 \le t \le 34$) is its age (in years). Use a graphing utility to graph the function and estimate the age of the tree when it is growing most rapidly. This point is called the *point of diminishing returns* because the increase in size will be less with each additional year. (*Hint:* Use a viewing window in which $-10 \le x \le 45$ and $-5 \le y \le 60$.)

Synthesis

True or False? **In Exercises 93 and 94, determine whether the statement is true or false. Justify your answer.**

93. A fifth-degree polynomial can have five turning points in its graph.

94. It is possible for a sixth-degree polynomial to have only one solution.

95. *Graphical Analysis* Describe a polynomial function that could represent the graph. (Indicate the degree of the function and the sign of its leading coefficient.)

(a)

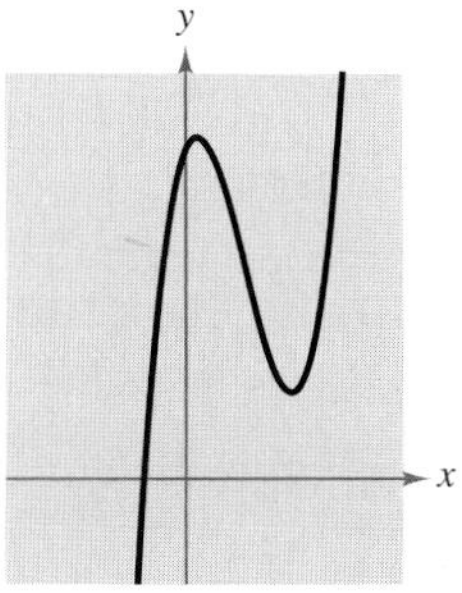

(b)

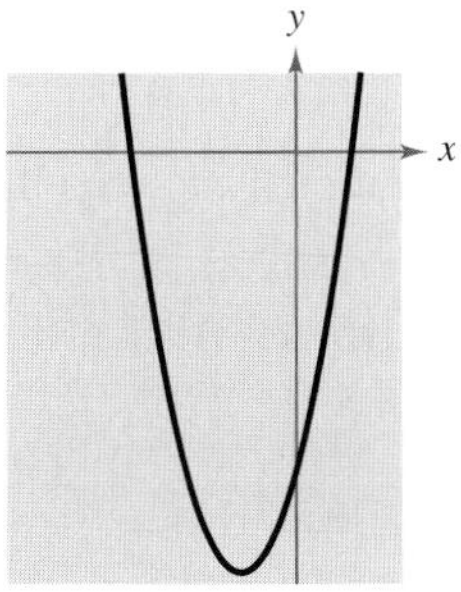

(c)

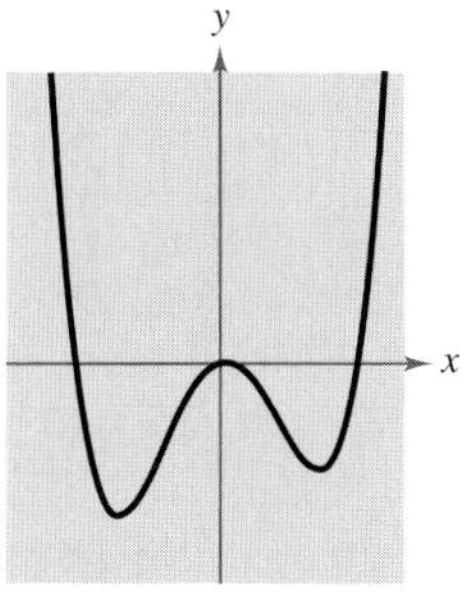

(d)

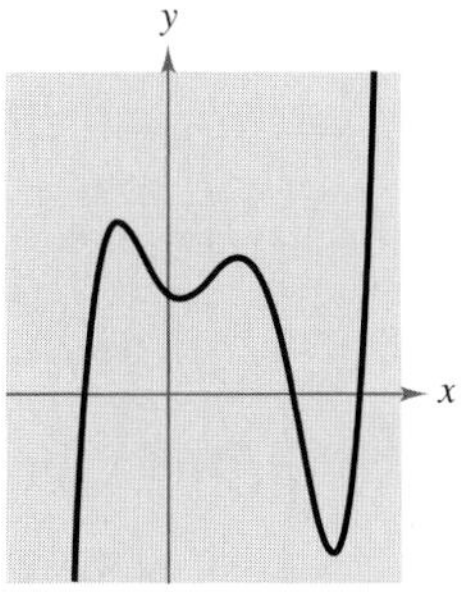

96. *Graphical Reasoning* Sketch a graph of the function $f(x) = x^4$. Explain how the graph of g differs (if it does) from the graph of f. Determine whether g is odd, even, or neither.

(a) $g(x) = f(x) + 2$ (b) $g(x) = f(x + 2)$

(c) $g(x) = f(-x)$ (d) $g(x) = -f(x)$

(e) $g(x) = f\left(\frac{1}{2}x\right)$ (f) $g(x) = \frac{1}{2}f(x)$

(g) $g(x) = f\left(x^{3/4}\right)$ (h) $g(x) = (f \circ f)(x)$

97. *Exploration* Explore the transformations of the form $g(x) = a(x - h)^5 + k$.

(a) Use a graphing utility to graph the functions

$$y_1 = -\frac{1}{3}(x - 2)^5 + 1$$

and

$$y_2 = \frac{3}{5}(x + 2)^5 - 3.$$

Determine whether the graphs are increasing or decreasing. Explain.

(b) Will the graph of g always be increasing or decreasing? If so, is this behavior determined by a, h, or k? Explain.

(c) Use a graphing utility to graph the function $H(x) = x^5 - 3x^3 + 2x + 1$.

Use the graph and the result of part (b) to determine whether H can be written in the form $H(x) = a(x - h)^5 + k$. Explain.

Review

In Exercises 98–101, solve the equation by factoring.

98. $2x^2 - x - 28 = 0$ **99.** $3x^2 - 22x - 16 = 0$

100. $12x^2 + 11x - 5 = 0$ **101.** $x^2 + 24x + 144 = 0$

In Exercises 102–105, solve the equation by completing the square.

102. $x^2 - 2x - 21 = 0$ **103.** $x^2 - 8x + 2 = 0$

104. $2x^2 + 5x - 20 = 0$ **105.** $3x^2 + 4x - 9 = 0$

In Exercises 106–109, factor the expression completely.

106. $5x^2 + 7x - 24$ **107.** $6x^3 - 61x^2 + 10x$

108. $4x^4 - 7x^3 - 15x^2$ **109.** $y^3 + 216$

110. Use the graphs of the functions f and g to answer each question.

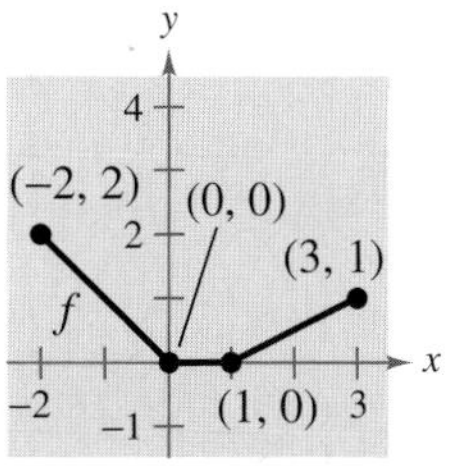

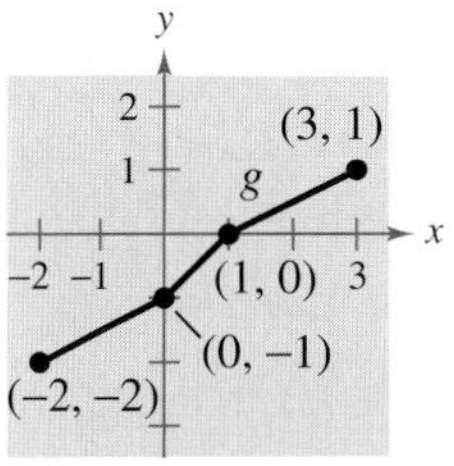

(a) Explain why f does not have an inverse.

(b) Find $g^{-1}(1)$.

2.3 Polynomial and Synthetic Division

▶ What you should learn

- How to use long division to divide polynomials by other polynomials
- How to use synthetic division to divide polynomials by binomials of the form $(x - k)$
- How to use the Remainder Theorem and the Factor Theorem
- How to use polynomial division to answer questions about real-life problems

▶ Why you should learn it

Polynomial division can help you rewrite polynomials that are used to model real-life problems. Rewriting a model can sometimes make the important properties of the model more apparent. For instance, Example 7 on page 232 shows how polynomial division can be used to rewrite the model for a person's take-home pay.

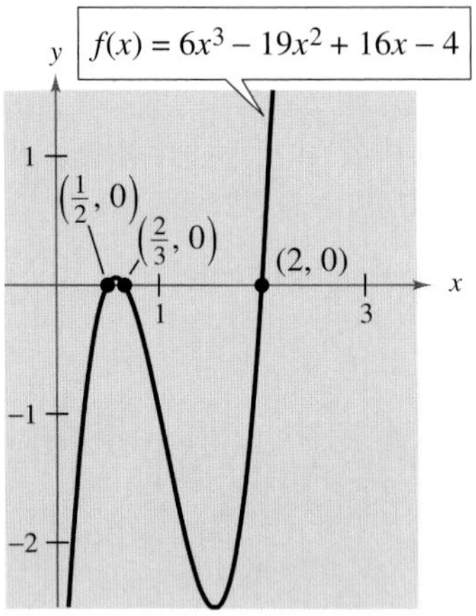

FIGURE 2.21

Long Division of Polynomials

In this section, you will study two procedures for *dividing* polynomials. These procedures are especially valuable in factoring and finding the zeros of polynomial functions. To begin, suppose you are given the graph of

$$f(x) = 6x^3 - 19x^2 + 16x - 4.$$

Notice that a zero of f occurs at $x = 2$, as shown in Figure 2.21. Because $x = 2$ is a zero of f, you know that $(x - 2)$ is a factor of $f(x)$. This means that there exists a second-degree polynomial $q(x)$ such that

$$f(x) = (x - 2) \cdot q(x).$$

To find $q(x)$, you can use **long division,** as illustrated in Example 1.

Example 1 ▶ Long Division of Polynomials

Divide $6x^3 - 19x^2 + 16x - 4$ by $x - 2$, and use the result to factor the polynomial completely.

Solution

Think $\frac{6x^3}{x} = 6x^2$.

Think $\frac{-7x^2}{x} = -7x$.

Think $\frac{2x}{x} = 2$.

$$\begin{array}{r}
6x^2 - 7x + 2 \\
x - 2 \,\overline{)\, 6x^3 - 19x^2 + 16x - 4} \\
\underline{6x^3 - 12x^2} \qquad\qquad \\
-7x^2 + 16x \qquad \\
\underline{-7x^2 + 14x} \qquad \\
2x - 4 \\
\underline{2x - 4} \\
0
\end{array}$$

$6x^3 - 12x^2$	Multiply: $6x^2(x - 2)$.
$-7x^2 + 16x$	Subtract.
$-7x^2 + 14x$	Multiply: $-7x(x - 2)$.
$2x - 4$	Subtract.
$2x - 4$	Multiply: $2(x - 2)$.
0	Subtract.

From this division, you can conclude that

$$6x^3 - 19x^2 + 16x - 4 = (x - 2)(6x^2 - 7x + 2)$$

and by factoring the quadratic $6x^2 - 7x + 2$, you have

$$6x^3 - 19x^2 + 16x - 4 = (x - 2)(2x - 1)(3x - 2).$$

Note that this factorization agrees with the graph shown in Figure 2.21 in that the three x-intercepts occur at $x = 2$, $x = \frac{1}{2}$, and $x = \frac{2}{3}$.

Note that one of the many uses of polynomial division is to write a function as a sum of terms to find slant asymptotes (see Section 2.6). This is a skill that is also used frequently in calculus.

In Example 1, $x - 2$ is a factor of the polynomial $6x^3 - 19x^2 + 16x - 4$, and the long division process produces a remainder of zero. Often, long division will produce a nonzero remainder. For instance, if you divide $x^2 + 3x + 5$ by $x + 1$, you obtain the following.

$$\begin{array}{rrl} & x + 2 & \leftarrow \text{Quotient} \\ \text{Divisor} \rightarrow x + 1 & \overline{)\,x^2 + 3x + 5} & \leftarrow \text{Dividend} \\ & \underline{x^2 + x \phantom{{}+5}} & \\ & 2x + 5 & \\ & \underline{2x + 2} & \\ & 3 & \leftarrow \text{Remainder} \end{array}$$

In fractional form, you can write this result as follows.

$$\underbrace{\frac{\overbrace{x^2 + 3x + 5}^{\text{Dividend}}}{x + 1}}_{\text{Divisor}} = \overbrace{x + 2}^{\text{Quotient}} + \underbrace{\frac{\overset{\text{Remainder}}{3}}{x + 1}}_{\text{Divisor}}$$

This implies that

$$x^2 + 3x + 5 = (x + 1)(x + 2) + 3 \qquad \text{Multiply each side by } (x + 1).$$

which illustrates the following theorem, called the **Division Algorithm.**

The Division Algorithm

If $f(x)$ and $d(x)$ are polynomials such that $d(x) \neq 0$, and the degree of $d(x)$ is less than or equal to the degree of $f(x)$, there exist unique polynomials $q(x)$ and $r(x)$ such that

$$\underset{\text{Dividend}}{f(x)} = \underset{\text{Divisor}}{d(x)}\,\underset{\text{Quotient}}{q(x)} + \underset{\text{Remainder}}{r(x)}$$

where $r(x) = 0$ *or* the degree of $r(x)$ is less than the degree of $d(x)$. If the remainder $r(x)$ is zero, $d(x)$ *divides evenly* into $f(x)$.

The Division Algorithm can also be written as

$$\frac{f(x)}{d(x)} = q(x) + \frac{r(x)}{d(x)}.$$

In the Division Algorithm, the rational expression $f(x)/d(x)$ is **improper** because the degree of $f(x)$ is greater than or equal to the degree of $d(x)$. On the other hand, the rational expression $r(x)/d(x)$ is **proper** because the degree of $r(x)$ is less than the degree of $d(x)$.

Example 2 ▶ Long Division of Polynomials

Divide $x^3 - 1$ by $x - 1$.

Solution

Because there is no x^2-term or x-term in the dividend, you need to line up the subtraction by using zero coefficients (or leaving spaces) for the missing terms.

$$\begin{array}{r}
x^2 + x + 1 \\
x - 1\,\overline{\smash{)}\,x^3 + 0x^2 + 0x - 1} \\
\underline{x^3 - x^2} \\
x^2 + 0x \\
\underline{x^2 - x} \\
x - 1 \\
\underline{x - 1} \\
0
\end{array}$$

So, $x - 1$ divides evenly into $x^3 - 1$ and you can write

$$\frac{x^3 - 1}{x - 1} = x^2 + x + 1, \quad x \neq 1.$$

You can check the result of a division problem by multiplying. For instance, in Example 2, try checking that

$$(x - 1)(x^2 + x + 1) = x^3 - 1.$$

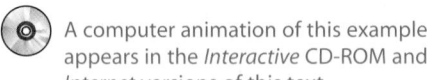

A computer animation of this example appears in the *Interactive* CD-ROM and *Internet* versions of this text.

Example 3 ▶ Long Division of Polynomials

Divide $2x^4 + 4x^3 - 5x^2 + 3x - 2$ by $x^2 + 2x - 3$.

Solution

$$\begin{array}{r}
2x^2 + 1 \\
x^2 + 2x - 3\,\overline{\smash{)}\,2x^4 + 4x^3 - 5x^2 + 3x - 2} \\
\underline{2x^4 + 4x^3 - 6x^2} \\
x^2 + 3x - 2 \\
\underline{x^2 + 2x - 3} \\
x + 1
\end{array}$$

Note that the first subtraction eliminated two terms from the dividend. When this happens, the quotient skips a term. You can write the result as

$$\frac{2x^4 + 4x^3 - 5x^2 + 3x - 2}{x^2 + 2x - 3} = 2x^2 + 1 + \frac{x + 1}{x^2 + 2x - 3}.$$

Remind students that when division yields a remainder, it is important that they write the remainder term correctly.

Synthetic Division

There is a nice shortcut for long division of polynomials when dividing by divisors of the form $x - k$. This shortcut is called **synthetic division.** The pattern for synthetic division of a cubic polynomial is summarized as follows. (The pattern for higher-degree polynomials is similar.)

Synthetic Division (for a Cubic Polynomial)

To divide $ax^3 + bx^2 + cx + d$ by $x - k$, use the following pattern.

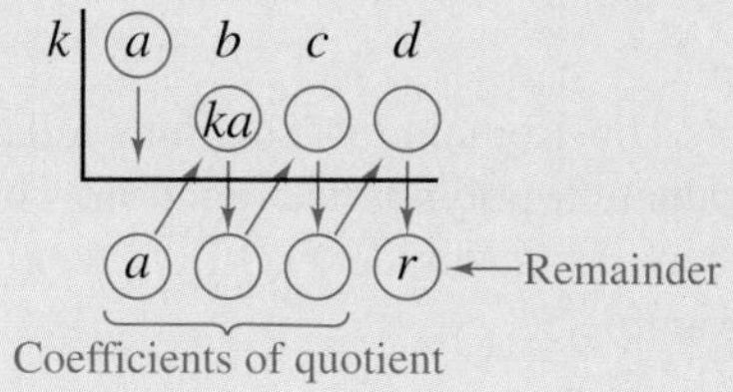

Vertical pattern: Add terms.
Diagonal pattern: Multiply by k.

Point out to students that a graphing utility can be used to check the answer to a polynomial division problem. When students graph both the original polynomial division problem and the answer in the same viewing window, the graphs should coincide.

Synthetic division works *only* for divisors of the form $x - k$. [Remember that $x + k = x - (-k)$.] You cannot use synthetic division to divide a polynomial by a quadratic such as $x^2 - 3$.

A computer animation of this example appears in the *Interactive* CD-ROM and *Internet* versions of this text.

Example 4 ▶ Using Synthetic Division

Use synthetic division to divide $x^4 - 10x^2 - 2x + 4$ by $x + 3$.

Solution

You should set up the array as follows. Note that a zero is included for each missing term in the dividend.

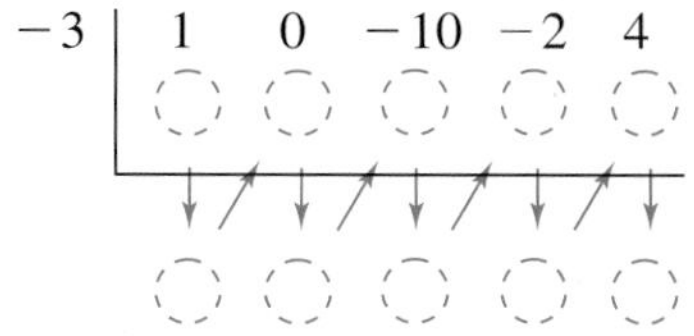

Then, use the synthetic division pattern by adding terms in columns and multiplying the results by -3.

Divisor: $x + 3$ Dividend: $x^4 - 10x^2 - 2x + 4$

-3	1	0	-10	-2	4
		-3	9	3	-3
	1	-3	-1	1	1 ← Remainder: 1

Quotient: $x^3 - 3x^2 - x + 1$

So, you have

$$\frac{x^4 - 10x^2 - 2x + 4}{x + 3} = x^3 - 3x^2 - x + 1 + \frac{1}{x + 3}.$$

The Remainder and Factor Theorems

The remainder obtained in the synthetic division process has an important interpretation, as described in the **Remainder Theorem.** (A proof is given in Appendix A.)

> **The Remainder Theorem**
>
> If a polynomial $f(x)$ is divided by $x - k$, then the remainder is
>
> $$r = f(k).$$

The Remainder Theorem tells you that synthetic division can be used to evaluate a polynomial function. That is, to evaluate a polynomial function $f(x)$ when $x = k$, divide $f(x)$ by $x - k$. The remainder will be $f(k)$, as illustrated in Example 5.

Example 5 ▶ Using the Remainder Theorem

Use the Remainder Theorem to evaluate the following function at $x = -2$.

$$f(x) = 3x^3 + 8x^2 + 5x - 7$$

Solution

Using synthetic division, you obtain the following.

$$\begin{array}{r|rrrr} -2 & 3 & 8 & 5 & -7 \\ & & -6 & -4 & -2 \\ \hline & 3 & 2 & 1 & -9 \end{array}$$

Because the remainder is $r = -9$, you can conclude that

$$f(-2) = -9.$$

This means that $(-2, -9)$ is a point on the graph of f. You can check this by substituting $x = -2$ in the original function.

Check

$$\begin{aligned} f(-2) &= 3(-2)^3 + 8(-2)^2 + 5(-2) - 7 \\ &= 3(-8) + 8(4) - 10 - 7 \\ &= -9 \end{aligned}$$

Additional Example

Use the Remainder Theorem to evaluate $f(x) = 4x^2 - 10x - 21$ when $x = 5$.

Solution

Using synthetic division, you obtain the following.

$$\begin{array}{r|rrr} 5 & 4 & -10 & -21 \\ & & 20 & 50 \\ \hline & 4 & 10 & 29 \end{array}$$

Because the remainder is 29, you can conclude that $f(5) = 29$.

Another important theorem is the **Factor Theorem,** which is stated below. This theorem states that you can test to see whether a polynomial has $(x - k)$ as a factor by evaluating the polynomial at $x = k$. If the result is 0, $(x - k)$ is a factor. (A proof is given in Appendix A.)

> **The Factor Theorem**
>
> A polynomial $f(x)$ has a factor $(x - k)$ if and only if $f(k) = 0$.

Example 6 ▶ Factoring a Polynomial: Repeated Division

Show that $(x - 2)$ and $(x + 3)$ are factors of

$$f(x) = 2x^4 + 7x^3 - 4x^2 - 27x - 18.$$

Then find the remaining factors of $f(x)$.

Solution

Using synthetic division with the factor $(x - 2)$, you obtain the following.

$$\begin{array}{r|rrrrr} 2 & 2 & 7 & -4 & -27 & -18 \\ & & 4 & 22 & 36 & 18 \\ \hline & 2 & 11 & 18 & 9 & 0 \end{array}$$

0 remainder, so $f(2) = 0$ and $(x - 2)$ is a factor.

Use the result of this division to perform synthetic division again with the factor $(x + 3)$.

$$\begin{array}{r|rrrr} -3 & 2 & 11 & 18 & 9 \\ & & -6 & -15 & -9 \\ \hline & 2 & 5 & 3 & 0 \end{array}$$

0 remainder, so $f(-3) = 0$ and $(x + 3)$ is a factor.

Because the resulting quadratic expression factors as

$$2x^2 + 5x + 3 = (2x + 3)(x + 1)$$

the complete factorization of $f(x)$ is

$$f(x) = (x - 2)(x + 3)(2x + 3)(x + 1).$$

Note that this factorization implies that f has four real zeros:

$$2, -3, -\tfrac{3}{2}, \text{ and } -1.$$

This is confirmed by the graph of f, which is shown in Figure 2.22.

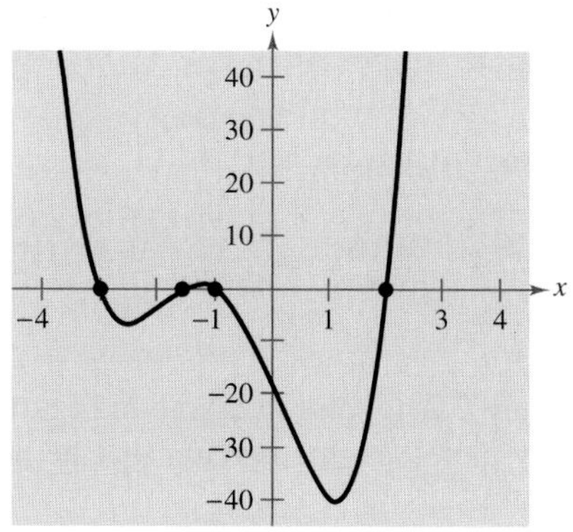

FIGURE 2.22

Using the Remainder in Synthetic Division

The remainder r, obtained in the synthetic division of $f(x)$ by $x - k$, provides the following information.

1. The remainder r gives the value of f at $x = k$. That is, $r = f(k)$.
2. If $r = 0$, $(x - k)$ is a factor of $f(x)$.
3. If $r = 0$, $(k, 0)$ is an x-intercept of the graph of f.

Throughout this text, we have emphasized the importance of developing several problem-solving strategies. In the exercises for this section, try using more than one strategy to solve several of the exercises. For instance, if you find that $x - k$ divides evenly into $f(x)$ (with no remainder), try sketching the graph of f. You should find that $(k, 0)$ is an x-intercept of the graph.

Activities

1. Use synthetic division to determine if $(x + 3)$ is a factor of $f(x) = 3x^3 + 4x^2 - 18x - 3$.
   Answer: No, it is not.
2. Divide using long division.
   $$\frac{4x^5 - x^3 + 2x^2 - x}{2x + 1}$$
   Answer: $2x^4 - x^3 + x - 1 + \dfrac{1}{2x + 1}$
3. Use the Remainder Theorem to evaluate $f(-3)$ for $f(x) = 2x^3 - 4x^2 + 1$.
   Answer: -89

Application

Example 7 ▶ Take-Home Pay

The 1999 monthly take-home pay for an employee who is single and claimed one deduction is given by the function

$$y = -0.00002436x^2 + 0.79337x + 42.096, \qquad 500 \le x \le 5000$$

where y represents the take-home pay and x represents the gross monthly salary. Find a function that gives the take-home pay as a *percent* of the gross monthly salary. (Source: UA Corporate Accounting Software, based on a state and local income tax rate of 3.8%)

Solution

Because the gross monthly salary is given by x and the take-home pay is given by y, the percent of gross monthly salary that the person takes home is

$$\begin{aligned} P &= \frac{y}{x} \\ &= \frac{-0.00002436x^2 + 0.79337x + 42.096}{x} \\ &= -0.00002436x + 0.79337 + \frac{42.096}{x}. \end{aligned}$$

The graphs of these functions are shown in Figure 2.23(a) and (b). Note in Figure 2.23(b) that as a person's gross monthly salary increases, the *percent* that he or she takes home decreases.

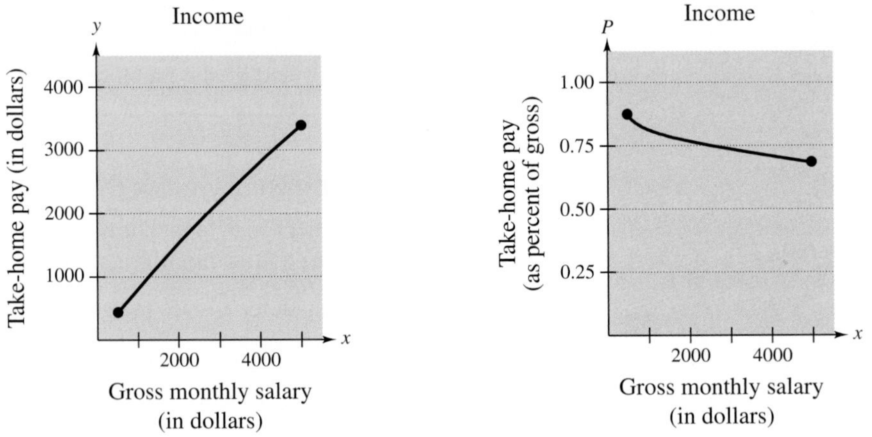

(a) (b)

FIGURE 2.23

In Example 7, the difference between gross pay and take-home pay results primarily from federal deductions for income tax, Social Security tax, and Medicare tax. Because the federal income tax is graduated, the percent that is deducted depends on a person's income. People who earn more do not simply pay more tax—they pay *much* more tax because the tax rate increases.

2.3 Exercises

Analytical Analysis **In Exercises 1–4, use long division to verify that $y_1 = y_2$.**

1. $y_1 = \dfrac{4x}{x-1}, \quad y_2 = 4 + \dfrac{4}{x-1}$

2. $y_1 = \dfrac{3x-5}{x-3}, \quad y_2 = 3 + \dfrac{4}{x-3}$

3. $y_1 = \dfrac{x^2}{x+2}, \quad y_2 = x - 2 + \dfrac{4}{x+2}$

4. $y_1 = \dfrac{x^4 - 3x^2 - 1}{x^2+5}, \quad y_2 = x^2 - 8 + \dfrac{39}{x^2+5}$

Graphical Analysis **In Exercises 5 and 6, use a graphing utility to graph the two equations in the same viewing window. Use the graphs to verify that the expressions are equivalent. Use long division to verify the results algebraically.**

5. $y_1 = \dfrac{x^5 - 3x^3}{x^2+1}, \quad y_2 = x^3 - 4x + \dfrac{4x}{x^2+1}$

6. $y_1 = \dfrac{x^3 - 2x^2 + 5}{x^2+x+1}, \quad y_2 = x - 3 + \dfrac{2(x+4)}{x^2+x+1}$

In Exercises 7–20, use long division to divide.

7. $(2x^2 + 10x + 12) \div (x + 3)$

8. $(5x^2 - 17x - 12) \div (x - 4)$

9. $(4x^3 - 7x^2 - 11x + 5) \div (4x + 5)$

10. $(6x^3 - 16x^2 + 17x - 6) \div (3x - 2)$

11. $(x^4 + 5x^3 + 6x^2 - x - 2) \div (x + 2)$

12. $(x^3 + 4x^2 - 3x - 12) \div (x - 3)$

13. $(7x + 3) \div (x + 2)$

14. $(8x - 5) \div (2x + 1)$

15. $(6x^3 + 10x^2 + x + 8) \div (2x^2 + 1)$

16. $(x^3 - 9) \div (x^2 + 1)$

17. $\dfrac{x^4 + 3x^2 + 1}{x^2 - 2x + 3}$

18. $\dfrac{x^5 + 7}{x^3 - 1}$

19. $\dfrac{x^4}{(x-1)^3}$

20. $\dfrac{2x^3 - 4x^2 - 15x + 5}{(x-1)^2}$

In Exercises 21–38, use synthetic division to divide.

21. $(3x^3 - 17x^2 + 15x - 25) \div (x - 5)$

22. $(5x^3 + 18x^2 + 7x - 6) \div (x + 3)$

23. $(4x^3 - 9x + 8x^2 - 18) \div (x + 2)$

24. $(9x^3 - 16x - 18x^2 + 32) \div (x - 2)$

25. $(-x^3 + 75x - 250) \div (x + 10)$

26. $(3x^3 - 16x^2 - 72) \div (x - 6)$

27. $(5x^3 - 6x^2 + 8) \div (x - 4)$

28. $(5x^3 + 6x + 8) \div (x + 2)$

29. $\dfrac{10x^4 - 50x^3 - 800}{x - 6}$

30. $\dfrac{x^5 - 13x^4 - 120x + 80}{x + 3}$

31. $\dfrac{x^3 + 512}{x + 8}$

32. $\dfrac{x^3 - 729}{x - 9}$

33. $\dfrac{-3x^4}{x - 2}$

34. $\dfrac{-3x^4}{x + 2}$

35. $\dfrac{180x - x^4}{x - 6}$

36. $\dfrac{5 - 3x + 2x^2 - x^3}{x + 1}$

37. $\dfrac{4x^3 + 16x^2 - 23x - 15}{x + \frac{1}{2}}$

38. $\dfrac{3x^3 - 4x^2 + 5}{x - \frac{3}{2}}$

In Exercises 39–46, express the function in the form $f(x) = (x - k)q(x) + r$ for the given value of k, and demonstrate that $f(k) = r$.

	Function	*Value of k*
39.	$f(x) = x^3 - x^2 - 14x + 11x$	$k = 4$
40.	$f(x) = x^3 - 5x^2 - 11x + 8$	$k = -2$
41.	$f(x) = 15x^4 + 10x^3 - 6x^2 + 14$	$k = -\frac{2}{3}$
42.	$f(x) = 10x^3 - 22x^2 - 3x + 4$	$k = \frac{1}{5}$
43.	$f(x) = x^3 + 3x^2 - 2x - 14$	$k = \sqrt{2}$
44.	$f(x) = x^3 + 2x^2 - 5x - 4$	$k = -\sqrt{5}$
45.	$f(x) = -4x^3 + 6x^2 + 12x + 4$	$k = 1 - \sqrt{3}$
46.	$f(x) = -3x^3 + 8x^2 + 10x - 8$	$k = 2 + \sqrt{2}$

In Exercises 47–50, use synthetic division to find each function value. Verify using another method.

47. $f(x) = 4x^3 - 13x + 10$

(a) $f(1)$ (b) $f(-2)$

(c) $f\left(\frac{1}{2}\right)$ (d) $f(8)$

48. $g(x) = x^6 - 4x^4 + 3x^2 + 2$

(a) $g(2)$ (b) $g(-4)$

(c) $g(3)$ (d) $g(-1)$

49. $h(x) = 3x^3 + 5x^2 - 10x + 1$
(a) $h(3)$ (b) $h\left(\frac{1}{3}\right)$
(c) $h(-2)$ (d) $h(-5)$

50. $f(x) = 0.4x^4 - 1.6x^3 + 0.7x^2 - 2$
(a) $f(1)$ (b) $f(-2)$
(c) $f(5)$ (d) $f(-10)$

In Exercises 51–58, use synthetic division to show that *x* is a solution of the third-degree polynomial equation, and use the result to factor the polynomial completely. List all real zeros of the function.

Polynomial Equation	*Value of x*
51. $x^3 - 7x + 6 = 0$	$x = 2$
52. $x^3 - 28x - 48 = 0$	$x = -4$
53. $2x^3 - 15x^2 + 27x - 10 = 0$	$x = \frac{1}{2}$
54. $48x^3 - 80x^2 + 41x - 6 = 0$	$x = \frac{2}{3}$
55. $x^3 + 2x^2 - 3x - 6 = 0$	$x = \sqrt{3}$
56. $x^3 + 2x^2 - 2x - 4 = 0$	$x = \sqrt{2}$
57. $x^3 - 3x^2 + 2 = 0$	$x = 1 + \sqrt{3}$
58. $x^3 - x^2 - 13x - 3 = 0$	$x = 2 - \sqrt{5}$

In Exercises 59–66, (a) verify the given factors of the function *f*, (b) find the remaining factors of *f*, (c) use your results to write the complete factorization of *f*, (d) list all real zeros of *f*, and (e) confirm your results by using a graphing utility to graph the function.

Function	*Factors*
59. $f(x) = 2x^3 + x^2 - 5x + 2$	$(x + 2), (x - 1)$
60. $f(x) = 3x^3 + 2x^2 - 19x + 6$	$(x + 3), (x - 2)$
61. $f(x) = x^4 - 4x^3 - 15x^2 + 58x - 40$	$(x - 5), (x + 4)$
62. $f(x) = 8x^4 - 14x^3 - 71x^2 - 10x + 24$	$(x + 2), (x - 4)$
63. $f(x) = 6x^3 + 41x^2 - 9x - 14$	$(2x + 1), (3x - 2)$
64. $f(x) = 10x^3 - 11x^2 - 72x + 45$	$(2x + 5), (5x - 3)$
65. $f(x) = 2x^3 - x^2 - 10x + 5$	$(2x - 1), (x + \sqrt{5})$
66. $f(x) = x^3 + 3x^2 - 48x - 144$	$(x + 4\sqrt{3}), (x + 3)$

Graphical Analysis **In Exercises 67–70, (a) use the root-finding capabilities of a graphing utility to approximate the zeros of the function accurate to three decimal places. (b) Determine one of the exact zeros and use synthetic division to verify your result. Then factor the polynomial completely.**

67. $f(x) = x^3 - 2x^2 - 5x + 10$

68. $g(x) = x^3 - 4x^2 - 2x + 8$

69. $h(t) = t^3 - 2t^2 - 7t + 2$

70. $f(s) = s^3 - 12s^2 + 40s - 24$

In Exercises 71–76, simplify the rational expression.

71. $\dfrac{4x^3 - 8x^2 + x + 3}{2x - 3}$

72. $\dfrac{x^3 + x^2 - 64x - 64}{x + 8}$

73. $\dfrac{x^3 + 3x^2 - x - 3}{x + 1}$

74. $\dfrac{2x^3 + 3x^2 - 3x - 2}{x - 1}$

75. $\dfrac{x^4 + 6x^3 + 11x^2 + 6x}{x^2 + 3x + 2}$

76. $\dfrac{x^4 + 9x^3 - 5x^2 - 36x + 4}{x^2 - 4}$

77. *Data Analysis* The average monthly basic rates R for cable television in the United States (in dollars) for the years 1988 through 1997 are shown in the table, where t represents the time (in years), with $t = 0$ corresponding to 1990. (Source: Paul Kagan Associates, Inc.)

t	-2	-1	0	1	2
R	13.86	15.21	16.78	18.10	19.08

t	3	4	5	6	7
R	19.39	21.62	23.07	24.41	26.48

(a) Use a graphing utility to sketch a scatter plot of the data.

(b) Use the regression feature of a graphing utility to find a cubic model for the data. Then graph the model in the same viewing window as the scatter plot. Compare the model with the data.

(c) Use the model to create a table of estimated values of R. Compare the estimated values with the actual data.

(d) Use synthetic division to evaluate the model for the year 2002. Even though the model is relatively accurate for estimating the given data, do you think it is accurate for predicting future cable rates? Explain.

78. *Data Analysis* The numbers of United States military personnel M (in thousands) on active duty for the years 1989 through 1996 are shown in the table, where t represents the time (in years), with $t = 0$ corresponding to 1990. (Source: U.S. Department of Defense)

t	-1	0	1	2
M	2130	2044	1986	1807

t	3	4	5	6
M	1705	1611	1518	1472

(a) Use a graphing utility to sketch a scatter plot of the data.

(b) Use the regression feature of a graphing utility to find a cubic model for the data. Then graph the model in the same viewing window as the scatter plot. Compare the model with the data.

(c) Use the model to create a table of estimated values of M. Compare the estimated values with the actual data.

(d) Use synthetic division to evaluate the model for the year 2001. Even though the model is relatively accurate for estimating the given data, would you use this model to predict the number of military personnel in the future? Explain.

Synthesis

True or False? **In Exercises 79 and 80, determine whether the statement is true or false. Justify your answer.**

79. If $(7x + 4)$ is a factor of some polynomial function f, then $\frac{4}{7}$ is a zero of f.

80. $(2x - 1)$ is a factor of the polynomial $6x^6 + x^5 - 92x^4 + 45x^3 + 184x^2 + 4x - 48$.

Think About It **In Exercises 81 and 82, perform the division by assuming that n is a positive integer.**

81. $\dfrac{x^{3n} + 9x^{2n} + 27x^n + 27}{x^n + 3}$

82. $\dfrac{x^{3n} - 3x^{2n} + 5x^n - 6}{x^n - 2}$

83. *Writing* Briefly explain what it means for a divisor to divide evenly into a dividend.

84. *Writing* Briefly explain how to check polynomial division, and justify your reasoning. Give an example.

Exploration **In Exercises 85 and 86, find the constant c such that the denominator will divide evenly into the numerator.**

85. $\dfrac{x^3 + 4x^2 - 3x + c}{x - 5}$

86. $\dfrac{x^5 - 2x^2 + x + c}{x + 2}$

Think About It **In Exercises 87 and 88, answer the questions about the division**

$$\frac{f(x)}{x - k}$$

where $f(x) = (x + 3)^2(x - 3)(x + 1)^3$.

87. What is the remainder when $k = -3$? Explain.

88. If it is necessary to find $f(2)$, is it easier to evaluate the function directly or to use synthetic division? Explain.

89. *Exploration* Use the form

$$f(x) = (x - k)q(x) + r$$

to create a cubic function that (a) passes through the point $(2, 5)$ and rises to the right, and (b) passes through the point $(-3, 1)$ and falls to the right. (There are many correct answers.)

Review

In Exercises 90–95, use any method to solve the quadratic equation.

90. $9x^2 - 25 = 0$

91. $16x^2 - 21 = 0$

92. $5x^2 - 3x - 14 = 0$

93. $8x^2 - 22x + 15 = 0$

94. $2x^2 + 6x + 3 = 0$

95. $x^2 + 3x - 3 = 0$

In Exercises 96–99, find a polynomial function that has the given zeros.

96. $0, 3, 4$

97. $-6, 1$

98. $-3, 1 + \sqrt{2}, 1 - \sqrt{2}$

99. $1, -2, 2 + \sqrt{3}, 2 - \sqrt{3}$

2.4 Complex Numbers

▶ **What you should learn**

- How to use the imaginary unit i to write complex numbers
- How to add, subtract, and multiply complex numbers
- How to use complex conjugates to divide complex numbers
- How to use the Quadratic Formula to find complex solutions of quadratic equations

▶ **Why you should learn it**

You can use complex numbers to model and solve real-life problems in electronics. For instance, in Exercise 84 on page 242, you will learn how to use complex numbers to find the impedance of an electrical circuit.

Chris Hamilton/The Stock Market

The Imaginary Unit i

Some quadratic equations have no real solutions. For instance, the quadratic equation

$$x^2 + 1 = 0 \qquad \text{Equation with no real solution}$$

has no real solution because there is no real number x that can be squared to produce -1. To overcome this deficiency, mathematicians created an expanded system of numbers using the **imaginary unit *i*,** defined as

$$i = \sqrt{-1} \qquad \text{Imaginary unit}$$

where $i^2 = -1$. By adding real numbers to real multiples of this imaginary unit, the set of **complex numbers** is obtained. Each complex number can be written in the **standard form $a + bi$.** The real number a is called the **real part** of the complex number $a + bi$, and the number bi (where b is a real number) is called the **imaginary part** of the complex number.

Definition of a Complex Number

If a and b are real numbers, the number $a + bi$ is a **complex number,** and it is said to be written in **standard form.** If $b = 0$, the number $a + bi = a$ is a real number. If $b \neq 0$, the number $a + bi$ is called an **imaginary number.** A number of the form bi, where $b \neq 0$, is called a **pure imaginary number.**

The set of real numbers is a subset of the set of complex numbers, as shown in Figure 2.24. This is true because every real number a can be written as a complex number using $b = 0$. That is, for every real number a, you can write $a = a + 0i$.

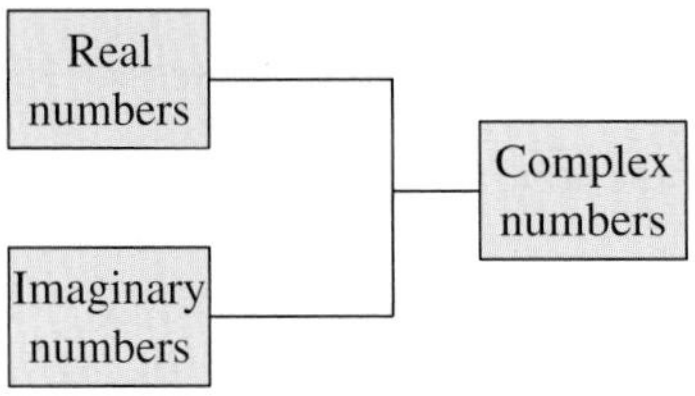

FIGURE 2.24

Equality of Complex Numbers

Two complex numbers $a + bi$ and $c + di$, written in standard form, are equal to each other

$$a + bi = c + di \qquad \text{Equality of two complex numbers}$$

if and only if $a = c$ and $b = d$.

Operations with Complex Numbers

To add (or subtract) two complex numbers, you add (or subtract) the real and imaginary parts of the numbers separately.

For each operation on complex numbers, you can show the parallel operation on polynomials.

Addition and Subtraction of Complex Numbers

If $a + bi$ and $c + di$ are two complex numbers written in standard form, their sum and difference are defined as follows.

Sum: $(a + bi) + (c + di) = (a + c) + (b + d)i$

Difference: $(a + bi) - (c + di) = (a - c) + (b - d)i$

The **additive identity** in the complex number system is zero (the same as in the real number system). Furthermore, the **additive inverse** of the complex number $a + bi$ is

$-(a + bi) = -a - bi.$ Additive inverse

So, you have

$(a + bi) + (-a - bi) = 0 + 0i = 0.$

Example 1 ▶ Adding and Subtracting Complex Numbers

a. $(3 - i) + (2 + 3i) = 3 - i + 2 + 3i$ Remove parentheses.

$= 3 + 2 - i + 3i$ Group like terms.

$= (3 + 2) + (-1 + 3)i$

$= 5 + 2i$ Write in standard form.

b. $2i + (-4 - 2i) = 2i - 4 - 2i$ Remove parentheses.

$= -4 + 2i - 2i$ Group like terms.

$= -4$ Write in standard form.

c. $3 - (-2 + 3i) + (-5 + i) = 3 + 2 - 3i - 5 + i$

$= 3 + 2 - 5 - 3i + i$

$= 0 - 2i$

$= -2i$

d. $(3 + 2i) + (4 - i) - (7 + i) = 3 + 2i + 4 - i - 7 - i$

$= 3 + 4 - 7 + 2i - i - i$

$= 0 + 0i$

$= 0$

Note in Example 1(b) that the sum of two complex numbers can be a real number.

Exploration

Complete the table:

$i^1 = i$	$i^7 =$
$i^2 = -1$	$i^8 =$
$i^3 = -i$	$i^9 =$
$i^4 = 1$	$i^{10} =$
$i^5 =$	$i^{11} =$
$i^6 =$	$i^{12} =$

What pattern do you see? Write a brief description of how you would find i raised to any positive integer power.

Many of the properties of real numbers are valid for complex numbers as well. Here are some examples.

Associative Properties of Addition and Multiplication

Commutative Properties of Addition and Multiplication

Distributive Property of Multiplication Over Addition

Notice below how these properties are used when two complex numbers are multiplied.

$$(a + bi)(c + di) = a(c + di) + bi(c + di) \quad \text{Distributive Property}$$
$$= ac + (ad)i + (bc)i + (bd)i^2 \quad \text{Distributive Property}$$
$$= ac + (ad)i + (bc)i + (bd)(-1) \quad i^2 = -1$$
$$= ac - bd + (ad)i + (bc)i \quad \text{Commutative Property}$$
$$= (ac - bd) + (ad + bc)i \quad \text{Associative Property}$$

Rather than trying to memorize this multiplication rule, you should simply remember how the Distributive Property is used to multiply two complex numbers. The procedure is similar to multiplying two polynomials and combining like terms (as in the FOIL Method).

Example 2 ▶ Multiplying Complex Numbers

a. $4(-2 + 3i) = 4(-2) + 4(3i)$ Distributive Property
$= -8 + 12i$ Simplify.

b. $(i)(-3i) = -3i^2$ Multiply.
$= -3(-1)$ $i^2 = -1$
$= 3$ Simplify.

c. $(2 - i)(4 + 3i) = 8 + 6i - 4i - 3i^2$ Product of binomials
$= 8 + 6i - 4i - 3(-1)$ $i^2 = -1$
$= (8 + 3) + (6i - 4i)$ Group like terms.
$= 11 + 2i$ Write in standard form.

d. $(3 + 2i)(3 - 2i) = 9 - 6i + 6i - 4i^2$ Product of binomials
$= 9 - 4(-1)$ $i^2 = -1$
$= 9 + 4$ Simplify.
$= 13$ Write in standard form.

e. $(3 + 2i)^2 = 9 + 6i + 6i + 4i^2$ Product of binomials
$= 9 + 4(-1) + 12i$ $i^2 = -1$
$= 9 - 4 + 12i$ Simplify.
$= 5 + 12i$ Write in standard form.

Complex Conjugates and Division

A comparison with the method of rationalizing denominators (Section P.2) may be helpful.

Notice in Example 2(d) that the product of two complex numbers can be a real number. This occurs with pairs of complex numbers of the form $a + bi$ and $a - bi$, called **complex conjugates.**

$$\begin{aligned}(a + bi)(a - bi) &= a^2 - abi + abi - b^2i^2 \\ &= a^2 - b^2(-1) \\ &= a^2 + b^2\end{aligned}$$

To find the quotient of $a + bi$ and $c + di$ where c and d are not both zero, multiply the numerator and denominator by the conjugate of the *denominator* to obtain

$$\begin{aligned}\frac{a + bi}{c + di} &= \frac{a + bi}{c + di}\left(\frac{c - di}{c - di}\right) \\ &= \frac{(ac + bd) + (bc - ad)i}{c^2 + d^2}.\end{aligned}$$

Example 3 ▶ Dividing Complex Numbers

$$\frac{1}{1 + i} = \frac{1}{1 + i}\left(\frac{1 - i}{1 - i}\right)$$ Multiply numerator and denominator by conjugate of denominator.

$$= \frac{1 - i}{1^2 - i^2}$$ Expand.

$$= \frac{1 - i}{1 - (-1)}$$ $i^2 = -1$

$$= \frac{1 - i}{2}$$ Simplify.

$$= \frac{1}{2} - \frac{1}{2}i$$ Write in standard form.

Example 4 ▶ Dividing Complex Numbers

$$\frac{2 + 3i}{4 - 2i} = \frac{2 + 3i}{4 - 2i}\left(\frac{4 + 2i}{4 + 2i}\right)$$ Multiply numerator and denominator by conjugate of denominator.

$$= \frac{8 + 4i + 12i + 6i^2}{16 - 4i^2}$$ Expand.

$$= \frac{8 - 6 + 16i}{16 + 4}$$ $i^2 = -1$

$$= \frac{2 + 16i}{20}$$ Simplify.

$$= \frac{1}{10} + \frac{4}{5}i$$ Write in standard form.

Complex Solutions of Quadratic Equations

When using the Quadratic Formula to solve a quadratic equation, you often obtain a result such as $\sqrt{-3}$, which you know is not a real number. By factoring out $i = \sqrt{-1}$, you can write this number in standard form.

$$\sqrt{-3} = \sqrt{3(-1)} = \sqrt{3}\sqrt{-1} = \sqrt{3}i$$

The number $\sqrt{3}i$ is called the *principal square root* of -3.

Principal Square Root of a Negative Number

If a is a positive number, the **principal square root** of the negative number $-a$ is defined as

$$\sqrt{-a} = \sqrt{a}i.$$

STUDY TIP

The definition of principal square root uses the rule

$$\sqrt{ab} = \sqrt{a}\sqrt{b}$$

for $a > 0$ and $b < 0$. This rule is not valid if *both* a and b are negative. For example,

$$\sqrt{-5}\sqrt{-5} = \sqrt{5(-1)}\sqrt{5(-1)}$$
$$= \sqrt{5}i\sqrt{5}i$$
$$= \sqrt{25i^2}$$
$$= 5i^2 = -5$$

whereas

$$\sqrt{(-5)(-5)} = \sqrt{25} = 5.$$

To avoid problems with multiplying square roots of negative numbers, be sure to convert to standard form *before* multiplying.

Example 5 ▶ Writing Complex Numbers in Standard Form

a. $\sqrt{-3}\sqrt{-12} = \sqrt{3}i\sqrt{12}i = \sqrt{36}i^2 = 6(-1) = -6$

b. $\sqrt{-48} - \sqrt{-27} = \sqrt{48}i - \sqrt{27}i = 4\sqrt{3}i - 3\sqrt{3}i = \sqrt{3}i$

c. $(-1 + \sqrt{-3})^2 = (-1 + \sqrt{3}i)^2$

$$= (-1)^2 - 2\sqrt{3}i + (\sqrt{3})^2(i^2)$$
$$= 1 - 2\sqrt{3}i + 3(-1)$$
$$= -2 - 2\sqrt{3}i$$

Example 6 ▶ Complex Solutions of a Quadratic Equation

Solve (a) $x^2 + 4 = 0$ and (b) $3x^2 - 2x + 5 = 0$.

Solution

a. $x^2 + 4 = 0$ — Write original equation.

$x^2 = -4$ — Subtract 4 from each side.

$x = \pm 2i$ — Extract square roots.

b. $3x^2 - 2x + 5 = 0$ — Write original equation.

$x = \dfrac{-(-2) \pm \sqrt{(-2)^2 - 4(3)(5)}}{2(3)}$ — Quadratic Formula

$= \dfrac{2 \pm \sqrt{-56}}{6}$ — Simplify.

$= \dfrac{2 \pm 2\sqrt{14}i}{6}$ — Write $\sqrt{-56}$ in standard form.

$= \dfrac{1}{3} \pm \dfrac{\sqrt{14}}{3}i$ — Write in standard form.

Activities

1. Perform the indicated operations and write the result in standard form.
   $(4 - \sqrt{-9})(2 + \sqrt{-9})$
   Answer: $17 + 6i$
2. Write $\dfrac{3 + i}{i}$ in standard form.
   Answer: $1 - 3i$
3. Use the Quadratic Formula to solve $2x^2 - 5x + 7 = 0$.
   Answer: $x = \dfrac{5}{4} \pm \dfrac{\sqrt{31}}{4}i$

2.4 Exercises

In Exercises 1–4, find real numbers *a* and *b* such that the equation is true.

1. $a + bi = -10 + 6i$
2. $a + bi = 13 + 4i$
3. $(a - 1) + (b + 3)i = 5 + 8i$
4. $(a + 6) + 2bi = 6 - 5i$

In Exercises 5–16, write the complex number in standard form.

5. $4 + \sqrt{-9}$
6. $3 + \sqrt{-16}$
7. $2 - \sqrt{-27}$
8. $1 + \sqrt{-8}$
9. $\sqrt{-75}$
10. $\sqrt{-4}$
11. 8
12. 45
13. $-6i + i^2$
14. $-4i^2 + 2i$
15. $\sqrt{-0.09}$
16. $\sqrt{-0.0004}$

In Exercises 17–26, perform the addition or subtraction and write the result in standard form.

17. $(5 + i) + (6 - 2i)$
18. $(13 - 2i) + (-5 + 6i)$
19. $(8 - i) - (4 - i)$
20. $(3 + 2i) - (6 + 13i)$
21. $(-2 + \sqrt{-8}) + (5 - \sqrt{-50})$
22. $(8 + \sqrt{-18}) - (4 + 3\sqrt{2}i)$
23. $13i - (14 - 7i)$
24. $22 + (-5 + 8i) + 10i$
25. $-\left(\frac{3}{2} + \frac{5}{2}i\right) + \left(\frac{5}{3} + \frac{11}{3}i\right)$
26. $(1.6 + 3.2i) + (-5.8 + 4.3i)$

In Exercises 27–40, perform the operation and write the result in standard form.

27. $\sqrt{-6} \cdot \sqrt{-2}$
28. $\sqrt{-5} \cdot \sqrt{-10}$
29. $(\sqrt{-10})^2$
30. $(\sqrt{-75})^2$
31. $(1 + i)(3 - 2i)$
32. $(6 - 2i)(2 - 3i)$
33. $6i(5 - 2i)$
34. $-8i(9 + 4i)$
35. $(\sqrt{14} + \sqrt{10}i)(\sqrt{14} - \sqrt{10}i)$
36. $(3 + \sqrt{-5})(7 - \sqrt{-10})$
37. $(4 + 5i)^2$
38. $(2 - 3i)^2$
39. $(2 + 3i)^2 + (2 - 3i)^2$
40. $(1 - 2i)^2 - (1 + 2i)^2$

In Exercises 41–48, write the conjugate of the complex number. Then multiply the number and its conjugate.

41. $6 + 3i$
42. $7 - 12i$
43. $-1 - \sqrt{5}i$
44. $-3 + \sqrt{2}i$
45. $\sqrt{-20}$
46. $\sqrt{-15}$
47. $\sqrt{8}$
48. $1 + \sqrt{8}$

In Exercises 49–62, perform the operation and write the result in standard form.

49. $\dfrac{5}{i}$
50. $-\dfrac{14}{2i}$
51. $\dfrac{2}{4 - 5i}$
52. $\dfrac{5}{1 - i}$
53. $\dfrac{3 + i}{3 - i}$
54. $\dfrac{6 - 7i}{1 - 2i}$
55. $\dfrac{6 - 5i}{i}$
56. $\dfrac{8 + 16i}{2i}$
57. $\dfrac{3i}{(4 - 5i)^2}$
58. $\dfrac{5i}{(2 + 3i)^2}$
59. $\dfrac{2}{1 + i} - \dfrac{3}{1 - i}$
60. $\dfrac{2i}{2 + i} + \dfrac{5}{2 - i}$
61. $\dfrac{i}{3 - 2i} + \dfrac{2i}{3 + 8i}$
62. $\dfrac{1 + i}{i} - \dfrac{3}{4 - i}$

In Exercises 63–72, use the Quadratic Formula to solve the quadratic equation.

63. $x^2 - 2x + 2 = 0$
64. $x^2 + 6x + 10 = 0$
65. $4x^2 + 16x + 17 = 0$
66. $9x^2 - 6x + 37 = 0$
67. $4x^2 + 16x + 15 = 0$
68. $16t^2 - 4t + 3 = 0$
69. $\frac{3}{2}x^2 - 6x + 9 = 0$
70. $\frac{7}{8}x^2 - \frac{3}{4}x + \frac{5}{16} = 0$
71. $1.4x^2 - 2x - 10 = 0$
72. $4.5x^2 - 3x + 12 = 0$

73. Express each of the following powers of i as i, $-i$, 1, or -1.

(a) i^{40} (b) i^{25} (c) i^{50} (d) i^{67}

In Exercises 74–81, simplify the complex number and write it in standard form.

74. $-6i^3 + i^2$

75. $4i^2 - 2i^3$

76. $-5i^5$

77. $(-i)^3$

78. $\left(\sqrt{-75}\right)^3$

79. $\left(\sqrt{-2}\right)^6$

80. $\dfrac{1}{i^3}$

81. $\dfrac{1}{(2i)^3}$

82. Cube each complex number.

(a) 2 (b) $-1 + \sqrt{3}i$ (c) $-1 - \sqrt{3}i$

83. Raise each complex number to the fourth power.

(a) 2 (b) -2 (c) $2i$ (d) $-2i$

84. *Impedance* The opposition to current in an electrical circuit is called its impedance. The impedance in a parallel circuit with two pathways satisfies the equation

$$\frac{1}{z} = \frac{1}{z_1} + \frac{1}{z_2}$$

where z_1 is the impedance (in ohms) of pathway 1 and z_2 is the impedance of pathway 2. Use the table to determine the impedance of each parallel circuit. The impedance of each pathway is found by adding the impedance of each component in the pathway.

	Resistor	Inductor	Capacitor
Symbol	$a\Omega$	$b\Omega$	$c\Omega$
Impedance	a	bi	$-ci$

(a)

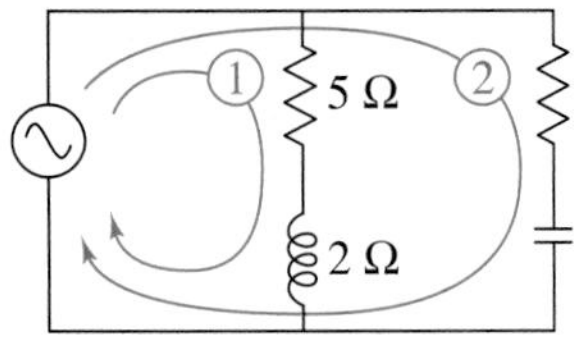

$z_1 = 5 + 2i, \quad z_2 = 3 - 4i$

(b)

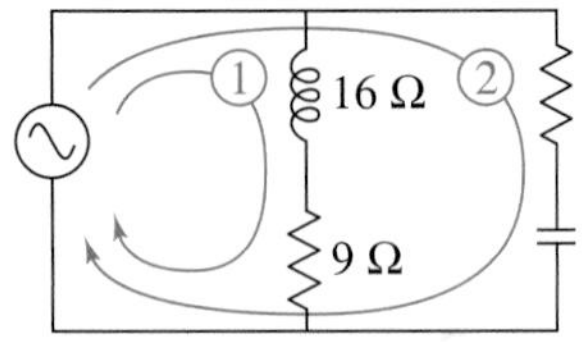

$z_1 = 9 + 16i, \quad z_2 = 20 - 10i$

Synthesis

True or False? **In Exercises 85–87, determine whether the statement is true or false. Justify your answer.**

85. There is no complex number that is equal to its conjugate.

86. $-i\sqrt{6}$ is a solution of $x^4 - x^2 + 14 = 56$.

87. $i^{44} + i^{150} - i^{74} - i^{109} + i^{61} = -1$

88. *Error Analysis* Describe the error.

$$\sqrt{-6}\sqrt{-6} = \sqrt{(-6)(-6)} = \sqrt{36} = 6 \quad ✗$$

89. Prove that the conjugate of the product of two complex numbers $a_1 + b_1 i$ and $a_2 + b_2 i$ is the product of their conjugates.

90. Prove that the conjugate of the sum of two complex numbers $a_1 + b_1 i$ and $a_2 + b_2 i$ is the sum of their conjugates.

Review

In Exercises 91–94, perform the operations and write the result in standard form.

91. $(4 + 3x) + (8 - 6x - x^2)$

92. $(x^3 - 3x^2) - (6 - 2x - 4x^2)$

93. $\left(3x - \frac{1}{2}\right)(x + 4)$

94. $(2x - 5)^2$

In Exercises 95–98, solve the equation and check your solution.

95. $-x - 12 = 19$

96. $8 - 3x = -34$

97. $4(5x - 6) - 3(6x + 1) = 0$

98. $5[x - (3x + 11)] = 20x - 15$

99. *Volume of an Oblate Spheroid*

Solve for a: $V = \frac{4}{3}\pi a^2 b$

100. *Newton's Law of Universal Gravitation*

Solve for r: $F = \alpha \dfrac{m_1 m_2}{r^2}$

101. *Mixture Problem* A 5-liter container contains a mixture with a concentration of 50%. How much of this mixture must be withdrawn and replaced by 100% concentrate to bring the mixture up to 60% concentration?

102. *Travel* A business executive drove at an average speed of 100 kilometers per hour on a 200-kilometer trip. Because of heavy traffic, the average speed on the return trip was 80 kilometers per hour. Find the average speed for the round trip.

Example 3 ▶ Rational Zero Test with Leading Coefficient of 1

Find the rational zeros of $f(x) = x^4 - x^3 + x^2 - 3x - 6$.

Solution

Because the leading coefficient is 1, the possible rational zeros are the factors of the constant term.

Possible rational zeros: $\pm 1, \pm 2, \pm 3, \pm 6$

A test of these possible zeros shows that $x = -1$ and $x = 2$ are the only two that work. Check the others to be sure.

Additional Example

List the possible rational zeros of $f(x) = x^3 + 8x^2 + 40x - 525$.

Solution

The leading coefficient is 1, so the possible rational zeros are $\pm 1, \pm 3, \pm 5, \pm 7, \pm 15, \pm 21, \pm 25, \pm 35, \pm 75, \pm 105$, and ± 175. To decide which possible rational zeros should be tested using synthetic division, graph the function. From the graph, you can see that the zero is positive and less than 10, so the only values of x that should be tested are $x = 1$, $x = 3$, $x = 5$, and $x = 7$.

If the leading coefficient of a polynomial is not 1, the list of possible rational zeros can increase dramatically. In such cases the search can be shortened in several ways: (1) a programmable calculator can be used to speed up the calculations; (2) a graph, drawn either by hand or with a graphing utility, can give a good estimate of the locations of the zeros; and (3) synthetic division can be used to test the possible rational zeros.

To see how to use synthetic division to test the possible rational zeros, take another look at the function $f(x) = x^4 - x^3 + x^2 - 3x - 6$ given in Example 3. To test that $x = -1$ and $x = 2$ are zeros of f, you can apply synthetic division successively, as follows.

$$\begin{array}{r|rrrrr} -1 & 1 & -1 & 1 & -3 & -6 \\ & & -1 & 2 & -3 & 6 \\ \hline & 1 & -2 & 3 & -6 & 0 \end{array}$$

$$\begin{array}{r|rrrr} 2 & 1 & -2 & 3 & -6 \\ & & 2 & 0 & 6 \\ \hline & 1 & 0 & 3 & 0 \end{array}$$

So, you have

$$f(x) = (x + 1)(x - 2)(x^2 + 3).$$

Because the factor $(x^2 + 3)$ produces no real zeros, you can conclude that $x = -1$ and $x = 2$ are the only *real* zeros of f, which is verified in Figure 2.26.

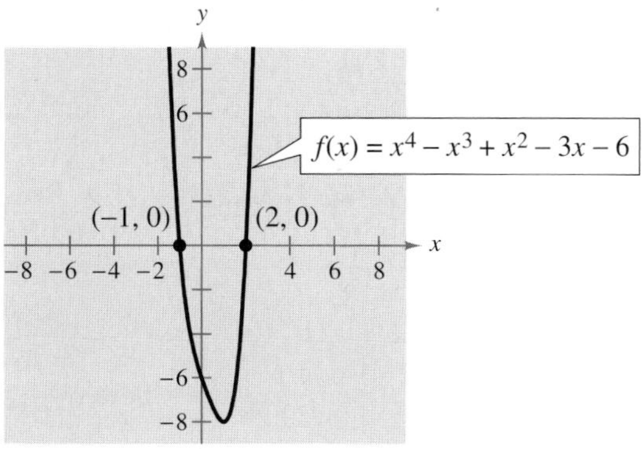

FIGURE 2.26

Finding the first zero is often the hardest part. After that, the search is simplified by using the lower-degree polynomial obtained in synthetic division.

Example 4 ▶ Using the Rational Zero Test

Find the rational zeros of $f(x) = 2x^3 + 3x^2 - 8x + 3$.

Solution

The leading coefficient is 2 and the constant term is 3.

$$\textit{Possible rational zeros: } \frac{\text{Factors of } 3}{\text{Factors of } 2} = \frac{\pm 1, \pm 3}{\pm 1, \pm 2} = \pm 1, \pm 3, \pm\frac{1}{2}, \pm\frac{3}{2}$$

By synthetic division, you can determine that $x = 1$ is a zero.

$$\begin{array}{r|rrrr} 1 & 2 & 3 & -8 & 3 \\ & & 2 & 5 & -3 \\ \hline & 2 & 5 & -3 & 0 \end{array}$$

So, $f(x)$ factors as

$$\begin{aligned} f(x) &= (x - 1)(2x^2 + 5x - 3) \\ &= (x - 1)(2x - 1)(x + 3) \end{aligned}$$

and you can conclude that the rational zeros of f are $x = 1$, $x = \frac{1}{2}$, and $x = -3$.

Example 5 ▶ Using the Rational Zero Test

Find all the real zeros of $f(x) = -10x^3 + 15x^2 + 16x - 12$.

Solution

The leading coefficient is -10 and the constant term is -12.

$$\textit{Possible rational zeros: } \frac{\text{Factors of } -12}{\text{Factors of } -10} = \frac{\pm 1, \pm 2, \pm 3, \pm 4, \pm 6, \pm 12}{\pm 1, \pm 2, \pm 5, \pm 10}$$

With so many possibilities (32, in fact), it is worth your time to stop and sketch a graph. From Figure 2.27, it looks like three reasonable choices would be $x = -\frac{6}{5}$, $x = \frac{1}{2}$, and $x = 2$. Testing these by synthetic division shows that only $x = 2$ works. So, you have

$$f(x) = (x - 2)(-10x^2 - 5x + 6).$$

Using the Quadratic Formula, you find that the two additional zeros are irrational numbers.

$$x = \frac{-(-5) + \sqrt{265}}{-20} \approx -1.0639$$

and

$$x = \frac{-(-5) - \sqrt{265}}{-20} \approx 0.5639$$

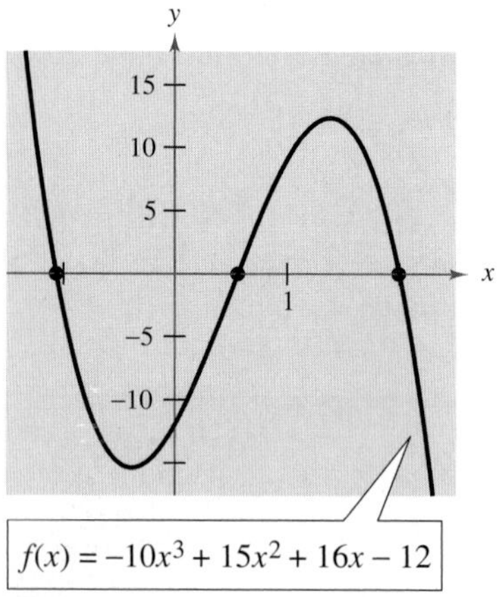

FIGURE 2.27

Conjugate Pairs

In Example 1(c) and (d), note that the pairs of complex zeros are **conjugates.** That is, they are of the form $a + bi$ and $a - bi$.

Complex Zeros Occur in Conjugate Pairs

Let $f(x)$ be a polynomial function that has *real coefficients*. If $a + bi$, where $b \neq 0$, is a zero of the function, the conjugate $a - bi$ is also a zero of the function.

Be sure you see that this result is true only if the polynomial function has *real coefficients*. For instance, the result applies to the function $f(x) = x^2 + 1$ but not to the function $g(x) = x - i$.

Example 6 ▶ Finding a Polynomial with Given Zeros

Find a fourth-degree polynomial function with real coefficients that has -1, -1, and $3i$ as zeros.

Solution

Because $3i$ is a zero *and* the polynomial is stated to have real coefficients, you know that the conjugate $-3i$ must also be a zero. So, from the Linear Factorization Theorem, $f(x)$ can be written as

$$f(x) = a(x + 1)(x + 1)(x - 3i)(x + 3i).$$

For simplicity, let $a = 1$ to obtain

$$\begin{aligned} f(x) &= (x^2 + 2x + 1)(x^2 + 9) \\ &= x^4 + 2x^3 + 10x^2 + 18x + 9. \end{aligned}$$

Factoring a Polynomial

The Linear Factorization Theorem shows that you can write any nth-degree polynomial as the product of n linear factors.

$$f(x) = a_n(x - c_1)(x - c_2)(x - c_3) \cdot \cdot \cdot (x - c_n)$$

However, this result includes the possibility that some of the values of c_i are complex. The following theorem says that even if you do not want to get involved with "complex factors," you can still write $f(x)$ as the product of linear and/or quadratic factors. A proof of this theorem is given in Appendix A.

Factors of a Polynomial

Every polynomial of degree $n > 0$ with real coefficients can be written as the product of linear and quadratic factors with real coefficients, where the quadratic factors have no real zeros.

You may want to remind students that a graphing calculator is helpful in determining real zeros, which in turn are useful in finding the complex zeros.

A quadratic factor with no real zeros is said to be *prime* or **irreducible over the reals.** Be sure you see that this is not the same as being *irreducible over the rationals*. For example, the quadratic

$$x^2 + 1 = (x - i)(x + i)$$

is irreducible over the reals (and therefore over the rationals). On the other hand, the quadratic

$$x^2 - 2 = \left(x - \sqrt{2}\right)\left(x + \sqrt{2}\right)$$

is irreducible over the rationals but *reducible* over the reals.

Example 7 ▶ Finding the Zeros of a Polynomial Function

Find all the zeros of

$$f(x) = x^4 - 3x^3 + 6x^2 + 2x - 60$$

given that $1 + 3i$ is a zero of f.

Solution

Because complex zeros occur in conjugate pairs, you know that $1 - 3i$ is also a zero of f. This means that both

$$[x - (1 + 3i)] \quad \text{and} \quad [x - (1 - 3i)]$$

are factors of f. Multiplying these two factors produces

$$\begin{aligned}[x - (1 + 3i)][x - (1 - 3i)] &= [(x - 1) - 3i][(x - 1) + 3i] \\ &= (x - 1)^2 - 9i^2 \\ &= x^2 - 2x + 1 - 9(-1) \\ &= x^2 - 2x + 10.\end{aligned}$$

Using long division, you can divide $x^2 - 2x + 10$ into f to obtain the following.

$$\begin{array}{r} x^2 - x - 6 \\ x^2 - 2x + 10 \overline{\smash{)}\, x^4 - 3x^3 + 6x^2 + 2x - 60} \\ \underline{x^4 - 2x^3 + 10x^2} \qquad\qquad \\ -x^3 - 4x^2 + 2x \qquad \\ \underline{-x^3 + 2x^2 - 10x} \qquad \\ -6x^2 + 12x - 60 \\ \underline{-6x^2 + 12x - 60} \\ 0 \end{array}$$

Therefore, you have

$$\begin{aligned}f(x) &= (x^2 - 2x + 10)(x^2 - x - 6) \\ &= (x^2 - 2x + 10)(x - 3)(x + 2)\end{aligned}$$

and you can conclude that the zeros of f are $1 + 3i$, $1 - 3i$, 3, and -2.

STUDY TIP

In Example 7, if you had not been told that $1 + 3i$ is a zero of f, you could still find all zeros of the function by using synthetic division to find the real zeros -2 and 3. Then you could factor the polynomial as $(x + 2)(x - 3)(x^2 - 2x + 10)$. Finally, by using the Quadratic Formula, you could determine that the zeros are $-2, 3, 1 + 3i$, and $1 - 3i$.

Technology

You can use the list editor feature of a graphing utility to help you determine which of the possible rational zeros are zeros of the polynomial in Example 8. Enter the possible rational zeros in the table. Then evaluate f at each of the possible rational zeros. When you do this, you will see that there are two rational zeros, -2 and 1, as shown in the table below.

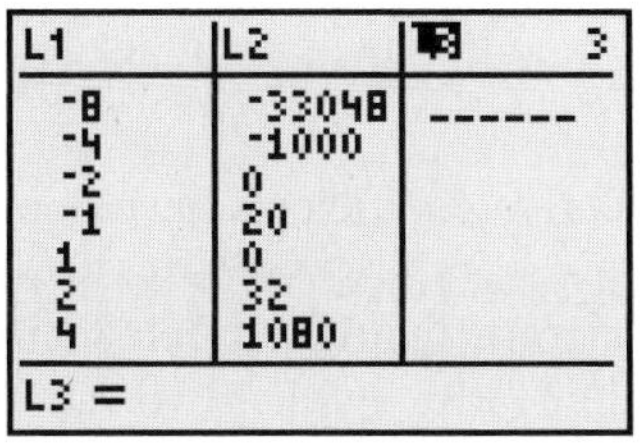

Example 8 shows how to find all the zeros of a polynomial function, including complex zeros.

Example 8 ▶ Finding the Zeros of a Polynomial Function

Write

$$f(x) = x^5 + x^3 + 2x^2 - 12x + 8$$

as the product of linear factors, and list all of its zeros.

Solution

The possible rational zeros are ±1, ±2, ±4, and ±8. Synthetic division produces the following.

$$\begin{array}{r|rrrrrr} 1 & 1 & 0 & 1 & 2 & -12 & 8 \\ & & 1 & 1 & 2 & 4 & -8 \\ \hline & 1 & 1 & 2 & 4 & -8 & 0 \end{array} \longrightarrow \text{1 is a zero.}$$

$$\begin{array}{r|rrrrr} 1 & 1 & 1 & 2 & 4 & -8 \\ & & 1 & 2 & 4 & 8 \\ \hline & 1 & 2 & 4 & 8 & 0 \end{array} \longrightarrow \text{1 is a repeated zero.}$$

$$\begin{array}{r|rrrr} -2 & 1 & 2 & 4 & 8 \\ & & -2 & 0 & -8 \\ \hline & 1 & 0 & 4 & 0 \end{array} \longrightarrow -2 \text{ is a zero.}$$

So, you have

$$\begin{aligned} f(x) &= x^5 + x^3 + 2x^2 - 12x + 8 \\ &= (x - 1)(x - 1)(x + 2)(x^2 + 4). \end{aligned}$$

By factoring $x^2 + 4$ as

$$\begin{aligned} x^2 - (-4) &= \left(x - \sqrt{-4}\right)\left(x + \sqrt{-4}\right) \\ &= (x - 2i)(x + 2i) \end{aligned}$$

you obtain

$$f(x) = (x - 1)(x - 1)(x + 2)(x - 2i)(x + 2i)$$

which gives the following five zeros of f.

$$1, \quad 1, \quad -2, \quad 2i, \quad \text{and} \quad -2i$$

Note from the graph of f shown in Figure 2.28 that the *real* zeros are the only ones that appear as x-intercepts.

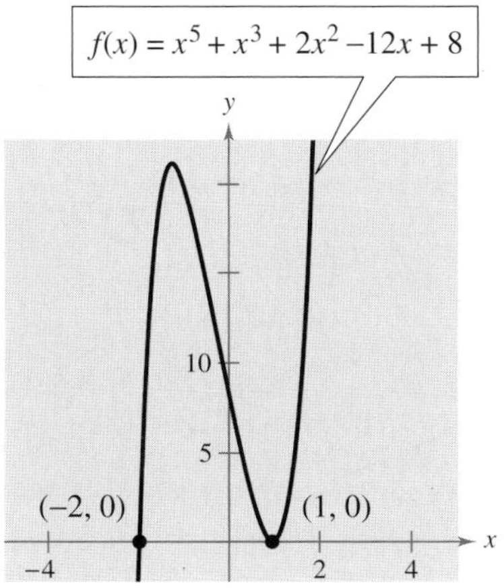

FIGURE 2.28

In Example 8, the fifth-degree polynomial function has three real zeros. In such cases, you can use the zoom and trace features or the zero or root feature of a graphing utility to approximate the real zeros. You can then use these real zeros to determine the complex zeros algebraically.

Other Tests for Zeros of Polynomials

You know that an nth-degree polynomial function can have *at most n* real zeros. Of course, many nth-degree polynomials do not have that many real zeros. For instance, $f(x) = x^2 + 1$ has no real zeros, and $f(x) = x^3 + 1$ has only one real zero. The following theorem, called **Descartes's Rule of Signs,** sheds more light on the number of real zeros of a polynomial.

Descartes's Rule of Signs

Let $f(x) = a_nx^n + a_{n-1}x^{n-1} + \cdots + a_2x^2 + a_1x + a_0$ be a polynomial with real coefficients and $a_0 \neq 0$.

1. The number of *positive real zeros* of f is either equal to the number of variations in sign of $f(x)$ *or* less than that number by an even integer.
2. The number of *negative real zeros* of f is either equal to the number of variations in sign of $f(-x)$ *or* less than that number by an even integer.

A **variation in sign** means that two consecutive coefficients have opposite signs.

When using Descartes's Rule of Signs, a zero of multiplicity m should be counted as m zeros. For instance, the polynomial $x^3 - 3x + 2$ has two variations in sign, and so has either two positive or no positive real zeros. Because

$$x^3 - 3x + 2 = (x - 1)(x - 1)(x + 2)$$

you can see that the two positive real zeros are $x = 1$ of multiplicity 2.

Example 9 ▶ Using Descartes's Rule of Signs

Describe the possible real zeros of $f(x) = 3x^3 - 5x^2 + 6x - 4$.

Solution

The original polynomial has *three* variations in sign.

+ to −   + to −

$$f(x) = 3x^3 - 5x^2 + 6x - 4$$

− to +

The polynomial

$$f(-x) = 3(-x)^3 - 5(-x)^2 + 6(-x) - 4$$
$$= -3x^3 - 5x^2 - 6x - 4$$

has no variations in sign. So, from Descartes's Rule of Signs, the polynomial $f(x) = 3x^3 - 5x^2 + 6x - 4$ has either three positive real zeros or one positive real zero, and has no negative real zeros. From the graph in Figure 2.29, you can see that the function has only one real zero (it is a positive number, near 1).

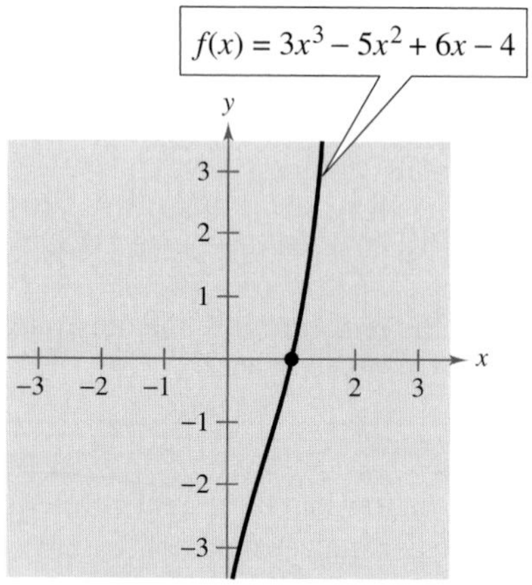

FIGURE 2.29

Another test for zeros of a polynomial function is related to the sign pattern in the last row of the synthetic division array. This test can give you an upper or lower bound of the real zeros of f. A real number b is an **upper bound** for the real zeros of f if no zeros are greater than b. Similarly, b is a **lower bound** if no real zeros of f are less than b.

Upper and Lower Bound Rules

Let $f(x)$ be a polynomial with real coefficients and a positive leading coefficient. Suppose $f(x)$ is divided by $x - c$, using synthetic division.

1. If $c > 0$ and each number in the last row is either positive or zero, c is an *upper bound* for the real zeros of f.
2. If $c < 0$ and the numbers in the last row are alternately positive and negative (zero entries count as positive or negative), c is a *lower bound* for the real zeros of f.

Example 10 ▶ Finding the Zeros of a Polynomial Function

Find the real zeros of

$$f(x) = 6x^3 - 4x^2 + 3x - 2.$$

Solution

The possible real zeros are as follows.

$$\frac{\text{Factors of } 2}{\text{Factors of } 6} = \frac{\pm 1, \pm 2}{\pm 1, \pm 2, \pm 3, \pm 6}$$

$$= \pm 1, \pm\frac{1}{2}, \pm\frac{1}{3}, \pm\frac{1}{6}, \pm\frac{2}{3}, \pm 2$$

Because $f(x)$ has three variations in sign and $f(-x)$ has none, you can apply Descartes's Rule of Signs to conclude that there are three positive real zeros or one positive real zero, and no negative zeros. Trying $x = 1$ produces the following.

In Example 10, notice how the Rational Zero Test, Descartes's Rule of Signs, and the Upper and Lower Bound Rules may be used together in a search for all real zeros of a polynomial function.

$$\begin{array}{r|rrrr} 1 & 6 & -4 & 3 & -2 \\ & & 6 & 2 & 5 \\ \hline & 6 & 2 & 5 & 3 \end{array}$$

So, $x = 1$ is not a zero, but because the last row has all positive entries, you know that $x = 1$ is an upper bound for the real zeros. So, you can restrict the search to zeros between 0 and 1. By trial and error, you can determine that $x = \frac{2}{3}$ is a zero. So,

$$f(x) = \left(x - \frac{2}{3}\right)(6x^2 + 3).$$

Because $6x^2 + 3$ has no real zeros, it follows that $x = \frac{2}{3}$ is the only real zero.

Before concluding this section, here are two additional hints that can help you find the real zeros of a polynomial.

1. If the terms of $f(x)$ have a common monomial factor, it should be factored out before applying the tests in this section. For instance, by writing

$$f(x) = x^4 - 5x^3 + 3x^2 + x$$
$$= x(x^3 - 5x^2 + 3x + 1)$$

you can see that $x = 0$ is a zero of f and that the remaining zeros can be obtained by analyzing the cubic factor.

2. If you are able to find all but two zeros of $f(x)$, you can always use the Quadratic Formula on the remaining quadratic factor. For instance, if you succeeded in writing

$$f(x) = x^4 - 5x^3 + 3x^2 + x$$
$$= x(x - 1)(x^2 - 4x - 1)$$

you can apply the Quadratic Formula to $x^2 - 4x - 1$ to conclude that the two remaining zeros are

$$x = 2 + \sqrt{5} \quad \text{and} \quad x = 2 - \sqrt{5}.$$

A computer animation of this concept appears in the *Interactive* CD-ROM and *Internet* versions of this text.

Example 11 ▶ Using a Polynomial Model

You are designing candle-making kits. Each kit will contain 25 cubic inches of candle wax and a mold for making a pyramid-shaped candle. You want the height of the candle to be 2 inches less than the length of each side of the candle's square base. What should the dimensions of your candle mold be?

Solution

The volume of a pyramid is $V = \frac{1}{3}Bh$, where B is the area of the base and h is the height. The area of the base is x^2 and the height is $(x - 2)$. So, the volume of the pyramid is

$$V = \frac{1}{3}Bh = \frac{1}{3}x^2(x - 2).$$

Substituting 25 for the volume yields the following.

$$25 = \frac{1}{3}x^2(x - 2) \qquad \text{Substitute 25 for } V.$$

$$75 = x^3 - 2x^2 \qquad \text{Multiply each side by 3.}$$

$$0 = x^3 - 2x^2 - 75 \qquad \text{Write in general form.}$$

The possible rational zeros are

$$x = \frac{\pm 1, \pm 3, \pm 5, \pm 15, \pm 25, \pm 75}{\pm 1}.$$

Using synthetic division, you can determine that $x = 5$ is a solution. The other two solutions, which satisfy $x^2 + 3x + 15 = 0$, are imaginary and can be discarded.

The base of the candle mold should be 5 inches by 5 inches. The height of the mold should be $5 - 2 = 3$ inches.

Activities

1. Write as a product of linear factors: $f(x) = x^4 - 16$.
   Answer: $(x - 2)(x + 2)(x - 2i)(x + 2i)$
2. Find a third-degree polynomial with integer coefficients that has $2, 3 + i$, and $3 - i$ as zeros.
   Answer: $x^3 - 8x^2 + 22x - 20$
3. Use the zero $x = 2i$ to find the zeros of $f(x) = x^4 - x^3 - 2x^2 - 4x - 24$.
   Answer: $-2, 3, 2i, -2i$

2.5 Exercises

In Exercises 1–6, find all the zeros of the function.

1. $f(x) = x(x - 6)^2$
2. $f(x) = x^2(x + 3)(x^2 - 1)$
3. $g(x) = (x - 2)(x + 4)^3$
4. $f(x) = (x + 5)(x - 8)^2$
5. $f(x) = (x + 6)(x + i)(x - i)$
6. $h(t) = (t - 3)(t - 2)(t - 3i)(t + 3i)$

In Exercises 7–10, use the Rational Zero Test to list all possible rational zeros of *f*. Verify that the zeros of *f* shown on the graph are contained in the list.

7. $f(x) = x^3 + 3x^2 - x - 3$

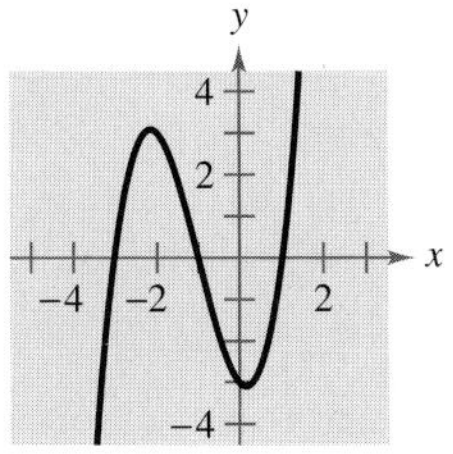

8. $f(x) = x^3 - 4x^2 - 4x + 16$

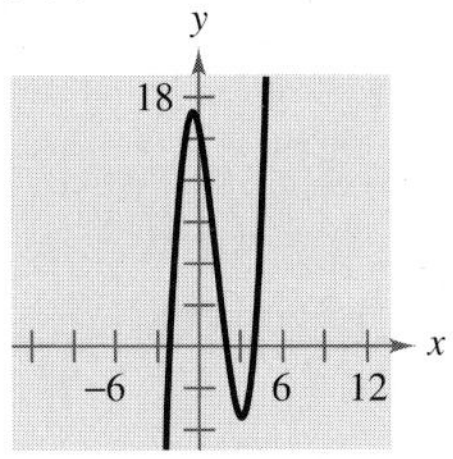

9. $f(x) = 2x^4 - 17x^3 + 35x^2 + 9x - 45$

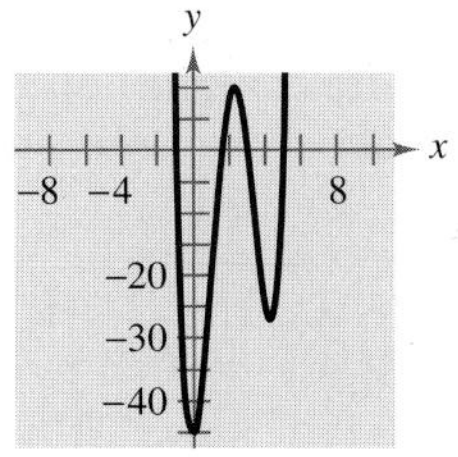

10. $f(x) = 4x^5 - 8x^4 - 5x^3 + 10x^2 + x - 2$

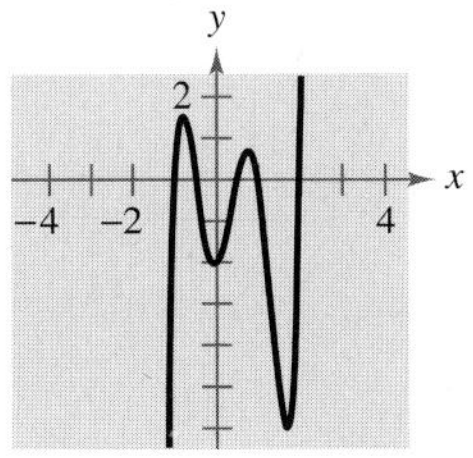

In Exercises 11–20, find all the real zeros of the function.

11. $f(x) = x^3 - 6x^2 + 11x - 6$
12. $f(x) = x^3 - 7x - 6$
13. $g(x) = x^3 - 4x^2 - x + 4$
14. $h(x) = x^3 - 9x^2 + 20x - 12$
15. $h(t) = t^3 + 12t^2 + 21t + 10$
16. $p(x) = x^3 - 9x^2 + 27x - 27$
17. $C(x) = 2x^3 + 3x^2 - 1$
18. $f(x) = 3x^3 - 19x^2 + 33x - 9$
19. $f(x) = 9x^4 - 9x^3 - 58x^2 + 4x + 24$
20. $f(x) = 2x^4 - 15x^3 + 23x^2 + 15x - 25$

In Exercises 21–24, find all real solutions of the polynomial equation.

21. $z^4 - z^3 - 2z - 4 = 0$
22. $x^4 - 13x^2 - 12x = 0$
23. $2y^4 + 7y^3 - 26y^2 + 23y - 6 = 0$
24. $x^5 - x^4 - 3x^3 + 5x^2 - 2x = 0$

In Exercises 25–28, (a) list the possible rational zeros of *f*, (b) sketch the graph of *f* so that some of the possible zeros in part (a) can be disregarded, and then (c) determine all real zeros of *f*.

25. $f(x) = x^3 + x^2 - 4x - 4$
26. $f(x) = -3x^3 + 20x^2 - 36x + 16$
27. $f(x) = -4x^3 + 15x^2 - 8x - 3$
28. $f(x) = 4x^3 - 12x^2 - x + 15$

In Exercises 29–32, (a) list the possible rational zeros of *f*, (b) use a graphing utility to graph *f* so that some of the possible zeros in part (a) can be disregarded, and then (c) determine all real zeros of *f*.

29. $f(x) = -2x^4 + 13x^3 - 21x^2 + 2x + 8$
30. $f(x) = 4x^4 - 17x^2 + 4$
31. $f(x) = 32x^3 - 52x^2 + 17x + 3$
32. $f(x) = 4x^3 + 7x^2 - 11x - 18$

Graphical Analysis **In Exercises 33–36, (a) use the root-finding capabilities of a graphing utility to approximate the zeros of the function accurate to three decimal places. (b) Determine one of the exact zeros and use synthetic division to verify your result, and then factor the polynomial completely.**

33. $f(x) = x^4 - 3x^2 + 2$

34. $P(t) = t^4 - 7t^2 + 12$

35. $h(x) = x^5 - 7x^4 + 10x^3 + 14x^2 - 24x$

36. $g(x) = 6x^4 - 11x^3 - 51x^2 + 99x - 27$

In Exercises 37–42, find a polynomial function with integer coefficients that has the given zeros. (There are many correct answers.)

37. $1, 5i, -5i$

38. $4, 3i, -3i$

39. $6, -5 + 2i, -5 - 2i$

40. $2, 4 + i, 4 - i$

41. $\frac{2}{3}, -1, 3 + \sqrt{2}i$

42. $-5, -5, 1 + \sqrt{3}i$

In Exercises 43–46, write the polynomial (a) as the product of factors that are irreducible over the *rationals*, (b) as the product of linear and quadratic factors that are irreducible over the *reals*, and (c) in completely factored form.

43. $f(x) = x^4 + 6x^2 - 27$

44. $f(x) = x^4 - 2x^3 - 3x^2 + 12x - 18$
(*Hint:* One factor is $x^2 - 6$.)

45. $f(x) = x^4 - 4x^3 + 5x^2 - 2x - 6$
(*Hint:* One factor is $x^2 - 2x - 2$.)

46. $f(x) = x^4 - 3x^3 - x^2 - 12x - 20$
(*Hint:* One factor is $x^2 + 4$.)

In Exercises 47–54, use the given zero to find all the zeros of the function.

	Function	*Zero*
47.	$f(x) = 2x^3 + 3x^2 + 50x + 75$	$5i$
48.	$f(x) = x^3 + x^2 + 9x + 9$	$3i$
49.	$f(x) = 2x^4 - x^3 + 7x^2 - 4x - 4$	$2i$
50.	$g(x) = x^3 - 7x^2 - x + 87$	$5 + 2i$
51.	$g(x) = 4x^3 + 23x^2 + 34x - 10$	$-3 + i$
52.	$h(x) = 3x^3 - 4x^2 + 8x + 8$	$1 - \sqrt{3}i$
53.	$f(x) = x^4 + 3x^3 - 5x^2 - 21x + 22$	$-3 + \sqrt{2}i$
54.	$f(x) = x^3 + 4x^2 + 14x + 20$	$-1 - 3i$

In Exercises 55–72, find all the zeros of the function and write the polynomial as a product of linear factors.

55. $f(x) = x^2 + 25$

56. $f(x) = x^2 - x + 56$

57. $h(x) = x^2 - 4x + 1$

58. $g(x) = x^2 + 10x + 23$

59. $f(x) = x^4 - 81$

60. $f(y) = y^4 - 625$

61. $f(z) = z^2 - 2z + 2$

62. $h(x) = x^3 - 3x^2 + 4x - 2$

63. $g(x) = x^3 - 6x^2 + 13x - 10$

64. $f(x) = x^3 - 2x^2 - 11x + 52$

65. $h(x) = x^3 - x + 6$

66. $h(x) = x^3 + 9x^2 + 27x + 35$

67. $f(x) = 5x^3 - 9x^2 + 28x + 6$

68. $g(x) = 3x^3 - 4x^2 + 8x + 8$

69. $g(x) = x^4 - 4x^3 + 8x^2 - 16x + 16$

70. $h(x) = x^4 + 6x^3 + 10x^2 + 6x + 9$

71. $f(x) = x^4 + 10x^2 + 9$

72. $f(x) = x^4 + 29x^2 + 100$

In Exercises 73–78, find all the zeros of the function. When there is an extended list of possible rational zeros, use a graphing utility to graph the function in order to discard any rational zeros that are obviously not zeros of the function.

73. $f(x) = x^3 + 24x^2 + 214x + 740$

74. $f(s) = 2s^3 - 5s^2 + 12s - 5$

75. $f(x) = 16x^3 - 20x^2 - 4x + 15$

76. $f(x) = 9x^3 - 15x^2 + 11x - 5$

77. $f(x) = 2x^4 + 5x^3 + 4x^2 + 5x + 2$

78. $g(x) = x^5 - 8x^4 + 28x^3 - 56x^2 + 64x - 32$

In Exercises 79–86, use Descartes's Rule of Signs to determine the possible number of positive and negative zeros of the function.

79. $g(x) = 5x^5 + 10x$

80. $h(x) = 4x^2 - 8x + 3$

81. $h(x) = 3x^4 + 2x^2 + 1$

82. $h(x) = 2x^4 - 3x + 2$

83. $g(x) = 2x^3 - 3x^2 - 3$

84. $f(x) = 4x^3 - 3x^2 + 2x - 1$

85. $f(x) = -5x^3 + x^2 - x + 5$

86. $f(x) = 3x^3 + 2x^2 + x + 3$

In Exercises 87–90, use synthetic division to verify the upper and lower bounds of the real zeros of *f*.

87. $f(x) = x^4 - 4x^3 + 15$

(a) Upper: $x = 4$ (b) Lower: $x = -1$

88. $f(x) = 2x^3 - 3x^2 - 12x + 8$

(a) Upper: $x = 4$ (b) Lower: $x = -3$

89. $f(x) = x^4 - 4x^3 + 16x - 16$

(a) Upper: $x = 5$ (b) Lower: $x = -3$

90. $f(x) = 2x^4 - 8x + 3$

(a) Upper: $x = 3$ (b) Lower: $x = -4$

In Exercises 91–94, find all the real zeros of the function.

91. $f(x) = 4x^3 - 3x - 1$

92. $f(z) = 12z^3 - 4z^2 - 27z + 9$

93. $f(y) = 4y^3 + 3y^2 + 8y + 6$

94. $g(x) = 3x^3 - 2x^2 + 15x - 10$

In Exercises 95–98, find all the rational zeros of the polynomial function.

95. $P(x) = x^4 - \frac{25}{4}x^2 + 9 = \frac{1}{4}(4x^4 - 25x^2 + 36)$

96. $f(x) = x^3 - \frac{3}{2}x^2 - \frac{23}{2}x + 6 = \frac{1}{2}(2x^3 - 3x^2 - 23x + 12)$

97. $f(x) = x^3 - \frac{1}{4}x^2 - x + \frac{1}{4} = \frac{1}{4}(4x^3 - x^2 - 4x + 1)$

98. $f(z) = z^3 + \frac{11}{6}z^2 - \frac{1}{2}z - \frac{1}{3} = \frac{1}{6}(6z^3 + 11z^2 - 3z - 2)$

In Exercises 99–102, match the cubic function with the numbers of rational and irrational zeros.

(a) Rational zeros: 0; Irrational zeros: 1

(b) Rational zeros: 3; Irrational zeros: 0

(c) Rational zeros: 1; Irrational zeros: 2

(d) Rational zeros: 1; Irrational zeros: 0

99. $f(x) = x^3 - 1$

100. $f(x) = x^3 - 2$

101. $f(x) = x^3 - x$

102. $f(x) = x^3 - 2x$

103. *Think About It* A third-degree polynomial function f has real zeros -2, $\frac{1}{2}$, and 3, and its leading coefficient is negative. Write an equation for f. Sketch the graph of f. How many different polynomial functions are possible for f?

104. *Think About It* Sketch the graph of a fifth-degree polynomial function, whose leading coefficient is positive, that has one root at $x = 3$ of multiplicity 2.

105. Use the information in the table.

Interval	Value of $f(x)$
$(-\infty, -2)$	Positive
$(-2, 1)$	Negative
$(1, 4)$	Negative
$(4, \infty)$	Positive

(a) What are the three zeros of the polynomial function f?

(b) What can be said about the behavior of the graph of f at $x = 1$?

(c) What is the least possible degree of f? Explain. Can the degree of f ever be odd? Explain.

(d) Is the leading coefficient of f positive or negative? Explain.

(e) Write an equation for f.

(f) Sketch a graph of the function you wrote in part (e).

106. Use the information in the table.

Interval	Value of $f(x)$
$(-\infty, -2)$	Negative
$(-2, 0)$	Positive
$(0, 2)$	Positive
$(2, \infty)$	Negative

(a) What are the three zeros of the polynomial function f?

(b) What can be said about the behavior of the graph of f at $x = 0$?

(c) What is the least possible degree of f? Explain. Can the degree of f ever be odd? Explain.

(d) Is the leading coefficient of f positive or negative? Explain.

(e) Write an equation for f.

(f) Sketch a graph of the function you wrote in part (e).

107. *Geometry* An open box is to be made from a rectangular piece of material, 15 centimeters by 9 centimeters, by cutting equal squares from the corners and turning up the sides.

(a) Let x represent the length of the sides of the squares removed. Draw a diagram showing the squares removed from the original piece of material and the resulting dimensions of the open box.

(b) Use the diagram to write the volume V of the box as a function of x. Determine the domain of the function.

(c) Sketch the graph of the function and approximate the dimensions of the box that yield a maximum volume.

(d) Find values of x such that $V = 56$. Which of these values is a physical impossibility in the construction of the box? Explain.

108. *Geometry* A rectangular package to be sent by a delivery service (see figure) can have a maximum combined length and girth (perimeter of a cross section) of 120 inches.

(a) Show that the volume of the package is

$$V(x) = 4x^2(30 - x).$$

(b) Use a graphing utility to graph the function and approximate the dimensions of the package that yield a maximum volume.

(c) Find values of x such that $V = 13{,}500$. Which of these values is a physical impossibility in the construction of the package? Explain.

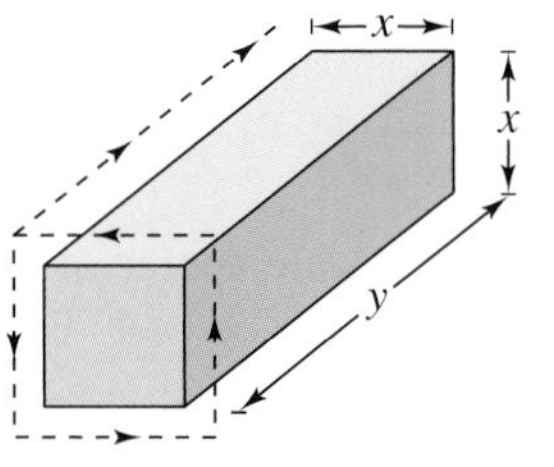

109. *Advertising Cost* A company that produces portable cassette players estimates that the profit for selling a particular model is

$$P = -76x^3 + 4830x^2 - 320{,}000, \quad 0 \le x \le 60$$

where P is the profit (in dollars) and x is the advertising expense (in tens of thousands of dollars). Using this model, find the smaller of two advertising amounts that yield a profit of \$2,500,000.

110. *Advertising Cost* A company that manufactures bicycles estimates that the profit for selling a particular model is

$$P = -45x^3 + 2500x^2 - 275{,}000, \quad 0 \le x \le 50$$

where P is the profit (in dollars) and x is the advertising expense (in tens of thousands of dollars). Using this model, find the smaller of two advertising amounts that yield a profit of \$800,000.

111. *Business* The ordering and transportation cost for the components used in manufacturing a certain product is

$$C = 100\left(\frac{200}{x^2} + \frac{x}{x + 30}\right), \quad 1 \le x$$

where C is the cost (in thousands of dollars) and x is the order size (in hundreds). In calculus, it can be shown that the cost is a minimum when

$$3x^3 - 40x^2 - 2400x - 36{,}000 = 0.$$

Use a calculator to approximate the optimal order size to the nearest hundred units.

112. *Foreign Trade* The values (in billions of dollars) of goods imported into the United States for the years 1988 through 1997 are shown in the table. (Source: U.S. International Trade Administration)

Year	1988	1989	1990	1991	1992
Imports	441.0	473.2	495.3	488.5	532.7

Year	1993	1994	1995	1996	1997
Imports	580.7	663.3	743.4	795.3	870.7

(a) A model for this data is

$$I = -0.222t^3 + 6.432t^2 + 23.328t + 473.991$$

where I is the annual value of goods imported (in billions of dollars) and t is the time (in years), with $t = 0$ corresponding to 1990. Use a graphing utility to sketch a scatter plot of the data and the model in the same viewing window. How do they compare?

(b) According to this model, when did the annual value of imports reach 750 billion dollars?

(c) According to the right-hand behavior of the model, will the value of imports continue to increase? Explain.

113. *Physics* A baseball is thrown upward from ground level with an initial velocity of 48 feet per second, and its height h (in feet) is

$$h = -16t^2 + 48t, \quad 0 \le t \le 3$$

where t is the time (in seconds). Suppose you are told the ball reaches a height of 64 feet. Is this possible?

114. *Economics* The demand equation for a certain product is $p = 140 - 0.0001x$, where p is the unit price (in dollars) of the product and x is the number of units produced and sold. The cost equation for the product is $C = 80x + 150{,}000$, where C is the total cost (in dollars) and x is the number of units produced. The total profit obtained by producing and selling x units is

$$P = R - C = xp - C.$$

You are working in the marketing department of the company that produces this product, and you are asked to determine a price p that will yield a profit of 9 million dollars. Is this possible? Explain.

Synthesis

True or False? **In Exercises 115 and 116, decide whether the statement is true or false. Justify your answer.**

115. It is possible for a third-degree polynomial function with integer coefficients to have no real zeros.

116. If $x = -i$ is a zero of the function $f(x) = x^3 + ix^2 + ix - 1$, then $x = i$ must also be a zero of f.

Think About It **In Exercises 117–122, determine (if possible) the zeros of the function g if the function f has zeros at $x = r_1$, $x = r_2$, and $x = r_3$.**

117. $g(x) = -f(x)$ **118.** $g(x) = 3f(x)$

119. $g(x) = f(x - 5)$ **120.** $g(x) = f(2x)$

121. $g(x) = 3 + f(x)$ **122.** $g(x) = f(-x)$

123. *Exploration* Use a graphing utility to graph the function $f(x) = x^4 - 4x^2 + k$ for different values of k. Find values of k such that the zeros of f satisfy the specified characteristics. (Some parts do not have unique answers.)

(a) Four real zeros

(b) Two real zeros, each of multiplicity 2

(c) Two real zeros and two complex roots

(d) Four complex zeros

124. *Think About It* Will the answers to Exercise 123 change for the function g?

(a) $g(x) = f(x - 2)$

(b) $g(x) = f(2x)$

125. (a) Find a quadratic function f (with integer coefficients) that has $\pm\sqrt{b}i$ as zeros. Assume that b is a positive integer.

(b) Find a quadratic function f (with integer coefficients) that has $a \pm bi$ as zeros. Assume that b is a positive integer.

126. *Graphical Reasoning* The graph of one of the following functions is shown below. Identify the function shown in the graph. Explain why each of the others is not the correct function. Use a graphing utility to verify your result.

(a) $f(x) = x^2(x + 2)(x - 3.5)$

(b) $g(x) = (x + 2)(x - 3.5)$

(c) $h(x) = (x + 2)(x - 3.5)(x^2 + 1)$

(d) $k(x) = (x + 1)(x + 2)(x - 3.5)$

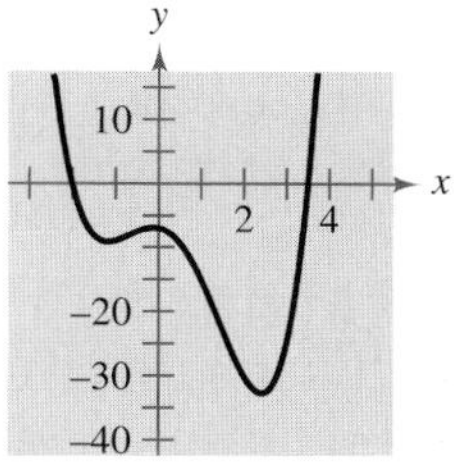

Review

In Exercises 127–132, perform the operation and simplify.

127. $(-3 + 6i) - (8 - 3i)$ **128.** $(12 - 5i) + 16i$

129. $(6 - 2i)(1 + 7i)$ **130.** $(9 - 5i)(9 + 5i)$

131. $\dfrac{1 + i}{1 - i}$ **132.** $(3 + i)^3$

In Exercises 133–138, use the graph of f to graph the function g.

133. $g(x) = f(x - 2)$

134. $g(x) = f(x) - 2$

135. $g(x) = 2f(x)$

136. $g(x) = f(-x)$

137. $g(x) = f(2x)$

138. $g(x) = f\left(\frac{1}{2}x\right)$

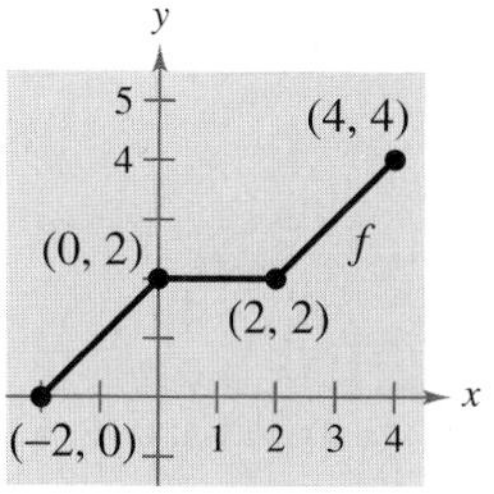

2.6 Rational Functions

▶ What you should learn

- How to find the domains of rational functions
- How to find the horizontal and vertical asymptotes of graphs of rational functions
- How to analyze and sketch graphs of rational functions
- How to sketch graphs of rational functions that have slant asymptotes
- How to use rational functions to model and solve real-life problems

▶ Why you should learn it

Rational functions can be used to model and solve real-life problems relating to the environment. For instance, Exercise 71 on page 269 shows how a rational function can be used to model the cost of removing pollutants from a river.

David Woodfall/Tony Stone Images

Introduction

A **rational function** can be written in the form

$$f(x) = \frac{N(x)}{D(x)}$$

where $N(x)$ and $D(x)$ are polynomials and $D(x)$ is not the zero polynomial. In this section it is assumed that $N(x)$ and $D(x)$ have no common factors.

In general, the *domain* of a rational function of x includes all real numbers except x-values that make the denominator zero. Much of the discussion of rational functions will focus on their graphical behavior near these x-values.

Example 1 ▶ Finding the Domain of a Rational Function

Find the domain of $f(x) = 1/x$ and discuss the behavior of f near any excluded x-values.

Solution

Because the denominator is zero when $x = 0$, the domain of f is all real numbers except $x = 0$. To determine the behavior of f near this excluded value, evaluate $f(x)$ to the left and right of $x = 0$, as indicated in the following tables.

x	-1	-0.5	-0.1	-0.01	-0.001	$\rightarrow 0$
$f(x)$	-1	-2	-10	-100	-1000	$\rightarrow -\infty$

x	$0 \leftarrow$	0.001	0.01	0.1	0.5	1
$f(x)$	$\infty \leftarrow$	1000	100	10	2	1

Note that as x approaches 0 *from the left*, $f(x)$ decreases without bound. In contrast, as x approaches 0 *from the right*, $f(x)$ increases without bound. The graph of f is shown in Figure 2.30.

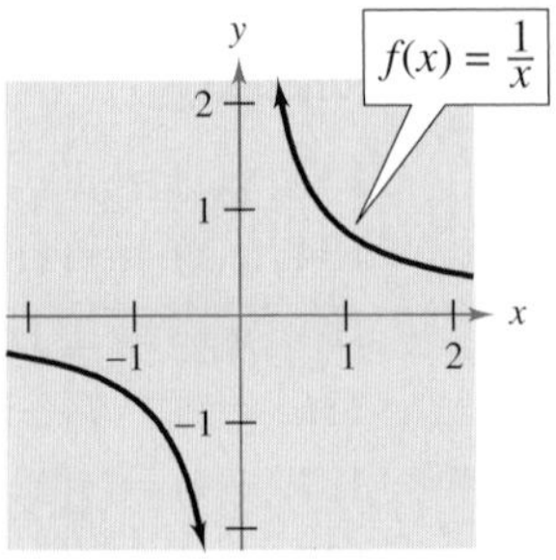

FIGURE 2.30

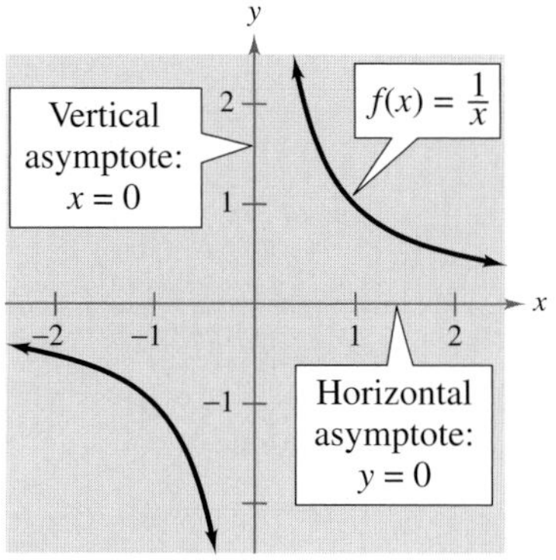

FIGURE 2.31

Horizontal and Vertical Asymptotes

In Example 1, the behavior of f near $x = 0$ is denoted as follows.

$$\underbrace{f(x) \longrightarrow -\infty \text{ as } x \longrightarrow 0^-}_{f(x) \text{ decreases without bound as } x \text{ approaches 0 from the left.}} \qquad \underbrace{f(x) \longrightarrow \infty \text{ as } x \longrightarrow 0^+}_{f(x) \text{ increases without bound as } x \text{ approaches 0 from the right.}}$$

The line $x = 0$ is a **vertical asymptote** of the graph of f, as shown in Figure 2.31. From this figure, you can see that the graph of f also has a **horizontal asymptote**—the line $y = 0$. This means that the values of $f(x) = 1/x$ approach zero as x increases or decreases without bound.

$$\underbrace{f(x) \longrightarrow 0 \text{ as } x \longrightarrow -\infty}_{f(x) \text{ approaches 0 as } x \text{ decreases without bound.}} \qquad \underbrace{f(x) \longrightarrow 0 \text{ as } x \longrightarrow \infty}_{f(x) \text{ approaches 0 as } x \text{ increases without bound.}}$$

Additional Examples

State the domain of the function.

a. $f(x) = \dfrac{3x}{x + 10}$

b. $f(x) = \dfrac{x + 1}{(x + 2)(x - 6)}$

Solution

a. The domain is all real numbers except $x = -10$.

b. The domain is all real numbers except $x = -2$ and $x = 6$.

Definition of Vertical and Horizontal Asymptotes

1. The line $x = a$ is a **vertical asymptote** of the graph of f if

$$f(x) \longrightarrow \infty \quad \text{or} \quad f(x) \longrightarrow -\infty$$

as $x \longrightarrow a$, either from the right or from the left.

2. The line $y = b$ is a **horizontal asymptote** of the graph of f if

$$f(x) \longrightarrow b$$

as $x \longrightarrow \infty$ or $x \longrightarrow -\infty$.

Eventually (as $x \longrightarrow \infty$ or $x \longrightarrow -\infty$), the distance between the horizontal asymptote and the points on the graph must approach zero. Figure 2.32 shows the horizontal and vertical asymptotes of the graphs of three rational functions.

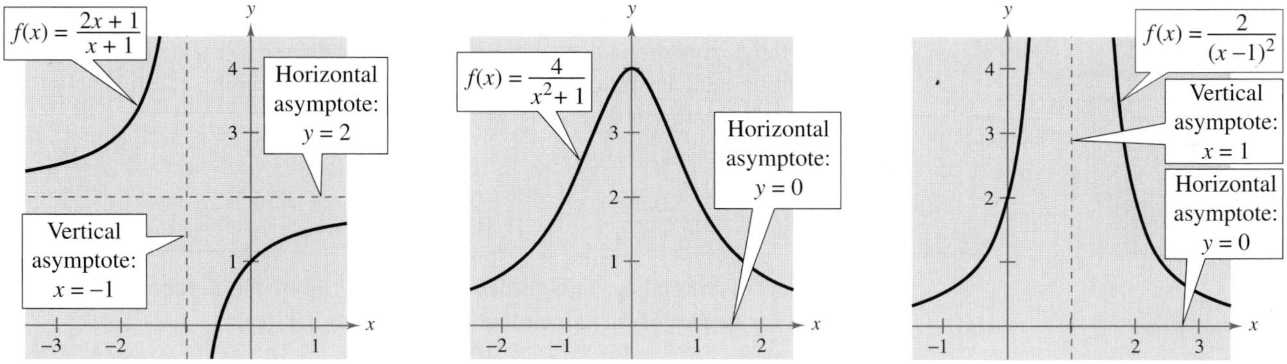

(a) (b) (c)

FIGURE 2.32

The graphs of $f(x) = 1/x$ in Figure 2.31 and $f(x) = (2x + 1)/(x + 1)$ in Figure 2.32(a) are **hyperbolas.** You will study hyperbolas in Section 10.4.

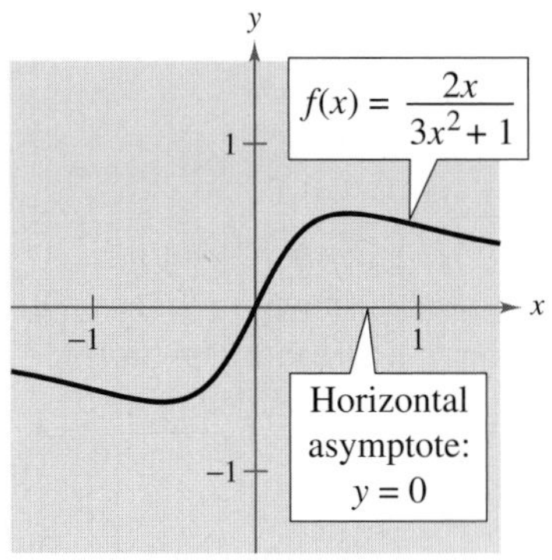

(a)

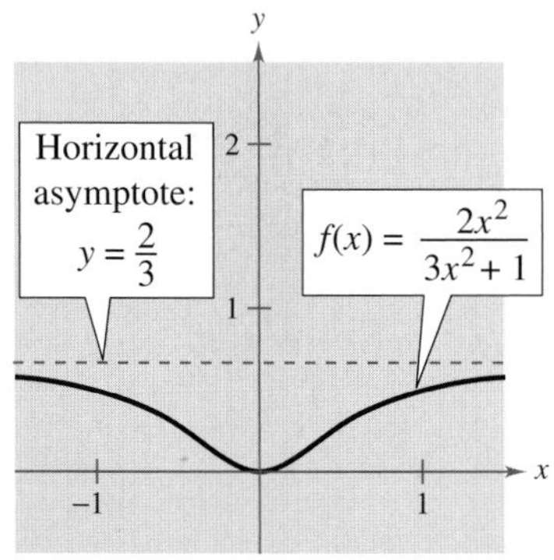

(b)

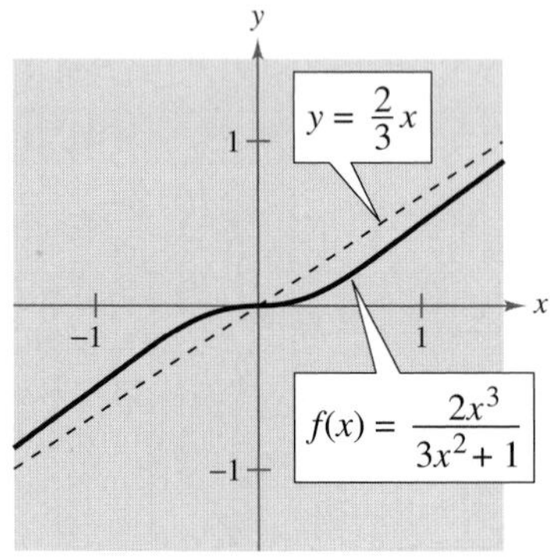

(c)

FIGURE 2.33

Asymptotes of a Rational Function

Let f be the rational function given by

$$f(x) = \frac{N(x)}{D(x)} = \frac{a_nx^n + a_{n-1}x^{n-1} + \cdot \cdot \cdot + a_1x + a_0}{b_mx^m + b_{m-1}x^{m-1} + \cdot \cdot \cdot + b_1x + b_0}$$

where $N(x)$ and $D(x)$ have no common factors.

1. The graph of f has *vertical* asymptotes at the zeros of $D(x)$.
2. The graph of f has one or no *horizontal* asymptote determined by comparing the degrees of $N(x)$ and $D(x)$.
   a. If $n < m$, the graph of f has the line $y = 0$ (the x-axis) as a horizontal asymptote.
   b. If $n = m$, the graph of f has the line $y = a_n/b_m$ as a horizontal asymptote.
   c. If $n > m$, the graph of f has no horizontal asymptote.

Example 2 ▶ Finding Horizontal Asymptotes

a. The graph of

$$f(x) = \frac{2x}{3x^2 + 1}$$

has the line $y = 0$ (the x-axis) as a horizontal asymptote, as shown in Figure 2.33(a). Note that the degree of the numerator is *less than* the degree of the denominator.

b. The graph of

$$f(x) = \frac{2x^2}{3x^2 + 1}$$

has the line $y = \frac{2}{3}$ as a horizontal asymptote, as shown in Figure 2.33(b). Note that the degree of the numerator is *equal to* the degree of the denominator, and the horizontal asymptote is given by the ratio of the leading coefficients of the numerator and denominator.

c. The graph of

$$f(x) = \frac{2x^3}{3x^2 + 1}$$

has no horizontal asymptote because the degree of the numerator is greater than the degree of the denominator. See Figure 2.33(c).

Although the graph of the function in Example 2(c) does not have a horizontal asymptote, it does have a *slant asymptote*—the line $y = \frac{2}{3}x$. You will study slant asymptotes later in this section.

Analyzing Graphs of Rational Functions

Guidelines for Analyzing Graphs of Rational Functions

Let $f(x) = N(x)/D(x)$, where $N(x)$ and $D(x)$ are polynomials with no common factors.

1. Find and plot the y-intercept (if any) by evaluating $f(0)$.
2. Find the zeros of the numerator (if any) by solving the equation $N(x) = 0$. Then plot the corresponding x-intercepts.
3. Find the zeros of the denominator (if any) by solving the equation $D(x) = 0$. Then sketch the corresponding vertical asymptotes.
4. Find and sketch the horizontal asymptote (if any) by using the rule for finding the horizontal asymptote of a rational function.
5. Plot at least one point *between* and one point *beyond* each x-intercept and vertical asymptote.
6. Use smooth curves to complete the graph between and beyond the vertical asymptotes.

Testing for symmetry can be useful, especially for simple rational functions. For example, the graph of $f(x) = 1/x$ is symmetric with respect to the origin, and the graph of $g(x) = 1/x^2$ is symmetric with respect to the y-axis.

Technology

Some graphing utilities have difficulty sketching graphs of rational functions that have vertical asymptotes. Often, the utility will connect parts of the graph that are not supposed to be connected. For instance, the screen below shows the graph of

$$f(x) = \frac{1}{x - 2}.$$

Notice that the graph should consist of two *separated* portions—one to the left of $x = 2$ and the other to the right of $x = 2$. To eliminate this problem, you can try changing the *mode* of the graphing utility to *dot mode*. The problem with this is that the graph is then represented as a collection of dots (as shown in the screen below on the right) rather than as a smooth curve.

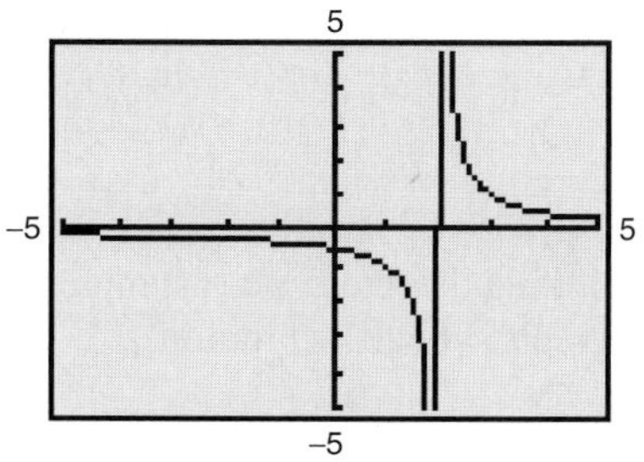

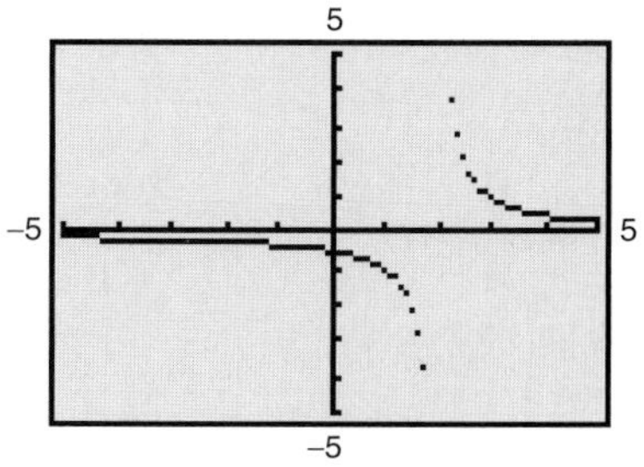

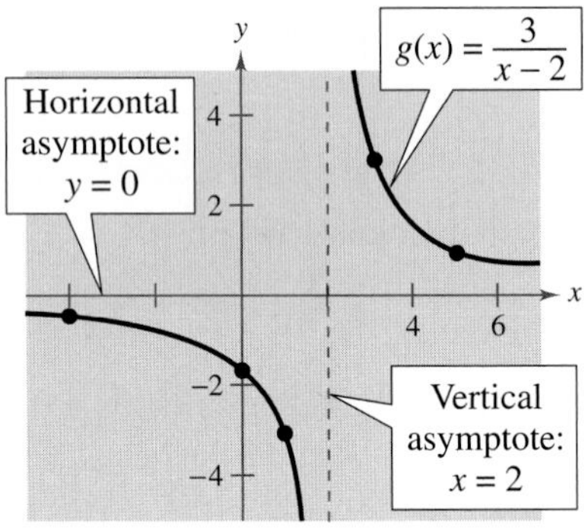

FIGURE 2.34

Example 3 ▶ Sketching the Graph of a Rational Function

Sketch the graph of the function and state its domain.

$$g(x) = \frac{3}{x-2}$$

Solution

y-Intercept: $\left(0, -\frac{3}{2}\right)$, from $g(0) = -\frac{3}{2}$

x-Intercept: None, because $3 \neq 0$

Vertical asymptote: $x = 2$, zero of denominator

Horizontal asymptote: $y = 0$, because degree of $N(x) <$ degree of $D(x)$

Additional points:

x	-4	1	3	5
$g(x)$	-0.5	-3	3	1

By plotting the intercepts, asymptotes, and a few additional points, you can obtain the graph shown in Figure 2.34. The domain of g is all real numbers except $x = 2$.

Additional Example

Identify all horizontal and vertical asymptotes of the graph of

$$f(x) = \frac{x}{x^2 - 9}.$$

Solution

The function has a horizontal asymptote at $y = 0$ because the degree of the numerator is less than the degree of the denominator. The function has two vertical asymptotes at $x = 3$ and $x = -3$, the zeros of the denominator.

The graph of g in Example 3 is a vertical stretch and a right shift of the graph of $f(x) = 1/x$ because

$$g(x) = \frac{3}{x-2} = 3\left(\frac{1}{x-2}\right) = 3f(x-2).$$

Example 4 ▶ Sketching the Graph of a Rational Function

Sketch the graph of the function and state its domain.

$$f(x) = \frac{2x-1}{x}$$

Solution

y-Intercept: None, because $x = 0$ is not in the domain

x-Intercept: $\left(\frac{1}{2}, 0\right)$, from $2x - 1 = 0$

Vertical asymptote: $x = 0$, zero of denominator

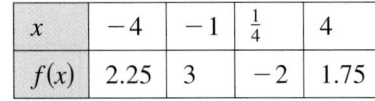

Horizontal asymptote: $y = 2$, because degree of $N(x) =$ degree of $D(x)$

Additional points:

x	-4	-1	$\frac{1}{4}$	4
$f(x)$	2.25	3	-2	1.75

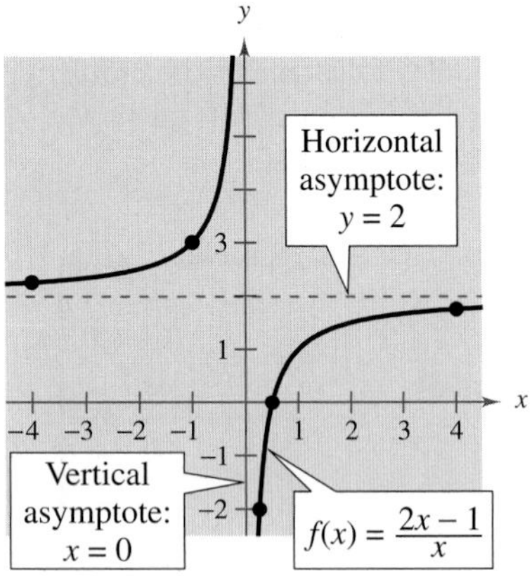

FIGURE 2.35

By plotting the intercepts, asymptotes, and a few additional points, you can obtain the graph shown in Figure 2.35. The domain of f is all real numbers except $x = 0$.

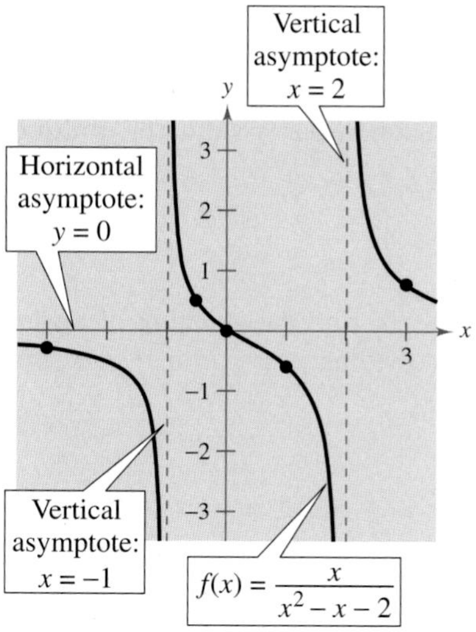

FIGURE 2.36

Example 5 ▶ Sketching the Graph of a Rational Function

Sketch the graph of

$$f(x) = \frac{x}{x^2 - x - 2}.$$

Solution

Factor the denominator to determine more easily the zeros of the denominator.

$$f(x) = \frac{x}{x^2 - x - 2} = \frac{x}{(x + 1)(x - 2)}$$

y-Intercept: (0, 0), because $f(0) = 0$

x-Intercept: (0, 0)

Vertical asymptotes: $x = -1$, $x = 2$, zeros of denominator

Horizontal asymptote: $y = 0$, because degree of $N(x) <$ degree of $D(x)$

Additional points:

x	-3	-0.5	1	3
$f(x)$	-0.3	0.4	-0.5	0.75

The graph is shown in Figure 2.36.

A computer animation of this example appears in the *Interactive* CD-ROM and *Internet* versions of this text.

Example 6 ▶ Sketching the Graph of a Rational Function

Sketch the graph of

$$f(x) = \frac{2(x^2 - 9)}{x^2 - 4}.$$

Solution

By factoring the numerator and denominator, you have

$$f(x) = \frac{2(x^2 - 9)}{x^2 - 4} = \frac{2(x - 3)(x + 3)}{(x - 2)(x + 2)}.$$

y-Intercept: $\left(0, \frac{9}{2}\right)$, because $f(0) = \frac{9}{2}$

x-Intercepts: $(-3, 0)$ and $(3, 0)$

Vertical asymptotes: $x = -2$, $x = 2$, zeros of denominator

Horizontal asymptote: $y = 2$, because degree of $N(x) =$ degree of $D(x)$

Symmetry: With respect to y-axis, because $f(-x) = f(x)$

Additional points:

x	0.5	2.5	6
$f(x)$	4.67	-2.44	1.69

The graph is shown in Figure 2.37.

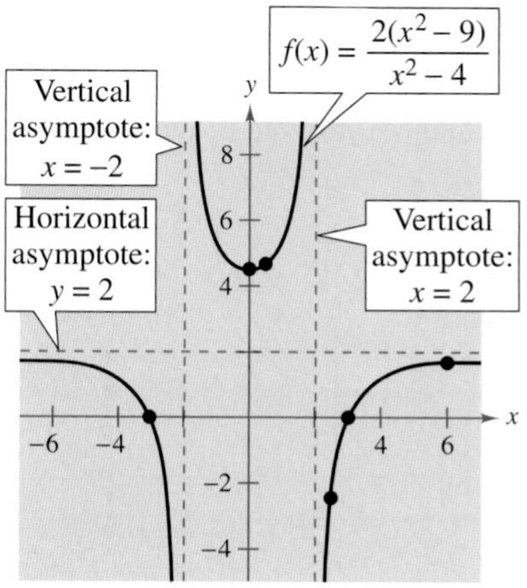

FIGURE 2.37

Slant Asymptotes

If the degree of the numerator of a rational function is exactly *one more* than the degree of the denominator, the graph of the function has a **slant** (or **oblique**) **asymptote.** For example, the graph of

$$f(x) = \frac{x^2 - x}{x + 1}$$

has a slant asymptote, as shown in Figure 2.38. To find the equation of a slant asymptote, use long division. For instance, by dividing $x + 1$ into $x^2 - x$, you obtain

$$f(x) = \frac{x^2 - x}{x + 1} = \underbrace{x - 2}_{\substack{\text{Slant asymptote} \\ (y = x - 2)}} + \frac{2}{x + 1}.$$

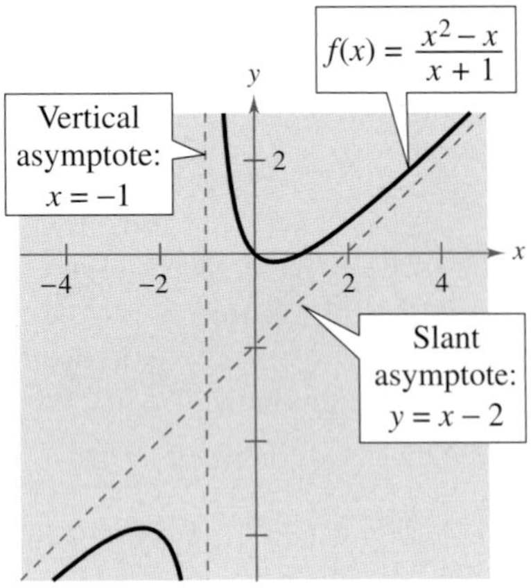

FIGURE 2.38

In Figure 2.38, notice that the graph of f approaches the line $y = x - 2$ as x moves to the right or left.

Example 7 ▶ A Rational Function with a Slant Asymptote

Sketch the graph of

$$f(x) = \frac{x^2 - x - 2}{x - 1}.$$

Solution

First write $f(x)$ in two different ways. Factoring the numerator

$$f(x) = \frac{x^2 - x - 2}{x - 1} = \frac{(x - 2)(x + 1)}{x - 1}$$

allows you to recognize the x-intercepts. Long division

$$f(x) = \frac{x^2 - x - 2}{x - 1} = x - \frac{2}{x - 1}$$

allows you to recognize that the line $y = x$ is a slant asymptote of the graph.

y-Intercept: $(0, 2)$, because $f(0) = 2$

x-Intercepts: $(-1, 0)$ and $(2, 0)$

Vertical asymptote: $x = 1$, zero of denominator

Slant asymptote: $y = x$

Additional points:

x	-2	0.5	1.5	3
$f(x)$	-1.33	4.5	-2.5	2

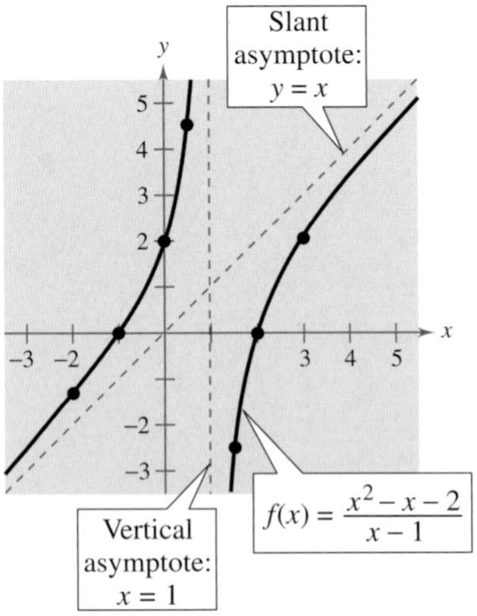

FIGURE 2.39

The graph is shown in Figure 2.39.

Applications

There are many examples of asymptotic behavior in real life. For instance, Example 8 shows how a vertical asymptote can be used to analyze the cost of removing pollutants from smokestack emissions.

Example 8 ▶ Cost-Benefit Model

A utility company burns coal to generate electricity. The cost of removing a certain *percent* of the pollutants from smokestack emissions is typically not a linear function. That is, if it costs C dollars to remove 25% of the pollutants, it would cost more than $2C$ dollars to remove 50% of the pollutants. As the percent of removed pollutants approaches 100%, the cost tends to increase without bound, becoming prohibitive. Suppose that the cost C (in dollars) of removing $p\%$ of the smokestack pollutants is $C = 80{,}000p/(100 - p)$ for $0 \le p < 100$. Sketch the graph of this function. Suppose you are a member of a state legislature considering a law that would require utility companies to remove 90% of the pollutants from their smokestack emissions. If the current law requires 85% removal, how much additional cost would the utility company incur as a result of the new law?

Solution

The graph of this function is shown in Figure 2.40. Note that the graph has a vertical asymptote at $p = 100$. Because the current law requires 85% removal, the current cost to the utility company is

$$C = \frac{80{,}000(85)}{100 - 85} \approx \$453{,}333. \qquad \text{Evaluate } C \text{ when } p = 85.$$

If the new law increases the percent removal to 90%, the cost to the utility company will be

$$C = \frac{80{,}000(90)}{100 - 90} = \$720{,}000. \qquad \text{Evaluate } C \text{ when } p = 90.$$

So, the new law would require the utility company to spend an additional

$$720{,}000 - 453{,}333 = \$266{,}667. \qquad \text{Subtract 85\% removal cost from 90\% removal cost.}$$

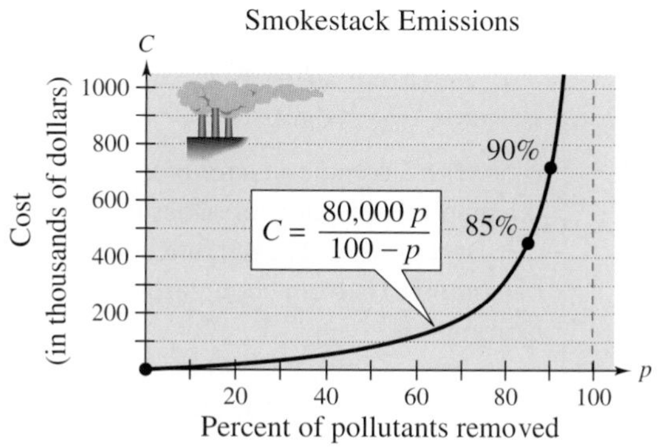

FIGURE 2.40

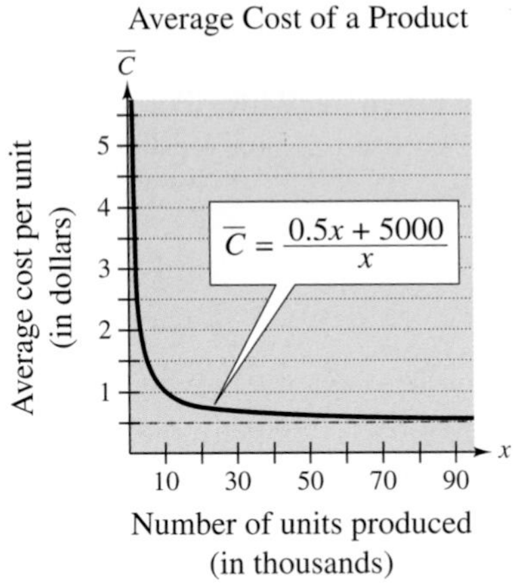

FIGURE 2.41

Example 9 ▶ Average Cost of Producing a Product

A business has a cost function of $C = 0.5x + 5000$, where C is measured in dollars and x is the number of units produced. The *average cost per unit* is

$$\overline{C} = \frac{C}{x} = \frac{0.5x + 5000}{x}.$$

Find the average cost per unit when $x = 1000$, 5000, 10,000, and 100,000. What is the horizontal asymptote for this function, and what does it represent?

Solution

When $x = 1000$, $\overline{C} = \dfrac{0.5(1000) + 5000}{1000} = \$5.50.$

When $x = 5000$, $\overline{C} = \dfrac{0.5(5000) + 5000}{5000} = \$1.50.$

When $x = 10{,}000$, $\overline{C} = \dfrac{0.5(10{,}000) + 5000}{10{,}000} = \$1.00.$

When $x = 100{,}000$, $\overline{C} = \dfrac{0.5(100{,}000) + 5000}{100{,}000} = \$0.55.$

As shown in Figure 2.41, the horizontal asymptote is the line $\overline{C} = 0.50$. This line represents the least possible unit cost for the product. This example points out one of the major problems of a small business. That is, it is difficult to have competitively low prices when the production level is low.

Activities

1. Which of the following functions has (have) $x = 0$ as a vertical asymptote? Discuss your answers.

   a. $f(x) = \dfrac{3 + x}{x}$ b. $f(x) = \dfrac{x}{4 - x^2}$

   c. $f(x) = \dfrac{3x^2 - x}{x^3 - x}$

   d. $f(x) = \dfrac{x^3 - x^2 + 1}{x}$

   Answer: (a) and (d). In (a) and (d), $x = 0$ is the zero of the denominator. In (b), x is not a factor of the denominator. So, $x = 0$ is not a vertical asymptote. In (c), the function is not reduced. The numerator and denominator have the common factor of x. So, $x = 0$ is not a vertical asymptote.

2. Identify any horizontal or vertical asymptotes.

   $f(x) = \dfrac{3x + 2}{5 - 2x}$

   Answer: Vertical asymptote: $x = \dfrac{5}{2}$

   Horizontal asymptote: $y = -\dfrac{3}{2}$

3. Find a function that has $x = 3$ as a vertical asymptote.

   Answers are not unique: $f(x) = \dfrac{5x}{x - 3}$

Writing ABOUT MATHEMATICS

Common Factors in the Numerator and Denominator When sketching the graph of a rational function, be sure that the rational function has no factor that is common to its numerator and denominator. To see why, consider the function

$$f(x) = \frac{2x^2 + x - 1}{x + 1}$$
$$= \frac{(x + 1)(2x - 1)}{x + 1}$$

which has a common factor of $x + 1$ in the numerator and denominator. Sketch the graph of this function. Does it have a vertical asymptote at $x = -1$?

Decide whether each function below has a vertical asymptote. Write a short paragraph to explain your reasoning. Include a graph of each function in your explanation.

a. $f(x) = \dfrac{x^2 - 4}{x + 2}$ **b.** $f(x) = \dfrac{x^2 - 4}{x}$ **c.** $f(x) = \dfrac{x^2 - 4}{2 - x}$

2.6 Exercises

In Exercises 1–6, (a) complete each table, (b) determine the vertical and horizontal asymptotes of the function, and (c) find the domain of the function.

x	$f(x)$
0.5	
0.9	
0.99	
0.999	

x	$f(x)$
1.5	
1.1	
1.01	
1.001	

x	$f(x)$
5	
10	
100	
1000	

1. $f(x) = \dfrac{1}{x-1}$

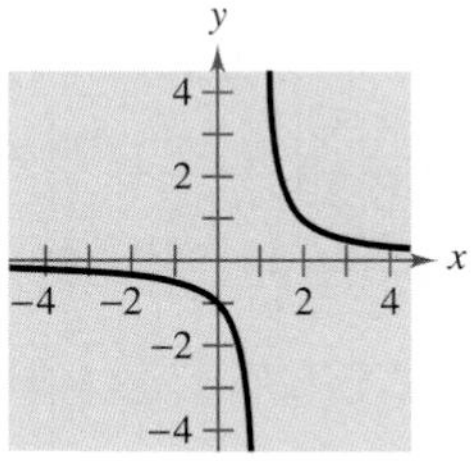

2. $f(x) = \dfrac{5x}{x-1}$

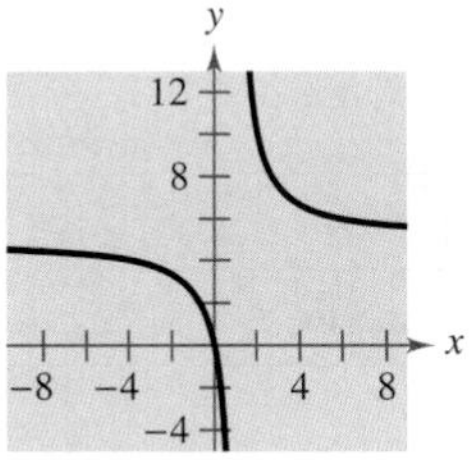

3. $f(x) = \dfrac{4x}{|x-1|}$

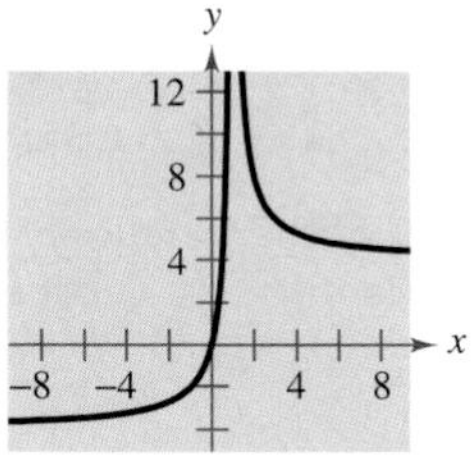

4. $f(x) = \dfrac{2}{|x-1|}$

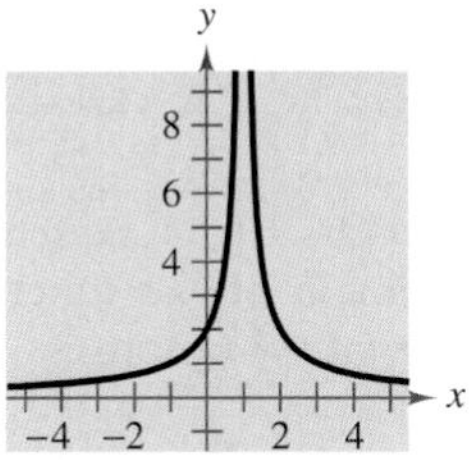

5. $f(x) = \dfrac{3x^2}{x^2-1}$

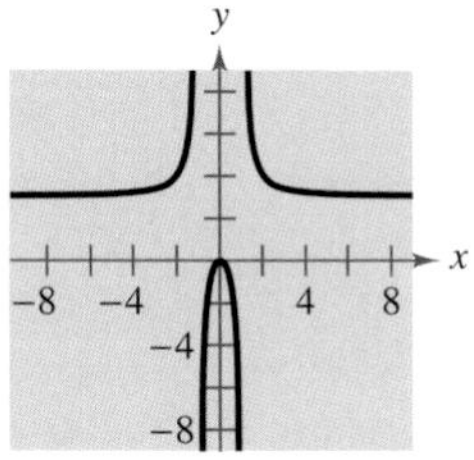

6. $f(x) = \dfrac{4x}{x^2-1}$

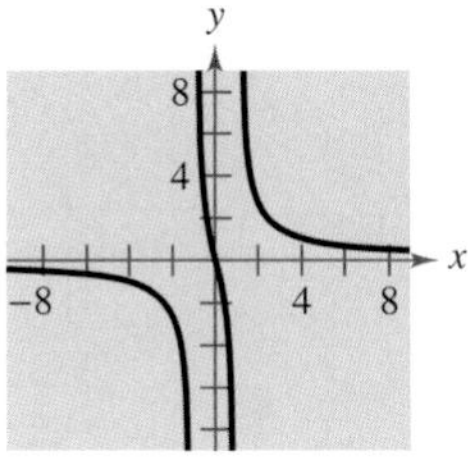

In Exercises 7–14, find the domain of the function and identify any horizontal and vertical asymptotes.

7. $f(x) = \dfrac{1}{x^2}$

8. $f(x) = \dfrac{4}{(x-2)^3}$

9. $f(x) = \dfrac{2+x}{2-x}$

10. $f(x) = \dfrac{1-5x}{1+2x}$

11. $f(x) = \dfrac{x^3}{x^2-1}$

12. $f(x) = \dfrac{2x^2}{x+1}$

13. $f(x) = \dfrac{3x^2+1}{x^2+x+9}$

14. $f(x) = \dfrac{3x^2+x-5}{x^2+1}$

In Exercises 15–20, match the rational function with its graph. [The graphs are labeled (a) through (f).]

(a)

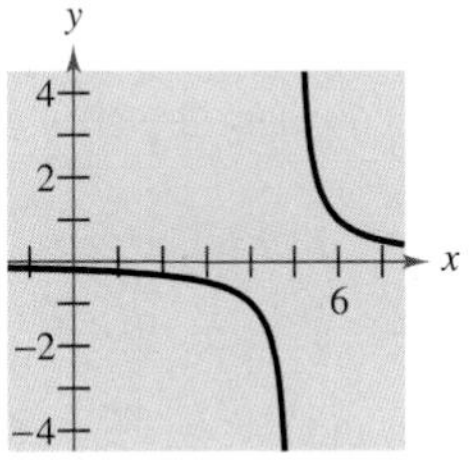

(b)

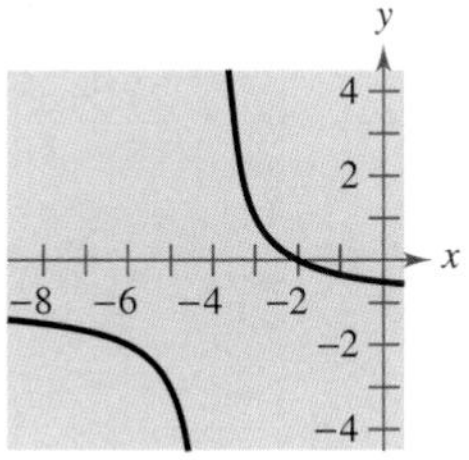

(c)

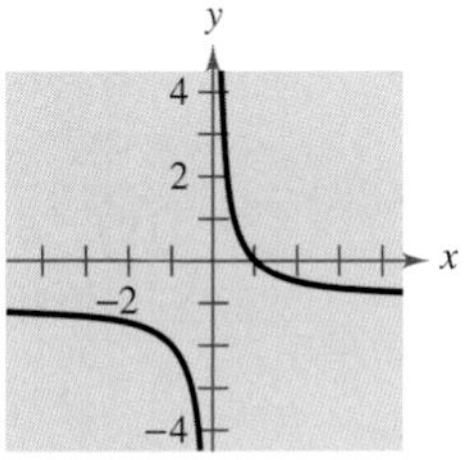

(d)

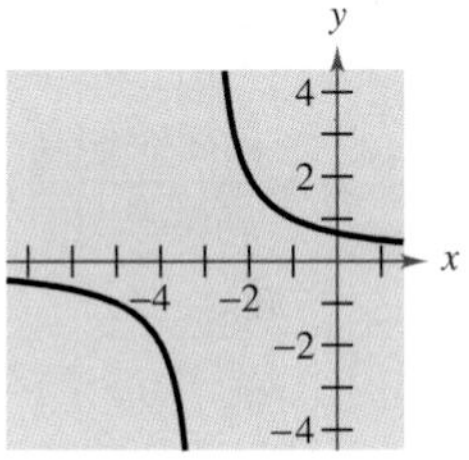

(e)

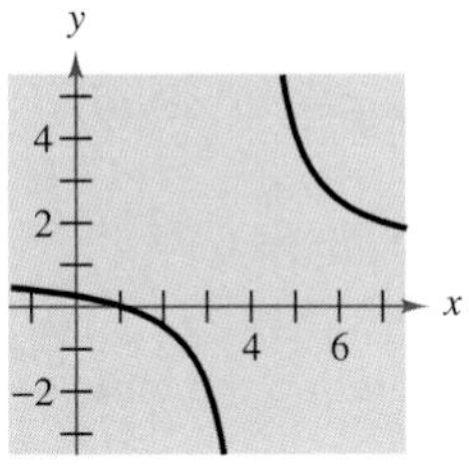

(f)

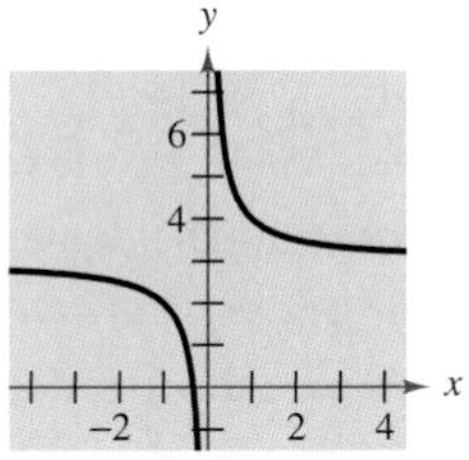

15. $f(x) = \dfrac{2}{x+3}$

16. $f(x) = \dfrac{1}{x-5}$

17. $f(x) = \dfrac{3x+1}{x}$

18. $f(x) = \dfrac{1-x}{x}$

19. $f(x) = \dfrac{x-1}{x-4}$

20. $f(x) = -\dfrac{x+2}{x+4}$

In Exercises 21–24, find the zeros (if any) of the rational function.

21. $g(x) = \dfrac{x^2 - 1}{x + 1}$

22. $h(x) = 2 + \dfrac{5}{x^2 + 2}$

23. $f(x) = 1 - \dfrac{3}{x - 3}$

24. $g(x) = \dfrac{x^3 - 8}{x^2 + 1}$

In Exercises 25–44, (a) identify all intercepts, (b) find any vertical and horizontal asymptotes, (c) check for symmetry, (d) plot additional solution points as needed, and (e) sketch the graph of the rational function.

25. $f(x) = \dfrac{1}{x + 2}$

26. $f(x) = \dfrac{1}{x - 3}$

27. $h(x) = \dfrac{-1}{x + 2}$

28. $g(x) = \dfrac{1}{3 - x}$

29. $C(x) = \dfrac{5 + 2x}{1 + x}$

30. $P(x) = \dfrac{1 - 3x}{1 - x}$

31. $g(x) = \dfrac{1}{x + 2} + 2$

32. $f(x) = 2 - \dfrac{3}{x^2}$

33. $f(x) = \dfrac{x^2}{x^2 + 9}$

34. $f(t) = \dfrac{1 - 2t}{t}$

35. $h(x) = \dfrac{x^2}{x^2 - 9}$

36. $g(x) = \dfrac{x}{x^2 - 9}$

37. $g(s) = \dfrac{s}{s^2 + 1}$

38. $f(x) = -\dfrac{1}{(x - 2)^2}$

39. $g(x) = \dfrac{4(x + 1)}{x(x - 4)}$

40. $h(x) = \dfrac{2}{x^2(x - 2)}$

41. $f(x) = \dfrac{3x}{x^2 - x - 2}$

42. $f(x) = \dfrac{2x}{x^2 + x - 2}$

43. $f(x) = \dfrac{6x}{x^2 - 5x - 14}$

44. $f(x) = \dfrac{3(x^2 + 1)}{x^2 + 2x - 15}$

Analytical, Numerical, and Graphical Analysis **In Exercises 45–48, do the following.**

(a) Determine the domains of *f* and *g*.

(b) Find any vertical asymptotes of *f*.

(c) Compare the functions by completing the table.

(d) Use a graphing utility to graph *f* and *g* in the same viewing window.

(e) Explain why the graphing utility may not show the difference in the domains of *f* and *g*.

45. $f(x) = \dfrac{x^2 - 1}{x + 1}, \quad g(x) = x - 1$

x	-3	-2	-1.5	-1	-0.5	0	1
$f(x)$							
$g(x)$							

46. $f(x) = \dfrac{x^2(x - 2)}{x^2 - 2x}, \quad g(x) = x$

x	-1	0	1	1.5	2	2.5	3
$f(x)$							
$g(x)$							

47. $f(x) = \dfrac{x - 2}{x^2 - 2x}, \quad g(x) = \dfrac{1}{x}$

x	-0.5	0	0.5	1	1.5	2	3
$f(x)$							
$g(x)$							

48. $f(x) = \dfrac{2x - 6}{x^2 - 7x + 12}, \quad g(x) = \dfrac{2}{x - 4}$

x	0	1	2	3	4	5	6
$f(x)$							
$g(x)$							

In Exercises 49–54, sketch the graph of the function. State the domain of the function and identify any vertical or horizontal asymptotes.

49. $h(t) = \dfrac{4}{t^2 + 1}$

50. $g(x) = -\dfrac{x}{(x - 2)^2}$

51. $f(t) = \dfrac{2t^2}{t^2 - 4}$

52. $f(x) = \dfrac{x + 4}{x^2 + x - 6}$

53. $f(x) = \dfrac{20x}{x^2 + 1} - \dfrac{1}{x}$

54. $f(x) = 5\left(\dfrac{1}{x - 4} - \dfrac{1}{x + 2}\right)$

In Exercises 55–62, (a) identify all intercepts, (b) find any vertical and slant asymptotes, (c) check for symmetry, (d) plot additional solution points as needed, and (e) sketch the graph of the rational function.

55. $f(x) = \dfrac{2x^2 + 1}{x}$ **56.** $f(x) = \dfrac{1 - x^2}{x}$

57. $g(x) = \dfrac{x^2 + 1}{x}$ **58.** $h(x) = \dfrac{x^2}{x - 1}$

59. $f(x) = \dfrac{x^3}{x^2 - 1}$ **60.** $g(x) = \dfrac{x^3}{2x^2 - 8}$

61. $f(x) = \dfrac{x^2 - x + 1}{x - 1}$ **62.** $f(x) = \dfrac{2x^2 - 5x + 5}{x - 2}$

In Exercises 63–66, use a graphing utility to graph the rational function. Give the domain of the function and identify any asymptotes. Then zoom out sufficiently far so that the graph appears as a line. Identify the line.

63. $f(x) = \dfrac{x^2 + 5x + 8}{x + 3}$ **64.** $f(x) = \dfrac{2x^2 + x}{x + 1}$

65. $g(x) = \dfrac{1 + 3x^2 - x^3}{x^2}$ **66.** $h(x) = \dfrac{12 - 2x - x^2}{2(4 + x)}$

Graphical Reasoning **In Exercises 67–70, (a) use the graph to determine any x-intercepts of the rational function, and (b) set $y = 0$ and solve the resulting equation to confirm your result in part (a).**

67. $y = \dfrac{x + 1}{x - 3}$

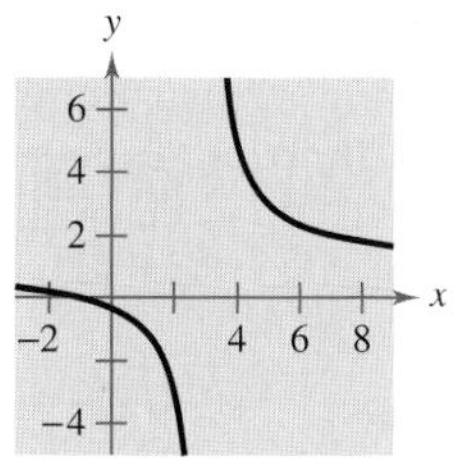

68. $y = \dfrac{2x}{x - 3}$

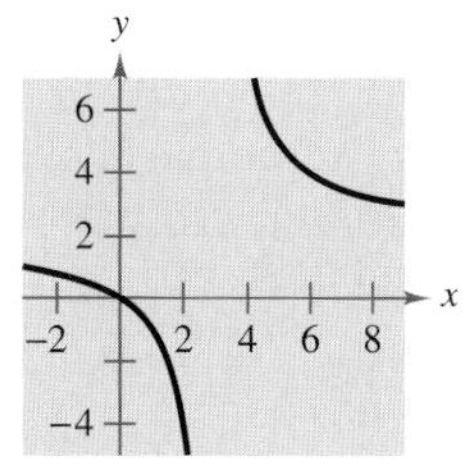

69. $y = \dfrac{1}{x} - x$

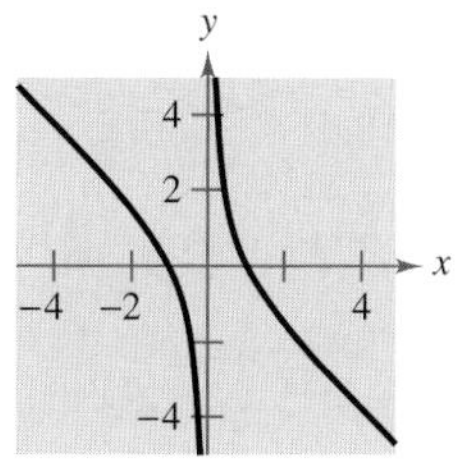

70. $y = x - 3 + \dfrac{2}{x}$

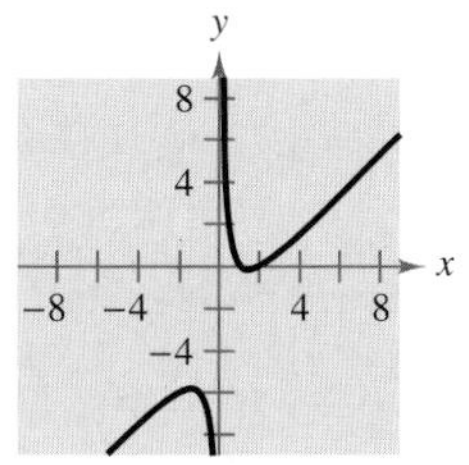

71. *Pollution* The cost (in millions of dollars) of removing $p\%$ of the industrial and municipal pollutants discharged into a river is

$$C = \frac{255p}{100 - p}, \qquad 0 \le p < 100.$$

(a) Find the cost of removing 10% of the pollutants.

(b) Find the cost of removing 40% of the pollutants.

(c) Find the cost of removing 75% of the pollutants.

(d) According to this model, would it be possible to remove 100% of the pollutants? Explain.

72. *Recycling* In a pilot project, a rural township is given recycling bins for separating and storing recyclable products. The cost (in dollars) for supplying bins to $p\%$ of the population is

$$C = \frac{25{,}000p}{100 - p}, \qquad 0 \le p < 100.$$

(a) Find the cost if 15% of the population gets bins.

(b) Find the cost if 50% of the population gets bins.

(c) Find the cost if 90% of the population gets bins.

(d) According to this model, would it be possible to supply bins to 100% of the residents? Explain.

73. *Population Growth* The game commission introduces 100 deer into newly acquired state game lands. The population N of the herd is

$$N = \frac{20(5 + 3t)}{1 + 0.04t}, \qquad 0 \le t$$

where t is the time in years.

(a) Find the population when t is 5, 10, and 25.

(b) What is the limiting size of the herd as time increases?

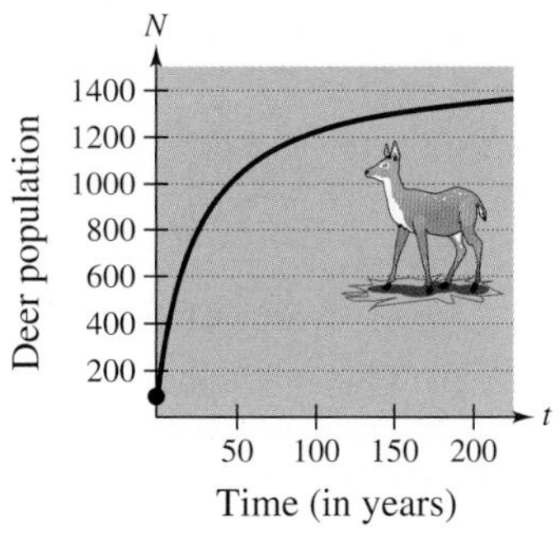

74. *Concentration of a Mixture* A 1000-liter tank contains 50 liters of a 25% brine solution. You add x liters of a 75% brine solution to the tank.

(a) Show that the concentration C, the proportion of brine to total solution, in the final mixture is

$$C = \frac{3x + 50}{4(x + 50)}.$$

(b) Determine the domain of the function based on the physical constraints of the problem.

(c) Graph the concentration function. As the tank is filled, what happens to the rate at which the concentration of brine is increasing? What percent does the concentration of brine appear to approach?

75. *Geometry* A rectangular region of length x and width y has an area of 500 square meters.

(a) Express the width y as a function of x.

(b) Determine the domain of the function based on the physical constraints of the problem.

(c) Sketch a graph of the function and determine the width of the rectangle when $x = 30$ meters.

76. *Minimum Area* A rectangular page is designed to contain 64 square inches of print. The margins at the top and bottom of the page are each 1 inch deep. The margins on each side are $1\frac{1}{2}$ inches wide. What should the dimensions of the page be so that the least amount of paper is used?

77. *Medicine* The concentration of a certain chemical in the bloodstream t hours after injection into muscle tissue is

$$C = \frac{3t^2 + t}{t^3 + 50}, \qquad t > 0.$$

(a) Determine the horizontal asymptote of the function and interpret its meaning in the context of the problem.

(b) Use a graphing utility to graph the function and approximate the time when the bloodstream concentration is greatest.

78. *Average Speed* A driver averaged 50 miles per hour on the round trip between home and a city 100 miles away. The average speeds for going and returning were x and y miles per hour, respectively.

(a) Show that $y = \dfrac{25x}{x - 25}$.

(b) Determine the vertical and horizontal asymptotes of the function.

Synthesis

True or False? **In Exercises 79 and 80, determine whether the statement is true or false. Justify your answer.**

79. A polynomial can have infinitely many vertical asymptotes.

80. A rational function can never cross one of its asymptotes.

Think About It **In Exercises 81–84, write a rational function f that has the specified characteristics.**

81. Vertical asymptotes: $x = -2, x = 1$

82. Vertical asymptote: None

Horizontal asymptote: $y = 0$

83. Vertical asymptote: None

Horizontal asymptote: $y = 2$

84. Vertical asymptotes: $x = 0, x = \frac{5}{2}$

Horizontal asymptote: $y = -3$

85. Give an example of a rational function whose domain is the set of all real numbers. Give an example of a rational function whose domain is the set of all real numbers except $x = 20$.

86. Describe what is meant by an asymptote of a graph.

87. *Think About It* Write a rational function satisfying the following criteria.

Vertical asymptote: $x = 2$

Slant asymptote: $y = x + 1$

Zero of the function: $x = -2$

Think About It **In Exercises 88 and 89, use a graphing utility to obtain the graph of the function. Explain why there is no vertical asymptote when a superficial examination of the function may indicate that there should be one.**

88. $h(x) = \dfrac{6 - 2x}{3 - x}$ **89.** $g(x) = \dfrac{x^2 + x - 2}{x - 1}$

Review

In Exercises 90–93, completely factor the expression.

90. $x^2 - 15x + 56$ **91.** $3x^2 + 23x - 36$

92. $x^3 - 5x^2 + 4x - 20$ **93.** $x^3 + 6x^2 - 2x - 12$

In Exercises 94–97, solve the inequality and show the solution on the real number line.

94. $10 - 3x \le 0$ **95.** $5 - 2x > 5(x + 1)$

96. $|4(x - 2)| < 20$ **97.** $\frac{1}{2}|2x + 3| \ge 5$

2.7 Partial Fractions

▶ **What you should learn**

- How to recognize partial fraction decompositions of rational expressions
- How to find partial fraction decompositions of rational expressions

▶ **Why you should learn it**

Partial fractions can help you analyze the behavior of a rational function. For instance, Exercise 57 on page 278 shows how partial fractions can help you analyze the exhaust temperatures of a diesel engine.

Michael Rosenfeld/Tony Stone Images

Introduction

In this section, you will learn to write a rational expression as the sum of two or more simpler rational expressions. For example, the rational expression

$$\frac{x+7}{x^2-x-6}$$

can be written as the sum of two fractions with first-degree denominators. That is,

$$\frac{x+7}{x^2-x-6} = \underbrace{\overbrace{\underbrace{\frac{2}{x-3}}_{\text{Partial fraction}} + \underbrace{\frac{-1}{x+2}}_{\text{Partial fraction}}}^{\text{Partial fraction decomposition of } \frac{x+7}{x^2-x-6}}}_{}.$$

Each fraction on the right side of the equation is a **partial fraction,** and together they make up the **partial fraction decomposition** of the left side.

Decomposition of $N(x)/D(x)$ into Partial Fractions

1. *Divide if improper:* If $N(x)/D(x)$ is an improper fraction [degree of $N(x) \geq$ degree of $D(x)$], divide the denominator into the numerator to obtain

$$\frac{N(x)}{D(x)} = (\text{polynomial}) + \frac{N_1(x)}{D(x)}$$

and apply Steps 2, 3, and 4 (below) to the proper rational expression $N_1(x)/D(x)$. Note that $N_1(x)$ is the remainder from the division of $N(x)$ by $D(x)$.

2. *Factor the denominator:* Completely factor the denominator into factors of the form

$$(px+q)^m \quad \text{and} \quad (ax^2+bx+c)^n$$

where (ax^2+bx+c) is irreducible.

3. *Linear factors:* For *each* factor of the form $(px+q)^m$, the partial fraction decomposition must include the following sum of m fractions.

$$\frac{A_1}{(px+q)} + \frac{A_2}{(px+q)^2} + \cdots + \frac{A_m}{(px+q)^m}$$

4. *Quadratic factors:* For *each* factor of the form $(ax^2+bx+c)^n$, the partial fraction decomposition must include the following sum of n fractions.

$$\frac{B_1x+C_1}{ax^2+bx+c} + \frac{B_2x+C_2}{(ax^2+bx+c)^2} + \cdots + \frac{B_nx+C_n}{(ax^2+bx+c)^n}$$

Partial Fraction Decomposition

Algebraic techniques for determining the constants in the numerators of partial fractions are demonstrated in the examples that follow. Note that the techniques vary slightly, depending on the type of factors of the denominator: linear or quadratic, distinct or repeated.

Technology

You can use a graphing utility to check *graphically* the decomposition found in Example 1. To do this, graph

$$y_1 = \frac{x+7}{x^2 - x - 6}$$

and

$$y_2 = \frac{2}{x-3} + \frac{-1}{x+2}$$

in the same viewing window. The graphs should be identical, as shown below.

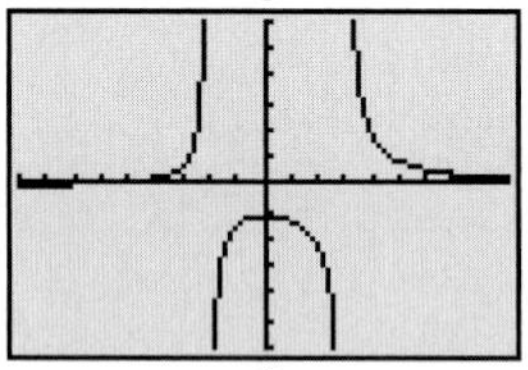

Example 1 ▶ Distinct Linear Factors

Write the partial fraction decomposition for $\dfrac{x+7}{x^2 - x - 6}$.

Solution

The expression is not improper, so factor the denominator. Because $x^2 - x - 6 = (x-3)(x+2)$, you should include one partial fraction with a constant numerator for each linear factor of the denominator. Write the form of the decomposition as follows.

$$\frac{x+7}{x^2 - x - 6} = \frac{A}{x-3} + \frac{B}{x+2}$$ Write form of decomposition.

Multiplying both sides of this equation by the least common denominator, $(x-3)(x+2)$, leads to the **basic equation**

$$x + 7 = A(x+2) + B(x-3).$$ Basic equation

Because this equation is true for all x, you can substitute any *convenient* values of x that will help determine the constants A and B. Values of x that are especially convenient are ones that make the factors $(x+2)$ and $(x-3)$ equal to zero. For instance, let $x = -2$. Then

$$-2 + 7 = A(-2+2) + B(-2-3)$$ Substitute -2 for x.

$$5 = A(0) + B(-5)$$

$$5 = -5B$$

$$-1 = B.$$

To solve for A, let $x = 3$ and obtain

$$3 + 7 = A(3+2) + B(3-3)$$ Substitute 3 for x.

$$10 = A(5) + B(0)$$

$$10 = 5A$$

$$2 = A.$$

So, the decomposition is

$$\frac{x+7}{x^2 - x - 6} = \frac{2}{x-3} + \frac{-1}{x+2}$$

as indicated at the beginning of this section. Check this result by combining the two partial fractions on the right side of the equation, or by using your graphing utility.

The next example shows how to find the partial fraction decomposition for a rational expression whose denominator has a repeated linear factor.

Example 2 ▶ Repeated Linear Factors

Write the partial fraction decomposition for $\dfrac{x^4 + 2x^3 + 6x^2 + 20x + 6}{x^3 + 2x^2 + x}$.

Solution

This rational expression is improper, so you should begin by dividing the numerator by the denominator to obtain

$$x + \frac{5x^2 + 20x + 6}{x^3 + 2x^2 + x}.$$

Have students check some of their answers or the answers to some of the examples by combining the partial fractions.

Because the denominator of the remainder factors as

$$x^3 + 2x^2 + x = x(x^2 + 2x + 1) = x(x + 1)^2$$

you should include one partial fraction with a constant numerator for each power of x and $(x + 1)$ and write the form of the decomposition as follows.

$$\frac{5x^2 + 20x + 6}{x(x + 1)^2} = \frac{A}{x} + \frac{B}{x + 1} + \frac{C}{(x + 1)^2}$$

Multiplying by the LCD, $x(x + 1)^2$, leads to the basic equation

$$5x^2 + 20x + 6 = A(x + 1)^2 + Bx(x + 1) + Cx. \qquad \text{Basic equation}$$

Letting $x = -1$ eliminates the A- and B-terms and yields

$$5(-1)^2 + 20(-1) + 6 = A(-1 + 1)^2 + B(-1)(-1 + 1) + C(-1)$$

$$5 - 20 + 6 = 0 + 0 - C$$

$$C = 9.$$

Letting $x = 0$ eliminates the B- and C-terms and yields

$$5(0)^2 + 20(0) + 6 = A(0 + 1)^2 + B(0)(0 + 1) + C(0)$$

$$6 = A(1) + 0 + 0$$

$$6 = A.$$

At this point, you have exhausted the most convenient choices for x, so to find the value of B, use *any other value* for x along with the known values of A and C. So, using $x = 1$, $A = 6$, and $C = 9$,

$$5(1)^2 + 20(1) + 6 = 6(1 + 1)^2 + B(1)(1 + 1) + 9(1)$$

$$31 = 6(4) + 2B + 9$$

$$-2 = 2B$$

$$-1 = B.$$

Therefore, the partial fraction decomposition is

$$\frac{x^4 + 2x^3 + 6x^2 + 20x + 6}{x^3 + 2x^2 + x} = x + \frac{6}{x} + \frac{-1}{x + 1} + \frac{9}{(x + 1)^2}.$$

The procedure used to solve for the constants in Examples 1 and 2 works well when the factors of the denominator are linear. However, when the denominator contains irreducible quadratic factors, you should use a different procedure, which involves writing the right side of the basic equation in polynomial form and *equating the coefficients* of like terms.

The Granger Collection

Historical Note

John Bernoulli (1667–1748), a Swiss mathematician, introduced the method of partial fractions and was instrumental in the early development of calculus. Bernoulli was a professor at the University of Basel and taught many outstanding students, the most famous of whom was Leonhard Euler.

Example 3 ▶ Distinct Linear and Quadratic Factors

Write the partial fraction decomposition for

$$\frac{3x^2 + 4x + 4}{x^3 + 4x}.$$

Solution

Because the denominator factors as

$$x^3 + 4x = x(x^2 + 4)$$

you should include one partial fraction with a constant numerator and one partial fraction with a linear numerator and write the form of the decomposition as follows.

$$\frac{3x^2 + 4x + 4}{x^3 + 4x} = \frac{A}{x} + \frac{Bx + C}{x^2 + 4}$$

Multiplying by the LCD, $x(x^2 + 4)$, yields the basic equation

$$3x^2 + 4x + 4 = A(x^2 + 4) + (Bx + C)x. \qquad \text{Basic equation}$$

Expanding this basic equation and collecting like terms produces

$$3x^2 + 4x + 4 = Ax^2 + 4A + Bx^2 + Cx$$

$$= (A + B)x^2 + Cx + 4A. \qquad \text{Polynomial form}$$

Finally, because two polynomials are equal if and only if the coefficients of like terms are equal,

$$3x^2 + 4x + 4 = (A + B)x^2 + Cx + 4A \qquad \text{Equate coefficients of like terms.}$$

you obtain the following equations.

$$3 = A + B, \qquad 4 = C, \qquad \text{and} \qquad 4 = 4A$$

So, $A = 1$ and $C = 4$. Moreover, substituting $A = 1$ in the equation $3 = A + B$ yields

$$3 = 1 + B$$

$$2 = B.$$

Therefore, the partial fraction decomposition is

$$\frac{3x^2 + 4x + 4}{x^3 + 4x} = \frac{1}{x} + \frac{2x + 4}{x^2 + 4}.$$

The next example shows how to find the partial fraction decomposition for a rational function whose denominator has a repeated quadratic factor.

Additional Examples

Partial fraction decomposition can be confusing to students. You may want to go over several additional examples.

a. $\frac{2x^2 + 7x + 4}{(x + 1)^3} = \frac{2}{x + 1} + \frac{3}{(x + 1)^2} - \frac{1}{(x + 1)^3}$

b. $\frac{-x^3 + 4x^2 - 2x + 6}{x^2(x^2 + 2)} = -\frac{1}{x} + \frac{3}{x^2} + \frac{1}{x^2 + 2}$

c. $\frac{3x^3 - x^2 + 5x + 1}{(x^2 + 1)^2} = \frac{3x - 1}{x^2 + 1} + \frac{2x + 2}{(x^2 + 1)^2}$

Example 4 ▶ Repeated Quadratic Factors

Write the partial fraction decomposition for

$$\frac{8x^3 + 13x}{(x^2 + 2)^2}.$$

Solution

You need to include one partial fraction with a linear numerator for each power of $(x^2 + 2)$.

$$\frac{8x^3 + 13x}{(x^2 + 2)^2} = \frac{Ax + B}{x^2 + 2} + \frac{Cx + D}{(x^2 + 2)^2}$$

Multiplying by the LCD, $(x^2 + 2)^2$, yields the basic equation

$$8x^3 + 13x = (Ax + B)(x^2 + 2) + Cx + D \qquad \text{Basic equation}$$

$$= Ax^3 + 2Ax + Bx^2 + 2B + Cx + D$$

$$= Ax^3 + Bx^2 + (2A + C)x + (2B + D). \qquad \text{Polynomial form}$$

Equating coefficients of like terms

$$8x^3 + 0x^2 + 13x + 0 = Ax^3 + Bx^2 + (2A + C)x + (2B + D)$$

produces

$$8 = A,\ 0 = B,\ 13 = 2A + C, \text{ and } 0 = 2B + D. \qquad \text{Equate coefficients.}$$

Finally, use the values $A = 8$ and $B = 0$ to obtain the following.

$$13 = 2A + C$$
$$= 2(8) + C$$
$$-3 = C$$

$$0 = 2B + D$$
$$= 2(0) + D$$
$$0 = D$$

Therefore,

$$\frac{8x^3 + 13x}{(x^2 + 2)^2} = \frac{8x}{x^2 + 2} + \frac{-3x}{(x^2 + 2)^2}.$$

Check students' understanding. Why do you use two different procedures? When is each procedure more effective? Is it mathematically correct (although not necessarily as efficient) to use either procedure regardless of whether the factors of the denominator are linear or quadratic?

By equating coefficients of like terms in Examples 3 and 4, you obtained several equations involving A, B, C, and D, which were solved by substitution. In a later chapter you will study a more general method for solving systems of equations.

Guidelines for Solving the Basic Equation

Linear Factors

1. Substitute the *zeros* of the distinct linear factors into the basic equation.
2. For repeated linear factors, use the coefficients determined above to rewrite the basic equation. Then substitute *other* convenient values of x and solve for the remaining coefficients.

Quadratic Factors

1. Expand the basic equation.
2. Collect terms according to powers of x.
3. Equate the coefficients of like terms to obtain equations involving A, B, C, and so on.
4. Use substitution to solve for $A, B, C, \ldots$.

Activities

1. Write the partial fraction decomposition.
$$\frac{5x - 10}{(x + 2)(2x - 1)}$$
Answer: $\dfrac{4}{x + 2} - \dfrac{3}{2x - 1}$

2. Write the partial fraction decomposition.
$$\frac{4x^3 + 9x^2 - 2x + 6}{x(x + 2)}$$
Answer: $4x + 1 + \dfrac{3}{x} - \dfrac{7}{x + 2}$

Keep in mind that for *improper* rational expressions such as

$$\frac{N(x)}{D(x)} = \frac{2x^3 + x^2 - 7x + 7}{x^2 + x - 2}$$

you must first divide before applying partial fraction decomposition.

Writing ABOUT MATHEMATICS

Error Analysis Suppose you are tutoring a student in algebra. In trying to find a partial fraction decomposition, your student writes the following.

$$\frac{x^2 + 1}{x(x - 1)} = \frac{A}{x} + \frac{B}{x - 1}$$

$$\frac{x^2 + 1}{x(x - 1)} = \frac{A(x - 1)}{x(x - 1)} + \frac{Bx}{x(x - 1)}$$

$$x^2 + 1 = A(x - 1) + Bx \qquad \text{Basic equation}$$

By substituting $x = 0$ and $x = 1$ into the basic equation, your student concludes that $A = -1$ and $B = 2$. However, in checking this solution, your student obtains the following. What has gone wrong?

$$\frac{-1}{x} + \frac{2}{x - 1} = \frac{(-1)(x - 1) + 2(x)}{x(x - 1)}$$

$$= \frac{x + 1}{x(x - 1)}$$

$$\neq \frac{x^2 + 1}{x(x - 1)}.$$

2.7 Exercises

In Exercises 1–4, match the rational expression with the form of its decomposition. [The decompositions are labeled (a), (b), (c), and (d).]

(a) $\frac{A}{x} + \frac{B}{x+2} + \frac{C}{x-2}$

(b) $\frac{A}{x} + \frac{B}{x-4}$

(c) $\frac{A}{x} + \frac{B}{x^2} + \frac{C}{x-4}$

(d) $\frac{A}{x} + \frac{Bx+C}{x^2+4}$

1. $\frac{3x-1}{x(x-4)}$

2. $\frac{3x-1}{x^2(x-4)}$

3. $\frac{3x-1}{x(x^2+4)}$

4. $\frac{3x-1}{x(x^2-4)}$

In Exercises 5–14, write the form of the partial fraction decomposition of the rational expression. Do not solve for the constants.

5. $\frac{7}{x^2-14x}$

6. $\frac{x-2}{x^2+4x+3}$

7. $\frac{12}{x^3-10x^2}$

8. $\frac{x^2-3x+2}{4x^3+11x^2}$

9. $\frac{4x^2+3}{(x-5)^3}$

10. $\frac{6x+5}{(x+2)^4}$

11. $\frac{2x-3}{x^3+10x}$

12. $\frac{x-6}{2x^3+8x}$

13. $\frac{x-1}{x(x^2+1)^2}$

14. $\frac{x+4}{x^2(3x-1)^2}$

In Exercises 15–38, write the partial fraction decomposition for the rational expression. Check your result algebraically.

15. $\frac{1}{x^2-1}$

16. $\frac{1}{4x^2-9}$

17. $\frac{1}{x^2+x}$

18. $\frac{3}{x^2-3x}$

19. $\frac{1}{2x^2+x}$

20. $\frac{5}{x^2+x-6}$

21. $\frac{3}{x^2+x-2}$

22. $\frac{x+1}{x^2+4x+3}$

23. $\frac{x^2+12x+12}{x^3-4x}$

24. $\frac{x+2}{x(x-4)}$

25. $\frac{4x^2+2x-1}{x^2(x+1)}$

26. $\frac{2x-3}{(x-1)^2}$

27. $\frac{3x}{(x-3)^2}$

28. $\frac{6x^2+1}{x^2(x-1)^2}$

29. $\frac{x^2-1}{x(x^2+1)}$

30. $\frac{x}{(x-1)(x^2+x+1)}$

31. $\frac{x}{x^3-x^2-2x+2}$

32. $\frac{x+6}{x^3-3x^2-4x+12}$

33. $\frac{x^2}{x^4-2x^2-8}$

34. $\frac{2x^2+x+8}{(x^2+4)^2}$

35. $\frac{x}{16x^4-1}$

36. $\frac{x+1}{x^3+x}$

37. $\frac{x^2+5}{(x+1)(x^2-2x+3)}$

38. $\frac{x^2-4x+7}{(x+1)(x^2-2x+3)}$

In Exercises 39–44, write the partial fraction decomposition for the improper rational expression.

39. $\frac{x^2-x}{x^2+x+1}$

40. $\frac{x^2-4x}{x^2+x+6}$

41. $\frac{2x^3-x^2+x+5}{x^2+3x+2}$

42. $\frac{x^3+2x^2-x+1}{x^2+3x-4}$

43. $\frac{x^4}{(x-1)^3}$

44. $\frac{16x^4}{(2x-1)^3}$

In Exercises 45–52, write the partial fraction decomposition for the rational expression. Use a graphing utility to check your result graphically.

45. $\frac{5-x}{2x^2+x-1}$

46. $\frac{3x^2-7x-2}{x^3-x}$

47. $\frac{x-1}{x^3+x^2}$

48. $\frac{4x^2-1}{2x(x+1)^2}$

49. $\frac{x^2+x+2}{(x^2+2)^2}$

50. $\frac{x^3}{(x+2)^2(x-2)^2}$

51. $\frac{2x^3-4x^2-15x+5}{x^2-2x-8}$

52. $\frac{x^3-x+3}{x^2+x-2}$

Graphical Analysis **In Exercises 53–56, write the partial fraction decomposition for the rational function. Identify the graph of the rational function and the graph of each term of its decomposition. State any relationship between the vertical asymptotes of the rational function and the vertical asymptotes of the terms of the decomposition.**

53. $y = \dfrac{x - 12}{x(x - 4)}$

54. $y = \dfrac{2(x + 1)^2}{x(x^2 + 1)}$

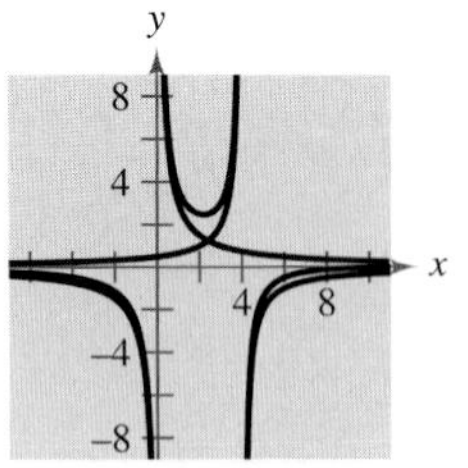

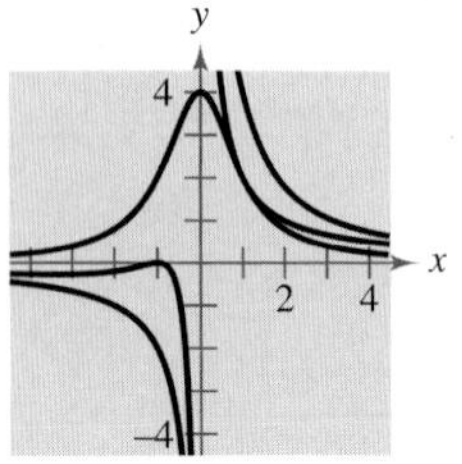

55. $y = \dfrac{2(4x - 3)}{x^2 - 9}$

56. $y = \dfrac{2(4x^2 - 15x + 39)}{x^2(x^2 - 10x + 26)}$

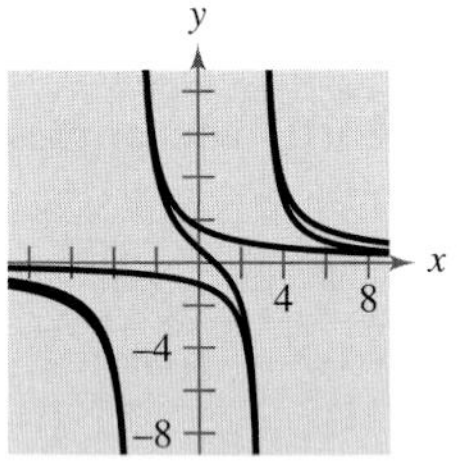

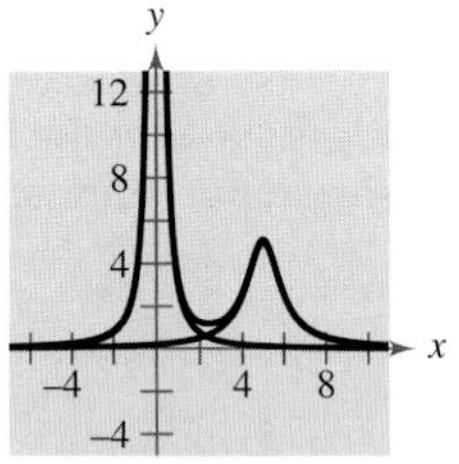

57. *Thermodynamics* The magnitude of the range of exhaust temperatures in degrees Fahrenheit in an experimental diesel engine is approximated by

$$R = \frac{2000(4 - 3x)}{(11 - 7x)(7 - 4x)}, \qquad 0 < x \le 1$$

where x is the relative load.

(a) Write the partial fraction decomposition of the expression.

(b) The decomposition in part (a) is the difference of two fractions. The absolute values of the terms give the expected maximum and minimum temperatures of the exhaust gases for different loads.

Ymax = |1st term|

Ymin = |2nd term|

Write the equations for Ymax and Ymin. Then use a graphing utility to graph each equation in the same viewing window.

Synthesis

True or False? **In Exercises 58 and 59, determine whether the statement is true or false. Justify your answer.**

58. For the rational expression

$$\frac{x}{(x + 10)(x - 10)^2}$$

the partial fraction decomposition is of the form

$$\frac{A}{x + 10} + \frac{B}{(x - 10)^2}.$$

59. When writing the partial fraction decomposition for the expression

$$\frac{x^3 + x - 2}{x^2 - 5x + 14}$$

the first step is to factor the denominator.

In Exercises 60–63, write the partial fraction decomposition for the rational expression. Check your result algebraically. Then assign a value to the constant *a* to check the result graphically.

60. $\dfrac{1}{a^2 - x^2}$

61. $\dfrac{1}{x(x + a)}$

62. $\dfrac{1}{y(a - y)}$

63. $\dfrac{1}{(x + 1)(a - x)}$

Review

In Exercises 64–71, sketch the graph of the function.

64. $f(x) = -3x + 7$

65. $f(x) = 6 - x$

66. $f(x) = -2x^2$

67. $f(x) = \frac{1}{4}x^2 + 1$

68. $f(x) = x^2 - 9x + 18$

69. $f(x) = 2x^2 - 9x - 5$

70. $f(x) = -x^2(x - 3)$

71. $f(x) = \frac{1}{2}x^3 - 1$

In Exercises 72–77, sketch the graph of the rational function.

72. $f(x) = \dfrac{6}{x - 1}$

73. $f(x) = \dfrac{1 - 4x}{x}$

74. $f(x) = \dfrac{x^2 + x - 6}{x + 5}$

75. $f(x) = \dfrac{3x - 1}{x^2 + 4x - 12}$

76. $f(x) = \dfrac{3(x^2 - 1)}{x^2 + 4x - 5}$

77. $f(x) = \dfrac{2x - 3}{x^2 - 16}$

Chapter Summary

What did you learn?

Section 2.1	Review Exercises
☐ How to analyze graphs of quadratic functions	1–6
☐ How to write quadratic functions in standard form and sketch their graphs	7–18
☐ How to use quadratic functions to model and solve real-life problems	19–22
Section 2.2	
☐ How to use transformations to sketch graphs of polynomial functions	23–28
☐ How to use the Leading Coefficient Test to determine the end behavior of graphs of polynomial functions	29–32
☐ How to use zeros of polynomial functions as sketching aids	33–38
☐ How to use the Intermediate Value Theorem to help locate zeros of polynomial functions	39–42
Section 2.3	
☐ How to use long division to divide polynomials by other polynomials	43–48
☐ How to use synthetic division to divide polynomials by binomials	49–56
☐ How to use the Remainder Theorem and the Factor Theorem	57–60
☐ How to use polynomial division to answer questions about real-life problems	61
Section 2.4	
☐ How to use the imaginary unit i to write complex numbers	62–65
☐ How to add, subtract, multiply, and divide complex numbers	66–75
☐ How to use the Quadratic Formula to find complex solutions of quadratic equations	76–79
Section 2.5	
☐ How to use the Fundamental Theorem of Algebra to determine the numbers of zeros of polynomial functions	80–85
☐ How to find rational zeros of polynomial functions	86–93
☐ How to find conjugate pairs of complex zeros	94, 95
☐ How to find zeros of polynomials by factoring and using Descartes's Rules of Signs	96–101
Section 2.6	
☐ How to find the domains of rational functions	102–105
☐ How to find the horizontal and vertical asymptotes of graphs of rational functions	106–109
☐ How to analyze and sketch graphs of rational functions	110–121
☐ How to sketch graphs of rational functions that have slant asymptotes	122–125
☐ How to use rational functions to model and solve real-life problems	126–129
Section 2.7	
☐ How to recognize partial fraction decompositions of rational expressions	130–133
☐ How to find partial fraction decompositions of rational expressions	134–141

Review Exercises

2.1 *Graphical Reasoning* **In Exercises 1 and 2, use a graphing utility to graph each equation in the same viewing window. Describe how each graph differs from the graph of $y = x^2$.**

1. (a) $y = 2x^2$ (b) $y = -2x^2$
(c) $y = x^2 + 2$ (d) $y = (x + 2)^2$

2. (a) $y = x^2 - 4$ (b) $y = 4 - x^2$
(c) $y = (x - 3)^2$ (d) $y = \frac{1}{2}x^2 - 1$

In Exercises 3–6, find the quadratic function that has the indicated vertex and whose graph passes through the given point.

3.

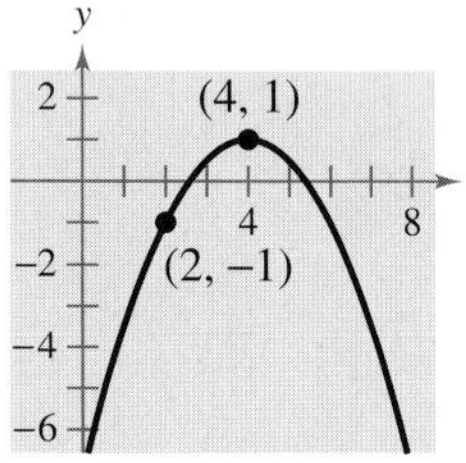

4.

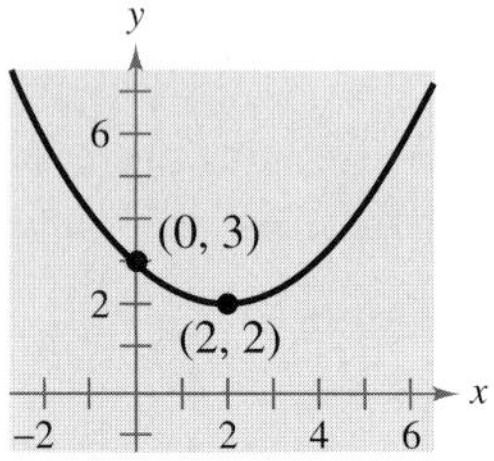

5. Vertex: $(1, -4)$; Point: $(2, -3)$

6. Vertex: $(2, 3)$; Point: $(-1, 6)$

In Exercises 7–18, write the quadratic function in standard form and sketch its graph.

7. $g(x) = x^2 - 2x$

8. $f(x) = 6x - x^2$

9. $f(x) = x^2 + 8x + 10$

10. $h(x) = 3 + 4x - x^2$

11. $f(t) = -2t^2 + 4t + 1$

12. $f(x) = x^2 - 8x + 12$

13. $h(x) = 4x^2 + 4x + 13$

14. $f(x) = x^2 - 6x + 1$

15. $h(x) = x^2 + 5x - 4$

16. $f(x) = 4x^2 + 4x + 5$

17. $f(x) = \frac{1}{3}(x^2 + 5x - 4)$

18. $f(x) = \frac{1}{2}(6x^2 - 24x + 22)$

19. *Geometry* The perimeter of a rectangle is 200 meters.

(a) Draw a rectangle that gives a visual representation of the problem. Label the length and width in terms of x and y, respectively.

(b) Write y as a function of x. Use the result to write the area as a function of x.

(c) Of all possible rectangles with perimeters of 200 meters, find the dimensions of the one with the maximum area.

20. *Maximum Profit* A real estate office handles 50 apartment units. When the rent is \$540 per month, all units are occupied. However, for each \$30 increase in rent, one unit becomes vacant. Each occupied unit requires an average of \$18 per month for service and repairs. What rent should be charged to obtain the maximum profit?

21. *Minimum Cost* A manufacturer has daily production costs of

$$C = 20{,}000 - 120x + 0.055x^2$$

where C is the total cost (in dollars) and x is the number of units produced. How many units should be produced each day to yield a minimum cost?

22. *Sociology* The average age of the groom at a wedding for a given age of the bride can be approximated by the model $y = -0.00428x^2 + 1.442x - 3.136$, $20 \le x \le 55$, where y is the age of the groom and x is the age of the bride. For what age of the bride is the average age of the groom 30? (Source: U.S. National Center for Health Statistics)

2.2 **In Exercises 23–28, sketch the graphs of $y = x^n$ and the transformation.**

23. $y = x^3$, $f(x) = -(x - 4)^3$

24. $y = x^3$, $f(x) = -4x^3$

25. $y = x^4$, $f(x) = 2 - x^4$

26. $y = x^4$, $f(x) = 2(x - 2)^4$

27. $y = x^5$, $f(x) = (x - 3)^5$

28. $y = x^5$, $f(x) = \frac{1}{2}x^5 + 3$

In Exercises 29–32, determine the right-hand and left-hand behavior of the graph of the polynomial function.

29. $f(x) = -x^2 + 6x + 9$

30. $f(x) = \frac{1}{2}x^3 + 2x$

31. $g(x) = \frac{3}{4}(x^4 + 3x^2 + 2)$

32. $h(x) = -x^5 - 7x^2 + 10x$

In Exercises 33–38, find the zeros of the function and sketch its graph.

33. $f(x) = 2x^2 + 11x - 21$

34. $f(x) = x(x + 3)^2$

35. $f(t) = t^3 - 3t$

36. $f(x) = x^3 - 8x^2$

37. $f(x) = -12x^3 + 20x^2$

38. $g(x) = x^4 - x^3 - 2x^2$

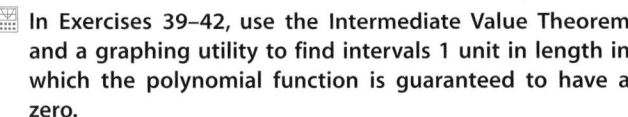

In Exercises 39–42, use the Intermediate Value Theorem and a graphing utility to find intervals 1 unit in length in which the polynomial function is guaranteed to have a zero.

39. $f(x) = 3x^3 - x^2 + 3$

40. $f(x) = 0.25x^3 - 3.65x + 6.12$

41. $f(x) = x^4 - 5x - 1$

42. $f(x) = 7x^4 + 3x^3 - 8x^2 + 2$

2.3 In Exercises 43–48, use long division to divide.

43. $\dfrac{24x^2 - x - 8}{3x - 2}$

44. $\dfrac{4x + 7}{3x - 2}$

45. $\dfrac{5x^3 - 13x^2 - x + 2}{x^2 - 3x + 1}$

46. $\dfrac{3x^4}{x^2 - 1}$

47. $\dfrac{x^4 - 3x^3 + 4x^2 - 6x + 3}{x^2 + 2}$

48. $\dfrac{6x^4 + 10x^3 + 13x^2 - 5x + 2}{2x^2 - 1}$

In Exercises 49–52, use synthetic division to divide.

49. $\dfrac{6x^4 - 4x^3 - 27x^2 + 18x}{x - \frac{2}{3}}$

50. $\dfrac{0.1x^3 + 0.3x^2 - 0.5}{x - 5}$

51. $\dfrac{2x^3 - 19x^2 + 38x + 24}{x - 4}$

52. $\dfrac{3x^3 + 20x^2 + 29x - 12}{x + 3}$

In Exercises 53 and 54, use synthetic division to determine whether the given values of *x* are zeros of the function.

53. $f(x) = 20x^4 + 9x^3 - 14x^2 - 3x$

(a) $x = -1$ (b) $x = \frac{3}{4}$

(c) $x = 0$ (d) $x = 1$

54. $f(x) = 3x^3 - 8x^2 - 20x + 16$

(a) $x = 4$ (b) $x = -4$

(c) $x = \frac{2}{3}$ (d) $x = -1$

In Exercises 55 and 56, use synthetic division to find the specified value of the function.

55. $f(x) = x^4 + 10x^3 - 24x^2 + 20x + 44$

(a) $f(-3)$ (b) $f(-1)$

56. $g(t) = 2t^5 - 5t^4 - 8t + 20$

(a) $g(-4)$ (b) $g(\sqrt{2})$

In Exercises 57–60, (a) verify the given factor(s) of the function *f*, (b) find the remaining factors of *f*, (c) use your results to write the complete factorization of *f*, (d) list all real zeros of *f*, and (e) confirm your results by using a graphing utility to graph the function.

	Function	*Factors*
57.	$f(x) = x^3 + 4x^2 - 25x - 28$	$(x - 4)$
58.	$f(x) = 2x^3 + 11x^2 - 21x - 90$	$(x + 6)$
59.	$f(x) = x^4 - 4x^3 - 7x^2 + 22x + 24$	$(x + 2)(x - 3)$
60.	$f(x) = x^4 - 11x^3 + 41x^2 - 61x + 30$	$(x - 2)(x - 5)$

61. *Data Analysis* The values V (in billions of dollars) of farm real estate in the United States for the years 1990 through 1997 are shown in the table. The variable t represents the year, with $t = 0$ corresponding to 1990. (Source: U.S. Department of Agriculture)

t	0	1	2	3
V	671.4	688	695.5	717.1

t	4	5	6	7
V	759.2	807	860.9	912.3

(a) Use a graphing utility to sketch a scatter plot of the data.

(b) Use the regression feature of a graphing utility to find a cubic model for the given data. Then graph the model in the same viewing window as the scatter plot. Compare the model with the data.

(c) Use the model to create a table of estimated values of V. Compare the estimated values with the actual data.

(d) Use synthetic division to evaluate the model for the year 2001. Do you think the model is accurate in predicting the future value of farm real estate? Explain.

2.4 In Exercises 62–65, write the complex number in standard form.

62. $6 + \sqrt{-4}$

63. $3 - \sqrt{-25}$

64. $i^2 + 3i$

65. $-5i + i^2$

In Exercises 66–75, perform the operations and write the result in standard form.

66. $(7 + 5i) + (-4 + 2i)$

67. $\left(\dfrac{\sqrt{2}}{2} - \dfrac{\sqrt{2}}{2}i\right) - \left(\dfrac{\sqrt{2}}{2} + \dfrac{\sqrt{2}}{2}i\right)$

68. $5i(13 - 8i)$

69. $(1 + 6i)(5 - 2i)$

70. $(10 - 8i)(2 - 3i)$

71. $i(6 + i)(3 - 2i)$

72. $\dfrac{6 + i}{4 - i}$

73. $\dfrac{3 + 2i}{5 + i}$

74. $\dfrac{4}{2 - 3i} + \dfrac{2}{1 + i}$

75. $\dfrac{1}{2 + i} - \dfrac{5}{1 + 4i}$

In Exercises 76–79, find all solutions of the equation.

76. $3x^2 + 1 = 0$

77. $2 + 8x^2 = 0$

78. $x^2 - 2x + 10 = 0$

79. $6x^2 + 3x + 27 = 0$

2.5 **In Exercises 80–85, use the Fundamental Theorem of Algebra to determine the number of zeros of the function. Find the zeros.**

80. $f(x) = 3x(x - 2)^2$

81. $f(x) = (x - 4)(x + 9)^2$

82. $f(x) = x^2 - 9x + 8$

83. $f(x) = x^3 + 6x$

84. $f(x) = (x + 4)(x - 6)(x - 2i)(x + 2i)$

85. $f(x) = (x - 8)(x - 5)^2(x - 3 + i)(x - 3 - i)$

In Exercises 86 and 87, use the Rational Zero Test to list all possible rational zeros of *f*.

86. $f(x) = -4x^3 + 8x^2 - 3x + 15$

87. $f(x) = 3x^4 + 4x^3 - 5x^2 - 8$

In Exercises 88–93, find all the real zeros of the function.

88. $f(x) = x^3 - 2x^2 - 21x - 18$

89. $f(x) = 3x^3 - 20x^2 + 7x + 30$

90. $f(x) = x^3 - 10x^2 + 17x - 8$

91. $f(x) = x^3 + 9x^2 + 24x + 20$

92. $f(x) = x^4 + x^3 - 11x^2 + x - 12$

93. $f(x) = 25x^4 + 25x^3 - 154x^2 - 4x + 24$

In Exercises 94 and 95, find a polynomial with real coefficients that has the given zeros.

94. $\frac{2}{3}, 4, \sqrt{3}i$

95. $2, -3, 1 - 2i$

In Exercises 96 and 97, write the polynomial (a) as the product of factors that are irreducible over the *rationals*, (b) as the product of linear and quadratic factors that are irreducible over the *reals*, and (c) in completely factored form.

96. $f(x) = x^4 - 2x^3 + 4x^2 + 2x - 5$

(*Hint:* One factor is $x^2 - 1$.)

97. $f(x) = x^4 - 2x^3 - 2x^2 - 2x - 3$

(*Hint:* One factor is $x^2 + 1$.)

In Exercises 98 and 99, use the given zero to find all the zeros of the function.

Function	*Zero*
98. $f(x) = x^3 - 12x^2 + x - 12$	i
99. $f(x) = x^3 - 2x^2 - 14x + 40$	$3 - i$

In Exercises 100 and 101, use Descartes's Rule of Signs to determine the possible numbers of positive and negative zeros of the function.

100. $g(x) = 5x^3 + 3x^2 - 6x + 9$

101. $h(x) = -2x^5 + 4x^3 - 2x^2 + 5$

2.6 **In Exercises 102–105, find the domain of the rational function.**

102. $f(x) = \dfrac{5x}{x + 12}$

103. $f(x) = \dfrac{3x^2}{1 + 3x}$

104. $f(x) = \dfrac{8}{x^2 - 10x + 24}$

105. $f(x) = \dfrac{x^2 + x - 2}{x^2 + 4}$

In Exercises 106–109, identify any horizontal or vertical asymptotes.

106. $f(x) = \dfrac{4}{x + 3}$

107. $f(x) = \dfrac{2x^2 + 5x - 3}{x^2 + 2}$

108. $g(x) = \dfrac{x^2}{x^2 - 4}$

109. $g(x) = \dfrac{1}{(x - 3)^2}$

In Exercises 110–121, identify intercepts, check for symmetry, identify any vertical or horizontal asymptotes, and sketch the graph of the rational function.

110. $f(x) = \dfrac{-5}{x^2}$

111. $f(x) = \dfrac{4}{x}$

112. $g(x) = \dfrac{2 + x}{1 - x}$

113. $h(x) = \dfrac{x - 3}{x - 2}$

114. $p(x) = \dfrac{x^2}{x^2 + 1}$

115. $f(x) = \dfrac{2x}{x^2 + 4}$

116. $f(x) = \dfrac{x}{x^2 + 1}$

117. $h(x) = \dfrac{4}{(x - 1)^2}$

118. $f(x) = \dfrac{-6x^2}{x^2 + 1}$

119. $y = \dfrac{2x^2}{x^2 - 4}$

120. $y = \dfrac{x}{x^2 - 1}$

121. $g(x) = \dfrac{-2}{(x + 3)^2}$

In Exercises 122–125, find the equation of the slant asymptote and sketch the graph of the rational function.

122. $f(x) = \dfrac{2x^3}{x^2 + 1}$

123. $f(x) = \dfrac{x^2 + 1}{x + 1}$

124. $f(x) = \dfrac{x^2 + 3x - 10}{x + 2}$

125. $f(x) = \dfrac{x^3}{x^2 - 4}$

126. *Average Cost* A business has a cost of $C = 0.5x + 500$ for producing x units. The average cost per unit is

$$\overline{C} = \frac{C}{x} = \frac{0.5x + 500}{x}, \qquad x > 0.$$

Determine the average cost per unit as x increases without bound. (Find the horizontal asymptote.)

127. *Seizure of Illegal Drugs* The cost (in millions of dollars) for the federal government to seize $p\%$ of a certain illegal drug as it enters the country is

$$C = \frac{528p}{100 - p}, \qquad 0 \le p < 100.$$

(a) Find the cost of seizing 25% of the drug.

(b) Find the cost of seizing 50% of the drug.

(c) Find the cost of seizing 75% of the drug.

(d) According to this model, would it be possible to seize 100% of the drug?

128. *Physics* The rise of distilled water in tubes of diameter x inches is approximated by the model

$$y = \left(\frac{0.80 - 0.54x}{1 + 2.72x}\right)^2, \qquad x > 0$$

where y is measured in inches. Approximate the diameter of the tube that will cause the water to rise 0.1 inch.

129. *Numerical and Graphical Analysis* A right triangle is formed in the first quadrant by the x- and y-axes and a line through the point $(2, 3)$.

(a) Draw a diagram that illustrates the problem. Label the known and unknown quantities.

(b) Verify that the area of the triangle is

$$A = \frac{3x^2}{2(x - 2)}, \qquad x > 2.$$

(c) Create a table that gives values of area for various values of x. Start the table with $x = 2.5$ and increment x in steps of 0.5. Continue until you can approximate the dimensions of the triangle of minimum area.

(d) Use a graphing utility to graph the area function. Use the graph to approximate the dimensions of the triangle of minimum area.

(e) Determine the slant asymptote of the area function. Explain its meaning.

2.7 **In Exercises 130–133, write the form of the partial fraction decomposition for the rational expression. Do not solve for the constants.**

130. $\dfrac{3}{x^2 + 20x}$

131. $\dfrac{x - 8}{x^2 - 3x - 28}$

132. $\dfrac{3x - 4}{x^3 - 5x^2}$

133. $\dfrac{x - 2}{x(x^2 + 2)^2}$

In Exercises 134–141, write the partial fraction decomposition for the rational expression.

134. $\dfrac{4 - x}{x^2 + 6x + 8}$

135. $\dfrac{-x}{x^2 + 3x + 2}$

136. $\dfrac{x^2}{x^2 + 2x - 15}$

137. $\dfrac{9}{x^2 - 9}$

138. $\dfrac{x^2 + 2x}{x^3 - x^2 + x - 1}$

139. $\dfrac{4x - 2}{3(x - 1)^2}$

140. $\dfrac{3x^3 + 4x}{(x^2 + 1)^2}$

141. $\dfrac{4x^2}{(x - 1)(x^2 + 1)}$

Synthesis

True or False? **In Exercises 142 and 143, determine whether the statement is true or false. Justify your answer.**

142. A fourth-degree polynomial can have -5, $-8i$, $4i$, and 5 as its zeros.

143. The domain of a rational function can never be the set of all real numbers.

144. Write quadratic equations that have (a) two distinct real solutions, (b) two complex solutions, and (c) no real solution.

145. Given the function $f(x) = a(x - h)^2 + k$, state the values of a, h, and k that give a reflection in the x-axis with either a shrink or stretch of the graph of the function $f(x) = x^2$.

146. What is the degree of a function that has exactly two real zeros and two complex zeros?

147. Because $i^2 = -1$, is the square of any complex number a real number? Explain.

Chapter Project ▶ Finding Points of Intersection Graphically

Example ▶ Approximating Points of Intersection

Approximate the points of intersection of the circle and parabola given by

$$x^2 + y^2 - 3x + 5y - 11 = 0 \quad \text{and} \quad y = x^2 - 4x + 5$$

using the zoom and trace features of a graphing utility.

Solution

Begin by writing the circle as the union of two functions.

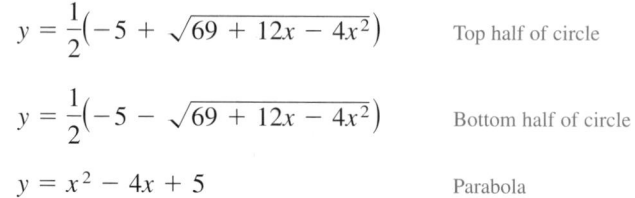

$$y = \frac{1}{2}\left(-5 + \sqrt{69 + 12x - 4x^2}\right) \qquad \text{Top half of circle}$$

$$y = \frac{1}{2}\left(-5 - \sqrt{69 + 12x - 4x^2}\right) \qquad \text{Bottom half of circle}$$

$$y = x^2 - 4x + 5 \qquad \text{Parabola}$$

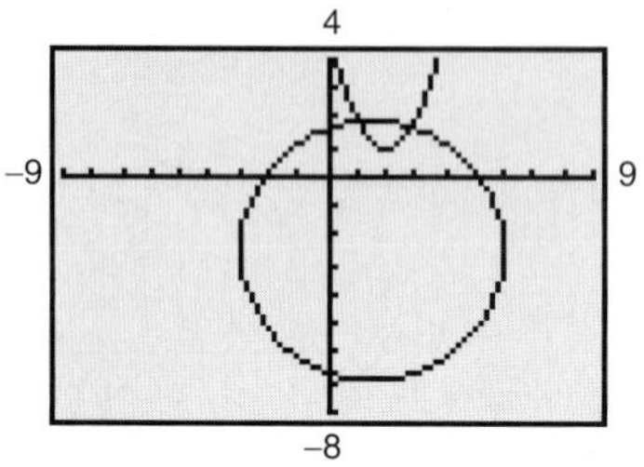

FIGURE 2.42

Next, graph all three functions in the same viewing window, as shown in Figure 2.42. Use the *zoom* and *trace* features of the graphing utility to estimate that the points of intersection are roughly (1, 1.9) and (2.8, 1.7).

Chapter Project Investigations

1. Using a setting of $1.05 \le x \le 1.06$ and $1.89 \le y \le 1.90$, graph the top half of the circle and the parabola in the example on the same screen. Then use the trace feature to approximate (accurate to three decimal places) the y-coordinate of the point of intersection that is shown on the screen.
2. Another method for finding the points of intersection is to substitute $x^2 - 4x + 5$ for y in the equation of the circle to get a fourth-degree polynomial equation. Graph this polynomial function.
   (a) Find a setting that allows you to approximate the solution $x \approx 1.055$ of the polynomial equation to two more decimal places.
   (b) Find a setting that allows you to approximate the solution $x \approx 2.841$ to two more decimal places.
3. Use a graphing utility to find the points of intersection of the circle and the parabola given by $x^2 + y^2 - 5x + 4y - 13 = 0$ and $y = x^2 - 3x + 2$.
4. The *market equilibrium* of a commodity is the quantity (and corresponding price) at which the supply of the commodity and the demand for the commodity are equal. The supply and demand curves for a business dealing with wheat are $p = 1.45 + 0.00014x^2$ and $p = (2.388 - 0.007x)^2$, respectively, where p is the price (in dollars per bushel) and x is the quantity (in bushels per day). Use a graphing utility to graph the supply and demand equations and find the market equilibrium. (*Hint:* The market equilibrium is the point of intersection of the graphs for $x > 0$.)

Chapter Test

Take this test as you would take a test in class. After you are done, check your work against the answers given in the back of the book.

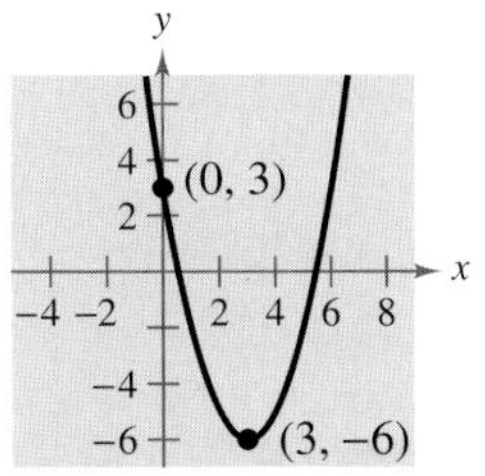

FIGURE FOR 2

The *Interactive* CD-ROM and *Internet* versions of this text provide answers to the Chapter Tests and Cumulative Tests. They also offer Chapter Pre-Tests (which test key skills and concepts covered in previous chapters) and Chapter Post-Tests, both of which have randomly generated exercises with diagnostic capabilities.

1. Describe how the graph of g differs from the graph of $f(x) = x^2$.
(a) $g(x) = 2 - x^2$ (b) $g(x) = \left(x - \frac{3}{2}\right)^2$

2. Write the equation in standard form of the parabola shown in the figure.

3. The path of a ball is given by $y = -\frac{1}{20}x^2 + 3x + 5$, where y is the height (in feet) of the ball and x is the horizontal distance (in feet) from where the ball was thrown.
(a) Find the maximum height of the ball.
(b) Find the distance the ball travels.

4. Determine the right-hand and left-hand behavior of the graph of the function $h(t) = -\frac{3}{4}t^5 + 2t^2$. Then sketch its graph.

5. Divide by long division.
$$\frac{3x^3 + 4x - 1}{x^2 + 1}$$

6. Divide by synthetic division.
$$\frac{2x^4 - 5x^2 - 3}{x - 2}$$

7. Use synthetic division to show that $x = \sqrt{3}$ is a solution of the equation $4x^3 - x^2 - 12x + 3 = 0$. Use the result to factor the polynomial completely and list all the real solutions of the equation.

8. Perform the operations and write the result in standard form.
(a) $10i - \left(3 + \sqrt{-25}\right)$ (b) $\left(2 + \sqrt{3}i\right)\left(2 - \sqrt{3}i\right)$ (c) $\dfrac{5}{2 + i}$

In Exercises 9 and 10, list all the possible rational zeros of the function. Use a graphing utility to graph the function and find all the rational zeros.

9. $g(t) = 2t^4 - 3t^3 + 16t - 24$ **10.** $h(x) = 3x^5 + 2x^4 - 3x - 2$

11. Find all zeros of $f(x) = x^4 - x^3 + 2x^2 - 4x - 8$ given that $f(2i) = 0$.

In Exercises 12 and 13, find a polynomial function with integer coefficients that has the given zeros.

12. $0, 3, 3 + i, 3 - i$ **13.** $1 + \sqrt{3}i, 1 - \sqrt{3}i, 2, 2$

In Exercises 14–16, find the domain of the function and identify any asymptotes.

14. $y = \dfrac{2}{4 - x}$ **15.** $f(x) = \dfrac{3 - x^2}{3 + x^2}$ **16.** $g(x) = \dfrac{x^2 + 2x - 3}{x - 2}$

In Exercises 17 and 18, graph the function. Identify any intercepts and asymptotes.

17. $h(x) = \dfrac{4}{x^2} - 1$ **18.** $g(x) = \dfrac{x^2 + 2}{x - 1}$

In Exercises 19–22, write the partial fraction decomposition for the rational expression.

19. $\dfrac{2x + 5}{x^2 - x - 2}$ **20.** $\dfrac{3x^2 - 2x + 4}{x^2(2 - x)}$ **21.** $\dfrac{x^2 + 5}{x^3 - x}$ **22.** $\dfrac{x^2 - 4}{x^3 + 2x}$

3.1 ▶ Exponential Functions and Their Graphs
3.2 ▶ Logarithmic Functions and Their Graphs
3.3 ▶ Properties of Logarithms
3.4 ▶ Exponential and Logarithmic Equations
3.5 ▶ Exponential and Logarithmic Models

Bill Aron/PhotoEdit

In the United States in 1997, personal savings as a percent of disposable income was 3.9% and the disposable per capita income was $21,969. So, the per capita personal savings in 1997 was $856.79. *(Source: U.S. Bureau of Economic Analysis)*

3 Exponential and Logarithmic Functions

▶ How to Study This Chapter

The Big Picture

In this chapter you will learn the following skills and concepts.

- How to recognize and evaluate exponential and logarithmic functions
- How to graph exponential and logarithmic functions
- How to rewrite logarithmic functions with a different base
- How to use properties of logarithms to evaluate, rewrite, expand, or condense logarithmic expressions
- How to solve exponential and logarithmic equations
- How to use exponential growth models, exponential decay models, Gaussian models, logistic growth models, and logarithmic models to solve real-life problems

Important Vocabulary

As you encounter each new vocabulary term in this chapter, add the term and its definition to your notebook glossary.

Algebraic functions (p. 288)
Transcendental functions (p. 288)
Exponential function f with base a (p. 288)
Natural base e (p. 292)
Natural exponential function (p. 292)
Continuous compounding (p. 293)
Logarithmic function with base a (p. 300)
Common logarithmic function (p. 301)
Natural logarithmic function (p. 304)
Exponential growth model (p. 328)
Exponential decay model (p. 328)
Gaussian model (p. 328)
Logistic growth model (p. 328)
Logarithmic models (p. 328)
Bell-shaped curve (p. 332)
Logistic curve (p. 333)
Sigmoidal curve (p. 333)

Study Tools

- Learning objectives at the beginning of each section
- Chapter Summary (p. 341)
- Review Exercises (pp. 342–345)
- Chapter Test (p. 347)
- Cumulative Test for Chapters 1–3 (pp. 348–349)

Additional Resources

- Study and Solutions Guide
- Interactive Precalculus
- Videotapes for Chapter 3
- Precalculus Website
- Student Success Organizer

STUDY T!P

Attend every class. Arrive on time with your text, a pen or pencil and paper for notes, and your calculator. If you must miss class, get the notes from another student, go to your tutor for help, or view the appropriate mathematics videotape.

3.1 Exponential Functions and Their Graphs

▶ **What you should learn**

- How to recognize and evaluate exponential functions with base *a*
- How to graph exponential functions
- How to recognize and evaluate exponential functions with base *e*
- How to use exponential functions to model and solve real-life applications

▶ **Why you should learn it**

Exponential functions can be used to model and solve real-life problems. For instance, Exercise 74 on page 298 shows how to use an exponential function to model the amount of defoliation caused by the gypsy moth.

Jenny Hager/The Image Works

Exponential Functions

So far, this book has dealt only with **algebraic functions,** which include polynomial functions and rational functions. In this chapter you will study two types of nonalgebraic functions—*exponential* functions and *logarithmic* functions. These functions are examples of **transcendental functions.**

Definition of Exponential Function

The **exponential function f with base a** is denoted by

$$f(x) = a^x$$

where $a > 0$, $a \neq 1$, and x is any real number.

The base $a = 1$ is excluded because it yields $f(x) = 1^x = 1$. This is a constant function, not an exponential function.

You already know how to evaluate a^x for integer and rational values of x. For example, you know that $4^3 = 64$ and $4^{1/2} = 2$. However, to evaluate 4^x for any real number x, you need to interpret forms with *irrational* exponents. For the purposes of this book, it is sufficient to think of

$$a^{\sqrt{2}} \quad (\text{where } \sqrt{2} \approx 1.41421356)$$

as the number that has the successively closer approximations

$$a^{1.4}, a^{1.41}, a^{1.414}, a^{1.4142}, a^{1.41421}, \ldots .$$

Example 1 shows how to use a calculator to evaluate exponential expressions.

Example 1 ▶ Evaluating Exponential Expressions

Use a calculator to evaluate each expression.

a. $2^{-3.1}$

b. $2^{-\pi}$

c. $12^{5/7}$

d. $(0.6)^{3/2}$

Solution

	Number	*Graphing Calculator Keystrokes*	*Display*
a.	$2^{-3.1}$	2 [^] [(−)] 3.1 [ENTER]	0.1166291
b.	$2^{-\pi}$	2 [^] [(−)] π [ENTER]	0.1133147
c.	$12^{5/7}$	12 [^] [(] 5 [÷] 7 [)] [ENTER]	5.8998877
d.	$(0.6)^{3/2}$	.6 [^] [(] 3 [÷] 2 [)] [ENTER]	0.4647580

Graphs of Exponential Functions

The graphs of all exponential functions have similar characteristics, as shown in Examples 2, 3, and 4.

Example 2 ▶ Graphs of $y = a^x$

In the same coordinate plane, sketch the graph of each function.

a. $f(x) = 2^x$ **b.** $g(x) = 4^x$

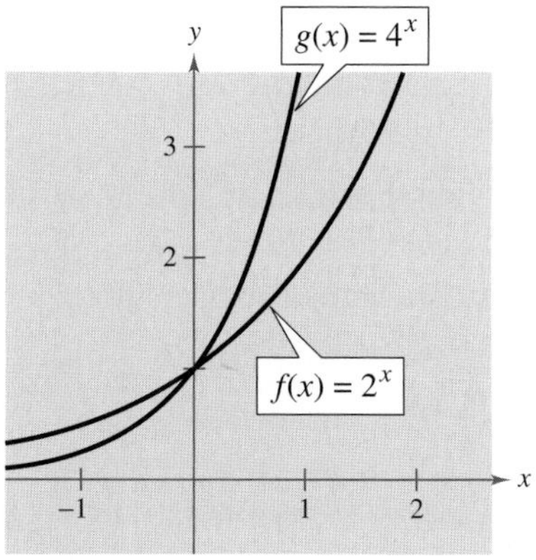

FIGURE 3.1

Solution

The table below lists some values for each function, and Figure 3.1 shows their graphs. Note that both graphs are increasing. Moreover, the graph of $g(x) = 4^x$ is increasing more rapidly than the graph of $f(x) = 2^x$.

x	-2	-1	0	1	2	3
2^x	$\frac{1}{4}$	$\frac{1}{2}$	1	2	4	8
4^x	$\frac{1}{16}$	$\frac{1}{4}$	1	4	16	64

The table in Example 2 was evaluated by hand. You could, of course, use a graphing utility to construct tables with even more values.

Example 3 ▶ Graphs of $y = a^{-x}$

In the same coordinate plane, sketch the graph of each function.

a. $F(x) = 2^{-x}$ **b.** $G(x) = 4^{-x}$

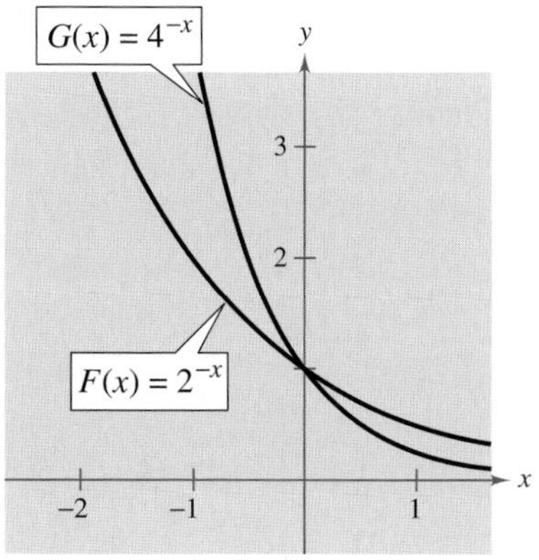

FIGURE 3.2

Solution

The table below lists some values for each function, and Figure 3.2 shows their graphs. Note that both graphs are decreasing. Moreover, the graph of $G(x) = 4^{-x}$ is decreasing more rapidly than the graph of $F(x) = 2^{-x}$.

x	-3	-2	-1	0	1	2
2^{-x}	8	4	2	1	$\frac{1}{2}$	$\frac{1}{4}$
4^{-x}	64	16	4	1	$\frac{1}{4}$	$\frac{1}{16}$

In Example 3, note that the functions $F(x) = 2^{-x}$ and $G(x) = 4^{-x}$ can be rewritten with positive exponents.

$$F(x) = 2^{-x} = \left(\frac{1}{2}\right)^x \quad \text{and} \quad G(x) = 4^{-x} = \left(\frac{1}{4}\right)^x$$

Comparing the functions in Examples 2 and 3, observe that

$$F(x) = 2^{-x} = f(-x) \quad \text{and} \quad G(x) = 4^{-x} = g(-x).$$

Consequently, the graph of F is a reflection (in the y-axis) of the graph of f. The graphs of G and g have the same relationship. The graphs in Figures 3.1 and 3.2 are typical of the exponential functions a^x and a^{-x}. They have one y-intercept and one horizontal asymptote (the x-axis), and they are continuous. The basic characteristics of these exponential functions are summarized in Figures 3.3 and 3.4.

STUDY TIP

Notice that the range of an exponential function is $(0, \infty)$, which means that $a^x > 0$ for all values of x.

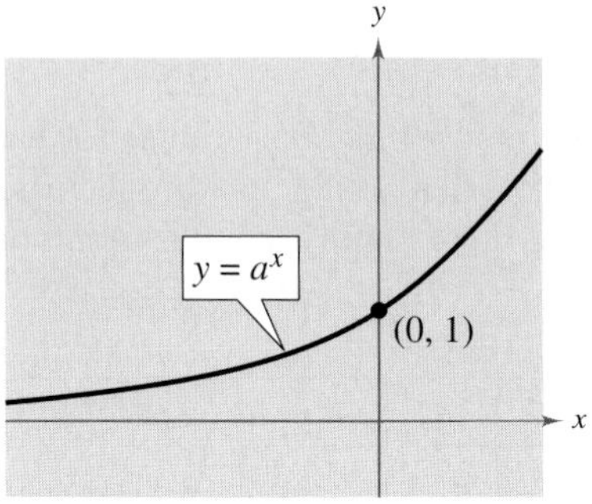

FIGURE 3.3

Graph of $y = a^x, a > 1$

- Domain: $(-\infty, \infty)$
- Range: $(0, \infty)$
- Intercept: $(0, 1)$
- Increasing
- x-Axis is a horizontal asymptote ($a^x \to 0$ as $x \to -\infty$).
- Continuous

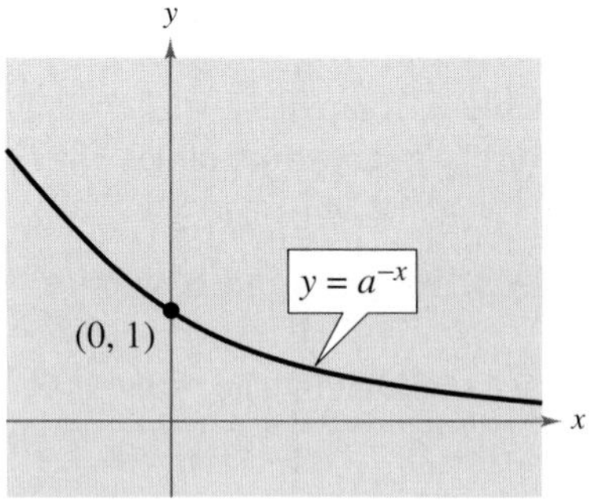

FIGURE 3.4

Graph of $y = a^{-x}, a > 1$

- Domain: $(-\infty, \infty)$
- Range: $(0, \infty)$
- Intercept: $(0, 1)$
- Decreasing
- x-Axis is a horizontal asymptote ($a^{-x} \to 0$ as $x \to \infty$).
- Continuous

The *Interactive* CD-ROM and *Internet* versions of this text offer a built-in graphing calculator, which can be used in the Examples, Explorations, Technology notes, and Exercises.

Exploration

Use a graphing utility to graph

$$y = a^x$$

for $a = 3$, 5, and 7 in the same viewing window. (Use a viewing window in which $-2 \le x \le 1$ and $0 \le y \le 2$.) For instance, the graph of

$$y = 3^x$$

is shown in Figure 3.5. How do the graphs compare with each other? Which graph is on the top in the interval $(-\infty, 0)$? Which is on the bottom? Which graph is on the top in the interval $(0, \infty)$? Which is on the bottom?

Repeat this experiment with the graphs of $y = b^x$ for $b = \frac{1}{3}, \frac{1}{5}$, and $\frac{1}{7}$. (Use a viewing window in which $-1 \le x \le 2$ and $0 \le y \le 2$.) What can you conclude about the shape of the graph of $y = b^x$ and the value of b?

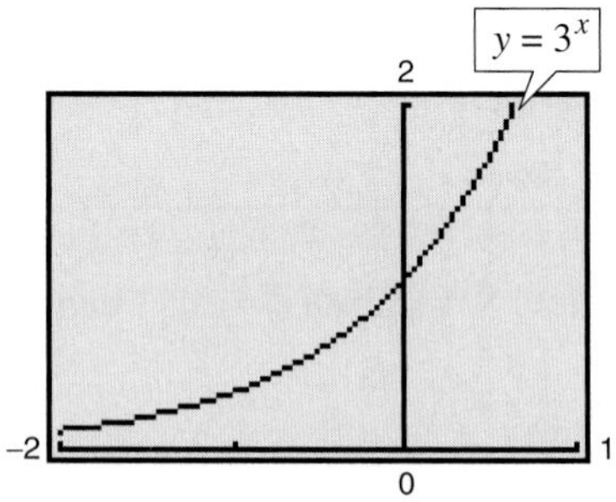

FIGURE 3.5

In the following example, notice how the graph of $y = a^x$ can be used to sketch the graphs of functions of the form $f(x) = b \pm a^{x+c}$.

Example 4 ▶ Transformations of Graphs of Exponential Functions

Each of the following graphs is a transformation of the graph of $f(x) = 3^x$, as shown in Figure 3.6.

a. Because $g(x) = 3^{x+1} = f(x + 1)$, the graph of g can be obtained by shifting the graph of f 1 unit to the left.

b. Because $h(x) = 3^x - 2 = f(x) - 2$, the graph of h can be obtained by shifting the graph of f down 2 units.

c. Because $k(x) = -3^x = -f(x)$, the graph of k can be obtained by reflecting the graph of f in the x-axis.

d. Because $j(x) = 3^{-x} = f(-x)$, the graph of j can be obtained by reflecting the graph of f in the y-axis.

A computer animation of this example appears in the *Interactive* CD-ROM and *Internet* versions of this text.

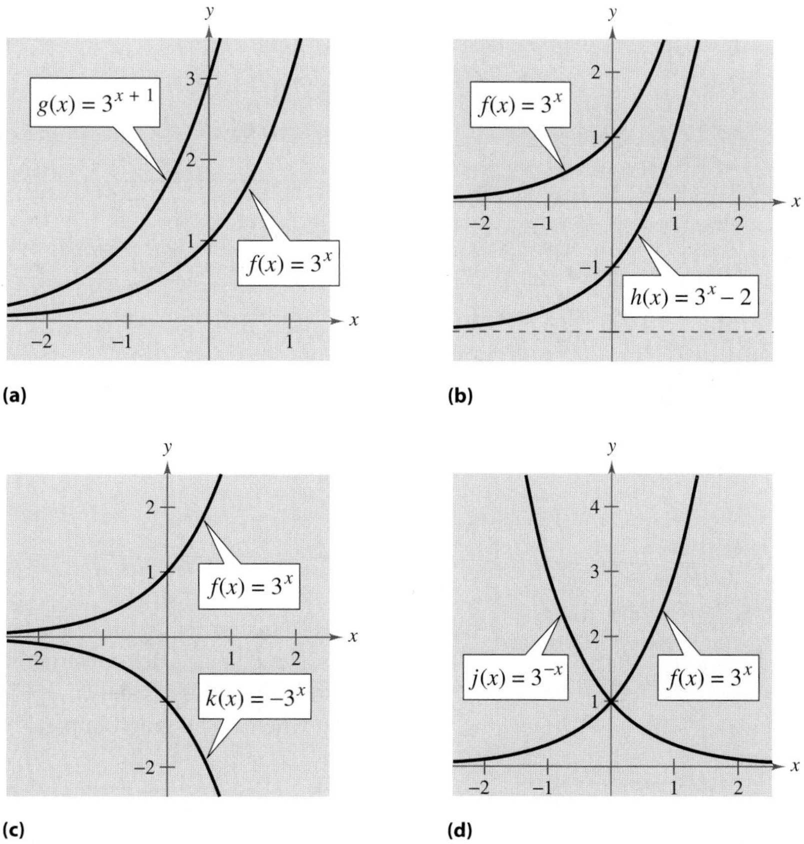

FIGURE 3.6

In Figure 3.6, notice that the transformations in parts (a), (c), and (d) keep the x-axis as a horizontal asymptote, but the transformation in part (b) yields a new horizontal asymptote of $y = -2$. Also, be sure to note how the y-intercept is affected by each transformation.

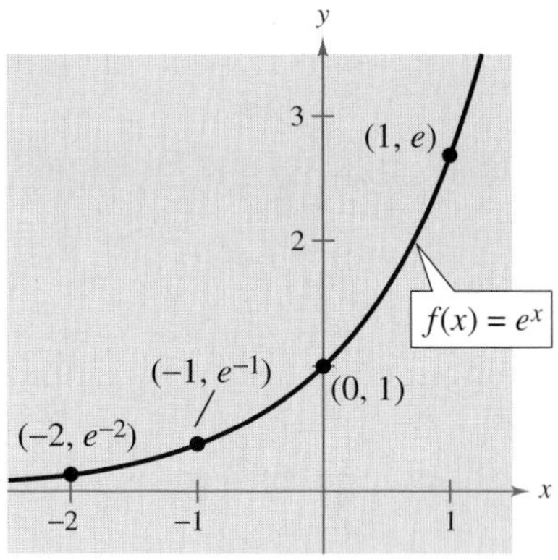

FIGURE 3.7

The Natural Base *e*

In many applications, the most convenient choice for a base is the irrational number

$$e \approx 2.718281828 \ . \ . \ . \ .$$

This number is called the **natural base.** The function $f(x) = e^x$ is called the **natural exponential function.** Its graph is shown in Figure 3.7. Be sure you see that for the exponential function $f(x) = e^x$, e is the constant 2.71828183 . . . , whereas x is the variable.

Example 5 ▶ Evaluating the Natural Exponential Function

Use a calculator to evaluate each expression.

a. e^{-2}

b. e^{-1}

c. e^1

d. e^2

Solution

Number	*Graphing Calculator Keystrokes*	*Display*
a. e^{-2}	[e^x] [(−)] 2 [ENTER]	0.1353353
b. e^{-1}	[e^x] [(−)] 1 [ENTER]	0.3678794
c. e^1	[e^x] 1 [ENTER]	2.7182818
d. e^2	[e^x] 2 [ENTER]	7.3890561

Example 6 ▶ Graphing Natural Exponential Functions

Sketch the graph of each natural exponential function.

a. $f(x) = 2e^{0.24x}$

b. $g(x) = \frac{1}{2}e^{-0.58x}$

Solution

To sketch these two graphs, you can use a graphing utility to construct a table of values, as shown below. After constructing the table, plot the points and connect them with smooth curves, as shown in Figure 3.8. Note that the graph in part (a) is increasing whereas the graph in part (b) is decreasing.

x	−3	−2	−1	0	1	2	3
$f(x)$	0.974	1.238	1.573	2.000	2.542	3.232	4.109
$g(x)$	2.849	1.595	0.893	0.500	0.280	0.157	0.088

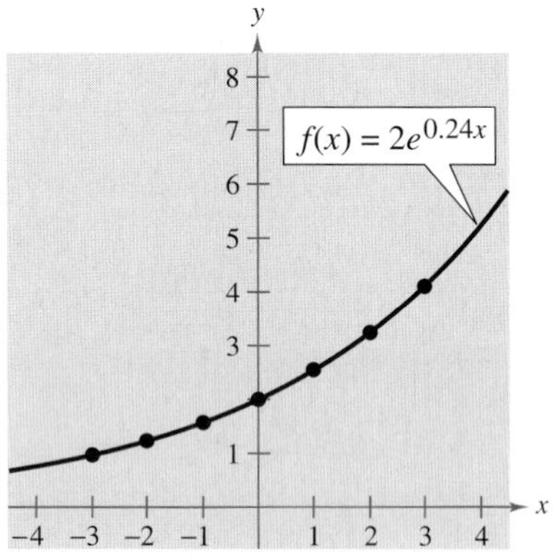

(a)

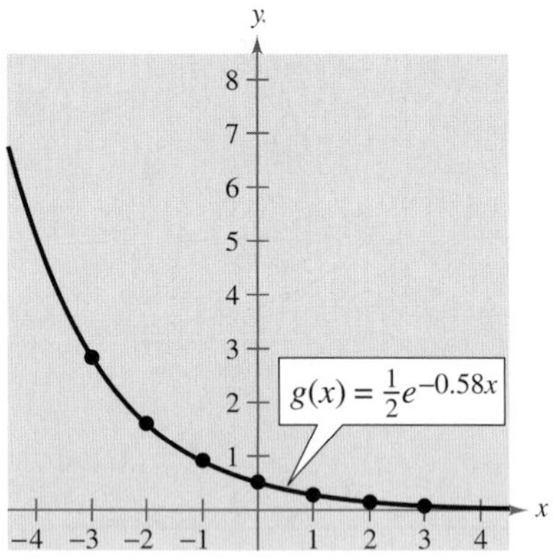

(b)

FIGURE 3.8

Applications

One of the most familiar examples of exponential growth is that of an investment earning *continuously compounded interest.* Using exponential functions, you can *develop* a formula for the balance in an account that pays compound interest, and show how it leads to continuous compounding.

Suppose a principal P is invested at an annual interest rate r, compounded once a year. If the interest is added to the principal at the end of the year, the new balance P_1 is

$$\begin{aligned} P_1 &= P + Pr \\ &= P(1 + r). \end{aligned}$$

This pattern of multiplying the previous principal by $1 + r$ is then repeated each successive year, as shown below.

Year	*Balance After Each Compounding*
0	$P = P$
1	$P_1 = P(1 + r)$
2	$P_2 = P_1(1 + r) = P(1 + r)(1 + r) = P(1 + r)^2$
3	$P_3 = P_2(1 + r) = P(1 + r)^2(1 + r) = P(1 + r)^3$
⋮	⋮
t	$P_t = P(1 + r)^t$

To accommodate more frequent (quarterly, monthly, or daily) compounding of interest, let n be the number of compoundings per year and let t be the number of years. Then the rate per compounding is r/n and the account balance after t years is

$$A = P\left(1 + \frac{r}{n}\right)^{nt}. \qquad \text{Amount (balance) with } n \text{ compoundings per year}$$

If you let the number of compoundings n increase without bound, the process approaches what is called **continuous compounding.** In the formula for n compoundings per year, let $m = n/r$. This produces

$$A = P\left(1 + \frac{r}{n}\right)^{nt} \qquad \text{Amount with } n \text{ compoundings per year}$$

$$= P\left(1 + \frac{r}{mr}\right)^{mrt} \qquad \text{Substitute } mr \text{ for } n.$$

$$= P\left(1 + \frac{1}{m}\right)^{mrt} \qquad \text{Simplify.}$$

$$= P\left[\left(1 + \frac{1}{m}\right)^{m}\right]^{rt}. \qquad \text{Property of exponents}$$

As m increases without bound, it can be shown that $[1 + (1/m)]^m$ approaches e. (Try the values $m = 10$, 10,000, and 10,000,000.) From this, you can conclude that the formula for continuous compounding is

$$A = Pe^{rt}. \qquad \text{Substitute } e \text{ for } (1 + 1/m)^m.$$

Technology

Use the formula

$$A = P\left(1 + \frac{r}{n}\right)^{nt}$$

to calculate the amount in an account when $P = \$3000$, $r = 6\%$, $t = 10$ years, and compounding is done (1) by the day, (2) by the hour, (3) by the minute, and (4) by the second. Use these results to present an argument that increasing the number of compoundings does not mean unlimited growth of the amount in the account.

A computer simulation of this concept appears in the *Interactive* CD-ROM and *Internet* versions of this text.

Activities

1. Sketch the graphs of the functions $f(x) = e^x$ and $g(x) = 1 + e^x$ on the same coordinate system.
2. Determine the balance A at the end of 20 years if \$1500 is invested at 6.5% interest and the interest is compounded (a) quarterly and (b) continuously.
   Answer: (a) \$5446.73 (b) \$5503.95
3. Determine the amount of money that should be invested at 9% interest, compounded monthly, to produce a final balance of \$30,000 in 15 years.
   Answer: \$7816.48

STUDY TIP

Be sure you see that the annual interest rate must be expressed in decimal form. For instance, 6% should be expressed as 0.06.

Formulas for Compound Interest

After t years, the balance A in an account with principal P and annual interest rate r (in decimal form) is given by the following formulas.

1. For n compoundings per year: $A = P\left(1 + \frac{r}{n}\right)^{nt}$
2. For continuous compounding: $A = Pe^{rt}$

Additional Example

The number of fruit flies in an experimental population after t hours is given by $Q(t) = 20e^{0.03t}$, $t \geq 0$.

a. Find the initial number of fruit flies in the population.

b. How large is the population of fruit flies after 72 hours?

Solution

a. To find the initial population, you evaluate $Q(t)$ at $t = 0$. $Q(0) = 20e^{0.03(0)} = 20e^0 = 20(1) = 20$ flies.

b. After 72 hours, the population size is $Q(72) = 20e^{0.03(72)} = 20e^{2.16} \approx 173$ flies.

Example 7 ▶ Compound Interest

A total of $12,000 is invested at an annual interest rate of 9%. Find the balance after 5 years if it is compounded

a. quarterly.

b. monthly.

c. continuously.

Solution

a. For quarterly compoundings, you have $n = 4$. So, in 5 years at 9%, the balance is

$$A = P\left(1 + \frac{r}{n}\right)^{nt} \quad \text{Formula for compound interest}$$

$$= 12{,}000\left(1 + \frac{0.09}{4}\right)^{4(5)} \quad \text{Substitute for } P, r, n, \text{ and } t.$$

$$\approx \$18{,}726.11. \quad \text{Use a calculator.}$$

b. For monthly compoundings, you have $n = 12$. So, in 5 years at 9%, the balance is

$$A = P\left(1 + \frac{r}{n}\right)^{nt} \quad \text{Formula for compound interest}$$

$$= 12{,}000\left(1 + \frac{0.09}{12}\right)^{12(5)} \quad \text{Substitute for } P, r, n, \text{ and } t.$$

$$\approx \$18{,}788.17. \quad \text{Use a calculator.}$$

c. For continuous compounding, the balance is

$$A = Pe^{rt} \quad \text{Formula for continuous compounding}$$

$$= 12{,}000e^{0.09(5)} \quad \text{Substitute for } P, r, n, \text{ and } t.$$

$$\approx \$18{,}819.75. \quad \text{Use a calculator.}$$

Group Activity

The sequence 3, 6, 9, 12, 15, ... is given by $f(n) = 3n$ and is an example of linear growth. The sequence 3, 9, 27, 81, 243, ... is given by $f(n) = 3^n$ and is an example of exponential growth. Explain the difference between these two types of growth. For each of the following sequences, indicate whether the sequence represents linear growth or exponential growth, and find a linear or exponential function that represents the sequence. Give several other examples of linear and exponential growth.

a. $\frac{1}{2}, \frac{1}{4}, \frac{1}{8}, \frac{1}{16}, \frac{1}{32}, \ldots$

b. 4, 8, 12, 16, 20, ...

c. $\frac{2}{3}, \frac{4}{3}, 2, \frac{8}{3}, \frac{10}{3}, 4, \ldots$

d. 5, 25, 125, 625, ...

In Example 7, note that continuous compounding yields more than quarterly or monthly compounding. This is typical of the two types of compounding. That is, for a given principal, interest rate, and time, continuous compounding will always yield a larger balance than compounding n times a year.

Example 8 ▶ Radioactive Decay

In 1986, a nuclear reactor accident occurred in Chernobyl in what was then the Soviet Union. The explosion spread highly toxic radioactive chemicals, such as plutonium, over hundreds of square miles, and the government evacuated the city and the surrounding area. To see why the city is now uninhabited, consider the model

$$P = 10e^{-0.00002845t}.$$

This model represents the amount of plutonium that remains (from an initial amount of 10 pounds) after t years. Sketch the graph of this function over the interval from $t = 0$ to $t = 100{,}000$. How much of the 10 pounds will remain in the year 2002? How much of the 10 pounds will remain after 100,000 years?

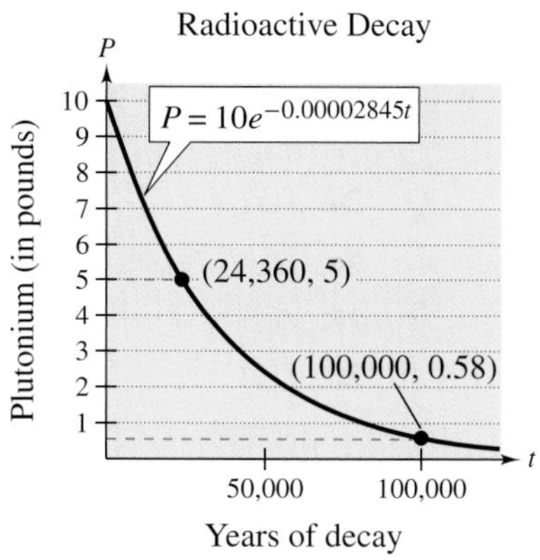

FIGURE 3.9

Solution

The graph of this function is shown in Figure 3.9. Note from this graph that plutonium has a *half-life* of about 24,360 years. That is, after 24,360 years, *half* of the original amount will remain. After another 24,360 years, one-quarter of the original amount will remain, and so on. In the year 2002 ($t = 16$), there will still be

$$P = 10e^{-0.00002845(16)}$$
$$= 10e^{-0.0004552}$$
$$\approx 9.995 \text{ pounds}$$

of plutonium remaining. After 100,000 years, there will still be

$$P = 10e^{-0.00002845(100{,}000)}$$
$$= 10e^{-2.845}$$
$$\approx 0.58 \text{ pound}$$

of plutonium remaining.

Writing ABOUT MATHEMATICS

Identifying Exponential Functions Which of the following functions generated the two tables below? Discuss how you were able to decide. What do these functions have in common? Are any the same? If so, explain why.

a. $f_1(x) = 2^{(x+3)}$ **b.** $f_2(x) = 8\left(\frac{1}{2}\right)^x$ **c.** $f_3(x) = \left(\frac{1}{2}\right)^{(x-3)}$

d. $f_4(x) = \left(\frac{1}{2}\right)^x + 7$ **e.** $f_5(x) = 7 + 2^x$ **f.** $f_6(x) = (8)2^x$

x	-1	0	1	2	3
$g(x)$	7.5	8	9	11	15

x	-2	-1	0	1	2
$h(x)$	32	16	8	4	2

Create two different exponential functions of the forms $y = a(b)^x$ and $y = c^x + d$ with y-intercepts of $(0, -3)$.

One way your students might approach this problem is to create a table, covering x-values from -2 through 3, for each of the functions and compare this table with the given tables. If this method is used, you might consider dividing your class into groups of three or six and having the groups assign one or two functions to each member. They should then pool their results and work cooperatively to determine that each function has a y-intercept of $(0, 8)$.

Another approach is a graphical one: the groups can create scatter plots of the data given in the table, and compare them with sketches of the graphs of the given functions. Consider assigning students to groups of four and giving the responsibility for sketching three graphs to each group member.

3.1 Exercises

The *Interactive* CD-ROM and *Internet* versions of this text contain step-by-step solutions to all odd-numbered Section and Review Exercises. They also provide Tutorial Exercises that link to Guided Examples for additional help.

In Exercises 1–10, evaluate the expression. Round your result to three decimal places.

1. $(3.4)^{5.6}$
2. $5000(2^{-1.5})$
3. $(1.005)^{400}$
4. $8^{2\pi}$
5. $5^{-\pi}$
6. $\sqrt[3]{4395}$
7. $100^{\sqrt{2}}$
8. $e^{1/2}$
9. $e^{-3/4}$
10. $e^{3.2}$

In Exercises 11–14, match the exponential function with its graph. [The graphs are labeled (a), (b), (c), and (d).]

(a)

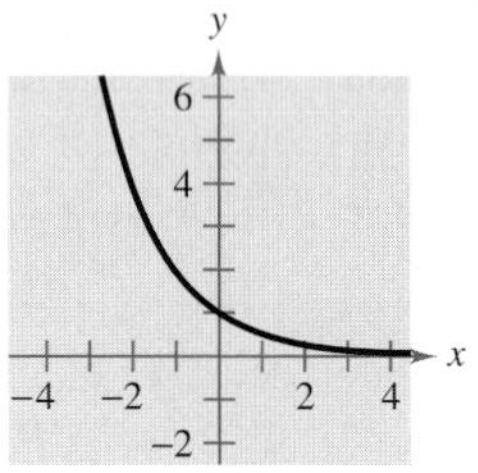

(b)

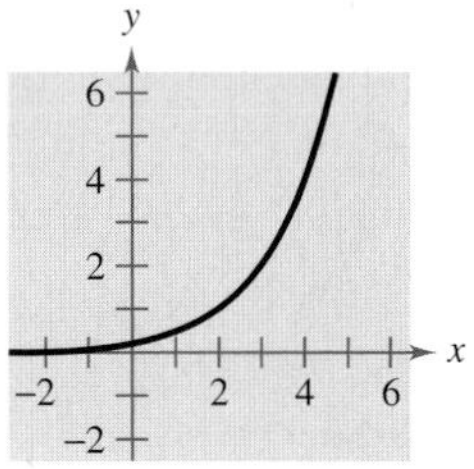

(c)

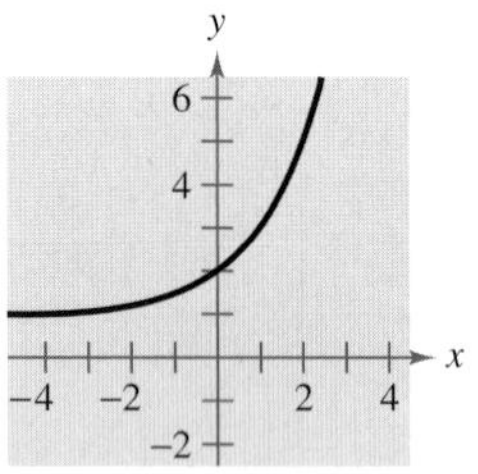

(d)

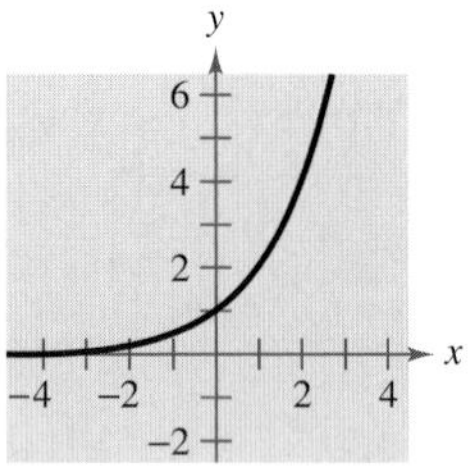

11. $f(x) = 2^x$
12. $f(x) = 2^x + 1$
13. $f(x) = 2^{-x}$
14. $f(x) = 2^{x-2}$

In Exercises 15–22, use the graph of *f* to describe the transformation that yields the graph of *g*.

15. $f(x) = 3^x, \quad g(x) = 3^{x-4}$
16. $f(x) = 4^x, \quad g(x) = 4^x + 1$
17. $f(x) = -2^x, \quad g(x) = 5 - 2^x$
18. $f(x) = 10^x, \quad g(x) = 10^{-x+3}$
19. $f(x) = \left(\frac{3}{5}\right)^x, \quad g(x) = -\left(\frac{3}{5}\right)^{x+4}$
20. $f(x) = \left(\frac{7}{2}\right)^x, \quad g(x) = -\left(\frac{7}{2}\right)^{-x+6}$
21. $f(x) = 0.3^x, \quad g(x) = -0.3^x + 5$
22. $f(x) = 3.6^x, \quad g(x) = -3.6^{-x} + 8$

In Exercises 23–36, use a graphing utility to construct a table of values for each function. Then sketch the graph of the function.

23. $f(x) = \left(\frac{1}{2}\right)^x$
24. $f(x) = 6^x$
25. $f(x) = \left(\frac{1}{2}\right)^{-x}$
26. $f(x) = 6^{-x}$
27. $f(x) = 2^{x-1}$
28. $f(x) = 3^{x+2}$
29. $f(x) = e^x$
30. $f(x) = e^{-x}$
31. $f(x) = 3e^{x+4}$
32. $f(x) = 2e^{-0.5x}$
33. $f(x) = 2e^{x-2} + 4$
34. $f(x) = 2 + e^{x-5}$
35. $f(x) = 4^{x-3} + 3$
36. $f(x) = -4^{x-3} - 3$

In Exercises 37–54, use a graphing utility to graph the exponential function.

37. $g(x) = 5^x$
38. $f(x) = \left(\frac{3}{2}\right)^x$
39. $f(x) = \left(\frac{1}{5}\right)^x = 5^{-x}$
40. $h(x) = \left(\frac{3}{2}\right)^{-x}$
41. $h(x) = 5^{x-2}$
42. $g(x) = \left(\frac{3}{2}\right)^{x+2}$
43. $g(x) = 5^{-x} - 3$
44. $f(x) = \left(\frac{3}{2}\right)^{-x} + 2$
45. $y = 2^{-x^2}$
46. $y = 3^{-|x|}$
47. $y = 3^{x-2} + 1$
48. $y = 4^{x+1} - 2$
49. $y = 1.08^{-5x}$
50. $y = 1.08^{5x}$
51. $s(t) = 2e^{0.12t}$
52. $s(t) = 3e^{-0.2t}$
53. $g(x) = 1 + e^{-x}$
54. $h(x) = e^{x-2}$

Finance **In Exercises 55–58, complete the table to determine the balance *A* for *P* dollars invested at rate *r* for *t* years and compounded *n* times per year.**

n	1	2	4	12	365	Continuous
A						

55. $P = \$2500, r = 8\%, t = 10$ years
56. $P = \$1000, r = 6\%, t = 10$ years
57. $P = \$2500, r = 8\%, t = 20$ years
58. $P = \$1000, r = 6\%, t = 40$ years

Finance **In Exercises 59–62, complete the table to determine the balance A for \$12,000 invested at rate r for t years, compounded continuously.**

t	1	10	20	30	40	50
A						

59. $r = 8\%$

60. $r = 6\%$

61. $r = 6.5\%$

62. $r = 7.5\%$

63. *Finance* On the day of a child's birth, a deposit of \$25,000 is made in a trust fund that pays 8.75% interest, compounded continuously. Determine the balance in this account on the child's 25th birthday.

64. *Finance* A deposit of \$5000 is made in a trust fund that pays 7.5% interest, compounded continuously. It is specified that the balance will be given to the college from which the donor graduated after the money has earned interest for 50 years. How much will the college receive?

65. *Graphical Reasoning* There are two options for investing \$500. The first earns 7% compounded annually and the second earns 7% simple interest. The figure shows the growth of each investment over a 30-year period.

(a) Identify which graph in the figure represents each type of investment. Explain your reasoning.

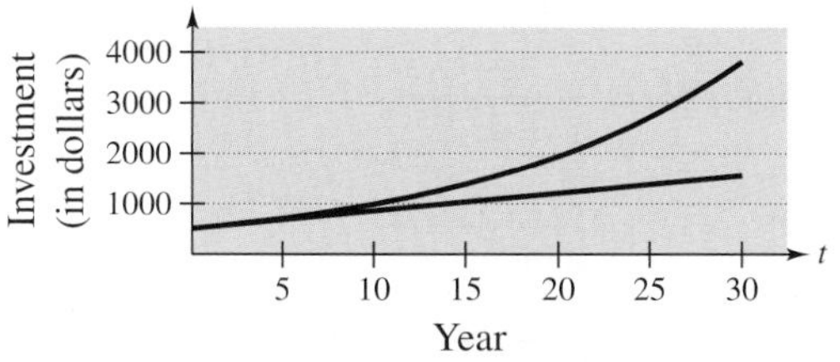

(b) Verify your answer in part (a) by finding the equations that model the investment growth and graphing the models.

66. *Depreciation* After t years, the value of a car that cost \$20,000 is modeled by

$$V(t) = 20{,}000\left(\frac{3}{4}\right)^t.$$

Graph the function and determine the value of the car 2 years after it was purchased.

67. *Inflation* If the annual rate of inflation averages 4% over the next 10 years, the approximate cost C of goods or services during any year in that decade will be modeled by

$$C(t) = P(1.04)^t$$

where t is the time in years and P is the present cost. If the price of an oil change for your car is presently \$23.95, estimate the price 10 years from now.

68. *Economics* The demand equation for a certain product is

$$p = 5000\left(1 - \frac{4}{4 + e^{-0.002x}}\right).$$

(a) Use a graphing utility to graph the demand function for $x > 0$ and $p > 0$.

(b) Find the price p for a demand of $x = 500$ units.

(c) Use the graph in part (a) to approximate the greatest price that will still yield a demand of at least 600 units.

69. *Population Growth* A certain type of bacterium increases according to the model $P(t) = 100e^{0.2197t}$, where t is the time in hours. Find (a) $P(0)$, (b) $P(5)$, and (c) $P(10)$.

70. *Population Growth* The population of a town increases according to the model $P(t) = 2500e^{0.0293t}$, where t is the time in years, with $t = 0$ corresponding to 1990. Use the model to estimate the population in (a) 2000 and (b) 2010.

71. *Radioactive Decay* Let Q (in grams) represent a mass of radioactive radium (^{226}Ra), whose half-life is 1620 years. The quantity of radium present after t years is

$$Q = 25\left(\frac{1}{2}\right)^{t/1620}.$$

(a) Determine the initial quantity (when $t = 0$).

(b) Determine the quantity present after 1000 years.

(c) Use a graphing utility to graph the function over the interval $t = 0$ to $t = 5000$.

72. *Radioactive Decay* Let Q (in grams) represent a mass of carbon 14 (^{14}C), whose half-life is 5730 years. The quantity of carbon 14 present after t years is

$$Q = 10\left(\frac{1}{2}\right)^{t/5730}.$$

(a) Determine the initial quantity (when $t = 0$).

(b) Determine the quantity present after 2000 years.

(c) Sketch the graph of this function over the interval $t = 0$ to $t = 10{,}000$.

73. *Data Analysis* A meteorologist measures the atmospheric pressure P (in pascals) at altitude h (in kilometers). The data is shown in the table.

h	0	5	10	15	20
P	101,293	54,735	23,294	12,157	5069

A model for the data is given by

$$P = 102{,}303e^{-0.137h}.$$

(a) Sketch a scatter plot of the data and graph the model on the same set of axes.

(b) Create a table that compares the model with the sample data.

(c) Estimate the atmospheric pressure at a height of 8 kilometers.

(d) Use the graph in part (a) to estimate the altitude at which the atmospheric pressure is 21,000 pascals.

74. *Data Analysis* To estimate the amount of defoliation caused by the gypsy moth during a given year, a forester counts the number x of egg masses on $\frac{1}{40}$ of an acre (circle of radius 18.6 feet) in the fall. The percent of defoliation y the next spring is shown in the table. (Source: USDA, Forest Service)

x	0	25	50	75	100
y	12	44	81	96	99

(a) A model for the data is

$$y = \frac{300}{3 + 17e^{-0.065x}}.$$

Use a graphing utility to create a scatter plot of the data and graph the model in the same viewing window.

(b) Create a table that compares the model with the sample data.

(c) Estimate the percent of defoliation if 36 egg masses are counted on $\frac{1}{40}$ acre.

(d) Use the graph in part (a) to estimate the number of egg masses per $\frac{1}{40}$ acre if you observe that $\frac{2}{3}$ of a forest is defoliated the following spring.

Synthesis

True or False? **In Exercises 75 and 76, determine whether the statement is true or false. Justify your answer.**

75. The x-axis is an asymptote for the graph of $f(x) = 10^x$.

76. $e = \dfrac{271{,}801}{99{,}990}$.

Think About It **In Exercises 77–80, use properties of exponents to determine which functions (if any) are the same.**

77. $f(x) = 3^{x-2}$
$g(x) = 3^x - 9$
$h(x) = \frac{1}{9}(3^x)$

78. $f(x) = 4^x + 12$
$g(x) = 2^{2x+6}$
$h(x) = 64(4^x)$

79. $f(x) = 16(4^{-x})$
$g(x) = \left(\frac{1}{4}\right)^{x-2}$
$h(x) = 16(2^{-2x})$

80. $f(x) = 5^{-x} + 3$
$g(x) = 5^{3-x}$
$h(x) = -5^{x-3}$

81. Graph the functions $y = 3^x$ and $y = 4^x$ and use the graphs to solve the inequalities.

(a) $4^x < 3^x$

(b) $4^x > 3^x$

82. Graph the functions $y = \left(\frac{1}{2}\right)^x$ and $y = \left(\frac{1}{4}\right)^x$ and use the graphs to solve the inequalities.

(a) $\left(\frac{1}{4}\right)^x < \left(\frac{1}{2}\right)^x$

(b) $\left(\frac{1}{4}\right)^x > \left(\frac{1}{2}\right)^x$

83. Use a graphing utility to graph each function. Use the graph to find any asymptotes of the function.

(a) $f(x) = \dfrac{8}{1 + e^{-0.5x}}$

(b) $g(x) = \dfrac{8}{1 + e^{-0.5/x}}$

84. Use a graphing utility to graph each function. Use the graph to find where the function is increasing and decreasing, and approximate any relative maximum or minimum values.

(a) $f(x) = x^2e^{-x}$

(b) $g(x) = x2^{3-x}$

85. Use a graphing utility to graph $y_1 = e^x$ and each of the functions $y_2 = x^2$, $y_3 = x^3$, $y_4 = \sqrt{x}$, and $y_5 = |x|$. Which function increases at the fastest rate as x approaches $+\infty$?

86. *Conjecture* Use the result of Exercise 85 to make a conjecture about the rate of growth of $y_1 = e^x$ and $y = x^n$, where n is a natural number and x approaches $+\infty$.

87. *Writing* Use the results of Exercises 85 and 86 to describe what is implied when it is stated that a quantity is growing exponentially.

88. *Graphical Analysis* Use a graphing utility to graph

$$f(x) = \left(1 + \frac{0.5}{x}\right)^x \quad \text{and} \quad g(x) = e^{0.5}$$

in the same viewing window. What is the relationship between f and g as x increases without bound?

89. *Conjecture* Use the result of Exercise 88 to make a conjecture about the value of $[1 + (r/x)]^x$ as x increases without bound. Create a table that illustrates your conjecture for $r = 1$.

90. *Think About It* Which functions are exponential?

(a) $3x$ (b) $3x^2$

(c) 3^x (d) 2^{-x}

91. *Think About It* Without using a calculator, why do you know that $2^{\sqrt{2}}$ is greater than 2, but less than 4?

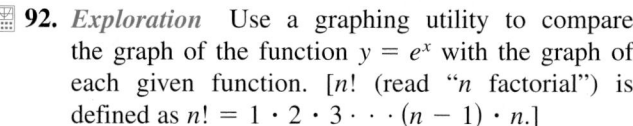

92. *Exploration* Use a graphing utility to compare the graph of the function $y = e^x$ with the graph of each given function. [$n!$ (read "n factorial") is defined as $n! = 1 \cdot 2 \cdot 3 \cdot \cdot \cdot (n - 1) \cdot n$.]

(a) $y_1 = 1 + \dfrac{x}{1!}$

(b) $y_2 = 1 + \dfrac{x}{1!} + \dfrac{x^2}{2!}$

(c) $y_3 = 1 + \dfrac{x}{1!} + \dfrac{x^2}{2!} + \dfrac{x^3}{3!}$

93. *Pattern Recognition* Identify the pattern of successive polynomials given in Exercise 92. Extend the pattern one more term and compare the graph of the resulting polynomial function with the graph of $y = e^x$. What do you think this pattern implies?

94. Given the exponential function

$$f(x) = a^x$$

show that

(a) $f(u + v) = f(u) \cdot f(v)$.

(b) $f(2x) = [f(x)]^2$.

Review

In Exercises 95–98, solve for *y*.

95. $2x - 7y + 14 = 0$

96. $x^2 + 3y = 4$

97. $x^2 + y^2 = 25$

98. $x - |y| = 2$

In Exercises 99–102, sketch the graph of the rational function.

99. $f(x) = \dfrac{2}{9 + x}$

100. $f(x) = \dfrac{4x - 3}{x}$

101. $f(x) = \dfrac{6}{x^2 + 5x - 24}$

102. $f(x) = \dfrac{x^2 - 7x + 12}{x + 2}$

3.2 Logarithmic Functions and Their Graphs

▶ **What you should learn**

- How to recognize and evaluate logarithmic functions with base *a*
- How to graph logarithmic functions
- How to recognize and evaluate natural logarithmic functions
- How to use logarithmic functions to model and solve real-life applications

▶ **Why you should learn it**

You can use logarithmic functions to model and solve real-life problems. For instance, Exercises 73 and 74 on page 308 show how to use a logarithmic function to model the minimum required ventilation rates in public school classrooms.

Mary Kate Kenny

Logarithmic Functions

In Section 1.7, you studied the concept of the inverse of a function. There, you learned that if a function has the property that no horizontal line intersects the graph of the function more than once, the function must have an inverse. By looking back at the graphs of the exponential functions introduced in Section 3.1, you will see that every function of the form

$$f(x) = a^x$$

passes the Horizontal Line Test and therefore must have an inverse. This inverse function is called the **logarithmic function with base *a*.**

Definition of Logarithmic Function with Base *a*

For $x > 0$ and $0 < a \neq 1$,

$$y = \log_a x \text{ if and only if } x = a^y.$$

The function given by

$$f(x) = \log_a x$$

is called the **logarithmic function with base *a*.**

The equations

$$y = \log_a x \qquad \text{and} \qquad x = a^y$$

are equivalent. The first equation is in logarithmic form and the second is in exponential form.

When evaluating logarithms, remember that *a logarithm is an exponent.* This means that $\log_a x$ is the exponent to which a must be raised to obtain x. For instance, $\log_2 8 = 3$ because 2 must be raised to the third power to get 8.

Example 1 ▶ Evaluating Logarithms

Use the definition of logarithmic function to evaluate the logarithms.

a. $\log_2 32$ **b.** $\log_3 27$ **c.** $\log_4 2$

d. $\log_{10} \frac{1}{100}$ **e.** $\log_3 1$ **f.** $\log_2 2$

Solution

a. $\log_2 32 = 5$ because $2^5 = 32.$

b. $\log_3 27 = 3$ because $3^3 = 27.$

c. $\log_4 2 = \frac{1}{2}$ because $4^{1/2} = \sqrt{4} = 2.$

d. $\log_{10} \frac{1}{100} = -2$ because $10^{-2} = \frac{1}{10^2} = \frac{1}{100}.$

e. $\log_3 1 = 0$ because $3^0 = 1.$

f. $\log_2 2 = 1$ because $2^1 = 2.$

Exploration

Complete the table for $f(x) = 10^x$.

x	-2	-1	0	1	2
$f(x)$					

Complete the table for $f(x) = \log_{10}x$.

x	$\frac{1}{100}$	$\frac{1}{10}$	1	10	100
$f(x)$					

Compare the two tables. What is the relationship between $f(x) = 10^x$ and $f(x) = \log_{10}x$?

The logarithmic function with base 10 is called the **common logarithmic function.** On most calculators, this function is denoted by [LOG]. Because $\log_a x$ is the inverse function of a^x, it follows that the domain of $\log_a x$ is the range of a^x, $(0, \infty)$. In other words, $\log_a x$ is defined only if x is positive.

Example 2 ▶ Evaluating Logarithms on a Calculator

Use a calculator to evaluate each expression.

a. $\log_{10} 10$

b. $2\log_{10} 2.5$

c. $\log_{10}(-2)$

Solution

Number	*Graphing Calculator Keystrokes*	*Display*
a. $\log_{10} 10$	[LOG] 10 [ENTER]	1
b. $2\log_{10} 2.5$	2 [×] [LOG] 2.5 [ENTER]	0.7958800
c. $\log_{10}(-2)$	[LOG] [(−)] 2 [ENTER]	ERROR

Note that the calculator displays an error message (or a complex number) when you try to evaluate $\log_{10}(-2)$. The reason for this is that the domain of every logarithmic function is the set of *positive real numbers.*

The following properties follow directly from the definition of the logarithmic function with base a.

The logarithmic function can be one of the most difficult concepts for students to understand. Remind students that a logarithm is an exponent. Converting back and forth from logarithmic form to exponential form supports this concept.

Properties of Logarithms

1. $\log_a 1 = 0$ because $a^0 = 1$.

2. $\log_a a = 1$ because $a^1 = a$.

3. $\log_a a^x = x$ and $a^{\log_a x} = x$ — Inverse Properties

4. If $\log_a x = \log_a y$, then $x = y$. — One-to-One Property

Example 3 ▶ Using Properties of Logarithms

Solve each equation for x.

a. $\log_2 x = \log_2 3$

b. $\log_4 4 = x$

Solution

a. Using the One-to-One Property (Property 4), you can conclude that $x = 3$.

b. Using Property 2, you can conclude that $x = 1$.

Graphs of Logarithmic Functions

To sketch the graph of

$$y = \log_a x$$

you can use the fact that the graphs of inverse functions are reflections of each other in the line $y = x$.

A computer animation of this example appears in the *Interactive* CD-ROM and *Internet* versions of this text.

Example 4 ▶ Graphs of Exponential and Logarithmic Functions

In the same coordinate plane, sketch the graph of each function.

a. $f(x) = 2^x$ **b.** $g(x) = \log_2 x$

Solution

a. For $f(x) = 2^x$, construct a table of values.

x	-2	-1	0	1	2	3
$f(x) = 2^x$	$\frac{1}{4}$	$\frac{1}{2}$	1	2	4	8

By plotting these points and connecting them with a smooth curve, you obtain the graph shown in Figure 3.10.

b. Because $g(x) = \log_2 x$ is the inverse of $f(x) = 2^x$, the graph of g is obtained by plotting the points $(f(x), x)$ and connecting them with a smooth curve. The graph of g is a reflection of the graph of f in the line $y = x$, as shown in Figure 3.10.

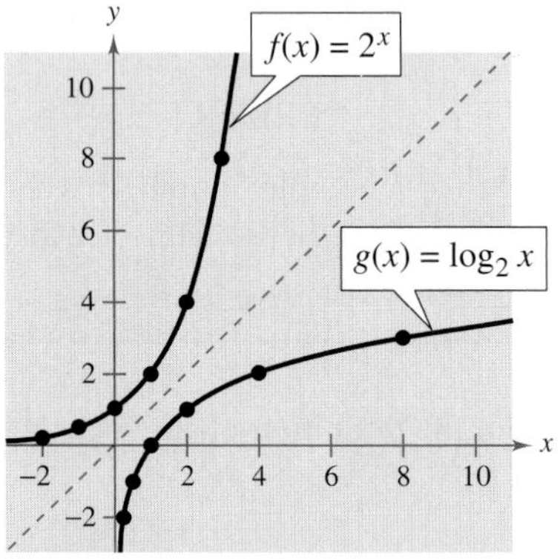

FIGURE 3.10

Example 5 ▶ Sketching the Graph of a Logarithmic Function

Sketch the graph of the common logarithmic function

$$f(x) = \log_{10} x.$$

Identify the x-intercept and the vertical asymptote.

Solution

Begin by constructing a table of values. Note that some of the values can be obtained without a calculator by using the Inverse Property of Logarithms. Others require a calculator. Next, plot the points and connect them with a smooth curve, as shown in Figure 3.11. The x-intercept of the graph is $(1, 0)$ and the vertical asymptote is $x = 0$ (y-axis).

	Without calculator				With calculator		
x	$\frac{1}{100}$	$\frac{1}{10}$	1	10	2	5	8
$\log_{10} x$	-2	-1	0	1	0.301	0.699	0.903

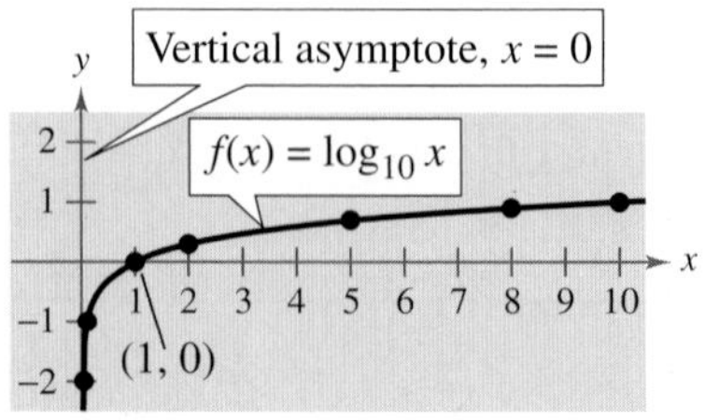

FIGURE 3.11

The nature of the graph in Figure 3.11 is typical of functions of the form $f(x) = \log_a x$, $a > 1$. They have one x-intercept and one vertical asymptote. Notice how slowly the graph rises for $x > 1$. In Figure 3.11 you would need to move out to $x = 1000$ before the graph rose to $y = 3$. The basic characteristics of logarithmic graphs are summarized in Figure 3.12.

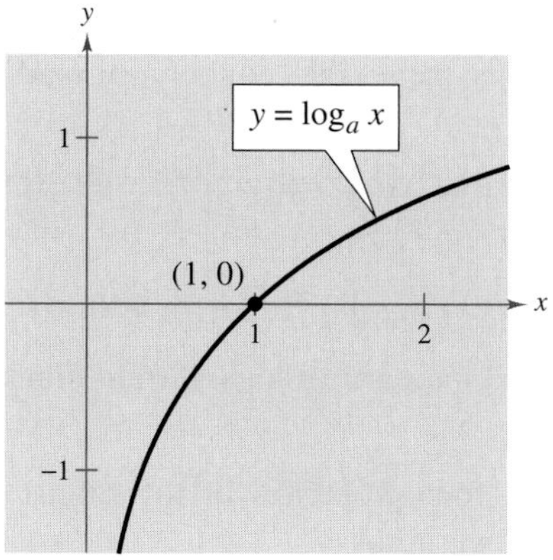

FIGURE 3.12

Graph of $y = \log_a x$, $a > 1$

- Domain: $(0, \infty)$
- Range: $(-\infty, \infty)$
- Intercept: $(1, 0)$
- Increasing
- y-axis is a vertical asymptote ($\log_a x \to -\infty$ as $x \to 0^+$)
- Continuous
- Reflection of graph of $y = a^x$ about the line $y = x$

In the next example, the graph of $\log_a x$ is used to sketch the graphs of functions of the form $y = b \pm \log_a(x + c)$. Notice how a horizontal shift of the graph results in a horizontal shift of the vertical asymptote.

STUDY TIP

You can use your understanding of transformations to identify vertical asymptotes of logarithmic functions. For instance, in Example 6(a) the graph of $g(x) = f(x - 1)$ shifts the graph of $f(x)$ 1 unit to the right. So the vertical asymptote of $g(x)$ is $x = 1$, 1 unit to the right of the asymptote of the graph of $f(x)$.

A computer animation of this example appears in the *Interactive* CD-ROM and *Internet* versions of this text.

Example 6 ▶ Shifting Graphs of Logarithmic Functions

The graph of each of the following functions is similar to the graph of $f(x) = \log_{10} x$, as shown in Figure 3.13.

a. Because

$$g(x) = \log_{10}(x - 1) = f(x - 1),$$

the graph of g can be obtained by shifting the graph of f 1 unit to the right.

b. Because

$$h(x) = 2 + \log_{10} x = 2 + f(x),$$

the graph of h can be obtained by shifting the graph of f 2 units up.

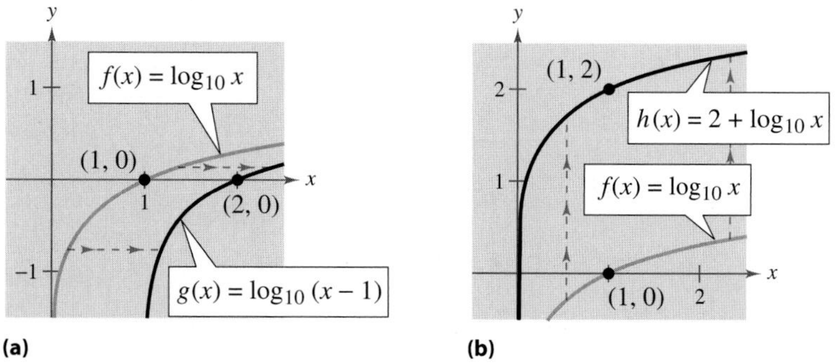

FIGURE 3.13

The Natural Logarithmic Function

As with exponential functions, the most widely used base for logarithmic functions is the number e, where

$$e \approx 2.718281828 \ldots .$$

The logarithmic function with base e is the **natural logarithmic function** and is denoted by the special symbol $\ln x$, read as "the natural log of x" or "el en of x."

The Natural Logarithmic Function

The function defined by

$$f(x) = \log_e x = \ln x, \quad x > 0$$

is called the **natural logarithmic function.**

The four properties of logarithms listed on page 301 are also valid for natural logarithms.

Properties of Natural Logarithms

1. $\ln 1 = 0$ because $e^0 = 1$.
2. $\ln e = 1$ because $e^1 = e$.
3. $\ln e^x = x$ and $e^{\ln x} = x$ — Inverse Properties
4. If $\ln x = \ln y$, then $x = y$. — One-to-One Property

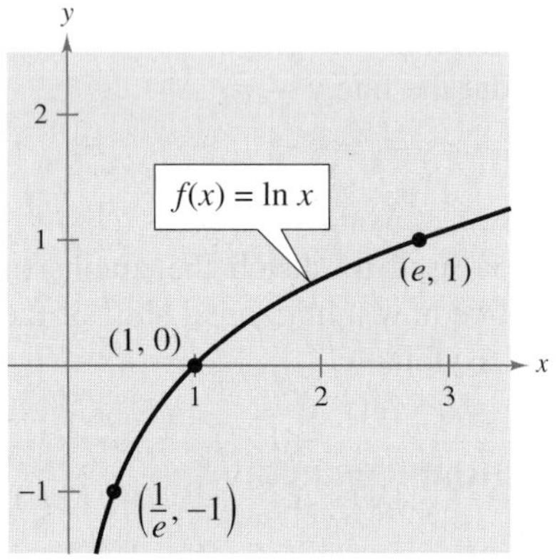

FIGURE 3.14

The graph of the natural logarithmic function is shown in Figure 3.14. Try using a graphing utility to confirm this graph. What is the domain of the natural logarithmic function?

Example 7 ▶ Using Properties of Natural Logarithms

Use the properties of natural logarithms to simplify the expression.

a. $\ln \dfrac{1}{e}$ **b.** $e^{\ln 5}$

c. $\dfrac{\ln 1}{3}$ **d.** $2 \ln e$

Solution

a. $\ln \dfrac{1}{e} = \ln e^{-1} = -1$ — Inverse Property

b. $e^{\ln 5} = 5$ — Inverse Property

c. $\dfrac{\ln 1}{3} = \dfrac{0}{3} = 0$ — Property 1

d. $2 \ln e = 2(1) = 2$ — Property 2

STUDY TIP

Some graphing utilities display a complex number instead of an ERROR message when evaluating an expression such as $\ln(-1)$.

On most calculators, the natural logarithm is denoted by LN, as illustrated in Example 8.

Example 8 ▶ Evaluating the Natural Logarithmic Function

Use a calculator to evaluate each expression.

a. $\ln 2$ **b.** $\ln 0.3$ **c.** $\ln e^2$ **d.** $\ln(-1)$ **e.** $\ln(1+\sqrt{2})$

Solution

Number	Graphing Calculator Keystrokes	Display
a. $\ln 2$	LN 2 ENTER	0.6931472
b. $\ln 0.3$	LN .3 ENTER	–1.2039728
c. $\ln e^2$	LN e^x 2 ENTER	2
d. $\ln(-1)$	LN (–) 1 ENTER	ERROR
e. $\ln(1+\sqrt{2})$	LN (1 + √ 2) ENTER	0.8813736

In Example 8, be sure you see that $\ln(-1)$ gives an error message on most calculators. This occurs because the domain of $\ln x$ is the set of positive real numbers (see Figure 3.14). So, $\ln(-1)$ is undefined.

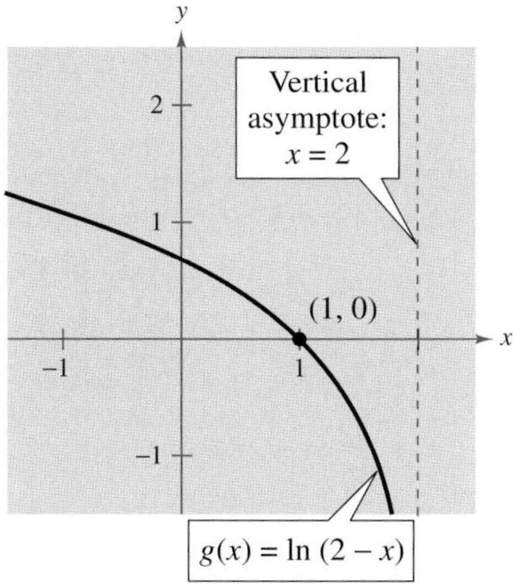

FIGURE 3.15

Example 9 ▶ Finding the Domains of Logarithmic Functions

Find the domain of each function.

a. $f(x) = \ln(x-2)$ **b.** $g(x) = \ln(2-x)$ **c.** $h(x) = \ln x^2$

Solution

a. Because $\ln(x-2)$ is defined only if

$$x - 2 > 0,$$

it follows that the domain of f is $(2, \infty)$.

b. Because $\ln(2-x)$ is defined only if

$$2 - x > 0,$$

it follows that the domain of g is $(-\infty, 2)$. The graph of g is shown in Figure 3.15.

c. Because $\ln x^2$ is defined only if

$$x^2 > 0,$$

it follows that the domain of h is all real numbers except $x = 0$.

In Example 9, suppose you had been asked to analyze the function given by $h(x) = \ln|x-2|$. How would the domain of this function compare with the domains of the functions given in parts (a) and (b) of the example?

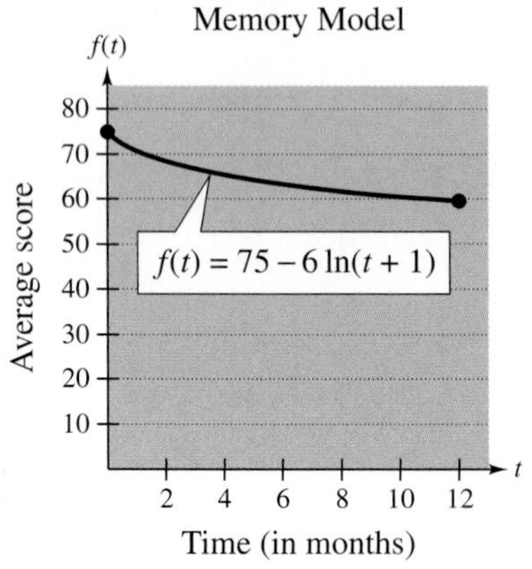

FIGURE 3.16

Application

Example 10 ▶ Human Memory Model

Students participating in a psychological experiment attended several lectures on a subject and were given an exam. Every month for a year after the exam, the students were retested to see how much of the material they remembered. The average scores for the group are given by the *human memory model*

$$f(t) = 75 - 6\ln(t + 1), \quad 0 \le t \le 12$$

where t is the time in months. The graph of f is shown in Figure 3.16.

a. What was the average score on the original $(t = 0)$ exam?

b. What was the average score at the end of $t = 2$ months?

c. What was the average score at the end of $t = 6$ months?

Solution

a. The original average score was

$$\begin{aligned} f(0) &= 75 - 6\ln(0 + 1) && \text{Substitute 0 for } t. \\ &= 75 - 6\ln 1 && \text{Simplify.} \\ &= 75 - 6(0) && \text{Property of natural logarithms} \\ &= 75. && \text{Solution} \end{aligned}$$

b. After 2 months, the average score was

$$\begin{aligned} f(2) &= 75 - 6\ln(2 + 1) \\ &= 75 - 6\ln 3 \\ &\approx 75 - 6(1.0986) \\ &\approx 68.4. \end{aligned}$$

c. After 6 months, the average score was

$$\begin{aligned} f(6) &= 75 - 6\ln(6 + 1) \\ &= 75 - 6\ln 7 \\ &\approx 75 - 6(1.9459) \\ &\approx 63.3. \end{aligned}$$

Alternate Writing About Mathematics
Use a graphing utility to graph $f(x) = \ln x$. How will the graphs of $h(x) = \ln x + 5$, $j(x) = \ln(x - 3)$, and $l(x) = \ln x - 4$ differ from the graph of f?

How will the basic graph of f be affected when a constant c is introduced: $g(x) = c \ln x$? Use a graphing utility to graph g with several different positive values of c, and summarize the effect of c.

Writing ABOUT MATHEMATICS

Analyzing a Human Memory Model Use a graphing utility to determine the time in months when the average score in Example 10 was 60. Explain your method of solving the problem. Describe another way that you can use a graphing utility to determine the answer.

3.2 Exercises

In Exercises 1–8, write the logarithmic equation in exponential form. For example, the exponential form of $\log_5 25 = 2$ is $5^2 = 25$.

1. $\log_4 64 = 3$
2. $\log_3 81 = 4$
3. $\log_7 \frac{1}{49} = -2$
4. $\log_{10} \frac{1}{1000} = -3$
5. $\log_{32} 4 = \frac{2}{5}$
6. $\log_{16} 8 = \frac{3}{4}$
7. $\ln 1 = 0$
8. $\ln 4 = 1.386\ldots$

In Exercises 9–18, write the exponential equation in logarithmic form. For example, the logarithmic form of $2^3 = 8$ is $\log_2 8 = 3$.

9. $5^3 = 125$
10. $8^2 = 64$
11. $81^{1/4} = 3$
12. $9^{3/2} = 27$
13. $6^{-2} = \frac{1}{36}$
14. $10^{-3} = 0.001$
15. $e^3 = 20.0855\ldots$
16. $e^x = 4$
17. $e^0 = 1$
18. $u^v = w$

In Exercises 19–32, evaluate the expression without using a calculator.

19. $\log_2 16$
20. $\log_2 \frac{1}{8}$
21. $\log_{16} 4$
22. $\log_{27} 9$
23. $\log_7 1$
24. $\log_{10} 1000$
25. $\log_{10} 0.01$
26. $\log_{10} 10$
27. $\log_8 32$
28. $\log_9 243$
29. $\ln e^3$
30. $\ln e^{-2}$
31. $\log_a a^2$
32. $\log_b b^{-3}$

In Exercises 33–44, use a calculator to evaluate the logarithm. Round to three decimal places.

33. $\log_{10} 345$
34. $\log_{10} 145$
35. $\log_{10} \frac{4}{5}$
36. $\log_{10} 12.5$
37. $\ln 18.42$
38. $\ln\sqrt{42}$
39. $3 \ln 0.32$
40. $2 \ln 0.75$
41. $\ln(1 + \sqrt{3})$
42. $\ln(\sqrt{5} - 2)$
43. $\ln \frac{2}{3}$
44. $\ln \frac{1}{2}$

In Exercises 45–50, use the graph of $y = \log_3 x$ to match the given function with its graph. [The graphs are labeled (a), (b), (c), (d), (e), and (f).]

(a)

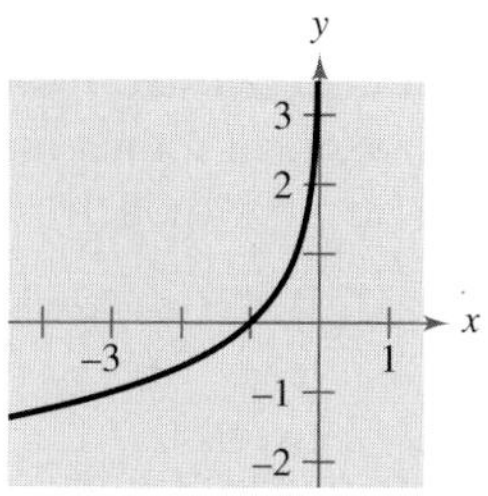

(b)

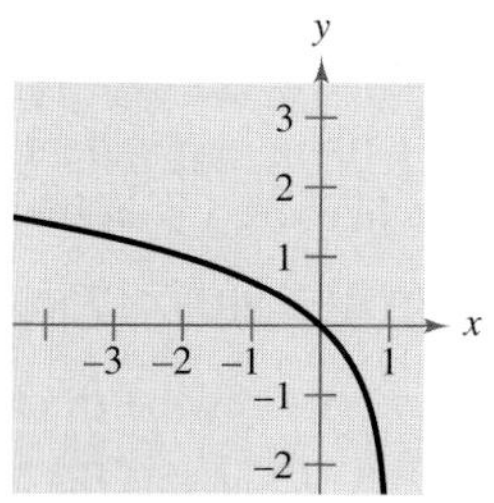

(c)

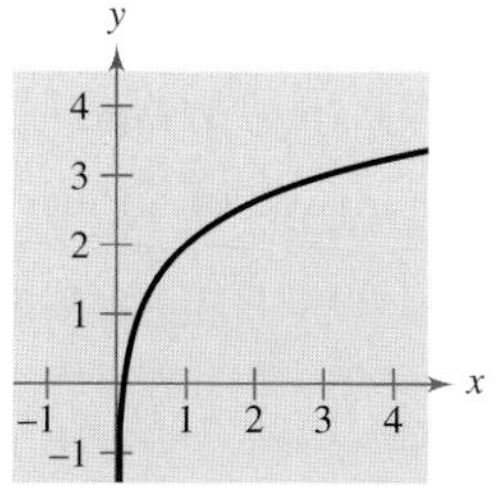

(d)

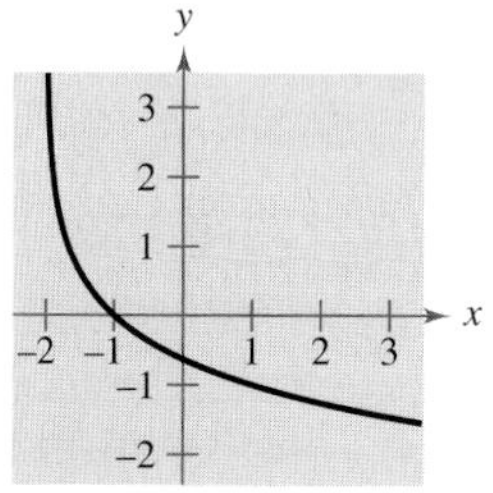

(e)

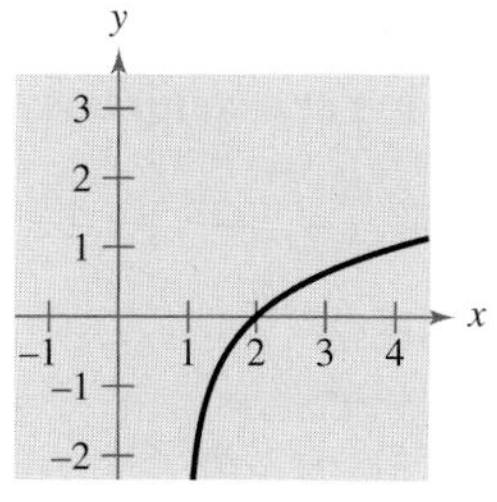

(f)

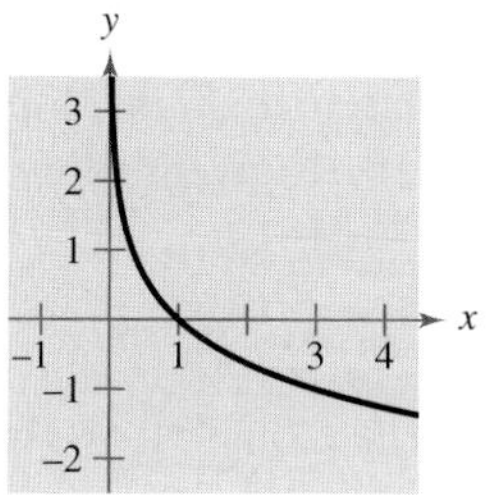

45. $f(x) = \log_3 x + 2$
46. $f(x) = -\log_3 x$
47. $f(x) = -\log_3(x + 2)$
48. $f(x) = \log_3(x - 1)$
49. $f(x) = \log_3(1 - x)$
50. $f(x) = -\log_3(-x)$

In Exercises 51–62, find the domain, x-intercept, and vertical asymptote of the logarithmic function and sketch its graph.

51. $f(x) = \log_4 x$
52. $g(x) = \log_6 x$
53. $y = -\log_3 x + 2$
54. $h(x) = \log_4(x - 3)$
55. $f(x) = -\log_6(x + 2)$
56. $y = \log_5(x - 1) + 4$
57. $y = \log_{10}\left(\frac{x}{5}\right)$
58. $y = \log_{10}(-x)$
59. $f(x) = \ln(x - 2)$
60. $h(x) = \ln(x + 1)$
61. $g(x) = \ln(-x)$
62. $f(x) = \ln(3 - x)$

In Exercises 63–68, use a graphing utility to graph the function. Be sure to use an appropriate viewing window.

63. $f(x) = \log_{10}(x + 1)$ **64.** $f(x) = \log_{10}(x - 1)$

65. $f(x) = \ln(x - 1)$ **66.** $f(x) = \ln(x + 2)$

67. $f(x) = \ln x + 2$ **68.** $f(x) = 3 \ln x - 1$

69. *Human Memory Model* Students in a mathematics class were given an exam and then retested monthly with an equivalent exam. The average scores for the class are given by the human memory model

$$f(t) = 80 - 17 \log_{10}(t + 1), \quad 0 \le t \le 12$$

where t is the time in months.

(a) What was the average score on the original exam $(t = 0)$?

(b) What was the average score after 4 months?

(c) What was the average score after 10 months?

70. *Population Growth* The population of a town will double in

$$t = \frac{10 \ln 2}{\ln 67 - \ln 50} \text{ years.}$$

Find t.

71. *Population* The time t in years for the world population to double if it is increasing at a continuous rate of r is given by

$$t = \frac{\ln 2}{r}.$$

(a) Complete the table.

r	0.005	0.01	0.015	0.02	0.025	0.03
t						

(b) Use a reference source to decide which value of r best approximates the actual rate of growth for the world population.

72. *Finance* A principal P, invested at $9\frac{1}{2}\%$ and compounded continuously, increases to an amount K times the original principal after t years, where t is given by

$$t = \frac{\ln K}{0.095}.$$

(a) Complete the table and interpret your results.

K	1	2	4	6	8	10	12
t							

(b) Sketch a graph of the function.

Ventilation **In Exercises 73 and 74, use the model**

$$y = 80.4 - 11 \ln x, \quad 100 \le x \le 1500$$

which approximates the minimum required ventilation rate in terms of the air space per child in a public school classroom. In the model, x is the air space per child in cubic feet and y is the ventilation rate in cubic feet per minute.

73. Use a graphing utility to graph the function and approximate the required ventilation rate if there is 300 cubic feet of air space per child.

74. A classroom is designed for 30 students. The air conditioning system in the room has the capacity of moving 450 cubic feet of air per minute.

(a) Determine the ventilation rate per child, assuming that the room is filled to capacity.

(b) Use the graph in Exercise 73 to estimate the air space required per child.

(c) Determine the minimum number of square feet of floor space required for the room if the ceiling height is 30 feet.

75. *Work* The work (in foot-pounds) done in compressing a volume of 9 cubic feet at a pressure of 15 pounds per square inch to a volume of 3 cubic feet is

$$W = 19{,}440(\ln 9 - \ln 3).$$

Find W.

76. *Sound Intensity* The relationship between the number of decibels β and the intensity of a sound I in watts per square meter is

$$\beta = 10 \log_{10}\left(\frac{I}{10^{-12}}\right).$$

(a) Determine the number of decibels of a sound with an intensity of 1 watt per square meter.

(b) Determine the number of decibels of a sound with an intensity of 10^{-2} watt per square meter.

(c) The intensity of the sound in part (a) is 100 times as great as that in part (b). Is the number of decibels 100 times as great? Explain.

Monthly Payment **In Exercises 77–80, use the model**

$$t = 12.542 \ln\left(\frac{x}{x - 1000}\right), \qquad x > 1000$$

which approximates the length of a home mortgage of \$150,000 at 8% in terms of the monthly payment. In the model, *t* is the length of the mortgage in years and *x* is the monthly payment in dollars (see figure).

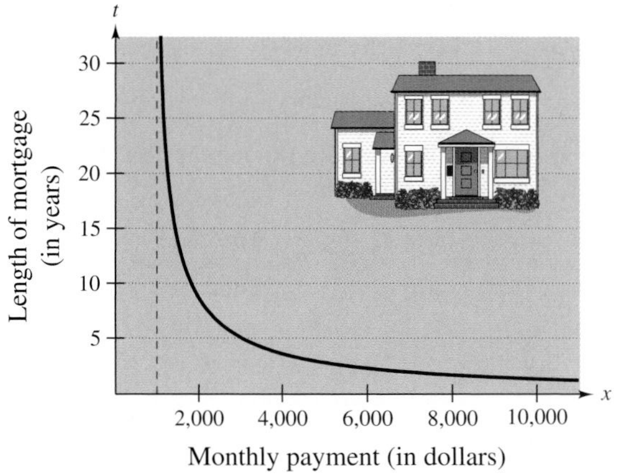

77. Use the model to approximate the length of a \$150,000 mortgage at 8% if the monthly payment is \$1100.65.

78. Use the model to approximate the length of a \$150,000 mortgage at 8% if the monthly payment is \$1254.68.

79. Approximate the total amount paid over the term of the mortgage in Exercise 77 with a monthly payment of \$1100.65. What is the total interest charge?

80. Approximate the total amount paid over the term of the mortgage in Exercise 78 with a monthly payment of \$1254.68. What is the total interest charge?

81. (a) Complete the table for the function

$$f(x) = \frac{\ln x}{x}.$$

x	1	5	10	10^2	10^4	10^6
$f(x)$						

(b) Use the table in part (a) to determine what value $f(x)$ approaches as x increases without bound.

(c) Use a graphing utility to confirm the result of part (b).

Synthesis

True or False? **In Exercises 82 and 83, determine whether the statement is true or false. Justify your answer.**

82. You can determine the graph of $f(x) = \log_6 x$ by graphing $g(x) = 6^x$ and reflecting it about the x-axis.

83. The graph of $f(x) = \log_3 x$ contains the point (27, 3).

In Exercises 84–87, describe the relationship between the graphs of *f* and *g*. What is the relationship between the functions *f* and *g*?

84. $f(x) = 3^x$
$g(x) = \log_3 x$

85. $f(x) = 5^x$
$g(x) = \log_5 x$

86. $f(x) = e^x$
$g(x) = \ln x$

87. $f(x) = 10^x$
$g(x) = \log_{10} x$

88. *Graphical Analysis* Use a graphing utility to graph f and g in the same viewing window and determine which is increasing at the greater rate as x approaches $+\infty$. What can you conclude about the rate of growth of the natural logarithmic function?

(a) $f(x) = \ln x, \quad g(x) = \sqrt{x}$

(b) $f(x) = \ln x, \quad g(x) = \sqrt[4]{x}$

89. *Exploration* The table of values was obtained by evaluating a function. Determine which of the statements may be true and which must be false.

x	1	2	8
y	0	1	3

(a) y is an exponential function of x.

(b) y is a logarithmic function of x.

(c) x is an exponential function of y.

(d) y is a linear function of x.

90. *Exploration* Use a graphing utility to compare the graph of the function $y = \ln x$ with the graph of each given function.

(a) $y_1 = x - 1$

(b) $y_2 = (x - 1) - \frac{1}{2}(x - 1)^2$

(c) $y_3 = (x - 1) - \frac{1}{2}(x - 1)^2 + \frac{1}{3}(x - 1)^3$

91. *Pattern Recognition* Identify the pattern of successive polynomials given in Exercise 90. Extend the pattern one more term and compare the graph of the resulting polynomial function with the graph of $y = \ln x$. What do you think the pattern implies?

92. *Exploration* Answer the following questions for the function $f(x) = \log_{10} x$. Do not use a calculator.

(a) What is the domain of f?

(b) What is f^{-1}?

(c) If x is a real number between 1000 and 10,000, in which interval will $f(x)$ be found?

(d) In which interval will x be found if $f(x)$ is negative?

(e) If $f(x)$ is increased by 1 unit, x must have been increased by what factor?

(f) If $f(x_1) = 3n$ and $f(x_2) = n$, what is the ratio of x_1 to x_2?

In Exercises 93–96, (a) use a graphing utility to graph the function, (b) use the graph to determine the intervals in which the function is increasing and decreasing, and (c) approximate any relative maximum or minimum values of the function.

93. $f(x) = |\ln x|$

94. $h(x) = \ln(x^2 + 1)$

95. $f(x) = \dfrac{x}{2} - \ln \dfrac{x}{4}$

96. $g(x) = \dfrac{12 \ln x}{x}$

Review

In Exercises 97–100, translate the statement into an algebraic expression.

97. The product of 8 and n is decreased by 3.

98. The total hourly wage for an employee is \$9.25 per hour plus 75 cents for each of q units produced per hour.

99. The total cost for auto repairs if the cost of parts was \$83.95 and there were t hours of labor at \$37.50 per hour

100. The area of a rectangle if the length is 10 units more than the width w

In Exercises 101–104, find the asymptotes of the rational function.

101. $f(x) = \dfrac{4}{-8 - x}$

102. $f(x) = \dfrac{2x^3 - 3}{x^2}$

103. $f(x) = \dfrac{x + 5}{2x^2 + x - 15}$

104. $f(x) = \dfrac{2x^2(x - 5)}{x - 7}$

In Exercises 105–108, evaluate the expression. Round your result to three decimal places.

105. e^6

106. $e^{3/2}$

107. e^{-4}

108. $4e^{-6}$

3.3 Properties of Logarithms

▶ **What you should learn**

- How to rewrite logarithmic functions with a different base
- How to use properties of logarithms to evaluate or rewrite logarithmic expressions
- How to use properties of logarithms to expand or condense logarithmic expressions
- How to use logarithmic functions to model and solve real-life applications

▶ **Why you should learn it**

Logarithmic functions are often used to model scientific observations. For instance, Exercise 85 on page 316 shows how to use a logarithmic function to model human memory.

Gary Conner/PhotoEdit

Change of Base

Most calculators have only two types of log keys, one for common logarithms (base 10) and one for natural logarithms (base e). Although common logs and natural logs are the most frequently used, you may occasionally need to evaluate logarithms to other bases. To do this, you can use the following *change-of-base formula.*

Change-of-Base Formula

Let a, b, and x be positive real numbers such that $a \neq 1$ and $b \neq 1$. Then $\log_a x$ can be converted to a different base as follows.

Base b	*Base 10*	*Base e*
$\log_a x = \dfrac{\log_b x}{\log_b a}$	$\log_a x = \dfrac{\log_{10} x}{\log_{10} a}$	$\log_a x = \dfrac{\ln x}{\ln a}$

One way to look at the change-of-base formula is that logarithms to base a are simply *constant multiples* of logarithms to base b. The constant multiplier is $1/(\log_b a)$.

Example 1 ▶ Changing Bases Using Common Logarithms

a. $\log_4 30 = \dfrac{\log_{10} 30}{\log_{10} 4}$ $\log_a b = \dfrac{\log_{10} b}{\log_{10} a}$

$\approx \dfrac{1.47712}{0.60206}$ Use a calculator.

≈ 2.4534 Use a calculator.

b. $\log_2 14 = \dfrac{\log_{10} 14}{\log_{10} 2} \approx \dfrac{1.14613}{0.30103} \approx 3.8074$

Example 2 ▶ Changing Bases Using Natural Logarithms

a. $\log_4 30 = \dfrac{\ln 30}{\ln 4}$ $\log_a b = \dfrac{\ln b}{\ln a}$

$\approx \dfrac{3.40120}{1.38629}$ Use a calculator.

≈ 2.4534 Use a calculator.

b. $\log_2 14 = \dfrac{\ln 14}{\ln 2} \approx \dfrac{2.63906}{0.693147} \approx 3.8074$

Properties of Logarithms

You know from the preceding section that the logarithmic function with base a is the *inverse* of the exponential function with base a. So, it makes sense that the properties of exponents should have corresponding properties involving logarithms. For instance, the exponential property $a^0 = 1$ has the corresponding logarithmic property $\log_a 1 = 0$.

Encourage your students to know these properties well. They will be used for solving logarithmic and exponential equations, as well as in calculus.

STUDY TIP

There is no general property that can be used to rewrite $\log_a(u \pm v)$. Specifically, $\log_a(x + y)$ is not equal to $\log_a x + \log_a y$.

Properties of Logarithms

Let a be a positive number such that $a \neq 1$, and let n be a real number. If u and v are positive real numbers, the following properties are true.

1. $\log_a(uv) = \log_a u + \log_a v$	**1.** $\ln(uv) = \ln u + \ln v$
2. $\log_a \dfrac{u}{v} = \log_a u - \log_a v$	**2.** $\ln \dfrac{u}{v} = \ln u - \ln v$
3. $\log_a u^n = n \log_a u$	**3.** $\ln u^n = n \ln u$

Remind your students to note the domain when applying the properties of logarithms to a logarithmic function. For example, the domain of $f(x) = \ln x^2$ is all real $x \neq 0$, whereas the domain of $g(x) = 2 \ln x$ is all real $x > 0$.

A proof of the first property listed above is given in Appendix A.

Example 3 ▶ Using Properties of Logarithms

Write the logarithm in terms of ln 2 and ln 3.

a. $\ln 6$ **b.** $\ln \dfrac{2}{27}$

Solution

a. $\ln 6 = \ln(2 \cdot 3)$ Rewrite 6 as $2 \cdot 3$.

$= \ln 2 + \ln 3$ Property 1

b. $\ln \dfrac{2}{27} = \ln 2 - \ln 27$ Property 2

$= \ln 2 - \ln 3^3$ Rewrite 27 as 3^3.

$= \ln 2 - 3 \ln 3$ Property 3

The Granger Collection

Historical Note

John Napier, a Scottish mathematician, developed logarithms as a way to simplify some of the tedious calculations of his day. Beginning in 1594, Napier worked about 20 years on the invention of logarithms. Napier was only partially successful in his quest to simplify tedious calculations. Nonetheless, the development of logarithms was a step forward and received immediate recognition.

Example 4 ▶ Using Properties of Logarithms

Use the properties of logarithms to verify that $-\log_{10} \frac{1}{100} = \log_{10} 100$.

Solution

$-\log_{10} \frac{1}{100} = -\log_{10}(100^{-1})$ Rewrite $\frac{1}{100}$ as 100^{-1}.

$= -(-1)\log_{10} 100$ Property 3

$= \log_{10} 100$ Simplify.

Try checking this result on your calculator.

A common error made in expanding logarithmic expressions is to rewrite $\log ax^n$ as $n \log ax$ instead of as $\log a + n \log x$.

Rewriting Logarithmic Expressions

The properties of logarithms are useful for rewriting logarithmic expressions in forms that simplify the operations of algebra. This is true because these properties convert complicated products, quotients, and exponential forms into simpler sums, differences, and products, respectively.

Example 5 ▶ Expanding Logarithmic Expressions

Expand the logarithmic expressions.

a. $\log_{10} 5x^3y$ **b.** $\ln \dfrac{\sqrt{3x - 5}}{7}$

A common error made in condensing logarithmic expressions is to rewrite $\log x - \log y$ as $\dfrac{\log x}{\log y}$ instead of as $\log \dfrac{x}{y}$.

Solution

a. $\log_{10} 5x^3y = \log_{10} 5 + \log_{10} x^3y$ Property 1

$= \log_{10} 5 + \log_{10} x^3 + \log_{10} y$ Property 1

$= \log_{10} 5 + 3 \log_{10} x + \log_{10} y$ Property 3

b. $\ln \dfrac{\sqrt{3x - 5}}{7} = \ln \dfrac{(3x - 5)^{1/2}}{7}$ Rewrite using rational exponent.

$= \ln(3x - 5)^{1/2} - \ln 7$ Property 2

$= \dfrac{1}{2} \ln(3x - 5) - \ln 7$ Property 3

In Example 5, the properties of logarithms were used to *expand* logarithmic expressions. In Example 6, this procedure is reversed and the properties of logarithms are used to *condense* logarithmic expressions.

Exploration

Use a graphing utility to graph the functions

$$y = \ln x - \ln(x - 3)$$

and

$$y = \ln \frac{x}{x - 3}$$

in the same viewing window. Does the graphing utility show the functions with the same domain? If so, should it? Explain your reasoning.

Example 6 ▶ Condensing Logarithmic Expressions

Condense the logarithmic expressions.

a. $\frac{1}{2} \log_{10} x + 3 \log_{10}(x + 1)$

b. $2 \ln(x + 2) - \ln x$

Solution

a. $\frac{1}{2} \log_{10} x + 3 \log_{10}(x + 1) = \log_{10} x^{1/2} + \log_{10}(x + 1)^3$ Property 3

$= \log_{10}\left[\sqrt{x} \cdot (x + 1)^3\right]$ Property 1

b. $2 \ln(x + 2) - \ln x = \ln(x + 2)^2 - \ln x$ Property 3

$= \ln \dfrac{(x + 2)^2}{x}$ Property 2

Application

One method of determining how the x- and y-values for a set of nonlinear data are related begins by taking the natural log of each of the x- and y-values. If the points are graphed and fall on a straight line, then you can determine that the x- and y-values are related by the equation

$$\ln y = m \ln x$$

where m is the slope of the straight line.

Example 7 ▶ Finding a Mathematical Model

The table gives the mean distance x and the period y of the six planets that are closest to the sun. In the table, the mean distance is given in terms of astronomical units (where the earth's mean distance is defined as 1.0), and the period is given in terms of years. Find an equation that expresses y as a function of x.

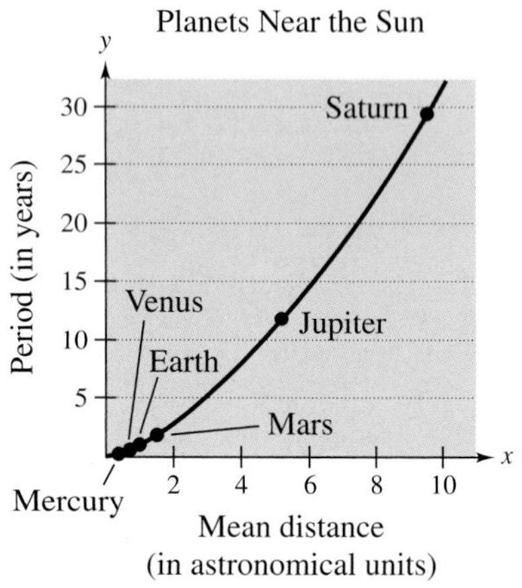

FIGURE 3.17

Planet	Mercury	Venus	Earth	Mars	Jupiter	Saturn
Period, y	0.241	0.615	1.0	1.881	11.862	29.458
Mean distance, x	0.387	0.723	1.0	1.524	5.203	9.539

Solution

The points in the table are plotted in Figure 3.17. From this figure it is not clear how to find an equation that relates y and x. To solve this problem, take the natural log of each of the x- and y-values given in the table. This produces the following results.

Planet	Mercury	Venus	Earth	Mars	Jupiter	Saturn
$\ln y$	−1.423	−0.486	0.0	0.632	2.473	3.383
$\ln x$	−0.949	−0.324	0.0	0.421	1.649	2.255

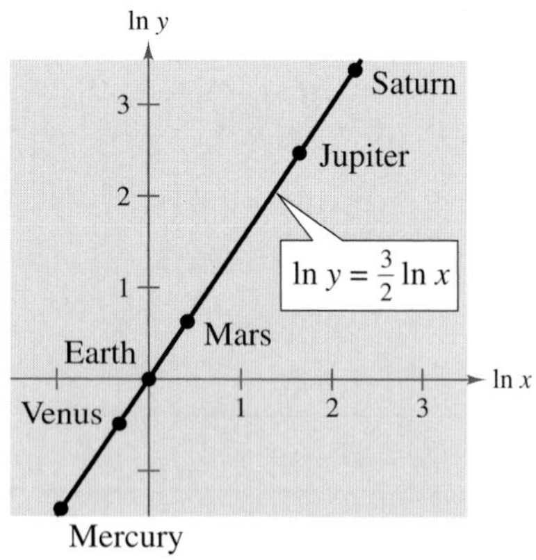

FIGURE 3.18

Now, by plotting the points in the second table, you can see that all six of the points appear to lie in a line (see Figure 3.18). You can use a graphical approach or the algebraic approach discussed in Section 1.8 to find that the slope of this line is $\frac{3}{2}$. You can therefore conclude that

$$\ln y = \frac{3}{2} \ln x.$$

Try to convert this to $y = f(x)$ form. You will get a function of the form $y = ax^b$, which is a power model (the variable x is raised to a power, b).

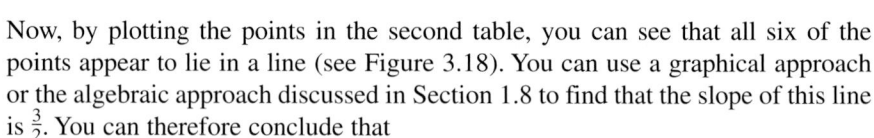

3.3 Exercises

In Exercises 1–8, evaluate the logarithm using the change-of-base formula. Round your result to three decimal places.

1. $\log_3 7$ **2.** $\log_7 4$

3. $\log_{1/2} 4$ **4.** $\log_{1/4} 5$

5. $\log_9 0.4$ **6.** $\log_{20} 0.125$

7. $\log_{15} 1250$ **8.** $\log_3 0.015$

In Exercises 9–16, rewrite the logarithm as a ratio of (a) common logarithms and (b) natural logarithms.

9. $\log_5 x$ **10.** $\log_3 x$

11. $\log_{1/5} x$ **12.** $\log_{1/3} x$

13. $\log_x \frac{3}{10}$ **14.** $\log_x \frac{3}{4}$

15. $\log_{2.6} x$ **16.** $\log_{7.1} x$

In Exercises 17–22, use the change-of-base formula to rewrite the logarithm as a ratio of logarithms. Then use a graphing utility to sketch the graph.

17. $f(x) = \log_2 x$ **18.** $f(x) = \log_4 x$

19. $f(x) = \log_{1/2} x$ **20.** $f(x) = \log_{1/4} x$

21. $f(x) = \log_{11.8} x$ **22.** $f(x) = \log_{12.4} x$

In Exercises 23–42, use the properties of logarithms to expand the expression as a sum, difference, and/or constant multiple of logarithms. (Assume all variables are positive.)

23. $\log_{10} 5x$ **24.** $\log_{10} 10z$

25. $\log_{10} \dfrac{5}{x}$ **26.** $\log_{10} \dfrac{y}{2}$

27. $\log_8 x^4$ **28.** $\log_6 z^{-3}$

29. $\ln \sqrt{z}$ **30.** $\ln \sqrt[3]{t}$

31. $\ln xyz$ **32.** $\ln \dfrac{xy}{z}$

33. $\ln \sqrt{a-1},\ a > 1$ **34.** $\ln\left(\dfrac{x^2-1}{x^3}\right),\ x > 1$

35. $\ln z(z-1)^2,\ z > 1$ **36.** $\ln \dfrac{x}{\sqrt{x^2+1}}$

37. $\ln \sqrt[3]{\dfrac{x}{y}}$ **38.** $\ln \sqrt{\dfrac{x^2}{y^3}}$

39. $\ln \dfrac{x^4\sqrt{y}}{z^5}$ **40.** $\ln \sqrt{x^2(x+2)}$

41. $\log_b \dfrac{x^2}{y^2z^3}$ **42.** $\log_b \dfrac{\sqrt{x}\,y^4}{z^4}$

In Exercises 43–62, condense the expression to the logarithm of a single quantity.

43. $\ln x + \ln 3$

44. $\ln y + \ln t$

45. $\log_4 z - \log_4 y$

46. $\log_5 8 - \log_5 t$

47. $2 \log_2(x+4)$

48. $-4 \log_6 2x$

49. $\frac{1}{4} \log_3 5x$

50. $\frac{2}{3} \log_7(z-2)$

51. $\ln x - 3 \ln(x+1)$

52. $2 \ln 8 + 5 \ln z$

53. $\ln(x-2) - \ln(x+2)$

54. $3 \ln x + 4 \ln y - 4 \ln z$

55. $\ln x - 4[\ln(x+2) + \ln(x-2)]$

56. $4[\ln z + \ln(z+5)] - 2 \ln(z-5)$

57. $\frac{1}{3}[2 \ln(x+3) + \ln x - \ln(x^2-1)]$

58. $2[\ln x - \ln(x+1) - \ln(x-1)]$

59. $\frac{1}{3}[\ln y + 2 \ln(y+4)] - \ln(y-1)$

60. $\frac{1}{2}[\ln(x+1) + 2 \ln(x-1)] + 6 \ln x$

61. $2 \ln 3 - \frac{1}{2} \ln(x^2+1)$

62. $\frac{3}{2} \ln 5t^6 - \frac{3}{4} \ln t^4$

In Exercises 63 and 64, compare the logarithmic quantities. If two are equal, explain why.

63. $\dfrac{\log_2 32}{\log_2 4}$, $\log_2 \dfrac{32}{4}$, $\log_2 32 - \log_2 4$

64. $\log_7 \sqrt{70}$, $\log_7 35$, $\frac{1}{2} + \log_7 \sqrt{10}$

In Exercises 65–78, find the exact value of the logarithm without using a calculator. (If this is not possible, state the reason.)

65. $\log_3 9$ **66.** $\log_6 \sqrt[3]{6}$

67. $\log_4 16^{1.2}$ **68.** $\log_5 \frac{1}{125}$

69. $\log_3(-9)$ **70.** $\log_2(-16)$

71. $\log_5 75 - \log_5 3$ **72.** $\log_4 2 + \log_4 32$

73. $\ln e^2 - \ln e^5$ **74.** $3 \ln e^4$

75. $\log_{10} 0$ **76.** $\ln 1$

77. $\ln e^{4.5}$ **78.** $\ln \sqrt[4]{e^3}$

In Exercises 79–84, use the properties of logarithms to rewrite and simplify the logarithmic expression.

79. $\log_4 8$

80. $\log_2(4^2 \cdot 3^4)$

81. $\log_5 \frac{1}{250}$

82. $\log_{10} \frac{9}{300}$

83. $\ln(5e^6)$

84. $\ln \dfrac{6}{e^2}$

85. *Human Memory Model* Students participating in a psychological experiment attended several lectures and were given an exam. Every month for a year after the exam, the students were retested to see how much of the material they remembered. The average scores for the group can be modeled by the memory model

$$f(t) = 90 - 15 \log_{10}(t + 1), \qquad 0 \le t \le 12$$

where t is the time in months.

(a) What was the average score on the original exam $(t = 0)$?

(b) What was the average score after 6 months?

(c) What was the average score after 12 months?

(d) When will the average score decrease to 75?

(e) Use the properties of logarithms to write the function in another form.

(f) Sketch the graph of the function over the specified domain.

86. *Sound Intensity* The relationship between the number of decibels β and the intensity of a sound I in watts per square meter is

$$\beta = 10 \log_{10}\left(\frac{I}{10^{-12}}\right).$$

Use the properties of logarithms to write the formula in simpler form, and determine the number of decibels of a sound with an intensity of 10^{-6} watt per square meter.

Synthesis

True or False? **In Exercises 87–92, determine whether the statement is true or false given that $f(x) = \ln x$. Justify your answer.**

87. $f(0) = 0$

88. $f(ax) = f(a) + f(x), \quad a > 0, x > 0$

89. $f(x - 2) = f(x) - f(2), \quad x > 2$

90. $\sqrt{f(x)} = \frac{1}{2}f(x)$

91. If $f(u) = 2f(v)$, then $v = u^2$.

92. If $f(x) < 0$, then $0 < x < 1$.

93. Prove that $\log_b \dfrac{u}{v} = \log_b u - \log_b v$.

94. Prove that $\log_b u^n = n \log_b u$.

In Exercises 95 and 96, use a graphing utility to graph the two functions in the same viewing window. Use the graphs to verify that the expressions are equivalent.

95. $f(x) = \log_{10} x$

$g(x) = \dfrac{\ln x}{\ln 10}$

96. $f(x) = \ln x$

$g(x) = \dfrac{\log_{10} x}{\log_{10} e}$

97. *Think About It* Sketch the graphs of

$$f(x) = \ln \frac{x}{2}, \quad g(x) = \frac{\ln x}{\ln 2}, \quad h(x) = \ln x - \ln 2$$

on the same set of axes. Which two functions have identical graphs? Explain why.

98. *Exploration* Approximate the natural logarithms of as many integers as possible between 1 and 20 given that $\ln 2 \approx 0.6931$, $\ln 3 \approx 1.0986$, and $\ln 5 \approx 1.6094$. (Do not use a calculator.)

Review

In Exercises 99–102, simplify the expression.

99. $\dfrac{24xy^{-2}}{16x^{-3}y}$

100. $\left(\dfrac{2x^2}{3y}\right)^{-3}$

101. $(18x^3y^4)^{-3}(18x^3y^4)^3$

102. $xy(x^{-1} + y^{-1})^{-1}$

In Exercises 103–108, use a calculator to evaluate the expression. Round your result to three decimal places.

103. $(2.8)^{7.6}$

104. $40(6^{-3.2})$

105. $7^{-\pi}$

106. $1.4^{3\pi}$

107. $\sqrt[4]{350}$

108. $96^{\sqrt{3}}$

In Exercises 109–112, use a calculator to evaluate the logarithm. Round your result to three decimal places.

109. $\log_{10} 26$

110. $\log_{10} \frac{5}{8}$

111. $\ln 10.6$

112. $\ln(\sqrt{7} + 1)$

3.4 Exponential and Logarithmic Equations

▶ **What you should learn**

- How to solve simple exponential and logarithmic equations
- How to solve more complicated exponential equations
- How to solve more complicated logarithmic equations
- How to use exponential and logarithmic equations to model and solve real-life applications

▶ **Why you should learn it**

Applications of exponential and logarithmic equations are found in consumer safety testing. For instance, Exercise 122 on page 327 shows how to use a logarithmic function to model crumple zones for automobile crash tests.

David Woods/The Stock Market

Introduction

So far in this chapter, you have studied the definitions, graphs, and properties of exponential and logarithmic functions. In this section, you will study procedures for *solving equations* involving these exponential and logarithmic functions.

There are two basic strategies for solving exponential or logarithmic equations. The first is based on the One-to-One Properties and the second is based on the Inverse Properties. For $a > 0$ and $a \neq 1$, the following properties are true for all x and y for which $\log_a x$ and $\log_a y$ are defined.

One-to-One Properties

$a^x = a^y$ if and only if $x = y$.

$\log_a x = \log_a y$ if and only if $x = y$.

Inverse Properties

$a^{\log_a x} = x$

$\log_a a^x = x$

Example 1 ▶ Solving Simple Equations

	Original Equation	*Rewritten Equation*	*Solution*	*Property*
a.	$2^x = 32$	$2^x = 2^5$	$x = 5$	One-to-One
b.	$\ln x - \ln 3 = 0$	$\ln x = \ln 3$	$x = 3$	One-to-One
c.	$\left(\frac{1}{3}\right)^x = 9$	$3^{-x} = 3^2$	$x = -2$	One-to-One
d.	$e^x = 7$	$\ln e^x = \ln 7$	$x = \ln 7$	Inverse
e.	$\ln x = -3$	$e^{\ln x} = e^{-3}$	$x = e^{-3}$	Inverse
f.	$\log_{10} x = -1$	$10^{\log_{10} x} = 10^{-1}$	$x = 10^{-1} = \frac{1}{10}$	Inverse

The strategies used in Example 1 are summarized as follows.

Strategies for Solving Exponential and Logarithmic Equations

1. Rewrite the given equation in a form that allows the use of the One-to-One Properties of exponential or logarithmic functions.
2. Rewrite an *exponential* equation in logarithmic form and apply the Inverse Property of logarithmic functions.
3. Rewrite a *logarithmic* equation in exponential form and apply the Inverse Property of exponential functions.

Before starting this section, have students review the logarithmic properties.

Solving Exponential Equations

Example 2 ▶ Solving Exponential Equations

Solve each equation and approximate the result to three decimal places.

a. $e^x = 72$ **b.** $3(2^x) = 42$

Solution

a.

$e^x = 72$	Write original equation.
$\ln e^x = \ln 72$	Take natural log of each side.
$x = \ln 72$	Inverse Property
$x \approx 4.277$	Use a calculator.

The solution is $\ln 72 \approx 4.277$. Check this in the original equation.

b.

$3(2^x) = 42$	Write original equation.
$2^x = 14$	Divide each side by 3.
$\log_2 2^x = \log_2 14$	Take log (base 2) of each side.
$x = \log_2 14$	Inverse Property
$x = \dfrac{\ln 14}{\ln 2}$	Change-of-base formula
$x \approx 3.807$	Use a calculator.

The solution is $\log_2 14 \approx 3.807$. Check this in the original equation.

Technology

When solving an exponential or logarithmic equation, remember that you can check your solution graphically by "graphing the left and right sides separately" and using the intersect feature of your graphing utility to determine the point of intersection. For instance, to check the solution of the equation in Example 2(a), you can graph

$$y = e^x \quad \text{and} \quad y = 72$$

in the same viewing window, as shown below. Using the intersect feature of your graphing utility, you can determine that the graphs intersect when $x \approx 4.277$, which confirms the solution found in Example 2(a).

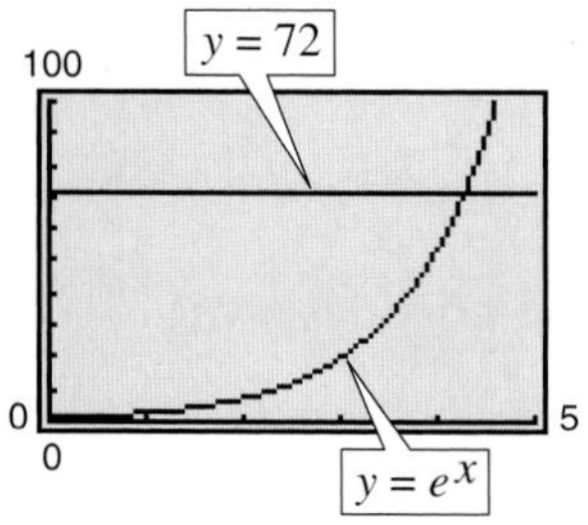

In Example 2(a), the exact solution is $x = \ln 72$ and the approximate solution is $x \approx 4.277$. An exact answer is preferred when the solution is an intermediate step in a larger problem. For a final answer, an approximate solution is easier to comprehend.

Example 3 ▶ Solving an Exponential Equation

Solve $e^x + 5 = 60$ and approximate the result to three decimal places.

Solution

$e^x + 5 = 60$	Write original equation.
$e^x = 55$	Subtract 5 from each side.
$\ln e^x = \ln 55$	Take natural log of each side.
$x = \ln 55$	Inverse Property
$x \approx 4.007$	Use a calculator.

The solution is $\ln 55 \approx 4.007$. Check this in the original equation.

Additional Example: A Base Other than *e*

$$2^x = 10$$
$$\ln 2^x = \ln 10$$
$$x \ln 2 = \ln 10$$
$$x = \frac{\ln 10}{\ln 2}$$

Note: Using the change-of-base formula, you could write this solution as $x = \log_2 10$.

Example 4 ▶ Solving an Exponential Equation

Solve $2(3^{2t-5}) - 4 = 11$ and approximate the result to three decimal places.

Solution

$2(3^{2t-5}) - 4 = 11$	Write original equation.
$2(3^{2t-5}) = 15$	Add 4 to each side.
$3^{2t-5} = \frac{15}{2}$	Divide each side by 2.
$\log_3 3^{2t-5} = \log_3 \frac{15}{2}$	Take log (base 3) of each side.
$2t - 5 = \log_3 \frac{15}{2}$	Inverse Property
$2t = 5 + \log_3 7.5$	Add 5 to each side.
$t = \frac{5}{2} + \frac{1}{2}\log_3 7.5$	Divide each side by 2.
$t \approx 3.417$	Use a calculator.

The solution is $\frac{5}{2} + \frac{1}{2}\log_3 7.5 \approx 3.417$. Check this in the original equation.

When an equation involves two or more exponential expressions, you can still use a procedure similar to that demonstrated in Examples 2, 3, and 4. However, the algebra is a bit more complicated.

Example 5 ▶ Solving an Exponential Equation of Quadratic Type

Solve $e^{2x} - 3e^x + 2 = 0$.

Solution

$e^{2x} - 3e^x + 2 = 0$	Write original equation.
$(e^x)^2 - 3e^x + 2 = 0$	Write in quadratic form.
$(e^x - 2)(e^x - 1) = 0$	Factor.
$e^x - 2 = 0$	Set 1st factor equal to 0.
$x = \ln 2$	Solution
$e^x - 1 = 0$	Set 2nd factor equal to 0.
$x = 0$	Solution

The solutions are ln 2 and 0. Check these in the original equation.

To ensure that students first solve for the unknown variable algebraically and then use their calculators, you can require both exact algebraic solutions and approximate numerical answers.

In Example 5, use a graphing utility to graph $y = e^{2x} - 3e^x + 2$. The graph should have two x-intercepts: one at $x = \ln 2 \approx 0.693$ and one at $x = 0$.

The *Interactive* CD-ROM and *Internet* versions of this text show every example with its solution; clicking on the *Try It!* button brings up similar problems. Guided Examples and Integrated Examples show step-by-step solutions to additional examples. Integrated Examples are related to several concepts in the section.

Solving Logarithmic Equations

To solve a logarithmic equation such as

$\ln x = 3$ Logarithmic form

write the equation in exponential form as follows.

$e^{\ln x} = e^3$ Exponentiate each side.

$x = e^3$ Exponential form

This procedure is called *exponentiating* both sides of an equation.

Example 6 ▶ Solving a Logarithmic Equation

a. Solve $\ln x = 2$. **b.** Solve $\log_3(5x - 1) = \log_3(x + 7)$.

Solution

a. $\ln x = 2$ Write original equation.

$e^{\ln x} = e^2$ Exponentiate each side.

$x = e^2$ Inverse Property

The solution is e^2. Check this in the original equation.

b. $\log_3(5x - 1) = \log_3(x + 7)$ Write original equation.

$5x - 1 = x + 7$ One-to-One Property

$4x = 8$ Add $-x$ and 1 to each side.

$x = 2$ Divide each side by 4.

The solution is 2. Check this in the original equation.

Example 7 ▶ Solving a Logarithmic Equation

Solve $5 + 2 \ln x = 4$ and approximate the result to three decimal places.

Solution

$5 + 2 \ln x = 4$ Write original equation.

$2 \ln x = -1$ Subtract 5 from each side.

$\ln x = -\frac{1}{2}$ Divide each side by 2.

$e^{\ln x} = e^{-1/2}$ Exponentiate each side.

$x = e^{-1/2}$ Inverse Property

$x \approx 0.607$ Use a calculator.

The solution is $e^{-1/2} \approx 0.607$. To check this result graphically, you can use a graphing utility to graph $y = 5 + 2 \ln x$ and $y = 4$ in the same viewing window, as shown in Figure 3.19.

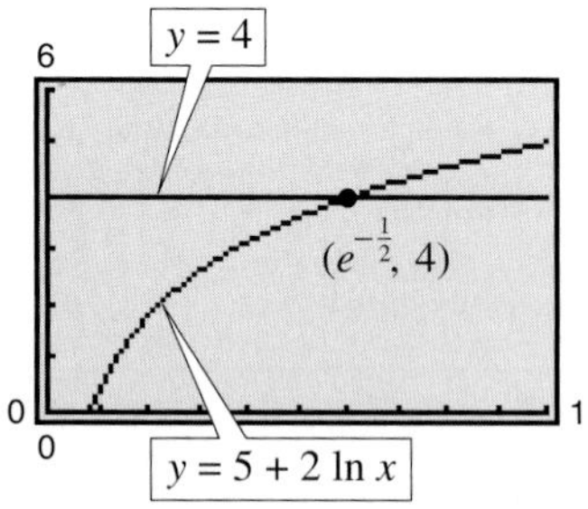

FIGURE 3.19

Activities

1. Write two or three sentences stating the general guidelines that you follow when solving (a) exponential equations and (b) logarithmic equations.

   Answer: (a) To solve exponential equations, it is useful first to isolate the exponential expression, then take the logarithms of both sides and solve for the variable. (b) To solve logarithmic equations, condense the logarithmic part into a single logarithm, then rewrite in exponential form and solve for the variable.

2. Solve for x: $7^x = 3$.

   Answer: $x = \log_7 3 = \dfrac{\ln 3}{\ln 7} \approx 0.5646$

3. Solve for x: $\log_{10}(x + 4) + \log_{10}(x + 1) = 1$.

   Answer: $x = 1$ ($x = -6$ is not in the domain.)

Example 8 ▶ Solving a Logarithmic Equation

Solve $2 \log_5 3x = 4$.

Solution

$2 \log_5 3x = 4$ — Write original equation.

$\log_5 3x = 2$ — Divide each side by 2.

$5^{\log_5 3x} = 5^2$ — Exponentiate each side (base 5).

$3x = 25$ — Inverse Property

$x = \dfrac{25}{3}$ — Divide each side by 3.

The solution is $\frac{25}{3}$. Check this in the original equation. Or try performing a graphical check by graphing the functions

$$y = 2 \log_5 3x \qquad \text{and} \qquad y = 4$$

on the same screen. The two graphs should intersect when $x = \frac{25}{3}$ and $y = 4$.

Because the domain of a logarithmic function generally does not include all real numbers, you should be sure to check for extraneous solutions of logarithmic equations.

STUDY TIP

In Example 9 the domain of $\log_{10} 5x$ is $x > 0$ and the domain of $\log_{10}(x - 1)$ is $x > 1$, so the domain of the original equation is $x > 1$. Because the domain is all real numbers greater than 1, the solution $x = -4$ is extraneous.

Example 9 ▶ Checking for Extraneous Solutions

Solve $\log_{10} 5x + \log_{10}(x - 1) = 2$.

Solution

$\log_{10} 5x + \log_{10}(x - 1) = 2$ — Write original equation.

$\log_{10}[5x(x - 1)] = 2$ — Product Property of Logarithms

$10^{\log_{10}(5x^2 - 5x)} = 10^2$ — Exponentiate each side (base 10).

$5x^2 - 5x = 100$ — Inverse Property

$x^2 - x - 20 = 0$ — Write in general form.

$(x - 5)(x + 4) = 0$ — Factor.

$x - 5 = 0$ — Set 1st factor equal to 0.

$x = 5$ — Solution

$x + 4 = 0$ — Set 2nd factor equal to 0.

$x = -4$ — Solution

The solutions appear to be 5 and -4. However, when you check these in the original equation or use a graphical check, you can see that $x = 5$ is the only solution.

Applications

Determine the amount of time it would take \$1000 to double in an account that pays 6.75% interest, compounded continuously. How does this compare to Example 10?

Answer: ≈ 10.27 years; it takes the same amount of time.

Example 10 ▶ Doubling an Investment

You have deposited \$500 in an account that pays 6.75% interest, compounded continuously. How long will it take your money to double?

Solution

Using the formula for continuous compounding, you can find that the balance in the account is

$$A = Pe^{rt}$$

$$A = 500e^{0.0675t}.$$

To find the time required for the balance to double, let $A = 1000$ and solve the resulting equation for t.

$$500e^{0.0675t} = 1000 \qquad \text{Let } A = 1000.$$

$$e^{0.0675t} = 2 \qquad \text{Divide each side by 500.}$$

$$\ln e^{0.0675t} = \ln 2 \qquad \text{Take natural log of each side.}$$

$$0.0675t = \ln 2 \qquad \text{Inverse Property}$$

$$t = \frac{\ln 2}{0.0675} \qquad \text{Divide each side by 0.0675.}$$

$$t \approx 10.27 \qquad \text{Use a calculator.}$$

The balance in the account will double after approximately 10.27 years. This result is demonstrated graphically in Figure 3.20.

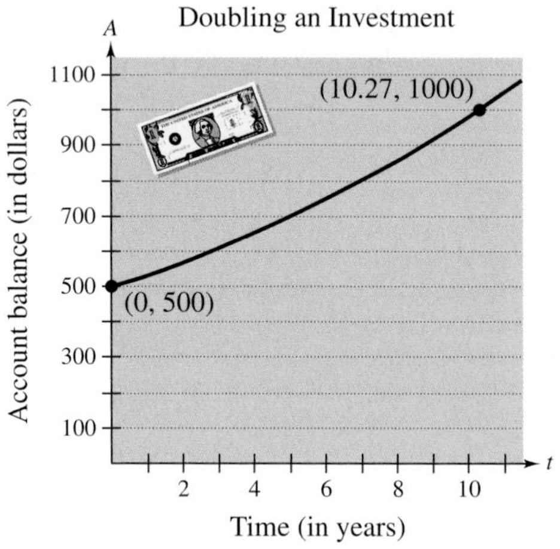

FIGURE 3.20

Exploration

The *effective yield* of a savings plan is the percent increase in the balance after 1 year. Find the effective yields for the following savings plans when \$1000 is deposited in a savings account.

a. 7% annual interest rate, compounded annually

b. 7% annual interest rate, compounded continuously

c. 7% annual interest rate, compounded quarterly

d. 7.25% annual interest rate, compounded quarterly

Which savings plan has the greatest effective yield? Which savings plan will have the highest balance after 5 years?

In Example 10, an approximate answer of 10.27 years is given. The exact solution, $(\ln 2)/0.0675$ years, does not make sense as an answer.

Example 11 ▶ Consumer Price Index for Sugar

From 1970 to 1997, the Consumer Price Index (CPI) value y for a fixed amount of sugar for the year t can be modeled by the equation

$$y = -171.8 + 87.1 \ln t$$

where $t = 10$ represents 1970 (see Figure 3.21). During which year did the price of sugar reach 4.5 times its 1970 price of 30.5 on the CPI? (Source: U.S. Bureau of Labor Statistics)

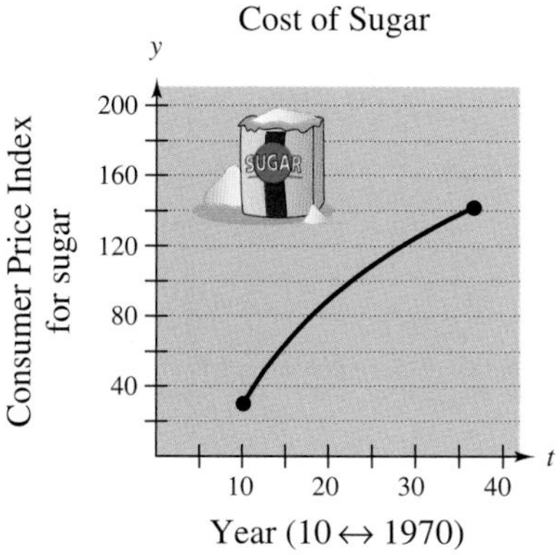

FIGURE 3.21

Solution

$$-171.8 + 87.1 \ln t = y \quad \text{Write original equation.}$$

$$-171.8 + 87.1 \ln t = 137.25 \quad \text{Let } y = (4.5)(30.5) = 137.25.$$

$$87.1 \ln t = 309.05 \quad \text{Add 171.8 to each side.}$$

$$\ln t \approx 3.548 \quad \text{Divide each side by 87.1.}$$

$$e^{\ln t} \approx e^{3.548} \quad \text{Exponentiate each side.}$$

$$t \approx e^{3.548} \quad \text{Inverse Property}$$

$$t \approx 35 \quad \text{Use a calculator.}$$

The solution is $t \approx 35$ years. Because $t = 10$ represents 1970, it follows that the price of sugar reached 4.5 times its 1970 price in 1995.

Writing ABOUT MATHEMATICS

Comparing Mathematical Models The table gives the numbers y (in millions) of single compact discs (CDs) shipped annually by manufacturers from 1993 through 1997, where $x = 3$ represents 1993. (Source: Recording Industry Association of America)

x	3	4	5	6	7
y	7.8	9.3	21.5	43.2	66.7

a. Create a scatter plot of the data. Find a linear model for the data, and add its graph to your scatter plot. According to this model, when will the annual shipment of single CDs reach 100 million?

b. Create a new table giving values for $\ln x$ and $\ln y$ and create a scatter plot of this transformed data. Use the method illustrated in Example 7 in Section 3.3 to find a model for the transformed data, and add its graph to your scatter plot. According to this model, when will the annual shipment of single CDs reach 100 million?

c. Solve the model in part (b) for y, and add its graph to your scatter plot in part (a). Which model better fits the original data? Which model will better predict future shipments? Explain.

3.4 Exercises

In Exercises 1–6, determine whether the *x*-values are solutions or approximate solutions of the equation.

1. $4^{2x-7} = 64$
 (a) $x = 5$
 (b) $x = 2$

2. $2^{3x+1} = 32$
 (a) $x = -1$
 (b) $x = 2$

3. $3e^{x+2} = 75$
 (a) $x = -2 + e^{25}$
 (b) $x = -2 + \ln 25$
 (c) $x \approx 1.219$

4. $5^{2x+3} = 812$
 (a) $x = -1.5 + \log_5 \sqrt{812}$
 (b) $x \approx 0.581$
 (c) $x = \frac{1}{2}\left(-3 + \frac{\ln 812}{\ln 5}\right)$

5. $\log_4(3x) = 3$
 (a) $x \approx 20.356$
 (b) $x = -4$
 (c) $x = \frac{64}{3}$

6. $\ln(x - 1) = 3.8$
 (a) $x = 1 + e^{3.8}$
 (b) $x \approx 45.701$
 (c) $x = 1 + \ln 3.8$

In Exercises 7–30, solve for *x*.

7. $4^x = 16$
8. $3^x = 243$
9. $5^x = 625$
10. $3^x = 729$
11. $7^x = \frac{1}{49}$
12. $8^x = 4$
13. $\left(\frac{1}{2}\right)^x = 32$
14. $\left(\frac{1}{4}\right)^x = 64$
15. $\left(\frac{3}{4}\right)^x = \frac{27}{64}$
16. $\left(\frac{2}{3}\right)^x = \frac{4}{9}$
17. $3^{x-1} = 27$
18. $2^{x-3} = 32$
19. $\ln x - \ln 2 = 0$
20. $\ln x - \ln 5 = 0$
21. $e^x = 2$
22. $e^x = 4$
23. $\ln x = -1$
24. $\ln x = -7$
25. $\log_4 x = 3$
26. $\log_x 625 = 4$
27. $\log_{10} x - 2 = 0$
28. $\log_{10} x + 3 = 0$
29. $\log_{10} x = -1$
30. $\ln(2x - 1) = 0$

In Exercises 31–34, approximate the point of intersection of the graphs of *f* and *g*. Then solve the equation $f(x) = g(x)$ algebraically.

31. $f(x) = 2^x$
 $g(x) = 8$

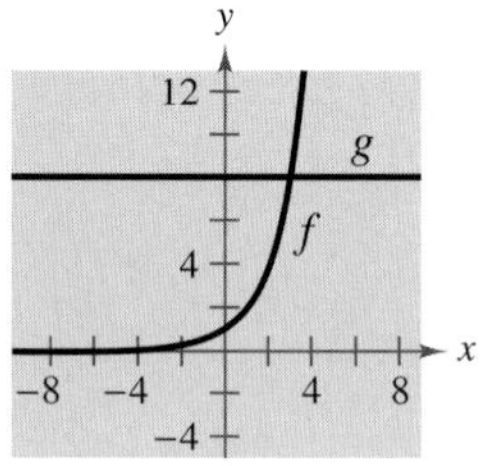

32. $f(x) = 27^x$
 $g(x) = 9$

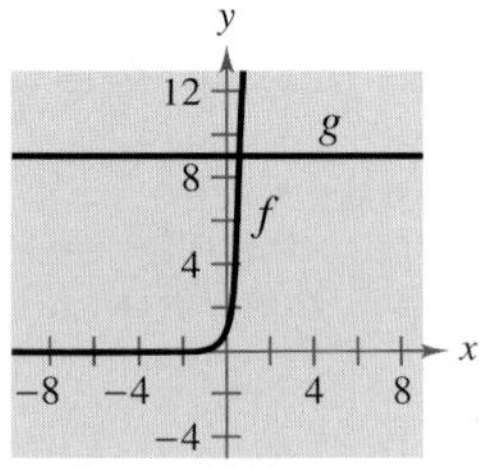

33. $f(x) = \log_3 x$
 $g(x) = 2$

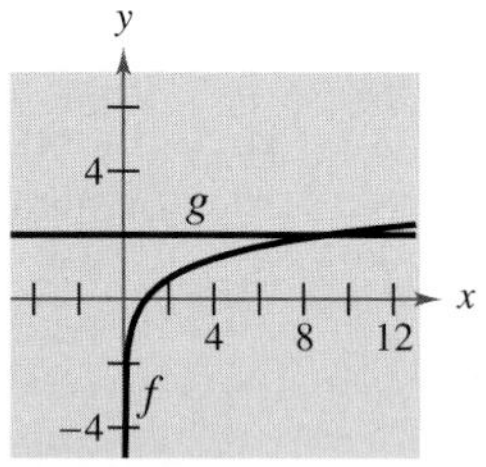

34. $f(x) = \ln(x - 4)$
 $g(x) = 0$

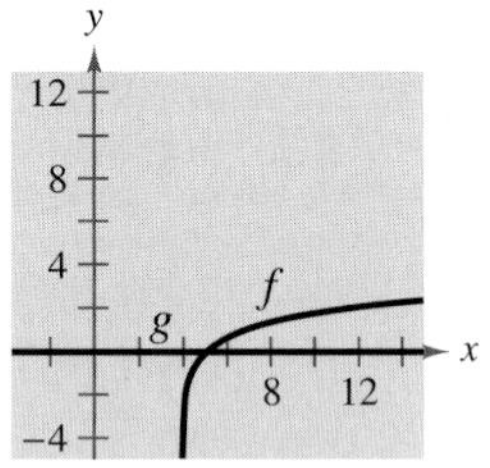

In Exercises 35–44, apply the inverse properties of ln *x* and e^x to simplify the expression.

35. $\log_{10} 10^{x^2}$
36. $\log_6 6^{2x-1}$
37. $8^{\log_8(x-2)}$
38. $4^{\log_4 x^3}$
39. $\ln e^{7x+2}$
40. $\ln e^{x^4}$
41. $e^{\ln(5x+2)}$
42. $e^{\ln x^2}$
43. $-1 + \ln e^{2x}$
44. $-8 + e^{\ln x^3}$

In Exercises 45–56, solve the exponential equation with base *e* algebraically. Approximate the result to three decimal places.

45. $e^x = 10$
46. $4e^x = 91$
47. $7 - 2e^x = 5$
48. $-14 + 3e^x = 11$
49. $e^{3x} = 12$
50. $e^{2x} = 50$
51. $500e^{-x} = 300$
52. $1000e^{-4x} = 75$
53. $e^{2x} - 4e^x - 5 = 0$
54. $e^{2x} - 5e^x + 6 = 0$
55. $20(100 - e^{x/2}) = 500$
56. $\frac{400}{1 + e^{-x}} = 350$

In Exercises 57–64, solve the exponential equation with base *a* algebraically. Approximate the result to three decimal places.

57. $10^x = 42$
58. $10^x = 570$
59. $3^{2x} = 80$
60. $6^{5x} = 3000$
61. $5^{-t/2} = 0.20$
62. $4^{-3t} = 0.10$
63. $2^{3-x} = 565$
64. $8^{-2-x} = 431$

In Exercises 65–72, use a graphing utility to graph the function. Approximate its zero to three decimal places.

65. $g(x) = 6e^{1-x} - 25$
66. $f(x) = -4e^{-x-1} + 15$
67. $f(x) = 3e^{3x/2} - 962$
68. $g(x) = 8e^{-2x/3} - 11$
69. $g(t) = e^{0.09t} - 3$
70. $f(x) = -e^{1.8x} + 7$
71. $h(t) = e^{0.125t} - 8$
72. $f(x) = e^{2.724x} - 29$

In Exercises 73–82, solve the exponential equation. Approximate the result to three decimal places.

73. $8(10^{3x}) = 12$
74. $5(10^{x-6}) = 7$
75. $3(5^{x-1}) = 21$
76. $8(3^{6-x}) = 40$
77. $\left(1 + \dfrac{0.065}{365}\right)^{365t} = 4$
78. $\left(4 - \dfrac{2.471}{40}\right)^{9t} = 21$
79. $\left(1 + \dfrac{0.10}{12}\right)^{12t} = 2$
80. $\left(16 - \dfrac{0.878}{26}\right)^{3t} = 30$
81. $\dfrac{3000}{2 + e^{2x}} = 2$
82. $\dfrac{119}{e^{6x} - 14} = 7$

In Exercises 83–96, solve the natural logarithmic equation algebraically. Approximate the result to three decimal places.

83. $\ln x = -3$
84. $\ln x = 2$
85. $\ln 2x = 2.4$
86. $\ln 4x = 1$
87. $3 \ln 5x = 10$
88. $2 \ln x = 7$
89. $\ln \sqrt{x + 2} = 1$
90. $\ln \sqrt{x - 8} = 5$
91. $\ln(x + 1)^2 = 2$
92. $\ln x + \ln(x + 1) = 1$
93. $\ln x + \ln(x - 2) = 1$
94. $\ln x + \ln(x + 3) = 1$
95. $\ln(x + 5) = \ln(x - 1) - \ln(x + 1)$
96. $\ln(x + 1) - \ln(x - 2) = \ln x^2$

In Exercises 97–106, solve the logarithmic equation algebraically. Approximate the result to three decimal places.

97. $\log_{10}(z - 3) = 2$
98. $\log_{10}x^2 = 6$
99. $6 \log_3(0.5x) = 11$
100. $5 \log_{10}(x - 2) = 11$
101. $\log_{10}(x + 4) - \log_{10}x = \log_{10}(x + 2)$
102. $\log_2 x + \log_2(x + 2) = \log_2(x + 6)$
103. $\log_4 x - \log_4(x - 1) = \frac{1}{2}$
104. $\log_3 x + \log_3(x - 8) = 2$
105. $\log_{10}8x - \log_{10}\left(1 + \sqrt{x}\right) = 2$
106. $\log_{10} 4x - \log_{10}\left(12 + \sqrt{x}\right) = 2$

In Exercises 107–110, use a graphing utility to approximate the point of intersection of the graphs. Approximate the result to three decimal places.

107. $y_1 = 7$
$y_2 = 2^x$
108. $y_1 = 500$
$y_2 = 1500e^{-x/2}$
109. $y_1 = 3$
$y_2 = \ln x$
110. $y_1 = 10$
$y_2 = 4 \ln(x - 2)$

Finance **In Exercises 111 and 112, find the time required for a \$1000 investment to double at interest rate *r*, compounded continuously.**

111. $r = 0.085$
112. $r = 0.12$

Finance **In Exercises 113 and 114, find the time required for a \$1000 investment to triple at interest rate *r*, compounded continuously.**

113. $r = 0.085$
114. $r = 0.12$

115. *Economics* The demand equation for a certain product is

$$p = 500 - 0.5(e^{0.004x}).$$

Find the demand x for a price of (a) $p = \$350$ and (b) $p = \$300$.

116. *Economics* The demand equation for a certain product is

$$p = 5000\left(1 - \frac{4}{4 + e^{-0.002x}}\right).$$

Find the demand x for a price of (a) $p = \$600$ and (b) $p = \$400$.

117. *Forest Yield* The yield V (in millions of cubic feet per acre) for a forest at age t years is

$$V = 6.7e^{-48.1/t}.$$

(a) Use a graphing utility to graph the function.

(b) Determine the horizontal asymptote of the function. Interpret its meaning in the context of the problem.

(c) Find the time necessary to obtain a yield of 1.3 million cubic feet.

118. *Trees per Acre* The number of trees per acre N of a certain species is approximated by the model

$$N = 68(10^{-0.04x}), \qquad 5 \le x \le 40$$

where x is the average diameter of the trees 3 feet above the ground. Use the model to approximate the average diameter of the trees in a test plot when $N = 21$.

119. *Average Heights* The percent of American males between the ages of 18 and 24 who are no more than x inches tall is

$$m(x) = \frac{100}{1 + e^{-0.6114(x - 69.71)}}.$$

The percent of American females between the ages of 18 and 24 who are no more than x inches tall is

$$f(x) = \frac{100}{1 + e^{-0.66607(x - 64.51)}}$$

where m and f are the percents and x is the height in inches. (Source: U.S. National Center for Health Statistics)

(a) Use the graph to determine any horizontal asymptotes of the functions. What do they mean?

(b) What is the average height of each sex?

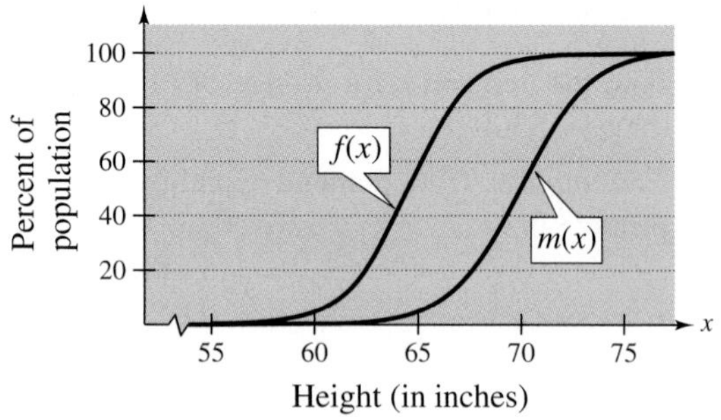

120. *Human Learning Model* In a group project in learning theory, a mathematical model for the proportion P of correct responses after n trials was found to be

$$P = \frac{0.83}{1 + e^{-0.2n}}.$$

(a) Use a graphing utility to graph the function.

(b) Use the graph to determine any horizontal asymptotes of the function. Interpret the meaning of the upper asymptote in the context of this problem.

(c) After how many trials will 60% of the responses be correct?

121. *Data Analysis* An object at a temperature of 160°C was removed from a furnace and placed in a room at 20°C. The temperature T of the object was measured each hour h and recorded in the table.

h	0	1	2	3	4	5
T	160°	90°	56°	38°	29°	24°

A model for this data is

$$T = 20[1 + 7(2^{-h})].$$

(a) The graph of this model is shown in the figure. Use the graph to identify the horizontal asymptote of the model and interpret the asymptote in the context of the problem.

(b) Use the model to approximate the time when the temperature of the object was 100°C.

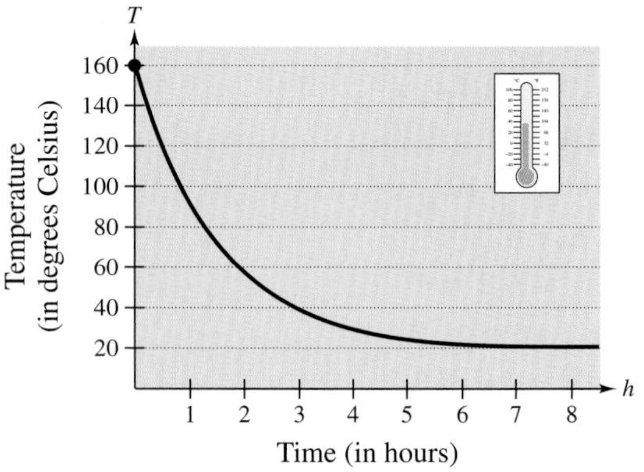

122. *Automobiles* Automobiles are designed with crumple zones that help protect their occupants in crashes. The crumple zones allow the occupants to move short distances when the automobiles come to abrupt stops. The greater the distance moved, the fewer g's the crash victims experience. (One g is equal to the acceleration due to gravity. For very short periods of time, humans have withstood as much as 40 g's.) In crash tests with vehicles moving at 90 kilometers per hour, analysts measured the numbers y of g's experienced during deceleration by crash dummies that were permitted to move x meters during impact. The data is shown in the table.

x	0.2	0.4	0.6	0.8	1
y	158	80	53	40	32

A model for this data is

$$y = -3.00 + 11.88 \ln x + \frac{36.94}{x}.$$

(a) Use a graphing utility to graph the data points and the model in the same viewing window. How do they compare?

(b) Use the model to estimate the distance traveled during impact if the passenger deceleration must not exceed 30 g's.

(c) Do you think it is practical to lower the number of g's experienced during impact to fewer than 23? Explain your reasoning.

Synthesis

True or False? **In Exercises 123–126, rewrite each verbal statement as an equation. Then decide whether the statement is true or false. Justify your answer.**

123. The logarithm of the product of two numbers is equal to the sum of the logarithms of the numbers.

124. The logarithm of the sum of two numbers is equal to the product of the logarithms of the numbers.

125. The logarithm of the difference of two numbers is equal to the difference of the logarithms of the numbers.

126. The logarithm of the quotient of two numbers is equal to the difference of the logarithms of the numbers.

127. *Finance* You are investing P dollars at an annual interest rate of r, compounded continuously, for t years. Which of the following would result in the highest value of the investment? Explain your reasoning.

(a) Double the amount you invest.

(b) Double your interest rate.

(c) Double the number of years.

128. *Think About It* Are the times required for the investments in Exercises 111 and 112 to quadruple twice as long as the times for them to double? Give a reason for your answer and verify your answer algebraically.

129. *Writing* Write a paragraph explaining whether or not the time required for an investment to double is dependent on the size of the investment.

Review

In Exercises 130–133, simplify the expression.

130. $\sqrt{48x^2y^5}$

131. $\sqrt{32} - 2\sqrt{25}$

132. $\sqrt[3]{25} \cdot \sqrt[3]{15}$

133. $\dfrac{3}{\sqrt{10} - 2}$

In Exercises 134–138, find a mathematical model for the verbal statement.

134. M varies directly as the cube of p.

135. t varies inversely as the cube of s.

136. d varies jointly as a and b.

137. x is inversely proportional to $b - 3$.

138. c varies inversely as the square root of w.

In Exercises 139–142, evaluate the logarithm using the change-of-base formula. Approximate your result to three decimal places.

139. $\log_6 9$

140. $\log_3 4$

141. $\log_{3/4} 5$

142. $\log_8 22$

3.5 Exponential and Logarithmic Models

▶ What you should learn

- How to recognize the five most common types of models involving exponential and logarithmic functions
- How to use exponential growth and decay functions to model and solve real-life problems
- How to use Gaussian functions to model and solve real-life problems
- How to use logistic growth functions to model and solve real-life problems
- How to use logarithmic functions to model and solve real-life problems

▶ Why you should learn it

Real-life applications of mathematics often involve deciding which mathematical model to use for a given situation. For instance, Exercise 45 on page 337 compares an exponential decay model and a linear model for the depreciation of a computer over 3 years.

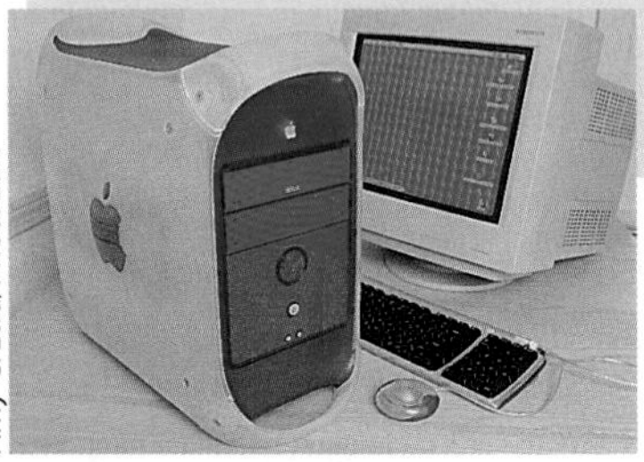

Amy C. Etra/PhotoEdit

Introduction

The five most common types of mathematical models involving exponential functions and logarithmic functions are as follows.

1. **Exponential growth model:** $y = ae^{bx}, \quad b > 0$
2. **Exponential decay model:** $y = ae^{-bx}, \quad b > 0$
3. **Gaussian model:** $y = ae^{-(x-b)^2/c}$
4. **Logistic growth model:** $y = \dfrac{a}{1 + be^{-rx}}$
5. **Logarithmic models:** $y = a + b\ln x, \quad y = a + b\log_{10} x$

The graphs of the basic forms of these functions are shown in Figure 3.22.

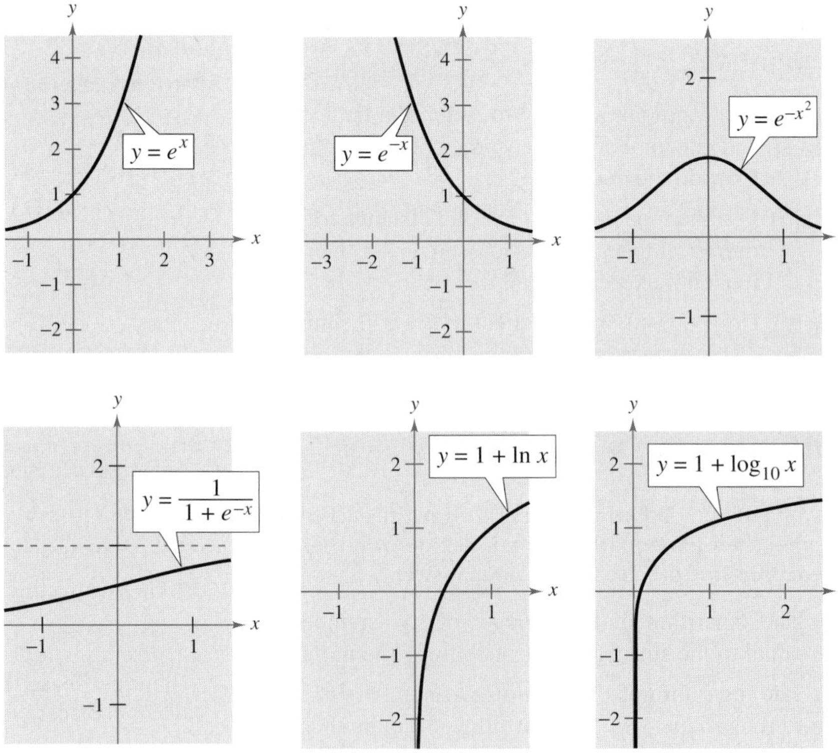

FIGURE 3.22

You can often gain quite a bit of insight into a situation modeled by an exponential or logarithmic function by identifying and interpreting the function's asymptotes. Use the graphs in Figure 3.22 to identify the asymptotes of each function.

This section shows students real-world applications for logarithmic and exponential functions.

Exponential Growth and Decay

Example 1 ▶ Population Increase

Estimates of the world population (in millions) from 1992 through 2000 are shown in the table. The scatter plot of the data is shown in Figure 3.23. (Source: U.S. Bureau of the Census, International Data Base)

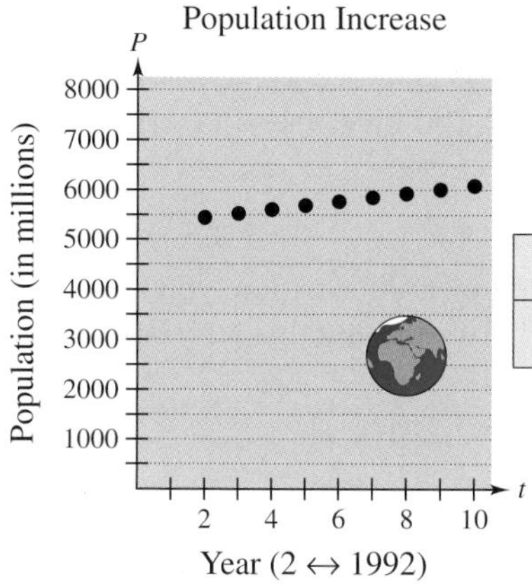

FIGURE 3.23

Year	1992	1993	1994	1995	1996	1997	1998	1999	2000
Population	5445	5527	5607	5688	5767	5847	5926	6005	6083

An exponential growth model that approximates this data is

$$P = 5304e^{0.013819t}, \quad 2 \le t \le 10$$

where P is the population (in millions) and $t = 2$ represents 1992. Compare the values given by the model with the estimates given by the U.S. Bureau of the Census. According to this model, when will the world population reach 6.5 billion?

Solution

The following table compares the two sets of population figures. The graph of the model is shown in Figure 3.24.

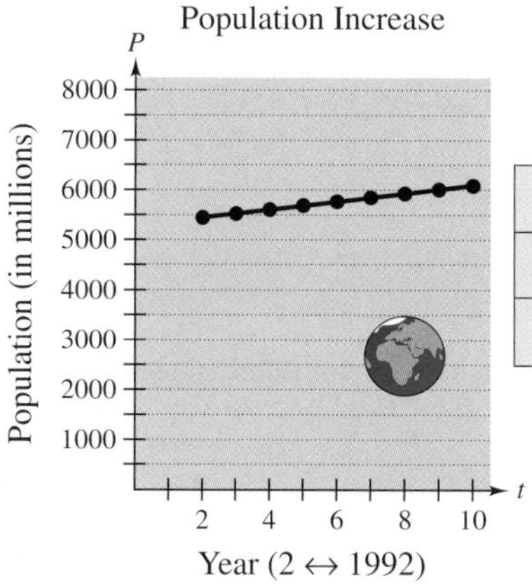

FIGURE 3.24

Year	1992	1993	1994	1995	1996	1997	1998	1999	2000
Population	5445	5527	5607	5688	5767	5847	5926	6005	6083
Model	5453	5529	5605	5683	5763	5843	5924	6006	6090

To find when the world population will reach 6.5 billion, let $P = 6500$ in the model and solve for t.

$$5304e^{0.013819t} = P \qquad \text{Write original model.}$$

$$5304e^{0.013819t} = 6500 \qquad \text{Let } P = 6500.$$

$$e^{0.013819t} \approx 1.22549 \qquad \text{Divide each side by 5304.}$$

$$\ln e^{0.013819t} \approx \ln 1.22549 \qquad \text{Take natural log of each side.}$$

$$0.013819t \approx 0.203341 \qquad \text{Inverse Property}$$

$$t \approx 14.71 \qquad \text{Divide each side by 0.013819.}$$

According to the model, the world population will reach 6.5 billion in 2004.

Technology

Some graphing utilities have curve-fitting capabilities that can be used to find models that represent data. If you have such a graphing utility, try using it to find a model for the data given in Example 1. How does your model compare with the model given in Example 1?

An exponential model increases (or decreases) by the same percent each year. What is the annual percent increase for the model in Example 1?

Additional Example

Radioactive iodine is a by-product of some types of nuclear reactors. Its half-life is 60 days. That is, after 60 days, a given amount of radioactive iodine will have decayed to half the original amount. Suppose a contained nuclear accident occurs and gives off an initial amount C of radioactive iodine.

a. Write an equation for the amount of radioactive iodine present at any time t following the accident.

b. How long will it take for the radioactive iodine to decay to a level of 20% of the original amount?

Solution

a. Knowing that half the original amount remains after 60 days, you can use the exponential decay model to obtain

$$y = ae^{-bt}$$
$$\frac{1}{2}C = Ce^{-b(60)}$$
$$\frac{1}{2} = e^{-60b}$$
$$-\ln 2 = -60b$$
$$b = \frac{\ln 2}{60} \approx 0.0116.$$

So, the exponential decay model is $y = Ce^{-0.0116t}$.

b. The time required for the radioactive iodine to decay to 20% of the original amount is

$$Ce^{-0.0116t} = (0.2)C$$
$$e^{-0.0116t} = 0.2$$
$$-0.0116t = \ln 0.2$$
$$t = \frac{\ln 0.2}{-0.0116} \approx 139 \text{ days.}$$

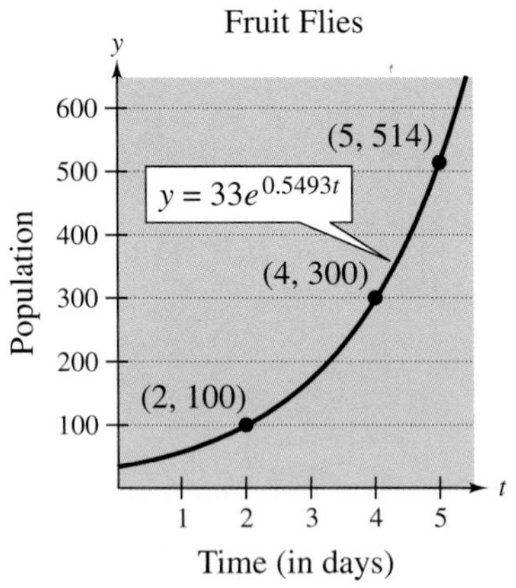

FIGURE 3.25

In Example 1, you were given the exponential growth model. But suppose this model were not given; how could you find such a model? One technique for doing this is demonstrated in Example 2.

Example 2 ▶ Modeling Population Growth

In a research experiment, a population of fruit flies is increasing according to the law of exponential growth. After 2 days there are 100 flies, and after 4 days there are 300 flies. How many flies will there be after 5 days?

Solution

Let y be the number of flies at time t. From the given information, you know that $y = 100$ when $t = 2$ and $y = 300$ when $t = 4$. Substituting this information into the model $y = ae^{bt}$ produces

$$100 = ae^{2b} \quad \text{and} \quad 300 = ae^{4b}.$$

To solve for b, solve for a in the first equation.

$$100 = ae^{2b} \quad \Longrightarrow \quad a = \frac{100}{e^{2b}}$$ Solve for a in the first equation.

Then substitute the result into the second equation.

$$300 = ae^{4b}$$ Write second equation.

$$300 = \left(\frac{100}{e^{2b}}\right)e^{4b}$$ Substitute $100/e^{2b}$ for a.

$$\frac{300}{100} = e^{2b}$$ Divide each side by 100.

$$\ln 3 = 2b$$ Take natural log of each side.

$$\frac{1}{2}\ln 3 = b$$ Solve for b.

Using $b = \frac{1}{2}\ln 3$ and the equation you found for a, you can determine that

$$a = \frac{100}{e^{2[(1/2)\ln 3]}}$$ Substitute $(1/2)\ln 3$ for b.

$$= \frac{100}{e^{\ln 3}}$$ Simplify.

$$= \frac{100}{3}$$ Inverse Property

$$\approx 33.$$ Simplify.

So, with $a \approx 33$ and $b = \frac{1}{2}\ln 3 \approx 0.5493$, the exponential growth model is

$$y = 33e^{0.5493t}$$

as shown in Figure 3.25. This implies that, after 5 days, the population is

$$y = 33e^{0.5493(5)} \approx 514 \text{ flies.}$$

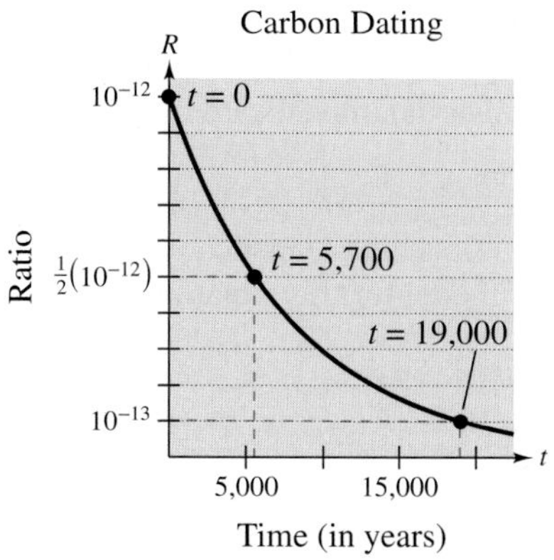

FIGURE 3.26

In living organic material, the ratio of the number of radioactive carbon isotopes (carbon 14) to the number of nonradioactive carbon isotopes (carbon 12) is about 1 to 10^{12}. When organic material dies, its carbon 12 content remains fixed, whereas its radioactive carbon 14 begins to decay with a half-life of about 5700 years. To estimate the age of dead organic material, scientists use the following formula, which denotes the ratio of carbon 14 to carbon 12 present at any time t (in years).

$$R = \frac{1}{10^{12}} e^{-t/8223} \qquad \text{Carbon dating model}$$

The graph of R is shown in Figure 3.26. Note that R decreases as t increases.

Example 3 ▶ Carbon Dating

The ratio of carbon 14 to carbon 12 in a newly discovered fossil is

$$R = \frac{1}{10^{13}}.$$

Estimate the age of the fossil.

Solution

In the carbon dating model, substitute the given value of R to obtain the following.

$$\frac{1}{10^{12}} e^{-t/8223} = R \qquad \text{Write original model.}$$

$$\frac{e^{-t/8223}}{10^{12}} = \frac{1}{10^{13}} \qquad \text{Let } R = \frac{1}{10^{13}}.$$

$$e^{-t/8223} = \frac{1}{10} \qquad \text{Multiply each side by } 10^{12}.$$

$$\ln e^{-t/8223} = \ln \frac{1}{10} \qquad \text{Take natural log of each side.}$$

$$-\frac{t}{8223} \approx -2.3026 \qquad \text{Inverse Property}$$

$$t \approx 18{,}934 \qquad \text{Multiply each side by } -8223.$$

So, to the nearest thousand years, you can estimate the age of the fossil to be 19,000 years.

STUDY TIP

To find how much of an initial 10 grams of ^{226}Ra isotope with a half-life of 1620 years is left after 500 years, you would use the exponential decay model.

$$y = ae^{-bt}$$

$$\tfrac{1}{2}(10) = 10e^{-b(1620)}$$

$$b = -\ln\left(\tfrac{1}{2}\right)/1620$$

Using the value of b found above and $a = 10$, the amount left is

$$y = 10e^{-[-\ln(1/2)/1620](500)}$$

$$y \approx 8.07 \text{ grams.}$$

The carbon dating model in Example 3 assumed that the carbon 14/carbon 12 ratio was one part in 10,000,000,000,000. Suppose an error in measurement occurred and the actual ratio was only one part in 8,000,000,000,000. The fossil age corresponding to the actual ratio would then be approximately 17,000 years. Try checking this result.

Gaussian Models

As mentioned at the beginning of this section, Gaussian models are of the form

$$y = ae^{-(x-b)^2/c}.$$

This type of model is commonly used in probability and statistics to represent populations that are **normally distributed.** One model for this situation takes the form

$$y = \frac{1}{\sigma\sqrt{2\pi}}e^{-x^2/(2\sigma^2)}$$

where σ is the standard deviation (σ is the lowercase Greek letter sigma). The graph of a Gaussian model is called a **bell-shaped curve.** Try assigning a value to σ and sketching a normal distribution curve with a graphing utility. Can you see why it is called a bell-shaped curve?

The average value for a population can be found from the bell-shaped curve by observing where the maximum y-value of the function occurs. The x-value corresponding to the maximum y-value of the function represents the average value of the independent variable—in this case, x.

Example 4 ▶ SAT Scores

In 1997, the Scholastic Aptitude Test (SAT) math scores for college-bound seniors roughly followed a normal distribution

$$y = 0.0036e^{-(x-511)^2/25{,}088}, \quad 200 \le x \le 800$$

where x is the SAT score for mathematics. Sketch the graph of this function. From the graph, estimate the average SAT score. (Source: College Board)

Solution

The graph of the function is given in Figure 3.27. From the graph, you can see that the average mathematics score for college-bound seniors in 1997 was 511.

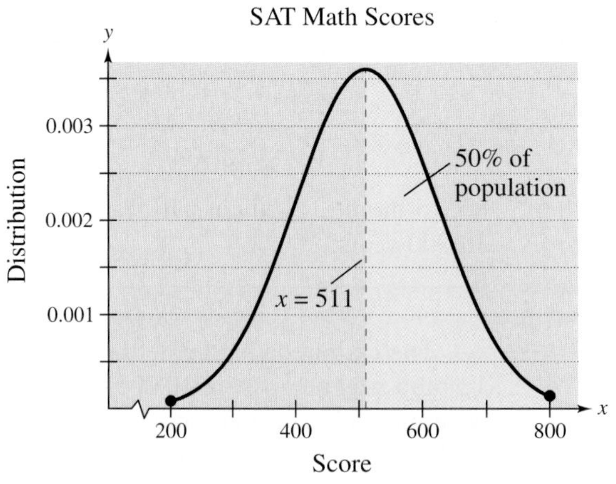

FIGURE 3.27

Logistic Growth Models

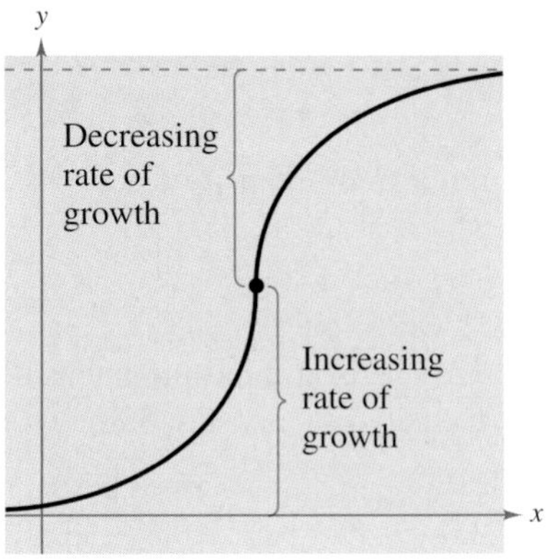

FIGURE 3.28

Some populations initially have rapid growth, followed by a declining rate of growth, as indicated by the graph in Figure 3.28. One model for describing this type of growth pattern is the **logistic curve** given by the function

$$y = \frac{a}{1 + be^{-rx}}$$

where y is the population size and x is the time. An example is a bacteria culture that is initially allowed to grow under ideal conditions, and then under less favorable conditions that inhibit growth. A logistic growth curve is also called a **sigmoidal curve.**

Example 5 ▶ Spread of a Virus

On a college campus of 5000 students, one student returns from vacation with a contagious and long-lasting flu virus. The spread of the virus is modeled by

$$y = \frac{5000}{1 + 4999e^{-0.8t}}, \quad 0 \le t$$

where y is the total number of students infected after t days. The college will cancel classes when 40% or more of the students are infected.

a. How many students are infected after 5 days?

b. After how many days will the college cancel classes?

Solution

a. After 5 days, the number of students infected is

$$y = \frac{5000}{1 + 4999e^{-0.8(5)}} = \frac{5000}{1 + 4999e^{-4}} \approx 54.$$

b. Classes are canceled when the number infected is $(0.40)(5000) = 2000$.

$$2000 = \frac{5000}{1 + 4999e^{-0.8t}}$$

$$1 + 4999e^{-0.8t} = 2.5$$

$$e^{-0.8t} \approx \frac{1.5}{4999}$$

$$\ln e^{-0.8t} \approx \ln \frac{1.5}{4999}$$

$$-0.8t \approx \ln \frac{1.5}{4999}$$

$$t = -\frac{1}{0.8} \ln \frac{1.5}{4999}$$

$$t \approx 10.1$$

So, after 10 days, at least 40% of the students will be infected, and classes will be canceled. The graph of the function is shown in Figure 3.29.

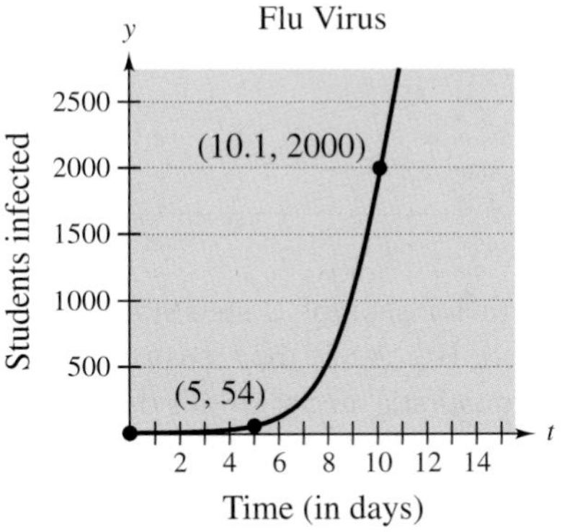

FIGURE 3.29

Twenty seconds of a 7.2 magnitude earthquake in Kobe, Japan on January 17, 1995 left damage approaching $60 billion.

Alternate Writing About Mathematics Use your school's library or some other reference source to find an application that fits one of the five models discussed in this section. After you have collected data for the model, plot the corresponding points and find an equation that describes the points you have plotted.

Logarithmic Models

Example 6 ▶ Magnitude of Earthquakes

On the Richter scale, the magnitude R of an earthquake of intensity I is

$$R = \log_{10} \frac{I}{I_0}$$

where $I_0 = 1$ is the minimum intensity used for comparison. Find the intensities per unit of area for the following earthquakes. (Intensity is a measure of the wave energy of an earthquake.)

a. Tokyo and Yokohama, Japan in 1923: $R = 8.3$.

b. Kobe, Japan in 1995: $R = 7.2$.

Solution

a. Because $I_0 = 1$ and $R = 8.3$, you have

$$8.3 = \log_{10} \frac{I}{1}$$ Substitute 1 for I_0 and 8.3 for R.

$$10^{8.3} = 10^{\log_{10} I}$$ Exponentiate each side.

$$I = 10^{8.3} \approx 199{,}526{,}000.$$ Inverse property of exponents and logs

b. For $R = 7.2$, you have

$$7.2 = \log_{10} \frac{I}{1}$$ Substitute 1 for I_0 and 7.2 for R.

$$10^{7.2} = 10^{\log_{10} I}$$ Exponentiate each side.

$$I = 10^{7.2} \approx 15{,}849{,}000.$$ Inverse property of exponents and logs

Note that an increase of 1.1 units on the Richter scale (from 7.2 to 8.3) represents an increase in intensity by a factor of

$$\frac{199{,}526{,}000}{15{,}849{,}000} \approx 13.$$

In other words, the earthquake in 1923 had an intensity about 13 times greater than that of the 1995 quake.

Writing ABOUT MATHEMATICS

Comparing Population Models The population (in millions) of the United States from 1810 to 1990 is given in the table. (Source: U.S. Bureau of the Census) Least squares regression analysis gives the best quadratic model for this data as $P = 0.6569t^2 + 0.305t + 6.12$ and the best exponential model for this data as $P = 8.325e^{0.195t}$. Which model better fits the data? Describe the method you used to reach your conclusion.

t	Year	Population
1	1810	7.23
3	1830	12.87
5	1850	23.19
7	1870	39.82
9	1890	62.95
11	1910	91.97
13	1930	122.78
15	1950	151.33
17	1970	203.30
19	1990	250.00

3.5 Exercises

In Exercises 1–6, match the function with its graph. [The graphs are labeled (a) through (f).]

(a)

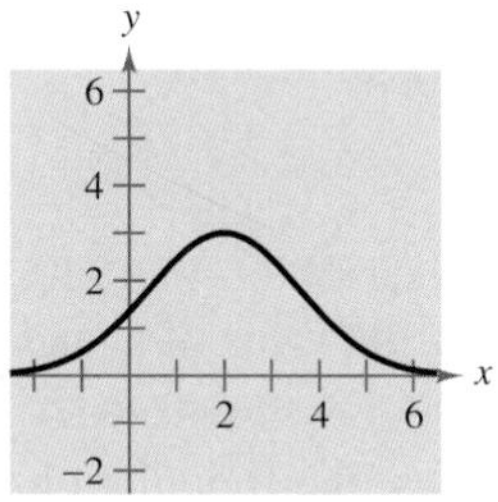

(b)

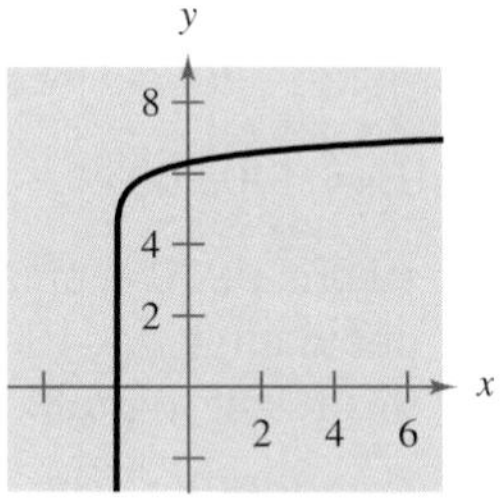

(c)

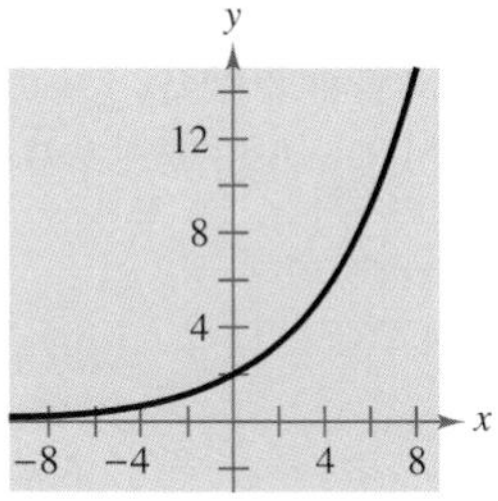

(d)

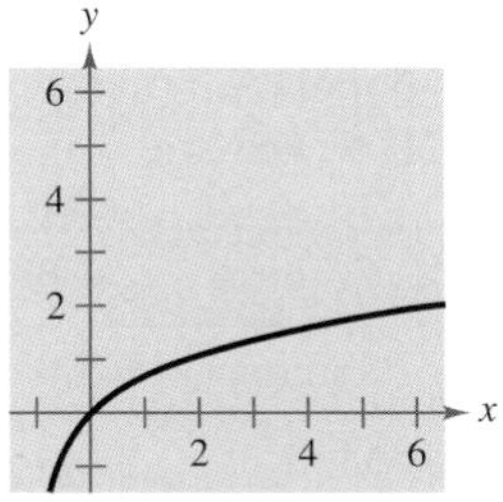

(e)

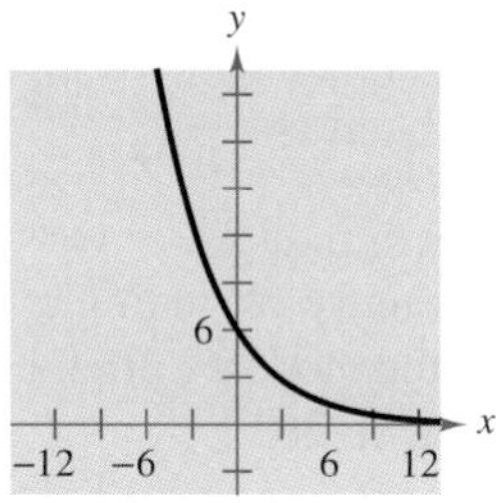

(f)

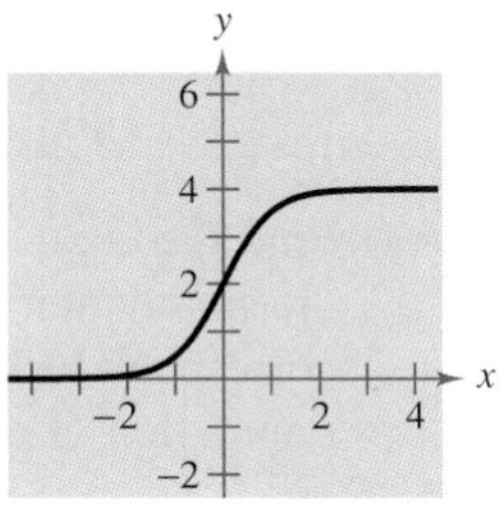

1. $y = 2e^{x/4}$

2. $y = 6e^{-x/4}$

3. $y = 6 + \log_{10}(x + 2)$

4. $y = 3e^{-(x-2)^2/5}$

5. $y = \ln(x + 1)$

6. $y = \dfrac{4}{1 + e^{-2x}}$

Finance **In Exercises 7–14, complete the table for a savings account in which interest is compounded continuously.**

	Initial Investment	*Annual % Rate*	*Time to Double*	*Amount After 10 Years*
7.	\$1000	12%		
8.	\$20,000	$10\frac{1}{2}\%$		
9.	\$750		$7\frac{3}{4}$ yr	
10.	\$10,000		12 yr	
11.	\$500			\$1505.00
12.	\$600			\$19,205.00
13.		4.5%		\$10,000.00
14.		8%		\$20,000.00

Finance **In Exercises 15 and 16, determine the principal P that must be invested at rate r, compounded monthly, so that \$500,000 will be available for retirement in t years.**

15. $r = 7\frac{1}{2}\%,\ t = 20$

16. $r = 12\%,\ t = 40$

Finance **In Exercises 17 and 18, determine the time necessary for \$1000 to double if it is invested at interest rate r compounded (a) annually, (b) monthly, (c) daily, and (d) continuously.**

17. $r = 11\%$

18. $r = 10\frac{1}{2}\%$

19. *Finance* Complete the table for the time t necessary for P dollars to triple if interest is compounded continuously at rate r.

r	2%	4%	6%	8%	10%	12%
t						

20. *Modeling Data* Draw a scatter plot of the data in Exercise 19. Use the curve-fitting capabilities of a graphing utility to find a model for the data.

21. *Finance* Complete the table for the time t necessary for P dollars to triple if interest is compounded annually at rate r.

r	2%	4%	6%	8%	10%	12%
t						

22. *Modeling Data* Draw a scatter plot of the data in Exercise 21. Use the curve-fitting capabilities of a graphing utility to find a model for the data.

23. *Finance* If \$1 is invested in an account over a 10-year period, the amount in the account, where t represents the time in years, is

$$A = 1 + 0.075[\![t]\!] \quad \text{or} \quad A = e^{0.07t}$$

depending on whether the account pays simple interest at $7\frac{1}{2}\%$ or continuous compound interest at 7%. Graph each function on the same set of axes. Which grows at the faster rate? (Remember that $[\![t]\!]$ is the greatest integer function discussed in Section 1.4.)

24. *Finance* If \$1 is invested in an account over a 10-year period, the amount in the account, where t represents the time in years, is

$$A = 1 + 0.06[\![t]\!] \quad \text{or} \quad A = \left(1 + \frac{0.055}{365}\right)^{[\![365t]\!]}$$

depending on whether the account pays simple interest at 6% or compound interest at $5\frac{1}{2}\%$ compounded daily. Use a graphing utility to graph each function in the same viewing window. Which grows at the faster rate?

In Exercises 25–30, complete the table for the radioactive isotope.

	Isotope	Half-life (years)	Initial Quantity	Amount After 1000 Years
25.	^{226}Ra	1620	10 g	
26.	^{226}Ra	1620		1.5 g
27.	^{14}C	5730		2 g
28.	^{14}C	5730	3 g	
29.	^{239}Pu	24,360		2.1 g
30.	^{239}Pu	24,360		0.4 g

In Exercises 31–34, find the exponential model $y = ae^{bx}$ that fits the points in the graph or table.

31.

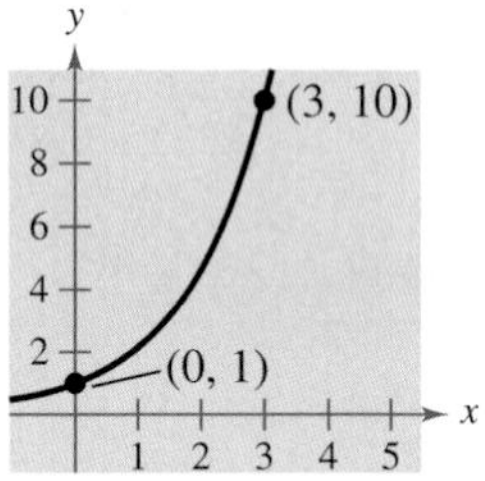

32.

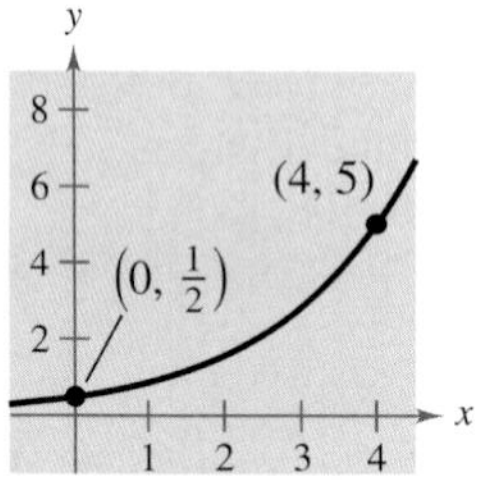

33.

x	0	4
y	5	1

34.

x	0	3
y	1	$\frac{1}{4}$

35. *Population* The population P of a city is

$$P = 105{,}300e^{0.015t}$$

where $t = 0$ represents the year 2000. According to this model, when will the population reach 150,000?

36. *Population* The population P of a city is

$$P = 240{,}360e^{0.012t}$$

where $t = 0$ represents the year 2000. According to this model, when will the population reach 275,000?

37. *Population* The population P of a city is

$$P = 2500e^{kt}$$

where $t = 0$ represents the year 2000. In 1945, the population was 1350. Find the value of k, and use this result to predict the population in the year 2010.

38. *Population* The population P of a city is

$$P = 140{,}500e^{kt}$$

where $t = 0$ represents the year 2000. In 1960, the population was 100,250. Find the value of k, and use this result to predict the population in the year 2020.

39. *Population* The table gives the population (in millions) of a country in 1997 and the projected population (in millions) for the year 2020. (Source: U.S. Bureau of the Census, International Data Base)

Country	1997	2020
Croatia	5.0	4.8
Mali	9.9	20.4
Singapore	3.5	4.3
Sweden	8.9	9.5

(a) Find the exponential growth model $y = ae^{bt}$ for the population in each country by letting $t = 0$ correspond to 1997. Use the model to predict the population of each country in 2030.

(b) You can see that the populations of Mali and Sweden are growing at different rates. What constant in the equation $y = ae^{bt}$ is determined by these different growth rates? Discuss the relationship between the different growth rates and the magnitude of the constant.

(c) You can see that the population of Singapore is increasing while the population of Croatia is decreasing. What constant in the equation $y = ae^{bt}$ reflects this difference? Explain.

40. *Bacteria Growth* The number of bacteria N in a culture is modeled by

$$N = 100e^{kt}$$

where t is the time in hours. If $N = 300$ when $t = 5$, estimate the time required for the population to double in size.

41. ***Bacteria Growth*** The number of bacteria N in a culture is modeled by

$N = 250e^{kt}$

where t is the time in hours. If $N = 280$ when $t = 10$, estimate the time required for the population to double in size.

42. ***Radioactive Decay*** The half-life of radioactive radium (^{226}Ra) is 1620 years. What percent of a present amount of radioactive radium will remain after 100 years?

43. ***Radioactive Decay*** Carbon 14 dating assumes that the carbon dioxide on earth today has the same radioactive content as it did centuries ago. If this is true, the amount of ^{14}C absorbed by a tree that grew several centuries ago should be the same as the amount of ^{14}C absorbed by a tree growing today. A piece of ancient charcoal contains only 15% as much radioactive carbon as a piece of modern charcoal. How long ago was the tree burned to make the ancient charcoal if the half-life of ^{14}C is 5730 years?

44. ***Depreciation*** A car that cost $22,000 new has a book value of $13,000 after 2 years.

(a) Find the straight-line model $V = mt + b$.

(b) Find the exponential model $V = ae^{kt}$.

(c) Use a graphing utility to graph the two models in the same viewing window. Which model depreciates faster in the first 2 years?

(d) Find the book values of the car after 1 year and after 3 years using each model.

(e) Interpret the slope of the straight-line model.

45. ***Depreciation*** A computer that costs $2000 new has a book value of $500 after 2 years.

(a) Find the straight-line model $V = mt + b$.

(b) Find the exponential model $V = ae^{kt}$.

(c) Use a graphing utility to graph the two models in the same viewing window. Which model depreciates faster in the first 2 years?

(d) Find the book values of the computer after 1 year and after 3 years using each model.

(e) Interpret the slope of the straight-line model.

46. ***Business*** The sales S (in thousands of units) of a new product after it has been on the market t years are modeled by

$S(t) = 100(1 - e^{kt})$.

Fifteen thousand units of the new product were sold the first year.

(a) Complete the model by solving for k.

(b) Sketch the graph of the model.

(c) Use the model to estimate the number of units sold after 5 years.

47. ***Business*** After discontinuing all advertising for a certain product in 1998, the manufacturer noted that sales began to drop according to the model

$$S = \frac{500{,}000}{1 + 0.6e^{kt}}$$

where S represents the number of units sold and $t = 0$ represents 1998. In 2000, the company sold 300,000 units.

(a) Complete the model by solving for k.

(b) Estimate sales in 2003.

48. ***Business*** The sales S (in thousands of units) of a product after x hundred dollars is spent on advertising are modeled by

$S = 10(1 - e^{kx})$.

When $500 is spent on advertising, 2500 units are sold.

(a) Complete the model by solving for k.

(b) Estimate the number of units that will be sold if advertising expenditures are raised to $700.

49. ***Business*** Because of a slump in the economy, a company finds that its annual profits have dropped from $742,000 in 1998 to $632,000 in 2000. If the profit follows an exponential pattern of decline, what is the expected profit for 2001? (Let $t = 0$ represent 1998.)

50. ***Learning Curve*** The management at a factory has found that the maximum number of units a worker can produce in a day is 30. The learning curve for the number of units N produced per day after a new employee has worked t days is

$N = 30(1 - e^{kt})$.

After 20 days on the job, a new employee produces 19 units.

(a) Find the learning curve for this employee (first, find the value of k).

(b) How many days should pass before this employee is producing 25 units per day?

(c) Is the employee's production increasing at a linear rate? Explain your reasoning.

51. *Population Growth* A conservation organization releases 100 animals of an endangered species into a game preserve. The organization believes that the preserve has a carrying capacity of 1000 animals and that the growth of the herd will be modeled by the logistic curve

$$p(t) = \frac{1000}{1 + 9e^{-0.1656t}}$$

where t is measured in months (see figure).

(a) Use a graphing utility to graph the function. Use the graph to determine the horizontal asymptotes, and interpret the meaning of the larger p-value in the context of the problem.

(b) Estimate the population after 5 months.

(c) After how many months will the population be 500?

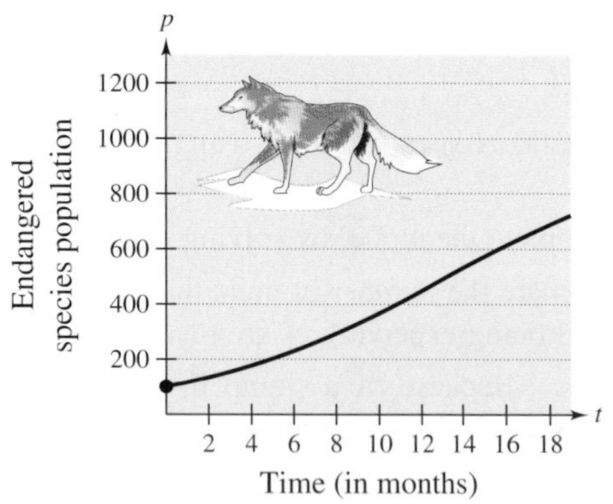

Geology **In Exercises 52 and 53, use the Richter scale for measuring the magnitudes of earthquakes.**

52. Find the magnitude R of an earthquake of intensity I (let $I_0 = 1$).

(a) $I = 80{,}500{,}000$

(b) $I = 48{,}275{,}000$

(c) $I = 251{,}200$

53. Find the intensity I of an earthquake measuring R on the Richter scale (let $I_0 = 1$).

(a) Chile in 1906, $R = 8.6$

(b) Los Angeles in 1971, $R = 6.7$

(c) Taiwan in 1999, $R = 7.7$

Physics **In Exercises 54–57, use the following information for determining sound intensity. The level of sound β, in decibels, with an intensity of I is**

$$\beta = 10 \log_{10} \frac{I}{I_0}$$

where I_0 is an intensity of 10^{-12} watt per square meter, corresponding roughly to the faintest sound that can be heard by the human ear. In Exercises 54 and 55, find the level of sound, β.

54. (a) $I = 10^{-10}$ watt per m^2 (faint whisper)

(b) $I = 10^{-5}$ watt per m^2 (busy street corner)

(c) $I = 10^{-2.5}$ watt per m^2 (air hammer)

(d) $I = 10^{0}$ watt per m^2 (threshold of pain)

55. (a) $I = 10^{-9}$ watt per m^2 (whisper)

(b) $I = 10^{-3.5}$ watt per m^2 (jet 4 miles from takeoff)

(c) $I = 10^{-3}$ watt per m^2 (diesel truck at 25 feet)

(d) $I = 10^{-0.5}$ watt per m^2 (auto horn at 3 feet)

56. Due to the installation of noise suppression materials, the noise level in an auditorium was reduced from 93 to 80 decibels. Find the percent decrease in the intensity level of the noise as a result of the installation of these materials.

57. Due to the installation of a muffler, the noise level of an engine was reduced from 88 to 72 decibels. Find the percent decrease in the intensity level of the noise as a result of the installation of the muffler.

Chemistry **In Exercises 58–63, use the acidity model given by $\text{pH} = -\log_{10}[H^+]$, where acidity (pH) is a measure of the hydrogen ion concentration $[H^+]$ (measured in moles of hydrogen per liter) of a solution.**

58. Find the pH if $[H^+] = 2.3 \times 10^{-5}$.

59. Find the pH if $[H^+] = 11.3 \times 10^{-6}$.

60. Compute $[H^+]$ for a solution in which pH $= 5.8$.

61. Compute $[H^+]$ for a solution in which pH $= 3.2$.

62. A certain fruit has a pH of 2.5 and an antacid tablet has a pH of 9.5. The hydrogen ion concentration of the fruit is how many times the concentration of the tablet?

63. If the pH of a solution is decreased by 1 unit, the hydrogen ion concentration is increased by what factor?

64. *Finance* A $120,000 home mortgage for 35 years at $7\frac{1}{2}\%$ has a monthly payment of $809.39. Part of the monthly payment goes for the interest charge on the unpaid balance, and the remainder of the payment is used to reduce the principal. The amount that goes for interest is

$$u = M - \left(M - \frac{Pr}{12}\right)\left(1 + \frac{r}{12}\right)^{12t}$$

and the amount that goes toward reduction of the principal is

$$v = \left(M - \frac{Pr}{12}\right)\left(1 + \frac{r}{12}\right)^{12t}.$$

In these formulas, P is the size of the mortgage, r is the interest rate, M is the monthly payment, and t is the time in years.

(a) Use a graphing utility to graph each function in the same viewing window. (The viewing window should show all 35 years of mortgage payments.)

(b) In the early years of the mortgage, the larger part of the monthly payment goes for what purpose? Approximate the time when the monthly payment is evenly divided between interest and principal reduction.

(c) Repeat parts (a) and (b) for a repayment period of 20 years ($M = \$966.71$). What can you conclude?

65. *Finance* The total interest u paid on a home mortgage of P dollars at interest rate r for t years is

$$u = P\left[\frac{rt}{1 - \left(\frac{1}{1 + r/12}\right)^{12t}} - 1\right].$$

Consider a $120,000 home mortgage at $7\frac{1}{2}\%$.

(a) Use a graphing utility to graph the total interest function.

(b) Approximate the length of the mortgage for which the total interest paid is the same as the size of the mortgage. Is it possible that some people are paying twice as much in interest charges as the size of the mortgage?

66. *Data Analysis* The table shows the time t (in seconds) required to attain a speed of s miles per hour from a standing start for a particular car.

s	30	40	50	60	70	80	90
t	3.4	5.0	7.0	9.3	12.0	15.8	20.0

Two models for this data are as follows.

$$t_1 = 40.757 + 0.556s - 15.817 \ln s$$

$$t_2 = 1.2259 + 0.0023s^2$$

(a) Use a graphing utility to fit a linear model t_3 and an exponential model t_4 to the data.

(b) Use a graphing utility to graph the data points and each model.

(c) Create a table comparing the data with estimates obtained from each model.

(d) Use the results of part (c) to find the sum of the absolute values of the differences between the data and estimated values given by each model. Based on the four sums, which model do you think better fits the data? Explain.

67. *Forensics* At 8:30 A.M., a coroner was called to the home of a person who had died during the night. In order to estimate the time of death, the coroner took the person's temperature twice. At 9:00 A.M. the temperature was 85.7°F, and at 9:30 A.M. the temperature was 82.8°F. From these two temperatures the coroner was able to determine that the time elapsed since death and the body temperature were related by the formula

$$t = -2.5 \ln \frac{T - 70}{98.6 - 70}$$

where t is the time in hours elapsed since the person died and T is the temperature (in degrees Fahrenheit) of the person's body. Assume that the person had a normal body temperature of 98.6°F at death, and that the room temperature was a constant 70°F. (This formula is derived from a general cooling principle called Newton's Law of Cooling.) Use the formula to estimate the time of death of the person.

Synthesis

True or False? **In Exercises 68–70, determine whether the statement is true or false. Justify your answer.**

68. The domain of a logistic growth function cannot be the set of real numbers.

69. A logistic growth function will always have an x-intercept.

70. The graph of $f(x) = \dfrac{4}{1 + 6e^{-2x}} + 5$ is the graph of $g(x) = \dfrac{4}{1 + 6e^{-2x}}$ shifted to the right 5 units.

71. Identify each model as linear, logarithmic, exponential, logistic, or none of the above. Explain your reasoning.

(a)

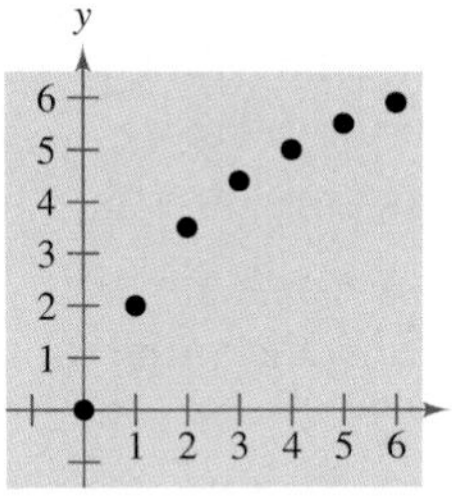

(b)

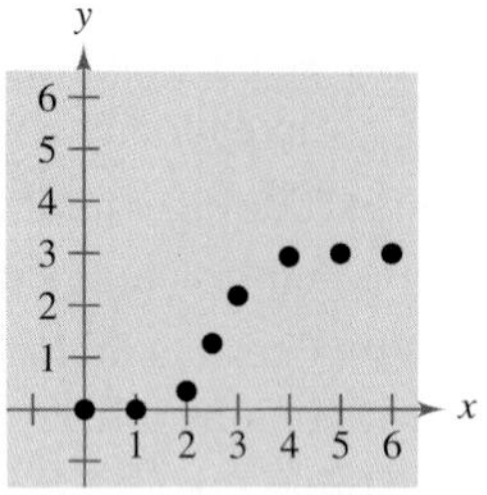

(c)

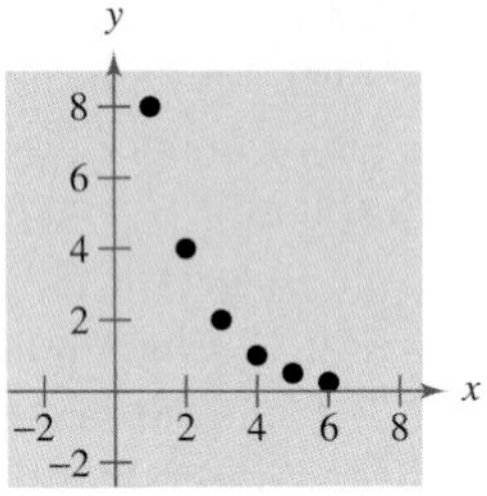

(d)

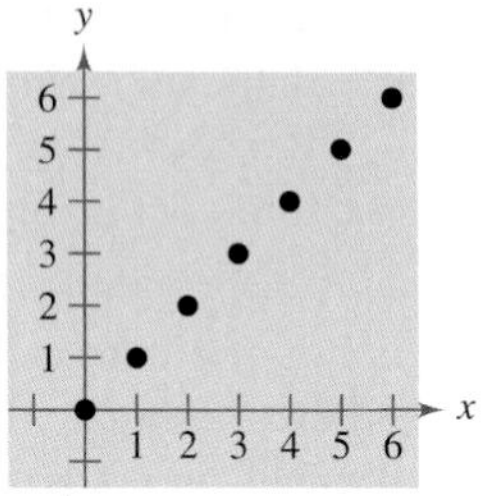

(e)

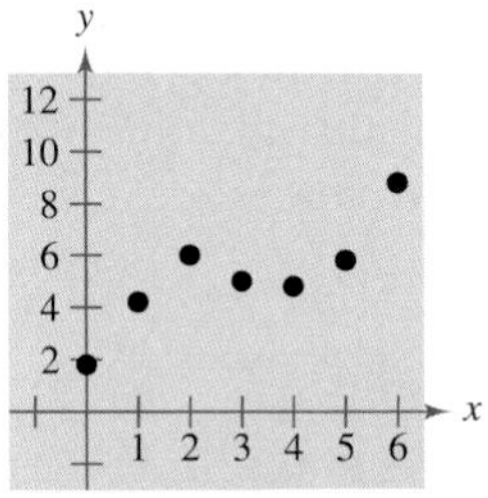

(f)

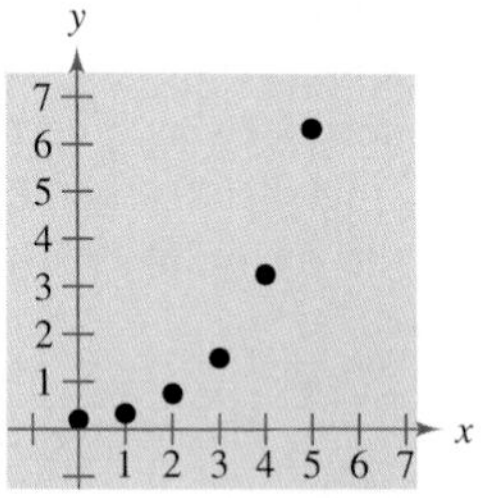

72. ***Writing*** Use your school's library or some other reference source to write a paper describing John Napier's work with logarithms.

73. ***Writing*** Before the development of electronic calculators and graphing utilities, some computations were done on slide rules. Use your school's library or some other reference source to write a paper describing the use of logarithmic scales on a slide rule.

Review

In Exercises 74–77, divide using synthetic division.

74. $\dfrac{4x^3 + 4x^2 - 39x + 36}{x + 4}$

75. $\dfrac{8x^3 - 36x^2 + 54x - 27}{x - \frac{3}{2}}$

76. $(2x^3 - 8x^2 + 3x - 9) \div (x - 4)$

77. $(x^4 - 3x + 1) \div (x + 5)$

In Exercises 78–87, sketch the graph of the equation.

78. $y = 10 - 3x$

79. $y = -4x - 1$

80. $y = -2x^2 - 3$

81. $y = 2x^2 - 7x - 30$

82. $3x^2 - 4y = 0$

83. $-x^2 - 8y = 0$

84. $y = \dfrac{4}{1 - 3x}$

85. $y = \dfrac{x^2}{-x - 2}$

86. $x^2 + (y - 8)^2 = 25$

87. $(x - 4)^2 + (y + 7) = 4$

In Exercises 88–91, graph the exponential function.

88. $f(x) = 2^{x-1} + 5$

89. $f(x) = -2^{-x-1} - 1$

90. $f(x) = 3^x - 4$

91. $f(x) = -3^x + 4$

Chapter Summary

What did you learn?

	Review Exercises
Section 3.1	
☐ How to recognize and evaluate exponential functions with base *a*	1–6
☐ How to graph exponential functions	7–18, 23–26
☐ How to recognize and evaluate exponential functions with base *e*	19–22
☐ How to use exponential functions to model and solve real-life applications	27–34
Section 3.2	
☐ How to recognize and evaluate logarithmic functions with base *a*	35–40
☐ How to graph logarithmic functions	41–46, 53–56
☐ How to recognize and evaluate natural logarithmic functions	47–52
☐ How to use logarithmic functions to model and solve real-life applications	57
Section 3.3	
☐ How to rewrite logarithmic functions with a different base	58–61
☐ How to use properties of logarithms to evaluate or rewrite logarithmic expressions	62–66
☐ How to use properties of logarithms to expand or condense logarithmic expressions	67–74
☐ How to use logarithmic functions to model and solve real-life applications	75
Section 3.4	
☐ How to solve simple exponential and logarithmic equations	76–81
☐ How to solve more complicated exponential equations	82–95
☐ How to solve more complicated logarithmic equations	96–111
☐ How to use exponential and logarithmic equations to model and solve real-life applications	112-114
Section 3.5	
☐ How to recognize the five most common types of models involving exponential and logarithmic functions	115–120
☐ How to use exponential growth and decay functions to model and solve real-life problems	121–125
☐ How to use Gaussian functions to model and solve real-life problems	126
☐ How to use logistic growth functions to model and solve real-life problems	127
☐ How to use logarithmic functions to model and solve real-life problems	128, 129

Review Exercises

3.1 In Exercises 1–6, evaluate the expression. Approximate your result to three decimal places.

1. $(6.1)^{2.4}$ **2.** $-14(5^{-0.8})$

3. $2^{-0.5\pi}$ **4.** $\sqrt[5]{1278}$

5. $60^{\sqrt{3}}$ **6.** $7^{-\sqrt{11}}$

In Exercises 7–10, match the function with its graph. [The graphs are labeled (a) through (d).]

(a)

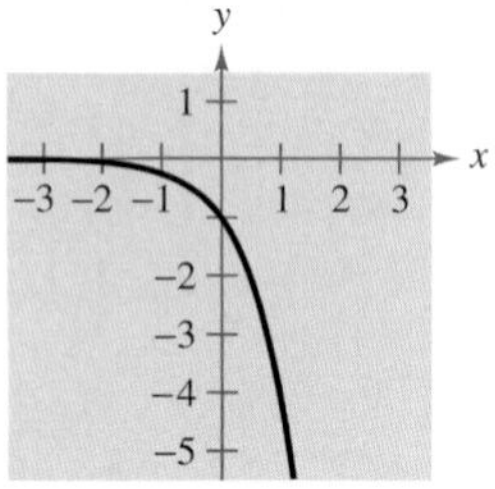

(b)

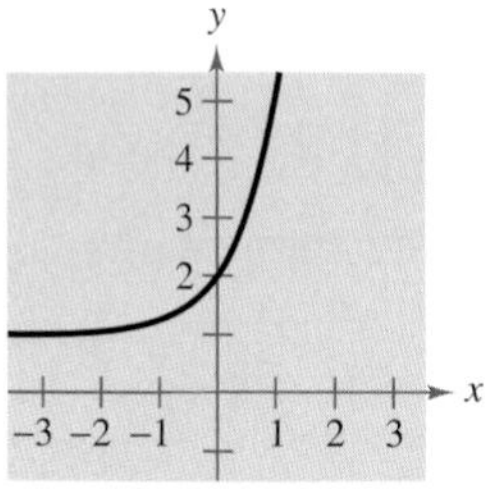

(c)

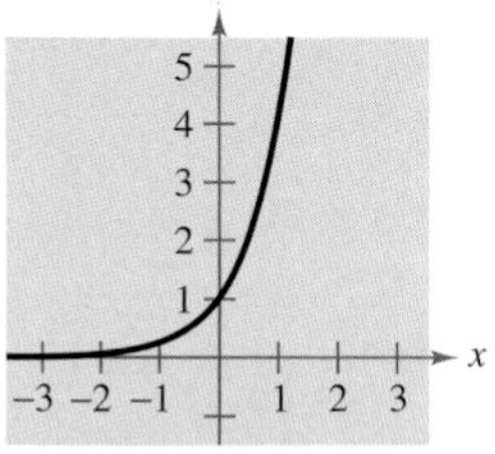

(d)

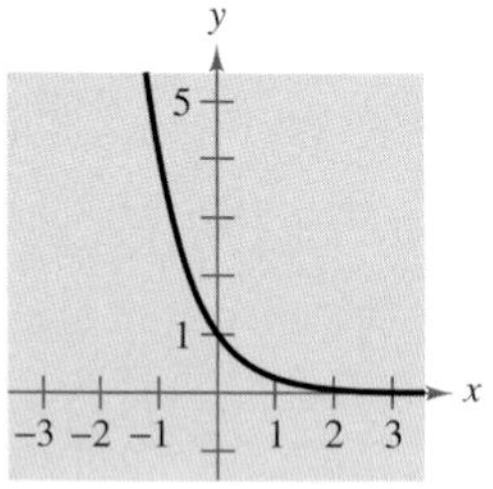

7. $f(x) = 4^x$ **8.** $f(x) = 4^{-x}$

9. $f(x) = -4^x$ **10.** $f(x) = 4^x + 1$

In Exercises 11–18, use a graphing utility to construct a table of values. Then sketch the graph of the function.

11. $f(x) = 4^{-x} + 4$ **12.** $f(x) = -4^x - 3$

13. $f(x) = -2.65^{x+1}$ **14.** $f(x) = 2.65^{x-1}$

15. $f(x) = 5^{x-2} + 4$ **16.** $f(x) = 2^{x-6} - 5$

17. $f(x) = \left(\frac{1}{2}\right)^{-x} + 3$ **18.** $f(x) = \left(\frac{1}{8}\right)^{x+2} - 5$

In Exercises 19–22, evaluate the expression. Approximate your result to three decimal places.

19. e^8 **20.** $e^{5/8}$

21. $e^{-1.7}$ **22.** $e^{0.278}$

In Exercises 23–26, use a graphing utility to construct a table of values. Then sketch the graph of the function.

23. $h(x) = e^{-x/2}$ **24.** $h(x) = 2 - e^{-x/2}$

25. $f(x) = e^{x+2}$ **26.** $s(t) = 4e^{-2/t}, \quad t > 0$

In Exercises 27 and 28, complete the table to determine the balance *A* for *P* dollars invested at rate *r* for *t* years and compounded *n* times per year.

n	1	2	4	12	365	Continuous
A						

27. $P = \$3500, \ r = 6.5\%, \ t = 10$ years

28. $P = \$2000, \ r = 5\%, \ t = 30$ years

In Exercises 29 and 30, complete the table to determine the amount *P* that should be invested at rate *r* to produce a balance of $200,000 in *t* years.

t	1	10	20	30	40	50
P						

29. $r = 8\%$, compounded continuously

30. $r = 6\%$, compounded monthly

31. *Waiting Times* The average time between incoming calls at a switchboard is 3 minutes. The probability of waiting less than t minutes until the next incoming call is approximated by the model

$$F(t) = 1 - e^{-t/3}.$$

If a call has just come in, find the probability that the next call will be within

(a) $\frac{1}{2}$ minute. (b) 2 minutes. (c) 5 minutes.

32. *Depreciation* After t years, the value of a car that cost $14,000 is

$$V(t) = 14{,}000\left(\tfrac{3}{4}\right)^t.$$

(a) Use a graphing utility to graph the function.

(b) Find the value of the car 2 years after it was purchased.

(c) According to the model, when does the car depreciate most rapidly? Is this realistic? Explain.

33. *Finance* On the day a person was born, a deposit of $50,000 was made in a trust fund that pays 8.75% interest, compounded continuously.

(a) Find the balance on the person's 35th birthday.

(b) How much longer would the person have to wait to get twice as much?

34. *Fuel Efficiency* A certain automobile gets 28 miles per gallon of gasoline for speeds up to 50 miles per hour. Over 50 miles per hour, the number of miles per gallon drops at a rate of 12% for each additional 10 miles per hour. If s is the speed and y is the number of miles per gallon, then

$$y = 28e^{0.6-0.012s}, \qquad s \geq 50.$$

Use this model to complete the table.

s	50	55	60	65	70
y					

3.2 **In Exercises 35 and 36, write the exponential equation in logarithmic form.**

35. $4^3 = 64$

36. $25^{3/2} = 125$

In Exercises 37–40, evaluate the expression by hand.

37. $\log_{10} 1000$

38. $\log_9 3$

39. $\log_2 \frac{1}{8}$

40. $\log_a \frac{1}{a}$

In Exercises 41–46, sketch the graph of the function. Identify any asymptotes.

41. $g(x) = \log_7 x$

42. $g(x) = \log_5 x$

43. $f(x) = \log_{10}\left(\frac{x}{3}\right)$

44. $f(x) = 6 + \log_{10} x$

45. $f(x) = 4 - \log_{10}(x + 5)$

46. $f(x) = \log_{10}(x - 3) + 1$

In Exercises 47–52, use your calculator to evaluate each expression. Approximate your result to three decimal places if necessary.

47. $\ln 22.6$

48. $\ln 0.98$

49. $\ln e^{-12}$

50. $\ln e^7$

51. $\ln(\sqrt{7} + 5)$

52. $\ln\left(\frac{\sqrt{3}}{8}\right)$

In Exercises 53–56, sketch the graph of the function. Identify any asymptotes.

53. $f(x) = \ln x + 3$

54. $f(x) = \ln(x - 3)$

55. $h(x) = \ln(x^2)$

56. $f(x) = \frac{1}{4} \ln x$

57. *Snow Removal* The number of miles s of roads cleared of snow is approximated by the model

$$s = 25 - \frac{13 \ln(h/12)}{\ln 3}, \qquad 2 \leq h \leq 15$$

where h is the depth of the snow in inches. Use this model to find s when $h = 10$ inches.

3.3 **In Exercises 58–61, evaluate the logarithm using the change-of-base formula. Do each problem twice, once with common logarithms and once with natural logarithms. Approximate the results to three decimal places.**

58. $\log_4 9$

59. $\log_{12} 200$

60. $\log_{1/2} 5$

61. $\log_3 0.28$

In Exercises 62–66, verify each statement using the properties of logarithms.

62. $\ln 8 + \ln 5 = \ln 40$

63. $-\ln\left(\frac{1}{12}\right) = \ln 12$

64. $\ln \sqrt[4]{\frac{x}{y}} = \frac{1}{4} \ln x - \frac{1}{4} \ln y$

65. $\log_8\left(\frac{\sqrt{x}}{y^3}\right) = \frac{1}{2} \log_8 x - 3 \log_8 y$

66. $\log_{10}\left(\frac{p^2q^3}{r}\right) = 2 \log_{10} p + 3 \log_{10} q - \log_{10} r$

In Exercises 67–70, use the properties of logarithms to write the expression as a sum, difference, and/or multiple of logarithms.

67. $\log_5 5x^2$

68. $\log_7 \frac{\sqrt{x}}{4}$

69. $\log_{10} \frac{5\sqrt{y}}{x^2}$

70. $\ln\left|\frac{x - 1}{x + 1}\right|$

In Exercises 71–74, write the expression as the logarithm of a single quantity.

71. $\log_2 5 + \log_2 x$

72. $\log_6 y - 2 \log_6 z$

73. $\frac{1}{2} \ln|2x - 1| - 2 \ln|x + 1|$

74. $5 \ln|x - 2| - \ln|x + 2| - 3 \ln|x|$

75. *Climb Rate* The time t, in minutes, for a small plane to climb to an altitude of h feet is modeled by

$$t = 50 \log_{10} \frac{18{,}000}{18{,}000 - h}$$

where 18,000 feet is the plane's absolute ceiling.

(a) Determine the domain of the function appropriate for the context of the problem.

(b) Use a graphing utility to graph the time function and identify any asymptotes.

(c) As the plane approaches its absolute ceiling, what can be said about the time required to increase its altitude further?

(d) Find the time for the plane to climb to an altitude of 4000 feet.

3.4 In Exercises 76–81, solve for x.

76. $8^x = 512$ **77.** $3^x = 729$

78. $6^x = \frac{1}{216}$ **79.** $6^{x-2} = 1296$

80. $\log_7 x = 4$ **81.** $\log_x 243 = 5$

In Exercises 82–91, solve the exponential equation. Approximate your result to three decimal places.

82. $e^x = 12$ **83.** $e^{3x} = 25$

84. $3e^{-5x} = 132$ **85.** $14e^{3x+2} = 560$

86. $e^x + 13 = 35$ **87.** $e^x - 28 = -8$

88. $-4(5^x) = -68$ **89.** $2(12^x) = 190$

90. $e^{2x} - 7e^x + 10 = 0$ **91.** $e^{2x} - 6e^x + 8 = 0$

In Exercises 92–95, use a graphing utility to solve the equation. Approximate the result to two decimal places.

92. $2^{0.6x} - 3x = 0$ **93.** $4^{-0.2x} + x = 0$

94. $25e^{-0.3x} = 12$

95. $4e^{1.2x} = 9$

In Exercises 96–107, solve the logarithmic equation. Approximate the result to three decimal places.

96. $\ln 3x = 8.2$ **97.** $\ln 5x = 7.2$

98. $2 \ln 4x = 15$ **99.** $4 \ln 3x = 15$

100. $\ln x - \ln 3 = 2$ **101.** $\ln\sqrt{x+8} = 3$

102. $\ln\sqrt{x+1} = 2$ **103.** $\ln x - \ln 5 = 4$

104. $\log_{10}(x - 1) = \log_{10}(x - 2) - \log_{10}(x + 2)$

105. $\log_{10}(x + 2) - \log_{10} x = \log_{10}(x + 5)$

106. $\log_{10}(1 - x) = -1$

107. $\log_{10}(-x - 4) = 2$

In Exercises 108–111, use a graphing utility to solve the equation. Approximate the result to two decimal places.

108. $2 \ln(x + 3) + 3x = 8$

109. $6 \log_{10}(x^2 + 1) - x = 0$

110. $4 \ln(x + 5) - x = 10$

111. $x - 2 \log_{10}(x + 4) = 0$

112. *Finance* \$7550 is deposited in an account that pays 7.25% interest, compounded continuously. How long will it take the money to triple?

113. *Finance* \$2440 is deposited in an account that pays 6.5% interest, compounded continuously. How long will it take the money to quadruple?

114. *Economics* The demand equation for a certain product is modeled by

$$p = 500 - 0.5e^{0.004x}.$$

Find the demand x for a price of (a) $p = \$450$ and (b) $p = \$400$.

3.5 In Exercises 115–120, match the function with its graph. [The graphs are labeled (a) through (f).]

(a)

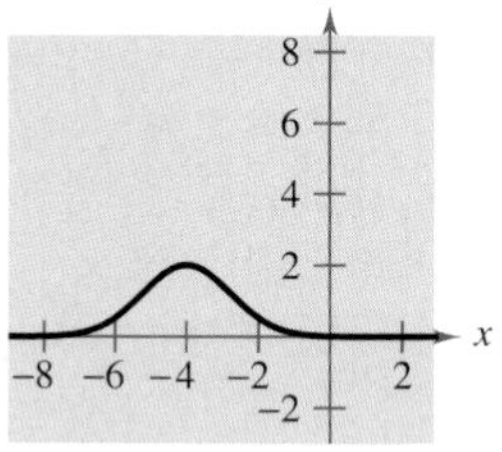

(b)

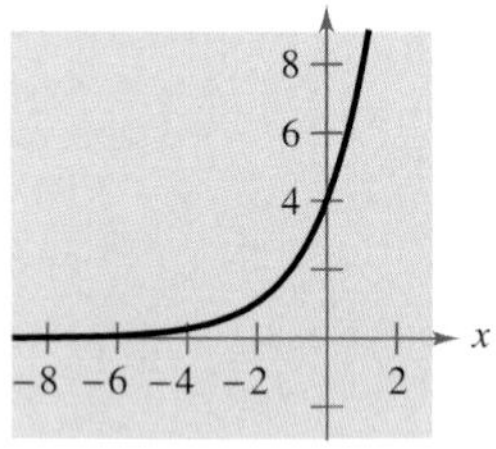

(c)

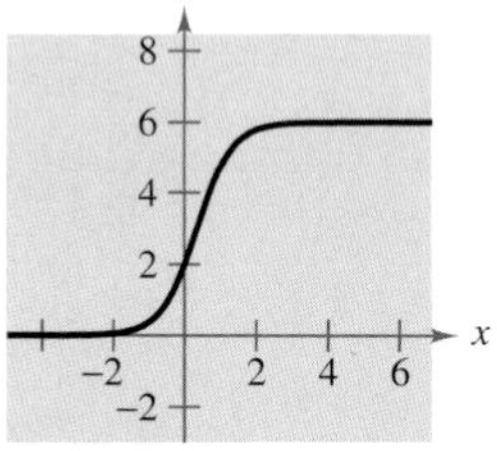

(d)

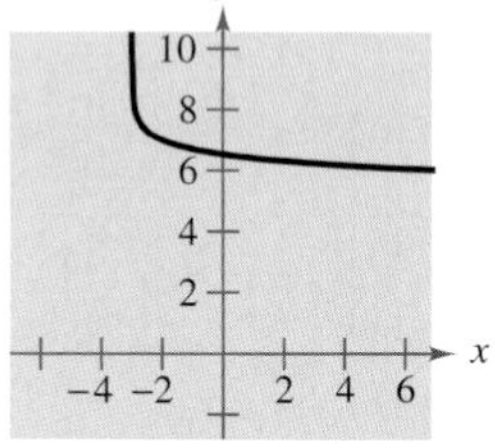

(e)

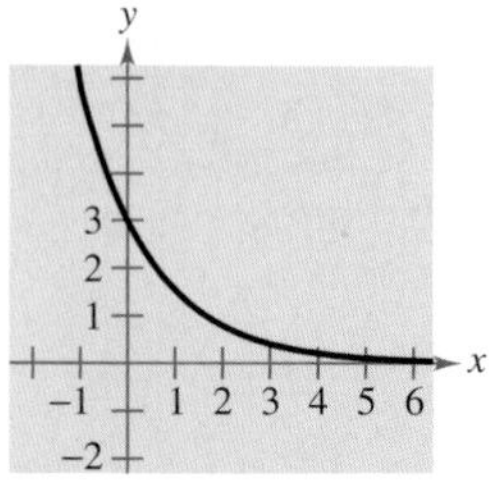

(f)

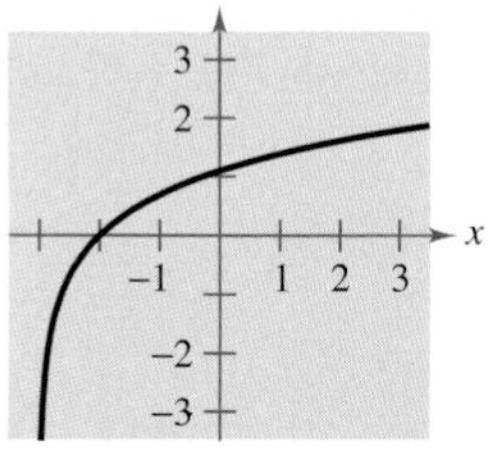

115. $y = 3e^{-2x/3}$ **116.** $y = 4e^{2x/3}$

117. $y = \ln(x + 3)$ **118.** $y = 7 - \log_{10}(x + 3)$

119. $y = 2e^{-(x+4)^2/3}$ **120.** $y = \dfrac{6}{1 + 2e^{-2x}}$

121. *Population* The population of a town is modeled by

$$P = 12{,}620e^{0.0118t}$$

where $t = 0$ represents the year 2000. According to this model, when will the population reach 17,000?

122. *Radioactive Decay* The half-life of radioactive uranium II (^{234}U) is 250,000 years. What percent of a present amount of radioactive uranium II will remain after 5000 years?

123. *Finance* A deposit of \$10,000 is made in a savings account for which the interest is compounded continuously. The balance will double in 5 years.

(a) What is the annual interest rate for this account?

(b) Find the balance after 1 year.

In Exercises 124 and 125, find the exponential function $y = ae^{bx}$ that passes through the points.

124. $(0, 2), (4, 3)$

125. $\left(0, \frac{1}{2}\right), (5, 5)$

126. *Test Scores* The test scores for a biology test follow a normal distribution modeled by

$$y = 0.0499e^{-(x-71)^2/128}, \quad 40 \le x \le 100$$

where x is the test score.

(a) Use a graphing utility to sketch the graph of the equation.

(b) From the graph, estimate the average test score.

127. *Typing Speed* In a typing class, the average number of words per minute typed after t weeks of lessons was found to be

$$N = \frac{157}{1 + 5.4e^{-0.12t}}.$$

Find the time necessary to type (a) 50 words per minute and (b) 75 words per minute.

128. *Physics* The relationship between the number of decibels β and the intensity of a sound I in watts per square centimeter is

$$\beta = 10 \log_{10}\left(\frac{I}{10^{-16}}\right).$$

Determine the intensity of a sound in watts per square centimeter if the decibel level is 125.

129. *Geology* On the Richter scale, the magnitude R of an earthquake of intensity I is

$$R = \log_{10} \frac{I}{I_0}$$

where $I_0 = 1$ is the minimum intensity used for comparison. Find the intensity per unit of area for the following values of R.

(a) $R = 8.4$

(b) $R = 6.85$

(c) $R = 9.1$

Synthesis

True or False? **In Exercises 130–135, determine whether the equation or statement is true or false. Justify your answer.**

130. $\log_b b^{2x} = 2x$

131. $e^{x-1} = \dfrac{e^x}{e}$

132. $\ln(x + y) = \ln x + \ln y$

133. $\ln(x + y) = \ln(x \cdot y)$

134. $\log_{10}\left(\dfrac{10}{x}\right) = 1 - \log_{10} x$

135. The domain of the function $f(x) = \ln x$ is the set of all real numbers.

136. The graphs of $y = e^{kt}$ are shown for $k = a, b, c,$ and d. Use the graphs to order $a, b, c,$ and d. Which of the four values are negative? Which are positive?

(a)

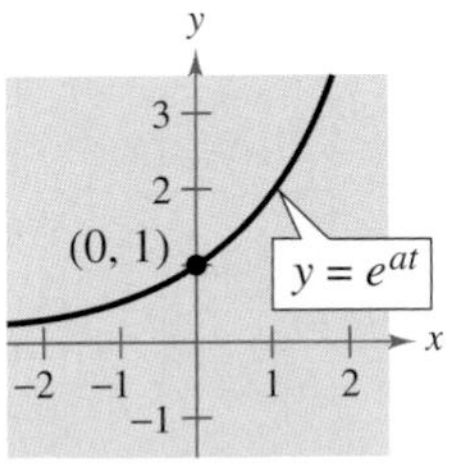

(b)

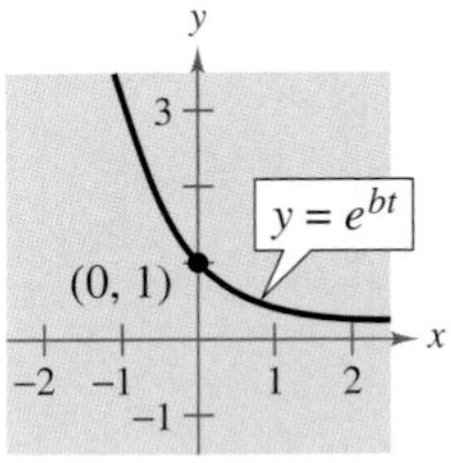

(c)

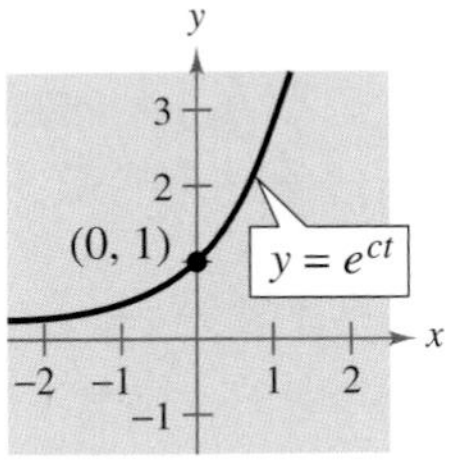

(d)

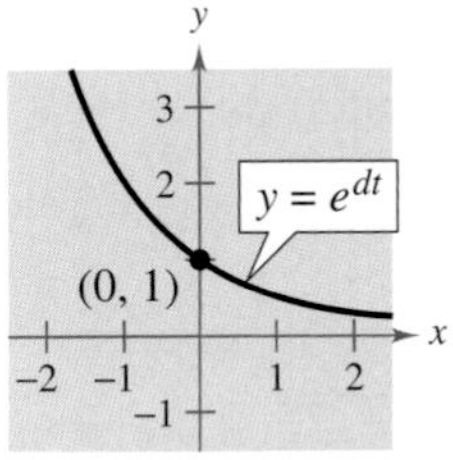

Chapter Project ▶ Graphical Approach to Compound Interest

A graphing utility can be used to investigate the rates of growth of different types of compound interest.

Example ▶ Comparing Balances

You are depositing \$1000 in a savings account. Which of the following will produce the largest balance?

a. 6% annual interest rate, compounded annually

b. 6% annual interest rate, compounded continuously

c. 6.25% annual interest rate, compounded quarterly

Solution

One way to compare all three options is to sketch their graphs in the same viewing window.

Option (a) $A = 1000(1 + 0.06)^t$

Option (b) $A = 1000e^{0.06t}$

Option (c) $A = 1000\left(1 + \frac{0.0625}{4}\right)^{4t}$

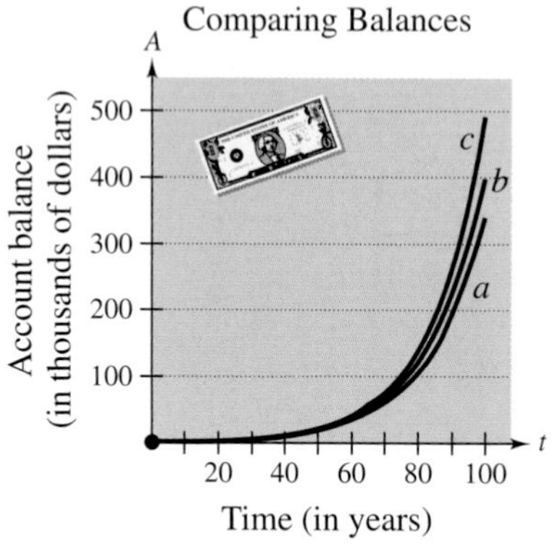

The graphs are shown at the left. From the graphs, you can conclude that option (c) is better than option (b), and option (b) is better than option (a). Note that for the first 50 years, there is little difference in the graphs. Between 50 and 100 years, however, the balances obtained begin to differ significantly. At the end of 100 years, the balances are (a) \$339,302, (b) \$403,429, and (c) \$493,575.

Chapter Project Investigations

1. Which would produce a larger balance: an annual interest rate of 8.05% compounded monthly or an annual interest rate of 8% compounded continuously? Explain.

2. You deposit \$1000 in each of two savings accounts. The interest for the accounts is paid according to the two options described in Question 1. How long would it take for the balance in one of the accounts to exceed the balance in the other account by \$100? By \$100,000?

3. No income tax is due on the interest earned in some types of investments. You deposit \$25,000 in an account. Which of the following plans is better? Explain.

(a) *Tax-free* The account pays 5% compounded annually. There is no income tax due on the earned interest.

(b) *Tax-deferred* The account pays 7% compounded annually. At maturity, the earned interest is taxable at a rate of 40%.

Chapter Test

The *Interactive* CD-ROM and *Internet* versions of this text provide answers to the Chapter Tests and Cumulative Tests. They also offer Chapter Pre-Tests (which test key skills and concepts covered in previous chapters) and Chapter Post-Tests, both of which have randomly generated exercises with diagnostic capabilities.

Take this test as you would take a test in class. After you are done, check your work against the answers given in the back of the book.

In Exercises 1–4, evaluate the expression. Approximate your result to three decimal places.

1. $12.4^{2.79}$ **2.** $4^{3\pi/2}$ **3.** $e^{-7/10}$ **4.** $e^{3.1}$

In Exercises 5–7, construct a table of values. Then sketch the graph of the function.

5. $f(x) = 10^{-x}$ **6.** $f(x) = -6^{x-2}$ **7.** $f(x) = 1 - e^{2x}$

8. Evaluate (a) $\log_7 7^{-0.89}$ and (b) $4.6 \ln e^2$.

In Exercises 9–11, construct a table of values. Then sketch the graph of the function. Identify any asymptotes.

9. $f(x) = -\log_{10} x - 6$ **10.** $f(x) = \ln(x - 4)$ **11.** $f(x) = 1 + \ln(x + 6)$

In Exercises 12–14, evaluate the expression. Approximate your result to three decimal places.

12. $\log_7 44$ **13.** $\log_{2/5} 0.9$ **14.** $\log_{24} 68$

In Exercises 15 and 16, use the properties of logarithms to write the expression as a sum, difference, and/or multiple of logarithms.

15. $\log_2 3a^4$ **16.** $\ln \dfrac{5\sqrt{x}}{6}$

In Exercises 17 and 18, rewrite the expression as the logarithm of a single quantity.

17. $\log_3 13 + \log_3 y$ **18.** $4 \ln x - 4 \ln y$

In Exercises 19 and 20, solve the equation algebraically. Approximate your result to three decimal places.

19. $\dfrac{1025}{8 + e^{4x}} = 5$ **20.** $\log_{10} x - \log_{10}(8 - 5x) = 2$

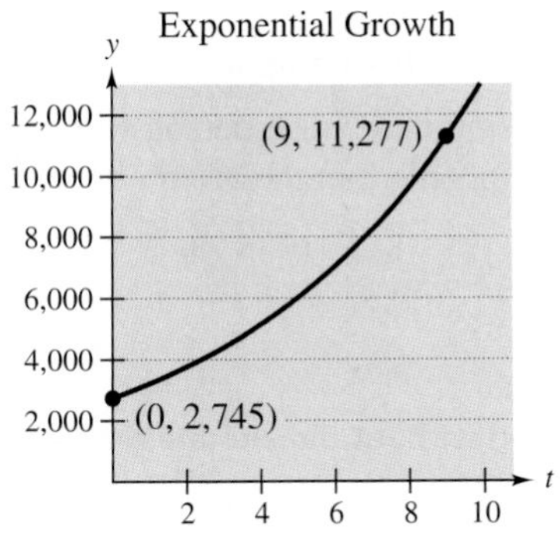

FIGURE FOR 21

21. Find an exponential growth model for the graph in the figure.

22. The half-life of radioactive actinium (^{227}Ac) is 22 years. What percent of a present amount of radioactive actinium will remain after 19 years?

23. A model that can be used for predicting the height H (in centimeters) of a child based on his or her age is

$$H = 70.228 + 5.104x + 9.222 \ln x, \qquad \tfrac{1}{4} \le x \le 6$$

where x is the age of the child in years. (Source: Snapshots of Applications in Mathematics)

(a) Construct a table of values. Then sketch the graph of the model.

(b) Use the graph from part (a) to estimate the height of a 4-year-old child. Then calculate the actual height using the model.

Cumulative Test for Chapters 1–3

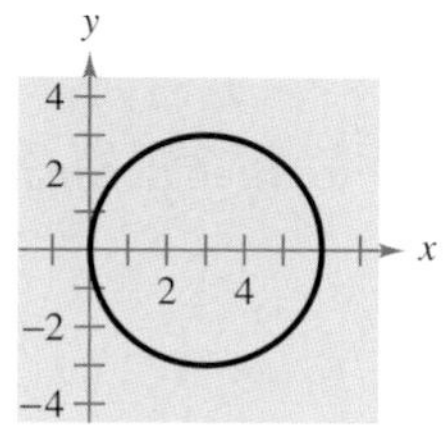

FIGURE FOR 5

Take this test to review the material from earlier chapters. After you are done, check your work against the answers given in the back of the book.

In Exercises 1–3, use intercepts and symmetry to sketch the graph of the equation.

1. $y = \sqrt{x - 5}$ **2.** $y = |x - 5|$ **3.** $y = x^3 - 4$

4. Find an equation for the line passing through $\left(-\frac{1}{2}, 1\right)$ and $(3, 8)$.

5. Explain why the graph at the left does not represent y as a function of x.

6. Evaluate (if possible) the function $f(x) = \dfrac{x}{x - 2}$ for each value.

(a) $f(6)$ (b) $f(2)$ (c) $f(s + 2)$

7. Describe how the graph of each function would differ from the graph of $y = \sqrt[3]{x}$. (*Note:* It is not necessary to sketch the graphs.)

(a) $r(x) = \frac{1}{2}\sqrt[3]{x}$ (b) $h(x) = \sqrt[3]{x} + 2$ (c) $g(x) = \sqrt[3]{x + 2}$

In Exercises 8 and 9, find (a) $(f + g)(x)$, (b) $(f - g)(x)$, (c) $(fg)(x)$, and (d) $(f/g)(x)$. What is the domain of f/g?

8. $f(x) = x - 3, \quad g(x) = 4x + 1$ **9.** $f(x) = \sqrt{x - 1}, \quad g(x) = x^2 + 1$

In Exercises 10 and 11, find (a) $f \circ g$ and (b) $g \circ f$.

10. $f(x) = 2x^2, \quad g(x) = \sqrt{x + 6}$ **11.** $f(x) = x - 2, \quad g(x) = |x|$

12. Determine whether $h(x) = 5x - 2$ has an inverse. If so, find it.

13. A group of n people decide to buy a \$36,000 minibus. Each person will pay an equal share of the cost. If three additional people join the group, the cost per person will decrease by \$1000. Find n.

14. Find the quadratic function $y = a(x - h)^2 + k$ whose graph has a vertex at $(-8, 5)$ and passes through the point $(-4, -7)$.

In Exercises 15–17, sketch the graph of the function without the aid of a graphing utility.

15. $h(x) = -(x^2 + 4x)$ **16.** $f(t) = \frac{1}{4}t(t - 2)^2$ **17.** $g(s) = s^2 + 4s + 10$

In Exercises 18 and 19, find all the zeros of the function.

18. $f(x) = x^3 + 2x^2 + 4x + 8$ **19.** $f(x) = x^4 + 4x^3 - 21x^2$

20. Divide: $\dfrac{6x^3 - 4x^2}{2x^2 + 1}$.

21. Use synthetic division to divide $2x^4 + 3x^3 - 6x + 5$ by $x + 2$.

22. Use a graphing utility to approximate the real zero of the function $g(x) = x^3 + 3x^2 - 6$ to the nearest hundredth.

23. Find a polynomial with integer coefficients that has -5, -2, and $2 + \sqrt{3}i$ as its zeros.

In Exercises 24–26, sketch the graph of the rational function by hand. Be sure to identify all intercepts and asymptotes.

24. $f(x) = \dfrac{2x}{x - 3}$ **25.** $f(x) = \dfrac{4x^2}{x - 5}$ **26.** $f(x) = \dfrac{2x}{x^2 - 9}$

In Exercises 27 and 28, write the partial fraction decomposition of the rational expression. Check your result algebraically.

27. $\dfrac{8}{x^2 - 4x - 21}$ **28.** $\dfrac{5x}{(x - 4)^2}$

In Exercises 29 and 30, use the graph of *f* to describe the transformation that yields the graph of *g*. Use a graphing utility to graph both equations in the same viewing window.

29. $f(x) = \left(\frac{2}{5}\right)^x, \quad g(x) = -\left(\frac{2}{5}\right)^{-x+3}$ **30.** $f(x) = 2.2^x, \quad g(x) = -2.2^x + 4$

In Exercises 31–34, use a calculator to evaluate each expression. Approximate your result to three decimal places.

31. $\log_{10} 98$ **32.** $\log_{10}\left(\frac{6}{7}\right)$ **33.** $\ln \sqrt{31}$ **34.** $\ln\left(\sqrt{40} - 5\right)$

In Exercises 35–37, evaluate the logarithm using the change-of-base formula. Approximate your answer to three decimal places.

35. $\log_7 1.8$ **36.** $\log_3 0.149$ **37.** $\log_{1/2} 17$

38. Use the logarithmic properties to expand $\ln\left(\dfrac{x^2 - 16}{x^4}\right)$, where $x > 4$.

39. Write $2 \ln x - \frac{1}{2} \ln(x + 5)$ as a logarithm of a single quantity.

In Exercises 40 and 41, solve the equation.

40. $6e^{2x} = 72$ **41.** $\log_2 x + \log_2 5 = 6$

42. Use a graphing utility to graph $f(x) = \dfrac{1000}{1 + 4e^{-0.2x}}$ and determine the horizontal asymptotes.

43. The numbers of used cars C (in millions) sold in the United States from 1990 through 1997 are shown in the table, where t represents the time in years, with $t = 0$ corresponding to 1990. (Source: National Automobile Dealers Association)

t	0	1	2	3	4	5	6	7
C	14.18	14.27	15.14	16.30	17.76	18.48	19.17	19.19

(a) Use a graphing utility to sketch a scatter plot of the data.

(b) A model for this data is given by $C = -0.025x^2 + 1.018x + 13.679$. Use a graphing utility to graph the model in the same viewing window as the scatter plot.

(c) Do you think this model could be used to predict the numbers of used cars sold in the future? Explain.

Photo Source Hawaii

The Mauna Loa Observatory in Hawaii conducts research to understand the global carbon cycle. It is located far from pollution sources that would affect the gases being measured. *(Source: NOAA/Climated Monitoring and Diagnostic Laboratory)*

4 Trigonometry

How to Study This Chapter

The Big Picture

In this chapter you will learn the following skills and concepts.

- How to describe an angle and convert between radian and degree measure
- How to identify a unit circle and its relationship to real numbers
- How to evaluate trigonometric functions of any angle
- How to use the fundamental trigonometric identities
- How to sketch the graphs of trigonometric functions and translations of graphs of sine and cosine functions
- How to evaluate the inverse trigonometric functions
- How to evaluate the compositions of trigonometric functions and inverse trigonometric functions

Important Vocabulary

As you encounter each new vocabulary term in this chapter, add the term and its definition to your notebook glossary.

Trigonometry (p. 352)
Angle (p. 352)
Initial side (p. 352)
Terminal side (p. 352)
Vertex (p. 352)
Standard position (p. 352)
Positive angles (p. 352)
Negative angles (p. 352)
Coterminal angles (p. 352)
Central angle (p. 353)
Radian (p. 353)
Acute angles (p. 353)
Obtuse angles (p. 353)
Complementary angles (p. 355)
Supplementary angles (p. 355)
Degree (p. 355)
Linear speed (p. 357)
Angular speed (p. 357)
Unit circle (p. 363)
Sine (pp. 364, 370)
Cosecant (pp. 364, 370)
Cosine (pp. 364, 370)
Secant (pp. 364, 370)
Tangent (pp. 364, 370)
Cotangent (pp. 364, 370)
Period (pp. 366, 393)
Hypotenuse (p. 370)
Opposite side (p. 370)
Adjacent side (p. 370)
Reference angles (p. 383)
Amplitude (p. 392)
Phase shift (p. 394)
Damping factor (p. 406)
Inverse sine function (p. 412)
Inverse cosine function (p. 414)
Inverse tangent function (p. 414)
Angle of elevation (p. 422)
Angle of depression (p. 422)
Bearings (p. 424)
Simple harmonic motion (pp. 425, 426)

Study Tools

- Learning objectives at the beginning of each section
- Chapter Summary (p. 433)
- Review Exercises (pp. 434–437)
- Chapter Test (p. 439)

Additional Resources

- Study and Solutions Guide
- Interactive Precalculus
- Videotapes for Chapter 4
- Precalculus Website
- Student Success Organizer

STUDY T!P

To prepare for a chapter test, review the learning objectives and work the review exercises. Take the sample Chapter Test and analyze the results.

4.1 Radian and Degree Measure

▶ **What you should learn**

- How to describe angles
- How to use radian measure
- How to use degree measure
- How to use angles to model and solve real-life problems

▶ **Why you should learn it**

You can use angles to model and solve real-life problems. For instance, in Exercises 91–94 on page 361, you are asked to use angles to find the distance between two cities on the same longitude.

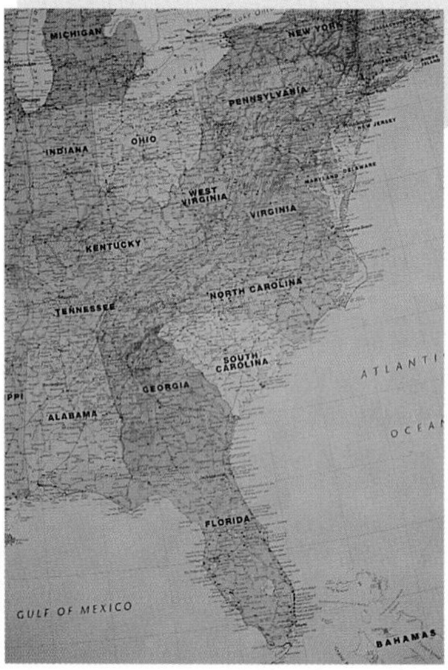

Ernest Jones

Angles

As derived from the Greek language, the word **trigonometry** means "measurement of triangles." Initially, trigonometry dealt with relationships among the sides and angles of triangles and was used in the development of astronomy, navigation, and surveying. With the development of calculus and the physical sciences in the 17th century, a different perspective arose—one that viewed the classic trigonometric relationships as *functions* with the set of real numbers as their domains. Consequently, the applications of trigonometry expanded to include a vast number of physical phenomena involving rotations and vibrations. These phenomena include sound waves, light rays, planetary orbits, vibrating strings, pendulums, and orbits of atomic particles.

The approach in this text incorporates *both* perspectives, starting with angles and their measure.

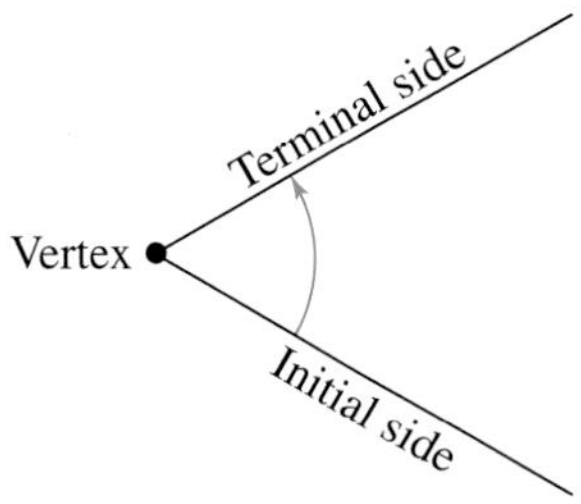

(a) Angle

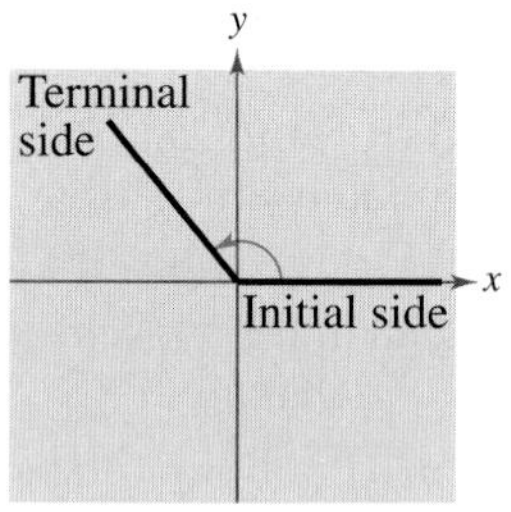

(b) Angle in Standard Position

FIGURE 4.1

An **angle** is determined by rotating a ray (half-line) about its endpoint. The starting position of the ray is the **initial side** of the angle, and the position after rotation is the **terminal side,** as shown in Figure 4.1(a). The endpoint of the ray is the **vertex** of the angle. This perception of an angle fits a coordinate system in which the origin is the vertex and the initial side coincides with the positive x-axis. Such an angle is in **standard position,** as shown in Figure 4.1(b). **Positive angles** are generated by counterclockwise rotation, and **negative angles** by clockwise rotation, as shown in Figure 4.2. Angles are labeled with Greek letters α (alpha), β (beta), and θ (theta), as well as uppercase letters A, B, and C. In Figure 4.3, note that angles α and β have the same initial and terminal sides. Such angles are **coterminal.**

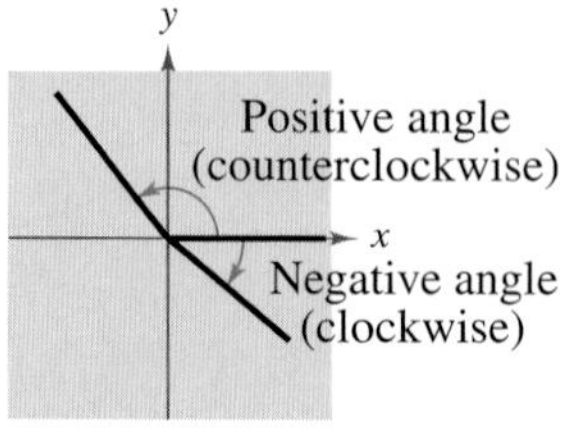

FIGURE 4.2

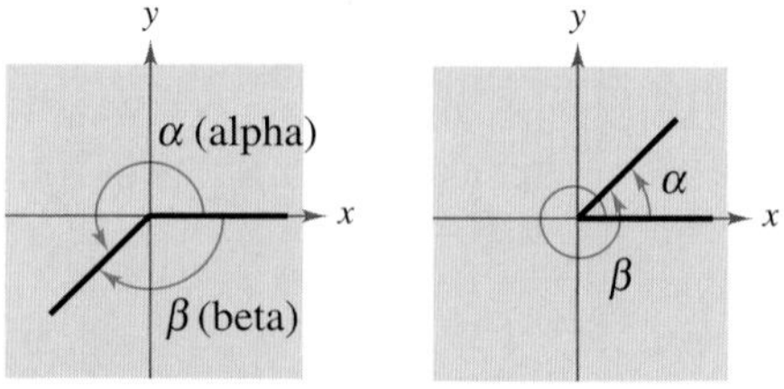

FIGURE 4.3

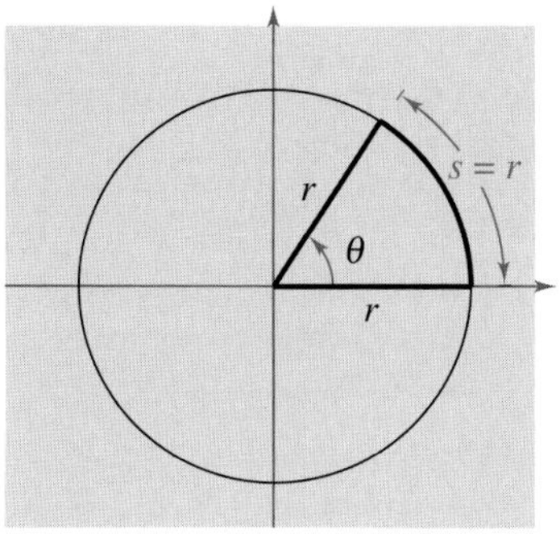

Arc length = radius when θ = 1 radian
FIGURE 4.4

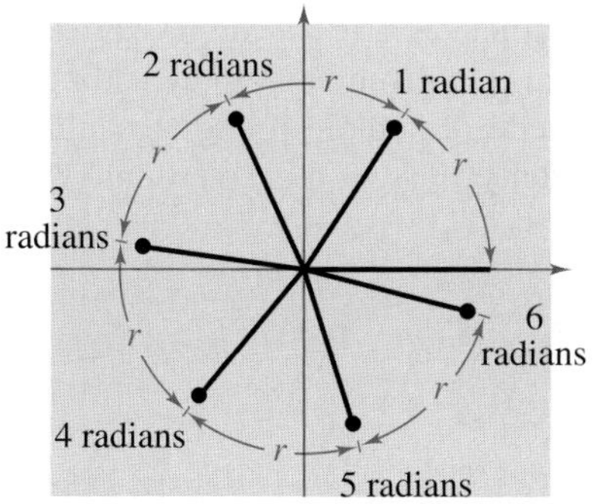

FIGURE 4.5

Radian Measure

The **measure of an angle** is determined by the amount of rotation from the initial side to the terminal side. One way to measure angles is in *radians*. This type of measure is especially useful in calculus. To define a radian, you can use a **central angle** of a circle, one whose vertex is the center of the circle, as shown in Figure 4.4.

Definition of Radian

One **radian** is the measure of a central angle θ that intercepts an arc s equal in length to the radius r of the circle. See Figure 4.4.

Because the circumference of a circle is $2\pi r$ units, it follows that a central angle of one full revolution (counterclockwise) corresponds to an arc length of

$$s = 2\pi r.$$

Moreover, because $2\pi \approx 6.28$, there are just over six radius lengths in a full circle, as shown in Figure 4.5. In general, the radian measure of a central angle θ is obtained by dividing the arc length s by r. That is, $s/r = \theta$, where *θ is measured in radians.* Because the units of measure for s and r are the same, this ratio is unitless—it is simply a real number.

Because the radian measure of an angle of one full revolution is 2π, you can obtain the following.

$$\frac{1}{2} \text{ revolution} = \frac{2\pi}{2} = \pi \text{ radians}$$

$$\frac{1}{4} \text{ revolution} = \frac{2\pi}{4} = \frac{\pi}{2} \text{ radians}$$

$$\frac{1}{6} \text{ revolution} = \frac{2\pi}{6} = \frac{\pi}{3} \text{ radians}$$

These and other common angles are shown in Figure 4.6.

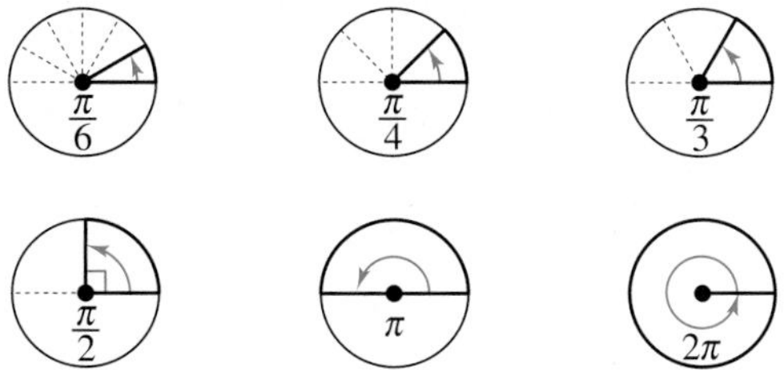

FIGURE 4.6

STUDY TIP

One revolution around a circle of radius r corresponds to an angle of 2π radians because

$$\frac{s}{r} = \frac{2\pi r}{r} = 2\pi \text{ radians.}$$

Recall that the four quadrants in a coordinate system are numbered I, II, III, and IV. Figure 4.7 shows which angles between 0 and 2π lie in each of the four quadrants. Note that angles between 0 and $\pi/2$ are **acute** and that angles between $\pi/2$ and π are **obtuse.**

STUDY TIP

The phrase "the terminal side of θ lies in a quadrant" is often abbreviated by simply saying that "θ lies in a quadrant." The terminal sides of the "quadrant angles" 0, $\pi/2$, π, and $3\pi/2$ do not lie within quadrants.

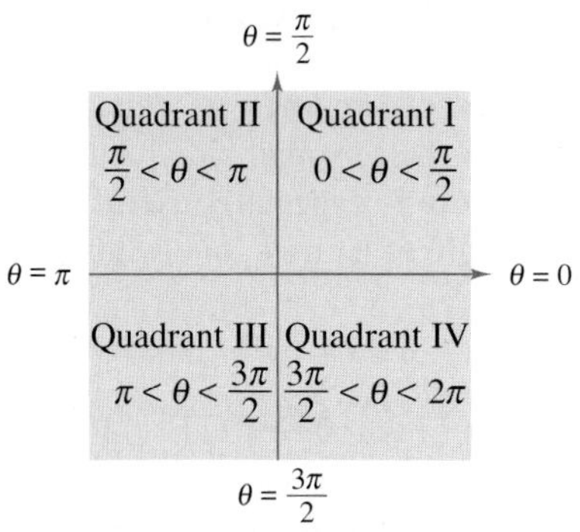

FIGURE 4.7

Two angles are coterminal if they have the same initial and terminal sides. For instance, the angles 0 and 2π are coterminal, as are the angles $\pi/6$ and $13\pi/6$. You can find an angle that is coterminal to a given angle θ by adding or subtracting 2π (one revolution), as demonstrated in Example 1. A given angle θ has infinitely many coterminal angles. For instance, $\theta = \pi/6$ is coterminal with

$$\frac{\pi}{6} + 2n\pi$$

where n is an integer.

Example 1 ▶ Sketching and Finding Coterminal Angles

Remind your students to work in radians.

a. For the positive angle $13\pi/6$, subtract 2π to obtain a coterminal angle

$$\frac{13\pi}{6} - 2\pi = \frac{\pi}{6}.$$ See Figure 4.8(a).

b. For the positive angle $3\pi/4$, subtract 2π to obtain a coterminal angle

$$\frac{3\pi}{4} - 2\pi = -\frac{5\pi}{4}.$$ See Figure 4.8(b).

c. For the negative angle $-2\pi/3$, add 2π to obtain a coterminal angle

$$-\frac{2\pi}{3} + 2\pi = \frac{4\pi}{3}.$$ See Figure 4.8(c).

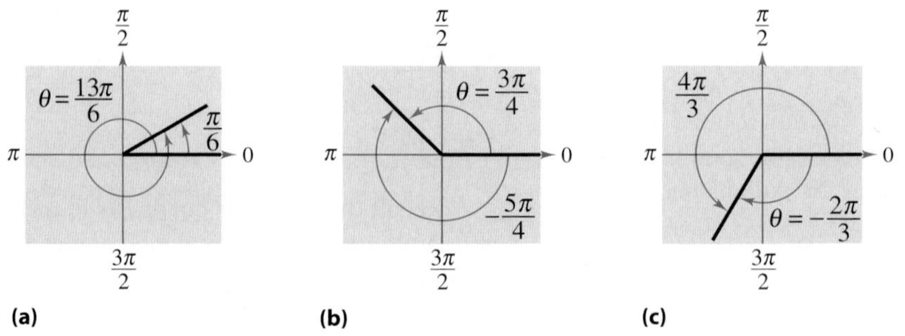

FIGURE 4.8

You might point out that complementary and supplementary angles do not necessarily share a common side. For example, the acute angles of a right triangle are complementary because the sum of their measures is $\pi/2$.

Two positive angles α and β are **complementary** (complements of each other) if their sum is $\pi/2$. Two positive angles are **supplementary** (supplements of each other) if their sum is π. See Figure 4.9.

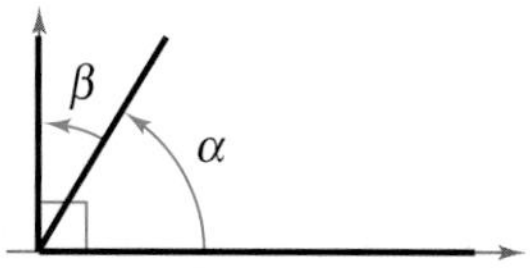

Complementary Angles

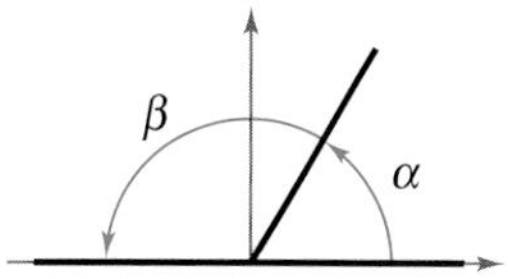

Supplementary Angles

FIGURE 4.9

Example 2 ▶ Complementary and Supplementary Angles

If possible, find the complement and the supplement of (a) $2\pi/5$ and (b) $4\pi/5$.

Solution

a. The complement of $2\pi/5$ is

$$\frac{\pi}{2} - \frac{2\pi}{5} = \frac{5\pi}{10} - \frac{4\pi}{10} = \frac{\pi}{10}.$$

The supplement of $2\pi/5$ is

$$\pi - \frac{2\pi}{5} = \frac{3\pi}{5}.$$

b. Because $4\pi/5$ is greater than $\pi/2$, it has no complement. (Remember to use only positive angles for complements.) The supplement is

$$\pi - \frac{4\pi}{5} = \frac{\pi}{5}.$$

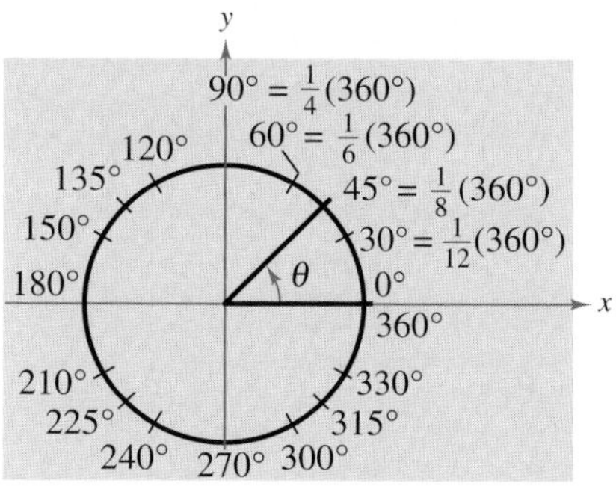

FIGURE 4.10

Degree Measure

A second way to measure angles is in terms of degrees. A measure of **1 degree** (1°) is equivalent to a rotation of $1/360$ of a complete revolution about the vertex. To measure angles, it is convenient to mark degrees on the circumference of a circle, as shown in Figure 4.10. So, a full revolution (counterclockwise) corresponds to 360°, a half revolution to 180°, a quarter revolution to 90°, and so on.

Because 2π radians corresponds to one complete revolution, degrees and radians are related by the equations

$$360° = 2\pi \text{ rad} \qquad \text{and} \qquad 180° = \pi \text{ rad}.$$

From the latter equation, you obtain

$$1° = \frac{\pi}{180} \text{ rad} \qquad \text{and} \qquad 1 \text{ rad} = \left(\frac{180°}{\pi}\right)$$

which lead to the conversion rules at the top of the next page.

Converting from degrees to radians and vice versa should help your students become familiar with radian measure.

The *Interactive* CD-ROM and *Internet* versions of this text show every example with its solution; clicking on the *Try It!* button brings up similar problems. Guided Examples and Integrated Examples show step-by-step solutions to additional examples. Integrated Examples are related to several concepts in the section.

Conversions Between Degrees and Radians

1. To convert degrees to radians, multiply degrees by $\dfrac{\pi \text{ rad}}{180°}$.

2. To convert radians to degrees, multiply radians by $\dfrac{180°}{\pi \text{ rad}}$.

To apply these two conversion rules, use the basic relationship π rad $= 180°$. (See Figure 4.11.)

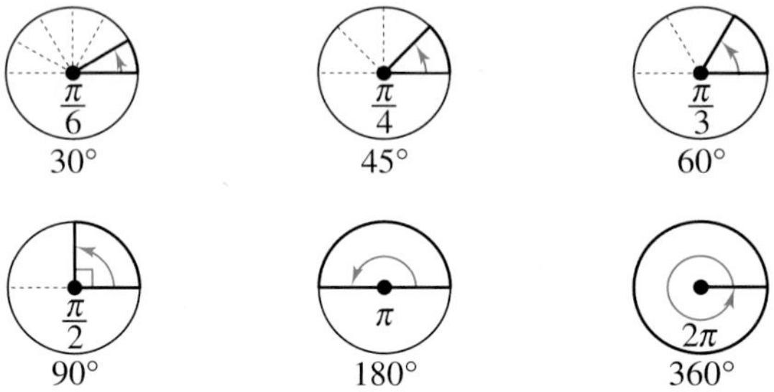

FIGURE 4.11

When no units of angle measure are specified, *radian measure is implied*. For instance, if you write $\theta = \pi$ or $\theta = 2$, you should mean $\theta = \pi$ radians or $\theta = 2$ radians.

Example 3 ▶ Converting from Degrees to Radians

a. $135° = (135 \text{ deg})\left(\dfrac{\pi \text{ rad}}{180 \text{ deg}}\right) = \dfrac{3\pi}{4}$ rad   Multiply by $\pi/180$.

b. $540° = (540 \text{ deg})\left(\dfrac{\pi \text{ rad}}{180 \text{ deg}}\right) = 3\pi$ rad   Multiply by $\pi/180$.

c. $-270° = (-270 \text{ deg})\left(\dfrac{\pi \text{ rad}}{180 \text{ deg}}\right) = -\dfrac{3\pi}{2}$ rad   Multiply by $\pi/180$.

Example 4 ▶ Converting from Radians to Degrees

a. $-\dfrac{\pi}{2} \text{ rad} = \left(-\dfrac{\pi}{2} \text{ rad}\right)\left(\dfrac{180 \text{ deg}}{\pi \text{ rad}}\right) = -90°$ Multiply by $180/\pi$.

b. $\dfrac{9\pi}{2} \text{ rad} = \left(\dfrac{9\pi}{2} \text{ rad}\right)\left(\dfrac{180 \text{ deg}}{\pi \text{ rad}}\right) = 810°$ Multiply by $180/\pi$.

c. $2 \text{ rad} = (2 \text{ rad})\left(\dfrac{180 \text{ deg}}{\pi \text{ rad}}\right) = \dfrac{360}{\pi} \approx 114.59°$ Multiply by $180/\pi$.

If you have a calculator with a "radian-to-degree" conversion key, try using it to verify the result shown in part (c) of Example 4.

Technology

With calculators it is convenient to use *decimal* degrees to denote fractional parts of degrees. Historically, however, fractional parts of degrees were expressed in *minutes* and *seconds*, using the prime (′) and double prime (″) notations, respectively. That is,

$1' = \text{one minute} = \frac{1}{60}(1°)$

$1'' = \text{one second} = \frac{1}{3600}(1°)$

Consequently, an angle of 64 degrees, 32 minutes, and 47 seconds is represented by $\theta = 64° \, 32' \, 47''$.

Many calculators have special keys for converting an angle in degrees, minutes, and seconds (D° M′ S″) into decimal degree form, and vice versa.

A computer simulation of this concept appears in the *Interactive* CD-ROM and *Internet* versions of this text.

Applications

The *radian measure* formula, $\theta = s/r$, can be used to measure arc length along a circle. Specifically, for a circle of radius r, a central angle θ intercepts an arc of length s given by

$$s = r\theta \qquad \text{Length of circular arc}$$

where θ is measured in radians.

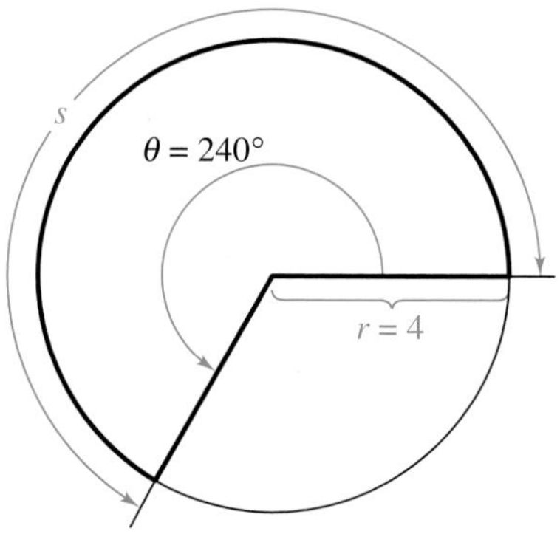

FIGURE 4.12

Example 5 ▶ Finding Arc Length

A circle has a radius of 4 inches. Find the length of the arc intercepted by a central angle of 240°, as shown in Figure 4.12.

Solution

To use the formula $s = r\theta$, first convert 240° to radian measure.

$$240° = (240 \text{ deg})\left(\frac{\pi \text{ rad}}{180 \text{ deg}}\right) \qquad \text{Convert from degrees to radians.}$$

$$= \frac{4\pi}{3} \text{ rad} \qquad \text{Simplify.}$$

Then, using a radius of $r = 4$ inches, you can find the arc length to be

$$s = r\theta \qquad \text{Length of circular arc}$$

$$= 4\left(\frac{4\pi}{3}\right) \qquad \text{Substitute for } r \text{ and } \theta.$$

$$= \frac{16\pi}{3} \qquad \text{Simplify.}$$

$$\approx 16.76 \text{ inches.} \qquad \text{Use a calculator.}$$

Note that the units for $r\theta$ are determined by the units for r because θ is given in radian measure and therefore has no units.

Because radian measure is so often used, you may want to encourage your students to be as familiar with the radian measure of angles as they are with degree measure. Measuring arc length along a circle is one of many applications where radian measure is used.

The formula for the length of a circular arc can be used to analyze the motion of a particle moving at a *constant speed* along a circular path.

Linear and Angular Speed

Consider a particle moving at a constant speed along a circular arc of radius r. If s is the length of the arc traveled in time t, then the **linear speed** of the particle is

$$\text{Linear speed} = \frac{\text{arc length}}{\text{time}} = \frac{s}{t}.$$

Moreover, if θ is the angle (in radian measure) corresponding to the arc length s, then the **angular speed** of the particle is

$$\text{Angular speed} = \frac{\text{central angle}}{\text{time}} = \frac{\theta}{t}.$$

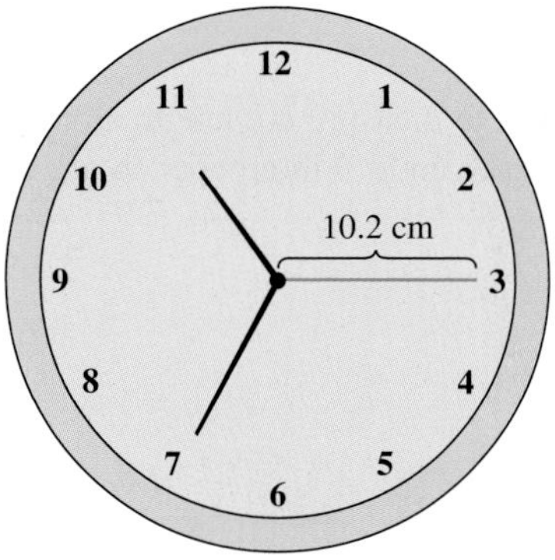

FIGURE 4.13

Example 6 ▶ Finding Linear Speed

The second hand of a clock is 10.2 centimeters long, as shown in Figure 4.13. Find the linear speed of the tip of this second hand as it passes around the clock face.

Solution

In one revolution, the arc length traveled is

$$s = 2\pi r$$
$$= 2\pi(10.2) \qquad \text{Substitute for } r.$$
$$= 20.4\pi \text{ centimeters.}$$

The time required for the second hand to travel this distance is

$$t = 1 \text{ minute} = 60 \text{ seconds.}$$

So, the linear speed of the tip of the second hand is

$$\text{Linear speed} = \frac{s}{t}$$
$$= \frac{20.4\pi \text{ centimeters}}{60 \text{ seconds}}$$
$$\approx 1.068 \text{ cm/sec.}$$

Example 7 ▶ Finding Angular and Linear Speed

A lawn roller with a 10-inch radius (see Figure 4.14) makes 1.2 revolutions per second.

a. Find the angular speed of the roller in radians per second.

b. Find the speed of the tractor that is pulling the roller.

Solution

a. Because each revolution generates 2π radians, it follows that the roller turns $(1.2)(2\pi) = 2.4\pi$ radians per second. In other words, the angular speed is

$$\text{Angular speed} = \frac{\theta}{t}$$
$$= \frac{2.4\pi \text{ radians}}{1 \text{ second}}$$
$$= 2.4\pi \text{ rad/sec.}$$

b. The linear speed is

$$\text{Linear speed} = \frac{s}{t}$$
$$= \frac{r\theta}{t}$$
$$= \frac{10(2.4\pi) \text{ inches}}{1 \text{ second}}$$
$$\approx 75.4 \text{ in./sec.}$$

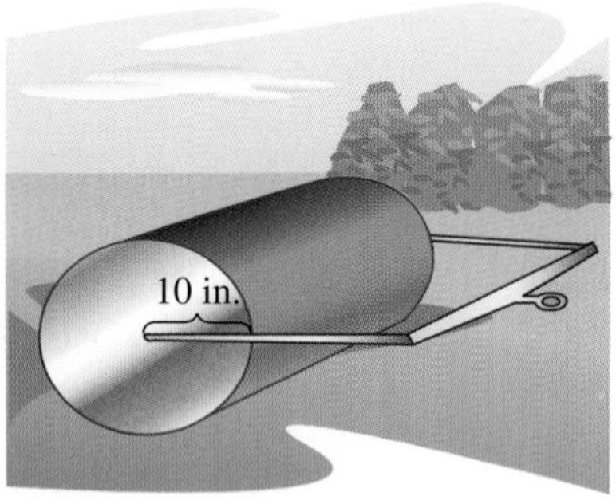

FIGURE 4.14

Activities

1. Find the supplement of the angle $\theta = \frac{5\pi}{7}$.

   Answer: $\frac{2\pi}{7}$

2. Convert 60° from degrees to radians.

   Answer: $\frac{\pi}{3}$

3. On a circle with a radius of 9 inches, find the length of the arc intercepted by a central angle of 140°.

   Answer: $7\pi \approx 22$ inches

4.1 Exercises

The *Interactive* CD-ROM and *Internet* versions of this text contain step-by-step solutions to all odd-numbered Section and Review Exercises. They also provide Tutorial Exercises that link to Guided Examples for additional help.

In Exercises 1–6, estimate the angle to the nearest one-half radian.

1.

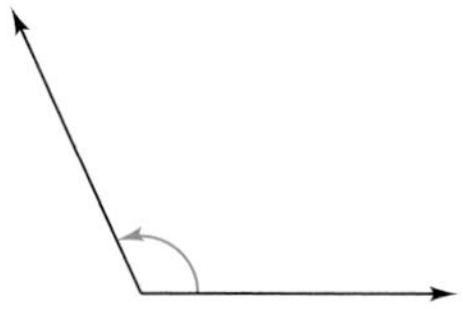

2.

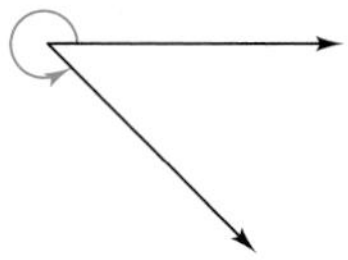

3.

4.

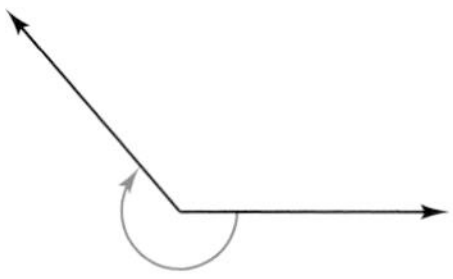

5.

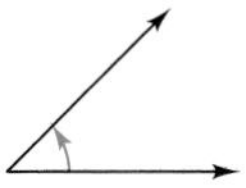

6.

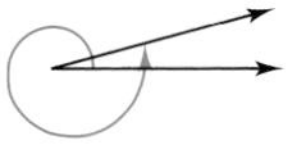

In Exercises 7–12, determine the quadrant in which each angle lies. (The angle measure is given in radians.)

7. (a) $\frac{\pi}{5}$ (b) $\frac{7\pi}{5}$

8. (a) $\frac{11\pi}{8}$ (b) $\frac{9\pi}{8}$

9. (a) $-\frac{\pi}{12}$ (b) $-\frac{11\pi}{9}$

10. (a) -1 (b) -2

11. (a) 3.5 (b) 2.25

12. (a) 6.02 (b) -4.25

In Exercises 13–16, sketch each angle in standard position.

13. (a) $\frac{5\pi}{4}$ (b) $\frac{2\pi}{3}$

14. (a) $-\frac{7\pi}{4}$ (b) $-\frac{5\pi}{2}$

15. (a) $\frac{11\pi}{6}$ (b) 7π

16. (a) 4 (b) -3

In Exercises 17–20, determine two coterminal angles (one positive and one negative) for each angle. Give your answers in radians.

17. (a)

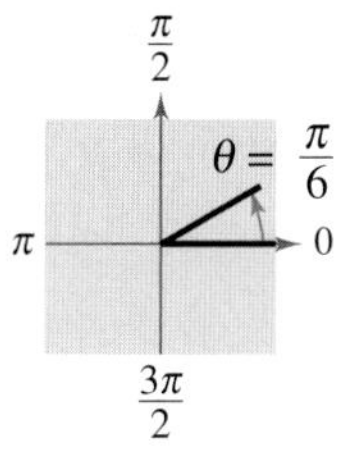

(b)

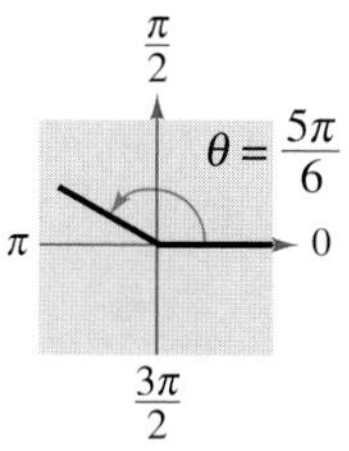

18. (a)

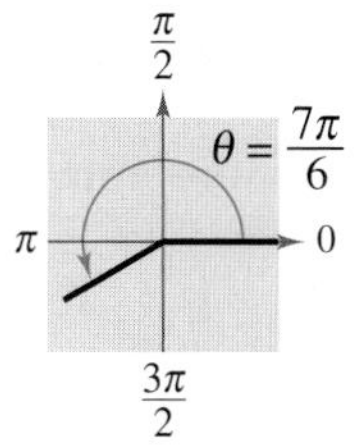

(b)

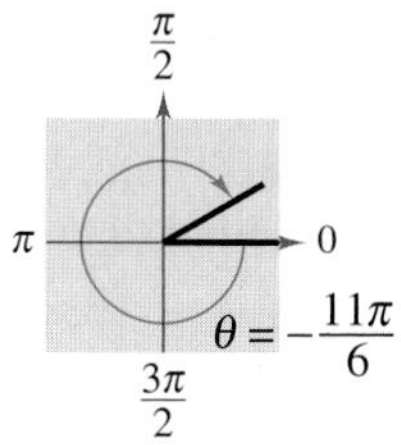

19. (a)

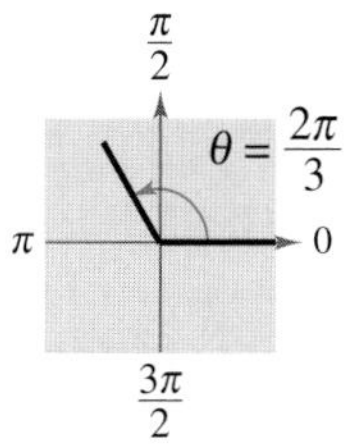

(b)

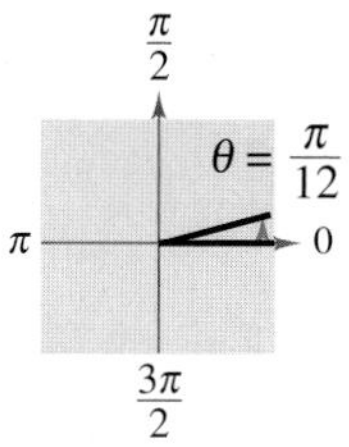

20. (a) $-\frac{9\pi}{4}$ (b) $-\frac{2\pi}{15}$

In Exercises 21–24, find (if possible) the complement and supplement of each angle.

21. (a) $\frac{\pi}{3}$ (b) $\frac{3\pi}{4}$

22. (a) $\frac{\pi}{12}$ (b) $\frac{11\pi}{12}$

23. (a) 1 (b) 2

24. (a) 3 (b) 1.5

In Exercises 25–28, express each angle in radian measure as a multiple of π. (Do not use a calculator.)

25. (a) 30° (b) 150°

26. (a) 315° (b) 120°

27. (a) $-20°$ (b) $-240°$

28. (a) $-270°$ (b) 144°

In Exercises 29–36, convert the measure from degrees to radians. Round to three decimal places.

29. $115°$ **30.** $87.4°$

31. $-216.35°$ **32.** $-48.27°$

33. $532°$ **34.** $345°$

35. $-0.83°$ **36.** $0.54°$

In Exercises 37–40, express each angle in degree measure. (Do not use a calculator.)

37. (a) $3\pi/2$ (b) $7\pi/6$

38. (a) $-7\pi/12$ (b) $\pi/9$

39. (a) $7\pi/3$ (b) $-11\pi/30$

40. (a) $11\pi/6$ (b) $34\pi/15$

In Exercises 41–48, convert the measure from radians to degrees. Round to three decimal places.

41. $\frac{\pi}{7}$ **42.** $\frac{5\pi}{11}$

43. $\frac{15\pi}{8}$ **44.** $\frac{13\pi}{2}$

45. -4.2π **46.** 4.8π

47. -2 **48.** -0.57

In Exercises 49–54, estimate the number of degrees in the angle.

49.
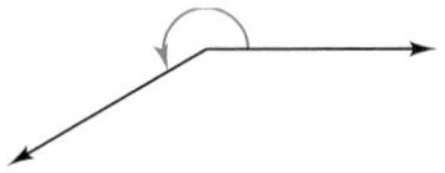

50.
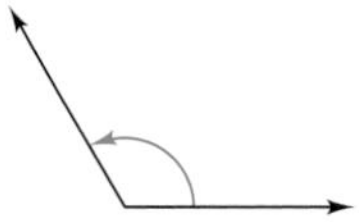

51.
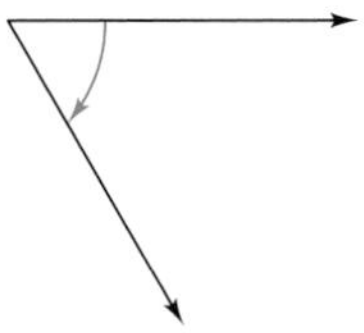

52.
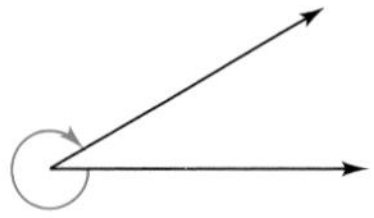

53.

54.
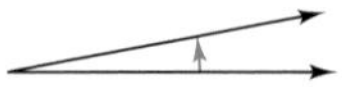

In Exercises 55–58, determine the quadrant in which each angle lies.

55. (a) $130°$ (b) $285°$

56. (a) $8.3°$ (b) $257° 30'$

57. (a) $-132° 50'$ (b) $-336°$

58. (a) $-260°$ (b) $-3.4°$

In Exercises 59–62, sketch each angle in standard position.

59. (a) $30°$ (b) $150°$

60. (a) $-270°$ (b) $-120°$

61. (a) $405°$ (b) $480°$

62. (a) $-750°$ (b) $-600°$

In Exercises 63–66, determine two coterminal angles (one positive and one negative) for each angle. Give your answers in degrees.

63. (a)
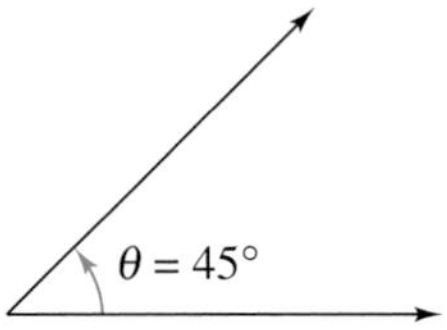

(b)
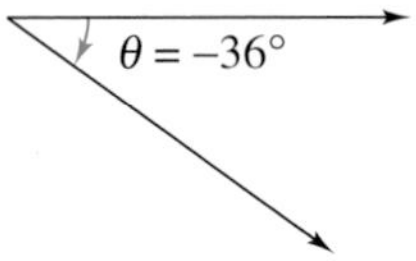

64. (a)
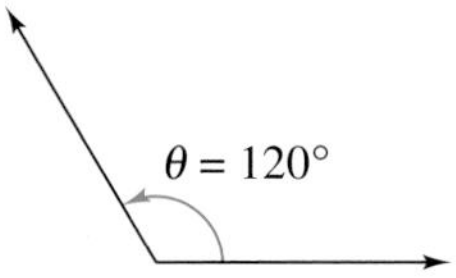

(b)
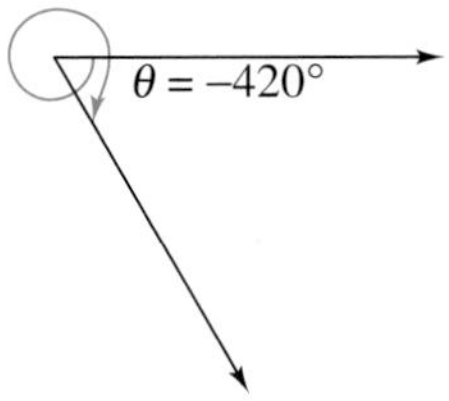

65. (a)
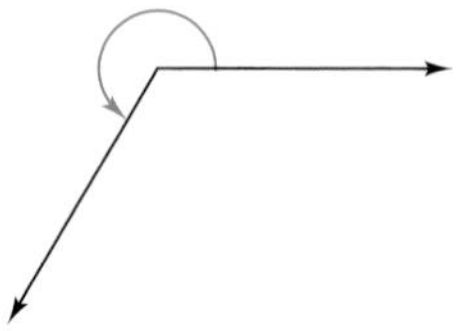

(b)
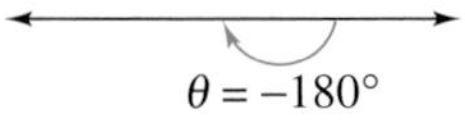

66. (a) $-420°$ (b) $230°$

In Exercises 67–70, find (if possible) the complement and supplement of each angle.

67. (a) $18°$ (b) $115°$

68. (a) $3°$ (b) $64°$

69. (a) $79°$ (b) $150°$

70. (a) $130°$ (b) $170°$

In Exercises 71–74, convert each angle measure to decimal degree form.

71. (a) $54° 45'$ (b) $-128° 30'$

72. (a) $245° 10'$ (b) $2° 12'$

73. (a) $85° 18' 30''$ (b) $330° 25''$

74. (a) $-135° 36''$ (b) $-408° 16' 20''$

In Exercises 75–78, convert each angle measure to D° M′ S″ form.

75. (a) $240.6°$ (b) $-145.8°$

76. (a) $-345.12°$ (b) $0.45°$

77. (a) $2.5°$ (b) $-3.58°$

78. (a) $-0.355°$ (b) $0.7865°$

In Exercises 79–82, find the angle in radians.

79.

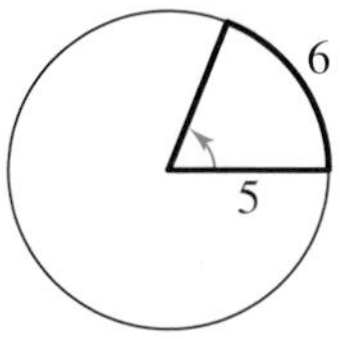

80.

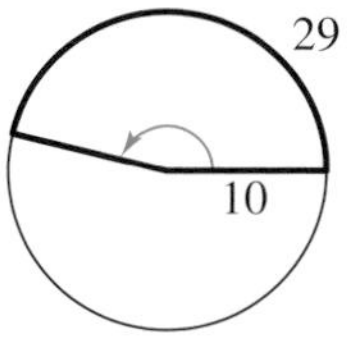

81.

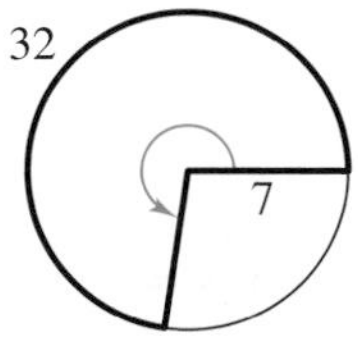

82.

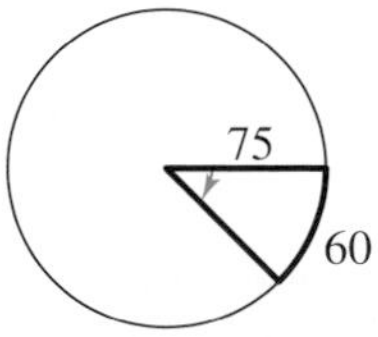

In Exercises 83–86, find the radian measure of the central angle of a circle of radius *r* that intercepts an arc of length *s*.

	Radius	*Arc Length*
83.	27 inches	6 inches
84.	14 feet	8 feet
85.	14.5 centimeters	25 centimeters
86.	80 kilometers	160 kilometers

In Exercises 87–90, find the length of the arc on a circle of radius *r* intercepted by a central angle θ.

	Radius	*Central Angle*
87.	15 inches	$180°$
88.	9 feet	$60°$
89.	3 meters	1 radian
90.	20 centimeters	$\pi/4$ radian

Distance Between Cities **In Exercises 91–94, find the distance between the cities. Assume that the earth is a sphere of radius 4000 miles and that the cities are on the same longitude (one city is due north of the other).**

	City	*Latitude*
91.	Dallas, Texas	32° 47′ 9″ N
	Omaha, Nebraska	41° 15′ 42″ N
92.	San Francisco, California	37° 46′ 39″ N
	Seattle, Washington	47° 36′ 32″ N
93.	Miami, Florida	25° 46′ 37″ N
	Erie, Pennsylvania	42° 7′ 15″ N
94.	Johannesburg, South Africa	26° 10′ S
	Jerusalem, Israel	31° 47′ N

95. *Difference in Latitudes* Assuming that the earth is a sphere of radius 6378 kilometers, what is the difference in latitude of two cities, one of which is 400 kilometers due north of the other?

96. *Difference in Latitudes* Assuming that the earth is a sphere of radius 6378 kilometers, what is the difference in latitude of two cities, one of which is 500 kilometers due north of the other?

97. *Instrumentation* The pointer on a voltmeter is 6 centimeters in length (see figure). Find the angle through which the pointer rotates when it moves 2.5 centimeters on the scale.

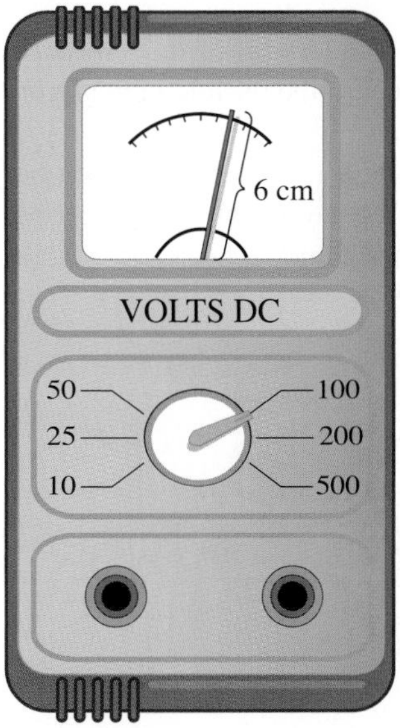

FIGURE FOR 97

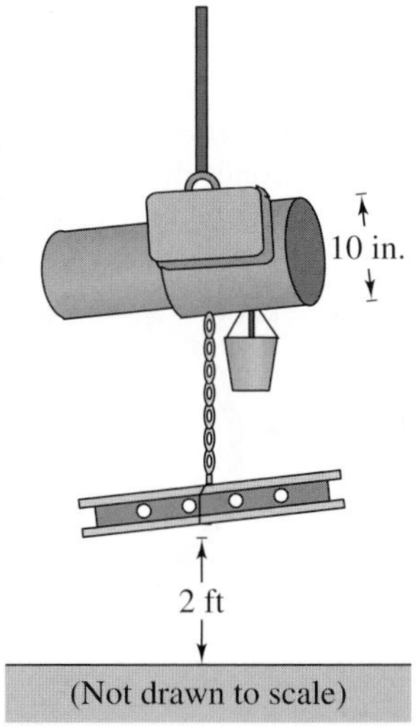

FIGURE FOR 98

98. *Electric Hoist* An electric hoist is being used to lift a beam (see figure). The diameter of the drum on the hoist is 10 inches, and the beam must be raised 2 feet. Find the number of degrees through which the drum must rotate.

99. *Angular Speed* A car is moving at a rate of 65 miles per hour, and the diameter of its wheels is 2.5 feet.

(a) Find the number of revolutions per minute the wheels are rotating.

(b) Find the angular speed of the wheels in radians per minute.

100. *Angular Speed* A 2-inch-diameter pulley on an electric motor that runs at 1700 revolutions per minute is connected by a belt to a 4-inch-diameter pulley on a saw arbor.

(a) Find the angular speed (in radians per minute) of each pulley.

(b) Find the revolutions per minute of the saw.

101. *Floppy Disk* The radius of the magnetic disk in a 3.5-inch diskette is 1.68 inches. Find the linear speed of a point on the circumference of the disk if it is rotating at a speed of 360 revolutions per minute.

102. *Speed of a Bicycle* The radii of the sprocket assemblies and the wheel of the bicycle in the figure are 4 inches, 2 inches, and 14 inches, respectively. If the cyclist is pedaling at a rate of 1 revolution per second, find the speed of the bicycle in (a) feet per second and (b) miles per hour.

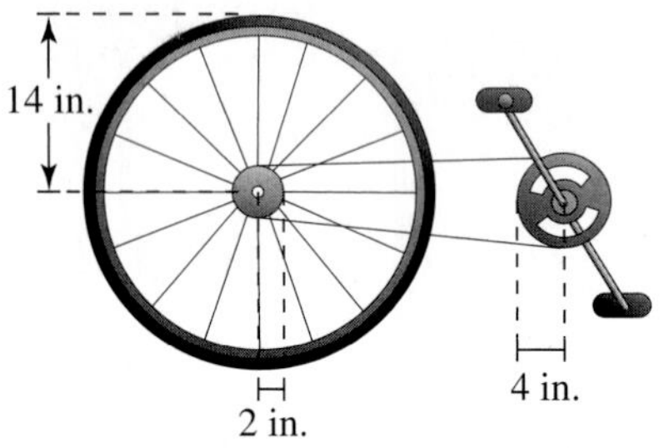

Synthesis

True or False? **In Exercises 103 and 104, determine whether the statement is true or false. Justify your answer.**

103. A measurement of 4 radians corresponds to two complete revolutions from the initial to the terminal side of an angle.

104. The difference of the measures of two coterminal angles is always a multiple of 360° if expressed in degrees and is always a multiple of 2π radians if expressed in radians.

105. In your own words, explain the meanings of (a) an angle in standard position, (b) a negative angle, (c) coterminal angles, and (d) an obtuse angle.

106. A fan motor turns at a given angular speed. How does the speed of the tips of the blades change if a fan of greater diameter is installed on the motor? Explain.

107. *Think About It* Is a degree or a radian the larger unit of measure? Explain.

108. *Writing* If the radius of a circle is increasing and the magnitude of a central angle is held constant, how is the length of the intercepted arc changing? Explain your reasoning.

109. *Geometry* Prove that the area of a circular sector of radius r with central angle θ is $A = \frac{1}{2}\theta r^2$, where θ is measured in radians.

Review

In Exercises 110–113, sketch the graphs of $y = x^5$ and the specified transformation.

110. $f(x) = (x - 2)^5$ **111.** $f(x) = x^5 - 4$

112. $f(x) = 2 - x^5$ **113.** $f(x) = -(x + 3)^5$

In Exercises 114–117, graph the exponential function.

114. $f(x) = 6^x$ **115.** $f(x) = 6^x - 2$

116. $f(x) = 6^{-x}$ **117.** $f(x) = 6^{x+1}$

In Exercises 118–123, graph the logarithmic function.

118. $f(x) = \log_4 x$ **119.** $f(x) = \log_4 x + 5$

120. $f(x) = -\log_4 x$ **121.** $f(x) = \log_4(x + 5)$

122. $f(x) = \log_4(-x)$ **123.** $f(x) = -\log_4(x + 5)$

In Exercises 124–131, simplify the radical expression.

124. $\frac{4}{4\sqrt{2}}$ **125.** $\frac{2}{\sqrt{3}}$

126. $\frac{2\sqrt{3}}{\sqrt{6}}$ **127.** $\frac{5\sqrt{5}}{2\sqrt{10}}$

128. $\sqrt{2^2 + 6^2}$ **129.** $\sqrt{18^2 + 12^2}$

130. $\sqrt{18^2 - 6^2}$ **131.** $\sqrt{17^2 - 9^2}$

4.2 Trigonometric Functions: The Unit Circle

▶ What you should learn

- How to identify a unit circle and its relationship to real numbers
- How to evaluate trigonometric functions using the unit circle
- How to use the domain and period to evaluate sine and cosine functions
- How to use a calculator to evaluate trigonometric functions

▶ Why you should learn it

Trigonometric functions are used to model the movement of an oscillating weight. For instance, in Exercises 59 and 60 on page 369, the displacement from equilibrium of an oscillating weight suspended by a spring is modeled as a function of time.

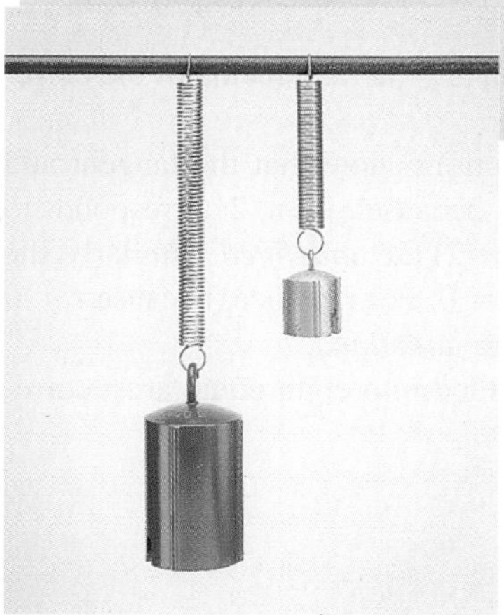

Richard Megna/Fundamental Photographs

A computer animation of this concept appears in the *Interactive* CD-ROM and *Internet* versions of this text.

Try demonstrating the wrapping function by using a spool and thread. Demonstrate the real number t as the length of the thread. Then wrap the thread around the spool to show the correspondence of t with (x, y) on the circle. For example, if $t = \pi/2$, the point on the unit circle to which it corresponds is $(0, 1)$.

The Unit Circle

The two historical perspectives of trigonometry incorporate different methods for introducing the trigonometric functions. Our first introduction to these functions is based on the unit circle.

Consider the **unit circle** given by

$$x^2 + y^2 = 1 \qquad \text{Unit circle}$$

as shown in Figure 4.15.

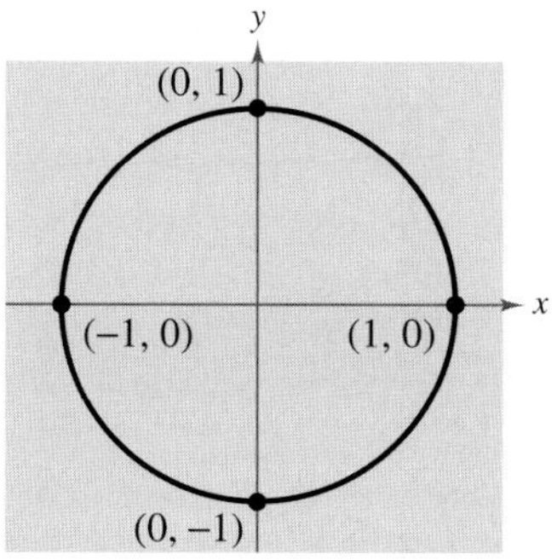

FIGURE 4.15

Imagine that the real number line is wrapped around this circle, with positive numbers corresponding to a counterclockwise wrapping and negative numbers corresponding to a clockwise wrapping, as shown in Figure 4.16.

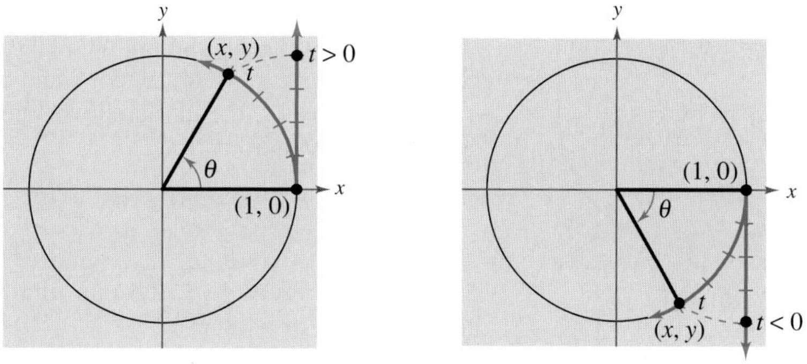

FIGURE 4.16

As the real number line is wrapped around the unit circle, each real number t corresponds to a point (x, y) on the circle. For example, the real number 0 corresponds to the point $(1, 0)$. Moreover, because the unit circle has a circumference of 2π, the real number 2π also corresponds to the point $(1, 0)$.

In general, each real number t also corresponds to a central angle θ (in standard position) whose radian measure is t. With this interpretation of t, the arc length formula $s = r\theta$ (with $r = 1$) indicates that the real number t is the length of the arc intercepted by the angle θ, given in radians.

The Trigonometric Functions

From the preceding discussion, it follows that the coordinates x and y are two functions of the real variable t. You can use these coordinates to define the six trigonometric functions of t.

sine	**cosecant**
cosine	**secant**
tangent	**cotangent**

These six functions are normally abbreviated sin, csc, cos, sec, tan, and cot, respectively.

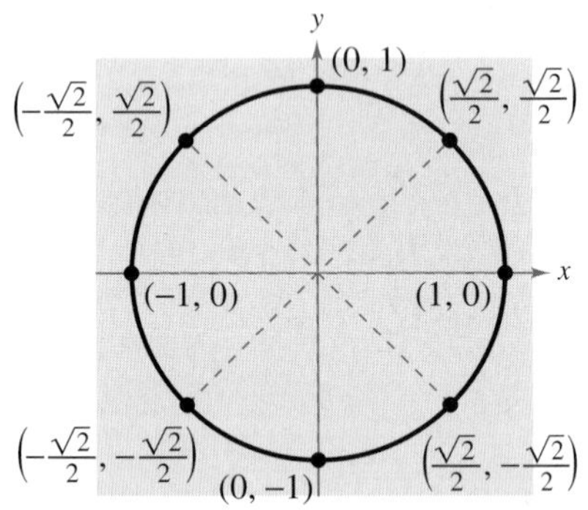

FIGURE 4.17

Definitions of Trigonometric Functions

Let t be a real number and let (x, y) be the point on the unit circle corresponding to t.

$$\sin t = y \qquad \csc t = \frac{1}{y}, \quad y \neq 0$$

$$\cos t = x \qquad \sec t = \frac{1}{x}, \quad x \neq 0$$

$$\tan t = \frac{y}{x}, \quad x \neq 0 \qquad \cot t = \frac{x}{y}, \quad y \neq 0$$

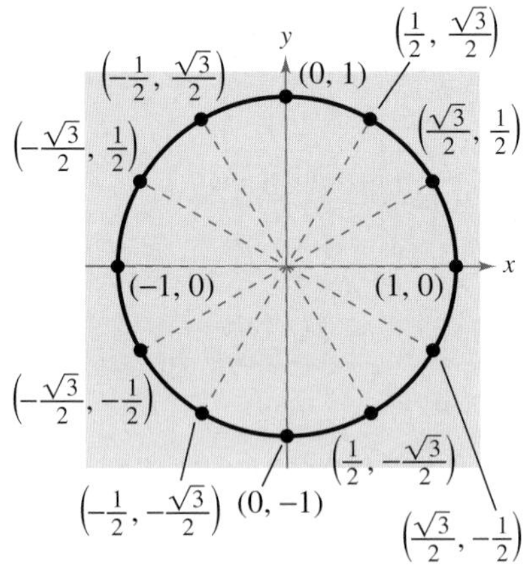

FIGURE 4.18

A computer animation of this concept appears in the *Interactive* CD-ROM and *Internet* versions of this text.

Note that the functions in the second column are the *reciprocals* of the corresponding functions in the first column.

In the definitions of the trigonometric functions, note that the tangent and secant are not defined when $x = 0$. For instance, because $t = \pi/2$ corresponds to $(x, y) = (0, 1)$, it follows that $\tan(\pi/2)$ and $\sec(\pi/2)$ are *undefined*. Similarly, the cotangent and cosecant are not defined when $y = 0$. For instance, because $t = 0$ corresponds to $(x, y) = (1, 0)$, cot 0 and csc 0 are *undefined*.

In Figure 4.17, the unit circle has been divided into eight equal arcs, corresponding to t-values of

$$0, \frac{\pi}{4}, \frac{\pi}{2}, \frac{3\pi}{4}, \pi, \frac{5\pi}{4}, \frac{3\pi}{2}, \frac{7\pi}{4}, \text{ and } 2\pi.$$

Similarly, in Figure 4.18, the unit circle has been divided into 12 equal arcs, corresponding to t-values of

$$0, \frac{\pi}{6}, \frac{\pi}{3}, \frac{\pi}{2}, \frac{2\pi}{3}, \frac{5\pi}{6}, \pi, \frac{7\pi}{6}, \frac{4\pi}{3}, \frac{3\pi}{2}, \frac{5\pi}{3}, \frac{11\pi}{6}, \text{ and } 2\pi.$$

Using the (x, y) coordinates in Figures 4.17 and 4.18, you can easily evaluate the trigonometric functions for common t-values. This procedure is demonstrated in Examples 1 and 2.

Example 1 ▶ Evaluating Trigonometric Functions

Evaluate the six trigonometric functions at each real number.

a. $t = \dfrac{\pi}{6}$ **b.** $t = \dfrac{5\pi}{4}$ **c.** $t = 0$ **d.** $t = \pi$

Solution

For each t-value, begin by finding the corresponding point (x, y) on the unit circle. Then use the definitions of trigonometric functions listed on page 364.

a. $t = \pi/6$ corresponds to the point $(x, y) = \left(\sqrt{3}/2, 1/2\right)$.

$$\sin\frac{\pi}{6} = y = \frac{1}{2} \qquad \csc\frac{\pi}{6} = \frac{1}{y} = 2$$

$$\cos\frac{\pi}{6} = x = \frac{\sqrt{3}}{2} \qquad \sec\frac{\pi}{6} = \frac{1}{x} = \frac{2}{\sqrt{3}} = \frac{2\sqrt{3}}{3}$$

$$\tan\frac{\pi}{6} = \frac{y}{x} = \frac{1/2}{\sqrt{3}/2} = \frac{1}{\sqrt{3}} \qquad \cot\frac{\pi}{6} = \frac{x}{y} = \sqrt{3}$$

b. $t = 5\pi/4$ corresponds to the point $(x, y) = \left(-\sqrt{2}/2, -\sqrt{2}/2\right)$.

$$\sin\frac{5\pi}{4} = y = -\frac{\sqrt{2}}{2} \qquad \csc\frac{5\pi}{4} = \frac{1}{y} = -\frac{2}{\sqrt{2}} = -\sqrt{2}$$

$$\cos\frac{5\pi}{4} = x = -\frac{\sqrt{2}}{2} \qquad \sec\frac{5\pi}{4} = \frac{1}{x} = -\frac{2}{\sqrt{2}} = -\sqrt{2}$$

$$\tan\frac{5\pi}{4} = \frac{y}{x} = \frac{-\sqrt{2}/2}{-\sqrt{2}/2} = 1 \qquad \cot\frac{5\pi}{4} = \frac{x}{y} = 1$$

c. $t = 0$ corresponds to the point $(x, y) = (1, 0)$.

$$\sin 0 = y = 0 \qquad \csc 0 = \frac{1}{y} \text{ is undefined.}$$

$$\cos 0 = x = 1 \qquad \sec 0 = \frac{1}{x} = 1$$

$$\tan 0 = \frac{y}{x} = \frac{0}{1} = 0 \qquad \cot 0 = \frac{x}{y} \text{ is undefined.}$$

d. $t = \pi$ corresponds to the point $(x, y) = (-1, 0)$.

$$\sin\pi = y = 0 \qquad \csc\pi = \frac{1}{y} \text{ is undefined.}$$

$$\cos\pi = x = -1 \qquad \sec\pi = \frac{1}{x} = -1$$

$$\tan\pi = \frac{y}{x} = \frac{0}{-1} = 0 \qquad \cot\pi = \frac{x}{y} \text{ is undefined.}$$

Quick Review

True or false:

1. $\csc t = \dfrac{1}{x}$
   Answer: False
2. cot 0 is undefined.
   Answer: True
3. $\cot\dfrac{\pi}{2} = 1$
   Answer: False

Additional Example

Evaluate the six trigonometric functions at $t = \dfrac{5\pi}{2}$.

Solution

Moving counterclockwise around the unit circle one and a quarter revolutions, you find that $t = 5\pi/2$ corresponds to the point $(x, y) = (0, 1)$.

$\sin\dfrac{5\pi}{2} = y = 1$ $\quad \csc\dfrac{5\pi}{2} = \dfrac{1}{y} = 1$

$\cos\dfrac{5\pi}{2} = x = 0$

$\sec\dfrac{5\pi}{2} = \dfrac{1}{x}$ is undefined.

$\tan\dfrac{5\pi}{2} = \dfrac{y}{x}$ is undefined.

$\cot\dfrac{5\pi}{2} = \dfrac{x}{y} = 0$

Exploration

With your graphing utility in radian and parametric modes, enter the equations

X1T = cos T and Y1T = sin T

and use the following settings.

Tmin = 0, Tmax = 6.3,
Tstep = 0.1
Xmin = -1.5, Xmax = 1.5,
Xscl = 1
Ymin = -1, Ymax = 1,
Yscl = 1

1. Graph the entered equations and describe the graph.
2. Use the trace key to move the cursor around the graph. What do the t-values represent? What do the x- and y-values represent?
3. What are the least and greatest values for x and y?

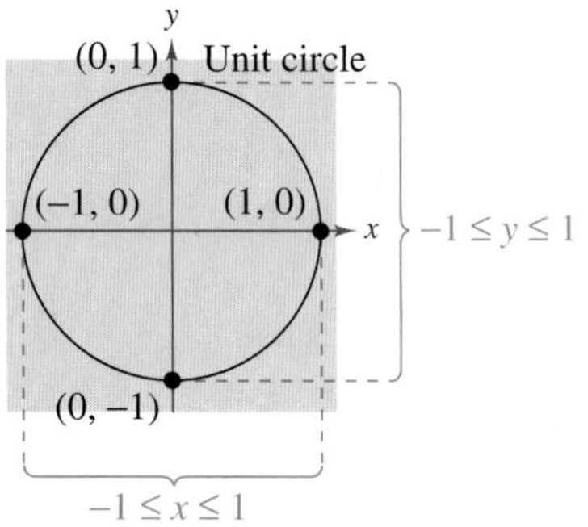

FIGURE 4.19

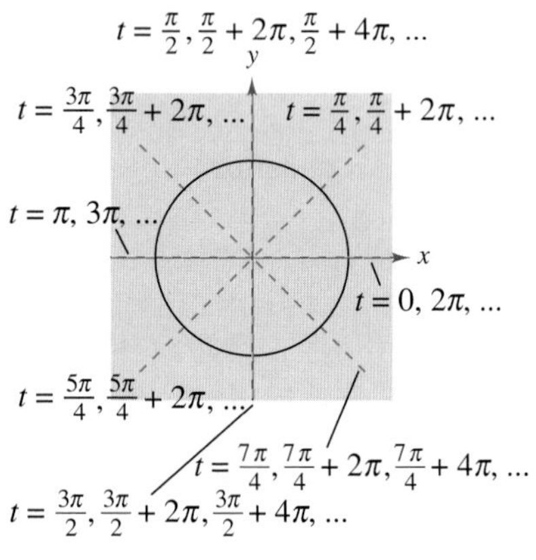

FIGURE 4.20

Example 2 ▶ Evaluating Trigonometric Functions

Evaluate the six trigonometric functions at $t = -\dfrac{\pi}{3}$.

Solution

Moving *clockwise* around the unit circle, it follows that $t = -\pi/3$ corresponds to the point $(x, y) = \left(1/2, -\sqrt{3}/2\right)$.

$$\sin\left(-\frac{\pi}{3}\right) = -\frac{\sqrt{3}}{2} \qquad \csc\left(-\frac{\pi}{3}\right) = -\frac{2}{\sqrt{3}}$$

$$\cos\left(-\frac{\pi}{3}\right) = \frac{1}{2} \qquad \sec\left(-\frac{\pi}{3}\right) = 2$$

$$\tan\left(-\frac{\pi}{3}\right) = -\sqrt{3} \qquad \cot\left(-\frac{\pi}{3}\right) = -\frac{1}{\sqrt{3}}$$

Domain and Period of Sine and Cosine

The *domain* of the sine and cosine functions is the set of all real numbers. To determine the *range* of these two functions, consider the unit circle shown in Figure 4.19. Because $r = 1$, it follows that $\sin t = y$ and $\cos t = x$. Moreover, because (x, y) is on the unit circle, you know that $-1 \le y \le 1$ and $-1 \le x \le 1$. So, the values of sine and cosine also range between -1 and 1.

$$-1 \le y \le 1 \qquad \text{and} \qquad -1 \le x \le 1$$
$$-1 \le \sin t \le 1 \qquad \qquad -1 \le \cos t \le 1$$

Adding 2π to each value of t in the interval $[0, 2\pi]$ completes a second revolution around the unit circle, as shown in Figure 4.20. The values of $\sin(t + 2\pi)$ and $\cos(t + 2\pi)$ correspond to those of $\sin t$ and $\cos t$. Similar results can be obtained for repeated revolutions (positive or negative) on the unit circle. This leads to the general result

$$\sin(t + 2\pi n) = \sin t$$

and

$$\cos(t + 2\pi n) = \cos t$$

for any integer n and real number t. Functions that behave in such a repetitive (or cyclic) manner are called **periodic.**

Definition of Periodic Function

A function f is **periodic** if there exists a positive real number c such that

$$f(t + c) = f(t)$$

for all t in the domain of f. The smallest number c for which f is periodic is called the **period** of f.

STUDY TIP

From the definition of periodic function it follows that the sine and cosine functions are periodic and have a period of 2π. The other four trigonometric functions are also periodic, and more will be said about this in Section 4.6.

Example 3 ▶ Using the Period to Evaluate the Sine and Cosine

a. Because $\dfrac{13\pi}{6} = 2\pi + \dfrac{\pi}{6}$, you have

$$\sin\frac{13\pi}{6} = \sin\left(2\pi + \frac{\pi}{6}\right) = \sin\frac{\pi}{6} = \frac{1}{2}.$$

b. Because $-\dfrac{7\pi}{2} = -4\pi + \dfrac{\pi}{2}$, you have

$$\cos\left(-\frac{7\pi}{2}\right) = \cos\left(-4\pi + \frac{\pi}{2}\right) = \cos\frac{\pi}{2} = 0.$$

Recall from Section 1.4 that a function f is *even* if $f(-t) = f(t)$, and is *odd* if $f(-t) = -f(t)$.

Students may have difficulty evaluating cosecant, secant, and/or cotangent functions using a calculator. Try having them rewrite the expression in terms of sine, cosine, or tangent before evaluating.

Example: $\csc 1.3 = \dfrac{1}{\sin 1.3} \approx 1.0378$

Even and Odd Trigonometric Functions

The cosine and secant functions are *even.*

$\cos(-t) = \cos t \qquad \sec(-t) = \sec t$

The sine, cosecant, tangent, and cotangent functions are *odd.*

$\sin(-t) = -\sin t \qquad \csc(-t) = -\csc t$

$\tan(-t) = -\tan t \qquad \cot(-t) = -\cot t$

Evaluating Trigonometric Functions with a Calculator

When evaluating a trigonometric function with a calculator, you need to set the calculator to the desired *mode* of measurement (degrees or radians).

Most calculators do not have keys for the cosecant, secant, and cotangent functions. To evaluate these functions, you can use the [x^{-1}] key with their respective reciprocal functions sine, cosine, and tangent. For example, to evaluate $\csc(\pi/8)$, use the fact that

$$\csc\frac{\pi}{8} = \frac{1}{\sin(\pi/8)}$$

and enter the following keystroke sequence in radian mode.

[(] [SIN] [(] [π] [÷] 8 [)] [)] [x^{-1}] [ENTER] Display 2.6131259

Technology

When evaluating trigonometric functions with a calculator, remember to enclose all fractional angle measures in parentheses. For instance, if you want to evaluate $\sin\theta$ for $\theta = \pi/6$, you should enter

[SIN] [(] [π] [÷] 6 [)] [ENTER]

rather than

[SIN] [π] [÷] 6 [ENTER].

The first set of keystrokes yields the correct value of 0.5. The second set yields the incorrect value of 0.

Example 4 ▶ Using a Calculator

	Function	*Mode*	*Calculator Keystrokes*	*Display*
a.	$\sin 2\pi/3$	Radian	[SIN] [(] 2 [π] [÷] 3 [)] [ENTER]	0.8660254
b.	$\cot 1.5$	Radian	[(] [TAN] 1.5 [)] [x^{-1}] [ENTER]	0.0709148

4.2 Exercises

In Exercises 1–4, determine the exact values of the six trigonometric functions of the angle θ.

1.
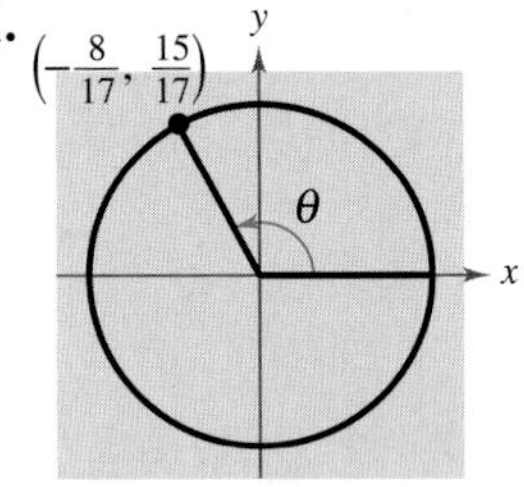

2.
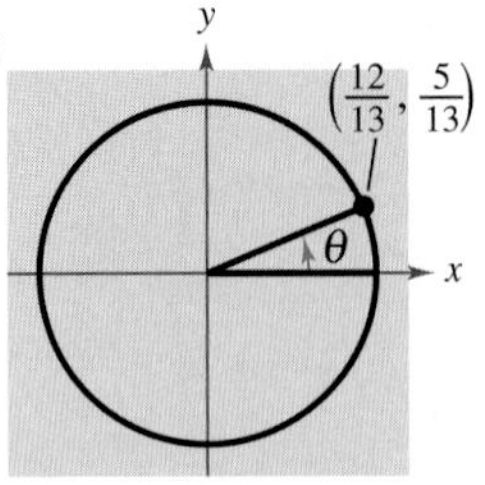

3.
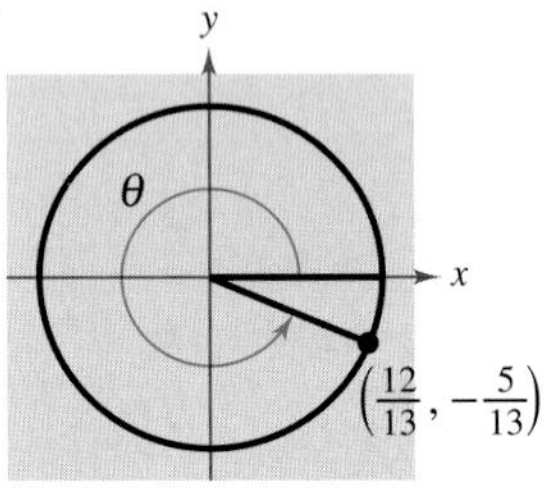

4.
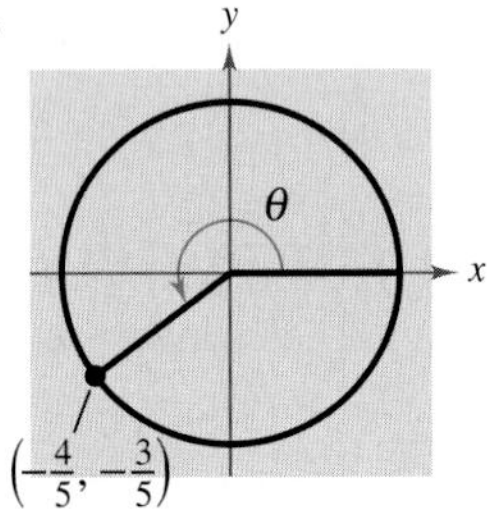

In Exercises 5–12, find the point (x, y) on the unit circle that corresponds to the real number t.

5. $t = \frac{\pi}{4}$
6. $t = \frac{\pi}{3}$
7. $t = \frac{7\pi}{6}$
8. $t = \frac{5\pi}{4}$
9. $t = \frac{4\pi}{3}$
10. $t = \frac{5\pi}{3}$
11. $t = \frac{3\pi}{2}$
12. $t = \pi$

In Exercises 13–22, evaluate (if possible) the sine, cosine, and tangent of the real number.

13. $t = \frac{\pi}{4}$
14. $t = \frac{\pi}{3}$
15. $t = -\frac{\pi}{6}$
16. $t = -\frac{\pi}{4}$
17. $t = -\frac{7\pi}{4}$
18. $t = -\frac{4\pi}{3}$
19. $t = \frac{11\pi}{6}$
20. $t = \frac{5\pi}{3}$
21. $t = -\frac{3\pi}{2}$
22. $t = -2\pi$

In Exercises 23–28, evaluate (if possible) the six trigonometric functions of the real number.

23. $t = \frac{3\pi}{4}$
24. $t = \frac{5\pi}{6}$
25. $t = \frac{\pi}{2}$
26. $t = \frac{3\pi}{2}$
27. $t = -\frac{\pi}{3}$
28. $t = -\frac{3\pi}{2}$

In Exercises 29–36, evaluate the trigonometric function using its period as an aid.

29. $\sin 5\pi$
30. $\cos 5\pi$
31. $\cos \frac{8\pi}{3}$
32. $\sin \frac{9\pi}{4}$
33. $\cos(-3\pi)$
34. $\sin(-3\pi)$
35. $\sin\left(-\frac{9\pi}{4}\right)$
36. $\cos\left(-\frac{8\pi}{3}\right)$

In Exercises 37–42, use the value of the trigonometric function to evaluate the indicated functions.

37. $\sin t = \frac{1}{3}$
 (a) $\sin(-t)$
 (b) $\csc(-t)$
38. $\sin(-t) = \frac{3}{8}$
 (a) $\sin t$
 (b) $\csc t$
39. $\cos(-t) = -\frac{1}{5}$
 (a) $\cos t$
 (b) $\sec(-t)$
40. $\cos t = -\frac{3}{4}$
 (a) $\cos(-t)$
 (b) $\sec(-t)$
41. $\sin t = \frac{4}{5}$
 (a) $\sin(\pi - t)$
 (b) $\sin(t + \pi)$
42. $\cos t = \frac{4}{5}$
 (a) $\cos(\pi - t)$
 (b) $\cos(t + \pi)$

In Exercises 43–52, use a calculator to evaluate the expression. Round to four decimal places.

43. $\sin \frac{\pi}{4}$
44. $\tan \frac{\pi}{3}$
45. $\csc 1.3$
46. $\cot 1$
47. $\cos(-1.7)$
48. $\cos(-2.5)$
49. $\csc 0.8$
50. $\sec 1.8$
51. $\sec 22.8$
52. $\sin(-0.9)$

Estimation **In Exercises 53 and 54, use the figure and a straightedge to approximate the value of each trigonometric function.**

53. (a) $\sin 5$ (b) $\cos 2$

54. (a) $\sin 0.75$ (b) $\cos 2.5$

Estimation **In Exercises 55 and 56, use the figure and a straightedge to approximate the solution of each equation, where $0 \le t < 2\pi$.**

55. (a) $\sin t = 0.25$ (b) $\cos t = -0.25$

56. (a) $\sin t = -0.75$ (b) $\cos t = 0.75$

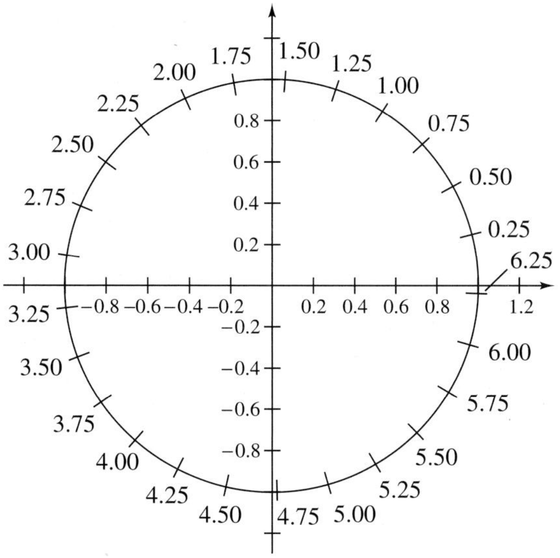

FIGURE FOR 53–56

57. Verify that $\cos 2t \neq 2 \cos t$ by approximating $\cos 1.5$ and $2 \cos 0.75$.

58. Verify that $\sin(t_1 + t_2) \neq \sin t_1 + \sin t_2$ by approximating $\sin 0.25$, $\sin 0.75$, and $\sin 1$.

59. *Harmonic Motion* The displacement from equilibrium of an oscillating weight suspended by a spring is

$$y(t) = \frac{1}{4} \cos 6t$$

where y is the displacement in feet and t is the time in seconds. Find the displacement when (a) $t = 0$, (b) $t = \frac{1}{4}$, and (c) $t = \frac{1}{2}$.

60. *Harmonic Motion* The displacement from equilibrium of an oscillating weight suspended by a spring and subject to the damping effect of friction is

$$y(t) = \frac{1}{4} e^{-t} \cos 6t$$

where y is the displacement in feet and t is the time in seconds. Find the displacement when (a) $t = 0$, (b) $t = \frac{1}{4}$, and (c) $t = \frac{1}{2}$.

Synthesis

True or False? **In Exercises 61 and 62, determine whether the statement is true or false. Justify your answer.**

61. Because $\sin(-t) = -\sin t$, it can be said that the sine of a negative angle is a negative number.

62. $\tan a = \tan(a - 6\pi)$

63. *Exploration* Let (x_1, y_1) and (x_2, y_2) be points on the unit circle corresponding to $t = t_1$ and $t = \pi - t_1$, respectively.

(a) Identify the symmetry of the points (x_1, y_1) and (x_2, y_2).

(b) Make a conjecture about any relationship between $\sin t_1$ and $\sin(\pi - t_1)$.

(c) Make a conjecture about any relationship between $\cos t_1$ and $\cos(\pi - t_1)$.

64. *Exploration* Let (x_1, y_1) and (x_2, y_2) be points on the unit circle corresponding to $t = t_1$ and $t = t_1 + \pi$, respectively.

(a) Identify the symmetry of the points (x_1, y_1) and (x_2, y_2).

(b) Make a conjecture about any relationship between $\sin t_1$ and $\sin(t_1 + \pi)$.

(c) Make a conjecture about any relationship between $\cos t_1$ and $\cos(t_1 + \pi)$.

65. Use the unit circle to verify that the cosine and secant functions are even and that the sine, cosecant, tangent, and cotangent functions are odd.

66. *Think About It* Because $f(t) = \sin t$ is an odd function and $g(t) = \cos t$ is an even function, what can be said about the function $h(t) = f(t)g(t)$?

67. *Think About It* Because $f(t) = \sin t$ and $g(t) = \tan t$ are odd functions, what can be said about the function $h(t) = f(t)g(t)$?

Review

In Exercises 68–71, find the inverse of the one-to-one function *f*.

68. $f(x) = \frac{1}{2}(3x - 2)$ **69.** $f(x) = \frac{1}{4}x^3 + 1$

70. $f(x) = \sqrt{x^2 - 4}, \quad x \ge 2$

71. $f(x) = \dfrac{2x}{x + 1}, \quad x > -1$

4.3 Right Triangle Trigonometry

▶ What you should learn

- How to evaluate trigonometric functions of acute angles
- How to use the fundamental trigonometric identities
- How to use a calculator to evaluate trigonometric functions
- How to use trigonometric functions to model and solve real-life problems

▶ Why you should learn it

Trigonometric functions are often used in mechanical calculations. For instance, Exercise 64 on page 379 shows you how trigonometric functions can be used to help find the width of a river.

Superstock

A computer animation of this concept appears in the *Interactive* CD-ROM and *Internet* versions of this text.

The Six Trigonometric Functions

Our second look at the trigonometric functions is from a *right triangle* perspective. Consider a right triangle, one of whose acute angles is labeled θ, as shown in Figure 4.21. Relative to the angle θ, the three sides of the triangle are the **hypotenuse,** the **opposite side** (the side opposite the angle θ), and the **adjacent side** (the side adjacent to the angle θ).

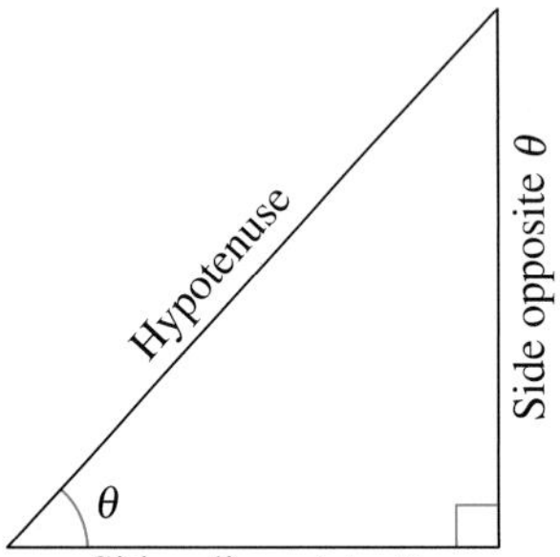

FIGURE 4.21

Using the lengths of these three sides, you can form six ratios that define the six trigonometric functions of the acute angle θ.

sine cosecant cosine secant tangent cotangent

In the following definition, it is important to see that $0° < \theta < 90°$ and that for such angles the value of each trigonometric function is *positive.*

Right Triangle Definitions of Trigonometric Functions

Let θ be an *acute* angle of a right triangle. The six trigonometric functions of the angle θ are defined as follows.

$$\sin\theta = \frac{\text{opp}}{\text{hyp}} \qquad \cos\theta = \frac{\text{adj}}{\text{hyp}} \qquad \tan\theta = \frac{\text{opp}}{\text{adj}}$$

$$\csc\theta = \frac{\text{hyp}}{\text{opp}} \qquad \sec\theta = \frac{\text{hyp}}{\text{adj}} \qquad \cot\theta = \frac{\text{adj}}{\text{opp}}$$

The abbreviations opp, adj, and hyp represent the lengths of the three sides of a right triangle.

opp = the length of the side *opposite* θ

adj = the length of the side *adjacent to* θ

hyp = the length of the *hypotenuse*

Note that the functions in the second row above are the *reciprocals* of the corresponding functions in the first row.

You may wish to review the Pythagorean Theorem before presenting the examples in this section.

Example 1 ▶ Evaluating Trigonometric Functions

Use the triangle in Figure 4.22 to find the values of the six trigonometric functions of θ.

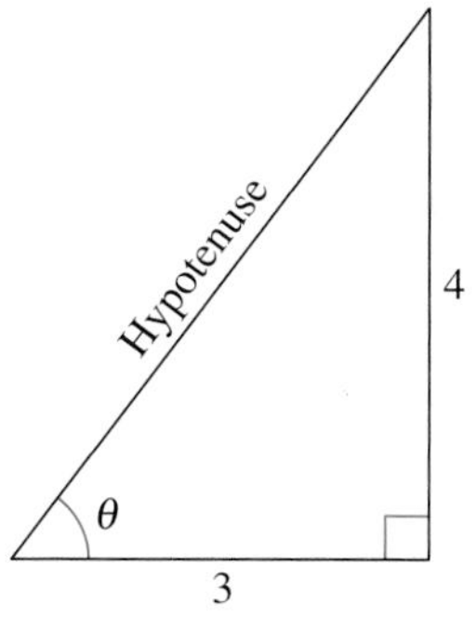

FIGURE 4.22

Solution

By the Pythagorean Theorem, $(\text{hyp})^2 = (\text{opp})^2 + (\text{adj})^2$, it follows that

$$\text{hyp} = \sqrt{4^2 + 3^2} = \sqrt{25} = 5.$$

So, the six trigonometric functions of θ are

$$\sin\theta = \frac{\text{opp}}{\text{hyp}} = \frac{4}{5} \qquad \cos\theta = \frac{\text{adj}}{\text{hyp}} = \frac{3}{5} \qquad \tan\theta = \frac{\text{opp}}{\text{adj}} = \frac{4}{3}$$

$$\csc\theta = \frac{\text{hyp}}{\text{opp}} = \frac{5}{4} \qquad \sec\theta = \frac{\text{hyp}}{\text{adj}} = \frac{5}{3} \qquad \cot\theta = \frac{\text{adj}}{\text{opp}} = \frac{3}{4}.$$

Historical Note
Georg Joachim Rhaeticus (1514–1576) was the leading Teutonic mathematical astronomer of the 16th century. He was the first to define the trigonometric functions as ratios of the sides of a right triangle.

In Example 1, you were given the lengths of two sides of the right triangle, but not the angle θ. It is more common in trigonometry to be asked to find the trigonometric functions of a *given* acute angle θ. To do this, you can construct a right triangle having θ as one of its angles.

Example 2 ▶ Evaluating Trigonometric Functions of 45°

Find the values of sin 45°, cos 45°, and tan 45°.

Solution

Construct a right triangle having 45° as one of its acute angles, as shown in Figure 4.23. Choose the length of the adjacent side to be 1. From geometry, you know that the other acute angle is also 45°. So, the triangle is isosceles and the length of the opposite side is also 1. Using the Pythagorean Theorem, you find the length of the hypotenuse to be $\sqrt{2}$.

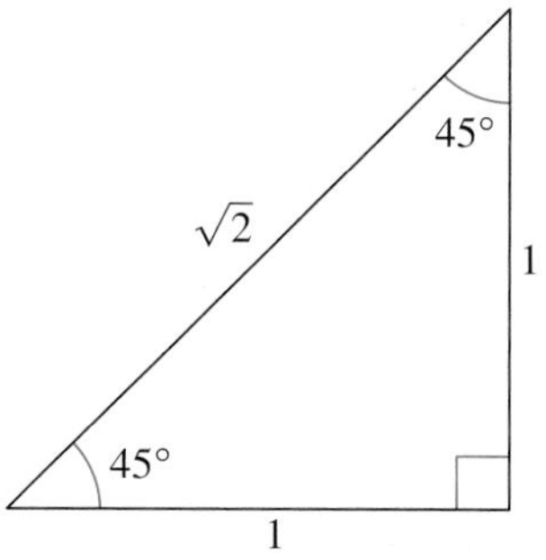

FIGURE 4.23

$$\sin 45^\circ = \frac{\text{opp}}{\text{hyp}} = \frac{1}{\sqrt{2}} = \frac{\sqrt{2}}{2} \qquad \cos 45^\circ = \frac{\text{adj}}{\text{hyp}} = \frac{1}{\sqrt{2}} = \frac{\sqrt{2}}{2}$$

$$\tan 45^\circ = \frac{\text{opp}}{\text{adj}} = \frac{1}{1} = 1$$

Example 3 ▶ Evaluating Trigonometric Functions of 30° and 60°

Use the equilateral triangle shown in Figure 4.24 to find the values of sin 60°, cos 60°, sin 30°, and cos 30°.

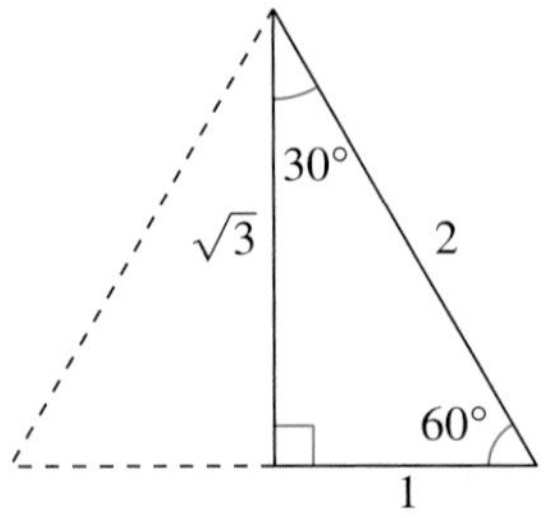

FIGURE 4.24

The triangles in Figures 4.22, 4.23, and 4.24 are useful problem-solving aids. Encourage your students to draw diagrams when they solve problems similar to Examples 1, 2, and 3.

Solution

Try using the Pythagorean Theorem and the equilateral triangle in Figure 4.24 to verify the lengths of the sides given in Figure 4.24. For $\theta = 60°$, you have adj $= 1$, opp $= \sqrt{3}$, and hyp $= 2$. Therefore,

$$\sin 60° = \frac{\text{opp}}{\text{hyp}} = \frac{\sqrt{3}}{2} \quad \text{and} \quad \cos 60° = \frac{\text{adj}}{\text{hyp}} = \frac{1}{2}.$$

For $\theta = 30°$, adj $= \sqrt{3}$, opp $= 1$, and hyp $= 2$. So,

$$\sin 30° = \frac{\text{opp}}{\text{hyp}} = \frac{1}{2} \quad \text{and} \quad \cos 30° = \frac{\text{adj}}{\text{hyp}} = \frac{\sqrt{3}}{2}.$$

Consider having your students construct the triangle in Figure 4.24 with angles in the corresponding radian measures, then find the six trigonometric functions for each of the acute angles.

Because the angles 30°, 45°, and 60° ($\pi/6$, $\pi/4$, and $\pi/3$) occur frequently in trigonometry, you should learn to construct the triangles shown in Figures 4.23 and 4.24.

A computer animation of this concept appears in the *Interactive* CD-ROM and *Internet* versions of this text.

Sines, Cosines, and Tangents of Special Angles

$$\sin 30° = \sin\frac{\pi}{6} = \frac{1}{2} \qquad \cos 30° = \cos\frac{\pi}{6} = \frac{\sqrt{3}}{2} \qquad \tan 30° = \tan\frac{\pi}{6} = \frac{\sqrt{3}}{3}$$

$$\sin 45° = \sin\frac{\pi}{4} = \frac{\sqrt{2}}{2} \qquad \cos 45° = \cos\frac{\pi}{4} = \frac{\sqrt{2}}{2} \qquad \tan 45° = \tan\frac{\pi}{4} = 1$$

$$\sin 60° = \sin\frac{\pi}{3} = \frac{\sqrt{3}}{2} \qquad \cos 60° = \cos\frac{\pi}{3} = \frac{1}{2} \qquad \tan 60° = \tan\frac{\pi}{3} = \sqrt{3}$$

In the box, note that $\sin 30° = \frac{1}{2} = \cos 60°$. This occurs because 30° and 60° are complementary angles. In general, it can be shown from the right triangle definitions that *cofunctions of complementary angles are equal.* That is, if θ is an acute angle, the following relationships are true.

$$\sin(90° - \theta) = \cos\theta \qquad \cos(90° - \theta) = \sin\theta$$

$$\tan(90° - \theta) = \cot\theta \qquad \cot(90° - \theta) = \tan\theta$$

$$\sec(90° - \theta) = \csc\theta \qquad \csc(90° - \theta) = \sec\theta$$

Trigonometric Identities

In trigonometry, a great deal of time is spent studying relationships between trigonometric functions (identities).

These identities will be used many times in trigonometry and later in calculus. Encourage your students to learn them well.

Fundamental Trigonometric Identities

Reciprocal Identities

$$\sin\theta = \frac{1}{\csc\theta} \qquad \cos\theta = \frac{1}{\sec\theta} \qquad \tan\theta = \frac{1}{\cot\theta}$$

$$\csc\theta = \frac{1}{\sin\theta} \qquad \sec\theta = \frac{1}{\cos\theta} \qquad \cot\theta = \frac{1}{\tan\theta}$$

Quotient Identities

$$\tan\theta = \frac{\sin\theta}{\cos\theta} \qquad \cot\theta = \frac{\cos\theta}{\sin\theta}$$

Pythagorean Identities

$$\sin^2\theta + \cos^2\theta = 1 \qquad 1 + \tan^2\theta = \sec^2\theta$$

$$1 + \cot^2\theta = \csc^2\theta$$

Note that $\sin^2\theta$ represents $(\sin\theta)^2$, $\cos^2\theta$ represents $(\cos\theta)^2$, and so on.

Example 4 ▶ Applying Trigonometric Identities

Let θ be an acute angle such that $\sin\theta = 0.6$. Find the values of (a) $\cos\theta$ and (b) $\tan\theta$ using trigonometric identities.

Solution

a. To find the value of $\cos\theta$, use the Pythagorean identity

$$\sin^2\theta + \cos^2\theta = 1.$$

So, you have

$$(0.6)^2 + \cos^2\theta = 1 \qquad \text{Substitute 0.6 for } \sin\theta.$$

$$\cos^2\theta = 1 - (0.6)^2 = 0.64 \qquad \text{Subtract } (0.6)^2 \text{ from each side.}$$

$$\cos\theta = \sqrt{0.64} = 0.8. \qquad \text{Extract the positive square root.}$$

b. Now, knowing the sine and cosine of θ, you can find the tangent of θ to be

$$\tan\theta = \frac{\sin\theta}{\cos\theta} = \frac{0.6}{0.8} = 0.75.$$

Try using the definitions of $\cos\theta$ and $\tan\theta$, and the triangle shown in Figure 4.25, to check these results.

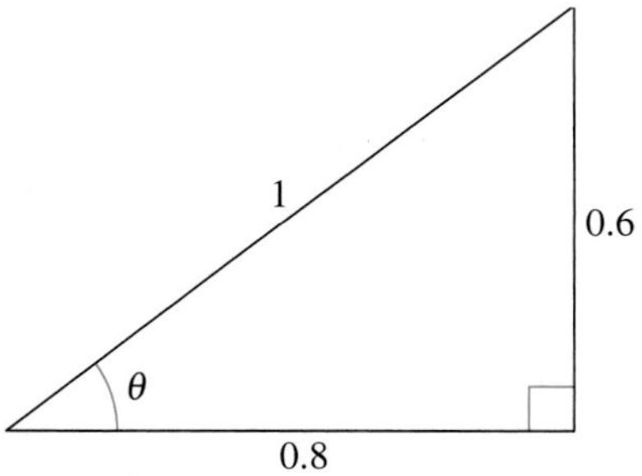

FIGURE 4.25

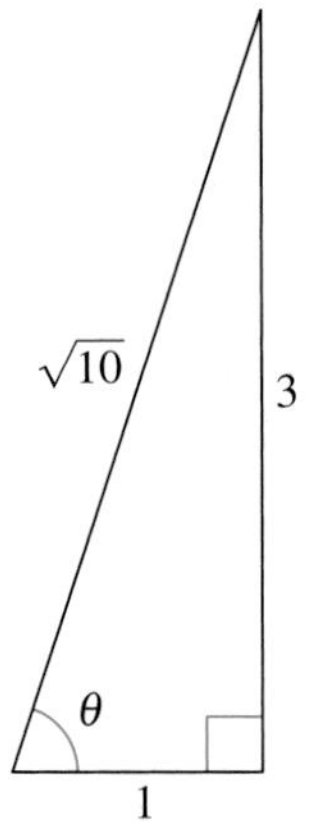

FIGURE 4.26

Example 5 ▶ Applying Trigonometric Identities

Let θ be an acute angle such that $\tan\theta = 3$. Find the values of each trigonometric function using trigonometric identities.

a. $\cot\theta$ **b.** $\sec\theta$

Solution

a. $\cot\theta = \dfrac{1}{\tan\theta} = \dfrac{1}{3}$ Reciprocal identity

b. $\sec^2\theta = 1 + \tan^2\theta$ Pythagorean identity

$\sec^2\theta = 1 + 3^2$

$\sec^2\theta = 10$

$\sec\theta = \sqrt{10}$

Try using the definitions of $\cot\theta$ and $\sec\theta$, and the triangle shown in Figure 4.26, to check these results.

Evaluating Trigonometric Functions with a Calculator

To use a calculator to evaluate trigonometric functions of angles measured in degrees, first set the calculator to degree mode and then proceed as demonstrated in Section 4.2. For instance, you can find values of cos 28° and sec 28° as follows.

Function	*Graphing Calculator Keystrokes*	*Display*
$\cos 28°$	[COS] 28 [ENTER]	0.8829476
$\sec 28°$	[(] [COS] 28 [)] [x^{-1}] [ENTER]	1.1325701

Throughout this text, angles are assumed to be measured in radians unless noted otherwise. For example, sin 1 means the sine of 1 radian and sin 1° means the sine of 1 degree.

One of the most common errors students make when they evaluate trigonometric functions with a calculator is not having their calculators set to the correct mode (radian vs. degree).

Example 6 ▶ Using a Calculator

Use a calculator to evaluate $\sec(5° \, 40' \, 12'')$.

Solution

Begin by converting to decimal form.

$$5° \, 40' \, 12'' = 5° + \left(\frac{40}{60}\right)^\circ + \left(\frac{12}{3600}\right)^\circ = 5.67°$$

Then, use a calculator to evaluate sec 5.67°.

$$\sec(5° \, 40' \, 12'') = \sec 5.67° = \frac{1}{\cos 5.67°} \approx 1.00492$$

The *Interactive* CD-ROM and *Internet* versions of this text offer a built-in graphing calculator, which can be used with the Examples, Explorations, Technology notes, and Exercises.

Applications Involving Right Triangles

Many applications of trigonometry involve a process called **solving right triangles.** In this type of application, you are usually given one side of a right triangle and one of the acute angles and asked to find one of the other sides, *or* you are given two sides and asked to find one of the acute angles.

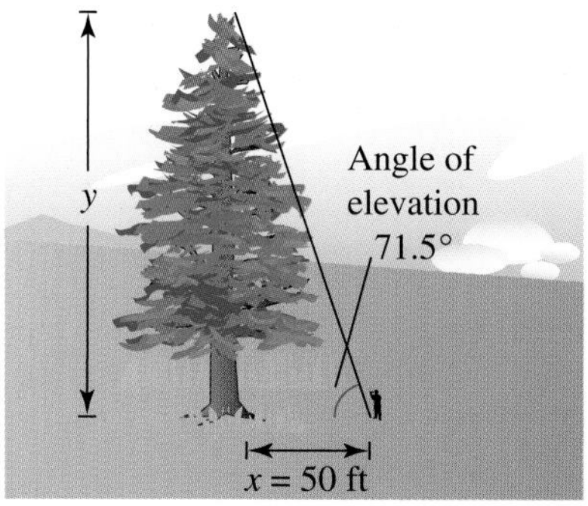

FIGURE 4.27

Example 7 ▶ Using Trigonometry to Solve a Right Triangle

A surveyor is standing 50 feet from the base of a large tree, as shown in Figure 4.27. The surveyor measures the angle of elevation to the top of the tree as 71.5°. How tall is the tree?

Solution

From Figure 4.27, you see that

$$\tan 71.5° = \frac{\text{opp}}{\text{adj}} = \frac{y}{x}$$

where $x = 50$ and y is the height of the tree. So, the height of the tree is

$$\begin{aligned} y &= x \tan 71.5° \\ &\approx 50(2.98868) \\ &\approx 149.4 \text{ feet.} \end{aligned}$$

Activities

1. Given the right triangle shown, find each of the six trigonometric functions of the angle θ.

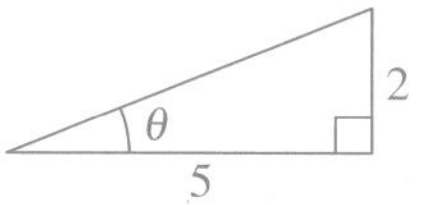

Answer:

$\sin\theta = \frac{2\sqrt{29}}{29}$, $\csc\theta = \frac{\sqrt{29}}{2}$

$\cos\theta = \frac{5\sqrt{29}}{29}$, $\sec\theta = \frac{\sqrt{29}}{5}$

$\tan\theta = \frac{2}{5}$, $\cot\theta = \frac{5}{2}$

2. A 10-foot ladder leans against the side of a house. The ladder makes an angle of 60° with the ground. How far up the side of the house does the ladder reach?

Answer: $5\sqrt{3} \approx 8.66$ feet

Example 8 ▶ Using Trigonometry to Solve a Right Triangle

A person is 200 yards from a river. Rather than walking directly to the river, the person walks 400 yards along a straight path to the river's edge. Find the acute angle θ between this path and the river's edge, as illustrated in Figure 4.28.

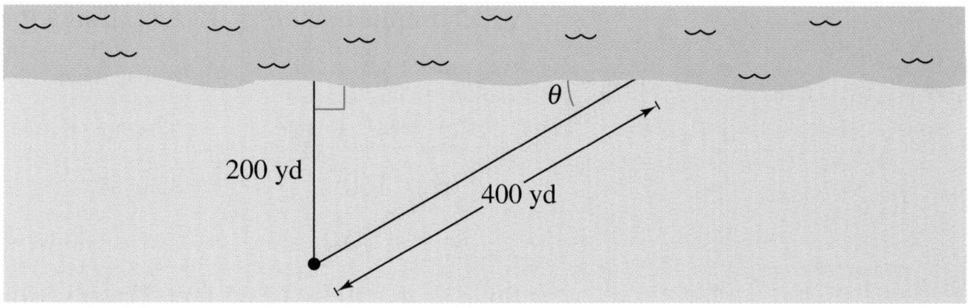

FIGURE 4.28

Solution

From Figure 4.28, you can see that the sine of the angle θ is

$$\sin\theta = \frac{\text{opp}}{\text{hyp}} = \frac{200}{400} = \frac{1}{2}.$$

Therefore, $\theta = 30°$.

In Example 8, you were able to recognize that the acute angle that satisfies the equation $\sin\theta = \frac{1}{2}$ is $\theta = 30°$. Suppose, however, that you were given the equation $\sin\theta = 0.6$ and asked to find the acute angle θ. Because

$$\sin 30° = \frac{1}{2}$$
$$= 0.5000$$

and

$$\sin 45° = \frac{1}{\sqrt{2}}$$
$$\approx 0.7071$$

you might guess that θ lies somewhere between 30° and 45°. A more precise value of θ can be found using the *inverse sine* key on a calculator. To do this, you can use the following keystroke sequence in degree mode.

[sin⁻¹] .6 [ENTER] Display 36.8699

So, you can conclude that if $\sin\theta = 0.6$, then $\theta \approx 36.87°$.

Example 9 ▶ Solving a Right Triangle

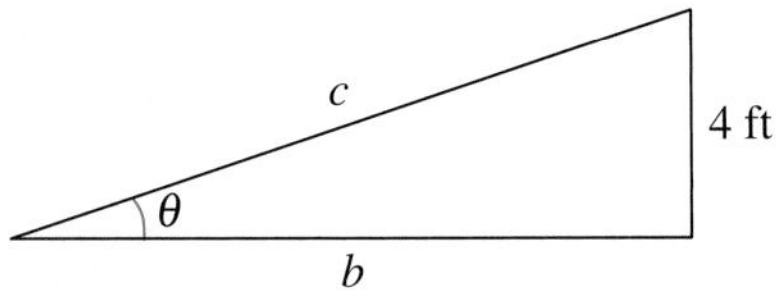

FIGURE 4.29

Specifications for a loading dock ramp require a rise of 1 foot for each 3 feet of horizontal length. In Figure 4.29, find the lengths of sides b and c and find the measure of θ.

Solution

From the given specifications, you can write

$$\frac{\text{rise}}{\text{run}} = \frac{1}{3} = \frac{4 \text{ ft}}{b \text{ ft}}$$

which implies that $b = 12$ feet. Using the Pythagorean Theorem, you can write

$c^2 = a^2 + b^2$ — Pythagorean Theorem

$c^2 = 4^2 + 12^2$ — Substitute for a and b.

$c^2 = 160$ — Simplify.

$c = 4\sqrt{10}.$ — Extract positive square root.

So, $c = 4\sqrt{10} \approx 12.65$ feet. To solve for θ, you can write

$$\tan\theta = \frac{4}{12}$$
$$= \frac{1}{3}.$$

Then, using the calculator keystrokes in degree mode

[tan⁻¹] [(] 1 [÷] 3 [)] [ENTER]

you obtain $\theta \approx 18.43°$.

Activity

Using your calculator, find the value of θ, $0 \le \theta \le \frac{\pi}{2}$, given that $\sin\theta = 0.2962$.

Answer: $\theta \approx 0.3007$

4.3 Exercises

In Exercises 1–4, find the exact values of the six trigonometric functions of the angle θ given in the figure. (Use the Pythagorean Theorem to find the third side of the triangle.)

1.

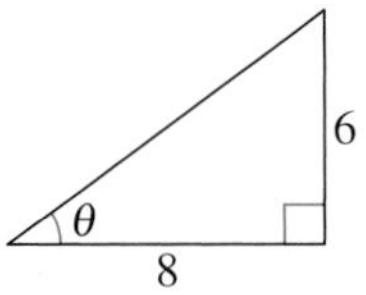

2.

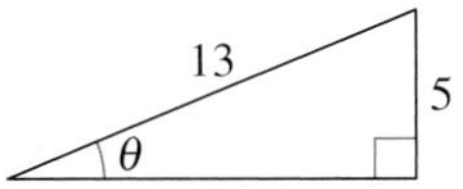

3.

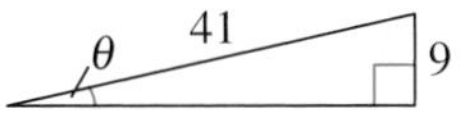

4.

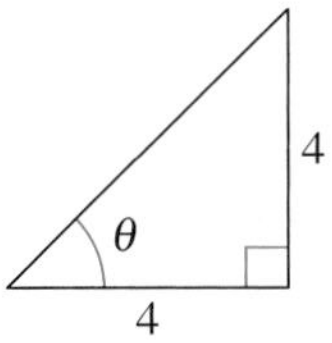

In Exercises 5–8, find the exact values of the six trigonometric functions of the angle θ for each of the two triangles. Explain why the function values are the same.

5.

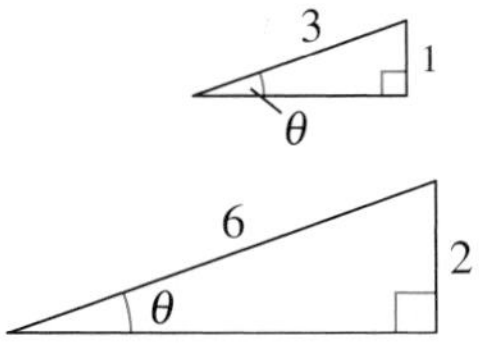

6.

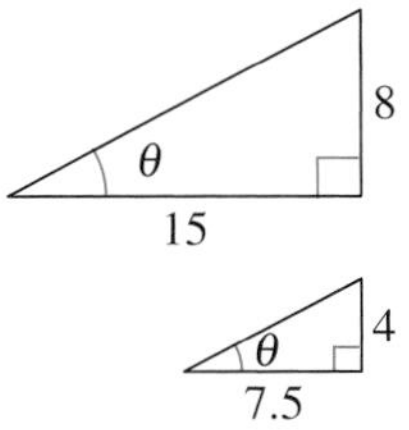

7.

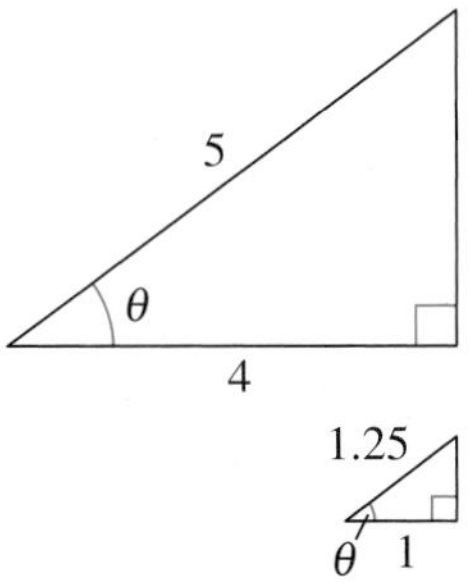

8.

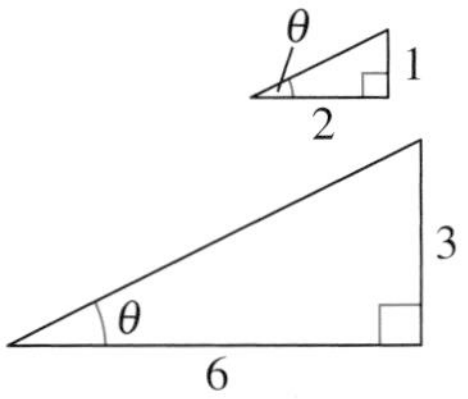

In Exercises 9–16, sketch a right triangle corresponding to the trigonometric function of the acute angle θ. Use the Pythagorean Theorem to determine the third side and then find the other five trigonometric functions of θ.

9. $\sin\theta = \frac{3}{4}$

10. $\cos\theta = \frac{5}{7}$

11. $\sec\theta = 2$

12. $\cot\theta = 5$

13. $\tan\theta = 3$

14. $\sec\theta = 6$

15. $\cot\theta = \frac{3}{2}$

16. $\csc\theta = \frac{17}{4}$

In Exercises 17–22, use the given function value(s), and trigonometric identities (including the cofunction identities), to find the indicated trigonometric functions.

17. $\sin 60^\circ = \dfrac{\sqrt{3}}{2}, \quad \cos 60^\circ = \dfrac{1}{2}$

(a) $\tan 60^\circ$ (b) $\sin 30^\circ$
(c) $\cos 30^\circ$ (d) $\cot 60^\circ$

18. $\sin 30^\circ = \dfrac{1}{2}, \quad \tan 30^\circ = \dfrac{\sqrt{3}}{3}$

(a) $\csc 30^\circ$ (b) $\cot 60^\circ$
(c) $\cos 30^\circ$ (d) $\cot 30^\circ$

19. $\csc\theta = \dfrac{\sqrt{13}}{2}, \quad \sec\theta = \dfrac{\sqrt{13}}{3}$

(a) $\sin\theta$ (b) $\cos\theta$
(c) $\tan\theta$ (d) $\sec(90^\circ - \theta)$

20. $\sec\theta = 5, \quad \tan\theta = 2\sqrt{6}$

(a) $\cos\theta$ (b) $\cot\theta$
(c) $\cot(90^\circ - \theta)$ (d) $\sin\theta$

21. $\cos\alpha = \frac{1}{3}$

(a) $\sec\alpha$ (b) $\sin\alpha$
(c) $\cot\alpha$ (d) $\sin(90^\circ - \alpha)$

22. $\tan\beta = 5$

(a) $\cot\beta$ (b) $\cos\beta$
(c) $\tan(90^\circ - \beta)$ (d) $\csc\beta$

In Exercises 23–26, evaluate the trigonometric function by memory or by constructing an appropriate triangle for the given special angle.

23. (a) $\cos 60^\circ$ (b) $\csc 30^\circ$ (c) $\tan 60^\circ$

24. (a) $\cot 45^\circ$ (b) $\cos 45^\circ$ (c) $\csc 45^\circ$

25. (a) $\sin 45^\circ$ (b) $\cos 30^\circ$ (c) $\tan 30^\circ$

26. (a) $\sin 60^\circ$ (b) $\tan 45^\circ$ (c) $\sec 30^\circ$

In Exercises 27–36, use a calculator to evaluate each function. Round your answers to four decimal places. (Be sure the calculator is in the correct angle mode.)

27. (a) $\sin 10^\circ$ (b) $\cos 80^\circ$

28. (a) $\tan 23.5^\circ$ (b) $\cot 66.5^\circ$

29. (a) $\sin 16.35^\circ$ (b) $\csc 16.35^\circ$

30. (a) $\cos 16° 18'$ (b) $\sin 73° 56'$

31. (a) $\sec 42° 12'$ (b) $\csc 48° 7'$

32. (a) $\cos 4° 50' 15''$ (b) $\sec 4° 50' 15''$

33. (a) $\cot 11° 15'$ (b) $\tan 11° 15'$

34. (a) $\sec 56° 8' 10''$ (b) $\cos 56° 8' 10''$

35. (a) $\csc 32° 40' 3''$ (b) $\tan 44° 28' 16''$

36. (a) $\sec\left(\frac{9}{5} \cdot 20 + 32\right)°$ (b) $\cot\left(\frac{9}{5} \cdot 30 + 32\right)°$

In Exercises 37–42, find the values of θ in degrees ($0° < \theta < 90°$) and radians ($0 < \theta < \pi/2$) without the aid of a calculator.

37. (a) $\sin \theta = \frac{1}{2}$ (b) $\csc \theta = 2$

38. (a) $\cos \theta = \frac{\sqrt{2}}{2}$ (b) $\tan \theta = 1$

39. (a) $\sec \theta = 2$ (b) $\cot \theta = 1$

40. (a) $\tan \theta = \sqrt{3}$ (b) $\cos \theta = \frac{1}{2}$

41. (a) $\csc \theta = \frac{2\sqrt{3}}{3}$ (b) $\sin \theta = \frac{\sqrt{2}}{2}$

42. (a) $\cot \theta = \frac{\sqrt{3}}{3}$ (b) $\sec \theta = \sqrt{2}$

In Exercises 43–46, find the values of θ in degrees ($0° < \theta < 90°$) and radians ($0 < \theta < \pi/2$) by using a calculator.

43. (a) $\sin \theta = 0.0145$ (b) $\sin \theta = 0.4565$

44. (a) $\cos \theta = 0.9848$ (b) $\cos \theta = 0.8746$

45. (a) $\tan \theta = 0.0125$ (b) $\tan \theta = 2.3545$

46. (a) $\sin \theta = 0.3746$ (b) $\cos \theta = 0.3746$

In Exercises 47–50, solve for *x*, *y*, or *r*, as indicated.

47. Solve for x.

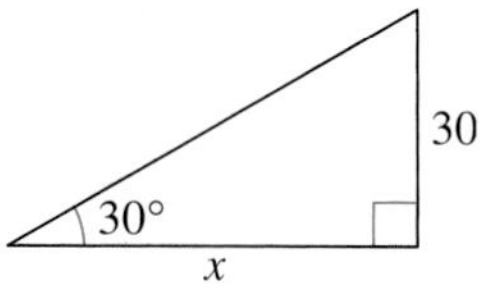

48. Solve for y.

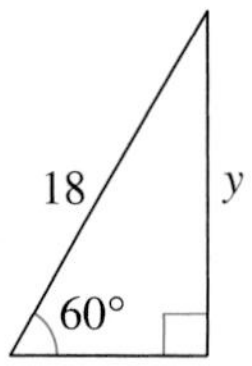

49. Solve for x.

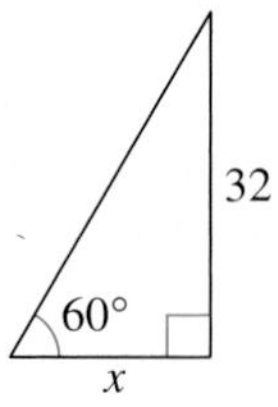

50. Solve for r.

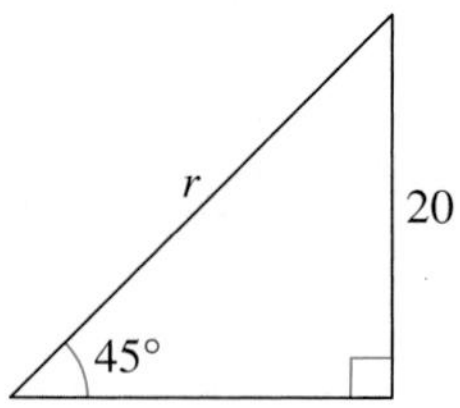

In Exercises 51–60, use trigonometric identities to transform the left side of the equation into the right side.

51. $\tan \theta \cot \theta = 1$

52. $\cos \theta \sec \theta = 1$

53. $\tan \alpha \cos \alpha = \sin \alpha$

54. $\cot \alpha \sin \alpha = \cos \alpha$

55. $(1 + \cos \theta)(1 - \cos \theta) = \sin^2 \theta$

56. $(1 + \sin \theta)(1 - \sin \theta) = \cos^2 \theta$

57. $(\sec \theta + \tan \theta)(\sec \theta - \tan \theta) = 1$

58. $\sin^2 \theta - \cos^2 \theta = 2 \sin^2 \theta - 1$

59. $\frac{\sin \theta}{\cos \theta} + \frac{\cos \theta}{\sin \theta} = \csc \theta \sec \theta$

60. $\frac{\tan \beta + \cot \beta}{\tan \beta} = \csc^2 \beta$

61. *Height* A 6-foot person walks from the base of a broadcasting tower directly toward the tip of the shadow cast by the tower. When the person is 132 feet from the tower and 3 feet from the tip of the shadow, the person's shadow starts to appear beyond the tower's shadow.

(a) Draw a right triangle that gives a visual representation of the problem. Show the known quantities of the triangle and use a variable to indicate the height of the tower.

(b) Use a trigonometric function to write an equation involving the unknown quantity.

(c) What is the height of the tower?

62. *Height* A 6-foot person standing 20 feet from a streetlight casts a 10-foot shadow (see figure). What is the height of the streetlight?

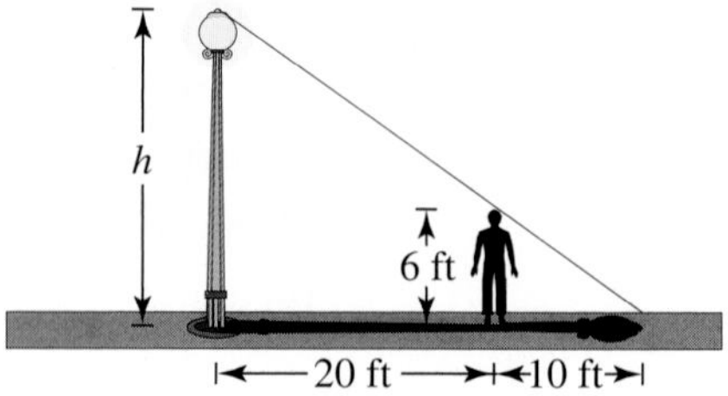

63. *Length* A 20-meter line is used to tether a helium-filled balloon. Because of a breeze, the line makes an angle of approximately 85° with the ground.

(a) Draw a right triangle that gives a visual representation of the problem. Show the known quantities on the triangle and use a variable to indicate the height of the balloon.

(b) Use a trigonometric function to write an equation involving the unknown quantity.

(c) What is the height of the balloon?

64. *Width of a River* A biologist wants to know the width w of a river in order to set instruments for studying the pollutants in the water. From point A, the biologist walks downstream 100 feet and sights to point C (see figure). From this sighting, it is determined that $\theta = 54°$. How wide is the river?

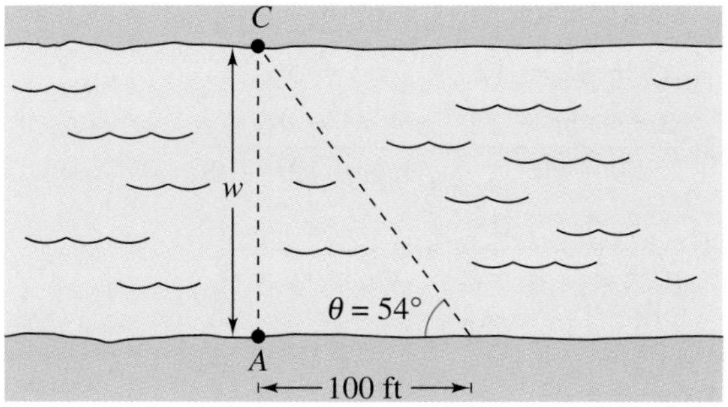

65. *Distance* From a 40-foot observation tower on the coast, a Coast Guard officer sights a boat in difficulty. The angle of depression of the boat is 4° (see figure). How far is the boat from the shoreline?

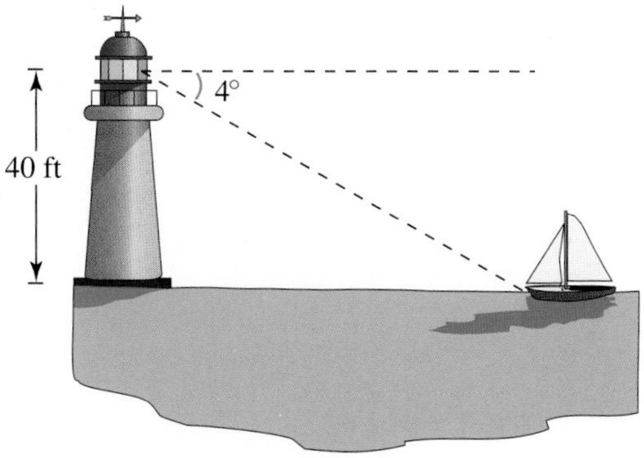

(Not drawn to scale)

66. *Angle of Elevation* A ramp 20 feet in length rises to a loading platform that is $3\frac{1}{3}$ feet off the ground.

(a) Draw a right triangle that gives a visual representation of the problem. Show the known quantities on the triangle and use a variable to indicate the angle of elevation of the ramp.

(b) Use a trigonometric function to write an equation involving the unknown quantity.

(c) What is the angle of elevation of the ramp?

67. *Machine Shop Calculations* A steel plate has the form of one-fourth of a circle with a radius of 60 centimeters. Two 2-centimeter holes are to be drilled in the plate positioned as shown in the figure. Find the coordinates of the center of each hole.

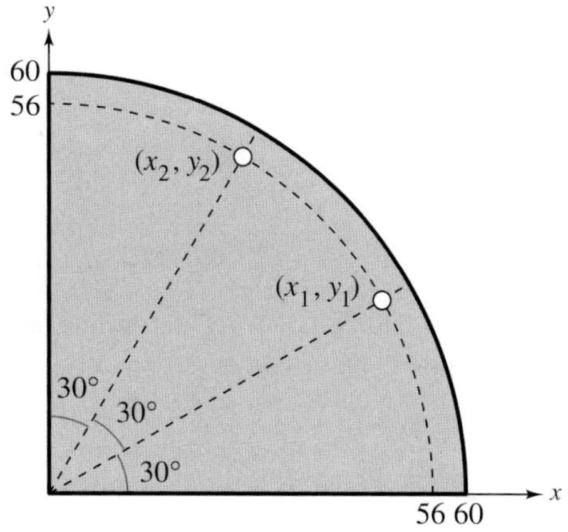

68. *Machine Shop Calculations* A tapered shaft has a diameter of 5 centimeters at the small end and is 15 centimeters long (see figure). If the taper is 3°, find the diameter d of the large end of the shaft.

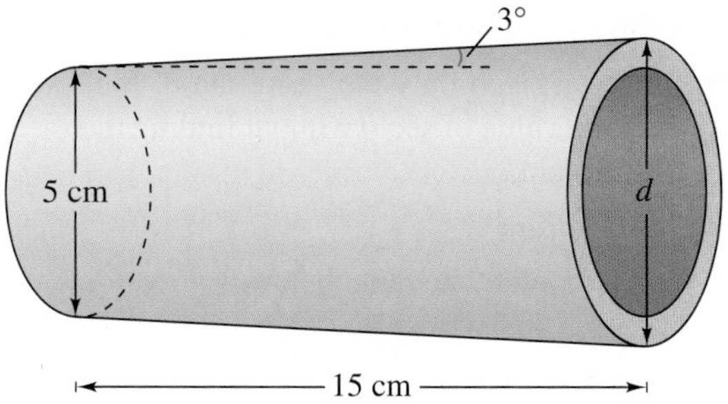

69. *Geometry* Use a compass to sketch a quarter of a circle of radius 10 centimeters. Using a protractor, construct an angle of 20° in standard position (see figure). Drop a perpendicular from the point of intersection of the terminal side of the angle and the arc of the circle. By actual measurement, calculate the coordinates (x, y) of the point of intersection and use these measurements to approximate the six trigonometric functions of a 20° angle.

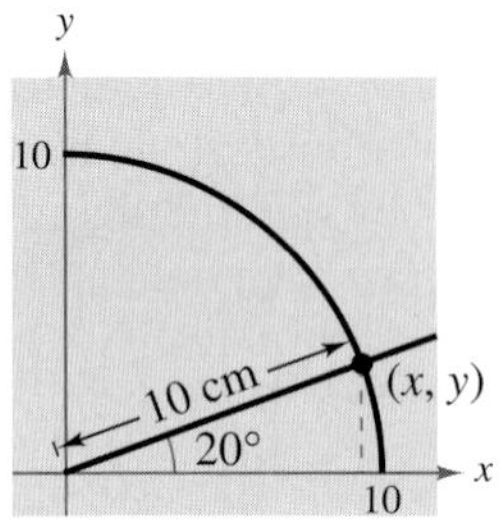

70. *Geometry* Repeat Exercise 69 using a 75° angle.

Synthesis

True or False? **In Exercises 71–76, determine whether the statement is true or false. Justify your answer.**

71. $\sin 60° \csc 60° = 1$

72. $\sec 30° = \csc 60°$

73. $\sin 45° + \cos 45° = 1$

74. $\cot^2 10° - \csc^2 10° = -1$

75. $\dfrac{\sin 60°}{\sin 30°} = \sin 2°$

76. $\tan[(5°)^2] = \tan^2(5°)$

77. In right triangle trigonometry, $\sin 30° = \frac{1}{2}$ regardless of the size of the triangle. Explain.

78. You are given only the value $\tan\theta$. Is it possible to find the value of $\sec\theta$ without finding the measure of θ? Explain.

79. *Exploration*

(a) Complete the table below.

θ	0.1	0.2	0.3	0.4	0.5
$\sin\theta$					

(b) As θ approaches 0, how do θ and $\sin\theta$ compare? Explain.

80. *Exploration*

(a) Complete the table.

θ	0°	18°	36°	54°	72°	90°
$\sin\theta$						
$\cos\theta$						

(b) Discuss the behavior of the sine function for θ in the range from 0° to 90°.

(c) Discuss the behavior of the cosine function for θ in the range from 0° to 90°.

(d) Use the definitions of the sine and cosine functions to explain the results of parts (b) and (c).

Review

In Exercises 81–84, perform the operations and simplify.

81. $\dfrac{x^2 - 6x}{x^2 + 4x - 12} \cdot \dfrac{x^2 + 12x + 36}{x^2 - 36}$

82. $\dfrac{2t^2 + 5t - 12}{9 - 4t^2} \div \dfrac{t^2 - 16}{4t^2 + 12t + 9}$

83. $\dfrac{3}{x + 2} - \dfrac{2}{x - 2} + \dfrac{x}{x^2 + 4x + 4}$

84. $\dfrac{\left(\dfrac{3}{x} - \dfrac{1}{4}\right)}{\left(\dfrac{12}{x} - 1\right)}$

In Exercises 85–88, solve for *x*.

85. $\dfrac{4}{x - 4} = \dfrac{12x}{24 - x}$

86. $\dfrac{6}{x} = \dfrac{72}{9x^2 - 6x}$

87. $\dfrac{2}{x + 3} + \dfrac{4}{x - 2} = \dfrac{12}{x^2 + x - 6}$

88. $\dfrac{3x + 2}{x^2 + x - 2} = \dfrac{4}{x + 2} - \dfrac{2}{1 - x}$

4.4 Trigonometric Functions of Any Angle

▶ What you should learn

- How to evaluate trigonometric functions of any angle
- How to use reference angles to evaluate trigonometric functions
- How to evaluate trigonometric functions of real numbers

▶ Why you should learn it

You can use trigonometric functions to model and solve real-life problems. For instance, Exercise 91 on page 388 shows how trigonometric functions can be used to model the average daily temperature in a city.

Henryk Kaiser/Leo de Wys

Introduction

In Section 4.3, the definitions of trigonometric functions were restricted to acute angles. In this section, the definitions are extended to cover *any* angle. If θ is an *acute* angle, these definitions coincide with those given in the preceding section.

Definitions of Trigonometric Functions of Any Angle

Let θ be an angle in standard position with (x, y) a point on the terminal side of θ and $r = \sqrt{x^2 + y^2} \neq 0$.

$$\sin\theta = \frac{y}{r} \qquad \cos\theta = \frac{x}{r}$$

$$\tan\theta = \frac{y}{x},\quad x \neq 0 \qquad \cot\theta = \frac{x}{y},\quad y \neq 0$$

$$\sec\theta = \frac{r}{x},\quad x \neq 0 \qquad \csc\theta = \frac{r}{y},\quad y \neq 0$$

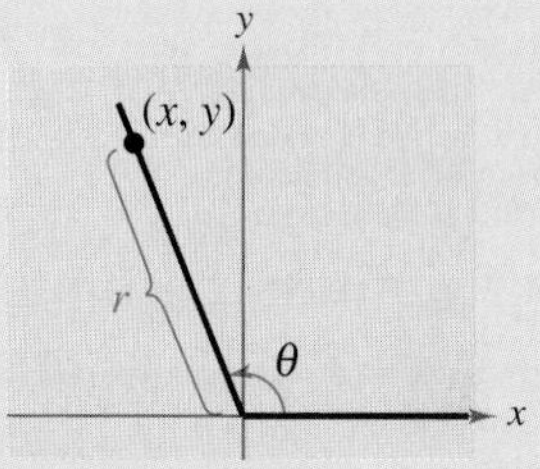

Because $r = \sqrt{x^2 + y^2}$ *cannot* be zero, it follows that the sine and cosine functions are defined for any real value of θ. However, if $x = 0$, the tangent and secant of θ are undefined. For example, the tangent of 90° is undefined. Similarly, if $y = 0$, the cotangent and cosecant of θ are undefined.

Example 1 ▶ Evaluating Trigonometric Functions

Let $(-3, 4)$ be a point on the terminal side of θ. Find the sine, cosine, and tangent of θ.

Solution

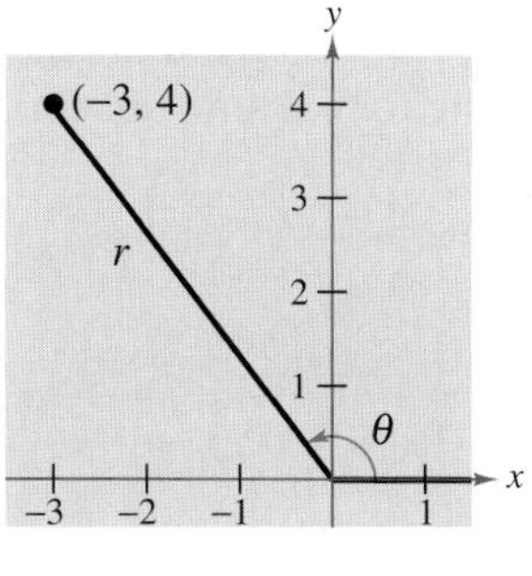

FIGURE 4.30

Referring to Figure 4.30, you see that $x = -3$, $y = 4$, and

$$r = \sqrt{x^2 + y^2} = \sqrt{(-3)^2 + 4^2} = \sqrt{25} = 5.$$

So, you have the following.

$$\sin\theta = y/r = 4/5, \qquad \cos\theta = x/r = -3/5, \qquad \tan\theta = y/x = -4/3$$

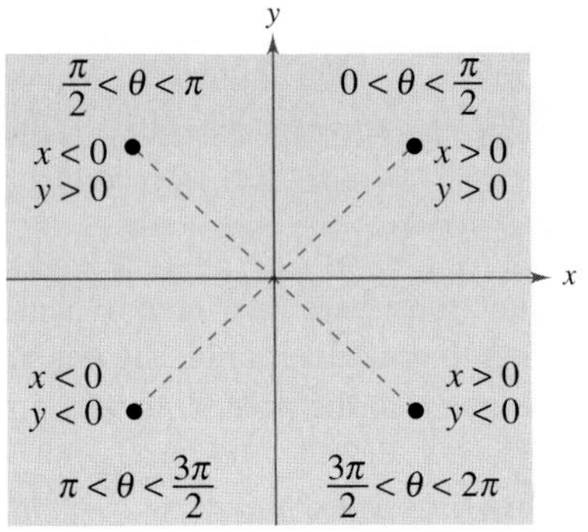

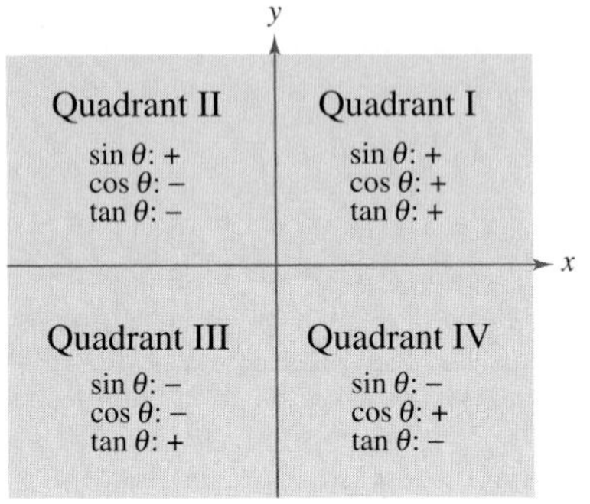

FIGURE 4.31

The *signs* of the trigonometric functions in the four quadrants can be determined easily from the definitions of the functions. For instance, because $\cos\theta = x/r$, it follows that $\cos\theta$ is positive wherever $x > 0$, which is in Quadrants I and IV. (Remember, r is always positive.) In a similar manner, you can verify the results shown in Figure 4.31.

Example 2 ▶ Evaluating Trigonometric Functions

Given $\tan\theta = -\frac{5}{4}$ and $\cos\theta > 0$, find $\sin\theta$ and $\sec\theta$.

Solution

Note that θ lies in Quadrant IV because that is the only quadrant in which the tangent is negative and the cosine is positive. Moreover, using

$$\tan\theta = \frac{y}{x}$$

$$= -\frac{5}{4}$$

and the fact that y is negative in Quadrant IV, you can let $y = -5$ and $x = 4$. So, $r = \sqrt{16 + 25} = \sqrt{41}$ and you have

$$\sin\theta = \frac{y}{r} = \frac{-5}{\sqrt{41}} \approx -0.7809$$

$$\sec\theta = \frac{r}{x} = \frac{\sqrt{41}}{4} \approx 1.6008.$$

Example 3 ▶ Trigonometric Functions of Quadrant Angles

Evaluate the sine function at the four quadrant angles $0, \frac{\pi}{2}, \pi$, and $\frac{3\pi}{2}$.

Solution

To begin, choose a point on the terminal side of each angle, as shown in Figure 4.32. For each of the four given points, $r = 1$, and you have

$$\sin 0 = \frac{y}{r} = \frac{0}{1} = 0 \qquad (x, y) = (1, 0)$$

$$\sin\frac{\pi}{2} = \frac{y}{r} = \frac{1}{1} = 1 \qquad (x, y) = (0, 1)$$

$$\sin\pi = \frac{y}{r} = \frac{0}{1} = 0 \qquad (x, y) = (-1, 0)$$

$$\sin\frac{3\pi}{2} = \frac{y}{r} = \frac{-1}{1} = -1. \qquad (x, y) = (0, -1)$$

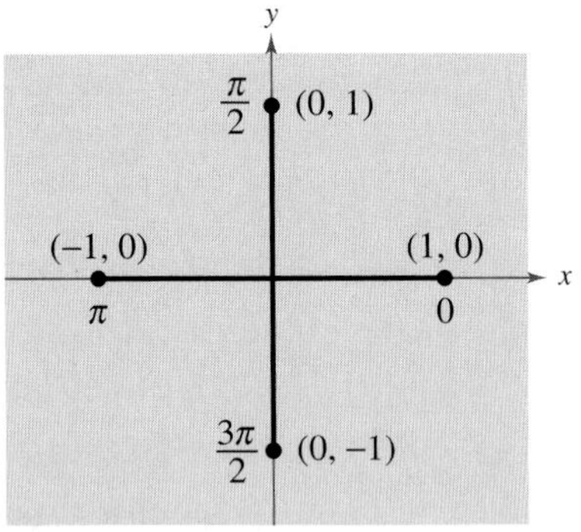

FIGURE 4.32

Sketching several angles with their reference angles helps reinforce the fact that the reference angle is the acute angle formed with the horizontal.

A computer simulation to accompany this concept appears in the *Interactive* CD-ROM and *Internet* versions of this text.

Reference Angles

The values of the trigonometric functions of angles greater than 90° (or less than 0°) can be determined from their values at corresponding acute angles called **reference angles.**

Definition of Reference Angle

Let θ be an angle in standard position. Its **reference angle** is the acute angle θ' formed by the terminal side of θ and the horizontal axis.

Figure 4.33 shows the reference angles for θ in Quadrants II, III, and IV.

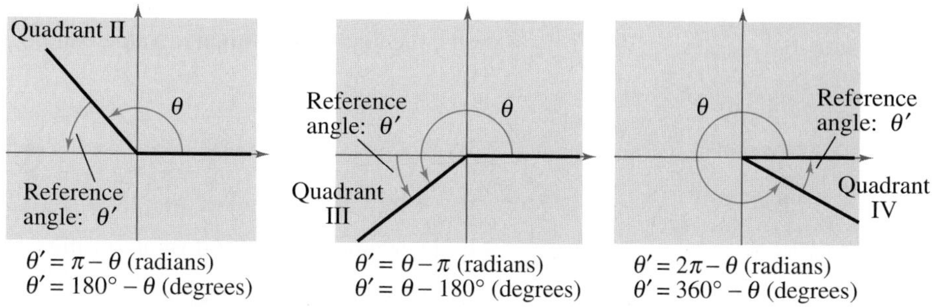

FIGURE 4.33

Example 4 ▶ Finding Reference Angles

Find the reference angle θ'.

a. $\theta = 300°$ **b.** $\theta = 2.3$ **c.** $\theta = -135°$

Solution

a. Because 300° lies in Quadrant IV, the angle it makes with the x-axis is

$$\theta' = 360° - 300° = 60°. \qquad \text{Degrees}$$

Figure 4.34(a) shows the angle $\theta = 300°$ and its reference angle $\theta' = 60°$.

b. Because 2.3 lies between $\pi/2 \approx 1.5708$ and $\pi \approx 3.1416$, it follows that it is in Quadrant II and its reference angle is

$$\theta' = \pi - 2.3 \approx 0.8416. \qquad \text{Radians}$$

Figure 4.34(b) shows the angle $\theta = 2.3$ and its reference angle $\theta' = \pi - 2.3$.

c. First, determine that $-135°$ is coterminal with 225°, which lies in Quadrant III. Hence, the reference angle is

$$\theta' = 225° - 180° = 45°. \qquad \text{Degrees}$$

Figure 4.34(c) shows the angle $\theta = -135°$ and its reference angle $\theta' = 45°$.

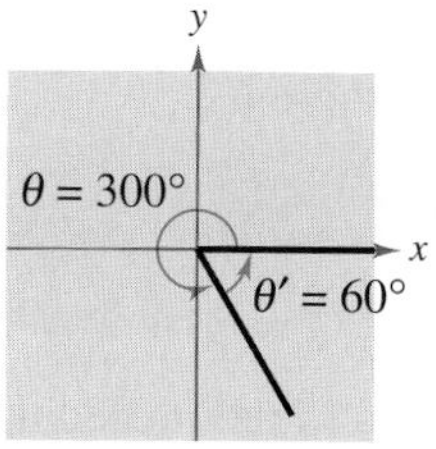

(a)

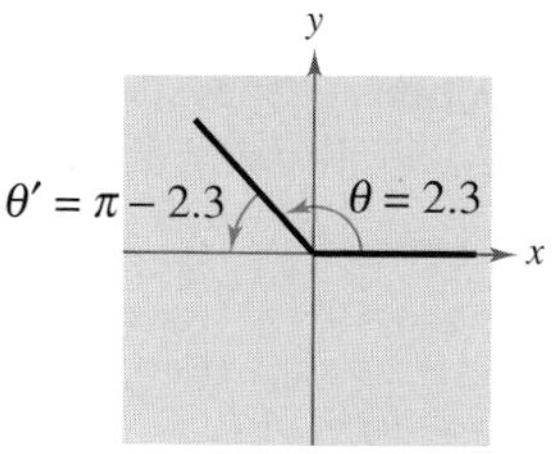

(b)

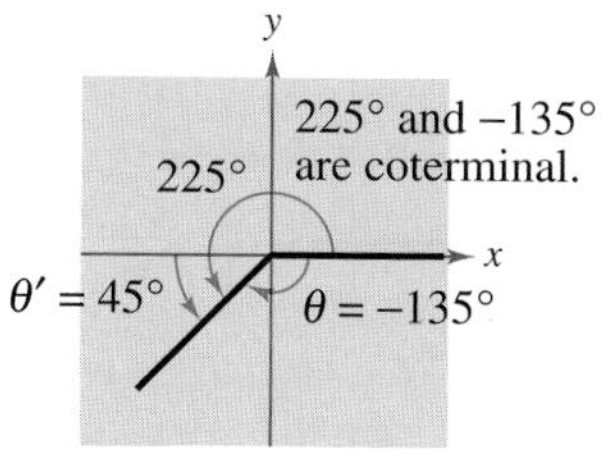

(c)

FIGURE 4.34

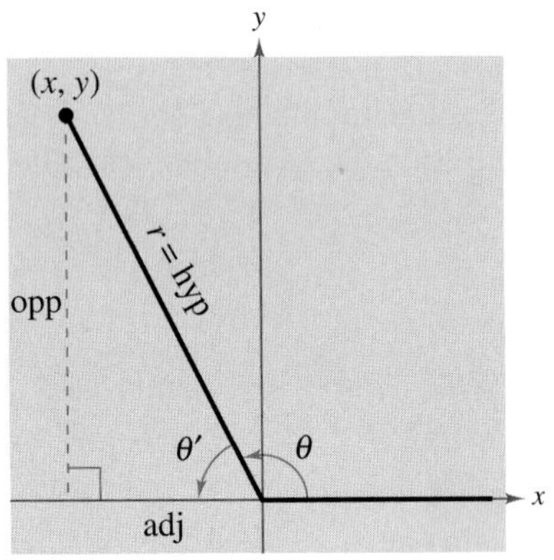

opp = |y|, adj = |x|

FIGURE 4.35

Trigonometric Functions of Real Numbers

To see how a reference angle is used to evaluate a trigonometric function, consider the point (x, y) on the terminal side of θ, as shown in Figure 4.35. By definition, you know that

$$\sin\theta = \frac{y}{r} \quad \text{and} \quad \tan\theta = \frac{y}{x}.$$

For the right triangle with acute angle θ' and sides of lengths $|x|$ and $|y|$, you have

$$\sin\theta' = \frac{\text{opp}}{\text{hyp}} = \frac{|y|}{r} \quad \text{and} \quad \tan\theta' = \frac{\text{opp}}{\text{adj}} = \frac{|y|}{|x|}.$$

So, it follows that $\sin\theta$ and $\sin\theta'$ are equal, *except possibly in sign.* The same is true for $\tan\theta$ and $\tan\theta'$ *and* for the other four trigonometric functions. In all cases, the sign of the function value can be determined by the quadrant in which θ lies.

Evaluating Trigonometric Functions of Any Angle

To find the value of a trigonometric function of any angle θ,

1. determine the function value for the associated reference angle θ';

2. depending on the quadrant in which θ lies, affix the appropriate sign to the function value.

By using reference angles and the special angles discussed in the preceding section, you can greatly extend the scope of *exact* trigonometric values. For instance, knowing the function values of 30° means that you know the function values of all angles for which 30° is a reference angle. For convenience, the table below gives the exact values of the trigonometric functions of special angles and quadrant angles.

Trigonometric Values of Common Angles

θ (degrees)	0°	30°	45°	60°	90°	180°	270°
θ (radians)	0	$\frac{\pi}{6}$	$\frac{\pi}{4}$	$\frac{\pi}{3}$	$\frac{\pi}{2}$	π	$\frac{3\pi}{2}$
$\sin\theta$	0	$\frac{1}{2}$	$\frac{\sqrt{2}}{2}$	$\frac{\sqrt{3}}{2}$	1	0	-1
$\cos\theta$	1	$\frac{\sqrt{3}}{2}$	$\frac{\sqrt{2}}{2}$	$\frac{1}{2}$	0	-1	0
$\tan\theta$	0	$\frac{\sqrt{3}}{3}$	1	$\sqrt{3}$	Undef.	0	Undef.

STUDY TIP

Learning the table of values at the right is worth the effort. Doing so will increase both your efficiency and your confidence. Here is a pattern for the sine function that may help you remember the values.

θ	0°	30°	45°	60°	90°
$\sin\theta$	$\frac{\sqrt{0}}{2}$	$\frac{\sqrt{1}}{2}$	$\frac{\sqrt{2}}{2}$	$\frac{\sqrt{3}}{2}$	$\frac{\sqrt{4}}{2}$

Reverse the order to get cosine values of the same angles.

Example 5 ▶ Trigonometric Functions of Nonacute Angles

Evaluate the trigonometric functions.

a. $\cos \dfrac{4\pi}{3}$ **b.** $\tan(-210°)$ **c.** $\csc \dfrac{11\pi}{4}$

Solution

a. Because $\theta = 4\pi/3$ lies in Quadrant III, the reference angle is $\theta' = (4\pi/3) - \pi = \pi/3$, as shown in Figure 4.36(a). Moreover, the cosine is negative in Quadrant III, so

$$\cos \frac{4\pi}{3} = (-) \cos \frac{\pi}{3}$$
$$= -\frac{1}{2}.$$

Emphasize the importance of reference angles in evaluating trigonometric functions of angles greater than 90°.

b. Because $-210° + 360° = 150°$, it follows that $-210°$ is coterminal with the second-quadrant angle 150°. Therefore, the reference angle is $\theta' = 180° - 150° = 30°$, as shown in Figure 4.36(b). Finally, because the tangent is negative in Quadrant II, you have

$$\tan(-210°) = (-) \tan 30°$$
$$= -\frac{\sqrt{3}}{3}.$$

c. Because $(11\pi/4) - 2\pi = 3\pi/4$, it follows that $11\pi/4$ is coterminal with the second-quadrant angle $3\pi/4$. Therefore, the reference angle is $\theta' = \pi - (3\pi/4) = \pi/4$, as shown in Figure 4.36(c). Because the cosecant is positive in Quadrant II, you have

$$\csc \frac{11\pi}{4} = (+) \csc \frac{\pi}{4}$$
$$= \frac{1}{\sin(\pi/4)}$$
$$= \sqrt{2}.$$

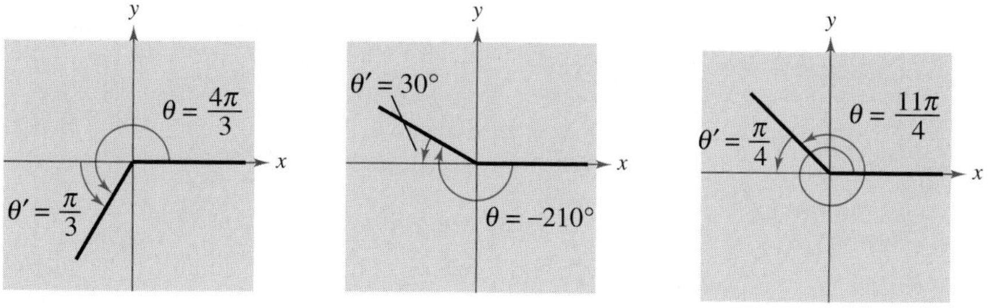

(a) (b) (c)

FIGURE 4.36

Example 6 ▶ Using Trigonometric Identities

Let θ be an angle in Quadrant II such that $\sin\theta = \frac{1}{3}$. Find (a) $\cos\theta$ and (b) $\tan\theta$ by using trigonometric identities.

Solution

a. Using the Pythagorean identity $\sin^2\theta + \cos^2\theta = 1$, you obtain

$$\left(\frac{1}{3}\right)^2 + \cos^2\theta = 1 \qquad \text{Substitute } \tfrac{1}{3} \text{ for } \sin\theta.$$

$$\cos^2\theta = 1 - \frac{1}{9} = \frac{8}{9}.$$

Because $\cos\theta < 0$ in Quadrant II, you can use the negative root to obtain

$$\cos\theta = -\frac{\sqrt{8}}{\sqrt{9}} = -\frac{2\sqrt{2}}{3}.$$

Students often have difficulty determining angles, especially when the functions given are csc, sec, and/or cot. Have your students rewrite the expression in terms of sin, cos, or tan, whichever is applicable, before evaluating.

b. Using the trigonometric identity $\tan\theta = \sin\theta/\cos\theta$, you obtain

$$\tan\theta = \frac{1/3}{-2\sqrt{2}/3} = -\frac{1}{2\sqrt{2}} = -\frac{\sqrt{2}}{4}.$$

Example 7 ▶ Using a Calculator

a. Use a calculator to evaluate $\cot 410^\circ$ and $\sin(-7)$.

b. Use a calculator to solve $\tan\theta = 4.812$, $0 \le \theta < 2\pi$.

Solution

	Function	*Mode*	*Calculator Keystrokes*	*Display*
a.	cot 410°	Degree	(TAN 410) x^{-1} ENTER	0.8390996
	sin(−7)	Radian	SIN ((−) 7) ENTER	−0.6569866

b. To solve the equation $\tan\theta = 4.812$, you can use the inverse tangent key, as follows.

Equation	*Mode*	*Calculator Keystrokes*	*Display*
$\tan\theta = 4.812$	Radian	TAN^{-1} 4.812 ENTER	1.365898912

The angle $\theta \approx 1.366$ lies in Quadrant I. A second value of θ lies in Quadrant III (tangent is positive) and is

$$\theta = \pi + 1.366 \approx 4.508.$$

Additional Examples

a. Solve the equation $\cot\theta = -4.3315$, $0^\circ \le \theta < 360^\circ$.

b. Solve the equation $\csc\theta = 1.1034$, $0 \le \theta < 2\pi$.

c. Solve the equation $\sec\theta = -1.3054$, $0 \le \theta < 2\pi$.

Solutions

a. $\tan\theta = \dfrac{1}{\cot\theta} = -0.230866905;$

$\theta \approx 167^\circ, 347^\circ$

b. $\sin\theta = \dfrac{1}{\csc\theta} = 0.906289650;$

$\theta \approx 1.13, 2.01$

c. $\cos\theta = \dfrac{1}{\sec\theta} = -0.766048721;$

$\theta = 2.44, 3.84$

4.4 Exercises

In Exercises 1–4, determine the exact values of the six trigonometric functions of the angle θ.

1. (a)

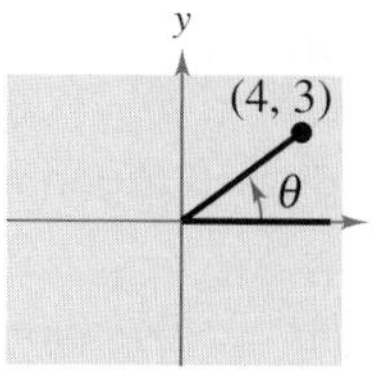

(b)

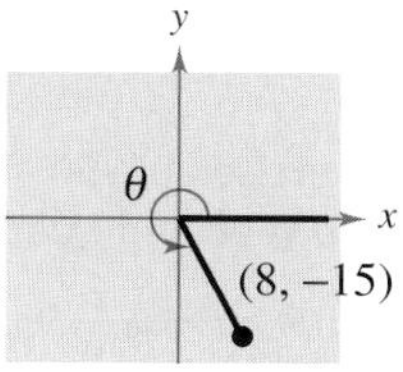

2. (a)

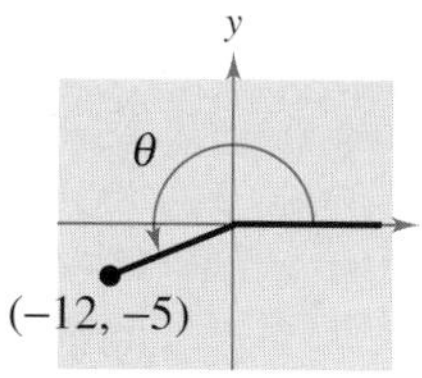

(b)

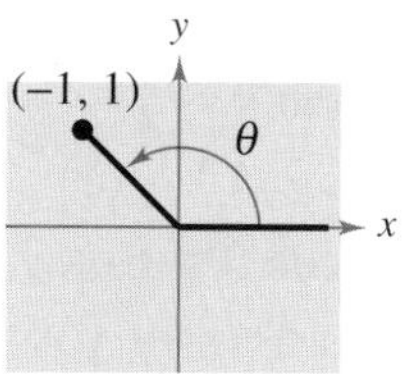

3. (a)

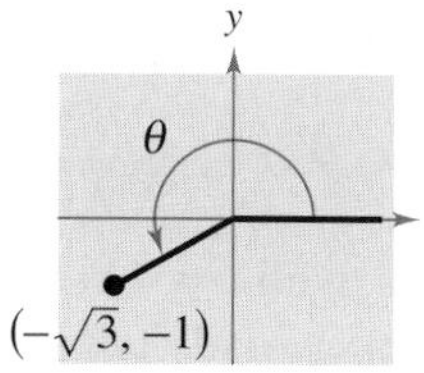

(b)

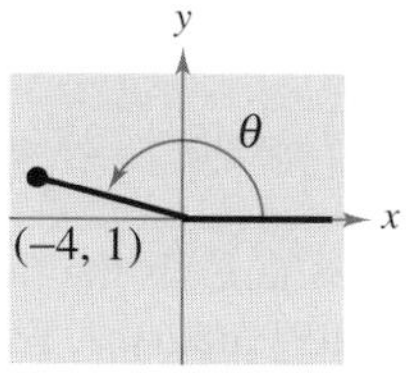

4. (a)

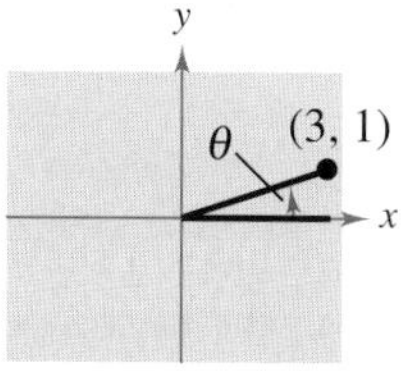

(b)

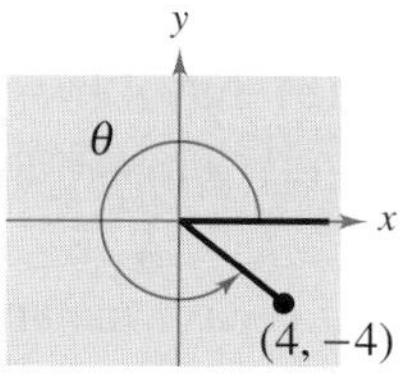

In Exercises 5–10, the point is on the terminal side of an angle in standard position. Determine the exact values of the six trigonometric functions of the angle.

5. $(7, 24)$

6. $(8, 15)$

7. $(-4, 10)$

8. $(-5, -2)$

9. $(-3.5, 6.8)$

10. $\left(3\frac{1}{2}, -7\frac{3}{4}\right)$

In Exercises 11–14, state the quadrant in which θ lies.

11. $\sin\theta < 0$ and $\cos\theta < 0$

12. $\sin\theta > 0$ and $\cos\theta > 0$

13. $\sin\theta > 0$ and $\tan\theta < 0$

14. $\sec\theta > 0$ and $\cot\theta < 0$

In Exercises 15–24, find the values of the six trigonometric functions of θ.

	Function Value	*Constraint*
15.	$\sin\theta = \frac{3}{5}$	θ lies in Quadrant II.
16.	$\cos\theta = -\frac{4}{5}$	θ lies in Quadrant III.
17.	$\tan\theta = -\frac{15}{8}$	$\sin\theta < 0$
18.	$\cos\theta = \frac{8}{17}$	$\tan\theta < 0$
19.	$\cot\theta = -3$	$\cos\theta > 0$
20.	$\csc\theta = 4$	$\cot\theta < 0$
21.	$\sec\theta = -2$	$\sin\theta > 0$
22.	$\sin\theta = 0$	$\sec\theta = -1$
23.	$\cot\theta$ is undefined.	$\pi/2 \le \theta \le 3\pi/2$
24.	$\tan\theta$ is undefined.	$\pi \le \theta \le 2\pi$

In Exercises 25–28, the terminal side of θ lies on the given line in the specified quadrant. Find the values of the six trigonometric functions of θ.

	Line	*Quadrant*
25.	$y = -x$	II
26.	$y = \frac{1}{3}x$	III
27.	$2x - y = 0$	III
28.	$4x + 3y = 0$	IV

In Exercises 29–36, evaluate the trigonometric function of the quadrant angle.

29. $\cos\pi$

30. $\cos\dfrac{3\pi}{2}$

31. $\sec\pi$

32. $\sec\dfrac{3\pi}{2}$

33. $\tan\dfrac{\pi}{2}$

34. $\tan\pi$

35. $\cot\dfrac{\pi}{2}$

36. $\csc\pi$

In Exercises 37–44, find the reference angle θ', and sketch θ and θ' in standard position.

37. $\theta = 203°$

38. $\theta = 309°$

39. $\theta = -245°$

40. $\theta = -145°$

41. $\theta = \dfrac{2\pi}{3}$

42. $\theta = \dfrac{7\pi}{4}$

43. $\theta = 3.5$

44. $\theta = \dfrac{11\pi}{3}$

In Exercises 45–58, evaluate the sine, cosine, and tangent of the angle without using a calculator.

45. $225°$ **46.** $300°$

47. $750°$ **48.** $-405°$

49. $-150°$ **50.** $-840°$

51. $\dfrac{4\pi}{3}$ **52.** $\dfrac{\pi}{4}$

53. $-\dfrac{\pi}{6}$ **54.** $-\dfrac{\pi}{2}$

55. $\dfrac{11\pi}{4}$ **56.** $\dfrac{10\pi}{3}$

57. $-\dfrac{3\pi}{2}$ **58.** $-\dfrac{25\pi}{4}$

In Exercises 59–68, use a calculator to evaluate the trigonometric function to four decimal places. (Be sure the calculator is set in the correct mode.)

59. $\sin 10°$ **60.** $\sec 225°$

61. $\cos(-110°)$ **62.** $\csc(-330°)$

63. $\tan 4.5$ **64.** $\cot 1.35$

65. $\tan \dfrac{\pi}{9}$ **66.** $\tan\left(-\dfrac{\pi}{9}\right)$

67. $\sin(-0.65)$ **68.** $\sin 0.65$

In Exercises 69–74, find two solutions of the equation. Give your answers in degrees ($0° \le \theta < 360°$) and radians ($0 \le \theta < 2\pi$). Do not use a calculator.

69. (a) $\sin \theta = \frac{1}{2}$ (b) $\sin \theta = -\frac{1}{2}$

70. (a) $\cos \theta = \dfrac{\sqrt{2}}{2}$ (b) $\cos \theta = -\dfrac{\sqrt{2}}{2}$

71. (a) $\csc \theta = \dfrac{2\sqrt{3}}{3}$ (b) $\cot \theta = -1$

72. (a) $\sec \theta = 2$ (b) $\sec \theta = -2$

73. (a) $\tan \theta = 1$ (b) $\cot \theta = -\sqrt{3}$

74. (a) $\sin \theta = \dfrac{\sqrt{3}}{2}$ (b) $\sin \theta = -\dfrac{\sqrt{3}}{2}$

In Exercises 75–78, use a calculator to approximate two values of θ ($0° \le \theta < 360°$) that satisfy the equation. Round the values to two decimal places.

75. $\sin \theta = 0.8191$ **76.** $\cos \theta = 0.8746$

77. $\cos \theta = -0.4367$ **78.** $\sin \theta = -0.6514$

In Exercises 79–84, use a calculator to approximate two values of θ ($0 \le \theta < 2\pi$) that satisfy the equation. Round the values to three decimal places.

79. $\cos \theta = 0.9848$ **80.** $\sin \theta = 0.0175$

81. $\tan \theta = 1.192$ **82.** $\cot \theta = 5.671$

83. $\sec \theta = -2.6667$ **84.** $\cos \theta = -0.3214$

In Exercises 85–90, find the indicated trigonometric value in the specified quadrant.

	Function	*Quadrant*	*Trigonometric Value*
85.	$\sin \theta = -\frac{3}{5}$	IV	$\cos \theta$
86.	$\cot \theta = -3$	II	$\sin \theta$
87.	$\tan \theta = \frac{3}{2}$	III	$\sec \theta$
88.	$\csc \theta = -2$	IV	$\cot \theta$
89.	$\cos \theta = \frac{5}{8}$	I	$\sec \theta$
90.	$\sec \theta = -\frac{9}{4}$	III	$\tan \theta$

91. *Average Temperature* The average daily temperature T (in degrees Fahrenheit) for a certain city is

$$T = 45 - 23\cos\left[\frac{2\pi}{365}(t - 32)\right]$$

where t is the time in days, with $t = 1$ corresponding to January 1. Find the average daily temperatures on the following days.

(a) January 1

(b) July 4 ($t = 185$)

(c) October 18 ($t = 291$)

92. *Sales* A company that produces a seasonal product forecasts monthly sales over the next 2 years to be

$$S = 23.1 + 0.442t + 4.3\sin\frac{\pi t}{6}$$

where S is measured in thousands of units and t is the time in months, with $t = 1$ representing January 2001. Predict sales for each of the following months.

(a) February 2001 (b) February 2002

(c) September 2001 (d) September 2002

93. *Harmonic Motion* The displacement from equilibrium of an oscillating weight suspended by a spring is

$$y(t) = 2\cos 6t$$

where y is the displacement in centimeters and t is the time in seconds (see figure). Find the displacement when (a) $t = 0$, (b) $t = \frac{1}{4}$, and (c) $t = \frac{1}{2}$.

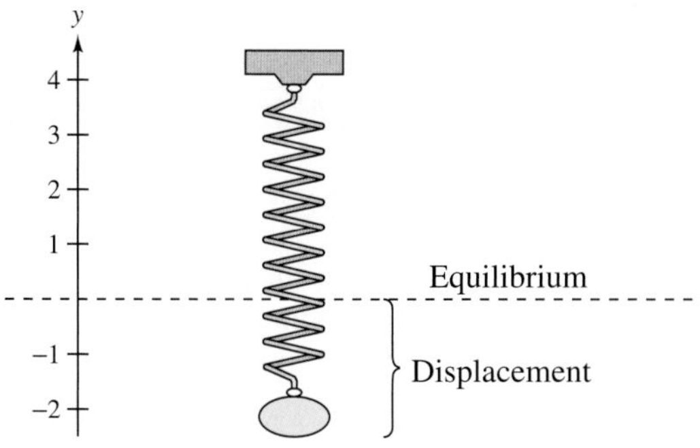

FIGURE FOR 93 AND 94

94. *Harmonic Motion* The displacement from equilibrium of an oscillating weight suspended by a spring and subject to the damping effect of friction is

$$y(t) = 2e^{-t} \cos 6t$$

where y is the displacement in centimeters and t is the time in seconds (see figure). Find the displacement when (a) $t = 0$, (b) $t = \frac{1}{4}$, and (c) $t = \frac{1}{2}$.

95. *Electric Circuits* The current I (in amperes) when 100 volts is applied to a circuit is

$$I = 5e^{-2t} \sin t$$

where t is the time in seconds after the voltage is applied (see figure). Approximate the current $t = 0.7$ second after the voltage is applied.

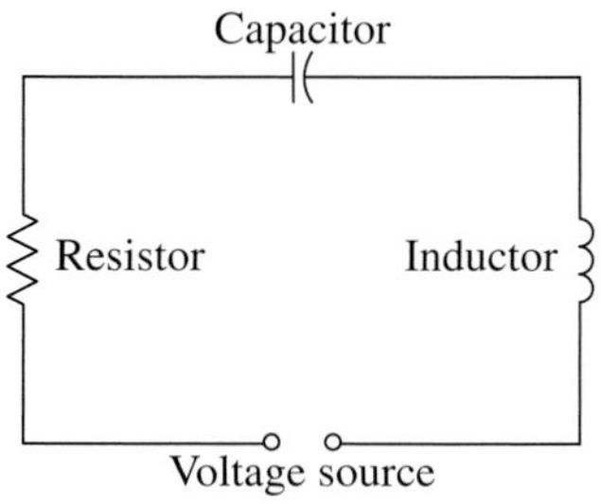

96. *Distance* An airplane, flying at an altitude of 6 miles, is on a flight path that passes directly over an observer (see figure). If θ is the angle of elevation from the observer to the plane, find the distance d from the observer to the plane when (a) $\theta = 30°$, (b) $\theta = 90°$, and (c) $\theta = 120°$.

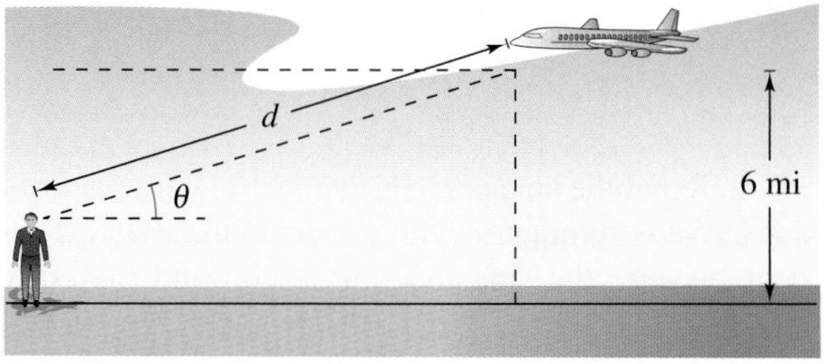

Synthesis

True or False? **In Exercises 97 and 98, determine whether the statement is true or false. Justify your answer.**

97. In each of the four quadrants, the sign of the secant function and sine function will be the same.

98. To find the reference angle for an angle θ (given in degrees), find the integer n such that $0 \le 360n - \theta \le 360$. The difference $360n - \theta$ is the reference angle.

99. *Writing* Consider an angle in standard position with $r = 12$ centimeters, as shown in the figure. Write a short paragraph describing the changes in the magnitudes of x, y, $\sin \theta$, $\cos \theta$, and $\tan \theta$ as θ increases continually from 0° to 90°.

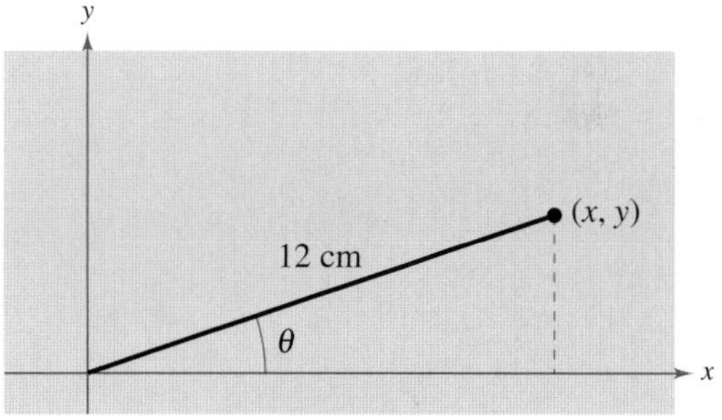

100. Explain how reference angles are used to find the trigonometric functions of obtuse angles.

101. Explain how reference angles are used to find the trigonometric functions of angles with negative measures.

Review

In Exercises 102–105, graph the exponential function. Identify the intercepts, asymptotes, domain, and range of the function.

102. $y = 2^{x-1}$

103. $y = 3^{x+1} + 2$

104. $y = 3^{-x/2}$

105. $y = 3^{(x+1)/2}$

In Exercises 106–109, graph the logarithmic function. Identify the intercepts, asymptotes, domain, and range of the function.

106. $y = \ln(x - 1)$

107. $y = \ln x^4$

108. $y = \log_{10}(x + 2)$

109. $y = \log_{10}(-3x)$

4.5 Graphs of Sine and Cosine Functions

▶ **What you should learn**

- How to sketch the graphs of basic sine and cosine functions
- How to use amplitude and period to help sketch the graphs of sine and cosine functions
- How to sketch translations of the graphs of sine and cosine functions
- How to use sine and cosine functions to model real-life data

▶ **Why you should learn it**

Many biological rhythms can be modeled with sine and cosine functions. For instance, Exercises 73 and 74 on pages 398 and 399 use a sine function to model the respiration of a person at rest and while exercising.

Bob Thomas/Tony Stone Images

Basic Sine and Cosine Curves

In this section you will study techniques for sketching the graphs of the sine and cosine functions. The graph of the sine function is a **sine curve.** In Figure 4.37, the black portion of the graph represents one period of the function and is called **one cycle** of the sine curve. The gray portion of the graph indicates that the basic sine wave repeats indefinitely to the right and left. The graph of the cosine function is shown in Figure 4.38.

Recall from Section 4.2 that the domain of the sine and cosine functions is the set of all real numbers. Moreover, the range of each function is the interval $[-1, 1]$, and each function has a period of 2π. Do you see how this information is consistent with the basic graphs given in Figures 4.37 and 4.38?

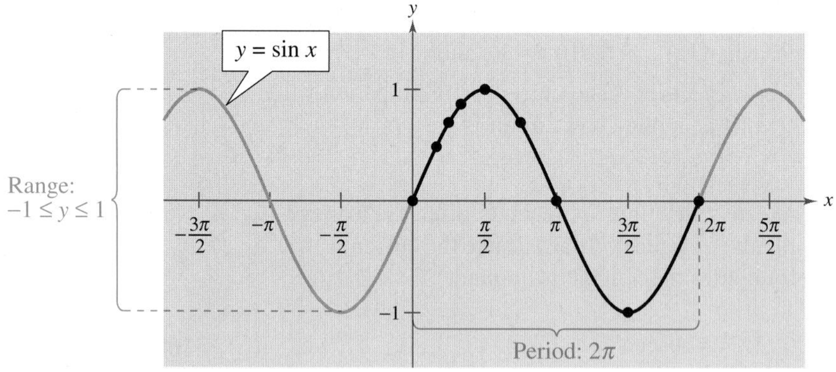

FIGURE 4.37

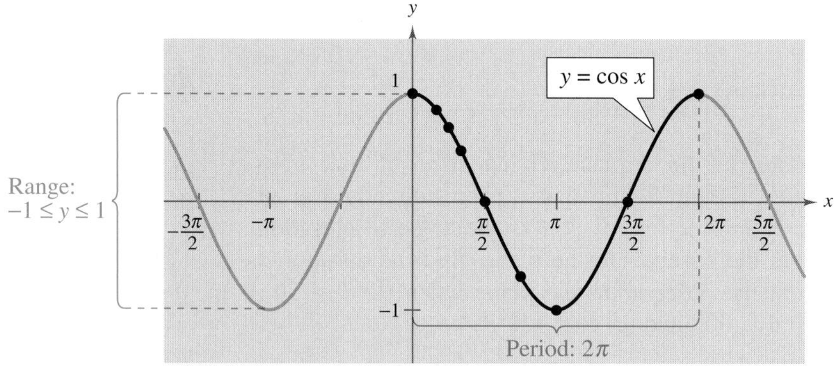

FIGURE 4.38

Note in Figures 4.37 and 4.38 that the sine curve is symmetric with respect to the *origin*, whereas the cosine curve is symmetric with respect to the *y-axis*. These properties of symmetry occur because the sine function is odd and the cosine function is even.

To sketch the graphs of the basic sine and cosine functions by hand, it helps to note five **key points** in one period of each graph: the *intercepts*, *maximum points*, and *minimum points* (see Figure 4.39).

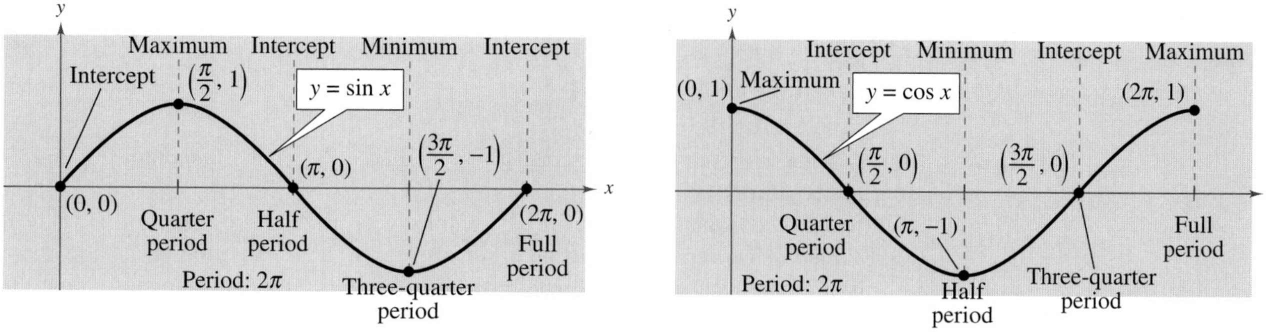

FIGURE 4.39

Example 1 ▶ Using Key Points to Sketch a Sine Curve

Sketch the graph of $y = 2 \sin x$ on the interval $[-\pi, 4\pi]$.

Solution

Note that

$$y = 2 \sin x = 2(\sin x)$$

indicates that the y-values for the key points will have twice the magnitude of those on the graph of $y = \sin x$. Divide the period 2π into four equal parts to get the key points for $y = 2 \sin x$.

$$(0, 0), \quad \left(\frac{\pi}{2}, 2\right), \quad (\pi, 0), \quad \left(\frac{3\pi}{2}, -2\right), \quad \text{and} \quad (2\pi, 0)$$

By connecting these key points with a smooth curve and extending the curve in both directions over the interval $[-\pi, 4\pi]$, you obtain the graph shown in Figure 4.40.

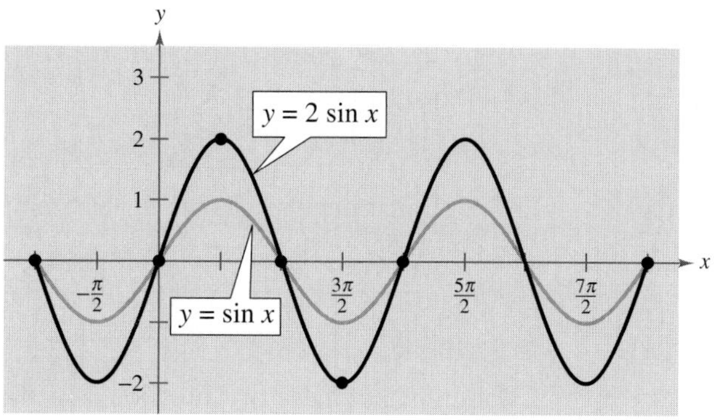

FIGURE 4.40

Technology

When using a graphing utility to graph trigonometric functions, pay special attention to the viewing window you use. For instance, try graphing $y = [\sin(10x)]/10$ in the standard viewing window in radian mode. What do you observe? Use the zoom feature to find a viewing window that displays a good view of the graph.

To graph the examples in this section, your students must know the basic graphs of $y = \sin x$ and $y = \cos x$. For example, to sketch the graph of $y = 3 \sin x$, your students must be able to identify that because $a = 3$, the amplitude is 3 times the amplitude of $y = \sin x$.

Amplitude and Period

In the remainder of this section you will study the graphic effect of each of the constants a, b, c, and d in equations of the forms

$$y = d + a \sin(bx - c)$$

and

$$y = d + a \cos(bx - c).$$

A quick review of the transformations studied in Section 1.5 should help in this investigation.

The constant factor a in $y = a \sin x$ acts as a *scaling factor*—a *vertical stretch* or *vertical shrink* of the basic sine curve. If $|a| > 1$, the basic sine curve is stretched, and if $|a| < 1$, the basic sine curve is shrunk. The result is that the graph of $y = a \sin x$ ranges between $-a$ and a instead of between -1 and 1. The absolute value of a is the **amplitude** of the function $y = a \sin x$. The range of the function $y = a \sin x$ is $-a \le y \le a$.

Definition of Amplitude of Sine and Cosine Curves

The **amplitude** of $y = a \sin x$ and $y = a \cos x$ represents half the distance between the maximum and minimum values of the function and is given by

$$\text{Amplitude} = |a|.$$

To help students learn how to determine and locate key points (intercepts, minimums, maximums), have them mark each of the points on their graphs and then check their graphs using a graphing utility.

Example 2 ▶ Scaling: Vertical Shrinking and Stretching

On the same coordinate axes, sketch the graphs of the functions.

a. $y = \dfrac{1}{2} \cos x$

b. $y = 3 \cos x$

Solution

a. Because the amplitude of $y = \frac{1}{2} \cos x$ is $\frac{1}{2}$, the maximum value is $\frac{1}{2}$ and the minimum value is $-\frac{1}{2}$. Divide one cycle, $0 \le x \le 2\pi$, into four equal parts to get the key points

$$\left(0, \frac{1}{2}\right), \quad \left(\frac{\pi}{2}, 0\right), \quad \left(\pi, -\frac{1}{2}\right), \quad \left(\frac{3\pi}{2}, 0\right), \quad \text{and} \quad \left(2\pi, \frac{1}{2}\right).$$

b. A similar analysis shows that the amplitude of $y = 3 \cos x$ is 3, and the key points are

$$(0, 3), \quad \left(\frac{\pi}{2}, 0\right), \quad (\pi, -3), \quad \left(\frac{3\pi}{2}, 0\right), \quad \text{and} \quad (2\pi, 3).$$

The graphs of these two functions are shown in Figure 4.41.

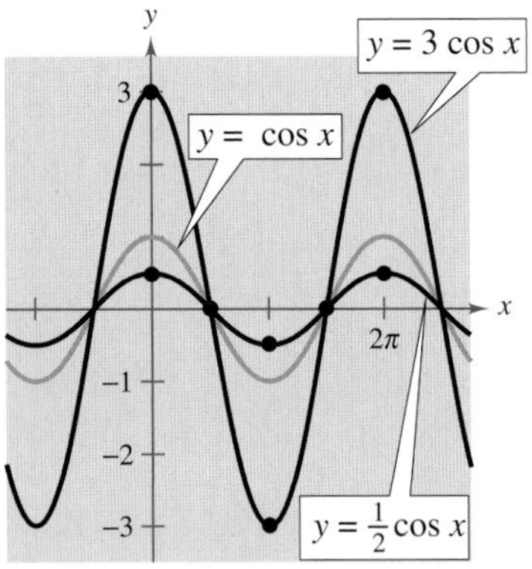

FIGURE 4.41

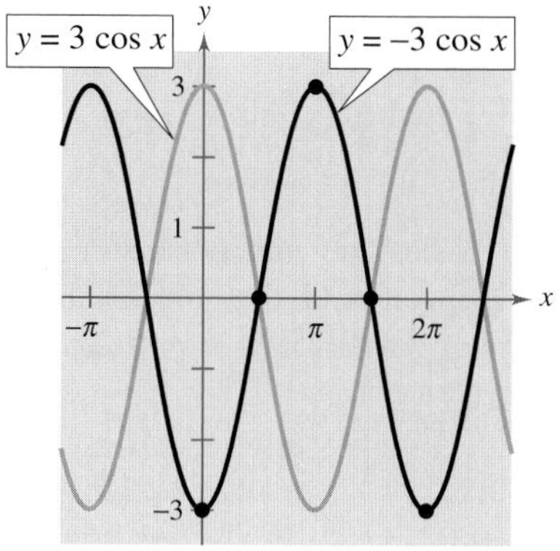

FIGURE 4.42

You know from Section 1.5 that the graph of $y = -f(x)$ is a **reflection** in the x-axis of the graph of $y = f(x)$. For instance, the graph of

$$y = -3 \cos x$$

is a reflection of the graph of

$$y = 3 \cos x,$$

as shown in Figure 4.42.

Because $y = a \sin x$ completes one cycle from $x = 0$ to $x = 2\pi$, it follows that $y = a \sin bx$ completes one cycle from $x = 0$ to $x = 2\pi/b$.

Period of Sine and Cosine Functions

Let b be a positive real number. The **period** of $y = a \sin bx$ and $y = a \cos bx$ is $2\pi/b$.

Note that if $0 < b < 1$, the period of $y = a \sin bx$ is greater than 2π and represents a *horizontal stretching* of the graph of $y = a \sin x$. Similarly, if $b > 1$, the period of $y = a \sin bx$ is less than 2π and represents a *horizontal shrinking* of the graph of $y = a \sin x$. If b is negative, the identities $\sin(-x) = -\sin x$ and $\cos(-x) = \cos x$ are used to rewrite the function.

Exploration

Sketch the graph of $y = \cos bx$ for $b = \frac{1}{2}$, 2, and 3. How does the value of b affect the graph? How many complete cycles occur between 0 and 2π for each value of b?

Example 3 ▶ Scaling: Horizontal Stretching

Sketch the graph of $y = \sin \dfrac{x}{2}$.

Solution

The amplitude is 1. Moreover, because $b = \frac{1}{2}$, the period is

$$\frac{2\pi}{b} = \frac{2\pi}{\frac{1}{2}} = 4\pi. \qquad \text{Substitute for } b.$$

Now, divide the period-interval $[0, 4\pi]$ into four equal parts with the values π, 2π, and 3π to obtain the key points on the graph.

$$(0, 0), \quad (\pi, 1), \quad (2\pi, 0), \quad (3\pi, -1), \quad \text{and} \quad (4\pi, 0)$$

The graph is shown in Figure 4.43.

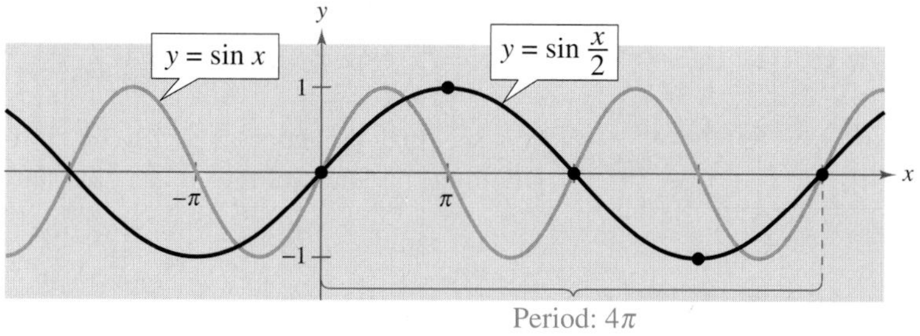

FIGURE 4.43

STUDY T!P

In general, to divide a period-interval into four equal parts, successively add "period/4," starting with the left endpoint of the interval. For instance, for the period-interval $[-\pi/6, \pi/2]$ of length $2\pi/3$, you would successively add

$$\frac{2\pi/3}{4} = \frac{\pi}{6}$$

to get $-\pi/6$, 0, $\pi/6$, $\pi/3$, and $\pi/2$.

Translations of Sine and Cosine Curves

The constant c in the general equations

$$y = a\sin(bx - c) \quad \text{and} \quad y = a\cos(bx - c)$$

creates a *horizontal translation* (shift) of the basic sine and cosine curves. Comparing $y = a\sin bx$ with $y = a\sin(bx - c)$, you find that the graph of $y = a\sin(bx - c)$ completes one cycle from $bx - c = 0$ to $bx - c = 2\pi$. By solving for x, you can find the interval for one cycle to be

$$\underbrace{\frac{c}{b}}_{\text{Left endpoint}} \le x \le \underbrace{\frac{c}{b} + \overbrace{\frac{2\pi}{b}}^{\text{Period}}}_{\text{Right endpoint}}.$$

This implies that the period of $y = a\sin(bx - c)$ is $2\pi/b$, and the graph of $y = a\sin bx$ is shifted by an amount c/b. The number c/b is the **phase shift.**

Exploration

Sketch the graph of

$$y = \sin(x + c)$$

where $c = -\pi/4$, 0, and $\pi/4$. How does the value of c affect the graph?

Graphs of Sine and Cosine Functions

The graphs of $y = a\sin(bx - c)$ and $y = a\cos(bx - c)$ have the following characteristics. (Assume $b > 0$.)

$$\text{Amplitude} = |a| \qquad \text{Period} = \frac{2\pi}{b}$$

The left and right endpoints of a one-cycle interval can be determined by solving the equations $bx - c = 0$ and $bx - c = 2\pi$.

Example 4 ▶ Horizontal Translation

Sketch the graph of $y = \frac{1}{2}\sin(x - \pi/3)$.

Solution

The amplitude is $\frac{1}{2}$ and the period is 2π. By solving the equations

$$x - \frac{\pi}{3} = 0 \quad \Rightarrow \quad x = \frac{\pi}{3}$$

and

$$x - \frac{\pi}{3} = 2\pi \quad \Rightarrow \quad x = \frac{7\pi}{3}$$

you see that the interval $[\pi/3, 7\pi/3]$ corresponds to one cycle of the graph. Dividing this interval into four equal parts produces the key points

$$\left(\frac{\pi}{3}, 0\right), \quad \left(\frac{5\pi}{6}, \frac{1}{2}\right), \quad \left(\frac{4\pi}{3}, 0\right), \quad \left(\frac{11\pi}{6}, -\frac{1}{2}\right), \quad \text{and} \quad \left(\frac{7\pi}{3}, 0\right).$$

The graph is shown in Figure 4.44.

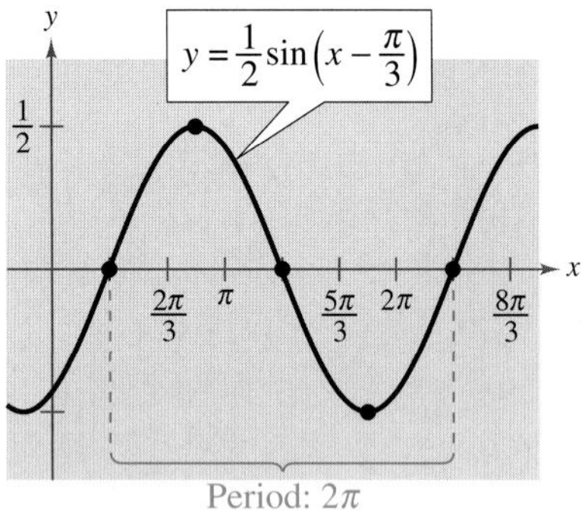

FIGURE 4.44

Example 5 ▶ Horizontal Translation

Sketch the graph of

$$y = -3\cos(2\pi x + 4\pi).$$

Solution

The amplitude is 3 and the period is $2\pi/2\pi = 1$. By solving the equations

$$2\pi x + 4\pi = 0$$
$$2\pi x = -4\pi$$
$$x = -2$$

and

$$2\pi x + 4\pi = 2\pi$$
$$2\pi x = -2\pi$$
$$x = -1$$

you see that the interval $[-2, -1]$ corresponds to one cycle of the graph. Dividing this interval into four equal parts produces the key points

$$(-2, -3), \quad \left(-\frac{7}{4}, 0\right), \quad \left(-\frac{3}{2}, 3\right), \quad \left(-\frac{5}{4}, 0\right), \quad \text{and} \quad (-1, -3).$$

The graph is shown in Figure 4.45.

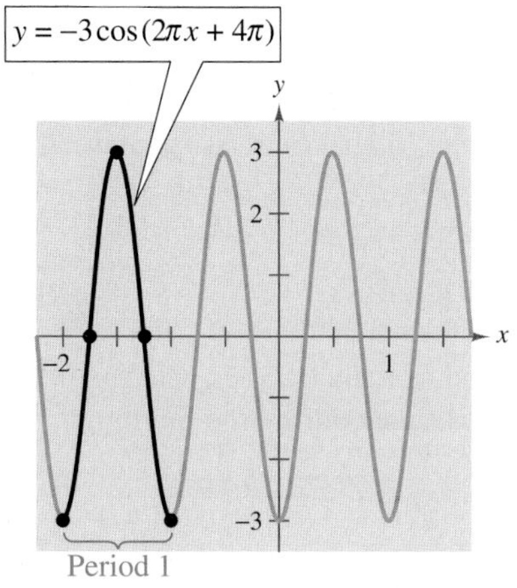

FIGURE 4.45

A computer animation of this example appears in the *Interactive* CD-ROM and *Internet* versions of this text.

Activities

1. Describe the relationship between the graphs of $f(x) = \sin x$ and $g(x) = 3\sin(2x + 1)$.
   Answer: The amplitude of the basic sine curve is 1, whereas the amplitude of $g(x)$ is 3. The period of the basic sine curve is 2π, whereas the period of $g(x)$ is π. Lastly, the graph of $g(x)$ has a phase shift $\frac{1}{2}$ unit to the left of the graph of $f(x) = \sin x$.
2. Determine the amplitude and period.
   $y = \frac{1}{2}\cos(\pi x - 1)$
   Answer: Amplitude, $\frac{1}{2}$; period, 2

The final type of transformation is the *vertical translation* caused by the constant d in the equations

$$y = d + a\sin(bx - c)$$

and

$$y = d + a\cos(bx - c).$$

The shift is d units upward for $d > 0$ and downward for $d < 0$. In other words, the graph oscillates about the horizontal line $y = d$ instead of the x-axis.

Example 6 ▶ Vertical Translation

Sketch the graph of

$$y = 2 + 3\cos 2x.$$

Solution

The amplitude is 3 and the period is π. The key points over the interval $[0, \pi]$ are

$$(0, 5), \quad \left(\frac{\pi}{4}, 2\right), \quad \left(\frac{\pi}{2}, -1\right), \quad \left(\frac{3\pi}{4}, 2\right), \quad \text{and} \quad (\pi, 5).$$

The graph is shown in Figure 4.46.

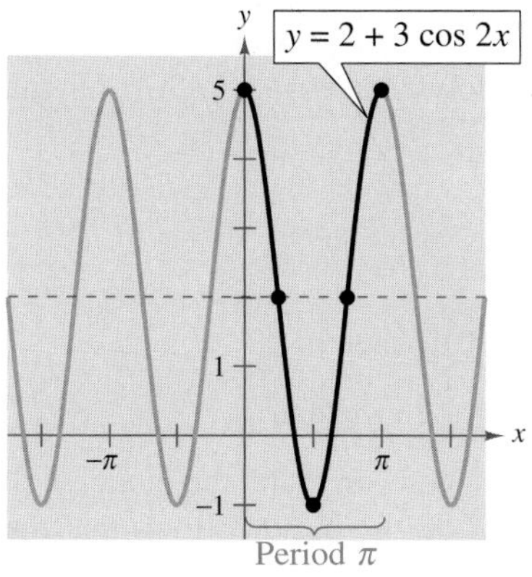

FIGURE 4.46

Mathematical Modeling

Sine and cosine functions can be used to model many real-life situations, including electric currents, musical tones, radio waves, tides, sunrises, and weather patterns.

Example 7 ▶ Finding a Trigonometric Model

Throughout the day, the depth of water at the end of a dock varies with the tides. The table shows the depths (in meters) at various times during the morning.

t (time)	Midnight	2 A.M.	4 A.M.	6 A.M.	8 A.M.	10 A.M.	Noon
y (depth)	2.55	3.80	4.40	3.80	2.55	1.80	2.27

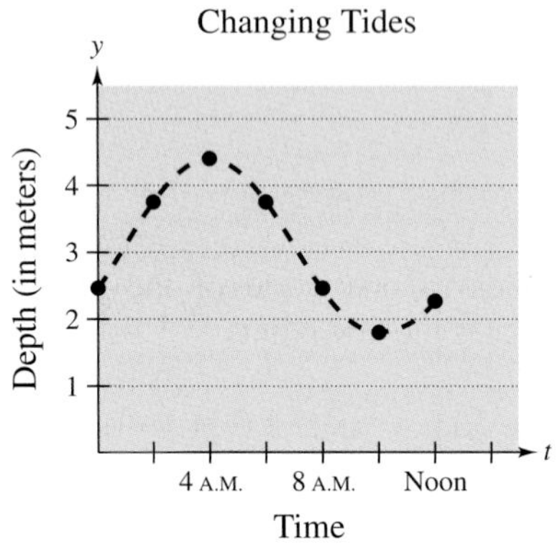

FIGURE 4.47

a. Use a trigonometric function to model this data.

b. Find the depths at 9 A.M. and 3 P.M.

c. A boat needs at least 3 meters of water to moor at the dock. During what times in the afternoon can it safely dock?

Solution

a. Begin by graphing the data, as shown in Figure 4.47. You can use either a sine or cosine model. Suppose you use a cosine model of the form

$$y = a\cos(bt - c) + d.$$

The amplitude is given by

$$a = \frac{1}{2}[(\text{high}) - (\text{low})] = \frac{1}{2}(4.4 - 1.8) = 1.3.$$

The period is

$$p = 2[(\text{low time}) - (\text{high time})] = 2(10 - 4) = 12$$

which implies that $b = 2\pi/p \approx 0.524$. Because high tide occurs 4 hours after midnight, you can conclude that $c/b = 4$, so $c \approx 2.094$. Moreover, because the average depth is $\frac{1}{2}(4.4 + 1.8) = 3.1$, it follows that $d = 3.1$. So, you can model the depth with the function

$$y = 1.3\cos(0.524t - 2.094) + 3.1.$$

b. The depths at 9 A.M. and 3 P.M. are as follows.

$$y = 1.3\cos(0.524 \cdot 9 - 2.094) + 3.1 \approx 1.97 \text{ meters} \qquad \text{9 A.M.}$$

$$y = 1.3\cos(0.524 \cdot 15 - 2.094) + 3.1 \approx 4.23 \text{ meters} \qquad \text{3 P.M.}$$

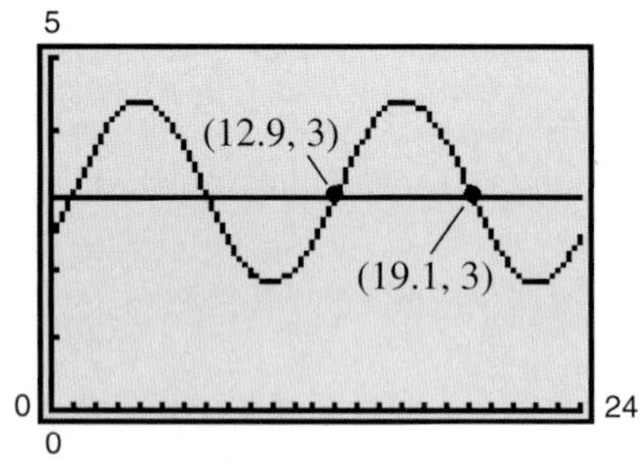

FIGURE 4.48

c. To find out when the depth y is at least 3 meters, you can graph the model with the line $y = 3$, as shown in Figure 4.48. From the graph, it follows that the depth is at least 3 meters between 12:54 P.M. ($t \approx 12.9$) and 7:06 P.M. ($t \approx 19.1$).

4.5 Exercises

In Exercises 1–14, find the period and amplitude.

1. $y = 3 \sin 2x$

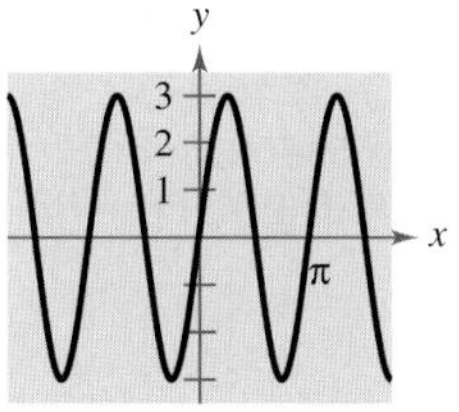

2. $y = 2 \cos 3x$

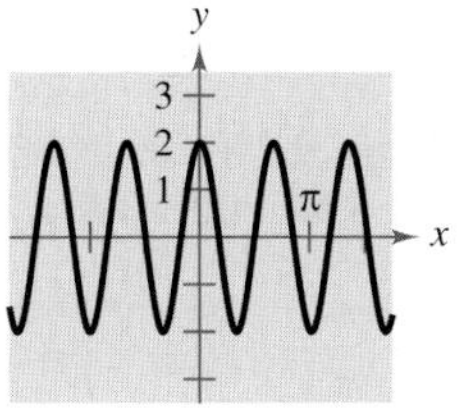

3. $y = \frac{5}{2} \cos \frac{x}{2}$

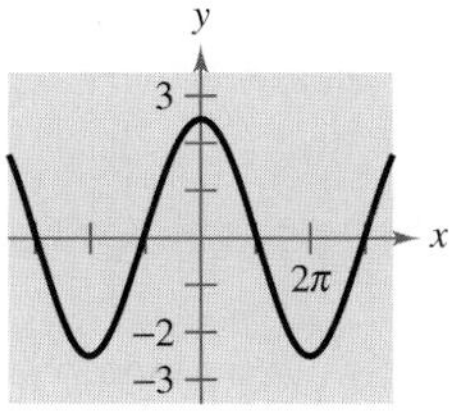

4. $y = -3 \sin \frac{x}{3}$

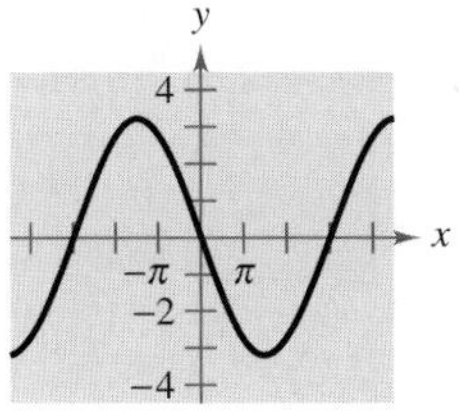

5. $y = \frac{1}{2} \sin \frac{\pi x}{3}$

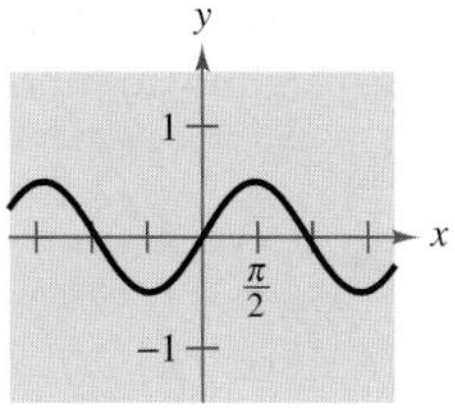

6. $y = \frac{3}{2} \cos \frac{\pi x}{2}$

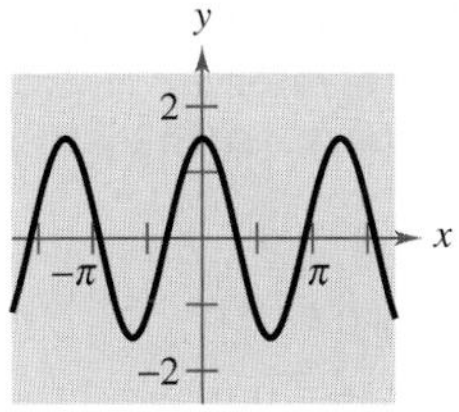

7. $y = -2 \sin x$

8. $y = -\cos \frac{2x}{3}$

9. $y = 3 \sin 10x$

10. $y = \frac{1}{3} \sin 8x$

11. $y = \frac{1}{2} \cos \frac{2x}{3}$

12. $y = \frac{5}{2} \cos \frac{x}{4}$

13. $y = \frac{1}{4} \sin 2\pi x$

14. $y = \frac{2}{3} \cos \frac{\pi x}{10}$

In Exercises 15–22, describe the relationship between the graphs of *f* and *g*.

15. $f(x) = \sin x$
$g(x) = \sin(x - \pi)$

16. $f(x) = \cos x$
$g(x) = \cos(x + \pi)$

17. $f(x) = \cos 2x$
$g(x) = -\cos 2x$

18. $f(x) = \sin 3x$
$g(x) = \sin(-3x)$

19. $f(x) = \cos x$
$g(x) = \cos 2x$

20. $f(x) = \sin x$
$g(x) = \sin 3x$

21. $f(x) = \sin 2x$
$g(x) = 3 + \sin 2x$

22. $f(x) = \cos 4x$
$g(x) = -2 + \cos 4x$

In Exercises 23–26, describe the relationship between the graphs of *f* and *g*.

23.

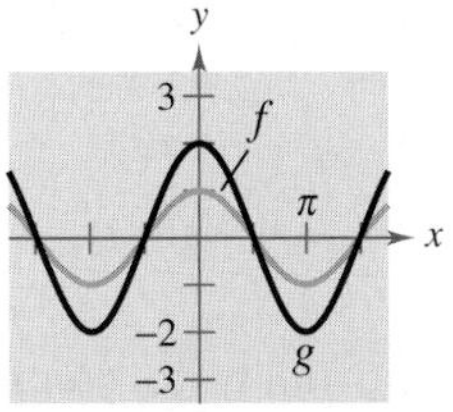

24.

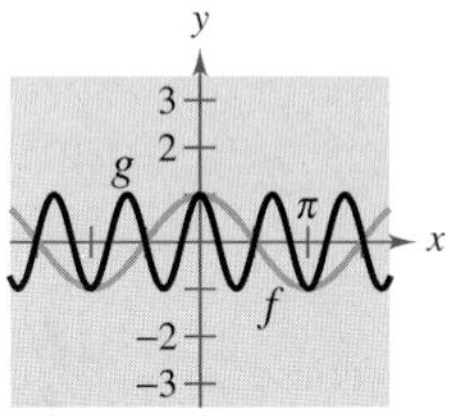

25.

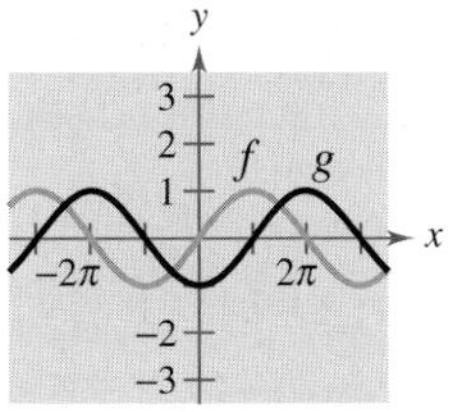

26.

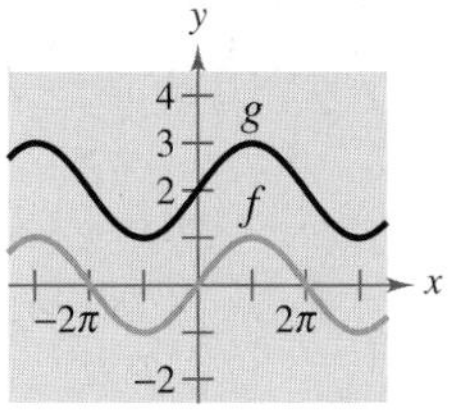

In Exercises 27–34, graph *f* and *g* on the same set of coordinate axes. (Include two full periods.)

27. $f(x) = -2 \sin x$
$g(x) = 4 \sin x$

28. $f(x) = \sin x$
$g(x) = \sin \frac{x}{3}$

29. $f(x) = \cos x$
$g(x) = 1 + \cos x$

30. $f(x) = 2 \cos 2x$
$g(x) = -\cos 4x$

31. $f(x) = -\frac{1}{2} \sin \frac{x}{2}$
$g(x) = 3 - \frac{1}{2} \sin \frac{x}{2}$

32. $f(x) = 4 \sin \pi x$
$g(x) = 4 \sin \pi x - 3$

33. $f(x) = 2 \cos x$
$g(x) = 2 \cos(x + \pi)$

34. $f(x) = -\cos x$
$g(x) = -\cos(x - \pi)$

In Exercises 35–52, sketch the graph of the function. (Include two full periods.)

35. $y = -2 \sin 6x$

36. $y = -3 \cos 4x$

37. $y = \cos 2\pi x$

38. $y = \sin \dfrac{\pi x}{4}$

39. $y = -\sin \dfrac{2\pi x}{3}$

40. $y = -10 \cos \dfrac{\pi x}{6}$

41. $y = \sin\left(x - \dfrac{\pi}{4}\right)$

42. $y = \sin(x - \pi)$

43. $y = 3 \cos(x + \pi)$

44. $y = 4 \cos\left(x + \dfrac{\pi}{4}\right)$

45. $y = 2 - \sin \dfrac{2\pi x}{3}$

46. $y = -3 + 5 \cos \dfrac{\pi t}{12}$

47. $y = 2 + \frac{1}{10} \cos 60\pi x$

48. $y = 2 \cos x - 3$

49. $y = 3 \cos(x + \pi) - 3$

50. $y = 4 \cos\left(x + \dfrac{\pi}{4}\right) + 4$

51. $y = \dfrac{2}{3} \cos\left(\dfrac{x}{2} - \dfrac{\pi}{4}\right)$

52. $y = -3 \cos(6x + \pi)$

In Exercises 53–60, use a graphing utility to graph the function. Include two full periods. Be sure to choose an appropriate viewing window.

53. $y = -2 \sin(4x + \pi)$

54. $y = -4 \sin\left(\dfrac{2}{3}x - \dfrac{\pi}{3}\right)$

55. $y = \cos\left(2\pi x - \dfrac{\pi}{2}\right) + 1$

56. $y = 3 \cos\left(\dfrac{\pi x}{2} + \dfrac{\pi}{2}\right) - 2$

57. $y = 5 \sin(\pi - 2x) + 10$

58. $y = 5 \cos(\pi - 2x) + 2$

59. $y = -0.1 \sin\left(\dfrac{\pi x}{10} + \pi\right)$

60. $y = \frac{1}{100} \sin 120\pi t$

Graphical Reasoning **In Exercises 61–64, find a and d for the function $f(x) = a \cos x + d$ such that the graph of f matches the figure.**

61.

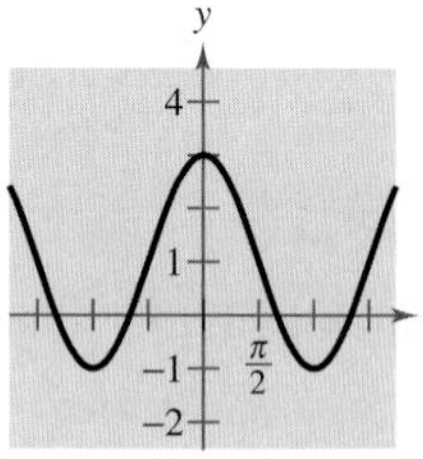

62.

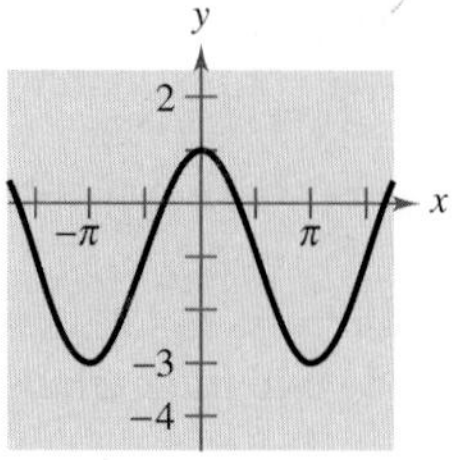

63.

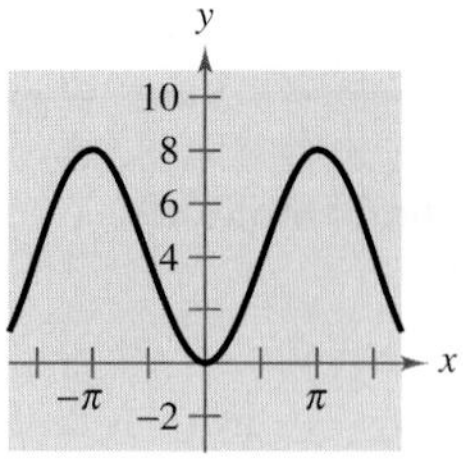

64.

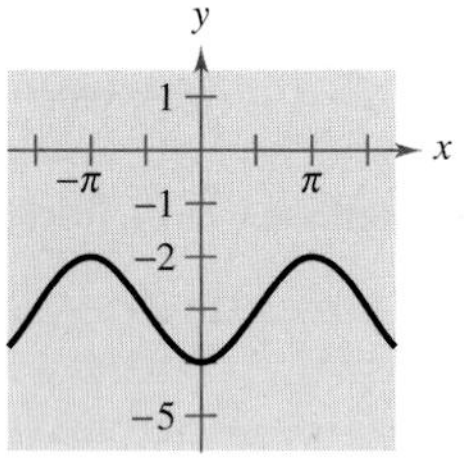

Graphical Reasoning **In Exercises 65–68, find a, b, and c for the function $f(x) = a \sin(bx - c)$ such that the graph of f matches the figure.**

65.

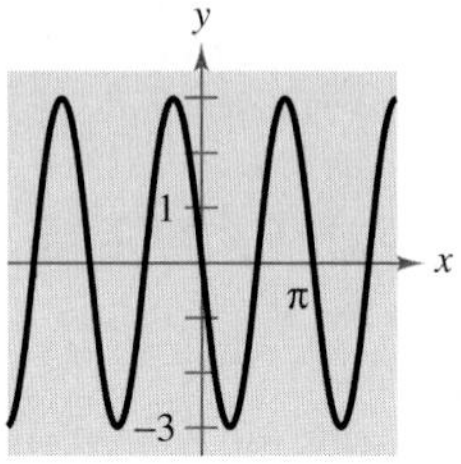

66.

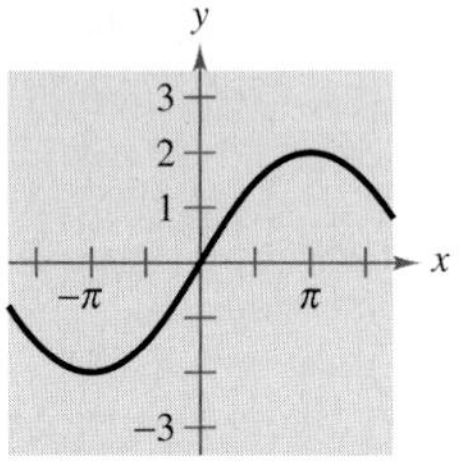

67.

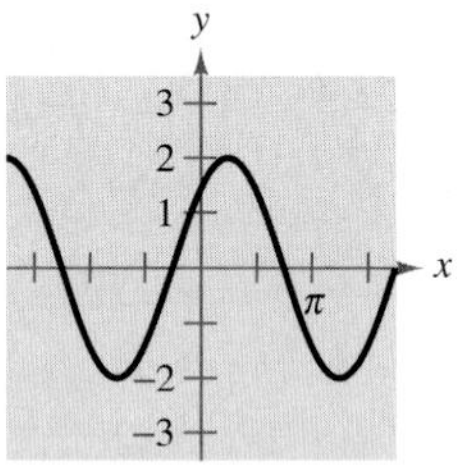

68.

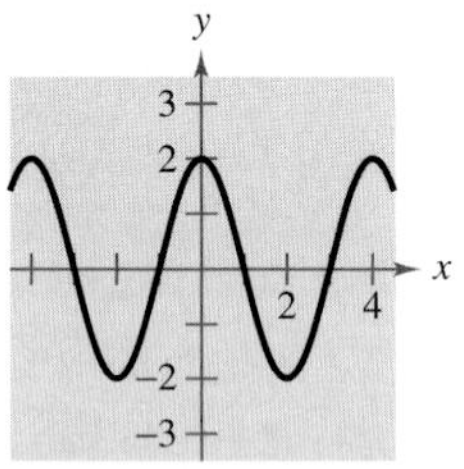

In Exercises 69–72, use a graphing utility to graph y_1 and y_2 in the interval $[-2\pi, 2\pi]$. Use the graphs to find real numbers x such that $y_1 = y_2$.

69. $y_1 = \sin x$
$y_2 = -\frac{1}{2}$

70. $y_1 = \cos x$
$y_2 = -1$

71. $y_1 = \cos x$
$y_2 = \dfrac{\sqrt{2}}{2}$

72. $y_1 = \sin x$
$y_2 = \dfrac{\sqrt{3}}{2}$

73. *Respiratory Cycle* For a person at rest, the velocity v (in liters per second) of air flow during a respiratory cycle (the time from the beginning of one breath to the beginning of the next) is

$$v = 0.85 \sin \frac{\pi t}{3}$$

where t is the time in seconds. (Inhalation occurs when $v > 0$, and exhalation occurs when $v < 0$.)

(a) Find the time for one full respiratory cycle.

(b) Find the number of cycles per minute.

(c) Sketch the graph of the velocity function.

74. *Respiratory Cycle* After exercising for a few minutes, a person has a respiratory cycle for which the velocity of air flow is approximated by $v = 1.75 \sin(\pi t/2)$, where t is the time in seconds. (Inhalation occurs when $v > 0$, and exhalation occurs when $v < 0$.)

(a) Find the time for one full respiratory cycle.

(b) Find the number of cycles per minute.

(c) Sketch the graph of the velocity function.

75. *Piano Tuning* When tuning a piano, a technician strikes a tuning fork for the A above middle C and sets up a wave motion that can be approximated by $y = 0.001 \sin 880\pi t$, where t is the time in seconds.

(a) What is the period of the function?

(b) The frequency f is given by $f = 1/p$. What is the frequency of the note?

76. *Biology* The function

$$P = 100 - 20 \cos \frac{5\pi t}{3}$$

approximates the blood pressure P in millimeters of mercury at time t in seconds for a person at rest.

(a) Find the period of the function.

(b) Find the number of heartbeats per minute.

77. *Data Analysis* The table gives the normal daily high temperatures for Honolulu H and Chicago C (in degrees Fahrenheit) for month t, with $t = 1$ corresponding to January. (Source: National Oceanic and Atmospheric Administration)

t	1	2	3	4	5	6
H	80.1	80.5	81.6	82.8	84.7	86.5
C	29.0	33.5	45.8	58.6	70.1	79.6

t	7	8	9	10	11	12
H	87.5	88.7	88.5	86.9	84.1	81.2
C	83.7	81.8	74.8	63.3	48.4	34.0

(a) A model for the temperature in Honolulu is

$$H(t) = 84.40 + 4.28 \sin\left(\frac{\pi t}{6} + 3.86\right).$$

Find a trigonometric model for Chicago.

(b) Use a graphing utility to graph the data points and the model for the temperatures in Honolulu. How well does the model fit the data?

(c) Use a graphing utility to graph the data points and the model for the temperatures in Chicago. How well does the model fit the data?

(d) Use the models to estimate the average annual temperature in each city. Which term of the models did you use? Explain.

(e) What is the period of each model? Are the periods what you expected? Explain.

(f) Which city has the greater variability in temperature throughout the year? Which factor of the models determines this variability? Explain.

Sales **In Exercises 78 and 79, use a graphing utility to graph the sales function over 1 year where S is the sales in thousands of units and t is the time in months, with $t = 1$ corresponding to January.**

78. $S = 22.3 - 3.4 \cos \dfrac{\pi t}{6}$

79. $S = 74.50 + 43.75 \sin \dfrac{\pi t}{6}$

80. *Fuel Consumption* The daily consumption C (in gallons) of diesel fuel on a farm is modeled by

$$C = 30.3 + 21.6 \sin\left(\frac{2\pi t}{365} + 10.9\right)$$

where t is the time in days, with $t = 1$ corresponding to January 1.

(a) What is the period of the model? Is it what you expected? Explain.

(b) What is the average daily fuel consumption? Which term of the model did you use? Explain.

(c) Use a graphing utility to graph the model. Use the graph to approximate the time of the year when consumption exceeds 40 gallons per day.

81. *Data Analysis* The percent y of the moon's face that is illuminated on day x of the year 2005, where $x = 70$ represents March 11, is given in the table. (Source: U.S. Naval Observatory)

x	76	84	91	98	106	114
y	0.5	1.0	0.5	0.0	0.5	1.0

(a) Create a scatter plot of the data.

(b) Find a trigonometric model that fits the data.

(c) Add the graph of your model in part (b) to the scatter plot. How well does the model fit the data?

(d) Estimate the moon's percent illumination for May 8, 2005.

Synthesis

True or False? **In Exercises 82 and 83, determine whether the statement is true or false. Justify your answer.**

82. The graph of the function $f(x) = \sin(x + 2\pi)$ translates the graph of $f(x) = \sin x$ exactly one period to the right so that the two graphs look identical.

83. The function $y = \frac{1}{2}\cos 2x$ has an amplitude that is twice that of the function $y = \cos x$.

84. *Writing* Use a graphing utility to graph the function $y = a \sin x$ for $a = \frac{1}{2}$, $a = \frac{3}{2}$, and $a = -3$. Write a paragraph describing the changes in the graph corresponding to the specified changes in a.

85. *Writing* Use a graphing utility to graph the function $y = d + \sin x$ for $d = 2$, $d = 3.5$, and $d = -2$. Write a paragraph describing the changes in the graph corresponding to the specified changes in d.

86. *Writing* Use a graphing utility to graph the function $y = \sin bx$ for $b = \frac{1}{2}$, $b = \frac{3}{2}$, and $b = 4$. Write a paragraph describing the changes in the graph corresponding to the specified changes in b.

87. *Writing* Use a graphing utility to graph the function $y = \sin(x - c)$ for $c = 1$, $c = 3$, and $c = -2$. Write a paragraph describing the changes in the graph corresponding to the specified changes in c.

Conjecture **In Exercises 88–91, graph f and g on the same set of coordinate axes. Include two full periods. Make a conjecture about the functions.**

88. $f(x) = \sin x, \quad g(x) = \cos\left(x - \frac{\pi}{2}\right)$

89. $f(x) = \sin x, \quad g(x) = -\cos\left(x + \frac{\pi}{2}\right)$

90. $f(x) = \cos x, \quad g(x) = -\sin\left(x - \frac{\pi}{2}\right)$

91. $f(x) = \cos x, \quad g(x) = -\cos(x - \pi)$

92. *Exploration* Using calculus, it can be shown that the sine and cosine functions can be approximated by the polynomials

$$\sin x \approx x - \frac{x^3}{3!} + \frac{x^5}{5!} \text{ and } \cos x \approx 1 - \frac{x^2}{2!} + \frac{x^4}{4!}$$

where x is in radians.

(a) Use a graphing utility to graph the sine function and its polynomial approximation in the same viewing window. How do the graphs compare?

(b) Use a graphing utility to graph the cosine function and its polynomial approximation in the same viewing window. How do the graphs compare?

(c) Study the patterns in the polynomial approximations of the sine and cosine functions and guess the next term in each. Then repeat parts (a) and (b). How did the accuracy of the approximations change when additional terms were added?

93. *Exploration* Use the polynomial approximations for the sine and cosine functions given in Exercise 92 to approximate the following functional values. Compare the results with those given by a calculator. Is the error in the approximation the same in each case? Explain.

(a) $\sin \frac{1}{2}$ (b) $\sin 1$ (c) $\sin \frac{\pi}{6}$

(d) $\cos(-0.5)$ (e) $\cos 1$ (f) $\cos \frac{\pi}{4}$

94. *Exploration* Use a graphing utility to graph h, and use the graph to decide whether h is even, odd, or neither.

(a) $h(x) = \cos^2 x$ (b) $h(x) = \sin^2 x$

95. *Conjecture* If f is an even function and g is an odd function, use the results of Exercise 94 to make a conjecture about h where

(a) $h(x) = [f(x)]^2$ (b) $h(x) = [g(x)]^2$.

Review

In Exercises 96–99, use the properties of logarithms to write the expression as a sum, difference, and/or constant multiple of a logarithm.

96. $\log_{10} \sqrt{x - 2}$

97. $\log_2[x^2(x - 3)]$

98. $\ln \frac{t^3}{t - 1}$

99. $\ln \sqrt{\frac{z}{z^2 + 1}}$

In Exercises 100–103, write the expression as the logarithm of a single quantity.

100. $\frac{1}{2}(\log_{10} x + \log_{10} y)$

101. $2 \log_2 x + \log_2(xy)$

102. $\ln 3x - 4 \ln y$

103. $\frac{1}{2}(\ln 2x - 2 \ln x) + 3 \ln x$

4.6 Graphs of Other Trigonometric Functions

▶ **What you should learn**

- How to sketch the graphs of tangent functions
- How to sketch the graphs of cotangent functions
- How to sketch the graphs of secant and cosecant functions
- How to sketch the graphs of damped trigonometric functions

▶ **Why you should learn it**

You can use tangent, cotangent, secant, and cosecant functions to model real-life data. For instance, Exercise 74 on page 410 shows you how a tangent function can be used to model and analyze the distance between a television camera and a parade unit.

A. Ramey/PhotoEdit

Graph of the Tangent Function

Recall that the tangent function is odd. That is, $\tan(-x) = -\tan x$. Consequently, the graph of $y = \tan x$ is symmetric with respect to the origin. You also know from the identity $\tan x = \sin x/\cos x$ that the tangent is undefined for values at which $\cos x = 0$. Two such values are $x = \pm\pi/2 \approx \pm 1.5708$.

x	$-\frac{\pi}{2}$	-1.57	-1.5	-1	0	1	1.5	1.57	$\frac{\pi}{2}$
$\tan x$	Undef.	-1255.8	-14.1	-1.56	0	1.56	14.1	1255.8	Undef.

As indicated in the table, $\tan x$ increases without bound as x approaches $\pi/2$ from the left, and decreases without bound as x approaches $-\pi/2$ from the right. So, the graph of $y = \tan x$ has *vertical asymptotes* at $x = \pi/2$ and $x = -\pi/2$, as shown in Figure 4.49. Moreover, because the period of the tangent function is π, vertical asymptotes also occur when $x = \pi/2 + n\pi$, where n is an integer. The domain of the tangent function is the set of all real numbers other than $x = \pi/2 + n\pi$, and the range is the set of all real numbers.

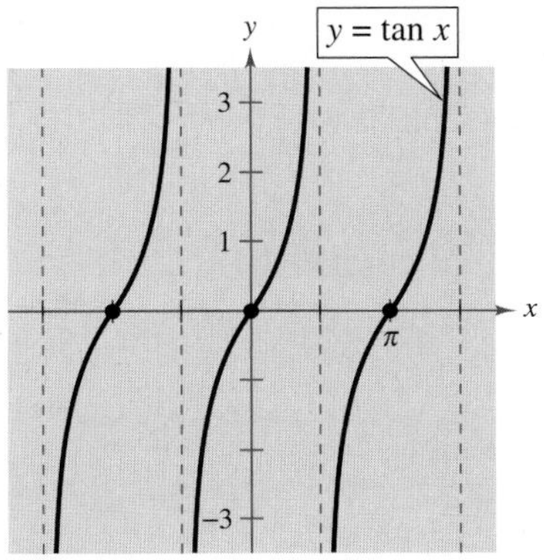

FIGURE 4.49

Period: π
Domain: all $x \neq \frac{\pi}{2} + n\pi$
Range: $(-\infty, \infty)$
Vertical asymptotes: $x = \frac{\pi}{2} + n\pi$

Sketching the graph of a function of the form $y = a\tan(bx - c)$ is similar to sketching the graph of $y = a\sin(bx - c)$ in that you locate key points that identify the intercepts and asymptotes. Two consecutive asymptotes can be found by solving the equations

$$bx - c = -\frac{\pi}{2} \quad \text{and} \quad bx - c = \frac{\pi}{2}.$$

The midpoint between two consecutive asymptotes is an x-intercept of the graph. The period of the function $y = a\tan(bx - c)$ is the distance between two consecutive asymptotes. The amplitude of a tangent function is not defined. After plotting the asymptotes and the x-intercept, plot a few additional points between the two asymptotes and sketch one cycle. Finally, sketch one or two additional cycles to the left and right.

Consider reviewing period, range, and domain for all six trigonometric functions, especially emphasizing the period difference in the tangent and cotangent functions.

Example 1 ▶ Sketching the Graph of a Tangent Function

Sketch the graph of $y = \tan \frac{x}{2}$.

Solution

By solving the equations

$$\frac{x}{2} = -\frac{\pi}{2} \quad \text{and} \quad \frac{x}{2} = \frac{\pi}{2}$$

$$x = -\pi \qquad\qquad x = \pi$$

you can see that two consecutive asymptotes occur at $x = -\pi$ and $x = \pi$. Between these two asymptotes, plot a few points, including the x-intercept, as shown in the table. Three cycles of the graph are shown in Figure 4.50.

x	$-\frac{\pi}{2}$	0	$\frac{\pi}{2}$
$\tan \frac{x}{2}$	-1	0	1

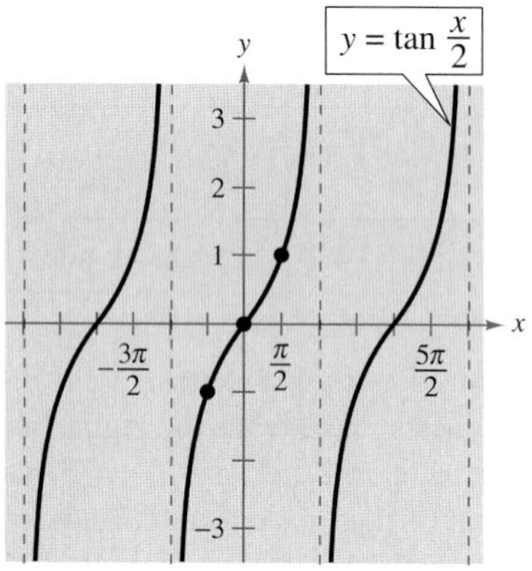

FIGURE 4.50

A computer animation of this example appears in the *Interactive* CD-ROM and *Internet* versions of this text.

Example 2 ▶ Sketching the Graph of a Tangent Function

Sketch the graph of $y = -3 \tan 2x$.

Solution

By solving the equations

$$2x = -\frac{\pi}{2} \quad \text{and} \quad 2x = \frac{\pi}{2}$$

$$x = -\frac{\pi}{4} \qquad\qquad x = \frac{\pi}{4}$$

you can see that two consecutive asymptotes occur at $x = -\pi/4$ and $x = \pi/4$. Between these two asymptotes, plot a few points, including the x-intercept, as shown in the table. Three cycles of the graph are shown in Figure 4.51.

x	$-\frac{\pi}{8}$	0	$\frac{\pi}{8}$
$-3 \tan 2x$	3	0	-3

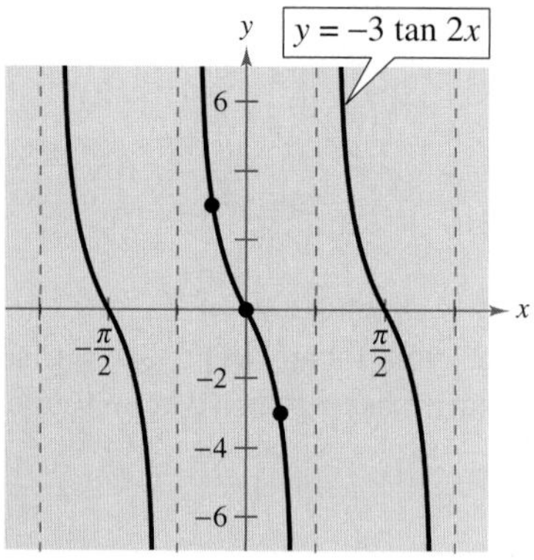

FIGURE 4.51

By comparing the graphs in Examples 1 and 2, you can see that the graph of $y = a \tan(bx - c)$ is increasing between consecutive vertical asymptotes if $a > 0$, and decreasing between consecutive vertical asymptotes if $a < 0$. In other words, the graph for $a < 0$ is a reflection in the x-axis of the graph for $a > 0$.

Graph of the Cotangent Function

The graph of the cotangent function is similar to the graph of the tangent function. It also has a period of π. However, from the identity

$$y = \cot x = \frac{\cos x}{\sin x}$$

you can see that the cotangent function has vertical asymptotes at $x = n\pi$, where n is an integer, because $\sin x$ is zero at these x-values. The graph of the cotangent function is shown in Figure 4.52.

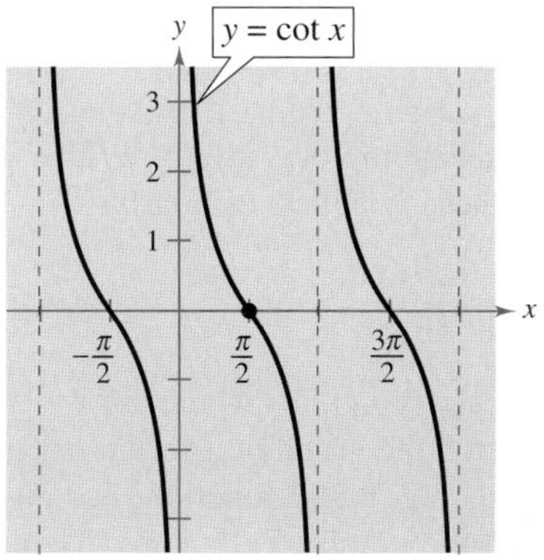

FIGURE 4.52

Period: π
Domain: *all* $x \neq n\pi$
Range: $(-\infty, \infty)$
Vertical asymptotes: $x = n\pi$

Example 3 ▶ Sketching the Graph of a Cotangent Function

Sketch the graph of $y = 2 \cot \dfrac{x}{3}$.

Solution

To locate two consecutive vertical asymptotes of the graph, solve the equations $x/3 = 0$ and $x/3 = \pi$, as follows.

$$\frac{x}{3} = 0 \quad \text{and} \quad \frac{x}{3} = \pi$$

$$x = 0 \qquad\qquad x = 3\pi$$

Then, between these two asymptotes, plot a few points, including the x-intercept, as shown in the table. Three cycles of the graph are shown in Figure 4.53. (Note that the period is 3π, the distance between consecutive asymptotes.)

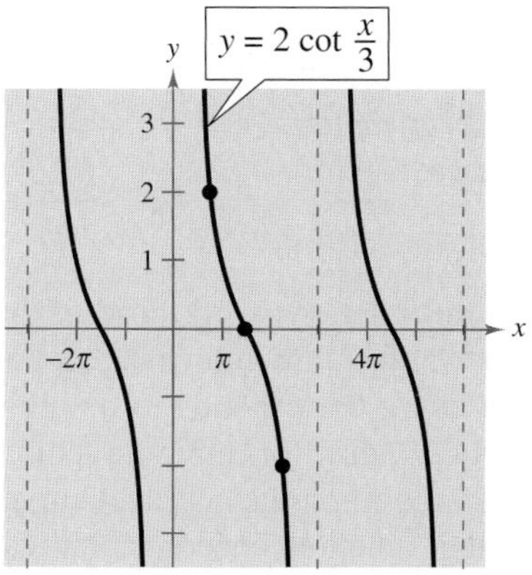

FIGURE 4.53

x	$\frac{3\pi}{4}$	$\frac{3\pi}{2}$	$\frac{9\pi}{4}$
$2 \cot \frac{x}{3}$	2	0	-2

Using a graphing utility to graph cosecant and secant functions can help reinforce the need to use the reciprocal functions.

Graphs of the Reciprocal Functions

The graphs of the two remaining trigonometric functions can be obtained from the graphs of the sine and cosine functions using the reciprocal identities

$$\csc x = \frac{1}{\sin x} \quad \text{and} \quad \sec x = \frac{1}{\cos x}.$$

For instance, at a given value of x, the y-coordinate of $\sec x$ is the reciprocal of the y-coordinate of $\cos x$. Of course, when $\cos x = 0$, the reciprocal does not exist. Near such values of x, the behavior of the secant function is similar to that of the tangent function. In other words, the graphs of

$$\tan x = \frac{\sin x}{\cos x} \quad \text{and} \quad \sec x = \frac{1}{\cos x}$$

have vertical asymptotes at $x = \pi/2 + n\pi$, where n is an integer, and the cosine is zero at these x-values. Similarly,

$$\cot x = \frac{\cos x}{\sin x} \quad \text{and} \quad \csc x = \frac{1}{\sin x}$$

have vertical asymptotes where $\sin x = 0$—that is, at $x = n\pi$.

To sketch the graph of a secant or cosecant function, you should first make a sketch of its reciprocal function. For instance, to sketch the graph of $y = \csc x$, first sketch the graph of $y = \sin x$. Then take reciprocals of the y-coordinates to obtain points on the graph of $y = \csc x$. This procedure is used to obtain the graphs shown in Figure 4.54.

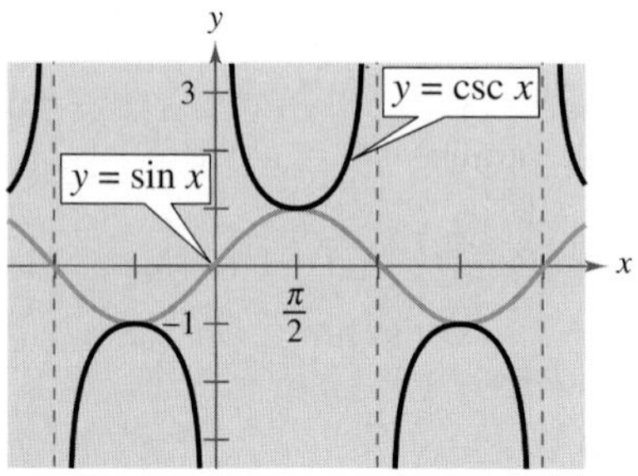

Period: 2π
Domain: all $x \neq n\pi$
Range: $(-\infty, -1]$ and $[1, \infty)$
Vertical asymptotes: $x = n\pi$
Symmetry: origin

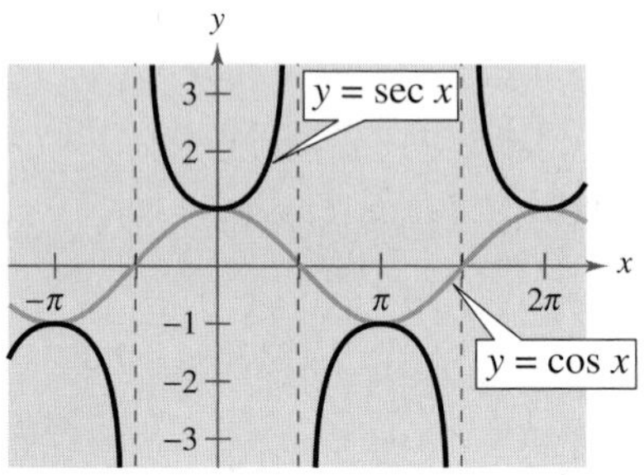

Period: 2π
Domain: all $x \neq \frac{\pi}{2} + n\pi$
Range: $(-\infty, -1]$ and $[1, \infty)$
Vertical asymptotes: $x = \frac{\pi}{2} + n\pi$
Symmetry: y-axis

FIGURE 4.54

In comparing the graphs of the secant and cosecant functions with those of the sine and cosine functions, note that the "hills" and "valleys" are interchanged. For example, a hill (or maximum point) on the sine curve corresponds to a valley (a local minimum) on the cosecant curve. Similarly, a valley (or minimum point) on the sine curve corresponds to a hill (a local maximum) on the cosecant curve. Additionally, x-intercepts of the sine and cosine functions become vertical asymptotes of the cosecant and secant functions, as shown in Figure 4.55.

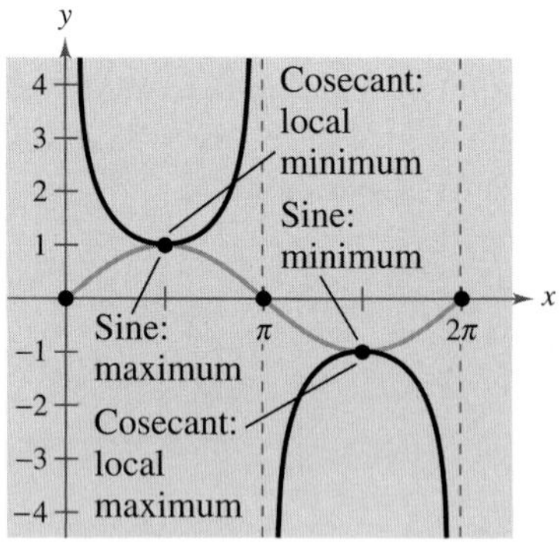

FIGURE 4.55

Example 4 ▶ Sketching the Graph of a Cosecant Function

Sketch the graph of $y = 2\csc\left(x + \frac{\pi}{4}\right)$.

Solution

Begin by sketching the graph of

$$y = 2\sin\left(x + \frac{\pi}{4}\right).$$

For this function, the amplitude is 2 and the period is 2π. By solving the equations

$$x + \frac{\pi}{4} = 0 \qquad \text{and} \qquad x + \frac{\pi}{4} = 2\pi$$

$$x = -\frac{\pi}{4} \qquad\qquad x = \frac{7\pi}{4}$$

you can see that one cycle of the sine function corresponds to the interval from $x = -\pi/4$ to $x = 7\pi/4$. The graph of this sine function is represented by the gray curve in Figure 4.56. Because the sine function is zero at the midpoint and endpoints of this interval, the corresponding cosecant function

$$y = 2\csc\left(x + \frac{\pi}{4}\right)$$

$$= 2\left(\frac{1}{\sin[x + (\pi/4)]}\right)$$

has vertical asymptotes at $x = -\pi/4$, $x = 3\pi/4$, $x = 7\pi/4$, etc. The graph of the cosecant function is represented by the black curve in Figure 4.56.

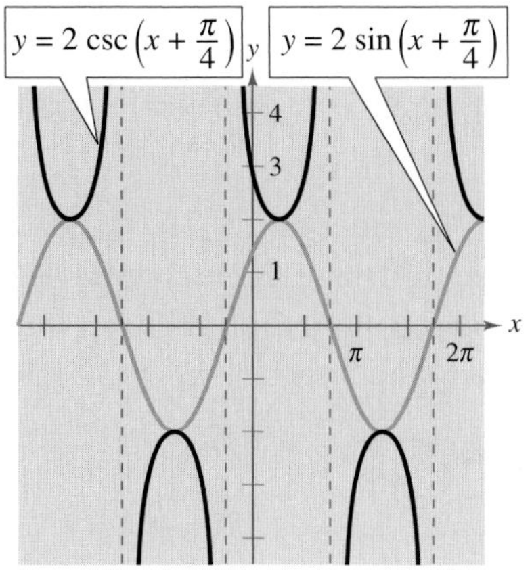

FIGURE 4.56

Example 5 ▶ Sketching the Graph of a Secant Function

Sketch the graph of $y = \sec 2x$.

Solution

Begin by sketching the graph of $y = \cos 2x$, as indicated by the gray curve in Figure 4.57. Then, form the graph of $y = \sec 2x$ as the black curve in the figure. Note that the x-intercepts of $y = \cos 2x$

$$\left(-\frac{\pi}{4}, 0\right), \quad \left(\frac{\pi}{4}, 0\right), \quad \left(\frac{3\pi}{4}, 0\right), \ldots$$

correspond to the vertical asymptotes

$$x = -\frac{\pi}{4}, \quad x = \frac{\pi}{4}, \quad x = \frac{3\pi}{4}, \ldots$$

of the graph of $y = \sec 2x$.

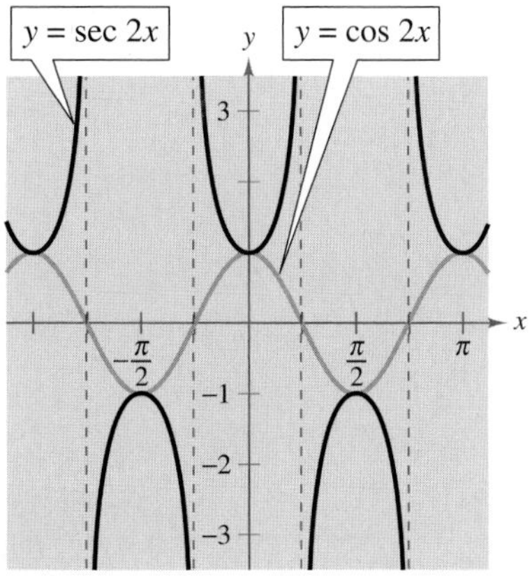

FIGURE 4.57

Damped Trigonometric Graphs

A *product* of two functions can be graphed using properties of the individual functions. For instance, consider the function

$$f(x) = x \sin x$$

as the product of the functions $y = x$ and $y = \sin x$. Using properties of absolute value and the fact that $|\sin x| \le 1$, you have $0 \le |x||\sin x| \le |x|$. Consequently,

$$-|x| \le x \sin x \le |x|$$

which means that the graph of $f(x) = x \sin x$ lies between the lines $y = -x$ and $y = x$. Furthermore, because

$$f(x) = x \sin x = \pm x \qquad \text{at} \qquad x = \frac{\pi}{2} + n\pi$$

and

$$f(x) = x \sin x = 0 \qquad \text{at} \qquad x = n\pi$$

the graph of f touches the line $y = -x$ or the line $y = x$ at $x = \pi/2 + n\pi$ and has x-intercepts at $x = n\pi$. A sketch of f is shown in Figure 4.58. In the function $f(x) = x \sin x$, the factor x is called the **damping factor.**

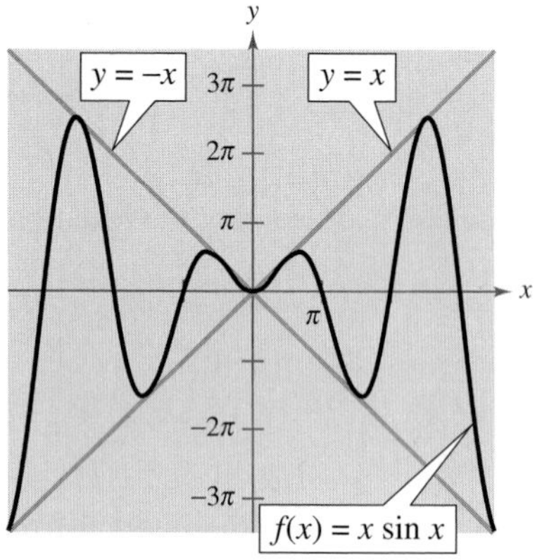

FIGURE 4.58

Example 6 ▶ Damped Sine Wave

Sketch the graph of

$$f(x) = e^{-x} \sin 3x.$$

Solution

Consider $f(x)$ as the product of the two functions

$$y = e^{-x} \qquad \text{and} \qquad y = \sin 3x$$

each of which has the set of real numbers as its domain. For any real number x, you know that $e^{-x} \ge 0$ and $|\sin 3x| \le 1$. Therefore, $e^{-x}|\sin 3x| \le e^{-x}$, which means that

$$-e^{-x} \le e^{-x} \sin 3x \le e^{-x}.$$

Furthermore, because

$$f(x) = e^{-x} \sin 3x = \pm e^{-x} \qquad \text{at} \qquad x = \frac{\pi}{6} + \frac{n\pi}{3}$$

and

$$f(x) = e^{-x} \sin 3x = 0 \qquad \text{at} \qquad x = \frac{n\pi}{3}$$

the graph of f touches the curves $y = -e^{-x}$ and $y = e^{-x}$ at $x = \pi/6 + n\pi/3$ and has intercepts at $x = n\pi/3$. A sketch is shown in Figure 4.59.

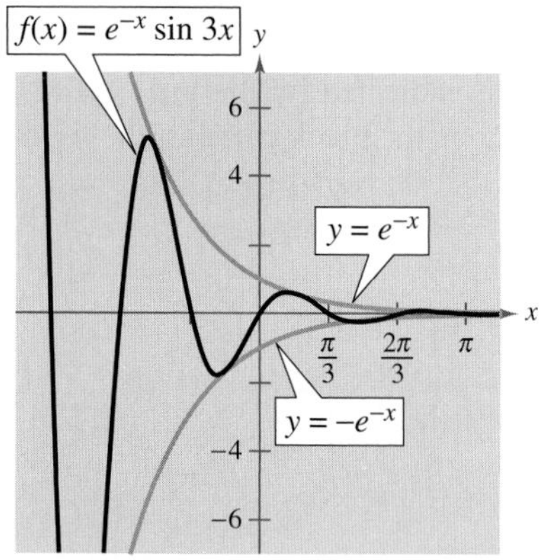

FIGURE 4.59

Figure 4.60 summarizes the six basic trigonometric functions.

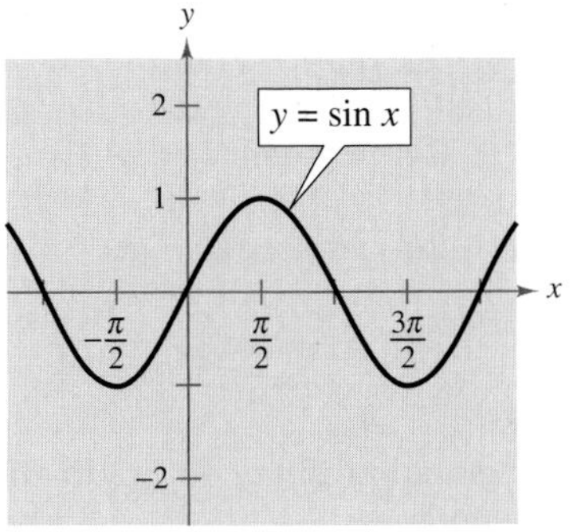

Domain: all reals
Range: $[-1, 1]$
Period: 2π

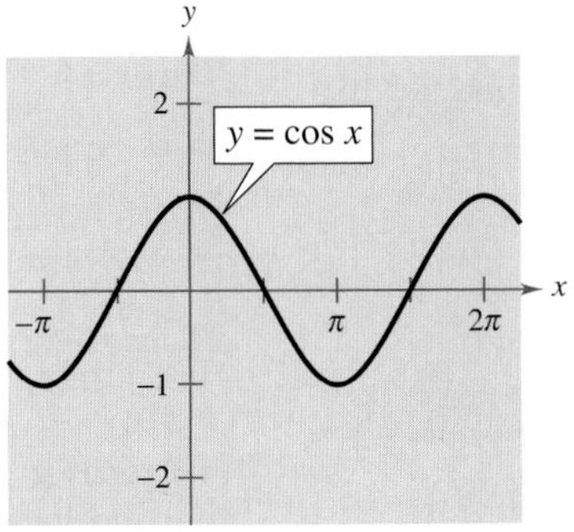

Domain: all reals
Range: $[-1, 1]$
Period: 2π

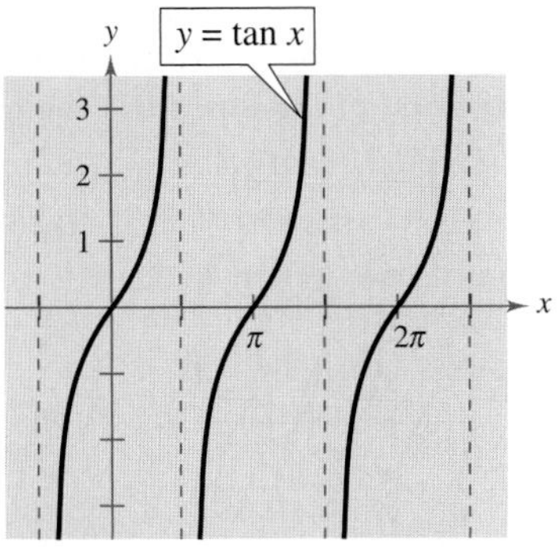

Domain: all $x \neq \frac{\pi}{2} + n\pi$
Range: $(-\infty, \infty)$
Period: π

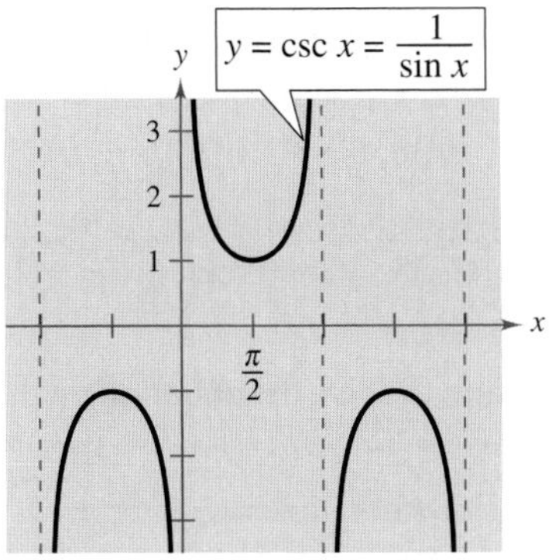

Domain: all $x \neq n\pi$
Range: $(-\infty, -1]$ *and* $[1, \infty)$
Period: 2π

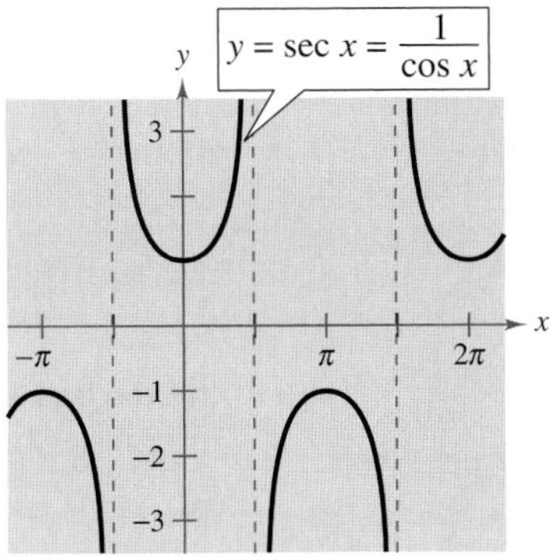

Domain: all $x \neq \frac{\pi}{2} + n\pi$
Range: $(-\infty, -1]$ *and* $[1, \infty)$
Period: 2π

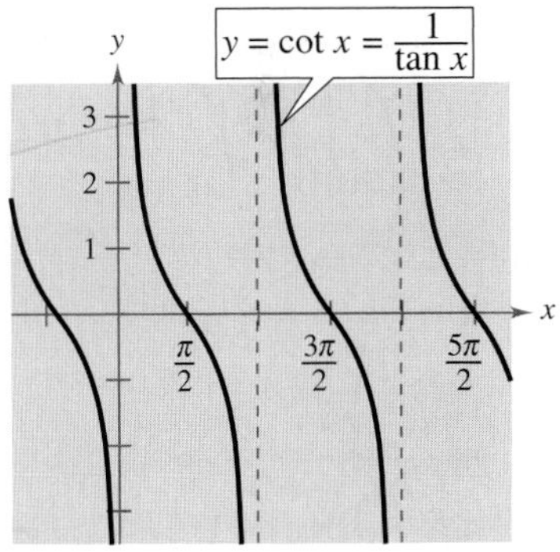

Domain: all $x \neq n\pi$
Range: $(-\infty, \infty)$
Period: π

FIGURE 4.60

Writing ABOUT MATHEMATICS

Combining Trigonometric Functions Recall from Section 1.6 that functions can be combined arithmetically. This also applies to trigonometric functions. For each of the functions

$$h(x) = x + \sin x \qquad \text{and} \qquad h(x) = \cos x - \sin 3x$$

(a) identify two simpler functions f and g that comprise the combination, (b) use a table to show how to obtain the numerical values of $h(x)$ from the numerical values of $f(x)$ and $g(x)$, and (c) use a graph of f and g to show how h may be formed.

Can you find functions

$$f(x) = d + a\sin(bx + c) \qquad \text{and} \qquad g(x) = d + a\cos(bx + c)$$

such that $f(x) + g(x) = 0$ for all x?

4.6 Exercises

In Exercises 1–6, match the function with its graph. State the period of the function. [The graphs are labeled (a), (b), (c), (d), (e), and (f).]

(a)

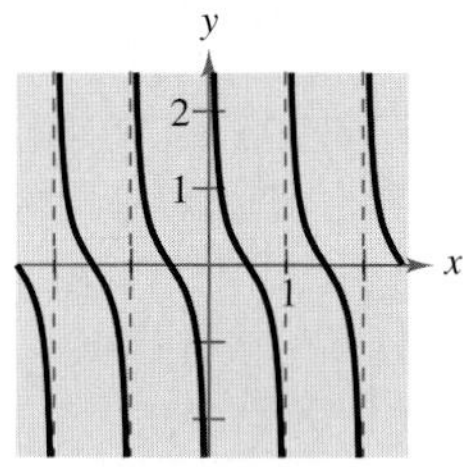

(b)

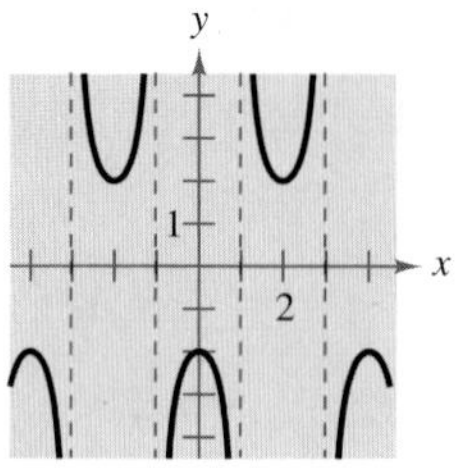

(c)

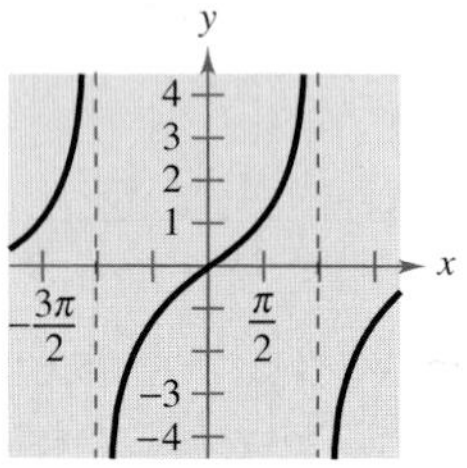

(d)

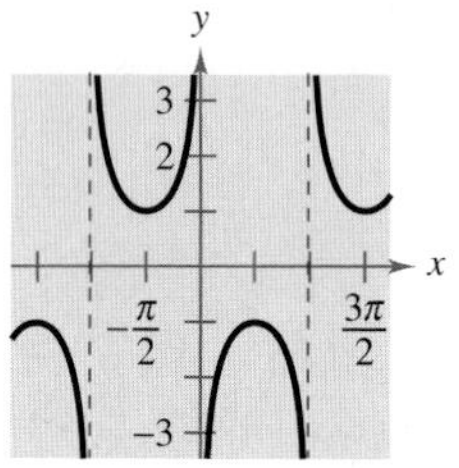

(e)

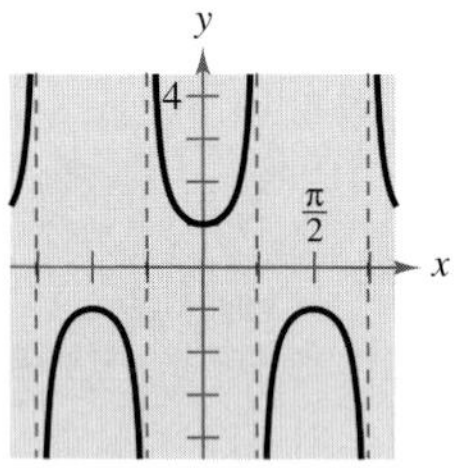

(f)

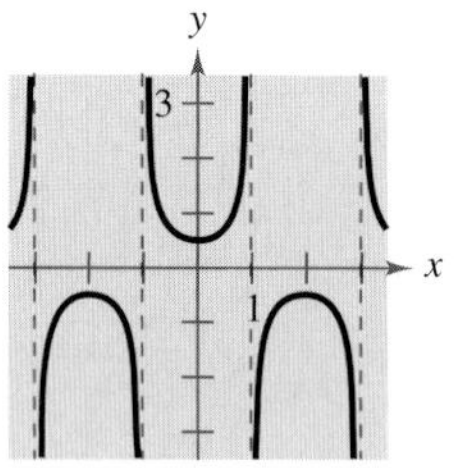

1. $y = \sec 2x$
2. $y = \tan \dfrac{x}{2}$
3. $y = \dfrac{1}{2} \cot \pi x$
4. $y = \dfrac{1}{2} \sec \dfrac{\pi x}{2}$
5. $y = -\csc x$
6. $y = -2 \sec \dfrac{\pi x}{2}$

In Exercises 7–28, sketch the graph of the function. Include two full periods.

7. $y = \frac{1}{3} \tan x$
8. $y = \frac{1}{4} \tan x$
9. $y = \tan 3x$
10. $y = -3 \tan \pi x$
11. $y = -\frac{1}{2} \sec x$
12. $y = \frac{1}{4} \sec x$
13. $y = \csc \pi x$
14. $y = 3 \csc 4x$
15. $y = \sec \pi x - 1$
16. $y = -2 \sec 4x + 2$
17. $y = \csc \dfrac{x}{2}$
18. $y = \csc \dfrac{x}{3}$
19. $y = \cot \dfrac{x}{2}$
20. $y = 3 \cot \dfrac{\pi x}{2}$
21. $y = \frac{1}{2} \sec 2x$
22. $y = -\frac{1}{2} \tan x$
23. $y = \tan \dfrac{\pi x}{4}$
24. $y = \tan(x + \pi)$
25. $y = \csc(\pi - x)$
26. $y = \sec(\pi - x)$
27. $y = \dfrac{1}{4} \csc\left(x + \dfrac{\pi}{4}\right)$
28. $y = 2 \cot\left(x + \dfrac{\pi}{2}\right)$

In Exercises 29–38, use a graphing utility to graph the function. Include two full periods.

29. $y = \tan \dfrac{x}{3}$
30. $y = -\tan 2x$
31. $y = -2 \sec 4x$
32. $y = \sec \pi x$
33. $y = \tan\left(x - \dfrac{\pi}{4}\right)$
34. $y = -\csc(4x - \pi)$
35. $y = \dfrac{1}{4} \cot\left(x - \dfrac{\pi}{2}\right)$
36. $y = 2 \sec(2x - \pi)$
37. $y = 0.1 \tan\left(\dfrac{\pi x}{4} + \dfrac{\pi}{4}\right)$
38. $y = \dfrac{1}{3} \sec\left(\dfrac{\pi x}{2} + \dfrac{\pi}{2}\right)$

In Exercises 39–46, use a graph to solve the equation on the interval $[-2\pi, 2\pi]$.

39. $\tan x = 1$
40. $\tan x = \sqrt{3}$
41. $\cot x = -\dfrac{\sqrt{3}}{3}$
42. $\cot x = 1$
43. $\sec x = -2$
44. $\sec x = 2$
45. $\csc x = \sqrt{2}$
46. $\csc x = -\dfrac{2\sqrt{3}}{3}$

In Exercises 47 and 48, use the graph of the function to determine whether the function is even, odd, or neither.

47. $f(x) = \sec x$
48. $f(x) = \tan x$

49. *Graphical Reasoning* Consider the functions

$$f(x) = 2 \sin x \quad \text{and} \quad g(x) = \frac{1}{2} \csc x$$

on the interval $(0, \pi)$.

(a) Graph f and g in the same coordinate plane.

(b) Approximate the interval where $f > g$.

(c) Describe the behavior of each of the functions as x approaches π. How is the behavior of g related to the behavior of f as x approaches π?

50. *Graphical Reasoning* Consider the functions

$$f(x) = \tan\frac{\pi x}{2} \quad \text{and} \quad g(x) = \frac{1}{2}\sec\frac{\pi x}{2}$$

on the interval $(-1, 1)$.

(a) Use a graphing utility to graph f and g in the same viewing window.

(b) Approximate the interval where $f < g$.

(c) Approximate the interval where $2f < 2g$. How does the result compare with that of part (b)? Explain.

In Exercises 51–54, use a graphing utility to graph the two equations in the same viewing window. Determine analytically whether the expressions are equivalent.

51. $y_1 = \sin x \csc x, \quad y_2 = 1$

52. $y_1 = \sin x \sec x, \quad y_2 = \tan x$

53. $y_1 = \dfrac{\cos x}{\sin x}, \quad y_2 = \cot x$

54. $y_1 = \sec^2 x - 1, \quad y_2 = \tan^2 x$

In Exercises 55–58, match the function with its graph. Describe the behavior of the function as x approaches zero. [The graphs are labeled (a), (b), (c), and (d).]

(a)

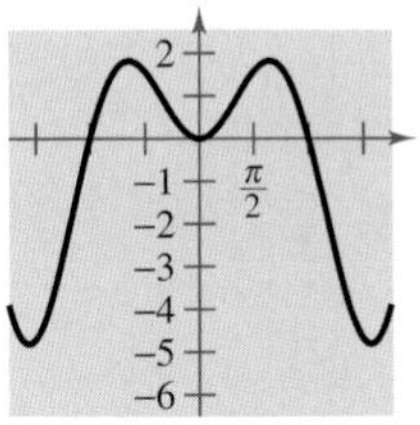

(b)

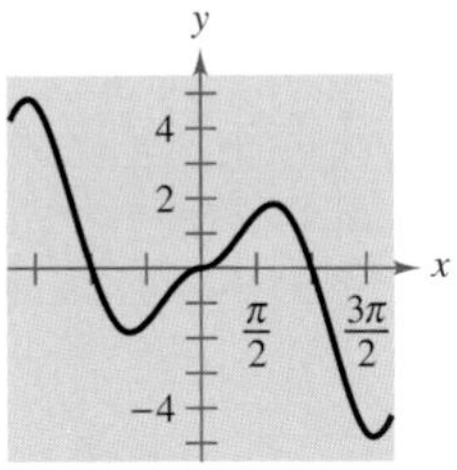

(c)

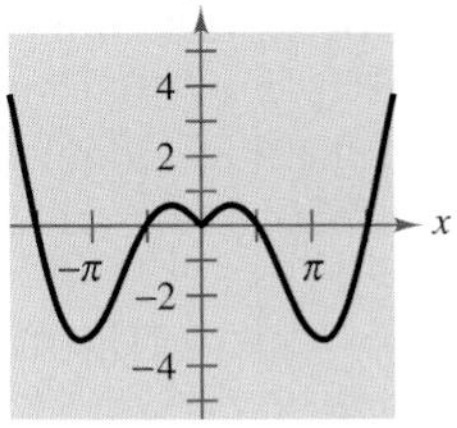

(d)

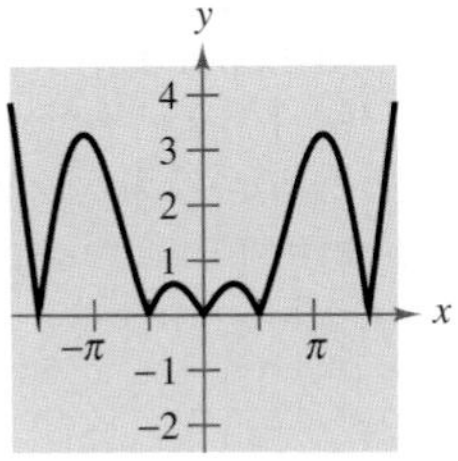

55. $f(x) = |x \cos x|$

56. $f(x) = x \sin x$

57. $g(x) = |x| \sin x$

58. $g(x) = |x| \cos x$

Conjecture **In Exercises 59–62, graph the functions f and g. Use the graphs to make a conjecture about the relationship between the functions.**

59. $f(x) = \sin x + \cos\left(x + \dfrac{\pi}{2}\right), \quad g(x) = 0$

60. $f(x) = \sin x - \cos\left(x + \dfrac{\pi}{2}\right), \quad g(x) = 2\sin x$

61. $f(x) = \sin^2 x, \quad g(x) = \frac{1}{2}(1 - \cos 2x)$

62. $f(x) = \cos^2 \dfrac{\pi x}{2}, \quad g(x) = \dfrac{1}{2}(1 + \cos \pi x)$

In Exercises 63–66, use a graphing utility to graph the function and the damping factor of the function in the same viewing window. Describe the behavior of the function as x increases without bound.

63. $f(x) = 2^{-x/4}\cos \pi x$

64. $f(x) = e^{-x}\cos x$

65. $g(x) = e^{-x^2/2}\sin x$

66. $h(x) = 2^{-x^2/4}\sin x$

Exploration **In Exercises 67–72, use a graphing utility to graph the function. Describe the behavior of the function as x approaches zero.**

67. $y = \dfrac{6}{x} + \cos x, \quad x > 0$

68. $y = \dfrac{4}{x} + \sin 2x, \quad x > 0$

69. $g(x) = \dfrac{\sin x}{x}$

70. $f(x) = \dfrac{1 - \cos x}{x}$

71. $f(x) = \sin\dfrac{1}{x}$

72. $h(x) = x\sin\dfrac{1}{x}$

73. *Distance* A plane flying at an altitude of 7 miles above a radar antenna will pass directly over the radar antenna (see figure). Let d be the ground distance from the antenna to the point directly under the plane and let x be the angle of elevation to the plane from the antenna. (d is positive as the plane approaches the antenna.) Write d as a function of x and graph the function over the interval $0 < x < \pi$.

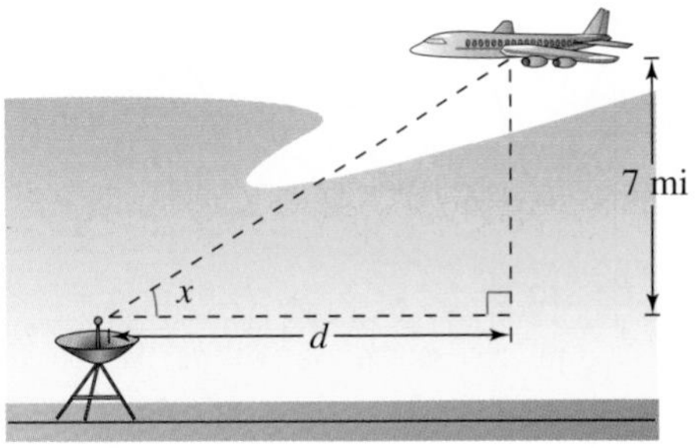

74. *Television Coverage* A television camera is on a reviewing platform 27 meters from the street on which a parade will be passing from left to right (see figure). Express the distance d from the camera to a particular unit in the parade as a function of the angle x, and graph the function over the interval $-\pi/2 < x < \pi/2$. (Consider x as negative when a unit in the parade approaches from the left.)

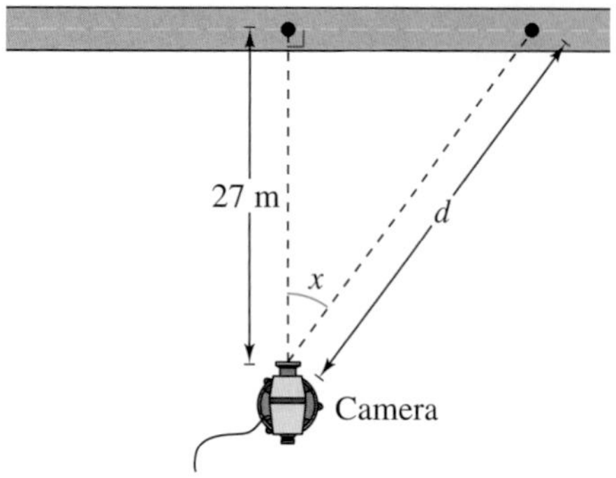

75. *Sales* The projected monthly sales S (in thousands of units) of a seasonal product are modeled by

$$S = 74 + 3t + 40 \sin \frac{\pi t}{6}$$

where t is the time in months, with $t = 1$ corresponding to January. Graph the sales function over 1 year.

76. *Predator-Prey Model* Suppose the population of a certain predator at time t (in months) in a given region is estimated to be

$$P = 10{,}000 + 3000 \sin \frac{2\pi t}{24}$$

and the population of its primary food source (its prey) is estimated to be

$$p = 15{,}000 + 5000 \cos \frac{2\pi t}{24}.$$

Use the graphs of the models to explain the oscillations in the size of each population.

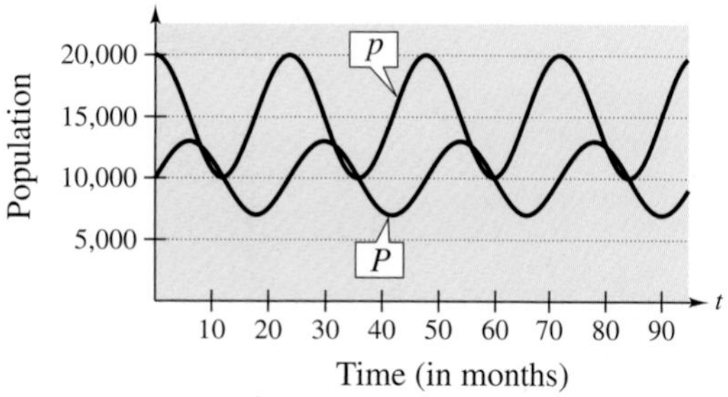

77. *Normal Temperatures* The normal monthly high temperatures in degrees Fahrenheit for Erie, Pennsylvania are approximated by

$$H(t) = 54.33 - 20.38 \cos \frac{\pi t}{6} - 15.69 \sin \frac{\pi t}{6}$$

and the normal monthly low temperatures are approximated by

$$L(t) = 39.36 - 15.70 \cos \frac{\pi t}{6} - 14.16 \sin \frac{\pi t}{6}$$

where t is the time in months, with $t = 1$ corresponding to January (see figure). (Source: National Oceanic and Atmospheric Administration)

(a) What is the period of each function?

(b) During what part of the year is the difference between the normal high and low temperatures greatest? When is it smallest?

(c) The sun is northernmost in the sky around June 21, but the graph shows the warmest temperatures at a later date. Approximate the lag time of the temperatures relative to the position of the sun.

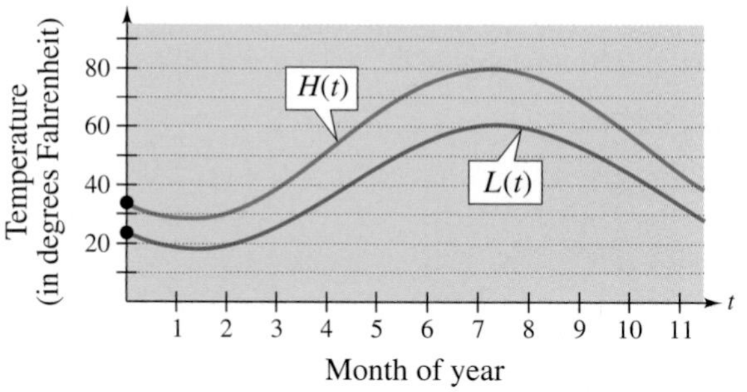

78. *Harmonic Motion* An object weighing W pounds is suspended from the ceiling by a steel spring (see figure). The weight is pulled downward (positive direction) from its equilibrium position and released. The resulting motion of the weight is described by the function

$$y = \frac{1}{2}e^{-t/4} \cos 4t, \qquad t > 0$$

where y is the distance in feet and t is the time in seconds.

(a) Use a graphing utility to graph the function.

(b) Describe the behavior of the displacement function for increasing values of time t.

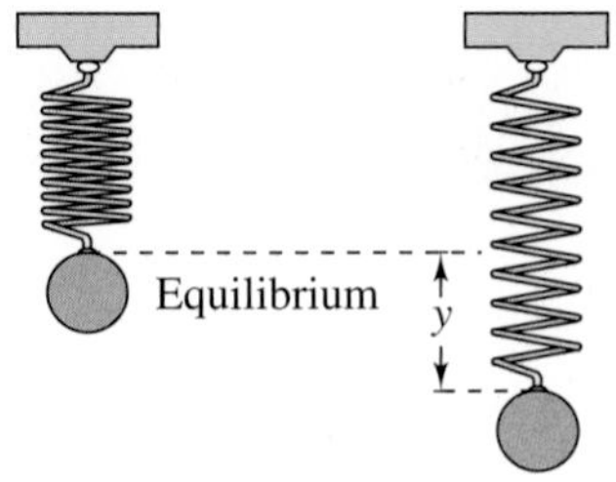

FIGURE FOR 78

Synthesis

True or False? **In Exercises 79 and 80, determine whether the statement is true or false. Justify your answer.**

79. The graph of $y = \csc x$ can be obtained on a calculator by graphing the reciprocal of $y = \sin x$.

80. The graph of $y = \sec x$ can be obtained on a calculator by graphing a translation of the reciprocal of $y = \sin x$.

81. *Writing* Describe the behavior of $f(x) = \tan x$ as x approaches $\pi/2$ from the left and from the right.

82. *Writing* Describe the behavior of $f(x) = \csc x$ as x approaches π from the left and from the right.

83. *Exploration* Consider the function

$$f(x) = x - \cos x.$$

(a) Use a graphing utility to graph the function and verify that there exists a zero between 0 and 1. Use the graph to approximate the zero.

(b) Starting with $x_0 = 1$, generate a sequence $x_1, x_2, x_3, \ldots$ where $x_n = \cos(x_{n-1})$. For example,

$$x_0 = 1$$
$$x_1 = \cos(x_0)$$
$$x_2 = \cos(x_1)$$
$$x_3 = \cos(x_2)$$
$$\vdots$$

Verify that the sequence approaches the zero of f.

84. *Approximation* Using calculus, it can be shown that the tangent function can be approximated by the polynomial

$$\tan x \approx x + \frac{2x^3}{3!} + \frac{16x^5}{5!}$$

where x is in radians. Use a graphing utility to graph the tangent function and its polynomial approximation in the same viewing window. How do the graphs compare?

85. *Approximation* Using calculus, it can be shown that the secant function can be approximated by the polynomial

$$\sec x \approx 1 + \frac{x^2}{2!} + \frac{5x^4}{4!}$$

where x is in radians. Use a graphing utility to graph the secant function and its polynomial approximation in the same viewing window. How do the graphs compare?

86. *Pattern Recognition*

(a) Use a graphing utility to graph each function.

$$y_1 = \frac{4}{\pi}\left(\sin \pi x + \frac{1}{3}\sin 3\pi x\right)$$

$$y_2 = \frac{4}{\pi}\left(\sin \pi x + \frac{1}{3}\sin 3\pi x + \frac{1}{5}\sin 5\pi x\right)$$

(b) Identify the pattern started in part (a) and find a function y_3 that continues the pattern one more term. Use a graphing utility to graph y_3.

(c) The graphs in parts (a) and (b) approximate the periodic function in the figure. Find a function y_4 that is a better approximation.

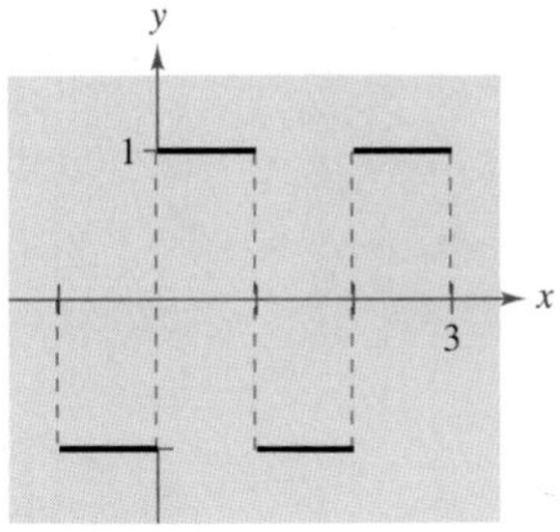

Review

In Exercises 87–90, solve the exponential equation. Round your answer to three decimal places.

87. $e^{2x} = 54$

88. $8^{3x} = 98$

89. $\dfrac{300}{1 + e^{-x}} = 100$

90. $\left(1 + \dfrac{0.15}{365}\right)^{365t} = 5$

In Exercises 91–96, solve the logarithmic equation. Round your answer to three decimal places.

91. $\ln(3x - 2) = 73$

92. $\ln(14 - 2x) = 68$

93. $\ln(x^2 + 1) = 3.2$

94. $\ln \sqrt{x + 4} = 5$

95. $\log_8 x + \log_8(x - 1) = \frac{1}{3}$

96. $\log_6 x + \log_6(x^2 - 1) = \log_6 64x$

4.7 Inverse Trigonometric Functions

▶ **What you should learn**

- How to evaluate the inverse sine function
- How to evaluate the other inverse trigonometric functions
- How to evaluate the compositions of trigonometric functions

▶ **Why you should learn it**

You can use inverse trigonometric functions to model and solve real-life problems. For instance, Exercise 91 on page 419 shows how an inverse trigonometric function can be used to model the angle of elevation from a boat to a winch and the length of the rope joining them.

Arlene Collins/The Image Works

Inverse Sine Function

Recall from Section 1.7 that, for a function to have an inverse, it must pass the Horizontal Line Test. From Figure 4.61 you can see that $y = \sin x$ does not pass the test because different values of x yield the same y-value.

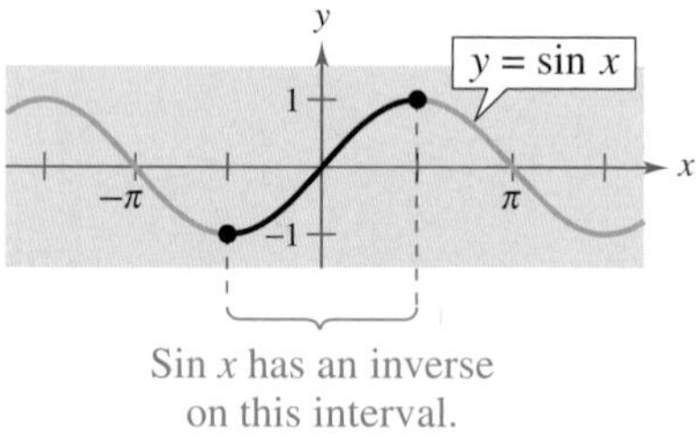

FIGURE 4.61

However, if you restrict the domain to the interval $-\pi/2 \le x \le \pi/2$ (corresponding to the black portion of the graph in Figure 4.61), the following properties hold.

1. On the interval $[-\pi/2, \pi/2]$, the function $y = \sin x$ is increasing.
2. On the interval $[-\pi/2, \pi/2]$, $y = \sin x$ takes on its full range of values, $-1 \le \sin x \le 1$.
3. On the interval $[-\pi/2, \pi/2]$, $y = \sin x$ passes the Horizontal Line Test.

So, on the restricted domain $-\pi/2 \le x \le \pi/2$, $y = \sin x$ has a unique inverse called the **inverse sine function.** It is denoted by

$$y = \arcsin x \qquad \text{or} \qquad y = \sin^{-1} x.$$

The notation $\sin^{-1} x$ is consistent with the inverse function notation $f^{-1}(x)$. The arcsin x notation (read as "the arcsine of x") comes from the association of a central angle with its subtended *arc length* on a unit circle. So, arcsin x means the angle (or arc) whose sine is x. Both notations, arcsin x and $\sin^{-1} x$, are commonly used in mathematics, so remember that $\sin^{-1} x$ denotes the *inverse* sine function rather than $1/\sin x$. The values of arcsin x lie in the interval $-\pi/2 \le \arcsin x \le \pi/2$. The graph of $y = \arcsin x$ is shown in Figure 4.62.

Definition of Inverse Sine Function

The **inverse sine function** is defined by

$$y = \arcsin x \qquad \text{if and only if} \qquad \sin y = x$$

where $-1 \le x \le 1$ and $-\pi/2 \le y \le \pi/2$. The domain of $y = \arcsin x$ is $[-1, 1]$, and the range is $[-\pi/2, \pi/2]$.

When evaluating the inverse sine function, it helps to remember the phrase "the arcsine of x is the angle (or number) whose sine is x."

STUDY TIP

As with the trigonometric functions, much of the work with the inverse trigonometric functions can be done by *exact* calculations rather than by calculator approximations. Exact calculations help to increase your understanding of the inverse functions by relating them to the triangle definitions of the trigonometric functions.

You may need to remind your students that the values of the inverse sine function are always in radians.

You may wish to illustrate the reflections of $y = \sin x$ and $y = \arcsin x$ about the line $y = x$. Consider using a graphing utility to do this.

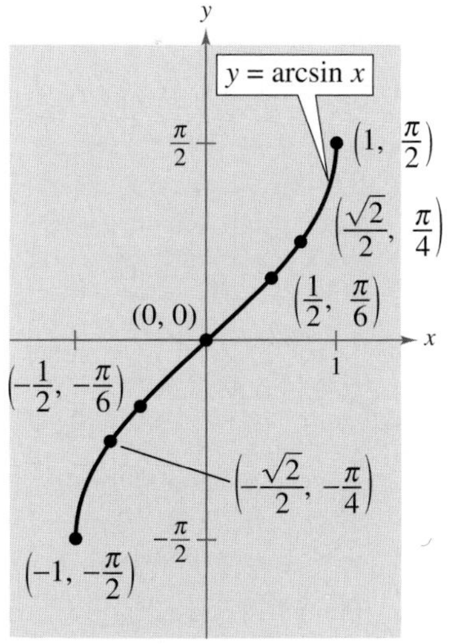

FIGURE 4.62

Example 1 ▶ Evaluating the Inverse Sine Function

If possible, find the exact value.

a. $\arcsin\left(-\frac{1}{2}\right)$ **b.** $\sin^{-1}\frac{\sqrt{3}}{2}$ **c.** $\sin^{-1}2$

Solution

a. Because $\sin(-\pi/6) = -\frac{1}{2}$ for $-\pi/2 \le y \le \pi/2$, it follows that

$$\arcsin\left(-\frac{1}{2}\right) = -\frac{\pi}{6}. \qquad \text{Angle whose sine is } -\tfrac{1}{2}$$

b. Because $\sin(\pi/3) = \sqrt{3}/2$ for $-\pi/2 \le y \le \pi/2$, it follows that

$$\sin^{-1}\frac{\sqrt{3}}{2} = \frac{\pi}{3}. \qquad \text{Angle whose sine is } \sqrt{3}/2$$

c. It is not possible to evaluate $y = \sin^{-1}x$ when $x = 2$ because there is no angle whose sine is 2. Remember that the domain of the inverse sine function is $[-1, 1]$.

Example 2 ▶ Graphing the Arcsine Function

Sketch a graph of $y = \arcsin x$.

Solution

By definition, the equations

$$y = \arcsin x \quad \text{and} \quad \sin y = x$$

are equivalent for $-\pi/2 \le y \le \pi/2$. So, their graphs are the same. From the interval $[-\pi/2, \pi/2]$, you can assign values to y in the second equation to make a table of values.

y	$-\frac{\pi}{2}$	$-\frac{\pi}{4}$	$-\frac{\pi}{6}$	0	$\frac{\pi}{6}$	$\frac{\pi}{4}$	$\frac{\pi}{2}$
$x = \sin y$	-1	$-\frac{\sqrt{2}}{2}$	$-\frac{1}{2}$	0	$\frac{1}{2}$	$\frac{\sqrt{2}}{2}$	1

The resulting graph for

$$y = \arcsin x$$

is shown in Figure 4.62. Note that it is the reflection (in the line $y = x$) of the black portion of the graph in Figure 4.61. Be sure you see that Figure 4.62 shows the *entire* graph of the inverse sine function. Remember that the range of $y = \arcsin x$ is the closed interval $[-\pi/2, \pi/2]$.

Other Inverse Trigonometric Functions

The cosine function is decreasing on the interval $0 \le x \le \pi$, as shown in Figure 4.63.

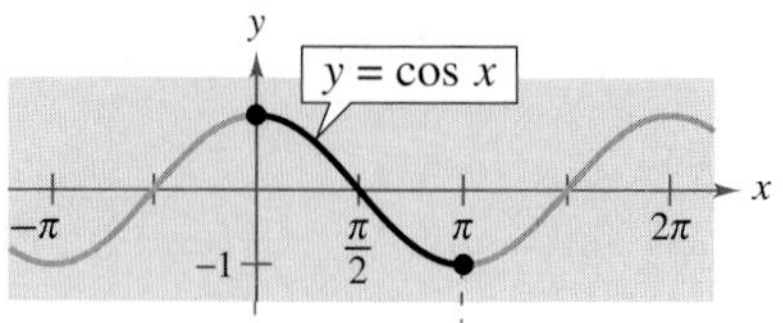

FIGURE 4.63

Consequently, on this interval the cosine function has an inverse function—the **inverse cosine function**—denoted by

$$y = \arccos x \qquad \text{or} \qquad y = \cos^{-1} x.$$

Similarly, you can define an **inverse tangent function** by restricting the domain of $y = \tan x$ to the interval $(-\pi/2, \pi/2)$. The following list summarizes the definitions of the three most common inverse trigonometric functions. The remaining three are defined in Exercises 99–101.

You may need to point out to your students that the range for each of these functions is different. Students should know these ranges well to ensure that their answers are within the correct range.

Definitions of the Inverse Trigonometric Functions

Function	*Domain*	*Range*
$y = \arcsin x$ if and only if $\sin y = x$	$-1 \le x \le 1$	$-\frac{\pi}{2} \le y \le \frac{\pi}{2}$
$y = \arccos x$ if and only if $\cos y = x$	$-1 \le x \le 1$	$0 \le y \le \pi$
$y = \arctan x$ if and only if $\tan y = x$	$-\infty < x < \infty$	$-\frac{\pi}{2} < y < \frac{\pi}{2}$

The graphs of these three inverse trigonometric functions are shown in Figure 4.64.

A computer animation of this concept appears in the *Interactive* CD-ROM and *Internet* versions of this text.

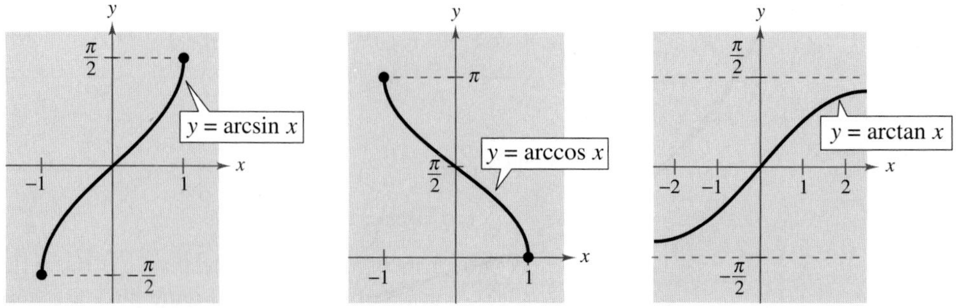

Domain: $[-1, 1]$
Range: $\left[-\frac{\pi}{2}, \frac{\pi}{2}\right]$

Domain: $[-1, 1]$
Range: $[0, \pi]$

Domain: $(-\infty, \infty)$
Range: $\left(-\frac{\pi}{2}, \frac{\pi}{2}\right)$

FIGURE 4.64

Example 3 ▶ Evaluating Inverse Trigonometric Functions

Find the exact value.

a. $\arccos \dfrac{\sqrt{2}}{2}$ **b.** $\arccos(-1)$ **c.** $\arctan 0$ **d.** $\arctan(-1)$

Solution

a. Because $\cos(\pi/4) = \sqrt{2}/2$, and $\pi/4$ lies in $[0, \pi]$, it follows that

$$\arccos \frac{\sqrt{2}}{2} = \frac{\pi}{4}. \qquad \text{Angle whose cosine is } \sqrt{2}/2$$

b. Because $\cos \pi = -1$, and π lies in $[0, \pi]$, it follows that

$$\arccos(-1) = \pi. \qquad \text{Angle whose cosine is } -1$$

c. Because $\tan 0 = 0$, and 0 lies in $(-\pi/2, \pi/2)$, it follows that

$$\arctan 0 = 0. \qquad \text{Angle whose tangent is } 0$$

d. Because $\tan(-\pi/4) = -1$, and $-\pi/4$ lies in $(-\pi/2, \pi/2)$, it follows that

$$\arctan(-1) = -\frac{\pi}{4}. \qquad \text{Angle whose tangent is } -1$$

Example 4 ▶ Calculators and Inverse Trigonometric Functions

Use a calculator to approximate the value (if possible).

a. $\arctan(-8.45)$

b. $\arcsin 0.2447$

c. $\arccos 2$

Solution

Function	*Mode*	*Calculator Keystrokes*
a. $\arctan(-8.45)$	Radian	[TAN⁻¹] [(] [(−)] 8.45 [)] [ENTER]

From the display, it follows that $\arctan(-8.45) \approx -1.453001$.

b. $\arcsin 0.2447$	Radian	[SIN⁻¹] 0.2447 [ENTER]

From the display, it follows that $\arcsin 0.2447 \approx 0.2472103$.

c. $\arccos 2$	Radian	[COS⁻¹] 2 [ENTER]

In real number mode, the calculator should display an *error message* because the domain of the inverse cosine function is $[-1, 1]$.

STUDY TIP

It is important to remember that the domain of the inverse sine function and the inverse cosine function is $[-1, 1]$, as shown in Example 4(c).

In Example 4, if you had set the calculator to degree mode, the display would have been in degrees rather than radians. This convention is peculiar to calculators. By definition, the values of inverse trigonometric functions are *always in radians.*

Compositions of Functions

Recall from Section 1.7 that inverse functions have the properties

$$f(f^{-1}(x)) = x \qquad \text{and} \qquad f^{-1}(f(x)) = x$$

for all x in the domain of f and f^{-1}.

Inverse Properties of Trigonometric Functions

If $-1 \le x \le 1$ and $-\pi/2 \le y \le \pi/2$, then

$$\sin(\arcsin x) = x \qquad \text{and} \qquad \arcsin(\sin y) = y.$$

If $-1 \le x \le 1$ and $0 \le y \le \pi$, then

$$\cos(\arccos x) = x \qquad \text{and} \qquad \arccos(\cos y) = y.$$

If x is a real number and $-\pi/2 < y < \pi/2$, then

$$\tan(\arctan x) = x \qquad \text{and} \qquad \arctan(\tan y) = y.$$

Activities

1. Evaluate $\arccos\left(-\frac{\sqrt{3}}{2}\right)$.

   Answer: $\frac{5\pi}{6}$

2. Use a calculator to evaluate arctan 3.2.

   Answer: 1.268

3. Write an algebraic expression that is equivalent to $\sin(\arctan 3x)$.

   Answer: $\frac{3x}{\sqrt{1 + 9x^2}}$

Keep in mind that these inverse properties do not apply for arbitrary values of x and y. For instance,

$$\arcsin\left(\sin\frac{3\pi}{2}\right) = \arcsin(-1) = -\frac{\pi}{2} \neq \frac{3\pi}{2}.$$

In other words, the property

$$\arcsin(\sin y) = y$$

is not valid for values of y outside the interval $[-\pi/2, \pi/2]$.

Example 5 ▶ Using Inverse Properties

If possible, find the exact value.

a. $\tan[\arctan(-5)]$ **b.** $\arcsin\left(\sin\frac{5\pi}{3}\right)$ **c.** $\cos(\cos^{-1}\pi)$

Solution

a. Because -5 lies in the domain of the arctan function, the inverse property applies, and you have

$$\tan[\arctan(-5)] = -5.$$

b. In this case, $5\pi/3$ does not lie within the range of the arcsine function, $-\pi/2 \le y \le \pi/2$. However, $5\pi/3$ is coterminal with

$$\frac{5\pi}{3} - 2\pi = -\frac{\pi}{3}$$

which does lie in the range of the arcsine function, and you have

$$\arcsin\left(\sin\frac{5\pi}{3}\right) = \arcsin\left[\sin\left(-\frac{\pi}{3}\right)\right] = -\frac{\pi}{3}.$$

c. The expression $\cos(\cos^{-1}\pi)$ is not defined because $\cos^{-1}\pi$ is not defined. Remember that the domain of the inverse cosine function is $[-1, 1]$.

Example 6 shows how to use right triangles to find exact values of compositions of inverse functions. Then, Example 7 shows how to use triangles to convert a trigonometric expression into an algebraic expression. This conversion technique is used frequently in calculus.

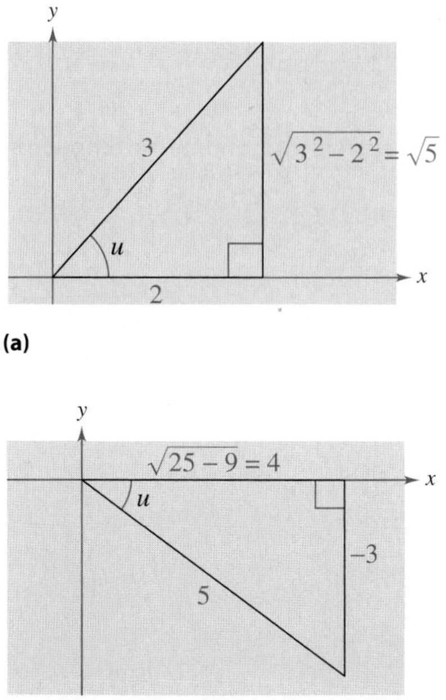

FIGURE 4.65

Example 6 ▶ Evaluating Compositions of Functions

Find the exact value.

a. $\tan\left(\arccos\dfrac{2}{3}\right)$ **b.** $\cos\left[\arcsin\left(-\dfrac{3}{5}\right)\right]$

Solution

a. If you let $u = \arccos\frac{2}{3}$, then $\cos u = \frac{2}{3}$. Because $\cos u$ is positive, u is a *first*-quadrant angle. You can sketch and label angle u as shown in Figure 4.65(a). Consequently,

$$\tan\left(\arccos\frac{2}{3}\right) = \tan u = \frac{\text{opp}}{\text{adj}} = \frac{\sqrt{5}}{2}.$$

b. If you let $u = \arcsin\left(-\frac{3}{5}\right)$, then $\sin u = -\frac{3}{5}$. Because $\sin u$ is negative, u is a *fourth*-quadrant angle. You can sketch and label angle u as shown in Figure 4.65(b). Consequently,

$$\cos\left[\arcsin\left(-\frac{3}{5}\right)\right] = \cos u = \frac{\text{adj}}{\text{hyp}} = \frac{4}{5}.$$

Example 7 ▶ Some Problems from Calculus

Write each of the following as an algebraic expression in x.

a. $\sin(\arccos 3x),\quad 0 \le x \le \dfrac{1}{3}$ **b.** $\cot(\arccos 3x),\quad 0 \le x \le \dfrac{1}{3}$

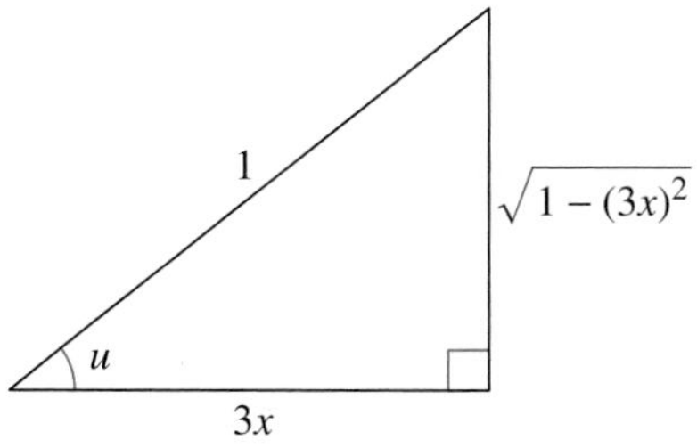

FIGURE 4.66

Solution

If you let $u = \arccos 3x$, then $\cos u = 3x$. Because

$$\cos u = \frac{3x}{1} = \frac{\text{adj}}{\text{hyp}}$$

you can sketch a right triangle with acute angle u, as shown in Figure 4.66. From this triangle, you can easily convert each expression to algebraic form.

a. $\sin(\arccos 3x) = \sin u = \dfrac{\text{opp}}{\text{hyp}} = \sqrt{1 - 9x^2},\quad 0 \le x \le \dfrac{1}{3}$

b. $\cot(\arccos 3x) = \cot u = \dfrac{\text{adj}}{\text{opp}} = \dfrac{3x}{\sqrt{1 - 9x^2}},\quad 0 \le x \le \dfrac{1}{3}$

In Example 7, similar arguments can be made for x-values lying in the interval $\left[-\frac{1}{3}, 0\right]$.

4.7 Exercises

In Exercises 1–16, evaluate the expression without the aid of a calculator.

1. $\arcsin \frac{1}{2}$
2. $\arcsin 0$
3. $\arccos \frac{1}{2}$
4. $\arccos 0$
5. $\arctan \dfrac{\sqrt{3}}{3}$
6. $\arctan(-1)$
7. $\arccos\left(-\dfrac{\sqrt{3}}{2}\right)$
8. $\arcsin\left(-\dfrac{\sqrt{2}}{2}\right)$
9. $\arctan(-\sqrt{3})$
10. $\arctan \sqrt{3}$
11. $\arccos\left(-\dfrac{1}{2}\right)$
12. $\arcsin \dfrac{\sqrt{2}}{2}$
13. $\arcsin \dfrac{\sqrt{3}}{2}$
14. $\arctan\left(-\dfrac{\sqrt{3}}{3}\right)$
15. $\arctan 0$
16. $\arccos 1$

In Exercises 17–32, use a calculator to approximate the expression. Round your result to two decimal places.

17. $\arccos 0.28$
18. $\arcsin 0.45$
19. $\arcsin(-0.75)$
20. $\arccos(-0.7)$
21. $\arctan(-3)$
22. $\arctan 15$
23. $\arcsin 0.31$
24. $\arccos 0.26$
25. $\arccos(-0.41)$
26. $\arcsin(-0.125)$
27. $\arctan 0.92$
28. $\arctan 2.8$
29. $\arcsin \frac{3}{4}$
30. $\arccos\left(-\frac{1}{3}\right)$
31. $\arctan \frac{7}{2}$
32. $\arctan\left(-\frac{95}{7}\right)$

In Exercises 33 and 34, determine the missing coordinates of the points on the graph of the function.

33.

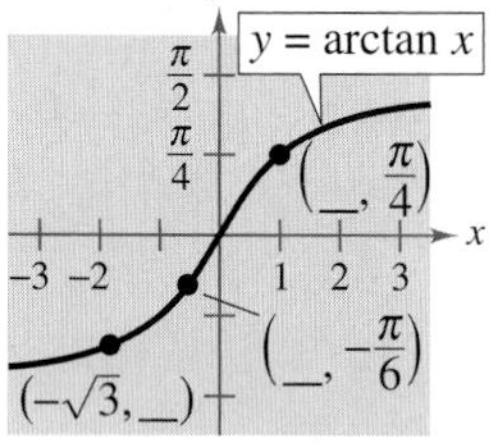

34.

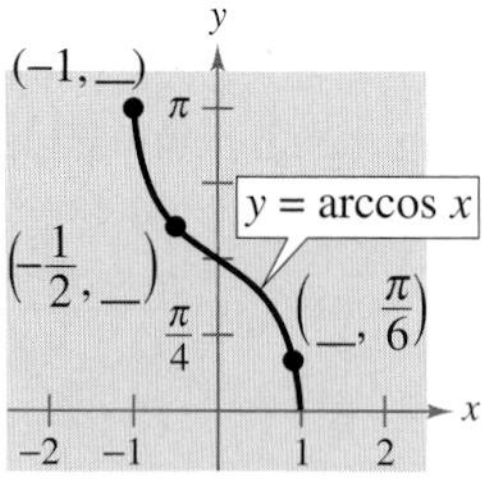

In Exercises 35 and 36, use a graphing utility to graph *f*, *g*, and *y* = *x* in the same viewing window to verify geometrically that *g* is the inverse of *f*. (Be sure to restrict the domain of *f* properly.)

35. $f(x) = \tan x, \quad g(x) = \arctan x$
36. $f(x) = \sin x, \quad g(x) = \arcsin x$

In Exercises 37–42, use an inverse trigonometric function to write θ as a function of *x*.

37.

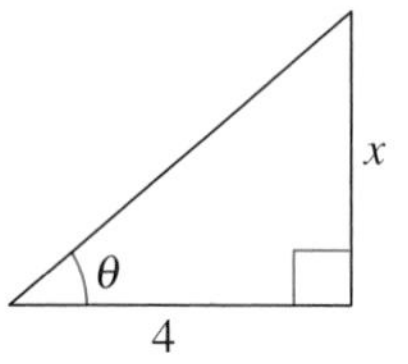

38.

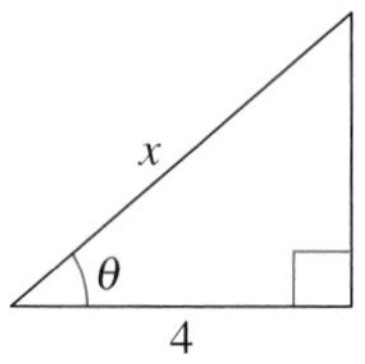

39.

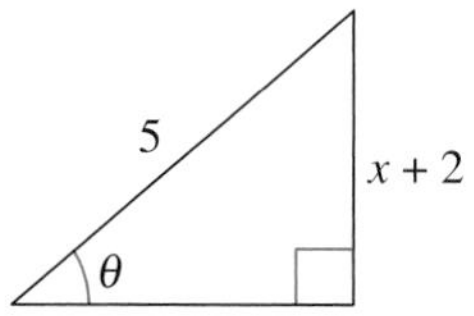

40.

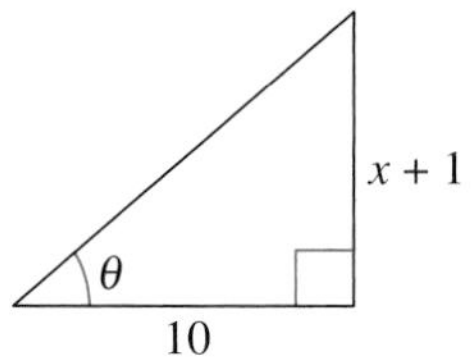

41.

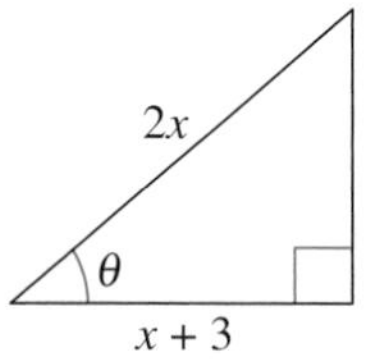

42.

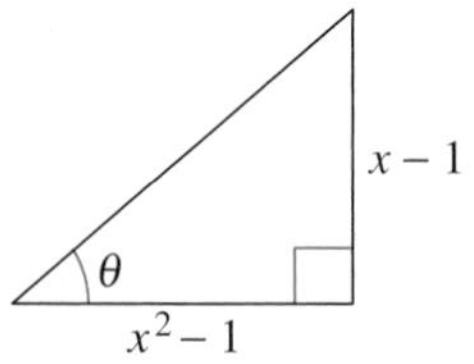

In Exercises 43–48, use the properties of inverse functions to evaluate the expression.

43. $\sin(\arcsin 0.3)$
44. $\tan(\arctan 25)$
45. $\cos[\arccos(-0.1)]$
46. $\sin[\arcsin(-0.2)]$
47. $\arcsin(\sin 3\pi)$
48. $\arccos\left(\cos \dfrac{7\pi}{2}\right)$

In Exercises 49–58, find the exact value of the expression. (*Hint:* Make a sketch of a right triangle.)

49. $\sin\left(\arctan \frac{3}{4}\right)$
50. $\sec\left(\arcsin \frac{4}{5}\right)$
51. $\cos(\arctan 2)$
52. $\sin\left(\arccos \dfrac{\sqrt{5}}{5}\right)$
53. $\cos\left(\arcsin \frac{5}{13}\right)$
54. $\csc\left[\arctan\left(-\frac{5}{12}\right)\right]$
55. $\sec\left[\arctan\left(-\frac{3}{5}\right)\right]$
56. $\tan\left[\arcsin\left(-\frac{3}{4}\right)\right]$
57. $\sin\left[\arccos\left(-\frac{2}{3}\right)\right]$
58. $\cot\left(\arctan \frac{5}{8}\right)$

In Exercises 59–68, write an algebraic expression that is equivalent to the expression. (*Hint:* Sketch a right triangle, as demonstrated in Example 7.)

59. $\cot(\arctan x)$

60. $\sin(\arctan x)$

61. $\cos(\arcsin 2x)$

62. $\sec(\arctan 3x)$

63. $\sin(\arccos x)$

64. $\sec[\arcsin(x - 1)]$

65. $\tan\left(\arccos \dfrac{x}{3}\right)$

66. $\cot\left(\arctan \dfrac{1}{x}\right)$

67. $\csc\left(\arctan \dfrac{x}{\sqrt{2}}\right)$

68. $\cos\left(\arcsin \dfrac{x - h}{r}\right)$

In Exercises 69 and 70, use a graphing utility to graph *f* and *g* in the same viewing window to verify that the two are equal. Explain why they are equal. Identify any asymptotes of the graphs.

69. $f(x) = \sin(\arctan 2x), \quad g(x) = \dfrac{2x}{\sqrt{1 + 4x^2}}$

70. $f(x) = \tan\left(\arccos \dfrac{x}{2}\right), \quad g(x) = \dfrac{\sqrt{4 - x^2}}{x}$

In Exercises 71–74, fill in the blank.

71. $\arctan \dfrac{9}{x} = \arcsin(\quad), \quad x \neq 0$

72. $\arcsin \dfrac{\sqrt{36 - x^2}}{6} = \arccos(\quad), \quad 0 \leq x \leq 6$

73. $\arccos \dfrac{3}{\sqrt{x^2 - 2x + 10}} = \arcsin(\quad)$

74. $\arccos \dfrac{x - 2}{2} = \arctan(\quad), \quad |x - 2| \leq 2$

In Exercises 75–82, sketch a graph of the function.

75. $y = 2 \arccos x$

76. $y = \arcsin \dfrac{x}{2}$

77. $f(x) = \arcsin(x - 1)$

78. $g(t) = \arccos(t + 2)$

79. $f(x) = \arctan 2x$

80. $f(x) = \dfrac{\pi}{2} + \arctan x$

81. $h(v) = \tan(\arccos v)$

82. $f(x) = \arccos \dfrac{x}{4}$

In Exercises 83–88, use a graphing utility to sketch a graph of the function.

83. $f(x) = 2 \arccos(2x)$

84. $f(x) = \pi \arcsin(4x)$

85. $f(x) = \arctan(2x - 3)$

86. $f(x) = -3 + \arctan(\pi x)$

87. $f(x) = \pi - \arcsin\left(\dfrac{2}{3}\right)$

88. $f(x) = \dfrac{\pi}{2} + \arccos\left(\dfrac{1}{\pi}\right)$

In Exercises 89 and 90, write the given function in terms of the sine function by using the identity

$$A \cos \omega t + B \sin \omega t = \sqrt{A^2 + B^2} \sin\left(\omega t + \arctan \frac{A}{B}\right).$$

Use a graphing utility to graph both forms of the function. What does the graph imply?

89. $f(t) = 3 \cos 2t + 3 \sin 2t$

90. $f(t) = 4 \cos \pi t + 3 \sin \pi t$

91. *Docking a Boat* A boat is pulled in by means of a winch located on a dock 5 feet above the deck of the boat (see figure). Let θ be the angle of elevation from the boat to the winch and let s be the length of the rope from the winch to the boat.

(a) Write θ as a function of s.

(b) Find θ when $s = 40$ feet and $s = 20$ feet.

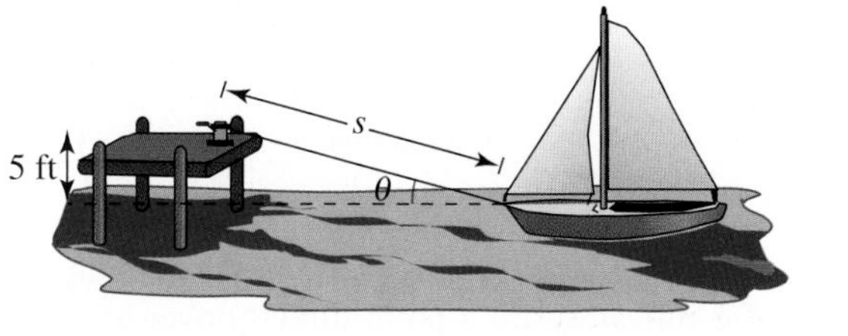

92. *Photography* A television camera at ground level is filming the lift-off of a space shuttle at a point 750 meters from the launch pad (see figure). Let θ be the angle of elevation to the shuttle and let s be the height of the shuttle.

(a) Write θ as a function of s.

(b) Find θ when $s = 300$ meters and $s = 1200$ meters.

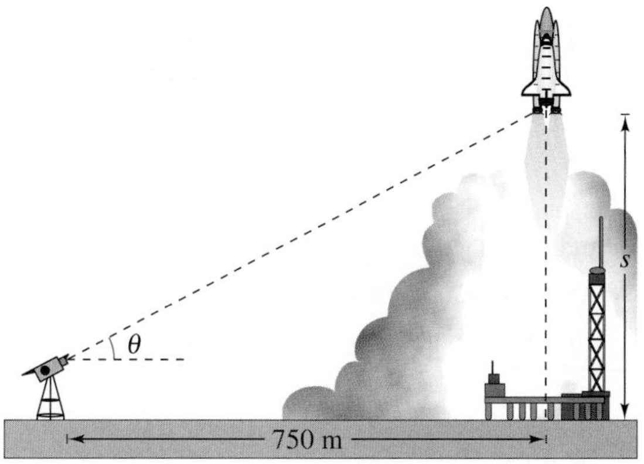

93. *Photography* A photographer is taking a picture of a 2-foot painting hung in an art gallery. The camera lens is 1 foot below the lower edge of the painting (see figure). The angle β subtended by the camera lens x feet from the painting is

$$\beta = \arctan \frac{3x}{x^2 + 4}, \qquad x > 0.$$

(a) Use a graphing utility to graph β as a function of x.

(b) Move the cursor along the graph to approximate the distance from the picture when β is maximum.

(c) Identify the asymptote of the graph and discuss its meaning in the context of the problem.

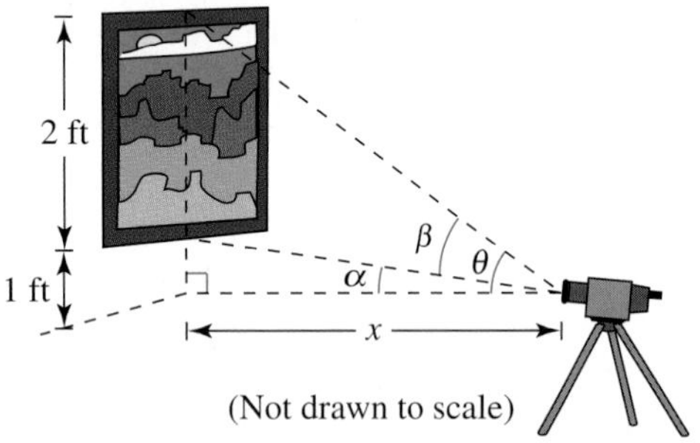

94. *Angle of Elevation* An airplane flies at an altitude of 6 miles toward a point directly over an observer. Consider θ and x as shown in the figure.

(a) Write θ as a function of x.

(b) Find θ when $x = 7$ miles and $x = 1$ mile.

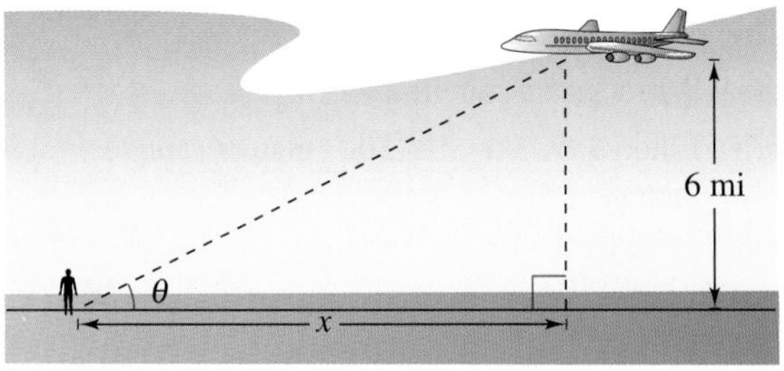

95. *Security Patrol* A security car with its spotlight on is parked 20 meters from a warehouse. Consider θ and x as shown in the figure.

(a) Write θ as a function of x.

(b) Find θ when $x = 5$ meters and $x = 12$ meters.

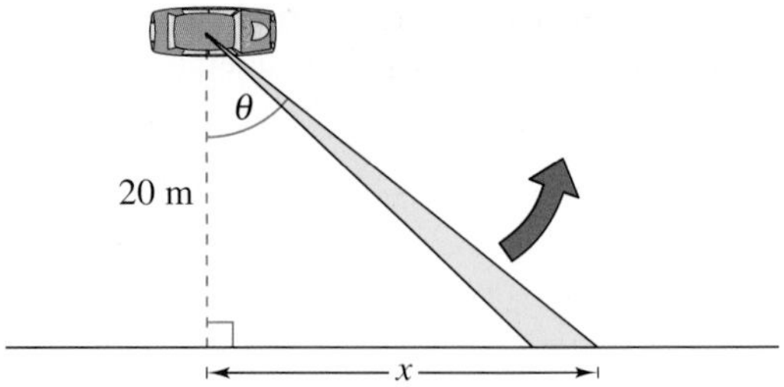

FIGURE FOR 95

Synthesis

True or False? **In Exercises 96 and 97, determine whether the statement is true or false. Justify your answer.**

96. $\sin \dfrac{5\pi}{6} = \dfrac{1}{2}$ ⇨ $\arcsin \dfrac{1}{2} = \dfrac{5\pi}{6}$

97. $\tan \dfrac{5\pi}{4} = 1$ ⇨ $\arctan 1 = \dfrac{5\pi}{4}$

98. *Area* In calculus, it is shown that the area of the region bounded by the graphs of $y = 0$, $y = 1/(x^2 + 1)$, $x = a$, and $x = b$ is given by

$$\text{Area} = \arctan b - \arctan a$$

(see figure). Find the area for the following values of a and b.

(a) $a = 0, b = 1$ (b) $a = -1, b = 1$

(c) $a = 0, b = 3$ (d) $a = -1, b = 3$

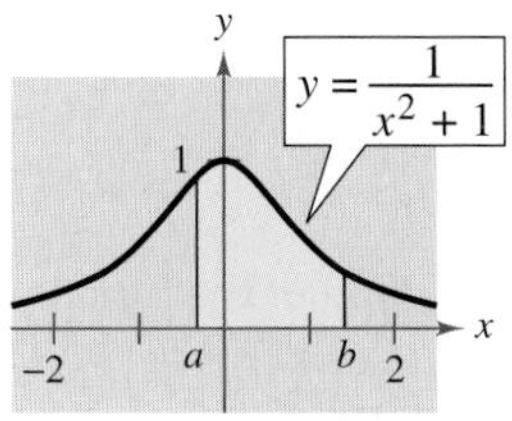

99. Define the inverse cotangent function by restricting the domain of the cotangent function to the interval $(0, \pi)$, and sketch its graph.

100. Define the inverse secant function by restricting the domain of the secant function to the intervals $[0, \pi/2)$ and $(\pi/2, \pi]$, and sketch its graph.

101. Define the inverse cosecant function by restricting the domain of the cosecant function to the intervals $[-\pi/2, 0)$ and $(0, \pi/2]$, and sketch its graph.

102. Use the results of Exercises 99–101 to evaluate the following without using a calculator.

(a) $\text{arcsec}\ \sqrt{2}$ (b) $\text{arcsec}\ 1$

(c) $\text{arccot}(-\sqrt{3})$ (d) $\text{arccsc}\ 2$

103. *Think About It* Use a graphing utility to graph the functions $f(x) = \sqrt{x}$ and $g(x) = 6 \arctan x$. For $x > 0$, it appears that $g > f$. Explain why you know that there exists a positive real number a such that $g < f$ for $x > a$. Approximate the number a.

104. *Think About It* Consider the functions

$$f(x) = \sin x \quad \text{and} \quad f^{-1}(x) = \arcsin x.$$

(a) Use a graphing utility to graph the composite functions $f \circ f^{-1}$ and $f^{-1} \circ f$.

(b) Explain why the graphs in part (a) are not the graph of the line $y = x$. Why do the graphs of $f \circ f^{-1}$ and $f^{-1} \circ f$ differ?

In Exercises 105–110, prove the identity.

105. $\arcsin(-x) = -\arcsin x$

106. $\arctan(-x) = -\arctan x$

107. $\arccos(-x) = \pi - \arccos x$

108. $\arctan x + \arctan \dfrac{1}{x} = \dfrac{\pi}{2}, \quad x > 0$

109. $\arcsin x + \arccos x = \dfrac{\pi}{2}$

110. $\arcsin x = \arctan \dfrac{x}{\sqrt{1 - x^2}}$

Review

In Exercises 111–114, sketch a right triangle corresponding to the trigonometric function of the acute angle θ. Use the Pythagorean Theorem to determine the third side.

111. $\sin \theta = \frac{3}{4}$

112. $\tan \theta = 2$

113. $\cos \theta = \frac{5}{6}$

114. $\sec \theta = 3$

In Exercises 115–118, evaluate the expression. Round your result to three decimal places.

115. $(8.2)^{3.4}$

116. $10(14)^{-2}$

117. $(1.1)^{50}$

118. $16^{-2\pi}$

119. *Partnership Costs* A group of people agree to share equally in the cost of a \$250,000 endowment to a college. If they could find two more people to join the group, each person's share of the cost would decrease by \$6250. How many people are presently in the group?

120. *Speed* A boat travels at a speed of 18 miles per hour in still water. It travels 35 miles upstream and then returns to the starting point in a total of 4 hours. Find the speed of the current.

4.8 Applications and Models

▶ What you should learn

- How to solve real-life problems involving right triangles
- How to solve real-life problems involving directional bearings
- How to solve real-life problems involving harmonic motion

▶ Why you should learn it

Trigonometric functions frequently model real-life problems involving wave motion. For instance, Exercise 60 on page 431 shows how a trigonometric function can be used to model the harmonic motion of a buoy.

Mary Kate Denny/PhotoEdit

Applications Involving Right Triangles

In this section the three angles of a right triangle are denoted by the letters A, B, and C (where C is the right angle), and the lengths of the sides opposite these angles by the letters a, b, and c (where c is the hypotenuse).

Example 1 ▶ Solving a Right Triangle

Solve the right triangle shown in Figure 4.67 for all unknown sides and angles.

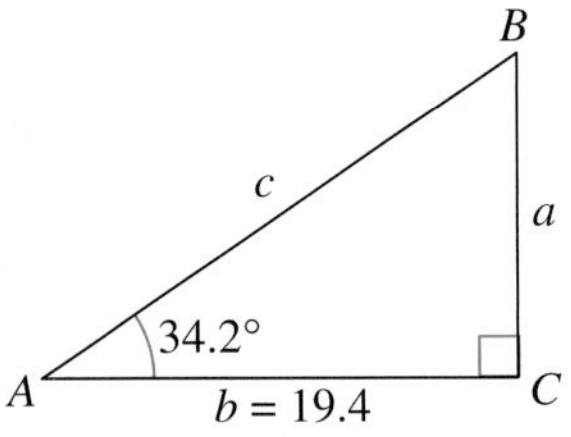

FIGURE 4.67

Solution

Because $C = 90°$, it follows that $A + B = 90°$ and $B = 90° - 34.2° = 55.8°$. To solve for a, use the fact that

$$\tan A = \frac{\text{opp}}{\text{adj}} = \frac{a}{b} \quad \Longrightarrow \quad a = b \tan A.$$

So, $a = 19.4 \tan 34.2° \approx 13.18$. Similarly, to solve for c, use the fact that

$$\cos A = \frac{\text{adj}}{\text{hyp}} = \frac{b}{c} \quad \Longrightarrow \quad c = \frac{b}{\cos A}.$$

So, $c = \dfrac{19.4}{\cos 34.2°} \approx 23.46$.

In the next three examples, the term **angle of elevation** represents the angle from the horizontal upward to an object. For objects that lie below the horizontal, it is common to use the term **angle of depression,** as shown in Figure 4.68.

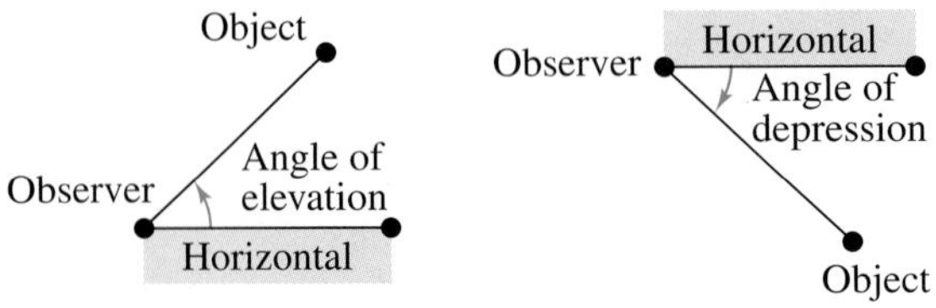

FIGURE 4.68

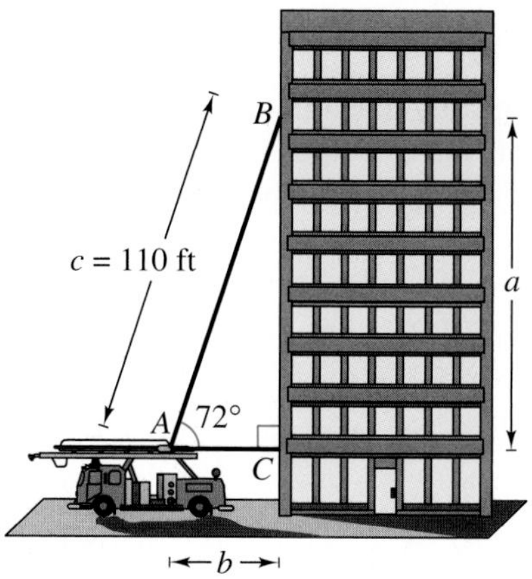

FIGURE 4.69

Example 2 ▶ Finding a Side of a Right Triangle

A safety regulation states that the maximum angle of elevation for a rescue ladder is 72°. If a fire department's longest ladder is 110 feet, what is the maximum safe rescue height?

Solution

A sketch is shown in Figure 4.69. From the equation $\sin A = a/c$, it follows that

$$a = c \sin A = 110 \sin 72° \approx 104.6.$$

So, the maximum safe rescue height is about 104.6 feet above the height of the fire truck.

Example 3 ▶ Finding a Side of a Right Triangle

At a point 200 feet from the base of a building, the angle of elevation to the *bottom* of a smokestack is 35°, whereas the angle of elevation to the *top* is 53°, as shown in Figure 4.70. Find the height s of the smokestack alone.

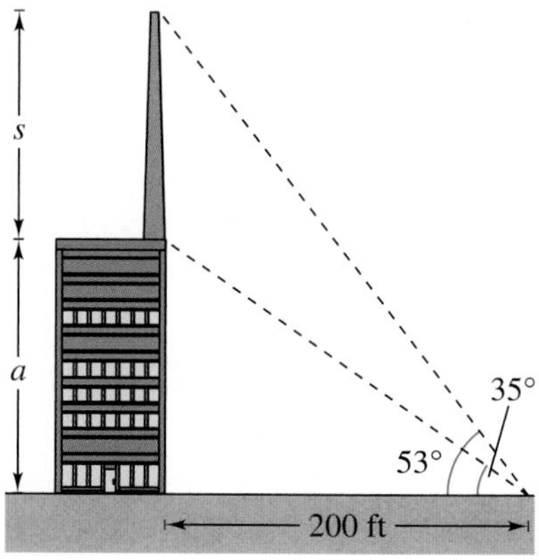

FIGURE 4.70

Solution

Note from Figure 4.70 that this problem involves two right triangles. In the smaller right triangle, use the fact that $\tan 35° = a/200$ to conclude that the height of the building is

$$a = 200 \tan 35°.$$

In the larger right triangle, use the equation

$$\tan 53° = \frac{a + s}{200}$$

to conclude that $a + s = 200 \tan 53°$. So, the height of the smokestack is

$$\begin{aligned} s &= 200 \tan 53° - a \\ &= 200 \tan 53° - 200 \tan 35° \\ &\approx 125.4 \text{ feet.} \end{aligned}$$

Example 4 ▶ Finding an Acute Angle of a Right Triangle

A swimming pool is 20 meters long and 12 meters wide. The bottom of the pool is slanted so that the water depth is 1.3 meters at the shallow end and 4 meters at the deep end, as shown in Figure 4.71. Find the angle of depression of the bottom of the pool.

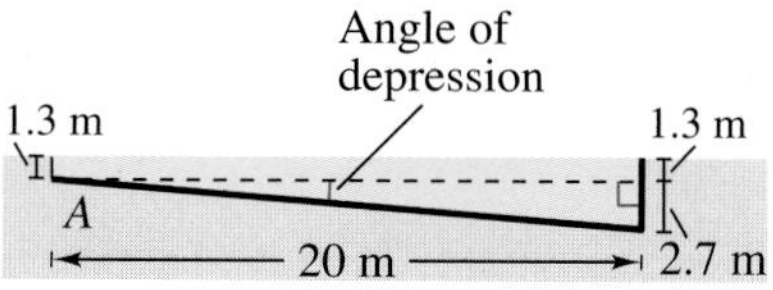

FIGURE 4.71

Solution

Using the tangent function, you see that

$$\tan A = \frac{\text{opp}}{\text{adj}} = \frac{2.7}{20} = 0.135.$$

So, the angle of depression is

$$A = \arctan 0.135 \approx 0.13419 \text{ radian} \approx 7.69°.$$

Trigonometry and Bearings

In surveying and navigation, directions are generally given in terms of **bearings.** A bearing measures the acute angle a path or line of sight makes with a fixed north-south line, as shown in Figure 4.72. For instance, the bearing S 35° E in Figure 4.72 means 35 degrees east of south.

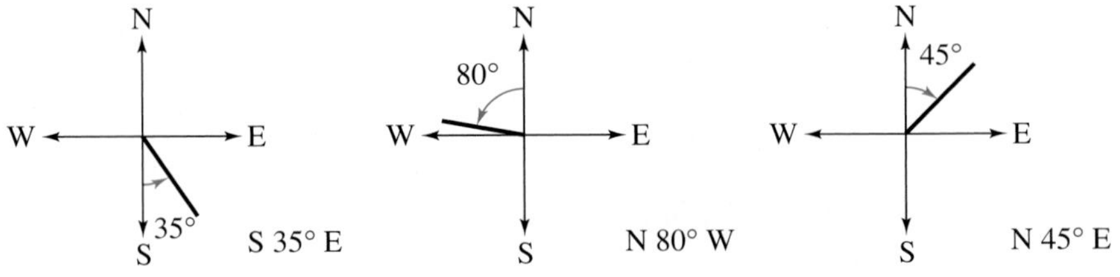

FIGURE 4.72

Example 5 ▶ Finding Directions in Terms of Bearings

A ship leaves port at noon and heads due west at 20 knots, or 20 nautical miles (nm) per hour. At 2 P.M. the ship changes course to N 54° W, as shown in Figure 4.73. Find the ship's bearing and distance from the port of departure at 3 P.M.

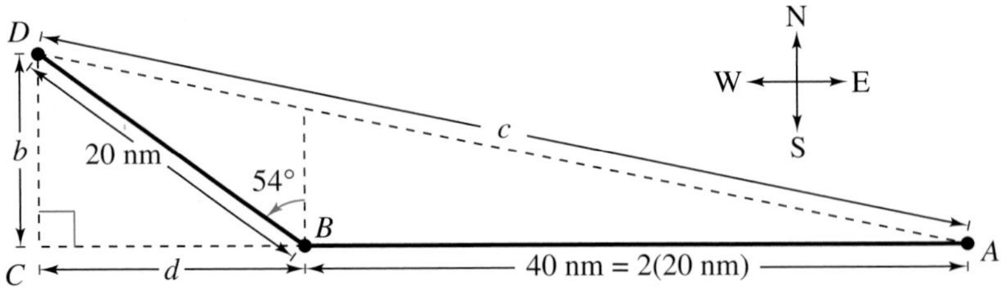

FIGURE 4.73

Solution

In triangle BCD, you have $B = 90° - 54° = 36°$. The two sides of this triangle can be determined to be

$$b = 20 \sin 36° \qquad \text{and} \qquad d = 20 \cos 36°.$$

In triangle ACD, you find angle A as follows.

$$\tan A = \frac{b}{d + 40} = \frac{20 \sin 36°}{20 \cos 36° + 40} \approx 0.2092494$$

$$A \approx \arctan 0.2092494 \approx 0.2062732 \text{ radian} \approx 11.82°$$

The angle with the north-south line is $90° - 11.82° = 78.18°$. Therefore, the bearing of the ship is

N 78.18° W. Bearing

Finally, from triangle ACD, you have $\sin A = b/c$, which yields

$$c = \frac{b}{\sin A} = \frac{20 \sin 36°}{\sin 11.82°}$$

$$\approx 57.4 \text{ nautical miles.}$$ Distance from port

Harmonic Motion

The periodic nature of the trigonometric functions is useful for describing the motion of a point on an object that vibrates, oscillates, rotates, or is moved by wave motion.

For example, consider a ball that is bobbing up and down on the end of a spring, as shown in Figure 4.74. Suppose that 10 centimeters is the maximum distance the ball moves vertically upward or downward from its equilibrium (at rest) position. Suppose further that the time it takes for the ball to move from its maximum displacement above zero to its maximum displacement below zero and back again is $t = 4$ seconds. Assuming the ideal conditions of perfect elasticity and no friction or air resistance, the ball would continue to move up and down in a uniform and regular manner.

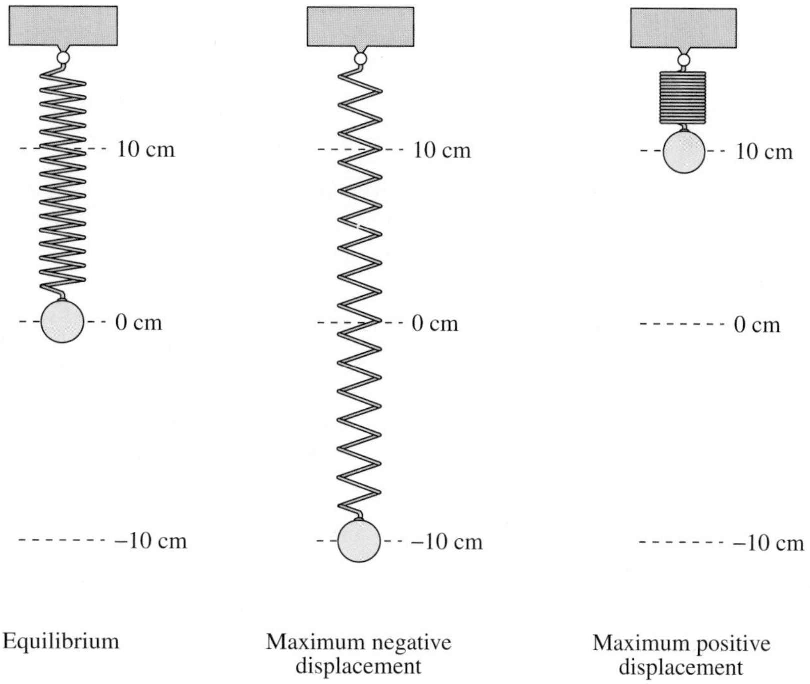

FIGURE 4.74

From this spring you can conclude that the period (time for one complete cycle) of the motion is

Period $=$ 4 seconds

and that its amplitude (maximum displacement from equilibrium) is

Amplitude $=$ 10 centimeters.

Motion of this nature can be described by a sine or cosine function, and is called **simple harmonic motion.**

Definition of Simple Harmonic Motion

A point that moves on a coordinate line is said to be in **simple harmonic motion** if its distance d from the origin at time t is given by either

$$d = a \sin \omega t \qquad \text{or} \qquad d = a \cos \omega t$$

where a and ω are real numbers such that $\omega > 0$. The motion has amplitude $|a|$, period $2\pi/\omega$, and frequency $\omega/2\pi$.

Example 6 ▶ Simple Harmonic Motion

Write the equation for the simple harmonic motion of the ball described in Figure 4.74, where the period is 4 seconds. What is the frequency of this harmonic motion?

Solution

Because the spring is at equilibrium ($d = 0$) when $t = 0$, you use the equation

$$d = a \sin \omega t.$$

Moreover, because the maximum displacement from zero is 10 and the period is 4, you have

$$\text{Amplitude} = |a| = 10$$

$$\text{Period} = \frac{2\pi}{\omega} = 4 \quad \Longrightarrow \quad \omega = \frac{\pi}{2}.$$

Consequently, the equation of motion is

$$d = 10 \sin \frac{\pi}{2} t.$$

Note that the choice of $a = 10$ or $a = -10$ depends on whether the ball initially moves up or down. The frequency is

$$\text{Frequency} = \frac{\omega}{2\pi} = \frac{\pi/2}{2\pi} = \frac{1}{4} \text{ cycle per second.}$$

One illustration of the relationship between sine waves and harmonic motion is seen in the wave motion resulting when a stone is dropped into a calm pool of water. The waves move outward in roughly the shape of sine (or cosine) waves, as shown in Figure 4.75. As an example, suppose you are fishing and your fishing bob is attached so that it does not move horizontally. As the waves move outward from the dropped stone, your fishing bob will move up and down in simple harmonic motion, as shown in Figure 4.76.

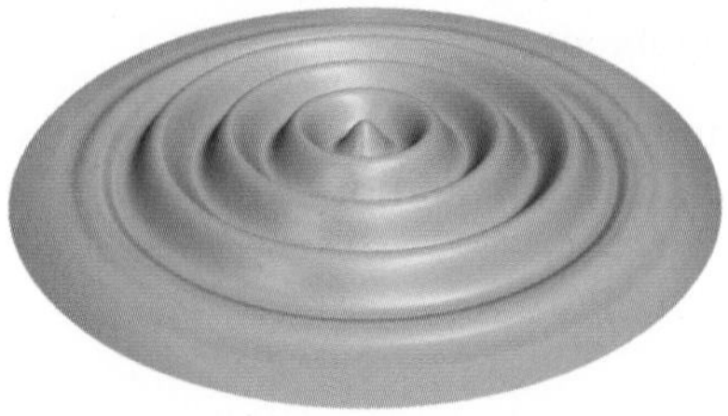

FIGURE 4.75

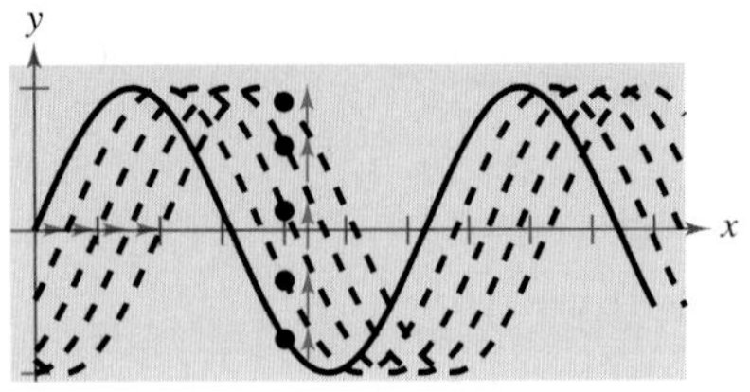

FIGURE 4.76

Example 7 ▶ Simple Harmonic Motion

Given the equation for simple harmonic motion

$$d = 6 \cos \frac{3\pi}{4} t$$

find (a) the maximum displacement, (b) the frequency, (c) the value of d when $t = 4$, and (d) the least positive value of t for which $d = 0$.

Solution

The given equation has the form $d = a \cos \omega t$, with $a = 6$ and $\omega = 3\pi/4$.

a. The maximum displacement (from the point of equilibrium) is given by the amplitude. So, the maximum displacement is 6.

b. $$\text{Frequency} = \frac{\omega}{2\pi} = \frac{3\pi/4}{2\pi} = \frac{3}{8} \text{ cycle per unit of time}$$

c. $$d = 6 \cos\left[\frac{3\pi}{4}(4)\right] = 6 \cos 3\pi = 6(-1) = -6$$

d. To find the least positive value of t for which $d = 0$, solve the equation

$$d = 6 \cos \frac{3\pi}{4} t = 0$$

to obtain

$$\frac{3\pi}{4} t = \frac{\pi}{2}, \frac{3\pi}{2}, \frac{5\pi}{2}, . . . \quad \Longrightarrow \quad t = \frac{2}{3}, 2, \frac{10}{3}, ...$$

So, the least positive value of t is $t = \frac{2}{3}$.

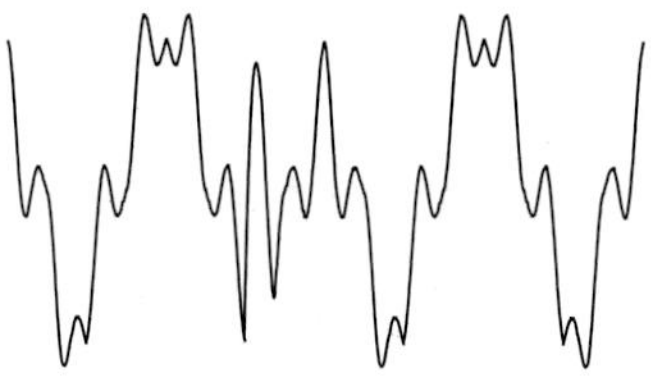

(a)

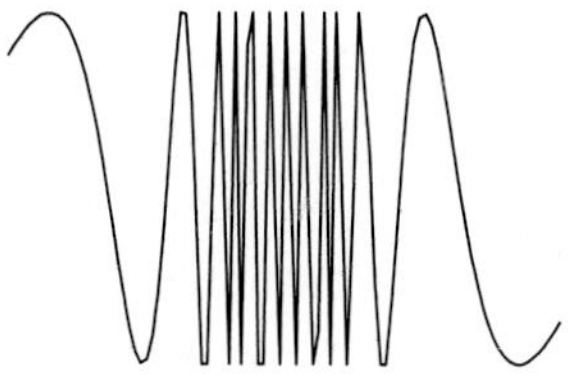

(b)

FIGURE 4.77

Writing ABOUT MATHEMATICS

Radio Waves Many different physical phenomena can be characterized by wave motion. These phenomena include electromagnetic waves such as radio waves, television waves, and microwaves. Radio waves transmit sound in two different ways. For an AM station, the *amplitude* of the wave is modified to carry sound. The letters AM stand for "amplitude modulation." An FM radio signal has its *frequency* modified in order to carry sound, hence the term "frequency modulation." Of the two graphs in Figure 4.77, one shows an AM wave and the other shows an FM wave. Which is which? Explain your reasoning.

4.8 Exercises

In Exercises 1–10, solve the right triangle shown in the figure. Round your answer to two decimal places.

1. $A = 20°,\ \ b = 10$
2. $B = 54°,\ \ c = 15$
3. $B = 71°,\ \ b = 24$
4. $A = 8.4°,\ \ a = 40.5$
5. $a = 6,\ \ b = 10$
6. $a = 25,\ \ c = 35$
7. $b = 16,\ \ c = 52$
8. $b = 1.32,\ \ c = 9.45$
9. $A = 12°15',\ \ c = 430.5$
10. $B = 65°12',\ \ a = 14.2$

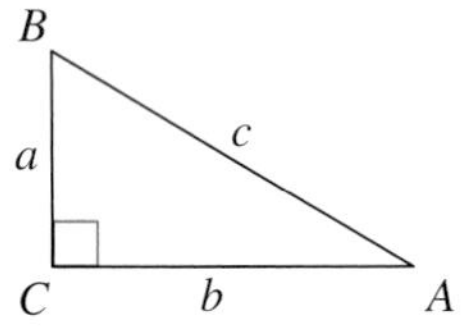

FIGURE FOR 1–10

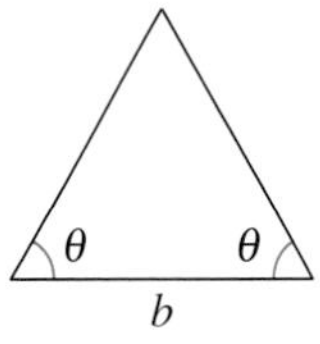

FIGURE FOR 11–14

In Exercises 11–14, find the altitude of the isosceles triangle shown in the figure. Round your answer to two decimal places.

11. $\theta = 52°,\ b = 4$ inches
12. $\theta = 18°,\ b = 10$ meters
13. $\theta = 41°,\ b = 46$ inches
14. $\theta = 27°,\ b = 11$ feet

15. *Length of a Shadow* If the sun is 25° above the horizon, find the length of a shadow cast by a silo that is 50 feet tall (see figure).

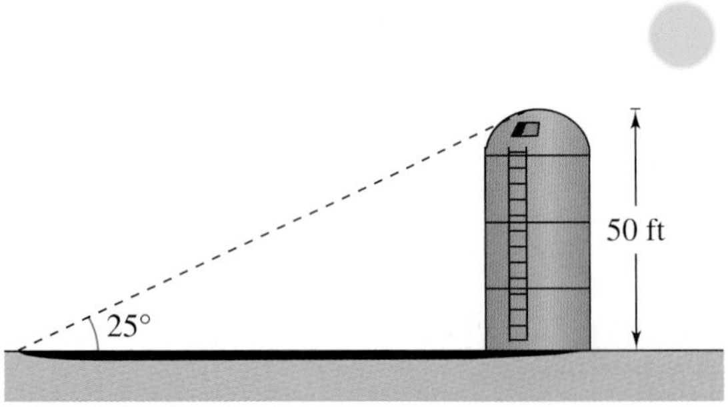

16. *Length of a Shadow* If the sun is 20° above the horizon, find the length of a shadow cast by a building that is 600 feet tall.

17. *Height* A ladder 20 feet long leans against the side of a house. Find the height h from the top of the ladder to the ground if the angle of elevation of the ladder is 80°.

18. *Height* The length of a shadow of a tree is 125 feet when the angle of elevation of the sun is 33°. Approximate the height h of the tree.

19. *Height* From a point 50 feet in front of a church, the angles of elevation to the base of the steeple and the top of the steeple are 35° and 47° 40′, respectively.
    (a) Draw right triangles that represent the problem. Label the known and unknown quantities.
    (b) Use a trigonometric function to write an equation involving the unknown quantity.
    (c) Find the height of the steeple.

20. *Height* You are standing 100 feet from the base of a platform from which people are bungee jumping. The angle of elevation from your position to the top of the platform from which they jump is 51°. From what height are the people jumping?

21. *Depth of a Submarine* The sonar of a navy cruiser detects a submarine that is 4000 feet from the cruiser. The angle between the water line and the submarine is 34° (see figure). How deep is the submarine?

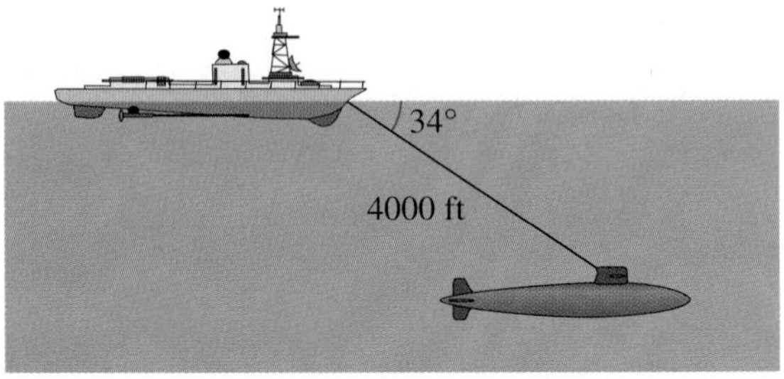

22. *Height of a Kite* A 75-foot line is attached to a kite. When the kite has pulled the line taut, the angle of elevation to the kite is approximately 60°. Approximate the height of the kite.

23. *Angle of Elevation* An amateur radio operator erects a 75-foot vertical tower for an antenna. Find the angle of elevation to the top of the tower at a point on level ground 50 feet from its base.

24. *Angle of Elevation* The height of an outdoor basketball backboard is $12\frac{1}{2}$ feet, and the backboard casts a shadow $17\frac{1}{3}$ feet long.
    (a) Draw a right triangle that represents the problem. Label the known and unknown quantities.
    (b) Use a trigonometric function to write an equation involving the unknown quantity.
    (c) Find the angle of elevation of the sun.

25. *Angle of Depression* A Global Positioning System satellite orbits 10,900 feet above earth's surface. Find the angle of depression from the satellite to the horizon. Assume the radius of earth is 4000 miles.

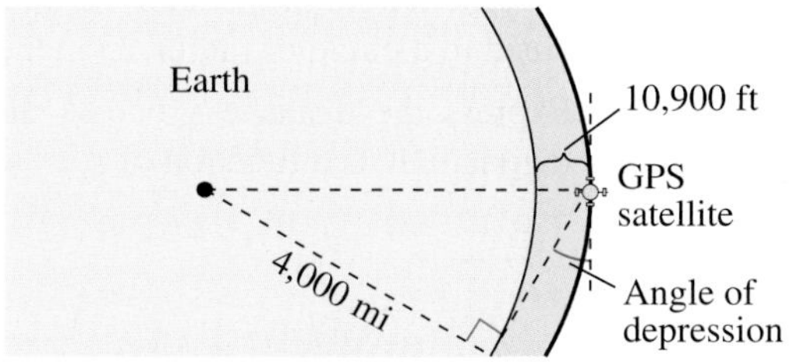

(Not drawn to scale)

26. *Angle of Depression* A cellular telephone tower that is 150 feet tall is placed on top of a mountain that is 1200 feet above sea level. What is the angle of depression from the top of the tower to a cell phone user who is 5 horizontal miles away and 400 feet above sea level?

27. *Airplane Ascent* During takeoff, an airplane's angle of climb is 18° and its speed is 275 feet per second. Find the plane's altitude after 1 minute.

28. *Airplane Ascent* How long will it take the plane in Exercise 27 to climb to an altitude of 10,000 feet?

29. *Mountain Descent* A sign on a roadway at the top of a mountain indicates that for the next 4 miles the grade is 10.5° (see figure). Find the change in elevation for a car descending the mountain.

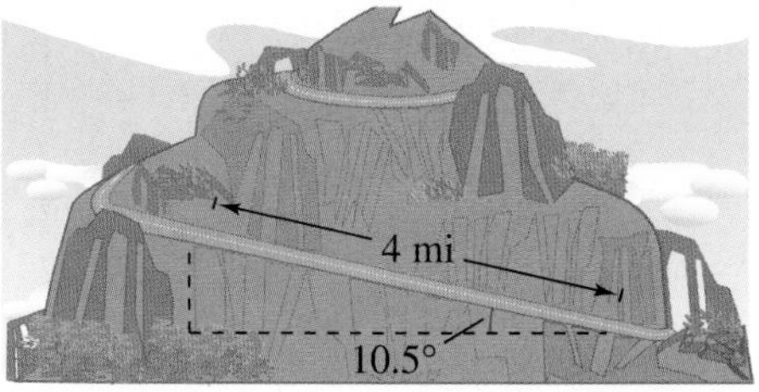

30. *Mountain Descent* A roadway sign at the top of a mountain indicates that for the next 4 miles the grade is 12%. Find the angle of the grade and the change in elevation for a car descending the mountain.

31. *Navigation* An airplane flying at 600 miles per hour has a bearing of N 52° E. After flying for 1.5 hours, how far north and how far east will the plane have traveled from its point of departure?

32. *Navigation* A ship leaves port at noon and has a bearing of S 27° W. If its speed is 20 knots, how many nautical miles south and how many nautical miles west will the ship have traveled by 6:00 P.M.?

33. *Surveying* A surveyor wishes to find the distance across a swamp (see figure). The bearing from A to B is N 32° W. The surveyor walks 50 meters from A, and at the point C the bearing to B is N 68° W. Find (a) the bearing from A to C and (b) the distance from A to B.

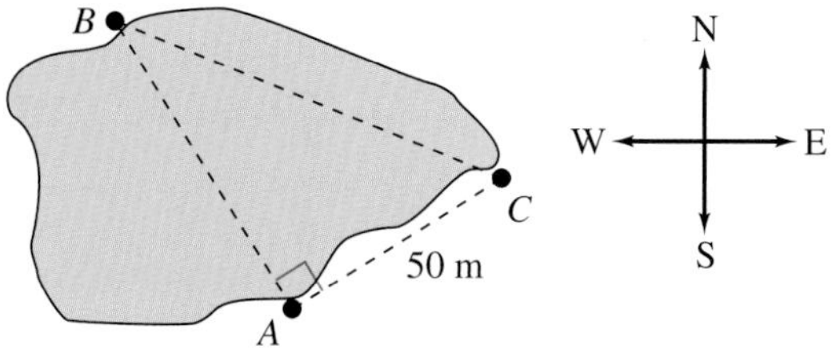

34. *Location of a Fire* Two fire towers are 30 kilometers apart, tower A being due west of tower B. A fire is spotted from the towers, and the bearings from A and B are E 14° N and W 34° N, respectively (see figure). Find the distance d of the fire from the line segment AB.

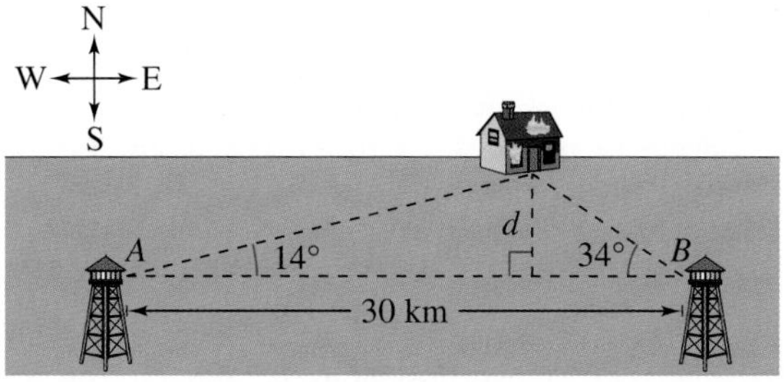

35. *Navigation* A ship is 45 miles east and 30 miles south of port. If the captain wants to sail directly to port, what bearing should be taken?

36. *Navigation* A plane is 120 miles north and 85 miles east of an airport. If the pilot wants to fly directly to the airport, what bearing should be taken?

37. *Distance Between Ships* An observer in a lighthouse 350 feet above sea level observes two ships directly offshore. The angles of depression to the ships are 4° and 6.5° (see figure). How far apart are the ships?

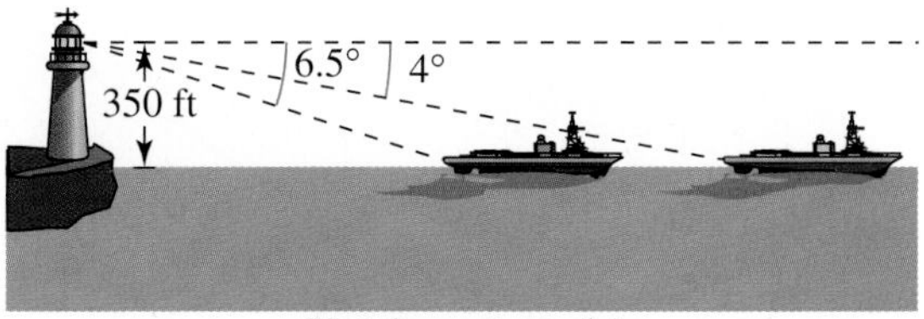

(Not drawn to scale)

38. *Distance Between Towns* A passenger in an airplane at an altitude of 10 kilometers sees two towns directly to the east of the plane. The angles of depression to the towns are 28° and 55° (see figure). How far apart are the towns?

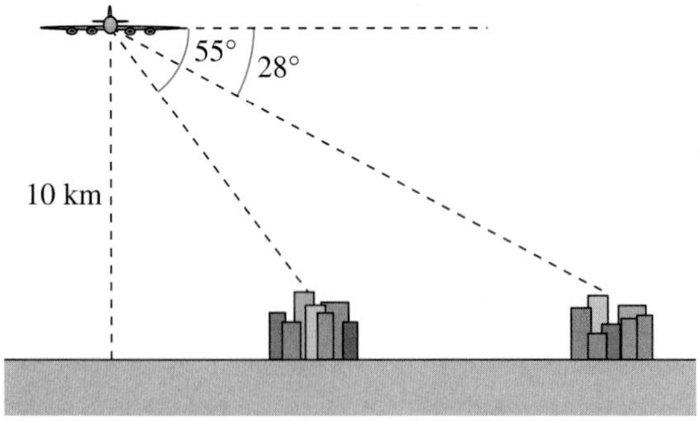

39. *Altitude of a Plane* A plane is observed approaching your home and you assume that its speed is 550 miles per hour. If the angle of elevation of the plane is 16° at one time and 57° one minute later, approximate the altitude of the plane.

40. *Height of a Mountain* While traveling across flat land, you notice a mountain directly in front of you. The angle of elevation to the peak is 2.5°. After you drive 17 miles closer to the mountain, the angle of elevation is 9°. Approximate the height of the mountain.

Geometry **In Exercises 41 and 42, find the angle α between two nonvertical lines L_1 and L_2. The angle α satisfies the equation**

$$\tan \alpha = \left| \frac{m_2 - m_1}{1 + m_2 m_1} \right|$$

where m_1 and m_2 are the slopes of L_1 and L_2, respectively. (Assume that $m_1 m_2 \neq -1$.)

41. L_1: $3x - 2y = 5$
L_2: $x + y = 1$

42. L_1: $2x - y = 8$
L_2: $x - 5y = -4$

43. *Geometry* Determine the angle between the diagonal of the cube and the diagonal of its base, as shown in the figure.

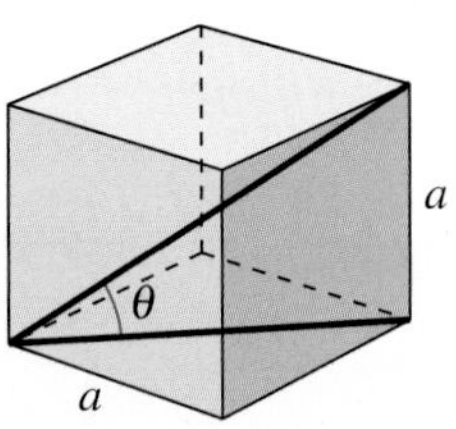

FIGURE FOR 43

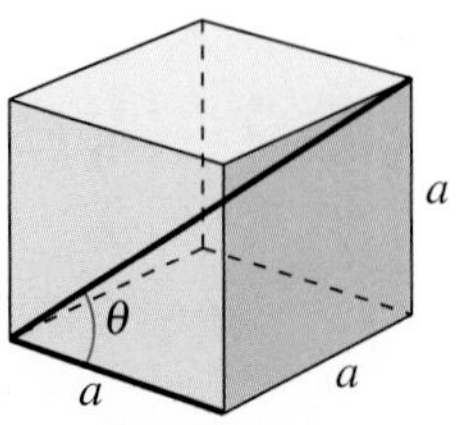

FIGURE FOR 44

44. *Geometry* Determine the angle between the diagonal of the cube and its edge, as shown in the figure.

45. *Geometry* Find the length of the sides of a regular pentagon inscribed in a circle of radius 25 inches.

46. *Geometry* Find the length of the sides of a regular hexagon inscribed in a circle of radius 25 inches.

47. *Hardware* Express the distance y across the flat sides of a hexagonal nut as a function of r, as shown in the figure.

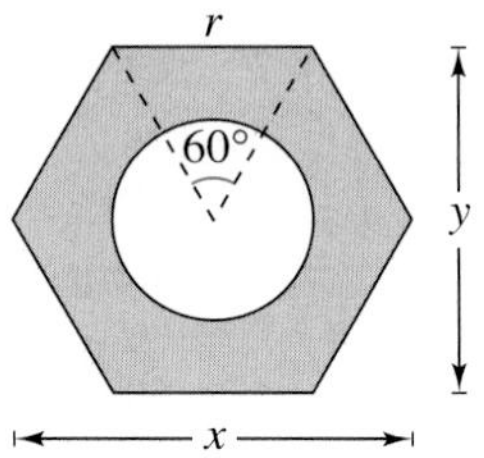

48. *Bolt Holes* The figure shows a circular piece of sheet metal that has a diameter of 40 centimeters and contains 12 equally spaced bolt holes. Determine the straight-line distance between the centers of consecutive bolt holes.

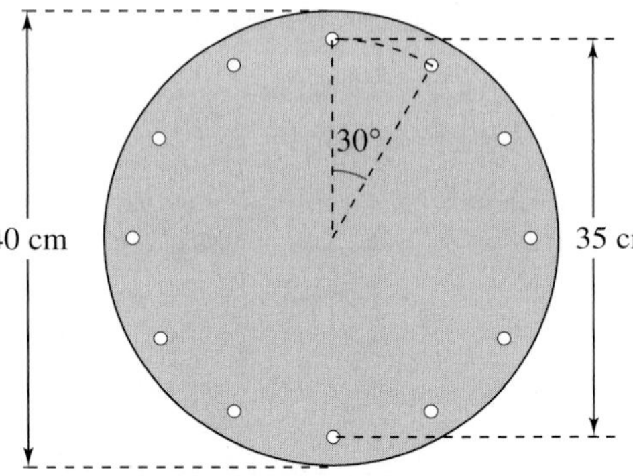

Trusses **In Exercises 49 and 50, find the lengths of all the unknown members of the truss.**

49.

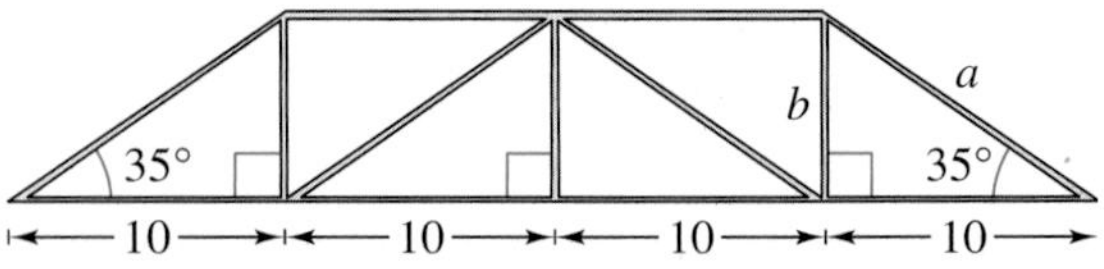

50.

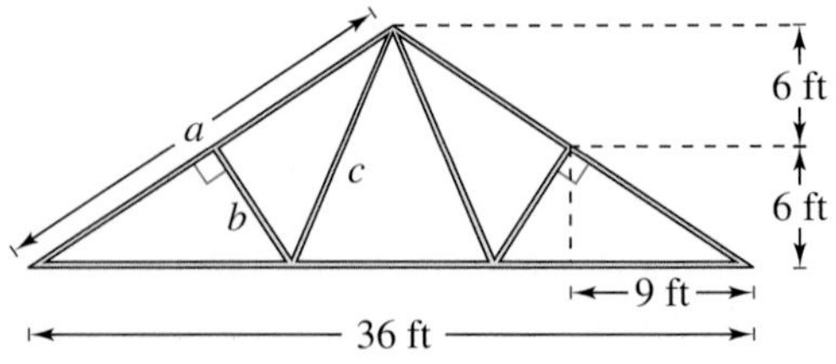

Harmonic Motion **In Exercises 51–54, for the simple harmonic motion described by the trigonometric function, find (a) the maximum displacement, (b) the frequency, and (c) the least positive value of t for which $d = 0$.**

51. $d = 4 \cos 8\pi t$

52. $d = \frac{1}{2} \cos 20\pi t$

53. $d = \frac{1}{16} \sin 120\pi t$

54. $d = \frac{1}{64} \sin 792\pi t$

Harmonic Motion **In Exercises 55–58, find a model for simple harmonic motion satisfying the specified conditions.**

	Displacement ($t = 0$)	*Amplitude*	*Period*
55.	0	4 cm	2 sec
56.	0	3 m	6 sec
57.	3 in.	3 in.	1.5 sec
58.	2 ft	2 ft	10 sec

59. *Tuning Fork* A point on the end of a tuning fork moves in simple harmonic motion described by $d = a \sin \omega t$. Find ω given that the tuning fork for middle C has a frequency of 264 vibrations per second.

60. *Wave Motion* A buoy oscillates in simple harmonic motion as waves go past. It is noted that the buoy moves a total of 3.5 feet from its low point to its high point (see figure), and that it returns to its high point every 10 seconds. Write an equation that describes the motion of the buoy if its high point is at $t = 0$.

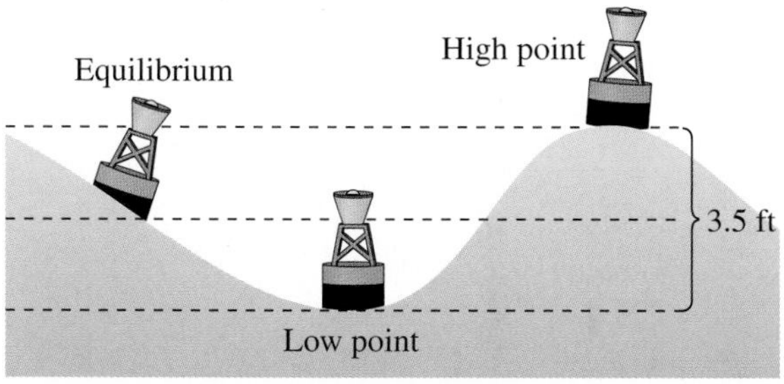

61. *Springs* A weight stretches a spring 1.5 inches. The weight is pushed 3 inches above its equilibrium position and released. Its motion is modeled by

$$y = \frac{1}{4} \cos 16t, \qquad t > 0$$

where y is in feet and t is in seconds.

(a) Graph the function.

(b) What is the period of the oscillations?

(c) Determine the first time the weight passes the point of equilibrium ($y = 0$).

Synthesis

True or False? **In Exercises 62 and 63, determine whether the statement is true or false. Justify your answer.**

62. A building that is famous for not being perfectly vertical is the Leaning Tower of Pisa. If you know the exact angle of elevation θ to the 191-foot tower when you stand near it, then you can determine the exact distance to the tower d by using the formula

$$\tan \theta = \frac{191}{d}.$$

63. For the harmonic motion of a ball bobbing up and down on the end of a spring, one period can be described as the length of one coil of the spring.

64. *Numerical and Graphical Analysis* A 2-meter-high fence is 3 meters from the side of a grain storage bin. A grain elevator must reach from ground level outside the fence to the storage bin (see figure). The objective is to determine the shortest elevator that meets the constraints.

(a) Complete four rows of the table.

θ	L_1	L_2	$L_1 + L_2$
0.1	$\frac{2}{\sin 0.1}$	$\frac{3}{\cos 0.1}$	23.0
0.2	$\frac{2}{\sin 0.2}$	$\frac{3}{\cos 0.2}$	13.1

(b) Use a graphing utility to generate additional rows of the table. Use the table to estimate the minimum length of the elevator.

(c) Write the length $L_1 + L_2$ as a function of θ.

(d) Use a graphing utility to graph the function. Use the graph to estimate the minimum length. How does your estimate compare with that of part (b)?

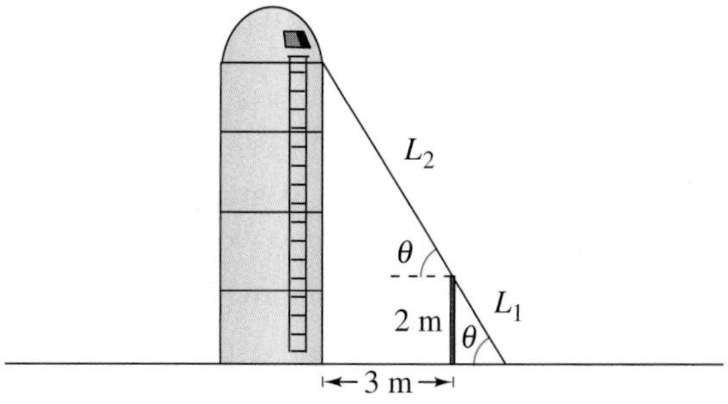

65. *Numerical and Graphical Analysis* The cross sections of an irrigation canal are isosceles trapezoids where the length of three of the sides is 8 feet (see figure). The objective is to find the angle θ that maximizes the area of the cross sections. [*Hint:* The area of a trapezoid is $(h/2)(b_1 + b_2)$.]

(a) Complete seven rows of the table.

Base 1	Base 2	Altitude	Area
8	$8 + 16 \cos 10°$	$8 \sin 10°$	22.1
8	$8 + 16 \cos 20°$	$8 \sin 20°$	42.5

(b) Use a graphing utility to generate additional rows of the table. Use the table to estimate the maximum cross-sectional area.

(c) Write the area A as a function of θ.

(d) Use a graphing utility to graph the function. Use the graph to estimate the maximum cross-sectional area. How does your estimate compare with that of part (b)?

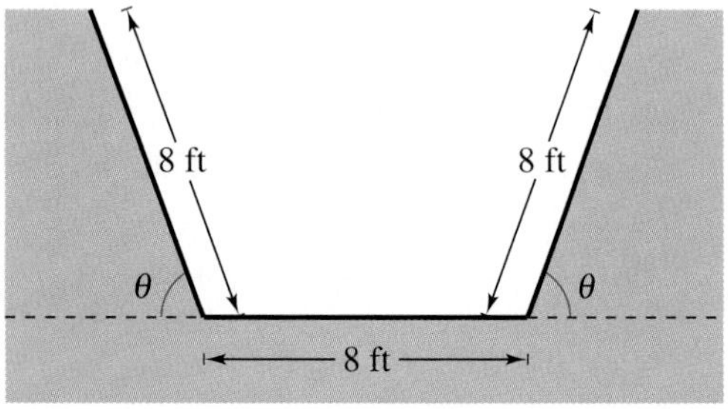

66. *Data Analysis* The times S of sunset (Greenwich Mean Time) at 40° north latitude on the 15th of each month are: 1(16:59), 2(17:35), 3(18:06), 4(18:38), 5(19:08), 6(19:30), 7(19:28), 8(18:57), 9(18:09), 10(17:21), 11(16:44), 12(16:36). The month is represented by t, with $t = 1$ corresponding to January. A model (where minutes have been converted to the decimal parts of an hour) for this data is

$$S(t) = 18.09 + 1.41 \sin\left(\frac{\pi t}{6} + 4.60\right).$$

(a) Use a graphing utility to graph the data points and the model in the same viewing window.

(b) What is the period of the model? Is it what you expected? Explain.

(c) What is the amplitude of the function? What does it represent in the model? Explain.

67. *Data Analysis* The table gives the average sales S (in millions of dollars) of an outerwear manufacturer for each month t, where $t = 1$ represents January.

t	1	2	3	4	5	6
S	13.46	11.15	8.00	4.85	2.54	1.70
t	7	8	9	10	11	12
S	2.54	4.85	8.00	11.15	13.46	14.30

(a) Create a scatter plot of the data.

(b) Find a trigonometric model that fits the data. Graph the model on your scatter plot. How well does the model fit the data?

(c) What is the period of the model? Do you think it is reasonable given the context? Explain your reasoning.

(d) Interpret the meaning of the model's amplitude in the context of the problem.

Review

In Exercises 68–77, graph the equation by hand.

68. $3x - 2y = 4$

69. $5y - 3x = 12$

70. $(y - 2)^2 = 8(x + 2)$

71. $(x + 3)^2 = 5y - 8$

72. $\frac{x^2}{4} + y^2 = 1$

73. $2x^2 + y^2 - 4 = 0$

74. $\frac{x^2}{4} - y^2 = 1$

75. $\frac{y^2}{4} - \frac{(x + 2)^2}{25} - 1 = 0$

76. $\frac{x^2}{4} + \frac{y^2}{4} = 1$

77. $(x - 2)^2 + y^2 = 25$

Chapter Summary

What did you learn?

Section 4.1	Review Exercises
☐ How to describe angles	1–4
☐ How to use radian and degree measure	5–20
☐ How to use angles to model and solve real-life problems	21, 22
Section 4.2	
☐ How to identify a unit circle and its relationship to real numbers	23–26
☐ How to evaluate trigonometric functions using the unit circle	27–30
☐ How to use the domain and period to evaluate sine and cosine functions	31–34
☐ How to use a calculator to evaluate trigonometric functions	35–38
Section 4.3	
☐ How to evaluate trigonometric functions of acute angles	39–42
☐ How to use the fundamental trigonometric identities	43–46
☐ How to use a calculator to evaluate trigonometric functions	47–52
☐ How to use trigonometric functions to model and solve real-life problems	53, 54
Section 4.4	
☐ How to evaluate trigonometric functions of any angle	55–68
☐ How to use reference angles to evaluate trigonometric functions	69–74
☐ How to evaluate trigonometric functions of real numbers	75–82
Section 4.5	
☐ How to use amplitude and period to sketch the graphs of sine and cosine functions	83–86
☐ How to sketch translations of graphs of sine and cosine functions	87–90
☐ How to use sine and cosine functions to model real-life data	91, 92
Section 4.6	
☐ How to sketch the graphs of tangent and cotangent functions	93–96
☐ How to sketch the graphs of secant and cosecant functions	97–100
☐ How to sketch the graphs of damped trigonometric functions	101, 102
Section 4.7	
☐ How to evaluate the inverse sine function	103–108
☐ How to evaluate the other inverse trigonometric functions	109–120
☐ How to evaluate the compositions of trigonometric functions	121–128
Section 4.8	
☐ How to solve real-life problems involving right triangles	129, 130
☐ How to solve real-life problems involving directional bearings	131
☐ How to solve real-life problems involving harmonic motion	132

Review Exercises

4.1 **In Exercises 1–4, estimate the angle to the nearest one-half radian.**

1. 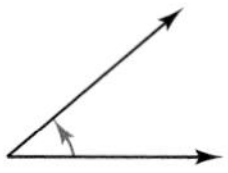

2. 

3. 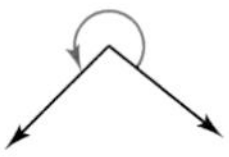

4. 

In Exercises 5–12, sketch the angle in standard position. List one positive and one negative coterminal angle.

5. $\frac{11\pi}{4}$ **6.** $\frac{2\pi}{9}$

7. $-\frac{4\pi}{3}$ **8.** $-\frac{23\pi}{3}$

9. $70°$ **10.** $280°$

11. $-110°$ **12.** $-405°$

In Exercises 13–16, convert the measure from radians to degrees. Round to two decimal places.

13. $\frac{5\pi}{7}$ **14.** $-\frac{11\pi}{6}$

15. -3.5 **16.** 5.7

In Exercises 17–20, convert the measure from degrees to radians. Round to four decimal places.

17. $480°$ **18.** $-127.5°$

19. $-33° 45'$ **20.** $196° 77'$

21. *Phonograph* Compact discs have all but replaced phonograph records. Phonograph records are vinyl discs that rotate on a turntable. A typical record album is 12 inches in diameter and plays at $33\frac{1}{3}$ revolutions per minute.

(a) What is the angular speed of a record album?

(b) What is the linear speed of the outer edge of a record album?

22. *Bicycle* At what speed is a bicyclist traveling if his 27-inch-diameter tires are rotating at an angular speed of 5π radians per second?

4.2 **In Exercises 23–26, find the point (x, y) on the unit circle that corresponds to the real number t.**

23. $t = \frac{2\pi}{3}$ **24.** $t = \frac{3\pi}{4}$

25. $t = \frac{5\pi}{6}$ **26.** $t = -\frac{4\pi}{3}$

In Exercises 27–30, evaluate (if possible) the six trigonometric functions of the real number.

27. $t = \frac{7\pi}{6}$ **28.** $t = \frac{\pi}{4}$

29. $t = -\frac{2\pi}{3}$ **30.** $t = 2\pi$

in Exercises 31–34, evaluate the trigonometric function using its period as an aid.

31. $\sin \frac{11\pi}{4}$ **32.** $\cos 4\pi$

33. $\sin\left(-\frac{17\pi}{6}\right)$ **34.** $\cos\left(-\frac{13\pi}{3}\right)$

In Exercises 35–38, use a calculator to evaluate the trigonometric function. Round to two decimal places.

35. $\tan 33$ **36.** $\csc 10.5$

37. $\sec \frac{12\pi}{5}$ **38.** $\sin\left(-\frac{\pi}{9}\right)$

4.3 **In Exercises 39–42, find the values of the six trigonometric functions of the angle θ in the figure.**

39. 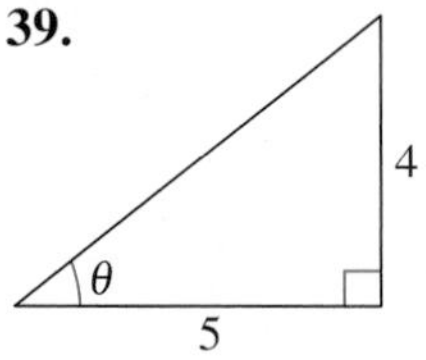

40. 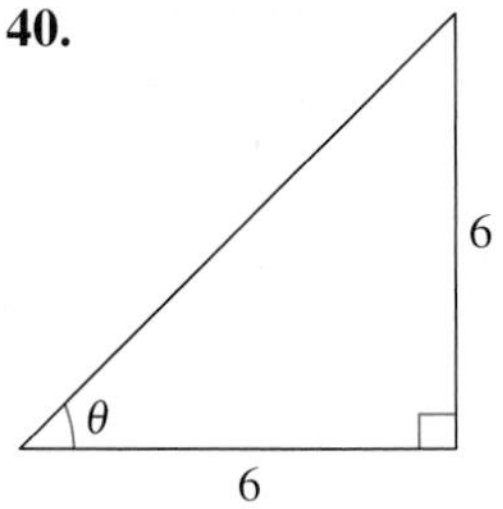

41. 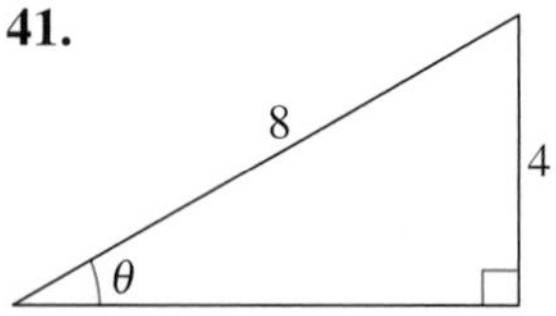

42. 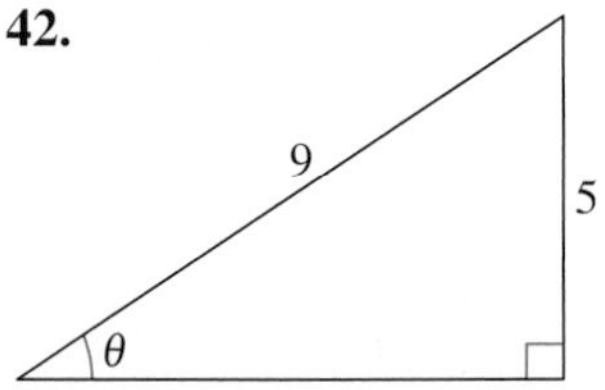

In Exercises 43–46, use the given function value and trigonometric identities (including the cofunction identities) to find the indicated trigonometric functions.

43. $\sin\theta = \frac{1}{3}$
(a) $\csc\theta$ (b) $\cos\theta$
(c) $\sec\theta$ (d) $\tan\theta$

44. $\tan\theta = 4$
(a) $\cot\theta$ (b) $\sec\theta$
(c) $\cos\theta$ (d) $\csc\theta$

45. $\csc\theta = 4$
(a) $\sin\theta$ (b) $\cos\theta$
(c) $\sec\theta$ (d) $\tan\theta$

46. $\csc\theta = 5$
(a) $\sin\theta$ (b) $\cot\theta$
(c) $\tan\theta$ (d) $\sec(90° - \theta)$

In Exercises 47–52, use a calculator to evaluate the trigonometric function. Round your answer to two decimal places.

47. $\tan 33°$ **48.** $\csc 11°$
49. $\sin 34.2°$ **50.** $\sec 79.3°$
51. $\cot 15°\ 14'$ **52.** $\cos 78°\ 11'58''$

53. *Railroad Grade* A train travels 3.5 kilometers on a straight track with a grade of $1°\ 10'$ (see figure). What is the vertical rise of the train in that distance?

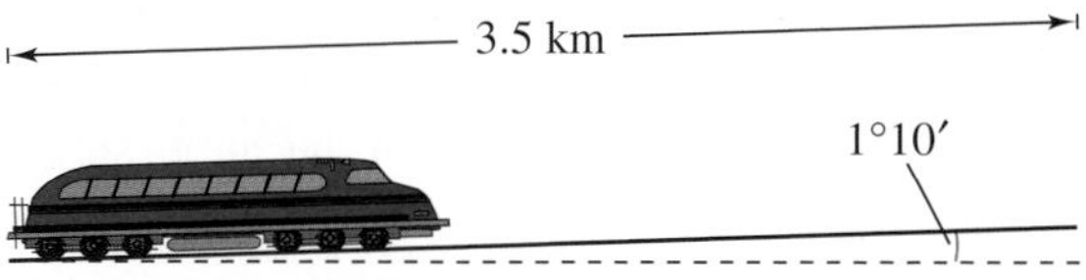

54. *Guy Wire* A guy wire runs from the ground to the top of a 25-foot telephone pole. The angle formed between the wire and the ground is 52°. How far from the base of the pole is the wire attached to the ground?

4.4 **In Exercises 55–62, find the six trigonometric functions of the angle θ (in standard position) whose terminal side passes through the point.**

55. $(12, 16)$ **56.** $(3, -4)$
57. $\left(\frac{2}{3}, \frac{5}{2}\right)$ **58.** $\left(-\frac{10}{3}, -\frac{2}{3}\right)$
59. $(-0.5, 4.5)$ **60.** $(0.3, 0.4)$
61. $(x, 4x),\ x > 0$ **62.** $(-2x, -3x),\ x > 0$

In Exercises 63–68, find the remaining five trigonometric functions of θ satisfying the condition.

63. $\sec\theta = \frac{6}{5},\ \tan\theta < 0$ **64.** $\csc\theta = \frac{3}{2},\ \cos\theta < 0$
65. $\sin\theta = \frac{3}{8},\ \cos\theta < 0$
66. $\tan\theta = \frac{5}{4},\ \cos\theta < 0$
67. $\cos\theta = -\frac{2}{5},\ \sin\theta > 0$
68. $\sin\theta = -\frac{2}{4},\ \cos\theta > 0$

In Exercises 69–74, evaluate the trigonometric function without using a calculator.

69. $\tan \pi/3$ **70.** $\sec \pi/4$
71. $\cos(-7\pi/3)$ **72.** $\cot(-5\pi/4)$
73. $\cos 495°$ **74.** $\sin(-150°)$

In Exercises 75–82, evaluate the trigonometric function of the real number. Round your answer to two decimal places.

75. $\sin 4$ **76.** $\tan 3$
77. $\sin(-3.2)$ **78.** $\cot(-4.8)$
79. $\sin 3\pi$ **80.** $\cot 1.5\pi$
81. $\sec 12\pi/5$ **82.** $\tan(-25\pi/7)$

4.5 **In Exercises 83–90, sketch a graph of the function. Include two full periods.**

83. $y = \sin x$ **84.** $y = \cos x$
85. $f(x) = 5 \sin \dfrac{2x}{5}$ **86.** $f(x) = 8 \cos\left(-\dfrac{x}{4}\right)$
87. $y = 2 + \sin x$ **88.** $y = -4 - \cos \pi x$
89. $g(t) = \frac{5}{2}\sin(t - \pi)$ **90.** $g(t) = 3\cos(t + \pi)$

91. *Sound Waves* Sound waves can be modeled by sine functions of the form $y = a \sin bx$, where x is measured in seconds.
(a) Write an equation of a sound wave whose amplitude is 2 and whose period is $\frac{1}{264}$ second.
(b) What is the frequency of the sound wave described in part (a)?

92. *Sound Waves* Use the cosine function $y = a \cos bx$ to model the sound wave described in Exercise 91.

4.6 **In Exercises 93–102, sketch a graph of the function. Include two full periods.**

93. $f(x) = \tan x$ **94.** $f(t) = \tan\left(t - \dfrac{\pi}{4}\right)$
95. $f(x) = \cot x$ **96.** $g(t) = 2 \cot 2t$
97. $f(x) = \sec x$ **98.** $h(t) = \sec\left(t - \dfrac{\pi}{4}\right)$
99. $f(x) = \csc x$ **100.** $f(t) = 3 \csc\left(2t + \dfrac{\pi}{4}\right)$

101. $f(x) = x \cos x$

102. $g(x) = e^x \cos x$

4.7 **In Exercises 103–108, evaluate the expression. If necessary, round your answer to two decimal places.**

103. $\arcsin\left(-\frac{1}{2}\right)$

104. $\arcsin(-1)$

105. $\arcsin 0.4$

106. $\arcsin 0.213$

107. $\sin^{-1}(-0.44)$

108. $\sin^{-1} 0.89$

In Exercises 109–112, evaluate the expression without the aid of a calculator.

109. $\arccos \dfrac{\sqrt{3}}{2}$

110. $\arccos \dfrac{\sqrt{2}}{2}$

111. $\cos^{-1}(-1)$

112. $\cos^{-1} \dfrac{\sqrt{3}}{2}$

In Exercises 113–120, use a calculator to approximate the value of the expression. Round your answer to two decimal places.

113. $\arccos 0.324$

114. $\arccos(-0.888)$

115. $\arctan 0.123$

116. $\arctan 2.34$

117. $\arctan 5.783$

118. $\arctan 99.1$

119. $\tan^{-1}(-1.5)$

120. $\tan^{-1} 8.2$

In Exercises 121–128, find the exact value of the expression.

121. $\sin(\arcsin 0.72)$

122. $\cos(\arccos 0.25)$

123. $\arctan(\tan \pi)$

124. $\arccos[\cos(-5\pi)]$

125. $\cos\left(\arctan \frac{3}{4}\right)$

126. $\tan\left(\arccos \frac{3}{5}\right)$

127. $\sec\left(\arctan \frac{12}{5}\right)$

128. $\cot\left[\arcsin\left(-\frac{12}{13}\right)\right]$

4.8 **129.** *Angle of Elevation* The height of a radio transmission tower is 70 meters, and it casts a shadow of length 30 meters (see figure). Find the angle of elevation of the sun.

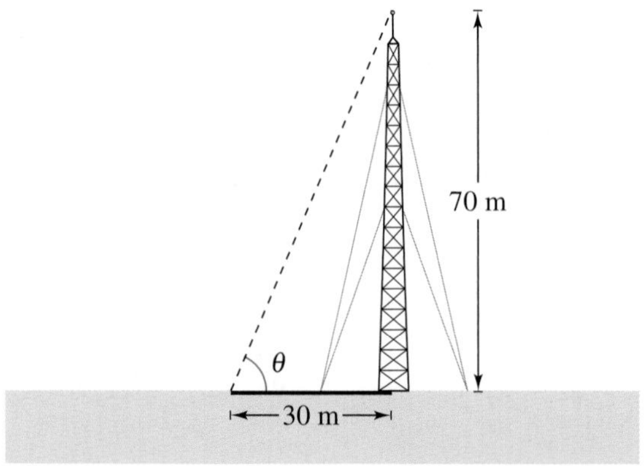

130. *Lost Ball* Your football has landed at the edge of the roof of your school building. When you are 25 feet from the base of the building, the angle of elevation to your football is 21°. How high off the ground is your football?

131. *Distance* From city A to city B, a plane flies 650 miles at a bearing of N 48° E. From city B to city C, the plane flies 810 miles at a bearing of S 65° E. Find the distance from A to C and the bearing from A to C.

132. *Wave Motion* Your fishing bobber oscillates in simple harmonic motion from the waves in the lake where you fish. Your bobber moves a total of 1.5 inches from its high point to its low point and returns to its high point every 3 seconds. Write an equation modeling the motion of your bobber if it is at its high point at time $t = 0$.

Synthesis

True or False? **In Exercises 133–136, determine whether the statement is true or false. Justify your answer.**

133. The tangent function is often useful for modeling simple harmonic motion.

134. The inverse sine function $y = \arcsin x$ cannot be defined as a function over any interval that is greater than the interval defined as $-\pi/2 \le y \le \pi/2$.

135. $y = \sin \theta$ is not a function because $\sin 30° = \sin 150°$.

136. Because $\tan 3\pi/4 = -1$, $\arctan(-1) = 3\pi/4$.

In Exercises 137–140, match the function $y = a \sin bx$ with its graph. Base your selection solely on your interpretation of the constants a and b. Explain your reasoning. [The graphs are labeled (a), (b), (c), and (d).]

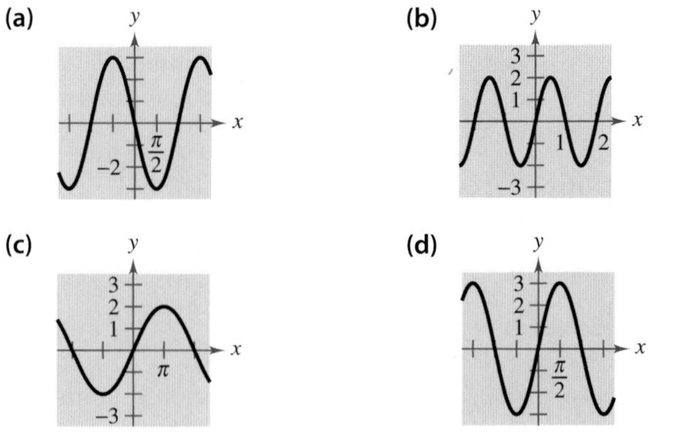

137. $y = 3 \sin x$

138. $y = -3 \sin x$

139. $y = 2 \sin \pi x$

140. $y = 2 \sin \dfrac{x}{2}$

141. Describe the behavior of $f(\theta) = \sec \theta$ at the zeros of $g(\theta) = \cos \theta$. Explain.

142. The function f is periodic, with period c. Therefore, $f(t + c) = f(t)$. Are the following equal? Explain.

(a) $f(t - 2c) \stackrel{?}{=} f(t)$

(b) $f\left(t + \frac{1}{2}c\right) \stackrel{?}{=} f\left(\frac{1}{2}t\right)$

(c) $f\left(\frac{1}{2}(t + c)\right) \stackrel{?}{=} f\left(\frac{1}{2}t\right)$

143. When graphing the sine and cosine functions, determining the amplitude is part of the analysis. Why is this not true for the other four trigonometric functions?

144. *Oscillation of a Spring* A weight is suspended from a ceiling by a steel spring. The weight is lifted (positive direction) from the equilibrium position and released. The resulting motion of the weight is modeled by

$$y = Ae^{-kt} \cos bt = \frac{1}{5}e^{-t/10} \cos 6t$$

where y is the distance in feet from equilibrium and t is the time in seconds. The graph of the function is given in the figure. For each of the following, describe the change in the system without graphing the resulting function.

(a) A is changed from $\frac{1}{5}$ to $\frac{1}{3}$.

(b) k is changed from $\frac{1}{10}$ to $\frac{1}{3}$.

(c) b is changed from 6 to 9.

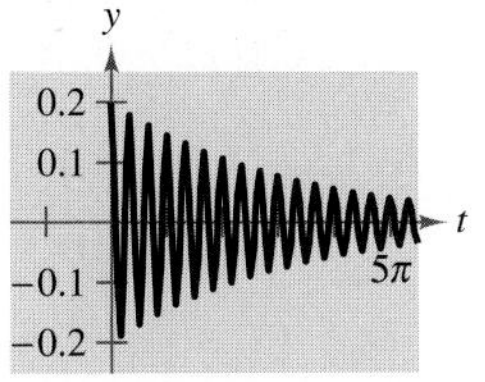

145. *Exploration* The base of the triangle in the figure is also the radius of a circular arc.

(a) Find the area A of the shaded region as a function of θ for $0 < \theta < \pi/2$.

(b) Use a graphing utility to graph the area function over the given domain. Interpret the graph in the context of the problem.

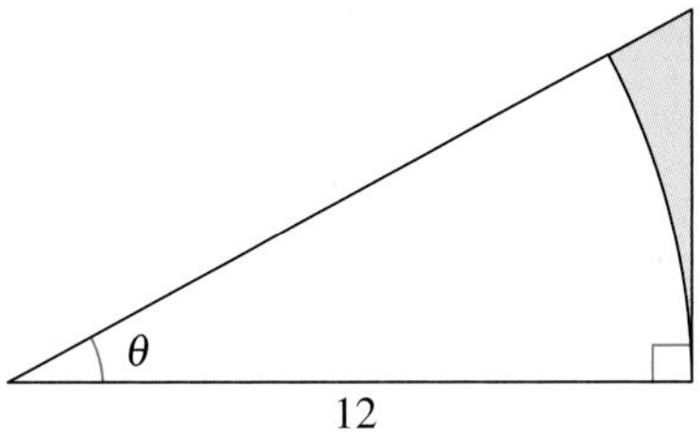

146. *Exploration* In calculus it can be shown that the arcsine and arctangent functions can be approximated by the polynomials

$$\arcsin x \approx x + \frac{x^3}{6} + \frac{3x^5}{40} + \frac{5x^7}{112}$$

$$\arctan x \approx x - \frac{x^3}{3} + \frac{x^5}{5} - \frac{x^7}{7}$$

where x is in radians.

(a) Use a graphing utility to graph the arcsine function and its polynomial approximation in the same viewing window. How do the graphs compare?

(b) Use a graphing utility to graph the arctangent function and its polynomial approximation in the same viewing window. How do the graphs compare?

(c) Study the pattern in the polynomial approximation of the arctangent function and guess the next term. Then repeat part (b). How did the accuracy of the approximation change when additional terms were added?

147. Describe a real-life application that can be represented by a simple harmonic motion model and is different from any that you've seen in this chapter. Explain which function you would use to model your application, and why. Explain how you would determine the amplitude, period, and frequency of the model for your application.

Chapter Project ▶ Analyzing a Graph

Graphs of functions that are combinations of algebraic functions and trigonometric functions can be difficult to sketch by hand. For such graphs, a graphing utility is helpful.

Example ▶ Sketching the Graph of a Function

Since 1974, the Mauna Loa Climate Observatory in Hawaii has been collecting data on the carbon dioxide level of the earth's atmosphere. A model that closely represents the data is

$$y = 323 + 1.5t + 0.001t^2 + 2.5 \sin 2\pi t$$

where y represents the monthly average of carbon dioxide concentration (in parts per million) and $t = 4$ represents January 1974, $t = 5$ represents January 1975, and so on. Sketch the graph of this function and explain the oscillations in the graph. (Source: National Oceanic and Atmospheric Administration, Climate Monitoring and Diagnostic Laboratory, Carbon Cycle-Greenhouse Gases)

Solution

The graph of the function is shown below. From the graph, you can see that the carbon dioxide level fluctuates each year. The low level each year, which occurs toward the end of the summer in the northern hemisphere, is caused by the intake of carbon dioxide in growing plants.

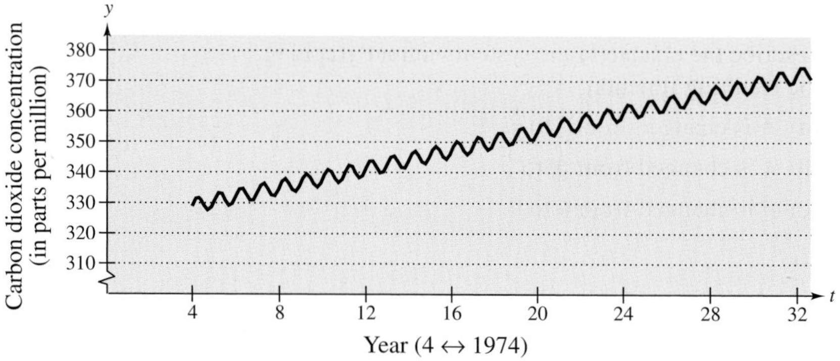

Chapter Project Investigations

1. Sketch the graph of the model given in the example for $27 \leq t \leq 29$. Between January 1997 and January 1999, what were the highest and lowest levels of carbon dioxide? When did each occur?

In Questions 2–4, use a graphing utility to graph the function. Choose a viewing window that you think produces a good representation of the important features of the graph.

2. $y = x^2 + \sin x$ **3.** $y = x^2 \sin x$ **4.** $y = \dfrac{\sin x}{x}$

Chapter Test

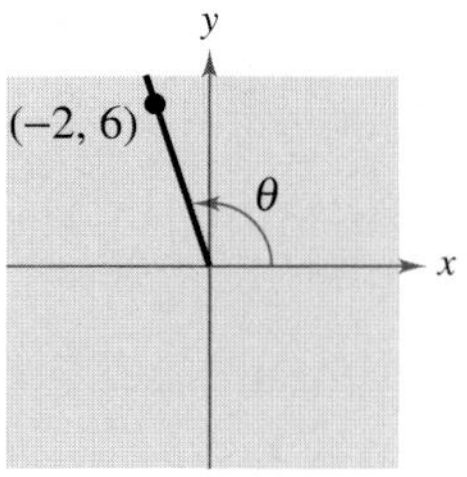

FIGURE FOR 3

The *Interactive* CD-ROM and *Internet* versions of this text provide answers to the Chapter Tests and Cumulative Tests. They also offer Chapter Pre-Tests (which test key skills and concepts covered in previous chapters) and Chapter Post-Tests, both of which have randomly generated exercises with diagnostic capabilities.

Take this test as you would take a test in class. After you are done, check your work against the answers given in the back of the book.

1. Consider the angle of magnitude $5\pi/4$ radians.
   (a) Sketch the angle in standard position.
   (b) Determine two coterminal angles (one positive and one negative).
   (c) Convert the angle to degree measure.
2. A truck is moving at a rate of 90 kilometers per hour, and the diameter of its wheels is 1 meter. Find the angular speed of the wheels in radians per minute.
3. Find the exact values of the six trigonometric functions of the angle θ shown in the figure.
4. Given that $\tan\theta = \frac{3}{2}$, find the other five trigonometric functions of θ.
5. Determine the reference angle θ' of the angle $\theta = 290°$ and sketch θ and θ' in standard position.
6. Determine the quadrant in which θ lies if $\sec\theta < 0$ and $\tan\theta > 0$.
7. Find two values of θ in degrees $(0 \le \theta < 360°)$ if $\cos\theta = -\sqrt{3}/2$. (Do not use a calculator.)
8. Use a calculator to approximate two values of θ in radians $(0 \le \theta < 2\pi)$ if $\csc\theta = 1.030$. Round the result to two decimal places.

In Exercises 9 and 10, find the remaining five trigonometric functions of θ satisfying the conditions.

9. $\cos\theta = \frac{3}{5},\ \tan\theta < 0$
10. $\sec\theta = -\frac{17}{8},\ \sin\theta > 0$

In Exercises 11 and 12, graph the function through two full periods without the aid of a graphing utility.

11. $g(x) = -2\sin\left(x - \frac{\pi}{4}\right)$
12. $f(\alpha) = \frac{1}{2}\tan 2\alpha$

In Exercises 13 and 14, use a graphing utility to graph the function. If the function is periodic, find its period.

13. $y = \sin 2\pi x + 2\cos\pi x$
14. $y = 6e^{-0.12t}\cos(0.25t),\quad 0 \le t \le 32$

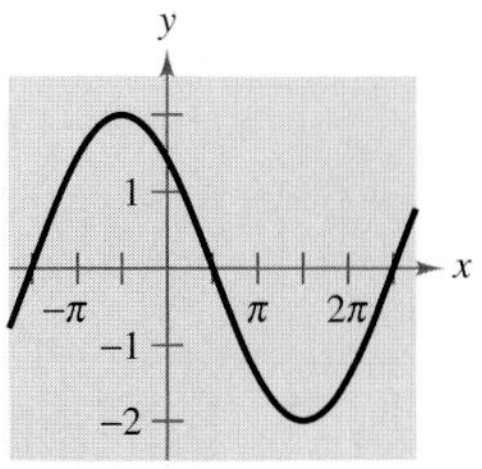

FIGURE FOR 15

15. Find a, b, and c for the function $f(x) = a\sin(bx + c)$ such that the graph of f matches the figure.
16. Find the exact value of $\tan\left(\arccos\frac{2}{3}\right)$ without the aid of a calculator.
17. Graph the function $f(x) = 2\arcsin\left(\frac{1}{2}x\right)$.
18. A plane is 80 miles south and 95 miles east of an airport. What bearing should be taken to fly directly to the airport?
19. Write the equation for the simple harmonic motion of a ball on a spring that starts at its lowest point of 6 inches below equilibrium, bounces to its maximum height of 6 inches above equilibrium, and returns to its lowest point in a total of 2 seconds.

Andy Lyons/Allsport

A football thrown by a quarterback follows a parabolic path. The horizontal distance the football travels depends not only on the speed of the throw, but on the angle at which the ball is thrown.

5 Analytic Trigonometry

▶ How to Study This Chapter

The Big Picture

In this chapter you will learn the following skills and concepts.

- How to use fundamental trigonometric identities to evaluate trigonometric functions and simplify trigonometric expressions
- How to verify trigonometric identities
- How to use standard algebraic techniques and inverse trigonometric functions to solve trigonometric equations
- How to use sum and difference formulas, multiple-angle formulas, power-reducing formulas, half-angle formulas, and product-to-sum formulas to rewrite and evaluate trigonometric functions

Important Vocabulary

As you encounter each new vocabulary term in this chapter, add the term and its definition to your notebook glossary.

Sum and difference formulas (p. 468)
Reduction formulas (p. 470)
Double-angle formulas (p. 475)
Power-reducing formulas (p. 477)
Half-angle formulas (p. 478)
Product-to-sum formulas (p. 479)
Sum-to-product formulas (p. 480)

Study Tools

- Learning objectives at the beginning of each section
- Chapter Summary (p. 486)
- Review Exercises (pp. 487–489)
- Chapter Test (p. 491)

Additional Resources

- Study and Solutions Guide
- Interactive Precalculus
- Videotapes for Chapter 5
- Precalculus Website
- Student Success Organizer

STUDY T!P

Avoid getting frustrated or spending too much time on one problem. Ask for help, take a break to clear your thoughts, sleep on it, rework the problem, or reread the section in the text.

5.1 Using Fundamental Identities

▶ **What you should learn**

- How to recognize and write the fundamental trigonometric identities
- How to use the fundamental trigonometric identities to evaluate trigonometric functions, simplify trigonometric expressions, and rewrite trigonometric expressions

▶ **Why you should learn it**

Fundamental trigonometric identities can be used to simplify trigonometric expressions. For instance, in Exercise 99 on page 449 you can use trigonometric identities to simplify an expression for the coefficient of friction.

Nancy Dudley/Stock Boston

Introduction

In Chapter 4, you studied the basic definitions, properties, graphs, and applications of the individual trigonometric functions. In this chapter, you will learn how to use the fundamental identities to

1. evaluate trigonometric functions.
2. simplify trigonometric expressions.
3. develop additional trigonometric identities.
4. solve trigonometric equations.

Fundamental Trigonometric Identities

Reciprocal Identities

$$\sin u = \frac{1}{\csc u} \qquad \cos u = \frac{1}{\sec u} \qquad \tan u = \frac{1}{\cot u}$$

$$\csc u = \frac{1}{\sin u} \qquad \sec u = \frac{1}{\cos u} \qquad \cot u = \frac{1}{\tan u}$$

Quotient Identities

$$\tan u = \frac{\sin u}{\cos u} \qquad \cot u = \frac{\cos u}{\sin u}$$

Pythagorean Identities

$$\sin^2 u + \cos^2 u = 1 \qquad 1 + \tan^2 u = \sec^2 u \qquad 1 + \cot^2 u = \csc^2 u$$

Cofunction Identities

$$\sin\left(\frac{\pi}{2} - u\right) = \cos u \qquad \cos\left(\frac{\pi}{2} - u\right) = \sin u$$

$$\tan\left(\frac{\pi}{2} - u\right) = \cot u \qquad \cot\left(\frac{\pi}{2} - u\right) = \tan u$$

$$\sec\left(\frac{\pi}{2} - u\right) = \csc u \qquad \csc\left(\frac{\pi}{2} - u\right) = \sec u$$

Even/Odd Identities

$$\sin(-u) = -\sin u, \qquad \cos(-u) = \cos u, \qquad \tan(-u) = -\tan u$$

$$\csc(-u) = -\csc u, \qquad \sec(-u) = \sec u, \qquad \cot(-u) = -\cot u$$

Pythagorean identities are sometimes used in radical form such as

$$\sin u = \pm\sqrt{1 - \cos^2 u}$$

or

$$\tan u = \pm\sqrt{\sec^2 u - 1}$$

where the sign depends on the choice of u.

The *Interactive* CD-ROM and *Internet* versions of this text offer a built-in graphing calculator, which can be used in the Examples, Explorations, Technology notes, and Exercises.

Encourage your students to learn the fundamental trigonometric identities well, because they are used frequently. They will also appear later in calculus. Also, stress that u can be an angle, a real number, or a variable.

Technology

You can use a graphing utility to check the result of Example 2. To do this, graph

$$y = \sin x \cos^2 x - \sin x$$

and

$$y = -\sin^3 x$$

in the same viewing window, as shown below. Because Example 2 shows the equivalence algebraically and the two graphs appear to coincide, you can conclude that the expressions are equivalent.

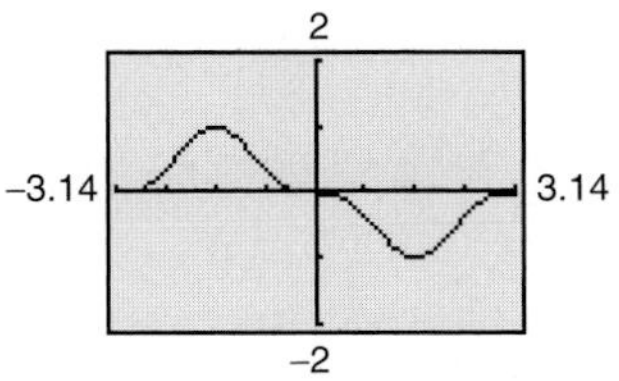

Remind students that they must use an algebraic approach to prove that two expressions are equivalent. A graphical approach can only confirm that the simplification found using algebraic techniques is correct.

Using the Fundamental Identities

One common use of trigonometric identities is to use given values of trigonometric functions to evaluate other trigonometric functions.

Example 1 ▶ Using Identities to Evaluate a Function

Use the values $\sec u = -\frac{3}{2}$ and $\tan u > 0$ to find the values of all six trigonometric functions.

Solution

Using a reciprocal identity, you have

$$\cos u = \frac{1}{\sec u} = \frac{1}{-3/2} = -\frac{2}{3}.$$

Using a Pythagorean identity, you have

$$\begin{aligned}\sin^2 u &= 1 - \cos^2 u \\ &= 1 - \left(-\frac{2}{3}\right)^2 \\ &= 1 - \frac{4}{9} \\ &= \frac{5}{9}.\end{aligned}$$

Because $\sec u < 0$ and $\tan u > 0$, it follows that u lies in Quadrant III. Moreover, because $\sin u$ is negative when u is in Quadrant III, you can choose the negative root and obtain $\sin u = -\sqrt{5}/3$. Now, knowing the values of the sine and cosine, you can find the values of all six trigonometric functions.

$$\sin u = -\frac{\sqrt{5}}{3} \qquad \csc u = \frac{1}{\sin u} = -\frac{3}{\sqrt{5}}$$

$$\cos u = -\frac{2}{3} \qquad \sec u = \frac{1}{\cos u} = -\frac{3}{2}$$

$$\tan u = \frac{\sin u}{\cos u} = \frac{-\sqrt{5}/3}{-2/3} = \frac{\sqrt{5}}{2} \qquad \cot u = \frac{1}{\tan u} = \frac{2}{\sqrt{5}}$$

Example 2 ▶ Simplifying a Trigonometric Expression

Simplify $\sin x \cos^2 x - \sin x$.

Solution

Factor the expression and then use a fundamental identity.

$$\begin{aligned}\sin x \cos^2 x - \sin x &= \sin x(\cos^2 x - 1) && \text{Monomial factor} \\ &= -\sin x(1 - \cos^2 x) && \text{Factor out } -1. \\ &= -\sin x(\sin^2 x) && \text{Pythagorean identity} \\ &= -\sin^3 x && \text{Multiply.}\end{aligned}$$

The *Interactive* CD-ROM and *Internet* versions of this text show every example with its solution; clicking on the *Try It!* button brings up similar problems. Guided Examples and Integrated Examples show step-by-step solutions to additional examples. Integrated Examples are related to several concepts in the section.

Example 3 ▶ Factoring Trigonometric Expressions

Factor each expression.

a. $\sec^2 \theta - 1$ **b.** $4 \tan^2 \theta + \tan \theta - 3$

Solution

a. Here you have the difference of two squares, which factors as

$$\sec^2 \theta - 1 = (\sec \theta - 1)(\sec \theta + 1).$$

b. This expression has the polynomial form $ax^2 + bx + c$, and it factors as

$$4 \tan^2 \theta + \tan \theta - 3 = (4 \tan \theta - 3)(\tan \theta + 1).$$

On occasion, factoring or simplifying can best be done by first rewriting the expression in terms of just *one* trigonometric function or in terms of *sine and cosine only.* These strategies are illustrated in Examples 4 and 5, respectively.

Example 4 ▶ Factoring a Trigonometric Expression

Factor $\csc^2 x - \cot x - 3$.

Solution

You can use the identity $\csc^2 x = 1 + \cot^2 x$ to rewrite the expression in terms of the cotangent.

$$\csc^2 x - \cot x - 3 = (1 + \cot^2 x) - \cot x - 3 \quad \text{Pythagorean identity}$$
$$= \cot^2 x - \cot x - 2 \quad \text{Combine like terms.}$$
$$= (\cot x - 2)(\cot x + 1) \quad \text{Factor.}$$

Example 5 ▶ Simplifying a Trigonometric Expression

Simplify $\sin t + \cot t \cos t$.

Solution

Begin by rewriting $\cot t$ in terms of sine and cosine.

$$\sin t + \cot t \cos t = \sin t + \left(\frac{\cos t}{\sin t}\right) \cos t \quad \text{Quotient identity}$$
$$= \frac{\sin^2 t + \cos^2 t}{\sin t} \quad \text{Add fractions.}$$
$$= \frac{1}{\sin t} \quad \text{Pythagorean identity}$$
$$= \csc t \quad \text{Reciprocal identity}$$

Example 6 ▶ Adding Trigonometric Expressions

Perform the addition and simplify.

$$\frac{\sin \theta}{1 + \cos \theta} + \frac{\cos \theta}{\sin \theta}$$

Solution

$$\frac{\sin \theta}{1 + \cos \theta} + \frac{\cos \theta}{\sin \theta} = \frac{(\sin \theta)(\sin \theta) + (\cos \theta)(1 + \cos \theta)}{(1 + \cos \theta)(\sin \theta)}$$

$$= \frac{\sin^2 \theta + \cos^2 \theta + \cos \theta}{(1 + \cos \theta)(\sin \theta)} \qquad \text{Multiply.}$$

$$= \frac{\cancel{1 + \cos \theta}}{\cancel{(1 + \cos \theta)}(\sin \theta)} \qquad \text{Pythagorean identity}$$

$$= \frac{1}{\sin \theta} \qquad \text{Divide out common factor.}$$

$$= \csc \theta \qquad \text{Reciprocal identity}$$

The last two examples in this section involve techniques for rewriting expressions in forms that are used in calculus.

Example 7 ▶ Rewriting a Trigonometric Expression

Rewrite $\frac{1}{1 + \sin x}$ so that it is *not* in fractional form.

Solution

From the Pythagorean identity $\cos^2 x = 1 - \sin^2 x = (1 - \sin x)(1 + \sin x)$, you can see that by multiplying both the numerator and the denominator by $(1 - \sin x)$ you produce a monomial denominator.

$$\frac{1}{1 + \sin x} = \frac{1}{1 + \sin x} \cdot \frac{1 - \sin x}{1 - \sin x} \qquad \text{Multiply numerator and denominator by } (1 - \sin x).$$

$$= \frac{1 - \sin x}{1 - \sin^2 x} \qquad \text{Multiply.}$$

$$= \frac{1 - \sin x}{\cos^2 x} \qquad \text{Pythagorean identity}$$

$$= \frac{1}{\cos^2 x} - \frac{\sin x}{\cos^2 x} \qquad \text{Separate fractions.}$$

$$= \frac{1}{\cos^2 x} - \frac{\sin x}{\cos x} \cdot \frac{1}{\cos x} \qquad \text{Separate fractions.}$$

$$= \sec^2 x - \tan x \sec x \qquad \text{Identities}$$

Example 8 ▶ Trigonometric Substitution

Use the substitution $x = 2 \tan \theta$, $0 < \theta < \pi/2$, to express

$$\sqrt{4 + x^2}$$

as a trigonometric function of θ.

Solution

Begin by letting $x = 2 \tan \theta$. Then, you can obtain

$$\sqrt{4 + x^2} = \sqrt{4 + (2 \tan \theta)^2} \quad \text{Substitute } 2 \tan \theta \text{ for } x.$$
$$= \sqrt{4 + 4 \tan^2 \theta} \quad \text{Rule of exponents}$$
$$= \sqrt{4(1 + \tan^2 \theta)} \quad \text{Factor.}$$
$$= \sqrt{4 \sec^2 \theta} \quad \text{Pythagorean identity}$$
$$= 2 \sec \theta. \quad \sec \theta > 0 \text{ for } 0 < \theta < \pi/2$$

Activities

1. Simplify, using the fundamental trigonometric identities.

   $$\frac{\cot^2 \theta}{\csc^2 \theta}$$

   Answer: $\cos^2 \theta$

2. Use the trigonometric substitution $x = 3 \cos \theta$ to rewrite the expression $\sqrt{9 - x^2}$ as a trigonometric function of θ, where

   $$0 < \theta < \frac{\pi}{2}.$$

   Answer: $3 \sin \theta$

Figure 5.1 shows the right triangle illustration of the trigonometric substitution in Example 8. For $0 < \theta < \pi/2$, you have

$$\text{opp} = x, \quad \text{adj} = 2, \quad \text{and} \quad \text{hyp} = \sqrt{4 + x^2}.$$

With these expressions, you can write the following.

$$\sec \theta = \frac{\sqrt{4 + x^2}}{2} \implies 2 \sec \theta = \sqrt{4 + x^2}$$

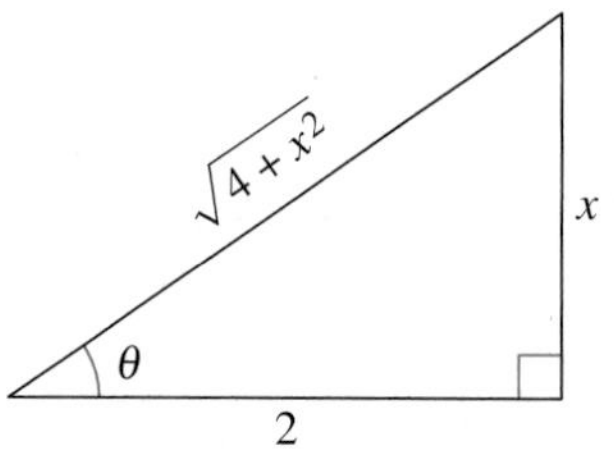

FIGURE 5.1

Writing ABOUT MATHEMATICS

Remembering Trigonometric Identities Most people find the Pythagorean identity involving sine and cosine to be fairly easy to remember: $\sin^2 u + \cos^2 u = 1$. The one involving tangent and secant, however, tends to give some people trouble. They can't remember if the identity is

$$1 + \tan^2 u \stackrel{?}{=} \sec^2 u \qquad \text{or} \qquad 1 + \sec^2 u \stackrel{?}{=} \tan^2 u.$$

Which of these two is the correct Pythagorean identity involving tangent and secant? Discuss how to remember (or derive) this identity. Can you think of easy ways to remember other fundamental trigonometric identities?

5.1 Exercises

The *Interactive* CD-ROM and *Internet* versions of this text contain step-by-step solutions to all odd-numbered Section and Review Exercises. They also provide Tutorial Exercises that link to Guided Examples for additional help.

In Exercises 1–14, use the given values to evaluate (if possible) the remaining trigonometric functions.

1. $\sin x = \dfrac{\sqrt{3}}{2}, \quad \cos x = -\dfrac{1}{2}$

2. $\tan x = \dfrac{\sqrt{3}}{3}, \quad \cos x = -\dfrac{\sqrt{3}}{2}$

3. $\sec \theta = \sqrt{2}, \quad \sin \theta = -\dfrac{\sqrt{2}}{2}$

4. $\csc \theta = \frac{5}{3}, \quad \tan \theta = \frac{3}{4}$

5. $\tan x = \frac{5}{12}, \quad \sec x = -\frac{13}{12}$

6. $\cot \phi = -3, \quad \sin \phi = \dfrac{\sqrt{10}}{10}$

7. $\sec \phi = \dfrac{3}{2}, \quad \csc \phi = -\dfrac{3\sqrt{5}}{5}$

8. $\cos\left(\dfrac{\pi}{2} - x\right) = \dfrac{3}{5}, \quad \cos x = \dfrac{4}{5}$

9. $\sin(-x) = -\dfrac{1}{3}, \quad \tan x = -\dfrac{\sqrt{2}}{4}$

10. $\sec x = 4, \quad \sin x > 0$

11. $\tan \theta = 2, \quad \sin \theta < 0$

12. $\csc \theta = -5, \quad \cos \theta < 0$

13. $\sin \theta = -1, \quad \cot \theta = 0$

14. $\tan \theta$ is undefined, $\quad \sin \theta > 0$

In Exercises 15–20, match the trigonometric expression with one of the following.

(a) $\sec x$ (b) -1 (c) $\cot x$
(d) 1 (e) $-\tan x$ (f) $\sin x$

15. $\sec x \cos x$

16. $\tan x \csc x$

17. $\cot^2 x - \csc^2 x$

18. $(1 - \cos^2 x)(\csc x)$

19. $\dfrac{\sin(-x)}{\cos(-x)}$

20. $\dfrac{\sin[(\pi/2) - x]}{\cos[(\pi/2) - x]}$

In Exercises 21–26, match the trigonometric expression with one of the following.

(a) $\csc x$ (b) $\tan x$ (c) $\sin^2 x$
(d) $\sin x \tan x$ (e) $\sec^2 x$ (f) $\sec^2 x + \tan^2 x$

21. $\sin x \sec x$

22. $\cos^2 x(\sec^2 x - 1)$

23. $\sec^4 x - \tan^4 x$

24. $\cot x \sec x$

25. $\dfrac{\sec^2 x - 1}{\sin^2 x}$

26. $\dfrac{\cos^2[(\pi/2) - x]}{\cos x}$

In Exercises 27–44, use the fundamental identities to simplify the expression. There is more than one correct form of each answer.

27. $\cot \theta \sec \theta$

28. $\cos \beta \tan \beta$

29. $\sin \phi(\csc \phi - \sin \phi)$

30. $\sec^2 x(1 - \sin^2 x)$

31. $\dfrac{\cot x}{\csc x}$

32. $\dfrac{\csc \theta}{\sec \theta}$

33. $\dfrac{1 - \sin^2 x}{\csc^2 x - 1}$

34. $\dfrac{1}{\tan^2 x + 1}$

35. $\sec \alpha \cdot \dfrac{\sin \alpha}{\tan \alpha}$

36. $\dfrac{\tan^2 \theta}{\sec^2 \theta}$

37. $\cos\left(\dfrac{\pi}{2} - x\right)\sec x$

38. $\cot\left(\dfrac{\pi}{2} - x\right)\cos x$

39. $\dfrac{\cos^2 y}{1 - \sin y}$

40. $\cos t(1 + \tan^2 t)$

41. $\sin \beta \tan \beta + \cos \beta$

42. $\csc \phi \tan \phi + \sec \phi$

43. $\cot u \sin u + \tan u \cos u$

44. $\sin \theta \sec \theta + \cos \theta \csc \theta$

In Exercises 45–56, factor the expression and use the fundamental identities to simplify. There is more than one correct form of each answer.

45. $\tan^2 x - \tan^2 x \sin^2 x$

46. $\sin^2 x \csc^2 x - \sin^2 x$

47. $\sin^2 x \sec^2 x - \sin^2 x$

48. $\cos^2 x + \cos^2 x \tan^2 x$

49. $\dfrac{\sec^2 x - 1}{\sec x - 1}$

50. $\dfrac{\cos^2 x - 4}{\cos x - 2}$

51. $\tan^4 x + 2\tan^2 x + 1$

52. $1 - 2\cos^2 x + \cos^4 x$

53. $\sin^4 x - \cos^4 x$

54. $\sec^4 x - \tan^4 x$

55. $\csc^3 x - \csc^2 x - \csc x + 1$

56. $\sec^3 x - \sec^2 x - \sec x + 1$

In Exercises 57–60, perform the multiplication and use the fundamental identities to simplify. There is more than one correct form of each answer.

57. $(\sin x + \cos x)^2$

58. $(\cot x + \csc x)(\cot x - \csc x)$

59. $(2 \csc x + 2)(2 \csc x - 2)$

60. $(3 - 3 \sin x)(3 + 3 \sin x)$

In Exercises 61–64, perform the addition or subtraction and use the fundamental identities to simplify. There is more than one correct form of each answer.

61. $\dfrac{1}{1 + \cos x} + \dfrac{1}{1 - \cos x}$

62. $\dfrac{1}{\sec x + 1} - \dfrac{1}{\sec x - 1}$

63. $\dfrac{\cos x}{1 + \sin x} + \dfrac{1 + \sin x}{\cos x}$

64. $\tan x - \dfrac{\sec^2 x}{\tan x}$

In Exercises 65–68, rewrite the expression so that it is not in fractional form. There is more than one correct form of each answer.

65. $\dfrac{\sin^2 y}{1 - \cos y}$

66. $\dfrac{5}{\tan x + \sec x}$

67. $\dfrac{3}{\sec x - \tan x}$

68. $\dfrac{\tan^2 x}{\csc x + 1}$

Numerical and Graphical Analysis **In Exercises 69–72, use a graphing utility to complete the table and graph the functions. Make a conjecture about y_1 and y_2.**

x	0.2	0.4	0.6	0.8	1.0	1.2	1.4
y_1							
y_2							

69. $y_1 = \cos\left(\dfrac{\pi}{2} - x\right), \quad y_2 = \sin x$

70. $y_1 = \sec x - \cos x, \quad y_2 = \sin x \tan x$

71. $y_1 = \dfrac{\cos x}{1 - \sin x}, \quad y_2 = \dfrac{1 + \sin x}{\cos x}$

72. $y_1 = \sec^4 x - \sec^2 x, \quad y_2 = \tan^2 x + \tan^4 x$

In Exercises 73–76, use a graphing utility to determine which of the six trigonometric functions is equal to the expression. Verify your answer algebraically.

73. $\cos x \cot x + \sin x$

74. $\sec x \csc x - \tan x$

75. $\dfrac{1}{\sin x}\left(\dfrac{1}{\cos x} - \cos x\right)$

76. $\dfrac{1}{2}\left(\dfrac{1 + \sin \theta}{\cos \theta} + \dfrac{\cos \theta}{1 + \sin \theta}\right)$

In Exercises 77–82, use the trigonometric substitution to write the algebraic expression as a trigonometric function of θ, where $0 < \theta < \pi/2$.

77. $\sqrt{9 - x^2}, \quad x = 3 \cos \theta$

78. $\sqrt{64 - 16x^2}, \quad x = 2 \cos \theta$

79. $\sqrt{x^2 - 9}, \quad x = 3 \sec \theta$

80. $\sqrt{x^2 - 4}, \quad x = 2 \sec \theta$

81. $\sqrt{x^2 + 25}, \quad x = 5 \tan \theta$

82. $\sqrt{x^2 + 100}, \quad x = 10 \tan \theta$

In Exercises 83–86, use the trigonometric substitution to write the algebraic equation as a trigonometric function of θ, where $-\pi/2 < \theta < \pi/2$. Then find $\sin \theta$ and $\cos \theta$.

83. $3 = \sqrt{9 - x^2}, \quad x = 3 \sin \theta$

84. $3 = \sqrt{36 - x^2}, \quad x = 6 \sin \theta$

85. $2\sqrt{2} = \sqrt{16 - 4x^2}, \quad x = 2 \cos \theta$

86. $-5\sqrt{3} = \sqrt{100 - x^2}, \quad x = 10 \cos \theta$

In Exercises 87–90, use a graphing utility to solve the equation for θ, where $0 \le \theta < 2\pi$.

87. $\sin \theta = \sqrt{1 - \cos^2 \theta}$

88. $\cos \theta = -\sqrt{1 - \sin^2 \theta}$

89. $\sec \theta = \sqrt{1 + \tan^2 \theta}$

90. $\csc \theta = \sqrt{1 + \cot^2 \theta}$

In Exercises 91–94, rewrite the expression as a single logarithm and simplify the result.

91. $\ln|\cos x| - \ln|\sin x|$

92. $\ln|\sec x| + \ln|\sin x|$

93. $\ln|\cot t| + \ln(1 + \tan^2 t)$

94. $\ln(\cos^2 t) + \ln(1 + \tan^2 t)$

In Exercises 95–98, use a calculator to demonstrate the identity for the given values of θ.

95. $\csc^2 \theta - \cot^2 \theta = 1$

(a) $\theta = 132°$, (b) $\theta = \dfrac{2\pi}{7}$

96. $\tan^2 \theta + 1 = \sec^2 \theta$

(a) $\theta = 346°$, (b) $\theta = 3.1$

97. $\cos\left(\dfrac{\pi}{2} - \theta\right) = \sin \theta$

(a) $\theta = 80°$, (b) $\theta = 0.8$

98. $\sin(-\theta) = -\sin \theta$

(a) $\theta = 250°$, (b) $\theta = \frac{1}{2}$

99. *Friction* The forces acting on an object weighing W units on an inclined plane positioned at an angle of θ with the horizontal (see figure) are modeled by $\mu W \cos \theta = W \sin \theta$, where μ is the coefficient of friction. Solve the equation for μ and simplify the result.

Synthesis

True or False? **In Exercises 100 and 101, determine whether the statement is true or false. Justify your answer.**

100. The even and odd trigonometric identities are helpful for determining whether the value of a trigonometric function is positive or negative.

101. A cofunction identity can be used to transform a tangent function so that it can be represented by a cosecant function.

Calculus **In Exercises 102–105, fill in the blanks. (*Note:* The notation $x \to c^+$ indicates that x approaches c from the right and $x \to c^-$ indicates that x approaches c from the left.)**

102. As $x \to \dfrac{\pi}{2}^-$, $\sin x \to$ ____ and $\csc x \to$ ____.

103. As $x \to 0^+$, $\cos x \to$ ____ and $\sec x \to$ ____.

104. As $x \to \dfrac{\pi}{2}^-$, $\tan x \to$ ____ and $\cot x \to$ ____.

105. As $x \to \pi^+$, $\sin x \to$ ____ and $\csc x \to$ ____.

In Exercises 106–111, determine whether or not the equation is an identity, and give a reason for your answer.

106. $\cos \theta = \sqrt{1 - \sin^2 \theta}$

107. $\cot \theta = \sqrt{\csc^2 \theta + 1}$

108. $(\sin k\theta)/(\cos k\theta) = \tan \theta$, k is a constant.

109. $1/(5 \cos \theta) = 5 \sec \theta$

110. $\sin \theta \csc \theta = 1$

111. $\sin \theta \csc \phi = 1$

112. Express each of the other trigonometric functions of θ in terms of $\sin \theta$.

113. Express each of the other trigonometric functions of θ in terms of $\cos \theta$.

Review

In Exercises 114–117, perform the operation and simplify.

114. $(\sqrt{x} + 5)(\sqrt{x} - 5)$

115. $\sqrt{v}(\sqrt{20} - \sqrt{5})$

116. $(2\sqrt{z} + 3)^2$

117. $50x/(\sqrt{30} - 5)$

In Exercises 118–121, sketch the graph of the function. Include two full periods.

118. $y = -4 \sin(2x - 2\pi)$

119. $y = 3 \sec(\pi x - \pi)$

120. $y = 2 - \frac{1}{2} \tan 2x$

121. $y = \frac{1}{4} \sec 2x$

5.2 Verifying Trigonometric Identities

▶ What you should learn

- How to plan a strategy for verifying trigonometric identities
- How to verify trigonometric identities

▶ Why you should learn it

You can use trigonometric identities to rewrite trigonometric expressions. For instance, in Exercise 65 on page 456 you can use trigonometric identities to simplify the expression for a rate of change.

A computer animation of this concept appears in the *Interactive* CD-ROM and *Internet* versions of this text.

You may want to review the distinctions among expressions, equations, and identities. Have your students look at some algebraic identities and conditional equations before starting this section. It is important for them to understand what it means to verify an identity and not try to solve it as an equation.

Introduction

In this section, you will study techniques for verifying trigonometric identities. In the next section, you will study techniques for solving trigonometric equations. The key to verifying identities *and* solving equations is the ability to use the fundamental identities and the rules of algebra to rewrite trigonometric expressions.

Remember that a *conditional equation* is an equation that is true for only some of the values in its domain. For example, the conditional equation

$$\sin x = 0 \qquad \text{Conditional equation}$$

is true only for $x = n\pi$, where n is an integer. When you find these values, you are *solving* the equation.

On the other hand, an equation that is true for all real values in the domain of the variable is an *identity*. For example, the familiar equation

$$\sin^2 x = 1 - \cos^2 x \qquad \text{Identity}$$

is true for all real numbers x. So, it is an identity.

Although there are similarities, verifying that a trigonometric equation is an identity is quite different from solving an equation. There is no well-defined set of rules to follow in verifying trigonometric identities, and the process is best learned by practice.

For instance, to verify that the trigonometric equation $\tan\theta\cos\theta = \sin\theta$ is an identity, begin by working with the more complicated, left side of the equation.

$$\tan\theta\cos\theta = \left(\frac{\sin\theta}{\cos\theta}\right)\cos\theta \qquad \text{Rewrite } \tan\theta \text{ as } \frac{\sin\theta}{\cos\theta}.$$

$$= \left(\frac{\sin\theta}{\cancel{\cos\theta}}\right)\cancel{\cos\theta} \qquad \text{Divide out } \cos\theta.$$

$$= \sin\theta \qquad \text{Simplify.}$$

The result shows that the left side of the equation is equal to the right side. Therefore, the identity has been verified.

Guidelines for Verifying Trigonometric Identities

1. Work with one side of the equation at a time. It is often better to work with the more complicated side first.
2. Look for opportunities to factor an expression, add fractions, square a binomial, or create a monomial denominator.
3. Look for opportunities to use the fundamental identities. Note which functions are in the final expression you want. Sines and cosines pair up well, as do secants and tangents, and cosecants and cotangents.
4. If the preceding guidelines do not help, try converting all terms to sines and cosines.
5. Always try *something*. Even paths that lead to dead ends give you insights.

Verifying Trigonometric Identities

Example 1 ▶ Verifying a Trigonometric Identity

Verify the identity $\dfrac{\sec^2\theta - 1}{\sec^2\theta} = \sin^2\theta$.

Solution

Because the left side is more complicated, start with it.

$$
\begin{aligned}
\frac{\sec^2\theta - 1}{\sec^2\theta} &= \frac{(\tan^2\theta + 1) - 1}{\sec^2\theta} && \text{Pythagorean identity}\\
&= \frac{\tan^2\theta}{\sec^2\theta} && \text{Simplify.}\\
&= \tan^2\theta(\cos^2\theta) && \text{Reciprocal identity}\\
&= \frac{\sin^2\theta}{\cancel{(\cos^2\theta)}}\cancel{(\cos^2\theta)} && \text{Quotient identity}\\
&= \sin^2\theta && \text{Simplify.}
\end{aligned}
$$

Here is another way to verify the identity in Example 1.

$$
\begin{aligned}
\frac{\sec^2\theta - 1}{\sec^2\theta} &= \frac{\sec^2\theta}{\sec^2\theta} - \frac{1}{\sec^2\theta} && \text{Rewrite as the difference of fractions.}\\
&= 1 - \cos^2\theta && \text{Reciprocal identity}\\
&= \sin^2\theta && \text{Pythagorean identity}
\end{aligned}
$$

As you can see, there can be more than one way to verify an identity. Your method may differ from that used by your instructor or fellow students. Here is a good chance to be creative and establish your own style, but try to be as efficient as possible.

Example 2 ▶ Combining Fractions Before Using Identities

Verify the identity $\dfrac{1}{1 - \sin\alpha} + \dfrac{1}{1 + \sin\alpha} = 2\sec^2\alpha$.

Solution

$$
\begin{aligned}
\frac{1}{1 - \sin\alpha} + \frac{1}{1 + \sin\alpha} &= \frac{1 + \sin\alpha + 1 - \sin\alpha}{(1 - \sin\alpha)(1 + \sin\alpha)} && \text{Add fractions.}\\
&= \frac{2}{1 - \sin^2\alpha} && \text{Simplify.}\\
&= \frac{2}{\cos^2\alpha} && \text{Pythagorean identity}\\
&= 2\sec^2\alpha && \text{Reciprocal identity}
\end{aligned}
$$

Example 3 ▶ Verifying a Trigonometric Identity

Verify the identity

$$(\tan^2 x + 1)(\cos^2 x - 1) = -\tan^2 x.$$

Solution

By applying identities before multiplying, you obtain the following.

$$(\tan^2 x + 1)(\cos^2 x - 1) = (\sec^2 x)(-\sin^2 x) \quad \text{Pythagorean identities}$$

$$= -\frac{\sin^2 x}{\cos^2 x} \quad \text{Reciprocal identity}$$

$$= -\left(\frac{\sin x}{\cos x}\right)^2 \quad \text{Rule of exponents}$$

$$= -\tan^2 x \quad \text{Quotient identity}$$

Example 4 ▶ Converting to Sines and Cosines

Verify the identity

$$\tan x + \cot x = \sec x \csc x.$$

Solution

In this case there appear to be no fractions to add, no products to find, and no opportunities to use the Pythagorean identities. So, try converting the left side into sines and cosines to see what happens.

$$\tan x + \cot x = \frac{\sin x}{\cos x} + \frac{\cos x}{\sin x} \quad \text{Quotient identities}$$

$$= \frac{\sin^2 x + \cos^2 x}{\cos x \sin x} \quad \text{Add fractions.}$$

$$= \frac{1}{\cos x \sin x} \quad \text{Pythagorean identity}$$

$$= \frac{1}{\cos x} \cdot \frac{1}{\sin x} \quad \text{Product of fractions}$$

$$= \sec x \csc x \quad \text{Reciprocal identities}$$

Recall from algebra that *rationalizing the denominator* using conjugates is, on occasion, a powerful simplification technique. A related form of this technique works for simplifying trigonometric expressions as well. For instance, to simplify $1/(1 - \cos x)$, multiply the numerator and the denominator by $1 + \cos x$.

$$\frac{1}{1 - \cos x} = \frac{1}{1 - \cos x}\left(\frac{1 + \cos x}{1 + \cos x}\right) = \frac{1 + \cos x}{1 - \cos^2 x} = \frac{1 + \cos x}{\sin^2 x}$$

$$= \csc^2 x(1 + \cos x)$$

This technique is demonstrated in the next example.

Example 5 ▶ Verifying Trigonometric Identities

Verify the identity $\sec y + \tan y = \dfrac{\cos y}{1 - \sin y}$.

Solution

Begin with the *right* side. Note that you can create a monomial denominator by multiplying the numerator and denominator by $(1 + \sin y)$.

$$\frac{\cos y}{1 - \sin y} = \frac{\cos y}{1 - \sin y}\left(\frac{1 + \sin y}{1 + \sin y}\right) \qquad \text{Multiply numerator and denominator by } (1 + \sin y).$$

$$= \frac{\cos y + \cos y \sin y}{1 - \sin^2 y} \qquad \text{Multiply.}$$

$$= \frac{\cos y + \cos y \sin y}{\cos^2 y} \qquad \text{Pythagorean identity}$$

$$= \frac{\cos y}{\cos^2 y} + \frac{\cos y \sin y}{\cos^2 y} \qquad \text{Separate fractions.}$$

$$= \frac{1}{\cos y} + \frac{\sin y}{\cos y} \qquad \text{Simplify.}$$

$$= \sec y + \tan y \qquad \text{Identities}$$

In Examples 1 through 5, you have been verifying trigonometric identities by working with one side of the equation and converting to the form given on the other side. On occasion it is practical to work with each side *separately*, to obtain one common form equivalent to both sides. This is illustrated in Example 6.

Example 6 ▶ Working with Each Side Separately

Verify the identity $\dfrac{\cot^2 \theta}{1 + \csc \theta} = \dfrac{1 - \sin \theta}{\sin \theta}$.

Solution

Working with the left side, you have

$$\frac{\cot^2 \theta}{1 + \csc \theta} = \frac{\csc^2 \theta - 1}{1 + \csc \theta} \qquad \text{Pythagorean identity}$$

$$= \frac{(\csc \theta - 1)\cancel{(\csc \theta + 1)}}{\cancel{1 + \csc \theta}} \qquad \text{Factor.}$$

$$= \csc \theta - 1. \qquad \text{Simplify.}$$

Now, simplifying the right side, you have

$$\frac{1 - \sin \theta}{\sin \theta} = \frac{1}{\sin \theta} - \frac{\sin \theta}{\sin \theta} \qquad \text{Separate fractions.}$$

$$= \csc \theta - 1. \qquad \text{Reciprocal identity}$$

The identity is verified because both sides are equal to $\csc \theta - 1$.

STUDY T!P

The technique of cross multiplication is not used when verifying trigonometric identities because you do not know that the expressions are equal. Cross multiplication is used in solving an equation, when you know that the left side equals the right side.

In Example 7, powers of trigonometric functions are rewritten as more complicated sums of products of trigonometric functions. This is a common procedure used in calculus.

Example 7 ▶ Three Examples from Calculus

Verify each identity.

a. $\tan^4 x = \tan^2 x \sec^2 x - \tan^2 x$

b. $\sin^3 x \cos^4 x = (\cos^4 x - \cos^6 x) \sin x$

c. $\csc^4 x \cot x = \csc^2 x(\cot x + \cot^3 x)$

Solution

a. $\tan^4 x = (\tan^2 x)(\tan^2 x)$ Separate factors.

$= \tan^2 x(\sec^2 x - 1)$ Pythagorean identity

$= \tan^2 x \sec^2 x - \tan^2 x$ Multiply.

b. $\sin^3 x \cos^4 x = \sin^2 x \cos^4 x \sin x$ Separate factors.

$= (1 - \cos^2 x)\cos^4 x \sin x$ Pythagorean identity

$= (\cos^4 x - \cos^6 x) \sin x$ Multiply.

c. $\csc^4 x \cot x = \csc^2 x \csc^2 x \cot x$ Separate factors.

$= \csc^2 x(1 + \cot^2 x) \cot x$ Pythagorean identity

$= \csc^2 x(\cot x + \cot^3 x)$ Multiply.

Alternate Writing About Mathematics

1. Ask students to assemble a list of techniques and strategies for rewriting trigonometric expressions such as those demonstrated in the examples of this section.
2. Ask students to work in pairs. Each student should create an identity equation from the fundamental trigonometric identities. Partners then trade and verify one another's identities. Then have students write a brief explanation of the techniques they used to create the identities.

Writing ABOUT MATHEMATICS

Error Analysis Suppose you are tutoring a student in trigonometry. One of the homework problems your student encounters asks whether the following statement is an identity.

$$\tan^2 x \sin^2 x \stackrel{?}{=} \frac{5}{6} \tan^2 x$$

Your student does not attempt to verify the equivalence algebraically, but mistakenly uses only a graphical approach. Using range settings of

Xmin $= -3\pi$

Xmax $= 3\pi$

Xscl $= \pi/2$

Ymin $= -20$

Ymax $= 20$

Yscl $= 1$

your student graphs both sides of the expression on a graphing utility and concludes that the statement is an identity.

What is wrong with your student's reasoning? Explain. Discuss the limitations of verifying identities graphically.

5.2 Exercises

In Exercises 1–44, verify the identity.

1. $\sin t \csc t = 1$
2. $\sec y \cos y = 1$
3. $(1 + \sin \alpha)(1 - \sin \alpha) = \cos^2 \alpha$
4. $\cot^2 y(\sec^2 y - 1) = 1$
5. $\cos^2 \beta - \sin^2 \beta = 1 - 2\sin^2 \beta$
6. $\cos^2 \beta - \sin^2 \beta = 2\cos^2 \beta - 1$
7. $\tan^2 \theta + 4 = \sec^2 \theta + 3$
8. $2 - \sec^2 z = 1 - \tan^2 z$
9. $\sin^2 \alpha - \sin^4 \alpha = \cos^2 \alpha - \cos^4 \alpha$
10. $\cos x + \sin x \tan x = \sec x$
11. $\dfrac{\csc^2 \theta}{\cot \theta} = \csc \theta \sec \theta$
12. $\dfrac{\cot^3 t}{\csc t} = \cos t(\csc^2 t - 1)$
13. $\dfrac{\cot^2 t}{\csc t} = \csc t - \sin t$
14. $\dfrac{1}{\tan \beta} + \tan \beta = \dfrac{\sec^2 \beta}{\tan \beta}$
15. $\sin^{1/2} x \cos x - \sin^{5/2} x \cos x = \cos^3 x\sqrt{\sin x}$
16. $\sec^6 x(\sec x \tan x) - \sec^4 x(\sec x \tan x) = \sec^5 x \tan^3 x$
17. $\dfrac{1}{\sec x \tan x} = \csc x - \sin x$
18. $\dfrac{\sec \theta - 1}{1 - \cos \theta} = \sec \theta$
19. $\cot \alpha + \tan \alpha = \csc \alpha \sec \alpha$
20. $\sec x - \cos x = \sin x \tan x$
21. $\sin x \cos x + \sin^3 x \sec x = \tan x$
22. $\dfrac{\sec x + \tan x}{\sec x - \tan x} = (\sec x + \tan x)^2$
23. $\dfrac{1}{\tan x} + \dfrac{1}{\cot x} = \tan x + \cot x$
24. $\dfrac{1}{\sin x} - \dfrac{1}{\csc x} = \csc x - \sin x$
25. $\dfrac{\cos \theta \cot \theta}{1 - \sin \theta} - 1 = \csc \theta$
26. $\dfrac{1 + \sin \theta}{\cos \theta} + \dfrac{\cos \theta}{1 + \sin \theta} = 2 \sec \theta$
27. $\dfrac{1}{\sin x + 1} + \dfrac{1}{\csc x + 1} = 1$
28. $\cos x - \dfrac{\cos x}{1 - \tan x} = \dfrac{\sin x \cos x}{\sin x - \cos x}$
29. $\tan\left(\dfrac{\pi}{2} - \theta\right) \tan \theta = 1$
30. $\dfrac{\cos[(\pi/2) - x]}{\sin[(\pi/2) - x]} = \tan x$
31. $\dfrac{\csc(-x)}{\sec(-x)} = -\cot x$
32. $(1 + \sin y)[1 + \sin(-y)] = \cos^2 y$
33. $\dfrac{\cos(-\theta)}{1 + \sin(-\theta)} = \sec \theta + \tan \theta$
34. $\dfrac{\csc(-\theta) + 1}{\cos(-\theta) + \cot(-\theta)} = \sec \theta$
35. $\dfrac{\sin x \cos y + \cos x \sin y}{\cos x \cos y - \sin x \sin y} = \dfrac{\tan x + \tan y}{1 - \tan x \tan y}$
36. $\dfrac{\tan x + \tan y}{1 - \tan x \tan y} = \dfrac{\cot x + \cot y}{\cot x \cot y - 1}$
37. $\dfrac{\tan x + \cot y}{\tan x \cot y} = \tan y + \cot x$
38. $\dfrac{\cos x - \cos y}{\sin x + \sin y} + \dfrac{\sin x - \sin y}{\cos x + \cos y} = 0$
39. $\sqrt{\dfrac{1 + \sin \theta}{1 - \sin \theta}} = \dfrac{1 + \sin \theta}{|\cos \theta|}$
40. $\sqrt{\dfrac{1 - \cos \theta}{1 + \cos \theta}} = \dfrac{1 - \cos \theta}{|\sin \theta|}$
41. $\cos^2 \beta + \cos^2\left(\dfrac{\pi}{2} - \beta\right) = 1$
42. $\sec^2 y - \cot^2\left(\dfrac{\pi}{2} - y\right) = 1$
43. $\sin t \csc\left(\dfrac{\pi}{2} - t\right) = \tan t$
44. $\sec^2\left(\dfrac{\pi}{2} - x\right) - 1 = \cot^2 x$

In Exercises 45–56, use a graphing utility to verify the identity graphically; then confirm it algebraically.

45. $2\sec^2 x - 2\sec^2 x \sin^2 x - \sin^2 x - \cos^2 x = 1$

46. $\csc x(\csc x - \sin x) + \dfrac{\sin x - \cos x}{\sin x} + \cot x = \csc^2 x$

47. $2 + \cos^2 x - 3\cos^4 x = \sin^2 x(2 + 3\cos^2 x)$

48. $4\tan^4 x + \tan^2 x - 3 = \sec^2 x(4\tan^2 x - 3)$

49. $\csc^4 x - 2\csc^2 x + 1 = \cot^4 x$

50. $(\sin^4 \beta - 2\sin^2 \beta + 1)\cos \beta = \cos^5 \beta$

51. $\sec^4 \theta - \tan^4 \theta = 1 + 2\tan^2 \theta$

52. $\csc^4 \theta - \cot^4 \theta = 2\csc^2 \theta - 1$

53. $\dfrac{\cos x}{1 + \sin x} = \dfrac{1 - \sin x}{\cos x}$

54. $\dfrac{\cot \alpha}{\csc \alpha - 1} = \dfrac{\csc \alpha + 1}{\cot \alpha}$

55. $\dfrac{\tan^3 \alpha - 1}{\tan \alpha - 1} = \tan^2 \alpha + \tan \alpha + 1$

56. $\dfrac{\sin^3 \beta + \cos^3 \beta}{\sin \beta + \cos \beta} = 1 - \sin \beta \cos \beta$

In Exercises 57–60, use the properties of logarithms and trigonometric identities to verify the identity.

57. $\ln|\tan \theta| = \ln|\sin \theta| - \ln|\cos \theta|$

58. $\ln|\sec \theta| = -\ln|\cos \theta|$

59. $-\ln(1 + \cos \theta) = \ln(1 - \cos \theta) - 2\ln|\sin \theta|$

60. $-\ln|\sec \theta + \tan \theta| = \ln|\sec \theta - \tan \theta|$

In Exercises 61–64, use the cofunction identities to evaluate the expression without the aid of a calculator.

61. $\sin^2 25° + \sin^2 65°$

62. $\cos^2 55° + \cos^2 35°$

63. $\cos^2 20° + \cos^2 52° + \cos^2 38° + \cos^2 70°$

64. $\sin^2 12° + \sin^2 40° + \sin^2 50° + \sin^2 78°$

65. *Rate of Change* The rate of change of the function

$f(x) = \sin x + \csc x$

with respect to change in the variable x is given by the expression

$\cos x - \csc x \cot x.$

Show that the expression for the rate of change can also be

$-\cos x \cot^2 x.$

Synthesis

True or False? **In Exercises 66 and 67, determine whether the statement is true or false. Justify your answer.**

66. The equation $\sin^2 \theta + \cos^2 \theta = 1 + \tan^2 \theta$ is an identity, because $\sin^2(0) + \cos^2(0) = 1$ and $1 + \tan^2(0) = 1$.

67. The equation $1 + \tan^2 \theta = 1 + \cot^2 \theta$ is *not* an identity, because it is true that $1 + \tan^2(\pi/6) = 1\frac{1}{3}$, and $1 + \cot^2(\pi/6) = 4$.

Think About It **In Exercises 68–71, explain why the equation is *not* an identity and find one value of the variable for which the equation is not true.**

68. $\sin \theta = \sqrt{1 - \cos^2 \theta}$

69. $\tan \theta = \sqrt{\sec^2 \theta - 1}$

70. $\sqrt{\tan^2 x} = \tan x$

71. $\sqrt{\sin^2 x + \cos^2 x} = \sin x + \cos x$

72. Verify that for all integers n, $\cos\left[\dfrac{(2n + 1)\pi}{2}\right] = 0$.

73. Verify that for all integers n, $\sin\left[\dfrac{(12n + 1)\pi}{6}\right] = \dfrac{1}{2}$.

Review

In Exercises 74–77, perform the operations and simplify.

74. $(2 + 3i) - \sqrt{-26}$

75. $(2 - 5i)^2$

76. $\sqrt{-16}\left(1 + \sqrt{-4}\right)$

77. $(3 + 2i)^3$

In Exercises 78–85, use the Quadratic Formula to solve the quadratic equation.

78. $x^2 - 6x + 12 = 0$

79. $x^2 + 5x + 7 = 0$

80. $3x^2 + 6x + 12 = 0$

81. $8x^2 - 4x + 3 = 0$

82. $-4x^2 + 3x - 12 = 0$

83. $14x^2 - 10x + 9 = 0$

84. $11x^2 - x + 22 = 0$

85. $13x^2 + 5x + 2 = 0$

5.3 Solving Trigonometric Equations

▶ **What you should learn**

- How to use standard algebraic techniques to solve trigonometric equations
- How to solve trigonometric equations of quadratic type
- How to solve trigonometric equations involving multiple angles
- How to use inverse trigonometric functions to solve trigonometric equations

▶ **Why you should learn it**

You can use trigonometric equations to solve a variety of real-life problems. For instance, in Exercise 76 on page 467 you solve a trigonometric equation to help answer questions about the unemployment rate in the United States.

M.Greenlar/The Image Works

Introduction

To solve a trigonometric equation, use standard algebraic techniques such as collecting like terms and factoring. Your preliminary goal in solving trigonometric equations is to isolate the trigonometric function involved in the equation.

Example 1 ▶ Solving a Trigonometric Equation

$2 \sin x - 1 = 0$ Write original equation.

$2 \sin x = 1$ Add 1 to each side.

$\sin x = \dfrac{1}{2}$ Divide each side by 2.

To solve for x, note in Figure 5.2 that the equation $\sin x = \frac{1}{2}$ has solutions $x = \pi/6$ and $x = 5\pi/6$ in the interval $[0, 2\pi)$. Moreover, because $\sin x$ has a period of 2π, there are infinitely many other solutions, which can be written as

$x = \pi/6 + 2n\pi$ and $x = 5\pi/6 + 2n\pi$ General solution

where n is an integer, as shown in Figure 5.2.

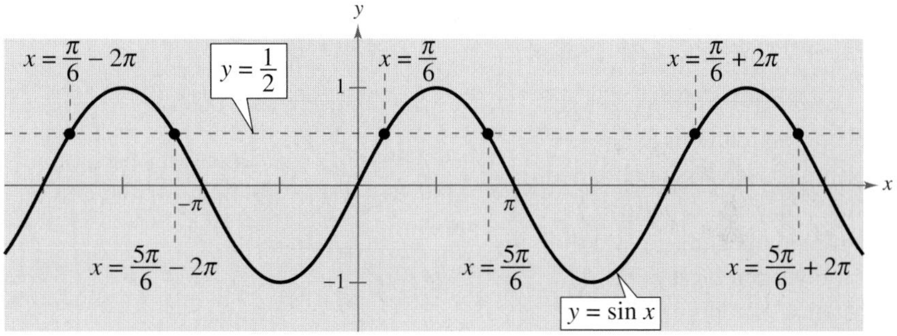

FIGURE 5.2

Another way to see that the equation $\sin x = \frac{1}{2}$ has infinitely many solutions is indicated in Figure 5.3. Any angles that are coterminal with $\pi/6$ or $5\pi/6$ will also be solutions of the equation.

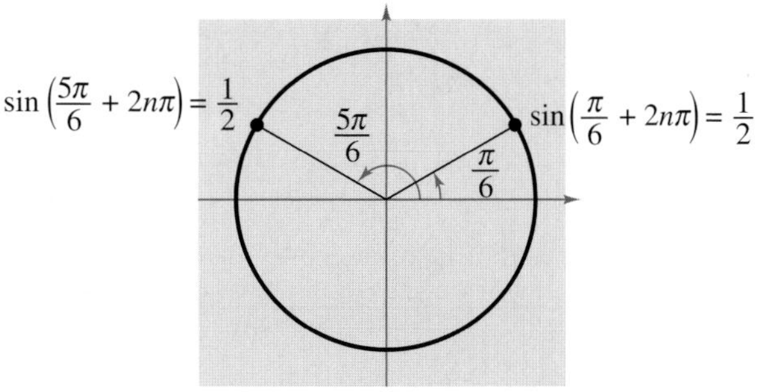

FIGURE 5.3

Encourage your students to give exact answers (rather than decimal approximations using a calculator) when solving these trigonometric equations.

Example 2 ▶ Collecting Like Terms

Solve $\sin x + \sqrt{2} = -\sin x$.

Solution

Begin by rewriting the equation so that $\sin x$ is isolated on one side of the equation.

$$\sin x + \sqrt{2} = -\sin x \qquad \text{Write original equation.}$$

$$\sin x + \sin x + \sqrt{2} = 0 \qquad \text{Add } \sin x \text{ to each side.}$$

$$\sin x + \sin x = -\sqrt{2} \qquad \text{Subtract } \sqrt{2} \text{ from each side.}$$

$$2 \sin x = -\sqrt{2} \qquad \text{Combine like terms.}$$

$$\sin x = -\frac{\sqrt{2}}{2} \qquad \text{Divide each side by 2.}$$

Because $\sin x$ has a period of 2π, first find all solutions in the interval $[0, 2\pi)$. These are $x = 5\pi/4$ and $x = 7\pi/4$. Finally, add $2n\pi$ to each of these solutions to get the general form

$$x = \frac{5\pi}{4} + 2n\pi \quad \text{and} \quad x = \frac{7\pi}{4} + 2n\pi \qquad \text{General solution}$$

where n is an integer.

Example 3 ▶ Extracting Square Roots

Solve $3 \tan^2 x - 1 = 0$.

Solution

Begin by rewriting the equation so that $\tan x$ is isolated on one side of the equation.

$$3 \tan^2 x - 1 = 0 \qquad \text{Write original equation.}$$

$$3 \tan^2 x = 1 \qquad \text{Add 1 to each side.}$$

$$\tan^2 x = \frac{1}{3} \qquad \text{Divide each side by 3.}$$

$$\tan x = \pm\frac{1}{\sqrt{3}} \qquad \text{Extract square roots.}$$

Because $\tan x$ has a period of π, first find all solutions in the interval $[0, \pi)$. These are $x = \pi/6$ and $x = 5\pi/6$. Finally, add $n\pi$ to each of these solutions to get the general form

$$x = \frac{\pi}{6} + n\pi \quad \text{and} \quad x = \frac{5\pi}{6} + n\pi \qquad \text{General solution}$$

where n is an integer.

Technology

The solutions in Examples 2 and 3 are obtained analytically. You can use a graphing utility to confirm the solutions graphically. For instance, to confirm the solutions found in Example 3, sketch the graph of

$$y = 3 \tan^2 x - 1$$

as shown below.

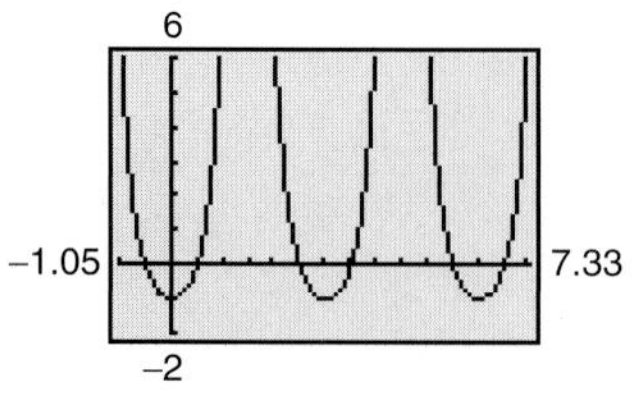

The equations in Examples 1, 2, and 3 involved only one trigonometric function. When two or more functions occur in the same equation, collect all terms on one side and try to separate the functions by factoring or by using appropriate identities. This may produce factors that yield no solutions, as illustrated in Example 4.

Example 4 ▶ Factoring

Solve $\cot x \cos^2 x = 2 \cot x$.

Solution

Begin by rewriting the equation so that all terms are collected on one side of the equation.

$$\cot x \cos^2 x = 2 \cot x \qquad \text{Write original equation.}$$

$$\cot x \cos^2 x - 2 \cot x = 0 \qquad \text{Subtract } 2 \cot x \text{ from each side.}$$

$$\cot x(\cos^2 x - 2) = 0 \qquad \text{Factor.}$$

By setting each of these factors equal to zero, you obtain

$$\cot x = 0 \qquad \text{and} \qquad \cos^2 x - 2 = 0$$

$$x = \frac{\pi}{2} \qquad\qquad \cos^2 x = 2$$

$$\cos x = \pm\sqrt{2}.$$

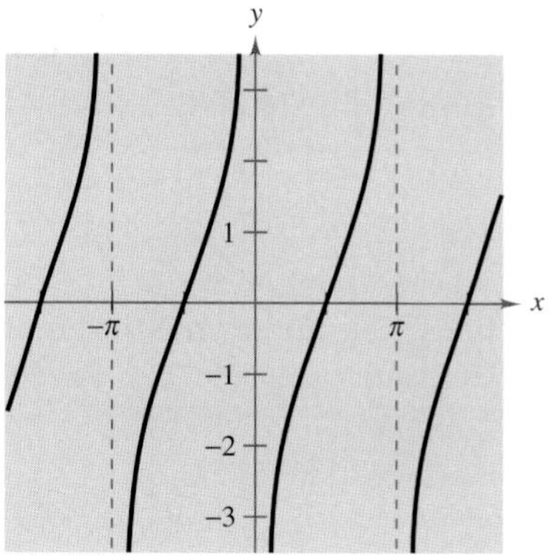

FIGURE 5.4

The equation $\cot x = 0$ has the solution $x = \pi/2$. No solution is obtained from $\cos x = \pm\sqrt{2}$ because $\pm\sqrt{2}$ are outside the range of the cosine function. Therefore, the general form of the solution is obtained by adding multiples of π to $x = \pi/2$, to get

$$x = \frac{\pi}{2} + n\pi \qquad \text{General solution}$$

where n is an integer. You can confirm this graphically by sketching the graph of $y = \cot x \cos^2 x - 2 \cot x$, as shown in Figure 5.4.

In Example 4, don't make the mistake of dividing each side of the equation by $\cot x$. If you do this, you lose the solutions. Can you see why?

Equations of Quadratic Type

Many trigonometric equations are of quadratic type. Here are a couple of examples.

Quadratic in sin x	*Quadratic in sec x*
$2 \sin^2 x - \sin x - 1 = 0$	$\sec^2 x - 3 \sec x - 2 = 0$
$2(\sin x)^2 - (\sin x) - 1 = 0$	$(\sec x)^2 - 3(\sec x) - 2 = 0$

To solve equations of this type, factor the quadratic or, if this is not possible, use the Quadratic Formula.

Example 5 ▶ Factoring an Equation of Quadratic Type

Find all solutions of $2\sin^2 x - \sin x - 1 = 0$ in the interval $[0, 2\pi)$.

Solution

Begin by treating the equation as a quadratic in $\sin x$ and factoring.

$$2\sin^2 x - \sin x - 1 = 0 \qquad \text{Write original equation.}$$

$$(2\sin x + 1)(\sin x - 1) = 0 \qquad \text{Factor.}$$

Setting each factor equal to zero, you can find the solutions in the interval $[0, 2\pi)$.

$$2\sin x + 1 = 0 \quad \text{and} \quad \sin x - 1 = 0$$

$$\sin x = -\frac{1}{2} \qquad \sin x = 1$$

$$x = \frac{7\pi}{6}, \frac{11\pi}{6} \qquad x = \frac{\pi}{2}$$

When working with an equation of quadratic type, be sure that the equation involves a *single* trigonometric function, as shown in the next example.

Example 6 ▶ Rewriting with a Single Trigonometric Function

Solve $2\sin^2 x + 3\cos x - 3 = 0$.

Solution

The algebraic model can be difficult for students to visualize. Encourage them to write or visualize the algebraic model, rewrite the corresponding trigonometric equation in terms of a single trigonometric function, and then solve for *x*. (See Exercises 43 and 44.)

This equation contains both sine and cosine functions. You can rewrite the equation so that it has only cosine functions by using the identity $\sin^2 x = 1 - \cos^2 x$.

$$2\sin^2 x + 3\cos x - 3 = 0 \qquad \text{Write original equation.}$$

$$2(1 - \cos^2 x) + 3\cos x - 3 = 0 \qquad \text{Pythagorean identity}$$

$$2\cos^2 x - 3\cos x + 1 = 0 \qquad \text{Multiply each side by } -1.$$

$$(2\cos x - 1)(\cos x - 1) = 0 \qquad \text{Factor.}$$

By setting each factor equal to zero, you can find the solutions in the interval $[0, 2\pi)$.

$$2\cos x - 1 = 0 \quad \text{and} \quad \cos x - 1 = 0$$

$$\cos x = \frac{1}{2} \qquad \cos x = 1$$

$$x = \frac{\pi}{3}, \frac{5\pi}{3} \qquad x = 0$$

The general solution is therefore

$$x = 2n\pi, \quad x = \frac{\pi}{3} + 2n\pi, \quad x = \frac{5\pi}{3} + 2n\pi \qquad \text{General solution}$$

where n is an integer.

Sometimes you must square both sides of an equation to obtain a quadratic, as demonstrated in the next example. Because this procedure can introduce extraneous solutions, you should check any solutions in the original equation to see whether they are valid or extraneous.

Example 7 ▶ Squaring and Converting to Quadratic Type

Find all solutions of $\cos x + 1 = \sin x$ in the interval $[0, 2\pi)$.

Solution

It is not clear how to rewrite this equation in terms of a single trigonometric function. See what happens when you square both sides of the equation.

$$\cos x + 1 = \sin x \quad \text{Write original equation.}$$

$$\cos^2 x + 2\cos x + 1 = \sin^2 x \quad \text{Square each side.}$$

$$\cos^2 x + 2\cos x + 1 = 1 - \cos^2 x \quad \text{Pythagorean identity}$$

$$\cos^2 x + \cos^2 x + 2\cos x + 1 - 1 = 0 \quad \text{Rewrite equation.}$$

$$2\cos^2 x + 2\cos x = 0 \quad \text{Combine like terms.}$$

$$2\cos x(\cos x + 1) = 0 \quad \text{Factor.}$$

Setting each factor equal to zero produces

$$2\cos x = 0 \qquad \text{and} \qquad \cos x + 1 = 0$$

$$\cos x = 0 \qquad\qquad \cos x = -1$$

$$x = \frac{\pi}{2}, \frac{3\pi}{2} \qquad\qquad x = \pi.$$

Because you squared the original equation, check for extraneous solutions.

Check for $x = \pi/2$

$$\cos\frac{\pi}{2} + 1 \stackrel{?}{=} \sin\frac{\pi}{2} \quad \text{Substitute } \pi/2 \text{ for } x.$$

$$0 + 1 = 1 \quad \text{Solution checks. ✓}$$

Check for $x = 3\pi/2$

$$\cos\frac{3\pi}{2} + 1 \stackrel{?}{=} \sin\frac{3\pi}{2} \quad \text{Substitute } 3\pi/2 \text{ for } x.$$

$$0 + 1 = -1 \quad \text{Solution does not check. ✗}$$

Check for $x = \pi$

$$\cos\pi + 1 \stackrel{?}{=} \sin\pi \quad \text{Substitute } \pi \text{ for } x.$$

$$-1 + 1 = 0 \quad \text{Solution checks. ✓}$$

Of the three possible solutions, $x = 3\pi/2$ is extraneous. So, in the interval $[0, 2\pi)$, the only two solutions are $x = \pi/2$ and $x = \pi$.

Exploration

Use a graphing utility to confirm the solutions found in Example 7 in two different ways. Do both methods produce the same x-values? Which method do you prefer? Why?

1. Graph both sides of the equation and find the x-coordinates of the points at which the graphs intersect.

   Left side: $y = \cos x + 1$

   Right side: $y = \sin x$

2. Graph the equation

   $y = \cos x + 1 - \sin x$

   and find the x-intercepts of the graph.

Functions Involving Multiple Angles

The next two examples involve trigonometric functions of multiple angles of the forms sin *ku* or cos *ku*.

Example 8 ▶ Functions of Multiple Angles

Find all solutions of $2\cos 3t - 1 = 0$.

Solution

$$2\cos 3t - 1 = 0 \quad \text{Write original equation.}$$

$$2\cos 3t = 1 \quad \text{Add 1 to each side.}$$

$$\cos 3t = \frac{1}{2} \quad \text{Divide each side by 2.}$$

In the interval $[0, 2\pi)$, you know that $3t = \pi/3$ and $3t = 5\pi/3$ are the only solutions so that, in general, you have

$$3t = \frac{\pi}{3} + 2n\pi \quad \text{and} \quad 3t = \frac{5\pi}{3} + 2n\pi.$$

Dividing these results by 3, you obtain the general solution

$$t = \frac{\pi}{9} + \frac{2n\pi}{3} \quad \text{and} \quad t = \frac{5\pi}{9} + \frac{2n\pi}{3} \quad \text{General solution}$$

where n is an integer.

Additional Examples

$(0 \le x < 2\pi)$

a. $2\sin^2 2x = 1$

$x = \frac{\pi}{8}, \frac{3\pi}{8}, \frac{5\pi}{8}, \frac{7\pi}{8}, \frac{9\pi}{8}, \frac{11\pi}{8}, \frac{13\pi}{8}, \frac{15\pi}{8}$

b. $\tan 4x = 1$

$x = \frac{\pi}{16}, \frac{5\pi}{16}, \frac{9\pi}{16}, \frac{13\pi}{16}, \frac{17\pi}{16}, \frac{21\pi}{16}, \frac{25\pi}{16}, \frac{29\pi}{16}$

Example 9 ▶ Functions of Multiple Angles

Find all solutions of $3\tan(x/2) + 3 = 0$.

Solution

$$3\tan\frac{x}{2} + 3 = 0 \quad \text{Write original equation.}$$

$$3\tan\frac{x}{2} = -3 \quad \text{Subtract 3 from each side.}$$

$$\tan\frac{x}{2} = -1 \quad \text{Divide each side by 3.}$$

In the interval $[0, \pi)$, you know that $x/2 = 3\pi/4$ is the only solution so that, in general, you have

$$\frac{x}{2} = \frac{3\pi}{4} + n\pi.$$

Multiplying this result by 2, you obtain the general solution

$$x = \frac{3\pi}{2} + 2n\pi \quad \text{General solution}$$

where n is an integer.

Consider having your students check their solutions with a graphing utility. This will help to reinforce the fact that there are many solutions, and will demonstrate why.

Using Inverse Functions

In the next example, you will see how inverse trigonometric functions are used to solve an equation.

Example 10 ▶ Using Inverse Functions

Find all solutions of $\sec^2 x - 2\tan x = 4$.

Solution

$\sec^2 x - 2\tan x = 4$	Write original equation.
$1 + \tan^2 x - 2\tan x - 4 = 0$	Pythagorean identity
$\tan^2 x - 2\tan x - 3 = 0$	Combine like terms.
$(\tan x - 3)(\tan x + 1) = 0$	Factor.

Setting each factor equal to zero, you obtain two solutions in the interval $(-\pi/2, \pi/2)$. [Recall that the range of the inverse tangent function is $(-\pi/2, \pi/2)$.]

$\tan x - 3 = 0$ and $\tan x + 1 = 0$

$\tan x = 3$ $\qquad$ $\tan x = -1$

$x = \arctan 3$ $\qquad$ $x = -\dfrac{\pi}{4}$

Finally, by adding multiples of π, you obtain the general solution

$$x = \arctan 3 + n\pi \quad \text{and} \quad x = -\frac{\pi}{4} + n\pi \qquad \text{General solution}$$

where n is an integer.

Activities

1. Find all solutions of the equation without using a calculator.
   $\tan^2 x = 1$
   Answer: $x = \dfrac{\pi}{4} + n\pi, \dfrac{3\pi}{4} + n\pi$
2. Solve for x, $0 \le x < 2\pi$, without using a calculator.
   $\sin^2 3x - 2\sin 3x + 1 = 0$
   Answer: $x = \dfrac{\pi}{6}, \dfrac{5\pi}{6}, \dfrac{3\pi}{2}$
3. Solve for x, $0 \le x < 2\pi$.
   $6\cos^2 x - \cos x - 2 = 0$
   Answer: $x = 0.841, \dfrac{2\pi}{3}, \dfrac{4\pi}{3}, 5.442$

Writing ABOUT MATHEMATICS

Equations with No Solutions One of the following equations has solutions and the other two don't. Which two equations do not have solutions?

a. $\sin^2 x - 5\sin x + 6 = 0$

b. $\sin^2 x - 4\sin x + 6 = 0$

c. $\sin^2 x - 5\sin x - 6 = 0$

Can you find conditions involving the constants b and c that will guarantee that the equation

$$\sin^2 x + b\sin x + c = 0$$

has at least one solution on some interval of length 2π?

5.3 Exercises

In Exercises 1–6, verify that the *x*-values are solutions.

1. $2\cos x - 1 = 0$

(a) $x = \frac{\pi}{3}$ (b) $x = \frac{5\pi}{3}$

2. $\sec x - 2 = 0$

(a) $x = \frac{\pi}{3}$ (b) $x = \frac{5\pi}{3}$

3. $3\tan^2 2x - 1 = 0$

(a) $x = \frac{\pi}{12}$ (b) $x = \frac{5\pi}{12}$

4. $2\cos^2 4x - 1 = 0$

(a) $x = \frac{\pi}{16}$ (b) $x = \frac{3\pi}{16}$

5. $2\sin^2 x - \sin x - 1 = 0$

(a) $x = \frac{\pi}{2}$ (b) $x = \frac{7\pi}{6}$

6. $\csc^4 x - 4\csc^2 x = 0$

(a) $x = \frac{\pi}{6}$ (b) $x = \frac{5\pi}{6}$

In Exercises 7–20, solve the equation.

7. $2\cos x + 1 = 0$

8. $2\sin x + 1 = 0$

9. $\sqrt{3}\csc x - 2 = 0$

10. $\tan x + \sqrt{3} = 0$

11. $3\sec^2 x - 4 = 0$

12. $3\cot^2 x - 1 = 0$

13. $\sin x(\sin x + 1) = 0$

14. $(3\tan^2 x - 1)(\tan^2 x - 3) = 0$

15. $4\cos^2 x - 1 = 0$

16. $\sin^2 x = 3\cos^2 x$

17. $2\sin^2 2x = 1$

18. $\tan^2 3x = 3$

19. $\tan 3x(\tan x - 1) = 0$

20. $\cos 2x(2\cos x + 1) = 0$

In Exercises 21–32, find all solutions of the equation in the interval $[0, 2\pi)$.

21. $\cos^3 x = \cos x$

22. $\sec^2 x - 1 = 0$

23. $3\tan^3 x = \tan x$

24. $2\sin^2 x = 2 + \cos x$

25. $\sec^2 x - \sec x = 2$

26. $\sec x \csc x = 2\csc x$

27. $2\sin x + \csc x = 0$

28. $\sec x + \tan x = 1$

29. $2\cos^2 x + \cos x - 1 = 0$

30. $2\sin^2 x + 3\sin x + 1 = 0$

31. $2\sec^2 x + \tan^2 x - 3 = 0$

32. $\cos x + \sin x \tan x = 2$

In Exercises 33–38, find all solutions of the equation.

33. $\cos 2x = \frac{1}{2}$

34. $\sin 2x = -\frac{\sqrt{3}}{2}$

35. $\tan 3x = 1$

36. $\sec 4x = 2$

37. $\cos\frac{x}{2} = \frac{\sqrt{2}}{2}$

38. $\sin\frac{x}{2} = -\frac{\sqrt{3}}{2}$

In Exercises 39–42, find the *x*-intercepts of the graph.

39. $y = \sin\frac{\pi x}{2} + 1$

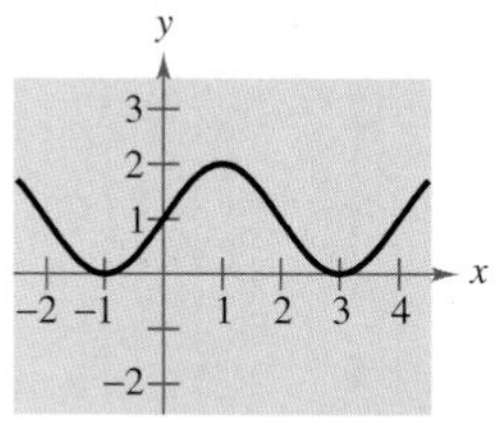

40. $y = \sin \pi x + \cos \pi x$

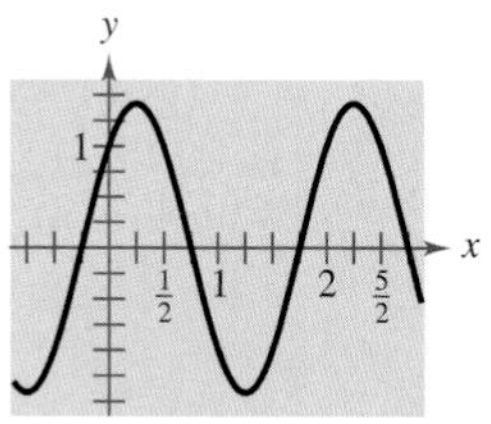

41. $y = \tan^2\left(\frac{\pi x}{6}\right) - 3$

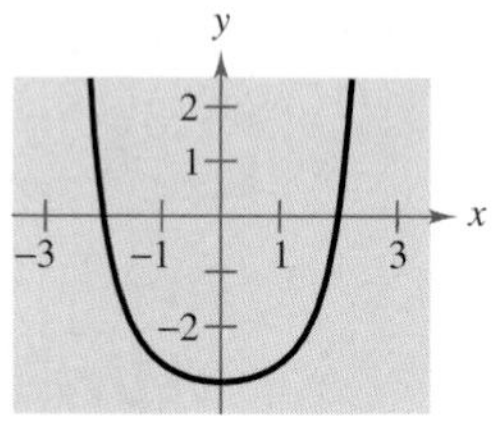

42. $y = \sec^4\left(\frac{\pi x}{8}\right) - 4$

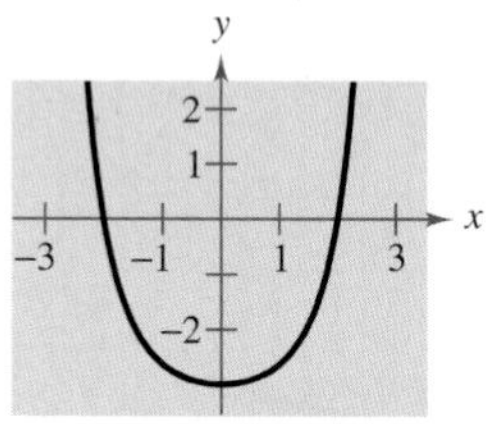

In Exercises 43 and 44, solve both equations. How do the solutions of the algebraic equation compare with the solutions of the trigonometric equation?

43. $6y^2 - 13y + 6 = 0$

$6\cos^2 x - 13\cos x + 6 = 0$

44. $y^2 + y - 20 = 0$

$\sin^2 x + \sin x - 20 = 0$

In Exercises 45–54, use a graphing utility to approximate the solutions of the equation in the interval $[0, 2\pi)$.

45. $2\sin x + \cos x = 0$

46. $4\sin^3 x + 2\sin^2 x - 2\sin x - 1 = 0$

47. $\dfrac{1 + \sin x}{\cos x} + \dfrac{\cos x}{1 + \sin x} = 4$

48. $\dfrac{\cos x \cot x}{1 - \sin x} = 3$

49. $x \tan x - 1 = 0$

50. $x \cos x - 1 = 0$

51. $\sec^2 x + 0.5 \tan x - 1 = 0$

52. $\csc^2 x + 0.5 \cot x - 5 = 0$

53. $2 \tan^2 x + 7 \tan x - 15 = 0$

54. $6 \sin^2 x - 7 \sin x + 2 = 0$

In Exercises 55–58, use the Quadratic Formula to solve the equation in the interval $[0, 2\pi)$. Then use a graphing utility to approximate the angle x.

55. $12 \sin^2 x - 13 \sin x + 3 = 0$

56. $3 \tan^2 x + 4 \tan x - 4 = 0$

57. $\tan^2 x + 3 \tan x + 1 = 0$

58. $4 \cos^2 x - 4 \cos x - 1 = 0$

In Exercises 59–62, use inverse functions where needed to find all solutions of the equation in the interval $[0, 2\pi)$.

59. $\tan^2 x - 6 \tan x + 5 = 0$

60. $\sec^2 x + \tan x - 3 = 0$

61. $2 \cos^2 x - 5 \cos x + 2 = 0$

62. $2 \sin^2 x - 7 \sin x + 3 = 0$

In Exercises 63 and 64, (a) use a graphing utility to graph the function and approximate the maximum and minimum points on the graph in the interval $[0, 2\pi)$, and (b) solve the trigonometric equation and demonstrate that its solutions are the x-coordinates of the maximum and minimum points of f. (Calculus is required to find the trigonometric equation.)

	Function	*Trigonometric Equation*
63.	$f(x) = \sin x + \cos x$	$\cos x - \sin x = 0$
64.	$f(x) = 2 \sin x + \cos 2x$	$2 \cos x - 4 \sin x \cos x = 0$

Fixed Point **In Exercises 65 and 66, find the smallest positive fixed point of the function f. [A fixed point of a function f is a real number c such that $f(c) = c$.]**

65. $f(x) = \tan \dfrac{\pi x}{4}$

66. $f(x) = \cos x$

67. *Graphical Reasoning* Consider the function

$$f(x) = \cos \frac{1}{x}$$

and its graph shown in the figure.

(a) What is the domain of the function?

(b) Identify any symmetry or asymptotes of the graph.

(c) Describe the behavior of the function as $x \to 0$.

(d) How many solutions does the equation

$$\cos \frac{1}{x} = 0$$

have in the interval $[-1, 1]$?

(e) Does the equation $\cos(1/x) = 0$ have a greatest solution? If so, approximate the solution. If not, explain why.

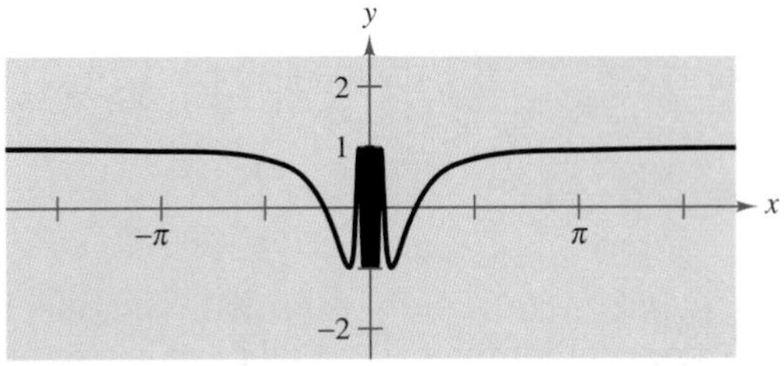

FIGURE FOR 67

68. *Graphical Reasoning* Consider the function

$$f(x) = \frac{\sin x}{x}$$

and its graph shown in the figure.

(a) What is the domain of the function?

(b) Identify any symmetry or asymptotes of the graph.

(c) Describe the behavior of the function as $x \to 0$.

(d) How many solutions does the equation

$$\frac{\sin x}{x} = 0$$

have in the interval $[-8, 8]$? Find the solutions.

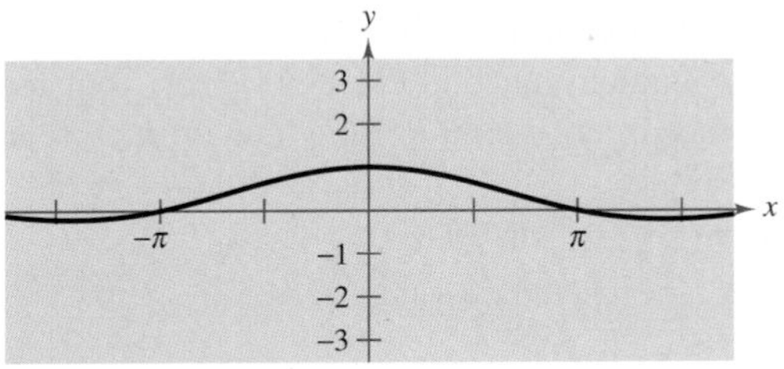

69. ***Harmonic Motion*** A weight is oscillating on the end of a spring (see figure). The position of the weight relative to the point of equilibrium is

$$y = \frac{1}{12}(\cos 8t - 3\sin 8t)$$

where y is the displacement in meters and t is the time in seconds. Find the times when the weight is at the point of equilibrium ($y = 0$) for $0 \le t \le 1$.

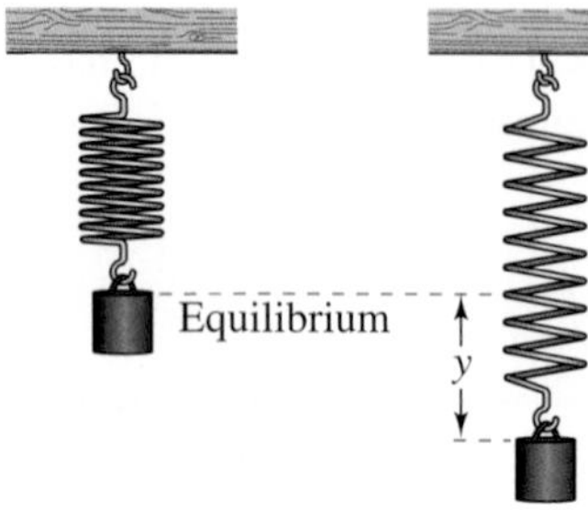

70. ***Damped Harmonic Motion*** The displacement from equilibrium of a weight oscillating on the end of a spring is

$$y = 1.56e^{-0.22t}\cos 4.9t$$

where y is the displacement in feet and t is the time in seconds. Use a graphing utility to graph the displacement function for $0 \le t \le 10$. Find the time beyond which the displacement does not exceed 1 foot from equilibrium.

71. ***Sales*** The monthly sales (in thousands of units) of a seasonal product are approximated by

$$S = 74.50 + 43.75\sin\frac{\pi t}{6}$$

where t is the time in months, with $t = 1$ corresponding to January. Determine the months when sales exceed 100,000 units.

72. ***Projectile Motion*** A batted baseball leaves the bat at an angle of θ with the horizontal and an initial velocity of $v_0 = 100$ feet per second. The ball is caught by an outfielder 300 feet from home plate (see figure). Find θ if the range r of a projectile is

$$r = \frac{1}{32}v_0^2\sin 2\theta.$$

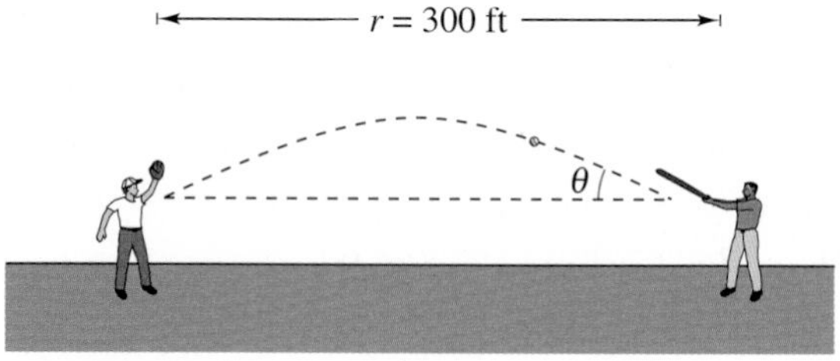

FIGURE FOR 72

73. ***Projectile Motion*** A sharpshooter intends to hit a target at a distance of 1000 yards with a gun that has a muzzle velocity of 1200 feet per second (see figure). Neglecting air resistance, determine the gun's minimum angle of elevation θ if the range r is

$$r = \frac{1}{32}v_0^2\sin 2\theta.$$

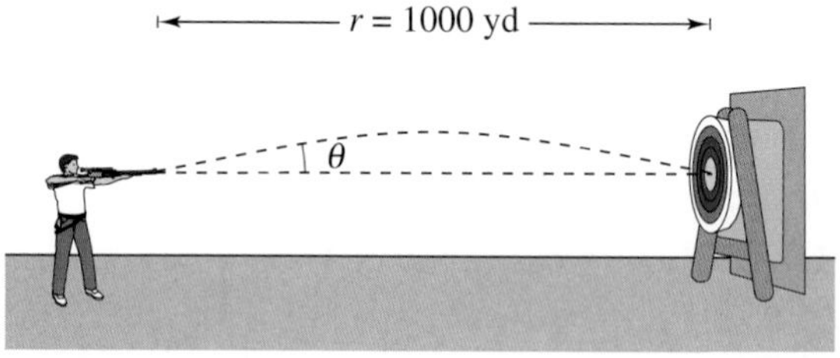

74. ***Area*** The area of a rectangle (see figure) inscribed in one arc of the graph of $y = \cos x$ is

$$A = 2x\cos x, \qquad 0 < x < \frac{\pi}{2}.$$

(a) Use a graphing utility to graph the area function, and approximate the area of the largest inscribed rectangle.

(b) Determine the values of x for which $A \ge 1$.

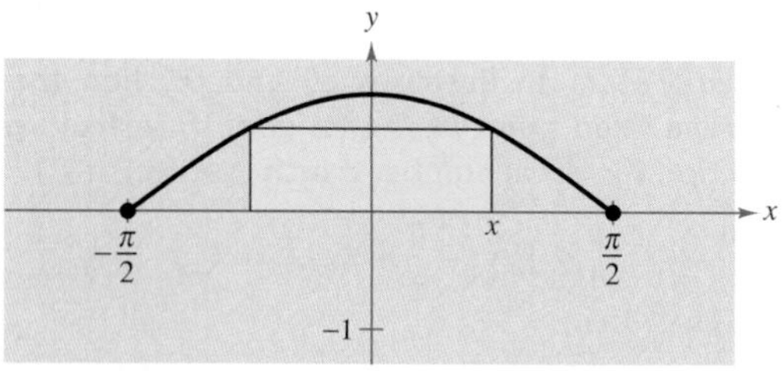

75. ***Quadratic Approximation*** Consider the function

$$f(x) = 3\sin(0.6x - 2).$$

(a) Approximate the zero of the function in the interval $[0, 6]$.

(b) A quadratic approximation agreeing with f at $x = 5$ is $g(x) = -0.45x^2 + 5.52x - 13.70$. Use a graphing utility to graph f and g in the same viewing window. Describe the result.

(c) Use the Quadratic Formula to find the zeros of g. Compare the zero in the interval $[0, 6]$ with the result of part (a).

76. *Data Analysis* The table gives the unemployment rate r for the years 1985 through 1996 in the United States. The time t is measured in years, with $t = 0$ corresponding to 1990. (Source: U.S. Bureau of Labor Statistics)

t	-5	-4	-3	-2	-1	0
r	7.2	7.0	6.2	5.5	5.3	5.6

t	1	2	3	4	5	6
r	6.8	7.5	6.9	6.1	5.6	5.4

(a) Create a scatter plot of the data.

(b) Which of the following models best represents the data? Explain your reasoning.

(1) $r = 1.5 \cos(t + 3.9) + 6.37$

(2) $r = 1.03 \sin(0.9t + 0.44) + 6.19$

(3) $r = \sin[0.91(t + 6.44)] + 6.26$

(4) $r = 1.5 \sin[0.5(t + 2.8)] + 6.25$

(c) What term in the model gives the average unemployment rate? What is the rate?

(d) Economists study the lengths of business cycles such as unemployment rates. Based on this short span of time, use the model to give the length of this cycle.

(e) Use the model to estimate the next time the unemployment rate will be 6% or less.

Synthesis

True or False? **In Exercises 77 and 78, determine whether the statement is true or false. Justify your answer.**

77. The equation $2 \sin 4t - 1 = 0$ has four times the number of solutions in the interval $[0, 2\pi)$ as the equation $2 \sin t - 1 = 0$.

78. If you correctly solve a trigonometric equation down to the statement $\sin x = 3.4$, then you can finish solving the equation by using an inverse function.

In Exercises 79 and 80, use the graph to approximate the number of points of intersection of the graphs of y_1 and y_2.

79. $y_1 = 2 \sin x$
$y_2 = 3x + 1$

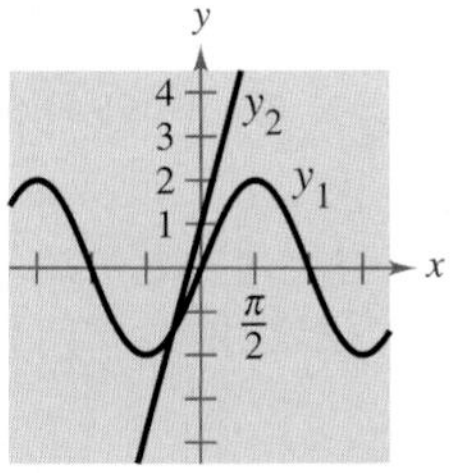

80. $y_1 = 2 \sin x$
$y_2 = \frac{1}{2}x + 1$

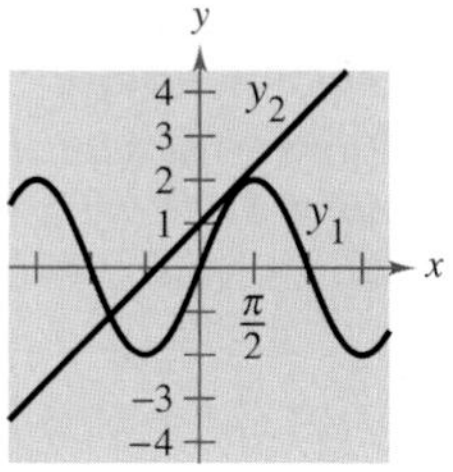

Review

In Exercises 81–84, solve triangle *ABC* by finding all missing angle measures and side lengths.

81.

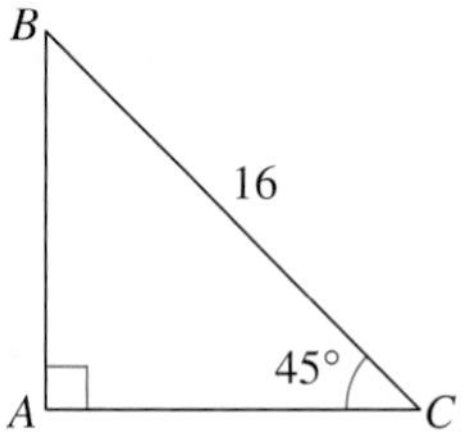

82.

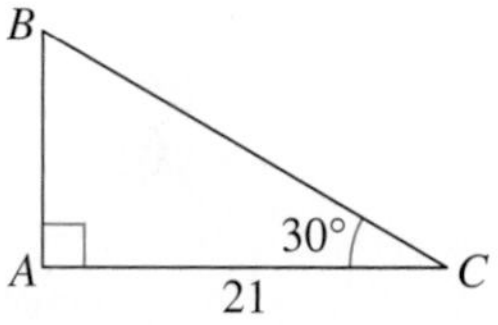

83.

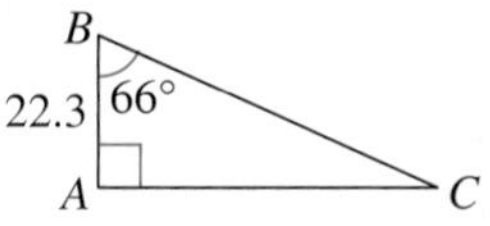

84.

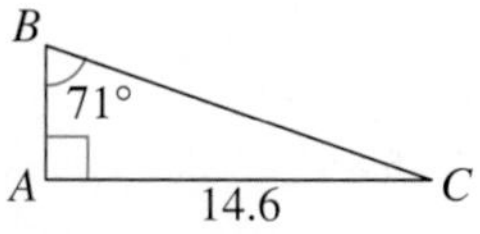

In Exercises 85–90, use reference angles to find the sine, cosine, and tangent of an angle with the given measure.

85. 390° **86.** 570°

87. 495° **88.** 855°

89. −1845° **90.** −1410°

91. *Height* From a point 100 feet in front of the public library, the angles of elevation to the base of the flagpole and the top of the pole are 28° and 39° 45′, respectively. The flagpole is mounted on the front of the library's roof. Find the height of the pole.

92. *Angle of Depression* Find the angle of depression from the top of a lighthouse 250 feet above water level to the water line of a ship 2 miles offshore.

5.4 Sum and Difference Formulas

▶ **What you should learn**

- How to use sum and difference formulas to evaluate trigonometric functions
- How to use sum and difference formulas to verify identities and solve trigonometric equations

▶ **Why you should learn it**

You can use sum and difference formulas to rewrite trigonometric expressions. For instance, in Exercise 76 on page 473 you use sum and difference formulas to rewrite a trigonometric expression in a form that helps you find the equation of a standing wave.

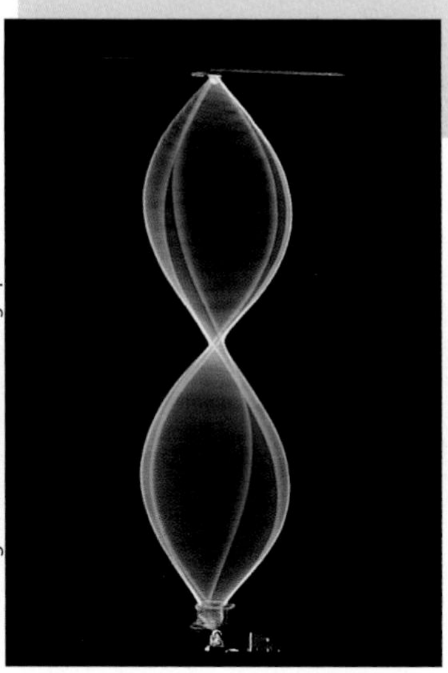

Richard Megna/Fundamental Photographs

Using Sum and Difference Formulas

In this and the following section, you will study the uses of several trigonometric identities and formulas. (Proofs of these formulas are given in Appendix A.)

Sum and Difference Formulas

$$\sin(u + v) = \sin u \cos v + \cos u \sin v$$

$$\sin(u - v) = \sin u \cos v - \cos u \sin v$$

$$\cos(u + v) = \cos u \cos v - \sin u \sin v$$

$$\cos(u - v) = \cos u \cos v + \sin u \sin v$$

$$\tan(u + v) = \frac{\tan u + \tan v}{1 - \tan u \tan v}$$

$$\tan(u - v) = \frac{\tan u - \tan v}{1 + \tan u \tan v}$$

Exploration

Use a graphing utility to graph $y = \cos(x + 2)$ and $y = \cos x + \cos 2$ in the same viewing window. What can you conclude about the graphs? Is it true that $\cos(x + 2) = \cos x + \cos 2$?

Use a graphing utility to graph $y = \sin(x + 4)$ and $y = \sin x + \sin 4$ in the same viewing window. What can you conclude about the graphs? Is it true that $\sin(x + 4) = \sin x + \sin 4$?

Examples 1 and 2 show how **sum and difference formulas** can be used to find exact values of trigonometric functions involving sums or differences of special angles.

Example 1 ▶ Evaluating a Trigonometric Function

Find the exact value of $\cos 75°$.

Solution

To find the *exact* value of $\cos 75°$, use the fact that $75° = 30° + 45°$. Consequently, the formula for $\cos(u + v)$ yields

$$\begin{aligned}\cos 75° &= \cos(30° + 45°)\\ &= \cos 30° \cos 45° - \sin 30° \sin 45°\\ &= \frac{\sqrt{3}}{2}\left(\frac{\sqrt{2}}{2}\right) - \frac{1}{2}\left(\frac{\sqrt{2}}{2}\right)\\ &= \frac{\sqrt{6} - \sqrt{2}}{4}.\end{aligned}$$

Try checking this result on your calculator. You will find that $\cos 75° \approx 0.259$.

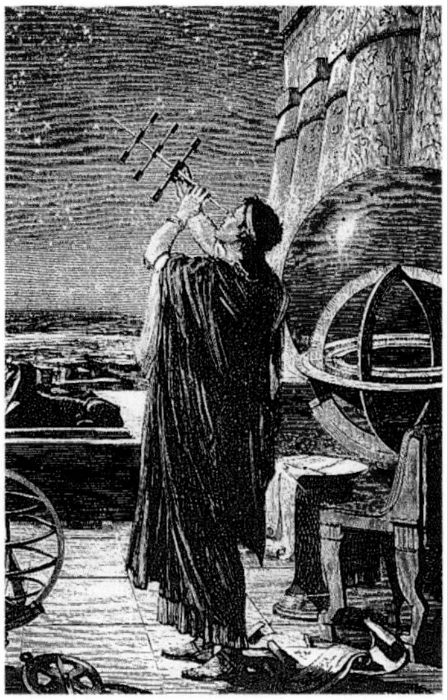

The Granger Collection, New York

Historical Note
Hipparchus, considered the most eminent of Greek astronomers, was born about 160 B.C. in Nicaea. He was credited with the invention of trigonometry. He also derived the sum and difference formulas for $\sin(A \pm B)$ and $\cos(A \pm B)$.

Example 2 ▶ Evaluating a Trigonometric Function

Find the exact value of $\sin \frac{\pi}{12}$.

Solution

Using the fact that

$$\frac{\pi}{12} = \frac{\pi}{3} - \frac{\pi}{4}$$

together with the formula for $\sin(u - v)$, you obtain

$$\begin{aligned}\sin \frac{\pi}{12} &= \sin\left(\frac{\pi}{3} - \frac{\pi}{4}\right)\\ &= \sin \frac{\pi}{3} \cos \frac{\pi}{4} - \cos \frac{\pi}{3} \sin \frac{\pi}{4}\\ &= \frac{\sqrt{3}}{2}\left(\frac{\sqrt{2}}{2}\right) - \frac{1}{2}\left(\frac{\sqrt{2}}{2}\right)\\ &= \frac{\sqrt{6} - \sqrt{2}}{4}.\end{aligned}$$

Example 3 ▶ Evaluating a Trigonometric Expression

Find the exact value of $\sin 42° \cos 12° - \cos 42° \sin 12°$.

Solution

Recognizing that this expression fits the formula for $\sin(u - v)$, you can write

$$\begin{aligned}\sin 42° \cos 12° - \cos 42° \sin 12° &= \sin(42° - 12°)\\ &= \sin 30°\\ &= \frac{1}{2}.\end{aligned}$$

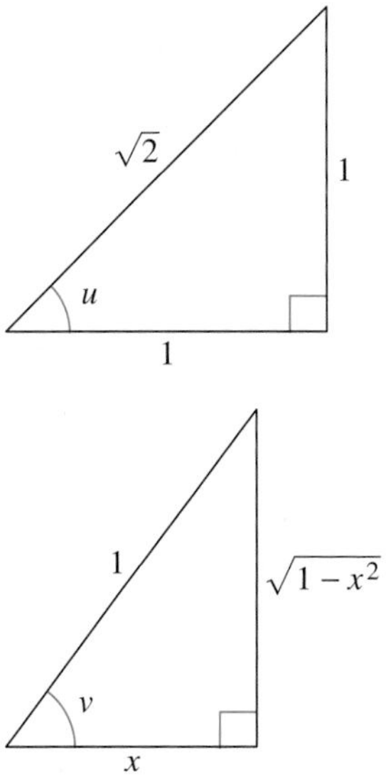

FIGURE 5.5

Example 4 ▶ An Application of a Sum Formula

Write $\cos(\arctan 1 + \arccos x)$ as an algebraic expression.

Solution

This expression fits the formula for $\cos(u + v)$. Angles $u = \arctan 1$ and $v = \arccos x$ are shown in Figure 5.5. So

$$\begin{aligned}\cos(u + v) &= \cos(\arctan 1) \cos(\arccos x) - \sin(\arctan 1) \sin(\arccos x)\\ &= \frac{1}{\sqrt{2}} \cdot x - \frac{1}{\sqrt{2}} \cdot \sqrt{1 - x^2}\\ &= \frac{x - \sqrt{1 - x^2}}{\sqrt{2}}.\end{aligned}$$

Example 5 ▶ Proving a Cofunction Identity

Prove the cofunction identity $\cos\left(\frac{\pi}{2} - x\right) = \sin x$.

Solution

Using the formula for $\cos(u - v)$, you have

$$\begin{aligned}\cos\left(\frac{\pi}{2} - x\right) &= \cos\frac{\pi}{2}\cos x + \sin\frac{\pi}{2}\sin x \\ &= (0)(\cos x) + (1)(\sin x) \\ &= \sin x.\end{aligned}$$

Sum and difference formulas can be used to derive **reduction formulas** involving expressions such as

$$\sin\left(\theta + \frac{n\pi}{2}\right) \quad \text{and} \quad \cos\left(\theta + \frac{n\pi}{2}\right), \quad \text{where } n \text{ is an integer.}$$

Example 6 ▶ Deriving Reduction Formulas

Simplify each expression.

a. $\cos\left(\theta - \frac{3\pi}{2}\right)$ **b.** $\tan(\theta + 3\pi)$

Solution

a. Using the formula for $\cos(u - v)$, you have

$$\begin{aligned}\cos\left(\theta - \frac{3\pi}{2}\right) &= \cos\theta\cos\frac{3\pi}{2} + \sin\theta\sin\frac{3\pi}{2} \\ &= (\cos\theta)(0) + (\sin\theta)(-1) \\ &= -\sin\theta.\end{aligned}$$

b. Using the formula for $\tan(u + v)$, you have

$$\begin{aligned}\tan(\theta + 3\pi) &= \frac{\tan\theta + \tan 3\pi}{1 - \tan\theta\tan 3\pi} \\ &= \frac{\tan\theta + 0}{1 - (\tan\theta)(0)} \\ &= \tan\theta.\end{aligned}$$

Activities

1. Use the sum and difference formulas to find the exact value of cos 15°.

   Answer: $\frac{\sqrt{2}}{4}(1 + \sqrt{3})$

2. Rewrite the expression using the sum and difference formulas.

   $\frac{\tan 40° + \tan 10°}{1 - \tan 40° \tan 10°}$

   Answer: $\tan(40° + 10°) = \tan 50°$

3. Verify the identity

   $\sin\left(\frac{\pi}{2} - \theta\right) = \cos\theta.$

   Answer:

   $$\begin{aligned}\sin\left(\frac{\pi}{2} - \theta\right) &= \sin\frac{\pi}{2}\cos\theta - \cos\frac{\pi}{2}\sin\theta \\ &= (1)\cos\theta - (0)\sin\theta \\ &= \cos\theta\end{aligned}$$

The next example was taken from calculus. It is used to derive the derivative of the sine function.

Example 7 ▶ An Application from Calculus

Verify that

$$\frac{\sin(x+h)-\sin x}{h} = (\cos x)\left(\frac{\sin h}{h}\right) - (\sin x)\left(\frac{1-\cos h}{h}\right)$$

where $h \neq 0$.

Solution

Using the formula for $\sin(u+v)$, you have

$$\frac{\sin(x+h)-\sin x}{h} = \frac{\sin x \cos h + \cos x \sin h - \sin x}{h}$$

$$= \frac{\cos x \sin h - \sin x(1-\cos h)}{h}$$

$$= (\cos x)\left(\frac{\sin h}{h}\right) - (\sin x)\left(\frac{1-\cos h}{h}\right).$$

Example 8 ▶ Solving a Trigonometric Equation

Find all solutions of

$$\sin\left(x+\frac{\pi}{4}\right) + \sin\left(x-\frac{\pi}{4}\right) = -1$$

in the interval $[0, 2\pi)$.

Solution

Using sum and difference formulas, rewrite the equation as

$$\sin x \cos\frac{\pi}{4} + \cos x \sin\frac{\pi}{4} + \sin x \cos\frac{\pi}{4} - \cos x \sin\frac{\pi}{4} = -1$$

$$2 \sin x \cos\frac{\pi}{4} = -1$$

$$2(\sin x)\left(\frac{\sqrt{2}}{2}\right) = -1$$

$$\sin x = -\frac{1}{\sqrt{2}}$$

$$\sin x = -\frac{\sqrt{2}}{2}.$$

Therefore, the only solutions in the interval $[0, 2\pi)$ are

$$x = \frac{5\pi}{4} \qquad \text{and} \qquad x = \frac{7\pi}{4}.$$

These solutions are checked graphically in Figure 5.6.

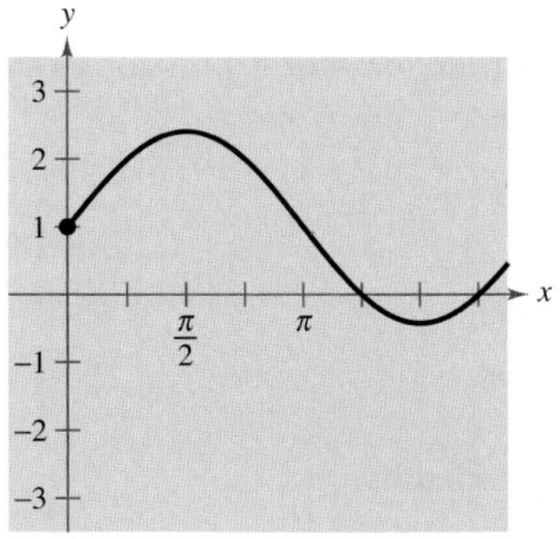

FIGURE 5.6

5.4 Exercises

In Exercises 1–6, find the exact value of each expression.

1. (a) $\cos\left(\frac{\pi}{4} + \frac{\pi}{3}\right)$ (b) $\cos\frac{\pi}{4} + \cos\frac{\pi}{3}$
2. (a) $\sin\left(\frac{3\pi}{4} + \frac{5\pi}{6}\right)$ (b) $\sin\frac{3\pi}{4} + \sin\frac{5\pi}{6}$
3. (a) $\sin\left(\frac{7\pi}{6} - \frac{\pi}{3}\right)$ (b) $\sin\frac{7\pi}{6} - \sin\frac{\pi}{3}$
4. (a) $\cos\left(\frac{2\pi}{3} - \frac{\pi}{6}\right)$ (b) $\cos\frac{2\pi}{3} + \cos\frac{\pi}{6}$
5. (a) $\cos(120° + 45°)$ (b) $\cos 120° + \cos 45°$
6. (a) $\sin(135° - 30°)$ (b) $\sin 135° - \cos 30°$

In Exercises 7–14, find the exact values of the sine, cosine, and tangent of the angle.

7. $105° = 60° + 45°$
8. $165° = 135° + 30°$
9. $195° = 225° - 30°$
10. $255° = 300° - 45°$
11. $\frac{11\pi}{12} = \frac{3\pi}{4} + \frac{\pi}{6}$
12. $\frac{7\pi}{12} = \frac{\pi}{3} + \frac{\pi}{4}$
13. $\frac{17\pi}{12} = \frac{9\pi}{4} - \frac{5\pi}{6}$
14. $-\frac{\pi}{12} = \frac{\pi}{6} - \frac{\pi}{4}$

In Exercises 15–22, find the exact values of the sine, cosine, and tangent of the angle.

15. $285°$
16. $-105°$
17. $-165°$
18. $15°$
19. $\frac{13\pi}{12}$
20. $-\frac{7\pi}{12}$
21. $-\frac{13\pi}{12}$
22. $\frac{5\pi}{12}$

In Exercises 23–30, write the expression as the sine, cosine, or tangent of an angle.

23. $\cos 25° \cos 15° - \sin 25° \sin 15°$
24. $\sin 140° \cos 50° + \cos 140° \sin 50°$
25. $\dfrac{\tan 325° - \tan 86°}{1 + \tan 325° \tan 86°}$
26. $\dfrac{\tan 140° - \tan 60°}{1 + \tan 140° \tan 60°}$
27. $\sin 3 \cos 1.2 - \cos 3 \sin 1.2$
28. $\cos\frac{\pi}{7}\cos\frac{\pi}{5} - \sin\frac{\pi}{7}\sin\frac{\pi}{5}$
29. $\dfrac{\tan 2x + \tan x}{1 - \tan 2x \tan x}$
30. $\cos 3x \cos 2y + \sin 3x \sin 2y$

In Exercises 31–36, find the exact value of the expression.

31. $\sin 330° \cos 30° - \cos 330° \sin 30°$
32. $\cos 15° \cos 60° + \sin 15° \sin 60°$
33. $\sin\frac{\pi}{12}\cos\frac{\pi}{4} + \cos\frac{\pi}{12}\sin\frac{\pi}{4}$
34. $\cos\frac{\pi}{16}\cos\frac{3\pi}{16} - \sin\frac{\pi}{16}\sin\frac{3\pi}{16}$
35. $\dfrac{\tan 25° + \tan 110°}{1 - \tan 25° \tan 110°}$
36. $\dfrac{\tan(5\pi/4) - \tan(\pi/12)}{1 + \tan(5\pi/4)\tan(\pi/12)}$

In Exercises 37–44, find the exact value of the trigonometric function given that $\sin u = \frac{5}{13}$ and $\cos v = -\frac{3}{5}$. (Both u and v are in Quadrant II.)

37. $\sin(u + v)$
38. $\cos(u - v)$
39. $\cos(u + v)$
40. $\sin(v - u)$
41. $\tan(u + v)$
42. $\csc(u - v)$
43. $\sec(v - u)$
44. $\cot(u + v)$

In Exercises 45–50, find the exact value of the trigonometric function given that $\sin u = -\frac{7}{25}$ and $\cos v = -\frac{4}{5}$. (Both u and v are in Quadrant III.)

45. $\cos(u + v)$
46. $\sin(u + v)$
47. $\tan(u - v)$
48. $\cot(v - u)$
49. $\sec(u + v)$
50. $\csc(u - v)$

In Exercises 51–54, write the trigonometric expression as an algebraic expression.

51. $\sin(\arcsin x + \arccos x)$
52. $\sin(\arctan 2x - \arccos x)$
53. $\cos(\arccos x + \arcsin x)$
54. $\cos(\arccos x - \arctan x)$

In Exercises 55–64, verify the identity.

55. $\sin(3\pi - x) = \sin x$
56. $\sin\left(\frac{\pi}{2} + x\right) = \cos x$

57. $\sin\left(\frac{\pi}{6} + x\right) = \frac{1}{2}(\cos x + \sqrt{3}\sin x)$

58. $\cos\left(\frac{5\pi}{4} - x\right) = -\frac{\sqrt{2}}{2}(\cos x + \sin x)$

59. $\cos(\pi - \theta) + \sin\left(\frac{\pi}{2} + \theta\right) = 0$

60. $\tan\left(\frac{\pi}{4} - \theta\right) = \frac{1 - \tan\theta}{1 + \tan\theta}$

61. $\cos(x + y)\cos(x - y) = \cos^2 x - \sin^2 y$

62. $\sin(x + y)\sin(x - y) = \sin^2 x - \sin^2 y$

63. $\sin(x + y) + \sin(x - y) = 2\sin x\cos y$

64. $\cos(x + y) + \cos(x - y) = 2\cos x\cos y$

In Exercises 65–68, simplify the expression algebraically and use a graphing utility to confirm your answer graphically.

65. $\cos\left(\frac{3\pi}{2} - x\right)$

66. $\cos(\pi + x)$

67. $\sin\left(\frac{3\pi}{2} + \theta\right)$

68. $\tan(\pi + \theta)$

In Exercises 69–72, find all solutions of the equation in the interval $[0, 2\pi)$.

69. $\sin\left(x + \frac{\pi}{3}\right) + \sin\left(x - \frac{\pi}{3}\right) = 1$

70. $\sin\left(x + \frac{\pi}{6}\right) - \sin\left(x - \frac{\pi}{6}\right) = \frac{1}{2}$

71. $\cos\left(x + \frac{\pi}{4}\right) - \cos\left(x - \frac{\pi}{4}\right) = 1$

72. $\tan(x + \pi) + 2\sin(x + \pi) = 0$

In Exercises 73 and 74, use a graphing utility to approximate the solutions in the interval $[0, 2\pi)$.

73. $\cos\left(x + \frac{\pi}{4}\right) + \cos\left(x - \frac{\pi}{4}\right) = 1$

74. $\tan(x + \pi) - \cos\left(x + \frac{\pi}{2}\right) = 0$

75. *Harmonic Motion* A weight is attached to a spring suspended vertically from a ceiling. When a driving force is applied to the system, the weight moves vertically from its equilibrium position, and this motion is modeled by

$$y = \frac{1}{3}\sin 2t + \frac{1}{4}\cos 2t$$

where y is the distance from equilibrium measured in feet and t is the time in seconds.

(a) Use the identity

$$a\sin B\theta + b\cos B\theta = \sqrt{a^2 + b^2}\sin(B\theta + C)$$

where $C = \arctan(b/a)$, $a > 0$, to write the model in the form

$$y = \sqrt{a^2 + b^2}\sin(Bt + C).$$

(b) Find the amplitude of the oscillations of the weight.

(c) Find the frequency of the oscillations of the weight.

76. *Standing Waves* The equation of a standing wave is obtained by adding the displacements of two waves traveling in opposite directions (see figure). Assume that each of the waves has amplitude A, period T, and wavelength λ. If the models for these waves are

$$y_1 = A\cos 2\pi\left(\frac{t}{T} - \frac{x}{\lambda}\right) \quad \text{and}$$

$$y_2 = A\cos 2\pi\left(\frac{t}{T} + \frac{x}{\lambda}\right)$$

show that

$$y_1 + y_2 = 2A\cos\frac{2\pi t}{T}\cos\frac{2\pi x}{\lambda}.$$

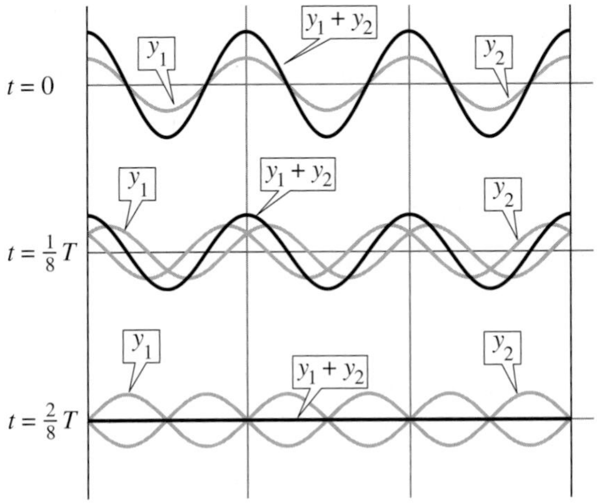

Synthesis

True or False? **In Exercises 77–80, determine whether the statement is true or false. Justify your answer.**

77. $\sin(u \pm v) = \sin u \pm \sin v$

78. $\cos(u \pm v) = \cos u \pm \cos v$

79. $\cos\left(x - \dfrac{\pi}{2}\right) = -\sin x$

80. $\sin\left(x - \dfrac{\pi}{2}\right) = -\cos x$

In Exercises 81–84, verify the identity.

81. $\cos(n\pi + \theta) = (-1)^n \cos \theta$, n is an integer.

82. $\sin(n\pi + \theta) = (-1)^n \sin \theta$, n is an integer.

83. $a \sin B\theta + b \cos B\theta = \sqrt{a^2 + b^2} \sin(B\theta + C)$, where $C = \arctan(b/a)$ and $a > 0$.

84. $a \sin B\theta + b \cos B\theta = \sqrt{a^2 + b^2} \cos(B\theta - C)$, where $C = \arctan(a/b)$ and $b > 0$.

In Exercises 85–88, use the formulas given in Exercises 83 and 84 to write the trigonometric expression in the following forms.

(a) $\sqrt{a^2 + b^2} \sin(B\theta + C)$

(b) $\sqrt{a^2 + b^2} \cos(B\theta - C)$

85. $\sin \theta + \cos \theta$

86. $3 \sin 2\theta + 4 \cos 2\theta$

87. $12 \sin 3\theta + 5 \cos 3\theta$

88. $\sin 2\theta - \cos 2\theta$

In Exercises 89 and 90, use the formulas given in Exercises 83 and 84 to write the trigonometric expression in the form $a \sin B\theta + b \cos B\theta$.

89. $2 \sin\left(\theta + \dfrac{\pi}{2}\right)$

90. $5 \cos\left(\theta + \dfrac{3\pi}{4}\right)$

In Exercises 91 and 92, use the figure, which shows two lines whose equations are

$y_1 = m_1 x + b_1$ and $y_2 = m_2 x + b_2$.

Assume that both lines have positive slopes. Derive a formula for the angle between the two lines. Then use your formula to find the angle between the given pair of lines.

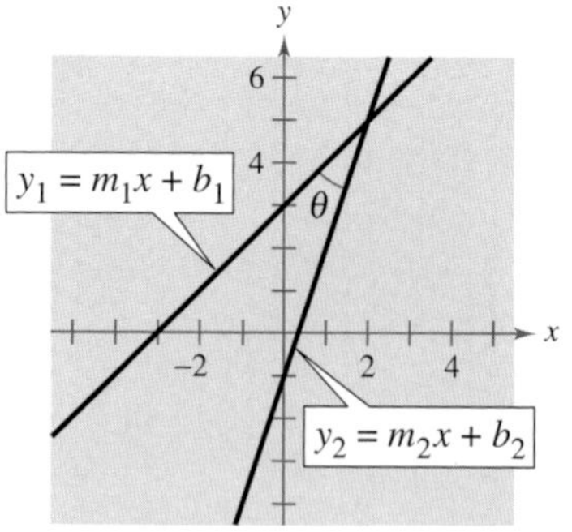

91. $y = x$ and $y = \sqrt{3}x$

92. $y = x$ and $y = \dfrac{1}{\sqrt{3}}x$

93. *Conjecture* Consider the function

$$f(\theta) = \sin^2\left(\theta + \frac{\pi}{4}\right) + \sin^2\left(\theta - \frac{\pi}{4}\right).$$

Use a graphing utility to graph the function and use the graph to create an identity. Prove your conjecture.

94. *Conjecture* Three squares of side s are placed side by side (see figure). Make a conjecture about the relationship between the sum $u + v$ and w. Prove your conjecture by using the identity for the tangent of the sum of two angles.

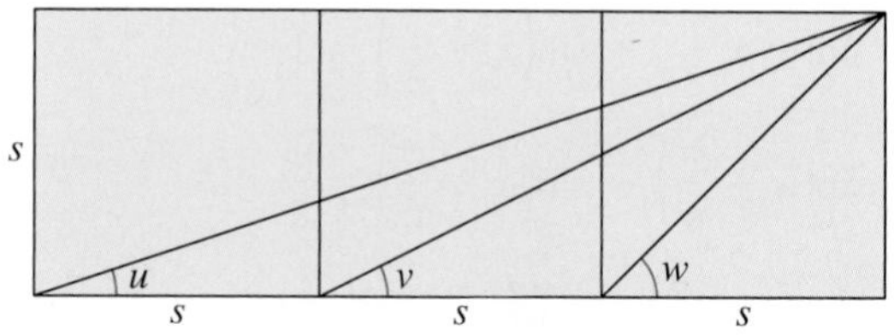

95. *Proof* Write a proof for the formula for $\sin(u + v)$.

96. *Proof* Write a proof for the formula for $\sin(u - v)$.

Review

In Exercises 97–100, find the inverse of f. Verify that $f(f^{-1}(x)) = x$ and $f^{-1}(f(x)) = x$.

97. $f(x) = 5(x - 3)$

98. $f(x) = \dfrac{7 - x}{8}$

99. $f(x) = x^2 - 8$

100. $f(x) = \sqrt{x - 16}$

In Exercises 101–104, apply the inverse properties of $\ln x$ and e^x to simplify the expression.

101. $\log_3 3^{4x-3}$

102. $\log_8 8^{3x^2}$

103. $e^{\ln(6x-3)}$

104. $12x + e^{\ln x(x-2)}$

5.5 Multiple-Angle and Product-to-Sum Formulas

▶ **What you should learn**

- How to use multiple-angle formulas to rewrite and evaluate trigonometric functions
- How to use power-reducing formulas to rewrite and evaluate trigonometric functions
- How to use half-angle formulas to rewrite and evaluate trigonometric functions
- How to use product-to-sum formulas to rewrite and evaluate trigonometric functions

▶ **Why you should learn it**

You can use a variety of trigonometric formulas to rewrite trigonometric functions in more convenient forms. For instance, in Exercise 121 on page 484 you use a half-angle formula to determine the apex angle of a sound wave cone from the speed of an airplane.

NASA-Liaison Agency

A computer animation of this concept appears in the *Interactive* CD-ROM and *Internet* versions of this text.

Consider having your students prove the double-angle formulas. Double-angle formulas are often used in calculus.

Multiple-Angle Formulas

In this section you will study four other categories of trigonometric identities.

1. The first category involves functions of multiple angles such as $\sin ku$ and $\cos ku$.
2. The second category involves squares of trigonometric functions such as $\sin^2 u$.
3. The third category involves functions of half-angles such as $\sin(u/2)$.
4. The fourth category involves products of trigonometric functions such as $\sin u \cos v$.

The most commonly used multiple-angle formulas are the **double-angle formulas.** They are used often, so you should learn them. (Proofs of the double-angle formulas are given in Appendix A.)

Double-Angle Formulas

$$\sin 2u = 2 \sin u \cos u \qquad \tan 2u = \frac{2 \tan u}{1 - \tan^2 u}$$

$$\cos 2u = \cos^2 u - \sin^2 u$$
$$= 2\cos^2 u - 1$$
$$= 1 - 2\sin^2 u$$

Example 1 ▶ Solving a Multiple-Angle Equation

Find all solutions of $2 \cos x + \sin 2x = 0$.

Solution

Begin by rewriting the equation so that it involves functions of x (rather than $2x$). Then factor and solve as usual.

$$2 \cos x + \sin 2x = 0 \qquad \text{Write original equation.}$$

$$2 \cos x + 2 \sin x \cos x = 0 \qquad \text{Double-angle formula}$$

$$2 \cos x(1 + \sin x) = 0 \qquad \text{Factor.}$$

$$\cos x = 0 \quad \text{and} \quad 1 + \sin x = 0 \qquad \text{Set factors equal to zero.}$$

$$x = \frac{\pi}{2}, \frac{3\pi}{2} \qquad x = \frac{3\pi}{2} \qquad \text{Solutions in } [0, 2\pi)$$

Therefore, the general solution is

$$x = \frac{\pi}{2} + 2n\pi \quad \text{and} \quad x = \frac{3\pi}{2} + 2n\pi$$

where n is an integer. Try verifying these solutions graphically.

Example 2 ▶ Using Double-Angle Formulas in Sketching Graphs

Use a double-angle formula to rewrite the equation

$$y = 4\cos^2 x - 2.$$

Then sketch the graph of the equation over the interval $[0, 2\pi]$.

Solution

Using a double-angle formula, you can rewrite the given function as

$$\begin{aligned} y &= 4\cos^2 x - 2 \\ &= 2(2\cos^2 x - 1) \\ &= 2\cos 2x. \end{aligned}$$

Using the techniques discussed in Section 4.5, you can recognize that the graph of this function has an amplitude of 2 and a period of π. The key points in the interval $[0, \pi]$ are as follows.

Maximum	Intercept	Minimum	Intercept	Maximum
$(0, 2)$	$\left(\frac{\pi}{4}, 0\right)$	$\left(\frac{\pi}{2}, -2\right)$	$\left(\frac{3\pi}{4}, 0\right)$	$(\pi, 2)$

Two cycles of the graph are shown in Figure 5.7.

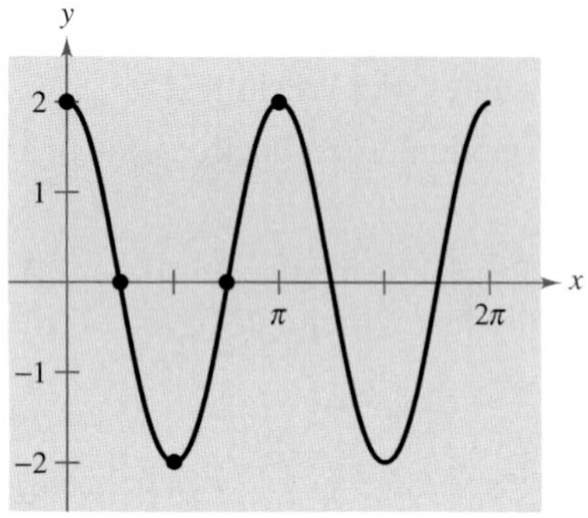

FIGURE 5.7

Example 3 ▶ Evaluating Functions Involving Double Angles

Use the following to find $\sin 2\theta$, $\cos 2\theta$, and $\tan 2\theta$.

$$\cos\theta = \frac{5}{13}, \qquad \frac{3\pi}{2} < \theta < 2\pi$$

Solution

From Figure 5.8, you can see that $\sin\theta = y/r = -12/13$. Consequently, you can write

$$\sin 2\theta = 2\sin\theta\cos\theta = 2\left(\frac{-12}{13}\right)\left(\frac{5}{13}\right) = -\frac{120}{169}$$

$$\cos 2\theta = 2\cos^2\theta - 1 = 2\left(\frac{25}{169}\right) - 1 = -\frac{119}{169}$$

$$\tan 2\theta = \frac{\sin 2\theta}{\cos 2\theta} = \frac{120}{119}.$$

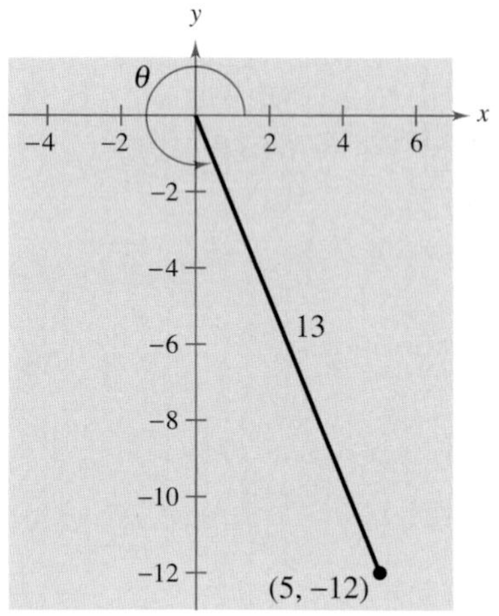

FIGURE 5.8

The double-angle formulas are not restricted to angles 2θ and θ. Other *double* combinations, such as 4θ and 2θ or 6θ and 3θ, are also valid. Here are two examples.

$$\sin 4\theta = 2\sin 2\theta\cos 2\theta \qquad \text{and} \qquad \cos 6\theta = \cos^2 3\theta - \sin^2 3\theta$$

By using double-angle formulas together with the sum formulas given in the preceding section, you can form other multiple-angle formulas.

Example 4 ▶ Deriving a Triple-Angle Formula

Express $\sin 3x$ in terms of $\sin x$.

Solution

$$\begin{aligned}
\sin 3x &= \sin(2x + x)\\
&= \sin 2x \cos x + \cos 2x \sin x\\
&= 2 \sin x \cos x \cos x + (1 - 2 \sin^2 x)\sin x\\
&= 2 \sin x \cos^2 x + \sin x - 2 \sin^3 x\\
&= 2 \sin x(1 - \sin^2 x) + \sin x - 2 \sin^3 x\\
&= 2 \sin x - 2 \sin^3 x + \sin x - 2 \sin^3 x\\
&= 3 \sin x - 4 \sin^3 x
\end{aligned}$$

Power-Reducing Formulas

The double-angle formulas can be used to obtain the following **power-reducing formulas.** (Proofs of the power-reducing formulas are given in Appendix A.) Example 5 shows a typical power reduction that is used in calculus.

Power-reducing formulas are often used in calculus.

Power-Reducing Formulas

$$\sin^2 u = \frac{1 - \cos 2u}{2} \qquad \cos^2 u = \frac{1 + \cos 2u}{2} \qquad \tan^2 u = \frac{1 - \cos 2u}{1 + \cos 2u}$$

A computer animation of this concept appears in the *Interactive* CD-ROM and *Internet* versions of this text.

Example 5 ▶ Reducing a Power

Rewrite $\sin^4 x$ as a sum of first powers of the cosines of multiple angles.

Solution

Note the repeated use of power-reducing formulas.

$$\begin{aligned}
\sin^4 x &= (\sin^2 x)^2\\
&= \left(\frac{1 - \cos 2x}{2}\right)^2\\
&= \frac{1}{4}(1 - 2 \cos 2x + \cos^2 2x)\\
&= \frac{1}{4}\left(1 - 2 \cos 2x + \frac{1 + \cos 4x}{2}\right)\\
&= \frac{1}{4} - \frac{1}{2}\cos 2x + \frac{1}{8} + \frac{1}{8}\cos 4x\\
&= \frac{1}{8}(3 - 4 \cos 2x + \cos 4x)
\end{aligned}$$

Half-Angle Formulas

You can derive some useful alternative forms of the power-reducing formulas by replacing u with $u/2$. The results are called **half-angle formulas.**

A computer animation of this concept appears in the *Interactive* CD-ROM and *Internet* versions of this text.

Half-Angle Formulas

$$\sin\frac{u}{2} = \pm\sqrt{\frac{1-\cos u}{2}}$$

$$\cos\frac{u}{2} = \pm\sqrt{\frac{1+\cos u}{2}}$$

$$\tan\frac{u}{2} = \frac{1-\cos u}{\sin u} = \frac{\sin u}{1+\cos u}$$

The signs of $\sin(u/2)$ and $\cos(u/2)$ depend on the quadrant in which $u/2$ lies.

Example 6 ▶ Using a Half-Angle Formula

Find the exact value of $\sin 105°$.

Solution

Begin by noting that 105° is half of 210°. Then, using the half-angle formula for $\sin(u/2)$ and the fact that 105° lies in Quadrant II, you have

$$\begin{aligned}\sin 105° &= \sqrt{\frac{1-\cos 210°}{2}}\\ &= \sqrt{\frac{1-(-\cos 30°)}{2}}\\ &= \sqrt{\frac{1+(\sqrt{3}/2)}{2}}\\ &= \frac{\sqrt{2+\sqrt{3}}}{2}.\end{aligned}$$

The positive square root is chosen because $\sin\theta$ is positive in Quadrant II.

Use your calculator to verify the result obtained in Example 6. That is, evaluate $\sin 105°$ and $\left(\sqrt{2+\sqrt{3}}\right)/2$.

$$\sin 105° \approx 0.9659258$$

$$\frac{\sqrt{2+\sqrt{3}}}{2} \approx 0.9659258$$

You can see that both values are approximately 0.9659258.

Example 7 ▶ Solving a Trigonometric Equation

Find all solutions of

$$2 - \sin^2 x = 2\cos^2 \frac{x}{2}$$

in the interval $[0, 2\pi)$.

Solution

A common error is to write $2\cos^2 \frac{x}{2}$ as $\cos^2 x$, rather than to correctly use the identity $2\cos^2 \frac{x}{2} = 2\left(\frac{1 + \cos x}{2}\right)$.

$$2 - \sin^2 x = 2\cos^2 \frac{x}{2} \quad \text{Write original equation.}$$

$$2 - \sin^2 x = 2\left(\pm\sqrt{\frac{1 + \cos x}{2}}\right)^2 \quad \text{Half-angle formula}$$

$$2 - \sin^2 x = 2\left(\frac{1 + \cos x}{2}\right) \quad \text{Simplify.}$$

$$2 - \sin^2 x = 1 + \cos x \quad \text{Simplify.}$$

$$2 - (1 - \cos^2 x) = 1 + \cos x \quad \text{Pythagorean identity}$$

$$\cos^2 x - \cos x = 0 \quad \text{Simplify.}$$

$$\cos x(\cos x - 1) = 0 \quad \text{Factor.}$$

By setting the factors $\cos x$ and $(\cos x - 1)$ equal to zero, you find that the solutions in the interval $[0, 2\pi)$ are

$$x = \frac{\pi}{2}, \quad x = \frac{3\pi}{2}, \quad \text{and} \quad x = 0.$$

Product-to-Sum Formulas

Each of the following **product-to-sum formulas** is easily verified using the sum and difference formulas discussed in the preceding section.

Product-to-Sum Formulas

$$\sin u \sin v = \frac{1}{2}[\cos(u - v) - \cos(u + v)]$$

$$\cos u \cos v = \frac{1}{2}[\cos(u - v) + \cos(u + v)]$$

$$\sin u \cos v = \frac{1}{2}[\sin(u + v) + \sin(u - v)]$$

$$\cos u \sin v = \frac{1}{2}[\sin(u + v) - \sin(u - v)]$$

Example 8 ▶ Writing Products as Sums

Rewrite the product as a sum or difference.

$\cos 5x \sin 4x$

Solution

$$\cos 5x \sin 4x = \frac{1}{2}[\sin(5x + 4x) - \sin(5x - 4x)]$$

$$= \frac{1}{2}\sin 9x - \frac{1}{2}\sin x$$

Occasionally, it is useful to reverse the procedure and write a sum of trigonometric functions as a product. This can be accomplished with the following **sum-to-product formulas.** (A proof of the first formula is given in Appendix A.)

Activities

1. Find the exact values of $\sin 2u$, $\cos 2u$, and $\tan 2u$ given $\cot u = -5$, $\frac{3\pi}{2} < u < 2\pi$.

   Answer: $\sin 2u = -\frac{5}{13}$

   $\cos 2u = \frac{12}{13}$

   $\tan 2u = -\frac{5}{12}$

2. Find all solutions of the equation in the interval $[0, 2\pi)$.

   $4 \sin x \cos x = 1$

   Answer: $x = \frac{\pi}{12}, \frac{5\pi}{12}, \frac{13\pi}{12}, \frac{17\pi}{12}$

3. Rewrite as a sum: $2 \sin \theta \sin \phi$.

   Answer: $\cos(\theta - \phi) - \cos(\theta + \phi)$

Sum-to-Product Formulas

$$\sin x + \sin y = 2 \sin\left(\frac{x+y}{2}\right) \cos\left(\frac{x-y}{2}\right)$$

$$\sin x - \sin y = 2 \cos\left(\frac{x+y}{2}\right) \sin\left(\frac{x-y}{2}\right)$$

$$\cos x + \cos y = 2 \cos\left(\frac{x+y}{2}\right) \cos\left(\frac{x-y}{2}\right)$$

$$\cos x - \cos y = -2 \sin\left(\frac{x+y}{2}\right) \sin\left(\frac{x-y}{2}\right)$$

Example 9 ▶ Using a Sum-to-Product Formula

Find the exact value of

$\cos 195° + \cos 105°.$

Solution

Using the appropriate sum-to-product formula, you obtain

$$\cos 195° + \cos 105° = 2 \cos\left(\frac{195° + 105°}{2}\right) \cos\left(\frac{195° - 105°}{2}\right)$$

$$= 2 \cos 150° \cos 45°$$

$$= 2\left(-\frac{\sqrt{3}}{2}\right)\left(\frac{\sqrt{2}}{2}\right)$$

$$= -\frac{\sqrt{6}}{2}.$$

Example 10 ▶ Solving a Trigonometric Equation

Find all solutions of $\sin 5x + \sin 3x = 0$.

Solution

$$\sin 5x + \sin 3x = 0 \qquad \text{Write original equation.}$$

$$2\sin\left(\frac{5x + 3x}{2}\right)\cos\left(\frac{5x - 3x}{2}\right) = 0 \qquad \text{Sum-to-product formula}$$

$$2\sin 4x \cos x = 0 \qquad \text{Simplify.}$$

By setting the factor $\sin 4x$ equal to zero, you can find that the solutions in the interval $[0, 2\pi)$ are

$$x = 0, \frac{\pi}{4}, \frac{\pi}{2}, \frac{3\pi}{4}, \pi, \frac{5\pi}{4}, \frac{3\pi}{2}, \frac{7\pi}{4}.$$

The equation $\cos x = 0$ yields no additional solutions, and you can conclude that the solutions are of the form

$$x = \frac{n\pi}{4}$$

where n is an integer. These solutions are verified graphically in Figure 5.9.

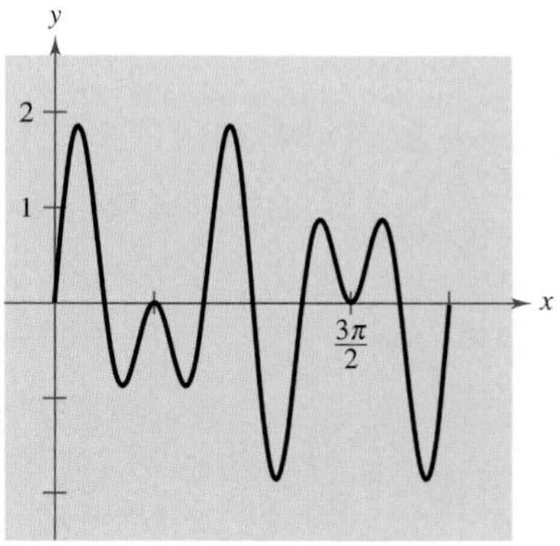

FIGURE 5.9

Example 11 ▶ Verifying a Trigonometric Identity

Verify the identity

$$\frac{\sin t + \sin 3t}{\cos t + \cos 3t} = \tan 2t.$$

Solution

Using appropriate sum-to-product formulas, you have

$$\frac{\sin t + \sin 3t}{\cos t + \cos 3t} = \frac{2\sin 2t\cos(-t)}{2\cos 2t\cos(-t)}$$

$$= \frac{\sin 2t}{\cos 2t}$$

$$= \tan 2t.$$

Writing ABOUT MATHEMATICS

Deriving an Area Formula Describe how you can use a double-angle formula or a half-angle formula to derive a formula for the area of an isosceles triangle. Use a labeled sketch to illustrate your derivation. Then write two examples that show how your formula can be used.

5.5 Exercises

In Exercises 1–8, use the figure to find the exact value of the trigonometric function.

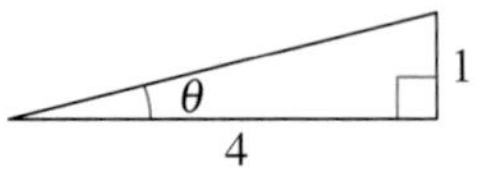

1. $\sin \theta$
2. $\tan \theta$
3. $\cos 2\theta$
4. $\sin 2\theta$
5. $\tan 2\theta$
6. $\sec 2\theta$
7. $\csc 2\theta$
8. $\cot 2\theta$

In Exercises 9–18, find the exact solutions of the equation in the interval $[0, 2\pi)$.

9. $\sin 2x - \sin x = 0$
10. $\sin 2x + \cos x = 0$
11. $4 \sin x \cos x = 1$
12. $\sin 2x \sin x = \cos x$
13. $\cos 2x - \cos x = 0$
14. $\cos 2x + \sin x = 0$
15. $\tan 2x - \cot x = 0$
16. $\tan 2x - 2 \cos x = 0$
17. $\sin 4x = -2 \sin 2x$
18. $(\sin 2x + \cos 2x)^2 = 1$

In Exercises 19–22, use a double-angle formula to rewrite the expression.

19. $6 \sin x \cos x$
20. $6 \cos^2 x - 3$
21. $4 - 8 \sin^2 x$
22. $(\cos x + \sin x)(\cos x - \sin x)$

In Exercises 23–28, find the exact values of sin 2*u*, cos 2*u*, and tan 2*u* using the double-angle formulas.

23. $\sin u = -\frac{4}{5}, \quad \pi < u < \frac{3\pi}{2}$
24. $\cos u = -\frac{2}{3}, \quad \frac{\pi}{2} < u < \pi$
25. $\tan u = \frac{3}{4}, \quad 0 < u < \frac{\pi}{2}$
26. $\cot u = -4, \quad \frac{3\pi}{2} < u < 2\pi$
27. $\sec u = -\frac{5}{2}, \quad \frac{\pi}{2} < u < \pi$
28. $\csc u = 3, \quad \frac{\pi}{2} < u < \pi$

In Exercises 29–34, use the power-reducing formulas to rewrite the expression in terms of the first power of the cosine.

29. $\cos^4 x$
30. $\sin^8 x$
31. $\sin^2 x \cos^2 x$
32. $\sin^4 x \cos^4 x$
33. $\sin^2 x \cos^4 x$
34. $\sin^4 x \cos^2 x$

In Exercises 35–40, use the figure to find the exact value of the trigonometric function.

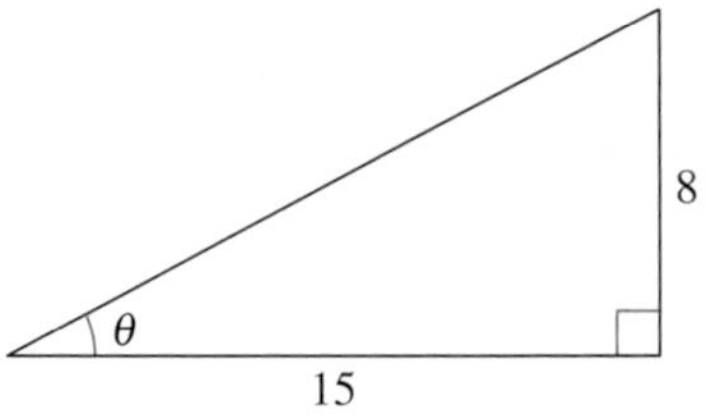

35. $\cos \frac{\theta}{2}$
36. $\sin \frac{\theta}{2}$
37. $\tan \frac{\theta}{2}$
38. $\sec \frac{\theta}{2}$
39. $\csc \frac{\theta}{2}$
40. $\cot \frac{\theta}{2}$

In Exercises 41–48, use the half-angle formulas to determine the exact values of the sine, cosine, and tangent of the angle.

41. 75°
42. 165°
43. $112^\circ\ 30'$
44. $67^\circ\ 30'$
45. $\frac{\pi}{8}$
46. $\frac{\pi}{12}$
47. $\frac{3\pi}{8}$
48. $\frac{7\pi}{12}$

In Exercises 49–54, find the exact values of sin(*u*/2), cos(*u*/2), and tan(*u*/2) using the half-angle formulas.

49. $\sin u = \frac{5}{13}, \quad \frac{\pi}{2} < u < \pi$
50. $\cos u = \frac{3}{5}, \quad 0 < u < \frac{\pi}{2}$

51. $\tan u = -\frac{5}{8}, \quad \frac{3\pi}{2} < u < 2\pi$

52. $\cot u = 3, \quad \pi < u < \frac{3\pi}{2}$

53. $\csc u = -\frac{5}{3}, \quad \pi < u < \frac{3\pi}{2}$

54. $\sec u = -\frac{7}{2}, \quad \frac{\pi}{2} < u < \pi$

In Exercises 55–58, use the half-angle formulas to simplify the expression.

55. $\sqrt{\frac{1 - \cos 6x}{2}}$

56. $\sqrt{\frac{1 + \cos 4x}{2}}$

57. $-\sqrt{\frac{1 - \cos 8x}{1 + \cos 8x}}$

58. $-\sqrt{\frac{1 - \cos(x - 1)}{2}}$

In Exercises 59–62, find all solutions in the interval $[0, 2\pi)$. Use a graphing utility to graph the function and verify the solutions.

59. $\sin \frac{x}{2} + \cos x = 0$

60. $\sin \frac{x}{2} + \cos x - 1 = 0$

61. $\cos \frac{x}{2} - \sin x = 0$

62. $\tan \frac{x}{2} - \sin x = 0$

In Exercises 63–74, use the product-to-sum formulas to write the product as a sum or difference.

63. $6 \sin \frac{\pi}{4} \cos \frac{\pi}{4}$

64. $4 \cos \frac{\pi}{3} \sin \frac{5\pi}{6}$

65. $\cos 4\theta \sin 6\theta$

66. $3 \sin 2\alpha \sin 3\alpha$

67. $5 \cos(-5\beta) \cos 3\beta$

68. $\cos 2\theta \cos 4\theta$

69. $\sin(x + y) \sin(x - y)$

70. $\sin(x + y) \cos(x - y)$

71. $\cos(\theta - \pi) \sin(\theta + \pi)$

72. $\sin(\theta + \pi) \sin(\theta - \pi)$

73. $10 \cos 75^\circ \cos 15^\circ$

74. $6 \sin 45^\circ \cos 15^\circ$

In Exercises 75–86, use the sum-to-product formulas to write the sum or difference as a product.

75. $\sin 60^\circ + \sin 30^\circ$

76. $\cos 120^\circ + \cos 30^\circ$

77. $\cos \frac{3\pi}{4} - \cos \frac{\pi}{4}$

78. $\sin \frac{5\pi}{4} - \sin \frac{3\pi}{4}$

79. $\sin 5\theta - \sin 3\theta$

80. $\sin 3\theta + \sin \theta$

81. $\cos 6x + \cos 2x$

82. $\sin x + \sin 5x$

83. $\sin(\alpha + \beta) - \sin(\alpha - \beta)$

84. $\cos(\phi + 2\pi) + \cos \phi$

85. $\cos\left(\theta + \frac{\pi}{2}\right) - \cos\left(\theta - \frac{\pi}{2}\right)$

86. $\sin\left(x + \frac{\pi}{2}\right) + \sin\left(x - \frac{\pi}{2}\right)$

In Exercises 87–90, find all solutions in the interval $[0, 2\pi)$. Use a graphing utility to graph the function and verify the solutions.

87. $\sin 6x + \sin 2x = 0$

88. $\cos 2x - \cos 6x = 0$

89. $\frac{\cos 2x}{\sin 3x - \sin x} - 1 = 0$

90. $\sin^2 3x - \sin^2 x = 0$

In Exercises 91–94, use the figure to find the exact value of the trigonometric function in two ways.

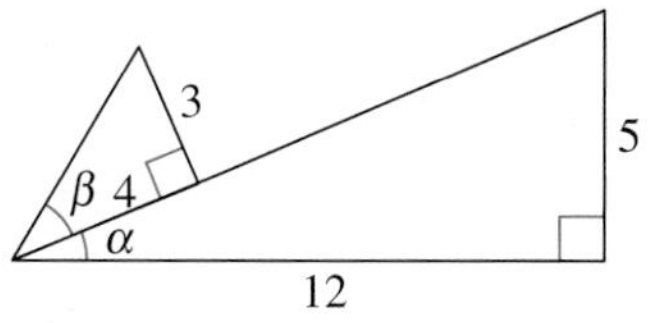

91. $\sin^2 \alpha$

92. $\cos^2 \alpha$

93. $\sin \alpha \cos \beta$

94. $\cos \alpha \sin \beta$

In Exercises 95–110, verify the identity.

95. $\csc 2\theta = \dfrac{\csc \theta}{2 \cos \theta}$

96. $\sec 2\theta = \dfrac{\sec^2 \theta}{2 - \sec^2 \theta}$

97. $\cos^2 2\alpha - \sin^2 2\alpha = \cos 4\alpha$

98. $\cos^4 x - \sin^4 x = \cos 2x$

99. $(\sin x + \cos x)^2 = 1 + \sin 2x$

100. $\sin \dfrac{\alpha}{3} \cos \dfrac{\alpha}{3} = \dfrac{1}{2} \sin \dfrac{2\alpha}{3}$

101. $1 + \cos 10y = 2 \cos^2 5y$

102. $\dfrac{\cos 3\beta}{\cos \beta} = 1 - 4 \sin^2 \beta$

103. $\sec \dfrac{u}{2} = \pm\sqrt{\dfrac{2 \tan u}{\tan u + \sin u}}$

104. $\tan \dfrac{u}{2} = \csc u - \cot u$

105. $\dfrac{\sin x \pm \sin y}{\cos x + \cos y} = \tan \dfrac{x \pm y}{2}$

106. $\dfrac{\sin x + \sin y}{\cos x - \cos y} = -\cot \dfrac{x - y}{2}$

107. $\dfrac{\cos 4x + \cos 2x}{\sin 4x + \sin 2x} = \cot 3x$

108. $\dfrac{\cos t + \cos 3t}{\sin 3t - \sin t} = \cot t$

109. $\sin\left(\dfrac{\pi}{6} + x\right) + \sin\left(\dfrac{\pi}{6} - x\right) = \cos x$

110. $\cos\left(\dfrac{\pi}{3} + x\right) + \cos\left(\dfrac{\pi}{3} - x\right) = \cos x$

In Exercises 111–114, use a graphing utility to verify the identity. Confirm that it is an identity algebraically.

111. $\cos 3\beta = \cos^3 \beta - 3 \sin^2 \beta \cos \beta$

112. $\sin 4\beta = 4 \sin \beta \cos \beta(1 - 2 \sin^2 \beta)$

113. $(\cos 4x - \cos 2x)/(2 \sin 3x) = -\sin x$

114. $(\cos 3x - \cos x)/(\sin 3x - \sin x) = -\tan 2x$

In Exercises 115 and 116, graph the function by hand in the interval $[0, 2\pi)$ by using the power-reducing formulas.

115. $f(x) = \sin^2 x$ **116.** $f(x) = \cos^2 x$

In Exercises 117 and 118, write the trigonometric expression as an algebraic expression.

117. $\sin(2 \arcsin x)$ **118.** $\cos(2 \arccos x)$

119. *Area* The length of each of the two equal sides of an isosceles triangle is 10 meters (see figure). The angle between the two sides is θ.

(a) Express the area of the triangle as a function of $\theta/2$.

(b) Express the area of the triangle as a function of θ. Determine the value of θ such that the area is a maximum.

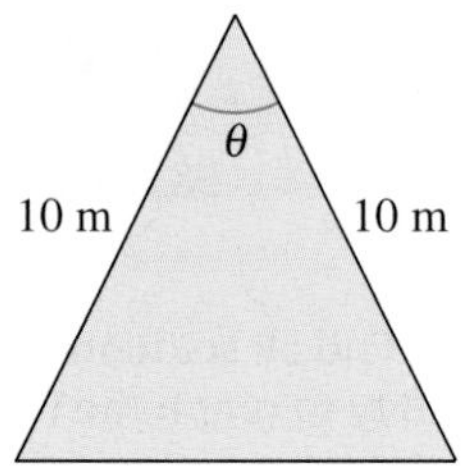

120. *Projectile Motion* The range of a projectile fired at an angle θ with the horizontal and with an initial velocity of v_0 feet per second is

$$r = \frac{1}{32} v_0^2 \sin 2\theta$$

where r is measured in feet. Determine the expression for the range in terms of θ.

121. *Mach Number* The mach number M of an airplane is the ratio of its speed to the speed of sound. When an airplane travels faster than the speed of sound, the sound waves form a cone behind the airplane. The mach number is related to the apex angle θ of the cone by

$$\sin \frac{\theta}{2} = \frac{1}{M}.$$

Find the angle θ that corresponds to a mach number of 4.5.

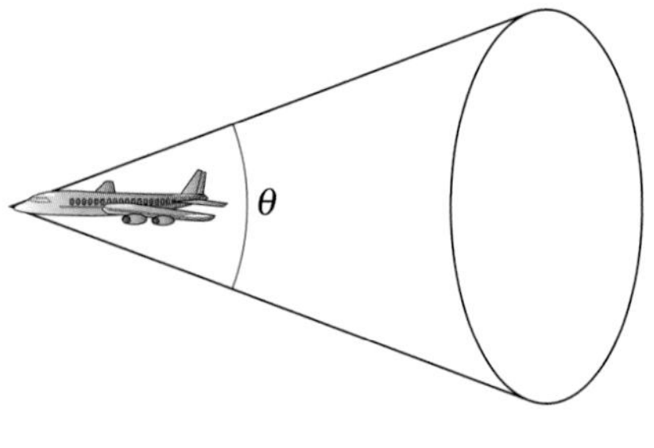

122. *Railroad Track* When two railroad tracks merge, the overlapping portions of the tracks are in the shapes of circular arcs. The radius of each arc r (in feet) and the angle θ are related by

$$\frac{x}{2} = 2r \sin^2 \frac{\theta}{2}.$$

Write a formula for x in terms of $\cos \theta$.

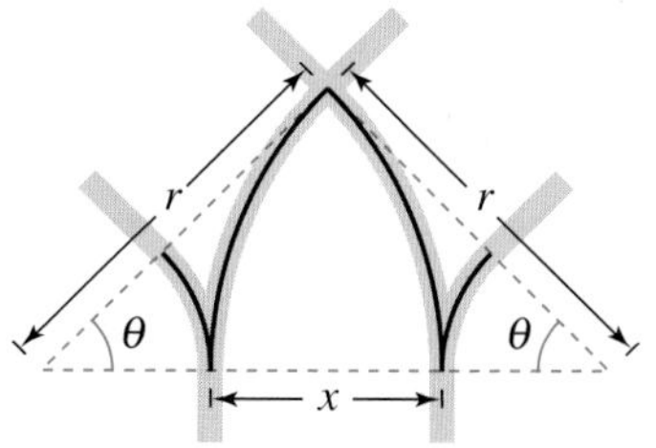

Synthesis

True or False? **In Exercises 123 and 124, determine whether the statement is true or false. Justify your answer.**

123. Because the sine function is an odd function, for a negative number u, $\sin 2u = -2 \sin u \cos u$.

124. $\sin \frac{u}{2} = -\sqrt{\frac{1 - \cos u}{2}}$ when u is in the second quadrant.

In Exercises 125 and 126, (a) use a graphing utility to graph the function and approximate the maximum and minimum points on the graph in the interval $[0, 2\pi)$ and (b) solve the trigonometric equation and verify that its solutions are the x-coordinates of the maximum and minimum points of f. (Calculus is required to find the trigonometric equation.)

	Function	*Trigonometric Equation*
125.	$f(x) = 4 \sin \frac{x}{2} + \cos x$	$2 \cos \frac{x}{2} - \sin x = 0$
126.	$f(x) = \cos 2x - 2 \sin x$	$-2 \cos x(2 \sin x + 1) = 0$

127. *Exploration* Consider the function

$$f(x) = \sin^4 x + \cos^4 x.$$

(a) Use the power-reducing formulas to write the function in terms of cosine to the first power.

(b) Determine another way of rewriting the function. Use a graphing utility to rule out incorrectly rewritten functions.

(c) Add a trigonometric term to the function so that it becomes a perfect square trinomial. Rewrite the function as a perfect square trinomial minus the term that you added. Use a graphing utility to rule out incorrectly rewritten functions.

(d) Rewrite the result of part (c) in terms of the sine of a double angle. Use a graphing utility to rule out incorrectly rewritten functions.

(e) When you rewrite a trigonometric expression, the result may not be the same as a friend's. Does this mean that one of you is wrong? Explain.

128. *Conjecture* Consider the function

$$f(x) = 2 \sin x[2 \cos^2(x/2) - 1].$$

(a) Use a graphing utility to graph the function.

(b) Make a conjecture about the function that is an identity with f.

(c) Verify your conjecture analytically.

Review

129. *Profit* The total profit for a company in October was 16% higher than it was in September. The total profit for the two months was \$507,600. Find the profit for each month.

130. *Average Speed* Two cars start at a given point and travel in the same direction at average speeds of 48 miles per hour and 56 miles per hour. How much time must elapse before the two cars are 12 miles apart?

131. *Mixture Problem* A 55-gallon barrel contains a mixture with a concentration of 30%. How much of this mixture must be withdrawn and replaced by 100% concentrate to bring the mixture up to 50% concentration?

132. *Baseball Diamond* A baseball diamond has the shape of a square where the distance between each of the consecutive bases is 90 feet. Approximate the distance from home plate to second base.

Chapter Summary

What did you learn?

	Review Exercises
Section 5.1	
☐ How to recognize and write the fundamental trigonometric identities	1–6
☐ How to use the fundamental trigonometric identities to evaluate trigonometric functions, simplify trigonometric expressions, and rewrite trigonometric expressions	7–23
Section 5.2	
☐ How to plan a strategy for verifying trigonometric identities	24–31
☐ How to verify trigonometric identities	24–31
Section 5.3	
☐ How to use standard algebraic techniques to solve trigonometric equations	32–37
☐ How to solve trigonometric equations of quadratic type	38–41
☐ How to solve trigonometric equations involving multiple angles	42–45
☐ How to use inverse trigonometric functions to solve trigonometric equations	46–49
Section 5.4	
☐ How to use sum and difference formulas to evaluate trigonometric functions	50–63
☐ How to use sum and difference formulas to verify identities and solve trigonometric equations	64–67
Section 5.5	
☐ How to use multiple-angle formulas to rewrite and evaluate trigonometric functions	68–72
☐ How to use power-reducing formulas to rewrite and evaluate trigonometric functions	73–76
☐ How to use half-angle formulas to rewrite and evaluate trigonometric functions	77–83
☐ How to use product-to-sum formulas to rewrite and evaluate trigonometric functions	84–92

Review Exercises

5.1 In Exercises 1–6, name the trigonometric function that is equivalent to the expression.

1. $\dfrac{1}{\cos x}$
2. $\dfrac{1}{\sin x}$
3. $\dfrac{1}{\sec x}$
4. $\dfrac{1}{\tan x}$
5. $\dfrac{\cos x}{\sin x}$
6. $\sqrt{1 + \tan^2 x}$

In Exercises 7–10, use the given values and trigonometric identities to evaluate (if possible) the other trigonometric functions of the angle.

7. $\sin x = \frac{3}{5}, \quad \cos x = \frac{4}{5}$
8. $\tan\theta = \dfrac{2}{3}, \quad \sec\theta = \dfrac{\sqrt{13}}{3}$
9. $\sin\left(\dfrac{\pi}{2} - x\right) = \dfrac{\sqrt{2}}{2}, \quad \sin x = -\dfrac{\sqrt{2}}{2}$
10. $\csc\left(\dfrac{\pi}{2} - \theta\right) = 9, \quad \sin\theta = \dfrac{4\sqrt{5}}{9}$

In Exercises 11–22, use the fundamental trigonometric identities to simplify the trigonometric expression.

11. $\dfrac{1}{\cot^2 x + 1}$
12. $\dfrac{\tan\theta}{1 - \cos^2\theta}$
13. $\tan^2 x(\csc^2 x - 1)$
14. $\cot^2 x(\sin^2 x)$
15. $\dfrac{\sin\left(\dfrac{\pi}{2} - \theta\right)}{\sin\theta}$
16. $\dfrac{\cot\left(\dfrac{\pi}{2} - u\right)}{\cos u}$
17. $\cos^2 x + \cos^2 x\cot^2 x$
18. $\tan^2\theta\csc^2\theta - \tan^2\theta$
19. $(\tan x + 1)^2\cos x$
20. $(\sec x - \tan x)^2$
21. $\dfrac{1}{\csc\theta + 1} - \dfrac{1}{\csc\theta - 1}$
22. $\dfrac{\cos^2 x}{1 - \sin x}$

23. *Rate of Change* The rate of change of the function $f(x) = 2\sqrt{\sin x}$ is the expression $\sin^{-1/2} x\cos x$. Show that this expression can also be written as $\cot x\sqrt{\sin x}$.

5.2 In Exercises 24–31, verify the identity.

24. $\cos x(\tan^2 x + 1) = \sec x$
25. $\sec^2 x\cot x - \cot x = \tan x$
26. $\cos\left(x + \dfrac{\pi}{2}\right) = -\sin x$
27. $\cot\left(\dfrac{\pi}{2} - x\right) = \tan x$
28. $\dfrac{1}{\tan\theta\csc\theta} = \cos\theta$
29. $\dfrac{1}{\tan x\csc x\sin x} = \cot x$
30. $\sin^5 x\cos^2 x = (\cos^2 x - 2\cos^4 x + \cos^6 x)\sin x$
31. $\cos^3 x\sin^2 x = (\sin^2 x - \sin^4 x)\cos x$

5.3 In Exercises 32–37, solve the equation.

32. $\sin x = \sqrt{3} - \sin x$
33. $4\cos\theta = 1 + 2\cos\theta$
34. $3\sqrt{3}\tan u = 3$
35. $\frac{1}{2}\sec x - 1 = 0$
36. $3\csc^2 x = 4$
37. $4\tan^2 u - 1 = \tan^2 u$

In Exercises 38–45, find all solutions of the equation in the interval $[0, 2\pi)$.

38. $2\cos^2 x - \cos x = 1$
39. $2\sin^2 x - 3\sin x = -1$
40. $\cos^2 x + \sin x = 1$
41. $\sin^2 x + 2\cos x = 2$
42. $2\sin 2x - \sqrt{2} = 0$
43. $\sqrt{3}\tan 3x = 0$
44. $\cos 4x(\cos x - 1) = 0$
45. $3\csc^2 5x = -4$

In Exercises 46–49, use inverse functions where needed to find all solutions of the equation in the interval $[0, 2\pi)$.

46. $\sin^2 x - 2\sin x = 0$
47. $2\cos^2 x + 3\cos x = 0$
48. $\tan^2\theta + \tan\theta - 12 = 0$
49. $\sec^2 x + 6\tan x + 4 = 0$

5.4 In Exercises 50–53, find the exact values of the sine, cosine, and tangent of the angle by using a sum or difference formula.

50. $285° = 315° - 30°$

51. $345° = 300° + 45°$

52. $\frac{25\pi}{12} = \frac{11\pi}{6} + \frac{\pi}{4}$

53. $\frac{19\pi}{12} = \frac{11\pi}{6} - \frac{\pi}{4}$

In Exercises 54–57, write the expression as the sine, cosine, or tangent of an angle.

54. $\sin 60° \cos 45° - \cos 60° \sin 45°$

55. $\cos 45° \cos 120° - \sin 45° \sin 120°$

56. $\frac{\tan 25° + \tan 10°}{1 - \tan 25° \tan 10°}$

57. $\frac{\tan 68° - \tan 115°}{1 + \tan 68° \tan 115°}$

In Exercises 58–63, find the exact value of the trigonometric function given that $\sin u = \frac{3}{4}$, $\cos v = -\frac{5}{13}$, and u and v are in Quadrant II.

58. $\sin(u + v)$

59. $\tan(u + v)$

60. $\cos(u - v)$

61. $\sin(u - v)$

62. $\cos(u + v)$

63. $\tan(u - v)$

In Exercises 64–67, find all solutions of the equation in the interval $[0, 2\pi)$.

64. $\sin\left(x + \frac{\pi}{4}\right) - \sin\left(x - \frac{\pi}{4}\right) = 1$

65. $\cos\left(x + \frac{\pi}{6}\right) - \cos\left(x - \frac{\pi}{6}\right) = 1$

66. $\sin\left(x + \frac{\pi}{2}\right) - \sin\left(x - \frac{\pi}{2}\right) = \sqrt{3}$

67. $\cos\left(x + \frac{3\pi}{4}\right) - \cos\left(x - \frac{3\pi}{4}\right) = 0$

5.5 In Exercises 68 and 69, use double-angle formulas to verify the identity algebraically and use a graphing utility to confirm it graphically.

68. $\sin 4x = 8 \cos^3 x \sin x - 4 \cos x \sin x$

69. $\tan^2 x = \frac{1 - \cos 2x}{1 + \cos 2x}$

In Exercises 70 and 71, find the exact values of $\sin 2u$, $\cos 2u$, and $\tan 2u$ using the double-angle formulas.

70. $\sin u = -\frac{4}{5}, \quad \pi < u < \frac{3\pi}{2}$

71. $\cos u = -\frac{2}{\sqrt{5}}, \quad \frac{\pi}{2} < u < \pi$

72. *Projectile Motion* A baseball leaves the hand of the person at first base at an angle of θ with the horizontal and an initial velocity of $v_0 = 80$ feet per second. The ball is caught by the person at second base 100 feet away. Find θ if the range r of a projectile is

$$r = \frac{1}{32} v_0^2 \sin 2\theta.$$

In Exercises 73–76, use the power-reducing formulas to rewrite the expression in terms of the first power of the cosine.

73. $\tan^2 2x$

74. $\cos^2 3x$

75. $\sin^2 x \tan^2 x$

76. $\cos^2 x \tan^2 x$

In Exercises 77–80, use the half-angle formulas to determine the exact values of the sine, cosine, and tangent of the angle.

77. $-75°$

78. $15°$

79. $\frac{19\pi}{12}$

80. $-\frac{17\pi}{12}$

In Exercises 81 and 82, use the half-angle formulas to simplify the expression.

81. $-\sqrt{\frac{1 + \cos 10x}{2}}$

82. $\frac{\sin 6x}{1 + \cos 6x}$

83. *Volume* A trough for feeding cattle is 4 meters long and its cross sections are isosceles triangles with the two equal sides being $\frac{1}{2}$ meter (see figure). The angle between the two sides is θ.

(a) Express the trough's volume as a function of $\frac{\theta}{2}$.

(b) Express the volume of the trough as a function of θ and determine the value of θ such that the volume is maximum.

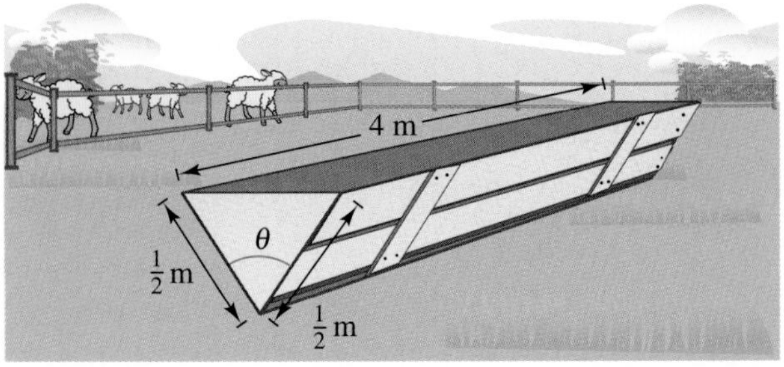

In Exercises 84–87, use the product-to-sum formulas to write the product as a sum or difference.

84. $\cos \frac{\pi}{6} \sin \frac{\pi}{6}$

85. $6 \sin 15^\circ \sin 45^\circ$

86. $\cos 5\theta \cos 3\theta$

87. $4 \sin 3\alpha \cos 2\alpha$

In Exercises 88–91, use the sum-to-product formulas to write the sum or difference as a product.

88. $\sin 60^\circ + \sin 90^\circ$

89. $\cos 3\theta + \cos 2\theta$

90. $\cos\left(x + \frac{\pi}{6}\right) - \cos\left(x - \frac{\pi}{6}\right)$

91. $\sin\left(x + \frac{\pi}{4}\right) - \sin\left(x - \frac{\pi}{4}\right)$

92. *Harmonic Motion* A weight is attached to a spring suspended vertically from a ceiling. When a driving force is applied to the system, the weight moves vertically from its equilibrium position, and this motion is described by the model

$$y = 1.5 \sin 8t - 0.5 \cos 8t$$

where y is the distance from equilibrium measured in feet and t is the time in seconds.

(a) Write the model in the form

$$y = \sqrt{a^2 + b^2} \sin(Bt + C).$$

(b) Find the amplitude of the oscillations of the weight.

(c) Find the frequency of the oscillations of the weight.

Synthesis

True or False? **In Exercises 93–96, determine whether the statement is true or false. Justify your answer.**

93. If $\frac{\pi}{2} < \theta < \pi$, then $\cos \frac{\theta}{2} < 0$.

94. $\sin(x + y) = \sin x + \sin y$

95. $4 \sin(-x) \cos(-x) = -2 \sin 2x$

96. $4 \sin 45^\circ \cos 15^\circ = 1 + \sqrt{3}$

97. List the reciprocal identities, quotient identities, and Pythagorean identities from memory.

98. Name the trigonometric functions that are considered to be odd.

99. *Think About It* If a trigonometric equation has an infinite number of solutions, is it true that the equation is an identity? Explain.

100. *Think About It* Explain why you know from observation that the equation $a \sin x - b = 0$ has no solution if $|a| < |b|$.

In Exercises 101 and 102, use the graphs of y_1 and y_2 to determine how to change one function to form the identity $y_1 = y_2$.

101. $y_1 = \sec^2\left(\frac{\pi}{2} - x\right)$

$y_2 = \cot^2 x$

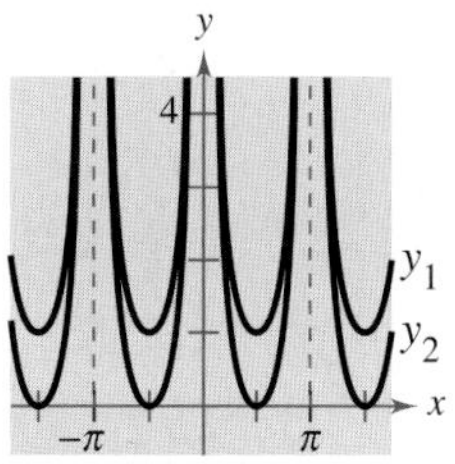

102. $y_1 = \frac{\cos 3x}{\cos x}$

$y_2 = (2 \sin x)^2$

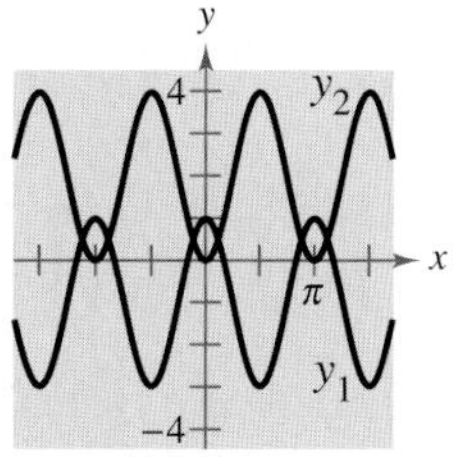

In Exercises 103 and 104, use a graphing utility with root- or zero-finding capabilities to approximate the zeros of the function.

103. $y = \sqrt{x + 3} + 4 \cos x$

104. $y = 2 - \frac{1}{2}x^2 + 3 \sin \frac{\pi x}{2}$

Chapter Project ▶ Projectile Motion

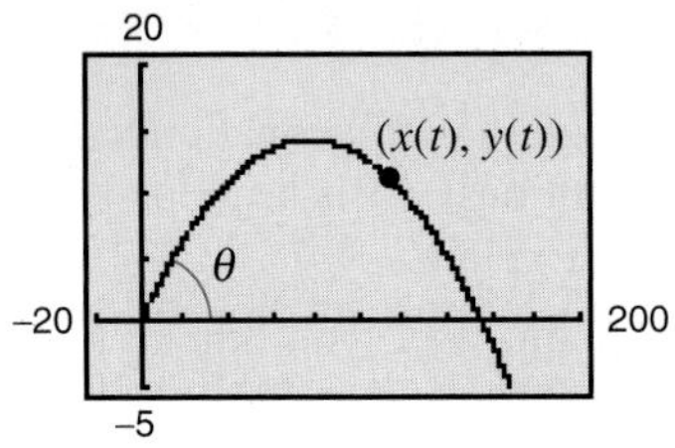

Tmin = 0
Tmax = 5
Tscl = .05
Xmin = -20
Xmax = 200
Xscl = 20
Ymin = -5
Ymax = 20
Yscl = 5

In this project, you will use parametric equations to model the path of a projectile. Parametric equations use a third variable t to represent time. The variables x and y are written as functions of t. For any time t, the horizontal position $x(t)$ and vertical position $y(t)$ of a projectile (ignoring air resistance) launched at ground level are given by the equations

$$x(t) = (v_0 \cos \theta)t$$

$$y(t) = (v_0 \sin \theta)t - 16t^2.$$

In these equations, θ is the angle with the horizontal and v_0 is the initial velocity in feet per second, as indicated in the figure at the left.

Set your graphing utility to parametric and degree modes, and use the viewing window shown at the left. Let $v_0 = 88$ feet per second and $\theta = 20°$ and graph the parametric equations

$$x(t) = (88 \cos 20°)t$$

$$y(t) = (88 \sin 20°)t - 16t^2.$$

Use the zoom and trace features to (a) find the maximum height attained by the projectile, (b) find the time at which the maximum height occurs, (c) determine the length of time that the projectile is in the air, and (d) determine the range of the projectile.

Chapter Project Investigations

1. Verify analytically the range of the projectile by solving the equation $(88 \sin 20°)t - 16t^2 = 0$ for t and then evaluating $x(t)$ at this t-value.
2. Use a graphing utility to find the maximum height and range of the projectile when $\theta = 30°$ and $v_0 = 132$ feet per second. What viewing window did you use?
3. Let $v_0 = 60$ feet per second and find the maximum range for the angles $\theta = 20°, 30°, 40°, 50°$, and $60°$. In general, what angle should you use to produce the maximum range?
4. What is the relationship between the time the projectile reaches its maximum height and the time it takes for the projectile to return to the ground? Explain.
5. Eliminate t from the parametric equations

   $x(t) = (v_0 \cos \theta)t$ and $y(t) = (v_0 \sin \theta)t - 16t^2$

   by solving for t in the first equation and substituting this value into the equation for y. Show in this case that the height of the projectile is given by the equation

   $$y = (\tan \theta)x - \frac{16 \sec^2 \theta}{v_0^2}x^2.$$

   Use this equation to find the angle θ corresponding to a maximum range of 200 feet and initial velocity of $v_0 = 80$ feet per second.

Chapter Test

The *Interactive* CD-ROM and *Internet* versions of this text provide answers to the Chapter Tests and Cumulative Tests. They also offer Chapter Pre-Tests (which test key skills and concepts covered in previous chapters) and Chapter Post-Tests, both of which have randomly generated exercises with diagnostic capabilities.

Take this test as you would take a test in class. After you are done, check your work against the answers given in the back of the book.

1. If $\tan\theta = \frac{3}{2}$ and $\cos\theta < 0$, use the fundamental identities to evaluate the other five trigonometric functions of θ.

2. Use the fundamental identities to simplify $\csc^2\beta(1 - \cos^2\beta)$.

3. Factor and simplify $\dfrac{\sec^4 x - \tan^4 x}{\sec^2 x + \tan^2 x}$.

4. Add and simplify $\dfrac{\cos\theta}{\sin\theta} + \dfrac{\sin\theta}{\cos\theta}$.

5. Determine the values of θ, $0 \le \theta < 2\pi$, for which $\tan\theta = -\sqrt{\sec^2\theta - 1}$ is true.

6. Use a graphing utility to graph the functions $y_1 = \cos x + \sin x \tan x$ and $y_2 = \sec x$. Make a conjecture about y_1 and y_2. Verify the result analytically.

In Exercises 7–12, verify the identity.

7. $\sin\theta\sec\theta = \tan\theta$

8. $\sec^2 x \tan^2 x + \sec^2 x = \sec^4 x$

9. $\dfrac{\csc\alpha + \sec\alpha}{\sin\alpha + \cos\alpha} = \cot\alpha + \tan\alpha$

10. $\cos\left(x + \dfrac{\pi}{2}\right) = -\sin x$

11. $\sin(n\pi + \theta) = (-1)^n \sin\theta$, n is an integer.

12. $(\sin x + \cos x)^2 = 1 + \sin 2x$

13. Rewrite $\sin^4 x \tan^2 x$ in terms of the first power of the cosine.

14. Use a half-angle formula to simplify the expression $\dfrac{\sin 4\theta}{1 + \cos 4\theta}$.

15. Write $4\cos 2\theta \sin 4\theta$ as a sum or difference.

16. Write $\sin 3\theta - \sin 4\theta$ as a product.

In Exercises 17–20, find all solutions of the equation in the interval $[0, 2\pi)$.

17. $\tan^2 x + \tan x = 0$

18. $\sin 2\alpha - \cos\alpha = 0$

19. $4\cos^2 x - 3 = 0$

20. $\csc^2 x - \csc x - 2 = 0$

21. Use a graphing utility to approximate the solutions of the equation $3\cos x - x = 0$ accurate to three decimal places.

22. Explain why the equation $\cos^2 x + \cos x - 6 = 0$ has no solution.

23. Find the exact value of $\cos 105°$ using the fact that $105° = 135° - 30°$.

24. Use the figure to find the exact values of $\sin 2u$ and $\tan 2u$.

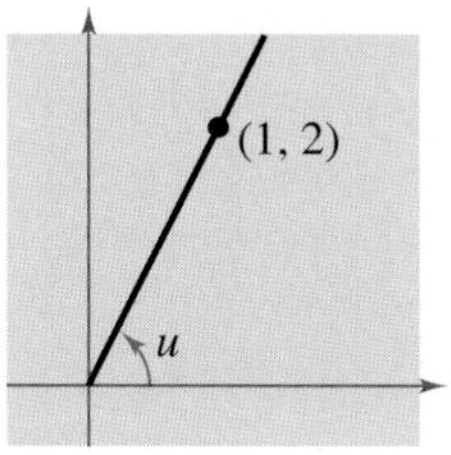

FIGURE FOR 24

25. The index of refraction n of a transparent material is the ratio of the speed of light in a vacuum to the speed of light in the material. For the glass triangular prism in the figure, $n = 1.5$ and $\alpha = 60°$. Find the angle θ for the glass prism.

$$n = \frac{\sin(\theta/2 + \alpha/2)}{\sin(\theta/2)}$$

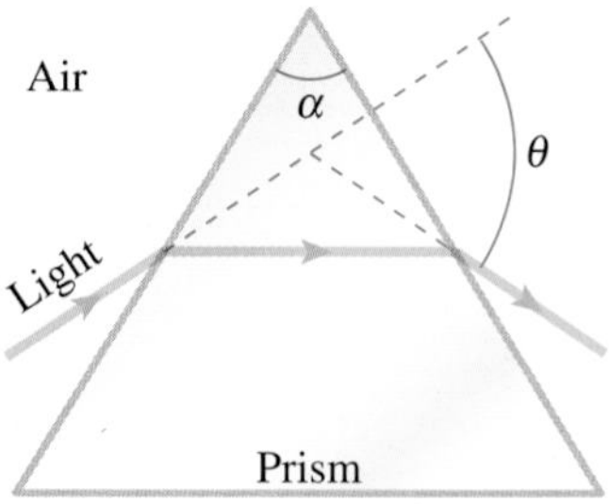

FIGURE FOR 25

Superstock

In November of 1999, 81.4% of U.S. airline flights arrived on time. Factors such as severe weather, aircraft maintenance, and air traffic control decisions can cause flight delays.
(Source: U.S. Department of Transportation)

6 Additional Topics in Trigonometry

▶ How to Study This Chapter

The Big Picture

In this chapter you will learn the following skills and concepts.

- ▶ How to use the Law of Sines and the Law of Cosines to solve oblique triangles
- ▶ How to find the areas of oblique triangles
- ▶ How to write the component forms of vectors and perform basic vector operations
- ▶ How to find the direction angles of vectors and the angle between two vectors
- ▶ How to multiply and divide complex numbers written in trigonometric form
- ▶ How to find powers and *n*th roots of complex numbers

Important Vocabulary

As you encounter each new vocabulary term in this chapter, add the term and its definition to your notebook glossary.

Oblique triangle (p. 494)
Law of Sines (p. 494)
Law of Cosines (p. 503)
Directed line segment (p. 511)
Initial point (p. 511)
Terminal point (p. 511)
Magnitude of directed line segment (p. 511)
Vector **v** in the plane (p. 511)
Standard position (p. 512)
Component form of vector **v** (p. 512)
Zero vector (p. 512)
Magnitude of **v** (p. 512)
Unit vector (p. 512)
Parallelogram law (p. 513)
Resultant (p. 513)
Standard unit vectors (p. 516)
Horizontal and vertical components of **v** (p. 516)
Linear combination of vectors (p. 516)
Direction angle (p. 517)
Dot product (p. 524)
Angle between two nonzero vectors (p. 525)
Orthogonal (p. 526)
Work (p. 529)
Complex plane (p. 532)
Real axis (p. 532)
Imaginary axis (p. 532)
Absolute value of a complex number (p. 532)
Trigonometric form of a complex number (p. 533)
Modulus (p. 533)
Argument (p. 533)
*n*th root of a complex number (p. 537)
*n*th roots of unity (p. 539)

Study Tools

- Learning objectives at the beginning of each section
- Chapter Summary (p. 543)
- Review Exercises (pp. 544–547)
- Chapter Test (p. 549)
- Cumulative Test for Chapters 4–6 (pp. 550–551)

Additional Resources

- Study and Solutions Guide
- Interactive Precalculus
- Videotapes for Chapter 6
- Precalculus Website
- Student Success Organizer

STUDY T!P

When your test is returned, review it carefully. Rework the problems you answered incorrectly. Discovering your mistakes will help you to avoid repeating them.

6.1 Law of Sines

▶ What you should learn

- How to use the Law of Sines to solve oblique triangles (AAS, ASA, or SSA)
- How to find the areas of oblique triangles
- How to use the Law of Sines to model and solve real-life problems

▶ Why you should learn it

You can use the Law of Sines to solve real-life problems involving oblique triangles. For instance, in Exercise 44 on page 501 you use the Law of Sines to determine the distance from a ranger station to a forest fire.

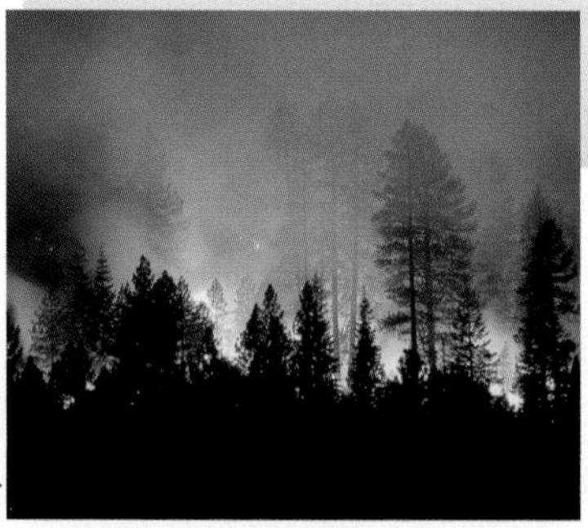

Superstock

You may want to illustrate each of the four cases (AAS or ASA, SSA, SSS, SAS) as a quick geometry review for your students.

Introduction

In Chapter 4 you looked at techniques for solving right triangles. In this section and the next, you will solve **oblique triangles**—triangles that have no right angles. As standard notation, the angles of a triangle are labeled as A, B, and C, and their opposite sides as a, b, and c, as shown in Figure 6.1.

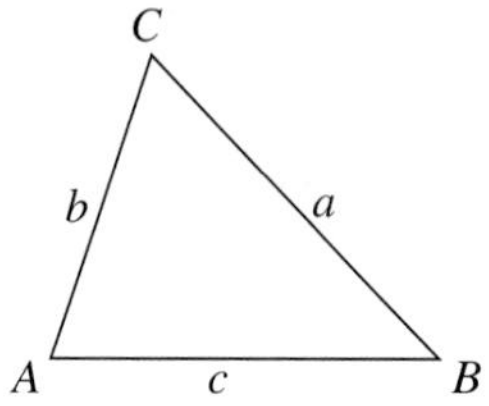

FIGURE 6.1

To solve an oblique triangle, you need to know the measure of at least one side and any two other parts of the triangle—either two sides, two angles, or one angle and one side. This breaks down into the following four cases.

1. Two angles and any side (AAS or ASA)
2. Two sides and an angle opposite one of them (SSA)
3. Three sides (SSS)
4. Two sides and their included angle (SAS)

The first two cases can be solved using the **Law of Sines,** whereas the last two cases require the Law of Cosines (Section 6.2). A proof of the Law of Sines is given in Appendix A.

Law of Sines

If ABC is a triangle with sides a, b, and c, then

$$\frac{a}{\sin A} = \frac{b}{\sin B} = \frac{c}{\sin C}.$$

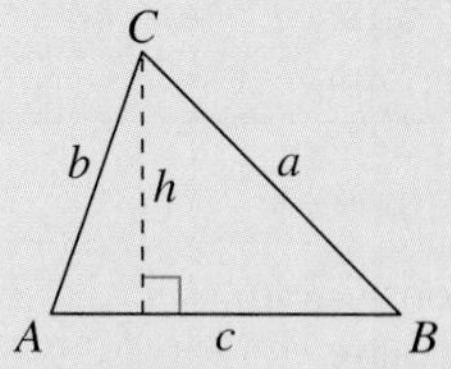

A is acute.

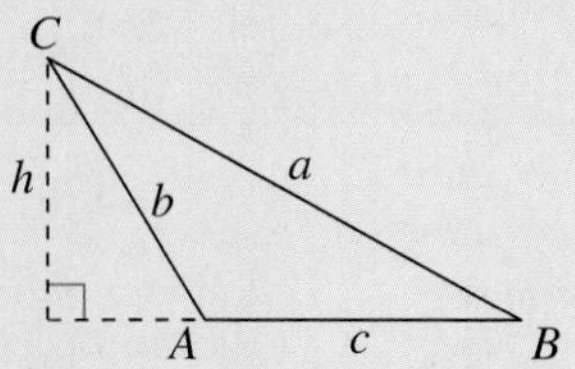

A is obtuse.

The Law of Sines can also be written in the reciprocal form

$$\frac{\sin A}{a} = \frac{\sin B}{b} = \frac{\sin C}{c}.$$

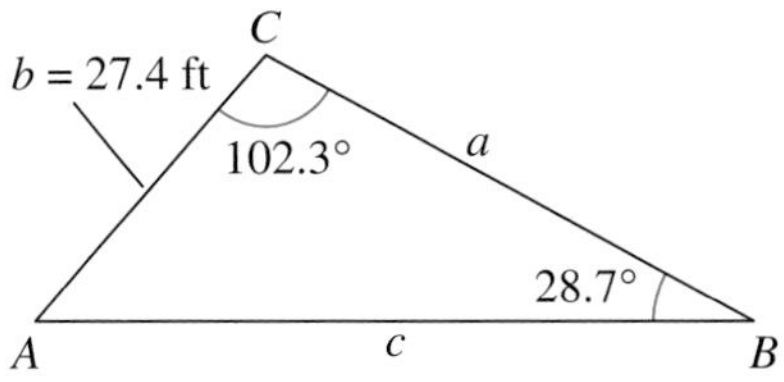

FIGURE 6.2

Example 1 ▶ Given Two Angles and One Side—AAS

For the triangle in Figure 6.2, $C = 102.3°$, $B = 28.7°$, and $b = 27.4$ feet. Find the remaining angle and sides.

Solution

The third angle of the triangle is

$$\begin{aligned} A &= 180° - B - C \\ &= 180° - 28.7° - 102.3° \\ &= 49.0°. \end{aligned}$$

By the Law of Sines, you have

$$\frac{a}{\sin 49°} = \frac{b}{\sin 28.7°} = \frac{c}{\sin 102.3°}.$$

Using $b = 27.4$ produces

$$a = \frac{b}{\sin B}(\sin A) = \frac{27.4}{\sin 28.7°}(\sin 49°) \approx 43.06 \text{ feet}$$

and

$$c = \frac{b}{\sin B}(\sin C) = \frac{27.4}{\sin 28.7°}(\sin 102.3°) \approx 55.75 \text{ feet.}$$

STUDY TIP

When solving triangles, a careful sketch is useful as a quick test for the feasibility of an answer. Remember that the longest side lies opposite the largest angle, and the shortest side lies opposite the smallest angle.

Example 2 ▶ Given Two Angles and One Side—ASA

A pole tilts *toward* the sun at an 8° angle from the vertical, and it casts a 22-foot shadow. The angle of elevation from the tip of the shadow to the top of the pole is 43°. How tall is the pole?

Solution

From Figure 6.3, note that $A = 43°$ and $B = 90° + 8° = 98°$. So, the third angle is

$$\begin{aligned} C &= 180° - A - B \\ &= 180° - 43° - 98° \\ &= 39°. \end{aligned}$$

By the Law of Sines, you have

$$\frac{a}{\sin 43°} = \frac{c}{\sin 39°}.$$

Because $c = 22$ feet, the length of the pole is

$$a = \frac{c}{\sin C}(\sin A) = \frac{22}{\sin 39°}(\sin 43°) \approx 23.84 \text{ feet.}$$

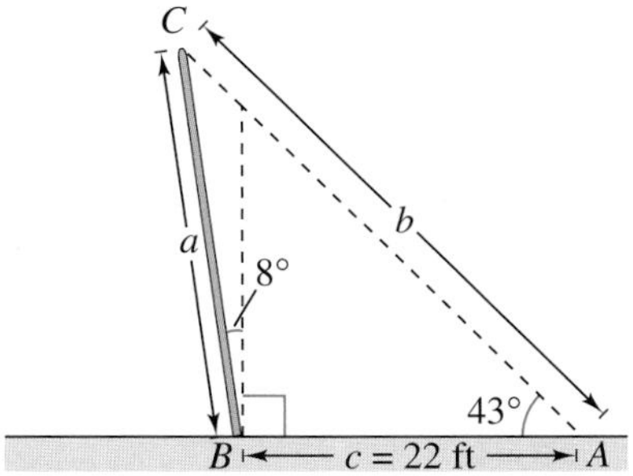

FIGURE 6.3

For practice, try reworking Example 2 for a pole that tilts *away from* the sun under the same conditions.

A computer simulation to accompany this concept appears in the *Interactive* CD-ROM and *Internet* versions of this text.

The Ambiguous Case (SSA)

In Examples 1 and 2 you saw that two angles and one side determine a unique triangle. However, if two sides and one opposite angle are given, three possible situations can occur: (1) no such triangle exists, (2) one such triangle exists, or (3) two distinct triangles may satisfy the conditions.

The Ambiguous Case (SSA)

Consider a triangle in which you are given a, b, and A. $(h = b \sin A)$

	A is acute.	A is acute.	A is acute.	A is acute.	A is obtuse.	A is obtuse.
Sketch	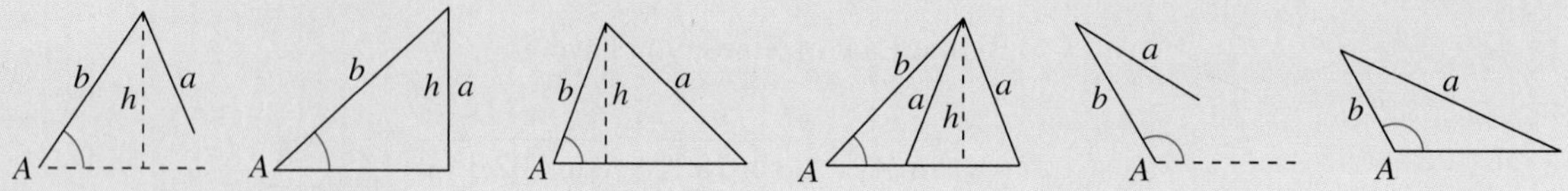					
Necessary condition	$a < h$	$a = h$	$a > b$	$h < a < b$	$a \le b$	$a > b$
Triangles possible	None	One	One	Two	None	One

Example 3 ▶ Single-Solution Case—SSA

For the triangle in Figure 6.4, $a = 22$ inches, $b = 12$ inches, and $A = 42°$. Find the remaining side and angles.

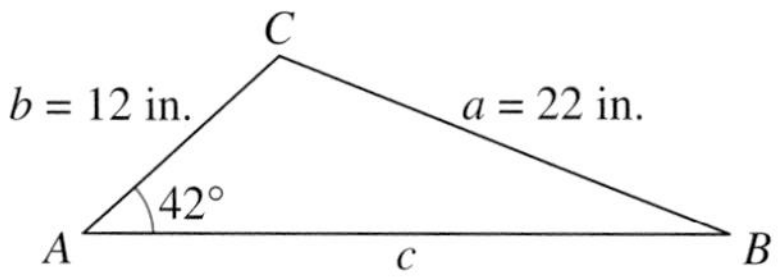

One solution: $a > b$

FIGURE 6.4

Solution

By the Law of Sines, you have

$$\frac{22}{\sin 42°} = \frac{12}{\sin B}$$

$$\sin B = b\left(\frac{\sin A}{a}\right) = 12\left(\frac{\sin 42°}{22}\right) \approx 0.3649803$$

$$B \approx 21.41°. \qquad B \text{ is acute.}$$

Now, you can determine that

$$C \approx 180° - 42° - 21.41°$$

$$= 116.59°.$$

Then the remaining side is

$$\frac{c}{\sin 116.59°} = \frac{22}{\sin 42°}$$

$$c = \sin C\left(\frac{a}{\sin A}\right) = \sin 116.59°\left(\frac{22}{\sin 42°}\right) \approx 29.40 \text{ inches.}$$

Encourage your students to sketch the triangle, keeping in mind that the longest side lies opposite the largest angle of the triangle. For practice, suggest that students also find h.

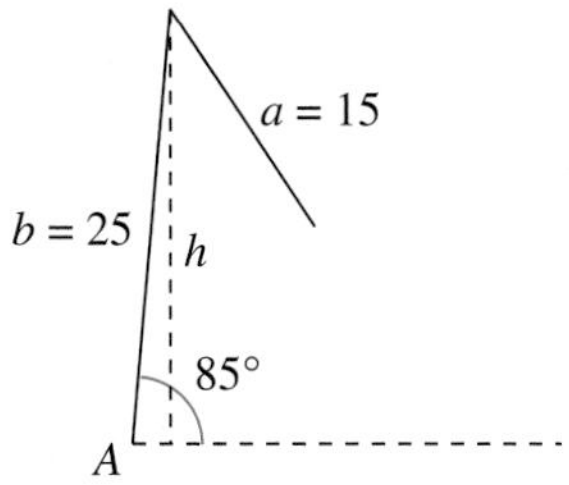

No solution: $a < h$
FIGURE 6.5

Example 4 ▶ No-Solution Case—SSA

Show that there is no triangle for which $a = 15$, $b = 25$, and $A = 85°$.

Solution

Begin by making the sketch shown in Figure 6.5. From this figure it appears that no triangle is formed. You can verify this using the Law of Sines.

$$\frac{a}{\sin A} = \frac{b}{\sin B}$$

$$\frac{15}{\sin 85°} = \frac{25}{\sin B}$$

$$\sin B = b\left(\frac{\sin A}{a}\right) = 25\left(\frac{\sin 85°}{15}\right) \approx 1.660 > 1$$

This contradicts the fact that $|\sin B| \le 1$. So, no triangle can be formed having sides $a = 15$ and $b = 25$ and an angle of $A = 85°$.

Activities
Have your students determine the number of triangles possible in each of the following cases.
1. $A = 62°$, $a = 10$, $b = 12$ (0 triangles)
2. $A = 98°$, $a = 10$, $b = 3$ (1 triangle)
3. $A = 54°$, $a = 7$, $b = 10$ (0 triangles)

Consider discussing several examples of the two-solution case.

Additional Example
Find two triangles for which $c = 29$, $b = 46$, and $C = 31°$.

Solution
$B = 54.8°$, $A = 94.2°$, $a = 56.2$
$B = 125.2°$, $A = 23.8°$, $a = 22.7$

Example 5 ▶ Two-Solution Case—SSA

Find two triangles for which $a = 12$ meters, $b = 31$ meters, and $A = 20.5°$.

Solution

By the Law of Sines, you have

$$\frac{a}{\sin A} = \frac{b}{\sin B}$$

$$\sin B = b\left(\frac{\sin A}{a}\right) = 31\left(\frac{\sin 20.5°}{12}\right) \approx 0.9047.$$

There are two angles $B_1 \approx 64.8°$ and $B_2 \approx 180° - 64.8° \approx 115.2°$ between 0° and 180° whose sine is 0.9047. For $B_1 \approx 64.8°$, you obtain

$$C \approx 180° - 20.5° - 64.8° = 94.7°$$

$$c = \frac{a}{\sin A}(\sin C) = \frac{12}{\sin 20.5°}(\sin 94.7°) \approx 34.15 \text{ meters.}$$

For $B_2 \approx 115.2°$, you obtain

$$C \approx 180° - 20.5° - 115.2° = 44.3°$$

$$c = \frac{a}{\sin A}(\sin C) = \frac{12}{\sin 20.5°}(\sin 44.3°) \approx 23.93 \text{ meters.}$$

The resulting triangles are shown in Figure 6.6.

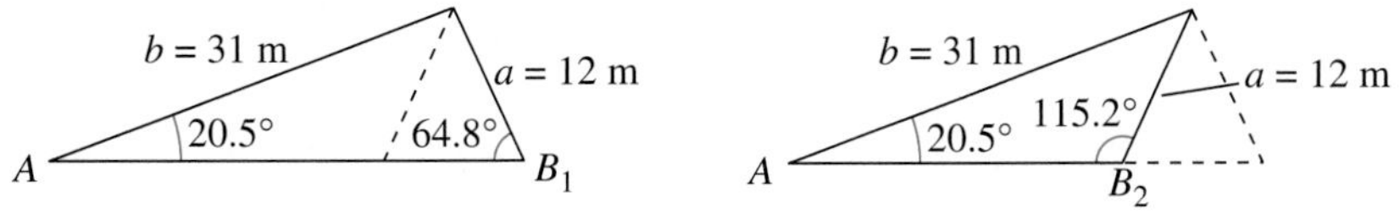

FIGURE 6.6

Activities

1. Use the given information to find (if possible) the remaining side and angles of the oblique triangle. If two solutions exist, find both.

   $A = 58°, \ a = 20, \ c = 10$

   Answer: $B = 97°, \ C = 25°, \ b = 23.4$

2. Use the given information to find (if possible) the remaining side and angles of the oblique triangle. If two solutions exist, find both.

   $B = 78°, \ b = 207, \ c = 210$

   Answer: Two solutions
   $A = 19.1° \ \ a = 69.2, \ C = 82.9°$
   $A = 4.9°, \ a = 18.1, \ C = 97.1°$

3. Find the area of the triangle with $B = 120°, \ a = 32, \ c = 50.$

   Answer: Area = 692.8 square units

Area of an Oblique Triangle

The procedure used to prove the Law of Sines leads to a simple formula for the area of an oblique triangle. Referring to Figure 6.7, note that each triangle has a height of $h = b \sin A$. Consequently, the area of each triangle is

$$\text{Area} = \frac{1}{2}(\text{base})(\text{height}) = \frac{1}{2}(c)(b \sin A) = \frac{1}{2}bc \sin A.$$

By similar arguments, you can develop the formulas

$$\text{Area} = \frac{1}{2}ab \sin C = \frac{1}{2}ac \sin B.$$

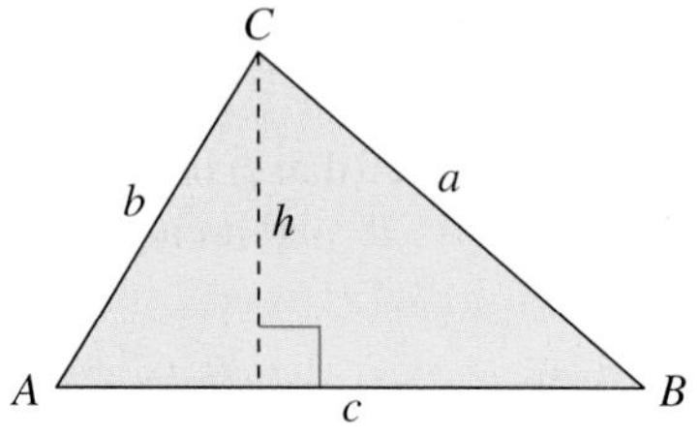

A is acute.

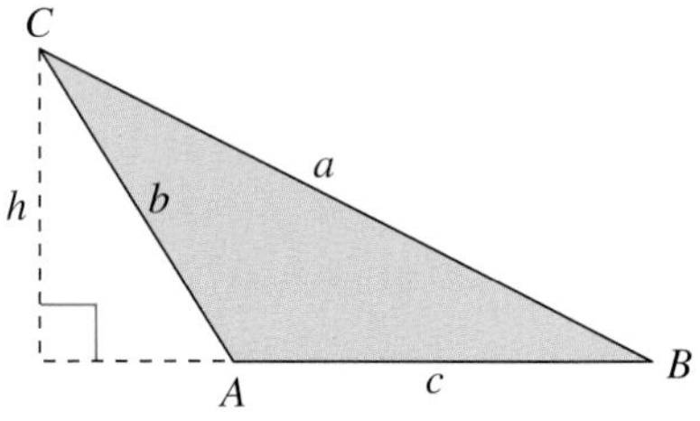

A is obtuse.

FIGURE 6.7

Area of an Oblique Triangle

The area of any triangle is one-half the product of the lengths of two sides times the sine of their included angle. That is,

$$\text{Area} = \frac{1}{2}bc \sin A = \frac{1}{2}ab \sin C = \frac{1}{2}ac \sin B.$$

Note that if angle A is 90°, the formula gives the area for a right triangle:

$$\text{Area} = \frac{1}{2}bc = \frac{1}{2}(\text{base})(\text{height}).$$

Similar results are obtained for angles C and B equal to 90°.

Example 6 ▶ Finding the Area of a Triangular Lot

Find the area of a triangular lot having two sides of lengths 90 meters and 52 meters and an included angle of 102°.

Solution

Consider $a = 90$ m, $b = 52$ m, and angle $C = 102°$, as shown in Figure 6.8. Then the area of the triangle is

$$\text{Area} = \frac{1}{2}ab \sin C = \frac{1}{2}(90)(52)(\sin 102°) \approx 2289 \text{ square meters.}$$

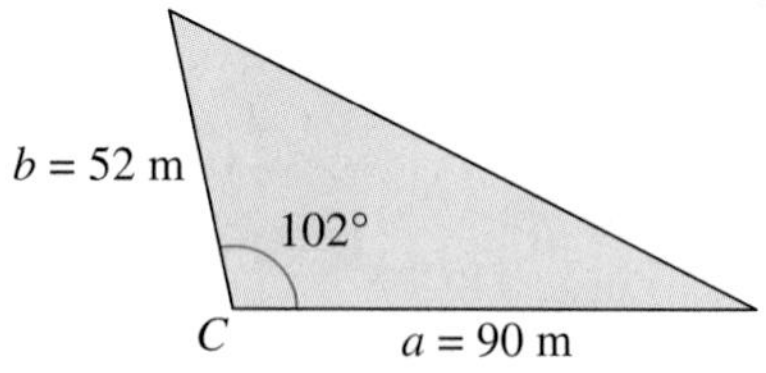

FIGURE 6.8

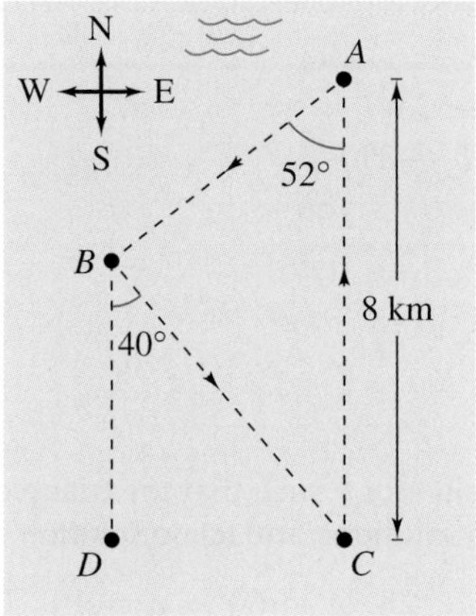

FIGURE 6.9

Application

Example 7 ▶ An Application of the Law of Sines

The course for a boat race starts at point A and proceeds in the direction S 52° W to point B, then in the direction S 40° E to point C, and finally back to A, as shown in Figure 6.9. Point C lies 8 kilometers directly south of point A. Approximate the total distance of the race course.

Solution

Because lines BD and AC are parallel, it follows that $\angle BCA \cong \angle DBC$. Consequently, triangle ABC has the measures shown in Figure 6.10. For angle B, you have $B = 180° - 52° - 40° = 88°$. Using the Law of Sines

$$\frac{a}{\sin 52°} = \frac{b}{\sin 88°} = \frac{c}{\sin 40°}$$

you can let $b = 8$ and obtain

$$a = \frac{8}{\sin 88°}(\sin 52°) \approx 6.308$$

and

$$c = \frac{8}{\sin 88°}(\sin 40°) \approx 5.145.$$

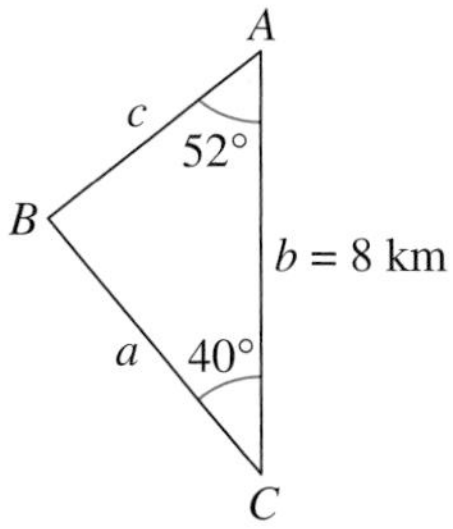

FIGURE 6.10

The total length of the course is approximately

$$\text{Length} \approx 8 + 6.308 + 5.145 = 19.453 \text{ kilometers.}$$

Writing ABOUT MATHEMATICS

Using the Law of Sines In this section, you have been using the Law of Sines to solve *oblique* triangles. Can the Law of Sines also be used to solve a right triangle? If so, write a short paragraph explaining how to use the Law of Sines to solve the following two triangles. Is there an easier way to solve these triangles?

a. (AAS)

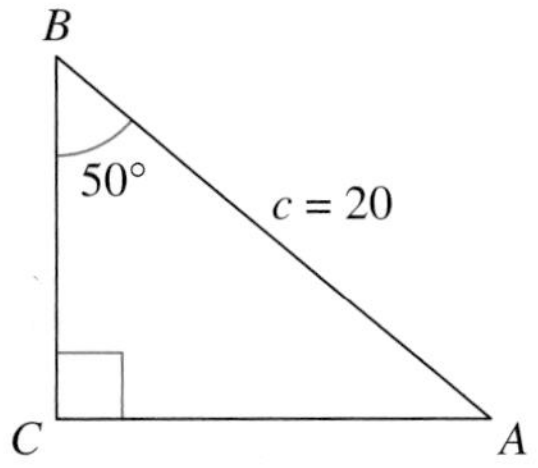

b. (ASA)

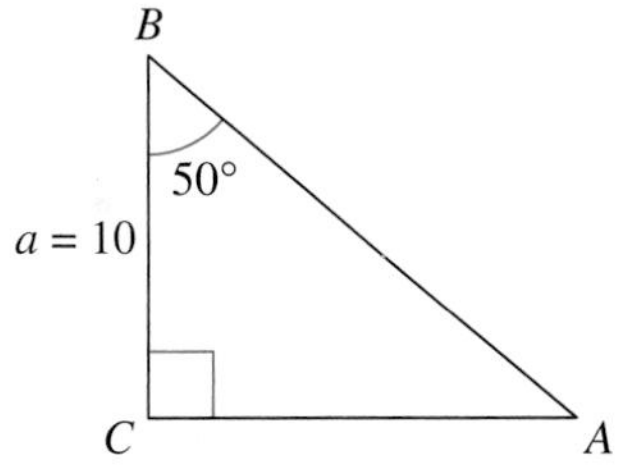

Alternate Writing About Mathematics: Error Analysis

Suppose you are a math instructor, and one of your students hands in the following solution. Discuss what is wrong with your student's solution. How could you help the student avoid making a similar mistake in the future? Use a diagram to illustrate your explanation.

Find side c in a triangle that has $a = 5.8$, $b = 7$, and $C = 82°$.

Student's Solution

Because this is an SSA situation, I can use the Law of Sines.

$$\frac{5.8}{7} = \frac{c}{\sin 82°} \quad \times$$

$$(\sin 82°)\frac{5.8}{7} = 0.821$$

6.1 Exercises

The *Interactive* CD-ROM and *Internet* versions of this text contain step-by-step solutions to all odd-numbered Section and Review Exercises. They also provide Tutorial Exercises that link to Guided Examples for additional help.

In Exercises 1–18, use the information to solve the triangle.

1.

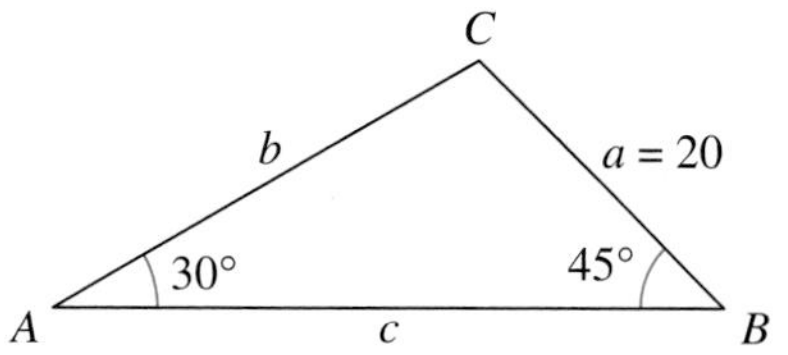

2.

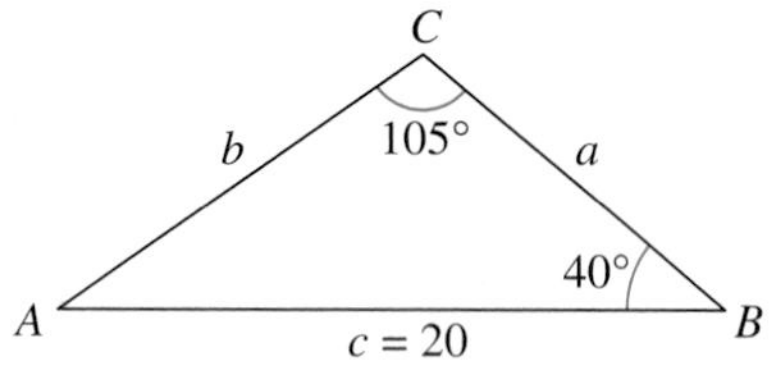

3.

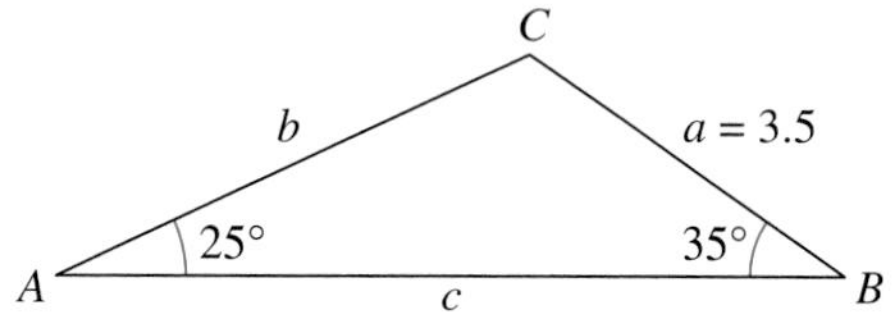

4.

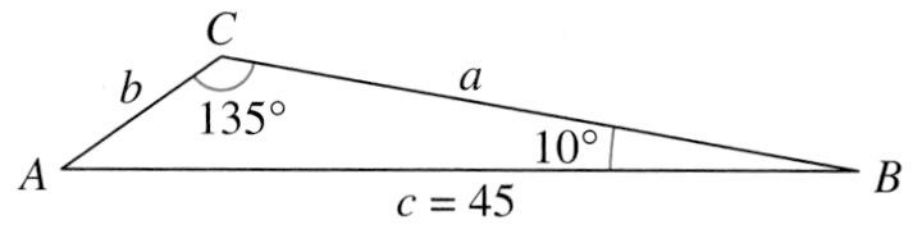

5. $A = 36°,\quad a = 8,\quad b = 5$
6. $A = 60°,\quad a = 9,\quad c = 10$
7. $A = 102.4°,\quad C = 16.7°,\quad a = 21.6$
8. $A = 24.3°,\quad C = 54.6°,\quad c = 2.68$
9. $A = 83° 20',\quad C = 54.6°,\quad c = 18.1$
10. $A = 5° 40',\quad B = 8° 15',\quad b = 4.8$
11. $B = 15° 30',\quad a = 4.5,\quad b = 6.8$
12. $B = 2° 45',\quad b = 6.2,\quad c = 5.8$
13. $C = 145°,\quad b = 4,\quad c = 14$
14. $A = 100°,\quad a = 125,\quad c = 10$
15. $A = 110° 15',\quad a = 48,\quad b = 16$
16. $C = 85° 20',\quad a = 35,\quad c = 50$
17. $A = 55°,\quad B = 42°,\quad c = \frac{3}{4}$
18. $B = 28°,\quad C = 104°,\quad a = 3\frac{5}{8}$

In Exercises 19–26, use the information to solve (if possible) the triangle. If two solutions exist, find both.

19. $A = 58°,\quad a = 4.5,\quad b = 12.8$
20. $A = 58°,\quad a = 11.4,\quad b = 12.8$
21. $A = 76°,\quad a = 18,\quad b = 20$
22. $A = 76°,\quad a = 34,\quad b = 21$
23. $A = 110°,\quad a = 125,\quad b = 200$
24. $A = 110°,\quad a = 125,\quad b = 100$
25. $A = 22°,\quad a = \frac{5}{12},\quad b = 1\frac{3}{8}$
26. $A = 22°,\quad a = \frac{5}{7},\quad b = \frac{5}{7}$

In Exercises 27–30, find a value for *b* such that the triangle has (a) one solution, (b) two solutions, and (c) no solution.

27. $A = 36°,\quad a = 5$
28. $A = 60°,\quad a = 10$
29. $A = 10°,\quad a = 10.8$
30. $A = 88°,\quad a = 315.6$

In Exercises 31–36, find the area of the triangle having the indicated sides and angle.

31. $C = 120°,\quad a = 4,\quad b = 6$
32. $B = 130°,\quad a = 62,\quad c = 20$
33. $A = 43° 45',\quad b = 57,\quad c = 85$
34. $A = 5° 15',\quad b = 4.5,\quad c = 22$
35. $B = 72° 30',\quad a = 105,\quad c = 64$
36. $C = 84° 30',\quad a = 16,\quad b = 20$

37. ***Height*** Because of prevailing winds, a tree grew so that it was leaning 4° from the vertical. At a point 35 meters from the tree, the angle of elevation to the top of the tree is 23° (see figure). Find the height h of the tree.

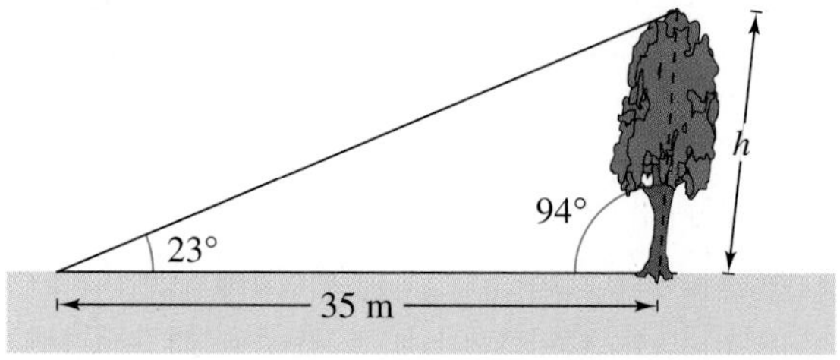

38. ***Height*** A flagpole at a right angle to the horizontal is located on a slope that makes an angle of 12° with the horizontal. The pole's shadow is 16 meters long and points directly up the slope. The angle of elevation from the tip of the shadow to the sun is 20°.
    (a) Draw a triangle that represents the problem. Show the known quantities on the triangle and use a variable to indicate the height of the flagpole.
    (b) Write an equation involving the unknown quantity.
    (c) Find the height of the flagpole.

39. *Angle of Elevation* A 10-meter telephone pole casts a 17-meter shadow directly down a slope when the angle of elevation of the sun is 42° (see figure). Find θ, the angle of elevation of the ground.

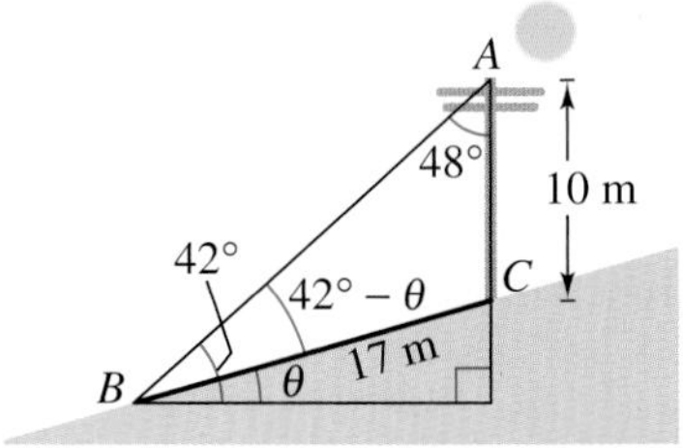

40. *Flight Path* A plane flies 500 kilometers with a bearing of N 44° W from Naples to Elgin (see figure). The plane then flies 720 kilometers from Elgin to Canton. Find the bearing of the flight from Elgin to Canton.

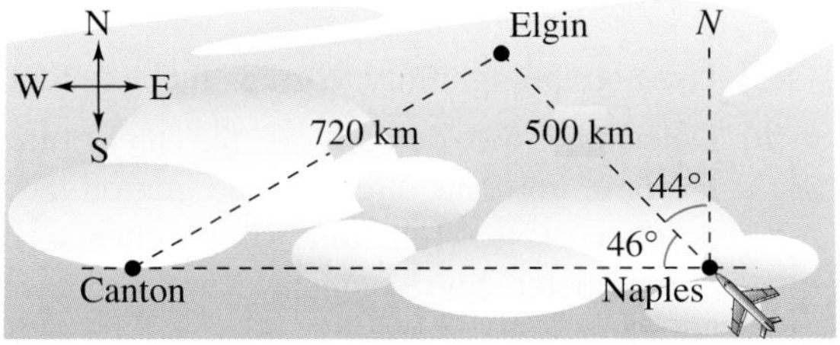

41. *Bridge Design* A bridge is to be built across a small lake from a gazebo to a dock (see figure). The bearing from the gazebo to the dock is S 41° W. From a tree 100 meters from the gazebo, the bearings to the gazebo and the dock are S 74° E and S 28° E, respectively. Find the distance from the gazebo to the dock.

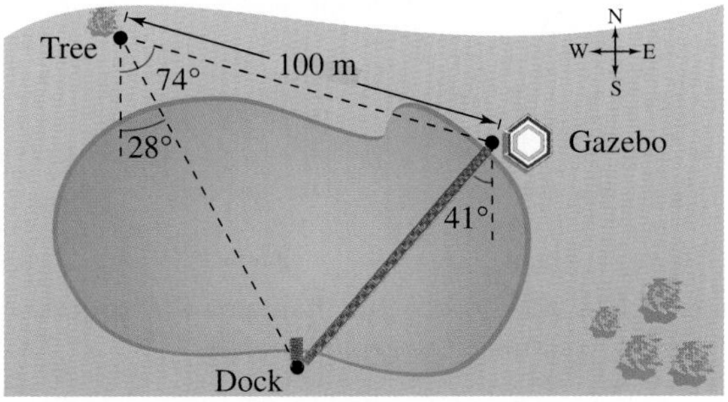

42. *Railroad Track Design* The circular arc of a railroad curve has a chord of length 3000 feet and a central angle of 40°.

(a) Draw a figure that visually represents the problem. Show the known quantities on the figure and use variables r and s to represent the radius of the arc and the length of the arc, respectively.

(b) Find the radius r of the circular arc.

(c) Find the length s of the circular arc.

43. *Glide Path* A pilot has just started on the glide path for landing at an airport where the length of the runway is 9000 feet. The angles of depression from the plane to the ends of the runway are 17.5° and 18.8°.

(a) Draw a figure that visually represents the problem.

(b) Find the air distance the plane must travel until touching down on the near end of the runway.

(c) Find the ground distance the plane must travel until touching down.

(d) Find the altitude of the plane when the pilot begins the descent.

44. *Locating a Fire* The bearing from the Pine Knob fire tower to the Colt Station fire tower is N 65° E and the two towers are 30 kilometers apart. A fire spotted by rangers in each tower has a bearing of N 80° E from Pine Knob and S 70° E from Colt Station. Find the distance of the fire from each tower.

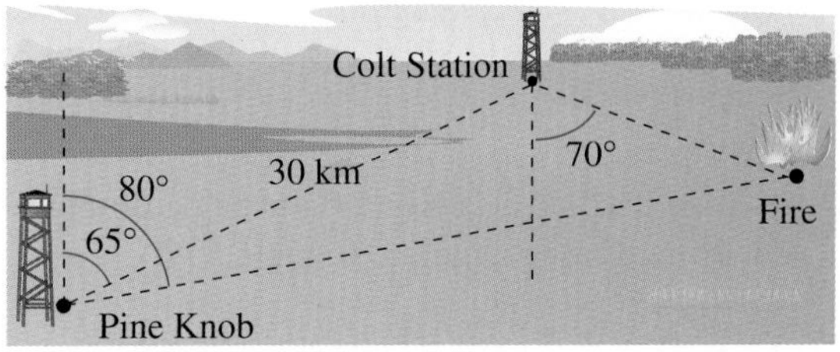

45. *Distance* A boat is sailing due east parallel to the shoreline at a speed of 10 miles per hour. At a given time the bearing to the lighthouse is S 70° E, and 15 minutes later the bearing is S 63° E (see figure). Find the distance from the boat to the shoreline if the lighthouse is at the shoreline.

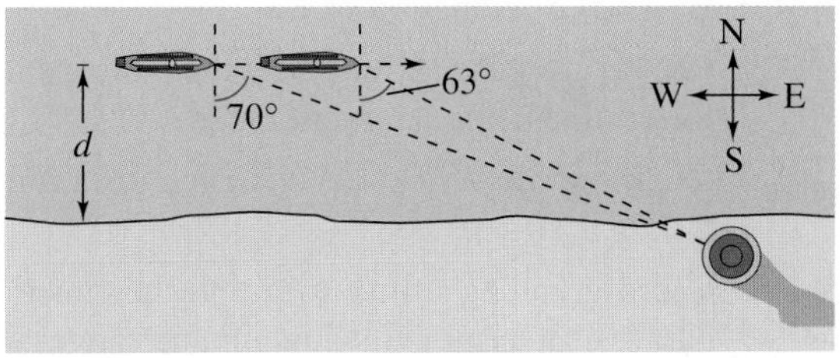

46. *Distance* A family is traveling due west on a road that passes a famous landmark. At a given time the bearing to the landmark is N 62° W, and after the family travels 5 miles farther the bearing is N 38° W. What is the closest the family will come to the landmark while on the road?

47. *Distance* The angles of elevation θ and ϕ to an airplane from the airport control tower and from an observation post 2 miles away are being continuously monitored (see figure). Write an equation giving the distance d between the plane and observation post in terms of θ and ϕ.

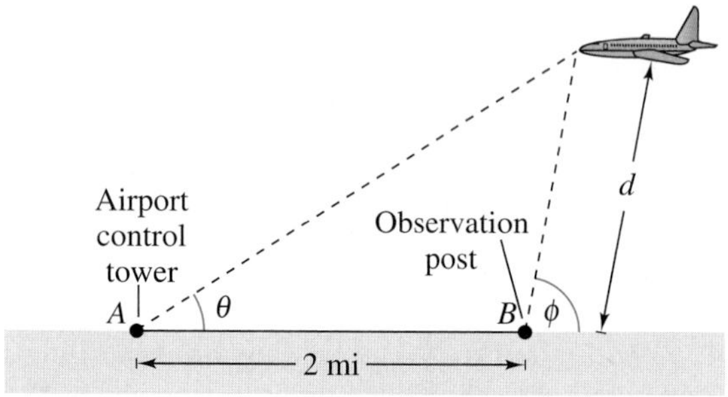

Synthesis

True or False? **In Exercises 48 and 49, determine whether the statement is true or false. Justify your answer.**

48. It is not possible to create an obtuse triangle whose longest side is one of the sides that forms its obtuse angle.

49. Two angles and one side of a triangle do not necessarily determine a unique triangle.

50. *Graphical and Numerical Analysis* In the figure, α and β are positive angles.

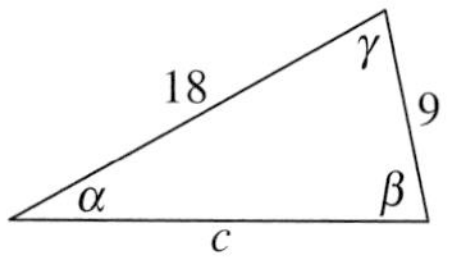

(a) Write α as a function of β.

(b) Use a graphing utility to graph the function. Determine its domain and range.

(c) Use the result of part (a) to write c as a function of β.

(d) Use a graphing utility to graph the function in part (c). Determine its domain and range.

(e) Complete the table. What can you infer?

β	0.4	0.8	1.2	1.6	2.0	2.4	2.8
α							
c							

51. *Graphical Analysis*

(a) Write the area A of the shaded region in the figure as a function of θ.

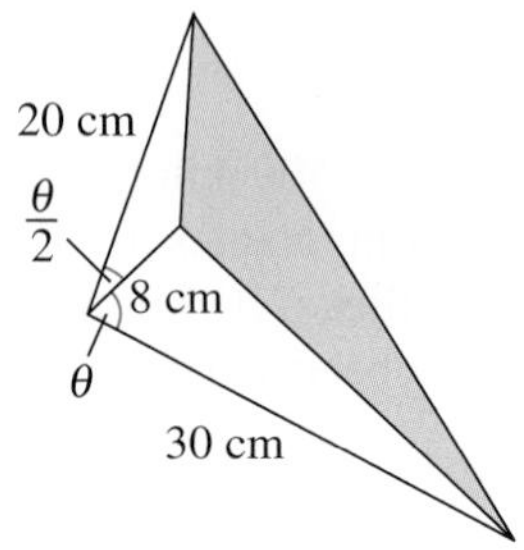

(b) Use a graphing utility to graph the area function.

(c) Determine the domain of the area function. Explain how the area of the region and the domain of the function would change if the 8-centimeter line segment were decreased in length.

Review

In Exercises 52–55, use the given values to evaluate the other five trigonometric functions.

52. $\sin x = \frac{2}{5}, \quad \frac{\pi}{2} < x < \pi$

53. $\cos x = \frac{1}{5}, \quad \frac{3\pi}{2} < x < 2\pi$

54. $\sec x = -4, \quad \pi < x < \frac{3\pi}{2}$

55. $\tan x = -5, \quad \frac{\pi}{2} < x < \pi$

In Exercises 56–59, use the fundamental trigonometric identities to simplify the expression.

56. $\sin x \cot x$

57. $\tan x \cos x \sec x$

58. $1 - \sin^2\left(\frac{\pi}{2} - x\right)$

59. $1 + \cot^2\left(\frac{\pi}{2} - x\right)$

6.2 Law of Cosines

▶ What you should learn

- How to use the Law of Cosines to solve oblique triangles (SSS or SAS)
- How to use the Law of Cosines to model and solve real-life problems
- How to use Heron's Area Formula to find the area of a triangle

▶ Why you should learn it

You can use the Law of Cosines to solve real-life problems involving oblique triangles. For instance, in Exercise 40 on page 508 you use the Law of Cosines to approximate the number of feet a baseball player must run to catch a ball.

Jed Jacobsohn/Allsport

In cases where the Law of Cosines must be used, encourage your students to solve for the largest angle first, then finish the problem using either the Law of Sines or the Law of Cosines.

Introduction

Two cases remain in the list of conditions needed to solve an oblique triangle—SSS and SAS. To use the Law of Sines, you must know at least one side and its opposite angle. If you are given three sides (SSS), or two sides and their included angle (SAS), none of the ratios would be complete. In such cases you can use the **Law of Cosines.** A proof of the Law of Cosines is given in Appendix A.

Law of Cosines

Standard Form	*Alternative Form*
$a^2 = b^2 + c^2 - 2bc \cos A$	$\cos A = \dfrac{b^2 + c^2 - a^2}{2bc}$
$b^2 = a^2 + c^2 - 2ac \cos B$	$\cos B = \dfrac{a^2 + c^2 - b^2}{2ac}$
$c^2 = a^2 + b^2 - 2ab \cos C$	$\cos C = \dfrac{a^2 + b^2 - c^2}{2ab}$

Example 1 ▶ Three Sides of a Triangle—SSS

Find the three angles of the triangle in Figure 6.11.

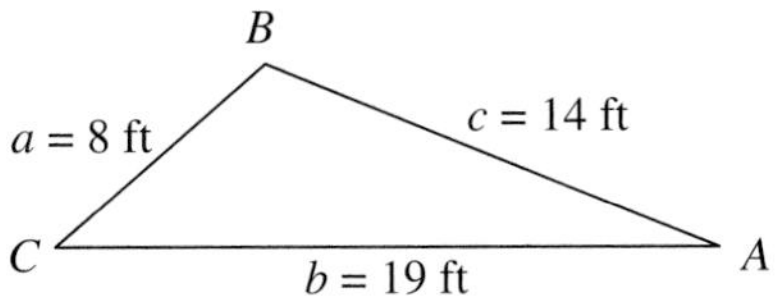

FIGURE 6.11

Solution

It is a good idea first to find the angle opposite the longest side—side b in this case (see Figure 6.11). Using the Law of Cosines, you find that

$$\cos B = \frac{a^2 + c^2 - b^2}{2ac} = \frac{8^2 + 14^2 - 19^2}{2(8)(14)} \approx -0.45089.$$

Because $\cos B$ is negative, you know that B is an *obtuse* angle given by $B \approx 116.80°$. At this point, knowing that $B \approx 116.80°$, it is simpler to use the Law of Sines to determine A.

$$\sin A = a\left(\frac{\sin B}{b}\right) \approx 8\left(\frac{\sin 116.80°}{19}\right) \approx 0.37582$$

Because B is obtuse, you know that A must be acute, because a triangle can have, at most, one obtuse angle. So, $A \approx 22.08°$ and

$$C \approx 180° - 22.08° - 116.80° = 41.12°.$$

Exploration

What familiar formula do you obtain when you use the third form of the Law of Cosines

$$c^2 = a^2 + b^2 - 2ab\cos C$$

and you let $C = 90°$? What is the relationship between the Law of Cosines and this formula?

The *Interactive* CD-ROM and *Internet* versions of this text show every example with its solution; clicking on the *Try It!* button brings up similar problems. Guided Examples and Integrated Examples show step-by-step solutions to additional examples. Integrated Examples are related to several concepts in the section.

Do you see why it was wise to find the largest angle *first* in Example 1? Knowing the cosine of an angle, you can determine whether the angle is acute or obtuse. That is,

$$\cos\theta > 0 \quad \text{for} \quad 0° < \theta < 90° \qquad \text{Acute}$$

$$\cos\theta < 0 \quad \text{for} \quad 90° < \theta < 180°. \qquad \text{Obtuse}$$

So, in Example 1, once you found that angle B was obtuse, you knew that angles A and C were both acute. If the largest angle is acute, the remaining two angles are acute also.

Example 2 ▶ Two Sides and the Included Angle—SAS

Find the remaining angles and side of the triangle in Figure 6.12.

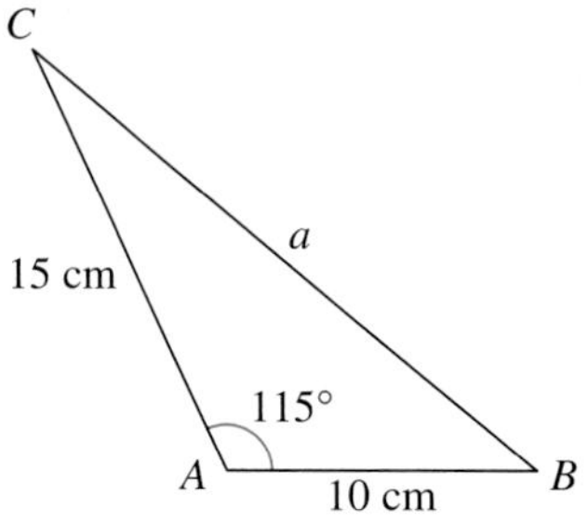

FIGURE 6.12

Solution

Use the Law of Cosines to find the unknown side a in the figure.

$$a^2 = b^2 + c^2 - 2bc\cos A$$

$$a^2 = 15^2 + 10^2 - 2(15)(10)\cos 115°$$

$$a^2 \approx 451.79$$

$$a \approx 21.26$$

Because $a \approx 21.26$ cm, you now know the ratio $a/\sin A$ and you can use the Law of Sines

$$\frac{a}{\sin A} = \frac{b}{\sin B}$$

to solve for B.

$$\sin B = b\left(\frac{\sin A}{a}\right)$$

$$= 15\left(\frac{\sin 115°}{21.26}\right)$$

$$\approx 0.63945$$

So, $B = \arcsin 0.63945 \approx 39.75°$ and $C \approx 180° - 115° - 39.75° = 25.25°$.

Applications

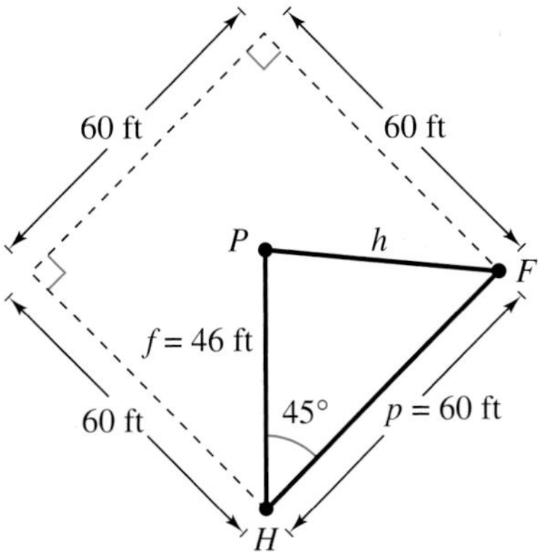

FIGURE 6.13

Example 3 ▶ An Application of the Law of Cosines

The pitcher's mound on a softball field is 46 feet from home plate and the distance between the bases is 60 feet, as shown in Figure 6.13. (The pitcher's mound is not halfway between home plate and second base.) How far is the pitcher's mound from first base?

Solution

In triangle HPF, $H = 45°$ (line HP bisects the right angle at H), $f = 46$, and $p = 60$. Using the Law of Cosines for this SAS case, you have

$$\begin{aligned} h^2 &= f^2 + p^2 - 2fp \cos H \\ &= 46^2 + 60^2 - 2(46)(60) \cos 45° \\ &\approx 1812.8. \end{aligned}$$

Therefore, the approximate distance from the pitcher's mound to first base is

$$\begin{aligned} h &\approx \sqrt{1812.8} \\ &\approx 42.58 \text{ feet.} \end{aligned}$$

Example 4 ▶ An Application of the Law of Cosines

The *Interactive* CD-ROM and *Internet* versions of this text offer a built-in graphing calculator, which can be used in the Examples, Explorations, Technology notes, and Exercises.

A ship travels 60 miles due east, then adjusts its course northward, as shown in Figure 6.14. After traveling 80 miles in that direction, the ship is 139 miles from its point of departure. Describe the bearing from point B to point C.

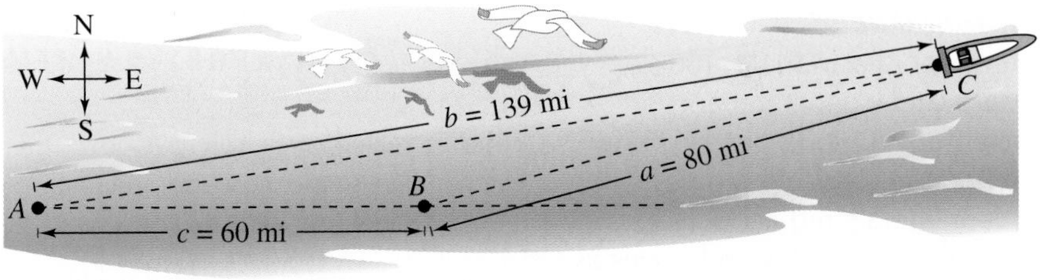

FIGURE 6.14

Solution

You have $a = 80$, $b = 139$, and $c = 60$; so, using the alternative form of the Law of Cosines, you have

$$\begin{aligned} \cos B &= \frac{a^2 + c^2 - b^2}{2ac} \\ &= \frac{80^2 + 60^2 - 139^2}{2(80)(60)} \\ &\approx -0.97094. \end{aligned}$$

Therefore, $B \approx \arccos(-0.97094) \approx 166.15°$. So, the bearing measured from due north from point B to point C is $166.15° - 90° = 76.15°$, or N 76.15° E.

Activities

1. Determine whether the Law of Sines or the Law of Cosines is needed to solve each of the triangles.
   (a) $A = 15°, B = 58°, c = 94$
   (b) $a = 96, b = 43, A = 105°$
   (c) $a = 24, b = 16, c = 29$
   (d) $a = 15, c = 42, B = 49°$
   Answer: (a) Law of Sines, (b) Law of Sines, (c) Law of Cosines, (d) Law of Cosines
2. Solve the triangle: $a = 31, b = 52, c = 28$.
   Answer: $A = 29.8°, B = 123.5°, C = 26.7°$.
3. Use Heron's Area Formula to find the area of a triangle with sides of lengths $a = 31$, $b = 52$, and $c = 28$.
   Answer: Area $= 361.8$ square units

Heron's Area Formula

The Law of Cosines can be used to establish the following formula for the area of a triangle. This formula is credited to the Greek mathematician Heron (c. 100 B.C.). A proof of this formula is given in Appendix A.

Heron's Area Formula

Given any triangle with sides of lengths a, b, and c, the area of the triangle is

$$\text{Area} = \sqrt{s(s-a)(s-b)(s-c)}$$

where $s = (a + b + c)/2$.

Example 5 ▶ Using Heron's Area Formula

Find the area of the triangular region having sides of lengths $a = 43$ meters, $b = 53$ meters, and $c = 72$ meters.

Solution

Because $s = (a + b + c)/2 = 168/2 = 84$, Heron's Area Formula yields

$$\begin{aligned}\text{Area} &= \sqrt{s(s-a)(s-b)(s-c)}\\ &= \sqrt{84(41)(31)(12)}\\ &\approx 1131.89 \text{ square meters.}\end{aligned}$$

Historical Note

Heron of Alexandria (c. 100 B.C.) was a Greek geometer and inventor. His works describe how to find the areas of triangles, quadrilaterals, regular polygons having 3 to 12 sides, and circles as well as the surface areas and volumes of three-dimensional objects.

Writing About Mathematics Suggestion:

You may want to ask students to draw the diagram of a triangle (labeling whatever sides and/or angles are necessary) that is well-suited to each area formula. Then ask students to exchange triangles with a partner to find the area of each.

Writing ABOUT MATHEMATICS

The Area of a Triangle You have now studied three different formulas for the area of a triangle. Use the most appropriate formula to find the area of each triangle below. Show your work and give your reasons for choosing each formula.

Standard Formula	$\text{Area} = \frac{1}{2}bh$
Oblique Triangle	$\text{Area} = \frac{1}{2}bc\sin A = \frac{1}{2}ab\sin C = \frac{1}{2}ac\sin B$
Heron's Area Formula	$\text{Area} = \sqrt{s(s-a)(s-b)(s-c)}$

a.

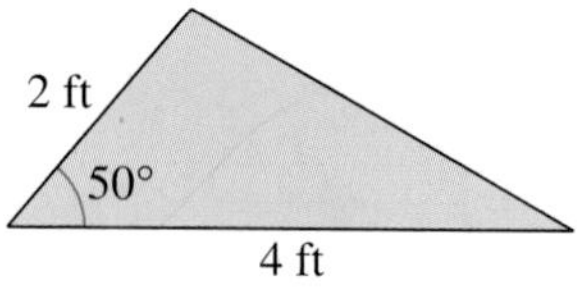

b.

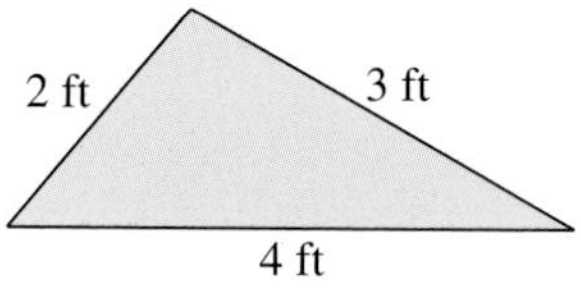

c.

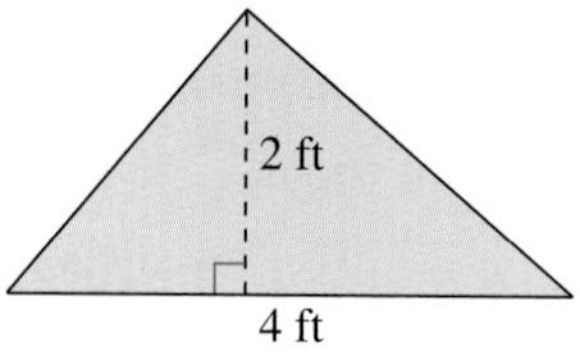

d.

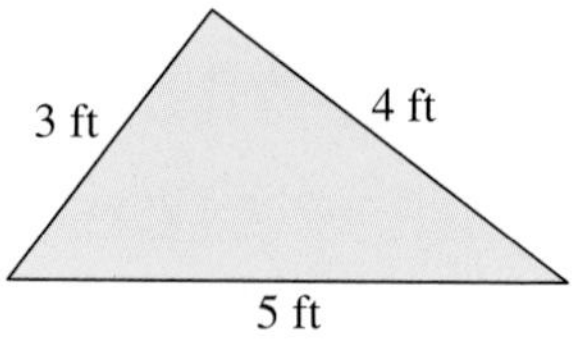

6.2 Exercises

In Exercises 1–16, use the Law of Cosines to solve the triangle.

1.
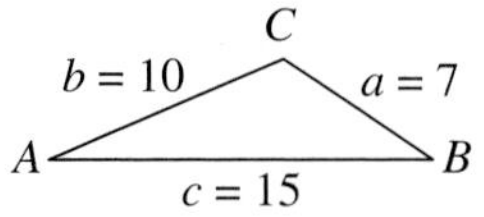

2.
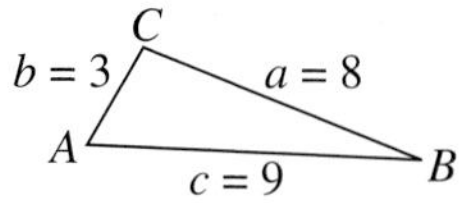

3.
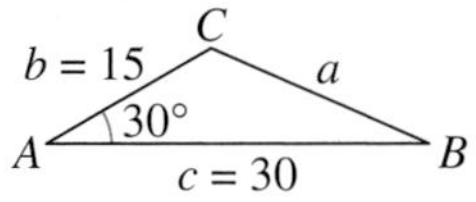

4.
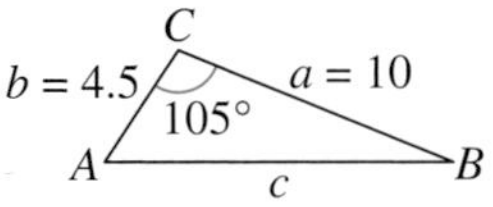

5. $a = 11,\quad b = 14,\quad c = 20$

6. $a = 55,\quad b = 25,\quad c = 72$

7. $a = 75.4,\quad b = 52,\quad c = 52$

8. $a = 1.42,\quad b = 0.75,\quad c = 1.25$

9. $A = 135°,\quad b = 4,\quad c = 9$

10. $A = 55°,\quad b = 3,\quad c = 10$

11. $B = 10° 35',\quad a = 40,\quad c = 30$

12. $B = 75° 20',\quad a = 6.2,\quad c = 9.5$

13. $B = 125° 40',\quad a = 32,\quad c = 32$

14. $C = 15° 15',\quad a = 6.25,\quad b = 2.15$

15. $C = 43°,\quad a = \frac{4}{9},\quad b = \frac{7}{9}$

16. $C = 103°,\quad a = \frac{3}{8},\quad b = \frac{3}{4}$

In Exercises 17–22, complete the table by solving the parallelogram shown in the figure. (The lengths of the diagonals are given by *c* and *d*.)

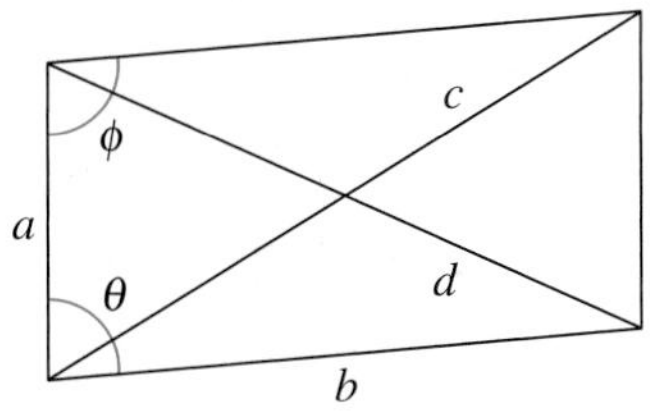

	a	b	c	d	θ	ϕ
17.	5	8			45°	
18.	25	35				120°
19.	10	14	20			
20.	40	60		80		
21.	15		25	20		
22.		25	50	35		

In Exercises 23–28, use Heron's Area Formula to find the area of the triangle.

23. $a = 5,\quad b = 7,\quad c = 10$

24. $a = 12,\quad b = 15,\quad c = 9$

25. $a = 2.5,\quad b = 10.2,\quad c = 9$

26. $a = 75.4,\quad b = 52,\quad c = 52$

27. $a = 12.32,\quad b = 8.46,\quad c = 15.05$

28. $a = 3.05,\quad b = 0.75,\quad c = 2.45$

29. *Navigation* A boat race runs along a triangular course marked by buoys *A*, *B*, and *C*. The race starts with the boats headed west for 3700 meters. The other two sides of the course lie to the north of the first side, and their lengths are 1700 meters and 3000 meters. Draw a figure that gives a visual representation of the problem, and find the bearings for the last two legs of the race.

30. *Navigation* A plane flies 810 miles from Niagara to Cuyahoga with a bearing of N 75° E. Then it flies 648 miles from Cuyahoga to Rosemount with a bearing of N 32° E. Draw a figure that visually represents the problem, and find the straight-line distance and bearing from Niagara to Rosemount.

31. *Surveying* To approximate the length of a marsh, a surveyor walks 250 meters from point *A* to point *B*, then turns 75° and walks 220 meters to point *C* (see figure). Approximate the length *AC* of the marsh.

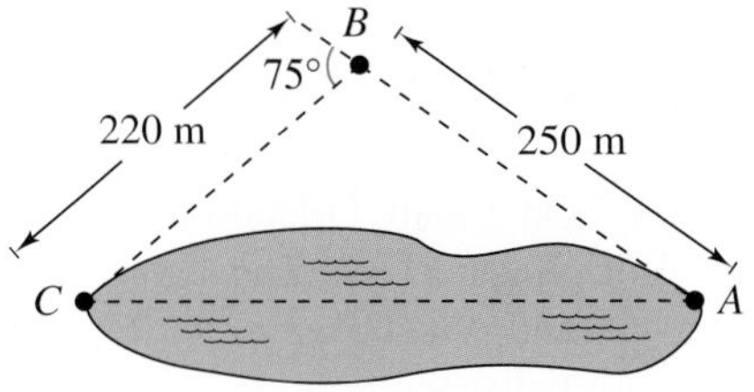

32. *Surveying* A triangular parcel of land has 115 meters of frontage, and the other boundaries have lengths of 76 meters and 92 meters. What angles does the frontage make with the two other boundaries?

33. *Surveying* A triangular parcel of ground has sides of lengths 725 feet, 650 feet, and 575 feet. Find the measure of the largest angle.

34. *Streetlight Design* Determine the angle θ in the design of the streetlight shown in the figure.

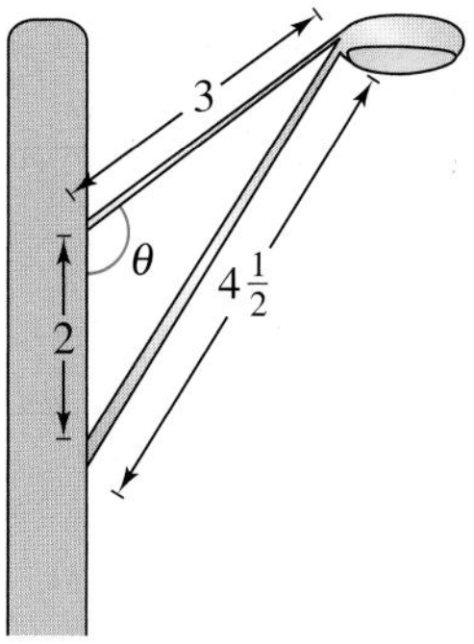

35. *Distance* Two ships leave a port at 9 A.M. One travels at a bearing of N 53° W at 12 miles per hour and the other travels at a bearing of S 67° W at 16 miles per hour. Approximate how far apart they are at noon that day.

36. *Distance* A 100-foot vertical tower is to be erected on the side of a hill that makes a 6° angle with the horizontal (see figure). Find the length of each of the two guy wires that will be anchored 75 feet uphill and downhill from the base of the tower.

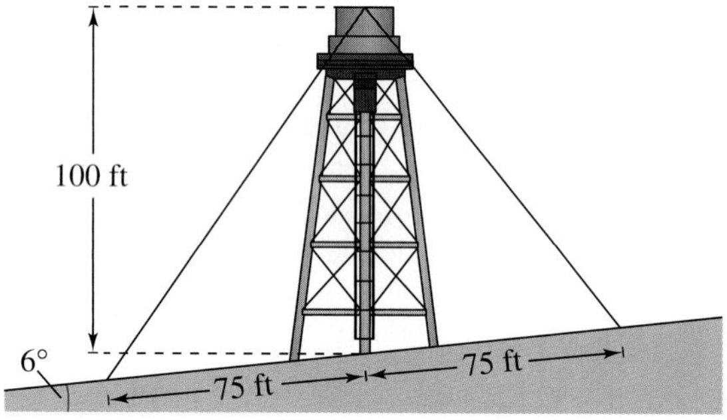

37. *Navigation* On a map, Orlando is 178 millimeters due south of Niagara Falls, Denver is 273 millimeters from Orlando, and Denver is 235 millimeters from Niagara Falls (see figure).

(a) Find the bearing of Denver from Orlando.

(b) Find the bearing of Denver from Niagara Falls.

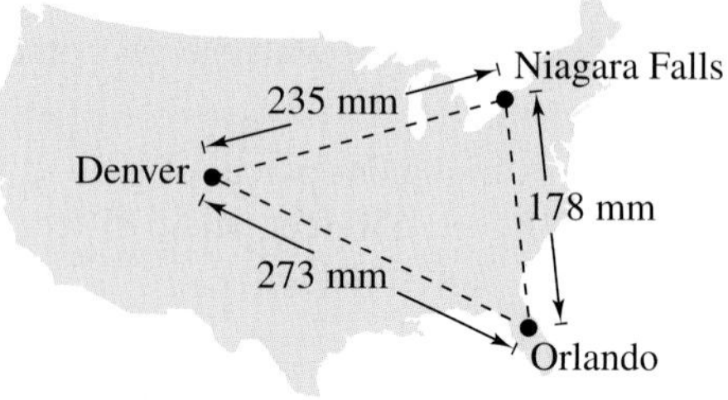

38. *Navigation* On a map, Minneapolis is 165 millimeters due west of Albany, Phoenix is 216 millimeters from Minneapolis, and Phoenix is 368 millimeters from Albany (see figure).

(a) Find the bearing of Minneapolis from Phoenix.

(b) Find the bearing of Albany from Phoenix.

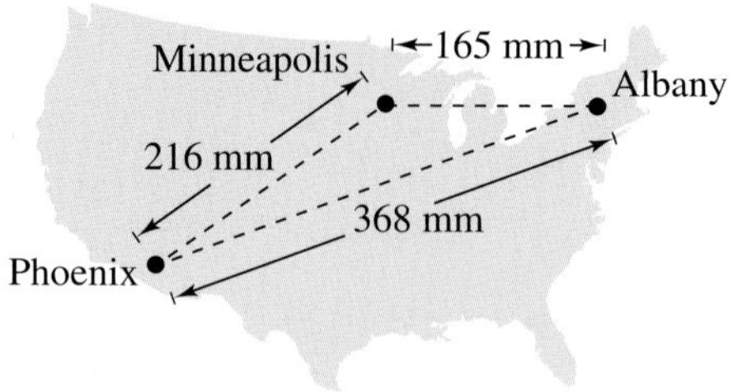

39. *Baseball* On a baseball diamond with 90-foot sides, the pitcher's mound is 60.5 feet from home plate. How far is it from the pitcher's mound to third base?

40. *Baseball* The baseball player in center field is playing approximately 330 feet from the television camera that is behind home plate. A batter hits a fly ball that goes to the wall 420 feet from the camera (see figure). Approximate the number of feet that the center fielder has to run to make the catch if the camera turns 8° to follow the play.

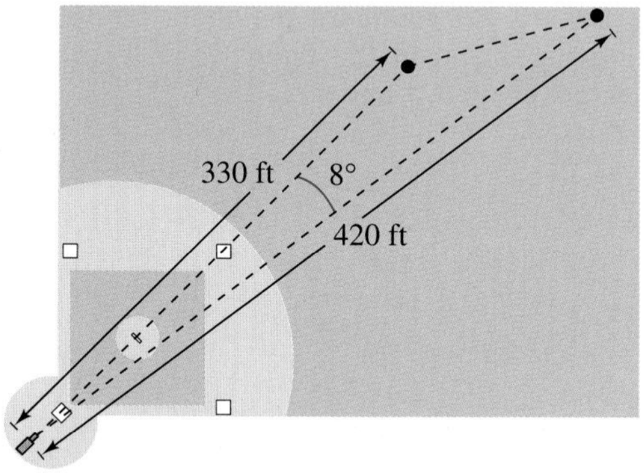

41. *Aircraft Tracking* To determine the distance between two aircraft, a tracking station continuously determines the distance to each aircraft and the angle A between them (see figure). Determine the distance a between the planes when $A = 42°$, $b = 35$ miles, and $c = 20$ miles.

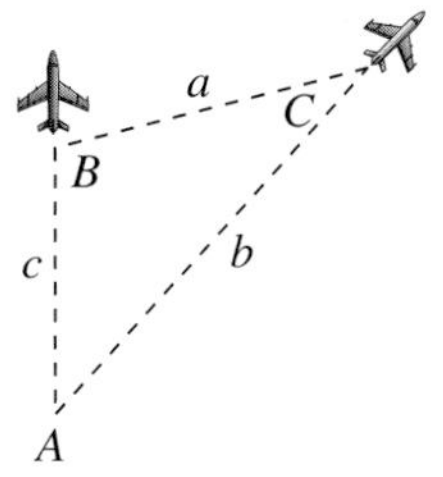

FIGURE FOR 41

42. *Aircraft Tracking* Use the figure for Exercise 41 to determine the distance a between the planes when $A = 11°$, $b = 20$ miles, and $c = 20$ miles.

43. *Engineering* If Q is the midpoint of the line segment $\overline{PR}$ in the truss rafter shown in the figure, find the lengths of the line segments $\overline{PQ}$, $\overline{QS}$, and $\overline{RS}$.

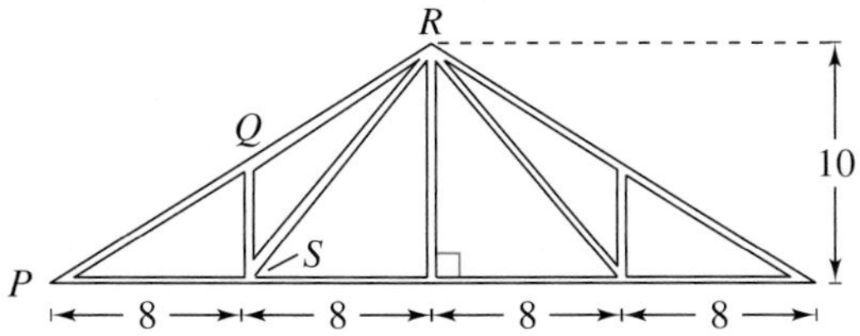

44. *Engine Design* An engine has a 7-inch connecting rod fastened to a crank (see figure).

(a) Use the Law of Cosines to write an equation giving the relationship between x and θ.

(b) Write x as a function of θ. (Select the sign that yields positive values of x.)

(c) Use a graphing utility to graph the function in part (b).

(d) Use the graph in part (c) to determine the maximum distance the piston moves in one cycle.

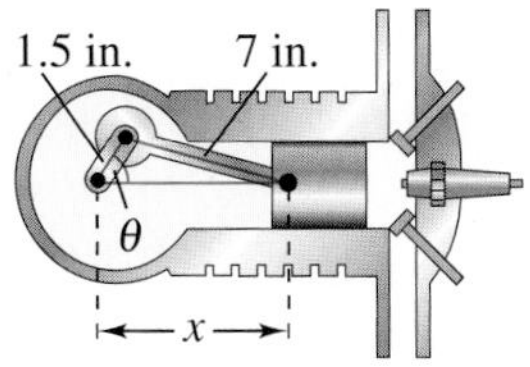

45. *Paper Manufacturing* In a certain process with continuous paper, the paper passes across three rollers of radii 3 inches, 4 inches, and 6 inches (see figure). The centers of the 3-inch and 6-inch rollers are d inches apart, and the length of the arc in contact with the paper on the 4-inch roller is s inches. Complete the following table.

d (inches)	9	10	12	13	14	15	16
θ (degrees)							
s (inches)							

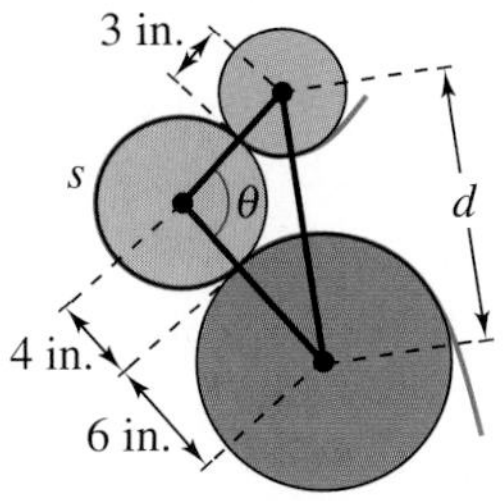

FIGURE FOR 45

46. *Awning Design* A retractable awning lowers at an angle of 50° from the top of a patio door that is 7 feet tall (see figure). Find the length x of the awning if no direct sunlight is to enter the door when the angle of elevation of the sun is greater than 70°.

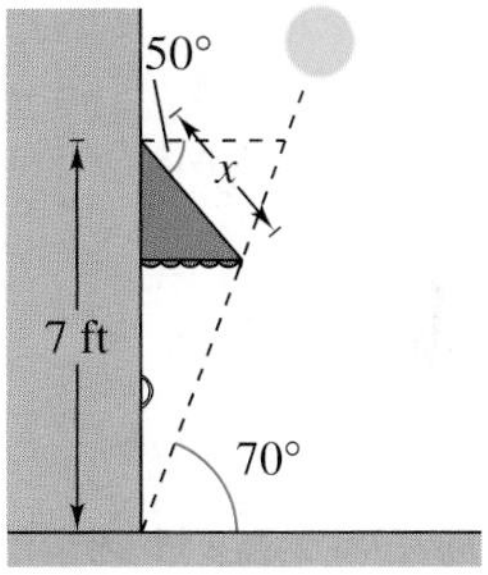

47. *Area* The lengths of the sides of a triangular parcel of land are approximately 200 feet, 500 feet, and 600 feet. Approximate the area of the parcel.

48. *Area* A parking lot has the shape of a parallelogram (see figure). The lengths of two adjacent sides are 70 meters and 100 meters. Find the area of the parking lot if the angle between the two sides is 70°.

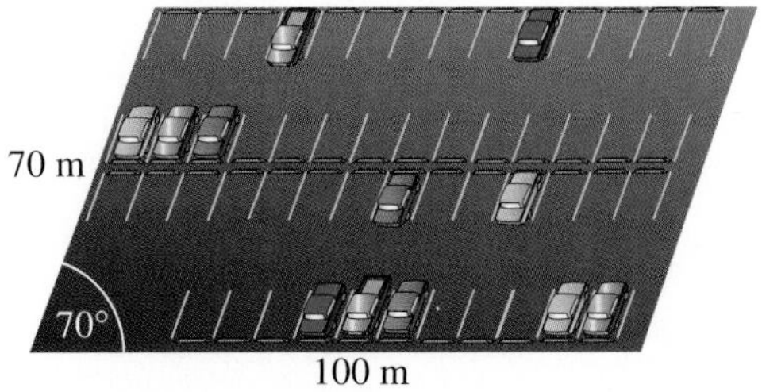

Synthesis

True or False? **In Exercises 49 and 50, determine whether the statement is true or false. Justify your answer.**

49. In Heron's Area Formula

$$\text{Area} = \sqrt{s(s-a)(s-b)(s-c)}$$

s is the average of the lengths of the three sides of the triangle.

50. In addition to SSS and SAS, the Law of Cosines can be used to solve triangles with SSA conditions.

51. *Circumscribed and Inscribed Circles* Let R and r be the radii of the circumscribed and inscribed circles of a triangle ABC, respectively (see figure), and let $s = (a + b + c)/2$.

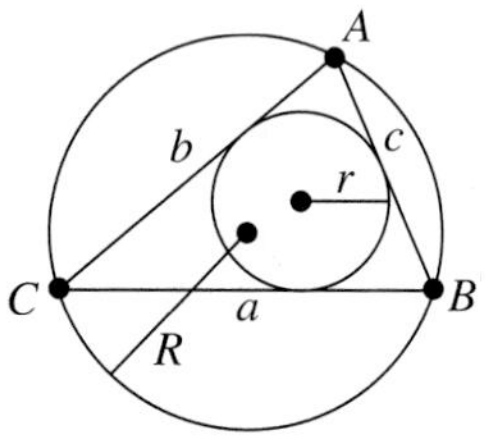

(a) Prove that $2R = \dfrac{a}{\sin A} = \dfrac{b}{\sin B} = \dfrac{c}{\sin C}$.

(b) Prove that $r = \sqrt{\dfrac{(s-a)(s-b)(s-c)}{s}}$.

Circumscribed and Inscribed Circles **In Exercises 52 and 53, use the results of Exercise 51.**

52. Given a triangle with $a = 25$, $b = 55$, and $c = 72$, find the area of (a) the triangle, (b) the circumscribed circle, and (c) the inscribed circle.

53. Find the length of the largest circular track that can be built on a triangular piece of property with sides of lengths 200 feet, 250 feet, and 325 feet.

54. Use the Law of Cosines to prove that

$$\frac{1}{2}bc(1 + \cos A) = \frac{a+b+c}{2} \cdot \frac{-a+b+c}{2}.$$

55. Use the Law of Cosines to prove that

$$\frac{1}{2}bc(1 - \cos A) = \frac{a-b+c}{2} \cdot \frac{a+b-c}{2}.$$

Review

In Exercises 56–61, evaluate the expression without the aid of a calculator.

56. $\arcsin(-1)$

57. $\arccos 0$

58. $\arctan \sqrt{3}$

59. $\arctan -\sqrt{3}$

60. $\arcsin\left(-\dfrac{\sqrt{3}}{2}\right)$

61. $\arccos\left(-\dfrac{\sqrt{3}}{2}\right)$

In Exercises 62–65, write an algebraic expression that is equivalent to the expression.

62. $\sec(\arcsin 2x)$

63. $\tan(\arccos 3x)$

64. $\cot[\arctan(x-2)]$

65. $\cos\left(\arcsin \dfrac{x-1}{2}\right)$

In Exercises 66–69, use trigonometric substitution to write the algebraic equation as a trigonometric function of θ, where $-\pi/2 < \theta < \pi/2$. Then find $\sec \theta$ and $\csc \theta$.

66. $5 = \sqrt{25 - x^2}, \quad x = 5 \sin \theta$

67. $-\sqrt{2} = \sqrt{4 - x^2}, \quad x = 2 \cos \theta$

68. $-\sqrt{3} = \sqrt{x^2 - 9}, \quad x = 3 \sec \theta$

69. $12 = \sqrt{36 + x^2}, \quad x = 6 \tan \theta$

6.3 Vectors in the Plane

▶ What you should learn

- How to represent vectors as directed line segments
- How to write the component forms of vectors
- How to perform basic vector operations and represent them graphically
- How to write vectors as linear combinations of unit vectors
- How to find the direction angles of vectors
- How to use vectors to model and solve real-life problems

▶ Why you should learn it

You can use vectors to model and solve real-life problems involving magnitude and direction. For instance, in Exercise 77 on page 522 you use vectors to determine the tension in the tow lines between a tugboat and a barge.

Introduction

Quantities such as force and velocity involve both *magnitude* and *direction* and cannot be completely characterized by a single real number. To represent such a quantity, you can use a **directed line segment,** as shown in Figure 6.15. The directed line segment $\overrightarrow{PQ}$ has **initial point** P and **terminal point** Q. Its **magnitude** (or length) is denoted by $\|\overrightarrow{PQ}\|$ and can be found using the distance formula.

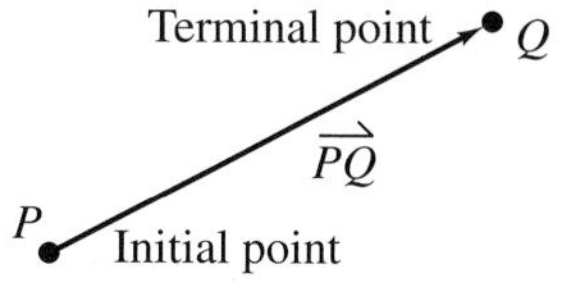

FIGURE 6.15

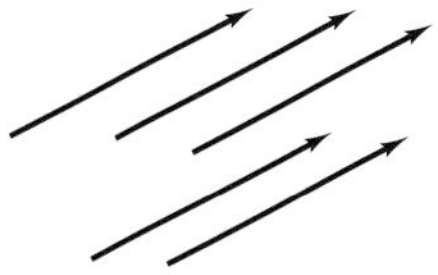

FIGURE 6.16

Two directed line segments that have the same magnitude and direction are equivalent. For example, the directed line segments in Figure 6.16 are all equivalent. The set of all directed line segments that are equivalent to given directed line segment $\overrightarrow{PQ}$ is a **vector v in the plane,** written $\mathbf{v} = \overrightarrow{PQ}$. Vectors are denoted by lowercase, boldface letters such as **u**, **v**, and **w**.

Example 1 ▶ Vector Representation by Directed Line Segments

Let **u** be represented by the directed line segment from $P = (0, 0)$ to $Q = (3, 2)$, and let **v** be represented by the directed line segment from $R = (1, 2)$ to $S = (4, 4)$, as shown in Figure 6.17. Show that $\mathbf{u} = \mathbf{v}$.

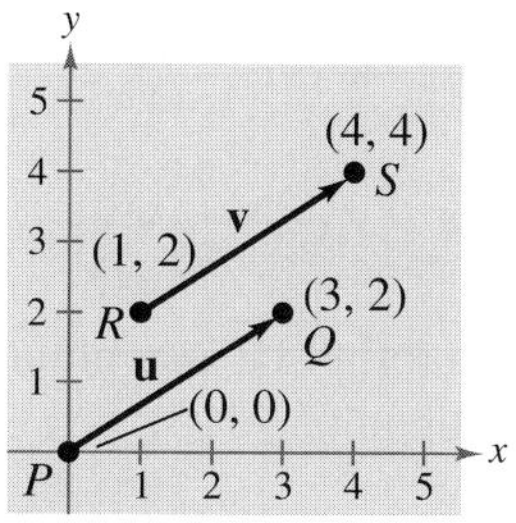

FIGURE 6.17

Solution

From the distance formula, it follows that $\overrightarrow{PQ}$ and $\overrightarrow{RS}$ have the *same magnitude.*

$$\|\overrightarrow{PQ}\| = \sqrt{(3-0)^2 + (2-0)^2} = \sqrt{13}$$

$$\|\overrightarrow{RS}\| = \sqrt{(4-1)^2 + (4-2)^2} = \sqrt{13}$$

Moreover, both line segments have the *same direction* because they are both directed toward the upper right on lines having a slope of $\frac{2}{3}$. So, $\overrightarrow{PQ}$ and $\overrightarrow{RS}$ have the same magnitude and direction, and it follows that $\mathbf{u} = \mathbf{v}$.

Component Form of a Vector

The directed line segment whose initial point is the origin is often the most convenient representative of a set of equivalent directed line segments. This representative of the vector **v** is in **standard position.**

A vector whose initial point is at the origin (0, 0) can be uniquely represented by the coordinates of its terminal point (v_1, v_2). This is the **component form of a vector v,** written

$$\mathbf{v} = \langle v_1, v_2 \rangle.$$

Help students see that $\mathbf{v} = \langle 1, 3 \rangle$ can be thought of as a vector with initial point (0, 0) and terminal point (1, 3), as well as a vector with initial point (0, −1) and terminal point (1, 2), and so on.

The coordinates v_1 and v_2 are the *components* of **v**. If both the initial point and the terminal point lie at the origin, **v** is the **zero vector** and is denoted by $\mathbf{0} = \langle 0, 0 \rangle$.

Component Form of a Vector

The component form of the vector with initial point $P = (p_1, p_2)$ and terminal point $Q = (q_1, q_2)$ is

$$\overrightarrow{PQ} = \langle q_1 - p_1, q_2 - p_2 \rangle = \langle v_1, v_2 \rangle = \mathbf{v}.$$

The **magnitude** (or length) of **v** is

$$\|\mathbf{v}\| = \sqrt{(q_1 - p_1)^2 + (q_2 - p_2)^2} = \sqrt{v_1^2 + v_2^2}.$$

If $\|\mathbf{v}\| = 1$, **v** is a **unit vector.** Moreover, $\|\mathbf{v}\| = 0$ if and only if **v** is the zero vector **0**.

Two vectors $\mathbf{u} = \langle u_1, u_2 \rangle$ and $\mathbf{v} = \langle v_1, v_2 \rangle$ are *equal* if and only if $u_1 = v_1$ and $u_2 = v_2$. For instance, in Example 1, the vector **u** from $P = (0, 0)$ to $Q = (3, 2)$ is

$$\mathbf{u} = \overrightarrow{PQ} = \langle 3 - 0, 2 - 0 \rangle = \langle 3, 2 \rangle$$

and the vector **v** from $R = (1, 2)$ to $S = (4, 4)$ is

$$\mathbf{v} = \overrightarrow{RS} = \langle 4 - 1, 4 - 2 \rangle = \langle 3, 2 \rangle.$$

Example 2 ▶ Finding the Component Form of a Vector

Find the component form and magnitude of the vector **v** that has initial point $(4, -7)$ and terminal point $(-1, 5)$.

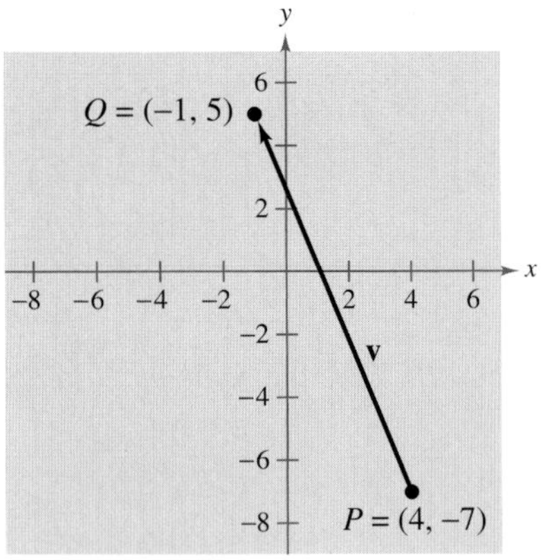

FIGURE 6.18

Solution

Let $P = (4, -7) = (p_1, p_2)$ and let $Q = (-1, 5) = (q_1, q_2)$, as shown in Figure 6.18. Then, the components of $\mathbf{v} = \langle v_1, v_2 \rangle$ are

$$v_1 = q_1 - p_1 = -1 - 4 = -5$$

$$v_2 = q_2 - p_2 = 5 - (-7) = 12.$$

So, $\mathbf{v} = \langle -5, 12 \rangle$ and the magnitude of **v** is

$$\|\mathbf{v}\| = \sqrt{(-5)^2 + 12^2} = \sqrt{169} = 13.$$

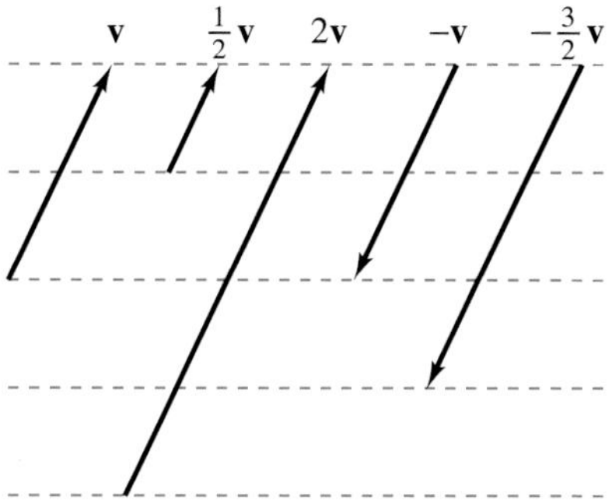

FIGURE 6.19

Vector Operations

The two basic vector operations are **scalar multiplication** and **vector addition.** In operations with vectors, numbers are usually referred to as **scalars.** In this text, scalars will always be real numbers. Geometrically, the product of a vector **v** and a scalar k is the vector that is $|k|$ times as long as **v**. If k is positive, k**v** has the same direction as **v**, and if k is negative, k**v** has the direction opposite that of **v**, as shown in Figure 6.19.

To add two vectors geometrically, position them (without changing length or direction) so that the initial point of one coincides with the terminal point of the other. The sum **u** + **v** is formed by joining the initial point of the second vector **v** with the terminal point of the first vector **u**, as shown in Figure 6.20. This technique is called the **parallelogram law** for vector addition because the vector **u** + **v**, often called the **resultant** of vector addition, is the diagonal of a parallelogram having **u** and **v** as its adjacent sides.

A computer animation of this concept appears in the *Interactive* CD-ROM and *Internet* versions of this text.

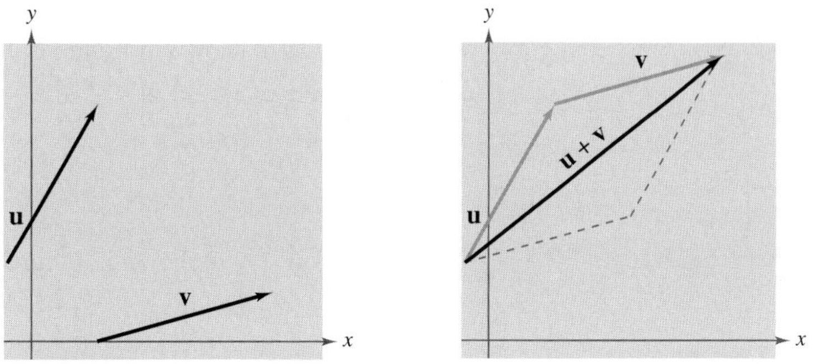

FIGURE 6.20

The graphical representation of the difference of two vectors may not be obvious to your students. You may want to go over this carefully.

Definitions of Vector Addition and Scalar Multiplication

Let $\mathbf{u} = \langle u_1, u_2 \rangle$ and $\mathbf{v} = \langle v_1, v_2 \rangle$ be vectors and let k be a scalar (a real number). Then the *sum* of **u** and **v** is the vector

$$\mathbf{u} + \mathbf{v} = \langle u_1 + v_1, u_2 + v_2 \rangle \qquad \text{Sum}$$

and the *scalar multiple* of k times **u** is the vector

$$k\mathbf{u} = k\langle u_1, u_2 \rangle = \langle ku_1, ku_2 \rangle. \qquad \text{Scalar multiple}$$

The *negative* of $\mathbf{v} = \langle v_1, v_2 \rangle$ is

$$-\mathbf{v} = (-1)\mathbf{v}$$

$$= \langle -v_1, -v_2 \rangle \qquad \text{Negative}$$

and the *difference* of **u** and **v** is

$$\mathbf{u} - \mathbf{v} = \mathbf{u} + (-\mathbf{v})$$

$$= \langle u_1 - v_1, u_2 - v_2 \rangle. \qquad \text{Difference}$$

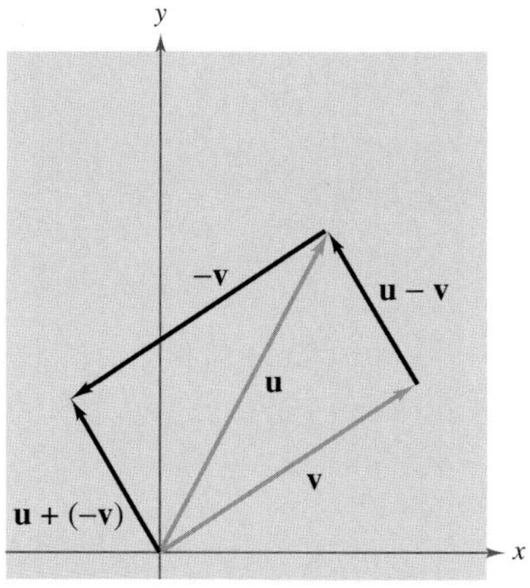

FIGURE 6.21

To represent **u** − **v** geometrically, you can use directed line segments with the *same* initial point. The difference **u** − **v** is the vector from the terminal point of **v** to the terminal point of **u**, as shown in Figure 6.21.

The component definitions of vector addition and scalar multiplication are illustrated in Example 3. In this example, notice that each of the vector operations can be interpreted geometrically.

A computer animation of this example appears in the *Interactive* CD-ROM and *Internet* versions of this text.

Example 3 ▶ Vector Operations

Let $\mathbf{v} = \langle -2, 5\rangle$ and $\mathbf{w} = \langle 3, 4\rangle$, and find each of the following vectors.

a. $2\mathbf{v}$ **b.** $\mathbf{w} - \mathbf{v}$ **c.** $\mathbf{v} + 2\mathbf{w}$

Solution

a. Because $\mathbf{v} = \langle -2, 5\rangle$, you have

$$\begin{aligned} 2\mathbf{v} &= 2\langle -2, 5\rangle \\ &= \langle 2(-2), 2(5)\rangle \\ &= \langle -4, 10\rangle. \end{aligned}$$

A sketch of $2\mathbf{v}$ is shown in Figure 6.22(a).

b. The difference of $\mathbf{w}$ and $\mathbf{v}$ is

$$\begin{aligned} \mathbf{w} - \mathbf{v} &= \langle 3 - (-2), 4 - 5\rangle \\ &= \langle 5, -1\rangle. \end{aligned}$$

A sketch of $\mathbf{w} - \mathbf{v}$ is shown in Figure 6.22(b).

c. The sum of $\mathbf{v}$ and $2\mathbf{w}$ is

$$\begin{aligned} \mathbf{v} + 2\mathbf{w} &= \langle -2, 5\rangle + 2\langle 3, 4\rangle \\ &= \langle -2, 5\rangle + \langle 2(3), 2(4)\rangle \\ &= \langle -2, 5\rangle + \langle 6, 8\rangle \\ &= \langle -2 + 6, 5 + 8\rangle \\ &= \langle 4, 13\rangle. \end{aligned}$$

A sketch of $\mathbf{v} + 2\mathbf{w}$ is shown in Figure 6.22(c).

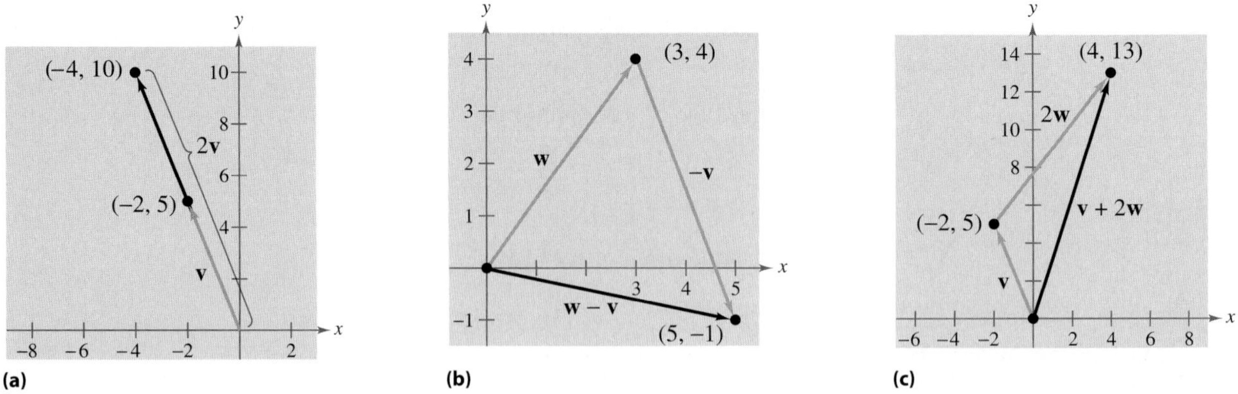

FIGURE 6.22

Vector addition and scalar multiplication share many of the properties of ordinary arithmetic.

Properties of Vector Addition and Scalar Multiplication

Let **u**, **v**, and **w** be vectors and let c and d be scalars. Then the following properties are true.

1. $\mathbf{u} + \mathbf{v} = \mathbf{v} + \mathbf{u}$
2. $(\mathbf{u} + \mathbf{v}) + \mathbf{w} = \mathbf{u} + (\mathbf{v} + \mathbf{w})$
3. $\mathbf{u} + \mathbf{0} = \mathbf{u}$
4. $\mathbf{u} + (-\mathbf{u}) = \mathbf{0}$
5. $c(d\mathbf{u}) = (cd)\mathbf{u}$
6. $(c + d)\mathbf{u} = c\mathbf{u} + d\mathbf{u}$
7. $c(\mathbf{u} + \mathbf{v}) = c\mathbf{u} + c\mathbf{v}$
8. $1(\mathbf{u}) = \mathbf{u}, 0(\mathbf{u}) = \mathbf{0}$
9. $\|c\mathbf{v}\| = |c|\,\|\mathbf{v}\|$

Property 9 can be stated as follows: the magnitude of the vector $c\mathbf{v}$ is the absolute value of c times the magnitude of **v**.

Unit Vectors

In many applications of vectors it is useful to find a unit vector that has the same direction as a given nonzero vector **v**. To do this, you can divide **v** by its magnitude to obtain

$$\mathbf{u} = \text{unit vector} = \frac{\mathbf{v}}{\|\mathbf{v}\|} = \left(\frac{1}{\|\mathbf{v}\|}\right)\mathbf{v}.$$

Note that **u** is a scalar multiple of **v**. The vector **u** has magnitude 1 and the same direction as **v**. The vector **u** is called a **unit vector in the direction of v.**

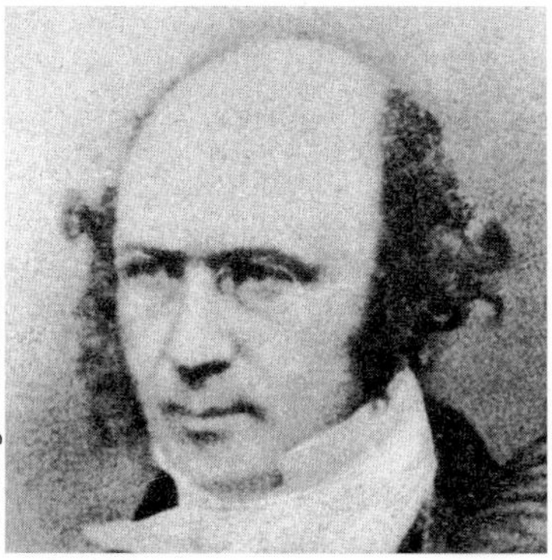
The Granger Collection

Historical Note

William Rowan Hamilton (1805–1865), an Irish mathematician, did some of the earliest work with vectors. Hamilton spent many years developing a system of vector-like quantities called quaternions. Although Hamilton was convinced of the benefits of quaternions, the operations he defined did not produce good models for physical phenomena. It wasn't until the latter half of the nineteenth century that the Scottish physicist James Maxwell (1831–1879) restructured Hamilton's quaternions in a form useful for representing physical quantities such as force, velocity, and acceleration.

Example 4 ▶ Finding a Unit Vector

Find a unit vector in the direction of $\mathbf{v} = \langle -2, 5\rangle$ and verify that the result has magnitude 1.

Solution

The unit vector in the direction of **v** is

$$\frac{\mathbf{v}}{\|\mathbf{v}\|} = \frac{\langle -2, 5\rangle}{\sqrt{(-2)^2 + (5)^2}}$$

$$= \frac{1}{\sqrt{29}}\langle -2, 5\rangle$$

$$= \left\langle \frac{-2}{\sqrt{29}}, \frac{5}{\sqrt{29}} \right\rangle.$$

This vector has magnitude 1 because

$$\sqrt{\left(\frac{-2}{\sqrt{29}}\right)^2 + \left(\frac{5}{\sqrt{29}}\right)^2} = \sqrt{\frac{4}{29} + \frac{25}{29}} = \sqrt{\frac{29}{29}} = 1.$$

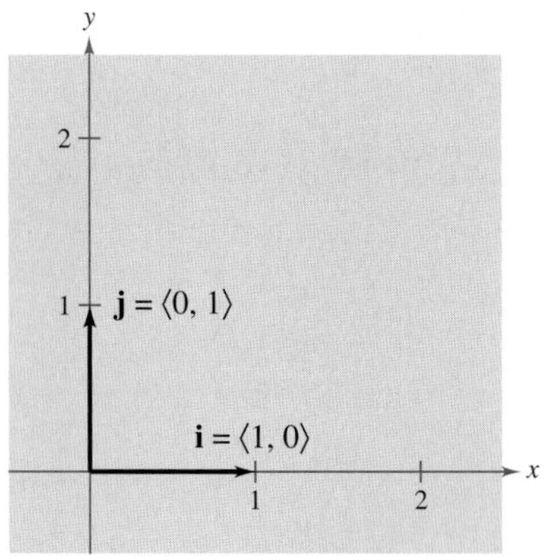

FIGURE 6.23

The unit vectors $\langle 1, 0\rangle$ and $\langle 0, 1\rangle$ are called the **standard unit vectors** and are denoted by

$$\mathbf{i} = \langle 1, 0\rangle \quad \text{and} \quad \mathbf{j} = \langle 0, 1\rangle,$$

as shown in Figure 6.23. (Note that the lowercase letter **i** is written in boldface to distinguish it from the imaginary number $i = \sqrt{-1}$.) These vectors can be used to represent any vector $\mathbf{v} = \langle v_1, v_2\rangle$ as follows.

$$\begin{aligned}\mathbf{v} &= \langle v_1, v_2\rangle \\ &= v_1\langle 1, 0\rangle + v_2\langle 0, 1\rangle \\ &= v_1\mathbf{i} + v_2\mathbf{j}\end{aligned}$$

The scalars v_1 and v_2 are called the **horizontal** and **vertical components of v,** respectively. The vector sum

$$v_1\mathbf{i} + v_2\mathbf{j}$$

is called a **linear combination** of the vectors **i** and **j**. Any vector in the plane can be expressed as a linear combination of the standard unit vectors **i** and **j**.

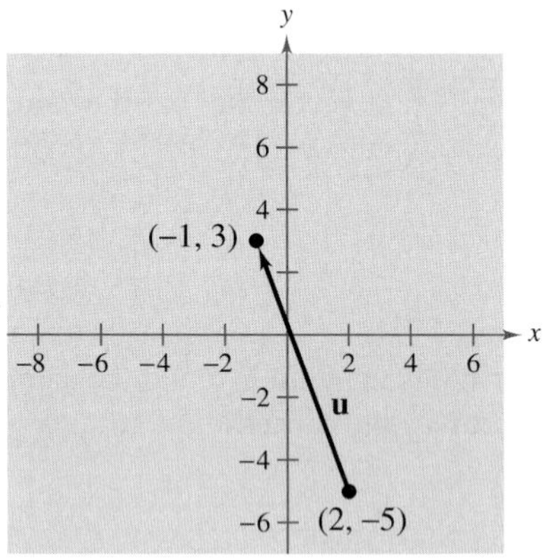

FIGURE 6.24

Example 5 ▶ Writing a Linear Combination of Unit Vectors

Let **u** be the vector with initial point $(2, -5)$ and terminal point $(-1, 3)$. Write **u** as a linear combination of the standard unit vectors **i** and **j**.

Solution

Begin by writing the component form of the vector **u**.

$$\begin{aligned}\mathbf{u} &= \langle -1 - 2, 3 + 5\rangle \\ &= \langle -3, 8\rangle \\ &= -3\mathbf{i} + 8\mathbf{j}\end{aligned}$$

This result is shown graphically in Figure 6.24.

Example 6 ▶ Vector Operations

Let $\mathbf{u} = -3\mathbf{i} + 8\mathbf{j}$ and let $\mathbf{v} = 2\mathbf{i} - \mathbf{j}$. Find $2\mathbf{u} - 3\mathbf{v}$.

Solution

You could solve this problem by converting **u** and **v** to component form. This, however, is not necessary. It is just as easy to perform the operations in unit vector form.

$$\begin{aligned}2\mathbf{u} - 3\mathbf{v} &= 2(-3\mathbf{i} + 8\mathbf{j}) - 3(2\mathbf{i} - \mathbf{j}) \\ &= -6\mathbf{i} + 16\mathbf{j} - 6\mathbf{i} + 3\mathbf{j} \\ &= -12\mathbf{i} + 19\mathbf{j}\end{aligned}$$

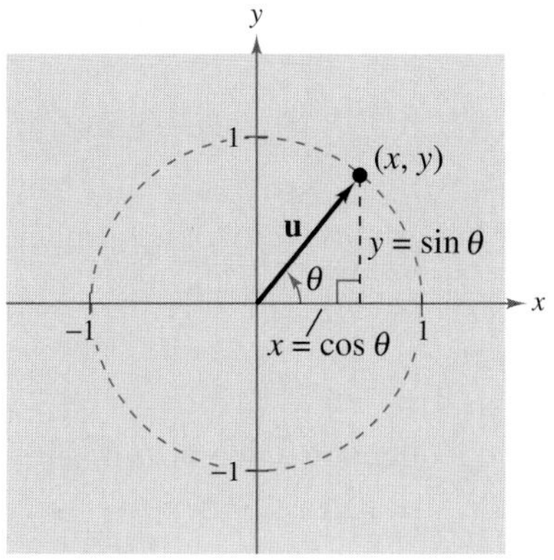

FIGURE 6.25

Direction Angles

If **u** is a *unit vector* such that θ is the angle (measured counterclockwise) from the positive x-axis to **u,** the terminal point of **u** lies on the unit circle and you have

$$\mathbf{u} = \langle x, y\rangle = \langle \cos\theta, \sin\theta\rangle = (\cos\theta)\mathbf{i} + (\sin\theta)\mathbf{j}$$

as shown in Figure 6.25. The angle θ is the **direction angle** of the vector **u**.

Suppose that **u** is a unit vector with direction angle θ. If **v** is any vector that makes an angle θ with the positive x-axis, it has the same direction as **u** and you can write

$$\begin{aligned}\mathbf{v} &= \|\mathbf{v}\|\langle \cos\theta, \sin\theta\rangle \\ &= \|\mathbf{v}\|(\cos\theta)\mathbf{i} + \|\mathbf{v}\|(\sin\theta)\mathbf{j}.\end{aligned}$$

Because $\mathbf{v} = a\mathbf{i} + b\mathbf{j} = \|\mathbf{v}\|(\cos\theta)\,\mathbf{i} + \|\mathbf{v}\|(\sin\theta)\,\mathbf{j}$, it follows that the direction angle θ for **v** is determined from

$$\begin{aligned}\tan\theta &= \frac{\sin\theta}{\cos\theta} \\ &= \frac{\|\mathbf{v}\|\sin\theta}{\|\mathbf{v}\|\cos\theta} \\ &= \frac{b}{a}.\end{aligned}$$

Example 7 ▶ Finding Direction Angles of Vectors

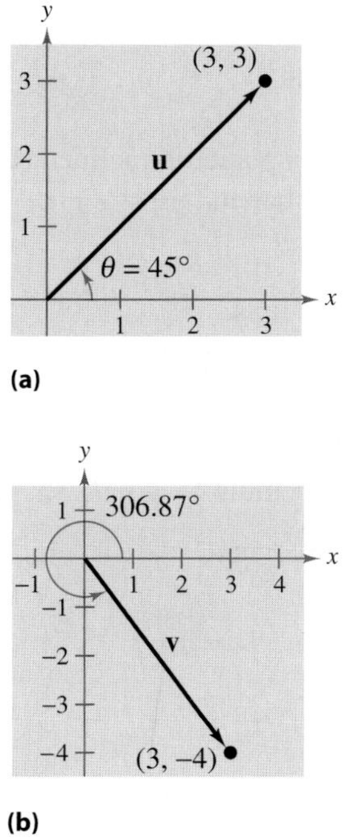

FIGURE 6.26

Find the direction angle of each vector.

a. $\mathbf{u} = 3\mathbf{i} + 3\mathbf{j}$ **b.** $\mathbf{v} = 3\mathbf{i} - 4\mathbf{j}$

Solution

a. The direction angle is

$$\tan\theta = \frac{b}{a} = \frac{3}{3} = 1.$$

Therefore, $\theta = 45°$, as shown in Figure 6.26(a).

b. The direction angle is

$$\tan\theta = \frac{b}{a} = \frac{-4}{3}.$$

Moreover, because $\mathbf{v} = 3\mathbf{i} - 4\mathbf{j}$ lies in Quadrant IV, θ lies in Quadrant IV and its reference angle is

$$\theta = \left|\arctan\left(-\frac{4}{3}\right)\right| \approx |-53.13°| = 53.13°.$$

Therefore, it follows that $\theta \approx 360° - 53.13° = 306.87°$, as shown in Figure 6.26(b).

Applications of Vectors

Example 8 ▶ Finding the Component Form of a Vector

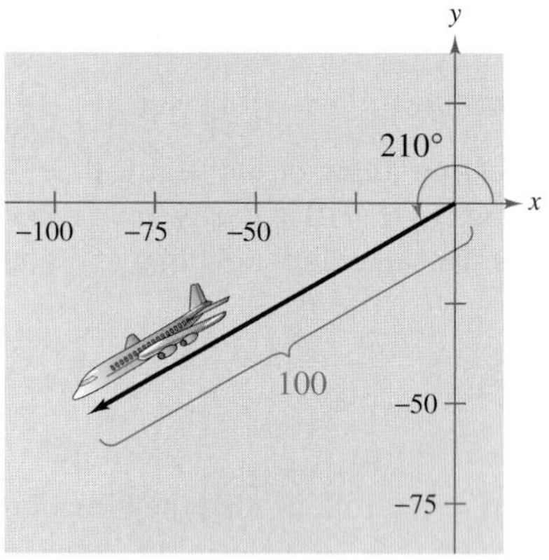

FIGURE 6.27

Find the component form of the vector that represents the velocity of an airplane descending at a speed of 100 miles per hour at an angle 30° below the horizontal, as shown in Figure 6.27.

Solution

The velocity vector **v** has a magnitude of 100 and a direction angle of $\theta = 210°$.

$$\begin{aligned}\mathbf{v} &= \|\mathbf{v}\|(\cos\theta)\mathbf{i} + \|\mathbf{v}\|(\sin\theta)\mathbf{j}\\ &= 100(\cos 210°)\mathbf{i} + 100(\sin 210°)\mathbf{j}\\ &= 100\left(\frac{-\sqrt{3}}{2}\right)\mathbf{i} + 100\left(\frac{-1}{2}\right)\mathbf{j}\\ &= -50\sqrt{3}\,\mathbf{i} - 50\mathbf{j}\\ &= \langle -50\sqrt{3}, -50\rangle\end{aligned}$$

You can check that **v** has a magnitude of 100.

$$\begin{aligned}\|\mathbf{v}\| &= \sqrt{\left(-50\sqrt{3}\right)^2 + (50)^2}\\ &= \sqrt{7500 + 2500}\\ &= \sqrt{10{,}000}\\ &= 100\end{aligned}$$

Example 9 ▶ Using Vectors to Determine Weight

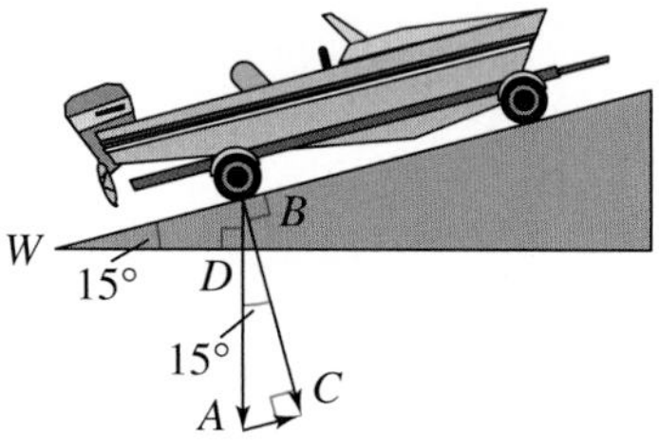

FIGURE 6.28

A force of 600 pounds is required to pull a boat and trailer up a ramp inclined at 15° from the horizontal. Find the combined weight of the boat and trailer.

Solution

Based on Figure 6.28, you can make the following observations.

$\|\overrightarrow{BA}\|$ = force of gravity = combined weight of boat and trailer

$\|\overrightarrow{BC}\|$ = force against ramp

$\|\overrightarrow{AC}\|$ = force required to move boat up ramp = 600 pounds

By construction, triangles *BWD* and *ABC* are similar. So, angle *ABC* is 15°. Therefore, in triangle *ABC* you have

$$\sin 15° = \frac{\|\overrightarrow{AC}\|}{\|\overrightarrow{BA}\|} = \frac{600}{\|\overrightarrow{BA}\|}$$

$$\|BA\| = \frac{600}{\sin 15°} \approx 2318.$$

Consequently, the combined weight is approximately 2318 pounds. (In Figure 6.28, note that $\overrightarrow{AC}$ is parallel to the ramp.)

Example 10 ▶ Using Vectors to Find Speed and Direction

An airplane is traveling at a fixed altitude with a negligible wind velocity. The airplane is headed N 30° W at a speed of 500 miles per hour, as shown in Figure 6.29. As the airplane reaches a certain point, it encounters a wind with a velocity of 70 miles per hour in the direction N 45° E. What are the resultant speed and direction of the airplane?

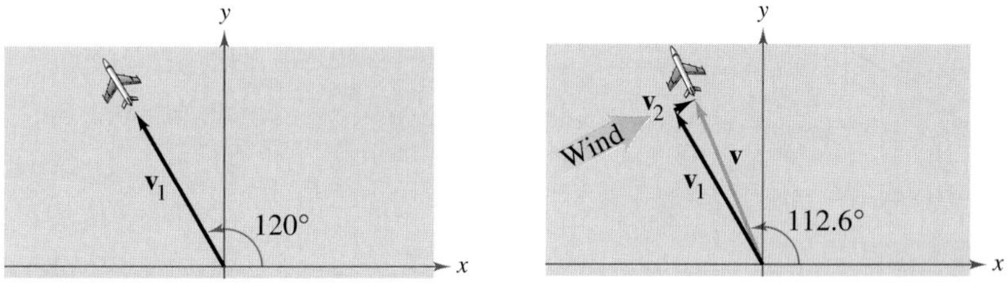

FIGURE 6.29

Solution

Using Figure 6.29, the velocity of the airplane (alone) is

$$\begin{aligned}\mathbf{v}_1 &= 500\langle \cos 120°, \sin 120° \rangle \\ &= \langle -250, 250\sqrt{3} \rangle\end{aligned}$$

and the velocity of the wind is

$$\begin{aligned}\mathbf{v}_2 &= 70\langle \cos 45°, \sin 45° \rangle \\ &= \langle 35\sqrt{2}, 35\sqrt{2} \rangle.\end{aligned}$$

So, the velocity of the airplane (in the wind) is

$$\begin{aligned}\mathbf{v} &= \mathbf{v}_1 + \mathbf{v}_2 \\ &= \langle -250 + 35\sqrt{2}, 250\sqrt{3} + 35\sqrt{2} \rangle \\ &\approx \langle -200.5, 482.5 \rangle\end{aligned}$$

and the speed of the airplane is

$$\begin{aligned}\|\mathbf{v}\| &= \sqrt{(-200.5)^2 + (482.5)^2} \\ &\approx 522.5 \text{ miles per hour.}\end{aligned}$$

Finally, if θ is the direction angle of the flight path, you have

$$\begin{aligned}\tan\theta &= \frac{482.5}{-200.5} \\ &\approx -2.4065\end{aligned}$$

which implies that

$$\begin{aligned}\theta &\approx 180° + \arctan(-2.4065) \\ &\approx 180° - 67.4° \\ &= 112.6°.\end{aligned}$$

Activities

1. Find the component form and the magnitude of the vector with initial point $(-3, 2)$ and terminal point $(1, 4)$.
   Answer: $\langle 4, 2 \rangle$; $2\sqrt{5}$
2. Vector $\mathbf{v}$ has direction angle $\theta = 30°$ and magnitude 6. Find $\mathbf{v}$.
   Answer: $\mathbf{v} = \langle 3\sqrt{3}, 3 \rangle$
3. $\mathbf{v} = 7\mathbf{i} - 2\mathbf{j}$, $\mathbf{w} = -2\mathbf{i} + \mathbf{j}$. Find $2\mathbf{v} + \mathbf{w}$.
   Answer: $12\mathbf{i} - 3\mathbf{j}$

6.3 Exercises

In Exercises 1–12, find the component form and the magnitude of the vector v.

1.

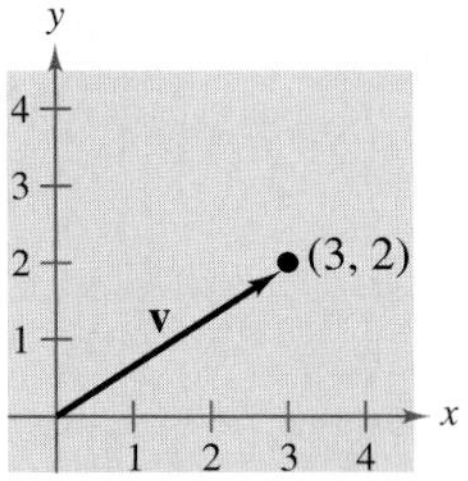

2.

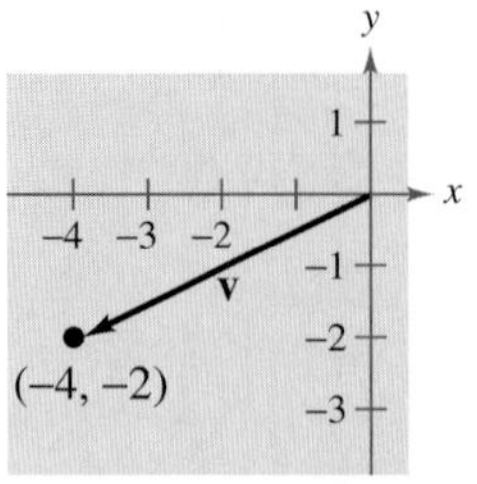

3.

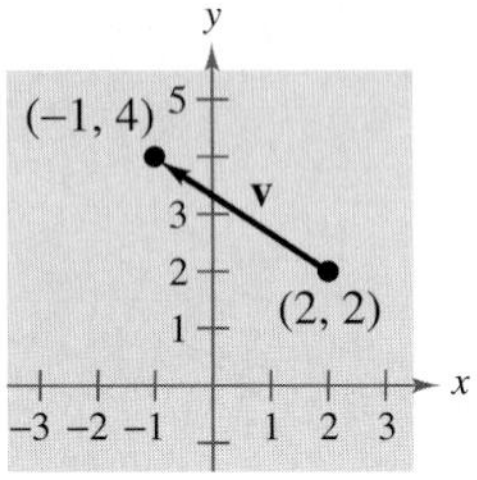

4.

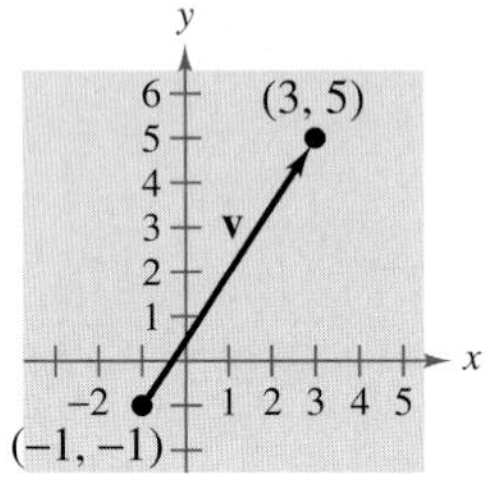

5.

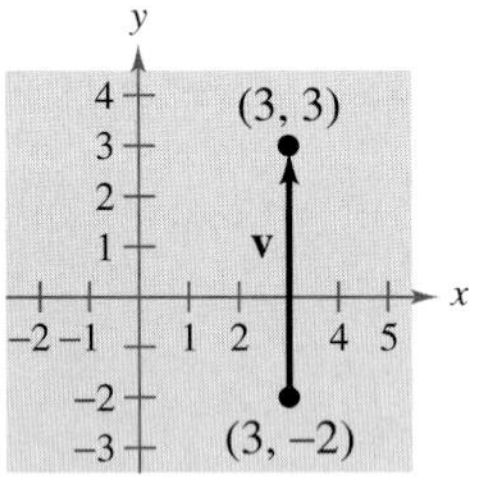

6.

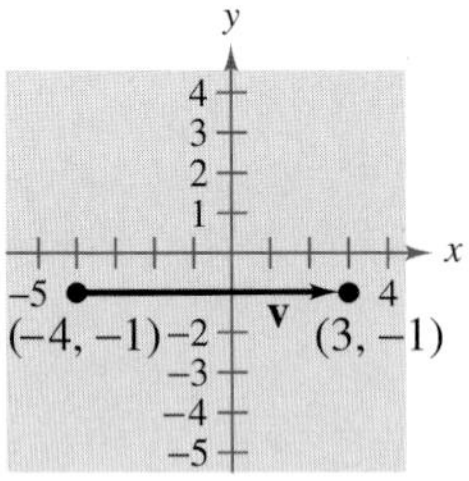

	Initial Point	*Terminal Point*
7.	$(-1, 5)$	$(15, 12)$
8.	$(1, 11)$	$(9, 3)$
9.	$(-3, -5)$	$(5, 1)$
10.	$(-3, 11)$	$(9, 40)$
11.	$(1, 3)$	$(-8, -9)$
12.	$(-2, 7)$	$(5, -17)$

In Exercises 13–18, use the figure to sketch a graph of the specified vector.

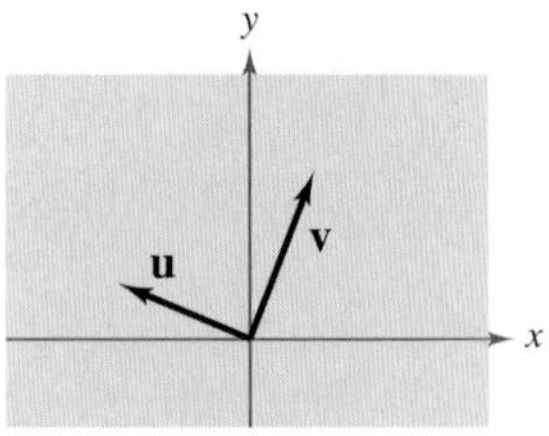

13. $-\mathbf{v}$
14. $5\mathbf{v}$
15. $\mathbf{u} + \mathbf{v}$
16. $\mathbf{u} - \mathbf{v}$
17. $\mathbf{u} + 2\mathbf{v}$
18. $\mathbf{v} - \frac{1}{2}\mathbf{u}$

In Exercises 19–26, find (a) u + v, (b) u − v, and (c) 2u − 3v. Then sketch your resultant vector.

19. $\mathbf{u} = \langle 2, 1 \rangle, \quad \mathbf{v} = \langle 1, 3 \rangle$
20. $\mathbf{u} = \langle 2, 3 \rangle, \quad \mathbf{v} = \langle 4, 0 \rangle$
21. $\mathbf{u} = \langle -5, 3 \rangle, \quad \mathbf{v} = \langle 0, 0 \rangle$
22. $\mathbf{u} = \langle 0, 0 \rangle, \quad \mathbf{v} = \langle 2, 1 \rangle$
23. $\mathbf{u} = \mathbf{i} + \mathbf{j}, \quad \mathbf{v} = 2\mathbf{i} - 3\mathbf{j}$
24. $\mathbf{u} = -2\mathbf{i} + \mathbf{j}, \quad \mathbf{v} = -\mathbf{i} + 2\mathbf{j}$
25. $\mathbf{u} = 2\mathbf{i}, \quad \mathbf{v} = \mathbf{j}$
26. $\mathbf{u} = 3\mathbf{j}, \quad \mathbf{v} = 2\mathbf{i}$

In Exercises 27–36, find a unit vector in the direction of the given vector.

27. $\mathbf{u} = \langle 3, 0 \rangle$
28. $\mathbf{u} = \langle 0, -2 \rangle$
29. $\mathbf{v} = \langle -2, 2 \rangle$
30. $\mathbf{v} = \langle 5, -12 \rangle$
31. $\mathbf{v} = 6\mathbf{i} - 2\mathbf{j}$
32. $\mathbf{v} = \mathbf{i} + \mathbf{j}$
33. $\mathbf{w} = 4\mathbf{j}$
34. $\mathbf{w} = -6\mathbf{i}$
35. $\mathbf{w} = \mathbf{i} - 2\mathbf{j}$
36. $\mathbf{w} = 7\mathbf{j} - 3\mathbf{i}$

In Exercises 37–40, find the vector v with the given magnitude and the same direction as u.

	Magnitude	*Direction*
37.	$\|\mathbf{v}\| = 5$	$\mathbf{u} = \langle 3, 3 \rangle$
38.	$\|\mathbf{v}\| = 6$	$\mathbf{u} = \langle -3, 3 \rangle$
39.	$\|\mathbf{v}\| = 9$	$\mathbf{u} = \langle 2, 5 \rangle$
40.	$\|\mathbf{v}\| = 10$	$\mathbf{u} = \langle -10, 0 \rangle$

In Exercises 41–46, find the component form of v and sketch the specified vector operations geometrically, where u = 2i − j and w = i + 2j.

41. $\mathbf{v} = \frac{3}{2}\mathbf{u}$
42. $\mathbf{v} = \frac{3}{4}\mathbf{w}$
43. $\mathbf{v} = \mathbf{u} + 2\mathbf{w}$
44. $\mathbf{v} = -\mathbf{u} + \mathbf{w}$
45. $\mathbf{v} = \frac{1}{2}(3\mathbf{u} + \mathbf{w})$
46. $\mathbf{v} = \mathbf{u} - 2\mathbf{w}$

In Exercises 47–50, find the magnitude and direction angle of the vector v.

47. $\mathbf{v} = 3(\cos 60^\circ \mathbf{i} + \sin 60^\circ \mathbf{j})$

48. $\mathbf{v} = 8(\cos 135°\mathbf{i} + \sin 135°\mathbf{j})$

49. $\mathbf{v} = 6\mathbf{i} - 6\mathbf{j}$

50. $\mathbf{v} = -5\mathbf{i} + 4\mathbf{j}$

In Exercises 51–58, find the component form of v given its magnitude and the angle it makes with the positive *x*-axis. Sketch v.

	Magnitude	*Angle*
51.	$\|\mathbf{v}\| = 3$	$\theta = 0°$
52.	$\|\mathbf{v}\| = 1$	$\theta = 45°$
53.	$\|\mathbf{v}\| = \frac{7}{2}$	$\theta = 150°$
54.	$\|\mathbf{v}\| = \frac{5}{2}$	$\theta = 45°$
55.	$\|\mathbf{v}\| = 3\sqrt{2}$	$\theta = 150°$
56.	$\|\mathbf{v}\| = 4\sqrt{3}$	$\theta = 90°$
57.	$\|\mathbf{v}\| = 2$	$\mathbf{v}$ in the direction $\mathbf{i} + 3\mathbf{j}$
58.	$\|\mathbf{v}\| = 3$	$\mathbf{v}$ in the direction $3\mathbf{i} + 4\mathbf{j}$

In Exercises 59–62, find the component form of the sum of u and v with direction angles θ_u and θ_v.

	Magnitude	*Angle*
59.	$\|\mathbf{u}\| = 5$	$\theta_u = 0°$
	$\|\mathbf{v}\| = 5$	$\theta_v = 90°$
60.	$\|\mathbf{u}\| = 4$	$\theta_u = 60°$
	$\|\mathbf{v}\| = 4$	$\theta_v = 90°$
61.	$\|\mathbf{u}\| = 20$	$\theta_u = 45°$
	$\|\mathbf{v}\| = 50$	$\theta_v = 180°$
62.	$\|\mathbf{u}\| = 50$	$\theta_u = 30°$
	$\|\mathbf{v}\| = 30$	$\theta_v = 110°$

In Exercises 63–66, use the Law of Cosines to find the angle α between the given vectors. (Assume $0° \le \alpha \le 180°$.)

63. $\mathbf{v} = \mathbf{i} + \mathbf{j}$, $\mathbf{w} = 2\mathbf{i} - 2\mathbf{j}$

64. $\mathbf{v} = 3\mathbf{i} - 2\mathbf{j}$, $\mathbf{w} = 2\mathbf{i} + 2\mathbf{j}$

65. $\mathbf{v} = \mathbf{i} + \mathbf{j}$, $\mathbf{w} = 3\mathbf{i} - \mathbf{j}$

66. $\mathbf{v} = \mathbf{i} + 2\mathbf{j}$, $\mathbf{w} = 2\mathbf{i} - \mathbf{j}$

In Exercises 67 and 68, find the angle between the forces given the magnitude of their resultant. (*Hint*: Write force 1 as a vector in the direction of the positive *x*-axis and force 2 as a vector at an angle θ with the positive *x*-axis.)

	Force 1	*Force 2*	*Resultant Force*
67.	45 pounds	60 pounds	90 pounds
68.	3000 pounds	1000 pounds	3750 pounds

69. *Resultant Force* Forces with magnitudes of 125 newtons and 300 newtons act on a hook (see figure). The angle between the two forces is 45°. Find the direction and magnitude of the resultant of these forces.

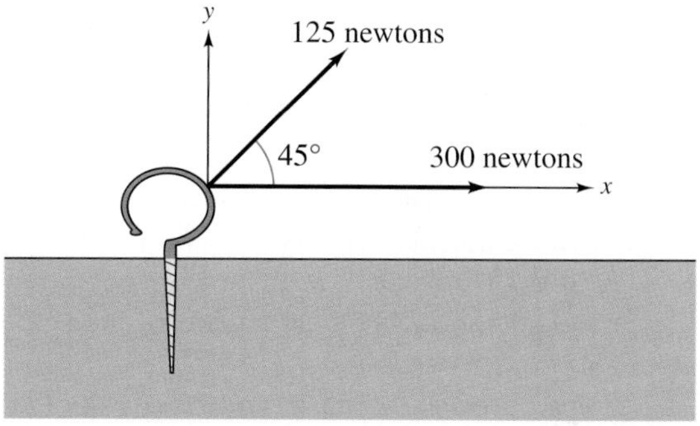

70. *Resultant Force* Forces with magnitudes of 2000 newtons and 900 newtons act on a machine part at angles of 30° and $-45°$, respectively, with the *x*-axis (see figure). Find the direction and magnitude of the resultant of these forces.

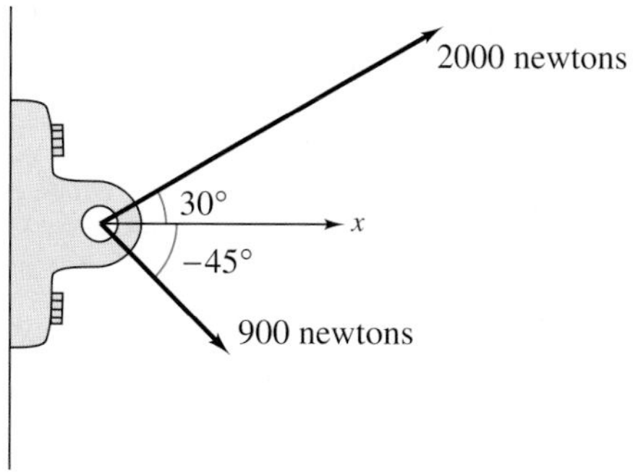

71. *Resultant Force* Three forces with magnitudes of 75 pounds, 100 pounds, and 125 pounds act on an object at angles of 30°, 45°, and 120°, respectively, with the positive *x*-axis. Find the direction and magnitude of the resultant of these forces.

72. *Resultant Force* Three forces with magnitudes of 70 pounds, 40 pounds, and 60 pounds act on an object at angles of $-30°$, 45°, and 135°, respectively, with the positive *x*-axis. Find the direction and magnitude of the resultant of these forces.

73. *Horizontal and Vertical Components of Velocity* A ball is thrown with an initial velocity of 70 feet per second, at an angle of 35° with the horizontal (see figure). Find the vertical and horizontal components of the velocity.

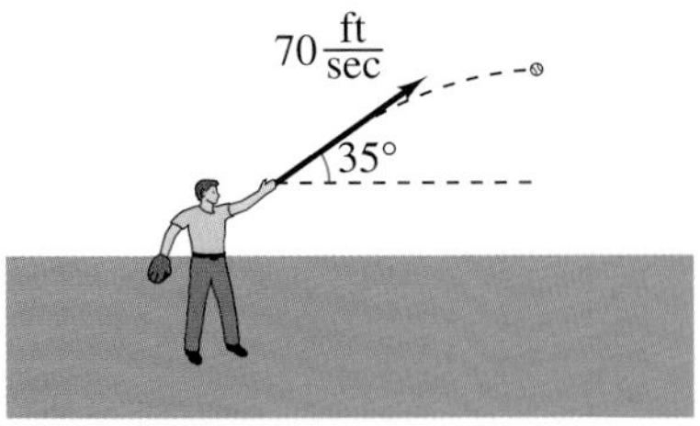

74. *Horizontal and Vertical Components of Velocity* A gun with a muzzle velocity of 1200 feet per second is fired at an angle of 6° with the horizontal. Find the vertical and horizontal components of the velocity.

Cable Tension **In Exercises 75 and 76, use the figure to determine the tension in each cable supporting the load.**

75.

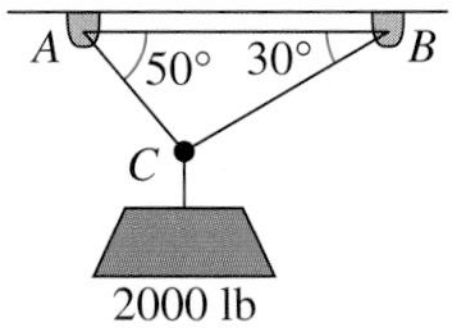

76.

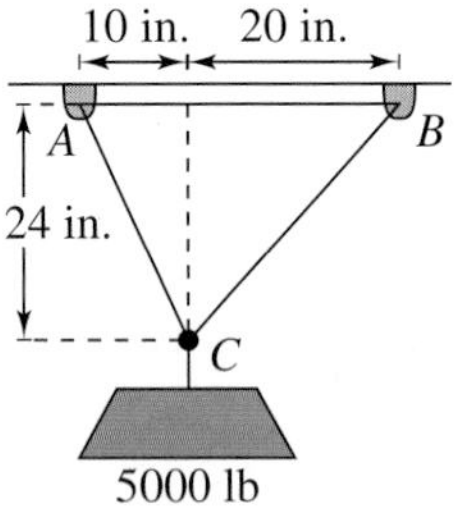

77. *Barge Towing* A loaded barge is being towed by two tugboats, and the magnitude of the resultant is 6000 pounds directed along the axis of the barge (see figure). Find the tension in the tow lines if they each make an 18° angle with the axis of the barge.

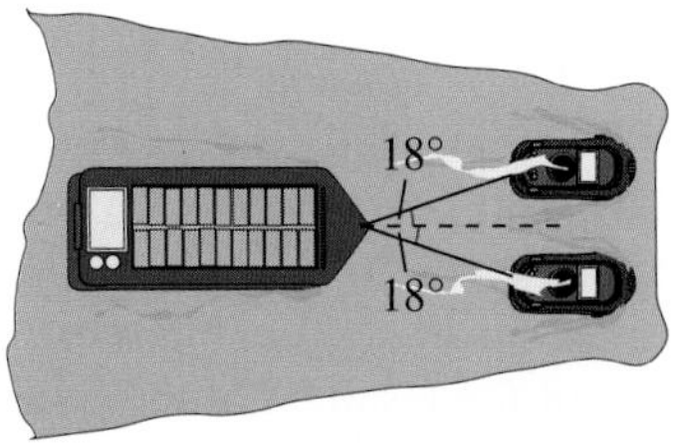

78. *Shared Load* To carry a 100-pound cylindrical weight, two people lift on the ends of short ropes that are tied to an eyelet on the top center of the cylinder. Each rope makes a 20° angle with the vertical. Draw a figure that gives a visual representation of the problem, and find the tension in the ropes.

79. *Navigation* An airplane is flying in the direction S 32° E, with an airspeed of 875 kilometers per hour. Because of the wind, its groundspeed and direction are 800 kilometers per hour and S 40° E, respectively (see figure). Find the direction and speed of the wind.

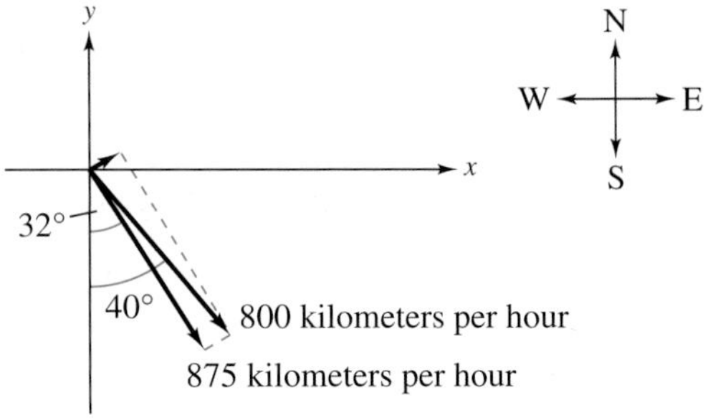

80. *Navigation* An airplane's velocity with respect to the air is 580 miles per hour, and it is heading N 60° W. The wind, at the altitude of the plane, is from the southwest and has a velocity of 60 miles per hour. Draw a figure that gives a visual representation of the problem. What is the true direction of the plane, and what is its speed with respect to the ground?

81. *Work* A heavy implement is pulled 30 feet across a floor, using a force of 100 pounds. Find the work done if the direction of the force is 50° above the horizontal (see figure). (Use the formula for work, $W = FD$, where F is the component of the force in the direction of motion and D is the distance.)

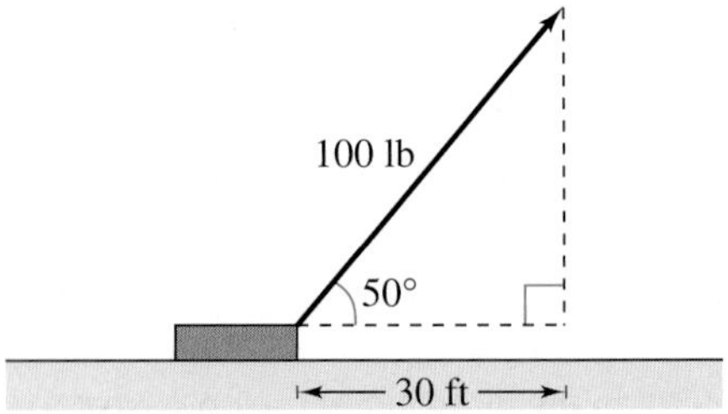

82. *Tetherball* A tetherball weighing 1 pound is pulled outward from the pole by a horizontal force **u** until the rope makes a 45° angle with the pole (see figure). Determine the resulting tension in the rope and the magnitude of **u**.

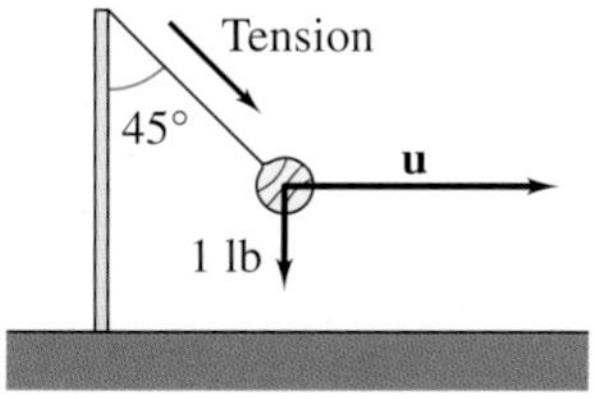

Synthesis

True or False? **In Exercises 83–86, decide whether the statement is true or false. Justify your answer.**

83. If **u** and **v** have the same magnitude and direction, then $\mathbf{u} = \mathbf{v}$.

84. If **u** is a unit vector in the direction of **v**, then $\mathbf{v} = \|\mathbf{v}\|\,\mathbf{u}$.

85. If $\mathbf{v} = a\mathbf{i} + b\mathbf{j} = \mathbf{0}$, then $a = -b$.

86. If $\mathbf{u} = a\mathbf{i} + b\mathbf{j}$ is a unit vector, then $a^2 + b^2 = 1$.

87. *Think About It* Consider two forces of equal magnitude acting on a point.

(a) If the magnitude of the resultant is the sum of the magnitudes of the two forces, make a conjecture about the angle between the forces.

(b) If the resultant of the forces is **0**, make a conjecture about the angle between the forces.

(c) Can the magnitude of the resultant be greater than the sum of the magnitudes of the two forces? Explain.

88. *Graphical Reasoning* Consider two forces

$\mathbf{F}_1 = \langle 10, 0\rangle$ and $\mathbf{F}_2 = 5\langle \cos\theta, \sin\theta\rangle$.

(a) Find $\|\mathbf{F}_1 + \mathbf{F}_2\|$ as a function of θ.

(b) Use a graphing utility to graph the function in part (a) for $0 \le \theta < 2\pi$.

(c) Use the graph in part (b) to determine the range of the function. What is its maximum, and for what value of θ does it occur? What is its minimum, and for what value of θ does it occur?

(d) Explain why the magnitude of the resultant is never 0.

89. Prove that $(\cos\theta)\mathbf{i} + (\sin\theta)\mathbf{j}$ is a unit vector for any value of θ.

90. *Technology* Write a program for your graphing utility that graphs two vectors and their difference given the vectors in component form.

In Exercises 91 and 92, use the program in Exercise 90 to find the difference of the vectors shown in the figure.

91.

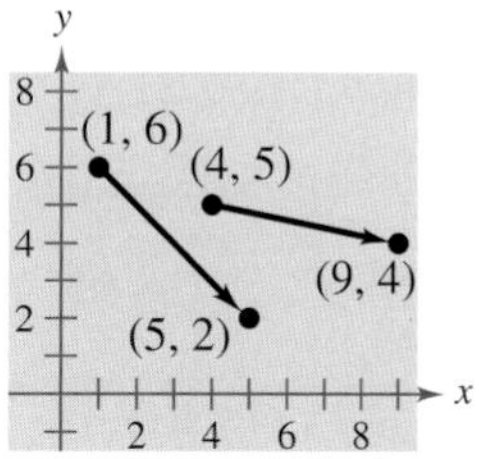

92.

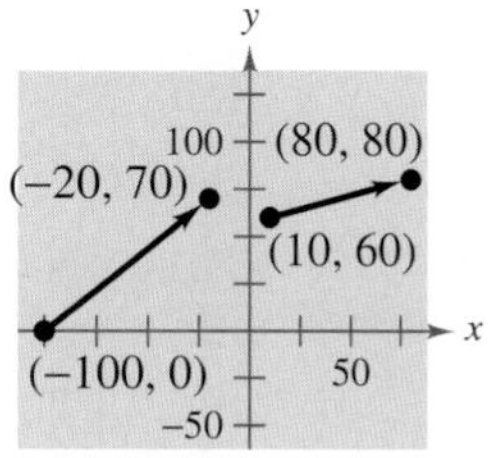

Review

In Exercises 93–96, use the specified trigonometric substitution to write the algebraic expression as a trigonometric function of θ, where $0 < \theta < \pi/2$.

93. $\sqrt{x^2 - 64}, \quad x = 8\sec\theta$

94. $\sqrt{64 - x^2}, \quad x = 8\sin\theta$

95. $\sqrt{x^2 + 36}, \quad x = 6\tan\theta$

96. $\sqrt{(x^2 - 25)^3}, \quad x = 5\sec\theta$

In Exercises 97–102, solve the equation.

97. $\cos x(\cos x + 1) = 0$

98. $\sin x\left(2\sin x + \sqrt{2}\right) = 0$

99. $3\sec x\sin x - 2\sqrt{3}\sin x = 0$

100. $\cos x\csc x + \cos x\sqrt{2} = 0$

101. $(\sin^2 x - 1)\sin^2 x = 0$

102. $(2\cos^2 x - 3)(\cos^2 x - 1) = 0$

6.4 Vectors and Dot Products

▶ **What you should learn**

- How to find the dot product of two vectors and use the Properties of the Dot Product
- How to find the angle between two vectors
- How to determine whether two vectors are orthogonal
- How to write a vector as the sum of two vector components
- How to use vectors to find the work done by a force

▶ **Why you should learn it**

You can use the dot product of two vectors to solve real-life problems involving two vector quantities. For instance, in Exercise 51 on page 531 you use the dot product to find the force necessary to keep a truck from rolling down a hill.

Alan Thornton/Tony Stone Images

The Dot Product of Two Vectors

So far you have studied two vector operations—vector addition and multiplication by a scalar—each of which yields another vector. In this section you will study a third vector operation, the **dot product.** This product yields a scalar, rather than a vector.

Definition of Dot Product

The **dot product** of $\mathbf{u} = \langle u_1, u_2 \rangle$ and $\mathbf{v} = \langle v_1, v_2 \rangle$ is

$$\mathbf{u} \cdot \mathbf{v} = u_1 v_1 + u_2 v_2.$$

Properties of the Dot Product

Let **u**, **v**, and **w** be vectors in the plane or in space and let c be a scalar.

1. $\mathbf{u} \cdot \mathbf{v} = \mathbf{v} \cdot \mathbf{u}$
2. $\mathbf{0} \cdot \mathbf{v} = 0$
3. $\mathbf{u} \cdot (\mathbf{v} + \mathbf{w}) = \mathbf{u} \cdot \mathbf{v} + \mathbf{u} \cdot \mathbf{w}$
4. $\mathbf{v} \cdot \mathbf{v} = \|\mathbf{v}\|^2$
5. $c(\mathbf{u} \cdot \mathbf{v}) = c\mathbf{u} \cdot \mathbf{v} = \mathbf{u} \cdot c\mathbf{v}$

Proofs of Properties 1, 4, and 5 are given in Appendix A.

Example 1 ▶ Finding Dot Products

Find each dot product.

a. $\langle 4, 5 \rangle \cdot \langle 2, 3 \rangle$

b. $\langle 2, -1 \rangle \cdot \langle 1, 2 \rangle$

c. $\langle 0, 3 \rangle \cdot \langle 4, -2 \rangle$

Solution

a. $\langle 4, 5 \rangle \cdot \langle 2, 3 \rangle = 4(2) + 5(3)$
$= 8 + 15$
$= 23$

b. $\langle 2, -1 \rangle \cdot \langle 1, 2 \rangle = 2(1) + (-1)(2) = 2 - 2 = 0$

c. $\langle 0, 3 \rangle \cdot \langle 4, -2 \rangle = 0(4) + 3(-2) = 0 - 6 = -6$

In Example 1, be sure you see that the dot product of two vectors is a scalar (a real number), not a vector. Moreover, notice that the dot product can be positive, zero, or negative.

Example 2 ▶ Using Properties of Dot Products

Let $\mathbf{u} = \langle -1, 3 \rangle$, $\mathbf{v} = \langle 2, -4 \rangle$, and $\mathbf{w} = \langle 1, -2 \rangle$. Find each dot product.

a. $(\mathbf{u} \cdot \mathbf{v})\mathbf{w}$ **b.** $\mathbf{u} \cdot 2\mathbf{v}$

Solution

Begin by finding the dot product of **u** and **v**.

$$\begin{aligned} \mathbf{u} \cdot \mathbf{v} &= \langle -1, 3 \rangle \cdot \langle 2, -4 \rangle \\ &= (-1)(2) + 3(-4) \\ &= -14 \end{aligned}$$

a. $$\begin{aligned} (\mathbf{u} \cdot \mathbf{v})\mathbf{w} &= -14\langle 1, -2 \rangle \\ &= \langle -14, 28 \rangle \end{aligned}$$

b. $$\begin{aligned} \mathbf{u} \cdot 2\mathbf{v} &= 2(\mathbf{u} \cdot \mathbf{v}) \\ &= 2(-14) \\ &= -28 \end{aligned}$$

Notice that the first product is a vector, whereas the second is a scalar. Can you see why?

Example 3 ▶ Dot Product and Magnitude

The dot product of **u** with itself is 5. What is the magnitude of **u**?

Solution

Because $\|\mathbf{u}\|^2 = \mathbf{u} \cdot \mathbf{u}$ and $\mathbf{u} \cdot \mathbf{u} = 5$, it follows that

$$\begin{aligned} \|\mathbf{u}\| &= \sqrt{\mathbf{u} \cdot \mathbf{u}} \\ &= \sqrt{5}. \end{aligned}$$

The Angle Between Two Vectors

The **angle between two nonzero vectors** is the angle θ, $0 \le \theta \le \pi$, between their respective standard position vectors, as shown in Figure 6.30 on page 526. This angle can be found using the dot product. (Note that the angle between the zero vector and another vector is not defined.) A proof is given in Appendix A.

Angle Between Two Vectors

If θ is the angle between two nonzero vectors **u** and **v**, then

$$\cos\theta = \frac{\mathbf{u} \cdot \mathbf{v}}{\|\mathbf{u}\| \, \|\mathbf{v}\|}.$$

Example 4 ▶ Finding the Angle Between Two Vectors

Find the angle between $\mathbf{u} = \langle 4, 3 \rangle$ and $\mathbf{v} = \langle 3, 5 \rangle$.

Solution

$$\cos\theta = \frac{\mathbf{u} \cdot \mathbf{v}}{\|\mathbf{u}\|\ \|\mathbf{v}\|}$$

$$= \frac{\langle 4, 3 \rangle \cdot \langle 3, 5 \rangle}{\|\langle 4, 3 \rangle\|\ \|\langle 3, 5 \rangle\|}$$

$$= \frac{27}{5\sqrt{34}}$$

This implies that the angle between the two vectors is

$$\theta = \arccos\frac{27}{5\sqrt{34}} \approx 22.2°$$

as shown in Figure 6.30.

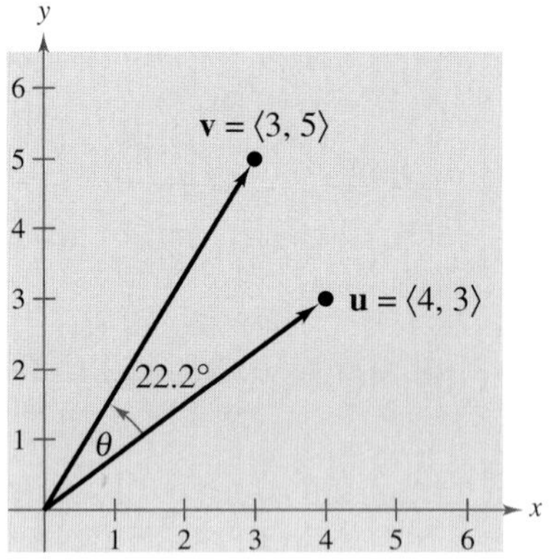

FIGURE 6.30

Rewriting the expression for the angle between two vectors in the form

$\mathbf{u} \cdot \mathbf{v} = \|\mathbf{u}\|\ \|\mathbf{v}\| \cos\theta$ Alternative form of dot product

produces an alternative way to calculate the dot product. From this form, you can see that because $\|\mathbf{u}\|$ and $\|\mathbf{v}\|$ are always positive, $\mathbf{u} \cdot \mathbf{v}$ and $\cos\theta$ will always have the same sign. Figure 6.31 shows the five possible orientations of two vectors.

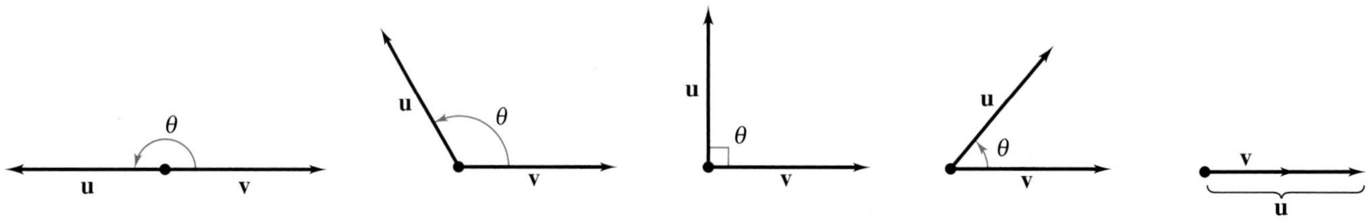

$\theta = \pi$	$\frac{\pi}{2} < \theta < \pi$	$\theta = \frac{\pi}{2}$	$0 < \theta < \frac{\pi}{2}$	$\theta = 0$
$\cos\theta = -1$	$-1 < \cos\theta < 0$	$\cos\theta = 0$	$0 < \cos\theta < 1$	$\cos\theta = 1$
Opposite Direction	*Obtuse Angle*	*90° Angle*	*Acute Angle*	*Same Direction*

FIGURE 6.31

Definition of Orthogonal Vectors

The vectors $\mathbf{u}$ and $\mathbf{v}$ are **orthogonal** if $\mathbf{u} \cdot \mathbf{v} = 0$.

The terms "orthogonal" and "perpendicular" mean essentially the same thing—meeting at right angles. Even though the angle between the zero vector and another vector is not defined, it is convenient to extend the definition of orthogonality to include the zero vector. In other words, the zero vector is orthogonal to every vector $\mathbf{u}$, because $\mathbf{0} \cdot \mathbf{u} = 0$.

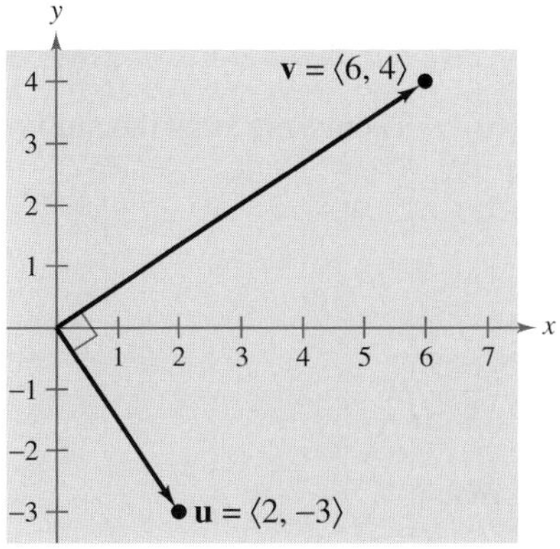

FIGURE 6.32

A computer simulation to accompany this concept appears in the *Interactive* CD-ROM and *Internet* versions of this text.

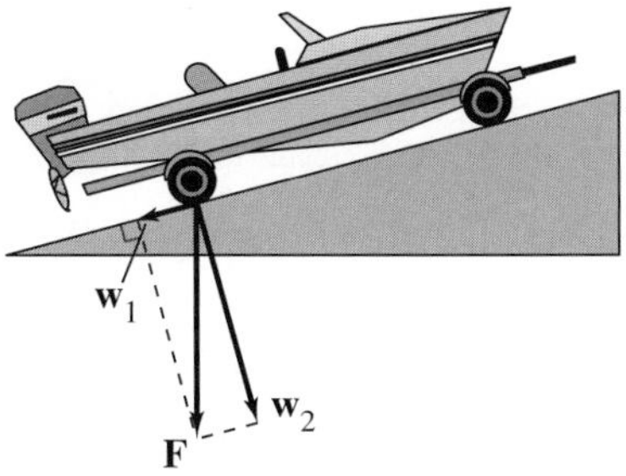

FIGURE 6.33

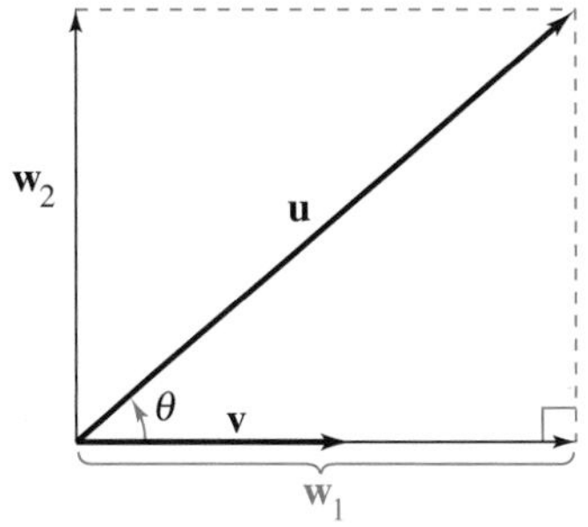

θ is acute.

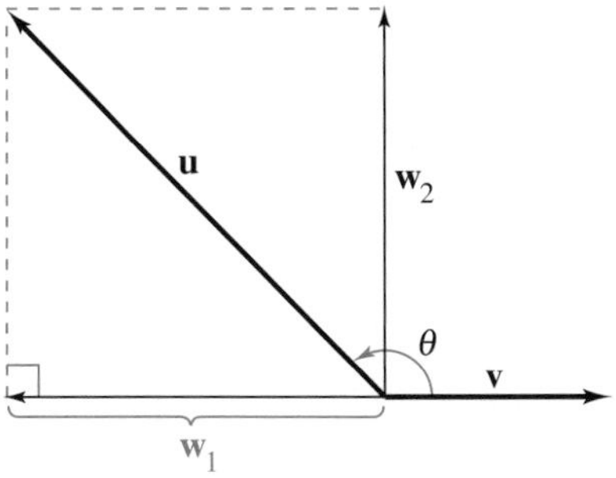

θ is obtuse.

FIGURE 6.34

Example 5 ▶ Determining Orthogonal Vectors

Are the vectors $\mathbf{u} = \langle 2, -3 \rangle$ and $\mathbf{v} = \langle 6, 4 \rangle$ orthogonal?

Solution

Begin by finding the dot product of the two vectors.

$$\mathbf{u} \cdot \mathbf{v} = \langle 2, -3 \rangle \cdot \langle 6, 4 \rangle = 2(6) + (-3)(4) = 0$$

Because the dot product is 0, the two vectors are orthogonal, as shown in Figure 6.32.

Finding Vector Components

You have already seen applications in which two vectors are added to produce a resultant vector. Many applications in physics and engineering pose the reverse problem—decomposing a given vector into the sum of two **vector components.**

Consider a boat on an inclined ramp, as shown in Figure 6.33. The force $\mathbf{F}$ due to gravity pulls the boat *down* the ramp and *against* the ramp. These two orthogonal forces, $\mathbf{w}_1$ and $\mathbf{w}_2$, are vector components of $\mathbf{F}$. That is,

$$\mathbf{F} = \mathbf{w}_1 + \mathbf{w}_2. \qquad \text{Vector components of } \mathbf{F}$$

The negative of component $\mathbf{w}_1$ represents the force needed to keep the boat from rolling down the ramp, whereas $\mathbf{w}_2$ represents the force that the tires must withstand against the ramp. A procedure for finding $\mathbf{w}_1$ and $\mathbf{w}_2$ is shown below.

Definition of Vector Components

Let $\mathbf{u}$ and $\mathbf{v}$ be nonzero vectors such that

$$\mathbf{u} = \mathbf{w}_1 + \mathbf{w}_2$$

where $\mathbf{w}_1$ and $\mathbf{w}_2$ are orthogonal and $\mathbf{w}_1$ is parallel to (or a scalar multiple of) $\mathbf{v}$, as shown in Figure 6.34. The vectors $\mathbf{w}_1$ and $\mathbf{w}_2$ are called vector components of $\mathbf{u}$. The vector $\mathbf{w}_1$ is the **projection** of $\mathbf{u}$ onto $\mathbf{v}$ and is denoted by

$$\mathbf{w}_1 = \text{proj}_{\mathbf{v}}\mathbf{u}.$$

The vector $\mathbf{w}_2$ is given by $\mathbf{w}_2 = \mathbf{u} - \mathbf{w}_1$.

From the definition of vector components, you can see that it is easy to find the component $\mathbf{w}_2$ once you have found the projection of $\mathbf{u}$ onto $\mathbf{v}$. To find the projection, you can use the dot product.

Projection of u onto v

Let $\mathbf{u}$ and $\mathbf{v}$ be nonzero vectors. The projection of $\mathbf{u}$ onto $\mathbf{v}$ is

$$\text{proj}_{\mathbf{v}}\mathbf{u} = \left(\frac{\mathbf{u} \cdot \mathbf{v}}{\|\mathbf{v}\|^2}\right)\mathbf{v}.$$

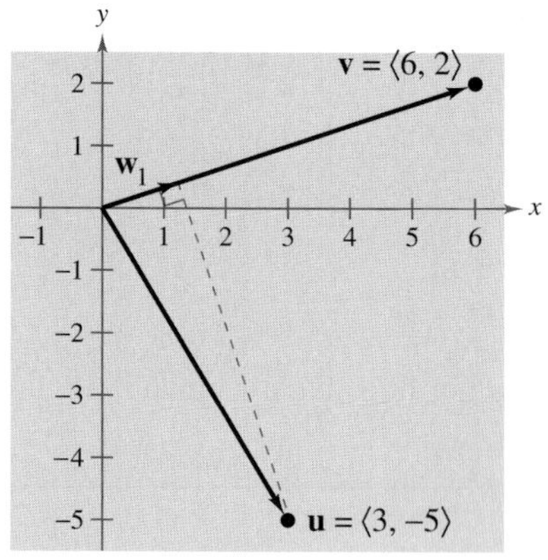

FIGURE 6.35

Example 6 ▶ Decomposing a Vector into Components

Find the projection of $\mathbf{u} = \langle 3, -5 \rangle$ onto $\mathbf{v} = \langle 6, 2 \rangle$. Then write $\mathbf{u}$ as the sum of two orthogonal vectors, one of which is $\text{proj}_{\mathbf{v}}\mathbf{u}$.

Solution

The projection of $\mathbf{u}$ onto $\mathbf{v}$ is

$$\mathbf{w}_1 = \text{proj}_{\mathbf{v}}\mathbf{u} = \left(\frac{\mathbf{u} \cdot \mathbf{v}}{\|\mathbf{v}\|^2}\right)\mathbf{v} = \left(\frac{8}{40}\right)\langle 6, 2 \rangle = \left\langle \frac{6}{5}, \frac{2}{5} \right\rangle$$

as shown in Figure 6.35. The other component, $\mathbf{w}_2$, is

$$\mathbf{w}_2 = \mathbf{u} - \mathbf{w}_1 = \langle 3, -5 \rangle - \left\langle \frac{6}{5}, \frac{2}{5} \right\rangle = \left\langle \frac{9}{5}, -\frac{27}{5} \right\rangle.$$

So,

$$\mathbf{u} = \mathbf{w}_1 + \mathbf{w}_2 = \left\langle \frac{6}{5}, \frac{2}{5} \right\rangle + \left\langle \frac{9}{5}, -\frac{27}{5} \right\rangle = \langle 3, -5 \rangle.$$

Example 7 ▶ Finding a Force

A 600-pound boat sits on a ramp inclined at 30°, as shown in Figure 6.36. What force is required to keep the boat from rolling down the ramp?

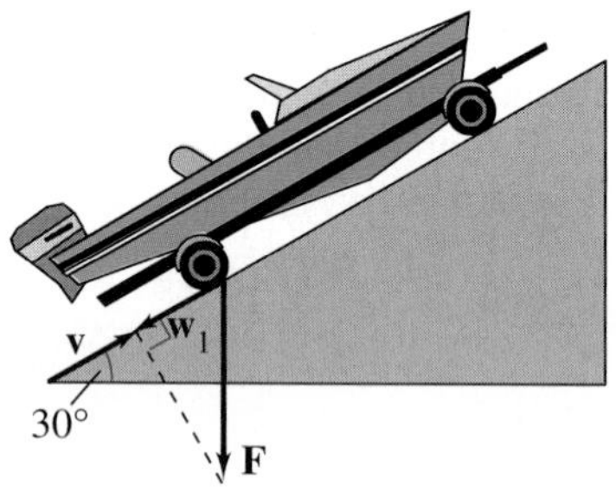

FIGURE 6.36

Solution

Because the force due to gravity is vertical and downward, you can represent the gravitational force by the vector

$$\mathbf{F} = -600\mathbf{j}.$$ Force due to gravity

To find the force required to keep the boat from rolling down the ramp, project $\mathbf{F}$ onto a unit vector $\mathbf{v}$ in the direction of the ramp, as follows.

$$\mathbf{v} = (\cos 30°)\mathbf{i} + (\sin 30°)\mathbf{j} = \frac{\sqrt{3}}{2}\mathbf{i} + \frac{1}{2}\mathbf{j}$$ Unit vector along ramp

Therefore, the projection of $\mathbf{F}$ onto $\mathbf{v}$ is

$$\begin{aligned}
\mathbf{w}_1 &= \text{proj}_{\mathbf{v}}\mathbf{F} \\
&= \left(\frac{\mathbf{F} \cdot \mathbf{v}}{\|\mathbf{v}\|^2}\right)\mathbf{v} \\
&= (\mathbf{F} \cdot \mathbf{v})\mathbf{v} \\
&= (-600)\left(\frac{1}{2}\right)\mathbf{v} \\
&= -300\left(\frac{\sqrt{3}}{2}\mathbf{i} + \frac{1}{2}\mathbf{j}\right).
\end{aligned}$$

The magnitude of this force is 300, and therefore a force of 300 pounds is required to keep the boat from rolling down the ramp.

Work

The work W done by a constant force $\mathbf{F}$ acting along the line of motion of an object is

$$W = (\text{magnitude of force})(\text{distance})$$
$$= \|\mathbf{F}\| \, \|\overrightarrow{PQ}\|$$

as shown in Figure 6.37(a). If the constant force $\mathbf{F}$ is not directed along the line of motion, as shown in Figure 6.37(b), the work W done by the force is

$$W = \|\text{proj}_{\overrightarrow{PQ}} \mathbf{F}\| \, \|\overrightarrow{PQ}\| = (\cos\theta)\|\mathbf{F}\| \, \|\overrightarrow{PQ}\| = \mathbf{F} \cdot \overrightarrow{PQ}.$$

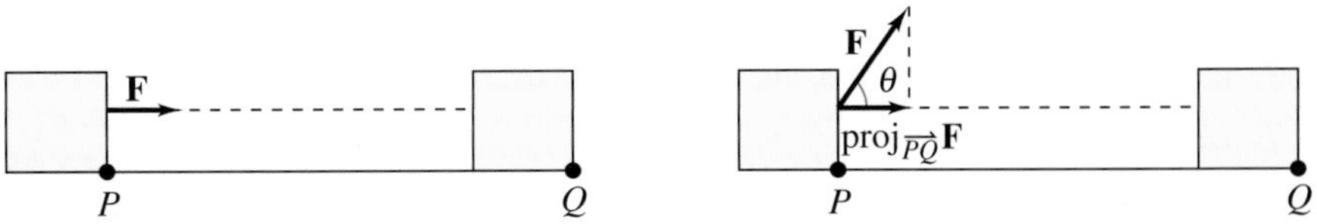

(a) **(b)**

FIGURE 6.37

This notion of work is summarized in the following definition.

Definition of Work

The **work** W done by a constant force $\mathbf{F}$ as its point of application moves along the vector $\overrightarrow{PQ}$ is given by either of the following.

1. $W = \|\text{proj}_{\overrightarrow{PQ}}\mathbf{F}\| \, \|\overrightarrow{PQ}\|$ Projection form
2. $W = \mathbf{F} \cdot \overrightarrow{PQ}$ Dot product form

Example 8 ▶ Finding Work

To close a sliding door, a person pulls on a rope with a constant force of 50 pounds at a constant angle of 60°, as shown in Figure 6.38. Find the work done in moving the door 12 feet to its closed position.

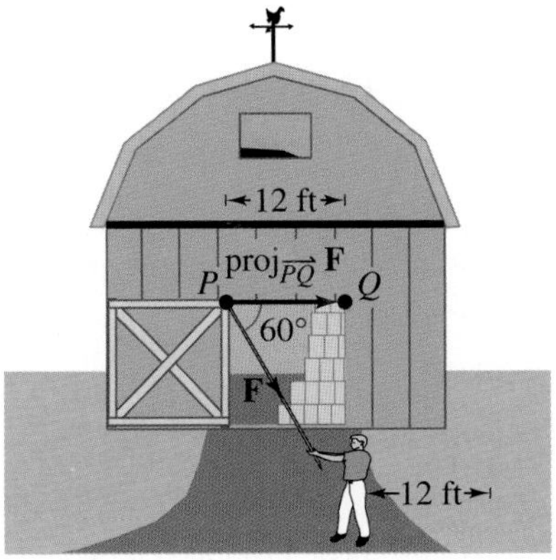

FIGURE 6.38

Solution

Using a projection, you can calculate the work as follows.

$$W = \|\text{proj}_{\overrightarrow{PQ}} \mathbf{F}\| \, \|\overrightarrow{PQ}\|$$
$$= (\cos 60°)\|\mathbf{F}\| \, \|\overrightarrow{PQ}\|$$
$$= \frac{1}{2}(50)(12)$$
$$= 300 \text{ foot-pounds}$$

So, the work done is 300 foot-pounds. You can verify this result by finding the vectors $\mathbf{F}$ and $\overrightarrow{PQ}$ and calculating their dot product.

6.4 Exercises

In Exercises 1–4, find the dot product of u and v.

1. $\mathbf{u} = \langle 6, 1 \rangle$
   $\mathbf{v} = \langle -2, 3 \rangle$
2. $\mathbf{u} = \langle 5, 12 \rangle$
   $\mathbf{v} = \langle -3, 2 \rangle$
3. $\mathbf{u} = 4\mathbf{i} - 2\mathbf{j}$
   $\mathbf{v} = \mathbf{i} - \mathbf{j}$
4. $\mathbf{u} = 3\mathbf{i} + 4\mathbf{j}$
   $\mathbf{v} = 7\mathbf{i} - 2\mathbf{j}$

In Exercises 5–8, use the vectors $\mathbf{u} = \langle 2, 2 \rangle$ and $\mathbf{v} = \langle -3, 4 \rangle$ to find the indicated quantity. State whether the result is a vector or a scalar.

5. $\mathbf{u} \cdot \mathbf{u}$
6. $\|\mathbf{v}\| + 3$
7. $(\mathbf{u} \cdot \mathbf{v})\mathbf{v}$
8. $3\mathbf{u} \cdot \mathbf{v}$

In Exercises 9–14, use the dot product to find the magnitude of u.

9. $\mathbf{u} = \langle -5, 12 \rangle$
10. $\mathbf{u} = \langle 2, -4 \rangle$
11. $\mathbf{u} = 20\mathbf{i} + 25\mathbf{j}$
12. $\mathbf{u} = 12\mathbf{i} - 16\mathbf{j}$
13. $\mathbf{u} = 6\mathbf{j}$
14. $\mathbf{u} = -21\mathbf{i}$

In Exercises 15–24, find the angle θ between the vectors.

15. $\mathbf{u} = \langle 1, 0 \rangle$
    $\mathbf{v} = \langle 0, -2 \rangle$
16. $\mathbf{u} = \langle 3, 2 \rangle$
    $\mathbf{v} = \langle 4, 0 \rangle$
17. $\mathbf{u} = 3\mathbf{i} + 4\mathbf{j}$
    $\mathbf{v} = -2\mathbf{j}$
18. $\mathbf{u} = 2\mathbf{i} - 3\mathbf{j}$
    $\mathbf{v} = \mathbf{i} - 2\mathbf{j}$
19. $\mathbf{u} = 2\mathbf{i} - \mathbf{j}$
    $\mathbf{v} = 6\mathbf{i} + 4\mathbf{j}$
20. $\mathbf{u} = -6\mathbf{i} - 3\mathbf{j}$
    $\mathbf{v} = -8\mathbf{i} + 4\mathbf{j}$
21. $\mathbf{u} = 5\mathbf{i} + 5\mathbf{j}$
    $\mathbf{v} = -6\mathbf{i} + 6\mathbf{j}$
22. $\mathbf{u} = 2\mathbf{i} - 3\mathbf{j}$
    $\mathbf{v} = 4\mathbf{i} + 3\mathbf{j}$
23. $\mathbf{u} = \cos\left(\frac{\pi}{3}\right)\mathbf{i} + \sin\left(\frac{\pi}{3}\right)\mathbf{j}$
    $\mathbf{v} = \cos\left(\frac{3\pi}{4}\right)\mathbf{i} + \sin\left(\frac{3\pi}{4}\right)\mathbf{j}$
24. $\mathbf{u} = \cos\left(\frac{\pi}{4}\right)\mathbf{i} + \sin\left(\frac{\pi}{4}\right)\mathbf{j}$
    $\mathbf{v} = \cos\left(\frac{\pi}{2}\right)\mathbf{i} + \sin\left(\frac{\pi}{2}\right)\mathbf{j}$

In Exercises 25–28, use vectors to find the interior angles of the triangle with the given vertices.

25. $(1, 2), (3, 4), (2, 5)$
26. $(-3, -4), (1, 7), (8, 2)$
27. $(-3, 0), (2, 2), (0, 6)$
28. $(-3, 5), (-1, 9), (7, 9)$

In Exercises 29–32, find $\mathbf{u} \cdot \mathbf{v}$, where θ is the angle between u and v.

29. $\|\mathbf{u}\| = 4, \|\mathbf{v}\| = 10, \theta = \frac{2\pi}{3}$
30. $\|\mathbf{u}\| = 100, \|\mathbf{v}\| = 250, \theta = \frac{\pi}{6}$
31. $\|\mathbf{u}\| = 81, \|\mathbf{v}\| = 64, \theta = \frac{\pi}{4}$
32. $\|\mathbf{u}\| = 9, \|\mathbf{v}\| = 144, \theta = \frac{\pi}{2}$

In Exercises 33–38, determine whether u and v are orthogonal, parallel, or neither.

33. $\mathbf{u} = \langle -12, 30 \rangle$
    $\mathbf{v} = \langle \frac{1}{2}, -\frac{5}{4} \rangle$
34. $\mathbf{u} = \langle 3, 15 \rangle$
    $\mathbf{v} = \langle -1, 5 \rangle$
35. $\mathbf{u} = \frac{1}{4}(3\mathbf{i} - \mathbf{j})$
    $\mathbf{v} = 5\mathbf{i} + 6\mathbf{j}$
36. $\mathbf{u} = \mathbf{i}$
    $\mathbf{v} = -2\mathbf{i} + 2\mathbf{j}$
37. $\mathbf{u} = 2\mathbf{i} - 2\mathbf{j}$
    $\mathbf{v} = -\mathbf{i} - \mathbf{j}$
38. $\mathbf{u} = \langle \cos\theta, \sin\theta \rangle$
    $\mathbf{v} = \langle \sin\theta, -\cos\theta \rangle$

In Exercises 39–42, find the projection of u onto v and the vector component of u orthogonal to v.

39. $\mathbf{u} = \langle 2, 2 \rangle$
    $\mathbf{v} = \langle 6, 1 \rangle$
40. $\mathbf{u} = \langle 4, 2 \rangle$
    $\mathbf{v} = \langle 1, -2 \rangle$
41. $\mathbf{u} = \langle 0, 3 \rangle$
    $\mathbf{v} = \langle 2, 15 \rangle$
42. $\mathbf{u} = \langle -3, -2 \rangle$
    $\mathbf{v} = \langle -4, -1 \rangle$

In Exercises 43–46, find two vectors in opposite directions that are orthogonal to the vector u. (The answers are not unique.)

43. $\mathbf{u} = \langle 3, 5 \rangle$
44. $\mathbf{u} = \langle -8, 3 \rangle$
45. $\mathbf{u} = \frac{1}{2}\mathbf{i} - \frac{2}{3}\mathbf{j}$
46. $\mathbf{u} = -\frac{5}{2}\mathbf{i} - 3\mathbf{j}$

Work **In Exercises 47 and 48, find the work done in moving a particle from P to Q if the magnitude and direction of the force are given by v.**

47. $P = (0, 0), \quad Q = (4, 7), \quad \mathbf{v} = \langle 1, 4 \rangle$
48. $P = (1, 3), \quad Q = (-3, 5), \quad \mathbf{v} = -2\mathbf{i} + 3\mathbf{j}$

49. *Revenue* The vector $\mathbf{u} = \langle 1650, 3200 \rangle$ gives the numbers of units of two products produced by a company. The vector $\mathbf{v} = \langle 15.25, 10.50 \rangle$ gives the price (in dollars) of each unit, respectively. Find the dot product $\mathbf{u} \cdot \mathbf{v}$ and explain what information it gives.

50. *Revenue* Repeat Exercise 49 after increasing the prices by 5%. Identify the vector operation used to increase the prices by 5%.

51. *Braking Load* A truck with a gross weight of 30,000 pounds is parked on a 5° slope (see figure). Assume that the only force to overcome is the force of gravity.

(a) Find the force required to keep the truck from rolling down the hill.

(b) Find the force perpendicular to the hill.

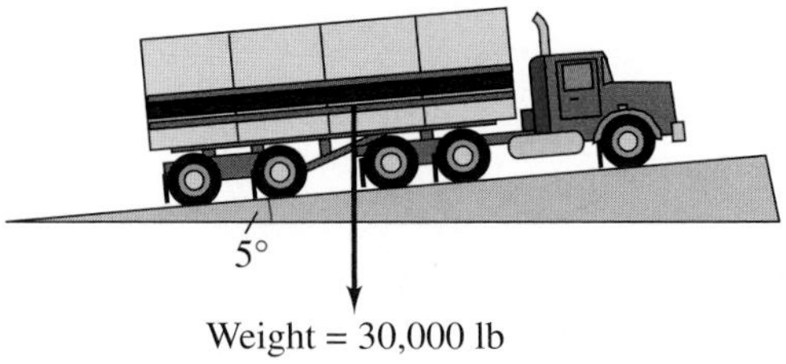

52. *Braking Load* Rework Exercise 51 for a truck that is parked on an 8° slope.

53. *Work* A 25-kilogram (245-newton) bag of sugar is lifted 3 meters. Determine the work done.

54. *Work* Determine the work done by a crane lifting a 2400-pound car 5 feet.

55. *Work* A force of 45 pounds in the direction of 30° above the horizontal is required to slide an implement across a floor (see figure). Find the work done if the implement is dragged 20 feet.

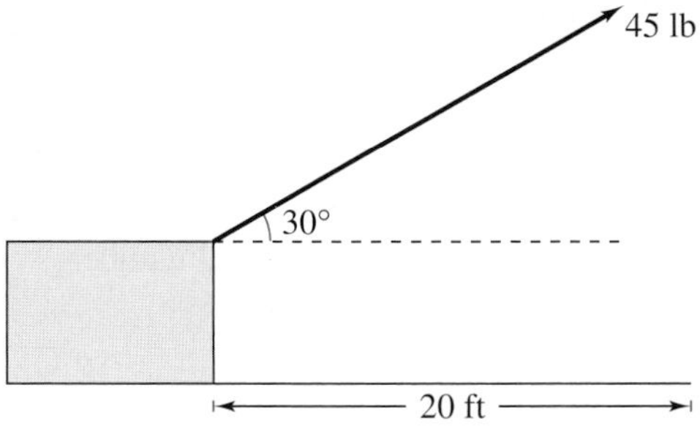

56. *Work* A tractor pulls a log 800 meters and the tension in the cable connecting the tractor and log is approximately 1600 kilograms (15,691 newtons). Approximate the work done if the direction of the force is 35° above the horizontal.

Synthesis

True or False? **In Exercises 57 and 58, determine whether the statement is true or false. Justify your answer.**

57. The work W done by a constant force $\mathbf{F}$ acting along the line of motion of an object is represented by a vector.

58. A sliding door moves along the line of vector $\overrightarrow{PQ}$. If a force is applied to the door along a vector that is orthogonal to $\overrightarrow{PQ}$, then no work is done.

59. *Think About It* What is known about θ, the angle between two nonzero vectors $\mathbf{u}$ and $\mathbf{v}$, under the following conditions?

(a) $\mathbf{u} \cdot \mathbf{v} = 0$ (b) $\mathbf{u} \cdot \mathbf{v} > 0$ (c) $\mathbf{u} \cdot \mathbf{v} < 0$

60. *Think About It* What can be said about the vectors $\mathbf{u}$ and $\mathbf{v}$ under the following conditions?

(a) The projection of $\mathbf{u}$ onto $\mathbf{v}$ equals $\mathbf{u}$.

(b) The projection of $\mathbf{u}$ onto $\mathbf{v}$ equals $\mathbf{0}$.

61. Use vectors to prove that the diagonals of a rhombus are perpendicular.

62. Prove the following.

$$\|\mathbf{u} - \mathbf{v}\|^2 = \|\mathbf{u}\|^2 + \|\mathbf{v}\|^2 - 2\mathbf{u} \cdot \mathbf{v}$$

63. Prove the following Properties of the Dot Product.

(a) $\mathbf{0} \cdot \mathbf{v} = 0$

(b) $\mathbf{u} \cdot (\mathbf{v} + \mathbf{w}) = \mathbf{u} \cdot \mathbf{v} + \mathbf{u} \cdot \mathbf{w}$

(c) $c(\mathbf{u} \cdot \mathbf{v}) = \mathbf{u} \cdot c\mathbf{v}$

64. Prove that if $\mathbf{u}$ is orthogonal to $\mathbf{v}$ and $\mathbf{w}$, then $\mathbf{u}$ is orthogonal to $c\mathbf{v} + d\mathbf{w}$ for any scalars c and d.

Review

In Exercises 65–68, find the exact solutions of the equation in the interval $[0, 2\pi)$.

65. $\sin 2x - \sqrt{3} \sin x = 0$ **66.** $\sin 2x + \sqrt{2} \cos x = 0$

67. $2 \tan x = \tan 2x$ **68.** $\cos 2x - 3 \sin x = 2$

In Exercises 69–72, find the exact value of the trigonometric function given that $\sin u = -\frac{12}{13}$ and $\cos v = \frac{24}{25}$. (Both u and v are in Quadrant IV.)

69. $\sin(u - v)$ **70.** $\sin(u + v)$

71. $\cos(v - u)$ **72.** $\tan(u - v)$

6.5 Trigonometric Form of a Complex Number

▶ What you should learn

- How to plot complex numbers in the complex plane
- How to write the trigonometric forms of complex numbers
- How to multiply and divide complex numbers written in trigonometric form
- How to use DeMoivre's Theorem to find powers of complex numbers
- How to find nth roots of complex numbers

▶ Why you should learn it

You can use the trigonometric form of a complex number to perform operations with complex numbers. For instance, in Exercises 107–114 on page 542 you use the trigonometric forms of complex numbers to help you solve polynomial equations.

The Complex Plane

Just as real numbers can be represented by points on the real number line, you can represent a complex number

$$z = a + bi$$

as the point (a, b) in a coordinate plane (the **complex plane**). The horizontal axis is called the **real axis** and the vertical axis is called the **imaginary axis,** as shown in Figure 6.39.

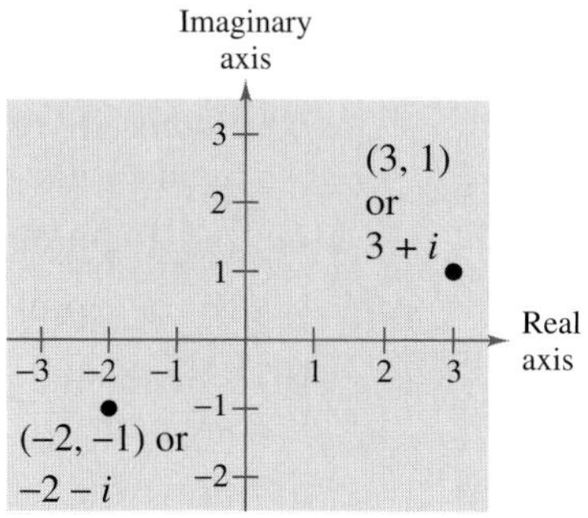

FIGURE 6.39

The **absolute value** of the complex number $a + bi$ is defined as the distance between the origin $(0, 0)$ and the point (a, b).

Definition of the Absolute Value of a Complex Number

The **absolute value** of the complex number $z = a + bi$ is

$$|a + bi| = \sqrt{a^2 + b^2}.$$

If the complex number $a + bi$ is a real number (that is, if $b = 0$), then this definition agrees with that given for the absolute value of a real number

$$|a + 0i| = \sqrt{a^2 + 0^2} = |a|.$$

Example 1 ▶ Finding the Absolute Value of a Complex Number

Plot $z = -2 + 5i$ and find its absolute value.

Solution

The number is plotted in Figure 6.40. It has an absolute value of

$$|z| = \sqrt{(-2)^2 + 5^2}$$
$$= \sqrt{29}.$$

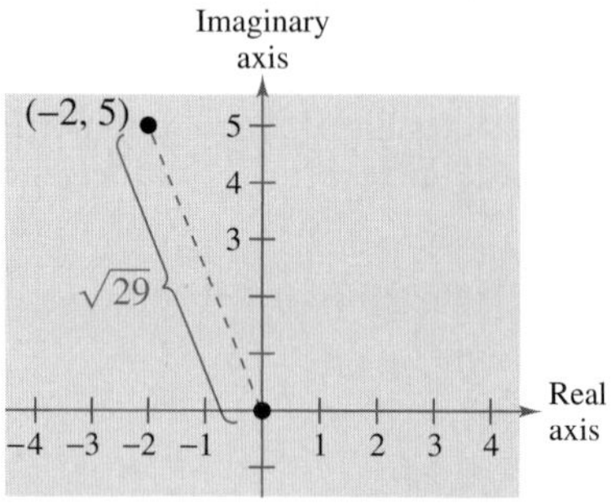

FIGURE 6.40

Trigonometric Form of a Complex Number

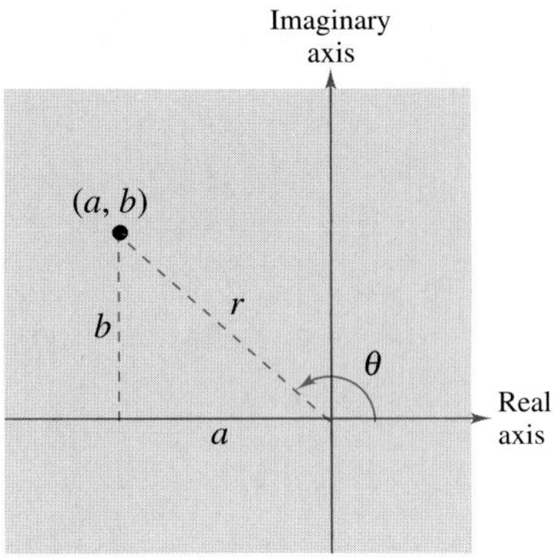

FIGURE 6.41

In Section 2.4 you learned how to add, subtract, multiply, and divide complex numbers. To work effectively with *powers* and *roots* of complex numbers, it is helpful to write complex numbers in trigonometric form. In Figure 6.41, consider the nonzero complex number $a + bi$. By letting θ be the angle from the positive real axis (measured counterclockwise) to the line segment connecting the origin and the point (a, b), you can write

$$a = r\cos\theta \qquad \text{and} \qquad b = r\sin\theta$$

where $r = \sqrt{a^2 + b^2}$. Consequently, you have

$$a + bi = (r\cos\theta) + (r\sin\theta)i$$

from which you can obtain the **trigonometric form of a complex number.**

Trigonometric Form of a Complex Number

The **trigonometric form** of the complex number $z = a + bi$ is

$$z = r(\cos\theta + i\sin\theta)$$

where $a = r\cos\theta$, $b = r\sin\theta$, $r = \sqrt{a^2 + b^2}$, and $\tan\theta = b/a$. The number r is the **modulus** of z, and θ is called an **argument** of z.

The trigonometric form of a complex number is also called the *polar form*. Because there are infinitely many choices for θ, the trigonometric form of a complex number is not unique. Normally, θ is restricted to the interval $0 \le \theta < 2\pi$, although on occasion it is convenient to use $\theta < 0$.

Example 2 ▶ Writing a Complex Number in Trigonometric Form

Write the complex number $z = -2 - 2\sqrt{3}i$ in trigonometric form.

Solution

The absolute value of z is

$$r = \left|-2 - 2\sqrt{3}i\right| = \sqrt{(-2)^2 + \left(-2\sqrt{3}\right)^2} = \sqrt{16} = 4$$

and the angle θ is

$$\tan\theta = \frac{b}{a} = \frac{-2\sqrt{3}}{-2} = \sqrt{3}.$$

Because $\tan(\pi/3) = \sqrt{3}$ and $z = -2 - 2\sqrt{3}i$ lies in Quadrant III, you choose θ to be $\theta = \pi + \pi/3 = 4\pi/3$. So, the trigonometric form is

$$z = r(\cos\theta + i\sin\theta)$$
$$= 4\left(\cos\frac{4\pi}{3} + i\sin\frac{4\pi}{3}\right).$$

See Figure 6.42.

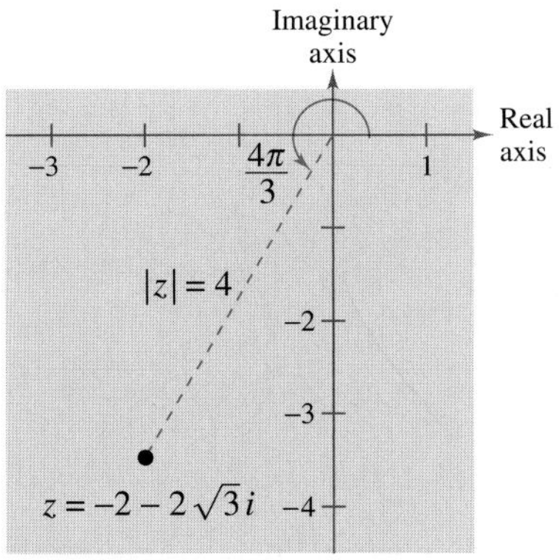

FIGURE 6.42

Example 3 ▶ Writing a Complex Number in Standard Form

Write the complex number in standard form $a + bi$.

$$z = \sqrt{8}\left[\cos\left(-\frac{\pi}{3}\right) + i\sin\left(-\frac{\pi}{3}\right)\right]$$

Technology

A graphing utility can be used to convert a complex number in trigonometric (or polar) form to rectangular form, and vice versa. For specific keystrokes, see the user's manual for your graphing utility.

Solution

Because $\cos(-\pi/3) = \frac{1}{2}$ and $\sin(-\pi/3) = -\sqrt{3}/2$, you can write

$$\begin{aligned} z &= \sqrt{8}\left[\cos\left(-\frac{\pi}{3}\right) + i\sin\left(-\frac{\pi}{3}\right)\right] \\ &= 2\sqrt{2}\left(\frac{1}{2} - \frac{\sqrt{3}}{2}i\right) \\ &= \sqrt{2} - \sqrt{6}i. \end{aligned}$$

Multiplication and Division of Complex Numbers

The trigonometric form adapts nicely to multiplication and division of complex numbers. Suppose you are given two complex numbers

$$z_1 = r_1(\cos\theta_1 + i\sin\theta_1) \qquad \text{and} \qquad z_2 = r_2(\cos\theta_2 + i\sin\theta_2).$$

The product of z_1 and z_2 is

$$\begin{aligned} z_1z_2 &= r_1r_2(\cos\theta_1 + i\sin\theta_1)(\cos\theta_2 + i\sin\theta_2) \\ &= r_1r_2[(\cos\theta_1\cos\theta_2 - \sin\theta_1\sin\theta_2) + i(\sin\theta_1\cos\theta_2 + \cos\theta_1\sin\theta_2)]. \end{aligned}$$

Using the sum and difference formulas for cosine and sine, you can rewrite this equation as

$$z_1z_2 = r_1r_2[\cos(\theta_1 + \theta_2) + i\sin(\theta_1 + \theta_2)].$$

This establishes the first part of the following rule. The second part is left to you (see Exercise 119).

Product and Quotient of Two Complex Numbers

Let $z_1 = r_1(\cos\theta_1 + i\sin\theta_1)$ and $z_2 = r_2(\cos\theta_2 + i\sin\theta_2)$ be complex numbers.

$$z_1z_2 = r_1r_2[\cos(\theta_1 + \theta_2) + i\sin(\theta_1 + \theta_2)] \qquad \text{Product}$$

$$\frac{z_1}{z_2} = \frac{r_1}{r_2}[\cos(\theta_1 - \theta_2) + i\sin(\theta_1 - \theta_2)], \quad z_2 \neq 0 \qquad \text{Quotient}$$

Note that this rule says that to *multiply* two complex numbers you multiply moduli and add arguments, whereas to *divide* two complex numbers you divide moduli and subtract arguments.

Example 4 ▶ Multiplying Complex Numbers

Find the product of the complex numbers.

$$z_1 = 2\left(\cos\frac{2\pi}{3} + i\sin\frac{2\pi}{3}\right) \qquad z_2 = 8\left(\cos\frac{11\pi}{6} + i\sin\frac{11\pi}{6}\right)$$

Solution

$$\begin{aligned}
z_1 z_2 &= 2\left(\cos\frac{2\pi}{3} + i\sin\frac{2\pi}{3}\right) \times 8\left(\cos\frac{11\pi}{6} + i\sin\frac{11\pi}{6}\right)\\
&= 16\left[\cos\left(\frac{2\pi}{3} + \frac{11\pi}{6}\right) + i\sin\left(\frac{2\pi}{3} + \frac{11\pi}{6}\right)\right]\\
&= 16\left(\cos\frac{5\pi}{2} + i\sin\frac{5\pi}{2}\right)\\
&= 16\left(\cos\frac{\pi}{2} + i\sin\frac{\pi}{2}\right)\\
&= 16[0 + i(1)]\\
&= 16i
\end{aligned}$$

You can check this result by first converting the complex numbers to the standard forms $z_1 = -1 + \sqrt{3}i$ and $z_2 = 4\sqrt{3} - 4i$ and then multiplying algebraically, as in Section 2.4.

$$\begin{aligned}
z_1 z_2 &= \left(-1 + \sqrt{3}i\right)\left(4\sqrt{3} - 4i\right)\\
&= -4\sqrt{3} + 4i + 12i + 4\sqrt{3}\\
&= 16i
\end{aligned}$$

Example 5 ▶ Dividing Complex Numbers

Find the quotient, z_1/z_2, of the complex numbers.

$$z_1 = 24(\cos 300^\circ + i\sin 300^\circ) \qquad z_2 = 8(\cos 75^\circ + i\sin 75^\circ)$$

Solution

$$\begin{aligned}
\frac{z_1}{z_2} &= \frac{24(\cos 300^\circ + i\sin 300^\circ)}{8(\cos 75^\circ + i\sin 75^\circ)}\\
&= \frac{24}{8}[\cos(300^\circ - 75^\circ) + i\sin(300^\circ - 75^\circ)]\\
&= 3(\cos 225^\circ + i\sin 225^\circ)\\
&= 3\left[\left(-\frac{\sqrt{2}}{2}\right) + i\left(-\frac{\sqrt{2}}{2}\right)\right]\\
&= -\frac{3\sqrt{2}}{2} - \frac{3\sqrt{2}}{2}i
\end{aligned}$$

Powers of Complex Numbers

To raise a complex number to a power, consider repeated use of the multiplication rule.

$$z = r(\cos\theta + i\sin\theta)$$
$$z^2 = r(\cos\theta + i\sin\theta)r(\cos\theta + i\sin\theta) = r^2(\cos 2\theta + i\sin 2\theta)$$
$$z^3 = r^2(\cos 2\theta + i\sin 2\theta)r(\cos\theta + i\sin\theta) = r^3(\cos 3\theta + i\sin 3\theta)$$
$$z^4 = r^4(\cos 4\theta + i\sin 4\theta)$$
$$z^5 = r^5(\cos 5\theta + i\sin 5\theta)$$
$$\vdots$$

This pattern leads to the following important theorem, which is named after the French mathematician Abraham DeMoivre (1667–1754).

Historical Note
Abraham DeMoivre (1667–1754) is remembered for his work in probability theory and DeMoivre's Theorem. His *The Doctrine of Chances* (published in 1718) includes the theory of recurring series and the theory of partial fractions.

DeMoivre's Theorem

If $z = r(\cos\theta + i\sin\theta)$ is a complex number and n is a positive integer, then

$$z^n = [r(\cos\theta + i\sin\theta)]^n$$
$$= r^n(\cos n\theta + i\sin n\theta).$$

Example 6 ▶ Finding Powers of a Complex Number

Use DeMoivre's Theorem to find $\left(-1 + \sqrt{3}i\right)^{12}$.

Solution

First convert the complex number to trigonometric form using

$$r = \sqrt{(-1)^2 + \left(\sqrt{3}\right)^2} = 2 \text{ and } \theta = \arctan\frac{\sqrt{3}}{-1} = \frac{2\pi}{3}.$$

$$-1 + \sqrt{3}i = 2\left(\cos\frac{2\pi}{3} + i\sin\frac{2\pi}{3}\right)$$

Then, by DeMoivre's Theorem, you have

$$(-1 + \sqrt{3}i)^{12} = \left[2\left(\cos\frac{2\pi}{3} + i\sin\frac{2\pi}{3}\right)\right]^{12}$$
$$= 2^{12}\left[\cos(12)\frac{2\pi}{3} + i\sin(12)\frac{2\pi}{3}\right]$$
$$= 4096(\cos 8\pi + i\sin 8\pi)$$
$$= 4096(1 + 0)$$
$$= 4096.$$

Roots of Complex Numbers

Recall that a consequence of the Fundamental Theorem of Algebra is that a polynomial equation of degree n has n solutions in the complex number system. So, the equation $x^6 = 1$ has six solutions, and in this particular case you can find the six solutions by factoring and using the Quadratic Formula.

$$\begin{aligned} x^6 - 1 &= (x^3 - 1)(x^3 + 1) \\ &= (x - 1)(x^2 + x + 1)(x + 1)(x^2 - x + 1) = 0 \end{aligned}$$

Consequently, the solutions are

$$x = \pm 1, \qquad x = \frac{-1 \pm \sqrt{3}i}{2}, \qquad \text{and} \qquad x = \frac{1 \pm \sqrt{3}i}{2}.$$

Each of these numbers is a sixth root of 1. In general, the ***n*th root** of a complex number is defined as follows.

Definition of *n*th Root of a Complex Number

The complex number $u = a + bi$ is an ***n*th root** of the complex number z if

$$z = u^n = (a + bi)^n.$$

Exploration

The nth roots of a complex number are useful for solving some polynomial equations. For instance, explain how you can use DeMoivre's Theorem to solve the polynomial equation

$$x^4 + 16 = 0.$$

[*Hint:* Write -16 as $16(\cos \pi + i \sin \pi)$.]

To find a formula for an nth root of a complex number, let u be an nth root of z, where

$$u = s(\cos \beta + i \sin \beta)$$

and

$$z = r(\cos \theta + i \sin \theta).$$

By DeMoivre's Theorem and the fact that $u^n = z$, you have

$$s^n (\cos n\beta + i \sin n\beta) = r(\cos \theta + i \sin \theta).$$

Taking the absolute values of both sides of this equation, it follows that $s^n = r$. Substituting back into the previous equation and dividing by r, you get

$$\cos n\beta + i \sin n\beta = \cos \theta + i \sin \theta.$$

So, it follows that

$$\cos n\beta = \cos \theta$$

and

$$\sin n\beta = \sin \theta.$$

Because both sine and cosine have a period of 2π, these last two equations have solutions if and only if the angles differ by a multiple of 2π. Consequently, there must exist an integer k such that

$$n\beta = \theta + 2\pi k$$

$$\beta = \frac{\theta + 2\pi k}{n}.$$

By substituting this value of β into the trigonometric form of u, you get the result stated on the following page.

*n*th Roots of a Complex Number

For a positive integer n, the complex number $z = r(\cos\theta + i\sin\theta)$ has exactly n distinct nth roots given by

$$\sqrt[n]{r}\left(\cos\frac{\theta + 2\pi k}{n} + i\sin\frac{\theta + 2\pi k}{n}\right)$$

where $k = 0, 1, 2, \ldots, n - 1$.

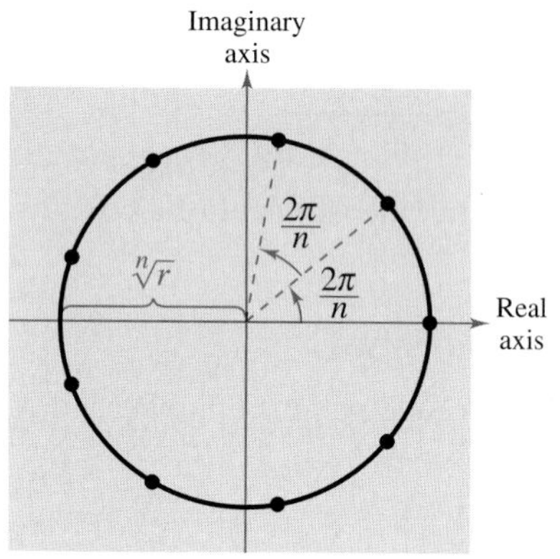

FIGURE 6.43

When k exceeds $n - 1$, the roots begin to repeat. For instance, if $k = n$, the angle

$$\frac{\theta + 2\pi n}{n} = \frac{\theta}{n} + 2\pi$$

is coterminal with θ/n, which is also obtained when $k = 0$.

The formula for the nth roots of a complex number z has a nice geometrical interpretation, as shown in Figure 6.43. Note that because the nth roots of z all have the same magnitude $\sqrt[n]{r}$, they all lie on a circle of radius $\sqrt[n]{r}$ with center at the origin. Furthermore, because successive nth roots have arguments that differ by $2\pi/n$, the n roots are equally spaced around the circle.

You have already found the sixth roots of 1 by factoring and by using the Quadratic Formula. Example 7 shows how you can solve the same problem with the formula for nth roots.

Example 7 ▶ Finding the *n*th Roots of a Real Number

Find all the sixth roots of 1.

Solution

First write 1 in the trigonometric form $1 = 1(\cos 0 + i\sin 0)$. Then, by the nth root formula, with $n = 6$ and $r = 1$, the roots have the form

$$\sqrt[6]{1}\left(\cos\frac{0 + 2\pi k}{6} + i\sin\frac{0 + 2\pi k}{6}\right)$$

or simply $\cos(\pi k/3) + i\sin(\pi k/3)$. So, for $k = 0, 1, 2, 3, 4,$ and 5, the sixth roots are as follows. (See Figure 6.44.)

$$\cos 0 + i\sin 0 = 1$$

$$\cos\frac{\pi}{3} + i\sin\frac{\pi}{3} = \frac{1}{2} + \frac{\sqrt{3}}{2}i \qquad \text{Increment by } \frac{2\pi}{n} = \frac{2\pi}{6} = \frac{\pi}{3}$$

$$\cos\frac{2\pi}{3} + i\sin\frac{2\pi}{3} = -\frac{1}{2} + \frac{\sqrt{3}}{2}i$$

$$\cos\pi + i\sin\pi = -1$$

$$\cos\frac{4\pi}{3} + i\sin\frac{4\pi}{3} = -\frac{1}{2} - \frac{\sqrt{3}}{2}i$$

$$\cos\frac{5\pi}{3} + i\sin\frac{5\pi}{3} = \frac{1}{2} - \frac{\sqrt{3}}{2}i$$

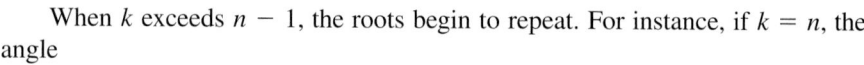

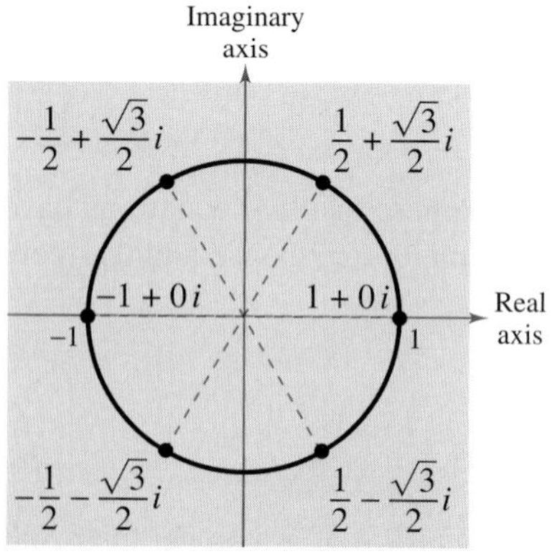

FIGURE 6.44

In Figure 6.44, notice that the roots obtained in Example 7 all have a magnitude of 1 and are equally spaced around the unit circle. Also notice that the complex roots occur in conjugate pairs, as discussed in Section 2.5. The n distinct nth roots of 1 are called the ***n*th roots of unity.**

Example 8 ▶ Finding the nth Roots of a Complex Number

Find the three cube roots of

$$z = -2 + 2i.$$

Solution

Because z lies in Quadrant II, the trigonometric form for z is

$$z = -2 + 2i$$
$$= \sqrt{8}\,(\cos 135° + i \sin 135°).$$

By the formula for nth roots, the cube roots have the form

$$\sqrt[6]{8}\left(\cos \frac{135° + 360°k}{3} + i \sin \frac{135° + 360°k}{3}\right).$$

Finally, for $k = 0$, 1, and 2, you obtain the roots

$$\sqrt[6]{8}\left(\cos \frac{135° + 360°(0)}{3} + i \sin \frac{135° + 360°(0)}{3}\right) = \sqrt{2}(\cos 45° + i \sin 45°)$$
$$= 1 + i$$

$$\sqrt[6]{8}\left(\cos \frac{135° + 360°(1)}{3} + i \sin \frac{135° + 360°(1)}{3}\right) = \sqrt{2}(\cos 165° + i \sin 165°)$$
$$\approx -1.3660 + 0.3660i$$

$$\sqrt[6]{8}\left(\cos \frac{135° + 360°(2)}{3} + i \sin \frac{135° + 360°(2)}{3}\right) = \sqrt{2}(\cos 285° + i \sin 285°)$$
$$\approx 0.3660 - 1.3660i.$$

Writing ABOUT MATHEMATICS

A Famous Mathematical Formula The famous formula

$$e^{a+bi} = e^a(\cos b + i \sin b)$$

is called Euler's Formula, after the German mathematician Leonhard Euler (1707–1783). Although the interpretation of this formula is beyond the scope of this text, we decided to include it because it gives rise to one of the most wonderful equations in mathematics.

$$e^{\pi i} + 1 = 0$$

This elegant equation relates the five most famous numbers in mathematics—0, 1, π, e, and i—in a single equation. Show how Euler's Formula can be used to derive this equation.

Activities

1. Use DeMoivre's Theorem to find $(-2 - 2\sqrt{3}i)^3$.

   Answer: $64(\cos 4\pi + i \sin 4\pi) = 64$

2. Find the cube roots of

   $8\left(-\frac{1}{2} + \frac{\sqrt{3}}{2}i\right) = 8\left(\cos \frac{2\pi}{3} + i \sin \frac{2\pi}{3}\right).$

   Answer: $2\left(\cos \frac{2\pi}{9} + i \sin \frac{2\pi}{9}\right)$

   $2\left(\cos \frac{8\pi}{9} + i \sin \frac{8\pi}{9}\right)$

   $2\left(\cos \frac{14\pi}{9} + i \sin \frac{14\pi}{9}\right)$

3. Find all of the solutions of the equation $x^4 + 1 = 0$.

   Answer: $\cos \frac{\pi}{4} + i \sin \frac{\pi}{4}$

   $\cos \frac{3\pi}{4} + i \sin \frac{3\pi}{4}$

   $\cos \frac{5\pi}{4} + i \sin \frac{5\pi}{4}$

   $\cos \frac{7\pi}{4} + i \sin \frac{7\pi}{4}$

6.5 Exercises

In Exercises 1–6, plot the complex number and find its absolute value.

1. $-7i$
2. -7
3. $-4 + 4i$
4. $5 - 12i$
5. $6 - 7i$
6. $-8 + 3i$

In Exercises 7–10, write the complex number in trigonometric form.

7. 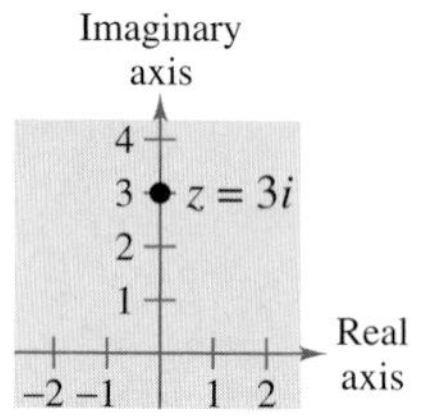

8. 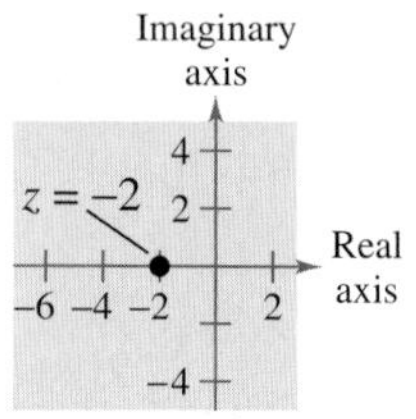

9. 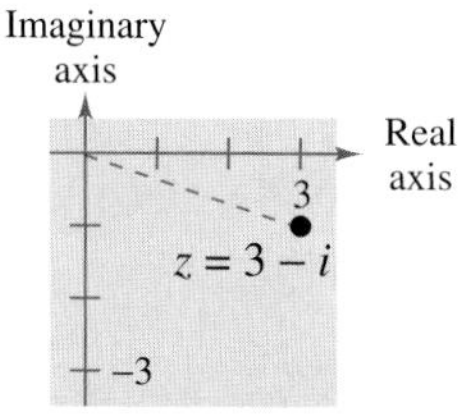

10. 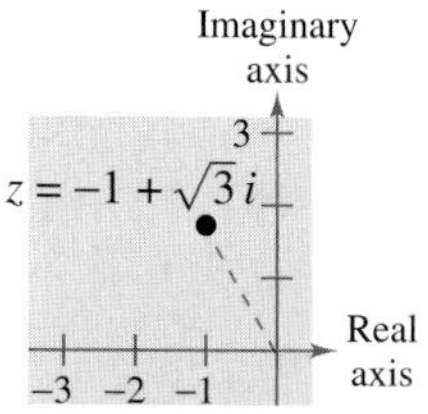

In Exercises 11–26, represent the complex number graphically, and find the trigonometric form of the number.

11. $3 - 3i$
12. $2 + 2i$
13. $\sqrt{3} + i$
14. $4 - 4\sqrt{3}i$
15. $-2(1 + \sqrt{3}i)$
16. $\frac{5}{2}(\sqrt{3} - i)$
17. $-5i$
18. $4i$
19. $-7 + 4i$
20. $3 - i$
21. 7
22. 4
23. $3 + \sqrt{3}i$
24. $2\sqrt{2} - i$
25. $-3 - i$
26. $1 + 3i$

In Exercises 27–34, use a graphing utility to represent the complex number in trigonometric form.

27. $5 + 2i$
28. $8 + 3i$
29. $-3 + i$
30. $-5 - i$
31. $3\sqrt{2} - 7i$
32. $4\sqrt{5} - 4i$
33. $-8 - 5\sqrt{3}i$
34. $-9 - 2\sqrt{10}i$

In Exercises 35–44, represent the complex number graphically, and find the standard form of the number.

35. $3(\cos 120° + i \sin 120°)$
36. $5(\cos 135° + i \sin 135°)$
37. $\frac{3}{2}(\cos 300° + i \sin 300°)$
38. $\frac{1}{4}(\cos 225° + i \sin 225°)$
39. $3.75\left(\cos \frac{3\pi}{4} + i \sin \frac{3\pi}{4}\right)$
40. $6\left(\cos \frac{5\pi}{12} + i \sin \frac{5\pi}{12}\right)$
41. $8\left(\cos \frac{\pi}{2} + i \sin \frac{\pi}{2}\right)$
42. $7(\cos 0 + i \sin 0)$
43. $3[\cos(18° 45') + i \sin(18° 45')]$
44. $6[\cos(230° 30') + i \sin(230° 30')]$

In Exercises 45–48, use a graphing utility to represent the complex number in standard form.

45. $5\left(\cos \frac{\pi}{9} + i \sin \frac{\pi}{9}\right)$
46. $10\left(\cos \frac{2\pi}{5} + i \sin \frac{2\pi}{5}\right)$
47. $3(\cos 165.5° + i \sin 165.5°)$
48. $9(\cos 58° + i \sin 58°)$

In Exercises 49–60, perform the operation and leave the result in trigonometric form.

49. $\left[2\left(\cos \frac{\pi}{4} + i \sin \frac{\pi}{4}\right)\right]\left[6\left(\cos \frac{\pi}{12} + i \sin \frac{\pi}{12}\right)\right]$
50. $\left[\frac{3}{4}\left(\cos \frac{\pi}{3} + i \sin \frac{\pi}{3}\right)\right]\left[4\left(\cos \frac{3\pi}{4} + i \sin \frac{3\pi}{4}\right)\right]$
51. $\left[\frac{5}{3}(\cos 140° + i \sin 140°)\right]\left[\frac{2}{3}(\cos 60° + i \sin 60°)\right]$
52. $[0.5(\cos 100° + i \sin 100°)] \times$
$[0.8(\cos 300° + i \sin 300°)]$
53. $[0.45(\cos 310° + i \sin 310°)] \times$
$[0.60(\cos 200° + i \sin 200°)]$
54. $(\cos 5° + i \sin 5°)(\cos 20° + i \sin 20°)$

55. $\dfrac{\cos 50° + i \sin 50°}{\cos 20° + i \sin 20°}$

56. $\dfrac{2(\cos 120° + i \sin 120°)}{4(\cos 40° + i \sin 40°)}$

57. $\dfrac{\cos(5\pi/3) + i \sin(5\pi/3)}{\cos \pi + i \sin \pi}$

58. $\dfrac{5(\cos 4.3 + i \sin 4.3)}{4(\cos 2.1 + i \sin 2.1)}$

59. $\dfrac{12(\cos 52° + i \sin 52°)}{3(\cos 110° + i \sin 110°)}$

60. $\dfrac{6(\cos 40° + i \sin 40°)}{7(\cos 100° + i \sin 100°)}$

In Exercises 61–68, (a) give the trigonometric form of the complex numbers, (b) perform the indicated operation using the trigonometric form, and (c) perform the indicated operation using the standard form, and check your result with that of part (b).

61. $(2 + 2i)(1 - i)$

62. $(\sqrt{3} + i)(1 + i)$

63. $-2i(1 + i)$

64. $4(1 - \sqrt{3}i)$

65. $\dfrac{3 + 4i}{1 - \sqrt{3}i}$

66. $\dfrac{1 + \sqrt{3}i}{6 - 3i}$

67. $\dfrac{5}{2 + 3i}$

68. $\dfrac{4i}{-4 + 2i}$

In Exercises 69–72, sketch the graphs of all complex numbers z satisfying the given condition.

69. $|z| = 2$

70. $|z| = 3$

71. $\theta = \dfrac{\pi}{6}$

72. $\theta = \dfrac{5\pi}{4}$

In Exercises 73–90, use DeMoivre's Theorem to find the indicated power of the complex number. Express the result in standard form.

73. $(1 + i)^5$

74. $(2 + 2i)^6$

75. $(-1 + i)^{10}$

76. $(3 - 2i)^8$

77. $2(\sqrt{3} + i)^7$

78. $4(1 - \sqrt{3}i)^3$

79. $[5(\cos 20° + i \sin 20°)]^3$

80. $[3(\cos 150° + i \sin 150°)]^4$

81. $\left(\cos \dfrac{\pi}{4} + i \sin \dfrac{\pi}{4}\right)^{12}$

82. $\left[2\left(\cos \dfrac{\pi}{2} + i \sin \dfrac{\pi}{2}\right)\right]^8$

83. $[5(\cos 3.2 + i \sin 3.2)]^4$

84. $(\cos 0 + i \sin 0)^{20}$

85. $(3 - 2i)^5$

86. $(\sqrt{5} - 4i)^3$

87. $[3(\cos 15° + i \sin 15°)]^4$

88. $[2(\cos 10° + i \sin 10°)]^8$

89. $\left[2\left(\cos \dfrac{\pi}{10} + i \sin \dfrac{\pi}{10}\right)\right]^5$

90. $\left[2\left(\cos \dfrac{\pi}{8} + i \sin \dfrac{\pi}{8}\right)\right]^6$

In Exercises 91–106, (a) use the formula on page 538 to find the indicated roots of the complex number, (b) represent each of the roots graphically, and (c) express each of the roots in standard form.

91. Square roots of $5(\cos 120° + i \sin 120°)$

92. Square roots of $16(\cos 60° + i \sin 60°)$

93. Cube roots of $8\left(\cos \dfrac{2\pi}{3} + i \sin \dfrac{2\pi}{3}\right)$

94. Fifth roots of $32\left(\cos \dfrac{5\pi}{6} + i \sin \dfrac{5\pi}{6}\right)$

95. Square roots of $-25i$

96. Fourth roots of $625i$

97. Cube roots of $-\frac{125}{2}(1 + \sqrt{3}i)$

98. Cube roots of $-4\sqrt{2}(1 - i)$

99. Fourth roots of 16

100. Fourth roots of i

101. Fifth roots of 1

102. Cube roots of 1000

103. Cube roots of -125

104. Fourth roots of -4

105. Fifth roots of $128(-1 + i)$

106. Sixth roots of $64i$

In Exercises 107–114, use the formula on page 538 to find all the solutions of the equation and represent the solutions graphically.

107. $x^4 - i = 0$

108. $x^3 + 1 = 0$

109. $x^5 + 243 = 0$

110. $x^3 - 27 = 0$

111. $x^4 + 16i = 0$

112. $x^6 - 64i = 0$

113. $x^3 - (1 - i) = 0$

114. $x^4 + (1 + i) = 0$

Synthesis

True or False? **In Exercises 115–118, determine whether the statement is true or false. Justify your answer.**

115. Although the square of the complex number bi is given by $(bi)^2 = -b^2$, the absolute value of the complex number $z = a + bi$ is defined as

$$|a + bi| = \sqrt{a^2 + b^2}.$$

116. Geometrically, the nth roots of any complex number z are all equally spaced around the unit circle centered at the origin.

117. The product of two complex numbers, $z_1 = r_1(\cos \theta_1 + i \sin \theta_1)$ and $z_2 = r_2(\cos \theta_2 + i \sin \theta_2)$, is zero only when $r_1 = 0$ and/or $r_2 = 0$.

118. By DeMoivre's Theorem,

$$\left(4 + \sqrt{6}i\right)^8 = \cos(32) + i \sin\left(8\sqrt{6}\right).$$

119. Given two complex numbers $z_1 = r_1(\cos \theta_1 + i \sin \theta_1)$ and $z_2 = r_2(\cos \theta_2 + i \sin \theta_2)$, $z_2 \neq 0$, show that

$$\frac{z_1}{z_2} = \frac{r_1}{r_2}[\cos(\theta_1 - \theta_2) + i \sin(\theta_1 - \theta_2)].$$

120. Show that $\bar{z} = r[\cos(-\theta) + i \sin(-\theta)]$ is the complex conjugate of $z = r(\cos \theta + i \sin \theta)$.

121. Use the trigonometric forms of z and $\bar{z}$ in Exercise 120 to find (a) $z\bar{z}$ and (b) $z/\bar{z}$, $\bar{z} \neq 0$.

122. Show that the negative of $z = r(\cos \theta + i \sin \theta)$ is $-z = r[\cos(\theta + \pi) + i \sin(\theta + \pi)]$.

123. Show that $-\frac{1}{2}\left(1 + \sqrt{3}i\right)$ is a sixth root of 1.

124. Show that $2^{-1/4}(1 - i)$ is a fourth root of -2.

Graphical Reasoning **In Exercises 125 and 126, use the graph of the roots of a complex number.**

(a) Write each of the roots in trigonometric form.

(b) Identify the complex number whose roots are given.

(c) Use a graphing utility to verify the results of part (b).

125.

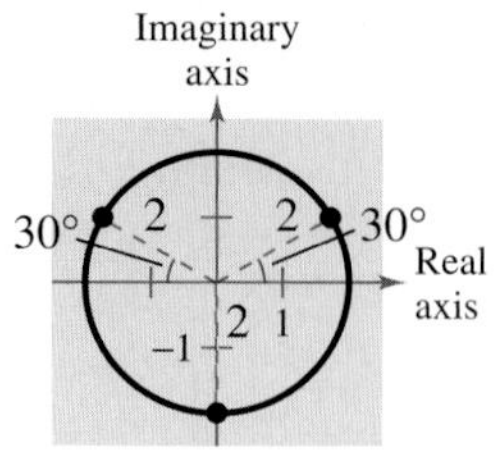

126.

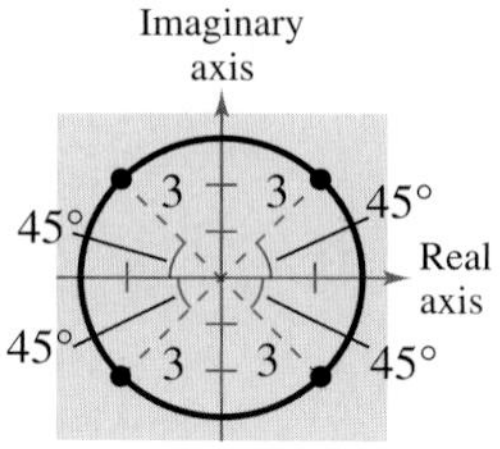

Review

In Exercises 127–132, solve the right triangle shown in the figure. Approximate the result to two decimal places.

127. $A = 22°$, $a = 8$

128. $B = 66°$, $a = 33.5$

129. $A = 30°$, $b = 112.6$

130. $B = 6°$, $b = 211.2$

131. $A = 42° 15'$, $c = 11.2$

132. $B = 81° 30'$, $c = 6.8$

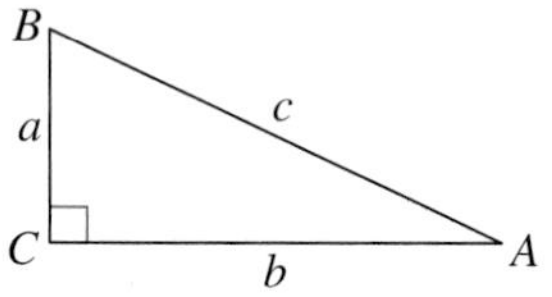

In Exercises 133–136, for the simple harmonic motion described by the trigonometric function, find the maximum displacement and the lowest possible value of t for which $d = 0$.

133. $d = 16 \cos \dfrac{\pi}{4}t$

134. $d = \dfrac{1}{8} \cos 12\pi t$

135. $d = \frac{1}{16} \sin \frac{5}{4}\pi t$

136. $d = \frac{1}{12} \sin 60\pi t$

Chapter Summary

What did you learn?

	Review Exercises
Section 6.1	
☐ How to use the Law of Sines to solve oblique triangles (AAS, ASA, or SSA)	1–12
☐ How to find the areas of oblique triangles	13–16
☐ How to use the Law of Sines to model and solve real-life problems	17–19
Section 6.2	
☐ How to use the Law of Cosines to solve oblique triangles (SSS or SAS)	20–27
☐ How to use the Law of Cosines to model and solve real-life problems	28, 29
☐ How to use Heron's Area Formula to find the area of a triangle	30–33
Section 6.3	
☐ How to represent vectors as directed line segments	34–37
☐ How to write the component forms of vectors	38–43
☐ How to perform basic vector operations and represent them graphically	44–47
☐ How to write vectors as linear combinations of unit vectors	48–53
☐ How to find the direction angles of vectors	54–59
☐ How to use vectors to model and solve real-life problems	60–62
Section 6.4	
☐ How to find the dot product of two vectors and use the Properties of the Dot Product	63–70
☐ How to find the angle between two vectors	71–74
☐ How to determine whether two vectors are orthogonal	75–78
☐ How to write a vector as the sum of two vector components	79–82
☐ How to use vectors to find the work done by a force	83, 84
Section 6.5	
☐ How to plot complex numbers in the complex plane	85–88
☐ How to write the trigonometric forms of complex numbers	89–92
☐ How to multiply and divide complex numbers written in trigonometric form	93, 94
☐ How to use DeMoivre's Theorem to find powers of complex numbers	95–98
☐ How to find *n*th roots of complex numbers	99–106

Review Exercises

6.1 In Exercises 1–12, use the Law of Sines to solve (if possible) the triangle. If two solutions exist, list both. Round your answers to two decimal places.

1.

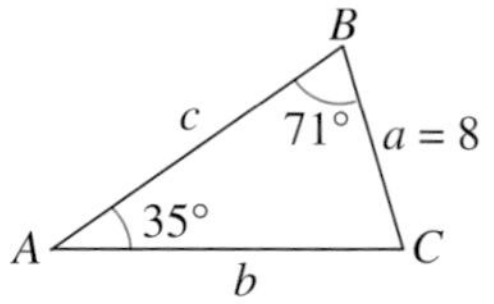

2.

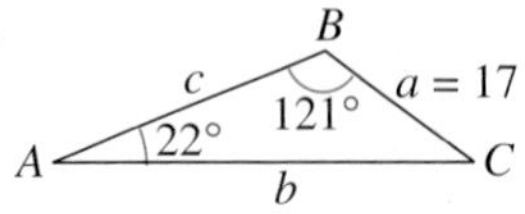

3. $B = 72°,\ C = 82°,\ b = 54$

4. $B = 10°,\ C = 20°,\ c = 33$

5. $A = 16°,\ B = 98°,\ c = 8.4$

6. $A = 95°,\ B = 45°,\ c = 104.8$

7. $A = 24°,\ C = 48°,\ b = 27.5$

8. $B = 64°,\ C = 36°,\ a = 367$

9. $B = 150°,\ b = 30,\ c = 10$

10. $B = 150°,\ a = 10,\ b = 3$

11. $A = 75°,\ a = 51.2,\ b = 33.7$

12. $B = 25°,\ a = 6.2,\ b = 4$

In Exercises 13–16, use the information to find the area of the triangle.

13. $A = 27°,\ b = 5,\ c = 7$

14. $B = 80°,\ a = 4,\ c = 8$

15. $C = 123°,\ a = 16,\ b = 5$

16. $A = 11°,\ b = 22,\ c = 21$

17. *Height* From a certain distance, the angle of elevation to the top of a building is 17°. At a point 50 meters closer to the building, the angle of elevation is 31° (see figure). Approximate the height of the building.

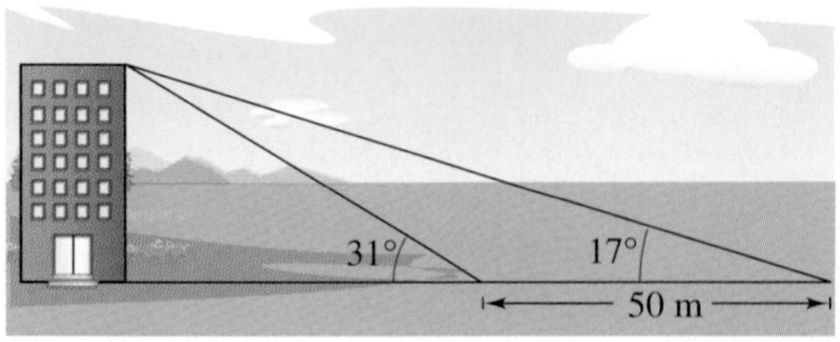

18. *Geometry* Find the length of the side w of the parallelogram.

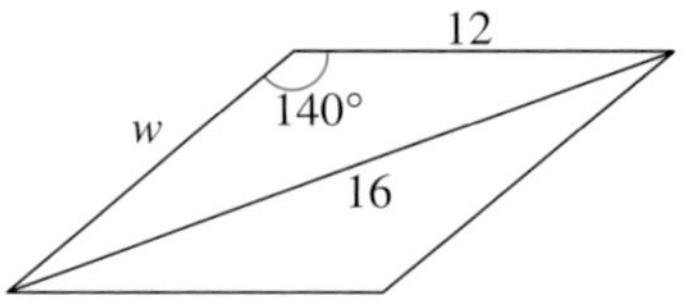

FIGURE FOR 18

19. *Height of a Tree* Find the height of a tree that stands on a hillside of slope 28°(from the horizontal) if from a point 75 feet down the hill the angle of elevation to the top of the tree is 45° (see figure).

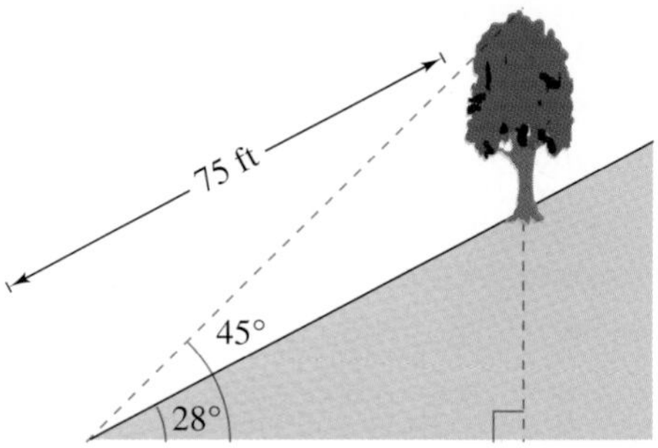

6.2 In Exercises 20–27, use the Law of Cosines to solve the triangle.

20. $a = 5,\ b = 8,\ c = 10$

21. $a = 80,\ b = 60,\ c = 100$

22. $a = 2.5,\ b = 5.0,\ c = 4.5$

23. $a = 16.4,\ b = 8.8,\ c = 12.2$

24. $B = 110°,\ a = 4,\ c = 4$

25. $B = 150°,\ a = 10,\ c = 20$

26. $C = 43°,\ a = 22.5,\ b = 31.4$

27. $A = 62°,\ b = 11.34,\ c = 19.52$

28. *Surveying* To approximate the length of a marsh, a surveyor walks 425 meters from point A to point B. Then the surveyor turns 65° and walks 300 meters to point C. Approximate the length AC of the marsh (see figure).

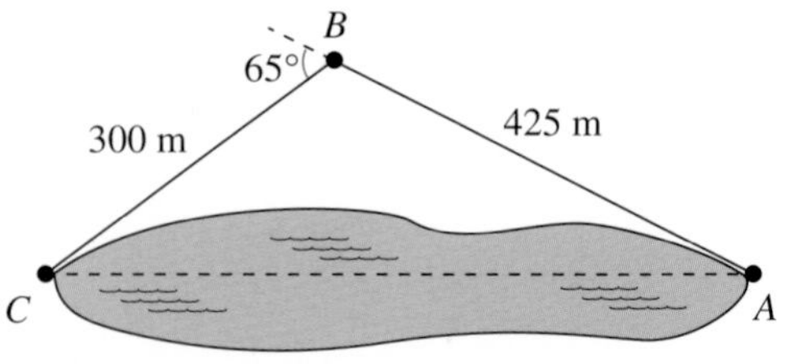

29. *Navigation* Two planes leave an airport at approximately the same time. One is flying 425 miles per hour at a bearing of N 5° W, and the other is flying 530 miles per hour at a bearing of N 67° E. Draw a figure that gives a visual representation of the problem and determine the distance between the planes after they have flown for 2 hours.

In Exercises 30–33, use Heron's Area Formula to find the area of the triangle.

30. $a = 4,\ b = 5,\ c = 7$

31. $a = 15,\ b = 8,\ c = 10$

32. $a = 12.3,\ b = 15.8,\ c = 3.7$

33. $a = 38.1,\ b = 26.7,\ c = 19.4$

6.3 In Exercises 34–37, graph the vector with the specified initial point and terminal point.

	Initial Point	*Terminal Point*
34.	(0, 0)	(8, 7)
35.	(3, 4)	(−5, −7)
36.	(−3, 9)	(8, −4)
37.	(−6, −8)	(8, 3)

In Exercises 38–43, find the component form of the vector v satisfying the conditions.

38.

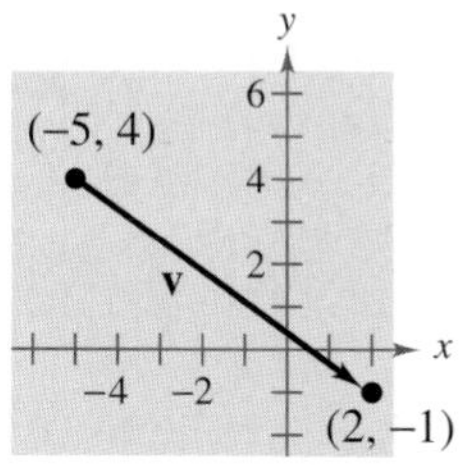

39.

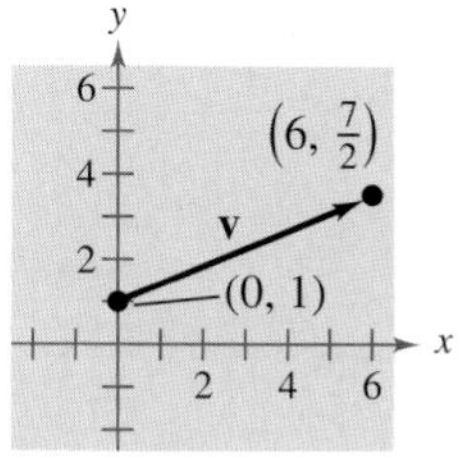

40. Initial point: (0, 10); Terminal point: (7, 3)

41. Initial point: (1, 5); Terminal point: (15, 9)

42. $\|\mathbf{v}\| = 8, \quad \theta = 120°$

43. $\|\mathbf{v}\| = \frac{1}{2}, \quad \theta = 225°$

In Exercises 44–47, find the component form of the specified vector given that u = 6i − 5j and v = 10i + 3j. Then sketch your result.

44. $2\mathbf{u} + \mathbf{v}$

45. $4\mathbf{u} - 5\mathbf{v}$

46. $3\mathbf{v}$

47. $\frac{1}{2}\mathbf{v}$

In Exercises 48–51, write vector u as a linear combination of the standard unit vectors i and j.

48. $\mathbf{u} = \langle -3, 4 \rangle$

49. $\mathbf{u} = \langle -6, -8 \rangle$

50. **u** has initial point (3, 4) and terminal point (9, 8).

51. **u** has initial point (−2, 7) and terminal point (5, −9).

In Exercises 52 and 53, write the vector v in the form $\|\mathbf{v}\|(\cos\theta\mathbf{i} + \sin\theta\mathbf{j})$.

52. $\mathbf{v} = -10\mathbf{i} + 10\mathbf{j}$

53. $\mathbf{v} = 4\mathbf{i} - \mathbf{j}$

In Exercises 54–59, find the magnitude and the direction angle of the vector v.

54. $\mathbf{v} = 7(\cos 60°\mathbf{i} + \sin 60°\mathbf{j})$

55. $\mathbf{v} = 3(\cos 150°\mathbf{i} + \sin 150°\mathbf{j})$

56. $\mathbf{v} = 5\mathbf{i} + 4\mathbf{j}$

57. $\mathbf{v} = -4\mathbf{i} + 7\mathbf{j}$

58. $\mathbf{v} = -3\mathbf{i} - 3\mathbf{j}$

59. $\mathbf{v} = 8\mathbf{i} - \mathbf{j}$

60. *Resultant Force* Forces of 85 pounds and 50 pounds act on a single point. The angle between the forces is 15°. Describe the resultant force.

61. *Rope Tension* A 180-pound weight is supported by two ropes, as shown in the figure. Find the tension exerted on each rope.

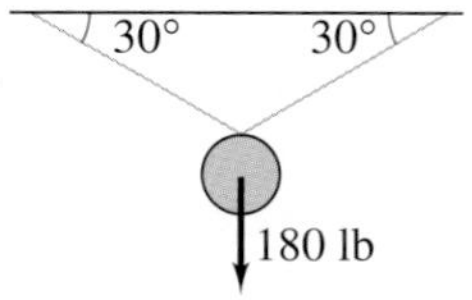

62. *Navigation* An airplane has an airspeed of 724 kilometers per hour at a bearing of N 30° E. If the wind velocity is 32 kilometers per hour from the west, find the groundspeed and the direction of the plane.

6.4 In Exercises 63–66, find the dot product of u and v.

63. $\mathbf{u} = \langle 6, 7 \rangle$
$\mathbf{v} = \langle -3, 9 \rangle$

64. $\mathbf{u} = \langle -7, 12 \rangle$
$\mathbf{v} = \langle -4, -14 \rangle$

65. $\mathbf{u} = 3\mathbf{i} + 7\mathbf{j}$
$\mathbf{v} = 11\mathbf{i} - 5\mathbf{j}$

66. $\mathbf{u} = -7\mathbf{i} + 2\mathbf{j}$
$\mathbf{v} = 16\mathbf{i} - 12\mathbf{j}$

In Exercises 67–70, use the vectors $\mathbf{u} = \langle -3, 4 \rangle$ and $\mathbf{v} = \langle 2, 1 \rangle$ to find the indicated quantity. State whether the result is a vector or a scalar.

67. $2\mathbf{u} \cdot \mathbf{u}$

68. $\|\mathbf{v}\|^2$

69. $\mathbf{u}(\mathbf{u} \cdot \mathbf{v})$

70. $3\mathbf{u} \cdot \mathbf{v}$

In Exercises 71–74, find the angle between u and v.

71. $\mathbf{u} = \cos \dfrac{7\pi}{4}\mathbf{i} + \sin \dfrac{7\pi}{4}\mathbf{j}$
$\mathbf{v} = \cos \dfrac{5\pi}{6}\mathbf{i} + \sin \dfrac{5\pi}{6}\mathbf{j}$

72. $\mathbf{u} = \cos 45°\mathbf{i} + \sin 45°\mathbf{j}$
$\mathbf{v} = \cos 300°\mathbf{i} + \sin 300°\mathbf{j}$

73. $\mathbf{u} = \langle 2\sqrt{2}, -4 \rangle, \quad \mathbf{v} = \langle -\sqrt{2}, 1 \rangle$

74. $\mathbf{u} = \langle 3, \sqrt{3} \rangle, \quad \mathbf{v} = \langle 4, 3\sqrt{3} \rangle$

In Exercises 75–78, determine whether u and v are orthogonal, parallel, or neither.

75. $\mathbf{u} = \langle -3, 8 \rangle$
$\mathbf{v} = \langle 8, 3 \rangle$

76. $\mathbf{u} = \langle \frac{1}{4}, -\frac{1}{2} \rangle$
$\mathbf{v} = \langle -2, 4 \rangle$

77. $\mathbf{u} = -\mathbf{i}$
$\mathbf{v} = \mathbf{i} + 2\mathbf{j}$

78. $\mathbf{u} = -2\mathbf{i} + \mathbf{j}$
$\mathbf{v} = 3\mathbf{i} + 6\mathbf{j}$

In Exercises 79–82, find $\text{proj}_{\mathbf{v}}\mathbf{u}$ and the vector component of u orthogonal to v.

79. $\mathbf{u} = \langle -4, 3 \rangle, \ \mathbf{v} = \langle -8, -2 \rangle$

80. $\mathbf{u} = \langle 5, 6 \rangle, \ \mathbf{v} = \langle 10, 0 \rangle$

81. $\mathbf{u} = \langle 2, 7 \rangle, \ \mathbf{v} = \langle 1, -1 \rangle$

82. $\mathbf{u} = \langle -3, 5 \rangle, \ \mathbf{v} = \langle -5, 2 \rangle$

In Exercises 83 and 84, find the work done in moving a particle from P to Q if the magnitude and direction of the force are given by v.

83. $P = (5, 3), Q = (8, 9), \mathbf{v} = \langle 2, 7 \rangle$

84. $P = (-2, -9), Q = (-12, 8), \mathbf{v} = 3\mathbf{i} - 6\mathbf{j}$

6.5 In Exercises 85–88, plot the complex number and find its absolute value.

85. $7i$

86. $-6i$

87. $5 + 3i$

88. $-10 - 4i$

In Exercises 89–92, write the trigonometric form of the complex number.

89. $5 - 5i$

90. $5 + 12i$

91. $-3\sqrt{3} + 3i$

92. -7

In Exercises 93 and 94, (a) express the two complex numbers in trigonometric form, and (b) use the trigonometric form to find z_1z_2 and z_1/z_2.

93. $z_1 = 2\sqrt{3} - 2i, \quad z_2 = -10i$

94. $z_1 = -3(1 + i), \quad z_2 = 2(\sqrt{3} + i)$

In Exercises 95–98, use DeMoivre's Theorem to find the indicated power of the complex number. Express the result in standard form.

95. $\left[5\left(\cos \dfrac{\pi}{12} + i \sin \dfrac{\pi}{12}\right)\right]^4$

96. $\left[2\left(\cos \dfrac{4\pi}{15} + i \sin \dfrac{4\pi}{15}\right)\right]^5$

97. $(2 + 3i)^6$

98. $(1 - i)^8$

Graphical Reasoning **In Exercises 99 and 100, use the graph of the roots of a complex number.**

(a) Write each of the roots in trigonometric form.

(b) Identify the complex number whose roots are given.

(c) Use a graphing utility to verify the results of part (b).

99.

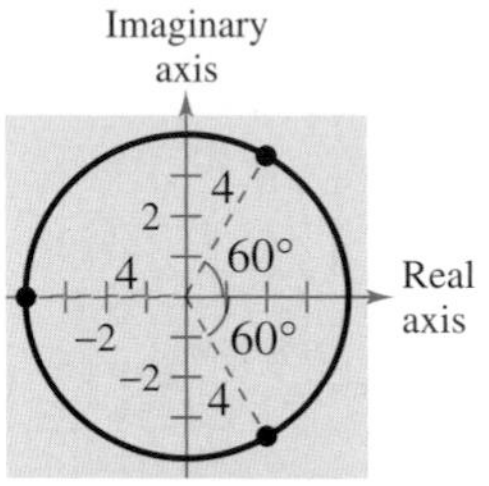

100.

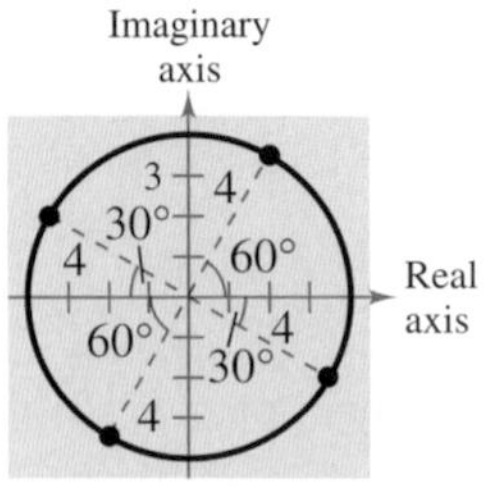

In Exercises 101 and 102, use the formula on page 538 to find the roots of the complex number.

101. Sixth roots of $-729i$

102. Fourth roots of 256

In Exercises 103–106, find all solutions of the equation and represent the solutions graphically.

103. $x^4 + 81 = 0$

104. $x^5 - 32 = 0$

105. $x^3 + 8i = 0$

106. $(x^3 - 1)(x^2 + 1) = 0$

Synthesis

True or False? **In Exercises 107 and 108, determine whether the statement is true or false. Justify your answer.**

107. The Law of Sines is true if one of the angles in the triangle is a right angle.

108. When the Law of Sines is used, the solution is always unique.

109. State the Law of Sines from memory.

110. State the Law of Cosines from memory.

111. If one of the angles in the triangle is a right angle, the Law of Cosines simplifies to what famous theorem?

112. What characterizes a vector in the plane?

113. Which vectors in the figure appear to be equivalent?

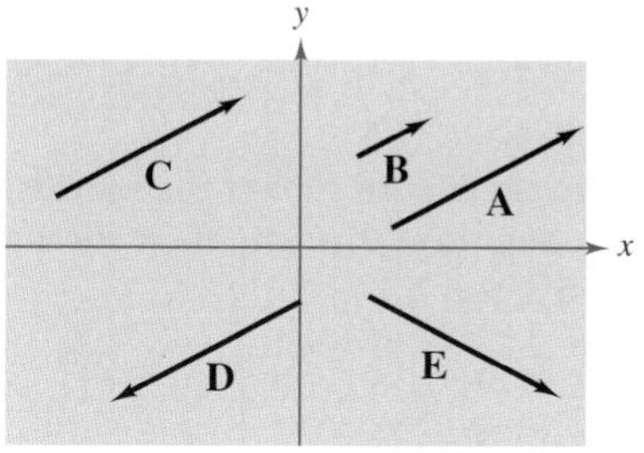

114. The vectors **u** and **v** have the same magnitudes in the two figures. In which figure will the magnitude of the sum be greater? Give a reason for your answer.

(a)

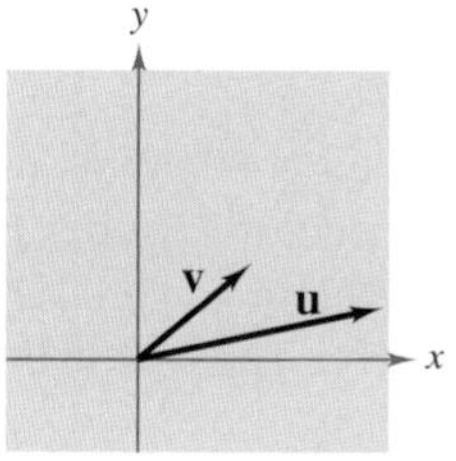

(b)

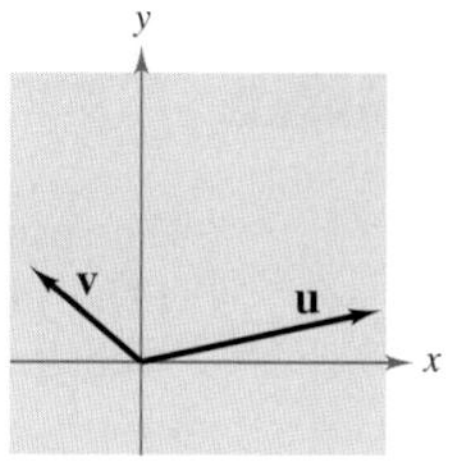

115. Give a geometric description of the scalar multiple $k\mathbf{u}$ of the vector **u**.

116. Give a geometric description of the sum of the vectors **u** and **v**.

117. The figure shows z_1 and z_2. Describe $z_1 z_2$ and z_1/z_2.

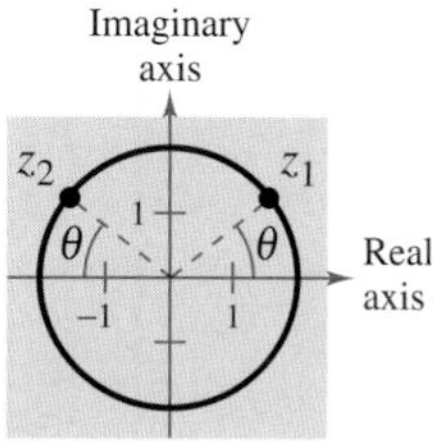

118. One of the fourth roots of a complex number z is shown in the figure.

(a) How many roots are not shown?

(b) Describe the other roots.

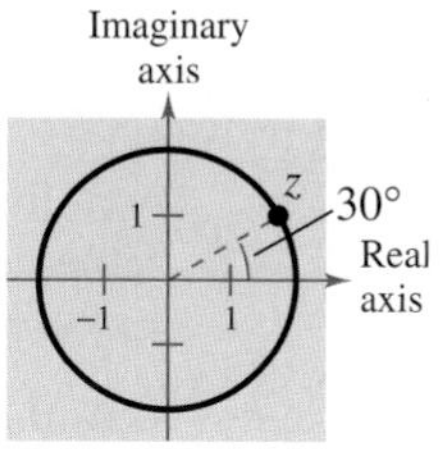

Chapter Project ▶ Adding Vectors Graphically

Program

- Input a
- Input b
- Input c
- Input d
- Draw a line from $(0, 0)$ to (a, b).
- Draw a line from $(0, 0)$ to (c, d).
- Add $a + c$ and store in e.
- Add $b + d$ and store in f.
- Draw a line from $(0, 0)$ to (e, f).
- Draw a line from (a, b) to (c, d).
- Draw a line from (c, d) to (e, f).
- Pause to view graph.
- End program

The psuedo-code at the left can be translated into a program for a graphing utility. (The program for several models of graphing calculators can be found at our website *college.hmco.com*.) The program sketches two vectors

$$\mathbf{u} = a\mathbf{i} + b\mathbf{j} \quad \text{and} \quad \mathbf{v} = c\mathbf{i} + d\mathbf{j}$$

in standard position. Then, using the parallelogram law for vector addition, the program also sketches the vector sum $\mathbf{u} + \mathbf{v}$. *Before* running the program, you should set values that produce an appropriate viewing window.

(a) Use the program to sketch the sum of the vectors $\mathbf{u} = 5\mathbf{i} + 2\mathbf{j}$ and $\mathbf{v} = -4\mathbf{i} + 3\mathbf{j}$. Set your viewing window as indicated in the figure below. Identify the vectors $\mathbf{u}$, $\mathbf{v}$, and $\mathbf{u} + \mathbf{v}$ in the graph.

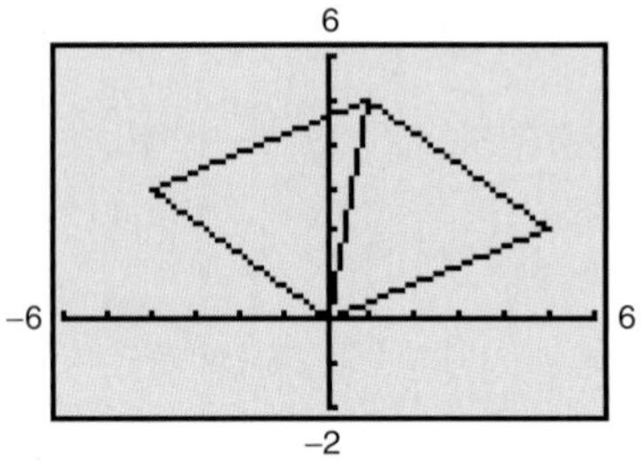

(b) An airplane is headed N 60° W at a speed of 400 miles per hour. The airplane encounters wind of velocity 75 miles per hour in the direction N 40° E. Use the program to find the resultant speed and direction of the airplane.

Chapter Project Investigations

In Questions 1–4, use the program above to sketch the sum of the vectors. Use the result to estimate graphically the components of the sum. Then check your result analytically. (Use $-9 \le x \le 9$ and $-6 \le y \le 6$.)

1. $\mathbf{u} = 3\mathbf{i} + 4\mathbf{j}, \quad \mathbf{v} = -5\mathbf{i} + \mathbf{j}$

2. $\mathbf{u} = 5\mathbf{i} - 4\mathbf{j}, \quad \mathbf{v} = 3\mathbf{i} + 2\mathbf{j}$

3. $\mathbf{u} = -4\mathbf{i} + 4\mathbf{j}, \quad \mathbf{v} = -2\mathbf{i} - 6\mathbf{j}$

4. $\mathbf{u} = 7\mathbf{i} + 3\mathbf{j}, \quad \mathbf{v} = -2\mathbf{i} - 6\mathbf{j}$

5. After encountering the wind, is the airplane in part (b) above traveling at a higher speed or a lower speed? Explain.

6. Consider the airplane described in part (b), headed N 60° W at a speed of 400 miles per hour. What wind velocity, in the direction of N 40° E, will produce a resultant direction of N 50° W? Explain how to use the program above to obtain the answer *experimentally*. Then explain how to obtain the answer analytically.

7. Consider the airplane described in part (b), headed N 60° W at a speed of 400 miles per hour. What wind direction, at a speed of 75 miles per hour, will produce a resultant direction of N 50° W? Explain how to use the program above to obtain the answer *experimentally*. Then explain how to obtain the answer analytically.

Chapter Test

The *Interactive* CD-ROM and *Internet* versions of this text provide answers to the Chapter Tests and Cumulative Tests. They also offer Chapter Pre-Tests (which test key skills and concepts covered in previous chapters) and Chapter Post-Tests, both of which have randomly generated exercises with diagnostic capabilities.

Take this test as you would take a test in class. After you are done, check your work against the answers given in the back of the book.

In Exercises 1–6, use the given information to solve the triangle. If two solutions exist, find both.

1. $A = 24°,\ B = 68°,\ a = 12.2$

2. $B = 104°,\ C = 33°,\ a = 18.1$

3. $A = 24°,\ a = 11.2,\ b = 13.4$

4. $a = 4.0,\ b = 7.3,\ c = 12.4$

5. $B = 100°,\ a = 15,\ b = 23$

6. $C = 123°,\ a = 41,\ b = 57$

7. A triangular parcel of land has borders of lengths 60 meters, 70 meters, and 82 meters. Find the area of the parcel of land.

8. A plane flies 370 miles from point A to point B with a bearing of N 24° E. It then flies 240 miles from point B to point C with a bearing of N 37° E (see figure). Find the distance and bearing from point A to point C.

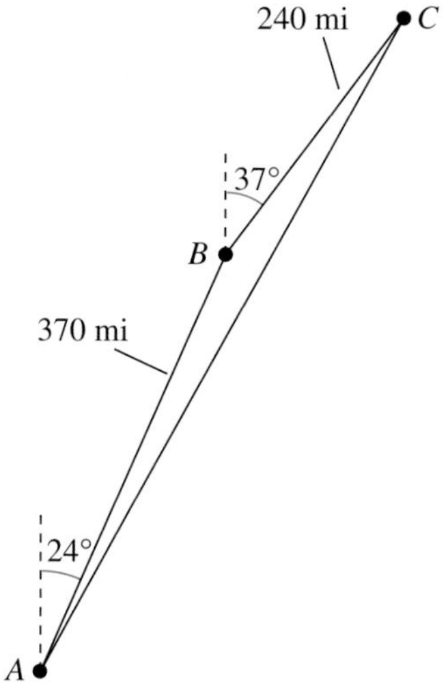

FIGURE FOR 8

In Exercises 9 and 10, find the component form of the vector v with the given components.

9. Initial point of **v**: $(-3, 7)$; Terminal point of **v**: $(11, -16)$

10. Magnitude of **v**: $\|\mathbf{v}\| = 12$; Direction of **v**: $\mathbf{u} = \langle 3, -5 \rangle$

In Exercises 11–13, $\mathbf{u} = \langle 3, 5 \rangle$ and $\mathbf{v} = \langle -7, 1 \rangle$. Find the resultant vector and sketch its graph.

11. $\mathbf{u} + \mathbf{v}$

12. $\mathbf{u} - \mathbf{v}$

13. $5\mathbf{u} - 3\mathbf{v}$

14. Find a unit vector in the direction of $\mathbf{u} = \langle 4, -3 \rangle$.

15. Forces with magnitudes of 250 pounds and 130 pounds act on an object at angles of 45° and $-60°$, respectively, with the x-axis. Find the direction and magnitude of the resultant of these forces.

16. Find the angle between the vectors $\mathbf{u} = \langle -1, 5 \rangle$ and $\mathbf{v} = \langle 3, -2 \rangle$.

17. Are the vectors $\mathbf{u} = \langle 6, 10 \rangle$ and $\mathbf{v} = \langle 2, 3 \rangle$ orthogonal?

18. Find the projection of $\mathbf{u} = \langle 6, 7 \rangle$ onto $\mathbf{v} = \langle -5, -1 \rangle$ and the vector component of **u** orthogonal to **v**.

19. Write the complex number $z = 5 - 5i$ in trigonometric form.

20. Write the complex number $z = 6(\cos 120° + i \sin 120°)$ in standard form.

In Exercises 21 and 22, use DeMoivre's Theorem to find the indicated power of the complex number.

21. $\left[3\left(\cos \frac{7\pi}{6} + i \sin \frac{7\pi}{6}\right)\right]^8$

22. $(3 - 3i)^6$

23. Find the fourth roots of $256\left(1 + \sqrt{3}i\right)$.

24. Find all solutions of the equation $x^3 - 27i = 0$ and represent the solutions graphically.

Cumulative Test for Chapters 4–6

Take this test to review the material from earlier chapters. After you are done, check your work against the answers given in the back of the book.

1. Consider the angle $\theta = -120°$.

(a) Sketch the angle in standard position.

(b) Determine a coterminal angle in the interval $[0°, 360°)$.

(c) Convert the angle to radian measure.

(d) Find the reference angle θ'.

(e) Find the exact values of the six trigonometric functions of θ.

2. Convert the angle of measure 2.35 radians to degrees. Round the answer to one decimal place.

3. Find $\cos \theta$ if $\tan \theta = -\frac{4}{3}$ and $\sin \theta < 0$.

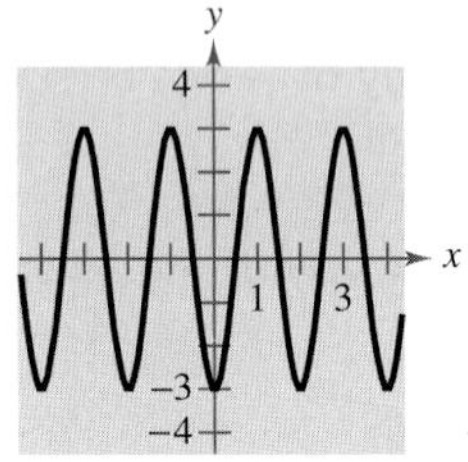

FIGURE FOR 6

In Exercises 4 and 5, find the period and amplitude, and sketch the graph of the trigonometric function.

4. $f(x) = 3 - 2 \sin \pi x$

5. $g(x) = \dfrac{1}{2} \tan\left(x - \dfrac{\pi}{2}\right)$

6. Find a, b, and c such that the graph of the function $h(x) = a \cos(bx + c)$ matches the graph in the figure.

7. Sketch the graph of the function $f(x) = \dfrac{x}{2} \sin x$ over the interval $-3\pi \le x \le 3\pi$.

In Exercises 8 and 9, find the exact value of the expression without the aid of a calculator.

8. $\tan(\arctan 6.7)$

9. $\tan\left(\arcsin \frac{3}{5}\right)$

10. Write an algebraic expression equivalent to $\sin(\arccos 2x)$.

11. Use the fundamental identities to simplify: $\cos\left(\dfrac{\pi}{2} - x\right) \csc x$.

12. Subtract and simplify: $\dfrac{\sin \theta - 1}{\cos \theta} - \dfrac{\cos \theta}{\sin \theta - 1}$.

In Exercises 13–15, prove the identity.

13. $\cot^2 \alpha(\sec^2 \alpha - 1) = 1$

14. $\sin(x + y) \sin(x - y) = \sin^2 x - \sin^2 y$

15. $\sin^2 x \cos^2 x = \frac{1}{8}(1 - \cos 4x)$

In Exercises 16 and 17, find all solutions of the equation in the interval $[0, 2\pi)$.

16. $2 \cos^2 \beta - \cos \beta = 0$

17. $3 \tan \theta - \cot \theta = 0$

18. Use the Quadratic Formula to solve the equation in the interval $[0, 2\pi)$: $\sin^2 x + 2 \sin x + 1 = 0$.

19. Given that $\sin u = \frac{12}{13}$, $\cos v = \frac{3}{5}$, and angles u and v are both in Quadrant I, find $\tan(u - v)$.

20. If $\tan\theta = \frac{1}{2}$, find the exact value of $\tan(2\theta)$.

21. If $\tan\theta = \dfrac{4}{3}$, find the exact value of $\sin\dfrac{\theta}{2}$.

22. Write the product $5\sin\dfrac{3\pi}{4}\cdot\cos\dfrac{7\pi}{4}$ as a sum or difference.

In Exercises 23–26, use the given information to solve the triangle shown in the figure.

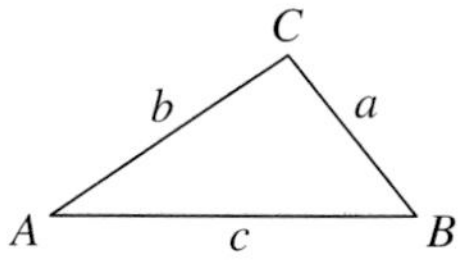

FIGURE FOR 23–26

23. $A = 30°,\ a = 9,\ b = 8$

24. $A = 30°, b = 8,\ c = 10$

25. $A = 30°,\ C = 90°,\ b = 10$

26. $a = 4,\ b = 8,\ c = 9$

27. Two sides of a triangle have lengths 7 and 12 inches. Their included angle measures 60°. Find the area of the triangle.

28. Find the area of a triangle with sides of lengths 11, 16, and 17 inches.

29. Write vector $\mathbf{u} = \langle 3, 5\rangle$ as a linear combination of the standard unit vectors $\mathbf{i}$ and $\mathbf{j}$.

30. Find $\mathbf{u}\cdot\mathbf{v}$ for $\mathbf{u} = 3\mathbf{i} + 4\mathbf{j}$ and $\mathbf{v} = \mathbf{i} - 2\mathbf{j}$.

31. Find the projection of $\mathbf{u} = \langle 8, -2\rangle$ onto $\mathbf{v} = \langle 1, 5\rangle$ and the vector component of $\mathbf{u}$ orthogonal to $\mathbf{v}$.

32. Find the trigonometric form of the complex number $-2 + 2i$.

33. Find the product of $[4(\cos 30° + i\sin 30°)][6(\cos 120° + i\sin 120°)]$. Write the answer in standard form.

34. Find the three cube roots of 1.

35. Write all the solutions of the equation $x^4 - 256i = 0$.

36. From a point 200 feet from a flagpole, the angles of elevation to the bottom and top of the flag are $16°\,45'$ and 18°, respectively. Approximate the height of the flag to the nearest foot.

37. A record single rotates on a turntable at 45 revolutions per minute. Find the angular speed of the record. Then find the speed of the groove that the needle is in when the needle is 3 inches from the center of the record.

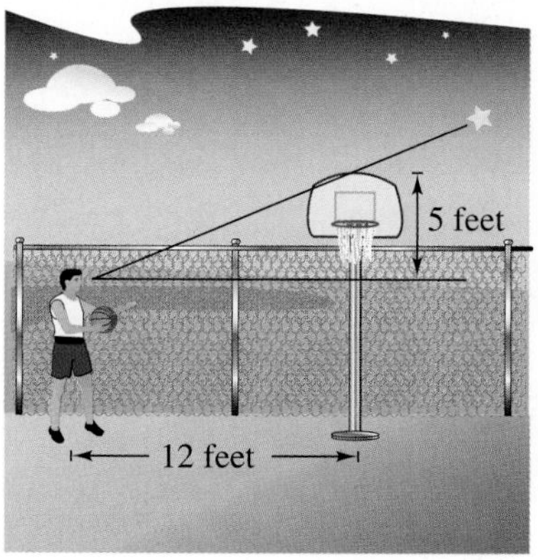

FIGURE FOR 38

38. To determine the angle of elevation of a star in the sky, you get the star in your line of vision with the backboard of a basketball hoop that is 5 feet higher than your eyes (see figure). Your horizontal distance from the backboard is 12 feet. What is the angle of elevation of the star?

39. Write a model for a particle in simple harmonic motion with a displacement of 4 inches and a period of 8 seconds.

40. An airplane's velocity with respect to the air is 500 kilometers per hour, with a bearing of N 30° E. The wind at the altitude of the plane has a velocity of 50 kilometers per hour with a bearing of N 60° E. What is the true direction of the plane, and what is its speed relative to the ground?

Lee Snider/The Image Works

In 1996, 57 million newspapers were printed daily in the United States. With a population of over 265 million, there were about 215 newspapers per 1000 people. (Source: U.S. Bureau of the Census and Editor & Publisher, Co.)

7 Systems of Equations and Inequalities

How to Study This Chapter

The Big Picture

In this chapter you will learn the following skills and concepts.

- How to solve systems of equations by substitution, by elimination, by Gaussian elimination, and by graphing
- How to recognize linear systems in row-echelon form and to use back-substitution to solve the systems
- How to solve nonsquare systems of equations
- How to sketch the graphs of inequalities in two variables and to solve systems of inequalities
- How to solve linear programming problems
- How to use systems of equations and inequalities to model and solve real-life problems

Important Vocabulary

As you encounter each new vocabulary term in this chapter, add the term and its definition to your notebook glossary.

System of equations (p. 554)
Solution of a system of equations (p. 554)
Solving a system of equations (p. 554)
Method of substitution (p. 554)
Graphical method (p. 558)
Points of intersection (p. 558)
Break-even point (p. 559)
Method of elimination (p. 565)
Equivalent systems (p. 566)
Consistent system (p. 568)
Inconsistent system (p. 568)
Row-echelon form (p. 577)
Ordered triple (p. 577)
Row operations (p. 578)
Gaussian elimination (p. 578)
Nonsquare system of equations (p. 582)
Position equation (p. 583)
Solution of an inequality (p. 590)
Graph of an inequality (p. 590)
Linear inequalities (p. 591)
Solution of a system of inequalities (p. 592)
Consumer surplus (p. 595)
Producer surplus (p. 595)
Optimization (p. 601)
Linear programming (p. 601)
Objective function (p. 601)
Constraints (p. 601)
Feasible solutions (p. 601)

Study Tools

- Learning objectives at the beginning of each section
- Chapter Summary (p. 611)
- Review Exercises (pp. 612–615)
- Chapter Test (p. 617)

Additional Resources

- Study and Solutions Guide
- Interactive Precalculus
- Videotapes for Chapter 7
- Precalculus Website
- Student Success Organizer

STUDY T!P

One way to check your work is to plug your answer into the equation or inequality, then solve to see if the numbers on each side are equal. Working on your "checking skills" should improve your test scores.

7.1 Solving Systems of Equations

▶ What you should learn

- How to use the method of substitution to solve systems of equations in two variables
- How to use a graphical approach to solve systems of equations in two variables
- How to use systems of equations to model and solve real-life problems

▶ Why you should learn it

Systems of equations help you solve real-life problems. For instance, Exercise 71 on page 563 shows how you can use a system of equations to compare the compensation plans of two different job offers.

Steve Smith/FPG International

The Method of Substitution

Up to this point in the book, most problems have involved either a function of one variable or a single equation in two variables. However, many problems in science, business, and engineering involve two or more equations in two or more variables. To solve such problems, you need to find solutions of a **system of equations.** Here is an example of a system of two equations in two unknowns.

$$\begin{cases} 2x + y = 5 & \text{Equation 1} \\ 3x - 2y = 4 & \text{Equation 2} \end{cases}$$

A **solution** of this system is an ordered pair that satisfies each equation in the system. Finding the set of all solutions is called **solving the system of equations.** For instance, the ordered pair $(2, 1)$ is a solution of this system. To check this, you can substitute 2 for x and 1 for y in *each* equation.

Check (2, 1) in Equation 1:

$$2x + y = 5 \qquad \text{Write Equation 1.}$$

$$2(2) + 1 \stackrel{?}{=} 5 \qquad \text{Substitute 2 for } x \text{ and 1 for } y.$$

$$4 + 1 = 5 \qquad \text{Solution checks in Equation 1. } \checkmark$$

Check (2, 1) in Equation 2:

$$3x - 2y = 4 \qquad \text{Write Equation 2.}$$

$$3(2) - 2(1) \stackrel{?}{=} 4 \qquad \text{Substitute 2 for } x \text{ and 1 for } y.$$

$$6 - 2 = 4 \qquad \text{Solution checks in Equation 2. } \checkmark$$

In this chapter you will study four ways to solve equations, beginning with the **method of substitution.**

	Method	*Section*	*Type of System*
1.	Substitution	7.1	Linear or nonlinear, two variables
2.	Graphical method	7.1	Linear or nonlinear, two variables
3.	Elimination	7.2	Linear, two variables
4.	Gaussian elimination	7.3	Linear, three or more variables

Method of Substitution

1. *Solve* one of the equations for one variable in terms of the other.
2. *Substitute* the expression found in Step 1 into the other equation to obtain an equation in one variable.
3. *Solve* the equation obtained in Step 2.
4. *Back-substitute* the value obtained in Step 3 into the expression obtained in Step 1 to find the value of the other variable.
5. *Check* that the solution satisfies *each* of the original equations.

Exploration

Use a graphing utility to graph $y = 4 - x$ and $y = x - 2$ in the same viewing window. Use the trace feature to find the coordinates of the point of intersection. Are the coordinates the same as the solution found in Example 1? Explain.

Example 1 ▶ Solving a System of Equations by Substitution

Solve the system of equations.

$$\begin{cases} x + y = 4 & \text{Equation 1} \\ x - y = 2 & \text{Equation 2} \end{cases}$$

Solution

Begin by solving for y in Equation 1.

$$y = 4 - x \qquad \text{Solve for } y \text{ in Equation 1.}$$

Next, substitute this expression for y into Equation 2 and solve the resulting single-variable equation for x.

$$\begin{aligned} x - y &= 2 && \text{Write Equation 2.} \\ x - (4 - x) &= 2 && \text{Substitute } 4 - x \text{ for } y. \\ x - 4 + x &= 2 && \text{Simplify.} \\ 2x &= 6 && \text{Combine like terms.} \\ x &= 3 && \text{Divide each side by 2.} \end{aligned}$$

The *Interactive* CD-ROM and *Internet* versions of this text offer a built-in graphing calculator, which can be used in the Examples, Explorations, Technology notes, and Exercises.

Finally, you can solve for y by *back-substituting* $x = 3$ into the equation $y = 4 - x$, to obtain

$$\begin{aligned} y &= 4 - x && \text{Write revised Equation 1.} \\ y &= 4 - 3 && \text{Substitute 3 for } x. \\ y &= 1. && \text{Solve for } y. \end{aligned}$$

The solution is the ordered pair (3, 1). You can check this as follows.

Check

Substitute (3, 1) into Equation 1:

$$\begin{aligned} x + y &= 4 && \text{Write Equation 1.} \\ 3 + 1 &\stackrel{?}{=} 4 && \text{Substitute for } x \text{ and } y. \\ 4 &= 4 && \text{Solution checks in Equation 1. } \checkmark \end{aligned}$$

Substitute (3, 1) into Equation 2:

$$\begin{aligned} x - y &= 2 && \text{Write Equation 2.} \\ 3 - 1 &\stackrel{?}{=} 2 && \text{Substitute for } x \text{ and } y. \\ 2 &= 2 && \text{Solution checks in Equation 2. } \checkmark \end{aligned}$$

Because (3, 1) satisfies both equations in the system, it is a solution of the system of equations.

STUDY TIP

Because many steps are required to solve a system of equations, it is very easy to make errors in arithmetic. So, we strongly suggest that you always check your solution by substituting it into each equation in the original system.

The term *back-substitution* implies that you work *backwards*. First you solve for one of the variables, and then you substitute that value *back* into one of the equations in the system to find the value of the other variable.

Example 2 ▶ Solving a System by Substitution

A total of \$12,000 is invested in two funds paying 9% and 11% simple interest. The yearly interest is \$1180. How much is invested at each rate?

Solution

Verbal Model: [9% fund] + [11% fund] = [Total investment]

[9% interest] + [11% interest] = [Total interest]

Labels:
Amount in 9% fund $= x$ (dollars)
Interest for 9% fund $= 0.09x$ (dollars)
Amount in 11% fund $= y$ (dollars)
Interest for 11% fund $= 0.11y$ (dollars)
Total investment $= \$12{,}000$ (dollars)
Total interest $= \$1180$ (dollars)

System:
$$\begin{cases} x + y = 12{,}000 & \text{Equation 1} \\ 0.09x + 0.11y = 1{,}180 & \text{Equation 2} \end{cases}$$

You may want to compare and contrast solving a problem using a system of two equations with solving a problem using one equation. Notice that for Example 2, the problem can be solved using either one equation

$0.09x + 0.11(12{,}000 - x) = 1180$

or a system of two equations

$$\begin{cases} x + y = 12{,}000 \\ 0.09x + 0.11y = 1180 \end{cases}.$$

To begin, it is convenient to multiply both sides of Equation 2 by 100. This eliminates the need to work with decimals.

$$100(0.09x + 0.11y) = 100(1180) \quad \text{Multiply each side by 100.}$$
$$9x + 11y = 118{,}000 \quad \text{Revised Equation 2}$$

To solve this system, you can solve for x in Equation 1.

$$x = 12{,}000 - y \quad \text{Revised Equation 1}$$

Then, substitute this expression for x into revised Equation 2 and solve the resulting equation for y.

$$9x + 11y = 118{,}000 \quad \text{Write revised Equation 2.}$$
$$9(12{,}000 - y) + 11y = 118{,}000 \quad \text{Substitute } 12{,}000 - y \text{ for } x.$$
$$108{,}000 - 9y + 11y = 118{,}000 \quad \text{Distributive Property}$$
$$2y = 10{,}000 \quad \text{Combine like terms.}$$
$$y = 5000 \quad \text{Divide each side by 2.}$$

Next, back-substitute the value $y = 5000$ to solve for x.

$$x = 12{,}000 - y \quad \text{Write revised Equation 1.}$$
$$x = 12{,}000 - 5000 \quad \text{Substitute 5000 for } y.$$
$$x = 7000 \quad \text{Simplify.}$$

The solution is (7000, 5000). So, \$7000 is invested at 9% and \$5000 is invested at 11%. Check this in the original problem.

Technology

One way to check the answers you obtain in this section is to use a graphing utility. For instance, enter the two equations in Example 2

$$y_1 = 12{,}000 - x$$
$$y_2 = \frac{1180 - 0.09x}{0.11}$$

and find an appropriate viewing window that shows where the lines intersect. Then use the zoom and trace features to find their point of intersection. Does this point agree with the solution obtained at the right?

A computer animation of this concept appears in the *Interactive* CD-ROM and *Internet* versions of this text.

The equations in Examples 1 and 2 are linear. Substitution can also be used to solve systems in which one or both of the equations are nonlinear.

Example 3 ▶ Substitution: Two-Solution Case

Solve the system of equations.

$$\begin{cases} x^2 + 4x - y = 7 & \text{Equation 1} \\ 2x - y = -1 & \text{Equation 2} \end{cases}$$

Solution

Begin by solving for y in Equation 2 to obtain

$y = 2x + 1.$ Solve for y in Equation 2.

Next, substitute this expression for y into Equation 1 and solve for x.

$x^2 + 4x - y = 7$ Write Equation 1.

$x^2 + 4x - (2x + 1) = 7$ Substitute $2x + 1$ for y.

$x^2 + 2x - 1 = 7$ Simplify.

$x^2 + 2x - 8 = 0$ General form

$(x + 4)(x - 2) = 0$ Factor.

$x = -4, 2$ Solve for x.

Back-substituting these values of x to solve for the corresponding values of y produces the solutions $(-4, -7)$ and $(2, 5)$. Check these in the original system.

Exploration

Use a graphing utility to graph the two equations in Example 3

$y_1 = x^2 + 4x - 7$

$y_2 = 2x + 1$

in the same viewing window. How many solutions do you think this system has?

Repeat this experiment for the equations in Example 4. How many solutions does this system have? Explain your reasoning.

Point out that it is not practical to solve Equation 2 for x (instead of for y) in Example 3.

Example 4 ▶ Substitution: No-Real-Solution Case

Solve the system of equations.

$$\begin{cases} -x + y = 4 & \text{Equation 1} \\ x^2 + y = 3 & \text{Equation 2} \end{cases}$$

Solution

Begin by solving for y in Equation 1 to obtain

$y = x + 4.$ Solve for y in Equation 1.

Next, substitute this expression for y into Equation 2 and solve for x.

$x^2 + y = 3$ Write Equation 2.

$x^2 + (x + 4) = 3$ Substitute $x + 4$ for y.

$x^2 + x + 1 = 0$ Simplify.

$x = \dfrac{-1 \pm \sqrt{1^2 - 4(1)(1)}}{2}$ Quadratic Formula

$x = \dfrac{-1 \pm \sqrt{-3}}{2}$ Simplify.

Because the discriminant is negative, the equation $x^2 + x + 1 = 0$ has no (real) solution. So, this system has no (real) solution.

Technology

Most graphing calculators have built-in programs that approximate the point(s) of intersection of two graphs. Typically, you must enter the equations of the graphs and visually locate a point of intersection before running the intersection program.

Use this feature to find the points of intersection for the graphs in Figure 7.1. Be sure to adjust your viewing window so that you see all the points of intersection.

The *Interactive* CD-ROM and *Internet* versions of this text show every example with its solution; clicking on the *Try It!* button brings up similar problems. Guided Examples and Integrated Examples show step-by-step solutions to additional examples. Integrated Examples are related to several concepts in the section.

Graphical Approach to Finding Solutions

From Examples 2, 3, and 4, you can see that a system of two equations in two unknowns can have exactly one solution, more than one solution, or no solution. By using a **graphical method,** you can gain insight about the number of solutions and the location(s) of the solution(s) of a system of equations by graphing each of the equations in the same coordinate plane. The solutions of the system correspond to the **points of intersection** of the graphs. For instance, the two equations in Figure 7.1(a) graph as two lines with *a single point* of intersection; the two equations in Figure 7.1(b) graph as a parabola and a line with *two points* of intersection; and the two equations in Figure 7.1(c) graph as a line and a parabola that have *no points* of intersection.

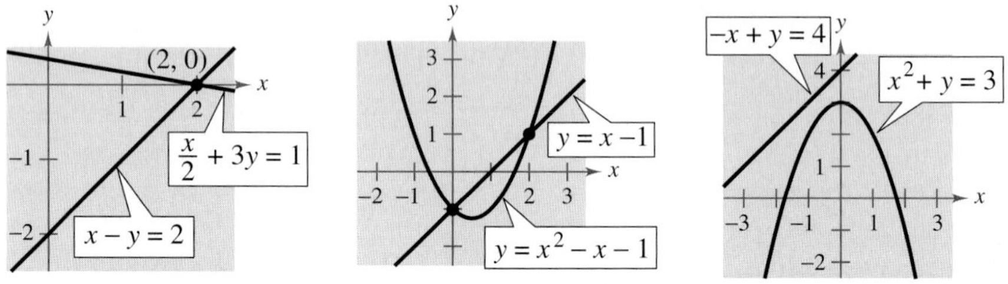

(a) One intersection point **(b) Two intersection points** **(c) No intersection points**

FIGURE 7.1

Example 5 ▶ Solving a System of Equations Graphically

Solve the system of equations.

$$\begin{cases} y = \ln x & \text{Equation 1} \\ x + y = 1 & \text{Equation 2} \end{cases}$$

Solution

Sketch the graphs of the two equations, as shown in Figure 7.2. From the graphs, it is clear that there is only one point of intersection and that $(1, 0)$ is the solution point. You can confirm this by substituting 1 for x and 0 for y in *both* equations.

Check (1, 0) in Equation 1:

$y = \ln x$ Write Equation 1.

$0 = \ln 1$ Equation 1 checks. ✓

Check (1, 0) in Equation 2:

$x + y = 1$ Write Equation 2.

$1 + 0 = 1$ Equation 2 checks. ✓

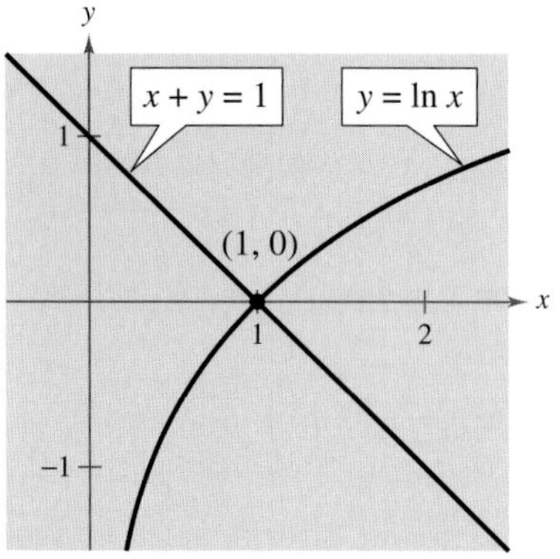

FIGURE 7.2

Example 5 shows the value of a graphical approach to solving systems of equations in two variables. Notice what would happen if you tried only the substitution method in Example 5. You would obtain the equation $x + \ln x = 1$. It would be difficult to solve this equation for x using standard algebraic techniques.

Activities:

1. Solve the system by the method of substitution.
$$\begin{cases} 3x + 2y = 16 \\ x - 2y = 0 \end{cases}$$
Answer: $(4, 2)$
2. Find all points of intersection.
$$\begin{cases} 4x - y - 5 = 0 \\ 4x^2 - 8x + y + 5 = 0 \end{cases}$$
Answer: $(0, -5), (1, -1)$
3. Solve the system graphically.
$$\begin{cases} 3x + 2y = 6 \\ y = \ln(x - 1) \end{cases}$$
Answer: $(2, 0)$

Applications

The total cost C of producing x units of a product typically has two components—the initial cost and the cost per unit. When enough units have been sold so that the total revenue R equals the total cost C, the sales are said to have reached the **break-even point.** You will find that the break-even point corresponds to the point of intersection of the cost and revenue curves.

Example 6 ▶ Break-Even Analysis

A small business invests \$10,000 in equipment to produce a product. Each unit of the product costs \$0.65 to produce and is sold for \$1.20. How many items must be sold before the business breaks even?

Solution

The total cost of producing x units is

Total cost = Cost per unit · Number of units + Initial cost

$$C = 0.65x + 10{,}000. \qquad \text{Equation 1}$$

The revenue obtained by selling x units is

Total revenue = Price per unit · Number of units

$$R = 1.20x. \qquad \text{Equation 2}$$

Because the break-even point occurs when $R = C$, you have $C = 1.20x$, and the system of equations to solve is

$$\begin{cases} C = 0.65x + 10{,}000 \\ C = 1.20x \end{cases}.$$

Now you can solve by substitution.

$$1.20x = 0.65x + 10{,}000 \qquad \text{Substitute } 1.2x \text{ for } C \text{ in Equation 1.}$$

$$0.55x = 10{,}000 \qquad \text{Subtract } 0.65x \text{ from each side.}$$

$$x = \frac{10{,}000}{0.55} \qquad \text{Divide each side by 0.55.}$$

$$x \approx 18{,}182 \text{ units} \qquad \text{Use a calculator.}$$

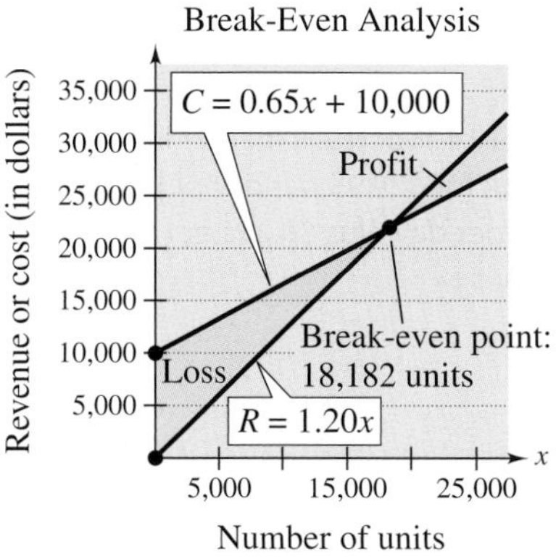

FIGURE 7.3

Note in Figure 7.3 that sales less than the break-even point correspond to an overall loss, whereas sales greater than the break-even point correspond to a profit.

Another way to view the solution in Example 6 is to consider the profit function

$$P = R - C.$$

The break-even point occurs when the profit is 0, which is the same as saying that $R = C$.

Example 7 ▶ State Population

From 1990 to 1997, the population of Arizona was increasing at a faster rate than the population of Alabama. Models that approximate the two populations P (in thousands) are

$$\begin{cases} P = 3631.2 + 131.7t & \text{Arizona} \\ P = 4055.6 + 39.9t & \text{Alabama} \end{cases}$$

where $t = 0$ represents 1990 (see Figure 7.4). According to these two models, when would you expect the population of Arizona to have exceeded the population of Alabama? (Source: U.S. Bureau of the Census)

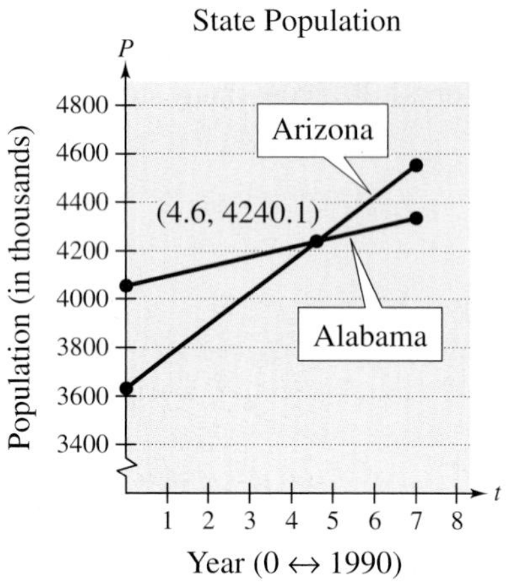

FIGURE 7.4

Solution

Because the first equation has already been solved for P in terms of t, substitute this value into the second equation and solve for t, as follows.

$$3631.2 + 131.7t = 4055.6 + 39.9t \qquad \text{Substitute for } P \text{ in Equation 2.}$$

$$131.7t - 39.9t = 4055.6 - 3631.2 \qquad \text{Subtract } 39.9t \text{ and } 3631.2 \text{ from each side.}$$

$$91.8t = 424.4 \qquad \text{Combine like terms.}$$

$$t \approx 4.6 \qquad \text{Divide each side by } 91.8.$$

So, from the given models, you would expect that the population of Arizona exceeded the population of Alabama after $t \approx 4.6$ years, which was sometime during 1994.

Writing About Mathematics Suggestion:
Consider asking your students to make up their own problems that involve a choice of solution methods. Students can then trade their problems with other members of their class. Make sure that students are able to interpret attributes such as slopes and points of intersection.

Writing ABOUT MATHEMATICS

Interpreting Points of Intersection You plan to rent a 14-foot truck for a 2-day local move. At truck rental agency A, you can rent a truck for \$29.95 per day plus \$0.49 per mile. At agency B, you can rent a truck for \$50 per day plus \$0.25 per mile. The total cost y (in dollars) for the truck from agency A is

$$y = (\$29.95 \text{ per day})(2 \text{ days}) + 0.49x$$

$$= 59.90 + 0.49x$$

where x is the total number of miles the truck is driven.

a. Write a total cost equation in terms of x and y for the total cost of the truck from agency B.

b. Use a graphing utility to graph the two equations and find the point of intersection. Interpret the meaning of the point of intersection in the context of the problem.

c. Which agency should you choose if you plan to travel a total of 100 miles over the 2-day move? Why?

d. How does the situation change if you plan to drive 200 miles over the 2-day move?

7.1 Exercises

The *Interactive* CD-ROM and *Internet* versions of this text contain step-by-step solutions to all odd-numbered Section and Review Exercises. They also provide Tutorial Exercises that link to Guided Examples for additional help.

Exercises containing systems with no solutions: 23, 24, 39, 40, 52, 55, 56

In Exercises 1–4, determine which ordered pairs are solutions of the system of equations.

1. $\begin{cases} 4x - y = 1 \\ 6x + y = -6 \end{cases}$ (a) $(0, -3)$ (b) $(-1, -4)$ (c) $\left(-\frac{3}{2}, -2\right)$ (d) $\left(-\frac{1}{2}, -3\right)$

2. $\begin{cases} 4x^2 + y = 3 \\ -x - y = 11 \end{cases}$ (a) $(2, -13)$ (b) $(2, -9)$ (c) $\left(-\frac{3}{2}, -\frac{31}{3}\right)$ (d) $\left(-\frac{7}{4}, -\frac{37}{4}\right)$

3. $\begin{cases} y = -2e^x \\ 3x - y = 2 \end{cases}$ (a) $(-2, 0)$ (b) $(0, -2)$ (c) $(0, -3)$ (d) $(-1, 2)$

4. $\begin{cases} -\log x + 3 = y \\ \frac{1}{9}x + y = \frac{28}{9} \end{cases}$ (a) $\left(9, \frac{37}{9}\right)$ (b) $(10, 2)$ (c) $(1, 3)$ (d) $(2, 4)$

In Exercises 5–14, solve the system by the method of substitution. Check your solution graphically.

5. $\begin{cases} 2x + y = 6 \\ -x + y = 0 \end{cases}$

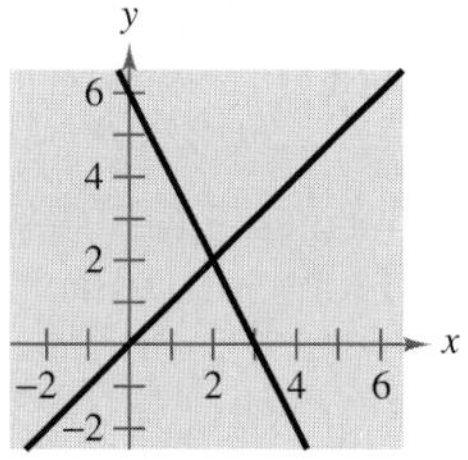

6. $\begin{cases} x - y = -4 \\ x + 2y = 5 \end{cases}$

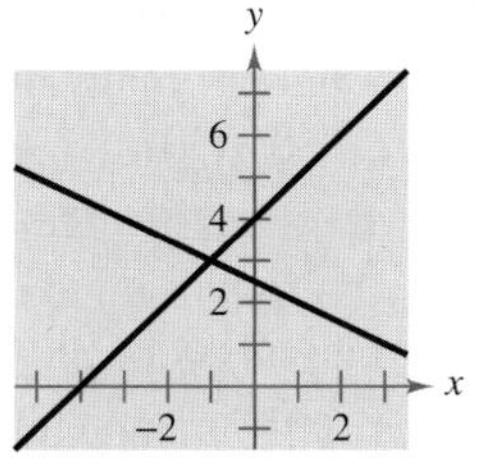

7. $\begin{cases} x - y = -4 \\ x^2 - y = -2 \end{cases}$

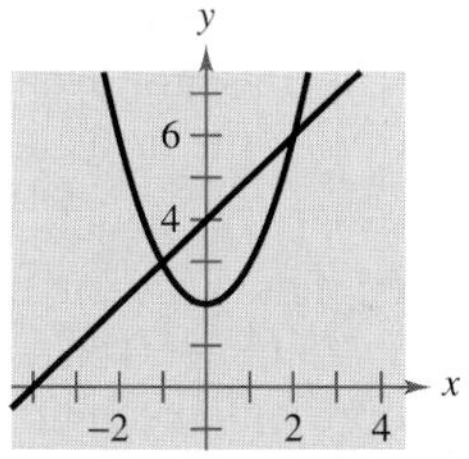

8. $\begin{cases} 3x + y = 2 \\ x^3 - 2 + y = 0 \end{cases}$

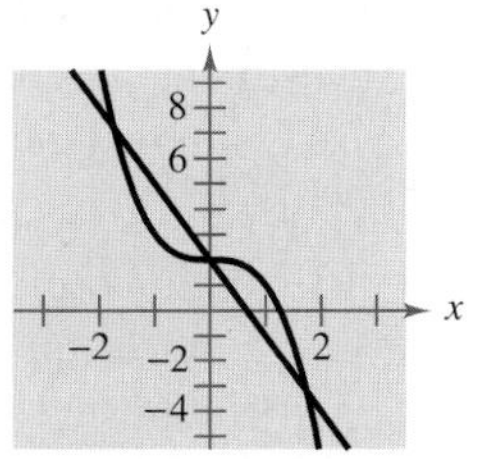

9. $\begin{cases} -2x + y = -5 \\ x^2 + y^2 = 25 \end{cases}$

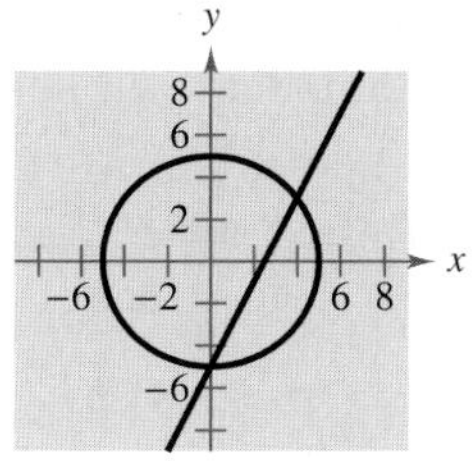

10. $\begin{cases} x + y = 0 \\ x^3 - 5x - y = 0 \end{cases}$

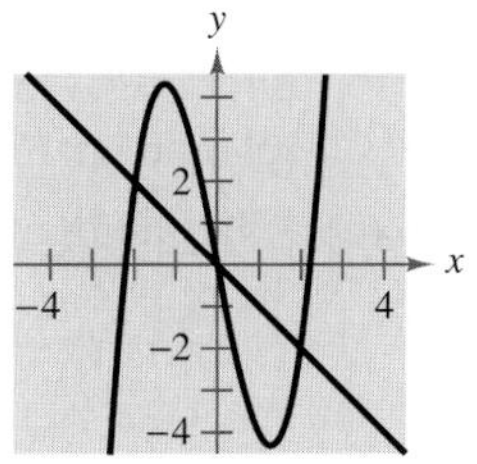

11. $\begin{cases} x^2 + y = 0 \\ x^2 - 4x - y = 0 \end{cases}$

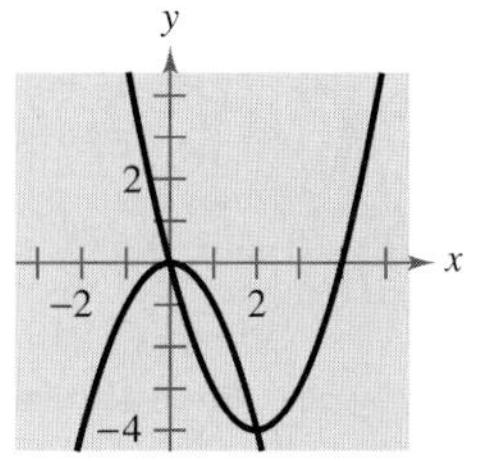

12. $\begin{cases} y = -2x^2 + 2 \\ y = 2(x^4 - 2x^2 + 1) \end{cases}$

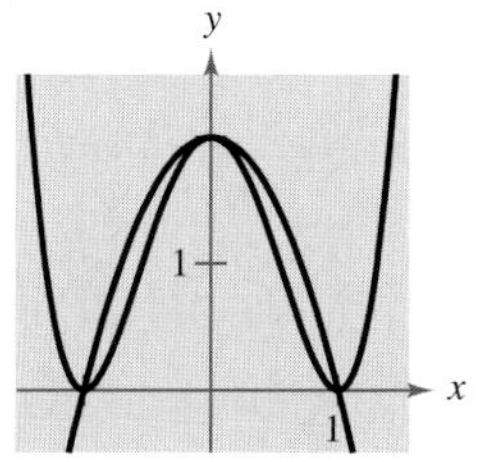

13. $\begin{cases} y = x^3 - 3x^2 + 1 \\ y = x^2 - 3x + 1 \end{cases}$

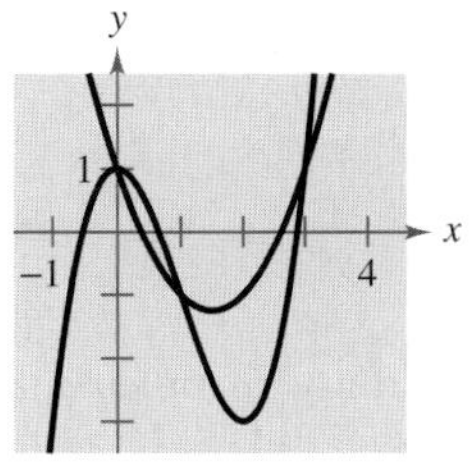

14. $\begin{cases} y = x^3 - 3x^2 + 4 \\ y = -2x + 4 \end{cases}$

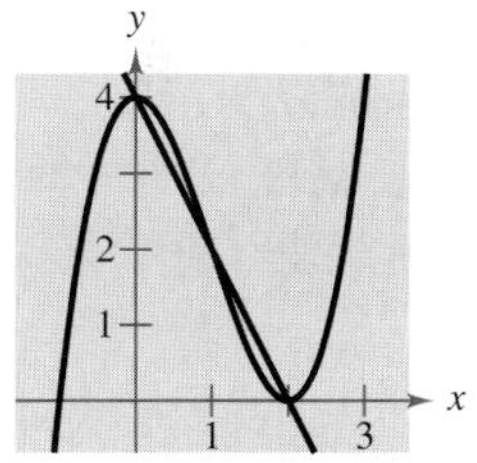

In Exercises 15–28, solve the system by the method of substitution.

15. $\begin{cases} x - y = 0 \\ 5x - 3y = 10 \end{cases}$

16. $\begin{cases} x + 2y = 1 \\ 5x - 4y = -23 \end{cases}$

17. $\begin{cases} 2x - y + 2 = 0 \\ 4x + y - 5 = 0 \end{cases}$

18. $\begin{cases} 6x - 3y - 4 = 0 \\ x + 2y - 4 = 0 \end{cases}$

19. $\begin{cases} 1.5x + 0.8y = 2.3 \\ 0.3x - 0.2y = 0.1 \end{cases}$

20. $\begin{cases} 0.5x + 3.2y = 9.0 \\ 0.2x - 1.6y = -3.6 \end{cases}$

21. $\begin{cases} \frac{1}{5}x + \frac{1}{2}y = 8 \\ x + y = 20 \end{cases}$

22. $\begin{cases} \frac{1}{2}x + \frac{3}{4}y = 10 \\ \frac{3}{4}x - y = 4 \end{cases}$

23. $\begin{cases} 6x + 5y = -3 \\ -x - \frac{5}{6}y = -7 \end{cases}$

24. $\begin{cases} -\frac{2}{3}x + y = 2 \\ 2x - 3y = 6 \end{cases}$

25. $\begin{cases} x^2 - y = 0 \\ 2x + y = 0 \end{cases}$

26. $\begin{cases} x - 2y = 0 \\ 3x - y^2 = 0 \end{cases}$

27. $\begin{cases} x^3 - y = 0 \\ x - y = 0 \end{cases}$

28. $\begin{cases} y = -x \\ y = x^3 + 3x^2 + 2x \end{cases}$

In Exercises 29–42, solve the system graphically.

29. $\begin{cases} -x + 2y = 2 \\ 3x + y = 15 \end{cases}$

30. $\begin{cases} x + y = 0 \\ 3x - 2y = 10 \end{cases}$

31. $\begin{cases} x - 3y = -2 \\ 5x + 3y = 17 \end{cases}$

32. $\begin{cases} -x + 2y = 1 \\ x - y = 2 \end{cases}$

33. $\begin{cases} x + y = 4 \\ x^2 + y^2 - 4x = 0 \end{cases}$

34. $\begin{cases} -x + y = 3 \\ x^2 - 6x - 27 + y^2 = 0 \end{cases}$

35. $\begin{cases} x - y + 3 = 0 \\ x^2 - 4x + 7 = y \end{cases}$

36. $\begin{cases} y^2 - 4x + 11 = 0 \\ -\frac{1}{2}x + y = -\frac{1}{2} \end{cases}$

37. $\begin{cases} 7x + 8y = 24 \\ x - 8y = 8 \end{cases}$

38. $\begin{cases} x - y = 0 \\ 5x - 2y = 6 \end{cases}$

39. $\begin{cases} 3x - 2y = 0 \\ x^2 - y^2 = 4 \end{cases}$

40. $\begin{cases} 2x - y + 3 = 0 \\ x^2 + y^2 - 4x = 0 \end{cases}$

41. $\begin{cases} x^2 + y^2 = 25 \\ 3x^2 - 16y = 0 \end{cases}$

42. $\begin{cases} x^2 + y^2 = 25 \\ (x - 8)^2 + y^2 = 41 \end{cases}$

In Exercises 43–50, use a graphing utility to approximate all points of intersection of the graphs.

43. $\begin{cases} y = e^x \\ x - y + 1 = 0 \end{cases}$

44. $\begin{cases} y = -4e^{-x} \\ y + 3x + 8 = 0 \end{cases}$

45. $\begin{cases} x + 2y = 8 \\ y = \log_2 x \end{cases}$

46. $\begin{cases} y = -2 + \ln(x - 1) \\ 3y + 2x = 9 \end{cases}$

47. $\begin{cases} y = \sqrt{x} \\ y = x \end{cases}$

48. $\begin{cases} x - y = 3 \\ x - y^2 = 1 \end{cases}$

49. $\begin{cases} x^2 + y^2 = 169 \\ x^2 - 8y = 104 \end{cases}$

50. $\begin{cases} x^2 + y^2 = 4 \\ 2x^2 - y = 2 \end{cases}$

In Exercises 51–62, solve the system graphically or algebraically. Explain your choice of method.

51. $\begin{cases} y = 2x \\ y = x^2 + 1 \end{cases}$

52. $\begin{cases} x + y = 4 \\ x^2 + y = 2 \end{cases}$

53. $\begin{cases} 3x - 7y + 6 = 0 \\ x^2 - y^2 = 4 \end{cases}$

54. $\begin{cases} x^2 + y^2 = 25 \\ 2x + y = 10 \end{cases}$

55. $\begin{cases} x - 2y = 4 \\ x^2 - y = 0 \end{cases}$

56. $\begin{cases} y = (x + 1)^3 \\ y = \sqrt{x - 1} \end{cases}$

57. $\begin{cases} y - e^{-x} = 1 \\ y - \ln x = 3 \end{cases}$

58. $\begin{cases} x^2 + y = 4 \\ e^x - y = 0 \end{cases}$

59. $\begin{cases} y = x^4 - 2x^2 + 1 \\ y = 1 - x^2 \end{cases}$

60. $\begin{cases} y = x^3 - 2x^2 + x - 1 \\ y = -x^2 + 3x - 1 \end{cases}$

61. $\begin{cases} xy - 1 = 0 \\ 2x - 4y + 7 = 0 \end{cases}$

62. $\begin{cases} x - 2y = 1 \\ y = \sqrt{x - 1} \end{cases}$

Break-Even Analysis **In Exercises 63–66, find the sales necessary to break even ($R = C$) for the cost C of x units and the revenue R obtained by selling x units. (Round to the nearest whole unit.)**

63. $C = 8650x + 250{,}000, \quad R = 9950x$

64. $C = 2.65x + 350{,}000, \quad R = 4.15x$

65. $C = 5.5\sqrt{x} + 10{,}000, \quad R = 3.29x$

66. $C = 7.8\sqrt{x} + 18{,}500, \quad R = 12.84x$

67. ***Break-Even Point*** A small business invests \$16,000 to produce an item that will sell for \$5.95. Each unit can be produced for \$3.45.

(a) How many units must be sold to break even?

(b) How many units must be sold to make a profit of \$6000?

68. ***Break-Even Point*** A small business invests \$5000 to produce an item that will sell for \$34.10. Each unit can be produced for \$21.60.

(a) How many units must be sold to break even?

(b) How many units must be sold to make a profit of \$8500?

69. ***Finance*** A total of \$25,000 is invested in two funds paying 6% and 8.5% simple interest. The 6% investment has a lower risk. The investor wants a yearly interest income of \$2000 from the investment.

(a) Write a system of equations in which one equation represents the total amount invested and the other equation represents the \$2000 required in interest. Let x and y represent the amounts invested at 6% and 8.5%, respectively.

(b) Use a graphing utility to graph the two equations. As the amount invested at 6% increases, how does the amount invested at 8.5% change? How does the amount of interest income change? Explain.

(c) What amount should be invested at 6% to meet the requirement of \$2000 per year in interest?

70. *Finance* A total of $20,000 is invested in two funds paying 6.5% and 8.5% simple interest. The 6.5% investment has a lower risk. The investor wants a yearly interest check of $1600 from the investments.

(a) Write a system of equations in which one equation represents the total amount invested and the other equation represents the $1600 required in interest. Let x and y represent the amounts invested at 6.5% and 8.5%, respectively.

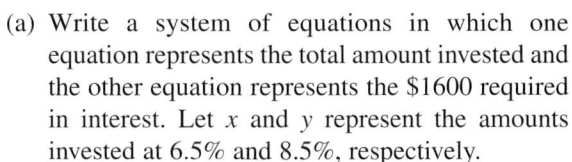

(b) Use a graphing utility to graph the two equations. As the amount invested at 6.5% increases, how does the amount invested at 8.5% change? How does the amount of interest change? Explain.

(c) What amount should be invested at 6.5% to meet the requirement of $1600 per year in interest?

71. *Choice of Two Jobs* You are offered two jobs selling dental supplies. One company offers a straight commission of 6% of sales. The other company offers a salary of $350 per week plus 3% of sales. How much would you have to sell in a week in order to make the straight commission offer better?

72. *Choice of Two Jobs* You are offered two different jobs selling college textbooks. One company offers an annual salary of $25,000 plus a year-end bonus of 2% of your total sales. The other company offers an annual salary of $20,000 plus a year-end bonus of 3% of your total sales. Determine the annual sales required to make the second offer better.

73. *Log Volume* You are offered two different rules for estimating the number of board feet in a 16-foot log. (A board foot is a unit of measure for lumber equal to a board 1 foot square and 1 inch thick.) The first rule is the *Doyle Log Rule* and is modeled by

$$V = (D - 4)^2, \qquad 5 \le D \le 40$$

and the other is the *Scribner Log Rule* and is modeled by

$$V = 0.79D^2 - 2D - 4, \qquad 5 \le D \le 40$$

where D is the diameter (in inches) of the log and V is its volume in board feet.

(a) Use a graphing utility to graph the two log rules in the same viewing window.

(b) For what diameter do the two scales agree?

(c) If you were selling large logs by the board foot, which scale would you use?

74. *Economics* The supply and demand curves for a business dealing with wheat are

Supply: $p = 1.45 + 0.00014x^2$

Demand: $p = (2.388 - 0.007x)^2$

where p is the price in dollars per bushel and x is the quantity in bushels per day. Use a graphing utility to graph the supply and demand equations and find the market equilibrium. (The *market equilibrium* is the point of intersection of the graphs for $x > 0$.)

Geometry **In Exercises 75–78, find the dimensions of the rectangle meeting the specified conditions.**

75. The perimeter is 30 meters and the length is 3 meters greater than the width.

76. The perimeter is 280 centimeters and the width is 20 centimeters less than the length.

77. The perimeter is 42 inches and the width is three-fourths the length.

78. The perimeter is 210 feet and the length is $1\frac{1}{2}$ times the width.

79. *Geometry* What are the dimensions of a rectangular tract of land if its perimeter is 40 kilometers and its area is 96 square kilometers?

80. *Geometry* What are the dimensions of an isosceles right triangle with a 2-inch hypotenuse and an area of 1 square inch?

81. *Data Analysis* The table gives the amount y, in millions of short tons, of paperboard produced in the years 1993 through 1996 in the United States. (Source: American Forest and Paper Association)

Year	1993	1994	1995	1996
y	43.1	45.7	46.6	47.9

(a) Use the regression features of a graphing utility to find a linear model f and a quadratic model g that represent the data in the interval from 1993 through 1996. Let $t = 3$ represent 1993.

(b) Use a graphing utility to graph the data and the two models in the same viewing window.

(c) Approximate the points of intersection of the graphs of the models.

Synthesis

True or False? **In Exercises 82 and 83, determine whether the statement is true or false. Justify your answer.**

82. In order to solve a system of equations by substitution, you must always solve for y in one of the two equations and then back-substitute.

83. If a system consists of a parabola and a circle, then the system can have at most two solutions.

84. What is meant by a solution of a system of equations in two variables?

85. *Think About It* When solving a system of equations by substitution, how do you recognize that the system has no solution?

86. *Writing* Write a brief paragraph describing any advantages of substitution over the graphical method of solving a system of equations.

87. *Exploration* Find an equation of a line whose graph intersects the graph of the parabola $y = x^2$ at (a) two points, (b) one point, and (c) no points. (There is more than one correct answer.)

88. *Conjecture* Consider the system of equations

$$\begin{cases} y = b^x \\ y = x^b \end{cases}.$$

(a) Use a graphing utility to graph the system for $b = 1, 2, 3,$ and 4.

(b) For a fixed even value of $b > 1$, make a conjecture about the number of points of intersection of the graphs in part (a).

Review

In Exercises 89–94, find the general form of the equation of the line through the two points.

89. $(-2, 7), (5, 5)$

90. $(3.5, 4), (10, 6)$

91. $(6, 3), (10, 3)$

92. $(4, -2), (4, 5)$

93. $\left(\frac{3}{5}, 0\right), (4, 6)$

94. $\left(-\frac{7}{3}, 8\right), \left(\frac{5}{2}, \frac{1}{2}\right)$

In Exercises 95–98, find the domain of the function and identify any horizontal or vertical asymptotes.

95. $f(x) = \dfrac{5}{x - 6}$

96. $f(x) = \dfrac{2x - 7}{3x + 2}$

97. $f(x) = \dfrac{x^2 + 2}{x^2 - 16}$

98. $f(x) = 3 - \dfrac{2}{x^2}$

In Exercises 99–102, sketch the graph of the equation.

99. $y = -2^{0.5x}$

100. $y = 2^{-0.5x}$

101. $y = 3e^{-x-4}$

102. $y = \dfrac{2 + e^{-x}}{5}$

7.2 Two-Variable Linear Systems

▶ **What you should learn**

- How to use the method of elimination to solve systems of linear equations in two variables
- How to interpret graphically the numbers of solutions of systems of linear equations in two variables
- How to use systems of equations in two variables to model and solve real-life problems

▶ **Why you should learn it**

You can use systems of equations in two variables to model and solve real-life problems. For instance, Exercise 64 on page 575 shows how to use a system of equations to analyze an investment portfolio.

Jon Riley/Tony Stone Images

The Method of Elimination

In Section 7.1, you studied two methods for solving a system of equations: substitution and graphing. Now you will study the **method of elimination.** The key step in this method is to obtain, for one of the variables, coefficients that differ only in sign so that *adding* the equations eliminates the variable.

$$3x + 5y = 7 \quad \text{Equation 1}$$
$$-3x - 2y = -1 \quad \text{Equation 2}$$
$$3y = 6 \quad \text{Add equations.}$$

Note that by adding the two equations, you eliminate the x-terms and obtain a single equation in y. Solving this equation for y produces $y = 2$, which you can then back-substitute into one of the original equations to solve for x.

Example 1 ▶ Solving a System of Equations by Elimination

Solve the system of linear equations.

$$\begin{cases} 3x + 2y = 4 & \text{Equation 1} \\ 5x - 2y = 8 & \text{Equation 2} \end{cases}$$

Solution

Because the coefficients of y differ only in sign, you can eliminate the y-terms by adding the two equations.

$$3x + 2y = 4 \quad \text{Write Equation 1.}$$
$$5x - 2y = 8 \quad \text{Write Equation 2.}$$
$$8x = 12 \quad \text{Add equations.}$$

Therefore, $x = \frac{3}{2}$. By back-substituting this value into Equation 1, you can solve for y.

$$3x + 2y = 4 \quad \text{Write Equation 1.}$$
$$3\left(\frac{3}{2}\right) + 2y = 4 \quad \text{Substitute } \tfrac{3}{2} \text{ for } x.$$
$$\frac{9}{2} + 2y = 4 \quad \text{Simplify.}$$
$$y = -\frac{1}{4} \quad \text{Solve for } y.$$

The solution is $\left(\frac{3}{2}, -\frac{1}{4}\right)$. Check this in the original system.

Try using the method of substitution to solve the system given in Example 1. Which method do you think is easier? Many people find that the method of elimination is more efficient.

To be efficient, students should know and understand both the substitution method and the elimination method. You may wish to point out the differences between the two methods and the advantages of one over the other.

STUDY T!P

To obtain coefficients (for one of the variables) that differ only in sign, you often need to multiply one or both of the equations by suitably chosen constants.

Remind your students to multiply the constant as well. A common error is to forget to multiply the constant terms. For instance, in Example 2, students may rewrite Equation 2 as $9x + 3y = -5$ rather than as $9x + 3y = -15$.

Example 2 ▶ Solving a System of Equations by Elimination

Solve the system of linear equations.

$$\begin{cases} 2x - 3y = -7 & \text{Equation 1} \\ 3x + y = -5 & \text{Equation 2} \end{cases}$$

Solution

For this system, you can obtain coefficients that differ only in sign by multiplying Equation 2 by 3.

$2x - 3y = -7$ ⇒	$2x - 3y = -7$	Write Equation 1.
$3x + y = -5$ ⇒	$9x + 3y = -15$	Multiply Equation 2 by 3.
	$11x = -22$	Add equations.

So, you can see that $x = -2$. By back-substituting this value of x into Equation 1, you can solve for y.

$$2x - 3y = -7 \qquad \text{Write Equation 1.}$$

$$2(-2) - 3y = -7 \qquad \text{Substitute } -2 \text{ for } x.$$

$$-3y = -3 \qquad \text{Collect like terms.}$$

$$y = 1 \qquad \text{Solve for } y.$$

The solution is $(-2, 1)$. Check this in the original system.

Check

$$2(-2) - 3(1) \stackrel{?}{=} -7 \qquad \text{Substitute into Equation 1.}$$

$$-4 - 3 = -7 \qquad \text{Equation 1 checks. ✓}$$

$$3(-2) + 1 \stackrel{?}{=} -5 \qquad \text{Substitute into Equation 2.}$$

$$-6 + 1 = -5 \qquad \text{Equation 2 checks. ✓}$$

In Example 2, the two systems of linear equations

$$\begin{cases} 2x - 3y = -7 \\ 3x + y = -5 \end{cases}$$

and

$$\begin{cases} 2x - 3y = -7 \\ 9x + 3y = -15 \end{cases}$$

are called **equivalent systems** because they have precisely the same solution set. The operations that can be performed on a system of linear equations to produce an equivalent system are (1) interchanging any two equations, (2) multiplying an equation by a nonzero constant, and (3) adding a multiple of one equation to any other equation in the system.

Method of Elimination

1. *Obtain coefficients* for x (or y) that differ only in sign by multiplying all terms of one or both equations by suitably chosen constants.
2. *Add* the equations to eliminate one variable and solve the resulting equation.
3. *Back-substitute* the value obtained in Step 2 into either of the original equations and solve for the other variable.
4. *Check* your solution in both of the original equations.

Exploration

Sketch the graph of each of the following systems of equations.

a. $\begin{cases} y = 5x + 1 \\ y - x = -5 \end{cases}$

b. $\begin{cases} 3y = 4x - 1 \\ -8x + 2 = -6y \end{cases}$

c. $\begin{cases} 2y = -x + 3 \\ -4 = y + \frac{1}{2}x \end{cases}$

Determine the number of solutions each system has. Explain your reasoning.

Consider having your students solve Example 3 by eliminating the x-terms instead of the y-terms to reinforce the fact that either variable can be eliminated first.

Example 3 ▶ Solving a System of Equations by Elimination

Solve the system of linear equations.

$$\begin{cases} 5x + 3y = 9 & \text{Equation 1} \\ 2x - 4y = 14 & \text{Equation 2} \end{cases}$$

Solution

You can obtain coefficients that differ only in sign by multiplying Equation 1 by 4 and multiplying Equation 2 by 3.

$$5x + 3y = 9 \quad \Rightarrow \quad 20x + 12y = 36 \qquad \text{Multiply Equation 1 by 4.}$$

$$2x - 4y = 14 \quad \Rightarrow \quad 6x - 12y = 42 \qquad \text{Multiply Equation 2 by 3.}$$

$$26x \qquad = 78 \qquad \text{Add equations.}$$

From this equation, you can see that $x = 3$. By back-substituting this value of x into Equation 2, you can solve for y.

$$2x - 4y = 14 \qquad \text{Write Equation 2.}$$

$$2(3) - 4y = 14 \qquad \text{Substitute 3 for } x.$$

$$-4y = 8 \qquad \text{Collect like terms.}$$

$$y = -2 \qquad \text{Solve for } y.$$

The solution is $(3, -2)$. Check this in the original system.

Remember that you can check the solution of a system of equations graphically. For instance, to check the solution found in Example 3, graph both equations in the same viewing window, as shown in Figure 7.5. Notice that the two lines intersect at $(3, -2)$.

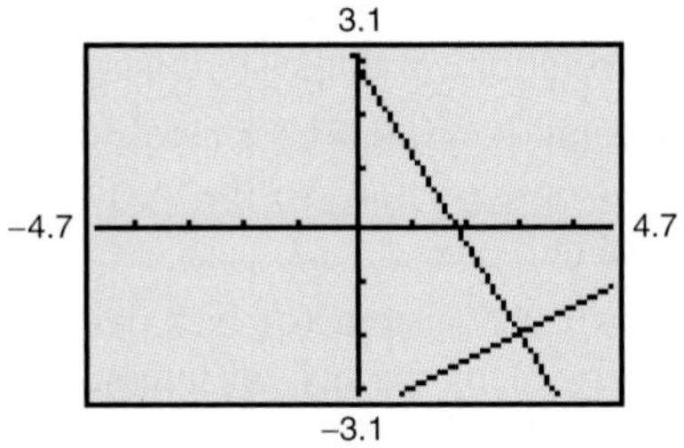

FIGURE 7.5

STUDY TIP

Keep in mind that the terminology and methods discussed in this section and the following section apply only to systems of *linear* equations.

Graphical Interpretation of Solutions

It is possible for a *general* system of equations to have exactly one solution, two or more solutions, or no solution. If a system of *linear* equations has two different solutions, it must have an *infinite* number of solutions. To see why this is true, consider the following graphical interpretations of a system of two linear equations in two variables.

Graphical Interpretations of Solutions

For a system of two linear equations in two variables, the number of solutions is one of the following.

Number of Solutions	*Graphical Interpretation*
1. Exactly one solution	The two lines intersect at one point.
2. Infinitely many solutions	The two lines are identical.
3. No solution	The two lines are parallel.

A system of linear equations is **consistent** if it has at least one solution. It is **inconsistent** if it has no solution.

Example 4 ▶ Recognizing Graphs of Linear Systems

A computer simulation of this example appears in the *Interactive* CD-ROM and *Internet* versions of this text.

Match the system of linear equations with its graph in Figure 7.6. State whether the system is consistent or inconsistent and describe the number of solutions.

a. $\begin{cases} 2x - 3y = 3 \\ -4x + 6y = 6 \end{cases}$ **b.** $\begin{cases} 2x - 3y = 3 \\ x + 2y = 5 \end{cases}$ **c.** $\begin{cases} 2x - 3y = 3 \\ -4x + 6y = -6 \end{cases}$

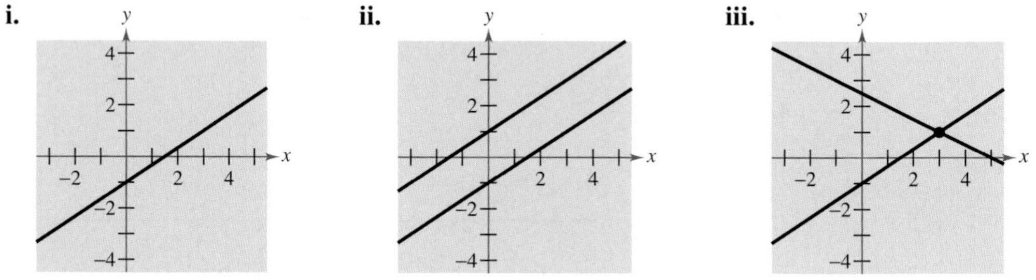

FIGURE 7.6

Solution

a. The graph of system (a) is a pair of parallel lines (ii). The lines have no point of intersection, so the system has no solution. The system is inconsistent.

b. The graph of system (b) is a pair of intersecting lines (iii). The lines have one point of intersection, so the system has exactly one solution. The system is consistent.

c. The graph of system (c) is a pair of lines that coincide (i). The lines have infinitely many points of intersection, so the system has infinitely many solutions. The system is consistent.

In Examples 5 and 6, note how you can use the method of elimination to determine that a system of linear equations has no solution or infinitely many solutions.

Example 5 ▶ No-Solution Case: Method of Elimination

Solve the system of linear equations.

$$\begin{cases} x - 2y = 3 & \text{Equation 1} \\ -2x + 4y = 1 & \text{Equation 2} \end{cases}$$

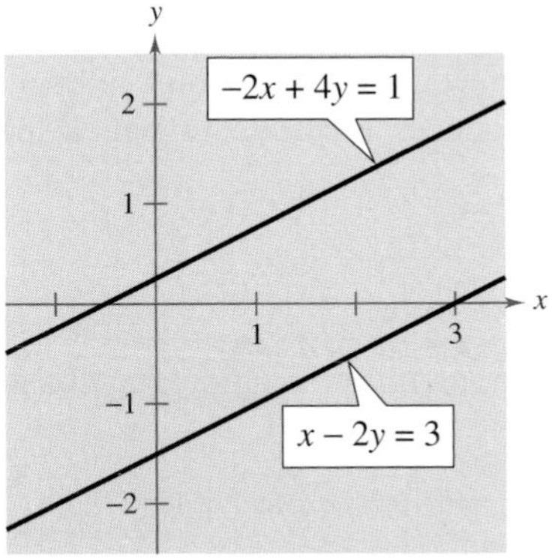

FIGURE 7.7

Solution

To obtain coefficients that differ only in sign, multiply Equation 1 by 2.

$x - 2y = 3$ ⟹ $2x - 4y = 6$ Multiply Equation 1 by 2.

$-2x + 4y = 1$ ⟹ $-2x + 4y = 1$ Write Equation 2.

$0 = 7$ False statement

Because there are no values of x and y for which $0 = 7$, you can conclude that the system is inconsistent and has no solution. The lines corresponding to the two equations in this system are shown in Figure 7.7. Note that the two lines are parallel and therefore have no point of intersection.

In Example 5, note that the occurrence of a false statement, such as $0 = 7$, indicates that the system has no solution. In the next example, note that the occurrence of a statement that is true for all values of the variables, such as $0 = 0$, indicates that the system has infinitely many solutions.

Example 6 ▶ Many-Solutions Case: Method of Elimination

Solve the system of linear equations.

$$\begin{cases} 2x - y = 1 & \text{Equation 1} \\ 4x - 2y = 2 & \text{Equation 2} \end{cases}$$

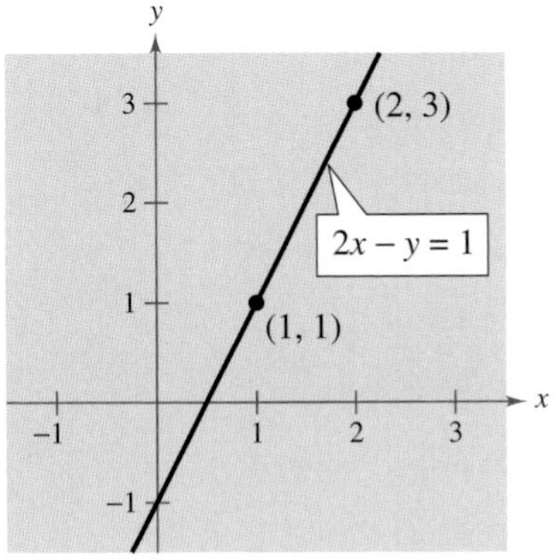

FIGURE 7.8

Solution

To obtain coefficients that differ only in sign, multiply Equation 2 by $-\frac{1}{2}$.

$2x - y = 1$ ⟹ $2x - y = 1$ Write Equation 1.

$4x - 2y = 2$ ⟹ $-2x + y = -1$ Multiply Equation 2 by $-\frac{1}{2}$.

$0 = 0$ Add equations.

Because the two equations turn out to be equivalent (have the same solution set), you can conclude that the system has infinitely many solutions. The solution set consists of all points (x, y) lying on the line $2x - y = 1$ as shown in Figure 7.8. Letting $x = a$, where a is any real number, you can see that the solutions to the system are $(a, 2a + 1)$.

Technology

The general solution of the linear system

$$\begin{cases} ax + by = c \\ dx + ey = f \end{cases}$$

is $x = (ce - bf)/(ae - db)$ and $y = (af - cd)/(ae - db)$. If $ae - db = 0$, the system does not have a unique solution. Graphing utility programs for solving such a system can be found at our website *college.hmco.com*. Try using the program for your graphing utility to solve the system in Example 7.

Example 7 illustrates a strategy for solving a system of linear equations that has decimal coefficients.

Example 7 ▶ A Linear System Having Decimal Coefficients

Solve the system of linear equations.

$$\begin{cases} 0.02x - 0.05y = -0.38 & \text{Equation 1} \\ 0.03x + 0.04y = 1.04 & \text{Equation 2} \end{cases}$$

Solution

Because the coefficients in this system have two decimal places, you can begin by multiplying each equation by 100. (This produces a system in which the coefficients are all integers.)

$$\begin{cases} 2x - 5y = -38 & \text{Revised Equation 1} \\ 3x + 4y = 104 & \text{Revised Equation 2} \end{cases}$$

Now, to obtain coefficients that differ only in sign, multiply Equation 1 by 3 and multiply Equation 2 by -2.

$$2x - 5y = -38 \;\Rightarrow\; 6x - 15y = -114 \quad \text{Multiply Equation 1 by 3.}$$

$$3x + 4y = 104 \;\Rightarrow\; -6x - 8y = -208 \quad \text{Multiply Equation 2 by } -2.$$

$$-23y = -322 \quad \text{Add equations.}$$

So, you can conclude that

$$y = \frac{-322}{-23} = 14.$$

Back-substituting this value into Equation 2 produces the following.

$$3x + 4y = 104 \quad \text{Write revised Equation 2.}$$

$$3x + 4(14) = 104 \quad \text{Substitute 14 for } y.$$

$$3x = 48 \quad \text{Collect like terms.}$$

$$x = 16 \quad \text{Solve for } x.$$

The solution is (16, 14). Check this in the original system.

Check

$$0.02x - 0.05y = -0.38 \quad \text{Write original Equation 1.}$$

$$0.02(16) - 0.05(14) \stackrel{?}{=} -0.38 \quad \text{Substitute into Equation 1.}$$

$$0.32 - 0.70 = -0.38 \quad \text{Equation 1 checks. ✓}$$

$$0.03x + 0.04y = 1.04 \quad \text{Write original Equation 2.}$$

$$0.03(16) + 0.04(14) \stackrel{?}{=} 1.04 \quad \text{Substitute into Equation 2.}$$

$$0.48 + 0.56 = 1.04 \quad \text{Equation 2 checks. ✓}$$

Activities

1. Solve the system by the elimination method.

   $$\begin{cases} 3x - 2y = 7 \\ 8x + 4y = 0 \end{cases}$$

   Answer: $(1, -2)$
2. How many milliliters of a 30% acid solution and a 55% acid solution must be mixed to obtain 100 milliliters of a 50% acid solution?
   Answer: 20 milliliters of the 30% acid solution and 80 milliliters of the 55% acid solution
3. Find the point of equilibrium if the demand equation is $p = 110 - 5x$ and the supply equation is $p = 60 + 20x$.
   Answer: $x = 2$, $p = \$100.00$.

Applications

At this point, you may be asking the question "How can I tell which application problems can be solved using a system of linear equations?" The answer comes from the following considerations.

1. Does the problem involve more than one unknown quantity?

2. Are there two (or more) equations or conditions to be satisfied?

If one or both of these conditions occur, the appropriate mathematical model for the problem may be a system of linear equations. Example 8 shows how to construct such a model.

Example 8 ▶ An Application of a Linear System

An airplane flying into a headwind travels the 2000-mile flying distance between two cities in 4 hours and 24 minutes. On the return flight, the same distance is traveled in 4 hours. Find the air speed of the plane and the speed of the wind, assuming that both remain constant.

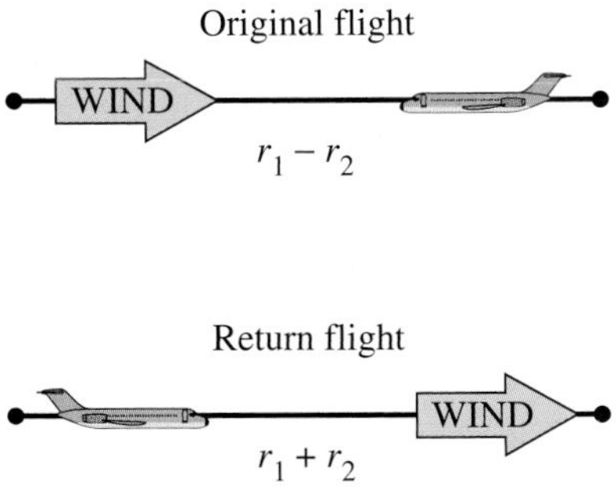

FIGURE 7.9

Solution

The two unknown quantities are the speeds of the wind and the plane. If r_1 is the speed of the plane and r_2 is the speed of the wind, then

$r_1 - r_2 =$ speed of the plane against the wind

$r_1 + r_2 =$ speed of the plane with the wind

as shown in Figure 7.9. Using the formula

Distance = (rate)(time)

for these two speeds, you obtain the following equations.

$$2000 = (r_1 - r_2)\left(4 + \frac{24}{60}\right)$$

$$2000 = (r_1 + r_2)(4)$$

These two equations simplify as follows.

$$\begin{cases} 5000 = 11r_1 - 11r_2 & \text{Equation 1} \\ 500 = r_1 + r_2 & \text{Equation 2} \end{cases}$$

By elimination, the solution is

$$r_1 = \frac{5250}{11} \approx 477.27 \text{ miles per hour} \qquad \text{Speed of plane}$$

$$r_2 = \frac{250}{11} \approx 22.73 \text{ miles per hour.} \qquad \text{Speed of wind}$$

Check this solution in the original statement of the problem.

In a free market, the demands for many products are related to the prices of the products. As the prices decrease, the demands by consumers increase and the amounts that producers are able or willing to supply decrease.

Example 9 ▶ Finding the Point of Equilibrium

The demand and supply functions for a certain type of calculator are

$$\begin{cases} p = 150 - 0.00001x & \text{Demand equation} \\ p = 60 + 0.00002x & \text{Supply equation} \end{cases}$$

where p is the price in dollars and x represents the number of units. Find the point of equilibrium for this market. The point of equilibrium is the price p and number of units x that satisfy both the demand and supply equations.

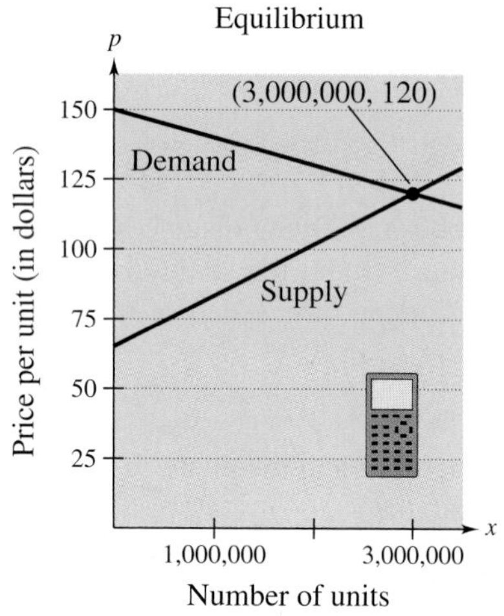

FIGURE 7.10

Solution

Begin by substituting the value of p given in the supply equation into the demand equation.

$$p = 150 - 0.00001x \qquad \text{Write demand equation.}$$

$$60 + 0.00002x = 150 - 0.00001x \qquad \text{Substitute } 60 + 0.00002x \text{ for } p.$$

$$0.00003x = 90 \qquad \text{Collect like terms.}$$

$$x = 3{,}000{,}000 \qquad \text{Solve for } x.$$

So, the point of equilibrium occurs when the demand and supply are each 3 million units. (See Figure 7.10.) The price that corresponds to this x-value is obtained by back-substituting $x = 3{,}000{,}000$ into either of the original equations. For instance, back-substituting into the demand equation produces

$$\begin{aligned} p &= 150 - 0.00001(3{,}000{,}000) \\ &= 150 - 30 \\ &= \$120. \end{aligned}$$

The solution is $(3{,}000{,}000,\ 120)$. You can check this as follows.

Check

Substitute $(3{,}000{,}000,\ 120)$ into the demand equation.

$$p = 150 - 0.00001x \qquad \text{Write demand equation.}$$

$$120 \stackrel{?}{=} 150 - 0.00001(3{,}000{,}000) \qquad \text{Substitute 120 for } p \text{ and 3,000,000 for } x.$$

$$120 = 120 \qquad \text{Solution checks in demand equation. ✓}$$

Substitute $(3{,}000{,}000,\ 120)$ into the supply equation.

$$p = 60 + 0.00002x \qquad \text{Write supply equation.}$$

$$120 \stackrel{?}{=} 60 + 0.00002(3{,}000{,}000) \qquad \text{Substitute 120 for } p \text{ and 3,000,000 for } x.$$

$$120 = 120 \qquad \text{Solution checks in supply equation. ✓}$$

Do your students understand why the elimination method is used in Example 8 and why the substitution method is used in Example 9? You may want to point out that if two equations are each written such that one variable is represented in terms of the other variable, substitution is usually more efficient.

7.2 Exercises

Exercises containing systems with no solutions: 5, 6, 19, 20, 33, 34, 37
Exercises containing systems with infinitely many solutions: 7, 8, 22, 23, 24, 36

In Exercises 1–10, solve by elimination. Match each line with its equation.

1. $\begin{cases} 2x + y = 5 \\ x - y = 1 \end{cases}$

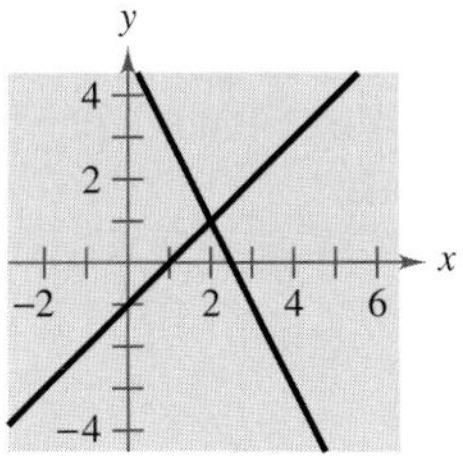

2. $\begin{cases} x + 3y = 1 \\ -x + 2y = 4 \end{cases}$

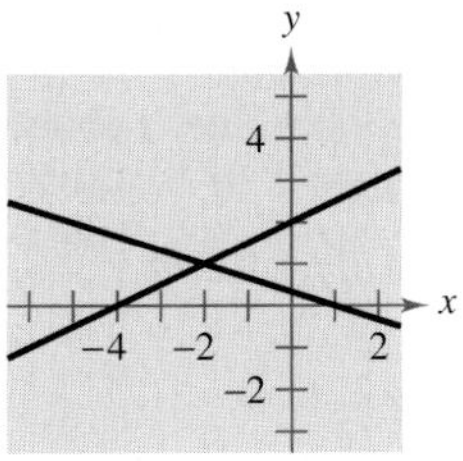

3. $\begin{cases} x + y = 0 \\ 3x + 2y = 1 \end{cases}$

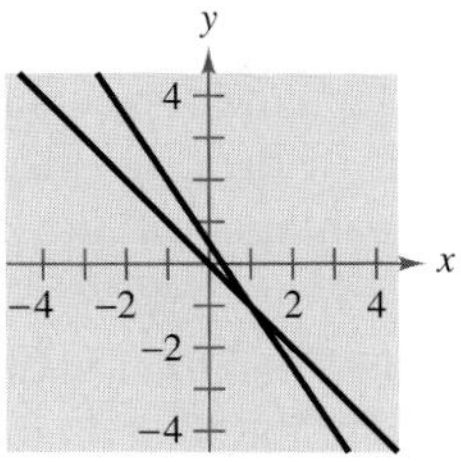

4. $\begin{cases} 2x - y = 3 \\ 4x + 3y = 21 \end{cases}$

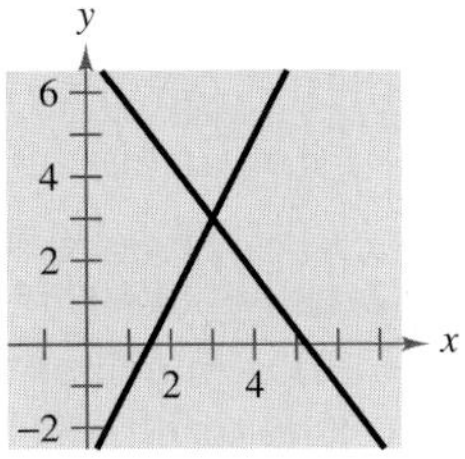

5. $\begin{cases} x - y = 2 \\ -2x + 2y = 5 \end{cases}$

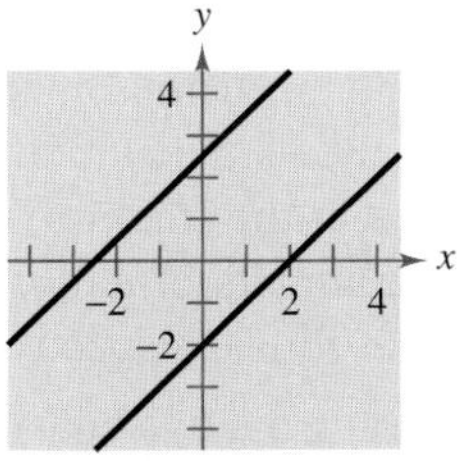

6. $\begin{cases} 3x + 2y = 3 \\ 6x + 4y = 14 \end{cases}$

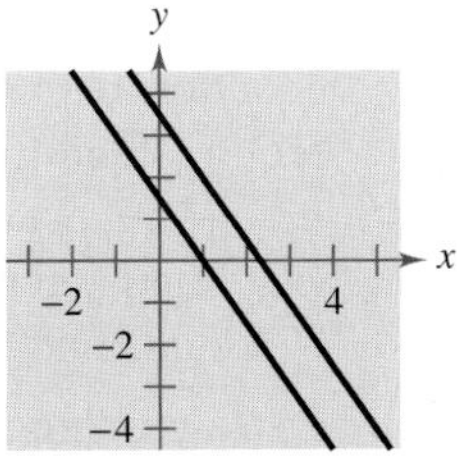

7. $\begin{cases} 3x - 2y = 5 \\ -6x + 4y = -10 \end{cases}$

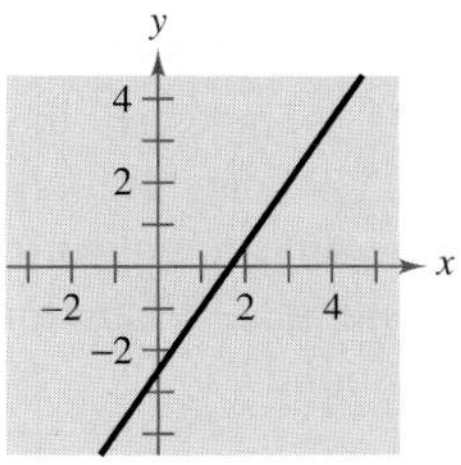

8. $\begin{cases} 9x - 3y = -15 \\ -3x + y = 5 \end{cases}$

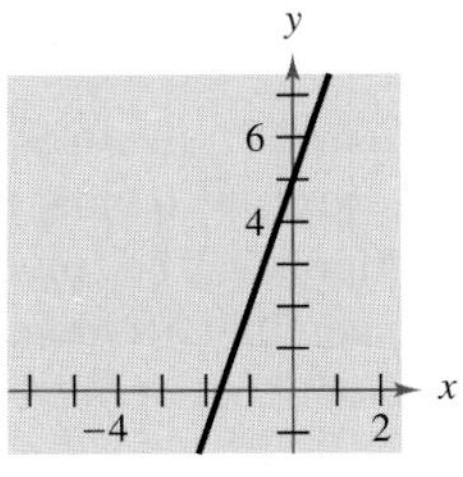

9. $\begin{cases} 9x + 3y = 1 \\ 3x - 6y = 5 \end{cases}$

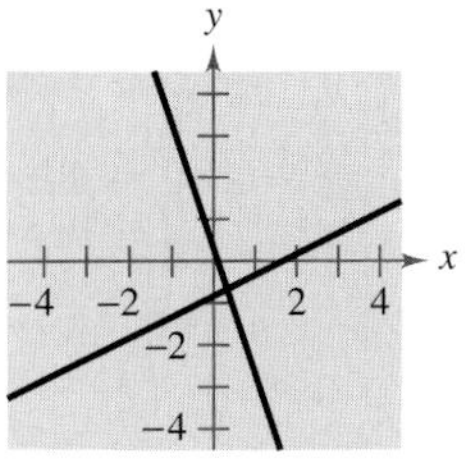

10. $\begin{cases} 5x + 3y = -18 \\ 2x - 6y = 1 \end{cases}$

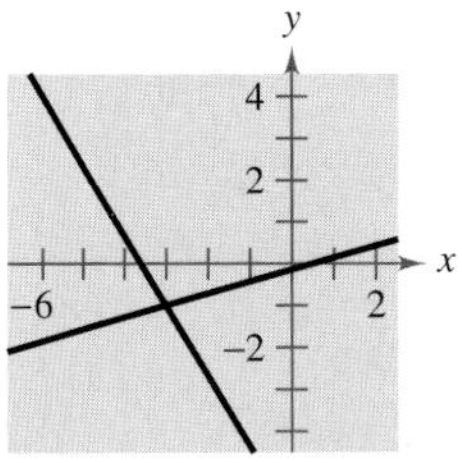

In Exercises 11–30, solve the system by elimination and check any solutions algebraically.

11. $\begin{cases} x + 2y = 4 \\ x - 2y = 1 \end{cases}$

12. $\begin{cases} 3x - 5y = 2 \\ 2x + 5y = 13 \end{cases}$

13. $\begin{cases} 2x + 3y = 18 \\ 5x - y = 11 \end{cases}$

14. $\begin{cases} x + 7y = 12 \\ 3x - 5y = 10 \end{cases}$

15. $\begin{cases} 3x + 2y = 10 \\ 2x + 5y = 3 \end{cases}$

16. $\begin{cases} 2r + 4s = 5 \\ 16r + 50s = 55 \end{cases}$

17. $\begin{cases} 5u + 6v = 24 \\ 3u + 5v = 18 \end{cases}$

18. $\begin{cases} 3x + 11y = 4 \\ -2x - 5y = 9 \end{cases}$

19. $\begin{cases} 1.8x + 1.2y = 4 \\ 9x + 6y = 3 \end{cases}$

20. $\begin{cases} 3.1x - 2.9y = -10.2 \\ 15.5x - 14.5y = 21 \end{cases}$

21. $\begin{cases} \dfrac{x}{4} + \dfrac{y}{6} = 1 \\ x - y = 3 \end{cases}$

22. $\begin{cases} \dfrac{2}{3}x + \dfrac{1}{6}y = \dfrac{2}{3} \\ 4x + y = 4 \end{cases}$

23. $\begin{cases} 2.5x - 3y = 1.5 \\ 2x - 2.4y = 1.2 \end{cases}$

24. $\begin{cases} 6.3x + 7.2y = 5.4 \\ 5.6x + 6.4y = 4.8 \end{cases}$

25. $\begin{cases} 0.05x - 0.03y = 0.21 \\ 0.07x + 0.02y = 0.16 \end{cases}$

26. $\begin{cases} 0.2x - 0.5y = -27.8 \\ 0.3x + 0.4y = 68.7 \end{cases}$

27. $\begin{cases} 4b + 3m = 3 \\ 3b + 11m = 13 \end{cases}$

28. $\begin{cases} 2x + 5y = 8 \\ 5x + 8y = 10 \end{cases}$

29. $\begin{cases} \dfrac{x+3}{4} + \dfrac{y-1}{3} = 1 \\ 2x - y = 12 \end{cases}$

30. $\begin{cases} \dfrac{x-1}{2} + \dfrac{y+2}{3} = 4 \\ x - 2y = 5 \end{cases}$

In Exercises 31–38, use a graphing utility to graph the lines in the system. Use the graphs to determine if the system is consistent or inconsistent. If the system is consistent, determine the number of solutions.

31. $\begin{cases} 2x - 5y = 0 \\ x - y = 3 \end{cases}$

32. $\begin{cases} 2x + y = 5 \\ x - 2y = -1 \end{cases}$

33. $\begin{cases} \frac{3}{5}x - y = 3 \\ -3x + 5y = 9 \end{cases}$

34. $\begin{cases} 4x - 6y = 9 \\ \frac{16}{3}x - 8y = 12 \end{cases}$

35. $\begin{cases} x + 7y = 2 \\ 4x - y = 9 \end{cases}$

36. $\begin{cases} 8x - 14y = 5 \\ 2x - 3.5y = 1.25 \end{cases}$

37. $\begin{cases} -x + 7y = 3 \\ -\frac{1}{7}x + y = 5 \end{cases}$

38. $\begin{cases} -7x + 6y = -4 \\ y + \frac{7}{6}x = -1 \end{cases}$

In Exercises 39–46, use a graphing utility to graph the two equations. Use the graphs to approximate the solutions of the system.

39. $\begin{cases} 8x + 9y = 42 \\ 6x - y = 16 \end{cases}$

40. $\begin{cases} 4y = -8 \\ 7x - 2y = 25 \end{cases}$

41. $\begin{cases} \frac{3}{2}x - \frac{1}{5}y = 8 \\ -2x + 3y = 3 \end{cases}$

42. $\begin{cases} \frac{3}{4}x - \frac{5}{2}y = -9 \\ -x + 6y = 28 \end{cases}$

43. $\begin{cases} 0.5x + 2.2y = 9 \\ 6x + 0.4y = -22 \end{cases}$

44. $\begin{cases} 2.4x + 3.8y = -17.6 \\ 4x - 0.2y = -3.2 \end{cases}$

45. $\begin{cases} 7x - 2y = 24 \\ 5x + 6y = -20 \end{cases}$

46. $\begin{cases} 10x - 13y = -20 \\ 8x + 11y = -16 \end{cases}$

In Exercises 47–54, use any method to solve the system.

47. $\begin{cases} 3x - 5y = 7 \\ 2x + y = 9 \end{cases}$

48. $\begin{cases} -x + 3y = 17 \\ 4x + 3y = 7 \end{cases}$

49. $\begin{cases} y = 2x - 5 \\ y = 5x - 11 \end{cases}$

50. $\begin{cases} 7x + 3y = 16 \\ y = x + 2 \end{cases}$

51. $\begin{cases} x - 5y = 21 \\ 6x + 5y = 21 \end{cases}$

52. $\begin{cases} y = -3x - 8 \\ y = 15 - 2x \end{cases}$

53. $\begin{cases} -2x + 8y = 19 \\ y = x - 3 \end{cases}$

54. $\begin{cases} 4x - 3y = 6 \\ -5x + 7y = -1 \end{cases}$

Supply and Demand **In Exercises 55–58, find the point of equilibrium of the given demand and supply equations.**

	Demand	*Supply*
55.	$p = 50 - 0.5x$	$p = 0.125x$
56.	$p = 100 - 0.05x$	$p = 25 + 0.1x$
57.	$p = 140 - 0.00002x$	$p = 80 + 0.00001x$
58.	$p = 400 - 0.0002x$	$p = 225 + 0.0005x$

59. *Airplane Speed* An airplane flying into a headwind travels the 1800-mile flying distance between two cities in 3 hours and 36 minutes. On the return flight, the distance is traveled in 3 hours. Find the air speed of the plane and the speed of the wind, assuming that both remain constant.

60. *Airplane Speed* Two planes start from the same airport and fly in opposite directions. The second plane starts $\frac{1}{2}$ hour after the first plane, but its speed is 80 kilometers per hour faster. Find the air speed of each plane if 2 hours after the first plane departs the planes are 3200 kilometers apart.

61. *Acid Mixture* Ten liters of a 30% acid solution is obtained by mixing a 20% solution with a 50% solution.

(a) Write a system of equations in which one equation represents the amount of final mixture required and the other represents the amount of acid in the final mixture. Let x and y represent the amounts of the 20% and 50% solutions, respectively.

(b) Use a graphing utility to graph the two equations in part (a). As the amount of the 20% solution increases, how does the amount of the 50% solution change?

(c) How much of each solution is required to obtain the specified concentration of the final mixture?

62. *Fuel Mixture* Five hundred gallons of 89 octane gasoline is obtained by mixing 87 octane gasoline with 92 octane gasoline.

(a) Write a system of equations in which one equation represents the amount of final mixture required and the other represents the amounts of 87 and 92 octane gasolines in the final mixture. Let x and y represent the numbers of gallons of 87 octane and 92 octane gasolines, respectively.

(b) Use a graphing utility to graph the two equations in part (a). As the amount of 87 octane gasoline increases, how does the amount of 92 octane gasoline change?

(c) How much of each type of gasoline is required to obtain the 500 gallons of 89 octane gasoline?

63. *Finance* A total of \$12,000 is invested in two corporate bonds that pay 7.5% and 9% simple interest. The investor wants an annual interest income of \$990 from the investments. What amount should be invested in the 7.5% bond?

64. *Finance* A total of \$32,000 is invested in two municipal bonds that pay 5.75% and 6.25% simple interest. The investor wants an annual interest income of \$1900 from the investments. What amount should be invested in the 5.75% bond?

65. *Ticket Sales* At a local championship basketball game, 1435 tickets were sold. A student admission ticket cost \$1.50 and an adult admission ticket cost \$5.00. The total ticket receipts for the ball game were \$3552.50. How many of each ticket were sold?

66. *Clearance Sale* A department store held a sale to sell all of the 214 winter jackets that remained after the season ended. Until noon, each jacket in the store was priced at \$31.95. At noon the price of the jackets was further reduced to \$18.95. After the last jacket was sold, total receipts for the clearance sale were \$5108.30. How many jackets were sold before noon and how many were sold after noon?

67. *Production* A plastics factory uses two different machines working continuously to produce deodorant containers. One machine produces the containers 1.8 times faster than the second machine. If 1764 containers are produced, how many are produced by each machine?

68. *Balloons* A child and his father blow up balloons together for a party. The child inflates two balloons for every three done by his father. How many balloons are inflated by each person to total 80?

Fitting a Line to Data **In Exercises 69–74, find the least squares regression line $y = ax + b$ for the points**

$(x_1, y_1), (x_2, y_2), \ldots, (x_n, y_n)$

by solving the system for a and b. Then use the linear regression capabilities of a graphing utility to confirm the result. (For an explanation of how the coefficients of a and b in the system are obtained, see Appendix B.)

69. $\begin{cases} 5b + 10a = 20.2 \\ 10b + 30a = 50.1 \end{cases}$

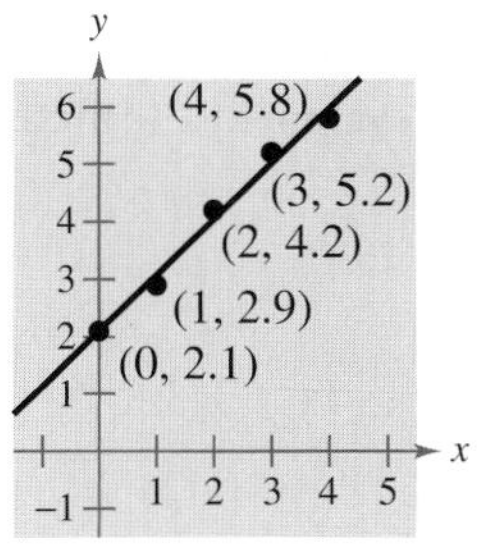

70. $\begin{cases} 5b + 10a = 11.7 \\ 10b + 30a = 25.6 \end{cases}$

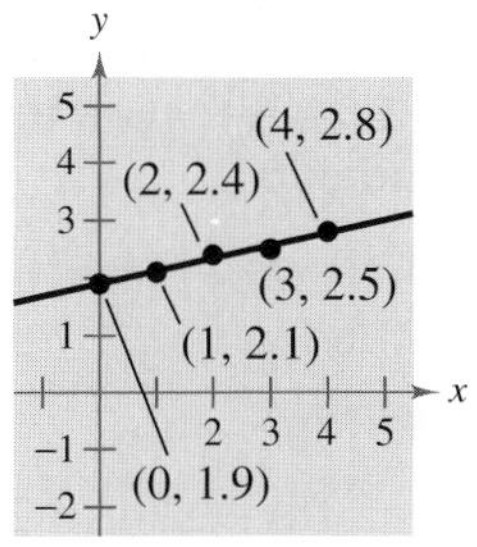

71. $\begin{cases} 7b + 21a = 35.1 \\ 21b + 91a = 114.2 \end{cases}$

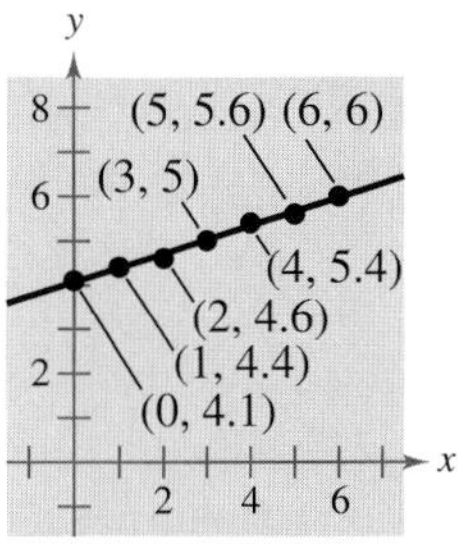

72. $\begin{cases} 6b + 15a = 23.6 \\ 15b + 55a = 48.8 \end{cases}$

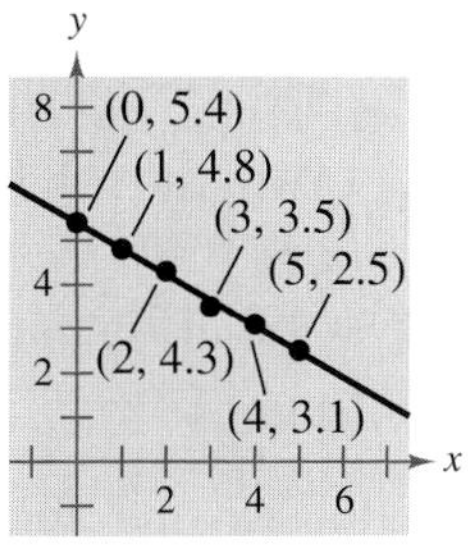

73. $\begin{cases} 4b + 4a = 8 \\ 4b + 6a = 4 \end{cases}$

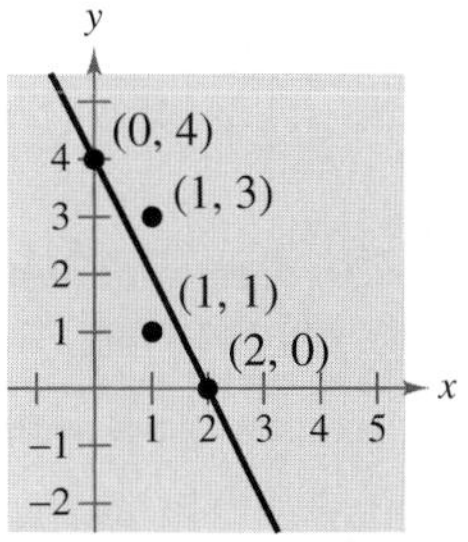

74. $\begin{cases} 8b + 28a = 8 \\ 28b + 116a = 37 \end{cases}$

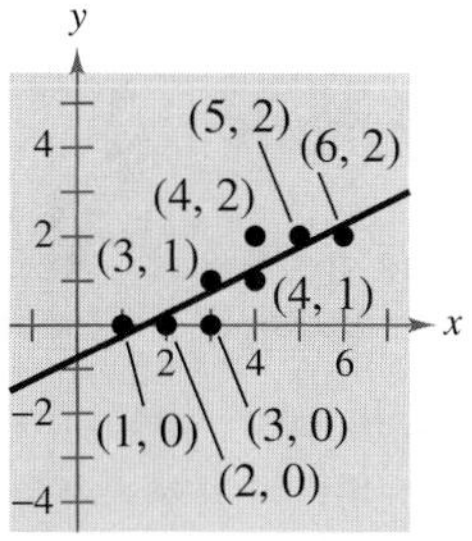

75. *Data Analysis* A store manager wants to know the demand for a certain product as a function of the price. The daily sales for the different prices of the product are given in the table.

Price (x)	\$1.00	\$1.25	\$1.50
Demand (y)	450	375	330

(a) Find the least squares regression line $y = ax + b$ for the data by solving the system for a and b.

$$\begin{cases} 3.00b + 3.7500a = 1155.00 \\ 3.75b + 4.8125a = 1413.75 \end{cases}$$

(b) Use the linear regression capabilities of a graphing utility to confirm the result.

(c) Plot the data and the linear regression equation.

(d) Use the line to predict the demand when the price is \$1.40.

76. *Data Analysis* A farmer used four test plots to determine the relationship between wheat yield in bushels per acre and the amount of fertilizer in hundreds of pounds per acre. The results are given in the table.

Fertilizer (x)	1.0	1.5	2.0	2.5
Yield (y)	32	41	48	53

(a) Find the least squares regression line $y = ax + b$ for the data by solving the system for a and b.

$$\begin{cases} 4b + 7.0a = 174 \\ 7b + 13.5a = 322 \end{cases}$$

(b) Use the linear regression capabilities of a graphing utility to confirm the result.

(c) Plot the data and the linear regression equation.

(d) Use the line to predict the yield for a fertilizer application of 160 pounds per acre.

Synthesis

True or False? **In Exercises 77–79, determine whether the statement is true or false. Justify your answer.**

77. If two lines do not have exactly one point of intersection, then they must be parallel.

78. Solving a system of equations graphically will always give an exact solution.

79. If a system of linear equations has no solution, then the lines must be parallel.

Exploration **In Exercises 80–83, find a system of linear equations that has the given solution. (There is more than one correct answer.)**

80. $(6, 3)$

81. $(8, -2)$

82. $\left(3, \frac{5}{2}\right)$

83. $\left(-\frac{2}{3}, -10\right)$

Think About It **In Exercises 84 and 85, the graphs of the two equations appear to be parallel. Yet, when the system is solved algebraically, you find that the system does have a solution. Find the solution and explain why it does not appear on the portion of the graph that is shown.**

84. $\begin{cases} 100y - x = 200 \\ 99y - x = -198 \end{cases}$

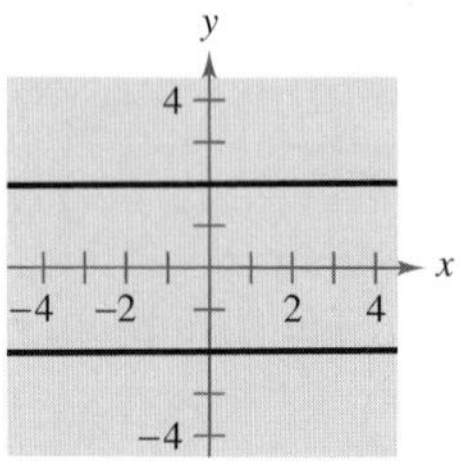

85. $\begin{cases} 21x - 20y = 0 \\ 13x - 12y = 120 \end{cases}$

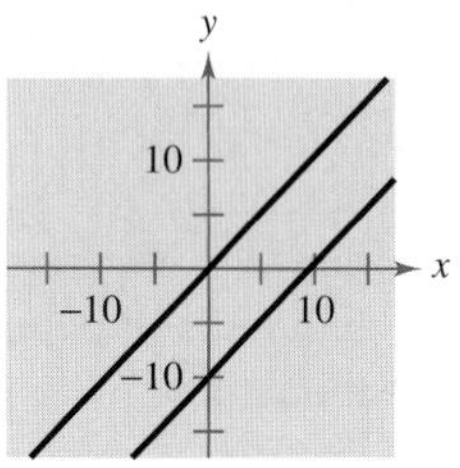

86. *Writing* Briefly explain whether or not it is possible for a consistent system of linear equations to have exactly two solutions.

87. *Think About It* Give examples of (a) a system of linear equations that has no solution and (b) a system that has an infinite number of solutions.

Exploration **In Exercises 88 and 89, find the value of k such that the system of linear equations is inconsistent.**

88. $\begin{cases} 4x - 8y = -3 \\ 2x + ky = 16 \end{cases}$

89. $\begin{cases} 15x + 3y = 6 \\ -10x + ky = 9 \end{cases}$

Review

In Exercises 90–97, solve the inequality and graph the solution on a real number line.

90. $-11 - 6x \geq 33$

91. $2(x - 3) > -5x + 1$

92. $8x - 15 \leq -4(2x - 1)$

93. $-6 \leq 3x - 10 < 6$

94. $|x - 8| < 10$

95. $|x + 10| \geq -3$

96. $2x^2 + 3x - 35 < 0$

97. $3x^2 + 12x > 0$

In Exercises 98 and 99, write the partial fraction decomposition for the rational expression.

98. $\dfrac{x - 1}{x^2 + 11x + 30}$

99. $\dfrac{3}{x(x^2 - 1)}$

In Exercises 100–103, write the expression as the logarithm of a single quantity.

100. $\ln x + \ln 6$

101. $\ln x - 5 \ln(x + 3)$

102. $\log_9 12 - \log_9 x$

103. $\frac{1}{4} \log_6 3x$

In Exercises 104 and 105, solve the system by the method of substitution.

104. $\begin{cases} 2x - y = 4 \\ -4x + 2y = -12 \end{cases}$

105. $\begin{cases} 30x - 40y - 33 = 0 \\ 10x + 20y - 21 = 0 \end{cases}$

7.3 Multivariable Linear Systems

▶ **What you should learn**

- How to recognize linear systems in row-echelon form and use back-substitution to solve the systems
- How to use Gaussian elimination to solve systems of linear equations
- How to solve nonsquare systems of linear equations
- How to use systems of linear equations in three or more variables to model and solve application problems

▶ **Why you should learn it**

Systems of linear equations in three or more variables can be used to model and solve real-life problems. For instance, Exercise 72 on page 588 shows how to use a system of linear equations to analyze the reproduction rates of deer in a wildlife preserve.

Jeanne Drake/Tony Stone Images

Row-Echelon Form and Back-Substitution

The method of elimination can be applied to a system of linear equations in more than two variables. In fact, this method easily adapts to computer use for solving linear systems with dozens of variables.

When elimination is used to solve a system of linear equations, the goal is to rewrite the system in a form to which back-substitution can be applied. To see how this works, consider the following two systems of linear equations.

System of Three Linear Equations in Three Variables: (See Example 3.)

$$\begin{cases} x - 2y + 3z = 9 \\ -x + 3y = -4 \\ 2x - 5y + 5z = 17 \end{cases}$$

Equivalent System in Row-Echelon Form: (See Example 1.)

$$\begin{cases} x - 2y + 3z = 9 \\ y + 3z = 5 \\ z = 2 \end{cases}$$

The second system is said to be in **row-echelon form,** which means that it has a "stair-step" pattern with leading coefficients of 1. After comparing the two systems, it should be clear that it is easier to solve the system in row-echelon form.

Example 1 ▶ Using Back-Substitution in Row-Echelon Form

Solve the system of linear equations.

$$\begin{cases} x - 2y + 3z = 9 & \text{Equation 1} \\ y + 3z = 5 & \text{Equation 2} \\ z = 2 & \text{Equation 3} \end{cases}$$

Solution

From Equation 3, you know the value of z. To solve for y, substitute $z = 2$ into Equation 2 to obtain

$$y + 3(2) = 5 \qquad \text{Substitute 2 for } z.$$

$$y = -1. \qquad \text{Solve for } y.$$

Finally, substitute $y = -1$ and $z = 2$ into Equation 1 to obtain

$$x - 2(-1) + 3(2) = 9 \qquad \text{Substitute } -1 \text{ for } y \text{ and 2 for } z.$$

$$x = 1. \qquad \text{Solve for } x.$$

The solution is $x = 1$, $y = -1$, and $z = 2$, which can be written as the **ordered triple** $(1, -1, 2)$. Check this in the original system of equations.

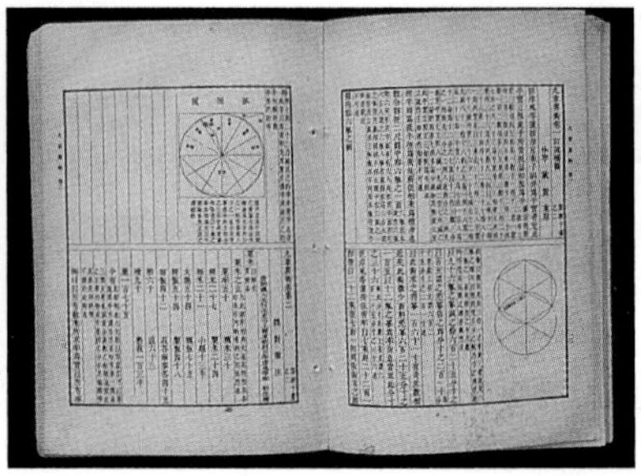

Christopher Lui/China Stock

Historical Note

One of the most influential Chinese mathematics books was the *Chui-chang suan-shu* or *Nine Chapters on the Mathematical Art* (written in approximately 250 B.C.). Chapter Eight of the *Nine Chapters* contained solutions of systems of linear equations using positive and negative numbers. One such system was

$$\begin{cases} 3x + 2y + z = 39 \\ 2x + 3y + z = 34. \\ x + 2y + 3z = 26 \end{cases}$$

This system was solved using column operations on a matrix. Matrices (plural for matrix) will be discussed in the next chapter.

Gaussian Elimination

Two systems of equations are *equivalent* if they have the same solution set. To solve a system that is not in row-echelon form, first convert it to an *equivalent* system that is in row-echelon form by using the following operations.

Operations That Produce Equivalent Systems

Each of the following **row operations** on a system of linear equations produces an *equivalent* system of linear equations.

1. Interchange two equations.
2. Multiply one of the equations by a nonzero constant.
3. Add a multiple of one of the equations to another equation to replace the latter equation.

To see how this is done, let's take another look at the method of elimination, as applied to a system of two linear equations.

Example 2 ▶ Using Gaussian Elimination to Solve a System

Solve the system of linear equations.

$$\begin{cases} 3x - 2y = -1 \\ x - y = 0 \end{cases}$$

Solution

There are two strategies that seem reasonable: eliminate the variable x or eliminate the variable y. The following steps show how to use the first strategy.

$$\begin{cases} x - y = 0 \\ 3x - 2y = -1 \end{cases}$$ Interchange two equations in the system.

$$-3x + 3y = 0$$ Multiply the first equation by -3.

$$\begin{array}{r} -3x + 3y = 0 \\ \underline{3x - 2y = -1} \\ y = -1 \end{array}$$ Add the multiple of the first equation to the second equation to obtain a new equation.

$$\begin{cases} x - y = 0 \\ y = -1 \end{cases}$$ New system in row-echelon form

Now, using back-substitution, you can determine that the solution is $y = -1$ and $x = -1$, which can be written as the ordered pair $(-1, -1)$. Check this in the original system of equations.

STUDY TIP

As demonstrated in the first step in the solution of Example 2, interchanging rows is the easiest way of obtaining a leading coefficient of 1.

As shown in Example 2, rewriting a system of linear equations in row-echelon form usually involves a chain of equivalent systems, each of which is obtained by using one of the three basic row operations listed above. This process is called **Gaussian elimination,** after the German mathematician Carl Friedrich Gauss (1777–1855).

Example 3 ▶ Using Gaussian Elimination to Solve a System

Solve the system of linear equations.

$$\begin{cases} x - 2y + 3z = 9 & \text{Equation 1} \\ -x + 3y = -4 & \text{Equation 2} \\ 2x - 5y + 5z = 17 & \text{Equation 3} \end{cases}$$

Solution

Because the leading coefficient of the first equation is 1, you can begin by saving the x at the upper left and eliminating the other x-terms from the first column.

Arithmetic errors are often made when performing elementary row operations. Have students note the operation performed in each step so that they can go back and check their work.

$$x - 2y + 3z = 9 \quad \text{Write Equation 1.}$$
$$-x + 3y = -4 \quad \text{Write Equation 2.}$$
$$y + 3z = 5 \quad \text{Add Equation 1 to Equation 2.}$$

$$\begin{cases} x - 2y + 3z = 9 \\ y + 3z = 5 \\ 2x - 5y + 5z = 17 \end{cases}$$

Adding the first equation to the second equation produces a new second equation.

$$-2x + 4y - 6z = -18 \quad \text{Multiply Equation 1 by } -2.$$
$$2x - 5y + 5z = 17 \quad \text{Write Equation 3.}$$
$$-y - z = -1 \quad \text{Add revised Equation 1 to Equation 3.}$$

$$\begin{cases} x - 2y + 3z = 9 \\ y + 3z = 5 \\ -y - z = -1 \end{cases}$$

Adding -2 times the first equation to the third equation produces a new third equation.

Now that all but the first x have been eliminated from the first column, go to work on the second column. (You need to eliminate y from the third equation.)

$$\begin{cases} x - 2y + 3z = 9 \\ y + 3z = 5 \\ 2z = 4 \end{cases}$$

Adding the second equation to the third equation produces a new third equation.

Finally, you need a coefficient of 1 for z in the third equation.

$$\begin{cases} x - 2y + 3z = 9 \\ y + 3z = 5 \\ z = 2 \end{cases}$$

Multiplying the third equation by $\frac{1}{2}$ produces a new third equation.

This is the same system that was solved in Example 1, and, as in that example, you can conclude that the solution is

$$x = 1, \qquad y = -1, \qquad \text{and} \qquad z = 2.$$

In Example 3, you can check the solution by substituting $x = 1$, $y = -1$, and $z = 2$ into each original equation, as follows.

Equation 1: $1 - 2(-1) + 3(2) = 9$ ✓

Equation 2: $-1 + 3(-1) = -4$ ✓

Equation 3: $2(1) - 5(-1) + 5(2) = 17$ ✓

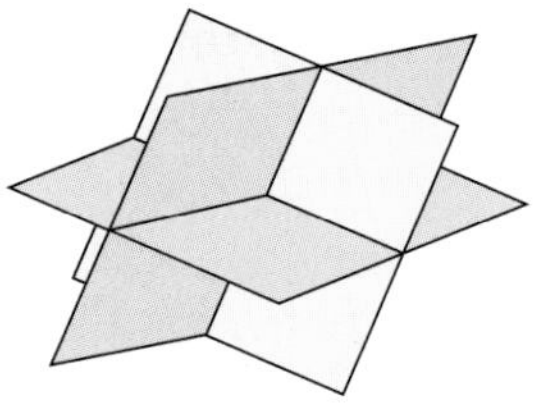

(a) Solution: one point

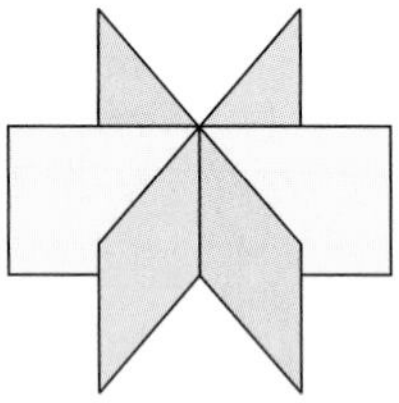

(b) Solution: one line

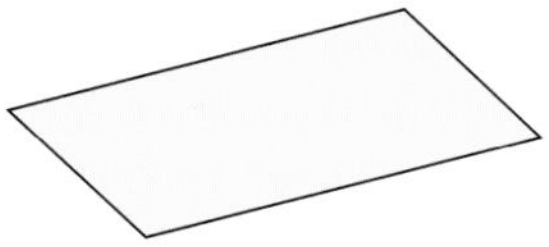

(c) Solution: one plane

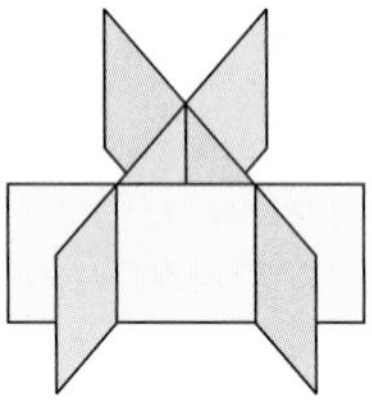

(d) Solution: none

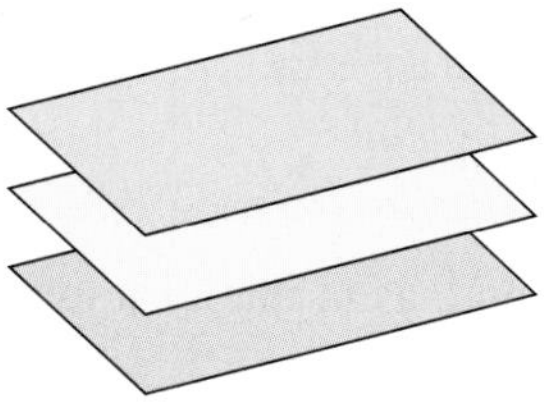

(e) Solution: none

FIGURE 7.11

The next example involves an inconsistent system—one that has no solution. The key to recognizing an inconsistent system is that at some stage in the elimination process you obtain a false statement such as $0 = -2$.

Example 4 ▶ An Inconsistent System

Solve the system of linear equations.

$$\begin{cases} x - 3y + z = 1 & \text{Equation 1} \\ 2x - y - 2z = 2 & \text{Equation 2} \\ x + 2y - 3z = -1 & \text{Equation 3} \end{cases}$$

Solution

$$\begin{cases} x - 3y + z = 1 \\ 5y - 4z = 0 \\ x + 2y - 3z = -1 \end{cases}$$

Adding -2 times the first equation to the second equation produces a new second equation.

$$\begin{cases} x - 3y + z = 1 \\ 5y - 4z = 0 \\ 5y - 4z = -2 \end{cases}$$

Adding -1 times the first equation to the third equation produces a new third equation.

$$\begin{cases} x - 3y + z = 1 \\ 5y - 4z = 0 \\ 0 = -2 \end{cases}$$

Adding -1 times the second equation to the third equation produces a new third equation.

Because the third "equation" is impossible, you can conclude that this system is inconsistent and therefore has no solution. Moreover, because this system is equivalent to the original system, you can conclude that the original system also has no solution.

As with a system of linear equations in two variables, the solution(s) of a system of linear equations in more than two variables must fall into one of three categories.

The Number of Solutions of a Linear System

For a system of linear equations, exactly one of the following is true.

1. There is exactly one solution.
2. There are infinitely many solutions.
3. There is no solution.

In Section 7.2, you learned that a system of two linear equations in two variables can be represented graphically as a pair of lines that are intersecting, coincident, or parallel. A system of three linear equations in three variables has a similar graphical representation—it can be represented as three planes in space that intersect in one point [see Figure 7.11(a)], intersect in a line or a plane [see Figures 7.11(b) and 7.11(c)], or have no points common to all three planes [see Figures 7.11(d) and 7.11(e)].

Example 5 ▶ A System with Infinitely Many Solutions

Solve the system of linear equations.

$$\begin{cases} x + y - 3z = -1 & \text{Equation 1} \\ y - z = 0 & \text{Equation 2} \\ -x + 2y = 1 & \text{Equation 3} \end{cases}$$

Solution

$$\begin{cases} x + y - 3z = -1 \\ y - z = 0 \\ 3y - 3z = 0 \end{cases}$$

Adding the first equation to the third equation produces a new third equation.

$$\begin{cases} x + y - 3z = -1 \\ y - z = 0 \\ 0 = 0 \end{cases}$$

Adding -3 times the second equation to the third equation produces a new third equation.

This means that Equation 3 depends on Equations 1 and 2 in the sense that it gives us no additional information about the variables. So, the original system is equivalent to the system

$$\begin{cases} x + y - 3z = -1 \\ y - z = 0. \end{cases}$$

In this last equation, solve for y in terms of z to obtain $y = z$. Back-substituting for y in the previous equation produces $x = 2z - 1$. Finally, letting $z = a$, where a is a real number, you can see that the solutions to the given system are all of the form

$$x = 2a - 1, \quad y = a, \quad \text{and} \quad z = a.$$

So, every ordered triple of the form

$$(2a - 1, a, a), \quad a \text{ is a real number}$$

is a solution of the system.

Look closely at the original equations in Example 5 and remind students how to recognize equivalent equations (see Section 7.2 for information on equivalent systems). Because Equations 2 and 3 are equivalent, there are infinitely many solutions because there are really only two equations and three unknowns.

STUDY TIP

In Example 5, x and y are solved in terms of the third variable z. To write a solution to the system that does not use any of the three variables of the system, let a represent any real number and let $z = a$. Then solve for x and y. The solution can then be written in terms of a, which is not one of the variables of the system.

In Example 5, there are other ways to write the same infinite set of solutions. For instance, the solutions could have been written as

$$\left(b, \tfrac{1}{2}(b + 1), \tfrac{1}{2}(b + 1)\right), \quad b \text{ is a real number.}$$

To convince yourself that this description produces the same set of solutions, consider the following.

Substitution	*Solution*	
$a = 0$	$(2(0) - 1, 0, 0) = (-1, 0, 0)$	Same solution
$b = -1$	$\left(-1, \tfrac{1}{2}(-1 + 1), \tfrac{1}{2}(-1 + 1)\right) = (-1, 0, 0)$	
$a = 1$	$(2(1) - 1, 1, 1) = (1, 1, 1)$	Same solution
$b = 1$	$\left(1, \tfrac{1}{2}(1 + 1), \tfrac{1}{2}(1 + 1)\right) = (1, 1, 1)$	
$a = 2$	$(2(2) - 1, 2, 2) = (3, 2, 2)$	Same solution
$b = 3$	$\left(3, \tfrac{1}{2}(3 + 1), \tfrac{1}{2}(3 + 1)\right) = (3, 2, 2)$	

STUDY TIP

When comparing descriptions of an infinite solution set, keep in mind that there is more than one way to describe the set.

Nonsquare Systems

So far, each system of linear equations you have looked at has been *square*, which means that the number of equations is equal to the number of variables. In a **nonsquare** system, the number of equations differs from the number of variables. A system of linear equations cannot have a unique solution unless there are at least as many equations as there are variables in the system.

Example 6 ▶ A System with Fewer Equations Than Variables

Solve the system of linear equations.

$$\begin{cases} x - 2y + z = 2 & \text{Equation 1} \\ 2x - y - z = 1 & \text{Equation 2} \end{cases}$$

Solution

Begin by rewriting the system in row-echelon form.

$$\begin{cases} x - 2y + z = 2 \\ 3y - 3z = -3 \end{cases}$$

Adding -2 times the first equation to the second equation produces a new second equation.

$$\begin{cases} x - 2y + z = 2 \\ y - z = -1 \end{cases}$$

Multiplying the second equation by $\frac{1}{3}$ produces a new second equation.

Solving for y in terms of z, you get

$$y = z - 1$$

and back-substitution into Equation 1 yields

$$x - 2(z - 1) + z = 2 \qquad \text{Substitute for } y \text{ in Equation 1.}$$

$$x - 2z + 2 + z = 2 \qquad \text{Distributive Property}$$

$$x = z. \qquad \text{Solve for } x.$$

Finally, by letting $z = a$, where a is a real number, you have the solution

$$x = a, \quad y = a - 1, \quad \text{and} \quad z = a.$$

So, every ordered triple of the form

$$(a, a - 1, a), \quad a \text{ is a real number}$$

is a solution of the system. Because there were originally three variables and only two equations, the system cannot have a unique solution.

In Example 6, try choosing some values of a to obtain different solutions of the system, such as $(1, 0, 1)$, $(2, 1, 2)$, and $(3, 2, 3)$. Then check each of the solutions in the original system.

Activities

Solve the systems of equations.

1. $\begin{cases} x - y + z = 4 \\ x + 3y - 2z = -3 \\ 3x + 2y + 2z = 6 \end{cases}$

   Answer: $(2, -1, 1)$

2. $\begin{cases} x - 2y - z = -5 \\ 2x + y + z = 5 \end{cases}$

   Answer: $(a, 3a, 5 - 5a)$

3. $\begin{cases} x - 2y + z = 4 \\ 3x - 6y + 3z = 7 \\ 2x + y + 4z = 2 \end{cases}$

   Answer: No solution

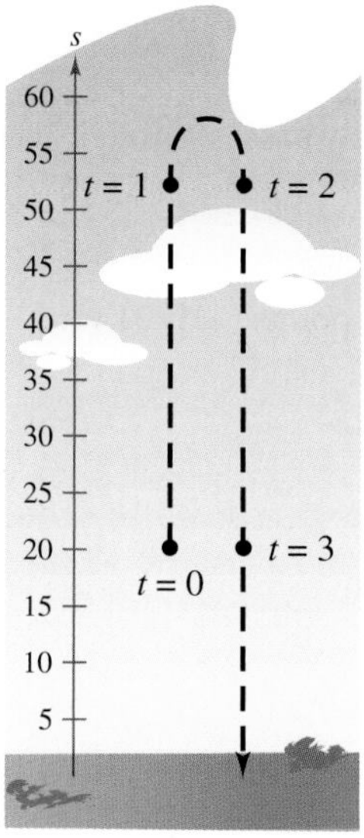

FIGURE 7.12

Applications

Example 7 ▶ Vertical Motion

The height at time t of an object that is moving in a (vertical) line with constant acceleration a is given by the **position equation**

$$s = \frac{1}{2}at^2 + v_0t + s_0.$$

The height s is measured in feet, t is measured in seconds, v_0 is the initial velocity (at $t = 0$), and s_0 is the initial height. Find the values of a, v_0, and s_0 if $s = 52$ at $t = 1$, $s = 52$ at $t = 2$, and $s = 20$ at $t = 3$. (See Figure 7.12.)

Solution

By substituting t and s into the position equation you can obtain three linear equations in a, v_0, and s_0.

When $t = 1$: $\frac{1}{2}a(1)^2 + v_0(1) + s_0 = 52$ ⇨ $a + 2v_0 + 2s_0 = 104$

When $t = 2$: $\frac{1}{2}a(2)^2 + v_0(2) + s_0 = 52$ ⇨ $2a + 2v_0 + s_0 = 52$

When $t = 3$: $\frac{1}{2}a(3)^2 + v_0(3) + s_0 = 20$ ⇨ $9a + 6v_0 + 2s_0 = 40$

Solving this system yields $a = -32$, $v_0 = 48$, and $s_0 = 20$. This solution results in a position equation of $s = -16t^2 + 48t + 20$ and implies that the object was thrown upward at a velocity of 48 feet per second from a height of 20 feet.

Example 8 ▶ Partial Fractions

Write the partial fraction decomposition for $\dfrac{3x + 4}{x^3 - 2x - 4}$.

Solution

Because $x^3 - 2x - 4 = (x - 2)(x^2 + 2x + 2)$, you can write

$$\frac{3x + 4}{x^3 - 2x - 4} = \frac{A}{x - 2} + \frac{Bx + C}{x^2 + 2x + 2}$$

$$3x + 4 = A(x^2 + 2x + 2) + (Bx + C)(x - 2)$$

$$3x + 4 = (A + B)x^2 + (2A - 2B + C)x + (2A - 2C).$$

By equating coefficients of like powers on both sides of the expanded equation, you obtain the following system in A, B, and C.

$$\begin{cases} A + B = 0 \\ 2A - 2B + C = 3 \\ 2A - 2C = 4 \end{cases}$$

You can solve this system to find that $A = 1$, $B = -1$, and $C = -1$. So, the partial fraction decomposition is

$$\frac{3x + 4}{x^3 - 2x - 4} = \frac{1}{x - 2} + \frac{-x - 1}{x^2 + 2x + 2} = \frac{1}{x - 2} - \frac{x + 1}{x^2 + 2x + 2}.$$

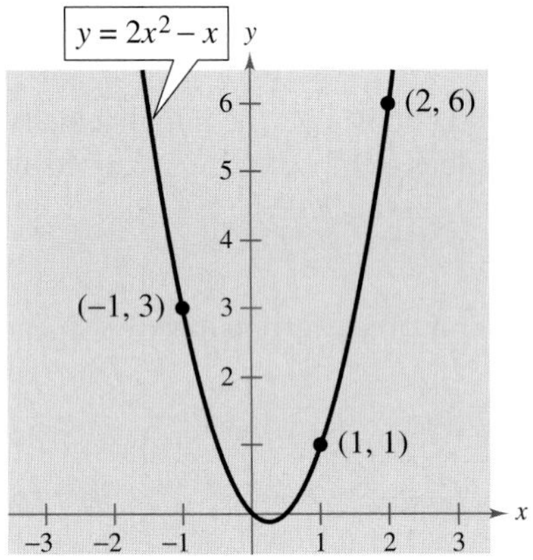

FIGURE 7.13

Example 9 ▶ Data Analysis: Curve-Fitting

Find a quadratic equation, $y = ax^2 + bx + c$, whose graph passes through the points $(-1, 3)$, $(1, 1)$, and $(2, 6)$.

Solution

Because the graph of $y = ax^2 + bx + c$ passes through the points $(-1, 3)$, $(1, 1)$, and $(2, 6)$, you can write the following.

When $x = -1$, $y = 3$: $a(-1)^2 + b(-1) + c = 3$

When $x = 1$, $y = 1$: $a(1)^2 + b(1) + c = 1$

When $x = 2$, $y = 6$: $a(2)^2 + b(2) + c = 6$

This produces the following system of linear equations.

$$\begin{cases} a - b + c = 3 & \text{Equation 1} \\ a + b + c = 1 & \text{Equation 2} \\ 4a + 2b + c = 6 & \text{Equation 3} \end{cases}$$

The solution of this system is $a = 2$, $b = -1$, and $c = 0$. So, the equation of the parabola is $y = 2x^2 - x$, as shown in Figure 7.13.

Example 10 ▶ Investment Analysis

An inheritance of \$12,000 was invested among three funds: a money-market fund that paid 5% annually, municipal bonds that paid 6% annually, and mutual funds that paid 12% annually. The amount invested in mutual funds was \$4000 more than the amount invested in municipal bonds. The total interest earned during the first year was \$1120. How much was invested in each type of fund?

Solution

Let x, y, and z represent the amounts invested in the money-market fund, municipal bonds, and mutual funds, respectively. From the given information, you can write the following equations.

$$\begin{cases} x + y + z = 12{,}000 & \text{Equation 1} \\ z = y + 4000 & \text{Equation 2} \\ 0.05x + 0.06y + 0.12z = 1120 & \text{Equation 3} \end{cases}$$

Rewriting this system in standard form without decimals produces the following.

$$\begin{cases} x + y + z = 12{,}000 & \text{Equation 1} \\ -y + z = 4{,}000 & \text{Equation 2} \\ 5x + 6y + 12z = 112{,}000 & \text{Equation 3} \end{cases}$$

Using Gaussian elimination to solve this system yields $x = 2000$, $y = 3000$, and $z = 7000$. So, \$2000 was invested in the money-market fund, \$3000 was invested in municipal bonds, and \$7000 was invested in mutual funds.

7.3 Exercises

Exercises containing systems with no solutions: 19, 20, 33
Exercises containing systems with infinitely many solutions: 23–30, 37, 38

In Exercises 1–4, determine which ordered triples are solutions of the system of equations.

1. $\begin{cases} 3x - y + z = 1 \\ 2x - 3z = -14 \\ 5y + 2z = 8 \end{cases}$

(a) $(2, 0, -3)$ (b) $(-2, 0, 8)$
(c) $(0, -1, 3)$ (d) $(-1, 0, 4)$

2. $\begin{cases} 3x + 4y - z = 17 \\ 5x - y + 2z = -2 \\ 2x - 3y + 7z = -21 \end{cases}$

(a) $(3, -1, 2)$ (b) $(1, 3, -2)$
(c) $(4, 1, -3)$ (d) $(1, -2, 2)$

3. $\begin{cases} 4x + y - z = 0 \\ -8x - 6y + z = -\frac{7}{4} \\ 3x - y = -\frac{9}{4} \end{cases}$

(a) $\left(\frac{1}{2}, -\frac{3}{4}, -\frac{7}{4}\right)$ (b) $\left(-\frac{3}{2}, \frac{5}{4}, -\frac{5}{4}\right)$
(c) $\left(-\frac{1}{2}, \frac{3}{4}, -\frac{5}{4}\right)$ (d) $\left(-\frac{1}{2}, \frac{1}{6}, -\frac{3}{4}\right)$

4. $\begin{cases} -4x - y - 8z = -6 \\ y + z = 0 \\ 4x - 7y = 6 \end{cases}$

(a) $(-2, -2, 2)$ (b) $\left(-\frac{33}{2}, -10, 10\right)$
(c) $\left(\frac{1}{8}, -\frac{1}{2}, \frac{1}{2}\right)$ (d) $\left(-\frac{11}{2}, -4, 4\right)$

In Exercises 5–10, use back-substitution to solve the system of linear equations.

5. $\begin{cases} 2x - y + 5z = 24 \\ y + 2z = 6 \\ z = 4 \end{cases}$

6. $\begin{cases} 4x - 3y - 2z = 21 \\ 6y - 5z = -8 \\ z = -2 \end{cases}$

7. $\begin{cases} 2x + y - 3z = 10 \\ y = 2 \\ y - z = 4 \end{cases}$

8. $\begin{cases} x = 8 \\ 2x + 3y = 10 \\ x - y + 2z = 22 \end{cases}$

9. $\begin{cases} 4x - 2y + z = 8 \\ 2z = 4 \\ -y + z = 4 \end{cases}$

10. $\begin{cases} 5x - 8z = 22 \\ 3y - 5z = 10 \\ z = -4 \end{cases}$

In Exercises 11 and 12, perform the row operation and write the equivalent system.

11. Add Equation 1 to Equation 2.

$\begin{cases} x - 2y + 3z = 5 & \text{Equation 1} \\ -x + 3y - 5z = 4 & \text{Equation 2} \\ 2x - 3z = 0 & \text{Equation 3} \end{cases}$

What did this operation accomplish?

12. Add -2 times Equation 1 to Equation 3.

$\begin{cases} x - 2y + 3z = 5 & \text{Equation 1} \\ -x + 3y - 5z = 4 & \text{Equation 2} \\ 2x - 3z = 0 & \text{Equation 3} \end{cases}$

What did this operation accomplish?

In Exercises 13–38, solve the system of linear equations and check any solution algebraically.

13. $\begin{cases} x + y + z = 6 \\ 2x - y + z = 3 \\ 3x - z = 0 \end{cases}$

14. $\begin{cases} x + y + z = 3 \\ x - 2y + 4z = 5 \\ 3y + 4z = 5 \end{cases}$

15. $\begin{cases} 2x + 2z = 2 \\ 5x + 3y = 4 \\ 3y - 4z = 4 \end{cases}$

16. $\begin{cases} 2x + 4y + z = 1 \\ x - 2y - 3z = 2 \\ x + y - z = -1 \end{cases}$

17. $\begin{cases} 6y + 4z = -12 \\ 3x + 3y = 9 \\ 2x - 3z = 10 \end{cases}$

18. $\begin{cases} 2x + 4y - z = 7 \\ 2x - 4y + 2z = -6 \\ x + 4y + z = 0 \end{cases}$

19. $\begin{cases} 2x + y - z = 7 \\ x - 2y + 2z = -9 \\ 3x - y + z = 5 \end{cases}$

20. $\begin{cases} 5x - 3y + 2z = 3 \\ 2x + 4y - z = 7 \\ x - 11y + 4z = 3 \end{cases}$

21. $\begin{cases} 3x - 5y + 5z = 1 \\ 5x - 2y + 3z = 0 \\ 7x - y + 3z = 0 \end{cases}$

22. $\begin{cases} 2x + y + 3z = 1 \\ 2x + 6y + 8z = 3 \\ 6x + 8y + 18z = 5 \end{cases}$

23. $\begin{cases} x + 2y - 7z = -4 \\ 2x + y + z = 13 \\ 3x + 9y - 36z = -33 \end{cases}$

24. $\begin{cases} 2x + y - 3z = 4 \\ 4x + 2z = 10 \\ -2x + 3y - 13z = -8 \end{cases}$

25. $\begin{cases} 3x - 3y + 6z = 6 \\ x + 2y - z = 5 \\ 5x - 8y + 13z = 7 \end{cases}$

26. $\begin{cases} x + 4z = 13 \\ 4x - 2y + z = 7 \\ 2x - 2y - 7z = -19 \end{cases}$

27. $\begin{cases} x - 2y + 5z = 2 \\ 4x - z = 0 \end{cases}$

28. $\begin{cases} x - 3y + 2z = 18 \\ 5x - 13y + 12z = 80 \end{cases}$

29. $\begin{cases} 2x - 3y + z = -2 \\ -4x + 9y = 7 \end{cases}$

30. $\begin{cases} 2x + 3y + 3z = 7 \\ 4x + 18y + 15z = 44 \end{cases}$

31. $\begin{cases} x + 3w = 4 \\ 2y - z - w = 0 \\ 3y - 2w = 1 \\ 2x - y + 4z = 5 \end{cases}$

32. $\begin{cases} x + y + z + w = 6 \\ 2x + 3y - w = 0 \\ -3x + 4y + z + 2w = 4 \\ x + 2y - z + w = 0 \end{cases}$

33. $\begin{cases} x + 4z = 1 \\ x + y + 10z = 10 \\ 2x - y + 2z = -5 \end{cases}$

34. $\begin{cases} 2x - 2y - 6z = -4 \\ -3x + 2y + 6z = 1 \\ x - y - 5z = -3 \end{cases}$

35. $\begin{cases} 2x + 3y = 0 \\ 4x + 3y - z = 0 \\ 8x + 3y + 3z = 0 \end{cases}$

36. $\begin{cases} 4x + 3y + 17z = 0 \\ 5x + 4y + 22z = 0 \\ 4x + 2y + 19z = 0 \end{cases}$

37. $\begin{cases} 12x + 5y + z = 0 \\ 23x + 4y - z = 0 \end{cases}$

38. $\begin{cases} 2x - y - z = 0 \\ -2x + 6y + 4z = 2 \end{cases}$

In Exercises 39–42, find the equation of the parabola

$$y = ax^2 + bx + c$$

that passes through the given points. To verify your result, use a graphing utility to plot the points and graph the parabola.

39. $(0, 0), (2, -2), (4, 0)$

40. $(0, 3), (1, 4), (2, 3)$

41. $(2, 0), (3, -1), (4, 0)$

42. $(1, 3), (2, 2), (3, -3)$

In Exercises 43–46, find the equation of the circle

$$x^2 + y^2 + Dx + Ey + F = 0$$

that passes through the given points. To verify your result, use a graphing utility to plot the points and graph the circle.

43. $(0, 0), (2, 2), (4, 0)$

44. $(0, 0), (0, 6), (3, 3)$

45. $(-3, -1), (2, 4), (-6, 8)$

46. $(0, 0), (0, -2), (3, 0)$

Vertical Motion **In Exercises 47–50, an object moving vertically is at the given heights at specified times. Find the position equation $s = \frac{1}{2}at^2 + v_0t + s_0$ for the object.**

47. At $t = 1$ second, $s = 128$ feet
At $t = 2$ seconds, $s = 80$ feet
At $t = 3$ seconds, $s = 0$ feet

48. At $t = 1$ second, $s = 48$ feet
At $t = 2$ seconds, $s = 64$ feet
At $t = 3$ seconds, $s = 48$ feet

49. At $t = 1$ second, $s = 452$ feet
At $t = 2$ seconds, $s = 372$ feet
At $t = 3$ seconds, $s = 260$ feet

50. At $t = 1$ second, $s = 132$ feet
At $t = 2$ seconds, $s = 100$ feet
At $t = 3$ seconds, $s = 36$ feet

51. *Football* Two teams playing in a football game scored a total of 72 points. The points came from a total of 20 different scoring plays, which were a combination of touchdowns, extra-point kicks, and field goals, worth 6 points, 1 point, and 3 points, respectively. The same number of extra points were scored as field goals were kicked. How many touchdowns, extra-point kicks, and field goals were scored?

52. *Basketball* The Aeros scored a total of 104 points in a basketball game. The scoring resulted from a combination of 3-point baskets, 2-point baskets, and 1-point free-throws. There were twice as many 2-point baskets as free-throws scored and twice as many free-throws as 3-point baskets. What combination of scoring accounted for the Aeros' 104 points?

53. *Finance* A small corporation borrowed \$775,000 to expand its product line. Some of the money was borrowed at 8%, some at 9%, and some at 10%. How much was borrowed at each rate if the annual interest owed was \$67,500 and the amount borrowed at 8% was four times the amount borrowed at 10%?

54. *Finance* A small corporation borrowed \$800,000 to expand its product line. Some of the money was borrowed at 8%, some at 9%, and some at 10%. How much was borrowed at each rate if the annual interest owed was \$67,000 and the amount borrowed at 8% was five times the amount borrowed at 10%?

Finance **In Exercises 55 and 56, consider an investor with a portfolio totaling \$500,000 that is invested in certificates of deposit, municipal bonds, blue-chip stocks, and growth or speculative stocks. How much is invested in each type of investment?**

55. The certificates of deposit pay 10% annually, and the municipal bonds pay 8% annually. Over a 5-year period, the investor expects the blue-chip stocks to return 12% annually and the growth stocks to return 13% annually. The investor wants a combined annual return of 10% and also wants to have only one-fourth of the portfolio invested in stocks.

56. The certificates of deposit pay 9% annually, and the municipal bonds pay 5% annually. Over a 5-year period, the investor expects the blue-chip stocks to return 12% annually and the growth stocks to return 14% annually. The investor wants a combined annual return of 10% and also wants to have only one-fourth of the portfolio invested in stocks.

57. *Agriculture* A mixture of 12 liters of chemical A, 16 liters of chemical B, and 26 liters of chemical C is required to kill a certain destructive crop insect. Commercial spray X contains 1, 2, and 2 parts, respectively, of these chemicals. Commercial spray Y contains only chemical C. Commercial spray Z contains only chemicals A and B in equal amounts. How much of each type of commercial spray is needed to get the desired mixture?

58. *Chemistry* A chemist needs 10 liters of a 25% acid solution. The solution is to be mixed from three solutions whose concentrations are 10%, 20%, and 50%. How many liters of each solution should the chemist use to satisfy the following?

(a) Use as little as possible of the 50% solution.

(b) Use as much as possible of the 50% solution.

(c) Use 2 liters of the 50% solution.

59. *Truck Scheduling* A small company that manufactures products A and B has an order for 15 units of product A and 16 units of product B. The company has trucks of three different sizes that can haul the products, as shown in the table.

Truck	Large	Medium	Small
Product A	6	4	0
Product B	3	4	3

How many trucks of each size are needed to deliver the order? Give two possible solutions.

60. *Electrical Network* Applying Kirchhoff's Laws to the electrical network in the figure, the currents I_1, I_2, and I_3 are the solution of the system

$$\begin{cases} I_1 - I_2 + I_3 = 0 \\ 3I_1 + 2I_2 = 7 \\ 2I_2 + 4I_3 = 8. \end{cases}$$

Find the currents.

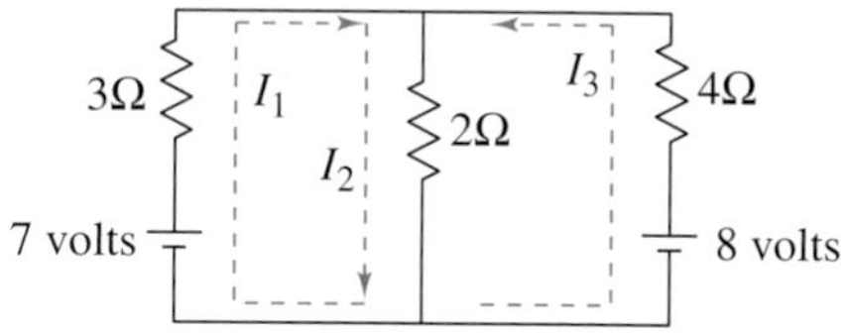

61. *Pulley System* A system of pulleys is loaded with 128-pound and 32-pound weights (see figure). The tensions t_1 and t_2 in the ropes and the acceleration a of the 32-pound weight are found by solving the system of equations

$$\begin{cases} t_1 - 2t_2 = 0 \\ t_1 - 2a = 128 \\ t_2 + a = 32 \end{cases}$$

where t_1 and t_2 are measured in pounds and a is measured in feet per second squared. Solve this system.

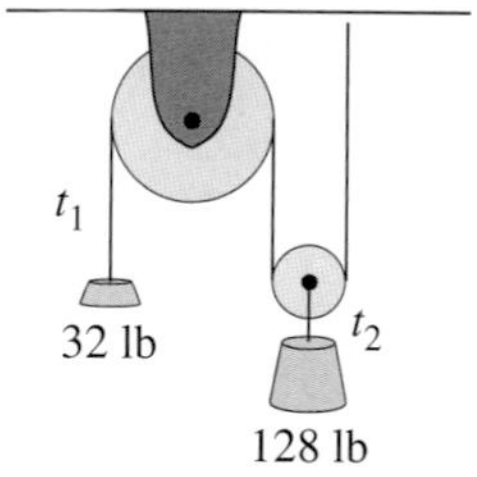

62. *Pulley System* If the 32-pound weight in the pulley system in Exercise 61 is replaced by a 64-pound weight, the new pulley system will be modeled by the following system of equations.

$$\begin{cases} t_1 - 2t_2 = 0 \\ t_1 - 2a = 128 \\ t_2 + 2a = 64 \end{cases}$$

Solve this system and use your answer for the acceleration to describe what (if anything) is happening in the pulley system.

Partial Fraction Decomposition **In Exercises 63–66, write the partial fraction decomposition for the rational expression.**

63. $\dfrac{1}{x^3 - x} = \dfrac{A}{x} + \dfrac{B}{x-1} + \dfrac{C}{x+1}$

64. $\dfrac{3}{x^2 + x - 2} = \dfrac{A}{x-1} + \dfrac{B}{x+2}$

65. $\dfrac{x^2 - 3x - 3}{x(x-2)(x+3)} = \dfrac{A}{x} + \dfrac{B}{x-2} + \dfrac{C}{x+3}$

66. $\dfrac{12}{x(x-2)(x+3)} = \dfrac{A}{x} + \dfrac{B}{x-2} + \dfrac{C}{x+3}$

Fitting a Parabola **In Exercises 67–70, find the least squares regression parabola $y = ax^2 + bx + c$ for the points $(x_1, y_1), (x_2, y_2), \ldots, (x_n, y_n)$ by solving the following system of linear equations for a, b, and c. Then use the least squares regression capabilities of a graphing utility to confirm the result. (For an explanation of how the coefficients of a, b, and c in the system are obtained, see Appendix B.)**

67. $\begin{cases} 4c \qquad\quad + 40a = 19 \\ \quad 40b \qquad\quad = -12 \\ 40c \qquad + 544a = 160 \end{cases}$

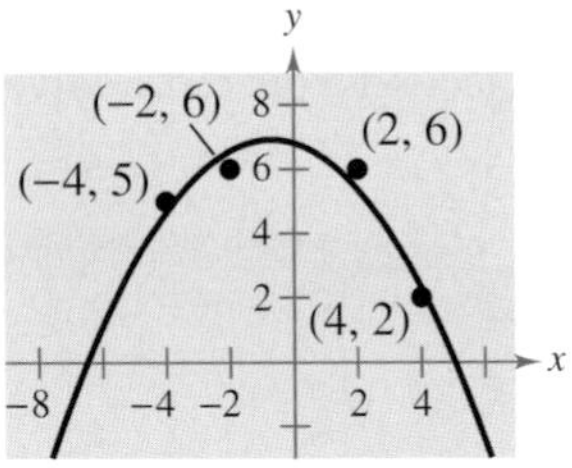

68. $\begin{cases} 5c \qquad + 10a = 8 \\ \quad 10b \qquad\quad = 12 \\ 10c \qquad + 34a = 22 \end{cases}$

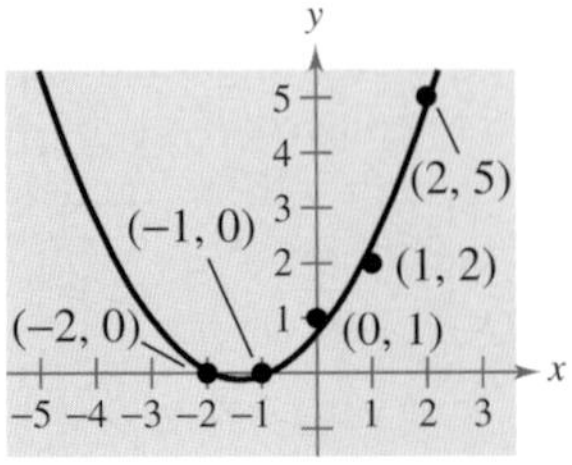

69. $\begin{cases} 4c + 9b + 29a = 20 \\ 9c + 29b + 99a = 70 \\ 29c + 99b + 353a = 254 \end{cases}$

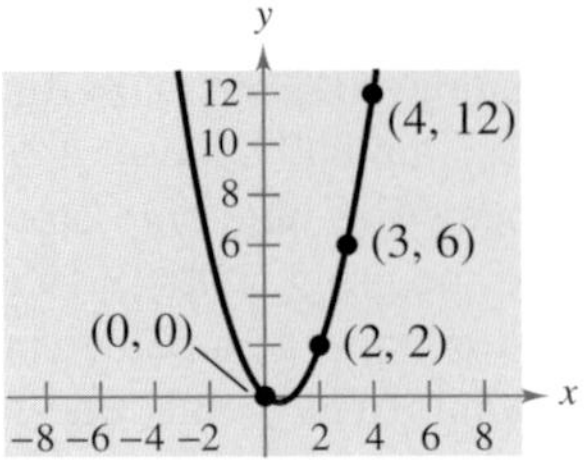

70. $\begin{cases} 4c + 6b + 14a = 25 \\ 6c + 14b + 36a = 21 \\ 14c + 36b + 98a = 33 \end{cases}$

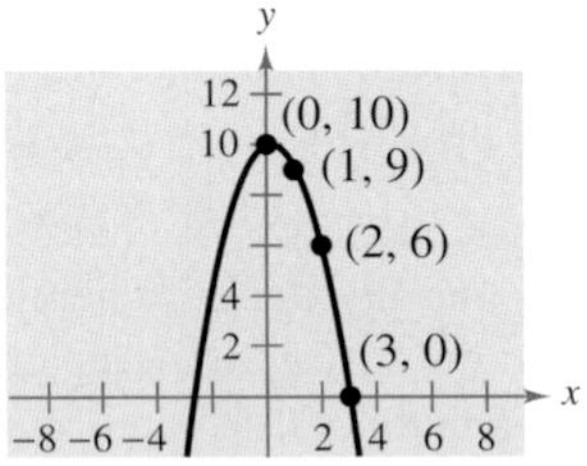

71. *Data Analysis* In testing a new braking system on an automobile, the speed in miles per hour and the stopping distance in feet were recorded in the table.

Speed (x)	30	40	50
Stopping distance (y)	55	105	188

(a) Find a quadratic equation that models the data.

(b) Graph the parabola and the data on the same set of axes.

(c) Use the model to estimate the stopping distance if the speed is 70 miles per hour.

72. *Data Analysis* A wildlife management team studied the reproduction rates of deer in three tracts of a wildlife preserve. Each tract contained 5 acres. In each tract the number of females and the percent of females that had offspring the following year were recorded. The results are given in the table.

Number (x)	100	120	140
Percent (y)	75	68	55

(a) Find a quadratic equation that models the data.

(b) Use a graphing utility to graph the parabola and the data in the same viewing window.

(c) Use the model to estimate the percent of females that had offspring if $x = 170$.

Advanced Applications **In Exercises 73–76, find x, y, and λ satisfying the system. These systems arise in certain optimization problems in calculus, and λ is called a Lagrange multiplier.**

73. $\begin{cases} y + \lambda = 0 \\ x + \lambda = 0 \\ x + y - 10 = 0 \end{cases}$

74. $\begin{cases} 2x + \lambda = 0 \\ 2y + \lambda = 0 \\ x + y - 4 = 0 \end{cases}$

75. $\begin{cases} 2x - 2x\lambda = 0 \\ -2y + \lambda = 0 \\ y - x^2 = 0 \end{cases}$

76. $\begin{cases} 2 + 2y + 2\lambda = 0 \\ 2x + 1 + \lambda = 0 \\ 2x + y - 100 = 0 \end{cases}$

Synthesis

True or False? **In Exercises 77 and 78, determine whether the statement is true or false. Justify your answer.**

77. The system $\begin{cases} x + 3y - 6z = -16 \\ 2y - z = -1 \\ z = 3 \end{cases}$ is in row-echelon form.

78. If a system of three linear equations is inconsistent, then its graph has no points common to all three equations.

79. *Think About It* Are the following two systems of equations equivalent? Give reasons for your answer.

$\begin{cases} x + 3y - z = 6 \\ 2x - y + 2z = 1 \\ 3x + 2y - z = 2 \end{cases}$ $\begin{cases} x + 3y - z = 6 \\ -7y + 4z = 1 \\ -7y - 4z = -16 \end{cases}$

80. *Think About It* One of the following systems is inconsistent and the other has one solution. How can you identify each by observation?

$\begin{cases} 3x - 5y = 3 \\ -12x + 20y = 8 \end{cases}$ $\begin{cases} 3x - 5y = 3 \\ 9x - 20y = 6 \end{cases}$

81. *Think About It* When using Gaussian elimination to solve a system of linear equations, how can you recognize that the system has no solution? Give an example that illustrates your answer.

Exploration **In Exercises 82–85, find a system of linear equations that has the ordered triple as a solution. (The answer is not unique.)**

82. $(4, -1, 2)$

83. $(-5, -2, 1)$

84. $\left(3, -\frac{1}{2}, \frac{7}{4}\right)$

85. $\left(-\frac{3}{2}, 4, -7\right)$

Review

In Exercises 86–89, solve the percent problem.

86. What is $7\frac{1}{2}\%$ of 85?

87. 225 is what percent of 150?

88. 0.5% of what number is 400?

89. 48% of what number is 132?

In Exercises 90–93, (a) determine the real zeros of f and (b) sketch the graph of f.

90. $f(x) = x^3 + x^2 - 12x$

91. $f(x) = -8x^4 + 32x^2$

92. $f(x) = 2x^3 + 5x^2 - 21x - 36$

93. $f(x) = 6x^3 - 29x^2 - 6x + 5$

In Exercises 94–97, use a graphing utility to construct a table of values. Then sketch the graph of the equation by hand.

94. $y = 4^{x-4} - 5$

95. $y = \left(\frac{5}{2}\right)^{-x+1} - 4$

96. $y = 1.9^{-0.8x} + 3$

97. $y = 3.5^{-x+2} + 6$

In Exercises 98 and 99, solve the system by elimination.

98. $\begin{cases} 2x + y = 120 \\ x + 2y = 120 \end{cases}$

99. $\begin{cases} 6x - 5y = 3 \\ 10x - 12y = 5 \end{cases}$

7.4 Systems of Inequalities

▶ **What you should learn**

- How to sketch the graphs of inequalities in two variables
- How to solve systems of inequalities
- How to use systems of inequalities in two variables to model and solve real-life problems

▶ **Why you should learn it**

You can use systems of inequalities in two variables to model and solve real-life problems. For instance, Exercise 72 on page 599 shows how to use a system of inequalities to analyze the compositions of dietary supplements.

Frank Siteman/PhotoEdit

The Graph of an Inequality

The statements $3x - 2y < 6$ and $2x^2 + 3y^2 \geq 6$ are inequalities in two variables. An ordered pair (a, b) is a **solution of an inequality** in x and y if the inequality is true when a and b are substituted for x and y, respectively. The **graph of an inequality** is the collection of all solutions of the inequality. To sketch the graph of an inequality, begin by sketching the graph of the *corresponding equation*. The graph of the equation will normally separate the plane into two or more regions. In each such region, one of the following must be true.

1. *All* points in the region are solutions of the inequality.
2. *No* point in the region is a solution of the inequality.

So, you can determine whether the points in an entire region satisfy the inequality by simply testing *one* point in the region.

Sketching the Graph of an Inequality in Two Variables

1. Replace the inequality sign by an equal sign, and sketch the graph of the resulting equation. (Use a dashed line for $<$ or $>$ and a solid line for $\leq$ or $\geq$.)
2. Test one point in each of the regions formed by the graph in Step 1. If the point satisfies the inequality, shade the entire region to denote that every point in the region satisfies the inequality.

Example 1 ▶ Sketching the Graph of an Inequality

To sketch the graph of $y \geq x^2 - 1$, begin by graphing the corresponding *equation* $y = x^2 - 1$, which is a parabola, as shown in Figure 7.14. By testing a point *above* the parabola (0, 0) and a point *below* the parabola (0, −2), you can see that the points that satisfy the inequality are those lying above (or on) the parabola.

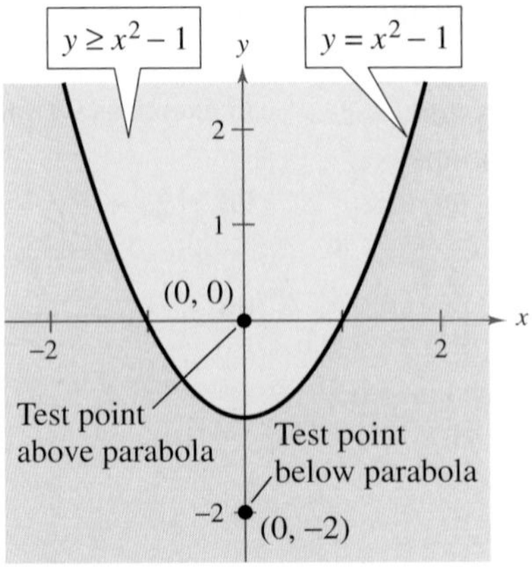

FIGURE 7.14

The inequality given in Example 1 is a nonlinear inequality in two variables. Most of the following examples involve **linear inequalities** such as $ax + by < c$. The graph of a linear inequality is a half-plane lying on one side of the line $ax + by = c$.

Example 2 ▶ Sketching the Graph of a Linear Inequality

Sketch the graph of each linear inequality.

a. $x > -2$ **b.** $y \leq 3$

Solution

a. The graph of the corresponding equation $x = -2$ is a vertical line. The points that satisfy the inequality $x > -2$ are those lying to the right of this line, as shown in Figure 7.15.

b. The graph of the corresponding equation $y = 3$ is a horizontal line. The points that satisfy the inequality $y \leq 3$ are those lying below (or on) this line, as shown in Figure 7.16.

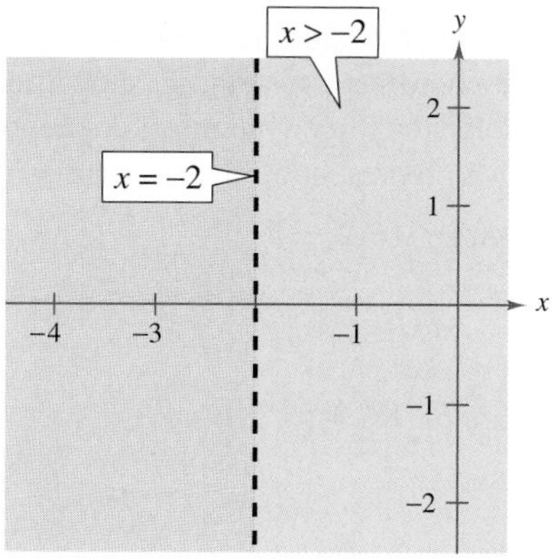

FIGURE 7.15

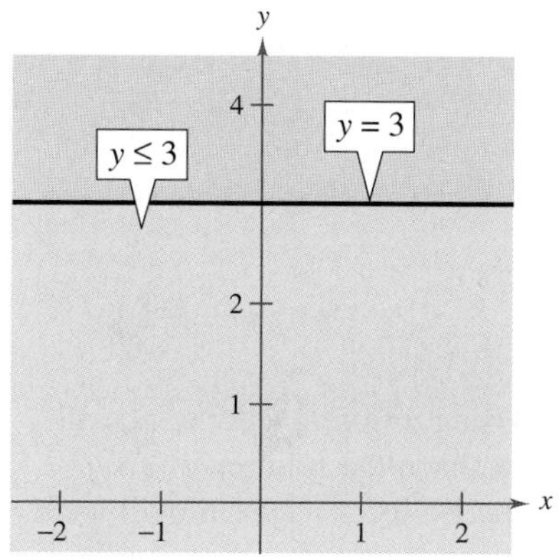

FIGURE 7.16

Technology

A graphing utility can be used to graph an inequality or a system of inequalities. Consult the user's guide for your graphing utility for keystrokes. The graph of the inequality from Example 3 is shown below.

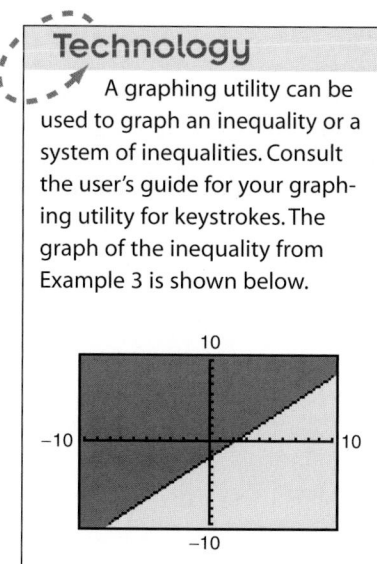

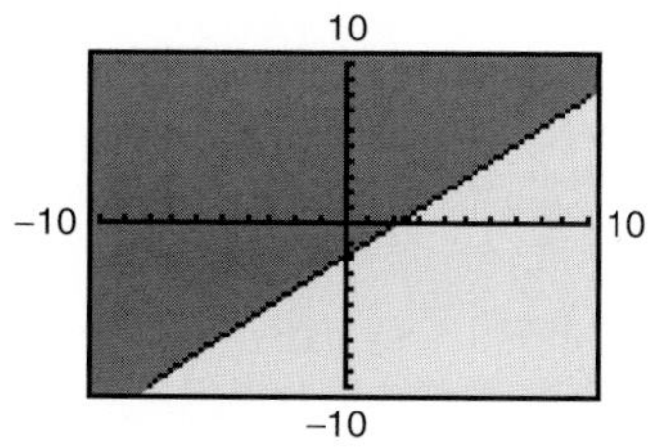

Example 3 ▶ Sketching the Graph of a Linear Inequality

Sketch the graph of $x - y < 2$.

Solution

The graph of the corresponding equation $x - y = 2$ is a line, as shown in Figure 7.17. Because the origin $(0, 0)$ satisfies the inequality, the graph consists of the half-plane lying above the line. (Try checking a point below the line. Regardless of which point you choose, you will see that it does not satisfy the inequality.)

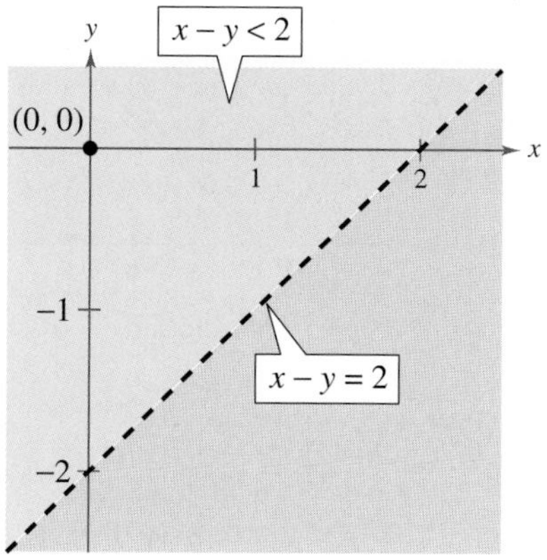

FIGURE 7.17

To graph a linear inequality, it can help to write the inequality in slope-intercept form. For instance, by writing $x - y < 2$ in the form

$$y > x - 2$$

you can see that the solution points lie *above* the line $x - y = 2$ (or $y = x - 2$), as shown in Figure 7.17.

Systems of Inequalities

Many practical problems in business, science, and engineering involve systems of linear inequalities. A **solution** of a system of inequalities in x and y is a point (x, y) that satisfies each inequality in the system.

To sketch the graph of a system of inequalities in two variables, first sketch the graph of each individual inequality (on the same coordinate system) and then find the region that is *common* to every graph in the system. For systems of *linear* inequalities, it is helpful to find the vertices of the solution region.

Example 4 ▶ Solving a System of Inequalities

Sketch the graph (and label the vertices) of the solution set of the system.

$$\begin{cases} x - y < 2 & \text{Inequality 1} \\ x > -2 & \text{Inequality 2} \\ y \le 3 & \text{Inequality 3} \end{cases}$$

Solution

The graphs of these inequalities are shown in Figures 7.15 to 7.17. The triangular region common to all three graphs can be found by superimposing the graphs on the same coordinate system, as shown in Figure 7.18. To find the vertices of the region, solve the three systems of corresponding equations obtained by taking *pairs* of equations representing the boundaries of the individual regions.

Vertex A: $(-2, -4)$

$$\begin{cases} x - y = 2 & \text{Boundary of Inequality 1} \\ x = -2 & \text{Boundary of Inequality 2} \end{cases}$$

Vertex B: $(5, 3)$

$$\begin{cases} x - y = 2 & \text{Boundary of Inequality 1} \\ y = 3 & \text{Boundary of Inequality 3} \end{cases}$$

Vertex C: $(-2, 3)$

$$\begin{cases} x = -2 & \text{Boundary of Inequality 2} \\ y = 3 & \text{Boundary of Inequality 3} \end{cases}$$

STUDY TIP

Using different colored pencils to shade the solution of each inequality in a system will make identifying the solution of the system of inequalities easier.

A computer animation of this example appears in the *Interactive* CD-ROM and *Internet* versions of this text.

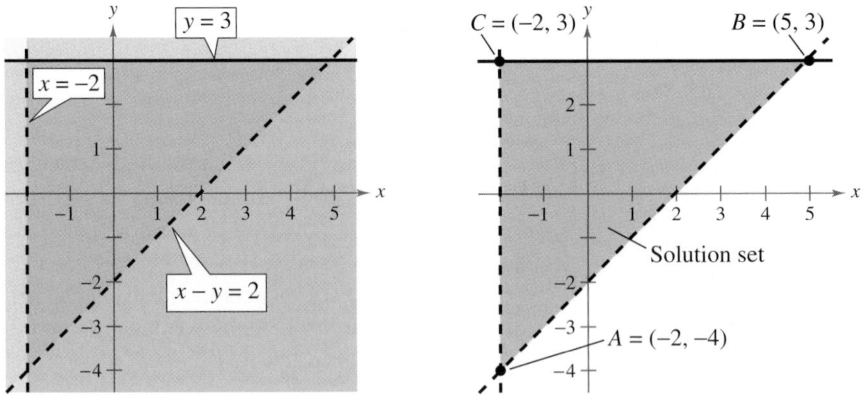

FIGURE 7.18

For the triangular region shown in Figure 7.18, each point of intersection of a pair of boundary lines corresponds to a vertex. With more complicated regions, two border lines can sometimes intersect at a point that is not a vertex of the region, as shown in Figure 7.19. To keep track of which points of intersection are actually vertices of the region, you should sketch the region and refer to your sketch as you find each point of intersection.

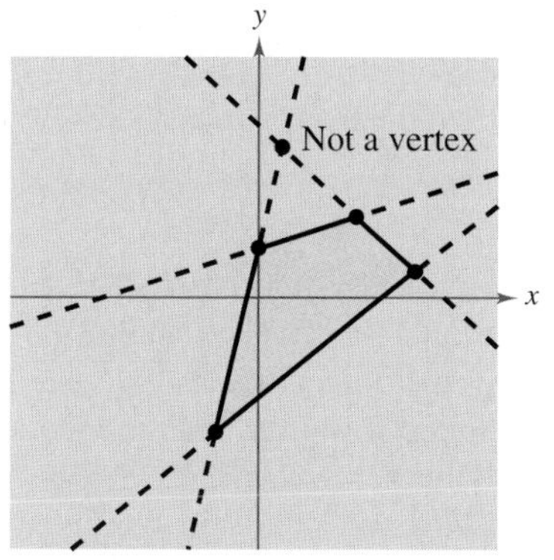

FIGURE 7.19

Example 5 ▶ Solving a System of Inequalities

Sketch the region containing all points that satisfy the system.

$$\begin{cases} x^2 - y \le 1 & \text{Inequality 1} \\ -x + y \le 1 & \text{Inequality 2} \end{cases}$$

Solution

As shown in Figure 7.20, the points that satisfy the inequality

$$x^2 - y \le 1 \qquad \text{Inequality 1}$$

are the points lying above (or on) the parabola given by

$$y = x^2 - 1. \qquad \text{Parabola}$$

The points satisfying the inequality

$$-x + y \le 1 \qquad \text{Inequality 2}$$

are the points lying below (or on) the line given by

$$y = x + 1. \qquad \text{Line}$$

To find the points of intersection of the parabola and the line, solve the system of corresponding equations.

$$\begin{cases} x^2 - y = 1 \\ -x + y = 1 \end{cases}$$

Using the method of substitution, you can find the solutions to be $(-1, 0)$ and $(2, 3)$, as shown in Figure 7.20.

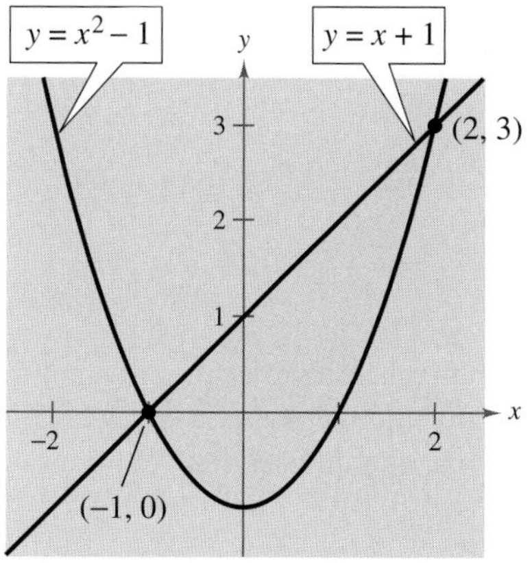

FIGURE 7.20

When solving a system of inequalities, you should be aware that the system might have no solution *or* it might be represented by an unbounded region in the plane. These two possibilities are shown in Examples 6 and 7.

Example 6 ▶ A System with No Solution

Sketch the solution set of the system.

$$\begin{cases} x + y > 3 & \text{Inequality 1} \\ x + y < -1 & \text{Inequality 2} \end{cases}$$

Solution

From the way the system is written, it is clear that the system has no solution, because the quantity $(x + y)$ cannot be both less than -1 and greater than 3. Graphically, the inequality $x + y > 3$ is represented by the half-plane lying above the line $x + y = 3$, and the inequality $x + y < -1$ is represented by the half-plane lying below the line $x + y = -1$, as shown in Figure 7.21. These two half-planes have no points in common. So the system of inequalities has no solution.

Activities

1. Sketch the graph of the inequality.

   $2x - 3y \le 6$

   Answer: The region above the solid line $2x - 3y = 6$.
2. Sketch the graph of the solution of the system of inequalities.

   $\begin{cases} x^2 + y^2 < 16 \\ y > x \end{cases}$

   Answer: The region above the line $y = x$, but in the interior of the circle.

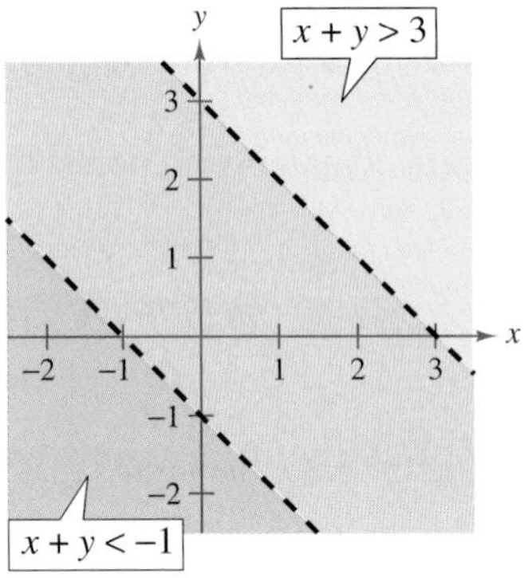

FIGURE 7.21

Example 7 ▶ An Unbounded Solution Set

Sketch the solution set of the system.

$$\begin{cases} x + y < 3 & \text{Inequality 1} \\ x + 2y > 3 & \text{Inequality 2} \end{cases}$$

Solution

The graph of the inequality $x + y < 3$ is the half-plane that lies below the line $x + y = 3$, as shown in Figure 7.22. The graph of the inequality $x + 2y > 3$ is the half-plane that lies above the line $x + 2y = 3$. The intersection of these two half-planes is an *infinite wedge* that has a vertex at $(3, 0)$.

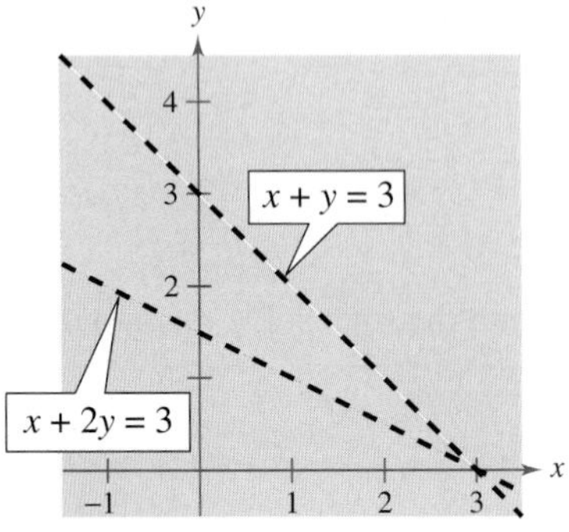

FIGURE 7.22

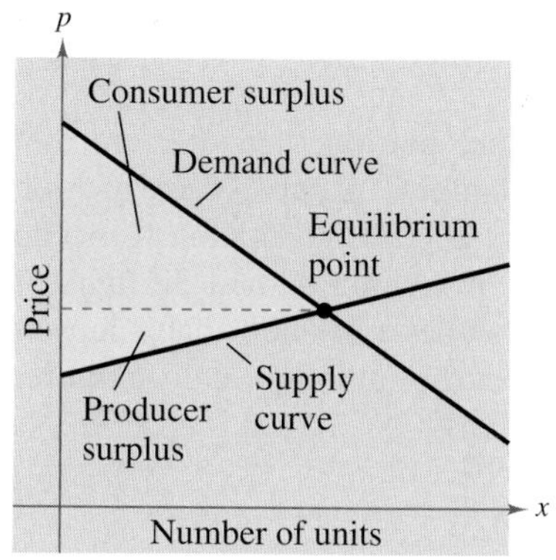

FIGURE 7.23

Applications

Example 9 in Section 7.2 discussed the *point of equilibrium* for a system of demand and supply functions. The next example discusses two related concepts that economists call **consumer surplus** and **producer surplus.** As shown in Figure 7.23, the consumer surplus is defined as the area of the region that lies *below* the demand curve, *above* the horizontal line passing through the equilibrium point, and to the right of the p-axis. Similarly, the producer surplus is defined as the area of the region that lies *above* the supply curve, *below* the horizontal line passing through the equilibrium point, and to the right of the p-axis. The consumer surplus is a measure of the amount that consumers would have been willing to pay *above what they actually paid,* whereas the producer surplus is a measure of the amount that producers would have been willing to receive *below what they actually received.*

Example 8 ▶ Consumer Surplus and Producer Surplus

The demand and supply functions for a certain type of calculator are given by

$$\begin{cases} p = 150 - 0.00001x & \text{Demand equation} \\ p = 60 + 0.00002x & \text{Supply equation} \end{cases}$$

where p is the price in dollars and x represents the number of units. Find the consumer surplus and producer surplus for these two equations.

Solution

Begin by finding the point of equilibrium (when supply and demand are equal) by solving the equation

$$60 + 0.00002x = 150 - 0.00001x.$$

In Example 9 in Section 7.2, you saw that the solution is $x = 3{,}000{,}000$, which corresponds to an equilibrium price of $p = \$120$. So, the consumer surplus and producer surplus are the areas of the following triangular regions.

Consumer Surplus

$$\begin{cases} p \le 150 - 0.00001x \\ p \ge 120 \\ x \ge 0 \end{cases}$$

Producer Surplus

$$\begin{cases} p \ge 60 + 0.00002x \\ p \le 120 \\ x \ge 0 \end{cases}$$

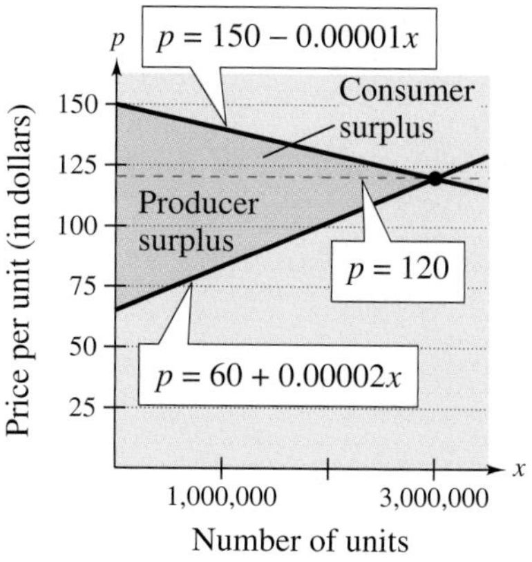

FIGURE 7.24

In Figure 7.24, you can see that the consumer and producer surpluses are defined as the areas of the shaded triangles.

$$\text{Consumer surplus} = \frac{1}{2}(\text{base})(\text{height})$$

$$= \frac{1}{2}(30)(3{,}000{,}000) = \$45{,}000{,}000$$

$$\text{Producer surplus} = \frac{1}{2}(\text{base})(\text{height})$$

$$= \frac{1}{2}(60)(3{,}000{,}000) = \$90{,}000{,}000$$

Example 9 ▶ Nutrition

The minimum daily requirements from the liquid portion of a diet are 300 calories, 36 units of vitamin A, and 90 units of vitamin C. A cup of dietary drink X provides 60 calories, 12 units of vitamin A, and 10 units of vitamin C. A cup of dietary drink Y provides 60 calories, 6 units of vitamin A, and 30 units of vitamin C. Set up a system of linear inequalities that describes how many cups of each drink should be consumed each day to meet the minimum daily requirements for calories and vitamins.

Solution

Begin by letting x and y represent the following.

$x =$ number of cups of dietary drink X

$y =$ number of cups of dietary drink Y

To meet the minimum daily requirements, the following inequalities must be satisfied.

$$\begin{cases} 60x + 60y \geq 300 & \text{Calories} \\ 12x + 6y \geq 36 & \text{Vitamin A} \\ 10x + 30y \geq 90 & \text{Vitamin C} \\ x \geq 0 \\ y \geq 0 \end{cases}$$

The last two inequalities are included because x and y cannot be negative. The graph of this system of inequalities is shown in Figure 7.25. (More is said about this application in Example 6 in Section 7.5.)

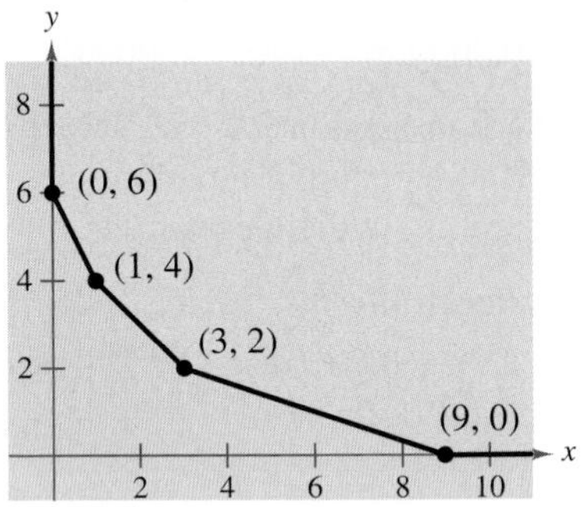

FIGURE 7.25

If you use graphing utilities in your course, you may want your students to work together on the following activity. Consider grouping students with the same model of graphing calculator together so that they can help one another with this activity.

Activity

Using the user's guide for your graphing utility, find how to graph a system of inequalities. Then use the graphing utility to graph the following systems.

$\begin{cases} y \leq 4 - x^2 \\ y \geq 2x - 3 \end{cases}$ $\begin{cases} y \leq 4 - x^2 \\ y \geq x^2 - 4 \end{cases}$

Writing ABOUT MATHEMATICS

Creating a System of Inequalities Plot the points (0, 0), (4, 0), (3, 2), and (0, 2) in a coordinate plane. Draw the quadrilateral that has these four points as its vertices. Write a system of linear inequalities that has the quadrilateral as its solution. Explain how you found the system of inequalities.

7.4 Exercises

Exercises containing systems with no solution: 39

In Exercises 1–16, sketch the graph of the inequality.

1. $x \geq 2$
2. $x \leq 4$
3. $y \geq -1$
4. $y \leq 3$
5. $y < 2 - x$
6. $y > 2x - 4$
7. $2y - x \geq 4$
8. $5x + 3y \geq -15$
9. $(x + 1)^2 + (y - 2)^2 < 9$
10. $y^2 - x < 0$
11. $y \leq \dfrac{1}{1 + x^2}$
12. $y > \dfrac{-15}{x^2 + x + 4}$
13. $y < \ln x$
14. $y \geq 6 - \ln(x + 5)$
15. $y < 3^{-x-4}$
16. $y \leq 2^{2x-0.5} - 7$

In Exercises 17–24, use a graphing utility to graph the inequality. Shade the region representing the solution.

17. $y \geq \frac{2}{3}x - 1$
18. $y \leq 6 - \frac{3}{2}x$
19. $y < -3.8x + 1.1$
20. $y \geq -20.74 + 2.66x$
21. $x^2 + 5y - 10 \leq 0$
22. $2x^2 - y - 3 > 0$
23. $\frac{5}{2}y - 3x^2 - 6 \geq 0$
24. $-\frac{1}{10}x^2 - \frac{3}{8}y < -\frac{1}{4}$

In Exercises 25–28, write an inequality for the shaded region shown in the figure.

25.

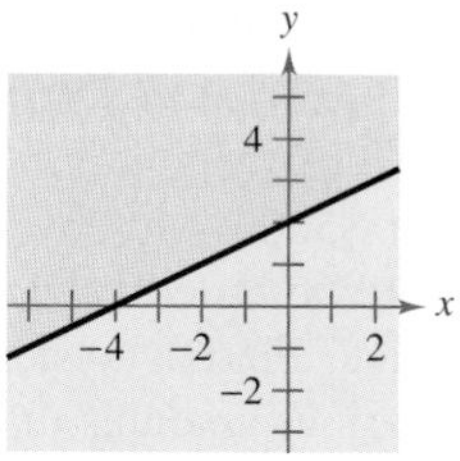

26.

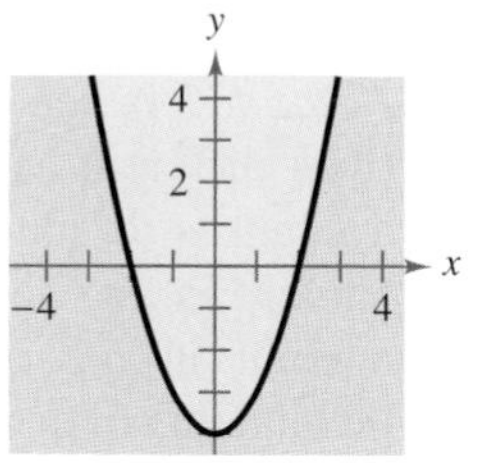

27.

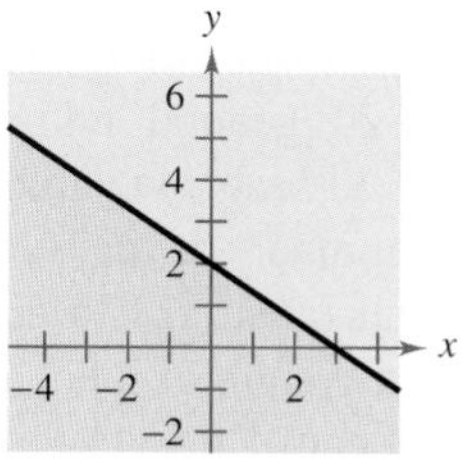

28.

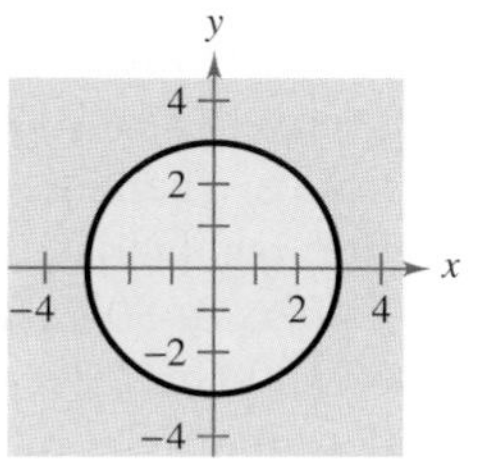

In Exercises 29–32, determine which ordered pairs are solutions of the system of linear inequalities.

29. $\begin{cases} x \geq -4 \\ y > -3 \\ y \leq -8x - 3 \end{cases}$
(a) $(0, 0)$ (b) $(-1, -3)$
(c) $(-4, 0)$ (d) $(-3, 11)$

30. $\begin{cases} -2x + 5y \geq 3 \\ y < 4 \\ -4x + 2y < 7 \end{cases}$
(a) $(0, 2)$ (b) $(-6, 4)$
(c) $(-8, -2)$ (d) $(-3, 2)$

31. $\begin{cases} 3x + y > 1 \\ -y - \frac{1}{2}x^2 \leq -4 \\ -15x + 4y > 0 \end{cases}$
(a) $(0, 10)$ (b) $(0, -1)$
(c) $(2, 9)$ (d) $(-1, 6)$

32. $\begin{cases} x^2 + y^2 \geq 36 \\ -3x + y \leq 10 \\ \frac{2}{3}x - y \geq 5 \end{cases}$
(a) $(-1, 7)$ (b) $(-5, 1)$
(c) $(6, 0)$ (d) $(4, -8)$

In Exercises 33–46, sketch the graph and label the vertices of the solution of the system of inequalities.

33. $\begin{cases} x + y \leq 1 \\ -x + y \leq 1 \\ y \geq 0 \end{cases}$

34. $\begin{cases} 3x + 2y < 6 \\ x > 0 \\ y > 0 \end{cases}$

35. $\begin{cases} x^2 + y \leq 5 \\ x \geq -1 \\ y \geq 0 \end{cases}$

36. $\begin{cases} 2x^2 + y \geq 2 \\ x \leq 2 \\ y \leq 1 \end{cases}$

37. $\begin{cases} -3x + 2y < 6 \\ x - 4y > -2 \\ 2x + y < 3 \end{cases}$

38. $\begin{cases} x - 7y > -36 \\ 5x + 2y > 5 \\ 6x - 5y > 6 \end{cases}$

39. $\begin{cases} 2x + y > 2 \\ 6x + 3y < 2 \end{cases}$

40. $\begin{cases} x - 2y < -6 \\ 5x - 3y > -9 \end{cases}$

41. $\begin{cases} x > y^2 \\ x < y + 2 \end{cases}$

42. $\begin{cases} x - y^2 > 0 \\ x - y > 2 \end{cases}$

43. $\begin{cases} x^2 + y^2 \leq 9 \\ x^2 + y^2 \geq 1 \end{cases}$

44. $\begin{cases} x^2 + y^2 \leq 25 \\ 4x - 3y \leq 0 \end{cases}$

45. $\begin{cases} 3x + 4 \geq y^2 \\ x - y < 0 \end{cases}$

46. $\begin{cases} x < 2y - y^2 \\ 0 < x + y \end{cases}$

In Exercises 47–52, use a graphing utility to graph the inequalities. Shade the region representing the solution of the system.

47. $\begin{cases} y \le \sqrt{3x} + 1 \\ y \ge x^2 + 1 \end{cases}$

48. $\begin{cases} y < -x^2 + 2x + 3 \\ y > x^2 - 4x + 3 \end{cases}$

49. $\begin{cases} y < x^3 - 2x + 1 \\ y > -2x \\ x \le 1 \end{cases}$

50. $\begin{cases} y \ge x^4 - 2x^2 + 1 \\ y \le 1 - x^2 \end{cases}$

51. $\begin{cases} x^2y \ge 1 \\ 0 < x \le 4 \\ y \le 4 \end{cases}$

52. $\begin{cases} y \le e^{-x^2/2} \\ y \ge 0 \\ -2 \le x \le 2 \end{cases}$

In Exercises 53–62, derive a set of inequalities to describe the region.

53.

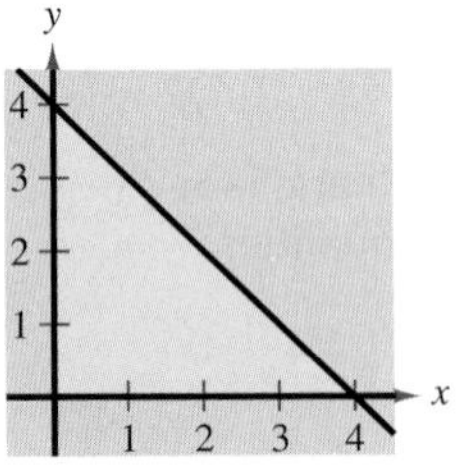

54.

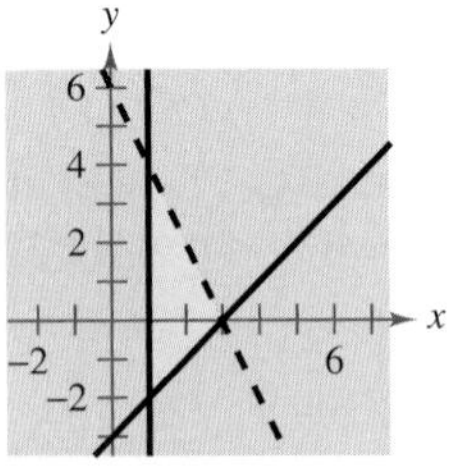

55.

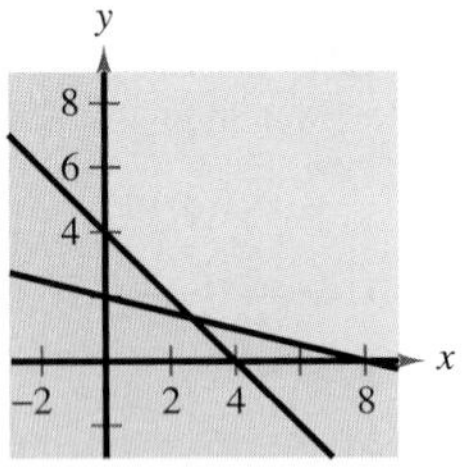

56.

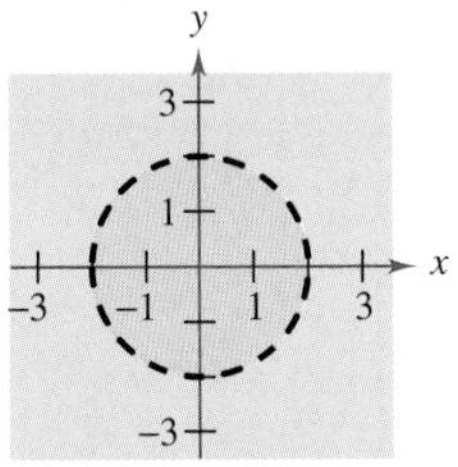

57.

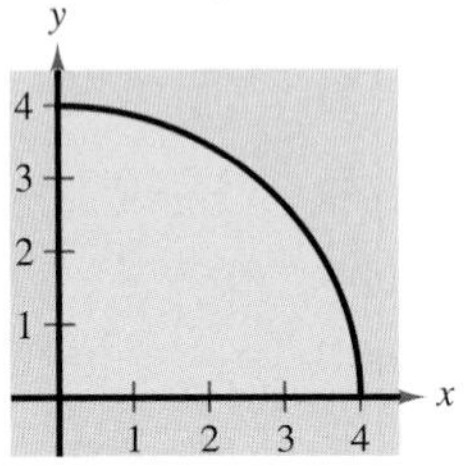

58.

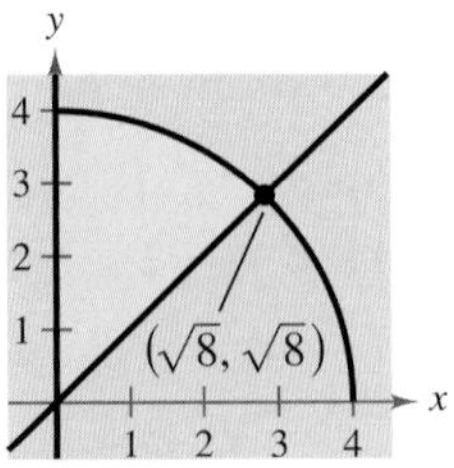

59. Rectangle: Vertices at (2, 1), (5, 1), (5, 7), (2, 7)

60. Parallelogram: Vertices at (0, 0), (4, 0), (1, 4), (5, 4)

61. Triangle: Vertices at (0, 0), (5, 0), (2, 3)

62. Triangle: Vertices at (−1, 0), (1, 0), (0, 1)

Economics **In Exercises 63–66, find the consumer surplus and producer surplus for the demand and supply equations.**

	Demand	*Supply*
63.	$p = 50 - 0.5x$	$p = 0.125x$
64.	$p = 100 - 0.05x$	$p = 25 + 0.1x$
65.	$p = 140 - 0.00002x$	$p = 80 + 0.00001x$
66.	$p = 400 - 0.0002x$	$p = 225 + 0.0005x$

67. *Business* A furniture company can sell all the tables and chairs it produces. Each table requires 1 hour in the assembly center and $1\frac{1}{3}$ hours in the finishing center. Each chair requires $1\frac{1}{2}$ hours in the assembly center and $1\frac{1}{2}$ hours in the finishing center. The company's assembly center is available 12 hours per day, and its finishing center is available 15 hours per day. Find and graph a system of inequalities describing all possible production levels.

68. *Business* A store sells two models of computers. Because of the demand, the store stocks at least twice as many units of model A as of model B. The costs to the store for the two models are \$800 and \$1200, respectively. The management does not want more than \$20,000 in computer inventory at any one time, and it wants at least four model A computers and two model B computers in inventory at all times. Devise a system of inequalities describing all possible inventory levels, and graph the system.

69. *Finance* A person plans to invest up to \$20,000 in two different interest-bearing accounts. Each account is to contain at least \$5000. Moreover, the amount in one account should be at least twice the amount in the other account. Find a system of inequalities to describe the various amounts that can be deposited in each account, and graph the system.

70. *Entertainment* For a concert event, there are \$30 reserved seat tickets and \$20 general admission tickets. There are 2000 reserved seats available, and fire regulations limit the number of paid ticket holders to 3000. The promoter must take in at least \$75,000 in ticket sales. Find a system of inequalities describing the different numbers of tickets that can be sold. Graph the system.

71. *Shipping* A warehouse supervisor is told to ship at least 50 packages of gravel that weigh 55 pounds each and at least 40 bags of stone that weigh 70 pounds each. The maximum weight capacity in the truck he is loading is 7500 pounds. Find a system of inequalities describing the numbers of bags of stone and gravel that he can send. Graph the system.

72. *Nutrition* A dietitian is asked to design a special dietary supplement using two different foods. Each ounce of food X contains 20 units of calcium, 15 units of iron, and 10 units of vitamin B. Each ounce of food Y contains 10 units of calcium, 10 units of iron, and 20 units of vitamin B. The minimum daily requirements of the diet are 300 units of calcium, 150 units of iron, and 200 units of vitamin B. Find and graph a system of inequalities describing the different amounts of food X and food Y that can be used.

73. *Physical Fitness Facility* An indoor running track is to be constructed with a space for body-building equipment inside the track (see figure). The track must be at least 125 meters long, and the body-building space must have an area of at least 500 square meters. Find a system of inequalities describing the requirements of the facility. Graph the system.

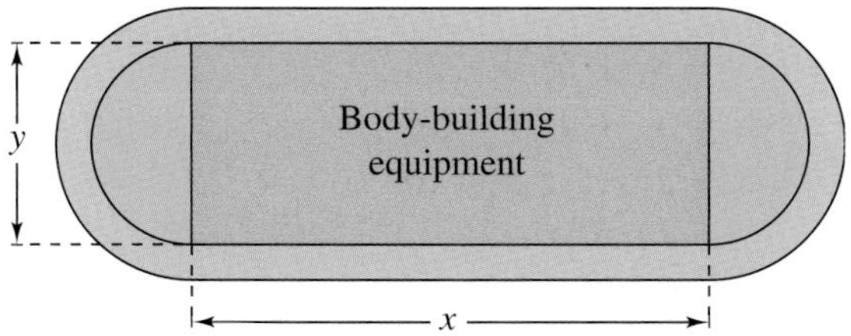

Synthesis

True or False? **In Exercises 74 and 75, determine whether the statement is true or false. Justify your answer.**

74. The area of the figure defined by the system

$$\begin{cases} x \ge -3 \\ x \le 6 \\ y \le 5 \\ y \ge -6 \end{cases}$$

is 99 square units.

75. The graph below shows the solution of the system

$$\begin{cases} y \le 6 \\ -4x - 9y > 6. \\ 3x + y^2 \ge 2 \end{cases}$$

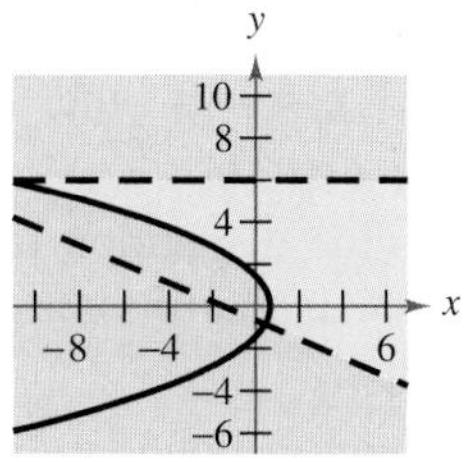

In Exercises 76–79, match the system of inequalities with the graph of its solution. [The graphs are labeled (a), (b), (c), and (d).]

(a)

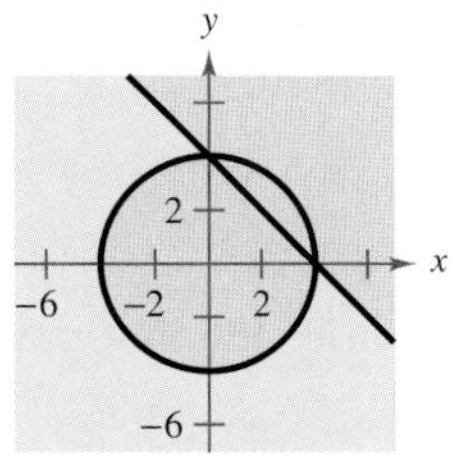

(b)

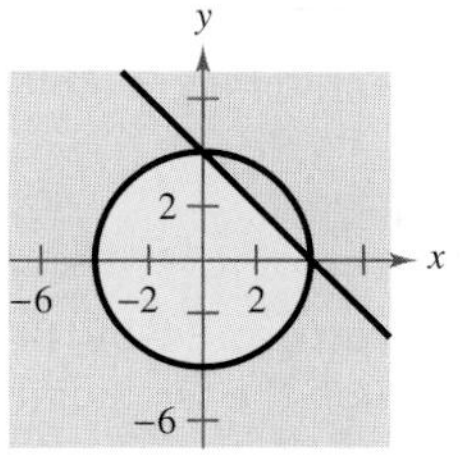

(c)

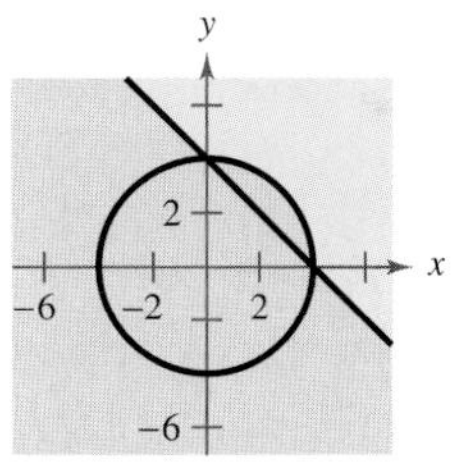

(d)

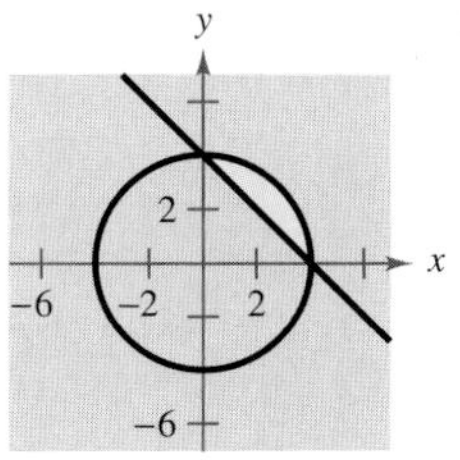

76. $\begin{cases} x^2 + y^2 \le 16 \\ x + y \ge 4 \end{cases}$

77. $\begin{cases} x^2 + y^2 \le 16 \\ x + y \le 4 \end{cases}$

78. $\begin{cases} x^2 + y^2 \ge 16 \\ x + y \ge 4 \end{cases}$

79. $\begin{cases} x^2 + y^2 \ge 16 \\ x + y \le 4 \end{cases}$

80. The graph of the solution of the inequality $x + 2y < 6$ is shown in the figure. Describe how the solution set would change for each of the following.

(a) $x + 2y \le 6$ (b) $x + 2y > 6$

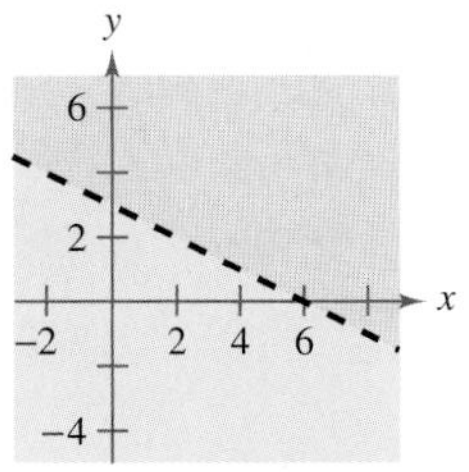

81. *Graphical Reasoning* Two concentric circles have radii x and y, where $y > x$. The area between the circles must be at least 10 square units.

(a) Find a system of inequalities describing the constraints on the circles.

(b) Use a graphing utility to graph the inequality in part (a). Graph the line $y = x$ in the same viewing window.

(c) Identify the graph of the line in relation to the boundary of the inequality. Explain its meaning in the context of the problem.

82. *Think About It* After graphing the boundary of an inequality in x and y, how do you decide on which side of the boundary the solution set of the inequality lies?

83. *Writing* Explain the difference between the graph of the inequality $x \le 4$ on the real number line and on the rectangular coordinate system.

90. *Mortgage Loans* The table shows the numbers of outstanding mortgage loans M (in millions) in the United States for the years 1992 to 1997. (Source: Mortgage Bankers Association of America)

Year	1992	1993	1994	1995	1996	1997
M	42.6	45.3	47.6	49.2	50.1	51.2

Use your graphing utility to find a linear model and a quadratic model that represent the data. Let $t = 0$ represent 1990. Use your graphing utility to plot the actual data and the models in the same viewing window. How closely do the models represent the data?

Review

In Exercises 84–89, find the equation of the line passing through the two points.

84. $(-2, 6), (4, -4)$

85. $(-8, 0), (3, -1)$

86. $\left(\frac{3}{4}, -2\right), \left(-\frac{7}{2}, 5\right)$

87. $\left(-\frac{1}{2}, 0\right), \left(\frac{11}{2}, 12\right)$

88. $(3.4, -5.2), (-2.6, 0.8)$

89. $(-4.1, -3.8), (2.9, 8.2)$

In Exercises 91–96, evaluate the expression. Round your result to three decimal places.

91. $(2.7)^{3.99}$

92. $150(4^{-2.6})$

93. $1.5^{-3\pi}$

94. $(3.815)^6$

95. $e^{-11/4}$

96. $e^{-\sqrt{13}}$

In Exercises 97 and 98, solve the system of linear equations and check any solution algebraically.

97. $\begin{cases} -x - 2y + 3z = -23 \\ 2x + 6y - z = 17 \\ 5y + z = 8 \end{cases}$

98. $\begin{cases} 7x - 3y + 5z = -28 \\ 4x + 4z = -16 \\ 7x + 2y - z = 0 \end{cases}$

7.5 Linear Programming

▶ **What you should learn**

- How to solve linear programming problems
- How to use linear programming to model and solve real-life problems

▶ **Why you should learn it**

Linear programming is often useful in making real-life economic decisions. For example, Exercise 35 on page 608 shows how a merchant could use linear programming to analyze the profitability of two models of compact disc players.

Jeff Greenberg/Unicorn Stock Photos

Linear Programming: A Graphical Approach

Many applications in business and economics involve a process called **optimization,** in which you are asked to find the minimum or maximum of a quantity. In this section you will study an optimization strategy called **linear programming.**

A two-dimensional linear programming problem consists of a linear **objective function** and a system of linear inequalities called **constraints.** The objective function gives the quantity that is to be maximized (or minimized), and the constraints determine the set of **feasible solutions.** For example, suppose you are asked to maximize the value of

$$z = ax + by \qquad \text{Objective function}$$

subject to a set of constraints that determines the shaded region in Figure 7.26.

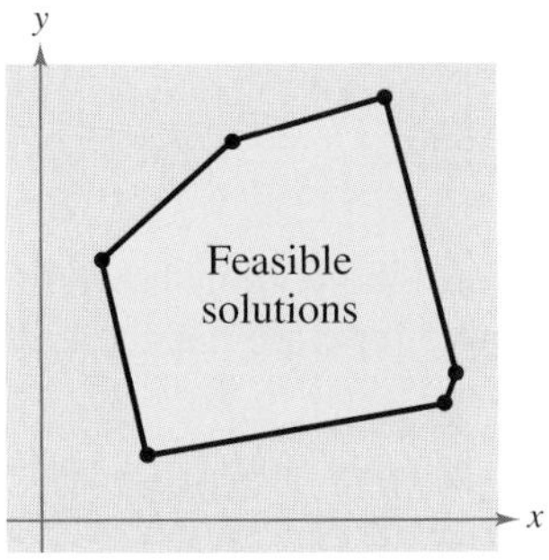

FIGURE 7.26

Because every point in the shaded region satisfies each constraint, it is not clear how you should find the point that yields a maximum value of z. Fortunately, it can be shown that if there is an optimal solution, it must occur at one of the vertices. This means that *you can find the maximum value of z by testing z at each of the vertices.*

Optimal Solution of a Linear Programming Problem

If a linear programming problem has a solution, it must occur at a vertex of the set of feasible solutions. If there is more than one solution, at least one of them must occur at such a vertex. In either case, the value of the objective function is unique.

Some guidelines for solving a linear programming problem in two variables are listed at the top of the next page.

It is important that students immediately become familiar with the terminology of this section. Emphasize that linear programming is powerful in that it identifies quickly those few points out of many that can be easily tested to find the point that gives the maximum or minimum value.

Solving a Linear Programming Problem

1. Sketch the region corresponding to the system of constraints. (The points inside or on the boundary of the region are *feasible solutions.*)
2. Find the vertices of the region.
3. Test the objective function at each of the vertices and select the values of the variables that optimize the objective function. For a bounded region, both a minimum and a maximum value will exist. (For an unbounded region, *if* an optimal solution exists, it will occur at a vertex.)

A computer animation of this example appears in the *Interactive* CD-ROM and *Internet* versions of this text.

Example 1 ▶ Solving a Linear Programming Problem

Find the maximum value of

$$z = 3x + 2y \qquad \text{Objective function}$$

subject to the following constraints.

$$\left.\begin{aligned} x &\ge 0 \\ y &\ge 0 \\ x + 2y &\le 4 \\ x - y &\le 1 \end{aligned}\right\} \qquad \text{Constraints}$$

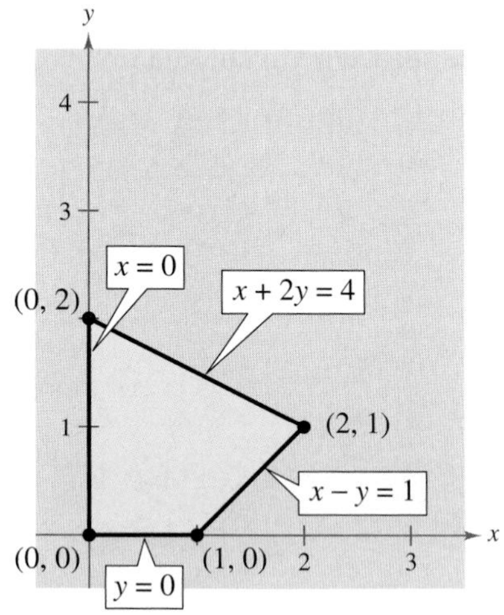

FIGURE 7.27

Solution

The constraints form the region shown in Figure 7.27. At the four vertices of this region, the objective function has the following values.

At (0, 0): $z = 3(0) + 2(0) = 0$

At (1, 0): $z = 3(1) + 2(0) = 3$

At (2, 1): $z = 3(2) + 2(1) = 8$ Maximum value of z

At (0, 2): $z = 3(0) + 2(2) = 4$

So, the maximum value of z is 8, and this occurs when $x = 2$ and $y = 1$.

In Example 1, try testing some of the *interior* points in the region. You will see that the corresponding values of z are less than 8. Here are some examples.

At (1, 1): $z = 3(1) + 2(1) = 5$

At $\left(\frac{1}{2}, \frac{3}{2}\right)$: $z = 3\left(\frac{1}{2}\right) + 2\left(\frac{3}{2}\right) = \frac{9}{2}$

To see why the maximum value of the objective function in Example 1 must occur at a vertex, consider writing the objective function in slope-intercept form

$$y = -\frac{3}{2}x + \frac{z}{2} \qquad \text{Family of lines}$$

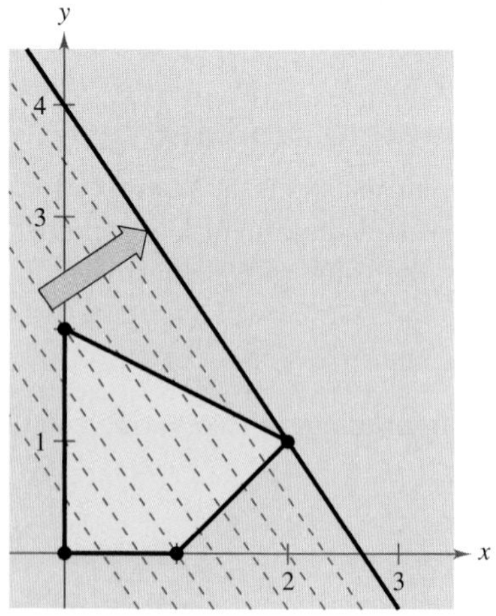

FIGURE 7.28

where $z/2$ is the y-intercept of the objective function. This equation represents a family of lines, each of slope $-\frac{3}{2}$. Of these infinitely many lines, you want the one that has the largest z-value while still intersecting the region determined by the constraints. In other words, of all the lines whose slope is $-\frac{3}{2}$, you want the one that has the largest y-intercept *and* intersects the given region, as shown in Figure 7.28. From the graph you can see that such a line will pass through one (or more) of the vertices of the region.

The next example shows that the same basic procedure can be used to solve a problem in which the objective function is to be *minimized*.

Example 2 ▶ Minimizing an Objective Function

Find the minimum value of

$$z = 5x + 7y \qquad \text{Objective function}$$

where $x \geq 0$ and $y \geq 0$, subject to the following constraints.

$$\left.\begin{aligned} 2x + 3y &\geq 6 \\ 3x - y &\leq 15 \\ -x + y &\leq 4 \\ 2x + 5y &\leq 27 \end{aligned}\right\} \qquad \text{Constraints}$$

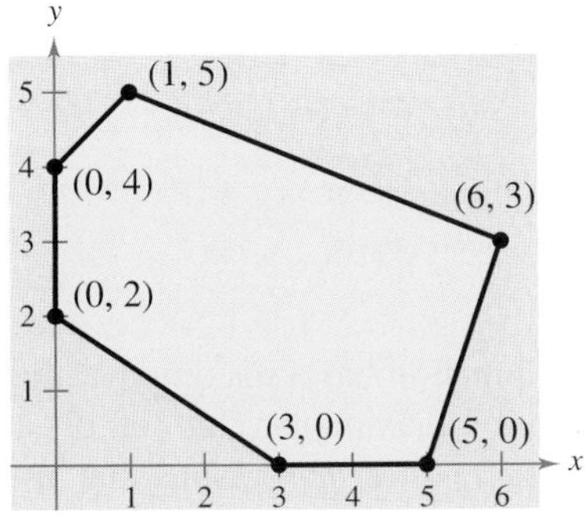

FIGURE 7.29

Solution

The region bounded by the constraints is shown in Figure 7.29. By testing the objective function at each vertex, you obtain the following.

At $(0, 2)$: $z = 5(0) + 7(2) = 14$ Minimum value of z
At $(0, 4)$: $z = 5(0) + 7(4) = 28$
At $(1, 5)$: $z = 5(1) + 7(5) = 40$
At $(6, 3)$: $z = 5(6) + 7(3) = 51$
At $(5, 0)$: $z = 5(5) + 7(0) = 25$
At $(3, 0)$: $z = 5(3) + 7(0) = 15$

So, the minimum value of z is 14, and this occurs when $x = 0$ and $y = 2$.

Edward W. Souza/News Service/ Stanford University

Historical Note
George Dantzig (1914–) was the first to propose the simplex method, or linear programming, in 1947. This technique defined the steps needed to find the optimal solution to a complex multivariable problem.

Example 3 ▶ Maximizing an Objective Function

Find the maximum value of

$$z = 5x + 7y \qquad \text{Objective function}$$

where $x \geq 0$ and $y \geq 0$, subject to the following constraints.

$$\left.\begin{aligned} 2x + 3y &\geq 6 \\ 3x - y &\leq 15 \\ -x + y &\leq 4 \\ 2x + 5y &\leq 27 \end{aligned}\right\} \qquad \text{Constraints}$$

Solution

This linear programming problem is identical to that given in Example 2 above, *except* that the objective function is maximized instead of minimized. Using the values of z at the vertices shown above, you can conclude that the maximum value of

$$z = 5(6) + 7(3) = 51$$

occurs when $x = 6$ and $y = 3$.

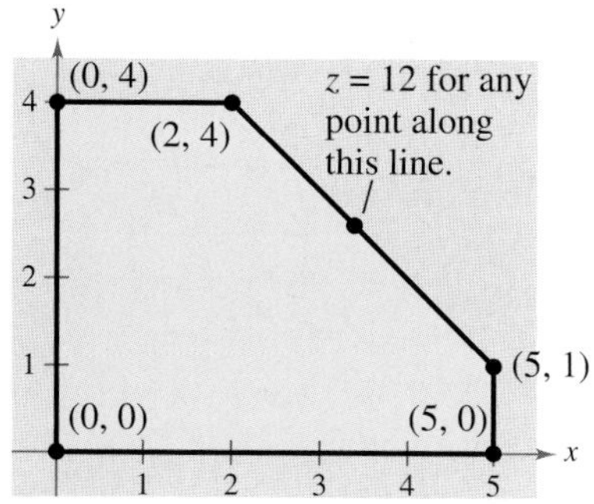

FIGURE 7.30

It is possible for the maximum (or minimum) value in a linear programming problem to occur at *two* different vertices. For instance, at the vertices of the region shown in Figure 7.30, the objective function

$$z = 2x + 2y$$ Objective function

has the following values.

At $(0, 0)$: $z = 2(0) + 2(0) = 0$

At $(0, 4)$: $z = 2(0) + 2(4) = 8$

At $(2, 4)$: $z = 2(2) + 2(4) = 12$ Maximum value of z

At $(5, 1)$: $z = 2(5) + 2(1) = 12$ Maximum value of z

At $(5, 0)$: $z = 2(5) + 2(0) = 10$

In this case, you can conclude that the objective function has a maximum value not only at the vertices $(2, 4)$ and $(5, 1)$; it also has a maximum value (of 12) at *any point on the line segment connecting these two vertices*. Note that the objective function in slope-intercept form

$$y = -x + \frac{1}{2}z$$

has the same slope as the line through the vertices $(2, 4)$ and $(5, 1)$.

Some linear programming problems have no optimal solutions. This can occur if the region determined by the constraints is *unbounded*. Example 4 illustrates such a problem.

Example 4 ▶ An Unbounded Region

Find the maximum value of

$$z = 4x + 2y$$ Objective function

where $x \ge 0$ and $y \ge 0$, subject to the following constraints.

$$\left.\begin{aligned} x + 2y &\ge 4 \\ 3x + y &\ge 7 \\ -x + 2y &\le 7 \end{aligned}\right\}$$ Constraints

Solution

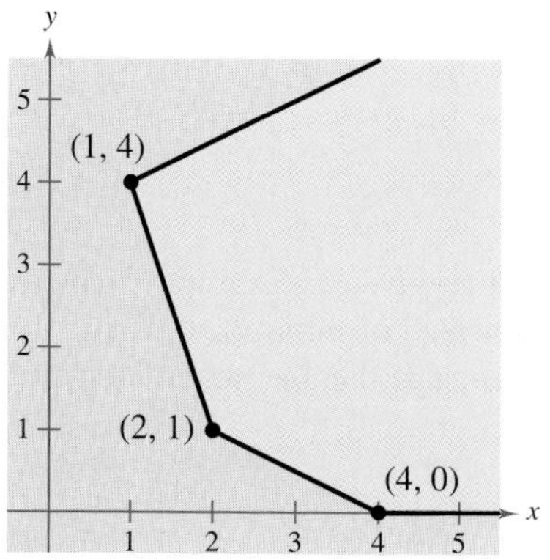

FIGURE 7.31

The region determined by the constraints is shown in Figure 7.31. For this unbounded region, there is no maximum value of z. To see this, note that the point $(x, 0)$ lies in the region for all values of $x \ge 4$. Substituting this point into the objective function, you get

$$z = 4(x) + 2(0) = 4x.$$

By choosing x to be large, you can obtain values of z that are as large as you want. So, there is no maximum value of z. For this problem, there *is* a minimum value of z.

At $(1, 4)$: $z = 4(1) + 2(4) = 12$

At $(2, 1)$: $z = 4(2) + 2(1) = 10$ Minimum value of z

At $(4, 0)$: $z = 4(4) + 2(0) = 16$

So, the minimum value of z is 10, and this occurs when $x = 2$ and $y = 1$.

Activities

1. Sketch the region determined by the constraints $x \geq 0$, $y \geq 0$, and $3x + 2y < 6$, and then find the minimum and maximum values of the objective function $z = 4x + 5y$ and where they occur, subject to the constraints.

   Answer: Minimum: (0, 0), 0; Maximum: (0, 3), 15

2. Find the maximum value of the objective function $z = 6x + 5y$ and where it occurs, subject to the constraints $3x + 2y \leq 16$, $x + 4y \leq 22$, $x \geq 0$, and $y \geq 0$.

   Answer: Maximum: (2, 5), 37

3. Find the maximum value and where it occurs, of $z = 2x + 5y$ where $x \geq 0$ and $y \geq 0$ subject to the constraints $x + 2y \geq 8$, $3x + y \geq 14$, and $-x + y \leq 10$.

   Answer: There is no maximum value of z.

Applications

Example 5 shows how linear programming can be used to find the maximum profit in a business application.

Example 5 ▶ Maximum Profit

A manufacturer wants to maximize the profit for two products. Product I yields a profit of \$1.50 per unit, and product II yields a profit of \$2.00 per unit. Market tests and available resources have indicated the following constraints.

1. The combined production level should not exceed 1200 units per month.
2. The demand for product II is no more than half the demand for product I.
3. The production level of product I should be less than or equal to 600 units plus three times the production level of product II.

Solution

If you let x be the number of units of product I and y be the number of units of product II, the objective function (for the combined profit) is given by

$$P = 1.5x + 2y. \quad \text{Objective function}$$

The three constraints translate into the following linear inequalities.

1. $x + y \leq 1200 \quad \Rightarrow \quad x + y \leq 1200$
2. $y \leq \frac{1}{2}x \quad \Rightarrow \quad -x + 2y \leq 0$
3. $x \leq 600 + 3y \quad \Rightarrow \quad x - 3y \leq 600$

Because neither x nor y can be negative, you also have the two additional constraints of $x \geq 0$ and $y \geq 0$. Figure 7.32 shows the region determined by the constraints. To find the maximum profit, test the values of P at the vertices of the region.

$$\begin{aligned} \text{At } (0, 0)\text{: } P &= 1.5(0) + 2(0) = 0 \\ \text{At } (800, 400)\text{: } P &= 1.5(800) + 2(400) = 2000 \quad \text{Maximum profit} \\ \text{At } (1050, 150)\text{: } P &= 1.5(1050) + 2(150) = 1875 \\ \text{At } (600, 0)\text{: } P &= 1.5(600) + 2(0) = 900 \end{aligned}$$

So, the maximum profit is \$2000, and it occurs when the monthly production consists of 800 units of product I and 400 units of product II.

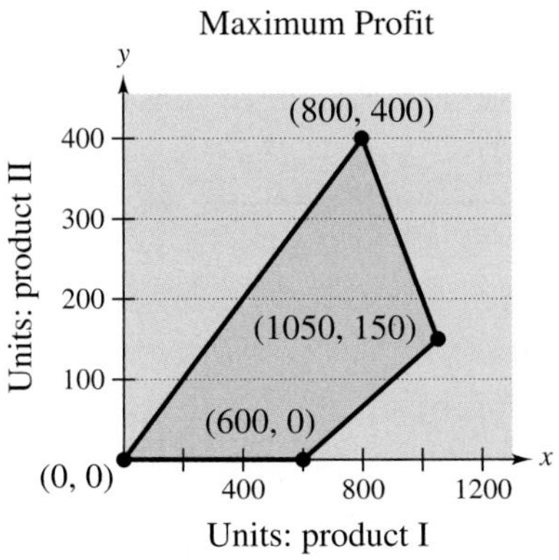

FIGURE 7.32

In Example 5, the manufacturer improved the production of product I so that it yielded a profit of \$2.50 per unit. The maximum profit can then be found using the objective function

$$P = 2.5x + 2y. \quad \text{Objective function}$$

By testing the values of P at the vertices of the region, you find the maximum profit is \$2925 and that it now occurs when $x = 1050$ and $y = 150$.

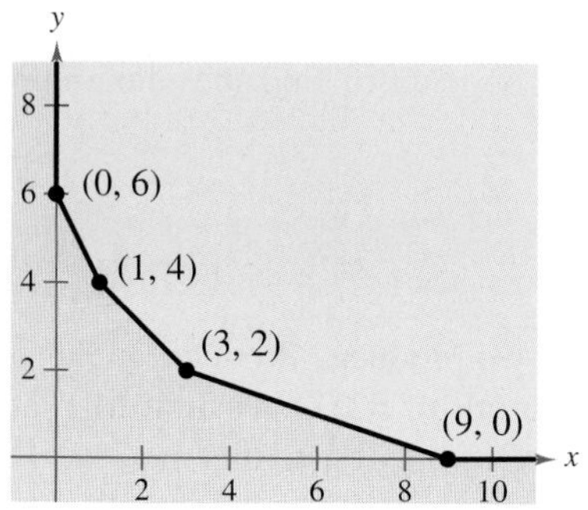

FIGURE 7.33

Example 6 ▶ Minimum Cost

The minimum daily requirements from the liquid portion of a diet are 300 calories, 36 units of vitamin A, and 90 units of vitamin C. A cup of dietary drink X costs \$0.12 and provides 60 calories, 12 units of vitamin A, and 10 units of vitamin C. A cup of dietary drink Y costs \$0.15 and provides 60 calories, 6 units of vitamin A, and 30 units of vitamin C. How many cups of each drink should be consumed each day to minimize the cost and still meet the daily requirements?

Solution

As in Example 9 on page 596, let x be the number of cups of dietary drink X and let y be the number of cups of dietary drink Y.

$$\left.\begin{array}{ll}\text{For calories:} & 60x + 60y \geq 300 \\ \text{For vitamin A:} & 12x + 6y \geq 36 \\ \text{For vitamin C:} & 10x + 30y \geq 90 \\ & x \geq 0 \\ & y \geq 0\end{array}\right\} \quad \text{Constraints}$$

The cost C is given by $C = 0.12x + 0.15y$. Objective function

The graph of the region corresponding to the constraints is shown in Figure 7.33. To determine the minimum cost, test C at each vertex of the region.

$$\begin{array}{ll}\text{At } (0, 6): & C = 0.12(0) + 0.15(6) = 0.90 \\ \text{At } (1, 4): & C = 0.12(1) + 0.15(4) = 0.72 \\ \text{At } (3, 2): & C = 0.12(3) + 0.15(2) = 0.66 \quad \text{Minimum value of } C \\ \text{At } (9, 0): & C = 0.12(9) + 0.15(0) = 1.08\end{array}$$

So, the minimum cost is \$0.66 per day, and this occurs when three cups of drink X and two cups of drink Y are consumed each day.

Activity

Explain what difficulties you might encounter with the following sets of linear programming constraints.

a. $$\left.\begin{array}{r}x - y < 0 \\ 3x + y > 9 \\ -4x + y > -2\end{array}\right\} \quad \text{Constraints}$$

b. $$\left.\begin{array}{r}2x + y > 11 \\ x - y > 0 \\ x \quad < 4 \\ y < 0\end{array}\right\} \quad \text{Constraints}$$

Writing ABOUT MATHEMATICS

Creating a Linear Programming Problem Sketch the region determined by the following constraints.

$$\left.\begin{array}{r}x + 2y \leq 8 \\ x + y \leq 5 \\ x \geq 0 \\ y \geq 0\end{array}\right\} \quad \text{Constraints}$$

Find, if possible, an objective function of the form $z = ax + by$ that has a maximum at the indicated vertex of the region.

a. $(0, 4)$ **b.** $(2, 3)$

c. $(5, 0)$ **d.** $(0, 0)$

Explain how you found each objective function.

7.5 Exercises

Exercises with no feasible solutions: 44
Exercises with unbounded regions:17, 19, 42

In Exercises 1–12, find the minimum and maximum values of the objective function and where they occur, subject to the indicated constraints. (For each exercise, the graph of the region determined by the constraints is provided.)

1. Objective function:

$z = 4x + 3y$

Constraints:

$x \geq 0$

$y \geq 0$

$x + y \leq 5$

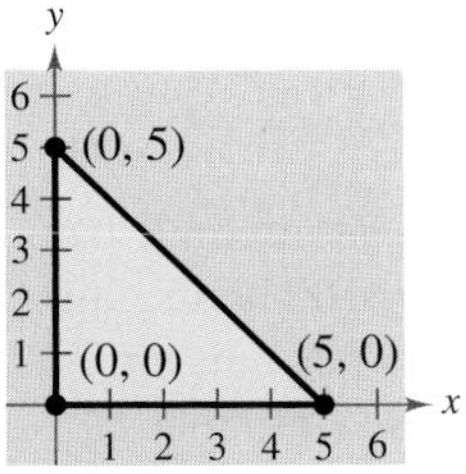

2. Objective function:

$z = 2x + 8y$

Constraints:

$x \geq 0$

$y \geq 0$

$2x + y \leq 4$

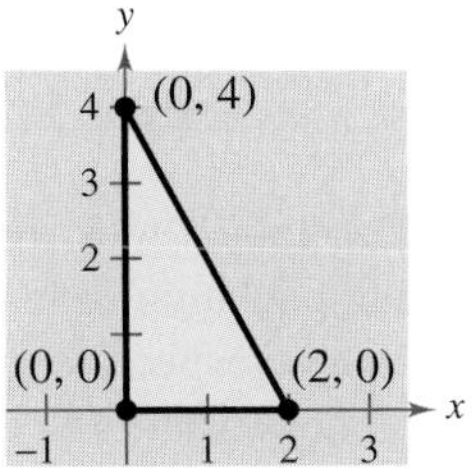

3. Objective function:

$z = 3x + 8y$

Constraints:
(See Exercise 1.)

4. Objective function:

$z = 7x + 3y$

Constraints:
(See Exercise 2.)

5. Objective function:

$z = 3x + 2y$

Constraints:

$x \geq 0$

$y \geq 0$

$x + 3y \leq 15$

$4x + y \leq 16$

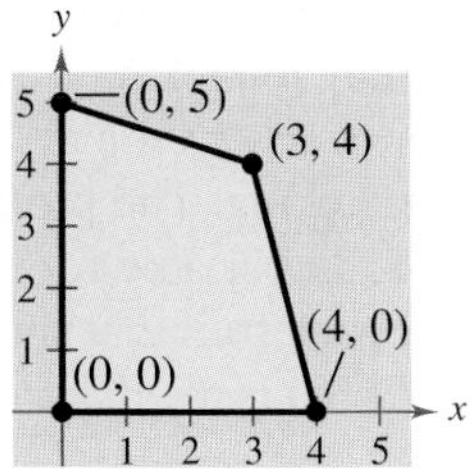

6. Objective function:

$z = 4x + 5y$

Constraints:

$x \geq 0$

$2x + 3y \geq 6$

$3x - y \leq 9$

$x + 4y \leq 16$

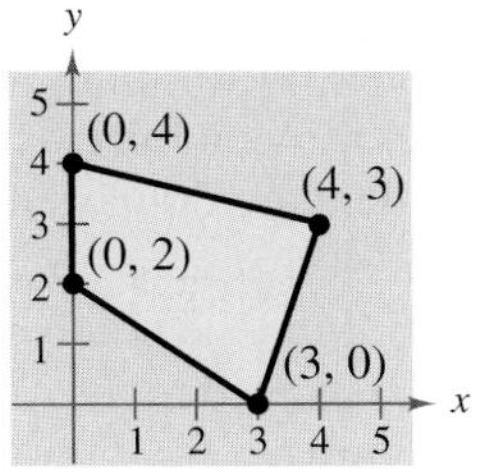

7. Objective function:

$z = 5x + 0.5y$

Constraints:
(See Exercise 5.)

8. Objective function:

$z = 2x + y$

Constraints:
(See Exercise 6.)

9. Objective function:

$z = 10x + 7y$

Constraints:

$0 \leq x \leq 60$

$0 \leq y \leq 45$

$5x + 6y \leq 420$

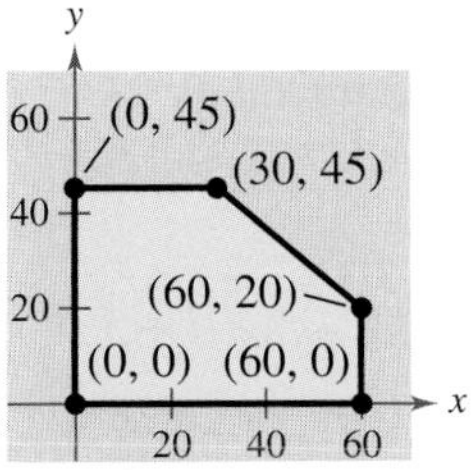

10. Objective function:

$z = 25x + 35y$

Constraints:

$x \geq 0$

$y \geq 0$

$8x + 9y \leq 7200$

$8x + 9y \geq 3600$

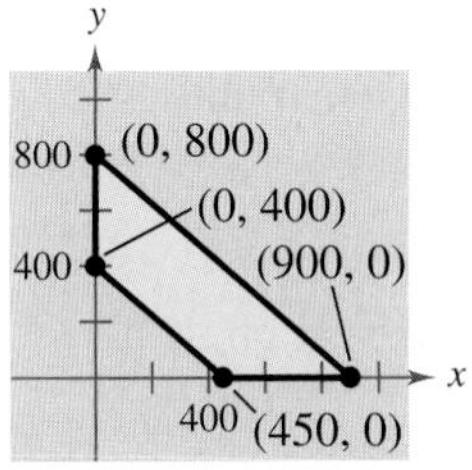

11. Objective function:

$z = 25x + 30y$

Constraints:
(See Exercise 9.)

12. Objective function:

$z = 15x + 20y$

Constraints:
(See Exercise 10.)

In Exercises 13–20, sketch the region determined by the constraints. Then find the minimum and maximum values of the objective function and where they occur, subject to the constraints.

13. Objective function:

$z = 6x + 10y$

Constraints:

$x \geq 0$

$y \geq 0$

$2x + 5y \leq 10$

14. Objective function:

$z = 7x + 8y$

Constraints:

$x \geq 0$

$y \geq 0$

$x + \frac{1}{2}y \leq 4$

15. Objective function:

$z = 9x + 24y$

Constraints:
(See Exercise 13.)

16. Objective function:

$z = 7x + 2y$

Constraints:
(See Exercise 14.)

17. Objective function:

$z = 4x + 5y$

Constraints:

$x \geq 0$

$y \geq 0$

$x + y \geq 8$

$3x + 5y \geq 30$

18. Objective function:

$z = 4x + 5y$

Constraints:

$x \geq 0$

$y \geq 0$

$2x + 2y \leq 10$

$x + 2y \leq 6$

19. Objective function:

$z = 2x + 7y$

Constraints:
(See Exercise 17.)

20. Objective function:

$z = 2x - y$

Constraints:
(See Exercise 18.)

In Exercises 21–24, use a graphing utility to sketch the region determined by the constraints. Then find the minimum and maximum values of the objective function and where they occur, subject to the constraints.

21. Objective function:

$z = 4x + y$

Constraints:

$x \geq 0$
$y \geq 0$
$x + 2y \leq 40$
$2x + 3y \geq 72$

22. Objective function:

$z = x$

Constraints:

$x \geq 0$
$y \geq 0$
$2x + 3y \leq 60$
$2x + y \leq 28$
$4x + y \leq 48$

23. Objective function:

$z = x + 4y$

Constraints:
(See Exercise 21.)

24. Objective function:

$z = y$

Constraints:
(See Exercise 22.)

In Exercises 25–28, find the maximum value of the objective function and where it occurs, subject to the constraints $x \geq 0$, $y \geq 0$, $3x + y \leq 15$, and $4x + 3y \leq 30$.

25. $z = 2x + y$

26. $z = 5x + y$

27. $z = x + y$

28. $z = 3x + y$

In Exercises 29–32, find the maximum value of the objective function, and where it occurs, subject to the constraints $x \geq 0$, $y \geq 0$, $x + 4y \leq 20$, $x + y \leq 18$, and $2x + 2y \leq 21$.

29. $z = x + 5y$

30. $z = 2x + 4y$

31. $z = 4x + 5y$

32. $z = 4x + y$

33. *Maximum Profit* A manufacturer produces two models of bicycles. The times (in hours) required for assembling, painting, and packaging each model are listed in the table.

Process	Model A	Model B
Assembling	2	2.5
Painting	4	1
Packaging	1	0.75

The total times available for assembling, painting, and packaging are 4000 hours, 4800 hours, and 1500 hours, respectively. The profits per unit are \$45 for model A and \$50 for model B. How many of each type should be produced to maximize profit? What is the maximum profit?

34. *Maximum Profit* A manufacturer produces two models of bicycles. The times (in hours) required for assembling, painting, and packaging each model are listed in the table.

Process	Model A	Model B
Assembling	2.5	3
Painting	2	1
Packaging	0.75	1.25

The total times available for assembling, painting, and packaging are 4000 hours, 2500 hours, and 1500 hours, respectively. The profits per unit are \$50 for model A and \$52 for model B. How many of each type should be produced to maximize profit? What is the maximum profit?

35. *Maximum Profit* A merchant plans to sell two models of compact disc players at costs of \$150 and \$200. The \$150 model yields a profit of \$25 per unit and the \$200 model yields a profit of \$40 per unit. The merchant estimates that the total monthly demand will not exceed 250 units. The merchant does not want to invest more than \$40,000 in inventory for these products. Find the number of units of each model that should be stocked in order to maximize profit. What is the maximum profit?

36. *Maximum Profit* A fruit grower has 150 acres of land available to raise two crops, A and B. It takes 1 day to trim an acre of crop A and 2 days to trim an acre of crop B, and there are 240 days per year available for trimming. It takes 0.3 day to pick an acre of crop A and 0.1 day to pick an acre of crop B, and there are 30 days available for picking. The profit is \$140 per acre for crop A and \$235 per acre for crop B. Find the number of acres of each fruit that should be planted to maximize profit. What is the maximum profit?

37. *Minimum Cost* A farming cooperative mixes two brands of cattle feed. Brand X costs \$25 per bag and contains 2 units of nutritional element A, 2 units of element B, and 2 units of element C. Brand Y costs \$20 per bag and contains 1 unit of nutritional element A, 9 units of element B, and 3 units of element C. The minimum requirements of nutrients A, B, and C are 12 units, 36 units, and 24 units, respectively. Find the number of bags of each brand that should be mixed to produce a mixture having a minimum cost. What is the minimum cost?

38. *Minimum Cost* Two gasolines, type A and type B, have octane ratings of 80 and 92, respectively. Type A costs \$1.13 per gallon and type B costs \$1.28 per gallon. Determine the blend of minimum cost with an octane rating of at least 90. What is the minimum cost? (*Hint:* Let x be the fraction of each gallon that is type A and let y be the fraction that is type B.)

39. *Maximum Revenue* An accounting firm has 800 hours of staff time and 96 hours of reviewing time available each week. The firm charges \$2000 for an audit and \$300 for a tax return. Each audit requires 100 hours of staff time and 8 hours of review time. Each tax return requires 12.5 hours of staff time and 2 hours of review time. What numbers of audits and tax returns will yield the maximum revenue? What is the maximum revenue?

40. *Maximum Revenue* The accounting firm in Exercise 39 lowers its charge for an audit to \$1000. What numbers of audits and tax returns will yield the maximum revenue? What is the maximum revenue?

In Exercises 41–46, the linear programming problem has an unusual characteristic. Sketch a graph of the solution region for the problem and describe the unusual characteristic. Find the maximum value of the objective function and where it occurs.

41. Objective function:

$z = 2.5x + y$

Constraints:

$x \geq 0$

$y \geq 0$

$3x + 5y \leq 15$

$5x + 2y \leq 10$

42. Objective function:

$z = x + y$

Constraints:

$x \geq 0$

$y \geq 0$

$-x + y \leq 1$

$-x + 2y \leq 4$

43. Objective function:

$z = -x + 2y$

Constraints:

$x \geq 0$

$y \geq 0$

$x \leq 10$

$x + y \leq 7$

44. Objective function:

$z = x + y$

Constraints:

$x \geq 0$

$y \geq 0$

$-x + y \leq 0$

$-3x + y \geq 3$

45. Objective function:

$z = 3x + 4y$

Constraints:

$x \geq 0$

$y \geq 0$

$x + y \leq 1$

$2x + y \leq 4$

46. Objective function:

$z = x + 2y$

Constraints:

$x \geq 0$

$y \geq 0$

$x + 2y \leq 4$

$2x + y \leq 4$

Synthesis

True or False? **In Exercises 47 and 48, determine whether the statement is true or false. Justify your answer.**

47. If an objective function has a maximum value at the vertices (4, 7) and (8, 3), you can conclude that it also has a maximum value at the points (4.5, 6.5) and (7.8, 3.2).

48. When solving a linear programming problem, if the objective function has a maximum value at more than one vertex, you can assume that there are an infinite number of points that will produce the maximum value.

In Exercises 49 and 50, determine values of t such that the objective function has a maximum value at the indicated vertex.

49. Objective function:

$z = 3x + ty$

Constraints:

$x \geq 0$

$y \geq 0$

$x + 3y \leq 15$

$4x + y \leq 16$

(a) (0, 5)

(b) (3, 4)

50. Objective function:

$z = 3x + ty$

Constraints:

$x \geq 0$

$y \geq 0$

$x + 2y \geq 4$

$x - y \leq 1$

(a) (2, 1)

(b) (0, 2)

Think About It **In Exercises 51–54, find an objective function that has a maximum or minimum value at the indicated vertex of the constraint region shown below. (There are many correct answers.)**

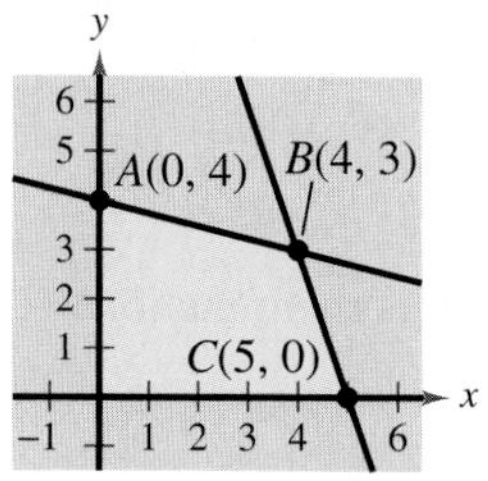

51. The maximum occurs at vertex A.

52. The maximum occurs at vertex B.

53. The maximum occurs at vertex C.

54. The minimum occurs at vertex C.

Review

In Exercises 55–58, simplify the compound fraction.

55. $\dfrac{\frac{9}{x}}{\left(\frac{6}{x}+2\right)}$

56. $\dfrac{\left(1+\frac{2}{x}\right)}{\left(x-\frac{4}{x}\right)}$

57. $\dfrac{\left(\frac{4}{x^2-9}+\frac{2}{x-2}\right)}{\left(\frac{1}{x+3}+\frac{1}{x-3}\right)}$

58. $\dfrac{\left(\frac{1}{x+1}+\frac{1}{2}\right)}{\left(\frac{3}{2x^2+4x+2}\right)}$

In Exercises 59–64, solve the equation algebraically. Round the result to three decimal places.

59. $e^{2x} + 2e^x - 15 = 0$

60. $e^{2x} - 10e^x + 24 = 0$

61. $8(62 - e^{x/4}) = 192$

62. $\dfrac{150}{e^{-x} - 4} = 75$

63. $7 \ln 3x = 12$

64. $\ln(x + 9)^2 = 2$

Chapter Summary

What did you learn?

	Review Exercises
Section 7.1	
☐ How to use the method of substitution to solve systems of equations in two variables	1–4
☐ How to use a graphical approach to solve systems of equations in two variables	5–10
☐ How to use systems of equations to model and solve real-life problems	11–13
Section 7.2	
☐ How to use the method of elimination to solve systems of linear equations in two variables	14–21
☐ How to interpret graphically the numbers of solutions of systems of equations in two variables	22–25
☐ How to use systems of equations in two variables to model and solve real-life problems	26–30
Section 7.3	
☐ How to recognize linear systems in row-echelon form and use back-substitution to solve the systems	31, 32
☐ How to use Gaussian elimination to solve systems of linear equations	33–42
☐ How to solve nonsquare systems of linear equations	43, 44
☐ How to use systems of linear equations in three or more variables to model and solve application problems	45–47
Section 7.4	
☐ How to sketch the graphs of inequalities in two variables	48–51
☐ How to solve systems of inequalities	52–59
☐ How to use systems of inequalities in two variables to model and solve real-life problems	60–63
Section 7.5	
☐ How to solve linear programming problems	64–67
☐ How to use linear programming to model and solve real-life problems	68–71

Review Exercises

Exercises containing systems with no solutions: 21, 23, 25
Exercises containing systems with infinitely many solutions: 20, 24, 35, 43, 44

7.1 In Exercises 1–4, solve the system by the method of substitution.

1. $\begin{cases} x^2 - y^2 = 9 \\ x - y = 1 \end{cases}$

2. $\begin{cases} x^2 + y^2 = 169 \\ 3x + 2y = 39 \end{cases}$

3. $\begin{cases} y = 2x^2 \\ y = x^4 - 2x^2 \end{cases}$

4. $\begin{cases} x = y + 3 \\ x = y^2 + 1 \end{cases}$

In Exercises 5–8, solve the system graphically.

5. $\begin{cases} 2x - y = 10 \\ x + 5y = -6 \end{cases}$

6. $\begin{cases} y^2 - 2y + x = 0 \\ x + y = 0 \end{cases}$

7. $\begin{cases} y = -2e^{-x} \\ 2e^x + y = 0 \end{cases}$

8. $\begin{cases} y = 2(6 - x) \\ y = 2^{x-2} \end{cases}$

In Exercises 9 and 10, use a graphing utility to solve the system of equations. Find the solution accurate to two decimal places.

9. $\begin{cases} y = 2x^2 - 4x + 1 \\ y = x^2 - 4x + 3 \end{cases}$

10. $\begin{cases} y = \ln(x - 1) - 3 \\ y = 4 - \frac{1}{2}x \end{cases}$

11. *Break-Even Point* You set up a business and make an initial investment of \$50,000. The unit cost of the product is \$2.15 and the selling price is \$6.95. How many units must you sell to break even?

12. *Choice of Two Jobs* You are offered two sales jobs. One company offers an annual salary of \$22,500 plus a year-end bonus of 1.5% of your total sales. The other company offers an annual salary of \$20,000 plus a year-end bonus of 2% of your total sales. What amount of sales will make the second offer better? Explain.

13. *Geometry* The perimeter of a rectangle is 480 meters and its length is 150% of its width. Find the dimensions of the rectangle.

7.2 In Exercises 14–21, solve the system by elimination.

14. $\begin{cases} 2x - y = 2 \\ 6x + 8y = 39 \end{cases}$

15. $\begin{cases} 40x + 30y = 24 \\ 20x - 50y = -14 \end{cases}$

16. $\begin{cases} 0.2x + 0.3y = 0.14 \\ 0.4x + 0.5y = 0.20 \end{cases}$

17. $\begin{cases} 12x + 42y = -17 \\ 30x - 18y = 19 \end{cases}$

18. $\begin{cases} 3x - 2y = 0 \\ 3x + 2(y + 5) = 10 \end{cases}$

19. $\begin{cases} 7x + 12y = 63 \\ 2x + 3(y + 2) = 21 \end{cases}$

20. $\begin{cases} 1.25x - 2y = 3.5 \\ 5x - 8y = 14 \end{cases}$

21. $\begin{cases} 1.5x + 2.5y = 8.5 \\ 6x + 10y = 24 \end{cases}$

In Exercises 22–25, use a graphing utility to graph the lines in the system. Use the graph to determine if the system is consistent or inconsistent. If the system is consistent, determine the number of solutions.

22. $\begin{cases} -3x - 5y = -1 \\ 6x + y = 4 \end{cases}$

23. $\begin{cases} \frac{1}{5}x = -4 + y \\ 5y = x \end{cases}$

24. $\begin{cases} 6x - 14.4y = 1.8 \\ 1.2x - 2.88y = 0.36 \end{cases}$

25. $\begin{cases} \frac{8}{5}x - y = 3 \\ -5y + 8x = -2 \end{cases}$

26. *Acid Mixture* Two hundred liters of a 75% acid solution is obtained by mixing a 90% solution with a 50% solution. How many liters of each must be used to obtain the desired mixture?

27. *Compact Disc Sales* Suppose you are the manager of a music store. At the end of one week you are going over receipts for the previous week's sales. Six hundred and fifty compact discs were sold. One type of compact disc sold for \$9.95 and another sold for \$14.95. The total compact disc receipts were \$7717.50. The cash register that was supposed to record the number of each type of compact disc sold malfunctioned. Can you recover the information? If so, how many of each type of compact disc were sold?

28. *Flying Speeds* Two planes leave Pittsburgh and Philadelphia at the same time, each going to the other city. One plane flies 25 miles per hour faster than the other. Find the air speed of each plane if the cities are 275 miles apart and the planes pass one another after 40 minutes of flying time.

Economics **In Exercises 29 and 30, find the point of equilibrium.**

	Demand Function	*Supply Function*
29.	$p = 37 - 0.0002x$	$p = 22 + 0.00001x$
30.	$p = 120 - 0.0001x$	$p = 45 + 0.0002x$

7.3 In Exercises 31 and 32, use back-substitution to solve the system.

31. $\begin{cases} x - 4y + 3z = 3 \\ -y + z = -1 \\ z = -5 \end{cases}$

32. $\begin{cases} x - 7y + 8z = 85 \\ y - 9z = -35 \\ z = 3 \end{cases}$

In Exercises 33–36, use Gaussian elimination to solve the system of equations.

33. $\begin{cases} x + 2y + 6z = 4 \\ -3x + 2y - z = -4 \\ 4x + 2z = 16 \end{cases}$

34. $\begin{cases} x + 3y - z = 13 \\ 2x - 5z = 23 \\ 4x - y - 2z = 14 \end{cases}$

35. $\begin{cases} x - 2y + z = -6 \\ 2x - 3y = -7 \\ -x + 3y - 3z = 11 \end{cases}$

36. $\begin{cases} 2x + 6z = -9 \\ 3x - 2y + 11z = -16 \\ 3x - y + 7z = -11 \end{cases}$

In Exercises 37 and 38, find the equation of the parabola $y = ax^2 + bx + c$ that passes through the points. Use a graphing utility to verify your result.

37.

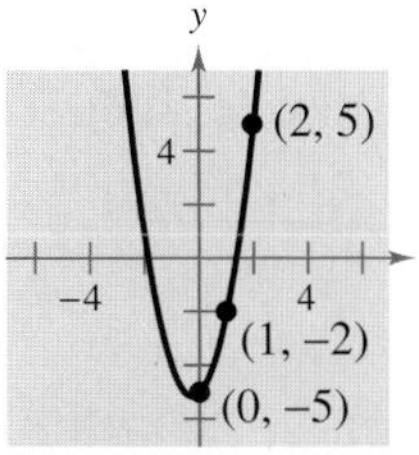

38.

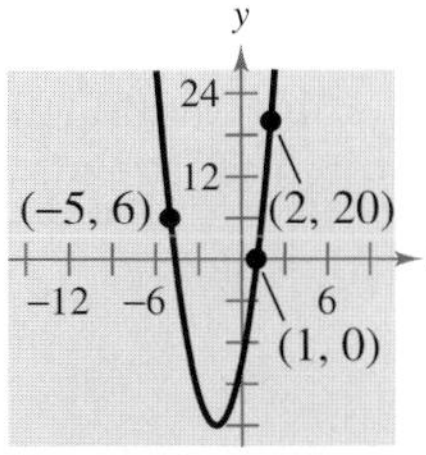

In Exercises 39 and 40, find the equation of the circle $x^2 + y^2 + Dx + Ey + F = 0$ that passes through the points. Use a graphing utility to verify your result.

39.

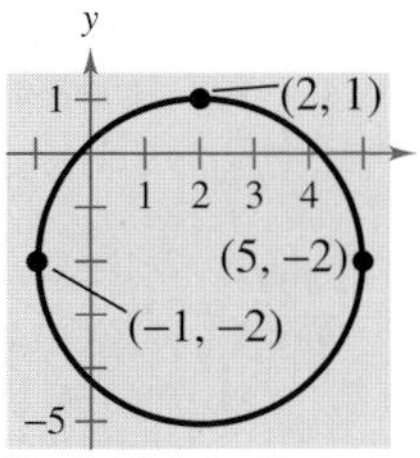

40.

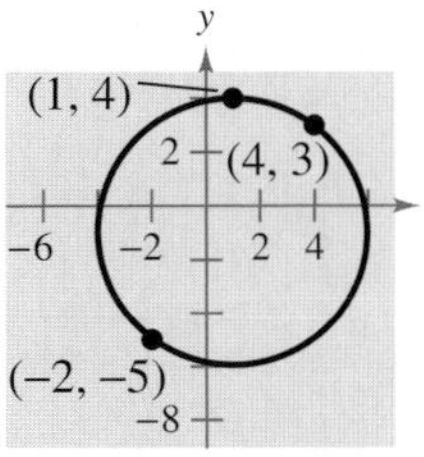

41. *Fitting a Line to Data* Solve the system of equations to find the least squares regression line $y = ax + b$ for the points in the figure. (For an explanation of how the coefficients of a and b in the system are obtained, see Appendix B.)

$$\begin{cases} 5b + 10a = 17.8 \\ 10b + 30a = 45.7 \end{cases}$$

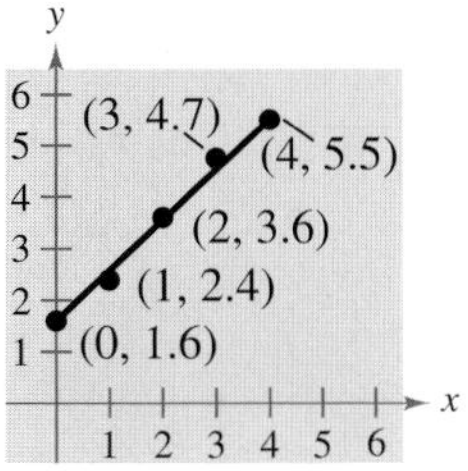

42. *Data Analysis* Let x and y represent the median ages at first marriage for women and men, respectively. In a sample of 6 years, these median ages are given by the following ordered pairs. (Source: U.S. Center for Health Statistics)

(23.0, 24.8), (23.3, 25.1), (23.6, 25.3),

(23.7, 25.5), (23.9, 25.9), (24.0, 25.9)

(a) Find the least squares regression line $y = ax + b$ for the data by solving the following system of linear equations.

$$\begin{cases} 6b + 141.5a = 152.5 \\ 141.5b + 3337.75a = 3597.27 \end{cases}$$

(b) Use a graphing utility to plot the data and graph the regression line in the same viewing window.

(c) Use the graph to determine whether the line is a good model for the data. Explain.

(d) What information is given by the slope of the regression line? Explain.

In Exercises 43 and 44, solve the nonsquare system of equations.

43. $\begin{cases} 5x - 12y + 7z = 16 \\ 3x - 7y + 4z = 9 \end{cases}$

44. $\begin{cases} 2x + 5y - 19z = 34 \\ 3x + 8y - 31z = 54 \end{cases}$

45. *Agriculture* A mixture of 6 gallons of chemical A, 8 gallons of chemical B, and 13 gallons of chemical C is required to kill a certain destructive crop insect. Commercial spray X contains 1, 2, and 2 parts, respectively, of these chemicals. Commercial spray Y contains only chemical C. Commercial spray Z contains chemicals A, B, and C in equal amounts. How much of each type of commercial spray is needed to get the desired mixture?

46. *Finance* An inheritance of \$40,000 was divided among three investments yielding \$3500 in interest per year. The interest rates for the three investments were 7%, 9%, and 11%. Find the amount placed in each investment if the second and third were \$3000 and \$5000 less than the first, respectively.

47. *Fitting a Parabola to Data* Solve the system of equations to find the least squares regression parabola $y = ax^2 + bx + c$ for the points in the figure. (For an explanation of how the coefficients of a, b, and c in the system are obtained, see Appendix B.)

$$\begin{cases} 5c + 10a = 9.1 \\ 10b = 8.0 \\ 10c + 34a = 19.8 \end{cases}$$

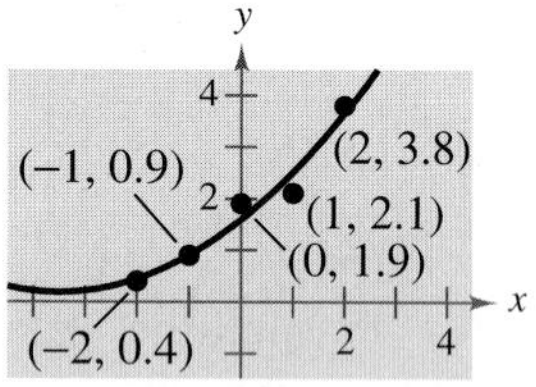

7.4 **In Exercises 48–51, sketch the graph of the inequality.**

48. $y \le 5 - \frac{1}{2}x$

49. $3y - x \ge 7$

50. $y - 4x^2 > -1$

51. $y \le 2 \ln x - 6$

In Exercises 52–59, sketch a graph and label the vertices of the solution set of the system of inequalities.

52. $\begin{cases} x + 2y \le 160 \\ 3x + y \le 180 \\ x \ge 0 \\ y \ge 0 \end{cases}$

53. $\begin{cases} 2x + 3y \le 24 \\ 2x + y \le 16 \\ x \ge 0 \\ y \ge 0 \end{cases}$

54. $\begin{cases} 3x + 2y \ge 24 \\ x + 2y \ge 12 \\ 2 \le x \le 15 \\ y \le 15 \end{cases}$

55. $\begin{cases} 2x + y \ge 16 \\ x + 3y \ge 18 \\ 0 \le x \le 25 \\ 0 \le y \le 25 \end{cases}$

56. $\begin{cases} y < x + 1 \\ y > x^2 - 1 \end{cases}$

57. $\begin{cases} y \le 6 - 2x - x^2 \\ y \ge x + 6 \end{cases}$

58. $\begin{cases} 2x - 3y \ge 0 \\ 2x - y \le 8 \\ y \ge 0 \end{cases}$

59. $\begin{cases} x^2 + y^2 \le 9 \\ (x - 3)^2 + y^2 \le 9 \end{cases}$

In Exercises 60 and 61, determine a system of inequalities that models the description. Use a graphing utility to graph and shade the solution of the system.

60. *Fruit Distribution* A Pennsylvania fruit grower has 1500 bushels of apples that are to be divided between markets in Harrisburg and Philadelphia. These two markets need at least 400 bushels and 600 bushels, respectively.

61. *Inventory Costs* A warehouse operator has 24,000 square feet of floor space in which to store two products. Each unit of product I requires 20 square feet of floor space and costs \$12 per day to store. Each unit of product II requires 30 square feet of floor space and costs \$8 per day to store. The total storage cost per day cannot exceed \$12,400.

In Exercises 62 and 63, find the consumer surplus and producer surplus for the demand and supply equations. Sketch the graph of the equations and shade the regions representing the consumer surplus and producer surplus.

	Demand	*Supply*
62.	$p = 160 - 0.0001x$	$p = 70 + 0.0002x$
63.	$p = 130 - 0.0002x$	$p = 30 + 0.0003x$

7.5 **In Exercises 64–67, find the required optimum value of the objective function and where it occurs, subject to the indicated constraints.**

64. Maximize:

$z = 3x + 4y$

Constraints:

$$\begin{aligned} x &\ge 0 \\ y &\ge 0 \\ 2x + 5y &\le 50 \\ 4x + y &\le 28 \end{aligned}$$

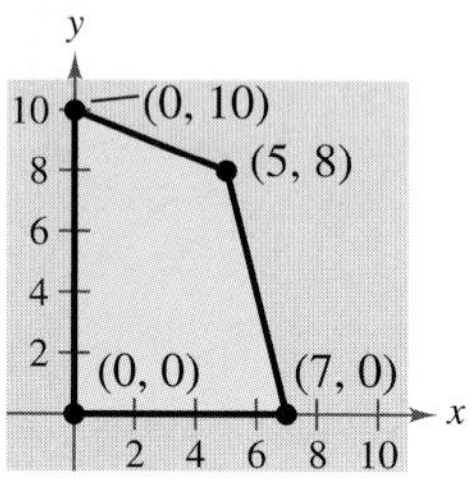

65. Minimize:

$z = 10x + 7y$

Constraints:

$$\begin{aligned} x &\ge 0 \\ y &\ge 0 \\ 2x + y &\ge 100 \\ x + y &\ge 75 \end{aligned}$$

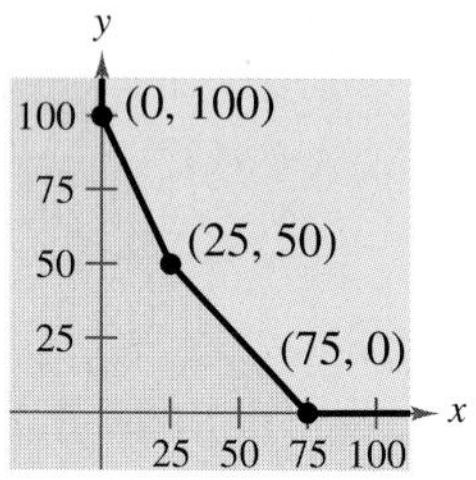

66. Minimize:

$z = 1.75x + 2.25y$

Constraints:

$$\begin{aligned} x &\ge 0 \\ y &\ge 0 \\ 2x + y &\ge 25 \\ 3x + 2y &\ge 45 \end{aligned}$$

67. Maximize:

$z = 50x + 70y$

Constraints:

$$\begin{aligned} x &\ge 0 \\ y &\ge 0 \\ x + 2y &\le 1500 \\ 5x + 2y &\le 3500 \end{aligned}$$

68. *Maximum Revenue* A student is working part time as a hairdresser to pay college expenses. The student may work no more than 24 hours per week. Haircuts cost \$25 and require an average of 20 minutes, and permanents cost \$70 and require an average of 1 hour and 10 minutes. What combination of haircuts and/or permanents will yield a maximum revenue? What is the maximum revenue?

69. *Maximum Profit* A manufacturer produces products A and B yielding profits of \$18 and \$24, respectively. Each product must go through three processes with the required times per unit shown in the table.

Process	Hours for Product A	Hours for Product B	Hours Available per Day
I	4	2	24
II	1	2	9
III	1	1	8

Find the daily production level for each unit to maximize the profit. What is the maximum profit?

70. *Minimum Cost* A pet supply company mixes two brands of dry dog food. Brand X costs \$15 per bag and contains 8 units of nutritional element A, 1 unit of nutritional element B, and 2 units of nutritional element C. Brand Y costs \$30 per bag and contains 2 units of nutritional element A, 1 unit of nutritional element B, and 7 units of nutritional element C. Each bag of mixed dog food must contain at least 16 units, 5 units, and 20 units of nutritional elements A, B, and C, respectively. Find the numbers of bags of brands X and Y that should be mixed to produce a mixture meeting the minimum nutritional requirements and having a minimum cost. What is the minimum cost?

71. *Minimum Cost* Two gasolines, type A and type B, have octane ratings of 80 and 92, respectively. Type A costs \$1.25 per gallon and type B costs \$1.55 per gallon. Determine the blend of minimum cost with an octane rating of at least 88. What is the minimum cost? (*Hint:* Let x be the fraction of each gallon that is type A and let y be the fraction that is type B.)

Synthesis

True or False? **In Exercises 72 and 73, determine whether the statement is true or false. Justify your answer.**

72. The system

$$\begin{cases} y \le 5 \\ y \ge -2 \\ y \ge \frac{7}{2}x - 9 \\ y \ge -\frac{7}{2}x + 26 \end{cases}$$

represents the region covered by an isosceles trapezoid.

73. It is possible for an objective function of a linear programming problem to have exactly 10 maximum value points.

Exploration **In Exercises 74–77, find a system of linear equations having the ordered pair as a solution. (There is more than one correct answer.)**

74. $(-6, 8)$

75. $(5, -4)$

76. $\left(\frac{4}{3}, 3\right)$

77. $\left(-1, \frac{9}{4}\right)$

Exploration **In Exercises 78–81, find a system of linear equations having the ordered triple as a solution. (There is more than one correct answer.)**

78. $(4, -1, 3)$

79. $(-3, 5, 6)$

80. $\left(5, \frac{3}{2}, 2\right)$

81. $\left(\frac{3}{4}, -2, 8\right)$

82. *Exploration* Find k_1 and k_2 such that the system of equations has an infinite number of solutions.

$$\begin{cases} 3x - 5y = 8 \\ 2x + k_1 y = k_2 \end{cases}$$

83. When solving a system of equations by substitution, how do you recognize that the system has no solution?

84. When solving a system of equations by elimination, how do you recognize that the system has no solution?

85. A system of two equations in two unknowns is solved and has a finite number of solutions. Determine the maximum number of solutions of the system satisfying each of the following.

(a) Both equations are linear.

(b) One equation is linear and the other is quadratic.

(c) Both equations are quadratic.

86. How can you tell graphically that a system of linear equations in two variables has no solution? Give an example.

Chapter Project ▶ Fitting Models to Data

Many of the models in this book were created with a statistical method called *least squares regression analysis.* This procedure can be performed easily with a computer or graphing calculator.

Example ▶ Fitting a Line to Data

The numbers of morning and evening newspapers published in the United States from 1990 through 1997 are shown in the table. Use the data to project the numbers of morning and evening newspapers that will be published in 2003. In the table, $t = 0$ represents 1990. (Source: Editor and Publisher Company)

Year, t	0	1	2	3	4	5	6	7
Morning	559	571	596	623	635	656	686	705
Evening	1084	1042	996	954	935	891	846	816

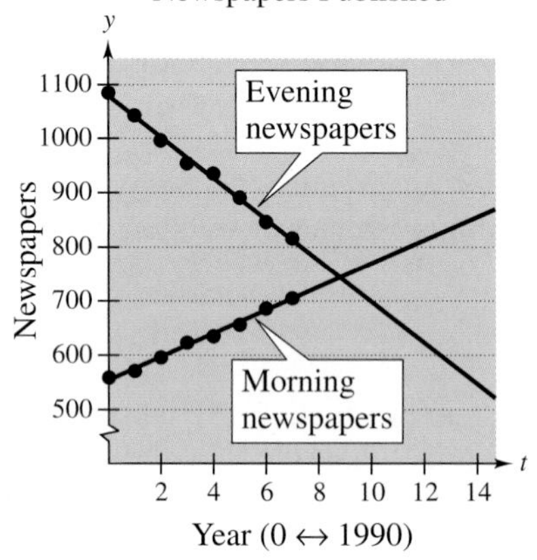

Solution

Begin by finding a computer or graphing calculator that will perform linear regression analysis. After entering the data and running the program, you should obtain the following models.

$$y = 554.3 + 21.3t \qquad \text{Morning newspapers}$$
$$y = 1078.4 - 38.0t \qquad \text{Evening newspapers}$$

With these models, you can project the numbers of newspapers in 2003. If the trend continues to follow the pattern from 1990 through 1997, the numbers of newspapers in 2003 should be about

$$y = 554.3 + 21.3(13) \approx 831 \qquad \text{Morning newspapers}$$
$$y = 1078.4 - 38.0(13) \approx 584. \qquad \text{Evening newspapers}$$

The graphs of the data and the models are shown at the left.

Year, t	Morning
0	41.3
1	41.5
2	42.4
3	43.1
4	43.4
5	44.3
6	44.8
7	45.4

Chapter Project Investigations

1. Use the models from the example to predict the year when the number of morning papers published will be equal to the number of evening papers published.
2. The total numbers (in millions) of *all* morning newspapers *sold* each day in the United States from 1990 to 1997 are shown in the table at the left. Find a linear model that represents this data. Use your model to project the number of morning papers that will be sold each day in 2003.
3. From 1990 through 1997, both the number of morning newspapers published and the total number of papers sold increased. Did the average circulation per morning paper (number sold ÷ number published) increase or decrease? Explain.

Chapter Test

The *Interactive* CD-ROM and *Internet* versions of this text provide answers to the Chapter Tests and Cumulative Tests. They also offer Chapter Pre-Tests (which test key skills and concepts covered in previous chapters) and Chapter Post-Tests, both of which have randomly generated exercises with diagnostic capabilities.

Take this test as you would take a test in class. After you are done, check your work against the answers given in the back of the book.

In Exercises 1–3, solve the system by the method of substitution.

1. $\begin{cases} x - y = -7 \\ 4x + 5y = 8 \end{cases}$ **2.** $\begin{cases} y = x - 1 \\ y = (x-1)^3 \end{cases}$ **3.** $\begin{cases} 2x - y^2 = 0 \\ x - y = 4 \end{cases}$

In Exercises 4–6, solve the system graphically.

4. $\begin{cases} 2x - 3y = 0 \\ 2x + 3y = 12 \end{cases}$ **5.** $\begin{cases} y = 9 - x^2 \\ y = x + 3 \end{cases}$ **6.** $\begin{cases} y - \ln x = 12 \\ 7x - 2y + 11 = -6 \end{cases}$

In Exercises 7 and 8, solve the linear system by elimination.

7. $\begin{cases} 2x + 3y = 17 \\ 5x - 4y = -15 \end{cases}$ **8.** $\begin{cases} x - 2y + 3z = 11 \\ 2x - z = 3 \\ 3y + z = -8 \end{cases}$

9. Find a system of linear equations that has the solution $\left(\frac{4}{3}, -5\right)$.

10. Find the equation of the parabola $y = ax^2 + bx + c$ passing through the points $(0, 6)$, $(-2, 2)$, and $\left(3, \frac{9}{2}\right)$.

11. Find a system of linear equations that has the solution $\left(-\frac{1}{2}, 6, -\frac{5}{4}\right)$.

In Exercises 12–14, sketch the graph and label the vertices of the solution of the system of inequalities.

12. $\begin{cases} 2x + y \le 4 \\ 2x - y \ge 0 \\ x \ge 0 \end{cases}$ **13.** $\begin{cases} y < -x^2 + x + 4 \\ y > 4x \end{cases}$ **14.** $\begin{cases} x^2 + y^2 \le 16 \\ x \ge 1 \\ y \ge -3 \end{cases}$

15. Find the maximum value of the objective function $z = 20x + 12y$ and where it occurs, subject to the constraints $x \ge 0$, $y \ge 0$, $x + 4y \le 32$, and $3x + 2y \le 36$.

16. A merchant plans to sell two models of compact disc players at costs of \$275 and \$400. The \$275 model yields a profit of \$55 and the \$400 model yields a profit of \$75. The merchant estimates that the total monthly demand will not exceed 300 units. The merchant does not want to invest more than \$100,000 in inventory for these products. Find the number of units of each model that should be stocked in order to maximize profit. What is the maximum profit?

17. A manufacturer produces two types of television stands. The amounts of time for assembling, staining, and packaging the two models are shown in the table at the left. The total amounts of time available for assembling, staining, and packaging are 4000, 8950, and 2650 hours, respectively. The profits per unit are \$30 (model I) and \$40 (model II). How many of each model should be produced to maximize the profit? What is the maximum profit?

	Model I	Model II
Assembling	0.5	0.75
Staining	2.0	1.5
Packaging	0.5	0.5

Peter Turnley/Corbis

In 1998, retail sales of jewelry and precious metals reached \$22.3 billion. *(Source: U.S. Bureau of the Census)*

8 Matrices and Determinants

▶ How to Study This Chapter

The Big Picture

In this chapter you will learn the following skills and concepts.

- ▶ How to use matrices, Gaussian elimination, and Gauss-Jordan elimination to solve systems of linear equations
- ▶ How to add and subtract matrices, multiply matrices by real numbers, and multiply two matrices
- ▶ How to find the inverses of matrices and use inverse matrices to solve systems of linear equations
- ▶ How to find minors, cofactors, and determinants of square matrices
- ▶ How to use Cramer's Rule to solve systems of linear equations
- ▶ How to use determinants and matrices to model and solve problems

Important Vocabulary

As you encounter each new vocabulary term in this chapter, add the term and its definition to your notebook glossary.

Matrix (p. 620)
Entry of a matrix (p. 620)
Order of a matrix (p. 620)
Square matrix (p. 620)
Main diagonal (p. 620)
Row matrix (p. 620)
Column matrix (p. 620)
Augmented matrix (p. 621)
Coefficient matrix (p. 621)
Elementary row operations (p. 622)
Row-equivalent matrices (p. 622)
Row-echelon form (p. 624)
Reduced row-echelon form (p. 624)
Gauss-Jordan elimination (p. 627)
Scalar (p. 636)
Scalar multiple (p.636)
Zero matrix (p. 639)
Matrix multiplication (p. 640)
Identity matrix of order *n* (p. 642)
Inverse of a matrix (p. 649)
Determinant (pp. 653, 658, 661)
Minors (p. 660)
Cofactors (p. 660)
Cramer's Rule (pp. 666, 668)

Study Tools

- Learning objectives at the beginning of each section
- Chapter Summary (p. 678)
- Review Exercises (pp. 679–683)
- Chapter Test (p. 685)

Additional Resources

- Study and Solutions Guide
- Interactive Precalculus
- Videotapes for Chapter 8
- Precalculus Website
- Student Success Organizer

STUDY T!P

Read the text before class. Write down any questions you may have about the material. Ask your instructor these questions during class. That way, you will understand the material better.

8.1 Matrices and Systems of Equations

▶ **What you should learn**

- How to write a matrix and identify its order
- How to perform elementary row operations on matrices
- How to use matrices and Gaussian elimination to solve systems of linear equations
- How to use matrices and Gauss-Jordan elimination to solve systems of linear equations

▶ **Why you should learn it**

You can use matrices to solve systems of linear equations in two or more variables. For instance, Exercise 88 on page 633 shows how a matrix can be used to help find a model for the parabolic path of a baseball.

Amwell/Tony Stone Images

Matrices

In this section you will study a streamlined technique for solving systems of linear equations. This technique involves the use of a rectangular array of real numbers called a **matrix.** The plural of matrix is *matrices.*

Definition of Matrix

If m and n are positive integers, an $m \times n$ matrix (read "m by n") is a rectangular array

$$\underbrace{\left.\begin{bmatrix} a_{11} & a_{12} & a_{13} & \cdots & a_{1n} \\ a_{21} & a_{22} & a_{23} & \cdots & a_{2n} \\ a_{31} & a_{32} & a_{33} & \cdots & a_{3n} \\ \vdots & \vdots & \vdots & & \vdots \\ a_{m1} & a_{m2} & a_{m3} & \cdots & a_{mn} \end{bmatrix}\right\} m \text{ rows}}_{n \text{ columns}}$$

in which each **entry,** a_{ij}, of the matrix is a number. An $m \times n$ matrix has m rows (horizontal lines) and n columns (vertical lines).

The entry in the ith row and jth column is denoted by the *double subscript* notation a_{ij}. A matrix having m rows and n columns is said to be of **order** $m \times n$. If $m = n$, the matrix is **square** of order n. For a square matrix, the entries $a_{11}, a_{22}, a_{33}, \ldots$ are the **main diagonal** entries.

Example 1 ▶ Order of Matrices

Determine the order of each matrix.

a. $[2]$

b. $\begin{bmatrix} 1 & -3 & 0 & \frac{1}{2} \end{bmatrix}$

c. $\begin{bmatrix} 0 & 0 \\ 0 & 0 \end{bmatrix}$

d. $\begin{bmatrix} 5 & 0 \\ 2 & -2 \\ -7 & 4 \end{bmatrix}$

Solution

a. This matrix has *one* row and *one* column. The order of the matrix is 1×1.

b. This matrix has *one* row and *four* columns. The order of the matrix is 1×4.

c. This matrix has *two* rows and *two* columns. The order of the matrix is 2×2.

d. This matrix has *three* rows and *two* columns. The order of the matrix is 3×2.

A matrix that has only one row is called a **row matrix,** and a matrix that has only one column is called a **column matrix.**

Encourage your students to become familiar with the terms in this chapter.

A matrix derived from a system of linear equations (each written in standard form with the constant term on the right) is the **augmented matrix** of the system. Moreover, the matrix derived from the coefficients of the system (but not including the constant terms) is the **coefficient matrix** of the system.

STUDY TIP

The vertical dots in an augmented matrix separate the coefficients of the linear system from the constant terms.

System
$$\begin{cases} x - 4y + 3z = 5 \\ -x + 3y - z = -3 \\ 2x - 4z = 6 \end{cases}$$

Augmented Matrix
$$\begin{bmatrix} 1 & -4 & 3 & \vdots & 5 \\ -1 & 3 & -1 & \vdots & -3 \\ 2 & 0 & -4 & \vdots & 6 \end{bmatrix}$$

Coefficient Matrix
$$\begin{bmatrix} 1 & -4 & 3 \\ -1 & 3 & -1 \\ 2 & 0 & -4 \end{bmatrix}$$

Note the use of 0 for the missing y-variable in the third equation, and also note the fourth column of constant terms in the augmented matrix.

When forming either the coefficient matrix or the augmented matrix of a system, you should begin by vertically aligning the variables in the equations and using zeros for the missing variables.

Example 2 ▶ Writing an Augmented Matrix

You may want to make sure that the way a linear system of equations matches up with a matrix is clear to your students.

Write the augmented matrix for the system of linear equations.

$$\begin{cases} x + 3y - w = 9 \\ -y + 4z + 2w = -2 \\ x - 5z - 6w = 0 \\ 2x + 4y - 3z = 4 \end{cases}$$

What is the order of the augmented matrix?

Solution

Begin by rewriting the linear system and aligning the variables.

$$\begin{cases} x + 3y \phantom{{}+4z} - w = 9 \\ \phantom{x+{}} -y + 4z + 2w = -2 \\ x \phantom{{}+3y} - 5z - 6w = 0 \\ 2x + 4y - 3z \phantom{{}-6w} = 4 \end{cases}$$

Next, use the coefficients as the matrix entries. Include zeros for the missing coefficients.

$$\begin{matrix} R_1 \\ R_2 \\ R_3 \\ R_4 \end{matrix} \begin{bmatrix} 1 & 3 & 0 & -1 & \vdots & 9 \\ 0 & -1 & 4 & 2 & \vdots & -2 \\ 1 & 0 & -5 & -6 & \vdots & 0 \\ 2 & 4 & -3 & 0 & \vdots & 4 \end{bmatrix}$$

The augmented matrix has 4 rows and 5 columns, so it is a 4-by-5 matrix. The notation R_n is used to designate each row in the matrix. For example, Row 1 is represented by R_1.

The *Interactive* CD-ROM and *Internet* versions of this text show every example with its solution; clicking on the *Try It!* button brings up similar problems. Guided Examples and Integrated Examples show step-by-step solutions to additional examples. Integrated Examples are related to several concepts in the section.

Elementary Row Operations

In Section 7.3, you studied three operations that can be used on a system of linear equations to produce an equivalent system.

1. Interchange two equations.
2. Multiply an equation by a nonzero constant.
3. Add a multiple of an equation to another equation.

In matrix terminology these three operations correspond to **elementary row operations.** An elementary row operation on an augmented matrix of a given system of linear equations produces a new augmented matrix corresponding to a new (but equivalent) system of linear equations. Two matrices are **row-equivalent** if one can be obtained from the other by a sequence of elementary row operations.

Students should recognize that the elementary row operations are essentially the same as the operations that led to equivalent systems of equations in Chapter 7.

Elementary Row Operations

1. Interchange two rows.
2. Multiply a row by a nonzero constant.
3. Add a multiple of a row to another row.

Although elementary row operations are simple to perform, they involve a lot of arithmetic. Because it is easy to make a mistake, you should get in the habit of noting the elementary row operations performed in each step so that you can go back and check your work.

Example 3 ▶ Elementary Row Operations

a. Interchange the first and second rows.

Original Matrix

$$\begin{bmatrix} 0 & 1 & 3 & 4 \\ -1 & 2 & 0 & 3 \\ 2 & -3 & 4 & 1 \end{bmatrix}$$

New Row-Equivalent Matrix

$$\begin{matrix} R_2 \\ R_1 \\ \end{matrix}\begin{bmatrix} -1 & 2 & 0 & 3 \\ 0 & 1 & 3 & 4 \\ 2 & -3 & 4 & 1 \end{bmatrix}$$

b. Multiply the first row by $\frac{1}{2}$.

Original Matrix

$$\begin{bmatrix} 2 & -4 & 6 & -2 \\ 1 & 3 & -3 & 0 \\ 5 & -2 & 1 & 2 \end{bmatrix}$$

New Row-Equivalent Matrix

$$\begin{matrix} \frac{1}{2}R_1 \rightarrow \\ \\ \end{matrix}\begin{bmatrix} 1 & -2 & 3 & -1 \\ 1 & 3 & -3 & 0 \\ 5 & -2 & 1 & 2 \end{bmatrix}$$

c. Add -2 times the first row to the third row.

Original Matrix

$$\begin{bmatrix} 1 & 2 & -4 & 3 \\ 0 & 3 & -2 & -1 \\ 2 & 1 & 5 & -2 \end{bmatrix}$$

New Row-Equivalent Matrix

$$\begin{matrix} \\ \\ -2R_1 + R_3 \rightarrow \end{matrix}\begin{bmatrix} 1 & 2 & -4 & 3 \\ 0 & 3 & -2 & -1 \\ 0 & -3 & 13 & -8 \end{bmatrix}$$

Note that the elementary row operation is written beside the row that is *changed.*

Technology

Most graphing utilities can perform elementary row operations on matrices. Consult the user's guide for your graphing utility for keystrokes.

After performing a row operation, the new row-equivalent matrix that is displayed on your graphing utility is stored in the *answer* variable. You should use the *answer* variable and not the original matrix for subsequent row operations.

In Example 3 in Section 7.3, you used Gaussian elimination with back-substitution to solve a system of linear equations. The next example demonstrates the matrix version of Gaussian elimination. The two methods are essentially the same. The basic difference is that with matrices you do not need to keep writing the variables.

Example 4 ▶ Comparing Linear Systems and Matrix Operations

Arithmetic errors are often made when elementary row operations are performed. Have students note the operation performed in each step so that they can go back and check their work.

Linear System	*Associated Augmented Matrix*
$\begin{cases} x - 2y + 3z = 9 \\ -x + 3y = -4 \\ 2x - 5y + 5z = 17 \end{cases}$	$\begin{bmatrix} 1 & -2 & 3 & \vdots & 9 \\ -1 & 3 & 0 & \vdots & -4 \\ 2 & -5 & 5 & \vdots & 17 \end{bmatrix}$
Add the first equation to the second equation.	Add the first row to the second row.
$\begin{cases} x - 2y + 3z = 9 \\ y + 3z = 5 \\ 2x - 5y + 5z = 17 \end{cases}$	$\begin{matrix} \\ R_1 + R_2 \rightarrow \\ \end{matrix} \begin{bmatrix} 1 & -2 & 3 & \vdots & 9 \\ 0 & 1 & 3 & \vdots & 5 \\ 2 & -5 & 5 & \vdots & 17 \end{bmatrix}$
Add -2 times the first equation to the third equation.	Add -2 times the first row to the third row.
$\begin{cases} x - 2y + 3z = 9 \\ y + 3z = 5 \\ -y - z = -1 \end{cases}$	$\begin{matrix} \\ \\ -2R_1 + R_3 \rightarrow \end{matrix} \begin{bmatrix} 1 & -2 & 3 & \vdots & 9 \\ 0 & 1 & 3 & \vdots & 5 \\ 0 & -1 & -1 & \vdots & -1 \end{bmatrix}$
Add the second equation to the third equation.	Add the second row to the third row.
$\begin{cases} x - 2y + 3z = 9 \\ y + 3z = 5 \\ 2z = 4 \end{cases}$	$\begin{matrix} \\ \\ R_2 + R_3 \rightarrow \end{matrix} \begin{bmatrix} 1 & -2 & 3 & \vdots & 9 \\ 0 & 1 & 3 & \vdots & 5 \\ 0 & 0 & 2 & \vdots & 4 \end{bmatrix}$
Multiply the third equation by $\frac{1}{2}$.	Multiply the third row by $\frac{1}{2}$.
$\begin{cases} x - 2y + 3z = 9 \\ y + 3z = 5 \\ z = 2 \end{cases}$	$\begin{matrix} \\ \\ \frac{1}{2}R_3 \rightarrow \end{matrix} \begin{bmatrix} 1 & -2 & 3 & \vdots & 9 \\ 0 & 1 & 3 & \vdots & 5 \\ 0 & 0 & 1 & \vdots & 2 \end{bmatrix}$

The *Interactive* CD-ROM and *Internet* versions of this text offer a built-in graphing calculator, which can be used in the Examples, Explorations, Technology notes, and Exercises.

At this point, you can use back-substitution to find x and y.

$$y + 3(2) = 5 \qquad \text{Substitute 2 for } z.$$

$$y = -1 \qquad \text{Solve for } y.$$

$$x - 2(-1) + 3(2) = 9 \qquad \text{Substitute } -1 \text{ for } y \text{ and 2 for } z.$$

$$x = 1 \qquad \text{Solve for } x.$$

The solution is $x = 1$, $y = -1$, and $z = 2$.

Remember that you can check a solution by substituting the values of x, y, and z into each equation in the original system.

The last matrix in Example 4 is said to be in **row-echelon form.** The term *echelon* refers to the stair-step pattern formed by the nonzero elements of the matrix. To be in this form, a matrix must have the following properties.

Row-Echelon Form and Reduced Row-Echelon Form

A matrix in **row-echelon form** has the following properties.

1. All rows consisting entirely of zeros occur at the bottom of the matrix.
2. For each row that does not consist entirely of zeros, the first nonzero entry is 1 (called a leading 1).
3. For two successive (nonzero) rows, the leading 1 in the higher row is farther to the left than the leading 1 in the lower row.

A matrix in *row-echelon form* is in **reduced row-echelon form** if every column that has a leading 1 has zeros in every position above and below its leading 1.

Example 5 ▶ Row-Echelon Form

The following matrices are in row-echelon form.

a. $\begin{bmatrix} 1 & 2 & -1 & 4 \\ 0 & 1 & 0 & 3 \\ 0 & 0 & 1 & -2 \end{bmatrix}$ **b.** $\begin{bmatrix} 0 & 1 & 0 & 5 \\ 0 & 0 & 1 & 3 \\ 0 & 0 & 0 & 0 \end{bmatrix}$

c. $\begin{bmatrix} 1 & -5 & 2 & -1 & 3 \\ 0 & 0 & 1 & 3 & -2 \\ 0 & 0 & 0 & 1 & 4 \\ 0 & 0 & 0 & 0 & 1 \end{bmatrix}$ **d.** $\begin{bmatrix} 1 & 0 & 0 & -1 \\ 0 & 1 & 0 & 2 \\ 0 & 0 & 1 & 3 \\ 0 & 0 & 0 & 0 \end{bmatrix}$

The matrices in (b) and (d) also happen to be in *reduced* row-echelon form. The following matrices are not in row-echelon form.

e. $\begin{bmatrix} 1 & 2 & -3 & 4 \\ 0 & 2 & 1 & -1 \\ 0 & 0 & 1 & -3 \end{bmatrix}$ First nonzero row entry in Row 2 is not 1.

f. $\begin{bmatrix} 1 & 2 & -1 & 2 \\ 0 & 0 & 0 & 0 \\ 0 & 1 & 2 & -4 \end{bmatrix}$ Row consisting entirely of zeros is not at the bottom of the matrix.

Every matrix is row-equivalent to a matrix in row-echelon form. For instance, in Example 5, you can change the matrix in part (e) to row-echelon form by multiplying its second row by $\frac{1}{2}$.

$$\frac{1}{2}R_2 \rightarrow \begin{bmatrix} 1 & 2 & -3 & 4 \\ 0 & 1 & \frac{1}{2} & -\frac{1}{2} \\ 0 & 0 & 1 & -3 \end{bmatrix}$$

What elementary row operation could you perform on the matrix in part (f) so that it would be in row-echelon form?

Gaussian Elimination with Back-Substitution

Gaussian elimination with back-substitution works well for solving systems of linear equations by hand or with a computer. For this algorithm, the order in which the elementary row operations are performed is important. We suggest operating from left to right by columns, using elementary row operations to obtain zeros in all entries directly below the leading 1's.

A computer animation of this example appears in the *Interactive* CD-ROM and *Internet* versions of this text.

Example 6 ▶ Gaussian Elimination with Back-Substitution

Solve the system
$$\begin{cases} y + z - 2w = -3 \\ x + 2y - z = 2 \\ 2x + 4y + z - 3w = -2 \\ x - 4y - 7z - w = -19 \end{cases}.$$

Solution

$$\begin{array}{r} R_2 \\ R_1 \\ \\ \\ \end{array}\left[\begin{array}{rrrr:r} 1 & 2 & -1 & 0 & 2 \\ 0 & 1 & 1 & -2 & -3 \\ 2 & 4 & 1 & -3 & -2 \\ 1 & -4 & -7 & -1 & -19 \end{array}\right]$$

Interchange R_1 and R_2 so first column has leading 1 in upper left corner.

$$\begin{array}{r} \\ \\ -2R_1 + R_3 \rightarrow \\ -R_1 + R_4 \rightarrow \end{array}\left[\begin{array}{rrrr:r} 1 & 2 & -1 & 0 & 2 \\ 0 & 1 & 1 & -2 & -3 \\ 0 & 0 & 3 & -3 & -6 \\ 0 & -6 & -6 & -1 & -21 \end{array}\right]$$

Perform operations on R_3 and R_4 so first column has zeros below its leading 1.

$$\begin{array}{r} \\ \\ \\ 6R_2 + R_4 \rightarrow \end{array}\left[\begin{array}{rrrr:r} 1 & 2 & -1 & 0 & 2 \\ 0 & 1 & 1 & -2 & -3 \\ 0 & 0 & 3 & -3 & -6 \\ 0 & 0 & 0 & -13 & -39 \end{array}\right]$$

Perform operations on R_4 so second column has zeros below its leading 1.

$$\begin{array}{r} \\ \\ \frac{1}{3}R_3 \rightarrow \\ \\ \end{array}\left[\begin{array}{rrrr:r} 1 & 2 & -1 & 0 & 2 \\ 0 & 1 & 1 & -2 & -3 \\ 0 & 0 & 1 & -1 & -2 \\ 0 & 0 & 0 & -13 & -39 \end{array}\right]$$

Perform operations on R_3 so third column has a leading 1.

$$\begin{array}{r} \\ \\ \\ -\frac{1}{13}R_4 \rightarrow \end{array}\left[\begin{array}{rrrr:r} 1 & 2 & -1 & 0 & 2 \\ 0 & 1 & 1 & -2 & -3 \\ 0 & 0 & 1 & -1 & -2 \\ 0 & 0 & 0 & 1 & 3 \end{array}\right]$$

Perform operations on R_4 so fourth column has a leading 1.

The matrix is now in row-echelon form, and the corresponding system is

$$\begin{cases} x + 2y - z = 2 \\ y + z - 2w = -3 \\ z - w = -2 \\ w = 3. \end{cases}$$

Using back-substitution, the solution is $x = -1$, $y = 2$, $z = 1$, and $w = 3$.

Gaussian Elimination with Back-Substitution

1. Write the augmented matrix of the system of linear equations.
2. Use elementary row operations to rewrite the augmented matrix in row-echelon form.
3. Write the system of linear equations corresponding to the matrix in row-echelon form, and use back-substitution to find the solution.

When solving a system of linear equations, remember that it is possible for the system to have no solution. If, in the elimination process, you obtain a row with zeros except for the last entry, it is unnecessary to continue the elimination process. You can simply conclude that the system has no solution, or is *inconsistent*.

Example 7 ▶ A System with No Solution

Solve the system.

$$\begin{cases} x - y + 2z = 4 \\ x + z = 6 \\ 2x - 3y + 5z = 4 \\ 3x + 2y - z = 1 \end{cases}$$

Solution

$$\begin{bmatrix} 1 & -1 & 2 & \vdots & 4 \\ 1 & 0 & 1 & \vdots & 6 \\ 2 & -3 & 5 & \vdots & 4 \\ 3 & 2 & -1 & \vdots & 1 \end{bmatrix}$$ Write augmented matrix.

$$\begin{matrix} \\ -R_1 + R_2 \to \\ -2R_1 + R_3 \to \\ -3R_1 + R_4 \to \end{matrix} \begin{bmatrix} 1 & -1 & 2 & \vdots & 4 \\ 0 & 1 & -1 & \vdots & 2 \\ 0 & -1 & 1 & \vdots & -4 \\ 0 & 5 & -7 & \vdots & -11 \end{bmatrix}$$ Perform row operations.

$$\begin{matrix} \\ \\ R_2 + R_3 \to \\ \\ \end{matrix} \begin{bmatrix} 1 & -1 & 2 & \vdots & 4 \\ 0 & 1 & -1 & \vdots & 2 \\ 0 & 0 & 0 & \vdots & -2 \\ 0 & 5 & -7 & \vdots & -11 \end{bmatrix}$$ Perform row operations.

Note that the third row of this matrix consists of zeros except for the last entry. This means that the original system of linear equations is inconsistent. You can see why this is true by converting back to a system of linear equations.

Reinforce the importance of recognizing systems with no solutions or systems with infinitely many solutions from row-echelon matrices.

$$\begin{cases} x - y + 2z = 4 \\ y - z = 2 \\ 0 = -2 \\ 5y - 7z = -11 \end{cases}$$

Because the third equation is not possible, the system has no solution.

Gauss-Jordan Elimination

With Gaussian elimination, elementary row operations are applied to a matrix to obtain a (row-equivalent) row-echelon form of the matrix. A second method of elimination, called **Gauss-Jordan elimination,** after Carl Friedrich Gauss and Wilhelm Jordan (1842–1899), continues the reduction process until a *reduced* row-echelon form is obtained. This procedure is demonstrated in Example 8.

Example 8 ▶ Gauss-Jordan Elimination

Use Gauss-Jordan elimination to solve the system $\begin{cases} x - 2y + 3z = 9 \\ -x + 3y = -4. \\ 2x - 5y + 5z = 17 \end{cases}$

Technology

For a demonstration of a graphical approach to Gauss-Jordan elimination on a 2×3 matrix, see the graphing calculator program available for several models of graphing calculators at our website *college.hmco.com.*

Solution

In Example 4, Gaussian elimination was used to obtain the row-echelon form of the linear system above.

$$\begin{bmatrix} 1 & -2 & 3 & \vdots & 9 \\ 0 & 1 & 3 & \vdots & 5 \\ 0 & 0 & 1 & \vdots & 2 \end{bmatrix}$$

Now, apply elementary row operations until you obtain zeros above each of the leading 1's, as follows.

$$\begin{array}{r} 2R_2 + R_1 \rightarrow \\ \\ \\ \end{array} \begin{bmatrix} 1 & 0 & 9 & \vdots & 19 \\ 0 & 1 & 3 & \vdots & 5 \\ 0 & 0 & 1 & \vdots & 2 \end{bmatrix}$$

Perform operations on R_1 so second column has zeros above its leading 1.

$$\begin{array}{r} -9R_3 + R_1 \rightarrow \\ -3R_3 + R_2 \rightarrow \\ \\ \end{array} \begin{bmatrix} 1 & 0 & 0 & \vdots & 1 \\ 0 & 1 & 0 & \vdots & -1 \\ 0 & 0 & 1 & \vdots & 2 \end{bmatrix}$$

Perform operations on R_1 and R_2 so third column has zeros above its leading 1.

The matrix is now in reduced row-echelon form. Now, converting back to a system of linear equations, you have

$$\begin{cases} x = 1 \\ y = -1. \\ z = 2 \end{cases}$$

Now you can simply read the solution.

The elimination procedures described in this section sometimes result in fractional coefficients. For instance, for the system

$$\begin{cases} 2x - 5y + 5z = 17 \\ 3x - 2y + 3z = 11 \\ -3x + 3y = -6 \end{cases}$$

the procedure would have required multiplication of the first row by $\frac{1}{2}$. You can sometimes avoid fractions by judiciously choosing the order in which you apply elementary row operations.

Group Activity Suggestion:
Because it can be difficult for students to find their own errors when applying Gauss-Jordan elimination to matrices, you might consider making up several more examples of common errors to give them additional practice.

Example 9 ▶ A System with an Infinite Number of Solutions

Solve the system.

$$\begin{cases} 2x + 4y - 2z = 0 \\ 3x + 5y \qquad = 1 \end{cases}$$

Solution

$$\begin{bmatrix} 2 & 4 & -2 & \vdots & 0 \\ 3 & 5 & 0 & \vdots & 1 \end{bmatrix}$$

$$\tfrac{1}{2}R_1 \rightarrow \begin{bmatrix} 1 & 2 & -1 & \vdots & 0 \\ 3 & 5 & 0 & \vdots & 1 \end{bmatrix}$$

$$-3R_1 + R_2 \rightarrow \begin{bmatrix} 1 & 2 & -1 & \vdots & 0 \\ 0 & -1 & 3 & \vdots & 1 \end{bmatrix}$$

$$-R_2 \rightarrow \begin{bmatrix} 1 & 2 & -1 & \vdots & 0 \\ 0 & 1 & -3 & \vdots & -1 \end{bmatrix}$$

$$-2R_2 + R_1 \rightarrow \begin{bmatrix} 1 & 0 & 5 & \vdots & 2 \\ 0 & 1 & -3 & \vdots & -1 \end{bmatrix}$$

The corresponding system of equations is

$$\begin{cases} x \quad + 5z = \ \ 2 \\ \quad y - 3z = -1 \end{cases}.$$

Solving for x and y in terms of z, you have

$$x = -5z + 2$$

and

$$y = 3z - 1.$$

To write a solution to the system that does not use any of the three variables of the system, let a represent any real number and let

$$z = a.$$

Now substitute a for z in the equations for x and y.

$$x = -5z + 2 = -5a + 2$$

$$y = 3z - 1 = 3a - 1$$

So, the solution set has the form

$$(-5a + 2, 3a - 1, a)$$

where a is any real number. Try substituting values for a to obtain a few solutions. Then check each solution in the original equation.

STUDY TIP

In Example 9, x and y are solved for in terms of the third variable z. To write a solution to the system that does not use any of the three variables of the system, let a represent any real number and let $z = a$. Then solve for x and y. The solution can then be written in terms of a, which is not one of the variables of the system.

It is worth noting that the row-echelon form of a matrix is not unique. That is, two different sequences of elementary row operations may yield different row-echelon forms. This is demonstrated in Example 10.

Example 10 ▶ Comparing Row-Echelon Forms

Compare the following row-echelon form with the one found in Example 4. Is it the same? Does it yield the same solution?

$$\begin{cases} x - 2y + 3z = 9 \\ -x + 3y = -4 \\ 2x - 5y + 5z = 17 \end{cases}$$

$$\begin{bmatrix} 1 & -2 & 3 & \vdots & 9 \\ -1 & 3 & 0 & \vdots & -4 \\ 2 & -5 & 5 & \vdots & 17 \end{bmatrix}$$

$$\begin{matrix} R_2 \\ R_1 \\ \\ \end{matrix} \begin{bmatrix} -1 & 3 & 0 & \vdots & -4 \\ 1 & -2 & 3 & \vdots & 9 \\ 2 & -5 & 5 & \vdots & 17 \end{bmatrix}$$

$$\begin{matrix} -R_1 \rightarrow \\ \\ \\ \end{matrix} \begin{bmatrix} 1 & -3 & 0 & \vdots & 4 \\ 1 & -2 & 3 & \vdots & 9 \\ 2 & -5 & 5 & \vdots & 17 \end{bmatrix}$$

$$\begin{matrix} \\ -R_1 + R_2 \rightarrow \\ -2R_1 + R_3 \rightarrow \end{matrix} \begin{bmatrix} 1 & -3 & 0 & \vdots & 4 \\ 0 & 1 & 3 & \vdots & 5 \\ 0 & 1 & 5 & \vdots & 9 \end{bmatrix}$$

$$\begin{matrix} \\ \\ -R_2 + R_3 \rightarrow \end{matrix} \begin{bmatrix} 1 & -3 & 0 & \vdots & 4 \\ 0 & 1 & 3 & \vdots & 5 \\ 0 & 0 & 2 & \vdots & 4 \end{bmatrix}$$

$$\begin{matrix} \\ \\ \frac{1}{2}R_3 \rightarrow \end{matrix} \begin{bmatrix} 1 & -3 & 0 & \vdots & 4 \\ 0 & 1 & 3 & \vdots & 5 \\ 0 & 0 & 1 & \vdots & 2 \end{bmatrix}$$

Solution

This row-echelon form is different from that obtained in Example 4. The corresponding system of linear equations for this row-echelon matrix is

$$\begin{cases} x - 3y = 4 \\ y + 3z = 5. \\ z = 2 \end{cases}$$

Using back-substitution on this system, you obtain the solution $x = 1$, $y = -1$, and $z = 2$, which is the same solution that was obtained in Example 4.

You have seen that the row-echelon form of a given matrix *is not* unique; however, the *reduced* row-echelon form of a given matrix *is* unique. Try applying Gauss-Jordan elimination to the row-echelon matrix in Example 10 to see that you obtain the same reduced row-echelon form as in Example 8.

Activities

1. Write two or three sentences comparing Gaussian elimination with back-substitution and Gauss-Jordan elimination.
2. Set up the augmented matrix needed to solve the system.

$$\begin{aligned} 2x + 5y - z + w &= 13 \\ x - 4y + 3w &= 7 \\ 5x + w &= 13 \\ x + 2y + 3z &= 1 \end{aligned}$$

Answer:

$$\begin{bmatrix} 2 & 5 & -1 & 1 & \vdots & 13 \\ 1 & -4 & 0 & 3 & \vdots & 7 \\ 5 & 0 & 0 & 1 & \vdots & 13 \\ 1 & 2 & 3 & 0 & \vdots & 1 \end{bmatrix}$$

3. Use Gauss-Jordan elimination to solve the system.

$$\begin{aligned} x + 2y + z &= -4 \\ 2x - y + z &= -4 \\ x + 3y - z &= -7 \end{aligned}$$

Answer: $(-3, -1, 1)$

8.1 Exercises

Exercises containing systems with no solutions: 57, 69, 70
Exercises containing systems with infinitely many solutions: 58, 65, 66, 67, 68, 71, 72, 75, 76

In Exercises 1–6, determine the order of the matrix.

1. $\begin{bmatrix} 7 & 0 \end{bmatrix}$

2. $\begin{bmatrix} 5 & -3 & 8 & 7 \end{bmatrix}$

3. $\begin{bmatrix} 2 \\ 36 \\ 3 \end{bmatrix}$

4. $\begin{bmatrix} -3 & 7 & 15 & 0 \\ 0 & 0 & 3 & 3 \\ 1 & 1 & 6 & 7 \end{bmatrix}$

5. $\begin{bmatrix} 33 & 45 \\ -9 & 20 \end{bmatrix}$

6. $\begin{bmatrix} -7 & 6 & 4 \\ 0 & -5 & 1 \end{bmatrix}$

In Exercises 7–12, write the augmented matrix for the system of linear equations.

7. $\begin{cases} 4x - 3y = -5 \\ -x + 3y = 12 \end{cases}$

8. $\begin{cases} 7x + 4y = 22 \\ 5x - 9y = 15 \end{cases}$

9. $\begin{cases} x + 10y - 2z = 2 \\ 5x - 3y + 4z = 0 \\ 2x + y = 6 \end{cases}$

10. $\begin{cases} -x - 8y + 5z = 8 \\ -7x - 15z = -38 \\ 3x - y + 8z = 20 \end{cases}$

11. $\begin{cases} 7x - 5y + z = 13 \\ 19x - 8z = 10 \end{cases}$

12. $\begin{cases} 9x + 2y - 3z = 20 \\ -25y + 11z = -5 \end{cases}$

In Exercises 13–18, write the system of linear equations represented by the augmented matrix. (Use variables *x*, *y*, *z*, and *w*.)

13. $\left[\begin{array}{cc:c} 1 & 2 & 7 \\ 2 & -3 & 4 \end{array}\right]$

14. $\left[\begin{array}{cc:c} 7 & -5 & 0 \\ 8 & 3 & -2 \end{array}\right]$

15. $\left[\begin{array}{ccc:c} 2 & 0 & 5 & -12 \\ 0 & 1 & -2 & 7 \\ 6 & 3 & 0 & 2 \end{array}\right]$

16. $\left[\begin{array}{ccc:c} 4 & -5 & -1 & 18 \\ -11 & 0 & 6 & 25 \\ 3 & 8 & 0 & -29 \end{array}\right]$

17. $\left[\begin{array}{cccc:c} 9 & 12 & 3 & 0 & 0 \\ -2 & 18 & 5 & 2 & 10 \\ 1 & 7 & -8 & 0 & -4 \\ 3 & 0 & 2 & 0 & -10 \end{array}\right]$

18. $\left[\begin{array}{cccc:c} 6 & 2 & -1 & -5 & -25 \\ -1 & 0 & 7 & 3 & 7 \\ 4 & -1 & -10 & 6 & 23 \\ 0 & 8 & 1 & -11 & -21 \end{array}\right]$

In Exercises 19–22, determine whether the matrix is in row-echelon form. If it is, determine if it is also in reduced row-echelon form.

19. $\begin{bmatrix} 1 & 0 & 0 & 0 \\ 0 & 1 & 1 & 5 \\ 0 & 0 & 0 & 0 \end{bmatrix}$

20. $\begin{bmatrix} 1 & 3 & 0 & 0 \\ 0 & 0 & 1 & 8 \\ 0 & 0 & 0 & 0 \end{bmatrix}$

21. $\begin{bmatrix} 2 & 0 & 4 & 0 \\ 0 & -1 & 3 & 6 \\ 0 & 0 & 1 & 5 \end{bmatrix}$

22. $\begin{bmatrix} 1 & 0 & 2 & 1 \\ 0 & 1 & -3 & 10 \\ 0 & 0 & 1 & 0 \end{bmatrix}$

In Exercises 23–26, fill in the blank(s) using elementary row operations to form a row-equivalent matrix.

23. $\begin{bmatrix} 1 & 4 & 3 \\ 2 & 10 & 5 \end{bmatrix}$

$\begin{bmatrix} 1 & 4 & 3 \\ 0 & \square & -1 \end{bmatrix}$

24. $\begin{bmatrix} 3 & 6 & 8 \\ 4 & -3 & 6 \end{bmatrix}$

$\begin{bmatrix} 1 & \square & \frac{8}{3} \\ 4 & -3 & 6 \end{bmatrix}$

25. $\begin{bmatrix} 1 & 1 & 4 & -1 \\ 3 & 8 & 10 & 3 \\ -2 & 1 & 12 & 6 \end{bmatrix}$

$\begin{bmatrix} 1 & 1 & 4 & -1 \\ 0 & 5 & \square & \square \\ 0 & 3 & \square & \square \end{bmatrix}$

$\begin{bmatrix} 1 & 1 & 4 & -1 \\ 0 & 1 & -\frac{2}{5} & \frac{6}{5} \\ 0 & 3 & \square & \square \end{bmatrix}$

26. $\begin{bmatrix} 2 & 4 & 8 & 3 \\ 1 & -1 & -3 & 2 \\ 2 & 6 & 4 & 9 \end{bmatrix}$

$\begin{bmatrix} 1 & \square & \square & \square \\ 1 & -1 & -3 & 2 \\ 2 & 6 & 4 & 9 \end{bmatrix}$

$\begin{bmatrix} 1 & 2 & 4 & \frac{3}{2} \\ 0 & \square & -7 & \frac{1}{2} \\ 0 & 2 & \square & \square \end{bmatrix}$

In Exercises 27-30, identify the elementary row operation(s) being performed to obtain the new row-equivalent matrix.

	Original Matrix	*New Row-Equivalent Matrix*
27.	$\begin{bmatrix} -2 & 5 & 1 \\ 3 & -1 & -8 \end{bmatrix}$	$\begin{bmatrix} 13 & 0 & -39 \\ 3 & -1 & -8 \end{bmatrix}$
28.	$\begin{bmatrix} 3 & -1 & -4 \\ -4 & 3 & 7 \end{bmatrix}$	$\begin{bmatrix} 3 & -1 & -4 \\ 5 & 0 & -5 \end{bmatrix}$
29.	$\begin{bmatrix} 0 & -1 & -5 & 5 \\ -1 & 3 & -7 & 6 \\ 4 & -5 & 1 & 3 \end{bmatrix}$	$\begin{bmatrix} -1 & 3 & -7 & 6 \\ 0 & -1 & -5 & 5 \\ 0 & 7 & -27 & 27 \end{bmatrix}$
30.	$\begin{bmatrix} -1 & -2 & 3 & -2 \\ 2 & -5 & 1 & -7 \\ 5 & 4 & -7 & 6 \end{bmatrix}$	$\begin{bmatrix} -1 & -2 & 3 & -2 \\ 0 & -9 & 7 & -11 \\ 0 & -6 & 8 & -4 \end{bmatrix}$

31. Perform the sequence of row operations on the matrix. What did the operations accomplish?

$$\begin{bmatrix} 1 & 2 & 3 \\ 2 & -1 & -4 \\ 3 & 1 & -1 \end{bmatrix}$$

(a) Add -2 times R_1 to R_2.

(b) Add -3 times R_1 to R_3.

(c) Add -1 times R_2 to R_3.

(d) Multiply R_2 by $-\frac{1}{5}$.

(e) Add -2 times R_2 to R_1.

32. Perform the sequence of row operations on the matrix. What did the operations accomplish?

$$\begin{bmatrix} 7 & 1 \\ 0 & 2 \\ -3 & 4 \\ 4 & 1 \end{bmatrix}$$

(a) Add R_3 to R_4.

(b) Interchange R_1 and R_4.

(c) Add 3 times R_1 to R_3.

(d) Add -7 times R_1 to R_4.

(e) Multiply R_2 by $\frac{1}{2}$.

(f) Add the appropriate multiples of R_2 to R_1, R_3, and R_4.

In Exercises 33–36, write the matrix in row-echelon form. Remember that the row-echelon form for a matrix is not unique.

33. $\begin{bmatrix} 1 & 1 & 0 & 5 \\ -2 & -1 & 2 & -10 \\ 3 & 6 & 7 & 14 \end{bmatrix}$

34. $\begin{bmatrix} 1 & 2 & -1 & 3 \\ 3 & 7 & -5 & 14 \\ -2 & -1 & -3 & 8 \end{bmatrix}$

35. $\begin{bmatrix} 1 & -1 & -1 & 1 \\ 5 & -4 & 1 & 8 \\ -6 & 8 & 18 & 0 \end{bmatrix}$

36. $\begin{bmatrix} 1 & -3 & 0 & -7 \\ -3 & 10 & 1 & 23 \\ 4 & -10 & 2 & -24 \end{bmatrix}$

In Exercises 37–42, use the matrix capabilities of a graphing utility to write the matrix in *reduced* row-echelon form.

37. $\begin{bmatrix} 3 & 3 & 3 \\ -1 & 0 & -4 \\ 2 & 4 & -2 \end{bmatrix}$ **38.** $\begin{bmatrix} 1 & 3 & 2 \\ 5 & 15 & 9 \\ 2 & 6 & 10 \end{bmatrix}$

39. $\begin{bmatrix} 1 & 2 & 3 & -5 \\ 1 & 2 & 4 & -9 \\ -2 & -4 & -4 & 3 \\ 4 & 8 & 11 & -14 \end{bmatrix}$

40. $\begin{bmatrix} -2 & 3 & -1 & -2 \\ 4 & -2 & 5 & 8 \\ 1 & 5 & -2 & 0 \\ 3 & 8 & -10 & -30 \end{bmatrix}$

41. $\begin{bmatrix} -3 & 5 & 1 & 12 \\ 1 & -1 & 1 & 4 \end{bmatrix}$

42. $\begin{bmatrix} 5 & 1 & 2 & 4 \\ -1 & 5 & 10 & -32 \end{bmatrix}$

In Exercises 43–46, write the system of linear equations represented by the augmented matrix. Then use back-substitution to solve. (Use variables *x*, *y*, and *z*.)

43. $\left[\begin{array}{cc|c} 1 & -2 & 4 \\ 0 & 1 & -3 \end{array}\right]$ **44.** $\left[\begin{array}{cc|c} 1 & 5 & 0 \\ 0 & 1 & -1 \end{array}\right]$

45. $\left[\begin{array}{ccc|c} 1 & -1 & 2 & 4 \\ 0 & 1 & -1 & 2 \\ 0 & 0 & 1 & -2 \end{array}\right]$

46. $\left[\begin{array}{ccc|c} 1 & 2 & -2 & -1 \\ 0 & 1 & 1 & 9 \\ 0 & 0 & 1 & -3 \end{array}\right]$

In Exercises 47–50, an augmented matrix that represents a system of linear equations (in variables *x*, *y*, and *z*) has been reduced using Gauss-Jordan elimination. Write the solution represented by the augmented matrix.

47. $\left[\begin{array}{cc|c} 1 & 0 & 3 \\ 0 & 1 & -4 \end{array}\right]$ **48.** $\left[\begin{array}{cc|c} 1 & 0 & -6 \\ 0 & 1 & 10 \end{array}\right]$

49. $\left[\begin{array}{ccc|c} 1 & 0 & 0 & -4 \\ 0 & 1 & 0 & -10 \\ 0 & 0 & 1 & 4 \end{array}\right]$

50. $\left[\begin{array}{ccc|c} 1 & 0 & 0 & 5 \\ 0 & 1 & 0 & -3 \\ 0 & 0 & 1 & 0 \end{array}\right]$

In Exercises 51–70, solve the system of equations if possible. Use Gaussian elimination with back-substitution or Gauss-Jordan elimination.

51. $\begin{cases} x + 2y = 7 \\ 2x + y = 8 \end{cases}$

52. $\begin{cases} 2x + 6y = 16 \\ 2x + 3y = 7 \end{cases}$

53. $\begin{cases} 3x - 2y = -27 \\ x + 3y = 13 \end{cases}$

54. $\begin{cases} -x + y = 4 \\ 2x - 4y = -34 \end{cases}$

55. $\begin{cases} -2x + 6y = -22 \\ x + 2y = -9 \end{cases}$

56. $\begin{cases} 5x - 5y = -5 \\ -2x - 3y = 7 \end{cases}$

57. $\begin{cases} -x + 2y = 1.5 \\ 2x - 4y = 3 \end{cases}$

58. $\begin{cases} x - 3y = 5 \\ -2x + 6y = -10 \end{cases}$

59. $\begin{cases} x - 3z = -2 \\ 3x + y - 2z = 5 \\ 2x + 2y + z = 4 \end{cases}$

60. $\begin{cases} 2x - y + 3z = 24 \\ 2y - z = 14 \\ 7x - 5y = 6 \end{cases}$

61. $\begin{cases} -x + y - z = -14 \\ 2x - y + z = 21 \\ 3x + 2y + z = 19 \end{cases}$

62. $\begin{cases} 2x + 2y - z = 2 \\ x - 3y + z = -28 \\ -x + y = 14 \end{cases}$

63. $\begin{cases} x + 2y - 3z = -28 \\ 4y + 2z = 0 \\ -x + y - z = -5 \end{cases}$

64. $\begin{cases} 3x - 2y + z = 15 \\ -x + y + 2z = -10 \\ x - y - 4z = 14 \end{cases}$

65. $\begin{cases} x + y - 5z = 3 \\ x - 2z = 1 \\ 2x - y - z = 0 \end{cases}$

66. $\begin{cases} 2x + 3z = 3 \\ 4x - 3y + 7z = 5 \\ 8x - 9y + 15z = 9 \end{cases}$

67. $\begin{cases} x + 2y + z + 2w = 8 \\ 3x + 7y + 6z + 9w = 26 \end{cases}$

68. $\begin{cases} 4x + 12y - 7z - 20w = 22 \\ 3x + 9y - 5z - 28w = 30 \end{cases}$

69. $\begin{cases} -x + y = -22 \\ 3x + 4y = 4 \\ 4x - 8y = 32 \end{cases}$

70. $\begin{cases} x + 2y = 0 \\ x + y = 6 \\ 3x - 2y = 8 \end{cases}$

In Exercises 71–76, use the matrix capabilities of a graphing utility to reduce the augmented matrix corresponding to the system of equations, and solve the system.

71. $\begin{cases} 3x + 3y + 12z = 6 \\ x + y + 4z = 2 \\ 2x + 5y + 20z = 10 \\ -x + 2y + 8z = 4 \end{cases}$

72. $\begin{cases} 2x + 10y + 2z = 6 \\ x + 5y + 2z = 6 \\ x + 5y + z = 3 \\ -3x - 15y - 3z = -9 \end{cases}$

73. $\begin{cases} 2x + y - z + 2w = -6 \\ 3x + 4y + w = 1 \\ x + 5y + 2z + 6w = -3 \\ 5x + 2y - z - w = 3 \end{cases}$

74. $\begin{cases} x + 2y + 2z + 4w = 11 \\ 3x + 6y + 5z + 12w = 30 \\ x + 3y - 3z + 2w = -5 \\ 6x - y - z + w = -9 \end{cases}$

75. $\begin{cases} x + y + z + w = 0 \\ 2x + 3y + z - 2w = 0 \\ 3x + 5y + z = 0 \end{cases}$

76. $\begin{cases} x + 2y + z + 3w = 0 \\ x - y + w = 0 \\ y - z + 2w = 0 \end{cases}$

In Exercises 77–80, determine whether the two systems of linear equations yield the same solutions. If so, find the solutions.

77. (a) $\begin{cases} x - 2y + z = -6 \\ y - 5z = 16 \\ z = -3 \end{cases}$ (b) $\begin{cases} x + y - 2z = 6 \\ y + 3z = -8 \\ z = -3 \end{cases}$

78. (a) $\begin{cases} x - 3y + 4z = -11 \\ y - z = -4 \\ z = 2 \end{cases}$

(b) $\begin{cases} x + 4y = -11 \\ y + 3z = 4 \\ z = 2 \end{cases}$

79. (a) $\begin{cases} x - 4y + 5z = 27 \\ y - 7z = -54 \\ z = 8 \end{cases}$

(b) $\begin{cases} x - 6y + z = 15 \\ y + 5z = 42 \\ z = 8 \end{cases}$

80. (a) $\begin{cases} x + 3y - z = 19 \\ y + 6z = -18 \\ z = -4 \end{cases}$

(b) $\begin{cases} x - y + 3z = -15 \\ y - 2z = 14 \\ z = -4 \end{cases}$

81. Use the system $\begin{cases} x + 3y + z = 3 \\ x + 5y + 5z = 1 \\ 2x + 6y + 3z = 8 \end{cases}$ to write two different matrices in row-echelon form that yield the same solutions.

82. *Electrical Network* The currents in an electrical network are given by the solution of the system

$$\begin{cases} I_1 - I_2 + I_3 = 0 \\ 3I_1 + 4I_2 = 18 \\ I_2 + 3I_3 = 6 \end{cases}$$

where I_1, I_2, and I_3 are measured in amperes. Solve the system of equations.

83. *Partial Fractions* Write the partial fraction decomposition for the rational expression

$$\frac{4x^2}{(x+1)^2(x-1)} = \frac{A}{x-1} + \frac{B}{x+1} + \frac{C}{(x+1)^2}.$$

84. *Finance* A small corporation borrowed \$1,500,000 to expand its product line. Some of the money was borrowed at 7%, some at 8%, and some at 10%. How much was borrowed at each rate if the annual interest was \$130,500 and the amount borrowed at 10% was 4 times the amount borrowed at 7%?

85. *Finance* A small corporation borrowed \$500,000 to expand its product line. Some of the money was borrowed at 9%, some at 10%, and some at 12%. How much was borrowed at each rate if the annual interest was \$52,000 and the amount borrowed at 10% was $2\frac{1}{2}$ times the amount borrowed at 9%?

In Exercises 86 and 87, find the specified equation that passes through the points. Use a graphing utility to verify your results.

86. Parabola: $y = ax^2 + bx + c$

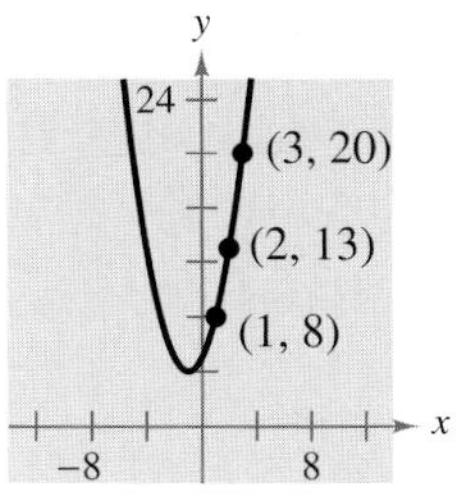

87. Parabola: $y = ax^2 + bx + c$

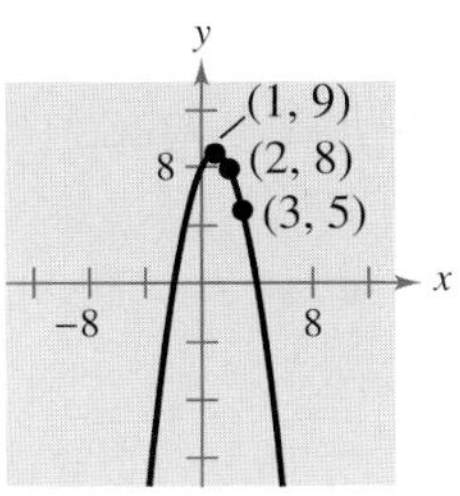

88. *Mathematical Modeling* A videotape of the path of a ball thrown by a baseball player was analyzed with a grid covering the TV screen (see figure). The tape was paused three times, and the position of the ball was measured each time. The coordinates were approximately (0, 5.0), (15, 9.6), and (30, 12.4). (The x-coordinate measures the horizontal distance from the player in feet, and the y-coordinate is the height of the ball in feet.)

(a) Find the equation of the parabola $y = ax^2 + bx + c$ that passes through the three points.

(b) Use a graphing utility to graph the parabola. Approximate the maximum height of the ball and the point at which the ball struck the ground.

(c) Find analytically the maximum height of the ball and the point at which it struck the ground.

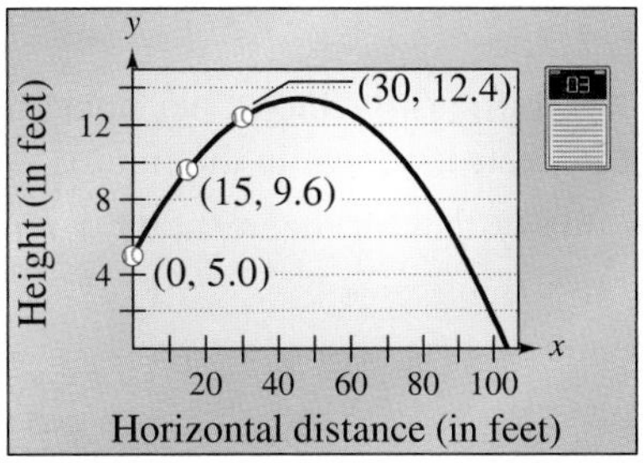

89. *Data Analysis* The bar graph gives the value y, in millions of dollars, for new orders of civil jet transport aircraft built by U.S. companies for the years 1995 through 1997. (Source: Aerospace Industries Association of America)

(a) Find the equation of the parabola that passes through the points. Let $t = 5$ represent 1995.

(b) Use a graphing utility to graph the parabola.

(c) Use the equation in part (a) to estimate y in the year 2000. Is the estimate reasonable? Explain.

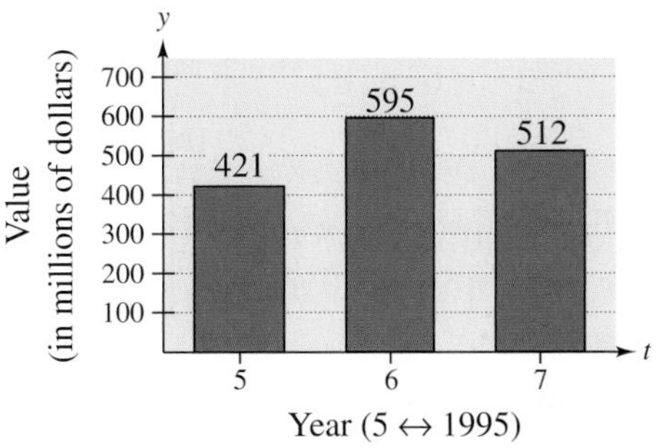

Network Analysis **In Exercises 90 and 91, answer the questions about the specified network. (In a network it is assumed that the total flow into each junction is equal to the total flow out of each junction.)**

90. Water flowing through a network of pipes (in thousands of cubic meters per hour) is shown in the figure.

(a) Solve this system for the water flow represented by x_i, $i = 1, 2, \ldots, 7$.

(b) Find the network flow pattern when $x_6 = x_7 = 0$.

(c) Find the network flow pattern when $x_5 = 1000$ and $x_6 = 0$.

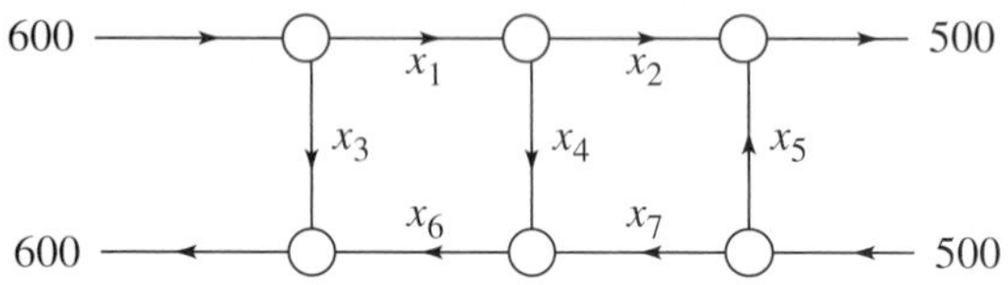

91. The flow of traffic (in vehicles per hour) through a network of streets is shown in the figure.

(a) Solve this system for the traffic flow represented by x_i, $i = 1, 2, \ldots, 5$.

(b) Find the traffic flow when $x_2 = 200$ and $x_3 = 50$.

(c) Find the traffic flow when $x_2 = 150$ and $x_3 = 0$.

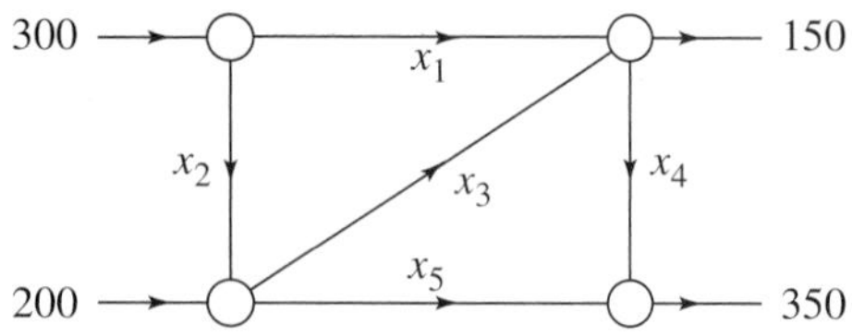

Synthesis

True or False? **In Exercises 92–94, determine whether the statement is true or false. Justify your answer.**

92. $\begin{bmatrix} 5 & 0 & -2 & 7 \\ -1 & 3 & -6 & 0 \end{bmatrix}$ is a 4×2 matrix.

93. The matrix $\begin{bmatrix} 0 & 0 & 0 & 0 \\ 0 & 0 & 1 & -4 \\ 0 & 1 & 0 & 2 \\ 1 & 0 & 0 & 5 \end{bmatrix}$ is in reduced row-echelon form.

94. Gaussian elimination reduces a matrix until a reduced row-echelon form is obtained.

95. *Think About It* The augmented matrix represents a system of linear equations (in variables x, y, and z) that has been reduced using Gauss-Jordan elimination. Write a system of equations with nonzero coefficients that is represented by the reduced matrix. (The answer is not unique.)

$$\left[\begin{array}{ccc:c} 1 & 0 & 3 & -2 \\ 0 & 1 & 4 & 1 \\ 0 & 0 & 0 & 0 \end{array}\right]$$

96. *Think About It*

(a) Describe the row-echelon form of an augmented matrix that corresponds to a system of linear equations that is inconsistent.

(b) Describe the row-echelon form of an augmented matrix that corresponds to a system of linear equations that has an infinite number of solutions.

97. Describe the three elementary row operations that can be performed on an augmented matrix.

98. What is the relationship between the three elementary row operations performed on an augmented matrix and the operations that lead to equivalent systems of equations?

99. *Writing* In your own words, describe the difference between a matrix in row-echelon form and a matrix in reduced row-echelon form.

Review

In Exercises 100 and 101, sketch the graph of the function. Identify any asymptotes.

100. $f(x) = \dfrac{7}{-x - 1}$

101. $f(x) = \dfrac{4x}{5x^2 + 2}$

In Exercises 102–105, sketch the graph of the function. Do not use a graphing utility.

102. $f(x) = 2^{x-1}$

103. $g(x) = 3^{-x+2}$

104. $h(x) = \ln(x - 1)$

105. $f(x) = 3 + \ln x$

8.2 Operations with Matrices

▶ **What you should learn**

- How to decide whether two matrices are equal
- How to add and subtract matrices and multiply matrices by real numbers
- How to multiply two matrices
- How to use matrix operations to model and solve real-life problems

▶ **Why you should learn it**

Matrix operations can be used to model and solve real-life problems. For instance, Exercise 75 on page 648 shows how to use matrix multiplication to help analyze the labor and wage requirements for a boat manufacturer.

Jim Brown/The Stock Market

Equality of Matrices

In Section 8.1, you used matrices to solve systems of linear equations. Matrices, however, can do much more than this. There is a rich mathematical theory of matrices, and its applications are numerous. This section and the next two introduce some fundamentals of matrix theory. It is standard mathematical convention to represent matrices in any of the following three ways.

1. A matrix can be denoted by an uppercase letter such as A, B, or C.
2. A matrix can be denoted by a representative element enclosed in brackets, such as $[a_{ij}]$, $[b_{ij}]$, or $[c_{ij}]$.
3. A matrix can be denoted by a rectangular array of numbers such as

$$A = [a_{ij}] = \begin{bmatrix} a_{11} & a_{12} & a_{13} & \cdots & a_{1n} \\ a_{21} & a_{22} & a_{23} & \cdots & a_{2n} \\ a_{31} & a_{32} & a_{33} & \cdots & a_{3n} \\ \vdots & \vdots & \vdots & & \vdots \\ a_{m1} & a_{m2} & a_{m3} & \cdots & a_{mn} \end{bmatrix}.$$

Two matrices $A = [a_{ij}]$ and $B = [b_{ij}]$ are equal if they have the same order $(m \times n)$ and $a_{ij} = b_{ij}$ for $1 \le i \le m$ and $1 \le j \le n$. In other words, two matrices are equal if their corresponding entries are equal.

Example 1 ▶ Equality of Matrices

Solve for a_{11}, a_{12}, a_{21}, and a_{22} in the following matrix equation.

$$\begin{bmatrix} a_{11} & a_{12} \\ a_{21} & a_{22} \end{bmatrix} = \begin{bmatrix} 2 & -1 \\ -3 & 0 \end{bmatrix}$$

Solution

Because two matrices are equal only if their corresponding entries are equal, you can conclude that

$$a_{11} = 2, \quad a_{12} = -1, \quad a_{21} = -3, \quad \text{and} \quad a_{22} = 0.$$

Be sure you see that for two matrices to be equal, they must have the same order *and* their corresponding entries must be equal. For instance,

$$\begin{bmatrix} 2 & -1 \\ \sqrt{4} & \frac{1}{2} \end{bmatrix} = \begin{bmatrix} 2 & -1 \\ 2 & 0.5 \end{bmatrix}$$

but

$$\begin{bmatrix} 2 & -1 \\ 3 & 4 \\ 0 & 0 \end{bmatrix} \neq \begin{bmatrix} 2 & -1 \\ 3 & 4 \end{bmatrix}.$$

The Granger Collection

Historical Note

Arthur Cayley (1821–1895), a British mathematician, invented matrices around 1858. Cayley was a Cambridge University graduate and a lawyer by profession. His ground-breaking work on matrices was begun as he studied the theory of transformations. Cayley also was instrumental in the development of determinants. Cayley and two American mathematicians, Benjamin Peirce (1809–1880) and his son Charles S. Peirce (1839–1914), are credited with developing "matrix algebra."

Matrix Addition and Scalar Multiplication

In this section, three basic matrix operations will be covered. The first two are matrix addition and scalar multiplication. With matrix addition, you can add two matrices (of the same order) by adding their corresponding entries.

Definition of Matrix Addition

If $A = [a_{ij}]$ and $B = [b_{ij}]$ are matrices of order $m \times n$, their sum is the $m \times n$ matrix given by

$$A + B = [a_{ij} + b_{ij}].$$

The sum of two matrices of different orders is undefined.

Example 2 ▶ Addition of Matrices

a. $\begin{bmatrix} -1 & 2 \\ 0 & 1 \end{bmatrix} + \begin{bmatrix} 1 & 3 \\ -1 & 2 \end{bmatrix} = \begin{bmatrix} -1+1 & 2+3 \\ 0-1 & 1+2 \end{bmatrix}$

$= \begin{bmatrix} 0 & 5 \\ -1 & 3 \end{bmatrix}$

b. $\begin{bmatrix} 0 & 1 & -2 \\ 1 & 2 & 3 \end{bmatrix} + \begin{bmatrix} 0 & 0 & 0 \\ 0 & 0 & 0 \end{bmatrix} = \begin{bmatrix} 0 & 1 & -2 \\ 1 & 2 & 3 \end{bmatrix}$

c. $\begin{bmatrix} 1 \\ -3 \\ -2 \end{bmatrix} + \begin{bmatrix} -1 \\ 3 \\ 2 \end{bmatrix} = \begin{bmatrix} 0 \\ 0 \\ 0 \end{bmatrix}$

d. The sum of

$$A = \begin{bmatrix} 2 & 1 & 0 \\ 4 & 0 & -1 \\ 3 & -2 & 2 \end{bmatrix} \quad \text{and}$$

$$B = \begin{bmatrix} 0 & 1 \\ -1 & 3 \\ 2 & 4 \end{bmatrix}$$

is undefined because A and B have different orders.

In operations with matrices, numbers are usually referred to as **scalars.** In this text, scalars will always be real numbers. You can multiply a matrix A by a scalar c by multiplying each entry in A by c.

Definition of Scalar Multiplication

If $A = [a_{ij}]$ is an $m \times n$ matrix and c is a scalar, the **scalar multiple** of A by c is the $m \times n$ matrix given by

$$cA = [ca_{ij}].$$

The symbol $-A$ represents the negation of A, or the scalar product $(-1)A$. Moreover, if A and B are of the same order, then $A - B$ represents the sum of A and $(-1)B$. That is,

$$A - B = A + (-1)B. \qquad \text{Subtraction of matrices}$$

Example 3 ▶ Scalar Multiplication and Matrix Subtraction

For the following matrices, find (a) $3A$, (b) $-B$, and (c) $3A - B$.

$$A = \begin{bmatrix} 2 & 2 & 4 \\ -3 & 0 & -1 \\ 2 & 1 & 2 \end{bmatrix} \quad \text{and} \quad B = \begin{bmatrix} 2 & 0 & 0 \\ 1 & -4 & 3 \\ -1 & 3 & 2 \end{bmatrix}$$

Solution

a. $$3A = 3\begin{bmatrix} 2 & 2 & 4 \\ -3 & 0 & -1 \\ 2 & 1 & 2 \end{bmatrix} \qquad \text{Scalar multiplication}$$

$$= \begin{bmatrix} 3(2) & 3(2) & 3(4) \\ 3(-3) & 3(0) & 3(-1) \\ 3(2) & 3(1) & 3(2) \end{bmatrix} \qquad \text{Multiply each entry by 3.}$$

$$= \begin{bmatrix} 6 & 6 & 12 \\ -9 & 0 & -3 \\ 6 & 3 & 6 \end{bmatrix} \qquad \text{Simplify.}$$

b. $$-B = (-1)\begin{bmatrix} 2 & 0 & 0 \\ 1 & -4 & 3 \\ -1 & 3 & 2 \end{bmatrix} \qquad \text{Definition of negation}$$

$$= \begin{bmatrix} -2 & 0 & 0 \\ -1 & 4 & -3 \\ 1 & -3 & -2 \end{bmatrix} \qquad \text{Multiply each entry by } -1.$$

c. $$3A - B = \begin{bmatrix} 6 & 6 & 12 \\ -9 & 0 & -3 \\ 6 & 3 & 6 \end{bmatrix} - \begin{bmatrix} 2 & 0 & 0 \\ 1 & -4 & 3 \\ -1 & 3 & 2 \end{bmatrix} \qquad \text{Matrix subtraction}$$

$$= \begin{bmatrix} 4 & 6 & 12 \\ -10 & 4 & -6 \\ 7 & 0 & 4 \end{bmatrix} \qquad \text{Subtract corresponding entries.}$$

STUDY TIP

The order of operations for matrix expressions is similar to that for real numbers. In particular, you perform scalar multiplication before matrix addition and subtraction, as shown in Example 3(c).

It is often convenient to rewrite the scalar multiple cA by factoring c out of every entry in the matrix. For instance, in the following example, the scalar $\frac{1}{2}$ has been factored out of the matrix.

$$\begin{bmatrix} \frac{1}{2} & -\frac{3}{2} \\ \frac{5}{2} & \frac{1}{2} \end{bmatrix} = \begin{bmatrix} \frac{1}{2}(1) & \frac{1}{2}(-3) \\ \frac{1}{2}(5) & \frac{1}{2}(1) \end{bmatrix}$$

$$= \frac{1}{2}\begin{bmatrix} 1 & -3 \\ 5 & 1 \end{bmatrix}$$

The properties of matrix addition and scalar multiplication are similar to those of addition and multiplication of real numbers.

Properties of Matrix Addition and Scalar Multiplication

Let A, B, and C be $m \times n$ matrices and let c and d be scalars.

1. $A + B = B + A$	Commutative Property of Matrix Addition
2. $A + (B + C) = (A + B) + C$	Associative Property of Matrix Addition
3. $(cd)A = c(dA)$	Associative Property of Scalar Multiplication
4. $1A = A$	Scalar Identity
5. $c(A + B) = cA + cB$	Distributive Property
6. $(c + d)A = cA + dA$	Distributive Property

Note that the Associative Property of Matrix Addition allows you to write expressions such as $A + B + C$ without ambiguity because the same sum occurs no matter how the matrices are grouped. In other words, you obtain the same sum whether you group $A + B + C$ as $(A + B) + C$ or as $A + (B + C)$. This same reasoning applies to sums of four or more matrices.

Example 4 ▶ Addition of More than Two Matrices

By adding corresponding entries, you obtain the following sum of four matrices.

$$\begin{bmatrix} 1 \\ 2 \\ -3 \end{bmatrix} + \begin{bmatrix} -1 \\ -1 \\ 2 \end{bmatrix} + \begin{bmatrix} 0 \\ 1 \\ 4 \end{bmatrix} + \begin{bmatrix} 2 \\ -3 \\ -2 \end{bmatrix} = \begin{bmatrix} 2 \\ -1 \\ 1 \end{bmatrix}$$

Example 5 ▶ Using the Distributive Property

$$3\left(\begin{bmatrix} -2 & 0 \\ 4 & 1 \end{bmatrix} + \begin{bmatrix} 4 & -2 \\ 3 & 7 \end{bmatrix}\right) = 3\begin{bmatrix} -2 & 0 \\ 4 & 1 \end{bmatrix} + 3\begin{bmatrix} 4 & -2 \\ 3 & 7 \end{bmatrix}$$

$$= \begin{bmatrix} -6 & 0 \\ 12 & 3 \end{bmatrix} + \begin{bmatrix} 12 & -6 \\ 9 & 21 \end{bmatrix}$$

$$= \begin{bmatrix} 6 & -6 \\ 21 & 24 \end{bmatrix}$$

STUDY TIP

In Example 5, you could add the two matrices first and then multiply the matrix by 3. The result would be the same.

Technology

Most graphing utilities can add and subtract matrices and multiply matrices by scalars. Try using a graphing utility to find the sum of the matrices

$$A = \begin{bmatrix} 2 & -3 \\ -1 & 0 \end{bmatrix} \quad \text{and} \quad B = \begin{bmatrix} -1 & 4 \\ 2 & -5 \end{bmatrix}.$$

One important property of addition of real numbers is that the number 0 is the additive identity. That is, $c + 0 = c$ for any real number c. For matrices, a similar property holds. That is, if A is an $m \times n$ matrix and O is the $m \times n$ **zero matrix** consisting entirely of zeros, then $A + O = A$.

In other words, O is the **additive identity** for the set of all $m \times n$ matrices. For example, the following matrices are the additive identities for the set of all 2×3 and 2×2 matrices.

$$O = \underbrace{\begin{bmatrix} 0 & 0 & 0 \\ 0 & 0 & 0 \end{bmatrix}}_{\text{Zero } 2 \times 3 \text{ matrix}} \quad \text{and} \quad O = \underbrace{\begin{bmatrix} 0 & 0 \\ 0 & 0 \end{bmatrix}}_{\text{Zero } 2 \times 2 \text{ matrix}}$$

The algebra of real numbers and the algebra of matrices have many similarities. For example, compare the following solutions.

Real Numbers (Solve for x.)	*$m \times n$ Matrices (Solve for X.)*
$x + a = b$	$X + A = B$
$x + a + (-a) = b + (-a)$	$X + A + (-A) = B + (-A)$
$x + 0 = b - a$	$X + O = B - A$
$x = b - a$	$X = B - A$

The algebra of real numbers and the algebra of matrices also have important differences, which will be discussed later.

Example 6 ▶ Solving a Matrix Equation

Solve for X in the equation $3X + A = B$, where

$$A = \begin{bmatrix} 1 & -2 \\ 0 & 3 \end{bmatrix} \quad \text{and} \quad B = \begin{bmatrix} -3 & 4 \\ 2 & 1 \end{bmatrix}.$$

Solution

Begin by solving the equation for X to obtain

$$3X = B - A$$

$$X = \frac{1}{3}(B - A).$$

Now, using the matrices A and B, you have

$$X = \frac{1}{3}\left(\begin{bmatrix} -3 & 4 \\ 2 & 1 \end{bmatrix} - \begin{bmatrix} 1 & -2 \\ 0 & 3 \end{bmatrix}\right)$$ Substitute the matrices.

$$= \frac{1}{3}\begin{bmatrix} -4 & 6 \\ 2 & -2 \end{bmatrix}$$ Subtract matrix B from matrix A.

$$= \begin{bmatrix} -\frac{4}{3} & 2 \\ \frac{2}{3} & -\frac{2}{3} \end{bmatrix}.$$ Multiply the matrix by $\frac{1}{3}$.

Matrix Multiplication

The third basic matrix operation is **matrix multiplication.** At first glance the definition may seem unusual. You will see later, however, that this definition of the product of two matrices has many practical applications.

Definition of Matrix Multiplication

If $A = [a_{ij}]$ is an $m \times n$ matrix and $B = [b_{ij}]$ is an $n \times p$ matrix, the product AB is an $m \times p$ matrix

$$AB = [c_{ij}]$$

where $c_{ij} = a_{i1}b_{1j} + a_{i2}b_{2j} + a_{i3}b_{3j} + \cdot \cdot \cdot + a_{in}b_{nj}$.

The definition of matrix multiplication indicates a *row-by-column* multiplication, where the entry in the ith row and jth column of the product AB is obtained by multiplying the entries in the ith row of A by the corresponding entries in the jth column of B and then adding the results. Example 7 illustrates this process.

A computer animation of this example appears in the *Interactive* CD-ROM and *Internet* versions of this text.

Example 7 ▶ Finding the Product of Two Matrices

Find the product AB where

$$A = \begin{bmatrix} -1 & 3 \\ 4 & -2 \\ 5 & 0 \end{bmatrix} \quad \text{and} \quad B = \begin{bmatrix} -3 & 2 \\ -4 & 1 \end{bmatrix}.$$

Solution

First, note that the product AB is defined because the number of columns of A is equal to the number of rows of B. Moreover, the product AB has order 3×2, and is of the form

$$\begin{bmatrix} -1 & 3 \\ 4 & -2 \\ 5 & 0 \end{bmatrix}\begin{bmatrix} -3 & 2 \\ -4 & 1 \end{bmatrix} = \begin{bmatrix} c_{11} & c_{12} \\ c_{21} & c_{22} \\ c_{31} & c_{32} \end{bmatrix}.$$

To find the entries of the product, multiply each row of A by each column of B, as follows. Use a graphing utility to check this result.

$$AB = \begin{bmatrix} -1 & 3 \\ 4 & -2 \\ 5 & 0 \end{bmatrix}\begin{bmatrix} -3 & 2 \\ -4 & 1 \end{bmatrix}$$

$$= \begin{bmatrix} (-1)(-3) + (3)(-4) & (-1)(2) + (3)(1) \\ (4)(-3) + (-2)(-4) & (4)(2) + (-2)(1) \\ (5)(-3) + (0)(-4) & (5)(2) + (0)(1) \end{bmatrix}$$

$$= \begin{bmatrix} -9 & 1 \\ -4 & 6 \\ -15 & 10 \end{bmatrix}$$

Technology

Some graphing utilities are able to add, subtract, and multiply matrices. If you have such a graphing utility, enter the matrices

$$A = \begin{bmatrix} 1 & 2 & 3 \\ 2 & -5 & 1 \end{bmatrix} \text{ and}$$

$$B = \begin{bmatrix} -3 & 2 & 1 \\ 4 & -2 & 0 \\ 1 & 2 & 3 \end{bmatrix}$$

and find their product AB. You should get:

$$\begin{bmatrix} 8 & 4 & 10 \\ -25 & 16 & 5 \end{bmatrix}.$$

This pattern for the order of the product of matrix multiplication is an effective memory device. You may want to demonstrate this pattern on several multiplication examples.

Be sure you understand that for the product of two matrices to be defined, the number of *columns* of the first matrix must equal the number of *rows* of the second matrix. That is, the middle two indices must be the same. The outside two indices give the order of the product, as shown below.

$$\underset{m \times n}{A} \quad \times \quad \underset{n \times p}{B} \quad = \quad \underset{m \times p}{AB}$$

Equal (middle indices n and n)

Order of AB (outside indices m and p)

Exploration

Use a graphing utility to multiply the matrices

$$A = \begin{bmatrix} 1 & 2 \\ 3 & 4 \end{bmatrix} \text{ and}$$

$$B = \begin{bmatrix} 0 & 1 \\ 2 & 3 \end{bmatrix}.$$

Do you obtain the same result for the product AB as for the product BA? What does this tell you about matrix multiplication and commutativity?

Example 8 ▶ Patterns in Matrix Multiplication

a. $$\underset{2 \times 3}{\begin{bmatrix} 1 & 0 & 3 \\ 2 & -1 & -2 \end{bmatrix}} \underset{3 \times 3}{\begin{bmatrix} -2 & 4 & 2 \\ 1 & 0 & 0 \\ -1 & 1 & -1 \end{bmatrix}} = \underset{2 \times 3}{\begin{bmatrix} -5 & 7 & -1 \\ -3 & 6 & 6 \end{bmatrix}}$$

b. $$\underset{2 \times 2}{\begin{bmatrix} 3 & 4 \\ -2 & 5 \end{bmatrix}} \underset{2 \times 2}{\begin{bmatrix} 1 & 0 \\ 0 & 1 \end{bmatrix}} = \underset{2 \times 2}{\begin{bmatrix} 3 & 4 \\ -2 & 5 \end{bmatrix}}$$

c. $$\underset{2 \times 2}{\begin{bmatrix} 1 & 2 \\ 1 & 1 \end{bmatrix}} \underset{2 \times 2}{\begin{bmatrix} -1 & 2 \\ 1 & -1 \end{bmatrix}} = \underset{2 \times 2}{\begin{bmatrix} 1 & 0 \\ 0 & 1 \end{bmatrix}}$$

d. $$\underset{3 \times 3}{\begin{bmatrix} 6 & 2 & 0 \\ 3 & -1 & 2 \\ 1 & 4 & 6 \end{bmatrix}} \underset{3 \times 1}{\begin{bmatrix} 1 \\ 2 \\ -3 \end{bmatrix}} = \underset{3 \times 1}{\begin{bmatrix} 10 \\ -5 \\ -9 \end{bmatrix}}$$

e. $$\underset{1 \times 3}{\begin{bmatrix} 1 & -2 & -3 \end{bmatrix}} \underset{3 \times 1}{\begin{bmatrix} 2 \\ -1 \\ 1 \end{bmatrix}} = \underset{1 \times 1}{\begin{bmatrix} 1 \end{bmatrix}}$$

f. $$\underset{3 \times 1}{\begin{bmatrix} 2 \\ -1 \\ 1 \end{bmatrix}} \underset{1 \times 3}{\begin{bmatrix} 1 & -2 & -3 \end{bmatrix}} = \underset{3 \times 3}{\begin{bmatrix} 2 & -4 & -6 \\ -1 & 2 & 3 \\ 1 & -2 & -3 \end{bmatrix}}$$

g. The product AB for the following matrices is not defined.

$$A = \underset{3 \times 2}{\begin{bmatrix} -2 & 1 \\ 1 & -3 \\ 1 & 4 \end{bmatrix}} \quad \text{and} \quad B = \underset{3 \times 4}{\begin{bmatrix} -2 & 3 & 1 & 4 \\ 0 & 1 & -1 & 2 \\ 2 & -1 & 0 & 1 \end{bmatrix}}$$

In parts (e) and (f) of Example 8, note that the two products are different. Matrix multiplication is not, in general, commutative. That is, for most matrices, $AB \neq BA$.

The general pattern for matrix multiplication is as follows. To obtain the entry in the ith row and the jth column of the product AB, use the ith row of A and the jth column of B.

$$\begin{bmatrix} a_{11} & a_{12} & a_{13} & \cdots & a_{1n} \\ a_{21} & a_{22} & a_{23} & \cdots & a_{2n} \\ a_{31} & a_{32} & a_{33} & \cdots & a_{3n} \\ \vdots & \vdots & \vdots & & \vdots \\ a_{i1} & a_{i2} & a_{i3} & \cdots & a_{in} \\ \vdots & \vdots & \vdots & & \vdots \\ a_{m1} & a_{m2} & a_{m3} & \cdots & a_{mn} \end{bmatrix} \begin{bmatrix} b_{11} & b_{12} & \cdots & b_{1j} & \cdots & b_{1p} \\ b_{21} & b_{22} & \cdots & b_{2j} & \cdots & b_{2p} \\ b_{31} & b_{32} & \cdots & b_{3j} & \cdots & b_{3p} \\ \vdots & \vdots & & \vdots & & \vdots \\ b_{n1} & b_{n2} & \cdots & b_{nj} & \cdots & b_{np} \end{bmatrix} = \begin{bmatrix} c_{11} & c_{12} & \cdots & c_{1j} & \cdots & c_{1p} \\ c_{21} & c_{22} & \cdots & c_{2j} & \cdots & c_{2p} \\ \vdots & \vdots & & \vdots & & \vdots \\ c_{i1} & c_{i2} & \cdots & c_{ij} & \cdots & c_{ip} \\ \vdots & \vdots & & \vdots & & \vdots \\ c_{m1} & c_{m2} & \cdots & c_{mj} & \cdots & c_{mp} \end{bmatrix}$$

$$a_{i1}b_{1j} + a_{i2}b_{2j} + a_{i3}b_{3j} + \cdots + a_{in}b_{nj} = c_{ij}$$

Properties of Matrix Multiplication

Let A, B, and C be matrices and let c be a scalar.

1. $A(BC) = (AB)C$ Associative Property of Multiplication
2. $A(B + C) = AB + AC$ Distributive Property
3. $(A + B)C = AC + BC$ Distributive Property
4. $c(AB) = (cA)B = A(cB)$

Using sigma notation, c_{ij} may also be written as

$$c_{ij} = \sum_{k=1}^{n} a_{ik} b_{kj}.$$

Sigma notation is introduced in Chapter 9.

Definition of Identity Matrix

The $n \times n$ matrix that consists of 1's on its main diagonal and 0's elsewhere is called the **identity matrix of order n** and is denoted by

$$I_n = \begin{bmatrix} 1 & 0 & 0 & \cdots & 0 \\ 0 & 1 & 0 & \cdots & 0 \\ 0 & 0 & 1 & \cdots & 0 \\ \vdots & \vdots & \vdots & & \vdots \\ 0 & 0 & 0 & \cdots & 1 \end{bmatrix}. \qquad \text{Identity matrix}$$

Note that an identity matrix must be *square*. When the order is understood to be n, you can denote I_n simply by I.

If A is an $n \times n$ matrix, the identity matrix has the property that $AI_n = A$ and $I_nA = A$. For example,

$$\begin{bmatrix} 3 & -2 & 5 \\ 1 & 0 & 4 \\ -1 & 2 & -3 \end{bmatrix} \begin{bmatrix} 1 & 0 & 0 \\ 0 & 1 & 0 \\ 0 & 0 & 1 \end{bmatrix} = \begin{bmatrix} 3 & -2 & 5 \\ 1 & 0 & 4 \\ -1 & 2 & -3 \end{bmatrix}$$

and

$$\begin{bmatrix} 1 & 0 & 0 \\ 0 & 1 & 0 \\ 0 & 0 & 1 \end{bmatrix} \begin{bmatrix} 3 & -2 & 5 \\ 1 & 0 & 4 \\ -1 & 2 & -3 \end{bmatrix} = \begin{bmatrix} 3 & -2 & 5 \\ 1 & 0 & 4 \\ -1 & 2 & -3 \end{bmatrix}.$$

Applications

Matrix multiplication can be used to represent a system of linear equations. Note how the system

$$\begin{cases} a_{11}x_1 + a_{12}x_2 + a_{13}x_3 = b_1 \\ a_{21}x_1 + a_{22}x_2 + a_{23}x_3 = b_2 \\ a_{31}x_1 + a_{32}x_2 + a_{33}x_3 = b_3 \end{cases}$$

can be written as the matrix equation $AX = B$, where A is the *coefficient matrix* of the system, and X and B are column matrices.

$$\underset{A}{\begin{bmatrix} a_{11} & a_{12} & a_{13} \\ a_{21} & a_{22} & a_{23} \\ a_{31} & a_{32} & a_{33} \end{bmatrix}} \times \underset{X}{\begin{bmatrix} x_1 \\ x_2 \\ x_3 \end{bmatrix}} = \underset{B}{\begin{bmatrix} b_1 \\ b_2 \\ b_3 \end{bmatrix}}$$

A computer animation of this example appears in the *Interactive* CD-ROM and *Internet* versions of this text.

Example 9 ▶ Solving a System of Linear Equations

Consider the following system of linear equations.

$$\begin{cases} x_1 - 2x_2 + x_3 = -4 \\ x_2 + 2x_3 = 4 \\ 2x_1 + 3x_2 - 2x_3 = 2 \end{cases}$$

a. Write this system as a matrix equation, $AX = B$.

b. Use Gauss-Jordan elimination on the augmented matrix $[A \vdots B]$ to solve for the matrix X.

Solution

a. In matrix form, $AX = B$, the system can be written as follows.

$$\begin{bmatrix} 1 & -2 & 1 \\ 0 & 1 & 2 \\ 2 & 3 & -2 \end{bmatrix}\begin{bmatrix} x_1 \\ x_2 \\ x_3 \end{bmatrix} = \begin{bmatrix} -4 \\ 4 \\ 2 \end{bmatrix}$$

b. The augmented matrix is formed by adjoining matrix B to matrix A.

$$[A \vdots B] = \left[\begin{array}{ccc:c} 1 & -2 & 1 & -4 \\ 0 & 1 & 2 & 4 \\ 2 & 3 & -2 & 2 \end{array}\right]$$

Using Gauss-Jordan elimination, you can rewrite this equation as

$$[I \vdots X] = \left[\begin{array}{ccc:c} 1 & 0 & 0 & -1 \\ 0 & 1 & 0 & 2 \\ 0 & 0 & 1 & 1 \end{array}\right].$$

So, the solution of the system of linear equations is $x_1 = -1$, $x_2 = 2$, and $x_3 = 1$, and the solution of the matrix equation is

$$X = \begin{bmatrix} x_1 \\ x_2 \\ x_3 \end{bmatrix} = \begin{bmatrix} -1 \\ 2 \\ 1 \end{bmatrix}.$$

Check this solution in the original system of equations.

STUDY TIP

The notation $[A \vdots B]$ represents the augmented matrix formed when matrix B is adjoined to matrix A.

Activities

1. Find $2A - 3B$.

$A = \begin{bmatrix} 2 & -1 \\ 3 & 6 \end{bmatrix}, B = \begin{bmatrix} -1 & 3 \\ 1 & 0 \end{bmatrix}$

Answer: $\begin{bmatrix} 7 & -11 \\ 3 & 12 \end{bmatrix}$

2. If possible, find AB.

$A = \begin{bmatrix} 3 & 1 & 0 \\ -1 & 0 & 2 \end{bmatrix}$

$B = \begin{bmatrix} 4 & 1 \\ -1 & 0 \\ 2 & -1 \end{bmatrix}$

Answer: $\begin{bmatrix} 11 & 3 \\ 0 & -3 \end{bmatrix}$

You may also wish to discuss some of the application problems in the exercise set. (See, for example, Exercises 69–76 on pages 647 and 648.)

Example 10 ▶ Softball Team Expenses

Two softball teams submit equipment lists to their sponsors.

	Women's Team	*Men's Team*
Bats	12	15
Balls	45	38
Gloves	15	17

Each bat costs \$48, each ball costs \$4, and each glove costs \$42. Use matrices to find the total cost of equipment for each team.

Solution

The equipment lists and the costs per item can be written in matrix form as

$$E = \begin{bmatrix} 12 & 15 \\ 45 & 38 \\ 15 & 17 \end{bmatrix}$$

and

$$C = \begin{bmatrix} 48 & 4 & 42 \end{bmatrix}.$$

The total cost of equipment for each team is given by the product

$$CE = \begin{bmatrix} 48 & 4 & 42 \end{bmatrix}\begin{bmatrix} 12 & 15 \\ 45 & 38 \\ 15 & 17 \end{bmatrix}$$

$$= \begin{bmatrix} 48(12) + 4(45) + 42(15) & 48(15) + 4(38) + 42(17) \end{bmatrix}$$

$$= \begin{bmatrix} 1386 & 1586 \end{bmatrix}.$$

So, the total cost of equipment for the women's team is \$1386 and the total cost of equipment for the men's team is \$1586.

Group Activity

Discuss the requirements for the order of the dimensions m and n for the multiplication of two $m \times n$ matrices. Determine which of the following matrix multiplications AB is (are) defined. For each case in which AB is defined, what is the order of the resulting matrix?

a. A is of order 1×3
   B is of order 2×1

b. A is of order 2×3
   B is of order 2×3

c. A is of order 3×4
   B is of order 4×2

d. A is of order 3×1
   B is of order 3×3

Discuss why matrix multiplication is not, in general, commutative. Work together to find an example of two 2×2 matrices such that $AB \neq BA$. Find an example of two 2×2 matrices such that $AB = BA$.

Writing ABOUT MATHEMATICS

Problem Posing Write a matrix multiplication application problem that uses the matrix

$$A = \begin{bmatrix} 20 & 42 & 33 \\ 17 & 30 & 50 \end{bmatrix}.$$

Exchange problems with another student in your class. Form the matrices that represent the problem, and solve the problem. Interpret your solution in the context of the problem. Check with the creator of the problem to see if you are correct. Discuss other ways to represent and/or approach the problem.

Exercises where the operation is not possible: 11a, 11b, 11d, 12a, 12b, 12d, 31c, 32c, 33, 34, 44, 45, 59, 60, 61, 64, 65

8.2 Exercises

In Exercises 1–4, find *x* and *y*.

1. $\begin{bmatrix} x & -2 \\ 7 & y \end{bmatrix} = \begin{bmatrix} -4 & -2 \\ 7 & 22 \end{bmatrix}$

2. $\begin{bmatrix} -5 & x \\ y & 8 \end{bmatrix} = \begin{bmatrix} -5 & 13 \\ 12 & 8 \end{bmatrix}$

3. $\begin{bmatrix} 16 & 4 & 5 & 4 \\ -3 & 13 & 15 & 6 \\ 0 & 2 & 4 & 0 \end{bmatrix} = \begin{bmatrix} 16 & 4 & 2x+1 & 4 \\ -3 & 13 & 15 & 3x \\ 0 & 2 & 3y-5 & 0 \end{bmatrix}$

4. $\begin{bmatrix} x+2 & 8 & -3 \\ 1 & 2y & 2x \\ 7 & -2 & y+2 \end{bmatrix} = \begin{bmatrix} 2x+6 & 8 & -3 \\ 1 & 18 & -8 \\ 7 & -2 & 11 \end{bmatrix}$

In Exercises 5–12, if possible, find (a) $A + B$, (b) $A - B$, (c) $3A$, and (d) $3A - 2B$.

5. $A = \begin{bmatrix} 1 & -1 \\ 2 & -1 \end{bmatrix}, \quad B = \begin{bmatrix} 2 & -1 \\ -1 & 8 \end{bmatrix}$

6. $A = \begin{bmatrix} 1 & 2 \\ 2 & 1 \end{bmatrix}, \quad B = \begin{bmatrix} -3 & -2 \\ 4 & 2 \end{bmatrix}$

7. $A = \begin{bmatrix} 6 & -1 \\ 2 & 4 \\ -3 & 5 \end{bmatrix}, \quad B = \begin{bmatrix} 1 & 4 \\ -1 & 5 \\ 1 & 10 \end{bmatrix}$

8. $A = \begin{bmatrix} 2 & 1 & 1 \\ -1 & -1 & 4 \end{bmatrix}, \quad B = \begin{bmatrix} 2 & -3 & 4 \\ -3 & 1 & -2 \end{bmatrix}$

9. $A = \begin{bmatrix} 2 & 2 & -1 & 0 & 1 \\ 1 & 1 & -2 & 0 & -1 \end{bmatrix},$

$B = \begin{bmatrix} 1 & 1 & -1 & 1 & 0 \\ -3 & 4 & 9 & -6 & -7 \end{bmatrix}$

10. $A = \begin{bmatrix} -1 & 4 & 0 \\ 3 & -2 & 2 \\ 5 & 4 & -1 \\ 0 & 8 & -6 \\ -4 & -1 & 0 \end{bmatrix}, \quad B = \begin{bmatrix} -3 & 5 & 1 \\ 2 & -4 & -7 \\ 10 & -9 & -1 \\ 3 & 2 & -4 \\ 0 & 1 & -2 \end{bmatrix}$

11. $A = \begin{bmatrix} 6 & 0 & 3 \\ -1 & -4 & 0 \end{bmatrix}, \quad B = \begin{bmatrix} 8 & -1 \\ 4 & -3 \end{bmatrix}$

12. $A = \begin{bmatrix} 3 \\ 2 \\ -1 \end{bmatrix}, \quad B = \begin{bmatrix} -4 & 6 & 2 \end{bmatrix}$

In Exercises 13–18, evaluate the expression.

13. $\begin{bmatrix} -5 & 0 \\ 3 & -6 \end{bmatrix} + \begin{bmatrix} 7 & 1 \\ -2 & -1 \end{bmatrix} + \begin{bmatrix} -10 & -8 \\ 14 & 6 \end{bmatrix}$

14. $\begin{bmatrix} 6 & 8 \\ -1 & 0 \end{bmatrix} + \begin{bmatrix} 0 & 5 \\ -3 & -1 \end{bmatrix} + \begin{bmatrix} -11 & -7 \\ 2 & -1 \end{bmatrix}$

15. $4\left(\begin{bmatrix} -4 & 0 & 1 \\ 0 & 2 & 3 \end{bmatrix} - \begin{bmatrix} 2 & 1 & -2 \\ 3 & -6 & 0 \end{bmatrix}\right)$

16. $\frac{1}{2}(\begin{bmatrix} 5 & -2 & 4 & 0 \end{bmatrix} + \begin{bmatrix} 14 & 6 & -18 & 9 \end{bmatrix})$

17. $-3\left(\begin{bmatrix} 0 & -3 \\ 7 & 2 \end{bmatrix} + \begin{bmatrix} -6 & 3 \\ 8 & 1 \end{bmatrix}\right) - 2\begin{bmatrix} 4 & -4 \\ 7 & -9 \end{bmatrix}$

18. $-1\begin{bmatrix} 4 & 11 \\ -2 & -1 \\ 9 & 3 \end{bmatrix} + \frac{1}{6}\left(\begin{bmatrix} -5 & -1 \\ 3 & 4 \\ 0 & 13 \end{bmatrix} + \begin{bmatrix} 7 & 5 \\ -9 & -1 \\ 6 & -1 \end{bmatrix}\right)$

In Exercises 19–22, use the matrix capabilities of a graphing utility to evaluate each expression. Round your results to three decimal places, if necessary.

19. $\frac{3}{7}\begin{bmatrix} 2 & 5 \\ -1 & -4 \end{bmatrix} + 6\begin{bmatrix} -3 & 0 \\ 2 & 2 \end{bmatrix}$

20. $55\left(\begin{bmatrix} 14 & -11 \\ -22 & 19 \end{bmatrix} + \begin{bmatrix} -22 & 20 \\ 13 & 6 \end{bmatrix}\right)$

21. $-\begin{bmatrix} 3.211 & 6.829 \\ -1.004 & 4.914 \\ 0.055 & -3.889 \end{bmatrix} - \begin{bmatrix} -1.630 & -3.090 \\ 5.256 & 8.335 \\ -9.768 & 4.251 \end{bmatrix}$

22. $-12\left(\begin{bmatrix} 6 & 20 \\ 1 & -9 \\ -2 & 5 \end{bmatrix} + \begin{bmatrix} 14 & -15 \\ -8 & -6 \\ 7 & 0 \end{bmatrix} + \begin{bmatrix} -31 & -19 \\ 16 & 10 \\ 24 & -10 \end{bmatrix}\right)$

In Exercises 23–26, solve for *X* when

$A = \begin{bmatrix} -2 & -1 \\ 1 & 0 \\ 3 & -4 \end{bmatrix}$ and $B = \begin{bmatrix} 0 & 3 \\ 2 & 0 \\ -4 & -1 \end{bmatrix}.$

23. $X = 3A - 2B$

24. $2X = 2A - B$

25. $2X + 3A = B$

26. $2A + 4B = -2X$

In Exercises 27–32, find (a) AB, (b) BA, and, if possible, (c) A^2. (*Note:* $A^2 = AA$.)

27. $A = \begin{bmatrix} 1 & 2 \\ 4 & 2 \end{bmatrix}, \quad B = \begin{bmatrix} 2 & -1 \\ -1 & 8 \end{bmatrix}$

28. $A = \begin{bmatrix} 2 & -1 \\ 1 & 4 \end{bmatrix}, \quad B = \begin{bmatrix} 0 & 0 \\ 3 & -3 \end{bmatrix}$

29. $A = \begin{bmatrix} 3 & -1 \\ 1 & 3 \end{bmatrix}, \quad B = \begin{bmatrix} 1 & -3 \\ 3 & 1 \end{bmatrix}$

30. $A = \begin{bmatrix} 1 & -1 \\ 1 & 1 \end{bmatrix}, \quad B = \begin{bmatrix} 1 & 3 \\ -3 & 1 \end{bmatrix}$

31. $A = \begin{bmatrix} 7 \\ 8 \\ -1 \end{bmatrix}$, $B = \begin{bmatrix} 1 & 1 & 2 \end{bmatrix}$

32. $A = \begin{bmatrix} 3 & 2 & 1 \end{bmatrix}$, $B = \begin{bmatrix} 2 \\ 3 \\ 0 \end{bmatrix}$

In Exercises 33–40, find *AB*, if possible.

33. $A = \begin{bmatrix} 2 & 1 \\ -3 & 4 \\ 1 & 6 \end{bmatrix}$, $B = \begin{bmatrix} 0 & -1 & 0 \\ 4 & 0 & 2 \\ 8 & -1 & 7 \end{bmatrix}$

34. $A = \begin{bmatrix} 1 & 0 & 3 & -2 \\ 6 & 13 & 8 & -17 \end{bmatrix}$, $B = \begin{bmatrix} 1 & 6 \\ 4 & 2 \end{bmatrix}$

35. $A = \begin{bmatrix} 0 & -1 & 0 \\ 4 & 0 & 2 \\ 8 & -1 & 7 \end{bmatrix}$, $B = \begin{bmatrix} 2 & 1 \\ -3 & 4 \\ 1 & 6 \end{bmatrix}$

36. $A = \begin{bmatrix} -1 & 3 \\ 4 & -5 \\ 0 & 2 \end{bmatrix}$, $B = \begin{bmatrix} 1 & 2 \\ 0 & 7 \end{bmatrix}$

37. $A = \begin{bmatrix} 1 & 0 & 0 \\ 0 & 4 & 0 \\ 0 & 0 & -2 \end{bmatrix}$, $B = \begin{bmatrix} 3 & 0 & 0 \\ 0 & -1 & 0 \\ 0 & 0 & 5 \end{bmatrix}$

38. $A = \begin{bmatrix} 5 & 0 & 0 \\ 0 & -8 & 0 \\ 0 & 0 & 7 \end{bmatrix}$, $B = \begin{bmatrix} \frac{1}{5} & 0 & 0 \\ 0 & -\frac{1}{8} & 0 \\ 0 & 0 & \frac{1}{2} \end{bmatrix}$

39. $A = \begin{bmatrix} 0 & 0 & 5 \\ 0 & 0 & -3 \\ 0 & 0 & 4 \end{bmatrix}$, $B = \begin{bmatrix} 6 & -11 & 4 \\ 8 & 16 & 4 \\ 0 & 0 & 0 \end{bmatrix}$

40. $A = \begin{bmatrix} 10 \\ 12 \end{bmatrix}$, $B = \begin{bmatrix} 6 & -2 & 1 & 6 \end{bmatrix}$

In Exercises 41–46, use the matrix capabilities of a graphing utility to find *AB*.

41. $A = \begin{bmatrix} 5 & 6 & -3 \\ -2 & 5 & 1 \\ 10 & -5 & 5 \end{bmatrix}$, $B = \begin{bmatrix} 1 & -1 & 2 \\ 8 & 1 & 4 \\ 4 & -2 & 9 \end{bmatrix}$

42. $A = \begin{bmatrix} 11 & -12 & 4 \\ 14 & 10 & 12 \\ 6 & -2 & 9 \end{bmatrix}$, $B = \begin{bmatrix} 12 & 10 \\ -5 & 12 \\ 15 & 16 \end{bmatrix}$

43. $A = \begin{bmatrix} -3 & 8 & -6 & 8 \\ -12 & 15 & 9 & 6 \\ 5 & -1 & 1 & 5 \end{bmatrix}$, $B = \begin{bmatrix} 3 & 1 & 6 \\ 24 & 15 & 14 \\ 16 & 10 & 21 \\ 8 & -4 & 10 \end{bmatrix}$

44. $A = \begin{bmatrix} -2 & 4 & 8 \\ 21 & 5 & 6 \\ 13 & 2 & 6 \end{bmatrix}$, $B = \begin{bmatrix} 2 & 0 \\ -7 & 15 \\ 32 & 14 \\ 0.5 & 1.6 \end{bmatrix}$

45. $A = \begin{bmatrix} 9 & 10 & -38 & 18 \\ 100 & -50 & 250 & 75 \end{bmatrix}$,

$B = \begin{bmatrix} 52 & -85 & 27 & 45 \\ 40 & -35 & 60 & 82 \end{bmatrix}$

46. $A = \begin{bmatrix} 15 & -18 \\ -4 & 12 \\ -8 & 22 \end{bmatrix}$, $B = \begin{bmatrix} -7 & 22 & 1 \\ 8 & 16 & 24 \end{bmatrix}$

In Exercises 47–50, use the matrix capabilities of a graphing utility to evaluate each expression.

47. $\begin{bmatrix} 3 & 1 \\ 0 & -2 \end{bmatrix}\begin{bmatrix} 1 & 0 \\ -2 & 2 \end{bmatrix}\begin{bmatrix} 1 & 0 \\ 2 & 4 \end{bmatrix}$

48. $-3\left(\begin{bmatrix} 6 & 5 & -1 \\ 1 & -2 & 0 \end{bmatrix}\begin{bmatrix} 0 & 3 \\ -1 & -3 \\ 4 & 1 \end{bmatrix}\right)$

49. $\begin{bmatrix} 0 & 2 & -2 \\ 4 & 1 & 2 \end{bmatrix}\left(\begin{bmatrix} 4 & 0 \\ 0 & -1 \\ -1 & 2 \end{bmatrix} + \begin{bmatrix} -2 & 3 \\ -3 & 5 \\ 0 & -3 \end{bmatrix}\right)$

50. $\begin{bmatrix} 3 \\ -1 \\ 5 \\ 7 \end{bmatrix}(\begin{bmatrix} 5 & -6 \end{bmatrix} + \begin{bmatrix} 7 & -1 \end{bmatrix} + \begin{bmatrix} -8 & 9 \end{bmatrix})$

In Exercises 51–58, (a) write each system of linear equations as a matrix equation, *AX* = *B*, and (b) use Gauss-Jordan elimination on the augmented matrix [*A* ⋮ *B*] to solve for the matrix *X*.

51. $\begin{cases} -x_1 + x_2 = 4 \\ -2x_1 + x_2 = 0 \end{cases}$

52. $\begin{cases} 2x_1 + 3x_2 = 5 \\ x_1 + 4x_2 = 10 \end{cases}$

53. $\begin{cases} -2x_1 - 3x_2 = -4 \\ 6x_1 + x_2 = -36 \end{cases}$

54. $\begin{cases} -4x_1 + 9x_2 = -13 \\ x_1 - 3x_2 = 12 \end{cases}$

55. $\begin{cases} x_1 - 2x_2 + 3x_3 = 9 \\ -x_1 + 3x_2 - x_3 = -6 \\ 2x_1 - 5x_2 + 5x_3 = 17 \end{cases}$

56. $\begin{cases} x_1 + x_2 - 3x_3 = 9 \\ -x_1 + 2x_2 = 6 \\ x_1 - x_2 + x_3 = -5 \end{cases}$

57. $\begin{cases} x_1 - 5x_2 + 2x_3 = -20 \\ -3x_1 + x_2 - x_3 = 8 \\ -2x_2 + 5x_3 = -16 \end{cases}$

58. $\begin{cases} x_1 - x_2 + 4x_3 = 17 \\ x_1 + 3x_2 = -11 \\ -6x_2 + 5x_3 = 40 \end{cases}$

Think About It **In Exercises 59–68, let matrices A, B, C, and D be of orders 2×3, 2×3, 3×2, and 2×2, respectively. Determine whether the matrices are of proper order to perform the operation(s). If so, give the order of the answer.**

59. $A + 2C$

60. $B - 3C$

61. AB

62. BC

63. $BC - D$

64. $CB - D$

65. $(CA)D$

66. $(BC)D$

67. $D(A - 3B)$

68. $(BC - D)A$

69. *Manufacturing* A certain corporation has three factories, each of which manufactures two products. The number of units of product i produced at factory j in one day is represented by a_{ij} in the matrix $A = \begin{bmatrix} 70 & 50 & 25 \\ 35 & 100 & 70 \end{bmatrix}$. Find the production levels if production is increased by 20%. (*Hint*: Because an increase of 20% corresponds to 100% + 20%, multiply the given matrix by 1.2.)

70. *Manufacturing* A certain corporation has four factories, each of which manufactures two products. The number of units of product i produced at factory j in one day is represented by a_{ij} in the matrix $A = \begin{bmatrix} 100 & 90 & 70 & 30 \\ 40 & 20 & 60 & 60 \end{bmatrix}$. Find the production levels if production is increased by 10%.

71. *Agriculture* A fruit grower raises two crops, which are shipped to three outlets. The number of units of product i that are shipped to outlet j is represented by a_{ij} in the matrix

$$A = \begin{bmatrix} 125 & 100 & 75 \\ 100 & 175 & 125 \end{bmatrix}.$$

The profit per unit is represented by the matrix

$$B = [\$3.50 \quad \$6.00].$$

Find the product BA, and state what each entry of the product represents.

72. *Revenue* A manufacturer produces three models of a product, which are shipped to two warehouses. The number of units of model i that are shipped to warehouse j is represented by a_{ij} in the matrix

$$A = \begin{bmatrix} 5{,}000 & 4{,}000 \\ 6{,}000 & 10{,}000 \\ 8{,}000 & 5{,}000 \end{bmatrix}.$$

The price per unit is represented by the matrix

$$B = [\$20.50 \quad \$26.50 \quad \$29.50].$$

Compute BA and interpret the result.

73. *Business* A company sells five models of computers through three retail outlets. The inventories are represented by S.

$$S = \overset{\text{Model}}{\overset{\begin{matrix} A & B & C & D & E \end{matrix}}{\begin{bmatrix} 3 & 2 & 2 & 3 & 0 \\ 0 & 2 & 3 & 4 & 3 \\ 4 & 2 & 1 & 3 & 2 \end{bmatrix}}} \begin{matrix} 1 \\ 2 \\ 3 \end{matrix} \Bigg\} \text{ Outlet}$$

The wholesale and retail prices are represented by T.

$$T = \overset{\text{Price}}{\overset{\begin{matrix} \text{Wholesale} & \text{Retail} \end{matrix}}{\begin{bmatrix} \$840 & \$1100 \\ \$1200 & \$1350 \\ \$1450 & \$1650 \\ \$2650 & \$3000 \\ \$3050 & \$3200 \end{bmatrix}}} \begin{matrix} A \\ B \\ C \\ D \\ E \end{matrix} \Bigg\} \text{ Model}$$

Compute ST and interpret the result.

74. *Voting Preferences* The matrix

$$P = \overset{\text{From}}{\overset{\begin{matrix} R & D & I \end{matrix}}{\begin{bmatrix} 0.6 & 0.1 & 0.1 \\ 0.2 & 0.7 & 0.1 \\ 0.2 & 0.2 & 0.8 \end{bmatrix}}} \begin{matrix} R \\ D \\ I \end{matrix} \Bigg\} \text{ To}$$

is called a stochastic matrix. Each entry $p_{ij}\,(i \neq j)$ represents the proportion of the voting population that changes from party i to party j, and p_{ii} represents the proportion that remains loyal to the party from one election to the next. Compute and interpret P^2.

75. *Labor/Wage Requirements* A company that manufactures boats has the following labor-hour and wage requirements.

Labor per boat

$$S = \overbrace{\begin{bmatrix} 1.0 \text{ hr} & 0.5 \text{ hr} & 0.2 \text{ hr} \\ 1.6 \text{ hr} & 1.0 \text{ hr} & 0.2 \text{ hr} \\ 2.5 \text{ hr} & 2.0 \text{ hr} & 0.4 \text{ hr} \end{bmatrix}}^{\text{Department: Cutting, Assembly, Packaging}} \left.\begin{matrix} \text{Small} \\ \text{Medium} \\ \text{Large} \end{matrix}\right\} \text{Boat size}$$

Wages per hour

$$T = \overbrace{\begin{bmatrix} \$12 & \$10 \\ \$9 & \$8 \\ \$8 & \$7 \end{bmatrix}}^{\text{Plant: A, B}} \left.\begin{matrix} \text{Cutting} \\ \text{Assembly} \\ \text{Packaging} \end{matrix}\right\} \text{Department}$$

Compute ST and interpret the result.

76. *Voting Preference* Use a graphing utility to find P^3, P^4, P^5, P^6, P^7, and P^8 for the matrix given in Exercise 74. Can you detect a pattern as P is raised to higher powers?

Synthesis

True or False? **In Exercises 77–79, determine whether the statement is true or false. Justify your answer.**

77. Two matrices can be added only if they have the same order.

78. $\begin{bmatrix} -6 & -2 \\ 2 & -6 \end{bmatrix}\begin{bmatrix} 4 & 0 \\ 0 & -1 \end{bmatrix} = \begin{bmatrix} 4 & 0 \\ 0 & -1 \end{bmatrix}\begin{bmatrix} -6 & -2 \\ 2 & -6 \end{bmatrix}$

79. $\begin{bmatrix} -2 & 4 \\ -3 & 0 \\ 6 & 1 \end{bmatrix}\begin{bmatrix} 1 & 1 \\ 1 & 1 \end{bmatrix} = \begin{bmatrix} -2 & 4 \\ -3 & 0 \\ 6 & 1 \end{bmatrix}$

80. *Think About It* If a, b, and c are real numbers such that $c \neq 0$ and $ac = bc$, then $a = b$. However, if A, B, and C are nonzero matrices such that $AC = BC$, then A is not necessarily equal to B. Illustrate this using the following matrices.

$$A = \begin{bmatrix} 0 & 1 \\ 0 & 1 \end{bmatrix}, \quad B = \begin{bmatrix} 1 & 0 \\ 1 & 0 \end{bmatrix}, \quad C = \begin{bmatrix} 2 & 3 \\ 2 & 3 \end{bmatrix}$$

81. *Think About It* If a and b are real numbers such that $ab = 0$, then $a = 0$ or $b = 0$. However, if A and B are matrices such that $AB = O$, it is *not* necessarily true that $A = O$ or $B = O$. Illustrate this using the following matrices.

$$A = \begin{bmatrix} 3 & 3 \\ 4 & 4 \end{bmatrix}, \quad B = \begin{bmatrix} 1 & -1 \\ -1 & 1 \end{bmatrix}$$

Exploration **In Exercises 82 and 83, let $i = \sqrt{-1}$.**

82. Consider the matrix

$$A = \begin{bmatrix} i & 0 \\ 0 & i \end{bmatrix}.$$

Find A^2, A^3, and A^4. Identify any similarities with i^2, i^3, and i^4.

83. Consider the matrix

$$A = \begin{bmatrix} 0 & -i \\ i & 0 \end{bmatrix}.$$

Find and identify A^2.

84. *Exploration* Let A and B be unequal diagonal matrices of the same order. (A diagonal matrix is a square matrix in which each entry not on the main diagonal is zero.) Determine the products AB for several pairs of such matrices. Make a conjecture about a quick rule for such products.

Review

In Exercises 85–90, solve the equation.

85. $3x^2 + 20x - 32 = 0$

86. $8x^2 - 10x - 3 = 0$

87. $4x^3 + 10x^2 - 3x = 0$

88. $3x^3 + 22x^2 - 45x = 0$

89. $3x^3 - 12x^2 + 5x - 20 = 0$

90. $2x^3 - 5x^2 - 12x + 30 = 0$

In Exercises 91–94, solve the system of linear equations both graphically and algebraically.

91. $\begin{cases} -x + 4y = -9 \\ 5x - 8y = 39 \end{cases}$

92. $\begin{cases} 8x - 3y = -17 \\ -6x + 7y = 27 \end{cases}$

93. $\begin{cases} -x + 2y = -5 \\ -3x - y = -8 \end{cases}$

94. $\begin{cases} 6x - 13y = 11 \\ 9x + 5y = 41 \end{cases}$

8.3 The Inverse of a Square Matrix

▶ **What you should learn**

- How to verify that two matrices are inverses of each other
- How to use Gauss-Jordan elimination to find the inverses of matrices
- How to use a formula to find the inverses of 2×2 matrices
- How to use inverse matrices to solve systems of linear equations

▶ **Why you should learn it**

You can use inverse matrices to model and solve real-life problems. For instance, Exercises 73 and 74 on page 657 show how to use an inverse matrix to help analyze an electrical circuit.

Bruce Forster/Tony Stone Images

The Inverse of a Matrix

This section further develops the algebra of matrices. To begin, consider the real number equation $ax = b$. To solve this equation for x, multiply each side of the equation by a^{-1} (provided that $a \neq 0$).

$$ax = b$$
$$(a^{-1}a)x = a^{-1}b$$
$$(1)x = a^{-1}b$$
$$x = a^{-1}b$$

The number a^{-1} is called the *multiplicative inverse of a* because $a^{-1}a = 1$. The definition of the multiplicative **inverse of a matrix** is similar.

Definition of the Inverse of a Square Matrix

Let A be an $n \times n$ matrix and let I_n be the $n \times n$ identity matrix. If there exists matrix A^{-1} such that

$$AA^{-1} = I_n = A^{-1}A$$

then A^{-1} is called the **inverse** of A. The symbol A^{-1} is read "A inverse."

Example 1 ▶ The Inverse of a Matrix

Show that B is the inverse of A, where

$$A = \begin{bmatrix} -1 & 2 \\ -1 & 1 \end{bmatrix}$$

and

$$B = \begin{bmatrix} 1 & -2 \\ 1 & -1 \end{bmatrix}.$$

Solution

To show that B is the inverse of A, show that $AB = I = BA$, as follows.

$$AB = \begin{bmatrix} -1 & 2 \\ -1 & 1 \end{bmatrix}\begin{bmatrix} 1 & -2 \\ 1 & -1 \end{bmatrix} = \begin{bmatrix} -1 + 2 & 2 - 2 \\ -1 + 1 & 2 - 1 \end{bmatrix} = \begin{bmatrix} 1 & 0 \\ 0 & 1 \end{bmatrix}$$

$$BA = \begin{bmatrix} 1 & -2 \\ 1 & -1 \end{bmatrix}\begin{bmatrix} -1 & 2 \\ -1 & 1 \end{bmatrix} = \begin{bmatrix} -1 + 2 & 2 - 2 \\ -1 + 1 & 2 - 1 \end{bmatrix} = \begin{bmatrix} 1 & 0 \\ 0 & 1 \end{bmatrix}$$

Recall that it is not always true that $AB = BA$, even if both products are defined. However, if A and B are both square matrices and $AB = I_n$, it can be shown that $BA = I_n$. So, in Example 1, you need only to check that $AB = I_2$.

Finding Inverse Matrices

If a matrix A has an inverse, A is called *invertible* (or *nonsingular*); otherwise, A is called *singular*. A nonsquare matrix cannot have an inverse. To see this, note that if A is of order $m \times n$ and B is of order $n \times m$ (where $m \neq n$), the products AB and BA are of different orders and so cannot be equal to each other. Not all square matrices have inverses (see the matrix at the bottom of page 652). If, however, a matrix does have an inverse, that inverse is unique. Example 2 shows how to use a system of equations to find the inverse of a matrix.

Example 2 ▶ Finding the Inverse of a Matrix

Find the inverse of

$$A = \begin{bmatrix} 1 & 4 \\ -1 & -3 \end{bmatrix}.$$

Solution

To find the inverse of A, try to solve the matrix equation $AX = I$ for X.

$$\overset{A}{\begin{bmatrix} 1 & 4 \\ -1 & -3 \end{bmatrix}} \overset{X}{\begin{bmatrix} x_{11} & x_{12} \\ x_{21} & x_{22} \end{bmatrix}} = \overset{I}{\begin{bmatrix} 1 & 0 \\ 0 & 1 \end{bmatrix}}$$

$$\begin{bmatrix} x_{11} + 4x_{21} & x_{12} + 4x_{22} \\ -x_{11} - 3x_{21} & -x_{12} - 3x_{22} \end{bmatrix} = \begin{bmatrix} 1 & 0 \\ 0 & 1 \end{bmatrix}$$

Equating corresponding entries, you obtain two systems of linear equations.

$$\begin{cases} x_{11} + 4x_{21} = 1 \\ -x_{11} - 3x_{21} = 0 \end{cases}$$ Linear system with two variables, x_{11} and x_{21}.

$$\begin{cases} x_{12} + 4x_{22} = 0 \\ -x_{12} - 3x_{22} = 1 \end{cases}$$ Linear system with two variables, x_{12} and x_{22}.

From the first system you can determine that $x_{11} = -3$ and $x_{21} = 1$, and from the second system you can determine that $x_{12} = -4$ and $x_{22} = 1$. Therefore, the inverse of A is

$$X = A^{-1}$$
$$= \begin{bmatrix} -3 & -4 \\ 1 & 1 \end{bmatrix}.$$

You can use matrix multiplication to check this result.

Check

$$AA^{-1} = \begin{bmatrix} 1 & 4 \\ -1 & -3 \end{bmatrix}\begin{bmatrix} -3 & -4 \\ 1 & 1 \end{bmatrix} = \begin{bmatrix} 1 & 0 \\ 0 & 1 \end{bmatrix} \checkmark$$

$$A^{-1}A = \begin{bmatrix} -3 & -4 \\ 1 & 1 \end{bmatrix}\begin{bmatrix} 1 & 4 \\ -1 & -3 \end{bmatrix} = \begin{bmatrix} 1 & 0 \\ 0 & 1 \end{bmatrix} \checkmark$$

In Example 2, note that the two systems of linear equations have the *same coefficient matrix A*. Rather than solve the two systems represented by

$$\begin{bmatrix} 1 & 4 & \vdots & 1 \\ -1 & -3 & \vdots & 0 \end{bmatrix}$$

and

$$\begin{bmatrix} 1 & 4 & \vdots & 0 \\ -1 & -3 & \vdots & 1 \end{bmatrix}$$

separately, you can solve them *simultaneously* by *adjoining* the identity matrix to the coefficient matrix to obtain

$$\overbrace{\begin{bmatrix} 1 & 4 \\ -1 & -3 \end{bmatrix}}^{A} \vdots \overbrace{\begin{bmatrix} 1 & 0 \\ 0 & 1 \end{bmatrix}}^{I}.$$

This "doubly augmented" matrix can be represented as $[A \ \vdots \ I]$. By applying Gauss-Jordan elimination to this matrix, you can solve *both* systems with a single elimination process.

$$\begin{bmatrix} 1 & 4 & \vdots & 1 & 0 \\ -1 & -3 & \vdots & 0 & 1 \end{bmatrix}$$

$$R_1 + R_2 \rightarrow \begin{bmatrix} 1 & 4 & \vdots & 1 & 0 \\ 0 & 1 & \vdots & 1 & 1 \end{bmatrix}$$

$$-4R_2 + R_1 \rightarrow \begin{bmatrix} 1 & 0 & \vdots & -3 & -4 \\ 0 & 1 & \vdots & 1 & 1 \end{bmatrix}$$

So, from the "doubly augmented" matrix $[A \ \vdots \ I]$, you obtained the matrix $[I \ \vdots \ A^{-1}]$.

$$\overbrace{\begin{bmatrix} 1 & 4 \\ -1 & -3 \end{bmatrix}}^{A} \vdots \overbrace{\begin{bmatrix} 1 & 0 \\ 0 & 1 \end{bmatrix}}^{I} \Rightarrow \overbrace{\begin{bmatrix} 1 & 0 \\ 0 & 1 \end{bmatrix}}^{I} \vdots \overbrace{\begin{bmatrix} -3 & -4 \\ 1 & 1 \end{bmatrix}}^{A^{-1}}$$

This procedure (or algorithm) works for an arbitrary square matrix that has an inverse.

Technology

You can find the inverse of a matrix with a graphing utility. Enter the matrix in the graphing utility, press the inverse key x^{-1}, and press ENTER. The inverse of the matrix will be displayed on the screen.

Finding an Inverse Matrix

Let A be a square matrix of order n.

1. Write the $n \times 2n$ matrix that consists of the given matrix A on the left and the $n \times n$ identity matrix I on the right to obtain $[A \ \vdots \ I]$.
2. If possible, row reduce A to I using elementary row operations on the *entire* matrix $[A \ \vdots \ I]$. The result will be the matrix $[I \ \vdots \ A^{-1}]$. If this is not possible, A is not invertible.
3. Check your work by multiplying to see that $AA^{-1} = I = A^{-1}A$.

A computer animation of this example appears in the *Interactive* CD-ROM and *Internet* versions of this text.

Example 3 ▶ Finding the Inverse of a Matrix

Find the inverse of

$$A = \begin{bmatrix} 1 & -1 & 0 \\ 1 & 0 & -1 \\ 6 & -2 & -3 \end{bmatrix}.$$

Solution

Begin by adjoining the identity matrix to A to form the matrix

$$[A \;\vdots\; I] = \left[\begin{array}{ccc|ccc} 1 & -1 & 0 & 1 & 0 & 0 \\ 1 & 0 & -1 & 0 & 1 & 0 \\ 6 & -2 & -3 & 0 & 0 & 1 \end{array}\right].$$

Use elementary row operations to obtain the form $[I \;\vdots\; A^{-1}]$, as follows.

$$\begin{array}{r} \\ -R_1 + R_2 \rightarrow \\ -6R_1 + R_3 \rightarrow \end{array} \left[\begin{array}{ccc|ccc} 1 & -1 & 0 & 1 & 0 & 0 \\ 0 & 1 & -1 & -1 & 1 & 0 \\ 0 & 4 & -3 & -6 & 0 & 1 \end{array}\right]$$

$$\begin{array}{r} R_2 + R_1 \rightarrow \\ \\ -4R_2 + R_3 \rightarrow \end{array} \left[\begin{array}{ccc|ccc} 1 & 0 & -1 & 0 & 1 & 0 \\ 0 & 1 & -1 & -1 & 1 & 0 \\ 0 & 0 & 1 & -2 & -4 & 1 \end{array}\right]$$

$$\begin{array}{r} R_3 + R_1 \rightarrow \\ R_3 + R_2 \rightarrow \\ \\ \end{array} \left[\begin{array}{ccc|ccc} 1 & 0 & 0 & -2 & -3 & 1 \\ 0 & 1 & 0 & -3 & -3 & 1 \\ 0 & 0 & 1 & -2 & -4 & 1 \end{array}\right]$$

Therefore, the matrix A is invertible and its inverse is

$$A^{-1} = \begin{bmatrix} -2 & -3 & 1 \\ -3 & -3 & 1 \\ -2 & -4 & 1 \end{bmatrix}.$$

Try confirming this result by multiplying A and A^{-1} to obtain I.

Check

$$AA^{-1} = \begin{bmatrix} 1 & -1 & 0 \\ 1 & 0 & -1 \\ 6 & -2 & -3 \end{bmatrix} \begin{bmatrix} -2 & -3 & 1 \\ -3 & -3 & 1 \\ -2 & -4 & 1 \end{bmatrix} = \begin{bmatrix} 1 & 0 & 0 \\ 0 & 1 & 0 \\ 0 & 0 & 1 \end{bmatrix} = I$$

You may want to show what happens when A^{-1} does not exist. For example,

$$A = \begin{bmatrix} 3 & 6 \\ 1 & 2 \end{bmatrix} \text{ or }$$

$$B = \begin{bmatrix} 3 & 2 & 1 \\ 1 & 0 & -1 \\ 0 & 1 & 2 \end{bmatrix}.$$

The process shown in Example 3 applies to any $n \times n$ matrix A. If A has an inverse, this process will find it. On the other hand, if A does not have an inverse (if A is *singular*), the process will tell you so. That is, matrix A will not reduce to the identity matrix. For instance, the following matrix has no inverse.

$$A = \begin{bmatrix} 1 & 2 & 0 \\ 3 & -1 & 2 \\ -2 & 3 & -2 \end{bmatrix}$$

Explain how the elimination process shows that this matrix is singular.

The Inverse of a 2 × 2 Matrix

Using Gauss-Jordan elimination to find the inverse of a matrix works well (even as a computer technique) for matrices of order 3 × 3 or greater. For 2 × 2 matrices, however, many people prefer to use a formula for the inverse rather than Gauss-Jordan elimination. This simple formula, which works *only* for 2 × 2 matrices, is explained as follows. If A is a 2 × 2 matrix given by

$$A = \begin{bmatrix} a & b \\ c & d \end{bmatrix}$$

then A is invertible if and only if $ad - bc \neq 0$. Moreover, if $ad - bc \neq 0$, the inverse is given by

$$A^{-1} = \frac{1}{ad - bc}\begin{bmatrix} d & -b \\ -c & a \end{bmatrix}.$$ Formula for inverse of matrix A

Try verifying this inverse by multiplication. The denominator $ad - bc$ is called the **determinant** of the 2 × 2 matrix A. You will study determinants in the next section.

Exploration

Use a graphing utility with matrix capabilities to find the inverse of the matrix

$$A = \begin{bmatrix} 1 & -3 \\ -2 & 6 \end{bmatrix}.$$

What message appears on the screen? Why does the graphing utility display this message?

Example 4 ▶ Finding the Inverse of a 2 × 2 Matrix

If possible, find the inverse of the matrix.

a. $A = \begin{bmatrix} 3 & -1 \\ -2 & 2 \end{bmatrix}$ **b.** $B = \begin{bmatrix} 3 & -1 \\ -6 & 2 \end{bmatrix}$

Solution

a. For the matrix A, apply the formula for the inverse of a 2 × 2 matrix to obtain

$$ad - bc = (3)(2) - (-1)(-2)$$
$$= 4.$$

Because this quantity is not zero, the inverse is formed by interchanging the entries on the main diagonal, changing the signs of the other two entries, and multiplying by the scalar $\frac{1}{4}$, as follows.

$$A^{-1} = \frac{1}{4}\begin{bmatrix} 2 & 1 \\ 2 & 3 \end{bmatrix}$$ Substitute for a, b, c, d, and the determinant.

$$= \begin{bmatrix} \frac{1}{2} & \frac{1}{4} \\ \frac{1}{2} & \frac{3}{4} \end{bmatrix}$$ Multiply by the scalar $\frac{1}{4}$.

b. For the matrix B, you have

$$ad - bc = (3)(2) - (-1)(-6)$$
$$= 0$$

which means that B is not invertible.

Activities

1. Find the inverse of the matrix.
   $A = \begin{bmatrix} 2 & 1 \\ -1 & -1 \end{bmatrix}$
   Answer: $A^{-1} = \begin{bmatrix} 1 & 1 \\ -1 & -2 \end{bmatrix}$
2. Use the inverse matrix
   $A^{-1} = \begin{bmatrix} -3 & -2 \\ -1 & -1 \end{bmatrix}$
   to solve the system of linear equations.
   $-x + 2y = -1$
   $x - 3y = 6$
   Answer: $x = -9, y = -5$

Systems of Linear Equations

You know that a system of linear equations can have exactly one solution, infinitely many solutions, or no solution. If the coefficient matrix A of a *square* system (a system that has the same number of equations as variables) is invertible, the system has a unique solution, which is defined as follows.

A System of Equations with a Unique Solution

If A is an invertible matrix, the system of linear equations represented by $AX = B$ has a unique solution given by

$$X = A^{-1}B.$$

Technology

To solve a system of equations with a graphing utility, enter the matrices A and B in the matrix editor. Then, using the inverse key, solve for X.

A [x^{-1}] B [ENTER]

The screen will display the solution, matrix X.

Example 5 ▶ Solving a System Using an Inverse

You are going to invest \$10,000 in AAA-rated bonds, AA-rated bonds, and B-rated bonds and want an annual return of \$730. The average yields are 6% on AAA bonds, 7.5% on AA bonds, and 9.5% on B bonds. You will invest twice as much in AAA bonds as in B bonds. Your investment can be represented as

$$\begin{cases} x + y + z = 10{,}000 \\ 0.06x + 0.075y + 0.095z = 730 \\ x - 2z = 0 \end{cases}$$

where x, y, and z represent the amounts invested in AAA, AA, and B bonds, respectively. Use an inverse matrix to solve the system.

Solution

Begin by writing the system in the matrix form $AX = B$.

$$\begin{bmatrix} 1 & 1 & 1 \\ 0.06 & 0.075 & 0.095 \\ 1 & 0 & -2 \end{bmatrix}\begin{bmatrix} x \\ y \\ z \end{bmatrix} = \begin{bmatrix} 10{,}000 \\ 730 \\ 0 \end{bmatrix}$$

Then, use Gauss-Jordan elimination to find A^{-1}.

$$A^{-1} = \begin{bmatrix} 15 & -200 & -2 \\ -21.5 & 300 & 3.5 \\ 7.5 & -100 & -1.5 \end{bmatrix}$$

Finally, multiply B by A^{-1} on the left to obtain the solution.

$$X = A^{-1}B$$

$$= \begin{bmatrix} 15 & -200 & -2 \\ -21.5 & 300 & 3.5 \\ 7.5 & -100 & -1.5 \end{bmatrix}\begin{bmatrix} 10{,}000 \\ 730 \\ 0 \end{bmatrix} = \begin{bmatrix} 4000 \\ 4000 \\ 2000 \end{bmatrix}$$

The solution to the system is $x = 4000$, $y = 4000$, and $z = 2000$. So, you will invest \$4000 in AAA bonds, \$4000 in AA bonds, and \$2000 in B bonds.

Group Activity

Create two linear equations that intersect and carefully graph them on graph paper. Do not write the equations on your graph. Exchange graphs with another student. Use the graph to reconstruct the equations of the two lines. Write the system as the matrix equation $AX = B$. Find A^{-1} and use it to find the point of intersection of the two lines. Does your solution agree with the graph you received? Compare results with the creator of the graph.

Exercises in which the inverse matrix does not exist: 17, 19, 20, 24, 32, 35, 41
Exercises containing systems with no solutions: 55
Exercises containing systems with infinitely many solutions: 63, 65

8.3 Exercises

In Exercises 1–10, show that B is the inverse of A.

1. $A = \begin{bmatrix} 2 & 1 \\ 5 & 3 \end{bmatrix}, \quad B = \begin{bmatrix} 3 & -1 \\ -5 & 2 \end{bmatrix}$

2. $A = \begin{bmatrix} 1 & -1 \\ -1 & 2 \end{bmatrix}, \quad B = \begin{bmatrix} 2 & 1 \\ 1 & 1 \end{bmatrix}$

3. $A = \begin{bmatrix} 1 & 2 \\ 3 & 4 \end{bmatrix}, \quad B = \begin{bmatrix} -2 & 1 \\ \frac{3}{2} & -\frac{1}{2} \end{bmatrix}$

4. $A = \begin{bmatrix} 1 & -1 \\ 2 & 3 \end{bmatrix}, \quad B = \begin{bmatrix} \frac{3}{5} & \frac{1}{5} \\ -\frac{2}{5} & \frac{1}{5} \end{bmatrix}$

5. $A = \begin{bmatrix} 2 & -17 & 11 \\ -1 & 11 & -7 \\ 0 & 3 & -2 \end{bmatrix},$

$B = \begin{bmatrix} 1 & 1 & 2 \\ 2 & 4 & -3 \\ 3 & 6 & -5 \end{bmatrix}$

6. $A = \begin{bmatrix} -4 & 1 & 5 \\ -1 & 2 & 4 \\ 0 & -1 & -1 \end{bmatrix},$

$B = \begin{bmatrix} -\frac{1}{2} & 1 & \frac{3}{2} \\ \frac{1}{4} & -1 & -\frac{11}{4} \\ -\frac{1}{4} & 1 & \frac{7}{4} \end{bmatrix}$

7. $A = \begin{bmatrix} 2 & 0 & 1 & 1 \\ 3 & 0 & 0 & 1 \\ -1 & 1 & -2 & 1 \\ 4 & -1 & 1 & 0 \end{bmatrix},$

$B = \begin{bmatrix} -1 & 2 & -1 & -1 \\ -4 & 9 & -5 & -6 \\ 0 & 1 & -1 & -1 \\ 3 & -5 & 3 & 3 \end{bmatrix}$

8. $A = \begin{bmatrix} -2 & 0 & 1 & 0 \\ 1 & -1 & -3 & 0 \\ -2 & -1 & 0 & -2 \\ 0 & 1 & 3 & -1 \end{bmatrix},$

$B = \begin{bmatrix} -3 & -3 & 1 & -2 \\ 12 & 14 & -5 & 10 \\ -5 & -6 & 2 & -4 \\ -3 & -4 & 1 & -3 \end{bmatrix}$

9. $A = \begin{bmatrix} -2 & 2 & 3 \\ 1 & -1 & 0 \\ 0 & 1 & 4 \end{bmatrix}, \quad B = \frac{1}{3}\begin{bmatrix} -4 & -5 & 3 \\ -4 & -8 & 3 \\ 1 & 2 & 0 \end{bmatrix}$

10. $A = \begin{bmatrix} -1 & 1 & 0 & -1 \\ 1 & -1 & 1 & 0 \\ -1 & 1 & 2 & 0 \\ 0 & -1 & 1 & 1 \end{bmatrix},$

$B = \frac{1}{3}\begin{bmatrix} -3 & 1 & 1 & -3 \\ -3 & -1 & 2 & -3 \\ 0 & 1 & 1 & 0 \\ -3 & -2 & 1 & 0 \end{bmatrix}$

In Exercises 11–26, find the inverse of the matrix (if it exists).

11. $\begin{bmatrix} 2 & 0 \\ 0 & 3 \end{bmatrix}$

12. $\begin{bmatrix} 1 & 2 \\ 3 & 7 \end{bmatrix}$

13. $\begin{bmatrix} 1 & -2 \\ 2 & -3 \end{bmatrix}$

14. $\begin{bmatrix} -7 & 33 \\ 4 & -19 \end{bmatrix}$

15. $\begin{bmatrix} -1 & 1 \\ -2 & 1 \end{bmatrix}$

16. $\begin{bmatrix} 11 & 1 \\ -1 & 0 \end{bmatrix}$

17. $\begin{bmatrix} 2 & 4 \\ 4 & 8 \end{bmatrix}$

18. $\begin{bmatrix} 2 & 3 \\ 1 & 4 \end{bmatrix}$

19. $\begin{bmatrix} 2 & 7 & 1 \\ -3 & -9 & 2 \end{bmatrix}$

20. $\begin{bmatrix} -2 & 5 \\ 6 & -15 \\ 0 & 1 \end{bmatrix}$

21. $\begin{bmatrix} 1 & 1 & 1 \\ 3 & 5 & 4 \\ 3 & 6 & 5 \end{bmatrix}$

22. $\begin{bmatrix} 1 & 2 & 2 \\ 3 & 7 & 9 \\ -1 & -4 & -7 \end{bmatrix}$

23. $\begin{bmatrix} 1 & 0 & 0 \\ 3 & 4 & 0 \\ 2 & 5 & 5 \end{bmatrix}$

24. $\begin{bmatrix} 1 & 0 & 0 \\ 3 & 0 & 0 \\ 2 & 5 & 5 \end{bmatrix}$

25. $\begin{bmatrix} -8 & 0 & 0 & 0 \\ 0 & 1 & 0 & 0 \\ 0 & 0 & 4 & 0 \\ 0 & 0 & 0 & -5 \end{bmatrix}$

26. $\begin{bmatrix} 1 & 3 & -2 & 0 \\ 0 & 2 & 4 & 6 \\ 0 & 0 & -2 & 1 \\ 0 & 0 & 0 & 5 \end{bmatrix}$

In Exercises 27–38, use the matrix capabilities of a graphing utility to find the inverse of the matrix (if it exists).

27. $\begin{bmatrix} 1 & 2 & -1 \\ 3 & 7 & -10 \\ -5 & -7 & -15 \end{bmatrix}$

28. $\begin{bmatrix} 10 & 5 & -7 \\ -5 & 1 & 4 \\ 3 & 2 & -2 \end{bmatrix}$

29. $\begin{bmatrix} 1 & 1 & 2 \\ 3 & 1 & 0 \\ -2 & 0 & 3 \end{bmatrix}$

30. $\begin{bmatrix} 3 & 2 & 2 \\ 2 & 2 & 2 \\ -4 & 4 & 3 \end{bmatrix}$

31. $\begin{bmatrix} -\frac{1}{2} & \frac{3}{4} & \frac{1}{4} \\ 1 & 0 & -\frac{3}{2} \\ 0 & -1 & \frac{1}{2} \end{bmatrix}$

32. $\begin{bmatrix} -\frac{5}{6} & \frac{1}{3} & \frac{11}{6} \\ 0 & \frac{2}{3} & 2 \\ 1 & -\frac{1}{2} & -\frac{5}{2} \end{bmatrix}$

33. $\begin{bmatrix} 0.1 & 0.2 & 0.3 \\ -0.3 & 0.2 & 0.2 \\ 0.5 & 0.4 & 0.4 \end{bmatrix}$ **34.** $\begin{bmatrix} 0.6 & 0 & -0.3 \\ 0.7 & -1 & 0.2 \\ 1 & 0 & -0.9 \end{bmatrix}$

35. $\begin{bmatrix} 1 & 0 & 3 & 0 \\ 0 & 2 & 0 & 4 \\ 1 & 0 & 3 & 0 \\ 0 & 2 & 0 & 4 \end{bmatrix}$ **36.** $\begin{bmatrix} 4 & 8 & -7 & 14 \\ 2 & 5 & -4 & 6 \\ 0 & 2 & 1 & -7 \\ 3 & 6 & -5 & 10 \end{bmatrix}$

37. $\begin{bmatrix} -1 & 0 & 1 & 0 \\ 0 & 2 & 0 & -1 \\ 2 & 0 & -1 & 0 \\ 0 & -1 & 0 & 1 \end{bmatrix}$

38. $\begin{bmatrix} 1 & -2 & -1 & -2 \\ 3 & -5 & -2 & -3 \\ 2 & -5 & -2 & -5 \\ -1 & 4 & 4 & 11 \end{bmatrix}$

In Exercises 39–44, use the formula on page 653 to find the inverse of the matrix.

39. $\begin{bmatrix} 5 & -2 \\ 2 & 3 \end{bmatrix}$ **40.** $\begin{bmatrix} 7 & 12 \\ -8 & -5 \end{bmatrix}$

41. $\begin{bmatrix} -4 & -6 \\ 2 & 3 \end{bmatrix}$ **42.** $\begin{bmatrix} -12 & 3 \\ 5 & -2 \end{bmatrix}$

43. $\begin{bmatrix} \frac{7}{2} & -\frac{3}{4} \\ \frac{1}{5} & \frac{4}{5} \end{bmatrix}$ **44.** $\begin{bmatrix} -\frac{1}{4} & \frac{9}{4} \\ \frac{5}{3} & \frac{8}{9} \end{bmatrix}$

In Exercises 45–48, use an inverse matrix to solve the system of linear equations. (Use the inverse matrix found in Exercise 13.)

45. $\begin{cases} x - 2y = 5 \\ 2x - 3y = 10 \end{cases}$ **46.** $\begin{cases} x - 2y = 0 \\ 2x - 3y = 3 \end{cases}$

47. $\begin{cases} x - 2y = 4 \\ 2x - 3y = 2 \end{cases}$ **48.** $\begin{cases} x - 2y = 1 \\ 2x - 3y = -2 \end{cases}$

In Exercises 49 and 50, use an inverse matrix to solve the system of linear equations. (Use the inverse matrix found in Exercise 21.)

49. $\begin{cases} x + y + z = 0 \\ 3x + 5y + 4z = 5 \\ 3x + 6y + 5z = 2 \end{cases}$ **50.** $\begin{cases} x + y + z = -1 \\ 3x + 5y + 4z = 2 \\ 3x + 6y + 5z = 0 \end{cases}$

In Exercises 51 and 52, use an inverse matrix to solve the system of linear equations. (Use the inverse matrix found in Exercise 38.)

51. $\begin{cases} x_1 - 2x_2 - x_3 - 2x_4 = 0 \\ 3x_1 - 5x_2 - 2x_3 - 3x_4 = 1 \\ 2x_1 - 5x_2 - 2x_3 - 5x_4 = -1 \\ -x_1 + 4x_2 + 4x_3 + 11x_4 = 2 \end{cases}$

52. $\begin{cases} x_1 - 2x_2 - x_3 - 2x_4 = 1 \\ 3x_1 - 5x_2 - 2x_3 - 3x_4 = -2 \\ 2x_1 - 5x_2 - 2x_3 - 5x_4 = 0 \\ -x_1 + 4x_2 + 4x_3 + 11x_4 = -3 \end{cases}$

In Exercises 53–62, use an inverse matrix to solve (if possible) the system of linear equations.

53. $\begin{cases} 3x + 4y = -2 \\ 5x + 3y = 4 \end{cases}$ **54.** $\begin{cases} 18x + 12y = 13 \\ 30x + 24y = 23 \end{cases}$

55. $\begin{cases} -0.4x + 0.8y = 1.6 \\ 2x - 4y = 5 \end{cases}$ **56.** $\begin{cases} 0.2x - 0.6y = 2.4 \\ -x + 1.4y = -8.8 \end{cases}$

57. $\begin{cases} 3x + 6y = 6 \\ 6x + 14y = 11 \end{cases}$ **58.** $\begin{cases} 3x + 2y = 1 \\ 2x + 10y = 6 \end{cases}$

59. $\begin{cases} -\frac{1}{4}x + \frac{3}{8}y = -2 \\ \frac{3}{2}x + \frac{3}{4}y = -12 \end{cases}$ **60.** $\begin{cases} \frac{5}{6}x - y = -20 \\ \frac{4}{3}x - \frac{7}{2}y = -51 \end{cases}$

61. $\begin{cases} 4x - y + z = -5 \\ 2x + 2y + 3z = 10 \\ 5x - 2y + 6z = 1 \end{cases}$ **62.** $\begin{cases} 4x - 2y + 3z = -2 \\ 2x + 2y + 5z = 16 \\ 8x - 5y - 2z = 4 \end{cases}$

In Exercises 63–68, use the matrix capabilities of a graphing utility to solve (if possible) the system of linear equations.

63. $\begin{cases} 5x - 3y + 2z = 2 \\ 2x + 2y - 3z = 3 \\ x - 7y + 8z = -4 \end{cases}$

64. $\begin{cases} 3x - 2y + z = -29 \\ -4x + y - 3z = 37 \\ x - 5y + z = -24 \end{cases}$

65. $\begin{cases} 2x + 3y + 5z = 4 \\ 3x + 5y + 9z = 7 \\ 5x + 9y + 17z = 13 \end{cases}$

66. $\begin{cases} -8x + 7y - 10z = -151 \\ 12x + 3y - 5z = 86 \\ 15x - 9y + 2z = 187 \end{cases}$

67. $\begin{cases} 7x - 3y + 2w = 41 \\ -2x + y - w = -13 \\ 4x + z - 2w = 12 \\ -x + y - w = -8 \end{cases}$

68. $\begin{cases} 2x + 5y + w = 11 \\ x + 4y + 2z - 2w = -7 \\ 2x - 2y + 5z + w = 3 \\ x - 3w = -1 \end{cases}$

Finance **In Exercises 69–72, consider a person who invests in AAA-rated bonds, A-rated bonds, and B-rated bonds. The average yields are 6.5% on AAA bonds, 7% on A bonds, and 9% on B bonds. The person invests twice as much in B bonds as in A bonds. Let x, y, and z represent the amounts invested in AAA, A, and B bonds, respectively.**

$$\begin{cases} x + y + z = \text{(total investment)} \\ 0.065x + 0.07y + 0.09z = \text{(annual return)} \\ 2y - z = 0 \end{cases}$$

Use the inverse of the coefficient matrix of this system to find the amount invested in each type of bond.

69. Total investment = \$10,000
Annual return = \$705

70. Total investment = \$10,000
Annual return = \$760

71. Total investment = \$12,000
Annual return = \$835

72. Total investment = \$500,000
Annual return = \$38,000

Circuit Analysis **In Exercises 73 and 74, consider the circuit in the figure. The currents I_1, I_2, and I_3, in amperes, are the solution of the system of linear equations**

$$\begin{cases} 2I_1 + 4I_3 = E_1 \\ I_2 + 4I_3 = E_2 \\ I_1 + I_2 - I_3 = 0 \end{cases}$$

where E_1 and E_2 are voltages. Use the inverse of the coefficient matrix of this system to find the unknown currents for the voltages.

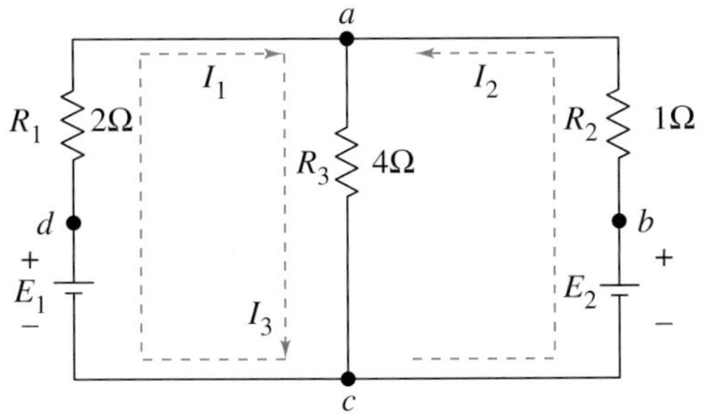

73. $E_1 = 14$ volts, $E_2 = 28$ volts

74. $E_1 = 24$ volts, $E_2 = 23$ volts

Synthesis

True or False? **In Exercises 75–77, determine whether the statement is true or false. Justify your answer.**

75. Multiplication of an invertible matrix and its inverse is commutative.

76. If you multiply two square matrices and obtain the identity matrix, you can assume the matrices are inverses of one another.

77. All nonsquare matrices do not have inverses.

78. If A is a 2×2 matrix

$$A = \begin{bmatrix} a & b \\ c & d \end{bmatrix}$$

then A is invertible if and only if $ad - bc \neq 0$. If $ad - bc \neq 0$, verify that the inverse is

$$A^{-1} = \frac{1}{ad - bc}\begin{bmatrix} d & -b \\ -c & a \end{bmatrix}.$$

79. *Writing* Write a brief paragraph explaining the advantage of using inverse matrices to solve the systems of linear equations in Exercises 45–52.

80. *Exploration* Consider matrices of the form

$$A = \begin{bmatrix} a_{11} & 0 & 0 & 0 & \cdots & 0 \\ 0 & a_{22} & 0 & 0 & \cdots & 0 \\ 0 & 0 & a_{33} & 0 & \cdots & 0 \\ \vdots & \vdots & \vdots & \vdots & & \vdots \\ 0 & 0 & 0 & 0 & \cdots & a_{nn} \end{bmatrix}.$$

(a) Write a 2×2 matrix and a 3×3 matrix in the form of A. Find the inverse of each.

(b) Use the result of part (a) to make a conjecture about the inverses of matrices in the form of A.

Review

In Exercises 81–84, solve the equation.

81. $3^{x/2} = 315$

82. $2000e^{-x/5} = 400$

83. $\log_2 x - 2 = 4.5$

84. $\ln x + \ln(x - 1) = 0$

In Exercises 85–88, perform the matrix operation.

85. $-3\begin{bmatrix} -4 & 6 \\ 2 & -8 \\ 1 & 12 \end{bmatrix}$

86. $\frac{1}{8}\begin{bmatrix} -6 & 2 & 0 \\ -4 & -24 & 18 \end{bmatrix}$

87. $\begin{bmatrix} 2 & 7 \\ -3 & -1 \end{bmatrix} - 4\begin{bmatrix} -1 & 2 \\ 6 & -5 \end{bmatrix}$

88. $8\begin{bmatrix} 2 & -3 \\ 1 & 0 \end{bmatrix} + \begin{bmatrix} 12 & 17 \\ -7 & 9 \end{bmatrix}$

8.4 The Determinant of a Square Matrix

▶ **What you should learn**

- How to find the determinants of 2 × 2 matrices
- How to find minors and cofactors of square matrices
- How to find the determinants of square matrices

▶ **Why you should learn it**

Determinants are often used in other branches of mathematics. For instance, Exercises 79–84 on page 665 show some types of determinants that are useful when changes in variables are made in calculus.

The Determinant of a 2 × 2 Matrix

Every *square* matrix can be associated with a real number called its **determinant.** Determinants have many uses, and several will be discussed in this and the next section. Historically, the use of determinants arose from special number patterns that occur when systems of linear equations are solved. For instance, the system

$$\begin{cases} a_1x + b_1y = c_1 \\ a_2x + b_2y = c_2 \end{cases}$$

has a solution

$$x = \frac{c_1b_2 - c_2b_1}{a_1b_2 - a_2b_1} \quad \text{and} \quad y = \frac{a_1c_2 - a_2c_1}{a_1b_2 - a_2b_1}$$

provided that $a_1b_2 - a_2b_1 \neq 0$. Note that the denominators of the two fractions are the same. This denominator is called the *determinant* of the coefficient matrix of the system.

Coefficient Matrix | *Determinant*

$$A = \begin{bmatrix} a_1 & b_1 \\ a_2 & b_2 \end{bmatrix} \qquad \det(A) = a_1b_2 - a_2b_1$$

The determinant of the matrix A can also be denoted by vertical bars on both sides of the matrix, as indicated in the following definition.

Definition of the Determinant of a 2 × 2 Matrix

The **determinant** of the matrix

$$A = \begin{bmatrix} a_1 & b_1 \\ a_2 & b_2 \end{bmatrix}$$

is given by

$$\det(A) = |A| = \begin{vmatrix} a_1 & b_1 \\ a_2 & b_2 \end{vmatrix} = a_1b_2 - a_2b_1.$$

In this book, $\det(A)$ and $|A|$ are used interchangeably to represent the determinant of A. Although vertical bars are also used to denote the absolute value of a real number, the context will show which use is intended.

A convenient method for remembering the formula for the determinant of a 2 × 2 matrix is shown in the diagram.

$$\det(A) = \begin{vmatrix} a_1 & b_1 \\ a_2 & b_2 \end{vmatrix} = a_1b_2 - a_2b_1$$

Note that the determinant is the difference of the products of the two diagonals of the matrix.

Point out to your students that a matrix is an array of numbers, but a determinant is a single numerical value.

Example 1 ▶ The Determinant of a 2 × 2 Matrix

Find the determinant of each matrix.

a. $A = \begin{bmatrix} 2 & -3 \\ 1 & 2 \end{bmatrix}$

b. $B = \begin{bmatrix} 2 & 1 \\ 4 & 2 \end{bmatrix}$

c. $C = \begin{bmatrix} 0 & \frac{3}{2} \\ 2 & 4 \end{bmatrix}$

Solution

a. $\det(A) = \begin{vmatrix} 2 & -3 \\ 1 & 2 \end{vmatrix}$

$= 2(2) - 1(-3)$

$= 4 + 3 = 7$

b. $\det(B) = \begin{vmatrix} 2 & 1 \\ 4 & 2 \end{vmatrix}$

$= 2(2) - 4(1)$

$= 4 - 4 = 0$

c. $\det(C) = \begin{vmatrix} 0 & \frac{3}{2} \\ 2 & 4 \end{vmatrix}$

$= 0(4) - 2\left(\frac{3}{2}\right)$

$= 0 = -3$

Exploration

Use a graphing utility to find the determinant of the following matrix.

$$A = \begin{bmatrix} 1 & 2 \\ -1 & 0 \\ 3 & -2 \end{bmatrix}$$

What message appears on the screen? Why does the graphing utility display this message?

Notice in Example 1 that the determinant of a matrix can be positive, zero, or negative.

The determinant of a matrix of order 1×1 is defined simply as the entry of the matrix. For instance, if $A = [-2]$, then $\det(A) = -2$.

Technology

Most graphing utilities can evaluate the determinant of a matrix. For instance, you can evaluate the determinant of

$$A = \begin{bmatrix} 2 & -3 \\ 1 & 2 \end{bmatrix}$$

by entering the matrix as $[A]$ and then choosing the determinant feature. The result should be 7, as in Example 1(a). Try evaluating determinants of other matrices.

Minors and Cofactors

To define the determinant of a square matrix of order 3×3 or higher, it is convenient to introduce the concepts of **minors** and **cofactors.**

Sign Pattern for Cofactors

$$\begin{bmatrix} + & - & + \\ - & + & - \\ + & - & + \end{bmatrix}$$

3×3 matrix

$$\begin{bmatrix} + & - & + & - \\ - & + & - & + \\ + & - & + & - \\ - & + & - & + \end{bmatrix}$$

4×4 matrix

$$\begin{bmatrix} + & - & + & - & + & \cdots \\ - & + & - & + & - & \cdots \\ + & - & + & - & + & \cdots \\ - & + & - & + & - & \cdots \\ + & - & + & - & + & \cdots \\ \vdots & \vdots & \vdots & \vdots & \vdots & \end{bmatrix}$$

$n \times n$ matrix

Minors and Cofactors of a Square Matrix

If A is a square matrix, the **minor** M_{ij} of the entry a_{ij} is the determinant of the matrix obtained by deleting the ith row and jth column of A. The **cofactor** C_{ij} of the entry a_{ij} is

$$C_{ij} = (-1)^{i+j}M_{ij}.$$

In the sign pattern for cofactors at the left, notice that *odd* positions (where $i + j$ is odd) have negative signs and *even* positions (where $i + j$ is even) have positive signs.

Example 2 ▶ Finding the Minors and Cofactors of a Matrix

Find all the minors and cofactors of

$$A = \begin{bmatrix} 0 & 2 & 1 \\ 3 & -1 & 2 \\ 4 & 0 & 1 \end{bmatrix}.$$

Solution

To find the minor M_{11}, delete the first row and first column of A and evaluate the determinant of the resulting matrix.

$$\begin{bmatrix} 0 & 2 & 1 \\ 3 & -1 & 2 \\ 4 & 0 & 1 \end{bmatrix}, \quad M_{11} = \begin{vmatrix} -1 & 2 \\ 0 & 1 \end{vmatrix} = -1(1) - 0(2) = -1$$

Similarly, to find M_{12}, delete the first row and second column.

$$\begin{bmatrix} 0 & 2 & 1 \\ 3 & -1 & 2 \\ 4 & 0 & 1 \end{bmatrix}, \quad M_{12} = \begin{vmatrix} 3 & 2 \\ 4 & 1 \end{vmatrix} = 3(1) - 4(2) = -5$$

Continuing this pattern, you obtain the minors.

$$\begin{array}{lll} M_{11} = -1 & M_{12} = -5 & M_{13} = 4 \\ M_{21} = 2 & M_{22} = -4 & M_{23} = -8 \\ M_{31} = 5 & M_{32} = -3 & M_{33} = -6 \end{array}$$

Now, to find the cofactors, combine the checkerboard pattern of signs for a 3×3 matrix (at left above) with these minors.

$$\begin{array}{lll} C_{11} = -1 & C_{12} = 5 & C_{13} = 4 \\ C_{21} = -2 & C_{22} = -4 & C_{23} = 8 \\ C_{31} = 5 & C_{32} = 3 & C_{33} = -6 \end{array}$$

The Determinant of a Square Matrix

The definition given below is called *inductive* because it uses determinants of matrices of order $n - 1$ to define determinants of matrices of order n.

Determinant of a Square Matrix

If A is a square matrix (of order 2×2 or greater), the determinant of A is the sum of the entries in any row (or column) of A multiplied by their respective cofactors. For instance, expanding along the first row yields

$$|A| = a_{11}C_{11} + a_{12}C_{12} + \cdot \cdot \cdot + a_{1n}C_{1n}.$$

Applying this definition to find a determinant is called *expanding by cofactors.*

Try checking that for a 2×2 matrix

$$A = \begin{bmatrix} a_1 & b_1 \\ a_2 & b_2 \end{bmatrix}$$

this definition of the determinant yields $|A| = a_1b_2 - a_2b_1$, as previously defined.

You may need to remind your students to use the appropriate signs when expanding a determinant by cofactors. (Refer them to the Sign Pattern for Cofactors chart on page 660.)

Example 3 ▶ The Determinant of a Matrix of Order 3 × 3

Find the determinant of

$$A = \begin{bmatrix} 0 & 2 & 1 \\ 3 & -1 & 2 \\ 4 & 0 & 1 \end{bmatrix}.$$

Solution

Note that this is the same matrix that was in Example 2. There you found the cofactors of the entries in the first row to be

$$C_{11} = -1, \quad C_{12} = 5, \quad \text{and} \quad C_{13} = 4.$$

Therefore, by the definition of a determinant, you have

$$\begin{aligned} |A| &= a_{11}C_{11} + a_{12}C_{12} + a_{13}C_{13} && \text{First-row expansion} \\ &= 0(-1) + 2(5) + 1(4) \\ &= 14. \end{aligned}$$

In Example 3 the determinant was found by expanding by the cofactors in the first row. You could have used any row or column. For instance, you could have expanded along the second row to obtain

$$\begin{aligned} |A| &= a_{21}C_{21} + a_{22}C_{22} + a_{23}C_{23} && \text{Second-row expansion} \\ &= 3(-2) + (-1)(-4) + 2(8) \\ &= 14. \end{aligned}$$

You might want to consider showing students the following alternative method that can be used to evaluate the determinant of a 3×3 matrix A. (Note that this method works only for 3×3 matrices.) Copy the first and second columns of A to form fourth and fifth columns. The determinant of A is obtained by adding the products of the three "downward diagonals" and subtracting the products of the three "upward diagonals" as shown below.

Subtract these products.

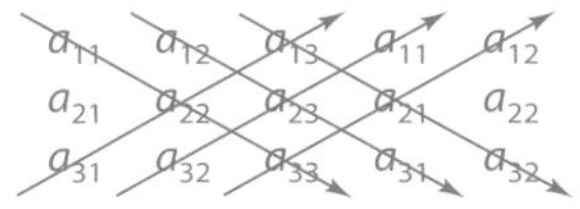

Add these products.

That is,

$$|A| = a_{11}a_{22}a_{33} + a_{12}a_{23}a_{31} + a_{13}a_{21}a_{32} - a_{31}a_{22}a_{13} - a_{32}a_{23}a_{11} - a_{33}a_{21}a_{12}.$$

Try using this method to find $|A|$ in Example 3.

When expanding by cofactors, you do not need to find cofactors of zero entries, because zero times its cofactor is zero.

$$a_{ij}C_{ij} = (0)C_{ij} = 0$$

So, the row (or column) containing the most zeros is usually the best choice for expansion by cofactors. This is demonstrated in the next example.

Example 4 ▶ The Determinant of a Matrix of Order 4 × 4

Find the determinant of

$$A = \begin{bmatrix} 1 & -2 & 3 & 0 \\ -1 & 1 & 0 & 2 \\ 0 & 2 & 0 & 3 \\ 3 & 4 & 0 & 2 \end{bmatrix}.$$

Solution

After inspecting this matrix, you can see that three of the entries in the third column are zeros. So, you can eliminate some of the work in the expansion by using the third column.

$$|A| = 3(C_{13}) + 0(C_{23}) + 0(C_{33}) + 0(C_{43})$$

Because C_{23}, C_{33}, and C_{43} have zero coefficients, you need only find the cofactor C_{13}. To do this, delete the first row and third column of A and evaluate the determinant of the resulting matrix.

$$C_{13} = (-1)^{1+3}\begin{vmatrix} -1 & 1 & 2 \\ 0 & 2 & 3 \\ 3 & 4 & 2 \end{vmatrix} \qquad \text{Delete 1st row and 3rd column.}$$

$$= \begin{vmatrix} -1 & 1 & 2 \\ 0 & 2 & 3 \\ 3 & 4 & 2 \end{vmatrix} \qquad \text{Simplify.}$$

Students may have difficulty choosing the best row or column about which to expand a determinant using cofactors. You might consider using the following matrices as exercises to give students practice in choosing the row or column that yields the most efficient expansion.

$$A = \begin{bmatrix} 0 & -1 & 3 \\ 4 & 0 & 3 \\ 2 & 0 & -3 \end{bmatrix}$$

$$B = \begin{bmatrix} -1 & 3 & 0 & 1 \\ 2 & 0 & 4 & -3 \\ -2 & 1 & 7 & 0 \\ 3 & 2 & 0 & 5 \end{bmatrix}$$

$$C = \begin{bmatrix} -6 & 0 & 1 & 3 & -2 \\ 2 & -1 & 4 & 0 & -3 \\ 7 & 0 & -1 & 0 & 0 \\ 1 & -3 & 0 & 2 & 2 \\ 0 & 4 & 1 & 3 & 2 \end{bmatrix}$$

Expanding by cofactors in the second row yields

$$C_{13} = 0(-1)^3\begin{vmatrix} 1 & 2 \\ 4 & 2 \end{vmatrix} + 2(-1)^4\begin{vmatrix} -1 & 2 \\ 3 & 2 \end{vmatrix} + 3(-1)^5\begin{vmatrix} -1 & 1 \\ 3 & 4 \end{vmatrix}$$

$$= 0 + 2(1)(-8) + 3(-1)(-7)$$

$$= 5.$$

So, you obtain

$$|A| = 3C_{13}$$

$$= 3(5)$$

$$= 15.$$

Try using a graphing utility to confirm the result of Example 4.

8.4 Exercises

The *Interactive* CD-ROM and *Internet* versions of this text contain step-by-step solutions to all odd-numbered Section and Review Exercises. They also provide Tutorial Exercises that link to Guided Examples for additional help.

In Exercises 1–16, find the determinant of the matrix.

1. $[5]$
2. $[-8]$
3. $\begin{bmatrix} 2 & 1 \\ 3 & 4 \end{bmatrix}$
4. $\begin{bmatrix} -3 & 1 \\ 5 & 2 \end{bmatrix}$
5. $\begin{bmatrix} 5 & 2 \\ -6 & 3 \end{bmatrix}$
6. $\begin{bmatrix} 2 & -2 \\ 4 & 3 \end{bmatrix}$
7. $\begin{bmatrix} -7 & 0 \\ 3 & 0 \end{bmatrix}$
8. $\begin{bmatrix} 4 & -3 \\ 0 & 0 \end{bmatrix}$
9. $\begin{bmatrix} 2 & 6 \\ 0 & 3 \end{bmatrix}$
10. $\begin{bmatrix} 2 & -3 \\ -6 & 9 \end{bmatrix}$
11. $\begin{bmatrix} -3 & -2 \\ -6 & -1 \end{bmatrix}$
12. $\begin{bmatrix} 4 & 7 \\ -2 & 5 \end{bmatrix}$
13. $\begin{bmatrix} 9 & 0 \\ 7 & 8 \end{bmatrix}$
14. $\begin{bmatrix} 0 & 6 \\ -3 & 2 \end{bmatrix}$
15. $\begin{bmatrix} -\frac{1}{2} & \frac{1}{3} \\ -6 & \frac{1}{3} \end{bmatrix}$
16. $\begin{bmatrix} \frac{2}{3} & \frac{4}{3} \\ -1 & -\frac{1}{3} \end{bmatrix}$

In Exercises 17–22, use the matrix capabilities of a graphing utility to find the determinant of the matrix.

17. $\begin{bmatrix} 0.3 & 0.2 & 0.2 \\ 0.2 & 0.2 & 0.2 \\ -0.4 & 0.4 & 0.3 \end{bmatrix}$
18. $\begin{bmatrix} 0.1 & 0.2 & 0.3 \\ -0.3 & 0.2 & 0.2 \\ 0.5 & 0.4 & 0.4 \end{bmatrix}$
19. $\begin{bmatrix} 0.9 & 0.7 & 0 \\ -0.1 & 0.3 & 1.3 \\ -2.2 & 4.2 & 6.1 \end{bmatrix}$
20. $\begin{bmatrix} 0.1 & 0.1 & -4.3 \\ 7.5 & 6.2 & 0.7 \\ 0.3 & 0.6 & -1.2 \end{bmatrix}$
21. $\begin{bmatrix} 1 & 4 & -2 \\ 3 & 6 & -6 \\ -2 & 1 & 4 \end{bmatrix}$
22. $\begin{bmatrix} 2 & 3 & 1 \\ 0 & 5 & -2 \\ 0 & 0 & -2 \end{bmatrix}$

In Exercises 23–30, find all (a) minors and (b) cofactors of the matrix.

23. $\begin{bmatrix} 3 & 4 \\ 2 & -5 \end{bmatrix}$
24. $\begin{bmatrix} 11 & 0 \\ -3 & 2 \end{bmatrix}$
25. $\begin{bmatrix} 3 & 1 \\ -2 & -4 \end{bmatrix}$
26. $\begin{bmatrix} -6 & 5 \\ 7 & -2 \end{bmatrix}$
27. $\begin{bmatrix} 4 & 0 & 2 \\ -3 & 2 & 1 \\ 1 & -1 & 1 \end{bmatrix}$
28. $\begin{bmatrix} 1 & -1 & 0 \\ 3 & 2 & 5 \\ 4 & -6 & 4 \end{bmatrix}$
29. $\begin{bmatrix} 3 & -2 & 8 \\ 3 & 2 & -6 \\ -1 & 3 & 6 \end{bmatrix}$
30. $\begin{bmatrix} -2 & 9 & 4 \\ 7 & -6 & 0 \\ 6 & 7 & -6 \end{bmatrix}$

In Exercises 31–36, find the determinant of the matrix by the method of expansion by cofactors. Expand using the indicated row or column.

31. $\begin{bmatrix} -3 & 2 & 1 \\ 4 & 5 & 6 \\ 2 & -3 & 1 \end{bmatrix}$
   (a) Row 1
   (b) Column 2
32. $\begin{bmatrix} -3 & 4 & 2 \\ 6 & 3 & 1 \\ 4 & -7 & -8 \end{bmatrix}$
   (a) Row 2
   (b) Column 3
33. $\begin{bmatrix} 5 & 0 & -3 \\ 0 & 12 & 4 \\ 1 & 6 & 3 \end{bmatrix}$
   (a) Row 2
   (b) Column 2
34. $\begin{bmatrix} 10 & -5 & 5 \\ 30 & 0 & 10 \\ 0 & 10 & 1 \end{bmatrix}$
   (a) Row 3
   (b) Column 1
35. $\begin{bmatrix} 6 & 0 & -3 & 5 \\ 4 & 13 & 6 & -8 \\ -1 & 0 & 7 & 4 \\ 8 & 6 & 0 & 2 \end{bmatrix}$
   (a) Row 2
   (b) Column 2
36. $\begin{bmatrix} 10 & 8 & 3 & -7 \\ 4 & 0 & 5 & -6 \\ 0 & 3 & 2 & 7 \\ 1 & 0 & -3 & 2 \end{bmatrix}$
   (a) Row 3
   (b) Column 1

In Exercises 37–52, find the determinant of the matrix. Expand by cofactors on the row or column that appears to make the computations easiest.

37. $\begin{bmatrix} 2 & -1 & 0 \\ 4 & 2 & 1 \\ 4 & 2 & 1 \end{bmatrix}$
38. $\begin{bmatrix} -2 & 2 & 3 \\ 1 & -1 & 0 \\ 0 & 1 & 4 \end{bmatrix}$
39. $\begin{bmatrix} 6 & 3 & -7 \\ 0 & 0 & 0 \\ 4 & -6 & 3 \end{bmatrix}$
40. $\begin{bmatrix} 1 & 1 & 2 \\ 3 & 1 & 0 \\ -2 & 0 & 3 \end{bmatrix}$
41. $\begin{bmatrix} -1 & 2 & -5 \\ 0 & 3 & 4 \\ 0 & 0 & 3 \end{bmatrix}$
42. $\begin{bmatrix} 1 & 0 & 0 \\ -4 & -1 & 0 \\ 5 & 1 & 5 \end{bmatrix}$
43. $\begin{bmatrix} 1 & 4 & -2 \\ 3 & 2 & 0 \\ -1 & 4 & 3 \end{bmatrix}$
44. $\begin{bmatrix} 2 & -1 & 3 \\ 1 & 4 & 4 \\ 1 & 0 & 2 \end{bmatrix}$
45. $\begin{bmatrix} 2 & 4 & 6 \\ 0 & 3 & 1 \\ 0 & 0 & -5 \end{bmatrix}$
46. $\begin{bmatrix} -3 & 0 & 0 \\ 7 & 11 & 0 \\ 1 & 2 & 2 \end{bmatrix}$

47. $\begin{bmatrix} 2 & 6 & 6 & 2 \\ 2 & 7 & 3 & 6 \\ 1 & 5 & 0 & 1 \\ 3 & 7 & 0 & 7 \end{bmatrix}$

48. $\begin{bmatrix} 3 & 6 & -5 & 4 \\ -2 & 0 & 6 & 0 \\ 1 & 1 & 2 & 2 \\ 0 & 3 & -1 & -1 \end{bmatrix}$

49. $\begin{bmatrix} 5 & 3 & 0 & 6 \\ 4 & 6 & 4 & 12 \\ 0 & 2 & -3 & 4 \\ 0 & 1 & -2 & 2 \end{bmatrix}$

50. $\begin{bmatrix} 1 & 4 & 3 & 2 \\ -5 & 6 & 2 & 1 \\ 0 & 0 & 0 & 0 \\ 3 & -2 & 1 & 5 \end{bmatrix}$

51. $\begin{bmatrix} 3 & 2 & 4 & -1 & 5 \\ -2 & 0 & 1 & 3 & 2 \\ 1 & 0 & 0 & 4 & 0 \\ 6 & 0 & 2 & -1 & 0 \\ 3 & 0 & 5 & 1 & 0 \end{bmatrix}$

52. $\begin{bmatrix} 5 & 2 & 0 & 0 & -2 \\ 0 & 1 & 4 & 3 & 2 \\ 0 & 0 & 2 & 6 & 3 \\ 0 & 0 & 3 & 4 & 1 \\ 0 & 0 & 0 & 0 & 2 \end{bmatrix}$

In Exercises 53–60, use the matrix capabilities of a graphing utility to evaluate the determinant.

53. $\begin{vmatrix} 3 & 8 & -7 \\ 0 & -5 & 4 \\ 8 & 1 & 6 \end{vmatrix}$

54. $\begin{vmatrix} 5 & -8 & 0 \\ 9 & 7 & 4 \\ -8 & 7 & 1 \end{vmatrix}$

55. $\begin{vmatrix} 7 & 0 & -14 \\ -2 & 5 & 4 \\ -6 & 2 & 12 \end{vmatrix}$

56. $\begin{vmatrix} 3 & 0 & 0 \\ -2 & 5 & 0 \\ 12 & 5 & 7 \end{vmatrix}$

57. $\begin{vmatrix} 1 & -1 & 8 & 4 \\ 2 & 6 & 0 & -4 \\ 2 & 0 & 2 & 6 \\ 0 & 2 & 8 & 0 \end{vmatrix}$

58. $\begin{vmatrix} 0 & -3 & 8 & 2 \\ 8 & 1 & -1 & 6 \\ -4 & 6 & 0 & 9 \\ -7 & 0 & 0 & 14 \end{vmatrix}$

59. $\begin{vmatrix} 3 & -2 & 4 & 3 & 1 \\ -1 & 0 & 2 & 1 & 0 \\ 5 & -1 & 0 & 3 & 2 \\ 4 & 7 & -8 & 0 & 0 \\ 1 & 2 & 3 & 0 & 2 \end{vmatrix}$

60. $\begin{vmatrix} -2 & 0 & 0 & 0 & 0 \\ 0 & 3 & 0 & 0 & 0 \\ 0 & 0 & -1 & 0 & 0 \\ 0 & 0 & 0 & 2 & 0 \\ 0 & 0 & 0 & 0 & -4 \end{vmatrix}$

In Exercises 61–68, find (a) $|A|$, (b) $|B|$, (c) AB, and (d) $|AB|$.

61. $A = \begin{bmatrix} -1 & 0 \\ 0 & 3 \end{bmatrix}, \quad B = \begin{bmatrix} 2 & 0 \\ 0 & -1 \end{bmatrix}$

62. $A = \begin{bmatrix} -2 & 1 \\ 4 & -2 \end{bmatrix}, \quad B = \begin{bmatrix} 1 & 2 \\ 0 & -1 \end{bmatrix}$

63. $A = \begin{bmatrix} 4 & 0 \\ 3 & -2 \end{bmatrix}, \quad B = \begin{bmatrix} -1 & 1 \\ -2 & 2 \end{bmatrix}$

64. $A = \begin{bmatrix} 5 & 4 \\ 3 & -1 \end{bmatrix}, \quad B = \begin{bmatrix} 0 & 6 \\ 1 & -2 \end{bmatrix}$

65. $A = \begin{bmatrix} 0 & 1 & 2 \\ -3 & -2 & 1 \\ 0 & 4 & 1 \end{bmatrix}, \quad B = \begin{bmatrix} 3 & -2 & 0 \\ 1 & -1 & 2 \\ 3 & 1 & 1 \end{bmatrix}$

66. $A = \begin{bmatrix} 3 & 2 & 0 \\ -1 & -3 & 4 \\ -2 & 0 & 1 \end{bmatrix}, \quad B = \begin{bmatrix} -3 & 0 & 1 \\ 0 & 2 & -1 \\ -2 & -1 & 1 \end{bmatrix}$

67. $A = \begin{bmatrix} -1 & 2 & 1 \\ 1 & 0 & 1 \\ 0 & 1 & 0 \end{bmatrix}, \quad B = \begin{bmatrix} -1 & 0 & 0 \\ 0 & 2 & 0 \\ 0 & 0 & 3 \end{bmatrix}$

68. $A = \begin{bmatrix} 2 & 0 & 1 \\ 1 & -1 & 2 \\ 3 & 1 & 0 \end{bmatrix}, \quad B = \begin{bmatrix} 2 & -1 & 4 \\ 0 & 1 & 3 \\ 3 & -2 & 1 \end{bmatrix}$

In Exercises 69–74, evaluate the determinant(s) to verify the equation.

69. $\begin{vmatrix} w & x \\ y & z \end{vmatrix} = -\begin{vmatrix} y & z \\ w & x \end{vmatrix}$

70. $\begin{vmatrix} w & cx \\ y & cz \end{vmatrix} = c\begin{vmatrix} w & x \\ y & z \end{vmatrix}$

71. $\begin{vmatrix} w & x \\ y & z \end{vmatrix} = \begin{vmatrix} w & x + cw \\ y & z + cy \end{vmatrix}$

72. $\begin{vmatrix} w & x \\ cw & cx \end{vmatrix} = 0$

73. $\begin{vmatrix} 1 & x & x^2 \\ 1 & y & y^2 \\ 1 & z & z^2 \end{vmatrix} = (y - x)(z - x)(z - y)$

74. $\begin{vmatrix} a + b & a & a \\ a & a + b & a \\ a & a & a + b \end{vmatrix} = b^2(3a + b)$

In Exercises 75–78, solve for x.

75. $\begin{vmatrix} x - 1 & 2 \\ 3 & x - 2 \end{vmatrix} = 0$

76. $\begin{vmatrix} x - 2 & -1 \\ -3 & x \end{vmatrix} = 0$

77. $\begin{vmatrix} x + 3 & 2 \\ 1 & x + 2 \end{vmatrix} = 0$

78. $\begin{vmatrix} x + 4 & -2 \\ 7 & x - 5 \end{vmatrix} = 0$

In Exercises 79–84, evaluate the determinant in which the entries are functions. Determinants of this type occur when changes in variables are made in calculus.

79. $\begin{vmatrix} 4u & -1 \\ -1 & 2v \end{vmatrix}$

80. $\begin{vmatrix} 3x^2 & -3y^2 \\ 1 & 1 \end{vmatrix}$

81. $\begin{vmatrix} e^{2x} & e^{3x} \\ 2e^{2x} & 3e^{3x} \end{vmatrix}$

82. $\begin{vmatrix} e^{-x} & xe^{-x} \\ -e^{-x} & (1-x)e^{-x} \end{vmatrix}$

83. $\begin{vmatrix} x & \ln x \\ 1 & 1/x \end{vmatrix}$

84. $\begin{vmatrix} x & x \ln x \\ 1 & 1 + \ln x \end{vmatrix}$

Synthesis

True or False? **In Exercises 85 and 86, determine whether the statement is true or false. Justify your answer.**

85. If a square matrix has an entire row of zeros, the determinant will always be zero.

86. If two columns of a square matrix are the same, then the determinant of the matrix will be zero.

87. *Exploration* Find square matrices A and B to demonstrate that

$|A + B| \neq |A| + |B|$.

88. *Exploration* Consider square matrices in which the entries are consecutive integers. An example of such a matrix is

$$\begin{bmatrix} 4 & 5 & 6 \\ 7 & 8 & 9 \\ 10 & 11 & 12 \end{bmatrix}.$$

(a) Use a graphing utility to evaluate the determinants of four matrices of this type. Make a conjecture based on the results.

(b) Verify your conjecture.

89. *Writing* Write a brief paragraph explaining the difference between a square matrix and its determinant.

90. *Think About It* If A is a matrix of order 3×3 such that $|A| = 5$, is it possible to find $|2A|$? Explain.

Review

In Exercises 91 and 92, sketch the graph of the system of inequalities.

91. $\begin{cases} x + y \le 8 \\ x \ge -3 \\ 2x - y < 5 \end{cases}$

92. $\begin{cases} -x - y > 4 \\ y \le 1 \\ 7x + 4y \le -10 \end{cases}$

In Exercises 93–96, find the inverse of the matrix (if it exists).

93. $\begin{bmatrix} -4 & 1 \\ 8 & -1 \end{bmatrix}$

94. $\begin{bmatrix} -5 & -8 \\ 3 & 6 \end{bmatrix}$

95. $\begin{bmatrix} -7 & 2 & 9 \\ 2 & -4 & -6 \\ 3 & 5 & 2 \end{bmatrix}$

96. $\begin{bmatrix} -6 & 2 & 0 \\ 1 & 3 & -2 \\ -2 & 0 & 1 \end{bmatrix}$

8.5 Applications of Matrices and Determinants

▶ What you should learn

- How to use Cramer's Rule to solve systems of linear equations
- How to use determinants to find the areas of triangles
- How to use a determinant to find an equation of a line passing through two points
- How to use matrices to code and decode messages

▶ Why you should learn it

You can use determinants and matrices to model and solve real-life problems. For instance, Exercise 27 on page 675 shows how matrices can be used to estimate the area of a region of land.

Layne Kennedy/Corbis

Cramer's Rule

So far, you have studied three methods for solving a system of linear equations: substitution, elimination with equations, and elimination with matrices. In this section, you will study one more method, **Cramer's Rule,** named after Gabriel Cramer (1704–1752). This rule uses determinants to write the solution of a system of linear equations. To see how Cramer's Rule works, take another look at the solution described at the beginning of Section 8.4. There, it was pointed out that the system

$$\begin{cases} a_1x + b_1y = c_1 \\ a_2x + b_2y = c_2 \end{cases}$$

has a solution

$$x = \frac{c_1b_2 - c_2b_1}{a_1b_2 - a_2b_1}$$

and

$$y = \frac{a_1c_2 - a_2c_1}{a_1b_2 - a_2b_1}$$

provided that $a_1b_2 - a_2b_1 \neq 0$. Each numerator and denominator in this solution can be expressed as a determinant, as follows.

$$x = \frac{c_1b_2 - c_2b_1}{a_1b_2 - a_2b_1} = \frac{\begin{vmatrix} c_1 & b_1 \\ c_2 & b_2 \end{vmatrix}}{\begin{vmatrix} a_1 & b_1 \\ a_2 & b_2 \end{vmatrix}}$$

$$y = \frac{a_1c_2 - a_2c_1}{a_1b_2 - a_2b_1} = \frac{\begin{vmatrix} a_1 & c_1 \\ a_2 & c_2 \end{vmatrix}}{\begin{vmatrix} a_1 & b_1 \\ a_2 & b_2 \end{vmatrix}}$$

Relative to the original system, the denominator for x and y is simply the determinant of the *coefficient* matrix of the system. This determinant is denoted by D. The numerators for x and y are denoted by D_x and D_y, respectively. They are formed by using the column of constants as replacements for the coefficients of x and y, as follows.

Coefficient Matrix	D	D_x	D_y
$\begin{bmatrix} a_1 & b_1 \\ a_2 & b_2 \end{bmatrix}$	$\begin{vmatrix} a_1 & b_1 \\ a_2 & b_2 \end{vmatrix}$	$\begin{vmatrix} c_1 & b_1 \\ c_2 & b_2 \end{vmatrix}$	$\begin{vmatrix} a_1 & c_1 \\ a_2 & c_2 \end{vmatrix}$

Example 1 ▶ Using Cramer's Rule for a 2 × 2 System

Your students should now be able to solve systems of linear equations by substitution (Section 7.1), elimination (Section 7.2), Gaussian elimination with back-substitution (Section 8.1), Gauss-Jordan elimination (Section 8.1), the inverse matrix method (Section 8.3), and now, after this section, Cramer's Rule.

Use Cramer's Rule to solve the system of linear equations.

$$\begin{cases} 4x - 2y = 10 \\ 3x - 5y = 11 \end{cases}$$

Solution

To begin, find the determinant of the coefficient matrix.

$$D = \begin{vmatrix} 4 & -2 \\ 3 & -5 \end{vmatrix} = -20 - (-6) = -14$$

Because this determinant is not zero, you can apply Cramer's Rule to find the solution, as follows.

$$x = \frac{D_x}{D} = \frac{\begin{vmatrix} 10 & -2 \\ 11 & -5 \end{vmatrix}}{-14} = \frac{(-50) - (-22)}{-14} = \frac{-28}{-14} = 2$$

$$y = \frac{D_y}{D} = \frac{\begin{vmatrix} 4 & 10 \\ 3 & 11 \end{vmatrix}}{-14} = \frac{44 - 30}{-14} = \frac{14}{-14} = -1$$

Therefore, the solution is $x = 2$ and $y = -1$. Check this in the original system.

Cramer's Rule generalizes easily to systems of n equations in n variables. The value of each variable is given as the quotient of two determinants. The denominator is the determinant of the coefficient matrix, and the numerator is the determinant of the matrix formed by replacing the column corresponding to the variable (being solved for) with the column representing the constants. For instance, the solution for x_3 in the system

$$\begin{aligned} a_{11}x_1 + a_{12}x_2 + a_{13}x_3 &= b_1 \\ a_{21}x_1 + a_{22}x_2 + a_{23}x_3 &= b_2 \\ a_{31}x_1 + a_{32}x_2 + a_{33}x_3 &= b_3 \end{aligned}$$

is given by

$$x_3 = \frac{|A_3|}{|A|} = \frac{\begin{vmatrix} a_{11} & a_{12} & b_1 \\ a_{21} & a_{22} & b_2 \\ a_{31} & a_{32} & b_3 \end{vmatrix}}{\begin{vmatrix} a_{11} & a_{12} & a_{13} \\ a_{21} & a_{22} & a_{23} \\ a_{31} & a_{32} & a_{33} \end{vmatrix}}.$$

STUDY TIP

When using Cramer's Rule, remember that the method does not apply if the determinant of the coefficient matrix is zero.

Cramer's Rule

If a system of n linear equations in n variables has a coefficient matrix A with a nonzero determinant $|A|$, the solution of the system is

$$x_1 = \frac{|A_1|}{|A|}, \quad x_2 = \frac{|A_2|}{|A|}, \quad . \; . \; . \; , \quad x_n = \frac{|A_n|}{|A|}$$

where the ith column of A_i is the column of constants in the system of equations. If the determinant of the coefficient matrix is zero, the system has either no solution or infinitely many solutions.

A computer animation of this example appears in the *Interactive* CD-ROM and *Internet* versions of this text.

Example 2 ▶ Using Cramer's Rule for a 3 × 3 System

Use Cramer's Rule to solve the system of linear equations.

$$\begin{cases} -x + 2y - 3z = 1 \\ 2x \qquad\;\; + z = 0 \\ 3x - 4y + 4z = 2 \end{cases}$$

Solution

The coefficient matrix

$$\begin{bmatrix} -1 & 2 & -3 \\ 2 & 0 & 1 \\ 3 & -4 & 4 \end{bmatrix}$$

can be expanded along the second row, as follows.

$$\begin{aligned} D &= 2(-1)^3\begin{vmatrix} 2 & -3 \\ -4 & 4 \end{vmatrix} + 0(-1)^4\begin{vmatrix} -1 & -3 \\ 3 & 4 \end{vmatrix} + 1(-1)^5\begin{vmatrix} -1 & 2 \\ 3 & -4 \end{vmatrix} \\ &= -2(-4) + 0 - 1(-2) \\ &= 10 \end{aligned}$$

Because this determinant is not zero, you can apply Cramer's Rule to find the solution, as follows.

$$x = \frac{D_x}{D} = \frac{\begin{vmatrix} 1 & 2 & -3 \\ 0 & 0 & 1 \\ 2 & -4 & 4 \end{vmatrix}}{10} = \frac{8}{10} = \frac{4}{5}$$

$$y = \frac{D_y}{D} = \frac{\begin{vmatrix} -1 & 1 & -3 \\ 2 & 0 & 1 \\ 3 & 2 & 4 \end{vmatrix}}{10} = \frac{-15}{10} = -\frac{3}{2}$$

$$z = \frac{D_z}{D} = \frac{\begin{vmatrix} -1 & 2 & 1 \\ 2 & 0 & 0 \\ 3 & -4 & 2 \end{vmatrix}}{10} = \frac{-16}{10} = -\frac{8}{5}$$

The solution is $\left(\frac{4}{5}, -\frac{3}{2}, -\frac{8}{5}\right)$. Check this in the original system.

Area of a Triangle

Another application of matrices and determinants is finding the area of a triangle whose vertices are given as points in a coordinate plane.

Area of a Triangle

The area of a triangle with vertices (x_1, y_1), (x_2, y_2), and (x_3, y_3) is

$$\text{Area} = \pm\frac{1}{2}\begin{vmatrix} x_1 & y_1 & 1 \\ x_2 & y_2 & 1 \\ x_3 & y_3 & 1 \end{vmatrix}$$

where the symbol $\pm$ indicates that the appropriate sign should be chosen to yield a positive area.

Example 3 ▶ Finding the Area of a Triangle

Find the area of a triangle whose vertices are $(1, 0)$, $(2, 2)$, and $(4, 3)$, as shown in Figure 8.1.

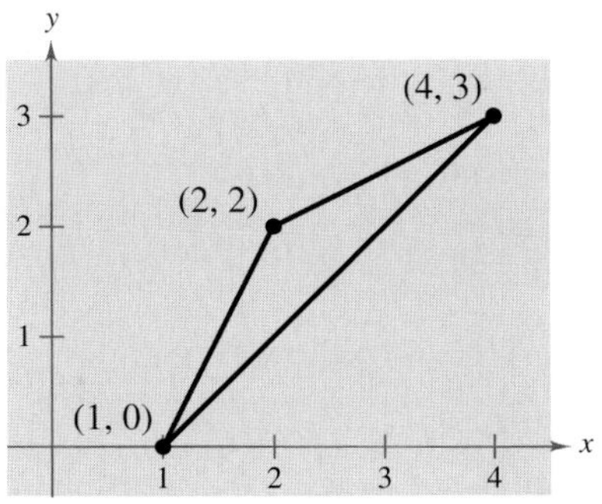

FIGURE 8.1

Solution

Let $(x_1, y_1) = (1, 0)$, $(x_2, y_2) = (2, 2)$, and $(x_3, y_3) = (4, 3)$. Then, to find the area of the triangle, evaluate the determinant.

$$\begin{vmatrix} x_1 & y_1 & 1 \\ x_2 & y_2 & 1 \\ x_3 & y_3 & 1 \end{vmatrix} = \begin{vmatrix} 1 & 0 & 1 \\ 2 & 2 & 1 \\ 4 & 3 & 1 \end{vmatrix}$$

$$= 1(-1)^2\begin{vmatrix} 2 & 1 \\ 3 & 1 \end{vmatrix} + 0(-1)^3\begin{vmatrix} 2 & 1 \\ 4 & 1 \end{vmatrix} + 1(-1)^4\begin{vmatrix} 2 & 2 \\ 4 & 3 \end{vmatrix}$$

$$= 1(-1) + 0 + 1(-2)$$

$$= -3.$$

Using this value, you can conclude that the area of the triangle is

$$\text{Area} = -\frac{1}{2}\begin{vmatrix} 1 & 0 & 1 \\ 2 & 2 & 1 \\ 4 & 3 & 1 \end{vmatrix} \qquad \text{Choose } (-) \text{ so that the area is positive.}$$

$$= -\frac{1}{2}(-3)$$

$$= \frac{3}{2}.$$

Try using determinants to find the area of a triangle with vertices $(3, -1)$, $(7, -1)$, and $(7, 5)$. Confirm your answer by plotting the points in a coordinate plane and using the formula

$$\text{Area} = \frac{1}{2}(\text{base})(\text{height}).$$

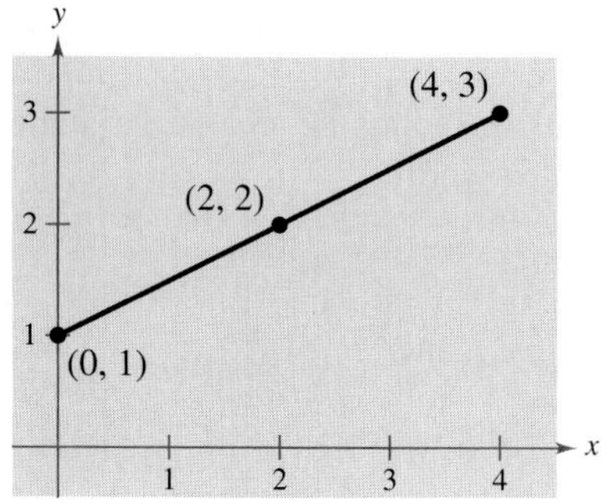

FIGURE 8.2

Lines in a Plane

What if the three points in Example 3 had been on the same line? What would have happened had the area formula been applied to three such points? The answer is that the determinant would have been zero. Consider, for instance, the three collinear points (0, 1), (2, 2), and (4, 3), as shown in Figure 8.2. The area of the "triangle" that has these three points as vertices is

$$\frac{1}{2}\begin{vmatrix} 0 & 1 & 1 \\ 2 & 2 & 1 \\ 4 & 3 & 1 \end{vmatrix} = \frac{1}{2}\left[0(-1)^2\begin{vmatrix} 2 & 1 \\ 3 & 1 \end{vmatrix} + 1(-1)^3\begin{vmatrix} 2 & 1 \\ 4 & 1 \end{vmatrix} + 1(-1)^4\begin{vmatrix} 2 & 2 \\ 4 & 3 \end{vmatrix}\right]$$

$$= \frac{1}{2}[0(-1) + 1(2) + 1(-2)]$$

$$= 0.$$

The result is generalized as follows.

Test for Collinear Points

Three points (x_1, y_1), (x_2, y_2), and (x_3, y_3) are collinear (lie on the same line) if and only if

$$\begin{vmatrix} x_1 & y_1 & 1 \\ x_2 & y_2 & 1 \\ x_3 & y_3 & 1 \end{vmatrix} = 0.$$

Example 4 ▶ Testing for Collinear Points

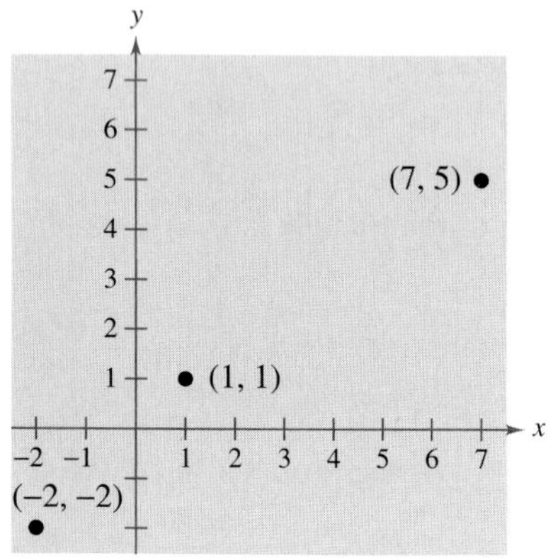

FIGURE 8.3

Determine whether the points $(-2, -2)$, $(1, 1)$, and $(7, 5)$ lie on the same line. (See Figure 8.3.)

Solution

Letting $(x_1, y_1) = (-2, -2)$, $(x_2, y_2) = (1, 1)$, and $(x_3, y_3) = (7, 5)$, you have

$$\begin{vmatrix} x_1 & y_1 & 1 \\ x_2 & y_2 & 1 \\ x_3 & y_3 & 1 \end{vmatrix} = \begin{vmatrix} -2 & -2 & 1 \\ 1 & 1 & 1 \\ 7 & 5 & 1 \end{vmatrix}$$

$$= -2(-1)^2\begin{vmatrix} 1 & 1 \\ 5 & 1 \end{vmatrix} + (-2)(-1)^3\begin{vmatrix} 1 & 1 \\ 7 & 1 \end{vmatrix} + 1(-1)^4\begin{vmatrix} 1 & 1 \\ 7 & 5 \end{vmatrix}$$

$$= -2(-4) + (-2)(6) + 1(-2)$$

$$= -6.$$

Because the value of this determinant is *not* zero, you can conclude that the three points do not lie on the same line.

The test for collinear points can be adapted to another use. That is, if you are given two points on a rectangular coordinate system, you can find an equation of the line passing through the two points, as follows.

Two-Point Form of the Equation of a Line

An equation of the line passing through the distinct points (x_1, y_1) and (x_2, y_2) is given by

$$\begin{vmatrix} x & y & 1 \\ x_1 & y_1 & 1 \\ x_2 & y_2 & 1 \end{vmatrix} = 0.$$

Example 5 ▶ Finding an Equation of a Line

Find an equation of the line passing through the two points $(2, 4)$ and $(-1, 3)$, as shown in Figure 8.4.

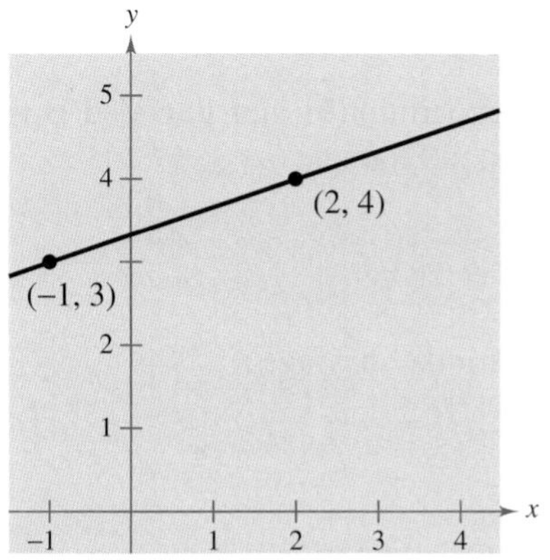

FIGURE 8.4

Solution

Applying the determinant formula for the equation of a line produces

$$\begin{vmatrix} x & y & 1 \\ 2 & 4 & 1 \\ -1 & 3 & 1 \end{vmatrix} = 0.$$

To evaluate this determinant, you can expand by cofactors along the first row to obtain the following.

$$x(-1)^2\begin{vmatrix} 4 & 1 \\ 3 & 1 \end{vmatrix} + y(-1)^3\begin{vmatrix} 2 & 1 \\ -1 & 1 \end{vmatrix} + 1(-1)^4\begin{vmatrix} 2 & 4 \\ -1 & 3 \end{vmatrix} = 0$$

$$x(1)(1) + y(-1)(3) + (1)(1)(10) = 0$$

$$x - 3y + 10 = 0$$

Therefore, an equation of the line is

$$x - 3y + 10 = 0.$$

Note that this method of finding the equation of a line works for all lines, including horizontal and vertical lines. For instance, the equation of the vertical line through $(2, 0)$ and $(2, 2)$ is

$$\begin{vmatrix} x & y & 1 \\ 2 & 0 & 1 \\ 2 & 2 & 1 \end{vmatrix} = 0$$

$$4 - 2x = 0$$

$$x = 2.$$

Cryptography

A **cryptogram** is a message written according to a secret code. (The Greek word *kryptos* means "hidden.") Matrix multiplication can be used to encode and decode messages. To begin, you need to assign a number to each letter in the alphabet (with 0 assigned to a blank space), as follows.

0 = _	9 = I	18 = R
1 = A	10 = J	19 = S
2 = B	11 = K	20 = T
3 = C	12 = L	21 = U
4 = D	13 = M	22 = V
5 = E	14 = N	23 = W
6 = F	15 = O	24 = X
7 = G	16 = P	25 = Y
8 = H	17 = Q	26 = Z

Then the message is converted to numbers and partitioned into **uncoded row matrices,** each having n entries, as demonstrated in Example 6.

Example 6 ▶ Forming Uncoded Row Matrices

Write the uncoded row matrices of order 1×3 for the message

MEET ME MONDAY.

Solution

Partitioning the message (including blank spaces, but ignoring punctuation) into groups of three produces the following uncoded row matrices.

$$\underset{\text{M E E}}{[13 \quad 5 \quad 5]} \quad \underset{\text{T \_ M}}{[20 \quad 0 \quad 13]} \quad \underset{\text{E \_ M}}{[5 \quad 0 \quad 13]} \quad \underset{\text{O N D}}{[15 \quad 14 \quad 4]} \quad \underset{\text{A Y \_}}{[1 \quad 25 \quad 0]}$$

Note that a blank space is used to fill out the last uncoded row matrix.

To encode a message, choose an $n \times n$ invertible matrix such as

$$A = \begin{bmatrix} 1 & -2 & 2 \\ -1 & 1 & 3 \\ 1 & -1 & -4 \end{bmatrix}$$

and multiply the uncoded row matrices by A (on the right) to obtain **coded row matrices.** Here is an example.

Uncoded Matrix *Encoding Matrix A* *Coded Matrix*

$$[13 \quad 5 \quad 5] \begin{bmatrix} 1 & -2 & 2 \\ -1 & 1 & 3 \\ 1 & -1 & -4 \end{bmatrix} = [13 \quad -26 \quad 21]$$

This technique is further illustrated in Example 7.

Activities

1. Solve for z using Cramer's Rule.

$$\begin{aligned} x - 2y - z &= 4 \\ 4x + y + z &= 7 \\ x + 3y - 4z &= -1 \end{aligned}$$

Answer: $z = 0$

2. Use a determinant to find the area of the triangle with vertices $(1, -3)$, $(2, 3)$, and $(3, 1)$.

Answer: 4 square units

3. Use a determinant to find an equation of the line through the points $(-1, 2)$ and $(3, 4)$.

Answer: $x - 2y + 5 = 0$

Example 7 ▶ Encoding a Message

Use the following matrix to encode the message MEET ME MONDAY.

$$A = \begin{bmatrix} 1 & -2 & 2 \\ -1 & 1 & 3 \\ 1 & -1 & -4 \end{bmatrix}$$

Solution

The coded row matrices are obtained by multiplying each of the uncoded row matrices found in Example 6 by the matrix A, as follows.

Uncoded Matrix Encoding Matrix A Coded Matrix

$$\begin{bmatrix} 13 & 5 & 5 \end{bmatrix}\begin{bmatrix} 1 & -2 & 2 \\ -1 & 1 & 3 \\ 1 & -1 & -4 \end{bmatrix} = \begin{bmatrix} 13 & -26 & 21 \end{bmatrix}$$

$$\begin{bmatrix} 20 & 0 & 13 \end{bmatrix}\begin{bmatrix} 1 & -2 & 2 \\ -1 & 1 & 3 \\ 1 & -1 & -4 \end{bmatrix} = \begin{bmatrix} 33 & -53 & -12 \end{bmatrix}$$

$$\begin{bmatrix} 5 & 0 & 13 \end{bmatrix}\begin{bmatrix} 1 & -2 & 2 \\ -1 & 1 & 3 \\ 1 & -1 & -4 \end{bmatrix} = \begin{bmatrix} 18 & -23 & -42 \end{bmatrix}$$

$$\begin{bmatrix} 15 & 14 & 4 \end{bmatrix}\begin{bmatrix} 1 & -2 & 2 \\ -1 & 1 & 3 \\ 1 & -1 & -4 \end{bmatrix} = \begin{bmatrix} 5 & -20 & 56 \end{bmatrix}$$

$$\begin{bmatrix} 1 & 25 & 0 \end{bmatrix}\begin{bmatrix} 1 & -2 & 2 \\ -1 & 1 & 3 \\ 1 & -1 & -4 \end{bmatrix} = \begin{bmatrix} -24 & 23 & 77 \end{bmatrix}$$

So, the sequence of coded row matrices is

$$[13 \quad -26 \quad 21]\,[33 \quad -53 \quad -12]\,[18 \quad -23 \quad -42]\,[5 \quad -20 \quad 56]\,[-24 \quad 23 \quad 77].$$

Finally, removing the matrix notation produces the following cryptogram.

$$13 \;\; -26 \;\; 21 \;\; 33 \;\; -53 \;\; -12 \;\; 18 \;\; -23 \;\; -42 \;\; 5 \;\; -20 \;\; 56 \;\; -24 \;\; 23 \;\; 77$$

For those who do not know the matrix A, decoding the cryptogram found in Example 7 is difficult. But for an authorized receiver who knows the matrix A, decoding is simple. The receiver need only multiply the coded row matrices by A^{-1} (on the right) to retrieve the uncoded row matrices. Here is an example.

$$\underbrace{[13 \quad -26 \quad 21]}_{\text{Coded}}A^{-1} = \underbrace{[13 \quad 5 \quad 5]}_{\text{Uncoded}}$$

Group Activity Suggestion:
Ask each group to work together to decide on their code's number scheme for assigning numbers to letters, to select a message to encode (some of the group members could work on the number scheme while the others are deciding on a message), and to find their invertible encoding matrix A. (A few group members could be asked to make up a matrix and the rest of the group could determine if it is invertible. You might also use this as an opportunity to discuss the creation of invertible matrices; suggest starting with the identity matrix and applying a series of elementary row operations.) Students can work together to partition their message into uncoded row matrices; each student can then be responsible for encoding several of the row matrices, and then the group can reassemble the coded message from each member's work. To decode the message they receive, the group can partition the message into coded row matrices, with each member being responsible for decoding several of these matrices, and then reassemble the message from each member's work. Alternatively, to save time, you might supply the entire class with a given number-to-letter code and an invertible matrix A. The groups must then only select a message to encode, divide up the message to encode with A, trade messages with another group, find A^{-1}, and decode the message they received.

Group Activity

Show how to use a graphing utility with matrix operations to decode the following cryptogram. (Use the matrix in Example 8.)

12 −25 15 28 −32 −89 10 −10 −49 12 −12 −51 17 −31 10 10 −28 55 4 −8 8

Example 8 ▶ Decoding a Message

Use the inverse of the matrix

$$A = \begin{bmatrix} 1 & -2 & 2 \\ -1 & 1 & 3 \\ 1 & -1 & -4 \end{bmatrix}$$

to decode the cryptogram

13 −26 21 33 −53 −12 18 −23 −42 5 −20 56 −24 23 77.

Solution

Partition the message into groups of three to form the coded row matrices. Then, multiply each coded row matrix by A^{-1} (on the right).

Coded Matrix *Decoding Matrix* A^{-1} *Decoded Matrix*

$$[13 \quad -26 \quad 21]\begin{bmatrix} -1 & -10 & -8 \\ -1 & -6 & -5 \\ 0 & -1 & -1 \end{bmatrix} = [13 \quad 5 \quad 5]$$

$$[33 \quad -53 \quad -12]\begin{bmatrix} -1 & -10 & -8 \\ -1 & -6 & -5 \\ 0 & -1 & -1 \end{bmatrix} = [20 \quad 0 \quad 13]$$

$$[18 \quad -23 \quad -42]\begin{bmatrix} -1 & -10 & -8 \\ -1 & -6 & -5 \\ 0 & -1 & -1 \end{bmatrix} = [5 \quad 0 \quad 13]$$

$$[5 \quad -20 \quad 56]\begin{bmatrix} -1 & -10 & -8 \\ -1 & -6 & -5 \\ 0 & -1 & -1 \end{bmatrix} = [15 \quad 14 \quad 4]$$

$$[-24 \quad 23 \quad 77]\begin{bmatrix} -1 & -10 & -8 \\ -1 & -6 & -5 \\ 0 & -1 & -1 \end{bmatrix} = [1 \quad 25 \quad 0]$$

So, the message is as follows.

[13	5	5]	[20	0	13]	[5	0	13]	[15	14	4]	[1	25	0]
M	E	E	T		M	E		M	O	N	D	A	Y	

Writing ABOUT MATHEMATICS

Cryptography Use your school's library or some other reference source to research information about another type of cryptography. Write a short paragraph describing how mathematics is used to code and decode messages.

8.5 Exercises

In Exercise 12, Cramer's Rule does not apply.
In Exercises 30, 31, and 34, the points are not collinear.

In Exercises 1–8, use Cramer's Rule to solve (if possible) the system of equations.

1. $\begin{cases} 3x + 4y = -2 \\ 5x + 3y = 4 \end{cases}$

2. $\begin{cases} -4x - 7y = 47 \\ -x + 6y = -27 \end{cases}$

3. $\begin{cases} -0.4x + 0.8y = 1.6 \\ 0.2x + 0.3y = 2.2 \end{cases}$

4. $\begin{cases} 2.4x - 1.3y = 14.63 \\ -4.6x + 0.5y = -11.51 \end{cases}$

5. $\begin{cases} 4x - y + z = -5 \\ 2x + 2y + 3z = 10 \\ 5x - 2y + 6z = 1 \end{cases}$

6. $\begin{cases} 4x - 2y + 3z = -2 \\ 2x + 2y + 5z = 16 \\ 8x - 5y - 2z = 4 \end{cases}$

7. $\begin{cases} x + 2y + 3z = -3 \\ -2x + y - z = 6 \\ 3x - 3y + 2z = -11 \end{cases}$

8. $\begin{cases} 5x - 4y + z = -14 \\ -x + 2y - 2z = 10 \\ 3x + y + z = 1 \end{cases}$

In Exercises 9–12, use a graphing utility and Cramer's Rule to solve (if possible) the system of equations.

9. $\begin{cases} 3x + 3y + 5z = 1 \\ 3x + 5y + 9z = 2 \\ 5x + 9y + 17z = 4 \end{cases}$

10. $\begin{cases} x + 2y - z = -7 \\ 2x - 2y - 2z = -8 \\ -x + 3y + 4z = 8 \end{cases}$

11. $\begin{cases} 2x + y + 2z = 6 \\ -x + 2y - 3z = 0 \\ 3x + 2y - z = 6 \end{cases}$

12. $\begin{cases} 2x + 3y + 5z = 4 \\ 3x + 5y + 9z = 7 \\ 5x + 9y + 17z = 13 \end{cases}$

In Exercises 13–22, use a determinant and the given vertices of a triangle to find the area of the triangle.

13.

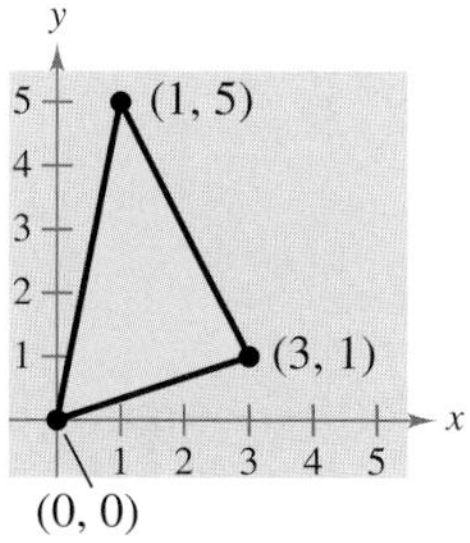

14.

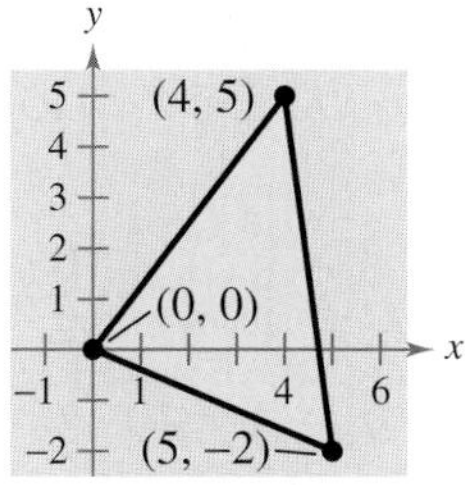

15.

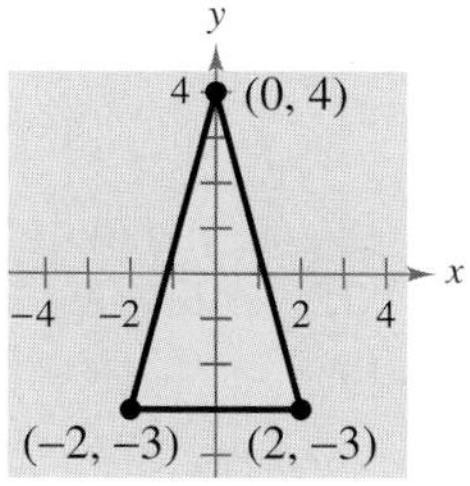

16.

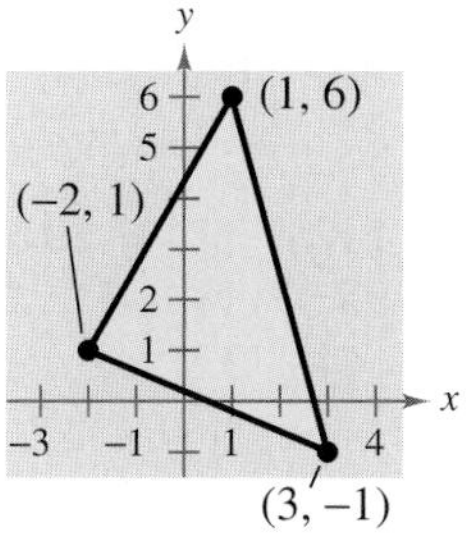

17.

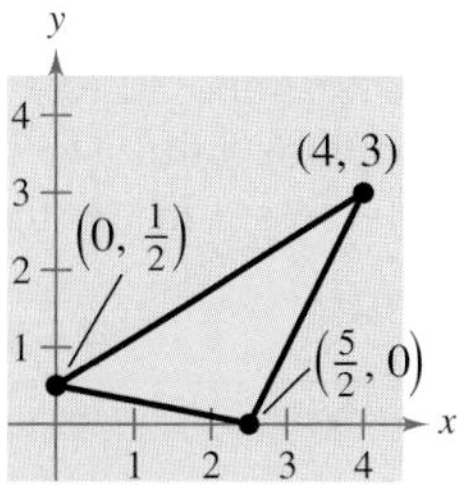

18.

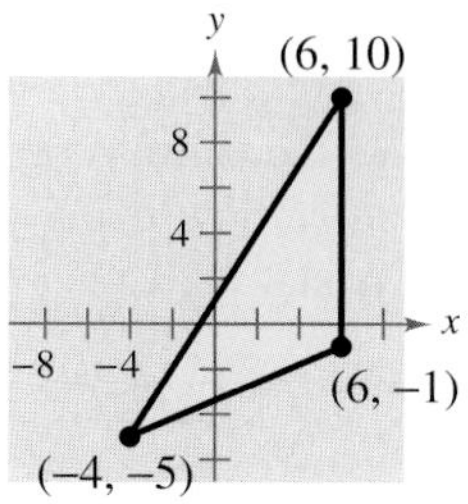

19. $(-2, 4), (2, 3), (-1, 5)$

20. $(0, -2), (-1, 4), (3, 5)$

21. $(-3, 5), (2, 6), (3, -5)$

22. $(-2, 4), (1, 5), (3, -2)$

In Exercises 23 and 24, find a value of x such that the triangle with the given vertices has an area of 4.

23. $(-5, 1), (0, 2), (-2, x)$

24. $(-4, 2), (-3, 5), (-1, x)$

In Exercises 25 and 26, find a value of x such that the triangle with the given vertices has an area of 6.

25. $(-2, -3), (1, -1), (-8, x)$

26. $(1, 0), (5, -3), (-3, x)$

27. *Area of a Region* A large region of forest has been infested with gypsy moths. The region is roughly triangular, as shown in the figure. From the northernmost vertex A of the region, the distances to the other vertices are 25 miles south and 10 miles east (for vertex B), and 20 miles south and 28 miles east (for vertex C). Use a graphing utility to approximate the number of square miles in this region.

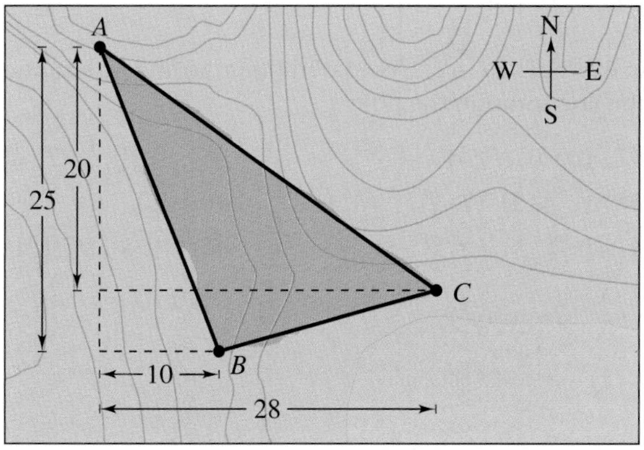

28. *Area of a Region* You own a triangular tract of land, as shown in the figure. To estimate the number of square feet in the tract, you start at one vertex, walk 65 feet east and 50 feet north to the second vertex, and then walk 85 feet west and 30 feet north to the third vertex. Use a graphing utility to determine how many square feet there are in the tract of land.

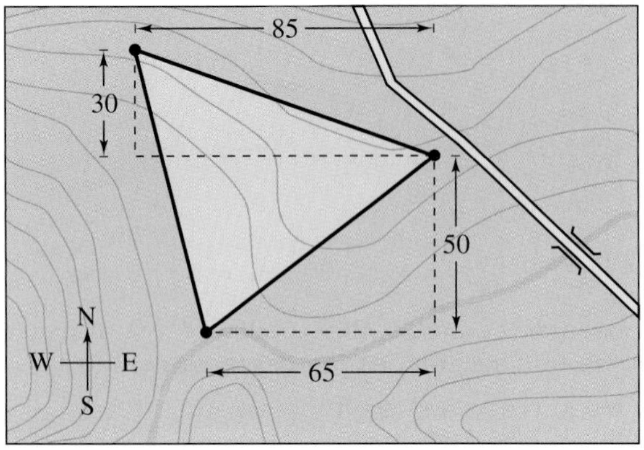

In Exercises 29–34, use a determinant to determine whether the points are collinear.

29. $(3, -1), (0, -3), (12, 5)$

30. $(-3, -5), (6, 1), (10, 2)$

31. $\left(2, -\frac{1}{2}\right), (-4, 4), (6, -3)$

32. $(0, 1), (4, -2), \left(-2, \frac{5}{2}\right)$

33. $(0, 2), (1, 2.4), (-1, 1.6)$

34. $(2, 3), (3, 3.5), (-1, 2)$

In Exercises 35 and 36, find x such that the points are collinear.

35. $(2, -5), (4, x), (5, -2)$

36. $(-6, 2), (-5, x), (-3, 5)$

In Exercises 37–42, use a determinant to find an equation of the line through the points.

37. $(0, 0), (5, 3)$

38. $(0, 0), (-2, 2)$

39. $(-4, 3), (2, 1)$

40. $(10, 7), (-2, -7)$

41. $\left(-\frac{1}{2}, 3\right), \left(\frac{5}{2}, 1\right)$

42. $\left(\frac{2}{3}, 4\right), (6, 12)$

In Exercises 43 and 44, find the uncoded 1 × 3 row matrices for the message. Then encode the message using the matrix.

Message *Matrix*

43. TROUBLE IN RIVER CITY $\begin{bmatrix} 1 & -1 & 0 \\ 1 & 0 & -1 \\ -6 & 2 & 3 \end{bmatrix}$

44. PLEASE SEND MONEY $\begin{bmatrix} 4 & 2 & 1 \\ -3 & -3 & -1 \\ 3 & 2 & 1 \end{bmatrix}$

In Exercises 45–48, write a cryptogram for the given message using the matrix A.

$$A = \begin{bmatrix} 1 & 2 & 2 \\ 3 & 7 & 9 \\ -1 & -4 & -7 \end{bmatrix}.$$

45. CALL AT NOON

46. ICEBERG DEAD AHEAD

47. HAPPY BIRTHDAY

48. OPERATION OVERLOAD

In Exercises 49–52, use A^{-1} to decode the cryptogram.

49. $A = \begin{bmatrix} 1 & 2 \\ 3 & 5 \end{bmatrix}$

11 21 64 112 25 50 29 53 23 46
40 75 55 92

50. $A = \begin{bmatrix} -5 & 2 \\ -7 & 3 \end{bmatrix}$

-136 58 -173 72 -120 51 -95 38
-178 73 -70 28 -242 101 -115 47
-90 36 -115 49 -199 82

51. $A = \begin{bmatrix} 1 & -1 & 0 \\ 1 & 0 & -1 \\ -6 & 2 & 3 \end{bmatrix}$

9 -1 -9 38 -19 -19 28 -9 -19
-80 25 41 -64 21 31 9 -5 -4

52. $A = \begin{bmatrix} 3 & -4 & 2 \\ 0 & 2 & 1 \\ 4 & -5 & 3 \end{bmatrix}$

112 -140 83 19 -25 13 72 -76 61
95 -118 71 20 21 38 35 -23 36 42
-48 32

In Exercises 53 and 54, decode the cryptogram by using the inverse of the matrix *A*.

$$A = \begin{bmatrix} 1 & 2 & 2 \\ 3 & 7 & 9 \\ -1 & -4 & -7 \end{bmatrix}$$

53. 20 17 −15 −12 −56 −104 1 −25 −65 62 143 181

54. 13 −9 −59 61 112 106 −17 −73 −131 11 24 29 65 144 172

55. The following cryptogram was encoded with a 2×2 matrix.

8 21 −15 −10 −13 −13 5 10 5 25 5 19 −1 6 20 40 −18 −18 1 16

The last word of the message is _RON. What is the message?

56. The following cryptogram was encoded with a 2×2 matrix.

5 2 25 11 −2 −7 −15 −15 32 14 −8 −13 38 19 −19 −19 37 16

The last word of the message is _SUE. What is the message?

Synthesis

True or False? **In Exercises 57 and 58, determine whether the statement is true or false. Justify your answer.**

57. You cannot use Cramer's Rule when solving a system of linear equations if the determinant of the coefficient matrix is zero.

58. In a system of linear equations, if the determinant of the coefficient matrix is zero, the system has no solution.

59. *Writing* At this point in the book, you have learned several methods for solving a system of linear equations. Briefly describe which method(s) you find easiest to use and which method(s) you find most difficult to use.

Review

In Exercises 60–63, use any method to solve the system of equations.

60. $\begin{cases} -x - 7y = -22 \\ 5x + y = -26 \end{cases}$

61. $\begin{cases} 3x + 8y = 11 \\ -2x + 12y = -16 \end{cases}$

62. $\begin{cases} -x - 3y + 5z = -14 \\ 4x + 2y - z = -1 \\ 5x - 3y + 2z = -11 \end{cases}$

63. $\begin{cases} 5x - y - z = 7 \\ -2x + 3y + z = -5 \\ 4x + 10y - 5z = -37 \end{cases}$

In Exercises 64 and 65, sketch the constraint region. Then find the minimum and maximum values of the objective function and where they occur, subject to the constraints.

64. Objective function:

$z = 6x + 4y$

Constraints:

$x \geq 0$

$y \geq 0$

$x + 6y \leq 30$

$6x + y \leq 40$

65. Objective function:

$z = 6x + 7y$

Constraints:

$x \geq 0$

$y \geq 0$

$4x + 3y \geq 24$

$x + 3y \geq 15$

In Exercises 66–69, find the determinant of the matrix.

66. $\begin{bmatrix} -8 & 5 \\ 7 & -9 \end{bmatrix}$

67. $\begin{bmatrix} 2.4 & -4.7 \\ -1.4 & -3 \end{bmatrix}$

68. $\begin{bmatrix} -6 & -1 & 5 \\ -3 & 0 & 4 \\ -2 & 4 & -5 \end{bmatrix}$

69. $\begin{bmatrix} 1 & 4 & -3 \\ 7 & -1 & 2 \\ 6 & 0 & 5 \end{bmatrix}$

Chapter Summary

What did you learn?

Section 8.1	Review Exercises
☐ How to write a matrix and identify its order	1–8
☐ How to perform elementary row operations on matrices	9, 10
☐ How to use matrices and Gaussian elimination to solve systems of linear equations	11–26
☐ How to use matrices and Gauss-Jordan elimination to solve systems of linear equations	27–32
Section 8.2	
☐ How to decide whether two matrices are equal	33–36
☐ How to add and subtract matrices and multiply matrices by real numbers	37–50
☐ How to multiply two matrices	51–66
☐ How to use matrix operations to model and solve real-life problems	67
Section 8.3	
☐ How to verify that two matrices are inverses of each other	68–71
☐ How to use Gauss-Jordan elimination to find the inverses of matrices	72–79
☐ How to use a formula to find the inverses of 2×2 matrices	80–83
☐ How to use inverse matrices to solve systems of linear equations	84–95
Section 8.4	
☐ How to find the determinants of 2×2 matrices	96–99
☐ How to find minors and cofactors of square matrices	100–103
☐ How to find the determinants of square matrices	104–107
Section 8.5	
☐ How to use Cramer's Rule to solve systems of linear equations	108–115
☐ How to use determinants to find the areas of triangles	116–119
☐ How to use a determinant to find an equation of a line passing through two points	120–123
☐ How to use matrices to code and decode messages	124–127

Review Exercises

Exercises containing systems with no solutions: 12, 22, 24, 32
Exercises containing systems with infinitely many solutions: 11, 14, 21
Exercises in which the operation is not possible: 39, 40, 42, 52, 56
Exercises in which the inverse matrix does not exist: 79

8.1 **In Exercises 1–4, determine the order of the matrix.**

1. $\begin{bmatrix} -4 \\ 0 \\ 5 \end{bmatrix}$

2. $\begin{bmatrix} 3 & -1 & 0 & 6 \\ -2 & 7 & 1 & 4 \end{bmatrix}$

3. $[3]$

4. $[6 \quad 2 \quad -5 \quad 8 \quad 0]$

In Exercises 5 and 6, form the augmented matrix for the system of linear equations.

5. $\begin{cases} 3x - 10y = 15 \\ 5x + 4y = 22 \end{cases}$

6. $\begin{cases} 8x - 7y + 4z = 12 \\ 3x - 5y + 2z = 20 \\ 5x + 3y - 3z = 26 \end{cases}$

In Exercises 7 and 8, write the system of linear equations represented by the augmented matrix. (Use variables *x*, *y*, *z*, and *w*.)

7. $\left[\begin{array}{ccc:c} 5 & 1 & 7 & -9 \\ 4 & 2 & 0 & 10 \\ 9 & 4 & 2 & 3 \end{array}\right]$

8. $\left[\begin{array}{cccc:c} 13 & 16 & 7 & 3 & 2 \\ 1 & 21 & 8 & 5 & 12 \\ 4 & 10 & -4 & 3 & -1 \end{array}\right]$

In Exercises 9 and 10, write the matrix in reduced row-echelon form.

9. $\begin{bmatrix} 0 & 1 & 1 \\ 1 & 2 & 3 \\ 2 & 2 & 2 \end{bmatrix}$

10. $\begin{bmatrix} 1 & 1 & 1 & 0 \\ 1 & 1 & 0 & 1 \\ 1 & 0 & 1 & 1 \\ 0 & 1 & 1 & 1 \end{bmatrix}$

In Exercises 11–14, the row-echelon form of an augmented matrix that corresponds to a system of linear equations is given. Use the matrix to determine whether the system is consistent or inconsistent, and if consistent, determine the number of solutions.

11. $\left[\begin{array}{ccc:c} 1 & 2 & 3 & 9 \\ 0 & 1 & -2 & 2 \\ 0 & 0 & 0 & 0 \end{array}\right]$

12. $\left[\begin{array}{ccc:c} 1 & 2 & 3 & 9 \\ 0 & 1 & -2 & 2 \\ 0 & 0 & 0 & 8 \end{array}\right]$

13. $\left[\begin{array}{ccc:c} 1 & 2 & 3 & 9 \\ 0 & 1 & -2 & 2 \\ 0 & 0 & 1 & -3 \end{array}\right]$

14. $\left[\begin{array}{ccccc:c} 1 & 2 & 3 & 10 & 6 & 0 \\ 0 & 1 & -5 & -2 & 0 & 5 \\ 0 & 0 & 1 & 12 & 0 & -2 \\ 0 & 0 & 0 & 1 & 1 & 0 \end{array}\right]$

In Exercises 15–24, use Gaussian elimination with back-substitution to solve the system of equations.

15. $\begin{cases} 5x + 4y = 2 \\ -x + y = -22 \end{cases}$

16. $\begin{cases} 2x - 5y = 2 \\ 3x - 7y = 1 \end{cases}$

17. $\begin{cases} 0.3x - 0.1y = -0.13 \\ 0.2x - 0.3y = -0.25 \end{cases}$

18. $\begin{cases} 0.2x - 0.1y = 0.07 \\ 0.4x - 0.5y = -0.01 \end{cases}$

19. $\begin{cases} 2x + 3y + z = 10 \\ 2x - 3y - 3z = 22 \\ 4x - 2y + 3z = -2 \end{cases}$

20. $\begin{cases} 2x + 3y + 3z = 3 \\ 6x + 6y + 12z = 13 \\ 12x + 9y - z = 2 \end{cases}$

21. $\begin{cases} 2x + y + 2z = 4 \\ 2x + 2y = 5 \\ 2x - y + 6z = 2 \end{cases}$

22. $\begin{cases} x + 2y + 6z = 1 \\ 2x + 5y + 15z = 4 \\ 3x + y + 3z = -6 \end{cases}$

23. $\begin{cases} 2x + y + z = 6 \\ -2y + 3z - w = 9 \\ 3x + 3y - 2z - 2w = -11 \\ x + z + 3w = 14 \end{cases}$

24. $\begin{cases} x + 2y + w = 3 \\ -3y + 3z = 0 \\ 4x + 4y + z + 2w = 0 \\ 2x + z = 3 \end{cases}$

Partial Fractions **In Exercises 25 and 26, write the partial fraction decomposition for the rational expression.**

25. $\dfrac{x + 9}{(x + 1)(x + 2)^2} = \dfrac{A}{x + 1} + \dfrac{B}{x + 2} + \dfrac{C}{(x + 2)^2}$

26. $\dfrac{3x - 2}{(x - 4)(x - 3)^2} = \dfrac{A}{x - 4} + \dfrac{B}{x - 3} + \dfrac{C}{(x - 3)^2}$

In Exercises 27–30, use Gauss-Jordan elimination to solve the system of equations.

27. $\begin{cases} -x + y + 2z = 1 \\ 2x + 3y + z = -2 \\ 5x + 4y + 2z = 4 \end{cases}$

28. $\begin{cases} 4x + 4y + 4z = 5 \\ 4x - 2y - 8z = 1 \\ 5x + 3y + 8z = 6 \end{cases}$

29. $\begin{cases} 2x - y + 9z = -8 \\ -x - 3y + 4z = -15 \\ 5x + 2y - z = 17 \end{cases}$

30. $\begin{cases} -3x + y + 7z = -20 \\ 5x - 2y - z = 34 \\ -x + y + 4z = -8 \end{cases}$

In Exercises 31 and 32, use the matrix capabilities of a graphing utility to reduce the augmented matrix corresponding to the system of equations, and solve the system.

31. $\begin{cases} 3x - y + 5z - 2w = -44 \\ x + 6y + 4z - w = 1 \\ 5x - y + z + 3w = -15 \\ 4y - z - 8w = 58 \end{cases}$

32. $\begin{cases} 4x + 12y + 2z = 20 \\ x + 6y + 4z = 12 \\ x + 6y + z = 8 \\ -2x - 10y - 2z = -10 \end{cases}$

8.2 In Exercises 33–36, find *x* and *y*.

33. $\begin{bmatrix} -1 & x \\ y & 9 \end{bmatrix} = \begin{bmatrix} -1 & 12 \\ -7 & 9 \end{bmatrix}$

34. $\begin{bmatrix} -1 & 0 \\ x & 5 \\ -4 & y \end{bmatrix} = \begin{bmatrix} -1 & 0 \\ 8 & 5 \\ -4 & 0 \end{bmatrix}$

35. $\begin{bmatrix} x+3 & 4 & -4y \\ 0 & -3 & 2 \\ -2 & y+5 & 6x \end{bmatrix} = \begin{bmatrix} 5x-1 & 4 & -44 \\ 0 & -3 & 2 \\ -2 & 16 & 6 \end{bmatrix}$

36. $\begin{bmatrix} -9 & 4 & 2 & -5 \\ 0 & -3 & 7 & -4 \\ 6 & -1 & 1 & 0 \end{bmatrix} = \begin{bmatrix} -9 & 4 & x-10 & -5 \\ 0 & -3 & 7 & 2y \\ \frac{1}{2}x & -1 & 1 & 0 \end{bmatrix}$

In Exercises 37–40, determine if the matrix operation $A + 3B$ can be performed. If not, state why.

37. $A = \begin{bmatrix} 2 & -2 \\ 3 & 5 \end{bmatrix}, \quad B = \begin{bmatrix} -3 & 10 \\ 12 & 8 \end{bmatrix}$

38. $A = \begin{bmatrix} 5 & 4 \\ -7 & 2 \\ 11 & 2 \end{bmatrix}, \quad B = \begin{bmatrix} 4 & 12 \\ 20 & 40 \\ 15 & 30 \end{bmatrix}$

39. $A = \begin{bmatrix} 5 & 4 \\ -7 & 2 \\ 11 & 2 \end{bmatrix}, \quad B = \begin{bmatrix} 4 & 12 \\ 20 & 40 \end{bmatrix}$

40. $A = \begin{bmatrix} 6 & -5 & 7 \end{bmatrix}, \quad B = \begin{bmatrix} -1 \\ 4 \\ 8 \end{bmatrix}$

In Exercises 41–44, perform the matrix operations. If it is not possible, explain why.

41. $\begin{bmatrix} 7 & 3 \\ -1 & 5 \end{bmatrix} + \begin{bmatrix} 10 & -20 \\ 14 & -3 \end{bmatrix}$

42. $\begin{bmatrix} -11 & 16 & 19 \\ -7 & -2 & 1 \end{bmatrix} - \begin{bmatrix} 6 & 0 \\ 8 & -4 \\ -2 & 10 \end{bmatrix}$

43. $-2\begin{bmatrix} 1 & 2 \\ 5 & -4 \\ 6 & 0 \end{bmatrix} + 8\begin{bmatrix} 7 & 1 \\ 1 & 2 \\ 1 & 4 \end{bmatrix}$

44. $-\begin{bmatrix} 8 & -1 & 8 \\ -2 & 4 & 12 \\ 0 & -6 & 0 \end{bmatrix} - 5\begin{bmatrix} -2 & 0 & -4 \\ 3 & -1 & 1 \\ 6 & 12 & -8 \end{bmatrix}$

In Exercises 45 and 46, use a graphing utility to perform the matrix operations.

45. $3\begin{bmatrix} 8 & -2 & 5 \\ 1 & 3 & -1 \end{bmatrix} + 6\begin{bmatrix} 4 & -2 & -3 \\ 2 & 7 & 6 \end{bmatrix}$

46. $-5\begin{bmatrix} 2 & 0 \\ 7 & -2 \\ 8 & 2 \end{bmatrix} + 4\begin{bmatrix} 4 & -2 \\ 6 & 11 \\ -1 & 3 \end{bmatrix}$

In Exercises 47–50, solve for *X* when

$$A = \begin{bmatrix} -4 & 0 \\ 1 & -5 \\ -3 & 2 \end{bmatrix} \quad \text{and} \quad B = \begin{bmatrix} 1 & 2 \\ -2 & 1 \\ 4 & 4 \end{bmatrix}.$$

47. $X = 3A - 2B$

48. $6X = 4A + 3B$

49. $3X + 2A = B$

50. $2A - 5B = 3X$

In Exercises 51–54, determine if the matrix operation *AB* can be performed. If not, state why.

51. $A = \begin{bmatrix} 2 & -2 \\ 3 & 5 \end{bmatrix}, \quad B = \begin{bmatrix} -3 & 10 \\ 12 & 8 \end{bmatrix}$

52. $A = \begin{bmatrix} 5 & 4 \\ -7 & 2 \\ 11 & 2 \end{bmatrix}, \quad B = \begin{bmatrix} 4 & 12 \\ 20 & 40 \\ 15 & 30 \end{bmatrix}$

53. $A = \begin{bmatrix} 5 & 4 \\ -7 & 2 \\ 11 & 2 \end{bmatrix}, \quad B = \begin{bmatrix} 4 & 12 \\ 20 & 40 \end{bmatrix}$

54. $A = [6 \quad -5 \quad 7]$, $B = \begin{bmatrix} -1 \\ 4 \\ 8 \end{bmatrix}$

In Exercises 55–62, perform the matrix operations. If it is not possible, explain why.

55. $\begin{bmatrix} 1 & 2 \\ 5 & -4 \\ 6 & 0 \end{bmatrix}\begin{bmatrix} 6 & -2 & 8 \\ 4 & 0 & 0 \end{bmatrix}$

56. $\begin{bmatrix} 1 & 5 & 6 \\ 2 & -4 & 0 \end{bmatrix}\begin{bmatrix} 6 & -2 & 8 \\ 4 & 0 & 0 \end{bmatrix}$

57. $\begin{bmatrix} 1 & 5 & 6 \\ 2 & -4 & 0 \end{bmatrix}\begin{bmatrix} 6 & 4 \\ -2 & 0 \\ 8 & 0 \end{bmatrix}$

58. $\begin{bmatrix} 1 & 3 & 2 \\ 0 & 2 & -4 \\ 0 & 0 & 3 \end{bmatrix}\begin{bmatrix} 4 & -3 & 2 \\ 0 & 3 & -1 \\ 0 & 0 & 2 \end{bmatrix}$

59. $\begin{bmatrix} 4 \\ 6 \end{bmatrix}[6 \quad -2]$

60. $[4 \quad -2 \quad 6]\begin{bmatrix} -2 & 1 \\ 0 & -3 \\ 2 & 0 \end{bmatrix}$

61. $\begin{bmatrix} 2 & 1 \\ 6 & 0 \end{bmatrix}\left(\begin{bmatrix} 4 & 2 \\ -3 & 1 \end{bmatrix} + \begin{bmatrix} -2 & 4 \\ 0 & 4 \end{bmatrix}\right)$

62. $-3\begin{bmatrix} 1 & -1 \\ 4 & 2 \end{bmatrix}\left(\begin{bmatrix} 0 & 3 \\ 1 & 2 \end{bmatrix}\begin{bmatrix} 1 & 0 \\ 5 & -3 \end{bmatrix}\right)$

In Exercises 63 and 64, use a graphing utility to perform the matrix operations.

63. $\begin{bmatrix} 4 & 1 \\ 11 & -7 \\ 12 & 3 \end{bmatrix}\begin{bmatrix} 3 & -5 & 6 \\ 2 & -2 & -2 \end{bmatrix}$

64. $\begin{bmatrix} -2 & 3 & 10 \\ 4 & -2 & 2 \end{bmatrix}\begin{bmatrix} 1 & 1 \\ -5 & 2 \\ 3 & 2 \end{bmatrix}$

65. Write the system of linear equations represented by the matrix equation

$$\begin{bmatrix} 5 & 4 \\ -1 & 1 \end{bmatrix}\begin{bmatrix} x \\ y \end{bmatrix} = \begin{bmatrix} 2 \\ -22 \end{bmatrix}.$$

66. Write the matrix equation $AX = B$ for the system of linear equations.

$$\begin{cases} 2x + 3y + z = 10 \\ 2x - 3y - 3z = 22 \\ 4x - 2y + 3z = -2 \end{cases}$$

67. *Manufacturing* A manufacturing company produces three models of a product that are shipped to two warehouses. The number of units of model i that are shipped to warehouse j is represented by a_{ij} in the matrix

$$A = \begin{bmatrix} 8200 & 7400 \\ 6500 & 9800 \\ 5400 & 4800 \end{bmatrix}.$$

The price per unit is represented by the matrix $B = [\$10.25 \quad \$14.50 \quad \$17.75]$.

(a) Use a graphing utility to compute BA and interpret the result.

(b) Suppose the numbers of units of each model shipped to the warehouses are increased by 25% for the following shipment. Use your graphing utility to find a new matrix A_n representing the number of units being shipped to the warehouses. Then calculate BA_n.

8.3 **In Exercises 68–71, show that *B* is the inverse of *A*.**

68. $A = \begin{bmatrix} -4 & -1 \\ 7 & 2 \end{bmatrix}$, $B = \begin{bmatrix} -2 & -1 \\ 7 & 4 \end{bmatrix}$

69. $A = \begin{bmatrix} 5 & -1 \\ 11 & -2 \end{bmatrix}$, $B = \begin{bmatrix} -2 & 1 \\ -11 & 5 \end{bmatrix}$

70. $A = \begin{bmatrix} 1 & 1 & 0 \\ 1 & 0 & 1 \\ 6 & 2 & 3 \end{bmatrix}$, $B = \begin{bmatrix} -2 & -3 & 1 \\ 3 & 3 & -1 \\ 2 & 4 & -1 \end{bmatrix}$

71. $A = \begin{bmatrix} 1 & -1 & 0 \\ -1 & 0 & -1 \\ 8 & -4 & 2 \end{bmatrix}$, $B = \begin{bmatrix} -2 & 1 & \frac{1}{2} \\ -3 & 1 & \frac{1}{2} \\ 2 & -2 & -\frac{1}{2} \end{bmatrix}$

In Exercises 72–75, use Gauss-Jordan elimination to find the inverse of the matrix (if it exists).

72. $\begin{bmatrix} -6 & 5 \\ -5 & 4 \end{bmatrix}$

73. $\begin{bmatrix} -3 & -5 \\ 2 & 3 \end{bmatrix}$

74. $\begin{bmatrix} -1 & -2 & -2 \\ 3 & 7 & 9 \\ 1 & 4 & 7 \end{bmatrix}$

75. $\begin{bmatrix} 0 & -2 & 1 \\ -5 & -2 & -3 \\ 7 & 3 & 4 \end{bmatrix}$

In Exercises 76–79, use a graphing utility to find the inverse of the matrix (if it exists).

76. $\begin{bmatrix} 2 & 6 \\ 3 & -6 \end{bmatrix}$

77. $\begin{bmatrix} 3 & -10 \\ 4 & 2 \end{bmatrix}$

78. $\begin{bmatrix} 2 & 0 & 3 \\ -1 & 1 & 1 \\ 2 & -2 & 1 \end{bmatrix}$

79. $\begin{bmatrix} 1 & 4 & 6 \\ 2 & -3 & 1 \\ -1 & 18 & 16 \end{bmatrix}$

In Exercises 80–83, let

$$A = \begin{bmatrix} a & b \\ c & d \end{bmatrix}.$$

Find the inverse of the above matrix

$$A^{-1} = \frac{1}{ad - bc}\begin{bmatrix} d & -b \\ -c & a \end{bmatrix}$$

(if it exists).

80. $\begin{bmatrix} -7 & 2 \\ -8 & 2 \end{bmatrix}$ **81.** $\begin{bmatrix} 10 & 4 \\ 7 & 3 \end{bmatrix}$

82. $\begin{bmatrix} -\frac{1}{2} & 20 \\ \frac{3}{10} & -6 \end{bmatrix}$ **83.** $\begin{bmatrix} -\frac{3}{4} & \frac{5}{2} \\ -\frac{4}{5} & -\frac{8}{3} \end{bmatrix}$

In Exercises 84–91, use an inverse matrix to solve (if possible) the system of linear equations.

84. $\begin{cases} -x + 4y = 8 \\ 2x - 7y = -5 \end{cases}$ **85.** $\begin{cases} 5x - y = 13 \\ -9x + 2y = -24 \end{cases}$

86. $\begin{cases} -3x + 10y = 8 \\ 5x - 17y = -13 \end{cases}$ **87.** $\begin{cases} 4x - 2y = -10 \\ -19x + 9y = 47 \end{cases}$

88. $\begin{cases} 3x + 2y - z = 6 \\ x - y + 2z = -1 \\ 5x + y + z = 7 \end{cases}$

89. $\begin{cases} -x + 4y - 2z = 12 \\ 2x - 9y + 5z = -25 \\ -x + 5y - 4z = 10 \end{cases}$

90. $\begin{cases} -2x + y + 2z = -13 \\ -x - 4y + z = -11 \\ -y - z = 0 \end{cases}$

91. $\begin{cases} 3x - y + 5z = -14 \\ -x + y + 6z = 8 \\ -8x + 4y - z = 44 \end{cases}$

In Exercises 92–95, use a graphing utility to solve (if possible) the system of linear equations using the inverse of the coefficient matrix.

92. $\begin{cases} x + 2y = -1 \\ 3x + 4y = -5 \end{cases}$ **93.** $\begin{cases} x + 3y = 23 \\ -6x + 2y = -18 \end{cases}$

94. $\begin{cases} -3x - 3y - 4z = 2 \\ y + z = -1 \\ 4x + 3y + 4z = -1 \end{cases}$

95. $\begin{cases} x - 3y - 2z = 8 \\ -2x + 7y + 3z = -19 \\ x - y - 3z = 3 \end{cases}$

8.4 In Exercises 96–99, find the determinant of the matrix.

96. $\begin{bmatrix} 8 & 5 \\ 2 & -4 \end{bmatrix}$ **97.** $\begin{bmatrix} -9 & 11 \\ 7 & -4 \end{bmatrix}$

98. $\begin{bmatrix} 50 & -30 \\ 10 & 5 \end{bmatrix}$ **99.** $\begin{bmatrix} 14 & -24 \\ 12 & -15 \end{bmatrix}$

In Exercises 100–103, find all (a) minors and (b) cofactors of the matrix.

100. $\begin{bmatrix} 2 & -1 \\ 7 & 4 \end{bmatrix}$ **101.** $\begin{bmatrix} 3 & 6 \\ 5 & -4 \end{bmatrix}$

102. $\begin{bmatrix} 3 & 2 & -1 \\ -2 & 5 & 0 \\ 1 & 8 & 6 \end{bmatrix}$ **103.** $\begin{bmatrix} 8 & 3 & 4 \\ 6 & 5 & -9 \\ -4 & 1 & 2 \end{bmatrix}$

In Exercises 104–107, find the determinant of the matrix. Expand by cofactors on the row or column that appears to make the computations easiest.

104. $\begin{bmatrix} -2 & 4 & 1 \\ -6 & 0 & 2 \\ 5 & 3 & 4 \end{bmatrix}$ **105.** $\begin{bmatrix} 4 & 7 & -1 \\ 2 & -3 & 4 \\ -5 & 1 & -1 \end{bmatrix}$

106. $\begin{bmatrix} 3 & 0 & -4 & 0 \\ 0 & 8 & 1 & 2 \\ 6 & 1 & 8 & 2 \\ 0 & 3 & -4 & 1 \end{bmatrix}$ **107.** $\begin{bmatrix} -5 & 6 & 0 & 0 \\ 0 & 1 & -1 & 2 \\ -3 & 4 & -5 & 1 \\ 1 & 6 & 0 & 3 \end{bmatrix}$

8.5 In Exercises 108–111, use Cramer's Rule to solve (if possible) the system of equations.

108. $\begin{cases} 5x - 2y = 6 \\ -11x + 3y = -23 \end{cases}$

109. $\begin{cases} 3x + 8y = -7 \\ 9x - 5y = 37 \end{cases}$

110. $\begin{cases} -2x + 3y - 5z = -11 \\ 4x - y + z = -3 \\ -x - 4y + 6z = 15 \end{cases}$

111. $\begin{cases} 5x - 2y + z = 15 \\ 3x - 3y - z = -7 \\ 2x - y - 7z = -3 \end{cases}$

In Exercises 112–115, use Cramer's Rule to solve.

112. *Mixture Problem* A florist wants to arrange a dozen flowers consisting of two varieties: carnations and roses. Carnations cost \$1.50 each and roses cost \$3.50 each. How many of each should the florist use so that the arrangement will cost \$30.00?

113. *Mixture Problem* One hundred liters of a 60% acid solution is obtained by mixing a 75% solution with a 50% solution. How many liters of each must be used to obtain the desired mixture?

114. *Fitting a Parabola to Three Points* Find an equation of the parabola $y = ax^2 + bx + c$ that passes through the points $(-1, 2)$, $(0, 3)$, and $(1, 6)$.

115. *Break-Even Point* A small business invests \$25,000 in equipment to produce a product. Each unit of the product costs \$3.75 to produce and is sold for \$5.25. How many items must be sold before the business breaks even?

In Exercises 116–119, use a determinant to find the area of the triangle with the given vertices.

116. $(1, 0), (5, 0), (5, 8)$ **117.** $(-4, 0), (4, 0), (0, 6)$

118. $\left(\frac{1}{2}, 1\right), \left(2, -\frac{5}{2}\right), \left(\frac{3}{2}, 1\right)$ **119.** $\left(\frac{3}{2}, 1\right), \left(4, -\frac{1}{2}\right), (4, 2)$

In Exercises 120–123, use a determinant to find an equation of the line through the points.

120. $(-4, 0), (4, 4)$ **121.** $(2, 5), (6, -1)$

122. $\left(-\frac{5}{2}, 3\right), \left(\frac{7}{2}, 1\right)$ **123.** $(-0.8, 0.2), (0.7, 3.2)$

In Exercises 124 and 125, find the uncoded 1 × 3 row matrices for the message. Then encode the message using the matrix.

	Message	*Matrix*
124.	LOOK OUT BELOW	$\begin{bmatrix} 2 & -2 & 0 \\ 3 & 0 & -3 \\ -6 & 2 & 3 \end{bmatrix}$
125.	RETURN TO BASE	$\begin{bmatrix} 2 & 1 & 0 \\ -6 & -6 & -2 \\ 3 & 2 & 1 \end{bmatrix}$

In Exercises 126 and 127, decode the cryptogram by using the inverse of the matrix

$$A = \begin{bmatrix} -5 & 4 & -3 \\ 10 & -7 & 6 \\ 8 & -6 & 5 \end{bmatrix}.$$

126. -5 11 -2 370 -265 225 -57 48 -33 32 -15 20 245 -171 147

127. 145 -105 92 264 -188 160 23 -16 15 129 -84 78 -9 8 -5 159 -118 100 219 -152 133 370 -265 225 -105 84 -63

Synthesis

True or False? **In Exercises 128 and 129, determine whether the statement is true or false. Justify your answer.**

128. It is possible to find the determinant of a 4×5 matrix.

129. $$\begin{vmatrix} a_{11} & a_{12} & a_{13} \\ a_{21} & a_{22} & a_{23} \\ a_{31} + c_1 & a_{32} + c_2 & a_{33} + c_3 \end{vmatrix} = \begin{vmatrix} a_{11} & a_{12} & a_{13} \\ a_{21} & a_{22} & a_{23} \\ a_{31} & a_{32} & a_{33} \end{vmatrix} + \begin{vmatrix} a_{11} & a_{12} & a_{13} \\ a_{21} & a_{22} & a_{23} \\ c_1 & c_2 & c_3 \end{vmatrix}$$

130. Under what conditions does a matrix have an inverse?

131. What is meant by the cofactor of an entry of a matrix? How are cofactors used to find the determinant of the matrix?

132. Three people were asked to solve a system of equations using an augmented matrix. Each person reduced the matrix to row-echelon form. The reduced matrices were

$$\left[\begin{array}{cc:c} 1 & 2 & 3 \\ 0 & 1 & 1 \end{array}\right]$$

$$\left[\begin{array}{cc:c} 1 & 0 & 1 \\ 0 & 1 & 1 \end{array}\right]$$

and

$$\left[\begin{array}{cc:c} 1 & 2 & 3 \\ 0 & 0 & 0 \end{array}\right].$$

Can all three be right? Explain.

133. *Think About It* Describe the row-echelon form of an augmented matrix that corresponds to a system of linear equations that has a unique solution.

134. Solve the equation $\begin{vmatrix} 2 - \lambda & 5 \\ 3 & -8 - \lambda \end{vmatrix} = 0.$

Chapter Project ▶ Solving Systems of Equations

Matrices have always been a powerful mathematical tool—especially for dealing with problems that involve a lot of data. With technology available to perform the calculations, matrices have also become a practical mathematical tool.

Example ▶ Solving a Metallurgy Problem

	Alloy X	Alloy Y	Alloy Z
Carbon	1%	1%	4%
Chromium	0%	15%	3%
Iron	99%	84%	93%

Three iron alloys contain different percents of carbon, chromium, and iron. Alloy X is a type of wrought iron, alloy Y is a type of stainless steel, and alloy Z is a type of cast iron. How much of each of the three alloys can you make with 15 tons of carbon, 39 tons of chromium, and 546 tons of iron?

Solution

Let x, y, and z represent the amounts of the three iron alloys. You can model the situation with the linear system

$$\begin{cases} 0.01x + 0.01y + 0.04z = 15 & \text{Carbon} \\ 0.15y + 0.03z = 39. & \text{Chromium} \\ 0.99x + 0.84y + 0.93z = 546 & \text{Iron} \end{cases}$$

The matrix equation $AX = B$ that represents this system is

$$\begin{bmatrix} 0.01 & 0.01 & 0.04 \\ 0 & 0.15 & 0.03 \\ 0.99 & 0.84 & 0.93 \end{bmatrix} \begin{bmatrix} x \\ y \\ z \end{bmatrix} = \begin{bmatrix} 15 \\ 39 \\ 546 \end{bmatrix}.$$

With a graphing utility or computer, you can solve the equation as follows.

$$X = A^{-1}B = \begin{bmatrix} -25.4 & -5.4 & 1.267 \\ -6.6 & 6.733 & 0.067 \\ 33 & -0.333 & -0.333 \end{bmatrix} \begin{bmatrix} 15 \\ 39 \\ 546 \end{bmatrix} = \begin{bmatrix} 100 \\ 200 \\ 300 \end{bmatrix}$$

So, you can make 100 tons of alloy X, 200 tons of alloy Y, and 300 tons of alloy Z.

Chapter Project Investigation

Three different gold alloys contain the percents of gold, copper, and silver shown in the matrix. You have 20,144 grams of gold, 766 grams of copper, and 1990 grams of silver. How much of each alloy can you make?

Percent by Weight

	Alloy X	Alloy Y	Alloy Z
Gold	94%	92%	80%
Copper	4%	2%	4%
Silver	2%	6%	16%

Chapter Test

The *Interactive* CD-ROM and *Internet* versions of this text provide answers to the Chapter Tests and Cumulative Tests. They also offer Chapter Pre-Tests (which test key skills and concepts covered in previous chapters) and Chapter Post-Tests, both of which have randomly generated exercises with diagnostic capabilities.

Take this test as you would take a test in class. After you are done, check your work against the answers given in the back of the book.

In Exercises 1 and 2, write the matrix in reduced row-echelon form.

1. $\begin{bmatrix} 1 & -1 & 5 \\ 6 & 2 & 3 \\ 5 & 3 & -3 \end{bmatrix}$ **2.** $\begin{bmatrix} 1 & 0 & -1 & 2 \\ -1 & 1 & 1 & -3 \\ 1 & 1 & -1 & 1 \\ 3 & 2 & -3 & 4 \end{bmatrix}$

3. Write the augmented matrix corresponding to the system of equations, and solve the system $\begin{cases} 4x + 3y - 2z = 14 \\ -x - y + 2z = -5. \\ 3x + y - 4z = 8 \end{cases}$

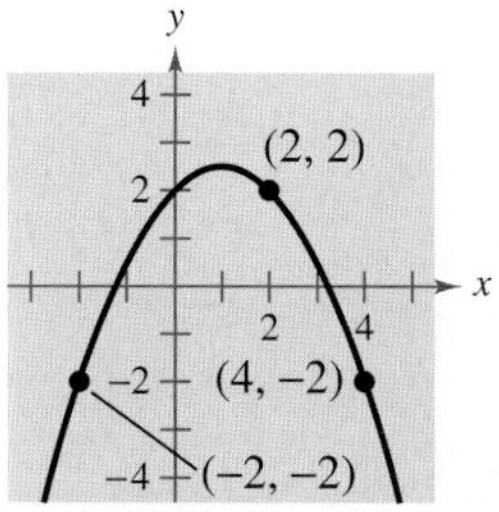

FIGURE FOR 4

4. Find the equation of the parabola $y = ax^2 + bx + c$ that passes through the points in the figure.

5. Find (a) $A - B$, (b) $3A$, (c) $3A - 2B$, and (d) AB (if possible).

$$A = \begin{bmatrix} 5 & 4 \\ -4 & -4 \end{bmatrix}, \quad B = \begin{bmatrix} 4 & -1 \\ -4 & 0 \end{bmatrix}$$

In Exercises 6 and 7, find the inverse of the matrix (if it exists).

6. $\begin{bmatrix} -6 & 4 \\ 10 & -5 \end{bmatrix}$ **7.** $\begin{bmatrix} -2 & 4 & -6 \\ 2 & 1 & 0 \\ 4 & -2 & 5 \end{bmatrix}$

8. Use the result of Exercise 6 to solve the system.

$$\begin{cases} -6x + 4y = 10 \\ 10x - 5y = 20 \end{cases}$$

In Exercises 9 and 10, evaluate the determinant of the matrix.

9. $\begin{bmatrix} -9 & 4 \\ 13 & 16 \end{bmatrix}$ **10.** $\begin{bmatrix} \frac{5}{2} & \frac{13}{4} \\ -8 & \frac{6}{5} \end{bmatrix}$

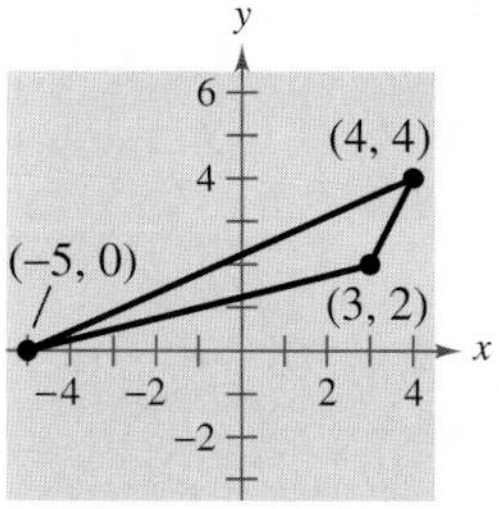

FIGURE FOR 13

In Exercises 11 and 12, use Cramer's Rule to solve (if possible) the system of equations.

11. $\begin{cases} 7x + 6y = 9 \\ -2x - 11y = -49 \end{cases}$ **12.** $\begin{cases} 6x - y + 2z = -4 \\ -2x + 3y - z = 10 \\ 4x - 4y + z = -18 \end{cases}$

13. Use a determinant to find the area of the triangle in the figure.

$$A = \begin{bmatrix} 1 & -1 & 0 \\ 1 & 0 & -1 \\ 6 & -2 & -3 \end{bmatrix}$$

MATRIX FOR 14

14. Find the uncoded 1×3 row matrices for the message KNOCK ON WOOD. Then encode the message using the matrix A at the left.

15. One hundred liters of a 50% solution is obtained by mixing a 60% solution with a 20% solution. How many liters of each must be used to obtain the desired mixture?

Siegfried Layda/Tony Stone Images

In 1997, 831.8 million trips were taken in the United States for pleasure or vacation. The average distance traveled per trip was 901 miles. (Source: Travel Industry Association of America)

9 Sequences, Series, and Probability

▶ How to Study This Chapter

The Big Picture

In this chapter you will learn the following skills and concepts.

▶ How to use sequence, factorial, and summation notation to write the terms and sum of a sequence

▶ How to recognize, write, and manipulate arithmetic sequences and geometric sequences

▶ How to use mathematical induction to prove a statement involving a positive integer n

▶ How to use the Binomial Theorem and Pascal's Triangle to calculate binomial coefficients and binomial expansions

▶ How to solve counting problems using the Fundamental Counting Principle, permutations, and combinations

▶ How to find the probabilities of events and their complements

Important Vocabulary

As you encounter each new vocabulary term in this chapter, add the term and its definition to your notebook glossary.

Infinite sequence (p. 688)
Terms of a sequence (p. 688)
Finite sequence (p. 688)
Recursive (p. 690)
Factorial (p. 690)
Summation or sigma notation (p. 692)
Infinite series (p. 693)
Finite series (p. 693)
nth partial sum (p. 693)
Arithmetic sequence (p. 699)
Common difference (p. 699)
Geometric sequence (p. 708)
Common ratio (p. 708)
Infinite geometric series (p. 712)
Geometric series (p. 712)
Mathematical induction (p. 718)
First differences (p. 725)
Second differences (p. 725)
Binomial coefficients (p. 728)
Binomial Theorem (p. 728)
Pascal's Triangle (p. 730)
Fundamental Counting Principle (p. 737)
Permutation (p. 738)
Distinguishable permutations (p. 740)
Combination (p. 741)
Experiment (p. 746)
Outcomes (p. 746)
Sample space (p. 746)
Event (p. 746)
Probability (p. 747)
Mutually exclusive (p. 750)
Independent events (p. 752)
Complement of an event (p. 753)

Study Tools

- Learning objectives at the beginning of each section
- Chapter Summary (p. 759)
- Review Exercises (pp. 760–763)
- Chapter Test (p. 765)
- Cumulative Test for Chapters 7–9 (pp. 766–767)

Additional Resources

- Study and Solutions Guide
- Interactive Precalculus
- Videotapes for Chapter 9
- Precalculus Website
- Student Success Organizer

STUDY T!P

Try working with a partner on assignments. You may find that teaching others is an excellent way to learn.

9.1 Sequences and Series

▶ **What you should learn**

- How to use sequence notation to write the terms of a sequence
- How to use factorial notation
- How to use summation notation to write sums
- How to find the sum of an infinite series
- How to use sequences and series to model and solve real-life problems

▶ **Why you should learn it**

Sequences and series can be used to model real-life problems. For instance, Exercise 109 on page 697 uses a sequence to model the net income of Wal-Mart from 1990 through 1998.

John Mailer, Jr./The Image Works

Sequences

In mathematics, the word *sequence* is used in much the same way as in ordinary English. Saying that a collection is listed *in sequence* means that it is ordered so that it has a first member, a second member, a third member, and so on.

Mathematically, you can think of a sequence as a *function* whose domain is the set of positive integers.

$$f(1) = a_1,\ f(2) = a_2,\ f(3) = a_3,\ f(4) = a_4, \ldots, f(n) = a_n, \ldots$$

Rather than using function notation, however, sequences are usually written using subscript notation, as indicated in the following definition.

Definition of a Sequence

An **infinite sequence** is a function whose domain is the set of positive integers. The function values

$$a_1, a_2, a_3, a_4, \ldots, a_n, \ldots$$

are the **terms** of the sequence. If the domain of the function consists of the first n positive integers only, the sequence is a **finite sequence.**

On occasion it is convenient to begin subscripting a sequence with 0 instead of 1 so that the terms of the sequence become

$$a_0, a_1, a_2, a_3, \ldots.$$

Example 1 ▶ Finding Terms of a Sequence

a. The first four terms of the sequence given by $a_n = 3n - 2$ are

$a_1 = 3(1) - 2 = 1$ 1st term

$a_2 = 3(2) - 2 = 4$ 2nd term

$a_3 = 3(3) - 2 = 7$ 3rd term

$a_4 = 3(4) - 2 = 10.$ 4th term

b. The first four terms of the sequence given by $a_n = 3 + (-1)^n$ are

$a_1 = 3 + (-1)^1 = 3 - 1 = 2$ 1st term

$a_2 = 3 + (-1)^2 = 3 + 1 = 4$ 2nd term

$a_3 = 3 + (-1)^3 = 3 - 1 = 2$ 3rd term

$a_4 = 3 + (-1)^4 = 3 + 1 = 4.$ 4th term

Exploration

Write out the first five terms of the sequence whose nth term is

$$a_n = \frac{(-1)^{n+1}}{2n-1}.$$

Are they the same as the first five terms of the sequence in Example 2? If not, how do they differ?

Additional Example

Write an expression for the apparent nth term a_n of the sequence

$\frac{2}{1}, \frac{3}{2}, \frac{4}{3}, \frac{5}{4}, \ldots$

n: 1 2 3 4 . . . n

Terms: $\frac{2}{1}$ $\frac{3}{2}$ $\frac{4}{3}$ $\frac{5}{4}$ . . . a_n

Apparent pattern: Each term has a numerator that is 1 greater than its denominator, which implies that

$$a_n = \frac{n+1}{n}.$$

The *Interactive* CD-ROM and *Internet* versions of this text offer a built-in graphing calculator, which can be used in the Examples, Explorations, Technology notes, and Exercises.

Technology

To graph a sequence using a graphing utility, set the mode to *seq* and *dot* and enter the sequence. The graph of the sequence in Example 3(a) is shown below.

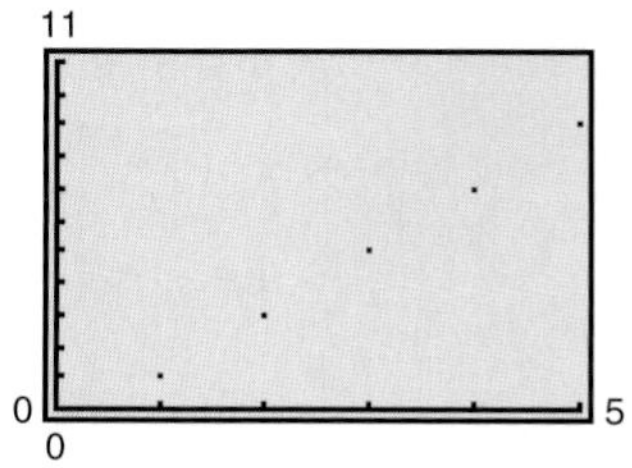

You can use the trace key to identify its terms.

Example 2 ▶ Finding Terms of a Sequence

The first five terms of the sequence given by

$$a_n = \frac{(-1)^n}{2n-1}$$

are as follows.

$$a_1 = \frac{(-1)^1}{2(1)-1} = \frac{-1}{2-1} = -1 \qquad \text{1st term}$$

$$a_2 = \frac{(-1)^2}{2(2)-1} = \frac{1}{4-1} = \frac{1}{3} \qquad \text{2nd term}$$

$$a_3 = \frac{(-1)^3}{2(3)-1} = \frac{-1}{6-1} = -\frac{1}{5} \qquad \text{3rd term}$$

$$a_4 = \frac{(-1)^4}{2(4)-1} = \frac{1}{8-1} = \frac{1}{7} \qquad \text{4th term}$$

$$a_5 = \frac{(-1)^5}{2(5)-1} = \frac{-1}{10-1} = -\frac{1}{9} \qquad \text{5th term}$$

Simply listing the first few terms is not sufficient to define a unique sequence—the nth term *must be given.* To see this, consider the following sequences, both of which have the same first three terms.

$$\frac{1}{2}, \frac{1}{4}, \frac{1}{8}, \frac{1}{16}, \ldots, \frac{1}{2^n}, \ldots$$

$$\frac{1}{2}, \frac{1}{4}, \frac{1}{8}, \frac{1}{15}, \ldots, \frac{6}{(n+1)(n^2-n+6)}, \ldots$$

Example 3 ▶ Finding the nth Term of a Sequence

Write an expression for the apparent nth term (a_n) of each sequence.

a. 1, 3, 5, 7, . . . **b.** 2, 5, 10, 17, . . .

Solution

a. *n:* 1 2 3 4 . . . n

Terms: 1 3 5 7 . . . a_n

Apparent pattern: Each term is 1 less than twice n, which implies that

$$a_n = 2n - 1.$$

b. *n:* 1 2 3 4 . . . n

Terms: 2 5 10 17 . . . a_n

Apparent pattern: Each term is 1 more than the square of n, which implies that

$$a_n = n^2 + 1.$$

Some sequences are defined **recursively.** To define a sequence recursively, you need to be given one or more of the first few terms. All other terms of the sequence are then defined using previous terms. A well-known example is the Fibonacci sequence shown in Example 4.

Example 4 ▶ The Fibonacci Sequence: A Recursive Sequence

The Fibonacci sequence is defined recursively, as follows.

$$a_0 = 1,\ a_1 = 1,\ a_k = a_{k-2} + a_{k-1}, \text{ where } k \geq 2$$

Write the first six terms of this sequence.

Solution

$a_0 = 1$ — 0th term is given.

$a_1 = 1$ — 1st term is given.

$a_{2-2} + a_{2-1} = a_0 + a_1 = 1 + 1 = 2$ — Use recursive formula.

$a_{3-2} + a_{3-1} = a_1 + a_2 = 1 + 2 = 3$ — Use recursive formula.

$a_{4-2} + a_{4-1} = a2 + a_3 = 2 + 3 = 5$ — Use recursive formula.

$a_{5-2} + a_{5-1} = a_3 + a_4 = 3 + 5 = 8$ — Use recursive formula.

STUDY TIP

Any variable can be used as a subscript. The most commonly used variable subscripts in sequence and series notation are i, j, k, and n.

Factorial Notation

Some very important sequences in mathematics involve terms that are defined with special types of products called **factorials.**

Definition of Factorial

If n is a positive integer, ***n* factorial** is defined by

$$n! = 1 \cdot 2 \cdot 3 \cdot 4 \cdot \cdot \cdot (n - 1) \cdot n.$$

As a special case, zero factorial is defined as $0! = 1$.

Here are some values of $n!$ for the first several nonnegative integers. Notice that $0!$ is 1 by definition.

$$0! = 1$$

$$1! = 1$$

$$2! = 1 \cdot 2 = 2$$

$$3! = 1 \cdot 2 \cdot 3 = 6$$

$$4! = 1 \cdot 2 \cdot 3 \cdot 4 = 24$$

$$5! = 1 \cdot 2 \cdot 3 \cdot 4 \cdot 5 = 120$$

The value of n does not have to be very large before the value of $n!$ becomes huge. For instance, $10! = 3{,}628{,}800$.

Factorials follow the same conventions for order of operations as do exponents. For instance,

$$2n! = 2(n!)$$
$$= 2(1 \cdot 2 \cdot 3 \cdot 4 \cdot \cdot \cdot n)$$

whereas $(2n)! = 1 \cdot 2 \cdot 3 \cdot 4 \cdot \cdot \cdot 2n$.

Example 5 ▶ Finding Terms of a Sequence Involving Factorials

List the first five terms of the sequence given by

$$a_n = \frac{2^n}{n!}.$$

Begin with $n = 0$. Then plot the points on a set of coordinate axes.

Solution

$$a_0 = \frac{2^0}{0!} = \frac{1}{1} = 1 \qquad \text{0th term}$$

$$a_1 = \frac{2^1}{1!} = \frac{2}{1} = 2 \qquad \text{1st term}$$

$$a_2 = \frac{2^2}{2!} = \frac{4}{2} = 2 \qquad \text{2nd term}$$

$$a_3 = \frac{2^3}{3!} = \frac{8}{6} = \frac{4}{3} \qquad \text{3rd term}$$

$$a_4 = \frac{2^4}{4!} = \frac{16}{24} = \frac{2}{3} \qquad \text{4th term}$$

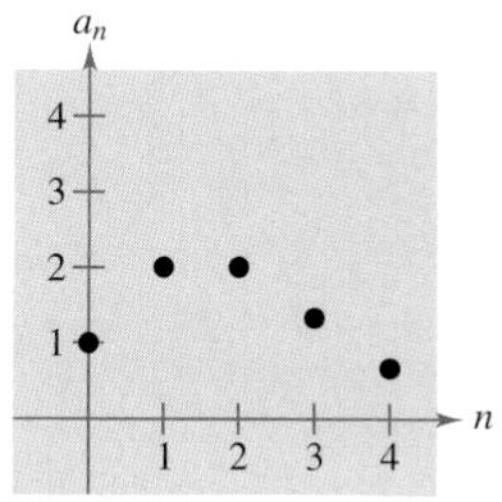

FIGURE 9.1

Figure 9.1 shows the first five terms of the sequence.

When working with fractions involving factorials, you will often find that the fractions can be reduced.

Additional Examples

a. $\frac{(2n+2)!}{(2n+4)!} = \frac{1}{(2n+4)(2n+3)}$

b. $\frac{2n!}{n!} = 2$

c. $\frac{(2n+1)!}{(2n)!} = 2n + 1$

You may want to point out to your students that $n! = n \cdot (n - 1)!$

Example 6 ▶ Evaluating Factorial Expressions

Evaluate each factorial expression.

a. $\frac{8!}{2! \cdot 6!}$ **b.** $\frac{2! \cdot 6!}{3! \cdot 5!}$ **c.** $\frac{n!}{(n-1)!}$

Solution

a. $\frac{8!}{2! \cdot 6!} = \frac{\cancel{1 \cdot 2 \cdot 3 \cdot 4 \cdot 5 \cdot 6} \cdot 7 \cdot 8}{1 \cdot 2 \cdot \cancel{1 \cdot 2 \cdot 3 \cdot 4 \cdot 5 \cdot 6}} = \frac{7 \cdot 8}{2} = 28$

b. $\frac{2! \cdot 6!}{3! \cdot 5!} = \frac{\cancel{1 \cdot 2} \cdot \cancel{1 \cdot 2 \cdot 3 \cdot 4 \cdot 5} \cdot 6}{\cancel{1 \cdot 2} \cdot 3 \cdot \cancel{1 \cdot 2 \cdot 3 \cdot 4 \cdot 5}} = \frac{6}{3} = 2$

c. $\frac{n!}{(n-1)!} = \frac{\cancel{1 \cdot 2 \cdot 3 \cdots (n-1)} \cdot n}{\cancel{1 \cdot 2 \cdot 3 \cdots (n-1)}} = n$

Technology

Most graphing utilities are able to sum the first n terms of a sequence. Check your user's guide for a "sum sequence" feature or a "series" feature.

Reading and writing the upper and lower limits of summation correctly will help with problems involving upper and lower limits in calculus.

The *Interactive* CD-ROM and *Internet* versions of this text show every example with its solution; clicking on the *Try It!* button brings up similar problems. Guided Examples and Integrated Examples show step-by-step solutions to additional examples. Integrated Examples are related to several concepts in the section.

Summation Notation

There is a convenient notation for the sum of the terms of a finite sequence. It is called **summation notation** or **sigma notation** because it involves the use of the uppercase Greek letter sigma, written as Σ.

Definition of Summation Notation

The sum of the first n terms of a sequence is represented by

$$\sum_{i=1}^{n} a_i = a_1 + a_2 + a_3 + a_4 + \cdot \cdot \cdot + a_n$$

where i is called the **index of summation,** n is the **upper limit of summation,** and 1 is the **lower limit of summation.**

Example 7 ▶ Summation Notation for Sums

Find each sum.

a. $\sum_{i=1}^{5} 3i$ **b.** $\sum_{k=3}^{6} (1 + k^2)$ **c.** $\sum_{i=0}^{8} \frac{1}{i!}$

Solution

a. $$\begin{aligned}\sum_{i=1}^{5} 3i &= 3(1) + 3(2) + 3(3) + 3(4) + 3(5)\\ &= 3(1 + 2 + 3 + 4 + 5)\\ &= 3(15)\\ &= 45\end{aligned}$$

b. $$\begin{aligned}\sum_{k=3}^{6} (1 + k^2) &= (1 + 3^2) + (1 + 4^2) + (1 + 5^2) + (1 + 6^2)\\ &= 10 + 17 + 26 + 37\\ &= 90\end{aligned}$$

c. $$\begin{aligned}\sum_{i=0}^{8} \frac{1}{i!} &= \frac{1}{0!} + \frac{1}{1!} + \frac{1}{2!} + \frac{1}{3!} + \frac{1}{4!} + \frac{1}{5!} + \frac{1}{6!} + \frac{1}{7!} + \frac{1}{8!}\\ &= 1 + 1 + \frac{1}{2} + \frac{1}{6} + \frac{1}{24} + \frac{1}{120} + \frac{1}{720} + \frac{1}{5040} + \frac{1}{40{,}320}\\ &\approx 2.71828\end{aligned}$$

For this summation, note that the sum is very close to the irrational number $e \approx 2.718281828$. It can be shown that as more terms of the sequence whose nth term is $1/n!$ are added, the sum becomes closer and closer to e.

In Example 7, note that the lower limit of a summation does not have to be 1. Also note that the index of summation does not have to be the letter i. For instance, in part (b), the letter k is the index of summation.

STUDY TIP

Variations in the upper and lower limits of summation can produce quite different-looking summation notations for *the same sum*. For example, the following two sums have the same terms.

$$\sum_{i=1}^{3} 3(2^i) = 3(2^1 + 2^2 + 2^3)$$

$$\sum_{i=0}^{2} 3(2^{i+1}) = 3(2^1 + 2^2 + 2^3)$$

Properties of Sums

1. $\displaystyle\sum_{i=1}^{n} ca_i = c\sum_{i=1}^{n} a_i,$ c is any constant.

2. $\displaystyle\sum_{i=1}^{n} (a_i + b_i) = \sum_{i=1}^{n} a_i + \sum_{i=1}^{n} b_i$

3. $\displaystyle\sum_{i=1}^{n} (a_i - b_i) = \sum_{i=1}^{n} a_i - \sum_{i=1}^{n} b_i$

A proof of Property 1 is given in Appendix A.

Series

Many applications involve the sum of the terms of a finite or infinite sequence. Such a sum is called a **series.**

Definition of a Series

Consider the infinite sequence $a_1, a_2, a_3, \ldots, a_i, \ldots$.

1. The sum of all terms of the infinite sequence is called an **infinite series** and is denoted by

$$a_1 + a_2 + a_3 + \cdots + a_i + \cdots = \sum_{i=1}^{\infty} a_i.$$

2. The sum of the first n terms of the sequence is called a **finite series** or the ***n*th partial sum** of the sequence and is denoted by

$$a_1 + a_2 + a_3 + \cdots + a_n = \sum_{i=1}^{n} a_i.$$

Example 8 ▶ Finding the Sum of a Series

For the series $\displaystyle\sum_{i=1}^{\infty} \frac{3}{10^i}$, find (a) the third partial sum and (b) the sum.

Solution

a. The third partial sum is

$$\sum_{i=1}^{3} \frac{3}{10^i} = \frac{3}{10^1} + \frac{3}{10^2} + \frac{3}{10^3} = 0.3 + 0.03 + 0.003 = 0.333.$$

b. The sum of the series is

$$\sum_{i=1}^{\infty} \frac{3}{10^i} = \frac{3}{10^1} + \frac{3}{10^2} + \frac{3}{10^3} + \frac{3}{10^4} + \frac{3}{10^5} + \cdots$$

$$= 0.3 + 0.03 + 0.003 + 0.0003 + 0.00003 + \cdots$$

$$= 0.33333\ldots = \frac{1}{3}.$$

Application

Sequences have many applications in business and science. One is illustrated in Example 9.

Example 9 ▶ Population of the United States

For the years 1960 to 1997, the resident population of the United States can be approximated by the model

$$a_n = \sqrt{33{,}282 + 801.3n + 6.12n^2} \qquad n = 0, 1, \ldots, 37$$

where a_n is the population in millions and n represents the calendar year, with $n = 0$ corresponding to 1960. Find the last five terms of this finite sequence, which represent the U.S. population for the years 1993 to 1997. (Source: U.S. Bureau of the Census)

Solution

The last five terms of this finite sequence are as follows.

$$a_{33} = \sqrt{33{,}282 + 801.3(33) + 6.12(33)^2} \approx 257.7 \qquad \text{1993 population}$$

$$a_{34} = \sqrt{33{,}282 + 801.3(34) + 6.12(34)^2} \approx 260.0 \qquad \text{1994 population}$$

$$a_{35} = \sqrt{33{,}282 + 801.3(35) + 6.12(35)^2} \approx 262.3 \qquad \text{1995 population}$$

$$a_{36} = \sqrt{33{,}282 + 801.3(36) + 6.12(36)^2} \approx 264.7 \qquad \text{1996 population}$$

$$a_{37} = \sqrt{33{,}282 + 801.3(37) + 6.12(37)^2} \approx 267.0 \qquad \text{1997 population}$$

The bar graph in Figure 9.2 graphically represents the population given by this sequence for the entire 38-year period from 1960 to 1997.

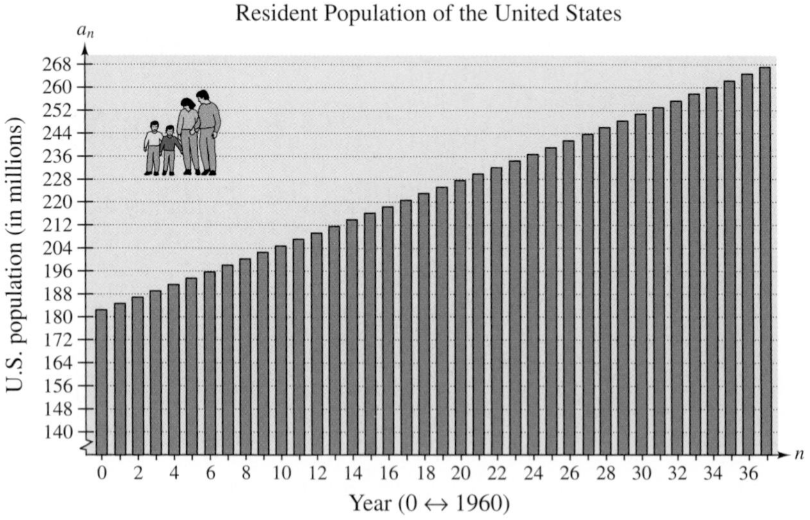

FIGURE 9.2

Activities

1. Write the first five terms of the sequence. (Assume that n begins with 1.)

   $a_n = \dfrac{2n - 1}{2n}$

   Answer: $\dfrac{1}{2}, \dfrac{3}{4}, \dfrac{5}{6}, \dfrac{7}{8}, \dfrac{9}{10}$

2. Write an expression for the most apparent nth term of the sequence

   $0, \dfrac{1}{2}, \dfrac{2}{6}, \dfrac{3}{24}, \dfrac{4}{120}.$

   Answer: $\dfrac{n-1}{n!}$

3. Find the sum.

   $\displaystyle\sum_{k=1}^{4} (-1)^k 2k$

   Answer: 4

9.1 Exercises

The *Interactive* CD-ROM and *Internet* versions of this text contain step-by-step solutions to all odd-numbered Section and Review Exercises. They also provide Tutorial Exercises that link to Guided Examples for additional help.

In Exercises 1–24, write the first five terms of the sequence. (Assume that *n* begins with 1.)

1. $a_n = 3n + 1$ **2.** $a_n = 5n - 3$

3. $a_n = 2^n$ **4.** $a_n = \left(\frac{1}{2}\right)^n$

5. $a_n = (-2)^n$ **6.** $a_n = \left(-\frac{1}{2}\right)^n$

7. $a_n = \dfrac{n+2}{n}$ **8.** $a_n = \dfrac{n}{n+2}$

9. $a_n = \dfrac{6n}{3n^2 - 1}$ **10.** $a_n = \dfrac{3n^2 - n + 4}{2n^2 + 1}$

11. $a_n = \dfrac{1 + (-1)^n}{n}$ **12.** $a_n = 1 + (-1)^n$

13. $a_n = 2 - \dfrac{1}{3^n}$ **14.** $a_n = \dfrac{2^n}{3^n}$

15. $a_n = \dfrac{1}{n^{3/2}}$ **16.** $a_n = \dfrac{10}{n^{2/3}}$

17. $a_n = \dfrac{3^n}{n!}$ **18.** $a_n = \dfrac{n!}{n}$

19. $a_n = \dfrac{(-1)^n}{n^2}$ **20.** $a_n = (-1)^n\left(\dfrac{n}{n+1}\right)$

21. $a_n = \frac{2}{3}$ **22.** $a_n = 0.3$

23. $a_n = n(n-1)(n-2)$ **24.** $a_n = n(n^2 - 6)$

In Exercises 25–30, find the indicated term of the sequence.

25. $a_n = (-1)^n(3n - 2)$

$a_{25} = \square$

26. $a_n = (-1)^{n-1}[n(n-1)]$

$a_{16} = \square$

27. $a_n = \dfrac{2^n}{n!}$ **28.** $a_n = \dfrac{n!}{2n}$

$a_{10} = \square$ $a_8 = \square$

29. $a_n = \dfrac{4n}{2n^2 - 3}$ **30.** $a_n = \dfrac{4n^2 - n + 3}{n(n-1)(n+2)}$

$a_{11} = \square$ $a_{13} = \square$

In Exercises 31–36, use a graphing utility to graph the first 10 terms of the sequence.

31. $a_n = \dfrac{3}{4}n$ **32.** $a_n = 2 - \dfrac{4}{n}$

33. $a_n = 16(-0.5)^{n-1}$ **34.** $a_n = 8(0.75)^{n-1}$

35. $a_n = \dfrac{2n}{n+1}$ **36.** $a_n = \dfrac{n^2}{n^2 + 2}$

In Exercises 37–40, match the sequence with the graph of its first 10 terms. [The graphs are labeled (a), (b), (c), and (d).]

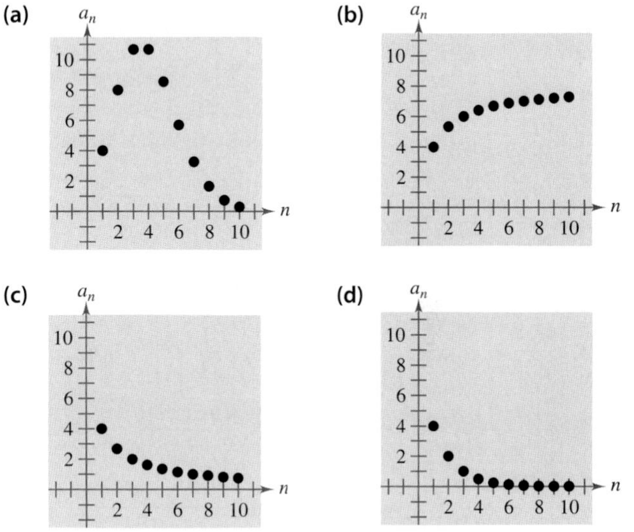

37. $a_n = \dfrac{8}{n+1}$ **38.** $a_n = \dfrac{8n}{n+1}$

39. $a_n = 4(0.5)^{n-1}$ **40.** $a_n = \dfrac{4^n}{n!}$

In Exercises 41–54, write an expression for the *most apparent* nth term of the sequence. (Assume that *n* begins with 1.)

41. $1, 4, 7, 10, 13, \ldots$ **42.** $3, 7, 11, 15, 19, \ldots$

43. $0, 3, 8, 15, 24, \ldots$ **44.** $2, -4, 6, -8, 10, \ldots$

45. $\dfrac{-2}{3}, \dfrac{3}{4}, \dfrac{-4}{5}, \dfrac{5}{6}, \dfrac{-6}{7}, \ldots$

46. $\dfrac{1}{2}, \dfrac{-1}{4}, \dfrac{1}{8}, \dfrac{-1}{16}, \ldots$ **47.** $\dfrac{2}{1}, \dfrac{3}{3}, \dfrac{4}{5}, \dfrac{5}{7}, \dfrac{6}{9}, \ldots$

48. $\frac{1}{3}, \frac{2}{9}, \frac{4}{27}, \frac{8}{81}, \ldots$ **49.** $1, \frac{1}{4}, \frac{1}{9}, \frac{1}{16}, \frac{1}{25}, \ldots$

50. $1, \frac{1}{2}, \frac{1}{6}, \frac{1}{24}, \frac{1}{120}, \ldots$ **51.** $1, -1, 1, -1, 1, \ldots$

52. $1, 2, \dfrac{2^2}{2}, \dfrac{2^3}{6}, \dfrac{2^4}{24}, \dfrac{2^5}{120}, \ldots$

53. $1 + \frac{1}{1}, 1 + \frac{1}{2}, 1 + \frac{1}{3}, 1 + \frac{1}{4}, 1 + \frac{1}{5}, \ldots$

54. $1 + \frac{1}{2}, 1 + \frac{3}{4}, 1 + \frac{7}{8}, 1 + \frac{15}{16}, 1 + \frac{31}{32}, \ldots$

In Exercises 55–58, write the first five terms of the sequence defined recursively.

55. $a_1 = 28, \quad a_{k+1} = a_k - 4$

56. $a_1 = 15, \quad a_{k+1} = a_k + 3$

57. $a_1 = 3, \quad a_{k+1} = 2(a_k - 1)$

58. $a_1 = 32, \quad a_{k+1} = \frac{1}{2}a_k$

In Exercises 59–62, write the first five terms of the sequence defined recursively. Use the pattern to write the *n*th term of the sequence as a function of *n*. (Assume that *n* begins with 1.)

59. $a_1 = 6, \quad a_{k+1} = a_k + 2$

60. $a_1 = 25, \quad a_{k+1} = a_k - 5$

61. $a_1 = 81, \quad a_{k+1} = \frac{1}{3}a_k$

62. $a_1 = 14, \quad a_{k+1} = (-2)a_k$

In Exercises 63–70, simplify the ratio of factorials.

63. $\dfrac{4!}{6!}$ **64.** $\dfrac{5!}{8!}$ **65.** $\dfrac{10!}{8!}$ **66.** $\dfrac{25!}{23!}$

67. $\dfrac{(n+1)!}{n!}$ **68.** $\dfrac{(n+2)!}{n!}$

69. $\dfrac{(2n-1)!}{(2n+1)!}$ **70.** $\dfrac{(3n+1)!}{(3n)!}$

In Exercises 71–82, find the sum.

71. $\sum_{i=1}^{5} (2i + 1)$ **72.** $\sum_{i=1}^{6} (3i - 1)$

73. $\sum_{k=1}^{4} 10$ **74.** $\sum_{k=1}^{5} 5$

75. $\sum_{i=0}^{4} i^2$ **76.** $\sum_{i=0}^{5} 2i^2$

77. $\sum_{k=0}^{3} \dfrac{1}{k^2 + 1}$ **78.** $\sum_{j=3}^{5} \dfrac{1}{j^2 - 3}$

79. $\sum_{k=2}^{5} (k+1)^2(k-3)$ **80.** $\sum_{i=1}^{4} [(i-1)^2 + (i+1)^3]$

81. $\sum_{i=1}^{4} 2^i$ **82.** $\sum_{j=0}^{4} (-2)^j$

In Exercises 83–86, use a calculator to find the sum.

83. $\sum_{j=1}^{6} (24 - 3j)$ **84.** $\sum_{j=1}^{10} \dfrac{3}{j+1}$

85. $\sum_{k=0}^{4} \dfrac{(-1)^k}{k+1}$ **86.** $\sum_{k=0}^{4} \dfrac{(-1)^k}{k!}$

In Exercises 87–96, use sigma notation to write the sum.

87. $\dfrac{1}{3(1)} + \dfrac{1}{3(2)} + \dfrac{1}{3(3)} + \cdots + \dfrac{1}{3(9)}$

88. $\dfrac{5}{1+1} + \dfrac{5}{1+2} + \dfrac{5}{1+3} + \cdots + \dfrac{5}{1+15}$

89. $\left[2\left(\frac{1}{8}\right) + 3\right] + \left[2\left(\frac{2}{8}\right) + 3\right] + \cdots + \left[2\left(\frac{8}{8}\right) + 3\right]$

90. $\left[1 - \left(\frac{1}{6}\right)^2\right] + \left[1 - \left(\frac{2}{6}\right)^2\right] + \cdots + \left[1 - \left(\frac{6}{6}\right)^2\right]$

91. $3 - 9 + 27 - 81 + 243 - 729$

92. $1 - \frac{1}{2} + \frac{1}{4} - \frac{1}{8} + \cdots - \frac{1}{128}$

93. $\dfrac{1}{1^2} - \dfrac{1}{2^2} + \dfrac{1}{3^2} - \dfrac{1}{4^2} + \cdots - \dfrac{1}{20^2}$

94. $\dfrac{1}{1 \cdot 3} + \dfrac{1}{2 \cdot 4} + \dfrac{1}{3 \cdot 5} + \cdots + \dfrac{1}{10 \cdot 12}$

95. $\frac{1}{4} + \frac{3}{8} + \frac{7}{16} + \frac{15}{32} + \frac{31}{64}$

96. $\frac{1}{2} + \frac{2}{4} + \frac{6}{8} + \frac{24}{16} + \frac{120}{32} + \frac{720}{64}$

In Exercises 97–100, find the indicated partial sum of the series.

97. $\sum_{i=1}^{\infty} 5\left(\frac{1}{2}\right)^i$, Fourth partial sum

98. $\sum_{i=1}^{\infty} 2\left(\frac{1}{3}\right)^i$, Fifth partial sum

99. $\sum_{n=1}^{\infty} 4\left(-\frac{1}{2}\right)^n$, Third partial sum

100. $\sum_{n=1}^{\infty} 8\left(-\frac{1}{4}\right)^n$, Fourth partial sum

In Exercises 101–104, find the sum of the infinite series.

101. $\sum_{i=1}^{\infty} 6\left(\frac{1}{10}\right)^i$ **102.** $\sum_{k=1}^{\infty} \left(\frac{1}{10}\right)^k$

103. $\sum_{k=1}^{\infty} 7\left(\frac{1}{10}\right)^k$ **104.** $\sum_{i=1}^{\infty} 2\left(\frac{1}{10}\right)^i$

105. *Compound Interest* A deposit of \$5000 is made in an account that earns 8% interest compounded quarterly. The balance in the account after n quarters is

$$A_n = 5000\left(1 + \frac{0.08}{4}\right)^n, \quad n = 1, 2, 3, \ldots .$$

(a) Compute the first eight terms of this sequence.

(b) Find the balance in this account after 10 years by computing the 40th term of the sequence.

106. *Compound Interest* A deposit of \$100 is made *each month* in an account that earns 12% interest compounded monthly. The balance in the account after n months is

$$A_n = 100(101)[(1.01)^n - 1], n = 1, 2, 3, \ldots.$$

(a) Compute the first six terms of this sequence.

(b) Find the balance in this account after 5 years by computing the 60th term of the sequence.

(c) Find the balance in this account after 20 years by computing the 240th term of the sequence.

107. *Per-Patient Hospital Care* The average cost to community hospitals per patient per day from 1989 to 1996 can be approximated by the model

$$a_n = 696 + 66.4n - 2.37n^2, \quad n = -1, 0, 1, \ldots, 6$$

where a_n is the per-patient cost (in dollars) and n is the year, with $n = 0$ corresponding to 1990. Find the terms of this finite sequence. Use a graphing utility to construct a bar graph that represents the sequence. (Source: American Hospital Association)

108. *Federal Debt* From 1989 to 1996, the federal debt of the United States rose from just under \$3 trillion to over \$5 trillion. The federal debt during the years from 1989 to 1996 is approximated by the model

$$a_n = \sqrt{10.9 + 2.8n}, \quad n = -1, 0, 1, \ldots, 6$$

where a_n is the debt in trillions of dollars and n is the year, with $n = 0$ corresponding to 1990. Find the terms of this finite sequence. Use a graphing utility to construct a bar graph that represents the sequence. (Source: Treasury Department, U.S. Bureau of the Census)

109. *Corporate Income* The net incomes a_n (in millions of dollars) of Wal-Mart for the years 1990 through 1998 are shown in the figure. These incomes can be approximated by the model

$$a_n = 1215 + 608.2n - 114.83n^2 + 11.00n^3,$$

$$n = 0, \ldots, 8$$

where $n = 0$ represents 1990. Use this model to approximate the total net income from 1990 through 1998. Compare this sum with the result of adding the incomes shown in the figure. (Source: Wal-Mart)

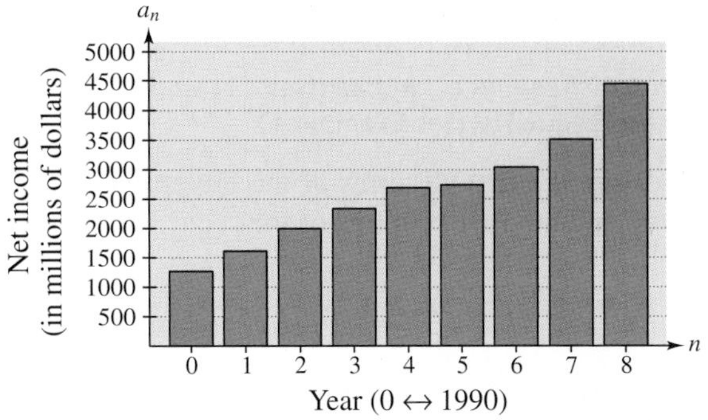

FIGURE FOR 109

110. *Corporate Dividends* The dividends a_n (in dollars) declared per share of common stock of Procter & Gamble Company for the years 1990 through 1998 are shown in the figure. These dividends can be approximated by the model

$$a_n = 4.27 + 0.294n - 2.934 \ln n,$$

$$n = 10, \ldots, 18$$

where $n = 10$ represents 1990. Use this model to approximate the total dividends per share of common stock from 1990 through 1998. Compare this sum with the result of adding the dividends shown in the figure. (Source: Procter & Gamble Company)

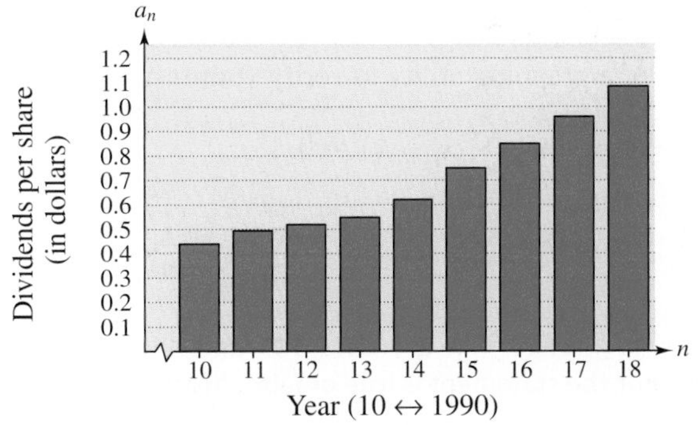

Synthesis

Fibonacci Sequence **In Exercises 111 and 112, use the Fibonacci sequence. (See Example 4.)**

111. Write the first 12 terms of the Fibonacci sequence a_n and the first 10 terms of the sequence given by

$$b_n = \frac{a_{n+1}}{a_n}, \quad n \geq 1.$$

112. Using the definition for b_n in Exercise 111, show that b_n can be defined recursively by

$$b_n = 1 + \frac{1}{b_{n-1}}.$$

Arithmetic Mean **In Exercises 113–116, use the following definition of the arithmetic mean $\bar{x}$ of a set of n measurements $x_1, x_2, x_3, \ldots, x_n$.**

$$\bar{x} = \frac{1}{n}\sum_{i=1}^{n} x_i$$

113. Find the arithmetic mean of the six checking account balances \$327.15, \$785.69, \$433.04, \$265.38, \$604.12, and \$590.30. Use the statistical capabilities of a graphing utility to verify your result.

114. Find the arithmetic mean of the following prices per gallon for regular unleaded gasoline at five gasoline stations in a city: \$1.279, \$1.259, \$1.289, \$1.329, and \$1.349. Use the statistical capabilities of a graphing utility to verify your result.

115. Prove that $\sum_{i=1}^{n} (x_i - \bar{x}) = 0$.

116. Prove that $\sum_{i=1}^{n} (x_i - \bar{x})^2 = \sum_{i=1}^{n} x_i^2 - \frac{1}{n}\left(\sum_{i=1}^{n} x_i\right)^2$.

True or False? **In Exercises 117 and 118, determine whether the statement is true or false. Justify your answer.**

117. $\sum_{i=1}^{4} (i^2 + 2i) = \sum_{i=1}^{4} i^2 + 2\sum_{i=1}^{4} i$

118. $\sum_{j=1}^{4} 2^j = \sum_{j=3}^{6} 2^{j-2}$

Review

In Exercises 119–122, find (a) $A - B$, (b) $4B - 3A$, (c) AB, and (d) BA.

119. $A = \begin{bmatrix} 6 & 5 \\ 3 & 4 \end{bmatrix}, \quad B = \begin{bmatrix} -2 & 4 \\ 6 & -3 \end{bmatrix}$

120. $A = \begin{bmatrix} 10 & 7 \\ -4 & 6 \end{bmatrix}, \quad B = \begin{bmatrix} 0 & -12 \\ 8 & 11 \end{bmatrix}$

121. $A = \begin{bmatrix} -2 & -3 & 6 \\ 4 & 5 & 7 \\ 1 & 7 & 4 \end{bmatrix}, \quad B = \begin{bmatrix} 1 & 4 & 2 \\ 0 & 1 & 6 \\ 0 & 3 & 1 \end{bmatrix}$

122. $A = \begin{bmatrix} -1 & 4 & 0 \\ 5 & 1 & 2 \\ 0 & -1 & 3 \end{bmatrix}, \quad B = \begin{bmatrix} 0 & 4 & 0 \\ 3 & 1 & -2 \\ -1 & 0 & 2 \end{bmatrix}$

In Exercises 123–126, find the determinant of the matrix.

123. $A = \begin{bmatrix} 3 & 5 \\ -1 & 7 \end{bmatrix}$

124. $A = \begin{bmatrix} -2 & 8 \\ 12 & 15 \end{bmatrix}$

125. $A = \begin{bmatrix} 3 & 4 & 5 \\ 0 & 7 & 3 \\ 4 & 9 & -1 \end{bmatrix}$

126. $A = \begin{bmatrix} 16 & 11 & 10 & 2 \\ 9 & 8 & 3 & 7 \\ -2 & -1 & 12 & 3 \\ -4 & 6 & 2 & 1 \end{bmatrix}$

9.2 Arithmetic Sequences and Partial Sums

▶ **What you should learn**

- How to recognize and write arithmetic sequences
- How to find an nth partial sum of an arithmetic sequence
- How to use arithmetic sequences to model and solve real-life problems

▶ **Why you should learn it**

Arithmetic sequences have practical real-life applications. For instance, Exercise 79 on page 706 uses an arithmetic sequence to find the number of bricks needed to lay a brick patio.

Frank Pedrick/The Image Works

Arithmetic Sequences

A sequence whose consecutive terms have a common difference is called an **arithmetic sequence.**

Definition of an Arithmetic Sequence

A sequence is arithmetic if the differences between consecutive terms are the same. So, the sequence

$$a_1, a_2, a_3, a_4, \ldots, a_n, \ldots$$

is arithmetic if there is a number d such that

$$a_2 - a_1 = a_3 - a_2 = a_4 - a_3 = \cdots = d$$

and so on. The number d is the **common difference** of the arithmetic sequence.

Example 1 ▶ Examples of Arithmetic Sequences

a. The sequence whose nth term is $4n + 3$ is arithmetic. For this sequence, the common difference between consecutive terms is 4.

$$\underbrace{7, 11}_{11 - 7 = 4}, 15, 19, \ldots, 4n + 3, \ldots$$

b. The sequence whose nth term is $7 - 5n$ is arithmetic. For this sequence, the common difference between consecutive terms is -5.

$$\underbrace{2, -3}_{-3 - 2 = -5}, -8, -13, \ldots, 7 - 5n, \ldots$$

c. The sequence whose nth term is $\frac{1}{4}(n + 3)$ is arithmetic. For this sequence, the common difference between consecutive terms is $\frac{1}{4}$.

$$\underbrace{1, \frac{5}{4}}_{\frac{5}{4} - 1 = \frac{1}{4}}, \frac{3}{2}, \frac{7}{4}, \ldots, \frac{n + 3}{4}, \ldots$$

The sequence 1, 4, 9, 16, . . . , whose nth term is n^2 is *not* arithmetic. The difference between the first two terms is

$$a_2 - a_1 = 4 - 1 = 3$$

but the difference between the second and third terms is

$$a_3 - a_2 = 9 - 4 = 5.$$

In Example 1, notice that each of the arithmetic sequences has an nth term that is of the form $dn + c$, where the common difference of the sequence is d. An arithmetic sequence may be thought of as a linear function whose domain is the set of natural numbers.

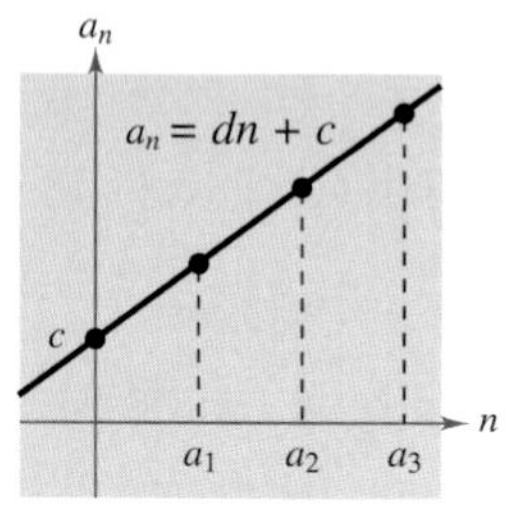

FIGURE 9.3

The nth Term of an Arithmetic Sequence

The nth term of an arithmetic sequence has the form

$$a_n = dn + c$$

where d is the common difference between consecutive terms of the sequence and $c = a_1 - d$. A graphical representation of this definition is shown in Figure 9.3.

Example 2 ▶ Finding the nth Term of an Arithmetic Sequence

Find a formula for the nth term of the arithmetic sequence whose common difference is 3 and whose first term is 2.

As an aid to learning the formula for the nth term of an arithmetic sequence, consider having your students intuitively find the nth term of each of the following sequences.

1. $a, a + 2, a + 4, a + 6, . . .$
   Answer: $2n + a - 2$
2. $5, 8, 11, 14, 17, . . .$
   Answer: $3n + 2$

Solution

Because the sequence is arithmetic, you know that the formula for the nth term is of the form $a_n = dn + c$. Moreover, because the common difference is $d = 3$, the formula must have the form

$$a_n = 3n + c. \qquad \text{Substitute 3 for } d.$$

Because $a_1 = 2$, it follows that

$$c = a_1 - d$$
$$= 2 - 3 \qquad \text{Substitute 2 for } a_1 \text{ and 3 for } d.$$
$$= -1.$$

So, the formula for the nth term is

$$a_n = 3n - 1.$$

The sequence therefore has the following form.

$$2, 5, 8, 11, 14, . . . , 3n - 1, . . .$$

Another way to find a formula for the nth term of the sequence in Example 2 is to begin by writing the terms of the sequence.

a_1	a_2	a_3	a_4	a_5	a_6	a_7	. . .
2	2 + 3	5 + 3	8 + 3	11 + 3	14 + 3	17 + 3	. . .
2	5	8	11	14	17	20	. . .

From these terms, you can reason that the nth term is of the form

$$a_n = dn + c = 3n - 1.$$

Example 3 ▶ Writing the Terms of an Arithmetic Sequence

The fourth term of an arithmetic sequence is 20, and the 13th term is 65. Write the first several terms of this sequence.

Solution

The fourth and 13th terms of the sequence are related by

$$a_{13} = a_4 + 9d.$$

Using $a_4 = 20$ and $a_{13} = 65$, you can conclude that $d = 5$, which implies that the sequence is as follows.

a_1	a_2	a_3	a_4	a_5	a_6	a_7	a_8	a_9	a_{10}	a_{11} . . .
5	10	15	20	25	30	35	40	45	50	55 . . .

If you know the nth term of an arithmetic sequence *and* you know the common difference of the sequence, you can find the $(n + 1)$th term by using the *recursive formula*

$$a_{n+1} = a_n + d. \qquad \text{Recursive formula}$$

With this formula, you can find any term of an arithmetic sequence, *provided* that you know the preceding term. For instance, if you know the first term, you can find the second term. Then, knowing the second term, you can find the third term, and so on.

If you substitute $a_1 - d$ for c in the formula $a_n = dn + c$, the nth term of an arithmetic sequence has the alternative recursive formula

$$a_n = a_1 + (n - 1)d. \qquad \text{Alternative recursive formula}$$

Use this formula to solve Example 4. You should get the same answer.

Example 4 ▶ Using a Recursive Formula

Find the ninth term of the arithmetic sequence that begins with 2 and 9.

Solution

For this sequence, the common difference is $d = 9 - 2 = 7$. There are two ways to find the ninth term. One way is simply to write out the first nine terms (by repeatedly adding 7).

2, 9, 16, 23, 30, 37, 44, 51, 58

Another way to find the ninth term is first to find a formula for the nth term. Because the first term is 2, it follows that

$$c = a_1 - d = 2 - 7 = -5.$$

Therefore, a formula for the nth term is

$$a_n = 7n - 5,$$

which implies that the ninth term is

$$a_9 = 7(9) - 5 = 58.$$

The Sum of a Finite Arithmetic Sequence

There is a simple formula for the *sum* of a finite arithmetic sequence. A proof is given in Appendix A.

STUDY T!P

Be sure you see that this formula works only for *arithmetic* sequences.

The Sum of a Finite Arithmetic Sequence

The sum of a finite arithmetic sequence with n terms is

$$S_n = \frac{n}{2}(a_1 + a_n).$$

Example 5 ▶ Finding the Sum of a Finite Arithmetic Sequence

Find the sum: $1 + 3 + 5 + 7 + 9 + 11 + 13 + 15 + 17 + 19$.

Solution

To begin, notice that the sequence is arithmetic (with a common difference of 2). Moreover, the sequence has 10 terms. So, the sum of the sequence is

$$S_n = 1 + 3 + 5 + 7 + 9 + 11 + 13 + 15 + 17 + 19$$

$$= \frac{n}{2}(a_1 + a_n)$$

$$= \frac{10}{2}(1 + 19) \qquad \text{Substitute 10 for } n, \text{ 1 for } a_1, \text{ 19 for } a_n.$$

$$= 5(20) = 100.$$

Example 6 ▶ Finding the Sum of a Finite Arithmetic Sequence

Find the sum of the integers (a) from 1 to 100 and (b) from 1 to N.

Solution

The integers from 1 to 100 form an arithmetic sequence that has 100 terms. So, you can use the formula for the sum of an arithmetic sequence, as follows.

a. $S_n = 1 + 2 + 3 + 4 + 5 + 6 + \cdots + 99 + 100$

$$= \frac{n}{2}(a_1 + a_n)$$

$$= \frac{100}{2}(1 + 100) \qquad \text{Substitute 100 for } n, \text{ 1 for } a_1, \text{ 100 for } a_n.$$

$$= 50(101) = 5050$$

b. $S_n = 1 + 2 + 3 + 4 + \cdots + N$

$$= \frac{n}{2}(a_1 + a_n)$$

$$= \frac{N}{2}(1 + N) \qquad \text{Substitute } N \text{ for } n, \text{ 1 for } a_1, N \text{ for } a_n.$$

Corbis-Bettmann

Historical Note

A teacher of Carl Friedrich Gauss (1777–1855) asked him to add all the integers from 1 to 100. When Gauss returned with the correct answer after only a few moments, the teacher could only look at him in astounded silence. This is what Gauss did:

$$\begin{array}{r} 1 + 2 + 3 + \cdots + 100 \\ 100 + 99 + 98 + \cdots + 1 \\ \hline 101 + 101 + 101 + \cdots + 101 \end{array}$$

$$\frac{100 \times 101}{2} = 5050$$

The sum of the first n terms of an infinite sequence is the nth partial sum.

Example 7 ▶ Finding a Partial Sum of an Arithmetic Sequence

Find the 150th partial sum of the arithmetic sequence

5, 16, 27, 38, 49, ...

Solution

For this arithmetic sequence, $a_1 = 5$ and $d = 16 - 5 = 11$. So,

$$c = a_1 - d = 5 - 11 = -6$$

and the nth term is $a_n = 11n - 6$. Therefore, $a_{150} = 11(150) - 6 = 1644$, and the sum of the first 150 terms is

$$S_n = \frac{n}{2}(a_1 + a_n)$$

$$= \frac{150}{2}(5 + 1644)$$

$$= 75(1649)$$

$$= 123{,}675.$$

Applications

Example 8 ▶ Seating Capacity

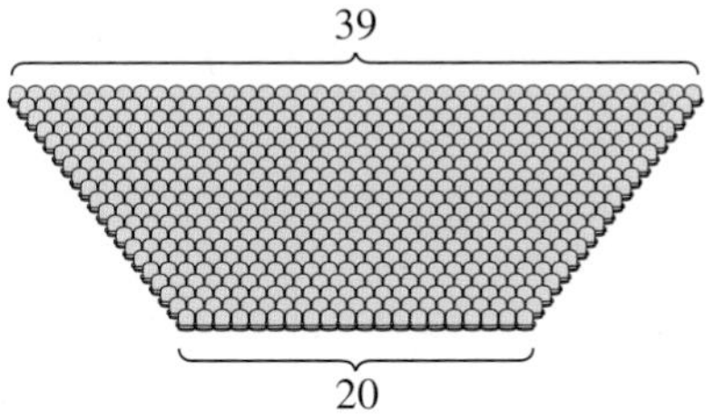

FIGURE 9.4

An auditorium has 20 rows of seats. There are 20 seats in the first row, 21 seats in the second row, 22 seats in the third row, and so on (see Figure 9.4). How many seats are there in all 20 rows?

Solution

The numbers of seats in the 20 rows form an arithmetic sequence in which the common difference is $d = 1$. Because

$$c = a_1 - d = 20 - 1 = 19$$

you can determine that the formula for the nth term of the sequence is $a_n = n + 19$. Therefore, the 20th term in the sequence is $a_{20} = 20 + 19 = 39$, and the total number of seats is

$$S_n = 20 + 21 + 22 + \cdot \cdot \cdot + 39$$

$$= \frac{n}{2}(a_1 + a_{20})$$

$$= \frac{20}{2}(20 + 39)$$

$$= 10(59)$$

$$= 590.$$

Activities

1. Determine which of the following are arithmetic sequences.
   (a) $3, 5, 7, 9, 11, \ldots$
   (b) $3, 6, 12, 24, 48, \ldots$
   (c) $-3, 6, -9, 12, -15, \ldots$
   (d) $5, 0, -5, -10, -15, \ldots$
   (e) $1, 3, 6, 10, 15, 21, \ldots$

   Answer: (a) and (d)
2. Find the first five terms of the arithmetic sequence $a_1 = 13, d = -4$.

   Answer: $13, 9, 5, 1, -3$
3. Find the sum.

   $$\sum_{n=1}^{100} (2 + 3n)$$

   Answer: 15,350

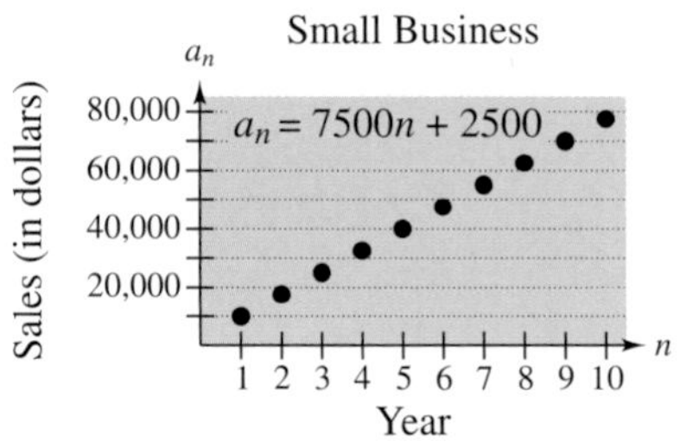

FIGURE 9.5

Example 9 ▶ Total Sales

A small business sells \$10,000 worth of products during its first year. The owner of the business has set a goal of increasing annual sales by \$7500 each year for 9 years. Assuming that this goal is met, find the total sales during the first 10 years this business is in operation.

Solution

The annual sales form an arithmetic sequence in which $a_1 = 10{,}000$ and $d = 7500$. So,

$$c = a_1 - d$$
$$= 10{,}000 - 7500$$
$$= 2500$$

and the nth term of the sequence is

$$a_n = 7500n + 2500.$$ See Figure 9.5.

This implies that the 10th term of the sequence is

$$a_{10} = 77{,}500.$$

The sum of the first 10 terms of the sequence is

$$S_n = \frac{n}{2}(a_1 + a_{10})$$
$$= \frac{10}{2}(10{,}000 + 77{,}500)$$
$$= 5(87{,}500)$$
$$= 437{,}500.$$

So, the total sales for the first 10 years are \$437,500.

Writing ABOUT MATHEMATICS

Numerical Relationships Decide whether it is possible to fill in the blanks in each of the sequences such that the resulting sequence is arithmetic. If so, find a recursive formula for the sequence.

a. -7, ___, ___, ___, ___, ___, 11

b. 17, ___, ___, ___, ___, ___, ___, ___, ___, 71

c. 2, 6, ___, ___, 162

d. 4, 7.5, ___, ___, ___, ___, ___, ___, ___, ___, 39

e. 8, 12, ___, ___, ___, 60.75

9.2 Exercises

In Exercises 1–10, determine whether the sequence is arithmetic. If it is, find the common difference.

1. $10, 8, 6, 4, 2, \ldots$

2. $4, 7, 10, 13, 16, \ldots$

3. $1, 2, 4, 8, 16, \ldots$

4. $80, 40, 20, 10, 5, \ldots$

5. $\frac{9}{4}, 2, \frac{7}{4}, \frac{3}{2}, \frac{5}{4}, \ldots$

6. $3, \frac{5}{2}, 2, \frac{3}{2}, 1, \ldots$

7. $\frac{1}{3}, \frac{2}{3}, 1, \frac{4}{3}, \frac{5}{6}, \ldots$

8. $5.3, 5.7, 6.1, 6.5, 6.9, \ldots$

9. $\ln 1, \ln 2, \ln 3, \ln 4, \ln 5, \ldots$

10. $1^2, 2^2, 3^2, 4^2, 5^2, \ldots$

In Exercises 11–18, write the first five terms of the sequence. Determine whether the sequence is arithmetic, and if it is, find the common difference.

11. $a_n = 5 + 3n$

12. $a_n = 100 - 3n$

13. $a_n = 3 - 4(n - 2)$

14. $a_n = 1 + (n - 1)4$

15. $a_n = (-1)^n$

16. $a_n = 2^{n-1}$

17. $a_n = \dfrac{(-1)^n 3}{n}$

18. $a_n = (2^n)n$

In Exercises 19–24, write the first five terms of the arithmetic sequence. Find the common difference and write the *n*th term of the sequence as a function of *n*.

19. $a_1 = 15, \quad a_{k+1} = a_k + 4$

20. $a_1 = 6, \quad a_{k+1} = a_k + 5$

21. $a_1 = 200, \quad a_{k+1} = a_k - 10$

22. $a_1 = 72, \quad a_{k+1} = a_k - 6$

23. $a_1 = \frac{5}{8}, \quad a_{k+1} = a_k - \frac{1}{8}$

24. $a_1 = 0.375, \quad a_{k+1} = a_k + 0.25$

In Exercises 25–32, write the first five terms of the arithmetic sequence.

25. $a_1 = 5, d = 6$

26. $a_1 = 5, d = -\frac{3}{4}$

27. $a_1 = -2.6, d = -0.4$

28. $a_1 = 16.5, d = 0.25$

29. $a_1 = 2, a_{12} = 46$

30. $a_4 = 16, a_{10} = 46$

31. $a_8 = 26, a_{12} = 42$

32. $a_3 = 19, a_{15} = -1.7$

In Exercises 33–44, find a formula for a_n for the arithmetic sequence.

33. $a_1 = 1, d = 3$

34. $a_1 = 15, d = 4$

35. $a_1 = 100, d = -8$

36. $a_1 = 0, d = -\frac{2}{3}$

37. $a_1 = x, d = 2x$

38. $a_1 = -y, d = 5y$

39. $4, \frac{3}{2}, -1, -\frac{7}{2}, \ldots$

40. $10, 5, 0, -5, -10, \ldots$

41. $a_1 = 5, a_4 = 15$

42. $a_1 = -4, a_5 = 16$

43. $a_3 = 94, a_6 = 85$

44. $a_5 = 190, a_{10} = 115$

In Exercises 45–48, match the sequence with its graph. [The graphs are labeled (a), (b), (c), and (d).]

(a)

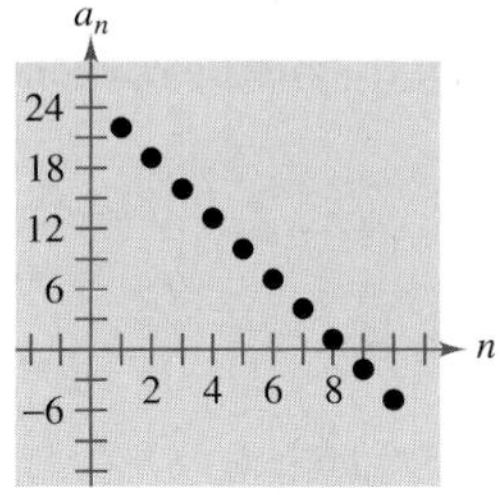

(b)

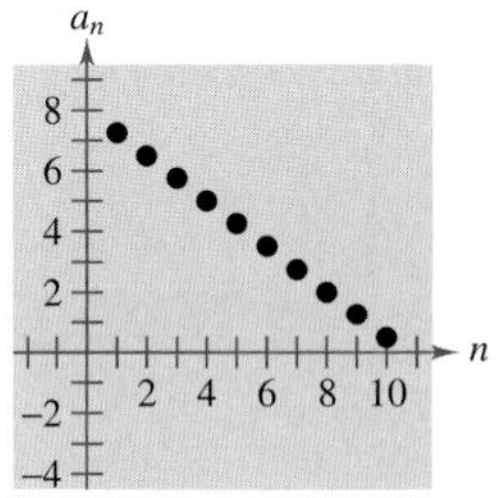

(c)

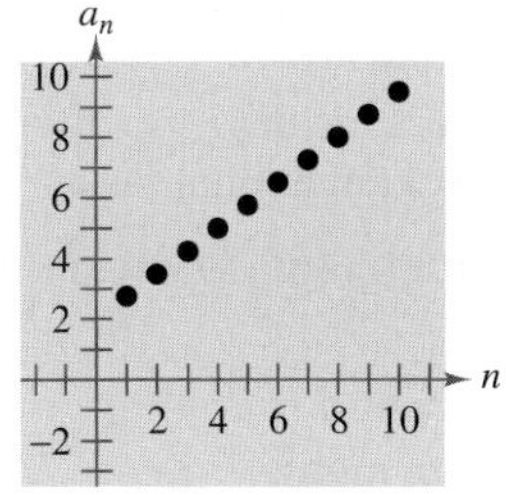

(d)

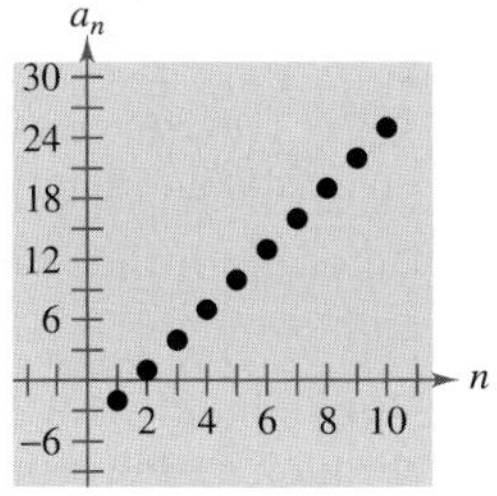

45. $a_n = -\frac{3}{4}n + 8$

46. $a_n = 3n - 5$

47. $a_n = 2 + \frac{3}{4}n$

48. $a_n = 25 - 3n$

In Exercises 49–52, use a graphing utility to graph the first 10 terms of the sequence.

49. $a_n = 15 - \frac{3}{2}n$

50. $a_n = -5 + 2n$

51. $a_n = 0.2n + 3$

52. $a_n = -0.3n + 8$

In Exercises 53–60, find the indicated *n*th partial sum of the arithmetic sequence.

53. $8, 20, 32, 44, \ldots, \quad n = 10$

54. $2, 8, 14, 20, \ldots, \quad n = 25$

55. $4.2, 3.7, 3.2, 2.7, \ldots, \quad n = 12$

56. $0.5, 0.9, 1.3, 1.7, \ldots, \quad n = 10$

57. $40, 37, 34, 31, \ldots, \quad n = 10$

58. $75, 70, 65, 60, \ldots, \quad n = 25$

59. $a_1 = 100, \quad a_{25} = 220, \quad n = 25$

60. $a_1 = 15, \quad a_{100} = 307, \quad n = 100$

In Exercises 61–68, find the partial sum.

61. $\sum_{n=1}^{50} n$

62. $\sum_{n=1}^{100} 2n$

63. $\sum_{n=10}^{100} 6n$

64. $\sum_{n=51}^{100} 7n$

65. $\sum_{n=11}^{30} n - \sum_{n=1}^{10} n$

66. $\sum_{n=51}^{100} n - \sum_{n=1}^{50} n$

67. $\sum_{n=1}^{400} (2n - 1)$

68. $\sum_{n=1}^{250} (1000 - n)$

In Exercises 69–74, use a calculator to find the partial sum.

69. $\sum_{n=1}^{20} (2n + 5)$

70. $\sum_{n=0}^{50} (1000 - 5n)$

71. $\sum_{n=1}^{100} \frac{n + 4}{2}$

72. $\sum_{n=0}^{100} \frac{8 - 3n}{16}$

73. $\sum_{i=1}^{60} \left(250 - \frac{8}{3}i\right)$

74. $\sum_{j=1}^{200} (4.5 + 0.025j)$

Job Offer **In Exercises 75 and 76, consider a job offer with the given starting salary and the given annual raise.**

(a) Determine the salary during the sixth year of employment.

(b) Determine the total compensation from the company through six full years of employment.

	Starting Salary	*Annual Raise*
75.	\$32,500	\$1500
76.	\$36,800	\$1750

77. *Seating Capacity* Determine the seating capacity of an auditorium with 30 rows of seats if there are 20 seats in the first row, 24 seats in the second row, 28 seats in the third row, and so on.

78. *Seating Capacity* Determine the seating capacity of an auditorium with 36 rows of seats if there are 15 seats in the first row, 18 seats in the second row, 21 seats in the third row, and so on.

79. *Brick Pattern* A brick patio has the approximate shape of a trapezoid (see figure). The patio has 18 rows of bricks. The first row has 14 bricks and the 18th row has 31 bricks. How many bricks are in the patio?

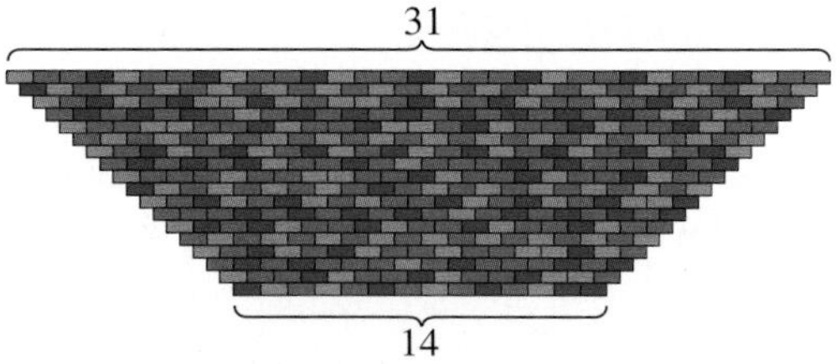

FIGURE FOR 79

80. *Brick Pattern* A triangular brick wall is made by cutting some bricks in half to use in the first column of every other row. The wall has 28 rows. The top row is one-half brick wide and the bottom row is 14 bricks wide. How many bricks are used in the finished wall?

81. *Falling Object* An object with negligible air resistance is dropped from a plane. During the first second of fall, the object falls 4.9 meters; during the second second, it falls 14.7 meters; during the third second, it falls 24.5 meters; during the fourth second, it falls 34.3 meters. If this arithmetic pattern continues, how many meters will the object fall in 10 seconds?

82. *Falling Object* An object with negligible air resistance is dropped from a height of 1500 meters. During the first second of fall, the object falls 4.9 meters; during the second second, it falls 14.7 meters; during the third second, it falls 24.5 meters; during the fourth second, it falls 34.3 meters. If this arithmetic pattern continues, how many meters will the object fall in 17 seconds?

Synthesis

True or False? **In Exercises 83 and 84, determine whether the statement is true or false. Justify your answer.**

83. Given an arithmetic sequence for which only the first and second terms are known, it is possible to find the nth term.

84. If the only known information about a finite arithmetic sequence is its first term and its last term, then it is possible to find the sum of the sequence.

85. *Writing* Describe the geometric pattern of the graph of an arithmetic sequence. Explain.

86. *Writing* Explain how to use the first two terms of an arithmetic sequence to find the nth term.

87. Find the sum of the first 100 odd integers.

88. Find the sum of the integers from -10 to 50.

89. *Pattern Recognition*

(a) Compute the following sums of positive odd integers.

$1 + 3 = \square$

$1 + 3 + 5 = \square$

$1 + 3 + 5 + 7 = \square$

$1 + 3 + 5 + 7 + 9 = \square$

$1 + 3 + 5 + 7 + 9 + 11 = \square$

(b) Use the sums in part (a) to make a conjecture about the sums of positive odd integers. Check your conjecture for the sum

$1 + 3 + 5 + 7 + 9 + 11 + 13 = \square$.

(c) Verify your conjecture analytically.

90. *Think About It* The following operations are performed on each term of an arithmetic sequence. Determine if the resulting sequence is arithmetic, and if so, state the common difference.

(a) A constant C is added to each term.

(b) Each term is multiplied by a nonzero constant C.

(c) Each term is squared.

91. *Think About It* The sum of the first 20 terms of an arithmetic sequence with a common difference of 3 is 650. Find the first term.

92. *Think About It* The sum of the first n terms of an arithmetic sequence with first term a_1 and common difference d is S_n. Determine the sum if each term is increased by 5. Explain.

Review

In Exercises 93–96, use Gaussian elimination to solve the system of equations.

93. $\begin{cases} 8x + 2y - 3z = 0 \\ 4x - 2y - 6z = 0 \\ -2x - 3y - 3z = 0 \end{cases}$

94. $\begin{cases} 2x + y - 3z = 0 \\ -x - y + 2z = 0 \\ 4x - 3y - z = 0 \end{cases}$

95. $\begin{cases} 9x + 2z = 0 \\ x - 3y + 2z = 0 \\ 7y - z = 0 \end{cases}$

96. $\begin{cases} 5x - 3y + 7z = 0 \\ -x + 6y + 4z = 0 \\ 6x + 3y + 4z = 0 \end{cases}$

In Exercises 97–100, use matrices to solve the system of equations.

97. $\begin{cases} 2x - 6y = 2 \\ y = 1 \end{cases}$

98. $\begin{cases} 7x + 3y = 0 \\ x + y = 4 \end{cases}$

99. $\begin{cases} 3x + y + z = 4 \\ x + 7y + 5z = 2 \\ 9x - 7y - 5z = 8 \end{cases}$

100. $\begin{cases} 3x + y = 0 \\ x - 2z = 0 \\ 2x - 3y + z = 0 \end{cases}$

9.3 Geometric Sequences and Series

▶ **What you should learn**

- How to recognize and write geometric sequences
- How to find the nth partial sum of a geometric sequence
- How to find the sum of an infinite geometric series
- How to use geometric sequences to model and solve real-life problems

▶ **Why you should learn it**

Geometric sequences can be used to model and solve real-life problems. For instance, Exercises 95 and 96 on page 716 uses a geometric sequence in retirement planning.

Ferguson/PhotoEdit

Geometric Sequences

In Section 9.2, you learned that a sequence whose consecutive terms have a common *difference* is an arithmetic sequence. In this section, you will study another important type of sequence called a **geometric sequence.** Consecutive terms of a geometric sequence have a common *ratio.*

Definition of a Geometric Sequence

A sequence is **geometric** if the ratios of consecutive terms are the same.

$$\frac{a_2}{a_1} = r, \quad \frac{a_3}{a_2} = r, \quad \frac{a_4}{a_3} = r, \ldots, \qquad r \neq 0$$

The number r is the **common ratio** of the sequence.

Example 1 ▶ Examples of Geometric Sequences

a. The sequence whose nth term is 2^n is geometric. For this sequence, the common ratio of consecutive terms is 2.

$$\underbrace{2, 4,}_{\frac{4}{2} = 2} 8, 16, \ldots, 2^n, \ldots$$

b. The sequence whose nth term is $4(3^n)$ is geometric. For this sequence, the common ratio of consecutive terms is 3.

$$\underbrace{12, 36,}_{\frac{36}{12} = 3} 108, 324, \ldots, 4(3^n), \ldots$$

c. The sequence whose nth term is $\left(-\frac{1}{3}\right)^n$ is geometric. For this sequence, the common ratio of consecutive terms is $-\frac{1}{3}$.

$$\underbrace{-\frac{1}{3}, \frac{1}{9},}_{\frac{1/9}{-1/3} = -\frac{1}{3}} -\frac{1}{27}, \frac{1}{81}, \ldots, \left(-\frac{1}{3}\right)^n, \ldots$$

The sequence 1, 4, 9, 16, . . . , whose nth term is n^2 is *not* geometric. The ratio of the second term to the first term is

$$\frac{a_2}{a_1} = \frac{4}{1} = 4$$

but the ratio of the third term to the second term is

$$\frac{a_3}{a_2} = \frac{9}{4}.$$

In Example 1, notice that each of the geometric sequences has an nth term that is of the form ar^n, where the common ratio of the sequence is r. A geometric sequence may be thought of as an exponential function whose domain is the set of natural numbers.

The nth Term of a Geometric Sequence

The nth term of a geometric sequence has the form

$$a_n = a_1 r^{n-1}$$

where r is the common ratio of consecutive terms of the sequence. So, every geometric sequence can be written in the following form.

$$\begin{array}{ccccccc} a_1, & a_2, & a_3, & a_4, & a_5, \ldots & a_n, \ldots \\ \downarrow & \downarrow & \downarrow & \downarrow & \downarrow & \downarrow \\ a_1, & a_1r, & a_1r^2, & a_1r^3, & a_1r^4, \ldots, & a_1r^{n-1}, \ldots \end{array}$$

Additional Example
Write the first five terms of the geometric sequence whose first term is $a_1 = 9$ and whose common ratio is $r = \frac{1}{3}$.
Solution
$9, 3, 1, \frac{1}{3}, \frac{1}{9}$

If you know the nth term of a geometric sequence, you can find the $(n + 1)$th term by multiplying by r. That is, $a_{n+1} = ra_n$.

Example 2 ▶ Finding the Terms of a Geometric Sequence

Write the first five terms of the geometric sequence whose first term is $a_1 = 3$ and whose common ratio is $r = 2$. Then plot the points on a set of coordinate axes.

Solution

Starting with 3, repeatedly multiply by 2 to obtain the following.

$a_1 = 3$ — 1st term

$a_2 = 3(2^1) = 6$ — 2nd term

$a_3 = 3(2^2) = 12$ — 3rd term

$a_4 = 3(2^3) = 24$ — 4th term

$a_5 = 3(2^4) = 48$ — 5th term

Figure 9.6 shows the first five terms of the geometric sequence.

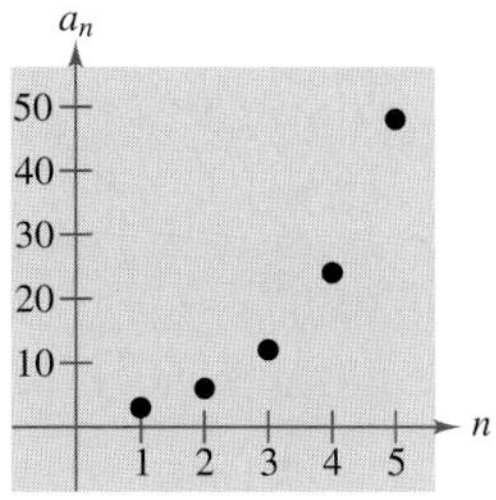

FIGURE 9.6

Example 3 ▶ Finding a Term of a Geometric Sequence

Find the 15th term of the geometric sequence whose first term is 20 and whose common ratio is 1.05.

Solution

$a_{15} = a_1 r^{n-1}$ — Formula for geometric sequence

$= 20(1.05)^{15-1}$ — Substitute for a_1, r, and n.

≈ 39.599 — Use a calculator.

Example 4 ▶ Finding a Term of a Geometric Sequence

Find the 12th term of the geometric sequence

5, 15, 45, ...

Solution

The common ratio of this sequence is

$$r = \frac{15}{5} = 3.$$

Because the first term is $a_1 = 5$, you can determine the 12th term $(n = 12)$ to be

$a_{12} = a_1 r^{n-1}$	Formula for geometric sequence
$= 5(3)^{12-1}$	Substitute for a_1, r, and n.
$= 5(177{,}147)$	Use a calculator.
$= 885{,}735.$	Simplify.

If you know any two terms of a geometric sequence, you can use that information to find a formula for the nth term of the sequence.

STUDY TIP

Remember that r is the common ratio of consecutive terms of a sequence. So in Example 5

$$\begin{aligned} a_{10} &= a_1 r^9 \\ &= a_1 \cdot r \cdot r \cdot r \cdot r^6 \\ &= a_1 \cdot \frac{a_2}{a_1} \cdot \frac{a_3}{a_2} \cdot \frac{a_4}{a_3} \cdot r^6 \\ &= a_4 r^6. \end{aligned}$$

Example 5 ▶ Finding a Term of a Geometric Sequence

The fourth term of a geometric sequence is 125, and the 10th term is 125/64. Find the 14th term. (Assume that the terms of the sequence are positive.)

Solution

The 10th term is related to the fourth term by the equation

$a_{10} = a_4 r^6.$	Multiply 4th term by r^{10-4}.

Because $a_{10} = 125/64$ and $a_4 = 125$, you can solve for r as follows.

$\frac{125}{64} = 125r^6$	Substitute $\frac{125}{64}$ for a_{10} and 125 for a_4.
$\frac{1}{64} = r^6$	Divide each side by 125.
$\frac{1}{2} = r$	Take the sixth root of each side.

You can obtain the 14th term by multiplying the 10th term by r^4.

$a_{14} = a_{10} r^4$	Multiply the 10th term by r^{14-10}.
$= \frac{125}{64}\left(\frac{1}{2}\right)^4$	Substitute $\frac{125}{64}$ for a_{10} and $\frac{1}{2}$ for r.
$= \frac{125}{1024}$	Simplify.

The Sum of a Finite Geometric Sequence

The formula for the sum of a *finite* geometric sequence is as follows. A proof is given in Appendix A.

The Sum of a Finite Geometric Sequence

The sum of the geometric sequence

$$a_1,\ a_1r,\ a_1r^2,\ a_1r^3,\ a_1r^4, \ldots, a_1r^{n-1}$$

with common ratio $r \neq 1$ is given by

$$S_n = a_1\left(\frac{1 - r^n}{1 - r}\right).$$

Example 6 ▶ Finding the Sum of a Finite Geometric Sequence

Find the sum $\displaystyle\sum_{n=1}^{12} 4(0.3)^n$.

Solution

By writing out a few terms, you have

$$\sum_{n=1}^{12} 4(0.3)^n = 4(0.3)^1 + 4(0.3)^2 + 4(0.3)^3 + \cdot \cdot \cdot + 4(0.3)^{12}.$$

Now, because $a_1 = 4(0.3)$, $r = 0.3$, and $n = 12$, you can apply the formula for the sum of a finite geometric sequence to obtain

$$\begin{aligned}\sum_{n=1}^{12} 4(0.3)^n &= a_1\left(\frac{1 - r^n}{1 - r}\right)\\ &= 4(0.3)\left[\frac{1 - (0.3)^{12}}{1 - 0.3}\right]\\ &\approx 1.714.\end{aligned}$$

When using the formula for the sum of a geometric sequence, be careful to check that the index begins at $i = 1$. If the index begins at $i = 0$, you must adjust the formula for the nth partial sum. For instance, if the index in Example 6 had begun with $n = 0$, the sum would have been

$$\begin{aligned}\sum_{n=0}^{12} 4(0.3)^n &= 4(0.3)^0 + \sum_{n=1}^{12} 4(0.3)^n\\ &= 4 + \sum_{n=1}^{12} 4(0.3)^n\\ &\approx 4 + 1.714\\ &= 5.714.\end{aligned}$$

Exploration

Use a graphing utility to graph

$$y = \left(\frac{1 - r^x}{1 - r}\right)$$

for $r = \frac{1}{2}, \frac{2}{3}$, and $\frac{4}{5}$. What happens as $x \to \infty$?

Use a graphing utility to graph

$$y = \left(\frac{1 - r^x}{1 - r}\right)$$

for $r = 1.5, 2$, and 3. What happens as $x \to \infty$?

A computer animation of this concept appears in the *Interactive* CD-ROM and *Internet* versions of this text.

Infinite geometric series and their sums have important uses in calculus. Refer to Exercise 101 of this section.

Geometric Series

The summation of the terms of an infinite geometric sequence is called an **infinite geometric series** or simply a **geometric series.**

The formula for the sum of a *finite* geometric sequence can, depending on the value of r, be extended to produce a formula for the sum of an *infinite* geometric series. Specifically, if the common ratio r has the property that $|r| < 1$, it can be shown that r^n becomes arbitrarily close to zero as n increases without bound. Consequently,

$$a_1\left(\frac{1 - r^n}{1 - r}\right) \longrightarrow a_1\left(\frac{1 - 0}{1 - r}\right) \quad \text{as} \quad n \longrightarrow \infty.$$

This result is summarized as follows.

The Sum of an Infinite Geometric Series

If $|r| < 1$, the infinite geometric series

$$a_1, a_1r, a_1r^2, a_1r^3, . . . , a_1r^{n-1}, . . .$$

has the sum

$$S = \frac{a_1}{1 - r}.$$

Example 7 ▶ Finding the Sum of an Infinite Geometric Sequence

Find the following sums.

a. $\displaystyle\sum_{n=1}^{\infty} 4(0.6)^{n-1}$

b. $3 + 0.3 + 0.03 + 0.003 + \cdot \cdot \cdot$

Solution

a. $\displaystyle\sum_{n=1}^{\infty} 4(0.6)^{n-1} = 4 + 4(0.6) + 4(0.6)^2 + 4(0.6)^3 + \cdot \cdot \cdot + 4(0.6)^{n-1} + \cdot \cdot \cdot$

$$= \frac{4}{1 - (0.6)} \qquad \frac{a_1}{1 - r}$$

$$= 10$$

b. $3 + 0.3 + 0.03 + 0.003 + \cdot \cdot \cdot = 3 + 3(0.1) + 3(0.1)^2 + 3(0.1)^3 + \cdot \cdot \cdot$

$$= \frac{3}{1 - (0.1)} \qquad \frac{a_1}{1 - r}$$

$$= \frac{10}{3}$$

$$\approx 3.33$$

Application

Example 8 ▶ Compound Interest

A deposit of \$50 is made on the first day of each month in a savings account that pays 6% compounded monthly. What is the balance of this annuity at the end of 2 years?

Solution

The first deposit will gain interest for 24 months, and its balance will be

$$A_{24} = 50\left(1 + \frac{0.06}{12}\right)^{24}$$
$$= 50(1.005)^{24}.$$

The second deposit will gain interest for 23 months, and its balance will be

$$A_{23} = 50\left(1 + \frac{0.06}{12}\right)^{23}$$
$$= 50(1.005)^{23}.$$

The last deposit will gain interest for only 1 month, and its balance will be

$$A_1 = 50\left(1 + \frac{0.06}{12}\right)^{1}$$
$$= 50(1.005).$$

The total balance in the annuity will be the sum of the balances of the 24 deposits. Using the formula for the sum of a finite geometric sequence, with $A_1 = 50(1.005)$ and $r = 1.005$, you have

$$S_{24} = 50(1.005)\left[\frac{1 - (1.005)^{24}}{1 - 1.005}\right] \qquad \text{Substitute for } A_1 \text{ and } r.$$
$$= \$1277.96.$$

Activities

1. Determine which of the following are geometric sequences.
   (a) $3, 6, 9, 12, 15, \ldots$
   (b) $2, 4, 8, 16, 32, \ldots$
   (c) $1, -1, 1, -1, 1, \ldots$
   (d) $4, 2, 1, \frac{1}{2}, \frac{1}{4}, \ldots$
   (e) $2, 4, 16, 64, 256, \ldots$
   Answer: (b), (c), (d)
2. Find the sum.
   $\sum_{n=1}^{10} 16\left(\frac{1}{2}\right)^n$
   Answer: ≈ 15.984
3. Find the sum.
   $\sum_{n=0}^{\infty} 16\left(\frac{1}{2}\right)^n$
   Answer: 32

Writing ABOUT MATHEMATICS

An Experiment You will need a piece of string or yarn, a pair of scissors, and a tape measure. Measure out any length of string at least 5 feet long. Double over the string and cut it in half. Take one of the resulting halves, double it over, and cut it in half. Continue this process until you are no longer able to cut a length of string in half. How many cuts were you able to make? Construct a sequence of the resulting string lengths after each cut, starting with the original length of the string. Find a formula for the *n*th term of this sequence. How many cuts could you theoretically make? Discuss why you were not able to make that many cuts.

Writing About Mathematics Suggestion: For the sake of simplicity, you may want to consider supplying each group with the tools they need for this activity rather than having them supply the materials themselves.

9.3 Exercises

In Exercises 1–10, determine whether the sequence is geometric. If it is, find the common ratio.

1. $5, 15, 45, 135, \dots$
2. $3, 12, 48, 192, \dots$
3. $3, 12, 21, 30, \dots$
4. $36, 27, 18, 9, \dots$
5. $1, -\frac{1}{2}, \frac{1}{4}, -\frac{1}{8}, \dots$
6. $5, 1, 0.2, 0.04, \dots$
7. $\frac{1}{8}, \frac{1}{4}, \frac{1}{2}, 1, \dots$
8. $9, -6, 4, -\frac{8}{3}, \dots$
9. $1, \frac{1}{2}, \frac{1}{3}, \frac{1}{4}, \dots$
10. $\frac{1}{5}, \frac{2}{7}, \frac{3}{9}, \frac{4}{11}, \dots$

In Exercises 11–20, write the first five terms of the geometric sequence.

11. $a_1 = 2, r = 3$
12. $a_1 = 6, r = 2$
13. $a_1 = 1, r = \frac{1}{2}$
14. $a_1 = 1, r = \frac{1}{3}$
15. $a_1 = 5, r = -\frac{1}{10}$
16. $a_1 = 6, r = -\frac{1}{4}$
17. $a_1 = 1, r = e$
18. $a_1 = 3, r = \sqrt{5}$
19. $a_1 = 2, r = \dfrac{x}{4}$
20. $a_1 = 5, r = 2x$

In Exercises 21–26, write the first five terms of the geometric sequence. Determine the common ratio and write the *n*th term of the sequence as a function of *n*.

21. $a_1 = 64, \quad a_{k+1} = \frac{1}{2}a_k$
22. $a_1 = 81, \quad a_{k+1} = \frac{1}{3}a_k$
23. $a_1 = 7, \quad a_{k+1} = 2a_k$
24. $a_1 = 5, \quad a_{k+1} = -2a_k$
25. $a_1 = 6, \quad a_{k+1} = -\frac{3}{2}a_k$
26. $a_1 = 48, \quad a_{k+1} = -\frac{1}{2}a_k$

In Exercises 27–38, find the *n*th term of the geometric sequence.

27. $a_1 = 4, r = \frac{1}{2}, n = 10$
28. $a_1 = 5, r = \frac{3}{2}, n = 8$
29. $a_1 = 6, r = -\frac{1}{3}, n = 12$
30. $a_1 = 64, r = -\frac{1}{4}, n = 10$
31. $a_1 = 100, r = e^x, n = 9$
32. $a_1 = 1, r = \sqrt{3}, n = 8$
33. $a_1 = 500, r = 1.02, n = 40$
34. $a_1 = 1000, r = 1.005, n = 60$
35. $a_1 = 16, a_4 = \frac{27}{4}, n = 3$
36. $a_2 = 3, a_5 = \frac{3}{64}, n = 1$
37. $a_4 = -18, a_7 = \frac{2}{3}, n = 6$
38. $a_3 = \frac{16}{3}, a_5 = \frac{64}{27}, n = 7$

In Exercises 39–42, match the sequence with its graph. [The graphs are labeled (a), (b), (c), and (d).]

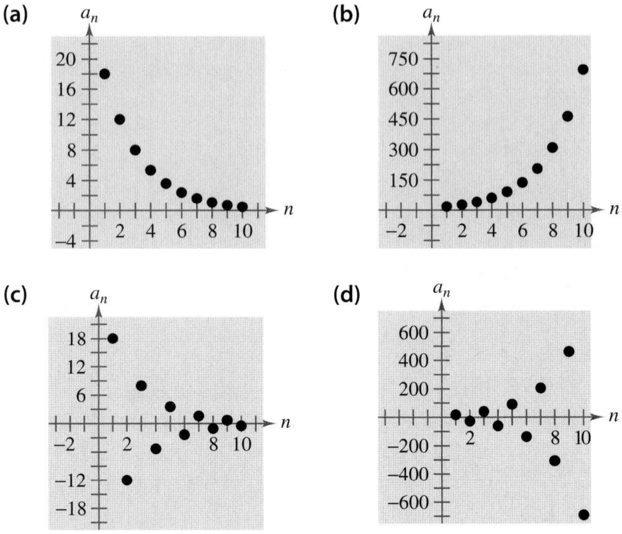

39. $a_n = 18\left(\frac{2}{3}\right)^{n-1}$
40. $a_n = 18\left(-\frac{2}{3}\right)^{n-1}$
41. $a_n = 18\left(\frac{3}{2}\right)^{n-1}$
42. $a_n = 18\left(-\frac{3}{2}\right)^{n-1}$

In Exercises 43–46, use a graphing utility to graph the first 10 terms of the sequence.

43. $a_n = 12(-0.75)^{n-1}$
44. $a_n = 12(-0.4)^{n-1}$
45. $a_n = 2(1.3)^{n-1}$
46. $a_n = 2(-1.4)^{n-1}$

In Exercises 47–56, find the sum of the finite geometric series.

47. $\displaystyle\sum_{n=1}^{9} 2^{n-1}$
48. $\displaystyle\sum_{n=1}^{9} (-2)^{n-1}$
49. $\displaystyle\sum_{i=1}^{7} 64\left(-\tfrac{1}{2}\right)^{i-1}$
50. $\displaystyle\sum_{i=1}^{6} 32\left(\tfrac{1}{4}\right)^{i-1}$

51. $\sum_{n=0}^{20} 3\left(\frac{3}{2}\right)^n$

52. $\sum_{n=0}^{15} 2\left(\frac{4}{3}\right)^n$

53. $\sum_{n=0}^{5} 300(1.06)^n$

54. $\sum_{n=0}^{6} 500(1.04)^n$

55. $\sum_{i=1}^{10} 8\left(-\frac{1}{4}\right)^{i-1}$

56. $\sum_{i=1}^{10} 5\left(-\frac{1}{3}\right)^{i-1}$

In Exercises 57–62, use summation notation to express the sum.

57. $5 + 15 + 45 + \cdot \cdot \cdot + 3645$

58. $7 + 14 + 28 + \cdot \cdot \cdot + 896$

59. $2 - \frac{1}{2} + \frac{1}{8} - \cdot \cdot \cdot + \frac{1}{2048}$

60. $15 - 3 + \frac{3}{5} - \cdot \cdot \cdot - \frac{3}{625}$

61. $0.1 + 0.4 + 1.6 + \cdot \cdot \cdot + 102.4$

62. $32 + 24 + 18 + \cdot \cdot \cdot + 10.125$

In Exercises 63–76, find the sum of the infinite geometric series.

63. $\sum_{n=0}^{\infty} \left(\frac{1}{2}\right)^n$

64. $\sum_{n=0}^{\infty} 2\left(\frac{2}{3}\right)^n$

65. $\sum_{n=0}^{\infty} \left(-\frac{1}{2}\right)^n$

66. $\sum_{n=0}^{\infty} 2\left(-\frac{2}{3}\right)^n$

67. $\sum_{n=0}^{\infty} 4\left(\frac{1}{4}\right)^n$

68. $\sum_{n=0}^{\infty} \left(\frac{1}{10}\right)^n$

69. $\sum_{n=0}^{\infty} (0.4)^n$

70. $\sum_{n=0}^{\infty} 4(0.2)^n$

71. $\sum_{n=0}^{\infty} -3(0.9)^n$

72. $\sum_{n=0}^{\infty} -10(0.2)^n$

73. $8 + 6 + \frac{9}{2} + \frac{27}{8} + \cdot \cdot \cdot$

74. $9 + 6 + 4 + \frac{8}{3} + \cdot \cdot \cdot$

75. $\frac{1}{9} - \frac{1}{3} + 1 - 3 + \cdot \cdot \cdot$

76. $-\frac{125}{36} + \frac{25}{6} - 5 + 6 - \cdot \cdot \cdot$

In Exercises 77–80, find the rational number representation of the repeating decimal.

77. $0.\overline{36}$

78. $0.\overline{297}$

79. $0.3\overline{18}$

80. $1.3\overline{8}$

Graphical Reasoning **In Exercises 81 and 82, use a graphing utility to graph the function. Identify the horizontal asymptote of the graph and determine its relationship to the sum.**

81. $f(x) = 6\left[\frac{1 - (0.5)^x}{1 - (0.5)}\right], \quad \sum_{n=0}^{\infty} 6\left(\frac{1}{2}\right)^n$

82. $f(x) = 2\left[\frac{1 - (0.8)^x}{1 - (0.8)}\right], \quad \sum_{n=0}^{\infty} 2\left(\frac{4}{5}\right)^n$

83. *Compound Interest* A principal of \$1000 is invested at 6% interest. Find the amount after 10 years if the interest is compounded (a) annually, (b) semiannually, (c) quarterly, (d) monthly, and (e) daily.

84. *Compound Interest* A principal of \$2500 is invested at 8% interest. Find the amount after 20 years if the interest is compounded (a) annually, (b) semiannually, (c) quarterly, (d) monthly, and (e) daily.

85. *Depreciation* A company buys a machine for \$135,000 and it depreciates at a rate of 30% per year. (In other words, at the end of each year the depreciated value is 70% of what it was at the beginning of the year.) Find the depreciated value of the machine after 5 full years.

86. *Population Growth* A city of 250,000 people is growing at a rate of 1.3% per year. Estimate the population of the city 30 years from now.

87. *Annuities* A deposit of \$100 is made at the beginning of each month in an account that pays 6%, compounded monthly. The balance A in the account at the end of 5 years is

$$A = 100\left(1 + \frac{0.06}{12}\right)^1 + \cdot \cdot \cdot + 100\left(1 + \frac{0.06}{12}\right)^{60}.$$

Find A.

88. *Annuities* A deposit of \$50 is made at the beginning of each month in an account that pays 8%, compounded monthly. The balance A in the account at the end of 5 years is

$$A = 50\left(1 + \frac{0.08}{12}\right)^1 + \cdot \cdot \cdot + 50\left(1 + \frac{0.08}{12}\right)^{60}.$$

Find A.

89. *Annuities* A deposit of P dollars is made at the beginning of each month in an account earning an annual interest rate r, compounded monthly. The balance A after t years is

$$A = P\left(1 + \frac{r}{12}\right) + P\left(1 + \frac{r}{12}\right)^2 + \cdots + P\left(1 + \frac{r}{12}\right)^{12t}.$$

Show that the balance is

$$A = P\left[\left(1 + \frac{r}{12}\right)^{12t} - 1\right]\left(1 + \frac{12}{r}\right).$$

90. *Annuities* A deposit of P dollars is made at the beginning of each month in an account earning an annual interest rate r, compounded continuously. The balance A after t years is

$$A = Pe^{r/12} + Pe^{2r/12} + \cdots + Pe^{12tr/12}.$$

Show that the balance is

$$A = \frac{Pe^{r/12}(e^{rt} - 1)}{e^{r/12} - 1}.$$

Annuities **In Exercises 91–94, consider making monthly deposits of P dollars in a savings account earning an annual interest rate r. Use the results of Exercises 89 and 90 to find the balance A after t years if the interest is compounded (a) monthly and (b) continuously.**

91. $P = \$50,\ r = 7\%,\ t = 20$ years

92. $P = \$75,\ r = 9\%,\ t = 25$ years

93. $P = \$100,\ r = 10\%,\ t = 40$ years

94. $P = \$20,\ r = 6\%,\ t = 50$ years

95. *Annuities* Consider an initial deposit of P dollars in an account earning an annual interest rate r, compounded monthly. At the end of each month a withdrawal of W dollars will occur and the account will be depleted in t years. The amount of the initial deposit required is

$$P = W\left(1 + \frac{r}{12}\right)^{-1} + W\left(1 + \frac{r}{12}\right)^{-2} + \cdots + W\left(1 + \frac{r}{12}\right)^{-12t}.$$

Show that the initial deposit is

$$P = W\left(\frac{12}{r}\right)\left[1 - \left(1 + \frac{r}{12}\right)^{-12t}\right].$$

96. *Annuities* Determine the amount required in a retirement account for an individual who retires at age 65 and wants an income of \$2000 from the account each month for 20 years. Use the result of Exercise 95 and assume that the account earns 9% compounded monthly.

97. *Geometry* The sides of a square are 16 inches in length. A new square is formed by connecting the midpoints of the sides of the original square, and two of the resulting triangles are shaded (see figure). If this process is repeated five more times, determine the total area of the shaded region.

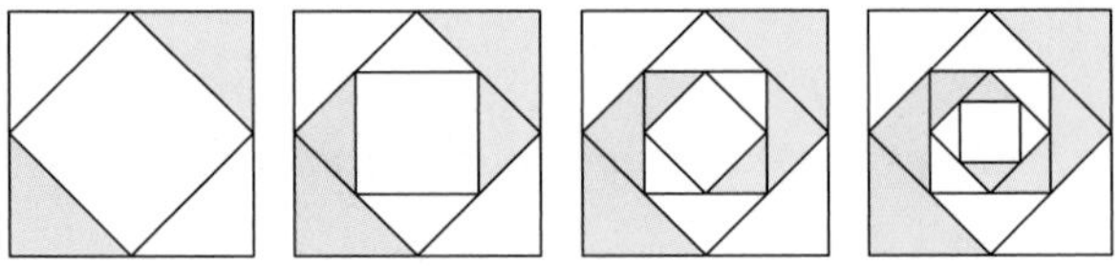

98. *Corporate Revenue* The annual revenues a_n (in billions of dollars) for Coca-Cola Enterprises for 1990 through 1996 can be approximated by the model

$$a_n = 3.978e^{0.11n}, \qquad n = 0, 1, \ldots, 6$$

where $n = 0$ represents 1990. Use this model and the formula for the sum of a finite geometric sequence to approximate the total revenue earned during this 7-year period. (Source: Coca-Cola Enterprises, Inc.)

99. *Think About It* Suppose you work for a company that pays \$0.01 the first day, \$0.02 the second day, \$0.04 the third day, and so on. If the daily wage keeps doubling, what would your total income be for working (a) 29 days, (b) 30 days, and (c) 31 days?

100. *Salary* A company has a job opening with a salary of \$30,000 for the first year. Suppose that during the next 39 years, there is a 5% raise each year. Find the total compensation over the 40-year period.

101. *Distance* A ball is dropped from a height of 16 feet. Each time it drops h feet, it rebounds $0.81h$ feet.

(a) Find the total distance traveled by the ball.

(b) The ball takes the following times for each fall.

$$\begin{aligned} s_1 &= -16t^2 + 16, & s_1 &= 0 \text{ if } t = 1 \\ s_2 &= -16t^2 + 16(0.81), & s_2 &= 0 \text{ if } t = 0.9 \\ s_3 &= -16t^2 + 16(0.81)^2, & s_3 &= 0 \text{ if } t = (0.9)^2 \\ s_4 &= -16t^2 + 16(0.81)^3, & s_4 &= 0 \text{ if } t = (0.9)^3 \\ &\vdots & &\vdots \\ s_n &= -16t^2 + 16(0.81)^{n-1}, & s_n &= 0 \text{ if } t = (0.9)^{n-1} \end{aligned}$$

Beginning with s_2, the ball takes the same amount of time to bounce up as it does to fall, and so the total time elapsed before it comes to rest is

$$t = 1 + 2\sum_{n=1}^{\infty} (0.9)^n.$$

Find this total.

Synthesis

True or False? **In Exercises 102 and 103, determine whether the statement is true or false. Justify your answer.**

102. A sequence is geometric if the ratios of consecutive differences of consecutive terms are the same.

103. You can find the *nth* term of a geometric sequence by multiplying its common ratio by the first term of the sequence raised to the $(n - 1)$th power.

104. *Writing* Write a brief paragraph explaining why the terms of a geometric sequence decrease in magnitude when $-1 < r < 1$.

105. *Writing* Write a brief paragraph explaining how to use the first two terms of a geometric sequence to find the nth term.

Review

In Exercises 106–109, evaluate the function for $f(x) = 3x + 1$ and $g(x) = x^2 - 1$.

106. $g(x + 1)$

107. $f(x + 1)$

108. $f(g(x + 1))$

109. $g(f(x + 1))$

In Exercises 110–113, completely factor the expression over the rational numbers.

110. $9x^3 - 64x$

111. $x^2 + 4x - 63$

112. $6x^2 - 13x - 5$

113. $16x^2 - 4x^4$

In Exercises 114–119, perform the indicated operation and simplify.

114. $\dfrac{3}{x + 3} \cdot \dfrac{x(x + 3)}{x - 3}$

115. $\dfrac{x - 2}{x + 7} \cdot \dfrac{2x(x + 7)}{6x(x - 2)}$

116. $\dfrac{x}{3} \div \dfrac{3x}{6x + 3}$

117. $\dfrac{x - 5}{x - 3} \div \dfrac{10 - 2x}{2(3 - x)}$

118. $5 + \dfrac{7}{x + 2} + \dfrac{2}{x - 2}$

119. $8 - \dfrac{x - 1}{x + 4} - \dfrac{4}{x - 1} - \dfrac{x + 4}{(x - 1)(x + 4)}$

9.4 Mathematical Induction

▶ **What you should learn**

- How to use mathematical induction to prove a statement
- How to find the sums of powers of integers
- How to recognize patterns and write the *n*th term of a sequence
- How to find finite differences of a sequence

▶ **Why you should learn it**

Mathematical induction can be used to prove statements involving a positive integer *n*. For instance, in Exercises 41–49 on pages 726 and 727, you are asked to use mathematical induction to prove properties of real numbers.

Introduction

In this section you will study a form of mathematical proof called **mathematical induction.** It is important that you see clearly the logical need for it, so let's take a closer look at the problem discussed in Example 5 on page 702.

$$S_1 = 1 = 1^2$$

$$S_2 = 1 + 3 = 2^2$$

$$S_3 = 1 + 3 + 5 = 3^2$$

$$S_4 = 1 + 3 + 5 + 7 = 4^2$$

$$S_5 = 1 + 3 + 5 + 7 + 9 = 5^2$$

Judging from the pattern formed by these first five sums, it appears that the sum of the first n odd integers is

$$S_n = 1 + 3 + 5 + 7 + 9 + \cdot \cdot \cdot + (2n - 1) = n^2.$$

Although this particular formula *is* valid, it is important for you to see that recognizing a pattern and then simply *jumping to the conclusion* that the pattern must be true for all values of n is *not* a logically valid method of proof. There are many examples in which a pattern appears to be developing for small values of n and then at some point the pattern fails. One of the most famous cases of this was the conjecture by the French mathematician Pierre de Fermat (1601–1665), who speculated that all numbers of the form

$$F_n = 2^{2^n} + 1, \quad n = 0, 1, 2, \ldots$$

are prime. For $n = 0, 1, 2, 3,$ and 4, the conjecture is true.

$$F_0 = 3$$

$$F_1 = 5$$

$$F_2 = 17$$

$$F_3 = 257$$

$$F_4 = 65{,}537$$

The size of the next Fermat number ($F_5 = 4{,}294{,}967{,}297$) is so great that it was difficult for Fermat to determine whether it was prime or not. However, another well-known mathematician, Leonhard Euler (1707–1783), later found the factorization

$$F_5 = 4{,}294{,}967{,}297$$

$$= 641(6{,}700{,}417)$$

which proved that F_5 is not prime and therefore Fermat's conjecture was false.

Just because a rule, pattern, or formula seems to work for several values of n, you cannot simply decide that it is valid for all values of n without going through a *legitimate proof.* Mathematical induction is one method of proof.

STUDY TIP

It is important to recognize that both parts of the Principle of Mathematical Induction are necessary.

The Principle of Mathematical Induction

Let P_n be a statement involving the positive integer n. If

1. P_1 is true, and
2. the truth of P_k implies the truth of P_{k+1} for every positive k,

then P_n must be true for all positive integers n.

To apply the Principle of Mathematical Induction, you need to be able to determine the statement P_{k+1} for a given statement P_k.

Example 1 ▶ A Preliminary Example

Find P_{k+1} for the following.

a. $P_k : S_k = \dfrac{k^2(k+1)^2}{4}$

b. $P_k : S_k = 1 + 5 + 9 + \cdot \cdot \cdot + [4(k-1) - 3] + (4k - 3)$

c. $P_k : 3^k \geq 2k + 1$

Solution

a. $P_{k+1} : S_{k+1} = \dfrac{(k+1)^2(k+1+1)^2}{4}$ Replace k by $k + 1$.

$= \dfrac{(k+1)^2(k+2)^2}{4}.$ Simplify.

b. $P_{k+1} : S_{k+1} = 1 + 5 + 9 + \cdot \cdot \cdot + \{4[(k+1) - 1] - 3\} + [4(k+1) - 3]$

$= 1 + 5 + 9 + \cdot \cdot \cdot + (4k - 3) + (4k + 1).$

c. $P_{k+1} : 3^{k+1} \geq 2(k+1) + 1$

$3^{k+1} \geq 2k + 3.$

A well-known illustration used to explain why the Principle of Mathematical Induction works is the unending line of dominoes shown in Figure 9.7.

FIGURE 9.7

If the line actually contains infinitely many dominoes, it is clear that you could not knock the entire line down by knocking down only *one domino* at a time. However, suppose it were true that each domino would knock down the next one as it fell. Then you could knock them all down simply by pushing the first one and starting a chain reaction. Mathematical induction works in the same way. If the truth of P_k implies the truth of P_{k+1} and if P_1 is true, the chain reaction proceeds as follows: P_1 implies P_2, P_2 implies P_3, P_3 implies P_4, and so on.

When using mathematical induction to prove a *summation* formula (such as the one in Example 2), it is helpful to think of S_{k+1} as

$$S_{k+1} = S_k + a_{k+1}$$

where a_{k+1} is the $(k + 1)$th term of the original sum.

Example 2 ▶ Using Mathematical Induction

Use mathematical induction to prove the following formula.

$$\begin{aligned} S_n &= 1 + 3 + 5 + 7 + \cdot \cdot \cdot + (2n - 1) \\ &= n^2 \end{aligned}$$

Your students may benefit from many demonstrations of proof by induction. Consider using proofs of the following.
$3 + 6 + 9 + 12 + \cdot \cdot \cdot + 3n = \frac{3}{2}n(n + 1)$
$5 + 7 + 9 + 11 + 13 + \cdot \cdot \cdot + (3 + 2n) = n(n + 4)$

Solution

Mathematical induction consists of two distinct parts. First, you must show that the formula is true when $n = 1$.

1. When $n = 1$, the formula is valid, because

$$S_1 = 1 = 1^2.$$

The second part of mathematical induction has two steps. The first step is to assume that the formula is valid for *some* integer k. The second step is to use this assumption to prove that the formula is valid for the next integer, $k + 1$.

2. Assuming that the formula

$$\begin{aligned} S_k &= 1 + 3 + 5 + 7 + \cdot \cdot \cdot + (2k - 1) \\ &= k^2 \end{aligned}$$

is true, you must show that the formula $S_{k+1} = (k + 1)^2$ is true.

$$\begin{aligned} S_{k+1} &= 1 + 3 + 5 + 7 + \cdot \cdot \cdot + (2k - 1) + [2(k + 1) - 1] \\ &= [1 + 3 + 5 + 7 + \cdot \cdot \cdot + (2k - 1)] + (2k + 2 - 1) \\ &= S_k + (2k + 1) && \text{Group terms to form } S_k. \\ &= k^2 + 2k + 1 && \text{Replace } S_k \text{ by } k^2. \\ &= (k + 1)^2 \end{aligned}$$

Combining the results of parts (1) and (2), you can conclude by mathematical induction that the formula is valid for *all* positive integer values of n.

It occasionally happens that a statement involving natural numbers is not true for the first $k - 1$ positive integers but is true for all values of $n \geq k$. In these instances, you use a slight variation of the Principle of Mathematical Induction in which you verify P_k rather than P_1. This variation is called the *extended principle of mathematical induction*. To see the validity of this, note from Figure 9.7 that all but the first $k - 1$ dominoes can be knocked down by knocking over the kth domino. This suggests that you can prove a statement P_n to be true for $n \geq k$ by showing that P_k is true and that P_k implies P_{k+1}. In Exercises 41–49 of this section you are asked to apply this extension of mathematical induction.

Example 3 ▶ Using Mathematical Induction

Use mathematical induction to prove the formula

$$S_n = 1^2 + 2^2 + 3^2 + 4^2 + \cdots + n^2$$
$$= \frac{n(n+1)(2n+1)}{6}.$$

Solution

1. When $n = 1$, the formula is valid, because

$$S_1 = 1^2 = \frac{1(2)(3)}{6}.$$

2. Assuming that

$$S_k = 1^2 + 2^2 + 3^2 + 4^2 + \cdots + k^2$$
$$= \frac{k(k+1)(2k+1)}{6}$$

you must show that

$$S_{k+1} = \frac{(k+1)(k+2)(2k+3)}{6}.$$

To do this, write the following.

$$S_{k+1} = S_k + a_{k+1}$$
$$= (1^2 + 2^2 + 3^2 + 4^2 + \cdots + k^2) + (k+1)^2$$
$$= \frac{k(k+1)(2k+1)}{6} + (k+1)^2$$
$$= \frac{k(k+1)(2k+1) + 6(k+1)^2}{6}$$
$$= \frac{(k+1)[k(2k+1) + 6(k+1)]}{6}$$
$$= \frac{(k+1)(2k^2 + 7k + 6)}{6}$$
$$= \frac{(k+1)(k+2)(2k+3)}{6}$$

Combining the results of parts (1) and (2), you can conclude by mathematical induction that the formula is valid for *all* $n \geq 1$.

When proving a formula using mathematical induction, the only statement that you *need* to verify is P_1. As a check, however, it is good to try verifying other statements. For instance, in Example 3, try verifying P_2 and P_3.

Sums of Powers of Integers

The formula in Example 3 is one of a collection of useful summation formulas. This and other formulas dealing with the sums of various powers of the first n positive integers are as follows.

Sums of Powers of Integers

1. $1 + 2 + 3 + 4 + \cdots + n = \dfrac{n(n+1)}{2}$

2. $1^2 + 2^2 + 3^2 + 4^2 + \cdots + n^2 = \dfrac{n(n+1)(2n+1)}{6}$

3. $1^3 + 2^3 + 3^3 + 4^3 + \cdots + n^3 = \dfrac{n^2(n+1)^2}{4}$

4. $1^4 + 2^4 + 3^4 + 4^4 + \cdots + n^4 = \dfrac{n(n+1)(2n+1)(3n^2+3n-1)}{30}$

5. $1^5 + 2^5 + 3^5 + 4^5 + \cdots + n^5 = \dfrac{n^2(n+1)^2(2n^2+2n-1)}{12}$

Example 4 ▶ Finding a Sum of Powers of Integers

Find the following sums.

a. $\displaystyle\sum_{n=1}^{7} n^3 = 1^3 + 2^3 + 3^3 + 4^3 + 5^3 + 6^3 + 7^3$ **b.** $\displaystyle\sum_{n=1}^{4} (6n - 4n^2)$

Solution

a. Using the formula for the sum of the cubes of the first n positive integers, you obtain

$$\sum_{n=1}^{7} n^3 = 1^3 + 2^3 + 3^3 + 4^3 + 5^3 + 6^3 + 7^3$$

$$= \frac{7^2(7+1)^2}{4} = \frac{49(64)}{4} = 784.$$

b.
$$\sum_{n=1}^{4} (6n - 4n^2) = \sum_{n=1}^{4} 6n - \sum_{n=1}^{4} 4n^2$$

$$= 6\sum_{n=1}^{4} n - 4\sum_{n=1}^{4} n^2$$

$$= 6\left[\frac{4(4+1)}{2}\right] - 4\left[\frac{4(4+1)(8+1)}{6}\right]$$

$$= 6(10) - 4(30)$$

$$= 60 - 120$$

$$= -60$$

Example 5 ▶ Proving an Inequality by Mathematical Induction

Prove that $n < 2^n$ for all positive integers n.

Solution

1. For $n = 1$ or 2, the statement is true, because

$$1 < 2^1 \quad \text{and} \quad 2 < 2^2.$$

2. Assuming that

$$k < 2^k$$

you need to show that $k + 1 < 2^{k+1}$. For $n = k$, you have

$$2^{k+1} = 2(2^k) > 2(k) = 2k. \qquad \text{By assumption}$$

Because $2k = k + k > k + 1$ for all $k > 1$, it follows that

$$2^{k+1} > 2k > k + 1$$

or

$$k + 1 < 2^{k+1}.$$

Therefore, $n < 2^n$ for all integers $n \geq 1$.

To check a result that you have proved by mathematical induction, it helps to list the statement for several values of n. For instance, in Example 5, you could list

$$1 < 2^1 = 2, \qquad 2 < 2^2 = 4, \qquad 3 < 2^3 = 8,$$
$$4 < 2^4 = 16, \qquad 5 < 2^5 = 32, \qquad 6 < 2^6 = 64.$$

From this list, your intuition confirms that the statement $n < 2^n$ is reasonable.

Pattern Recognition

Although choosing a formula on the basis of a few observations does *not* guarantee the validity of the formula, pattern recognition *is* important. Once you have a pattern or formula that you think works, you can try using mathematical induction to prove your formula.

Finding a Formula for the *n*th Term of a Sequence

To find a formula for the nth term of a sequence, consider these guidelines.

1. Calculate the first several terms of the sequence. It is often a good idea to write the terms in both simplified and factored forms.

2. Try to find a recognizable pattern for the terms and write a formula for the nth term of the sequence. This is your *hypothesis* or *conjecture*. You might try computing one or two more terms in the sequence to test your hypothesis.

3. Use mathematical induction to prove your hypothesis.

Example 6 ▶ Finding a Formula for a Finite Sum

Find a formula for the finite sum and prove its validity.

$$\frac{1}{1 \cdot 2} + \frac{1}{2 \cdot 3} + \frac{1}{3 \cdot 4} + \frac{1}{4 \cdot 5} + \cdot \cdot \cdot + \frac{1}{n(n + 1)}$$

Solution

Begin by writing out the first few sums.

$$S_1 = \frac{1}{1 \cdot 2} = \frac{1}{2} = \frac{1}{1 + 1}$$

$$S_2 = \frac{1}{1 \cdot 2} + \frac{1}{2 \cdot 3} = \frac{4}{6} = \frac{2}{3} = \frac{2}{2 + 1}$$

$$S_3 = \frac{1}{1 \cdot 2} + \frac{1}{2 \cdot 3} + \frac{1}{3 \cdot 4} = \frac{9}{12} = \frac{3}{4} = \frac{3}{3 + 1}$$

$$S_4 = \frac{1}{1 \cdot 2} + \frac{1}{2 \cdot 3} + \frac{1}{3 \cdot 4} + \frac{1}{4 \cdot 5} = \frac{48}{60} = \frac{4}{5} = \frac{4}{4 + 1}$$

From this sequence, it appears that the formula for the kth sum is

$$S_k = \frac{1}{1 \cdot 2} + \frac{1}{2 \cdot 3} + \frac{1}{3 \cdot 4} + \frac{1}{4 \cdot 5} + \cdot \cdot \cdot + \frac{1}{k(k + 1)} = \frac{k}{k + 1}.$$

To prove the validity of this hypothesis, use mathematical induction, as follows. Note that you have already verified the formula for $n = 1$, so you can begin by assuming that the formula is valid for $n = k$ and trying to show that it is valid for $n = k + 1$.

$$\begin{aligned} S_{k+1} &= \left[\frac{1}{1 \cdot 2} + \frac{1}{2 \cdot 3} + \frac{1}{3 \cdot 4} + \frac{1}{4 \cdot 5} + \cdot \cdot \cdot + \frac{1}{k(k + 1)}\right] + \frac{1}{(k + 1)(k + 2)} \\ &= \frac{k}{k + 1} + \frac{1}{(k + 1)(k + 2)} \\ &= \frac{k(k + 2) + 1}{(k + 1)(k + 2)} \\ &= \frac{k^2 + 2k + 1}{(k + 1)(k + 2)} \\ &= \frac{(k + 1)^2}{(k + 1)(k + 2)} \\ &= \frac{k + 1}{k + 2} \end{aligned}$$

So, the hypothesis is valid.

STUDY TIP

For a linear model, the first differences should be the same nonzero number. For a quadratic model, the second differences are the same nonzero number.

Finite Differences

The **first differences** of a sequence are found by subtracting consecutive terms. The **second differences** are found by subtracting consecutive first differences. The first and second differences of the sequence 3, 5, 8, 12, 17, 23, . . . are as follows.

n:	1		2		3		4		5		6
a_n:	3		5		8		12		17		23
First differences:		2		3		4		5		6	
Second differences:			1		1		1		1		

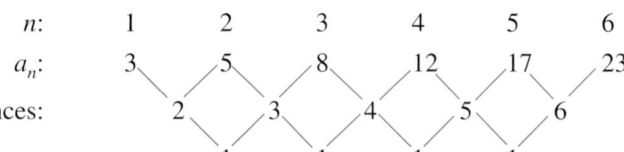

For this sequence, the second differences are all the same. When this happens, the sequence has a perfect *quadratic* model. If the first differences are all the same, the sequence has a *linear* model. That is, it is arithmetic.

Group Activity

A regular n-sided polygon is a polygon that has n equal sides and n equal angles. For instance, an equilateral triangle is a regular three-sided polygon. Each angle of an equilateral triangle measures 60°, and the sum of all three angles is 180°. Similarly, the sum of the four angles of a four-sided polygon (a square) is 360°.

Polygon	*Number of Sides*	*Sum of Angles*
Equilateral triangle	3	180°
Square	4	360°
Regular pentagon	5	540°
Regular hexagon	6	720°

a. The list above gives the sums of the angles of four regular polygons. Use this data to write a conjecture about the sum of the angles of any regular n-sided polygon.

b. Discuss how you could *prove* that your formula is valid.

Example 7 ▶ Finding a Quadratic Model

Find the quadratic model for the sequence

3, 5, 8, 12, 17, 23, ...

Solution

You know from the second differences shown above that the model is quadratic and has the form

$$a_n = an^2 + bn + c.$$

By substituting 1, 2, and 3 for n, you can obtain a system of three linear equations in three variables.

$$a_1 = a(1)^2 + b(1) + c = 3 \qquad \text{Substitute 1 for } n.$$

$$a_2 = a(2)^2 + b(2) + c = 5 \qquad \text{Substitute 2 for } n.$$

$$a_3 = a(3)^2 + b(3) + c = 8 \qquad \text{Substitute 3 for } n.$$

You now have a system of three equations in a, b, and c.

$$\begin{cases} a + b + c = 3 & \text{Equation 1} \\ 4a + 2b + c = 5 & \text{Equation 2} \\ 9a + 3b + c = 8 & \text{Equation 3} \end{cases}$$

Using the techniques discussed in Chapter 7, you can find the solution to be $a = \frac{1}{2}$, $b = \frac{1}{2}$, and $c = 2$. So, the quadratic model is

$$a_n = \frac{1}{2}n^2 + \frac{1}{2}n + 2.$$

Try checking the values of a_1, a_2, and a_3.

9.4 Exercises

In Exercises 1–4, find P_{k+1} for the given P_k.

1. $P_k = \dfrac{5}{k(k+1)}$

2. $P_k = \dfrac{1}{2(k+2)}$

3. $P_k = \dfrac{k^2(k+1)^2}{4}$

4. $P_k = \dfrac{k}{3}(2k+1)$

In Exercises 5–18, use mathematical induction to prove the formula for every positive integer *n*.

5. $2 + 4 + 6 + 8 + \cdots + 2n = n(n+1)$

6. $3 + 7 + 11 + 15 + \cdots + (4n-1) = n(2n+1)$

7. $2 + 7 + 12 + 17 + \cdots + (5n-3) = \dfrac{n}{2}(5n-1)$

8. $1 + 4 + 7 + 10 + \cdots + (3n-2) = \dfrac{n}{2}(3n-1)$

9. $1 + 2 + 2^2 + 2^3 + \cdots + 2^{n-1} = 2^n - 1$

10. $2(1 + 3 + 3^2 + 3^3 + \cdots + 3^{n-1}) = 3^n - 1$

11. $1 + 2 + 3 + 4 + \cdots + n = \dfrac{n(n+1)}{2}$

12. $1^2 + 2^2 + 3^2 + 4^2 + \cdots + n^2 = \dfrac{n(n+1)(2n+1)}{6}$

13. $1^3 + 2^3 + 3^3 + 4^3 + \cdots + n^3 = \dfrac{n^2(n+1)^2}{4}$

14. $\left(1 + \dfrac{1}{1}\right)\left(1 + \dfrac{1}{2}\right)\left(1 + \dfrac{1}{3}\right) \cdots \left(1 + \dfrac{1}{n}\right) = n + 1$

15. $\displaystyle\sum_{i=1}^{n} i^5 = \frac{n^2(n+1)^2(2n^2+2n-1)}{12}$

16. $\displaystyle\sum_{i=1}^{n} i^4 = \frac{n(n+1)(2n+1)(3n^2+3n-1)}{30}$

17. $\displaystyle\sum_{i=1}^{n} i(i+1) = \frac{n(n+1)(n+2)}{3}$

18. $\displaystyle\sum_{i=1}^{n} \frac{1}{(2i-1)(2i+1)} = \frac{n}{2n+1}$

In Exercises 19–28, find the sum using the formulas for the sums of powers of integers.

19. $\displaystyle\sum_{n=1}^{15} n$

20. $\displaystyle\sum_{n=1}^{30} n$

21. $\displaystyle\sum_{n=1}^{6} n^2$

22. $\displaystyle\sum_{n=1}^{10} n^3$

23. $\displaystyle\sum_{n=1}^{5} n^4$

24. $\displaystyle\sum_{n=1}^{8} n^5$

25. $\displaystyle\sum_{n=1}^{6} (n^2 - n)$

26. $\displaystyle\sum_{n=1}^{20} (n^3 - n)$

27. $\displaystyle\sum_{i=1}^{6} (6i - 8i^3)$

28. $\displaystyle\sum_{j=1}^{10} \left(3 - \tfrac{1}{2}j + \tfrac{1}{2}j^2\right)$

In Exercises 29–34, find a formula for the sum of the first *n* terms of the sequence.

29. $1, 5, 9, 13, \ldots$

30. $25, 22, 19, 16, \ldots$

31. $1, \frac{9}{10}, \frac{81}{100}, \frac{729}{1000}, \ldots$

32. $3, -\frac{9}{2}, \frac{27}{4}, -\frac{81}{8}, \ldots$

33. $\dfrac{1}{4}, \dfrac{1}{12}, \dfrac{1}{24}, \dfrac{1}{40}, \ldots, \dfrac{1}{2n(n+1)}, \ldots$

34. $\dfrac{1}{2 \cdot 3}, \dfrac{1}{3 \cdot 4}, \dfrac{1}{4 \cdot 5}, \dfrac{1}{5 \cdot 6}, \ldots, \dfrac{1}{(n+1)(n+2)}, \ldots$

In Exercises 35–40, prove the inequality for the indicated integer values of *n*.

35. $n! > 2^n, \quad n \geq 4$

36. $\left(\frac{4}{3}\right)^n > n, \quad n \geq 7$

37. $\dfrac{1}{\sqrt{1}} + \dfrac{1}{\sqrt{2}} + \dfrac{1}{\sqrt{3}} + \cdots + \dfrac{1}{\sqrt{n}} > \sqrt{n}, \quad n \geq 2$

38. $\left(\dfrac{x}{y}\right)^{n+1} < \left(\dfrac{x}{y}\right)^n, \quad n \geq 1$ and $0 < x < y$.

39. $(1 + a)^n \geq na, \quad n \geq 1$ and $a > 0$

40. $2n^2 > (n+1)^2, \quad n \geq 3$

In Exercises 41–49, use mathematical induction to prove the given property for all positive integers *n*.

41. $(ab)^n = a^n b^n$

42. $\left(\dfrac{a}{b}\right)^n = \dfrac{a^n}{b^n}$

43. If $x_1 \neq 0, x_2 \neq 0, \ldots, x_n \neq 0$, then

$(x_1 x_2 x_3 \cdots x_n)^{-1} = x_1^{-1} x_2^{-1} x_3^{-1} \cdots x_n^{-1}$.

44. If $x_1 > 0, x_2 > 0, \ldots, x_n > 0$, then

$\ln(x_1 x_2 x_3 \cdots x_n) = \ln x_1 + \ln x_2 + \ln x_3 + \cdots + \ln x_n$.

45. Generalized Distributive Law:

$$x(y_1 + y_2 + \cdots + y_n) = xy_1 + xy_2 + \cdots + xy_n$$

46. $(a + bi)^n$ and $(a - bi)^n$ are complex conjugates for all $n \geq 1$.

47. A factor of $(n^3 + 3n^2 + 2n)$ is 3.

48. A factor of $(2^{2n-1} + 3^{2n-1})$ is 5.

49. A factor of $(9^n - 8n - 1)$ is 64 for all $n \geq 2$.

In Exercises 50–53, write the first five terms of the sequence.

50. $a_0 = 1$, $a_n = a_{n-1} + 2$

51. $a_0 = 10$, $a_n = 4a_{n-1}$

52. $a_0 = 4$, $a_1 = 2$, $a_n = a_{n-1} - a_{n-2}$

53. $a_0 = 0$, $a_1 = 2$, $a_n = a_{n-1} + 2a_{n-2}$

In Exercises 54–63, write the first five terms of the sequence. Then calculate the first and second differences of the sequence. Does the sequence have a linear model, a quadratic model, or neither?

54. $a_1 = 0$, $a_n = a_{n-1} + 3$

55. $a_1 = 2$, $a_n = n - a_{n-1}$

56. $a_1 = 3$, $a_n = a_{n-1} - n$

57. $a_2 = -3$, $a_n = -2a_{n-1}$

58. $a_0 = 0$, $a_n = a_{n-1} + n$

59. $a_0 = 2$, $a_n = (a_{n-1})^2$

60. $a_1 = 2$, $a_n = a_{n-1} + 2$

61. $a_1 = 0$, $a_n = a_{n-1} + 2n$

62. $a_0 = 1$, $a_n = a_{n-1} + n^2$

63. $a_0 = 0$, $a_n = a_{n-1} - 1$

In Exercises 64–67, find a quadratic model for the sequence with the indicated terms.

64. $a_0 = 3,\ a_1 = 3,\ a_4 = 15$

65. $a_0 = 7,\ a_1 = 6,\ a_3 = 10$

66. $a_0 = -3,\ a_2 = 1,\ a_4 = 9$

67. $a_0 = 3,\ a_2 = 0,\ a_6 = 36$

Synthesis

True or False? **In Exercises 68–71, determine whether the statement is true or false. Justify your answer.**

68. If the statement P_1 is true but the true statement P_6 does *not* imply that the statement P_7 is true, then P_n is not necessarily true for all positive integers n.

69. If the statement P_k is true and P_k implies P_{k+1}, then P_1 is also true.

70. If the second differences of a sequence are all zero, then the sequence is arithmetic.

71. A sequence with n terms has $n - 1$ second differences.

72. *Writing* In your own words, explain what is meant by a proof by mathematical induction.

73. *Think About It* What conclusion can be drawn from the information about the sequence of statements P_n?

(a) P_3 is true and P_k implies P_{k+1}.

(b) $P_1, P_2, P_3, \ldots, P_{50}$ are all true.

(c) P_1, P_2, and P_3 are all true, but the truth of P_k does not imply that P_{k+1} is true.

(d) P_2 is true and P_{2k} implies P_{2k+2}.

Review

In Exercises 74–77, solve the system of equations.

74. $\begin{cases} y = x^2 \\ -3x + 2y = 2 \end{cases}$

75. $\begin{cases} x - y^3 = 0 \\ x - 2y^2 = 0 \end{cases}$

76. $\begin{cases} x - y = -1 \\ x + 2y - 2z = 3 \\ 3x - y + 2z = 3 \end{cases}$

77. $\begin{cases} 2x + y - 2z = 1 \\ x - z = 1 \\ 3x + 3y + z = 12 \end{cases}$

In Exercises 78–81, find the product.

78. $(2x^2 - 1)^2$

79. $(2x - y)^2$

80. $(5 - 4x)^3$

81. $(2x - 4y)^3$

9.5 The Binomial Theorem

▶ What you should learn

- How to use the Binomial Theorem to calculate binomial coefficients
- How to use Pascal's Triangle to calculate binomial coefficients
- How to use binomial coefficients to write binomial expansions

▶ Why you should learn it

You can use binomial coefficients to model and solve real-life problems. For instance, in Exercise 69 on page 734, you are asked to use binomial coefficients to find the probability that a baseball player will get three hits during the next 10 times at bat.

Jonathan Daniel/Allsport

Binomial Coefficients

Recall that a binomial is a polynomial that has two terms. In this section, you will study a formula that gives a quick method of raising a binomial to a power. To begin, let's look at the expansion of $(x + y)^n$ for several values of n.

$$(x + y)^0 = 1$$

$$(x + y)^1 = x + y$$

$$(x + y)^2 = x^2 + 2xy + y^2$$

$$(x + y)^3 = x^3 + 3x^2y + 3xy^2 + y^3$$

$$(x + y)^4 = x^4 + 4x^3y + 6x^2y^2 + 4xy^3 + y^4$$

$$(x + y)^5 = x^5 + 5x^4y + 10x^3y^2 + 10x^2y^3 + 5xy^4 + y^5$$

There are several observations you can make about these expansions.

1. In each expansion, there are $n + 1$ terms.
2. In each expansion, x and y have symmetrical roles. The powers of x decrease by 1 in successive terms, whereas the powers of y increase by 1.
3. The sum of the powers of each term is n. For instance, in the expansion of $(x + y)^5$, the sum of the powers of each term is 5.

$$(x + y)^5 = x^5 + 5\overbrace{x^4y^1}^{4+1=5} + 10\overbrace{x^3y^2}^{3+2=5} + 10x^2y^3 + 5x^1y^4 + y^5$$

4. The coefficients increase and then decrease in a symmetric pattern.

The coefficients of a binomial expansion are called **binomial coefficients.** To find them, you can use the **Binomial Theorem.** A proof of this theorem is given in Appendix A.

The Binomial Theorem

In the expansion of $(x + y)^n$

$$(x + y)^n = x^n + nx^{n-1}y + \cdots + {}_nC_r\, x^{n-r}y^r + \cdots + nxy^{n-1} + y^n$$

the coefficient of $x^{n-r}y^r$ is

$${}_nC_r = \frac{n!}{(n-r)!r!}.$$

The symbol $\binom{n}{r}$ is often used in place of ${}_nC_r$ to denote binomial coefficients.

Technology

Most graphing calculators are programmed to evaluate ${}_nC_r$. Consult your user's manual and then evaluate ${}_8C_5$. You should get an answer of 56.

Example 1 ▶ Finding Binomial Coefficients

Find the binomial coefficients.

a. ${}_8C_2$ **b.** $\binom{10}{3}$ **c.** ${}_7C_0$ **d.** $\binom{8}{8}$

Solution

a. $${}_8C_2 = \frac{8!}{6! \cdot 2!} = \frac{(8 \cdot 7) \cdot \cancel{6!}}{\cancel{6!} \cdot 2!} = \frac{8 \cdot 7}{2 \cdot 1} = 28$$

b. $$\binom{10}{3} = \frac{10!}{7! \cdot 3!} = \frac{(10 \cdot 9 \cdot 8) \cdot \cancel{7!}}{\cancel{7!} \cdot 3!} = \frac{10 \cdot 9 \cdot 8}{3 \cdot 2 \cdot 1} = 120$$

c. $${}_7C_0 = \frac{\cancel{7!}}{\cancel{7!} \cdot 0!} = 1$$ **d.** $$\binom{8}{8} = \frac{\cancel{8!}}{0! \cdot \cancel{8!}} = 1$$

When $r \neq 0$ and $r \neq n$, as in parts (a) and (b) above, there is a simple pattern for evaluating binomial coefficients.

$${}_8C_2 = \frac{\overbrace{8 \cdot 7}^{2\text{ factors}}}{\underbrace{2 \cdot 1}_{2\text{ factors}}} \quad \text{and} \quad \binom{10}{3} = \frac{\overbrace{10 \cdot 9 \cdot 8}^{3\text{ factors}}}{\underbrace{3 \cdot 2 \cdot 1}_{3\text{ factors}}}$$

Example 2 ▶ Finding Binomial Coefficients

Find the binomial coefficients.

a. ${}_7C_3$ **b.** $\binom{7}{4}$ **c.** ${}_{12}C_1$ **d.** $\binom{12}{11}$

Solution

a. $${}_7C_3 = \frac{7 \cdot \cancel{6} \cdot 5}{\cancel{3} \cdot \cancel{2} \cdot 1} = 35$$

b. $$\binom{7}{4} = \frac{7 \cdot \cancel{6} \cdot 5 \cdot \cancel{4}}{\cancel{4} \cdot \cancel{3} \cdot \cancel{2} \cdot 1} = 35$$

c. $${}_{12}C_1 = \frac{12}{1} = 12$$

d. $$\binom{12}{11} = \frac{12!}{1! \cdot 11!} = \frac{(12) \cdot \cancel{11!}}{1! \cdot \cancel{11!}} = \frac{12}{1} = 12$$

It is not a coincidence that the results in parts (a) and (b) of Example 2 are the same and that the results in parts (c) and (d) are the same. In general, it is true that

$${}_nC_r = {}_nC_{n-r}.$$

This shows the symmetric property of binomial coefficients that was identified earlier.

Pascal's Triangle

There is a convenient way to remember the pattern for binomial coefficients. By arranging the coefficients in a triangular pattern, you obtain **Pascal's Triangle.** This triangle is named after the famous French mathematician Blaise Pascal (1623–1662).

$$
\begin{array}{ccccccccccccccc}
 & & & & & & & 1 & & & & & & & \\
 & & & & & & 1 & & 1 & & & & & & \\
 & & & & & 1 & & 2 & & 1 & & & & & \\
 & & & & 1 & & 3 & & 3 & & 1 & & & & \\
 & & & 1 & & 4 & & 6 & & 4 & & 1 & & & \\
 & & 1 & & 5 & & 10 & & 10 & & 5 & & 1 & & \\
 & 1 & & 6 & & 15 & & 20 & & 15 & & 6 & & 1 & \\
1 & & 7 & & 21 & & 35 & & 35 & & 21 & & 7 & & 1
\end{array}
$$

$4 + 6 = 10$

$15 + 6 = 21$

> **Exploration**
>
> Complete the following table and describe the result.
>
> | n | r | ${}_nC_r$ | ${}_nC_{n-r}$ |
> |---|---|---|---|
> | 9 | 5 | | |
> | 7 | 1 | | |
> | 12 | 4 | | |
> | 6 | 0 | | |
> | 10 | 7 | | |
>
> What characteristic of Pascal's Triangle is illustrated by this table?

The first and last numbers in each row of Pascal's Triangle are 1. Every other number in each row is formed by adding the two numbers immediately above the number. Pascal noticed that numbers in this triangle are precisely the same numbers that are the coefficients of binomial expansions, as follows.

$$(x + y)^0 = 1$$

$$(x + y)^1 = 1x + 1y$$

$$(x + y)^2 = 1x^2 + 2xy + 1y^2$$

$$(x + y)^3 = 1x^3 + 3x^2y + 3xy^2 + 1y^3$$

$$(x + y)^4 = 1x^4 + 4x^3y + 6x^2y^2 + 4xy^3 + 1y^4$$

$$(x + y)^5 = 1x^5 + 5x^4y + 10x^3y^2 + 10x^2y^3 + 5xy^4 + 1y^5$$

$$(x + y)^6 = 1x^6 + 6x^5y + 15x^4y^2 + 20x^3y^3 + 15x^2y^4 + 6xy^5 + 1y^6$$

$$(x + y)^7 = 1x^7 + 7x^6y + 21x^5y^2 + 35x^4y^3 + 35x^3y^4 + 21x^2y^5 + 7xy^6 + 1y^7$$

The top row in Pascal's Triangle is called the *zero row* because it corresponds to the binomial expansion

$$(x + y)^0 = 1.$$

Similarly, the next row is called the *first row* because it corresponds to the binomial expansion $(x + y)^1 = 1(x) + 1(y)$. In general, the *nth row* in Pascal's Triangle gives the coefficients of $(x + y)^n$.

Example 3 ▶ Using Pascal's Triangle

Use the seventh row of Pascal's Triangle to find the binomial coefficients.

$${}_8C_0, {}_8C_1, {}_8C_2, {}_8C_3, {}_8C_4, {}_8C_5, {}_8C_6, {}_8C_7, {}_8C_8$$

Solution

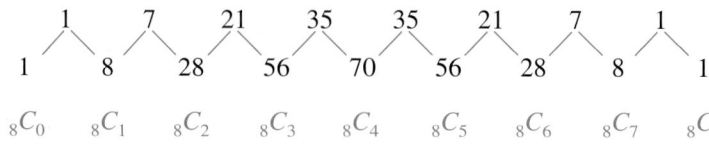

$$
\begin{array}{ccccccccccccccccc}
 & 1 & & 7 & & 21 & & 35 & & 35 & & 21 & & 7 & & 1 & \\
1 & & 8 & & 28 & & 56 & & 70 & & 56 & & 28 & & 8 & & 1 \\
{}_8C_0 & & {}_8C_1 & & {}_8C_2 & & {}_8C_3 & & {}_8C_4 & & {}_8C_5 & & {}_8C_6 & & {}_8C_7 & & {}_8C_8
\end{array}
$$

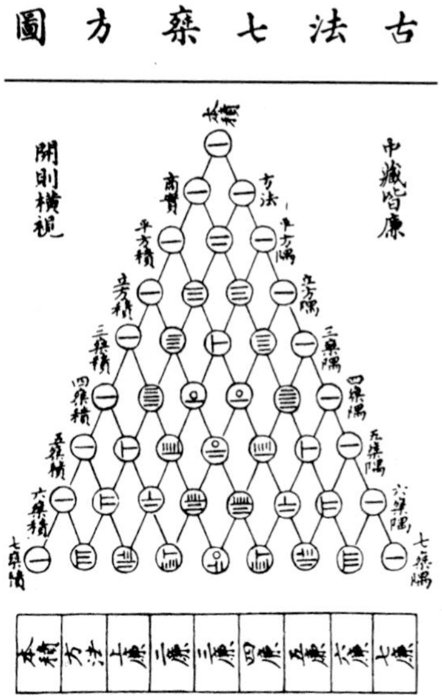

Historical Note

Precious Mirror "Pascal's" Triangle and forms of the Binomial Theorem were known in Eastern cultures prior to the Western "discovery" of the theorem. A Chinese text entitled *Precious Mirror* contains a triangle of binomial expansions through the eighth power.

A computer animation of this concept appears in the *Interactive* CD-ROM and *Internet* versions of this text.

Binomial Expansions

As mentioned at the beginning of this section, when you write out the coefficients for a binomial that is raised to a power, you are **expanding a binomial.** The formulas for binomial coefficients give you an easy way to expand binomials, as demonstrated in the next four examples.

Example 4 ▶ Expanding a Binomial

Write the expansion for the expression

$(x + 1)^3$.

Solution

The binomial coefficients from the third row of Pascal's Triangle are

1, 3, 3, 1.

Therefore, the expansion is as follows.

$$(x + 1)^3 = (1)x^3 + (3)x^2(1) + (3)x(1^2) + (1)(1^3)$$
$$= x^3 + 3x^2 + 3x + 1$$

To expand binomials representing *differences* rather than sums, you alternate signs. Here are two examples.

$$(x - 1)^3 = x^3 - 3x^2 + 3x - 1$$
$$(x - 1)^4 = x^4 - 4x^3 + 6x^2 - 4x + 1$$

Example 5 ▶ Expanding a Binomial

Write the expansion for each expression.

a. $(2x + 3)^4$ **b.** $(2x - 3)^4$

Solution

a. The binomial coefficients from the fourth row of Pascal's Triangle are

1, 4, 6, 4, 1.

Therefore, the expansion is as follows.

$$(2x + 3)^4 = (1)(2x)^4 + (4)(2x)^3(3) + (6)(2x)^2(3^2) + (4)(2x)(3^3) + (1)(3^4)$$
$$= 16x^4 + 96x^3 + 216x^2 + 216x + 81$$

b. The binomial coefficients from the fourth row of Pascal's Triangle are

1, 4, 6, 4, 1.

Therefore, the expansion is as follows.

$$(2x - 3)^4 = (1)(2x)^4 - (4)(2x)^3(3) + (6)(2x)^2(3^2) - (4)(2x)(3^3) + (1)(3^4)$$
$$= 16x^4 - 96x^3 + 216x^2 - 216x + 81$$

Activities

1. Expand and simplify $(x - 3)^4$.
   Answer:
   $x^4 - 12x^3 + 54x^2 - 108x + 81$
2. Find the fifth term in the expansion of $(2x + 1)^9$.
   Answer:
   $\frac{9 \cdot 8 \cdot 7 \cdot 6 \cdot 5}{5 \cdot 4 \cdot 3 \cdot 2 \cdot 1}(2x)^5(1)^4 = 4032x^5$
3. Evaluate ${}_8C_3$.
   Answer: $\frac{8 \cdot 7 \cdot 6}{3 \cdot 2 \cdot 1} = 56$

Example 6 ▶ Expanding a Binomial

Write the expansion for $(x - 2y)^4$.

Solution

Use the fourth row of Pascal's Triangle, as follows.

$$\begin{aligned}(x-2y)^4 &= (1)x^4 - (4)x^3(2y) + (6)x^2(2y)^2 - (4)x(2y)^3 + (1)(2y)^4 \\ &= x^4 - 8x^3y + 24x^2y^2 - 32xy^3 + 16y^4\end{aligned}$$

Example 7 ▶ Expanding a Binomial

Write the expansion for $(x^2 + 4)^3$.

Solution

Use the third row of Pascal's Triangle, as follows.

$$\begin{aligned}(x^2+4)^3 &= (1)(x^2)^3 + (3)(x^2)^2(4) + (3)x^2(4^2) + (1)(4^3) \\ &= x^6 + 12x^4 + 48x^2 + 64\end{aligned}$$

Group Activity

Add the binomial coefficients in each of the first five rows of Pascal's Triangle. What pattern do you see? Work together to use the pattern to find the sums of the terms in the 10th, 15th, and 20th rows of Pascal's Triangle. Check your answer by actually adding the terms of the 10th, 15th, and 20th rows.

Example 8 ▶ Finding a Term in a Binomial Expansion

Find the sixth term of $(a + 2b)^8$.

Solution

For the first term of the binomial expansion, you would use $n = 8$ and $r = 0$ to get ${}_8C_0a^{8-0}(2b)^0$. For the second term of the binomial expansion, you would use $n = 8$ and $r = 1$ to get ${}_8C_1a^{8-1}(2b)^1$. So, for the sixth term of this binomial expansion, use $n = 8$ and $r = 5$ to get

$$\begin{aligned}{}_8C_5a^{8-5}(2b)^5 &= 56 \cdot a^3 \cdot (2b)^5 \\ &= 56(2^5)a^3b^5 \\ &= 1792a^3b^5.\end{aligned}$$

Writing ABOUT MATHEMATICS

Error Analysis You are a math instructor and receive the following solutions from one of your students on a quiz. Find the error(s) in each solution. Discuss ways that your student could avoid the error(s) in the future.

a. Find the second term in the expansion of $(2x - 3y)^5$.

$$5(2x)^4(3y)^2 = 720x^4y^2 \quad \times$$

b. Find the fourth term in the expansion of $\left(\frac{1}{2}x + 7y\right)^6$.

$${}_6C_4\left(\tfrac{1}{2}x\right)^2(7y)^4 = 9003.75x^2y^4 \quad \times$$

9.5 Exercises

In Exercises 1–10, find the binomial coefficients.

1. ${}_5C_3$
2. ${}_8C_6$
3. ${}_{12}C_0$
4. ${}_{20}C_{20}$
5. ${}_{20}C_{15}$
6. ${}_{12}C_5$
7. $\binom{10}{4}$
8. $\binom{10}{6}$
9. $\binom{100}{98}$
10. $\binom{100}{2}$

In Exercises 11–14, evaluate using Pascal's Triangle.

11. $\binom{8}{5}$
12. $\binom{8}{7}$
13. ${}_7C_4$
14. ${}_6C_3$

In Exercises 15–34, use the Binomial Theorem to expand and simplify the expression.

15. $(x + 1)^4$
16. $(x + 1)^6$
17. $(a + 6)^4$
18. $(a + 5)^5$
19. $(y - 4)^3$
20. $(y - 2)^5$
21. $(x + y)^5$
22. $(c + d)^3$
23. $(r + 3s)^6$
24. $(x + 2y)^4$
25. $(3a - b)^5$
26. $(2x - y)^5$
27. $(1 - 2x)^3$
28. $(5 - 3y)^3$
29. $(x^2 + 5)^4$
30. $(x^2 + y^2)^6$
31. $\left(\frac{1}{x} + y\right)^5$
32. $\left(\frac{1}{x} + 2y\right)^6$
33. $2(x - 3)^4 + 5(x - 3)^2$
34. $3(x + 1)^5 - 4(x + 1)^3$

In Exercises 35–38, expand the binomial using Pascal's Triangle to determine the coefficients.

35. $(2t - s)^5$
36. $(3 - 2z)^4$
37. $(x + 2y)^5$
38. $(2v + 3)^6$

In Exercises 39–46, find the coefficient a of the term in the expansion of the binomial.

	Binomial	*Term*
39.	$(x + 3)^{12}$	ax^5
40.	$(x^2 + 3)^{12}$	ax^8
41.	$(x - 2y)^{10}$	ax^8y^2
42.	$(4x - y)^{10}$	ax^2y^8
43.	$(3x - 2y)^9$	ax^4y^5
44.	$(2x - 3y)^8$	ax^6y^2
45.	$(x^2 + y)^{10}$	ax^8y^6
46.	$(z^2 - t)^{10}$	az^4t^8

In Exercises 47–50, use the Binomial Theorem to expand and simplify the expression.

47. $\left(\sqrt{x} + 3\right)^4$
48. $\left(2\sqrt{t} - 1\right)^3$
49. $(x^{2/3} - y^{1/3})^3$
50. $(u^{3/5} + 2)^5$

In Exercises 51–54, expand the binomial in the difference quotient and simplify.

$$\frac{f(x + h) - f(x)}{h} \qquad \text{Difference quotient}$$

51. $f(x) = x^3$
52. $f(x) = x^4$
53. $f(x) = \sqrt{x}$
54. $f(x) = \frac{1}{x}$

In Exercises 55–60, use the Binomial Theorem to expand the complex number. Simplify your result.

55. $(1 + i)^4$
56. $(2 - i)^5$
57. $(2 - 3i)^6$
58. $\left(5 + \sqrt{-9}\right)^3$
59. $\left(-\frac{1}{2} + \frac{\sqrt{3}}{2}i\right)^3$
60. $\left(5 - \sqrt{3}i\right)^4$

Approximation **In Exercises 61–64, use the Binomial Theorem to approximate the given quantity accurate to three decimal places. For example, in Exercise 61, use the expansion**

$$(1.02)^8 = (1 + 0.02)^8 = 1 + 8(0.02) + 28(0.02)^2 + \cdots.$$

61. $(1.02)^8$
62. $(2.005)^{10}$
63. $(2.99)^{12}$
64. $(1.98)^9$

Graphical Reasoning **In Exercises 65 and 66, use a graphing utility to graph f and g in the same viewing window. What is the relationship between the two graphs? Use the Binomial Theorem to write the polynomial function g in standard form.**

65. $f(x) = x^3 - 4x, \quad g(x) = f(x + 4)$
66. $f(x) = -x^4 + 4x^2 - 1, \quad g(x) = f(x - 3)$

67. *Graphical Reasoning* Use a graphing utility to graph the functions in the given order and in the same viewing window. Compare the graphs. Which two functions have identical graphs, and why?

(a) $f(x) = (1 - x)^3$

(b) $g(x) = 1 - 3x$

(c) $h(x) = 1 - 3x + 3x^2$

(d) $p(x) = 1 - 3x + 3x^2 - x^3$

Probability **In Exercises 68–71, consider n independent trials of an experiment in which each trial has two possible outcomes: "success" or "failure." The probability of a success on each trial is p, and the probability of a failure is $q = 1 - p$. In this context, the term ${}_nC_k\, p^k q^{n-k}$ in the expansion of $(p + q)^n$ gives the probability of k successes in the n trials of the experiment.**

68. A fair coin is tossed seven times. To find the probability of obtaining four heads, evaluate the term

$$ {}_7C_4\left(\frac{1}{2}\right)^4\left(\frac{1}{2}\right)^3 $$

in the expansion of $\left(\frac{1}{2} + \frac{1}{2}\right)^7$.

69. The probability of a baseball player getting a hit during any given time at bat is $\frac{1}{4}$. To find the probability that the player gets three hits during the next 10 times at bat, evaluate the term

$$ {}_{10}C_3\left(\frac{1}{4}\right)^3\left(\frac{3}{4}\right)^7 $$

in the expansion of $\left(\frac{1}{4} + \frac{3}{4}\right)^{10}$.

70. The probability of a sales representative making a sale with any one customer is $\frac{1}{3}$. The sales representative makes eight contacts a day. To find the probability of making four sales, evaluate the term

$$ {}_8C_4\left(\frac{1}{3}\right)^4\left(\frac{2}{3}\right)^4 $$

in the expansion of $\left(\frac{1}{3} + \frac{2}{3}\right)^8$.

71. To find the probability that the sales representative in Exercise 70 makes four sales if the probability of a sale with any one customer is $\frac{1}{2}$, evaluate the term

$$ {}_8C_4\left(\frac{1}{2}\right)^4\left(\frac{1}{2}\right)^4 $$

in the expansion of $\left(\frac{1}{2} + \frac{1}{2}\right)^8$.

72. *Life Insurance* The average amount of life insurance per household $f(t)$ (in thousands of dollars) from 1980 through 1996 can be approximated by

$$ f(t) = 0.0348t^2 + 5.1083t + 41.0250, \quad 0 \le t \le 16 $$

where $t = 0$ represents 1980 (see figure). You want to adjust this model so that $t = 0$ corresponds to 1990 rather than 1980. To do this, you shift the graph of f 10 units *to the left* and obtain

$$ g(t) = f(t + 10). $$

(Source: American Council of Life Insurance)

(a) Write $g(t)$ in standard form.

(b) Use a graphing utility to graph f and g in the same viewing window.

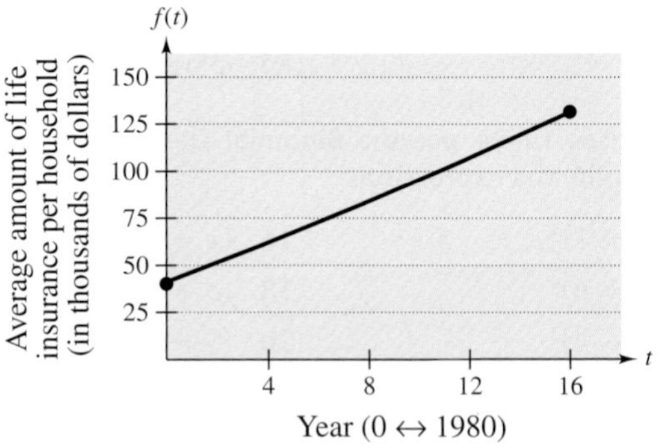

73. *Health Maintenance Organizations* The number of people $f(t)$ (in millions) enrolled in health maintenance organizations in the United States from 1976 through 1997 can be approximated by the model

$$ f(t) = 0.0834t^2 + 0.07657t + 5.3680, \quad 0 \le t \le 21 $$

where $t = 0$ represents 1976 (see figure). You want to adjust this model so that $t = 0$ corresponds to 1980 rather than 1976. To do this, you shift the graph of f 4 units *to the left* and obtain

$$ g(t) = f(t + 4). $$

(Source: Group Health Association of America, Interstudy)

(a) Write $g(t)$ in standard form.

(b) Use a graphing utility to graph f and g in the same viewing window.

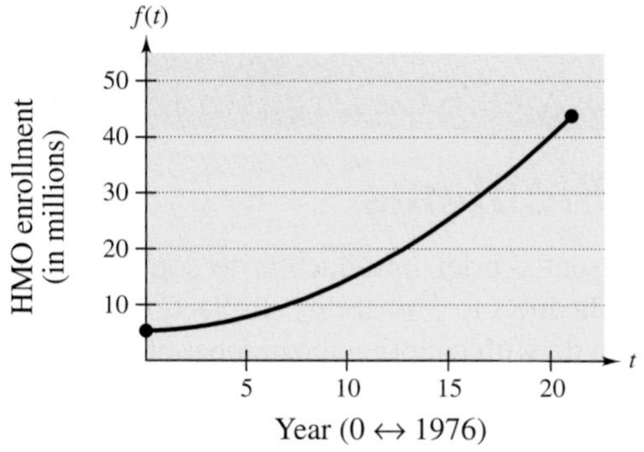

FIGURE FOR 73

Synthesis

True or False? **In Exercises 74 and 75, determine whether the statement is true or false. Justify your answer.**

74. The Binomial Theorem could be used to produce each row of Pascal's Triangle.

75. A binomial that represents a difference cannot always be accurately expanded using the Binomial Theorem.

76. *Writing* In your own words, explain how to form the rows of Pascal's Triangle.

77. Form the first nine rows of Pascal's Triangle.

78. *Think About It* How many terms are in the expansion of $(x + y)^n$?

79. *Think About It* How do the expansions of $(x + y)^n$ and $(x - y)^n$ differ?

In Exercises 80–83, prove the given property for all integers *r* and *n* where $0 \le r \le n$.

80. ${}_nC_r = {}_nC_{n-r}$

81. ${}_nC_0 - {}_nC_1 + {}_nC_2 - \cdot \cdot \cdot \pm {}_nC_n = 0$

82. ${}_{n+1}C_r = {}_nC_r + {}_nC_{r-1}$

83. The sum of the numbers in the nth row of Pascal's Triangle is 2^n.

Review

In Exercises 84–87, describe the relationship between the graphs of *f* and *g*.

84. $g(x) = f(x) + 8$

85. $g(x) = f(x - 3)$

86. $g(x) = f(-x)$

87. $g(x) = -f(x)$

In Exercises 88–91, the graph of $y = g(x)$ is shown. Graph *f* and use the graph to write an equation for the graph of *g*.

88. $f(x) = x^2$

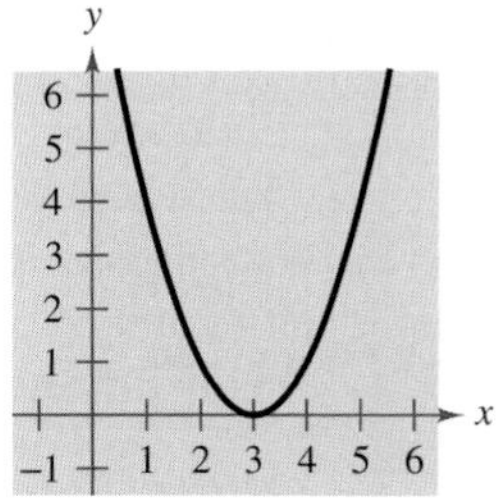

89. $f(x) = x^2$

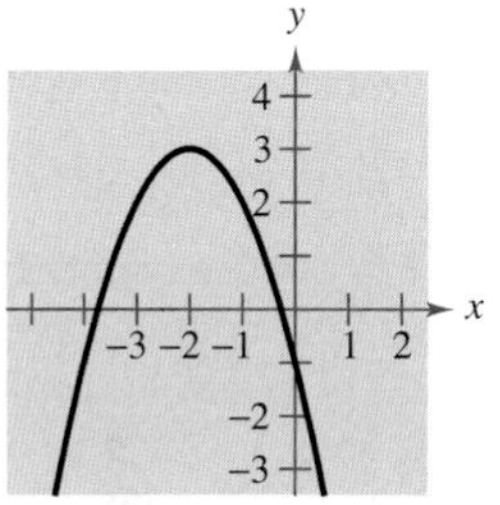

90. $f(x) = \sqrt{x}$

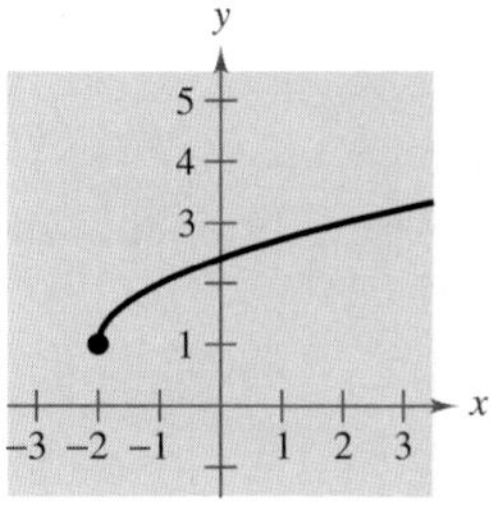

91. $f(x) = \sqrt{x}$

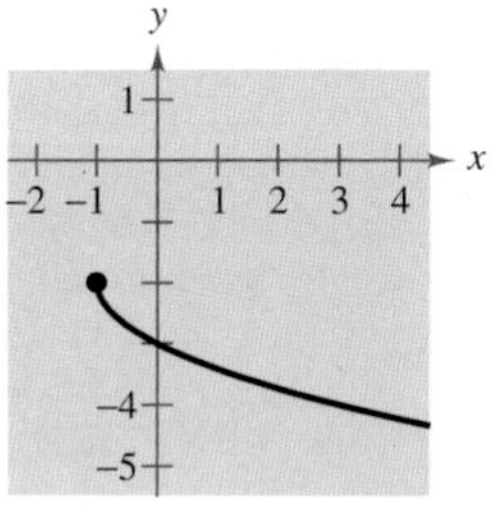

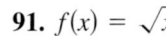

9.6 Counting Principles

▶ **What you should learn**

- How to solve simple counting problems
- How to use the Fundamental Counting Principle to solve counting problems
- How to use permutations to solve counting problems
- How to use combinations to solve counting problems

▶ **Why you should learn it**

You can use counting principles to solve counting problems that occur in real life. For instance, in Exercises 53 and 54 on page 744, you are asked to use counting principles to determine the number of possible ways of selecting the numbers on a lottery ticket.

Dion Ogust/The Image Works

Simple Counting Problems

This section and Section 9.7 present a brief introduction to some of the basic counting principles and their application to probability. In Section 9.7 you will see that much of probability has to do with counting the number of ways an event can occur.

Example 1 ▶ Selecting Pairs of Numbers at Random

Eight pieces of paper are numbered from 1 to 8 and placed in a box. One piece of paper is drawn from the box, its number is written down, and the piece of paper is *replaced in the box*. Then, a second piece of paper is drawn from the box, and its number is written down. Finally, the two numbers are added together. How many different ways can a total of 12 be obtained?

Solution

To solve this problem, count the different ways that a total of 12 can be obtained using two numbers from 1 to 8.

First number	4	5	6	7	8
Second number	8	7	6	5	4

From this list, you can see that a total of 12 can occur in five different ways.

Example 2 ▶ Selecting Pairs of Numbers at Random

Eight pieces of paper are numbered from 1 to 8 and placed in a box. Two pieces of paper are drawn from the box *at the same time*, and the numbers on the pieces of paper are written down and totaled. How many different ways can a total of 12 be obtained?

Solution

To solve this problem, count the different ways that a total of 12 can be obtained *using two different numbers* from 1 to 8.

First number	4	5	7	8
Second number	8	7	5	4

So, a total of 12 can be obtained in four different ways.

The difference between the counting problems in Examples 1 and 2 can be expressed by saying that the random selection in Example 1 occurs **with replacement,** whereas the random selection in Example 2 occurs **without replacement,** which eliminates the possibility of choosing two 6's.

You may want to consider opening class by asking students if they can predict which offers more choices for license plates: (a) a plate with three different letters of the alphabet in any order or (b) a plate with four different nonzero digits in any order. You may want to end class by verifying the answers: (a) offers 26 · 25 · 24 = 15,600 choices and (b) offers 9 · 8 · 7 · 6 = 3024 choices.

The Fundamental Counting Principle

Examples 1 and 2 describe simple counting problems in which you can *list* each possible way that an event can occur. When it is possible, this is always the best way to solve a counting problem. However, some events can occur in so many different ways that it is not feasible to write out the entire list. In such cases, you must rely on formulas and counting principles. The most important of these is the **Fundamental Counting Principle.**

Fundamental Counting Principle

Let E_1 and E_2 be two events. The first event E_1 can occur in m_1 different ways. After E_1 has occurred, E_2 can occur in m_2 different ways. The number of ways that the two events can occur is $m_1 \cdot m_2$.

The Fundamental Counting Principle can be extended to three or more events. For instance, the number of ways that three events E_1, E_2, and E_3 can occur is $m_1 \cdot m_2 \cdot m_3$.

Example 3 ▶ Using the Fundamental Counting Principle

How many different pairs of letters from the English alphabet are possible?

Solution

There are two events in this situation. The first event is the choice of the first letter, and the second event is the choice of the second letter. Because the English alphabet contains 26 letters, it follows that the number of two-letter pairs is $26 \cdot 26 = 676$.

Example 4 ▶ Using the Fundamental Counting Principle

Telephone numbers in the United States currently have 10 digits. The first three are the *area code* and the next seven are the *local telephone number.* How many different telephone numbers are possible within each area code? (Note that at this time, a local telephone number cannot begin with 0 or 1.)

Solution

Because the first digit cannot be 0 or 1, there are only eight choices for the first digit. For each of the other six digits, there are 10 choices.

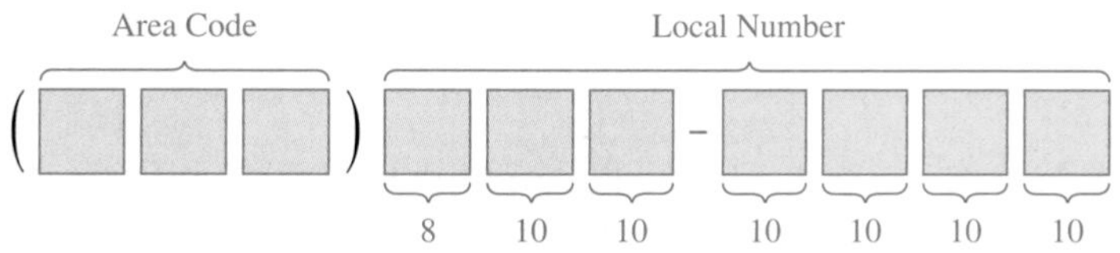

So, the number of local telephone numbers that are possible within each area code is $8 \cdot 10 \cdot 10 \cdot 10 \cdot 10 \cdot 10 \cdot 10 = 8{,}000{,}000$.

Permutations

One important application of the Fundamental Counting Principle is in determining the number of ways that n elements can be arranged (in order). An ordering of n elements is called a **permutation** of the elements.

Definition of Permutation

A **permutation** of n different elements is an ordering of the elements such that one element is first, one is second, one is third, and so on.

Example 5 ▶ Finding the Number of Permutations of n Elements

How many permutations are possible for the letters A, B, C, D, E, and F?

Solution

Consider the following reasoning.

First position: Any of the *six* letters
Second position: Any of the remaining *five* letters
Third position: Any of the remaining *four* letters
Fourth position: Any of the remaining *three* letters
Fifth position: Any of the remaining *two* letters
Sixth position: The *one* remaining letter

So, the numbers of choices for the six positions are as follows.

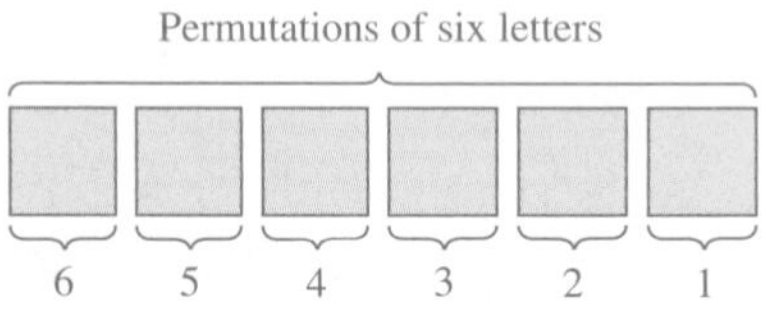

The total number of permutations of the six letters is

$$6! = 6 \cdot 5 \cdot 4 \cdot 3 \cdot 2 \cdot 1$$
$$= 720.$$

Additional Example
How many permutations are possible for the numbers 0, 1, 2, and 3?
Solution
$4! = 24$

Number of Permutations of n Elements

The number of permutations of n elements is

$$n \cdot (n-1) \cdot \cdot \cdot 4 \cdot 3 \cdot 2 \cdot 1 = n!.$$

In other words, there are $n!$ different ways that n elements can be ordered.

Bob Coglianese Photos, Inc.

Eleven thoroughbred racehorses hold the title of Triple Crown winner for winning the Kentucky Derby, the Preakness, and the Belmont Stakes in the same year. Forty horses have won two out of the three races.

Occasionally, you are interested in ordering a *subset* of a collection of elements rather than the entire collection. For example, you might want to choose (and order) r elements out of a collection of n elements. Such an ordering is called a **permutation of n elements taken r at a time.**

Example 6 ▶ Counting Horse Race Finishes

Eight horses are running in a race. In how many different ways can these horses come in first, second, and third? (Assume that there are no ties.)

Solution

Here are the different possibilities.

Win (first position): *Eight* choices
Place (second position): *Seven* choices
Show (third position): *Six* choices

Using the Fundamental Counting Principle, multiply these three numbers together to obtain the following.

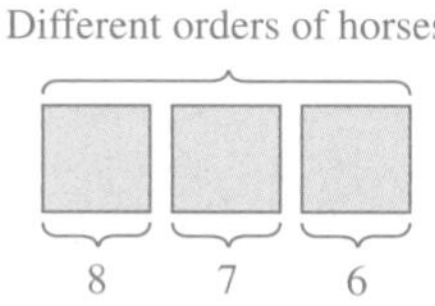

So, there are $8 \cdot 7 \cdot 6 = 336$ different orders.

Technology

Most graphing calculators are programmed to evaluate $_nP_r$. Consult your user's manual and then evaluate $_8P_5$. You should get an answer of 6720.

Permutations of *n* Elements Taken *r* at a Time

The number of permutations of n elements taken r at a time is

$$_nP_r = \frac{n!}{(n-r)!}$$
$$= n(n-1)(n-2) \cdot \cdot \cdot (n-r+1).$$

Using this formula, you can rework Example 6 to find that the number of permutations of eight horses taken three at a time is

$$_8P_3 = \frac{8!}{5!}$$
$$= \frac{8 \cdot 7 \cdot 6 \cdot \cancel{5!}}{\cancel{5!}}$$
$$= 336$$

which is the same answer obtained in the example.

Remember that for permutations, order is important. So, if you are looking at the possible permutations of the letters A, B, C, and D taken three at a time, the permutations (A, B, D) and (B, A, D) are counted as different because the *order* of the elements is different.

Suppose, however, that you are asked to find the possible permutations of the letters A, A, B, and C. The total number of permutations of the four letters would be ${}_4P_4 = 4!$. However, not all of these arrangements would be *distinguishable* because there are two A's in the list. To find the number of distinguishable permutations, you can use the following formula.

Distinguishable Permutations

Suppose a set of n objects has n_1 of one kind of object, n_2 of a second kind, n_3 of a third kind, and so on, with

$$n = n_1 + n_2 + n_3 + \cdot \cdot \cdot + n_k .$$

Then the number of **distinguishable permutations** of the n objects is

$$\frac{n!}{n_1! \cdot n_2! \cdot n_3! \cdot \cdot \cdot \cdot \cdot n_k!}.$$

Example 7 ▶ Distinguishable Permutations

In how many distinguishable ways can the letters in BANANA be written?

Solution

This word has six letters, of which three are A's, two are N's, and one is a B. So, the number of distinguishable ways the letters can be written is

$$\frac{n!}{n_1! \cdot n_2! \cdot n_3!} = \frac{6!}{3! \cdot 2! \cdot 1!}$$

$$= \frac{6 \cdot 5 \cdot 4 \cdot \cancel{3!}}{\cancel{3!} \cdot 2!}$$

$$= 60.$$

Additional Example
In how many different ways can the letters in INDIANA be written?
Solution
630

The 60 different distinguishable permutations are as follows.

AAABNN	AAANBN	AAANNB	AABANN	AABNAN	AABNNA
AANABN	AANANB	AANBAN	AANBNA	AANNAB	AANNBA
ABAANN	ABANAN	ABANNA	ABNAAN	ABNANA	ABNNAA
ANAABN	ANAANB	ANABAN	ANABNA	ANANAB	ANANBA
ANBAAN	ANBANA	ANBNAA	ANNAAB	ANNABA	ANNBAA
BAAANN	BAANAN	BAANNA	BANAAN	BANANA	BANNAA
BNAAAN	BNAANA	BNANAA	BNNAAA	NAAABN	NAAANB
NAABAN	NAABNA	NAANAB	NAANBA	NABAAN	NABANA
NABNAA	NANAAB	NANABA	NANBAA	NBAAAN	NBAANA
NBANAA	NBNAAA	NNAAAB	NNAABA	NNABAA	NNBAAA

Combinations

When you count the number of possible permutations of a set of elements, *order* is important. As a final topic in this section, you will look at a method of selecting subsets of a larger set in which order is *not* important. Such subsets are called **combinations of n elements taken r at a time.** For instance, the combinations

$$\{A, B, C\} \quad \text{and} \quad \{B, A, C\}$$

are equivalent because both sets contain the same three elements, and the order in which the elements are listed is not important. So, you would count only one of the two sets. A common example of how a combination occurs is a card game in which the player is free to reorder the cards after they have been dealt.

Careful attention to detail and numerous examples will help students understand when to count different orders and when not to. You may want to discuss the misnaming of the "combination lock" according to the definitions presented in this section.

Example 8 ▶ Combinations of n Elements Taken r at a Time

In how many different ways can three letters be chosen from the letters A, B, C, D, and E? (The order of the three letters is not important.)

Solution

The following subsets represent the different combinations of three letters that can be chosen from the five letters.

{A, B, C}　　{A, B, D}
{A, B, E}　　{A, C, D}
{A, C, E}　　{A, D, E}
{B, C, D}　　{B, C, E}
{B, D, E}　　{C, D, E}

From this list, you can conclude that there are 10 different ways that three letters can be chosen from five letters.

Combinations of n Elements Taken r at a Time

The number of combinations of n elements taken r at a time is

$${}_nC_r = \frac{n!}{(n-r)!r!}.$$

Note that the formula for ${}_nC_r$ is the same one given for binomial coefficients. To see how this formula is used, let's solve the counting problem in Example 8. In that problem, you are asked to find the number of combinations of five elements taken three at a time. So, $n = 5$, $r = 3$, and the number of combinations is

$${}_5C_3 = \frac{5!}{2!3!} = \frac{5 \cdot \overset{2}{\cancel{4}} \cdot \cancel{3!}}{\cancel{2} \cdot 1 \cdot \cancel{3!}} = 10$$

which is the same answer obtained in the example.

Example 9 ▶ Counting Card Hands

A standard poker hand consists of five cards dealt from a deck of 52. How many different poker hands are possible? (After the cards are dealt, the player may reorder them, and therefore order is not important.)

Solution

You can find the number of different poker hands by using the formula for the number of combinations of 52 elements taken five at a time, as follows.

$$\begin{aligned} {}_{52}C_5 &= \frac{52!}{47!5!} \\ &= \frac{52 \cdot 51 \cdot 50 \cdot 49 \cdot 48 \cdot 47!}{5 \cdot 4 \cdot 3 \cdot 2 \cdot 1 \cdot 47!} \\ &= 2{,}598{,}960 \end{aligned}$$

Activities

1. Evaluate (a) ${}_6P_2$ and (b) ${}_6C_2$.
   Answers: (a) 30 and (b) 15
2. A local building supply company is hiring extra summer help. They need four additional employees to work outside in the lumber yard and three more to work inside the store. In how many ways can these positions be filled if there are 10 applicants for outside work and five for inside work?
   Answer: 2100
3. In how many distinguishable ways can the letters C A L C U L U S be written?
   Answer: 5040

Example 10 ▶ The Number of Subsets of a Set

Find the total number of subsets of a set that has 10 elements.

Solution

Begin by considering the number of subsets with 0 elements, the number with 1 element, the number with 2 elements, and so on.

$$\begin{aligned} \text{Number of subsets} &= \text{Subsets with 0 elements} + \text{Subsets with 1 element} + \text{Subsets with 2 elements} + \cdots + \text{Subsets with 10 elements} \\ &= \binom{10}{0} + \binom{10}{1} + \binom{10}{2} + \cdots + \binom{10}{10} \end{aligned}$$

By comparing this expression with the binomial expansion of $(1 + 1)^{10}$, you see that they are the same.

$$\begin{aligned} (1+1)^{10} &= \binom{10}{0}1^{10}1^0 + \binom{10}{1}1^9 1^1 + \binom{10}{2}1^8 1^2 + \cdots + \binom{10}{10}1^0 1^{10} \\ &= \binom{10}{0} + \binom{10}{1} + \binom{10}{2} + \cdots + \binom{10}{10} \end{aligned}$$

This implies that the total number of subsets of a set of 10 elements is

$$\begin{aligned} (1+1)^{10} &= 2^{10} \\ &= 1024. \end{aligned}$$

The result of Example 10 can be generalized to conclude that the total number of subsets of a set of n elements is 2^n.

9.6 Exercises

Random Selection **In Exercises 1–8, determine the number of ways a computer can randomly generate one or more such integers from 1 through 12.**

1. An odd integer

2. An even integer

3. A prime integer

4. An integer that is greater than 9

5. An integer that is divisible by 4

6. An integer that is divisible by 3

7. Two integers whose sum is 8

8. Two *distinct* integers whose sum is 8

9. *Entertainment Systems* A customer can choose one of three amplifiers, one of two compact disc players, and one of five speaker models for an entertainment system. Determine the number of possible system configurations.

10. *Computer Systems* A customer in a computer store can choose one of four monitors, one of three keyboards, and one of five computers. If all the choices are compatible, determine the number of possible system configurations.

11. *Job Applicants* A college needs two additional faculty members: a chemist and a statistician. In how many ways can these positions be filled if there are five applicants for the chemistry position and three applicants for the statistics position?

12. *Course Schedule* A college student is preparing a course schedule for the next semester. The student may select one of two mathematics courses, one of three science courses, and one of five courses from the social sciences and humanities. How many schedules are possible?

13. *True-False Exam* In how many ways can a six-question true-false exam be answered? (Assume that no questions are omitted.)

14. *True-False Exam* In how many ways can a 12-question true-false exam be answered? (Assume that no questions are omitted.)

15. *Toboggan Ride* Three people are lining up for a ride on a toboggan, but only two of the three are willing to take the first position. With that constraint, in how many ways can the three people be seated on the toboggan?

16. *Aircraft Boarding* Eight people are boarding an aircraft. Two have tickets for first class and board before those in the economy class. In how many ways can the eight people board the aircraft?

17. *License Plate Numbers* In a certain state, each automobile license plate number consists of three letters followed by a four-digit number. How many distinct license plate numbers can be formed?

18. *License Plate Numbers* In a certain state, each automobile license plate number consists of two letters followed by a four-digit number. To avoid confusion between "O" and "zero" and between "I" and "one," the letters "O" and "I" are not used. How many distinct license plate numbers can be formed?

19. *Three-Digit Numbers* How many three-digit numbers can be formed under the following conditions?

(a) The leading digit cannot be zero.

(b) The leading digit cannot be zero and no repetition of digits is allowed.

(c) The leading digit cannot be zero and the number must be a multiple of 5.

(d) The number is at least 400.

20. *Four-Digit Numbers* How many four-digit numbers can be formed under the following conditions?

(a) The leading digit cannot be zero.

(b) The leading digit cannot be zero and no repetition of digits is allowed.

(c) The leading digit cannot be zero and the number must be less than 5000.

(d) The leading digit cannot be zero and the number must be even.

21. *Combination Lock* A combination lock will open when the right choice of three numbers (from 1 to 40, inclusive) is selected. How many different lock combinations are possible?

22. *Combination Lock* A combination lock will open when the right choice of three numbers (from 1 to 50, inclusive) is selected. How many different lock combinations are possible?

23. *Concert Seats* Four couples have reserved seats in a given row for a concert. In how many different ways can they be seated if

(a) there are no seating restrictions?

(b) the two members of each couple wish to sit together?

24. *Single File* In how many orders can four girls and four boys walk through a doorway single file if

(a) there are no restrictions?

(b) the girls walk through before the boys?

In Exercises 25–30, evaluate $_nP_r$.

25. $_4P_4$

26. $_5P_5$

27. $_8P_3$

28. $_{20}P_2$

29. $_5P_4$

30. $_7P_4$

In Exercises 31 and 32, solve for n.

31. $14 \cdot {_nP_3} = {_{n+2}P_4}$

32. $_nP_5 = 18 \cdot {_{n-2}P_4}$

In Exercises 33–38, evaluate using a calculator.

33. $_{20}P_5$

34. $_{100}P_5$

35. $_{100}P_3$

36. $_{10}P_8$

37. $_{20}C_5$

38. $_{10}C_7$

In Exercises 39–42, find the number of distinguishable permutations of the group of letters.

39. A, A, G, E, E, E, M

40. B, B, B, T, T, T, T, T

41. A, L, G, E, B, R, A

42. M, I, S, S, I, S, S, I, P, P, I

43. Write all permutations of the letters A, B, C, and D.

44. Write all the permutations of the letters A, B, C, and D if the letters B and C must remain between the letters A and D.

45. Write all the possible selections of two letters that can be formed from the letters A, B, C, D, E, and F. (The order of the two letters is not important.)

46. Write all the possible selections of three letters that can be formed from the letters A, B, C, D, E, and F. (The order of the three letters is not important.)

47. *Posing for a Photograph* In how many ways can five children line up in a row?

48. *Riding in a Car* In how many ways can six people sit in a six-passenger car?

49. *Choosing Officers* From a pool of 12 candidates, the offices of president, vice-president, secretary, and treasurer will be filled. In how many different ways can the offices be filled?

50. *Assembly Line Production* There are four processes involved in assembling a certain product, and these processes can be performed in any order. The management wants to test each order to determine which is the least time-consuming. How many different orders will have to be tested?

51. *Forming an Experimental Group* In order to conduct a certain experiment, five students are randomly selected from a class of 20. How many different groups of five students are possible?

52. *Test Questions* You can answer any 10 questions from a total of 12 questions on an exam. In how many different ways can you select the questions?

53. *Lottery Choices* There are 40 numbers in a particular state lottery. In how many ways can a player select six of the numbers?

54. *Lottery Choices* There are 50 numbers in a particular state lottery. In how many ways can a player select six of the numbers?

55. *Number of Subsets* How many subsets of four elements can be formed from a set of 100 elements?

56. *Number of Subsets* How many subsets of five elements can be formed from a set of 80 elements?

57. *Geometry* Three points that are not on a line determine three lines. How many lines are determined by seven points, no three of which are on a line?

58. *Defective Units* A shipment of 10 microwave ovens contains three defective units. In how many ways can a vending company purchase four of these units and receive (a) all good units, (b) two good units, and (c) at least two good units?

59. *Job Applicants* An employer interviews eight people for four openings in the company. Three of the eight people are women. If all eight are qualified, in how many ways can the employer fill the four positions if (a) the selection is random and (b) exactly two selections are women?

60. *Poker Hand* You are dealt five cards from an ordinary deck of 52 playing cards. In how many ways can you get a full house? (A full house consists of three of one kind and two of another. For example, A-A-A-5-5 and K-K-K-10-10 are full houses.)

61. *Forming a Committee* Four people are to be selected at random from a group of four couples. In how many ways can this be done under the following conditions?

(a) There are no restrictions.

(b) The group must have at least one couple.

(c) Each couple must be represented in the group.

62. *Interpersonal Relationships* The complexity of the interpersonal relationships increases dramatically as the size of a group increases. Determine the number of different two-person relationships in a group of people of size (a) 3, (b) 8, (c) 12, and (d) 20.

In Exercises 63–66, find the number of diagonals of the polygon. (A line segment connecting any two nonadjacent vertices is called a diagonal of the polygon.)

63. Pentagon

64. Hexagon

65. Octagon

66. Decagon (10 sides)

Synthesis

True or False? **In Exercises 67 and 68, determine whether the statement is true or false. Justify your answer.**

67. The number of letter pairs that can be formed from any of the first 13 letters in the alphabet (A–M) is an example of a permutation.

68. The number of permutations of n elements can be determined by using the Fundamental Counting Principle.

69. What is the relationship between ${}_nC_r$ and ${}_nC_{n-r}$?

70. Without calculating the numbers, determine which of the following is greater. Explain.

(a) The combinations of 10 elements taken six at a time

(b) The permutations of 10 elements taken six at a time

In Exercises 71–74, prove the identity.

71. ${}_nP_{n-1} = {}_nP_n$

72. ${}_nC_n = {}_nC_0$

73. ${}_nC_{n-1} = {}_nC_1$

74. ${}_nC_r = \dfrac{{}_nP_r}{r!}$

75. *Think About It* Can your calculator evaluate ${}_{100}P_{80}$? If not, explain why.

76. *Writing* Explain in words the meaning of ${}_nP_r$.

Review

In Exercises 77–80, evaluate the function at the specified values of the independent variable.

77. $f(x) = 3x^2 + 8$

(a) $f(3)$ (b) $f(0)$ (c) $f(-5)$

78. $g(x) = \sqrt{x - 3} + 2$

(a) $g(3)$ (b) $g(7)$ (c) $g(x + 3)$

79. $f(x) = -|x - 5| + 6$

(a) $f(-5)$ (b) $f(-1)$ (c) $f(11)$

80. $f(x) = \begin{cases} x^2 - 2x + 5, & x \le -4 \\ -x^2 - 2, & x > -4 \end{cases}$

(a) $f(-4)$ (b) $f(-1)$ (c) $f(-20)$

In Exercises 81–84, use the Binomial Theorem to expand and simplify the expression.

81. $(x + 1)^5$

82. $(y - 2)^6$

83. $(x^2 + 2y)^5$

84. $(x^2 - y^2)^4$

9.7 Probability

▶ **What you should learn**

- How to find the probability of an event
- How to find the probabilities of mutually exclusive events
- How to find the probabilities of independent events
- How to find the probability of the complement of an event

▶ **Why you should learn it**

You can use probability to solve a variety of problems that occur in real life. For instance, in Exercise 33 on page 754, you are asked to use probability to help analyze the age distribution of employees who find work through temporary help agencies.

Telegraph Colour Library/FPG

The Probability of an Event

Any happening for which the result is uncertain is called an **experiment.** The possible results of the experiment are **outcomes,** the set of all possible outcomes of the experiment is the **sample space** of the experiment, and any subcollection of a sample space is an **event.**

For instance, when a six-sided die is tossed, the sample space can be represented by the numbers 1 through 6. For this experiment, each of the outcomes is *equally likely.*

To describe sample spaces in such a way that each outcome is equally likely, you must sometimes distinguish between or among various outcomes in ways that appear artificial. Example 1 illustrates such a situation.

Example 1 ▶ Finding the Sample Space

Find the sample space for each of the following.

a. One coin is tossed.

b. Two coins are tossed.

c. Three coins are tossed.

Solution

a. Because the coin will land either heads up (denoted by H) or tails up (denoted by T), the sample space is

$$S = \{H, T\}.$$

b. Because either coin can land heads up or tails up, the possible outcomes are as follows.

HH = heads up on both coins

HT = heads up on first coin and tails up on second coin

TH = tails up on first coin and heads up on second coin

TT = tails up on both coins

So, the sample space is

$$S = \{HH, HT, TH, TT\}.$$

Note that this list distinguishes between the two cases HT and TH, even though these two outcomes appear to be similar.

c. Following the notation of part (b), the sample space is

$$S = \{HHH, HHT, HTH, HTT, THH, THT, TTH, TTT\}.$$

To calculate the probability of an event, count the number of outcomes in the event and in the sample space. The *number of outcomes* in event E is denoted by $n(E)$, and the number of outcomes in the sample space S is denoted by $n(S)$. The probability that event E will occur is given by $n(E)/n(S)$.

The Probability of an Event

If an event E has $n(E)$ equally likely outcomes and its sample space S has $n(S)$ equally likely outcomes, the **probability** of event E is

$$P(E) = \frac{n(E)}{n(S)}.$$

Because the number of outcomes in an event must be less than or equal to the number of outcomes in the sample space, the probability of an event must be a number between 0 and 1. That is,

$$0 \leq P(E) \leq 1.$$

If $P(E) = 0$, event E *cannot occur*, and E is called an impossible event. If $P(E) = 1$, event E *must occur*, and E is called a certain event.

Example 2 ▶ Finding the Probability of an Event

a. Two coins are tossed. What is the probability that both land heads up?

b. A card is drawn from a standard deck of playing cards. What is the probability that it is an ace?

Solution

a. Following the procedure in Example 1(b), let

$$E = \{HH\}$$

and

$$S = \{HH, HT, TH, TT\}.$$

The probability of getting two heads is

$$P(E) = \frac{n(E)}{n(S)} = \frac{1}{4}.$$

b. Because there are 52 cards in a standard deck of playing cards and there are four aces (one in each suit), the probability of drawing an ace is

$$P(E) = \frac{n(E)}{n(S)} = \frac{4}{52} = \frac{1}{13}.$$

Additional Example

Two coins are tossed. What is the probability that one coin lands heads up and the other coin lands tails up?

Solution

$\frac{1}{2}$

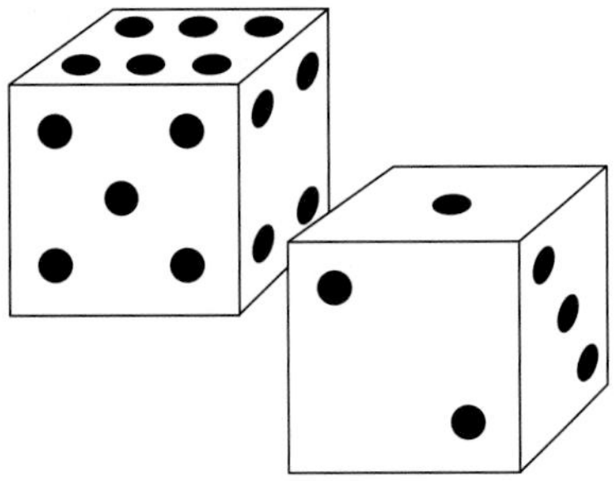

FIGURE 9.8

A computer simulation of this example appears in the *Interactive* CD-ROM and *Internet* versions of this text.

Example 3 ▶ Finding the Probability of an Event

Two six-sided dice are tossed. What is the probability that the total of the two dice is 7? (See Figure 9.8.)

Solution

Because there are six possible outcomes on each die, you can use the Fundamental Counting Principle to conclude that there are $6 \cdot 6$ or 36 different outcomes when two dice are tossed. To find the probability of rolling a total of 7, you must first count the number of ways in which this can occur.

First die	1	2	3	4	5	6
Second die	6	5	4	3	2	1

So, a total of 7 can be rolled in six ways, which means that the probability of rolling a 7 is

$$P(E) = \frac{n(E)}{n(S)} = \frac{6}{36} = \frac{1}{6}.$$

Additional Example
What is the probability that the total of the two dice is 8?
Solution
$\frac{5}{36}$

You could have written out each sample space in Examples 2 and 3 and simply counted the outcomes in the desired events. For larger sample spaces, however, you should use the counting principles discussed in Section 9.6.

Example 4 ▶ Finding the Probability of an Event

Twelve-sided dice, as shown in Figure 9.9, can be constructed (in the shape of regular dodecahedrons) such that each of the numbers from 1 to 6 appears twice on each die. Prove that these dice can be used in any game requiring ordinary six-sided dice without changing the probabilities of different outcomes.

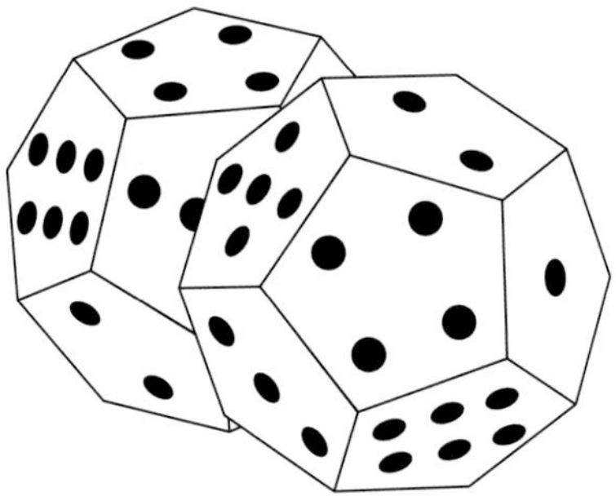

FIGURE 9.9

Solution

For an ordinary six-sided die, each of the numbers 1, 2, 3, 4, 5, and 6 occurs only once, so the probability of any particular number coming up is

$$P(E) = \frac{n(E)}{n(S)} = \frac{1}{6}.$$

For one of the 12-sided dice, each number occurs twice, so the probability of any particular number coming up is

$$P(E) = \frac{n(E)}{n(S)} = \frac{2}{12} = \frac{1}{6}.$$

Courtesy of Corel Professional Photos

Although popular in the early 1800s, lotteries were banned in more and more states until, by 1894, no state allowed lotteries. In 1964, New Hampshire became the first state to reinstitute a state lottery. Today, lotteries are conducted by almost all states.

Example 5 ▶ The Probability of Winning a Lottery

In a state lottery, a player chooses six different numbers from 1 to 40. If these six numbers match the six numbers drawn by the lottery commission, the player wins (or shares) the top prize. What is the probability of winning?

Solution

To find the number of elements in the sample space, use the formula for the number of combinations of 40 elements taken six at a time.

$$\begin{aligned} n(S) &= {}_{40}C_6 \\ &= \frac{40 \cdot 39 \cdot 38 \cdot 37 \cdot 36 \cdot 35}{6 \cdot 5 \cdot 4 \cdot 3 \cdot 2 \cdot 1} \\ &= 3{,}838{,}380 \end{aligned}$$

If a person buys only one ticket, the probability of winning is

$$\begin{aligned} P(E) &= \frac{n(E)}{n(S)} \\ &= \frac{1}{3{,}838{,}380}. \end{aligned}$$

Example 6 ▶ Random Selection

The numbers of colleges and universities in various regions of the United States in 1996 are shown in Figure 9.10. One institution is selected at random. What is the probability that the institution is in one of the three southern regions? (Source: U.S. National Center for Education Statistics)

Solution

From the figure, the total number of colleges and universities is 3696. Because there are $607 + 265 + 298 = 1170$ colleges and universities in the three southern regions, the probability that the institution is in one of these regions is

$$P(E) = \frac{n(E)}{n(S)} = \frac{1170}{3696} \approx 0.317.$$

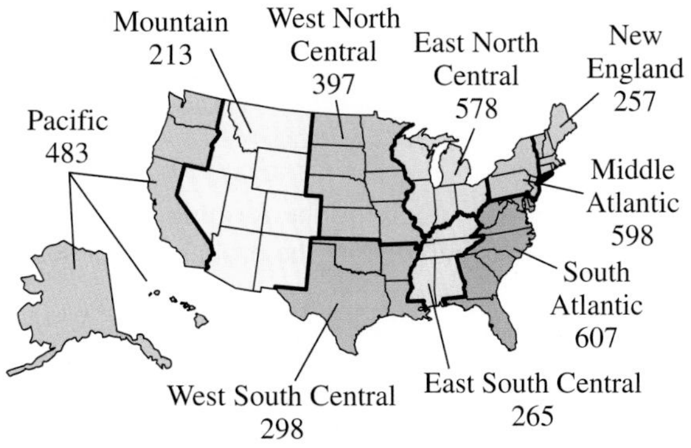

FIGURE 9.10

Mutually Exclusive Events

Two events A and B (from the same sample space) are **mutually exclusive** if A and B have no outcomes in common. In the terminology of sets, the intersection of A and B is the empty set, which is expressed as

$$P(A \cap B) = 0.$$

For instance, if two dice are tossed, the event A of rolling a total of 6 and the event B of rolling a total of 9 are mutually exclusive. To find the probability that one or the other of two mutually exclusive events will occur, you can *add* their individual probabilities.

Probability of the Union of Two Events

If A and B are events in the same sample space, the probability of A *or* B occurring is given by

$$P(A \cup B) = P(A) + P(B) - P(A \cap B).$$

If A and B are mutually exclusive, then

$$P(A \cup B) = P(A) + P(B).$$

Example 7 ▶ The Probability of a Union of Events

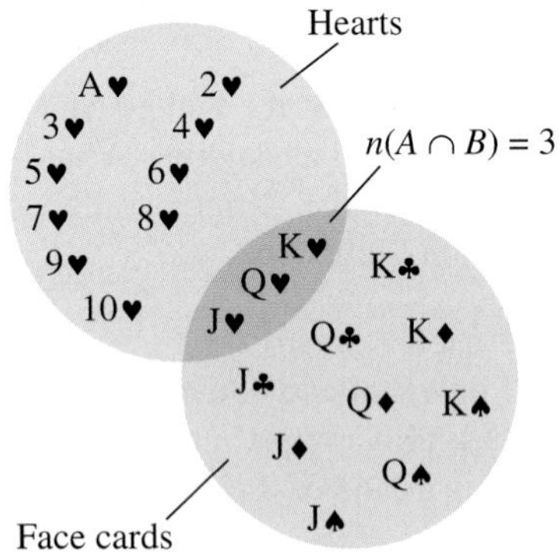

FIGURE 9.11

One card is selected from a standard deck of 52 playing cards. What is the probability that the card is either a heart or a face card?

Solution

Because the deck has 13 hearts, the probability of selecting a heart (event A) is

$$P(A) = \frac{13}{52}.$$

Similarly, because the deck has 12 face cards, the probability of selecting a face card (event B) is

$$P(B) = \frac{12}{52}.$$

Because three of the cards are hearts *and* face cards (see Figure 9.11), it follows that

$$P(A \cap B) = \frac{3}{52}.$$

Finally, applying the formula for the probability of the union of two events, you can conclude that the probability of selecting a heart or a face card is

$$P(A \cup B) = P(A) + P(B) - P(A \cap B)$$

$$= \frac{13}{52} + \frac{12}{52} - \frac{3}{52} = \frac{22}{52} \approx 0.423.$$

Example 8 ▶ Probability of Mutually Exclusive Events

The personnel department of a company has compiled data on the numbers of employees who have been with the company for various periods of time. The results are shown in the table.

Years of service	Number of employees
0–4	157
5–9	89
10–14	74
15–19	63
20–24	42
25–29	38
30–34	37
35–39	21
40–44	8

If an employee is chosen at random, what is the probability that the employee has 9 or fewer years of service?

Solution

To begin, add the number of employees and find that the total is 529. Next, let event A represent choosing an employee with 0 to 4 years of service and let event B represent choosing an employee with 5 to 9 years of service. Then

$$P(A) = \frac{157}{529} \quad \text{and} \quad P(B) = \frac{89}{529}.$$

Because A and B have no outcomes in common, you can conclude that these two events are mutually exclusive and that

$$\begin{aligned} P(A \cup B) &= P(A) + P(B) \\ &= \frac{157}{529} + \frac{89}{529} \\ &= \frac{246}{529} \\ &\approx 0.465. \end{aligned}$$

So, the probability of choosing an employee who has 9 or fewer years of service is about 0.465.

Additional Examples

a. If an employee from Example 8 is chosen at random, what is the probability that the employee has 30 or more years of service?

b. If an employee from Example 8 is chosen at random, what is the probability that the employee has either less than 15 years or more than 34 years of service?

Solutions

a. ≈ 0.125

b. ≈ 0.660

Independent Events

Two events are **independent** if the occurrence of one has no effect on the occurrence of the other. For instance, rolling a total of 12 with two six-sided dice has no effect on the outcome of future rolls of the dice. To find the probability that two independent events will occur, *multiply* the probabilities of each.

Probability of Independent Events

If A and B are independent events, the probability that both A and B will occur is

$$P(A \text{ and } B) = P(A) \cdot P(B).$$

Example 9 ▶ Probability of Independent Events

A random number generator on a computer selects three integers from 1 to 20. What is the probability that all three numbers are less than or equal to 5?

Solution

The probability of selecting a number from 1 to 5 is

$$P(A) = \frac{5}{20} = \frac{1}{4}.$$

So, the probability that all three numbers are less than or equal to 5 is

$$P(A) \cdot P(A) \cdot P(A) = \left(\frac{1}{4}\right)\left(\frac{1}{4}\right)\left(\frac{1}{4}\right) = \frac{1}{64}.$$

Example 10 ▶ Probability of Independent Events

In 1997, 58% of the population of the United States was 30 years old or older. Suppose that in a survey, 10 people were chosen at random from the population. What is the probability that all 10 were 30 years old or older? (Source: U.S. Bureau of the Census)

Solution

Let A represent choosing a person who was 30 years old or older. Because the probability of choosing a person who was 30 years old or older was 0.58, you can conclude that the probability that all 10 people were 30 years old or older is

$$[P(A)]^{10} = (0.58)^{10} \approx 0.0043.$$

Activities

1. You have two bags, each of which contains a red marble and a green marble. You select one marble from each bag. Write the sample space.
   Answer: $S = \{RR, RG, GR, GG\}$
2. A person draws two cards from a deck of 52 cards. What is the probability that the two cards drawn will both be face cards if the deck contains 12 face cards?
   Answer: $\frac{11}{221}$
3. A bag contains four red, three yellow, and five green marbles. One marble is selected from the bag. What is the probability that it is a red marble? (Assume that the marbles are of the same size, shape, etc.)
   Answer: $\frac{1}{3}$
4. Using the situation in today's class,
   (a) if two students are chosen at random, what is the probability that both will be female?
   (b) if one student is chosen at random, what is the probability that this student was born in a month beginning with the letter "J"?
   Answers will vary depending on the class.

Exploration

You are in a class with 22 other people. What is the probability that at least two out of the 23 people will have a birthday on the same day of the year?

The complement of the probability that at least two people have the same birthday is the probability that all 23 birthdays are different. So, first find the probability that all 23 people have different birthdays and then find the complement.

Now, determine the probability that in a room with 50 people at least two people have the same birthday.

The Complement of an Event

The **complement of an event** A is the collection of all outcomes in the sample space that are *not* in A. The complement of event A is denoted by A'. Because $P(A \text{ or } A') = 1$ and because A and A' are mutually exclusive, it follows that $P(A) + P(A') = 1$. Therefore, the probability of A' is

$$P(A') = 1 - P(A).$$

For instance, if the probability of *winning* a certain game is

$$P(A) = \frac{1}{4}$$

the probability of *losing* the game is

$$P(A') = 1 - \frac{1}{4}$$

$$= \frac{3}{4}.$$

Probability of a Complement

Let A be an event and let A' be its complement. If the probability of A is $P(A)$, the probability of the complement is

$$P(A') = 1 - P(A).$$

Example 11 ▶ Finding the Probability of a Complement

A manufacturer has determined that a certain machine averages one faulty unit for every 1000 it produces. What is the probability that an order of 200 units will have one or more faulty units?

Solution

To solve this problem as stated, you would need to find the probabilities of having exactly one faulty unit, exactly two faulty units, exactly three faulty units, and so on. However, using complements, you can simply find the probability that all units are perfect and then subtract this value from 1. Because the probability that any given unit is perfect is 999/1000, the probability that all 200 units are perfect is

$$P(A) = \left(\frac{999}{1000}\right)^{200}$$

$$\approx 0.8186.$$

Therefore, the probability that at least one unit is faulty is

$$P(A') = 1 - P(A)$$

$$\approx 0.1814.$$

Activity

Consider having students create a book or pamphlet on probability by taking each type of probability situation discussed in this section, defining the terms, giving the formulas needed, and then creating their own examples to illustrate how to calculate probabilities.

9.7 Exercises

In Exercises 1–6, determine the sample space for the experiment.

1. A coin and a six-sided die are tossed.

2. A six-sided die is tossed twice and the sum of the points is recorded.

3. A taste tester has to rank three varieties of yogurt, A, B, and C, according to preference.

4. Two marbles are selected from a sack containing two red marbles, two blue marbles, and one black marble. The color of each marble is recorded.

5. Two county supervisors are selected from five supervisors, A, B, C, D, and E, to study a recycling plan.

6. A sales representative makes presentations of a product in three homes per day. In each home there may be a sale (denote by S) or there may be no sale (denote by F).

Heads or Tails **In Exercises 7–10, find the probability for the experiment of tossing a coin three times. Use the sample space $S = \{HHH, HHT, HTH, HTT, THH, THT, TTH, TTT\}$.**

7. The probability of getting exactly one tail

8. The probability of getting a head on the first toss

9. The probability of getting at least one head

10. The probability of getting at least two heads

Drawing a Card **In Exercises 11–14, find the probability for the experiment of selecting one card from a standard deck of 52 playing cards.**

11. The card is a face card.

12. The card is not a face card.

13. The card is a red face card.

14. The card is a 6 or less.

Tossing a Die **In Exercises 15–20, find the probability for the experiment of tossing a six-sided die twice.**

15. The sum is 4.

16. The sum is at least 7.

17. The sum is less than 11.

18. The sum is 2, 3, or 12.

19. The sum is odd and no more than 7.

20. The sum is odd or prime.

Drawing Marbles **In Exercises 21–24, find the probability for the experiment of drawing two marbles (without replacement) from a bag containing one green, two yellow, and three red marbles.**

21. Both marbles are red.

22. Both marbles are yellow.

23. Neither marble is yellow.

24. The marbles are of different colors.

In Exercises 25–28, you are given the probability that an event *will* happen. Find the probability that the event *will not* happen.

25. $p = 0.7$

26. $p = 0.36$

27. $p = \frac{1}{3}$

28. $p = \frac{5}{6}$

In Exercises 29–32, you are given the probability that an event *will not* happen. Find the probability that the event *will* happen.

29. $p = 0.15$

30. $p = 0.84$

31. $p = \frac{13}{20}$

32. $p = \frac{87}{100}$

33. *Graphical Reasoning* In 1997 there were approximately 1.3 million temporary help agency workers in the United States. The figure gives the age profile of these workers. (Source: U.S. Bureau of Labor Statistics)

Age of U.S. Workers

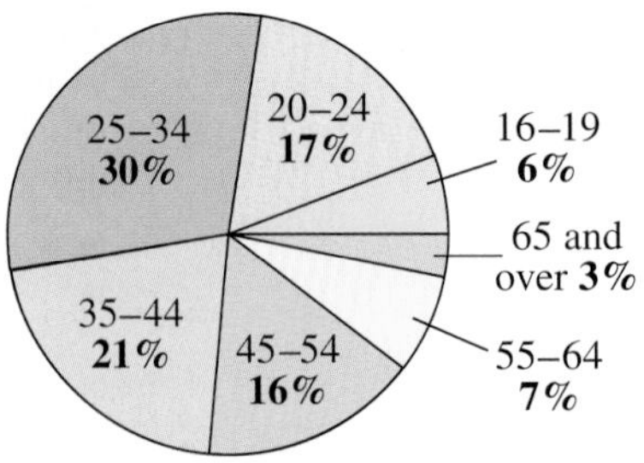

(a) Estimate the number of temporary help agency workers in the age category 16–19.

(b) What is the probability that a person selected at random from the population of temporary agency workers is in the 25–34 age group?

(c) What is the probability that a person selected at random from the population of temporary agency workers is in the 35–54 age group?

34. *Graphical Reasoning* In 1997 there were approximately 111 million workers in the civilian labor force in the United States. The figure gives the educational attainment of these workers. (Source: U.S. Bureau of Labor Statistics)

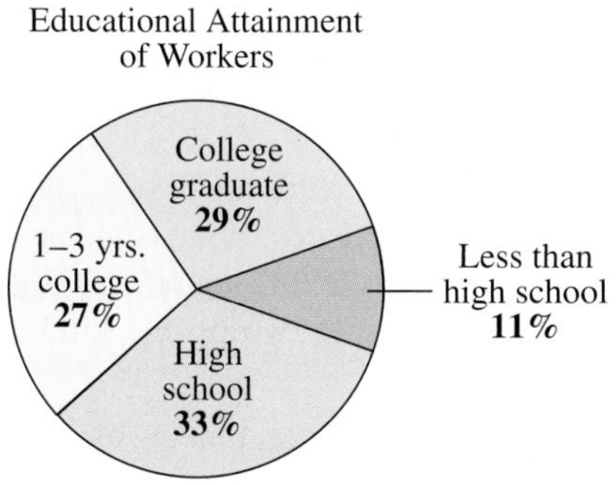

(a) Estimate the number of workers whose highest education level was a high school education.

(b) A person is selected at random from the civilian work force. What is the probability that the person has 1 to 3 years of college education?

(c) A person is selected at random from the civilian work force. What is the probability that the person has more than a high school education?

35. *Data Analysis* A study of the effectiveness of a flu vaccine was conducted with a sample of 500 people. Some in the study were given no vaccine, some were given one injection, and others were given two injections. The results of the study are listed in the table.

	No vaccine	One injection	Two injections	Total
Flu	7	2	13	22
No flu	149	52	277	478
Total	156	54	290	500

A person is selected at random from the sample. Find the specified probability.

(a) The person had two injections.

(b) The person did not get the flu.

(c) The person got the flu and had one injection.

36. *Data Analysis* One hundred college students were interviewed to determine their political party affiliations and whether they favored a balanced-budget amendment to the Constitution. The results of the study are listed in the table.

	Favor	Not favor	Unsure	Total
Democrat	23	25	7	55
Republican	32	9	4	45
Total	55	34	11	100

A person is selected at random from the sample. Find the probability that the described person is selected.

(a) A person who doesn't favor the amendment

(b) A Republican

(c) A Democrat who favors the amendment

37. *Alumni Association* A college sends a survey to selected members of the class of 2000. Of the 1254 people who graduated that year, 672 are women, of whom 124 went on to graduate school. Of the 582 male graduates, 198 went on to graduate school. If an alumnus member is selected at random, what is the probability that the person is (a) female, (b) male, and (c) female and did not attend graduate school?

38. *Post–High School Education* In a high school graduating class of 72 students, 28 are on the honor roll. Of these, 18 are going on to college, and of the other 44 students, 12 are going on to college. If a student is selected at random from the class, what is the probability that the person chosen is (a) going to college, (b) not going to college, and (c) on the honor roll, but not going to college?

39. *Winning an Election* Taylor, Moore, and Jenkins are candidates for public office. It is estimated that Moore and Jenkins have about the same probability of winning, and Taylor is believed to be twice as likely to win as either of the others. Find the probability of each candidate winning the election.

40. *Winning an Election* Three people have been nominated for president of a class. From a poll, it is estimated that the first has a 37% chance of winning and the second has a 44% chance of winning. What is the probability that the third candidate will win?

In Exercises 41–52, the sample spaces are large and you should use the counting principles discussed in Section 9.6.

41. *Preparing for a Test* A class is given a list of 20 study problems, from which 10 will be part of an upcoming exam. If a given student knows how to solve 15 of the problems, find the probability that the student will be able to answer (a) all 10 questions on the exam, (b) exactly eight questions on the exam, and (c) at least nine questions on the exam.

42. *Preparing for a Test* A class is given a list of eight study problems, from which five will be part of an upcoming exam. If a given student knows how to solve six of the problems, find the probability that the student will be able to answer (a) all five questions on the exam, (b) exactly four questions on the exam, and (c) at least four questions on the exam.

43. *Letter Mix-Up* Four letters and envelopes are addressed to four different people. If the letters are randomly inserted into the envelopes, what is the probability that (a) exactly one will be inserted in the correct envelope and (b) at least one will be inserted in the correct envelope?

44. *Payroll Mix-Up* Five paychecks and envelopes are addressed to five different people. If the paychecks are randomly inserted into the envelopes, what is the probability that (a) exactly one will be inserted in the correct envelope and (b) at least one will be inserted in the correct envelope?

45. *Game Show* On a game show you are given five digits to arrange in the proper order to form the price of a car. If you are correct, you win the car. What is the probability of winning, given the following conditions?

(a) You guess the position of each digit.

(b) You know the first digit and guess the positions of the others.

46. *Game Show* On a game show you are given five digits in the price of a car. The first digit is 1, and you are given the other four digits to arrange in the correct order to win the car. What is your probability of winning given the following conditions?

(a) You guess the position of each digit.

(b) You know the second digit but guess the others.

47. *Drawing Cards from a Deck* Two cards are selected at random from an ordinary deck of 52 playing cards. Find the probability that two aces are selected, given the following conditions.

(a) The cards are drawn in sequence, with the first card being replaced and the deck reshuffled prior to the second drawing.

(b) The two cards are drawn consecutively, without replacement.

48. *Poker Hand* Five cards are drawn from an ordinary deck of 52 playing cards. What is the probability that the hand drawn is a full house? (A full house is a hand that consists of two of one kind and three of another kind.)

49. *Defective Units* A shipment of 12 microwave ovens contains three defective units. A vending company has ordered four of these units, and because each is identically packaged, the selection will be random. What is the probability that (a) all four units are good, (b) exactly two units are good, and (c) at least two units are good?

50. *Defective Units* A shipment of 20 compact disc players contains four defective units. A retail outlet has ordered five of these units. What is the probability that (a) all five units are good, (b) exactly four units are good, and (c) at least one unit is defective?

51. *Random Number Generator* Two integers from 1 through 30 are chosen by a random number generator. What is the probability that (a) the numbers are both even, (b) one number is even and one is odd, (c) both numbers are less than 10, and (d) the same number is chosen twice?

52. *Random Number Generator* Two integers from 1 through 40 are chosen by a random number generator. What is the probability that (a) the numbers are both even, (b) one number is even and one is odd, (c) both numbers are less than 30, and (d) the same number is chosen twice?

53. *Backup System* A space vehicle has an independent backup system for one of its communication networks. The probability that either system will function satisfactorily during a flight is 0.985. What is the probability that during a given flight (a) both systems function satisfactorily, (b) at least one system functions satisfactorily, and (c) both systems fail?

54. *Backup Vehicle* A fire company keeps two rescue vehicles. Because of the demand on the vehicles and the chance of mechanical failure, the probability that a specific vehicle is available when needed is 90%. If the availability of one vehicle is *independent* of the availability of the other, find the probability that (a) both vehicles are available at a given time, (b) neither vehicle is available at a given time, and (c) at least one vehicle is available at a given time.

55. *Making a Sale* A sales representative makes a sale at approximately one-fourth of all calls. If, on a given day, the representative contacts five potential clients, what is the probability that a sale will be made with (a) each of the five contacts, (b) none of the contacts, and (c) at least one contact?

56. *A Boy or a Girl?* Assume that the probability of the birth of a child of a particular sex is 50%. In a family with four children, what is the probability that (a) all the children are boys, (b) all the children are the same sex, and (c) there is at least one boy?

57. *Flexible Work Hours* In a survey, people were asked if they would prefer to work flexible hours—even if it meant slower career advancement—so they could spend more time with their families. The results of the survey are shown in the figure. Suppose that three people from the survey were chosen at random. What is the probability that all three people would prefer flexible work hours?

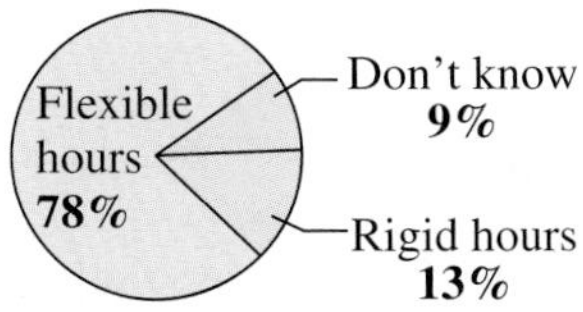

58. *Will That Be Cash or Charge?* Suppose that the methods used by shoppers to pay for merchandise are as shown in the circle graph. If two shoppers are chosen at random, what is the probability that both shoppers paid for their purchases only in cash?

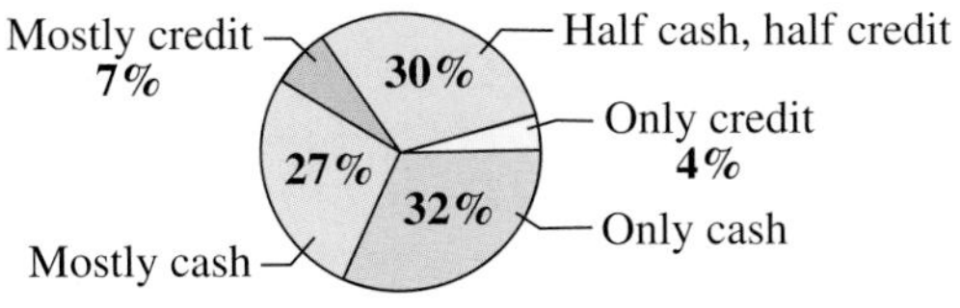

59. *Geometry* You and a friend agree to meet at your favorite fast-food restaurant between 5:00 and 6:00 P.M. The one who arrives first will wait 15 minutes for the other, and then will leave (see figure). What is the probability that the two of you will actually meet, assuming that your arrival times are random within the hour?

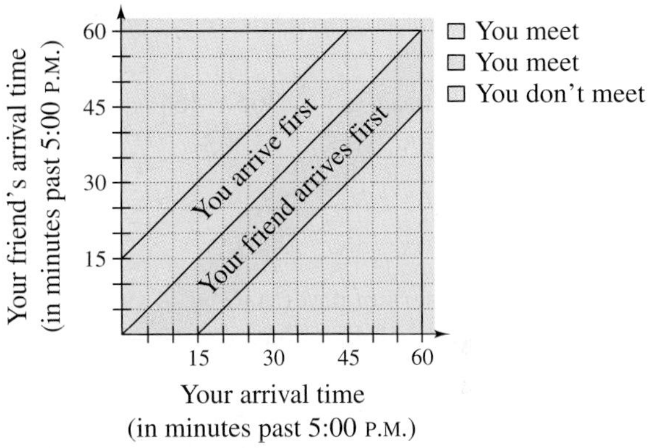

60. *Estimating π* A coin of diameter d is dropped onto a paper that contains a grid of squares d units on a side (see figure).

(a) Find the probability that the coin covers a vertex of one of the squares on the grid.

(b) Perform the experiment 100 times and use the results to approximate π.

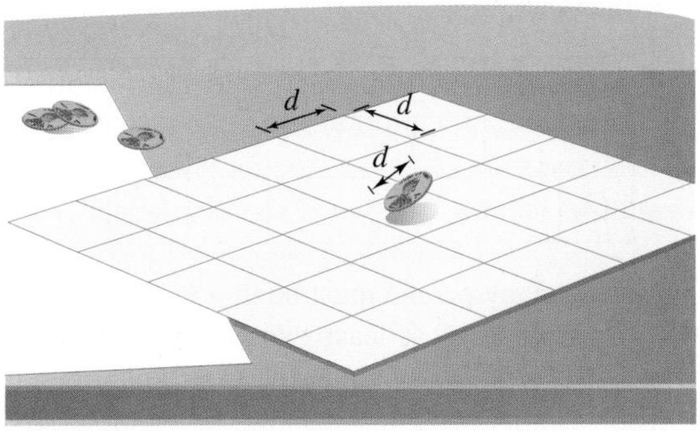

Synthesis

True or False? **In Exercises 61 and 62, determine whether the statement is true or false. Justify your answer.**

61. If A and B are independent events with nonzero probabilities, then A can occur when B occurs.

62. Rolling a number less than 3 on a normal six-sided die has a probability of $\frac{1}{3}$. The complement of this event is to roll a number greater than 3, and its probability is $\frac{1}{2}$.

63. *Pattern Recognition and Exploration* Consider a group of n people.

(a) Explain why the following pattern gives the probabilities that the n people have distinct birthdays.

$$n = 2: \quad \frac{365}{365} \cdot \frac{364}{365} = \frac{365 \cdot 364}{365^2}$$

$$n = 3: \quad \frac{365}{365} \cdot \frac{364}{365} \cdot \frac{363}{365} = \frac{365 \cdot 364 \cdot 363}{365^3}$$

(b) Use the pattern in part (a) to write an expression for the probability that $n = 4$ people have distinct birthdays.

(c) Let P_n be the probability that the n people have distinct birthdays. Verify that this probability can be obtained recursively by

$$P_1 = 1 \text{ and } P_n = \frac{365 - (n - 1)}{365} P_{n-1}.$$

(d) Explain why $Q_n = 1 - P_n$ gives the probability that at least two people in a group of n people have the same birthday.

(e) Use the results of parts (c) and (d) to complete the table.

n	10	15	20	23	30	40	50
P_n							
Q_n							

(f) How many people must be in a group so that the probability of at least two of them having the same birthday is greater than $\frac{1}{2}$? Explain.

64. *Think About It* A weather forecast indicates that the probability of rain is 40%. What does this mean?

Review

In Exercises 65–68, find all real solutions of the polynomial equation.

65. $6x^2 + 8 = 0$

66. $4x^2 + 6x - 12 = 0$

67. $x^3 - x^2 - 3x = 0$

68. $x^5 + x^3 - 2x = 0$

In Exercises 69–74, find all real solutions of the rational equation.

69. $\dfrac{12}{x} = -3$

70. $\dfrac{32}{x} = 2x$

71. $\dfrac{2}{x - 5} = 4$

72. $\dfrac{3}{2x + 3} - 4 = \dfrac{-1}{2x + 3}$

73. $\dfrac{3}{x - 2} + \dfrac{x}{x + 2} = 1$

74. $\dfrac{2}{x} - \dfrac{5}{x - 2} = \dfrac{-13}{x^2 - 2x}$

In Exercises 75–80, find all real solutions of the exponential equation.

75. $e^x = 27$

76. $e^x + 7 = 35$

77. $e^{2x} - 4e^x + 3 = 0$

78. $e^{2x} - 7e^x + 12 = 0$

79. $200e^{-x} = 75$

80. $800e^{-x} = 250$

In Exercises 81–84, find all real solutions of the logarithmic equation.

81. $\ln x = 8$

82. $3 - 4 \ln x = 6$

83. $4 \ln 6x = 16$

84. $5 \ln 2x - 4 = 11$

Chapter Summary

What did you learn?

Section 9.1	Review Exercises
☐ How to use sequence notation to write the terms of a sequence	1–4
☐ How to use factorial notation	5–8
☐ How to use summation notation to write sums	9–16
☐ How to find the sum of an infinite series	17–20
☐ How to use sequences and series to model and solve real-life problems	21, 22
Section 9.2	
☐ How to recognize and write arithmetic sequences	23–30
☐ How to find an nth partial sum of an arithmetic sequence	31–36
☐ How to use arithmetic sequences to model and solve real-life problems	37, 38
Section 9.3	
☐ How to recognize and write geometric sequences	39–46
☐ How to find the nth partial sum of a geometric sequence	47–56
☐ How to find the sum of an infinite geometric series	57–62
☐ How to use geometric sequences to model and solve real-life problems	63, 64
Section 9.4	
☐ How to use mathematical induction to prove a statement	65–68
☐ How to find the sums of powers of integers	69–72
☐ How to recognize patterns and write the nth term of a sequence	73–76
☐ How to find finite differences of a sequence	77–80
Section 9.5	
☐ How to use the Binomial Theorem to calculate binomial coefficients	81–84
☐ How to use Pascal's Triangle to calculate binomial coefficients	85–88
☐ How to use binomial coefficients to write binomial expansions	89–94
Section 9.6	
☐ How to solve simple counting problems	95, 96
☐ How to use the Fundamental Counting Principle to solve counting problems	97, 98
☐ How to use permutations to solve counting problems	99, 100
☐ How to use combinations to solve counting problems	101, 102
Section 9.7	
☐ How to find the probability of an event	103, 104
☐ How to find the probabilities of mutually exclusive events	105, 106
☐ How to find the probabilities of independent events	107, 108
☐ How to find the probability of the complement of an event	109, 110

Review Exercises

9.1 In Exercises 1–4, write the first five terms of the sequence. (Assume that n begins with 1.)

1. $a_n = 2 + \dfrac{6}{n}$
2. $a_n = \dfrac{5n}{2n - 1}$
3. $a_n = \dfrac{72}{n!}$
4. $a_n = n(n - 1)$

In Exercises 5–8, evaluate the factorial expression.

5. $5!$
6. $3! \cdot 2!$
7. $\dfrac{3! \cdot 5!}{6!}$
8. $\dfrac{7! \cdot 6!}{6! \cdot 8!}$

In Exercises 9–14, find the sum.

9. $\sum_{i=1}^{6} 5$
10. $\sum_{k=2}^{5} 4k$
11. $\sum_{j=1}^{4} \dfrac{6}{j^2}$
12. $\sum_{i=1}^{8} \dfrac{i}{i + 1}$
13. $\sum_{k=1}^{10} 2k^3$
14. $\sum_{j=0}^{4} (j^2 + 1)$

In Exercises 15 and 16, use sigma notation to write the sum.

15. $\dfrac{1}{2(1)} + \dfrac{1}{2(2)} + \dfrac{1}{2(3)} + \cdot \cdot \cdot + \dfrac{1}{2(20)}$
16. $\dfrac{1}{2} + \dfrac{2}{3} + \dfrac{3}{4} + \cdot \cdot \cdot + \dfrac{9}{10}$

In Exercises 17–20, find the sum of the infinite series.

17. $\sum_{i=1}^{\infty} \dfrac{5}{10^i}$
18. $\sum_{i=1}^{\infty} \dfrac{3}{10^i}$
19. $\sum_{k=1}^{\infty} \dfrac{2}{100^k}$
20. $\sum_{k=2}^{\infty} \dfrac{9}{10^k}$

21. *Job Offer* The starting salary for a job is \$34,000 with a guaranteed salary increase of \$2250 per year. Determine (a) the salary during the fifth year and (b) the total compensation through 5 full years of employment.

22. *Baling Hay* In the first two trips baling hay around a large field, a farmer obtains 123 bales and 112 bales, respectively. Because each round gets shorter, the farmer estimates that the same pattern will continue. Estimate the total number of bales made if there are another six trips around the field.

9.2 In Exercises 23–26, determine whether the sequence is arithmetic. If it is, find the common difference.

23. $5, 3, 1, -1, -3, . . .$
24. $0, 1, 3, 6, 10, . . .$
25. $\frac{1}{2}, 1, \frac{3}{2}, 2, \frac{5}{2}, . . .$
26. $\frac{9}{9}, \frac{8}{9}, \frac{7}{9}, \frac{6}{9}, \frac{5}{9}, . . .$

In Exercises 27–30, find a formula for a_n for the arithmetic sequence.

27. $a_1 = 7, d = 12$
28. $a_1 = 25, d = -3$
29. $a_1 = y, d = 3y$
30. $a_1 = -2x, d = x$

In Exercises 31–34, find the sum.

31. $\sum_{j=1}^{10} (2j - 3)$
32. $\sum_{j=1}^{8} (20 - 3j)$
33. $\sum_{k=1}^{11} \left(\frac{2}{3}k + 4\right)$
34. $\sum_{k=1}^{25} \left(\dfrac{3k + 1}{4}\right)$

35. Find the sum of the first 100 positive multiples of 5.

36. Find the sum of the integers from 20 to 80 (inclusive).

37. *Running* The first time you run a 5-mile distance course, it takes you 49 minutes. If you run the same course 30 seconds faster each week, how fast can you run the 5-mile course after 12 weeks?

38. *Running* On the first day of a new training schedule you run 2 miles. If you increase your distance by one-half mile every day, how many total miles will you run in 14 days?

9.3 In Exercises 39–42, write the first five terms of the geometric sequence.

39. $a_1 = 4, \ r = -\frac{1}{4}$
40. $a_1 = 2, \ r = 2$
41. $a_1 = 9, \ a_3 = 4$
42. $a_1 = 2, \ a_3 = 12$

In Exercises 43–46, write an expression for the *n*th term of the geometric sequence and find the sum of the first 20 terms of the sequence.

43. $a_1 = 16, a_2 = -8$ **44.** $a_3 = 6, a_4 = 1$

45. $a_1 = 100, r = 1.05$ **46.** $a_1 = 5, r = 0.2$

In Exercises 47–52, find the sum.

47. $\sum_{i=1}^{7} 2^{i-1}$ **48.** $\sum_{i=1}^{5} 3^{i-1}$

49. $\sum_{i=1}^{4} \left(\frac{1}{2}\right)^i$ **50.** $\sum_{i=1}^{6} \left(\frac{1}{3}\right)^{i-1}$

51. $\sum_{i=1}^{5} (2)^{i-1}$ **52.** $\sum_{i=1}^{4} 6(3)^i$

In Exercises 53–56, use a graphing utility to find the sum.

53. $\sum_{i=1}^{10} 10\left(\frac{3}{5}\right)^{i-1}$ **54.** $\sum_{i=1}^{15} 20(0.2)^{i-1}$

55. $\sum_{i=1}^{25} 100(1.06)^{i-1}$ **56.** $\sum_{i=1}^{20} 8\left(\frac{6}{5}\right)^{i-1}$

In Exercises 57–62, find the sum of the infinite series.

57. $\sum_{i=1}^{\infty} \left(\frac{7}{8}\right)^{i-1}$ **58.** $\sum_{i=1}^{\infty} \left(\frac{1}{3}\right)^{i-1}$

59. $\sum_{i=1}^{\infty} (0.1)^{i-1}$ **60.** $\sum_{i=1}^{\infty} (0.5)^{i-1}$

61. $\sum_{k=1}^{\infty} 4\left(\frac{2}{3}\right)^{k-1}$ **62.** $\sum_{k=1}^{\infty} 1.3\left(\frac{1}{10}\right)^{k-1}$

63. *Depreciation* A company buys a machine for \$120,000. During the next 5 years it will depreciate at a rate of 30% per year. (That is, at the end of each year the depreciated value will be 70% of what it was at the beginning of the year.)

(a) Find the formula for the *n*th term of a geometric sequence that gives the value of the machine *t* full years after it was purchased.

(b) Find the depreciated value of the machine at the end of 5 full years.

64. *Total Compensation* A job pays a salary of \$32,000 the first year. During the next 39 years, suppose there is a 5.5% raise each year. What would be the total amount earned over the 40-year period?

9.4 In Exercises 65–68, use mathematical induction to prove the formula for every positive integer *n*.

65. $1 + 4 + \cdots + (3n - 2) = \frac{n}{2}(3n - 1)$

66. $1 + \frac{3}{2} + 2 + \frac{5}{2} + \cdots + \frac{1}{2}(n + 1) = \frac{n}{4}(n + 3)$

67. $\sum_{i=0}^{n-1} ar^i = \frac{a(1 - r^n)}{1 - r}$

68. $\sum_{k=0}^{n-1} (a + kd) = \frac{n}{2}[2a + (n - 1)d]$

In Exercises 69–72, find the sum using the formulas for the sums of powers of integers.

69. $\sum_{n=1}^{30} n$ **70.** $\sum_{n=1}^{10} n^2$

71. $\sum_{n=1}^{7} n^4$ **72.** $\sum_{n=1}^{6} n^5$

In Exercises 73–76, find a formula for the sum of the first *n* terms of the sequence.

73. $9, 13, 17, 21, \ldots$ **74.** $68, 60, 52, 44, \ldots$

75. $1, \frac{3}{5}, \frac{9}{25}, \frac{27}{125}, \ldots$ **76.** $12, -1, \frac{1}{12}, -\frac{1}{144}, \ldots$

In Exercises 77–80, find the first five terms of the sequence. Then calculate the first and second differences of the sequence. Does the sequence have a linear model, a quadratic model, or neither?

77. $a_1 = 5$
$a_n = a_{n-1} + 5$

78. $a_1 = -3$
$a_n = a_{n-1} - 2n$

79. $a_1 = 16$
$a_n = a_{n-1} - 1$

80. $a_0 = 0$
$a_n = n - a_{n-1}$

9.5 In Exercises 81–84, use the Binomial Theorem to calculate the binomial coefficient.

81. ${}_6C_4$ **82.** ${}_{10}C_7$

83. ${}_8C_5$ **84.** ${}_{12}C_3$

In Exercises 85–88, use Pascal's Triangle to calculate the binomial coefficient.

85. $\binom{7}{3}$ **86.** $\binom{9}{4}$

87. $\binom{8}{6}$

88. $\binom{5}{3}$

In Exercises 89–94, use the Binomial Theorem to expand the binomial. Simplify your answer. (Remember that $i = \sqrt{-1}$.)

89. $\left(\frac{x}{2} + y\right)^4$

90. $\left(\frac{2}{x} - 3x\right)^6$

91. $(a - 3b)^5$

92. $(3x + y^2)^7$

93. $(5 + 2i)^4$

94. $(4 - 5i)^3$

9.6 **95.** *Dice* In how many different ways can a pair of dice be rolled to obtain a total of 10?

96. *Numbers in a Hat* If slips of paper numbered 1 through 14 are placed in a hat, in how many ways could you draw two numbers with replacement that total 12?

97. *Telephone Numbers* The same three-digit prefix is used for all of the telephone numbers in a small town. How many different telephone numbers are possible by changing only the last four digits?

98. *Telephone Numbers* A telephone number in another town can use any one of five different three-digit prefixes. How many different telephone numbers are possible in this town?

99. *Bike Race* There are 10 bicyclists entered in a race. In how many different orders could these 10 bicyclists finish?

100. *Bike Race* If 10 bicyclists are entered in a race, in how many different ways could the top three places be decided?

101. *Travel Wardrobe* You have eight different suits to choose from to take on a trip. How many combinations of three suits could you take?

102. *Neckties* If you have 20 different neckties in your wardrobe, how many combinations of three ties could you choose?

9.7 **103.** *Matching Socks* A man has five pairs of socks of which no two pairs are the same color. If he randomly selects two socks from a drawer, what is the probability that he gets a matched pair?

104. *Bookshelf Order* A child returns a five-volume set of books to a bookshelf. The child is not able to read, and so cannot distinguish one volume from another. What is the probability that the books are shelved in the correct order?

105. *Students by Class* At a particular high school, the numbers of students in each class are broken down by the following percents.

Freshmen	31%
Sophomores	26%
Juniors	25%
Seniors	18%

If a single student is picked randomly by lottery for a cash scholarship, what is the probability that the scholarship winner is

(a) a junior or senior?

(b) a freshman, sophomore, or junior?

106. *Data Analysis* A sample of college students, faculty, and administration were asked whether they favored a proposed increase in the annual activity fee to enhance student life on campus. The results of the study are listed in the table.

	Students	Faculty	Admin.	Total
Favor	237	37	18	292
Oppose	163	38	7	208
Total	400	75	25	500

A person is selected at random from the sample. Find the specified probability.

(a) The person is not in favor of the proposal.

(b) The person is a student.

(c) The person is a faculty member and is in favor of the proposal.

107. *Roll of the Dice* If a six-sided die is rolled three times, what is the probability of a 6 on each roll?

108. *Roll of the Dice* A six-sided die is rolled six times. What is the probability that each side appears exactly once?

109. *Picking a Card* You randomly select a card from a 52-card deck. What is the probability that the card is *not* a club?

110. *Tossing a Coin* Find the probability of obtaining at least one tail when a coin is tossed five times.

Synthesis

True or False? **In Exercises 111–114, determine whether the statement is true or false. Justify your answer.**

111. $\dfrac{(n+2)!}{n!} = (n+2)(n+1)$

112. $\sum_{i=1}^{5}(i^3 + 2i) = \sum_{i=1}^{5} i^3 + \sum_{i=1}^{5} 2i$

113. $\sum_{k=1}^{8} 3k = 3\sum_{k=1}^{8} k$

114. $\sum_{j=1}^{6} 2^j = \sum_{j=3}^{8} 2^{j-2}$

115. An infinite sequence is a function. What is the domain of the function?

116. How do the two sequences differ?

(a) $a_n = \dfrac{(-1)^n}{n}$

(b) $a_n = \dfrac{(-1)^{n+1}}{n}$

117. In your own words, explain what makes a sequence (a) arithmetic and (b) geometric.

118. The graphs of two sequences are shown below. Identify each sequence as arithmetic or geometric. Explain your reasoning.

(a)

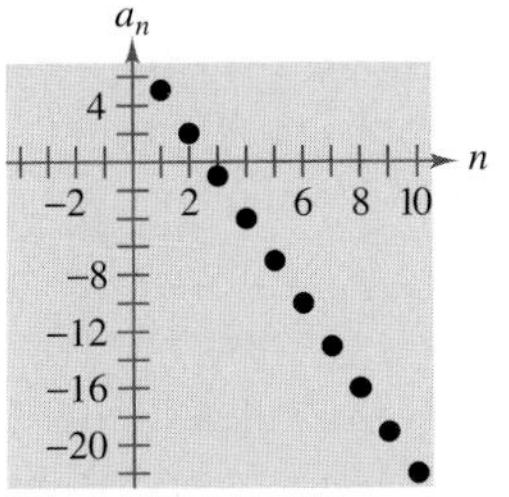

(b)

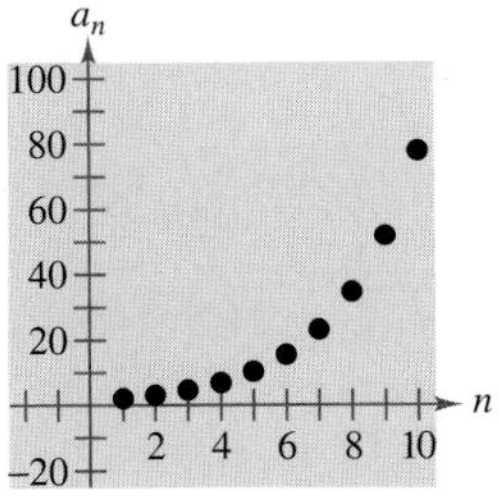

119. Explain what a recursion formula is.

120. Explain why the terms of a geometric sequence decrease when $0 < r < 1$.

In Exercises 121–124, match the sequence or sum of a sequence with its graph without doing any calculations. Explain your reasoning. [The graphs are labeled (a), (b), (c), and (d).]

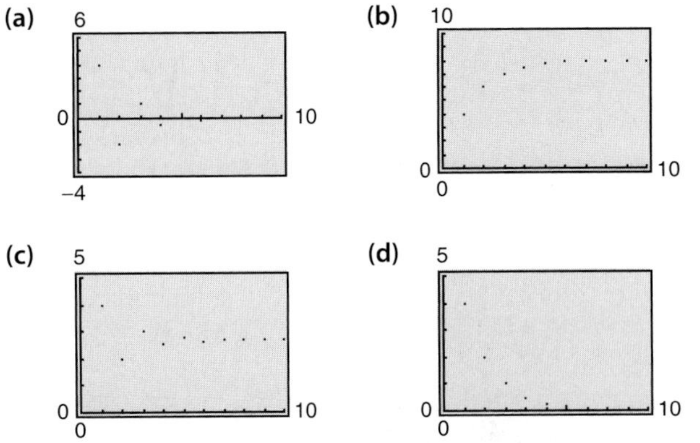

121. $a_n = 4\left(\frac{1}{2}\right)^{n-1}$

122. $a_n = 4\left(-\frac{1}{2}\right)^{n-1}$

123. $a_n = \sum_{k=1}^{n} 4\left(\frac{1}{2}\right)^{k-1}$

124. $a_n = \sum_{k=1}^{n} 4\left(-\frac{1}{2}\right)^{k-1}$

125. How do the expansions of $(x+y)^n$ and $(x-y)^n$ differ?

126. The probability of an event must be a real number in what interval? Is the interval open or closed?

127. The probability of an event is $\frac{2}{3}$. What is the probability that the event does not occur? Explain.

128. A weather forecast indicates that the probability of rain is 60%. Explain what this means.

Chapter Project ▶ Exploring Difference Quotients

In this project, you will explore difference quotients and see how they can be used to find average rates.

Example ▶ Finding an Average Speed

You are driving from Atlanta to Miami. The trip is about 700 miles and takes you 12 hours. Your average speed is

$$\text{Average speed} = \frac{\text{distance}}{\text{time}} = \frac{700}{12} \approx 58.3 \text{ miles per hour.}$$

The concept of average rate of change can be generalized as follows. Let f be a function defined on the interval $[a, b]$. The **average rate of change of f** from a to b is

$$\text{Average rate of change} = \frac{f(b) - f(a)}{b - a}.$$

This expression is called a **difference quotient.**

Chapter Project Investigations

1. (a) Calculate the average rate of change of $f(x) = 3x + 4$ on the interval $[2, 6]$. Select any other interval and show that you obtain the same average rate of change.

(b) Calculate the average rates of change of $f(x) = x^2$ on the intervals $[1, 3]$ and $[4, 6]$. Are they equal? Explain.

(c) During a 1-hour trip, your average speed is 50 miles per hour. Discuss the relationship between this speed and the speeds shown on your speedometer.

2. The data in the table gives the costs of first-class postage in the United States for selected years from 1971 to 1999, where $t = 71$ corresponds to 1971. (Source: U.S. Postal Service)

t	71	74	75	78	81
Postage	\$0.08	\$0.10	\$0.13	\$0.15	\$0.20

t	85	88	91	95	99
Postage	\$0.22	\$0.25	\$0.29	\$0.32	\$0.33

(a) Find the average rate of change of the cost of postage between each two adjacent time intervals.

(b) When was the average rate of change largest? When was it smallest?

(c) Are there intervals over which the average rate of change was zero? What does this mean?

Chapter Test

The *Interactive* CD-ROM and *Internet* versions of this text provide answers to the Chapter Tests and Cumulative Tests. They also offer Chapter Pre-Tests (which test key skills and concepts covered in previous chapters) and Chapter Post-Tests, both of which have randomly generated exercises with diagnostic capabilities.

Take this test as you would take a test in class. After you are done, check your work against the answers given in the back of the book.

1. Write the first five terms of the sequence $a_n = \dfrac{(-1)^n}{3n + 2}$.

2. Write an expression for the nth term of the sequence.

$\dfrac{3}{1!}, \dfrac{4}{2!}, \dfrac{5}{3!}, \dfrac{6}{4!}, \dfrac{7}{5!}, \ldots$

3. Find the fifth partial sum of the series.

$6 + 17 + 28 + 39 + \cdots$

4. The fifth term of an arithmetic series is 5.4, and the 12th term is 11.0. Find the 30th term.

5. Write the first five terms of the sequence $a_n = 5(2)^{n-1}$.

6. Find the sum of the finite series $\displaystyle\sum_{i=1}^{50} (2i^2 + 5)$.

7. Find the sum of the infinite series $\displaystyle\sum_{i=1}^{\infty} 4\left(\tfrac{1}{2}\right)^i$.

8. Use mathematical induction to prove the formula.

$$5 + 10 + 15 + \cdots + 5n = \frac{5n(n + 1)}{2}$$

9. Use the Binomial Theorem to expand the expression $(x + 2y)^4$.

10. To dress for a party, you can choose from six different pairs of slacks, 10 different shirts, and three different pairs of shoes. How many possible outfit combinations do you have?

In Exercises 11 and 12, evaluate the expression.

11. (a) ${}_9P_2$ (b) ${}_{70}P_3$

12. (a) ${}_{11}C_4$ (b) ${}_{66}C_4$

13. Eight people are going for a ride in a boat that seats eight people. The owner of the boat will drive, and only three of the remaining people are willing to ride in the two bow seats. How many seating arrangements are possible?

14. You attend a karaoke night and hope to hear your favorite song. The karaoke song book has 300 different songs. Assuming that the singers are equally likely to pick any song and no song is repeated, what is the probability that your favorite song is one of the 20 that you hear?

15. You are with seven of your friends at a party. Names of all of the 60 guests are placed in a hat and drawn randomly to award eight door prizes. If each guest is limited to one prize, what is the probability that you and your friends win all eight of the prizes?

16. The weather report calls for a 75% chance of rain. According to this report, what is the probability that it will not rain?

Cumulative Test for Chapters 7–9

Take this test to review the material from earlier chapters. After you are done, check your work against the answers given in the back of the book.

In Exercises 1–4, solve the system by the specified method.

1. Substitution

$$\begin{cases} y = 3 - x^2 \\ 2(y - 2) = x - 1 \end{cases}$$

2. Elimination

$$\begin{cases} x + 3y = -1 \\ 2x + 4y = 0 \end{cases}$$

3. Elimination

$$\begin{cases} -2x + 4y - z = 3 \\ x - 2y + 2z = -6 \\ x - 3y - z = 1 \end{cases}$$

4. Gauss-Jordan Elimination

$$\begin{cases} x + 3y - 2z = -7 \\ -2x + y - z = -5 \\ 4x + y + z = 3 \end{cases}$$

In Exercises 5 and 6, sketch the graph of the solution set of the system of inequalities.

5. $\begin{cases} 2x + y \geq -3 \\ x - 3y \leq 2 \end{cases}$

6. $\begin{cases} x - y > 6 \\ 5x + 2y < 10 \end{cases}$

7. Sketch a graph of the solution of the constraints and maximize the objective function $z = 3x + 2y$ subject to the constraints.

$$\begin{aligned} x + 4y &\leq 20 \\ 2x + y &\leq 12 \\ x &\geq 0 \\ y &\geq 0 \end{aligned}$$

8. A custom-blend bird seed is to be mixed from seed mixtures costing \$0.75 per pound and \$1.25 per pound. How many pounds of each seed mixture are used to make 200 pounds of custom-blend bird seed costing \$0.95 per pound?

9. Find the equation of the parabola $y = ax^2 + bx + c$ passing through the points (0, 4), (3, 1), and (6, 4).

$$\begin{cases} -x + 2y - z = 9 \\ 2x - y + 2z = -9 \\ 3x + 3y - 4z = 7 \end{cases}$$

SYSTEM FOR 10 AND 11

In Exercises 10 and 11, use the system of equations at the left.

10. Write the augmented matrix corresponding to the system of equations.

11. Solve the system using the matrix and Gauss-Jordan elimination.

In Exercises 12–15, use the following matrices.

$$A = \begin{bmatrix} 4 & 0 \\ -1 & 2 \end{bmatrix}, \quad B = \begin{bmatrix} -1 & 3 \\ 1 & 0 \end{bmatrix}$$

12. Find $A - B$.

13. Find $-2B$.

14. Find $A - 2B$.

15. Find AB, if possible.

16. Find the inverse (if it exists): $\begin{bmatrix} 1 & 2 & -1 \\ 3 & 7 & -10 \\ -5 & -7 & -15 \end{bmatrix}$.

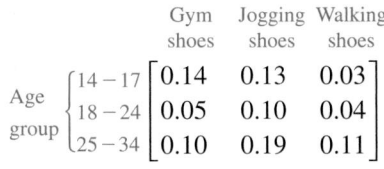

		Gym shoes	Jogging shoes	Walking shoes
Age group	14 – 17	0.14	0.13	0.03
	18 – 24	0.05	0.10	0.04
	25 – 34	0.10	0.19	0.11

MATRIX FOR 17

17. The percents (by age group) of the total amounts spent on three types of shoes in 1996 are shown in the matrix. The total amounts (in millions) spent by each age group on the three types of shoes were \$518.97 (14–17 age group), \$336.16 (18–24 age group), and \$753.37 (25–34 age group). How many dollars worth of gym shoes, jogging shoes, and walking shoes were sold in 1996? (Source: National Sporting Goods Association)

In Exercises 18 and 19, use Cramer's Rule to solve the system of equations.

18. $\begin{cases} 8x - 3y = -52 \\ 3x + 5y = 5 \end{cases}$

19. $\begin{cases} 5x + 4y + 3z = 7 \\ -3x - 8y + 7z = -9 \\ 7x - 5y - 6z = -53 \end{cases}$

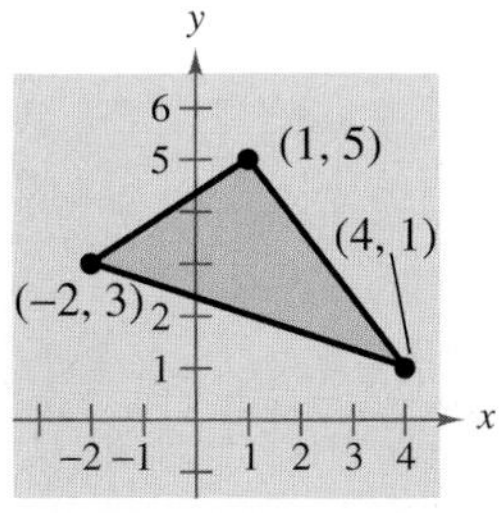

FIGURE FOR 20

20. Find the area of the triangle in the figure.

21. Write the first five terms of the sequence $a_n = \dfrac{(-1)^{n+1}}{2n + 3}$.

22. Write an expression for the nth term of the sequence.

$$\frac{2!}{4}, \frac{3!}{5}, \frac{4!}{6}, \frac{5!}{7}, \frac{6!}{8}, \ldots$$

23. Sum the first 20 terms of the arithmetic sequence 8, 12, 16, 20, ...

24. The sixth term of an arithmetic series is 20.6, and the ninth term is 30.2. Find the 20th term.

25. Write the first five terms of the sequence $a_n = 3(2)^{n-1}$.

26. Find the sum: $\sum_{i=0}^{\infty} 3\left(\frac{1}{2}\right)^i$.

27. Use mathematical induction to prove the formula

$$3 + 7 + 11 + 15 + \cdots + (4n - 1) = n(2n + 1).$$

28. Use the Binomial Theorem to expand and simplify $(z - 3)^4$.

In Exercises 29–32, evaluate the expression.

29. ${}_7P_3$ 30. ${}_{25}P_2$ 31. $\binom{8}{4}$ 32. ${}_{10}C_3$

33. A personnel manager has 10 applicants to fill three different positions. In how many ways can this be done, assuming that all the applicants are qualified for any of the three positions?

34. On a game show, the digits 3, 4, and 5 must be arranged in the proper order to form the price of an appliance. If the digits are arranged correctly, the contestant wins the appliance. What is the probability of winning if the contestant knows that the price is at least \$400?

VGC/FPG International

The space shuttle orbits earth at about 200 miles. By comparison, geosynchronous satellites orbit at about 24,000 miles, and Global Positioning System (GPS) satellites orbit at about 11,000 miles. *(Source: Howstuffworks.com, Inc.)*

10 Topics in Analytic Geometry

▶ How to Study This Chapter

The Big Picture

In this chapter you will learn the following skills and concepts.

- ▶ How to find the inclination of a line, the angle between two lines, and the distance between a point and a line
- ▶ How to write the standard form of the equation of a parabola, an ellipse, and a hyperbola
- ▶ How to eliminate the *xy*-term in the equation of a conic and use the discriminant to identify a conic
- ▶ How to rewrite a set of parametric equations as a rectangular equation and find a set of parametric equations for a graph
- ▶ How to write equations in polar form and graph polar equations

Important Vocabulary

As you encounter each new vocabulary term in this chapter, add the term and its definition to your notebook glossary.

Inclination (p. 770)
Angle between two lines (p. 771)
Conic section or conic (pp. 777, 834)
Parabola (pp. 778, 834)
Directrix (p. 778)
Focus or foci (pp. 778, 786)
Focal chord (p. 780)
Latus rectum (p. 780)
Tangent (p. 780)
Ellipse (pp. 786, 834)
Vertices (pp. 786, 795)
Major axis (p. 786)
Minor axis (p. 786)
Center (pp. 786, 795)
Eccentricity (pp. 790, 799, 834)
Hyperbola (pp. 795, 834)
Branches (p. 795)
Transverse axis (p. 795)
Asymptotes (p. 797)
Conjugate axis (p. 797)
Invariant under rotation (p. 808)
Discriminant (p. 808)
Parameter (p. 812)
Parametric equations (p. 812)
Plane curve (p. 812)
Orientation (p. 813)
Polar coordinate system (p. 820)
Pole or origin (p. 820)
Polar axis (p. 820)
Polar coordinates (p. 820)

Study Tools

- Learning objectives at the beginning of each section
- Chapter Summary (p. 841)
- Review Exercises (pp. 842–845)
- Chapter Test (p. 847)

Additional Resources

- Study and Solutions Guide
- Interactive Precalculus
- Videotapes for Chapter 10
- Precalculus Website
- Student Success Organizer

STUDY T!P

Once a test has begun, read the directions carefully and work at a reasonable pace. If you finish early, take a few moments to clear your thoughts and then go over your work.

10.1 Lines

▶ **What you should learn**

- How to find the inclination of a line
- How to find the angle between two lines
- How to find the distance between a point and a line

▶ **Why you should learn it**

The inclination of a line can be used to measure heights indirectly. For instance, Exercise 56 on page 776 shows how the inclination of a line can be used to determine the heights of two mountain peaks.

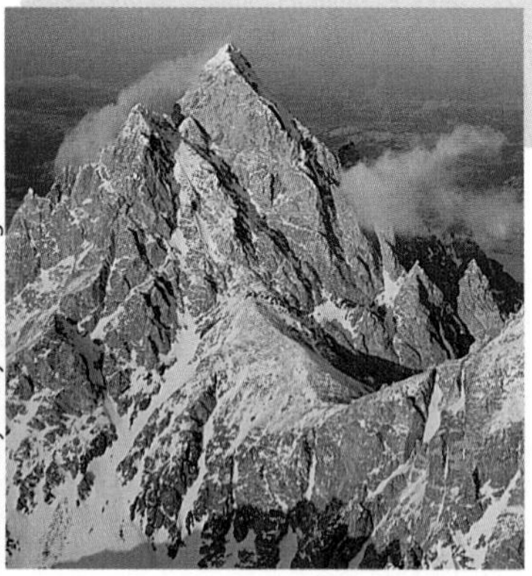

Paul Chesley/Tony Stone Images

Inclination of a Line

In Section 1.2 you learned that the graph of the linear equation

$$y = mx + b$$

is a nonvertical line with slope m and y-intercept at $(0, b)$. There, the slope of a line was described as the rate of change in y with respect to x. In this section, you will look at the slope of a line in terms of the angle of inclination of the line.

Every nonhorizontal line must intersect the x-axis. The angle formed by such an intersection determines the **inclination** of the line, as specified in the following definition.

Definition of Inclination

The **inclination** of a nonhorizontal line is the positive angle θ (less than π) measured counterclockwise from the x-axis to the line. (See Figure 10.1.)

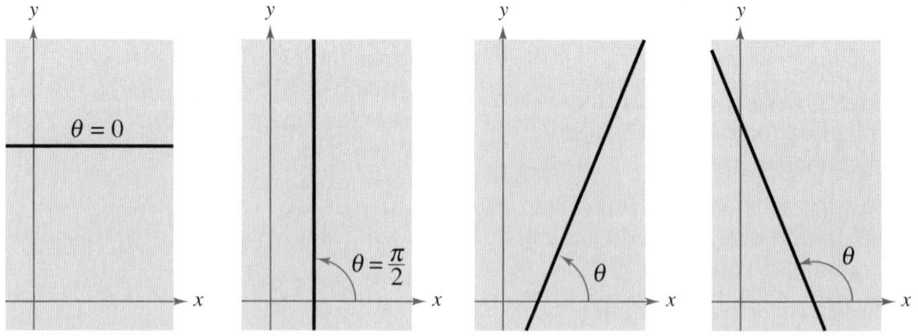

Horizontal line *Vertical line* *Acute angle* *Obtuse angle*

FIGURE 10.1

The inclination of a line is related to its slope in the following manner. A proof is given in Appendix A.

Inclination and Slope

If a nonvertical line has inclination θ and slope m, then

$$m = \tan \theta.$$

If the line has a positive slope, it will intersect the x-axis. Call this point $(x_1, 0)$. If (x_2, y_2) is a second point on the line, the slope is

$$m = \frac{y_2 - 0}{x_2 - x_1} = \frac{y_2}{x_2 - x_1}$$
$$= \tan \theta.$$

Try proving the case in which the line has a negative slope.

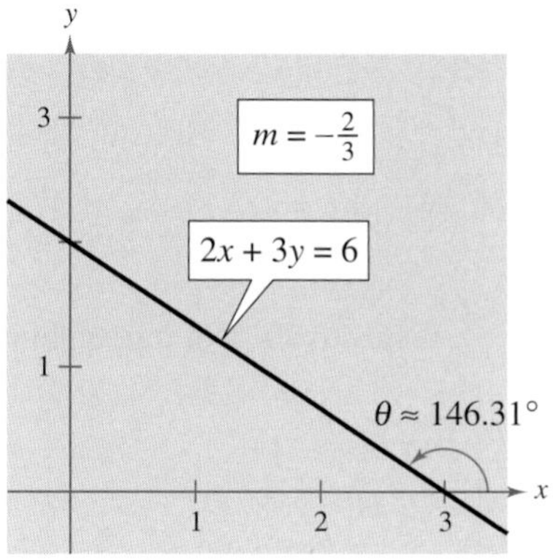

FIGURE 10.2

A computer simulation of this example appears in the *Interactive* CD-ROM and *Internet* versions of this text.

Example 1 ▶ Finding the Inclination of a Line

Find the inclination of the line given by $2x + 3y = 6$.

Solution

The slope of this line is $m = -\frac{2}{3}$. So, its inclination is determined from the equation

$$\tan \theta = -\frac{2}{3}.$$

From Figure 10.2, it follows that $\frac{\pi}{2} < \theta < \pi$. This means that

$$\begin{aligned}\theta &= \pi + \arctan\left(-\frac{2}{3}\right)\\ &\approx \pi + (-0.588)\\ &= \pi - 0.588\\ &\approx 2.554.\end{aligned}$$

The angle of inclination is about 2.554 radians or about 146.3°.

The Angle Between Two Lines

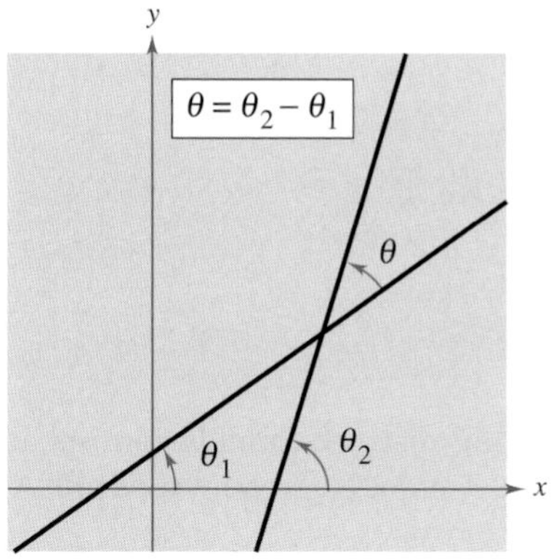

FIGURE 10.3

Two distinct lines in a plane are either parallel or intersecting. If they intersect, their intersection forms two pairs of opposite angles. One pair is acute and the other pair is obtuse. The smaller of these angles is called the **angle between the two lines.** As shown in Figure 10.3, you can use the inclinations of the two lines to find the angle between the two lines. Specifically, if two lines have inclinations θ_1 and θ_2, the angle between the two lines is

$$\theta = \theta_2 - \theta_1$$

where $\theta_1 < \theta_2$. You can use the formula for the tangent of the difference of two angles

$$\begin{aligned}\tan \theta &= \tan(\theta_2 - \theta_1)\\ &= \frac{\tan \theta_2 - \tan \theta_1}{1 + \tan \theta_1 \tan \theta_2}\end{aligned}$$

to obtain the formula for the angle between two lines.

Angle Between Two Lines

If two nonperpendicular lines have slopes m_1 and m_2, the angle between the two lines is

$$\tan \theta = \left|\frac{m_2 - m_1}{1 + m_1 m_2}\right|.$$

A computer simulation of this example appears in the *Interactive* CD-ROM and *Internet* versions of this text.

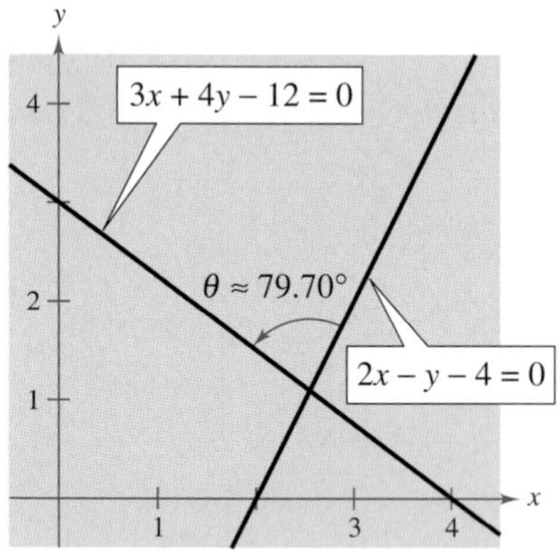

FIGURE 10.4

Example 2 ▶ Finding the Angle Between Two Lines

Find the angle between the two lines.

Line 1: $2x - y - 4 = 0$ *Line 2:* $3x + 4y - 12 = 0$

Solution

The two lines have slopes of $m_1 = 2$ and $m_2 = -\frac{3}{4}$, respectively. So, the tangent of the angle between the two lines is

$$\tan\theta = \left|\frac{m_2 - m_1}{1 + m_1 m_2}\right| = \left|\frac{(-3/4) - 2}{1 + (2)(-3/4)}\right| = \left|\frac{-11/4}{-2/4}\right| = \frac{11}{2}.$$

Finally, you can conclude that the angle is

$$\theta = \arctan\frac{11}{2} \approx 1.391 \text{ radians} \approx 79.70°$$

as shown in Figure 10.4.

The Distance Between a Point and a Line

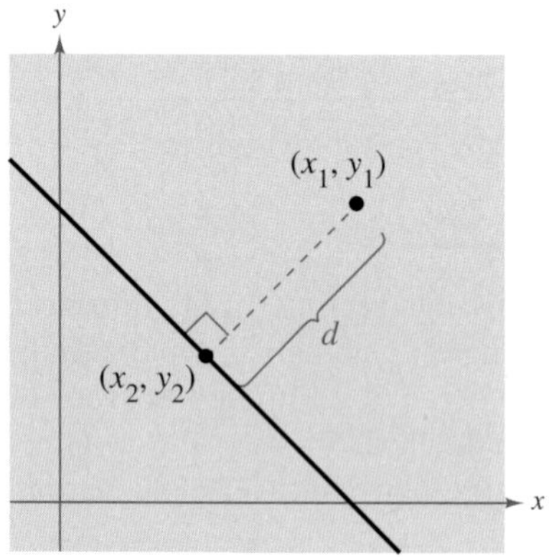

FIGURE 10.5

Finding the distance between a line and a point not on the line is an application of perpendicular lines. This distance is defined to be the length of the perpendicular line segment joining the point to the given line, as shown in Figure 10.5.

Distance Between a Point and a Line

The distance between the point (x_1, y_1) and the line $Ax + By + C = 0$ is

$$d = \frac{|Ax_1 + By_1 + C|}{\sqrt{A^2 + B^2}}.$$

Remember that the values of A, B, and C in this distance formula correspond to the general equation of a line, $Ax + By + C = 0$. A proof is given in Appendix A.

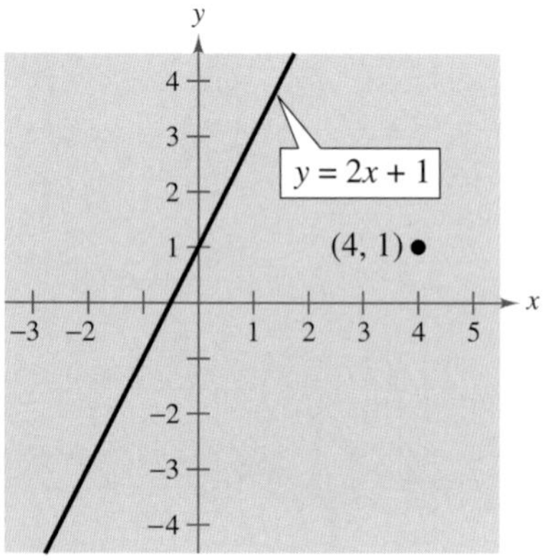

FIGURE 10.6

Example 3 ▶ Finding the Distance Between a Point and a Line

Find the distance between the point (4, 1) and the line $y = 2x + 1$.

Solution

The general form of the equation is

$$-2x + y - 1 = 0.$$

So, the distance between the point and the line is

$$d = \frac{|-2(4) + 1(1) - 1|}{\sqrt{(-2)^2 + 1^2}} = \frac{8}{\sqrt{5}} \approx 3.58.$$

The line and the point are shown in Figure 10.6.

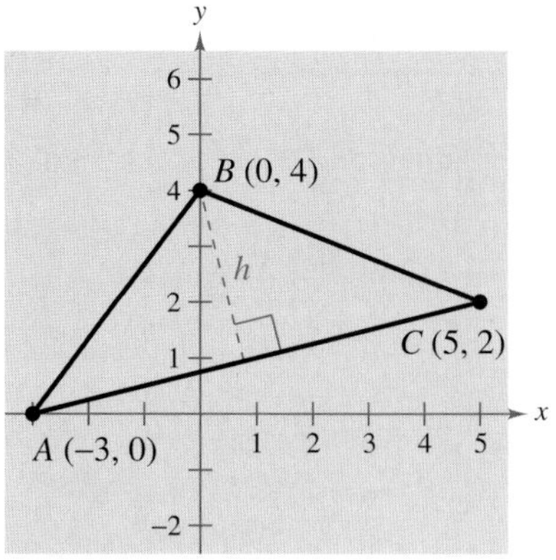

FIGURE 10.7

Activities

1. Find the inclination of the line $5x - 4y = 20$.
   Answer: $\theta = 0.896$ radian or $51.34°$
2. Find the angle θ between the lines $x + 2y = 5$ and $3x - y = 6$.
   Answer: $\theta = 1.429$ radians or $81.87°$
3. Find the distance between the point $(3, -1)$ and the line $4x - 3y - 12 = 0$.
   Answer: $d = \frac{3}{5}$

The *Interactive* CD-ROM and *Internet* versions of this text show every example with its solution; clicking on the *Try It!* button brings up similar problems. Guided Examples and Integrated Examples show step-by-step solutions to additional examples. Integrated Examples are related to several concepts in the section.

Group Activity: Graphing Utility
Put your students in groups of two. Ask each group to write a graphing calculator program that finds the angle between two lines based on user input. Ask each group to demonstrate the program on an example or exercise in this section.

Example 4 ▶ An Application of Two Distance Formulas

Figure 10.7 shows a triangle with vertices $A(-3, 0)$, $B(0, 4)$, and $C(5, 2)$.

a. Find the altitude from vertex B to side AC.

b. Find the area of the triangle.

Solution

a. To find the altitude, use the formula for the distance between line AC and the point $(0, 4)$. The equation of line AC is obtained as follows.

$$\textit{Slope:}\quad m = \frac{2-0}{5+3} = \frac{1}{4}$$

$$\textit{Equation:}\quad y - 0 = \frac{1}{4}(x+3)$$

$$4y = x + 3$$

$$x - 4y + 3 = 0$$

Therefore, the distance between this line and the point $(0, 4)$ is

$$\text{Altitude} = h = \frac{|1(0) - 4(4) + 3|}{\sqrt{1^2 + (-4)^2}} = \frac{13}{\sqrt{17}}.$$

b. Using the formula for the distance between two points, you can find the length of the base AC to be

$$\begin{aligned} b &= \sqrt{(5+3)^2 + (2-0)^2} \\ &= \sqrt{68} \\ &= 2\sqrt{17}. \end{aligned}$$

Finally, the area of the triangle in Figure 10.7 is

$$\begin{aligned} A &= \frac{1}{2}bh \\ &= \frac{1}{2}\left(2\sqrt{17}\right)\left(\frac{13}{\sqrt{17}}\right) \\ &= 13 \text{ square units.} \end{aligned}$$

Writing ABOUT MATHEMATICS

Inclination and the Angle Between Two Lines Discuss why the inclination of a line can be an angle that is larger than $\pi/2$, but the angle between two lines cannot be larger than $\pi/2$. Decide whether the following statement is true or false: "The inclination of a line is the angle between the line and the *x*-axis." Explain.

10.1 Exercises

The *Interactive* CD-ROM and *Internet* versions of this text contain step-by-step solutions to all odd-numbered Section and Review Exercises. They also provide Tutorial Exercises that link to Guided Examples for additional help.

In Exercises 1–8, find the slope of the line with inclination θ.

1.

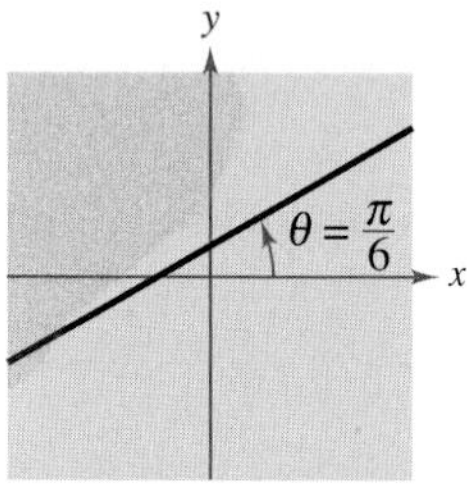

2.

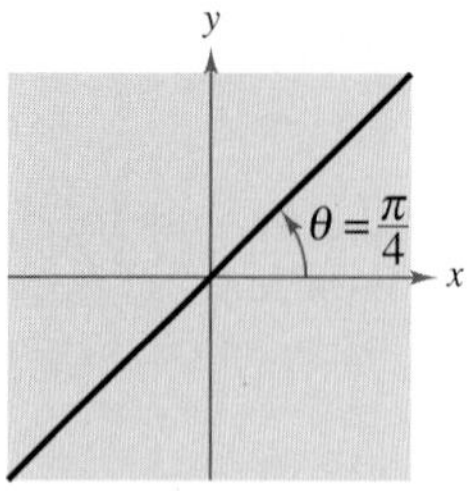

3.

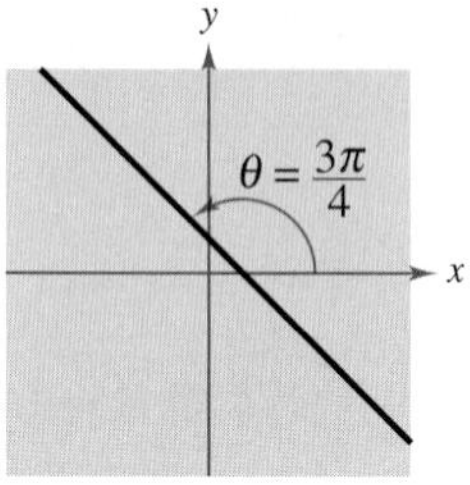

4.

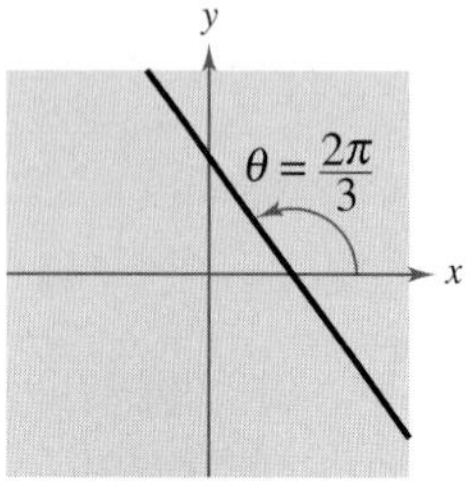

5. $\theta = \frac{\pi}{3}$ radians

6. $\theta = \frac{5\pi}{6}$ radians

7. $\theta = 1.27$ radians

8. $\theta = 2.88$ radians

In Exercises 9–14, find, in radians and degrees, the inclination θ of the line with a slope of m.

9. $m = -1$

10. $m = -2$

11. $m = 1$

12. $m = 2$

13. $m = \frac{3}{4}$

14. $m = -\frac{5}{2}$

In Exercises 15–18, find, in radians and degrees, the inclination θ of the line passing through the points.

15. $(6, 1), (10, 8)$

16. $(12, 8), (-4, -3)$

17. $(-2, 20), (10, 0)$

18. $(0, 100), (50, 0)$

In Exercises 19–22, find, in radians and degrees, the inclination θ of the line.

19. $6x - 2y + 8 = 0$

20. $4x + 5y - 9 = 0$

21. $5x + 3y = 0$

22. $x - y - 10 = 0$

In Exercises 23–32, find, in radians and degrees, the angle θ between the lines.

23. $3x + y = 3$
$x - y = 2$

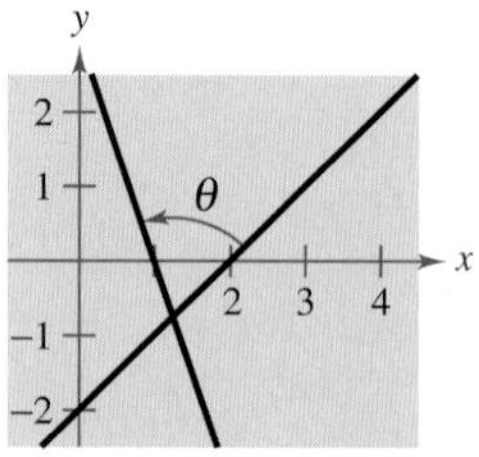

24. $x + 3y = 2$
$x - 2y = -3$

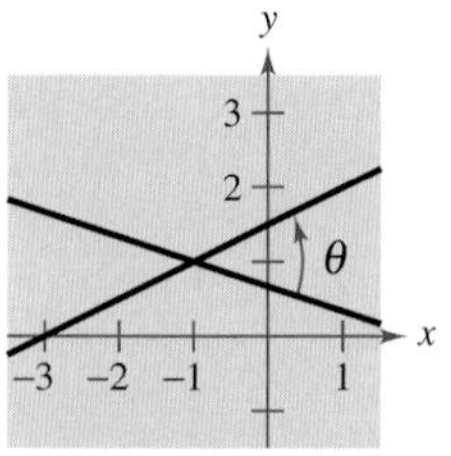

25. $x - y = 0$
$3x - 2y = -1$

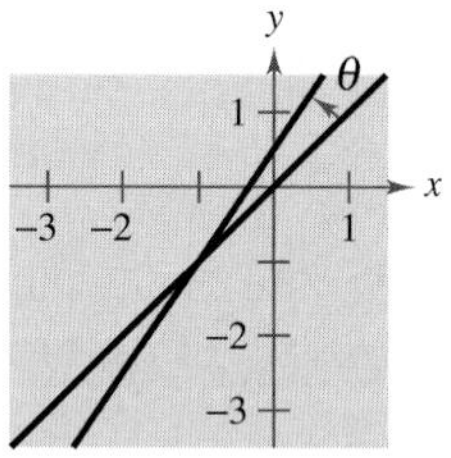

26. $2x - y = 2$
$4x + 3y = 24$

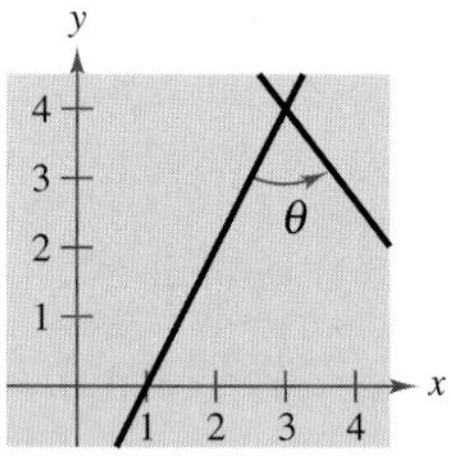

27. $x - 2y = 7$
$6x + 2y = 5$

28. $5x + 2y = 16$
$3x - 5y = -1$

29. $x + 2y = 8$
$x - 2y = 2$

30. $3x - 5y = 3$
$3x + 5y = 12$

31. $0.05x - 0.03y = 0.21$
$0.07x + 0.02y = 0.16$

32. $0.02x - 0.05y = -0.19$
$0.03x + 0.04y = 0.52$

Angle Measurement **In Exercises 33–36, find the slope of each side of the triangle and use the slopes to find the measures of the interior angles.**

33.

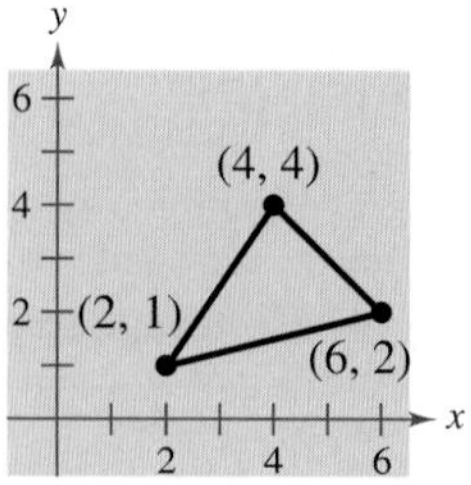

34.

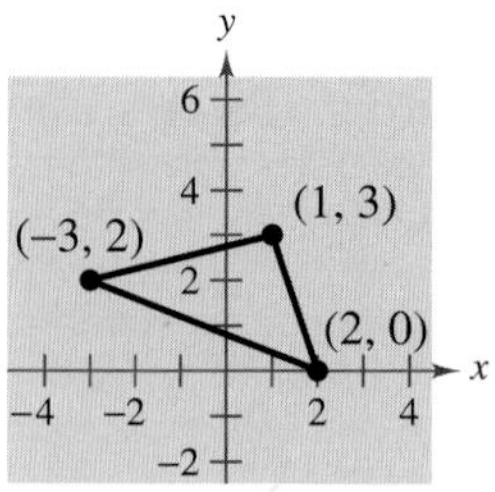

35.

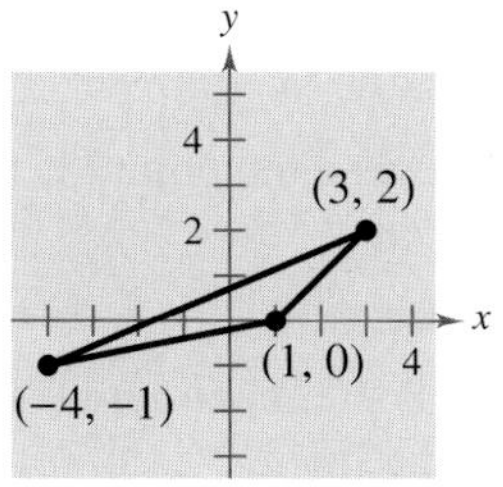

36.

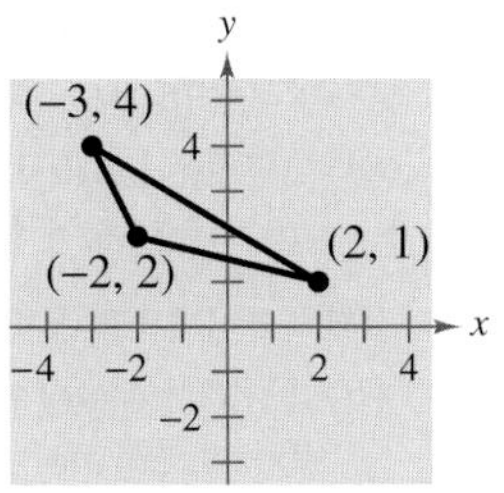

In Exercises 37–44, find the distance between the point and the line.

	Point	*Line*
37.	$(0, 0)$	$4x + 3y = 0$
38.	$(0, 0)$	$2x - y = 4$
39.	$(2, 3)$	$4x + 3y = 10$
40.	$(-2, 1)$	$x - y = 2$
41.	$(6, 2)$	$x + 1 = 0$
42.	$(10, 8)$	$y - 4 = 0$
43.	$(0, 8)$	$6x - y = 0$
44.	$(4, 2)$	$x - y = 20$

Area **In Exercises 45–48, (a) find the altitude from vertex *B* of the triangle to side *AC*, and (b) find the area of the triangle.**

45. $A = (0, 0),\ B = (1, 4),\ C = (4, 0)$

46. $A = (0, 0),\ B = (4, 5),\ C = (5, -2)$

47. $A = \left(-\frac{1}{2}, \frac{1}{2}\right),\ B = (2, 3),\ C = \left(\frac{5}{2}, 0\right)$

48. $A = (-4, -5),\ B = (3, 10),\ C = (6, 12)$

In Exercises 49 and 50, find the distance between the parallel lines.

49. $x + y = 1$
$x + y = 5$

50. $3x - 4y = 1$
$3x - 4y = 10$

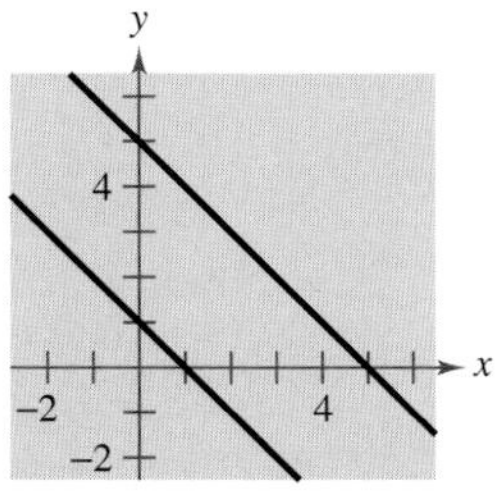

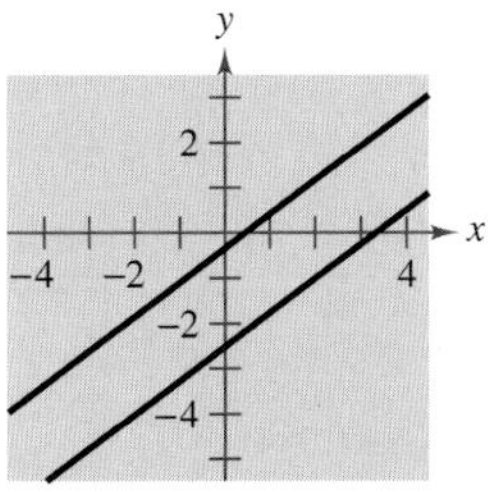

51. *Grade of a Road* A straight road rises with an inclination of 0.10 radian from the horizontal. Find the slope of the road and the change in elevation over a 2-mile stretch of the road.

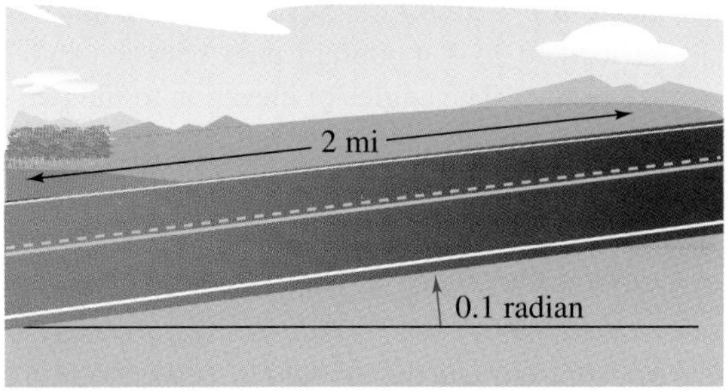

FIGURE FOR 51

52. *Grade of a Road* A straight road rises with an inclination of 0.20 radian from the horizontal. Find the slope of the road and the change in elevation over a 1-mile stretch of the road.

53. *Conveyor Design* A moving conveyor is built so that it rises 1 meter for each 3 meters of horizontal travel.

(a) Find the inclination of the conveyor.

(b) The conveyor runs between two floors in a factory. The distance between the floors is 5 meters. Find the length of the conveyor.

54. *Pitch of a Roof* A roof has a rise of 3 feet for every horizontal change of 5 feet (see figure). Find the inclination of the roof.

55. *Truss* Find the angles α and β shown in the drawing of the roof truss.

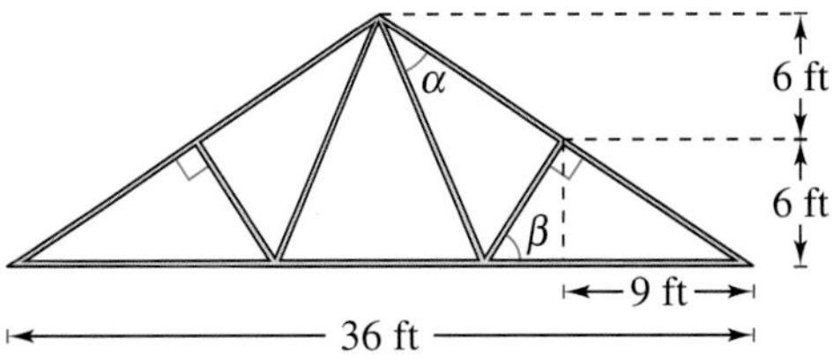

56. *Mountain Climbing* Several mountain climbers are located in a mountain pass between two peaks (see figure). The angles of elevation to the two peaks are 0.84 radian and 1.10 radians. A range finder shows that the distances to the peaks are 3250 feet and 6700 feet, respectively.

(a) Find the angle between the two lines of sight to the peaks.

(b) Approximate the amount of vertical climb that is necessary to reach the summit of each peak.

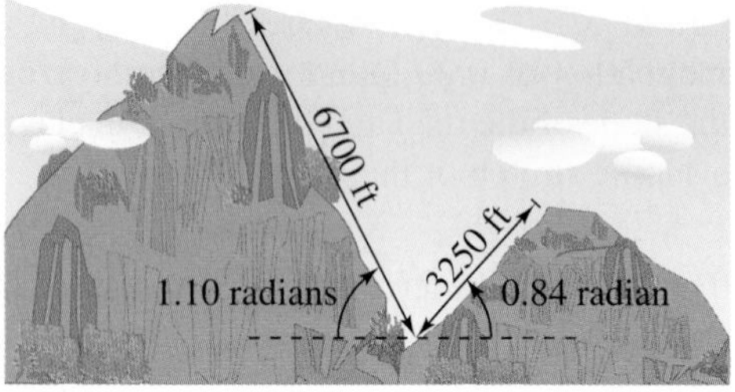

Synthesis

True or False? **In Exercises 57 and 58, determine whether the statement is true or false. Justify your answer.**

57. A line that has an inclination greater than $\pi/2$ radians has a negative slope.

58. To find the angle between two lines whose angles of inclination θ_1 and θ_2 are known, substitute θ_1 and θ_2 for m_1 and m_2, respectively, in the formula for the angle between two lines.

59. *Exploration* Consider a line with slope m and y-intercept $(0, 4)$.

(a) Write the distance d between the origin and the line as a function of m.

(b) Graph the function in part (a).

(c) Find the slope that yields the maximum distance between the origin and the line.

(d) Find the asymptote of the graph in part (b) and interpret its meaning in the context of the problem.

60. *Exploration* Consider a line with slope m and y-intercept $(0, 4)$.

(a) Write the distance d between the point $(3, 1)$ and the line as a function of m.

(b) Graph the function in part (a).

(c) Find the slope that yields the maximum distance between the point and the line.

(d) Is it possible for the distance to be 0? If so, what is the slope of the line that yields a distance of 0?

(e) Find the asymptote of the graph in part (b) and interpret its meaning in the context of the problem.

Review

In Exercises 61–66, find all x-intercepts and y-intercepts of the quadratic function.

61. $f(x) = (x - 7)^2$

62. $f(x) = (x + 9)^2$

63. $f(x) = (x - 5)^2 - 5$

64. $f(x) = (x + 11)^2 + 12$

65. $f(x) = x^2 - 7x - 1$

66. $f(x) = x^2 + 9x - 22$

In Exercises 67–72, write the quadratic function in standard form by completing the square. Identify the vertex of the function.

67. $f(x) = 3x^2 + 2x - 16$

68. $f(x) = 2x^2 - x - 21$

69. $f(x) = 5x^2 + 34x - 7$

70. $f(x) = -x^2 - 8x - 15$

71. $f(x) = 6x^2 - x - 12$

72. $f(x) = -8x^2 - 34x - 21$

10.2 Introduction to Conics: Parabolas

▶ **What you should learn**

- How to recognize a conic as the intersection of a plane and a double-napped cone
- How to write the standard form of the equation of a parabola
- How to use the reflective property of parabolas to solve real-life problems

▶ **Why you should learn it**

Parabolas can be used to model and solve many types of real-life problems. For instance, in Exercise 62 on page 783, a parabola is used to model the cables of a suspension bridge.

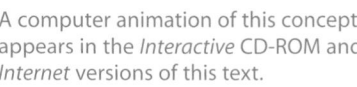

Adam Woolfitt/CORBIS

A computer animation of this concept appears in the *Interactive* CD-ROM and *Internet* versions of this text.

Conics

Conic sections were discovered during the classical Greek period, 600 to 300 B.C. The early Greeks were concerned largely with the geometric properties of conics. In the early 17th century the broad applicability of conics became apparent, and they then played a prominent role in the early development of calculus.

Each **conic section** (or simply **conic**) is the intersection of a plane and a double-napped cone. Notice in Figure 10.8(a) that in the formation of the four basic conics, the intersecting plane does not pass through the vertex of the cone. When the plane does pass through the vertex, the resulting figure is a *degenerate conic*, as shown in Figure 10.8(b).

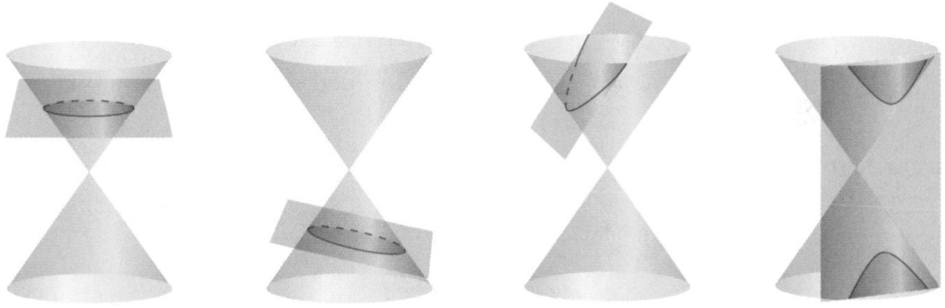

Circle *Ellipse* *Parabola* *Hyperbola*

(a) Conic Sections

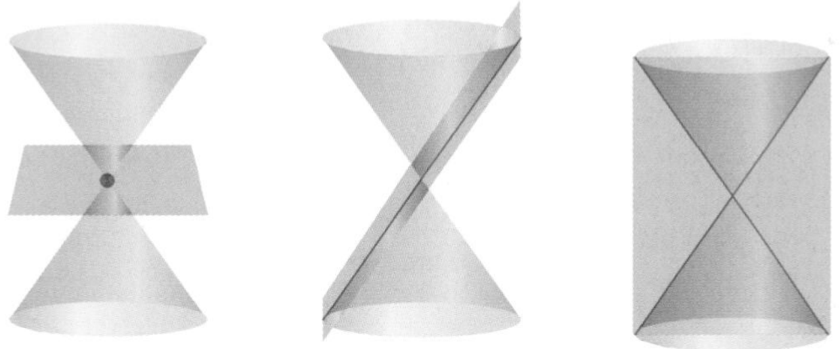

Point *Line* *Two Intersecting Lines*

(b) Degenerate Conics

FIGURE 10.8

There are several ways to approach a study of conics. You could define conics in terms of the intersections of planes and cones, as the Greeks did, or you could define them algebraically in terms of the general second-degree equation

$$Ax^2 + Bxy + Cy^2 + Dx + Ey + F = 0.$$

However, you will study a third approach, in which each of the conics is defined as a **locus** (collection) of points satisfying a geometric property. For example, a circle is defined as the collection of all points (x, y) that are equidistant from a fixed point (h, k). This leads to the standard equation of a circle

$$(x - h)^2 + (y - k)^2 = r^2. \qquad \text{Equation of circle}$$

This study of conics is from a locus-of-points approach, which leads to the development of the standard equation for each conic. Your students should know the standard equations of all conics well. Make sure they understand the relationship of *h* and *k* to the horizontal and vertical shifts.

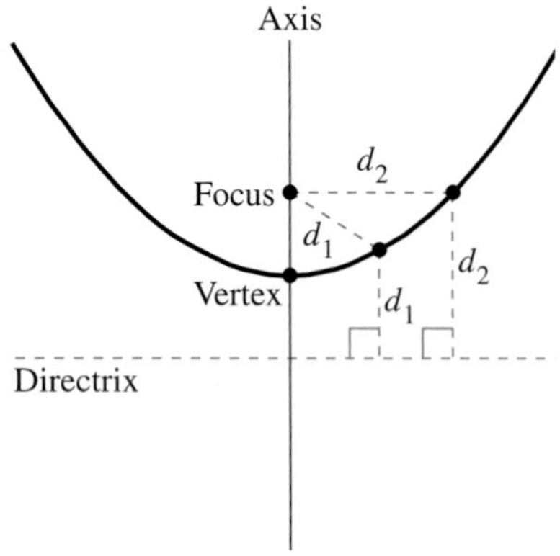

FIGURE 10.9

Parabolas

The first type of conic is called a **parabola,** and is defined as follows.

Definition of Parabola

A **parabola** is the set of all points (x, y) that are equidistant from a fixed line **(directrix)** and a fixed point **(focus)** not on the line.

The midpoint between the focus and the directrix is called the **vertex,** and the line passing through the focus and the vertex is called the **axis** of the parabola. Note in Figure 10.9 that a parabola is symmetric with respect to its axis. Using the definition of a parabola, you can derive the following **standard form** of the equation of a parabola whose directrix is parallel to the x-axis or to the y-axis. A proof is given in Appendix A.

Standard Equation of a Parabola

The **standard form** of the equation of a parabola with vertex at (h, k) is as follows.

$(x - h)^2 = 4p(y - k),\ p \neq 0$ Vertical axis, directrix: $y = k - p$

$(y - k)^2 = 4p(x - h),\ p \neq 0$ Horizontal axis, directrix: $x = h - p$

The focus lies on the axis p units (*directed distance*) from the vertex. If the vertex is at the origin $(0, 0)$, the equation takes one of the following forms.

$x^2 = 4py$ Vertical axis

$y^2 = 4px$ Horizontal axis

See Figure 10.10.

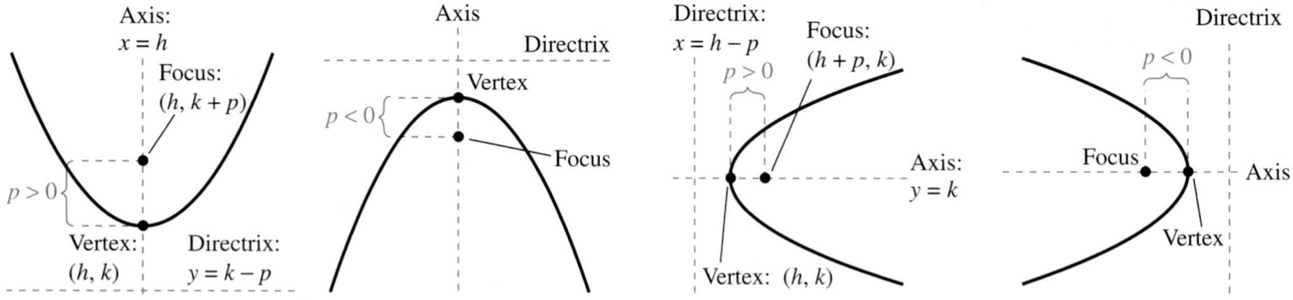

(a) $(x - h)^2 = 4p(y - k)$ **Vertical axis:** $p > 0$

(b) $(x - h)^2 = 4p(y - k)$ **Vertical axis:** $p < 0$

(c) $(y - k)^2 = 4p(x - h)$ **Horizontal axis:** $p > 0$

(d) $(y - k)^2 = 4p(x - h)$ **Horizontal axis:** $p < 0$

FIGURE 10.10

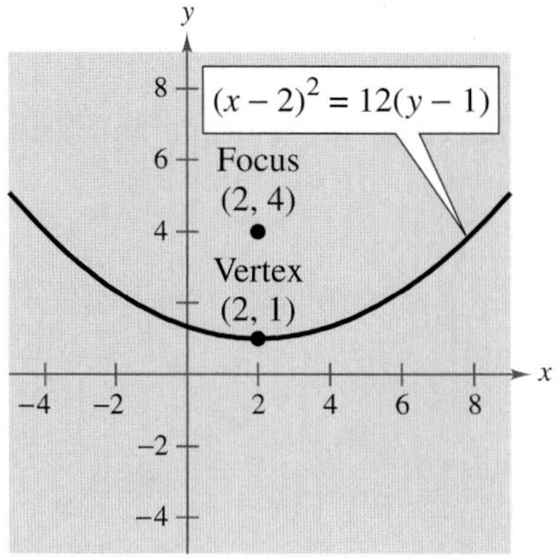

FIGURE 10.11

STUDY TIP

You may want to review the technique of completing the square found in Section P.5, which will be used to rewrite each of the conics in standard form.

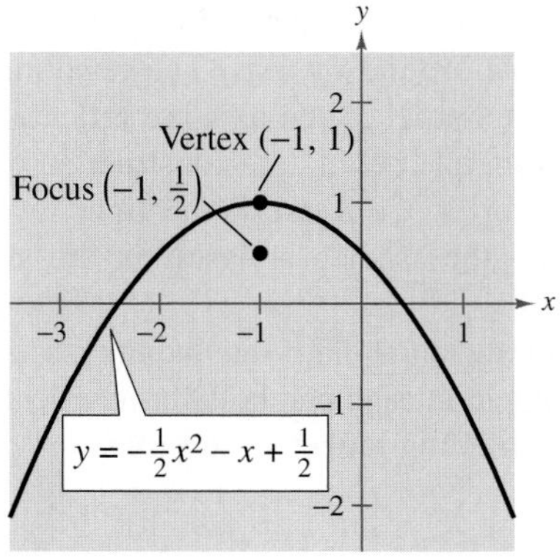

FIGURE 10.12

Example 1 ▶ Finding the Standard Equation of a Parabola

Find the standard form of the equation of the parabola with vertex $(2, 1)$ and focus $(2, 4)$.

Solution

Because the axis of the parabola is vertical, consider the equation

$$(x - h)^2 = 4p(y - k)$$

where $h = 2$, $k = 1$, and $p = 4 - 1 = 3$. So, the standard form is

$$(x - 2)^2 = 12(y - 1).$$

You can obtain the more common quadratic form as follows.

$$x^2 - 4x + 4 = 12y - 12 \qquad \text{Multiply.}$$

$$x^2 - 4x + 16 = 12y \qquad \text{Add 12 to each side.}$$

$$\frac{1}{12}(x^2 - 4x + 16) = y \qquad \text{Divide each side by 12.}$$

The graph of this parabola is shown in Figure 10.11.

Example 2 ▶ Finding the Focus of a Parabola

Find the focus of the parabola

$$y = -\frac{1}{2}x^2 - x + \frac{1}{2}.$$

Solution

To find the focus, convert to standard form by completing the square.

$$y = -\frac{1}{2}x^2 - x + \frac{1}{2} \qquad \text{Write original equation.}$$

$$-2y = x^2 + 2x - 1 \qquad \text{Multiply each side by } -2.$$

$$1 - 2y = x^2 + 2x \qquad \text{Add 1 to each side.}$$

$$1 + 1 - 2y = x^2 + 2x + 1 \qquad \text{Complete the square.}$$

$$2 - 2y = x^2 + 2x + 1 \qquad \text{Combine like terms.}$$

$$-2(y - 1) = (x + 1)^2 \qquad \text{Standard form}$$

Comparing this equation with

$$(x - h)^2 = 4p(y - k),$$

you can conclude that $h = -1$, $k = 1$, and $p = -\frac{1}{2}$. Because p is negative, the parabola opens downward, as shown in Figure 10.12. Therefore, the focus of the parabola is

$$(h, k + p) = \left(-1, \frac{1}{2}\right). \qquad \text{Focus}$$

Technology

Try using a graphing utility to confirm the equation found in Example 3. To do this, it helps to split the equation into two parts:

$y_1 = \sqrt{8x}$ (upper part)

and

$y_2 = -\sqrt{8x}$ (lower part).

Example 3 ▶ Vertex at the Origin

Find the standard equation of the parabola with vertex at the origin and focus $(2, 0)$.

Solution

The axis of the parabola is horizontal, passing through $(0, 0)$ and $(2, 0)$, as shown in Figure 10.13.

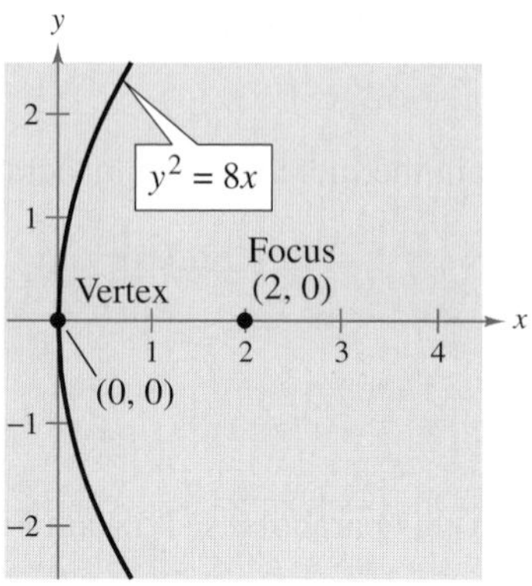

FIGURE 10.13

So, the standard form is $y^2 = 4px$, where $h = k = 0$ and $p = 2$. Therefore, the equation is $y^2 = 8x$.

Application

A line segment that passes through the focus of a parabola and has endpoints on the parabola is called a **focal chord.** The specific focal chord perpendicular to the axis of the parabola is called the **latus rectum.**

Parabolas occur in a wide variety of applications. For instance, a parabolic reflector can be formed by revolving a parabola around its axis. The resulting surface has the property that all incoming rays parallel to the axis are reflected through the focus of the parabola; this is the principle behind the construction of the parabolic mirrors used in reflecting telescopes. Conversely, the light rays emanating from the focus of a parabolic reflector used in a flashlight are all parallel to one another, as shown in Figure 10.14.

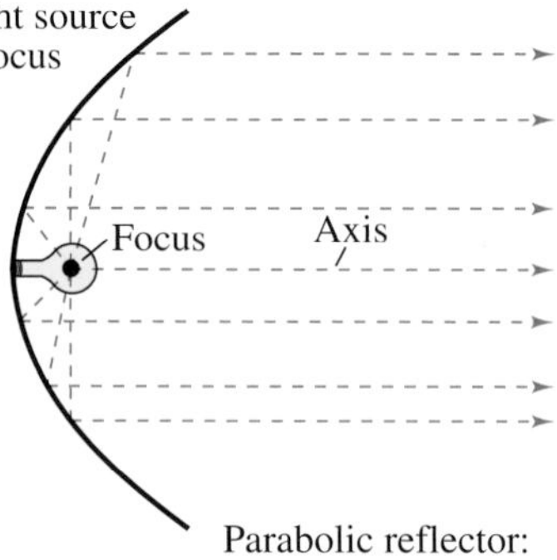

FIGURE 10.14

A line is **tangent** to a parabola at a point on the parabola if the line intersects, but does not cross, the parabola at the point. Tangent lines to parabolas have special properties related to the use of parabolas in constructing reflective surfaces.

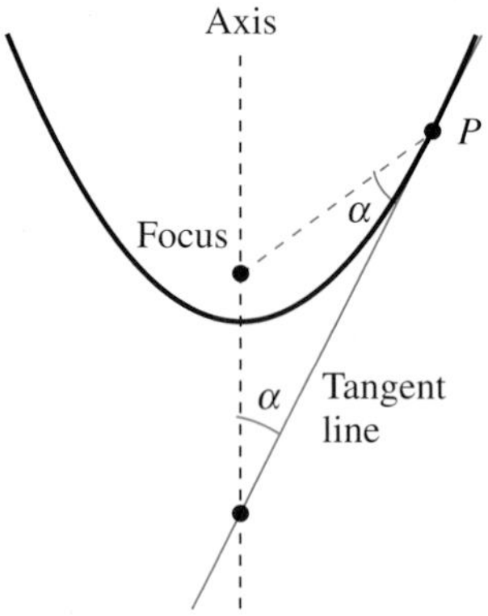

FIGURE 10.15

Reflective Property of a Parabola

The tangent line to a parabola at a point P makes equal angles with the following two lines (see Figure 10.15).

1. The line passing through P and the focus
2. The axis of the parabola

Your students may question why $d_1 = \frac{1}{4} - b$ rather than $b - \frac{1}{4}$. You may need to explain that the distance must be positive, so the order of subtraction is important.

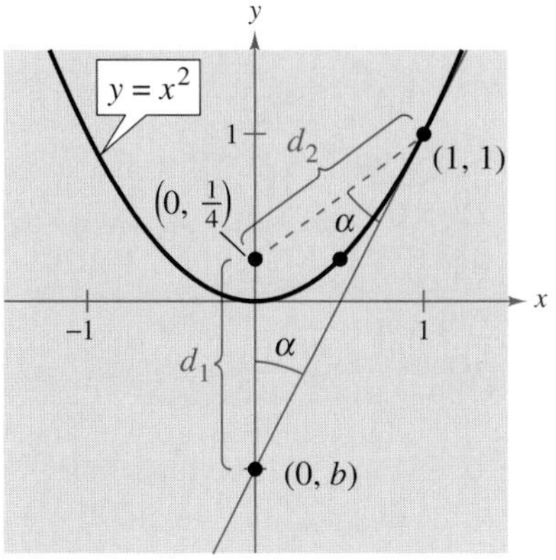

FIGURE 10.16

Technology

Try using a graphing utility to confirm the result of Example 4. By graphing

$$y = x^2 \quad \text{and} \quad y = 2x - 1$$

in the same viewing window, you should be able to see that the line touches the parabola at the point (1, 1).

The *Interactive* CD-ROM and *Internet* versions of this text offer a built-in graphing calculator, which can be used with the Examples, Explorations, Technology notes, and Exercises.

Activities

1. Find the vertex, focus, and directrix of the parabola $x^2 - 6x - 4y + 5 = 0$.
   Answer: Vertex $(3, -1)$; Focus $(3, 0)$; Directrix $y = -2$
2. Find an equation of the parabola with vertex $(4, 0)$ and directrix $x = 5$.
   Answer: $y^2 + 4x - 16 = 0$
3. Find an equation of the tangent line to the parabola $y = 2x^2$ at the point $(1, 2)$.
   Answer: $y = 4x - 2$

Example 4 ▶ Finding the Tangent Line at a Point on a Parabola

Find the equation of the tangent line to the parabola given by $y = x^2$ at the point $(1, 1)$.

Solution

For this parabola, $p = \frac{1}{4}$ and the focus is $\left(0, \frac{1}{4}\right)$, as shown in Figure 10.16. You can find the y-intercept $(0, b)$ of the tangent line by equating the lengths of the two sides of the isosceles triangle shown in Figure 10.16:

$$d_1 = \frac{1}{4} - b$$

and

$$d_2 = \sqrt{(1 - 0)^2 + [1 - (1/4)]^2} = \frac{5}{4}.$$

Setting $d_1 = d_2$ produces

$$\frac{1}{4} - b = \frac{5}{4}$$

$$b = -1.$$

So, the slope of the tangent line is

$$m = \frac{1 - (-1)}{1 - 0} = 2$$

and its slope-intercept equation is

$$y = 2x - 1.$$

Writing ABOUT MATHEMATICS

Television Antenna Dishes Cross sections of television antenna dishes are parabolic in shape. Write a paragraph explaining why these dishes are parabolic.

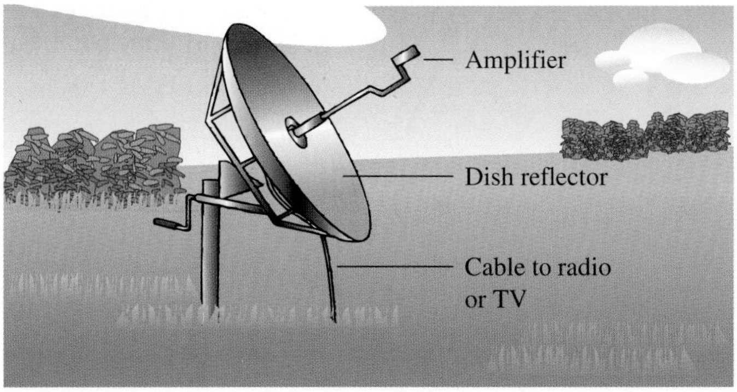

10.2 Exercises

In Exercises 1–4, describe in words how a plane could intersect with the double-napped cone shown to form the conic section.

1. Circle
2. Ellipse
3. Parabola
4. Hyperbola

In Exercises 5–10, match the equation with its graph. [The graphs are labeled (a), (b), (c), (d), (e), and (f).]

(a)

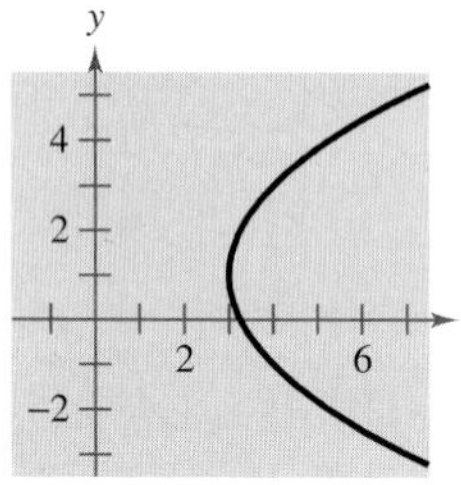

(b)

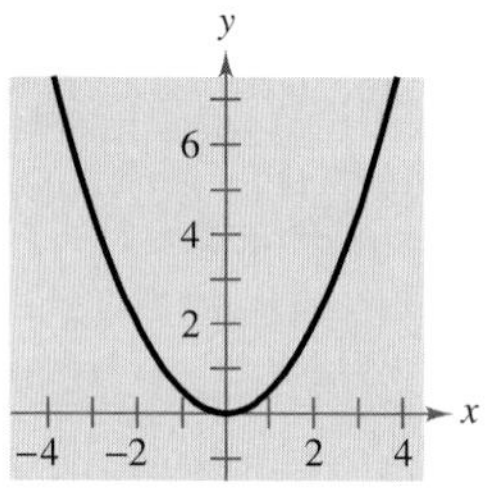

(c)

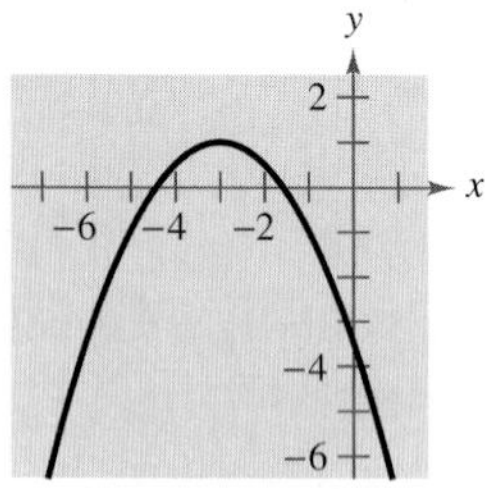

(d)

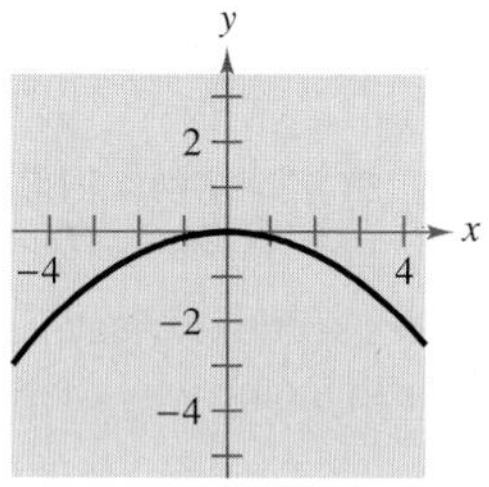

(e)

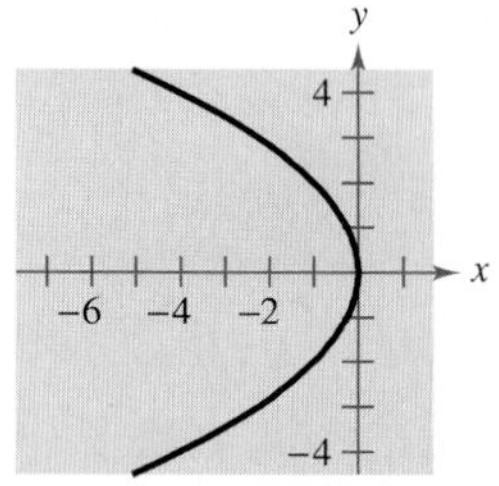

(f)

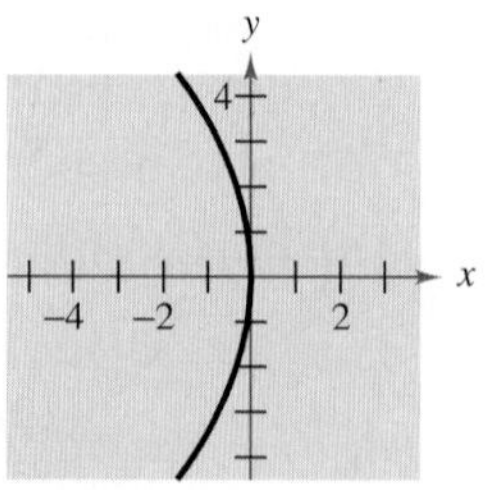

5. $y^2 = -4x$
6. $x^2 = 2y$
7. $x^2 = -8y$
8. $y^2 = -12x$
9. $(y - 1)^2 = 4(x - 3)$
10. $(x + 3)^2 = -2(y - 1)$

In Exercises 11–24, find the vertex, focus, and directrix of the parabola and sketch its graph.

11. $y = \frac{1}{2}x^2$
12. $y = -2x^2$
13. $y^2 = -6x$
14. $y^2 = 3x$
15. $x^2 + 6y = 0$
16. $x + y^2 = 0$
17. $(x - 1)^2 + 8(y + 2) = 0$
18. $(x + 5) + (y - 1)^2 = 0$
19. $\left(x + \frac{3}{2}\right)^2 = 4(y - 2)$
20. $\left(x + \frac{1}{2}\right)^2 = 4(y - 1)$
21. $y = \frac{1}{4}(x^2 - 2x + 5)$
22. $x = \frac{1}{4}(y^2 + 2y + 33)$
23. $y^2 + 6y + 8x + 25 = 0$
24. $y^2 - 4y - 4x = 0$

In Exercises 25–28, find the vertex, focus, and directrix of the parabola. Use a graphing utility to graph the parabola.

25. $x^2 + 4x + 6y - 2 = 0$
26. $x^2 - 2x + 8y + 9 = 0$
27. $y^2 + x + y = 0$
28. $y^2 - 4x - 4 = 0$

In Exercises 29 and 30, the equations of a parabola and a tangent line to the parabola are given. Use a graphing utility to graph both equations in the same viewing window. Determine the coordinates of the point of tangency.

	Parabola	*Tangent Line*
29.	$y^2 - 8x = 0$	$x - y + 2 = 0$
30.	$x^2 + 12y = 0$	$x + y - 3 = 0$

In Exercises 31–42, find the standard form of the equation of the parabola with its vertex at the origin.

31.

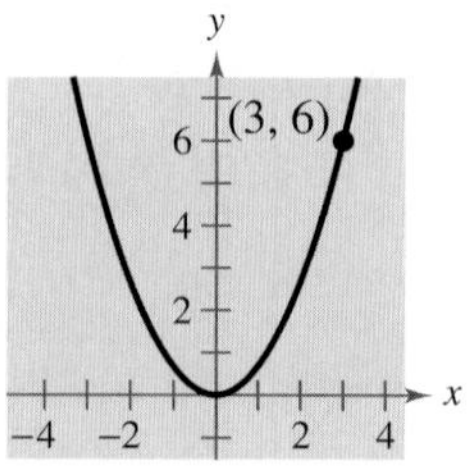

32.

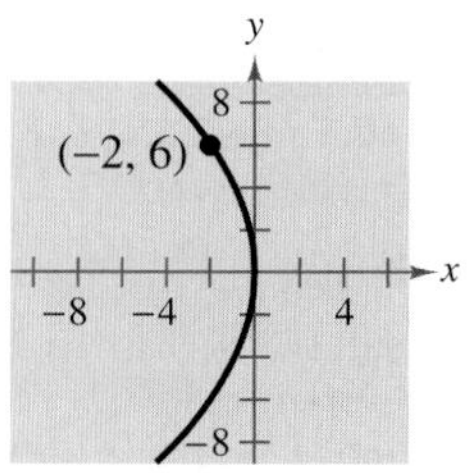

33. Focus: $\left(0, -\frac{3}{2}\right)$
34. Focus: $(2, 0)$
35. Focus: $(-2, 0)$

36. Focus: $(0, -2)$

37. Directrix: $y = -1$

38. Directrix: $y = 3$

39. Directrix: $x = 2$

40. Directrix: $x = -3$

41. Horizontal axis and passes through the point $(4, 6)$

42. Vertical axis and passes through the point $(-3, -3)$

In Exercises 43–52, find the standard form of the equation of the parabola.

43.

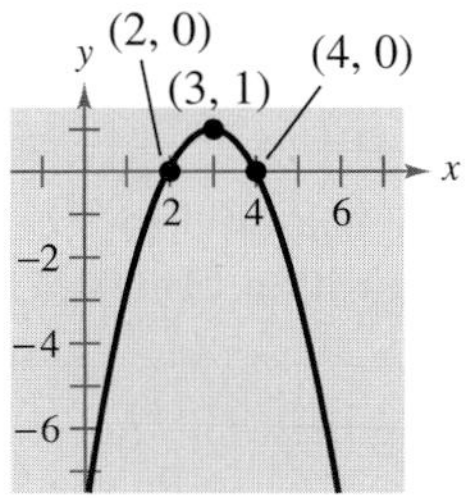

44.

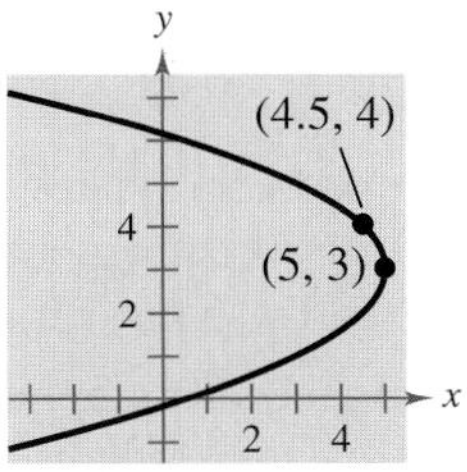

45.

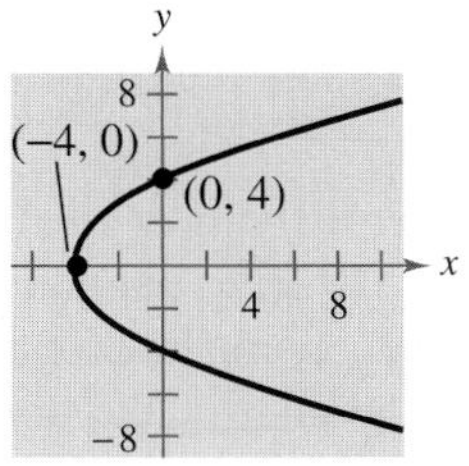

46.

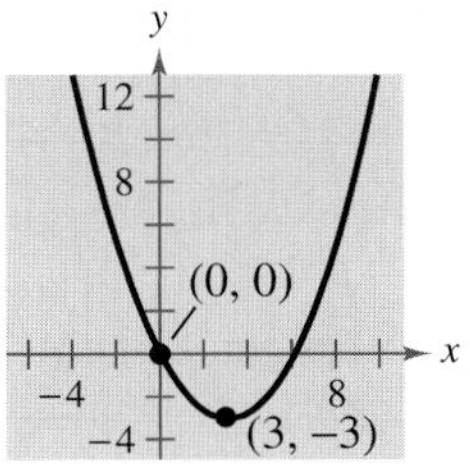

47. Vertex: $(5, 2)$; Focus: $(3, 2)$

48. Vertex: $(-1, 2)$; Focus: $(-1, 0)$

49. Vertex: $(0, 4)$; Directrix: $y = 2$

50. Vertex: $(-2, 1)$; Directrix: $x = 1$

51. Focus: $(2, 2)$; Directrix: $x = -2$

52. Focus: $(0, 0)$; Directrix: $y = 8$

In Exercises 53 and 54, change the equation so that its graph matches the description.

53. $(y - 3)^2 = 6(x + 1)$; upper half of parabola

54. $(y + 1)^2 = 2(x - 4)$; lower half of parabola

In Exercises 55–58, find an equation of the tangent line to the parabola at the given point, and find the x-intercept of the line.

55. $x^2 = 2y, \ (4, 8)$

56. $x^2 = 2y, \ \left(-3, \frac{9}{2}\right)$

57. $y = -2x^2, \ (-1, -2)$

58. $y = -2x^2, \ (2, -8)$

59. ***Revenue*** The revenue R generated by the sale of x units of a product is

$$R = 265x - \frac{5}{4}x^2.$$

Use a graphing utility to graph the function and approximate the number of sales that will maximize revenue.

60. ***Revenue*** The revenue R generated by the sale of x units of a product is

$$R = 378x - \frac{7}{5}x^2.$$

Use a graphing utility to graph the function and approximate the number of sales that will maximize revenue.

61. ***Satellite Antenna*** The receiver in a parabolic television dish antenna is 4.5 feet from the vertex and is located at the focus (see figure). Find an equation of a cross section of the reflector. (Assume that the dish is directed upward and the vertex is at the origin.)

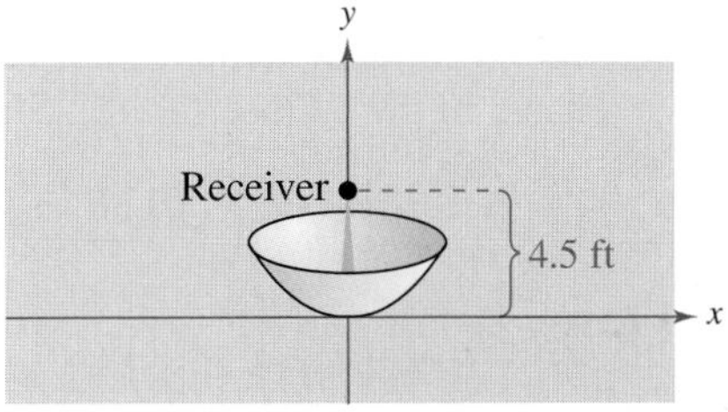

62. ***Suspension Bridge*** Each cable of a suspension bridge is suspended (in the shape of a parabola) between two towers that are 120 meters apart and whose tops are 20 meters above the roadway. The cables touch the roadway midway between the towers.

(a) Create a sketch of the bridge. Draw a rectangular coordinate system on the bridge with the center of the bridge at the origin. Identify the coordinates of the known points.

(b) Find an equation for the parabolic shape of each cable.

(c) Complete the table by finding the heights y of the suspension cables over the roadway at distances of x meters from the center of the bridge.

x	0	20	40	60
y				

63. *Road Design* Roads are often designed with parabolic surfaces to allow rain to drain off. A particular road that is 32 feet wide is 0.4 foot higher in the center than it is on the sides (see figure).

(a) Find an equation of the parabola that models the road surface. (Assume that the origin is at the center of the road.)

(b) How far from the center of the road is the road surface 0.1 foot lower than in the middle?

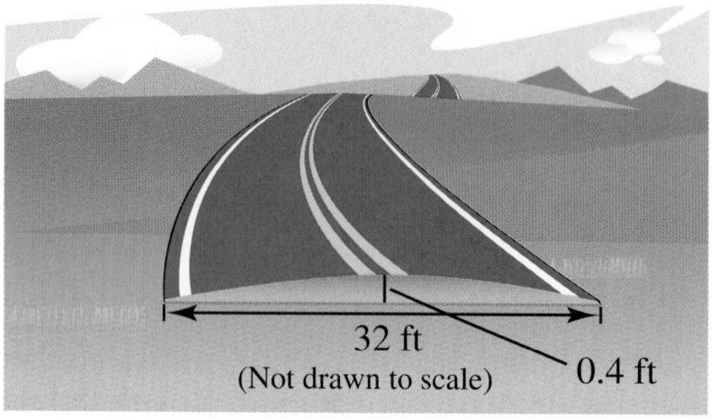

Cross section of road surface

64. *Highway Design* Highway engineers design a parabolic curve for an entrance ramp from a straight street to an interstate highway (see figure). Find an equation of the parabola.

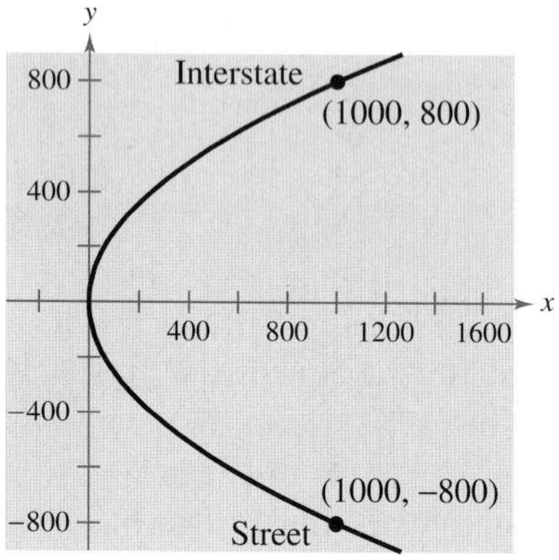

65. *Satellite Orbit* An earth satellite in a 100-mile-high circular orbit around the earth has a velocity of approximately 17,500 miles per hour. If this velocity is multiplied by $\sqrt{2}$, the satellite will have the minimum velocity necessary to escape the earth's gravity and it will follow a parabolic path with the center of the earth as the focus (see figure).

(a) Find the escape velocity of the satellite.

(b) Find an equation of its path (assume that the radius of the earth is 4000 miles).

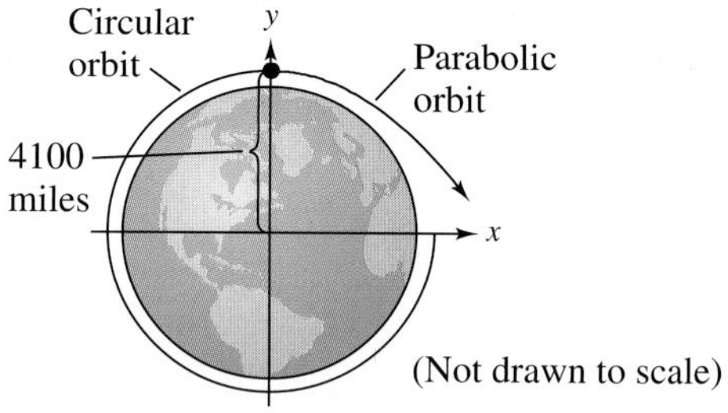

FIGURE FOR 65

66. *Path of a Projectile* The path of a softball is given by the equation

$$y = -0.08x^2 + x + 4.$$

The coordinates x and y are measured in feet, with $x = 0$ corresponding to the position from which the ball was thrown.

(a) Use a graphing utility to graph the trajectory of the softball.

(b) Move the cursor along the path to approximate the highest point. Approximate the range of the trajectory.

67. *Projectile Motion* A bomber is flying at an altitude of 30,000 feet and a speed of 540 miles per hour (792 feet per second). When should a bomb be dropped so that it will hit the target if the path of the bomb is modeled by

$$y = 30{,}000 - \frac{x^2}{39{,}204}?$$

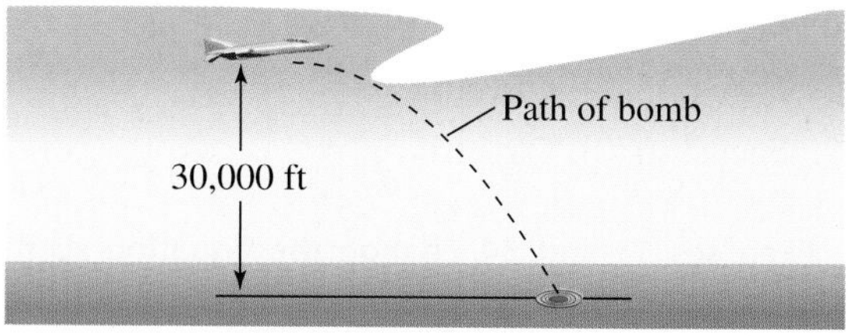

Synthesis

True or False? **In Exercises 68 and 69, determine whether the statement is true or false. Justify your answer.**

68. It is possible for a parabola to intersect its directrix.

69. If the vertex and focus of a parabola are on a horizontal line, then the directrix of the parabola is vertical.

70. *Exploration* Consider the parabola

$x^2 = 4py.$

(a) Use a graphing utility to graph the parabola for $p = 1$, $p = 2$, $p = 3$,and $p = 4$. Describe the effect on the graph when p increases.

(b) Locate the focus for each parabola in part (a).

(c) For each parabola in part (a), find the length of the chord passing through the focus and parallel to the directrix. How can the length of this chord be determined directly from the standard form of the equation of the parabola?

(d) Explain how the result of part (c) can be used as a sketching aid when graphing parabolas.

71. *Area* The area of the shaded region in the figure is

$$A = \frac{8}{3}p^{1/2}b^{3/2}.$$

(a) Find the area if $p = 2$ and $b = 4$.

(b) Give a geometric explanation of why the area approaches 0 as p approaches 0.

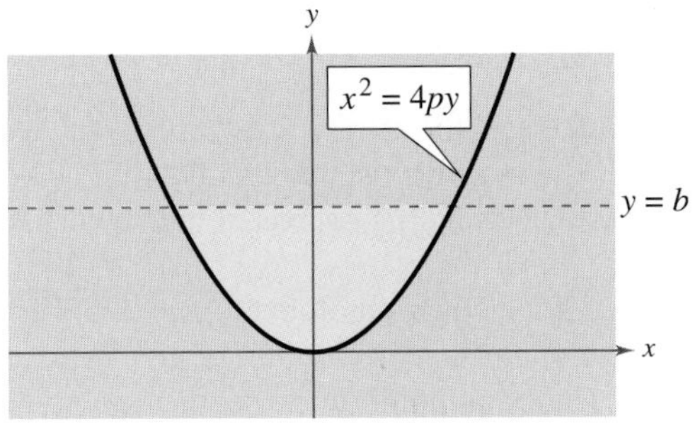

72. *Exploration* Let (x_1, y_1) be the coordinates of a point on the parabola $x^2 = 4py$. The equation of the line tangent to the parabola at the point is

$$y - y_1 = \frac{x_1}{2p}(x - x_1).$$

What is the slope of the tangent line?

Review

In Exercises 73–76, use Descartes's Rule of Signs to determine the possible numbers of positive and negative zeros of the function.

73. $f(x) = 2x^3 + 3x^2 - 2$

74. $f(x) = 3x^4 - 2x^3 + 6x - 3$

75. $f(x) = 5x^5 - 2x^2 + 3$

76. $f(x) = x^5 - 2x^4 + x^3 + x^2 - 3x + 1$

In Exercises 77–80, list the possible rational zeros given by the Rational Zero Test.

77. $f(x) = x^3 - 2x^2 + 2x - 4$

78. $f(x) = 2x^3 + 4x^2 - 3x + 10$

79. $f(x) = 2x^5 + x^2 + 16$

80. $f(x) = 3x^3 - 12x + 22$

In Exercises 81–84, solve as a review of the skills and problem-solving techniques you learned in previous sections.

81. Find a polynomial with integer coefficients that has the zeros 3, $2 + i$, and $2 - i$.

82. Find all the zeros of

$$f(x) = 2x^3 - 3x^2 + 50x - 75$$

if one of the zeros is $x = \frac{3}{2}$.

83. Find all the zeros of the function

$$g(x) = 6x^4 + 7x^3 - 29x^2 - 28x + 20$$

if two of the zeros are $x = \pm 2$.

84. Use a graphing utility to graph the function

$$h(x) = 2x^4 + x^3 - 19x^2 - 9x + 9.$$

Use the graph to approximate the zeros of h.

10.3 Ellipses

▶ **What you should learn**

- How to write the standard form of the equation of an ellipse
- How to use properties of ellipses to model and solve real-life problems
- How to find the eccentricity of an ellipse

▶ **Why you should learn it**

Ellipses can be used to model and solve many types of real-life problems. For instance, in Exercise 52 on page 793, an ellipse is used to model the orbit of Halley's comet.

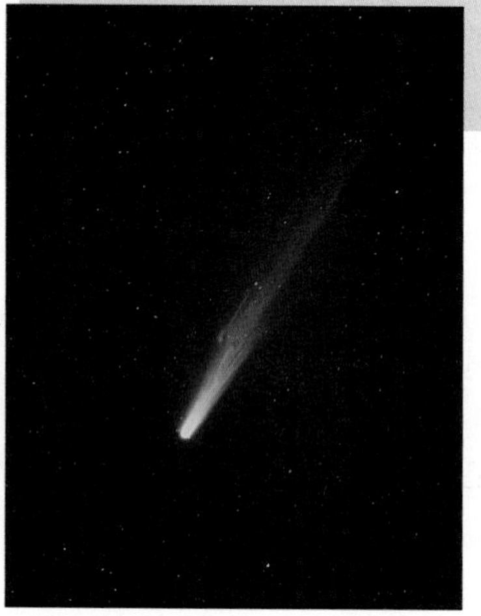

Introduction

The second type of conic is called an **ellipse,** and is defined as follows.

Definition of Ellipse

An **ellipse** is the set of all points (x, y) the sum of whose distances from two distinct fixed points (**foci**) is constant. (See Figure 10.17.)

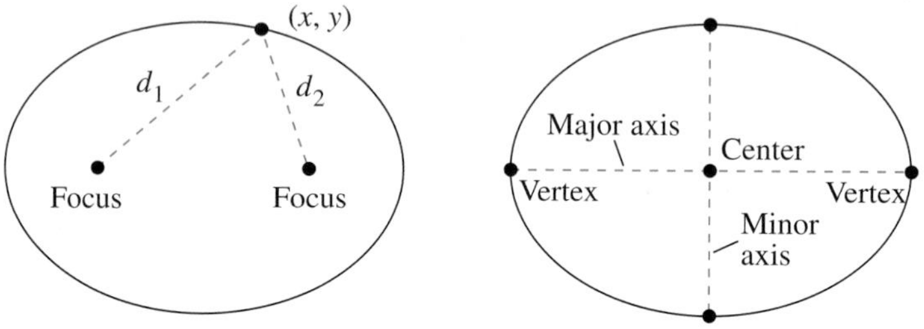

$d_1 + d_2$ is constant.

FIGURE 10.17

The line through the foci intersects the ellipse at two points called **vertices.** The chord joining the vertices is the **major axis,** and its midpoint is the **center** of the ellipse. The chord perpendicular to the major axis at the center is the **minor axis** of the ellipse.

To derive the standard form of the equation of an ellipse, consider the ellipse in Figure 10.18 with the following points: center, (h, k); vertices, $(h \pm a, k)$; foci, $(h \pm c, k)$. Note that the center is the midpoint of the segment joining the foci.

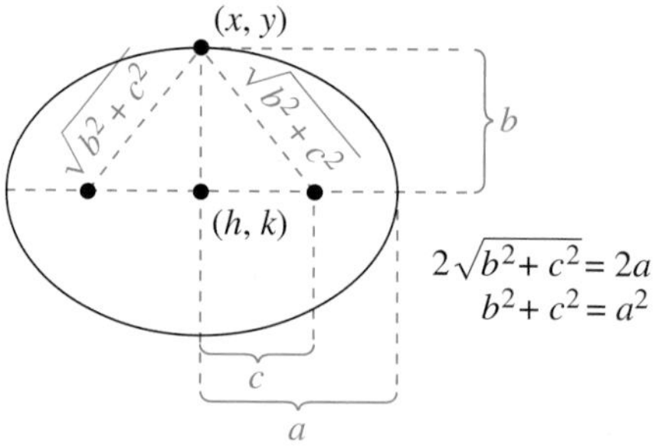

FIGURE 10.18

The sum of the distances from any point on the ellipse to the two foci is constant. Using a vertex point, this constant sum is

$$(a + c) + (a - c) = 2a \qquad \text{Length of major axis}$$

or simply the length of the major axis. Now, if you let (x, y) be *any* point on the ellipse, the sum of the distances between (x, y) and the two foci must also be $2a$.

When discussing ellipses, you might also choose to discuss the latera recta as background for Exercises 55–59.

That is,

$$\sqrt{[x-(h-c)]^2+(y-k)^2}+\sqrt{[x-(h+c)]^2+(y-k)^2}=2a.$$

Finally, in Figure 10.18, you can see that $b^2 = a^2 - c^2$, which implies that the equation of the ellipse is

$$b^2(x-h)^2+a^2(y-k)^2=a^2b^2$$

$$\frac{(x-h)^2}{a^2}+\frac{(y-k)^2}{b^2}=1.$$

You would obtain a similar equation in the derivation by starting with a vertical major axis. Both results are summarized as follows.

Standard Equation of an Ellipse

The standard form of the equation of an ellipse, with center (h, k) and major and minor axes of lengths $2a$ and $2b$, where $0 < b < a$, is

$$\frac{(x-h)^2}{a^2}+\frac{(y-k)^2}{b^2}=1 \qquad \text{Major axis is horizontal.}$$

$$\frac{(x-h)^2}{b^2}+\frac{(y-k)^2}{a^2}=1. \qquad \text{Major axis is vertical.}$$

The foci lie on the major axis, c units from the center, with $c^2 = a^2 - b^2$. If the center is at the origin $(0, 0)$, the equation takes one of the following forms.

$$\frac{x^2}{a^2}+\frac{y^2}{b^2}=1 \qquad \text{Major axis is horizontal.}$$

$$\frac{x^2}{b^2}+\frac{y^2}{a^2}=1 \qquad \text{Major axis is vertical.}$$

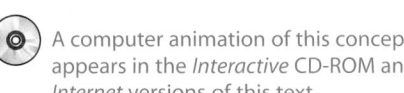

A computer animation of this concept appears in the *Interactive* CD-ROM and *Internet* versions of this text.

Figure 10.19 shows both the vertical and horizontal orientations for an ellipse.

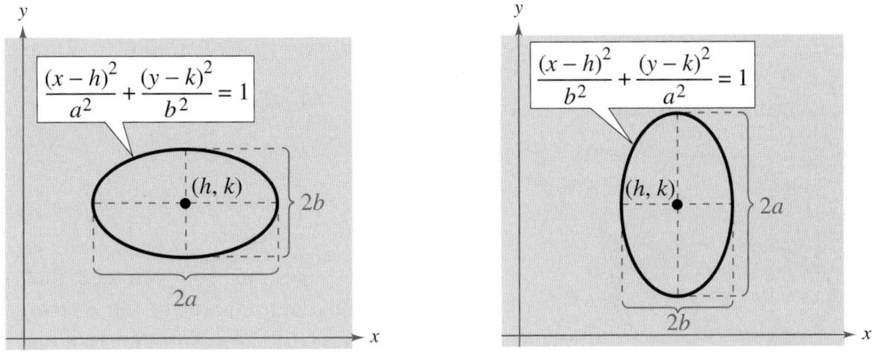

FIGURE 10.19

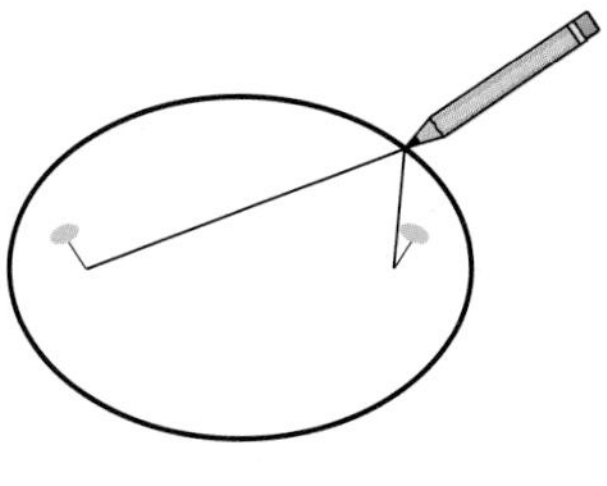

FIGURE 10.20

You can visualize the definition of an ellipse by imagining two thumbtacks placed at the foci, as shown in Figure 10.20. If the ends of a fixed length of string are fastened to the thumbtacks and the string is drawn taut with a pencil, the path traced by the pencil will be an ellipse.

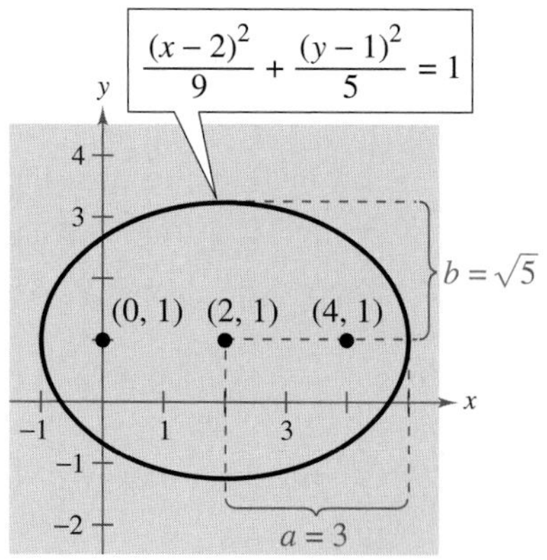

FIGURE 10.21

Example 1 ▶ Finding the Standard Equation of an Ellipse

Find the standard form of the equation of the ellipse having foci at (0, 1) and (4, 1) and a major axis of length 6, as shown in Figure 10.21.

Solution

Because the foci occur at (0, 1) and (4, 1), the center of the ellipse is (2, 1) and the distance from the center to one of the foci is $c = 2$. Because $2a = 6$, you know that $a = 3$. Now, from $c^2 = a^2 - b^2$, you have

$$\begin{aligned} b^2 &= a^2 - c^2 \\ &= 3^2 - 2^2 \\ &= 5. \end{aligned}$$

Because the major axis is horizontal, the standard equation is

$$\frac{(x-2)^2}{9} + \frac{(y-1)^2}{5} = 1.$$

In Example 1, note the use of the equation $c^2 = a^2 - b^2$. Don't confuse this equation with the Pythagorean Theorem—there is a difference in sign.

Remind your students that completing the square must be performed twice to write the equation of the ellipse in standard form in Example 2.

Example 2 ▶ Writing an Equation in Standard Form

Sketch the graph of the ellipse whose equation is

$$x^2 + 4y^2 + 6x - 8y + 9 = 0.$$

Solution

Begin by writing the given equation in standard form. In the fourth step, note that 9 and 4 are added to *both* sides of the equation when completing the squares.

$$x^2 + 4y^2 + 6x - 8y + 9 = 0 \qquad \text{Write original equation.}$$

$$(x^2 + 6x + \square) + (4y^2 - 8y + \square) = -9 \qquad \text{Group terms.}$$

$$(x^2 + 6x + \square) + 4(y^2 - 2y + \square) = -9 \qquad \text{Factor 4 out of } y\text{-terms.}$$

$$(x^2 + 6x + 9) + 4(y^2 - 2y + 1) = -9 + 9 + 4(1)$$

$$(x+3)^2 + 4(y-1)^2 = 4 \qquad \text{Completed square form}$$

$$\frac{(x+3)^2}{4} + \frac{(y-1)^2}{1} = 1 \qquad \text{Standard form}$$

From the equation, you can see that the center occurs at $(h, k) = (-3, 1)$. Because the denominator of the x-term is $a^2 = 2^2$, you can locate the endpoints of the major axis 2 units to the right and left of the center. Similarly, because the denominator of the y-term is $b^2 = 1^2$, you can locate the endpoints of the minor axis 1 unit up and down from the center. The graph of this ellipse is shown in Figure 10.22.

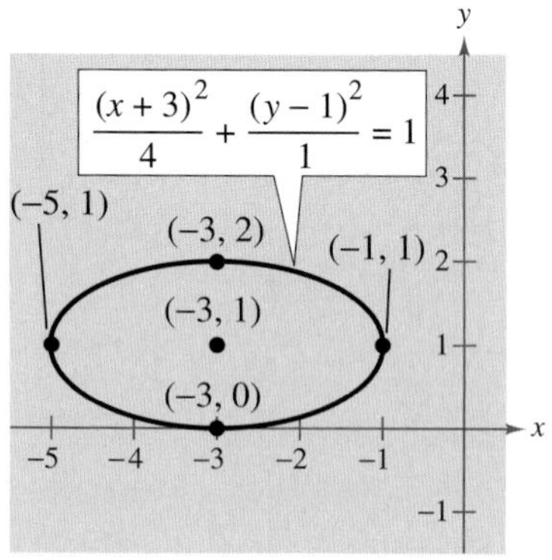

FIGURE 10.22

Example 3 ▶ Analyzing an Ellipse

Find the center, vertices, and foci of the ellipse

$$4x^2 + y^2 - 8x + 4y - 8 = 0.$$

Solution

By completing the square, you can write the given equation in standard form.

$$4x^2 + y^2 - 8x + 4y - 8 = 0$$

$$(4x^2 - 8x + \quad) + (y^2 + 4y + \quad) = 8$$

$$4(x^2 - 2x + \quad) + (y^2 + 4y + \quad) = 8$$

$$4(x^2 - 2x + 1) + (y^2 + 4y + 4) = 8 + 4(1) + 4$$

$$4(x - 1)^2 + (y + 2)^2 = 16$$

$$\frac{(x - 1)^2}{4} + \frac{(y + 2)^2}{16} = 1$$

So, the major axis is vertical, where $h = 1$, $k = -2$, $a = 4$, $b = 2$, and

$$c = \sqrt{16 - 4} = \sqrt{12} = 2\sqrt{3}.$$

Therefore, you have the following.

Center: $(1, -2)$ Vertices: $(1, -6)$, $(1, 2)$ Foci: $(1, -2 - 2\sqrt{3})$, $(1, -2 + 2\sqrt{3})$

The graph of the ellipse is shown in Figure 10.23.

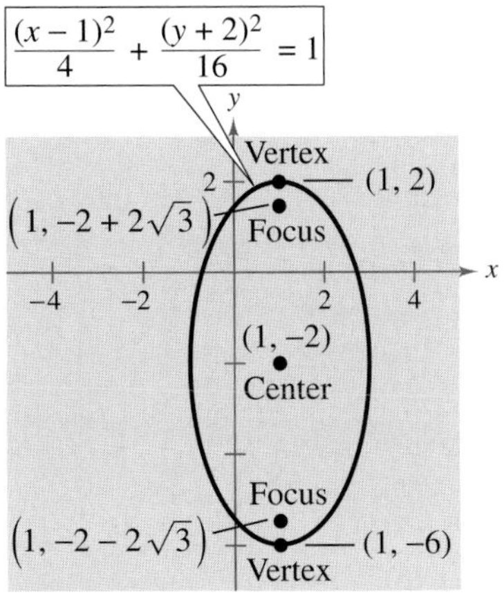

FIGURE 10.23

Technology

You can use a graphing utility to graph an ellipse by graphing the upper and lower portions in the same viewing window. For instance, to graph the ellipse in Example 3, first solve for y to get

$$y_1 = -2 + 4\sqrt{1 - \frac{(x - 1)^2}{4}}$$

and

$$y_2 = -2 - 4\sqrt{1 - \frac{(x - 1)^2}{4}}.$$

Use a viewing window in which $-6 \le x \le 9$ and $-7 \le y \le 3$. You should obtain the graph shown below.

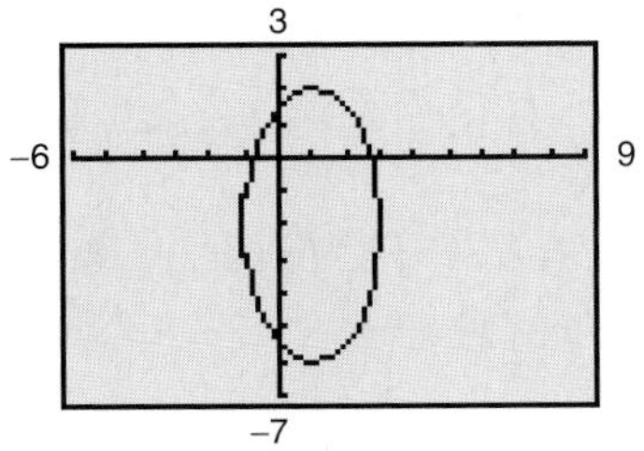

Application

Ellipses have many practical and aesthetic uses. For instance, machine gears, supporting arches, and acoustic designs often involve elliptical shapes. The orbits of satellites and planets are also ellipses. Example 4 investigates the elliptical orbit of the moon about the earth.

Example 4 ▶ An Application Involving an Elliptical Orbit

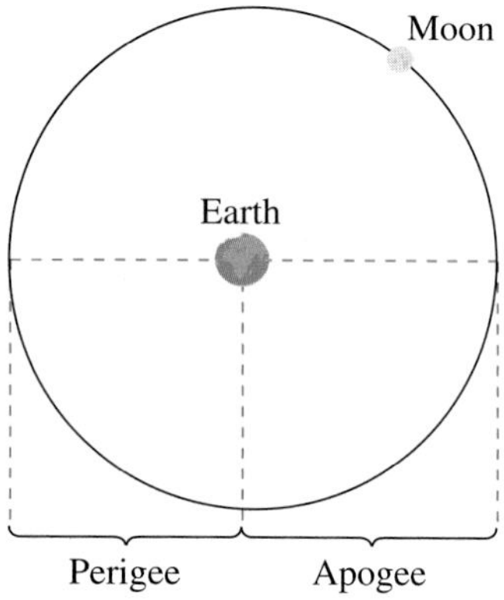

FIGURE 10.24

The moon travels about the earth in an elliptical orbit with the earth at one focus, as shown in Figure 10.24. The major and minor axes of the orbit have lengths of 768,806 kilometers and 767,746 kilometers, respectively. Find the greatest and smallest distances (the *apogee* and *perigee*) from the earth's center to the moon's center.

Solution

Because $2a = 768{,}806$ and $2b = 767{,}746$, you have $a = 384{,}403$ and $b = 383{,}873$, which implies that

$$\begin{aligned} c &= \sqrt{a^2 - b^2} \\ &= \sqrt{384{,}403^2 - 383{,}873^2} \\ &\approx 20{,}179. \end{aligned}$$

Therefore, the greatest distance between the center of the earth and the center of the moon is

$$a + c \approx 404{,}582 \text{ kilometers}$$

and the smallest distance is

$$a - c \approx 364{,}224 \text{ kilometers.}$$

Eccentricity

One of the reasons it was difficult for early astronomers to detect that the orbits of the planets are ellipses is that the foci of the planetary orbits are relatively close to their centers, and so the orbits are nearly circular. To measure the ovalness of an ellipse, you can use the concept of **eccentricity.**

Ask students to make a conjecture about the eccentricity of a circle before going further.

Definition of Eccentricity

The **eccentricity** e of an ellipse is given by the ratio

$$e = \frac{c}{a}.$$

Note that $0 < e < 1$ for *every* ellipse.

Activities

1. Find the center, foci, vertices, and eccentricity of the ellipse $\frac{(x-2)^2}{25} + \frac{(y+1)^2}{9} = 1.$

   Answer: Center $(2, -1)$; Foci $(-2, -1), (6, -1)$; Vertices $(-3, -1), (7, -1)$; $e = \frac{4}{5}$

2. Rewrite in standard form: $9x^2 + 5y^2 + 36x - 30y + 36 = 0.$

   Answer: $\frac{(x+2)^2}{5} + \frac{(y-3)^2}{9} = 1$

To see how this ratio is used to describe the shape of an ellipse, note that because the foci of an ellipse are located along the major axis between the vertices and the center, it follows that

$$0 < c < a.$$

For an ellipse that is nearly circular, the foci are close to the center and the ratio c/a is small, as shown in Figure 10.25(a). On the other hand, for an elongated ellipse, the foci are close to the vertices, and the ratio c/a is close to 1, as shown in Figure 10.25(b).

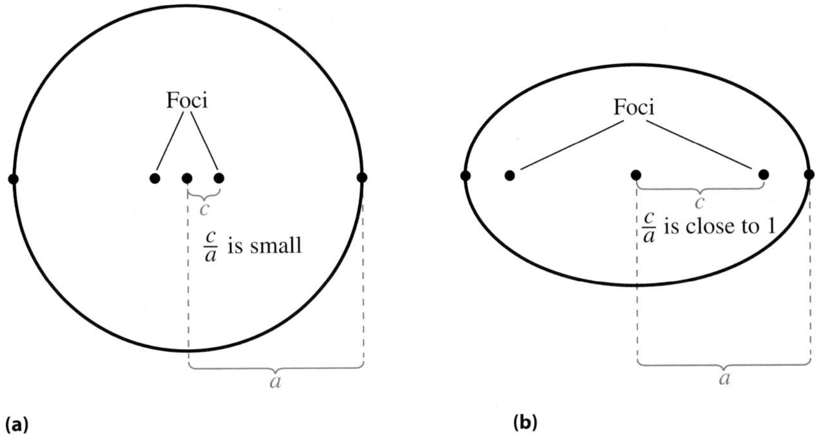

FIGURE 10.25

The orbit of the moon has an eccentricity of $e \approx 0.0525$, and the eccentricities of the nine planetary orbits are as follows.

Mercury: $e \approx 0.2056$	Saturn: $e \approx 0.0543$
Venus: $e \approx 0.0068$	Uranus: $e \approx 0.0460$
Earth: $e \approx 0.0167$	Neptune: $e \approx 0.0082$
Mars: $e \approx 0.0934$	Pluto: $e \approx 0.2481$
Jupiter: $e \approx 0.0484$	

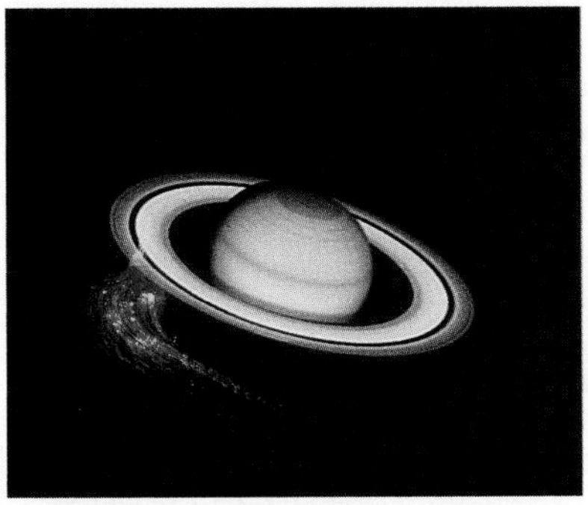

NASA

The time it takes Saturn to orbit the sun is equal to 29.5 earth years.

Writing ABOUT MATHEMATICS

Ellipses and Circles

a. Show that the equation of an ellipse can be written as

$$\frac{(x-h)^2}{a^2} + \frac{(y-k)^2}{a^2(1-e^2)} = 1.$$

b. For the equation in part (a), let $a = 4$, $h = 1$, and $k = 2$, and use a graphing utility to graph the ellipse for $e = 0.95$, $e = 0.75$, $e = 0.5$, $e = 0.25$, and $e = 0.1$. Discuss the changes in the shape of the ellipse as e approaches 0.

c. Make a conjecture about the shape of the graph in part (b) when $e = 0$. What is the equation of this ellipse? What is another name for an ellipse with an eccentricity of 0?

10.3 Exercises

In Exercises 1–6, match the equation with its graph. [The graphs are labeled (a), (b), (c), (d), (e), and (f).]

(a)

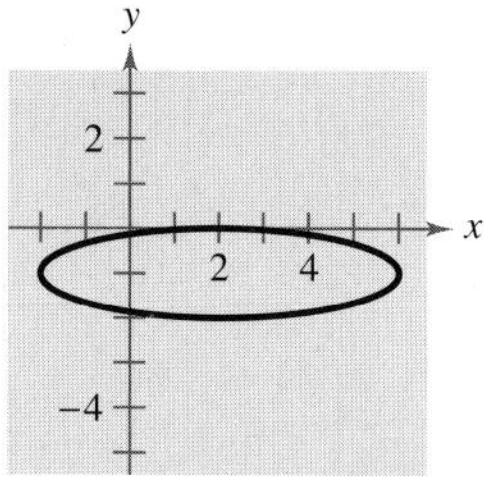

(b)

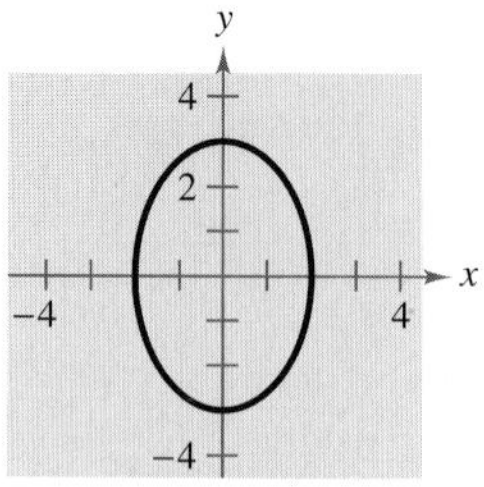

(c)

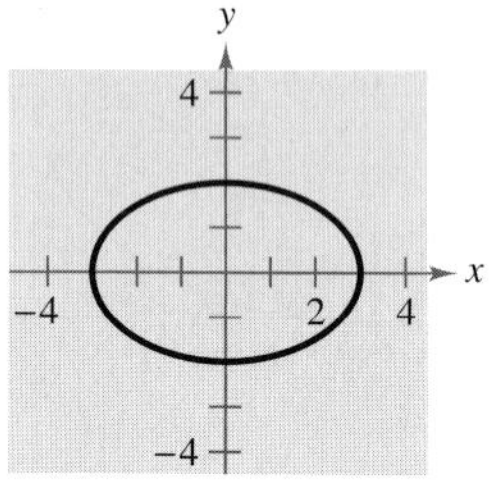

(d)

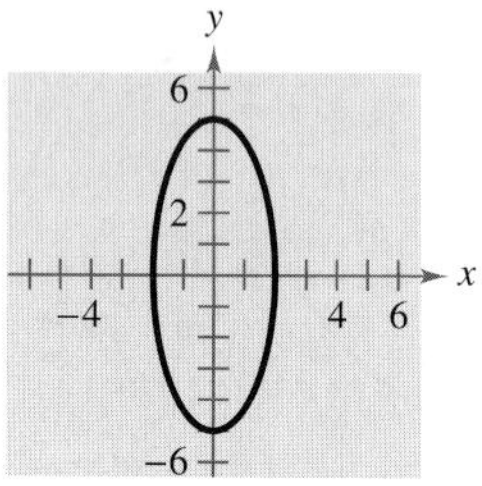

(e)

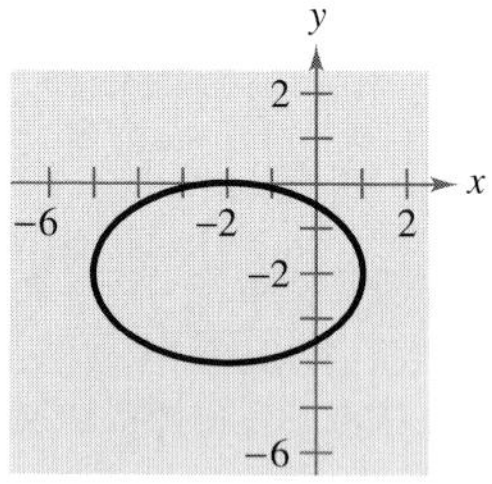

(f)

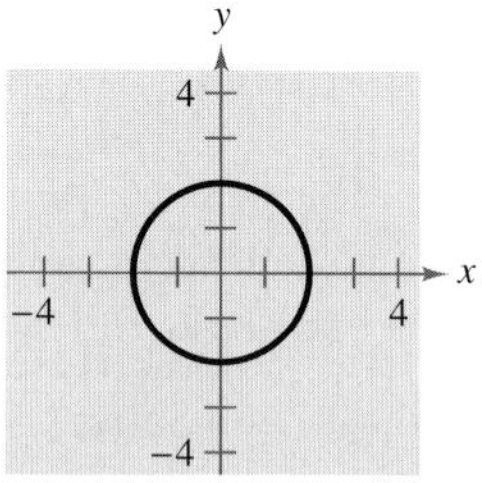

1. $\dfrac{x^2}{4} + \dfrac{y^2}{9} = 1$

2. $\dfrac{x^2}{9} + \dfrac{y^2}{4} = 1$

3. $\dfrac{x^2}{4} + \dfrac{y^2}{25} = 1$

4. $\dfrac{x^2}{4} + \dfrac{y^2}{4} = 1$

5. $\dfrac{(x-2)^2}{16} + (y+1)^2 = 1$

6. $\dfrac{(x+2)^2}{9} + \dfrac{(y+2)^2}{4} = 1$

In Exercises 7–22, find the center, vertices, foci, and eccentricity of the ellipse, then sketch its graph.

7. $\dfrac{x^2}{25} + \dfrac{y^2}{16} = 1$

8. $\dfrac{x^2}{81} + \dfrac{y^2}{144} = 1$

9. $\dfrac{x^2}{5} + \dfrac{y^2}{9} = 1$

10. $\dfrac{x^2}{64} + \dfrac{y^2}{28} = 1$

11. $\dfrac{(x+3)^2}{16} + \dfrac{(y-5)^2}{25} = 1$

12. $\dfrac{(x-4)^2}{12} + \dfrac{(y+3)^2}{16} = 1$

13. $\dfrac{(x+5)^2}{9/4} + (y-1)^2 = 1$

14. $(x+2)^2 + \dfrac{(y+4)^2}{1/4} = 1$

15. $9x^2 + 4y^2 + 36x - 24y + 36 = 0$

16. $9x^2 + 4y^2 - 54x + 40y + 37 = 0$

17. $x^2 + 5y^2 - 8x - 30y - 39 = 0$

18. $3x^2 + y^2 + 18x - 2y - 8 = 0$

19. $6x^2 + 2y^2 + 18x - 10y + 2 = 0$

20. $x^2 + 4y^2 - 6x + 20y - 2 = 0$

21. $16x^2 + 25y^2 - 32x + 50y + 16 = 0$

22. $9x^2 + 25y^2 - 36x - 50y + 60 = 0$

In Exercises 23–26, use a graphing utility to graph the ellipse. Find the center, foci, and vertices. (Recall that it may be necessary to solve the equation for *y* and obtain two functions.)

23. $5x^2 + 3y^2 = 15$

24. $3x^2 + 4y^2 = 12$

25. $12x^2 + 20y^2 - 12x + 40y - 37 = 0$

26. $36x^2 + 9y^2 + 48x - 36y - 72 = 0$

In Exercises 27–34, find the standard form of the equation of the ellipse with center at the origin.

27.

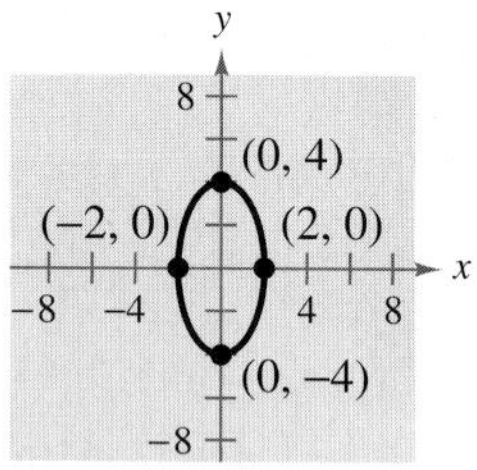

28.

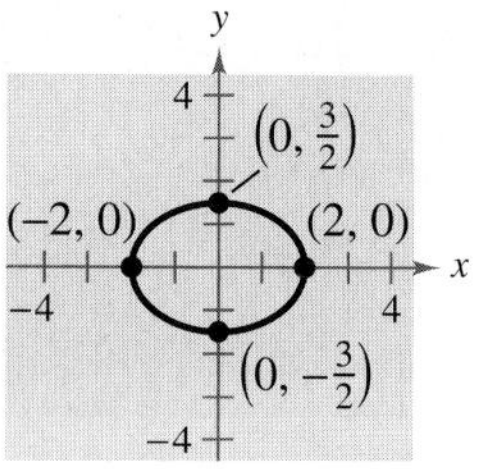

29. Vertices: $(\pm 6, 0)$; Foci: $(\pm 2, 0)$

30. Vertices: $(0, \pm 8)$; Foci: $(0, \pm 4)$

31. Foci: $(\pm 5, 0)$; Major axis of length 12

32. Foci: $(\pm 2, 0)$; Major axis of length 8

33. Vertices: $(0, \pm 5)$; Passes through the point $(4, 2)$

34. Major axis vertical; Passes through the points $(0, 4)$ and $(2, 0)$

In Exercises 35–46, find the standard form of the equation of the specified ellipse.

35.

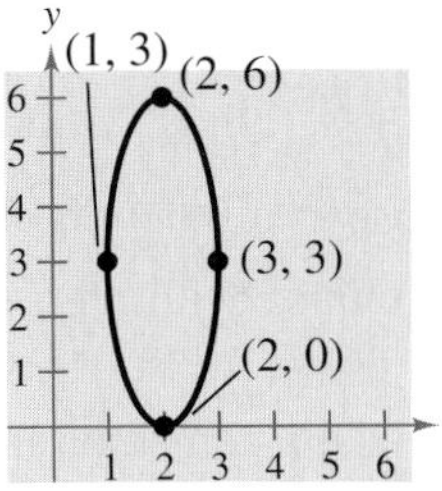

36.

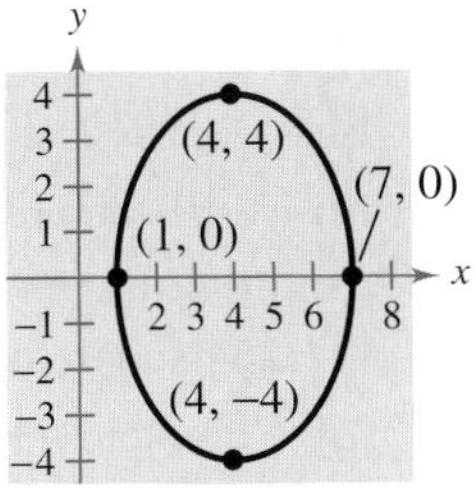

37.

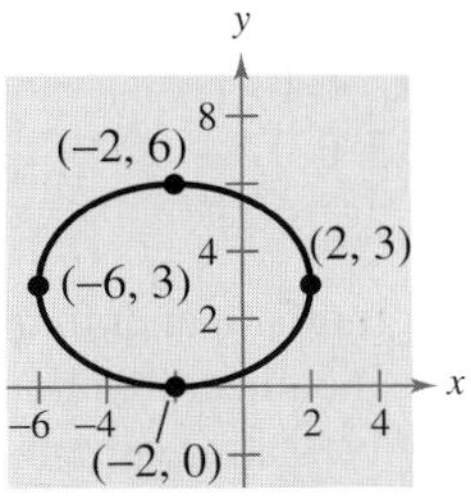

38.

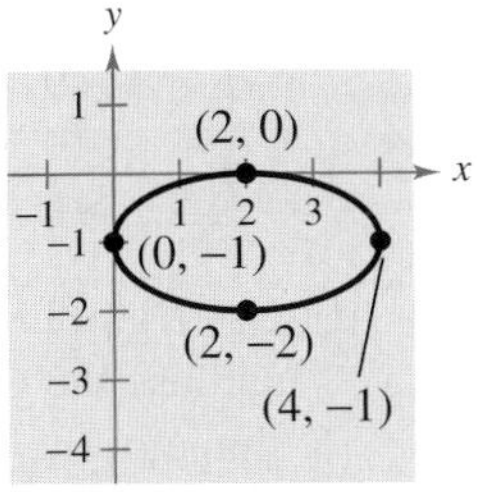

39. Vertices: $(0, 4), (4, 4)$; Minor axis of length 2

40. Foci: $(0, 0), (4, 0)$; Major axis of length 8

41. Foci: $(0, 0), (0, 8)$; Major axis of length 16

42. Center: $(2, -1)$; Vertex: $\left(2, \frac{1}{2}\right)$; Minor axis of length 2

43. Vertices: $(3, 1), (3, 9)$; Minor axis of length 6

44. Center: $(3, 2)$; $a = 3c$; Foci: $(1, 2), (5, 2)$

45. Center: $(0, 4)$; $a = 2c$; Vertices: $(-4, 4), (4, 4)$

46. Vertices: $(5, 0), (5, 12)$; Endpoints of the minor axis: $(1, 6), (9, 6)$

47. Find an equation of the ellipse with vertices $(\pm 5, 0)$ and eccentricity $e = \frac{3}{5}$.

48. Find an equation of the ellipse with vertices $(0, \pm 8)$ and eccentricity $e = \frac{1}{2}$.

49. *Fireplace Arch* A fireplace arch is to be constructed in the shape of a semiellipse. The opening is to have a height of 2 feet at the center and a width of 6 feet along the base (see figure). The contractor draws the outline of the ellipse using tacks as described at the beginning of this section. Give the required positions of the tacks and the length of the string.

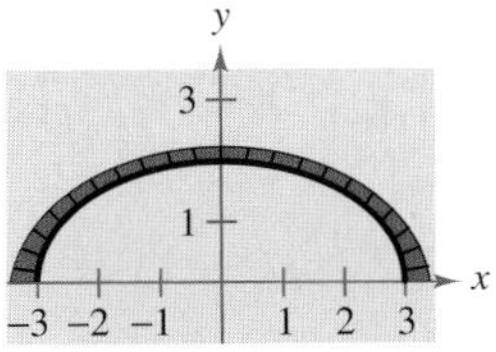

50. *Mountain Tunnel* A semielliptical arch over a tunnel for a road through a mountain has a major axis of 80 feet and a height at the center of 30 feet.

(a) Draw a rectangular coordinate system on a sketch of the tunnel with the center of the road entering the tunnel at the origin. Identify the coordinates of the known points.

(b) Find an equation of the elliptical tunnel.

(c) Determine the height of the arch 5 feet from the edge of the tunnel.

51. *Geometry* The area of the ellipse in the figure is twice the area of the circle. What is the length of the major axis? (*Hint:* $A = \pi ab$ for an ellipse.)

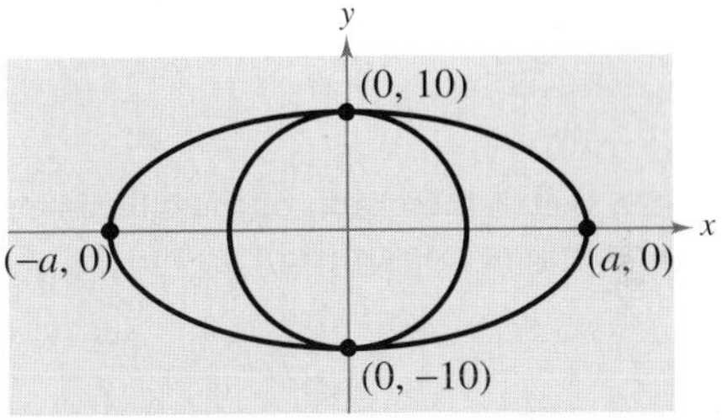

52. *Comet Orbit* Halley's comet has an elliptical orbit, with the sun at one focus. The eccentricity of the orbit is approximately 0.97. The length of the major axis of the orbit is approximately 36.18 astronomical units. (An astronomical unit is about 93 million miles.) Find an equation of the orbit. Place the center of the orbit at the origin, and place the major axis on the x-axis.

53. *Comet Orbit* The comet Encke has an elliptical orbit, with the sun at one focus. Encke ranges from 0.34 to 4.08 astronomical units from the sun. Find an equation of the orbit. Place the center of the orbit at the origin, and place the major axis on the x-axis.

54. *Satellite Orbit* The first artificial satellite to orbit earth was Sputnik I (launched by Russia in 1957). Its highest point above earth's surface was 938 kilometers, and its lowest point was 212 kilometers (see figure). The radius of earth is 6378 kilometers. Find the eccentricity of the orbit.

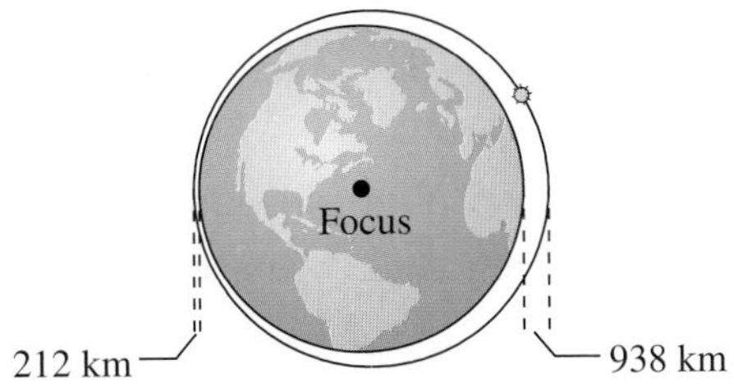

55. *Geometry* A line segment through a focus of an ellipse with endpoints on the ellipse and perpendicular to the major axis is called a **latus rectum** of the ellipse. Therefore, an ellipse has two latera recta. Knowing the length of the latera recta is helpful in sketching an ellipse because it yields other points on the curve (see figure). Show that the length of each latus rectum is $2b^2/a$.

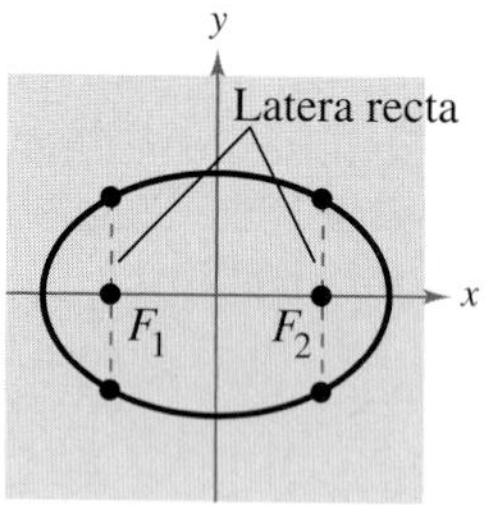

In Exercises 56–59, sketch the graph of the ellipse, making use of the latera recta (see Exercise 55).

56. $\frac{x^2}{4} + \frac{y^2}{1} = 1$

57. $\frac{x^2}{9} + \frac{y^2}{16} = 1$

58. $9x^2 + 4y^2 = 36$

59. $5x^2 + 3y^2 = 15$

Synthesis

True or False? **In Exercises 60–63, determine whether the statement is true or false. Justify your answer.**

60. The graph of $(x^2/4) + y^4 = 1$ is an ellipse.

61. It is easier to distinguish the graph of an ellipse from the graph of a circle if the eccentricity of the ellipse is large (close to 1).

62. The area of a circle with diameter $d = 2r = 8$ is greater than the area of an ellipse with major axis $2a = 8$.

63. It is possible for the foci of an ellipse to occur outside the ellipse.

64. *Think About It* At the beginning of this section it was noted that an ellipse can be drawn using two thumbtacks, a string of fixed length (greater than the distance between the two tacks), and a pencil. If the ends of the string are fastened at the tacks and the string is drawn taut with a pencil, the path traced by the pencil is an ellipse.

(a) What is the length of the string in terms of a?

(b) Explain why the path is an ellipse.

65. *Exploration* The area A of the ellipse

$$\frac{x^2}{a^2} + \frac{y^2}{b^2} = 1$$

is $A = \pi ab$. Let $a + b = 20$.

(a) Write the area of the ellipse as a function of a.

(b) Find the equation of an ellipse with an area of 264 square centimeters.

(c) Complete the table and make a conjecture about the shape of the ellipse with a maximum area.

a	8	9	10	11	12	13
A						

(d) Use a graphing utility to graph the area function, and use the graph to make a conjecture about the shape of the ellipse that yields a maximum area.

Review

In Exercises 66–69, determine whether the sequence is arithmetic, geometric, or neither.

66. 66, 55, 44, 33, 22, . . .

67. 80, 40, 20, 10, 5, . . .

68. $\frac{1}{4}, \frac{1}{2}, 1, 2, 4, \ldots$

69. $-\frac{1}{2}, \frac{1}{2}, \frac{3}{2}, \frac{5}{2}, \frac{7}{2}, \ldots$

In Exercises 70–73, find a formula for a_n for the arithmetic sequence.

70. $a_1 = 13,\ d = 3$

71. $a_1 = 0,\ d = -\frac{1}{4}$

72. $a_1 = 5,\ a_4 = 9.5$

73. $a_3 = 27,\ a_8 = 72$

In Exercises 74–77, find the sum.

74. $\sum_{n=0}^{6} 3^n$

75. $\sum_{n=0}^{6} (-3)^n$

76. $\sum_{n=1}^{10} 4\left(\frac{3}{4}\right)^{n-1}$

77. $\sum_{n=0}^{10} 5\left(\frac{4}{3}\right)^n$

10.4 Hyperbolas

▶ What you should learn

- How to write the standard form of the equation of a hyperbola
- How to find the asymptotes of a hyperbola
- How to use properties of hyperbolas to solve real-life problems
- How to classify a conic from its general equation

▶ Why you should learn it

Hyperbolas can be used to model and solve many types of real-life problems. For instance, in Exercise 39 on page 803, hyperbolas are used to locate the position of an explosion that was recorded by three listening stations.

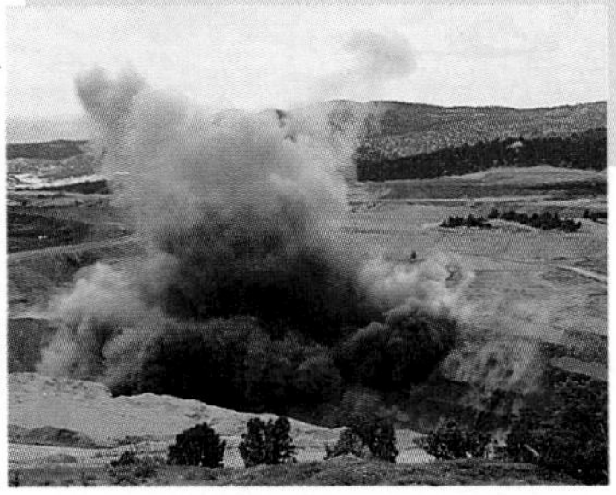

James Foote/Photo Researchers, Inc.

Introduction

The definition of a hyperbola parallels that of an ellipse. The difference is that for an ellipse the *sum* of the distances between the foci and a point on the ellipse is fixed, whereas for a hyperbola the *difference* of these distances is fixed.

Definition of Hyperbola

A **hyperbola** is the set of all points (x, y) the difference of whose distances from two distinct fixed points (foci) is a positive constant. (See Figure 10.26.)

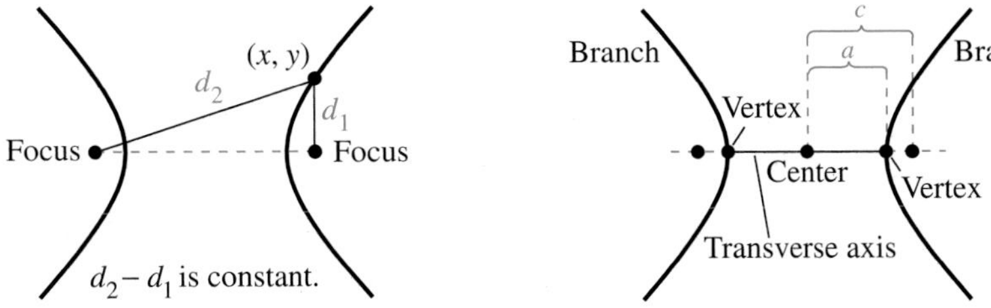

FIGURE 10.26

Every hyperbola has two disconnected **branches.** The line through the two foci intersects a hyperbola at its two **vertices.** The line segment connecting the vertices is called the **transverse axis,** and the midpoint of the transverse axis is called the **center** of the hyperbola. The development of the standard form of the equation of a hyperbola is similar to that of an ellipse.

Standard Equation of a Hyperbola

The standard form of the equation of a hyperbola with center at (h, k) is

$$\frac{(x-h)^2}{a^2} - \frac{(y-k)^2}{b^2} = 1 \qquad \text{Transverse axis is horizontal.}$$

$$\frac{(y-k)^2}{a^2} - \frac{(x-h)^2}{b^2} = 1. \qquad \text{Transverse axis is vertical.}$$

The vertices are a units from the center, and the foci are c units from the center. Moreover, $c^2 = a^2 + b^2$. If the center of the hyperbola is at the origin $(0, 0)$, the equation takes one of the following forms.

$$\frac{x^2}{a^2} - \frac{y^2}{b^2} = 1 \qquad \text{Transverse axis is horizontal.} \qquad \frac{y^2}{a^2} - \frac{x^2}{b^2} = 1 \qquad \text{Transverse axis is vertical.}$$

Note that a, b, and c are related differently for hyperbolas than for ellipses.

Figure 10.27 shows both the horizontal and vertical orientations for a hyperbola.

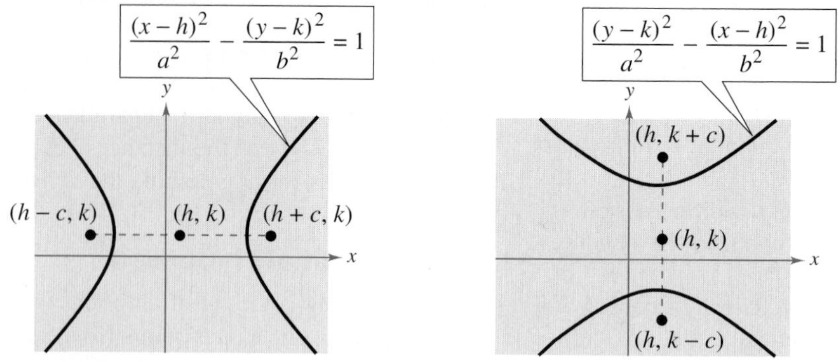

FIGURE 10.27

Example 1 ▶ Finding the Standard Equation of a Hyperbola

Find the standard form of the equation of the hyperbola with foci at $(-1, 2)$ and $(5, 2)$ and vertices at $(0, 2)$ and $(4, 2)$.

Solution

By the Midpoint Formula, the center of the hyperbola occurs at the point $(2, 2)$. Furthermore, $c = 3$ and $a = 2$, and it follows that

$$\begin{aligned} b^2 &= c^2 - a^2 \\ &= 3^2 - 2^2 \\ &= 9 - 4 \\ &= 5. \end{aligned}$$

So, the equation of the hyperbola is

$$\frac{(x-2)^2}{4} - \frac{(y-2)^2}{5} = 1.$$

Figure 10.28 shows the hyperbola.

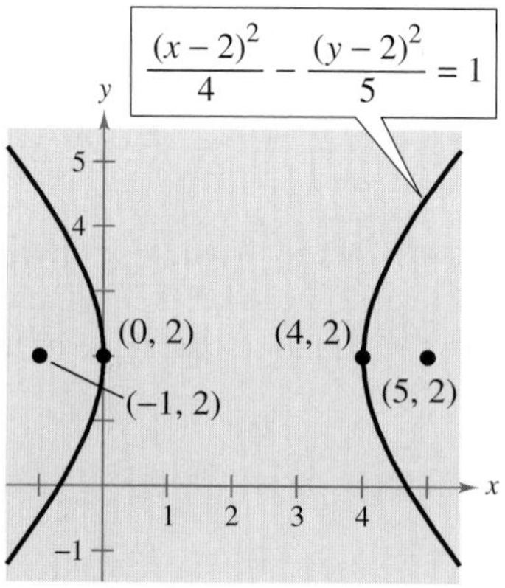

FIGURE 10.28

Exploration

Most graphing utilities have a parametric mode. Try using parametric mode to graph the hyperbola $x = 2 + 2 \sec t$ and $y = 2 + \sqrt{5} \tan t$. How does the result compare with the graph given in Figure 10.28? (Let the parameter vary from Tmin $= 0$ to Tmax $= 6.28$ with Tstep $= 0.13$.)

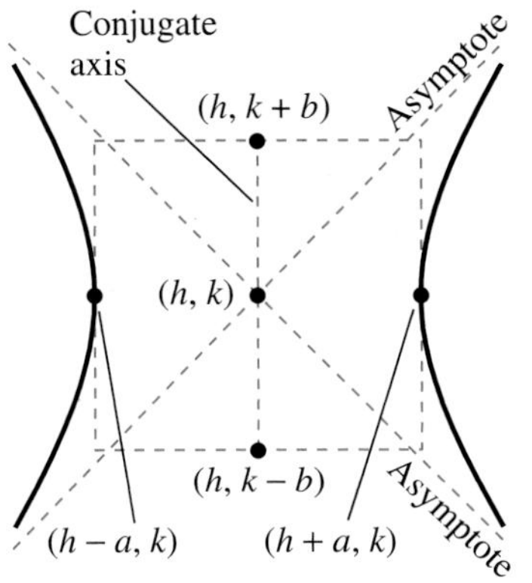

FIGURE 10.29

Asymptotes of a Hyperbola

Each hyperbola has two **asymptotes** that intersect at the center of the hyperbola, as shown in Figure 10.29. The asymptotes pass through the vertices of a rectangle of dimensions $2a$ by $2b$, with its center at (h, k). The line segment of length $2b$ joining $(h, k + b)$ and $(h, k - b)$ [or $(h + b, k)$ and $(h - b, k)$] is the **conjugate axis** of the hyperbola.

Asymptotes of a Hyperbola

The equations for the asymptotes of a hyperbola are

$$y = k \pm \frac{b}{a}(x - h) \qquad \text{Asymptotes for horizontal transverse axis}$$

$$y = k \pm \frac{a}{b}(x - h). \qquad \text{Asymptotes for vertical transverse axis}$$

Example 2 ▶ Using Asymptotes to Sketch a Hyperbola

Sketch the hyperbola whose equation is $4x^2 - y^2 = 16$.

Solution

Divide both sides of the original equation by 16, and rewrite the equation.

$$\frac{x^2}{4} - \frac{y^2}{16} = 1 \qquad \text{Standard form}$$

From this, you can conclude that $a = 2$, $b = 4$, and the transverse axis is horizontal. So, the vertices occur at $(-2, 0)$ and $(2, 0)$, and the ends of the conjugate axis occur at $(0, -4)$ and $(0, 4)$. Using these four points, you are able to sketch the rectangle shown in Figure 10.30(a). Finally, after drawing the asymptotes through the corners of this rectangle, you can complete the sketch, as shown in Figure 10.30(b).

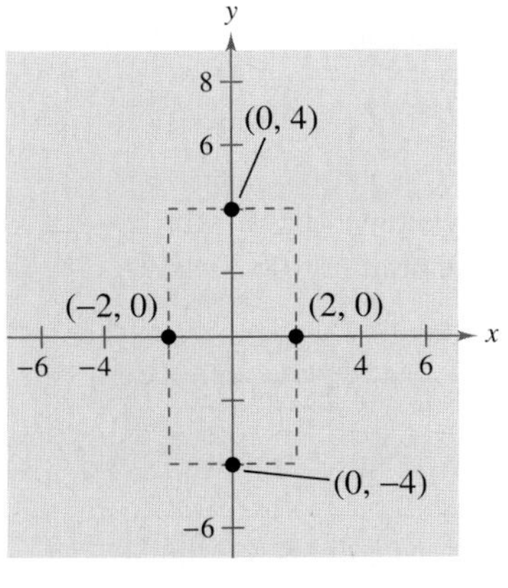

(a)

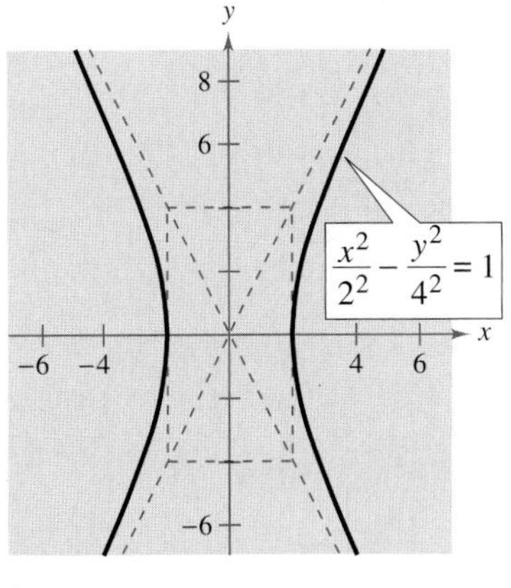

(b)

FIGURE 10.30

Example 3 ▶ Finding the Asymptotes of a Hyperbola

Sketch the hyperbola given by $4x^2 - 3y^2 + 8x + 16 = 0$ and find the equations of its asymptotes.

Solution

$4x^2 - 3y^2 + 8x + 16 = 0$	Write original equation.
$4(x^2 + 2x) - 3y^2 = -16$	Subtract 16 from each side and factor.
$4(x^2 + 2x + 1) - 3y^2 = -16 + 4$	Add 4 to each side.
$4(x + 1)^2 - 3y^2 = -12$	Complete the square.
$\frac{y^2}{4} - \frac{(x+1)^2}{3} = 1$	Standard form

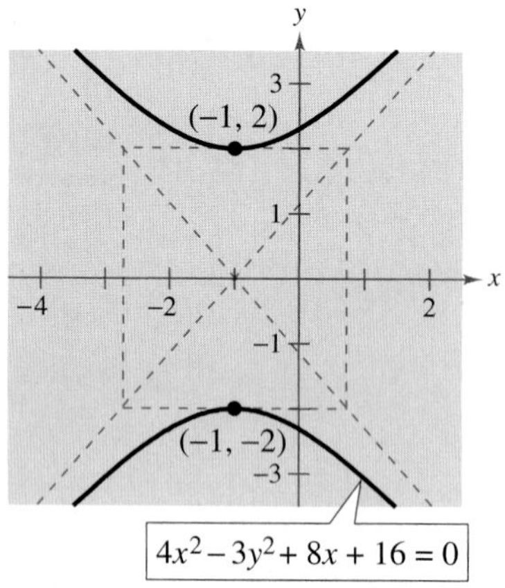

FIGURE 10.31

From this equation you can conclude that the hyperbola is centered at $(-1, 0)$, has vertices at $(-1, 2)$ and $(-1, -2)$, and has a conjugate axis with ends at $(-1 - \sqrt{3}, 0)$ and $(-1 + \sqrt{3}, 0)$. To sketch the hyperbola, draw a rectangle through these four points. The asymptotes are the lines passing through the corners of the rectangle, as shown in Figure 10.31. Finally, using $a = 2$ and $b = \sqrt{3}$, you can conclude that the equations of the asymptotes are

$$y = \frac{2}{\sqrt{3}}(x + 1) \quad \text{and} \quad y = -\frac{2}{\sqrt{3}}(x + 1).$$

If the constant term F in the equation in Example 3 had been 4 instead of 16, you would have obtained the following degenerate case.

Two Intersecting Lines: $\frac{y^2}{4} - \frac{(x+1)^2}{3} = 0$

Technology

You can use a graphing utility to graph a hyperbola by graphing the upper and lower portions in the same viewing window. For instance, to graph the hyperbola in Example 3, first solve for y to get

$$y_1 = 2\sqrt{1 + \frac{(x+1)^2}{3}} \quad \text{and} \quad y_2 = -2\sqrt{1 + \frac{(x+1)^2}{3}}.$$

Use a viewing window in which $-9 \le x \le 9$ and $-6 \le y \le 6$. You should obtain the graph shown below.

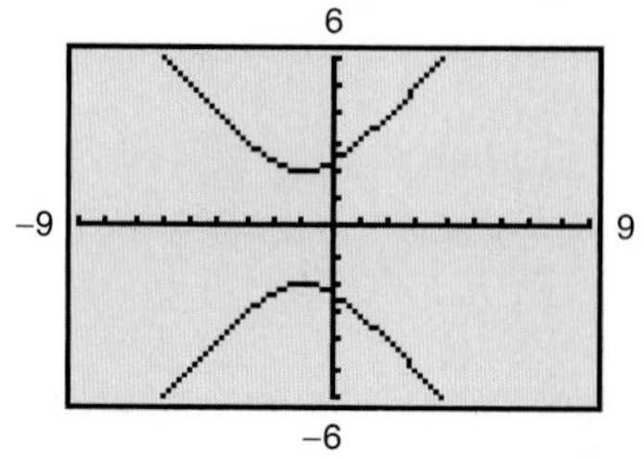

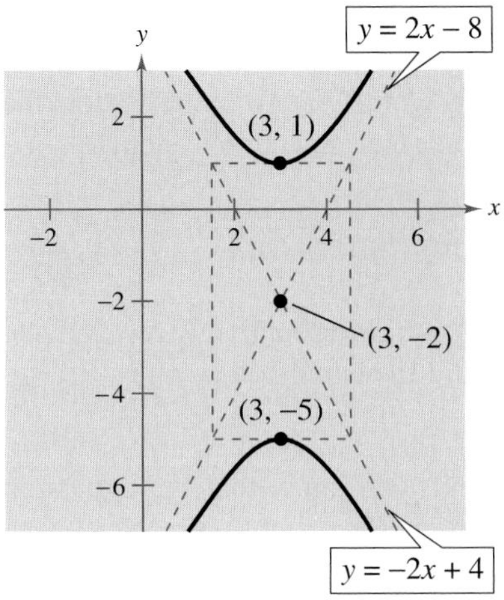

FIGURE 10.32

Example 4 ▶ Using Asymptotes to Find the Standard Equation

Find the standard form of the equation of the hyperbola having vertices at $(3, -5)$ and $(3, 1)$ and having asymptotes

$$y = 2x - 8 \qquad \text{and} \qquad y = -2x + 4$$

as shown in Figure 10.32.

Solution

According to the Midpoint Formula, the center of the hyperbola is at $(3, -2)$. Furthermore, the hyperbola has a vertical transverse axis with $a = 3$. From the given equations, you can determine the slopes of the asymptotes to be

$$m_1 = 2 = \frac{a}{b} \qquad \text{and} \qquad m_2 = -2 = -\frac{a}{b}$$

and, because $a = 3$,

$$2 = \frac{a}{b} \quad \Rightarrow \quad 2 = \frac{3}{b} \quad \Rightarrow \quad b = \frac{3}{2}.$$

So, the standard equation is

$$\frac{(y+2)^2}{9} - \frac{(x-3)^2}{9/4} = 1.$$

As with ellipses, the **eccentricity** of a hyperbola is

$$e = \frac{c}{a} \qquad \text{Eccentricity}$$

and because $c > a$ it follows that $e > 1$. If the eccentricity is large, the branches of the hyperbola are nearly flat, as shown in Figure 10.33(a). If the eccentricity is close to 1, the branches of the hyperbola are more pointed, as shown in Figure 10.33(b).

Quick Review
True or false:

1. $\dfrac{(x-3)^2}{16} = 1 - \dfrac{(y+2)^2}{25}$ is the equation of a hyperbola.
   Answer: False
2. For an ellipse, $c < a$.
   Answer: True
3. A parabola has no focus.
   Answer: False

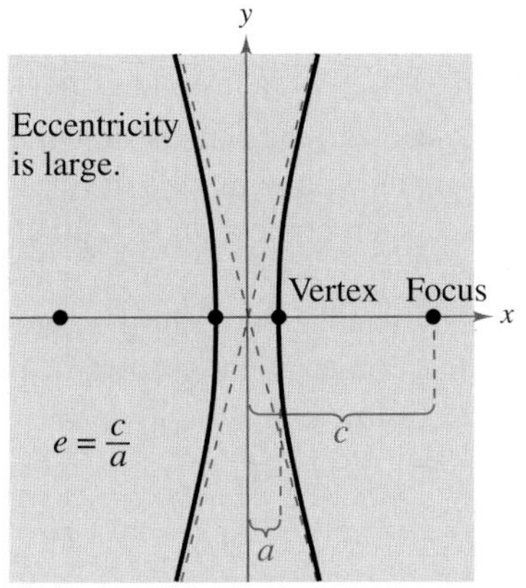

(a)

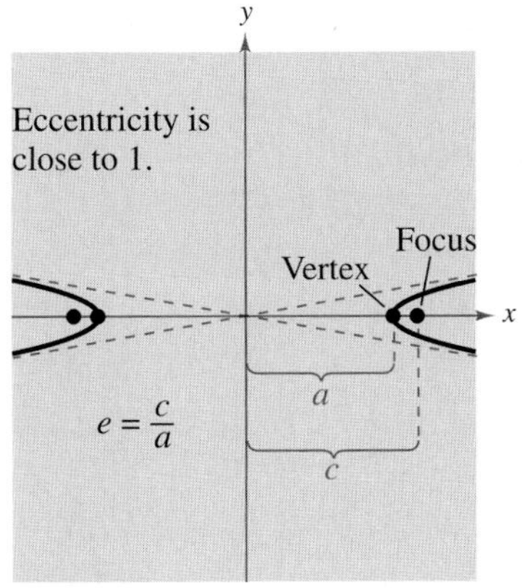

(b)

FIGURE 10.33

Applications

The following application was developed during World War II. It shows how the properties of hyperbolas can be used in radar and other detection systems.

Example 5 ▶ An Application Involving Hyperbolas

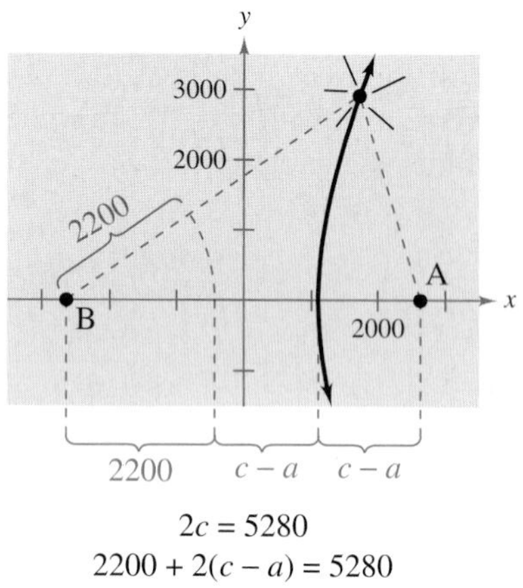

FIGURE 10.34

Two microphones, 1 mile apart, record an explosion. Microphone A receives the sound 2 seconds before microphone B. Where did the explosion occur?

Solution

Assuming sound travels at 1100 feet per second, you know that the explosion took place 2200 feet farther from B than from A, as shown in Figure 10.34. The locus of all points that are 2200 feet closer to A than to B is one branch of the hyperbola

$$\frac{x^2}{a^2} - \frac{y^2}{b^2} = 1$$

where

$$c = \frac{5280}{2} = 2640$$

and

$$a = \frac{2200}{2} = 1100.$$

So, $b^2 = c^2 - a^2 = 5{,}759{,}600$, and you conclude that the explosion occurred somewhere on the right branch of the hyperbola

$$\frac{x^2}{1{,}210{,}000} - \frac{y^2}{5{,}759{,}600} = 1.$$

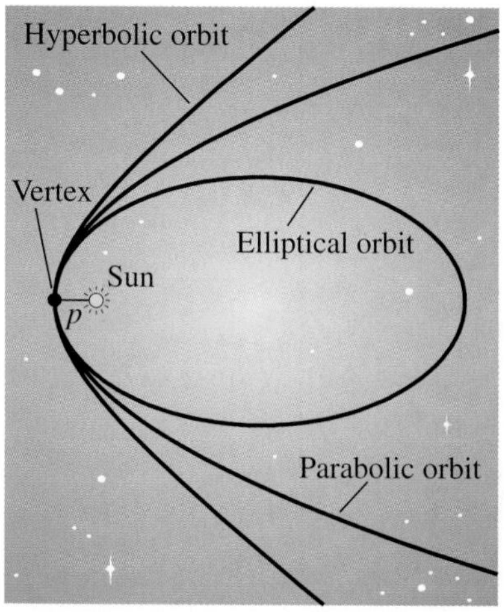

FIGURE 10.35

Another interesting application of conic sections involves the orbits of comets in our solar system. Of the 610 comets identified prior to 1970, 245 have elliptical orbits, 295 have parabolic orbits, and 70 have hyperbolic orbits. The center of the sun is a focus of each of these orbits, and each orbit has a vertex at the point where the comet is closest to the sun, as shown in Figure 10.35. Undoubtedly, there have been many comets with parabolic or hyperbolic orbits that were not identified. We only get to see such comets *once.* Comets with elliptical orbits, such as Halley's comet, are the only ones that remain in our solar system.

If p is the distance between the vertex and the focus in meters, and v is the velocity of the comet at the vertex in meters per second, the type of orbit is determined as follows.

1. *Ellipse:* $v < \sqrt{2GM/p}$
2. *Parabola:* $v = \sqrt{2GM/p}$
3. *Hyperbola:* $v > \sqrt{2GM/p}$

In each of these equations, $M \approx 1.991 \times 10^{30}$ kilograms (the mass of the sun) and $G \approx 6.67 \times 10^{-11}$ cubic meters per kilogram-second squared.

General Equations of Conics

Classifying a Conic from Its General Equation

The graph of $Ax^2 + Cy^2 + Dx + Ey + F = 0$ is one of the following.

1. *Circle:*	$A = C$	
2. *Parabola:*	$AC = 0$	$A = 0$ or $C = 0$, but not both.
3. *Ellipse:*	$AC > 0$	A and C have like signs.
4. *Hyperbola:*	$AC < 0$	A and C have unlike signs.

The test above is valid if the graph is a conic. The test does not apply to equations such as $x^2 + y^2 = -1$, which is not a conic.

Example 6 ▶ Classifying Conics from General Equations

Classify each graph.

a. $4x^2 - 9x + y - 5 = 0$

b. $4x^2 - y^2 + 8x - 6y + 4 = 0$

c. $2x^2 + 4y^2 - 4x + 12y = 0$

d. $2x^2 + 2y^2 - 8x + 12y + 26 = 0$

Solution

a. For the equation $4x^2 - 9x + y - 5 = 0$, you have

$$AC = 4(0) = 0. \qquad \text{Parabola}$$

So, the graph is a parabola.

b. For the equation $4x^2 - y^2 + 8x - 6y + 4 = 0$, you have

$$AC = 4(-1) < 0. \qquad \text{Hyperbola}$$

So, the graph is a hyperbola.

c. For the equation $2x^2 + 4y^2 - 4x + 12y = 0$, you have

$$AC = 2(4) > 0. \qquad \text{Ellipse}$$

So, the graph is an ellipse.

d. For the equation $2x^2 + 2y^2 - 8x + 12y + 2 = 0$, you have

$$A = C = 2. \qquad \text{Circle}$$

So, the graph is a circle.

Writing ABOUT MATHEMATICS

Sketching Conics Sketch each of the conics described in Example 6. Write a paragraph describing the procedures that allow you to sketch the conics efficiently.

Activities

1. Classify each equation.
   a. $3x^2 - 2y^2 + 4y - 3 = 0$
   b. $2y^2 - 3x + 2 = 0$
   c. $x^2 + 4y^2 - 2x - 3 = 0$
   d. $x^2 - 2x + 4y - 1 = 0$

   Answer: (a) Hyperbola, (b) Parabola, (c) Ellipse, (d) Parabola
2. Find an equation of the hyperbola with asymptotes $y = \pm 2x$ and vertices $(0, \pm 2)$.

   Answer: $\dfrac{y^2}{4} - \dfrac{x^2}{1} = 1$

The Granger Collection

Historical Note

Caroline Herschel (1750–1848) was the first woman to be credited with detecting a new comet. During her long life, this English astronomer discovered a total of eight new comets.

10.4 Exercises

In Exercises 1–4, match the equation with its graph. [The graphs are labeled (a), (b), (c), and (d).]

(a)

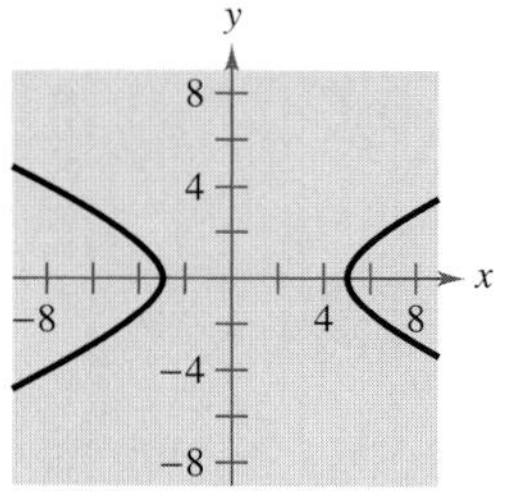

(b)

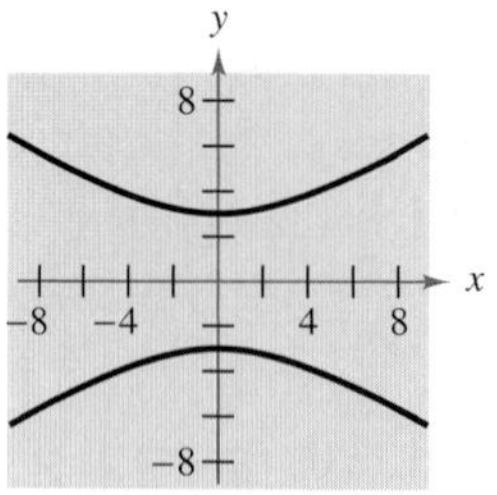

(c)

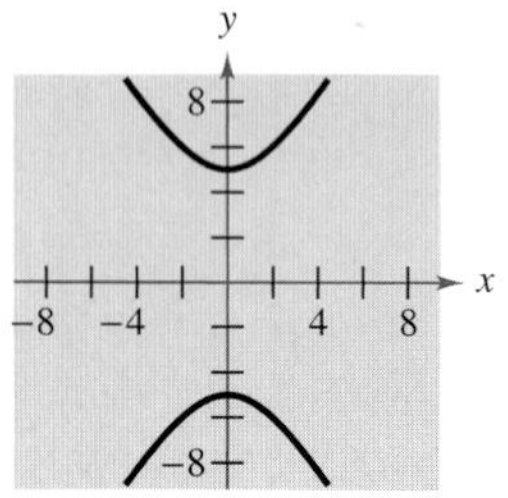

(d)

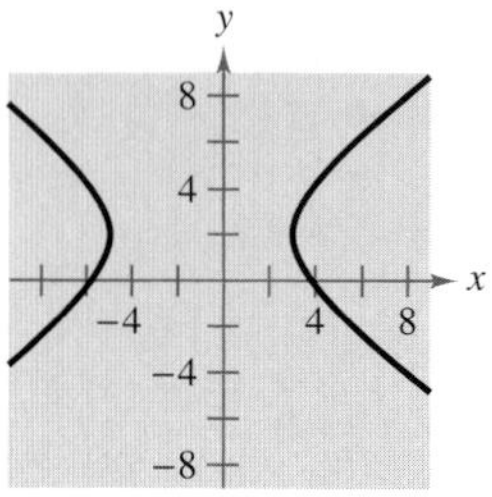

1. $\dfrac{y^2}{9} - \dfrac{x^2}{25} = 1$

2. $\dfrac{y^2}{25} - \dfrac{x^2}{9} = 1$

3. $\dfrac{(x-1)^2}{16} - \dfrac{y^2}{4} = 1$

4. $\dfrac{(x+1)^2}{16} - \dfrac{(y-2)^2}{9} = 1$

In Exercises 5–16, find the center, vertices, foci, and the equations of the asymptotes of the hyperbola, and sketch its graph.

5. $x^2 - y^2 = 1$

6. $\dfrac{x^2}{9} - \dfrac{y^2}{25} = 1$

7. $\dfrac{y^2}{25} - \dfrac{x^2}{81} = 1$

8. $\dfrac{x^2}{36} - \dfrac{y^2}{4} = 1$

9. $\dfrac{(x-1)^2}{4} - \dfrac{(y+2)^2}{1} = 1$

10. $\dfrac{(x+3)^2}{144} - \dfrac{(y-2)^2}{25} = 1$

11. $\dfrac{(y+6)^2}{1/9} - \dfrac{(x-2)^2}{1/4} = 1$

12. $\dfrac{(y-1)^2}{1/4} - \dfrac{(x+3)^2}{1/16} = 1$

13. $9x^2 - y^2 - 36x - 6y + 18 = 0$

14. $x^2 - 9y^2 + 36y - 72 = 0$

15. $x^2 - 9y^2 + 2x - 54y - 80 = 0$

16. $16y^2 - x^2 + 2x + 64y + 63 = 0$

In Exercises 17–20, find the center, vertices, foci, and the equations of the asymptotes of the hyperbola. Use a graphing utility to graph the hyperbola and its asymptotes.

17. $2x^2 - 3y^2 = 6$

18. $6y^2 - 3x^2 = 18$

19. $9y^2 - x^2 + 2x + 54y + 62 = 0$

20. $9x^2 - y^2 + 54x + 10y + 55 = 0$

In Exercises 21–26, find the standard form of the equation of the specified hyperbola with center at the origin.

21. Vertices: $(0, \pm 2)$; Foci: $(0, \pm 4)$

22. Vertices: $(\pm 4, 0)$; Foci: $(\pm 6, 0)$

23. Vertices: $(\pm 1, 0)$; Asymptotes: $y = \pm 5x$

24. Vertices: $(0, \pm 3)$; Asymptotes: $y = \pm 3x$

25. Foci: $(0, \pm 8)$; Asymptotes: $y = \pm 4x$

26. Foci: $(\pm 10, 0)$; Asymptotes: $y = \pm \frac{3}{4}x$

In Exercises 27–38, find the standard form of the equation of the specified hyperbola.

27. Vertices: $(2, 0), (6, 0)$; Foci: $(0, 0), (8, 0)$

28. Vertices: $(2, 3), (2, -3)$; Foci: $(2, 6), (2, -6)$

29. Vertices: $(4, 1), (4, 9)$; Foci: $(4, 0), (4, 10)$

30. Vertices: $(-2, 1), (2, 1)$; Foci: $(-3, 1), (3, 1)$

31. Vertices: $(2, 3), (2, -3)$;
    Passes through the point $(0, 5)$

32. Vertices: $(-2, 1), (2, 1)$;
    Passes through the point $(5, 4)$

33. Vertices: $(0, 4), (0, 0)$;
    Passes through the point $(\sqrt{5}, -1)$

34. Vertices: $(1, 2), (1, -2)$;
    Passes through the point $(0, \sqrt{5})$

35. Vertices: $(1, 2), (3, 2)$
    Asymptotes: $y = x$, $y = 4 - x$

36. Vertices: $(3, 0), (3, 6)$
    Asymptotes: $y = 6 - x$, $y = x$

37. Vertices: $(0, 2), (6, 2)$;
    Asymptotes: $y = \frac{2}{3}x$, $y = 4 - \frac{2}{3}x$

38. Vertices: $(3, 0), (3, 4)$;
    Asymptotes: $y = \frac{2}{3}x$, $y = 4 - \frac{2}{3}x$

39. *Sound Location* Three listening stations located at (3300, 0), (3300, 1100), and (−3300, 0) monitor an explosion. If the last two stations detect the explosion 1 second and 4 seconds after the first, respectively, determine the coordinates of the explosion. (Assume that the coordinate system is measured in feet and that sound travels at 1100 feet per second.)

40. *LORAN* Long distance radio navigation for aircraft and ships uses synchronized pulses transmitted by widely separated transmitting stations. These pulses travel at the speed of light (186,000 miles per second). The difference in the times of arrival of these pulses at an aircraft or ship is constant on a hyperbola having the transmitting stations as foci. Assume that two stations, 300 miles apart, are positioned on the rectangular coordinate system at points with coordinates (−150, 0) and (150, 0), and that a ship is traveling on a path with coordinates $(x, 75)$ (see figure). Find the x-coordinate of the position of the ship if the time difference between the pulses from the transmitting stations is 1000 microseconds (0.001 second).

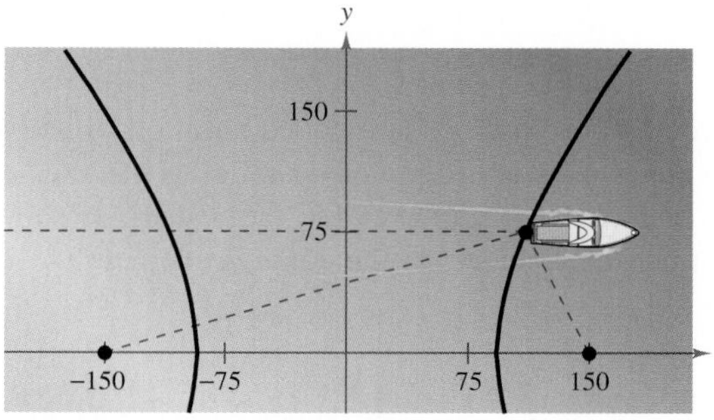

41. *Hyperbolic Mirror* A hyperbolic mirror (used in some telescopes) has the property that a light ray directed at a focus will be reflected to the other focus (see figure). The focus of a hyperbolic mirror has coordinates (24, 0). Find the vertex of the mirror if its mount has coordinates (24, 24).

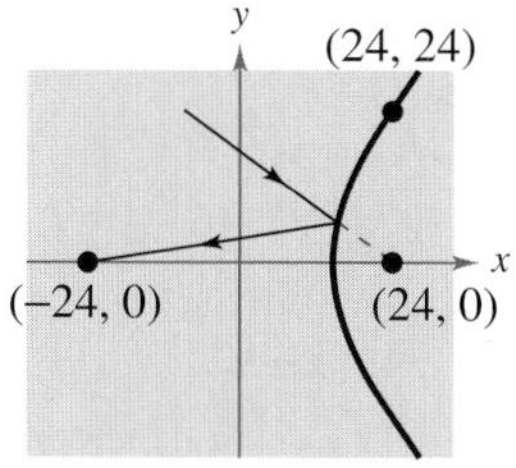

In Exercises 42–49, classify the graph of the equation as a circle, a parabola, an ellipse, or a hyperbola.

42. $x^2 + y^2 - 6x + 4y + 9 = 0$

43. $x^2 + 4y^2 - 6x + 16y + 21 = 0$

44. $4x^2 - y^2 - 4x - 3 = 0$

45. $y^2 - 6y - 4x + 21 = 0$

46. $4x^2 + 3y^2 + 8x - 24y + 51 = 0$

47. $4y^2 - 2x^2 - 4y - 8x - 15 = 0$

48. $25x^2 - 10x - 200y - 119 = 0$

49. $4y^2 + 4x^2 - 24x + 35 = 0$

Synthesis

True or False? **In Exercises 50 and 51, determine whether the statement is true or false. Justify your answer.**

50. In the standard form of the equation of a hyperbola, the larger the ratio of b to a, the larger the eccentricity of the hyperbola.

51. In the standard form of the equation of a hyperbola, the trivial solution of two intersecting lines occurs when $b = 0$.

52. Consider a hyperbola centered at the origin with a horizontal transverse axis. Use the definition of a hyperbola to derive its standard form.

53. Explain how the central rectangle of a hyperbola can be used to sketch its asymptotes.

Review

In Exercises 54–57, perform the indicated polynomial operation.

54. Subtract: $(x^3 - 3x^2) - (6 - 2x - 4x^2)$

55. Multiply: $\left(3x - \frac{1}{2}\right)(x + 4)$

56. Divide: $\dfrac{x^3 - 3x + 4}{x + 2}$

57. Expand: $[(x + y) + 3]^2$

In Exercises 58–63, factor the given polynomial.

58. $x^3 - 16x$

59. $x^2 + 14x + 49$

60. $2x^3 - 24x^2 + 72x$

61. $6x^3 - 11x^2 - 10x$

62. $16x^3 + 54$

63. $4 - x + 4x^2 - x^3$

10.5 Rotation of Conics

▶ What you should learn

- How to rotate the coordinate axes to eliminate the *xy*-term in the equation of a conic
- How to use the discriminant to classify a conic

▶ Why you should learn it

As illustrated in Exercises 7–18 on page 810, rotation of the coordinate axes can help you identify the graph of a general second-degree equation.

Rotation

In the preceding section you learned that the equation of a conic with axes parallel to one of the coordinate axes has a standard form that can be written in the general form

$$Ax^2 + Cy^2 + Dx + Ey + F = 0. \qquad \text{Horizontal or vertical axis}$$

In this section you will study the equations of conics whose axes are rotated so that they are not parallel to either the x-axis or the y-axis. The general equation for such conics contains an xy-term.

$$Ax^2 + Bxy + Cy^2 + Dx + Ey + F = 0 \qquad \text{Equation in } xy\text{-plane}$$

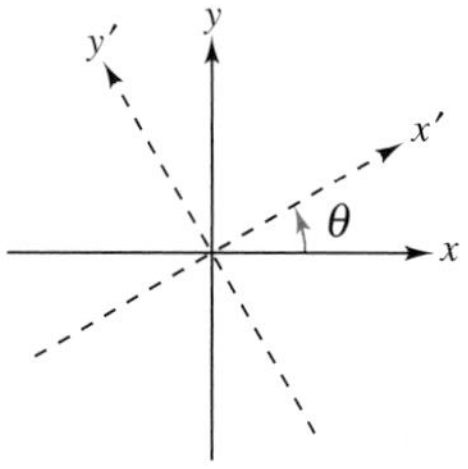

FIGURE 10.36

To eliminate this xy-term, you can use a procedure called **rotation of axes.** The objective is to rotate the x- and y-axes until they are parallel to the axes of the conic. The rotated axes are denoted as the x'-axis and the y'-axis, as shown in Figure 10.36. After the rotation, the equation of the conic in the new $x'y'$-plane will have the form

$$A'(x')^2 + C'(y')^2 + D'x' + E'y' + F' = 0. \qquad \text{Equation in } x'y'\text{-plane}$$

Because this equation has no xy-term, you can obtain a standard form by completing the square. The following theorem identifies how much to rotate the axes to eliminate the xy-term and also the equations for determining the new coefficients A', C', D', E', and F'.

Rotation of Axes to Eliminate an *xy*-Term

The general second-degree equation $Ax^2 + Bxy + Cy^2 + Dx + Ey + F = 0$ can be rewritten as

$$A'(x')^2 + C'(y')^2 + D'x' + E'y' + F' = 0$$

by rotating the coordinate axes through an angle θ, where

$$\cot 2\theta = \frac{A - C}{B}.$$

The coefficients of the new equation are obtained by making the substitutions

$$x = x'\cos\theta - y'\sin\theta \qquad \text{and} \qquad y = x'\sin\theta + y'\cos\theta.$$

STUDY TIP

Remember that the substitutions

$$x = x'\cos\theta - y'\sin\theta$$

and

$$y = x'\sin\theta + y'\cos\theta$$

were developed to eliminate the $x'y'$-term in the rotated system. You can use this as a check on your work. In other words, if your final equation contains an $x'y'$-term, you know that you made a mistake.

Example 1 ▶ Rotation of Axes for a Hyperbola

Write the equation $xy - 1 = 0$ in standard form.

Solution

Because $A = 0$, $B = 1$, and $C = 0$, you have

$$\cot 2\theta = \frac{A - C}{B} = 0 \quad \Rightarrow \quad 2\theta = \frac{\pi}{2} \quad \Rightarrow \quad \theta = \frac{\pi}{4}$$

which implies that

$$x = x'\cos\frac{\pi}{4} - y'\sin\frac{\pi}{4} = x'\left(\frac{\sqrt{2}}{2}\right) - y'\left(\frac{\sqrt{2}}{2}\right) = \frac{x' - y'}{\sqrt{2}}$$

and

$$y = x'\sin\frac{\pi}{4} + y'\cos\frac{\pi}{4} = x'\left(\frac{\sqrt{2}}{2}\right) + y'\left(\frac{\sqrt{2}}{2}\right) = \frac{x' + y'}{\sqrt{2}}.$$

The equation in the $x'y'$-system is obtained by substituting these expressions in the equation $xy - 1 = 0$.

$$\left(\frac{x' - y'}{\sqrt{2}}\right)\left(\frac{x' + y'}{\sqrt{2}}\right) - 1 = 0$$

$$\frac{(x')^2 - (y')^2}{2} - 1 = 0$$

$$\frac{(x')^2}{(\sqrt{2})^2} - \frac{(y')^2}{(\sqrt{2})^2} = 1 \qquad \text{Standard form}$$

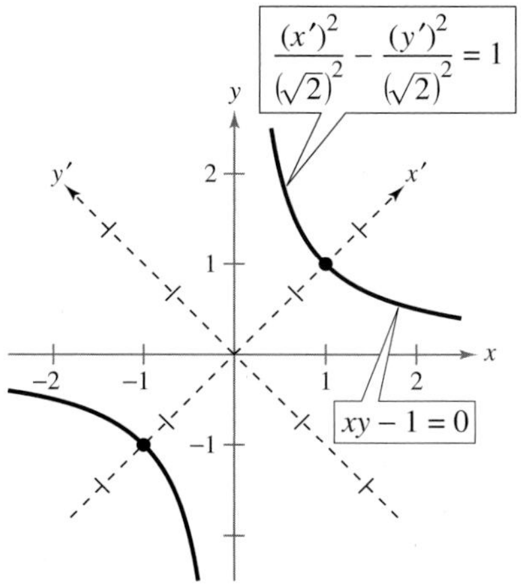

Vertices:
In x'y'-system: $(\sqrt{2}, 0), (-\sqrt{2}, 0)$
In xy-system: $(1, 1), (-1, -1)$

FIGURE 10.37

In the $x'y'$-system, this is a hyperbola centered at the origin with vertices at $(\pm\sqrt{2}, 0)$, as shown in Figure 10.37. To find the coordinates of the vertices in the xy-system, substitute the coordinates $(\pm\sqrt{2}, 0)$ in the equations

$$x = \frac{x' - y'}{\sqrt{2}} \quad \text{and} \quad y = \frac{x' + y'}{\sqrt{2}}.$$

This substitution yields the vertices $(1, 1)$ and $(-1, -1)$ in the xy-system. Note also that the asymptotes of the hyperbola have equations $y' = \pm x'$, which correspond to the original x- and y-axes.

Example 2 ▶ Rotation of Axes for an Ellipse

Sketch the graph of $7x^2 - 6\sqrt{3}xy + 13y^2 - 16 = 0$.

Solution

Because $A = 7$, $B = -6\sqrt{3}$, and $C = 13$, you have

$$\cot 2\theta = \frac{A - C}{B} = \frac{7 - 13}{-6\sqrt{3}} = \frac{1}{\sqrt{3}}$$

which implies that $\theta = \pi/6$. The equation in the $x'y'$-system is obtained by making the substitutions

$$x = x' \cos\frac{\pi}{6} - y' \sin\frac{\pi}{6} = x'\left(\frac{\sqrt{3}}{2}\right) - y'\left(\frac{1}{2}\right) = \frac{\sqrt{3}x' - y'}{2}$$

and

$$y = x' \sin\frac{\pi}{6} + y' \cos\frac{\pi}{6} = x'\left(\frac{1}{2}\right) + y'\left(\frac{\sqrt{3}}{2}\right) = \frac{x' + \sqrt{3}y'}{2}$$

in the original equation. So, you have

$$7x^2 - 6\sqrt{3}xy + 13y^2 - 16 = 0$$

$$7\left(\frac{\sqrt{3}x' - y'}{2}\right)^2 - 6\sqrt{3}\left(\frac{\sqrt{3}x' - y'}{2}\right)\left(\frac{x' + \sqrt{3}y'}{2}\right) + 13\left(\frac{x' + \sqrt{3}y'}{2}\right)^2 - 16 = 0$$

which simplifies to

$$4(x')^2 + 16(y')^2 - 16 = 0$$

$$4(x')^2 + 16(y')^2 = 16$$

$$\frac{(x')^2}{4} + \frac{(y')^2}{1} = 1. \qquad \text{Standard form}$$

This is the equation of an ellipse centered at the origin with vertices $(\pm 2, 0)$ in the $x'y'$-system, as shown in Figure 10.38.

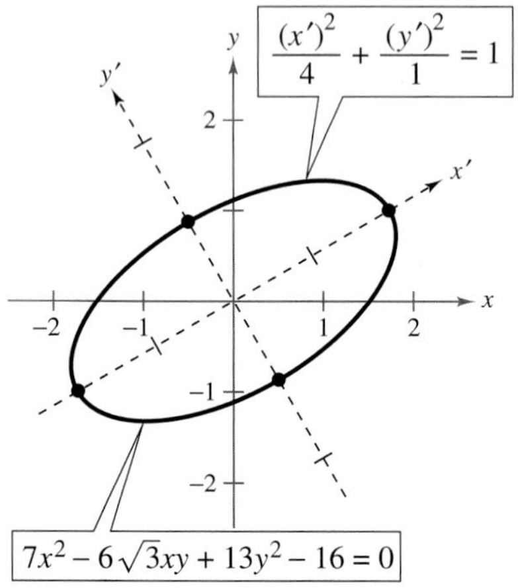

Vertices:

In $x'y'$-system: $(\pm 2, 0)$, $(0, \pm 1)$

In xy-system: $(\sqrt{3}, 1)$, $(-\sqrt{3}, -1)$, $\left(\frac{1}{2}, -\frac{\sqrt{3}}{2}\right)$, $\left(-\frac{1}{2}, \frac{\sqrt{3}}{2}\right)$

FIGURE 10.38

Example 3 ▶ Rotation of Axes for a Parabola

Sketch the graph of $x^2 - 4xy + 4y^2 + 5\sqrt{5}y + 1 = 0$.

Solution

Because $A = 1$, $B = -4$, and $C = 4$, you have

$$\cot 2\theta = \frac{A - C}{B} = \frac{1 - 4}{-4} = \frac{3}{4}.$$

Using the identity $\cot 2\theta = (\cot^2 \theta - 1)/(2 \cot \theta)$ produces

$$\cot 2\theta = \frac{3}{4} = \frac{\cot^2 \theta - 1}{2 \cot \theta}$$

from which you obtain the equation

$$4 \cot^2 \theta - 4 = 6 \cot \theta$$

$$4 \cot^2 \theta - 6 \cot \theta - 4 = 0$$

$$(2 \cot \theta - 4)(2 \cot \theta + 1) = 0.$$

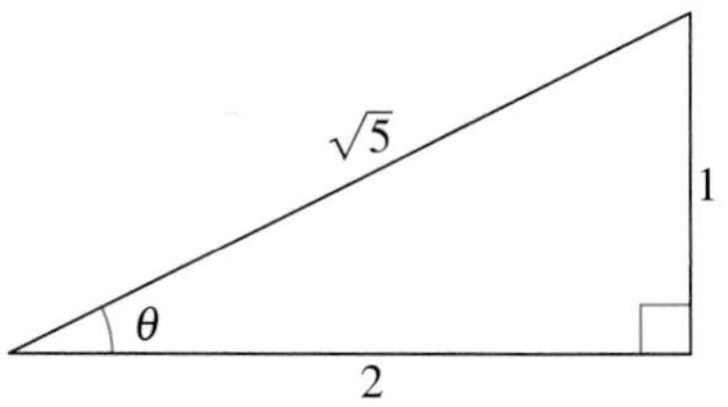

FIGURE 10.39

Considering $0 < \theta < \pi/2$, you have $2 \cot \theta = 4$. So,

$$\cot \theta = 2 \quad \Rightarrow \quad \theta \approx 26.6°.$$

From the triangle in Figure 10.39, you obtain $\sin \theta = 1/\sqrt{5}$ and $\cos \theta = 2/\sqrt{5}$. Consequently, you use the substitutions

$$x = x' \cos \theta - y' \sin \theta$$

$$= x'\left(\frac{2}{\sqrt{5}}\right) - y'\left(\frac{1}{\sqrt{5}}\right) = \frac{2x' - y'}{\sqrt{5}}$$

$$y = x' \sin \theta + y' \cos \theta$$

$$= x'\left(\frac{1}{\sqrt{5}}\right) + y'\left(\frac{2}{\sqrt{5}}\right) = \frac{x' + 2y'}{\sqrt{5}}.$$

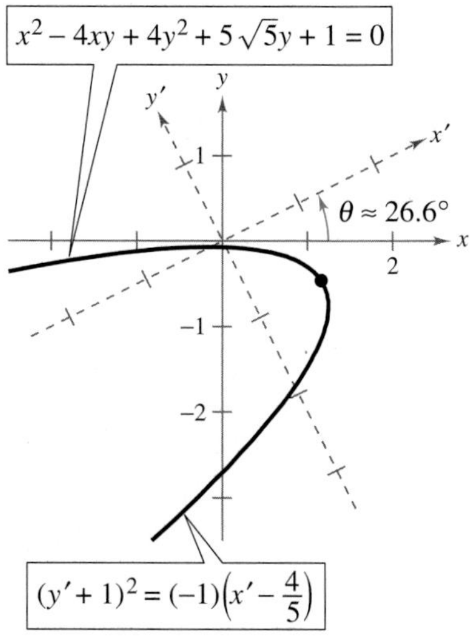

Vertex:
In $x'y'$-system: $\left(\frac{4}{5}, -1\right)$
In xy-system: $\left(\frac{13}{5\sqrt{5}}, -\frac{6}{5\sqrt{5}}\right)$

FIGURE 10.40

Substituting these expressions in the original equation, you have

$$x^2 - 4xy + 4y^2 + 5\sqrt{5}y + 1 = 0$$

$$\left(\frac{2x' - y'}{\sqrt{5}}\right)^2 - 4\left(\frac{2x' - y'}{\sqrt{5}}\right)\left(\frac{x' + 2y'}{\sqrt{5}}\right) + 4\left(\frac{x' + 2y'}{\sqrt{5}}\right)^2 + 5\sqrt{5}\left(\frac{x' + 2y'}{\sqrt{5}}\right) + 1 = 0$$

which simplifies as follows.

$$5(y')^2 + 5x' + 10y' + 1 = 0$$

$$5(y' + 1)^2 = -5x' + 4 \qquad \text{Complete the square.}$$

$$(y' + 1)^2 = (-1)\left(x' - \frac{4}{5}\right) \qquad \text{Standard form}$$

The graph of this equation is a parabola with vertex at $\left(\frac{4}{5}, -1\right)$. Its axis is parallel to the x'-axis in the $x'y'$-system, as shown in Figure 10.40.

Invariants Under Rotation

In the rotation of axes theorem listed at the beginning of this section, note that the constant term is the same in both equations, $F' = F$. Such quantities are **invariant under rotation.** The next theorem lists some other rotation invariants.

Rotation Invariants

The rotation of the coordinate axes through an angle θ that transforms the equation $Ax^2 + Bxy + Cy^2 + Dx + Ey + F = 0$ into the form

$$A'(x')^2 + C'(y')^2 + D'x' + E'y' + F' = 0$$

has the following rotation invariants.

1. $F = F'$

2. $A + C = A' + C'$

3. $B^2 - 4AC = (B')^2 - 4A'C'$

You can use the results of this theorem to classify the graph of a second-degree equation *with* an xy-term in much the same way you do for a second-degree equation *without* an xy-term. Note that because $B' = 0$, the invariant $B^2 - 4AC$ reduces to

$$B^2 - 4AC = -4A'C'. \qquad \text{Discriminant}$$

This quantity is called the **discriminant** of the equation

$$Ax^2 + Bxy + Cy^2 + Dx + Ey + F = 0.$$

Now, from the classification procedure given in Section 10.4, you know that the sign of $A'C'$ determines the type of graph for the equation

$$A'(x')^2 + C'(y')^2 + D'x' + E'y' + F' = 0.$$

Consequently, the sign of $B^2 - 4AC$ will determine the type of graph for the original equation, as given in the following classification.

Classification of Conics by the Discriminant

The graph of the equation $Ax^2 + Bxy + Cy^2 + Dx + Ey + F = 0$ is, except in degenerate cases, determined by its discriminant as follows.

1. *Ellipse or circle:* $B^2 - 4AC < 0$

2. *Parabola:* $B^2 - 4AC = 0$

3. *Hyperbola:* $B^2 - 4AC > 0$

If the *xy*-term exists, your students should be able to recognize immediately the fact that rotation occurs. They should then use the discriminant to classify the conic before rotating the axes.

For example, in the general equation

$$3x^2 + 7xy + 5y^2 - 6x - 7y + 15 = 0$$

you have $A = 3$, $B = 7$, and $C = 5$. So the discriminant is

$$B^2 - 4AC = 7^2 - 4(3)(5) = 49 - 60 = -11.$$

Because $-11 < 0$, the graph of the equation is an ellipse or a circle.

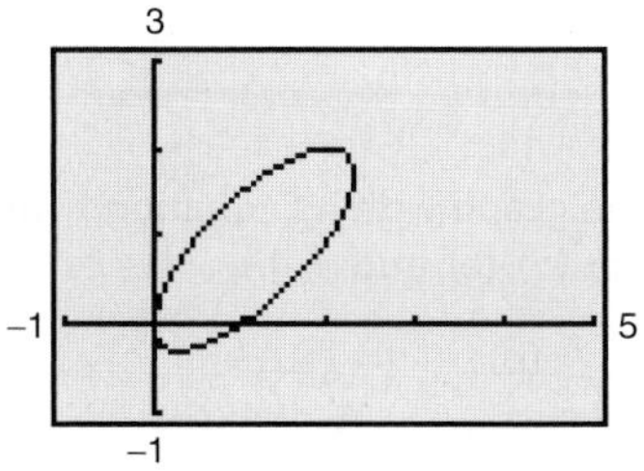

FIGURE 10.41

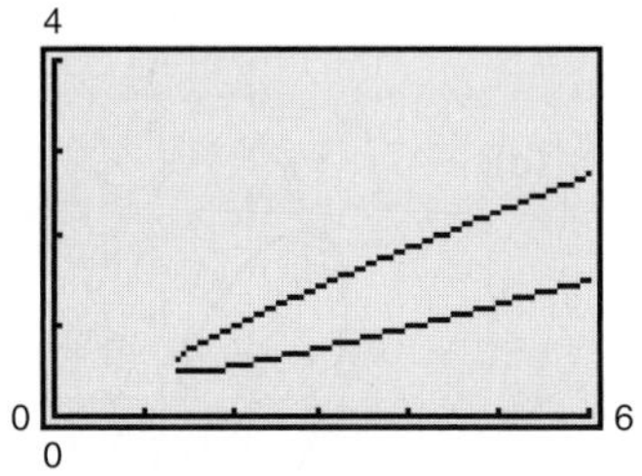

FIGURE 10.42

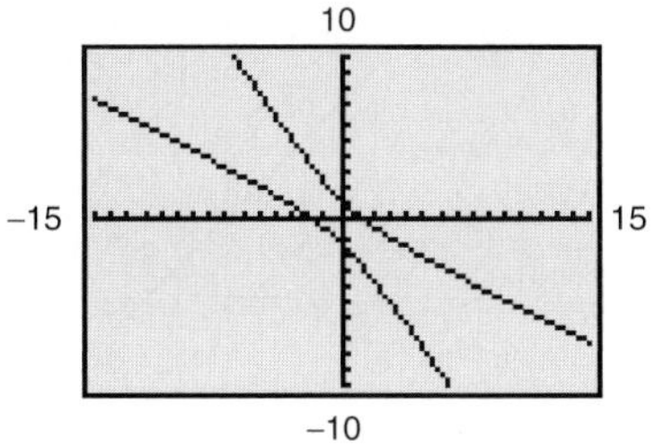

FIGURE 10.43

Example 4 ▶ Rotations and Graphing Utilities

For each of the following, classify the graph, use the Quadratic Formula to solve for y, and then use a graphing utility to graph the equation.

a. $2x^2 - 3xy + 2y^2 - 2x = 0$ **b.** $x^2 - 6xy + 9y^2 - 2y + 1 = 0$

c. $3x^2 + 8xy + 4y^2 - 7 = 0$

Solution

a. Because $B^2 - 4AC = 9 - 16 < 0$, the graph is a circle or an ellipse. Solve for y as follows.

$$2x^2 - 3xy + 2y^2 - 2x = 0 \quad \text{Write original equation.}$$

$$2y^2 - 3xy + (2x^2 - 2x) = 0 \quad \text{Quadratic form } ay^2 + by + c = 0$$

$$y = \frac{-(-3x) \pm \sqrt{(-3x)^2 - 4(2)(2x^2 - 2x)}}{2(2)}$$

Graph both of the equations to obtain the ellipse in Figure 10.41.

$$y = \frac{3x + \sqrt{9x^2 - 16(x^2 - x)}}{4} \quad \text{Top half of ellipse}$$

$$y = \frac{3x - \sqrt{9x^2 - 16(x^2 - x)}}{4} \quad \text{Bottom half of ellipse}$$

b. Because $B^2 - 4AC = 36 - 36 = 0$, the graph is a parabola.

$$x^2 - 6xy + 9y^2 - 2y + 1 = 0 \quad \text{Write original equation.}$$

$$9y^2 - (6x + 2)y + (x^2 + 1) = 0 \quad \text{Quadratic form } ay^2 + by + c = 0$$

$$y = \frac{(6x + 2) \pm \sqrt{(6x + 2)^2 - 4(9)(x^2 + 1)}}{18}$$

Graphing the resulting two equations gives the parabola in Figure 10.42.

c. Because $B^2 - 4AC = 64 - 48 > 0$, the graph is a hyperbola.

$$3x^2 + 8xy + 4y^2 - 7 = 0 \quad \text{Write original equation.}$$

$$4y^2 + 8xy + (3x^2 - 7) = 0 \quad \text{Quadratic form } ay^2 + by + c = 0$$

$$y = \frac{-8x \pm \sqrt{(8x)^2 - 4(4)(3x^2 - 7)}}{8}$$

The graphs of the resulting two equations yield the hyperbola in Figure 10.43.

Activities

1. Use the discriminant to classify each equation.

a. $x^2 - 4xy + y^2 + 2x - 3y + 1 = 0$
b. $4x^2 - xy + 4y^2 + 2x - 3y + 1 = 0$
c. $x^2 - xy + 4y^2 + 2x - 3y + 1 = 0$
d. $x^2 - 4xy + 4y^2 + 2x - 3y + 1 = 0$

Answer: (a) Hyperbola, (b) Ellipse or circle, (c) Ellipse or circle, (d) Parabola

2. Find the angle of rotation necessary to eliminate the xy-term in the equation $3x^2 + 2xy - y^2 + x - 1 = 0$.

Answer: $\theta \approx 13.28°$

3. Rotate the axes to eliminate the xy-term in the equation $x^2 + 2xy + y^2 - \sqrt{2}y = 0$.

Answer: $2(x')^2 - x' - y' = 0, \theta = \frac{\pi}{4}$

Writing ABOUT MATHEMATICS

Classifying a Graph as a Hyperbola In Section 2.6, it was mentioned that the graph of $f(x) = 1/x$ is a hyperbola. Discuss how you could use the techniques in this section to verify this, and then do so. Compare your statement with that of another student.

10.5 Exercises

In Exercises 1–6, the $x'y'$-coordinate system has been rotated θ degrees from the xy-coordinate system. The coordinates of a point in the xy-coordinate system are given. Find the coordinates of the point in the rotated coordinate system.

1. $\theta = 90°, (0, 3)$

2. $\theta = 45°, (3, 3)$

3. $\theta = 30°, (1, 3)$

4. $\theta = 60°, (3, 1)$

5. $\theta = 45°, (2, 1)$

6. $\theta = 30°, (2, 4)$

In Exercises 7–18, rotate the axes to eliminate the xy-term. Sketch the graph of the resulting equation, showing both sets of axes.

7. $xy + 1 = 0$

8. $xy - 2 = 0$

9. $x^2 - 2xy + y^2 - 1 = 0$

10. $xy + x - 2y + 3 = 0$

11. $xy - 2y - 4x = 0$

12. $2x^2 - 3xy - 2y^2 + 10 = 0$

13. $5x^2 - 6xy + 5y^2 - 12 = 0$

14. $13x^2 + 6\sqrt{3}xy + 7y^2 - 16 = 0$

15. $3x^2 - 2\sqrt{3}xy + y^2 + 2x + 2\sqrt{3}y = 0$

16. $16x^2 - 24xy + 9y^2 - 60x - 80y + 100 = 0$

17. $9x^2 + 24xy + 16y^2 + 90x - 130y = 0$

18. $9x^2 + 24xy + 16y^2 + 80x - 60y = 0$

In Exercises 19–26, use a graphing utility to graph the conic. Determine the angle θ through which the axes are rotated. Explain how you used the graphing utility to obtain the graph.

19. $x^2 + 2xy + y^2 = 20$

20. $x^2 - 4xy + 2y^2 = 6$

21. $17x^2 + 32xy - 7y^2 = 75$

22. $40x^2 + 36xy + 25y^2 = 52$

23. $32x^2 + 48xy + 8y^2 = 50$

24. $24x^2 + 18xy + 12y^2 = 34$

25. $4x^2 - 12xy + 9y^2 + (4\sqrt{13} - 12)x - (6\sqrt{13} + 8)y = 91$

26. $6x^2 - 4xy + 8y^2 + (5\sqrt{5} - 10)x - (7\sqrt{5} + 5)y = 80$

In Exercises 27–32, match the graph with its equation. [The graphs are labeled (a), (b), (c), (d), (e), and (f).]

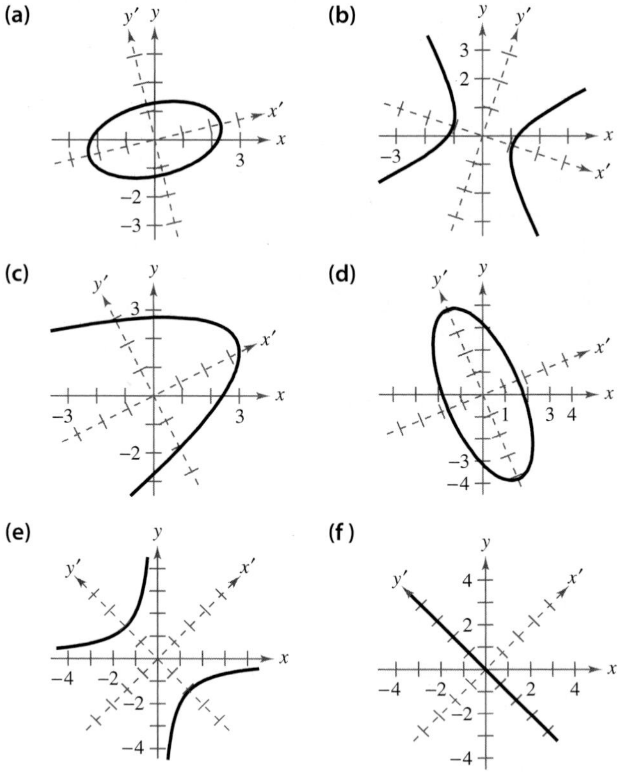

27. $xy + 2 = 0$

28. $x^2 + 2xy + y^2 = 0$

29. $-2x^2 + 3xy + 2y^2 + 3 = 0$

30. $x^2 - xy + 3y^2 - 5 = 0$

31. $3x^2 + 2xy + y^2 - 10 = 0$

32. $x^2 - 4xy + 4y^2 + 10x - 30 = 0$

In Exercises 33–40, use the discriminant to classify the graph, use the Quadratic Formula to solve for y, and then use a graphing utility to graph the equation.

33. $16x^2 - 8xy + y^2 - 10x + 5y = 0$

34. $x^2 - 4xy - 2y^2 - 6 = 0$

35. $12x^2 - 6xy + 7y^2 - 45 = 0$

36. $2x^2 + 4xy + 5y^2 + 3x - 4y - 20 = 0$

37. $x^2 - 6xy - 5y^2 + 4x - 22 = 0$

38. $36x^2 - 60xy + 25y^2 + 9y = 0$

39. $x^2 + 4xy + 4y^2 - 5x - y - 3 = 0$

40. $x^2 + xy + 4y^2 + x + y - 4 = 0$

In Exercises 41–44, sketch (if possible) the graph of the degenerate conic.

41. $y^2 - 9x^2 = 0$

42. $x^2 + y^2 - 2x + 6y + 10 = 0$

43. $x^2 + 2xy + y^2 - 1 = 0$

44. $x^2 - 10xy + y^2 = 0$

In Exercises 45–58, find any points of intersection of the graphs algebraically and then verify using a graphing utility.

45. $-x^2 + y^2 + 4x - 6y + 4 = 0$
$x^2 + y^2 - 4x - 6y + 12 = 0$

46. $-x^2 - y^2 - 8x + 20y - 7 = 0$
$x^2 + 9y^2 + 8x + 4y + 7 = 0$

47. $-4x^2 - y^2 - 16x + 24y - 16 = 0$
$4x^2 + y^2 + 40x - 24y + 208 = 0$

48. $x^2 - 4y^2 - 20x - 64y - 172 = 0$
$16x^2 + 4y^2 - 320x + 64y + 1600 = 0$

49. $x^2 - y^2 - 12x + 16y - 64 = 0$
$x^2 + y^2 - 12x - 16y + 64 = 0$

50. $x^2 + 4y^2 - 2x - 8y + 1 = 0$
$-x^2 + 2x - 4y - 1 = 0$

51. $-16x^2 - y^2 + 24y - 80 = 0$
$16x^2 + 25y^2 - 400 = 0$

52. $16x^2 - y^2 + 16y - 128 = 0$
$y^2 - 48x - 16y - 32 = 0$

53. $x^2 + y^2 - 4 = 0$
$3x - y^2 = 0$

54. $4x^2 + 9y^2 - 36y = 0$
$x^2 + 9y - 27 = 0$

55. $x^2 + 2y^2 - 4x + 6y - 5 = 0$
$-x + y - 4 = 0$

56. $x^2 + 2y^2 - 4x + 6y - 5 = 0$
$x^2 - 4x - y + 4 = 0$

57. $xy + x - 2y + 3 = 0$
$x^2 + 4y^2 - 9 = 0$

58. $5x^2 - 2xy + 5y^2 - 12 = 0$
$x + y - 1 = 0$

Synthesis

True or False? **In Exercises 59 and 60, determine whether the statement is true or false. Justify your answer.**

59. The graph of the equation

$$x^2 + xy + ky^2 + 6x + 10 = 0$$

where k is any constant less than $\frac{1}{4}$, is a hyperbola.

60. After a rotation of axes is used to eliminate the xy-term from an equation of the form

$$Ax^2 + Bxy + Cy^2 + Dx + Ey + F = 0$$

the coefficients of the x^2- and y^2-terms remain A and C, respectively.

61. Show that the equation $x^2 + y^2 = r^2$ is invariant under rotation of axes.

62. Find the lengths of the major and minor axes of the ellipse graphed in Exercise 14.

Review

In Exercises 63–68, find the zeros (if any) of the rational function.

63. $f(x) = \dfrac{x^2 - 9}{x + 1}$

64. $f(x) = \dfrac{3}{x^2 + 1}$

65. $f(x) = 4 - \dfrac{8}{x^2 - 2}$

66. $f(x) = \dfrac{x^3 - 27}{x^2 - 1}$

67. $f(x) = \dfrac{x^2 + 4x + 4}{x^2 + 3}$

68. $f(x) = \dfrac{x^2 - 1}{2x^2 + 3x - 2}$

In Exercises 69–76, graph the rational function. Identify all intercepts and asymptotes.

69. $g(x) = \dfrac{2}{2 - x}$

70. $f(x) = \dfrac{2x}{2 - x}$

71. $g(t) = \dfrac{2}{1 + t}$

72. $h(t) = \dfrac{2t}{2 + t}$

73. $g(x) = \dfrac{x^2 - 2x - 3}{x - 2}$

74. $f(x) = \dfrac{x^2}{2 - x}$

75. $h(s) = \dfrac{2}{4 - s^2}$

76. $f(t) = \dfrac{t}{t^2 - t - 6}$

10.6 Parametric Equations

▶ **What you should learn**

- How to evaluate a set of parametric equations for a given value of the parameter
- How to sketch the curve that is represented by a set of parametric equations
- How to rewrite a set of parametric equations as a single rectangular equation
- How to find a set of parametric equations for a graph

▶ **Why you should learn it**

Parametric equations are useful for modeling the path of an object. For instance, in Exercise 60 on page 819, you will use a set of parametric equations to model the path of an arrow.

Dennis O'Clair/Tony Stone Images

A computer animation of this concept appears in the *Interactive* CD-ROM and *Internet* versions of this text.

Plane Curves

Up to this point you have been representing a graph by a single equation involving the *two* variables x and y. In this section, you will study situations in which it is useful to introduce a *third* variable to represent a curve in the plane.

To see the usefulness of this procedure, consider the path followed by an object that is propelled into the air at an angle of 45°. If the initial velocity of the object is 48 feet per second, it can be shown that the object follows the parabolic path

$$y = -\frac{x^2}{72} + x \qquad \text{Rectangular equation}$$

as shown in Figure 10.44. However, this equation does not tell the whole story. Although it does tell you *where* the object has been, it doesn't tell you *when* the object was at a given point (x, y) on the path. To determine this time, you can introduce a third variable t, called a **parameter.** It is possible to write both x and y as functions of t to obtain the **parametric equations**

$$x = 24\sqrt{2}t \qquad \text{Parametric equation for } x$$

$$y = -16t^2 + 24\sqrt{2}t. \qquad \text{Parametric equation for } y$$

From this set of equations you can determine that at time $t = 0$, the object is at the point $(0, 0)$. Similarly, at time $t = 1$, the object is at the point $\left(24\sqrt{2}, 24\sqrt{2} - 16\right)$, and so on.

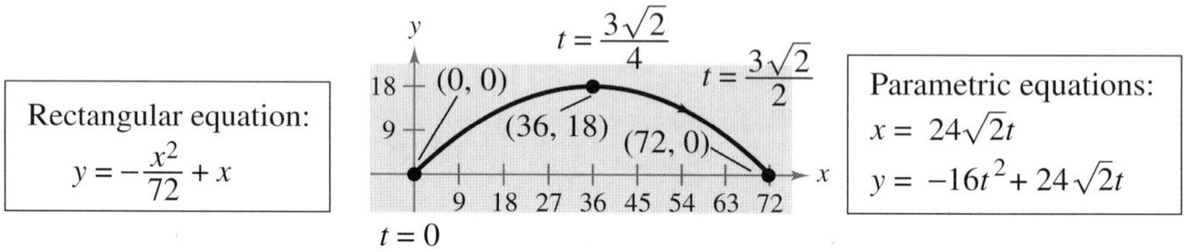

FIGURE 10.44 *Curvilinear Motion: Two Variables for Position, One Variable for Time*

For this particular motion problem, x and y are continuous functions of t, and the resulting path is a **plane curve.** (For this text, it is sufficient to think of a *continuous function* as one whose graph can be traced without lifting the pencil from the paper.)

Definition of Plane Curve

If f and g are continuous functions of t on an interval I, the set of ordered pairs $(f(t), g(t))$ is a **plane curve** C. The equations

$$x = f(t) \qquad \text{and} \qquad y = g(t)$$

are **parametric equations** for C, and t is the **parameter.**

Point out to your students the importance of knowing the orientation of a curve, and thus the usefulness of parametric equations.

Sketching a Plane Curve

When sketching a curve represented by a pair of parametric equations, you still plot points in the xy-plane. Each set of coordinates (x, y) is determined from a value chosen for the parameter t. Plotting the resulting points in the order of *increasing* values of t traces the curve in a specific direction. This is called the **orientation** of the curve.

A computer animation of this example appears in the *Interactive* CD-ROM and *Internet* versions of this text.

Example 1 ▶ Sketching a Curve

Sketch the curve described by the parametric equations

$$x = t^2 - 4 \quad \text{and} \quad y = \frac{t}{2}, \quad -2 \le t \le 3.$$

Solution

Using values of t in the given interval, the parametric equations yield the points (x, y) shown in the table.

t	-2	-1	0	1	2	3
x	0	-3	-4	-3	0	5
y	-1	$-1/2$	0	$1/2$	1	$3/2$

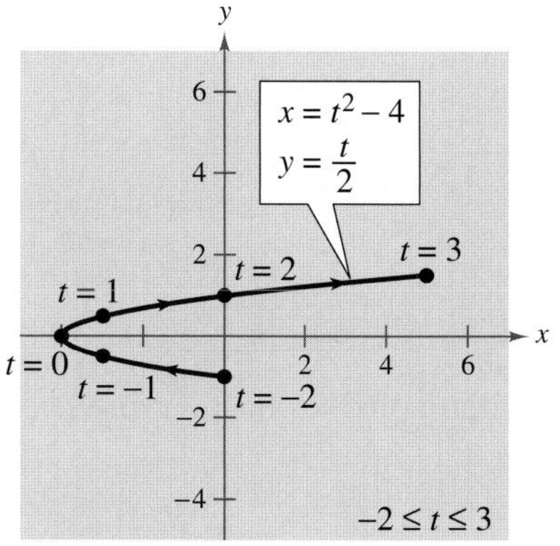

FIGURE 10.45

By plotting these points in the order of increasing t, you obtain the curve C shown in Figure 10.45. Note that the arrows on the curve indicate its orientation as t increases from -2 to 3.

Note that the graph shown in Figure 10.45 does not define y as a function of x. This points out one benefit of parametric equations—they can be used to represent graphs that are more general than graphs of functions.

It often happens that two different sets of parametric equations have the same graph. For example, the set of parametric equations

$$x = 4t^2 - 4 \quad \text{and} \quad y = t, \quad -1 \le t \le \frac{3}{2}$$

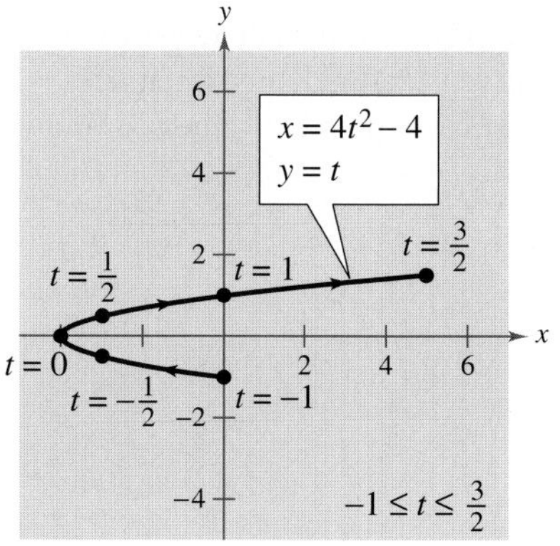

FIGURE 10.46

has the same graph as the set given in Example 1. However, by comparing the values of t in Figures 10.45 and 10.46, you see that this second graph is traced out more *rapidly* (considering t as time) than the first graph. So, in applications, different parametric representations can be used to represent various *speeds* at which objects travel along a given path.

Eliminating the Parameter

Example 1 uses simple point plotting to sketch the given curve. This tedious process can sometimes be simplified by finding a rectangular equation (in x and y) that has the same graph. This process is called **eliminating the parameter.**

Parametric equations	⇨	Solve for t in one equation.	⇨	Substitute in other equation.	⇨	Rectangular equation
$x = t^2 - 4$ $y = t/2$		$t = 2y$		$x = (2y)^2 - 4$		$x = 4y^2 - 4$

Now you can recognize that the equation $x = 4y^2 - 4$ represents a parabola with a horizontal axis and a vertex at $(-4, 0)$.

When converting equations from parametric to rectangular form, you may need to alter the domain of the rectangular equation so that its graph matches the graph of the parametric equations. Such a situation is demonstrated in Example 2.

Example 2 ▶ Eliminating the Parameter

Emphasize that converting equations from parametric to rectangular form is primarily an aid in graphing.

Sketch the curve represented by the equations $x = 1/\sqrt{t+1}$ and $y = t/(t+1)$ by eliminating the parameter and adjusting the domain of the resulting rectangular equation.

Solution

Solving for t in the equation for x, you have

$$x = \frac{1}{\sqrt{t+1}} \quad \Rightarrow \quad x^2 = \frac{1}{t+1}$$

which implies that $t = (1 - x^2)/x^2$. Now, substituting in the equation for y, you obtain

$$y = \frac{t}{t+1} = \frac{\dfrac{(1-x^2)}{x^2}}{\left[\dfrac{(1-x^2)}{x^2}\right] + 1} = \frac{\dfrac{1-x^2}{x^2}}{\dfrac{1-x^2}{x^2} + 1} \cdot \frac{x^2}{x^2} = 1 - x^2.$$

The rectangular equation, $y = 1 - x^2$, is defined for all values of x, but from the parametric equation for x you can see that the curve is defined only when $t > -1$. This implies that you should restrict the domain of x to positive values, as shown in Figure 10.47.

Exploration

Most graphing utilities have a parametric graphing mode. If yours does, try entering the parametric equations given in Example 2. Over what values should you let t vary to obtain the graph shown in Figure 10.47?

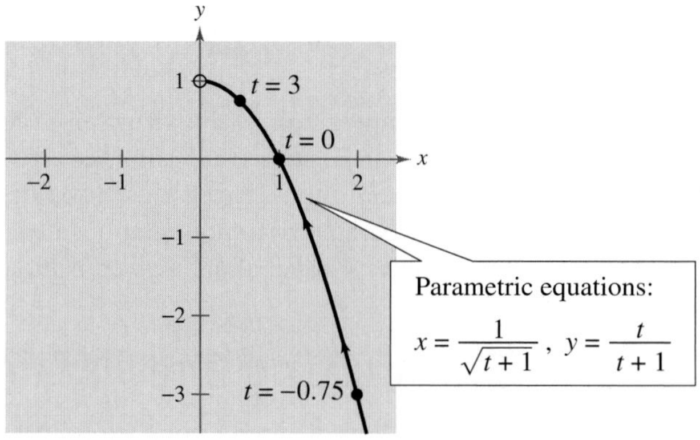

FIGURE 10.47

STUDY TIP

To eliminate the parameter in equations involving trigonometric functions, try using the identities

$$\sin^2 \theta + \cos^2 \theta = 1$$

$$\sec^2 \theta - \tan^2 \theta = 1$$

as done in Example 3.

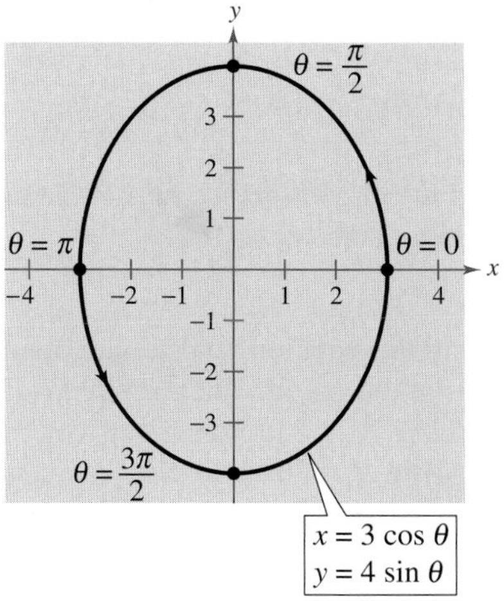

FIGURE 10.48

It is not necessary for the parameter in a set of parametric equations to represent time. The next example uses an *angle* as the parameter.

Example 3 ▶ Eliminating the Parameter

Sketch the curve represented by

$$x = 3 \cos \theta \qquad \text{and} \qquad y = 4 \sin \theta, \qquad 0 \le \theta \le 2\pi$$

by eliminating the parameter.

Solution

Begin by solving for $\cos \theta$ and $\sin \theta$ in the equations.

$$\cos \theta = \frac{x}{3} \qquad \text{and} \qquad \sin \theta = \frac{y}{4} \qquad \text{Solve for } \cos \theta \text{ and } \sin \theta$$

Make use of the identity $\sin^2 \theta + \cos^2 \theta = 1$ to form an equation involving only x and y.

$$\cos^2 \theta + \sin^2 \theta = 1 \qquad \text{Trigonometric identity}$$

$$\left(\frac{x}{3}\right)^2 + \left(\frac{y}{4}\right)^2 = 1 \qquad \text{Substitute } \frac{x}{3} \text{ for } \cos \theta \text{ and } \frac{y}{4} \text{ for } \sin \theta.$$

$$\frac{x^2}{9} + \frac{y^2}{16} = 1 \qquad \text{Rectangular equation}$$

From this rectangular equation, you can see that the graph is an ellipse centered at $(0, 0)$, with vertices at $(0, 4)$ and $(0, -4)$ and minor axis of length $2b = 6$, as shown in Figure 10.48. Note that the elliptic curve is traced out *counterclockwise* as θ varies from 0 to 2π.

In Examples 2 and 3 it is important to realize that eliminating the parameter is primarily an *aid to curve sketching.* If the parametric equations represent the path of a moving object, the graph alone is not sufficient to describe the object's motion. You still need the parametric equations to tell you the *position, direction,* and *speed* at a given time.

Finding Parametric Equations for a Graph

You have been studying techniques for sketching the graph represented by a set of parametric equations. Now consider the reverse problem—that is, how can you find a set of parametric equations for a given graph or a given physical description? From the discussion following Example 1, you know that such a representation is not unique. That is, the equations

$$x = 4t^2 - 4 \qquad \text{and} \qquad y = t, \ -1 \le t \le \frac{3}{2}$$

produced the same graph as the equations

$$x = t^2 - 4 \qquad \text{and} \qquad y = \frac{t}{2}, \ -2 \le t \le 3.$$

This is further demonstrated in Example 4.

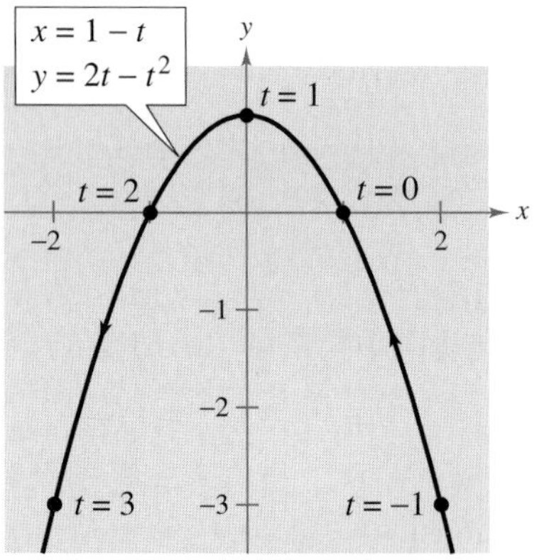

FIGURE 10.49

Point out that a single rectangular equation can have many different parametric representations. To reinforce this, demonstrate along with parts (a) and (b) of Example 4 the parametric equation representations of the graph of $y = 1 - x^2$ using the parameters $t = 2x$ and $t = 2 - 3x$. A graphing utility can be a helpful tool in demonstrating that each of these representations yields the same graph.

Example 4 ▶ Finding Parametric Equations for a Given Graph

Find a set of parametric equations to represent the graph of $y = 1 - x^2$, using the following parameters.

a. $t = x$ **b.** $t = 1 - x$

Solution

a. Letting $t = x$, you obtain the parametric equations

$$x = t \quad \text{and} \quad y = 1 - x^2 = 1 - t^2.$$

b. Letting $t = 1 - x$, you obtain

$$x = 1 - t \quad \text{and} \quad y = 1 - (1 - t)^2 = 2t - t^2.$$

In Figure 10.49, note how the resulting curve is oriented by the increasing values of t. For part (a), the curve would have the opposite orientation.

Example 5 ▶ Parametric Equations for a Cycloid

Describe the **cycloid** traced out by a point P on the circumference of a circle of radius a as the circle rolls along a straight line in a plane.

Solution

As the parameter, let θ be the measure of the circle's rotation, and let the point $P = (x, y)$ begin at the origin. When $\theta = 0$, P is at the origin; when $\theta = \pi$, P is at a maximum point $(\pi a, 2a)$; and when $\theta = 2\pi$, P is back on the x-axis at $(2\pi a, 0)$. From Figure 10.50, you can see that $\angle APC = 180° - \theta$. So, you have

$$\sin\theta = \sin(180° - \theta) = \sin(\angle APC) = \frac{AC}{a} = \frac{BD}{a}$$

$$\cos\theta = -\cos(180° - \theta) = -\cos(\angle APC) = \frac{AP}{-a}$$

which implies that $AP = -a\cos\theta$ and $BD = a\sin\theta$. Because the circle rolls along the x-axis, you know that $OD = \overset{\frown}{PD} = a\theta$. Furthermore, because $BA = DC = a$, you have

$$x = OD - BD = a\theta - a\sin\theta \quad \text{and} \quad y = BA + AP = a - a\cos\theta.$$

Therefore, the parametric equations are

$$x = a(\theta - \sin\theta) \quad \text{and} \quad y = a(1 - \cos\theta).$$

STUDY TIP

In Example 5, $\overset{\frown}{PD}$ represents the arc of the circle between points P and D.

Technology

Use a graphing utility in parametric mode to obtain a graph similar to Figure 10.50 by graphing the following equations.

$X_{1T} = T - \sin T$

$Y_{1T} = 1 - \cos T$

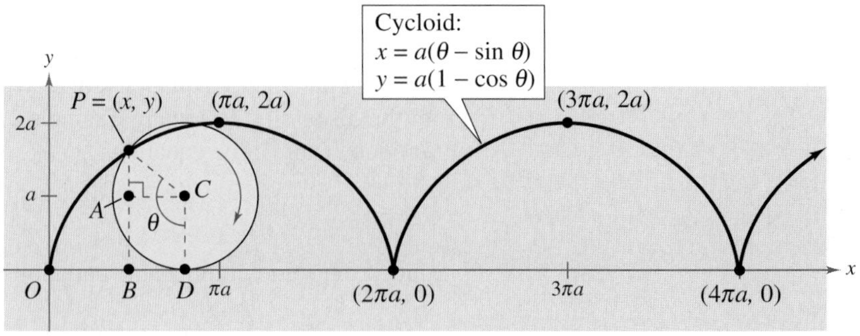

FIGURE 10.50

A computer animation of this example appears in the *Interactive* CD-ROM and *Internet* versions of this text.

10.6 Exercises

1. Consider the parametric equations $x = \sqrt{t}$ and $y = 3 - t$.
   (a) Create a table of x- and y-values using $t = 0, 1, 2, 3,$ and 4.
   (b) Plot the points (x, y) generated in part (a), and sketch a graph of the parametric equations.
   (c) Find the rectangular equation by eliminating the parameter. Sketch its graph. How do the graphs differ?

2. Consider the parametric equations $x = 4\cos^2\theta$ and $y = 2\sin\theta$.
   (a) Create a table of x- and y-values using $\theta = -\pi/2, -\pi/4, 0, \pi/4,$ and $\pi/2$.
   (b) Plot the points (x, y) generated in part (a), and sketch a graph of the parametric equations.
   (c) Find the rectangular equation by eliminating the parameter. Sketch its graph. How do the graphs differ?

In Exercises 3–22, sketch the curve represented by the parametric equations (indicate the direction of the curve) by eliminating the parameter and adjusting the domain of the resulting rectangular equation.

3. $x = 3t - 3$, $y = 2t + 1$
4. $x = 3 - 2t$, $y = 2 + 3t$
5. $x = \frac{1}{4}t$, $y = t^2$
6. $x = t$, $y = t^3$
7. $x = t + 2$, $y = t^2$
8. $x = \sqrt{t}$, $y = 1 - t$
9. $x = t + 1$, $y = t/(t + 1)$
10. $x = t - 1$, $y = t/(t - 1)$
11. $x = 2(t + 1)$, $y = |t - 2|$
12. $x = |t - 1|$, $y = t + 2$
13. $x = 3\cos\theta$, $y = 3\sin\theta$
14. $x = 2\cos\theta$, $y = 3\sin\theta$
15. $x = 4\sin 2\theta$, $y = 2\cos 2\theta$
16. $x = \cos\theta$, $y = 2\sin 2\theta$
17. $x = 4 + 2\cos\theta$, $y = -1 + \sin\theta$
18. $x = 4 + 2\cos\theta$, $y = 2 + 3\sin\theta$
19. $x = e^{-t}$, $y = e^{3t}$
20. $x = e^{2t}$, $y = e^t$
21. $x = t^3$, $y = 3\ln t$
22. $x = \ln 2t$, $y = 2t^2$

In Exercises 23 and 24, determine how the plane curves differ from each other.

23. (a) $x = t$, $y = 2t + 1$ (b) $x = \cos\theta$, $y = 2\cos\theta + 1$
    (c) $x = e^{-t}$, $y = 2e^{-t} + 1$ (d) $x = e^t$, $y = 2e^t + 1$

24. (a) $x = t$, $y = t^2 - 1$ (b) $x = t^2$, $y = t^4 - 1$
    (c) $x = \sin t$, $y = \sin^2 t - 1$ (d) $x = e^t$, $y = e^{2t} - 1$

In Exercises 25–28, eliminate the parameter and obtain the standard form of the rectangular equation.

25. Line through (x_1, y_1) and (x_2, y_2):
    $x = x_1 + t(x_2 - x_1),\ y = y_1 + t(y_2 - y_1)$
26. Circle:
    $x = h + r\cos\theta,\ y = k + r\sin\theta$
27. Ellipse:
    $x = h + a\cos\theta,\ y = k + b\sin\theta$
28. Hyperbola:
    $x = h + a\sec\theta,\ y = k + b\tan\theta$

In Exercises 29–36, use the results of Exercises 25–28 to find a set of parametric equations for the line or conic.

29. Line: Passes through $(0, 0)$ and $(6, -3)$
30. Line: Passes through $(2, 3)$ and $(6, -3)$
31. Circle: Center: $(3, 2)$; Radius: 4
32. Circle: Center: $(-3, 2)$; Radius: 5
33. Ellipse: Vertices: $(\pm 4, 0)$; Foci: $(\pm 3, 0)$
34. Ellipse: Vertices: $(4, 7), (4, -3)$; Foci: $(4, 5), (4, -1)$
35. Hyperbola: Vertices: $(\pm 4, 0)$; Foci: $(\pm 5, 0)$
36. Hyperbola: Vertices: $(0, \pm 2)$; Foci: $(0, \pm 4)$

In Exercises 37–44, find a set of parametric equations for the given rectangular equation using (a) $t = x$ and (b) $t = 2 - x$.

37. $y = 3x - 2$

38. $x = 3y - 2$

39. $y = x^2$

40. $y = x^3$

41. $y = x^2 + 1$

42. $y = 2 - x$

43. $y = \dfrac{1}{x}$

44. $y = \dfrac{1}{2x}$

In Exercises 45–52, use a graphing utility to obtain a graph of the curve represented by the parametric equations.

45. Cycloid: $x = 4(\theta - \sin\theta),\ y = 4(1 - \cos\theta)$

46. Cycloid: $x = \theta + \sin\theta,\ y = 1 - \cos\theta$

47. Prolate cycloid: $x = \theta - \frac{3}{2}\sin\theta,\ y = 1 - \frac{3}{2}\cos\theta$

48. Prolate cycloid: $x = 2\theta - 4\sin\theta,\ y = 2 - 4\cos\theta$

49. Hypocycloid: $x = 3\cos^3\theta,\ y = 3\sin^3\theta$

50. Curtate cycloid: $x = 8\theta - 4\sin\theta,\ y = 8 - 4\cos\theta$

51. Witch of Agnesi: $x = 2\cot\theta,\ y = 2\sin^2\theta$

52. Folium of Descartes: $x = \dfrac{3t}{1 + t^3},\ y = \dfrac{3t^2}{1 + t^3}$

In Exercises 53–56, match the parametric equations with the correct graph and describe the domain and range. [The graphs are labeled (a), (b), (c), and (d).]

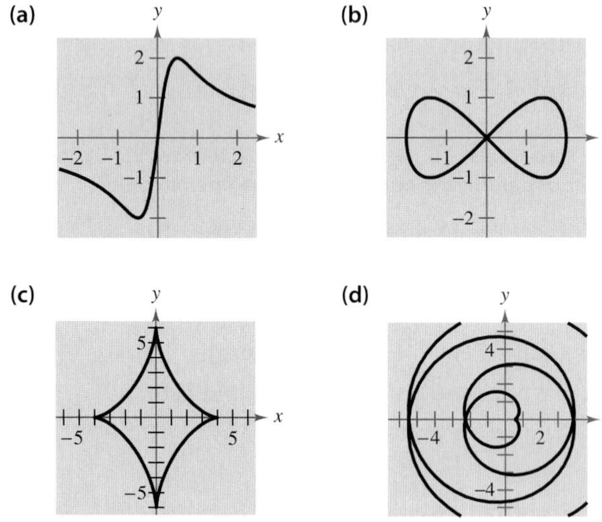

53. Lissajous curve: $x = 2\cos\theta$
$y = \sin 2\theta$

54. Evolute of ellipse: $x = 4\cos^3\theta$
$y = 6\sin^3\theta$

55. Involute of circle: $x = \frac{1}{2}(\cos\theta + \theta\sin\theta)$
$y = \frac{1}{2}(\sin\theta - \theta\cos\theta)$

56. Serpentine curve: $x = \frac{1}{2}\cot\theta$
$y = 4\sin\theta\cos\theta$

Projectile Motion **A projectile is launched at a height of h feet above the ground and at an angle θ with the horizontal. If the initial velocity is v_0 feet per second, the path of the projectile is modeled by the parametric equations**

$$x = (v_0 \cos\theta)t \quad \text{and} \quad y = h + (v_0 \sin\theta)t - 16t^2.$$

In Exercises 57 and 58, use a graphing utility to graph the paths of a projectile launched from ground level at the specified values of θ and v_0. For each case, use the graph to approximate the maximum height and the range of the projectile.

57. (a) $\theta = 60°$, $v_0 = 88$ ft/sec
(b) $\theta = 60°$, $v_0 = 132$ ft/sec
(c) $\theta = 45°$, $v_0 = 88$ ft/sec
(d) $\theta = 45°$, $v_0 = 132$ ft/sec

58. (a) $\theta = 15°$, $v_0 = 60$ ft/sec
(b) $\theta = 15°$, $v_0 = 100$ ft/sec
(c) $\theta = 30°$, $v_0 = 60$ ft/sec
(d) $\theta = 30°$, $v_0 = 100$ ft/sec

59. ***Baseball*** The center field fence in a ballpark is 10 feet high and 400 feet from home plate. The baseball is hit 3 feet above the ground. It leaves the bat at an angle of θ degrees with the horizontal at a speed of 100 miles per hour (see figure).

(a) Write a set of parametric equations for the path of the baseball.

(b) Use a graphing utility to sketch the path of the baseball if $\theta = 15°$. Is the hit a home run?

(c) Use a graphing utility to sketch the path of the baseball if $\theta = 23°$. Is the hit a home run?

(d) Find the minimum angle required for the hit to be a home run.

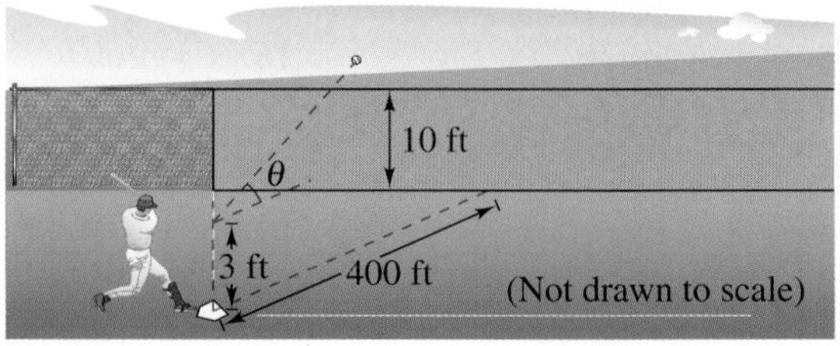

60. *Archery* An archer releases an arrow from a bow 5 feet above the ground. The arrow leaves the bow at an angle of 10° with the horizontal and at an initial speed of 240 feet per second.

(a) Write a set of parametric equations for the path of the arrow.

(b) Assuming the ground is level, find the distance the arrow travels before it hits the ground. (Ignore air resistance.)

(c) Use a graphing utility to graph the path of the arrow and approximate its maximum height.

(d) Find the time the arrow is in the air.

61. *Projectile Motion* Eliminate the parameter t from the position function for the motion of a projectile to show that the rectangular equation is

$$y = -\frac{16\sec^2\theta}{v_0^2}x^2 + (\tan\theta)x + h.$$

62. *Path of a Projectile* The path of a projectile is given by the rectangular equation

$$y = 7 + x - 0.02x^2.$$

(a) Use the result of Exercise 61 to find h, v_0, and θ. Find the parametric equations of the path.

(b) Use a graphing utility to graph the rectangular equation for the path of the projectile. Confirm your answer in part (a) by sketching the curve represented by the parametric equations.

(c) Use a graphing utility to approximate the maximum height of the projectile and its range.

63. *Curtate Cycloid* A wheel of radius a rolls along a straight line without slipping. The curve traced by a point P that is b units from the center $(b < a)$ is called a **curtate cycloid** (see figure). Use the angle θ shown in the figure to find a set of parametric equations for the curve.

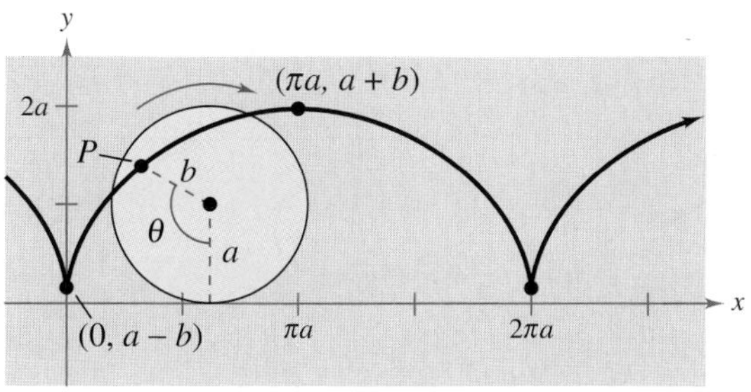

64. *Epicycloid* A circle of radius 1 rolls around the outside of a circle of radius 2 without slipping. The curve traced by a point on the circumference of the smaller circle is called an **epicycloid** (see figure). Use the angle θ shown in the figure to find a set of parametric equations for the curve.

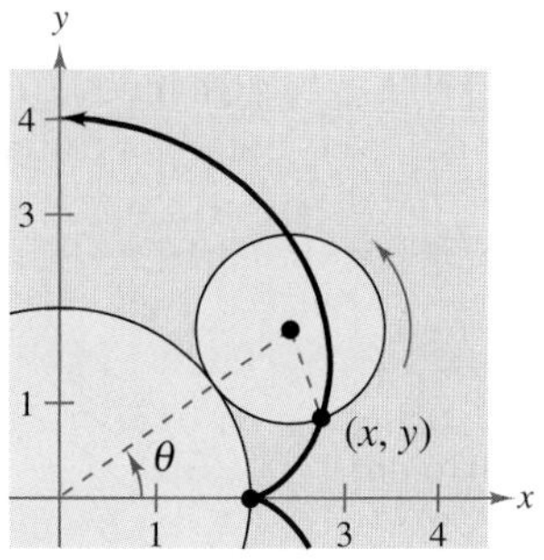

Synthesis

True or False? **In Exercises 65 and 66, determine whether the statement is true or false. Justify your answer.**

65. The two sets of parametric equations $x = t$, $y = t^2 + 1$ and $x = 3t$, $y = 9t^2 + 1$ have the same rectangular equation.

66. The graph of the parametric equations $x = t^2$ and $y = t^2$ is the line $y = x$.

Review

In Exercises 67–70, solve the system of equations.

67. $\begin{cases} 5x - 7y = 11 \\ -3x + y = -13 \end{cases}$

68. $\begin{cases} 3x + 5y = 9 \\ 4x - 2y = -14 \end{cases}$

69. $\begin{cases} 3a - 2b + c = 8 \\ 2a + b - 3c = -3 \\ a - 3b + 9c = 16 \end{cases}$

70. $\begin{cases} 5u + 7v + 9w = 4 \\ u - 2v - 3w = 7 \\ 8u - 2v + w = 20 \end{cases}$

In Exercises 71–74, find the equation of the parabola

$$y = ax^2 + bx + c$$

that passes through the given points.

71. $(0, 0), (3, -3), (6, 0)$

72. $(0, -6), (2, -2), (4, -6)$

73. $(4, 12), (8, -2), (12, 12)$

74. $(-8, -6), (-7, -8), (-6, -6)$

10.7 Polar Coordinates

▶ **What you should learn**

- How to plot points in the polar coordinate system
- How to convert points from rectangular to polar form and vice versa
- How to convert equations from rectangular to polar form and vice versa

▶ **Why you should learn it**

Polar coordinates offer a different mathematical perspective on graphing. For instance, in Exercises 1–8 on page 824, you see that a polar coordinate can be written in more than one way.

A computer animation of this concept appears in the *Interactive* CD-ROM and *Internet* versions of this text.

Introduction

So far, you have been representing graphs of equations as collections of points (x, y) on the rectangular coordinate system, where x and y represent the directed distances from the coordinate axes to the point (x, y). In this section you will study a different system called the **polar coordinate system.**

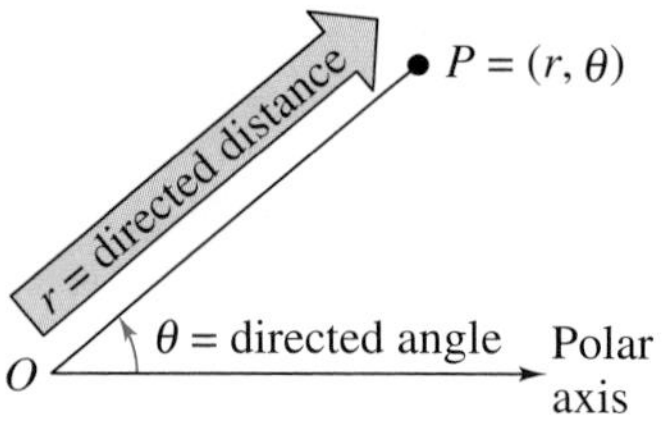

FIGURE 10.51

To form the polar coordinate system in the plane, fix a point O, called the **pole** (or **origin**), and construct from O an initial ray called the **polar axis,** as shown in Figure 10.51. Then each point P in the plane can be assigned **polar coordinates** (r, θ) as follows.

1. r = *directed distance* from O to P
2. θ = *directed angle*, counterclockwise from polar axis to segment $\overline{OP}$

Example 1 ▶ Plotting Points on the Polar Coordinate System

a. The point $(r, \theta) = (2, \pi/3)$ lies 2 units from the pole on the terminal side of the angle $\theta = \pi/3$, as shown in Figure 10.52(a).

b. The point $(r, \theta) = (3, -\pi/6)$ lies 3 units from the pole on the terminal side of the angle $\theta = -\pi/6$, as shown in Figure 10.52(b).

c. The point $(r, \theta) = (3, 11\pi/6)$ coincides with the point $(3, -\pi/6)$, as shown in Figure 10.52(c).

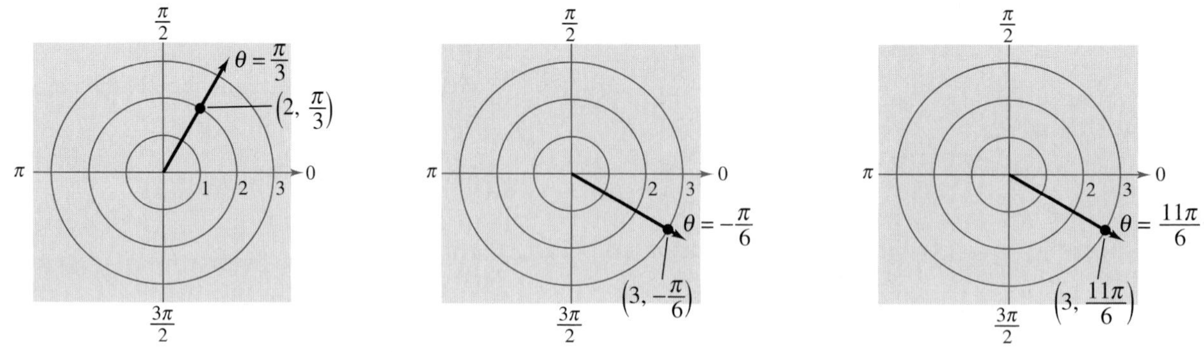

(a) (b) (c)

FIGURE 10.52

Exploration

Most graphing calculators have a *polar* graphing mode. If yours does, try graphing the equation $r = 3$. (Use a setting of $-6 \le x \le 6$ and $-4 \le y \le 4$.) You should obtain a circle of radius 3.

a. Use the trace feature to cursor around the circle. Can you locate the point $(3, 5\pi/4)$?

b. Can you find other polar representations of the point $(3, 5\pi/4)$? If so, explain how you did it.

In rectangular coordinates, each point (x, y) has a unique representation. This is not true for polar coordinates. For instance, the coordinates (r, θ) and $(r, \theta + 2\pi)$ represent the same point, as illustrated in Example 1. Another way to obtain multiple representations of a point is to use negative values for r. Because r is a *directed distance*, the coordinates (r, θ) and $(-r, \theta + \pi)$ represent the same point. In general, the point (r, θ) can be represented as

$$(r, \theta) = (r, \theta \pm 2n\pi) \qquad \text{or} \qquad (r, \theta) = (-r, \theta \pm (2n + 1)\pi)$$

where n is any integer. Moreover, the pole is represented by $(0, \theta)$, where θ is any angle.

Example 2 ▶ Multiple Representation of Points

Plot the point $(3, -3\pi/4)$ and find three additional polar representations of this point, using $-2\pi < \theta < 2\pi$.

Solution

The point is shown in Figure 10.53. Three other representations are as follows.

$$\left(3, -\frac{3\pi}{4} + 2\pi\right) = \left(3, \frac{5\pi}{4}\right) \qquad \text{Add } 2\pi \text{ to } \theta.$$

$$\left(-3, -\frac{3\pi}{4} - \pi\right) = \left(-3, -\frac{7\pi}{4}\right) \qquad \text{Replace } r \text{ by } -r; \text{ subtract } \pi \text{ from } \theta.$$

$$\left(-3, -\frac{3\pi}{4} + \pi\right) = \left(-3, \frac{\pi}{4}\right) \qquad \text{Replace } r \text{ by } -r; \text{ add } \pi \text{ to } \theta.$$

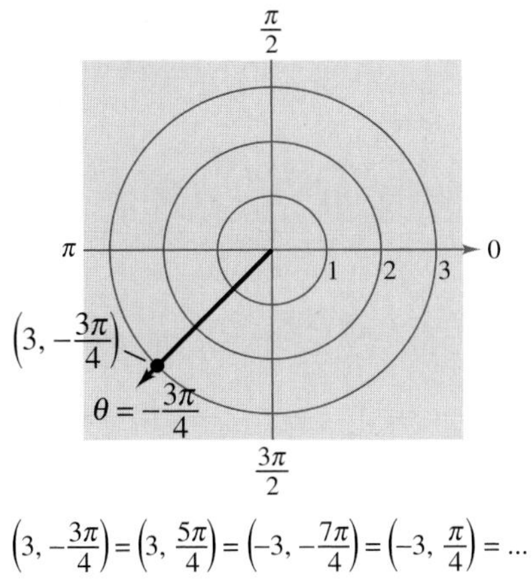

FIGURE 10.53

Coordinate Conversion

To establish the relationship between polar and rectangular coordinates, let the polar axis coincide with the positive x-axis and the pole with the origin, as shown in Figure 10.54. Because (x, y) lies on a circle of radius r, it follows that $r^2 = x^2 + y^2$. Moreover, for $r > 0$, the definitions of the trigonometric functions imply that

$$\tan\theta = \frac{y}{x}, \qquad \cos\theta = \frac{x}{r}, \qquad \text{and} \qquad \sin\theta = \frac{y}{r}.$$

If $r < 0$, you can show that the same relationships hold.

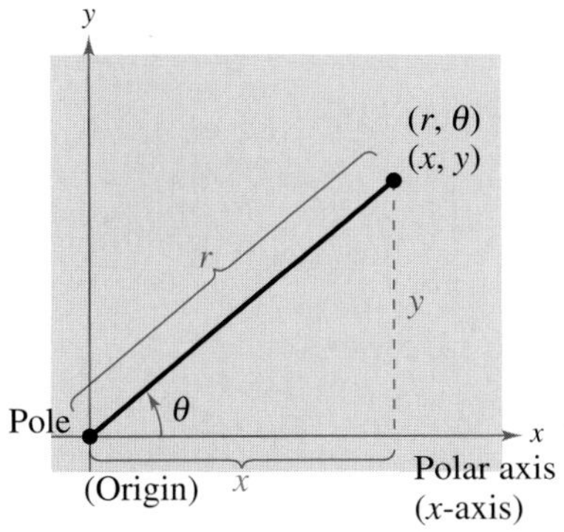

FIGURE 10.54

Coordinate Conversion

The polar coordinates (r, θ) are related to the rectangular coordinates (x, y) as follows.

$$x = r\cos\theta \qquad \text{and} \qquad \tan\theta = \frac{y}{x}$$

$$y = r\sin\theta \qquad\qquad r^2 = x^2 + y^2$$

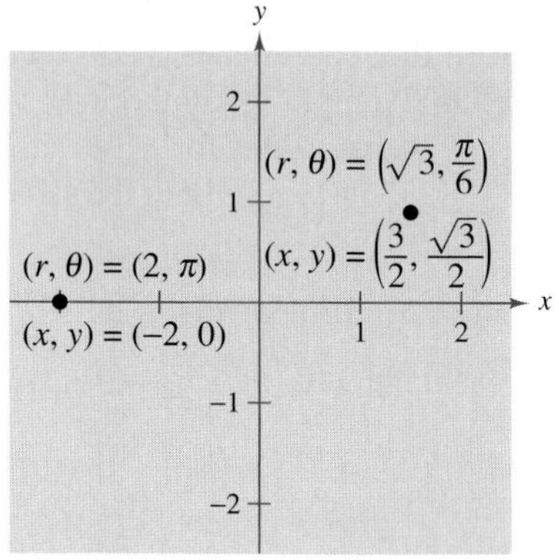

FIGURE 10.55

Additional Examples

a. Three polar representations of the point $\left(1, -\frac{\pi}{4}\right)$ are $\left(1, \frac{7\pi}{4}\right)$, $\left(-1, \frac{3\pi}{4}\right)$, and $\left(-1, -\frac{5\pi}{4}\right)$.

b. Three polar representations of the point $\left(-2, \frac{\pi}{3}\right)$ are $\left(2, \frac{4\pi}{3}\right)$, $\left(2, -\frac{2\pi}{3}\right)$, and $\left(-2, -\frac{5\pi}{3}\right)$.

Example 3 ▶ Polar-to-Rectangular Conversion

Convert the points to rectangular coordinates. (See Figure 10.55.)

a. $(2, \pi)$ **b.** $\left(\sqrt{3}, \pi/6\right)$

Solution

a. For the point $(r, \theta) = (2, \pi)$, you have

$$x = r\cos\theta = 2\cos\pi = -2$$

and

$$y = r\sin\theta = 2\sin\pi = 0.$$

The rectangular coordinates are $(x, y) = (-2, 0)$.

b. For the point $(r, \theta) = \left(\sqrt{3}, \pi/6\right)$, you have

$$x = \sqrt{3}\cos\frac{\pi}{6} = \sqrt{3}\left(\frac{\sqrt{3}}{2}\right) = \frac{3}{2}$$

and

$$y = \sqrt{3}\sin\frac{\pi}{6} = \sqrt{3}\left(\frac{1}{2}\right) = \frac{\sqrt{3}}{2}.$$

The rectangular coordinates are $(x, y) = \left(3/2, \sqrt{3}/2\right)$.

Example 4 ▶ Rectangular-to-Polar Conversion

Convert the points to polar coordinates.

a. $(-1, 1)$ **b.** $(0, 2)$

Solution

a. For the second-quadrant point $(x, y) = (-1, 1)$, you have

$$\tan\theta = \frac{y}{x}$$

$$\tan\theta = -1$$

$$\theta = \frac{3\pi}{4}.$$

Because θ lies in the same quadrant as (x, y), use positive r.

$$r = \sqrt{x^2 + y^2} = \sqrt{(-1)^2 + (1)^2} = \sqrt{2}$$

So, *one* set of polar coordinates is $(r, \theta) = \left(\sqrt{2}, 3\pi/4\right)$, as shown in Figure 10.56(a).

b. Because the point $(x, y) = (0, 2)$ lies on the positive y-axis, choose

$$\theta = \frac{\pi}{2} \quad \text{and} \quad r = 2.$$

This implies that *one* set of polar coordinates is $(r, \theta) = (2, \pi/2)$, as shown in Figure 10.56(b).

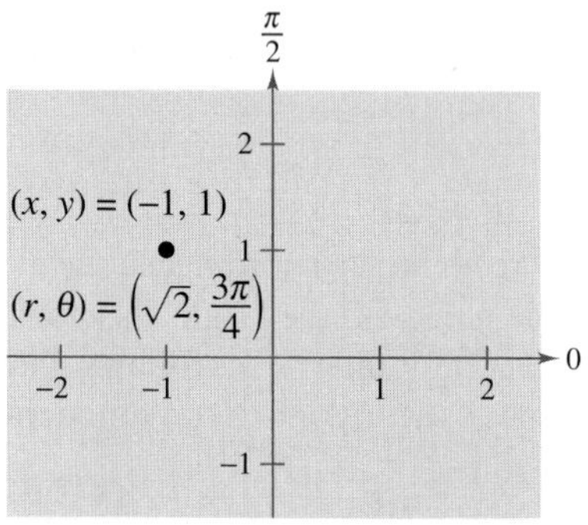

(a)

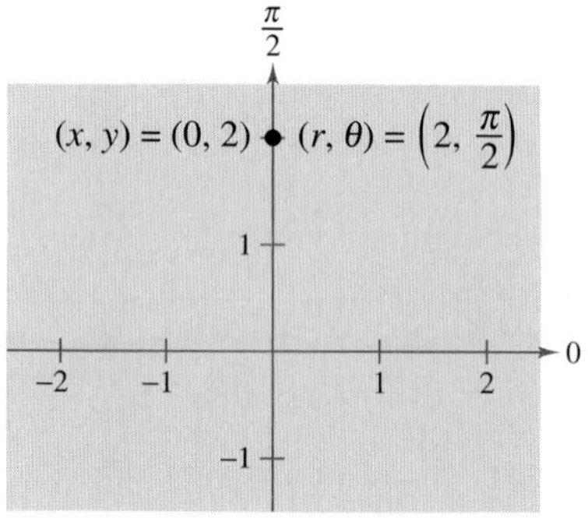

(b)

FIGURE 10.56

Multiple representation of points and equations in the polar system can cause confusion. You may want to discuss several examples.

Activities

1. Convert the point $(-4, 2)$ from rectangular to polar form.
   Answer: $(2\sqrt{5}, -0.4636)$ or $(-2\sqrt{5}, 2.6779)$
2. Convert the polar equation $r = 3 \cos \theta$ to rectangular form.
   Answer: $x^2 + y^2 - 3x = 0$
3. Convert the rectangular equation $x = 4$ to polar form.
   Answer: $r = 4 \sec \theta$

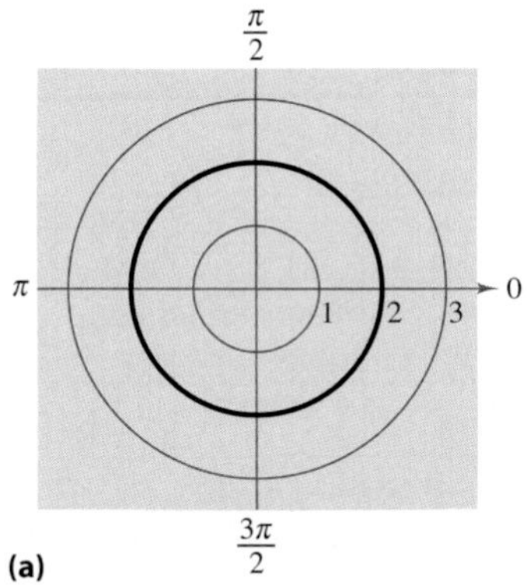

(a)

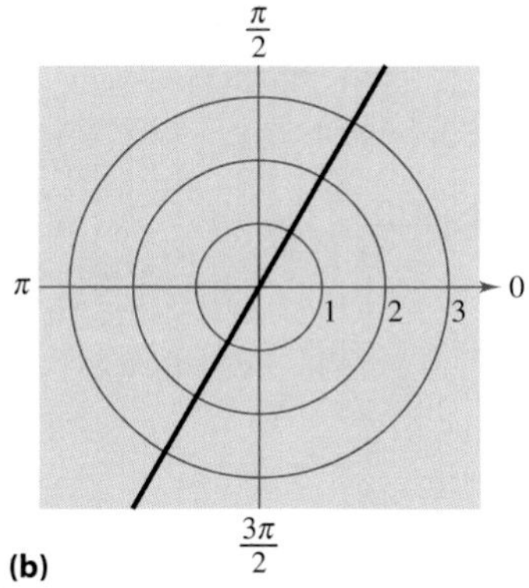

(b)

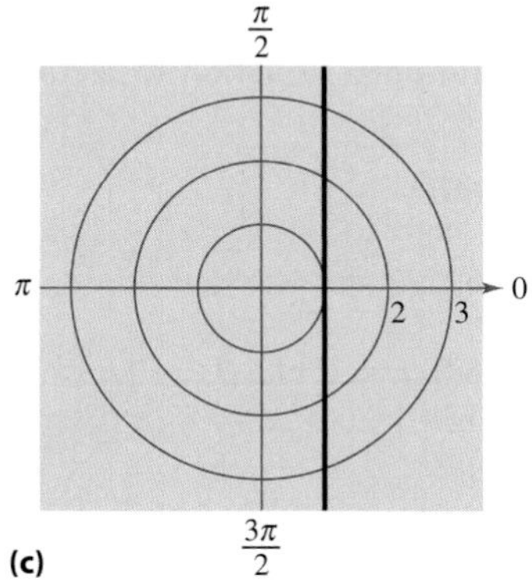

(c)

FIGURE 10.57

Equation Conversion

By comparing Examples 3 and 4, you can see that point conversion from the polar to the rectangular system is straightforward, whereas point conversion from the rectangular to the polar system is more involved. For equations, the opposite is true. To convert a rectangular equation to polar form, you simply replace x by $r \cos \theta$ and y by $r \sin \theta$. For instance, the rectangular equation $y = x^2$ can be written in polar form as follows.

$$y = x^2 \qquad \text{Rectangular equation}$$

$$r \sin \theta = (r \cos \theta)^2 \qquad \text{Polar equation}$$

$$r = \sec \theta \tan \theta \qquad \text{Simplest form}$$

On the other hand, converting a polar equation to rectangular form requires considerable ingenuity.

Example 5 demonstrates several polar-to-rectangular conversions that enable you to sketch the graphs of some polar equations.

Example 5 ▶ Converting Polar Equations to Rectangular Form

Describe the graph of each polar equation and find the corresponding rectangular equation.

a. $r = 2$ **b.** $\theta = \dfrac{\pi}{3}$ **c.** $r = \sec \theta$

Solution

a. The graph of the polar equation $r = 2$ consists of all points that are 2 units from the pole. In other words, this graph is a circle centered at the origin with a radius of 2, as shown in Figure 10.57(a). You can confirm this by converting to rectangular form, using the relationship $r^2 = x^2 + y^2$.

$$\underbrace{r = 2}_{\text{Polar equation}} \quad \Rightarrow \quad r^2 = 2^2 \quad \Rightarrow \quad \underbrace{x^2 + y^2 = 2^2}_{\text{Rectangular equation}}$$

b. The graph of the polar equation $\theta = \pi/3$ consists of all points on the line that makes an angle of $\pi/3$ with the positive polar axis, as shown in Figure 10.57(b). To convert to rectangular form, make use of the relationship $\tan \theta = y/x$.

$$\underbrace{\theta = \frac{\pi}{3}}_{\text{Polar equation}} \quad \Rightarrow \quad \tan \theta = \sqrt{3} \quad \Rightarrow \quad \underbrace{y = \sqrt{3}x}_{\text{Rectangular equation}}$$

c. The graph of the polar equation $r = \sec \theta$ is not evident by simple inspection, so you convert to rectangular form by using the relationship $r \cos \theta = x$.

$$\underbrace{r = \sec \theta}_{\text{Polar equation}} \quad \Rightarrow \quad r \cos \theta = 1 \quad \Rightarrow \quad \underbrace{x = 1}_{\text{Rectangular equation}}$$

Now you see that the graph is a vertical line, as shown in Figure 10.57(c).

10.7 Exercises

In Exercises 1–8, plot the point given in polar coordinates and find two additional polar representations.

1. $(4, -\pi/3)$
2. $(-1, -3\pi/4)$
3. $(0, -7\pi/6)$
4. $(16, 5\pi/2)$
5. $(\sqrt{2}, 2.36)$
6. $(-3, -1.57)$
7. $(2\sqrt{2}, 4.71)$
8. $(-5, -2.36)$

In Exercises 9–16, a point in polar coordinates is given. Convert the point to rectangular coordinates.

9. $\left(3, \dfrac{\pi}{2}\right)$

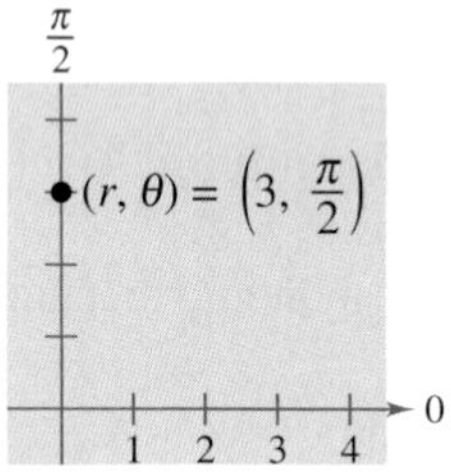

10. $\left(3, \dfrac{3\pi}{2}\right)$

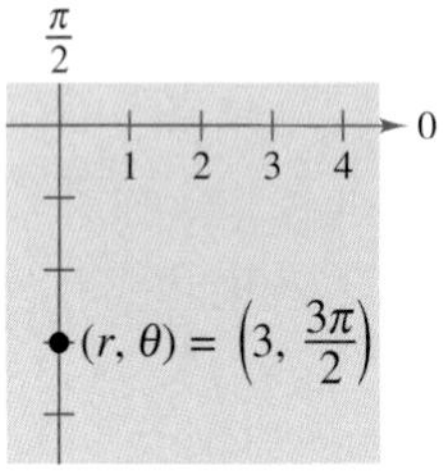

11. $\left(-1, \dfrac{5\pi}{4}\right)$

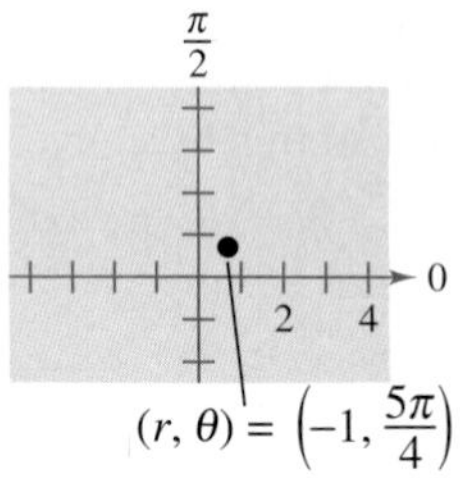

12. $(0, -\pi)$

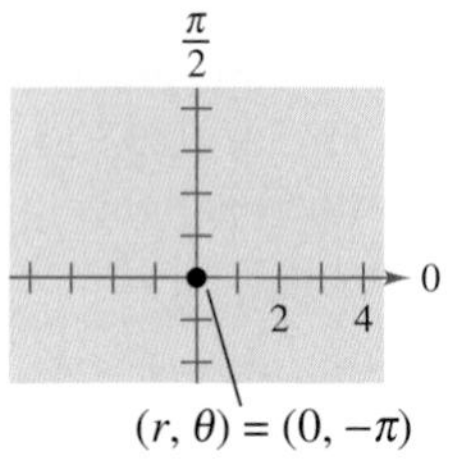

13. $(2, 3\pi/4)$
14. $(-2, 7\pi/6)$
15. $(-2.5, 1.1)$
16. $(8.25, 3.5)$

In Exercises 17–26, a point in rectangular coordinates is given. Convert the point to polar coordinates.

17. $(1, 1)$
18. $(-3, -3)$
19. $(-6, 0)$
20. $(0, -5)$
21. $(-3, 4)$
22. $(3, -1)$
23. $(-\sqrt{3}, -\sqrt{3})$
24. $(\sqrt{3}, -1)$
25. $(6, 9)$
26. $(5, 12)$

In Exercises 27–32, use a graphing utility to find one set of polar coordinates of the point given in rectangular coordinates.

27. $(3, -2)$
28. $(-5, 2)$
29. $(\sqrt{3}, 2)$
30. $(3\sqrt{2}, 3\sqrt{2})$
31. $\left(\frac{5}{2}, \frac{4}{3}\right)$
32. $\left(\frac{7}{4}, \frac{3}{2}\right)$

In Exercises 33–48, convert the rectangular equation to polar form.

33. $x^2 + y^2 = 9$
34. $x^2 + y^2 = 16$
35. $y = 4$
36. $x = 10$
37. $3x - y + 2 = 0$
38. $3x + 5y - 2 = 0$
39. $xy = 16$
40. $y = x$
41. $y^2 - 8x - 16 = 0$
42. $(x^2 + y^2)^2 = 9(x^2 - y^2)$
43. $x^2 + y^2 = a^2$
44. $x^2 + y^2 = 9a^2$
45. $y = b$
46. $x = 4a$
47. $x^2 + y^2 - 2ax = 0$
48. $x^2 + y^2 - 2ay = 0$

In Exercises 49–58, convert the polar equation to rectangular form.

49. $r = 4 \sin \theta$
50. $r = 3 \cos \theta$
51. $\theta = \dfrac{2\pi}{3}$
52. $r = 4$
53. $r = 2 \csc \theta$
54. $r^2 = \sin 2\theta$
55. $r = 2 \sin 3\theta$
56. $r = \dfrac{1}{1 - \cos \theta}$
57. $r = \dfrac{6}{2 - 3 \sin \theta}$
58. $r = \dfrac{6}{2 \cos \theta - 3 \sin \theta}$

In Exercises 59–64, convert the polar equation to rectangular form and sketch its graph.

59. $r = 6$
60. $r = 8$
61. $\theta = \dfrac{\pi}{6}$
62. $\theta = \dfrac{3\pi}{4}$
63. $r = 3 \sec \theta$
64. $r = 2 \csc \theta$

Synthesis

True or False? **In Exercises 65 and 66, determine whether the statement is true or false. Justify your answer.**

65. If $\theta_1 = \theta_2 + 2\pi n$ for some integer n, then (r, θ_1) and (r, θ_2) represent the same point on the polar coordinate system.

66. If $|r_1| = |r_2|$, then (r_1, θ) and (r_2, θ) represent the same point on the polar coordinate system.

67. Convert the polar equation

$$r = 2(h\cos\theta + k\sin\theta)$$

to rectangular form and verify that it is the equation of a circle. Find the radius and the rectangular coordinates of the center of the circle.

68. Convert the polar equation $r = \cos\theta + 3\sin\theta$ to rectangular form and identify the graph.

69. *Think About It*

(a) Show that the distance between the points (r_1, θ_1) and (r_2, θ_2) is

$$\sqrt{r_1^{\,2} + r_2^{\,2} - 2r_1r_2\cos(\theta_1 - \theta_2)}.$$

(b) Describe the positions of the points relative to each other if $\theta_1 = \theta_2$. Simplify the Distance Formula for this case. Is the simplification what you expected? Explain.

(c) Simplify the Distance Formula if $\theta_1 - \theta_2 = 90°$. Is the simplification what you expected? Explain.

(d) Choose two points on the polar coordinate system and find the distance between them. Then choose different polar representations of the same two points and apply the Distance Formula again. Discuss the result.

70. *Exploration*

(a) Set the window format of your graphing utility on rectangular coordinates and locate the cursor at any position off the coordinate axes. Move the cursor horizontally and observe any changes in the displayed coordinates of the points. Explain the changes. Now repeat the process moving the cursor vertically.

(b) Set the window format of your graphing utility on polar coordinates and locate the cursor at any position off the coordinate axes. Move the cursor horizontally and observe any changes in the displayed coordinates of the points. Explain the changes. Now repeat the process moving the cursor vertically.

(c) Explain why the results of parts (a) and (b) are not the same.

Review

In Exercises 71–76, use determinants to solve the system of equations.

71. $\begin{cases} 5x - 7y = -11 \\ -3x + y = -3 \end{cases}$

72. $\begin{cases} 3x + 5y = 10 \\ 4x - 2y = -5 \end{cases}$

73. $\begin{cases} 3a - 2b + c = 0 \\ 2a + b - 3c = 0 \\ a - 3b + 9c = 8 \end{cases}$

74. $\begin{cases} 5u + 7v + 9w = 15 \\ u - 2v - 3w = 7 \\ 8u - 2v + w = 0 \end{cases}$

75. $\begin{cases} x + y + z - 3w = -8 \\ 3x - y - 2z + w = 7 \\ -x + y - z + 2w = -2 \\ 2y + w = -6 \end{cases}$

76. $\begin{cases} 2y + 5z + 6w = 32 \\ 2x + 4y - 5z - w = -7 \\ 3x - 6y + z + 5w = 6 \\ 4x - 2y - z = -12 \end{cases}$

In Exercises 77–80, use a determinant to determine whether the points are collinear.

77. $(4, -3), (6, -7), (-2, -1)$

78. $(-2, 4), (0, 1), (4, -5)$

79. $(-6, -4), (-1, -3), (1.5, -2.5)$

80. $(-2.3, 5), (-0.5, 0), (1.5, -3)$

10.8 Graphs of Polar Equations

▶ **What you should learn**

- How to graph a polar equation by point plotting
- How to use symmetry, zeros, and maximum r-values as graphing aids
- How to recognize special polar graphs

▶ **Why you should learn it**

Equations of several common figures are simpler in polar form than in rectangular form. For instance, Exercise 6 on page 832 shows the graph of a circle and its polar equation.

Introduction

In previous chapters you spent a lot of time learning how to sketch graphs on rectangular coordinate systems. You began with the basic point-plotting method, which was then enhanced by sketching aids such as symmetry, intercepts, asymptotes, periods, and shifts. This section approaches curve sketching on the polar coordinate system similarly, beginning with a demonstration of point plotting.

Example 1 ▶ Graphing a Polar Equation by Point Plotting

Sketch the graph of the polar equation $r = 4 \sin \theta$.

Solution

The sine function is periodic, so you can get a full range of r-values by considering values of θ in the interval $0 \le \theta \le 2\pi$, as shown in the following table.

θ	0	$\frac{\pi}{6}$	$\frac{\pi}{3}$	$\frac{\pi}{2}$	$\frac{2\pi}{3}$	$\frac{5\pi}{6}$	π	$\frac{7\pi}{6}$	$\frac{3\pi}{2}$	$\frac{11\pi}{6}$	2π
r	0	2	$2\sqrt{3}$	4	$2\sqrt{3}$	2	0	-2	-4	-2	0

If you plot these points as shown in Figure 10.58, it appears that the graph is a circle of radius 2 whose center is at the point $(x, y) = (0, 2)$. Try confirming this by squaring both sides of the polar equation and converting the result to rectangular form.

Emphasize setting up the table of θ values. Your students will benefit from labeling the points as they plot them.

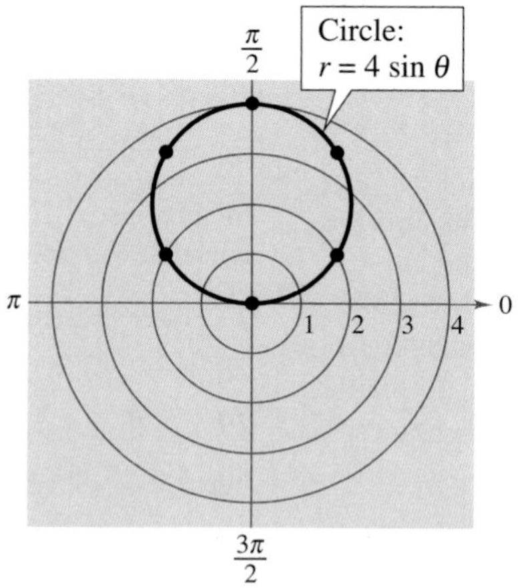

FIGURE 10.58

Symmetry

In Figure 10.58, note that as θ increases from 0 to 2π the graph is traced out twice. Moreover, note that the graph is *symmetric with respect to the line* $\theta = \pi/2$. Had you known about this symmetry and retracing ahead of time, you could have used fewer points.

Symmetry with respect to the line $\theta = \pi/2$ is one of three important types of symmetry to consider in polar curve sketching. (See Figure 10.59.)

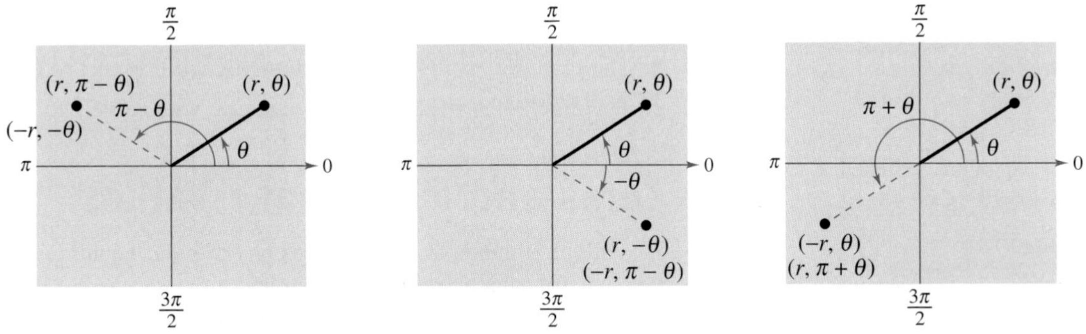

Symmetry with Respect to the Line $\theta = \dfrac{\pi}{2}$

Symmetry with Respect to the Polar Axis

Symmetry with Respect to the Pole

FIGURE 10.59

Point out to your students that these tests are sufficient for showing symmetry; however, they are not necessary. A polar graph can exhibit symmetry even when the tests fail to indicate symmetry.

Tests for Symmetry in Polar Coordinates

The graph of a polar equation is symmetric with respect to the following if the given substitution yields an equivalent equation.

1. *The line* $\theta = \pi/2$: Replace (r, θ) by $(r, \pi - \theta)$ or $(-r, -\theta)$.
2. *The polar axis:* Replace (r, θ) by $(r, -\theta)$ or $(-r, \pi - \theta)$.
3. *The pole:* Replace (r, θ) by $(r, \pi + \theta)$ or $(-r, \theta)$.

Example 2 ▶ Using Symmetry to Sketch a Polar Graph

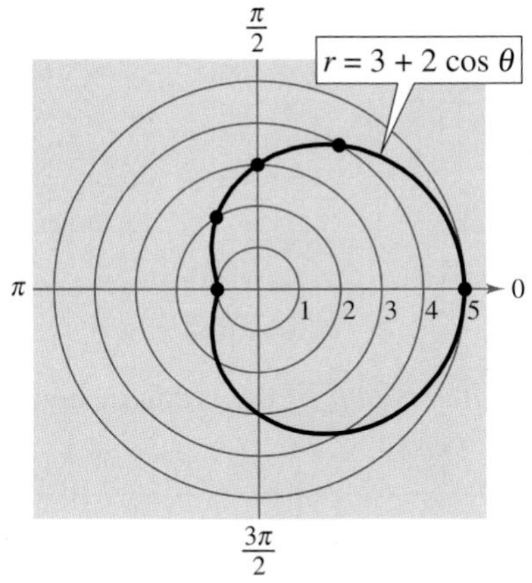

FIGURE 10.60

Use symmetry to sketch the graph of $r = 3 + 2\cos\theta$.

Solution

Replacing (r, θ) by $(r, -\theta)$ produces $r = 3 + 2\cos(-\theta) = 3 + 2\cos\theta$. So, you can conclude that the curve is symmetric with respect to the polar axis. Plotting the points in the table and using polar axis symmetry, you obtain the graph of a **limaçon,** as shown in Figure 10.60.

θ	0	$\frac{\pi}{3}$	$\frac{\pi}{2}$	$\frac{2\pi}{3}$	π
r	5	4	3	2	1

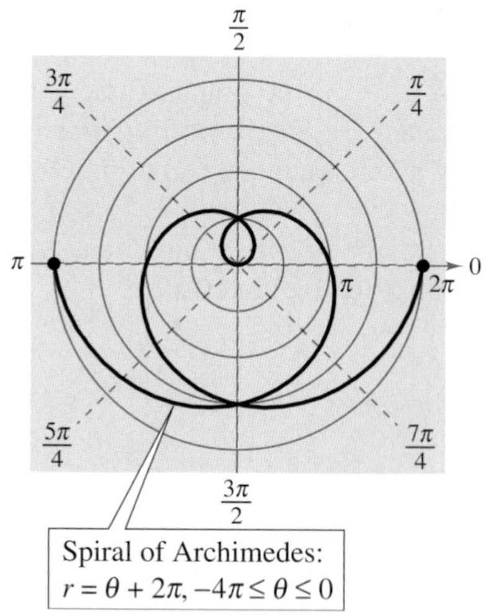

FIGURE 10.61

The three tests for symmetry in polar coordinates listed on page 827 are sufficient to guarantee symmetry, but they are not necessary. For instance, Figure 10.61 shows the graph of $r = \theta + 2\pi$ to be symmetric with respect to the line $\theta = \pi/2$, and yet the tests on page 827 fail to indicate symmetry.

The equations discussed in Examples 1 and 2 are of the form

$$r = 4 \sin \theta = f(\sin \theta) \qquad \text{and} \qquad r = 3 + 2 \cos \theta = g(\cos \theta).$$

The graph of the first equation is symmetric with respect to the line $\theta = \pi/2$, and the graph of the second equation is symmetric with respect to the polar axis. This observation can be generalized to yield the following *quick tests for symmetry*.

1. The graph of $r = f(\sin \theta)$ is symmetric with respect to the line $\theta = \pi/2$.
2. The graph of $r = g(\cos \theta)$ is symmetric with respect to the polar axis.

Zeros and Maximum *r*-Values

Two additional aids to sketching graphs of polar equations involve knowing the θ-values for which $|r|$ is maximum and knowing the θ-values for which $r = 0$. For instance, in Example 1, the maximum value of $|r|$ for $r = 4 \sin \theta$ is $|r| = 4$, and this occurs when $\theta = \pi/2$, as shown in Figure 10.58. Moreover, $r = 0$ when $\theta = 0$.

Example 3 ▶ Sketching a Polar Graph

Sketch the graph of

$$r = 1 - 2 \cos \theta.$$

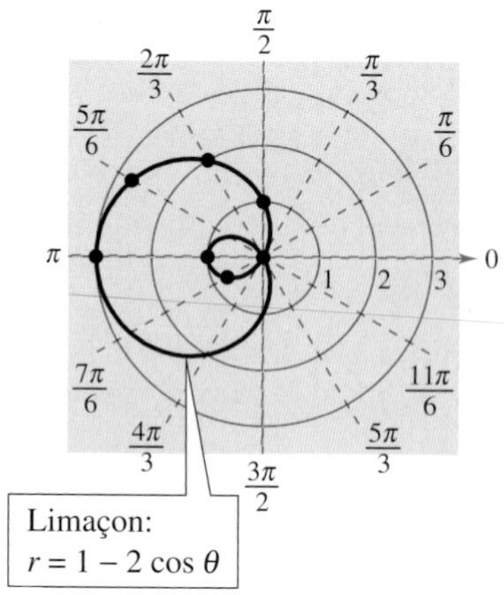

FIGURE 10.62

Solution

From the equation $r = 1 - 2 \cos \theta$, you can obtain the following.

Symmetry: With respect to the polar axis

Maximum value of $|r|$: $r = 3$ when $\theta = \pi$

Zero of r: $r = 0$ when $\theta = \pi/3$

The table shows several θ-values in the interval $[0, \pi]$. By plotting the corresponding points, you can sketch the graph shown in Figure 10.62.

θ	0	$\frac{\pi}{6}$	$\frac{\pi}{3}$	$\frac{\pi}{2}$	$\frac{2\pi}{3}$	$\frac{5\pi}{6}$	π
r	-1	-0.73	0	1	2	2.73	3

Note how the negative r-values determine the *inner loop* of the graph in Figure 10.62. This graph, like the one in Figure 10.60, is a limaçon.

Some curves reach their zeros and maximum r-values at more than one point. Example 4 shows how to handle this situation.

Example 4 ▶ Sketching a Polar Graph

Sketch the graph of $r = 2 \cos 3\theta$.

Solution

Symmetry: With respect to the polar axis

Maximum value of $|r|$: $|r| = 2$ when $3\theta = 0, \pi, 2\pi, 3\pi$ or
$\theta = 0, \pi/3, 2\pi/3, \pi$

Zeros of r: $r = 0$ when $3\theta = \pi/2, 3\pi/2, 5\pi/2$ or
$\theta = \pi/6, \pi/2, 5\pi/6$

A computer animation of this example appears in the *Interactive* CD-ROM and *Internet* versions of this text.

θ	0	$\frac{\pi}{12}$	$\frac{\pi}{6}$	$\frac{\pi}{4}$	$\frac{\pi}{3}$	$\frac{5\pi}{12}$	$\frac{\pi}{2}$
r	2	$\sqrt{2}$	0	$-\sqrt{2}$	-2	$-\sqrt{2}$	0

By plotting these points and using the specified symmetry, zeros, and maximum values, you can obtain the graph shown in Figure 10.63. This graph is called a **rose curve,** and each of the loops on the graph is called a *petal* of the rose curve. Note how the entire curve is generated as θ increases from 0 to π.

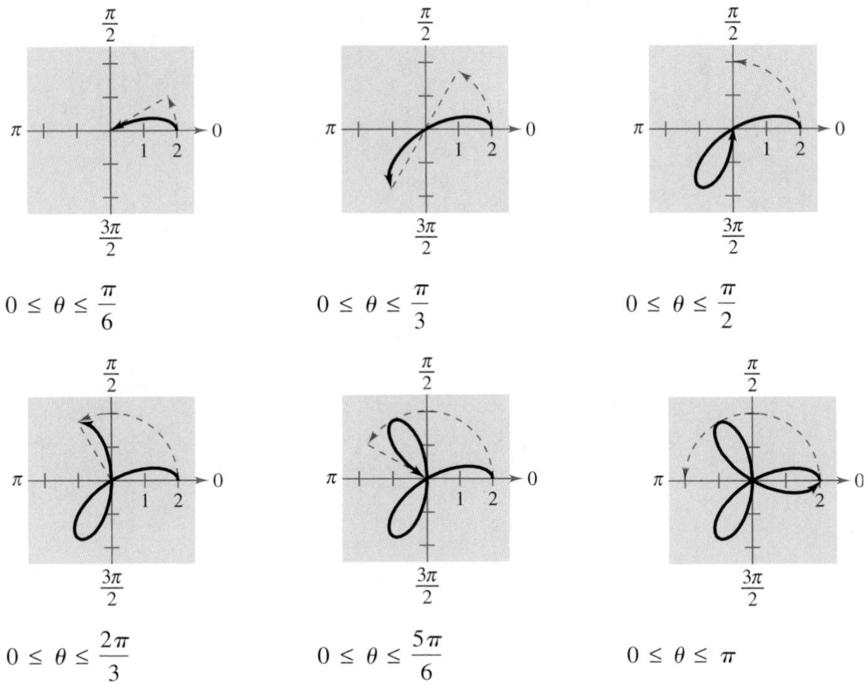

FIGURE 10.63

Technology

Use a graphing utility in polar mode to verify the graph of $r = 2 \cos 3\theta$ shown in Figure 10.63.

Special Polar Graphs

Several important types of graphs have equations that are simpler in polar form than in rectangular form. For example, the circle

$$r = 4 \sin \theta$$

in Example 1 has the more complicated rectangular equation

$$x^2 + (y - 2)^2 = 4.$$

Several other types of graphs that have simple polar equations are shown below.

Limaçons
$r = a \pm b \cos \theta$
$r = a \pm b \sin \theta$
$(a > 0, b > 0)$

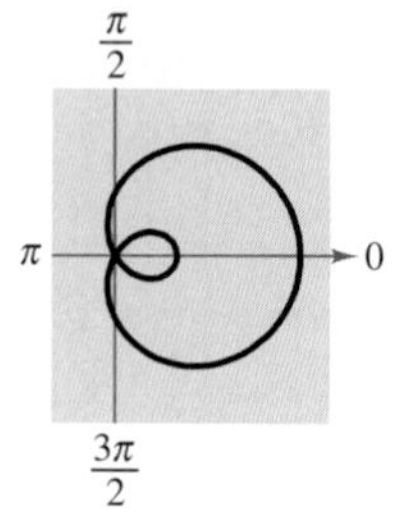

$\frac{a}{b} < 1$

Limaçon with inner loop

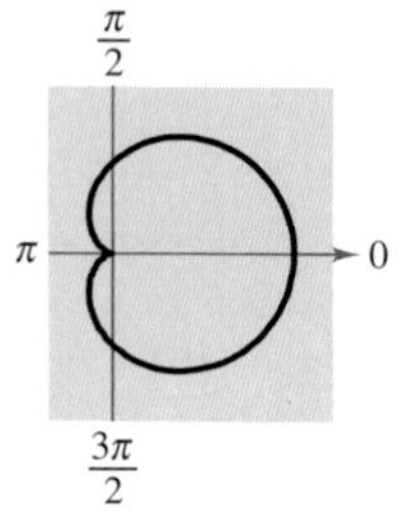

$\frac{a}{b} = 1$

Cardioid (heart-shaped)

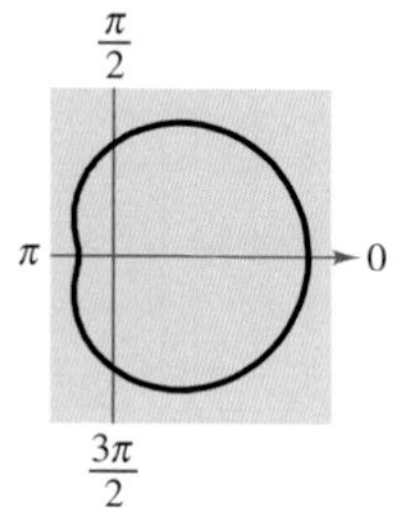

$1 < \frac{a}{b} < 2$

Dimpled limaçon

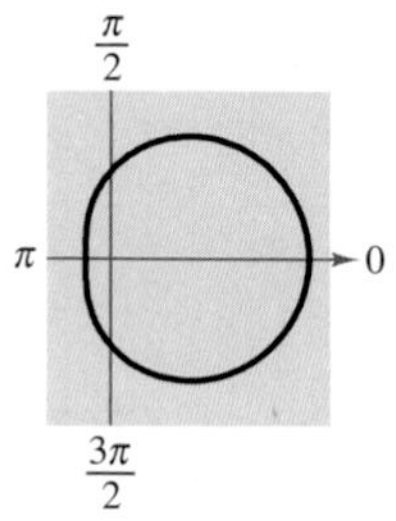

$\frac{a}{b} \geq 2$

Convex limaçon

Rose Curves
n petals if n is odd
$2n$ petals if n is even
$(n \geq 2)$

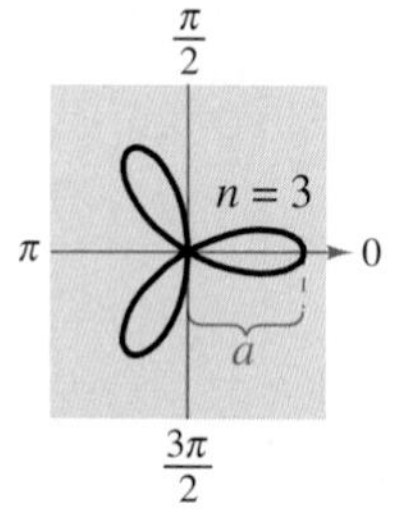

$r = a \cos n\theta$

Rose curve

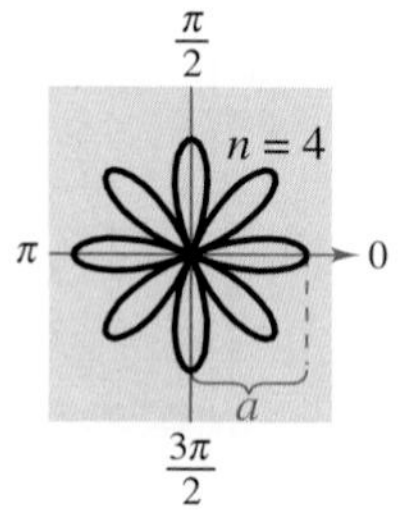

$r = a \cos n\theta$

Rose curve

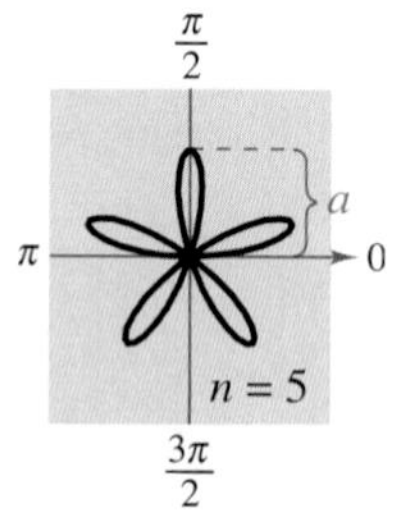

$r = a \sin n\theta$

Rose curve

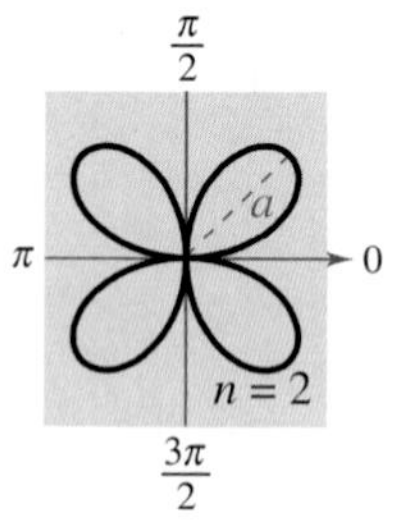

$r = a \sin n\theta$

Rose curve

Circles and Lemniscates

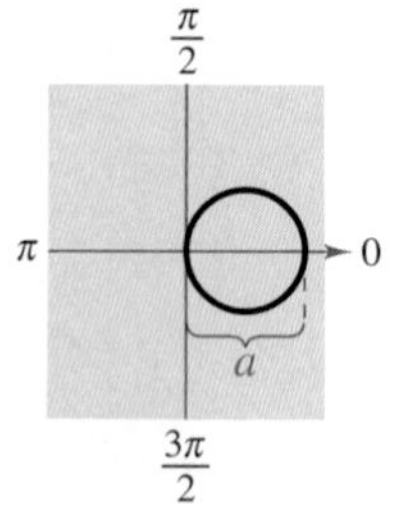

$r = a \cos \theta$

Circle

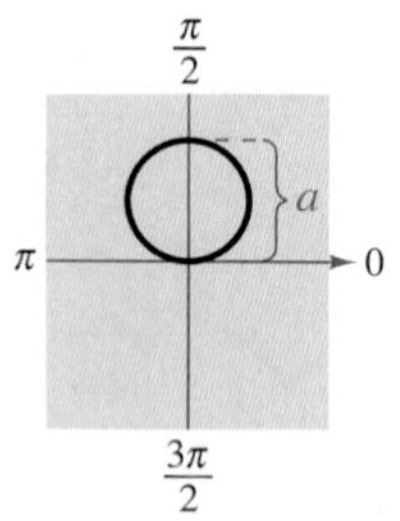

$r = a \sin \theta$

Circle

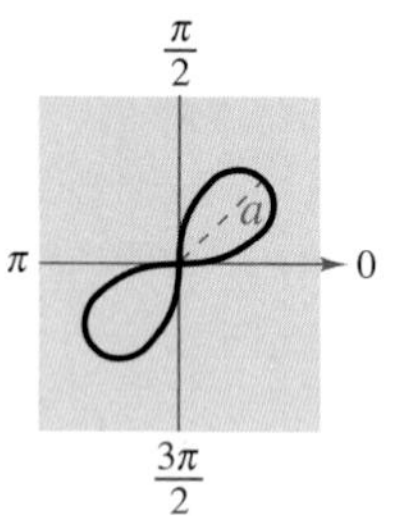

$r^2 = a^2 \sin 2\theta$

Lemniscate

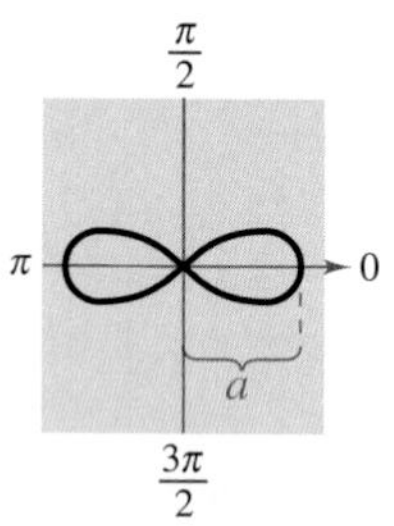

$r^2 = a^2 \cos 2\theta$

Lemniscate

The quick tests for symmetry presented in this section can be especially useful when graphing these special polar graphs.

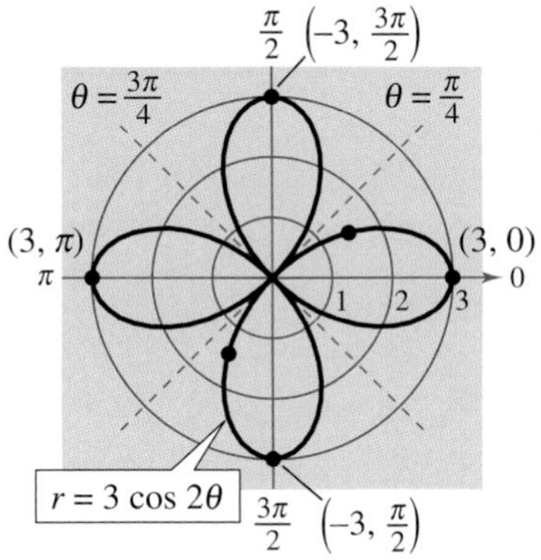

FIGURE 10.64

Example 5 ▶ Sketching a Rose Curve

Sketch the graph of

$$r = 3\cos 2\theta.$$

Solution

Type of curve:	Rose curve with $2n = 4$ petals
Symmetry:	With respect to polar axis, the line $\theta = \pi/2$, and the pole
Maximum value of $\lvert r\rvert$:	$\lvert r\rvert = 3$ when $\theta = 0, \pi/2, \pi, 3\pi/2$
Zeros of r:	$r = 0$ when $\theta = \pi/4, 3\pi/4$

Using this information together with the additional points shown in the following table, you obtain the graph shown in Figure 10.64.

θ	0	$\frac{\pi}{6}$	$\frac{\pi}{4}$	$\frac{\pi}{3}$
r	3	$\frac{3}{2}$	0	$-\frac{3}{2}$

Activities

1. Test $r^2 = 3\sin\theta$ for symmetry.
   Answer: Symmetric with respect to the pole
2. Find the maximum value of $|r|$ and any zeros of r in the polar equation $r = 2 + 2\cos\theta$.
   Answer: Maximum value of $|r| = 4$ when $\theta = 0$ and $r = 0$ when $\theta = \pi$.
3. Identify the shape of the graph of the polar equation $r = 2\sin 3\theta$.
   Answer: Rose curve with 3 petals

Example 6 ▶ Sketching a Lemniscate

Sketch the graph of

$$r^2 = 9\sin 2\theta.$$

Solution

Type of curve:	Lemniscate
Symmetry:	With respect to the pole
Maximum value of $\lvert r\rvert$:	$\lvert r\rvert = 3$ when $\theta = \pi/4$
Zeros of r:	$r = 0$ when $\theta = 0, \pi/2$

If $\sin 2\theta < 0$, this equation has no solution points. So, you restrict the values of θ to those for which $\sin 2\theta \geq 0$.

$$0 \leq \theta \leq \frac{\pi}{2} \quad \text{or} \quad \pi \leq \theta \leq \frac{3\pi}{2}$$

Moreover, using symmetry, you need to consider only the first of these two intervals. By finding a few additional points, you can obtain the graph shown in Figure 10.65.

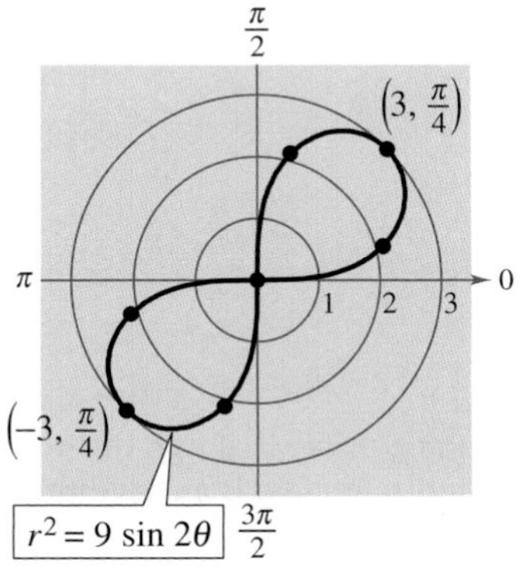

FIGURE 10.65

θ	0	$\frac{\pi}{12}$	$\frac{\pi}{4}$	$\frac{5\pi}{12}$	$\frac{\pi}{2}$
$r = \pm 3\sqrt{\sin 2\theta}$	0	$\frac{\pm 3}{\sqrt{2}}$	± 3	$\frac{\pm 3}{\sqrt{2}}$	0

10.8 Exercises

In Exercises 1–6, identify the type of polar graph.

1. 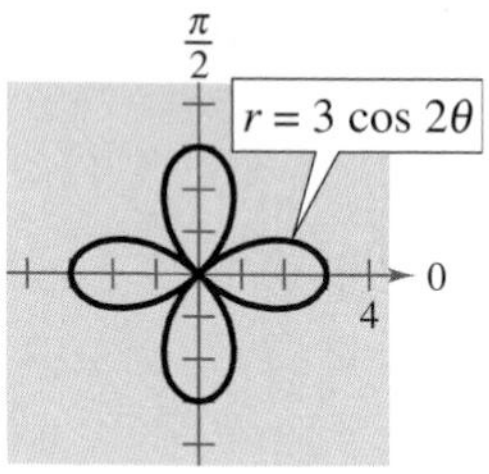

2. 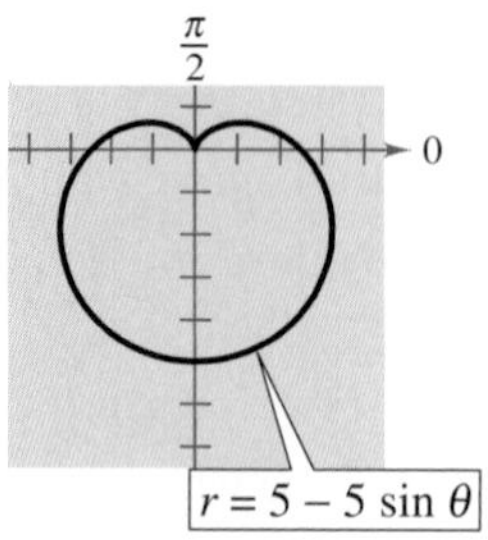

3. 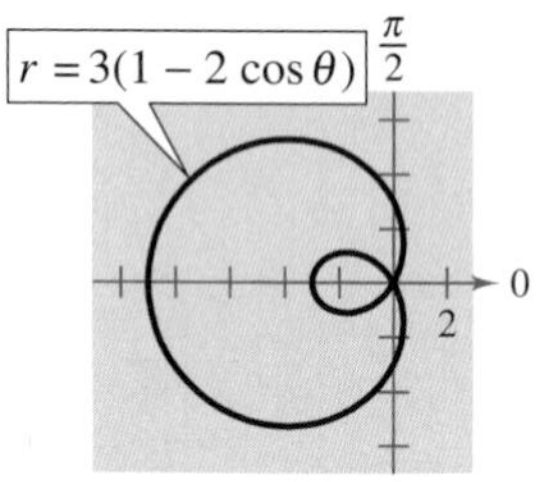

4. 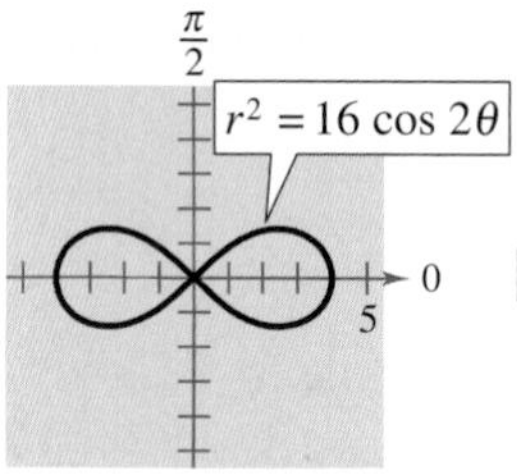

5. 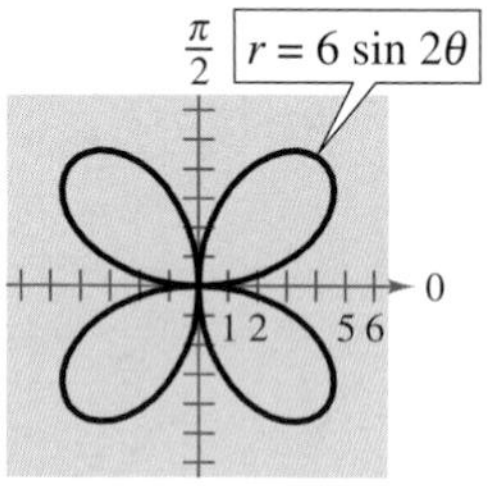

6. 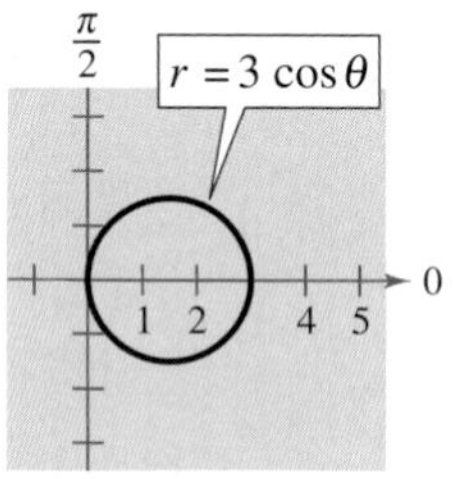

In Exercises 7–12, test for symmetry with respect to $\theta = \pi/2$, the polar axis, and the pole.

7. $r = 5 + 4\cos\theta$
8. $r = 16\cos 3\theta$
9. $r = \dfrac{2}{1 + \sin\theta}$
10. $r = \dfrac{3}{2 + \cos\theta}$
11. $r^2 = 16\cos 2\theta$
12. $r^2 = 36\sin 2\theta$

In Exercises 13–16, find the maximum value of $|r|$ and any zeros of r.

13. $r = 10(1 - \sin\theta)$
14. $r = 6 + 12\cos\theta$
15. $r = 4\cos 3\theta$
16. $r = 3\sin 2\theta$

In Exercises 17–40, sketch the graph of the polar equation.

17. $r = 5$
18. $r = 2$
19. $r = \dfrac{\pi}{6}$
20. $r = -\dfrac{3\pi}{4}$
21. $r = 3\sin\theta$
22. $r = 4\cos\theta$
23. $r = 3(1 - \cos\theta)$
24. $r = 4(1 - \sin\theta)$
25. $r = 4(1 + \sin\theta)$
26. $r = 2(1 + \cos\theta)$
27. $r = 3 + 6\sin\theta$
28. $r = 4 - 3\sin\theta$
29. $r = 1 - 2\sin\theta$
30. $r = 1 - 2\cos\theta$
31. $r = 3 - 4\cos\theta$
32. $r = 4 + 3\cos\theta$
33. $r = 5\sin 2\theta$
34. $r = 3\cos 2\theta$
35. $r = 2\sec\theta$
36. $r = 5\csc\theta$
37. $r = \dfrac{3}{\sin\theta - 2\cos\theta}$
38. $r = \dfrac{6}{2\sin\theta - 3\cos\theta}$
39. $r^2 = 9\cos 2\theta$
40. $r^2 = 4\sin\theta$

In Exercises 41–46, use a graphing utility to graph the polar equation.

41. $r = 8\cos\theta$
42. $r = \cos 2\theta$
43. $r = 3(2 - \sin\theta)$
44. $r = 2\cos(3\theta - 2)$
45. $r = 8\sin\theta\cos^2\theta$
46. $r = 2\csc\theta + 5$

In Exercises 47–52, use a graphing utility to graph the polar equation. Find an interval for θ for which the graph is traced *only once*.

47. $r = 3 - 4\cos\theta$
48. $r = 5 + 4\cos\theta$
49. $r = 2\cos\left(\dfrac{3\theta}{2}\right)$
50. $r = 3\sin\left(\dfrac{5\theta}{2}\right)$
51. $r^2 = 9\sin 2\theta$
52. $r^2 = \dfrac{1}{\theta}$

In Exercises 53–56, use a graphing utility to graph the polar equation and show that the indicated line is an asymptote of the graph.

Name of Graph	*Polar Equation*	*Asymptote*
53. Conchoid	$r = 2 - \sec\theta$	$x = -1$
54. Conchoid	$r = 2 + \csc\theta$	$y = 1$
55. Hyperbolic spiral	$r = \dfrac{3}{\theta}$	$y = 3$
56. Strophoid	$r = 2\cos 2\theta\sec\theta$	$x = -2$

Synthesis

True or False? **In Exercises 57 and 58, determine whether the statement is true or false. Justify your answer.**

57. In the polar coordinate system, if a graph that has symmetry with respect to the polar axis were folded on the line $\theta = 0$, the portion of the graph above the polar axis would coincide with the portion of the graph below the polar axis.

58. In the polar coordinate system, if a graph that has symmetry with respect to the pole were folded on the line $\theta = 3\pi/4$, the portion of the graph on one side of the fold would coincide with the portion of the graph on the other side of the fold.

59. *Exploration* Sketch the graph of $r = 6 \cos \theta$ over each interval. Describe the part of the graph obtained in each case.

(a) $0 \le \theta \le \dfrac{\pi}{2}$ (b) $\dfrac{\pi}{2} \le \theta \le \pi$

(c) $-\dfrac{\pi}{2} \le \theta \le \dfrac{\pi}{2}$ (d) $\dfrac{\pi}{4} \le \theta \le \dfrac{3\pi}{4}$

60. *Graphical Reasoning* Use a graphing utility to graph the polar equation

$$r = 6[1 + \cos(\theta - \phi)]$$

for (a) $\phi = 0$, (b) $\phi = \pi/4$, and (c) $\phi = \pi/2$. Use the graphs to describe the effect of the angle ϕ. Write the equation as a function of $\sin \theta$ for part (c).

61. The graph of $r = f(\theta)$ is rotated about the pole through an angle ϕ. Show that the equation of the rotated graph is $r = f(\theta - \phi)$.

62. Consider the graph of $r = f(\sin \theta)$.

(a) Show that if the graph is rotated counterclockwise $\pi/2$ radians about the pole, the equation of the rotated graph is $r = f(-\cos \theta)$.

(b) Show that if the graph is rotated counterclockwise π radians about the pole, the equation of the rotated graph is $r = f(-\sin \theta)$.

(c) Show that if the graph is rotated counterclockwise $3\pi/2$ radians about the pole, the equation of the rotated graph is $r = f(\cos \theta)$.

In Exercises 63–66, use the results of Exercises 61 and 62.

63. Write an equation for the limaçon $r = 2 - \sin \theta$ after it has been rotated by the given amount.

(a) $\dfrac{\pi}{4}$ (b) $\dfrac{\pi}{2}$ (c) π (d) $\dfrac{3\pi}{2}$

64. Write an equation for the rose curve $r = 2 \sin 2\theta$ after it has been rotated by the given amount.

(a) $\dfrac{\pi}{6}$ (b) $\dfrac{\pi}{2}$ (c) $\dfrac{2\pi}{3}$ (d) π

65. Sketch the graph of each equation.

(a) $r = 1 - \sin \theta$ (b) $r = 1 - \sin\left(\theta - \dfrac{\pi}{4}\right)$

66. Sketch the graph of each equation.

(a) $r = 3 \sec \theta$ (b) $r = 3 \sec\left(\theta - \dfrac{\pi}{4}\right)$

(c) $r = 3 \sec\left(\theta + \dfrac{\pi}{3}\right)$ (d) $r = 3 \sec\left(\theta - \dfrac{\pi}{2}\right)$

67. *Exploration* Use a graphing utility to graph and identify $r = 2 + k \sin \theta$ for $k = 0, 1, 2,$ and 3.

68. *Exploration* Consider the equation $r = 3 \sin k\theta$.

(a) Use a graphing utility to graph the equation for $k = 1.5$. Find the interval for θ over which the graph is traced only once.

(b) Use a graphing utility to graph the equation for $k = 2.5$. Find the interval for θ over which the graph is traced only once.

(c) Is it possible to find an interval for θ over which the graph is traced only once for any rational number k? Explain.

Review

In Exercises 69–74, solve the equation algebraically. Round the result to three decimal places.

69. $e^x = 19$ **70.** $6e^x = 47$

71. $10^x = 84$ **72.** $5^{2x} = 60$

73. $\ln x = 4$ **74.** $4 \ln 4x = 18$

In Exercises 75–78, find the zeros (if any) of the rational function.

75. $y = \dfrac{x^2 - 9}{x + 1}$ **76.** $y = 6 + \dfrac{4}{x^2 + 4}$

77. $y = 5 - \dfrac{3}{x - 2}$ **78.** $y = \dfrac{x^3 - 27}{x^2 + 4}$

10.9 Polar Equations of Conics

▶ **What you should learn**

- How to define a conic in terms of eccentricity
- How to write equations of conics in polar form
- How to use equations of conics in polar form to model real-life problems

▶ **Why you should learn it**

The ellipitical orbits of planets and satellites can be modeled with polar equations. For instance, Exercise 49 on page 839 uses a polar equation to model a planet's orbit.

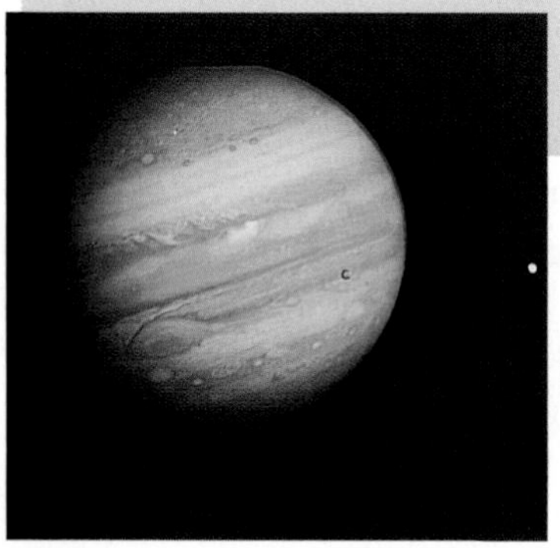

NASA

Alternative Definition of Conics

In Sections 10.3 and 10.4, you learned that the rectangular equations of ellipses and hyperbolas take simple forms when the origin lies at their *centers*. As it happens, there are many important applications of conics in which it is more convenient to use one of the *foci* as the origin of the coordinate system. For example, the sun lies at a focus of the earth's orbit. In this section you will learn that polar equations of conics take simple forms if one of the foci lies at the pole.

To begin, consider the following alternative definition of conic that uses the concept of eccentricity.

Alternative Definition of Conic

The locus of a point in the plane that moves so that its distance from a fixed point (focus) is in a constant ratio to its distance from a fixed line (directrix) is a **conic.** The constant ratio is the **eccentricity** of the conic and is denoted by e. Moreover, the conic is an **ellipse** if $e < 1$, a **parabola** if $e = 1$, and a **hyperbola** if $e > 1$.

In Figure 10.66, note that for each type of conic, the pole corresponds to the fixed point (focus) given in the definition.

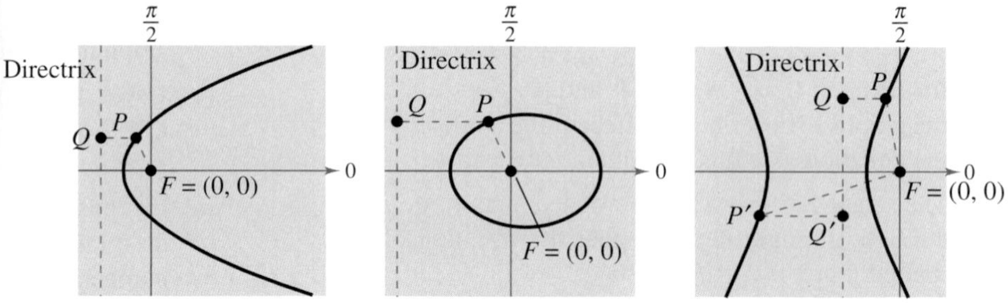

Parabola: $e = 1$ *Ellipse:* $0 < e < 1$, $\frac{PF}{PQ} < 1$ *Hyperbola:* $e > 1$, $\frac{PF}{PQ} = \frac{P'F}{P'Q'} > 1$

FIGURE 10.66

A computer simulation of this concept appears in the *Interactive* CD-ROM and *Internet* versions of this text.

Polar Equations of Conics

The benefit of locating a focus of a conic at the pole is that the equation of the conic takes on a simpler form. A proof of the polar form is given in Appendix A.

Polar Equations of Conics

The graph of a polar equation of the form

1. $r = \dfrac{ep}{1 \pm e\cos\theta}$ **2.** $r = \dfrac{ep}{1 \pm e\sin\theta}$

is a conic, where $e > 0$ is the eccentricity and $|p|$ is the distance between the focus (pole) and the directrix.

The equations

$$r = \frac{ep}{1 \pm e \cos \theta}$$ Vertical directrix

correspond to conics with vertical directrices, and the equations

$$r = \frac{ep}{1 \pm e \sin \theta}$$ Horizontal directrix

correspond to conics with horizontal directrices. Moreover, the converse is also true—that is, any conic with a focus at the pole and having a horizontal or vertical directrix can be represented by one of the given equations.

Example 1 ▶ Determining a Conic from Its Equation

Identify the conic and sketch its graph.

$$r = \frac{15}{3 - 2 \cos \theta}$$

Solution

To identify the type of conic, rewrite the equation as

$$r = \frac{15}{3 - 2 \cos \theta} = \frac{5}{1 - (2/3) \cos \theta}.$$ Divide numerator and denominator by 3.

From this form you can conclude that the graph is an ellipse with $e = \frac{2}{3}$. You can sketch the upper half of the ellipse by plotting points from $\theta = 0$ to $\theta = \pi$, as shown in Figure 10.67. Using symmetry with respect to the polar axis, you can sketch the lower half.

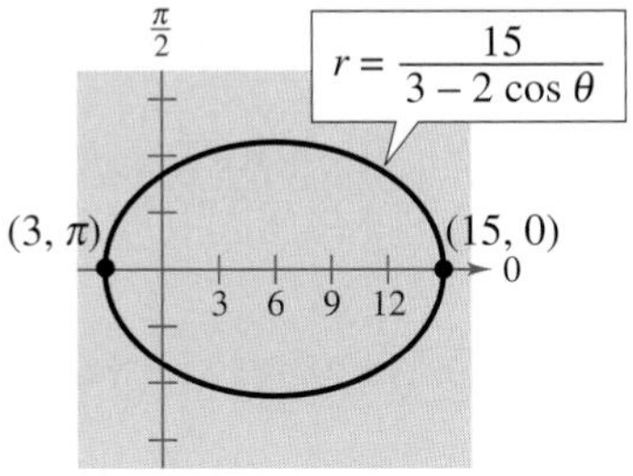

FIGURE 10.67

For the ellipse in Figure 10.67, the major axis is horizontal and the vertices lie at $(15, 0)$ and $(3, \pi)$. So, the length of the *major* axis is $2a = 18$. To find the length of the *minor* axis, you can use the equations $e = c/a$ and $b^2 = a^2 - c^2$ to conclude that

$$b^2 = a^2 - c^2 = a^2 - (ea)^2 = a^2(1 - e^2).$$ Ellipse

Because $e = \frac{2}{3}$, you have $b^2 = 9^2\left[1 - \left(\frac{2}{3}\right)^2\right] = 45$, which implies that $b = \sqrt{45} = 3\sqrt{5}$. So, the length of the minor axis is $2b = 6\sqrt{5}$. A similar analysis for hyperbolas yields

$$b^2 = c^2 - a^2 = (ea)^2 - a^2 = a^2(e^2 - 1).$$ Hyperbola

Example 2 ▶ Sketching a Conic from Its Polar Equation

Identify the conic $r = 32/(3 + 5 \sin \theta)$ and sketch its graph.

Solution

Dividing each term by 3, you have

$$r = \frac{32/3}{1 + (5/3) \sin \theta}.$$

Because $e = \frac{5}{3} > 1$, the graph is a hyperbola. The transverse axis of the hyperbola lies on the line $\theta = \pi/2$, and the vertices occur at $(4, \pi/2)$ and $(-16, 3\pi/2)$. Because the length of the transverse axis is 12, you can see that $a = 6$. To find b, write

$$b^2 = a^2(e^2 - 1) = 6^2\left[\left(\frac{5}{3}\right)^2 - 1\right] = 64.$$

Therefore, $b = 8$. Finally, you can use a and b to determine the asymptotes of the hyperbola and obtain the sketch shown in Figure 10.68.

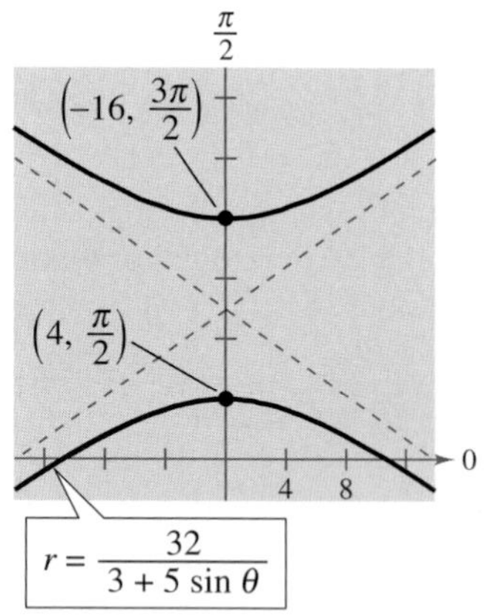

FIGURE 10.68

In the next example you are asked to find a polar equation of a specified conic. To do this, let p be the distance between the pole and the directrix.

1. *Horizontal directrix above the pole:* $r = \dfrac{ep}{1 + e \sin \theta}$
2. *Horizontal directrix below the pole:* $r = \dfrac{ep}{1 - e \sin \theta}$
3. *Vertical directrix to the right of the pole:* $r = \dfrac{ep}{1 + e \cos \theta}$
4. *Vertical directrix to the left of the pole:* $r = \dfrac{ep}{1 - e \cos \theta}$

Technology

Most graphing utilities have a polar mode. Try using a graphing utility set in polar mode to verify the four orientations shown at the right. Remember that e must be positive, but p can be positive or negative.

Example 3 ▶ Finding the Polar Equation of a Conic

Find the polar equation of the parabola whose focus is the pole and whose directrix is the line $y = 3$.

Solution

From Figure 10.69, you can see that the directrix is horizontal and above the pole. So, you can choose an equation of the form

$$r = \frac{ep}{1 + e \sin \theta}.$$

Moreover, because the eccentricity of a parabola is $e = 1$ and the distance between the pole and the directrix is $p = 3$, you have the equation

$$r = \frac{3}{1 + \sin \theta}.$$

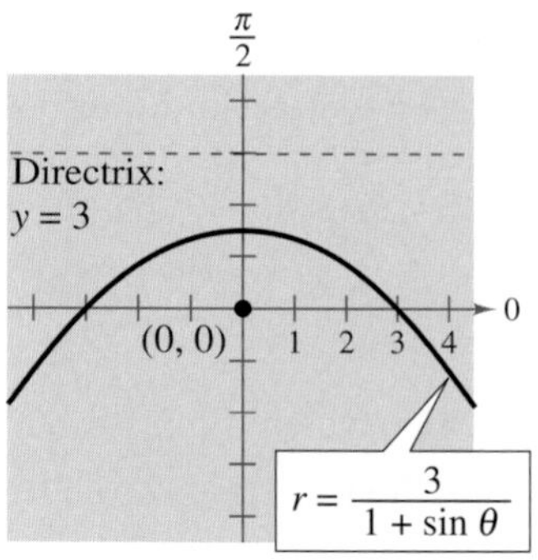

FIGURE 10.69

Activities

1. Find a polar equation of the parabola with focus at the pole and directrix $y = 2$.

   Answer: $r = \dfrac{2}{1 + \sin\theta}$

2. Identify the graph of the conic $r = \dfrac{4}{3 - 2\cos\theta}$

   Answer: Ellipse

Quick Review

True or false:

1. For converting rectangular to polar coordinates, $x = r\cos\theta$.
   Answer: True
2. The graph of $2r = 3\sec\theta$ is symmetric with respect to the polar axis.
   Answer: True
3. The graph of $r = \frac{5}{4}\sin\theta$ is a lemniscate.
   Answer: False
4. The directrix of the polar equation $r = \dfrac{0.5}{1 - 0.25\cos\theta}$ is a conic with a vertical directrix to the right of the pole.
   Answer: False

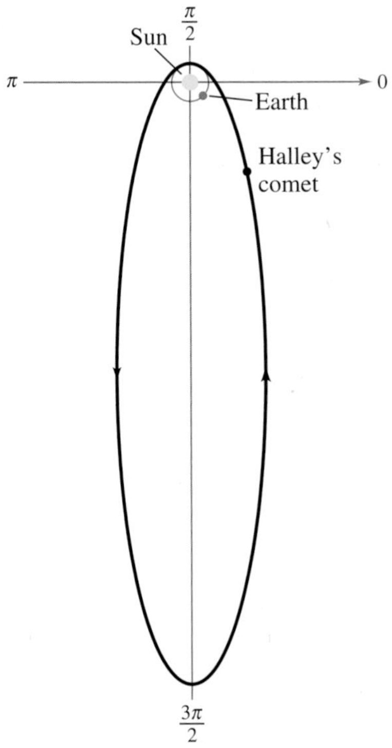

FIGURE 10.70

Applications

Kepler's Laws (listed below), named after the German astronomer Johannes Kepler (1571–1630), can be used to describe the orbits of the planets about the sun.

1. Each planet moves in an elliptical orbit with the sun at one focus.
2. A ray from the sun to the planet sweeps out equal areas of the ellipse in equal times.
3. The square of the period is proportional to the cube of the mean distance between the planet and the sun.

Although Kepler simply stated these laws on the basis of observation, they were later validated by Isaac Newton (1642–1727). In fact, Newton was able to show that each law can be deduced from a set of universal laws of motion and gravitation that govern the movement of all heavenly bodies, including comets and satellites. This is illustrated in the next example, which involves the comet named after the English mathematician and physicist Edmund Halley (1656–1742).

If you use earth as a reference with a period of 1 year and a distance of 1 astronomical unit, the proportionality constant in Kepler's third law is 1. For example, because Mars has a mean distance to the sun of $d = 1.523$ astronomical units, its period P is given by $d^3 = P^2$. So, the period of Mars is $P = 1.88$ years.

Example 4 ▶ Halley's Comet

Halley's comet has an elliptical orbit with an eccentricity of $e \approx 0.97$. The length of the major axis of the orbit is approximately 36.18 astronomical units. (An *astronomical unit* is defined as the mean distance between earth and the sun, or about 93 million miles.) Find a polar equation for the orbit. How close does Halley's comet come to the sun?

Solution

Using a vertical axis, as shown in Figure 10.70, choose an equation of the form $r = ep/(1 + e\sin\theta)$. Because the vertices of the ellipse occur when $\theta = \pi/2$ and $\theta = 3\pi/2$, you can determine the length of the major axis to be the sum of the r-values of the vertices. That is,

$$2a = \frac{0.97p}{1 + 0.97} + \frac{0.97p}{1 - 0.97} \approx 32.83p \approx 36.18.$$

So, $p \approx 1.102$ and $ep \approx (0.97)(1.102) \approx 1.069$. Using this value of ep in the equation, you have

$$r = \frac{1.069}{1 + 0.97\sin\theta}$$

where r is measured in astronomical units. To find the closest point to the sun (the focus), substitute $\theta = \pi/2$ in this equation to obtain

$$r = \frac{1.069}{1 + 0.97\sin(\pi/2)} \approx 0.54 \text{ astronomical unit} \approx 50{,}000{,}000 \text{ miles.}$$

10.9 Exercises

In Exercises 1–4, use a graphing utility to graph the polar equation for $e = 1$, $e = 0.5$, and $e = 1.5$. What can you conclude?

1. $r = \dfrac{4e}{1 + e \cos \theta}$

2. $r = \dfrac{4e}{1 - e \cos \theta}$

3. $r = \dfrac{4e}{1 - e \sin \theta}$

4. $r = \dfrac{4e}{1 + e \sin \theta}$

In Exercises 5–10, match the polar equation with its graph. [The graphs are labeled (a), (b), (c), (d), (e), and (f).]

(a)

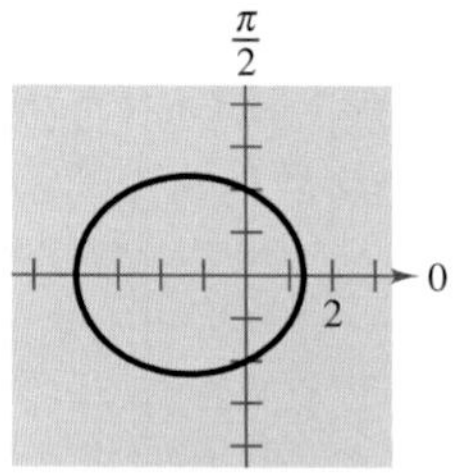

(b)

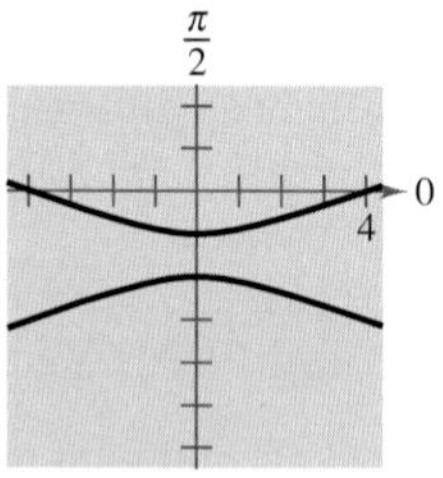

(c)

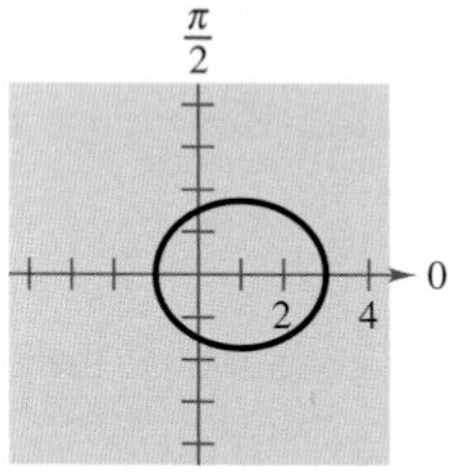

(d)

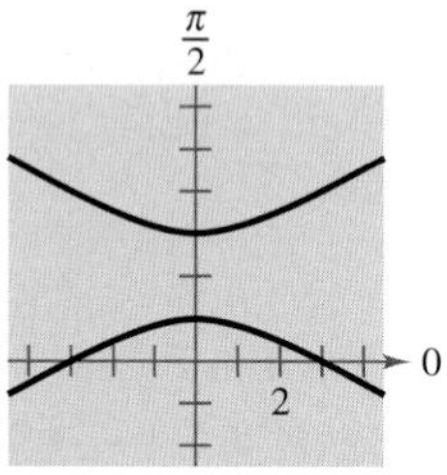

(e)

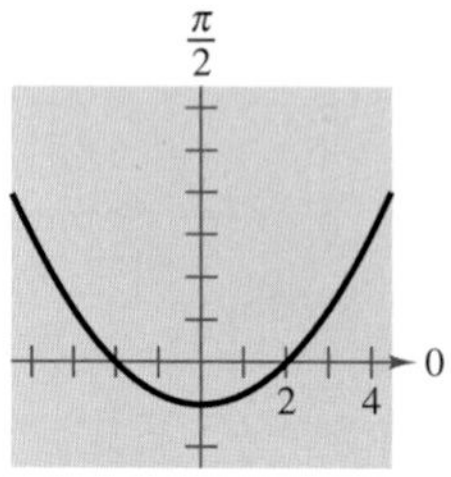

(f)

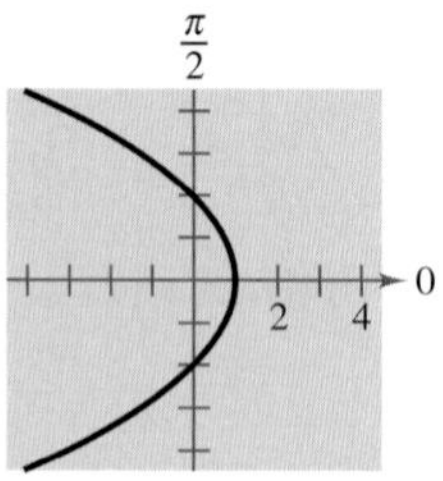

5. $r = \dfrac{2}{1 + \cos \theta}$

6. $r = \dfrac{3}{2 - \cos \theta}$

7. $r = \dfrac{3}{1 + 2 \sin \theta}$

8. $r = \dfrac{2}{1 - \sin \theta}$

9. $r = \dfrac{4}{2 + \cos \theta}$

10. $r = \dfrac{4}{1 - 3 \sin \theta}$

In Exercises 11–24, identify the conic and sketch its graph.

11. $r = \dfrac{2}{1 - \cos \theta}$

12. $r = \dfrac{3}{1 + \sin \theta}$

13. $r = \dfrac{5}{1 + \sin \theta}$

14. $r = \dfrac{6}{1 + \cos \theta}$

15. $r = \dfrac{2}{2 - \cos \theta}$

16. $r = \dfrac{3}{3 + \sin \theta}$

17. $r = \dfrac{6}{2 + \sin \theta}$

18. $r = \dfrac{9}{3 - 2 \cos \theta}$

19. $r = \dfrac{3}{2 + 4 \sin \theta}$

20. $r = \dfrac{5}{-1 + 2 \cos \theta}$

21. $r = \dfrac{3}{2 - 6 \cos \theta}$

22. $r = \dfrac{3}{2 + 6 \sin \theta}$

23. $r = \dfrac{4}{2 - \cos \theta}$

24. $r = \dfrac{2}{2 + 3 \sin \theta}$

In Exercises 25–28, use a graphing utility to graph the polar equation. Identify the graph.

25. $r = \dfrac{-1}{1 - \sin \theta}$

26. $r = \dfrac{-5}{2 + 4 \sin \theta}$

27. $r = \dfrac{3}{-4 + 2 \cos \theta}$

28. $r = \dfrac{4}{1 - 2 \cos \theta}$

In Exercises 29–32, use a graphing utility to graph the rotated conic.

29. $r = \dfrac{2}{1 - \cos(\theta - \pi/4)}$ (See Exercise 11.)

30. $r = \dfrac{3}{3 + \sin(\theta - \pi/3)}$ (See Exercise 16.)

31. $r = \dfrac{6}{2 + \sin(\theta + \pi/6)}$ (See Exercise 17.)

32. $r = \dfrac{5}{-1 + 2 \cos(\theta + 2\pi/3)}$ (See Exercise 20.)

In Exercises 33–48, find a polar equation of the conic with its focus at the pole.

	Conic	*Eccentricity*	*Directrix*
33.	Parabola	$e = 1$	$x = -1$
34.	Parabola	$e = 1$	$y = -2$
35.	Ellipse	$e = \frac{1}{2}$	$y = 1$
36.	Ellipse	$e = \frac{3}{4}$	$y = -3$
37.	Hyperbola	$e = 2$	$x = 1$
38.	Hyperbola	$e = \frac{3}{2}$	$x = -1$

	Conic	*Vertex or Vertices*
39.	Parabola	$(1, -\pi/2)$
40.	Parabola	$(6, 0)$
41.	Parabola	$(5, \pi)$
42.	Parabola	$(10, \pi/2)$
43.	Ellipse	$(2, 0), (10, \pi)$
44.	Ellipse	$(2, \pi/2), (4, 3\pi/2)$
45.	Ellipse	$(20, 0), (4, \pi)$
46.	Hyperbola	$(2, 0), (8, 0)$
47.	Hyperbola	$(1, 3\pi/2), (9, 3\pi/2)$
48.	Hyperbola	$(4, \pi/2), (-1, 3\pi/2)$

49. *Planetary Motion* The planets travel in elliptical orbits with the sun at one focus. Assume that the focus is at the pole, the major axis lies on the polar axis, and the length of the major axis is $2a$ (see figure). Show that the polar equation of the orbit is

$$r = \frac{(1 - e^2)a}{1 - e \cos \theta}$$

where e is the eccentricity.

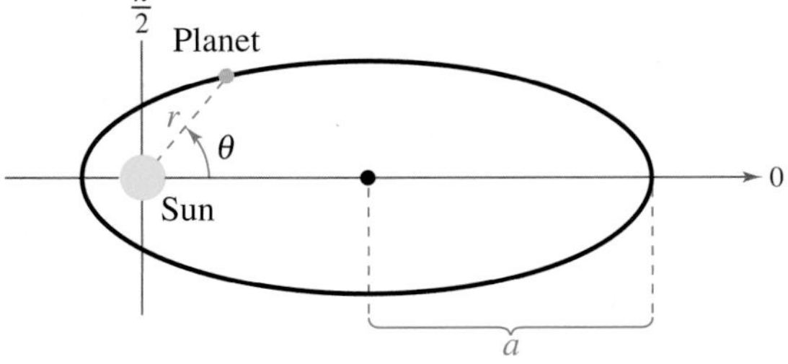

50. *Planetary Motion* Use the result of Exercise 49 to show that the minimum distance (*perihelion distance*) from the sun to the planet is $r = a(1 - e)$ and the maximum distance (*aphelion distance*) is $r = a(1 + e)$.

In Exercises 51–56, use the results of Exercises 49 and 50 to find the polar equation of the planet's orbit and the perihelion and aphelion distances.

51. Earth $a = 92.960 \times 10^6$ miles
$e = 0.0167$

52. Saturn $a = 1.429 \times 10^9$ kilometers
$e = 0.0543$

53. Pluto $a = 5.900 \times 10^9$ kilometers
$e = 0.2481$

54. Mercury $a = 35.98 \times 10^6$ miles
$e = 0.2056$

55. Mars $a = 141.00 \times 10^6$ miles
$e = 0.0934$

56. Jupiter $a = 778.40 \times 10^6$ kilometers
$e = 0.0484$

57. *Satellite Tracking* A satellite in a 100-mile-high circular orbit around earth has a velocity of approximately 17,500 miles per hour. If this velocity is multiplied by $\sqrt{2}$, the satellite will have the minimum velocity necessary to escape the earth's gravity and it will follow a parabolic path with the center of earth as the focus (see figure). Find a polar equation of the parabolic path of the satellite (assume the radius of earth is 4000 miles). Find the distance between the surface of the earth and the satellite when $\theta = 30°$.

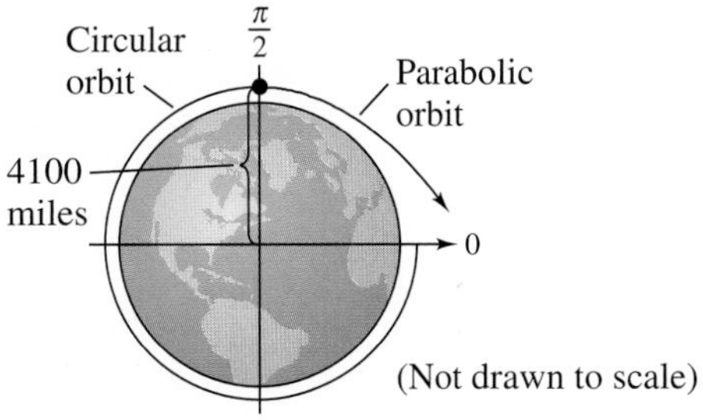

58. Use the equation found in Exercise 57. Find the distance between the surface of the earth and the satellite when $\theta = 60°$.

Synthesis

True or False? **In Exercises 59 and 60, determine whether the statement is true or false. Justify your answer.**

59. If $r = \dfrac{ep}{1 \pm e \sin \theta}$ is the equation of an ellipse with $e < 1$, changing the value of p will affect the length of the major axis, but will not affect the length of the minor axis.

60. For a given value of $e > 1$ over the interval $\theta = 0$ to $\theta = 2\pi$, the graph of $r = \dfrac{ex}{1 - e \cos \theta}$ is the same as the graph of $r = \dfrac{e(-x)}{1 + e \cos \theta}$.

61. Show that the polar equation of the ellipse

$$\frac{x^2}{a^2} + \frac{y^2}{b^2} = 1 \quad \text{is} \quad r^2 = \frac{b^2}{1 - e^2 \cos^2 \theta}.$$

62. Show that the polar equation of the hyperbola

$$\frac{x^2}{a^2} - \frac{y^2}{b^2} = 1 \quad \text{is} \quad r^2 = \frac{-b^2}{1 - e^2 \cos^2 \theta}.$$

In Exercises 63–68, use the results of Exercises 61 and 62 to write the polar form of the equation of the conic.

63. $\dfrac{x^2}{169} + \dfrac{y^2}{144} = 1$

64. $\dfrac{x^2}{25} + \dfrac{y^2}{16} = 1$

65. $\dfrac{x^2}{9} - \dfrac{y^2}{16} = 1$

66. $\dfrac{x^2}{36} - \dfrac{y^2}{4} = 1$

67. Hyperbola One focus: $(5, \pi/2)$
Vertices: $(4, \pi/2), (4, -\pi/2)$

68. Ellipse One focus: $(4, 0)$
Vertices: $(5, 0), (5, \pi)$

Review

In Exercises 69–74, solve the trigonometric equation.

69. $4\sqrt{3} \tan \theta - 3 = 1$

70. $6 \cos x - 2 = 1$

71. $12 \sin^2 \theta = 9$

72. $9 \csc^2 x - 10 = 2$

73. $2 \cot x = 5 \cos \dfrac{\pi}{2}$

74. $\sqrt{2} \sec \theta = 2 \csc \dfrac{\pi}{4}$

In Exercises 75–78, find the value of the trigonometric function given that u and v are in Quadrant IV and $\sin u = -\frac{3}{5}$ and $\cos v = 1/\sqrt{2}$.

75. $\cos(u + v)$

76. $\sin(u + v)$

77. $\cos(u - v)$

78. $\sin(u - v)$

In Exercises 79 and 80, find the exact values of $\sin 2u$, $\cos 2u$, and $\tan 2u$ using the double-angle formulas.

79. $\sin u = \dfrac{4}{5}, \ \dfrac{\pi}{2} < u < \pi$

80. $\tan u = -\sqrt{3}, \ \dfrac{3\pi}{2} < u < 2\pi$

Chapter Summary

What did you learn?

Section 10.1	Review Exercises
☐ How to find the inclination of a line and the angle between two lines	1–8
☐ How to find the distance between a point and a line	9, 10
Section 10.2	
☐ How to recognize a conic as the intersection of a plane and a cone	11, 12
☐ How to write the standard form of the equation of a parabola	13–16
☐ How to use the reflective property of parabolas to solve real-life problems	17–20
Section 10.3	
☐ How to write the standard form of the equation of an ellipse	21–24
☐ How to use properties of ellipses to model and solve real-life problems	25, 26
☐ How to find the eccentricity of an ellipse	27–30
Section 10.4	
☐ How to write the standard form of the equation of a hyperbola	31–34
☐ How to find the asymptotes of a hyperbola	35–38
☐ How to use properties of hyperbolas to solve real-life problems	39, 40
☐ How to classify a conic from its general equation	41, 42
Section 10.5	
☐ How to rotate coordinate axes to eliminate the *xy*-term of a conic	43–46
☐ How to use the discriminant to classify a conic	47–50
Section 10.6	
☐ How to evaluate a set of parametric equations for a given value of the parameter	51–54
☐ How to sketch the curve that is represented by a set of parametric equations and rewrite the equations as a single rectangular equation	55–60
☐ How to find a set of parametric equations for a graph	61–64
Section 10.7	
☐ How to plot points in the polar coordinate system	65–68
☐ How to convert points from rectangular to polar form and vice versa	69–72
☐ How to convert equations from rectangular to polar form and vice versa	73–78
Section 10.8	
☐ How to graph a polar equation by point plotting	79–84
☐ How to use symmetry, zeros, and maximum *r*-values as graphing aids	85–88
☐ How to recognize special polar graphs	89–92
Section 10.9	
☐ How to define a conic in terms of eccentricity	93–96
☐ How to write equations of conics in polar form	97–100
☐ How to use equations of conics in polar form to model real-life problems	101, 102

Review Exercises

10.1 In Exercises 1–4, find the inclination θ of the line with the given characteristics.

1. Slope: $m = \frac{3}{5}$

2. Passes through the points $(3, 4)$ and $(-2, 7)$

3. Equation: $y = 2x + 4$

4. Equation: $6x - 7y - 5 = 0$

In Exercises 5–8, find the angle θ between the lines.

5. $4x + y = 2$
$-5x + y = -1$

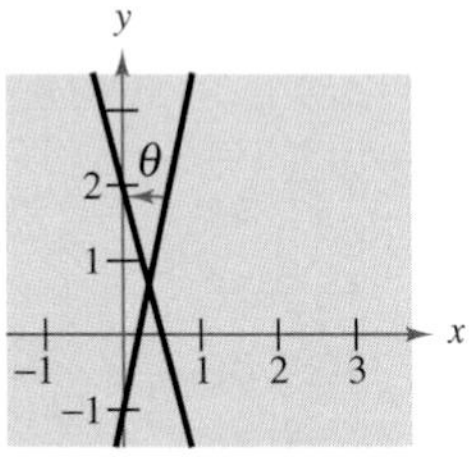

6. $-5x + 3y = 3$
$-2x + 3y = 1$

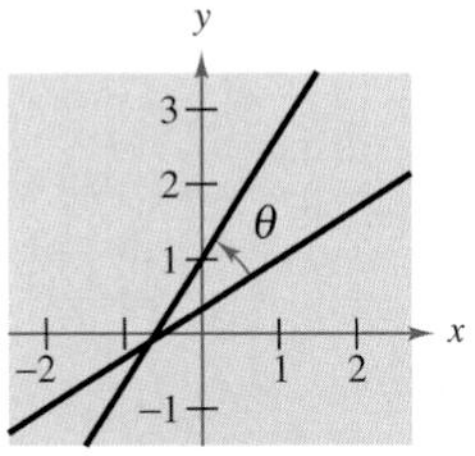

7. $2x - 7y = 8$
$0.4x + y = 0$

8. $0.02x + 0.07y = 0.18$
$0.09x - 0.04y = 0.17$

In Exercises 9 and 10, find the distance between the point and the line.

	Point	*Line*
9.	$(1, 2)$	$x - y - 3 = 0$
10.	$(0, 4)$	$x + 2y - 2 = 0$

10.2 In Exercises 11 and 12, state what type of conic is formed by the intersection of the plane and the double-napped cone.

11.

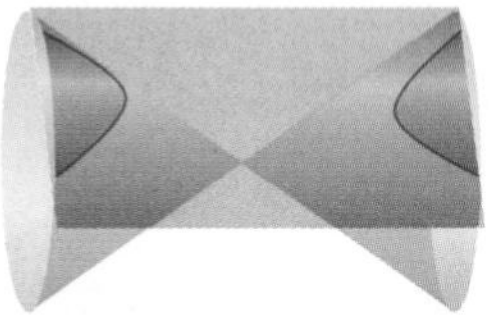

12.

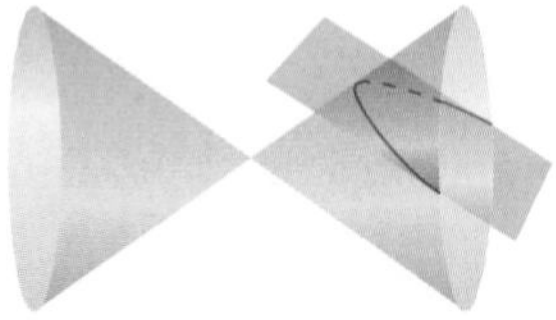

In Exercises 13–16, find the standard form of the equation of the parabola.

13. Vertex: $(4, 2)$
Focus: $(4, 0)$

14. Vertex: $(2, 0)$
Focus: $(0, 0)$

15. Vertex: $(0, 2)$
Directrix: $x = -3$

16. Vertex: $(2, 2)$
Directrix: $y = 0$

In Exercises 17 and 18, find an equation of a tangent line to the parabola at the given point, and find the x-intercept of the line.

17. $x^2 = -2y,\ (2, -2)$

18. $x^2 = -2y,\ (-4, -8)$

19. *Parabolic Archway* A parabolic archway is 12 meters high at the vertex. At a height of 10 meters, the width of the archway is 8 meters (see figure). How wide is the archway at ground level?

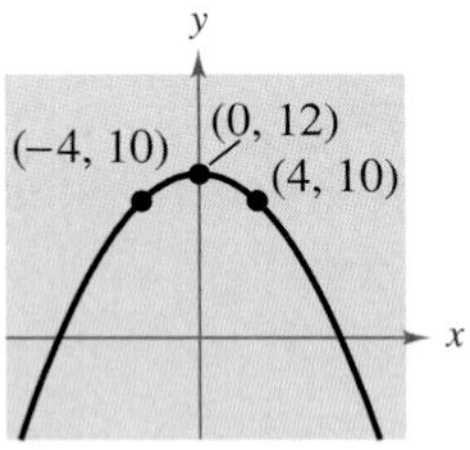

FIGURE FOR 19

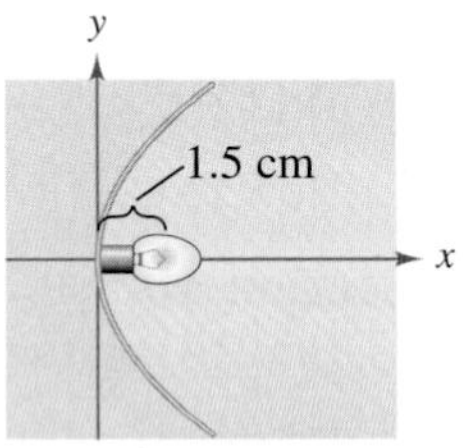

FIGURE FOR 20

20. *Flashlight* The light bulb in a flashlight is at the focus of its parabolic reflector, 1.5 centimeters from the vertex of the reflector (see figure). Write an equation for a cross section of the flashlight's reflector with its focus on the positive x-axis and its vertex at the origin.

10.3 In Exercises 21–24, find the standard form of the equation of the ellipse.

21. Vertices: $(-3, 0), (7, 0)$; Foci: $(0, 0), (4, 0)$

22. Vertices: $(2, 0), (2, 4)$; Foci: $(2, 1), (2, 3)$

23. Vertices: $(0, \pm 6)$; Passes through $(2, 2)$

24. Vertices: $(0, 1), (4, 1)$;
Endpoints of the minor axis: $(2, 0), (2, 2)$

25. *Semielliptical Archway* A semielliptical archway is set on pillars that are 10 feet apart. Its height (atop the pillars) is 4 feet. Where should the foci be placed in order to sketch the semielliptical arch?

26. *Wading Pool* You are building a wading pool that is in the shape of an ellipse. Your plans give an equation for the elliptical shape of the pool measured in feet as

$$\frac{x^2}{324} + \frac{y^2}{196} = 1.$$

Find the longest distance across the pool, the shortest distance, and the distance between the foci.

In Exercises 27–30, find the center, vertices, foci, and eccentricity of the ellipse.

27. $16x^2 + 9y^2 - 32x + 72y + 16 = 0$

28. $4x^2 + 25y^2 + 16x - 150y + 141 = 0$

29. $\dfrac{(x + 2)^2}{81} + \dfrac{(y - 1)^2}{100} = 1$

30. $\dfrac{(x - 5)^2}{1} + \dfrac{(y + 3)^2}{36} = 1$

10.4 In Exercises 31–34, find the standard form of the equation of the hyperbola.

31. Vertices: $(0, \pm1)$; Foci: $(0, \pm3)$

32. Vertices: $(2, 2), (-2, 2)$; Foci: $(4, 2), (-4, 2)$

33. Foci: $(0, 0), (8, 0)$; Asymptotes: $y = \pm2(x - 4)$

34. Foci: $(3, \pm2)$; Asymptotes: $y = \pm2(x - 3)$

In Exercises 35–38, find the center, vertices, foci, and the equations of the asymptotes of the hyperbola. Then sketch its graph.

35. $9x^2 - 16y^2 - 18x - 32y - 151 = 0$

36. $-4x^2 + 25y^2 - 8x + 150y + 121 = 0$

37. $\dfrac{(x - 3)^2}{16} - \dfrac{(y + 5)^2}{4} = 1$

38. $\dfrac{(y - 1)^2}{4} - x^2 = 1$

39. *Loran* Radio transmitting station A is located 200 miles east of transmitting station B. A ship is in an area to the north and 40 miles west of station A. Synchronized radio pulses transmitted at 186,000 miles per second by the two stations are received 0.0005 second sooner from station A than from station B. How far north is the ship?

40. *Locating an Explosion* Two of your friends live 4 miles apart and on the same "east-west" street, and you live halfway between them. You are talking on a three-way phone call when you hear an explosion. Six seconds later your friend to the east hears the explosion, and your friend to the west hears it 8 seconds after you do. Find equations of two hyperbolas that would locate the explosion. (Sound travels at a rate of 1100 feet per second.)

In Exercises 41 and 42, classify the conic from its general equation.

41. $5x^2 - 2y^2 + 10x - 4y + 17 = 0$

42. $-4y^2 + 5x + 3y + 7 = 0$

10.5 In Exercises 43–46, rotate the axes to eliminate the *xy*-term. Sketch the graph of the resulting equation, showing both sets of axes.

43. $xy - 4 = 0$

44. $x^2 - 10xy + y^2 + 1 = 0$

45. $5x^2 - 2xy + 5y^2 - 12 = 0$

46. $4x^2 + 8xy + 4y^2 + 7\sqrt{2}x + 9\sqrt{2}y = 0$

In Exercises 47–50, use the discriminant to classify the graph, use the Quadratic Formula to solve for *y*, and then use a graphing utility to graph the equation.

47. $16x^2 - 24xy + 9y^2 - 30x - 40y = 0$

48. $13x^2 - 8xy + 7y^2 - 45 = 0$

49. $x^2 + y^2 + 2xy + 2\sqrt{2}x - 2\sqrt{2}y + 2 = 0$

50. $x^2 - 10xy + y^2 + 1 = 0$

10.6 In Exercises 51–54, evaluate the parametric equations $x = 3\cos\theta$ and $y = 2\sin^2\theta$ for the given value of θ.

51. $\theta = 0$

52. $\theta = \dfrac{\pi}{3}$

53. $\theta = \dfrac{\pi}{6}$

54. $\theta = -\dfrac{\pi}{4}$

In Exercises 55–60, sketch the curve represented by the parametric equations and, where possible, write the corresponding rectangular equation by eliminating the parameter. Verify your result with a graphing utility.

55. $x = 2t$, $y = 4t$

56. $x = 1 + 4t$, $y = 2 - 3t$

57. $x = t^2$, $y = \sqrt{t}$

58. $x = t + 4$, $y = t^2$

59. $x = 6\cos\theta$
$y = 6\sin\theta$

60. $x = 3 + 3\cos\theta$
$y = 2 + 5\sin\theta$

61. Find a parametric representation of the ellipse with center at $(-3, 4)$, major axis horizontal and 8 units in length, and minor axis 6 units in length.

62. Find a parametric representation of the hyperbola with vertices $(0, \pm 4)$ and foci $(0, \pm 5)$.

63. *Rotary Engine* The rotary engine was developed by Felix Wankel in the 1950s. It features a rotor that is basically a modified equilateral triangle. The rotor moves in a chamber that, in two dimensions, is an epitrochoid. Use a graphing utility to graph the chamber modeled by the parametric equations $x = \cos 3\theta + 5\cos\theta$ and $y = \sin 3\theta + 5\sin\theta$.

64. *Involute of a Circle* The *involute* of a circle is described by the endpoint P of a string that is held taut as it is unwound from a spool (see figure). The spool does not rotate. Show that a parametric representation of the involute of a circle is

$$x = r(\cos\theta + \theta\sin\theta)$$

$$y = r(\sin\theta - \theta\cos\theta).$$

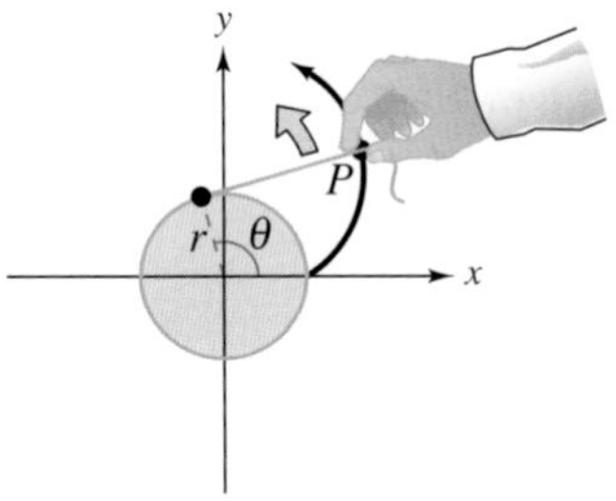

10.7 **In Exercises 65–68, plot the point given in polar coordinates and find the corresponding rectangular coordinates of the point.**

65. $\left(2, \frac{\pi}{4}\right)$

66. $\left(-5, -\frac{\pi}{3}\right)$

67. $(-7, 4.19)$

68. $\left(\sqrt{3}, 2.62\right)$

In Exercises 69–72, the rectangular coordinates of a point are given. Find two sets of polar coordinates of the point for $0 \le \theta \le 2\pi$.

69. $(0, 2)$

70. $\left(-\sqrt{5}, \sqrt{5}\right)$

71. $(4, 6)$

72. $(3, -4)$

In Exercises 73–76, convert the polar equation to rectangular form.

73. $r = 3\cos\theta$

74. $r = 10$

75. $r = \dfrac{2}{1 + \sin\theta}$

76. $r^2 = \cos 2\theta$

In Exercises 77 and 78, convert the rectangular equation to polar form.

77. $(x^2 + y^2)^2 = ax^2y$

78. $x^2 + y^2 - 4x = 0$

10.8 **In Exercises 79–84, identify and sketch the graph of the polar equation.**

79. $r = 4$

80. $r = 2\theta$

81. $r = 4\sin 2\theta$

82. $r = \cos 5\theta$

83. $r = -2(1 + \cos\theta)$

84. $r = 3 - 4\cos\theta$

In Exercises 85–88, determine the symmetry of r, the maximum value of $|r|$, and any zeros of r. Then sketch the graph of the equation.

85. $r = 2 + 6\sin\theta$

86. $r = 5 - 5\cos\theta$

87. $r = -3\cos 2\theta$

88. $r^2 = \cos 2\theta$

In Exercises 89–92, identify the type of polar graph.

89. $r = 3(2 - \cos\theta)$

90. $r = 3(1 - 2\cos\theta)$

91. $r = 4\cos 3\theta$

92. $r^2 = 9\cos 2\theta$

10.9 **In Exercises 93–96, state the eccentricity of the conic and identify the conic from its eccentricity. Sketch a graph of the conic.**

93. $r = \dfrac{1}{1 + 2\sin\theta}$

94. $r = \dfrac{2}{1 - \sin\theta}$

95. $r = \dfrac{4}{5 - 3\cos\theta}$

96. $r = \dfrac{16}{4 + 5\cos\theta}$

In Exercises 97–100, find a polar equation of the conic.

97. Parabola — Vertex: $(2, \pi)$; Focus: $(0, 0)$

98. Parabola — Vertex: $(2, \pi/2)$; Focus: $(0, 0)$

99. Ellipse — Vertices: $(5, 0), (1, \pi)$; One focus: $(0, 0)$

100. Hyperbola — Vertices: $(1, 0), (7, 0)$; One focus: $(0, 0)$

101. *Explorer 18* On November 26, 1963, the United States launched Explorer 18. Its low and high points above the surface of earth were 119 miles and 122,000 miles, respectively (see figure). The center of earth is at one focus of the orbit. Find the polar equation of the orbit and find the distance between the surface of the earth (assume a radius of 4000 miles) and the satellite when $\theta = \pi/3$ radians.

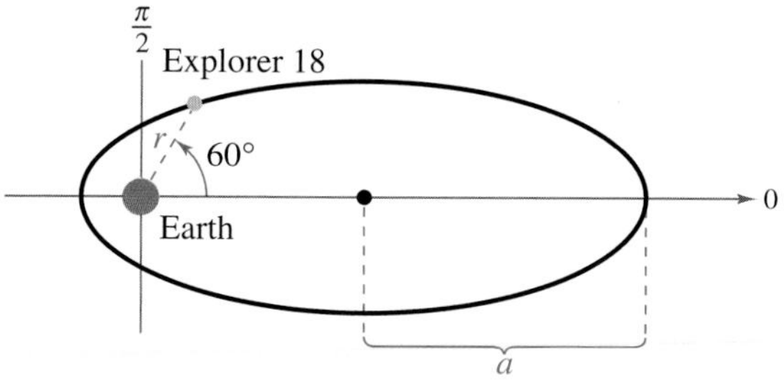

102. *Asteroid* An asteroid takes a parabolic path with earth as its focus. It is about 6,000,000 miles from earth at its closest approach. Write the polar equation of the path of the asteroid with its vertex at $\theta = \pi/2$. Find the distance between the asteroid and earth when $\theta = -\pi/3$.

Synthesis

True or False? **In Exercises 103–106, determine whether the statement is true or false. Justify your answer.**

103. When $B = 0$ in an equation of the form $Ax^2 + Bxy + Cy^2 + Dx + Ey + F = 0$, the graph of the equation can be a parabola only if $C = 0$ also.

104. The graph of $(x^2/4) - y^4 = 1$ is a hyperbola.

105. Only one set of parametric equations can represent the line $y = 3 - 2x$.

106. There is a unique polar coordinate representation of each point in the plane.

107. Consider an ellipse with the major axis horizontal and 10 units in length. The number b in the standard form of the equation of the ellipse must be less than what real number? Explain the change in the shape of the ellipse as b approaches this number.

108. The graph of the parametric equations $x = 2 \sec t$ and $y = 3 \tan t$ is given in the figure. Would the graph change for the equations $x = 2 \sec(-t)$ and $y = 3 \tan(-t)$? If so, how would it change?

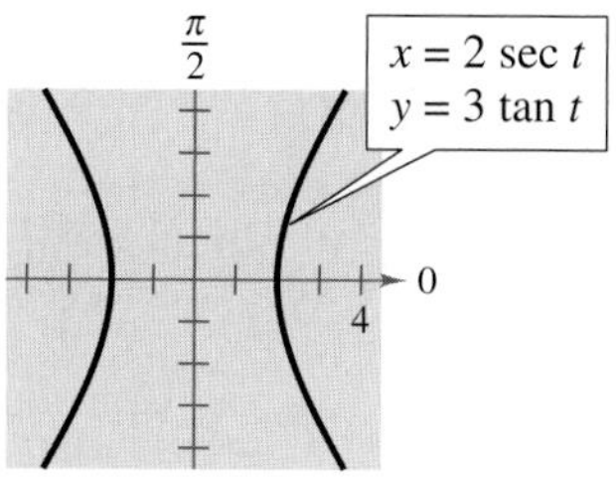

FIGURE FOR 108

109. A moving object is modeled by the parametric equations $x = 4 \cos t$ and $y = 3 \sin t$, where t is time (see figure). How would the orbit change for the following?

(a) $x = 4 \cos 2t, \quad y = 3 \sin 2t$

(b) $x = 5 \cos t, \quad y = 3 \sin t$

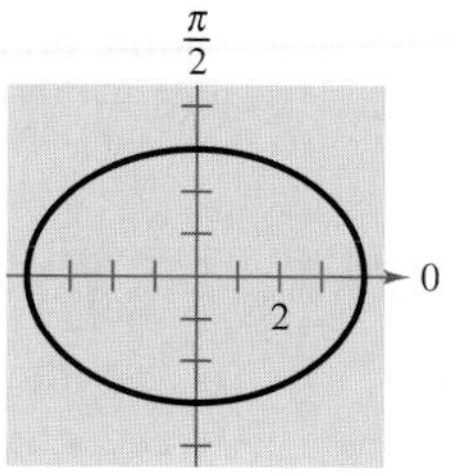

110. Identify the type of symmetry each of the following polar points has with the point in the figure.

(a) $\left(-4, \frac{\pi}{6}\right)$

(b) $\left(4, -\frac{\pi}{6}\right)$

(c) $\left(-4, -\frac{\pi}{6}\right)$

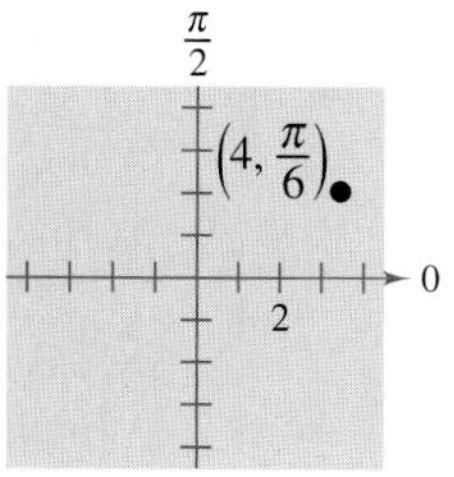

111. What is the relationship between the graphs of the rectangular and polar equations?

(a) $x^2 + y^2 = 25, \quad r = 5$

(b) $x - y = 0, \quad \theta = \frac{\pi}{4}$

Chapter Project ▶ Polar, Rectangular, and Parametric Forms

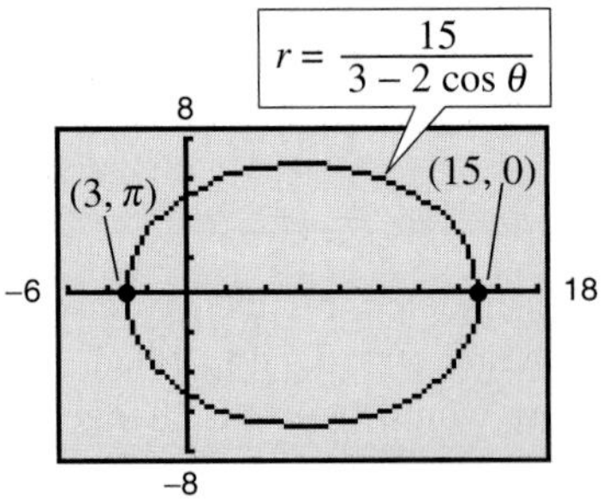

In this project, you will compare the polar, rectangular, and parametric forms of equations for conics.

(a) Consider the polar equation

$$r = \frac{15}{3 - 2\cos\theta}.$$

In Section 10.9, you learned that the graph of this polar equation is an ellipse and that one of the ellipse's foci is at the pole, as shown in the graph at the left. To find the rectangular equation of this ellipse, begin by rewriting the equation as

$$3r - 2r\cos\theta = 15.$$

Then use the substitutions $r = \sqrt{x^2 + y^2}$ and $r\cos\theta = x$ to find the rectangular equation. After you find the rectangular equation, write it in the standard form

$$\frac{(x-h)^2}{a^2} + \frac{(y-k)^2}{b^2} = 1.$$

(b) Use the standard form of the ellipse in part (a) to find the center, foci, and eccentricity of the ellipse. Compare your results with those obtained in Example 1 in Section 10.9. Use the function mode of a graphing utility to graph the ellipse.

(c) Use the standard form of the ellipse and the identity $\sin^2\theta + \cos^2\theta = 1$ to write parametric equations for the ellipse. Then use the parametric mode of a graphing utility to graph the ellipse.

Chapter Project Investigations

1. Sketch the graph of the polar equation

$$r = \frac{32}{3 - 5\sin\theta}.$$

Which type of conic is this?

2. Write the standard form of the rectangular equation of the conic in Question 1. Then sketch the graph using the *function* mode of a graphing utility. Does your graph agree with the graph obtained in Question 1? Which of the two graphs is easier to obtain? Explain your reasoning.

3. Write parametric equations for the conic given in Question 1. Then use the *parametric* mode of a graphing utility to graph the conic. Does your graph agree with the graph obtained in Question 1?

4. Consider the parametric equations

$$x = 4 + 5\cos t \quad \text{and} \quad y = 3\sin t.$$

With appropriate scaling of the coordinate system, could these equations represent the motion of a comet about the sun? Explain your reasoning.

Chapter Test

The *Interactive* CD-ROM and *Internet* versions of this text provide answers to the Chapter Tests and Cumulative Tests. They also offer Chapter Pre-Tests (which test key skills and concepts covered in previous chapters) and Chapter Post-Tests, both of which have randomly generated exercises with diagnostic capabilities.

Take this test as you would take a test in class. After you are done, check your work against the answers given in the back of the book.

1. Find the inclination of the line $2x - 7y + 3 = 0$.

2. Find the angle between the lines $3x + 2y - 4 = 0$ and $4x - y + 6 = 0$.

3. Find the distance between the point $(7, 5)$ and the line $y = 5 - x$.

In Exercises 4–7, classify the conic and write the equation in standard form. Identify the center, vertices, foci, and asymptotes (if any). Then sketch the graph of the conic.

4. $y^2 - 4x + 4 = 0$

5. $x^2 - 4y^2 - 4x = 0$

6. $9x^2 + 16y^2 + 54x - 32y - 47 = 0$

7. $2x^2 + 2y^2 - 8x - 4y + 9 = 0$

8. Find the standard form of the equation of the parabola with vertex $(3, -2)$, with vertical axis, and passing through the point $(0, 4)$.

9. Find the standard form of the equation of the hyperbola with foci $(0, 0)$ and $(0, 4)$ and asymptotes $y = \pm\frac{1}{2}x + 2$.

10. (a) Determine the number of degrees the axis must be rotated to eliminate the xy-term of the conic $x^2 + 6xy + y^2 - 6 = 0$.

(b) Graph the conic and use a graphing utility to confirm your result.

11. Sketch the curve represented by the parametric equations $x = 2 + 3\cos\theta$ and $y = 2\sin\theta$. Eliminate the parameter and write the corresponding rectangular equation.

12. Find a set of parametric equations of the line passing through the points $(2, -3)$ and $(6, 4)$. (The answer is not unique.)

13. Convert the polar coordinate $(-2, 5\pi/6)$ to rectangular form.

14. Convert the rectangular coordinate $(2, -2)$ to polar form and find two additional representations for this point.

15. Convert the rectangular equation $x^2 + y^2 - 4y = 0$ to polar form.

In Exercises 16–19, sketch the graph of the polar equation. Identify the type of graph.

16. $r = \dfrac{4}{1 + \cos\theta}$

17. $r = \dfrac{4}{2 + \cos\theta}$

18. $r = 2 + 3\sin\theta$

19. $r = 3\sin 2\theta$

20. A straight road rises with an inclination of 0.15 radian from the horizontal. Find the slope of the road and the change in elevation over a 1-mile stretch of the road.

21. A baseball is hit 3 feet above the ground toward the left field fence. The fence is 10 feet high and 375 feet from home plate. The path of the baseball can be modeled by the parametric equations $x = (115\cos\theta)t$ and $y = 3 + (115\sin\theta)t - 16t^2$. Does the baseball go over the fence when the baseball is hit at an angle of $\theta = 30°$? Does the baseball go over the fence when $\theta = 35°$?

Appendix A Proofs of Selected Theorems

SECTION P.8, PAGE 85

The Midpoint Formula

The midpoint of the segment joining the points (x_1, y_1) and (x_2, y_2) is given by the Midpoint Formula

$$\text{Midpoint} = \left(\frac{x_1 + x_2}{2}, \frac{y_1 + y_2}{2}\right).$$

Proof

Using the figure, you must show that

$$d_1 = d_2 \quad \text{and} \quad d_1 + d_2 = d_3.$$

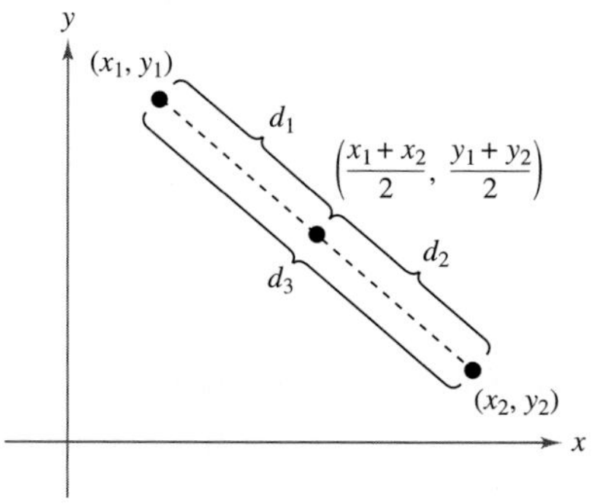

Midpoint Formula

By the Distance Formula, you obtain

$$d_1 = \sqrt{\left(\frac{x_1 + x_2}{2} - x_1\right)^2 + \left(\frac{y_1 + y_2}{2} - y_1\right)^2} = \frac{1}{2}\sqrt{(x_2 - x_1)^2 + (y_2 - y_1)^2}$$

$$d_2 = \sqrt{\left(x_2 - \frac{x_1 + x_2}{2}\right)^2 + \left(y_2 - \frac{y_1 + y_2}{2}\right)^2} = \frac{1}{2}\sqrt{(x_2 - x_1)^2 + (y_2 - y_1)^2}$$

$$d_3 = \sqrt{(x_2 - x_1)^2 + (y_2 - y_1)^2}.$$

So, it follows that $d_1 = d_2$ and $d_1 + d_2 = d_3$.

SECTION 2.3, PAGE 230

The Remainder Theorem

If a polynomial $f(x)$ is divided by $x - k$, the remainder is

$$r = f(k).$$

Proof

From the Division Algorithm, you have

$$f(x) = (x - k)q(x) + r(x)$$

and because either $r(x) = 0$ or the degree of $r(x)$ is less than the degree of $x - k$, you know that $r(x)$ must be a constant. That is, $r(x) = r$. Now, by evaluating $f(x)$ at $x = k$, you have

$$f(k) = (k - k)q(k) + r = (0)q(k) + r = r.$$

SECTION 2.3, PAGE 230

The Factor Theorem

A polynomial $f(x)$ has a factor $(x - k)$ if and only if $f(k) = 0$.

Proof

Using the Division Algorithm with the factor $(x - k)$, you have

$$f(x) = (x - k)q(x) + r(x).$$

By the Remainder Theorem, $r(x) = r = f(k)$, and you have

$$f(x) = (x - k)q(x) + f(k)$$

where $q(x)$ is a polynomial of lesser degree than $f(x)$. If $f(k) = 0$, then

$$f(x) = (x - k)q(x)$$

and you see that $(x - k)$ is a factor of $f(x)$. Conversely, if $(x - k)$ is a factor of $f(x)$, division of $f(x)$ by $(x - k)$ yields a remainder of 0. So, by the Remainder Theorem, you have $f(k) = 0$.

SECTION 2.5, PAGE 243

Linear Factorization Theorem

If $f(x)$ is a polynomial of degree n, where $n > 0$, then f has precisely n linear factors

$$f(x) = a_n(x - c_1)(x - c_2) \cdot \cdot \cdot (x - c_n)$$

where $c_1, c_2, \ldots, c_n$ are complex numbers.

Proof

Using the Fundamental Theorem of Algebra, you know that f must have at least one zero, c_1. Consequently, $(x - c_1)$ is a factor of $f(x)$, and you have

$$f(x) = (x - c_1)f_1(x).$$

If the degree of $f_1(x)$ is greater than zero, you again apply the Fundamental Theorem to conclude that f_1 must have a zero c_2, which implies that

$$f(x) = (x - c_1)(x - c_2)f_2(x).$$

It is clear that the degree of $f_1(x)$ is $n - 1$, that the degree of $f_2(x)$ is $n - 2$, and that you can repeatedly apply the Fundamental Theorem n times until you obtain

$$f(x) = a_n(x - c_1)(x - c_2) \cdot \cdot \cdot (x - c_n)$$

where a_n is the leading coefficient of the polynomial $f(x)$.

SECTION 2.5, PAGE 247

Factors of a Polynomial

Every polynomial of degree $n > 0$ with real coefficients can be written as the product of linear and quadratic factors with real coefficients, where the quadratic factors have no real zeros.

Proof

To begin, you use the Linear Factorization Theorem to conclude that $f(x)$ can be *completely* factored in the form

$$f(x) = d(x - c_1)(x - c_2)(x - c_3) \cdot \cdot \cdot (x - c_n).$$

If each c_i is real, there is nothing more to prove. If any c_i is complex ($c_i = a + bi$, $b \neq 0$), then, because the coefficients of $f(x)$ are real, you know that the conjugate $c_j = a - bi$ is also a zero. By multiplying the corresponding factors, you obtain

$$\begin{aligned}(x - c_i)(x - c_j) &= [x - (a + bi)][x - (a - bi)] \\ &= x^2 - 2ax + (a^2 + b^2)\end{aligned}$$

where each coefficient is real.

SECTION 3.3, PAGE 312

Properties of Logarithms

Let a be a positive number such that $a \neq 1$, and let n be a real number. If u and v are positive real numbers, the following properties are true.

1. $\log_a(uv) = \log_a u + \log_a v$	**1.** $\ln(uv) = \ln u + \ln v$
2. $\log_a \dfrac{u}{v} = \log_a u - \log_a v$	**2.** $\ln \dfrac{u}{v} = \ln u - \ln v$
3. $\log_a u^n = n \log_a u$	**3.** $\ln u^n = n \ln u$

Proof

To prove Property 1, let $x = \log_a u$ and $y = \log_a v$. The corresponding exponential forms of these two equations are

$$a^x = u \quad \text{and} \quad a^y = v.$$

Multiplying u and v produces $uv = a^x a^y = a^{x+y}$. The corresponding logarithmic form of $uv = a^{x+y}$ is $\log_a(uv) = x + y$. So, $\log_a(uv) = \log_a u + \log_a v$.

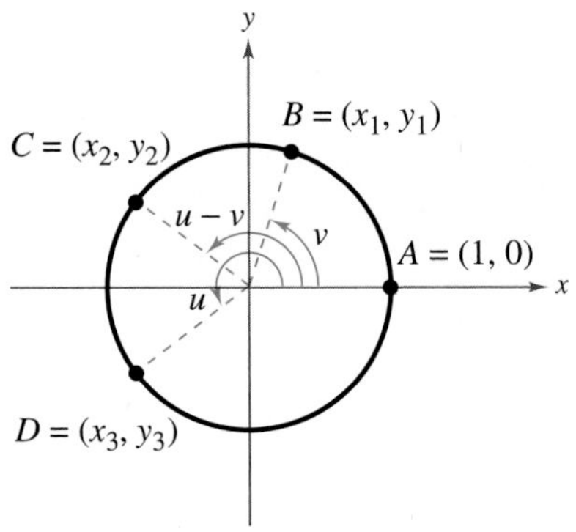

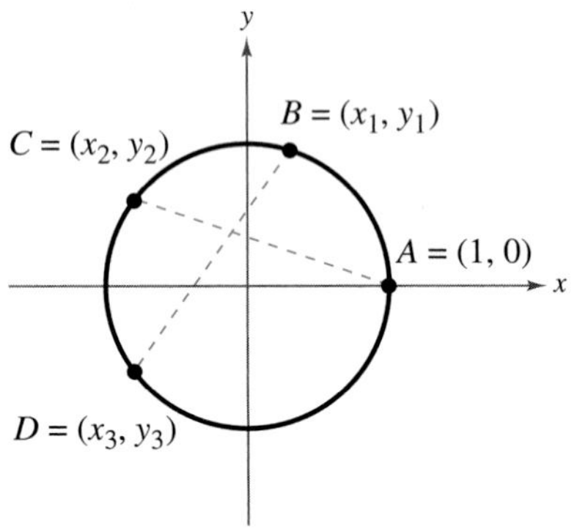

SECTION 5.4, PAGE 468

Sum and Difference Formulas

$\sin(u + v) = \sin u \cos v + \cos u \sin v$

$\sin(u - v) = \sin u \cos v - \cos u \sin v$

$\cos(u + v) = \cos u \cos v - \sin u \sin v$

$\cos(u - v) = \cos u \cos v + \sin u \sin v$

$\tan(u + v) = \dfrac{\tan u + \tan v}{1 - \tan u \tan v}$

$\tan(u - v) = \dfrac{\tan u - \tan v}{1 + \tan u \tan v}$

Proof

Here are proofs for the formulas for $\cos(u \pm v)$. In the top figure, let A be the point $(1, 0)$ and then use u and v to locate the points $B = (x_1, y_1)$, $C = (x_2, y_2)$, and $D = (x_3, y_3)$ on the unit circle. So, $x_i^2 + y_i^2 = 1$ for $i = 1, 2,$ and 3. For convenience, assume that $0 < v < u < 2\pi$. In the bottom figure, note that arcs AC and BD have the same length. So, line segments AC and BD are also equal in length, which implies that

$$\sqrt{(x_2 - 1)^2 + (y_2 - 0)^2} = \sqrt{(x_3 - x_1)^2 + (y_3 - y_1)^2}$$

$$x_2^2 - 2x_2 + 1 + y_2^2 = x_3^2 - 2x_1x_3 + x_1^2 + y_3^2 - 2y_1y_3 + y_1^2$$

$$(x_2^2 + y_2^2) + 1 - 2x_2 = (x_3^2 + y_3^2) + (x_1^2 + y_1^2) - 2x_1x_3 - 2y_1y_3$$

$$1 + 1 - 2x_2 = 1 + 1 - 2x_1x_3 - 2y_1y_3$$

$$x_2 = x_3x_1 + y_3y_1.$$

Finally, by substituting the values $x_2 = \cos(u - v)$, $x_3 = \cos u$, $x_1 = \cos v$, $y_3 = \sin u$, and $y_1 = \sin v$, you obtain $\cos(u - v) = \cos u \cos v + \sin u \sin v$.

The formula for $\cos(u + v)$ can be established by considering $u + v = u - (-v)$ and using the formula just derived to obtain

$$\begin{aligned}\cos(u + v) &= \cos[u - (-v)] \\ &= \cos u \cos(-v) + \sin u \sin(-v) \\ &= \cos u \cos v - \sin u \sin v.\end{aligned}$$

SECTION 5.5, PAGE 475

Double-Angle Formulas

$\sin 2u = 2 \sin u \cos u$

$\tan 2u = \dfrac{2 \tan u}{1 - \tan^2 u}$

$$\begin{aligned}\cos 2u &= \cos^2 u - \sin^2 u \\ &= 2\cos^2 u - 1 = 1 - 2\sin^2 u\end{aligned}$$

Proof

To prove the first formula, let $v = u$ in the formula for $\sin(u + v)$.

$$\begin{aligned}\sin 2u &= \sin(u + u) \\ &= \sin u \cos u + \cos u \sin u \\ &= 2 \sin u \cos u\end{aligned}$$

To prove the second formula, let $v = u$ in the formula for $\cos(u + v)$.

$$\begin{aligned}\cos 2u &= \cos(u + u) \\ &= \cos u \cos u - \sin u \sin u \\ &= \cos^2 u - \sin^2 u\end{aligned}$$

The tangent double-angle formula can be proven in a similar way.

SECTION 5.5, PAGE 477

Power-Reducing Formulas

$\sin^2 u = \dfrac{1 - \cos 2u}{2}$ $\qquad$ $\cos^2 u = \dfrac{1 + \cos 2u}{2}$ $\qquad$ $\tan^2 u = \dfrac{1 - \cos 2u}{1 + \cos 2u}$

Proof

The first two formulas can be verified by solving for $\sin^2 u$ and $\cos^2 u$, respectively, in the double-angle formulas

$$\cos 2u = 1 - 2\sin^2 u \quad \text{and} \quad \cos 2u = 2\cos^2 u - 1.$$

The third formula can be verified using the fact that

$$\tan^2 u = \frac{\sin^2 u}{\cos^2 u}.$$

SECTION 5.5, PAGE 480

Sum-to-Product Formulas

$$\sin x + \sin y = 2\sin\left(\frac{x+y}{2}\right)\cos\left(\frac{x-y}{2}\right)$$

$$\sin x - \sin y = 2\cos\left(\frac{x+y}{2}\right)\sin\left(\frac{x-y}{2}\right)$$

$$\cos x + \cos y = 2\cos\left(\frac{x+y}{2}\right)\cos\left(\frac{x-y}{2}\right)$$

$$\cos x - \cos y = -2\sin\left(\frac{x+y}{2}\right)\sin\left(\frac{x-y}{2}\right)$$

Proof

To prove the first formula, let $x = u + v$ and $y = u - v$. Then substitute $u = (x + y)/2$ and $v = (x - y)/2$ in the product-to-sum formula.

$$\sin u \cos v = \frac{1}{2}[\sin(u+v) + \sin(u-v)]$$

$$\sin\left(\frac{x+y}{2}\right)\cos\left(\frac{x-y}{2}\right) = \frac{1}{2}(\sin x + \sin y)$$

$$2\sin\left(\frac{x+y}{2}\right)\cos\left(\frac{x-y}{2}\right) = \sin x + \sin y$$

SECTION 6.1, PAGE 494

Law of Sines

If ABC is a triangle with sides a, b, and c, then

$$\frac{a}{\sin A} = \frac{b}{\sin B} = \frac{c}{\sin C}.$$

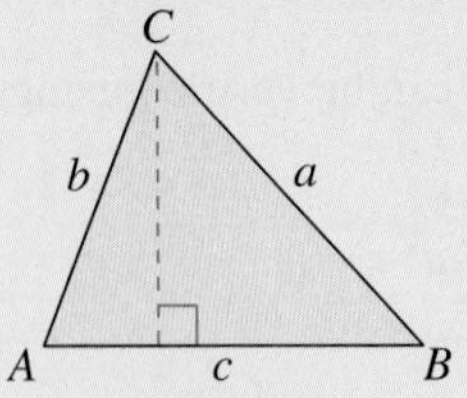

A is acute.

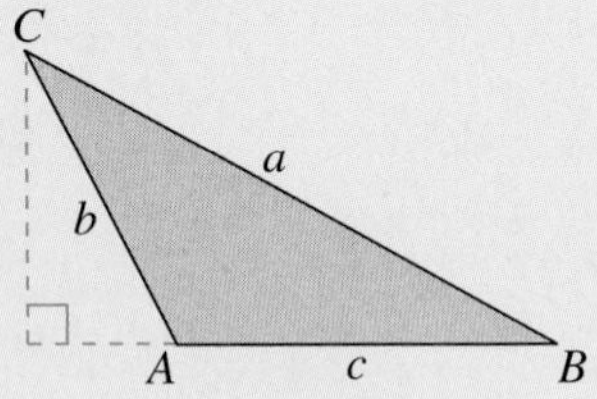

A is obtuse.

Proof

Let h be the altitude of either triangle found in the figure on the preceding page. Then you have

$$\sin A = \frac{h}{b} \quad \text{or} \quad h = b \sin A$$

$$\sin B = \frac{h}{a} \quad \text{or} \quad h = a \sin B.$$

Equating these two values of h, you have

$$a \sin B = b \sin A \quad \text{or} \quad \frac{a}{\sin A} = \frac{b}{\sin B}.$$

Note that $\sin A \neq 0$ and $\sin B \neq 0$ because no angle of a triangle can have a measure of $0°$ or $180°$. In a similar manner, by constructing an altitude from vertex B to side AC (extended), you can show that

$$\frac{a}{\sin A} = \frac{c}{\sin C}.$$

So, the Law of Sines is established.

SECTION 6.2, PAGE 503

Law of Cosines

Standard Form	*Alternative Form*
$a^2 = b^2 + c^2 - 2bc \cos A$	$\cos A = \dfrac{b^2 + c^2 - a^2}{2bc}$
$b^2 = a^2 + c^2 - 2ac \cos B$	$\cos B = \dfrac{a^2 + c^2 - b^2}{2ac}$
$c^2 = a^2 + b^2 - 2ab \cos C$	$\cos C = \dfrac{a^2 + b^2 - c^2}{2ab}$

Proof

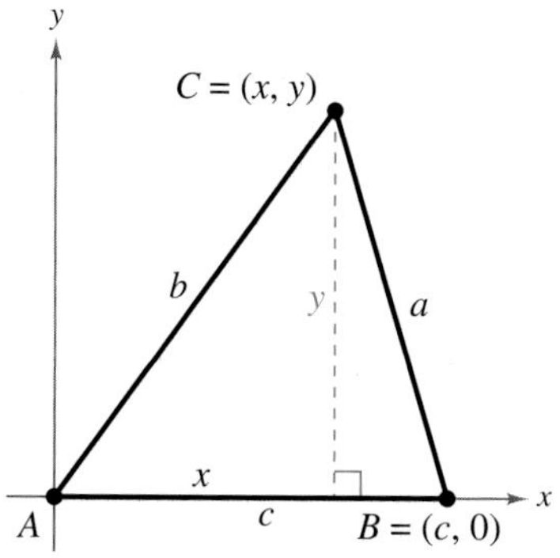

Consider a triangle that has three acute angles, as shown in the figure. Note that vertex B has coordinates $(c, 0)$. Furthermore, C has coordinates (x, y), where $x = b \cos A$ and $y = b \sin A$. Because a is the distance from vertex C to vertex B, it follows that

$$a = \sqrt{(x - c)^2 + (y - 0)^2}$$

$$a^2 = (b \cos A - c)^2 + (b \sin A)^2$$

$$a^2 = b^2 \cos^2 A - 2bc \cos A + c^2 + b^2 \sin^2 A$$

$$a^2 = b^2(\sin^2 A + \cos^2 A) + c^2 - 2ab \cos A$$

$$a^2 = b^2 + c^2 - 2bc \cos A. \qquad \sin^2 A + \cos^2 A = 1$$

Similar arguments can be used to establish the other two equations.

SECTION 6.2, PAGE 506

Heron's Area Formula

Given any triangle with sides of lengths a, b, and c, the area of the triangle is

$$\text{Area} = \sqrt{s(s-a)(s-b)(s-c)}$$

where $s = (a + b + c)/2$.

Proof

From Section 6.1, you know that

$$\begin{aligned}
\text{Area} &= \frac{1}{2}bc \sin A \\
&= \sqrt{\frac{1}{4}b^2c^2 \sin^2 A} \\
&= \sqrt{\frac{1}{4}b^2c^2(1 - \cos^2 A)} \\
&= \sqrt{\left[\frac{1}{2}bc(1 + \cos A)\right]\left[\frac{1}{2}bc(1 - \cos A)\right]}.
\end{aligned}$$

Using the Law of Cosines, you can show that

$$\frac{1}{2}bc(1 + \cos A) = \frac{a + b + c}{2} \cdot \frac{-a + b + c}{2}$$

and

$$\frac{1}{2}bc(1 - \cos A) = \frac{a - b + c}{2} \cdot \frac{a + b - c}{2}.$$

Letting $s = (a + b + c)/2$, these two equations can be rewritten as

$$\frac{1}{2}bc(1 + \cos A) = s(s - a)$$

and

$$\frac{1}{2}bc(1 - \cos A) = (s - b)(s - c).$$

So, you can conclude that

$$\text{Area} = \sqrt{s(s-a)(s-b)(s-c)}.$$

SECTION 6.4, PAGE 524

Properties of the Dot Product

Let $\mathbf{u}$, $\mathbf{v}$, and $\mathbf{w}$ be vectors in the plane or in space and let c be a scalar.

1. $\mathbf{u} \cdot \mathbf{v} = \mathbf{v} \cdot \mathbf{u}$
2. $\mathbf{0} \cdot \mathbf{v} = 0$
3. $\mathbf{u} \cdot (\mathbf{v} + \mathbf{w}) = \mathbf{u} \cdot \mathbf{v} + \mathbf{u} \cdot \mathbf{w}$
4. $\mathbf{v} \cdot \mathbf{v} = \|\mathbf{v}\|^2$
5. $c(\mathbf{u} \cdot \mathbf{v}) = c\mathbf{u} \cdot \mathbf{v} = \mathbf{u} \cdot c\mathbf{v}$

To prove Property 1, let $\mathbf{u} = \langle u_1, u_2 \rangle$ and $\mathbf{v} = \langle v_1, v_2 \rangle$. Then

$$\begin{aligned}\mathbf{u} \cdot \mathbf{v} &= u_1v_1 + u_2v_2\\ &= v_1u_1 + v_2u_2\\ &= \mathbf{v} \cdot \mathbf{u}.\end{aligned}$$

To prove Property 4, let $\mathbf{v} = \langle v_1, v_2 \rangle$. Then

$$\begin{aligned}\mathbf{v} \cdot \mathbf{v} &= v_1^2 + v_2^2\\ &= \left(\sqrt{v_1^2 + v_2^2}\right)^2\\ &= \|\mathbf{v}\|^2.\end{aligned}$$

To prove part of Property 5, let $\mathbf{u} = \langle u_1, u_2 \rangle$ and $\mathbf{v} = \langle v_1, v_2 \rangle$ and let c be a scalar. Then,

$$\begin{aligned}c(\mathbf{u} \cdot \mathbf{v}) &= c(\langle u_1, u_2 \rangle \cdot \langle v_1, v_2 \rangle)\\ &= c(u_1v_1 + u_2v_2)\\ &= (cu_1)v_1 + (cu_2)v_2\\ &= \langle cu_1, cu_2 \rangle \cdot \langle v_1, v_2 \rangle\\ &= c\mathbf{u} \cdot \mathbf{v}.\end{aligned}$$

SECTION 6.4, PAGE 525

Angle Between Two Vectors

If θ is the angle between two nonzero vectors $\mathbf{u}$ and $\mathbf{v}$, then

$$\cos \theta = \frac{\mathbf{u} \cdot \mathbf{v}}{\|\mathbf{u}\| \, \|\mathbf{v}\|}.$$

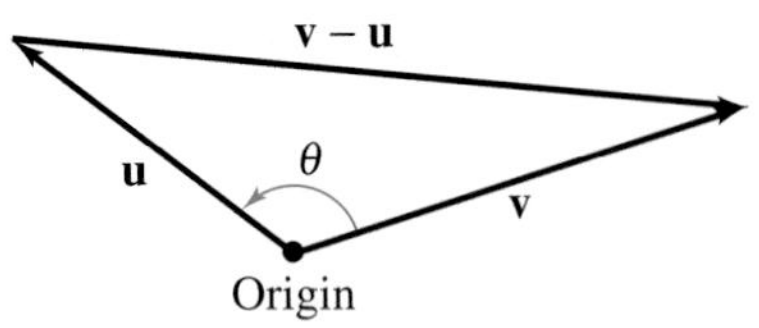

Proof

Consider the triangle determined by vectors $\mathbf{u}$, $\mathbf{v}$, and $\mathbf{v} - \mathbf{u}$, as shown in the figure. By the Law of Cosines, you can write

$$\begin{aligned}
\|\mathbf{v} - \mathbf{u}\|^2 &= \|\mathbf{u}\|^2 + \|\mathbf{v}\|^2 - 2\|\mathbf{u}\|\,\|\mathbf{v}\| \cos \theta \\
(\mathbf{v} - \mathbf{u}) \cdot (\mathbf{v} - \mathbf{u}) &= \|\mathbf{u}\|^2 + \|\mathbf{v}\|^2 - 2\|\mathbf{u}\|\,\|\mathbf{v}\| \cos \theta \\
(\mathbf{v} - \mathbf{u}) \cdot \mathbf{v} - (\mathbf{v} - \mathbf{u}) \cdot \mathbf{u} &= \|\mathbf{u}\|^2 + \|\mathbf{v}\|^2 - 2\|\mathbf{u}\|\,\|\mathbf{v}\| \cos \theta \\
\mathbf{v} \cdot \mathbf{v} - \mathbf{u} \cdot \mathbf{v} - \mathbf{v} \cdot \mathbf{u} + \mathbf{u} \cdot \mathbf{u} &= \|\mathbf{u}\|^2 + \|\mathbf{v}\|^2 - 2\|\mathbf{u}\|\,\|\mathbf{v}\| \cos \theta \\
\|\mathbf{v}\|^2 - 2\mathbf{u} \cdot \mathbf{v} + \|\mathbf{u}\|^2 &= \|\mathbf{u}\|^2 + \|\mathbf{v}\|^2 - 2\|\mathbf{u}\|\,\|\mathbf{v}\| \cos \theta \\
-2\mathbf{u} \cdot \mathbf{v} &= -2\|\mathbf{u}\|\,\|\mathbf{v}\| \cos \theta \\
\cos \theta &= \frac{\mathbf{u} \cdot \mathbf{v}}{\|\mathbf{u}\|\,\|\mathbf{v}\|}.
\end{aligned}$$

SECTION 9.1, PAGE 693

Properties of Sums

1. $\displaystyle\sum_{i=1}^{n} ca_i = c\sum_{i=1}^{n} a_i,$ c is any constant.

2. $\displaystyle\sum_{i=1}^{n} (a_i + b_i) = \sum_{i=1}^{n} a_i + \sum_{i=1}^{n} b_i$

3. $\displaystyle\sum_{i=1}^{n} (a_i - b_i) = \sum_{i=1}^{n} a_i - \sum_{i=1}^{n} b_i$

Proof

Each of these properties follows directly from the Associative Property of Addition, the Commutative Property of Addition, and the Distributive Property of multiplication over addition. For example, note the use of the Distributive Property in the proof of Property 1.

$$\begin{aligned}
\sum_{i=1}^{n} ca_i &= ca_1 + ca_2 + ca_3 + \cdots + ca_n \\
&= c(a_1 + a_2 + a_3 + \cdots + a_n) = c\sum_{i=1}^{n} a_i
\end{aligned}$$

SECTION 9.2, PAGE 702

The Sum of a Finite Arithmetic Sequence

The sum of a finite arithmetic sequence with n terms is

$$S_n = \frac{n}{2}(a_1 + a_n).$$

Proof

Begin by generating the terms of the arithmetic sequence in two ways. In the first way, repeatedly add d to the first term to obtain

$$\begin{aligned} S_n &= a_1 + a_2 + a_3 + \cdots + a_{n-2} + a_{n-1} + a_n \\ &= a_1 + [a_1 + d] + [a_1 + 2d] + \cdots + [a_1 + (n-1)d]. \end{aligned}$$

In the second way, repeatedly subtract d from the nth term to obtain

$$\begin{aligned} S_n &= a_n + a_{n-1} + a_{n-2} + \cdots + a_3 + a_2 + a_1 \\ &= a_n + [a_n - d] + [a_n - 2d] + \cdots + [a_n - (n-1)d]. \end{aligned}$$

If you add these two versions of S_n, the multiples of d cancel and you obtain

$$\begin{aligned} 2S_n &= \overbrace{(a_1 + a_n) + (a_1 + a_n) + (a_1 + a_n) + \cdots + (a_1 + a_n)}^{n \text{ terms}} \\ &= n(a_1 + a_n). \end{aligned}$$

So, you have $S_n = \frac{n}{2}(a_1 + a_n)$.

SECTION 9.3, PAGE 711

The Sum of a Finite Geometric Sequence

The sum of the geometric sequence

$$a_1,\ a_1r,\ a_1r^2,\ a_1r^3,\ a_1r^4,\ \ldots,\ a_1r^{n-1}$$

with common ratio $r \neq 1$ is given by $S_n = a_1\left(\frac{1 - r^n}{1 - r}\right)$.

Proof

Begin by writing out the nth partial sum.

$$S_n = a_1 + a_1r + a_1r^2 + \cdots + a_1r^{n-2} + a_1r^{n-1}$$

Multiplication by r yields

$$rS_n = a_1r + a_1r^2 + a_1r^3 + \cdots + a_1r^{n-1} + a_1r^n.$$

Subtracting the second equation from the first yields

$$S_n - rS_n = a_1 - a_1r^n.$$

So, $S_n(1 - r) = a_1(1 - r^n)$, and, because $r \neq 1$, you have $S_n = a_1\left(\frac{1 - r^n}{1 - r}\right)$.

SECTION 9.5, PAGE 728

The Binomial Theorem

In the expansion of $(x + y)^n$

$$(x + y)^n = x^n + nx^{n-1}y + \cdot \cdot \cdot + {}_nC_r\, x^{n-r}y^r + \cdot \cdot \cdot + nxy^{n-1} + y^n$$

the coefficient of $x^{n-r}y^r$ is

$${}_nC_r = \frac{n!}{(n-r)!r!}.$$

Proof

The Binomial Theorem can be proved quite nicely using mathematical induction. The steps are straightforward but look a little messy, so only an outline of the proof is presented.

1. If $n = 1$, you have

$$(x + y)^1 = x^1 + y^1 = {}_1C_0x + {}_1C_1y$$

and the formula is valid.

2. Assuming that the formula is true for $n = k$, the coefficient of $x^{k-r}y^r$ is

$${}_kC_r = \frac{k!}{(k-r)!r!} = \frac{k(k-1)(k-2)\cdot \cdot \cdot (k-r+1)}{r!}.$$

To show that the formula is true for $n = k + 1$, look at the coefficient of $x^{k+1-r}y^r$ in the expansion of

$$(x + y)^{k+1} = (x + y)^k(x + y).$$

From the right-hand side, you can determine that the term involving $x^{k+1-r}y^r$ is the sum of two products.

$$\begin{aligned}
&({}_kC_r x^{k-r}y^r)(x) + ({}_kC_{r-1}x^{k+1-r}y^{r-1})(y)\\
&= \left[\frac{k!}{(k-r)!r!} + \frac{k!}{(k-r+1)!(r-1)!}\right]x^{k+1-r}y^r\\
&= \left[\frac{(k+1-r)k!}{(k+1-r)!r!} + \frac{k!r}{(k+1-r)!r!}\right]x^{k+1-r}y^r\\
&= \left[\frac{k!(k+1-r+r)}{(k+1-r)!r!}\right]x^{k+1-r}y^r\\
&= \left[\frac{(k+1)!}{(k+1-r)!r!}\right]x^{k+1-r}y^r\\
&= {}_{k+1}C_r x^{k+1-r}y^r
\end{aligned}$$

So, by mathematical induction, the Binomial Theorem is valid for all positive integers n.

SECTION 10.1, PAGE 770

Inclination and Slope

If a nonvertical line has inclination θ and slope m, then $m = \tan \theta$.

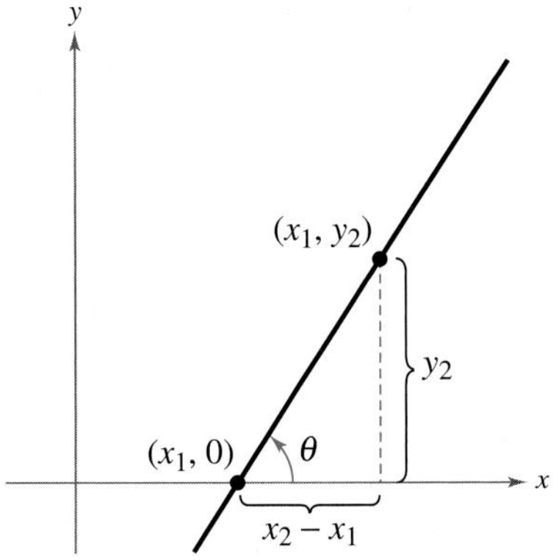

Proof

If $m = 0$, the line is horizontal and $\theta = 0$. So, the result is true for horizontal lines because $m = 0 = \tan 0$.

If the line has a positive slope, it will intersect the x-axis. Label this point $(x_1, 0)$, as shown. If (x_2, y_2) is a second point on the line, the slope is

$$m = \frac{y_2 - 0}{x_2 - x_1} = \frac{y_2}{x_2 - x_1} = \tan \theta.$$

The case in which the line has a negative slope is left for your to prove.

SECTION 10.1, PAGE 772

Distance Between a Point and a Line

The distance between the point (x_1, y_1) and the line $Ax + By + C = 0$ is

$$d = \frac{|Ax_1 + By_1 + C|}{\sqrt{A^2 + B^2}}.$$

Proof

For simplicity's sake, assume that the given line is neither horizontal nor vertical. By writing the equation $Ax + By + C = 0$ in slope-intercept form

$$y = -\frac{A}{B}x - \frac{C}{B}$$

you can see that the line has a slope of $m = -A/B$. So, the slope of the line passing through (x_1, y_1) and perpendicular to the given line is B/A, and its equation is $y - y_1 = (B/A)(x - x_1)$. These two lines intersect at the point (x_2, y_2), where

$$x_2 = \frac{B(Bx_1 - Ay_1) - AC}{A^2 + B^2} \quad \text{and} \quad y_2 = \frac{A(-Bx_1 + Ay_1) - BC}{A^2 + B^2}.$$

Finally, the distance between (x_1, y_1) and (x_2, y_2) is

$$\begin{aligned} d &= \sqrt{(x_2 - x_1)^2 + (y_2 - y_1)^2} \\ &= \sqrt{\left(\frac{B^2x_1 - ABy_1 - AC}{A^2 + B^2} - x_1\right)^2 + \left(\frac{-ABx_1 + A^2y_1 - BC}{A^2 + B^2} - y_1\right)^2} \\ &= \sqrt{\frac{A^2(Ax_1 + By_1 + C)^2 + B^2(Ax_1 + By_1 + C)^2}{(A^2 + B^2)^2}} \\ &= \frac{|Ax_1 + By_1 + C|}{\sqrt{A^2 + B^2}}. \end{aligned}$$

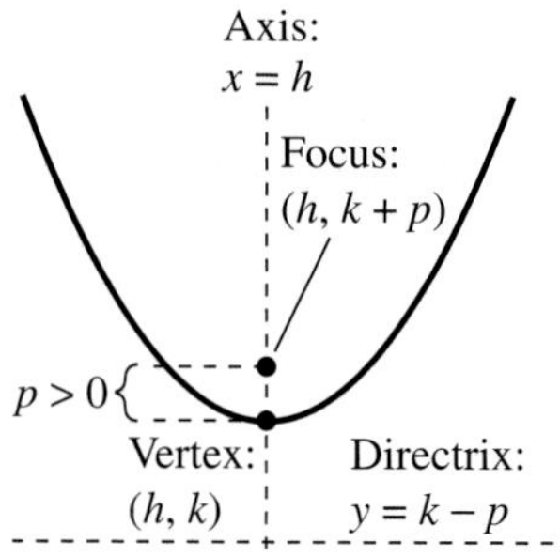

SECTION 10.2, PAGE 778

Standard Equation of a Parabola

The standard form of the equation of a parabola with vertex at (h, k) is as follows.

$(x - h)^2 = 4p(y - k),\ p \neq 0$ Vertical axis, directrix: $y = k - p$

$(y - k)^2 = 4p(x - h),\ p \neq 0$ Horizontal axis, directrix: $x = h - p$

The focus lies on the axis p units (*directed distance*) from the vertex. If the vertex is at the origin $(0, 0)$, the equation takes one of the following forms.

$x^2 = 4py$ Vertical axis

$y^2 = 4px$ Horizontal axis

Proof

The case for which the directrix is parallel to the x-axis and the focus lies above the vertex, as shown in the figure, is proven here. If (x, y) is any point on the parabola, then, by definition, it is equidistant from the focus $(h, k + p)$ and the directrix $y = k - p$, and you have

$$\sqrt{(x - h)^2 + [y - (k + p)]^2} = y - (k - p)$$

$$(x - h)^2 + [y - (k + p)]^2 = [y - (k - p)]^2$$

$$(x - h)^2 + y^2 - 2y(k + p) + (k + p)^2 = y^2 - 2y(k - p) + (k - p)^2$$

$$(x - h)^2 - 2py + 2pk = 2py - 2pk$$

$$(x - h)^2 = 4p(y - k).$$

The case in which the directrix is parallel to the y-axis and the focus lies to the right of the vertex is left for you to prove.

SECTION 10.9, PAGE 834

Polar Equations of Conics

The graph of a polar equation of the form

1. $r = \dfrac{ep}{1 \pm e \cos \theta}$

2. $r = \dfrac{ep}{1 \pm e \sin \theta}$

is a conic, where $e > 0$ is the eccentricity and $|p|$ is the distance between the focus (pole) and the directrix.

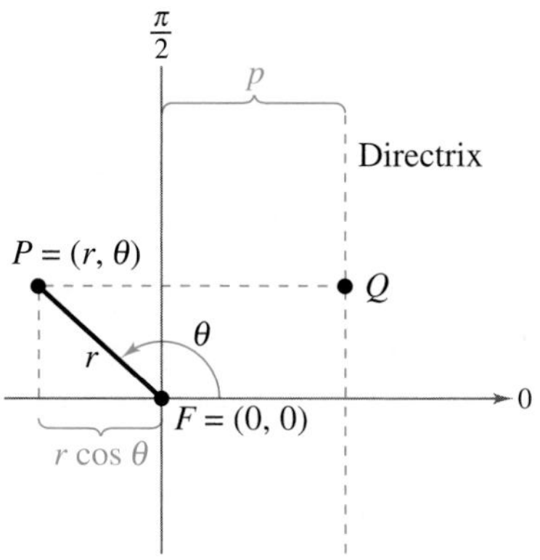

Proof

A proof for $r = ep/(1 + e \cos \theta)$ with $p > 0$ is listed here. The proofs of the other cases are similar. In the figure, consider a vertical directrix, p units to the right of the focus $F = (0, 0)$. If $P = (r, \theta)$ is a point on the graph of

$$r = \frac{ep}{1 + e \cos \theta}$$

the distance between P and the directrix is

$$\begin{aligned} PQ &= |p - x| \\ &= |p - r \cos \theta| \\ &= \left| p - \left(\frac{ep}{1 + e \cos \theta} \right) \cos \theta \right| \\ &= \left| p \left(1 - \frac{e \cos \theta}{1 + e \cos \theta} \right) \right| \\ &= \left| \frac{p}{1 + e \cos \theta} \right| \\ &= \left| \frac{r}{e} \right|. \end{aligned}$$

Moreover, because the distance between P and the pole is simply $PF = |r|$, the ratio of PF to PQ is

$$\begin{aligned} \frac{PF}{PQ} &= \frac{|r|}{|r/e|} \\ &= |e| \\ &= e \end{aligned}$$

and, by definition, the graph of the equation must be a conic.

Appendix B Concepts in Statistics

B.1 Representing Data

▶ What you should learn

- How to use line plots to order and analyze data
- How to use stem-and-leaf plots to organize and compare data
- How to use histograms to represent frequency distributions

▶ Why you should learn it

Line plots, stem-and-leaf plots, and histograms are quick methods for determining which elements in a set of data occur with the greatest frequency. For instance, in Exercise 9 on page A16, you are asked to construct a frequency distribution and a histogram of employees' monthly salary contributions to a retirement plan.

Line Plots

Statistics is the branch of mathematics that studies techniques for collecting, organizing, and interpreting data. In this section, you will study several ways to organize data. The first is a **line plot,** which uses a portion of a real number line to order numbers. Line plots are especially useful for ordering small sets of numbers (about 50 or less) by hand.

Example 1 ▶ Constructing a Line Plot

Use a line plot to organize the following test scores. Which number occurs with the greatest frequency?

93, 70, 76, 67, 86, 93, 82, 78, 83, 86, 64, 78, 76, 66, 83,
83, 96, 74, 69, 76, 64, 74, 79, 76, 88, 76, 81, 82, 74, 70

Solution

Begin by scanning the data to find the smallest and largest numbers. For this data, the smallest number is 64 and the largest is 96. Next, draw a portion of a real number line that includes the interval [64, 96]. To create the line plot, start with the first number, 93, and enter an × above 93 on the number line. Continue recording ×'s for each number in the list until you obtain the line plot shown in Figure B.1. From the line plot, you can see that 76 had the greatest frequency.

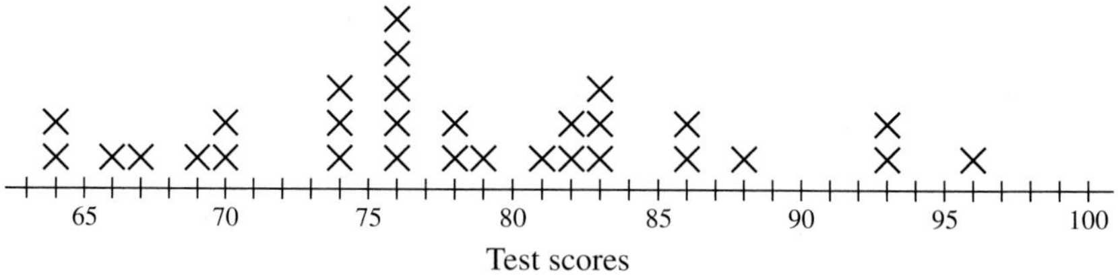

FIGURE B.1

Stem-and-Leaf Plots

Another type of plot that can be used to organize sets of numbers by hand is a **stem-and-leaf plot.** A set of test scores and the corresponding stem-and-leaf plot are shown at the left.

Test Scores

93, 70, 76, 58, 86, 93, 82, 78,
83, 86, 64, 78, 76, 66, 83, 83,
96, 74, 69, 76, 64, 74, 79, 76,
88, 76, 81, 82, 74, 70

Stems	*Leaves*
5	8
6	4 4 6 9
7	0 0 4 4 4 6 6 6 6 6 8 8 9
8	1 2 2 3 3 3 6 6 8
9	3 3 6

Note that the *leaves* represent the units digits of the numbers and the *stems* represent the tens digits. Stem-and-leaf plots can also be used to compare two sets of data, as shown in the following example.

Example 2 ▶ Comparing Two Sets of Data

Use a stem-and-leaf plot to compare the test scores given on page A13 with the following test scores. Which set of test scores is better?

90, 81, 70, 62, 64, 73, 81, 92, 73, 81, 92, 93, 83, 75, 76,
83, 94, 96, 86, 77, 77, 86, 96, 86, 77, 86, 87, 87, 79, 88

Solution

Begin by ordering the second set of scores.

62, 64, 70, 73, 73, 75, 76, 77, 77, 77, 79, 81, 81, 81, 83,
83, 86, 86, 86, 86, 87, 87, 88, 90, 92, 92, 93, 94, 96, 96

Now that the data has been ordered, you can construct a *double* stem-and-leaf plot by letting the leaves to the right of the stems represent the units digits for the first group of test scores and letting the leaves to the left of the stems represent the units digits for the second group of test scores.

Leaves (2nd Group)	*Stems*	*Leaves (1st Group)*
	5	8
4 2	6	4 4 6 9
9 7 7 7 6 5 3 3 0	7	0 0 4 4 4 6 6 6 6 6 8 8 9
8 7 7 6 6 6 6 3 3 1 1 1	8	1 2 2 3 3 3 6 6 8
6 6 4 3 2 2 0	9	3 3 6

By comparing the two sets of leaves, you can see that the second group of test scores is better than the first group.

AK	5.3	MT	13.2
AL	13.0	NC	12.5
AR	14.3	ND	14.4
AZ	13.2	NE	13.7
CA	11.1	NH	12.1
CO	10.1	NJ	13.7
CT	14.4	NM	11.2
DC	13.9	NV	11.5
DE	12.9	NY	13.4
FL	18.5	OH	13.4
GA	9.9	OK	13.4
HI	13.2	OR	13.3
IA	15.0	PA	15.8
ID	11.3	RI	15.8
IL	12.5	SC	12.1
IN	12.5	SD	14.3
KS	13.5	TN	12.5
KY	12.5	TX	10.1
LA	11.4	UT	8.7
MA	14.1	VA	11.2
MD	11.5	VT	12.3
ME	13.9	WA	11.5
MI	12.4	WI	13.2
MN	12.3	WV	15.1
MO	13.7	WY	11.3
MS	12.2		

Example 3 ▶ Using a Stem-and-Leaf Plot

The table at the left above shows the percent of the population of each state and the District of Columbia that was at least 65 years old in 1997. Use a stem-and-leaf plot to organize the data. (Source: U.S. Bureau of the Census)

Solution

Begin by ordering the numbers, as shown below.

5.3, 8.7, 9.9, 10.1, 10.1, 11.1, 11.2, 11.2, 11.3, 11.3, 11.4, 11.5, 11.5, 11.5,
12.1, 12.1, 12.2, 12.3, 12.3, 12.4, 12.5, 12.5, 12.5, 12.5, 12.5, 12.9, 13.0,
13.2, 13.2, 13.2, 13.2, 13.3, 13.4, 13.4, 13.4, 13.5, 13.7, 13.7, 13.7, 13.9,
13.9, 14.1, 14.3, 14.3, 14.4, 14.4, 15.0, 15.1, 15.8, 15.8, 18.5

Next construct the stem-and-leaf plot using the leaves to represent the digits to the right of the decimal points, as shown at the left. From the stem-and-leaf plot, you can see that Alaska has the lowest percent and Florida has the highest percent.

Stems	*Leaves*
5.	3
6.	
7.	
8.	7
9.	9
10.	1 1
11.	1 2 2 3 3 4 5 5 5
12.	1 1 2 3 3 4 5 5 5 5 5 9
13.	0 2 2 2 2 3 4 4 4 5 7 7 7 9 9
14.	1 3 3 4 4
15.	0 1 8 8
16.	
17.	
18.	5

Histograms and Frequency Distributions

With data such as that given in Example 3, it is useful to group the numbers into intervals and plot the frequency of the data in each interval. For instance, the **frequency distribution** and **histogram** shown in Figure B.2 represent the data given in Example 3.

Frequency Distribution

Interval	*Tally*
[5, 7)	|
[7, 9)	|
[9, 11)	|||
[11, 13)	𝍸 𝍸 𝍸 𝍸 |
[13, 15)	𝍸 𝍸 𝍸 𝍸
[15, 17)	||||
[17, 19)	|

Histogram

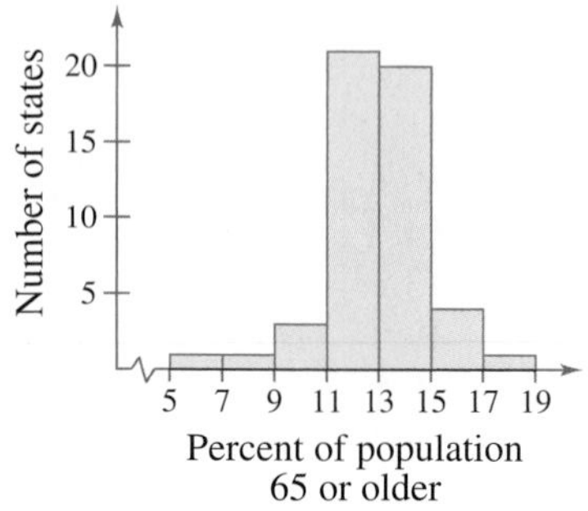

FIGURE B.2

A histogram has a portion of a real number line as its horizontal axis. A histogram is similar to a bar graph, except that the rectangles (bars) in a bar graph can be either horizontal or vertical and the labels of the bars are not necessarily numbers.

Another difference between a bar graph and a histogram is that the bars in a bar graph are usually separated by spaces, whereas the bars in a histogram are not separated by spaces.

Interval	*Tally*
100–109	𝍸 |||
110–119	|
120–129	|||
130–139	|||
140–149	𝍸 ||
150–159	𝍸
160–169	𝍸 |||
170–179	𝍸 |
180–189	||
190–199	𝍸

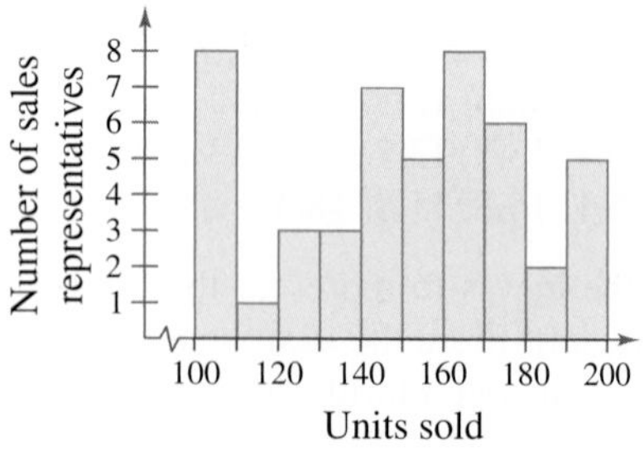

FIGURE B.3

Example 4 ▶ Constructing a Histogram

A company has 48 sales representatives who sold the following numbers of units during the first quarter of 2000. Construct a grouped frequency distribution for this data.

107	162	184	170	177	102	145	141
105	193	167	149	195	127	193	191
150	153	164	167	171	163	141	129
109	171	150	138	100	164	147	153
171	163	118	142	107	144	100	132
153	107	124	162	192	134	187	177

Solution

To begin constructing a grouped frequency distribution, you must first decide on the number of groups. There are several ways to group this data. However, because the smallest number is 100 and the largest is 195, it seems that 10 groups of 10 each would be appropriate. The first group would be 100–109, the second group would be 110–119, and so on. By tallying the data into the 10 groups, you obtain the distribution shown at the left above. A histogram for the distribution is shown in Figure B.3.

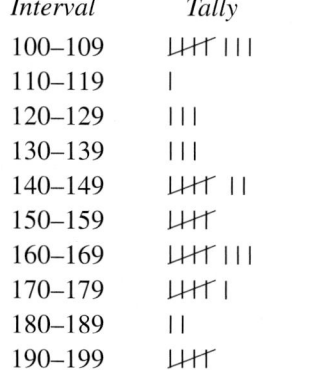

B.1 Exercises

1. *Gasoline Prices* The line plot shows a sample of prices of unleaded regular gasoline from 25 different cities.

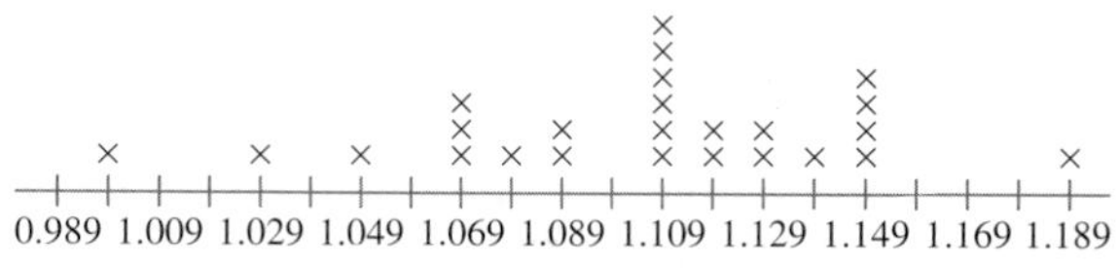

(a) What price occurred with the greatest frequency?

(b) What is the range of prices?

2. *Livestock Weights* The line plot shows the weights (to the nearest hundred pounds) of 30 head of cattle sold by a rancher.

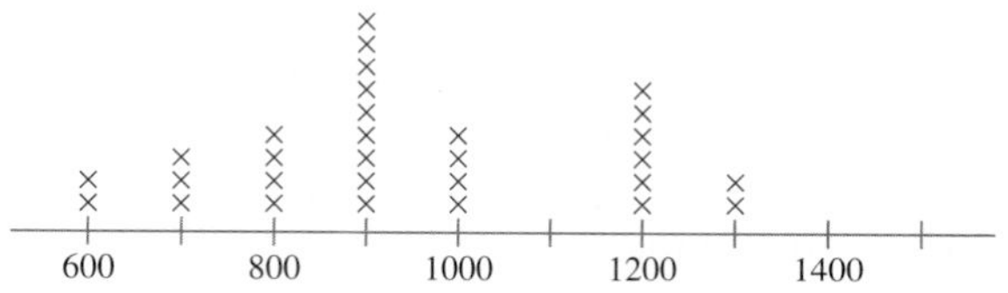

(a) What weight occurred with the greatest frequency?

(b) What is the range of weights?

Quiz and Exam Scores **In Exercises 3–6, use the following scores from a math class of 30 students. The scores are for two 25-point quizzes and two 100-point exams.**

Quiz #1 20, 15, 14, 20, 16, 19, 10, 21, 24, 15, 15, 14, 15, 21, 19, 15, 20, 18, 18, 22, 18, 16, 18, 19, 21, 19, 16, 20, 14, 12

Quiz #2 22, 22, 23, 22, 21, 24, 22, 19, 21, 23, 23, 25, 24, 22, 22, 23, 23, 23, 23, 22, 24, 23, 22, 24, 21, 24, 16, 21, 16, 14

Exam #1 77, 100, 77, 70, 83, 89, 87, 85, 81, 84, 81, 78, 89, 78, 88, 85, 90, 92, 75, 81, 85, 100, 98, 81, 78, 75, 85, 89, 82, 75

Exam #2 76, 78, 73, 59, 70, 81, 71, 66, 66, 73, 68, 67, 63, 67, 77, 84, 87, 71, 78, 78, 90, 80, 77, 70, 80, 64, 74, 68, 68, 68

3. Construct a line plot for each quiz. For each quiz, which score occurred with the greatest frequency?

4. Construct a line plot for each exam. For each exam, which score occurred with the greatest frequency?

5. Construct a stem-and-leaf plot for Exam #1.

6. Construct a double stem-and-leaf plot to compare the scores for Exam #1 and Exam #2. Which set of scores is higher?

7. *Insurance Coverage* The table shows the total numbers of persons (in thousands) without health insurance coverage in the 50 states and the District of Columbia in 1996. Use a stem-and-leaf plot to organize the data. (Source: U.S. Bureau of the Census)

AK	89	AL	550	AR	566	AZ	1159	CA	6514
CO	644	CT	368	DC	80	DE	98	FL	2722
GA	1319	HI	101	IA	335	ID	196	IL	1337
IN	600	KS	292	KY	601	LA	890	MA	766
MD	581	ME	146	MI	857	MN	480	MO	700
MS	518	MT	124	NC	1160	ND	62	NE	190
NH	109	NJ	1317	NM	412	NV	255	NY	3132
OH	1292	OK	570	OR	496	PA	1133	RI	93
SC	634	SD	67	TN	841	TX	4680	UT	240
VA	811	VT	65	WA	761	WI	438	WV	261
WY	66								

8. *Snowfall* The data below shows the seasonal snowfall (in inches) at Lincoln, Nebraska for the years 1968 through 1997 (the amounts are listed in order by year). How would you organize this data? Explain your reasoning. (Source: University of Nebraska–Lincoln)

39.8, 26.2, 49.0, 21.6, 29.2, 33.6, 42.1, 21.1, 21.8, 31.0, 34.4, 23.3, 13.0, 32.3, 38.0, 47.5, 21.5, 18.9, 15.7, 13.0, 19.1, 18.7, 25.8, 23.8, 32.1, 21.3, 21.8, 30.7, 29.0, 44.6

9. *Retirement Contributions* The employees of a company must contribute 7% of their monthly salaries to a company-sponsored retirement plan. The contributed amounts (in dollars) for the company's 35 employees are as follows.

100, 200, 130, 136, 161, 156, 209, 126, 135, 98, 114, 117, 168, 133, 140, 124, 172, 127, 143, 157, 124, 152, 104, 126, 155, 92, 194, 115, 120, 136, 148, 112, 116, 146, 96

(a) Construct a frequency distribution using groups of 20. The first group should be 90–109.

(b) Construct a histogram for this frequency distribution.

B.2 Measures of Central Tendency and Dispersion

▶ **What you should learn**

- How to find and interpret the mean, median, and mode of a set of data
- How to determine the measure of central tendency that best represents a set of data
- How to find the standard deviation of a set of data
- How to use box-and-whisker plots

▶ **Why you should learn it**

Measures of central tendency and dispersion provide a convenient way to describe and compare sets of data. For instance, in Exercise 36 on page A25, the mean and standard deviation are used to analyze the price of gold for the years 1978 through 1997.

Mean, Median, and Mode

In many real-life situations, it is helpful to describe data by a single number that is most representative of the entire collection of numbers. Such a number is called a **measure of central tendency.** The most commonly used measures are as follows.

1. The **mean,** or **average,** of n numbers is the sum of the numbers divided by n.
2. The **median** of n numbers is the middle number when the numbers are written in order. If n is even, the median is the average of the two middle numbers.
3. The **mode** of n numbers is the number that occurs most frequently. If two numbers tie for most frequent occurrence, the collection has two modes and is called **bimodal.**

Example 1 ▶ Comparing Measures of Central Tendency

On an interview for a job, the interviewer tells you that the average annual income of the company's 25 employees is \$60,849. The actual annual incomes of the 25 employees are shown below. What are the mean, median, and mode of the incomes? Was the person telling you the truth?

\$17,305,	\$478,320,	\$45,678,	\$18,980,	\$17,408,
\$25,676,	\$28,906,	\$12,500,	\$24,540,	\$33,450,
\$12,500,	\$33,855,	\$37,450,	\$20,432,	\$28,956,
\$34,983,	\$36,540,	\$250,921,	\$36,853,	\$16,430,
\$32,654,	\$98,213,	\$48,980,	\$94,024,	\$35,671

Solution

The mean of the incomes is

$$\text{Mean} = \frac{17{,}305 + 478{,}320 + 45{,}678 + 18{,}980 + \cdot \cdot \cdot + 35{,}671}{25}$$

$$= \frac{1{,}521{,}225}{25} = \$60{,}849.$$

To find the median, order the incomes as follows.

\$12,500,	\$12,500,	\$16,430,	\$17,305,	\$17,408,
\$18,980,	\$20,432,	\$24,540,	\$25,676,	\$28,906,
\$28,956,	\$32,654,	\$33,450,	\$33,855,	\$34,983,
\$35,671,	\$36,540,	\$36,853,	\$37,450,	\$45,678,
\$48,980,	\$94,024,	\$98,213,	\$250,921,	\$478,320

From this list, you can see that the median (the middle number) is \$33,450. From the same list, you can see that \$12,500 is the only income that occurs more than once. So, the mode is \$12,500. Technically, the person was telling the truth because the average is (generally) defined to be the mean. However, of the three measures of central tendency *Mean:* \$60,849 *Median:* \$33,450 *Mode:* \$12,500 it seems clear that the median is most representative. The mean is inflated by the two highest salaries.

Choosing a Measure of Central Tendency

Which of the three measures of central tendency is the most representative? The answer is that it depends on the distribution of the data *and* the way in which you plan to use the data.

For instance, in Example 1, the mean salary of \$60,849 does not seem very representative to a potential employee. To a city income tax collector who wants to estimate 1% of the total income of the 25 employees, however, the mean is precisely the right measure.

Example 2 ▶ Choosing a Measure of Central Tendency

Which measure of central tendency is the most representative of the data given in each of the following frequency distributions?

a.

Number	*Tally*
1	7
2	20
3	15
4	11
5	8
6	3
7	2
8	0
9	15

b.

Number	*Tally*
1	9
2	8
3	7
4	6
5	5
6	6
7	7
8	8
9	9

c.

Number	*Tally*
1	6
2	1
3	2
4	3
5	5
6	5
7	4
8	3
9	0

Solution

a. For this data, the mean is 4.23, the median is 3, and the mode is 2. Of these, the mode is probably the most representative.

b. For this data, the mean and median are each 5 and the modes are 1 and 9 (the distribution is bimodal). Of these, the mean or median is the most representative.

c. For this data, the mean is 4.59, the median is 5, and the mode is 1. Of these, the mean or median is the most representative.

Variance and Standard Deviation

Very different sets of numbers can have the same mean. You will now study two **measures of dispersion,** which give you an idea of how much the numbers in a set differ from the mean of the set. These two measures are called the *variance* of the set and the *standard deviation* of the set.

Definitions of Variance and Standard Deviation

Consider a set of numbers $\{x_1, x_2, \ldots, x_n\}$ with a mean of $\bar{x}$. The **variance** of the set is

$$v = \frac{(x_1 - \bar{x})^2 + (x_2 - \bar{x})^2 + \cdots + (x_n - \bar{x})^2}{n}$$

and the **standard deviation** of the set is $\sigma = \sqrt{v}$ (σ is the lowercase Greek letter *sigma*).

The standard deviation of a set is a measure of how much a typical number in the set differs from the mean. The greater the standard deviation, the more the numbers in the set *vary* from the mean. For instance, each of the following sets has a mean of 5.

$$\{5, 5, 5, 5\}, \quad \{4, 4, 6, 6\}, \quad \text{and} \quad \{3, 3, 7, 7\}$$

The standard deviations of the sets are 0, 1, and 2.

$$\sigma_1 = \sqrt{\frac{(5-5)^2 + (5-5)^2 + (5-5)^2 + (5-5)^2}{4}}$$
$$= 0$$
$$\sigma_2 = \sqrt{\frac{(4-5)^2 + (4-5)^2 + (6-5)^2 + (6-5)^2}{4}}$$
$$= 1$$
$$\sigma_3 = \sqrt{\frac{(3-5)^2 + (3-5)^2 + (7-5)^2 + (7-5)^2}{4}}$$
$$= 2$$

Example 3 ▶ Estimations of Standard Deviation

Consider the three sets of data represented by the bar graphs in Figure B.4. Which set has the smallest standard deviation? Which has the largest?

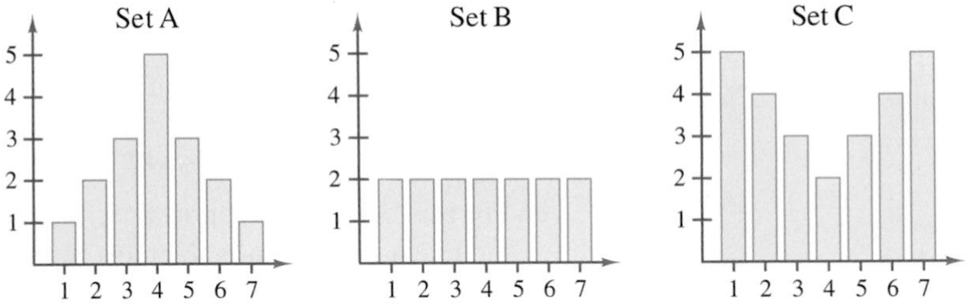

FIGURE B.4

Solution

Of the three sets, the numbers in set A are grouped most closely to the center and the numbers in set C are the most dispersed. So, set A has the smallest standard deviation and set C has the largest standard deviation.

Example 4 ▶ Finding Standard Deviation

Find the standard deviation of each set shown in Example 3.

Solution

Because of the symmetry of each bar graph, you can conclude that each has a mean of $\bar{x} = 4$. The standard deviation of set A is

$$\sigma = \sqrt{\frac{(-3)^2 + 2(-2)^2 + 3(-1)^2 + 5(0)^2 + 3(1)^2 + 2(2)^2 + (3)^2}{17}}$$

$$\approx 1.53.$$

The standard deviation of set B is

$$\sigma = \sqrt{\frac{2(-3)^2 + 2(-2)^2 + 2(-1)^2 + 2(0)^2 + 2(1)^2 + 2(2)^2 + 2(3)^2}{14}}$$

$$= 2.$$

The standard deviation of set C is

$$\sigma = \sqrt{\frac{5(-3)^2 + 4(-2)^2 + 3(-1)^2 + 2(0)^2 + 3(1)^2 + 4(2)^2 + 5(3)^2}{26}}$$

$$\approx 2.22.$$

These values confirm the results of Example 3. That is, set A has the smallest standard deviation and set C has the largest.

The following alternative formula provides a more efficient way to compute the standard deviation.

Alternative Formula for Standard Deviation

The standard deviation of $\{x_1, x_2, . . . , x_n\}$ is

$$\sigma = \sqrt{\frac{x_1^2 + x_2^2 + \cdot \cdot \cdot + x_n^2}{n} - \bar{x}^2}.$$

Because of messy computations, this formula is difficult to verify. Conceptually, however, the process is straightforward. It consists of showing that the expressions

$$\sqrt{\frac{(x_1 - \bar{x})^2 + (x_2 - \bar{x})^2 + \cdot \cdot \cdot + (x_n - \bar{x})^2}{n}}$$

and

$$\sqrt{\frac{x_1^2 + x_2^2 + \cdot \cdot \cdot + x_n^2}{n} - \bar{x}^2}$$

are equivalent. Try verifying this equivalence for the set $\{x_1, x_2, x_3\}$ with $\bar{x} = (x_1 + x_2 + x_3)/3$.

AK	66	MT	62
AL	40	NC	42
AR	39	ND	47
AZ	51	NE	63
CA	62	NH	59
CO	69	NJ	77
CT	80	NM	45
DC	94	NV	49
DE	44	NY	73
FL	50	OH	55
GA	46	OK	47
HI	80	OR	70
IA	55	PA	61
ID	53	RI	56
IL	61	SC	41
IN	47	SD	49
KS	51	TN	53
KY	53	TX	47
LA	45	UT	66
MA	74	VA	54
MD	68	VT	57
ME	47	WA	68
MI	62	WI	65
MN	67	WV	43
MO	53	WY	52
MS	37		

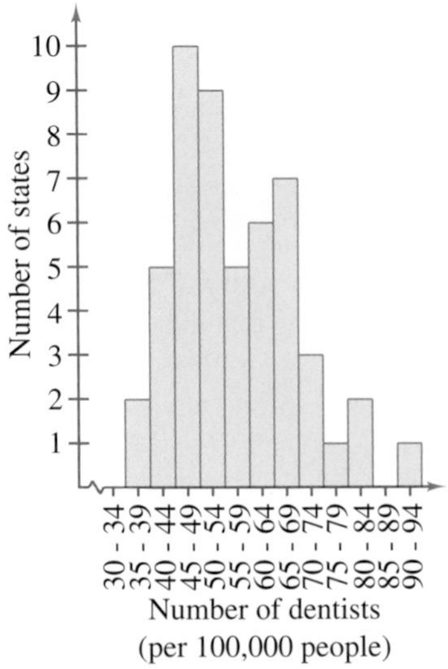

FIGURE B.5

Example 5 ▶ Using the Alternative Formula

Use the alternative formula for standard deviation to find the standard deviation of the following set of numbers.

5, 6, 6, 7, 7, 8, 8, 8, 9, 10

Solution

Begin by finding the mean of the set, which is 7.4. So, the standard deviation is

$$\sigma = \sqrt{\frac{5^2 + 2(6^2) + 2(7^2) + 3(8^2) + 9^2 + 10^2}{10} - (7.4)^2}$$

$$= \sqrt{\frac{568}{10} - 54.76}$$

$$= \sqrt{2.04}$$

$$\approx 1.43.$$

You can use the statistical features of a graphing utility to check this result.

A well-known theorem in statistics, called *Chebychev's Theorem*, states that at least

$$1 - \frac{1}{k^2}$$

of the numbers in a distribution must lie within k standard deviations of the mean. So, 75% of the numbers in a collection must lie within two standard deviations of the mean, and at least 88.9% of the numbers must lie within three standard deviations of the mean. For most distributions, these percentages are low. For instance, in all three distributions shown in Example 3, 100% of the numbers lie within two standard deviations of the mean.

Example 6 ▶ Describing a Distribution

The table at the left above shows the number of dentists (per 100,000 people) in each state and the District of Columbia. Find the mean and standard deviation of the numbers. What percent of the numbers lie within two standard deviations of the mean? (Source: American Dental Association)

Solution

Begin by entering the numbers into a graphing utility that has a standard deviation program. After running the program, you should obtain

$$\bar{x} \approx 56.76 \quad \text{and} \quad \sigma = 12.14.$$

The interval that contains all numbers that lie within two standard deviations of the mean is

$$[56.76 - 2(12.14), 56.76 + 2(12.14)] \quad \text{or} \quad [32.48, 81.04].$$

From the histogram in Figure B.5, you can see that all but one of the numbers (98%) lie in this interval—all but the number that corresponds to the number of dentists (per 100,000 people) in Washington, DC.

Box-and-Whisker Plots

Standard deviation is the measure of dispersion that is associated with the mean. **Quartiles** measure dispersion associated with the median.

Definition of Quartiles

Consider an ordered set of numbers whose median is m. The **lower quartile** is the median of the numbers that occur before m. The **upper quartile** is the median of the numbers that occur after m.

Example 7 ▶ Finding Quartiles of a Set

Find the lower and upper quartiles for the following set.

34, 14, 24, 16, 12, 18, 20, 24, 16, 26, 13, 27

Solution

Begin by ordering the set.

$$\underbrace{12, 13, 14,}_{\text{1st }25\%}\ \underbrace{16, 16, 18,}_{\text{2nd }25\%}\ \underbrace{20, 24, 24,}_{\text{3rd }25\%}\ \underbrace{26, 27, 34}_{\text{4th }25\%}$$

The median of the entire set is 19. The median of the six numbers that are less than 19 is 15. So, the lower quartile is 15. The median of the six numbers that are greater than 19 is 25. So, the upper quartile is 25.

Quartiles are represented graphically by a **box-and-whisker plot,** as shown in Figure B.6. In the plot, notice that five numbers are listed: the smallest number, the lower quartile, the median, the upper quartile, and the largest number. Also notice that the numbers are spaced proportionally, as though they were on a real number line.

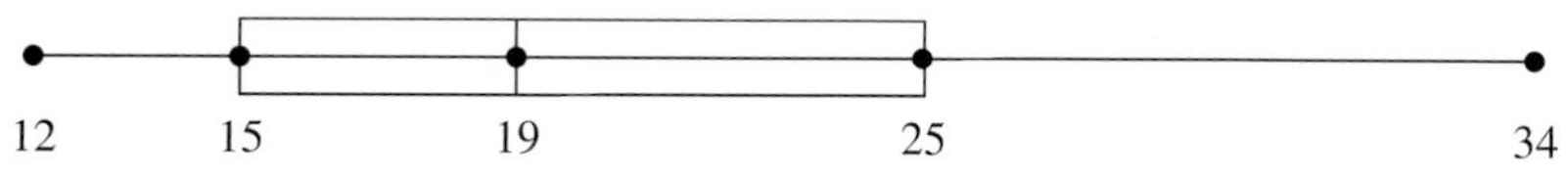

FIGURE B.6

The next example shows how to find quartiles when the number of elements in a set is not divisible by 4.

Example 8 ▶ Sketching Box-and-Whisker Plots

Sketch a box-and-whisker plot for each of the given sets.

a. 27, 28, 30, 42, 45, 50, 50, 61, 62, 64, 66

b. 82, 82, 83, 85, 87, 89, 90, 94, 95, 95, 96, 98, 99

c. 11, 13, 13, 15, 17, 18, 20, 24, 24, 27

Solution

a. This set has 11 numbers. The median is 50 (the sixth number). The lower quartile is 30 (the median of the first five numbers). The upper quartile is 62 (the median of the last five numbers).

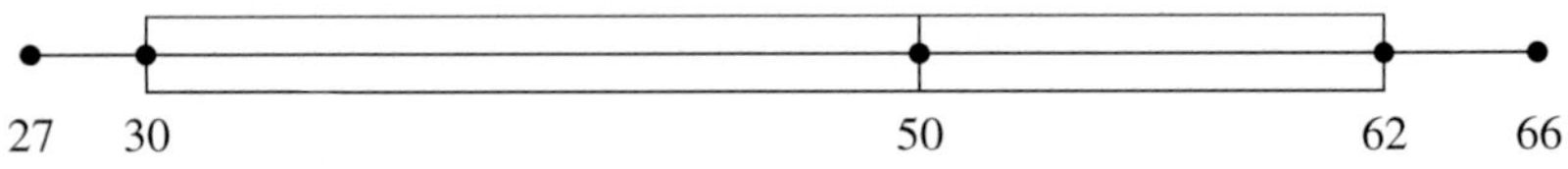

b. This set has 13 numbers. The median is 90 (the seventh number). The lower quartile is 84 (the median of the first six numbers). The upper quartile is 95.5 (the median of the last six numbers).

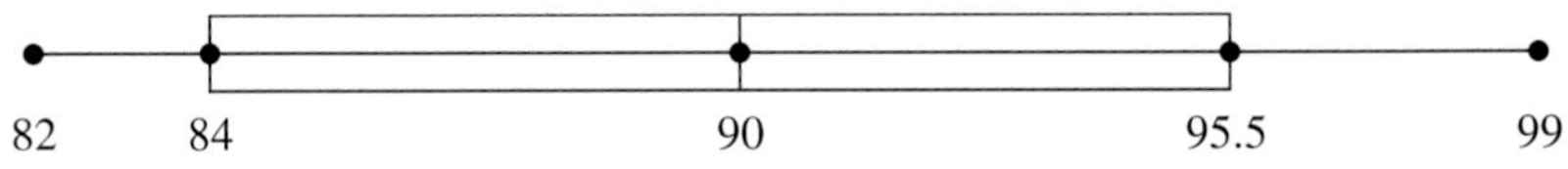

c. This set has 10 numbers. The median is 17.5 (the average of the fifth and sixth numbers). The lower quartile is 13 (the median of the first five numbers). The upper quartile is 24 (the median of the last five numbers).

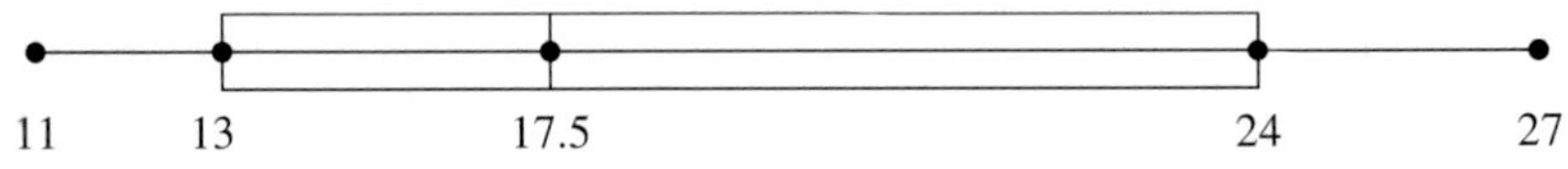

B.2 Exercises

In Exercises 1–6, find the mean, median, and mode of the set of measurements.

1. 5, 12, 7, 14, 8, 9, 7

2. 30, 37, 32, 39, 33, 34, 32

3. 5, 12, 7, 24, 8, 9, 7

4. 20, 37, 32, 39, 33, 34, 32

5. 5, 12, 7, 14, 9, 7

6. 30, 37, 32, 39, 34, 32

7. *Reasoning* Compare your answers for Exercises 1 and 3 with those for Exercises 2 and 4. Which of the measures of central tendency is sensitive to extreme measurements? Explain your reasoning.

8. *Reasoning*

(a) Add 6 to each measurement in Exercise 1 and calculate the mean, median, and mode of the revised measurements. How are the measures of central tendency changed?

(b) If a constant k is added to each measurement in a set of data, how will the measures of central tendency change?

9. *Electric Bills* A person had the following monthly bills for electricity. What are the mean and median of the collection of bills?

January	\$67.92	February	\$59.84
March	\$52.00	April	\$52.50
May	\$57.99	June	\$65.35
July	\$81.76	August	\$74.98
September	\$87.82	October	\$83.18
November	\$65.35	December	\$57.00

10. *Car Rental* A car rental company kept the following record of the numbers of miles a rental car was driven. What are the mean, median, and mode of this data?

Monday	410	Tuesday	260
Wednesday	320	Thursday	320
Friday	460	Saturday	150

11. *Six-Child Families* A study was done on families having six children. The table gives the numbers of families in the study with the indicated numbers of girls. Determine the mean, median, and mode of this set of data.

Number of girls	0	1	2	3	4	5	6
Frequency	1	24	45	54	50	19	7

12. *Baseball* A baseball fan examined the records of a favorite baseball player's performance during his last 50 games. The numbers of games in which the player had 0, 1, 2, 3, and 4 hits are recorded in the table.

Number of hits	0	1	2	3	4
Frequency	14	26	7	2	1

(a) Determine the average number of hits per game.

(b) Determine the player's batting average if he had 200 at-bats during the 50-game series.

13. *Think About It* Construct a collection of numbers that has the following properties. If this is not possible, explain why it is not.

Mean = 6, median = 4, mode = 4

14. *Think About It* Construct a collection of numbers that has the following properties. If this is not possible, explain why it is not.

Mean = 6, median = 6, mode = 4

15. *Test Scores* A professor records the following scores for a 100-point exam.

99, 64, 80, 77, 59, 72, 87, 79, 92, 88,
90, 42, 20, 89, 42, 100, 98, 84, 78, 91

Which measure of central tendency best describes these test scores?

16. *Shoe Sales* A salesman sold eight pairs of a certain style of men's shoes. The sizes of the eight pairs were as follows: $10\frac{1}{2}$, 8, 12, $10\frac{1}{2}$, 10, $9\frac{1}{2}$, 11, and $10\frac{1}{2}$. Which measure (or measures) of central tendency best describes the typical shoe size for this data?

In Exercises 17–24, find the mean ($\bar{x}$), variance (v), and standard deviation (σ) of the numbers.

17. 4, 10, 8, 2

18. 3, 15, 6, 9, 2

19. 0, 1, 1, 2, 2, 2, 3, 3, 4

20. 2, 2, 2, 2, 2, 2

21. 1, 2, 3, 4, 5, 6, 7

22. 1, 1, 1, 5, 5, 5

23. 49, 62, 40, 29, 32, 70

24. 1.5, 0.4, 2.1, 0.7, 0.8

In Exercises 25–30, use the alternative formula to find the standard deviation of the numbers.

25. 2, 4, 6, 6, 13, 5

26. 10, 25, 50, 26, 15, 33, 29, 4

27. 246, 336, 473, 167, 219, 359

28. 6.0, 9.1, 4.4, 8.7, 10.4

29. 8.1, 6.9, 3.7, 4.2, 6.1

30. 9.0, 7.5, 3.3, 7.4, 6.0

In Exercises 31 and 32, line plots of sets of data are given. Determine the mean and standard deviation of each set.

31. (a)

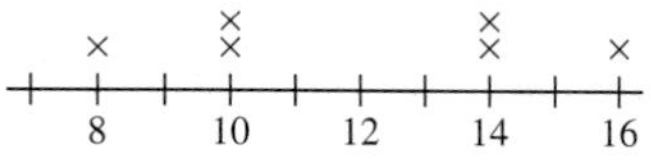

(b)

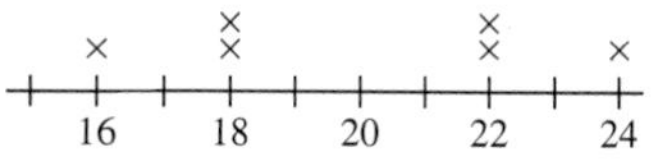

(c)

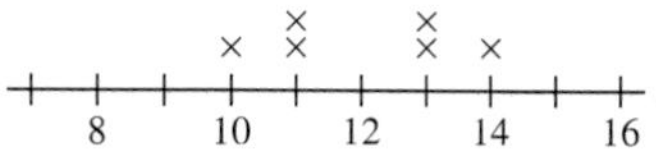

(d)

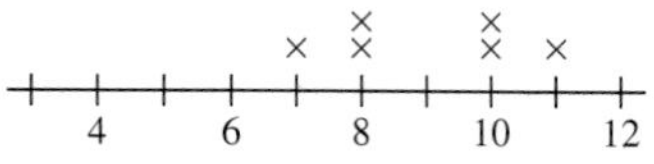

32. (a)
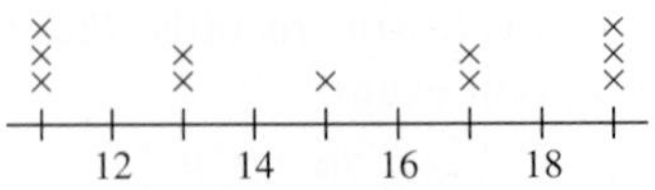

(b)
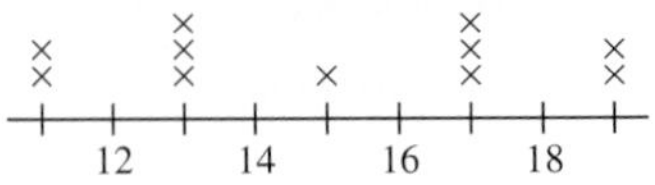

(c)
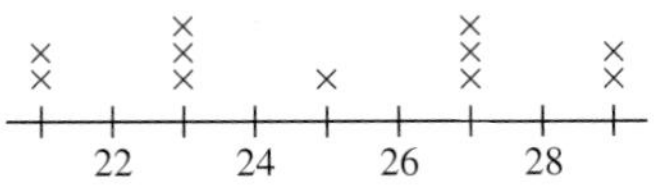

(d)
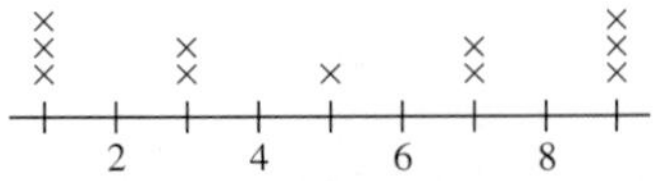

33. *Reasoning* Without calculating the standard deviation, explain why the set $\{4, 4, 20, 20\}$ has a standard deviation of 8.

34. *Reasoning* If the standard deviation of a set of numbers is 0, what does this imply about the set?

35. *Test Scores* An instructor adds five points to each student's exam score. Will this change the mean or standard deviation of the exam scores? Explain.

36. *Price of Gold* The following data represents the average prices of gold (in dollars per fine ounce) for the years 1978 to 1997. Use a computer or calculator to find the mean, variance, and standard deviation of the data. What percent of the data lies within two standard deviations of the mean? (Source: U.S. Bureau of Mines)

194,	308,	613,	460,	376,
424,	361,	318,	368,	448,
438,	383,	385,	363,	345,
361,	385,	386,	389,	333

37. *Think About It* The histograms represent the test scores of two classes of a college course in mathematics. Which histogram has the smaller standard deviation?

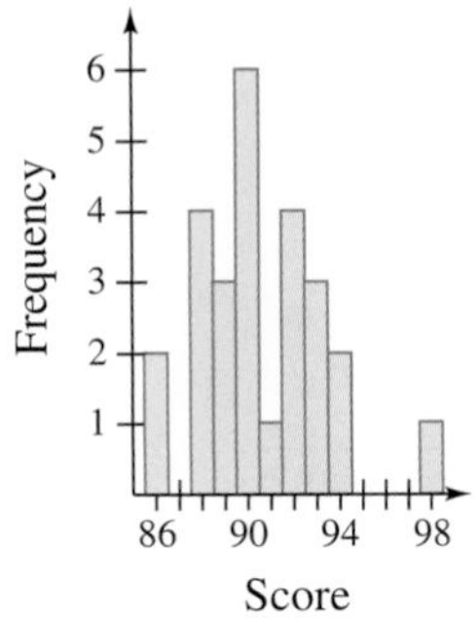

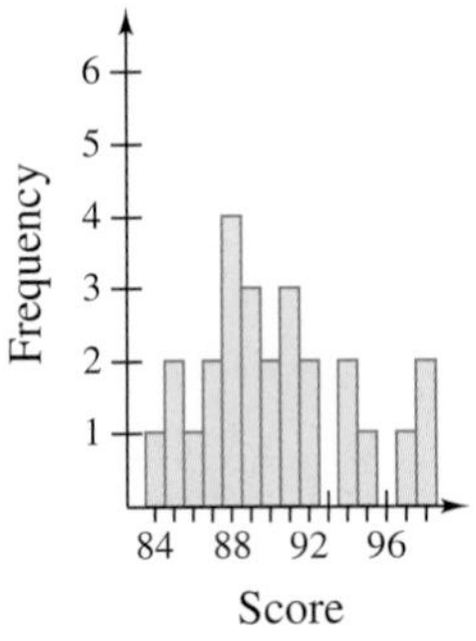

In Exercises 38–41, sketch a box-and-whisker plot for the data without the aid of a graphing utility.

38. 23, 15, 14, 23, 13, 14, 13, 20, 12

39. 11, 10, 11, 14, 17, 16, 14, 11, 8, 14, 20

40. 46, 48, 48, 50, 52, 47, 51, 47, 49, 53

41. 25, 20, 22, 28, 24, 28, 25, 19, 27, 29, 28, 21

In Exercises 42–45, use a graphing utility to create a box-and-whisker plot for the data.

42. 19, 12, 14, 9, 14, 15, 17, 13, 19, 11, 10, 19

43. 9, 5, 5, 5, 6, 5, 4, 12, 7, 10, 7, 11, 8, 9, 9

44. 20.1, 43.4, 34.9, 23.9, 33.5, 24.1, 22.5, 42.4, 25.7, 17.4, 23.8, 33.3, 17.3, 36.4, 21.8

45. 78.4, 76.3, 107.5, 78.5, 93.2, 90.3, 77.8, 37.1, 97.1, 75.5, 58.8, 65.6

46. *Product Lifetime* A company has redesigned a product in an attempt to increase the lifetime of the product. The two sets of data list the lifetimes (in months) of 20 units with the original design and 20 units with the new design. Create a box-and-whisker plot for each set of data, and then comment on the differences between the plots.

Original Design

15.1	78.3	56.3	68.9	30.6
27.2	12.5	42.7	72.7	20.2
53.0	13.5	11.0	18.4	85.2
10.8	38.3	85.1	10.0	12.6

New Design

55.8	71.5	25.6	19.0	23.1
37.2	60.0	35.3	18.9	80.5
46.7	31.1	67.9	23.5	99.5
54.0	23.2	45.5	24.8	87.8

B.3 Least Squares Regression

▶ What you should learn

- How to use the sum of squared differences to determine a least squares regression line
- How to find a least squares regression line for a set of data
- How to find a least squares regression parabola for a set of data

▶ Why you should learn it

The method of least squares provides a way of creating mathematical models for a set of data, which can then be analyzed.

In many of the examples and exercises in this text, you have been asked to use the regression capabilities of a graphing utility to find mathematical models for sets of data. Another way to find a mathematical model for a set of data is to use the **method of least squares.** As a measure of how well a model fits a set of data points

$$\{(x_1, y_1), (x_2, y_2), (x_3, y_3), . . . , (x_n, y_n)\}$$

you can add the squares of the differences between the actual y-values and the values given by the model to obtain the **sum of the squared differences.** For instance, the table shows the heights x (in feet) and the diameters y (in inches) of eight trees. The table also shows the values of a linear model $y^* = 0.69x - 40$ for each x-value. The sum of squared differences for the model is 53.3.

x	70	72	75	76	85	78	77	80
y	8.3	10.5	11.0	11.4	12.9	14.0	16.3	18.0
y^*	8.3	9.68	11.75	12.44	18.65	13.82	13.13	15.2

The model that has the least sum of squared differences is the **least squares regression line** for the data. The least squares regression line for the data in the table is $y \approx 0.43x - 20.3$. The sum of squared differences is 43.32.

To find the least squares regression line $y = ax + b$ for the points $\{(x_1, y_1), (x_2, y_2), (x_3, y_3), . . . , (x_n, y_n)\}$ algebraically, you need to solve the following system for a and b.

$$\begin{cases} nb + \left(\sum_{i=1}^{n} x_i\right)a = \sum_{i=1}^{n} y_i \\ \left(\sum_{i=1}^{n} x_i\right)b + \left(\sum_{i=1}^{n} x_i^2\right)a = \sum_{i=1}^{n} x_i y_i \end{cases}$$

In the system,

$$\sum_{i=1}^{n} x_i = x_1 + x_2 + \cdot \cdot \cdot + x_n$$

$$\sum_{i=1}^{n} y_i = y_1 + y_2 + \cdot \cdot \cdot + y_n$$

$$\sum_{i=1}^{n} x_i^2 = x_1^2 + x_2^2 + \cdot \cdot \cdot + x_n^2$$

$$\sum_{i=1}^{n} x_i y_i = x_1 y_1 + x_2 y_2 + \cdot \cdot \cdot + x_n y_n.$$

Example 1 ▶ Finding a Least Squares Regression Line

Find the least squares regression line for the points $(-3, 0)$, $(-1, 1)$, $(0, 2)$, and $(2, 3)$.

Solution

Begin by constructing a table like that shown below.

x	y	xy	x^2
-3	0	0	9
-1	1	-1	1
0	2	0	0
2	3	6	4
$\sum_{i=1}^{n} x_i = -2$	$\sum_{i=1}^{n} y_i = 6$	$\sum_{i=1}^{n} x_i y_i = 5$	$\sum_{i=1}^{n} x_i^2 = 14$

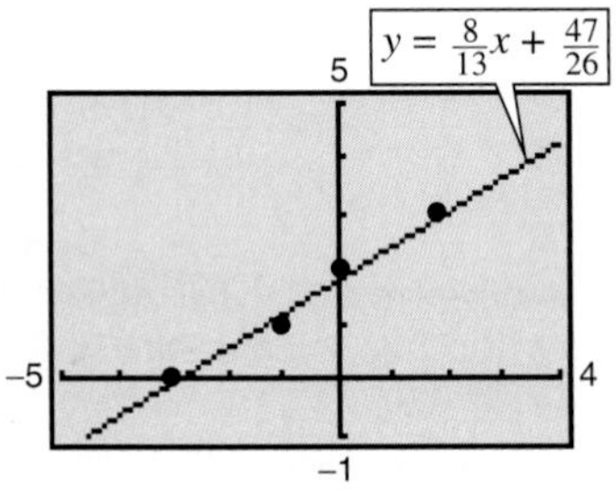

FIGURE B.7

Applying the system for the least squares regression line with $n = 4$ produces

$$\begin{cases} nb + \left(\sum_{i=1}^{n} x_i\right)a = \sum_{i=1}^{n} y_i \\ \left(\sum_{i=1}^{n} x_i\right)b + \left(\sum_{i=1}^{n} x_i^2\right)a = \sum_{i=1}^{n} x_i y_i \end{cases} \Longrightarrow \begin{cases} 4b - 2a = 6 \\ -2b + 14a = 5 \end{cases}.$$

Solving this system of equations produces $a = \frac{8}{13}$ and $b = \frac{47}{26}$. So, the least squares regression line is $y = \frac{8}{13}x + \frac{47}{26}$, as shown in Figure B.7.

The least squares regression parabola $y = ax^2 + bx + c$ for the points

$$\{(x_1, y_1), (x_2, y_2), (x_3, y_3), \ldots, (x_n, y_n)\}$$

is obtained in a similar manner by solving the following system of three equations in three unknowns for a, b, and c.

$$\begin{cases} nc + \left(\sum_{i=1}^{n} x_i\right)b + \left(\sum_{i=1}^{n} x_i^2\right)a = \sum_{i=1}^{n} y_i \\ \left(\sum_{i=1}^{n} x_i\right)c + \left(\sum_{i=1}^{n} x_i^2\right)b + \left(\sum_{i=1}^{n} x_i^3\right)a = \sum_{i=1}^{n} x_i y_i \\ \left(\sum_{i=1}^{n} x_i^2\right)c + \left(\sum_{i=1}^{n} x_i^3\right)b + \left(\sum_{i=1}^{n} x_i^4\right)a = \sum_{i=1}^{n} x_i^2 y_i \end{cases}$$

Fortunately, graphing utilities have built-in least squares regression capabilities.

Answers to Odd-Numbered Exercises and Tests

Chapter P

Section P.1 *(page 9)*

1. (a) 5, 1, 2 (b) $-9, 5, 0, 1, -4, 2, -11$
(c) $-9, -\frac{7}{2}, 5, \frac{2}{3}, 0, 1, -4, 2, -11$ (d) $\sqrt{2}$

3. (a) 1 (b) $-13, 1, -6$
(c) $2.01, 0.666\ldots, -13, 1, -6$ (d) $0.010110111\ldots$

5. (a) $\frac{6}{3}, 8$ (b) $\frac{6}{3}, -1, 8, -22$
(c) $-\frac{1}{3}, \frac{6}{3}, -7.5, -1, 8, -22$ (d) $-\pi, \frac{1}{2}\sqrt{2}$

7. 0.625 **9.** $0.\overline{123}$ **11.** $\frac{41}{10}$

13. $\frac{92}{9}$ **15.** $-\frac{1811}{900}$ **17.** $-1 < 2.5$

19. $-4 > -8$

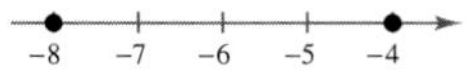

21. $\frac{3}{2} < 7$

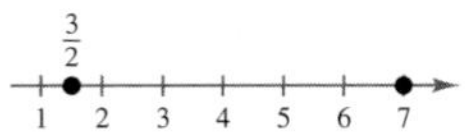

23. $\frac{5}{6} > \frac{2}{3}$

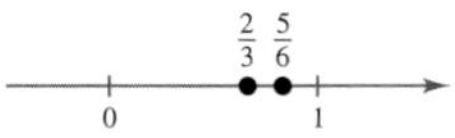

25. $x \le 5$ denotes the set of all real numbers less than or equal to 5. Unbounded

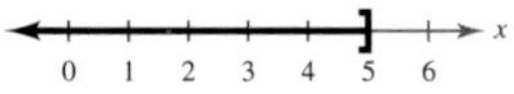

27. $x < 0$ denotes the set of all negative real numbers. Unbounded

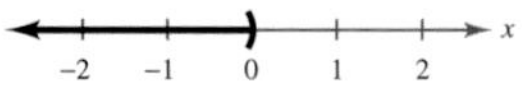

29. $x \ge 4$ denotes the set of all real numbers greater than or equal to 4. Unbounded

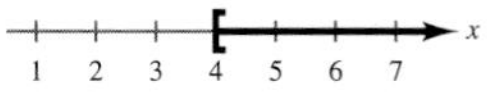

31. $-2 < x < 2$ denotes the set of all real numbers greater than -2 and less than 2. Bounded

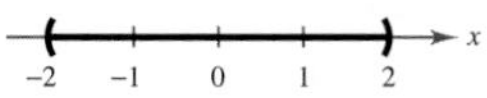

33. $-1 \le x < 0$ denotes the set of all negative real numbers greater than or equal to -1. Bounded

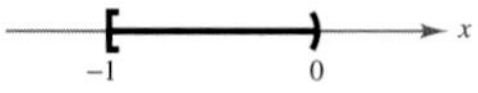

35. $\frac{127}{90}, \frac{584}{413}, \frac{7071}{5000}, \sqrt{2}, \frac{47}{33}$ **37.** $-2 < x \le 4$ **39.** $y \ge 0$

41. $10 \le t \le 22$ **43.** $W > 65$

45. This interval consists of all real numbers greater than or equal to 0 and less than 8.

47. This interval consists of all real numbers greater than -6.

49. 10 **51.** If $x \le 3$, $3 - x$. If $x > 3$, $x - 3$. **53.** -1

55. -1 **57.** -1 **59.** $|-3| > -|-3|$

61. $-5 = -|5|$ **63.** $-|-2| = -|2|$ **65.** 4

67. 51 **69.** $\frac{5}{2}$ **71.** $\frac{128}{75}$ **73.** (a) Negative (b) Negative

75. $|7 - 18| = 11$ miles **77.** $|60 - 23| = 37°$

79. $|x - 5| \le 3$ **81.** $|y| \ge 6$

83. $|\$113{,}356 - \$112{,}700| = \$656 > \$500$
$0.05(\$112{,}700) = \$5635$
Because the actual expenses differ from the budget by more than \$500, there is failure to meet the "budget variance test."

85. $|\$37{,}335 - \$37{,}640| = \$305 < \$500$
$0.05(\$37{,}640) = \$1882$
Because the difference between the actual expenses and the budget is less than \$500 and less than 5% of the budgeted amount, there is compliance with the "budget variance test."

87. $|92.5 - 92.2| = 0.3$
There was a surplus of \$0.3 billion.

89. $|1032.0 - 1253.2| = 221.2$
There was a deficit of \$221.2 billion.

91. $7x$ and 4 are the terms; 7 is the coefficient.

93. $\sqrt{3}x^2, -8x$, and -11 are the terms; $\sqrt{3}$ and -8 are the coefficients.

95. $4x^3, x/2$, and -5 are the terms; 4 and $\frac{1}{2}$ are the coefficients.

97. (a) -10 (b) -6 **99.** (a) 14 (b) 2

101. (a) Division by 0 is undefined. (b) 0

103. Commutative Property of Addition

105. Multiplicative Inverse Property

107. Distributive Property

109. Multiplicative Identity Property

111. Associative and Commutative Properties of Multiplication

113. $\frac{1}{2}$ **115.** $\frac{3}{8}$ **117.** 48 **119.** $\frac{5x}{12}$

121. -2.57 **123.** 1.56

125. (a)

n	1	0.5	0.01	0.0001	0.000001
$5/n$	5	10	500	50,000	5,000,000

(b) The value of $5/n$ approaches infinity as n approaches 0.

127. (a) No. If one variable is negative while the other is positive, the expressions are unequal.

(b) $|u + v| \leq |u| + |v|$

The expressions are equal when u and v have the same sign. If u and v differ in sign, $|u + v|$ is less than $|u| + |v|$.

129. The only even prime number is 2, because its only factors are itself and 1.

131. False. The denominators cannot be added when adding fractions.

133. Yes. $|a| = -a$ if $a < 0$.

Section P.2 *(page 21)*

1. $(8 \times 8 \times 8 \times 8 \times 8)$

3. $-(0.4 \times 0.4 \times 0.4 \times 0.4 \times 0.4 \times 0.4)$ **5.** 4.9^6

7. $(-10)^5$ **9.** (a) 27 (b) 81

11. (a) 729 (b) -9 **13.** (a) $\frac{243}{64}$ (b) -1

15. (a) $\frac{5}{6}$ (b) 4 **17.** -1600 **19.** 2.125

21. -24 **23.** 6 **25.** -54 **27.** 1

29. (a) $-125z^3$ (b) $5x^6$ **31.** (a) $24y^{10}$ (b) $3x^2$

33. (a) $\dfrac{7}{x}$ (b) $\dfrac{4}{3}(x + y)^2$ **35.** (a) 1 (b) $\dfrac{1}{4x^4}$

37. (a) $-2x^3$ (b) $\dfrac{10}{x}$ **39.** (a) $\dfrac{a^6}{64b^9}$ (b) $\dfrac{1}{625x^8y^8}$

41. (a) 3^{3n} (b) $\dfrac{b^5}{a^5}$ **43.** $9^{1/2} = 3$ **45.** $\sqrt[5]{32} = 2$

47. $\sqrt{196} = 14$ **49.** $(-216)^{1/3} = -6$ **51.** $\sqrt[3]{27^2} = 9$

53. $81^{3/4} = 27$

55. (a) 3 (b) 2 **57.** (a) 3 (b) $\frac{1}{2}$

59. (a) -125 (b) 3 **61.** (a) $\frac{1}{8}$ (b) $\frac{27}{8}$

63. (a) -4 (b) 2 **65.** (a) 7.550 (b) -7.225

67. (a) 14.499 (b) 0.528

69. (a) -0.011 (b) 0.005

71. (a) $2\sqrt{2}$ (b) $2\sqrt[3]{3}$ **73.** (a) $6x\sqrt{2x}$ (b) $\dfrac{18}{z\sqrt{z}}$

75. (a) $2x\sqrt[3]{2x^2}$ (b) $\dfrac{5|x|\sqrt{3}}{y^2}$ **77.** 625 **79.** $\dfrac{2}{x}$

81. $\dfrac{1}{x^3}$, $x > 0$ **83.** (a) $\dfrac{\sqrt{3}}{3}$ (b) $4\sqrt[3]{4}$

85. (a) $\dfrac{x(5 + \sqrt{3})}{11}$ (b) $3(\sqrt{6} - \sqrt{5})$

87. (a) $\dfrac{2}{\sqrt{2}}$ (b) $\dfrac{3}{\sqrt[3]{75}}$

89. (a) $\dfrac{2}{3(\sqrt{5} - \sqrt{3})}$ (b) $-\dfrac{1}{2(\sqrt{7} + 3)}$

91. (a) $\sqrt{3}$ (b) $\sqrt[3]{(x + 1)^2}$

93. (a) $2\sqrt[4]{2}$ (b) $\sqrt[8]{2x}$

95. (a) $34\sqrt{2}$ (b) $22\sqrt{2}$ **97.** (a) $2\sqrt{x}$ (b) $4\sqrt{y}$

99. (a) $13\sqrt{x + 1}$ (b) $18\sqrt{5x}$

101. $\sqrt{5} + \sqrt{3} > \sqrt{5 + 3}$

103. $5 > \sqrt{3^2 + 2^2}$ **105.** 5.73×10^7 **107.** 8.99×10^{-5}

109. 604,800,000 **111.** 0.0000000000000000001602

113. (a) 50,000 (b) 200,000

115. (a) 954.448 (b) 3.077×10^{10}

117. (a) 67,082.039 (b) 39.791

119. When any positive integer is squared, the units digit is 0, 1, 4, 5, 6, or 9. Therefore, $\sqrt{5233}$ is not an integer.

121. $\dfrac{\pi}{2} \approx 1.57$ seconds **123.** 13.29 seconds **125.** 0.280

127. True. When dividing variables, you subtract exponents.

129. $a^0 = 1, a \neq 0$, using the property $\dfrac{a^m}{a^n} = a^{m-n}$: $\dfrac{a^m}{a^m} = a^{m-m} = a^0 = 1$.

131. No. A number written in scientific notation has the form $c \times 10^n$, where $1 \leq c < 10$ and n is an integer. In true scientific notation, the number 52.7×10^5 is 5.27×10^6.

Section P.3 *(page 33)*

1. d **3.** b **5.** f

7. $-2x^3 + 4x^2 - 3x + 20$ (Answers will vary.) **9.** $-15x^4 + 1$ (Answers will vary.)

11. Degree: 1; Leading coefficient: 2

13. Degree: 5; Leading coefficient: -4

15. Degree: 5; Leading coefficient: 1

17. Polynomial: $-3x^3 + 2x + 8$

19. Not a polynomial because of the operation of division

21. Polynomial: $-y^4 + y^3 + y^2$ **23.** $x^2 + 2x$

25. $8.3x^3 + 29.7x^2 + 11$ **27.** $12z + 8$

29. $3x^3 - 6x^2 + 3x$ **31.** $-15z^2 + 5z$

33. $-4x^4 + 4x$ **35.** $7.5x^3 + 9x$ **37.** $-\frac{1}{2}x^2 - 12x$

39. $x^2 + 7x + 12$ **41.** $6x^2 - 7x - 5$

43. $4x^2 + 12x + 9$ **45.** $4x^2 - 20xy + 25y^2$

47. $x^2 - 100$ **49.** $x^2 - 4y^2$ **51.** $m^2 - n^2 - 6m + 9$

53. $x^2 + 2xy + y^2 - 6x - 6y + 9$ **55.** $4r^4 - 25$

57. $x^3 + 3x^2 + 3x + 1$ **59.** $8x^3 - 12x^2y + 6xy^2 - y^3$

61. $\frac{1}{4}x^2 - 3x + 9$ **63.** $\frac{1}{9}x^2 - 4$

65. $1.44x^2 + 7.2x + 9$ **67.** $2.25x^2 - 16$

69. $2x^2 + 2x$ **71.** $u^4 - 16$ **73.** $x - y$

75. $x^2 - 2\sqrt{5}x + 5$ **77.** Yes

79. Yes **81.** $3(x + 2)$

83. $2x(x^2 - 3)$ **85.** $(x - 1)(x + 6)$

87. $(x + 3)(x - 1)$ **89.** $\frac{1}{2}x(x^2 + 4x - 10)$

91. $\frac{2}{3}(x - 6)(x - 3)$ **93.** $(4y + 3)(4y - 3)$

95. $\left(4x + \frac{1}{3}\right)\left(4x - \frac{1}{3}\right)$ **97.** $(x + 1)(x - 3)$

99. $(3u + 2v)(3u - 2v)$ **101.** $(x - 2)^2$

103. $(6y - 9)^2$ **105.** $(3u + 4v)^2$ **107.** $\left(x - \frac{2}{3}\right)^2$

109. $(x + 2)(x - 1)$ **111.** $(s - 3)(s - 2)$

113. $-(5 + y)(y - 4)$ **115.** $(3x - 2)(x - 1)$

117. $(5x + 1)(x + 5)$ **119.** $-(3z - 2)(3z + 1)$

121. $(x - 2)(x^2 + 2x + 4)$ **123.** $(y + 4)(y^2 - 4y + 16)$

125. $(2t - 1)(4t^2 + 2t + 1)$

127. $(u + 3v)(u^2 - 3uv + 9v^2)$ **129.** $(x - 1)(x^2 + 2)$

131. $(2x - 1)(x^2 - 3)$ **133.** $(3x^2 - 1)(2x + 1)$

135. $(x + 2)(3x + 4)$ **137.** $(3x - 1)(5x - 2)$

139. $6(x + 3)(x - 3)$ **141.** $x^2(x - 4)$

143. $-2x(x + 1)(x - 2)$ **145.** $(x^2 + 5)(3x + 1)$

147. $\frac{1}{81}(x - 18)(x + 36)$ **149.** $x(x - 4)(x^2 + 1)$

151. $(x + 1)^2(x - 1)^2$ **153.** $2(t - 2)(t^2 + 2t + 4)$

155. $(2x - 1)(6x - 1)$ **157.** $5(1 - x)^2(3x + 2)(4x + 3)$

159. $(x - 2)^2(x + 1)^3(7x - 5)$ **161.** $-14, 14, -2, 2$

163. $-11, 11, -4, 4, -1, 1$ **165.** Answers will vary.

167. Answers will vary. **169.** $P = \$85{,}000$

171. (a) $500r^2 + 1000r + 500$

(b)

r	$2\frac{1}{2}\%$	3%	4%
$500(1 + r)^2$	\$525.31	\$530.45	\$540.80

r	$4\frac{1}{2}\%$	5%
$500(1 + r)^2$	\$546.01	\$551.25

(c) The amount increases with increasing r.

173. $V = x(26 - 2x)(18 - 2x)$

$= 4x(x - 13)(x - 9)$

x (cm)	1	2	3
V (cm^3)	384	616	720

175. (a) $6x^2 - 3x$ (b) $42x^2$

177. $2x^2 + 46x + 252$

179.

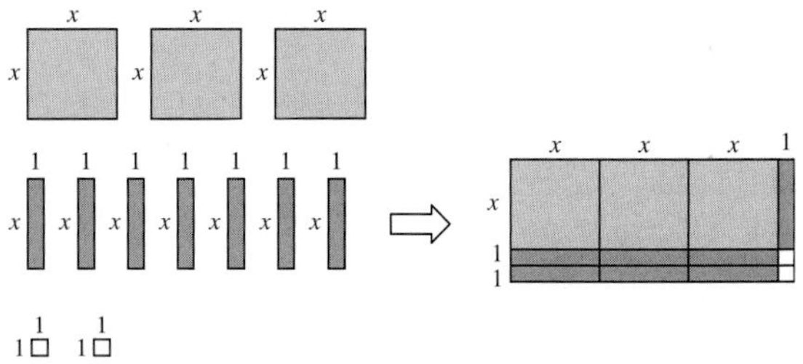

181. $4\pi(r + 1)$ **183.** $4(6 - x)(6 + x)$

185. (a) $\pi h(R - r)(R + r)$ (b) $2\pi\left[\left(\frac{R + r}{2}\right)(R - r)\right]h$

187. False. $(4x^2 + 1)(3x + 1) = 12x^3 + 4x^2 + 3x + 1$

189. True. $(x^2 - 16) = (x + 4)(x - 4)$ **191.** $m + n$

193. $(3 + 4)^2 = 49 \neq 25 = 3^2 + 4^2$

The only values of x and y that can make the equation true are 0. No other values work, because adding two numbers and then squaring produces larger values than squaring the two numbers separately and then adding.

195. $(x^n + y^n)(x^n - y^n)$ **197.** $x^{3n} - y^{2n}$

Section P.4 *(page 45)*

1. All real numbers **3.** All nonnegative real numbers

5. All real numbers x such that $x \neq 2$

7. All real numbers x such that $x \geq -1$

9. $3x,\ x \neq 0$ **11.** $x - 2,\ x \neq 2$ **13.** $x,\ x \neq 0$

15. $\frac{3x}{2},\ x \neq 0$ **17.** $\frac{3y}{y + 1},\ x \neq 0$ **19.** $\frac{-4y}{5},\ y \neq \frac{1}{2}$

21. $-\frac{1}{2},\ x \neq 5$ **23.** $y - 4,\ y \neq -4$

25. $\frac{x(x + 3)}{x - 2},\ x \neq -2$ **27.** $\frac{y - 4}{y + 6},\ y \neq 3$

29. $\frac{-(x^2 + 1)}{(x + 2)},\ x \neq 2$ **31.** $z - 2$

33.

x	0	1	2	3	4	5	6
$\frac{x^2 - 2x - 3}{x - 3}$	1	2	3	Undef.	5	6	7
$x + 1$	1	2	3	4	5	6	7

The expressions are equivalent except at $x = 3$.

35. The expression cannot be simplified.

37. $\frac{\pi}{4},\ r \neq 0$ **39.** $\frac{1}{5(x - 2)},\ x \neq 1$

41. $\frac{x-3}{(x+2)^2}$, $x \neq -5$ **43.** $\frac{r+1}{r}$, $r \neq 1$

45. $\frac{t-3}{(t+3)(t-2)}$, $t \neq -2$ **47.** $\frac{x-y}{x(x+y)^2}$, $x \neq -2y$

49. $\frac{3}{2}$, $x \neq -y$ **51.** $\frac{(x^2+16)(x+2)(x-2)}{(x+4)^2}$

53. $\frac{(x+6)(x+1)}{x^2}$, $x \neq 6$ **55.** $\frac{x+5}{x-1}$ **57.** $\frac{6x+13}{x+3}$

59. $-\frac{2}{x-2}$ **61.** $\frac{x-4}{(x+2)(x-2)(x-1)}$

63. $-\frac{x^2+3}{(x+1)(x-2)(x-3)}$ **65.** $\frac{2-x}{x^2+1}$, $x \neq 0$

67. $\frac{x^7-2}{x^2}$ **69.** $\frac{3x^2-2}{x^{1/2}}$ **71.** $\frac{-1}{(x^2+1)^5}$

73. $\frac{2x^3-2x^2-5}{(x-1)^{1/2}}$ **75.** $\frac{-2(x-6)}{x+2}$ **77.** $\frac{1}{2}$, $x \neq 2$

79. $x(x+1)$, $x \neq -1, 0$ **81.** $\frac{1}{x}$, $x \neq -1$

83. $\frac{(x+3)^3}{2x(x-3)}$, $x \neq -3$ **85.** $-\frac{2x+h}{x^2(x+h)^2}$, $h \neq 0$

87. $\frac{2x-1}{2x}$, $x > 0$ **89.** $\frac{3x-1}{3}$

91. $-\frac{1}{x^2(x+1)^{3/4}}$ **93.** $\frac{1}{\sqrt{x+2}+\sqrt{x}}$

95. (a) $\frac{1}{16}$ minute (b) $\frac{x}{16}$ minute(s)

(c) $\frac{60}{16} = \frac{15}{4}$ minutes

97. $\frac{11x}{30}$ **99.** (a) 9.09% (b) $\frac{288(MN-P)}{N(MN+12P)}$, 9.09%

101. (a)

t	0	2	4	6	8	10
T	75	55.9	48.3	45	43.3	42.3

t	12	14	16	18	20	22
T	41.7	41.3	41.1	40.9	40.7	40.6

(b) The model is approaching a T-value of 40.

103. $\frac{x}{3(x+3)}$ **105.** $\frac{7}{6x(x+5)}$

107. False. In order for the simplified expression to be equivalent to the original expression, the domain of the simplified expression needs to be restricted. If n is even, $x \neq -1, 1$. If n is odd, $x \neq 1$.

109. False. The least common denominator of several fractions consists of the product of all *prime factors* in the denominators, with each factor given the highest power of its occurrence in any denominator.

111. No. $\frac{ax-b}{b-ax} = \frac{\cancel{ax-b}}{-(\cancel{ax-b})} = -1$, $x \neq \frac{b}{a}$

Section P.5 *(page 58)*

1. Identity **3.** Conditional equation **5.** Identity

7. Conditional equation **9.** Conditional equation

11. 4 **13.** -9 **15.** 5 **17.** 9 **19.** No solution

21. -4 **23.** $-\frac{6}{5}$ **25.** 9

27. No solution. The x-terms sum to zero. **29.** 10

31. 4 **33.** 3

35. No solution. The variable is divided out. **37.** $\frac{5}{3}$

39. No solution. The solution is extraneous. **41.** 5

43. No solution. The solution is extraneous. **45.** 0

47. All real numbers **49.** $2x^2+8x-3=0$

51. $x^2-6x+6=0$ **53.** $3x^2-90x-10=0$

55. $0, -\frac{1}{2}$ **57.** $4, -2$ **59.** -5 **61.** $3, -\frac{1}{2}$

63. $2, -6$ **65.** $-\frac{20}{3}, -4$ **67.** $-a$ **69.** $\pm 7; \pm 7.00$

71. $\pm\sqrt{11}; \pm 3.32$ **73.** $\pm 3\sqrt{3}; \pm 5.20$

75. 8, 16; 8.00, 16.00 **77.** $-2 \pm \sqrt{14}$; 1.74, -5.74

79. $\frac{1 \pm 3\sqrt{2}}{2}$; 2.62, -1.62 **81.** 2; 2.00 **83.** 0, 2

85. $4, -8$ **87.** $-3 \pm \sqrt{7}$ **89.** $1 \pm \frac{\sqrt{6}}{3}$

91. $2 \pm 2\sqrt{3}$ **93.** $\frac{1}{2}, -1$ **95.** $\frac{1}{4}, -\frac{3}{4}$ **97.** $1 \pm \sqrt{3}$

99. $-7 \pm \sqrt{5}$ **101.** $-4 \pm 2\sqrt{5}$ **103.** $\frac{2}{3} \pm \frac{\sqrt{7}}{3}$

105. $-\frac{4}{3}$ **107.** $-\frac{1}{2} \pm \sqrt{2}$ **109.** $\frac{2}{7}$ **111.** $2 \pm \frac{\sqrt{6}}{2}$

113. $6 \pm \sqrt{11}$ **115.** $-\frac{3}{8} \pm \frac{\sqrt{265}}{8}$

117. 0.976, -0.643 **119.** 1.355, -14.071

121. 1.687, -0.488 **123.** $-0.290, -2.200$

125. $1 \pm \sqrt{2}$ **127.** $6, -12$ **129.** $\frac{1}{2} \pm \sqrt{3}$

131. $-\frac{1}{2}$ **133.** $\frac{3}{4} \pm \frac{\sqrt{97}}{4}$ **135.** $0, \pm\frac{3\sqrt{2}}{2}$

137. ± 3 **139.** -6 **141.** $-3, 0$ **143.** $3, 1, -1$

145. ± 1 **147.** $\pm\sqrt{3}, \pm 1$ **149.** $\pm\frac{1}{2}, \pm 4$

151. $1, -2$ **153.** 50 **155.** 26 **157.** -16

159. $2, -5$ **161.** 0 **163.** 9

165. $-3 \pm 16\sqrt{2}$ **167.** $\pm\sqrt{14}$

169. 1 **171.** 4, -5 **173.** $\dfrac{-3 \pm \sqrt{21}}{6}$

175. 2, $-\frac{3}{2}$ **177.** 1, -3 **179.** 3, -2

181. $\sqrt{3}, -3$ **183.** 3, $\dfrac{-1 - \sqrt{17}}{2}$

185. 61.2 inches **187.** 23,437.5 miles

189. (a)

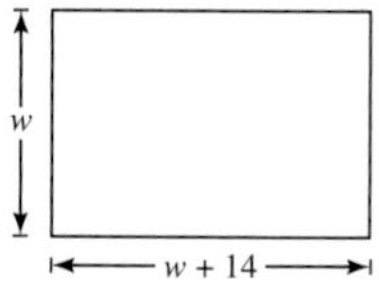

(b) $w(w + 14) = 1632$

(c) $w = 34$ feet, $l = 48$ feet

191. $\dfrac{5\sqrt{2}}{2} \approx 3.54$ centimeters

193. $\approx$550 miles per hour and 600 miles per hour

195. 50,000 units **197.** 24.725 pounds per square inch

199. False. The product must equal zero for the Zero-Factor Property to be used.

201. The student should have subtracted $15x$ from both sides so that the equation was equal to zero. By factoring out an x, there are two solutions, $x = 0$ and $x = 6$.

203. Remove symbols of grouping, combine like terms, reduce fractions.

Add (or subtract) the same quantity to (from) both sides of the equation.

Multiply (or divide) both sides of the equation by the same nonzero quantity.

Interchange the two sides of the equation.

205. (a) $0, -\dfrac{b}{a}$ (b) 0, 1

207. Isolate the absolute value by subtracting x from both sides of the equation. The expression inside the absolute value signs can be positive or negative, so two separate equations must be solved.

Section P.6 *(page 70)*

1. $-1 \le x \le 5$. Bounded **3.** $11 < x < \infty$. Unbounded

5. $-\infty < x < -2$. Unbounded **7.** b **9.** d **11.** e

13. (a) Yes (b) No (c) Yes (d) No

15. (a) Yes (b) No (c) No (d) Yes

17. (a) Yes (b) Yes (c) Yes (d) No

19. $x < 3$

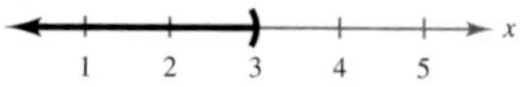

21. $x > \frac{3}{2}$

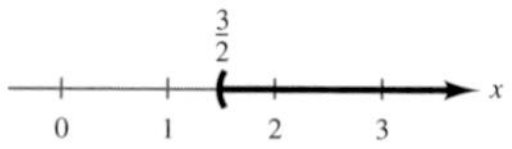

23. $x \ge 12$

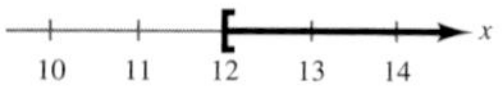

25. $x > 2$

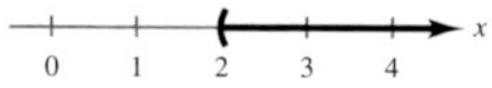

27. $x \ge \frac{2}{7}$

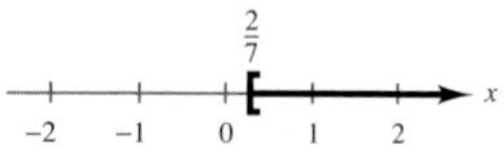

29. $x < 5$

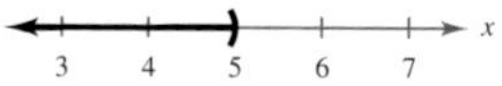

31. $x \ge 4$

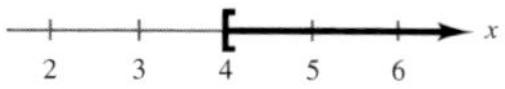

33. $x \ge 2$

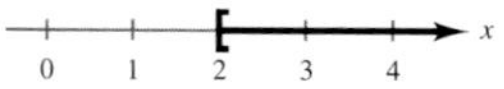

35. $x \ge -4$

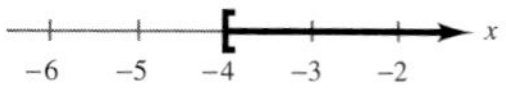

37. $-1 < x < 3$

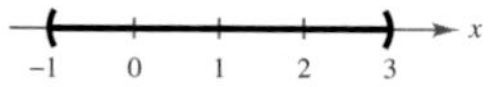

39. $-\frac{9}{2} < x < \frac{15}{2}$

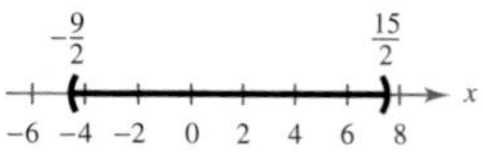

41. $-\frac{3}{4} < x < -\frac{1}{4}$

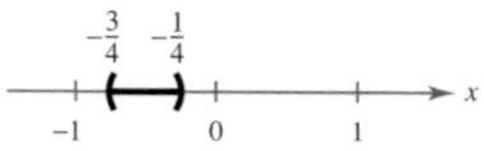

43. $10.5 \le x \le 13.5$

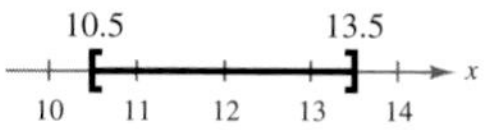

45. $-6 < x < 6$

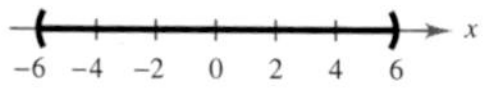

47. $x < -10, x > 10$

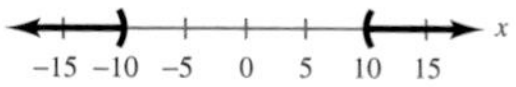

49. No solution

51. $14 \le x \le 26$

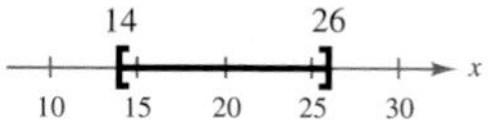

53. $x \le -\frac{3}{2}, x \ge 3$

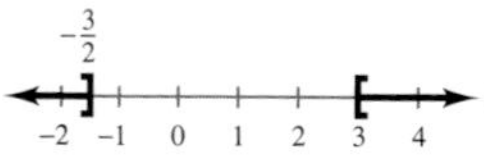

55. $x \le -7, x \ge 13$

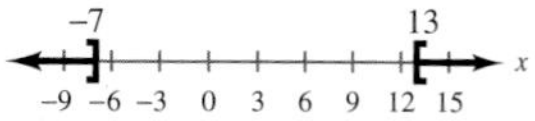

57. $4 < x < 5$

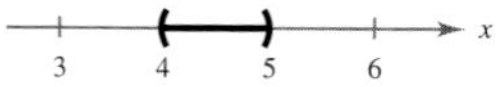

59. $x \le -\frac{29}{2}, x \ge -\frac{11}{2}$

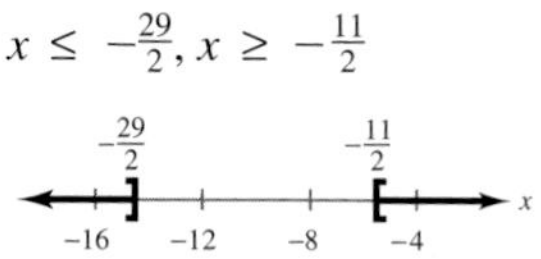

61.

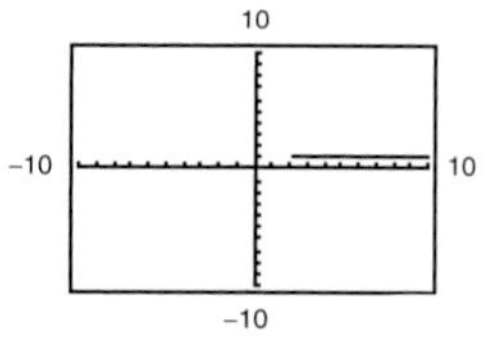

$x > 2$

63.

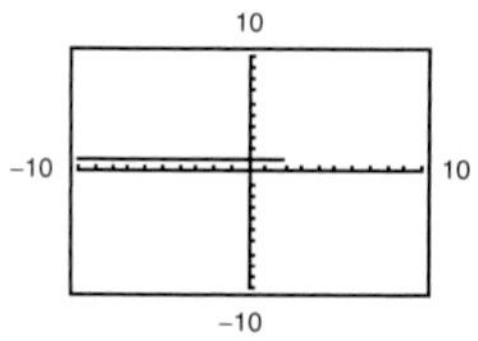

$x \le 2$

65.

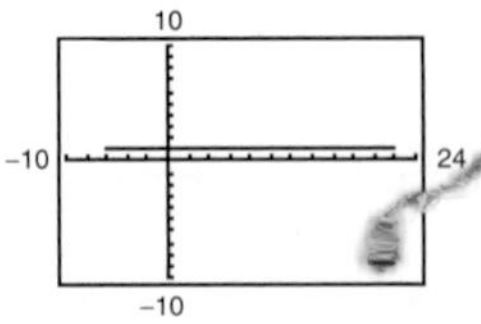

$-6 \le x \le 22$

67.

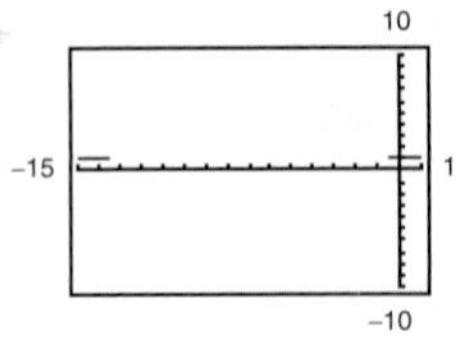

$x \le -\frac{27}{2}, x \ge -\frac{1}{2}$

69. $[5, \infty)$ **71.** $[-3, \infty)$ **73.** $\left(-\infty, \frac{7}{2}\right]$

75. All real numbers within 8 units of 10

77. $|x| \le 3$ **79.** $|x - 7| \ge 3$

81. $|x - 12| < 10$ **83.** $|x + 3| > 5$

85. (a) No (b) Yes (c) Yes (d) No

87. (a) Yes (b) No (c) No (d) Yes

89. $2, -\frac{3}{2}$ **91.** $\frac{7}{2}, 5$

93. $[-3, 3]$

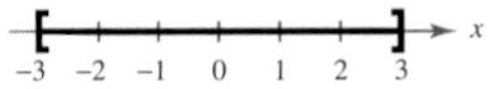

95. $(-7, 3)$

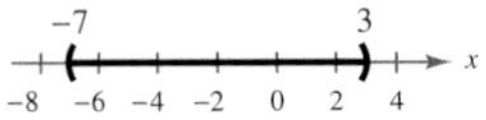

97. $(-\infty, -5], [1, \infty)$

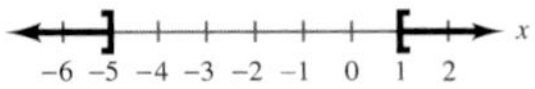

99. $(-3, 2)$

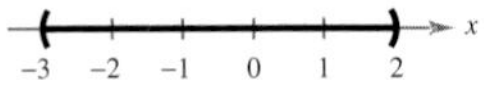

101. $(-3, 1)$

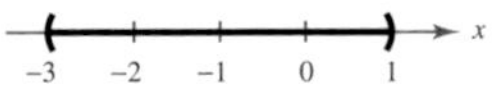

103. $\left(-\infty, -4 - \sqrt{21}\right], \left[-4 + \sqrt{21}, \infty\right)$

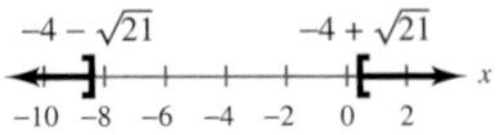

105. $(-1, 1), (3, \infty)$

107. $[-3, 2], [3, \infty)$

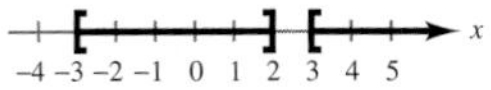

109. $(-\infty, 0), \left(0, \frac{3}{2}\right)$ **111.** $[-2, 0], [2, \infty)$

113. $[-2, \infty)$

115. $(-\infty, -1), (0, 1)$

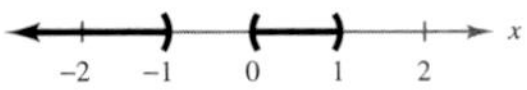

117. $(-\infty, -1), (4, \infty)$

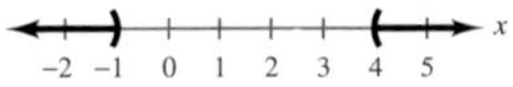

119. $(5, 15)$

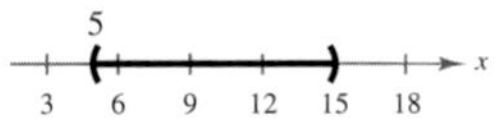

121. $\left(-5, -\frac{3}{2}\right), (-1, \infty)$

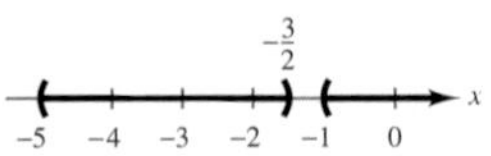

123. $\left(-\frac{3}{4}, 3\right), [6, \infty)$

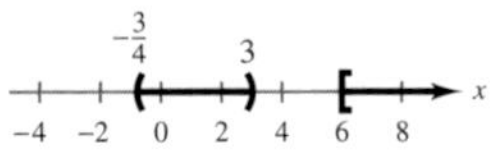

125. $(-3, -2], [0, 3)$

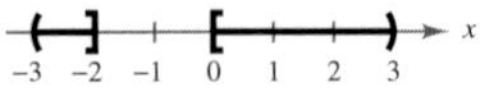

127. $(-\infty, -1), \left(-\frac{2}{3}, 1\right), (3, \infty)$ **129.** $[-2, 2]$

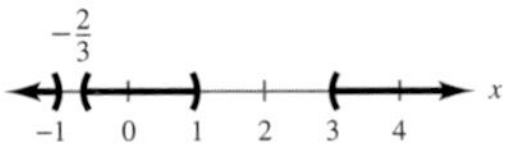

131. $(-\infty, 3], [4, \infty)$ **133.** $(-5, 0], (7, \infty)$

135. $(-3.51, 3.51)$ **137.** $(-0.13, 25.13)$

139. $(2.26, 2.39)$ **141.** More than 400 miles

143. $r > 3.1\%$ **145.** $65.8 \le h \le 71.2$

147. Between 13.8 and 36.2 meters **149.** $r > 4.88\%$

151. $R_1 \ge 2$ **153.** False. c has to be greater than zero.

155. True. The test intervals are $(-\infty, -3), (-3, 1), (1, 4)$, and $(4, \infty)$.

157. b **159.** $(-\infty, -4] \cup [4, \infty)$

161. $\left(-\infty, -2\sqrt{30}\right] \cup \left[2\sqrt{30}, \infty\right)$

163. If $a > 0$ and $c \le 0$, b can be any real number. If $a > 0$ and $c > 0$, $b < -2\sqrt{ac}$ or $b > 2\sqrt{ac}$.

165. (a) $x = a, \quad x = b$

(b)

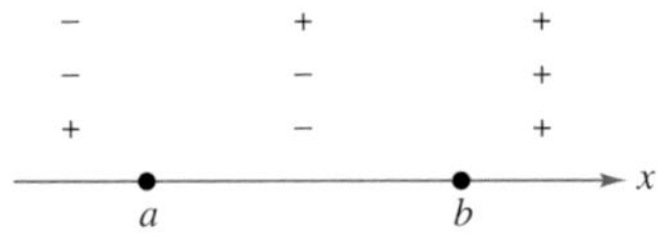

(c) The real zeros of the polynomial

Section P.7 *(page 79)*

1. Change all signs when distributing the minus sign.

$2x - (3y + 4) = 2x - 3y - 4$

3. Change all signs when distributing the minus sign.

$$\frac{4}{16x - (2x + 1)} = \frac{4}{14x - 1}$$

5. z occurs twice as a factor.

$(5z)(6z) = 30z^2$

7. The fraction as a whole is multiplied by a, not the numerator and denominator separately.

$$a\left(\frac{x}{y}\right) = \frac{ax}{y}$$

9. The exponent applies to the denominator also.

$$\left(\frac{x}{y}\right)^3 = \frac{x \cdot x \cdot x}{y \cdot y \cdot y} = \frac{x^3}{y^3}$$

11. $\sqrt{x + 9}$ cannot be simplified.

13. $\dfrac{6x + y}{6x - y}$ cannot be simplified.

15. The negative exponent is on a term of the denominator, not a factor.

$$\frac{1}{x + y^{-1}} = \frac{y}{xy + 1}, \quad y \ne 0$$

17. Exponents are applied before multiplying.

$x(2x - 1)^2 = x(4x^2 - 4x + 1) = 4x^3 - 4x^2 + x$

19. Radicals apply to every factor of the radicand.

$\sqrt[3]{x^3 + 7x^2} = \sqrt[3]{x^2(x + 7)} = \left(\sqrt[3]{x^2}\right)\left(\sqrt[3]{x + 7}\right)$

21. To add fractions, first find a common denominator.

$\frac{3}{x} + \frac{4}{y} = \frac{3y + 4x}{xy}$

23. $3x + 2$ **25.** $2x^2 + x + 15$ **27.** $\frac{1}{4}$ **29.** $-\frac{1}{4}$

31. $\frac{1}{2}$ **33.** $\frac{1}{2x^2}$ **35.** $\frac{25}{9}, \frac{49}{16}$ **37.** 1, 2

39. $1 - 5x$ **41.** $1 - 7x$ **43.** $3x - 1$ **45.** $\frac{16}{x} - 5 - x$

47. $4x^{8/3} - 7x^{5/3} + \frac{1}{x^{1/3}}$ **49.** $\frac{3}{\sqrt{x}} - 5x^{3/2} - x^{7/2}$

51. $\frac{-7x^2 - 4x + 9}{(x^2 - 3)^3(x + 1)^4}$ **53.** $\frac{27x^2 - 24x + 2}{(6x + 1)^4}$

55. $\frac{-1}{(x + 3)^{2/3}(x + 2)^{7/4}}$ **57.** $\frac{4x - 3}{(3x - 1)^{4/3}}$ **59.** $\frac{x}{x^2 + 4}$

61. $\frac{(3x - 2)^{1/2}(15x^2 - 4x + 45)}{2(x^2 + 5)^{1/2}}$

63. (a) Answers will vary.

(b)

x	-2	-1	$-\frac{1}{2}$	0	1	2	$\frac{5}{2}$
y_1	-8.7	-2.9	-1.1	0	2.9	8.7	12.5
y_2	-8.7	-2.9	-1.1	0	2.9	8.7	12.5

65. $y_1(0) = 0$ and $y_2(0) = 2$ so, $y_1 \neq y_2$.

$y_2 = \frac{2x - 3x^3}{\sqrt{1 - x^2}}$. (Answers will vary.)

67. False. Cannot move term-by-term from denominator to numerator.

69. False. $x^2 - 9$ does not factor into $(\sqrt{x} + 3)(\sqrt{x} - 3)$.

71. There is no error. **73.** There is no error.

Section P.8 *(page 87)*

1.

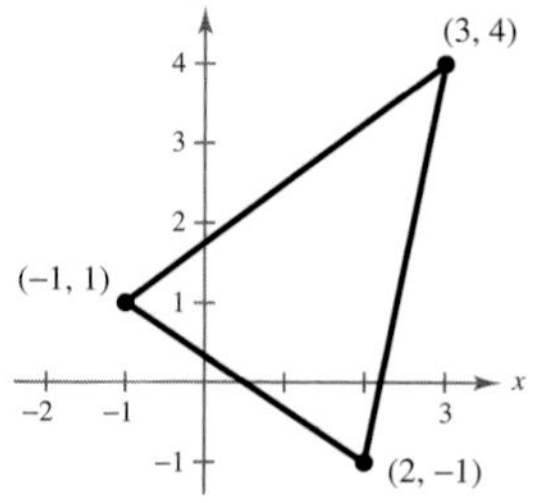

3.

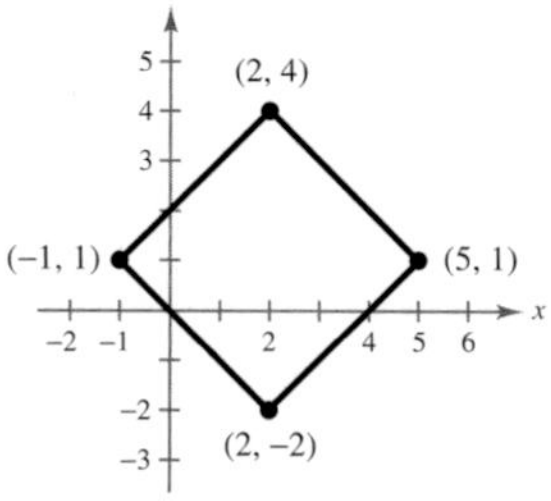

5. A: (2, 6), B: $(-6, -2)$, C: $(4, -4)$, D: $(-3, 2)$

7. $(-3, 4)$ **9.** $(-5, -5)$

11. Quadrant IV **13.** Quadrant II

15. Quadrants III and IV **17.** Quadrant III

19. Quadrants I and III **21.** (0, 1), (4, 2), (1, 4)

23. $(-3, 6)$, (2, 10), (2, 4), $(-3, 4)$

25.

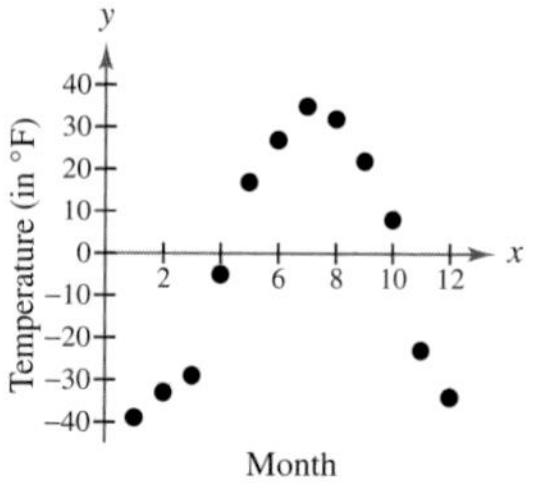

27. 1996 : \$1.65 per one-half gallon **29.** 166.67%

31. 1990s **33.** 65 **35.** 8 **37.** 5

39. (a) 4, 3, 5 (b) $4^2 + 3^2 = 5^2$

41. (a) 10, 3, $\sqrt{109}$ (b) $10^2 + 3^2 = (\sqrt{109})^2$

43. (a)

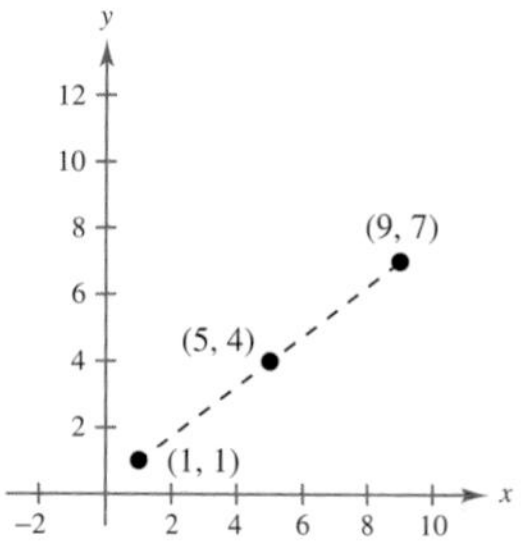

(b) 10

(c) (5, 4)

45. (a)

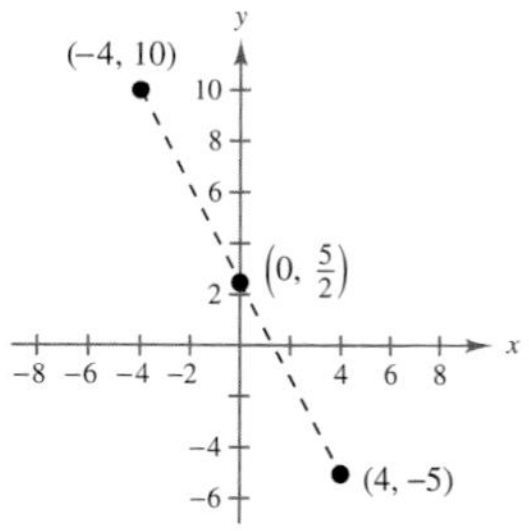

(b) 17

(c) $(0, \frac{5}{2})$

47. (a)

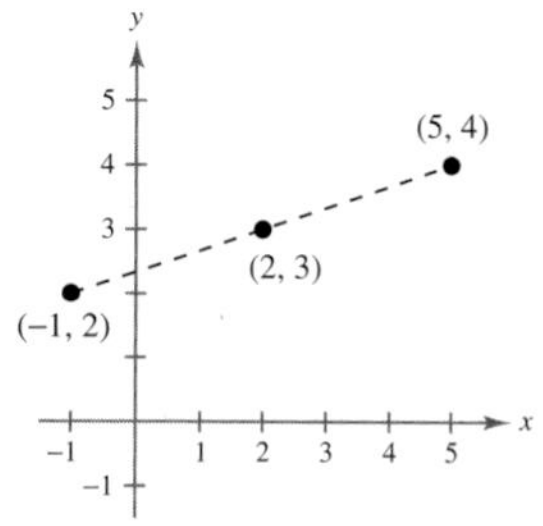

(b) $2\sqrt{10}$

(c) (2, 3)

49. (a)

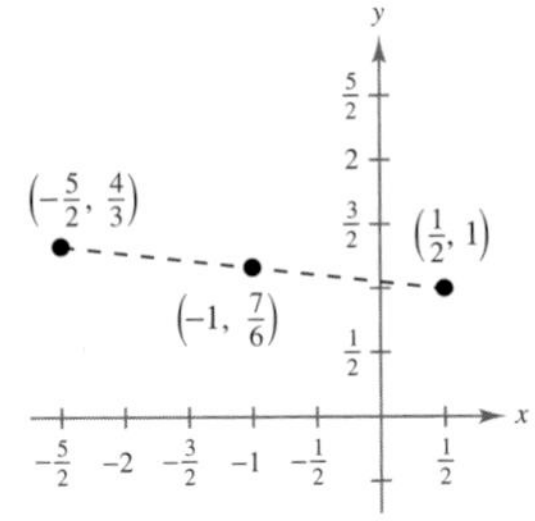

(b) $\frac{\sqrt{82}}{3}$

(c) $\left(-1, \frac{7}{6}\right)$

51. (a)

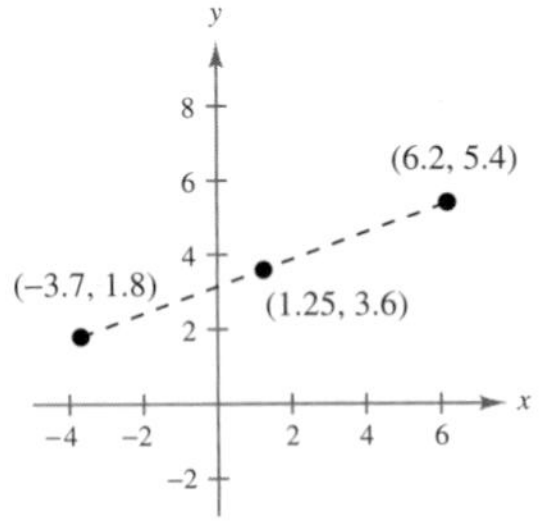

(b) $\sqrt{110.97}$

(c) $(1.25, 3.6)$

53. (a)

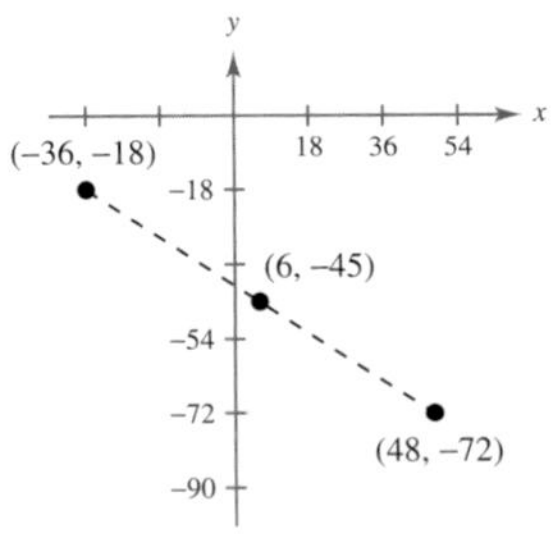

(b) $6\sqrt{277}$

(c) $(6, -45)$

55. $630,000

57. $(\sqrt{5})^2 + (\sqrt{45})^2 = (\sqrt{50})^2$

59. Opposite sides have equal lengths of $2\sqrt{5}$ and $\sqrt{85}$.

61. $(2x_m - x_1, 2y_m - y_1)$

63. $\left(\frac{3x_1 + x_2}{4}, \frac{3y_1 + y_2}{4}\right), \left(\frac{x_1 + x_2}{2}, \frac{y_1 + y_2}{2}\right),$
$\left(\frac{x_1 + 3x_2}{4}, \frac{y_1 + 3y_2}{4}\right)$

65. $5\sqrt{74} \approx 43$ yards

67.

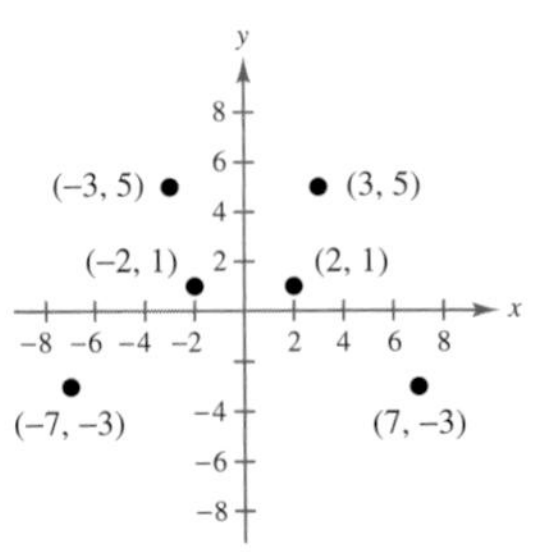

(a) The point is reflected through the y-axis.

(b) The point is reflected through the x-axis.

(c) The point is reflected through the origin.

69. $1002.6 million

71. False. The Midpoint Formula would be used 15 times.

73. Point on x-axis: $y = 0$; Point on y-axis: $x = 0$

75. Use the Midpoint Formula to prove that the diagonals of the parallelogram bisect each other.

$$\left(\frac{b + a}{2}, \frac{c + 0}{2}\right) = \left(\frac{a + b}{2}, \frac{c}{2}\right)$$

$$\left(\frac{a + b + 0}{2}, \frac{c + 0}{2}\right) = \left(\frac{a + b}{2}, \frac{c}{2}\right)$$

77. c **79.** a

Review Exercises *(page 92)*

1. (a) 11 (b) $11, -14$ (c) $11, -14, -\frac{8}{9}, \frac{5}{2}, 0.4$
(d) $\sqrt{6}$

3. (a) $.8\overline{3}$ (b) 0.875

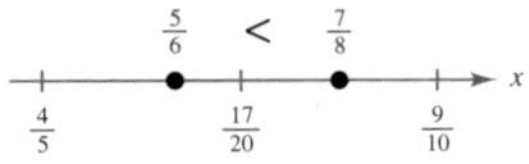

5. The set consists of all real numbers less than or equal to 7.

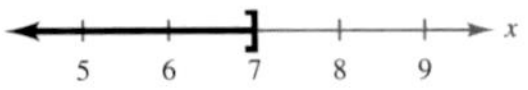

7. 155 **9.** 24.6°F change **11.** 0.017 **13.** -35

15. (a) $\frac{1}{2}$ (b) Undefined **17.** -1 **19.** 2

21. -100 **23.** (a) $\frac{y}{xy + 1}$ (b) $\frac{1}{x^4}$

25. 3.048×10^{-1} **27.** 32 **29.** $2u, u \neq 0$

31. (a) $(2x + 1)\sqrt{2x}$ (b) $(3x^2 - 2x)\sqrt{2x}$

33. $\frac{x - 1}{2(\sqrt{x} + 1)}$ **35.** $\frac{1}{16}$ **37.** $(x - 1)^{1/12}$

39. $\sqrt[4]{16}$ **41.** $-5x^5 + 3x^3 + x - 4$

43. $-7x^2 + 12x + 6$ **45.** $-2y^2 + 11y - 8$

47. $x^2 + 2x - 1 - \frac{2}{x}$ **49.** $36x^2 - 25$

51. $x^3 - 12x^2 + 48x - 64$ **53.** $x(x + 1)(x - 1)$

55. $(5x + 7)(5x - 7)$ **57.** $(x - 4)(x^2 + 4x + 16)$

59. $(x + 10)(2x + 1)$ **61.** $(x - 1)(x^2 + 2)$

63. All real numbers except $x = -6$

65. $\frac{x - 8}{15}, x \neq -8$ **67.** $\frac{1}{x^2}, x \neq \pm 2$

69. $\frac{4x^2 - 16x + 3}{2(x - 4)}$ **71.** $\frac{3x}{(x - 1)(x^2 + x + 1)}$

73. $\dfrac{3ax^2}{(a^2 - x)(a - x)}$ **75.** Identity **77.** Identity

79. 20 **81.** $-\frac{1}{2}$ **83.** $2\frac{6}{7}$ liters **85.** $-\frac{7}{2}, 4$

87. $\pm\frac{5}{4}$ **89.** $8 \pm \sqrt{15}$ **91.** $-3 \pm 2\sqrt{3}$

93. $\dfrac{1}{2} \pm \dfrac{\sqrt{249}}{6}$ **95.** $0, \dfrac{3}{2}$ **97.** $0, -3, \pm\dfrac{2}{3}$ **99.** 66

101. 2 **103.** 79 **105.** $\pm 2, \pm\frac{2}{3}$ **107.** $2, -5$

109. 2, 3 **111.** (a) Yes (b) No (c) Yes (d) No

113. $(-\infty, 12]$ **115.** $\left[\frac{32}{15}, \infty\right)$ **117.** $\left(-\frac{2}{3}, 17\right]$

119. $[-4, 4]$ **121.** $(-\infty, -1), (7, \infty)$

123. $x = 37$ units **125.** $(-3, 9)$ **127.** $\left(-\frac{4}{3}, \frac{1}{2}\right)$

129. $[-5, -1), (1, \infty)$ **131.** $[-4, -3], (0, \infty)$

133. $r > 4.88\%$

135. The multiplication in parentheses comes first.
$10(4 \cdot 7) = 10(28) = 280$

137. The exponent applies to the coefficient also.
$(2x)^4 = 16x^4$

139. Multiply exponents when raising a power to a power.
$(3^4)^4 = 3^{16}$

141. Add what is in parentheses first before squaring.
$(5 + 8)^2 = 13^2 = 169$

143. $16x^2 - 9x + 20$ **145.** $-5x^2 + 2x + 15$

147. $\dfrac{2(x + 1)}{(x + 2)^{1/2}}$ **149.** $x - 4 + 2x^{-1}$

151. $2x^{5/2} - 4x^{3/2} + 3x^{-1/2}$

153.

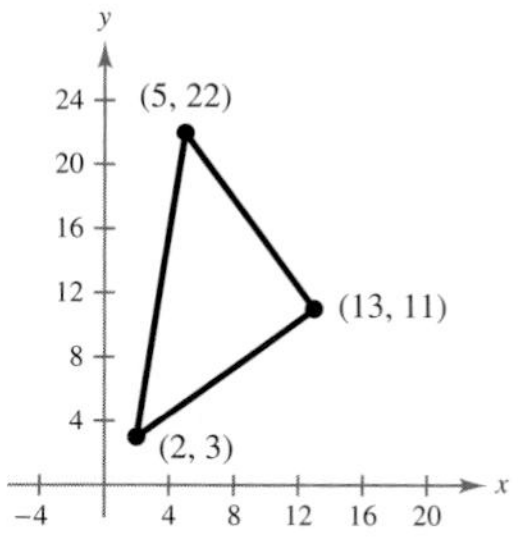

$(\sqrt{185})^2 + (\sqrt{185})^2 = (\sqrt{370})^2$

155. Quadrant IV **157.** Quadrant IV

159. (a)

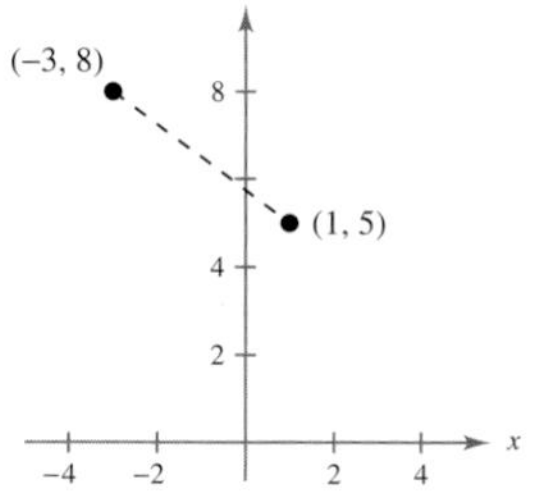

(b) 5

161. (a)

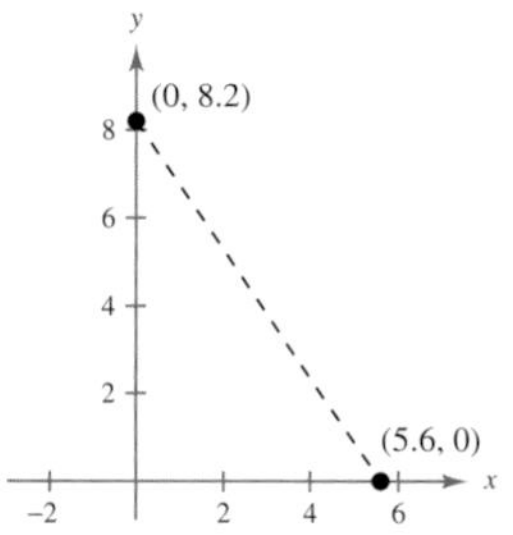

(b) 9.9

163. (a)

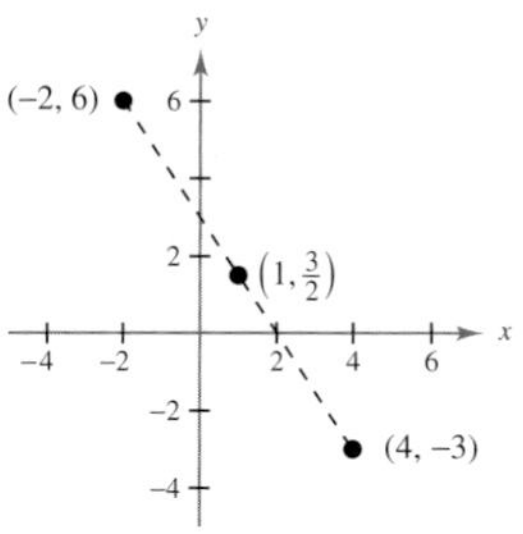

(b) $\left(1, \frac{3}{2}\right)$

165. (a)

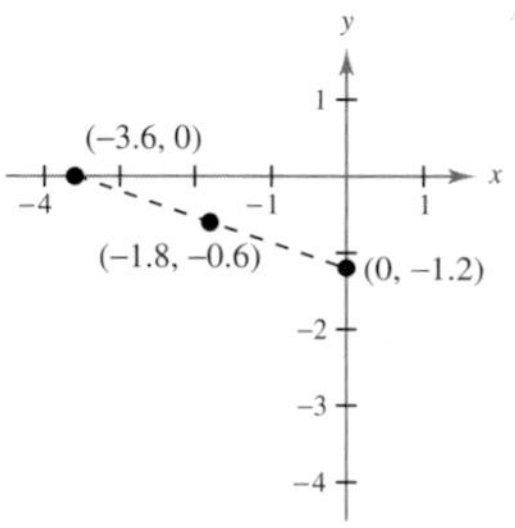

(b) $(-1.8, -0.6)$

167. 80°F

169. False. There is also a cross-product term when a binomial sum is squared. $(x + a)^2 = x^2 + 2ax + a^2$

171. True. If a quadratic equation cannot be factored, then the Quadratic Formula can be used to solve the equation.

173. (a) Neither (b) Both (c) Quadratic (d) Neither

175. Radicals cannot be combined unless the index and the radicand are the same.

177. Answers will vary.

Chapter Test *(page 97)*

1. $-\frac{10}{3} > -|-4|$ **2.** 9.15 **3.** (a) -18 (b) $\frac{4}{27}$

4. (a) $-\frac{27}{125}$ (b) $\frac{8}{729}$ **5.** (a) 25 (b) 6

6. (a) 1.8×10^5 (b) 2.7×10^{13}

7. (a) $12z^8$ (b) $(u - 2)^{-7}$ (c) $\dfrac{3x^2}{y^2}$

8. (a) $15z\sqrt{2z}$ (b) $-10\sqrt{y}$ (c) $\dfrac{2}{v}\sqrt[3]{\dfrac{2}{v^2}}$

9. $2x^2 - 3x - 5$ **10.** $x^2 - 5$

11. (a) $x^2(2x + 1)(x - 2)$ (b) $(x - 2)(x + 2)^2$

12. (a) $4\sqrt[3]{4}$ (b) $-3(1 + \sqrt{3})$ **13.** $\frac{128}{11}$ **14.** $-4, 5$

15. No solution **16.** $\pm\sqrt{2}$ **17.** 4

18. $-2, \frac{8}{3}$

19. $-\frac{11}{2} \le x < 3$

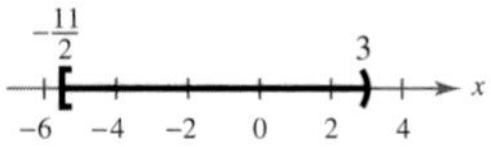

20. $x < -6$ or $0 < x < 4$

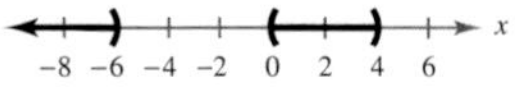

21. $x < -4$ or $x > \frac{3}{2}$

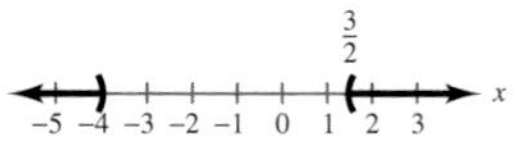

22. $x \le 10$ or $x \ge 20$

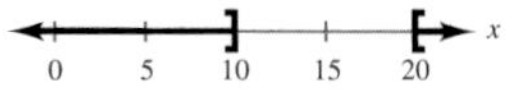

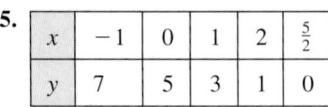

23. $\dfrac{3x - 1}{2x - 3}, x \ne -2$ **24.** $\dfrac{15}{28}x; \dfrac{7}{15}$

25. Distance: $\sqrt{89}$ miles; Midpoint: $\left(2, \frac{5}{2}\right)$

Chapter 1

Section 1.1 *(page 107)*

1. (a) Yes (b) Yes **3.** (a) No (b) Yes

5.

x	-1	0	1	2	$\frac{5}{2}$
y	7	5	3	1	0

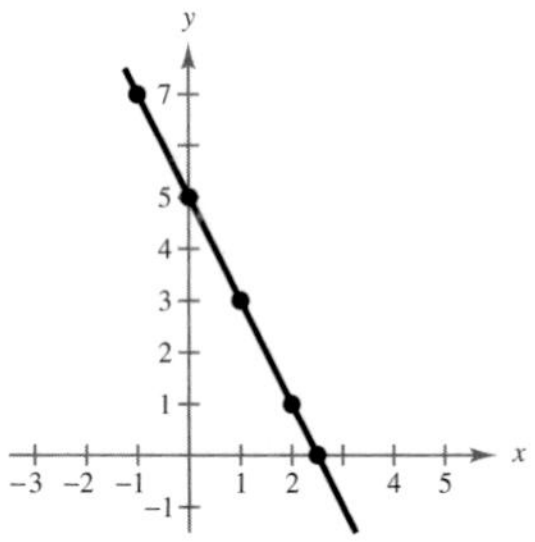

7.

x	-1	0	1	2	3
y	4	0	-2	-2	0

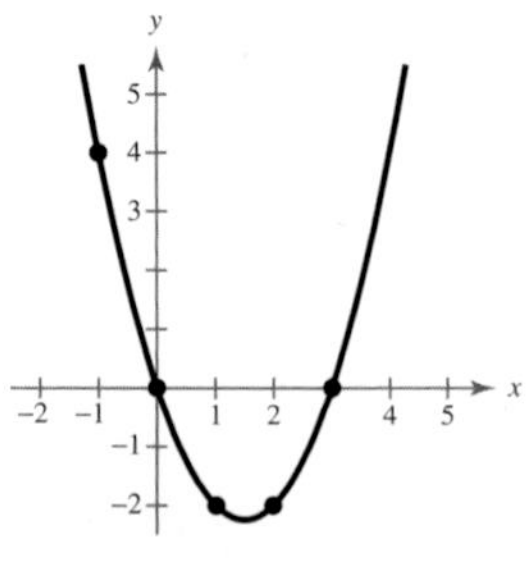

9. y-axis symmetry **11.** Origin symmetry

13. Origin symmetry **15.** x-axis symmetry

17.

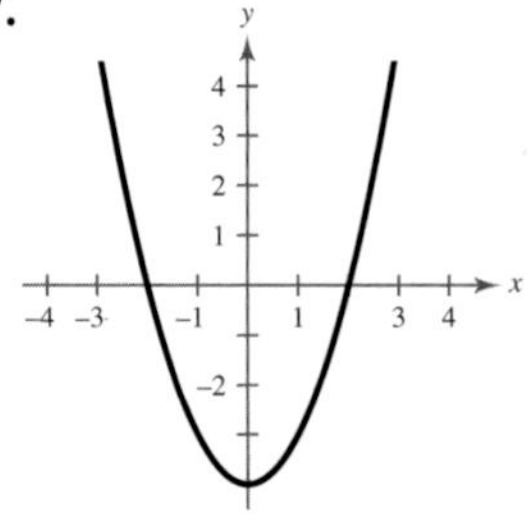

19.

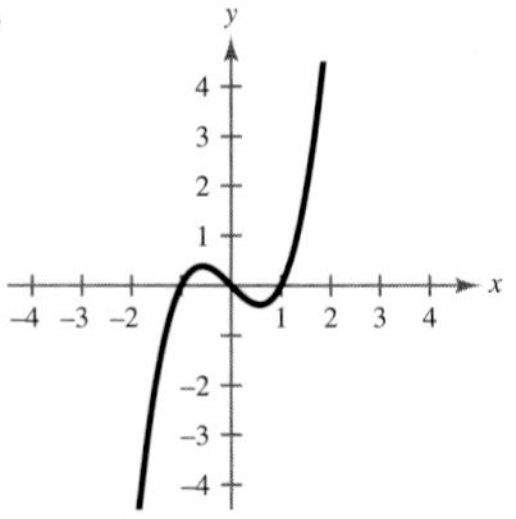

21. c **23.** b

25. x-intercepts: $(\pm 2, 0)$

y-intercept: $(0, 16)$

27. x-intercepts: $(0, 0), \left(\frac{5}{2}, 0\right)$

y-intercept: $(0, 0)$

29.

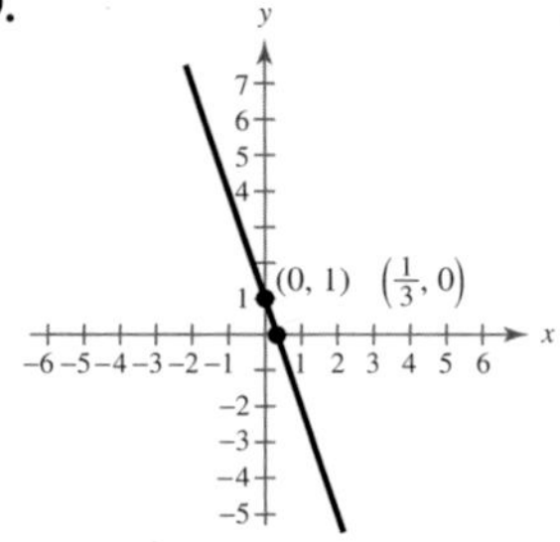

31.

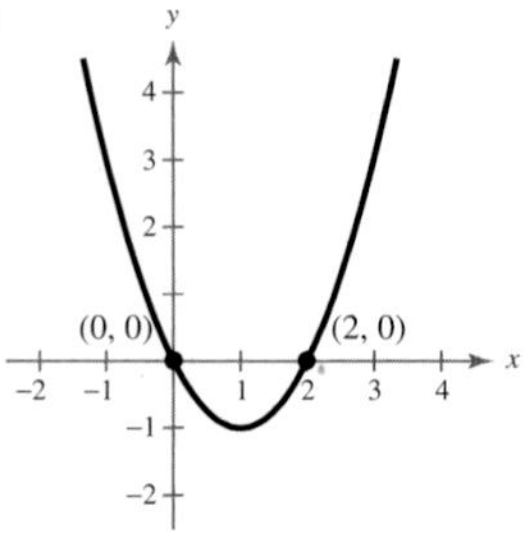

33.

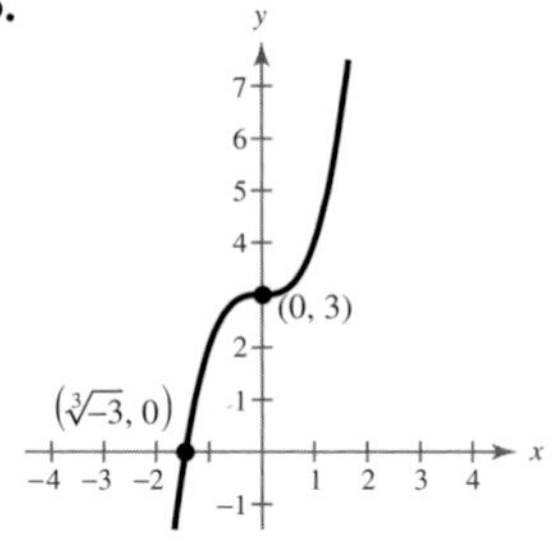

35.

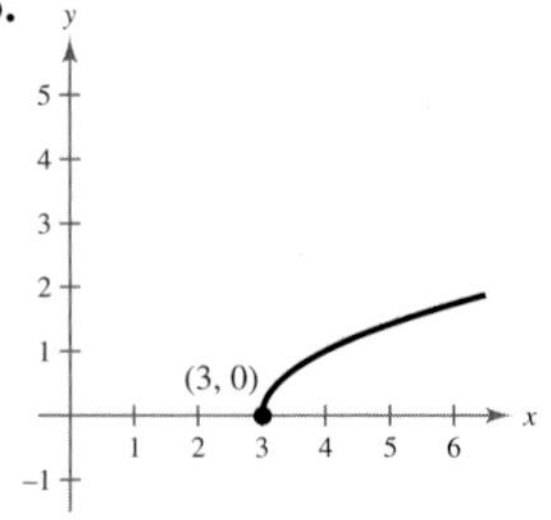

37.

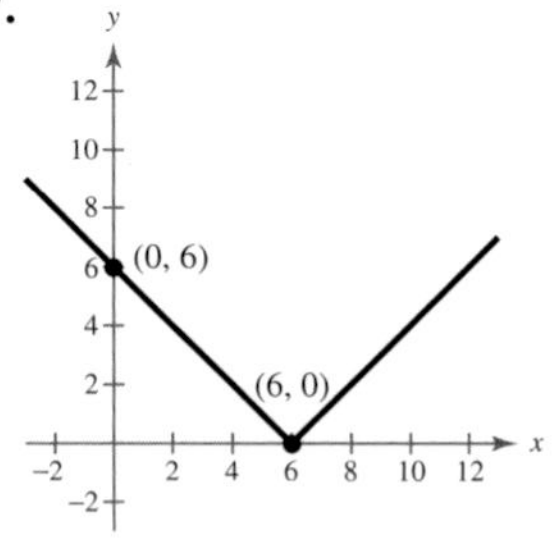

39.

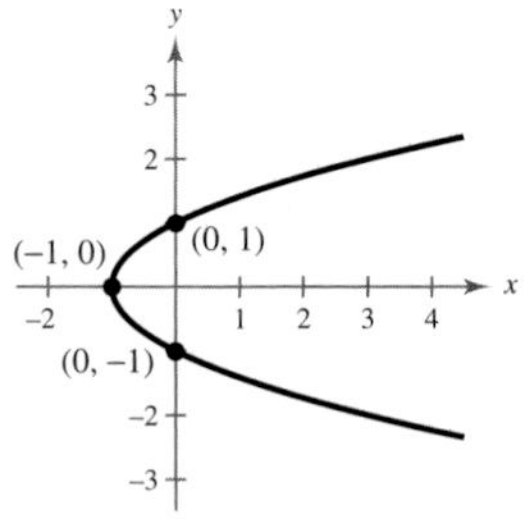

41.

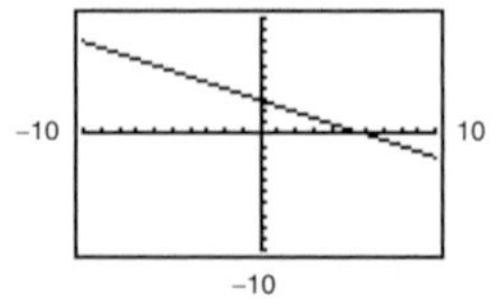

Intercepts: $(6, 0)$, $(0, 3)$

43.

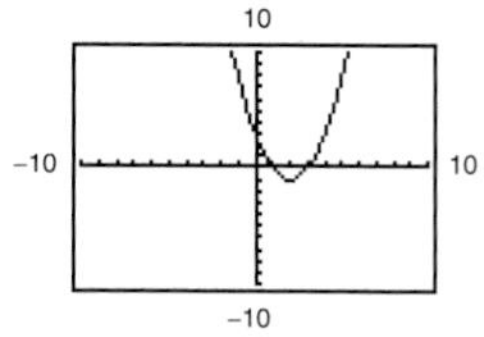

Intercepts: $(3, 0)$, $(1, 0)$, $(0, 3)$

45.

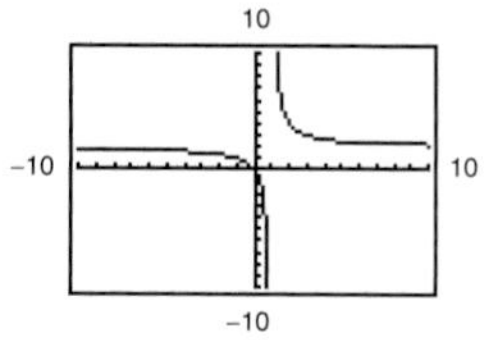

Intercept: $(0, 0)$

47.

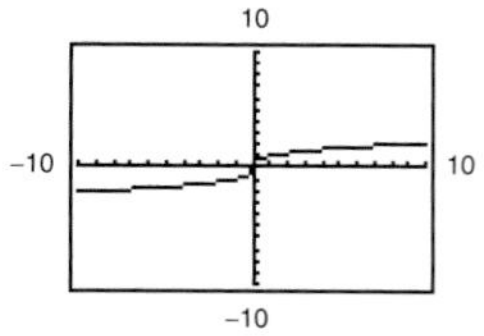

Intercept: $(0, 0)$

49.

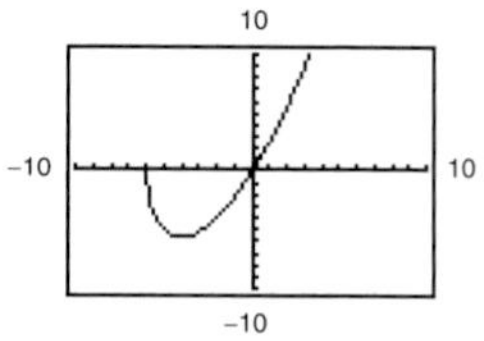

Intercepts: $(0, 0)$, $(-6, 0)$

51.

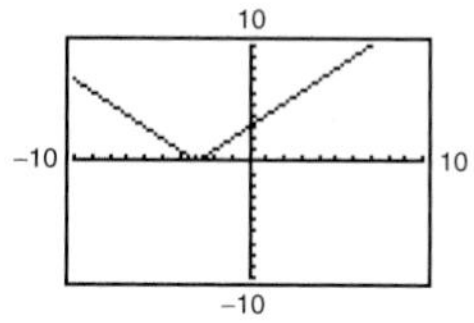

Intercepts: $(-3, 0)$, $(0, 3)$

53. $x^2 + y^2 = 16$

55. $(x - 2)^2 + (y + 1)^2 = 16$

57. $(x + 1)^2 + (y - 2)^2 = 5$

59. $(x - 3)^2 + (y - 4)^2 = 25$

61. Center: $(0, 0)$; Radius: 5

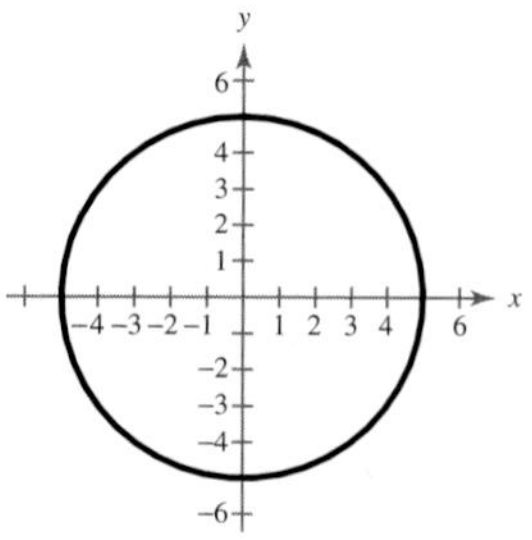

63. Center: $(1, -3)$; Radius: 3

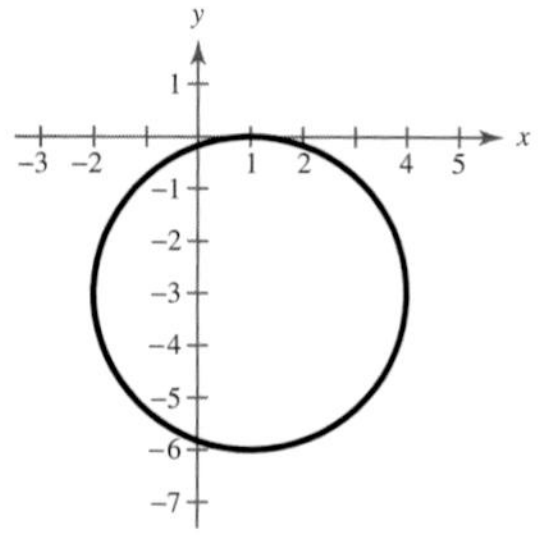

65. Center: $\left(\frac{1}{2}, \frac{1}{2}\right)$; Radius: $\frac{3}{2}$

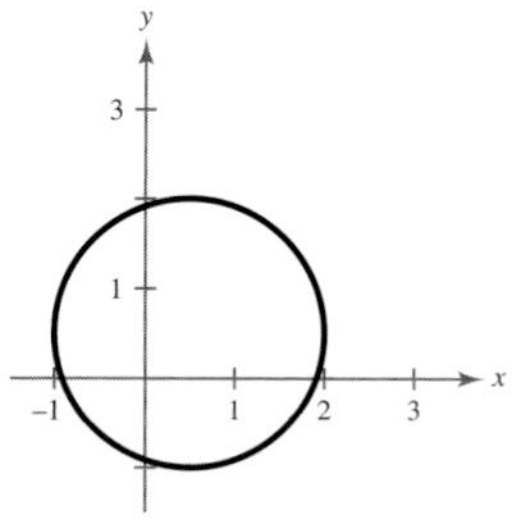

67.

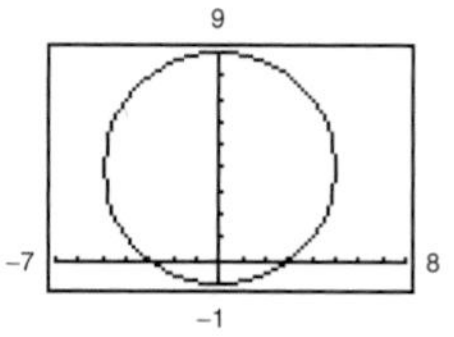

Circle

69.

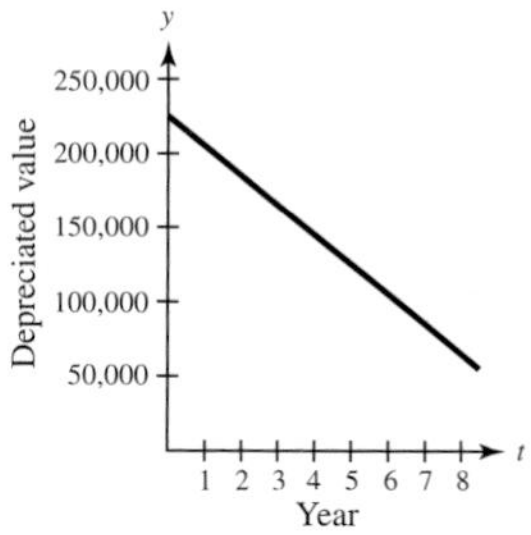

71. (a)

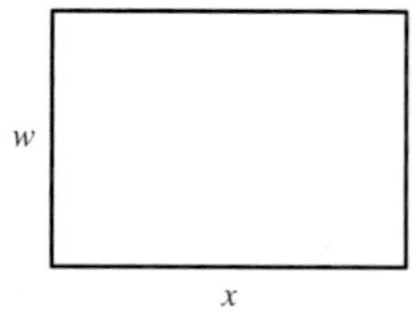

(b) Answers will vary.

(c)

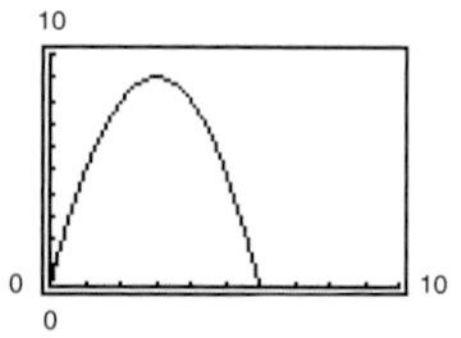

(d) $x = 3$, $w = 3$

73. (a) and (b)

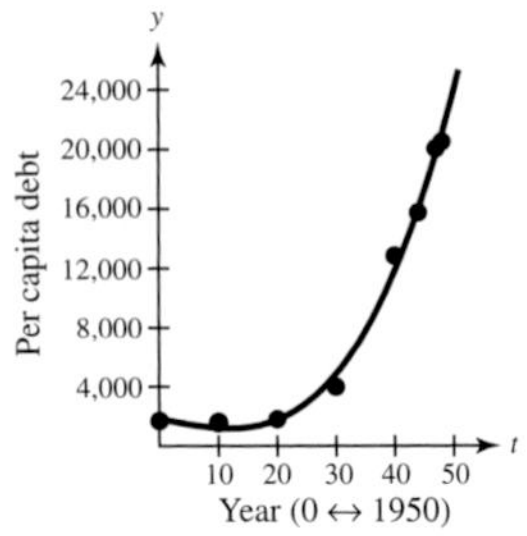

(b) The curve seems to be a good fit for the data.

(c) 2002: \$27,142; 2004: \$30,589

75. 3.9 ohms

77. True. All linear equations of the form $y = mx + b$, which excludes vertical lines, cross the y-axis one time.

79. (a) $a = 1, b = 0$ (b) $a = 0, b = 1$

81. False. $\dfrac{1}{3 \cdot 4^{-1}} = \dfrac{4}{3}$ **83.** $9x^5,\ 4x^3,\ -7$

85. $2\sqrt{2x}$ **87.** $\dfrac{10\sqrt{7x}}{x}$ **89.** $\sqrt[3]{|t|}$

Section 1.2 *(page 119)*

1. (a) L_2 (b) L_3 (c) L_1

3.

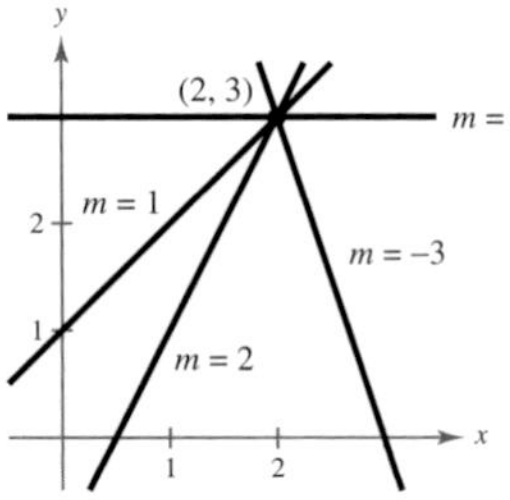

5. $\frac{8}{5}$ **7.** 0 **9.** -4

11. $m = 2$

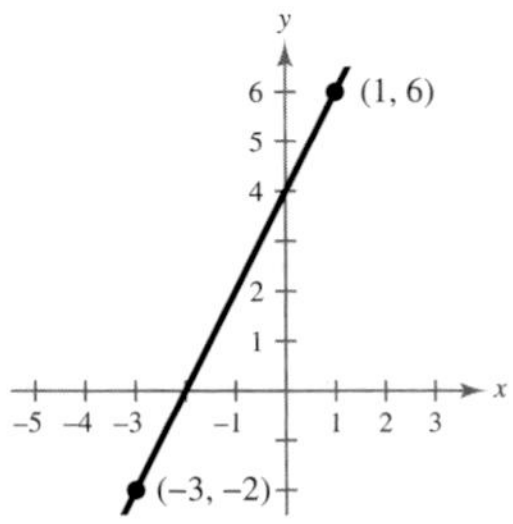

13. m is undefined.

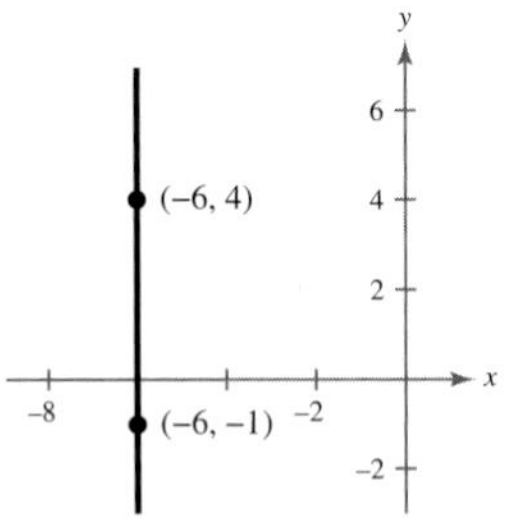

15. $m = -\frac{1}{7}$

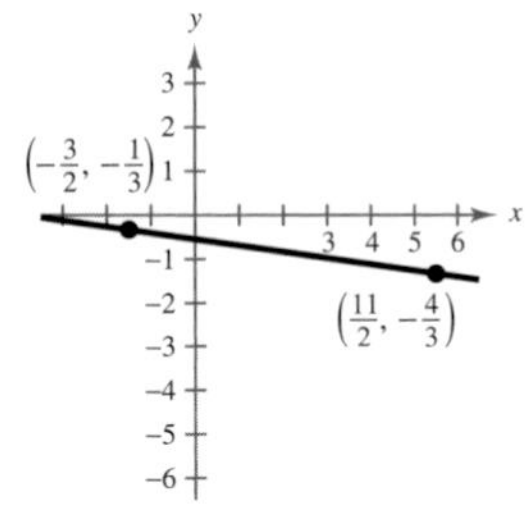

17. $m = 0.15$

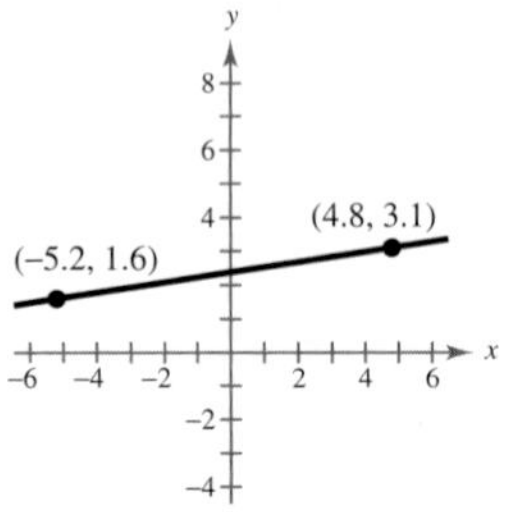

19. $(0, 1),\ (3, 1),\ (-1, 1)$ **21.** $(6, -5),\ (7, -4),\ (8, -3)$

23. $(-8, 0),\ (-8, 2),\ (-8, 3)$

25. $(-4, 6), (-3, 8), (-2, 10)$

27. $(9, -1),\ (11, 0),\ (13, 1)$ **29.** Perpendicular

31. Parallel

33. (a) Sales increasing 135 units per year

(b) No change in sales

(c) Sales decreasing 40 units per year

35. (a) Greatest increase: 1990 to 1991 and 1996 to 1997

Greatest decrease: 1997 to 1998

(b) $m = 0.037$

(c) Each year, the earnings per share increase by \$0.037.

37. (a) and (b)

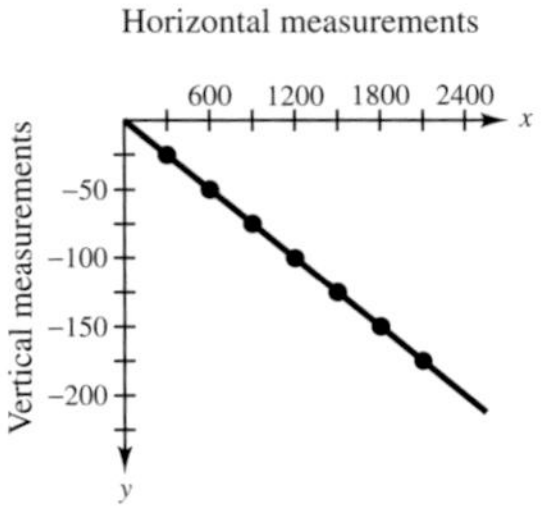

(c) $y = -\frac{1}{12}x$

(d) For every 12 horizontal measurements, the vertical measurement decreases by 1.

(e) "8.3% grade"

39. 12 feet

41. $m = 1$; Intercept: $(0, -10)$

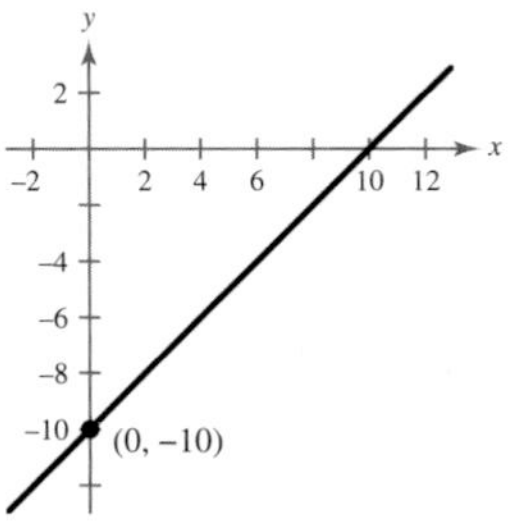

43. $m = 0$; Intercept: $\left(0, -\frac{5}{3}\right)$ **45.** $m = -\frac{2}{3}$; Intercept: $(0, 3)$

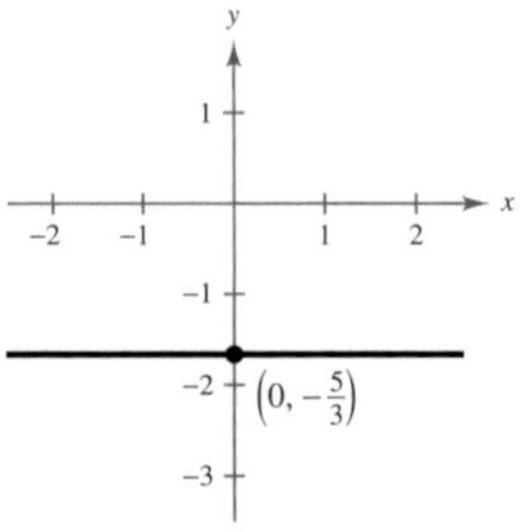

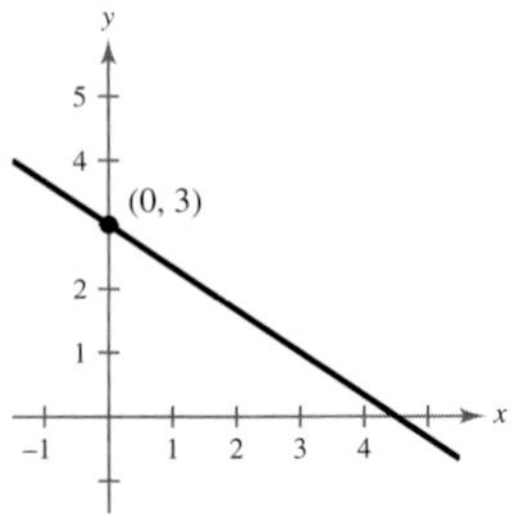

47. $y = -x + 10$

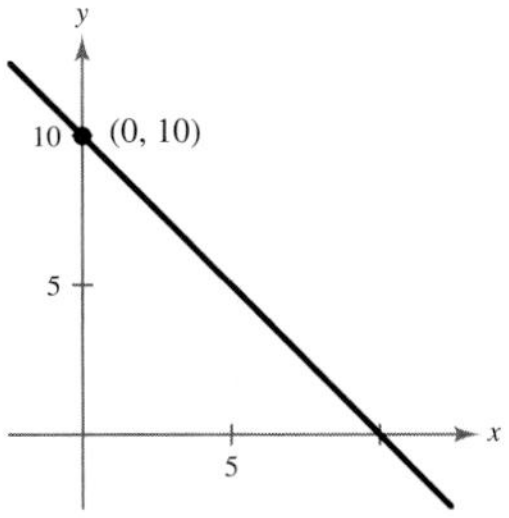

49. $y = 4x$

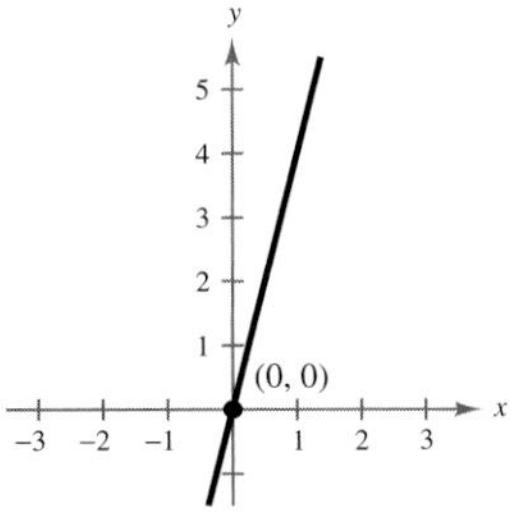

51. $y = \frac{3}{4}x - \frac{7}{2}$

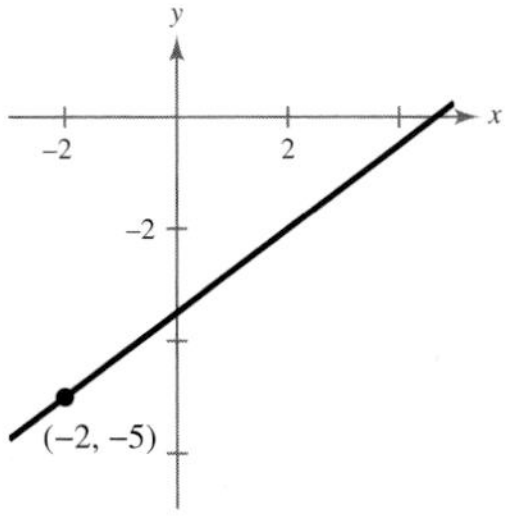

53. $y = 4$

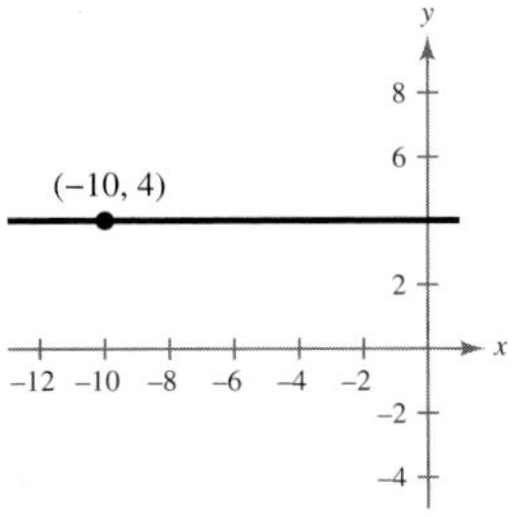

55. $y = -3x$

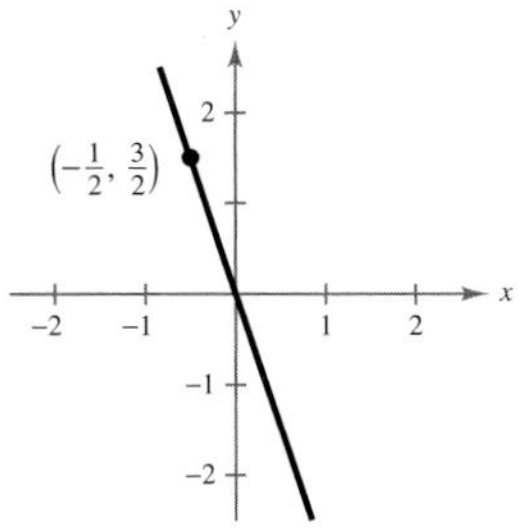

57. $y = -2.5x - 2.75$

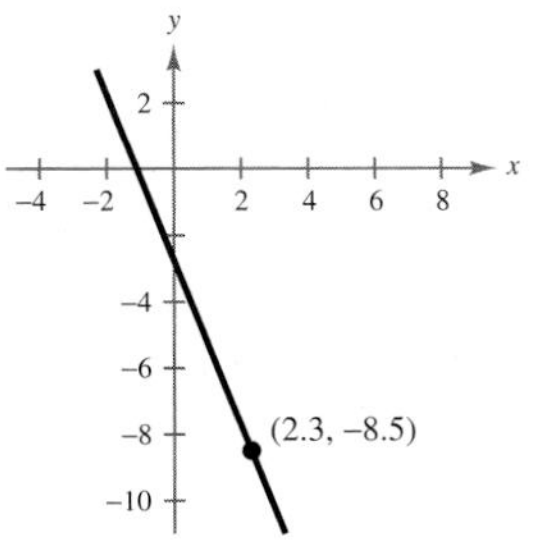

59. $y = \frac{7}{8}x - \frac{1}{2}$

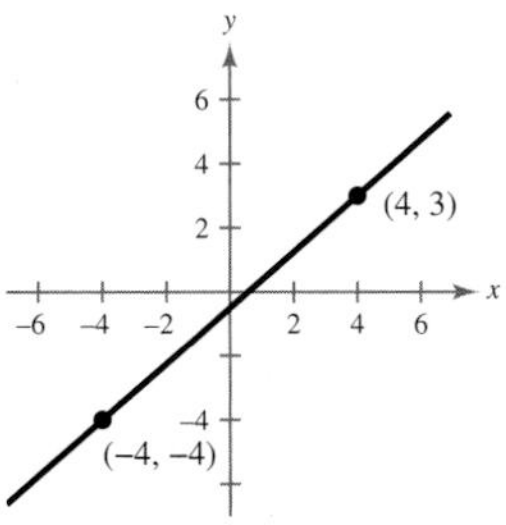

61. $y = 4$

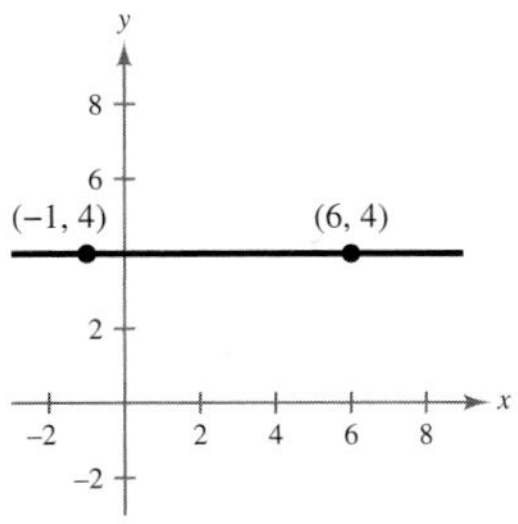

63. $y = -\frac{1}{3}x + \frac{4}{3}$

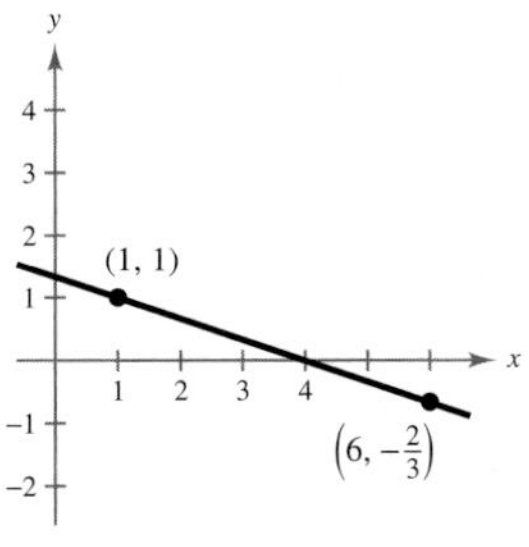

65. $y = -\frac{3}{25}x + \frac{159}{100}$

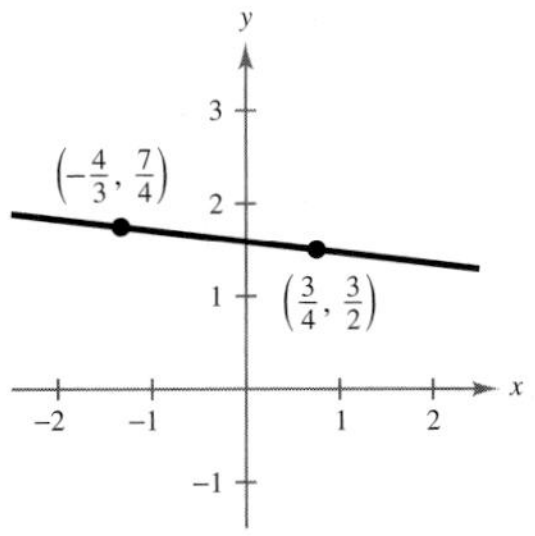

67. $y = 0.3x - 1.8$

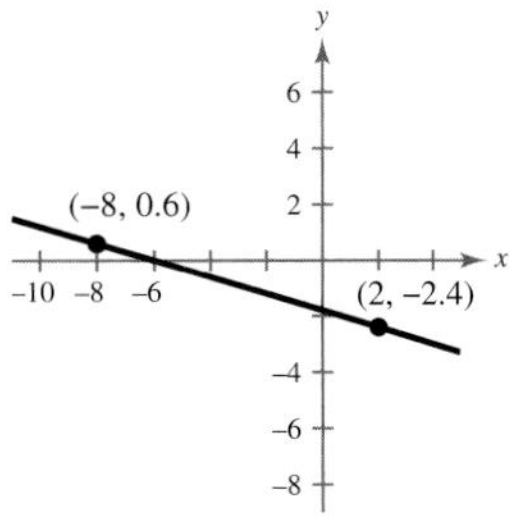

69. $4x - 3y + 12 = 0$ **71.** $3x - y - 2 = 0$

73. $x + y - 1 = 0$

75. (a) $y = -x - 1$ (b) $y = x + 5$

77. (a) $y = -\frac{5}{3}x + \frac{53}{24}$ (b) $y = \frac{3}{5}x + \frac{9}{40}$

79. (a) $x = 2$ (b) $y = 5$

81. (a) $y = -3x - 13.1$ (b) $y = \frac{1}{3}x - 0.1$

83. (a) is parallel to (c). (b) is perpendicular to (a) and (c).

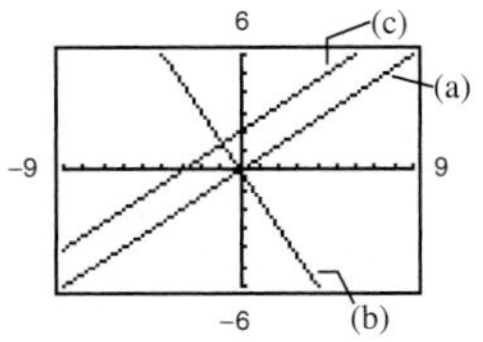

85. (a) is parallel to (b). (c) is perpendicular to (a) and (b).

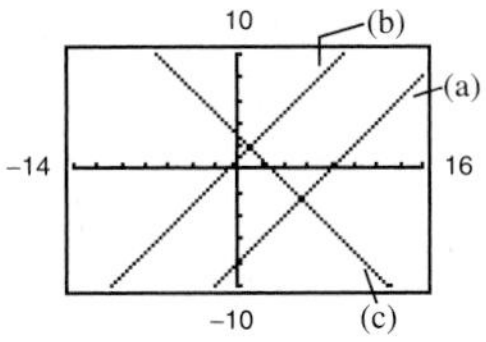

87. $V = 4.5t + 151.5$

89. c; The slope is 2, which represents the hourly wage per unit produced.

91. d; The slope is -100, which represents the decrease in the value of the word processor each year.

93. $y = -\frac{5}{13}x - \frac{2}{13}$ **95.** $y = -\frac{16}{21}x - \frac{13}{56}$

97. $y = 54t + 3927$; 1999: 4089 stores, 2000: 4143 stores

99.

C	$-17.8°$	$-10°$	$10°$	$20°$	$32.2°$	$177°$
F	$0°$	$14°$	$50°$	$68°$	$90°$	$350.6°$

101. 3014 students **103.** $V = 25{,}000 - 2300t$

105. $W = 0.75x + 11.50$

107. (a) $x = -\frac{1}{15}p + \frac{266}{3}$ (b) 45 (c) 49

109. $W = 0.07S + 2500$

111. (a) $12{,}000 - x$ (b) $y = -0.015x + 480$

(c) 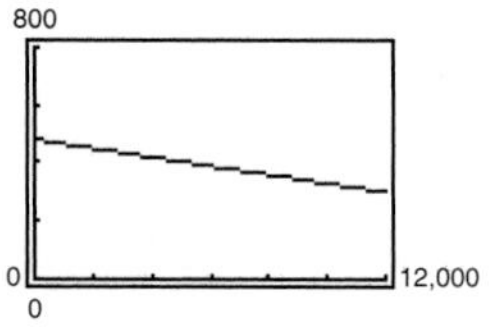

(d) As the amount invested at the lower interest rate increases, the annual interest decreases.

113. (a) and (b)

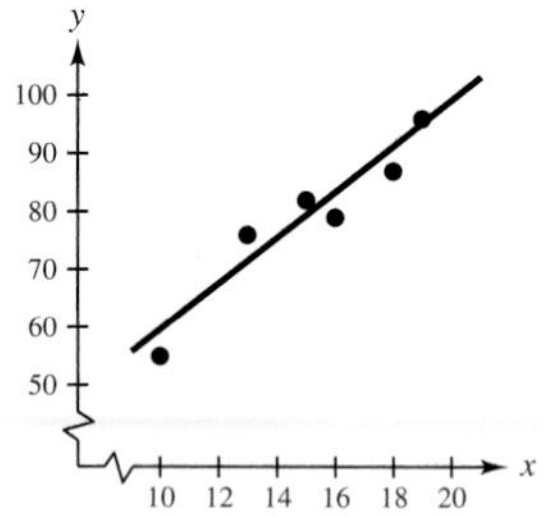

(c) $y = 4x + 19$ (d) 87

(e) Vertical shift 4 units upward

115. False. The slope of the first line is $\frac{2}{7}$ and the slope of the second line is $-\frac{11}{7}$.

117. The slope of a vertical line is undefined because division by zero is undefined.

119. The line with a slope of -4 is steeper. The slope with the greatest magnitude corresponds to the steepest line.

121. Yes. The rate of change remains the same on a line.

123. d **125.** a **127.** -1 **129.** $\frac{7}{2}, 7$

131. No solution

Section 1.3 *(page 133)*

1. Yes **3.** No

5. Yes, each input value has exactly one output value.

7. No, the input value of 7 has two output values, 6 and 12.

9. (a) Function

(b) Not a function, because the element 1 in A corresponds to two elements, -2 and 1, in B.

(c) Function

(d) Not a function, because not every element in A is matched with an element in B.

11. Each is a function. For each year there corresponds one and only one circulation.

13. Not a function **15.** Function **17.** Function

19. Not a function **21.** Function

23. (a) 4 (b) 0 (c) $4x$ (d) $(x + c)$

25. (a) -1 (b) -9 (c) $2x - 5$

27. (a) 36π (b) $\frac{9}{2}\pi$ (c) $\frac{32}{3}\pi r^3$

29. (a) 1 (b) 2.5 (c) $3 - 2|x|$

31. (a) $-\frac{1}{9}$ (b) Undefined (c) $\dfrac{1}{y^2 + 6y}$

33. (a) 1 (b) -1 (c) $\dfrac{|x - 1|}{x - 1}$

35. (a) -1 (b) 2 (c) 6

37.

x	-2	-1	0	1	2
$f(x)$	1	-2	-3	-2	1

39.

t	-5	-4	-3	-2	-1
$h(t)$	1	$\frac{1}{2}$	0	$\frac{1}{2}$	1

41.

x	-2	-1	0	1	2
$f(x)$	5	$\frac{9}{2}$	4	1	0

43. 5 **45.** $\frac{4}{3}$ **47.** ± 3 **49.** $0, \pm 1$

51. $2, -1$ **53.** $3, 0$ **55.** All real numbers x

57. All real numbers $t \neq 0$

59. $y \geq 10$ **61.** $-1 \leq x \leq 1$

63. All real numbers $x \neq 0, -2$ **65.** $s \geq 1,\ s \neq 4$

67. All real numbers $x \neq 0$

69. $\{(-2, 4), (-1, 1), (0, 0), (1, 1), (2, 4)\}$

71. $\{(-2, 0), (-1, 1), (0, \sqrt{2}), (1, \sqrt{3}), (2, 2)\}$

73. $g(x) = -2x^2$ **75.** $r(x) = \dfrac{32}{x}$ **77.** $3 + h, h \neq 0$

79. $3x^2 + 3xc + c^2, c \neq 0$

81. $3, x \neq 3$ **83.** $\dfrac{\sqrt{5x} - 5}{x - 5}$ **85.** $A = \dfrac{p^2}{16}$

87. $A = 8\sqrt{s^2 - 64}$

89. (a)

Height x	Width	Volume V
1	$24 - 2(1)$	$1[24 - 2(1)]^2 = 484$
2	$24 - 2(2)$	$2[24 - 2(2)]^2 = 800$
3	$24 - 2(3)$	$3[24 - 2(3)]^2 = 972$
4	$24 - 2(4)$	$4[24 - 2(4)]^2 = 1024$
5	$24 - 2(5)$	$5[24 - 2(5)]^2 = 980$
6	$24 - 2(6)$	$6[24 - 2(6)]^2 = 864$

Maximum when $x = 4$

(b)

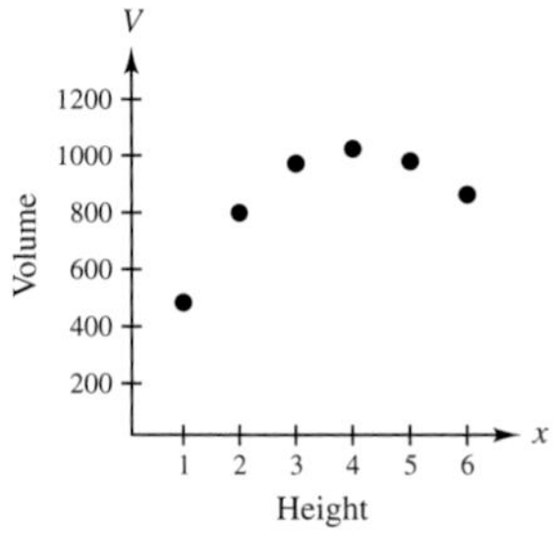

V is a function of x.

(c) $V = x(24 - 2x)^2, 0 < x < 12$

91. $A = \dfrac{x^2}{2(x-2)}, x > 2$

93. 1978: \$15,198; 1988: \$25,558; 1993: \$30,800; 1997: \$41,008

95. (a) $C = 12.30x + 98{,}000$

(b) $R = 17.98x$

(c) $P = 5.68x - 98{,}000$

97. (a) $R = \dfrac{240n - n^2}{20}, n \geq 80$

(b)

n	90	100	110	120	130	140	150
$R(n)$	\$675	\$700	\$715	\$720	\$715	\$700	\$675

The revenue is maximum when 120 people take the trip.

99. (a)

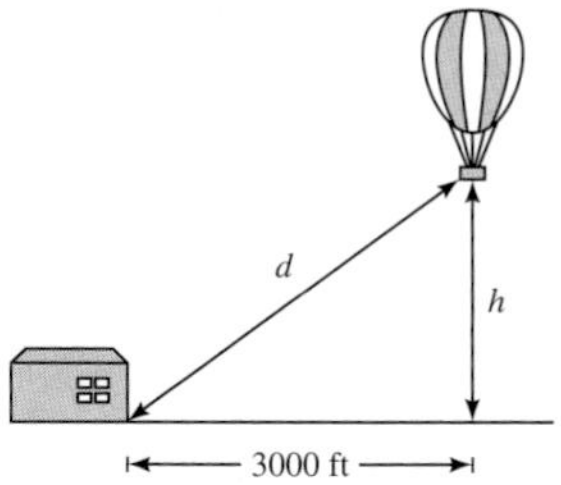

(b) $h = \sqrt{d^2 - 3000^2}, d \geq 3000$

101. Yes, the ball will be at a height of 6 feet.

103. True. Each x-value corresponds to one y-value.

105. The domain is the set of inputs of the function, and the range is the set of outputs.

107. $\frac{15}{8}$ **109.** $-\frac{1}{5}$ **111.** $2x - 3y - 11 = 0$

113. $10x + 9y + 15 = 0$

Section 1.4 *(page 147)*

1. Domain: all real numbers
Range: all real numbers

3. Domain: all real numbers
Range: $(-\infty, 1]$

5. Domain: $(-\infty, -1], [1, \infty)$
Range: $[0, \infty)$

7. Domain: $[-4, 4]$
Range: $[0, 4]$

9. Function **11.** Not a function **13.** Function

15. $-\frac{5}{2}, 6$ **17.** 0 **19.** $0, \pm\sqrt{2}$ **21.** $\pm\frac{1}{2}, 6$

23. $-\frac{5}{3}$

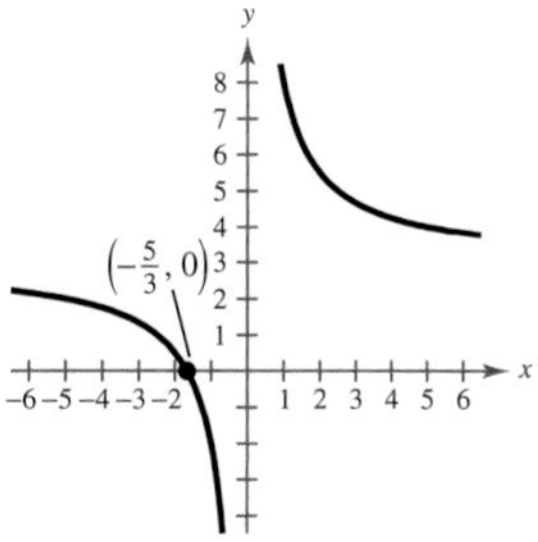

25. $-\frac{11}{2}$

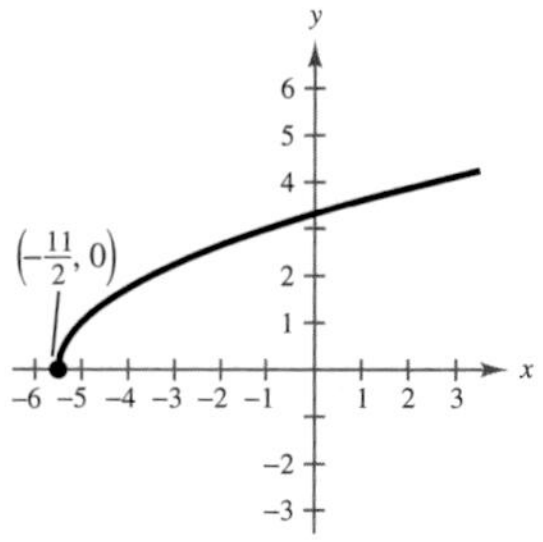

27. $\frac{1}{3}$

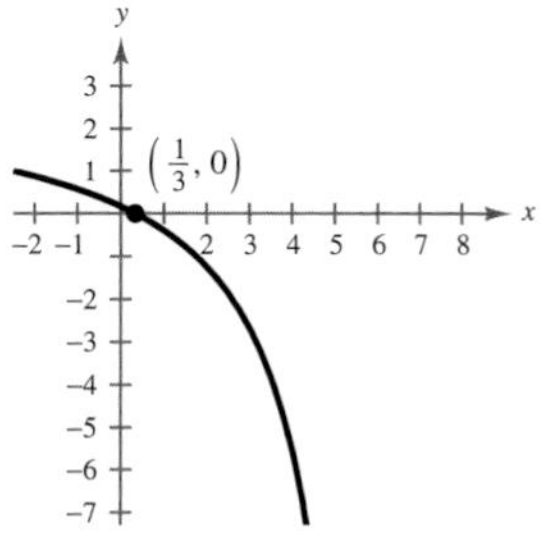

29. (a) Increasing on $(-\infty, \infty)$ (b) Odd

31. (a) Increasing on $(-\infty, 0)$ and $(2, \infty)$

Decreasing on $(0, 2)$

(b) Neither even nor odd

33. (a)

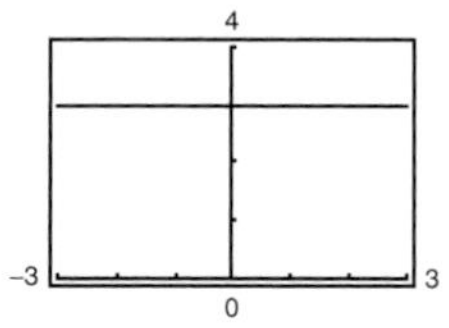

Constant on $(-\infty, \infty)$

(b)

x	-2	-1	0	1	2
$f(x)$	3	3	3	3	3

35. (a)

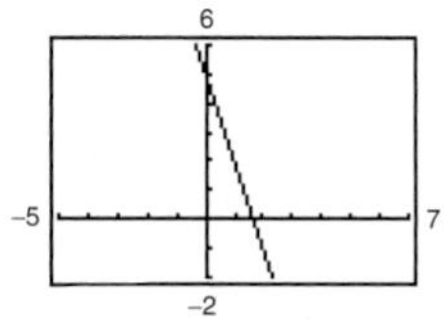

Decreasing on $(-\infty, \infty)$

(b)

x	-2	-1	0	1	2
$f(x)$	11	8	5	2	-1

37. (a)

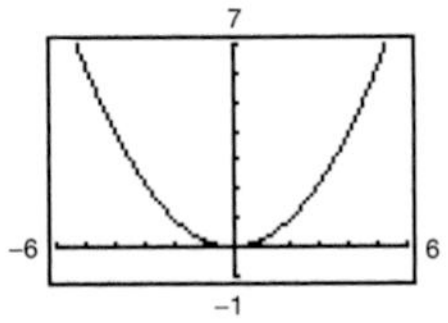

Decreasing on $(-\infty, 0)$; Increasing on $(0, \infty)$

(b)

s	-4	-2	0	2	4
$g(s)$	4	1	0	1	4

39. (a)

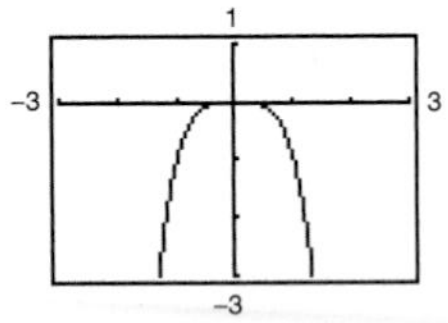

Increasing on $(-\infty, 0)$; Decreasing on $(0, \infty)$

(b)

t	-2	-1	0	1	2
$f(t)$	-16	-1	0	-1	-16

41. (a)

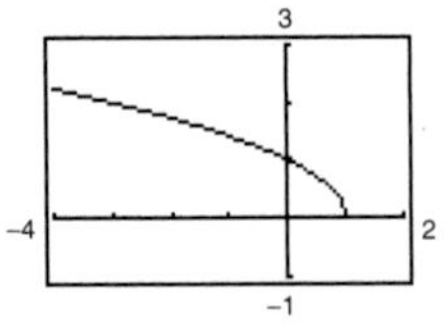

Decreasing on $(-\infty, 1)$

(b)

x	-3	-2	-1	0	1
$f(x)$	2	$\sqrt{3}$	$\sqrt{2}$	1	0

43. (a)

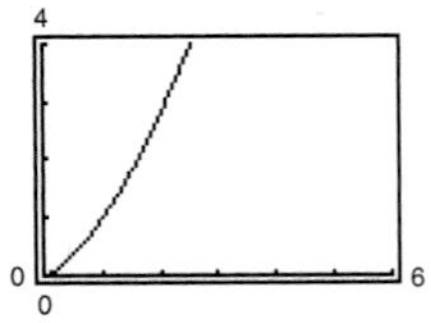

Increasing on $(0, \infty)$

(b)

x	0	1	2	3	4
$f(x)$	0	1	2.82	5.2	8

45. (a)

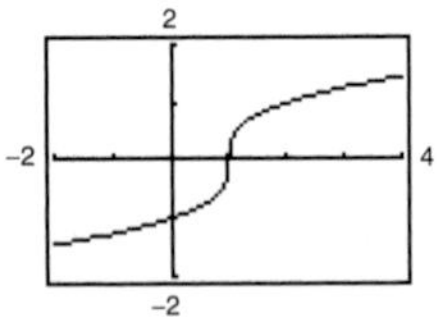

Increasing on $(-\infty, \infty)$

(b)

t	-2	-1	0	1	2
$g(t)$	-1.44	-1.26	-1	0	1

47. (a)

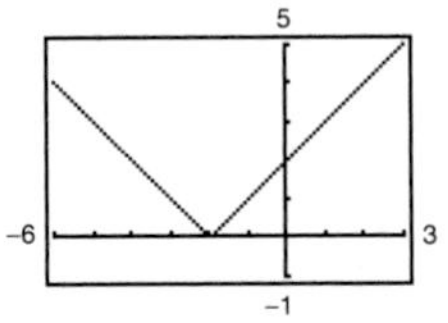

Decreasing on $(-\infty, -2)$; Increasing on $(-2, \infty)$

(b)

x	-6	-4	-2	0	2
$f(x)$	4	2	0	2	4

49. (a)

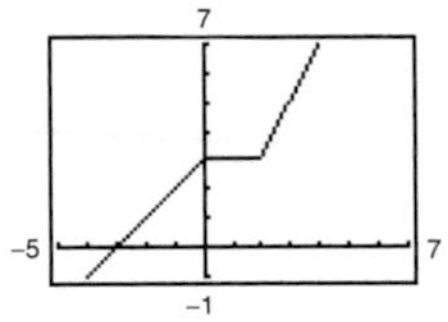

Increasing on $(-\infty, 0)$ and $(2, \infty)$; Constant on $(0, 2)$

(b)

x	-2	-1	0	1	2	3	4
$f(x)$	1	2	3	3	3	5	7

51.

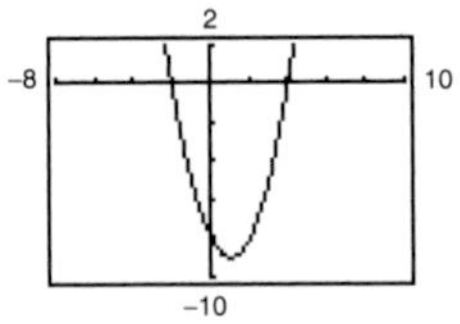

Relative minimum: $(1, -9)$

53.

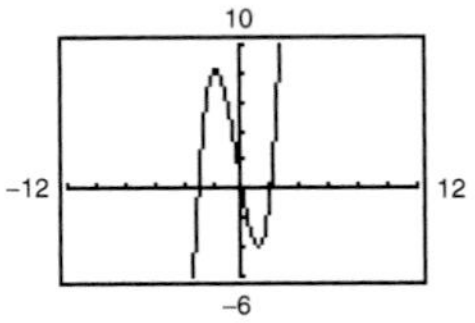

Relative maximum: $(-1.79, 8.21)$

Relative minimum: $(1.12, -4.06)$

55.

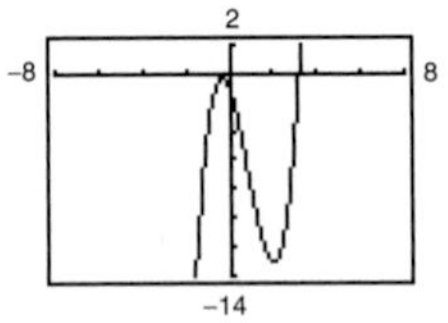

Relative maximum: $(-0.33, -0.30)$

Relative minimum: $(2, -13)$

57.

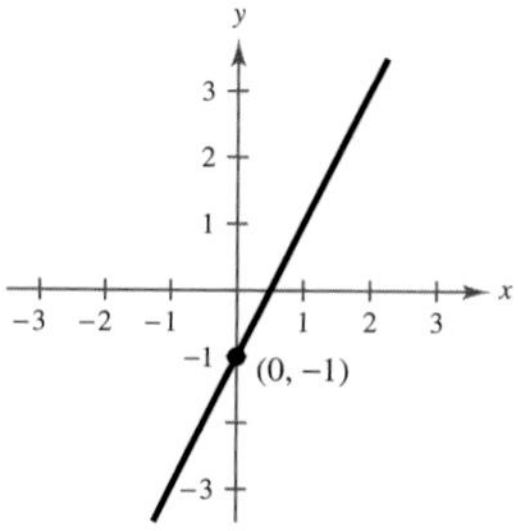

59.

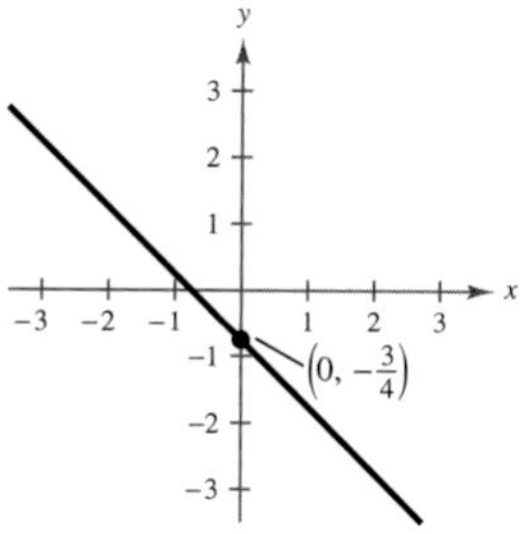

61.

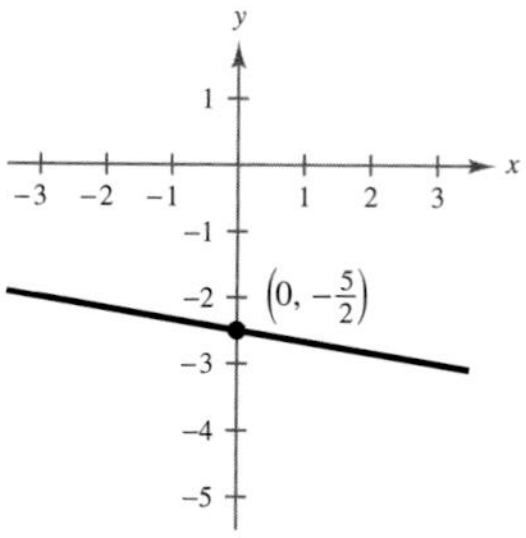

63.

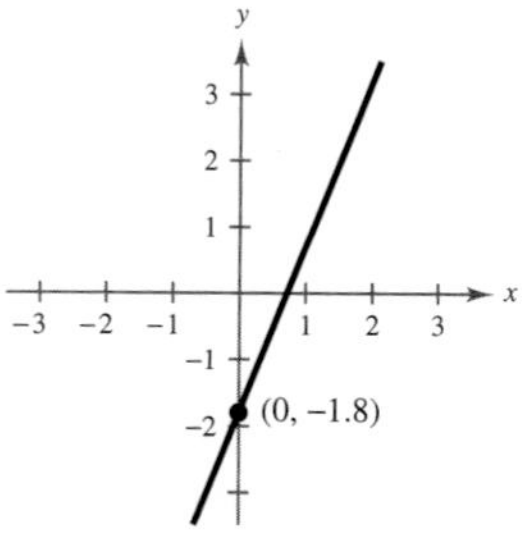

65. $f(x) = -2x + 6$ **67.** $f(x) = -3x + 11$

69. $f(x) = \frac{2}{5}x - 3$ **71.** $f(x) = \frac{6}{7}x - \frac{45}{7}$

73.

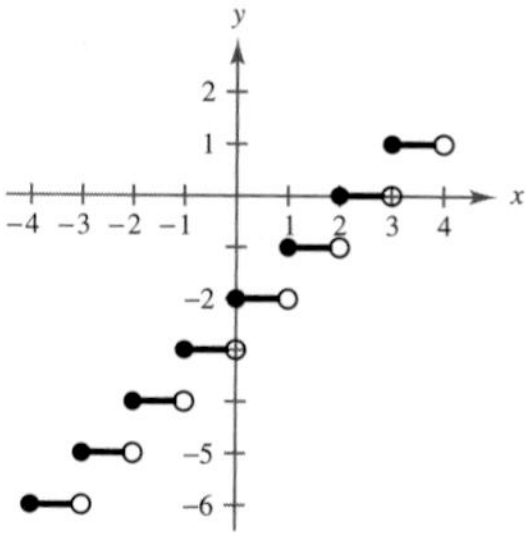

Vertical shift 2 units downward

75.

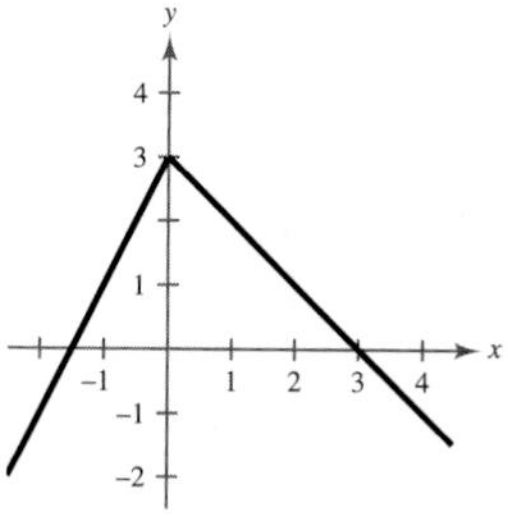

77.

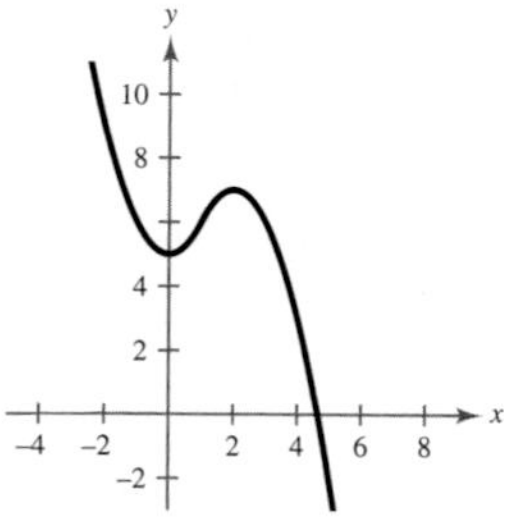

79. $(-\infty, 4]$

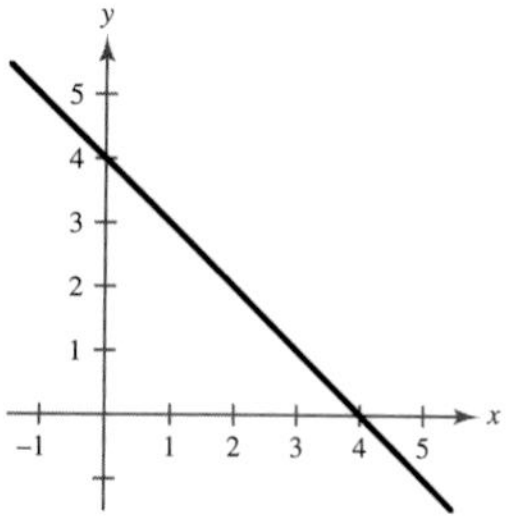

81. $(-\infty, -3], [3, \infty)$

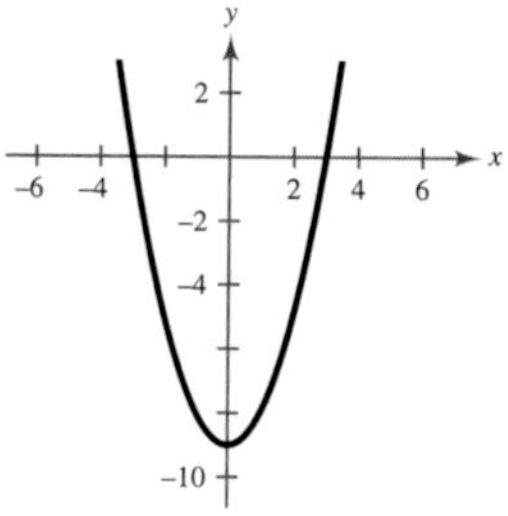

83. $[-1, 1]$

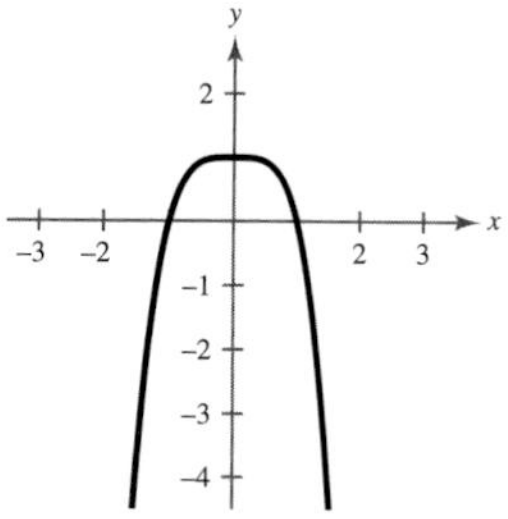

85. $(-\infty, \infty)$

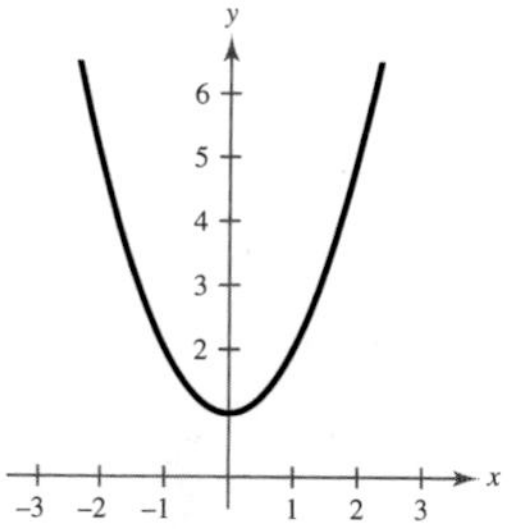

87. $f(x) < 0$ for all x.

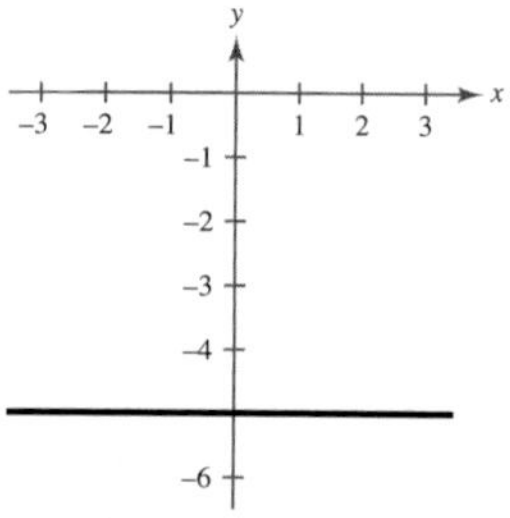

89. $(-2, 8]$

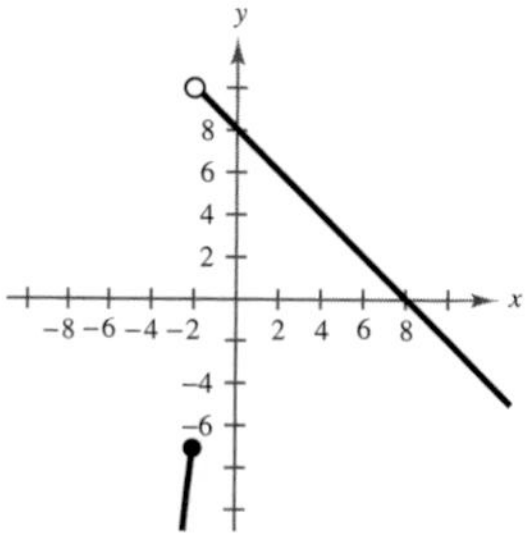

91.

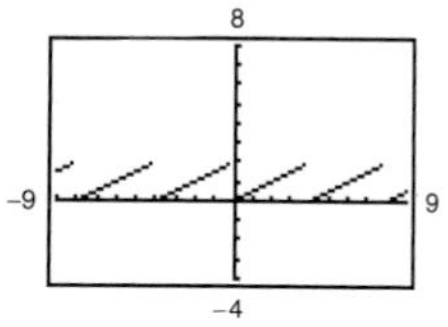

Domain: $(-\infty, \infty)$

Range: $[0, 2)$

Sawtooth pattern

93. Even **95.** Odd **97.** Neither even nor odd

99. (a) $\left(\frac{3}{2}, 4\right)$ (b) $\left(\frac{3}{2}, -4\right)$

101. (a) $(-4, 9)$ (b) $(-4, -9)$

103. (a) C_2 is the appropriate model, because the cost does not increase until after the next minute of conversation has started.

(b)

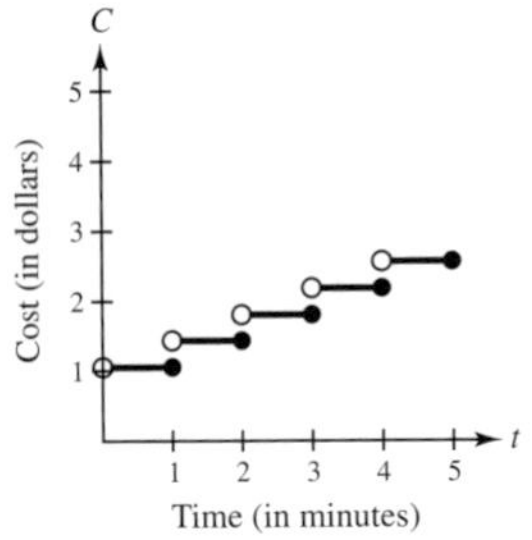

$7.89

105.

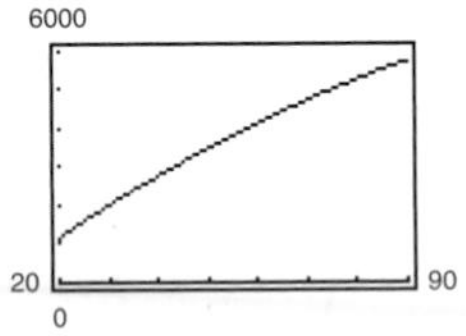

30 watts

107. $h = 3 - 4x + x^2$ **109.** $h = 2 - \sqrt[3]{x}$

111. $L = 2 - \sqrt[3]{2y}$ **113.** $L = \dfrac{2}{y}$

115. (a) $A = 64 - 2x^2,\ 0 \le x \le 4$

(b)

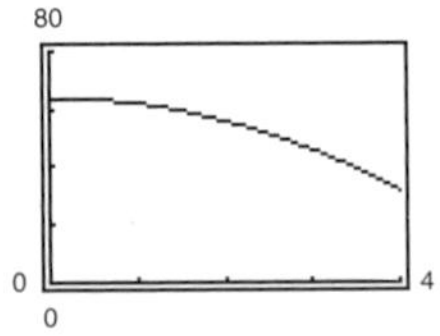

$32 \le A \le 64$

(c) Square with sides $4\sqrt{2}$ m

117.

Interval	*Inside Pipe*	*Drainpipe 1*	*Drainpipe 2*
[0, 5]	Open	Closed	Closed
[5, 10]	Open	Open	Closed
[10, 20]	Closed	Closed	Closed
[20, 30]	Closed	Closed	Open
[30, 40]	Open	Open	Open
[40, 45]	Open	Closed	Open
[45, 50]	Open	Open	Open
[50, 60]	Open	Open	Closed

119. False. A piecewise-defined function is a function that is defined by two or more equations over a specified domain. That domain may or may not include x- and y-intercepts.

121. Answers will vary.

123. Yes. For each value of y there corresponds one and only one value of x.

125. (a)

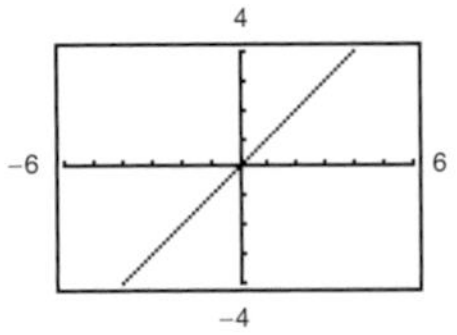

(b)

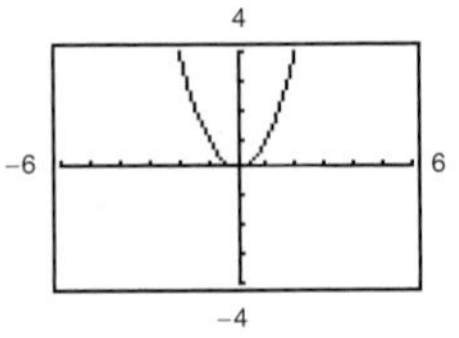

(c)

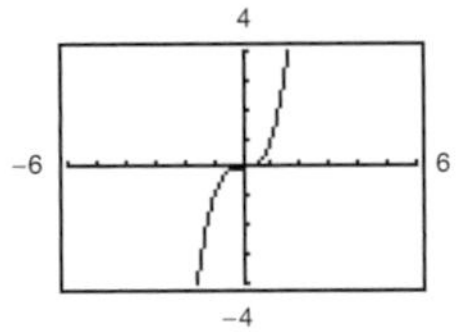

(d)

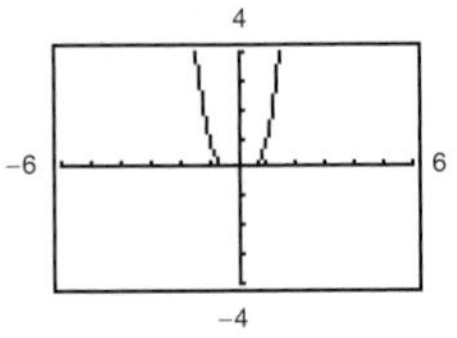

(e)

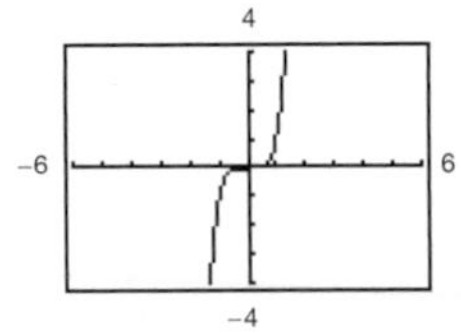

(f)

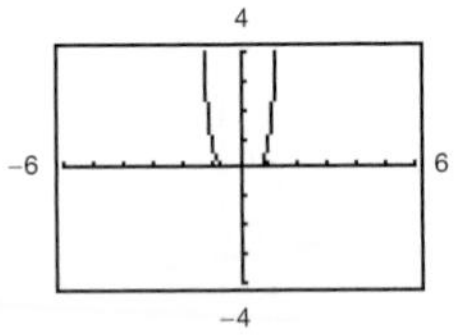

All the graphs pass through the origin. The graphs of the odd powers of x are symmetric with respect to the origin, and the graphs of the even powers are symmetric with respect to the y-axis. As the powers increase, the graphs become flatter in the interval $-1 < x < 1$.

127. 0, 10 **129.** $0, \pm 1$

131. (a) 37 (b) -28 (c) $5x - 43$

133. (a) -9 (b) $2\sqrt{7} - 9$ (c) $-9 + 3\sqrt{2}i$

135. $h + 4,\ h \neq 0$

Section 1.5 *(page 158)*

1. (a)

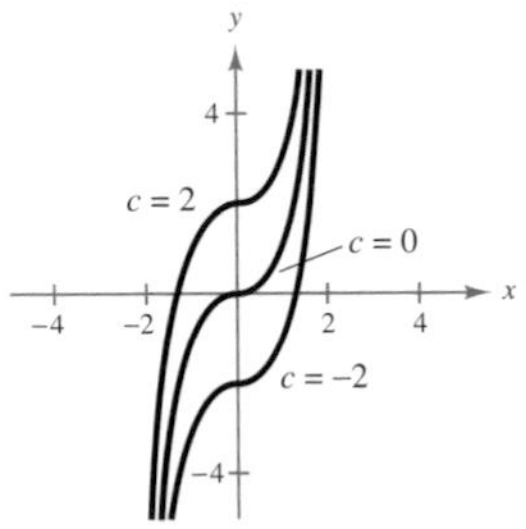

(b)

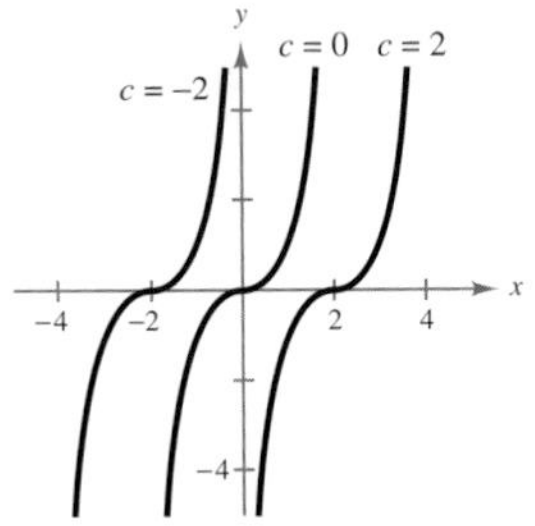

3. (a)

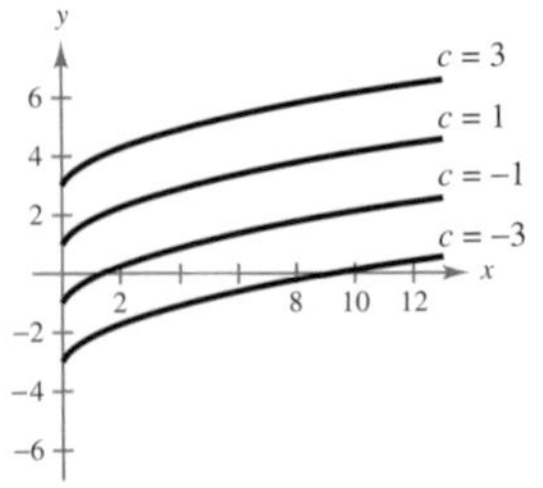

(b)

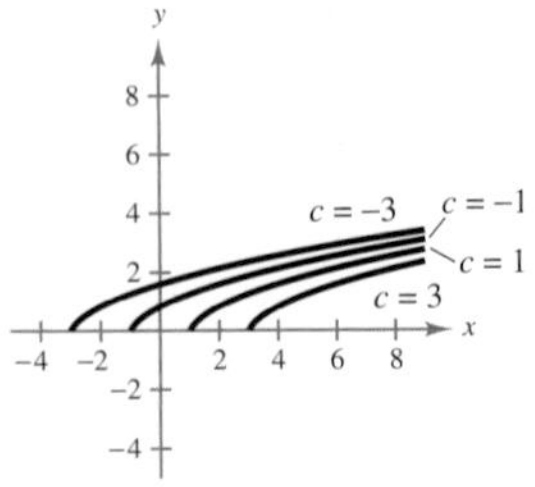

(c)

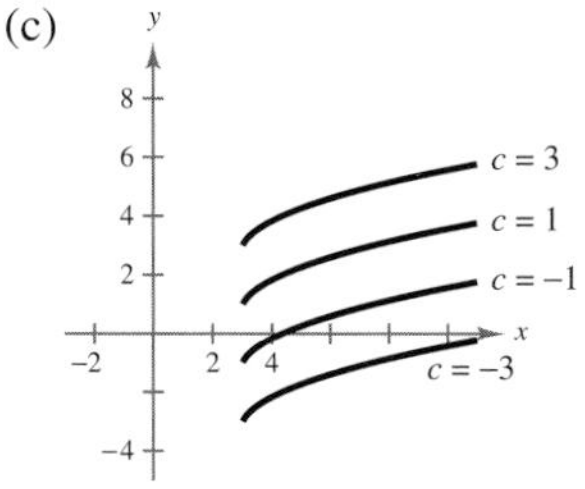

5. (a)

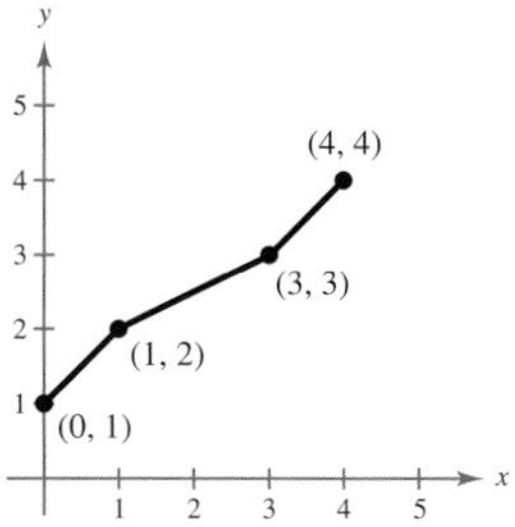

(b)

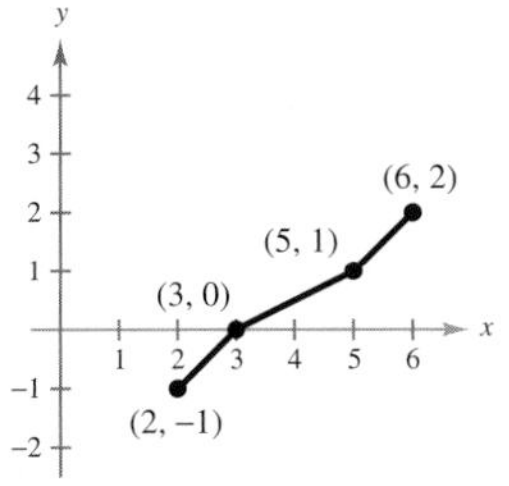

(c)

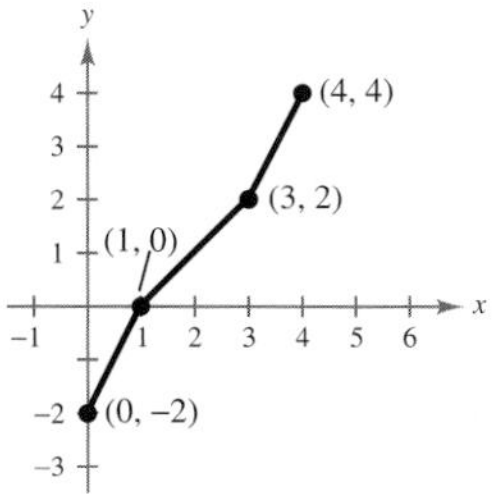

(d)

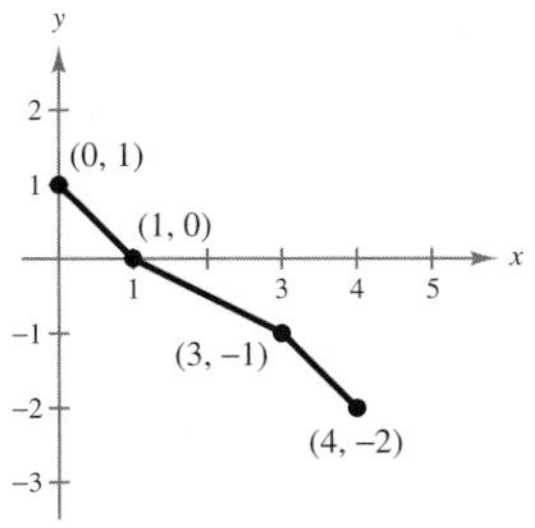

(e)

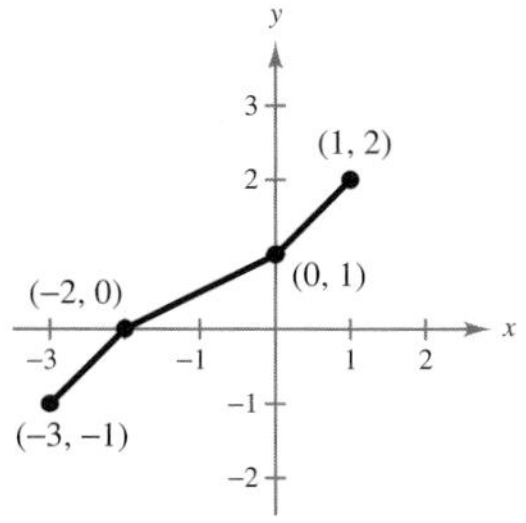

(f)

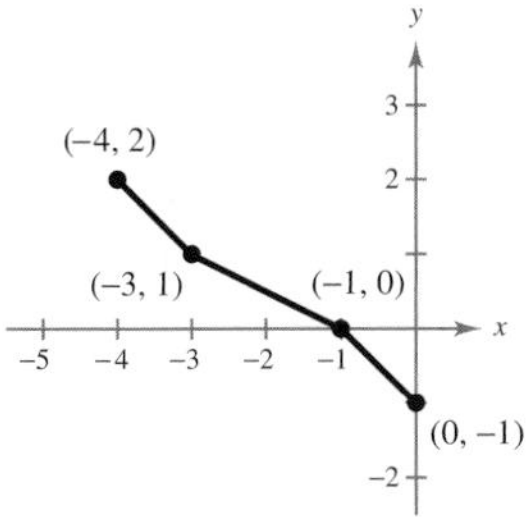

7. (a)

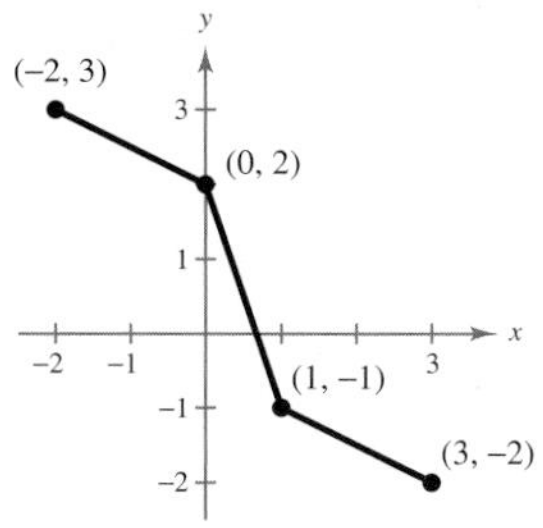

(b)

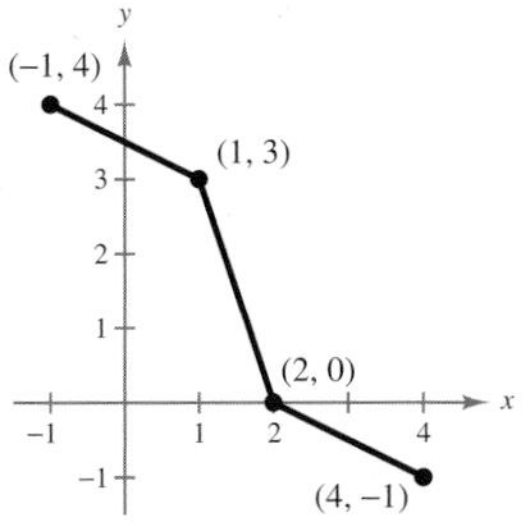

(c)

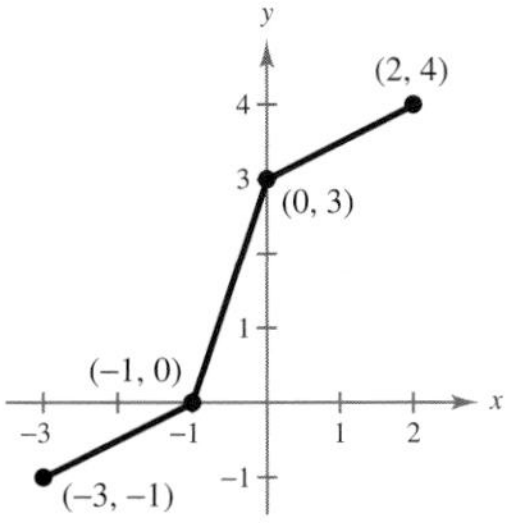

(d)

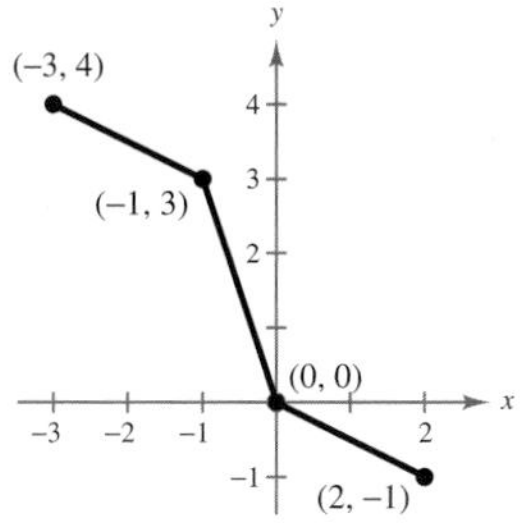

(e)

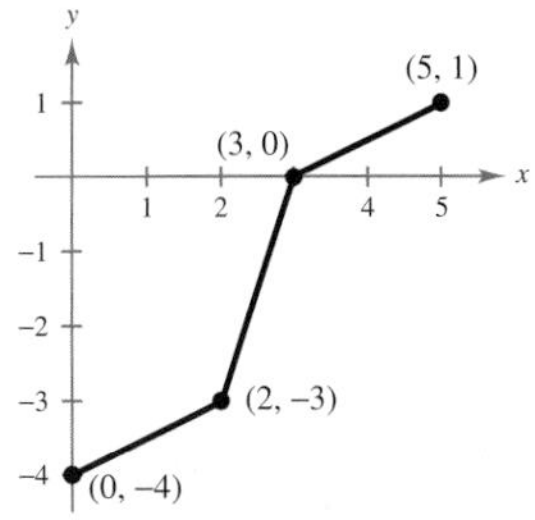

(f)

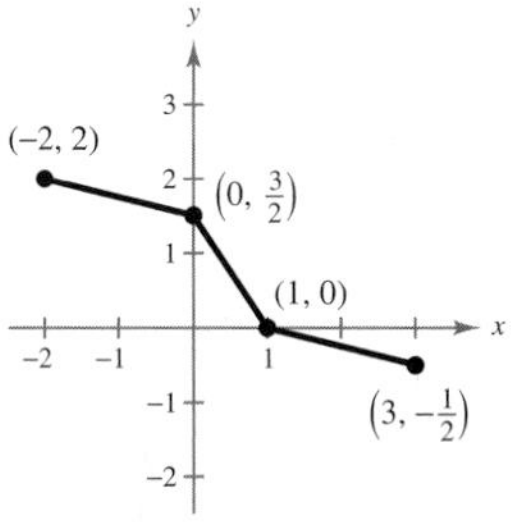

9. (a) $y = x^2 - 1$ (b) $y = 1 - (x + 1)^2$
(c) $y = -(x - 2)^2 + 6$ (d) $y = (x - 5)^2 - 3$

11. (a) $y = |x| + 5$ (b) $y = -|x + 3|$
(c) $y = |x - 2| - 4$ (d) $y = -|x - 6| - 1$

13. Horizontal shift of $y = x^3$; $y = (x - 2)^3$

15. Reflection in the x-axis of $y = x^2$; $y = -x^2$

17. Reflection in the x-axis and vertical shift of $y = \sqrt{x}$; $y = 1 - \sqrt{x}$

19. Reflection of $f(x) = x^2$ in x-axis and vertical shift of 12 units upward

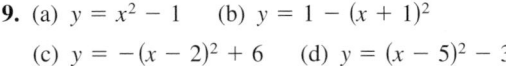

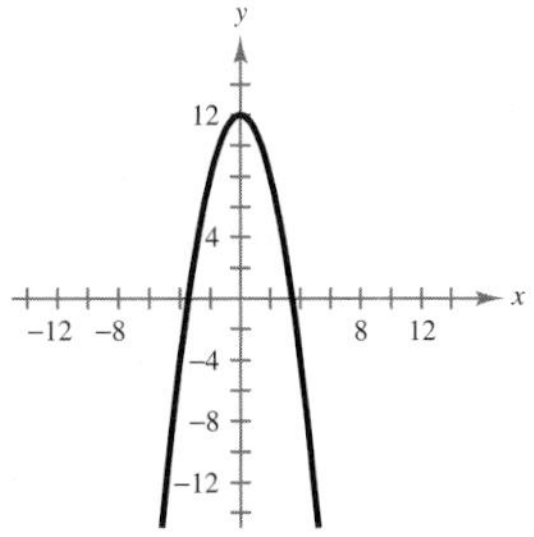

21. Vertical shift of $f(x) = x^3$ seven units upward

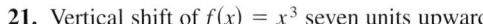

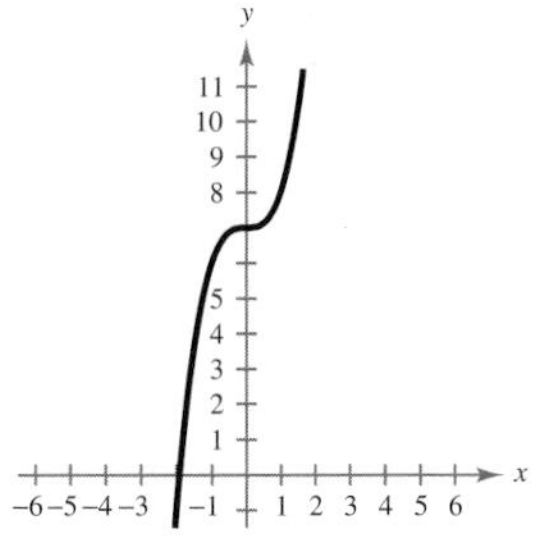

23. Reflection of $f(x) = x^2$ in x-axis, vertical shift of 2 units upward, and horizontal shift of 5 units to the left

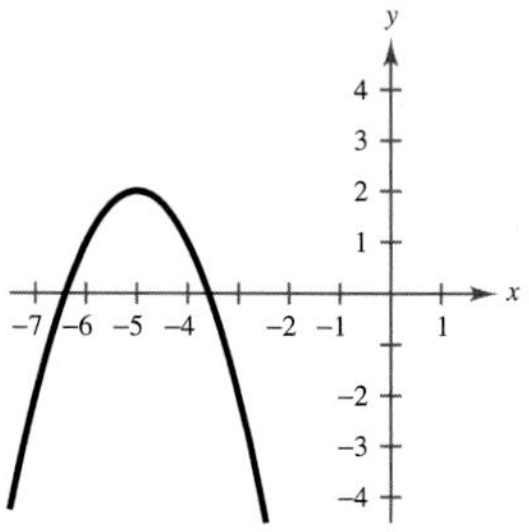

25. Vertical shift of $f(x) = x^3$ 2 units upward and horizontal shift of 1 unit to the right

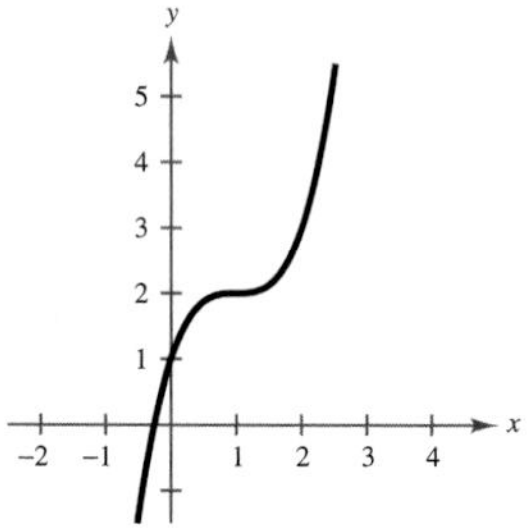

27. Reflection of $f(x) = |x|$ in x-axis and vertical shift of 2 units downward

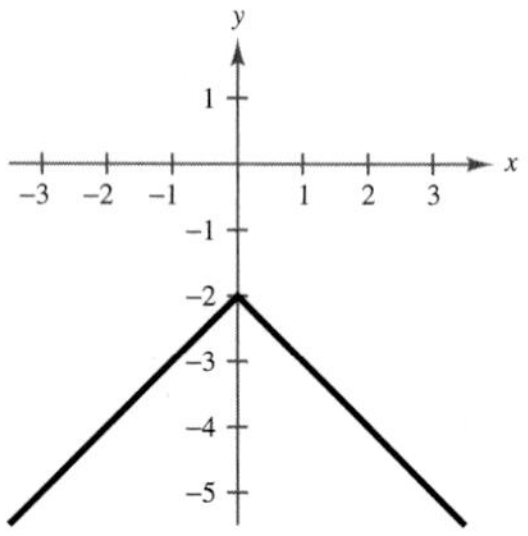

29. Reflection of $f(x) = |x|$ in x-axis, vertical shift of 8 units upward, and horizontal shift of 4 units to the left

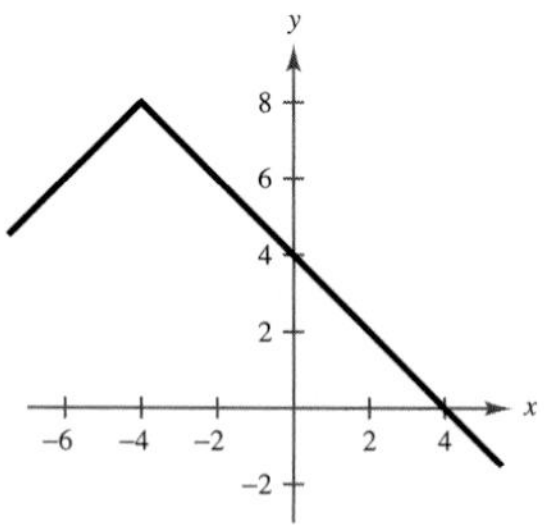

31. Horizontal shift of $f(x) = \sqrt{x}$ 9 units to the right

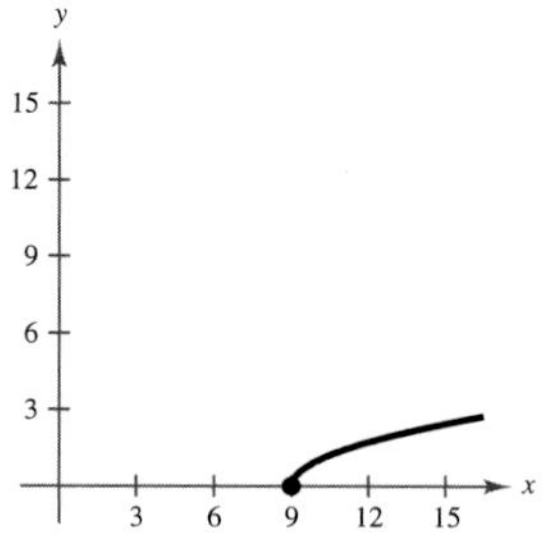

33. Reflection of $f(x) = \sqrt{x}$ in y-axis, vertical shift of 2 units downward, and horizontal shift of 7 units to the right

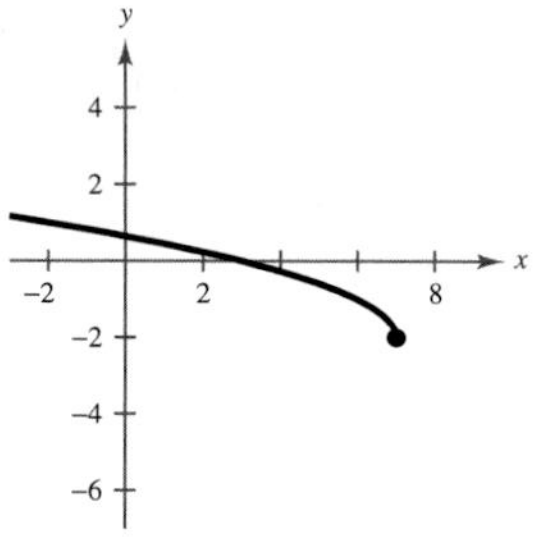

35. $f(x) = (x - 2)^2 - 8$ **37.** $f(x) = (x - 13)^3$

39. $f(x) = -|x| - 10$ **41.** $f(x) = -\sqrt{-x + 6}$

43. (a) $y = -3x^2$ (b) $y = 4x^2 + 3$

45. (a) $y = -\frac{1}{2}|x|$ (b) $y = 3|x| - 3$

47. Vertical stretch of $y = x^3$; $y = 2x^3$

49. Reflection in x-axis and vertical shrink of $y = x^2$; $y = -\frac{1}{2}x^2$

51. Reflection in y-axis and vertical shrink of $y = \sqrt{x}$; $y = \frac{1}{2}\sqrt{-x}$

53. $y = -(x - 2)^3 + 2$ **55.** $y = -\sqrt{x} - 3$

57. (a) (b)

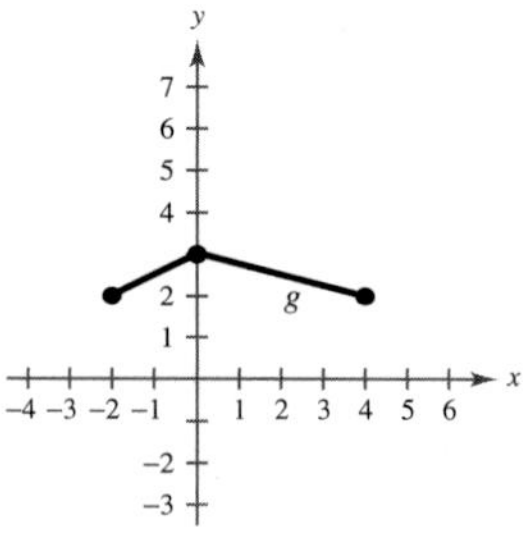

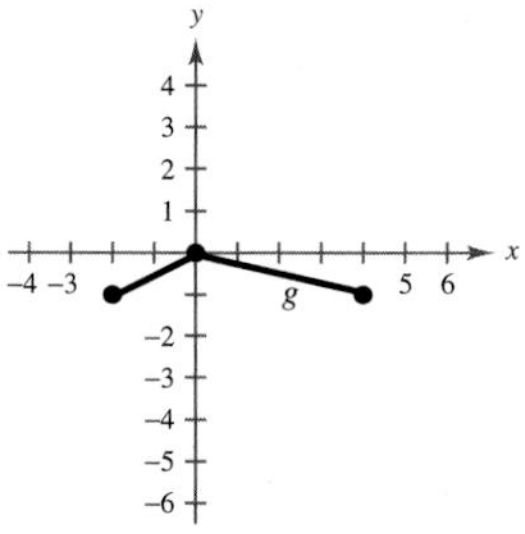

(c)

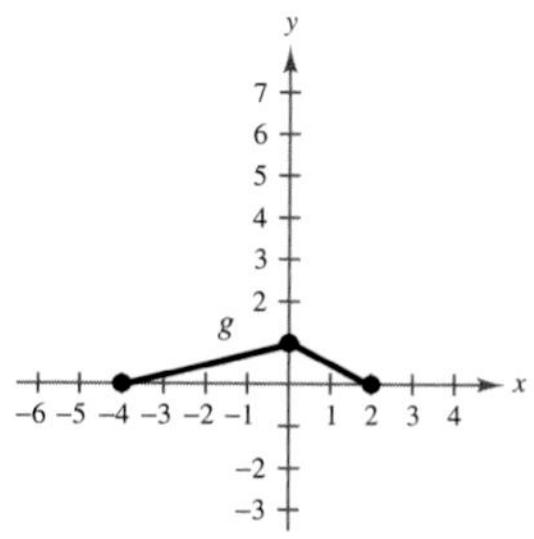

(d)

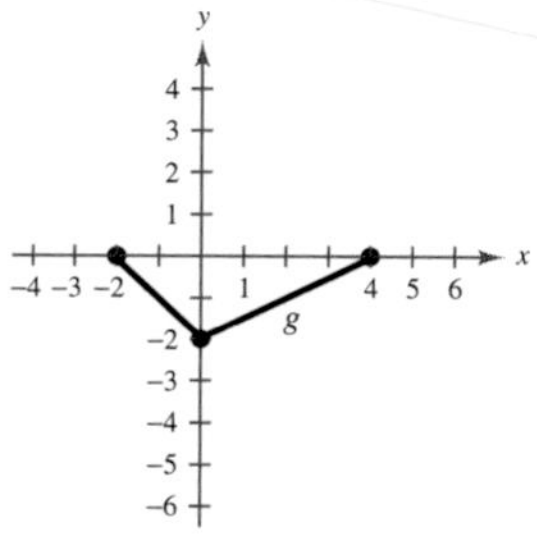

59. (a) Vertical shrink of 0.04 and vertical shift of 20.46 units upward

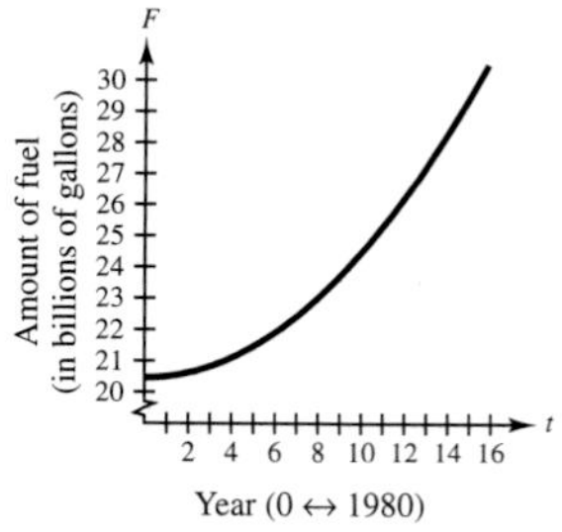

(b) $f(t) = 20.46 + 0.04(t + 10)^2$. By shifting the graph 10 units to the left, you obtain $t = 0$ represents 1990.

61. True. $|-x| = |x|$

63. (a) $g(t) = \frac{3}{4}f(t)$ (b) $g(t) = f(t) + 10{,}000$

(c) $g(t) = f(t - 2)$

65. $(-2, 0), (-1, 1), (0, 2)$

67. $\dfrac{4}{x(1 - x)}$ **69.** $\dfrac{3x - 2}{x(x - 1)}$ **71.** $\dfrac{(x - 4)\sqrt{x^2 - 4}}{x^2 - 4}$

73. $5(x - 3), x \neq -3$

75. (a) 38 (b) $\frac{57}{4}$ (c) $x^2 - 12x + 38$

77. All real numbers $x \neq 11$ **79.** $-9 \leq x \leq 9$

Section 1.6 *(page 168)*

1.

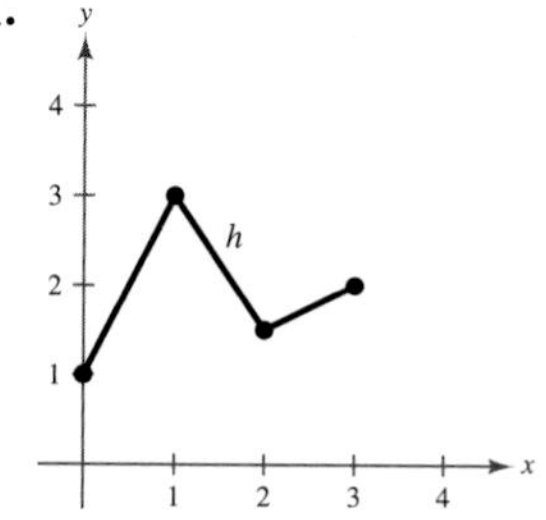

3.

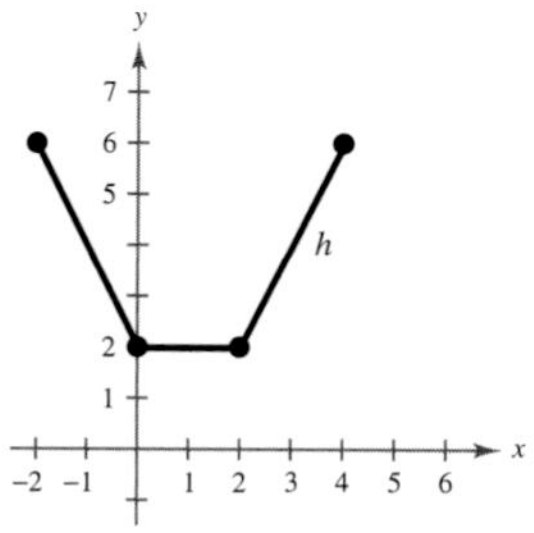

5. (a) $2x$ (b) 4 (c) $x^2 - 4$ (d) $\dfrac{x + 2}{x - 2}$; $x \neq 2$

7. (a) $x^2 - x + 2$ (b) $x^2 + x - 2$

(c) $2x^2 - x^3$ (d) $\dfrac{x^2}{2 - x}$; $x \neq 2$

9. (a) $x^2 + 6 + \sqrt{1 - x}$ (b) $x^2 + 6 - \sqrt{1 - x}$

(c) $(x^2 + 6)\sqrt{1 - x}$ (d) $\dfrac{(x^2 + 6)\sqrt{1 - x}}{1 - x}$; $x < 1$

11. (a) $\dfrac{x + 1}{x^2}$ (b) $\dfrac{x - 1}{x^2}$ (c) $\dfrac{1}{x^3}$ (d) x; $x \neq 0$

13. 3 **15.** 5 **17.** $9t^2 - 3t + 5$ **19.** 74

21. 26 **23.** $\frac{3}{5}$

25.

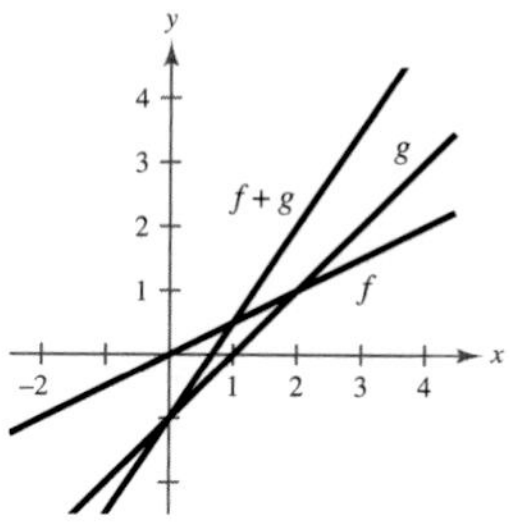

27.

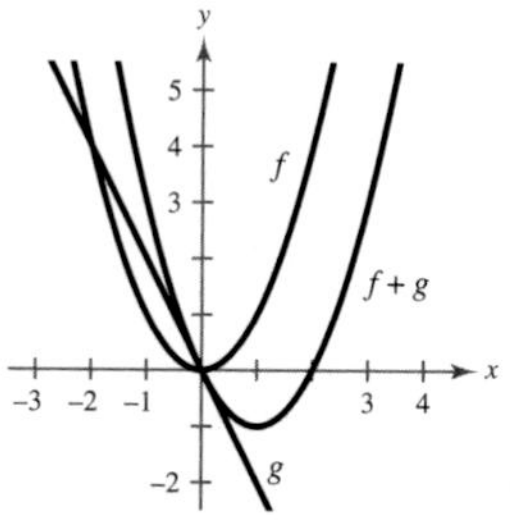

29.

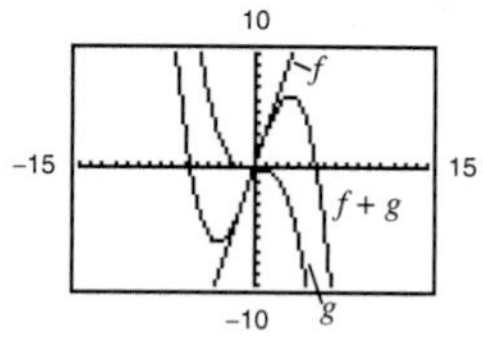

$f(x), g(x)$

31. $T = \frac{3}{4}x + \frac{1}{15}x^2$

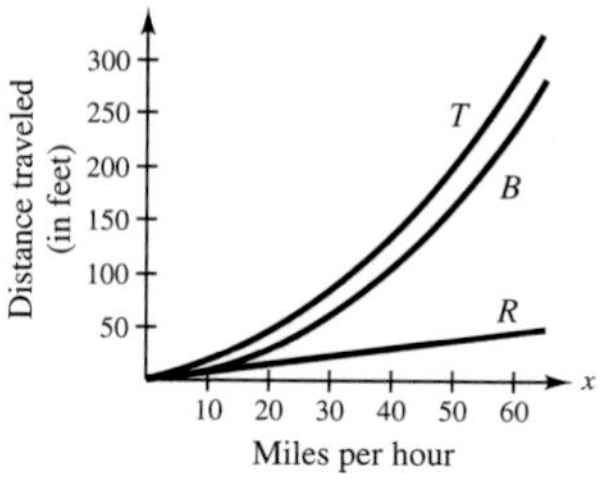

33. $y_1 = -0.59t^2 + 7.66t - 144.90$

$y_2 = 16.58t + 245.06$

$y_3 = 1.85t + 21.88$

35. (a) For each time t there corresponds one and only one temperature T.

(b) 60°, 72°

(c) All the temperature changes would occur 1 hour later.

(d) The temperature would be decreased by 1 degree.

37. (a) $(x - 1)^2$ (b) $x^2 - 1$ (c) x^4

39. (a) $20 - 3x$ (b) $-3x$ (c) $9x + 20$

41. (a) $\sqrt{x^2 + 4}$ (b) $x + 4$

Domain of f and $g \circ f$: $x \geq -4$;

Domain of g and $f \circ g$: all real numbers

43. (a) $x - \frac{8}{3}$ (b) $x - 8$

Domain of f, g, $f \circ g$, and $g \circ f$: all real numbers

45. (a) x^{16} (b) x^{16}

Domain of f, g, $f \circ g$, and $g \circ f$: all real numbers

47. (a) $|x + 6|$ (b) $|x| + 6$

Domain of f, g, $f \circ g$, and $g \circ f$: all real numbers

49. (a) $\dfrac{1}{x + 3}$ (b) $\dfrac{1}{x} + 3$

Domain of f and $g \circ f$: all real numbers $x \neq 0$

Domain of g: all real numbers

Domain of $f \circ g$: all real numbers $x \neq -3$

51. (a) 3 (b) 0 **53.** (a) 0 (b) 4

55. $f(x) = x^2,\ g(x) = 2x + 1$

57. $f(x) = \sqrt[3]{x},\ g(x) = x^2 - 4$

59. $f(x) = \dfrac{1}{x},\ g(x) = x + 2$

61. $f(x) = \dfrac{x + 3}{4 + x},\ g(x) = -x^2$

63. (a) $r(x) = \dfrac{x}{2}$ (b) $A(r) = \pi r^2$

(c) $(A \circ r)(x) = \pi\left(\dfrac{x}{2}\right)^2$; $(A \circ r)(x)$ represents the area of the circular base of the tank on the square foundation with side length x.

65. $(C \circ x)(t) = 3000t + 750$; $(C \circ x)(t)$ represents the cost after t production hours.

67. True. The range of g must be a subset of the domain of f for $(f \circ g)(x)$ to be defined.

69. Answers will vary.

71. 3 **73.** $\dfrac{-4}{x(x + h)}$

75. $3x - y - 10 = 0$ **77.** $3x + 2y - 22 = 0$

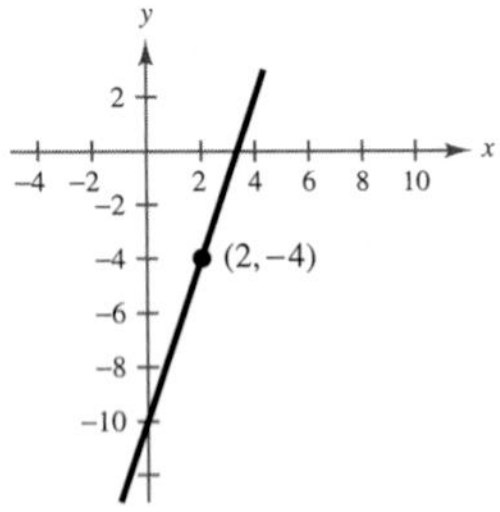

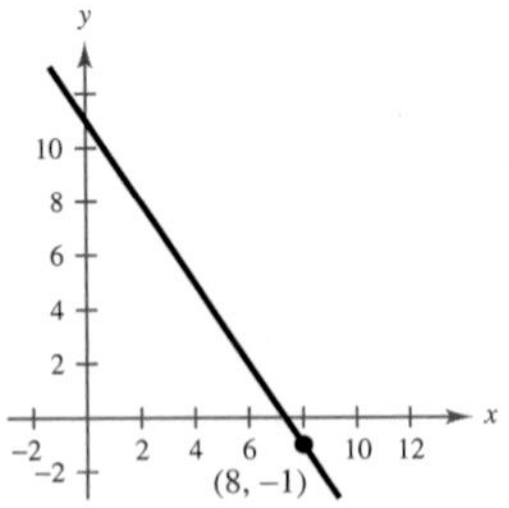

Section 1.7 *(page 177)*

1. c **3.** a **5.** $f^{-1}(x) = \frac{1}{6}x$ **7.** $f^{-1}(x) = x - 9$

9. $f^{-1}(x) = \dfrac{x - 1}{3}$ **11.** $f^{-1}(x) = x^3$

13. (a) $f(g(x)) = f\left(\dfrac{x}{2}\right) = 2\left(\dfrac{x}{2}\right) = x$

$g(f(x)) = g(2x) = \dfrac{(2x)}{2} = x$

(b)

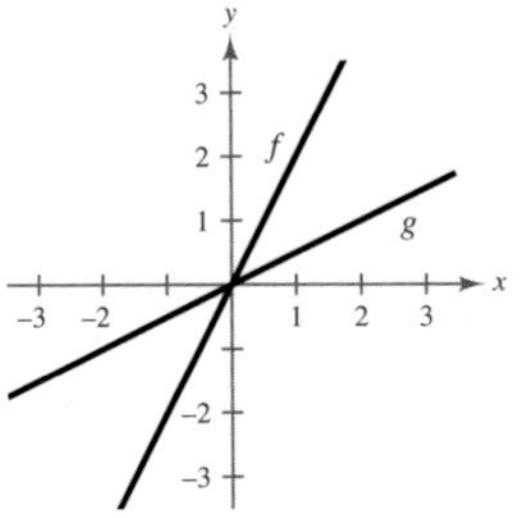

15. (a) $f(g(x)) = f\left(\dfrac{x - 1}{5}\right) = 5\left(\dfrac{x - 1}{5}\right) + 1 = x$

$g(f(x)) = g(5x + 1) = \dfrac{(5x + 1) - 1}{5} = x$

(b)

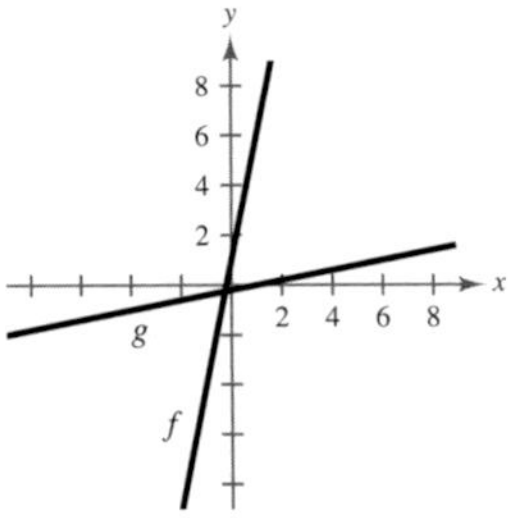

17. (a) $f(g(x)) = f\left(\sqrt[3]{x}\right) = \left(\sqrt[3]{x}\right)^3 = x$

$g(f(x)) = g(x^3) = \sqrt[3]{x^3} = x$

(b)

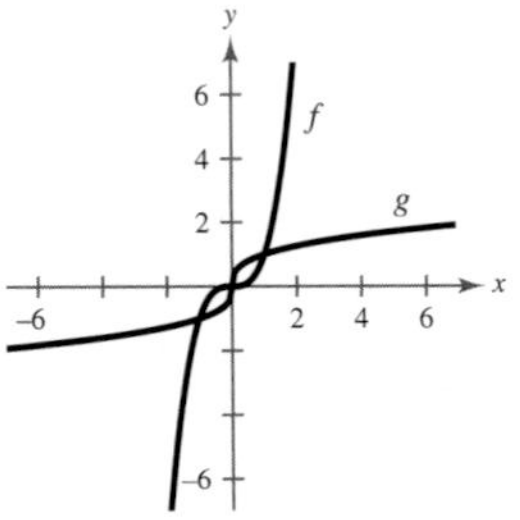

19. (a) $f(g(x)) = f(x^2 + 4),\ x \geq 0$

$= \sqrt{(x^2 + 4) - 4} = x$

$g(f(x)) = g\left(\sqrt{x - 4}\right)$

$= \left(\sqrt{x - 4}\right)^2 + 4 = x$

(b)

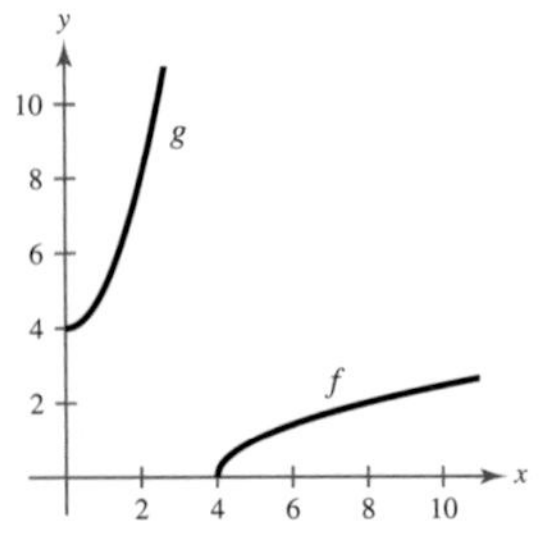

21. (a) $f(g(x)) = f\left(\sqrt{9-x}\right),\ x \le 9$

$= 9 - \left(\sqrt{9-x}\right)^2 = x$

$g(f(x)) = g(9 - x^2),\ x \ge 0$

$= \sqrt{9 - (9 - x^2)} = x$

(b)

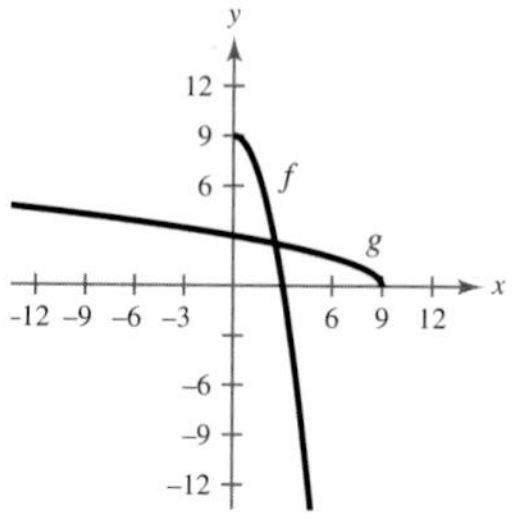

23. (a) $f(g(x)) = f\left(-\dfrac{5x+1}{x-1}\right) = \dfrac{-\left(\dfrac{5x+1}{x-1}\right) - 1}{-\left(\dfrac{5x+1}{x-1}\right) + 5}$

$= \dfrac{-5x - 1 - x + 1}{-5x - 1 + 5x - 5} = x$

$g(f(x)) = g\left(\dfrac{x-1}{x+5}\right) = \dfrac{-5\left(\dfrac{x-1}{x+5}\right) - 1}{\dfrac{x-1}{x+5} - 1}$

$= \dfrac{-5x + 5 - x - 5}{x - 1 - x - 5} = x$

(b)

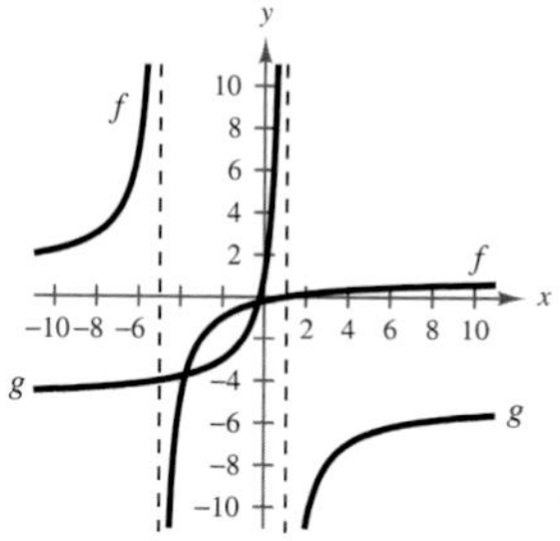

25. No

27.

x	-2	0	2	4	6	8
$f^{-1}(x)$	-2	-1	0	1	2	3

29. Yes **31.** No

33.

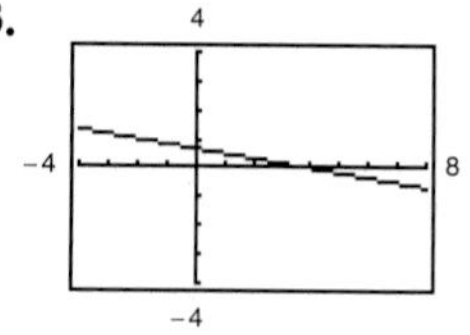

The function has an inverse.

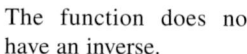

35.

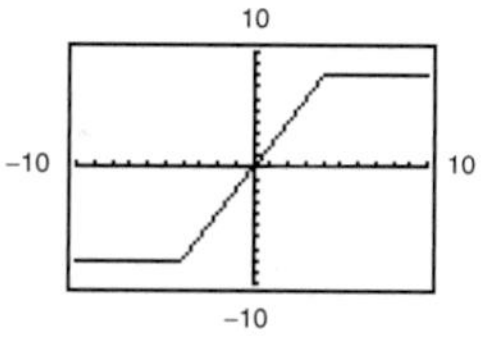

The function does not have an inverse.

37.

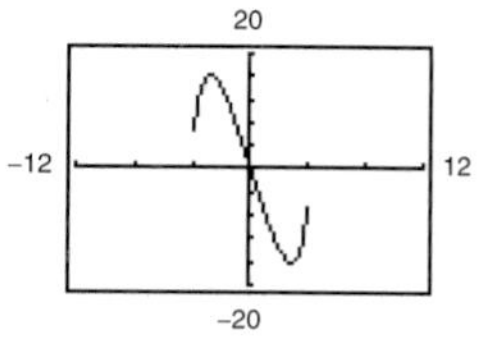

The function does not have an inverse.

39. $f^{-1}(x) = \dfrac{x+3}{2}$

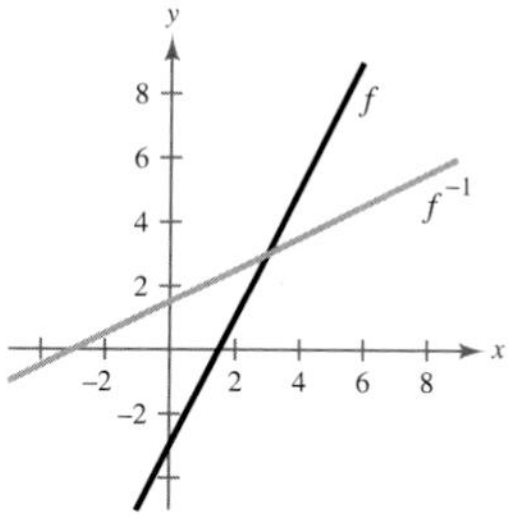

41. $f^{-1}(x) = \sqrt[5]{x+2}$

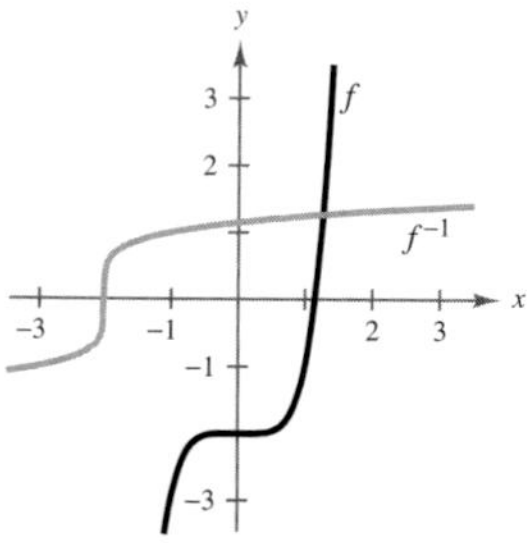

43. $f^{-1}(x) = x^2,\ x \ge 0$

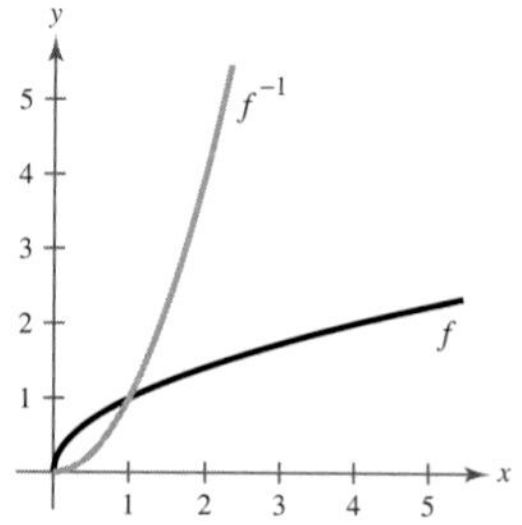

45. $f^{-1}(x) = \sqrt{4 - x^2},\ 0 \le x \le 2$

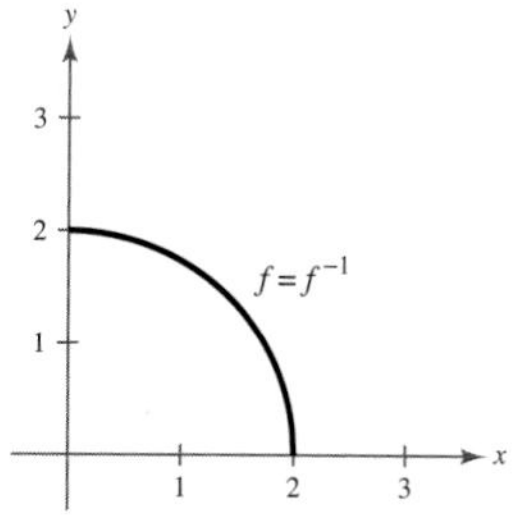

47. $f^{-1}(x) = \dfrac{4}{x}$

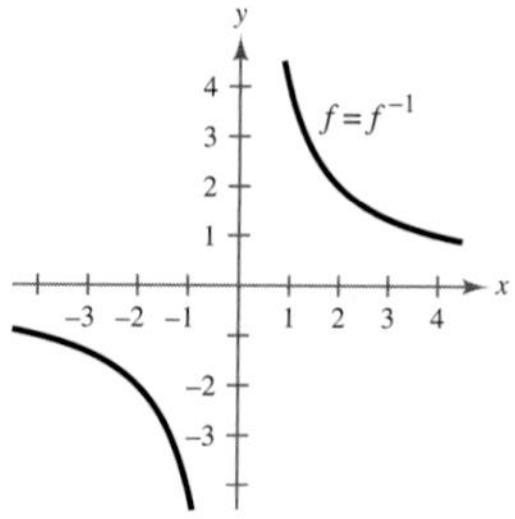

49. $f^{-1}(x) = \dfrac{2x+1}{x-1}$

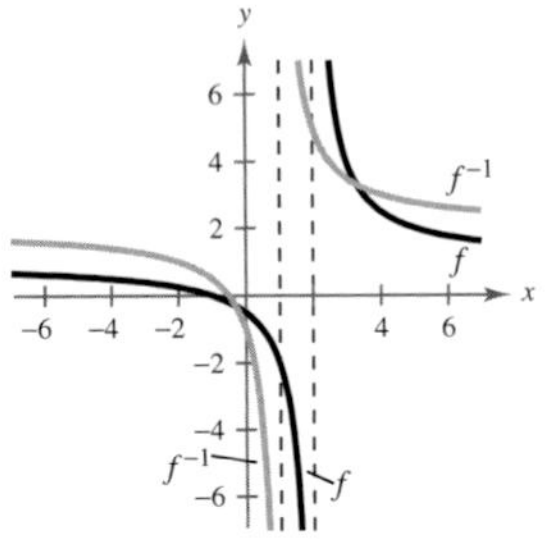

51. $f^{-1}(x) = x^3 + 1$

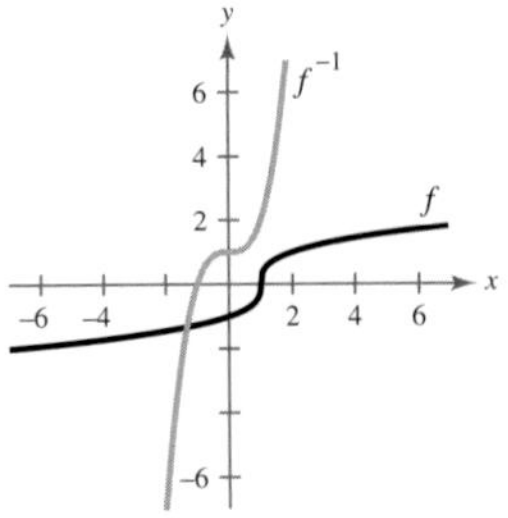

53. $f^{-1}(x) = \dfrac{5x-4}{6-4x}$

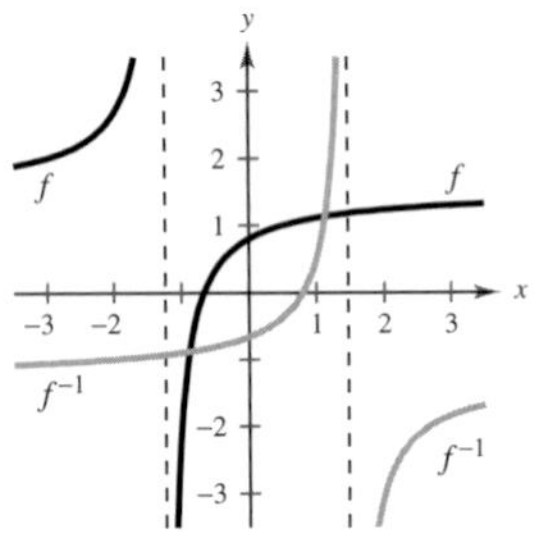

55. No inverse **57.** $g^{-1}(x) = 8x$ **59.** No inverse

61. $f^{-1}(x) = \sqrt{x} - 3$ **63.** No inverse

65. $h^{-1}(x) = \dfrac{1}{x}$ **67.** $f^{-1}(x) = \dfrac{x^2-3}{2},\ x \ge 0$ **69.** 32

71. 600 **73.** $2\sqrt[3]{x+3}$ **75.** $\dfrac{x+1}{2}$ **77.** $\dfrac{x+1}{2}$

79. (a) $y = \dfrac{x-8}{0.75}$

(b) $y =$ number of units produced; $x =$ hourly wage

(c) 19 units

81. (a) $y = \sqrt{\dfrac{x - 245.50}{0.03}},\ 245.5 < x < 545.5$

$x =$ degrees Fahrenheit; $y =$ % load

(b)

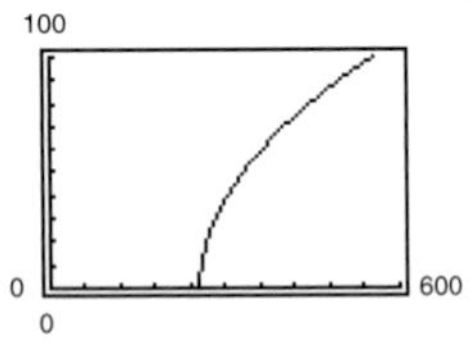

(c) $0 < x < 92.11$

83. No. The function would not pass the Horizontal Line Test.

85. (a) Yes. f^{-1} yields the year for a given per capita regular soft drink consumption.

(b) 5

87. True. If $f(x) = x - 6$ and $f^{-1}(x) = x + 6$, then the y-intercept of f is $(0, -6)$ and the x-intercept of f^{-1} is $(-6, 0)$.

89. False. $f(x) = \dfrac{1}{x} = f^{-1}(x)$.

91.

x	-2	-1	1	3
y	-5	-2	2	3

x	-5	-2	2	3
$f^{-1}(x)$	-2	-1	1	3

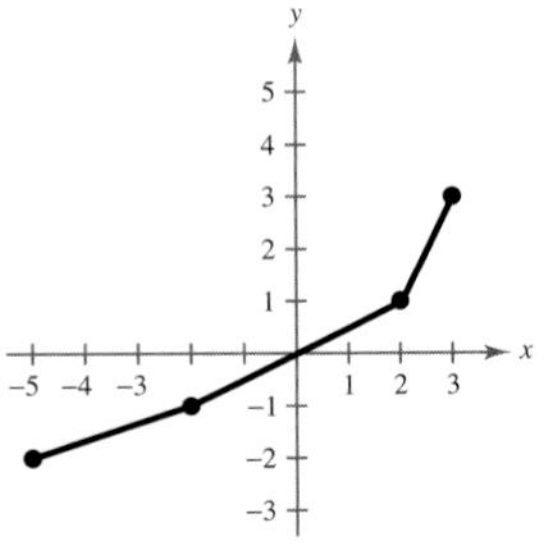

93.

x	-4	-2	0	3
y	3	4	0	-1

The graph of f does not pass the Horizontal Line Test, so $f^{-1}(x)$ does not exist.

95. ± 8 **97.** $\frac{3}{2}$ **99.** $3 \pm \sqrt{5}$ **101.** $5, -\frac{10}{3}$

103. All real numbers **105.** All real numbers $x \ne 0, 4$

107. 16, 18 **109.** $b = h = 2\sqrt{5}$ feet

Section 1.8 *(page 187)*

1.

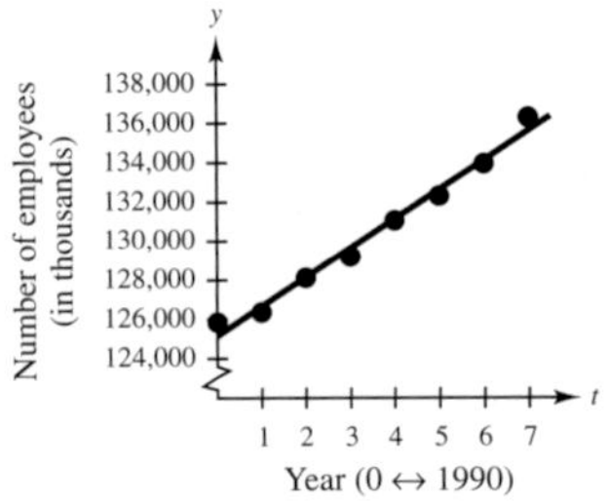

The model is a "good fit" for the actual data.

3. Inversely

5.

x	2	4	6	8	10
$y = kx^2$	4	16	36	64	100

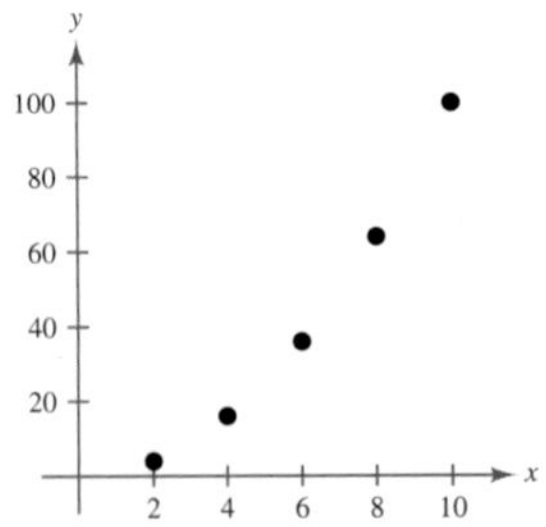

7.

x	2	4	6	8	10
$y = kx^2$	2	8	18	32	50

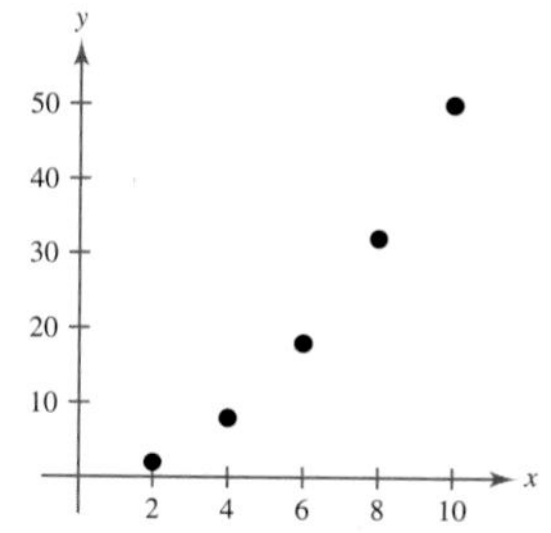

9.

x	2	4	6	8	10
$y = k/x^2$	$\frac{1}{2}$	$\frac{1}{8}$	$\frac{1}{18}$	$\frac{1}{32}$	$\frac{1}{50}$

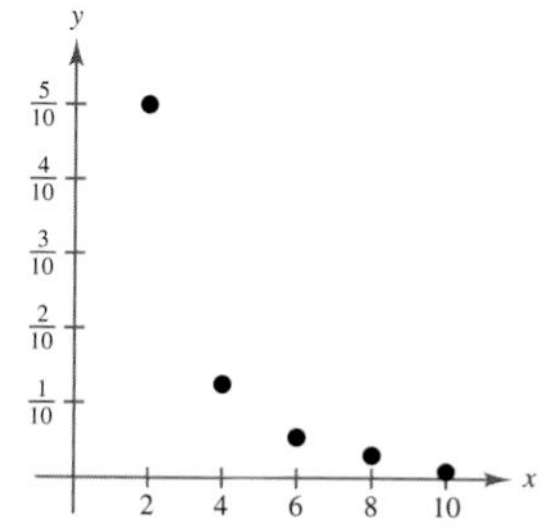

11.

x	2	4	6	8	10
$y = k/x^2$	$\frac{5}{2}$	$\frac{5}{8}$	$\frac{5}{18}$	$\frac{5}{32}$	$\frac{1}{10}$

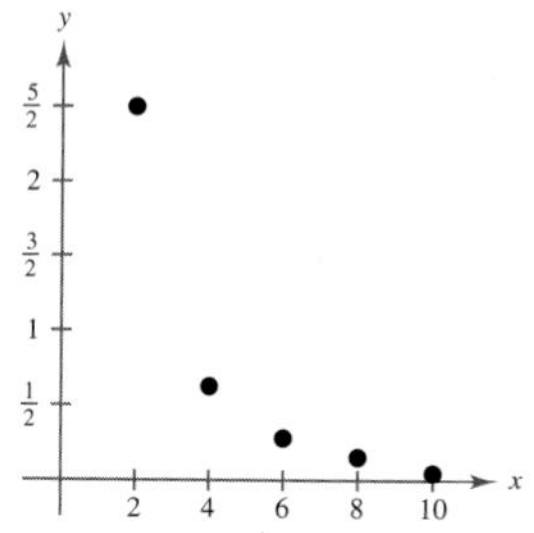

13. $y = \dfrac{5}{x}$ **15.** $y = -\dfrac{7}{10}x$ **17.** $y = \dfrac{12}{5}x$

19. $y = 205x$ **21.** $I = 0.035P$

23. Model: $y = \frac{33}{13}x$

Inches	5	10	20	25	30
Centimeters	12.7	25.4	50.8	63.5	76.2

25. $y = 0.0368x$; \$7360

27. (a) 0.05 meter (b) $176\frac{2}{3}$ newtons **29.** 39.47 pounds

31. $A = kr^2$ **33.** $y = \dfrac{k}{x^2}$ **35.** $F = \dfrac{kg}{r^2}$

37. $P = \dfrac{k}{V}$ **39.** $F = \dfrac{km_1m_2}{r^2}$

41. The area of a triangle is jointly proportional to its base and height.

43. The volume of a sphere varies directly as the cube of its radius.

45. Average speed is directly proportional to the distance and inversely proportional to the time.

47. $A = \pi r^2$ **49.** $y = \dfrac{28}{x}$ **51.** $F = 14rs^3$

53. $z = \dfrac{2x^2}{3y}$ **55.** ≈ 0.61 mile per hour **57.** 506 feet

59. 400 feet **61.** No. The 15-inch pizza is the best buy.

63. (a) The velocity is increased by one-third.

(b) The velocity is decreased by one-fourth.

65. (a)

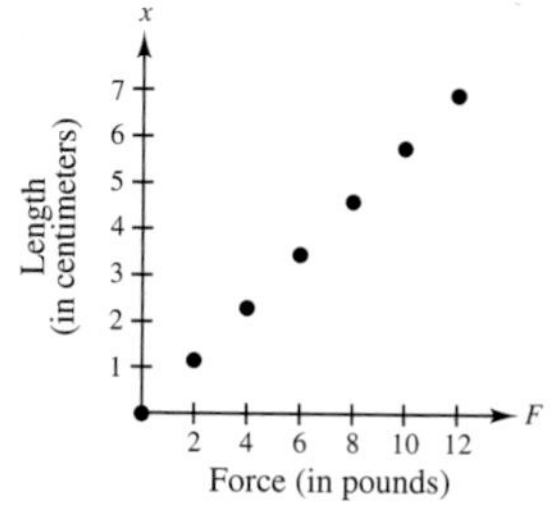

(b) Yes. $k = 0.575$

(c) 5 pounds

67. (a)

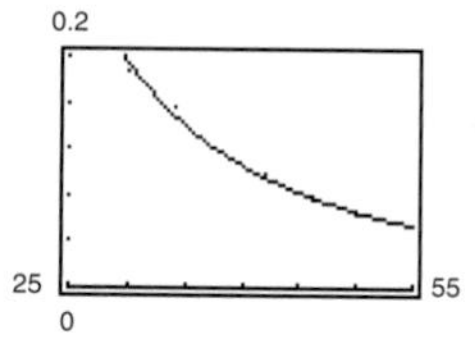

(b) 0.2857 microwatt per square centimeter

69. (a) $y = 127.4t + 218.4$

(b)

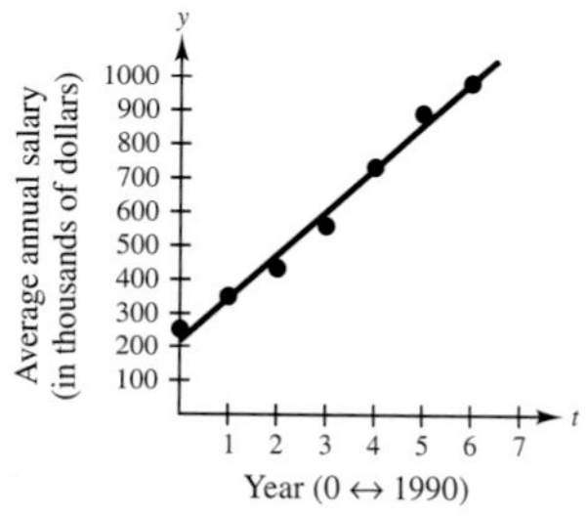

(c) 1997: \$1110 thousand
1998: \$1238 thousand
1999: \$1365 thousand

(d) Answers will vary.

71. (a) $y = 489.58t + 5628.4$

(b)

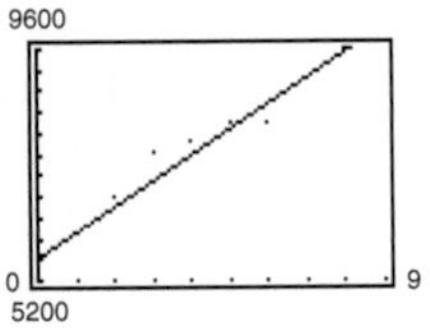

(c) \$10,035 million (d) Answers will vary.

73. (a) $y = -0.0186x + 689$

(b)

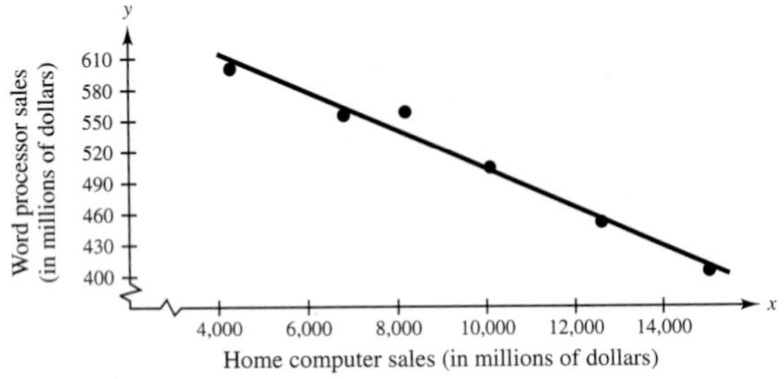

(c) $\approx$\$354 million

75. False. E is jointly proportional to the mass of an object and the square of its velocity.

77. Poor approximation **79.** Good approximation

81.

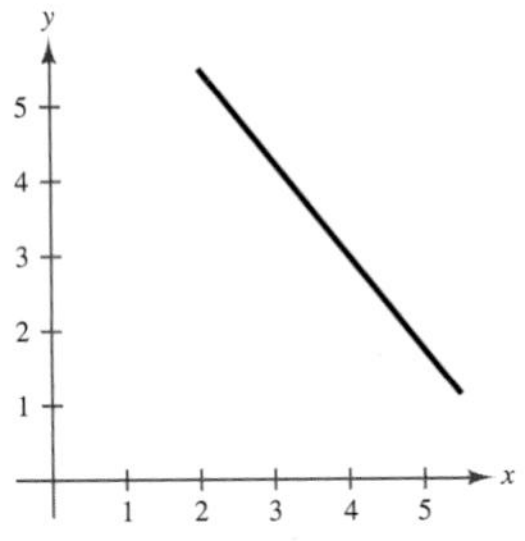

$y = -\frac{5}{4}x + 8$

83.

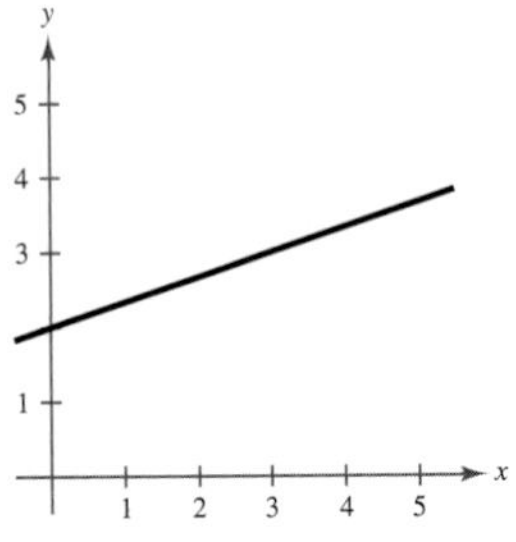

$y = \frac{1}{3}x + 2$

85. $x \le 4, x \ge 6$

87. $x > 5$

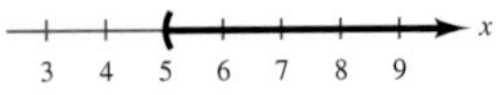

89. (a) $-\frac{5}{3}$ (b) $-\frac{7}{3}$ (c) 21 **91.** All real numbers

93. All real numbers $x \neq -7$

Review Exercises *(page 194)*

1.

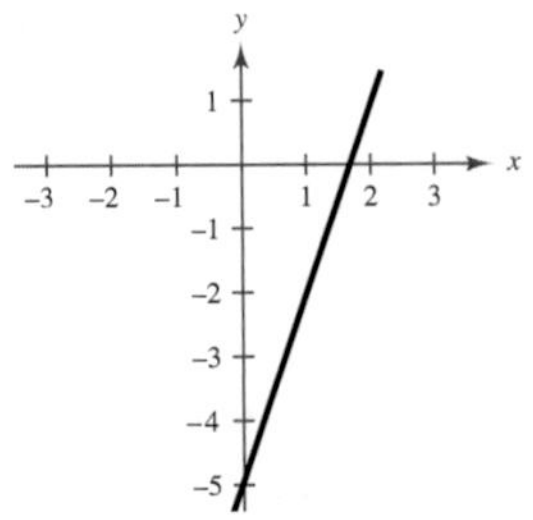

3.

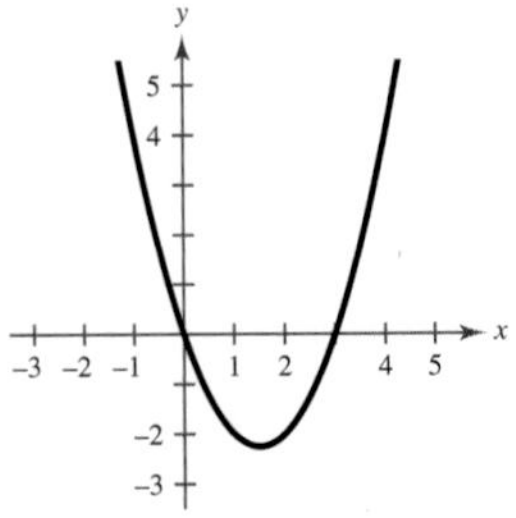

5. x-intercept: $\left(\frac{9}{2}, 0\right)$

y-intercept: $(0, -9)$

7. x-intercept: $(-1, 0)$

y-intercept: $(0, 1)$

9.

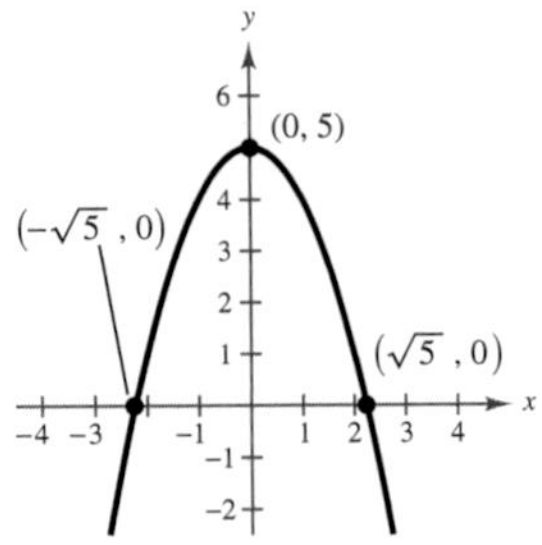

11.

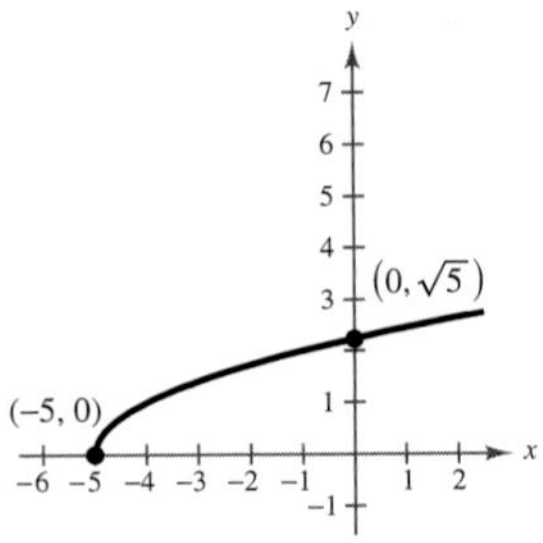

13. Center: $(0, 0)$

Radius: 5

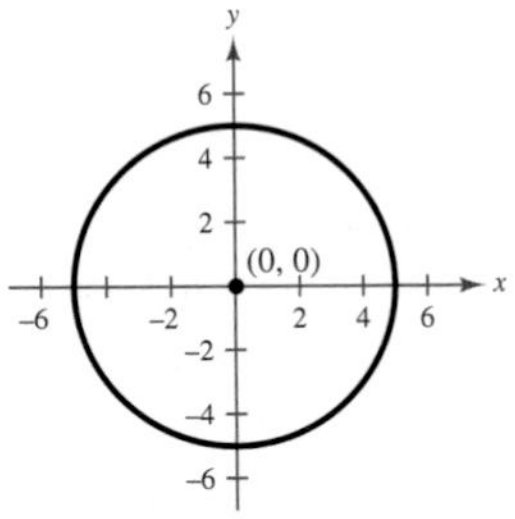

15. Center: $(-2, 0)$

Radius: 4

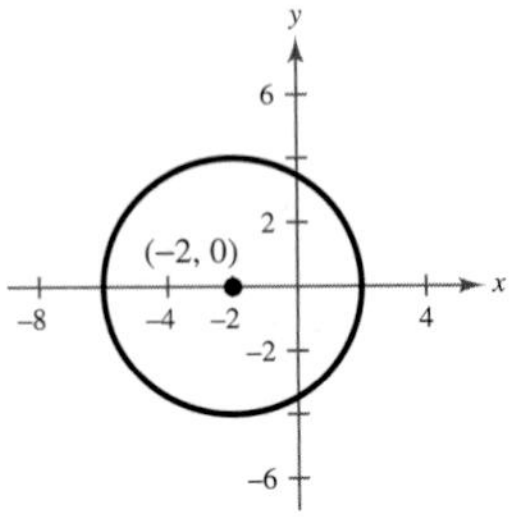

17. $(x - 2)^2 + (y + 3)^2 = 13$

19.

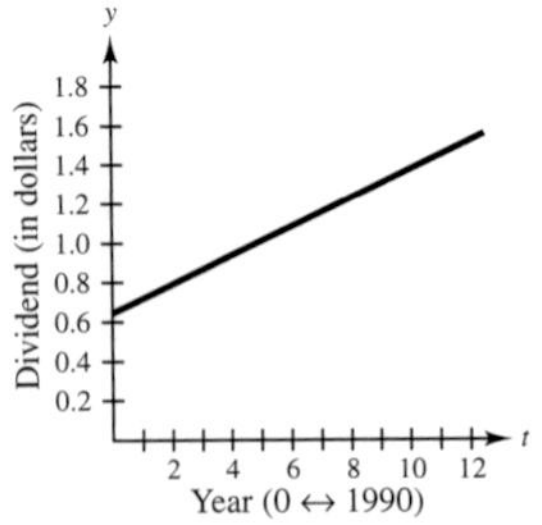

21. Slope: 0
y-intercept: $(0, 6)$

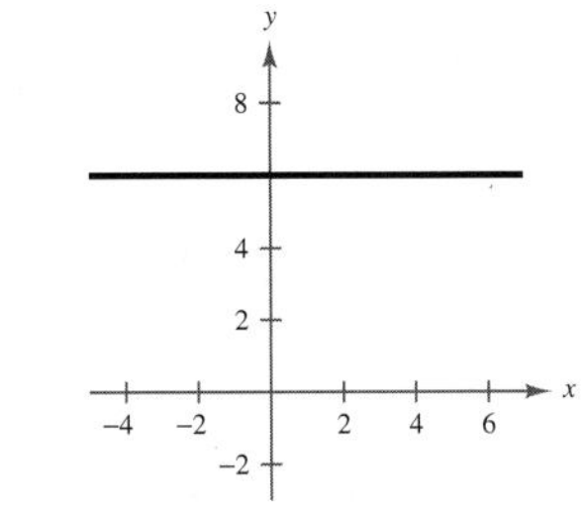

23. Slope: 3
y-intercept: $(0, 13)$

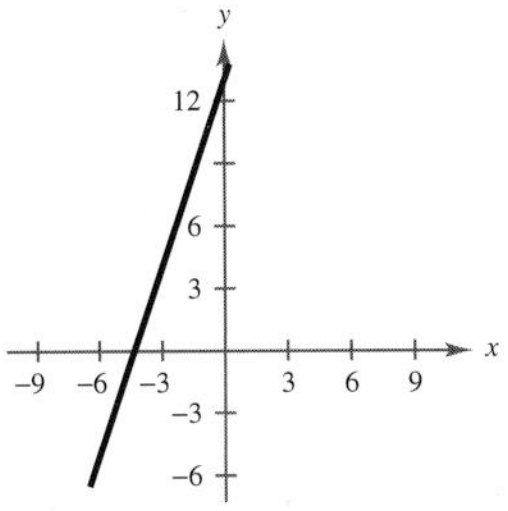

25. (a) L_2 (b) L_3 (c) L_1 (d) L_4

27.

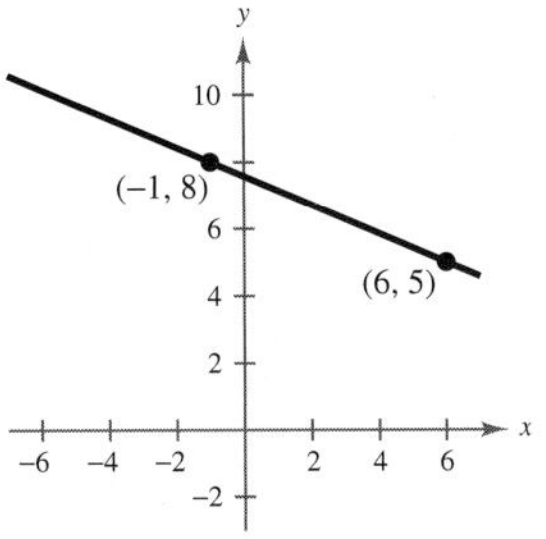

$m = -\frac{3}{7}$

29.

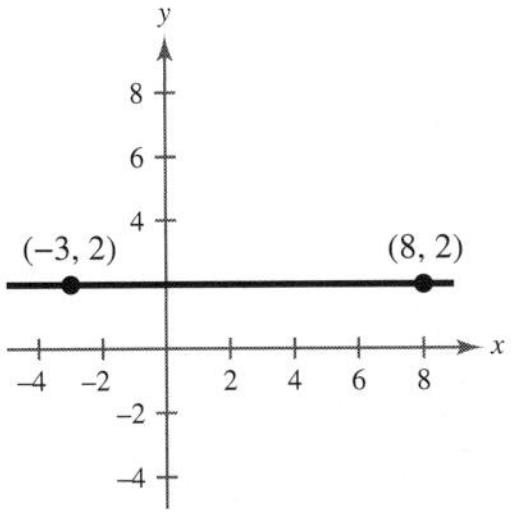

$m = 0$

31. $3x - 2y + 4 = 0$ **33.** $x + 5y - 1 = 0$

35. $y - 6 = 0$

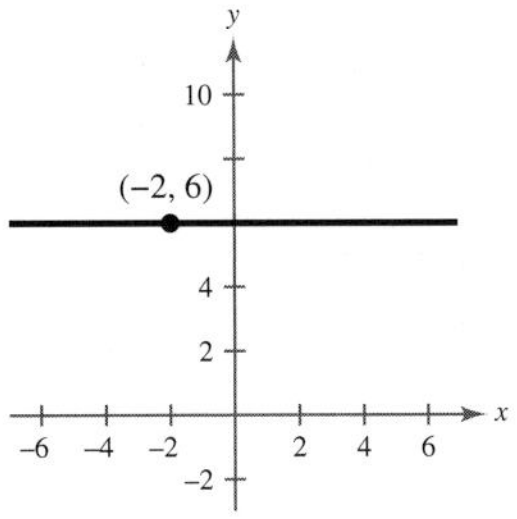

37. $x + 8 = 0$

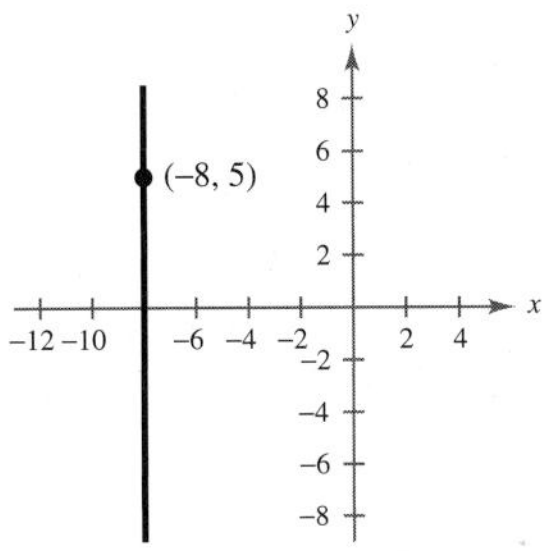

39. (a) $2x + 3y + 7 = 0$ (b) $3x - 2y + 30 = 0$

41. $V = 5.15t + 72.95$ **43.** Yes **45.** No

47. (a) 16 (b) $(t + 1)^{4/3}$ (c) $\frac{15}{7}$ (d) $x^{4/3}$

49. All real numbers

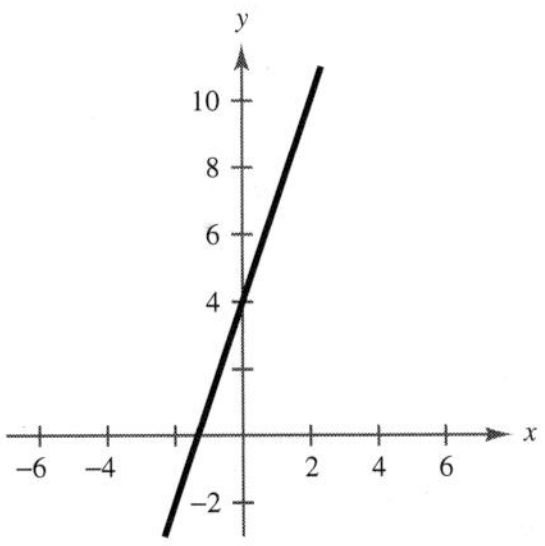

51. All real numbers

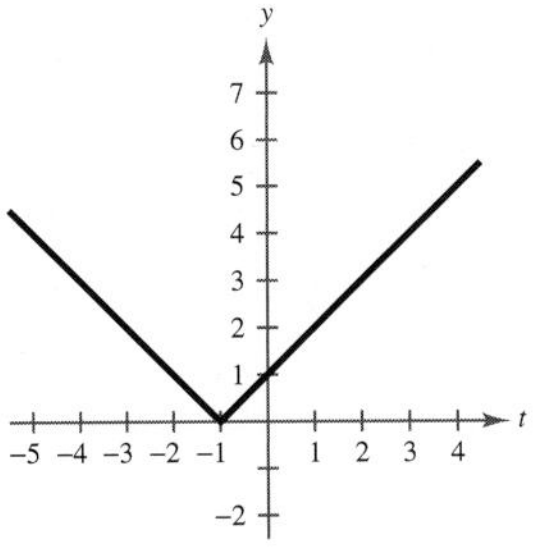

53. (a) $f(x) = 0.4(50 - x) + x = 20 + 0.6x$
(b) Domain: $0 \le x \le 50$; Range: $20 \le y \le 50$
(c) $8\frac{1}{3}$ liters

55. Function **57.** Not a function **59.** $-1, \frac{1}{5}$ **61.** $1, \pm 5$

63.

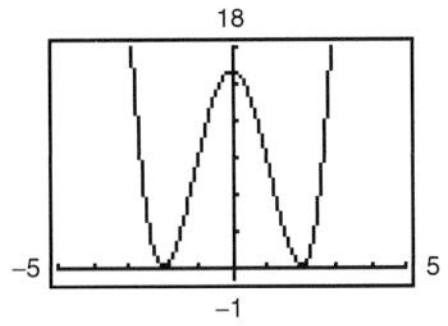

Increasing on $(-2, 0)$ and $(2, \infty)$
Decreasing on $(-\infty, -2)$ and $(0, 2)$

65.

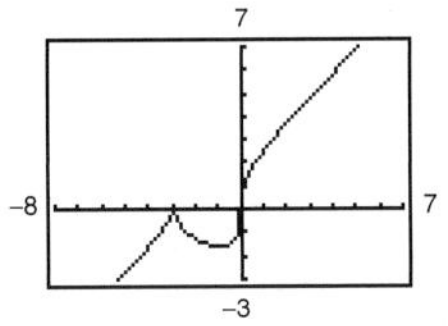

Increasing on $(-\infty, -3)$ and $(-1, \infty)$
Decreasing on $(-3, -1)$

67. $f(x) = -\frac{3}{4}x - 5$

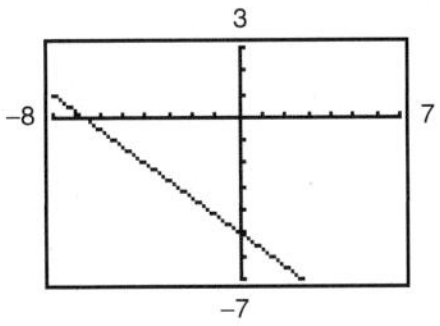

69. $f(x) = \frac{7}{8}x + \frac{217}{80}$

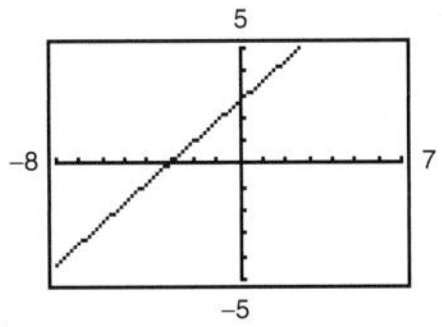

71.

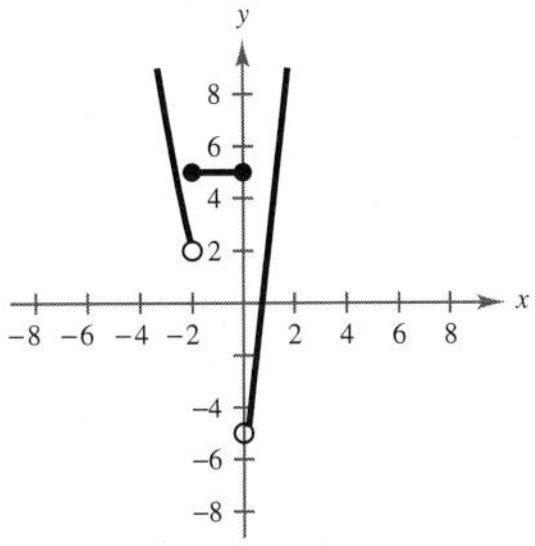

73. Even **75.** Even

77. Reflection in x-axis and horizontal shift of 3 units to the left of $y = x^2$

79. Horizontal shift of 7 units to the right

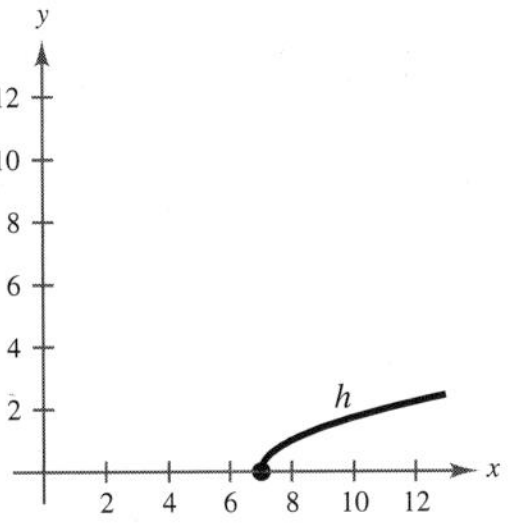

81. Reflection in x-axis, horizontal shift of 3 units to the left, and vertical shift of 1 unit upward

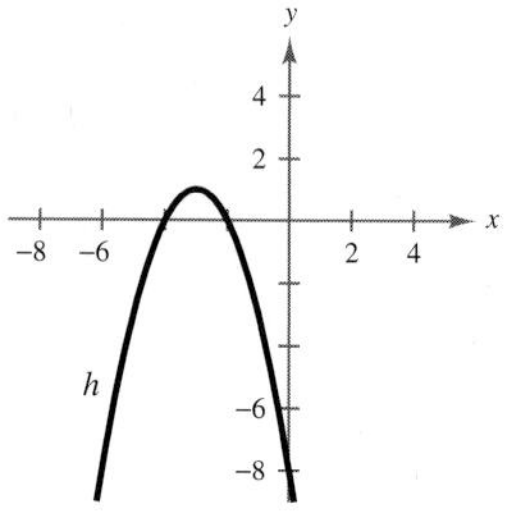

83. Reflection in x-axis and vertical shrink

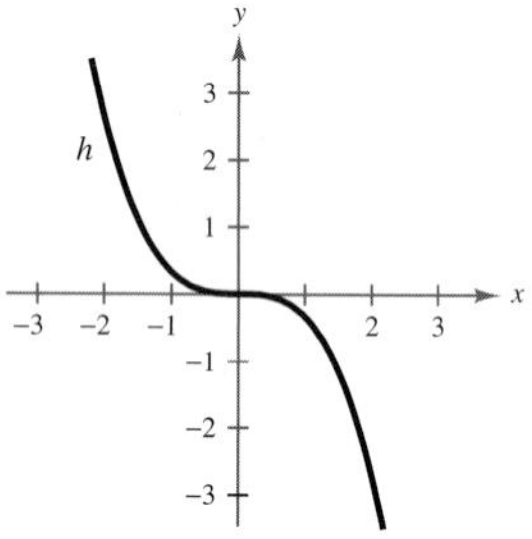

85. $\frac{3}{2}$ **87.** $\sqrt{7}$

89.

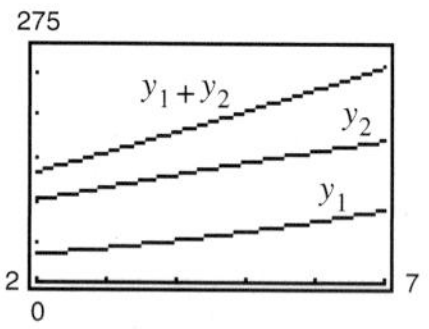

$397.76 billion

91. $f^{-1}(x) = 12x$

$f(f^{-1}(x)) = \frac{1}{12}(12x) = x$

$f^{-1}(f(x)) = 12\left(\frac{1}{12}x\right) = x$

93. $f^{-1}(x) = x - 5$

$f(f^{-1}(x)) = x - 5 + 5 = x$

$f^{-1}(f(x)) = x + 5 - 5 = x$

95.

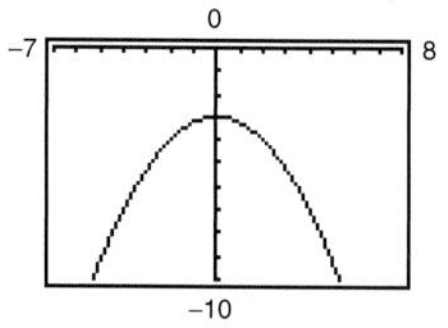

No

97.

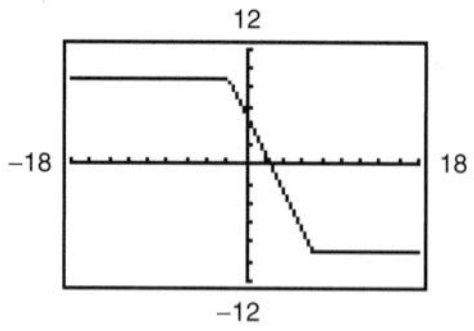

No

99. (a) $f^{-1}(x) = \dfrac{x + 7}{5}$

(b)

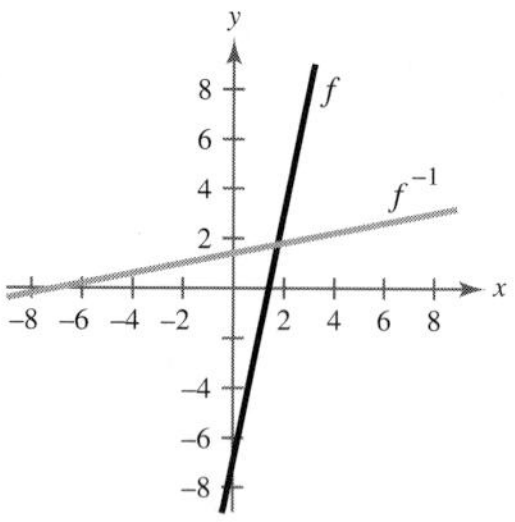

(c) $f^{-1}(f(x)) = f^{-1}(5x - 7)$

$= \dfrac{5x - 7 + 7}{5}$

$= x$

$f(f^{-1}(x)) = f\left(\dfrac{x + 7}{5}\right)$

$= 5\left(\dfrac{x + 7}{5}\right) - 7$

$= x$

101. (a) $f^{-1}(x) = \sqrt[3]{x - 2}$

(b)

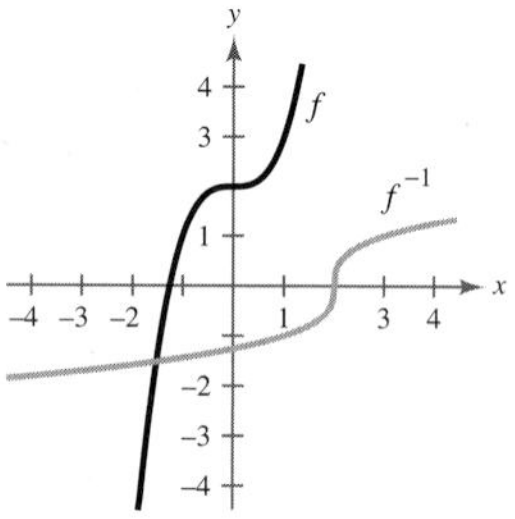

(c) $f^{-1}(f(x)) = f^{-1}(x^3 + 2)$

$= \sqrt[3]{x^3 + 2 - 2}$

$= x$

$f(f^{-1}(x)) = f\left(\sqrt[3]{x - 2}\right)$

$= (x - 2) + 2$

$= x$

103. $x \geq 2;\ f^{-1}(x) = x + 2,\ x \geq 0$

105.

Miles	2	5	10	12
Kilometers	3.2	8	16	19.2

107. A factor of 4 **109.** $F = \frac{1}{3}x\sqrt{y}$

111. False. The graph is reflected in the x-axis, shifted 9 units to the left, and then shifted 13 units down.

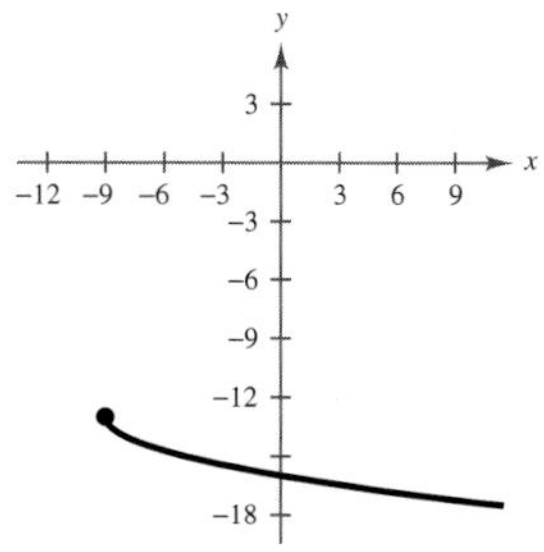

113. A function from a set A to a set B is a relation that assigns to each element x in the set A exactly one element y in the set B.

115. The y-intercept is zero.

Chapter Test *(page 199)*

1.

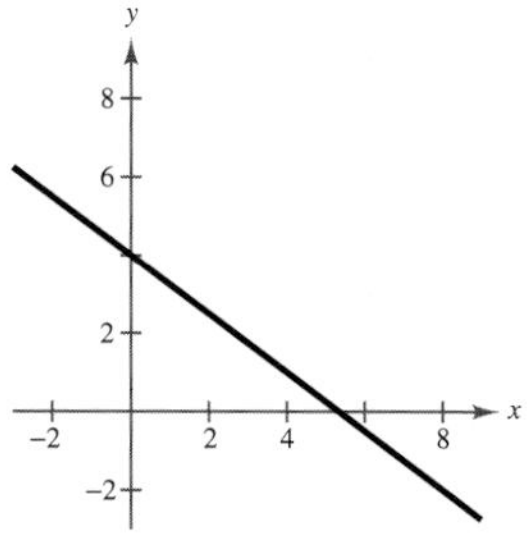

2.

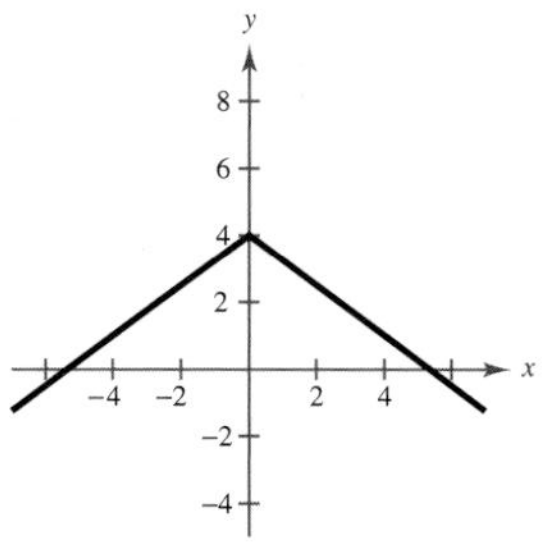

3.

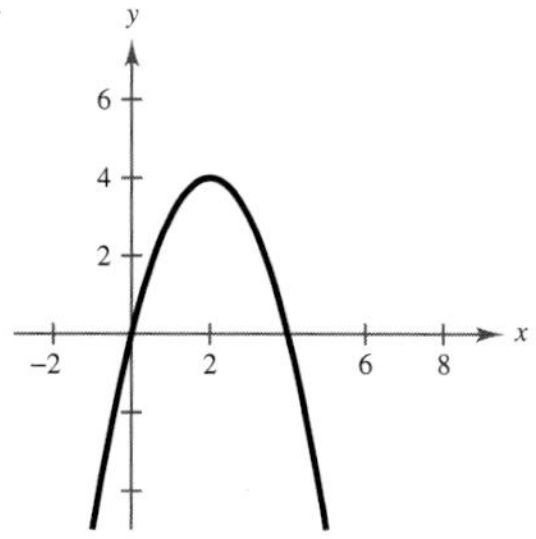

4. $2x + y - 1 = 0$ **5.** $17x + 10y - 59 = 0$

6. (a) $4x - 7y + 44 = 0$ (b) $7x + 4y - 53 = 0$

7. (a) $-\dfrac{1}{8}$ (b) $-\dfrac{1}{28}$ (c) $\dfrac{\sqrt{x}}{x^2 - 18x}$

8. $-10 \le x \le 10$ **9.** All real numbers

10. (a) 0

(b)

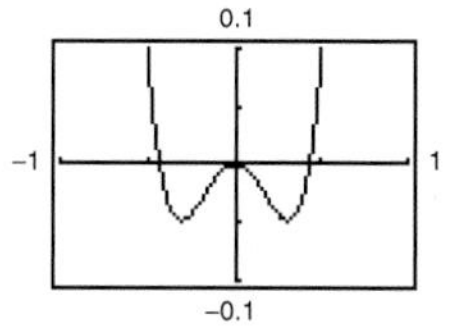

(c) Increasing on $(-0.31, 0)$, $(0.31, \infty)$

Decreasing on $(-\infty, -0.31)$, $(0, 0.31)$

(d) Even

11. (a) 0, 3

(b)

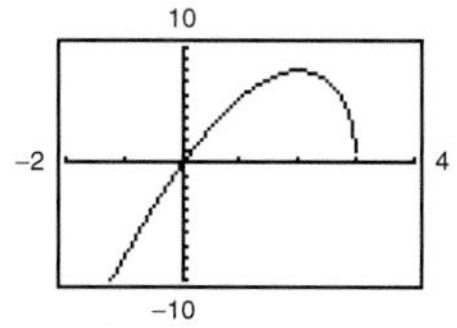

(c) Increasing on $(-\infty, 2)$

Decreasing on $(2, 3]$

(d) Neither

12. (a) -5

(b)

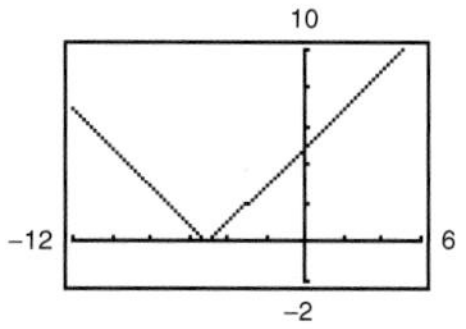

(c) Increasing on $(-5, \infty)$

Decreasing on $(-\infty, -5)$

(d) Neither

13.

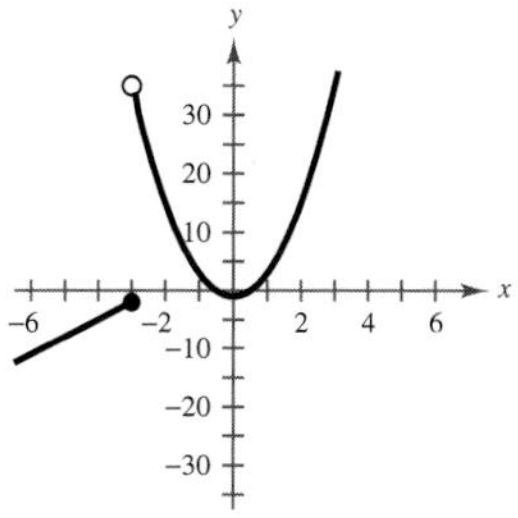

14.

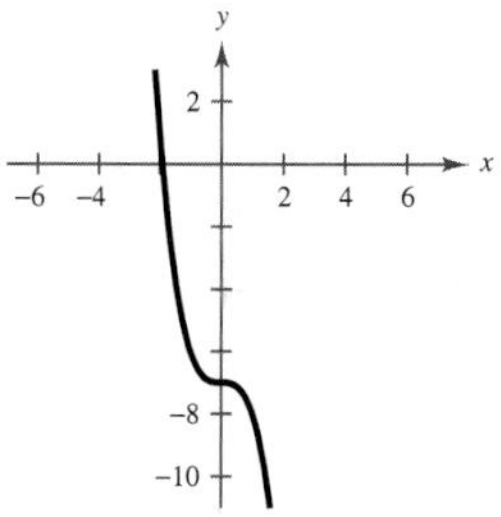

15.

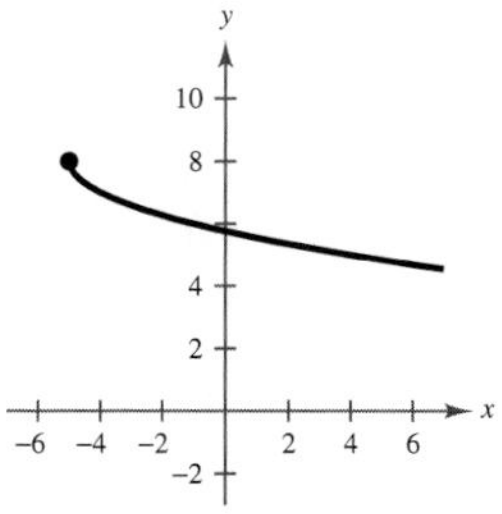

16.

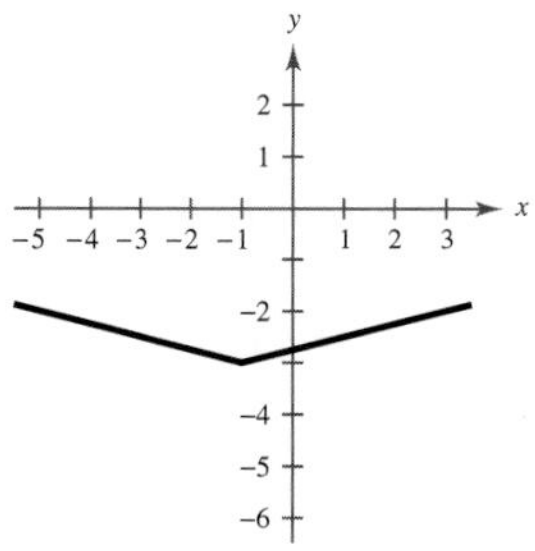

17. -2 **18.** 12 **19.** -35 **20.** 5

21. $f^{-1}(x) = \sqrt[3]{x - 8}$ **22.** No inverse

23. $f^{-1}(x) = \left(\frac{8}{3}x\right)^{2/3}$ **24.** $v = 6\sqrt{5}$ **25.** $A = \frac{25}{6}xy$

26. $b = \dfrac{48}{a}$ **27.** $153

Chapter 2

Section 2.1 *(page 208)*

1. g **3.** b **5.** f **7.** e

9. (a)

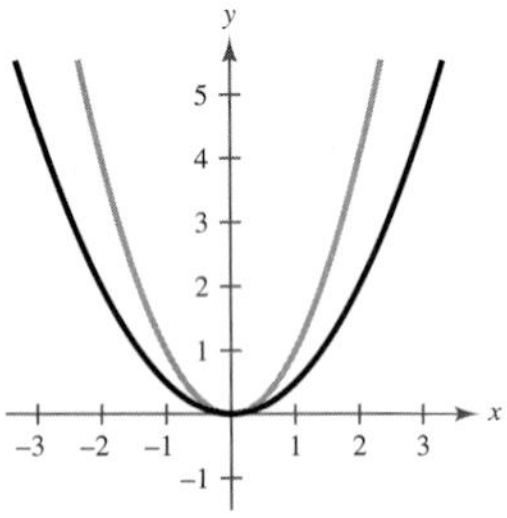

Vertical shrink

(b)

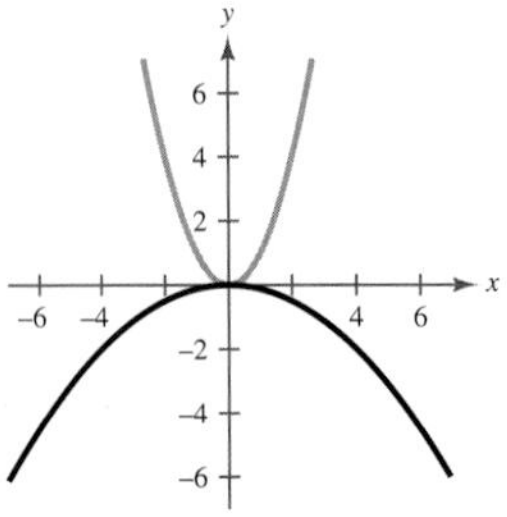

Vertical shrink and reflection in the x-axis

(c)

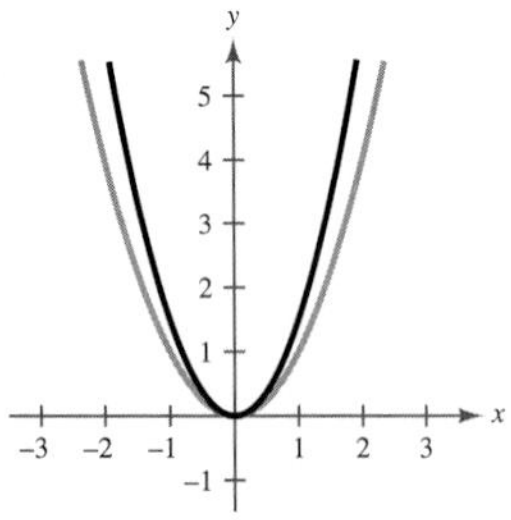

Vertical stretch

(d)

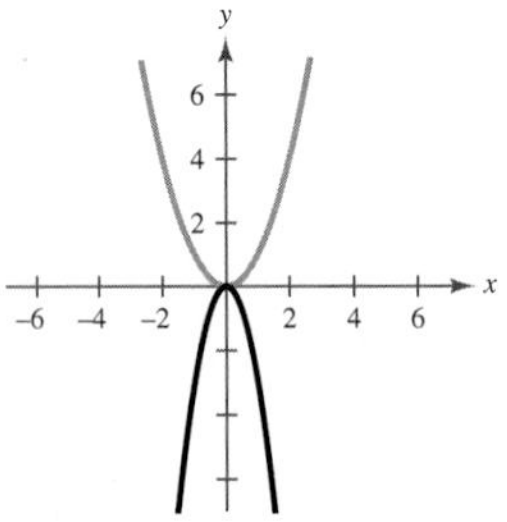

Vertical stretch and reflection in the x-axis

11. (a)

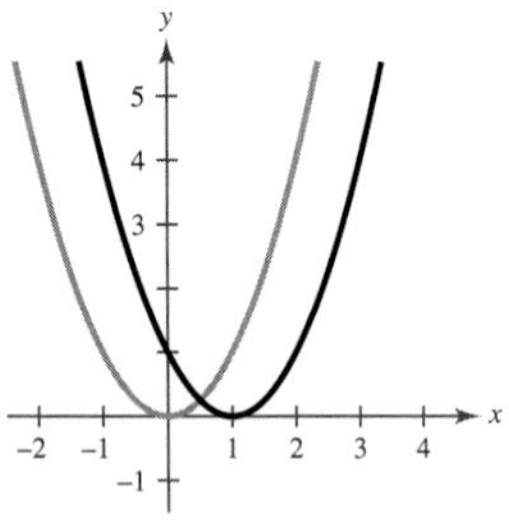

Horizontal translation

(b)

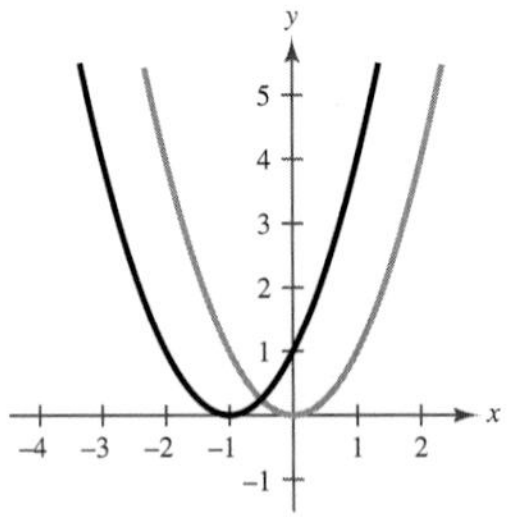

Horizontal translation

(c)

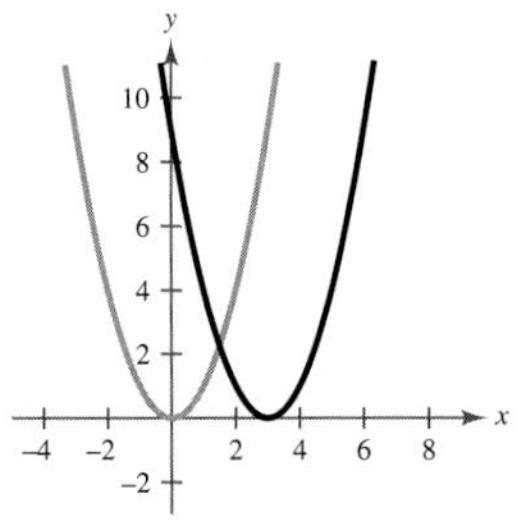

Horizontal translation

(d)

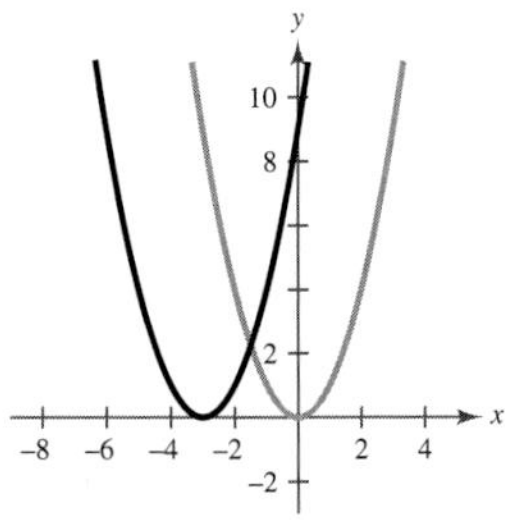

Horizontal translation

13.

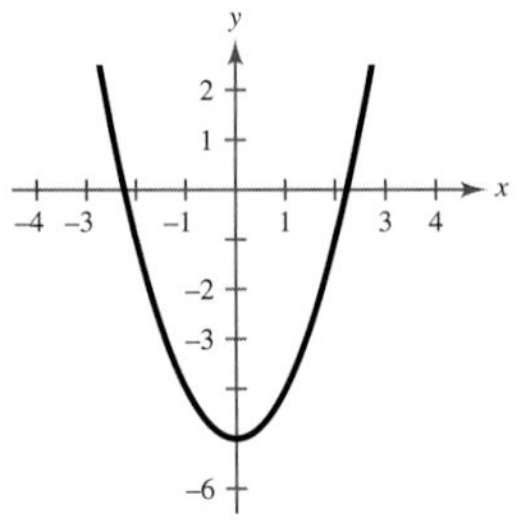

Vertex: $(0, -5)$

x-intercepts: $\left(\pm\sqrt{5}, 0\right)$

15.

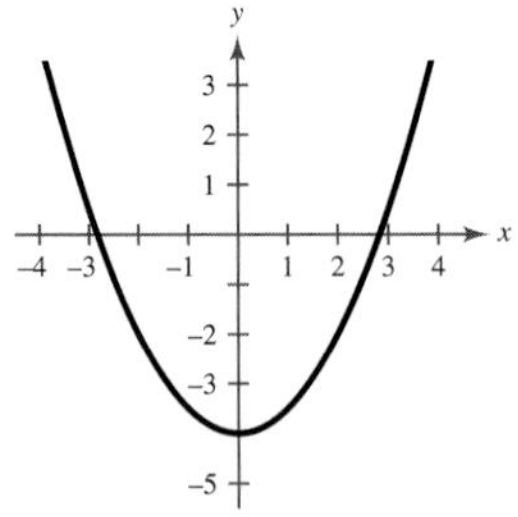

Vertex: $(0, -4)$

x-intercepts: $\left(\pm 2\sqrt{2}, 0\right)$

17.

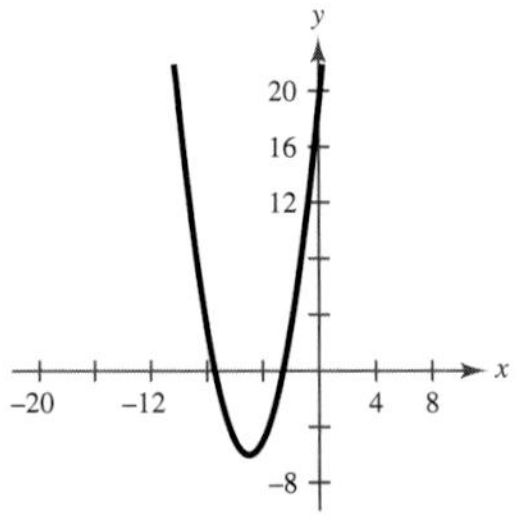

Vertex: $(-5, -6)$

x-intercepts: $\left(-5 \pm \sqrt{6}, 0\right)$

19. 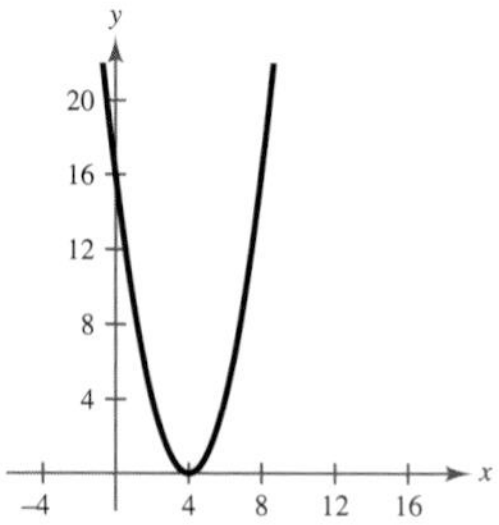

Vertex: $(4, 0)$

x-intercept: $(4, 0)$

21. 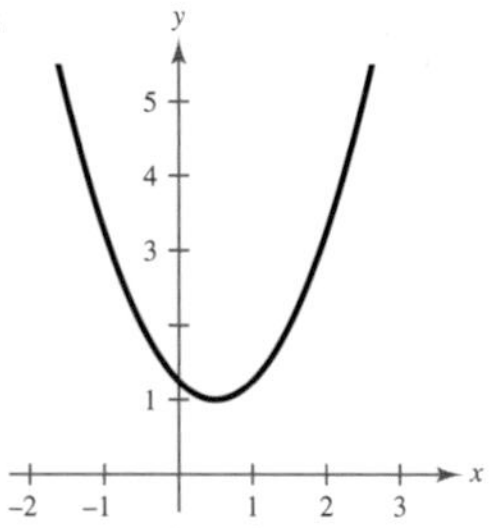

Vertex: $\left(\frac{1}{2}, 1\right)$

No x-intercept

23. 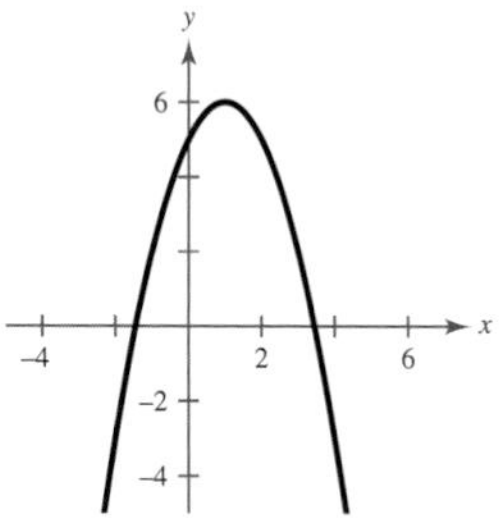

Vertex: $(1, 6)$

x-intercepts: $\left(1 \pm \sqrt{6}, 0\right)$

25. 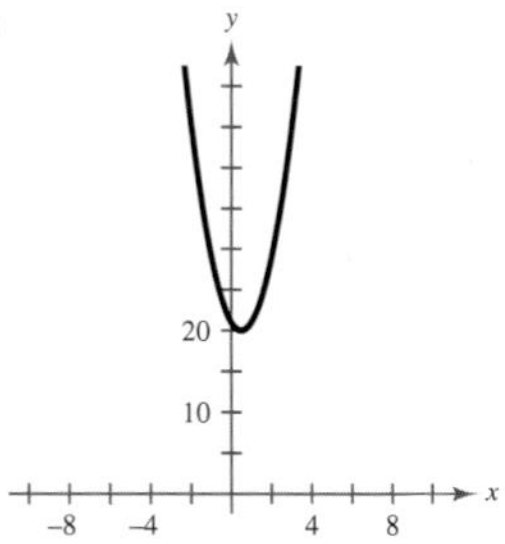

Vertex: $\left(\frac{1}{2}, 20\right)$

No x-intercept

27. 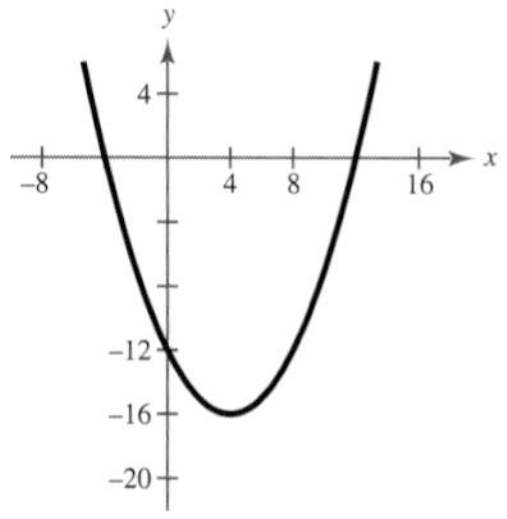

Vertex: $(4, -16)$

x-intercepts: $(-4, 0), (12, 0)$

29. 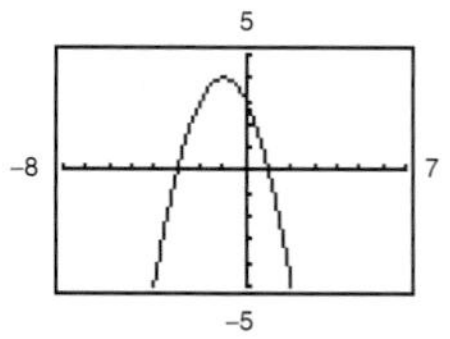

Vertex: $(-1, 4)$

x-intercepts: $(1, 0), (-3, 0)$

31. 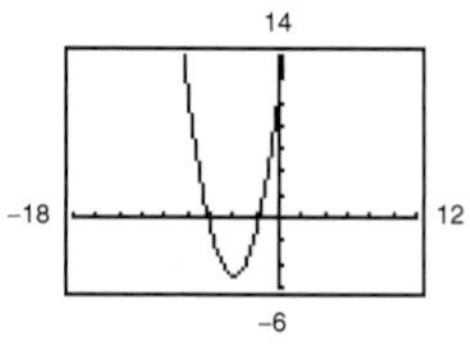

Vertex: $(-4, -5)$

x-intercepts: $\left(-4 \pm \sqrt{5}, 0\right)$

33. 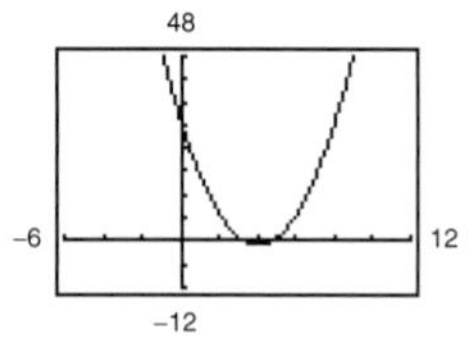

Vertex: $(4, -1)$

x-intercepts: $\left(4 \pm \frac{1}{2}\sqrt{2}, 0\right)$

35. 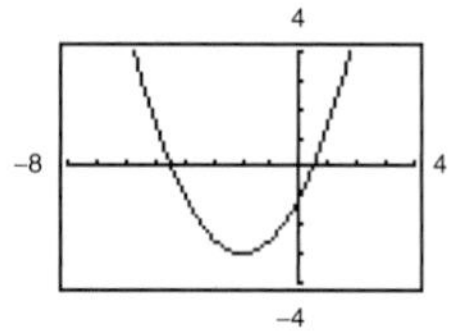

Vertex: $(-2, -3)$

x-intercepts: $\left(-2 \pm \sqrt{6}, 0\right)$

37. $y = (x - 1)^2$ **39.** $y = -(x + 1)^2 + 4$

41. $y = -2(x + 2)^2 + 2$ **43.** $f(x) = (x + 2)^2 + 5$

45. $f(x) = -\frac{1}{2}(x - 3)^2 + 4$ **47.** $f(x) = \frac{3}{4}(x - 5)^2 + 12$

49. $f(x) = -\frac{24}{49}\left(x + \frac{1}{4}\right)^2 + \frac{3}{2}$ **51.** $f(x) = -\frac{16}{3}\left(x + \frac{5}{2}\right)^2$

53. $(\pm 4, 0)$. The x-intercepts and solutions of the equation are the same.

55. $(5, 0), (-1, 0)$. The x-intercepts and solutions of the equation are the same.

57. 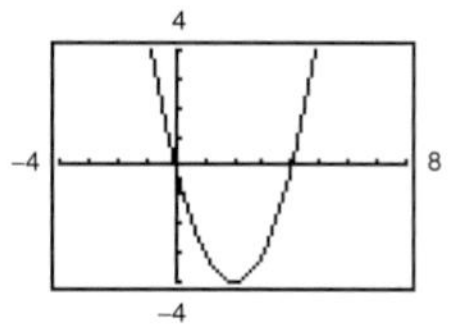

$(0, 0), (4, 0)$

59. 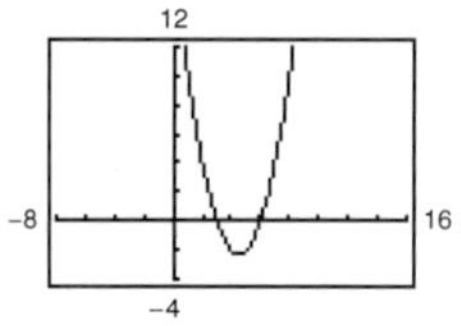

$(3, 0), (6, 0)$

61. 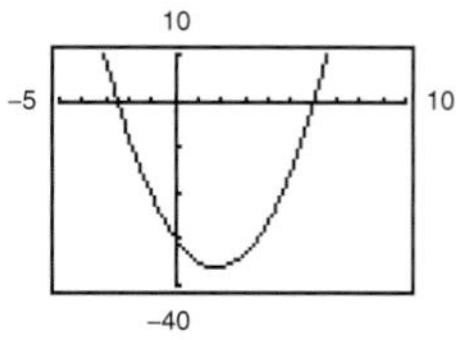

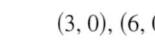

$\left(-\frac{5}{2}, 0\right), (6, 0)$

63. 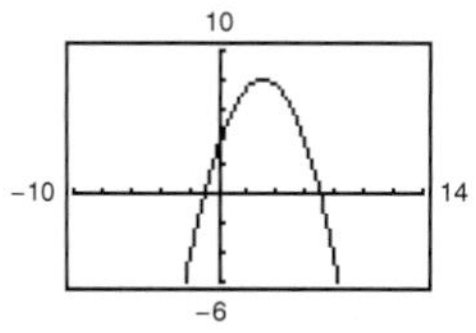

$(7, 0), (-1, 0)$

65. $f(x) = x^2 - 2x - 3$
$g(x) = -x^2 + 2x + 3$

67. $f(x) = x^2 - 10x$
$g(x) = -x^2 + 10x$

69. $f(x) = 2x^2 + 7x + 3$
$g(x) = -2x^2 - 7x - 3$

71. 55, 55 **73.** 12, 6

75. (a) $A = x(50 - x),\ 0 < x < 50$

(b)

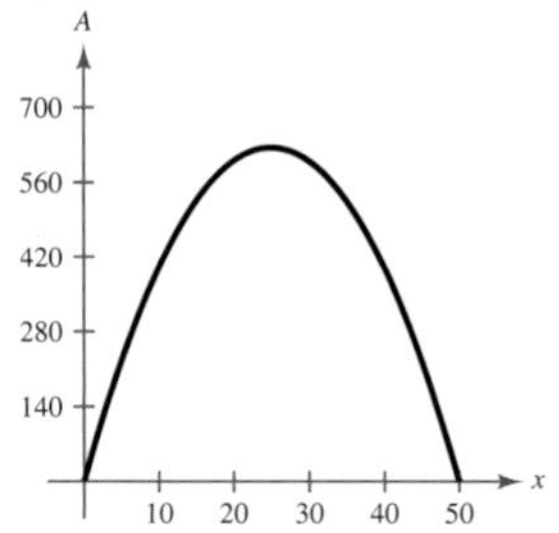

(c) 25 feet × 25 feet

77. (a)

x	y	Area
2	$\frac{1}{3}[200 - 4(2)]$	$(2)(2)(\frac{1}{3})[200 - 4(2)] = 256$
4	$\frac{1}{3}[200 - 4(4)]$	$(2)(4)(\frac{1}{3})[200 - 4(4)] \approx 491$
6	$\frac{1}{3}[200 - 4(6)]$	$(2)(6)(\frac{1}{3})[200 - 4(6)] = 704$
8	$\frac{1}{3}[200 - 4(8)]$	$(2)(8)(\frac{1}{3})[200 - 4(8)] = 896$
10	$\frac{1}{3}[200 - 4(10)]$	$(2)(10)(\frac{1}{3})[200 - 4(10)] \approx 1067$
12	$\frac{1}{3}[200 - 4(12)]$	$(2)(12)(\frac{1}{3})[200 - 4(12)] = 1216$

(b)

x	y	Area
20	$\frac{1}{3}[200 - 4(20)]$	$(2)(20)(\frac{1}{3})[200 - 4(20)] = 1600$
22	$\frac{1}{3}[200 - 4(22)]$	$(2)(22)(\frac{1}{3})[200 - 4(22)] \approx 1643$
24	$\frac{1}{3}[200 - 4(24)]$	$(2)(24)(\frac{1}{3})[200 - 4(24)] = 1664$
26	$\frac{1}{3}[200 - 4(26)]$	$(2)(26)(\frac{1}{3})[200 - 4(26)] = 1664$
28	$\frac{1}{3}[200 - 4(28)]$	$(2)(28)(\frac{1}{3})[200 - 4(28)] \approx 1643$
30	$\frac{1}{3}[200 - 4(30)]$	$(2)(30)(\frac{1}{3})[200 - 4(30)] = 1600$

$x = 25$ ft, $y = 33\frac{1}{3}$ ft

(c) $A = \dfrac{8x(50 - x)}{3}$

(d)

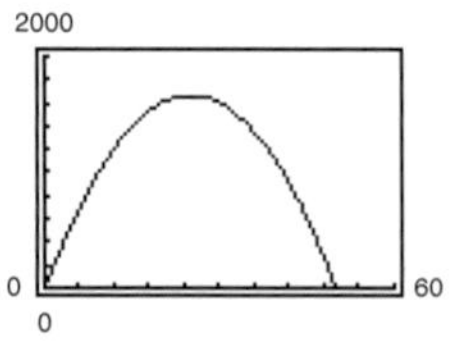

$x = 25$ feet; $y = 33\frac{1}{3}$ feet

(e) $A = -\frac{8}{3}(x - 25)^2 + \frac{5000}{3}$

79. 4500 units **81.** 20 fixtures **83.** 350,000 units

85. (a) 4 feet (b) 16 feet (c) 25.86 feet

87. 16 feet

89. (a)

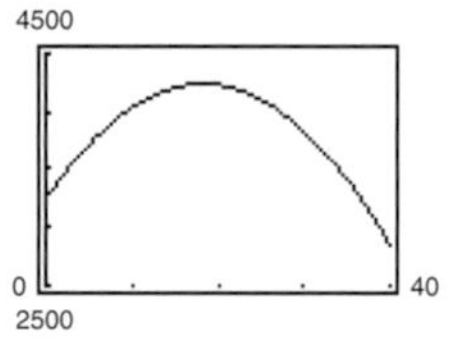

(b) 4242; Yes (c) 9703 annually; 27 daily

91. (a) and (c)

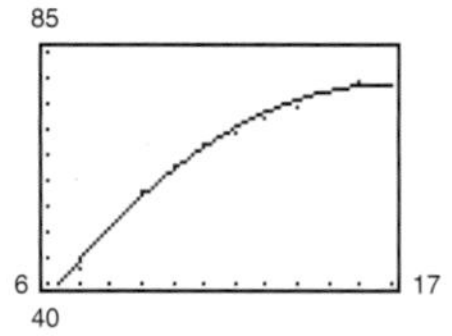

(b) $y = -0.35t^2 + 11.8t - 21$

(d) No. The model decreases and eventually becomes negative.

93. True. The vertex of $f(x)$ is $\left(-\frac{5}{4}, \frac{53}{4}\right)$ and the vertex of $g(x)$ is $\left(-\frac{5}{4}, -\frac{71}{4}\right)$.

95. Conditions (a) and (d) are preferable because profits would be increasing.

97. Answers will vary. **99.** $y = -\frac{1}{3}x + \frac{5}{3}$

101. $y = \frac{5}{4}x + 3$ **103.** 27 **105.** $-\frac{1408}{49}$ **107.** 109

Section 2.2 *(page 222)*

1. c **3.** h **5.** a **7.** d

9. (a)

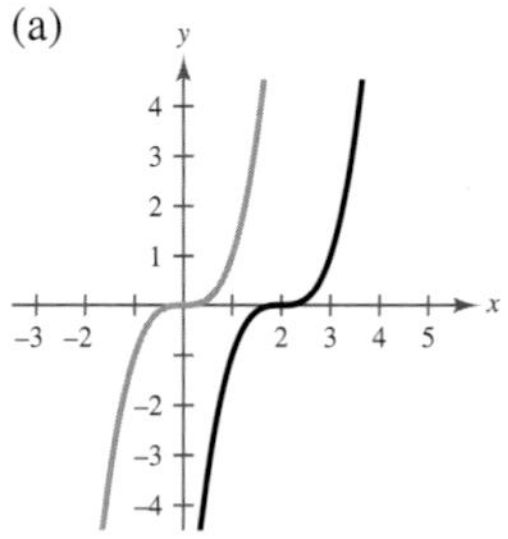

(b)

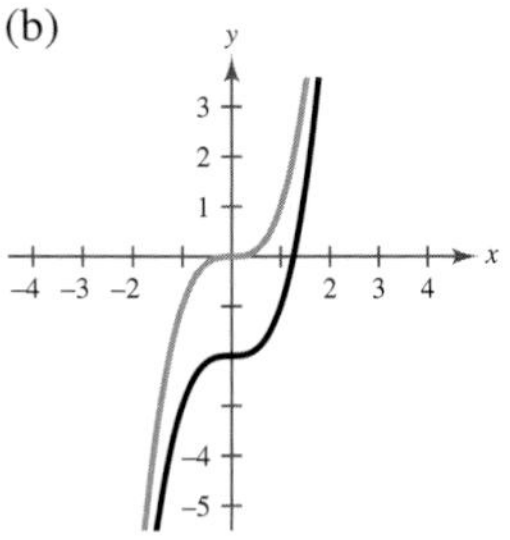

(c)

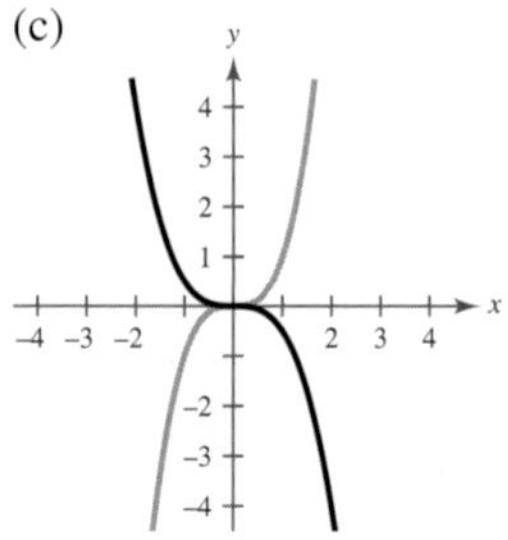

(d)

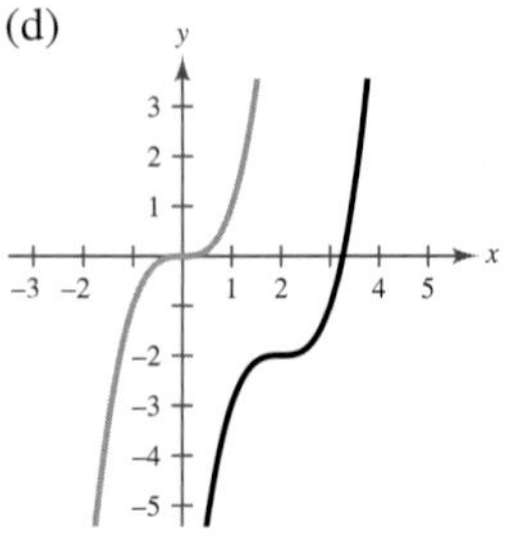

11. (a)

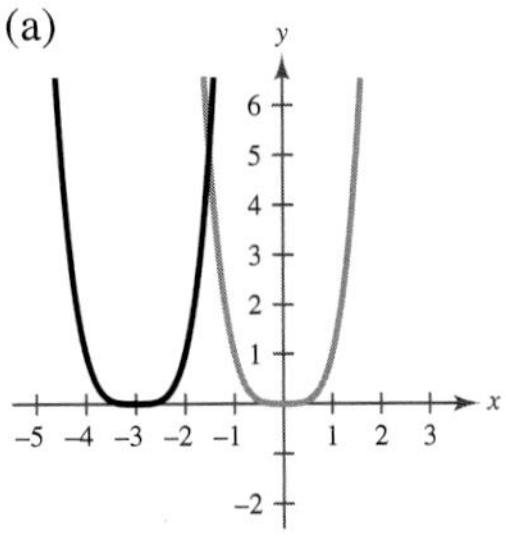

(b)

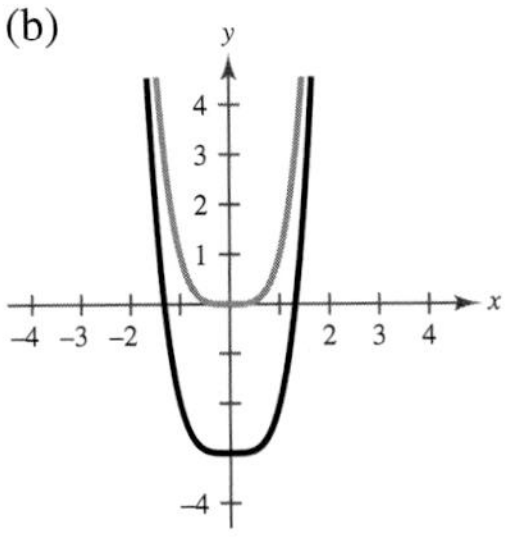

(c)

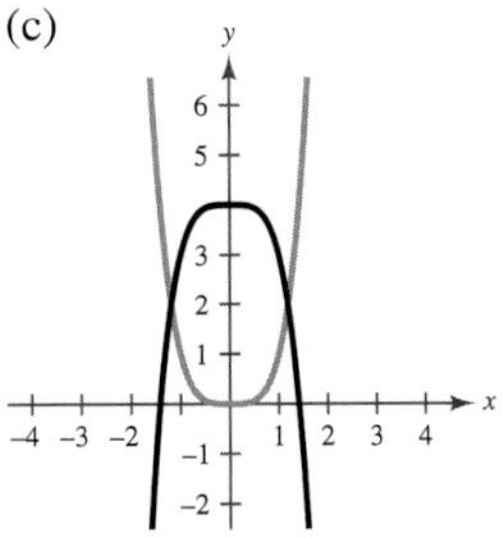

(d)

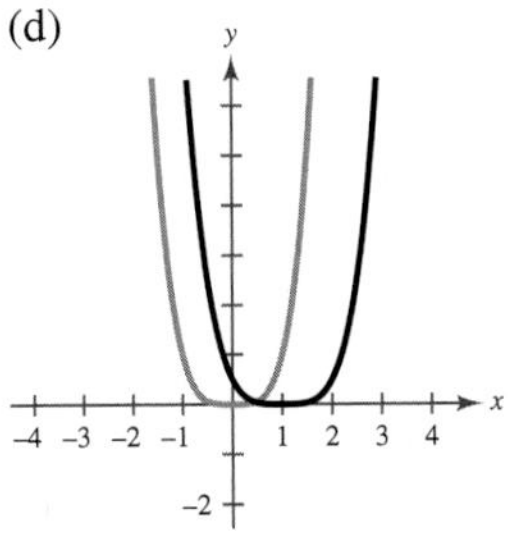

13. Falls to the left.
Rises to the right.

15. Falls to the left.
Falls to the right.

17. Rises to the left.
Falls to the right.

19. Rises to the left.
Falls to the right.

21. Falls to the left.
Falls to the right.

23.

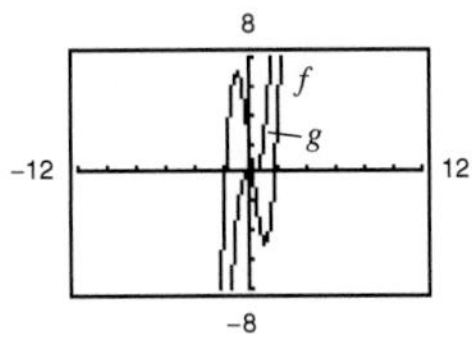

25.

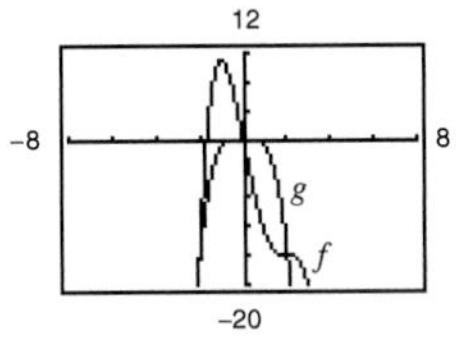

27. ± 5 **29.** 3 **31.** $1, -2$ **33.** $2 \pm \sqrt{3}$

35. 2, 0 **37.** ± 1 **39.** $0, \pm\sqrt{3}$ **41.** No real zeros

43.

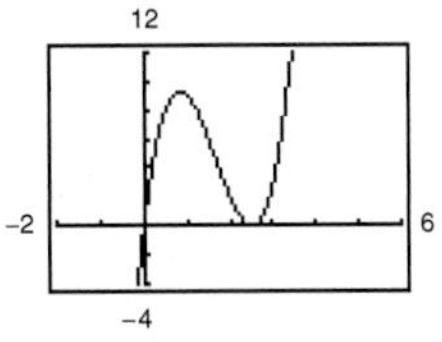

$(0, 0), \left(\frac{5}{2}, 0\right)$

45.

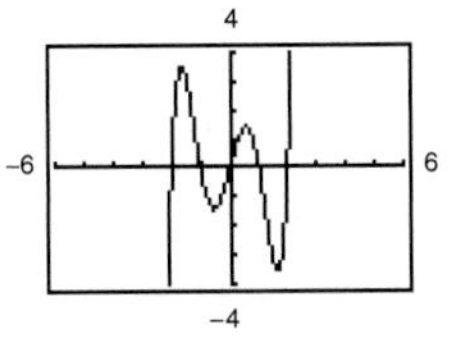

$(0, 0), (\pm 1, 0), (\pm 2, 0)$

47. $f(x) = x^2 - 10x$ **49.** $f(x) = x^2 + 4x - 12$

51. $f(x) = x^3 + 5x^2 + 6x$

53. $f(x) = x^4 - 4x^3 - 9x^2 + 36x$

55. $f(x) = x^2 - 2x - 2$ **57.** $f(x) = x^2 + 4x + 4$

59. $f(x) = x^3 + 2x^2 - 3x$ **61.** $f(x) = x^3 - 3x$

63. $f(x) = x^4 + x^3 - 15x^2 + 23x - 10$

65. $f(x) = x^5 + 16x^4 + 96x^3 + 256x^2 + 256x$

67.

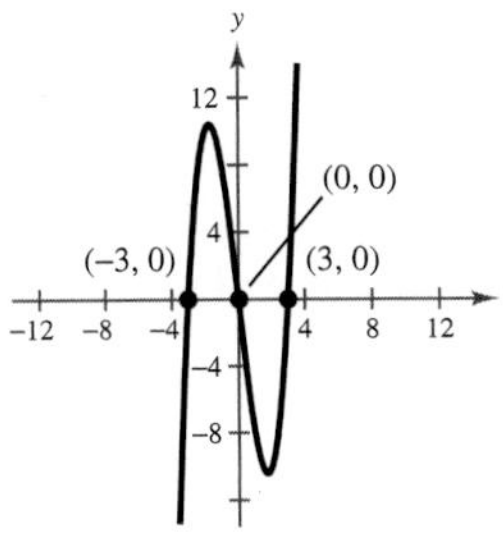

69.

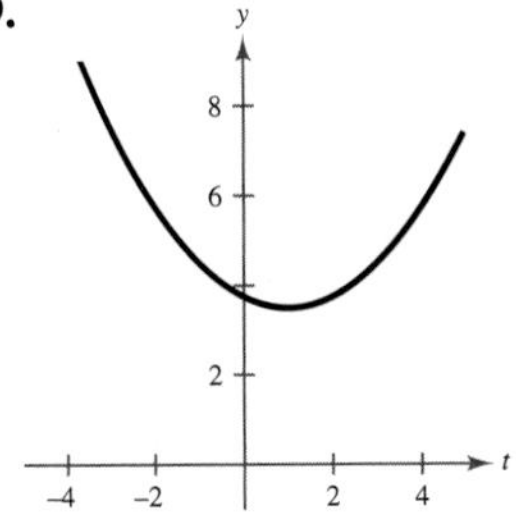

71.

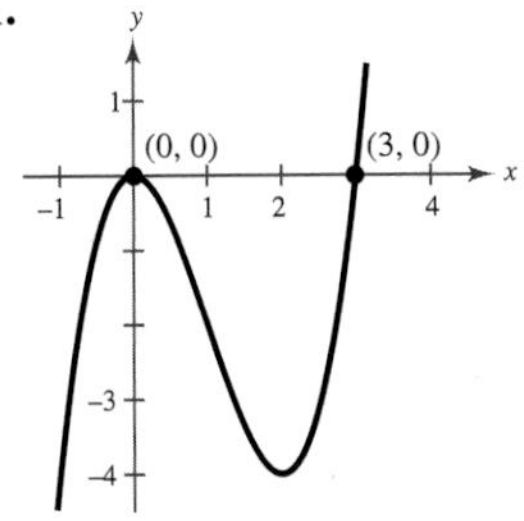

73.

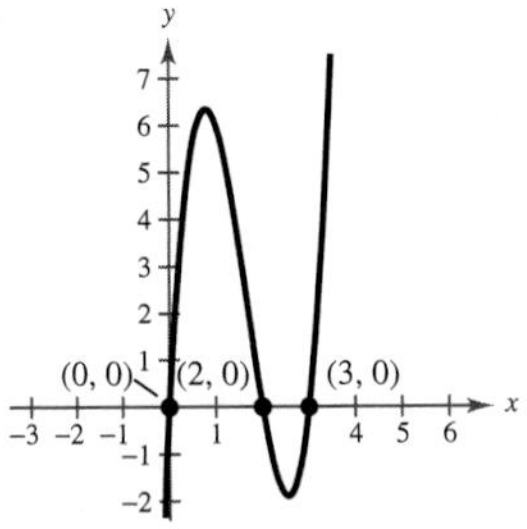

75.

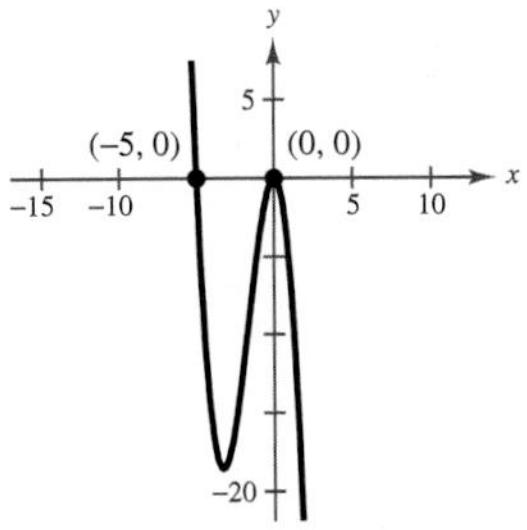

77.

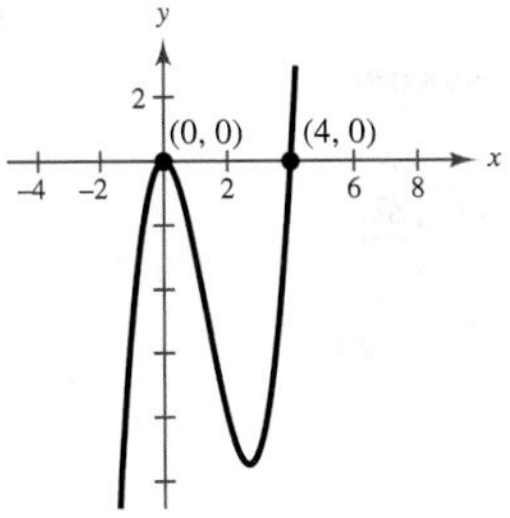

79.

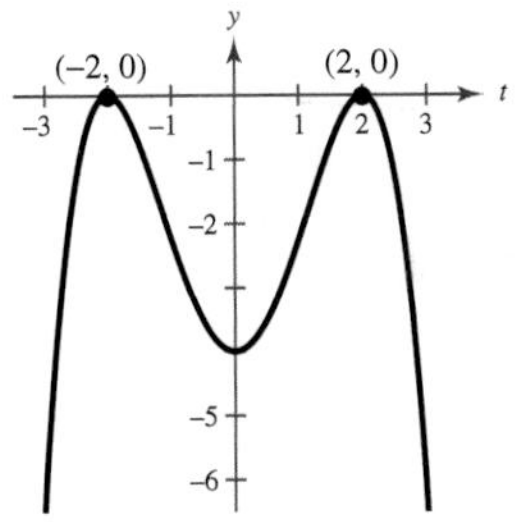

81.

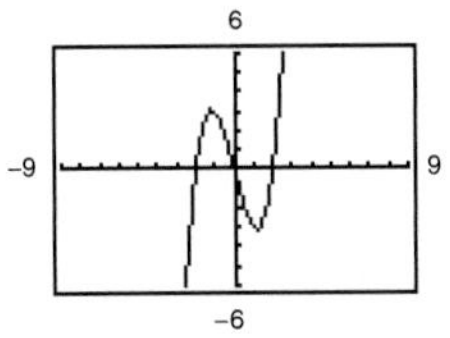

Zeros: $0, \pm 2$, all of odd multiplicity

83.

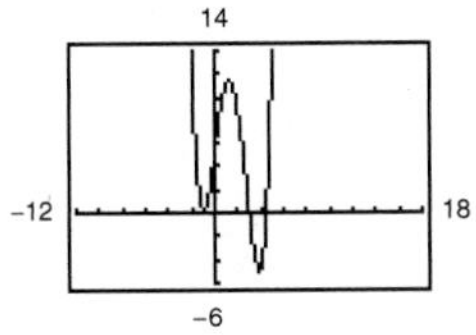

Zeros: -1, even multiplicity; $3, \frac{9}{2}$, odd multiplicity

85. $(-1, 0), (1, 2), (2, 3)$ **87.** $(-2, -1), (0, 1)$

89. (a) and (b)

Box height	Box width	Box volume
1	$36 - 2(1)$	$1[36 - 2(1)]^2 = 1156$
2	$36 - 2(2)$	$2[36 - 2(2)]^2 = 2048$
3	$36 - 2(3)$	$3[36 - 2(3)]^2 = 2700$
4	$36 - 2(4)$	$4[36 - 2(4)]^2 = 3136$
5	$36 - 2(5)$	$5[36 - 2(5)]^2 = 3380$
6	$36 - 2(6)$	$6[36 - 2(6)]^2 = 3456$
7	$36 - 2(7)$	$7[36 - 2(7)]^2 = 3388$

6 in. × 24 in. × 24 in.

(c) Domain: $0 < x < 18$

(d)

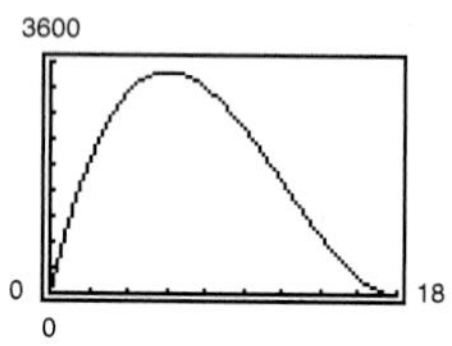

$x = 6$

91. $x = 200$

93. False. A fifth-degree polynomial can have at most four turning points.

95. (a) Degree: 3; Leading coefficient: positive

(b) Degree: 2; Leading coefficient: positive

(c) Degree: 4; Leading coefficient: positive

(d) Degree: 5; Leading coefficient: positive

97. (a)

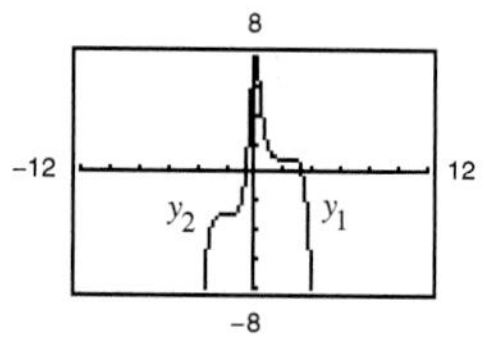

y_1 is decreasing. y_2 is increasing.

(b) Either always increasing or always decreasing. The behavior is determined by a.

(c)

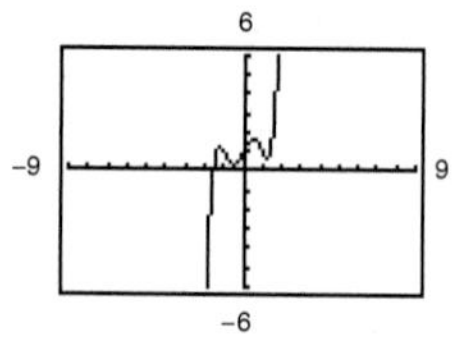

Because $H(x)$ is not always increasing or always decreasing, H cannot be written in the form $H(x) = a(x - h)^5 + k$.

99. $-\frac{2}{3}, 8$ **101.** -12 **103.** $4 \pm \sqrt{14}$

105. $\dfrac{-2 \pm \sqrt{31}}{3}$ **107.** $x(6x - 1)(x - 10)$

109. $(y + 6)(y^2 - 6y + 36)$

Section 2.3 *(page 233)*

1. Answers will vary. **3.** Answers will vary.

5.

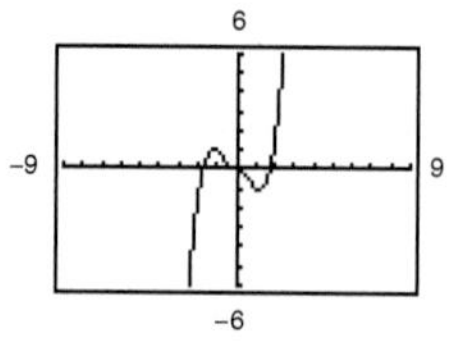

7. $2x + 4$ **9.** $x^2 - 3x + 1$ **11.** $x^3 + 3x^2 - 1$

13. $7 - \dfrac{11}{x + 2}$ **15.** $3x + 5 - \dfrac{2x - 3}{2x^2 + 1}$

17. $x^2 + 2x + 4 + \dfrac{2x - 11}{x^2 - 2x + 3}$

19. $x + 3 + \dfrac{6x^2 - 8x + 3}{(x - 1)^3}$ **21.** $3x^2 - 2x + 5$

23. $4x^2 - 9$ **25.** $-x^2 + 10x - 25$

27. $5x^2 + 14x + 56 + \dfrac{232}{x - 4}$

29. $10x^3 + 10x^2 + 60x + 360 + \dfrac{1360}{x - 6}$

31. $x^2 - 8x + 64$

33. $-3x^3 - 6x^2 - 12x - 24 - \dfrac{48}{x - 2}$

35. $-x^3 - 6x^2 - 36x - 36 - \dfrac{216}{x - 6}$

37. $4x^2 + 14x - 30$

39. $f(x) = (x - 4)(x^2 + 3x - 2) + 3,\ \ f(4) = 3$

41. $f(x) = \left(x + \frac{2}{3}\right)(15x^3 - 6x + 4) + \frac{34}{3}$, $f\left(-\frac{2}{3}\right) = \frac{34}{3}$

43. $f(x) = \left(x - \sqrt{2}\right)\left[x^2 + \left(3 + \sqrt{2}\right)x + 3\sqrt{2}\right] - 8$, $f\left(\sqrt{2}\right) = -8$

45. $f(x) = \left(x - 1 + \sqrt{3}\right)\left[-4x^2 + \left(2 + 4\sqrt{3}\right)x + \left(2 + 2\sqrt{3}\right)\right]$, $f\left(1 - \sqrt{3}\right) = 0$

47. (a) 1 (b) 4 (c) 4 (d) 1954

49. (a) 97 (b) $-\frac{5}{3}$ (c) 17 (d) -199

51. $(x - 2)(x + 3)(x - 1)$; Zeros: $2, -3, 1$

53. $(2x - 1)(x - 5)(x - 2)$; Zeros: $\frac{1}{2}, 5, 2$

55. $\left(x + \sqrt{3}\right)\left(x - \sqrt{3}\right)(x + 2)$; Zeros: $-\sqrt{3}, \sqrt{3}, -2$

57. $(x-1)(x-1-\sqrt{3})(x-1+\sqrt{3})$;
Zeros: $1,\ 1+\sqrt{3},\ 1-\sqrt{3}$

59. (a) Answers will vary. (b) $2x-1$
(c) $f(x)=(2x-1)(x+2)(x-1)$ (d) $\frac{1}{2}, -2, 1$
(e)

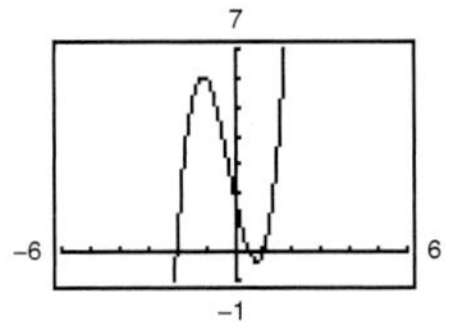

61. (a) Answers will vary. (b) $(x-1), (x-2)$
(c) $f(x)=(x-1)(x-2)(x-5)(x+4)$
(d) $1, 2, 5, -4$
(e)

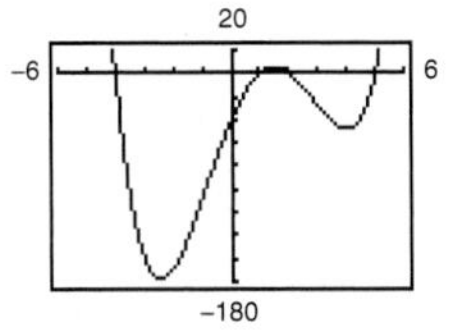

63. (a) Answers will vary. (b) $x+7$
(c) $f(x)=(x+7)(2x+1)(3x-2)$ (d) $-7, -\frac{1}{2}, \frac{2}{3}$
(e)

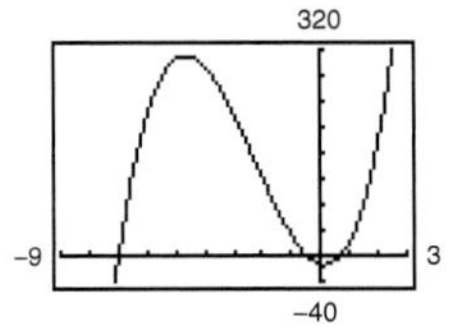

65. (a) Answers will vary. (b) $(x-\sqrt{5})$
(c) $f(x)=(x-\sqrt{5})(x+\sqrt{5})(2x-1)$ (d) $\pm\sqrt{5}, \frac{1}{2}$
(e)

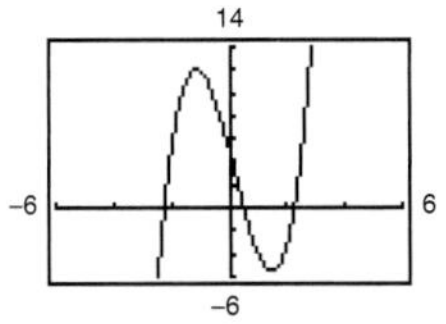

67. (a) Zeros are 2 and $\approx\pm 2.236$.
(b) $f(x)=(x-2)(x-\sqrt{5})(x+\sqrt{5})$

69. (a) Zeros are -2, ≈ 0.268, and ≈ 3.732.
(b) $h(t)=(t+2)[t-(2+\sqrt{3})][t-(2-\sqrt{3})]$

71. $2x^2-x-1,\ x\neq\frac{3}{2}$ **73.** $x^2+2x-3,\ x\neq -1$

75. $x^2+3x,\ x\neq -2, -1$

77. (a) and (b)

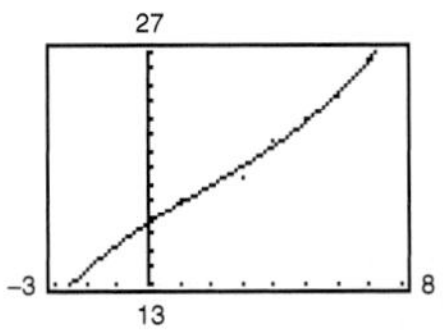

$R=0.01326t^3-0.0677t^2+1.231t+16.68$

(c)

t	-2	-1	0	1	2
R	13.84	15.37	16.68	17.86	18.98

t	3	4	5	6	7
R	20.12	21.37	22.80	24.49	26.53

(d) $44.62. No, because the model will approach infinity quickly.

79. False. $-\frac{4}{7}$ is a zero of f. **81.** $x^{2n}+6x^n+9$

83. The remainder is 0. **85.** $c=-210$

87. 0; $x+3$ is a factor of f.

89. (a) $f(x)=(x-2)x^2+5=x^3-2x^2+5$
(b) $f(x)=-(x+3)x^2+1=-x^3-3x^2+1$

91. $\pm\dfrac{\sqrt{21}}{4}$ **93.** $\dfrac{5}{4}, \dfrac{3}{2}$ **95.** $\dfrac{-3\pm\sqrt{21}}{2}$

97. $f(x)=x^2+5x-6$
(Answer is not unique.)

99. $f(x)=x^4-3x^3-5x^2+9x-2$
(Answer is not unique.)

Section 2.4 *(page 241)*

1. $a=-10,\ b=6$ **3.** $a=6,\ b=5$ **5.** $4+3i$
7. $2-3\sqrt{3}i$ **9.** $5\sqrt{3}i$ **11.** 8 **13.** $-1-6i$
15. $0.3i$ **17.** $11-i$ **19.** 4 **21.** $3-3\sqrt{2}i$
23. $-14+20i$ **25.** $\frac{1}{6}+\frac{7}{6}i$ **27.** $-2\sqrt{3}$ **29.** -10
31. $5+i$ **33.** $12+30i$ **35.** 24 **37.** $-9+40i$
39. -10 **41.** $6-3i$, 45 **43.** $-1+\sqrt{5}i$, 6
45. $-2\sqrt{5}i$, 20 **47.** $\sqrt{8}$, 8 **49.** $-5i$ **51.** $\frac{8}{41}+\frac{10}{41}i$
53. $\frac{4}{5}+\frac{3}{5}i$ **55.** $-5-6i$ **57.** $-\frac{120}{1681}-\frac{27}{1681}i$
59. $-\frac{1}{2}-\frac{5}{2}i$ **61.** $\frac{62}{949}+\frac{297}{949}i$ **63.** $1\pm i$
65. $-2\pm\frac{1}{2}i$ **67.** $-\frac{3}{2}, -\frac{5}{2}$ **69.** $2\pm\sqrt{2}i$
71. $\dfrac{5}{7}\pm\dfrac{5\sqrt{15}}{7}$ **73.** (a) 1 (b) i (c) -1 (d) $-i$
75. $-4+2i$ **77.** i **79.** -8 **81.** $\frac{1}{8}i$
83. (a) 16 (b) 16 (c) 16 (d) 16

85. False. If the complex number is real, the number equals its conjugate.

87. False.
$i^{44}+i^{150}-i^{74}-i^{109}+i^{61}=1-1+1-i+i=1$

89. Answers will vary. **91.** $-x^2-3x+12$

93. $3x^2+\frac{23}{2}x-2$ **95.** -31 **97.** $\frac{27}{2}$

99. $a=\dfrac{\sqrt{3V\pi b}}{2\pi b}$ **101.** 1 liter

Section 2.5 *(page 253)*

1. 0, 6 **3.** 2, -4 **5.** $-6, \pm i$ **7.** $\pm 1, \pm 3$

9. $\pm 1, \pm 3, \pm 5, \pm 9, \pm 15, \pm 45, \pm\frac{1}{2}, \pm\frac{3}{2}, \pm\frac{5}{2}, \pm\frac{9}{2}, \pm\frac{15}{2}, \pm\frac{45}{2}$

11. 1, 2, 3 **13.** 1, -1, 4 **15.** $-1, -10$

17. $\frac{1}{2}, -1$ **19.** $-2, 3, \pm\frac{2}{3}$ **21.** $-1, 2$ **23.** $-6, \frac{1}{2}, 1$

25. (a) $\pm 1, \pm 2, \pm 4$

(b)

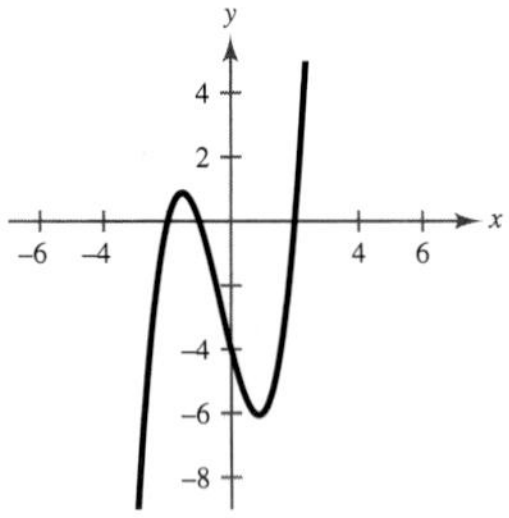

(c) $-2, -1, 2$

27. (a) $\pm 1, \pm 3, \pm\frac{1}{2}, \pm\frac{3}{2}, \pm\frac{1}{4}, \pm\frac{3}{4}$

(b)

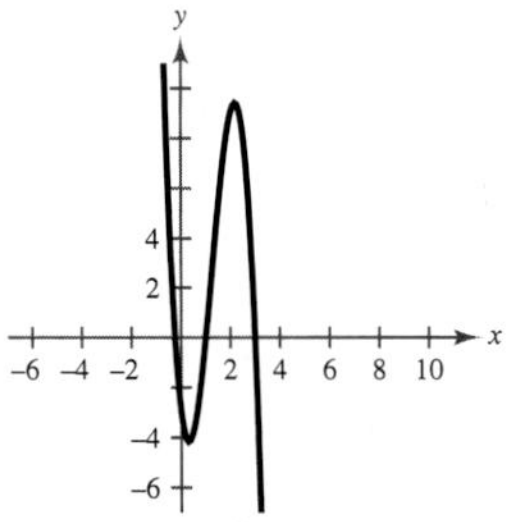

(c) $-\frac{1}{4}, 1, 3$

29. (a) $\pm 1, \pm 2, \pm 4, \pm 8, \pm\frac{1}{2}$

(b)

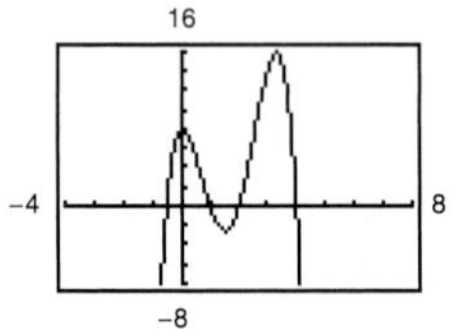

(c) $-\frac{1}{2}, 1, 2, 4$

31. (a) $\pm 1, \pm 3, \pm\frac{1}{2}, \pm\frac{3}{2}, \pm\frac{1}{4}, \pm\frac{3}{4}, \pm\frac{1}{8}, \pm\frac{3}{8}, \pm\frac{1}{16}, \pm\frac{3}{16}, \pm\frac{1}{32}, \pm\frac{3}{32}$

(b)

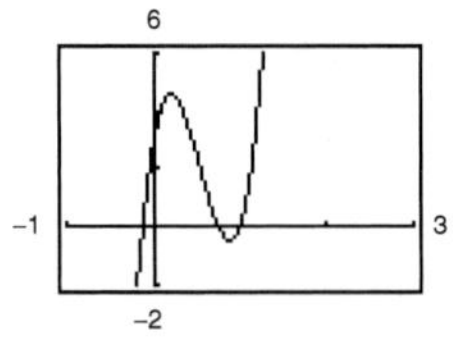

(c) $1, \frac{3}{4}, -\frac{1}{8}$

33. (a) $\pm 1, \approx \pm 1.414$

(b) $f(x) = (x + 1)(x - 1)(x + \sqrt{2})(x - \sqrt{2})$

35. (a) $0, 3, 4, \approx \pm 1.414$

(b) $h(x) = x(x - 3)(x - 4)(x + \sqrt{2})(x - \sqrt{2})$

37. $x^3 - x^2 + 25x - 25$ **39.** $x^3 + 4x^2 - 31x - 174$

41. $3x^4 - 17x^3 + 25x^2 + 23x - 22$

43. (a) $(x^2 + 9)(x^2 - 3)$

(b) $(x^2 + 9)(x + \sqrt{3})(x - \sqrt{3})$

(c) $(x + 3i)(x - 3i)(x + \sqrt{3})(x - \sqrt{3})$

45. (a) $(x^2 - 2x - 2)(x^2 - 2x + 3)$

(b) $(x - 1 + \sqrt{3})(x - 1 - \sqrt{3})(x^2 - 2x + 3)$

(c) $(x - 1 + \sqrt{3})(x - 1 - \sqrt{3})(x - 1 + \sqrt{2}i)(x - 1 - \sqrt{2}i)$

47. $-\frac{3}{2}, \pm 5i$ **49.** $\pm 2i, 1, -\frac{1}{2}$ **51.** $-3 \pm i, \frac{1}{4}$

53. $2, -3 \pm \sqrt{2}i, 1$ **55.** $\pm 5i$; $(x + 5i)(x - 5i)$

57. $2 \pm \sqrt{3}$; $(x - 2 - \sqrt{3})(x - 2 + \sqrt{3})$

59. $\pm 3, \pm 3i$; $(x + 3)(x - 3)(x + 3i)(x - 3i)$

61. $1 \pm i$; $(z - 1 + i)(z - 1 - i)$

63. $2, 2 \pm i$; $(x - 2)(x - 2 + i)(x - 2 - i)$

65. $-2, 1 \pm \sqrt{2}i$; $(x + 2)(x - 1 + \sqrt{2}i)(x - 1 - \sqrt{2}i)$

67. $-\frac{1}{5}, 1 \pm \sqrt{5}i$; $(5x + 1)(x - 1 + \sqrt{5}i)(x - 1 - \sqrt{5}i)$

69. $2, \pm 2i$; $(x - 2)^2(x + 2i)(x - 2i)$

71. $\pm i, \pm 3i$; $(x + i)(x - i)(x + 3i)(x - 3i)$

73. $-10, -7 \pm 5i$ **75.** $-\frac{3}{4}, 1 \pm \frac{1}{2}i$ **77.** $-2, -\frac{1}{2}, \pm i$

79. No real zeros **81.** No real zeros

83. One positive zero **85.** One or three positive zeros

87. Answers will vary. **89.** Answers will vary.

91. $1, -\frac{1}{2}$ **93.** $-\frac{3}{4}$ **95.** $\pm 2, \pm\frac{3}{2}$ **97.** $\pm 1, \frac{1}{4}$

99. d **101.** b

103. $f(x) = -2x^3 + 3x^2 + 11x - 6$

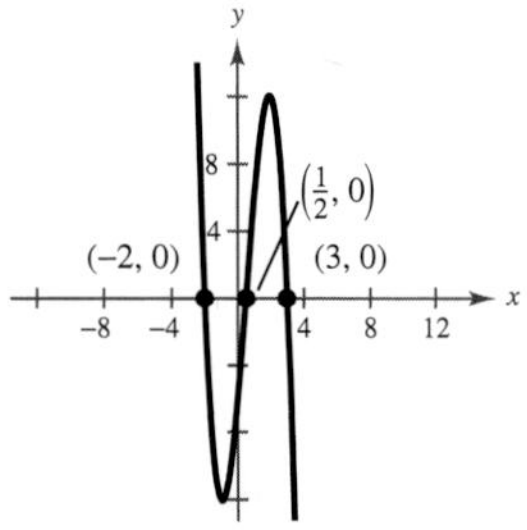

(Equations and graphs will vary.) There are infinitely many possible functions for f.

105. (a) $-2, 1, 4$

(b) The graph touches the x-axis at $x = 1$.

(c) The least possible degree of the function is 4 because there are at least four real zeros (1 is repeated) and a function can have at most the number of real zeros equal to the degree of the function. The degree cannot be odd by the definition of multiplicity.

(d) Positive. From the information in the table, you can conclude that the graph will eventually rise to the left and to the right.

(e) $f(x) = x^4 - 4x^3 - 3x^2 + 14x - 8$
(Answer is not unique.)

(f)

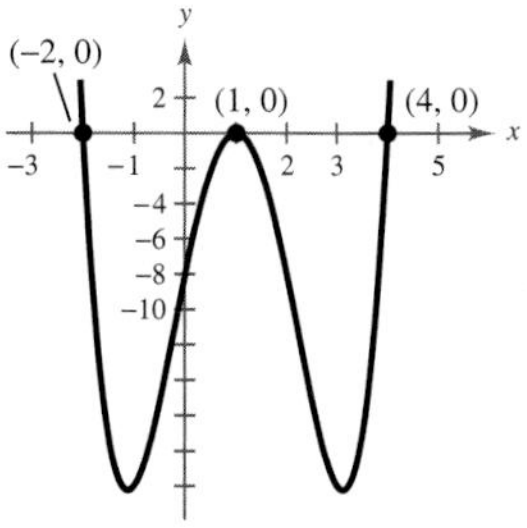

107. (a)

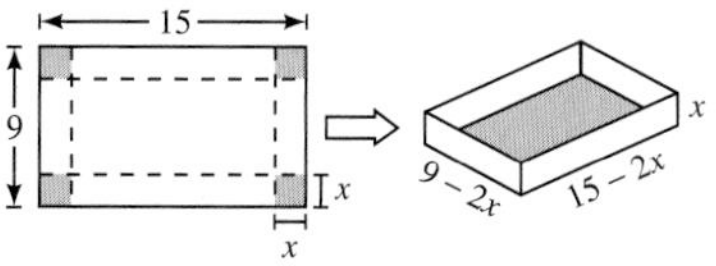

(b) $V = x(9 - 2x)(15 - 2x)$

Domain: $0 < x < \frac{9}{2}$

(c)

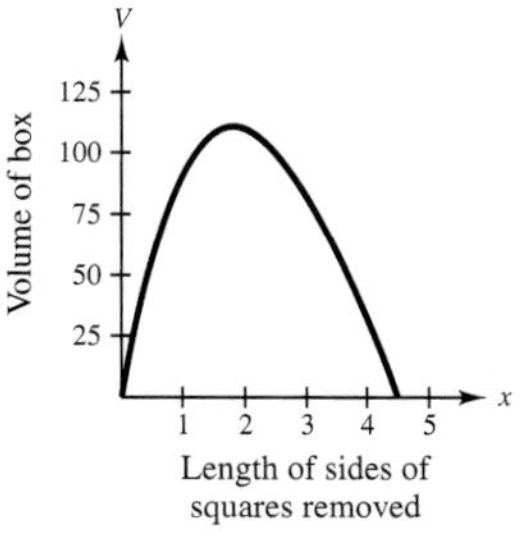

1.82 cm × 5.36 cm × 11.36 cm

(d) $\frac{1}{2}, \frac{7}{2}, 8$; 8 is not in the domain of V.

109. $x \approx 38.4$, or $384,000 **111.** $x \approx 40$

113. No. Setting $h = 64$ and solving the resulting equation yields imaginary roots.

115. False. The most complex zeros it can have is two and the Linear Factorization Theorem guarantees that there are three linear factors, so one zero must be real.

117. r_1, r_2, r_3 **119.** $5 + r_1, 5 + r_2, 5 + r_3$

121. The zeros cannot be determined.

123. (a) $0 < k < 4$ (b) $k = 4$ (c) $k < 0$ (d) $k > 4$

125. (a) $x^2 + b$ (b) $x^2 - 2ax + a^2 + b^2$

127. $-11 + 9i$ **129.** $20 + 40i$ **131.** i

133.

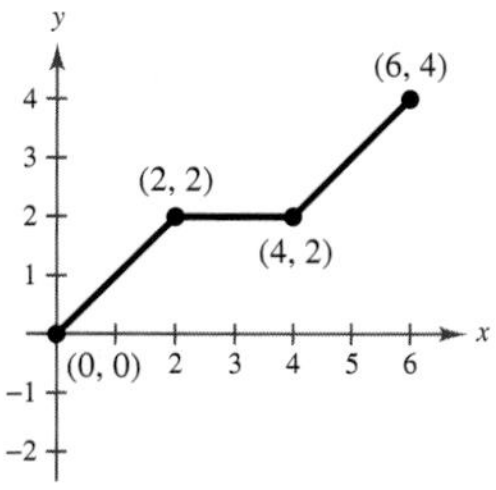

135.

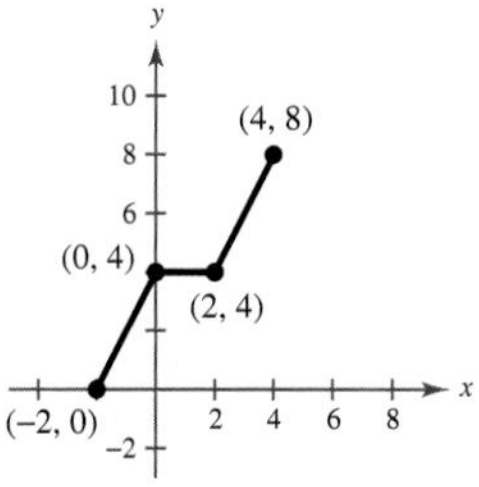

137.

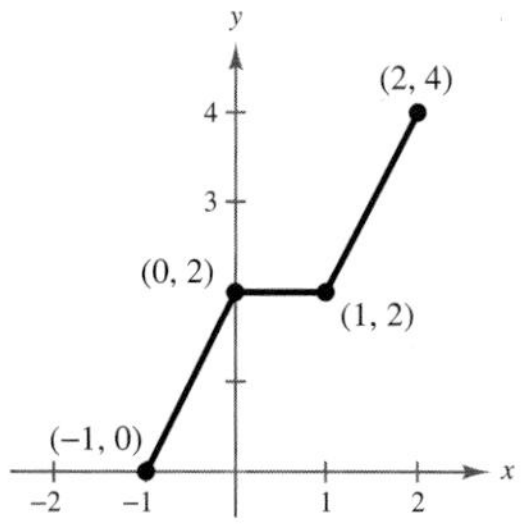

Section 2.6 *(page 267)*

1. (a)

x	$f(x)$
0.5	−2
0.9	−10
0.99	−100
0.999	−1000

x	$f(x)$
1.5	2
1.1	10
1.01	100
1.001	1000

x	$f(x)$
5	0.25
10	$0.\overline{1}$
100	$0.\overline{01}$
1000	$0.\overline{001}$

(b) Vertical asymptote: $x = 1$

Horizontal asymptote: $y = 0$

(c) Domain: all real numbers $x \neq 1$

3. (a)

x	$f(x)$
0.5	4
0.9	36
0.99	396
0.999	3996

x	$f(x)$
1.5	12
1.1	44
1.01	404
1.001	4004

x	$f(x)$
5	5
10	$4.\overline{4}$
100	$4.\overline{04}$
1000	$4.\overline{004}$

(b) Vertical asymptote: $x = 1$

Horizontal asymptotes: $y = \pm 4$

(c) Domain: all real numbers $x \neq 1$

5. (a)

x	$f(x)$
0.5	−1
0.9	−12.79
0.99	−147.8
0.999	−1498

x	$f(x)$
1.5	5.4
1.1	17.29
1.01	152.3
1.001	1502

x	$f(x)$
5	3.125
10	$3.\overline{03}$
100	$3.\overline{0003}$
1000	3

(b) Vertical asymptotes: $x = \pm 1$

Horizontal asymptote: $y = 3$

(c) Domain: all real numbers $x \neq \pm 1$

7. Domain: all real numbers $x \neq 0$

Vertical asymptote: $x = 0$

Horizontal asymptote: $y = 0$

9. Domain: all real numbers $x \neq 2$

Vertical asymptote: $x = 2$

Horizontal asymptote: $y = -1$

11. Domain: all real numbers $x \neq \pm 1$

Vertical asymptotes: $x = \pm 1$

13. Domain: all real numbers

Horizontal asymptote: $y = 3$

15. d **17.** f **19.** e **21.** 1 **23.** 6

25. (a) Intercept: $\left(0, \frac{1}{2}\right)$

(b) Vertical asymptote: $x = -2$
Horizontal asymptote: $y = 0$

(c) No symmetry

(d) and (e)

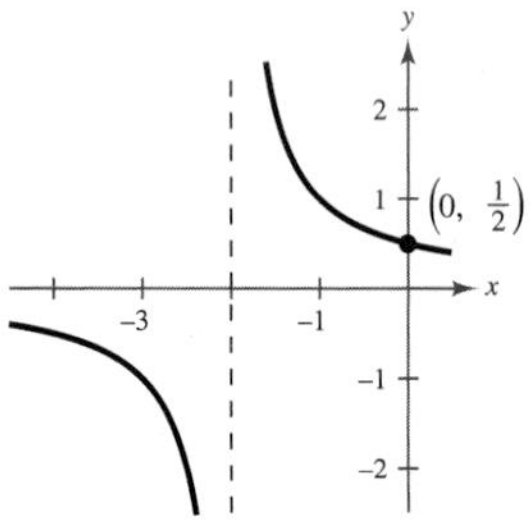

27. (a) Intercept: $\left(0, -\frac{1}{2}\right)$

(b) Vertical asymptote: $x = -2$
Horizontal asymptote: $y = 0$

(c) No symmetry

(d) and (e)

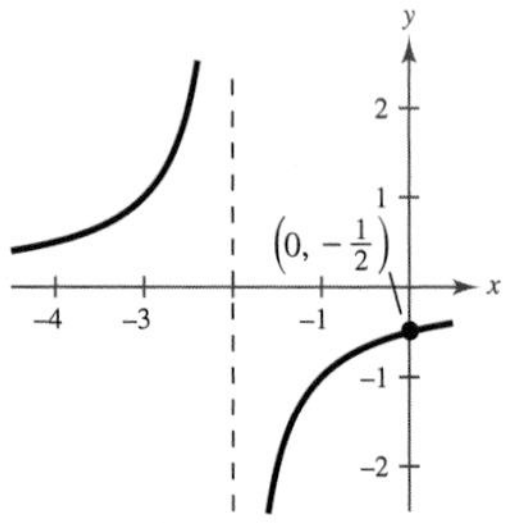

29. (a) Intercepts: $(0, 5)$, $\left(-\frac{5}{2}, 0\right)$

(b) Vertical asymptote: $x = -1$
Horizontal asymptote: $y = 2$

(c) No symmetry

(d) and (e)

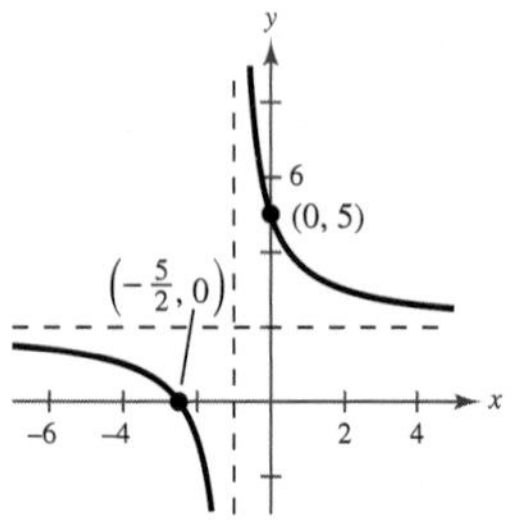

31. (a) Intercepts: $\left(0, \frac{5}{2}\right)$, $\left(-\frac{5}{2}, 0\right)$

(b) Vertical asymptote: $x = -2$
Horizontal asymptote: $y = 2$

(c) No symmetry

(d) and (e)

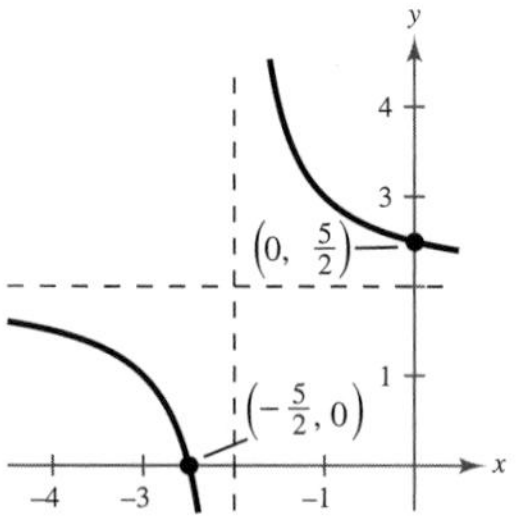

33. (a) Intercept: $(0, 0)$

(b) Horizontal asymptote: $y = 1$

(c) y-axis

(d) and (e)

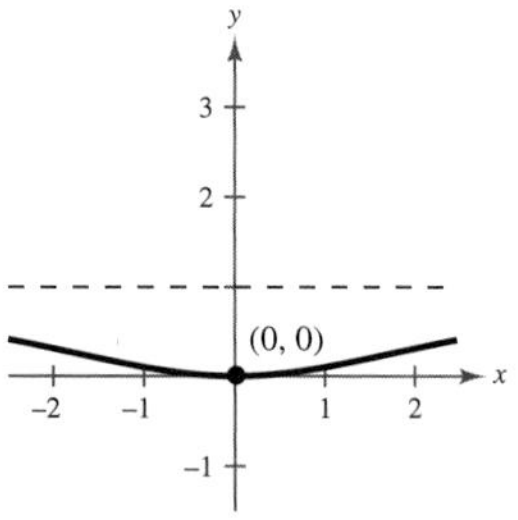

35. (a) Intercept: $(0, 0)$

(b) Vertical asymptotes: $x = \pm 3$
Horizontal asymptote: $y = 1$

(c) y-axis

(d) and (e)

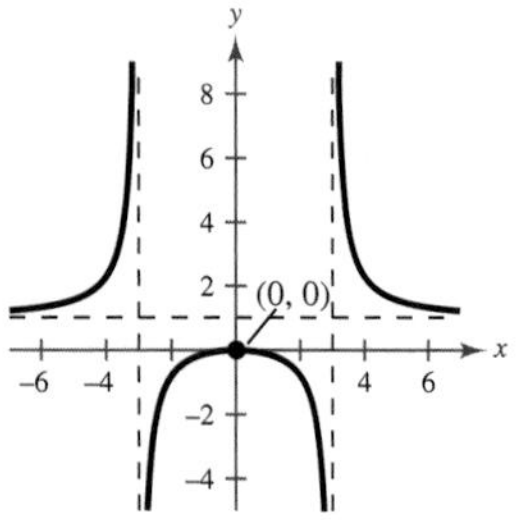

37. (a) Intercept: $(0, 0)$

(b) Horizontal asymptote: $y = 0$

(c) Origin

(d) and (e)

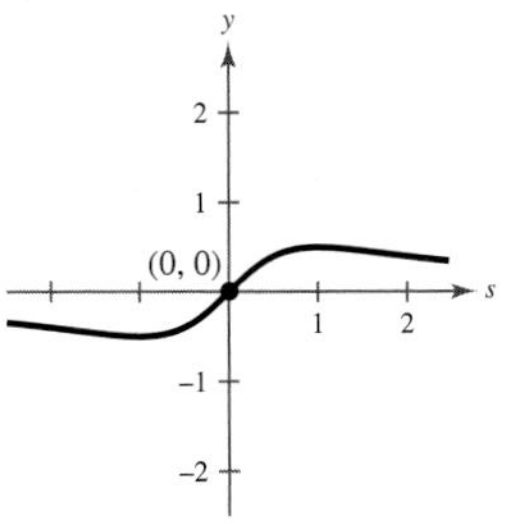

39. (a) Intercept: $(-1, 0)$

(b) Vertical asymptotes: $x = 0, x = 4$
Horizontal asymptote: $y = 0$

(c) No symmetry

(d) and (e)

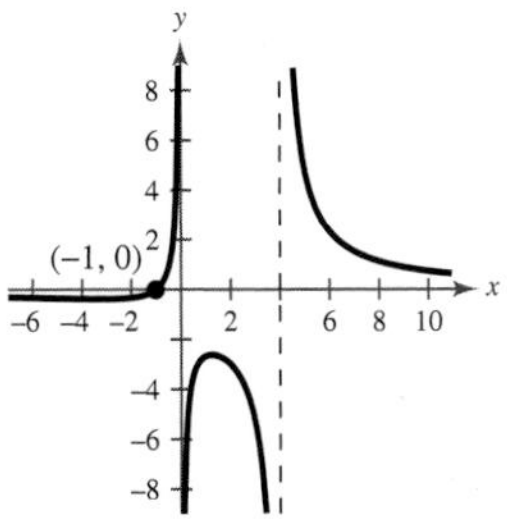

41. (a) Intercept: $(0, 0)$

(b) Vertical asymptotes: $x = -1, x = 2$
Horizontal asymptote: $y = 0$

(c) No symmetry

(d) and (e)

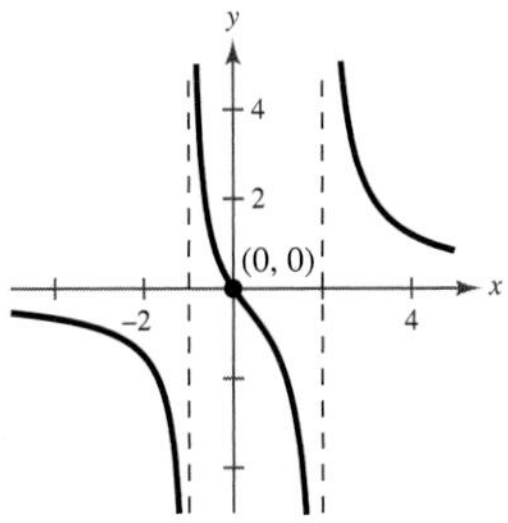

43. (a) Intercept: $(0, 0)$

(b) Vertical asymptotes: $x = -2, x = 7$
Horizontal asymptote: $y = 0$

(c) No symmetry

(d) and (e)

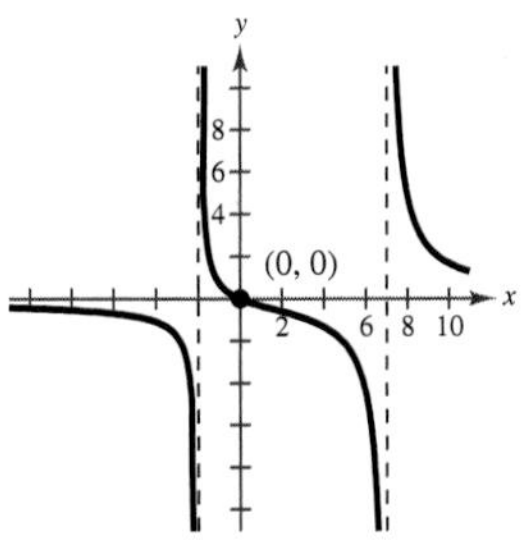

45. (a) Domain of f: all real numbers $x \neq -1$

Domain of g: all real numbers

(b) Vertical asymptote: none

(c)

x	-3	-2	-1.5	-1	-0.5	0	1
$f(x)$	-4	-3	-2.5	Undef.	-1.5	-1	0
$g(x)$	-4	-3	-2.5	-2	-1.5	-1	0

(d)

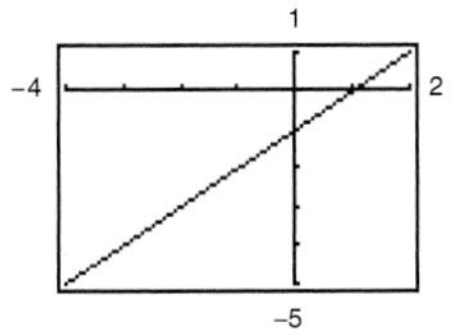

(e) Because there are only a finite number of pixels, the utility may not attempt to evaluate the function where it does not exist.

47. (a) Domain of f: all real numbers $x \neq 0, 2$

Domain of g: all real numbers $x \neq 0$

(b) Vertical asymptote: $x = 0$

(c)

x	-0.5	0	0.5	1	1.5	2	3
$f(x)$	-2	Undef.	2	1	$\frac{2}{3}$	Undef.	$\frac{1}{3}$
$g(x)$	-2	Undef.	2	1	$\frac{2}{3}$	$\frac{1}{2}$	$\frac{1}{3}$

(d)

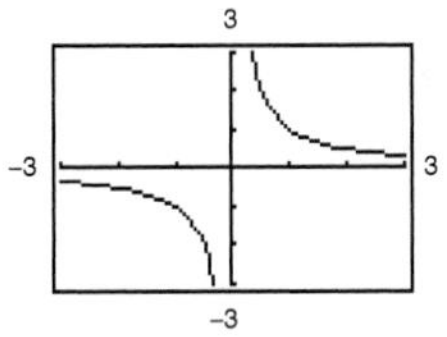

(e) Because there are only a finite number of pixels, the utility may not attempt to evaluate the function where it does not exist.

49.

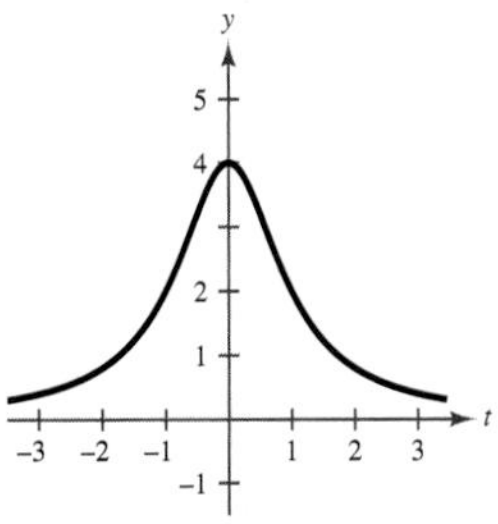

Domain: all real numbers

$y = 0$

51.

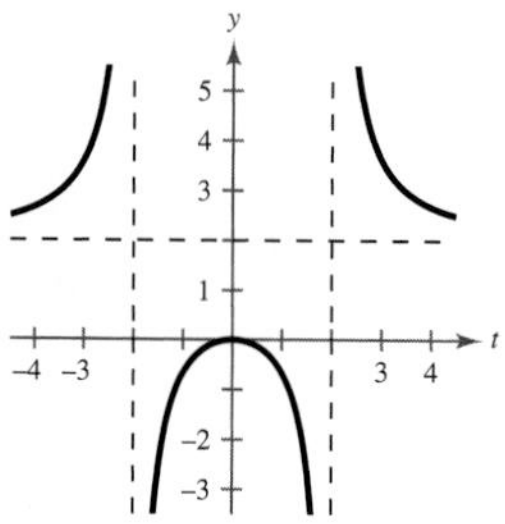

Domain: all real numbers $t \neq \pm 2$

$t = \pm 2,\ y = 2$

53.

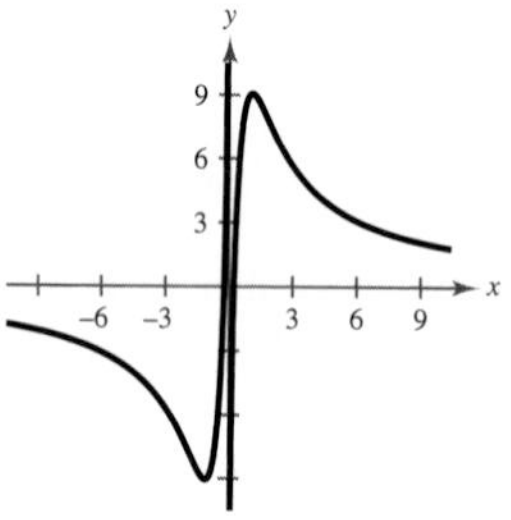

Domain: all real numbers $x \neq 0$

$x = 0,\ y = 0$

55. (a) No intercepts

(b) Vertical asymptote: $x = 0$
Slant asymptote: $y = 2x$

(c) Origin

(d) and (e)

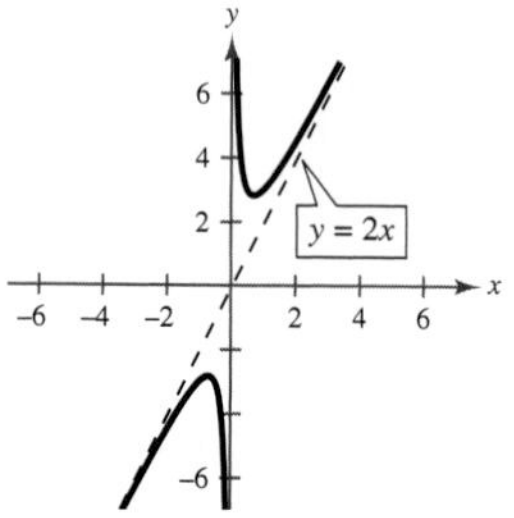

57. (a) No intercepts

(b) Vertical asymptote: $x = 0$
Slant asymptote: $y = x$

(c) Origin

(d) and (e)

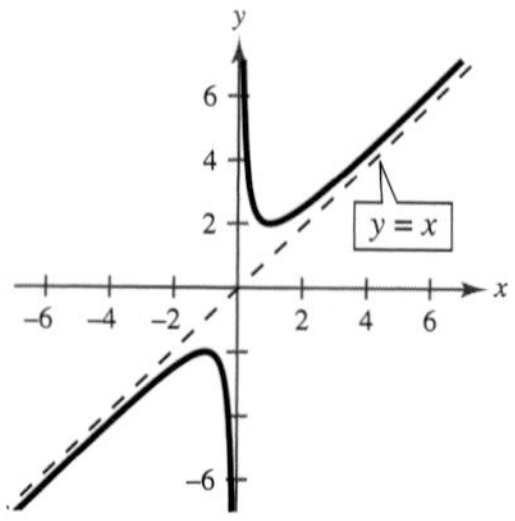

59. (a) Intercept: $(0, 0)$

(b) Vertical asymptotes: $x = \pm 1$
Slant asymptote: $y = x$

(c) Origin

(d) and (e)

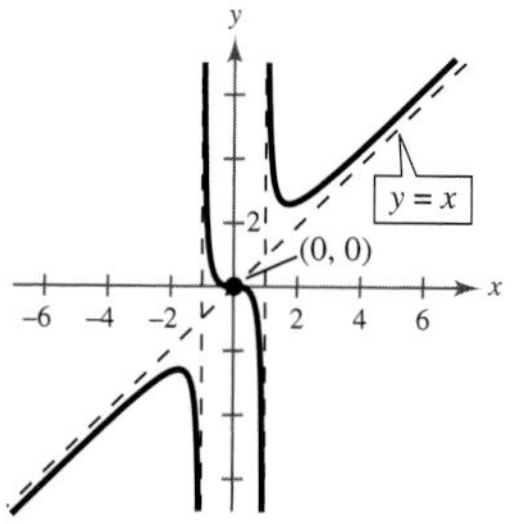

61. (a) Intercept: $(0, -1)$

(b) Vertical asymptote: $x = 1$
Slant asymptote: $y = x$

(c) No symmetry

(d) and (e)

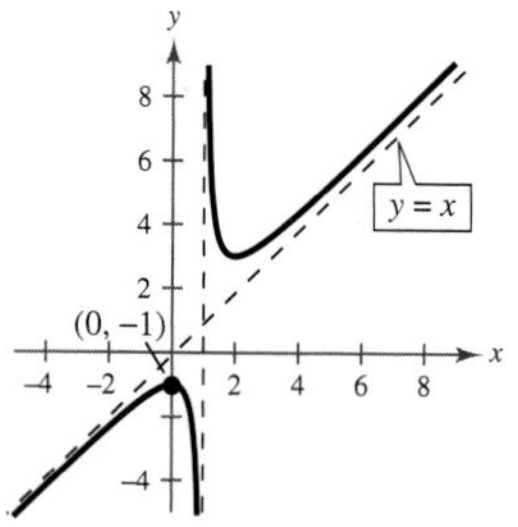

63.

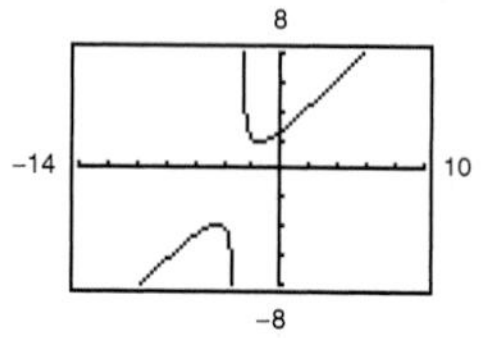

Domain: all real numbers $x \neq -3$

Vertical asymptote: $x = -3$

Slant asymptote: $y = x + 2$

$y = x + 2$

65.

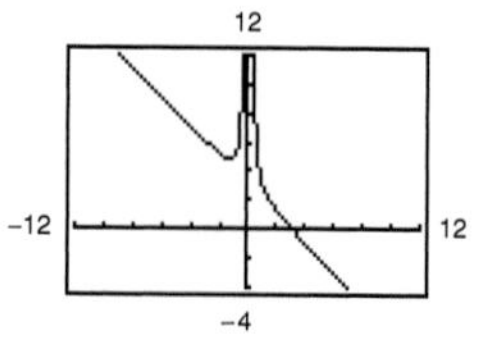

Domain: all real numbers $x \neq 0$

Vertical asymptote: $x = 0$

Slant asymptote: $y = -x + 3$

$y = -x + 3$

67. (a) and (b) $(-1, 0)$

69. (a) and (b) $(1, 0), (-1, 0)$

71. (a) 28.33 million dollars
(b) 170 million dollars
(c) 765 million dollars
(d) No. The function is undefined at $p = 100$.

73. (a) 333 deer, 500 deer, 800 deer (b) 1500

75. (a) $y = \dfrac{500}{x}$
(b) $x > 0$
(c)

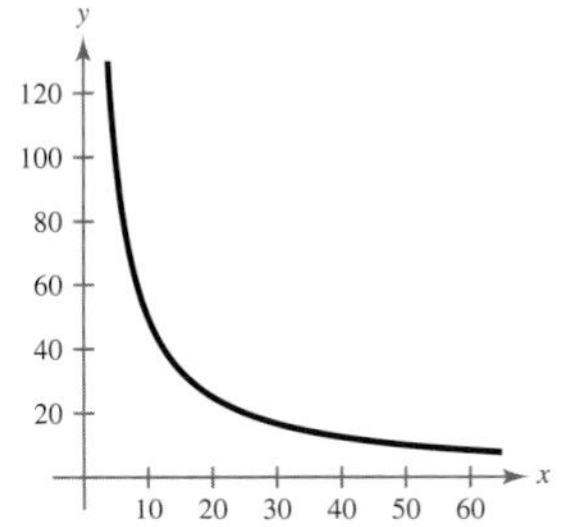

$16\frac{2}{3}$ meters

77. (a) $C = 0$; the chemical will eventually dissipate.
(b)

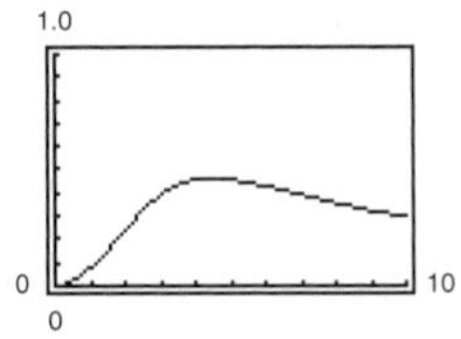

$t \approx 4.5$

79. False. Polynomials do not have vertical asymptotes.

81. $f(x) = \dfrac{1}{x^2 + x - 2}$ **83.** $f(x) = \dfrac{2x^2}{1 + x^2}$

85. $f(x) = \dfrac{1}{x^2 + 2}$; $f(x) = \dfrac{1}{x - 20}$
(Answers are not unique.)

87. $f(x) = \dfrac{x^2 - x - 6}{x - 2}$

89.

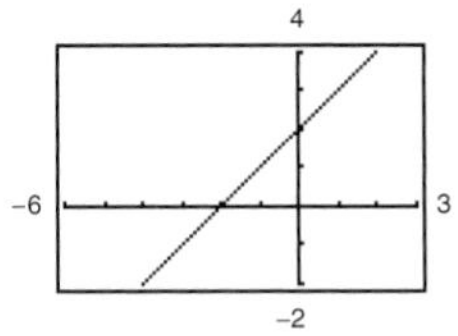

The fraction is not reduced.

91. $(3x - 4)(x + 9)$ **93.** $(x + 6)(x^2 - 2)$

95. $x < 0$

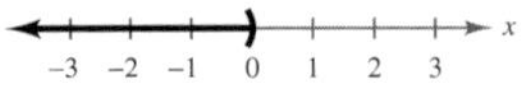

97. $x \le -\frac{13}{2}, x \ge \frac{7}{2}$

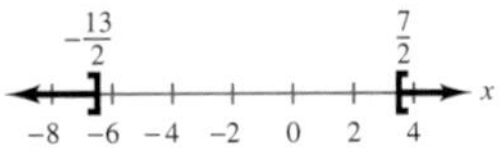

Section 2.7 *(page 277)*

1. b **3.** d **5.** $\dfrac{A}{x} + \dfrac{B}{x - 14}$

7. $\dfrac{A}{x} + \dfrac{B}{x^2} + \dfrac{C}{x - 10}$ **9.** $\dfrac{A}{x - 5} + \dfrac{B}{(x - 5)^2} + \dfrac{C}{(x - 5)^3}$

11. $\dfrac{A}{x} + \dfrac{Bx + C}{x^2 + 10}$ **13.** $\dfrac{A}{x} + \dfrac{Bx + C}{x^2 + 1} + \dfrac{Dx + E}{(x^2 + 1)^2}$

15. $\dfrac{1}{2}\left(\dfrac{1}{x - 1} - \dfrac{1}{x + 1}\right)$ **17.** $\dfrac{1}{x} - \dfrac{1}{x + 1}$

19. $\dfrac{1}{x} - \dfrac{2}{2x + 1}$ **21.** $\dfrac{1}{x - 1} - \dfrac{1}{x + 2}$

23. $-\dfrac{3}{x} - \dfrac{1}{x + 2} + \dfrac{5}{x - 2}$ **25.** $\dfrac{3}{x} - \dfrac{1}{x^2} + \dfrac{1}{x + 1}$

27. $\dfrac{3}{x - 3} + \dfrac{9}{(x - 3)^2}$ **29.** $-\dfrac{1}{x} + \dfrac{2x}{x^2 + 1}$

31. $\dfrac{-1}{x - 1} + \dfrac{x + 2}{x^2 - 2}$

33. $\dfrac{1}{3(x^2 + 2)} - \dfrac{1}{6(x + 2)} + \dfrac{1}{6(x - 2)}$

35. $\dfrac{1}{8(2x + 1)} + \dfrac{1}{8(2x - 1)} - \dfrac{x}{2(4x^2 + 1)}$

37. $\dfrac{1}{x + 1} + \dfrac{2}{x^2 - 2x + 3}$ **39.** $1 - \dfrac{2x + 1}{x^2 + x + 1}$

41. $2x - 7 + \dfrac{17}{x + 2} + \dfrac{1}{x + 1}$

43. $x + 3 + \dfrac{6}{x - 1} + \dfrac{4}{(x - 1)^2} + \dfrac{1}{(x - 1)^3}$

45. $\dfrac{3}{2x - 1} - \dfrac{2}{x + 1}$ **47.** $\dfrac{2}{x} - \dfrac{1}{x^2} - \dfrac{2}{x + 1}$

49. $\dfrac{1}{x^2 + 2} + \dfrac{x}{(x^2 + 2)^2}$ **51.** $2x + \dfrac{1}{2}\left(\dfrac{3}{x - 4} - \dfrac{1}{x + 2}\right)$

53. $\dfrac{3}{x} - \dfrac{2}{x - 4}$

$y = \dfrac{x - 12}{x(x - 4)}$ $y = \dfrac{3}{x}, y = -\dfrac{2}{x - 4}$

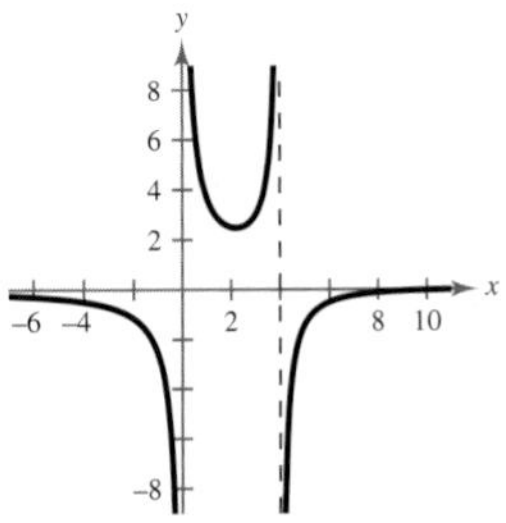

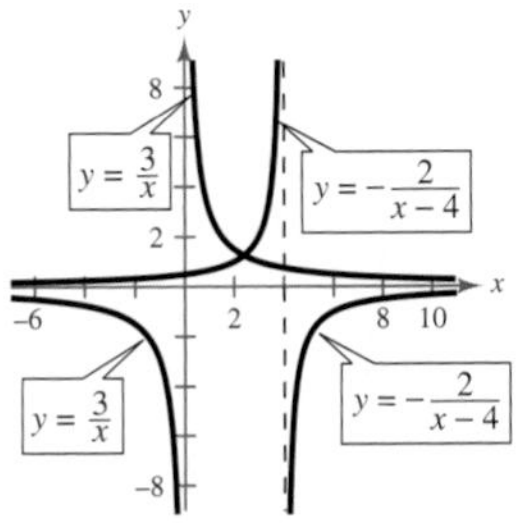

The vertical asymptotes are the same.

55. $\dfrac{3}{x-3} + \dfrac{5}{x+3}$

$y = \dfrac{2(4x-3)}{x^2-9}$ $\qquad$ $y = \dfrac{3}{x-3},\ y = \dfrac{5}{x+3}$

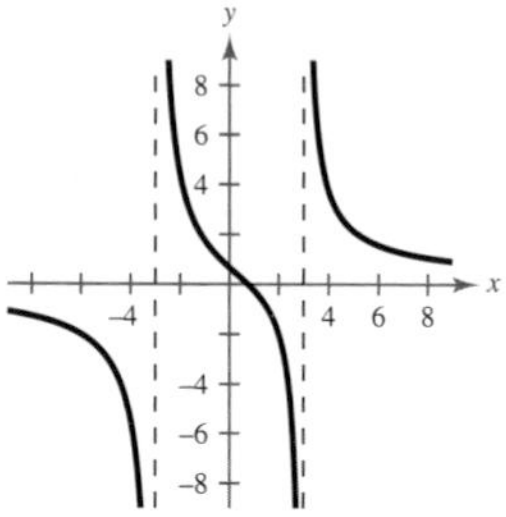

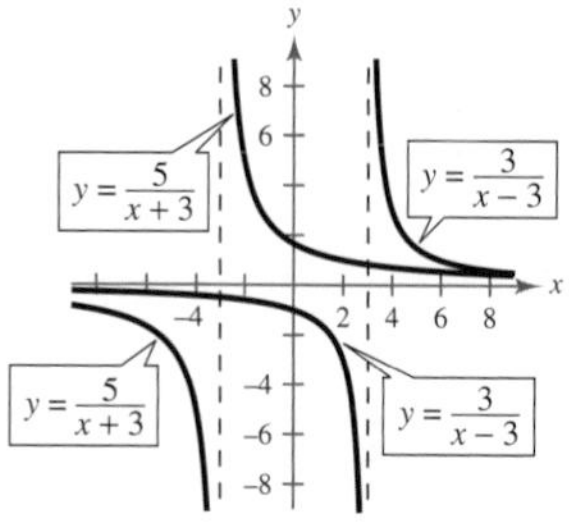

The vertical asymptotes are the same.

57. (a) $\dfrac{2000}{7-4x} - \dfrac{2000}{11-7x}, \quad 0 < x \le 1$

(b) $\text{Ymax} = \left|\dfrac{2000}{7-4x}\right|$

$\text{Ymin} = \left|\dfrac{-2000}{11-7x}\right|$

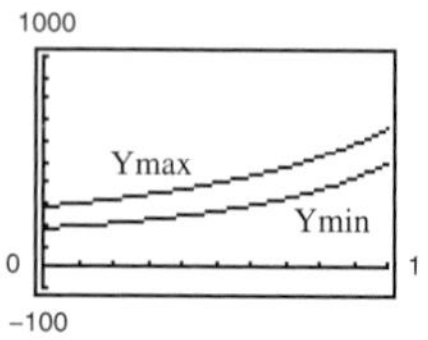

59. False. The expression is an improper rational expression, so you must first divide before applying partial fraction decomposition.

61. $\dfrac{1}{a}\left(\dfrac{1}{x} - \dfrac{1}{x+a}\right)$ $\qquad$ **63.** $\dfrac{1}{a+1}\left(\dfrac{1}{x+1} + \dfrac{1}{a-x}\right)$

65.

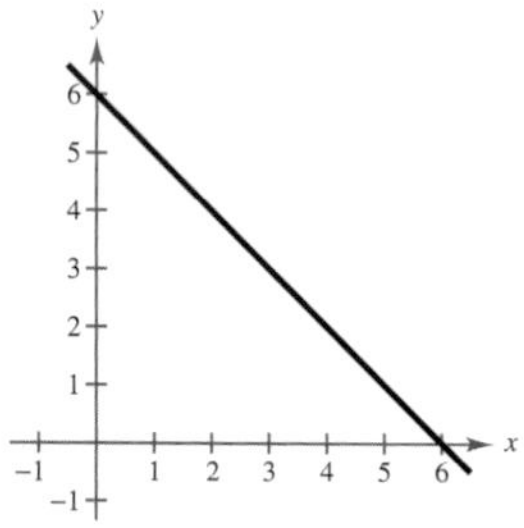

67.

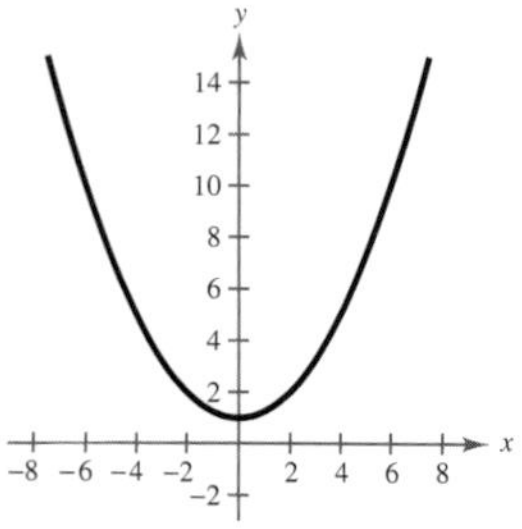

69.

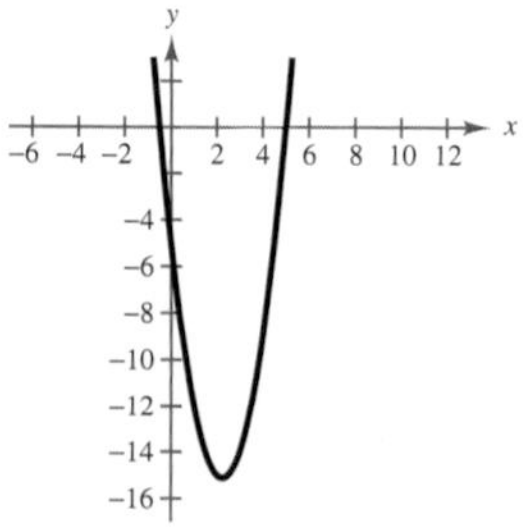

71.

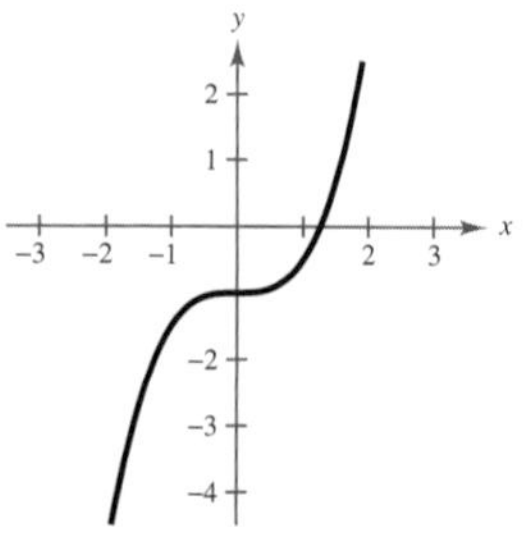

73.

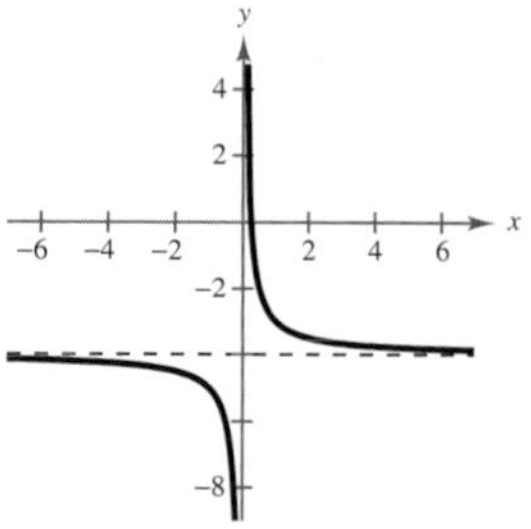

75.

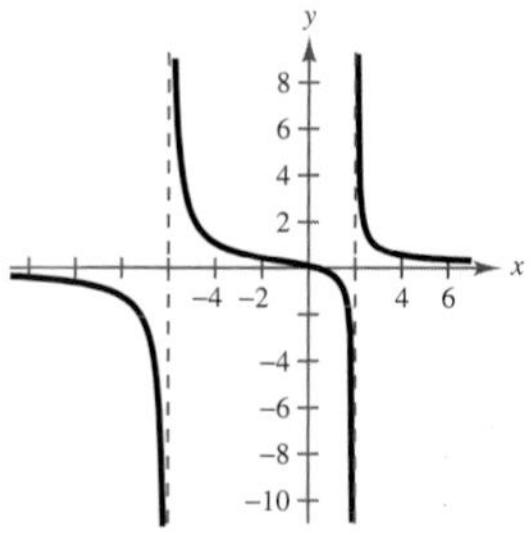

77.

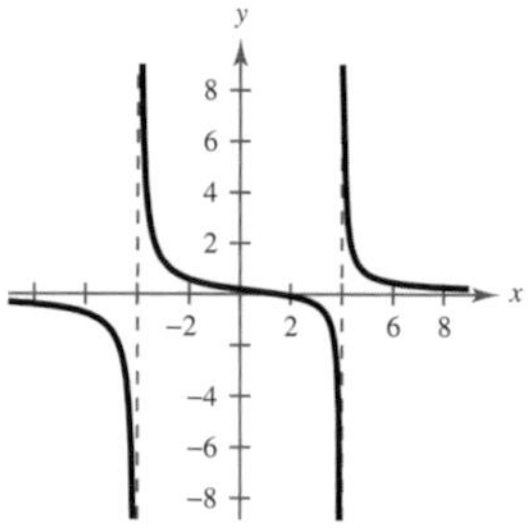

Review Exercises *(page 280)*

1. (a)

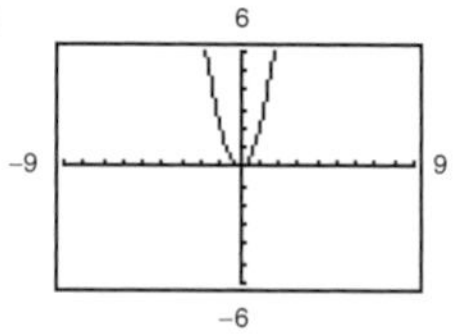

Vertical stretch

(b)

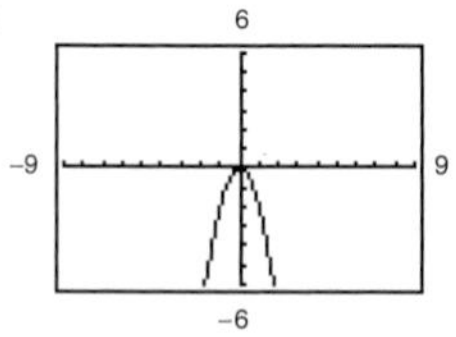

Vertical stretch and reflection in the x-axis

(c)

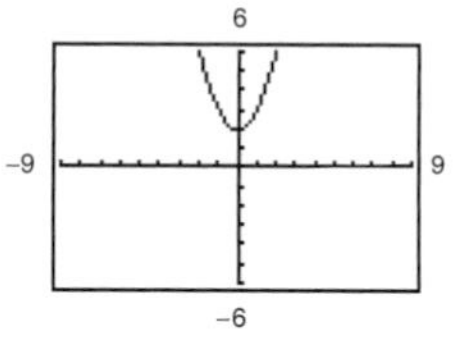

Vertical translation

(d)

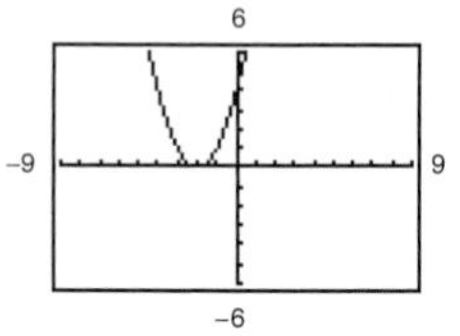

Horizontal translation

3. $f(x) = -\frac{1}{2}(x-4)^2 + 1$ $\qquad$ **5.** $f(x) = (x-1)^2 - 4$

7. $g(x) = (x-1)^2 - 1$ $\qquad$ **9.** $f(x) = (x+4)^2 - 6$

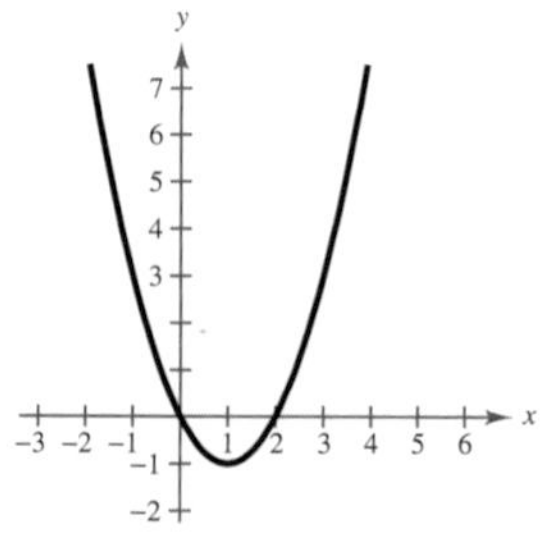

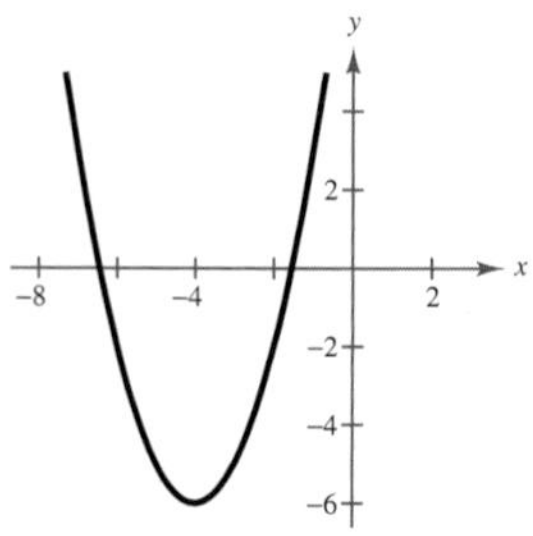

11. $f(t) = -2(t-1)^2 + 3$

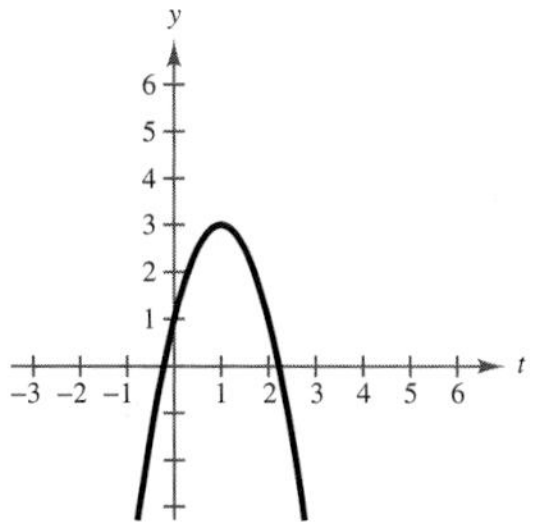

13. $h(x) = 4\left(x + \frac{1}{2}\right)^2 + 12$

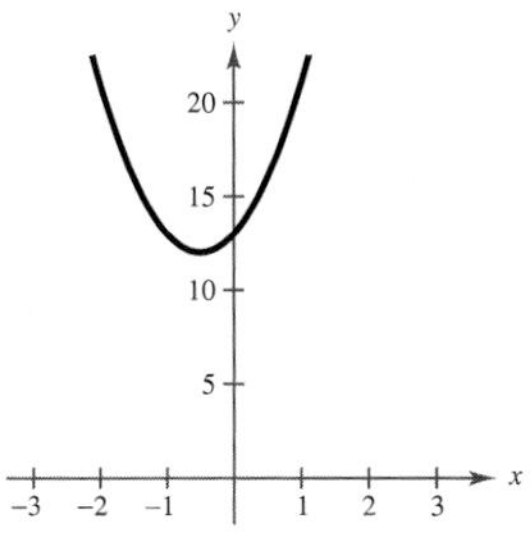

15. $h(x) = \left(x + \frac{5}{2}\right)^2 - \frac{41}{4}$

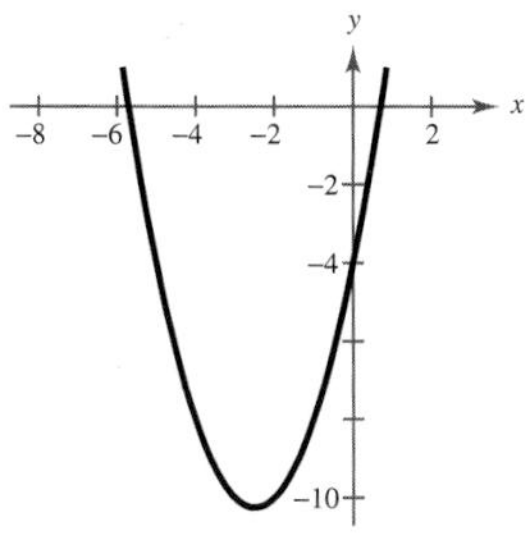

17. $f(x) = \frac{1}{3}\left(x + \frac{5}{2}\right)^2 - \frac{41}{12}$

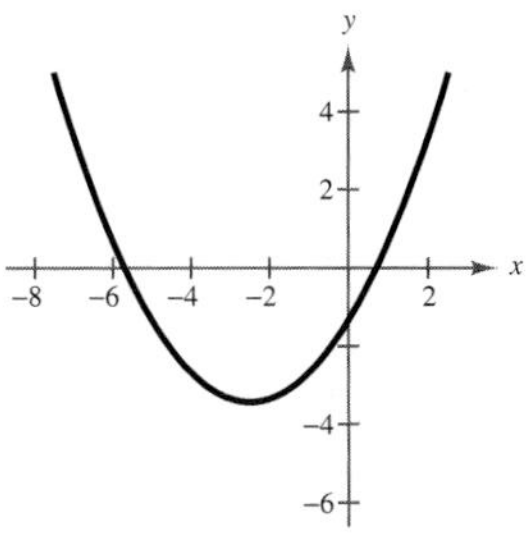

19. (a)

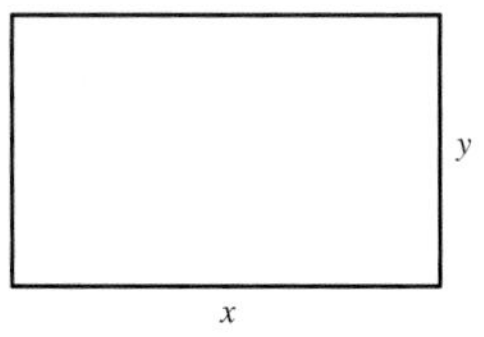

(b) $y = 100 - x$

$A = 100x - x^2$

(c) $x = 50,\ y = 50$

21. 1091 units

23.

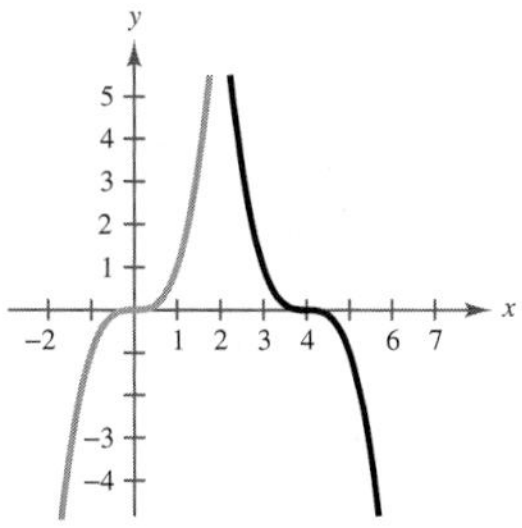

25.

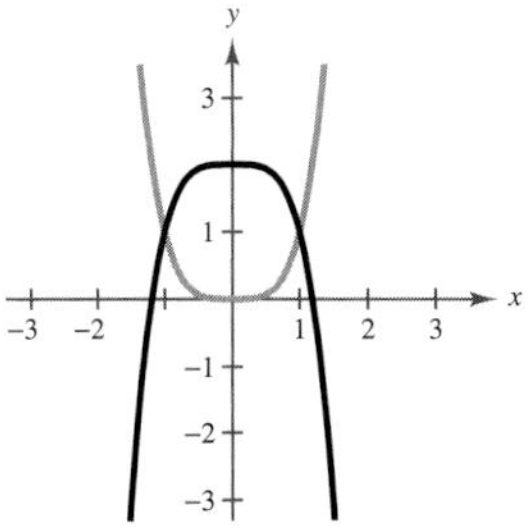

27.

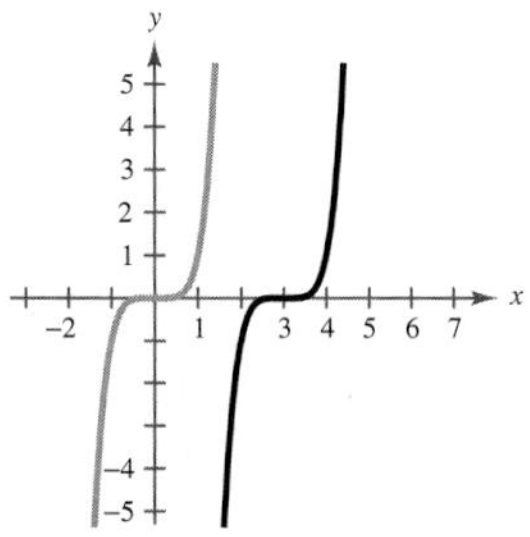

29. Falls to the left.
Falls to the right.

31. Rises to the left.
Rises to the right.

33. $-7, \frac{3}{2}$

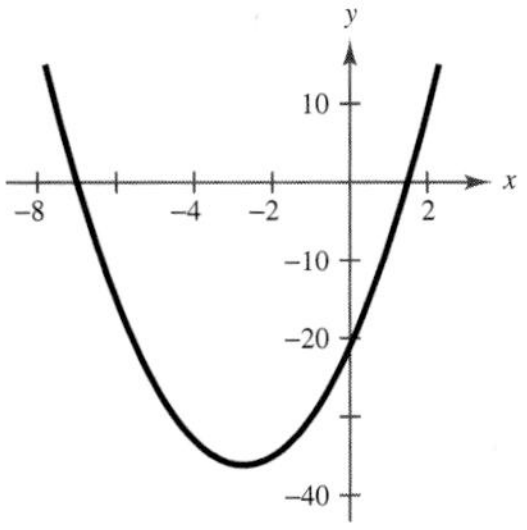

35. $0, \pm\sqrt{3}$

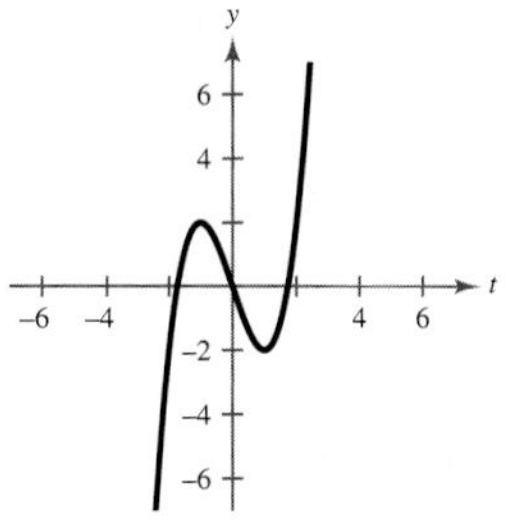

37. $0, \frac{5}{3}$

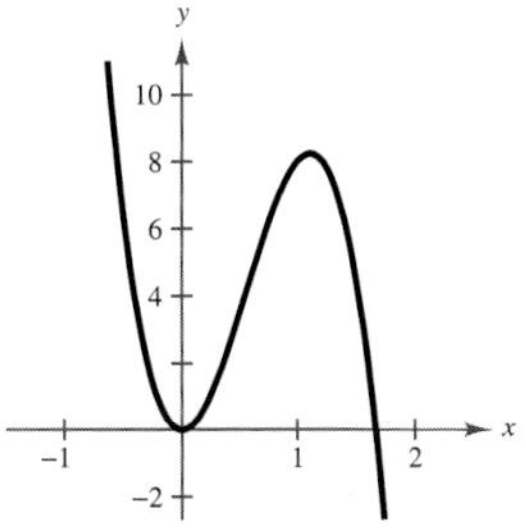

39. $(-1, 0)$ **41.** $(-1, 0), (1, 2)$

43. $8x + 5 + \dfrac{2}{3x - 2}$ **45.** $5x + 2$

47. $x^2 - 3x + 2 - \dfrac{1}{x^2 + 2}$ **49.** $6x^3 - 27x$

51. $2x^2 - 11x - 6$

53. (a) Yes (b) Yes (c) Yes (d) No

55. (a) -421 (b) -9

57. (a) Answers will vary. (b) $(x + 7), (x + 1)$
(c) $f(x) = (x + 7)(x + 1)(x - 4)$ (d) $-7, -1, 4$
(e)

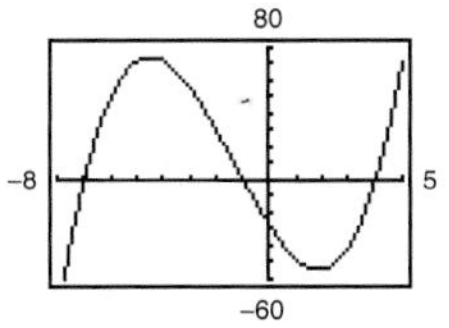

59. (a) Answers will vary. (b) $(x + 1), (x - 4)$
(c) $f(x) = (x + 1)(x - 4)(x + 2)(x - 3)$
(d) $-2, -1, 3, 4$
(e)

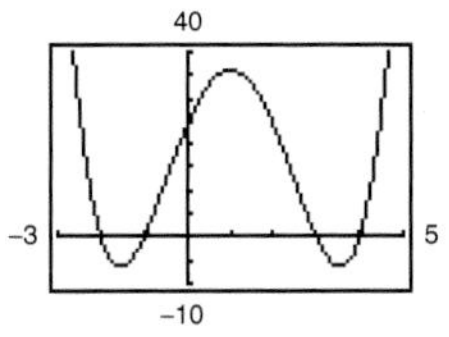

61. (a) and (b)

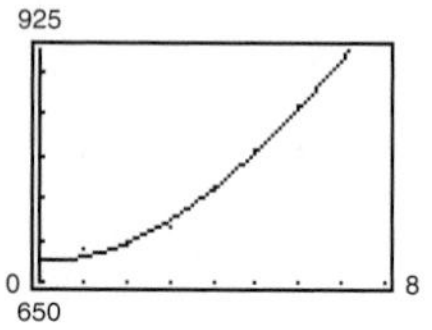

$V = -0.215t^3 + 6.693t^2 - 2.128t + 675.279$

The model closely fits the data.

(c)

t	0	1	2	3
V	675.3	679.6	696.1	723.3

t	4	5	6	7
V	760.1	805.1	857	914.6

The estimated values are close to the actual data.

(d) $1175.6 billion. No. According to the model, the value of farm real estate will eventually become negative.

63. $3 - 5i$ **65.** $-1 - 5i$ **67.** $-\sqrt{2}i$

69. $17 + 28i$ **71.** $9 + 20i$ **73.** $\frac{17}{26} + \frac{7}{26}i$

75. $\dfrac{9}{85} + \dfrac{83}{85}i$ **77.** $\pm\dfrac{1}{2}i$ **79.** $-\dfrac{1}{4} \pm \dfrac{\sqrt{71}}{4}i$

81. $-9, -9, 4$ **83.** $0, \pm\sqrt{6}i$ **85.** $5, 5, 8, 3 \pm i$

87. $\pm 1, \pm 2, \pm 4, \pm 8, \pm\frac{1}{3}, \pm\frac{2}{3}, \pm\frac{4}{3}, \pm\frac{8}{3}$

89. $-1, \frac{5}{3}, 6$ **91.** $-5, -2$ **93.** $-3, 2, \pm\frac{2}{5}$

95. $x^4 - x^3 - 3x^2 + 17x - 30$

97. (a) $(x^2 + 1)(x + 1)(x - 3)$

(b) $(x^2 + 1)(x + 1)(x - 3)$

(c) $(x + i)(x - i)(x + 1)(x - 3)$

99. $-4, 3 - i, 3 + i$ **101.** One or three positive zeros, two or no negative zeros

103. Domain: all real numbers $x \neq -\frac{1}{3}$

105. Domain: all real numbers

107. Horizontal asymptote: $y = 2$

109. Vertical asymptote: $x = 3$

Horizontal asymptote: $y = 0$

111. No intercepts

Origin

Vertical asymptote: $x = 0$

Horizontal asymptote: $y = 0$

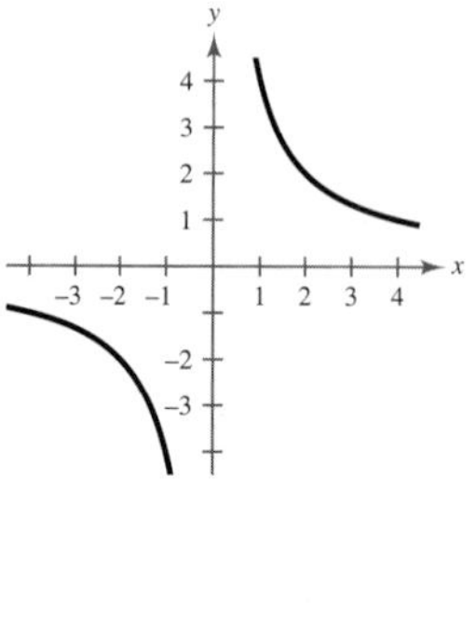

113. Intercepts: $(0, \frac{3}{2}), (3, 0)$

No symmetry

Vertical asymptote: $x = 2$

Horizontal asymptote: $y = 1$

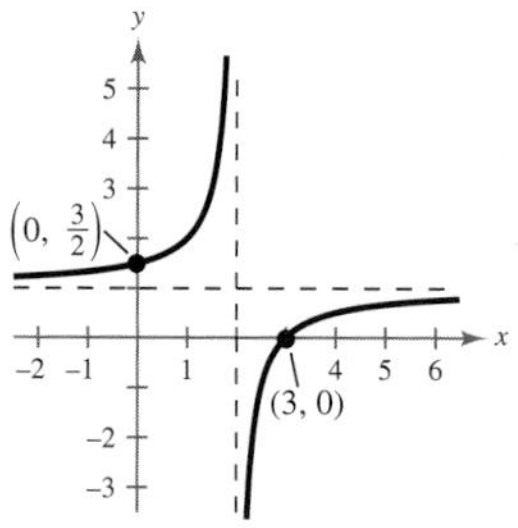

115. Intercept: $(0, 0)$

Origin

Horizontal asymptote: $y = 0$

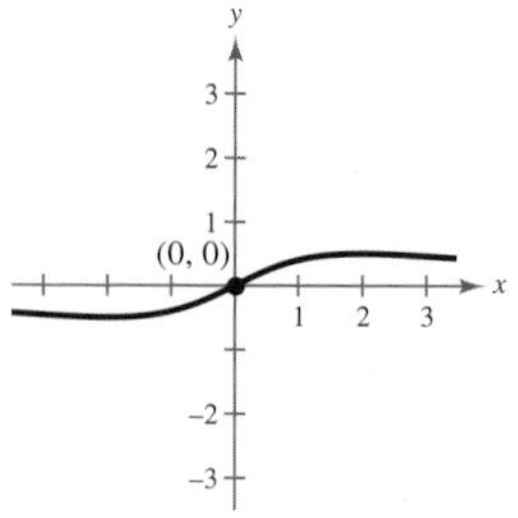

117. Intercept: $(0, 4)$

No symmetry

Vertical asymptote: $x = 1$

Horizontal asymptote: $y = 0$

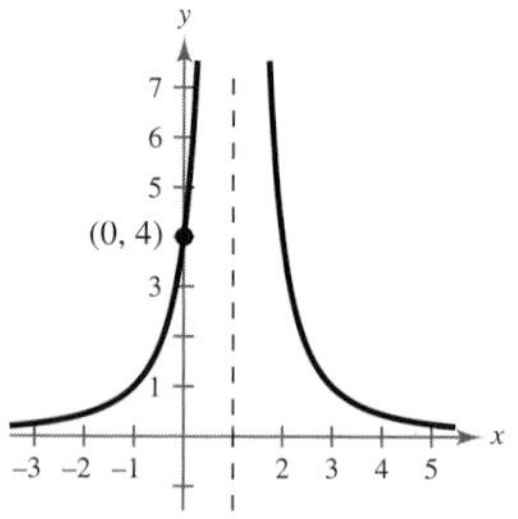

119. Intercept: $(0, 0)$

y-axis

Vertical asymptotes: $x = \pm 2$

Horizontal asymptote: $y = 2$

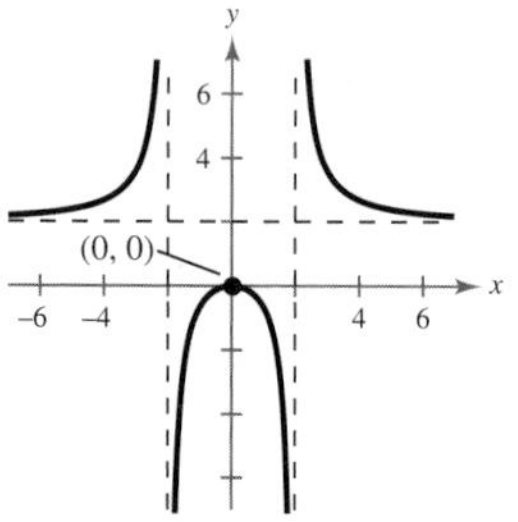

121. Intercept: $(0, -\frac{2}{9})$

No symmetry

Vertical asymptote: $x = -3$

Horizontal asymptote: $y = 0$

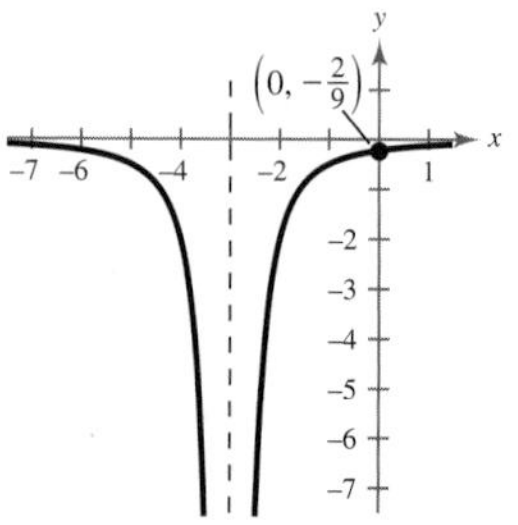

123. $y = x - 1$

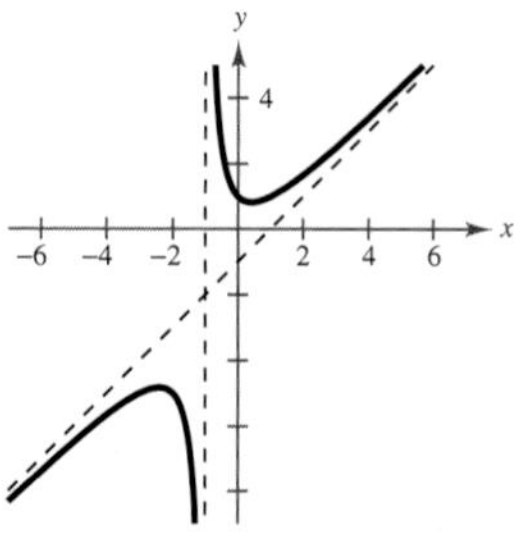

125. $y = x$

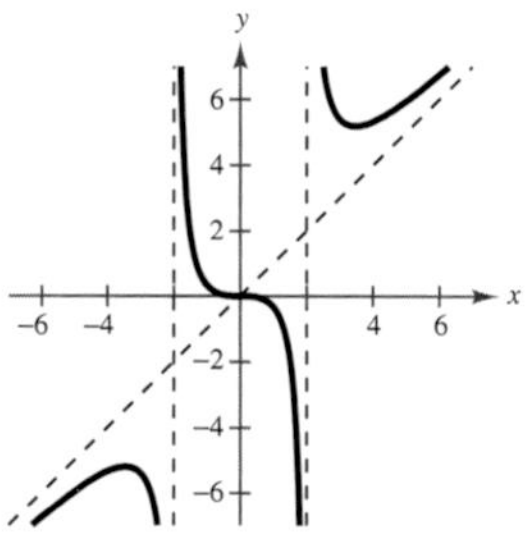

127. (a) \$176 million (b) \$528 million (c) \$1584 million (d) No

129. (a)

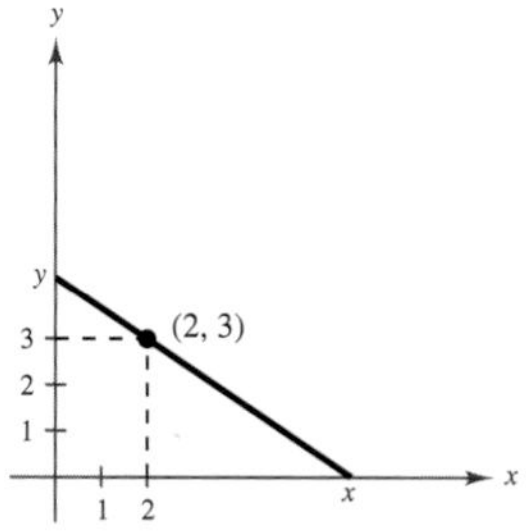

(b) Answers will vary.

(c)

x	2.5	3.0	3.5	4.0	4.5	5.0
A	18.75	13.5	12.25	12.0	12.15	12.5

Base: 4; height: 6

(d)

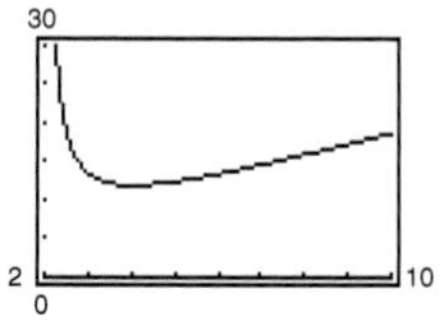

(e) $y = \frac{3}{2}x + 3$. The ratio of the area of the triangle to the side x approaches $\frac{3}{2}$ as x increases.

131. $\dfrac{A}{x - 7} + \dfrac{B}{x + 4}$

133. $\dfrac{A}{x} + \dfrac{Bx + C}{x^2 + 2} + \dfrac{Dx + E}{(x^2 + 2)^2}$

135. $\dfrac{1}{x + 1} - \dfrac{2}{x + 2}$

137. $\dfrac{1}{2}\left(\dfrac{3}{x - 3} - \dfrac{3}{x + 3}\right)$

139. $\dfrac{4}{3(x - 1)} + \dfrac{2}{3(x - 1)^2}$

141. $2\left(\dfrac{1}{x - 1} + \dfrac{x + 1}{x^2 + 1}\right)$

143. False. The domain of $f(x) = \dfrac{1}{x^2 + 1}$ is the set of all real numbers.

145. a is negative and h and k are zero.

147. No. For a complex number $a + bi$, if $a = 0$ or $b = 0$, then the square of the complex number is a real number.

Chapter Test *(page 285)*

1. (a) Reflection in the x-axis followed by a vertical translation

(b) Horizontal translation

2. $y = (x - 3)^2 - 6$

3. (a) 50 feet

(b) 61.62 feet

4. Rises to the left.

Falls to the right.

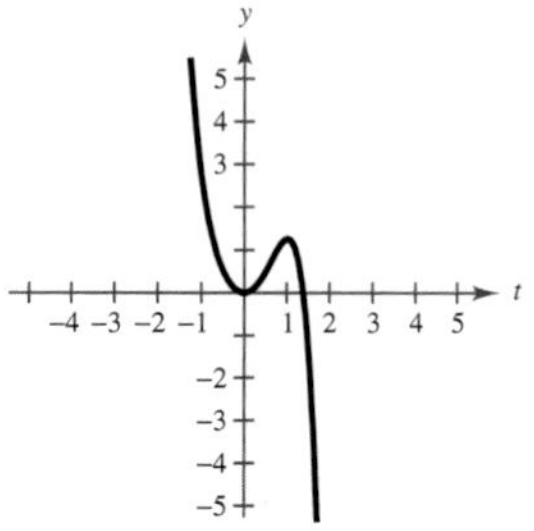

5. $3x + \dfrac{x - 1}{x^2 + 1}$

6. $2x^3 + 4x^2 + 3x + 6 + \dfrac{9}{x - 2}$

7. $(4x - 1)(x - \sqrt{3})(x + \sqrt{3})$; $\frac{1}{4}, \pm\sqrt{3}$

8. (a) $-3 + 5i$ (b) 7 (c) $2 - i$

9. $\pm 1, \pm 2, \pm 3, \pm 4, \pm 6, \pm 8, \pm 12, \pm 24, \pm\frac{1}{2}, \pm\frac{3}{2}$

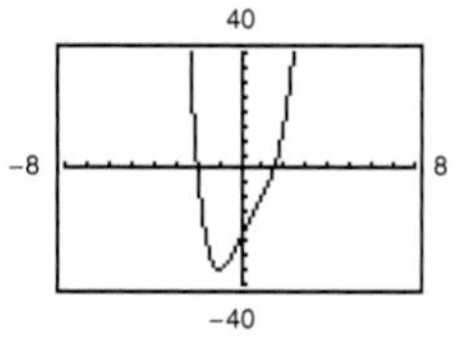

$-2, \frac{3}{2}$

10. $\pm 1, \pm 2, \pm\frac{1}{3}, \pm\frac{2}{3}$

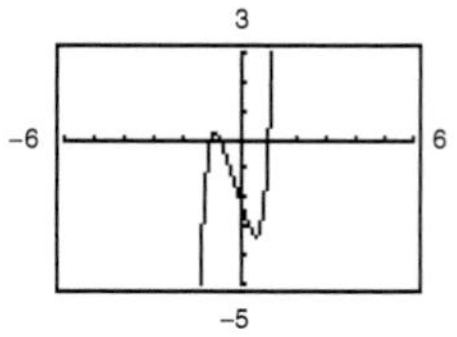

$\pm 1, -\frac{2}{3}$

11. $-1, 2, \pm 2i$

12. $f(x) = x^4 - 9x^3 + 28x^2 - 30x$

13. $f(x) = x^4 - 6x^3 + 16x^2 - 24x + 16$

14. Domain: all real numbers except $x \neq 4$

Vertical asymptote: $x = 4$

Horizontal asymptote: $y = 0$

15. Domain: all real numbers

Horizontal asymptote: $y = -1$

16. Domain: all real numbers except $x \neq 2$

Vertical asymptote: $x = 2$

Slant asymptote: $y = x + 4$

17. Intercepts: $(-2, 0), (2, 0)$

Vertical asymptote: $x = 0$

Horizontal asymptote: $y = -1$

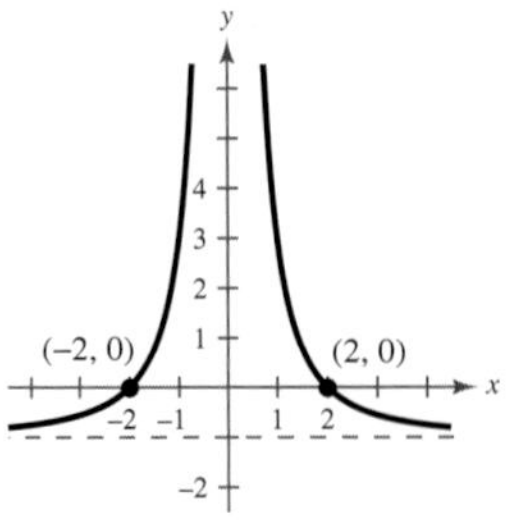

18. Intercept: $(0, -2)$

Vertical asymptote: $x = 1$

Slant asymptote: $y = x + 1$

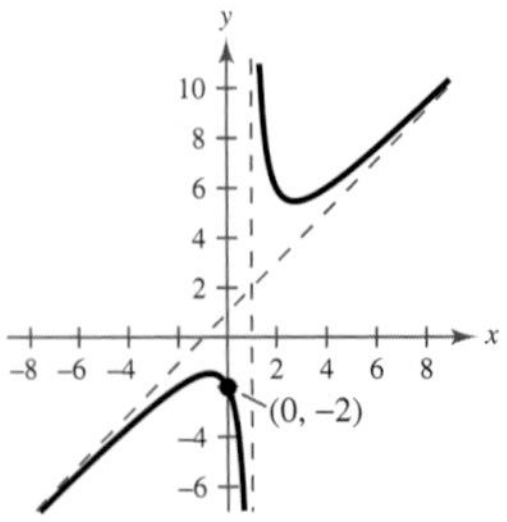

19. $\dfrac{3}{x-2} - \dfrac{1}{x+1}$ **20.** $\dfrac{2}{x^2} - \dfrac{3}{x-2}$

21. $-\dfrac{5}{x} + \dfrac{3}{x-1} + \dfrac{3}{x+1}$ **22.** $-\dfrac{2}{x} + \dfrac{3x}{x^2+2}$

Chapter 3

Section 3.1 *(page 296)*

1. 946.852 **3.** 7.352 **5.** 0.006 **7.** 673.639

9. 0.472 **11.** d **13.** a

15. Shift the graph of f 4 units to the right.

17. Shift the graph of f 5 units upward.

19. Reflect f in the x-axis and shift f 4 units to the left.

21. Reflect f in the x-axis and shift f 5 units upward.

23.

x	-2	-1	0	1	2
$f(x)$	4	2	1	0.5	0.25

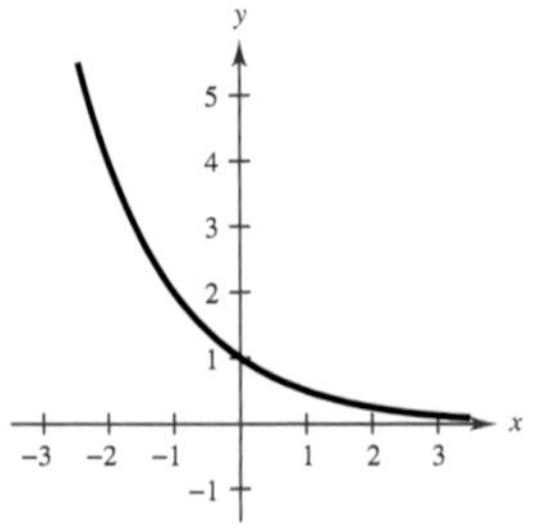

25.

x	-2	-1	0	1	2
$f(x)$	0.25	0.5	1	2	4

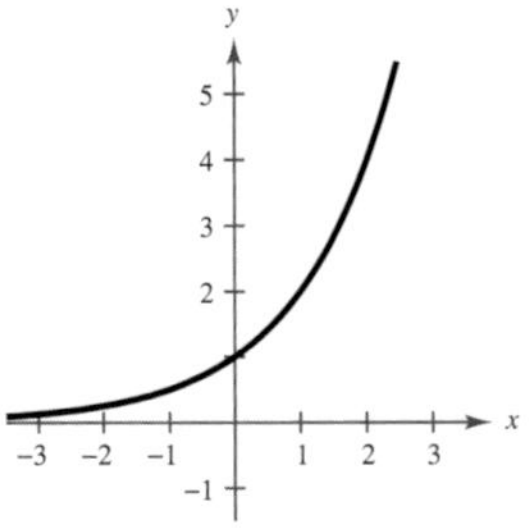

27.

x	-2	-1	0	1	2
$f(x)$	0.125	0.25	0.5	1	2

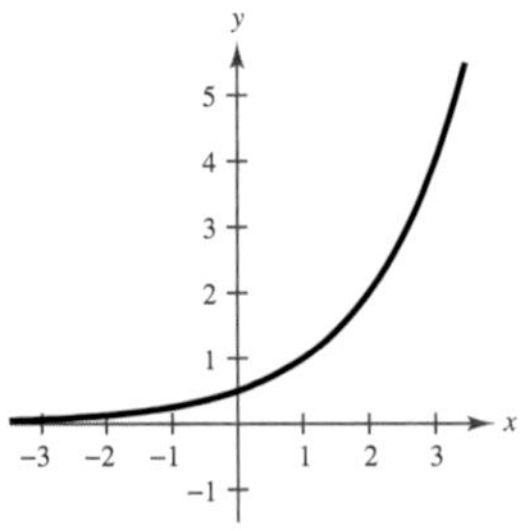

29.

x	-2	-1	0	1	2
$f(x)$	0.135	0.368	1	2.718	7.389

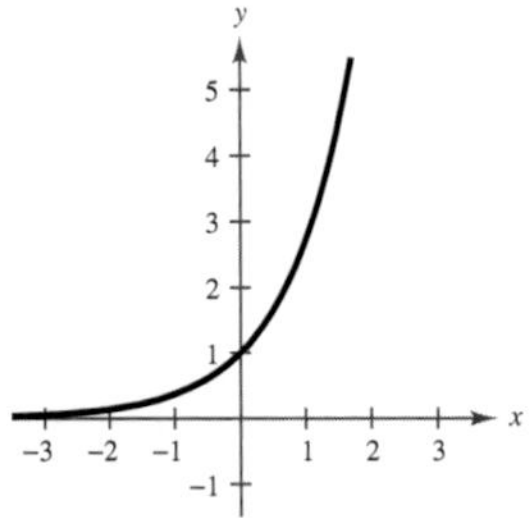

31.

x	-8	-7	-6	-5	-4
$f(x)$	0.055	0.149	0.406	1.104	3

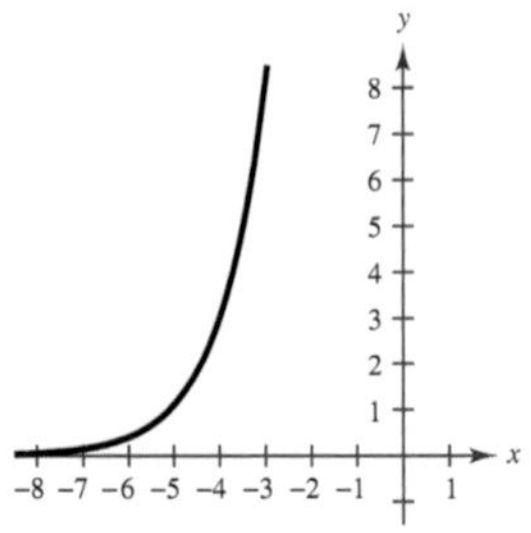

33.

x	-2	-1	0	1	2
$f(x)$	4.037	4.100	4.271	4.736	6

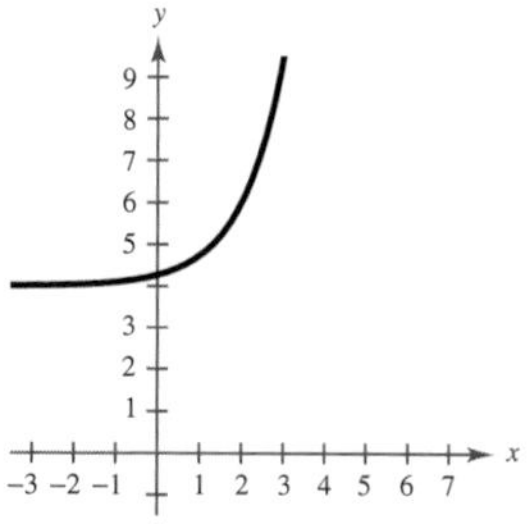

35.

x	-1	0	1	2	3
$f(x)$	3.003	3.016	3.063	3.25	4

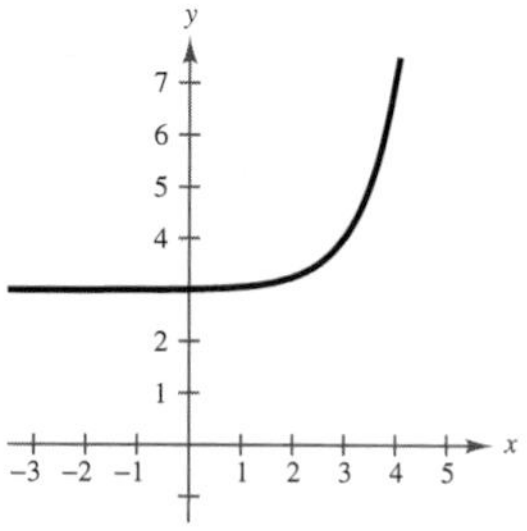

37.

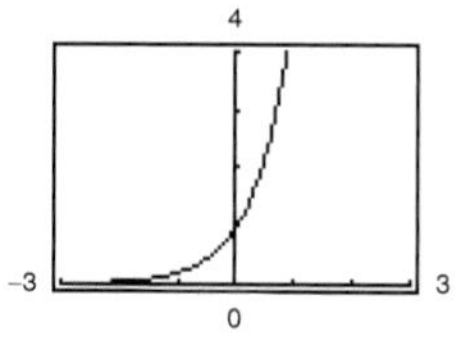

39.

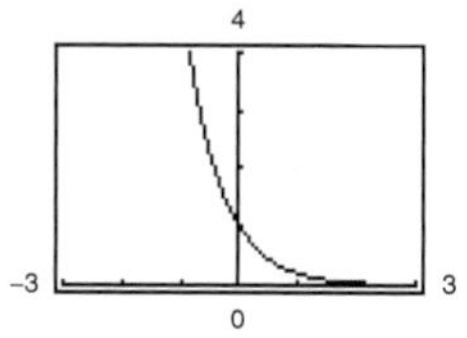

41.

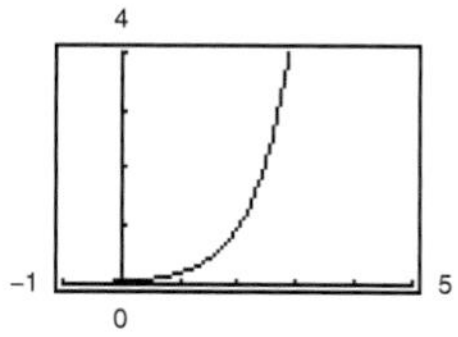

43.

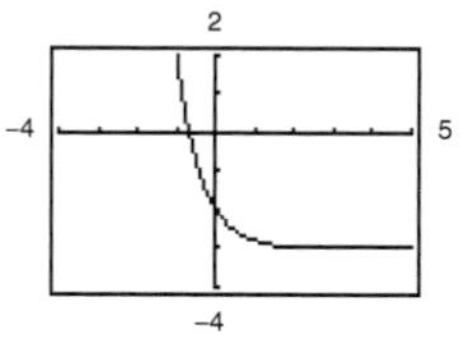

45.

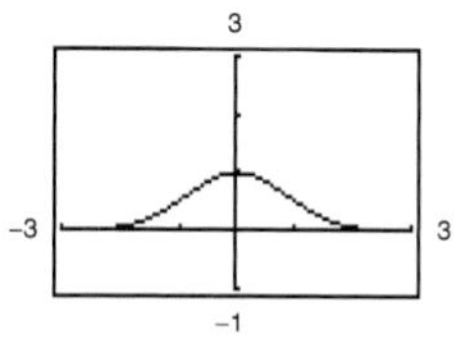

47.

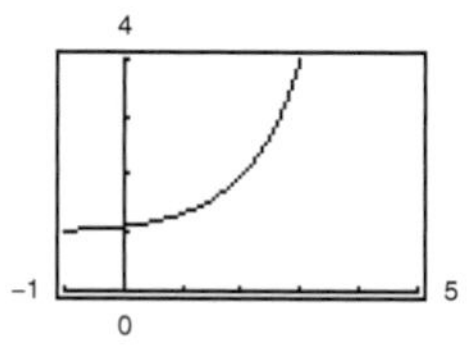

49.

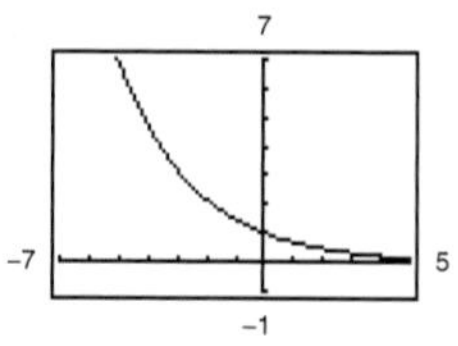

51.

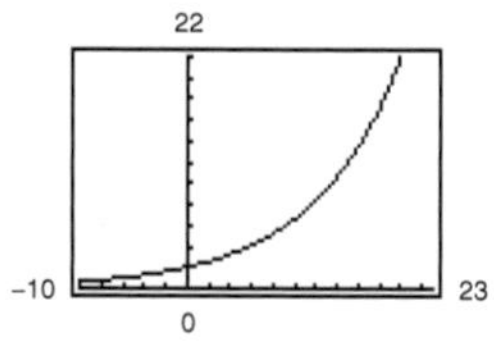

53.

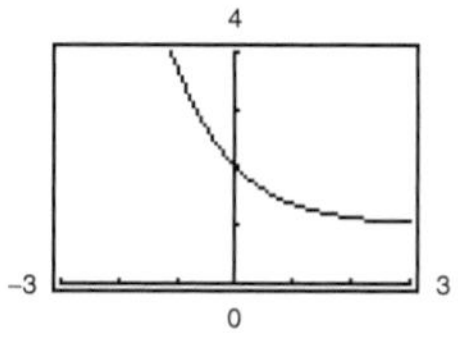

55.

n	1	2	4
A	\$5397.31	\$5477.81	\$5520.10

n	12	365	Continuous
A	\$5549.10	\$5563.36	\$5563.85

57.

n	1	2	4
A	\$11,652.39	\$12,002.55	\$12,188.60

n	12	365	Continuous
A	\$12,317.01	\$12,380.41	\$12,382.58

59.

t	1	10	20
A	\$12,999.44	\$26,706.49	\$59,436.39

t	30	40	50
A	\$132,278.12	\$294,390.36	\$655,177.80

61.

t	1	10	20
A	\$12,805.91	\$22,986.49	\$44,031.56

t	30	40	50
A	\$84,344.25	\$161,564.86	\$309,484.08

63. \$222,822.57

65. (a) The steeper curve represents the investment earning compound interest, because compound interest earns more than simple interest.

(b) $A = 500(1.07)^t$

$A = 500(0.07)t + 500$

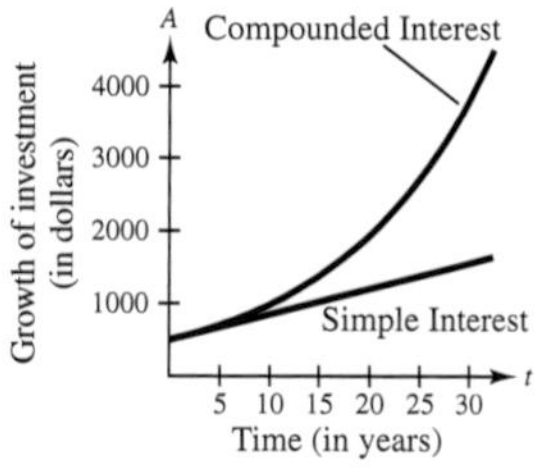

67. \$35.45 **69.** (a) 100 (b) 300 (c) 900

71. (a) 25 units (b) 16.30 units

(c)

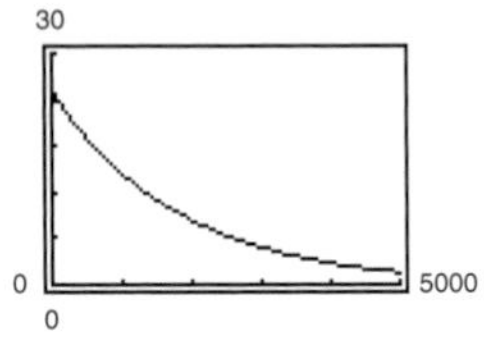

73. (a)

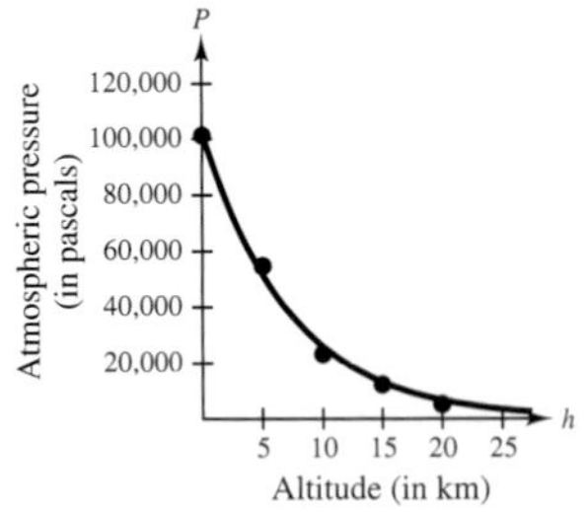

(b)

h	0	5	10	15	20
P	102,303	51,570	25,996	13,104	6606

(c) 34,190 pascals (d) ≈ 11.6 kilometers

75. True. As $x \to -\infty$, $f(x) \to 0$ but never reaches 0.

77. $f(x) = h(x)$ **79.** $f(x) = g(x) = h(x)$

81. (a) $x < 0$ (b) $x > 0$

83. (a)

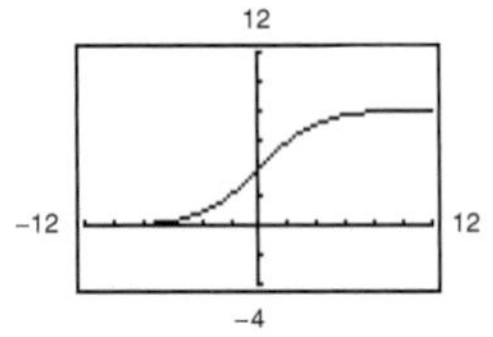

Horizontal asymptotes: $y = 0$, $y = 8$

(b)

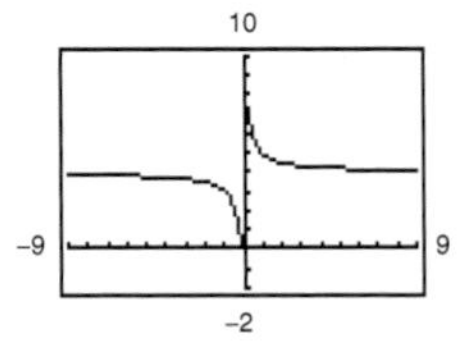

Horizontal asymptote: $y = 4$

Vertical asymptote: $x = 0$

85.

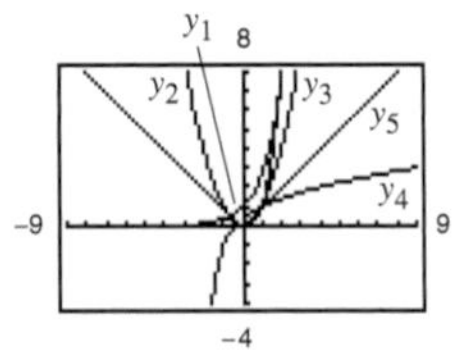

$y = e^x$

87. It usually implies rapid growth.

89. The value of $f(x)$ approaches e^r.

x	1	10	100	200	500
$[1 + (1/x)]^x$	2	2.5937	2.7048	2.7115	2.7156

x	1100	10,000
$[1 + (1/x)]^x$	2.7170	2.718

91. $1 < \sqrt{2} < 2$

$2^1 < 2^{\sqrt{2}} < 2^2$

93. $y_4 = 1 + \dfrac{x}{1!} + \dfrac{x^2}{2!} + \dfrac{x^3}{3!} + \dfrac{x^4}{4!}$

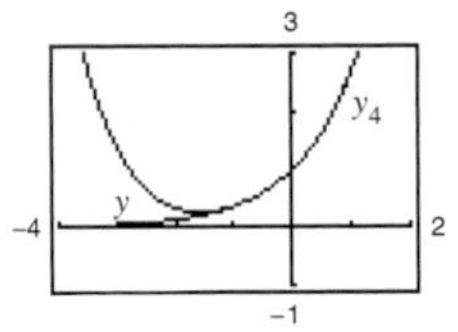

As more terms are added, the polynomial approaches e^x.

95. $y = \frac{1}{7}(2x + 14)$ **97.** $y = \pm\sqrt{25 - x^2}$

99.

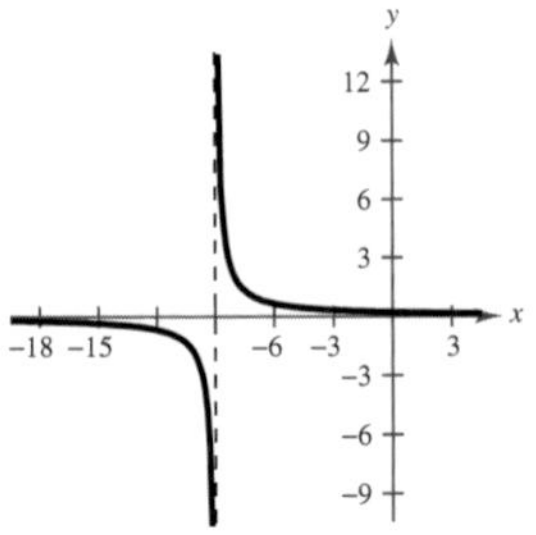

101.

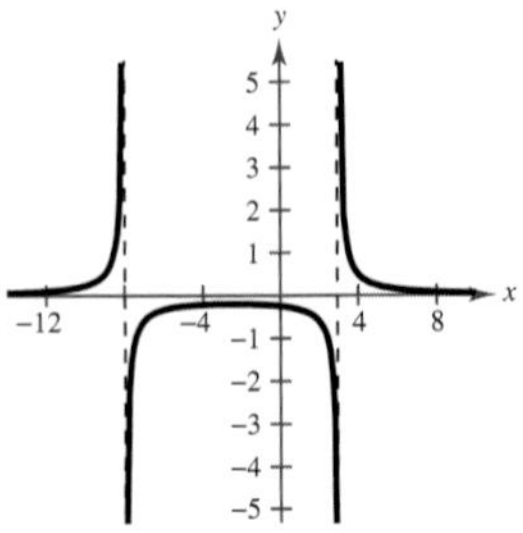

Section 3.2 *(page 307)*

1. $4^3 = 64$ **3.** $7^{-2} = \frac{1}{49}$ **5.** $32^{2/5} = 4$

7. $e^0 = 1$ **9.** $\log_5 125 = 3$ **11.** $\log_{81} 3 = \frac{1}{4}$

13. $\log_6 \frac{1}{36} = -2$ **15.** $\ln 20.0855 \ldots = 3$

17. $\ln 1 = 0$ **19.** 4 **21.** $\frac{1}{2}$ **23.** 0

25. -2 **27.** $\frac{5}{3}$ **29.** 3 **31.** 2

33. 2.538 **35.** -0.097 **37.** 2.913 **39.** -3.418

41. 1.005 **43.** -0.405 **45.** c **47.** d **49.** b

51. Domain: $(0, \infty)$

Intercept: $(1, 0)$

Vertical asymptote: $x = 0$

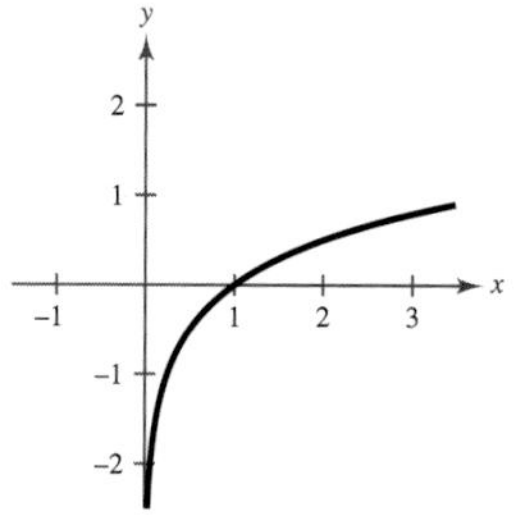

53. Domain: $(0, \infty)$

Intercept: $(9, 0)$

Vertical asymptote: $x = 0$

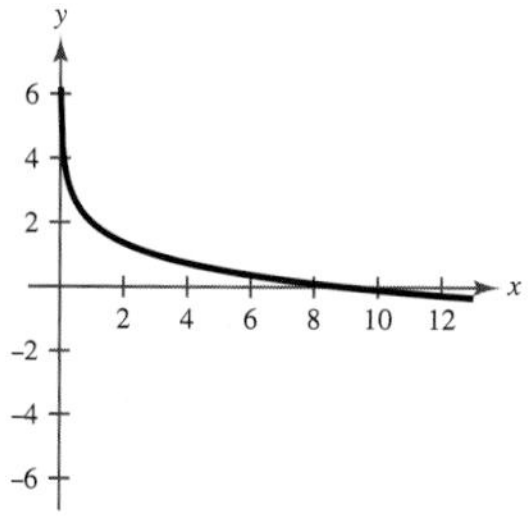

55. Domain: $(-2, \infty)$

Intercept: $(-1, 0)$

Vertical asymptote: $x = -2$

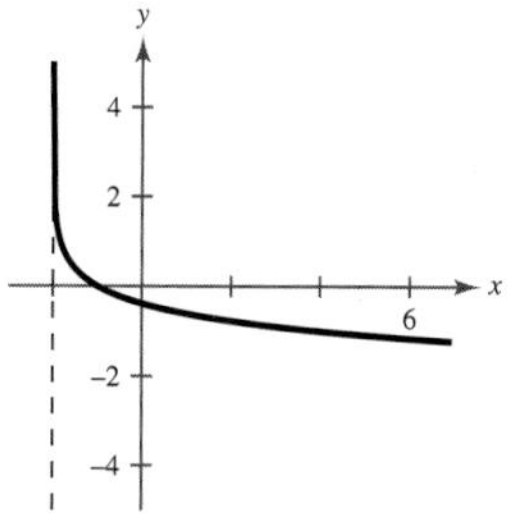

57. Domain: $(0, \infty)$

Intercept: $(5, 0)$

Vertical asymptote: $x = 0$

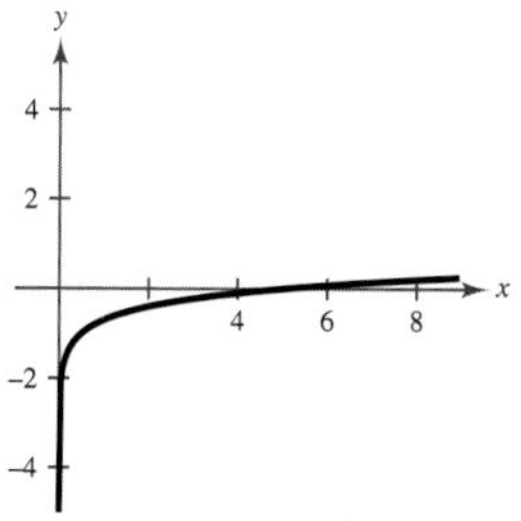

59. Domain: $(2, \infty)$

Intercept: $(3, 0)$

Vertical asymptote: $x = 2$

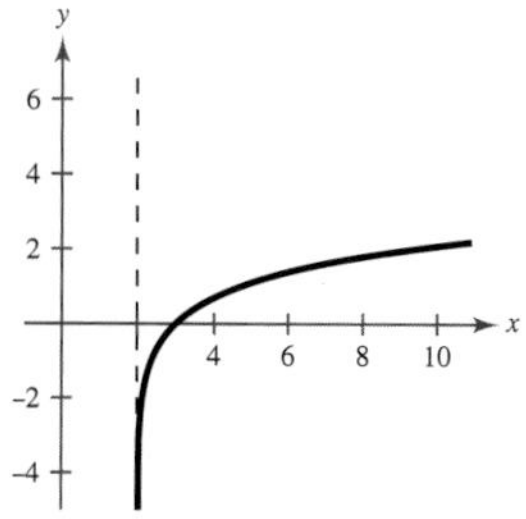

61. Domain: $(-\infty, 0)$

Intercept: $(-1, 0)$

Vertical asymptote: $x = 0$

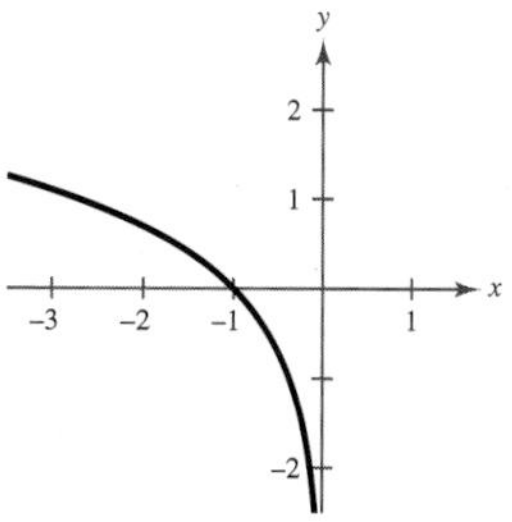

63.

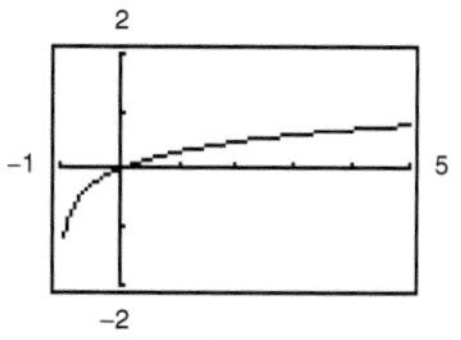

65.

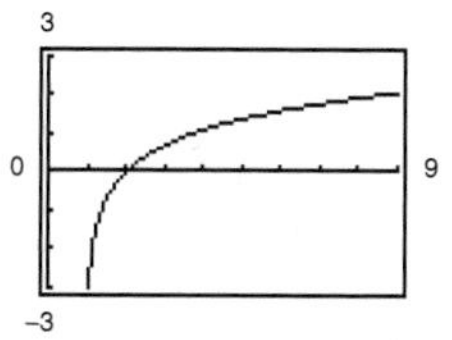

67.

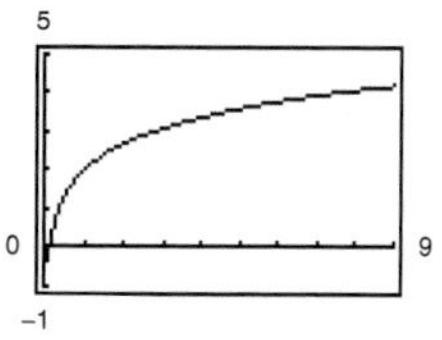

69. (a) 80 (b) 68.1 (c) 62.3

71. (a)

r	0.005	0.01	0.015	0.02	0.025	0.03
t	138.6	69.3	46.2	34.7	27.7	23.1

(b) Answers will vary.

73.

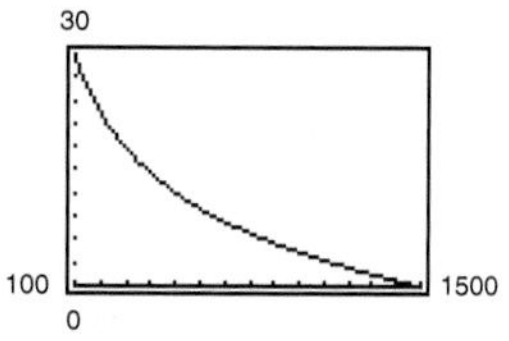

17.66 cubic feet per minute

75. 21,357 foot-pounds **77.** 30 years

79. Total amount: \$396,234

Interest: \$246,234

81. (a)

x	1	5	10	10^2
$f(x)$	0	0.322	0.230	0.046

x	10^4	10^6
$f(x)$	0.00092	0.0000138

(b) 0

(c)

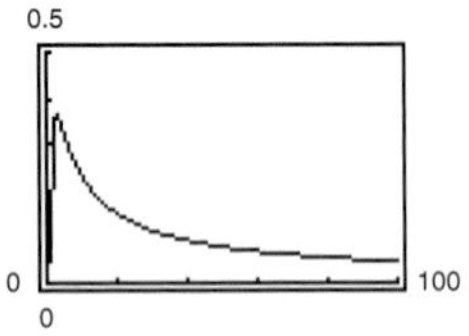

83. True. $\log_3 27 = 3 \Rightarrow 3^3 = 27$

85.

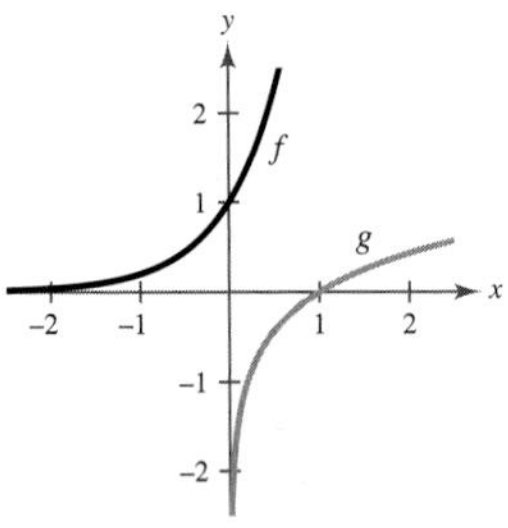

$g = f^{-1}$

87.

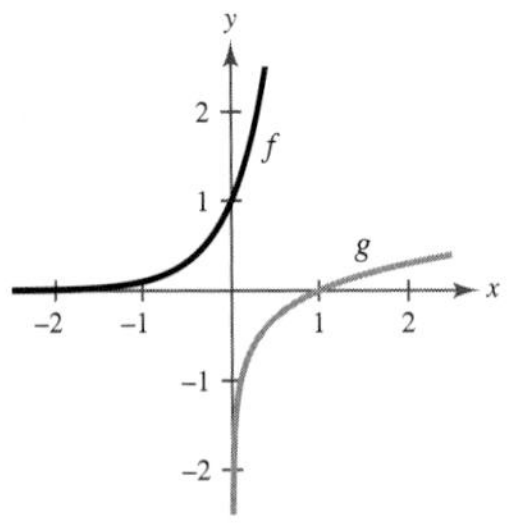

$g = f^{-1}$

89. (a) False (b) True (c) True (d) False

91. $y_4 = (x-1) - \frac{1}{2}(x-1)^2 + \frac{1}{3}(x-1)^3 - \frac{1}{4}(x-1)^4$

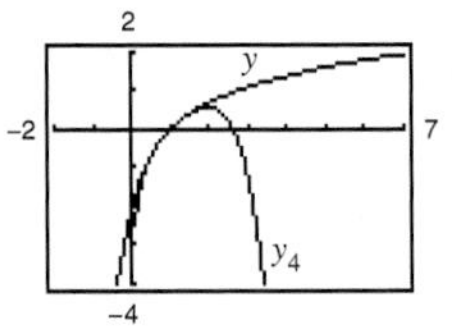

93. (a)

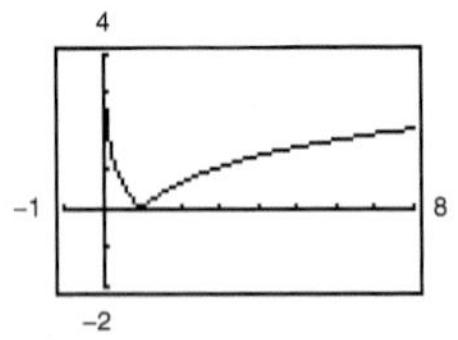

(b) Increasing: $(1, \infty)$

Decreasing: $(0, 1)$

(c) Relative minimum: $(1, 0)$

95. (a)

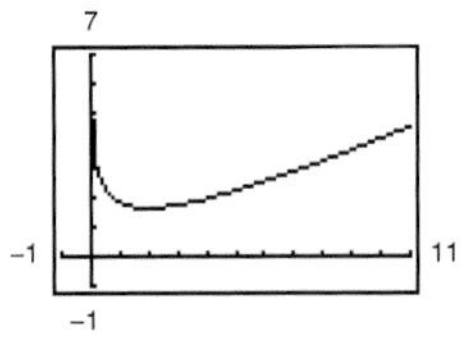

(b) Increasing: $(2, \infty)$

Decreasing: $(0, 2)$

(c) Relative minimum: $\left(2, 1 - \ln \frac{1}{2}\right)$

97. $8n - 3$ **99.** $83.95 + 37.50t$

101. Vertical asymptote: $x = -8$

Horizontal asymptote: $y = 0$

103. Vertical asymptotes: $x = -3, x = \frac{5}{2}$

Horizontal asymptote: $y = 0$

105. 403.429 **107.** 0.018

Section 3.3 *(page 315)*

1. 1.771 **3.** -2.000 **5.** -0.417 **7.** 2.633

9. (a) $\dfrac{\log_{10} x}{\log_{10} 5}$ (b) $\dfrac{\ln x}{\ln 5}$ **11.** (a) $\dfrac{\log_{10} x}{\log_{10} \frac{1}{5}}$ (b) $\dfrac{\ln x}{\ln \frac{1}{5}}$

13. (a) $\dfrac{\log_{10} \frac{3}{10}}{\log_{10} x}$ (b) $\dfrac{\ln \frac{3}{10}}{\ln x}$

15. (a) $\dfrac{\log_{10} x}{\log_{10} 2.6}$ (b) $\dfrac{\ln x}{\ln 2.6}$

17. $f(x) = \dfrac{\log_{10} x}{\log_{10} 2} = \dfrac{\ln x}{\ln 2}$

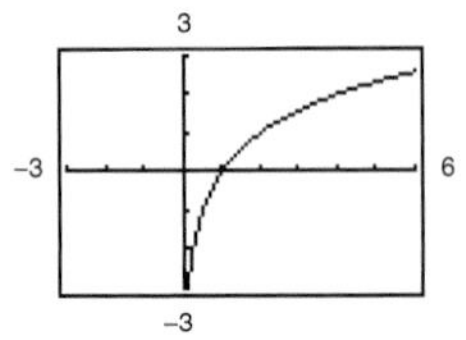

19. $f(x) = \dfrac{\log_{10} x}{\log_{10} \frac{1}{2}} = \dfrac{\ln x}{\ln \frac{1}{2}}$

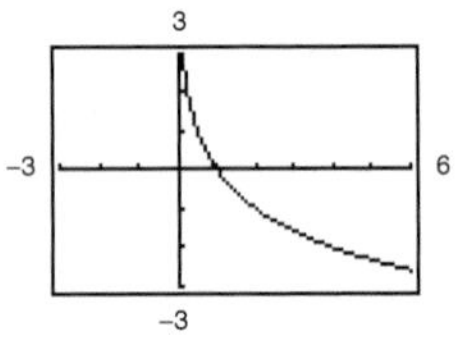

21. $f(x) = \dfrac{\log_{10} x}{\log_{10} 11.8} = \dfrac{\ln x}{\ln 11.8}$

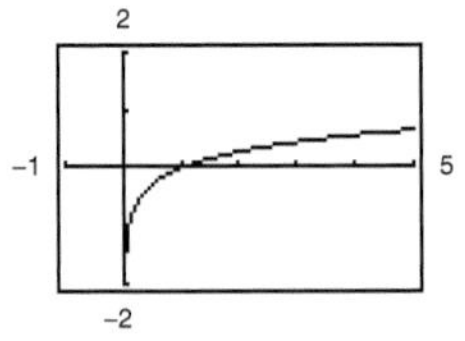

23. $\log_{10} 5 + \log_{10} x$ **25.** $\log_{10} 5 - \log_{10} x$

27. $4 \log_8 x$ **29.** $\frac{1}{2} \ln z$ **31.** $\ln x + \ln y + \ln z$

33. $\frac{1}{2} \ln(a - 1)$ **35.** $\ln z + 2 \ln(z - 1)$

37. $\frac{1}{3} \ln x - \frac{1}{3} \ln y$ **39.** $4 \ln x + \frac{1}{2} \ln y - 5 \ln z$

41. $2 \log_b x - 2 \log_b y - 3 \log_b z$

43. $\ln 3x$ **45.** $\log_4 \dfrac{z}{y}$ **47.** $\log_2 (x + 4)^2$

49. $\log_3 \sqrt[4]{5x}$ **51.** $\ln \dfrac{x}{(x+1)^3}$ **53.** $\ln \dfrac{x-2}{x+2}$

55. $\ln \dfrac{x}{(x^2-4)^4}$ **57.** $\ln \sqrt[3]{\dfrac{x(x+3)^2}{x^2-1}}$

59. $\ln \dfrac{\sqrt[3]{y(y+4)^2}}{y-1}$ **61.** $\ln \dfrac{9}{\sqrt{x^2+1}}$

63. $\log_2 \frac{32}{4} = \log_2 32 - \log_2 4$; Property 2

65. 2 **67.** 2.4 **69.** -9 is not in the domain of $\log_3 x$.

71. 2 **73.** -3 **75.** 0 is not in the domain of $\log_{10} x$.

77. 4.5 **79.** $\frac{3}{2}$ **81.** $-3 - \log_5 2$ **83.** $6 + \ln 5$

85. (a) 90 (b) 77 (c) 73 (d) 9 months

(e) $90 - \log_{10}(t+1)^{15}$

(f)

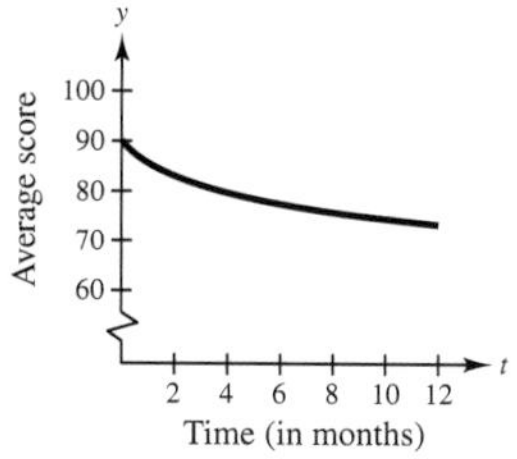

87. False. $\ln 1 = 0$ **89.** False. $\ln(x-2) \neq \ln x - \ln 2$

91. False. $u = v^2$ **93.** Answers will vary.

95.

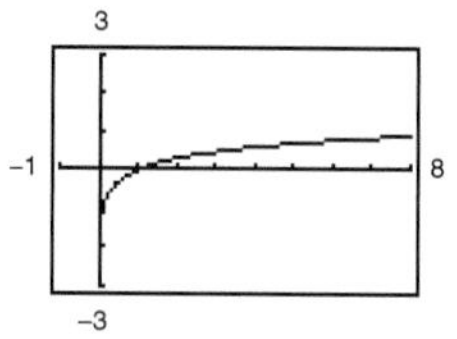

97.

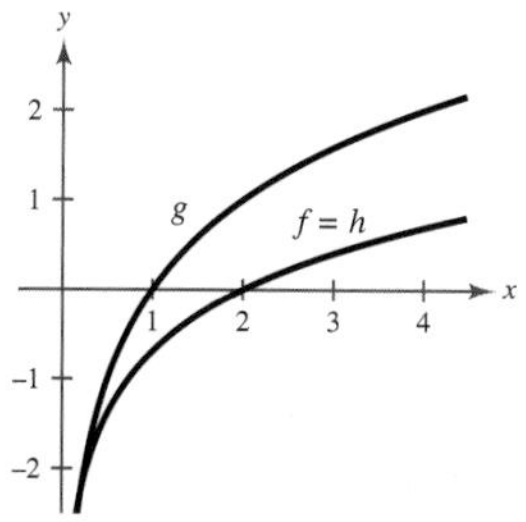

$f(x) = h(x)$; Property 2

99. $\dfrac{3x^4}{2y^3}$, $x \neq 0$ **101.** $1, x \neq 0, y \neq 0$

103. 2502.655 **105.** 0.002 **107.** 4.325

109. 1.415 **111.** 2.361

Section 3.4 *(page 324)*

1. (a) Yes (b) No

3. (a) No (b) Yes (c) Yes, approximate

5. (a) Yes, approximate (b) No (c) Yes **7.** 2

9. 4 **11.** -2 **13.** -5 **15.** 3 **17.** 4

19. 2 **21.** $\ln 2 \approx 0.693$ **23.** $e^{-1} \approx 0.368$

25. 64 **27.** 100 **29.** $\frac{1}{10}$ **31.** (3, 8) **33.** (9, 2)

35. x^2 **37.** $x - 2, x > 2$ **39.** $7x + 2$

41. $5x + 2, x > -\frac{2}{5}$ **43.** $2x - 1$ **45.** $\ln 10 \approx 2.303$

47. 0 **49.** $\dfrac{\ln 12}{3} \approx 0.828$ **51.** $\ln \dfrac{5}{3} \approx 0.511$

53. $\ln 5 \approx 1.609$ **55.** $2 \ln 75 \approx 8.635$

57. $\log_{10} 42 \approx 1.623$ **59.** $\dfrac{\ln 80}{2 \ln 3} \approx 1.994$

61. 2 **63.** $\dfrac{\ln 8 - \ln 565}{\ln 2} \approx -6.142$

65.

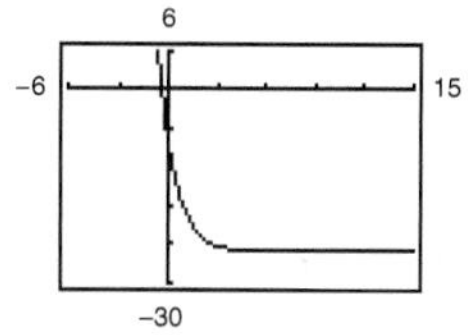

-0.427

67.

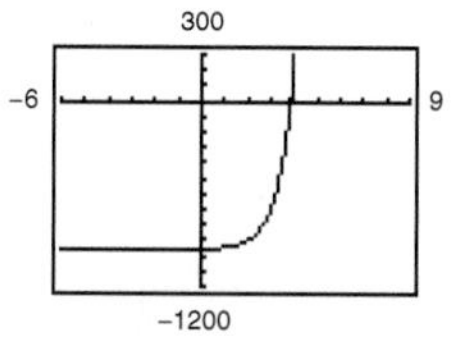

3.847

69.

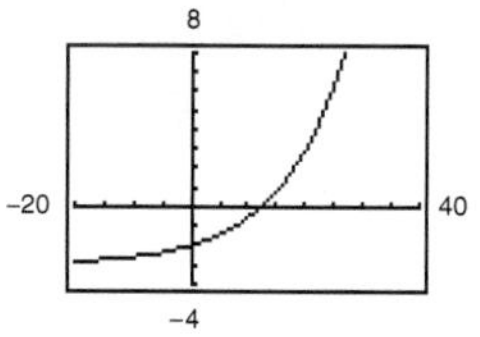

12.207

71.

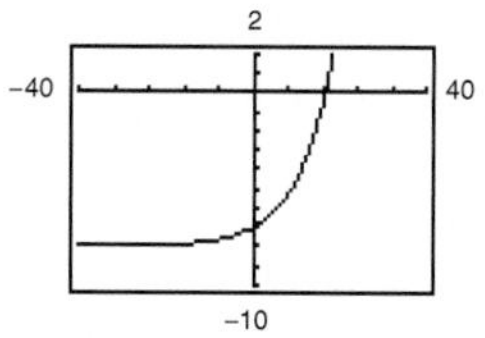

16.636

73. $\dfrac{1}{3} \log_{10}\left(\dfrac{3}{2}\right) \approx 0.059$ **75.** $1 + \dfrac{\ln 7}{\ln 5} \approx 2.209$

77. $\dfrac{\ln 4}{365 \ln\left(1 + \frac{0.065}{365}\right)} \approx 21.330$

79. $\dfrac{\ln 2}{12 \ln\left(1 + \frac{0.10}{12}\right)} \approx 6.960$ **81.** $\dfrac{\ln 1498}{2} \approx 3.656$

83. $e^{-3} \approx 0.050$ **85.** $\dfrac{e^{2.4}}{2} \approx 5.512$

87. $\dfrac{e^{10/3}}{5} \approx 5.606$ **89.** $e^2 - 2 \approx 5.389$

91. $e - 1 \approx 1.718$ and $-e - 1 \approx -3.718$

93. $1 + \sqrt{1 + e} \approx 2.928$ **95.** No solution

97. $10^2 + 3 = 103$ **99.** $2(3^{11/6}) \approx 14.988$

101. $\dfrac{-1 + \sqrt{17}}{2} \approx 1.562$ **103.** 2

105. $\dfrac{725 + 125\sqrt{33}}{8} \approx 180.384$

107.

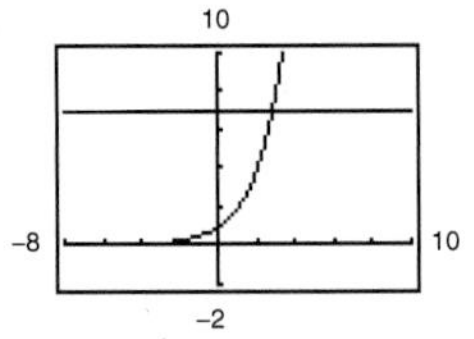

(2.807, 7)

109.

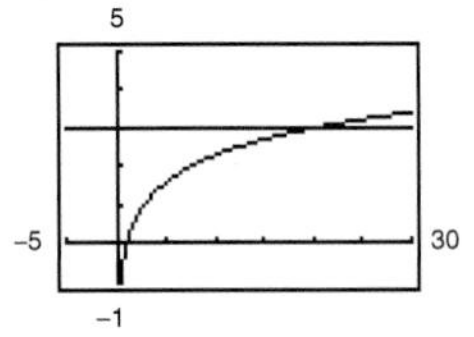

(20.086, 3)

111. 8.2 years **113.** 12.9 years

115. (a) 1426 units (b) 1498 units

117. (a)

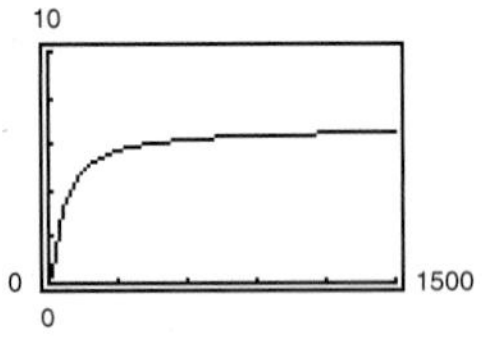

(b) $V = 6.7$. Yield will approach 6.7 million cubic feet per acre.

(c) 29.3 years

119. (a) $y = 100$ and $y = 0$; the range falls between 0% and 100%.

(b) Males: 69.71 inches Females: 64.51 inches

121. (a) $T = 20$; Room temperature (b) ≈ 0.81 hour

123. $\log_b uv = \log_b u + \log_b v$

True by Property 1 in Section 3.3.

125. $\log_b(u - v) = \log_b u - \log_b v$

False. $1.95 \approx \log_{10}(100 - 10) \neq \log_{10}100 - \log_{10}10 = 1$

127. For $rt < \ln 2$ years, double the amount you invest. For $rt > \ln 2$ years, double the interest rate or double the number of years, because either of these will double the exponent in the exponential function.

129. No. It is dependent on the interest rate.

131. $4\sqrt{2} - 10$ **133.** $\frac{1}{2}\sqrt{10} + 1$ **135.** $t = \frac{k}{s^3}$

137. $x = \frac{k}{b - 3}$ **139.** 1.226 **141.** −5.595

Section 3.5 *(page 335)*

1. c **3.** b **5.** d

	Initial Investment	*Annual % Rate*	*Time to Double*	*Amount after 10 Years*
7.	\$1000	12%	5.78 yr	\$3,320.12
9.	\$750	8.9438%	7.75 yr	\$1,834.37
11.	\$500	11.0%	6.3 yr	\$1,505.00
13.	\$6376.28	4.5%	15.4 yr	\$10,000.00

15. \$112,087.09

17. (a) 6.642 years (b) 6.330 years (c) 6.302 years (d) 6.301 years

19.

r	2%	4%	6%	8%	10%	12%
t	54.93	27.47	18.31	13.73	10.99	9.16

21.

r	2%	4%	6%	8%	10%	12%
t	55.48	28.01	18.85	14.27	11.53	9.69

23.

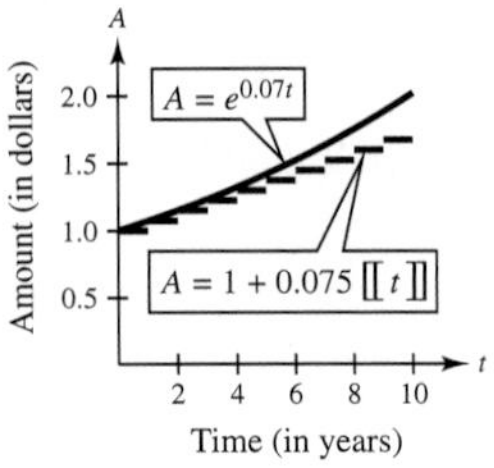

Continuous compounding

	Isotope	*Half-life (Years)*	*Initial Quantity*	*Amount After 1000 Years*
25.	^{226}Ra	1620	10 g	6.52 g
27.	^{14}C	5730	2.26 g	2 g
29.	^{239}Pu	24,360	2.16 g	2.1 g

31. $y = e^{0.7675x}$ **33.** $y = 5e^{-0.4024x}$

35. 2023 **37.** $k = 0.0112$; 2796

39. (a) Croatia: $y = 5e^{-0.0018t}$; 4.7 million

Mali: $y = 9.9e^{0.0314t}$; 27.9 million

Singapore: $y = 3.5e^{0.0090t}$; 4.7 million

Sweden: $y = 8.9e^{0.0028t}$; 9.8 million

(b) The greater the rate of growth, the greater the value of b.

(c) b determines whether the population is increasing ($b > 0$) or decreasing ($b < 0$).

41. 61.16 hours **43.** 15,683 years

45. (a) $V = -750t + 2000$ (b) $V = 2000e^{-0.6931t}$

(c)

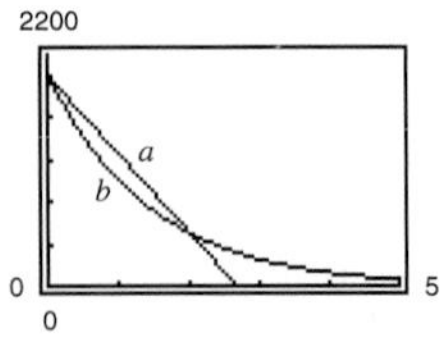

Exponential

(d) 1 year:
Straight-line, \$1250; Exponential, \$1000
3 years:
Straight-line, −\$250; Exponential, \$250

(e) Decreases \$750 per year

47. (a) $S = \dfrac{500{,}000}{1 + 0.6e^{0.053t}}$ (b) 280,771 units

49. \$583,275

51. (a)

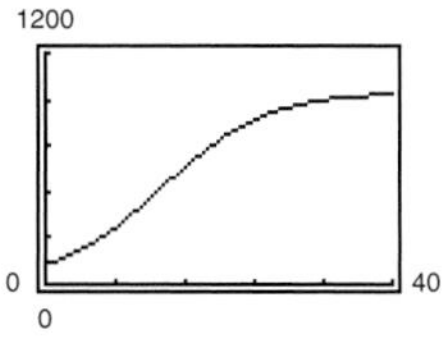

Asymptotes: $y = 0, y = 1000$. The population size will approach 1000 as time increases.

(b) 203 animals

(c) 13 months

53. (a) 398,107,171 (b) 5,011,872 (c) 50,118,723

55. (a) 30 decibels (b) 85 decibels

(c) 90 decibels (d) 115 decibels

57. 97% **59.** 4.95

61. $10^{-3.2} \approx 6.3 \times 10^{-4}$ moles per liter **63.** 10

65. (a)

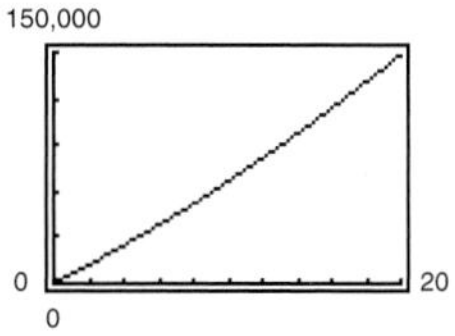

(b) ≈ 21 years; Yes

67. 7:30 A.M.

69. False. A logistics growth function never has an x-intercept.

71. (a) Logarithmic (b) Logistic (c) Exponential (d) Linear (e) None of the above (f) Exponential

73. Answers will vary.

75. $8x^2 - 24x + 18$

77. $x^3 - 5x^2 + 25x - 128 + \dfrac{641}{x + 5}$

79.

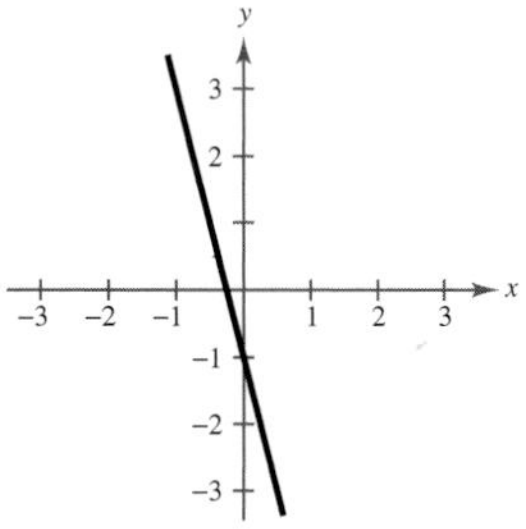

81.

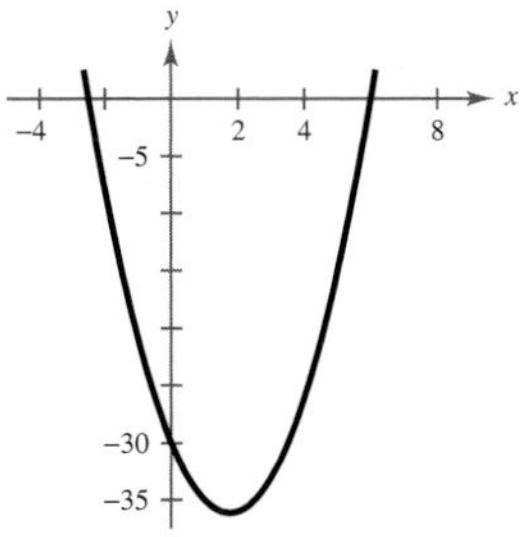

83.

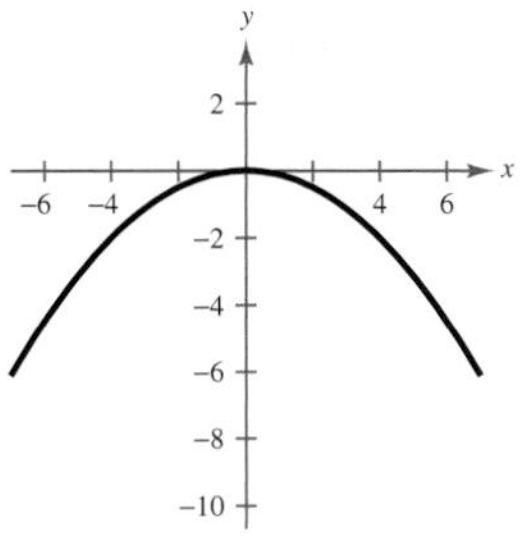

85.

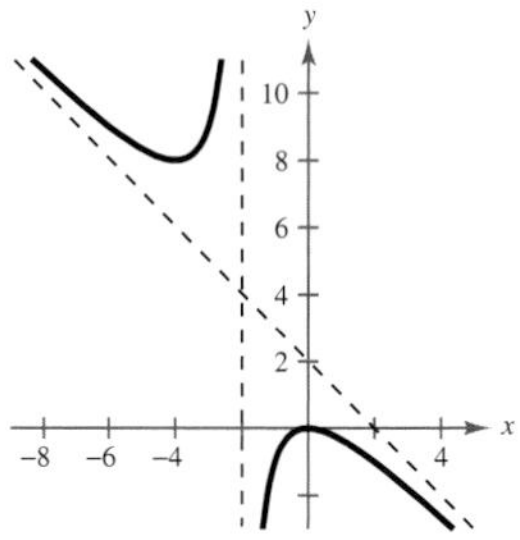

87.

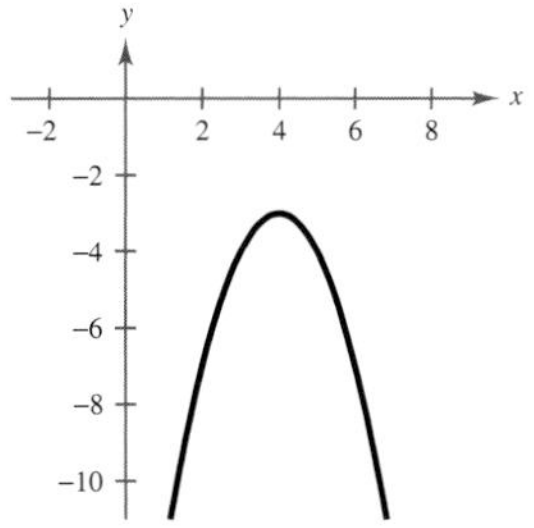

89.

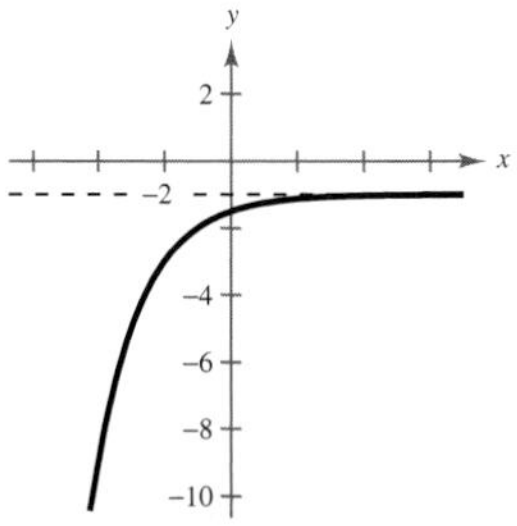

91.

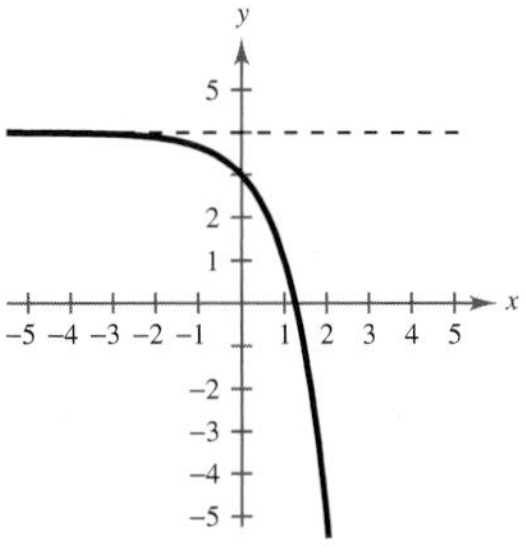

Review Exercises *(page 342)*

1. 76.699 **3.** 0.337 **5.** 1201.845 **7.** c **9.** a

11.

x	-1	0	1	2	3
$f(x)$	8	5	4.25	4.063	4.016

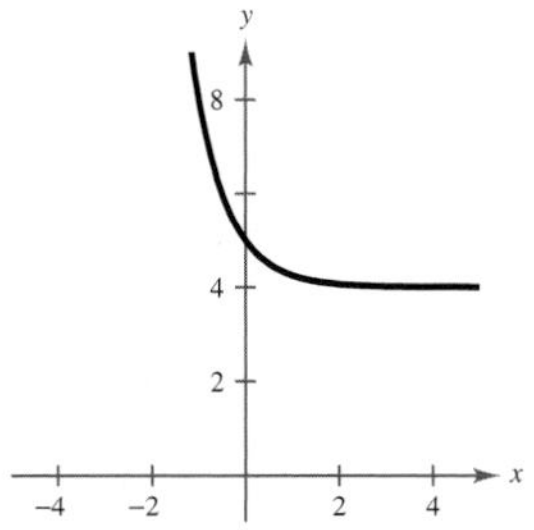

13.

x	-2	-1	0	1	2
$f(x)$	-0.377	-1	-2.65	-7.023	-18.61

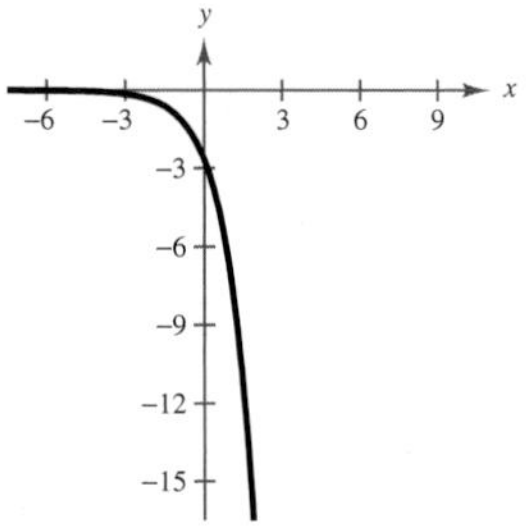

15.

x	-1	0	1	2	3
$f(x)$	4.008	4.04	4.2	5	9

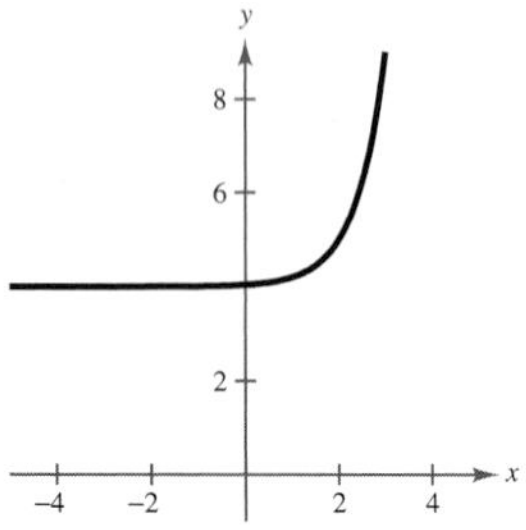

17.

x	-2	-1	0	1	2
$f(x)$	3.25	3.5	4	5	7

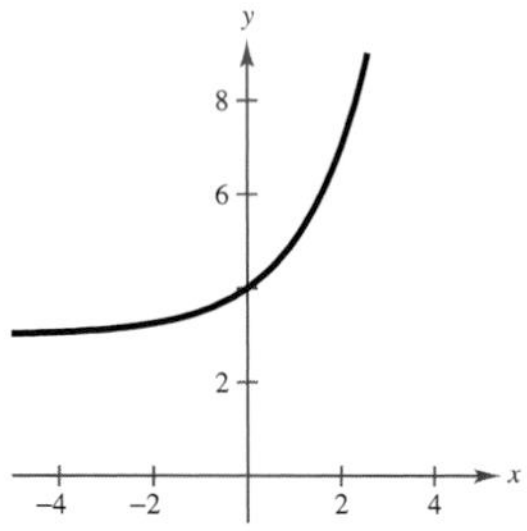

19. 2980.958 **21.** 0.183

23.

x	-2	-1	0	1	2
$h(x)$	2.72	1.65	1	0.61	0.37

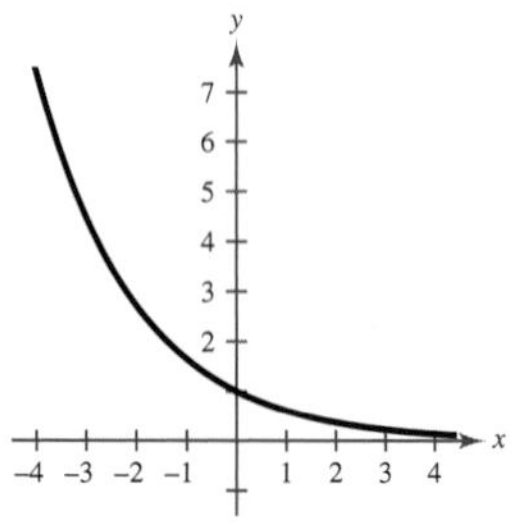

25.

x	-3	-2	-1	0	1
$f(x)$	0.37	1	2.72	7.39	20.09

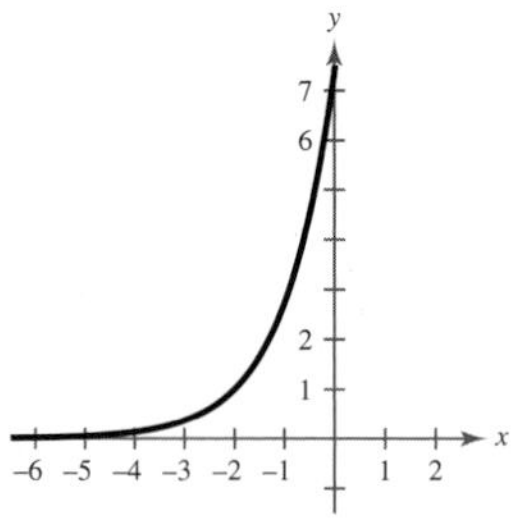

27.

n	1	2	4	12
A	\$6569.98	\$6635.43	\$6669.46	\$6692.64

n	365	Continuous
A	\$6704.00	\$6704.39

29.

t	1	10	20
P	\$184,623.27	\$89,865.79	\$40,379.30

t	30	40	50
P	\$18,143.59	\$8152.44	\$3663.13

31. (a) 0.154 (b) 0.487 (c) 0.811

33. (a) \$1,069,047.14 (b) 7.9 years

35. $\log_4 64 = 3$ **37.** 3 **39.** -3

41.

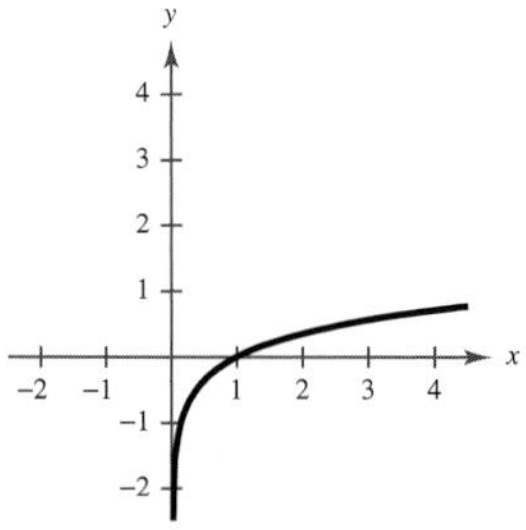

Vertical asymptote: $x = 0$

43.

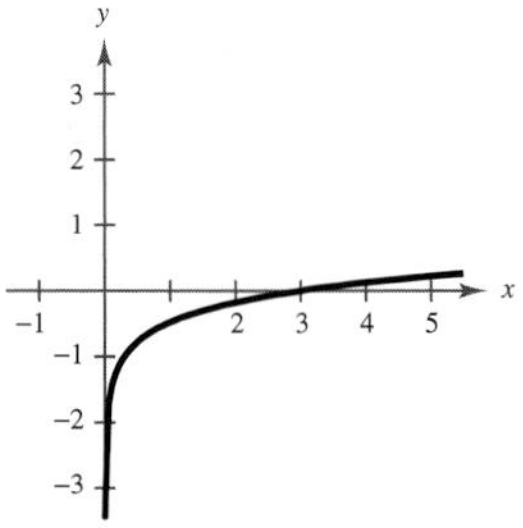

Vertical asymptote: $x = 0$

45.

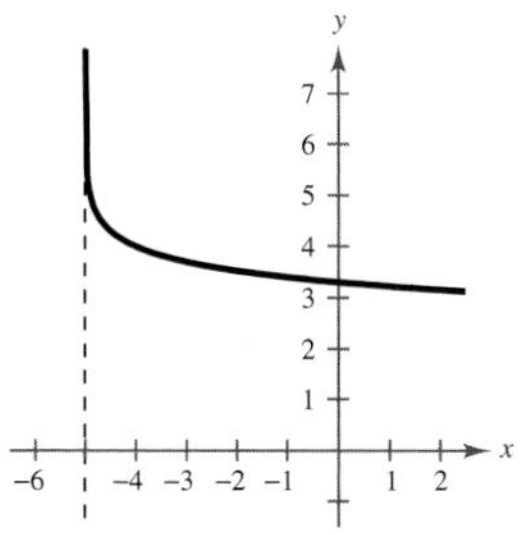

Vertical asymptote: $x = -5$

47. 3.118 **49.** -12 **51.** 2.034

53.

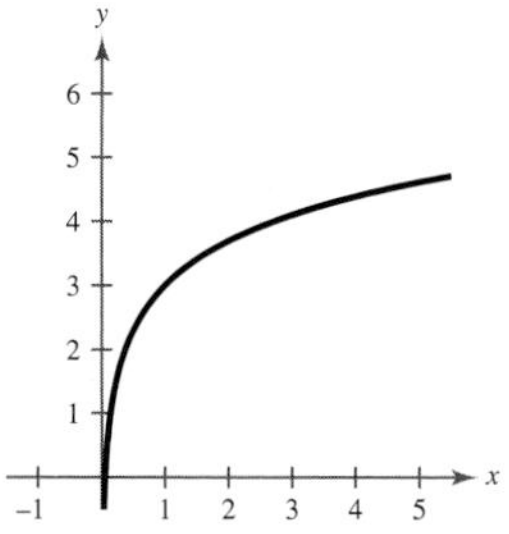

Vertical asymptote: $x = 0$

55.

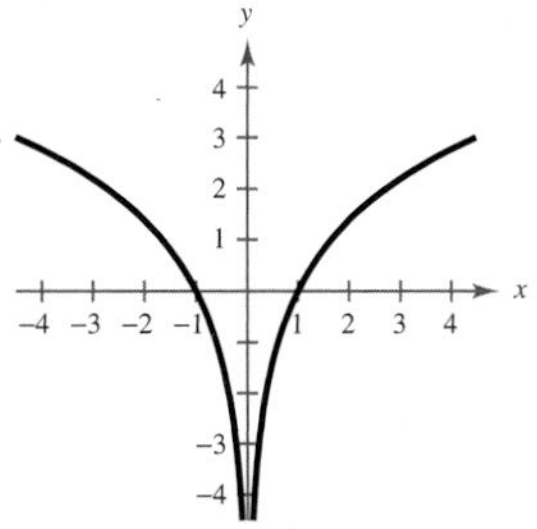

Vertical asymptote: $x = 0$

57. 27.16 miles **59.** 2.132 **61.** -1.159

63. and 65. Answers will vary.

67. $1 + 2\log_5|x|$ **69.** $\log_{10} 5 + \frac{1}{2}\log_{10} y - 2\log_{10}|x|$

71. $\log_2 5x$ **73.** $\ln \dfrac{\sqrt{|2x-1|}}{(x+1)^2}$

75. (a) $0 \le h < 18{,}000$

(b)

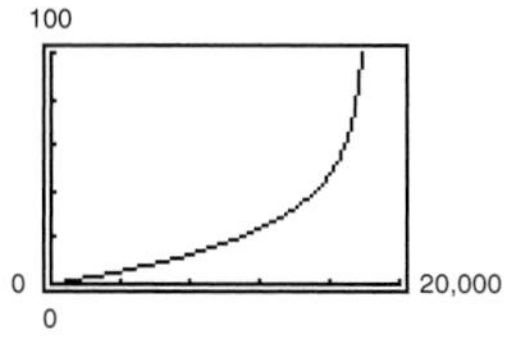

Vertical asymptote: $h = 18{,}000$

(c) Climbing at a slower rate, so the time required increases.

(d) 5.46 minutes

77. 6 **79.** 6 **81.** 3 **83.** $\dfrac{\ln 25}{3} \approx 1.073$

85. $\frac{1}{3}(\ln 40 - 2) \approx 0.563$ **87.** $\ln 20 \approx 2.996$

89. $\dfrac{\ln 95}{\ln 12} \approx 1.833$ **91.** $\ln 2 \approx 0.693, \ln 4 \approx 1.386$

93. $-7.04, -1.53$ **95.** 0.68 **97.** $\frac{1}{5}e^{7.2} \approx 267.886$

99. $\frac{1}{3}e^{15/4} \approx 14.174$ **101.** $e^6 - 8 \approx 395.429$

103. $5e^4 \approx 272.991$ **105.** $-2 + \sqrt{6} \approx 0.449$

107. -104 **109.** 0, 0.42, 13.63 **111.** $-3.99, 1.48$

113. 21.3 years **115.** e **117.** f **119.** a

121. 2025 **123.** (a) 13.8629% (b) $11,486.98

125. $y = \frac{1}{2}e^{0.4605x}$

127. (a) 7.7 weeks (b) 13.3 weeks

129. (a) 251,188,643 (b) 7,079,458 (c) 1,258,925,412

131. True by properties of exponents.

$$e^{x-1} = e^x \cdot e^{-1} = \frac{e^x}{e}$$

133. False. $\ln(x \cdot y) = \ln x + \ln y \neq \ln(x + y)$

135. False. The domain of $f(x) = \ln x$ is $(0, \infty)$.

Chapter Test *(page 347)*

1. 1123.690 **2.** 687.291 **3.** 0.497 **4.** 22.198

5.

x	-1	$-\frac{1}{2}$	0	$\frac{1}{2}$	1
$f(x)$	10	3.162	1	0.316	0.1

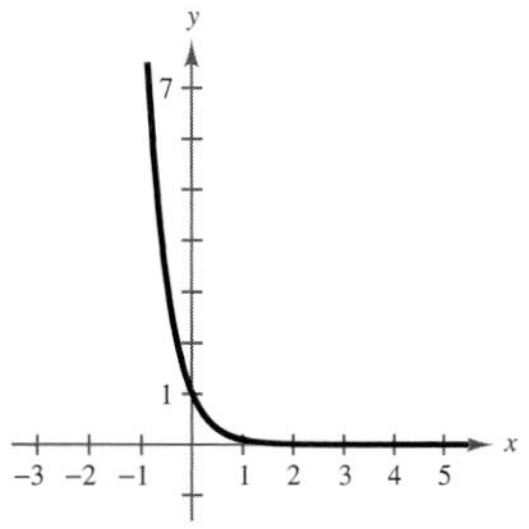

6.

x	-1	0	1	2	3
$f(x)$	-0.005	-0.028	-0.167	-1	-6

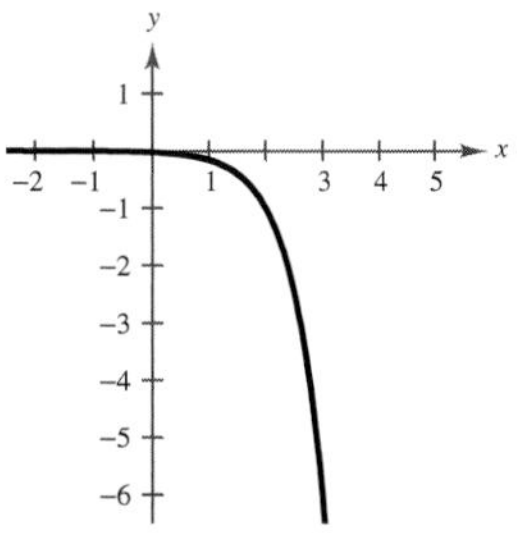

7.

x	-1	$-\frac{1}{2}$	0	$\frac{1}{2}$	1
$f(x)$	0.865	0.632	0	-1.718	-6.389

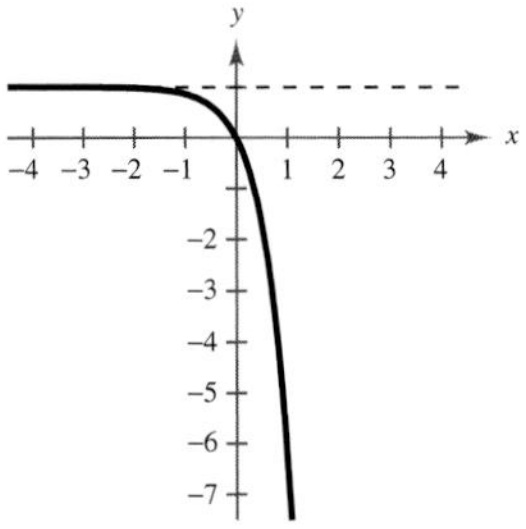

8. (a) -0.89 (b) 9.2

9.

x	$\frac{1}{2}$	1	$\frac{3}{2}$	2	4
$f(x)$	-5.699	-6	-6.176	-6.301	-6.602

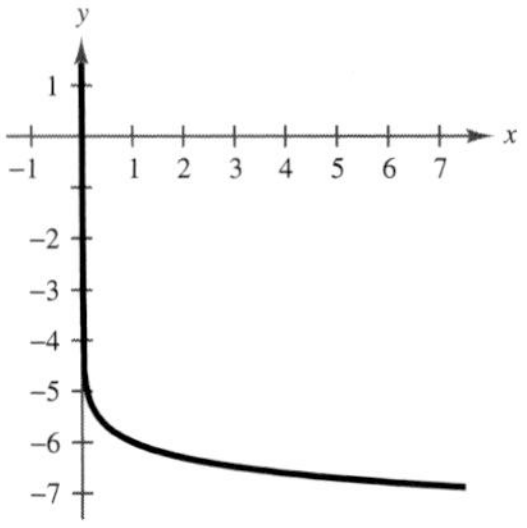

Vertical asymptote: $x = 0$

10.

x	5	7	9	11	13
$f(x)$	0	1.099	1.609	1.946	2.197

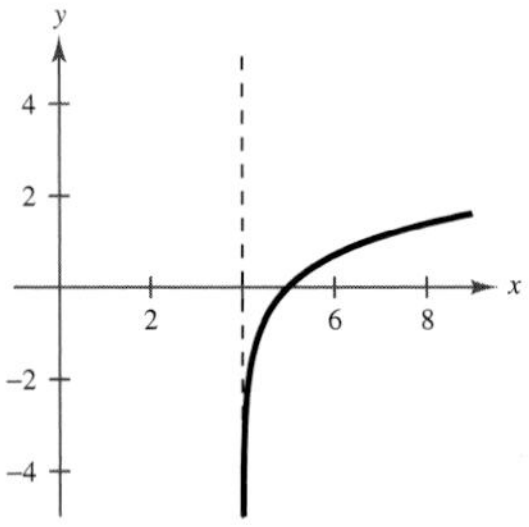

Vertical asymptote: $x = 4$

11.

x	-5	-3	-1	0	1
$f(x)$	1	2.099	2.609	2.792	2.946

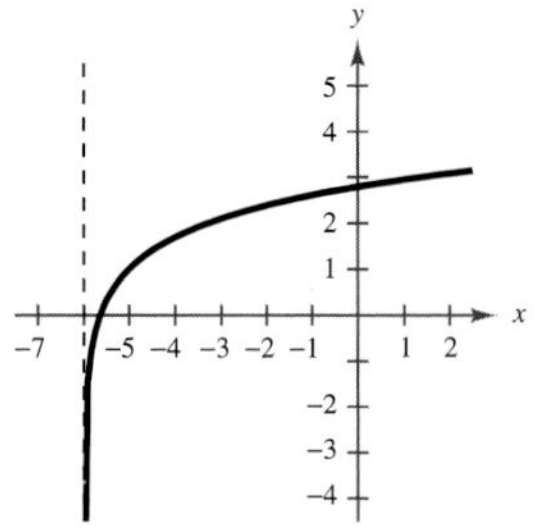

Vertical asymptote: $x = -6$

12. 1.945 **13.** 0.115 **14.** 1.328

15. $\log_2 3 + 4 \log_2 a$ **16.** $\ln 5 + \frac{1}{2} \ln x - \ln 6$

17. $\log_3 13y$ **18.** $\ln \frac{x^4}{y^4}$ **19.** $\frac{\ln 197}{4} \approx 1.321$

20. $\frac{800}{501} \approx 1.597$ **21.** $y = 2745e^{0.1570x}$ **22.** 55%

23. (a)

x	$\frac{1}{4}$	1	2	4	5	6
H	58.720	75.332	86.828	103.43	110.59	117.38

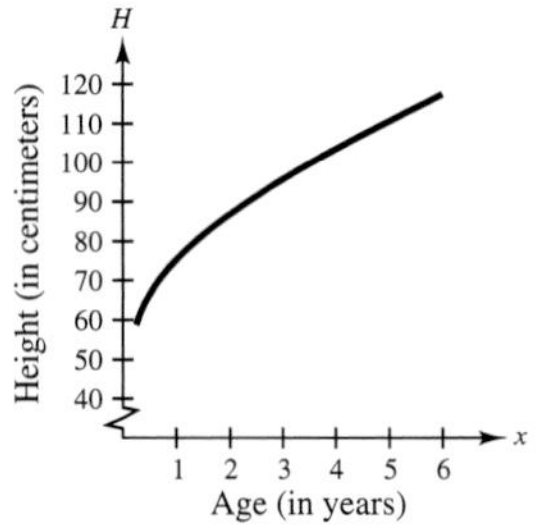

(b) 103 centimeters; 103.43 centimeters

Cumulative Test for Chapters 1–3
(page 348)

1.

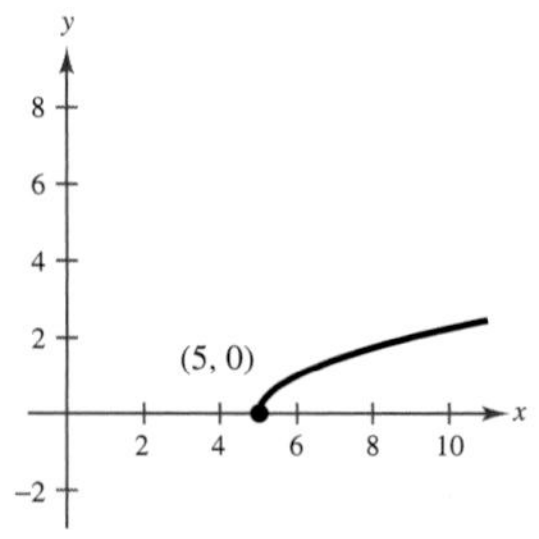

2.

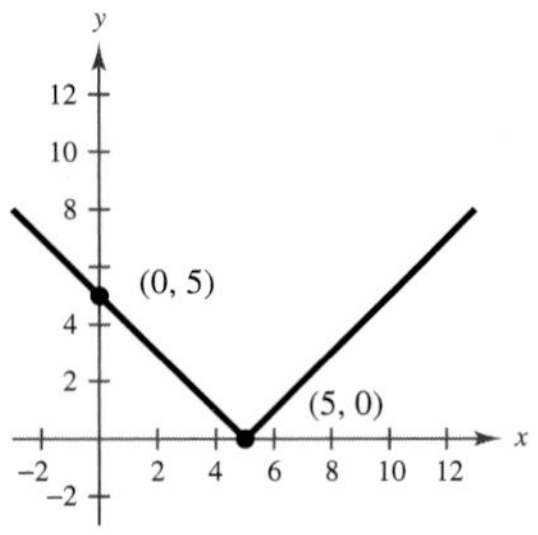

3.

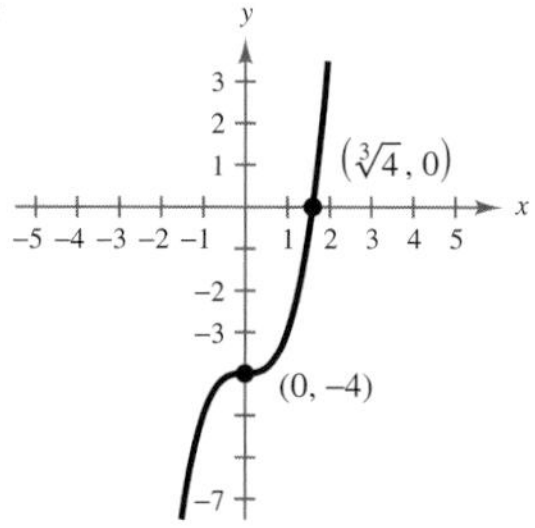

4. $2x - y + 2 = 0$

5. For some values of x there correspond two values of y.

6. (a) $\frac{3}{2}$ (b) Division by 0 is undefined. (c) $\frac{s + 2}{s}$

7. (a) Vertical shrink by $\frac{1}{2}$

(b) Vertical shift of 2 units upward

(c) Horizontal shift of 2 units to the left

8. (a) $5x - 2$ (b) $-3x - 4$

(c) $4x^2 - 11x - 3$ (d) $\frac{x - 3}{4x + 1}$

All real numbers $x \neq -\frac{1}{4}$

9. (a) $\sqrt{x - 1} + x^2 + 1$ (b) $\sqrt{x - 1} - x^2 - 1$

(c) $x^2\sqrt{x - 1} + \sqrt{x - 1}$ (d) $\frac{\sqrt{x - 1}}{x^2 + 1}$

All real numbers $x \geq 1$

10. (a) $2x + 12$ (b) $\sqrt{2x^2 + 6}$

11. (a) $|x| - 2$ (b) $|x - 2|$ **12.** $h^{-1}(x) = \frac{1}{5}(x + 2)$

13. $n = 9$ **14.** $y = -\frac{3}{4}(x + 8)^2 + 5$

15.

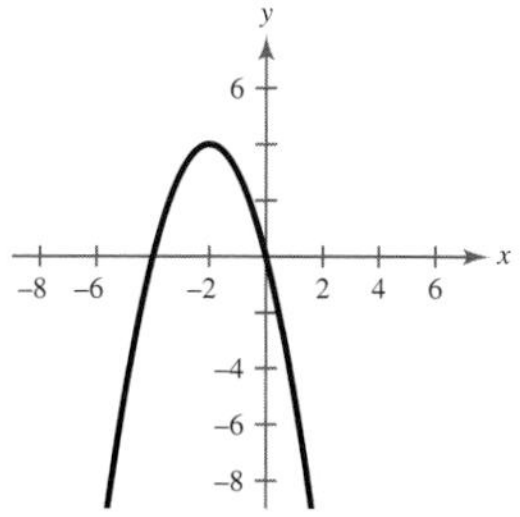

16.

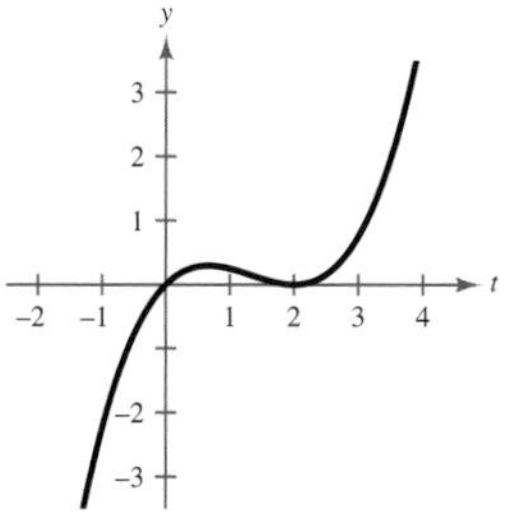

17.

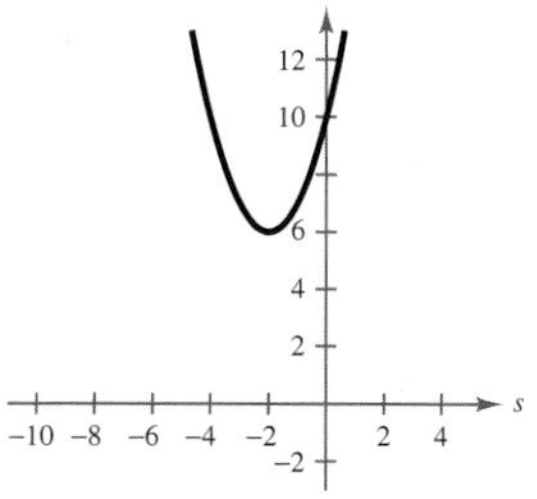

18. $-2, \pm 2i$ **19.** $-7, 0, 3$ **20.** $3x - 2 - \frac{3x - 2}{2x^2 + 1}$

21. $2x^3 - x^2 + 2x - 10 + \frac{25}{x + 2}$ **22.** 1.20

23. $x^4 + 3x^3 - 11x^2 + 9x + 70$

24. Intercept: (0, 0)

Vertical asymptote: $x = 3$

Horizontal asymptote: $y = 2$

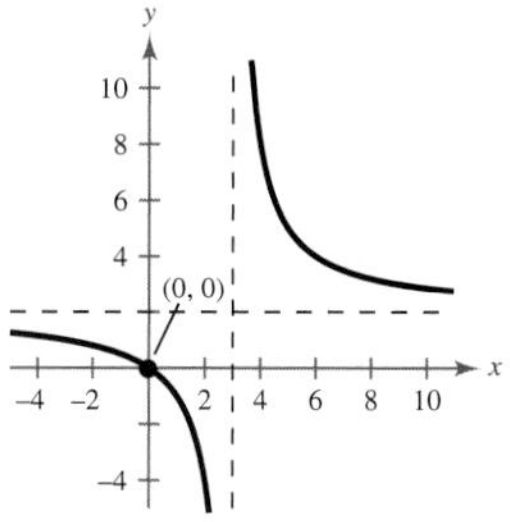

25. Intercept: (0, 0)

Vertical asymptote: $x = 5$

Slant asymptote:
$y = 4x + 20$

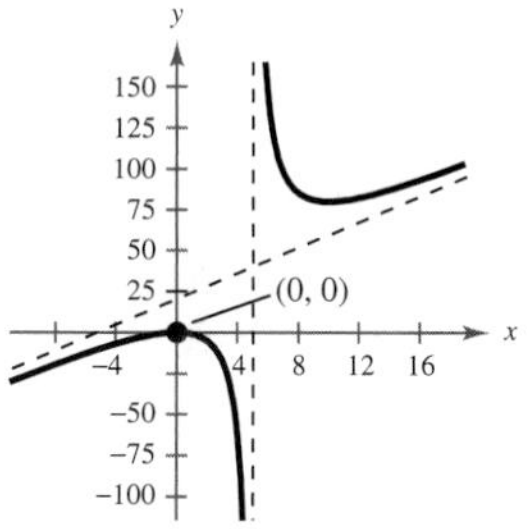

26. Intercept: (0, 0)

Vertical asymptotes: $x = \pm 3$

Horizontal asymptote: $y = 0$

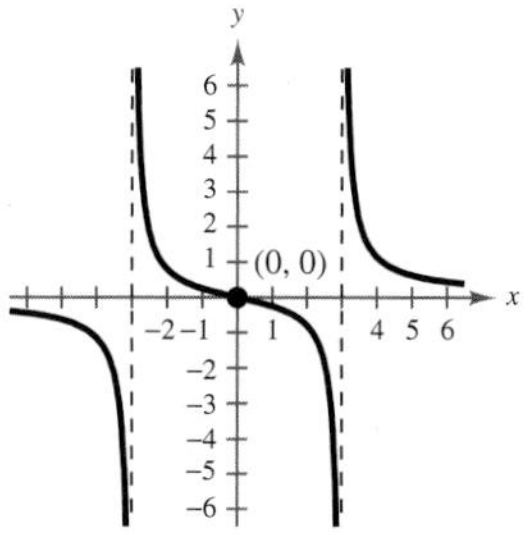

27. $\frac{1}{5}\left(\frac{4}{x - 7} - \frac{4}{x + 3}\right)$ **28.** $\frac{5}{x - 4} + \frac{20}{(x - 4)^2}$

29. Reflect f in x-axis and y-axis and shift f 3 units to the right.

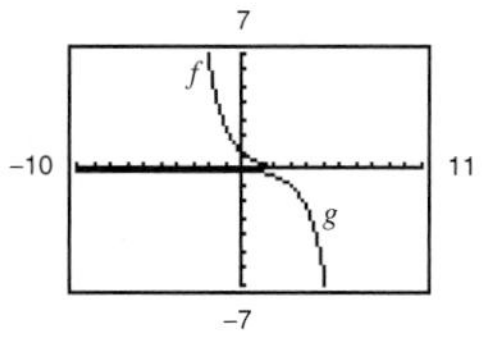

30. Reflect f in x-axis and shift f 4 units upward.

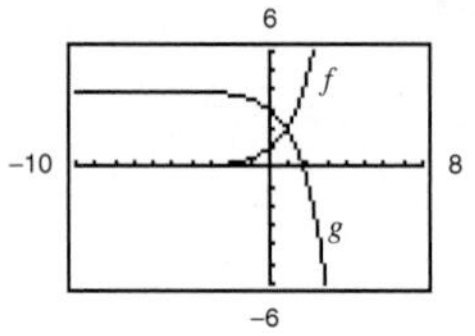

31. 1.991 **32.** -0.067 **33.** 1.717 **34.** 0.281

35. 0.302 **36.** -1.733 **37.** -4.087

38. $\ln(x + 4) + \ln(x - 4) - 4 \ln x,\ x > 4$

39. $\ln \dfrac{x^2}{\sqrt{x + 5}},\ x > -5, x \neq 0$

40. $\dfrac{\ln 12}{2} \approx 1.242$ **41.** $\dfrac{64}{5}$

42.

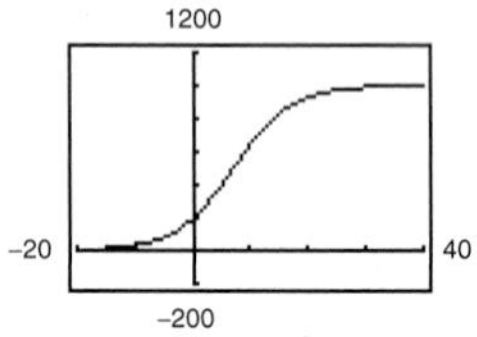

$y = 0, y = 1000$

43. (a) and (b)

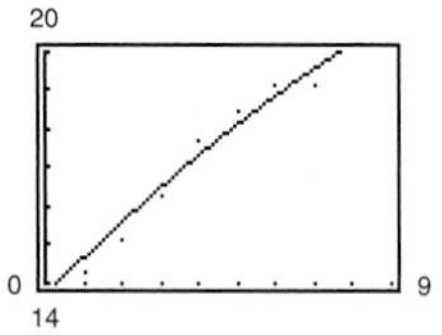

(c) No, because the model will eventually become negative.

Chapter 4

Section 4.1 *(page 359)*

1. 2 **3.** -3 **5.** 1

7. (a) Quadrant I (b) Quadrant III

9. (a) Quadrant IV (b) Quadrant II

11. (a) Quadrant III (b) Quadrant II

13. (a) (b)

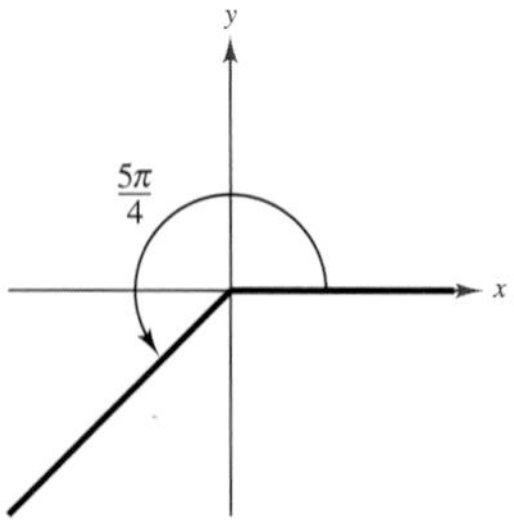

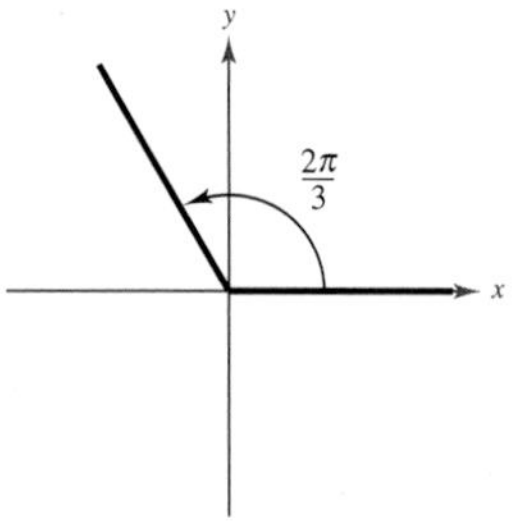

15. (a) (b)

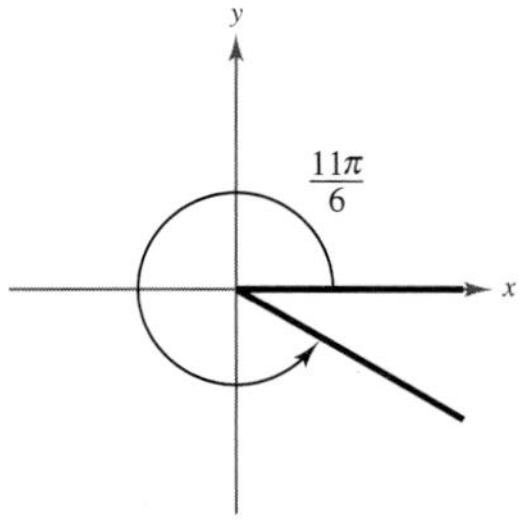

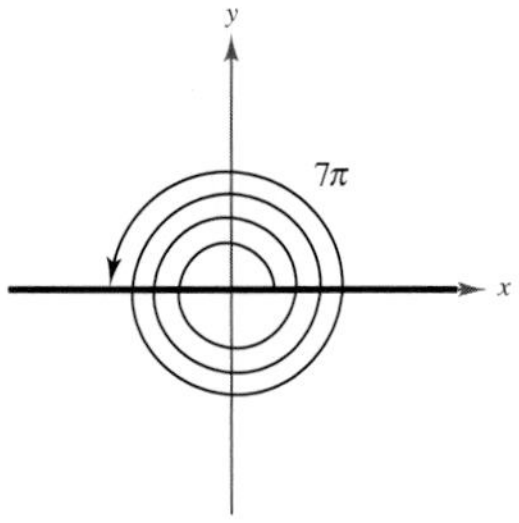

17. (a) $\dfrac{13\pi}{6}, -\dfrac{11\pi}{6}$ (b) $\dfrac{17\pi}{6}, -\dfrac{7\pi}{6}$

19. (a) $\dfrac{8\pi}{3}, -\dfrac{4\pi}{3}$ (b) $\dfrac{25\pi}{12}, -\dfrac{23\pi}{12}$

21. (a) Complement: $\dfrac{\pi}{6}$; Supplement: $\dfrac{2\pi}{3}$

(b) Complement: none; Supplement: $\dfrac{\pi}{4}$

23. (a) Complement: $\dfrac{\pi}{2} - 1 \approx 0.57$;

Supplement: $\pi - 1 \approx 2.14$

(b) Complement: none; Supplement: $\pi - 2 \approx 1.14$

25. (a) $\dfrac{\pi}{6}$ (b) $\dfrac{5\pi}{6}$ **27.** (a) $-\dfrac{\pi}{9}$ (b) $-\dfrac{4\pi}{3}$

29. 2.007 **31.** -3.776 **33.** 9.285 **35.** -0.014

37. (a) 270° (b) 210° **39.** (a) 420° (b) $-66°$

41. 25.714° **43.** 337.5° **45.** $-756°$

47. $-114.592°$ **49.** 210° **51.** $-60°$ **53.** 165°

55. (a) Quadrant II (b) Quadrant IV

57. (a) Quadrant III (b) Quadrant I

59. (a) (b)

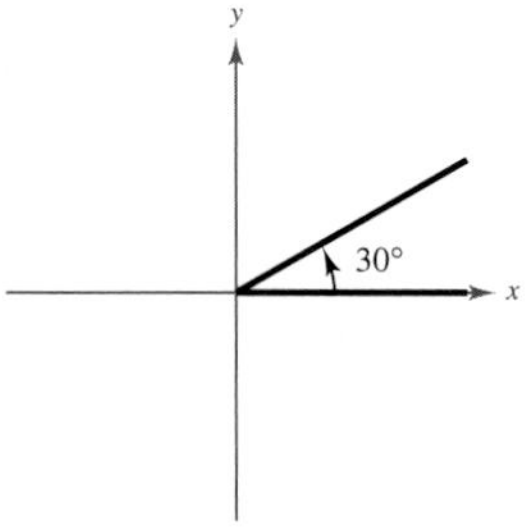

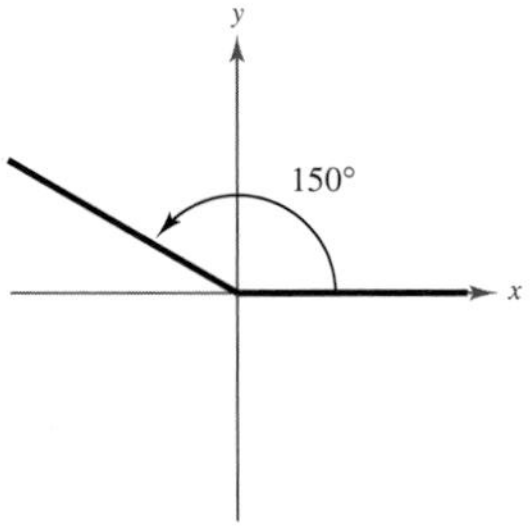

61. (a)

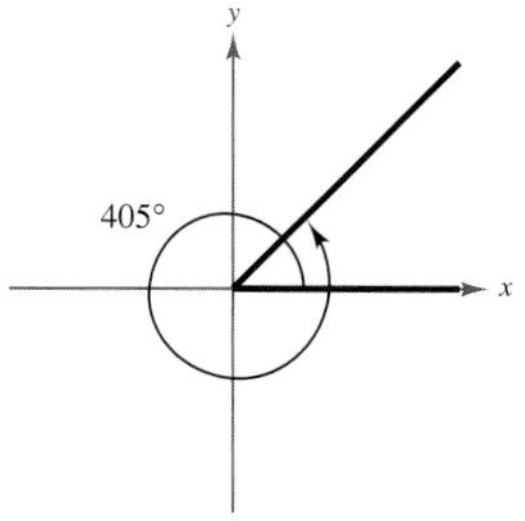

(b)

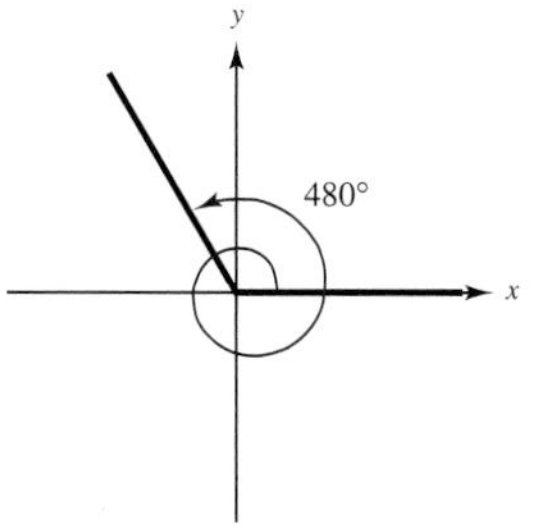

63. (a) 405°, −315° (b) 324°, −396°

65. (a) 600°, −120° (b) 180°, −540°

67. (a) Complement: 72°; Supplement: 162°
(b) Complement: none; Supplement: 65°

69. (a) Complement: 11°; Supplement: 101°
(b) Complement: none; Supplement: 30°

71. (a) 54.75° (b) −128.5°

73. (a) 85.308° (b) 330.007°

75. (a) 240° 36′ (b) −145° 48′

77. (a) 2° 30′ (b) −3° 34′48″ **79.** $\frac{6}{5}$ radians

81. $\frac{32}{7}$ radians **83.** $\frac{2}{9}$ radian **85.** $\frac{50}{29}$ radians

87. 15π inches ≈ 47.12 inches **89.** 3 meters

91. 591.72 miles **93.** 1141.02 miles

95. 0.063 radian ≈ 3.59° **97.** $\frac{5}{12}$ radian

99. (a) 728.3 revolutions per minute
(b) 4576 radians per minute

101. 20.16π inches per second

103. False. A measurement of 4π radians corresponds to two complete revolutions from the initial to the terminal side of an angle.

105. (a) The vertex is at the origin and the initial side is on the positive x-axis.
(b) Clockwise rotation of the terminal side
(c) Two angles in standard position where the terminal sides coincide
(d) The magnitude of the angle is between 90° and 180°.

107. Radian. 1 radian ≈ 57.3° **109.** Answers will vary.

111.

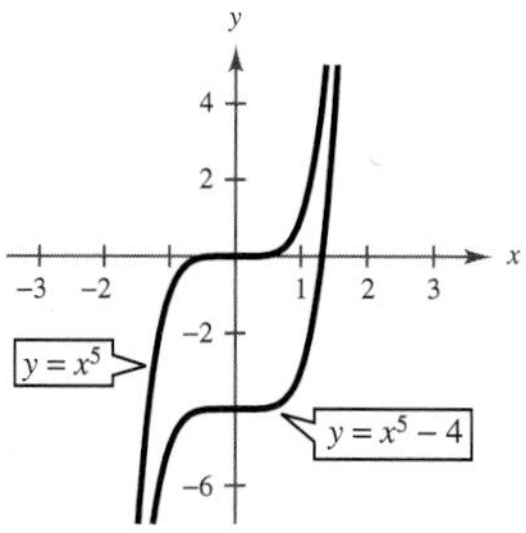

113.

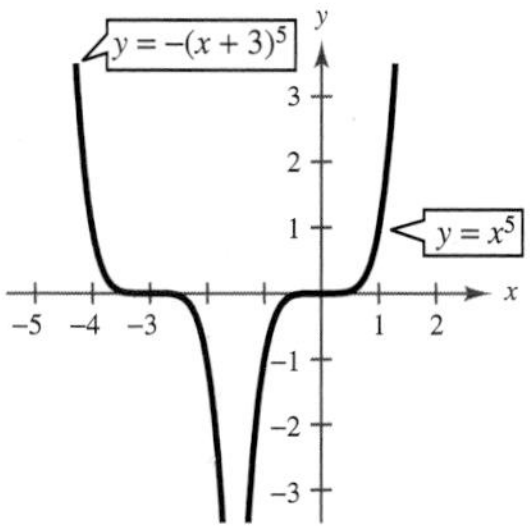

115.

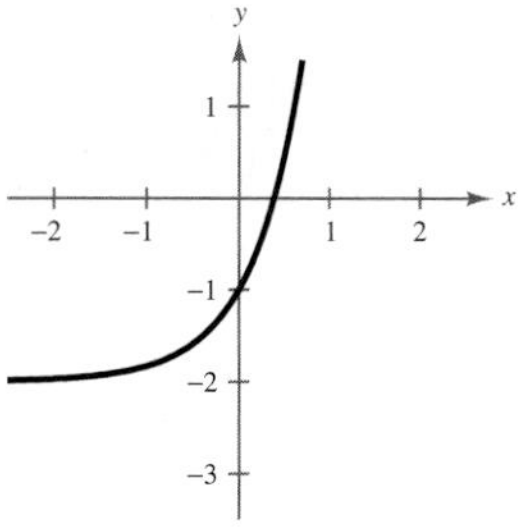

117.

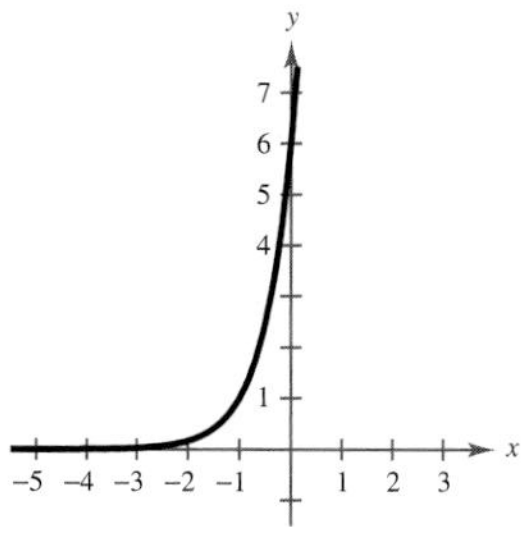

119.

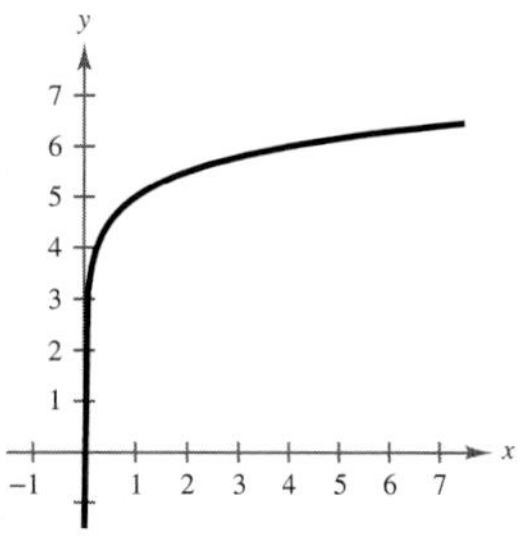

121.

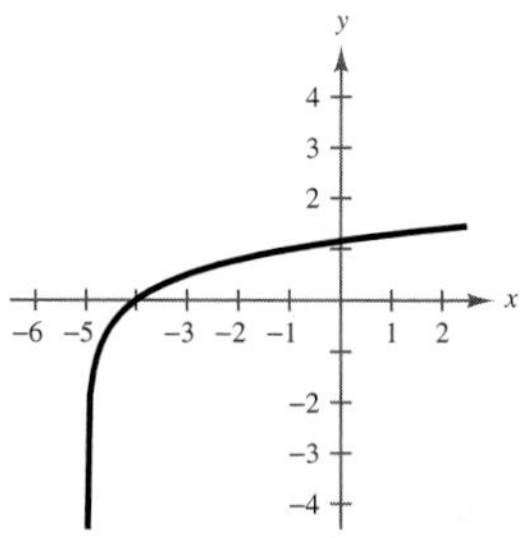

123.

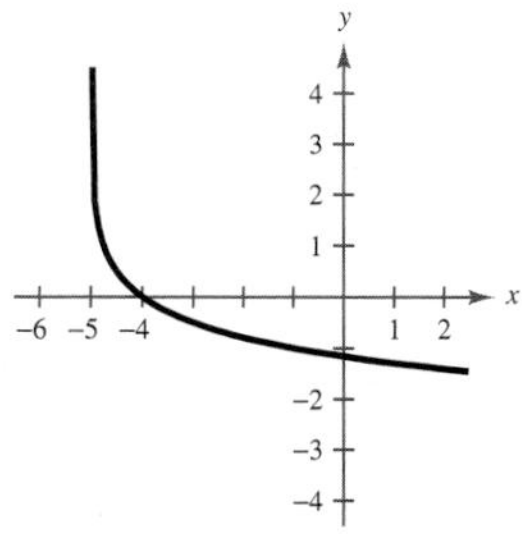

125. $\frac{2\sqrt{3}}{3}$ **127.** $\frac{5\sqrt{2}}{4}$ **129.** $6\sqrt{13}$ **131.** $4\sqrt{13}$

Section 4.2 *(page 368)*

1. $\sin\theta = \frac{15}{17}$
$\cos\theta = -\frac{8}{17}$
$\tan\theta = -\frac{15}{8}$
$\csc\theta = \frac{17}{15}$
$\sec\theta = -\frac{17}{8}$
$\cot\theta = -\frac{8}{15}$

3. $\sin\theta = -\frac{5}{13}$
$\cos\theta = \frac{12}{13}$
$\tan\theta = -\frac{5}{12}$
$\csc\theta = -\frac{13}{5}$
$\sec\theta = \frac{13}{12}$
$\cot\theta = -\frac{12}{5}$

5. $\left(\frac{\sqrt{2}}{2}, \frac{\sqrt{2}}{2}\right)$ **7.** $\left(-\frac{\sqrt{3}}{2}, -\frac{1}{2}\right)$ **9.** $\left(-\frac{1}{2}, -\frac{\sqrt{3}}{2}\right)$

11. (0, −1)

13. $\sin\frac{\pi}{4} = \frac{\sqrt{2}}{2}$
$\cos\frac{\pi}{4} = \frac{\sqrt{2}}{2}$
$\tan\frac{\pi}{4} = 1$

15. $\sin\left(-\frac{\pi}{6}\right) = -\frac{1}{2}$
$\cos\left(-\frac{\pi}{6}\right) = \frac{\sqrt{3}}{2}$
$\tan\left(-\frac{\pi}{6}\right) = -\frac{\sqrt{3}}{3}$

17. $\sin\left(-\frac{7\pi}{4}\right) = \frac{\sqrt{2}}{2}$

$\cos\left(-\frac{7\pi}{4}\right) = \frac{\sqrt{2}}{2}$

$\tan\left(-\frac{7\pi}{4}\right) = 1$

19. $\sin\frac{11\pi}{6} = -\frac{1}{2}$

$\cos\frac{11\pi}{6} = \frac{\sqrt{3}}{2}$

$\tan\frac{11\pi}{6} = -\frac{\sqrt{3}}{3}$

21. $\sin\left(-\frac{3\pi}{2}\right) = 1$

$\cos\left(-\frac{3\pi}{2}\right) = 0$

$\tan\left(-\frac{3\pi}{2}\right)$ is undefined.

23. $\sin\frac{3\pi}{4} = \frac{\sqrt{2}}{2}$ $\quad\csc\frac{3\pi}{4} = \sqrt{2}$

$\cos\frac{3\pi}{4} = -\frac{\sqrt{2}}{2}$ $\quad\sec\frac{3\pi}{4} = -\sqrt{2}$

$\tan\frac{3\pi}{4} = -1$ $\quad\cot\frac{3\pi}{4} = -1$

25. $\sin\frac{\pi}{2} = 1$ $\quad\csc\frac{\pi}{2} = 1$

$\cos\frac{\pi}{2} = 0$ $\quad\sec\frac{\pi}{2}$ is undefined.

$\tan\frac{\pi}{2}$ is undefined. $\quad\cot\frac{\pi}{2} = 0$

27. $\sin\left(-\frac{\pi}{3}\right) = -\frac{\sqrt{3}}{2}$ $\quad\csc\left(-\frac{\pi}{3}\right) = -\frac{2\sqrt{3}}{3}$

$\cos\left(-\frac{\pi}{3}\right) = \frac{1}{2}$ $\quad\sec\left(-\frac{\pi}{3}\right) = 2$

$\tan\left(-\frac{\pi}{3}\right) = -\sqrt{3}$ $\quad\cot\left(-\frac{\pi}{3}\right) = -\frac{\sqrt{3}}{3}$

29. $\sin 5\pi = \sin \pi = 0$ **31.** $\cos\frac{8\pi}{3} = \cos\frac{2\pi}{3} = -\frac{1}{2}$

33. $\cos(-3\pi) = \cos(-\pi) = -1$

35. $\sin\left(-\frac{9\pi}{4}\right) = \sin\frac{7\pi}{4} = -\frac{\sqrt{2}}{2}$

37. (a) $-\frac{1}{3}$ (b) -3 **39.** (a) $-\frac{1}{5}$ (b) -5

41. (a) $\frac{4}{5}$ (b) $-\frac{4}{5}$ **43.** 0.7071 **45.** 1.0378

47. -0.1288 **49.** 1.3940 **51.** -1.4486

53. (a) -1 (b) -0.4

55. (a) 0.25, 2.89 (b) 1.82, 4.46

57. $0.0707 = \cos 1.5 \neq 2\cos 0.75 = 1.4634$

59. (a) 0.2500 foot (b) 0.0177 foot (c) -0.2475 foot

61. False. $\sin(-t) = -\sin t$ means that the function is odd, not that the sine of a negative angle is a negative number.

63. (a) y-axis (b) $\sin t_1 = \sin(\pi - t_1)$
(c) $\cos(\pi - t_1) = -\cos t_1$

65. Answers will vary. **67.** Even

69. $f^{-1}(x) = \sqrt[3]{4(x-1)}$ **71.** $f^{-1}(x) = \frac{x}{2-x}, \ x < 2$

Section 4.3 *(page 377)*

1. $\sin\theta = \frac{3}{5}$

$\cos\theta = \frac{4}{5}$

$\tan\theta = \frac{3}{4}$

$\csc\theta = \frac{5}{3}$

$\sec\theta = \frac{5}{4}$

$\cot\theta = \frac{4}{3}$

3. $\sin\theta = \frac{9}{41}$

$\cos\theta = \frac{40}{41}$

$\tan\theta = \frac{9}{40}$

$\csc\theta = \frac{41}{9}$

$\sec\theta = \frac{41}{40}$

$\cot\theta = \frac{40}{9}$

5. $\sin\theta = \frac{1}{3}$ $\quad\csc\theta = 3$

$\cos\theta = \frac{2\sqrt{2}}{3}$ $\quad\sec\theta = \frac{3\sqrt{2}}{4}$

$\tan\theta = \frac{\sqrt{2}}{4}$ $\quad\cot\theta = 2\sqrt{2}$

The triangles are similar, and corresponding sides are proportional.

7. $\sin\theta = \frac{3}{5}$ $\quad\csc\theta = \frac{5}{3}$

$\cos\theta = \frac{4}{5}$ $\quad\sec\theta = \frac{5}{4}$

$\tan\theta = \frac{3}{4}$ $\quad\cot\theta = \frac{4}{3}$

The triangles are similar, and corresponding sides are proportional.

9.

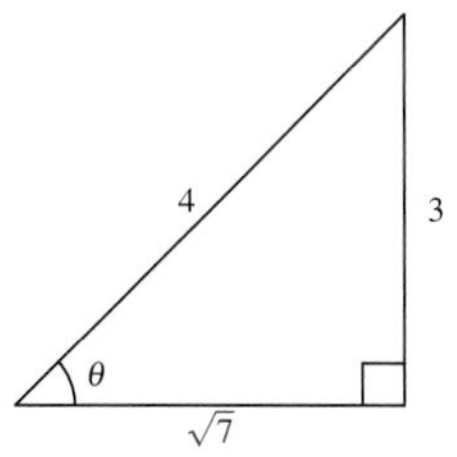

$\cos\theta = \frac{\sqrt{7}}{4}$

$\tan\theta = \frac{3\sqrt{7}}{7}$

$\csc\theta = \frac{4}{3}$

$\sec\theta = \frac{4\sqrt{7}}{7}$

$\cot\theta = \frac{\sqrt{7}}{3}$

11.

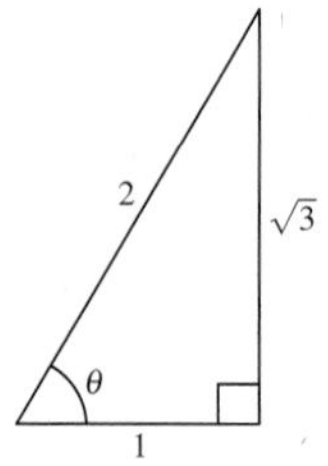

$\sin\theta = \frac{\sqrt{3}}{2}$

$\cos\theta = \frac{1}{2}$

$\tan\theta = \sqrt{3}$

$\csc\theta = \frac{2\sqrt{3}}{3}$

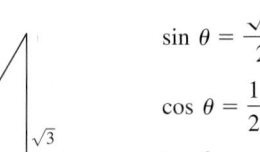

$\cot\theta = \frac{\sqrt{3}}{3}$

13.

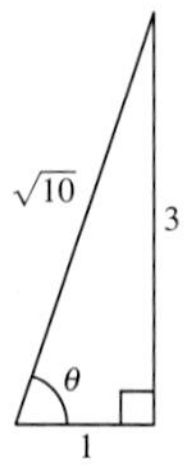

$\sin \theta = \dfrac{3\sqrt{10}}{10}$

$\cos \theta = \dfrac{\sqrt{10}}{10}$

$\csc \theta = \dfrac{\sqrt{10}}{3}$

$\sec \theta = \sqrt{10}$

$\cot \theta = \dfrac{1}{3}$

15.

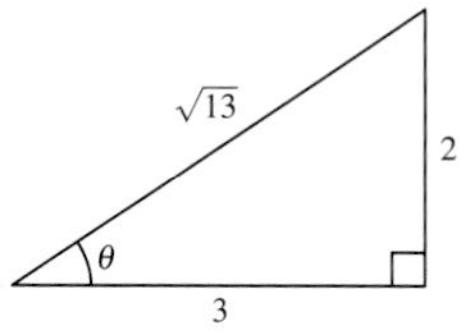

$\sin \theta = \dfrac{2\sqrt{13}}{13}$

$\cos \theta = \dfrac{3\sqrt{13}}{13}$

$\tan \theta = \dfrac{2}{3}$

$\csc \theta = \dfrac{\sqrt{13}}{2}$

$\sec \theta = \dfrac{\sqrt{13}}{3}$

17. (a) $\sqrt{3}$ (b) $\dfrac{1}{2}$ (c) $\dfrac{\sqrt{3}}{2}$ (d) $\dfrac{\sqrt{3}}{3}$

19. (a) $\dfrac{2\sqrt{13}}{13}$ (b) $\dfrac{3\sqrt{13}}{13}$ (c) $\dfrac{2}{3}$ (d) $\dfrac{\sqrt{13}}{2}$

21. (a) 3 (b) $\dfrac{2\sqrt{2}}{3}$ (c) $\dfrac{\sqrt{2}}{4}$ (d) $\dfrac{1}{3}$

23. (a) $\frac{1}{2}$ (b) 2 (c) $\sqrt{3}$

25. (a) $\dfrac{\sqrt{2}}{2}$ (b) $\dfrac{\sqrt{3}}{2}$ (c) $\dfrac{\sqrt{3}}{3}$

27. (a) 0.1736 (b) 0.1736 **29.** (a) 0.2815 (b) 3.5523

31. (a) 1.3499 (b) 1.3432 **33.** (a) 5.0273 (b) 0.1989

35. (a) 1.8527 (b) 0.9817

37. (a) $30° = \dfrac{\pi}{6}$ (b) $30° = \dfrac{\pi}{6}$

39. (a) $60° = \dfrac{\pi}{3}$ (b) $45° = \dfrac{\pi}{4}$

41. (a) $60° = \dfrac{\pi}{3}$ (b) $45° = \dfrac{\pi}{4}$

43. (a) $0.83° \approx 0.015$ (b) $27° \approx 0.474$

45. (a) $0.72° \approx 0.012$ (b) $67° \approx 1.169$

47. $30\sqrt{3}$ **49.** $\dfrac{32\sqrt{3}}{3}$ **51.–59.** Answers will vary.

61. (a)

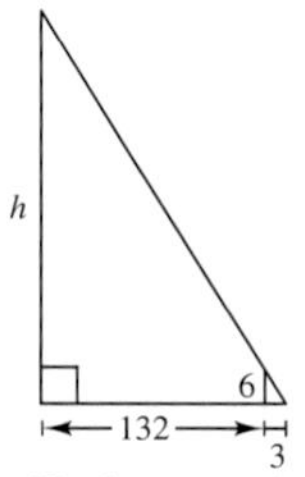

Not drawn to scale

(b) $\dfrac{6}{3} = \dfrac{h}{135}$

(c) 270 feet

63. (a)

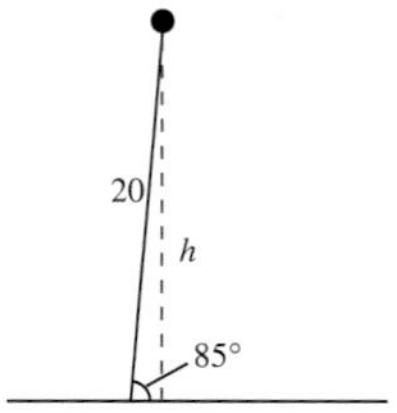

(b) $\sin 85° = \dfrac{h}{20}$

(c) 19.9 meters

65. 572 feet **67.** $(x_1, y_1) = (28\sqrt{3}, 28)$
$(x_2, y_2) = (28, 28\sqrt{3})$

69. $\sin 20° \approx 0.34$ $\csc 20° \approx 2.92$
$\cos 20° \approx 0.94$ $\sec 20° \approx 1.06$
$\tan 20° \approx 0.36$ $\cot 20° \approx 2.75$

71. True, $\csc x = \dfrac{1}{\sin x}$ **73.** False, $\dfrac{\sqrt{2}}{2} + \dfrac{\sqrt{2}}{2} \neq 1$

75. False, $1.7321 \neq 0.0349$

77. Corresponding sides of similar triangles are proportional.

79. (a)

θ	0.1	0.2	0.3	0.4	0.5
$\sin \theta$	0.0998	0.1987	0.2955	0.3894	0.4794

(b) As θ approaches 0, $\sin \theta$ approaches θ.

81. $\dfrac{x}{x-2}$, $x \neq \pm 6$ **83.** $\dfrac{2(x^2 - 5x - 10)}{(x-2)(x+2)^2}$

85. $x = \dfrac{11 \pm \sqrt{409}}{6}$ **87.** $x = \dfrac{2}{3}$

Section 4.4 *(page 387)*

1. (a) $\sin \theta = \frac{3}{5}$ (b) $\sin \theta = -\frac{15}{17}$
$\cos \theta = \frac{4}{5}$ $\cos \theta = \frac{8}{17}$
$\tan \theta = \frac{3}{4}$ $\tan \theta = -\frac{15}{8}$
$\csc \theta = \frac{5}{3}$ $\csc \theta = -\frac{17}{15}$
$\sec \theta = \frac{5}{4}$ $\sec \theta = \frac{17}{8}$
$\cot \theta = \frac{4}{3}$ $\cot \theta = -\frac{8}{15}$

3. (a) $\sin\theta = -\dfrac{1}{2}$ (b) $\sin\theta = \dfrac{\sqrt{17}}{17}$

$\cos\theta = -\dfrac{\sqrt{3}}{2}$ $\cos\theta = -\dfrac{4\sqrt{17}}{17}$

$\tan\theta = \dfrac{\sqrt{3}}{3}$ $\tan\theta = -\dfrac{1}{4}$

$\csc\theta = -2$ $\csc\theta = \sqrt{17}$

$\sec\theta = -\dfrac{2\sqrt{3}}{3}$ $\sec\theta = -\dfrac{\sqrt{17}}{4}$

$\cot\theta = \sqrt{3}$ $\cot\theta = -4$

5. $\sin\theta = \frac{24}{25}$ $\csc\theta = \frac{25}{24}$

$\cos\theta = \frac{7}{25}$ $\sec\theta = \frac{25}{7}$

$\tan\theta = \frac{24}{7}$ $\cot\theta = \frac{7}{24}$

7. $\sin\theta = \dfrac{5\sqrt{29}}{29}$ $\csc\theta = \dfrac{\sqrt{29}}{5}$

$\cos\theta = -\dfrac{2\sqrt{29}}{29}$ $\sec\theta = -\dfrac{\sqrt{29}}{2}$

$\tan\theta = -\dfrac{5}{2}$ $\cot\theta = -\dfrac{2}{5}$

9. $\sin\theta = \dfrac{68\sqrt{5849}}{5849} \approx 0.9$ $\csc\theta = \dfrac{\sqrt{5849}}{68} \approx 1.1$

$\cos\theta = -\dfrac{35\sqrt{5849}}{5849} \approx -0.5$ $\sec\theta = -\dfrac{\sqrt{5849}}{35} \approx -2.2$

$\tan\theta = -\dfrac{68}{35} \approx -1.9$ $\cot\theta = -\dfrac{35}{68} \approx -0.5$

11. Quadrant III **13.** Quadrant II

15. $\sin\theta = \frac{3}{5}$ $\csc\theta = \frac{5}{3}$

$\cos\theta = -\frac{4}{5}$ $\sec\theta = -\frac{5}{4}$

$\tan\theta = -\frac{3}{4}$ $\cot\theta = -\frac{4}{3}$

17. $\sin\theta = -\frac{15}{17}$ $\csc\theta = -\frac{17}{15}$

$\cos\theta = \frac{8}{17}$ $\sec\theta = \frac{17}{8}$

$\tan\theta = -\frac{15}{8}$ $\cot\theta = -\frac{8}{15}$

19. $\sin\theta = -\dfrac{\sqrt{10}}{10}$ $\csc\theta = -\sqrt{10}$

$\cos\theta = \dfrac{3\sqrt{10}}{10}$ $\sec\theta = \dfrac{\sqrt{10}}{3}$

$\tan\theta = -\dfrac{1}{3}$ $\cot\theta = -3$

21. $\sin\theta = \dfrac{\sqrt{3}}{2}$ $\csc\theta = \dfrac{2\sqrt{3}}{3}$

$\cos\theta = -\dfrac{1}{2}$ $\sec\theta = -2$

$\tan\theta = -\sqrt{3}$ $\cot\theta = -\dfrac{\sqrt{3}}{3}$

23. $\sin\theta = 0$ $\csc\theta$ is undefined.

$\cos\theta = -1$ $\sec\theta = -1$

$\tan\theta = 0$ $\cot\theta$ is undefined.

25. $\sin\theta = \dfrac{\sqrt{2}}{2}$ $\csc\theta = \sqrt{2}$

$\cos\theta = -\dfrac{\sqrt{2}}{2}$ $\sec\theta = -\sqrt{2}$

$\tan\theta = -1$ $\cot\theta = -1$

27. $\sin\theta = -\dfrac{2\sqrt{5}}{5}$ $\csc\theta = -\dfrac{\sqrt{5}}{2}$

$\cos\theta = -\dfrac{\sqrt{5}}{5}$ $\sec\theta = -\sqrt{5}$

$\tan\theta = 2$ $\cot\theta = \dfrac{1}{2}$

29. -1 **31.** -1 **33.** Undefined **35.** 0

37. $\theta' = 23°$

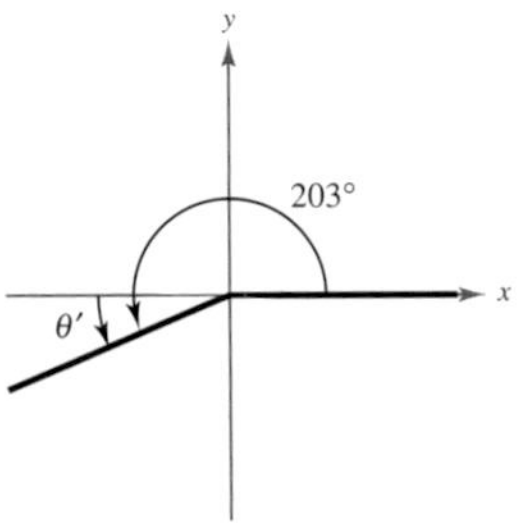

39. $\theta' = 65°$

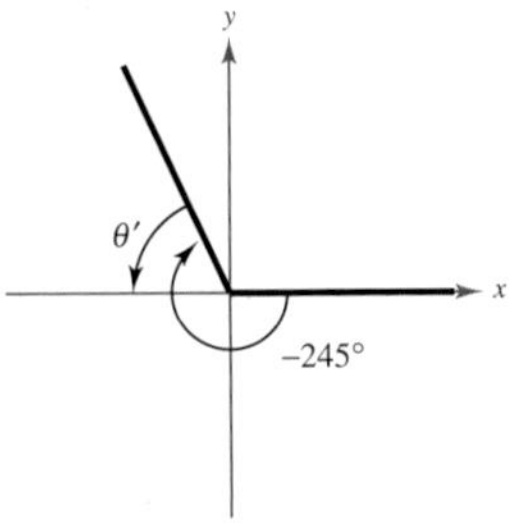

41. $\theta' = \dfrac{\pi}{3}$

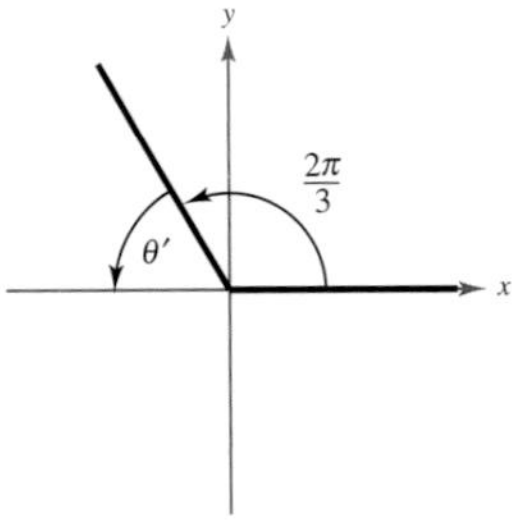

43. $\theta' = 3.5 - \pi$

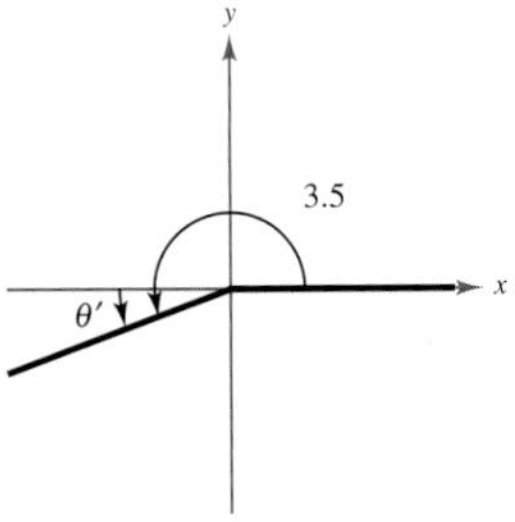

45. $\sin 225° = -\dfrac{\sqrt{2}}{2}$

$\cos 225° = -\dfrac{\sqrt{2}}{2}$

$\tan 225° = 1$

47. $\sin 750° = \dfrac{1}{2}$

$\cos 750° = \dfrac{\sqrt{3}}{2}$

$\tan 750° = \dfrac{\sqrt{3}}{3}$

49. $\sin(-150°) = -\dfrac{1}{2}$

$\cos(-150°) = -\dfrac{\sqrt{3}}{2}$

$\tan(-150°) = \dfrac{\sqrt{3}}{3}$

51. $\sin\dfrac{4\pi}{3} = -\dfrac{\sqrt{3}}{2}$

$\cos\dfrac{4\pi}{3} = -\dfrac{1}{2}$

$\tan\dfrac{4\pi}{3} = \sqrt{3}$

53. $\sin\left(-\frac{\pi}{6}\right) = -\frac{1}{2}$

$\cos\left(-\frac{\pi}{6}\right) = \frac{\sqrt{3}}{2}$

$\tan\left(-\frac{\pi}{6}\right) = -\frac{\sqrt{3}}{3}$

55. $\sin \frac{11\pi}{4} = \frac{\sqrt{2}}{2}$

$\cos \frac{11\pi}{4} = -\frac{\sqrt{2}}{2}$

$\tan \frac{11\pi}{4} = -1$

57. $\sin\left(-\frac{3\pi}{2}\right) = 1$

$\cos\left(-\frac{3\pi}{2}\right) = 0$

$\tan\left(-\frac{3\pi}{2}\right)$ is undefined.

59. 0.1736 **61.** -0.3420

63. 4.6373 **65.** 0.3640 **67.** -0.6052

69. (a) $30° = \frac{\pi}{6}$, $150° = \frac{5\pi}{6}$ (b) $210° = \frac{7\pi}{6}$, $330° = \frac{11\pi}{6}$

71. (a) $60° = \frac{\pi}{3}$, $120° = \frac{2\pi}{3}$ (b) $135° = \frac{3\pi}{4}$, $315° = \frac{7\pi}{4}$

73. (a) $45° = \frac{\pi}{4}$, $225° = \frac{5\pi}{4}$ (b) $150° = \frac{5\pi}{6}$, $330° = \frac{11\pi}{6}$

75. 54.99°, 125.01° **77.** 115.89°, 244.11°

79. 0.175, 6.109 **81.** 0.873, 4.014 **83.** 1.955, 4.328

85. $\frac{4}{5}$ **87.** $-\frac{\sqrt{13}}{2}$ **89.** $\frac{8}{5}$

91. (a) 25.2°F
(b) 65.1°F
(c) 50.8°F

93. (a) 2 centimeters
(b) 0.14 centimeter
(c) -1.98 centimeters

95. 0.79

97. False. In each of the four quadrants, the signs of the secant function and cosine function will be the same, because these functions are reciprocals of each other.

99. As θ increases from 0° to 90°, x decreases from 12 cm to 0 cm and y increases from 0 cm to 12 cm. Therefore, $\sin \theta = y/12$ increases from 0 to 1 and $\cos \theta = x/12$ decreases from 1 to 0. Thus, $\tan \theta = y/x$ and increases without bound. When $\theta = 90°$, the tangent is undefined.

101. First, determine a positive coterminal angle. Then determine the trigonometric function of the reference angle and prefix the appropriate sign.

103.

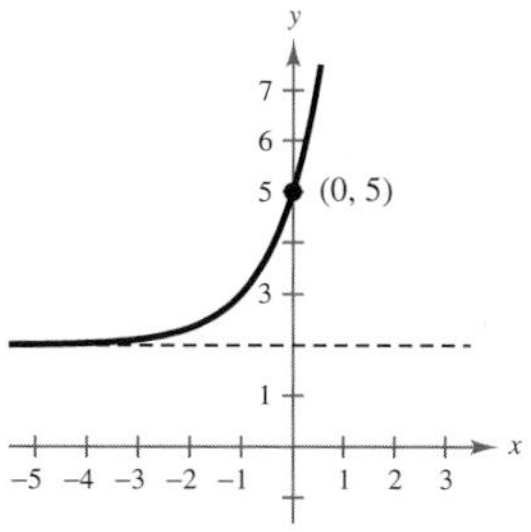

Intercept: $(0, 5)$
Asymptote: $y = 2$
Domain: all real numbers
Range: $y > 2$

105.

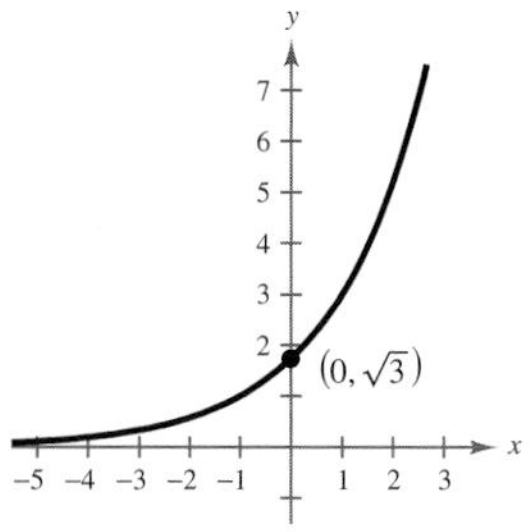

Intercept: $(0, \sqrt{3})$
Asymptote: $y = 0$
Domain: all real numbers
Range: $y > 0$

107.

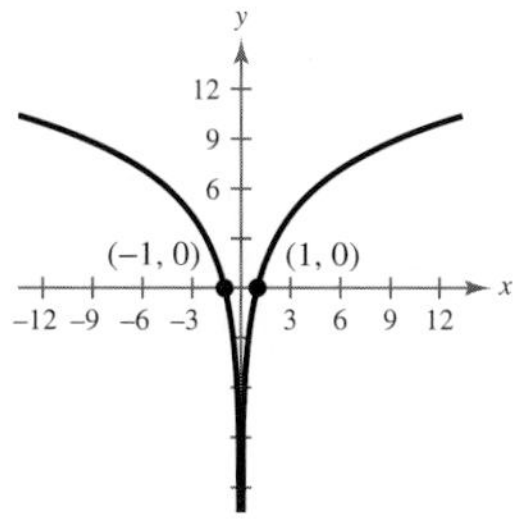

Intercepts: $(\pm 1, 0)$
Asymptote: $x = 0$
Domain: all real numbers $x \neq 0$
Range: all real numbers

109.

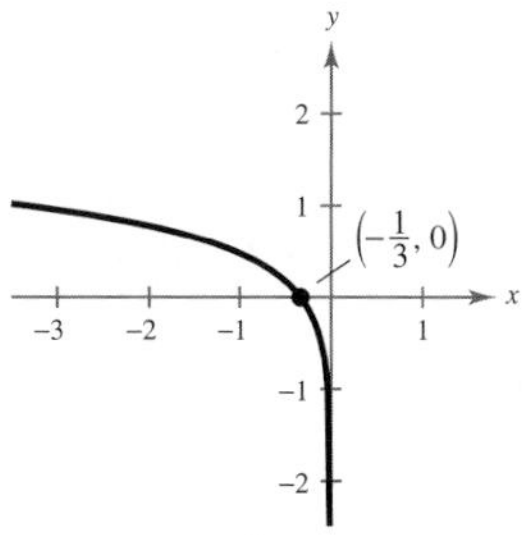

Intercept: $\left(-\frac{1}{3}, 0\right)$
Asymptote: $x = 0$
Domain: $x < 0$
Range: all real numbers

Section 4.5 *(page 397)*

1. Period: π
Amplitude: 3

3. Period: 4π
Amplitude: $\frac{5}{2}$

5. Period: 6
Amplitude: $\frac{1}{2}$

7. Period: 2π
Amplitude: 2

9. Period: $\frac{\pi}{5}$
Amplitude: 3

11. Period: 3π
Amplitude: $\frac{1}{2}$

13. Period: 1
Amplitude: $\frac{1}{4}$

15. g is a shift of f π units to the right.

17. g is a reflection of f in the x-axis.

19. The period of f is twice the period of g.

21. g is a shift of f 3 units upward.

23. The graph of g has twice the amplitude of the graph of f.

25. The graph of g is a horizontal shift of the graph of f π units to the right.

27.

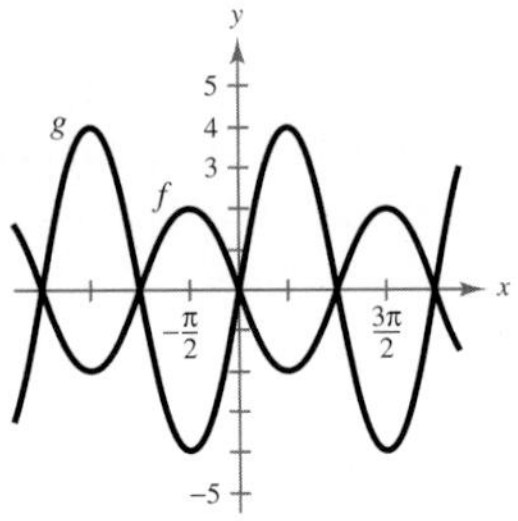

29.

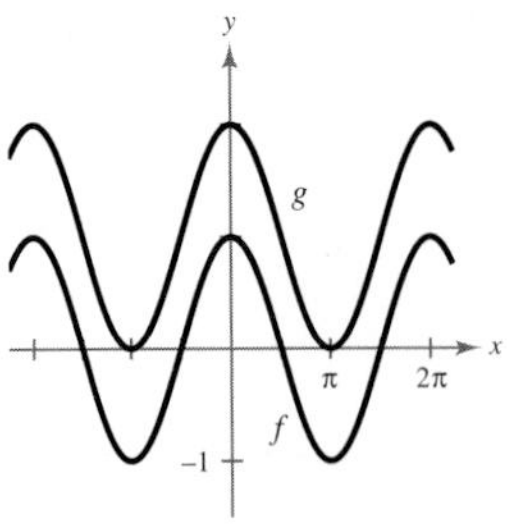

31.

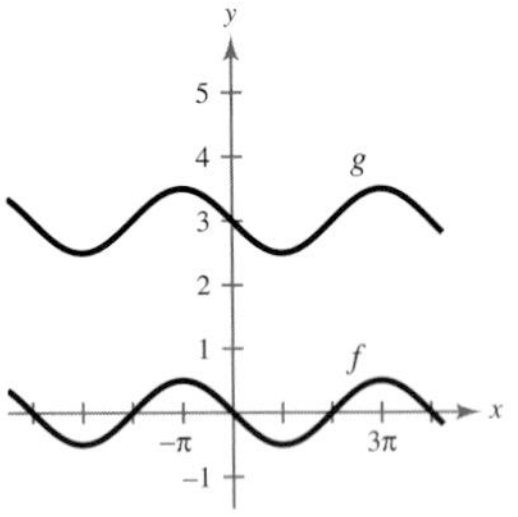

33.

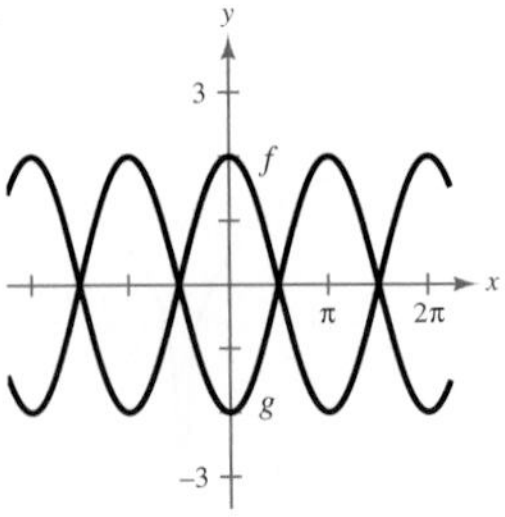

35.

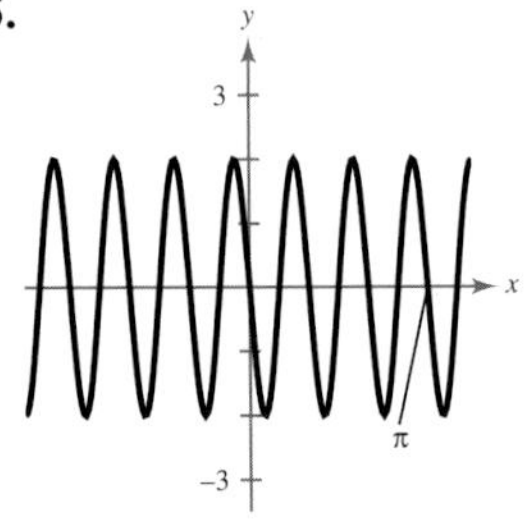

37.

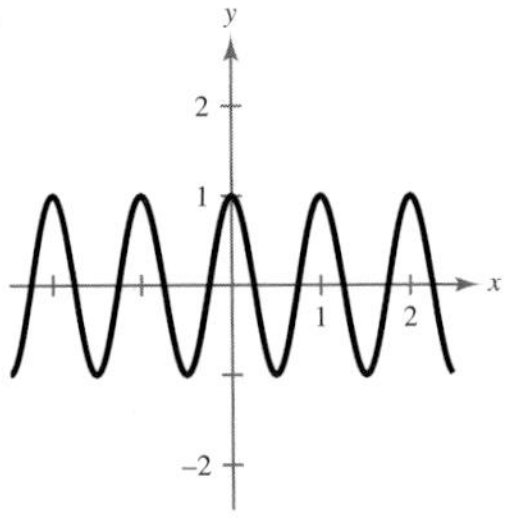

39.

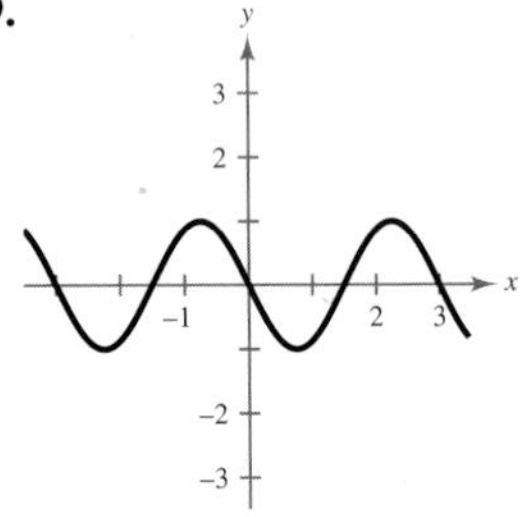

41.

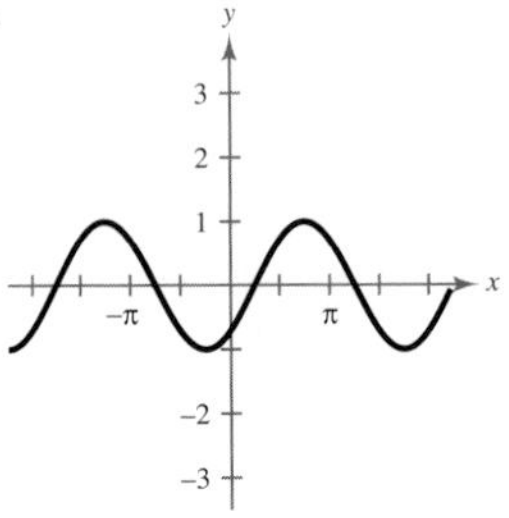

43.

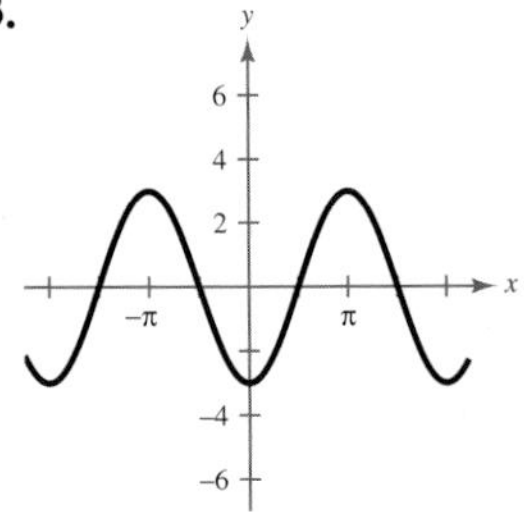

45.

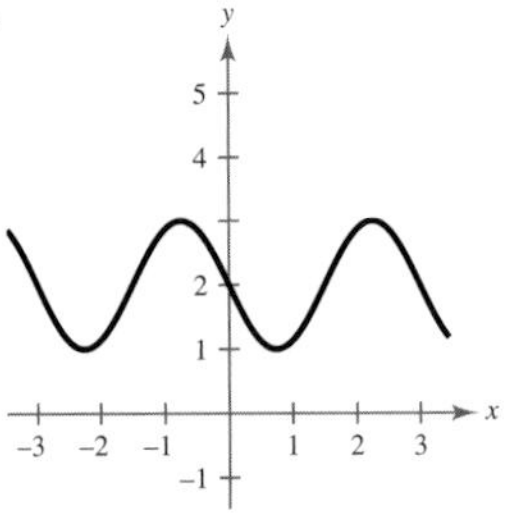

47.

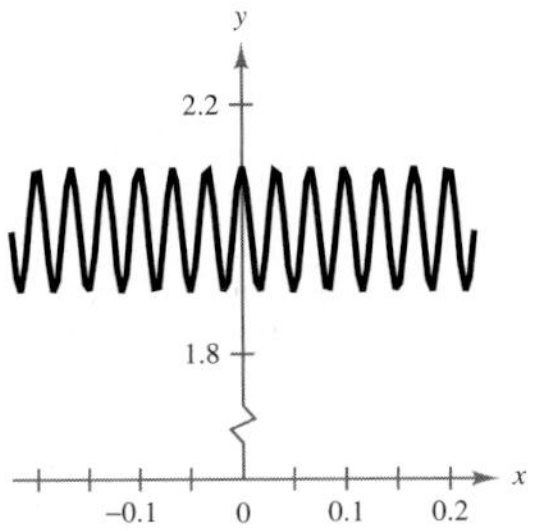

49.

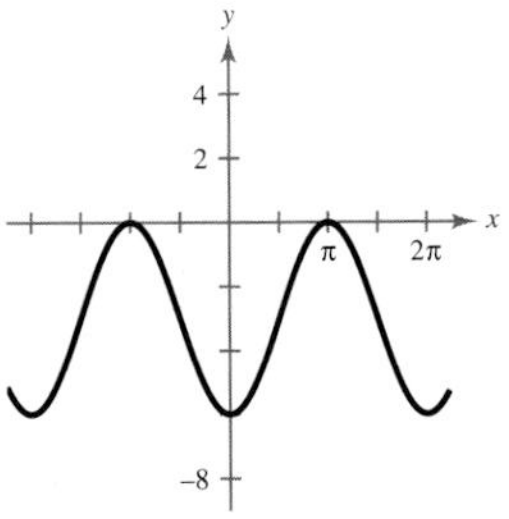

51.

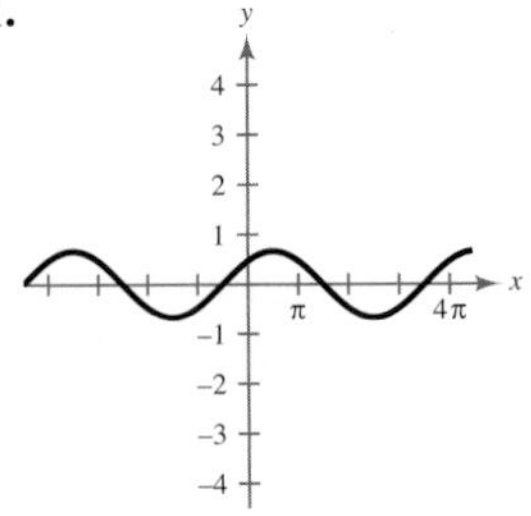

53.

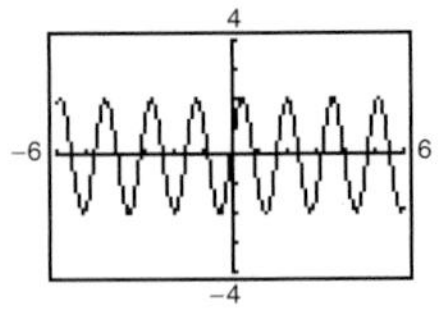

55.

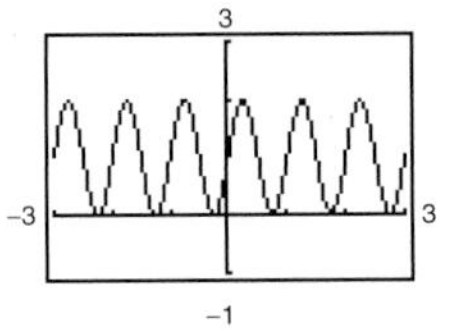

57.

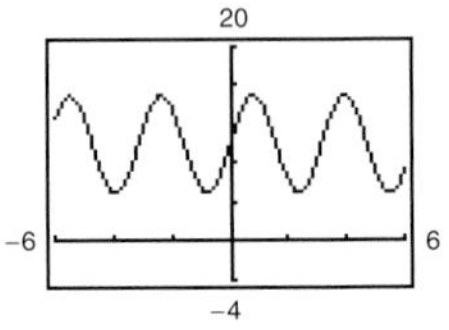

59.

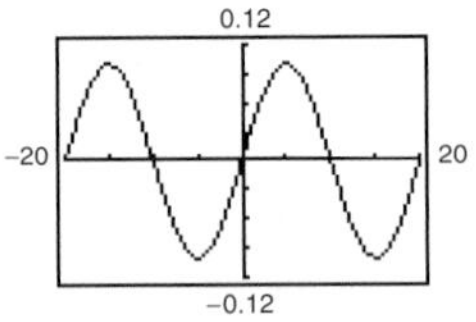

61. $a = 2, d = 1$ **63.** $a = -4, d = 4$

65. $a = -3, b = 2, c = 0$ **67.** $a = 2, b = 1, c = -\frac{\pi}{4}$

69.

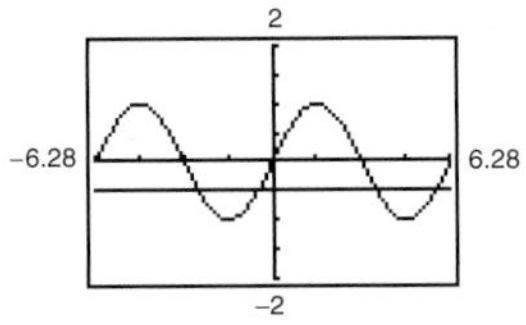

$x = -\frac{\pi}{6}, -\frac{5\pi}{6}, \frac{7\pi}{6}, \frac{11\pi}{6}$

71.

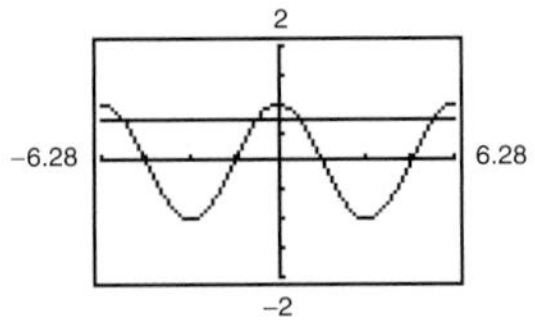

$x = \pm\frac{\pi}{4}, \pm\frac{7\pi}{4}$

73. (a) 6 seconds (b) 10 cycles per minute

(c)

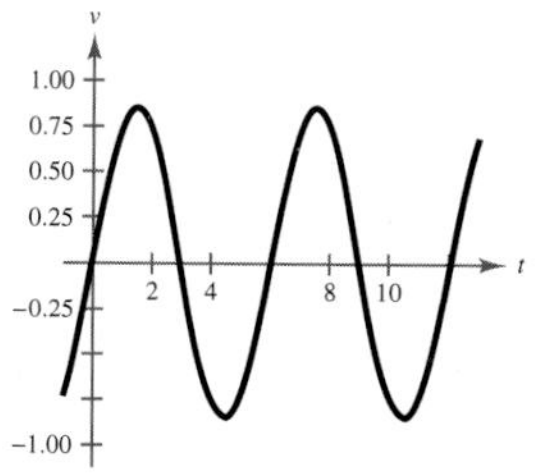

75. (a) $\frac{1}{440}$ second (b) 440 cycles per second

77. (a) $C(t) = 56.35 + 27.35 \sin\left(\frac{\pi t}{6} + 4.19\right)$

(b)

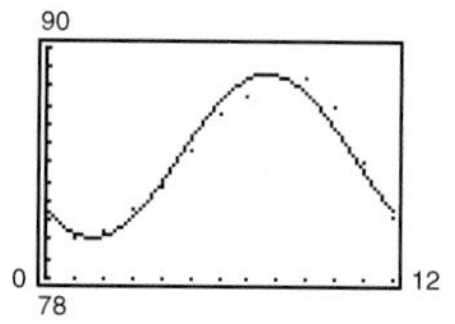

The model is a good fit.

(c)

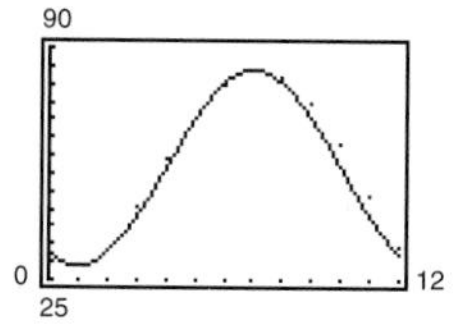

The model is a good fit.

(d) Honolulu: 84.40°; Chicago: 56.35°.
The constant term gives annual average temperature.

(e) 12. Yes. One full period is 1 year.

(f) Chicago; amplitude

79.

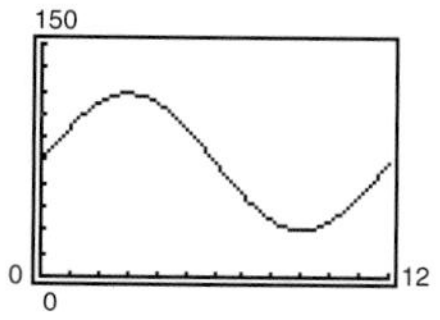

81. (a) and (c)

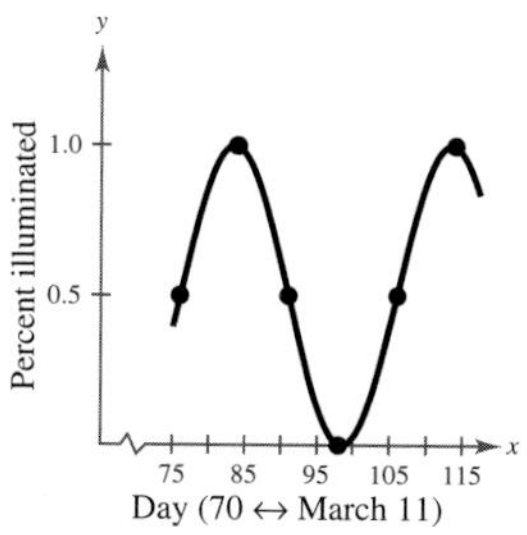

The model is a good fit.

(b) $y = \frac{1}{2} + \frac{1}{2}\sin\left[\frac{\pi}{15}(x - 76)\right]$ (d) 0

83. False. The function $y = \frac{1}{2}\cos 2x$ has an amplitude that is one-half that of $y = \cos x$. For $y = a \cos bx$, the amplitude is $|a|$.

85.

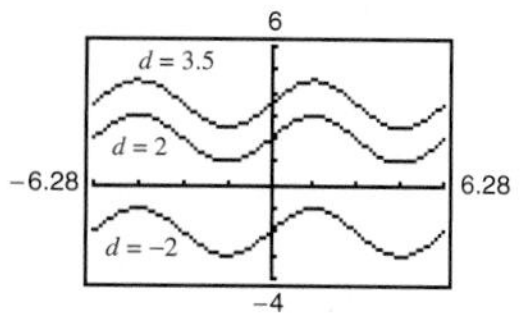

Vertical translations

87.

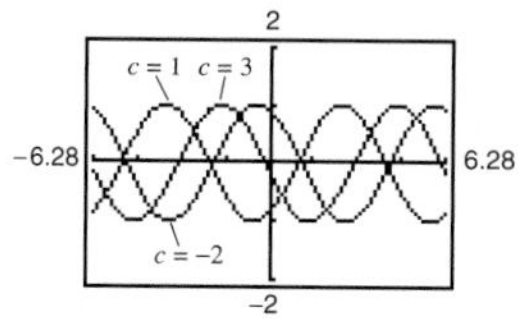

Horizontal shifts

89.

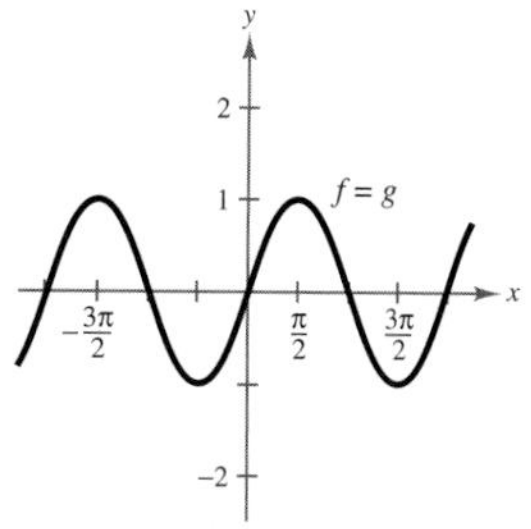

Conjecture: $\sin x = -\cos\left(x + \frac{\pi}{2}\right)$

91.

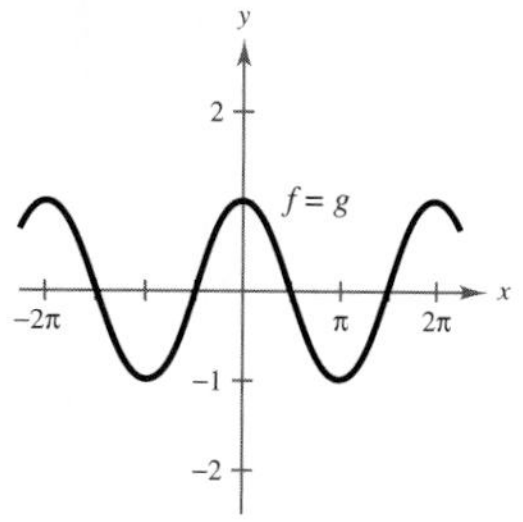

Conjecture: $\cos x = -\cos(x - \pi)$

93. (a) 0.4794, 0.4794

(b) 0.8417, 0.8415

(c) 0.5, 0.5

(d) 0.8776, 0.8776

(e) 0.5417, 0.5403

(f) 0.7074, 0.7071.

The error increases as x moves farther away from 0.

95. (a) Even (b) Even

97. $2 \log_2 x + \log_2(x - 3)$

99. $\frac{1}{2}\ln z - \frac{1}{2}\ln(z^2 + 1)$

101. $\log_2(x^3 y)$

103. $\ln\left(x^3\sqrt{\frac{2x}{x^2}}\right) = \ln\left(x^2\sqrt{2x}\right)$

Section 4.6 *(page 408)*

1. e, π **3.** a, 1 **5.** d, 2π

7.
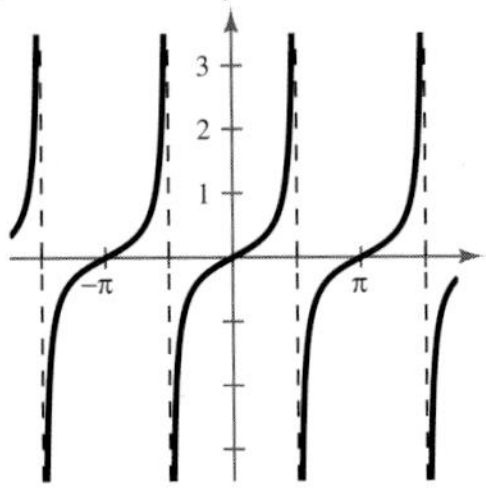

9.
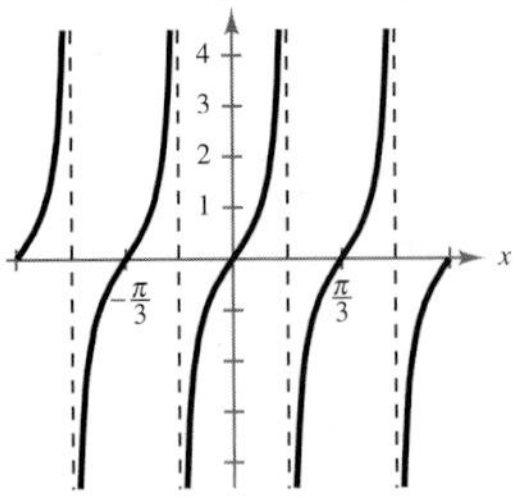

11.
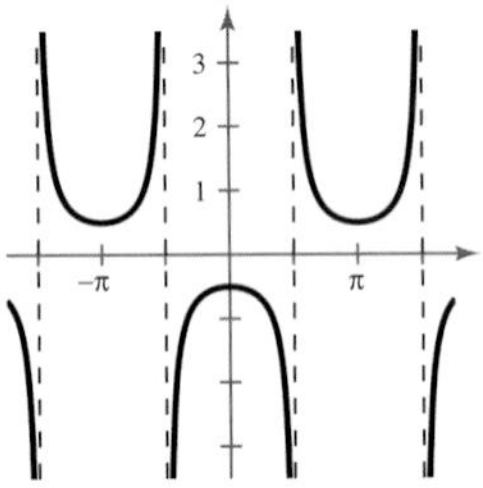

13.
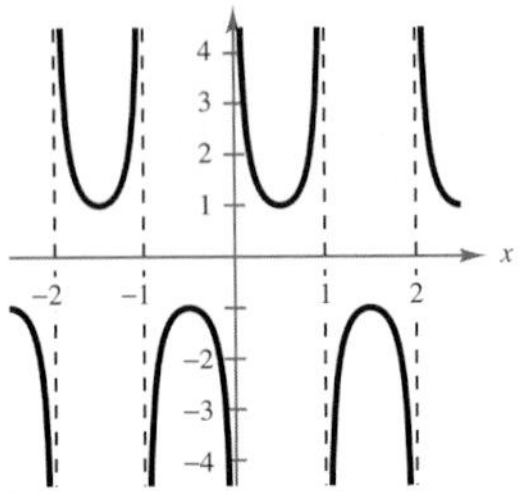

15.
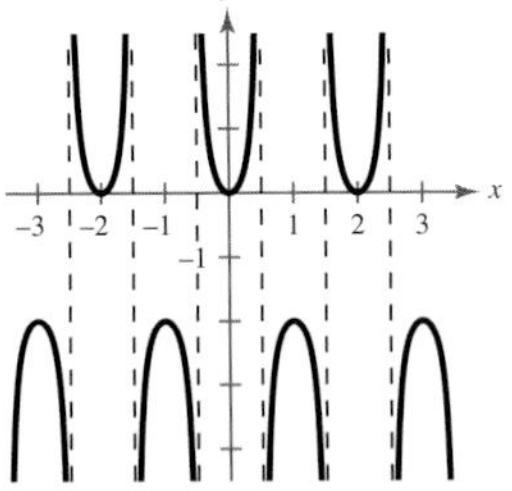

17.
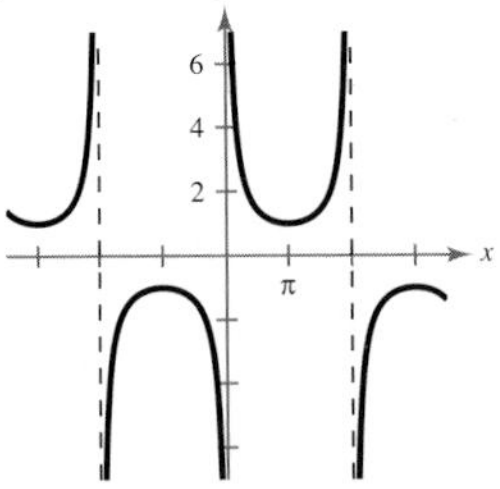

19.
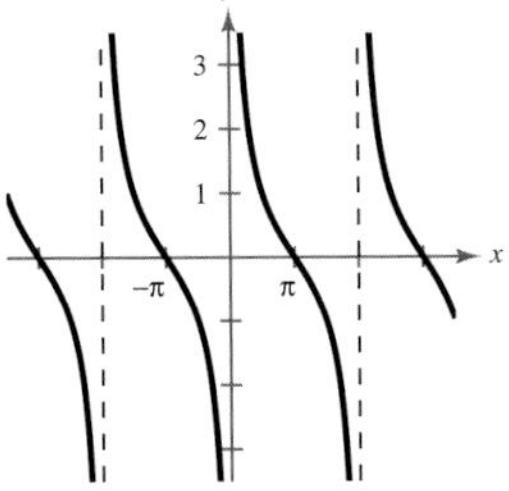

21.
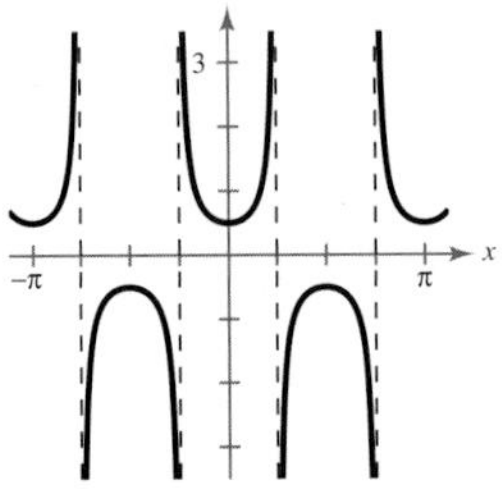

23.
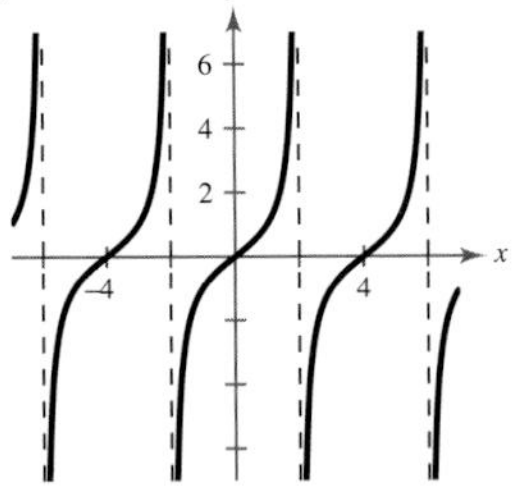

25.
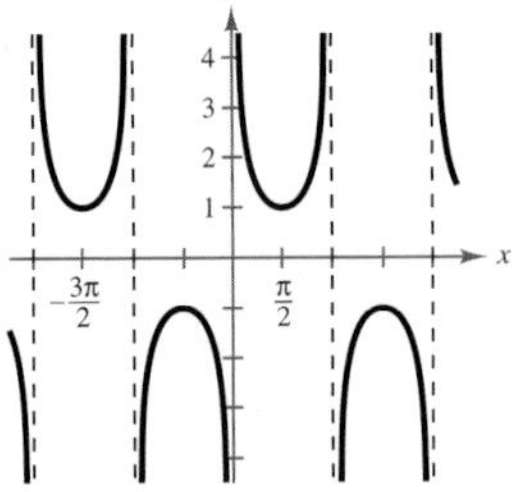

27.
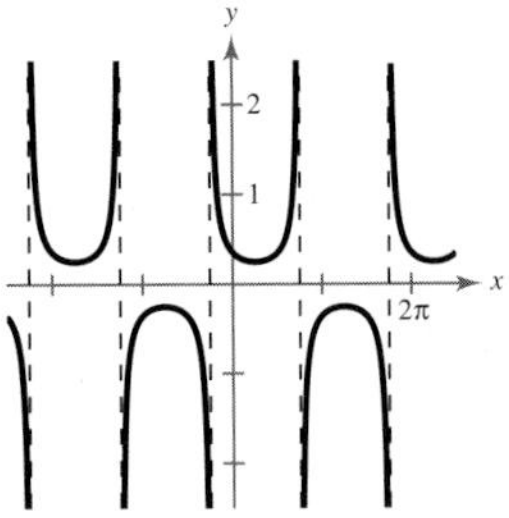

29.
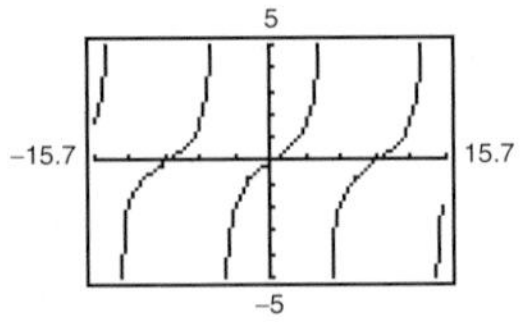

31.
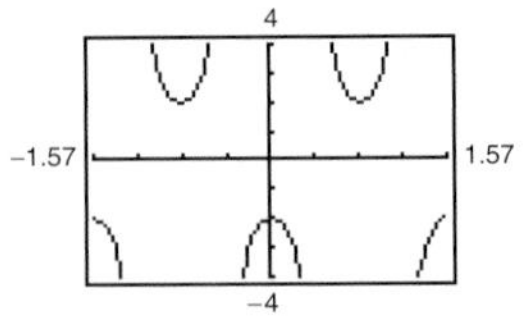

33.
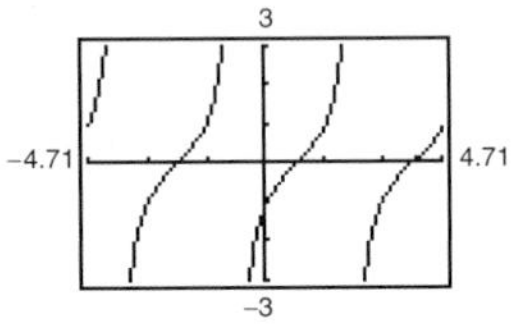

35.
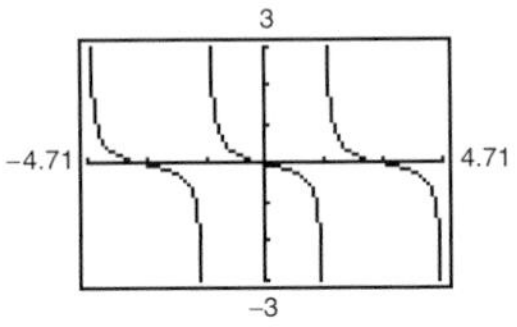

37.
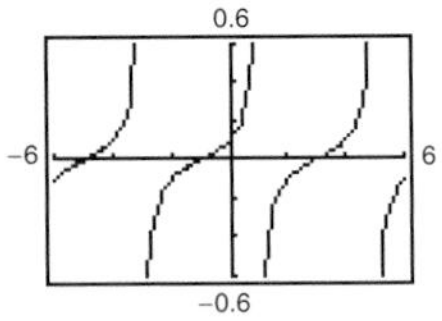

39. $-\frac{7\pi}{4}, -\frac{3\pi}{4}, \frac{\pi}{4}, \frac{5\pi}{4}$

41. $-\frac{4\pi}{3}, -\frac{\pi}{3}, \frac{2\pi}{3}, \frac{5\pi}{3}$

43. $-\frac{4\pi}{3}, -\frac{2\pi}{3}, \frac{2\pi}{3}, \frac{4\pi}{3}$

45. $-\frac{7\pi}{4}, -\frac{5\pi}{4}, \frac{\pi}{4}, \frac{3\pi}{4}$

47. Even

49. (a)
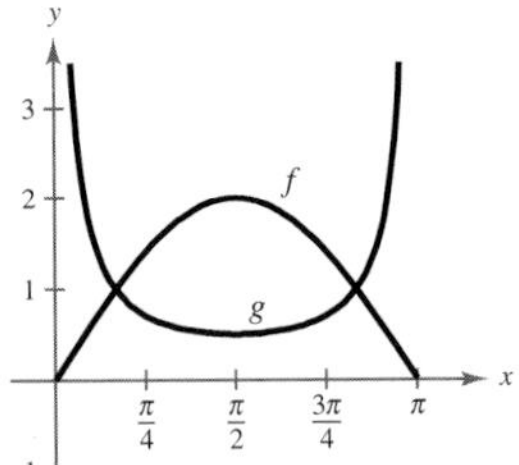

(b) $\frac{\pi}{6} < x < \frac{5\pi}{6}$

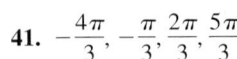

(c) f approaches 0 and g approaches $+\infty$ because the cosecant is the reciprocal of the sine.

51.
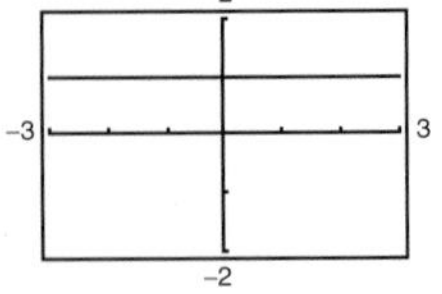

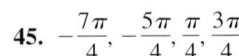

The expressions are equivalent except that when $x = 0$, y_1 is undefined.

53.

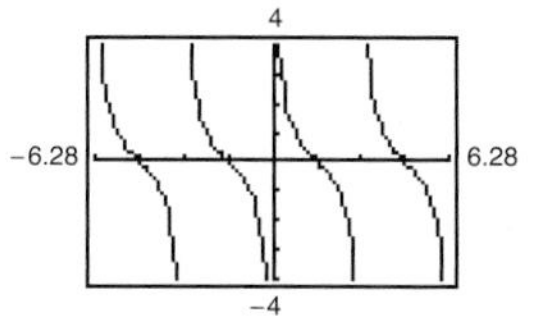

The expressions are equivalent.

55. d, $f \to 0$ as $x \to 0$. **57.** b, $g \to 0$ as $x \to 0$.

59.

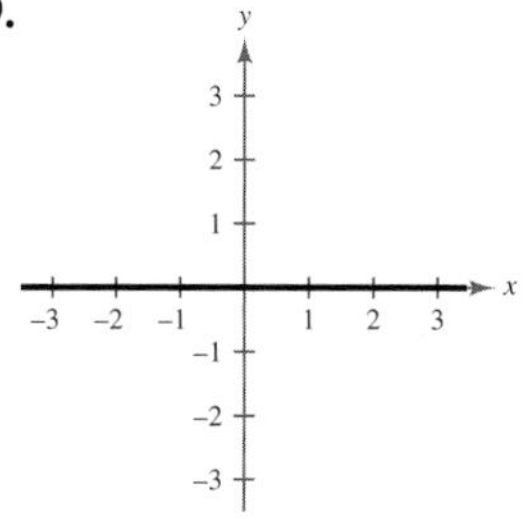

The functions are equal.

61.

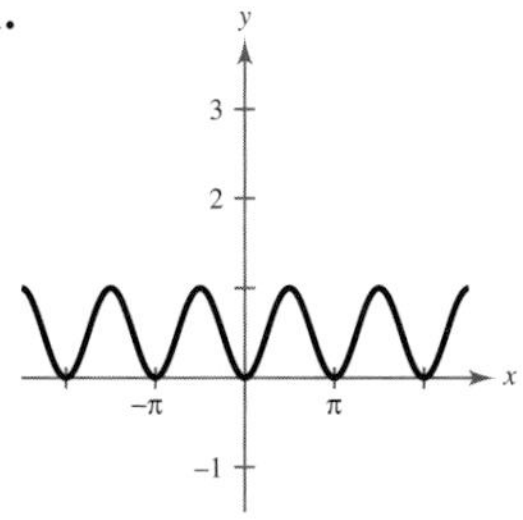

The functions are equal.

63.

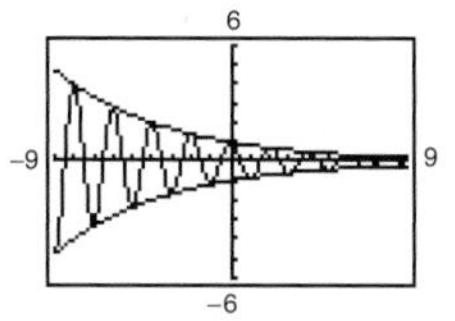

As $x \to \infty$, $f(x) \to 0$.

65.

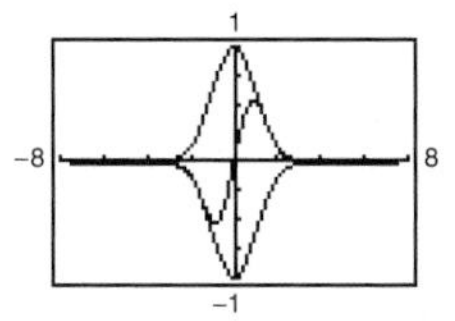

As $x \to \infty$, $g(x) \to 0$.

67.

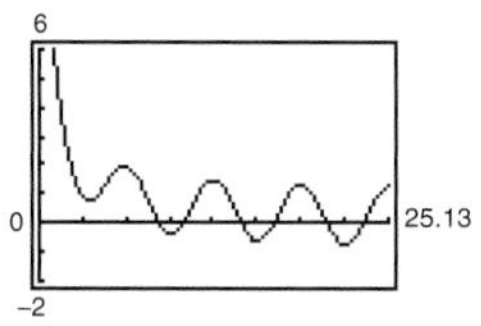

As $x \to 0$, $y \to \infty$.

69.

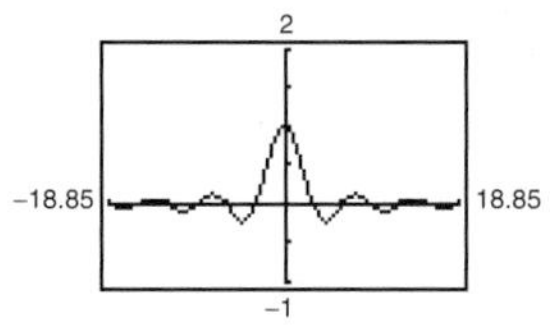

As $x \to 0$, $g(x) \to 1$.

71.

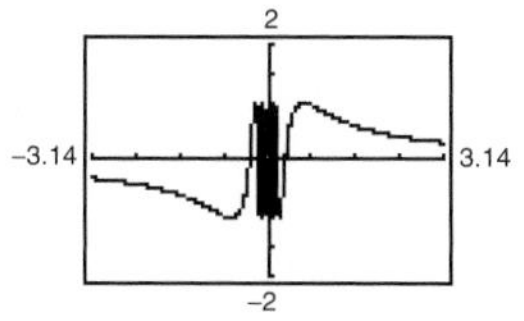

As $x \to 0$, $f(x)$ oscillates between 1 and -1.

73. $d = 7 \cot x$

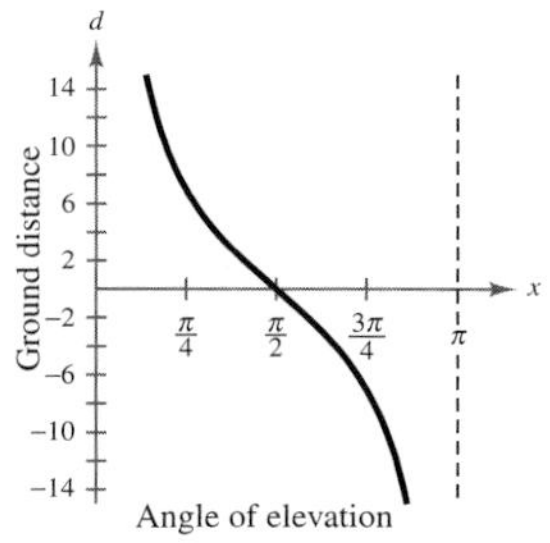

75.

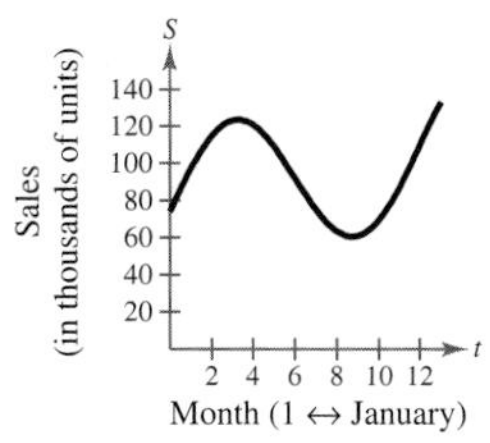

77. (a) 12
(b) Summer; winter
(c) 1 month

79. True. For a given value of x, the y-coordinate of $\csc x$ is the reciprocal of the y-coordinate of $\sin x$.

81. As x approaches $\pi/2$ from the left, f approaches ∞. As x approaches $\pi/2$ from the right, f approaches $-\infty$.

83. (a)

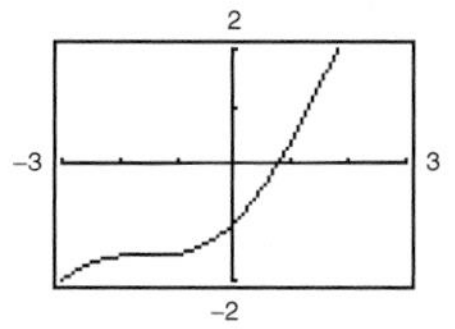

0.739

(b) 1, 0.5403, 0.8576, 0.6543, 0.7935, 0.7014, 0.7640, 0.7221, 0.7504, 0.7314, . . .

85.

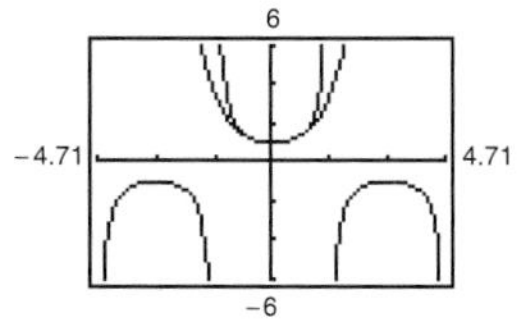

The graphs appear to coincide on the interval $-1.1 \le x \le 1.1$.

87. $\dfrac{\ln 54}{2} \approx 1.994$ **89.** $-\ln 2 \approx -0.693$

91. $\dfrac{2 + e^{73}}{3} \approx 1.684 \times 10^{31}$ **93.** $\pm\sqrt{e^{3.2} - 1} \approx \pm 4.851$

95. 2

Section 4.7 *(page 418)*

1. $\dfrac{\pi}{6}$ **3.** $\dfrac{\pi}{3}$ **5.** $\dfrac{\pi}{6}$ **7.** $\dfrac{5\pi}{6}$ **9.** $-\dfrac{\pi}{3}$ **11.** $\dfrac{2\pi}{3}$

13. $\dfrac{\pi}{3}$ **15.** 0 **17.** 1.29 **19.** -0.85 **21.** -1.25

23. 0.32 **25.** 1.99 **27.** 0.74 **29.** 0.85

31. 1.29 **33.** $-\dfrac{\pi}{3}, -\dfrac{\sqrt{3}}{3}, 1$

35.

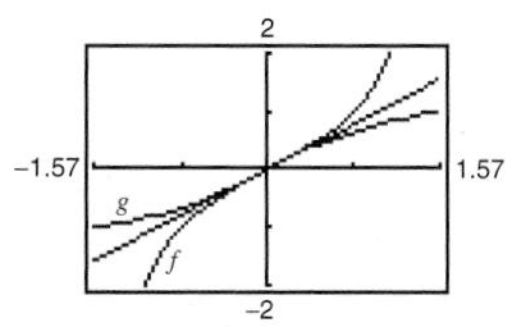

37. $\theta = \arctan \dfrac{x}{4}$

39. $\theta = \arcsin \dfrac{x+2}{5}$ **41.** $\theta = \arccos \dfrac{x+3}{2x}$ **43.** 0.3

45. -0.1 **47.** 0 **49.** $\dfrac{3}{5}$ **51.** $\dfrac{\sqrt{5}}{5}$ **53.** $\dfrac{12}{13}$

55. $\dfrac{\sqrt{34}}{5}$ **57.** $\dfrac{\sqrt{5}}{3}$ **59.** $\dfrac{1}{x}$ **61.** $\sqrt{1-4x^2}$

63. $\sqrt{1-x^2}$ **65.** $\dfrac{\sqrt{9-x^2}}{x}$ **67.** $\dfrac{\sqrt{x^2+2}}{x}$

69.

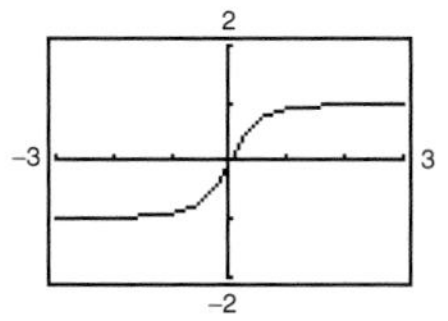

Asymptotes: $y = \pm 1$

71. $\dfrac{9}{\sqrt{x^2+81}},\ x > 0;\ \dfrac{-9}{\sqrt{x^2+81}},\ x < 0$

73. $\dfrac{|x-1|}{\sqrt{x^2-2x+10}}$

75.

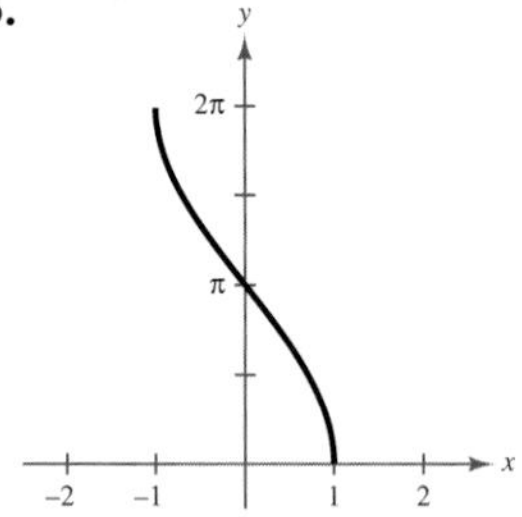

77.

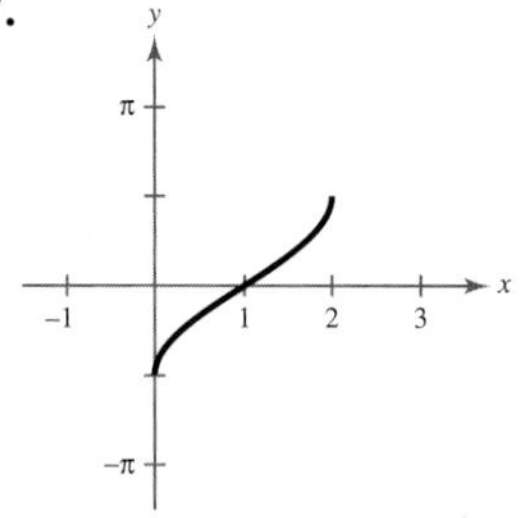

79.

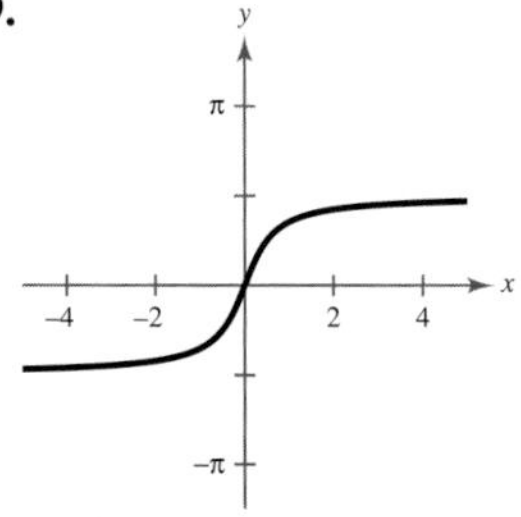

81.

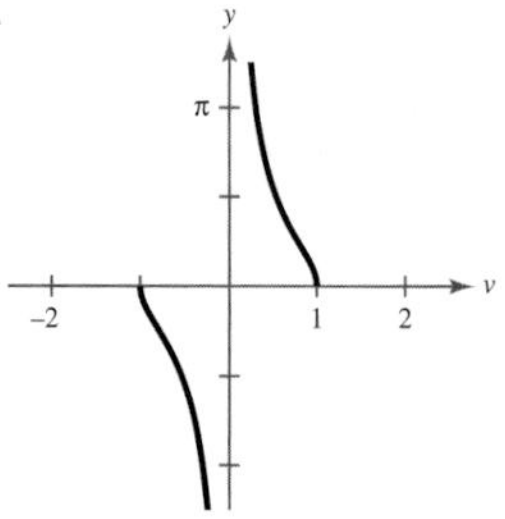

83.

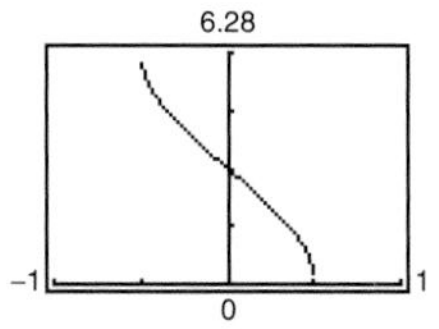

85.

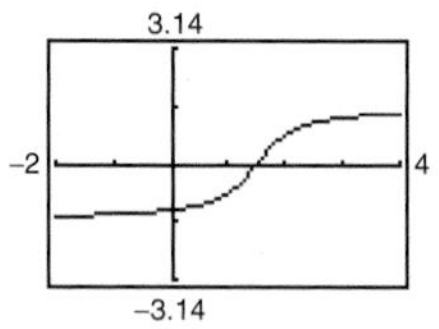

87.

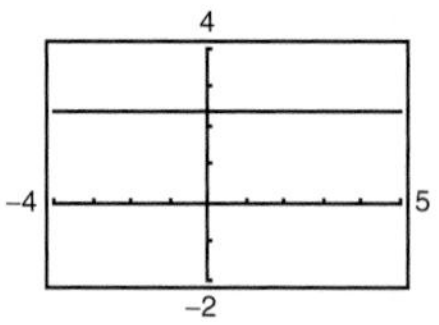

89. $3\sqrt{2}\sin\left(2t + \dfrac{\pi}{4}\right)$

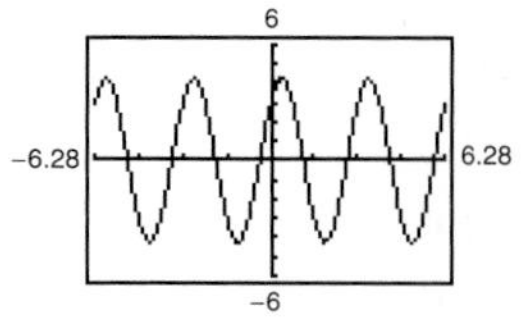

The graph implies that the identity is true.

91. (a) $\theta = \arcsin \dfrac{5}{s}$ (b) 0.13, 0.25

93. (a)

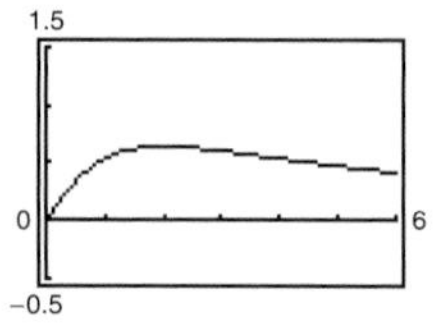

(b) 2 feet

(c) $\beta = 0$; As x increases, β approaches 0.

95. (a) $\theta = \arctan \dfrac{x}{20}$ (b) 0.24, 0.54

97. False. $\dfrac{5\pi}{4}$ is not in the range of the arctangent.

99. Domain: $(-\infty, \infty)$

Range: $(0, \pi)$

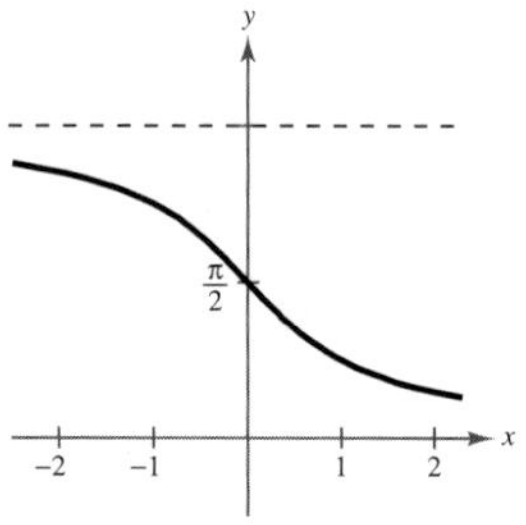

101. Domain: $(-\infty, -1] \cup [1, \infty)$

Range: $[-\pi/2, 0) \cup (0, \pi/2]$

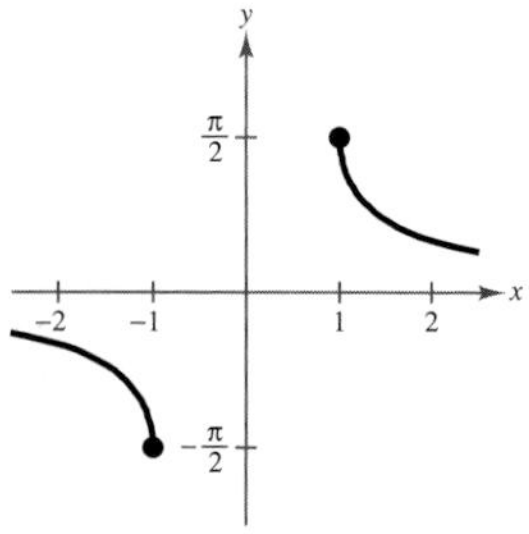

103.

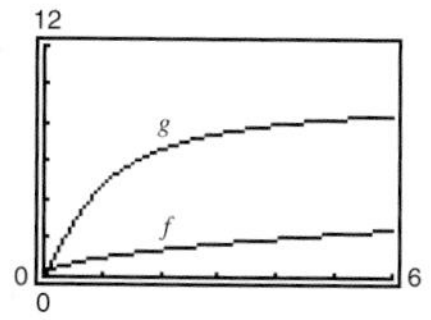

As x increases to infinity, g approaches 3π, but f has no maximum.

$a \approx 87.54$

105.–109. Answers will vary.

111.

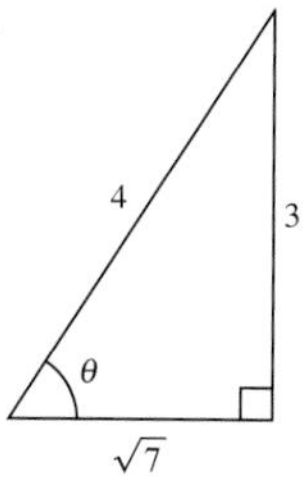

113.

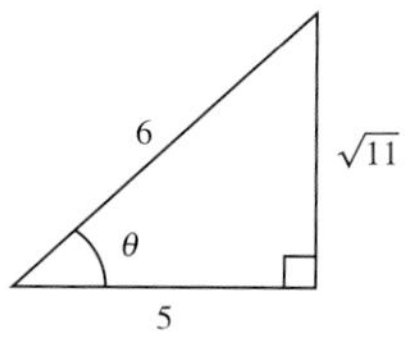

115. 1279.284 **117.** 117.391 **119.** 8 people

Section 4.8 *(page 428)*

1. $a \approx 3.64$, $c \approx 10.64$, $B = 70°$

3. $a \approx 8.26$, $c \approx 25.38$, $A = 19°$

5. $c \approx 11.66$, $A \approx 30.96°$, $B \approx 59.04°$

7. $a \approx 49.48$, $A \approx 72.08°$, $B = 17.92°$

9. $a \approx 91.34$, $b \approx 420.70$, $B = 77°45'$

11. 2.56 inches

13. 19.99 inches **15.** 107.2 feet **17.** 19.7 feet

19. (a)

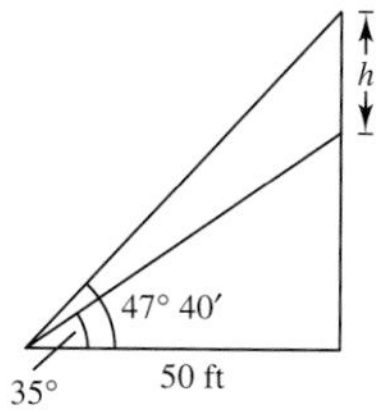

(b) $h = 50(\tan 47° 40' - \tan 35°)$ (c) 19.9 feet

21. 2236.8 feet **23.** 56.3° **25.** 1.84°

27. 5099 feet **29.** 0.73 mile

31. 554 miles north; 709 miles east

33. (a) N 58° E (b) 68.82 meters **35.** N 56.3° W

37. 1933.3 feet **39.** ≈ 3.23 miles or $\approx 17{,}054$ feet

41. 78.7° **43.** 35.3° **45.** 29.4 inches

47. $y = \sqrt{3}r$ **49.** $a \approx 12.2, b \approx 7$

51. (a) 4 (b) 4 (c) $\frac{1}{16}$

53. (a) $\frac{1}{16}$ (b) 60 (c) $\frac{1}{120}$ **55.** $d = 4\sin(\pi t)$

57. $d = 3\cos\left(\frac{4\pi t}{3}\right)$ **59.** $\omega = 528\pi$

61. (a)

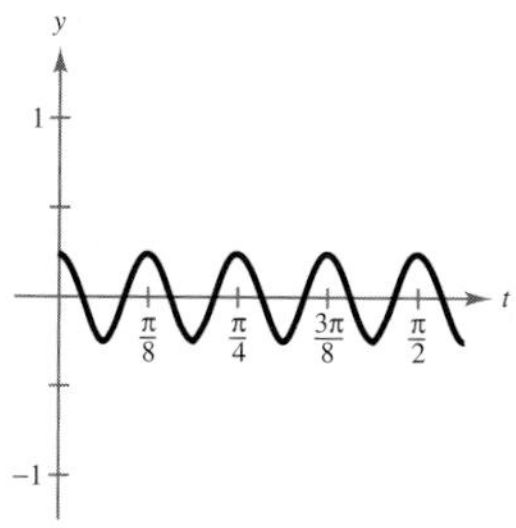

(b) $\frac{\pi}{8}$

(c) $\frac{\pi}{32}$

63. False. One period is the time for one complete cycle of the motion.

65. (a) and (b)

Base 1	Base 2	Altitude	Area
8	8 + 16 cos 10°	8 sin 10°	22.1
8	8 + 16 cos 20°	8 sin 20°	42.5
8	8 + 16 cos 30°	8 sin 30°	59.7
8	8 + 16 cos 40°	8 sin 40°	72.7
8	8 + 16 cos 50°	8 sin 50°	80.5
8	8 + 16 cos 60°	8 sin 60°	83.1
8	8 + 16 cos 70°	8 sin 70°	80.7

≈ 83.1 square feet when $\theta = 60°$

(c) $A = 64(1 + \cos\theta)(\sin\theta)$

(d)

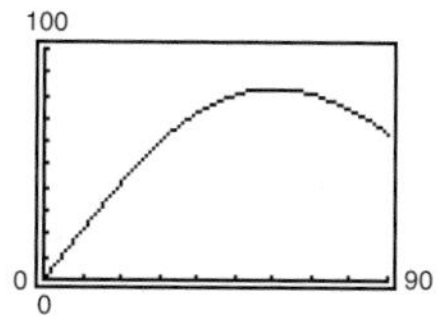

≈ 83.1 square feet when $\theta = 60°$

67. (a)

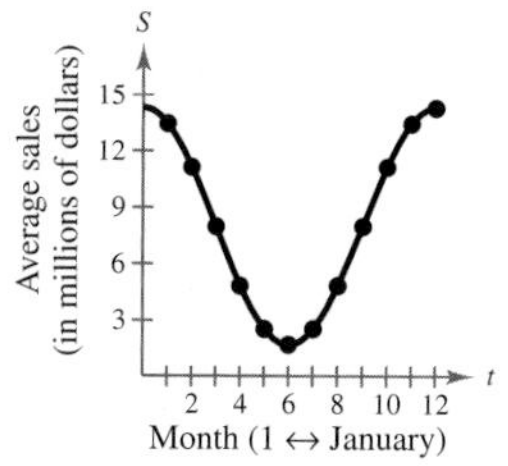

(b) $S = 8 + 6.3\cos\left(\frac{\pi}{6}t\right)$ or $S = 8 + 6.3\sin\left(\frac{\pi}{6}t + \frac{\pi}{2}\right)$

The model is a good fit.

(c) 12. Yes, sales of outerwear are seasonal.

(d) Maximum displacement from average sales of \$8 million.

69.

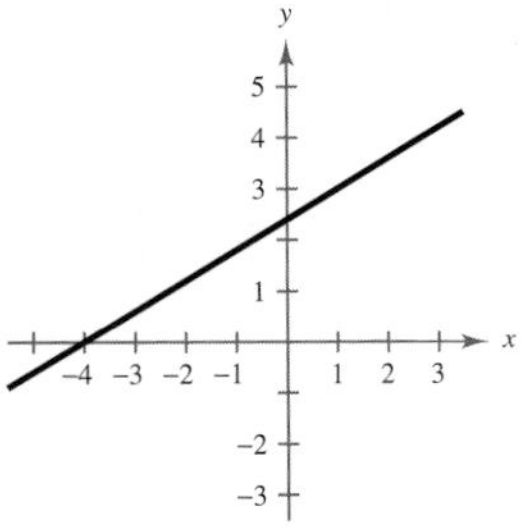

71.

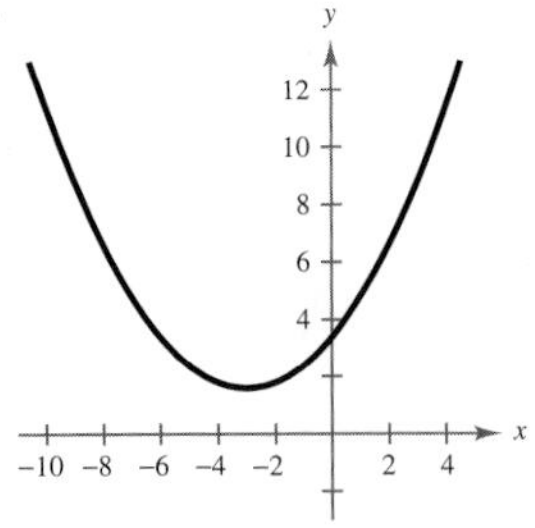

73.

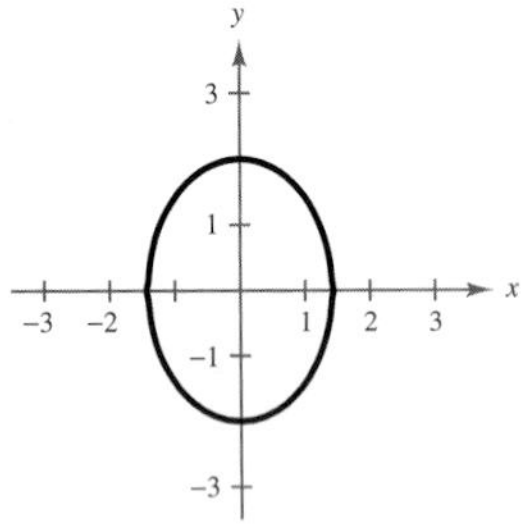

75.

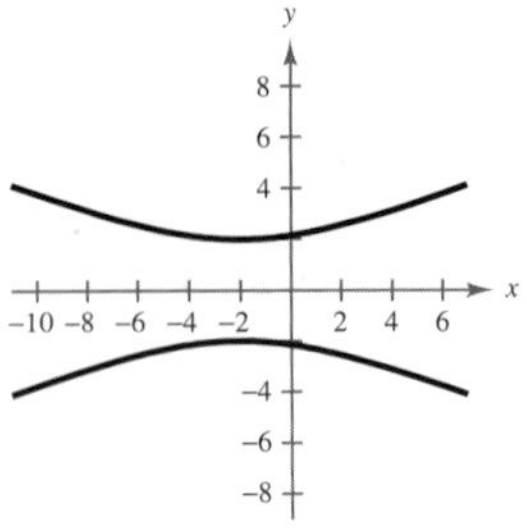

77.

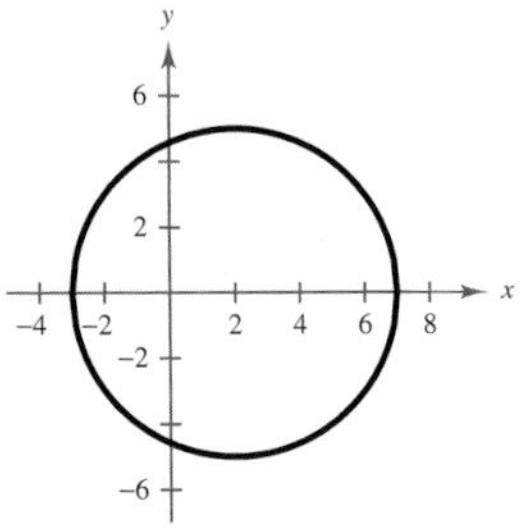

Review Exercises *(page 434)*

1. 0.5 **3.** 4.5

5.

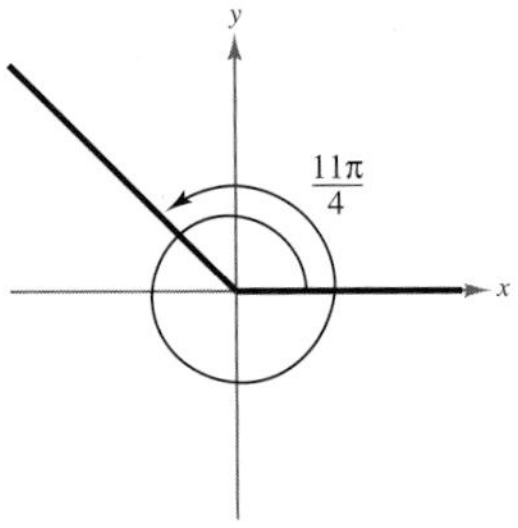

$\frac{3\pi}{4}, -\frac{5\pi}{4}$

7.

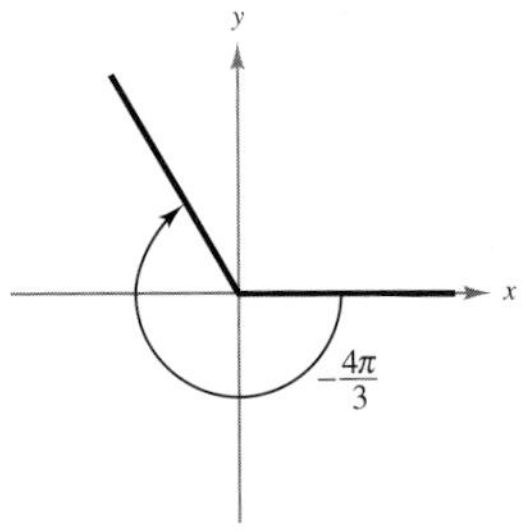

$\frac{2\pi}{3}, -\frac{10\pi}{3}$

9.

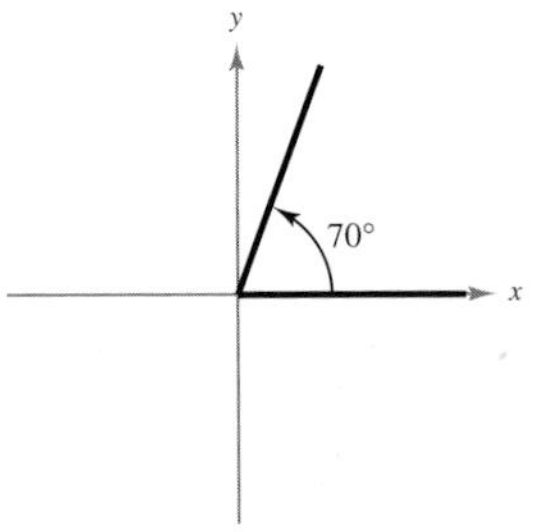

430°, −290°

11.

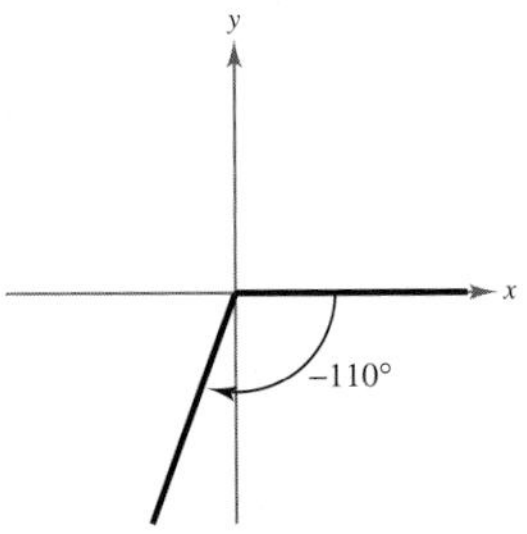

250°, −470°

13. 128.57° **15.** −200.54°

17. 8.3776 **19.** −0.5890

21. (a) $66\frac{2}{3}\pi$ radians per minute

(b) 400π inches per minute

23. $\left(-\frac{1}{2}, \frac{\sqrt{3}}{2}\right)$ **25.** $\left(-\frac{\sqrt{3}}{2}, \frac{1}{2}\right)$

27. $\sin\frac{7\pi}{6} = -\frac{1}{2}$

$\cos\frac{7\pi}{6} = -\frac{\sqrt{3}}{2}$

$\tan\frac{7\pi}{6} = \frac{\sqrt{3}}{3}$

$\csc\frac{7\pi}{6} = -2$

$\sec\frac{7\pi}{6} = -\frac{2\sqrt{3}}{3}$

$\cot\frac{7\pi}{6} = \sqrt{3}$

29. $\sin\left(-\frac{2\pi}{3}\right) = -\frac{\sqrt{3}}{2}$

$\cos\left(-\frac{2\pi}{3}\right) = -\frac{1}{2}$

$\tan\left(-\frac{2\pi}{3}\right) = \sqrt{3}$

$\csc\left(-\frac{2\pi}{3}\right) = -\frac{2\sqrt{3}}{3}$

$\sec\left(-\frac{2\pi}{3}\right) = -2$

$\cot\left(-\frac{2\pi}{3}\right) = \frac{\sqrt{3}}{3}$

31. $\sin\frac{11\pi}{4} = \sin\frac{3\pi}{4} = \frac{\sqrt{2}}{2}$

33. $\sin\left(-\frac{17\pi}{6}\right) = \sin\left(-\frac{5\pi}{6}\right) = -\frac{1}{2}$

35. −75.31 **37.** 3.24

39. $\sin\theta = \frac{4\sqrt{41}}{41}$

$\cos\theta = \frac{5\sqrt{41}}{41}$

$\tan\theta = \frac{4}{5}$

$\csc\theta = \frac{\sqrt{41}}{4}$

$\sec\theta = \frac{\sqrt{41}}{5}$

$\cot\theta = \frac{5}{4}$

41. $\sin\theta = \frac{1}{2}$

$\cos\theta = \frac{\sqrt{3}}{2}$

$\tan\theta = \frac{\sqrt{3}}{3}$

$\csc\theta = 2$

$\sec\theta = \frac{2\sqrt{3}}{3}$

$\cot\theta = \sqrt{3}$

43. (a) 3 (b) $\frac{2\sqrt{2}}{3}$ (c) $\frac{3\sqrt{2}}{4}$ (d) $\frac{\sqrt{2}}{4}$

45. (a) $\frac{1}{4}$ (b) $\frac{\sqrt{15}}{4}$ (c) $\frac{4\sqrt{15}}{15}$ (d) $\frac{\sqrt{15}}{15}$

47. 0.65 **49.** 0.56 **51.** 3.67 **53.** 0.07 kilometer

55. $\sin\theta = \frac{4}{5}$ $\csc\theta = \frac{5}{4}$
$\cos\theta = \frac{3}{5}$ $\sec\theta = \frac{5}{3}$
$\tan\theta = \frac{4}{3}$ $\cot\theta = \frac{3}{4}$

57. $\sin\theta = \frac{15\sqrt{241}}{241}$ $\csc\theta = \frac{\sqrt{241}}{15}$
$\cos\theta = \frac{4\sqrt{241}}{241}$ $\sec\theta = \frac{\sqrt{241}}{4}$
$\tan\theta = \frac{15}{4}$ $\cot\theta = \frac{4}{15}$

59. $\sin\theta \approx 1$ $\csc\theta \approx 1$
$\cos\theta \approx -0.1$ $\sec\theta \approx -9$
$\tan\theta = -9$ $\cot\theta \approx -0.1$

61. $\sin\theta = \frac{4\sqrt{17}}{17}$ $\csc\theta = \frac{\sqrt{17}}{4}$
$\cos\theta = \frac{\sqrt{17}}{17}$ $\sec\theta = \sqrt{17}$
$\tan\theta = 4$ $\cot\theta = \frac{1}{4}$

63. $\sin\theta = -\frac{\sqrt{11}}{6}$
$\cos\theta = \frac{5}{6}$
$\tan\theta = -\frac{\sqrt{11}}{5}$
$\csc\theta = -\frac{6\sqrt{11}}{11}$
$\cot\theta = -\frac{5\sqrt{11}}{11}$

65. $\cos\theta = -\frac{\sqrt{55}}{8}$
$\tan\theta = -\frac{3\sqrt{55}}{55}$
$\csc\theta = \frac{8}{3}$
$\sec\theta = -\frac{8\sqrt{55}}{55}$
$\cot\theta = -\frac{\sqrt{55}}{3}$

67. $\sin\theta = \frac{\sqrt{21}}{5}$
$\tan\theta = -\frac{\sqrt{21}}{2}$
$\csc\theta = \frac{5\sqrt{21}}{21}$
$\sec\theta = -\frac{5}{2}$
$\cot\theta = -\frac{2\sqrt{21}}{21}$

69. $\sqrt{3}$ **71.** $\frac{1}{2}$ **73.** $-\frac{\sqrt{2}}{2}$ **75.** -0.76

77. 0.06 **79.** 0 **81.** 3.24

83.

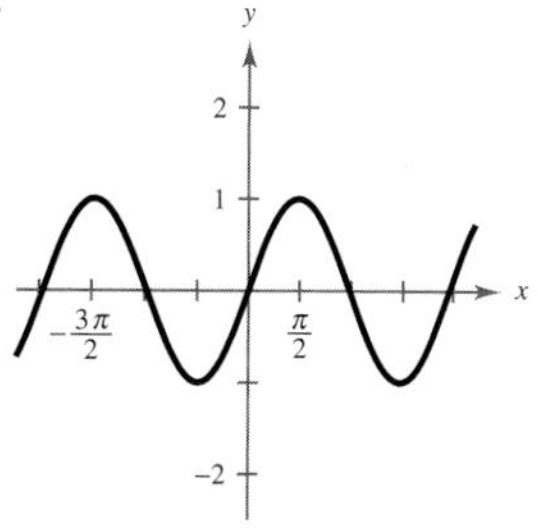

85.

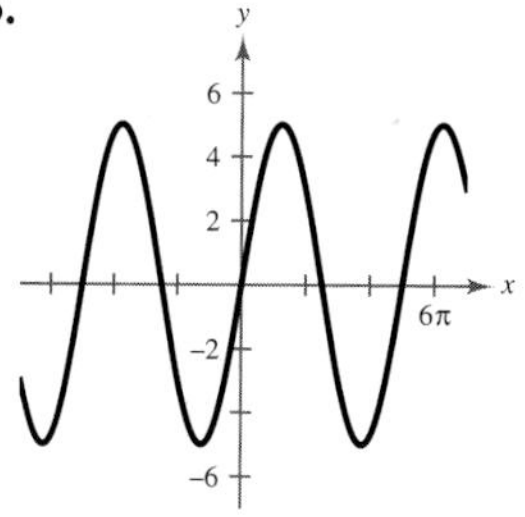

87.

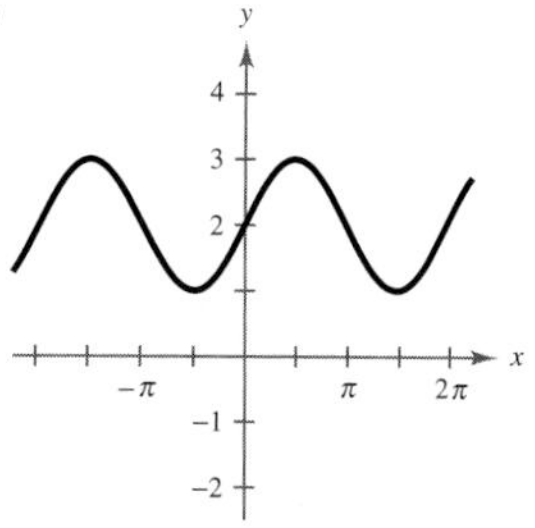

89.

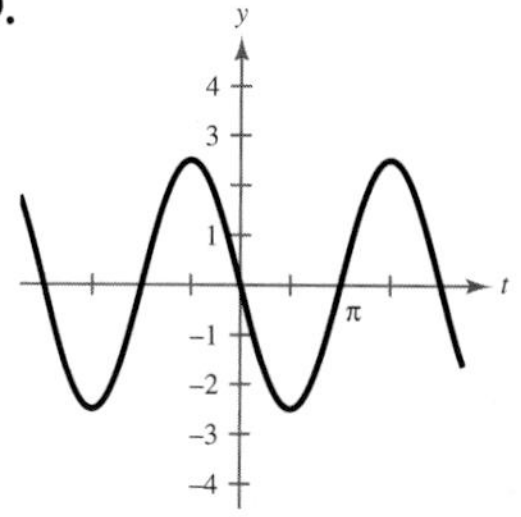

91. (a) $y = 2\sin 528\pi x$
(b) 264 cycles per second

93.

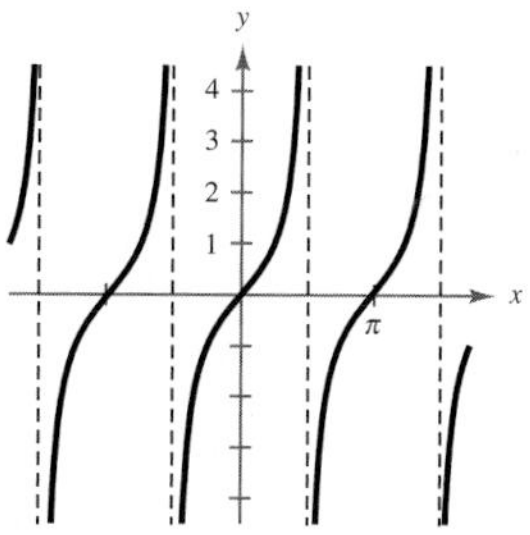

95.

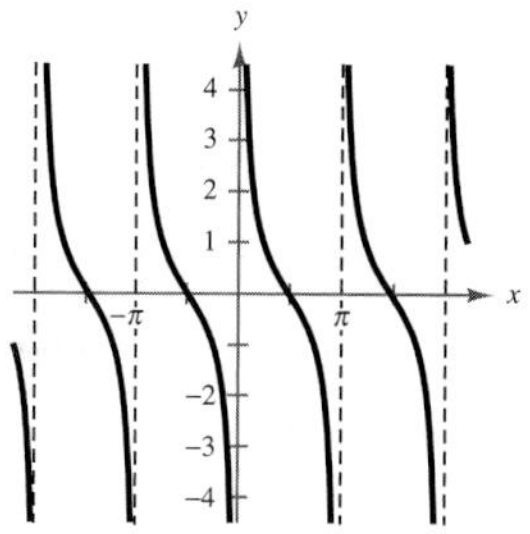

97.

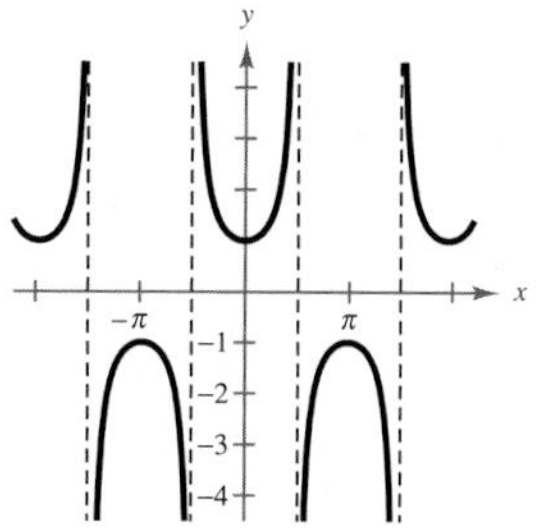

99.

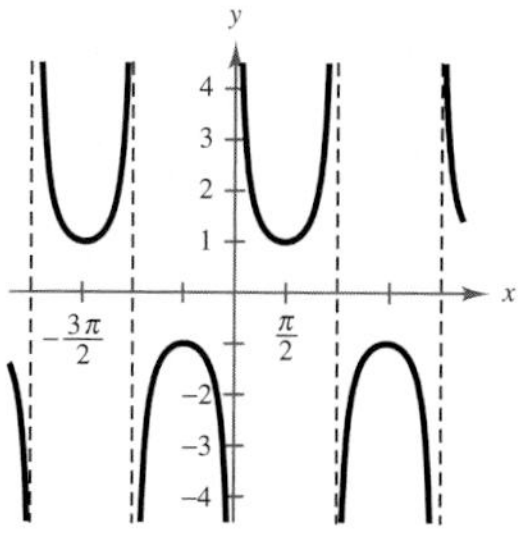

101.

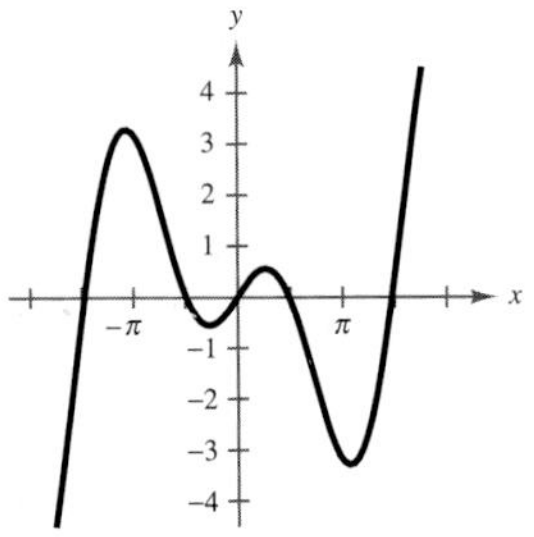

103. $-\dfrac{\pi}{6}$ **105.** 0.41

107. -0.46 **109.** $\dfrac{\pi}{6}$ **111.** π **113.** 1.24

115. 0.12 **117.** 1.40 **119.** -0.98 **121.** 0.72

123. 0 **125.** $\frac{4}{5}$ **127.** $\frac{13}{5}$ **129.** 66.8°

131. 1221 miles; N 85.6° E

133. False. The sine or cosine function is often useful for modeling simple harmonic motion.

135. False. For each θ there corresponds exactly one value of y.

137. d; the period is 2π and the amplitude is 3.

139. b; the period is 2 and the amplitude is 2.

141. Undefined because $\sec\theta = 1/\cos\theta$

143. Their ranges are $(-\infty, \infty)$ or $(-\infty, -1] \cup [1, \infty)$.

145. (a) $A = 72(\tan\theta - \theta)$

(b)

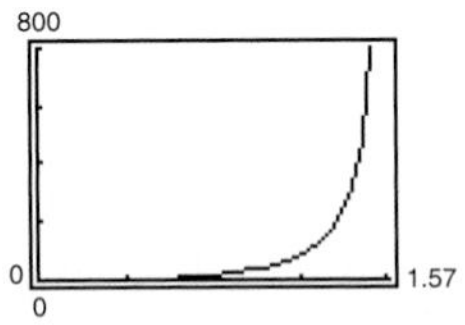

Area increases without bound as θ approaches $\pi/2$.

147. Answers will vary.

Chapter Test *(page 439)*

1. (a)

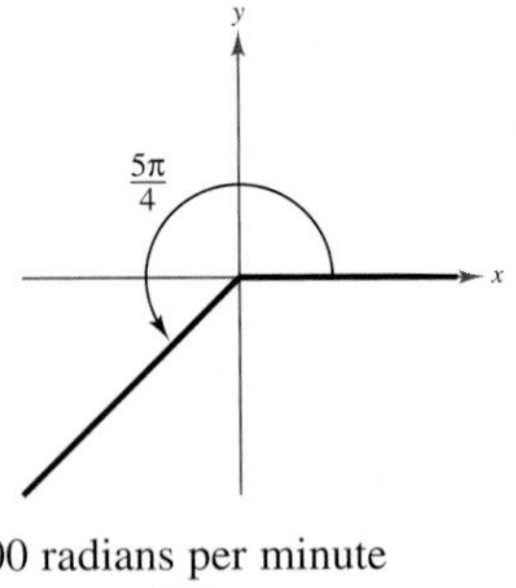

(b) $\dfrac{13\pi}{4}, -\dfrac{3\pi}{4}$

(c) 225°

2. 3000 radians per minute

3. $\sin\theta = \dfrac{3\sqrt{10}}{10}$ $\csc\theta = \dfrac{\sqrt{10}}{3}$

$\cos\theta = -\dfrac{\sqrt{10}}{10}$ $\sec\theta = -\sqrt{10}$

$\tan\theta = -3$ $\cot\theta = -\dfrac{1}{3}$

4. For $0 \le \theta < \dfrac{\pi}{2}$:

$\sin\theta = \dfrac{3\sqrt{13}}{13}$

$\cos\theta = \dfrac{2\sqrt{13}}{13}$

$\csc\theta = \dfrac{\sqrt{13}}{3}$

$\sec\theta = \dfrac{\sqrt{13}}{2}$

$\cot\theta = \dfrac{2}{3}$

For $\pi \le \theta < \dfrac{3\pi}{2}$:

$\sin\theta = -\dfrac{3\sqrt{13}}{13}$

$\cos\theta = -\dfrac{2\sqrt{13}}{13}$

$\csc\theta = -\dfrac{\sqrt{13}}{3}$

$\sec\theta = -\dfrac{\sqrt{13}}{2}$

$\cot\theta = \dfrac{2}{3}$

5. $\theta' = 70°$

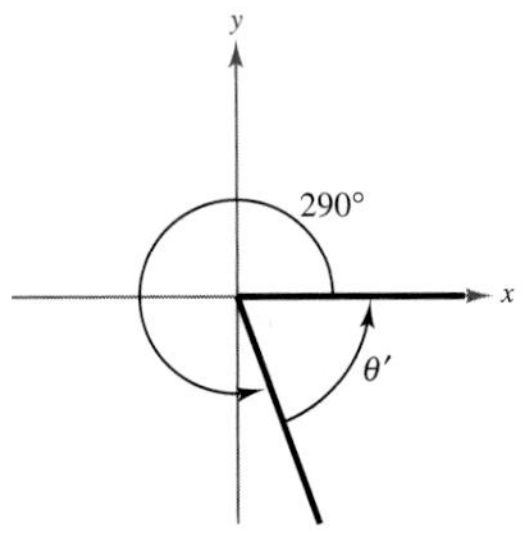

6. Quadrant III **7.** 150°, 210° **8.** 1.33, 1.81

9. $\sin\theta = -\frac{4}{5}$

$\tan\theta = -\frac{4}{3}$

$\csc\theta = -\frac{5}{4}$

$\sec\theta = \frac{5}{3}$

$\cot\theta = -\frac{3}{4}$

10. $\sin\theta = \frac{15}{17}$

$\cos\theta = -\frac{8}{17}$

$\tan\theta = -\frac{15}{8}$

$\csc\theta = \frac{17}{15}$

$\cos\theta = -\frac{8}{15}$

11.

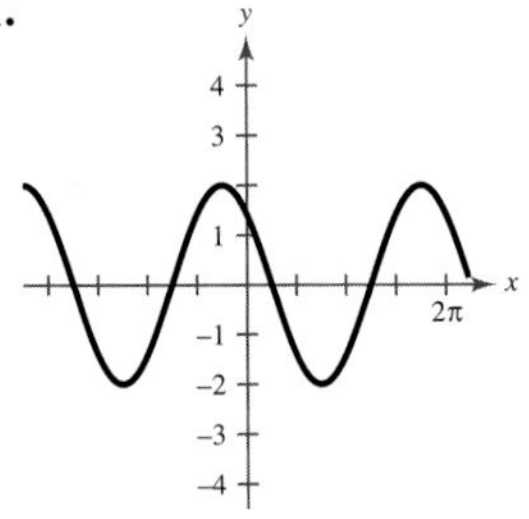

12.

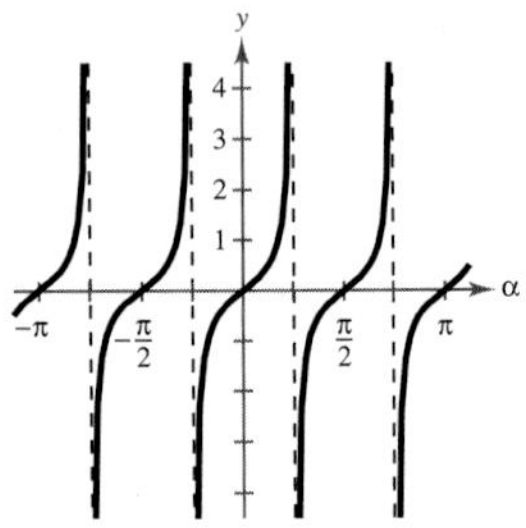

13.

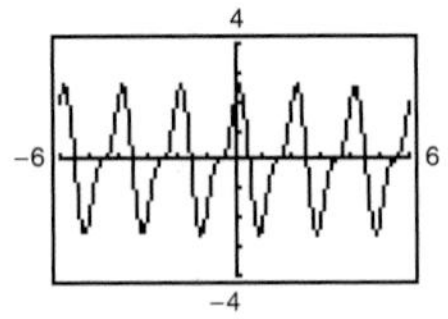

Period: 2

14.

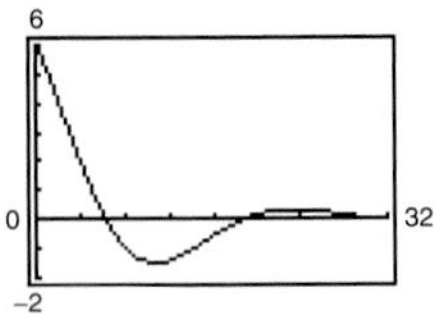

Not periodic

15. $y = -2\sin\left(\frac{x}{2} - \frac{\pi}{4}\right)$ **16.** $\frac{\sqrt{5}}{2}$

17.

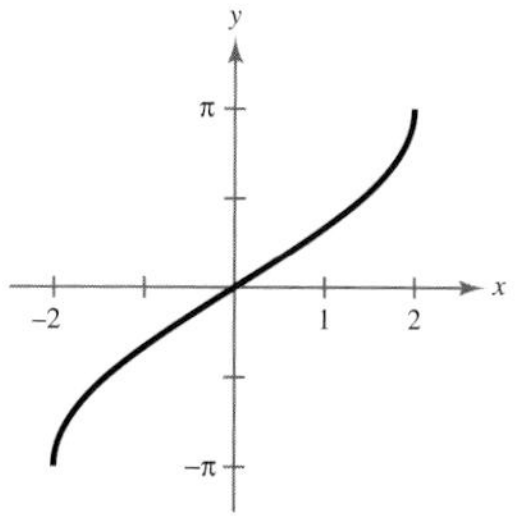

18. N 49.9° W **19.** $d = -6\cos \pi t$

Chapter 5

Section 5.1 *(page 447)*

1. $\tan x = -\sqrt{3}$

$\csc x = \frac{2\sqrt{3}}{3}$

$\sec x = -2$

$\cot x = -\frac{\sqrt{3}}{3}$

3. $\cos\theta = \frac{\sqrt{2}}{2}$

$\tan\theta = -1$

$\csc\theta = -\sqrt{2}$

$\cot\theta = -1$

5. $\sin x = -\frac{5}{13}$

$\cos x = -\frac{12}{13}$

$\csc x = -\frac{13}{5}$

$\cot x = \frac{12}{5}$

7. $\sin\phi = -\frac{\sqrt{5}}{3}$

$\cos\phi = \frac{2}{3}$

$\tan\phi = -\frac{\sqrt{5}}{2}$

$\cot\phi = -\frac{2\sqrt{5}}{5}$

9. $\sin x = \frac{1}{3}$

$\cos x = -\frac{2\sqrt{2}}{3}$

$\csc x = 3$

$\sec x = -\frac{3\sqrt{2}}{4}$

$\cot x = -2\sqrt{2}$

11. $\sin\theta = -\frac{2\sqrt{5}}{5}$

$\cos\theta = -\frac{\sqrt{5}}{5}$

$\csc\theta = -\frac{\sqrt{5}}{2}$

$\sec\theta = -\sqrt{5}$

$\cot\theta = \frac{1}{2}$

13. $\cos\theta = 0$

$\tan\theta$ is undefined.

$\csc\theta = -1$

$\sec\theta$ is undefined.

15. d **17.** b **19.** e **21.** b **23.** f **25.** e

27. $\csc\theta$ **29.** $\cos^2\phi$ **31.** $\cos x$ **33.** $\sin^2 x$

35. 1 **37.** $\tan x$ **39.** $1 + \sin y$ **41.** $\sec\beta$

43. $\cos u + \sin u$ **45.** $\sin^2 x$ **47.** $\sin^2 x \tan^2 x$

49. $\sec x + 1$ **51.** $\sec^4 x$ **53.** $\sin^2 x - \cos^2 x$

55. $\cot^2 x(\csc x - 1)$ **57.** $1 + 2\sin x\cos x$

59. $4\cot^2 x$ **61.** $2\csc^2 x$ **63.** $2\sec x$

65. $1 + \cos y$ **67.** $3(\sec x + \tan x)$

69.

x	0.2	0.4	0.6	0.8	1.0
y_1	0.1987	0.3894	0.5646	0.7174	0.8415
y_2	0.1987	0.3894	0.5646	0.7174	0.8415

x	1.2	1.4
y_1	0.9320	0.9854
y_2	0.9320	0.9854

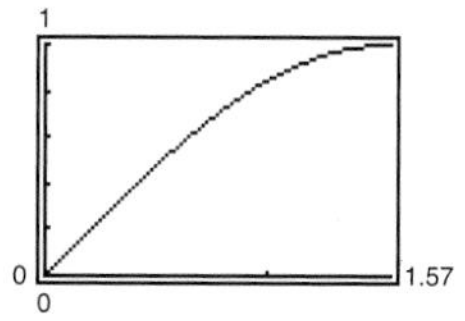

$y_1 = y_2$

71.

x	0.2	0.4	0.6	0.8	1.0
y_1	1.2230	1.5085	1.8958	2.4650	3.4082
y_2	1.2230	1.5085	1.8958	2.4650	3.4082

x	1.2	1.4
y_1	5.3319	11.6814
y_2	5.3319	11.6814

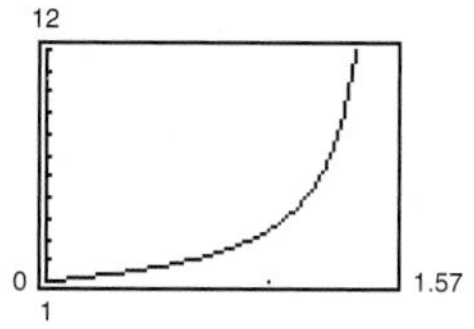

$y_1 = y_2$

73. $\csc x$ **75.** $\tan x$ **77.** $3\sin\theta$ **79.** $3\tan\theta$

81. $5\sec\theta$ **83.** $3\cos\theta = 3$; $\sin\theta = 0$; $\cos\theta = 1$

85. $4\sin\theta = 2\sqrt{2}$; $\sin\theta = \frac{\sqrt{2}}{2}$; $\cos\theta = \frac{\sqrt{2}}{2}$

87. $0 \le \theta \le \pi$ **89.** $0 \le \theta < \frac{\pi}{2}, \frac{3\pi}{2} < \theta < 2\pi$

91. $\ln|\cot \theta|$ **93.** $\ln|\csc t \sec t|$

95. (a) $\csc^2 132° - \cot^2 132° \approx 1.8107 - 0.8107 = 1$

(b) $\csc^2 \dfrac{2\pi}{7} - \cot^2 \dfrac{2\pi}{7} \approx 1.6360 - 0.6360 = 1$

97. (a) $\cos(90° - 80°) = \sin 80° \approx 0.9848$

(b) $\cos\left(\dfrac{\pi}{2} - 0.8\right) = \sin 0.8 \approx 0.7174$

99. $\tan \theta$

101. False. A cofunction identity can be used to transform a tangent function so that it can be represented by a cotangent function.

103. 1, 1 **105.** $0, -\infty$

107. Not an identity because $\cot \theta = \pm\sqrt{\csc^2 \theta - 1}$

109. Not an identity because $\dfrac{5}{\cos \theta} \neq \dfrac{1}{5 \cos \theta}$

111. Not an identity because θ and ϕ may not be equal

113. $\sin \theta = \pm\sqrt{1 - \cos^2 \theta}$

$\tan \theta = \pm\dfrac{\sqrt{1 - \cos^2 \theta}}{\cos \theta}$

$\csc \theta = \pm\dfrac{1}{\sqrt{1 - \cos^2 \theta}}$

$\sec \theta = \dfrac{1}{\cos \theta}$

$\cot \theta = \pm\dfrac{\cos \theta}{\sqrt{1 - \cos^2 \theta}}$

115. $\sqrt{5v}$ **117.** $10x(\sqrt{30} + 5)$

119.

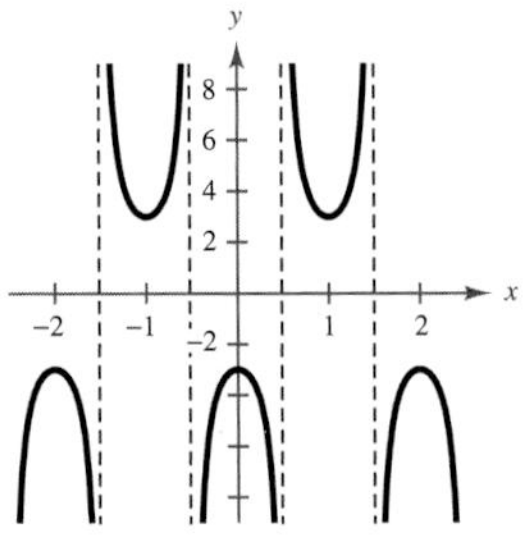

121.

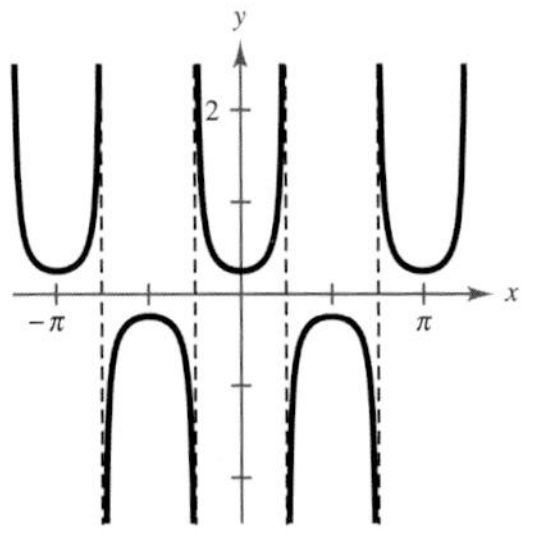

Section 5.2 *(page 455)*

1.–59. Answers will vary.

61. 1 **63.** 2 **65.** Answers will vary.

67. True. An identity is an equation that is true for all real values in the domain of the variable.

69. Not an identity because $\tan \theta = \pm\sqrt{\sec^2 \theta - 1}$

Possible answer: $\dfrac{3\pi}{4}$

71. Not an identity because $\sqrt{\sin^2 x + \cos^2 x} \neq \sin x + \cos x$

The left side is 1 for any x, but the right side is not necessarily 1.

Possible answer: $\dfrac{\pi}{4}$

73. Answers will vary. **75.** $-21 - 20i$

77. $-9 + 46i$ **79.** $\dfrac{-5 \pm \sqrt{3}i}{2}$ **81.** $\dfrac{1 \pm \sqrt{5}i}{4}$

83. $\dfrac{5 \pm \sqrt{101}i}{14}$ **85.** $\dfrac{-5 \pm \sqrt{79}i}{26}$

Section 5.3 *(page 464)*

1.–5. Answers will vary. **7.** $\dfrac{2\pi}{3} + 2n\pi, \dfrac{4\pi}{3} + 2n\pi$

9. $\dfrac{\pi}{3} + 2n\pi, \dfrac{2\pi}{3} + 2n\pi$ **11.** $\dfrac{\pi}{6} + n\pi, \dfrac{5\pi}{6} + n\pi$

13. $n\pi, \dfrac{3\pi}{2} + 2n\pi$ **15.** $\dfrac{\pi}{3} + n\pi, \dfrac{2\pi}{3} + n\pi$

17. $\dfrac{\pi}{8} + n\pi, \dfrac{3\pi}{8} + n\pi, \dfrac{5\pi}{8} + n\pi, \dfrac{7\pi}{8} + n\pi$

19. $\dfrac{n\pi}{3}, \dfrac{\pi}{4} + n\pi$ **21.** $0, \dfrac{\pi}{2}, \pi, \dfrac{3\pi}{2}$

23. $0, \pi, \dfrac{\pi}{6}, \dfrac{5\pi}{6}, \dfrac{7\pi}{6}, \dfrac{11\pi}{6}$ **25.** $\dfrac{\pi}{3}, \dfrac{5\pi}{3}, \pi$

27. No solution **29.** $\pi, \dfrac{\pi}{3}, \dfrac{5\pi}{3}$ **31.** $\dfrac{\pi}{6}, \dfrac{5\pi}{6}, \dfrac{7\pi}{6}, \dfrac{11\pi}{6}$

33. $\dfrac{\pi}{6} + n\pi, \dfrac{5\pi}{6} + n\pi$ **35.** $\dfrac{\pi}{12} + \dfrac{n\pi}{3}$

37. $\dfrac{\pi}{2} + 4n\pi, \dfrac{7\pi}{2} + 4n\pi$ **39.** $-1, 3$ **41.** ± 2

43. $\frac{2}{3}, \frac{3}{2}$; $0.8411 + 2n\pi, 5.4421 + 2n\pi$

45. 2.6779, 5.8195

47. 1.0472, 5.2360

49. 0.8603, 3.4256

51. 0, 2.6779, 3.1416, 5.8195

53. 0.9828, 1.7682, 4.1244, 4.9098

55. 0.3398, 0.8481, 2.2935, 2.8018

57. 1.9357, 2.7767, 5.0773, 5.9183

59. $\dfrac{\pi}{4}, \dfrac{5\pi}{4}$, arctan 5, arctan 5 + π **61.** $\dfrac{\pi}{3}, \dfrac{5\pi}{3}$

63. (a)

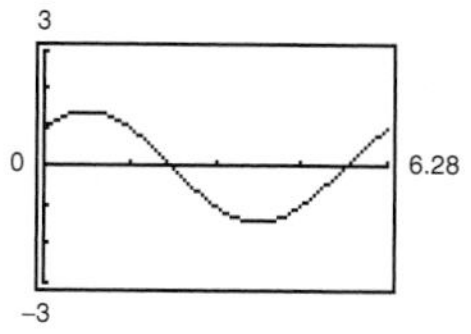

(b) $\frac{\pi}{4}, \frac{5\pi}{4}$

Maximum: $\left(\frac{\pi}{4}, \sqrt{2}\right)$

Minimum: $\left(\frac{5\pi}{4}, -\sqrt{2}\right)$

65. 1

67. (a) All real numbers except $x = 0$

(b) y-axis symmetry; horizontal asymptote: $y = 1$

(c) Oscillates

(d) Infinitely many solutions

(e) Yes, 0.6366

69. 0.04 second, 0.43 second, 0.83 second

71. February, March, and April **73.** 1.9°

75. (a) $\frac{10}{3}$

(b)

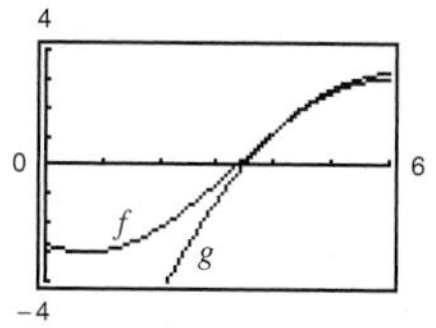

For $3.5 \le x \le 6$, the approximation appears to be good.

(c) 3.46, 8.81

3.46 is close to the zero of f in the interval $[0, 6]$.

77. True. The first equation has a smaller period than the second equation, so it will have more solutions in the interval $[0, 2\pi)$.

79. 1 **81.** $B = 45°$, $b \approx 11.31$, $c \approx 11.31$ **83.** $C = 24°$, $a \approx 54.8$, $b \approx 50.1$

85. $\sin\theta = \frac{1}{2}$, $\cos\theta = \frac{\sqrt{3}}{2}$, $\tan\theta = \frac{\sqrt{3}}{3}$

87. $\sin\theta = \frac{\sqrt{2}}{2}$, $\cos\theta = -\frac{\sqrt{2}}{2}$, $\tan\theta = -1$

89. $\sin\theta = -\frac{\sqrt{2}}{2}$, $\cos\theta = \frac{\sqrt{2}}{2}$, $\tan\theta = -1$

91. 30 feet

Section 5.4 *(page 472)*

1. (a) $\frac{\sqrt{2} - \sqrt{6}}{4}$ (b) $\frac{\sqrt{2} + 1}{2}$

3. (a) $\frac{1}{2}$ (b) $\frac{-\sqrt{3} - 1}{2}$

5. (a) $\frac{-\sqrt{2} - \sqrt{6}}{4}$ (b) $\frac{-1 + \sqrt{2}}{2}$

7. $\sin 105° = \frac{\sqrt{2}}{4}\left(\sqrt{3} + 1\right)$

$\cos 105° = \frac{\sqrt{2}}{4}\left(1 - \sqrt{3}\right)$

$\tan 105° = -2 - \sqrt{3}$

9. $\sin 195° = \frac{\sqrt{2}}{4}\left(1 - \sqrt{3}\right)$

$\cos 195° = -\frac{\sqrt{2}}{4}\left(\sqrt{3} + 1\right)$

$\tan 195° = 2 - \sqrt{3}$

11. $\sin\frac{11\pi}{12} = \frac{\sqrt{2}}{4}\left(\sqrt{3} - 1\right)$

$\cos\frac{11\pi}{12} = -\frac{\sqrt{2}}{4}\left(\sqrt{3} + 1\right)$

$\tan\frac{11\pi}{12} = -2 + \sqrt{3}$

13. $\sin\frac{17\pi}{12} = -\frac{\sqrt{2}}{4}\left(\sqrt{3} + 1\right)$

$\cos\frac{17\pi}{12} = \frac{\sqrt{2}}{4}\left(1 - \sqrt{3}\right)$

$\tan\frac{17\pi}{12} = 2 + \sqrt{3}$

15. $\sin 285° = -\frac{\sqrt{2}}{4}\left(\sqrt{3} + 1\right)$

$\cos 285° = \frac{\sqrt{2}}{4}\left(\sqrt{3} - 1\right)$

$\tan 285° = -\left(2 + \sqrt{3}\right)$

17. $\sin(-165°) = -\frac{\sqrt{2}}{4}\left(\sqrt{3} - 1\right)$

$\cos(-165°) = -\frac{\sqrt{2}}{4}\left(1 + \sqrt{3}\right)$

$\tan(-165°) = 2 - \sqrt{3}$

19. $\sin\frac{13\pi}{12} = \frac{\sqrt{2}}{4}(1 - \sqrt{3})$

$\cos\frac{13\pi}{12} = -\frac{\sqrt{2}}{4}(1 + \sqrt{3})$

$\tan\frac{13\pi}{12} = 2 - \sqrt{3}$

21. $\sin\left(-\frac{13\pi}{12}\right) = \frac{\sqrt{2}}{4}(\sqrt{3} - 1)$

$\cos\left(-\frac{13\pi}{12}\right) = -\frac{\sqrt{2}}{4}(\sqrt{3} + 1)$

$\tan\left(-\frac{13\pi}{12}\right) = -2 + \sqrt{3}$

23. $\cos 40°$ **25.** $\tan 239°$ **27.** $\sin 1.8$ **29.** $\tan 3x$

31. $-\frac{\sqrt{3}}{2}$ **33.** $\frac{\sqrt{3}}{2}$ **35.** -1 **37.** $-\frac{63}{65}$

39. $\frac{16}{65}$ **41.** $-\frac{63}{16}$ **43.** $\frac{65}{56}$ **45.** $\frac{3}{5}$ **47.** $-\frac{44}{117}$

49. $\frac{5}{3}$ **51.** 1 **53.** 0 **55.–63.** Answers will vary.

65. $-\sin x$ **67.** $-\cos\theta$ **69.** $\frac{\pi}{2}$ **71.** $\frac{5\pi}{4}, \frac{7\pi}{4}$

73. $\frac{\pi}{4}, \frac{7\pi}{4}$ **75.** (a) $y = \frac{5}{12}\sin(2t + 0.6435)$

(b) $\frac{5}{12}$ feet (c) $\frac{1}{\pi}$ cycle per second

77. False. $\sin(u \pm v) = \sin u \cos v \pm \cos u \sin v$

79. False.

$\cos\left(x - \frac{\pi}{2}\right) = \cos x \cos\frac{\pi}{2} + \sin x \sin\frac{\pi}{2} = \sin x$

81. and 83. Answers will vary.

85. (a) $\sqrt{2}\sin\left(\theta + \frac{\pi}{4}\right)$ (b) $\sqrt{2}\cos\left(\theta - \frac{\pi}{4}\right)$

87. (a) $13\sin(3\theta + 0.3948)$ (b) $13\cos(3\theta - 1.1760)$

89. $2\cos\theta$ **91.** 15°

93.

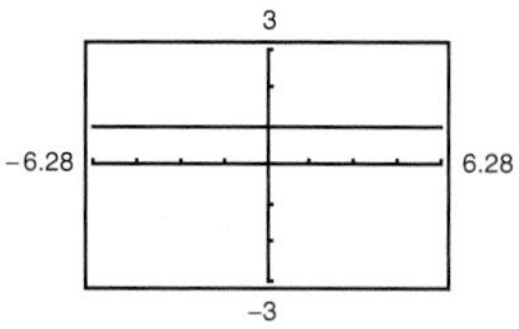

$\sin^2\left(\theta + \frac{\pi}{4}\right) + \sin^2\left(\theta - \frac{\pi}{4}\right) = 1$

95. Answers will vary. **97.** $f^{-1}(x) = \frac{x + 15}{5}$

99. Because f is not one-to-one, f^{-1} does not exist.

101. $4x - 3$ **103.** $6x - 3$

Section 5.5 *(page 482)*

1. $\frac{\sqrt{17}}{17}$ **3.** $\frac{15}{17}$ **5.** $\frac{8}{15}$ **7.** $\frac{17}{8}$

9. $0, \frac{\pi}{3}, \pi, \frac{5\pi}{3}$ **11.** $\frac{\pi}{12}, \frac{5\pi}{12}, \frac{13\pi}{12}, \frac{17\pi}{12}$

13. $0, \frac{2\pi}{3}, \frac{4\pi}{3}$ **15.** $\frac{\pi}{2}, \frac{\pi}{6}, \frac{5\pi}{6}, \frac{7\pi}{6}, \frac{3\pi}{2}, \frac{11\pi}{6}$

17. $0, \frac{\pi}{2}, \pi, \frac{3\pi}{2}$ **19.** $3\sin 2x$ **21.** $4\cos 2x$

23. $\sin 2u = \frac{24}{25}$, $\cos 2u = -\frac{7}{25}$, $\tan 2u = -\frac{24}{7}$

25. $\sin 2u = \frac{24}{25}$, $\cos 2u = \frac{7}{25}$, $\tan 2u = \frac{24}{7}$

27. $\sin 2u = -\frac{4\sqrt{21}}{25}$

$\cos 2u = -\frac{17}{25}$

$\tan 2u = \frac{4\sqrt{21}}{17}$

29. $\frac{1}{8}(3 + 4\cos 2x + \cos 4x)$ **31.** $\frac{1}{8}(1 - \cos 4x)$

33. $\frac{1}{16}(1 + \cos 2x - \cos 4x - \cos 2x \cos 4x)$

35. $\frac{4\sqrt{17}}{17}$ **37.** $\frac{1}{4}$ **39.** $\sqrt{17}$

41. $\sin 75° = \frac{1}{2}\sqrt{2 + \sqrt{3}}$

$\cos 75° = \frac{1}{2}\sqrt{2 - \sqrt{3}}$

$\tan 75° = 2 + \sqrt{3}$

43. $\sin 112° 30' = \frac{1}{2}\sqrt{2 + \sqrt{2}}$

$\cos 112° 30' = -\frac{1}{2}\sqrt{2 - \sqrt{2}}$

$\tan 112° 30' = -1 - \sqrt{2}$

45. $\sin\frac{\pi}{8} = \frac{1}{2}\sqrt{2 - \sqrt{2}}$

$\cos\frac{\pi}{8} = \frac{1}{2}\sqrt{2 + \sqrt{2}}$

$\tan\frac{\pi}{8} = \sqrt{2} - 1$

47. $\sin\frac{3\pi}{8} = \frac{1}{2}\sqrt{2 + \sqrt{2}}$

$\cos\frac{3\pi}{8} = \frac{1}{2}\sqrt{2 - \sqrt{2}}$

$\tan\frac{3\pi}{8} = \sqrt{2} + 1$

49. $\sin\frac{u}{2} = \frac{5\sqrt{26}}{26}$

$\cos\frac{u}{2} = \frac{\sqrt{26}}{26}$

$\tan\frac{u}{2} = 5$

51. $\sin\frac{u}{2} = \sqrt{\frac{89 - 8\sqrt{89}}{178}}$

$\cos\frac{u}{2} = -\sqrt{\frac{89 + 8\sqrt{89}}{178}}$

$\tan\frac{u}{2} = \frac{8 - \sqrt{89}}{5}$

53. $\sin \frac{u}{2} = \frac{3\sqrt{10}}{10}$

$\cos \frac{u}{2} = -\frac{\sqrt{10}}{10}$

$\tan \frac{u}{2} = -3$

55. $|\sin 3x|$ **57.** $-|\tan 4x|$

59. π **61.** $\frac{\pi}{3}, \pi, \frac{5\pi}{3}$ **63.** $3\left(\sin \frac{\pi}{2} + \sin 0\right)$

65. $\frac{1}{2}[\sin 10\theta - \sin(-2\theta)]$ **67.** $\frac{5}{2}(\cos 8\beta + \cos 2\beta)$

69. $\frac{1}{2}(\cos 2y - \cos 2x)$ **71.** $\frac{1}{2}[\sin 2\theta - \sin(-2\pi)]$

73. $5(\cos 60^\circ + \cos 90^\circ)$ **75.** $2 \sin 45^\circ \cos 15^\circ$

77. $-2 \sin \frac{\pi}{2} \sin \frac{\pi}{4}$ **79.** $2 \cos 4\theta \sin \theta$

81. $2 \cos 4x \cos 2x$ **83.** $2 \cos \alpha \sin \beta$

85. $-2 \sin \theta \sin \frac{\pi}{2}$ **87.** $0, \frac{\pi}{4}, \frac{\pi}{2}, \frac{3\pi}{4}, \pi, \frac{5\pi}{4}, \frac{3\pi}{2}, \frac{7\pi}{4}$

89. $\frac{\pi}{6}, \frac{5\pi}{6}$ **91.** $\frac{25}{169}$ **93.** $\frac{4}{13}$

95.–109. Answers will vary.

111.

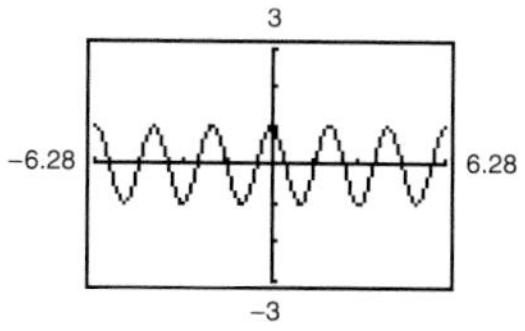

113.

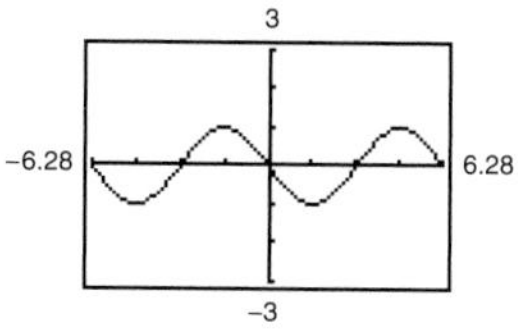

115.

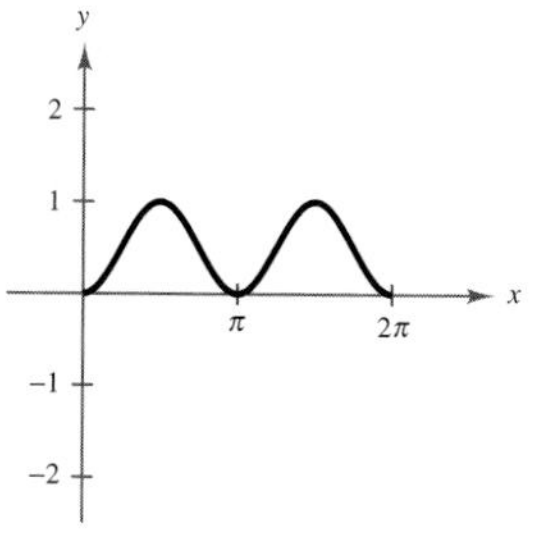

117. $2x\sqrt{1 - x^2}$

119. (a) $A = 100 \sin \frac{\theta}{2} \cos \frac{\theta}{2}$

(b) $A = 50 \sin \theta$

The area is maximum when $\theta = \pi/2$.

121. 0.4482

123. False. For $u < 0$,

$\sin 2u = -\sin(-2u)$

$= -2 \sin(-u) \cos(-u)$

$= -2(-\sin u) \cos u$

$= 2 \sin u \cos u.$

125. (a)

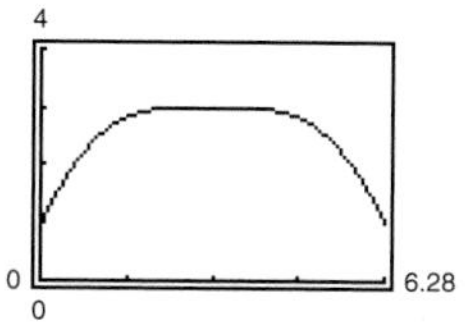

Maximum: $(\pi, 3)$

(b) π

127. (a) $\frac{1}{4}(3 + \cos 4x)$

(b) $2 \cos^4 x - 2 \cos^2 x + 1$

(c) $1 - 2 \sin^2 x \cos^2 x$

(d) $1 - \frac{1}{2} \sin^2 2x$

(e) No. There is often more than one way to rewrite a trigonometric expression.

129. September: \$235,000 **131.** 15.7 gallons

October: \$272,600

Review Exercises *(page 487)*

1. $\sec x$ **3.** $\cos x$ **5.** $\cot x$

7. $\tan x = \frac{3}{4}$

$\csc x = \frac{5}{3}$

$\sec x = \frac{5}{4}$

$\cot x = \frac{4}{3}$

9. $\cos x = \frac{\sqrt{2}}{2}$

$\tan x = -1$

$\csc x = -\sqrt{2}$

$\sec x = \sqrt{2}$

$\cot x = -1$

11. $\sin^2 x$ **13.** 1 **15.** $\cot \theta$ **17.** $\cot^2 x$

19. $\sec x + 2 \sin x$ **21.** $-2 \tan^2 \theta$

23.–31. Answers will vary. **33.** $\frac{\pi}{3} + 2n\pi, \frac{5\pi}{3} + 2n\pi$

35. $\frac{\pi}{3} + 2n\pi, \frac{5\pi}{3} + 2n\pi$ **37.** $\frac{\pi}{6} + n\pi, \frac{5\pi}{6} + n\pi$

39. $\frac{\pi}{6}, \frac{\pi}{2}, \frac{5\pi}{6}$ **41.** 0 **43.** $0, \frac{\pi}{3}, \frac{2\pi}{3}, \pi, \frac{4\pi}{3}, \frac{5\pi}{3}$

45. No solution **47.** $\frac{\pi}{2}, \frac{3\pi}{2}$

49. $\frac{3\pi}{4}, \frac{7\pi}{4}, \arctan(-5) + \pi, \arctan(-5) + 2\pi$

51. $\sin 345° = \frac{\sqrt{2}}{4}(1 - \sqrt{3})$

$\cos 345° = \frac{\sqrt{2}}{4}(1 + \sqrt{3})$

$\tan 345° = -2 + \sqrt{3}$

53. $\sin \frac{19\pi}{12} = -\frac{\sqrt{2}}{4}(\sqrt{3} + 1)$

$\cos \frac{19\pi}{12} = \frac{\sqrt{2}}{4}(\sqrt{3} - 1)$

$\tan \frac{19\pi}{12} = -2 - \sqrt{3}$

55. $\cos 165°$ **57.** $\tan(-47°)$ **59.** $\frac{960 + 507\sqrt{7}}{1121}$

61. $\frac{12\sqrt{7} - 15}{52}$ **63.** $\frac{-960 + 507\sqrt{7}}{1121}$ **65.** $\frac{3\pi}{2}$

67. $0, \pi$

69.

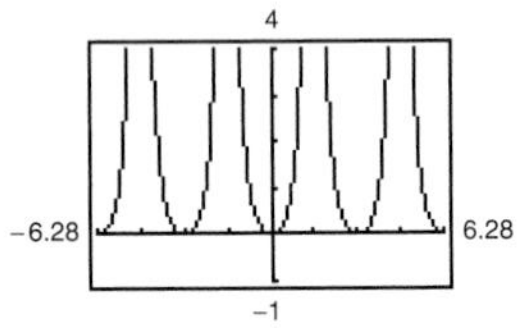

71. $\sin 2u = -\frac{4}{5}$

$\cos 2u = \frac{3}{5}$

$\tan 2u = -\frac{4}{3}$

73. $\frac{1 - \cos 4x}{1 + \cos 4x}$ **75.** $\frac{3 - 4\cos 2x + \cos 4x}{4(1 + \cos 2x)}$

77. $\sin(-75°) = -\frac{1}{2}\sqrt{2 + \sqrt{3}}$

$\cos(-75°) = \frac{1}{2}\sqrt{2 - \sqrt{3}}$

$\tan(-75°) = -2 - \sqrt{3}$

79. $\sin \frac{19\pi}{12} = -\frac{1}{2}\sqrt{2 + \sqrt{3}}$

$\cos \frac{19\pi}{12} = \frac{1}{2}\sqrt{2 - \sqrt{3}}$

$\tan \frac{19\pi}{12} = -2 - \sqrt{3}$

81. $-|\cos 5x|$

83. (a) $V = \sin\frac{\theta}{2}\cos\frac{\theta}{2}$ cubic meters

(b) $V = \sin\theta$ cubic meters

Volume is maximum when $\theta = \pi/2$.

85. $3[\cos(-30°) - \cos 60°]$ **87.** $2(\sin 5\alpha + \sin\alpha)$

89. $2\cos\frac{5\theta}{2}\cos\frac{\theta}{2}$ **91.** $2\cos x \sin\frac{\pi}{4}$

93. False. If $\frac{\pi}{2} < \theta < \pi$, then $\cos\frac{\theta}{2} > 0$. The sign of $\cos\frac{\theta}{2}$ depends on the quadrant in which $\frac{\theta}{2}$ lies.

95. True. $4\sin(-x)\cos(-x) = 4(-\sin x)\cos x$

$= -4\sin x\cos x$

$= -2(2\sin x\cos x)$

$= -2\sin 2x$

97. Reciprocal identities: $\sin\theta = \frac{1}{\csc\theta}$, $\cos\theta = \frac{1}{\sec\theta}$,

$\tan\theta = \frac{1}{\cot\theta}$, $\csc\theta = \frac{1}{\sin\theta}$, $\sec\theta = \frac{1}{\cos\theta}$,

$\cot\theta = \frac{1}{\tan\theta}$

Quotient identities: $\tan\theta = \frac{\sin\theta}{\cos\theta}$, $\cot\theta = \frac{\cos\theta}{\sin\theta}$

Pythagorean identities: $\sin^2\theta + \cos^2\theta = 1$, $1 + \tan^2\theta = \sec^2\theta$, $1 + \cot^2\theta = \csc^2\theta$

99. No. For an equation to be an identity, the equation must be true for all real numbers x. $\sin\theta = \frac{1}{2}$ has an infinite number of solutions but is not an identity.

101. $y_1 = y_2 + 1$

103. $-1.8431, 2.1758, 3.9903, 8.8935, 9.8820$

Chapter Test *(page 491)*

1. $\sin\theta = -\frac{3\sqrt{13}}{13}$

$\cos\theta = -\frac{2\sqrt{13}}{13}$

$\csc\theta = -\frac{\sqrt{13}}{3}$

$\sec\theta = -\frac{\sqrt{13}}{2}$

$\cot\theta = \frac{2}{3}$

2. 1 **3.** 1 **4.** $\csc\theta\sec\theta$

5. $\theta = 0, \frac{\pi}{2} < \theta \le \pi, \frac{3\pi}{2} < \theta < 2\pi$

6.

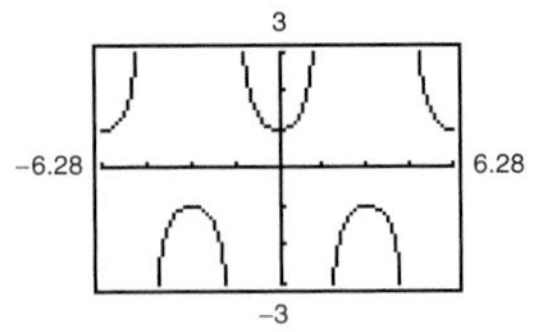

$y_1 = y_2$

7.–12. Answers will vary.

13. $\frac{1}{16}\left(\frac{10 - 15\cos 2x + 6\cos 4x - \cos 6x}{1 + \cos 2x}\right)$ **14.** $\tan 2\theta$

15. $2(\sin 6\theta + \sin 2\theta)$ **16.** $-2\cos\frac{7\theta}{2}\sin\frac{\theta}{2}$

17. $0, \frac{3\pi}{4}, \pi, \frac{7\pi}{4}$ **18.** $\frac{\pi}{6}, \frac{\pi}{2}, \frac{5\pi}{6}, \frac{3\pi}{2}$

19. $\frac{\pi}{6}, \frac{5\pi}{6}, \frac{7\pi}{6}, \frac{11\pi}{6}$ **20.** $\frac{\pi}{6}, \frac{5\pi}{6}, \frac{3\pi}{2}$

21. $-2.938, -2.663, 1.170$

22. $|\cos^2 x + \cos x| \le 2$ for all x

23. $\frac{\sqrt{2} - \sqrt{6}}{4}$ **24.** $\sin 2u = \frac{4}{5}, \tan 2u = -\frac{4}{3}$

25. $\theta \approx 76.5°$

Chapter 6

Section 6.1 *(page 500)*

1. $C = 105°, b \approx 28.28, c \approx 38.64$

3. $C = 120°, b \approx 4.8, c \approx 7.2$

5. $B \approx 21.55°, C \approx 122.45°, c \approx 11.49$

7. $B = 60.9°, b \approx 19.3, c \approx 6.4$

9. $B = 42°4', a \approx 22.05, b \approx 14.88$

11. $A \approx 10°11', C \approx 154°19', c \approx 11.03$

13. $A \approx 25.57°, B \approx 9.43°, a \approx 10.53$

15. $B \approx 18°13', C \approx 51°32', c \approx 40.05$

17. $C = 83°, a \approx 0.62, b \approx 0.51$ **19.** No solution

21. No solution **23.** No solution **25.** No solution

27. (a) $b \le 5,\ b = \frac{5}{\sin 36°}$

(b) $5 < b < \frac{5}{\sin 36°}$

(c) $b > \frac{5}{\sin 36°}$

29. (a) $b \le 10.8, b = \frac{10.8}{\sin 10°}$

(b) $10.8 < b < \frac{10.8}{\sin 10°}$

(c) $b > \frac{10.8}{\sin 10°}$

31. 10.4 **33.** 1675.2 **35.** 3204.5 **37.** 15.3 meters

39. 16.1° **41.** 77 meters

43. (a)

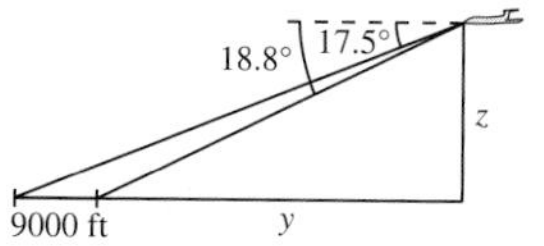

(b) 22.6 miles

(c) 21.4 miles

(d) 38,443 feet

45. 3.2 miles **47.** $d = \frac{2 \sin \theta}{\sin(\phi - \theta)}$

49. False. Two sides and one opposite angle do not necessarily determine a unique triangle.

51. (a) $A = 20\left[15 \sin \frac{3\theta}{2} - 4 \sin \frac{\theta}{2} - 6 \sin \theta\right]$

(b)

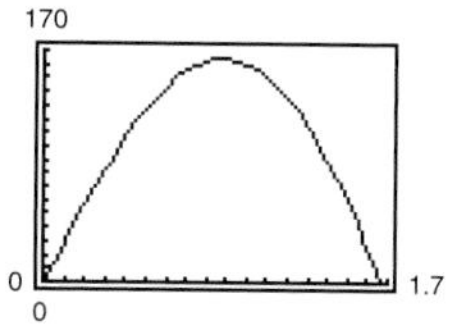

(c) Domain: $0 \le \theta \le 1.6690$

The area would increase and the domain would increase in length.

53. $\sin x = -\frac{2\sqrt{6}}{5}$

$\tan x = -2\sqrt{6}$

$\csc x = -\frac{5\sqrt{6}}{12}$

$\sec x = 5$

$\cot x = -\frac{\sqrt{6}}{12}$

55. $\sin x = \frac{5\sqrt{26}}{26}$

$\cos x = -\frac{\sqrt{26}}{26}$

$\csc x = \frac{\sqrt{26}}{5}$

$\sec x = -\sqrt{26}$

$\cot x = -\frac{1}{5}$

57. $\tan x$ **59.** $\sec^2 x$

Section 6.2 *(page 507)*

1. $A \approx 23.07°, B \approx 34.05°, C \approx 122.88°$

3. $B \approx 23.8°, C \approx 126.2°, a \approx 18.6$

5. $A \approx 31.98°, B \approx 42.38°, C \approx 105.63°$

7. $A \approx 92.94°, B \approx 43.53°, C \approx 43.53°$

9. $B \approx 13.45°, C = 31.55°, a = 12.16$

11. $A \approx 141°45', C \approx 27°40', b \approx 11.9$

13. $A = 27°10', C = 27°10', b \approx 56.9$

15. $A \approx 33.8°, B \approx 103.2°, c \approx 0.5448$

	a	b	c	d	θ	ϕ
17.	5	8	12.07	5.69	45°	135°
19.	10	14	20	13.86	68.2°	111.8°
21.	15	16.96	25	20	77.2°	102.8°

23. 16.25 **25.** 10.44 **27.** 52.11

29.

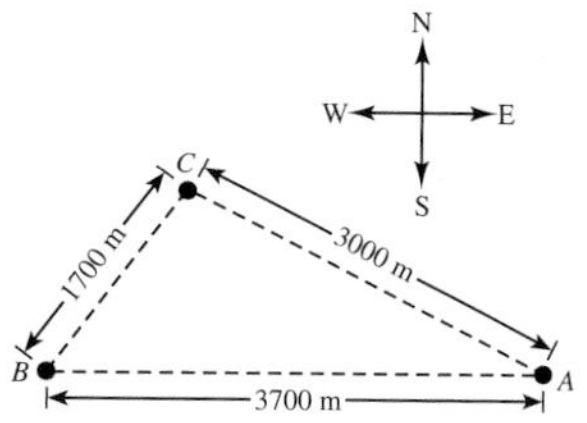

N 37.1° E, S 63.1° E

31. 373.3 meters **33.** 72.3° **35.** 43.3 miles

37. (a) N 58.4° W (b) S 81.5° W **39.** 63.7 feet

41. 24.2 miles **43.** $\overline{PQ} \approx 9.4$, $\overline{QS} = 5$, $\overline{RS} \approx 12.8$

45.

d (inches)	9	10	12	13	14
θ (degrees)	60.9°	69.5°	88.0°	98.2°	109.6°
s (inches)	20.88	20.28	18.99	18.28	17.48

d (inches)	15	16
θ (degrees)	122.9°	139.8°
s (inches)	16.55	15.37

47. 46,837.5 square feet

49. False. For s to be the average of the lengths of the three sides of the triangle, s would be equal to $(a + b + c)/3$.

51. Answers will vary. **53.** 405.2 feet

55. Answers will vary. **57.** $\frac{\pi}{2}$ **59.** $-\frac{\pi}{3}$ **61.** $\frac{5\pi}{6}$

63. $\frac{\sqrt{1 - 9x^2}}{3x}$ **65.** $\frac{\sqrt{4 - (x - 1)^2}}{2}$

67. $\sin \theta = -\frac{\sqrt{2}}{2}$

$\sec \theta = \sqrt{2}$

$\csc \theta = -\sqrt{2}$

69. $12 = 6 \sec \theta$

$\sec \theta = 2$

$\csc \theta = \pm\frac{2\sqrt{3}}{3}$

Section 6.3 *(page 520)*

1. $\mathbf{v} = \langle 3, 2\rangle$; $\|\mathbf{v}\| = \sqrt{13}$ **3.** $\mathbf{v} = \langle -3, 2\rangle$; $\|\mathbf{v}\| = \sqrt{13}$

5. $\mathbf{v} = \langle 0, 5\rangle$; $\|\mathbf{v}\| = 5$ **7.** $\mathbf{v} = \langle 16, 7\rangle$; $\|\mathbf{v}\| = \sqrt{305}$

9. $\mathbf{v} = \langle 8, 6\rangle$; $\|\mathbf{v}\| = 10$ **11.** $\mathbf{v} = \langle -9, -12\rangle$; $\|\mathbf{v}\| = 15$

13.

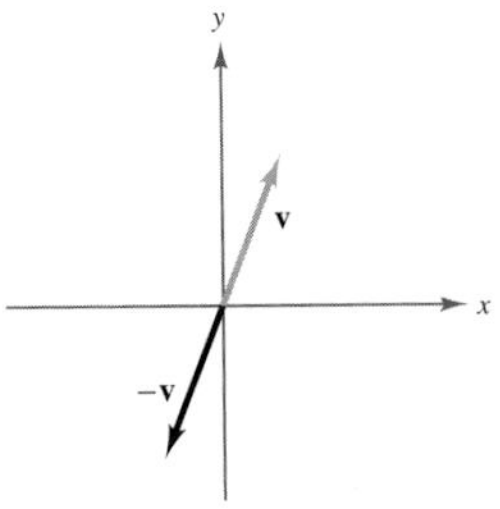

15.

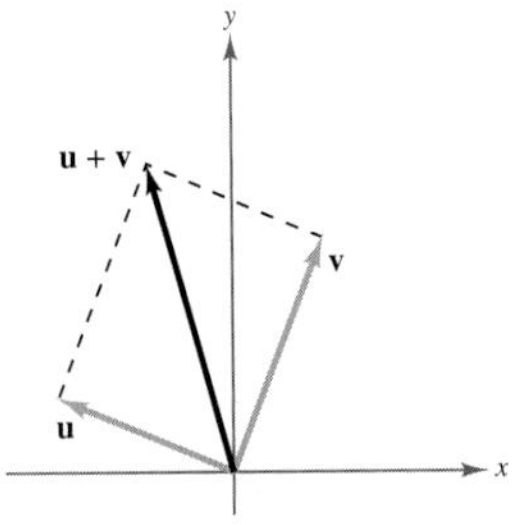

17.

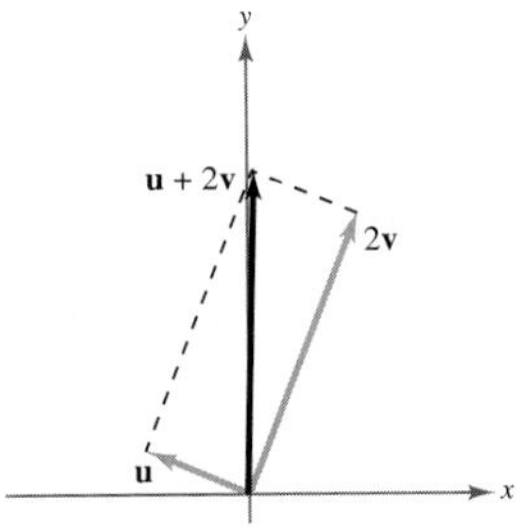

19. (a) $\langle 3, 4\rangle$

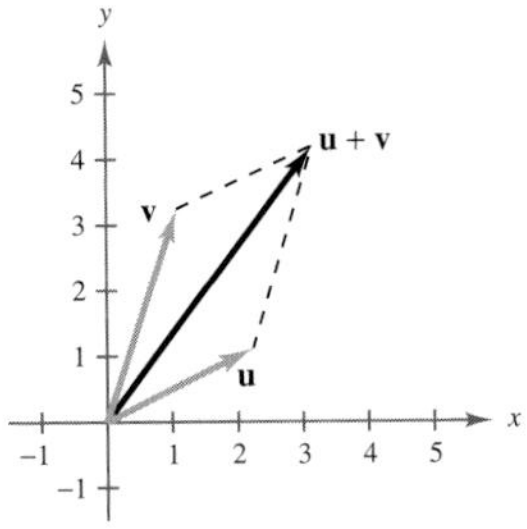

(b) $\langle 1, -2\rangle$

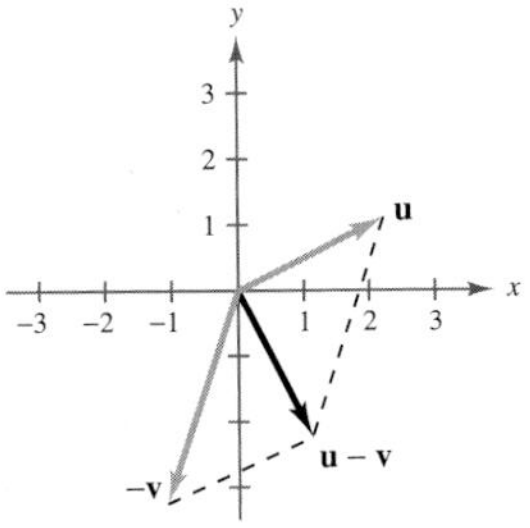

(c) $\langle 1, -7\rangle$

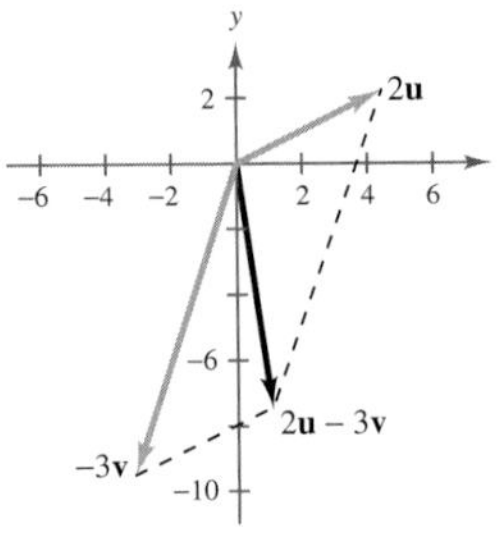

21. (a) $\langle -5, 3\rangle$

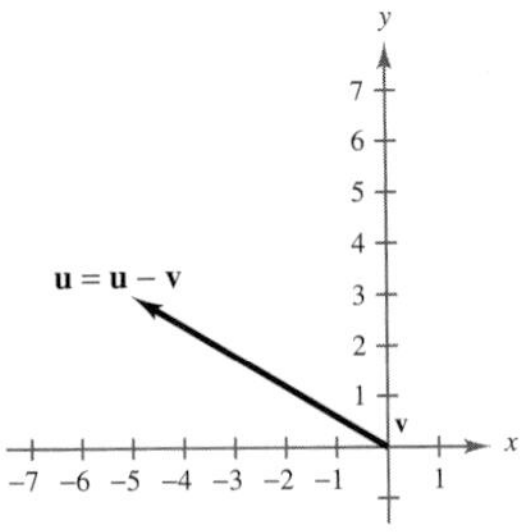

(b) $\langle -5, 3\rangle$

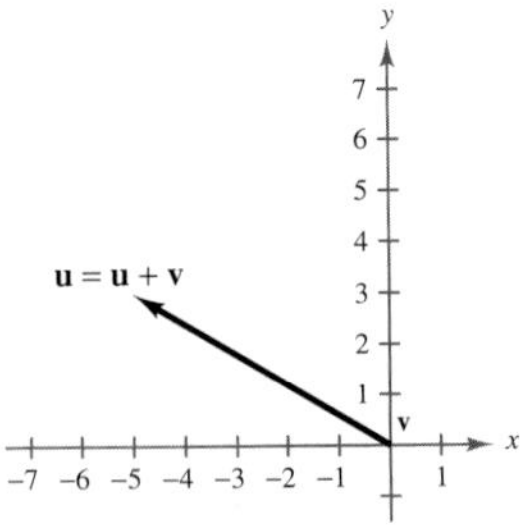

(c) $\langle -10, 6\rangle$

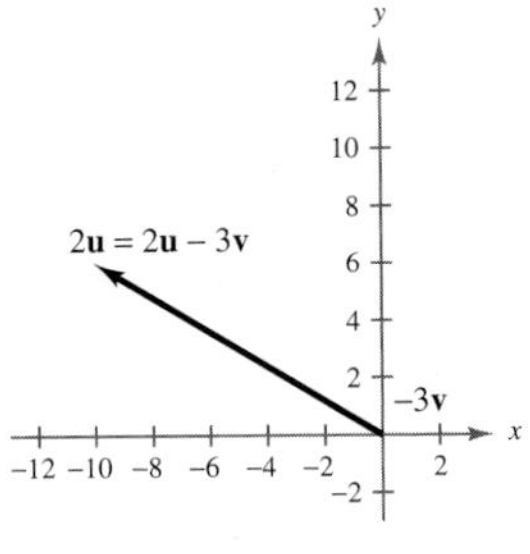

23. (a) $3\mathbf{i} - 2\mathbf{j}$

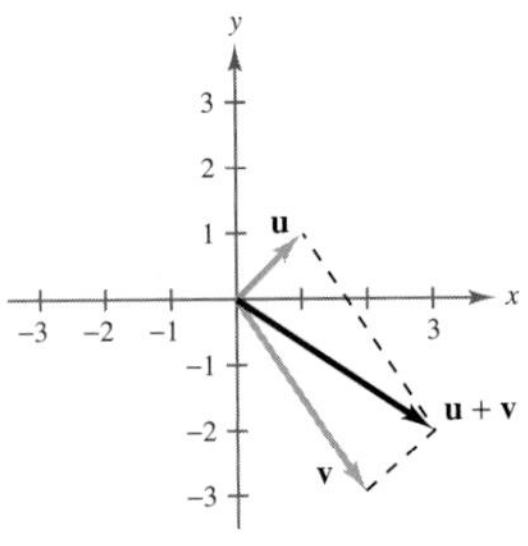

(b) $-\mathbf{i} + 4\mathbf{j}$

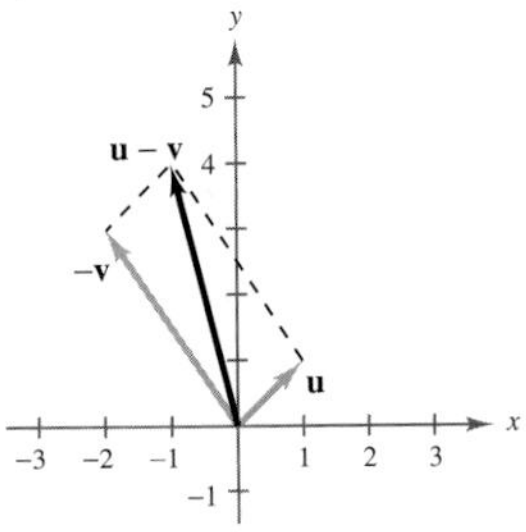

(c) $-4\mathbf{i} + 11\mathbf{j}$

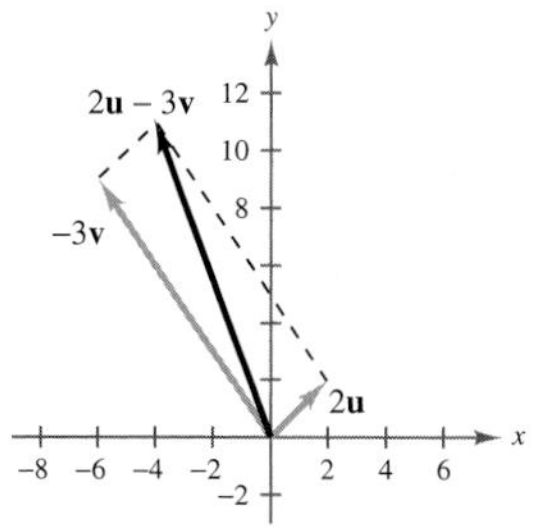

25. (a) $2\mathbf{i} + \mathbf{j}$

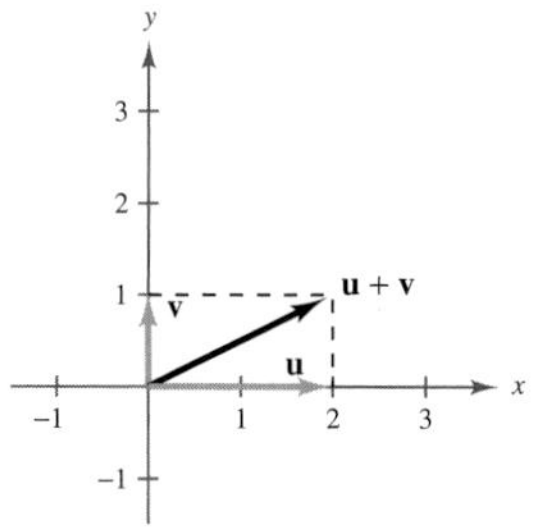

(b) $2\mathbf{i} - \mathbf{j}$

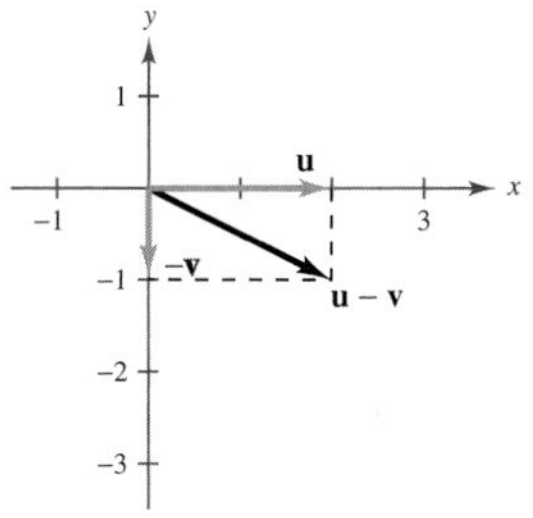

(c) $4\mathbf{i} - 3\mathbf{j}$

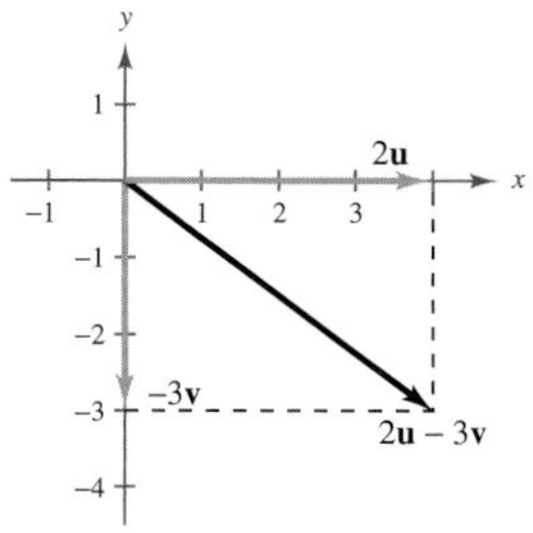

27. $\langle 1, 0 \rangle$ **29.** $\left\langle -\frac{1}{\sqrt{2}}, \frac{1}{\sqrt{2}} \right\rangle$ **31.** $\frac{3}{\sqrt{10}}\mathbf{i} - \frac{1}{\sqrt{10}}\mathbf{j}$

33. $\mathbf{j}$ **35.** $\frac{1}{\sqrt{5}}\mathbf{i} - \frac{2}{\sqrt{5}}\mathbf{j}$ **37.** $\left\langle \frac{5}{\sqrt{2}}, \frac{5}{\sqrt{2}} \right\rangle$

39. $\left\langle \frac{18}{\sqrt{29}}, \frac{45}{\sqrt{29}} \right\rangle$

41. $\mathbf{v} = \langle 3, -\frac{3}{2} \rangle$

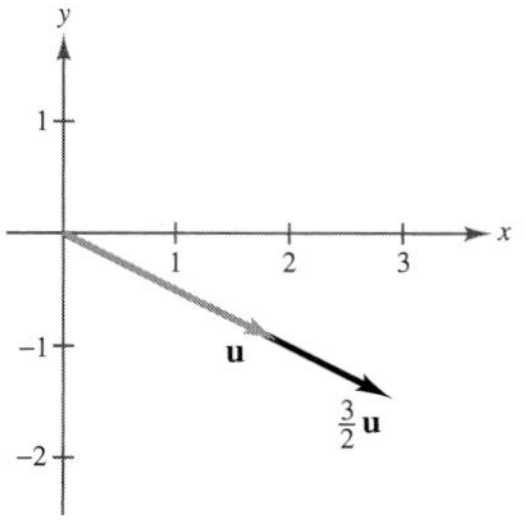

43. $\mathbf{v} = \langle 4, 3 \rangle$

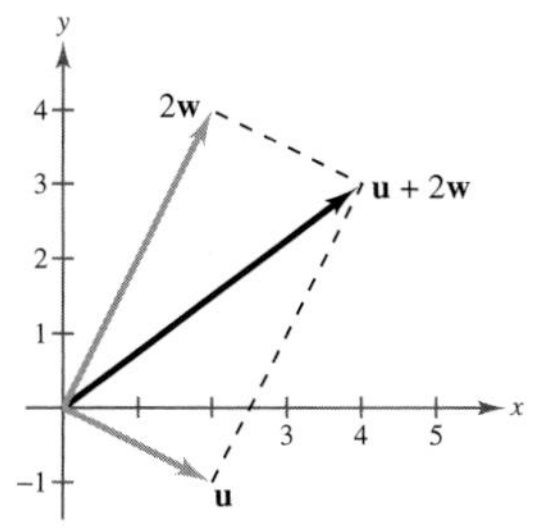

45. $\mathbf{v} = \langle \frac{7}{2}, -\frac{1}{2} \rangle$

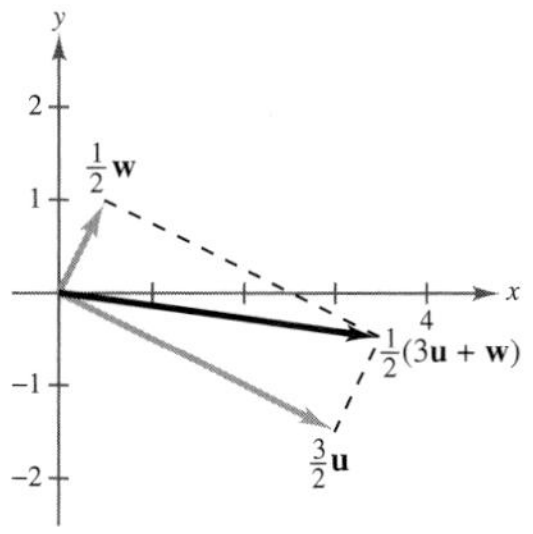

47. $\|\mathbf{v}\| = 3$; $\theta = 60°$ **49.** $\|\mathbf{v}\| = 6\sqrt{2}$; $\theta = 315°$

51. $\mathbf{v} = \langle 3, 0 \rangle$

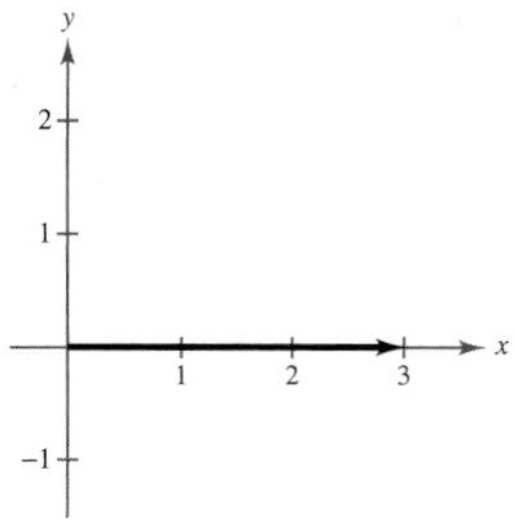

53. $\mathbf{v} = \left\langle -\frac{7\sqrt{3}}{4}, \frac{7}{4} \right\rangle$

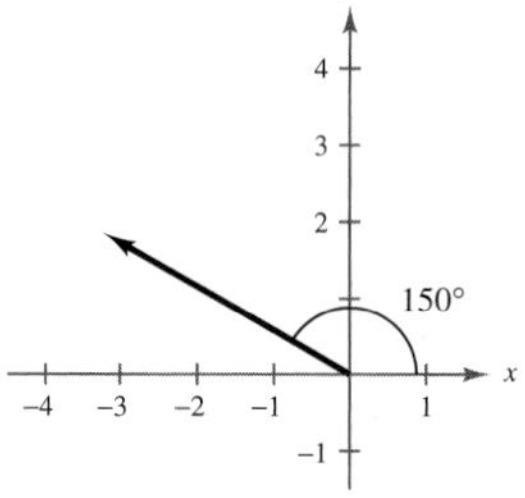

55. $\mathbf{v} = \left\langle -\frac{3\sqrt{6}}{2}, \frac{3\sqrt{2}}{2} \right\rangle$

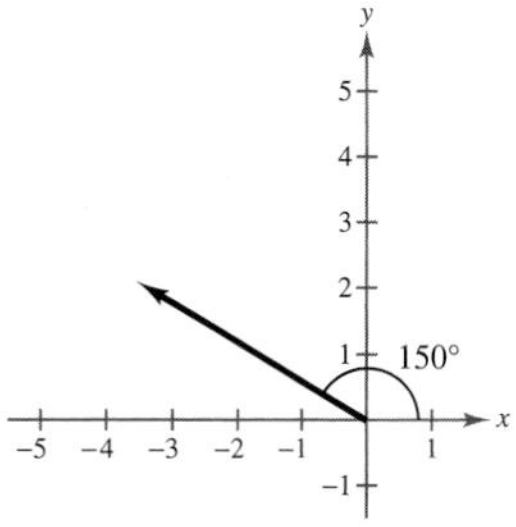

57. $\mathbf{v} = \left\langle \frac{\sqrt{10}}{5}, \frac{3\sqrt{10}}{5} \right\rangle$

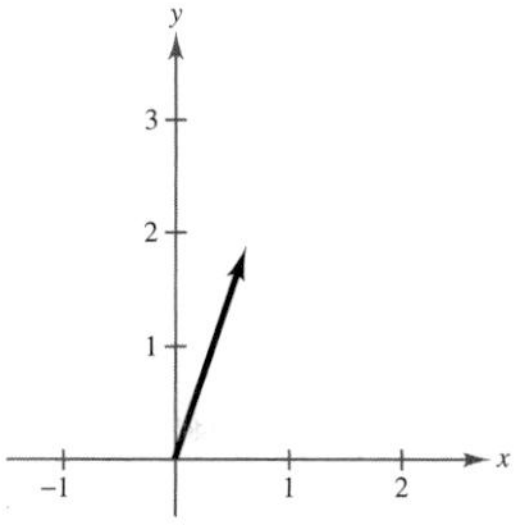

59. $\langle 5, 5 \rangle$ **61.** $\langle 10\sqrt{2} - 50, 10\sqrt{2} \rangle$ **63.** 90°

65. 63.4° **67.** 62.7° **69.** 12.8°; 398.32 newtons

71. 71.3°; 228.5 pounds

73. Vertical component: $70 \sin 35° \approx 40.15$ feet per second

Horizontal component: $70 \cos 35° \approx 57.34$ feet per second

75. $T_{AC} \approx 1758.8$ pounds **77.** 3154.4 pounds

$T_{BC} \approx 1305.4$ pounds

79. N 21.4° E; 138.7 kilometers per hour

81. 1928.4 foot-pounds **83.** True. See Example 1.

85. False. $a = b = 0$

87. (a) 0° (b) 180°

(c) No. The magnitude is at most equal to the sum when the angle between the vectors is 0°.

89. Answers will vary. **91.** $\langle 1, 3\rangle$ or $\langle -1, -3\rangle$

93. $8 \tan \theta$ **95.** $6 \sec \theta$

97. $\dfrac{\pi}{2} + n\pi, \pi + 2n\pi$

99. $n\pi, \dfrac{\pi}{6} + 2n\pi, \dfrac{11\pi}{6} + 2n\pi$ **101.** $\dfrac{n\pi}{2}$

Section 6.4 *(page 530)*

1. -9 **3.** 6 **5.** 8; scalar **7.** $\langle -6, 8\rangle$; vector

9. 13 **11.** $5\sqrt{41}$ **13.** 6 **15.** $90°$ **17.** $143.13°$

19. $60.26°$ **21.** $90°$ **23.** $\dfrac{5\pi}{12}$ **25.** $26.6°, 63.4°, 90°$

27. $41.63°, 53.13°, 85.24°$ **29.** -20 **31.** $2592\sqrt{2}$

33. Parallel **35.** Neither **37.** Orthogonal

39. $\frac{1}{37}\langle 84, 14\rangle, \frac{1}{37}\langle -10, 60\rangle$ **41.** $\frac{45}{229}\langle 2, 15\rangle, \frac{6}{229}\langle -15, 2\rangle$

43. $\langle -5, 3\rangle, \langle 5, -3\rangle$ **45.** $\frac{2}{3}\mathbf{i} + \frac{1}{2}\mathbf{j}, -\frac{2}{3}\mathbf{i} - \frac{1}{2}\mathbf{j}$ **47.** 32

49. \$58,762.50

This value gives the total revenue that can be earned by selling all of the units.

51. (a) 2614.7 pounds (b) 29,885.9 pounds

53. 735 newton-meters **55.** 779.4 foot-pounds

57. False. Work is represented by a scalar.

59. (a) $\theta = \dfrac{\pi}{2}$ (b) $0 \le \theta < \dfrac{\pi}{2}$ (c) $\dfrac{\pi}{2} < \theta \le \pi$

61. and 63. Answers will vary. **65.** $0, \dfrac{\pi}{6}, \pi, \dfrac{11\pi}{6}$

67. $0, \pi$ **69.** $-\frac{253}{325}$ **71.** $\frac{204}{325}$

Section 6.5 *(page 540)*

1.

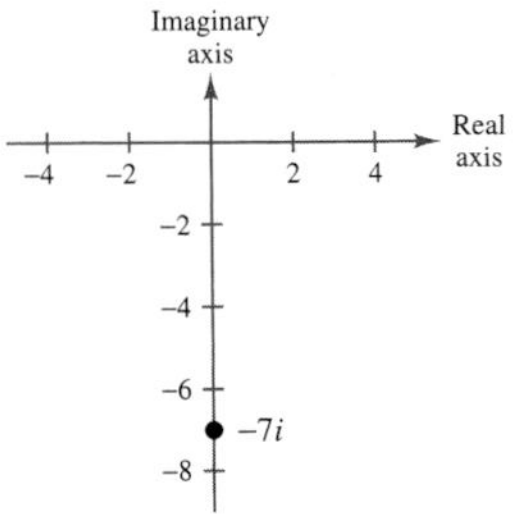

7

3.

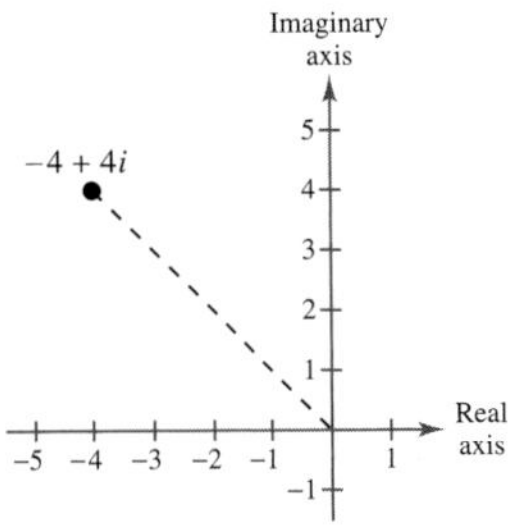

$4\sqrt{2}$

5.

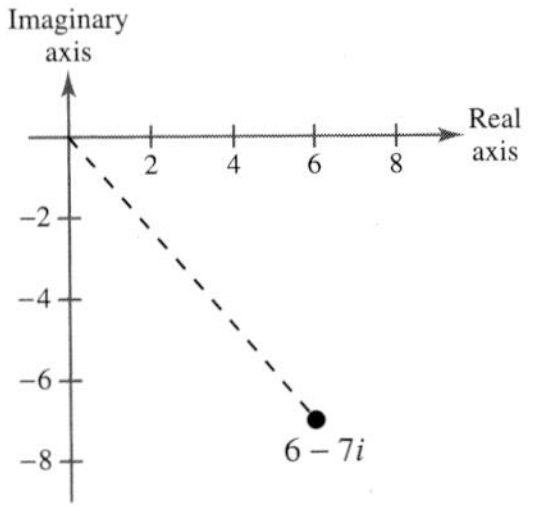

$\sqrt{85}$

7. $3\left(\cos \dfrac{\pi}{2} + i \sin \dfrac{\pi}{2}\right)$ **9.** $\sqrt{10}(\cos 5.96 + i \sin 5.96)$

11.

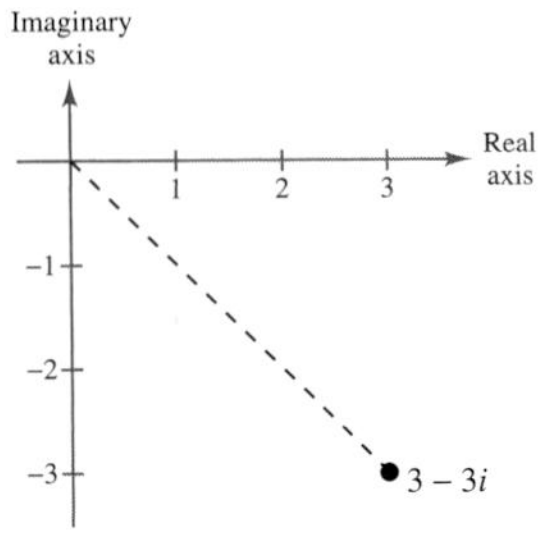

$3\sqrt{2}\left(\cos \dfrac{7\pi}{4} + i \sin \dfrac{7\pi}{4}\right)$

13.

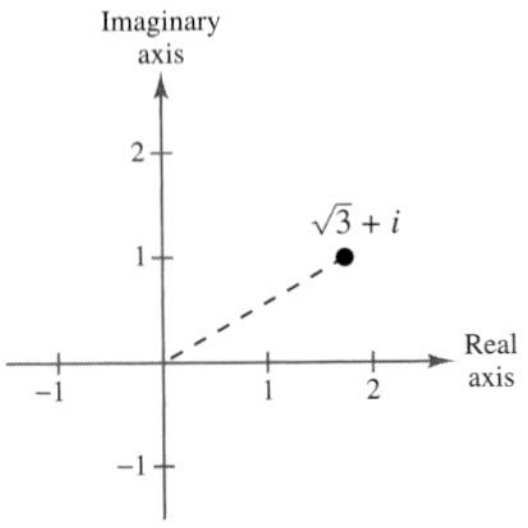

$2\left(\cos \dfrac{\pi}{6} + i \sin \dfrac{\pi}{6}\right)$

15.

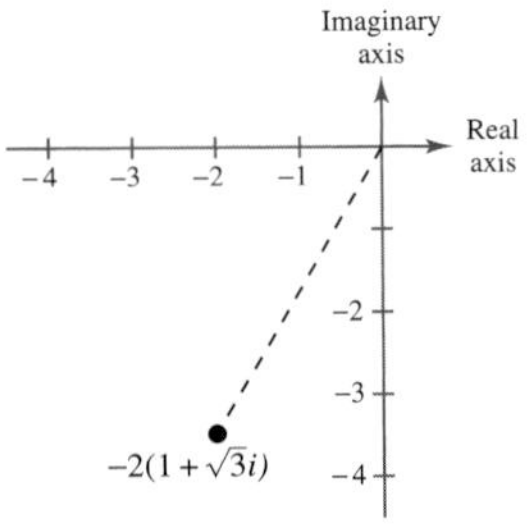

$4\left(\cos \dfrac{4\pi}{3} + i \sin \dfrac{4\pi}{3}\right)$

17.

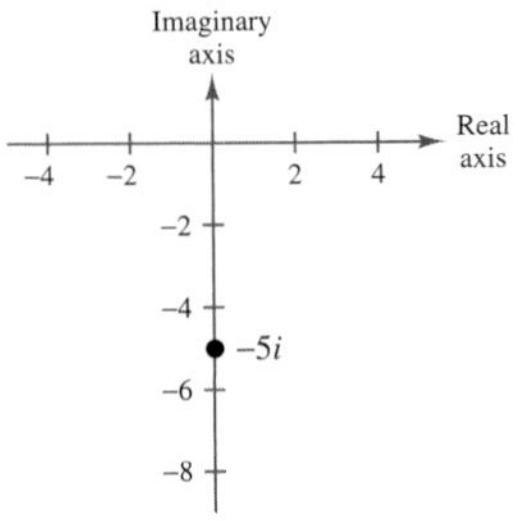

$5\left(\cos \dfrac{3\pi}{2} + i \sin \dfrac{3\pi}{2}\right)$

19.

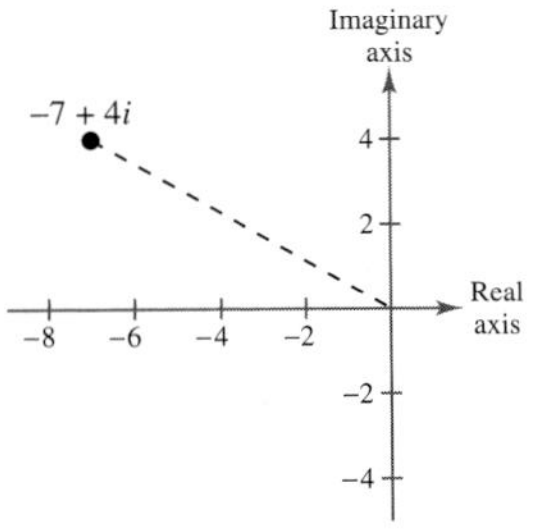

$\sqrt{65}\,(\cos 2.62 + i \sin 2.62)$

21.

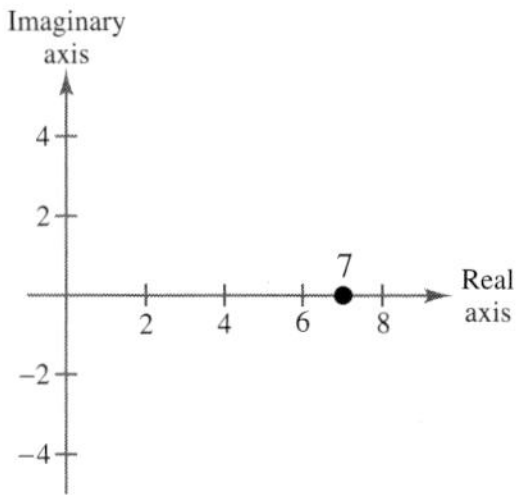

$7(\cos 0 + i \sin 0)$

23.

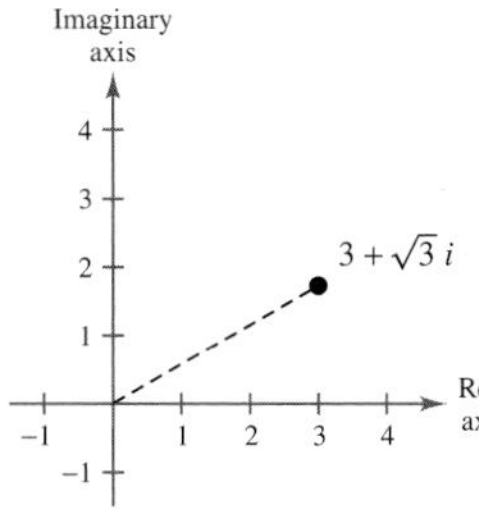

$2\sqrt{3}\left(\cos \frac{\pi}{6} + i \sin \frac{\pi}{6}\right)$

25.

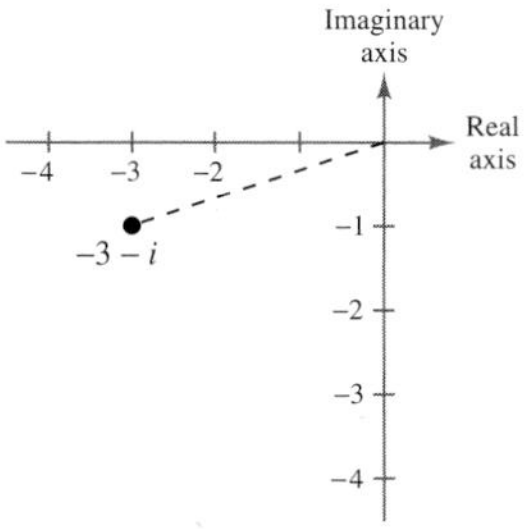

$\sqrt{10}(\cos 3.46 + i \sin 3.46)$

27. $5.39(\cos 0.38 + i \sin 0.38)$

29. $3.16(\cos 2.82 + i \sin 2.82)$

31. $8.19(\cos 5.26 + i \sin 5.26)$

33. $11.79(\cos 3.97 + i \sin 3.97)$

35.

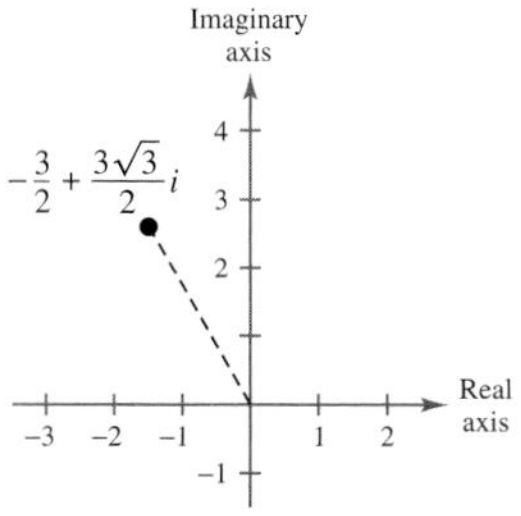

$-\frac{3}{2} + \frac{3\sqrt{3}}{2}i$

37.

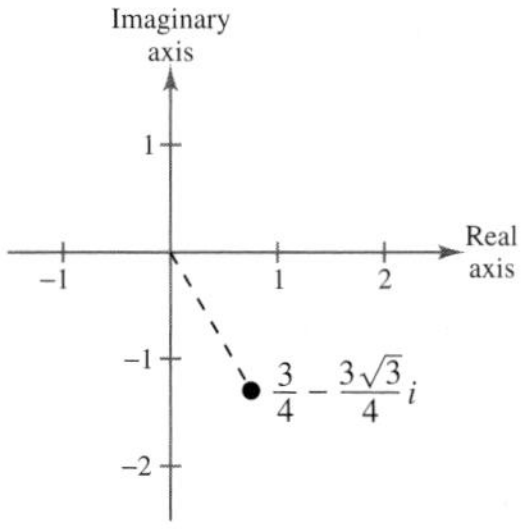

$\frac{3}{4} - \frac{3\sqrt{3}}{4}i$

39.

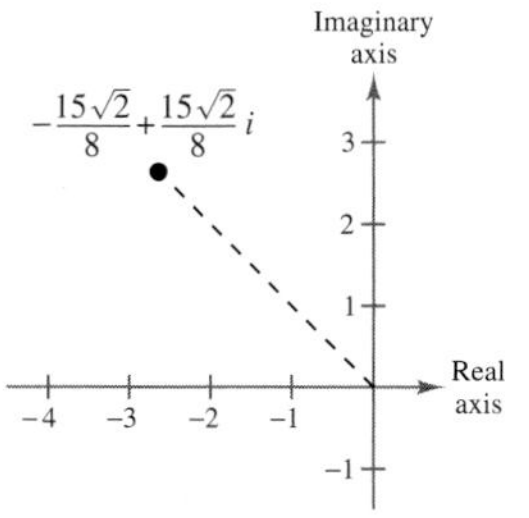

$-\frac{15\sqrt{2}}{8} + \frac{15\sqrt{2}}{8}i$

41.

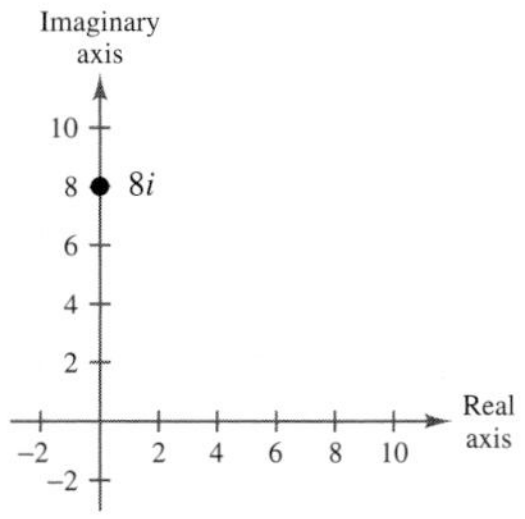

$8i$

43.

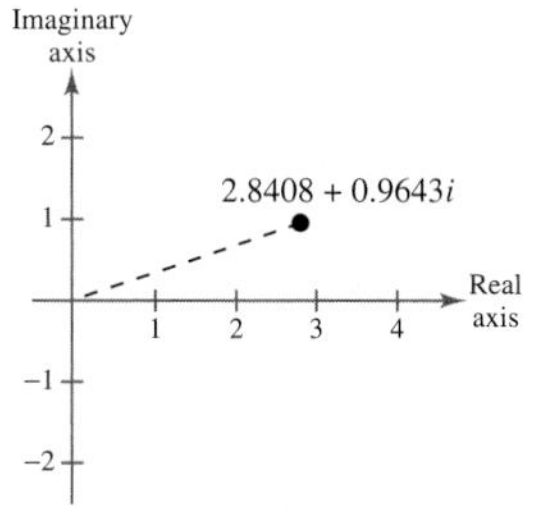

$2.8408 + 0.9643i$

45. $4.70 + 1.71i$ **47.** $-2.90 + 0.75i$

49. $12\left(\cos \frac{\pi}{3} + i \sin \frac{\pi}{3}\right)$ **51.** $\frac{10}{9}(\cos 200^\circ + i \sin 200^\circ)$

53. $0.27(\cos 150^\circ + i \sin 150^\circ)$ **55.** $\cos 30^\circ + i \sin 30^\circ$

57. $\cos \frac{2\pi}{3} + i \sin \frac{2\pi}{3}$ **59.** $4[\cos(302^\circ) + i \sin(302^\circ)]$

61. (a) $\left[2\sqrt{2}\left(\cos \frac{\pi}{4} + i \sin \frac{\pi}{4}\right)\right]\left\{\sqrt{2}\left[\cos\left(\frac{7\pi}{4}\right) + i \sin\left(\frac{7\pi}{4}\right)\right]\right\}$

(b) $4(\cos 0 + i \sin 0) = 4$

(c) 4

63. (a) $2\left[\cos\left(\frac{3\pi}{2}\right) + i \sin\left(\frac{3\pi}{2}\right)\right]\left[\sqrt{2}\left(\cos \frac{\pi}{4} + i \sin \frac{\pi}{4}\right)\right]$

(b) $2\sqrt{2}\left[\cos\left(\frac{7\pi}{4}\right) + i \sin\left(\frac{7\pi}{4}\right)\right] = 2 - 2i$

(c) $-2i - 2i^2 = -2i + 2 = 2 - 2i$

65. (a) $[5(\cos 0.93 + i \sin 0.93)] \div \left[2\left(\cos \frac{5\pi}{3} + i \sin \frac{5\pi}{3}\right)\right]$

(b) $\frac{5}{2}[\cos(1.97) + i \sin(1.97)] = -0.982 + 2.299i$

(c) $\approx -0.982 + 2.299i$

67. (a) $[5(\cos 0 + i \sin 0)] \div \left[\sqrt{13}(\cos 0.98 + i \sin 0.98)\right]$

(b) $\frac{5}{\sqrt{13}}[\cos(5.30) + i \sin(5.30)] \approx 0.769 - 1.154\text{i}$

(c) $\frac{10}{13} - \frac{15}{13}i \approx 0.769 - 1.154i$

69.

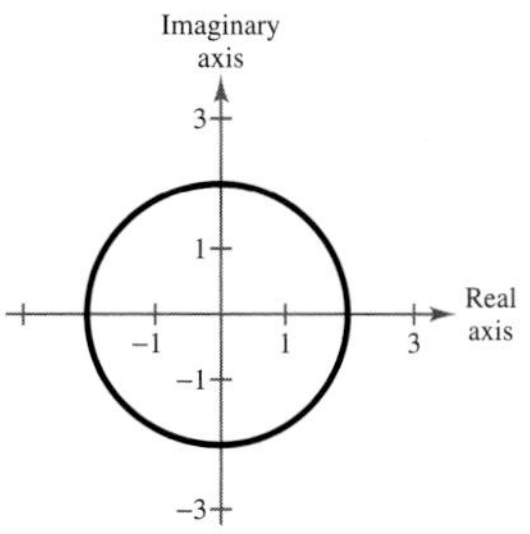

71.

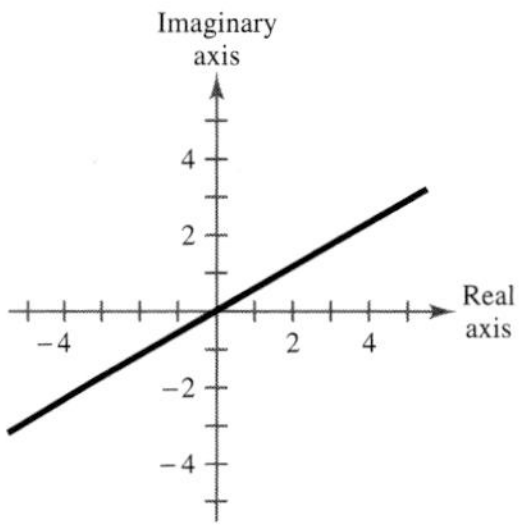

73. $-4 - 4i$ **75.** $-32i$ **77.** $-128\sqrt{3} - 128i$

79. $\frac{125}{2} + \frac{125\sqrt{3}}{2}i$ **81.** -1 **83.** $608.02 + 144.69i$

85. $-597 - 122i$ **87.** $\frac{81}{2} + \frac{81\sqrt{3}}{2}i$ **89.** $32i$

91. (a) $\sqrt{5}(\cos 60° + i \sin 60°)$

$\sqrt{5}(\cos 240° + i \sin 240°)$

(b)

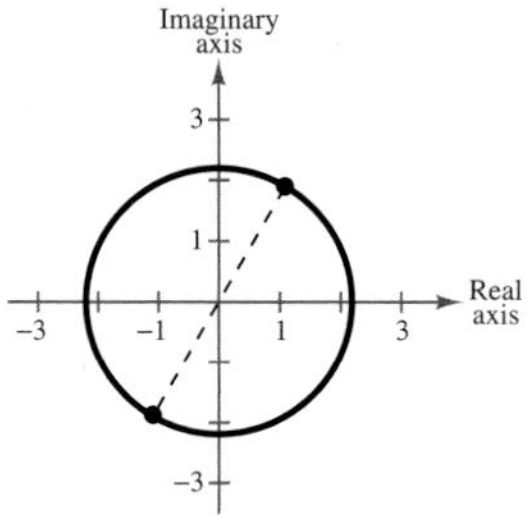

(c) $\frac{\sqrt{5}}{2} + \frac{\sqrt{15}}{2}i, \ -\frac{\sqrt{5}}{2} - \frac{\sqrt{15}}{2}i$

93. (a) $2\left(\cos \frac{2\pi}{9} + i \sin \frac{2\pi}{9}\right)$

$2\left(\cos \frac{8\pi}{9} + i \sin \frac{8\pi}{9}\right)$

$2\left(\cos \frac{14\pi}{9} + i \sin \frac{14\pi}{9}\right)$

(b)

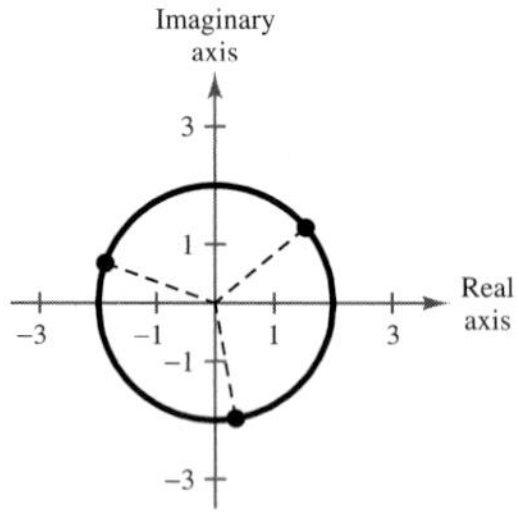

(c) $1.5321 + 1.2856i, \ -1.8794 + 0.6840i,$

$0.3473 - 1.9696i$

95. (a) $5\left(\cos \frac{3\pi}{4} + i \sin \frac{3\pi}{4}\right)$

$5\left(\cos \frac{7\pi}{4} + i \sin \frac{7\pi}{4}\right)$

(b)

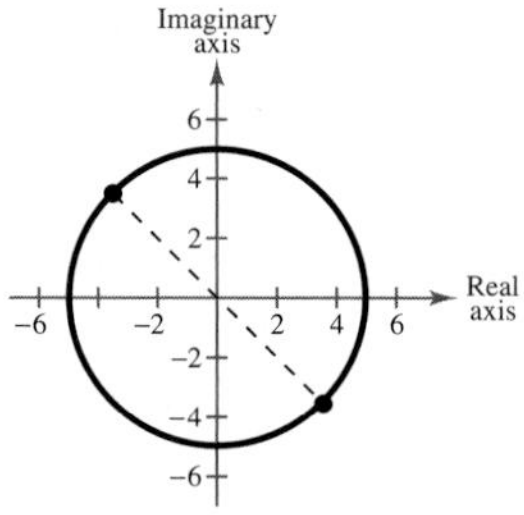

(c) $-\frac{5\sqrt{2}}{2} + \frac{5\sqrt{2}}{2}i, \ \frac{5\sqrt{2}}{2} - \frac{5\sqrt{2}}{2}i$

97. (a) $5\left(\cos \frac{4\pi}{9} + i \sin \frac{4\pi}{9}\right)$

$5\left(\cos \frac{10\pi}{9} + i \sin \frac{10\pi}{9}\right)$

$5\left(\cos \frac{16\pi}{9} + i \sin \frac{16\pi}{9}\right)$

(b)

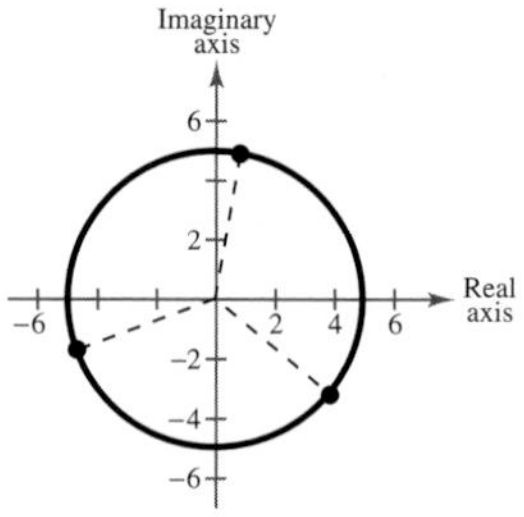

(c) $0.8682 + 4.924i, \ -4.6985 - 1.7101i,$

$3.8302 - 3.214i$

99. (a) $2(\cos 0 + i \sin 0)$

$2\left(\cos \frac{\pi}{2} + i \sin \frac{\pi}{2}\right)$

$2(\cos \pi + i \sin \pi)$

$2\left(\cos \frac{3\pi}{2} + i \sin \frac{3\pi}{2}\right)$

(b)

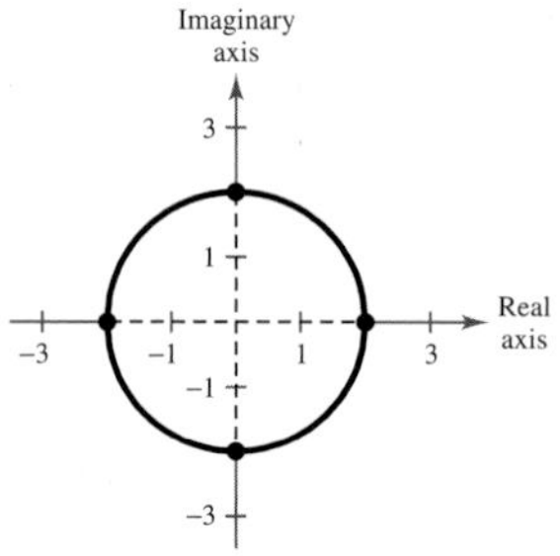

(c) $2, 2i, -2, -2i$

101. (a) $\cos 0 + i \sin 0$

$\cos \frac{2\pi}{5} + i \sin \frac{2\pi}{5}$

$\cos \frac{4\pi}{5} + i \sin \frac{4\pi}{5}$

$\cos \frac{6\pi}{5} + i \sin \frac{6\pi}{5}$

$\cos \frac{8\pi}{5} + i \sin \frac{8\pi}{5}$

(b)

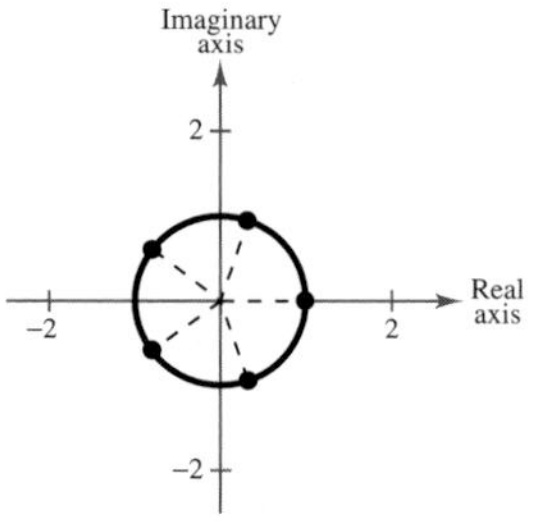

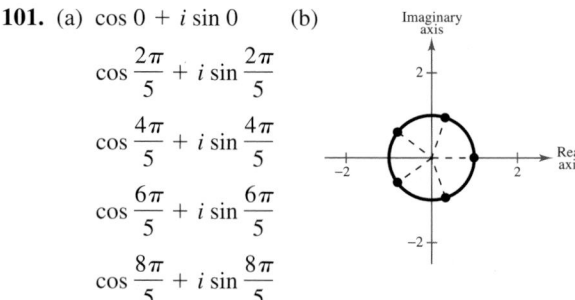

(c) $1, 0.3090 + 0.9511i, \ -0.8090 + 0.5878i,$

$-0.8090 - 0.5878i, 0.3090 - 0.9511i$

103. (a) $5\left(\cos\frac{\pi}{3} + i\sin\frac{\pi}{3}\right)$

$5(\cos\pi + i\sin\pi)$

$5\left(\cos\frac{5\pi}{3} + i\sin\frac{5\pi}{3}\right)$

(b)

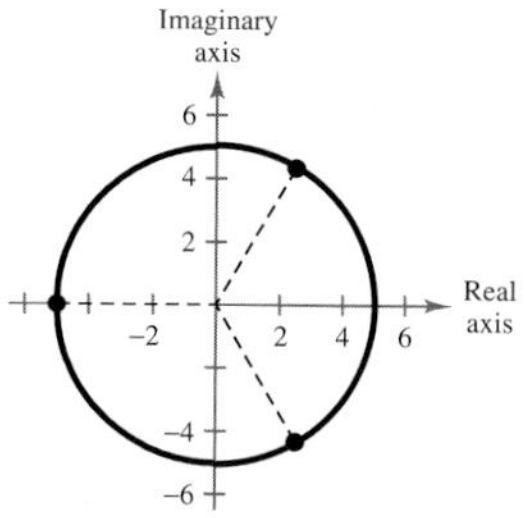

(c) $\frac{5}{2} + \frac{5\sqrt{3}}{2}i,\ -5,\ \frac{5}{2} - \frac{5\sqrt{3}}{2}i$

105. (a) $2\sqrt[5]{4\sqrt{2}}\left(\cos\frac{3\pi}{20} + i\sin\frac{3\pi}{20}\right)$

$2\sqrt[5]{4\sqrt{2}}\left(\cos\frac{11\pi}{20} + i\sin\frac{11\pi}{20}\right)$

$2\sqrt[5]{4\sqrt{2}}\left(\cos\frac{19\pi}{20} + i\sin\frac{19\pi}{20}\right)$

$2\sqrt[5]{4\sqrt{2}}\left(\cos\frac{27\pi}{20} + i\sin\frac{27\pi}{20}\right)$

$2\sqrt[5]{4\sqrt{2}}\left(\cos\frac{7\pi}{4} + i\sin\frac{7\pi}{4}\right)$

(b)

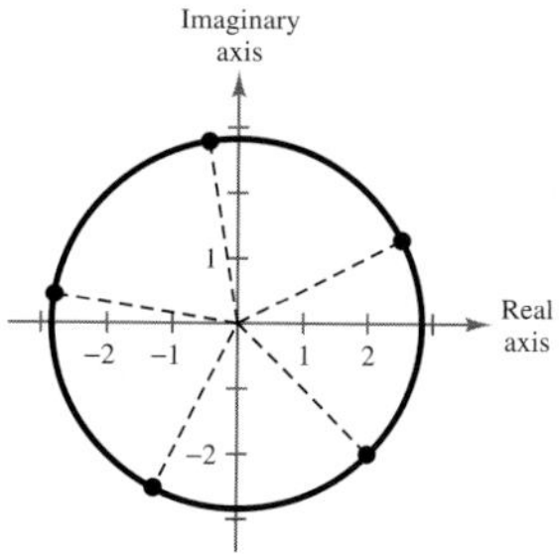

(c) $2.5201 + 1.2841i,\ -0.4425 + 2.7936i,$

$-2.7936 + 0.4425i,\ -1.2841 - 2.5201i,\ 2 - 2i$

107. $\cos\frac{\pi}{8} + i\sin\frac{\pi}{8}$

$\cos\frac{5\pi}{8} + i\sin\frac{5\pi}{8}$

$\cos\frac{9\pi}{8} + i\sin\frac{9\pi}{8}$

$\cos\frac{13\pi}{8} + i\sin\frac{13\pi}{8}$

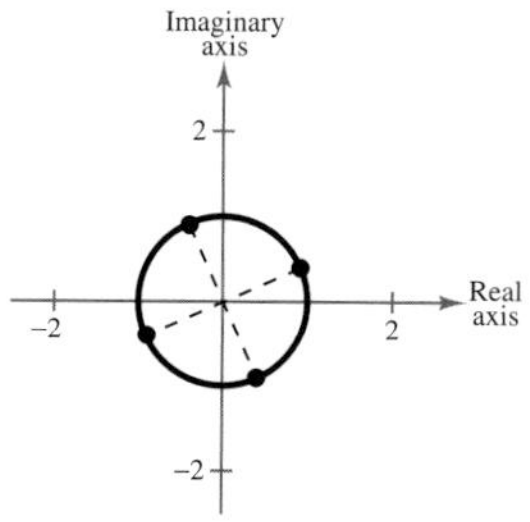

109. $3\left(\cos\frac{\pi}{5} + i\sin\frac{\pi}{5}\right)$

$3\left(\cos\frac{3\pi}{5} + i\sin\frac{3\pi}{5}\right)$

$3(\cos\pi + i\sin\pi)$

$3\left(\cos\frac{7\pi}{5} + i\sin\frac{7\pi}{5}\right)$

$3\left(\cos\frac{9\pi}{5} + i\sin\frac{9\pi}{5}\right)$

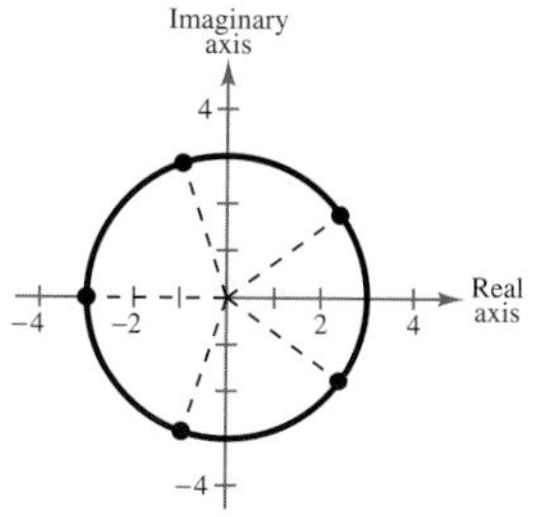

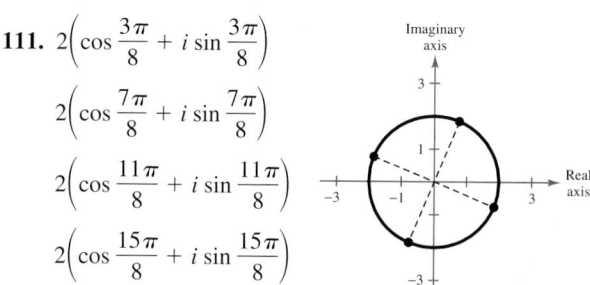

111. $2\left(\cos\frac{3\pi}{8} + i\sin\frac{3\pi}{8}\right)$

$2\left(\cos\frac{7\pi}{8} + i\sin\frac{7\pi}{8}\right)$

$2\left(\cos\frac{11\pi}{8} + i\sin\frac{11\pi}{8}\right)$

$2\left(\cos\frac{15\pi}{8} + i\sin\frac{15\pi}{8}\right)$

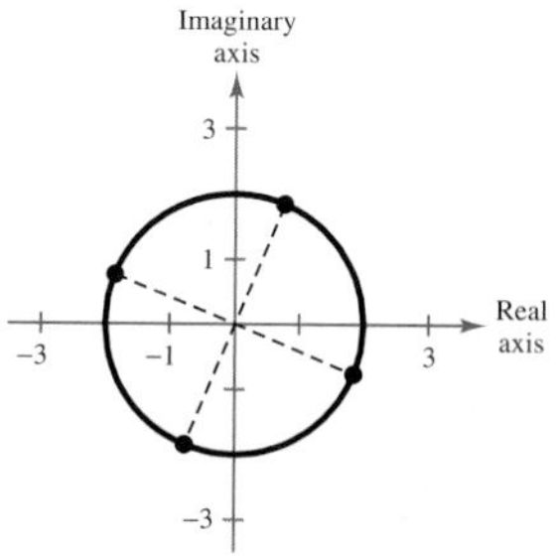

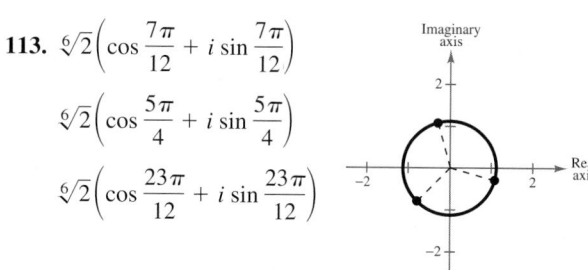

113. $\sqrt[6]{2}\left(\cos\frac{7\pi}{12} + i\sin\frac{7\pi}{12}\right)$

$\sqrt[6]{2}\left(\cos\frac{5\pi}{4} + i\sin\frac{5\pi}{4}\right)$

$\sqrt[6]{2}\left(\cos\frac{23\pi}{12} + i\sin\frac{23\pi}{12}\right)$

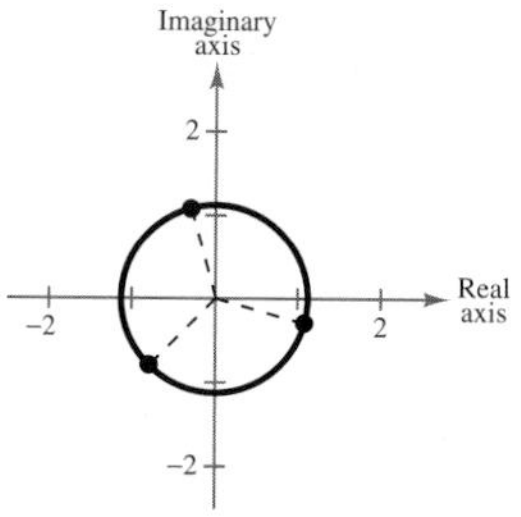

115. True, by the definition of the absolute value of a complex number.

117. True. $z_1 z_2 = r_1 r_2[\cos(\theta_1 + \theta_2) + i\sin(\theta_1 + \theta_2)] = 0$ if and only if $r_1 = 0$ and/or $r_2 = 0$.

119. Answers will vary.

121. (a) r^2 (b) $\cos 2\theta + i\sin 2\theta$

123. Answers will vary.

125. (a) $2(\cos 30° + i\sin 30°)$

$2(\cos 150° + i\sin 150°)$

$2(\cos 270° + i\sin 270°)$

(b) $8i$

127. $B = 68°,\ b \approx 19.80,\ c \approx 21.36$

129. $B = 60°,\ a \approx 65.01,\ c \approx 130.02$

131. $B = 47°45',\ a \approx 7.53,\ b \approx 8.29$

133. 16; 2 **135.** $\frac{1}{16}$; 0

Review Exercises *(page 544)*

1. $C = 74°,\ b \approx 13.19,\ c \approx 13.41$

3. $A = 26°,\ a \approx 24.89,\ c \approx 56.23$

5. $C = 66°, a \approx 2.53, b \approx 9.11$

7. $B = 108°, a \approx 11.76, c \approx 21.49$

9. $A \approx 20.41°, C \approx 9.59°, a \approx 20.92$

11. $B \approx 39.48°, C \approx 65.52°, c \approx 48.24$

13. 7.945 **15.** 33.547 **17.** 31.1 meters

19. 31.01 feet **21.** $A \approx 53.13°, B \approx 36.87°, C \approx 90°$

23. $A \approx 101.47°, B \approx 31.73°, C \approx 46.8°$

25. $A \approx 9.90°, C \approx 20.10°, b \approx 29.09$

27. $B \approx 35.20°, C \approx 82.8°, a \approx 17.37$

29.

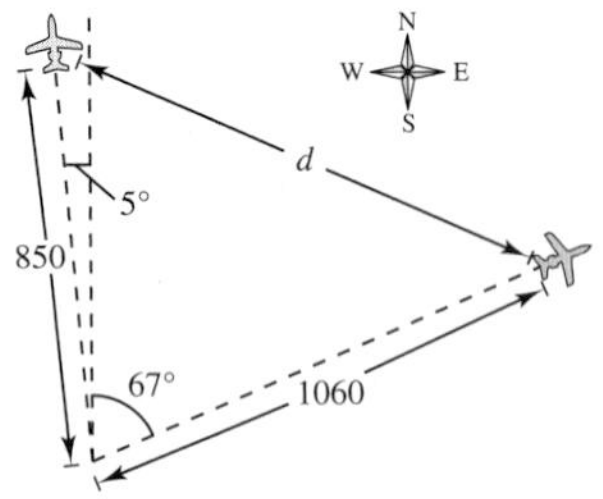

1135 miles

31. 36.979 **33.** 242.630

35.

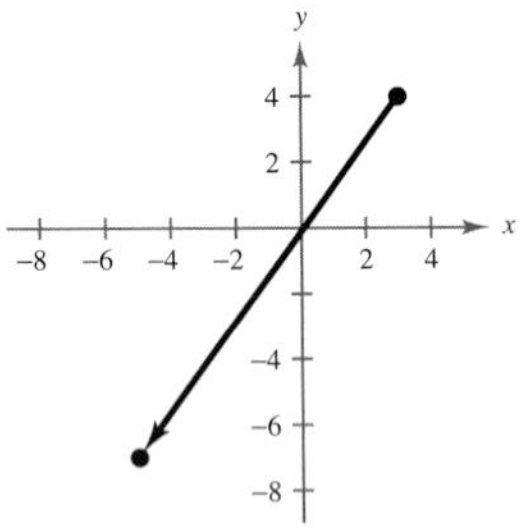

37.

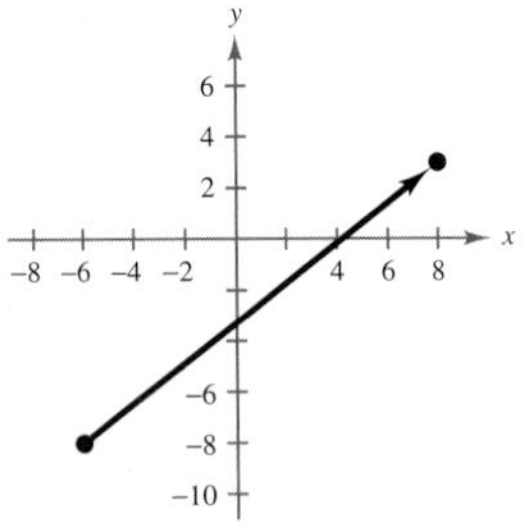

39. $\left\langle 6, \frac{5}{2} \right\rangle$ **41.** $\langle 14, 4 \rangle$ **43.** $\left\langle -\frac{\sqrt{2}}{4}, -\frac{\sqrt{2}}{4} \right\rangle$

45. $\langle -26, -35 \rangle$ **47.** $\left\langle 5, \frac{3}{2} \right\rangle$

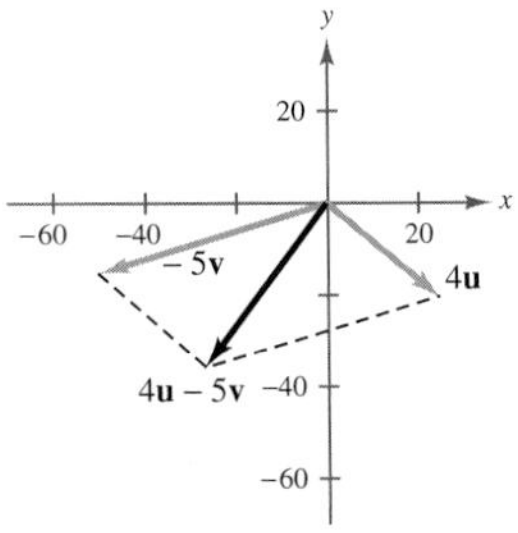

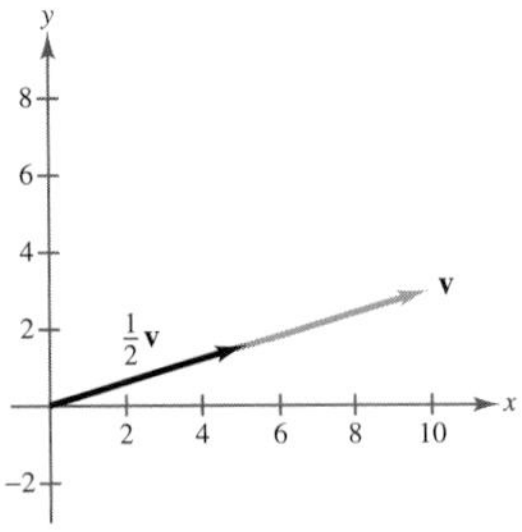

49. $-6\mathbf{i} - 8\mathbf{j}$ **51.** $7\mathbf{i} - 16\mathbf{j}$

53. $\sqrt{17}(\mathbf{i} \cos 346° + \mathbf{j} \sin 346°)$ **55.** $\|\mathbf{v}\| = 3;\ \theta = 150°$

57. $\|\mathbf{v}\| = \sqrt{65};\ \theta = 119.7°$ **59.** $\|\mathbf{v}\| = \sqrt{65};\ \theta = 352.9°$

61. 180 pounds **63.** 45 **65.** -2 **67.** 50; scalar

69. $\langle 6, -8 \rangle$; vector **71.** $\frac{11\pi}{12}$ **73.** 160.5°

75. Orthogonal **77.** Neither **79.** $-\frac{13}{17}\langle 4, 1 \rangle, \frac{16}{17}\langle -1, 4 \rangle$

81. $\frac{5}{2}\langle -1, 1 \rangle, \frac{9}{2}\langle 1, 1 \rangle$ **83.** 48

85.

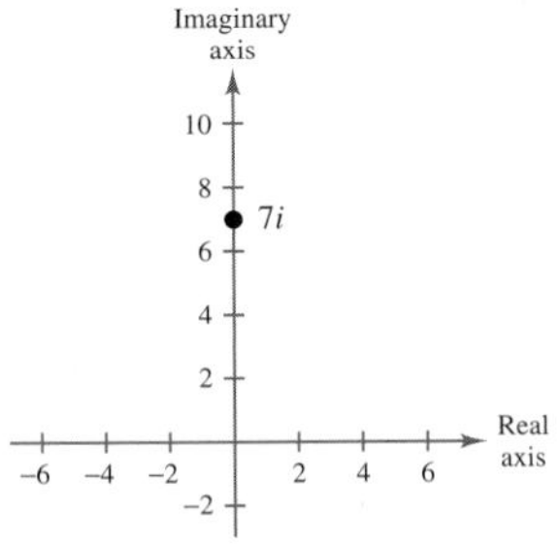

7

87.

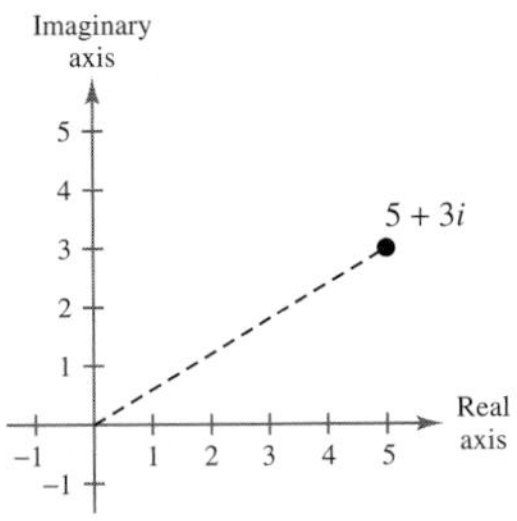

$\sqrt{34}$

89. $5\sqrt{2}\left(\cos \frac{7\pi}{4} + i \sin \frac{7\pi}{4}\right)$

91. $6\left(\cos \frac{5\pi}{6} + i \sin \frac{5\pi}{6}\right)$

93. (a) $z_1 = 4\left(\cos \frac{11\pi}{6} + i \sin \frac{11\pi}{6}\right)$

$z_2 = 10\left(\cos \frac{3\pi}{2} + i \sin \frac{3\pi}{2}\right)$

(b) $z_1 z_2 = 40\left(\cos \frac{10\pi}{3} + i \sin \frac{10\pi}{3}\right)$

$\frac{z_1}{z_2} = \frac{2}{5}\left(\cos \frac{\pi}{3} + i \sin \frac{\pi}{3}\right)$

95. $\frac{625}{2} + \frac{625\sqrt{3}}{2}i$ **97.** $2035 - 828i$

99. (a) $4(\cos 60° + i \sin 60°)$

$4(\cos 180° + i \sin 180°)$

$4(\cos 300° + i \sin 300°)$

(b) -64

101. $3\left(\cos \frac{\pi}{4} + i \sin \frac{\pi}{4}\right)$

$3\left(\cos \frac{7\pi}{12} + i \sin \frac{7\pi}{12}\right)$

$3\left(\cos \frac{11\pi}{12} + i \sin \frac{11\pi}{12}\right)$

$3\left(\cos \frac{5\pi}{4} + i \sin \frac{5\pi}{4}\right)$

$3\left(\cos \frac{19\pi}{12} + i \sin \frac{19\pi}{12}\right)$

$3\left(\cos \frac{23\pi}{12} + i \sin \frac{23\pi}{12}\right)$

103. $3\left(\cos\frac{\pi}{4} + i\sin\frac{\pi}{4}\right) = \frac{3\sqrt{2}}{2} + \frac{3\sqrt{2}}{2}i$

$3\left(\cos\frac{3\pi}{4} + i\sin\frac{3\pi}{4}\right) = -\frac{3\sqrt{2}}{2} + \frac{3\sqrt{2}}{2}i$

$3\left(\cos\frac{5\pi}{4} + i\sin\frac{5\pi}{4}\right) = -\frac{3\sqrt{2}}{2} - \frac{3\sqrt{2}}{2}i$

$3\left(\cos\frac{7\pi}{4} + i\sin\frac{7\pi}{4}\right) = \frac{3\sqrt{2}}{2} - \frac{3\sqrt{2}}{2}i$

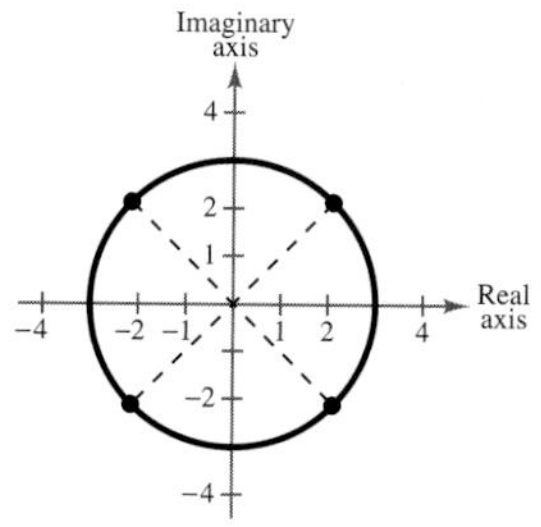

105. $2\left(\cos\frac{\pi}{2} + i\sin\frac{\pi}{2}\right) = 2i$

$2\left(\cos\frac{7\pi}{6} + i\sin\frac{7\pi}{6}\right) = -\sqrt{3} - i$

$2\left(\cos\frac{11\pi}{6} + i\sin\frac{11\pi}{6}\right) = \sqrt{3} - i$

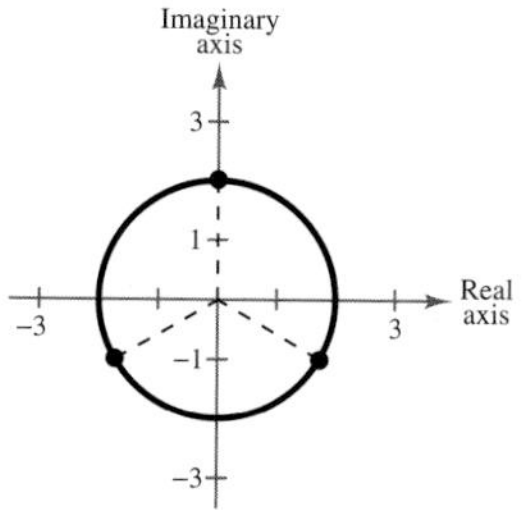

107. True. sin 90° is defined in the Law of Sines.

109. $\frac{a}{\sin A} = \frac{b}{\sin B} = \frac{c}{\sin C}$ **111.** Pythagorean Theorem

113. **A** and **C**

115. If $k > 0$, the direction is the same and the magnitude is k times as great.

If $k < 0$, the result is a vector in the opposite direction and the magnitude is k times as great.

117. $z_1z_2 = -4; \frac{z_1}{z_2} = \cos(2\theta - \pi) + i\sin(2\theta - \pi)$

$= -\cos 2\theta - i\sin 2\theta$

Chapter Test *(page 549)*

1. $C = 88°, b \approx 27.81, c \approx 29.98$

2. $A = 43°, b \approx 25.75, c \approx 14.45$

3. Two solutions:

$B \approx 29.12°, C \approx 126.88°, c \approx 22.03$

$B \approx 150.88°, C \approx 5.12°, c \approx 2.46$

4. No solution **5.** $A \approx 39.96°, C \approx 40.04°, c \approx 15.02$

6. $A \approx 23.44°, B \approx 33.56°, c \approx 86.46$

7. 2052.5 square meters **8.** 606.3 miles; N 29° E

9. $\langle 14, -23\rangle$ **10.** $\left\langle \frac{18\sqrt{34}}{17}, -\frac{30\sqrt{34}}{17}\right\rangle$

11. $\langle -4, 6\rangle$

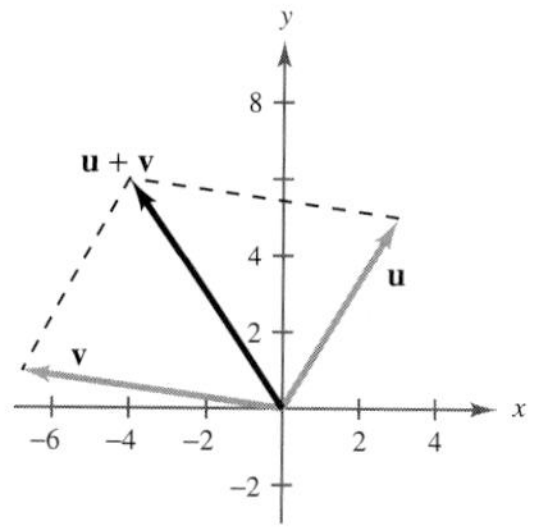

12. $\langle 10, 4\rangle$

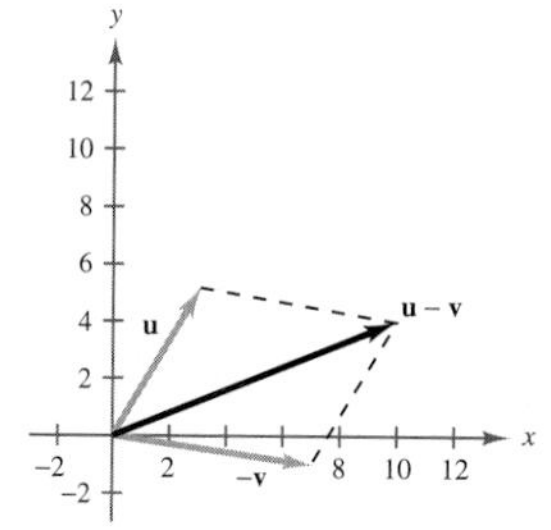

13. $\langle 36, 22\rangle$

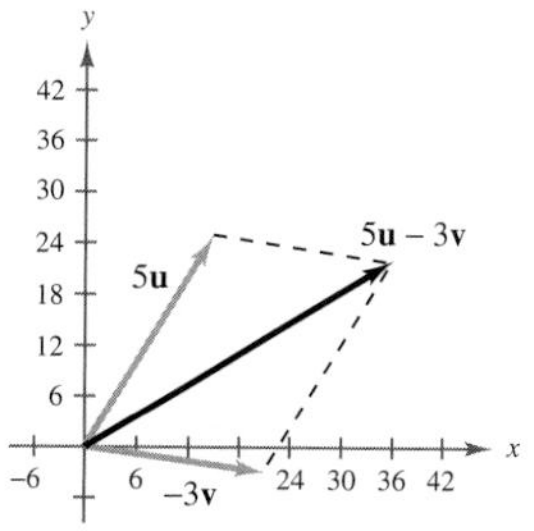

14. $\left\langle \frac{4}{5}, -\frac{3}{5}\right\rangle$

15. 14.9°; 250.15 pounds **16.** 135° **17.** No

18. $\frac{37}{26}\langle 5, 1\rangle; \frac{29}{26}\langle -1, 5\rangle$ **19.** $5\sqrt{2}\left(\cos\frac{7\pi}{4} + i\sin\frac{7\pi}{4}\right)$

20. $-3 + 3\sqrt{3}i$ **21.** $-\frac{6561}{2} - \frac{6561\sqrt{3}}{2}i$ **22.** $5832i$

23. $4\sqrt[4]{2}\left(\cos\frac{\pi}{12} + i\sin\frac{\pi}{12}\right)$

$4\sqrt[4]{2}\left(\cos\frac{7\pi}{12} + i\sin\frac{7\pi}{12}\right)$

$4\sqrt[4]{2}\left(\cos\frac{13\pi}{12} + i\sin\frac{13\pi}{12}\right)$

$4\sqrt[4]{2}\left(\cos\frac{19\pi}{12} + i\sin\frac{19\pi}{12}\right)$

24. $3\left(\cos\frac{\pi}{6} + i\sin\frac{\pi}{6}\right)$

$3\left(\cos\frac{5\pi}{6} + i\sin\frac{5\pi}{6}\right)$

$3\left(\cos\frac{3\pi}{2} + i\sin\frac{3\pi}{2}\right)$

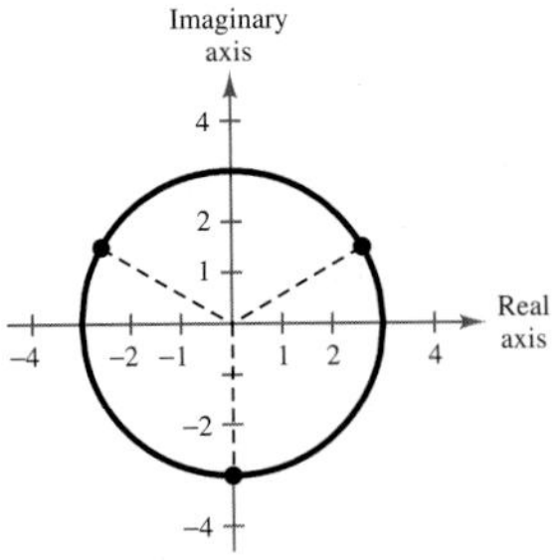

Cumulative Test for Chapters 4–6 *(page 550)*

1. (a)

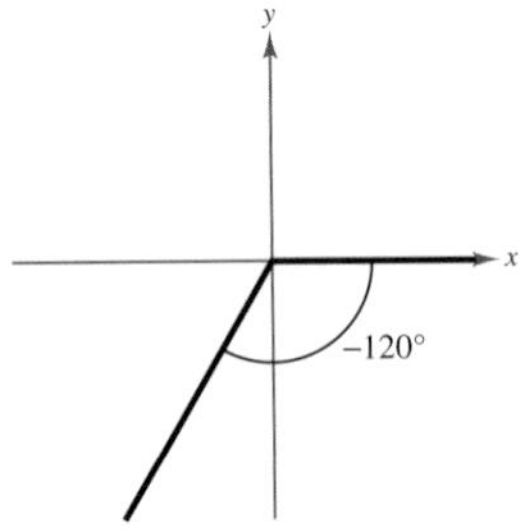

(b) 240°

(c) $-\frac{2\pi}{3}$

(d) 60°

(e) $\sin(-120°) = -\frac{\sqrt{3}}{2}$ $\qquad \csc(-120°) = -\frac{2\sqrt{3}}{3}$

$\cos(-120°) = -\frac{1}{2}$ $\qquad \sec(-120°) = -2$

$\tan(-120°) = \sqrt{3}$ $\qquad \cot(-120°) = \frac{\sqrt{3}}{3}$

2. 134.6° **3.** $\frac{3}{5}$

4. Period: 2; Amplitude: 2

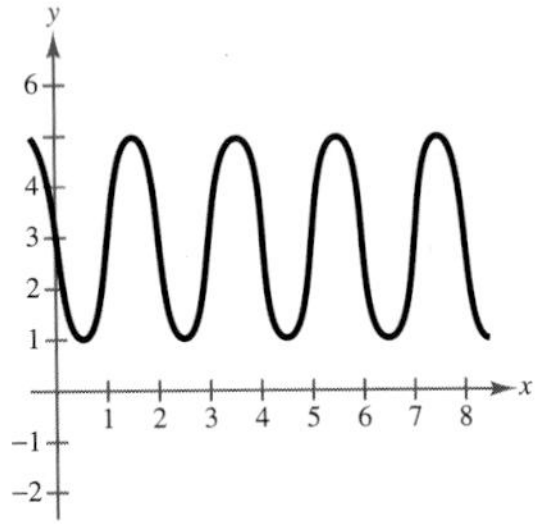

5. Period: π

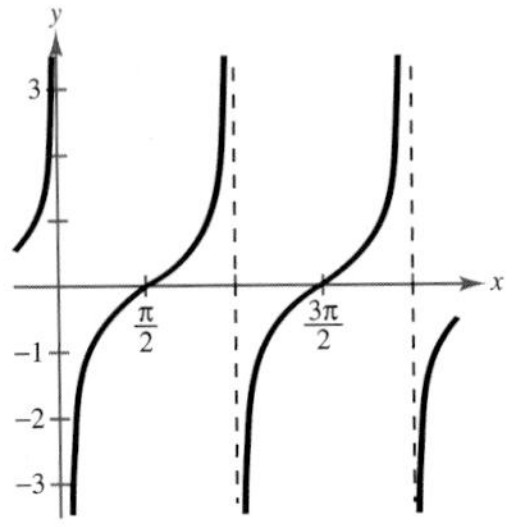

6. $a = -3, b = \pi, c = 0$

7.

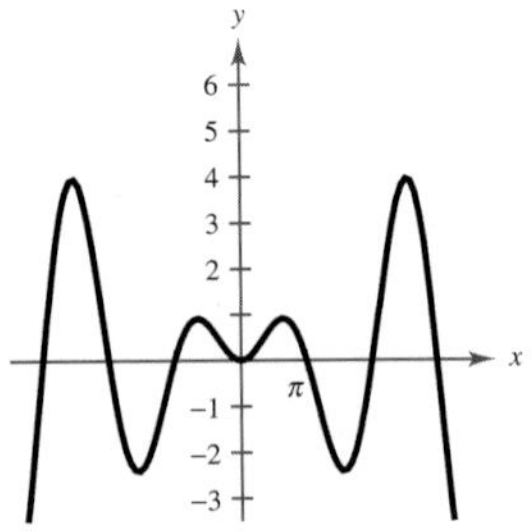

8. 6.7 **9.** $\frac{3}{4}$ **10.** $\sqrt{1 - 4x^2}$ **11.** 1

12. $2\tan\theta$ **13.–15.** Answers will vary.

16. $\frac{\pi}{3}, \frac{\pi}{2}, \frac{3\pi}{2}, \frac{5\pi}{3}$ **17.** $\frac{\pi}{6}, \frac{5\pi}{6}, \frac{7\pi}{6}, \frac{11\pi}{6}$ **18.** $\frac{3\pi}{2}$

19. $\frac{16}{63}$ **20.** $\frac{4}{3}$ **21.** $\frac{\sqrt{5}}{5}, \frac{2\sqrt{5}}{5}$

22. $\frac{5}{2}\left(\sin\frac{5\pi}{2} - \sin\pi\right)$

23. $B \approx 26.4°, C \approx 123.6°, c \approx 15.0$

24. $B \approx 52.5°, C \approx 97.5°, a \approx 5.0$

25. $B = 60°, a \approx 5.8, c \approx 11.5$

26. $A = 26.4°, B \approx 62.7°, C \approx 90.9°$

27. 36.4 square inches **28.** 85.2 square inches

29. $3\mathbf{i} + 5\mathbf{j}$ **30.** -5 **31.** $-\frac{1}{13}\langle 1, 5\rangle; \frac{21}{13}\langle 5, -1\rangle$

32. $2\sqrt{2}\left(\cos\frac{3\pi}{4} + i\sin\frac{3\pi}{4}\right)$ **33.** $-12\sqrt{3} + 12i$

34. $\cos 0 + i\sin 0 = 1$

$\cos\frac{2\pi}{3} + i\sin\frac{2\pi}{3} = -\frac{1}{2} + \frac{\sqrt{3}}{2}i$

$\cos\frac{4\pi}{3} + i\sin\frac{4\pi}{3} = -\frac{1}{2} - \frac{\sqrt{3}}{2}i$

35. $4\left(\cos\frac{\pi}{8} + i\sin\frac{\pi}{8}\right)$

$4\left(\cos\frac{5\pi}{8} + i\sin\frac{5\pi}{8}\right)$

$4\left(\cos\frac{9\pi}{8} + i\sin\frac{9\pi}{8}\right)$

$4\left(\cos\frac{13\pi}{8} + i\sin\frac{13\pi}{8}\right)$

36. 5 feet

37. 90π radians per minute; 848.23 inches per minute

38. 22.6° **39.** $d = 4\cos\frac{\pi}{4}t$

40. N 32.6° E; 543.9 kilometers per hour

Chapter 7

Section 7.1 *(page 561)*

1. d **3.** b **5.** (2, 2) **7.** (2, 6), (−1, 3)

9. (0, −5), (4, 3) **11.** (0, 0), (2, −4)

13. (0, 1), (1, −1), (3, 1) **15.** (5, 5) **17.** $\left(\frac{1}{2}, 3\right)$

19. (1, 1) **21.** $\left(\frac{20}{3}, \frac{40}{3}\right)$ **23.** No solution

25. (−2, 4), (0, 0) **27.** (0, 0), (−1, −1), (1, 1)

29. (4, 3) **31.** $\left(\frac{5}{2}, \frac{3}{2}\right)$ **33.** (2, 2), (4, 0)

35. (1, 4), (4, 7) **37.** $\left(4, -\frac{1}{2}\right)$

39. No solution **41.** $(4, 3), (-4, 3)$

43.

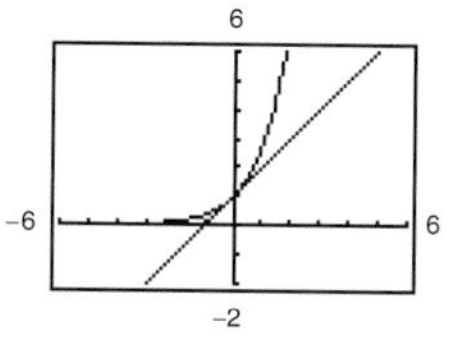

$(0, 1)$

45.

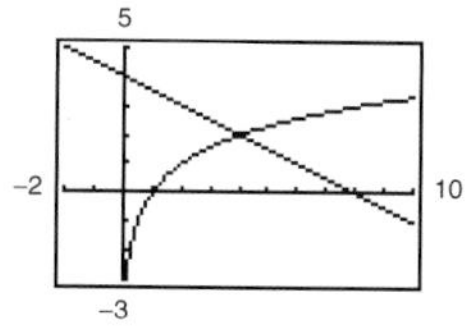

$(4, 2)$

47.

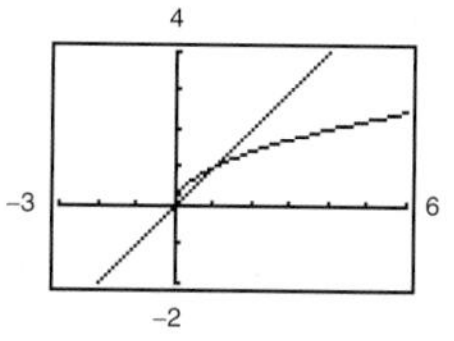

$(0, 0), (1, 1)$

49.

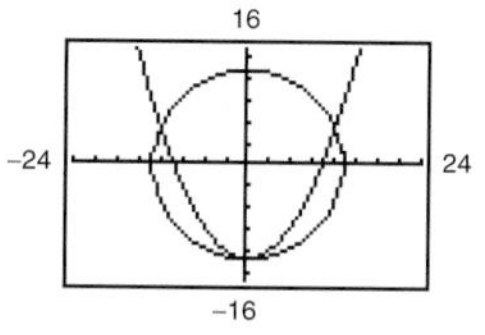

$(0, -13), (\pm 12, 5)$

51. $(1, 2)$ **53.** $(-2, 0), \left(\frac{29}{10}, \frac{21}{10}\right)$ **55.** No solution

57. $(0.287, 1.75)$ **59.** $(-1, 0), (0, 1), (1, 0)$

61. $\left(\frac{1}{2}, 2\right), \left(-4, -\frac{1}{4}\right)$ **63.** 192 units **65.** 3133 units

67. (a) 6400 units (b) 8800 units

69. (a) $\begin{cases} x + y = 25{,}000 \\ 0.06x + 0.085y = 2{,}000 \end{cases}$

(b)

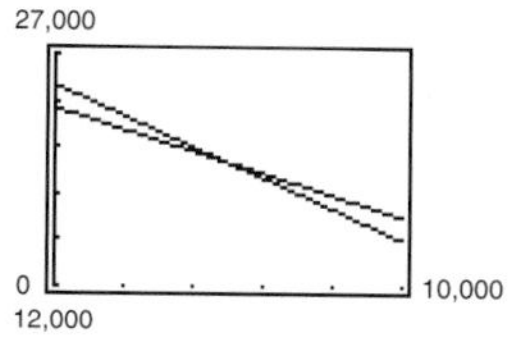

Decreases; interest is fixed.

(c) \$5000

71. More than \$11,666.67

73. (a)

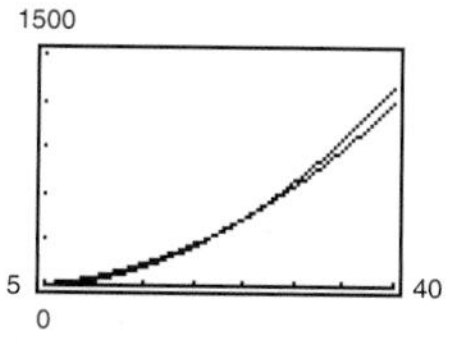

(b) 24.7 inches (c) Doyle Log Rule

75. 6×9 meters **77.** 9×12 inches

79. 8×12 kilometers

81. (a) $f(t) = 1.53t + 38.94$

$g(t) = -0.325t^2 + 4.455t + 32.765$

(b)

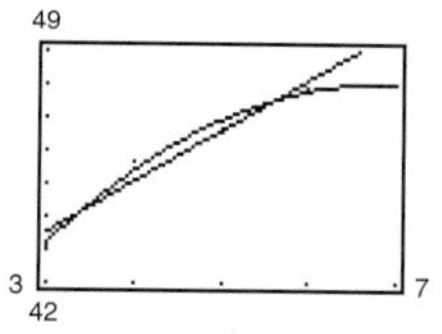

(c) $(3.382, 44.114), (5.618, 47.536)$

83. False. The system can have at most four solutions because a parabola and a circle can intersect at most four times.

85. For a linear system the result will be a contradictory equation such as $0 = N$, where N is a nonzero real number. For a nonlinear system there may be an equation with imaginary solutions.

87. (a) $y = 2x$ (b) $y = 0$ (c) $y = x - 2$

89. $2x + 7y - 45 = 0$ **91.** $y - 3 = 0$

93. $30x - 17y - 18 = 0$

95. Domain: All real numbers $x \neq 6$

Horizontal asymptote: $y = 0$

Vertical asymptote: $x = 6$

97. Domain: All real numbers $x \neq \pm 4$

Horizontal asymptote: $y = 1$

Vertical asymptotes: $x = \pm 4$

99.

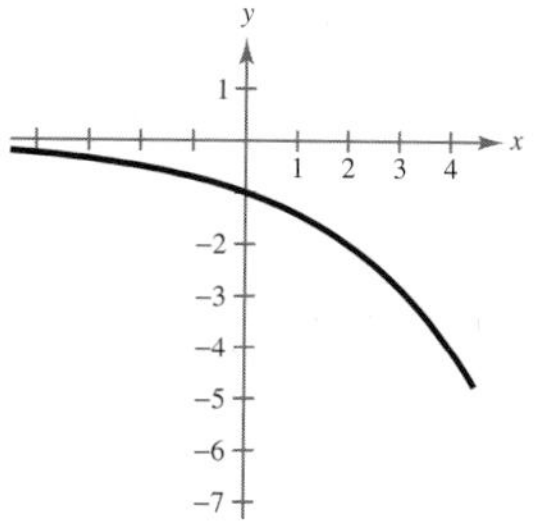

101.

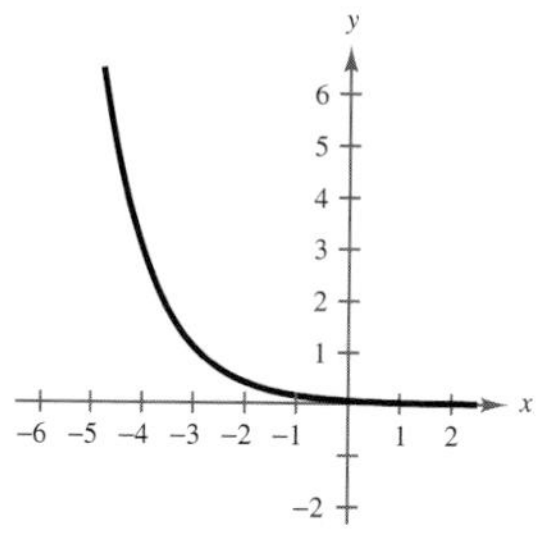

Section 7.2 *(page 573)*

1. $(2, 1)$ **3.** $(1, -1)$ **5.** No solution

7. $\left(a, \frac{3}{2}a - \frac{5}{2}\right)$ **9.** $\left(\frac{1}{3}, -\frac{2}{3}\right)$ **11.** $\left(\frac{5}{2}, \frac{3}{4}\right)$

13. $(3, 4)$ **15.** $(4, -1)$ **17.** $\left(\frac{12}{7}, \frac{18}{7}\right)$

19. No solution **21.** $\left(\frac{18}{5}, \frac{3}{5}\right)$ **23.** $\left(a, \frac{5}{6}a - \frac{1}{2}\right)$

25. $\left(\frac{90}{31}, -\frac{67}{31}\right)$ **27.** $\left(-\frac{6}{35}, \frac{43}{35}\right)$ **29.** $(5, -2)$

31.

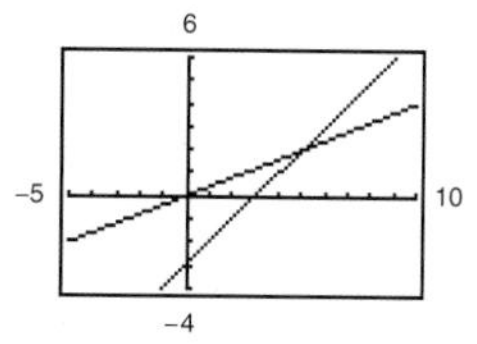

Consistent, one solution

33.

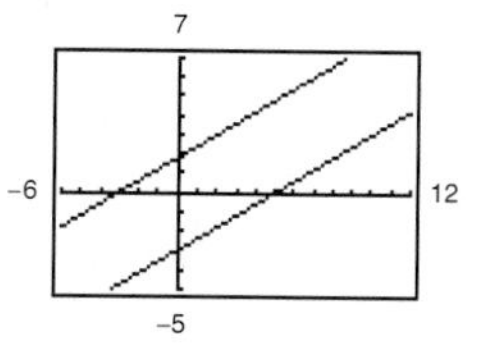

Inconsistent

35.

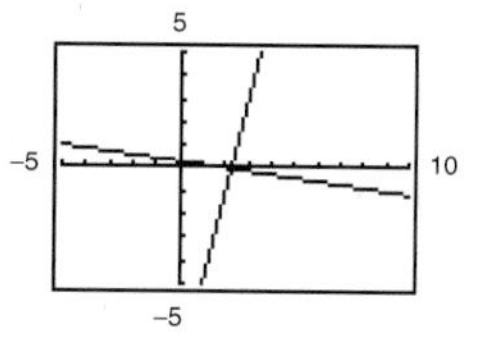

Consistent, one solution

37.

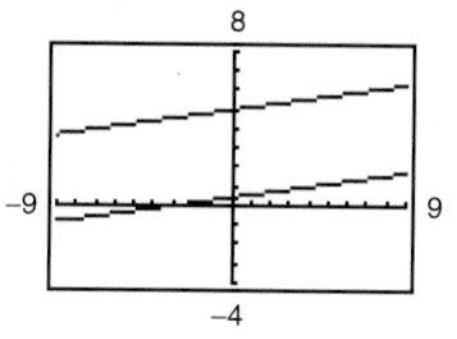

Inconsistent

39.
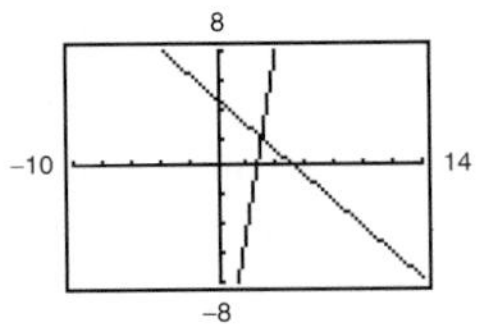

$(3, 2)$

41.
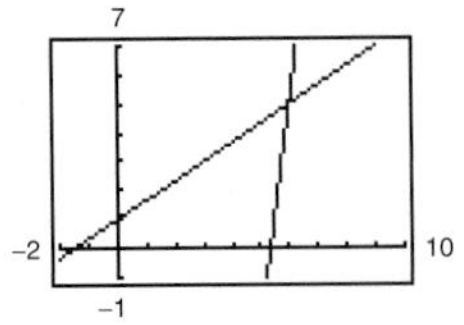

$(6, 5)$

43.
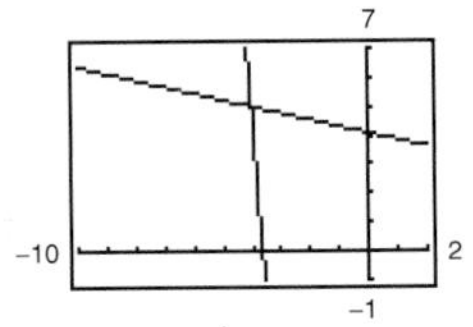

$(-4, 5)$

45.
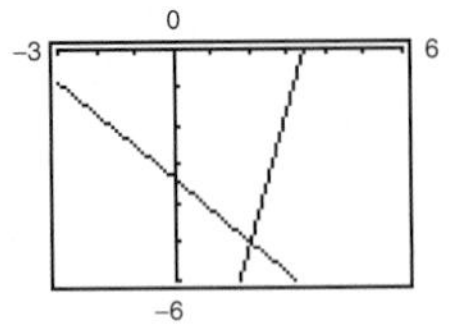

$(2, -5)$

47. $(4, 1)$ **49.** $(2, -1)$ **51.** $(6, -3)$ **53.** $\left(\frac{43}{6}, \frac{25}{6}\right)$

55. $(80, 10)$ **57.** $(2{,}000{,}000, 100)$

59. 550 miles per hour, 50 miles per hour

61. (a) $\begin{cases} x + y = 10 \\ 0.2x + 0.5y = 3 \end{cases}$

(b)
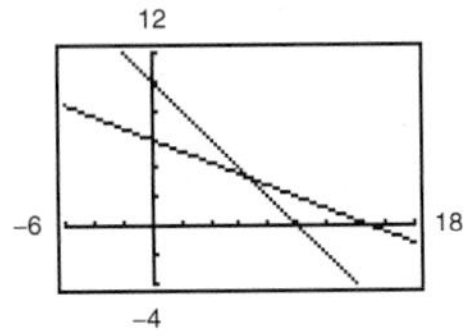

Decreases

(c) 20% solution: $6\frac{2}{3}$ liters

50% solution: $3\frac{1}{3}$ liters

63. \$6000 **65.** 400 adults, 1035 students

67. Machine 1: 1134 containers

Machine 2: 630 containers

69. $y = 0.97x + 2.10$ **71.** $y = 0.318x + 4.061$

73. $y = -2x + 4$

75. (a) and (b) $y = -240x + 685$

(c)
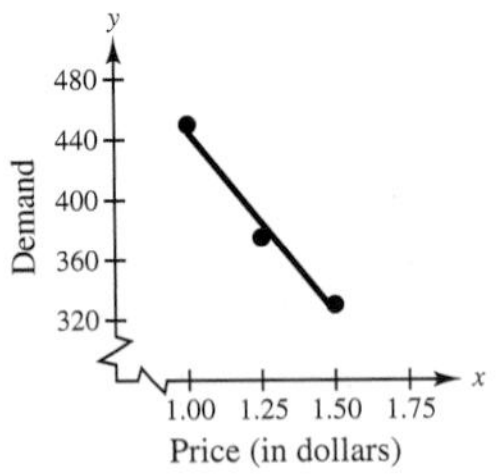

(d) 349 units

77. False. Two lines that coincide have infinitely many points of intersection.

79. True

81. $\begin{cases} x + y = 6 \\ 2x - y = 18 \end{cases}$ **83.** $\begin{cases} 3x + 3y = -32 \\ 3x + 6y = -62 \end{cases}$

85. $(300, 315)$. It is necessary to change the scale on the axes to see the point of intersection.

87. (a) $\begin{cases} x + y = 10 \\ x + y = 20 \end{cases}$ (b) $\begin{cases} x + y = 3 \\ 2x + 2y = 6 \end{cases}$

89. $k = -2$

91. $x > 1$

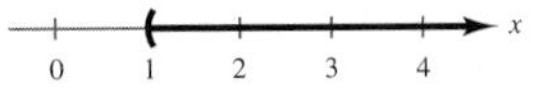

93. $\frac{4}{3} \le x < \frac{16}{3}$

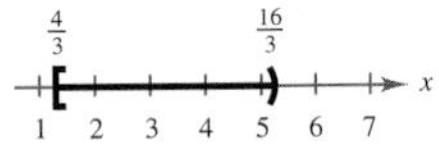

95. All real numbers x

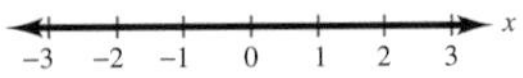

97. $x < -4, x > 0$

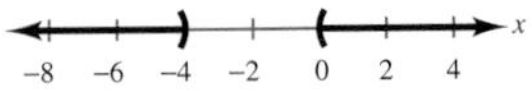

99. $\frac{1}{2}\left(-\frac{6}{x} + \frac{3}{x+1} + \frac{3}{x-1}\right)$ **101.** $\ln \frac{x}{(x+3)^5}$

103. $\log_6 \sqrt[4]{3x}$ **105.** $\left(\frac{3}{2}, \frac{3}{10}\right)$

Section 7.3 *(page 585)*

1. d **3.** c **5.** $(1, -2, 4)$ **7.** $(1, 2, -2)$

9. $\left(\frac{1}{2}, -2, 2\right)$

11. $\begin{cases} x - 2y + 3z = 5 \\ y - 2z = 9 \\ 2x - 3z = 0 \end{cases}$

First step in putting the system in row-echelon form

13. $(1, 2, 3)$ **15.** $(-4, 8, 5)$ **17.** $(5, -2, 0)$

19. No solution **21.** $\left(-\frac{1}{2}, 1, \frac{3}{2}\right)$

23. $(-3a + 10, 5a - 7, a)$ **25.** $(-a + 3, a + 1, a)$

27. $(2a, 21a - 1, 8a)$ **29.** $\left(-\frac{3}{2}a + \frac{1}{2}, -\frac{2}{3}a + 1, a\right)$

31. $(1, 1, 1, 1)$ **33.** No solution **35.** $(0, 0, 0)$

37. $(9a, -35a, 67a)$

39. $y = \frac{1}{2}x^2 - 2x$

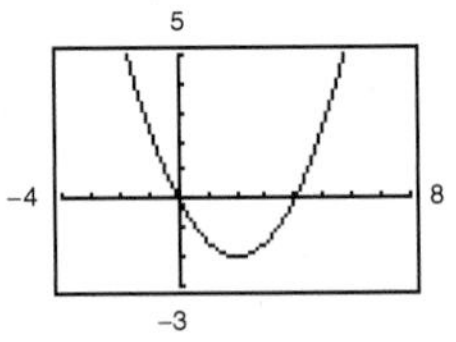

41. $y = x^2 - 6x + 8$

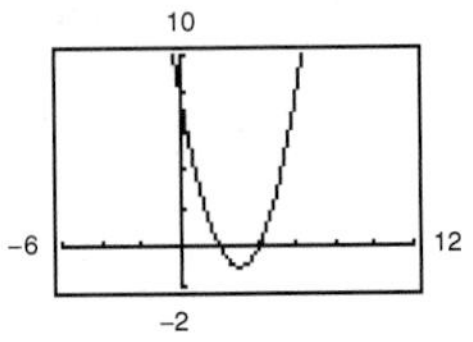

43. $x^2 + y^2 - 4x = 0$

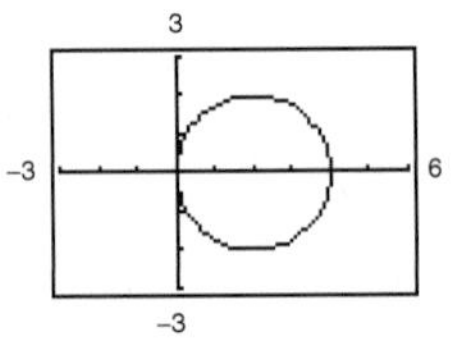

45. $x^2 + y^2 + 6x - 8y = 0$

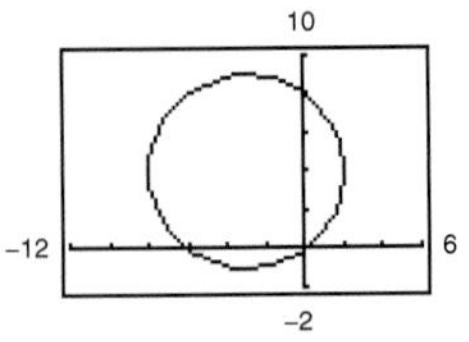

47. $s = -16t^2 + 144$ **49.** $s = -16t^2 - 32t + 500$

51. 8 touchdowns
6 extra points
6 field goals

53. \$300,000 at 8%
\$400,000 at 9%
\$75,000 at 10%

55. $250{,}000 - \frac{1}{2}s$ in certificates of deposit,
$125{,}000 + \frac{1}{2}s$ in municipal bonds,
$125{,}000 - s$ in blue-chip stocks,
s in growth stocks

57. 20 liters of spray X,
18 liters of spray Y,
16 liters of spray Z

59. Use four medium trucks or use two large, one medium, and two small trucks. Other answers are possible.

61. $t_1 = 96$ pounds
$t_2 = 48$ pounds
$a = -16$ feet per second squared

63. $\frac{1}{2}\left(-\frac{2}{x} + \frac{1}{x-1} + \frac{1}{x+1}\right)$ **65.** $\frac{1}{2}\left(\frac{1}{x} - \frac{1}{x-2} + \frac{2}{x+3}\right)$

67. $y = -\frac{5}{24}x^2 - \frac{3}{10}x + \frac{41}{6}$ **69.** $y = x^2 - x$

71. (a) $y = 0.165x^2 - 6.55x + 103$
(b)

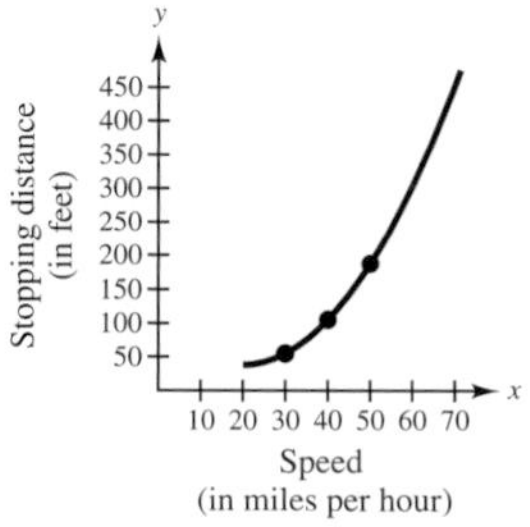

(c) 453 feet

73. $x = 5$, $y = 5$, $\lambda = -5$ **75.** $x = \pm\sqrt{2}/2$, $y = \frac{1}{2}$, $\lambda = 1$ or $x = 0$, $y = 0$, $\lambda = 0$

77. False. Equation 2 does not have a leading coefficient of 1.

79. No. Answers will vary.

81. There will be a row representing a contradictory equation such as $0 = N$, where N is a nonzero real number.

83. $$\begin{cases} x + y + z = -6 \\ -2x - y + 3z = 15 \\ x + 4y - z = -14 \end{cases}$$

85. $$\begin{cases} 2x - y + 3z = -28 \\ -6x + 4y + z = 18 \\ -4x - 2y - 3z = 19 \end{cases}$$

87. 150% **89.** 275

91. (a) $\pm 2, 0$
(b)

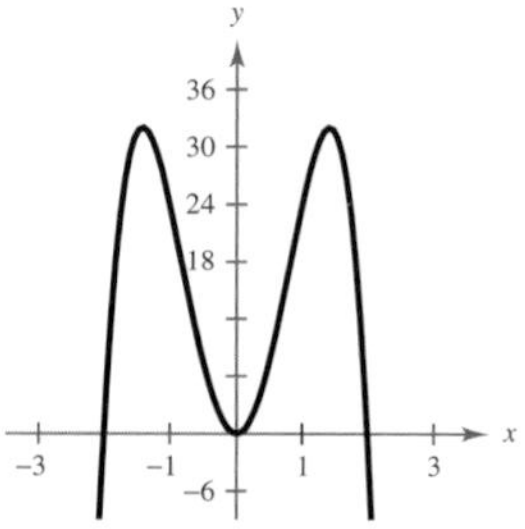

93. (a) $-\frac{1}{2}, \frac{1}{3}, 5$
(b)

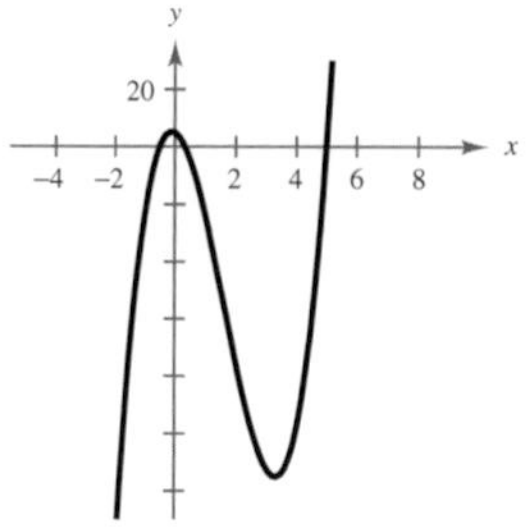

95.

x	-2	-1	0	1	2
y	11.625	2.25	-1.5	-3	-3.6

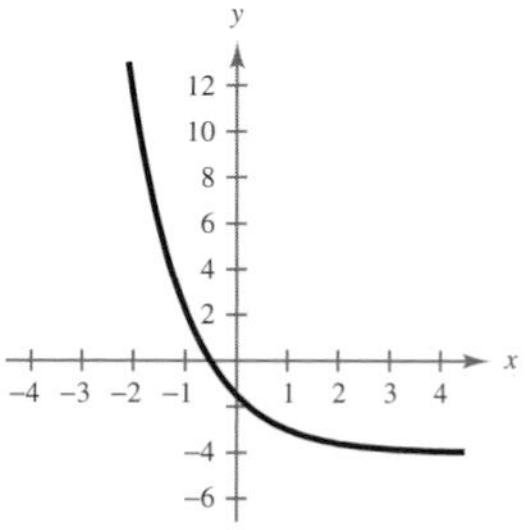

97.

x	$-\frac{1}{2}$	0	$\frac{1}{2}$	1	2
y	28.918	18.25	12.548	9.5	7

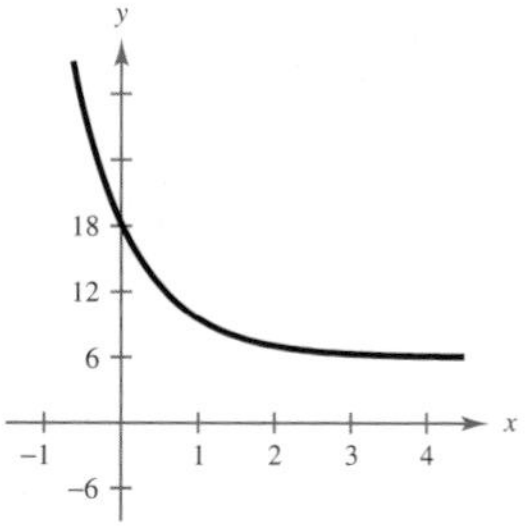

99. $\left(\frac{1}{2}, 0\right)$

Section 7.4 *(page 597)*

1.

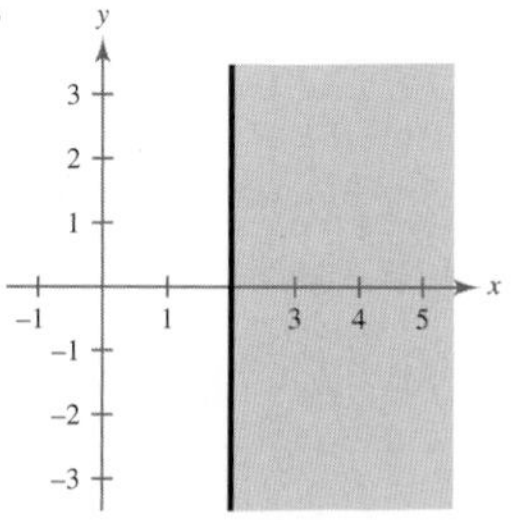

3.

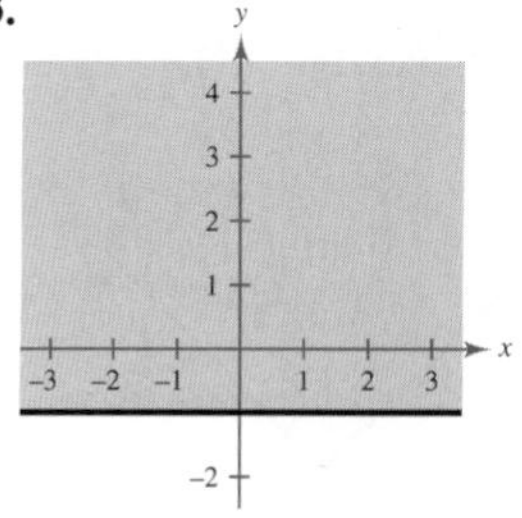

5.

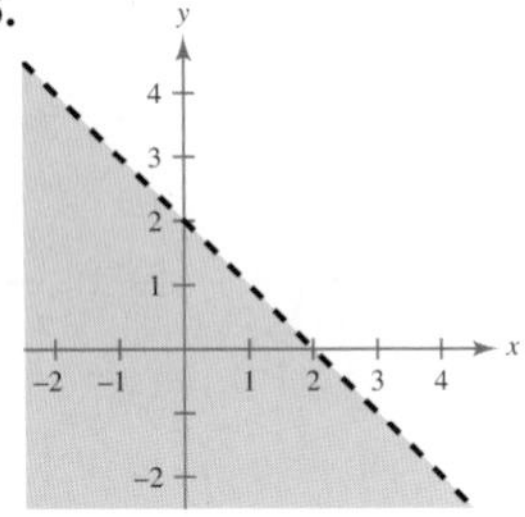

7.

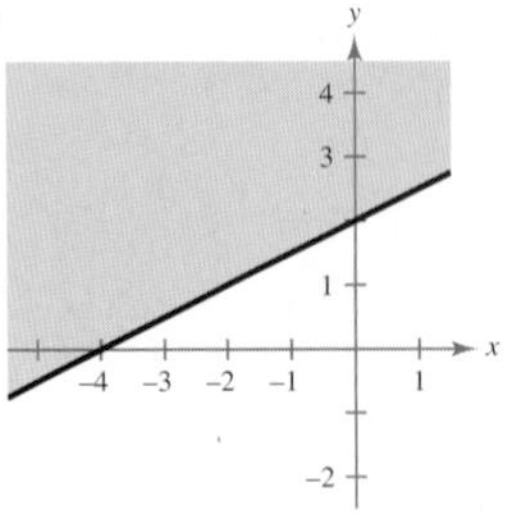

9.

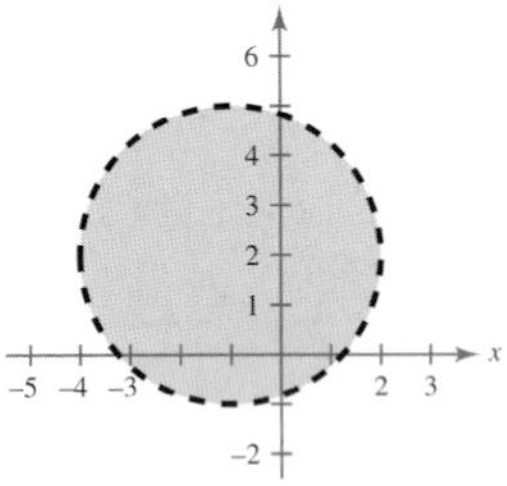

11.

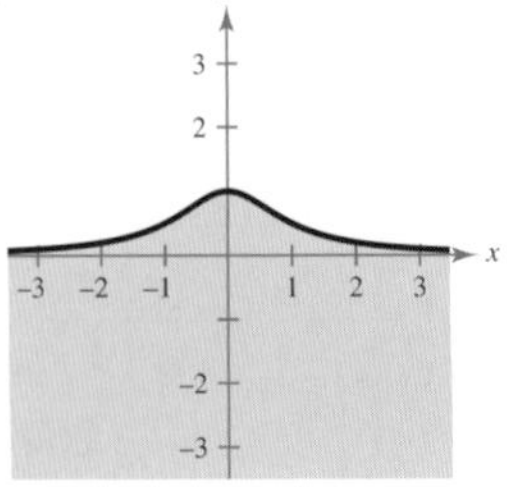

13.

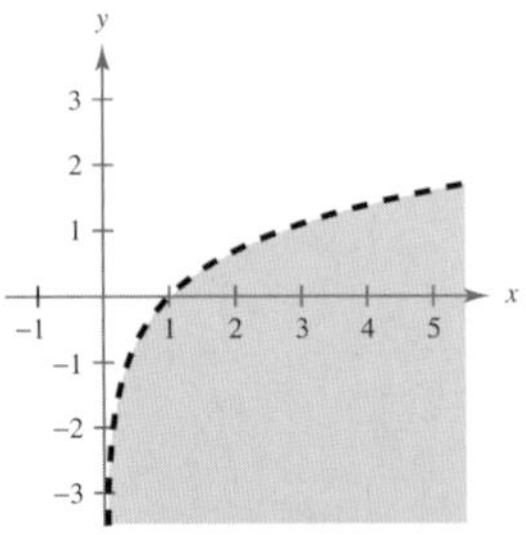

15.

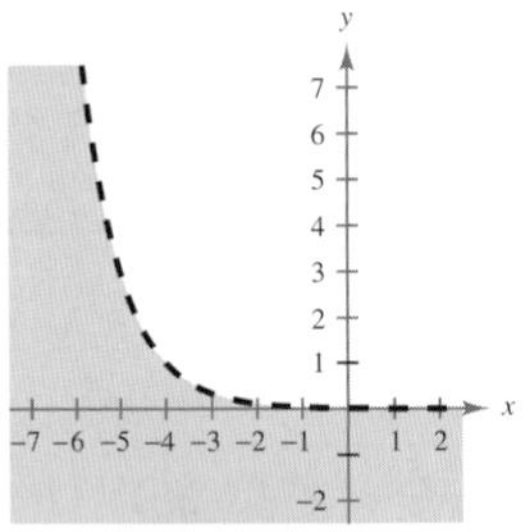

17.

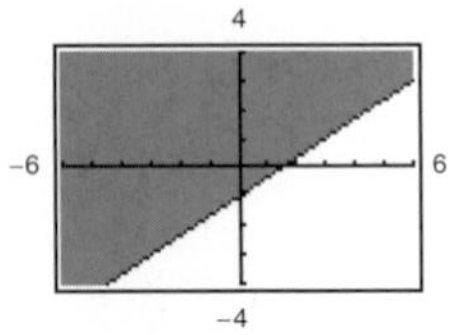

19.

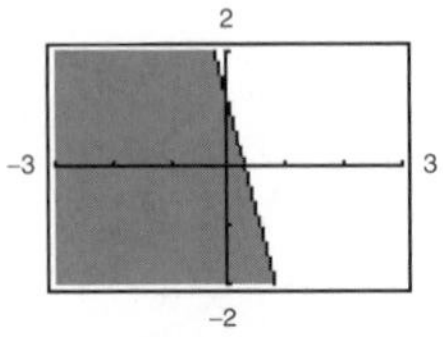

21.

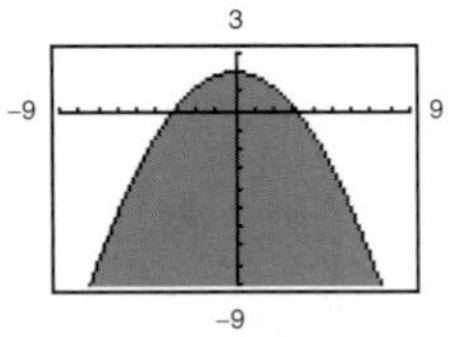

23.

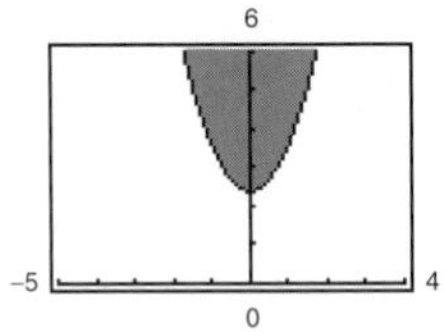

25. $y \le \frac{1}{2}x + 2$ **27.** $y \ge -\frac{2}{3}x + 2$

29. c and d **31.** a, c, and d

33.

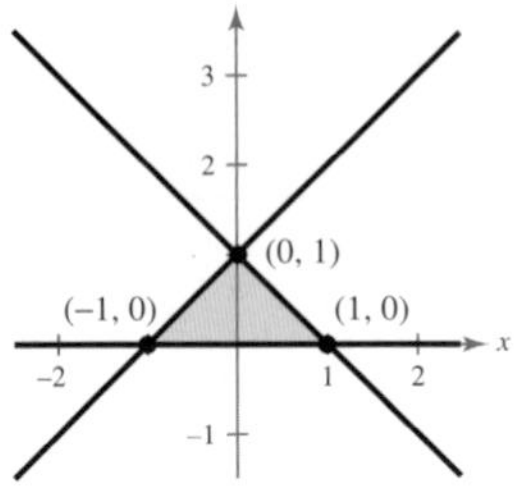

35.

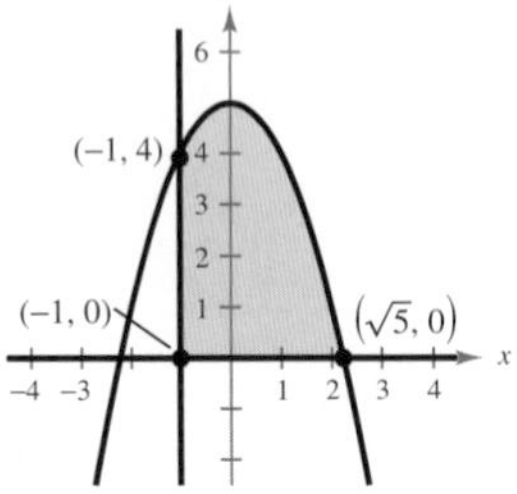

37.

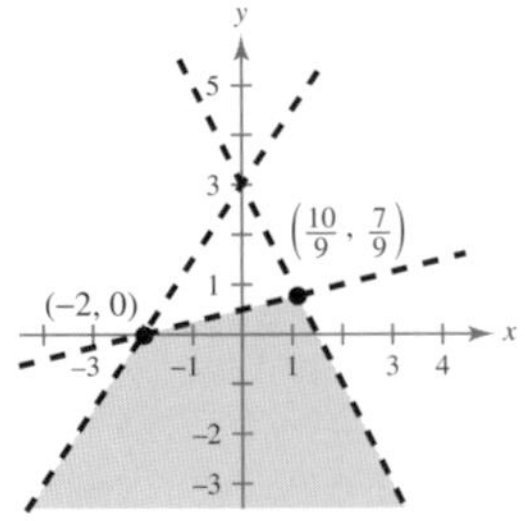

39. No solution

41.

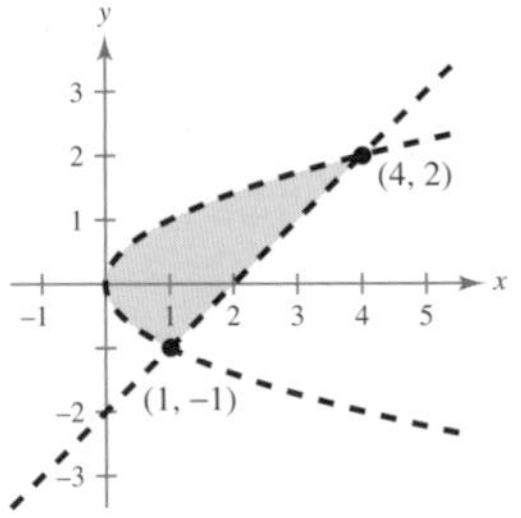

43.

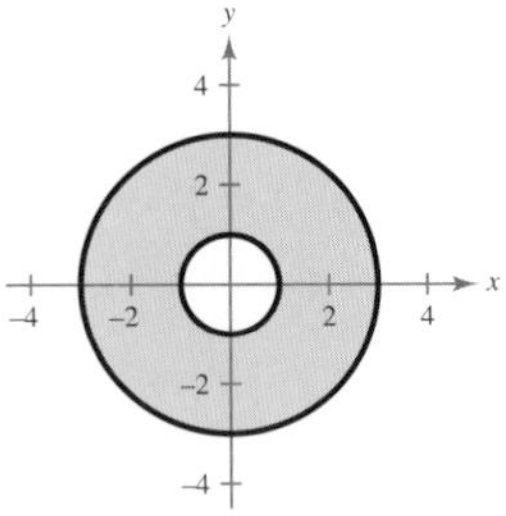

45.

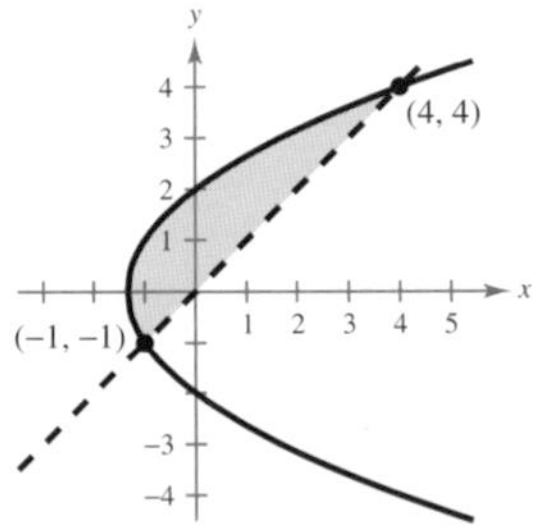

47.

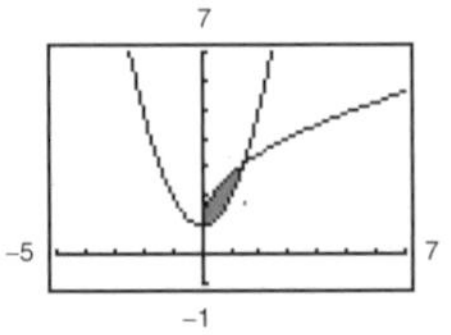

49.

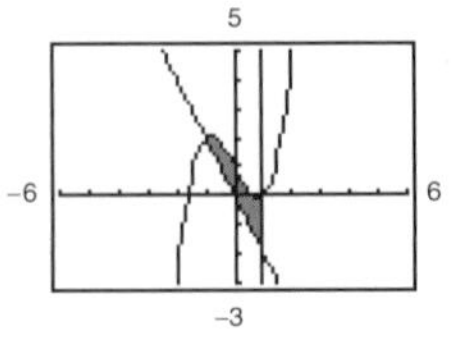

51.

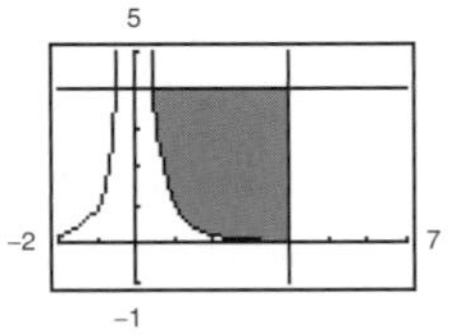

53. $\begin{cases} y \le 4 - x \\ x \ge 0 \\ y \ge 0 \end{cases}$ **55.** $\begin{cases} y \ge 4 - x \\ y \ge 2 - \frac{1}{4}x \\ x \ge 0,\ y \ge 0 \end{cases}$

57. $\begin{cases} x^2 + y^2 \le 16 \\ x \ge 0 \\ y \ge 0 \end{cases}$ **59.** $\begin{cases} 2 \le x \le 5 \\ 1 \le y \le 7 \end{cases}$ **61.** $\begin{cases} y \le \frac{3}{2}x \\ y \le -x + 5 \\ y \ge 0 \end{cases}$

63. Consumer surplus: 1600
Producer surplus: 400

65. Consumer surplus: 40,000,000
Producer surplus: 20,000,000

67. $\begin{cases} x + \frac{3}{2}y \le 12 \\ \frac{4}{3}x + \frac{3}{2}y \le 15 \\ x \ge 0 \\ y \ge 0 \end{cases}$

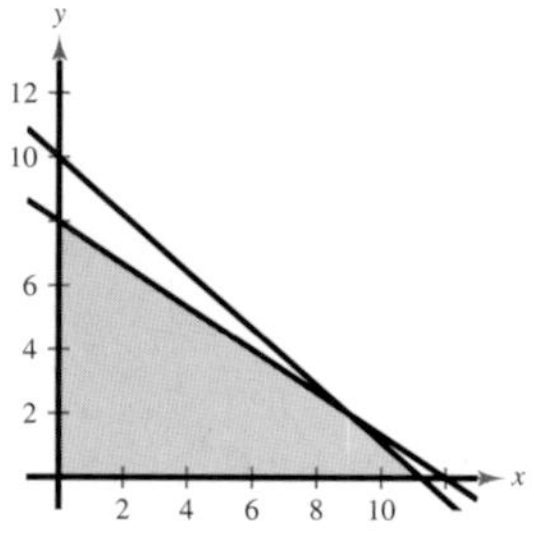

69. $\begin{cases} x + y \le 20{,}000 \\ y \ge 2x \\ x \ge 5{,}000 \\ y \ge 5{,}000 \end{cases}$

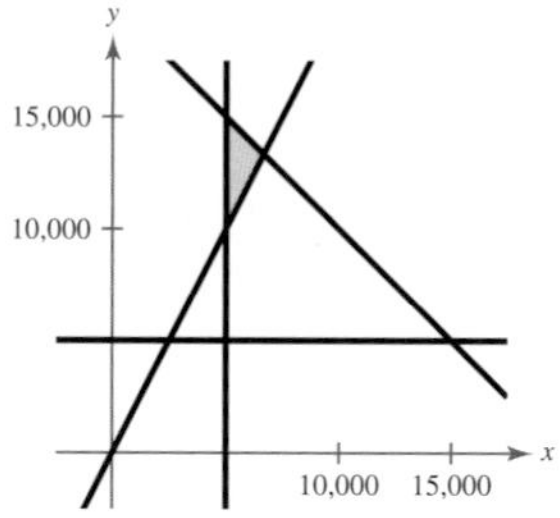

71. $\begin{cases} 55x + 70y \le 7500 \\ x \ge 50 \\ y \ge 40 \end{cases}$

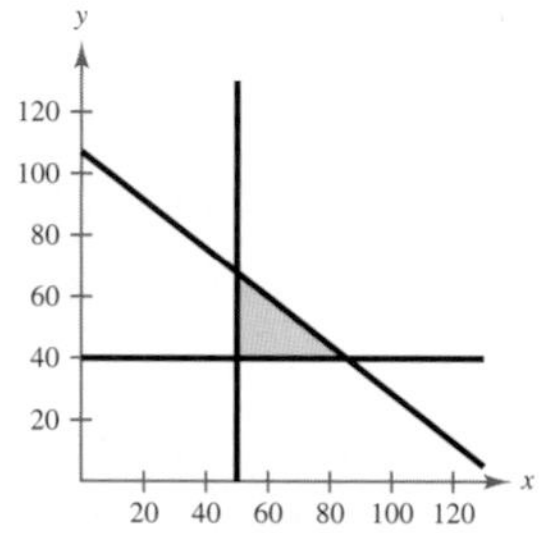

73. $\begin{cases} xy \ge 500 \\ 2x + \pi y \ge 125 \\ x \ge 0 \\ y \ge 0 \end{cases}$

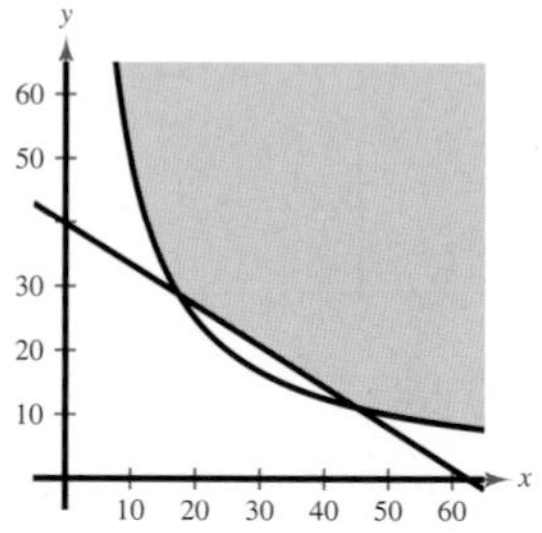

75. False. The graph shows the solution of the system

$$\begin{cases} y < 6 \\ -4x - 9y < 6. \\ 3x + y^2 \ge 2 \end{cases}$$

77. b **79.** a

81. (a) $\begin{cases} \pi y^2 - \pi x^2 \ge 10 \\ y > x \\ x > 0 \end{cases}$

(b)

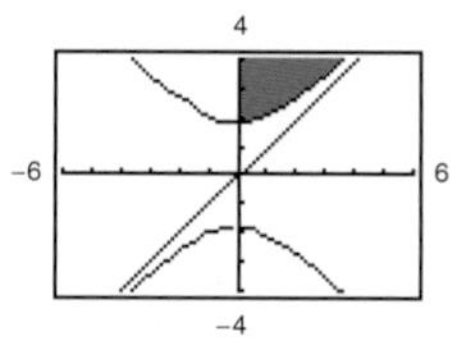

(c) The line is an asymptote to the boundary. The larger the circles, the closer the radii can be and the constraint still be satisfied.

83. The graph is a half-line on the real number line; on the rectangular coordinate system, the graph is a half-plane.

85. $x + 11y + 8 = 0$ **87.** $2x - y + 1 = 0$

89. $60x - 35y + 113 = 0$ **91.** 52.619 **93.** 0.022

95. 0.064 **97.** $(-4, 3, -7)$

Section 7.5 *(page 607)*

1. Minimum at $(0, 0)$: 0
Maximum at $(5, 0)$: 20

3. Minimum at $(0, 0)$: 0
Maximum at $(0, 5)$: 40

5. Minimum at $(0, 0)$: 0
Maximum at $(3, 4)$: 17

7. Minimum at $(0, 0)$: 0
Maximum at $(4, 0)$: 20

9. Minimum at $(0, 0)$: 0
Maximum at $(60, 20)$: 740

11. Minimum at $(0, 0)$: 0
Maximum at any point on the line segment connecting $(60, 20)$ and $(30, 45)$: 2100

13.

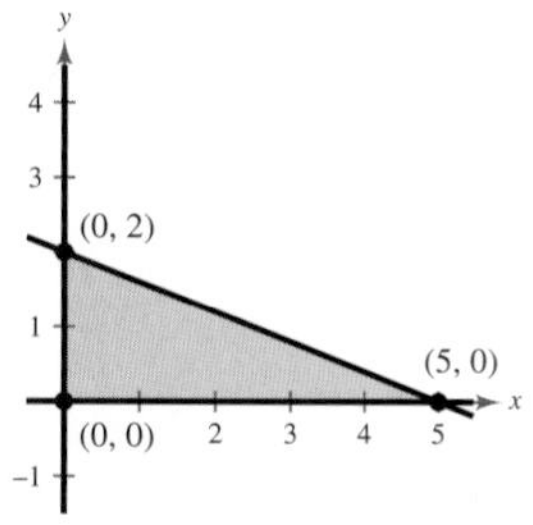

Minimum at $(0, 0)$: 0
Maximum at $(5, 0)$: 30

15.

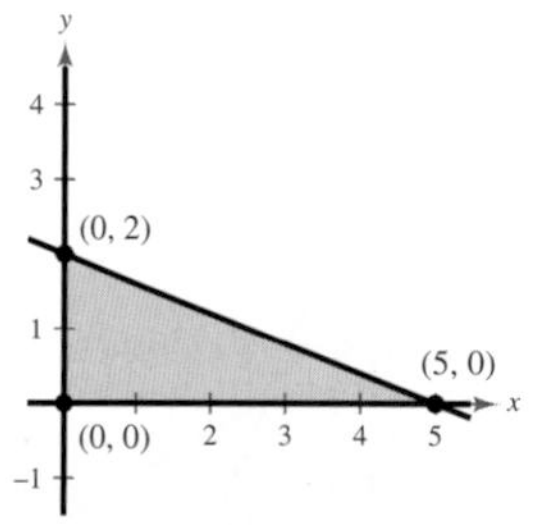

Minimum at $(0, 0)$: 0
Maximum at $(0, 2)$: 48

17.

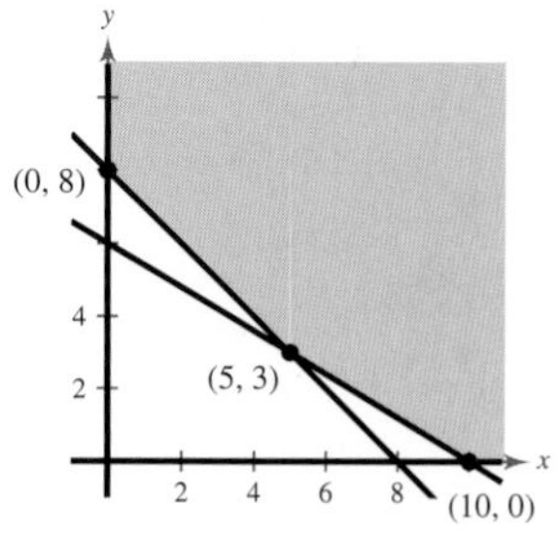

Minimum at $(5, 3)$: 35
No maximum

19.

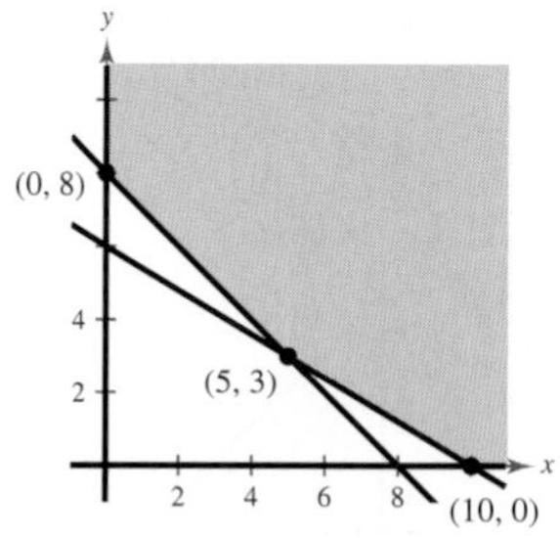

Minimum at $(10, 0)$: 20
No maximum

21.

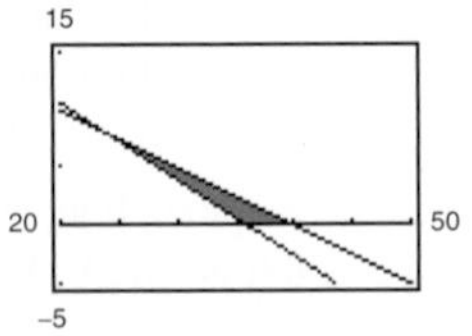

Minimum at (24, 8): 104

Maximum at (40, 0): 160

23.

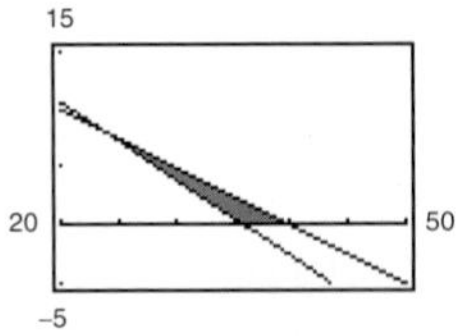

Minimum at (36, 0): 36

Maximum at (24, 8): 56

25. Maximum at (3, 6): 12 **27.** Maximum at (0, 10): 10

29. Maximum at (0, 5): 25 **31.** Maximum at $\left(\frac{22}{3}, \frac{19}{6}\right)$: $\frac{271}{6}$

33. 750 units of model A

1000 units of model B

Maximum profit: \$83,750

35. 0 units of the \$150 model

200 units of the \$200 model

Maximum profit: \$8000

37. Three bags of brand X

Six bags of brand Y

Minimum cost: \$195

39. 4 audits

32 tax returns

Maximum revenue: \$17,600

41.

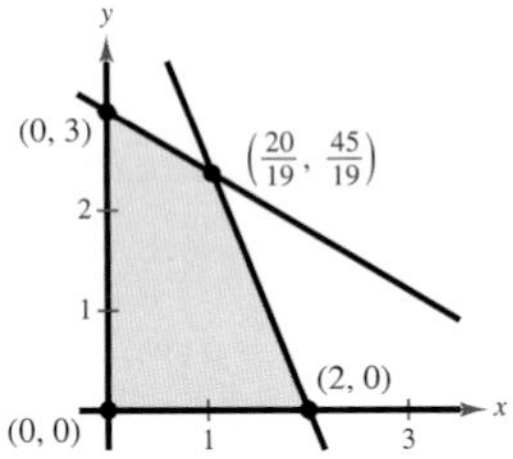

The maximum, 5, occurs at any point on the line segment connecting (2, 0) and $\left(\frac{20}{19}, \frac{45}{19}\right)$.

43.

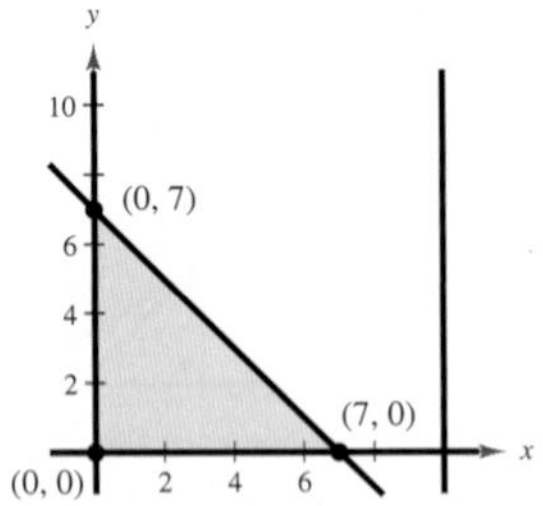

The constraint $x \le 10$ is extraneous. Maximum at (0, 7): 14

45.

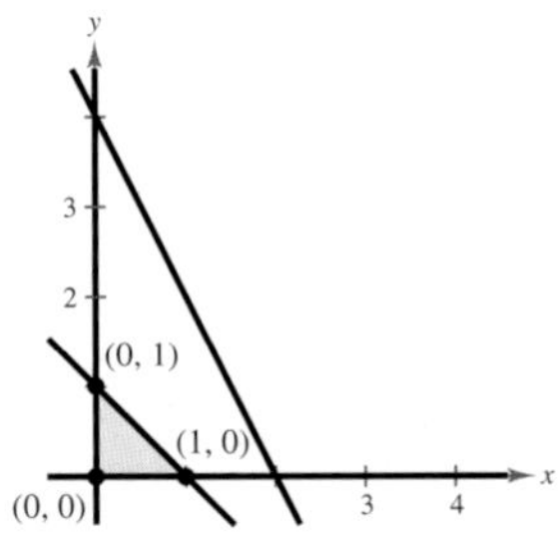

The constraint $2x + y \le 4$ is extraneous. Maximum at (0, 1): 4

47. True. The objective function has a maximum value at any point on the line segment connecting the two vertices.

49. (a) $t \ge 9$ (b) $\frac{3}{4} \le t \le 9$ **51.** $z = x + 5y$

53. $z = 4x + y$

55. $\dfrac{9}{2(x + 3)}, \ x \ne 0$ **57.** $\dfrac{x^2 + 2x - 13}{x(x - 2)}, \ x \ne \pm 3$

59. $\ln 3 \approx 1.099$ **61.** $4 \ln 38 \approx 14.550$

63. $\frac{1}{3}e^{12/7} \approx 1.851$

Review Exercises *(page 612)*

1. (5, 4) **3.** (0, 0), (2, 8), (−2, 8) **5.** (4, −2)

7. (0, −2) **9.** (1.41, −0.66), (−1.41, 10.66)

11. 10,417 units **13.** 96 × 144 meters **15.** $\left(\frac{3}{10}, \frac{2}{5}\right)$

17. $\left(\frac{1}{3}, -\frac{1}{2}\right)$ **19.** (−3, 7) **21.** No solution

23.

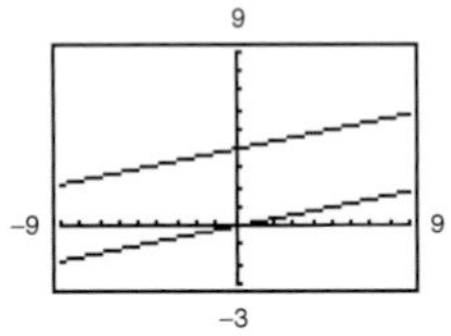

Inconsistent

25.

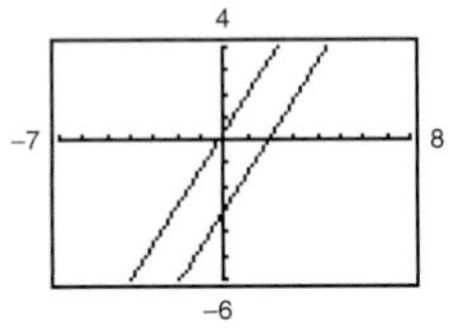

Inconsistent

27. 400 at \$9.95; 250 at \$14.95 **29.** $\left(\dfrac{500{,}000}{7}, \dfrac{159}{7}\right)$

31. (2, −4, −5) **33.** $\left(\frac{24}{5}, \frac{22}{5}, -\frac{8}{5}\right)$

35. $(3a + 4, 2a + 5, a)$ **37.** $y = 2x^2 + x - 5$

39. $x^2 + y^2 - 4x + 4y - 1 = 0$ **41.** $y = 1.01x + 1.54$

43. $(a - 1, a, a + 3)$

45. 10 gallons of spray X

5 gallons of spray Y

12 gallons of spray Z

47. $y = 0.114x^2 + 0.800x + 1.591$

49.

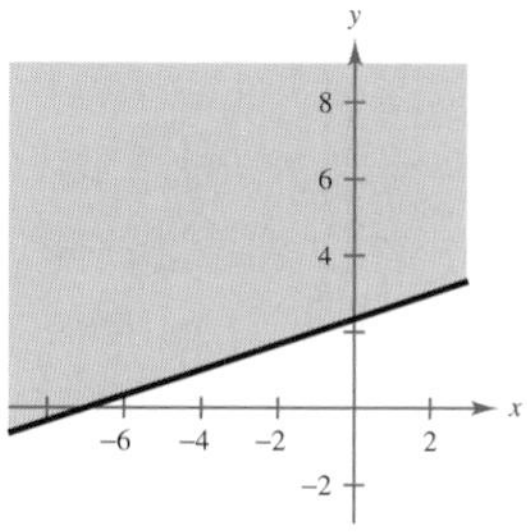

51.

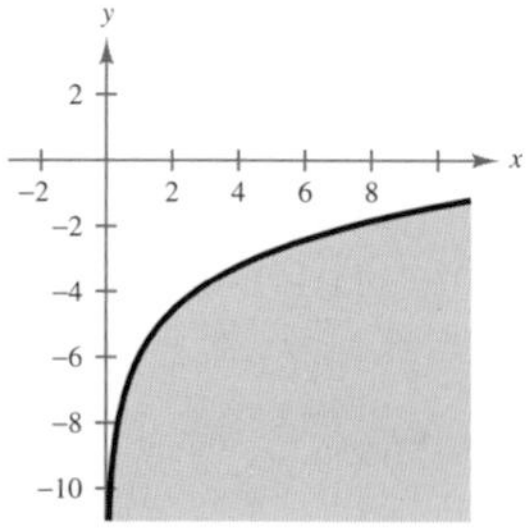

53.

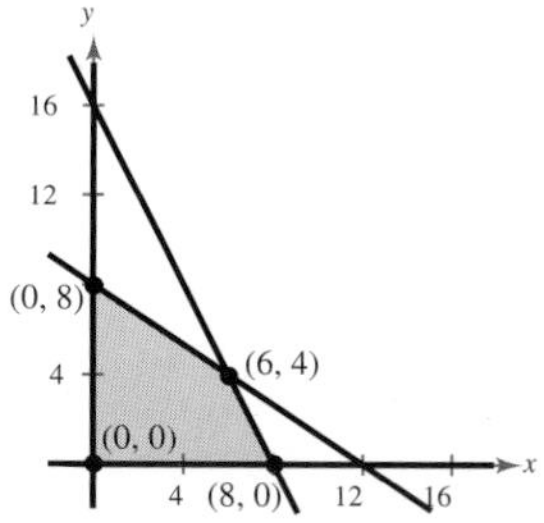

55.

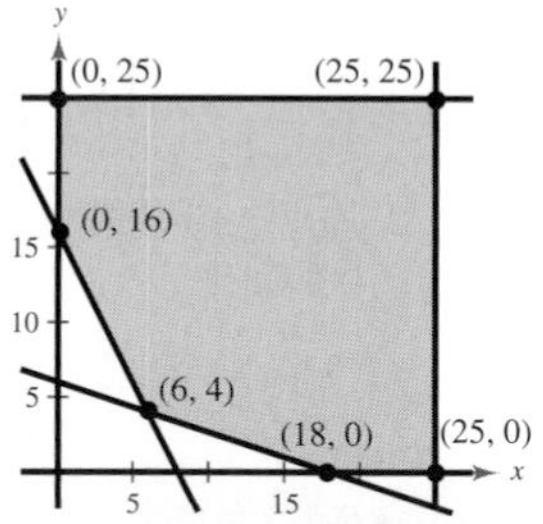

57.

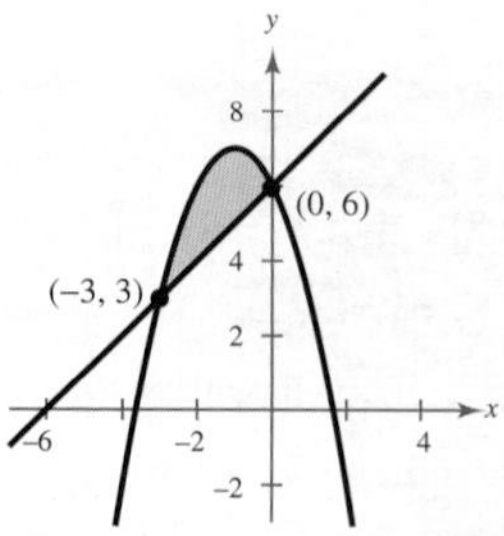

59.

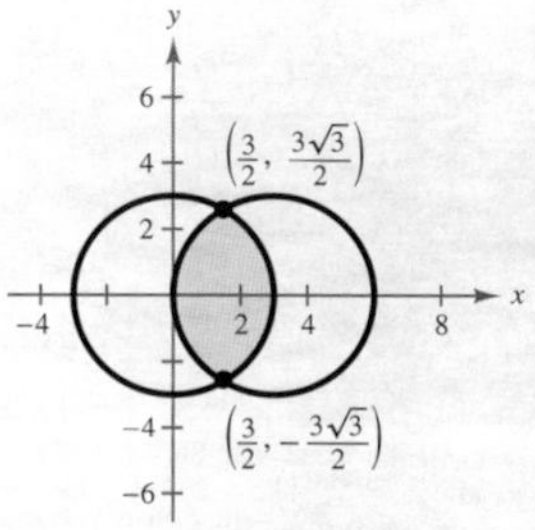

61. $\begin{cases} 20x + 30y \le 24{,}000 \\ 12x + 8y \le 12{,}400 \\ x \ge 0 \\ y \ge 0 \end{cases}$

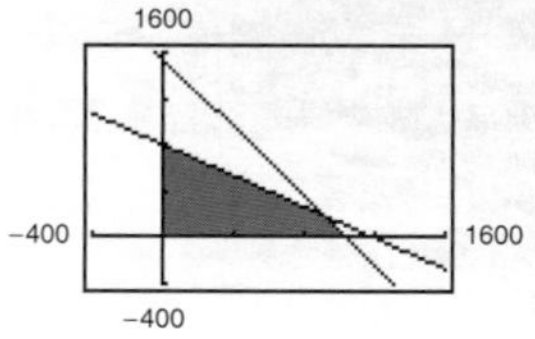

63. Consumer surplus: \$4,000,000

Producer surplus: \$6,000,000

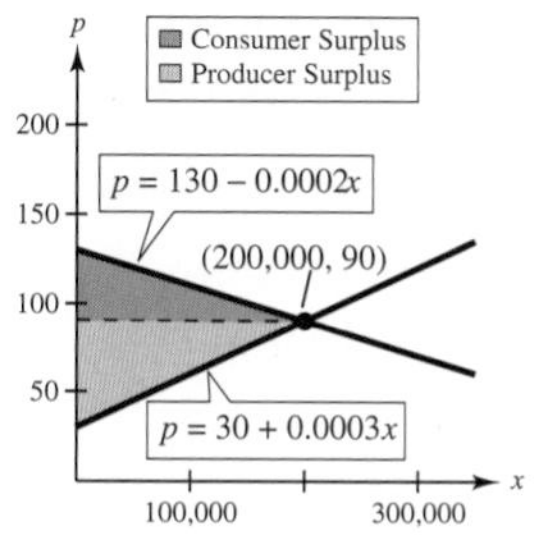

65. Minimum at (25, 50): 600

67. Maximum at (500, 500): 60,000

69. Five units of product A
Two units of product B
Maximum profit: \$138

71. $\frac{1}{3}$ gallon of type A
$\frac{2}{3}$ gallon of type B
Minimum cost: \$1.45

73. False. An objective function can have no maximum value, one point where a maximum value occurs, or an infinite number of points where a maximum value occurs.

75. $\begin{cases} x - y = 9 \\ 3x + y = 11 \end{cases}$ **77.** $\begin{cases} -x + 4y = 10 \\ 3x - 8y = -21 \end{cases}$

79. $\begin{cases} x - 2y + z = -7 \\ 2x + y - 4z = -25 \\ -x + 3y - z = 12 \end{cases}$ **81.** $\begin{cases} 4x + y - z = -7 \\ 8x + 3y + 2z = 16 \\ 4x - 2y + 3z = 31 \end{cases}$

83. For a linear system, the result will be a contradictory equation such as $0 = N$, where N is a nonzero real number. For a nonlinear system, there may be an equation with imaginary roots.

85. (a) One (b) Two (c) Four

Chapter Test *(page 617)*

1. (−3, 4) **2.** (0, −1), (1, 0), (2, 1)

3. (8, 4), (2, −2) **4.** (3, 2) **5.** (−3, 0), (2, 5)

6. (1, 12), (0.034, 8.619) **7.** (1, 5) **8.** (2, −3, 1)

9. $\begin{cases} 3x - y = 9 \\ 6x + y = 3 \end{cases}$

(Answer is not unique.)

10. $y = -\frac{1}{2}x^2 + x + 6$

11. $\begin{cases} 2x - y - 4z = -2 \\ 4x + 3y + 8z = 6 \\ -6x + y - 12z = 24 \end{cases}$

(Answer is not unique.)

12.

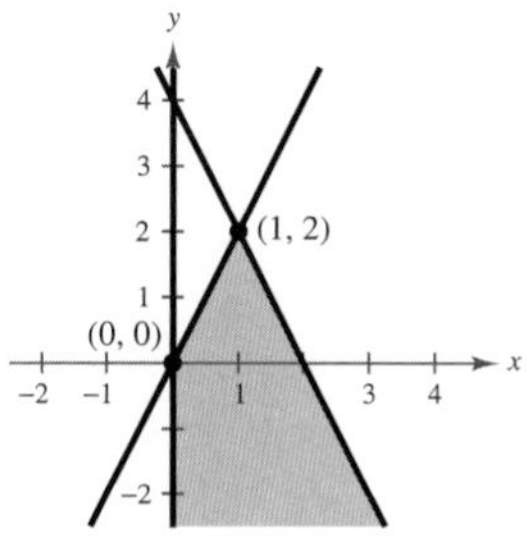

13.

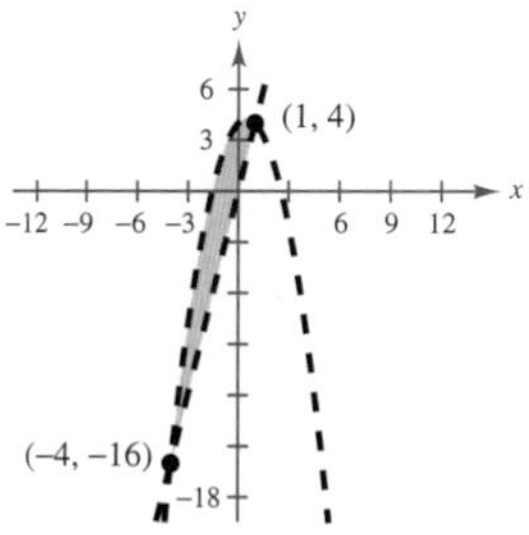

14.

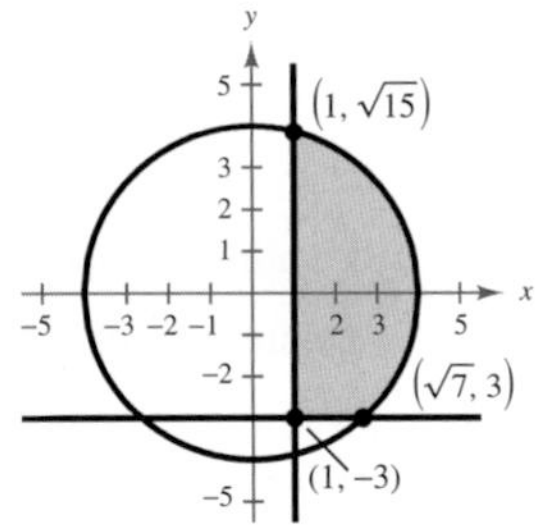

15. $(12, 0)$; $z = 240$

16. \$275 model: 160 units
\$400 model: 140 units
Maximum profit: \$19,300

17. 0 units of model I
5300 units of model II
Maximum profit: \$212,000

Chapter 8

Section 8.1 *(page 630)*

1. 1×2 **3.** 3×1 **5.** 2×2

7. $\begin{bmatrix} 4 & -3 & \vdots & -5 \\ -1 & 3 & \vdots & 12 \end{bmatrix}$ **9.** $\begin{bmatrix} 1 & 10 & -2 & \vdots & 2 \\ 5 & -3 & 4 & \vdots & 0 \\ 2 & 1 & 0 & \vdots & 6 \end{bmatrix}$

11. $\begin{bmatrix} 7 & -5 & 1 & \vdots & 13 \\ 19 & 0 & -8 & \vdots & 10 \end{bmatrix}$

13. $\begin{cases} x + 2y = 7 \\ 2x - 3y = 4 \end{cases}$ **15.** $\begin{cases} 2x + 5z = -12 \\ y - 2z = 7 \\ 6x + 3y = 2 \end{cases}$

17. $\begin{cases} 9x + 12y + 3z = 0 \\ -2x + 18y + 5z + 2w = 10 \\ x + 7y - 8z = -4 \\ 3x + 2z = -10 \end{cases}$

19. Reduced row-echelon form

21. Not in row-echelon form

23. $\begin{bmatrix} 1 & 4 & 3 \\ 0 & 2 & -1 \end{bmatrix}$ **25.** $\begin{bmatrix} 1 & 1 & 4 & -1 \\ 0 & 5 & -2 & 6 \\ 0 & 3 & 20 & 4 \end{bmatrix}$
$\begin{bmatrix} 1 & 1 & 4 & -1 \\ 0 & 1 & -\frac{2}{5} & \frac{6}{5} \\ 0 & 3 & 20 & 4 \end{bmatrix}$

27. Add 5 times Row 2 to Row 1.

29. Interchange Row 1 and Row 2. Add 4 times new Row 1 to Row 3.

31. (a) $\begin{bmatrix} 1 & 2 & 3 \\ 0 & -5 & -10 \\ 3 & 1 & -1 \end{bmatrix}$ (b) $\begin{bmatrix} 1 & 2 & 3 \\ 0 & -5 & -10 \\ 0 & -5 & -10 \end{bmatrix}$

(c) $\begin{bmatrix} 1 & 2 & 3 \\ 0 & -5 & -10 \\ 0 & 0 & 0 \end{bmatrix}$ (d) $\begin{bmatrix} 1 & 2 & 3 \\ 0 & 1 & 2 \\ 0 & 0 & 0 \end{bmatrix}$

(e) $\begin{bmatrix} 1 & 0 & -1 \\ 0 & 1 & 2 \\ 0 & 0 & 0 \end{bmatrix}$

The matrix is in reduced row-echelon form.

33. $\begin{bmatrix} 1 & 1 & 0 & 5 \\ 0 & 1 & 2 & 0 \\ 0 & 0 & 1 & -1 \end{bmatrix}$ **35.** $\begin{bmatrix} 1 & -1 & -1 & 1 \\ 0 & 1 & 6 & 3 \\ 0 & 0 & 0 & 0 \end{bmatrix}$

37. $\begin{bmatrix} 1 & 0 & 0 \\ 0 & 1 & 0 \\ 0 & 0 & 1 \end{bmatrix}$ **39.** $\begin{bmatrix} 1 & 2 & 0 & 0 \\ 0 & 0 & 1 & 0 \\ 0 & 0 & 0 & 1 \\ 0 & 0 & 0 & 0 \end{bmatrix}$

41. $\begin{bmatrix} 1 & 0 & 3 & 16 \\ 0 & 1 & 2 & 12 \end{bmatrix}$

43. $\begin{cases} x - 2y = 4 \\ y = -3 \end{cases}$ **45.** $\begin{cases} x - y + 2z = 4 \\ y - z = 2 \\ z = -2 \end{cases}$

$(-2, -3)$ $(8, 0, -2)$

47. $(3, -4)$ **49.** $(-4, -10, 4)$ **51.** $(3, 2)$

53. $(-5, 6)$ **55.** $(-1, -4)$ **57.** Inconsistent

59. $(4, -3, 2)$ **61.** $(7, -3, 4)$ **63.** $(-4, -3, 6)$

65. $(2a + 1, 3a + 2, a)$

67. $(4 + 5b + 4a, 2 - 3b - 3a, b, a)$ **69.** Inconsistent

71. $(0, 2 - 4a, a)$ **73.** $(1, 0, 4, -2)$

75. $(-2a, a, a, 0)$ **77.** Yes; $(-1, 1, -3)$ **79.** No

81. $\begin{bmatrix} 1 & 3 & \frac{3}{2} & \vdots & 4 \\ 0 & 1 & \frac{7}{4} & \vdots & -\frac{3}{2} \\ 0 & 0 & 1 & \vdots & 2 \end{bmatrix}, \begin{bmatrix} 1 & 3 & 1 & \vdots & 3 \\ 0 & 1 & 2 & \vdots & -1 \\ 0 & 0 & 1 & \vdots & 2 \end{bmatrix}$

83. $\dfrac{4x^2}{(x + 1)^2(x - 1)} = \dfrac{1}{x - 1} + \dfrac{3}{x + 1} - \dfrac{2}{(x + 1)^2}$

85. \$100,000 at 9%
\$250,000 at 10%
\$150,000 at 12%

87. $y = -x^2 + 2x + 8$

89. (a) $y = -128.5t^2 + 1587.5t - 4304.0$

(b)

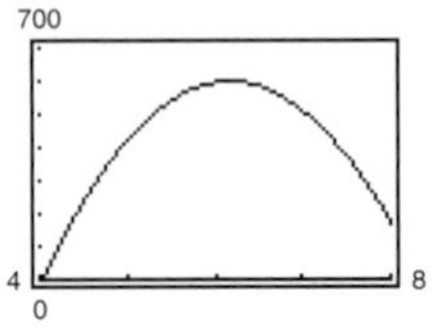

(c) -1279. The estimate is not reasonable because it is a negative number.

91. (a) $x_1 = 500 - s - t, x_2 = -200 + s + t, x_3 = s,$
$x_4 = 350 - t, x_5 = t$

(b) $x_1 = 100, x_2 = 200, x_3 = 50, x_4 = 0, x_5 = 350$

(c) $x_1 = 150, x_2 = 150, x_3 = 0, x_4 = 0, x_5 = 350$

93. False. A matrix is in reduced row-echelon form if (1) all rows consisting entirely of zeros occur at the bottom of the matrix, (2) for each row that does not consist entirely of zeros, the first nonzero entry is 1, (3) for two successive nonzero rows, the leading 1 in the higher row is farther to the left than the leading 1 in the lower row, and (4) every column that has a leading 1 has zeros in every position above and below its leading 1.

95. $\begin{cases} x + y + 7z = -1 \\ x + 2y + 11z = 0 \\ 2x + y + 10z = -3 \end{cases}$

97. Interchange two rows.
Multiply a row by a nonzero constant.
Add a multiple of a row to another row.

99. A matrix in reduced row-echelon form has zeros above the leading 1s. A matrix in row-echelon form may have any real number above the leading 1s.

101.

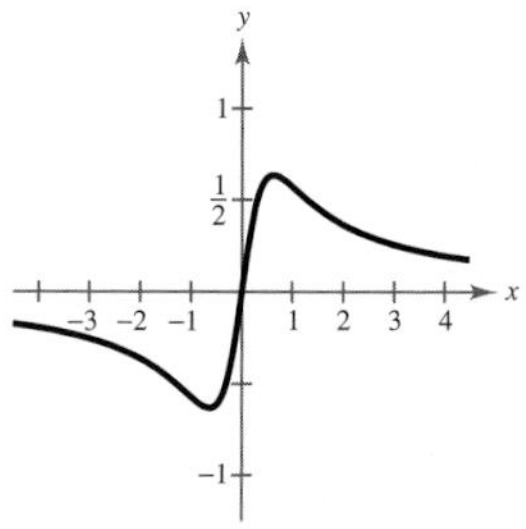

Horizontal asymptote: $y = 0$

103.

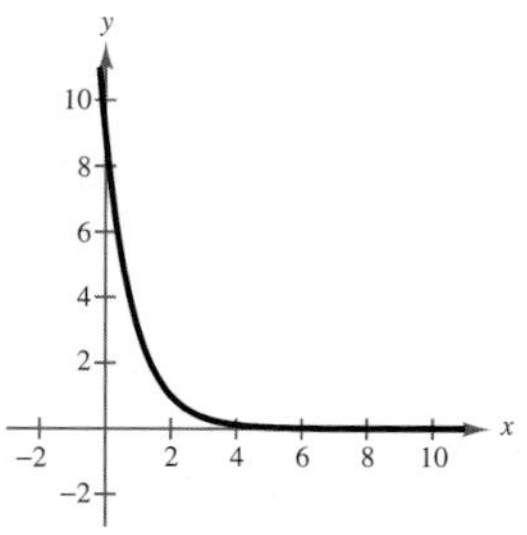

105.

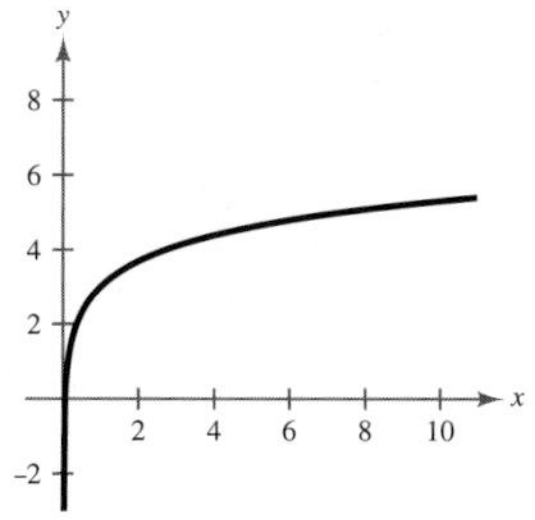

Section 8.2 *(page 645)*

1. $x = -4, y = 22$ **3.** $x = 2, y = 3$

5. (a) $\begin{bmatrix} 3 & -2 \\ 1 & 7 \end{bmatrix}$ (b) $\begin{bmatrix} -1 & 0 \\ 3 & -9 \end{bmatrix}$

(c) $\begin{bmatrix} 3 & -3 \\ 6 & -3 \end{bmatrix}$ (d) $\begin{bmatrix} -1 & -1 \\ 8 & -19 \end{bmatrix}$

7. (a) $\begin{bmatrix} 7 & 3 \\ 1 & 9 \\ -2 & 15 \end{bmatrix}$ (b) $\begin{bmatrix} 5 & -5 \\ 3 & -1 \\ -4 & -5 \end{bmatrix}$ (c) $\begin{bmatrix} 18 & -3 \\ 6 & 12 \\ -9 & 15 \end{bmatrix}$

(d) $\begin{bmatrix} 16 & -11 \\ 8 & 2 \\ -11 & -5 \end{bmatrix}$

9. (a) $\begin{bmatrix} 3 & 3 & -2 & 1 & 1 \\ -2 & 5 & 7 & -6 & -8 \end{bmatrix}$

(b) $\begin{bmatrix} 1 & 1 & 0 & -1 & 1 \\ 4 & -3 & -11 & 6 & 6 \end{bmatrix}$

(c) $\begin{bmatrix} 6 & 6 & -3 & 0 & 3 \\ 3 & 3 & -6 & 0 & -3 \end{bmatrix}$

(d) $\begin{bmatrix} 4 & 4 & -1 & -2 & 3 \\ 9 & -5 & -24 & 12 & 11 \end{bmatrix}$

11. (a), (b), and (d) not possible

(c) $\begin{bmatrix} 18 & 0 & 9 \\ -3 & -12 & 0 \end{bmatrix}$

13. $\begin{bmatrix} -8 & -7 \\ 15 & -1 \end{bmatrix}$ **15.** $\begin{bmatrix} -24 & -4 & 12 \\ -12 & 32 & 12 \end{bmatrix}$

17. $\begin{bmatrix} 10 & 8 \\ -59 & 9 \end{bmatrix}$ **19.** $\begin{bmatrix} -17.143 & 2.143 \\ 11.571 & 10.286 \end{bmatrix}$

21. $\begin{bmatrix} -1.581 & -3.739 \\ -4.252 & -13.249 \\ 9.713 & -0.362 \end{bmatrix}$

23. $\begin{bmatrix} -6 & -9 \\ -1 & 0 \\ 17 & -10 \end{bmatrix}$ **25.** $\begin{bmatrix} 3 & 3 \\ -\frac{1}{2} & 0 \\ -\frac{13}{2} & \frac{11}{2} \end{bmatrix}$

27. (a) $\begin{bmatrix} 0 & 15 \\ 6 & 12 \end{bmatrix}$ (b) $\begin{bmatrix} -2 & 2 \\ 31 & 14 \end{bmatrix}$ (c) $\begin{bmatrix} 9 & 6 \\ 12 & 12 \end{bmatrix}$

29. (a) $\begin{bmatrix} 0 & -10 \\ 10 & 0 \end{bmatrix}$ (b) $\begin{bmatrix} 0 & -10 \\ 10 & 0 \end{bmatrix}$ (c) $\begin{bmatrix} 8 & -6 \\ 6 & 8 \end{bmatrix}$

31. (a) $\begin{bmatrix} 7 & 7 & 14 \\ 8 & 8 & 16 \\ -1 & -1 & -2 \end{bmatrix}$ (b) $[13]$ (c) Not possible

33. Not possible **35.** $\begin{bmatrix} 3 & -4 \\ 10 & 16 \\ 26 & 46 \end{bmatrix}$ **37.** $\begin{bmatrix} 3 & 0 & 0 \\ 0 & -4 & 0 \\ 0 & 0 & -10 \end{bmatrix}$

39. $\begin{bmatrix} 0 & 0 & 0 \\ 0 & 0 & 0 \\ 0 & 0 & 0 \end{bmatrix}$ **41.** $\begin{bmatrix} 41 & 7 & 7 \\ 42 & 5 & 25 \\ -10 & -25 & 45 \end{bmatrix}$

43. $\begin{bmatrix} 151 & 25 & 48 \\ 516 & 279 & 387 \\ 47 & -20 & 87 \end{bmatrix}$ **45.** Not possible

47. $\begin{bmatrix} 5 & 8 \\ -4 & -16 \end{bmatrix}$ **49.** $\begin{bmatrix} -4 & 10 \\ 3 & 14 \end{bmatrix}$

51. (a) $\begin{bmatrix} -1 & 1 \\ -2 & 1 \end{bmatrix}\begin{bmatrix} x_1 \\ x_2 \end{bmatrix} = \begin{bmatrix} 4 \\ 0 \end{bmatrix}$ (b) $\begin{bmatrix} 4 \\ 8 \end{bmatrix}$

53. (a) $\begin{bmatrix} -2 & -3 \\ 6 & 1 \end{bmatrix}\begin{bmatrix} x_1 \\ x_2 \end{bmatrix} = \begin{bmatrix} -4 \\ -36 \end{bmatrix}$ (b) $\begin{bmatrix} -7 \\ 6 \end{bmatrix}$

55. (a) $\begin{bmatrix} 1 & -2 & 3 \\ -1 & 3 & -1 \\ 2 & -5 & 5 \end{bmatrix}\begin{bmatrix} x_1 \\ x_2 \\ x_3 \end{bmatrix} = \begin{bmatrix} 9 \\ -6 \\ 17 \end{bmatrix}$ (b) $\begin{bmatrix} 1 \\ -1 \\ 2 \end{bmatrix}$

57. (a) $\begin{bmatrix} 1 & -5 & 2 \\ -3 & 1 & -1 \\ 0 & -2 & 5 \end{bmatrix}\begin{bmatrix} x_1 \\ x_2 \\ x_3 \end{bmatrix} = \begin{bmatrix} -20 \\ 8 \\ -16 \end{bmatrix}$ (b) $\begin{bmatrix} -1 \\ 3 \\ -2 \end{bmatrix}$

59. Not possible **61.** Not possible **63.** 2×2

65. Not possible **67.** 2×3 **69.** $\begin{bmatrix} 84 & 60 & 30 \\ 42 & 120 & 84 \end{bmatrix}$

71. $BA = [\$1037.50 \quad \$1400 \quad \$1012.50]$

The entries represent the profits from both products at each of the three outlets.

73. $\begin{bmatrix} \$15{,}770 & \$18{,}300 \\ \$26{,}500 & \$29{,}250 \\ \$21{,}260 & \$24{,}150 \end{bmatrix}$

The entries are the wholesale and retail inventory values of the inventories at the three outlets.

75. $\begin{bmatrix} \$18.10 & \$15.40 \\ \$29.80 & \$25.40 \\ \$51.20 & \$43.80 \end{bmatrix}$

The entries are labor costs at each plant for each size of boat.

77. True. The sum of two matrices of different orders is undefined.

79. False.

$$\begin{bmatrix} -2 & 4 \\ -3 & 0 \\ 6 & 1 \end{bmatrix}\begin{bmatrix} 1 & 1 \\ 1 & 1 \end{bmatrix} = \begin{bmatrix} 2 & 2 \\ -3 & -3 \\ 7 & 7 \end{bmatrix}$$

81. $AB = \begin{bmatrix} 0 & 0 \\ 0 & 0 \end{bmatrix}$

$AB = O$ and neither A nor B is O.

83. $A^2 = \begin{bmatrix} 1 & 0 \\ 0 & 1 \end{bmatrix} = I$, the identity matrix.

85. $-8, \dfrac{4}{3}$ **87.** $0, \dfrac{-5 \pm \sqrt{37}}{4}$ **89.** $4, \pm\dfrac{\sqrt{15}}{3}i$

91. $\left(7, -\frac{1}{2}\right)$ **93.** $(3, -1)$

Section 8.3 *(page 655)*

1.–9. $AB = I$ and $BA = I$

11. $\begin{bmatrix} \frac{1}{2} & 0 \\ 0 & \frac{1}{3} \end{bmatrix}$ **13.** $\begin{bmatrix} -3 & 2 \\ -2 & 1 \end{bmatrix}$ **15.** $\begin{bmatrix} 1 & -1 \\ 2 & -1 \end{bmatrix}$

17. Does not exist **19.** Does not exist

21. $\begin{bmatrix} 1 & 1 & -1 \\ -3 & 2 & -1 \\ 3 & -3 & 2 \end{bmatrix}$ **23.** $\begin{bmatrix} 1 & 0 & 0 \\ -\frac{3}{4} & \frac{1}{4} & 0 \\ \frac{7}{20} & -\frac{1}{4} & \frac{1}{5} \end{bmatrix}$

25. $\begin{bmatrix} -\frac{1}{8} & 0 & 0 & 0 \\ 0 & 1 & 0 & 0 \\ 0 & 0 & \frac{1}{4} & 0 \\ 0 & 0 & 0 & -\frac{1}{5} \end{bmatrix}$ **27.** $\begin{bmatrix} -175 & 37 & -13 \\ 95 & -20 & 7 \\ 14 & -3 & 1 \end{bmatrix}$

29. $\begin{bmatrix} -1.5 & 1.5 & 1 \\ 4.5 & -3.5 & -3 \\ -1 & 1 & 1 \end{bmatrix}$ **31.** $\begin{bmatrix} -12 & -5 & -9 \\ -4 & -2 & -4 \\ -8 & -4 & -6 \end{bmatrix}$

33. $\begin{bmatrix} 0 & -1.\overline{81} & 0.\overline{90} \\ -10 & 5 & 5 \\ 10 & -2.\overline{72} & -3.\overline{63} \end{bmatrix}$ **35.** Does not exist

37. $\begin{bmatrix} 1 & 0 & 1 & 0 \\ 0 & 1 & 0 & 1 \\ 2 & 0 & 1 & 0 \\ 0 & 1 & 0 & 2 \end{bmatrix}$ **39.** $\begin{bmatrix} \frac{3}{19} & \frac{2}{19} \\ -\frac{2}{19} & \frac{5}{19} \end{bmatrix}$

41. Does not exist **43.** $\begin{bmatrix} \frac{16}{59} & \frac{15}{59} \\ -\frac{4}{59} & \frac{70}{59} \end{bmatrix}$ **45.** $(5, 0)$

47. $(-8, -6)$ **49.** $(3, 8, -11)$ **51.** $(2, 1, 0, 0)$

53. $(2, -2)$ **55.** No solution **57.** $\left(3, -\frac{1}{2}\right)$

59. $(-4, -8)$ **61.** $(-1, 3, 2)$

63. $\left(\frac{5}{16}a + \frac{13}{16}, \frac{19}{16}a + \frac{11}{16}, a\right)$ **65.** $(2a - 1, -3a + 2, a)$

67. $(5, 0, -2, 3)$

69. \$7000 in AAA-rated bonds
\$1000 in A-rated bonds
\$2000 in B-rated bonds

71. \$9000 in AAA-rated bonds
\$1000 in A-rated bonds
\$2000 in B-rated bonds

73. $I_1 = -3$ amperes
$I_2 = 8$ amperes
$I_3 = 5$ amperes

75. True. If B is the inverse of A, then $AB = I = BA$.

77. True. If A is of order $m \times n$ and B is of order $n \times m$ (where $m \neq n$), the products AB and BA are of different orders and so cannot be equal to each other.

79. The inverse matrix can be calculated once and used for more than one exercise.

81. $x = \dfrac{2 \ln 315}{\ln 3} \approx 10.47$ **83.** $x = 2^{6.5} \approx 90.51$

85. $\begin{bmatrix} 12 & -18 \\ -6 & 24 \\ -3 & -36 \end{bmatrix}$ **87.** $\begin{bmatrix} 6 & -1 \\ -27 & 19 \end{bmatrix}$

Section 8.4 *(page 663)*

1. 5 **3.** 5 **5.** 27 **7.** 0 **9.** 6 **11.** -9

13. 72 **15.** $\frac{11}{6}$ **17.** -0.002 **19.** -4.842 **21.** 0

23. (a) $M_{11} = -5, M_{12} = 2, M_{21} = 4, M_{22} = 3$
(b) $C_{11} = -5, C_{12} = -2, C_{21} = -4, C_{22} = 3$

25. (a) $M_{11} = -4, M_{12} = -2, M_{21} = 1, M_{22} = 3$
(b) $C_{11} = -4, C_{12} = 2, C_{21} = -1, C_{22} = 3$

27. (a) $M_{11} = 3, M_{12} = -4, M_{13} = 1, M_{21} = 2, M_{22} = 2, M_{23} = -4, M_{31} = -4, M_{32} = 10, M_{33} = 8$
(b) $C_{11} = 3, C_{12} = 4, C_{13} = 1, C_{21} = -2, C_{22} = 2, C_{23} = 4, C_{31} = -4, C_{32} = -10, C_{33} = 8$

29. (a) $M_{11} = 30, M_{12} = 12, M_{13} = 11, M_{21} = -36, M_{22} = 26, M_{23} = 7, M_{31} = -4, M_{32} = -42, M_{33} = 12$
(b) $C_{11} = 30, C_{12} = -12, C_{13} = 11, C_{21} = 36, C_{22} = 26, C_{23} = -7, C_{31} = -4, C_{32} = 42, C_{33} = 12$

31. (a) -75 (b) -75 **33.** (a) 96 (b) 96

35. (a) 170 (b) 170 **37.** 0 **39.** 0

41. -9 **43.** -58 **45.** -30 **47.** -168 **49.** 0

51. 412 **53.** -126 **55.** 0 **57.** -336 **59.** 410

61. (a) -3 (b) -2 (c) $\begin{bmatrix} -2 & 0 \\ 0 & -3 \end{bmatrix}$ (d) 6

63. (a) -8 (b) 0 (c) $\begin{bmatrix} -4 & 4 \\ 1 & -1 \end{bmatrix}$ (d) 0

65. (a) -21 (b) -19 (c) $\begin{bmatrix} 7 & 1 & 4 \\ -8 & 9 & -3 \\ 7 & -3 & 9 \end{bmatrix}$ (d) 399

67. (a) 2 (b) -6 (c) $\begin{bmatrix} 1 & 4 & 3 \\ -1 & 0 & 3 \\ 0 & 2 & 0 \end{bmatrix}$ (d) -12

69.–73. Answers will vary. **75.** $-1, 4$ **77.** $-1, -4$

79. $8uv - 1$ **81.** e^{5x} **83.** $1 - \ln x$

85. True. If an entire row is zero, then each cofactor in the expansion is multiplied by zero.

87. Answers will vary.

89. A square matrix is a square array of numbers. The determinant of a square matrix is a real number.

91.

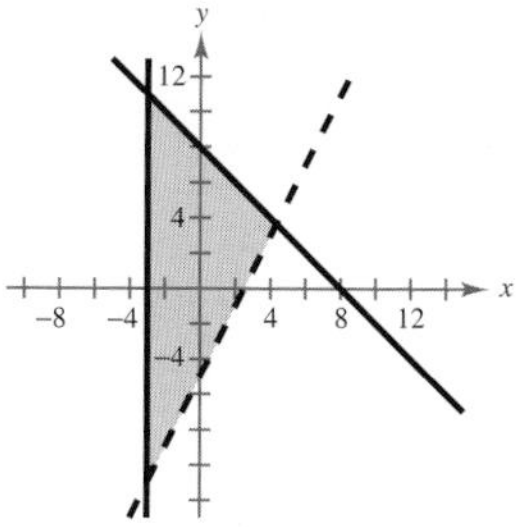

93. $\begin{bmatrix} \frac{1}{4} & \frac{1}{4} \\ 2 & 1 \end{bmatrix}$ **95.** Does not exist

Section 8.5 *(page 675)*

1. $(2, -2)$ **3.** $\left(\frac{32}{7}, \frac{30}{7}\right)$ **5.** $(-1, 3, 2)$

7. $(-2, 1, -1)$ **9.** $\left(0, -\frac{1}{2}, \frac{1}{2}\right)$ **11.** $(1, 2, 1)$ **13.** 7

15. 14 **17.** $\frac{33}{8}$ **19.** $\frac{5}{2}$ **21.** 28 **23.** $\frac{16}{5}$ or 0

25. -3 or -11 **27.** 250 square miles **29.** Collinear

31. Not collinear **33.** Collinear **35.** $x = -3$

37. $3x - 5y = 0$ **39.** $x + 3y - 5 = 0$

41. $2x + 3y - 8 = 0$

43. Uncoded: [20 18 15], [21 2 12], [5 0 9], [14 0 18], [9 22 5], [18 0 3], [9 20 25]
Encoded: -52 10 27 -49 3 34 -49 13 27 -94 22 54 1 1 -7 0 -12 9 -121 41 55

45. -6 -35 -69 11 20 17 6 -16 -58 46 79 67

47. -5 -41 -87 91 207 257 11 -5 -41 40 80 84 76 177 227

49. HAPPY NEW YEAR **51.** CLASS IS CANCELED

53. SEND PLANES **55.** MEET ME TONIGHT RON

57. True. If the determinant of the coefficient matrix is zero, then the solution of the system would result in division by zero, which is undefined.

59. Answers will vary. **61.** $\left(5, -\frac{1}{2}\right)$ **63.** $(2, -2, 5)$

65.

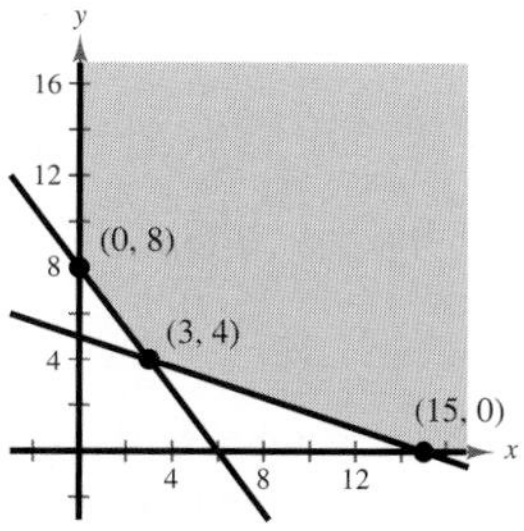

Minimum at (3, 4): 46
No maximum

67. -13.78 **69.** -115

Review Exercises *(page 679)*

1. 3×1 **3.** 1×1 **5.** $\left[\begin{array}{rr|r} 3 & -10 & 15 \\ 5 & 4 & 22 \end{array}\right]$

7. $\begin{cases} 5x + y + 7z = -9 \\ 4x + 2y = 10 \\ 9x + 4y + 2z = 3 \end{cases}$

9. $\begin{bmatrix} 1 & 0 & 0 \\ 0 & 1 & 0 \\ 0 & 0 & 1 \end{bmatrix}$

11. Consistent, infinitely many solutions

13. Consistent, one solution **15.** $(10, -12)$

17. $(-0.2, 0.7)$ **19.** $(5, 2, -6)$

21. $\left(-2a + \frac{3}{2}, 2a + 1, a\right)$ **23.** $(1, 0, 4, 3)$

25. $\dfrac{8}{x+1} - \dfrac{8}{x+2} - \dfrac{7}{(x+2)^2}$ **27.** $(2, -3, 3)$

29. $(2, 3, -1)$ **31.** $(2, 6, -10, -3)$

33. $x = 12, y = -7$ **35.** $x = 1, y = 11$ **37.** Yes

39. Not possible because matrices are not of the same order.

41. $\begin{bmatrix} 17 & -17 \\ 13 & 2 \end{bmatrix}$ **43.** $\begin{bmatrix} 54 & 4 \\ -2 & 24 \\ -4 & 32 \end{bmatrix}$

45. $\begin{bmatrix} 48 & -18 & -3 \\ 15 & 51 & 33 \end{bmatrix}$ **47.** $\begin{bmatrix} -14 & -4 \\ 7 & -17 \\ -17 & -2 \end{bmatrix}$

49. $\frac{1}{3}\begin{bmatrix} 9 & 2 \\ -4 & 11 \\ 10 & 0 \end{bmatrix}$ **51.** Yes **53.** Yes

55. $\begin{bmatrix} 14 & -2 & 8 \\ 14 & -10 & 40 \\ 36 & -12 & 48 \end{bmatrix}$ **57.** $\begin{bmatrix} 44 & 4 \\ 20 & 8 \end{bmatrix}$

59. $\begin{bmatrix} 24 & -8 \\ 36 & -12 \end{bmatrix}$ **61.** $\begin{bmatrix} 1 & 17 \\ 12 & 36 \end{bmatrix}$

63. $\begin{bmatrix} 14 & -22 & 22 \\ 19 & -41 & 80 \\ 42 & -66 & 66 \end{bmatrix}$

65. $\begin{cases} 5x + 4y = 2 \\ -x + y = -22 \end{cases}$

67. (a) [\$274,150 \$303,150]

The merchandise shipped to warehouse 1 is worth \$274,150 and the merchandise shipped to warehouse 2 is worth \$303,150.

(b) $A_n = \begin{bmatrix} 10{,}250 & 9{,}250 \\ 8{,}125 & 12{,}250 \\ 6{,}750 & 6{,}000 \end{bmatrix}$

$BA_n =$ [\$342,687.5 \$378,937.5]

69. and 71. $AB = I$ and $BA = I$

73. $\begin{bmatrix} 3 & 5 \\ -2 & -3 \end{bmatrix}$ **75.** $\begin{bmatrix} 1 & 11 & 8 \\ -1 & -7 & -5 \\ -1 & -14 & -10 \end{bmatrix}$

77. $\begin{bmatrix} 0.0434\ldots & 0.2173\ldots \\ -0.0869\ldots & 0.0652\ldots \end{bmatrix}$ **79.** Does not exist

81. $\begin{bmatrix} \frac{3}{2} & -2 \\ -\frac{7}{2} & 5 \end{bmatrix}$ **83.** $\begin{bmatrix} -\frac{2}{3} & -\frac{5}{8} \\ \frac{1}{5} & -\frac{3}{16} \end{bmatrix}$ **85.** $(2, -3)$

87. $(-2, 1)$ **89.** $(-2, 4, 3)$ **91.** $(-3, 5, 0)$

93. $(5, 6)$ **95.** $(4, -2, 1)$ **97.** -41 **99.** 78

101. (a) $M_{11} = -4, M_{12} = 5, M_{21} = 6, M_{22} = 3$

(b) $C_{11} = -4, C_{12} = -5, C_{21} = -6, C_{22} = 3$

103. (a) $M_{11} = 19, M_{12} = -24, M_{13} = 26, M_{21} = 2,$ $M_{22} = 32, M_{23} = 20, M_{31} = -47, M_{32} = -96,$ $M_{33} = 22$

(b) $C_{11} = 19, C_{12} = 24, C_{13} = 26, C_{21} = -2,$ $C_{22} = 32, C_{23} = -20, C_{31} = -47, C_{32} = 96,$ $C_{33} = 22$

105. -117 **107.** -255 **109.** $(3, -2)$

111. $(6, 8, 1)$

113. 40 liters of 75% solution; 60 liters of 50% solution

115. 16,667 units **117.** 24 **119.** $\frac{25}{8}$

121. $3x + 2y - 16 = 0$ **123.** $10x - 5y + 9 = 0$

125. Uncoded: [18 5 20], [21 18 14], [0 20 15], [0 2 1], [19 5 0]

Encoded: 66 28 10 −24 −59 −22 −75 −90 −25 −9 −10 −3 8 −11 −10

127. MAY THE FORCE BE WITH YOU

129. True. Answers will vary.

131. If A is a square matrix, the cofactor C_{ij} of the entry a_{ij} is $(-1)^{i+j}M_{ij}$, where M_{ij} is the determinant obtained by deleting the ith row and jth column of A. The determinant of A is the sum of the entries of any row or column of A multiplied by their respective cofactors.

133. The part of the matrix corresponding to the coefficients of the system reduces to a matrix in which the number of rows with nonzero entries is the same as the number of variables.

Chapter Test *(page 685)*

1. $\begin{bmatrix} 1 & 0 & 0 \\ 0 & 1 & 0 \\ 0 & 0 & 1 \end{bmatrix}$ **2.** $\begin{bmatrix} 1 & 0 & -1 & 2 \\ 0 & 1 & 0 & -1 \\ 0 & 0 & 0 & 0 \\ 0 & 0 & 0 & 0 \end{bmatrix}$

3. $\left[\begin{array}{ccc:c} 4 & 3 & -2 & 14 \\ -1 & -1 & 2 & -5 \\ 3 & 1 & -4 & 8 \end{array}\right]$, $\left(1, 3, -\frac{1}{2}\right)$

4. $y = -\frac{1}{2}x^2 + x + 2$

5. (a) $\begin{bmatrix} 1 & 5 \\ 0 & -4 \end{bmatrix}$ (b) $\begin{bmatrix} 15 & 12 \\ -12 & -12 \end{bmatrix}$

(c) $\begin{bmatrix} 7 & 14 \\ -4 & -12 \end{bmatrix}$ (d) $\begin{bmatrix} 4 & -5 \\ 0 & 4 \end{bmatrix}$

6. $\begin{bmatrix} \frac{1}{2} & \frac{2}{5} \\ 1 & \frac{3}{5} \end{bmatrix}$ **7.** $\begin{bmatrix} -\frac{5}{2} & 4 & -3 \\ 5 & -7 & 6 \\ 4 & -6 & 5 \end{bmatrix}$

8. (13, 22) **9.** −196 **10.** 29 **11.** (−3, 5)

12. (−2, 4, 6) **13.** 7

14. Uncoded: [11 14 15], [3 11 0], [15 14 0], [23 15 15], [4 0 0]

Encoded: 115 −41 −59 14 −3 −11 29 −15 −14 128 −53 −60 4 −4 0

15. 75 liters of 60% solution

25 liters of 20% solution

Chapter 9

Section 9.1 *(page 695)*

1. 4, 7, 10, 13, 16 **3.** 2, 4, 8, 16, 32

5. −2, 4, −8, 16, −32 **7.** 3, 2, $\frac{5}{3}$, $\frac{3}{2}$, $\frac{7}{5}$

9. 3, $\frac{12}{11}$, $\frac{9}{13}$, $\frac{24}{47}$, $\frac{15}{37}$ **11.** 0, 1, 0, $\frac{1}{2}$, 0 **13.** $\frac{5}{3}$, $\frac{17}{9}$, $\frac{53}{27}$, $\frac{161}{81}$, $\frac{485}{243}$

15. 1, $\frac{1}{2^{3/2}}$, $\frac{1}{3^{3/2}}$, $\frac{1}{8}$, $\frac{1}{5^{3/2}}$ **17.** 3, $\frac{9}{2}$, $\frac{9}{2}$, $\frac{27}{8}$, $\frac{81}{40}$

19. −1, $\frac{1}{4}$, $-\frac{1}{9}$, $\frac{1}{16}$, $-\frac{1}{25}$ **21.** $\frac{2}{3}$, $\frac{2}{3}$, $\frac{2}{3}$, $\frac{2}{3}$, $\frac{2}{3}$

23. 0, 0, 6, 24, 60 **25.** −73

27. 0.000282 **29.** $\frac{44}{239}$

31.

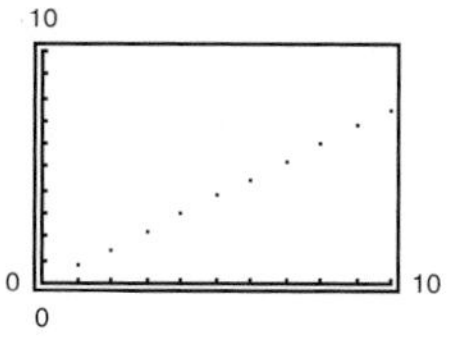

33.

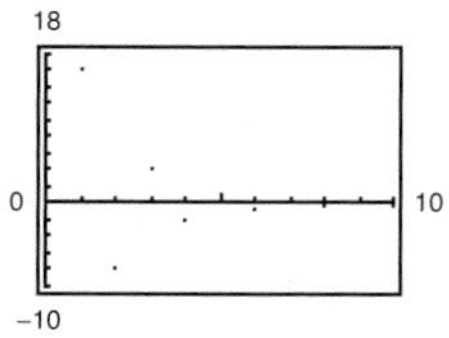

35.

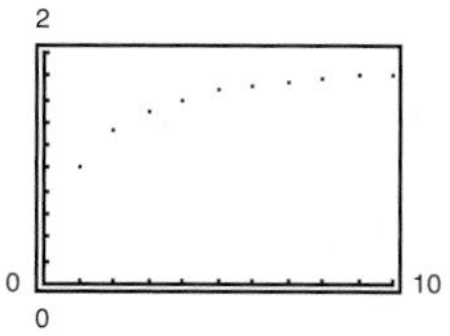

37. c **39.** d **41.** $a_n = 3n - 2$ **43.** $a_n = n^2 - 1$

45. $\frac{(-1)^n(n+1)}{n+2}$ **47.** $a_n = \frac{n+1}{2n-1}$ **49.** $a_n = \frac{1}{n^2}$

51. $a_n = (-1)^{n+1}$ **53.** $a_n = 1 + \frac{1}{n}$

55. 28, 24, 20, 16, 12 **57.** 3, 4, 6, 10, 18

59. 6, 8, 10, 12, 14 **61.** 81, 27, 9, 3, 1

$a_n = 2n + 4$ $a_n = \frac{243}{3^n}$

63. $\frac{1}{30}$ **65.** 90 **67.** $n + 1$ **69.** $\frac{1}{2n(2n+1)}$

71. 35 **73.** 40 **75.** 30 **77.** $\frac{9}{5}$ **79.** 88

81. 30 **83.** 81 **85.** $\frac{47}{60}$

87. $\sum_{i=1}^{9} \frac{1}{3i}$ **89.** $\sum_{i=1}^{8} \left[2\left(\frac{i}{8}\right) + 3\right]$ **91.** $\sum_{i=1}^{6} (-1)^{i+1}3^i$

93. $\sum_{i=1}^{20} \frac{(-1)^{i+1}}{i^2}$ **95.** $\sum_{i=1}^{5} \frac{2^i - 1}{2^{i+1}}$ **97.** $\frac{75}{16}$ **99.** $-\frac{3}{2}$

101. $\frac{2}{3}$ **103.** $\frac{7}{9}$

105. (a) A_1 = \$5100.00, A_2 = \$5202.00, A_3 = \$5306.04, A_4 = \$5412.16, A_5 = \$5520.40, A_6 = \$5630.81, A_7 = \$5743.43, A_8 = \$5858.30

(b) \$11,040.20

107. a_{-1} = \$627.23, a_0 = \$696.00, a_1 = \$760.03, a_2 = \$819.32, a_3 = \$873.87, a_4 = \$923.68, a_5 = \$968.75, a_6 = \$1009.08

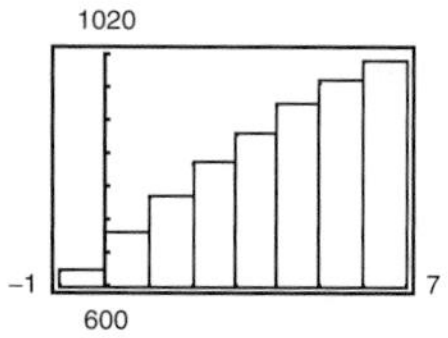

109. \$23,660.88 million

111. 1, 1, 2, 3, 5, 8, 13, 21, 34, 55, 89, 144

1, 2, $\frac{3}{2}$, $\frac{5}{3}$, $\frac{8}{5}$, $\frac{13}{8}$, $\frac{21}{13}$, $\frac{34}{21}$, $\frac{55}{34}$, $\frac{89}{55}$

113. \$500.95 **115.** Answers will vary.

117. True by the Properties of Sums

119. (a) $\begin{bmatrix} 8 & 1 \\ -3 & 7 \end{bmatrix}$ (b) $\begin{bmatrix} -26 & 1 \\ 15 & -24 \end{bmatrix}$

(c) $\begin{bmatrix} 18 & 9 \\ 18 & 0 \end{bmatrix}$ (d) $\begin{bmatrix} 0 & 6 \\ 27 & 18 \end{bmatrix}$

121. (a) $\begin{bmatrix} -3 & -7 & 4 \\ 4 & 4 & 1 \\ 1 & 4 & 3 \end{bmatrix}$ (b) $\begin{bmatrix} 10 & 25 & -10 \\ -12 & -11 & 3 \\ -3 & -9 & -8 \end{bmatrix}$

(c) $\begin{bmatrix} -2 & 7 & -16 \\ 4 & 42 & 45 \\ 1 & 23 & 48 \end{bmatrix}$ (d) $\begin{bmatrix} 16 & 31 & 42 \\ 10 & 47 & 31 \\ 13 & 22 & 25 \end{bmatrix}$

123. 26 **125.** −194

Section 9.2 *(page 705)*

1. Arithmetic sequence, $d = -2$

3. Not an arithmetic sequence

5. Arithmetic sequence, $d = -\frac{1}{4}$

7. Not an arithmetic sequence

9. Not an arithmetic sequence

11. 8, 11, 14, 17, 20

Arithmetic sequence, $d = 3$

13. $7, 3, -1, -5, -9$

Arithmetic sequence, $d = -4$

15. $-1, 1, -1, 1, -1,$

Not an arithmetic sequence

17. $-3, \frac{3}{2}, -1, \frac{3}{4}, -\frac{3}{5}$

Not an arithmetic sequence

19. 15, 19, 23, 27, 31; $d = 4$; $a_n = 4n + 11$

21. 200, 190, 180, 170, 160; $d = -10$; $a_n = -10n + 210$

23. $\frac{5}{8}, \frac{1}{2}, \frac{3}{8}, \frac{1}{4}, \frac{1}{8}$; $d = -\frac{1}{8}$; $a_n = -\frac{1}{8}n + \frac{3}{4}$

25. 5, 11, 17, 23, 29 **27.** $-2.6, -3.0, -3.4, -3.8, -4.2$

29. 2, 6, 10, 14, 18 **31.** $-2, 2, 6, 10, 14$

33. $a_n = 3n - 2$ **35.** $a_n = -8n + 108$

37. $a_n = 2xn - x$ **39.** $a_n = -\frac{5}{2}n + \frac{13}{2}$

41. $a_n = \frac{10}{3}n + \frac{5}{3}$ **43.** $a_n = -3n + 103$

45. b **47.** c

49.

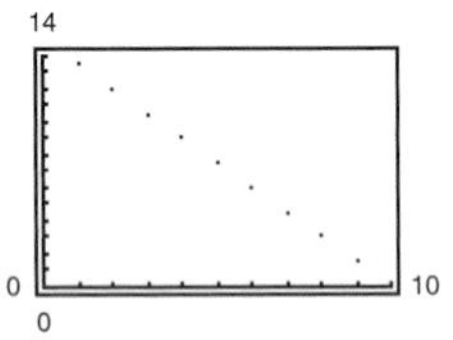

51.

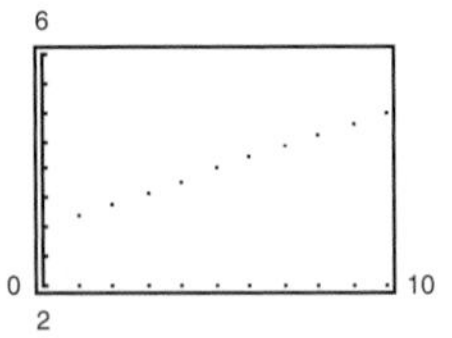

53. 620 **55.** 17.4 **57.** 265 **59.** 4000

61. 1275 **63.** 30,030 **65.** 355 **67.** 160,000

69. 520 **71.** 2725 **73.** 10,120

75. (a) \$40,000 (b) \$217,500

77. 2340 seats **79.** 405 bricks **81.** 490 meters

83. True. Given a_1 and a_2, $d = a_2 - a_1$ and $a_n = a_1 + (n - 1)d$.

85. Because $a_n = dn + c$, the geometric pattern is linear.

87. 10,000

89. (a) 4, 9, 16, 25, 36 (b) n^2

(c) $\frac{n}{2}[1 + (2n - 1)] = n^2$

91. 4 **93.** $\left(\frac{3}{4}a, -\frac{3}{2}a, a\right)$ **95.** (0, 0, 0)

97. (4, 1) **99.** $(1, -2, 3)$

Section 9.3 *(page 714)*

1. Geometric sequence, $r = 3$

3. Not a geometric sequence

5. Geometric sequence, $r = -\frac{1}{2}$

7. Geometric sequence, $r = 2$

9. Not a geometric sequence

11. 2, 6, 18, 54, 162 **13.** $1, \frac{1}{2}, \frac{1}{4}, \frac{1}{8}, \frac{1}{16}$

15. $5, -\frac{1}{2}, \frac{1}{20}, -\frac{1}{200}, \frac{1}{2000}$ **17.** $1, e, e^2, e^3, e^4$

19. $2, \frac{x}{2}, \frac{x^2}{8}, \frac{x^3}{32}, \frac{x^4}{128}$

21. 64, 32, 16, 8, 4; $r = \frac{1}{2}$; $a_n = 128\left(\frac{1}{2}\right)^n$

23. 7, 14, 28, 56, 112; $r = 2$; $a_n = \frac{7}{2}(2)^n$

25. $6, -9, \frac{27}{2}, -\frac{81}{4}, \frac{243}{8}$; $r = -\frac{3}{2}$; $a_n = -4\left(-\frac{3}{2}\right)^n$

27. $\frac{1}{128}$ **29.** $-\frac{2}{3^{10}}$ **31.** $100e^{8x}$ **33.** 1082.372

35. 9 **37.** -2 **39.** a **41.** b

43.

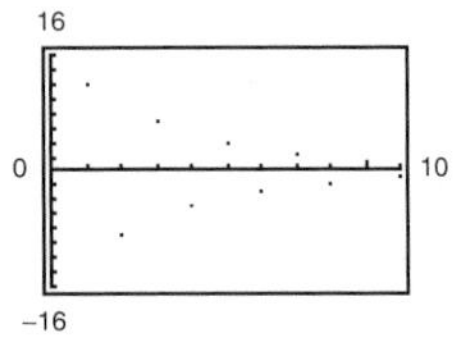

45.

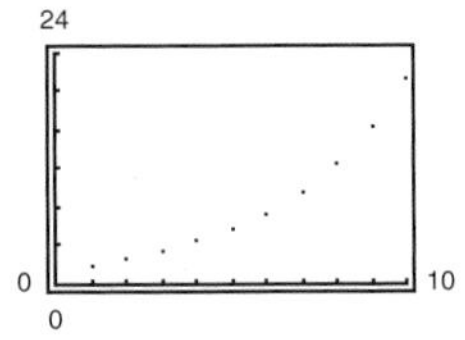

47. 511 **49.** 43 **51.** 29,921.311 **53.** 2092.596

55. 6.400 **57.** $\sum_{n=1}^{7} 5(3)^{n-1}$ **59.** $\sum_{n=1}^{7} 2\left(-\frac{1}{4}\right)^{n-1}$

61. $\sum_{n=1}^{6} 0.1(4)^{n-1}$ **63.** 2 **65.** $\frac{2}{3}$ **67.** $\frac{16}{3}$ **69.** $\frac{5}{3}$

71. -30 **73.** 32 **75.** Undefined **77.** $\frac{4}{11}$ **79.** $\frac{7}{22}$

81.

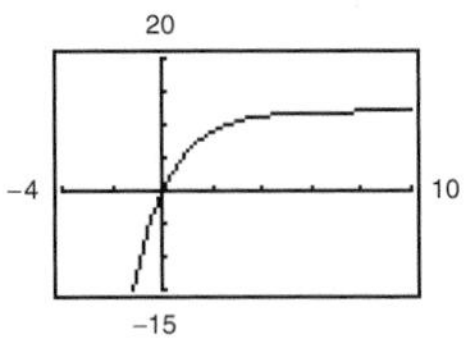

Horizontal asymptote: $y = 12$

Corresponds to the sum of the series

83. (a) \$1790.85

(b) \$1806.11

(c) \$1814.02

(d) \$1819.40

(e) \$1822.03

85. \$22,689.45 **87.** \$7011.89

89. Answers will vary.

91. (a) \$26,198.27

(b) \$26,263.88

93. (a) \$637,678.02

(b) \$645,861.43

95. Answers will vary. **97.** 126 square inches

99. (a) \$5,368,709.11

(b) \$10,737,418.23

(c) \$21,474,836.47

101. (a) 152.42 feet

(b) 19 seconds

103. False. $a_n = a_1r^{n-1}$

105. Given a_1 and a_2, $r = a_2/a_1$ and $a_n = a_1r^{n-1}$.

107. $3x + 4$ **109.** $9x^2 + 24x + 15$

111. Does not factor **113.** $4x^2(2 + x)(2 - x)$

115. $\frac{1}{3}, x \neq 2, -7$ **117.** $1, x \neq 3, 5$

119. $\dfrac{7x^2 + 21x - 53}{(x - 1)(x + 4)}$

Section 9.4 *(page 726)*

1. $\dfrac{5}{(k + 1)(k + 2)}$ **3.** $\dfrac{(k + 1)^2(k + 2)^2}{4}$

5.–17. Answers will vary.

19. 120 **21.** 91 **23.** 979 **25.** 70 **27.** -3402

29. $S_n = n(2n - 1)$ **31.** $S_n = 10 - 10\left(\frac{9}{10}\right)^n$

33. $S_n = \dfrac{n}{2(n + 1)}$ **35.–49.** Answers will vary.

51. 10, 40, 160, 640, 2560 **53.** 0, 2, 2, 6, 10

55. 2, 0, 3, 1, 4

First differences: $-2, 3, -2, 3$

Second differences: $5, -5, 5$

Neither

57. $\frac{3}{2}, -3, 6, -12, 24$

First differences: $-\frac{9}{2}, 9, -18, 36$

Second differences: $\frac{27}{2}, -27, 54$

Neither

59. 2, 4, 16, 256, 65,536

First differences: 2, 12, 240, 65,280

Second differences: 10, 228, 65,040

Neither

61. 0, 4, 10, 18, 28

First differences: 4, 6, 8, 10

Second differences: 2, 2, 2

Quadratic

63. $0, -1, -2, -3, -4$

First differences: $-1, -1, -1, -1$

Second differences: 0, 0, 0

Linear

65. $a_n = n^2 - 2n + 7$ **67.** $a_n = \frac{7}{4}n^2 - 5n + 3$

69. False. P_1 must be proven to be true.

71. False. A sequence with n terms has $n - 2$ second differences.

73. (a) P_n is true for integers $n \geq 3$.

(b) P_n is true for integers $1 \leq n \leq 50$.

(c) P_1, P_2, and P_3 are true.

(d) P_{2n} is true for any positive integer n.

75. (8, 2) and (0, 0) **77.** $(4, -1, 3)$

79. $4x^2 - 4xy + y^2$ **81.** $8x^3 - 48x^2y + 96xy^2 - 64y^3$

Section 9.5 *(page 733)*

1. 10 **3.** 1 **5.** 15,504 **7.** 210 **9.** 4950

11. 56 **13.** 35 **15.** $x^4 + 4x^3 + 6x^2 + 4x + 1$

17. $a^4 + 24a^3 + 216a^2 + 864a + 1296$

19. $y^3 - 12y^2 + 48y - 64$

21. $x^5 + 5x^4y + 10x^3y^2 + 10x^2y^3 + 5xy^4 + y^5$

23. $r^6 + 18r^5s + 135r^4s^2 + 540r^3s^3 + 1215r^2s^4 + 1458rs^5 + 729s^6$

25. $243a^5 - 405a^4b + 270a^3b^2 - 90a^2b^3 + 15ab^4 - b^5$

27. $1 - 6x + 12x^2 - 8x^3$

29. $x^8 + 20x^6 + 150x^4 + 500x^2 + 625$

31. $\dfrac{1}{x^5} + \dfrac{5y}{x^4} + \dfrac{10y^2}{x^3} + \dfrac{10y^3}{x^2} + \dfrac{5y^4}{x} + y^5$

33. $2x^4 - 24x^3 + 113x^2 - 246x + 207$

35. $32t^5 - 80t^4s + 80t^3s^2 - 40t^2s^3 + 10ts^4 - s^5$

37. $x^5 + 10x^4y + 40x^3y^2 + 80x^2y^3 + 80xy^4 + 32y^5$

39. 1,732,104 **41.** 180 **43.** $-326,592$ **45.** 210

47. $x^2 + 12x^{3/2} + 54x + 108x^{1/2} + 81$

49. $x^2 - 3x^{4/3}y^{1/3} + 3x^{2/3}y^{2/3} - y$ **51.** $3x^2 + 3xh + h^2$

53. $\dfrac{1}{\sqrt{x + h} + \sqrt{x}}$ **55.** -4 **57.** $2035 + 828i$

59. 1 **61.** 1.172 **63.** 510,568.785

65.

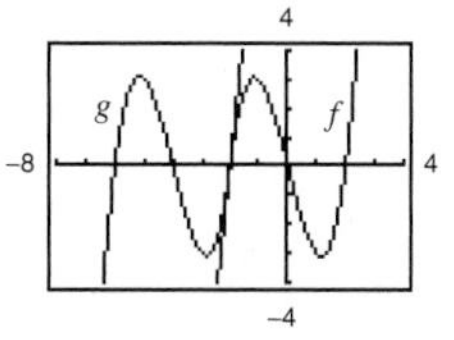

g is shifted 4 units to the left of f.

$g(x) = x^3 + 12x^2 + 44x + 48$

67.

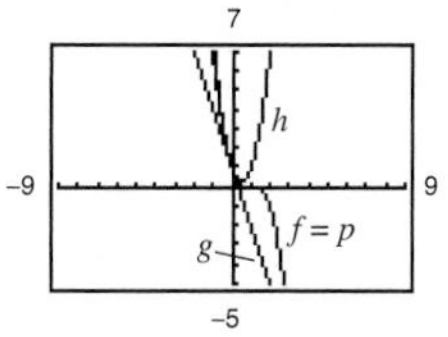

$p(x)$ is the expansion of $f(x)$.

69. 0.2503 **71.** 0.273

73. (a) $g(t) = 0.0834t^2 + 0.74377t + 7.00868$

(b)

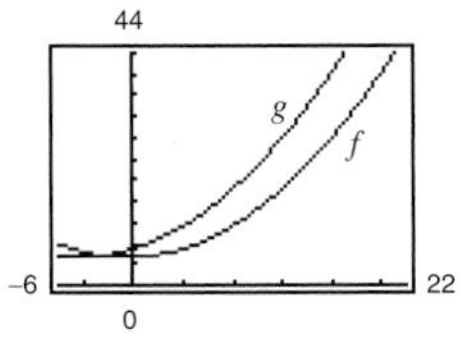

75. False. Expanding binomials that represent differences is accurate. The coefficients have alternating signs.

77.

1
1 1
1 2 1
1 3 3 1
1 4 6 4 1
1 5 10 10 5 1
1 6 15 20 15 6 1
1 7 21 35 35 21 7 1
1 8 28 56 70 56 28 8 1

79. The signs of the terms in the expansion of $(x - y)^n$ alternate from positive to negative.

81. and 83. Answers will vary.

85. $g(x)$ is shifted 3 units to the right of $f(x)$.

87. $g(x)$ is the reflection of $f(x)$ in the x-axis.

89.

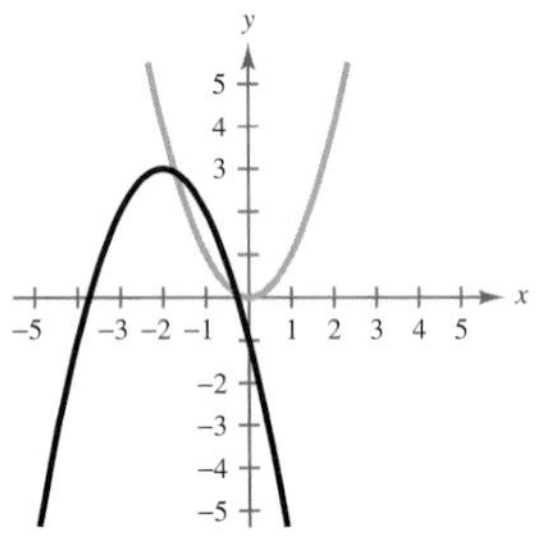

$g(x) = -(x + 2)^2 + 3$

91.

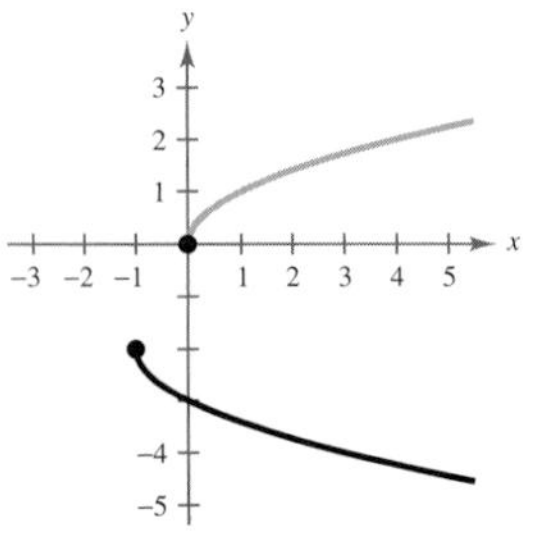

$g(x) = -\sqrt{x + 1} - 2$

Section 9.6 *(page 743)*

1. 6 **3.** 5 **5.** 3 **7.** 7 **9.** 30 **11.** 15

13. 64 **15.** 4 **17.** 175,760,000

19. (a) 900 (b) 648 (c) 180 (d) 600

21. 64,000 **23.** (a) 40,320 (b) 384

25. 24 **27.** 336 **29.** 120 **31.** $n = 5$ or $n = 6$

33. 1,860,480 **35.** 970,200 **37.** 15,504

39. 420 **41.** 2520

43. ABCD, ABDC, ACBD, ACDB, ADBC, ADCB, BACD, BADC, CABD, CADB, DABC, DACB, BCAD, BDAC, CBAD, CDAB, DBAC, DCAB, BCDA, BDCA, CBDA, CDBA, DBCA, DCBA

45. AB, AC, AD, AE, AF, BC, BD, BE, BF, CD, CE, CF, DE, DF, EF

47. 120 **49.** 11,880 **51.** 15,504 **53.** 3,838,380

55. 3,921,225 **57.** 21

59. (a) 70 (b) 30

61. (a) 70 (b) 54 (c) 16

63. 5 **65.** 20

67. False. It is an example of a combination.

69. They are the same. **71. and 73.** Answers will vary.

75. No. For some calculators the number is too great.

77. (a) 35 (b) 8 (c) 83

79. (a) -4 (b) 0 (c) 0

81. $x^5 + 5x^4 + 10x^3 + 10x^2 + 5x + 1$

83. $x^{10} + 10x^8y + 40x^6y^2 + 80x^4y^3 + 80x^2y^4 + 32y^5$

Section 9.7 *(page 754)*

1. $\{(H, 1), (H, 2), (H, 3), (H, 4), (H, 5), (H, 6), (T, 1), (T, 2), (T, 3), (T, 4), (T, 5), (T, 6)\}$

3. {ABC, ACB, BAC, BCA, CAB, CBA}

5. {(A, B), (A, C), (A, D), (A, E), (B, C), (B, D), (B, E), (C, D), (C, E), (D, E)}

7. $\frac{3}{8}$ **9.** $\frac{7}{8}$ **11.** $\frac{3}{13}$ **13.** $\frac{3}{26}$ **15.** $\frac{1}{12}$

17. $\frac{11}{12}$ **19.** $\frac{1}{3}$ **21.** $\frac{1}{5}$ **23.** $\frac{2}{5}$ **25.** 0.3

27. $\frac{2}{3}$ **29.** 0.85 **31.** $\frac{7}{20}$

33. (a) 78,000 (b) 30% (c) 37%

35. (a) 58% (b) 95.6% (c) 0.4%

37. (a) $\frac{112}{209}$ (b) $\frac{97}{209}$ (c) $\frac{274}{627}$

39. $P(\{\text{Taylor wins}\}) = \frac{1}{2}$

$P(\{\text{Moore wins}\}) = P(\{\text{Jenkins wins}\}) = \frac{1}{4}$

41. (a) $\frac{21}{1292}$ (b) $\frac{225}{646}$ (c) $\frac{49}{323}$

43. (a) $\frac{1}{3}$ (b) $\frac{5}{8}$

45. (a) $\frac{1}{120}$ (b) $\frac{1}{24}$

47. (a) $\frac{1}{169}$ (b) $\frac{1}{221}$

49. (a) $\frac{14}{55}$ (b) $\frac{12}{55}$ (c) $\frac{54}{55}$

51. (a) $\frac{1}{4}$ (b) $\frac{1}{2}$ (c) $\frac{9}{100}$ (d) $\frac{1}{30}$

53. (a) 0.9702 (b) 0.9998 (c) 0.0002

55. (a) $\frac{1}{1024}$ (b) $\frac{243}{1024}$ (c) $\frac{781}{1024}$

57. 0.4746 **59.** $\frac{7}{16}$

61. True. Two events are independent if the occurrence of one has no effect on the occurrence of the other.

63. (a) As you consider successive people with distinct birthdays, the probabilities must decrease to take into account the birth dates already used. Because the birth dates of people are independent events, multiply the respective probabilities of distinct birthdays.

(b) $\frac{365}{365} \cdot \frac{364}{365} \cdot \frac{363}{365} \cdot \frac{362}{365}$

(c) Answers will vary.

(d) Q_n is the probability that the birthdays are *not* distinct, which is equivalent to at least two people having the same birthday.

(e)

n	10	15	20	23	30	40	50
P_n	0.88	0.75	0.59	0.49	0.29	0.11	0.03
Q_n	0.12	0.25	0.41	0.51	0.71	0.89	0.97

(f) 23

65. No real solution **67.** $0, \frac{1 \pm \sqrt{13}}{2}$ **69.** -4

71. $\frac{11}{2}$ **73.** -10 **75.** $\ln 27 \approx 3.296$

77. $\ln 1 = 0, \ln 3 \approx 1.099$ **79.** $\ln \frac{8}{3} \approx 0.981$

81. $e^8 \approx 2980.958$ **83.** $\frac{e^4}{6} \approx 9.100$

Review Exercises *(page 760)*

1. $8, 5, 4, \frac{7}{2}, \frac{16}{5}$ **3.** $72, 36, 12, 3, \frac{3}{5}$ **5.** 120 **7.** 1

9. 30 **11.** $\frac{205}{24}$ **13.** 6050 **15.** $\sum_{k=1}^{20} \frac{1}{2k}$

17. $\frac{5}{9}$ **19.** $\frac{2}{99}$

21. (a) \$43,000 (b) \$192,500

23. Arithmetic sequence, $d = -2$

25. Arithmetic sequence, $d = \frac{1}{2}$ **27.** $a_n = 12n - 5$

29. $a_n = 3ny - 2y$ **31.** 80 **33.** 88 **35.** 25,250

37. 43 minutes **39.** $4, -1, \frac{1}{4}, -\frac{1}{16}, \frac{1}{64}$

41. $9, 6, 4, \frac{8}{3}, \frac{16}{9}$ or $9, -6, 4, -\frac{8}{3}, \frac{16}{9}$

43. $a_n = 16\left(-\frac{1}{2}\right)^{n-1}$, 10.67

45. $a_n = 100(1.05)^{n-1}$, 3306.60 **47.** 127 **49.** $\frac{15}{16}$

51. 31 **53.** 24.849 **55.** 5486.45 **57.** 8

59. $\frac{10}{9}$ **61.** 12

63. (a) $a_t = 120{,}000(0.7)^t$

(b) \$20,168.40

65. and 67. Answers will vary.

69. 465 **71.** 4676

73. $S_n = n(2n + 7)$ **75.** $S_n = \frac{5}{2}\left[1 - \left(\frac{3}{5}\right)^n\right]$

77. 5, 10, 15, 20, 25

First differences: 5, 5, 5, 5

Second differences: 0, 0, 0

Linear

79. 16, 15, 14, 13, 12

First differences: $-1, -1, -1, -1$

Second differences: 0, 0, 0

Linear

81. 15 **83.** 56 **85.** 35 **87.** 28

89. $\frac{x^4}{16} + \frac{x^3y}{2} + \frac{3x^2y^2}{2} + 2xy^3 + y^4$

91. $a^5 - 15a^4b + 90a^3b^2 - 270a^2b^3 + 405ab^4 - 243b^5$

93. $41 + 840i$ **95.** 3 **97.** 10,000

99. 3,628,800 **101.** 56 **103.** $\frac{1}{9}$

105. (a) 43% (b) 82% **107.** $\frac{1}{216}$ **109.** $\frac{3}{4}$

111. True. $\frac{(n+2)!}{n!} = \frac{(n+2)(n+1)n!}{n!} = (n+2)(n+1)$

113. True by Properties of Sums **115.** Natural numbers

117. (a) Each term is obtained by adding the same constant (common difference) to the previous term.

(b) Each term is obtained by multiplying the same constant (common ratio) by the previous term.

119. Each term of the sequence is defined in terms of the previous term.

121. d **123.** b

125. The terms of the expansion of $(x + y)^n$ are positive; the signs of the terms in the expansion of $(x - y)^n$ alternate.

127. $\frac{1}{3}$. The probability that an event does not occur is 1 minus the probability that it does occur.

Chapter Test *(page 765)*

1. $-\frac{1}{5}, \frac{1}{8}, -\frac{1}{11}, \frac{1}{14}, -\frac{1}{17}$ **2.** $a_n = \frac{n+2}{n!}$ **3.** 140

4. 25.4 **5.** 5, 10, 20, 40, 80 **6.** 86,100 **7.** 4

8. Answers will vary.

9. $x^4 + 8x^3y + 24x^2y^2 + 32xy^3 + 16y^4$ **10.** 180

11. (a) 72 (b) 328,440 **12.** (a) 330 (b) 720,720

13. 720 **14.** $\frac{1}{15}$ **15.** 3.908×10^{-10} **16.** 25%

Cumulative Test for Chapters 7–9 *(page 766)*

1. $(1, 2), \left(-\frac{3}{2}, \frac{3}{4}\right)$ **2.** $(2, -1)$

3. $(4, 2, -3)$ **4.** $(1, -2, 1)$

5.

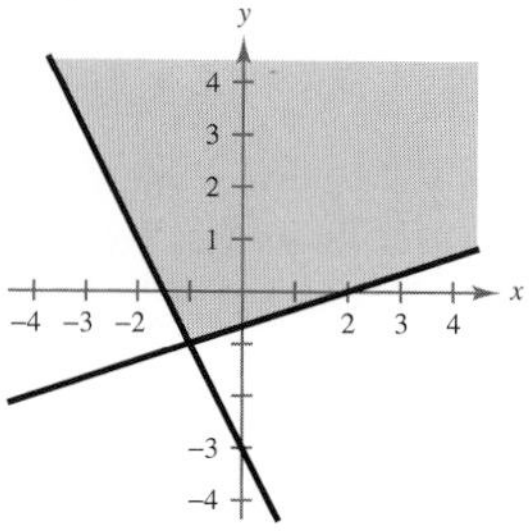

6.

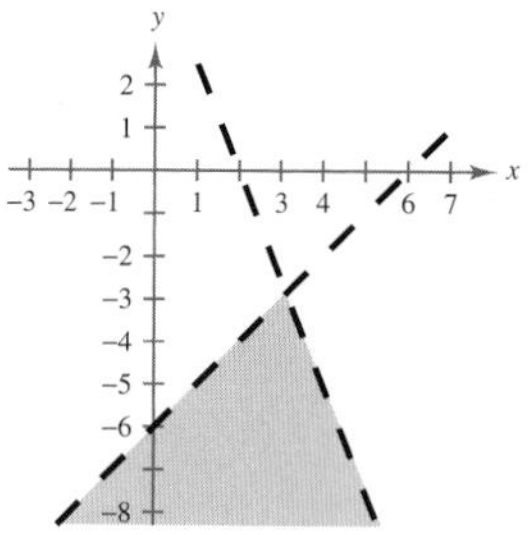

7.

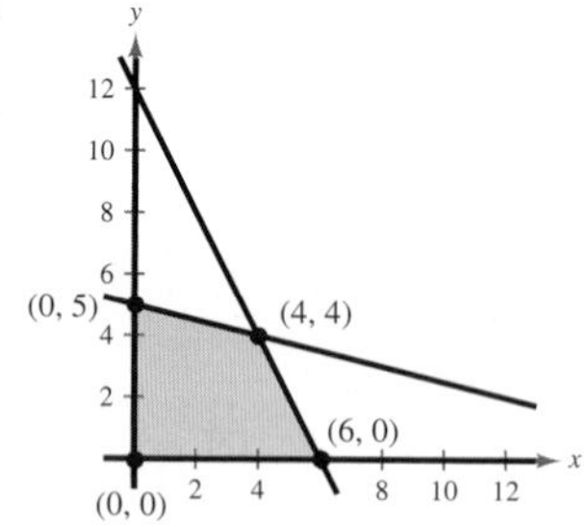

Maximum at (4, 4): $z = 20$

8. 120 pounds, 80 pounds **9.** $y = \frac{1}{3}x^2 - 2x + 4$

10. $\begin{bmatrix} -1 & 2 & -1 & \vdots & 9 \\ 2 & -1 & 2 & \vdots & -9 \\ 3 & 3 & -4 & \vdots & 7 \end{bmatrix}$ **11.** $(-2, 3, -1)$

12. $\begin{bmatrix} 5 & -3 \\ -2 & 2 \end{bmatrix}$ **13.** $\begin{bmatrix} 2 & -6 \\ -2 & 0 \end{bmatrix}$ **14.** $\begin{bmatrix} 6 & -6 \\ -3 & 2 \end{bmatrix}$

15. $\begin{bmatrix} -4 & 12 \\ 3 & -3 \end{bmatrix}$ **16.** $\begin{bmatrix} -175 & 37 & -13 \\ 95 & -20 & 7 \\ 14 & -3 & 1 \end{bmatrix}$

17. Gym shoes: \$164.80 million

Jogging shoes: \$244.22 million

Walking shoes: \$111.89 million

18. $(-5, 4)$ **19.** $(-3, 4, 2)$ **20.** 9

21. $\frac{1}{5}, -\frac{1}{7}, \frac{1}{9}, -\frac{1}{11}, \frac{1}{13}$ **22.** $a_n = \frac{(n+1)!}{n+3}$

23. 920 **24.** 65.4 **25.** 3, 6, 12, 24, 48 **26.** 6

27. Answers will vary.

28. $z^4 - 12z^3 + 54z^2 - 108z + 81$ **29.** 210 **30.** 600

31. 70 **32.** 120 **33.** 720 **34.** $\frac{1}{4}$

Chapter 10

Section 10.1 *(page 774)*

1. $\frac{\sqrt{3}}{3}$ **3.** -1 **5.** $\sqrt{3}$ **7.** 3.2236

9. $\frac{3\pi}{4}$ radians, 135° **11.** $\frac{\pi}{4}$ radian, 45°

13. 0.6435 radian, 36.9° **15.** 1.0517 radians, 60.3°

17. 2.1112 radians, 121.0° **19.** 1.2490 radians, 71.6°

21. 2.1112 radians, 121.0° **23.** 1.1071 radians, 63.4°

25. 0.1974 radian, 11.3° **27.** 1.4289 radians, 81.9°

29. 0.9273 radian, 53.1° **31.** 0.8187 radian, 46.9°

33. (2, 1), 42.3°; (4, 4), 78.7°; (6, 2), 59.0°

35. $(-4, -1)$, 11.9°; (3, 2), 21.8°; (1, 0), 146.3° **37.** 0

39. $\frac{7}{5}$ **41.** 7 **43.** $\frac{8\sqrt{37}}{37} \approx 1.3152$

45. (a) 4 (b) 8 **47.** (a) $\frac{35\sqrt{37}}{74}$ (b) $\frac{35}{8}$

49. $2\sqrt{2}$ **51.** 0.1003, 1054 feet

53. (a) 18.4° (b) 15.8 meters

55. $\alpha = 33.69°$; $\beta = 56.31°$

57. True. The inclination of a line is related to its slope by $m = \tan\theta$. If the angle is greater than $\pi/2$ but less than π, then the angle is in the second quadrant, where the tangent function is negative.

59. (a) $d = \frac{4}{\sqrt{m^2 + 1}}$

(b)

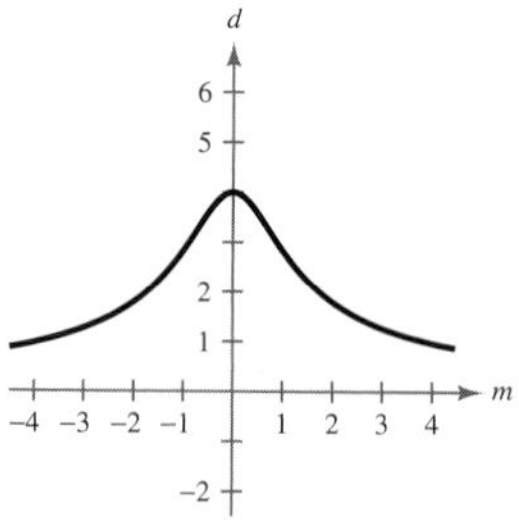

(c) $m = 0$

(d) $d = 0$. As the line approaches the vertical, the distance approaches 0.

61. x-intercept: (7, 0)

y-intercept: (0, 49)

63. x-intercepts: $(5 \pm \sqrt{5}, 0)$

y-intercept: (0, 20)

65. x-intercepts: $\left(\frac{7 \pm \sqrt{53}}{2}, 0\right)$

y-intercept: $(0, -1)$

67. $f(x) = 3\left(x + \frac{1}{3}\right)^2 - \frac{49}{3}$

Vertex: $\left(-\frac{1}{3}, -\frac{49}{3}\right)$

69. $f(x) = 5\left(x + \frac{17}{5}\right)^2 - \frac{324}{5}$

Vertex: $\left(-\frac{17}{5}, -\frac{324}{5}\right)$

71. $f(x) = 6\left(x - \frac{1}{12}\right)^2 - \frac{289}{24}$

Vertex: $\left(\frac{1}{12}, -\frac{289}{24}\right)$

Section 10.2 *(page 782)*

1. A circle is formed when a plane intersects the top or bottom half of a double-napped cone and is perpendicular to the axis of the cone.

3. A parabola is formed when a plane intersects the top or bottom half of a double-napped cone, is parallel to the side of the cone, and doesn't intersect the vertex.

5. e **7.** d **9.** a

11. Vertex: $(0, 0)$
Focus: $\left(0, \frac{1}{2}\right)$
Directrix: $y = -\frac{1}{2}$

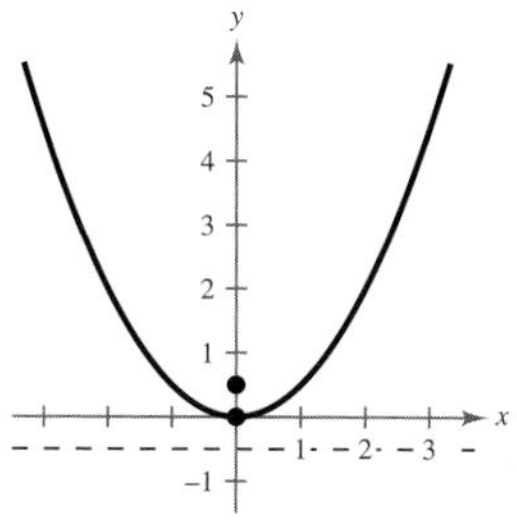

13. Vertex: $(0, 0)$
Focus: $\left(-\frac{3}{2}, 0\right)$
Directrix: $x = \frac{3}{2}$

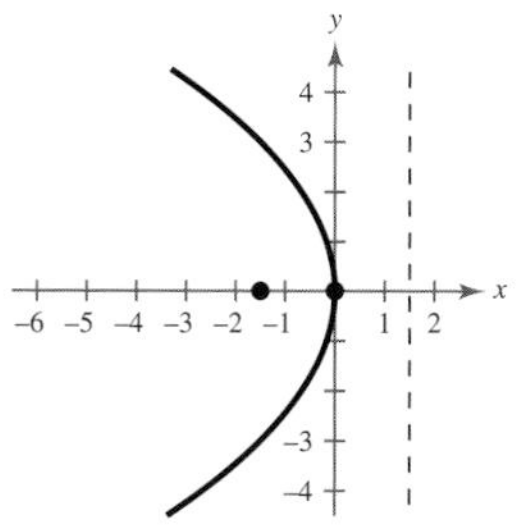

15. Vertex: $(0, 0)$
Focus: $\left(0, -\frac{3}{2}\right)$
Directrix: $y = \frac{3}{2}$

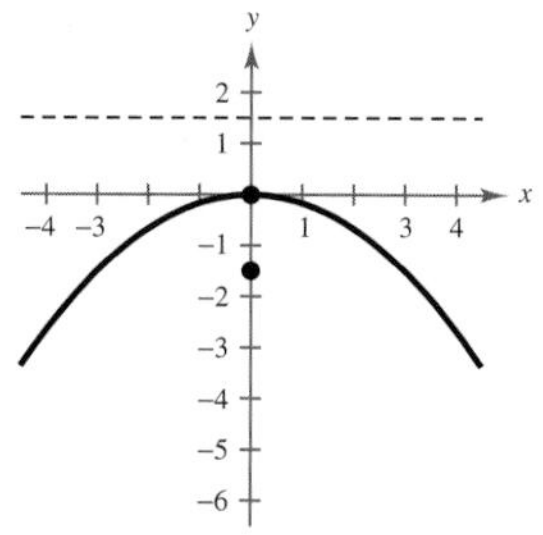

17. Vertex: $(1, -2)$
Focus: $(1, -4)$
Directrix: $y = 0$

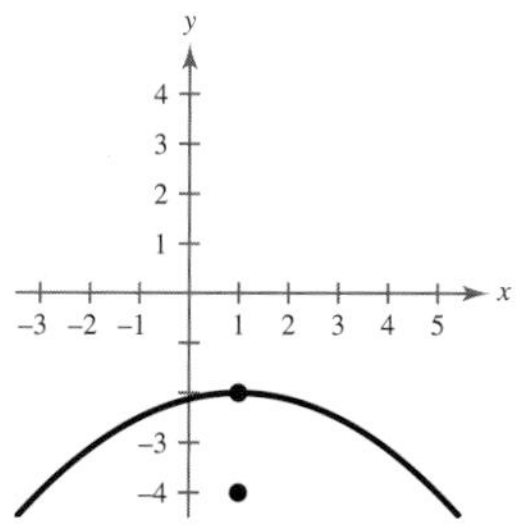

19. Vertex: $\left(-\frac{3}{2}, 2\right)$
Focus: $\left(-\frac{3}{2}, 3\right)$
Directrix: $y = 1$

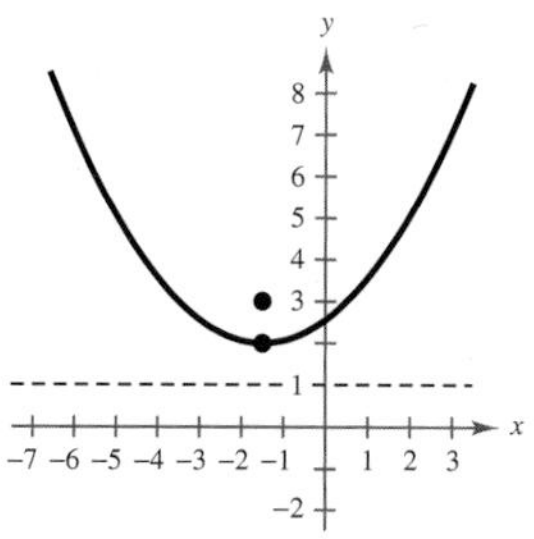

21. Vertex: $(1, 1)$
Focus: $(1, 2)$
Directrix: $y = 0$

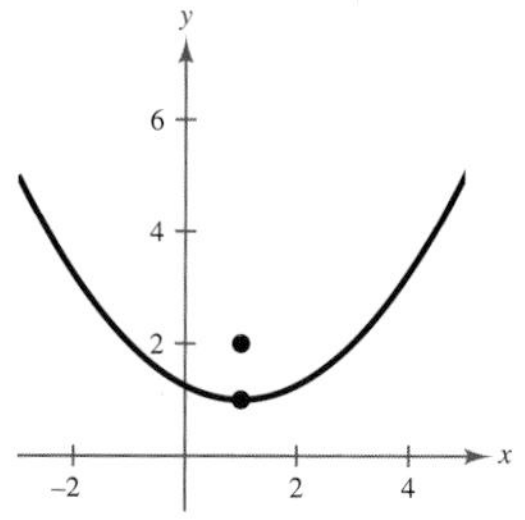

23. Vertex: $(-2, -3)$
Focus: $(-4, -3)$
Directrix: $x = 0$

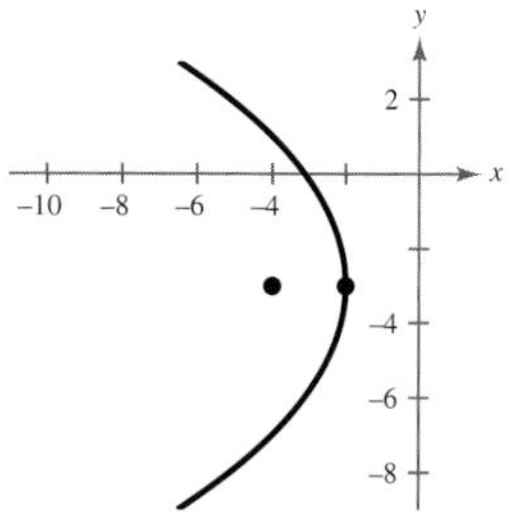

25. Vertex: $(-2, 1)$
Focus: $\left(-2, -\frac{1}{2}\right)$
Directrix: $y = \frac{5}{2}$

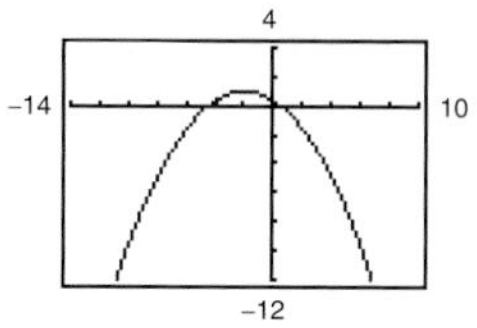

27. Vertex: $\left(\frac{1}{4}, -\frac{1}{2}\right)$
Focus: $\left(0, -\frac{1}{2}\right)$
Directrix: $x = \frac{1}{2}$

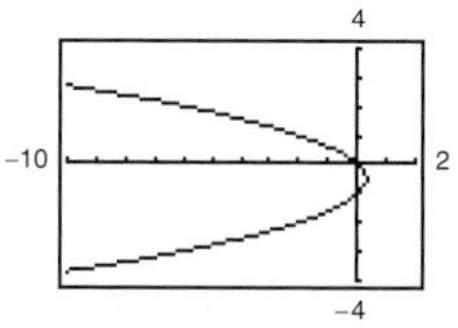

29.

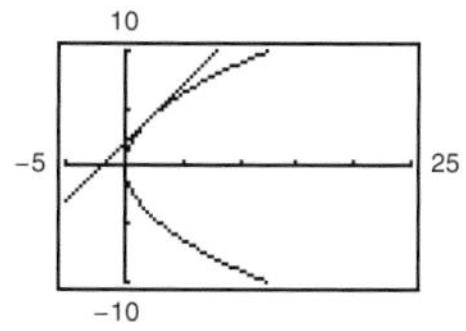

$(2, 4)$

31. $x^2 = \frac{3}{2}y$ **33.** $x^2 = -6y$ **35.** $y^2 = -8x$

37. $x^2 = 4y$ **39.** $y^2 = -8x$ **41.** $y^2 = 9x$

43. $(x - 3)^2 = -(y - 1)$ **45.** $y^2 = 4(x + 4)$

47. $(y - 2)^2 = -8(x - 5)$ **49.** $x^2 = 8(y - 4)$

51. $(y - 2)^2 = 8x$ **53.** $y = \sqrt{6(x + 1)} + 3$

55. $4x - y - 8 = 0$; $(2, 0)$ **57.** $4x - y + 2 = 0$; $\left(-\frac{1}{2}, 0\right)$

59.

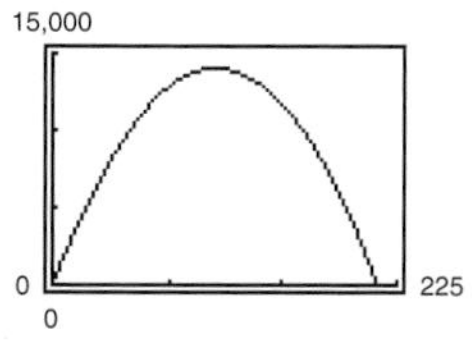

$x = 106$ units

61. $y = \frac{1}{18}x^2$ **63.** (a) $y = -\frac{1}{640}x^2$ (b) 8 feet

65. (a) $17{,}500\sqrt{2}$ miles per hour
(b) $x^2 = -16{,}400(y - 4100)$

67. 43.3 seconds prior to being over the target

69. True. A parabola is the set of all points that are equidistant from the directrix and the focus.

71. (a) $\dfrac{64\sqrt{2}}{3}$
(b) The parabola becomes narrower for $0 \le y \le b$.

73. One positive real zero, two or no negative real zeros

75. Two or no positive real zeros, one negative real zero

77. $\pm 1, \pm 2, \pm 4$ **79.** $\pm 1, \pm \frac{1}{2}, \pm 2, \pm 4, \pm 8, \pm 16$

81. $y = x^3 - 7x^2 + 17x - 15$ **83.** $\pm 2, \frac{1}{2}, -\frac{5}{3}$

Section 10.3 *(page 792)*

1. b **3.** d **5.** a

7. Center: $(0, 0)$
Vertices: $(\pm 5, 0)$
Foci: $(\pm 3, 0)$
Eccentricity: $\frac{3}{5}$

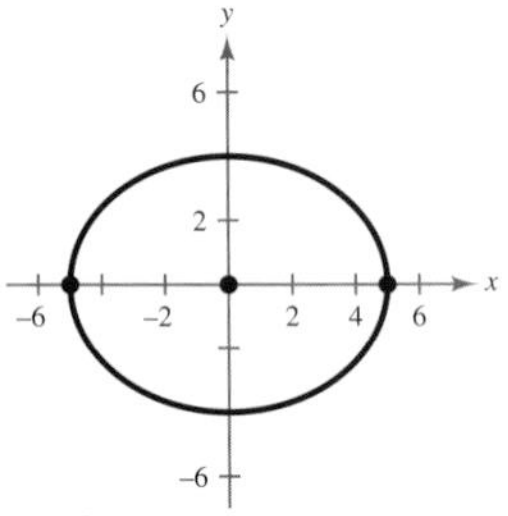

9. Center: $(0, 0)$
Vertices: $(0, \pm 3)$
Foci: $(0, \pm 2)$
Eccentricity: $\frac{2}{3}$

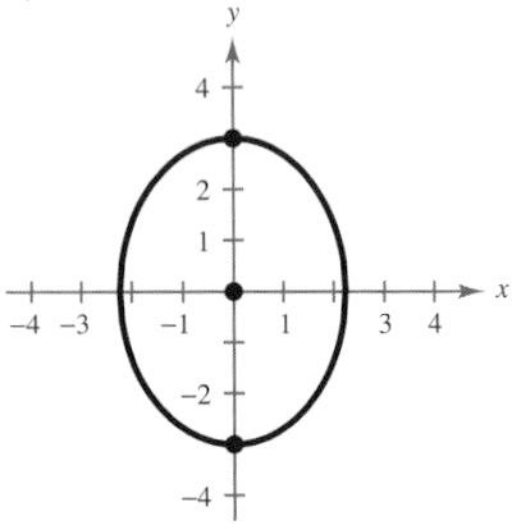

11. Center: $(-3, 5)$
Vertices: $(-3, 10), (-3, 0)$
Foci: $(-3, 8), (-3, 2)$
Eccentricity: $\frac{3}{5}$

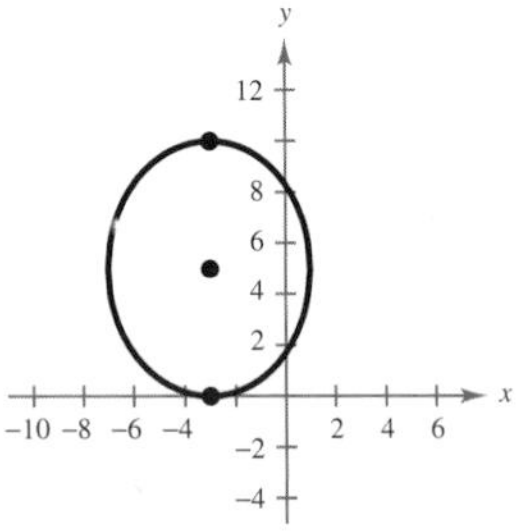

13. Center: $(-5, 1)$
Vertices: $\left(-\frac{7}{2}, 1\right), \left(-\frac{13}{2}, 1\right)$
Foci: $\left(-5 + \frac{\sqrt{5}}{2}, 1\right), \left(-5 - \frac{\sqrt{5}}{2}, 1\right)$
Eccentricity: $\frac{\sqrt{5}}{3}$

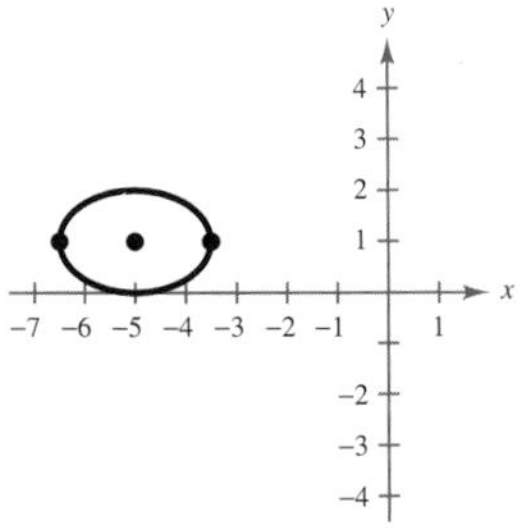

15. Center: $(-2, 3)$
Vertices: $(-2, 6), (-2, 0)$
Foci: $\left(-2, 3 \pm \sqrt{5}\right)$
Eccentricity: $\frac{\sqrt{5}}{3}$

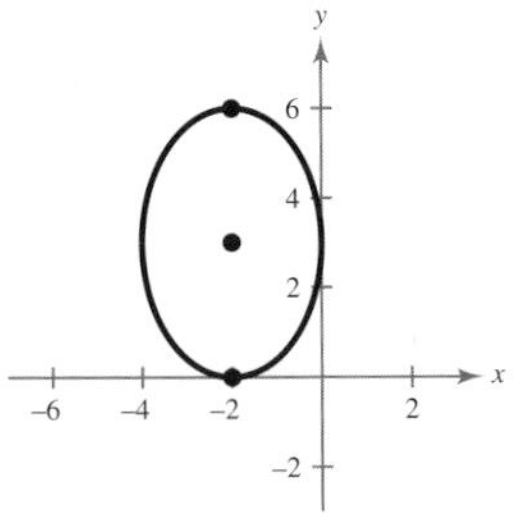

17. Center: $(4, 3)$
Vertices: $(14, 3), (-6, 3)$
Foci: $\left(4 \pm 4\sqrt{5}, 3\right)$
Eccentricity: $\frac{2\sqrt{5}}{5}$

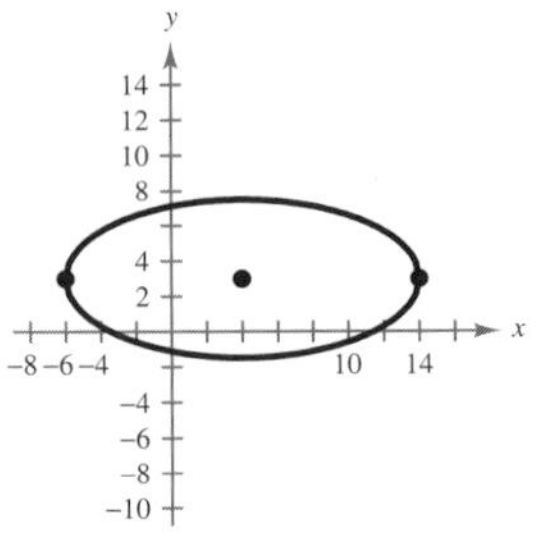

19. Center: $\left(-\frac{3}{2}, \frac{5}{2}\right)$
Vertices: $\left(-\frac{3}{2}, \frac{5}{2} \pm 2\sqrt{3}\right)$
Foci: $\left(-\frac{3}{2}, \frac{5}{2} \pm 2\sqrt{2}\right)$
Eccentricity: $\frac{\sqrt{6}}{3}$

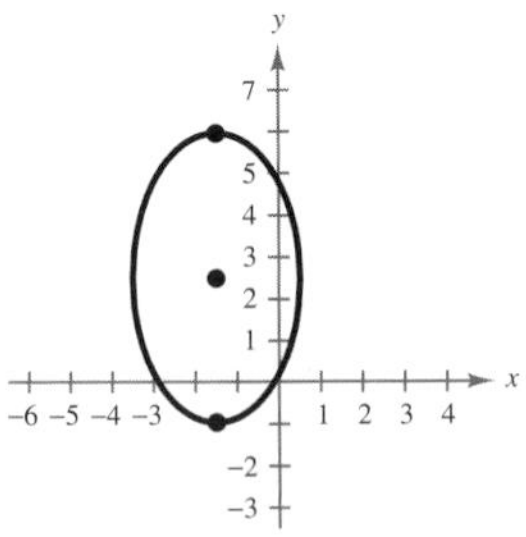

21. Center: $(1, -1)$
Vertices: $\left(\frac{9}{4}, -1\right), \left(-\frac{1}{4}, -1\right)$
Foci: $\left(\frac{7}{4}, -1\right), \left(\frac{1}{4}, -1\right)$
Eccentricity: $\frac{3}{5}$

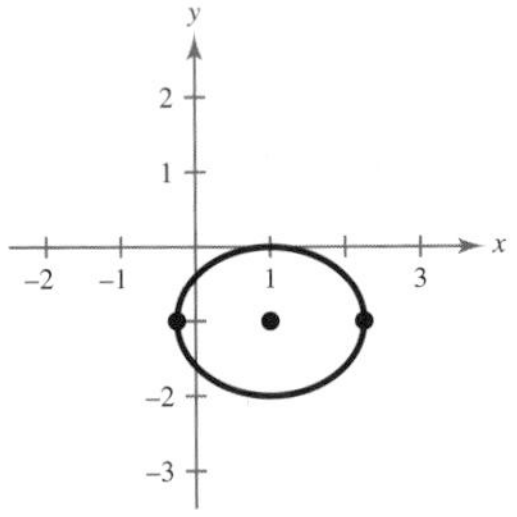

23.

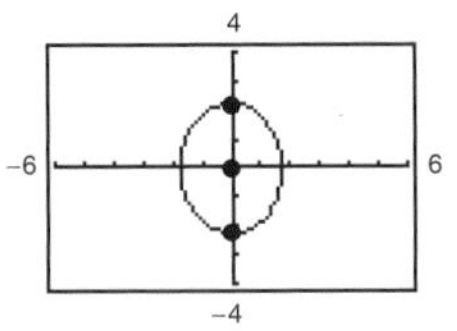

Center: $(0, 0)$
Vertices: $\left(0, \pm\sqrt{5}\right)$
Foci: $\left(0, \pm\sqrt{2}\right)$

25.

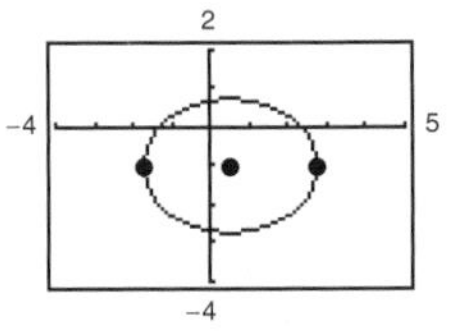

Center: $\left(\frac{1}{2}, -1\right)$
Vertices: $\left(\frac{1}{2} \pm \sqrt{5}, -1\right)$
Foci: $\left(\frac{1}{2} \pm \sqrt{2}, -1\right)$

27. $\frac{x^2}{4} + \frac{y^2}{16} = 1$ **29.** $\frac{x^2}{36} + \frac{y^2}{32} = 1$

31. $\frac{x^2}{36} + \frac{y^2}{11} = 1$ **33.** $\frac{21x^2}{400} + \frac{y^2}{25} = 1$

35. $\frac{(x-2)^2}{1} + \frac{(y-3)^2}{9} = 1$

37. $\frac{(x+2)^2}{16} + \frac{(y-3)^2}{9} = 1$

39. $\frac{(x-2)^2}{4} + \frac{(y-4)^2}{1} = 1$ **41.** $\frac{x^2}{48} + \frac{(y-4)^2}{64} = 1$

43. $\frac{(x-3)^2}{9} + \frac{(y-5)^2}{16} = 1$ **45.** $\frac{x^2}{16} + \frac{(y-4)^2}{12} = 1$

47. $\frac{x^2}{25} + \frac{y^2}{16} = 1$

49. Positions: $(\pm\sqrt{5}, 0)$; Length of string: 6 feet

51. 40 **53.** $\frac{x^2}{4.88} + \frac{y^2}{1.39} = 1$ **55.** Answers will vary.

57.

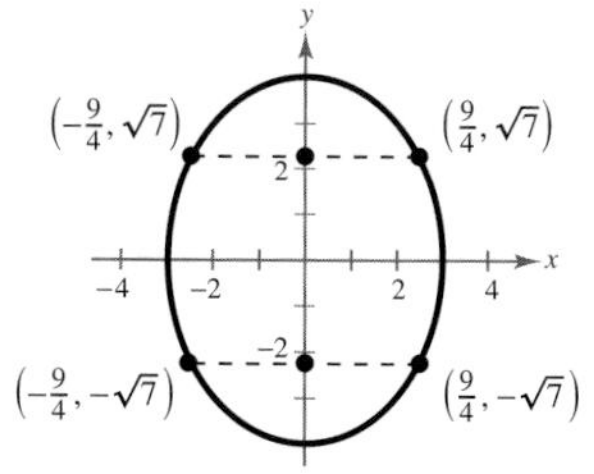

59.

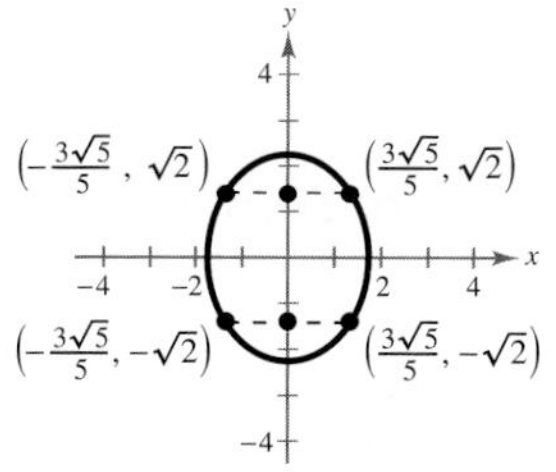

61. True. The ellipse is elongated and the foci are close to the vertices.

63. False. The foci of an ellipse cannot occur outside the ellipse because $0 < c < a$.

65. (a) $A = \pi a(20 - a)$ (b) $\frac{x^2}{196} + \frac{y^2}{36} = 1$

(c)

a	8	9	10	11	12	13
A	301.6	311.0	314.2	311.0	301.6	285.9

$a = 10$, circle

(d)

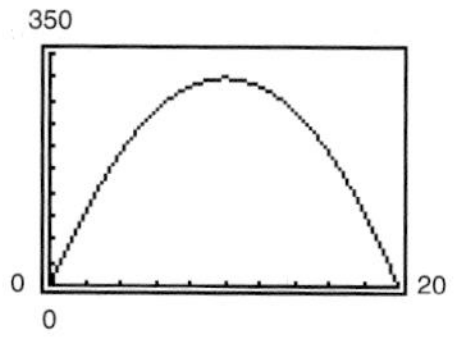

The shape of an ellipse with a maximum area is a circle.

67. Geometric **69.** Arithmetic **71.** $a_n = -\frac{1}{4}n + \frac{1}{4}$

73. $a_n = 9n$ **75.** 547 **77.** 340.15

Section 10.4 *(page 802)*

1. b **3.** a

5. Center: $(0, 0)$
Vertices: $(\pm 1, 0)$
Foci: $(\pm\sqrt{2}, 0)$
Asymptotes: $y = \pm x$

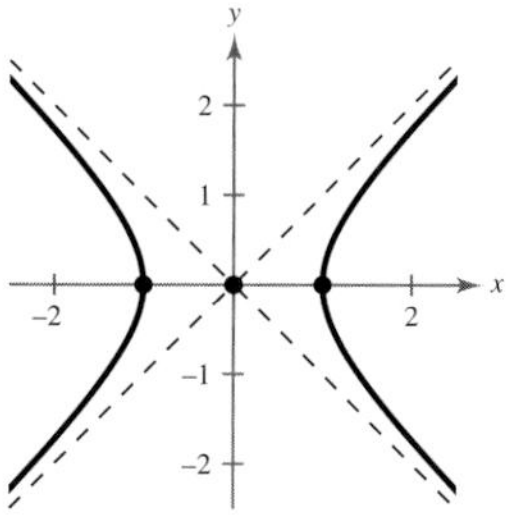

7. Center: $(0, 0)$
Vertices: $(0, \pm 5)$
Foci: $(0, \pm\sqrt{106})$
Asymptotes: $y = \pm\frac{5}{9}x$

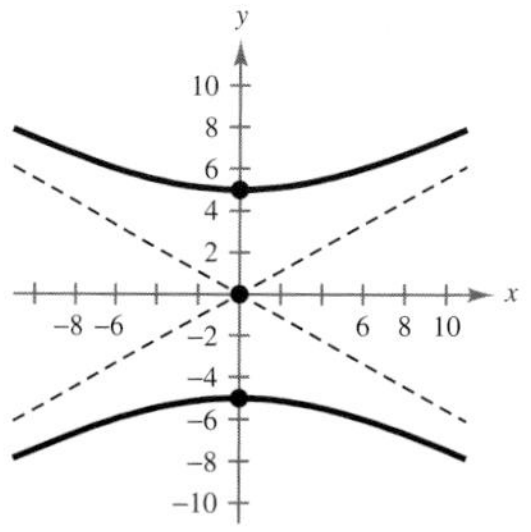

9. Center: $(1, -2)$
Vertices: $(3, -2), (-1, -2)$
Foci: $(1 \pm \sqrt{5}, -2)$
Asymptotes: $y = -2 \pm \frac{1}{2}(x - 1)$

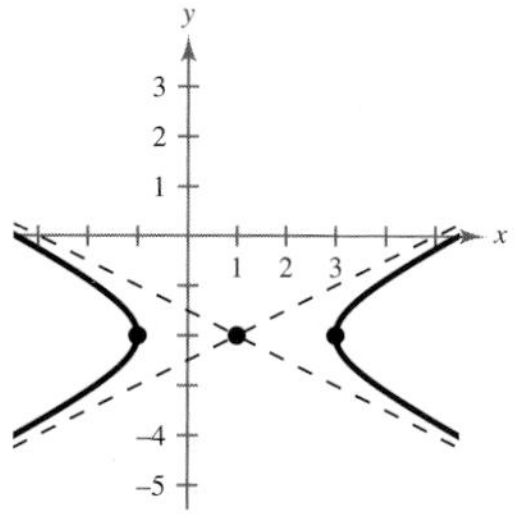

11. Center: $(2, -6)$
Vertices: $\left(2, -\frac{17}{3}\right), \left(2, -\frac{19}{3}\right)$
Foci: $\left(2, -6 \pm \frac{\sqrt{13}}{6}\right)$
Asymptotes: $y = -6 \pm \frac{2}{3}(x - 2)$

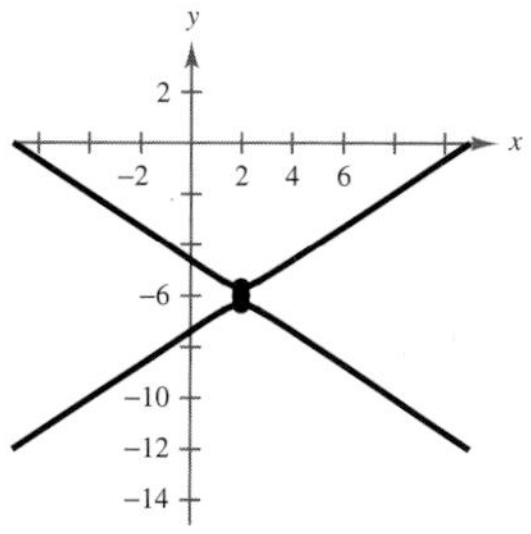

13. Center: $(2, -3)$

Vertices: $(3, -3), (1, -3)$

Foci: $\left(2 \pm \sqrt{10}, -3\right)$

Asymptotes: $y = -3 \pm 3(x - 2)$

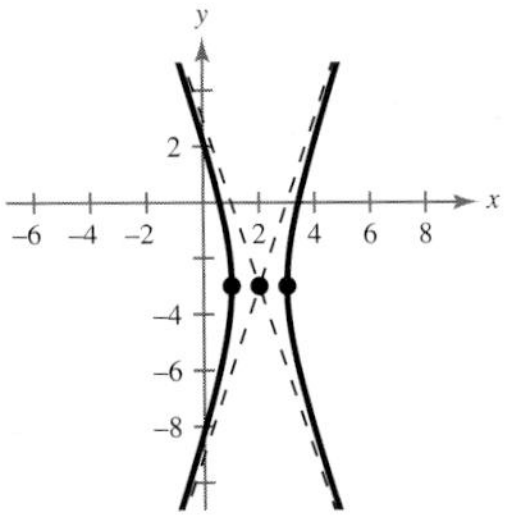

15. The graph of this equation is two lines intersecting at $(-1, -3)$.

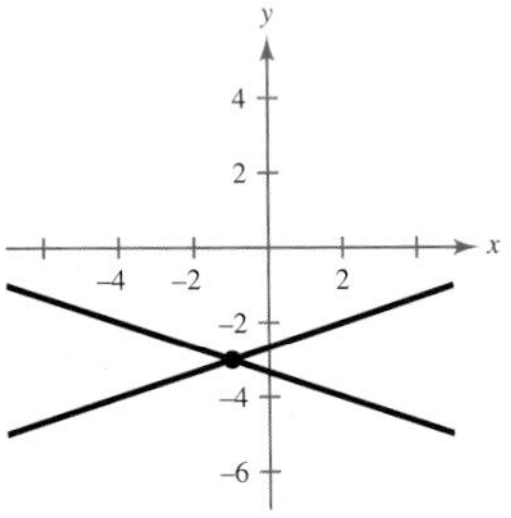

17. Center: $(0, 0)$

Vertices: $\left(\pm\sqrt{3}, 0\right)$

Foci: $\left(\pm\sqrt{5}, 0\right)$

Asymptotes: $y = \pm\dfrac{\sqrt{6}}{3}x$

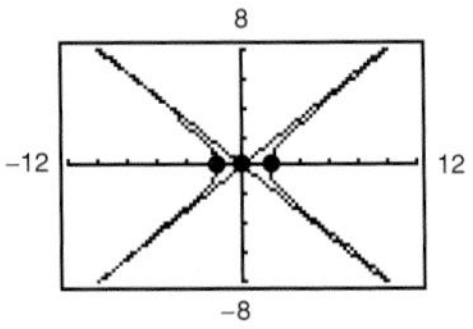

19. Center: $(1, -3)$

Vertices: $\left(1, -3 \pm \sqrt{2}\right)$

Foci: $\left(1, -3 \pm 2\sqrt{5}\right)$

Asymptotes: $y = -3 \pm \frac{1}{3}(x - 1)$

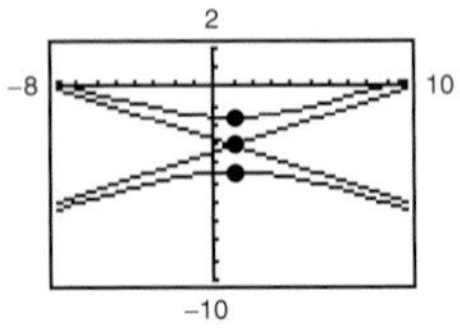

21. $\dfrac{y^2}{4} - \dfrac{x^2}{12} = 1$ **23.** $\dfrac{x^2}{1} - \dfrac{y^2}{25} = 1$

25. $\dfrac{17y^2}{1024} - \dfrac{17x^2}{64} = 1$ **27.** $\dfrac{(x - 4)^2}{4} - \dfrac{y^2}{12} = 1$

29. $\dfrac{(y - 5)^2}{16} - \dfrac{(x - 4)^2}{9} = 1$ **31.** $\dfrac{y^2}{9} - \dfrac{4(x - 2)^2}{9} = 1$

33. $\dfrac{(y - 2)^2}{4} - \dfrac{x^2}{4} = 1$ **35.** $\dfrac{(x - 2)^2}{1} - \dfrac{(y - 2)^2}{1} = 1$

37. $\dfrac{(x - 3)^2}{9} - \dfrac{(y - 2)^2}{4} = 1$ **39.** $(3300, -2750)$

41. $\left(12\left(\sqrt{5} - 1\right), 0\right) \approx (14.83, 0)$ **43.** Ellipse

45. Parabola **47.** Hyperbola **49.** Circle

51. False. For the trivial solution of two intersecting lines to occur, the standard form of the equation of the hyperbola would be equal to zero,

$$\frac{(x - h)^2}{a^2} - \frac{(y - k)^2}{b^2} = 0 \text{ or } \frac{(y - k)^2}{a^2} - \frac{(x - h)^2}{b^2} = 0.$$

53. The extended diagonals of the central rectangle are asymptotes of the hyperbola.

55. $3x^2 + \frac{23}{2}x - 2$ **57.** $x^2 + 2xy + y^2 + 6x + 6y + 9$

59. $(x + 7)^2$ **61.** $x(3x + 2)(2x - 5)$

63. $(x^2 + 1)(4 - x)$

Section 10.5 *(page 810)*

1. $(3, 0)$ **3.** $\left(\dfrac{3 + \sqrt{3}}{2}, \dfrac{3\sqrt{3} - 1}{2}\right)$

5. $\left(\dfrac{3\sqrt{2}}{2}, -\dfrac{\sqrt{2}}{2}\right)$

7. $\dfrac{(y')^2}{2} - \dfrac{(x')^2}{2} = 1$ **9.** $y' = \pm\dfrac{\sqrt{2}}{2}$

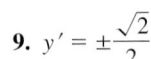

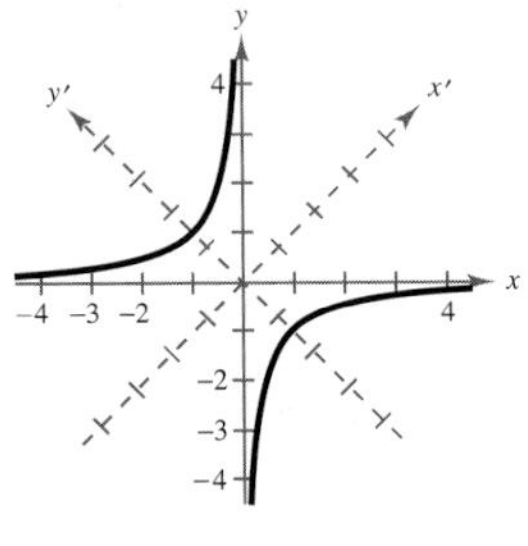

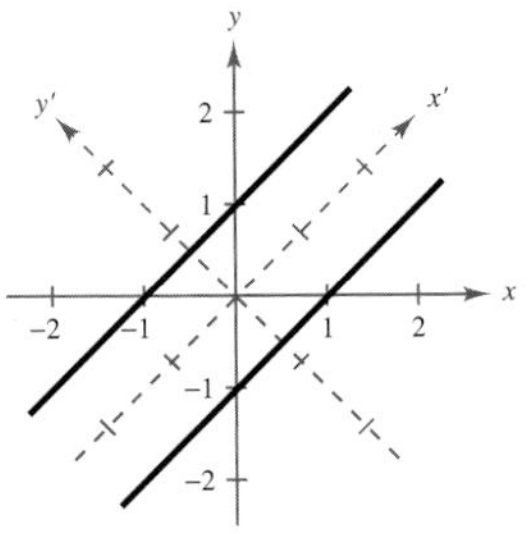

11. $\dfrac{\left(x' - 3\sqrt{2}\right)^2}{16} - \dfrac{\left(y' - \sqrt{2}\right)^2}{16} = 1$

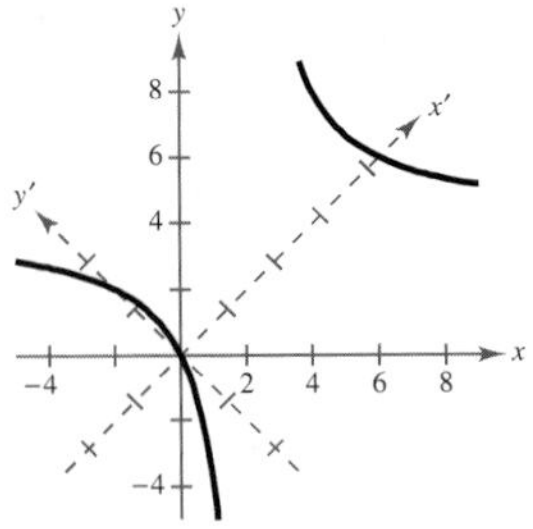

13. $\dfrac{(x')^2}{6} + \dfrac{(y')^2}{\frac{3}{2}} = 1$

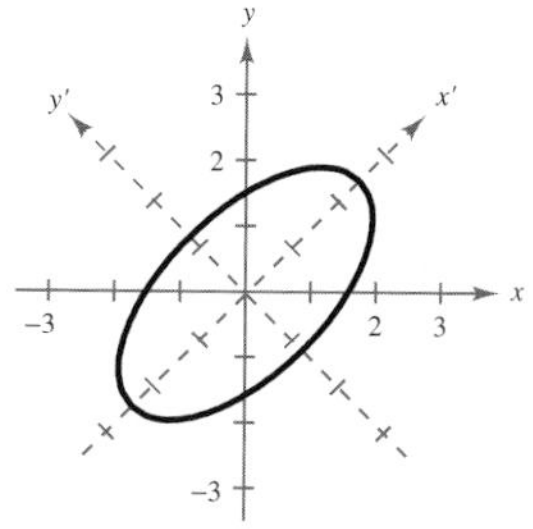

15. $x' = -(y')^2$

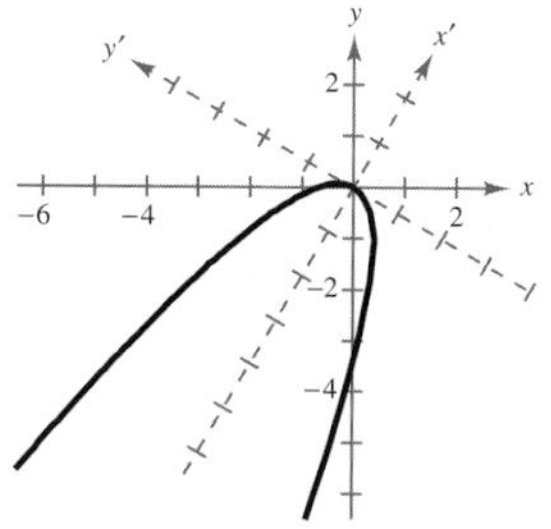

17. $y' = \frac{1}{6}(x')^2 - \frac{1}{3}x'$

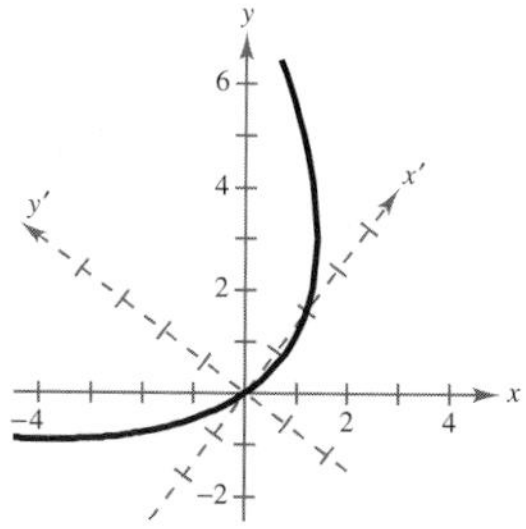

19.

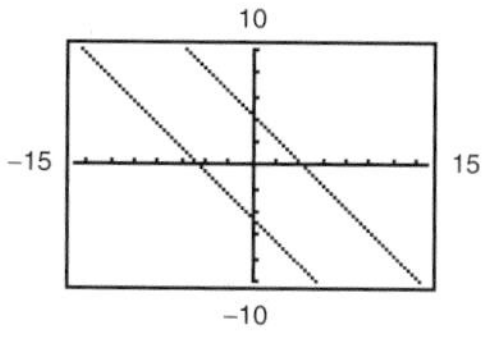

$\theta = 45°$

21.

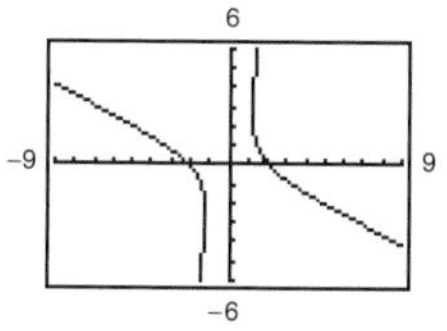

$\theta = 26.57°$

23.

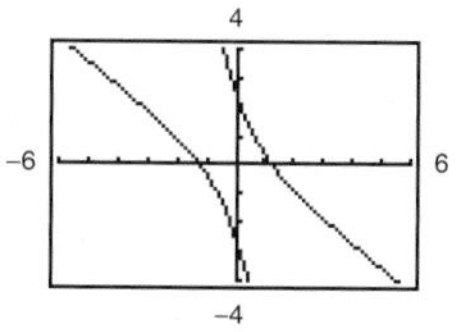

$\theta = 31.72°$

25.

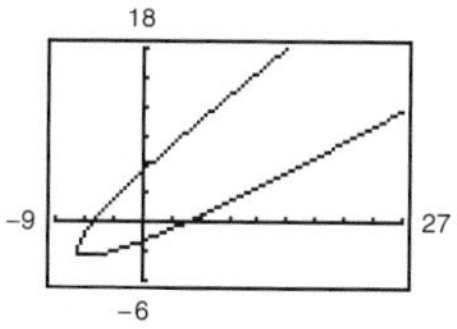

$\theta = 33.69°$

27. e **29.** b **31.** d

33. Parabola

$$y = \frac{(8x - 5) \pm \sqrt{(8x - 5)^2 - 4(16x^2 - 10x)}}{2}$$

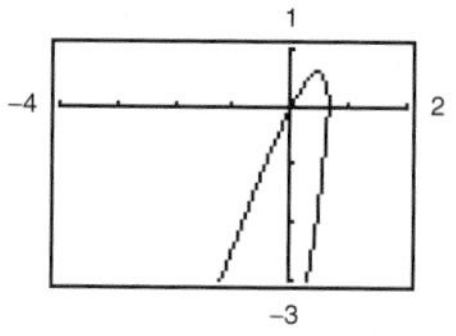

35. Ellipse

$$y = \frac{6x \pm \sqrt{36x^2 - 28(12x^2 - 45)}}{14}$$

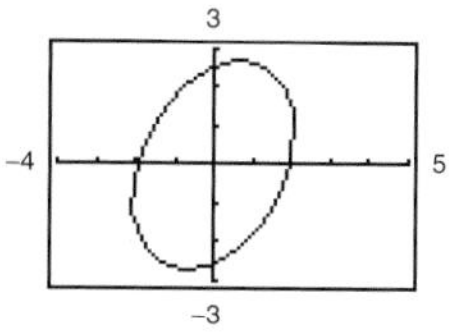

37. Hyperbola

$$y = \frac{6x \pm \sqrt{36x^2 + 20(x^2 + 4x - 22)}}{-10}$$

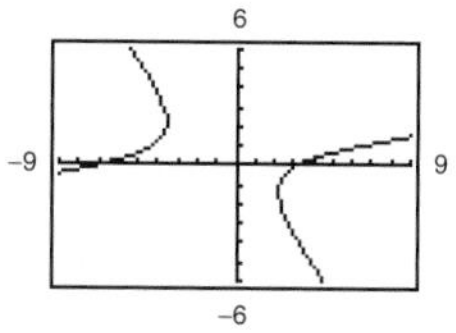

39. Parabola

$$y = \frac{-(4x - 1) \pm \sqrt{(4x - 1)^2 - 16(x^2 - 5x - 3)}}{8}$$

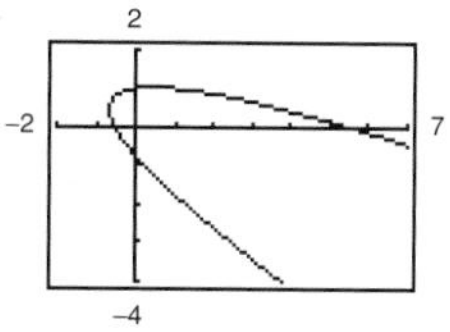

41.

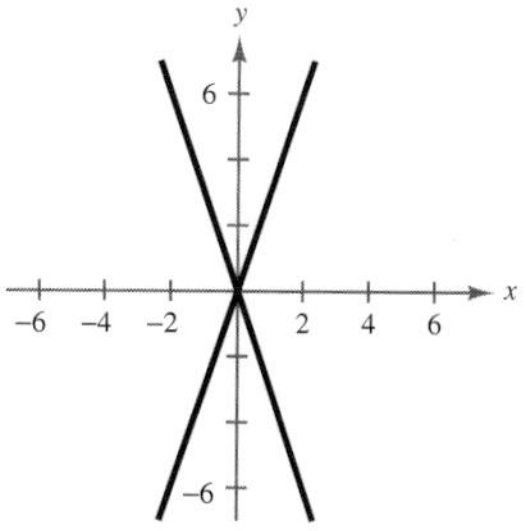

43.

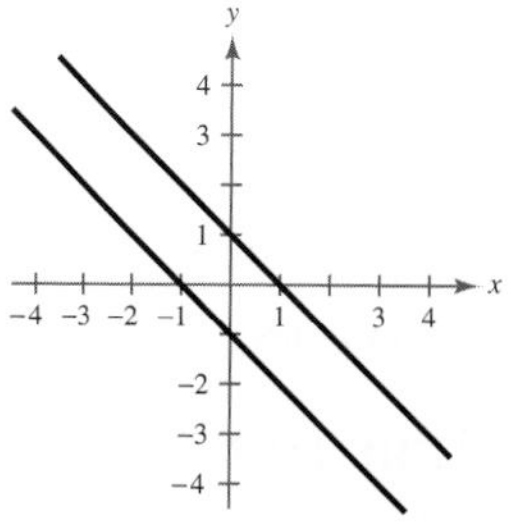

45. (2, 2), (2, 4) **47.** (−8, 12) **49.** (0, 8), (12, 8)

51. (0, 4) **53.** $(1, \sqrt{3}), (1, -\sqrt{3})$

55. No solution **57.** $(0, \frac{3}{2}), (-3, 0)$

59. True. The graph of the equation can be classified by finding the discriminant. For a graph to be a hyperbola, the discriminant must be greater than zero. If $k \geq \frac{1}{4}$, then the discriminant would be less than or equal to zero.

61. Answers will vary. **63.** ±3 **65.** ±2 **67.** −2

69.

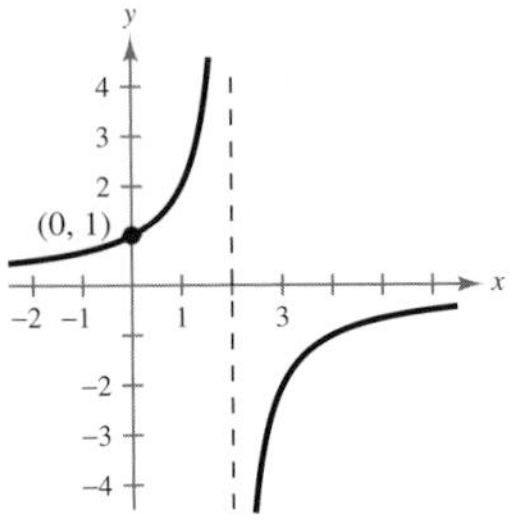

Intercept: $(0, 1)$

Asymptotes: $x = 2, y = 0$

71.

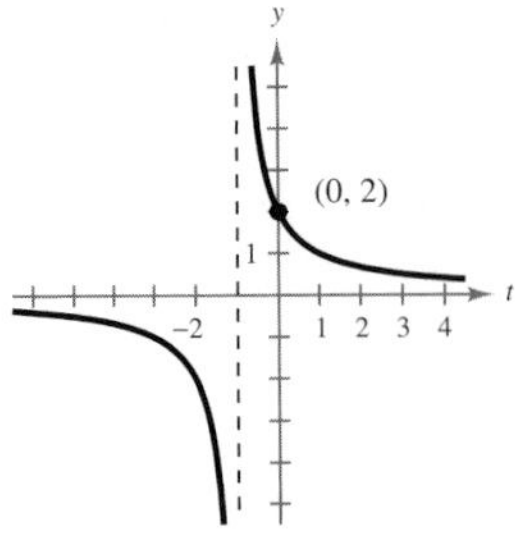

Intercept: $(0, 2)$

Asymptotes: $t = -1, y = 0$

73.

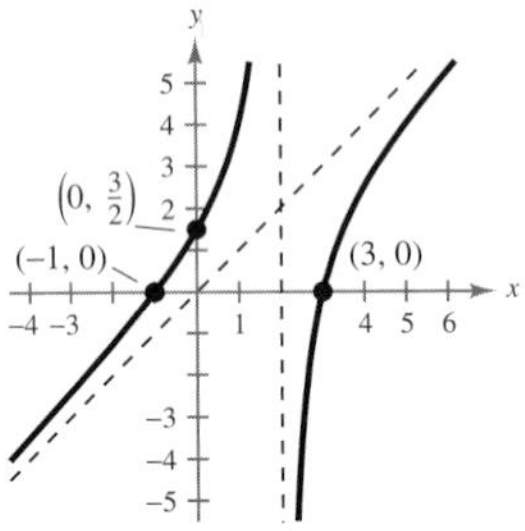

Intercepts: $(-1, 0), (3, 0), \left(0, \frac{3}{2}\right)$

Asymptotes: $x = 2, y = x$

75.

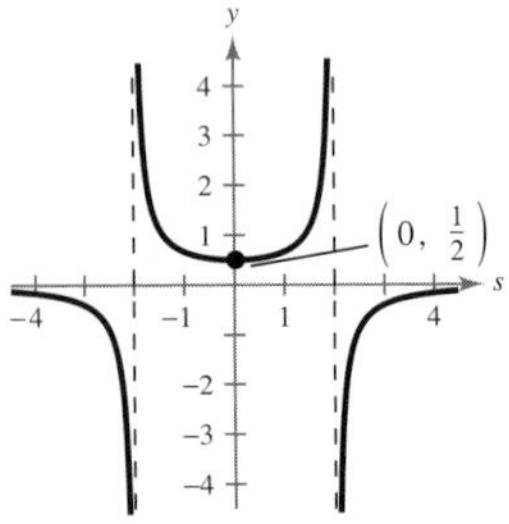

Intercept: $\left(0, \frac{1}{2}\right)$

Asymptotes: $s = \pm 2, y = 0$

Section 10.6 *(page 817)*

1. (a)

t	0	1	2	3	4
x	0	1	$\sqrt{2}$	$\sqrt{3}$	2
y	3	2	1	0	-1

(b)

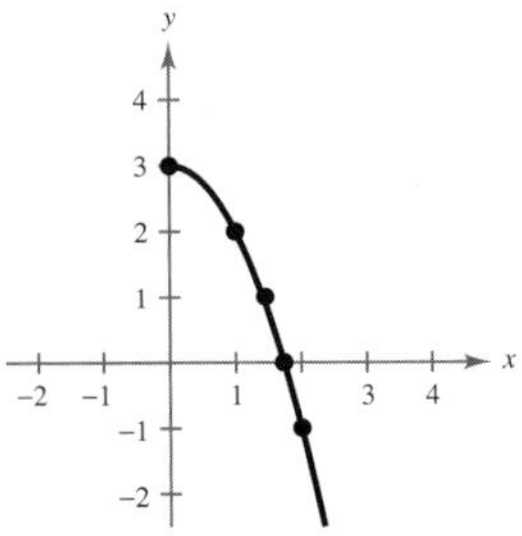

(c) $y = 3 - x^2$

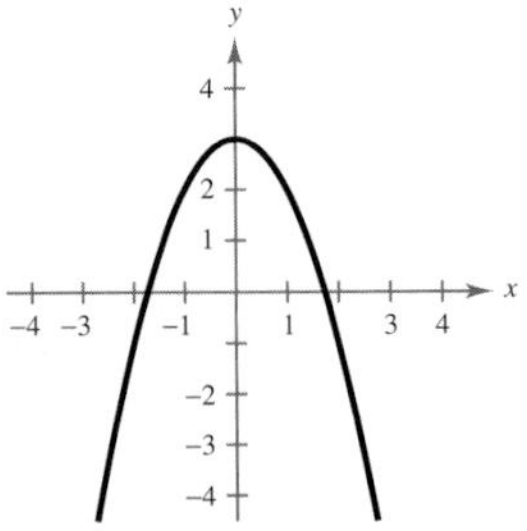

The graph of the rectangular equation shows the entire parabola rather than just the right half.

3.

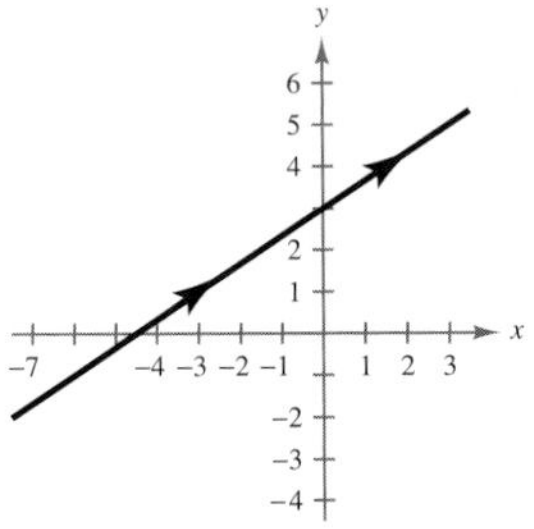

5.

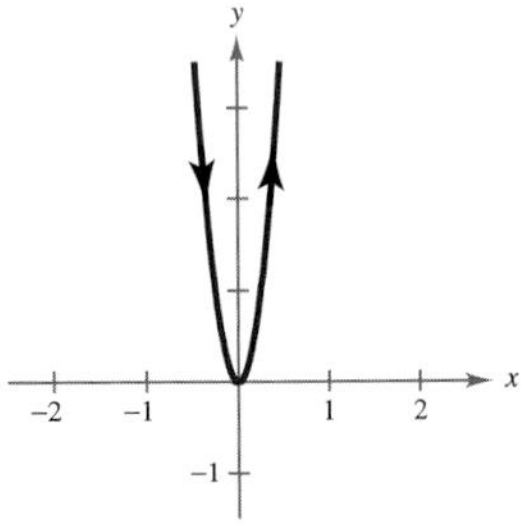

7.

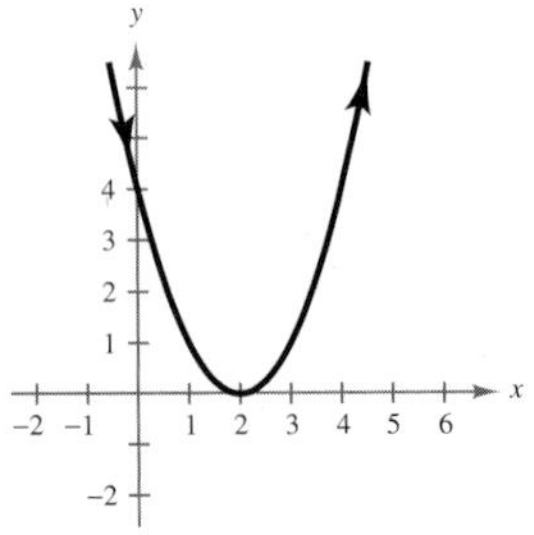

9.

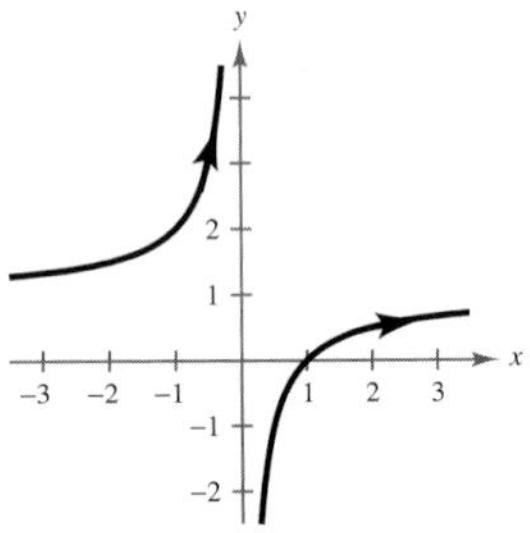

11.

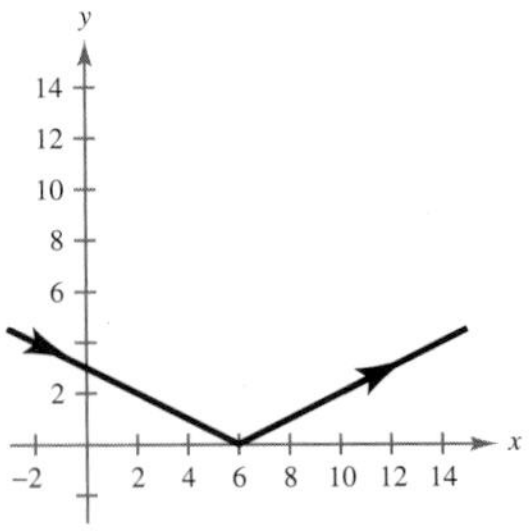

13.

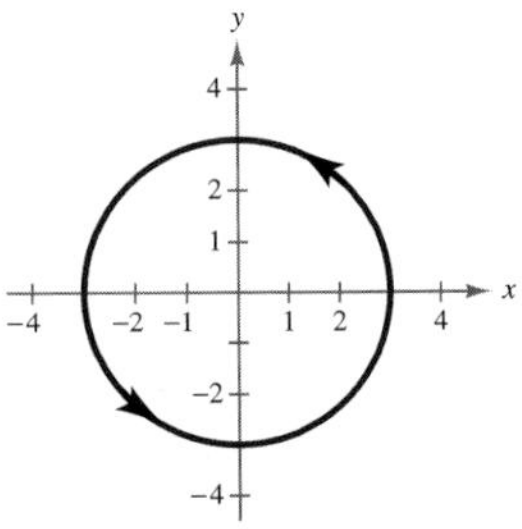

15.

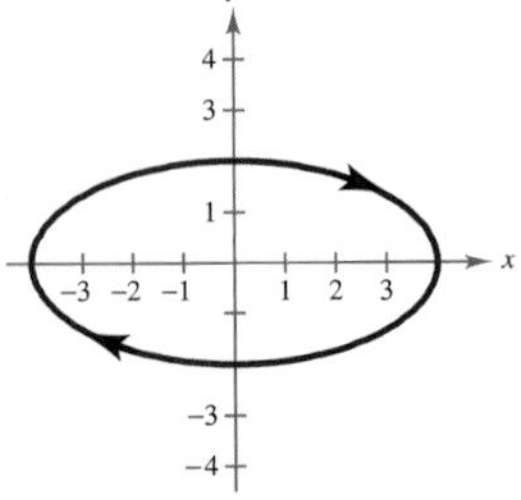

17.

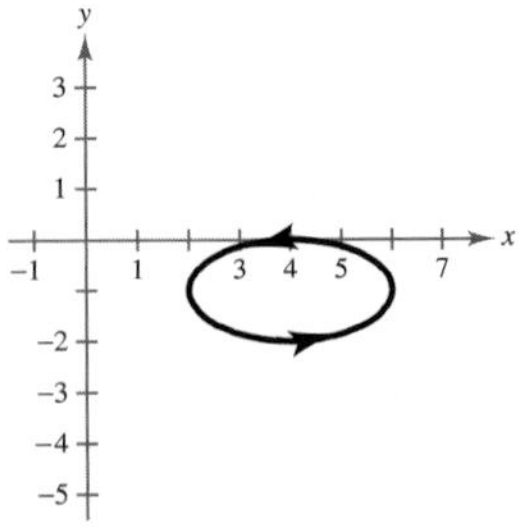

19.

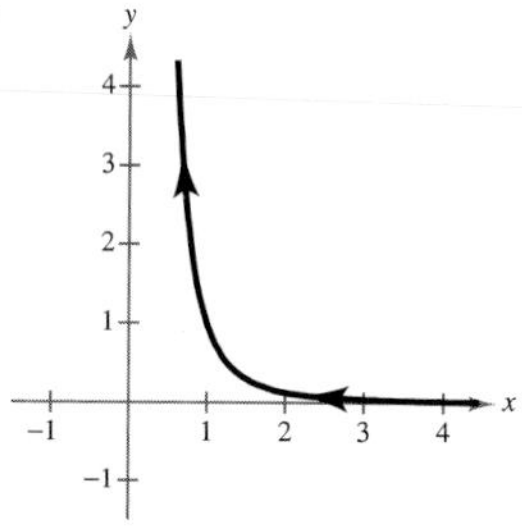

21.

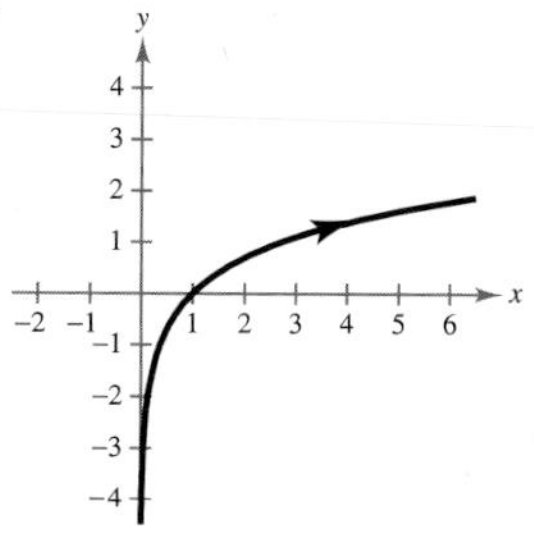

23. Each curve represents a portion of the line $y = 2x + 1$.

	Domain	*Orientation*
(a)	$(-\infty, \infty)$	Left to right
(b)	$[-1, 1]$	Depends on θ
(c)	$(0, \infty)$	Right to left
(d)	$(0, \infty)$	Left to right

25. $y - y_1 = m(x - x_1)$ **27.** $\dfrac{(x-h)^2}{a^2} + \dfrac{(y-k)^2}{b^2} = 1$

29. $x = 6t$, $y = -3t$ **31.** $x = 3 + 4\cos\theta$, $y = 2 + 4\sin\theta$ **33.** $x = 4\cos\theta$, $y = \sqrt{7}\sin\theta$

35. $x = 4\sec\theta$, $y = 3\tan\theta$ **37.** (a) $x = t,\ y = 3t - 2$ (b) $x = -t + 2,\ y = -3t + 4$

39. (a) $x = t, y = t^2$

(b) $x = -t + 2, y = t^2 - 4t + 4$

41. (a) $x = t, y = t^2 + 1$

(b) $x = -t + 2, y = t^2 - 4t + 5$

43. (a) $x = t, y = \dfrac{1}{t}$

(b) $x = -t + 2, y = -\dfrac{1}{t - 2}$

45.

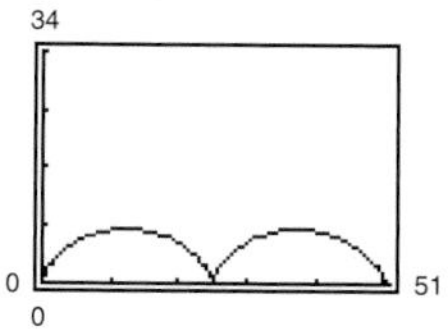

47.

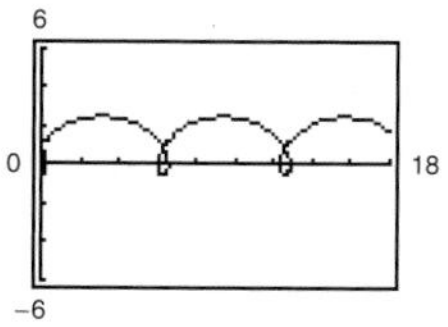

49.

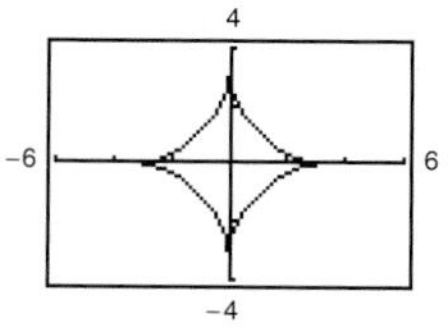

51.

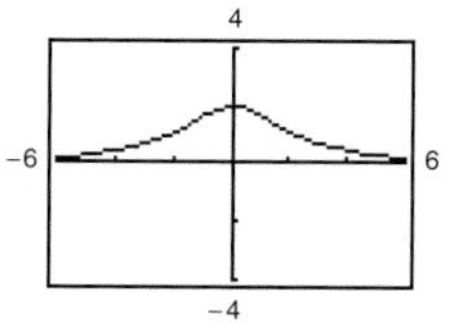

53. b

Domain: $[-2, 2]$

Range: $[-1, 1]$

55. d

Domain: $(-\infty, \infty)$

Range: $(-\infty, \infty)$

57. (a)

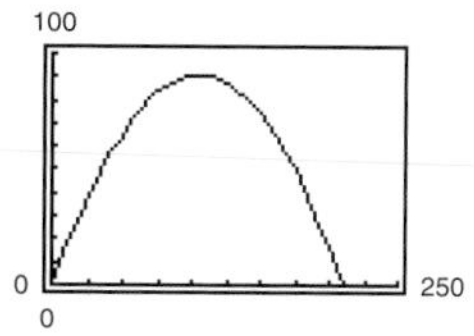

Maximum height: 90.7 feet

Range: 209.6 feet

(b)

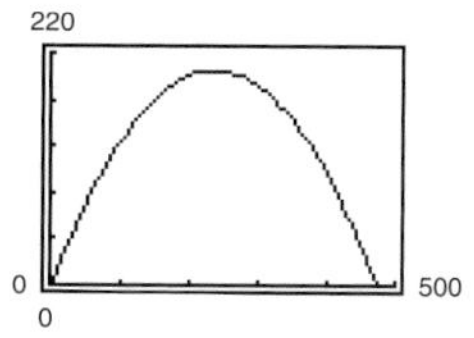

Maximum height: 204.2 feet

Range: 471.6 feet

(c)

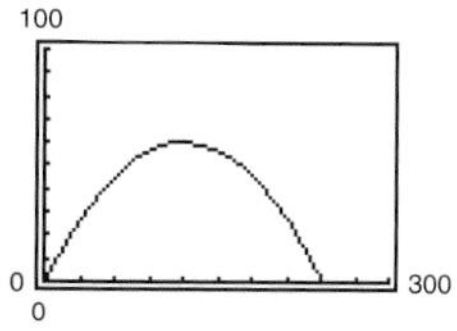

Maximum height: 60.5 feet

Range: 242.0 feet

(d)

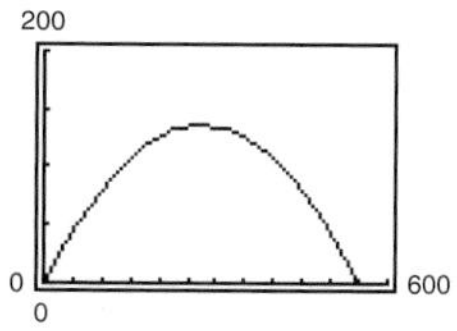

Maximum height: 136.1 feet

Range: 544.5 feet

59. (a) $x = (146.67\cos\theta)t$

$y = 3 + (146.67\sin\theta)t - 16t^2$

(b)

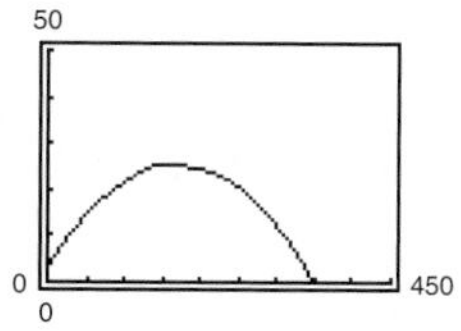

No

(c)

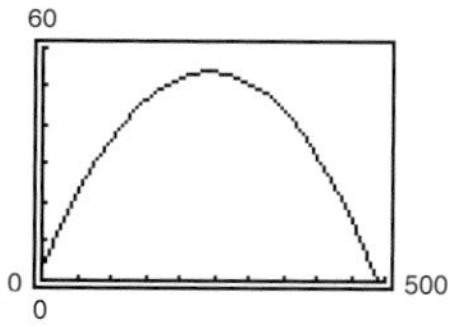

Yes

(d) 19.4°

61. Answers will vary. **63.** $x = a\theta - b \sin \theta$
$y = a - b \cos \theta$

65. True

$x = t$

$y = t^2 + 1 \Rightarrow y = x^2 + 1$

$x = 3t$

$y = 9t^2 + 1 \Rightarrow y = x^2 + 1$

67. $(5, 2)$ **69.** $(1, -2, 1)$ **71.** $y = \frac{1}{3}x^2 - 2x$

73. $y = \frac{7}{8}x^2 - 14x + 54$

Section 10.7 *(page 824)*

1.

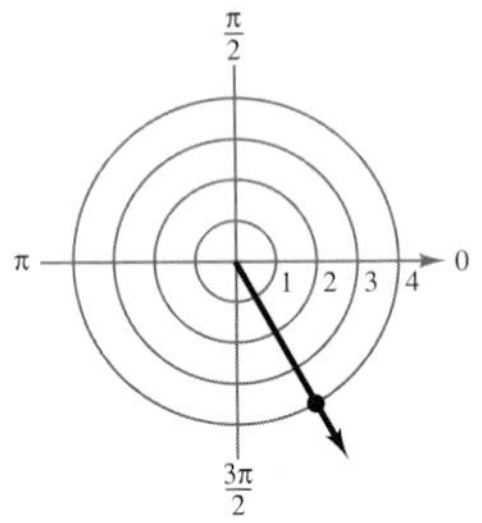

$\left(4, \frac{5\pi}{3}\right), \left(-4, -\frac{4\pi}{3}\right)$

3.

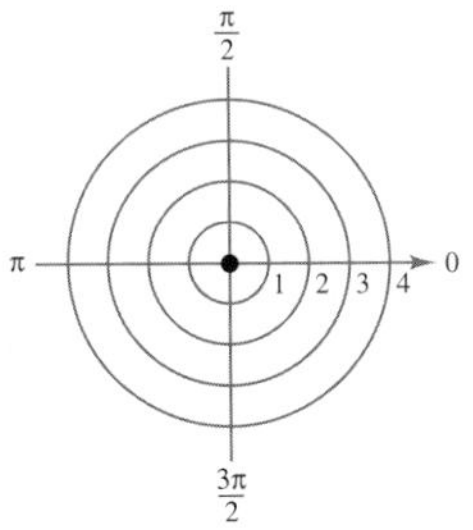

$\left(0, \frac{5\pi}{6}\right), \left(0, -\frac{13\pi}{6}\right)$

5.

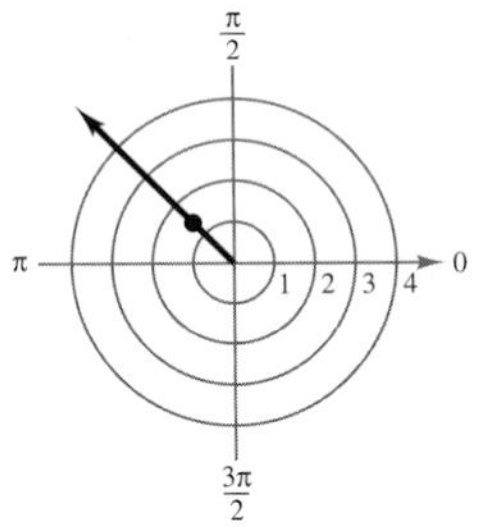

$(\sqrt{2}, 8.64), (-\sqrt{2}, -0.78)$

7.

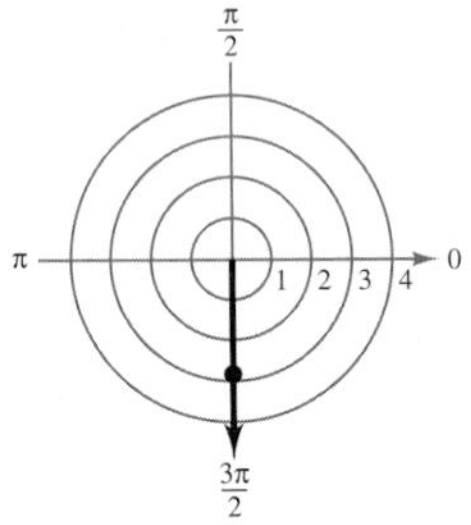

$(2\sqrt{2}, 10.99), (-2\sqrt{2}, 7.85)$

9. $(0, 3)$ **11.** $\left(\frac{\sqrt{2}}{2}, \frac{\sqrt{2}}{2}\right)$ **13.** $(-\sqrt{2}, \sqrt{2})$

15. $(-1.1, -2.2)$ **17.** $\left(\sqrt{2}, \frac{\pi}{4}\right)$ **19.** $(6, \pi)$

21. $(5, 2.2143)$ **23.** $\left(\sqrt{6}, \frac{5\pi}{4}\right)$ **25.** $(3\sqrt{13}, 0.9828)$

27. $(\sqrt{13}, -0.5880)$ **29.** $(\sqrt{7}, 0.8571)$

31. $\left(\frac{17}{6}, 0.4900\right)$ **33.** $r = 3$ **35.** $r = 4 \csc \theta$

37. $r = \dfrac{-2}{3 \cos \theta - \sin \theta}$

39. $r^2 = 16 \sec \theta \csc \theta = 32 \csc 2\theta$

41. $r = \dfrac{4}{1 - \cos \theta}$ or $-\dfrac{4}{1 + \cos \theta}$ **43.** $r = a$

45. $r = b \csc \theta$ **47.** $r = 2a \cos \theta$

49. $x^2 + y^2 - 4y = 0$ **51.** $\sqrt{3}x + y = 0$

53. $y = 2$ **55.** $(x^2 + y^2)^2 = 6x^2y - 2y^3$

57. $4x^2 - 5y^2 - 36y - 36 = 0$

59. $x^2 + y^2 = 36$

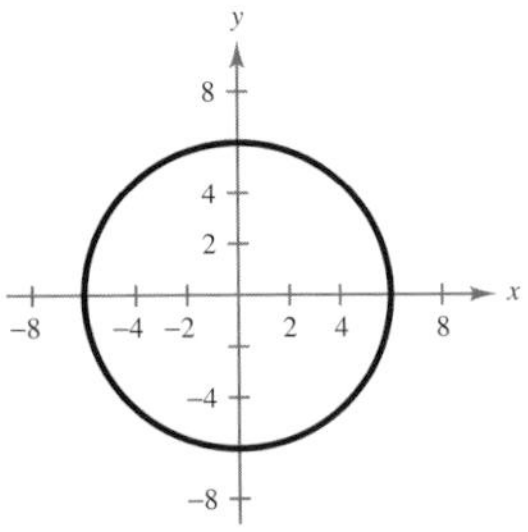

61. $-\sqrt{3}x + 3y = 0$

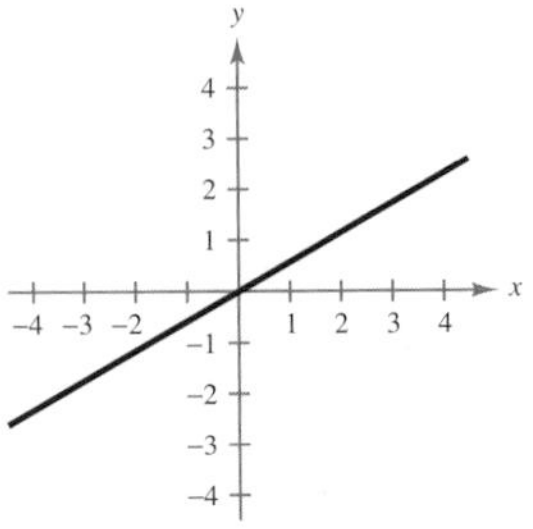

63. $x - 3 = 0$

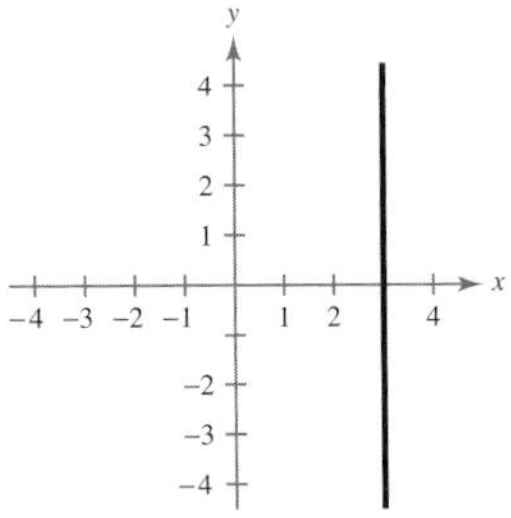

65. True. Because r is a directed distance, the point (r, θ) can be represented as $(r, \theta \pm 2n\pi)$.

67. $(x - h)^2 + (y - k)^2 = h^2 + k^2$

Radius: $\sqrt{h^2 + k^2}$

Center: (h, k)

69. (a) Answers will vary.

(b) (r_1, θ_1), (r_2, θ_2) and the pole are collinear.

$d = \sqrt{r_1^2 + r_2^2 - 2r_1r_2} = |r_1 - r_2|$

The distance between two points on the line $\theta = \theta_1 = \theta_2$.

(c) $d = \sqrt{r_1^2 + r_2^2}$

Pythagorean Theorem

(d) Example: Points: $(3, \pi/6)$, $(4, \pi/3)$

Distance: 2.053

Points: $(-3, 7\pi/6)$, $(-4, 4\pi/3)$

Distance: 2.053

71. $(2, 3)$ **73.** $\left(\frac{8}{7}, \frac{88}{35}, \frac{8}{5}\right)$ **75.** $(1, -4, 1, 2)$

77. Not collinear **79.** Collinear

Section 10.8 *(page 832)*

1. Rose curve **3.** Limaçon **5.** Rose curve

7. Polar axis **9.** $\theta = \dfrac{\pi}{2}$ **11.** $\theta = \dfrac{\pi}{2}$, polar axis, pole

13. Maximum: $|r| = 20$ when $\theta = \dfrac{3\pi}{2}$

Zero: $r = 0$ when $\theta = \dfrac{\pi}{2}$

15. Maximum: $|r| = 4$ when $\theta = 0, \dfrac{\pi}{3}, \dfrac{2\pi}{3}$

Zero: $r = 0$ when $\theta = \dfrac{\pi}{6}, \dfrac{\pi}{2}, \dfrac{5\pi}{6}$

17.

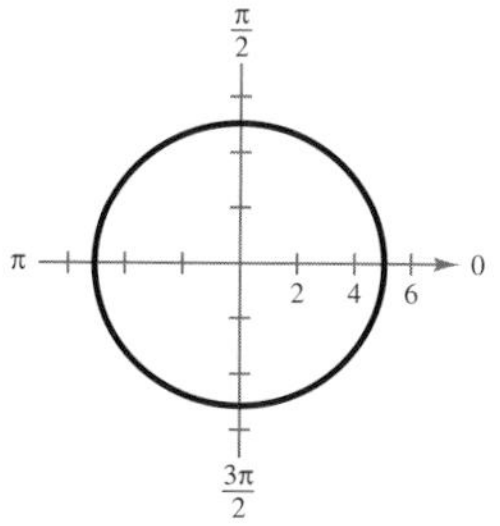

19.

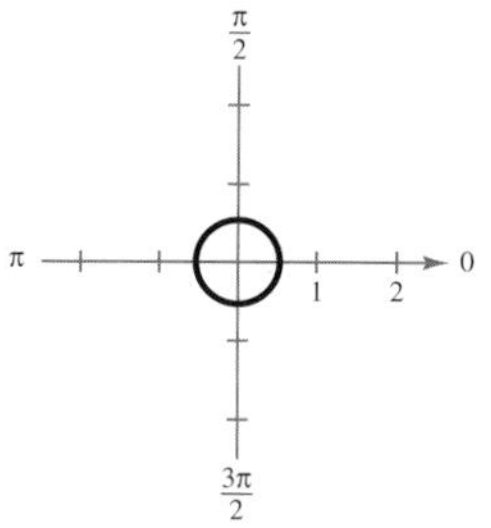

21.

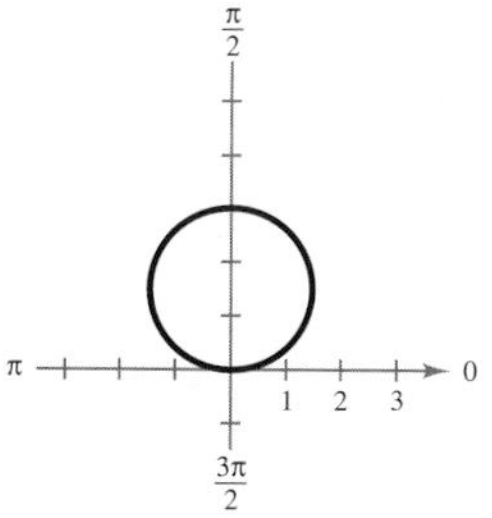

23.

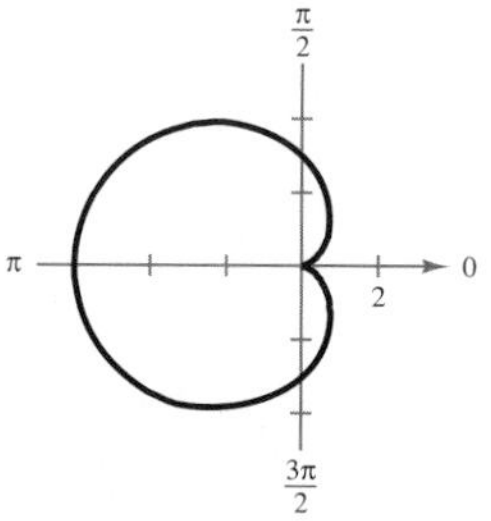

25.

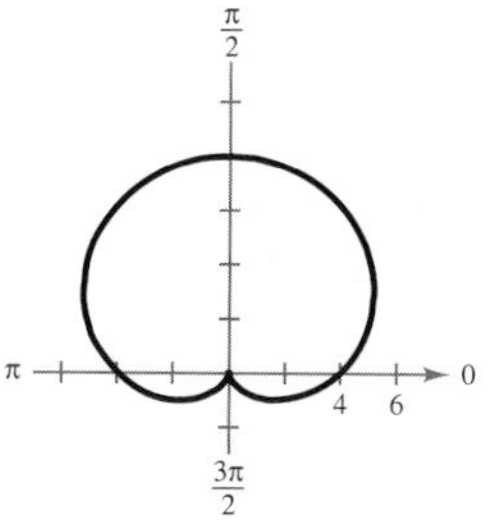

27.

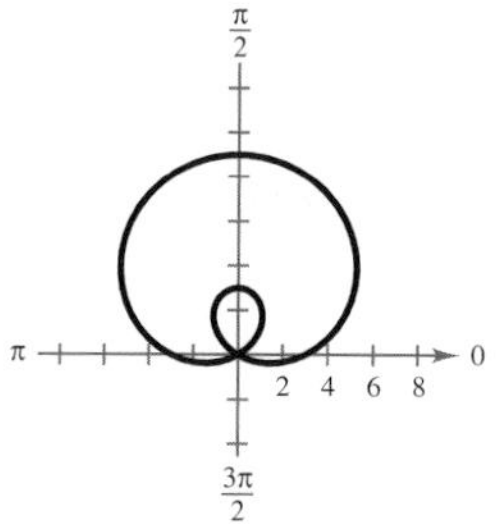

29.

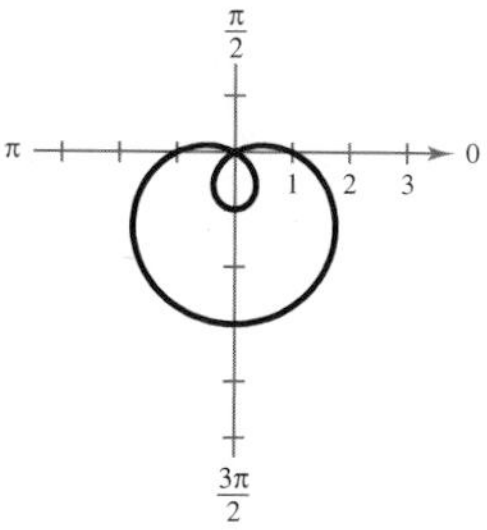

31.

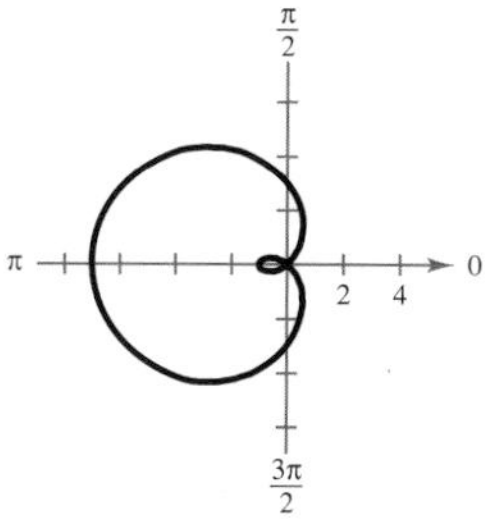

33.

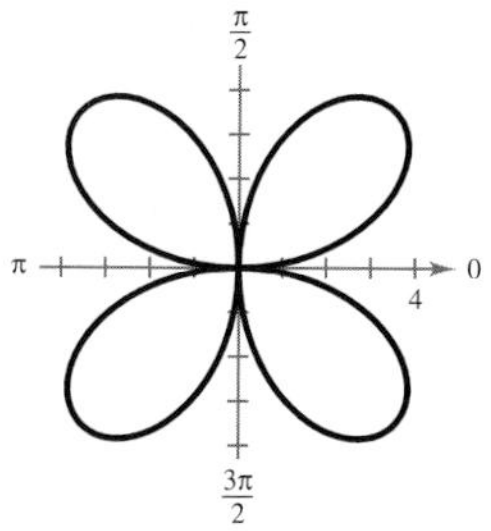

35.

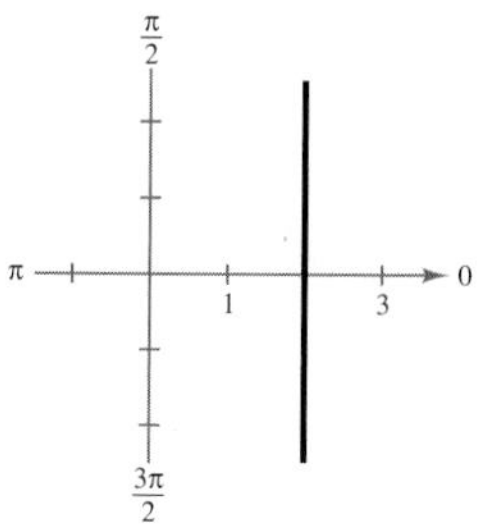

37.

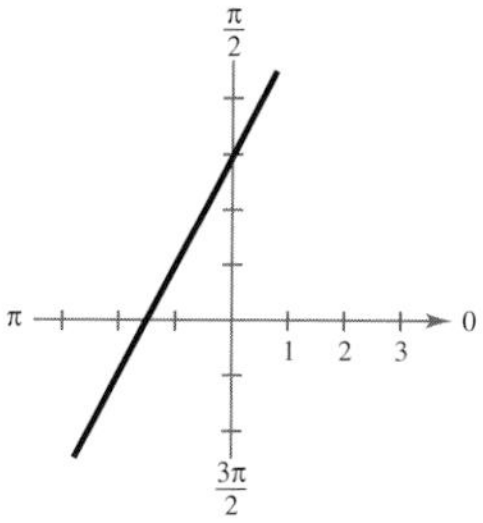

39.

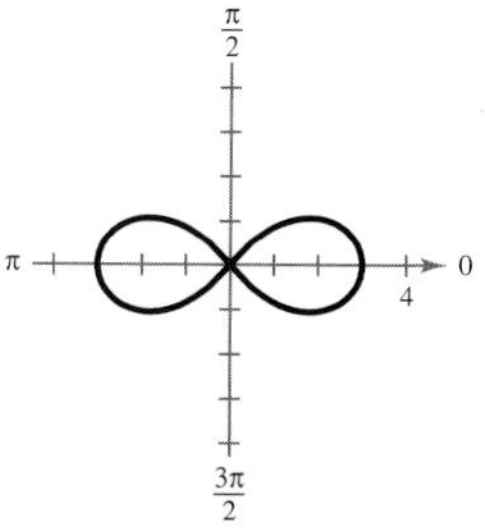

41.

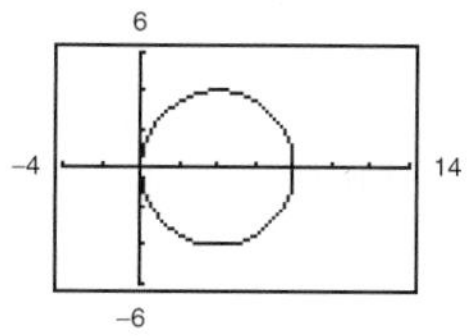

43.

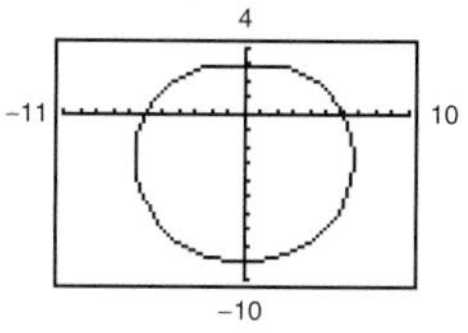

45.

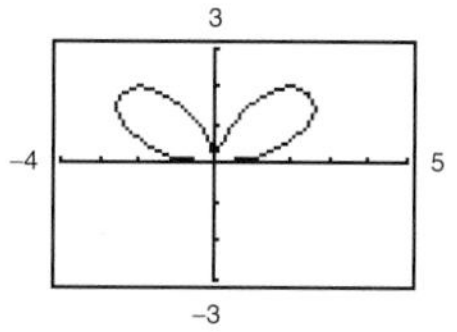

47.

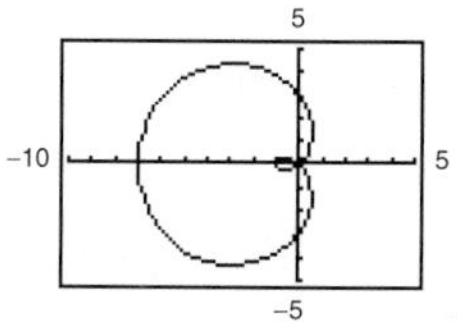

$0 \le \theta < 2\pi$

49.

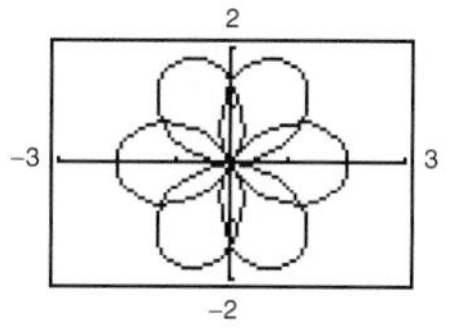

$0 \le \theta < 4\pi$

51.

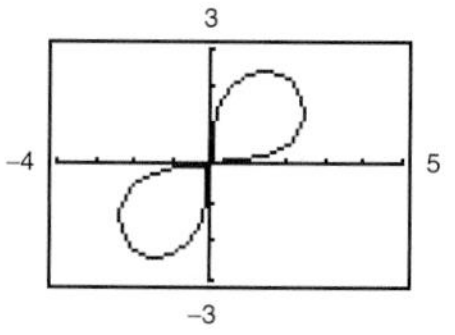

$0 \le \theta < \pi$

53.

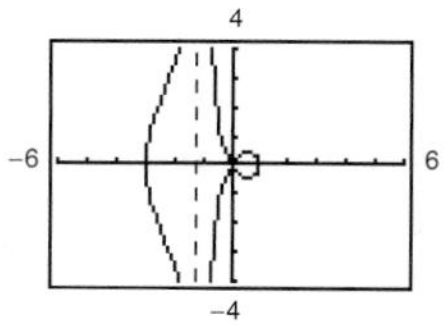

55.

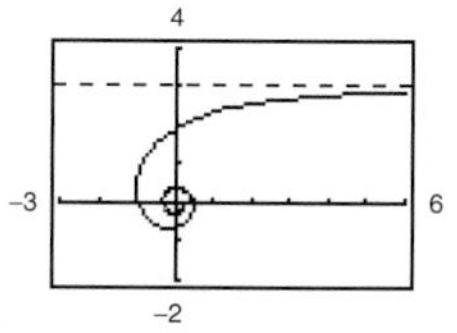

57. True. For a graph to have polar axis symmetry, replace (r, θ) by $(r, -\theta)$ or $(-r, \pi - \theta)$.

59. (a)

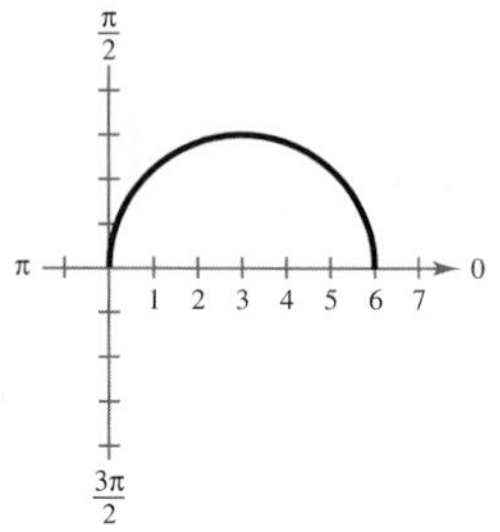

Upper half of circle

(b)

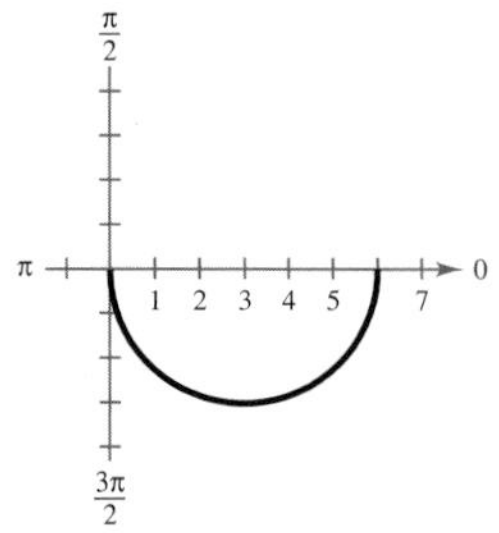

Lower half of circle

(c)

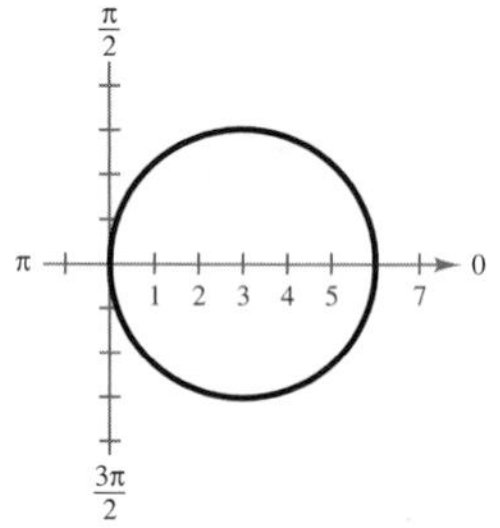

Full circle

(d)

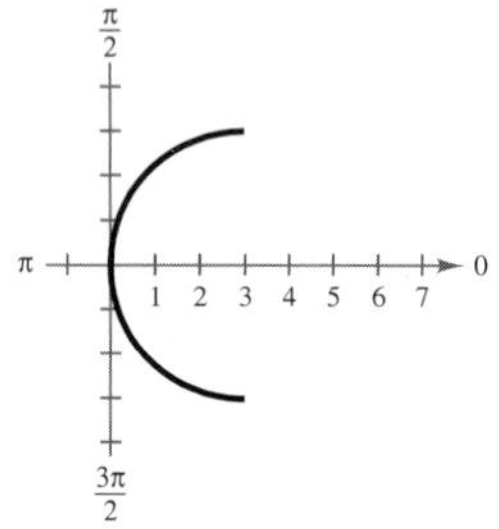

Left half of circle

61. Answers will vary.

63. (a) $r = 2 - \sin\left(\theta - \dfrac{\pi}{4}\right)$

(b) $r = 2 + \cos\theta$

(c) $r = 2 + \sin\theta$

(d) $r = 2 - \cos\theta$

65. (a)

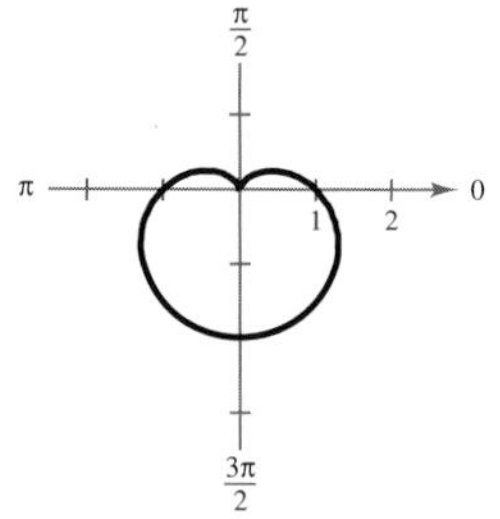

(b)

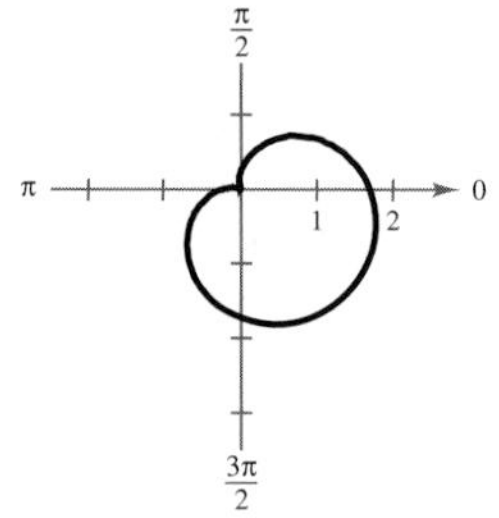

67.

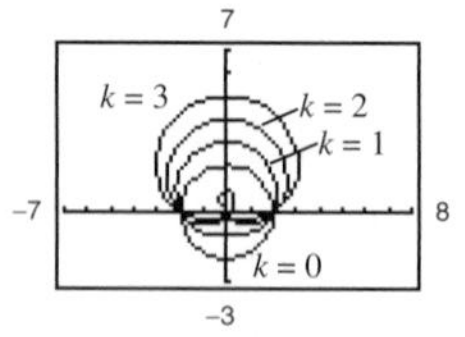

$k = 0$, circle

$k = 1$, limaçon

$k = 2$, cardioid

$k = 3$, limaçon

69. $\ln 19 \approx 2.944$ **71.** $\log_{10} 84 \approx 1.924$

73. $e^4 \approx 54.598$ **75.** ± 3 **77.** $\frac{13}{5}$

Section 10.9 *(page 838)*

1.

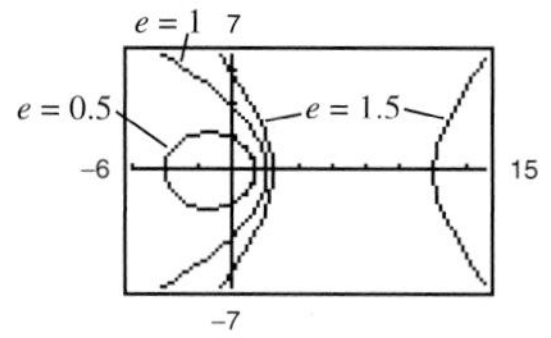

3.

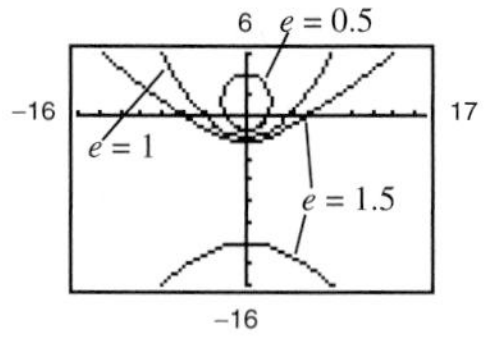

5. f **7.** d **9.** a

11. Parabola

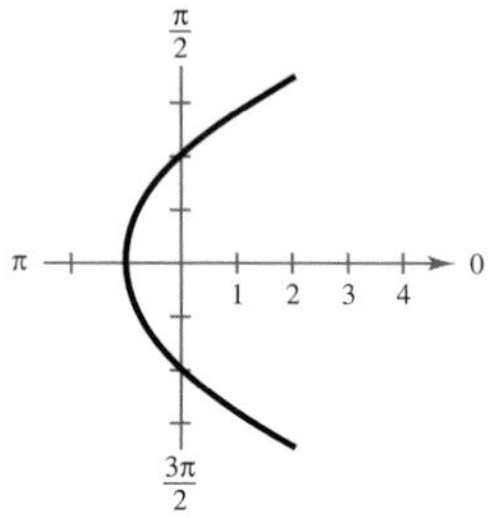

13. Parabola

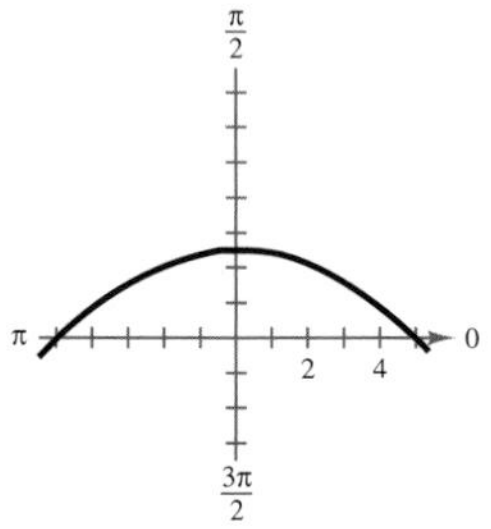

15. Ellipse

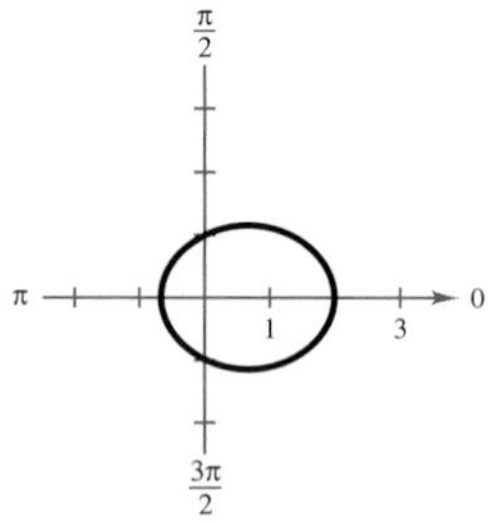

17. Ellipse

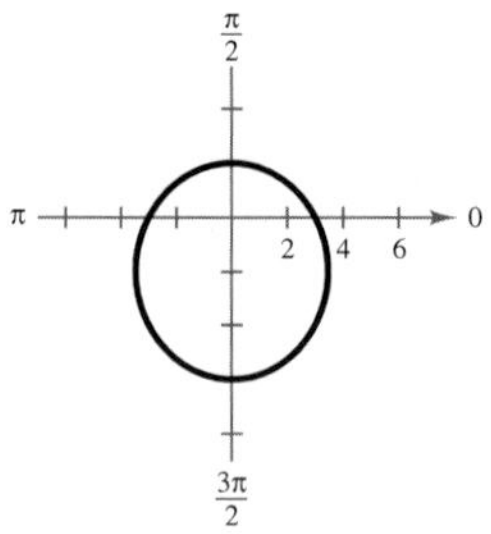

19. Hyperbola

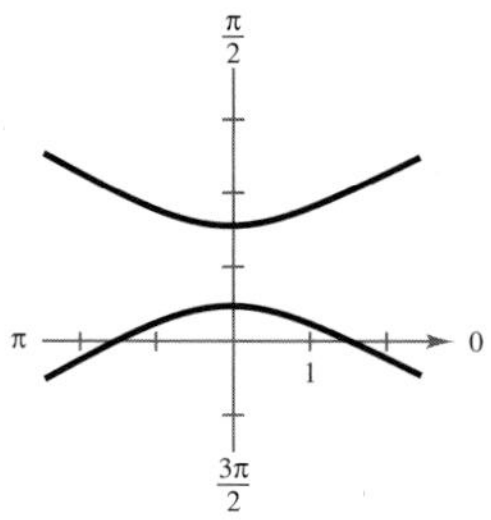

21. Hyperbola

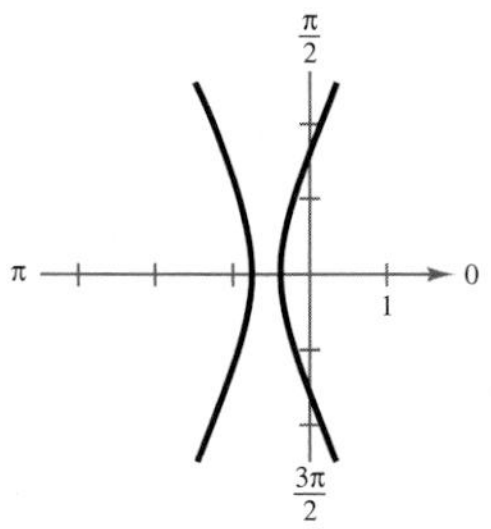

23. Ellipse

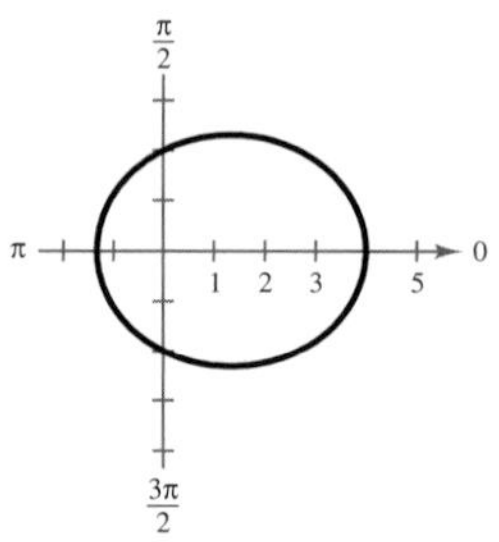

25.

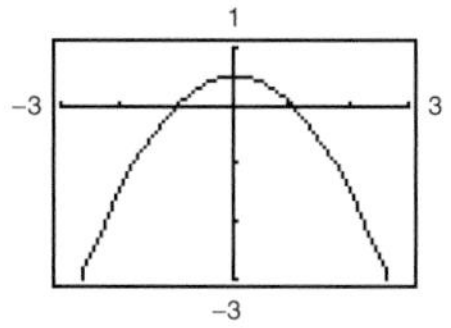

Parabola

27.

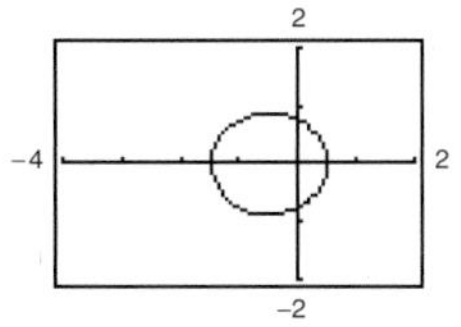

Ellipse

29.

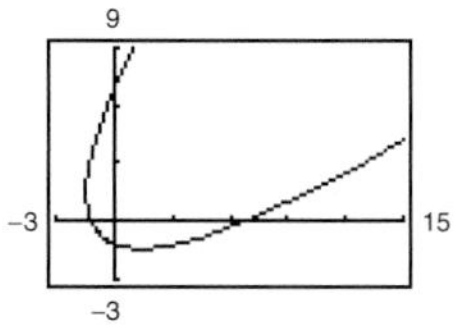

31.

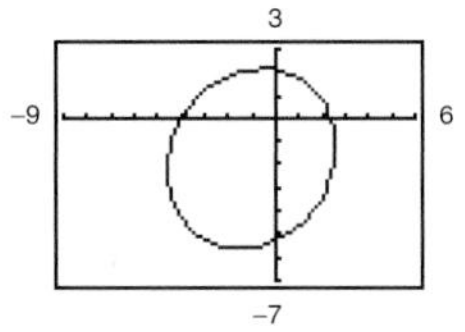

33. $r = \dfrac{1}{1 - \cos\theta}$ **35.** $r = \dfrac{1}{2 + \sin\theta}$

37. $r = \dfrac{2}{1 + 2\cos\theta}$ **39.** $r = \dfrac{2}{1 - \sin\theta}$

41. $r = \dfrac{10}{1 - \cos\theta}$ **43.** $r = \dfrac{10}{3 + 2\cos\theta}$

45. $r = \dfrac{20}{3 - 2\cos\theta}$ **47.** $r = \dfrac{9}{4 - 5\sin\theta}$

49. Answers will vary.

51. $r = \dfrac{9.2934 \times 10^7}{1 - 0.0167\cos\theta}$

Perihelion: 9.1408×10^7 miles

Aphelion: 9.4512×10^7 miles

53. $r = \dfrac{5.5368 \times 10^9}{1 - 0.2481\cos\theta}$

Perihelion: 4.4362×10^9 km

Aphelion: 7.3638×10^9 km

55. $r = \dfrac{1.3977 \times 10^8}{1 - 0.0934\cos\theta}$

Perihelion: 1.2783×10^8 miles

Aphelion: 1.5417×10^8 miles

57. $r = \dfrac{8200}{1 + \sin\theta}$

1467 miles

59. False. If e remains fixed and p changes, then the lengths of both the major axis and the minor axis change. For example, graph

$r = \dfrac{5}{1 - \frac{2}{3}\cos\theta}$, with $e = \frac{2}{3}$ and $p = \frac{15}{2}$, and graph

$r = \dfrac{6}{1 - \frac{2}{3}\cos\theta}$, with $e = \frac{2}{3}$ and $p = 9$, on the same set of coordinate axes.

61. Answers will vary. **63.** $r^2 = \dfrac{24{,}336}{169 - 25\cos^2\theta}$

65. $r^2 = \dfrac{144}{25\cos^2\theta - 9}$ **67.** $r^2 = \dfrac{144}{25\cos^2\theta - 16}$

69. $\dfrac{\pi}{6} + n\pi$

71. $\dfrac{\pi}{3} + n\pi, \dfrac{2\pi}{3} + n\pi$

73. $\dfrac{\pi}{2} + n\pi$ **75.** $\dfrac{\sqrt{2}}{10}$ **77.** $\dfrac{7\sqrt{2}}{10}$

79. $\sin 2u = -\frac{24}{25}$

$\cos 2u = -\frac{7}{25}$

$\tan 2u = \frac{24}{7}$

Review Exercises *(page 842)*

1. 30.96° **3.** 63.43° **5.** 25.35° **7.** 37.75°

9. $2\sqrt{2}$ **11.** Hyperbola **13.** $(x - 4)^2 = -8(y - 2)$

15. $(y - 2)^2 = 12x$ **17.** $y = -2x + 2$; $(1, 0)$

19. $8\sqrt{6}$ meters **21.** $\dfrac{(x - 2)^2}{25} + \dfrac{y^2}{21} = 1$

23. $\dfrac{2x^2}{9} + \dfrac{y^2}{36} = 1$

25. The foci occur 3 feet from the center of the arch on a line connecting the tops of the pillars.

27. Center: $(1, -4)$

Vertices: $(1, 0), (1, -8)$

Foci: $\left(1, -4 \pm \sqrt{7}\right)$

Eccentricity: $\dfrac{\sqrt{7}}{4}$

29. Center: $(-2, 1)$

Vertices: $(-2, 11), (-2, -9)$

Foci: $\left(-2, 1 \pm \sqrt{19}\right)$

Eccentricity: $\dfrac{\sqrt{19}}{10}$

31. $y^2 - \dfrac{x^2}{8} = 1$ **33.** $\dfrac{5(x - 4)^2}{16} - \dfrac{5y^2}{64} = 1$

35. Center: $(1, -1)$

Vertices: $(5, -1), (-3, -1)$

Foci: $(6, -1), (-4, -1)$

Asymptotes: $y = -1 \pm \frac{3}{4}(x - 1)$

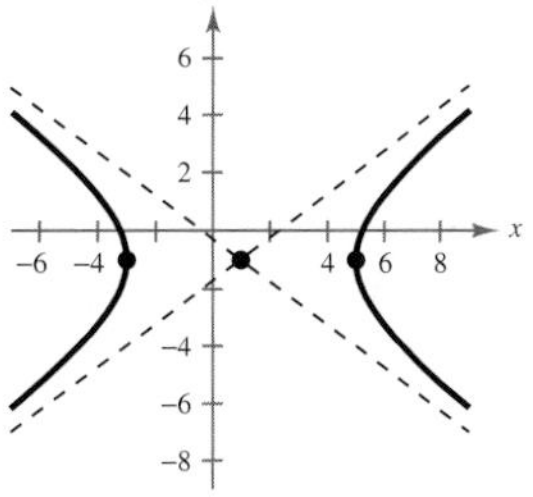

37. Center: $(3, -5)$

Vertices: $(7, -5), (-1, -5)$

Foci: $\left(3 \pm 2\sqrt{5}, -5\right)$

Asymptotes: $y = -5 \pm \frac{1}{2}(x - 3)$

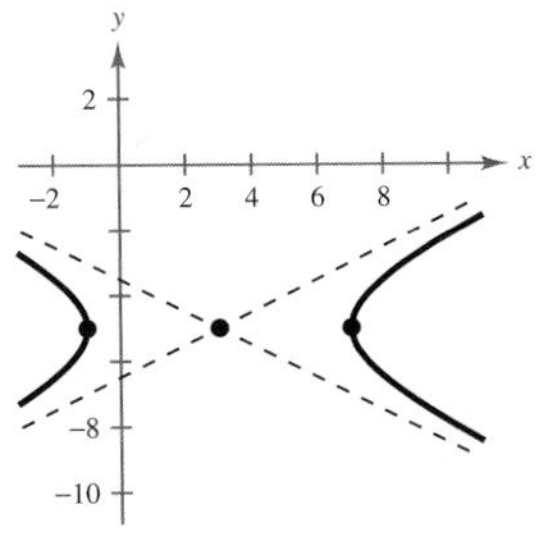

39. 72 miles **41.** Hyperbola

43. $\dfrac{(x')^2}{8} - \dfrac{(y')^2}{8} = 1$ **45.** $\dfrac{(x')^2}{3} + \dfrac{(y')^2}{2} = 1$

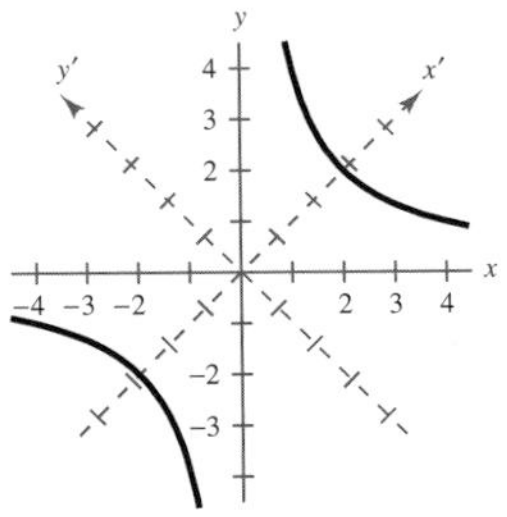

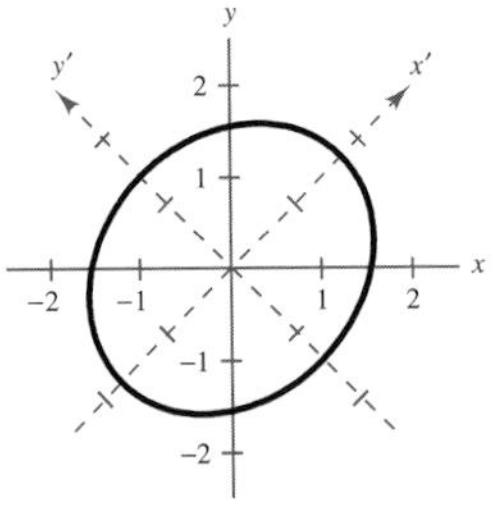

47. Parabola

$$y = \frac{24x + 40 \pm \sqrt{(24x + 40)^2 - 36(16x^2 - 30x)}}{18}$$

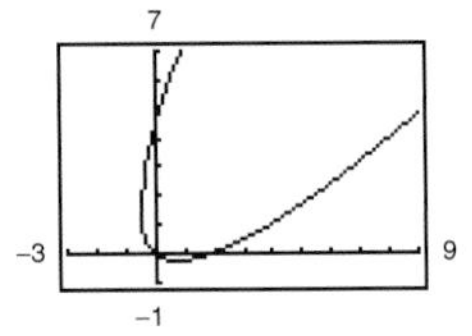

49. Parabola

$$y = \frac{-\left(2x - 2\sqrt{2}\right) \pm \sqrt{\left(2x - 2\sqrt{2}\right)^2 - 4\left(x^2 + 2\sqrt{2}x + 2\right)}}{2}$$

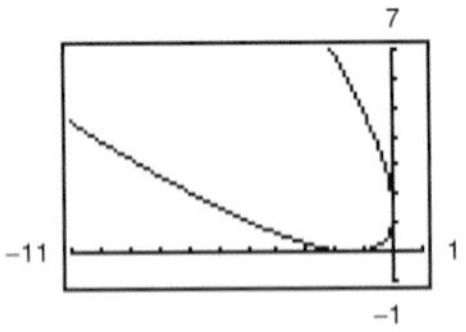

51. $x = 3, y = 0$ **53.** $x = \dfrac{3\sqrt{3}}{2}, y = \dfrac{1}{2}$

55.

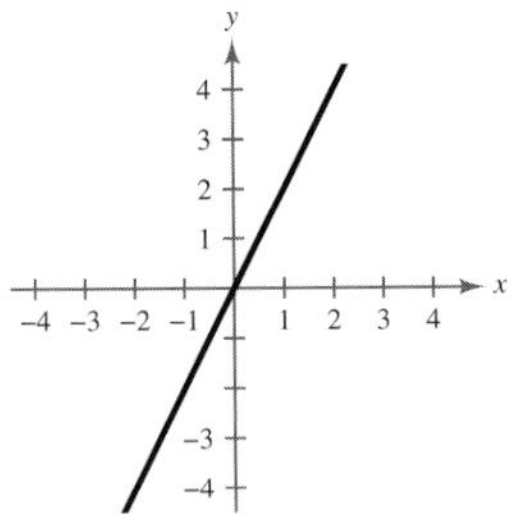

$y = 2x$

57.

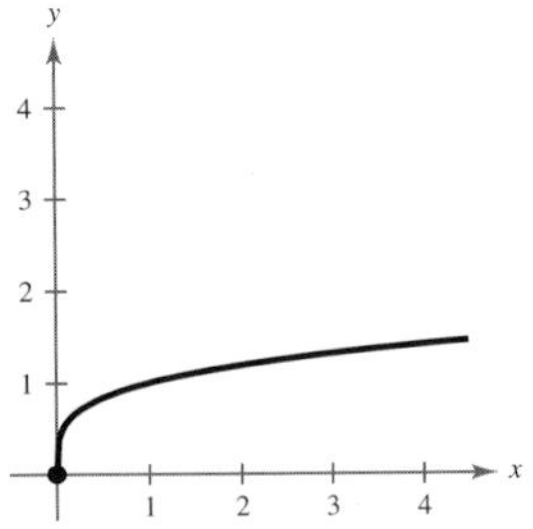

$y = \sqrt[4]{x}$

59.

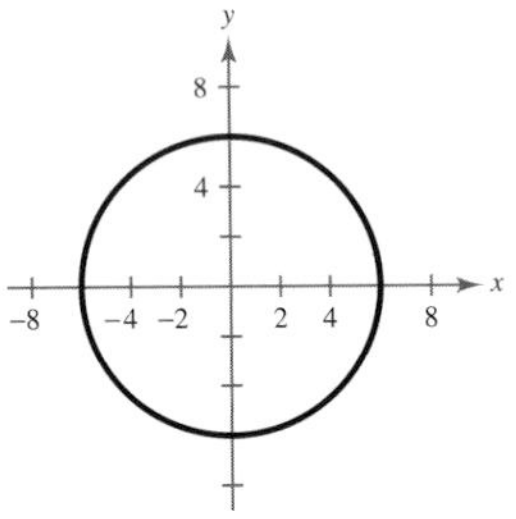

$x^2 + y^2 = 36$

61. $x = -3 + 4\cos\theta$

$y = 4 + 3\sin\theta$

63.

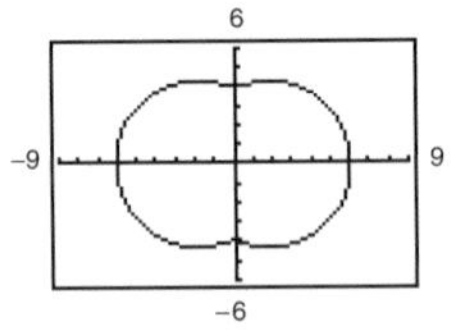

65.

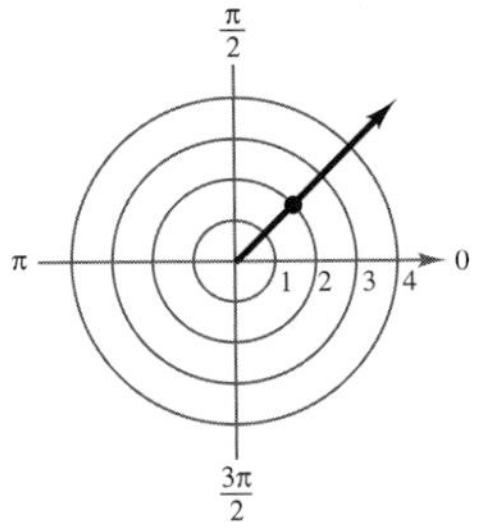

$\left(\sqrt{2}, \sqrt{2}\right)$

67.

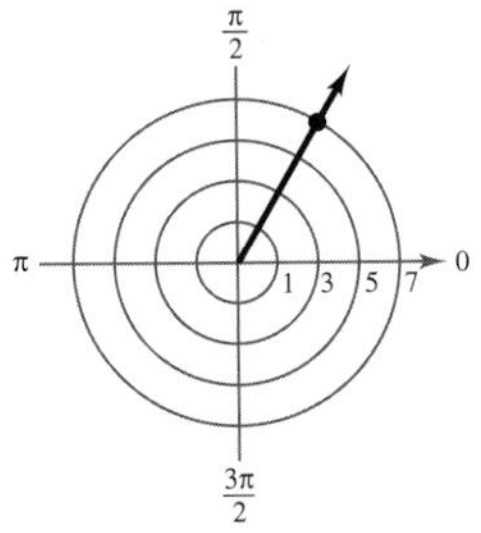

$(3.4927, 6.0664)$

69. $\left(2, \dfrac{\pi}{2}\right), \left(-2, \dfrac{3\pi}{2}\right)$

71. $(7.2111, 0.9828), (-7.211, 4.124)$ **73.** $x^2 + y^2 = 3x$

75. $x^2 + 4y - 4 = 0$ **77.** $r = a\cos^2\theta\sin\theta$

79. Circle

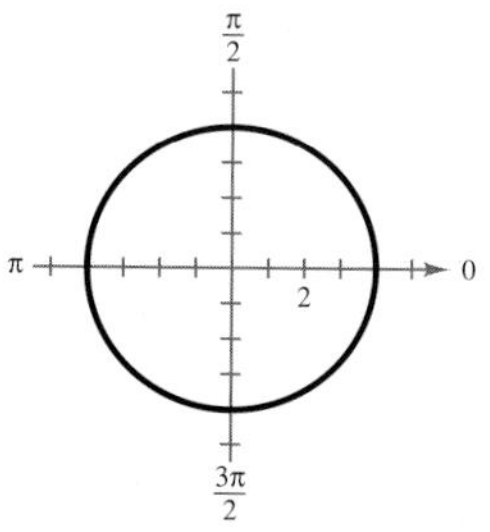

81. Rose curve

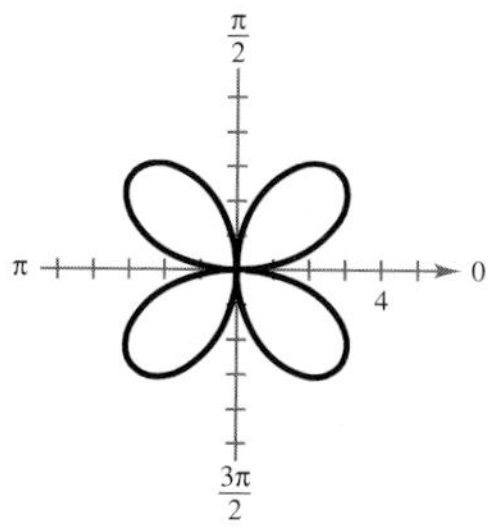

83. Cardioid

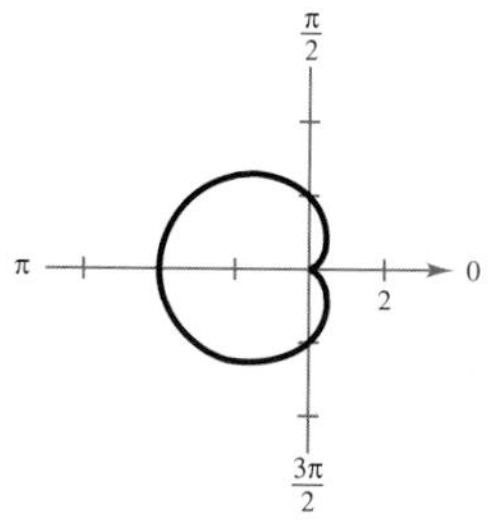

85. Symmetry: $\theta = \frac{\pi}{2}$

Maximum value of $|r|$: $|r| = 8$ when $\theta = \frac{\pi}{2}$

Zeros of r: $r = 0$ when $\theta = 3.4814, 5.9433$

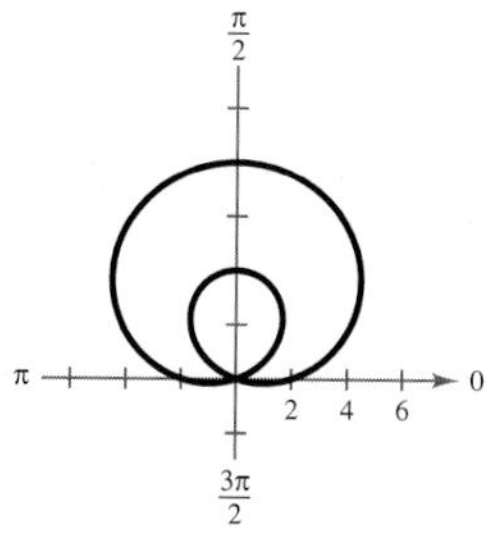

87. Symmetry: $\theta = \frac{\pi}{2}$, polar axis

Maximum value of $|r|$: $|r| = 3$ when $\theta = 0, \frac{\pi}{2}, \pi, \frac{3\pi}{2}$

Zeros of r: $r = 0$ when $\theta = \frac{\pi}{4}, \frac{3\pi}{4}$

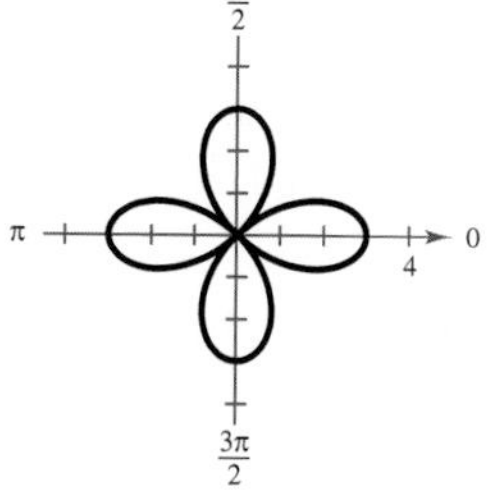

89. Limaçon **91.** Rose curve

93. Eccentricity: 2; Hyperbola

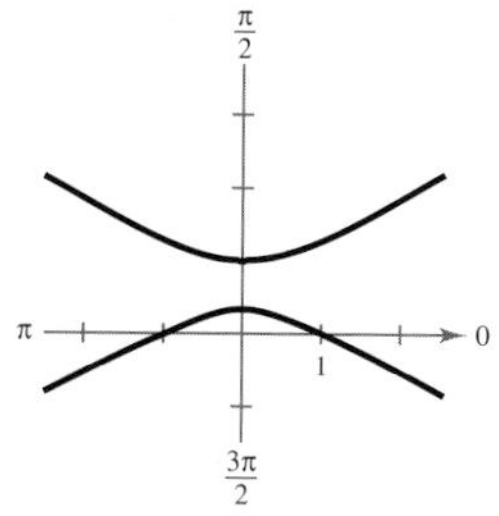

95. Eccentricity: $\frac{3}{5}$; Ellipse

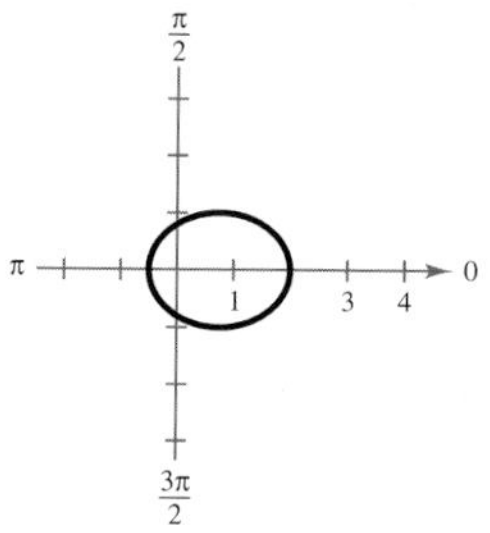

97. $r = \dfrac{4}{1 - \cos\theta}$ **99.** $r = \dfrac{5}{3 - 2\cos\theta}$

101. $r = \dfrac{7977.2}{1 - 0.937\cos\theta}$; 11,008 miles

103. False. When classifying an equation of the form $Ax^2 + Bxy + Cy^2 + Dx + Ey + F = 0$, its graph can be determined by its discriminant. For a graph to be a parabola, its discriminant, $B^2 - 4AC$, must equal zero. So, if $B = 0$, then A or C equals 0.

105. False. The following are two sets of parametric equations for the line.

$x = t,\ y = 3 - 2t$

$x = 3t,\ y = 3 - 6t$

107. 5. The ellipse becomes more circular and approaches a circle of radius 5.

109. (a) The speed would double.

(b) The elliptical orbit would be flatter; the length of the major axis would be greater.

111. (a) The graphs are the same.

(b) The graphs are the same.

Chapter Test *(page 847)*

1. 15.9° **2.** 47.7° **3.** $\frac{7\sqrt{2}}{2}$

4. Parabola: $y^2 = 4(x - 1)$

Vertex: $(1, 0)$

Focus: $(2, 0)$

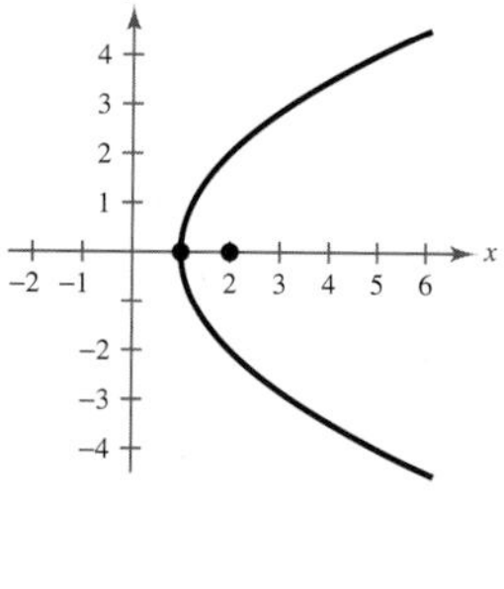

5. Hyperbola: $\dfrac{(x-2)^2}{4} - y^2 = 1$

Center: $(2, 0)$

Vertices: $(0, 0)$, $(4, 0)$

Foci: $\left(2 \pm \sqrt{5}, 0\right)$

Asymptotes: $y = \pm\dfrac{1}{2}(x - 2)$

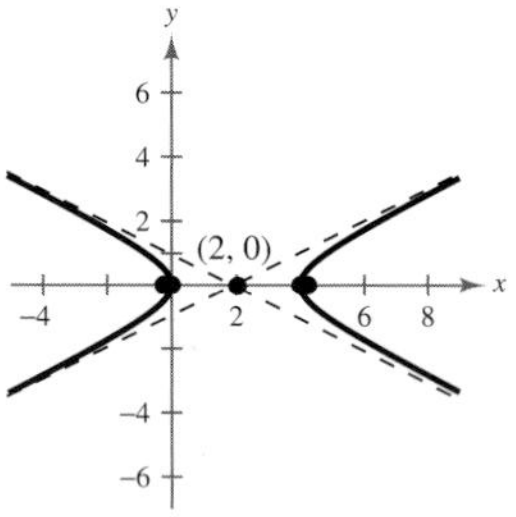

6. Ellipse: $\dfrac{(x+3)^2}{16} + \dfrac{(y-1)^2}{9} = 1$

Center: $(-3, 1)$

Vertices: $(1, 1)$, $(-7, 1)$

Foci: $\left(-3 \pm \sqrt{7}, 1\right)$

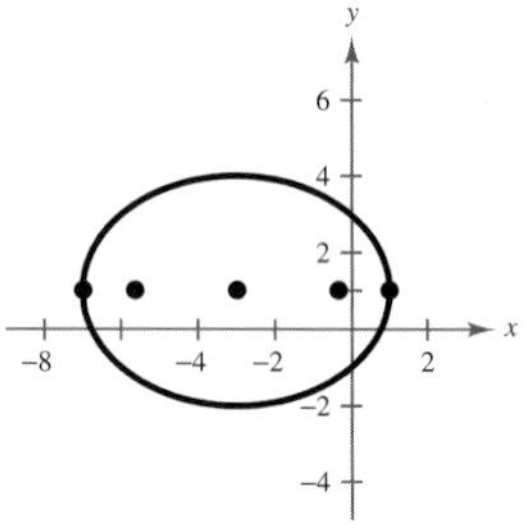

7. Circle: $(x-2)^2 + (y-1)^2 = \frac{1}{2}$

Center: $(2, 1)$

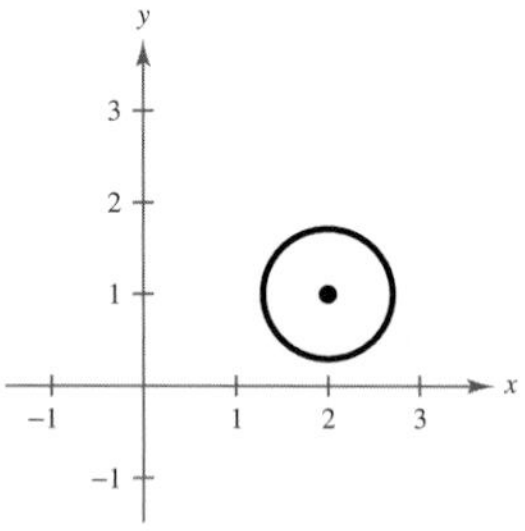

8. $(x-3)^2 = \dfrac{3}{2}(y+2)$ **9.** $\dfrac{5(y-2)^2}{4} - \dfrac{5x^2}{16} = 1$

10. (a) $45°$

(b)

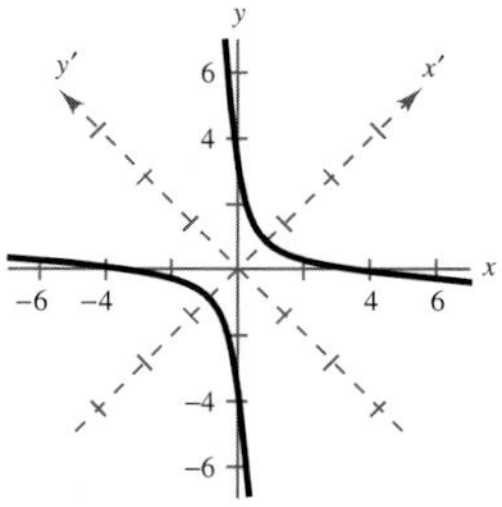

11.

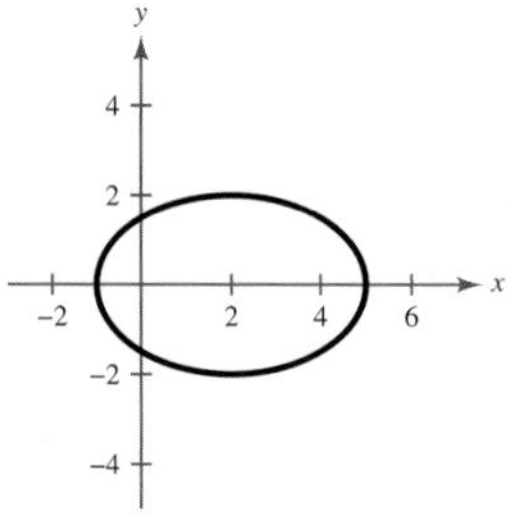

$\dfrac{(x-2)^2}{9} + \dfrac{y^2}{4} = 1$

12. $x = 6 + 4t$
$y = 4 + 7t$

13. $\left(\sqrt{3}, -1\right)$

14. $\left(2\sqrt{2}, \dfrac{7\pi}{4}\right), \left(-2\sqrt{2}, \dfrac{3\pi}{4}\right), \left(2\sqrt{2}, -\dfrac{\pi}{4}\right)$

15. $r = 4 \sin \theta$

16.

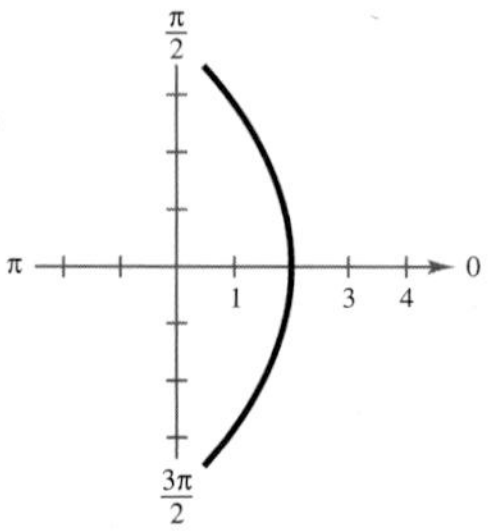

Parabola

17.

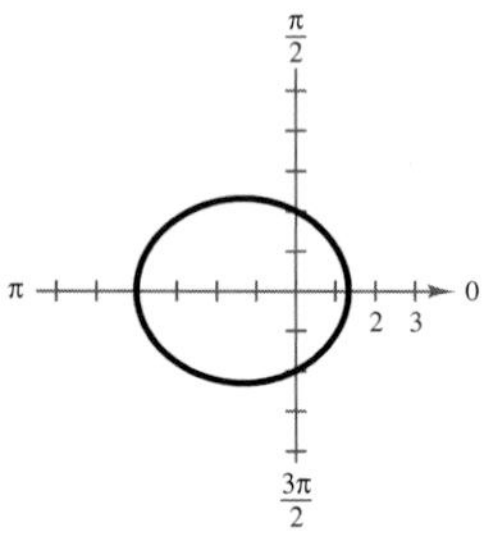

Ellipse

18.

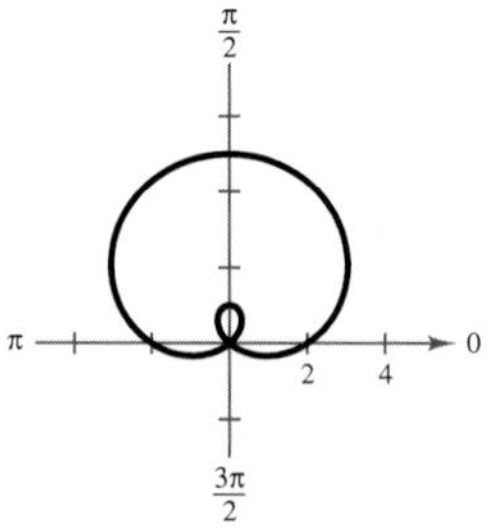

Limaçon

19.

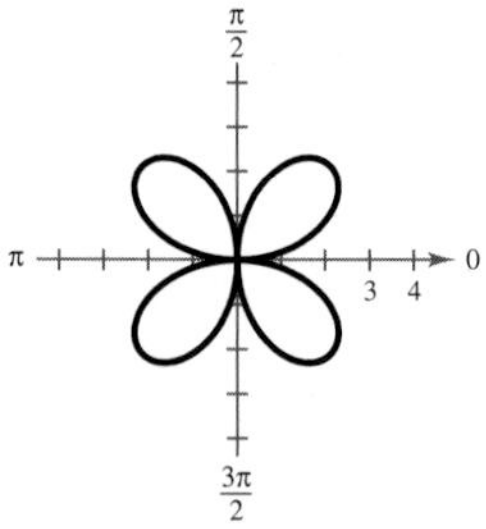

Rose curve

20. Slope: 0.1511; Change in elevation: 789.03 feet

21. No; Yes

Appendix B

Section B.1 *(page A16)*

1. (a) \$1.109 (b) [\$0.999, \$1.189]

3. Quiz 1:

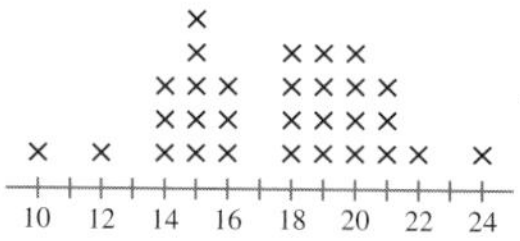

15

Quiz 2:

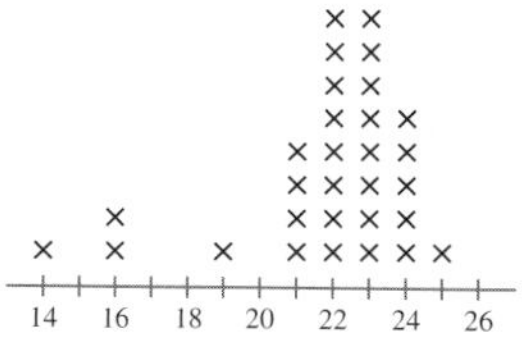

22 and 23

5.

Stems	*Leaves*
7	0 5 5 5 7 7 8 8 8
8	1 1 1 1 2 3 4 5 5 5 5 7 8 9 9 9
9	0 2 8
10	0 0

7.

Stems	*Leaves*
0	89 66 67 65 80 98 62 93
1	09 01 46 24 96 90
2	92 55 40 61
3	68 35
4	12 96 80 38
5	81 18 50 70 66
6	44 00 34 01
7	61 66 00
8	11 57 41 90
9	
10	
11	60 59 33
12	92
13	19 17 37
27	22
31	32
46	80
65	14

9. (a)

Interval	*Tally*
90–109	𝍸
110–129	𝍸 𝍸 |
130–149	𝍸 ||||
150–169	𝍸 |
170–189	|
190–209	|||

(b)

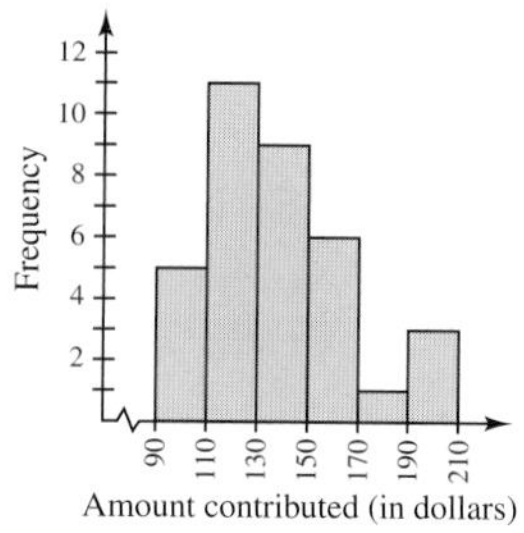

Section B.2 *(page A23)*

1. Mean: 8.86; median: 8; mode: 7

3. Mean: 10.29; median: 8; mode: 7

5. Mean: 9; median: 8; mode: 7

7. The mean is sensitive to extreme values.

9. Mean: \$67.14; median: \$65.35

11. Mean: 3.07; median: 3; mode: 3

13. One possibility: {4, 4, 10}

15. The median gives the most representative description.

17. $\bar{x} = 6,\ v = 10,\ \sigma = 3.16$

19. $\bar{x} = 2,\ v = \frac{4}{3},\ \sigma = 1.15$ **21.** $\bar{x} = 4,\ v = 4,\ \sigma = 2$

23. $\bar{x} = 47,\ v = 226,\ \sigma = 15.03$ **25.** 3.42

27. 101.55 **29.** 1.65

31. (a) $\bar{x} = 12;\ \sigma = 2.83$ (b) $\bar{x} = 20;\ \sigma = 2.83$
(c) $\bar{x} = 12;\ \sigma = 1.41$ (d) $\bar{x} = 9;\ \sigma = 1.41$

33. $\bar{x} = 12$ and $|x_i - 12| = 8$ for all x_i

35. It will increase the mean by 5, but the standard deviation will not change.

37. First histogram

39.

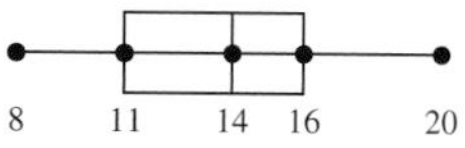

41.

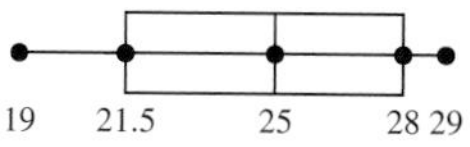

43.

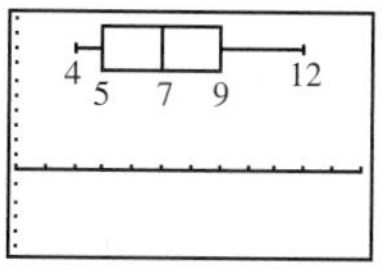

45.

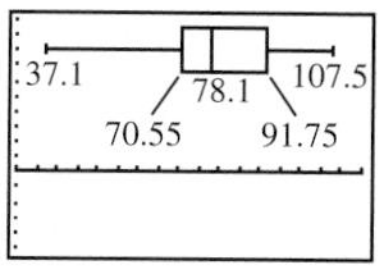

Instructor's Answers

Chapter P

Section P.1 *(page 9)*

2. (a) 12, 5 (b) $-7, 0, -3, 12, 5$
(c) $-7, -\frac{7}{3}, 0, 3.12, \frac{5}{4}, -3, 12, 5$ (d) $\sqrt{5}$

4. (a) 15, 31 (b) $-75, 15, 31$
(c) $0.7575, -4.63, -75, 15, 31$
(d) $2.3030030003 \ldots, \sqrt{10}$

6. (a) $25, \sqrt{9}, 7, 13$ (b) $25, -17, \sqrt{9}, 7, 13$
(c) $25, -17, -\frac{12}{5}, \sqrt{9}, 3.12, 7, -11.1, 13$ (d) $\frac{1}{2}\pi$

8. $0.\overline{3}$ **10.** $0.\overline{54}$ **12.** $\frac{17}{2}$ **14.** $\frac{60}{11}$ **16.** $-\frac{149}{90}$

18. $-6 < -2.5$

20. $-3.5 < 1$

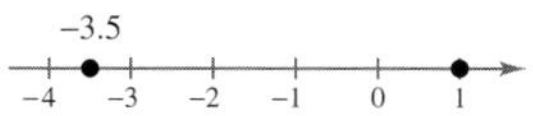

22. $1 < \frac{16}{3}$

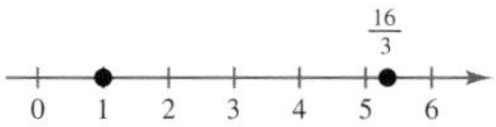

24. $-\frac{8}{7} < -\frac{3}{7}$

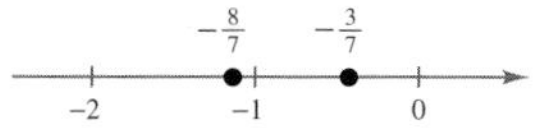

26. $x \geq -2$ denotes the set of all real numbers greater than or equal to -2. Unbounded

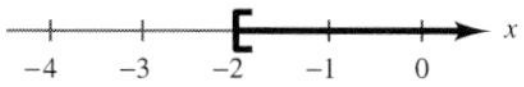

28. $x > 3$ denotes the set of all real numbers greater than 3. Unbounded

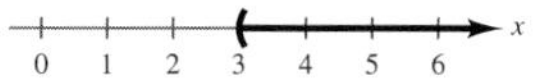

30. $x < 2$ denotes the set of all real numbers less than 2. Unbounded

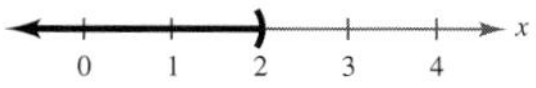

32. $0 \leq x \leq 5$ denotes the set of all real numbers greater than or equal to zero and less than or equal to 5. Bounded

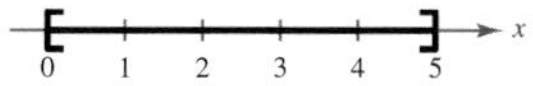

34. $0 < x \leq 6$ denotes the set of positive real numbers less than or equal to 6. Bounded

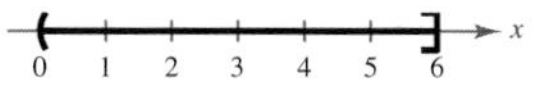

36. $\frac{381}{220}, 1.7320, \sqrt{3}, \frac{2103}{1214}, \frac{26}{15}$ **38.** $-6 \leq y < 0$

40. $y \leq 25$ **42.** $-3 \leq k < 5$ **44.** $2.5\% \leq r \leq 5\%$

46. This interval consists of all real numbers greater than or equal to -5 and less than or equal to 7.

48. This interval consists of all real numbers less than or equal to 4.

50. 0 **52.** If $x \leq 4, 4 - x$. If $x > 4, x - 4$.

54. -6 **56.** -9 **58.** 1 **60.** $|-4| = |4|$

62. $-|-6| < |-6|$ **64.** $-(-2) > -2$ **66.** $\frac{5}{2}$

68. 51 **70.** $\frac{5}{2}$ **72.** 14.99

74. (a) Positive (b) Positive **76.** $|103 - 86| = 17$

78. $|48 - 82| = 34°$ **80.** $|x + 10| \geq 6$

82. $|y - a| \leq 2$

84. $|\$9772 - \$9400| = \$372 < \$500$
$0.05(\$9400) = \$470$
Because the difference between the actual expenses and the budget is less than \$500 and less than 5% of the budgeted amount, there is compliance with the "budget variance test."

86. $|\$2613 - \$2575| = \$38 < \$500$
$0.05(\$2575) = \$128.75$
Because the difference between the actual expenses and the budget is less than \$500 and less than 5% of the budgeted amount, there is compliance with the "budget variance test."

88. $|517.1 - 590.9| = 73.8$
There was a deficit of \$73.8 billion.

90. $|1657.9 - 1667.8| = 9.9$
There was a deficit of \$9.9 billion.

92. $6x^3$ and $-5x$ are the terms; 6 and -5 are the coefficients.

94. $3\sqrt{3}x^2$ and 1 are the terms; $3\sqrt{3}$ is the coefficient.

96. $3x^4$ and $-x^2/4$ are the terms; 3 and $-\frac{1}{4}$ are the coefficients.

98. (a) 30 (b) -12 **100.** (a) -10 (b) 0

102. (a) $\frac{1}{2}$ (b) Division by 0 is undefined.

104. Multiplicative Inverse Property

106. Additive Inverse Property

108. Additive Identity Property

110. Associative Property of Addition

112. $\frac{1}{7}(7 \cdot 12) = \left(\frac{1}{7} \cdot 7\right)12$ Associative Property of Multiplication
$= 1 \cdot 12$ Multiplicative Inverse Property
$= 12$ Multiplicative Identity Property

114. $\frac{2}{7}$ **116.** $\frac{59}{66}$ **118.** -3

120. $\dfrac{5x}{27}$ **122.** -0.13 **124.** 10.20

126. (a)

n	1	10	100	10,000	100,000
$5/n$	5	0.5	0.05	0.0005	0.00005

(b) The value of $5/n$ approaches 0 as n increases without bound.

128. Yes. y is nonnegative if $y \geq 0$. y is positive if $y > 0$.

130. False. If $a < b$, then $\dfrac{1}{a} > \dfrac{1}{b}$, where $a \neq b \neq 0$.

132.

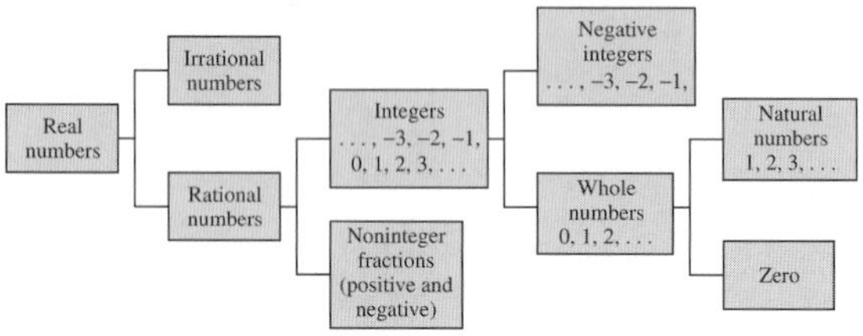

Section P.2 *(page 21)*

2. $(-2) \times (-2) \times (-2) \times (-2) \times (-2) \times (-2) \times (-2)$

4. $11.3 \times 11.3 \times 11.3 \times 11.3$ **6.** $(2\sqrt{5})^4$

8. $-\left(\frac{3}{2}\right)^4$ **10.** (a) 125 (b) $\frac{1}{9}$

12. (a) 5184 (b) $-\frac{3}{5}$ **14.** (a) $\frac{16}{3}$ (b) 1

16. (a) $\frac{7}{12}$ (b) $\frac{1}{81}$ **18.** 0.244 **20.** 5184

22. $\frac{7}{16}$ **24.** -135 **26.** -48 **28.** $-\frac{5}{27}$

30. (a) $9x^2$ (b) $16x^6$ **32.** (a) $-3z^7$ (b) $\frac{5}{2}y^4$

34. (a) $\dfrac{1}{r^2}$ (b) $\dfrac{5184}{y^7}$ **36.** (a) 1 (b) $\dfrac{1}{(z+2)^4}$

38. (a) $32y^2$ (b) $\dfrac{125x^9}{y^{12}}$ **40.** (a) $\dfrac{x^2}{y^2}$ (b) 1

42. (a) $\dfrac{1}{x}$ (b) 1 **44.** $64^{1/3} = 4$

46. $-\sqrt{144} = -12$ **48.** $(614.125)^{1/3} = 8.5$

50. $\sqrt[5]{-243} = -3$ **52.** $81^{3/4} = 27$ **54.** $\sqrt[4]{16^5} = 32$

56. (a) 7 (b) $\frac{3}{2}$ **58.** (a) 0 (b) 1

60. (a) 562 (b) 216 **62.** (a) $\frac{1}{1000}$ (b) $\frac{2}{3}$

64. (a) $-\frac{3}{5}$ (b) -625 **66.** (a) 12.651 (b) 2.236

68. (a) 0.022 (b) 21.316 **70.** (a) 3.495 (b) 5.906

72. (a) $\dfrac{2\sqrt[3]{2}}{3}$ (b) $\dfrac{5\sqrt{3}}{2}$

74. (a) $3y^2\sqrt{6x}$ (b) $\dfrac{4a^2}{|b|}\sqrt{2}$

76. (a) $3x^2$ (b) $2x\sqrt[5]{3}$ **78.** 64 **80.** $xy^{1/3}$

82. $\dfrac{x}{5}$, $x > 0$ **84.** (a) $\dfrac{\sqrt{10}}{2}$ (b) $\dfrac{\sqrt[3]{5x}}{x}$

86. (a) $\dfrac{\sqrt{14}+2}{2}$ (b) $\dfrac{\sqrt{10}+5}{-3}$

88. (a) $\dfrac{2}{3\sqrt{2}}$ (b) $\dfrac{5}{\sqrt[4]{500}}$

90. (a) $\dfrac{1}{2\left(\sqrt{3}+\sqrt{2}\right)}$ (b) $\dfrac{3}{\sqrt{3}}$

92. (a) $\sqrt{x}$ (b) $3x^2$

94. (a) $3\sqrt[4]{3(x+1)}$ (b) $a\sqrt[6]{10ab}$

96. (a) $7\sqrt{3}$ (b) $11\sqrt[3]{2}$

98. (a) $-84\sqrt{x}$ (b) $23x\sqrt{3}$

100. (a) $4\sqrt{x^3-7}$ (b) $50x\sqrt{5x}$

102. $\sqrt{\dfrac{3}{11}} = \dfrac{\sqrt{3}}{\sqrt{11}}$ **104.** $5 = \sqrt{3^2 + 4^2}$

106. 9.46×10^{15} **108.** 3.937×10^{-5}

110. $15{,}000{,}000$ **112.** 0.00009

114. (a) $60{,}000$ (b) 2.0×10^{11}

116. (a) 4.907×10^{17} (b) 1.479

118. (a) 0.064 (b) 0.030

120. No. Rationalizing the denominator produces a number equivalent to the original fraction; squaring does not.

122. 0.026 inch **124.** $\frac{25}{3}$ minutes

126. Paper: 7.99×10^7 tons; Metals: 1.61×10^7 tons; Glass: 1.24×10^7 tons; Plastics: 1.97×10^7 tons; Yard waste: 2.81×10^7 tons; Other: 5.35×10^7 tons

128. False. When a power is raised to a power, you multiply the exponents: $(a^n)^k = a^{nk}$.

130. (a) 3 is also raised to the -1 power, so $(3x)^{-1} = \dfrac{1}{3x}$.

(b) When two powers have the same base, the exponents are added: $y^3 \cdot y^2 = y^5$.

(c) When a power is raised to a power, the exponents are multiplied: $(a^2b^3)^4 = a^8b^{12}$.

(d) The square of a binomial contains a cross-product term: $(a+b)^2 = a^2 + 2ab + b^2$.

(e) If $x < 0$, then $\sqrt{4x^2} > 0$ but $2x < 0$: $\sqrt{4x^2} = 2|x|$.

(f) Radicals can only be added together if they have the same radicand and index: $\sqrt{2} + \sqrt{2} = 2\sqrt{2}$.

132.

Planet	Mercury	Venus	Earth	Mars	Jupiter
x	0.387	0.723	1.0	1.523	5.203
$\sqrt{x}$	0.622	0.850	1.0	1.234	2.281
y	0.241	0.615	1.0	1.881	11.861
$\sqrt[3]{y}$	0.622	0.850	1.0	1.234	2.281

$y = x^{3/2}$. The time it takes for a planet to orbit the sun is equal to the average distance of a planet from the sun raised to the $\frac{3}{2}$ power.

Section P.3 *(page 33)*

2. e **4.** a **6.** c

8. $6x^5 + 3x + 1$ (Answers will vary.) **10.** $2x^3 + 1$ (Answers will vary.)

12. Degree: 4; Leading coefficient: -3

14. Degree: 0; Leading coefficient: 3

16. Degree: 6; Leading coefficient: -1

18. Not a polynomial because of the negative exponent

20. Polynomial: $\frac{1}{2}x^2 + x - \frac{3}{2}$

22. Not a polynomial because of the square root

24. $-2x^2 - 4$ **26.** $1.3x^4 - 8.4x - 34.1$

28. $y^3 - y^2 - 3y + 7$ **30.** $4y^4 + 2y^3 - 3y^2$

32. $-15x^2 - 6x$ **34.** $4x^4 - 12x$

36. $-7y^4 + 4y^3$ **38.** $-\frac{7}{4}y^2 + 8y$ **40.** $x^2 + 5x - 50$

42. $28x^2 - 29x + 6$ **44.** $16x^2 + 40x + 25$

46. $25 - 80x + 64x^2$ **48.** $4x^2 - 9$

50. $4x^2 - 9y^2$ **52.** $x^2 + 2xy + y^2 - 1$

54. $x^2 - 2xy + y^2 + 2x - 2y + 1$

56. $9a^6 - 16b^4$ **58.** $x^3 - 6x^2 + 12x - 8$

60. $16x^6 - 24x^3 + 9$ **62.** $\frac{4}{9}t^2 + \frac{20}{3}t + 25$

64. $4x^2 - \frac{1}{25}$ **66.** $2.25y^2 - 9y + 9$ **68.** $6.25y^2 - 9$

70. $2x^2 + 8x + 6$ **72.** $x^4 - y^4$ **74.** $25 - x$

76. $x^2 + 2\sqrt{3}x + 3$ **78.** No. $(x + 3)(x + 3)(x - 3)$

80. No. $x(x + 3)(2x + 1)(2x - 1)$ **82.** $5(y - 6)$

84. $2x(2x^2 - 3x + 6)$ **86.** $(x + 2)(3x - 4)$

88. $3x(3x - 1)$ **90.** $\frac{1}{3}y(y^3 - 15y + 6)$

92. $\frac{2}{5}(y + 1)(2y - 5)$ **94.** $(7 + 3y)(7 - 3y)$

96. $\left(\frac{2}{5}y + 8\right)\left(\frac{2}{5}y - 8\right)$ **98.** $-z(z + 10)$

100. $(5x + 4y)(5x - 4y)$ **102.** $(x + 5)^2$

104. $(3x - 2)^2$ **106.** $(2x - y)^2$ **108.** $\left(z + \frac{1}{2}\right)^2$

110. $(x + 2)(x + 3)$ **112.** $(t + 2)(t - 3)$

114. $(8 - z)(3 + z)$ **116.** $(2x + 1)(x - 1)$

118. $(3x + 1)(4x + 1)$ **120.** $(-5u + 2)(u + 3)$

122. $(x - 3)(x^2 + 3x + 9)$ **124.** $(z + 5)(z^2 - 5z + 25)$

126. $(3x + 2)(9x^2 - 6x + 4)$

128. $(4x - y)(16x^2 + 4xy + y^2)$ **130.** $(x + 5)(x^2 - 5)$

132. $(3 + x)(2 - x^3)$ **134.** $(4x^3 - 3)(2x^2 + 3)$

136. $(x + 3)(2x + 3)$ **138.** $(x - 1)(12x - 1)$

140. $12(x + 2)(x - 2)$ **142.** $x(x + 3)(x - 3)$

144. $y(2y + 3)(y - 5)$ **146.** $(5x + 3)(x + 2)$

148. $\frac{1}{8}\left(x - \frac{3}{4}\right)\left(x + \frac{2}{3}\right)$ **150.** $(u + 2)(3 - u^2)$

152. $(x + 2)(x + 4)(x - 2)(x - 4)$

154. $5(x + 2)(x^2 - 2x + 4)$ **156.** $(3 - 4x)(23 - 60x)$

158. $7(x^2 + 1)(3x^2 - 1)$

160. $3(x^2 + 1)^4(x^4 - x^2 + 1)^4(3x + 2)^2(33x^6 + 20x^5 + 3)$

162. $-51, 51, -15, 15, -27, 27$

164. $25, -25, 14, -14, 11, -11, 10, -10$ **166.** $3, -8$

168. 9, 7 **170.** $P = \$548$

172. (a) $1200r^3 + 3600r^2 + 3600r + 1200$

(b)

r	2%	3%	$3\frac{1}{2}$%
$1200(1 + r)^3$	\$1273.45	\$1311.27	\$1330.46

r	4%	$4\frac{1}{2}$%
$1200(1 + r)^3$	\$1349.84	\$1369.40

(c) The amount increases with increasing r.

174. (a) $3x^2 + 8x$ (b) $30x^2$ **176.** $44x + 308$

178. (a) $T = 0.14x^2 - 3.33x + 58.40$

(b)

x (mi/hr)	30	40	55
T (ft)	84.50	149.20	298.75

(c) Stopping distance increases at an accelerating rate as speed increases.

180.

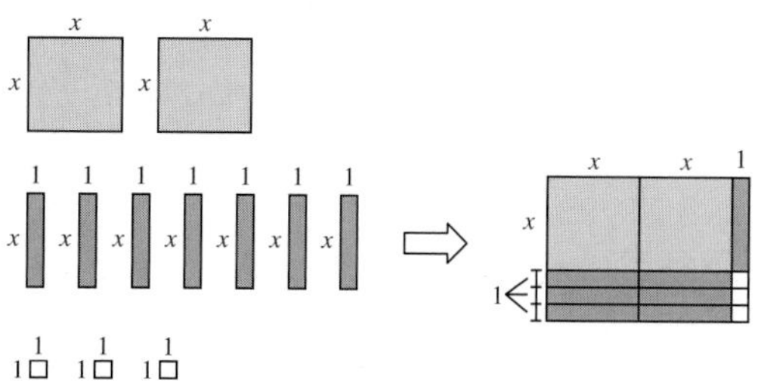

182. $r^2(4 - \pi)$ **184.** $\frac{5}{8}(x + 7)(x - 1)$ **186.** $kx(Q - x)$

188. False.

$$\begin{aligned}(4x+3)+(-4x+6) &= 4x+3-4x+6\\ &= 3+6\\ &= 9\end{aligned}$$

190. False. A perfect square trinomial can be factored as the binomial sum squared.

192. n

194. (a) x^2-1 (b) x^3-1 (c) x^4-1

The result is x^5-1.

196. $(x^n+y^n)(x^{2n}-x^ny^n+y^{2n})$

Section P.4 *(page 45)*

2. All real numbers **4.** All positive real numbers

6. All real numbers x such that $x \neq -\frac{1}{2}$

8. All real numbers x such that $x \leq 6$

10. $(x+1),\ x \neq -1$ **12.** $(y-1),\ y \neq 1$

14. $(1+z),\ z \neq -1$ **16.** $\dfrac{3}{10y^3}$ **18.** $\dfrac{2x^2}{x-1}$

20. $\dfrac{9x}{2},\ x \neq -1$ **22.** $-4, x \neq 3$

24. $-(x+5),\ x \neq 5$ **26.** $\dfrac{x-2}{x+1},\ x \neq -10$

28. $\dfrac{(x-6)(x-1)}{(x+10)(x+1)}$ **30.** $\dfrac{1}{x+1},\ x \neq \pm 3$

32. $\dfrac{y(y-3)}{y^2-y+1},\ y \neq -1$

34.

x	0	1	2	3	4	5	6
$\dfrac{x-3}{x^2-x-6}$	$\frac{1}{2}$	$\frac{1}{3}$	$\frac{1}{4}$	Undef.	$\frac{1}{6}$	$\frac{1}{7}$	$\frac{1}{8}$
$\dfrac{1}{x+2}$	$\frac{1}{2}$	$\frac{1}{3}$	$\frac{1}{4}$	$\frac{1}{5}$	$\frac{1}{6}$	$\frac{1}{7}$	$\frac{1}{8}$

The expressions are equivalent except at $x = 3$.

36. The expression cannot be simplified.

38. $\dfrac{x+5}{4(2x+3)}$ **40.** $-\dfrac{x+13}{5x^2},\ x \neq 3$

42. $-\dfrac{x(x+7)}{x+1},\ x \neq 9$ **44.** $-\dfrac{8}{5},\ y \neq -3, 4$

46. $\dfrac{2(y^2+2y+4)}{y^2(y-3)},\ y \neq 2$

48. $\dfrac{(x^2+x+1)(x^2+1)}{(x+1)^2},\ x \neq 1$ **50.** $\dfrac{x+2}{x-2},\ x \neq 3$

52. $\dfrac{3}{x^2},\ x \neq -6$ **54.** $\dfrac{1}{3},\ x \neq \pm 7$ **56.** $\dfrac{x}{x+3}$

58. $\dfrac{8-5x}{x-1}$ **60.** $\dfrac{2x+5}{x-5}$ **62.** $\dfrac{1}{(x+2)(x-1)}$

64. $\dfrac{6(2x+3)}{(x-2)(x+1)(x+4)}$ **66.** $\dfrac{4x+1}{(x-1)(x+1)}$

68. $\dfrac{x^8-5}{x^3}$ **70.** $\dfrac{5x^{13/2}-3}{x^{3/2}}$ **72.** $\dfrac{-2x(x+5)}{(x-5)^4}$

74. $\dfrac{4x^3(2x-1)^2-2x}{(2x-1)^{1/2}}$ **76.** $\dfrac{2(5x+6)}{x^2(x+2)}$

78. $\dfrac{4x}{x+4},\ x \neq 0, 4$ **80.** $\dfrac{x+1}{x-1},\ x \neq 0$

82. $\dfrac{1}{2y+1},\ y \neq 0, -\frac{5}{4}$ **84.** $\dfrac{1}{(x+5)(x+1)}$

86. $\dfrac{1}{(x+h+1)(x+1)},\ h \neq 0$ **88.** $\dfrac{-1}{t^2\sqrt{t^2+1}}$

90. $\dfrac{x^2-2}{x^3(1-x^2)^{1/2}}$ **92.** $\dfrac{2x+3}{3(2x+1)^{4/3}}$

94. $\dfrac{-1}{\sqrt{z-3}+\sqrt{z}}$ **96.** $\dfrac{8}{15}t$ **98.** $\dfrac{7x}{16}, \dfrac{13x}{24}, \dfrac{31x}{48}$

100. (a) 7.27% (b) $\dfrac{288(MN-P)}{N(MN+12P)}$, 7.27%

102. (a)

Year	1992	1993	1994	1995	1996	1997
Gold	\$345	\$364	\$383	\$399	\$404	\$366
Silver	\$3.91	\$4.01	\$4.71	\$5.13	\$4.97	\$4.81

The estimates are fairly close to the actual values.

(b) Ratio =

$$\frac{-2.156t^3+34.691t^2-178.045t+310.1}{0.0017t^4-0.056t^3+0.598t^2-2.568t+4.02}$$

Year	1992	1993	1994	1995	1996	1997
Ratio	88.31	92.59	89.67	91.69	101.17	109.86

Increase

104. $\dfrac{x}{2(2x+1)}$ **106.** $\dfrac{8(x+2)}{(x+4)^2}$

108. False. The two expressions are equivalent for all values of x such that $x \neq 1$.

110. Factor the numerator and the denominator and cancel all common factors.

Section P.5 *(page 58)*

2. Conditional equation **4.** Identity **6.** Identity

8. Identity **10.** Conditional equation **12.** -12

14. 3 **16.** -5 **18.** $-\frac{5}{8}$ **20.** All real numbers

22. -5 **24.** 6 **26.** 50

28. No solution. The x-terms sum to zero. **30.** $\frac{1}{2}$ **32.** 0

34. $\frac{9}{7}$ **36.** 0 **38.** $\frac{11}{6}$ **40.** $-\frac{13}{3}$ **42.** $\frac{7}{4}$

44. No solution. The solution is extraneous. **46.** $\frac{1}{5}$

48. No solution. The x-terms sum to zero.

50. $x^2 - 16x = 0$ **52.** $-3x^2 - 42x - 134 = 0$

54. $4x^2 - 2x + 1 = 0$ **56.** $-\frac{1}{3}, \frac{1}{3}$ **58.** 9, 1

60. $-\frac{3}{2}$ **62.** $-\frac{3}{2}, 11$ **64.** 2, 6 **66.** $-8, 16$

68. $-a - b, -a + b$ **70.** $\pm 13; \pm 13.00$

72. $\pm 4\sqrt{2}; \pm 5.66$ **74.** $\pm 2; \pm 2.00$

76. $-18, -8; -18.00, -8.00$

78. $5 \pm \sqrt{30}$; 10.48, -0.48

80. $-\dfrac{7}{4} \pm \dfrac{\sqrt{11}}{2}; -0.09, -3.41$ **82.** $-\dfrac{9}{2}; -4.50$

84. $0, -4$ **86.** $3, -1$ **88.** $-4 \pm \sqrt{2}$

90. $\frac{2}{3} \pm \sqrt{2}$ **92.** $\frac{11}{2}, -\frac{9}{2}$ **94.** $1, -\frac{1}{2}$ **96.** $\frac{3}{5}, \frac{1}{5}$

98. $5 \pm \sqrt{3}$ **100.** $-3 \pm \sqrt{13}$ **102.** $\dfrac{1}{2} \pm \dfrac{\sqrt{5}}{2}$

104. $\dfrac{5}{4} \pm \dfrac{\sqrt{3}}{4}$ **106.** $-\dfrac{1}{3} \pm \dfrac{\sqrt{11}}{6}$ **108.** $\dfrac{5}{4} \pm \dfrac{\sqrt{5}}{2}$

110. $-\dfrac{3}{2} \pm \dfrac{\sqrt{13}}{2}$ **112.** $-\dfrac{8}{5} \pm \dfrac{\sqrt{3}}{5}$ **114.** $-7 \pm \sqrt{13}$

116. $\dfrac{686 \pm 196\sqrt{6}}{25}$ **118.** 1.400, -0.150

120. 2.137, 18.063 **122.** 0.672, -0.968

124. -2.995, 2.971 **126.** $0, -3$ **128.** 7

130. $-\dfrac{3}{2} \pm \sqrt{3}$ **132.** $-\dfrac{b}{a}, \dfrac{b}{a}$ **134.** ± 1 **136.** $0, \pm\dfrac{5}{2}$

138. ± 2 **140.** $\frac{8}{3}$ **142.** $0, \frac{4}{3}$ **144.** -2 **146.** ± 2

148. ± 2 **150.** $\pm\dfrac{\sqrt{7}}{6}$ **152.** $-\sqrt[3]{2}, -1$ **154.** $\frac{9}{16}$

156. -4 **158.** $\frac{124}{3}$ **160.** $3, -2$ **162.** No solution

164. 1 **166.** $-29, 25$ **168.** $\dfrac{1 \pm 5\sqrt{5}}{2}$

170. $0, 1, \dfrac{3}{5}$ **172.** $2, -12$ **174.** $\dfrac{1 \pm \sqrt{31}}{3}$

176. $\frac{3}{4}, -1$ **178.** ± 1 **180.** $\frac{5}{3}, -3$

182. $-6, -3, 3$ **184.** $10, -1$

186. Yes. The estimated height of a male with a 19-inch thigh bone is 69.4 inches.

188. $y = -3t + 8$; about 2.33 hours

190. 6 inches $\times$ 6 inches **192.** $\dfrac{20\sqrt{3}}{3} \approx 11.55$ inches

194. 26,250 passengers **196.** 500 units

198. False.

$x(3 - x) = 10$

$3x - x^2 = 10$

The equation cannot be written in the form $ax + b = 0$.

200. False. $|x| = 0$ has only one solution to check, 0.

202. The equations have the same solution set, and one is derived from the other by steps for generating equivalent equations.

$2x = 5, 2x + 3 = 8$

204. (a) and (b) $x = -5, -\frac{10}{3}$

(c) The method of part (a) reduces the number of algebraic steps.

206. $a = 9, b = 9$

Section P.6 *(page 70)*

2. $2 < x \le 10$. Bounded **4.** $-5 \le x < \infty$. Unbounded

6. $-\infty < x \le 7$. Unbounded **8.** f **10.** c **12.** a

14. (a) No (b) No (c) Yes (d) No

16. (a) No (b) No (c) Yes (d) No

18. (a) No (b) Yes (c) No (d) Yes

20. $x > -4$

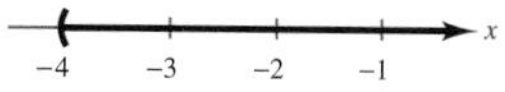

22. $x < -\frac{5}{2}$

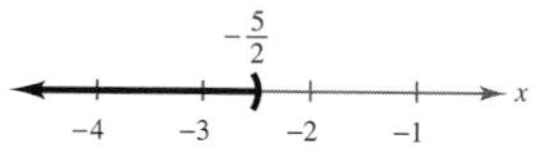

24. $x \le 5$

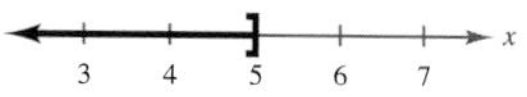

26. $x \ge \frac{1}{2}$

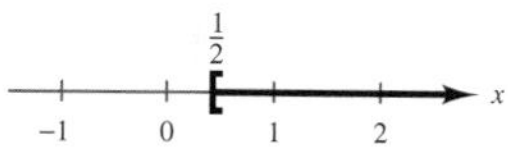

28. $x \ge -3$

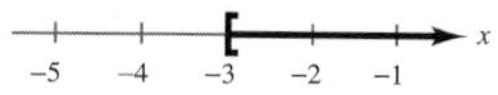

30. $x < -\frac{1}{2}$

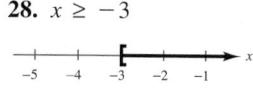

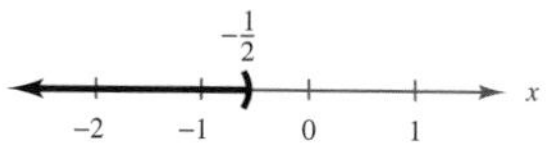

32. $x < 7$

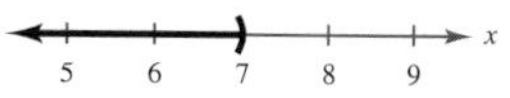

34. $x > \frac{1}{6}$

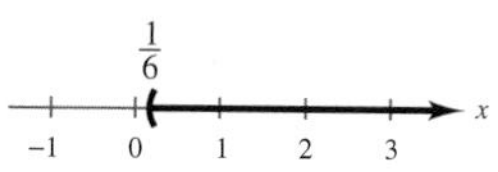

36. $x > 16$

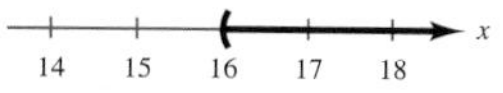

38. $-6 < x \le 1$

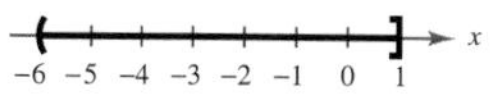

40. $-3 \le x < 7$

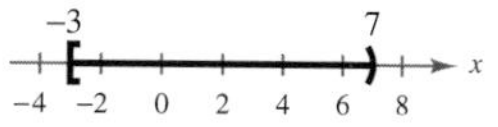

42. $3 < x < 9$

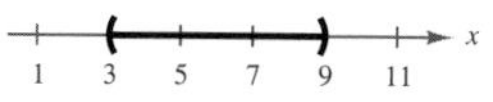

CHAPTER P

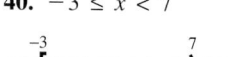

44. No solution

46. $x < -4, x > 4$

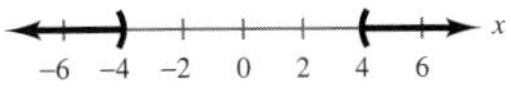

48. $x < -15, x > 15$

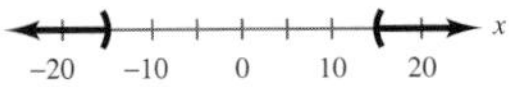

50. All real numbers x

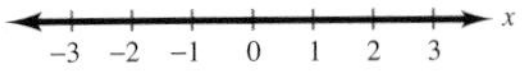

52. No solution

54. $-2 < x < 3$

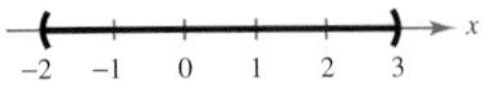

56. $0 < x < 3$

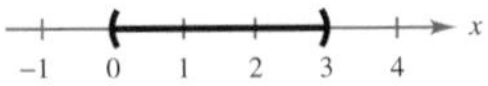

58. $x < -28, x > 0$

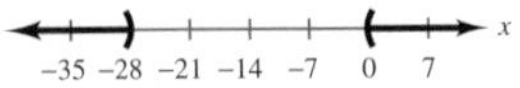

60. $\frac{1}{5} \leq x \leq \frac{7}{5}$

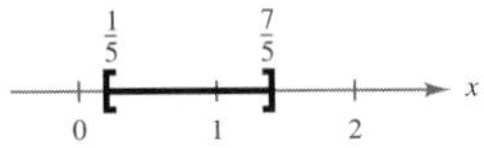

62.

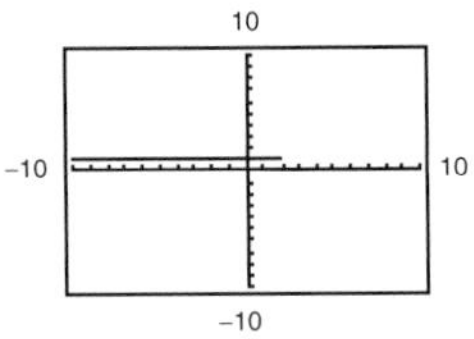

$x \leq 2$

64.

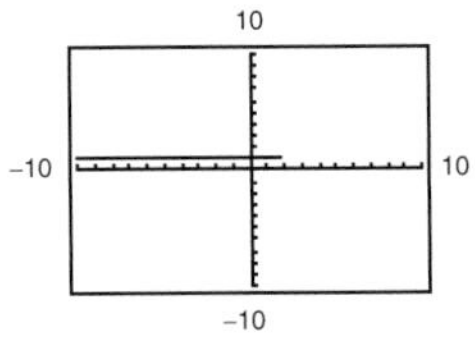

$x < 2$

66.

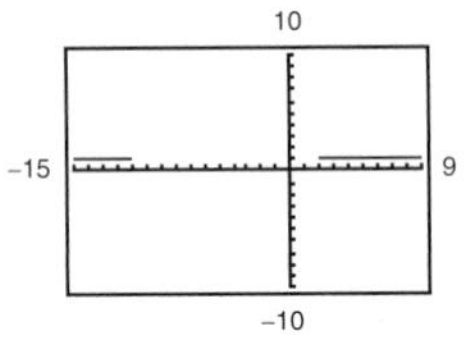

$x < -11, x > 2$

68.

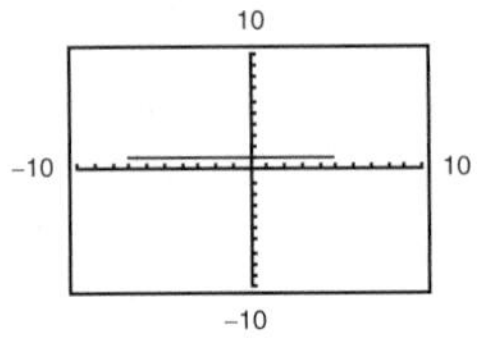

$-7 \leq x \leq 5$

70. $[10, \infty)$ **72.** $(-\infty, 3]$ **74.** $\left[-\frac{5}{2}, \infty\right)$

76. All real numbers that are more than 4 units from 8

78. $|x| > 3$ **80.** $|x + 1| \leq 4$ **82.** $|x - 8| \geq 5$

84. $|x + 6| \leq 7$

86. (a) Yes (b) No (c) Yes (d) Yes

88. (a) No (b) Yes (c) Yes (d) No

90. $0, \frac{25}{9}$ **92.** $-2, -1, 1, 4$

94. $\left(-\sqrt{5}, \sqrt{5}\right)$

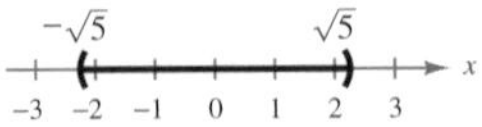

96. $(-\infty, 2], [4, \infty)$

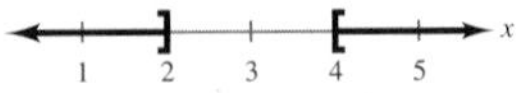

98. $(-1, 7)$

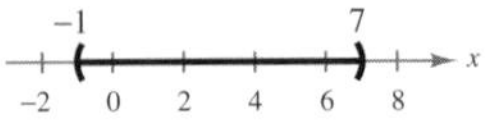

100. $(-\infty, -3), (1, \infty)$

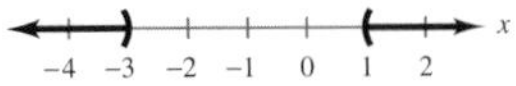

102. $\left(-\infty, 2 - \sqrt{5}\right), \left(2 + \sqrt{5}, \infty\right)$

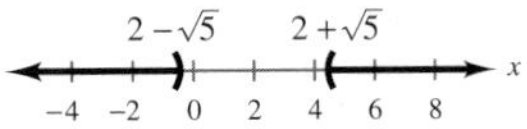

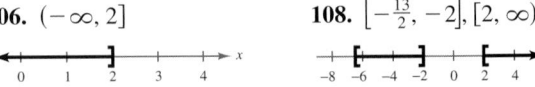

104. $\left(-\infty, \frac{3}{2} - \frac{\sqrt{39}}{2}\right], \left[\frac{3}{2} + \frac{\sqrt{39}}{2}, \infty\right)$

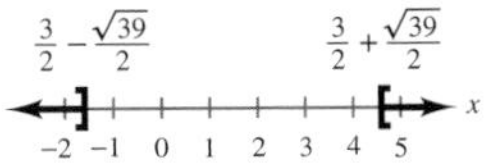

106. $(-\infty, 2]$

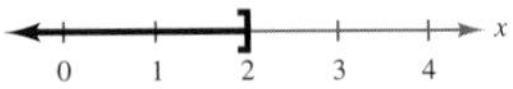

108. $\left[-\frac{13}{2}, -2\right], [2, \infty)$

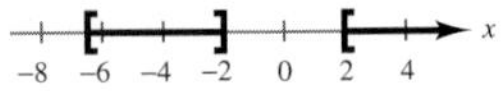

110. $(3, \infty)$ **112.** $(-\infty, 0], [2, \infty)$ **114.** $(-\infty, 3]$

116. $(-\infty, 0), \left(\frac{1}{4}, \infty\right)$

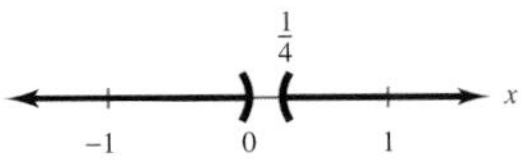

118. $(-2, 3]$

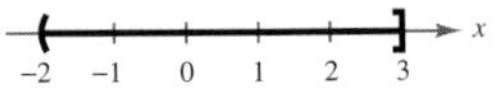

120. $\left(-\infty, -\frac{1}{2}\right), (1, \infty)$

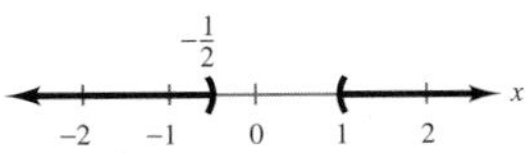

122. $(-14, -2), (6, \infty)$

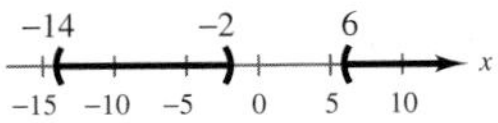

124. $(-\infty, -3), (0, \infty)$

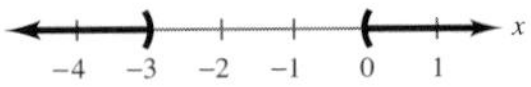

126. $[-3, 0), [2, \infty)$

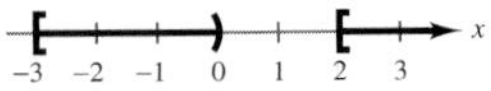

128. $(-\infty, -4), [-2, 1), [6, \infty)$

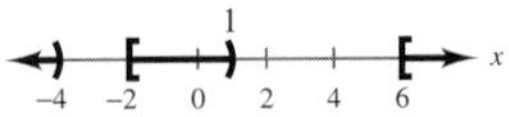

130. $(-\infty, -2], [2, \infty)$ **132.** $[-4, 4]$

134. $(-3, 0], (3, \infty)$ **136.** $(-1.13, 1.13)$

138. $(-4.42, 0.42)$ **140.** $(1.19, 1.30)$

142. More than 42,857 copies **144.** $x \geq 187$

146. $20 \leq h \leq 80$

148. Between 45.97 feet and 174.03 feet

150. Between 90,000 and 100,000 units

152. (a) 2223.9, 5593.9, 10,312, 16,378, 23,792

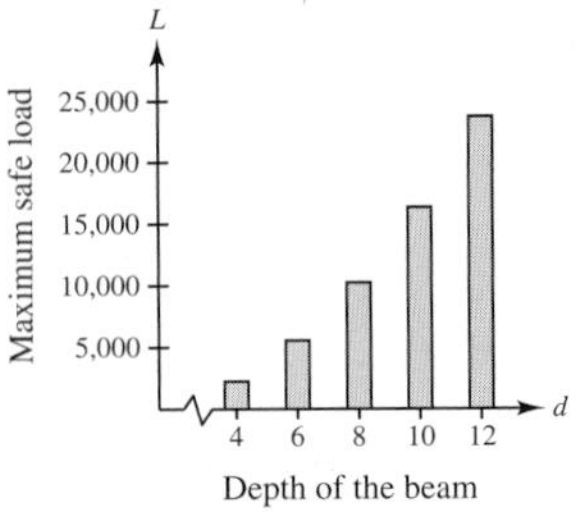

(b) 3.83 inches

154. False. If $-10 \le x \le 8$, then $10 \ge -x$ and $-x \ge 8$.

156. True. The y-values are greater than zero for all values of x.

158. $a = 1, b = 5, c = 5$ **160.** $(-\infty, \infty)$

162. $\left(-\infty, -2\sqrt{10}\right] \cup \left[2\sqrt{10}, \infty\right)$ **164.** 0

Section P.7 *(page 79)*

2. The 3 is distributed to both terms.

$5z + 3(x - 2) = 5z + 3x - 6$

4. The expression on the right should be negative.

$$\frac{1 - x}{(5 - x)(-x)} = -\frac{x - 1}{x(x - 5)}$$

6. yz is one term, not two. $x(yz) = xyz$

8. The exponent applies to the coefficient also. $(4x)^2 = 16x^2$

10. Do not apply radicals term-by-term.

$\sqrt{25 - x^2} = \sqrt{(5 + x)(5 - x)}$

12. Divide out common factors, not common terms.

$\dfrac{2x^2 + 1}{5x}$ cannot be simplified.

14. To get rid of negative exponents:

$$\frac{1}{a^{-1} + b^{-1}} = \frac{1}{a^{-1} + b^{-1}} \cdot \frac{ab}{ab} = \frac{ab}{b + a}.$$

16. Factor within grouping symbols before applying exponent to each factor.

$(x^2 + 5x)^{1/2} = [x(x + 5)]^{1/2} = x^{1/2}(x + 5)^{1/2}$

18. Factor within grouping symbols before applying exponent to each factor.

$(3x^2 - 6x)^3 = [3x(x - 2)]^3 = 27x^3(x - 2)^3$

20. The 5 needs to be distributed before the addition can take place.

$$\frac{7 + 5(x + 3)}{x + 3} = \frac{7 + 5x + 15}{x + 3} = \frac{5x + 22}{x + 3}$$

22. Be careful when using a slash to denote division.

$$(1/2)y = \frac{1}{2} \cdot y = \frac{y}{2}$$

24. x^2 **26.** $3x + 2$ **28.** $\frac{1}{3}$ **30.** 2 **32.** $x - 1$

34. $x - 1$ **36.** $\frac{4}{3}, \frac{16}{9}$ **38.** 4, 7 **40.** $10x + 3$

42. $1 + 10x^2 - 20x^3$ **44.** $4t - 3$ **46.** $x - 5 + \dfrac{4}{x^2}$

48. $2x^{7/2} - 3x^{3/2} + 5x^{-1/2} - \dfrac{1}{x^{3/2}}$ **50.** $\dfrac{x}{3} - \dfrac{5x^2}{3}$

52. $\dfrac{-11x^2 - 5}{x^6(x^2 + 1)^4}$ **54.** $\dfrac{-6(2x - 3)}{(4x^2 + 9)^{3/2}}$ **56.** $\dfrac{x - 3}{(2x - 1)^{1/2}}$

58. $\dfrac{1 - 4x}{(x + 1)^2(2x - 3x^2)^{1/2}}$ **60.** $\dfrac{2(3x^2 + 5x - 6)}{(x^2 - 6)(2x + 5)}$

62. $\dfrac{3(x - 6)^{1/2}[6x + 4 - (x - 6)^{5/2}]}{2(3x + 2)^{3/2}}$

64. (a) Answers will vary.

(b)

x	-2	-1	$-\frac{1}{2}$	$\frac{1}{4}$
y_1	-1.01	-3.18	-12.17	-48.17
y_2	-1.01	-3.18	-12.17	-48.17

x	1	2	$\frac{5}{2}$
y_1	-3.18	-1.01	-0.87
y_2	-3.18	-1.01	-0.87

66. True. $x^{-1} + y^{-2} = \dfrac{1}{x} + \dfrac{1}{y^2}$

$$= \frac{y^2 + x}{xy^2}$$

68. True. $\dfrac{1}{\sqrt{x} + 4} = \dfrac{1}{\sqrt{x} + 4} \cdot \dfrac{\sqrt{x} - 4}{\sqrt{x} - 4}$

$$= \frac{\sqrt{x} - 4}{x - 16}$$

70. Add exponents when multiplying powers with like bases.

$x^n \cdot x^{3n} = x^{4n}$

72. When squaring binomials, there is also a middle term.

$(x^n + y^n)^2 = x^{2n} + 2x^n y^n + y^{2n} \ne x^{2n} + y^{2n}$

74. The two answers are equivalent and can be obtained by factoring.

$\frac{1}{10}(2x - 1)^{5/2} + \frac{1}{6}(2x - 1)^{3/2}$

$= \frac{1}{60}(2x - 1)^{3/2}[6(2x - 1) + 10]$

$= \frac{1}{60}(2x - 1)^{3/2}(12x + 4)$

$= \frac{4}{60}(2x - 1)^{3/2}(3x + 1)$

$= \frac{1}{15}(2x - 1)^{3/2}(3x + 1)$

Section P.8 *(page 87)*

2.

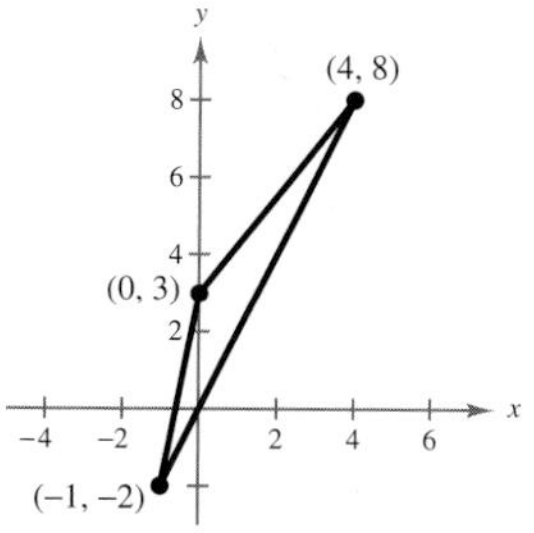

4.

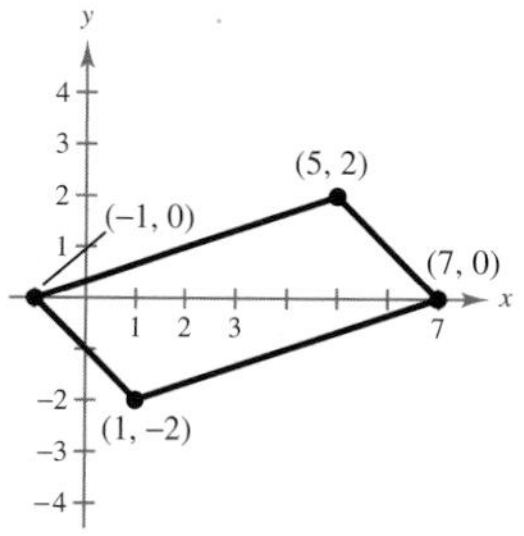

6. $A: \left(\frac{3}{2}, -4\right)$, $B: (0, -2)$, $C: \left(-3, \frac{5}{2}\right)$, $D: (-6, 0)$

8. $(4, -8)$ **10.** $(-12, 0)$

12. Quadrant III **14.** Quadrant I

16. Quadrants I and IV **18.** Quadrant III

20. Quadrants II and IV **22.** $(3, 3), (1, 0), (3, -3), (5, 0)$

24. $(-5, 2), (-7, 0), (-3, 0), (-5, -4)$

26.

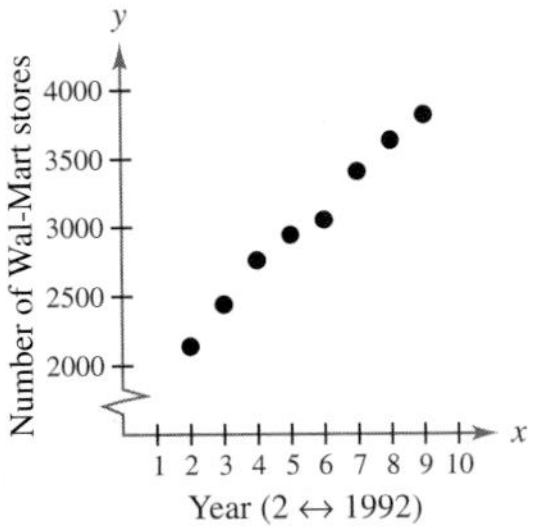

28. 19% **30.** (a) $250,000 (b) $750,000

32. (a) 12% (b) 467%

34. No. There are many variables that will affect the final exam score.

36. 7 **38.** 10

40. (a) 5, 12, 13 (b) $5^2 + 12^2 = 13^2$

42. (a) 4, 7, $\sqrt{65}$ (b) $4^2 + 7^2 = (\sqrt{65})^2$

44. (a)

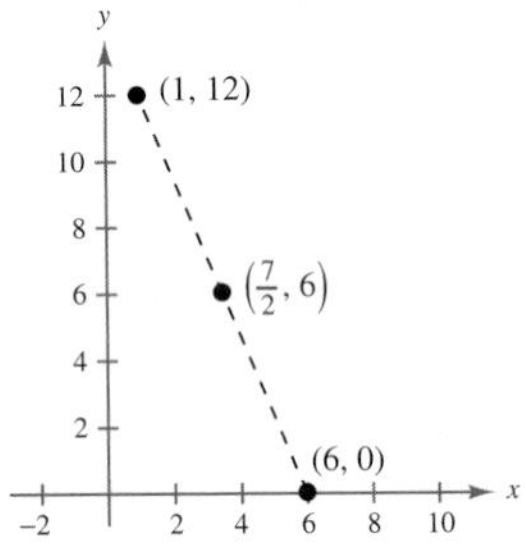

(b) 13 (c) $(\frac{7}{2}, 6)$

46. (a)

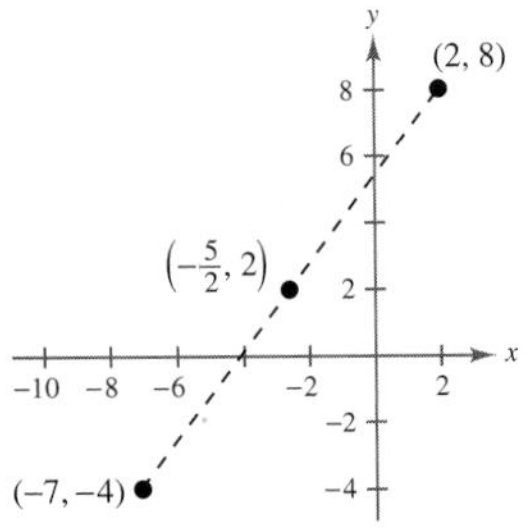

(b) 15 (c) $(-\frac{5}{2}, 2)$

48. (a)

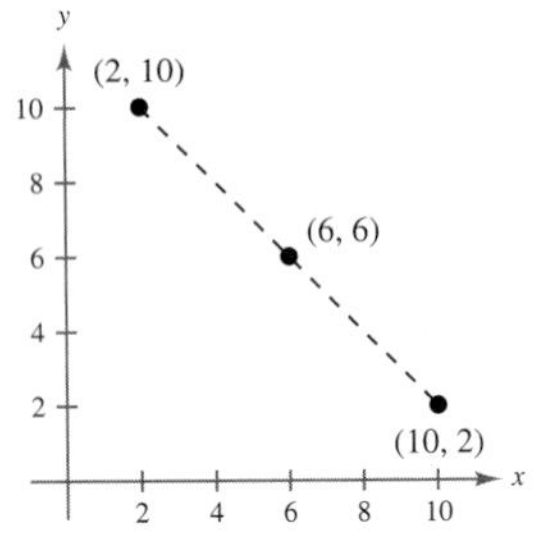

(b) $8\sqrt{2}$ (c) (6, 6)

50. (a)

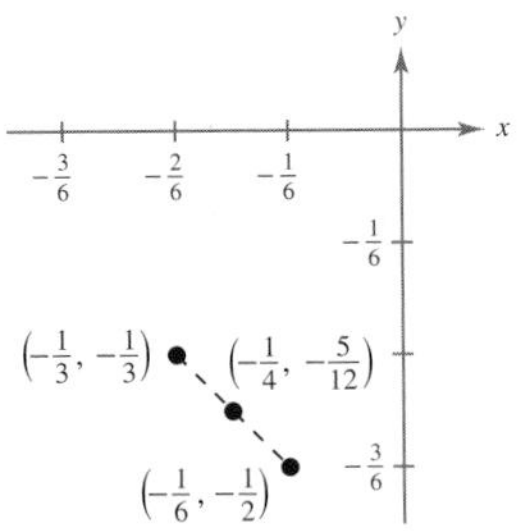

(b) $\frac{\sqrt{2}}{6}$ (c) $(-\frac{1}{4}, -\frac{5}{12})$

52. (a)

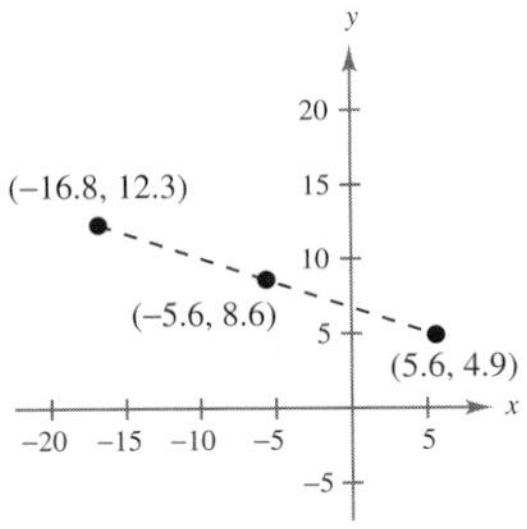

(b) $\sqrt{556.52}$ (c) $(-5.6, 8.6)$

54. (a)

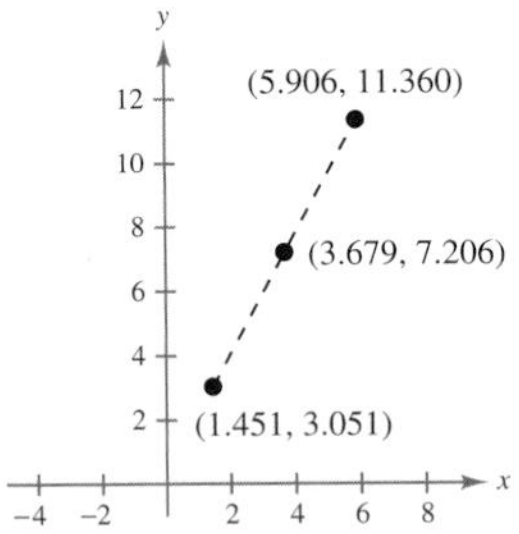

(b) $\sqrt{88.887}$ (c) (3.679, 7.206)

56. $4,925,000 **58.** $\sqrt{29}, \sqrt{58}, \sqrt{29}$

60. Opposite sides have equal lengths of $3\sqrt{5}$ and $\sqrt{10}$.

62. (a) (7, 0) (b) $(9, -3)$

64. (a) $(\frac{7}{4}, -\frac{7}{4}), (\frac{5}{2}, -\frac{3}{2}), (\frac{13}{4}, -\frac{5}{4})$
(b) $(-\frac{3}{2}, -\frac{9}{4}), (-1, -\frac{3}{2}), (-\frac{1}{2}, -\frac{3}{4})$

66. $50\sqrt{13} \approx 180.28$ kilometers

68. (a) The number of artists elected each year seems to be nearly steady except for the first few years. Between six and eight artists will be elected in 2001.
(b) Elections for inclusion in the Rock and Roll Hall of Fame began in 1986.

70. $1245.05 million

72. True. Two sides of the triangle have lengths of $\sqrt{149}$, and the third side has a length of $\sqrt{18}$.

74. No. It depends on the magnitude of the quantities measured.

76. b **78.** d

Review Exercises *(page 92)*

2. (a) None (b) $-22, 0$ (c) $-22, -\frac{10}{3}, 0, 5.2, \frac{3}{7}$
(d) $\sqrt{15}$

4. (a) 0.36 (b) $0.\overline{714285}$

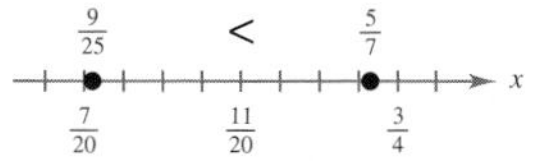

6. The set consists of all real numbers greater than 1.

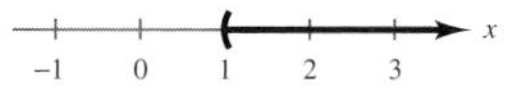

8. 14.73 **10.** 37.234 **12.** 27.2

14. (a) -1 (b) -3 **16.** -11 **18.** $\frac{1}{12}$

20. -144 **22.** (a) $\dfrac{3u^5}{v^4}$ (b) $\dfrac{1}{m^2}$ **24.** 3.3674×10^7

26. 216 **28.** $6u$ **30.** (a) $2\sqrt{2}$ (b) $26\sqrt{2}$

32. $2 + \sqrt{3}$ **34.** 64 **36.** $6x^{9/10}$ **38.** $16^{1/2}$

40. $-11x^2 + 3$ **42.** $-12x^2 - 4$ **44.** $-3x^2 - 7x + 1$

46. $15x^2 - 27x - 6$ **48.** $4x^2 - 12x + 9$ **50.** 41

52. (a) \$300,000 (b) \$600 **54.** $(x - 3)(x + 4)$

56. $(x - 6)^2$ **58.** $(2x + 3)(4x^2 - 6x + 9)$

60. $(x + 4)(3x + 2)$ **62.** $(x^2 + 2)(x - 4)$

64. All real numbers greater than or equal to -4

66. $\dfrac{x^2 - 3x + 9}{x - 2},\ x \neq -3$ **68.** $\dfrac{2(x + 3)}{x(x - 1)},\ x \neq -3, \dfrac{3}{2}$

70. $\dfrac{x + 1}{x(x^2 + 1)}$ **72.** $\dfrac{2x^2 - 3x + 5}{(x + 2)(2x - 1)}$

74. $\dfrac{4x}{2x - 3},\ x \neq -\dfrac{3}{2}, 0$

76. Conditional equation **78.** Conditional equation

80. $-\frac{9}{2}$ **82.** $-\frac{17}{3}$ **84.** $-\frac{5}{2}, 3$ **86.** $\pm\sqrt{2}$

88. $-4 \pm 3\sqrt{2}$ **90.** $6 \pm \sqrt{6}$ **92.** $-\dfrac{5}{4} \pm \dfrac{\sqrt{241}}{4}$

94. $0, \frac{12}{5}$ **96.** $\pm\sqrt{2}, \pm\sqrt{3}$ **98.** 5 **100.** No solution

102. $-124, 126$ **104.** $-2 \pm \dfrac{\sqrt{95}}{5}, -4$ **106.** $-5, 15$

108. 1, 3 **110.** 143,203 units

112. (a) Yes (b) Yes (c) No (d) No

114. $(-2, \infty)$ **116.** $\left(-\frac{5}{3}, \infty\right)$ **118.** $[-2, 10)$

120. $(1, 3)$ **122.** $(-\infty, 0], [3, \infty)$

124. $353.44 \text{ cm}^2 \leq \text{area} \leq 392.04 \text{ cm}^2$

126. $(-\infty, -1], [3, \infty)$ **128.** $(-\infty, -3], \left[\frac{5}{2}, \infty\right)$

130. $(-\infty, 3), (5, \infty)$ **132.** $(-\infty, 0), (2, \infty)$

134. $t \geq 9$ days

136. $\frac{1}{3}$ occurs twice as a factor. $\left(\frac{1}{3}x\right)\left(\frac{1}{3}y\right) = \frac{1}{9}xy$

138. The negative sign is also raised to the 6th power.

$(-x)^6 = x^6$

140. Do not apply radicals term-by-term.

$\sqrt{3^2 + 4^2} = \sqrt{9 + 16} = \sqrt{25} = 5$

142. When factoring, apply exponents to all factors.

$$(9x + 12)^2 = [3(3x + 4)]^2 = 9(3x + 4)^2$$

144. -1 **146.** -1 **148.** $-\dfrac{2}{15}\left[\dfrac{x^2 + 3x + 16}{(4 + x)^{1/2}}\right]$

150. $3x + 4 + x^{-1} - 5x^{-2}$ **152.** $\frac{8}{5}x^2 + x - 2x^{-1} + x^{-2}$

154.

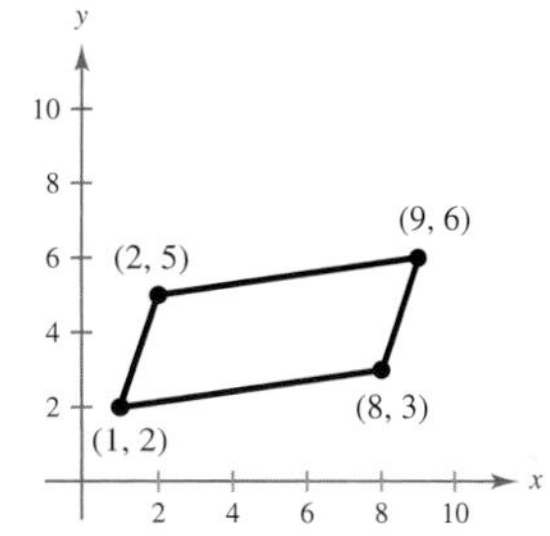

Opposite sides have equal lengths of $\sqrt{10}$ and $5\sqrt{2}$.

156. Quadrants I and II **158.** Quadrants I and III

160. (a)

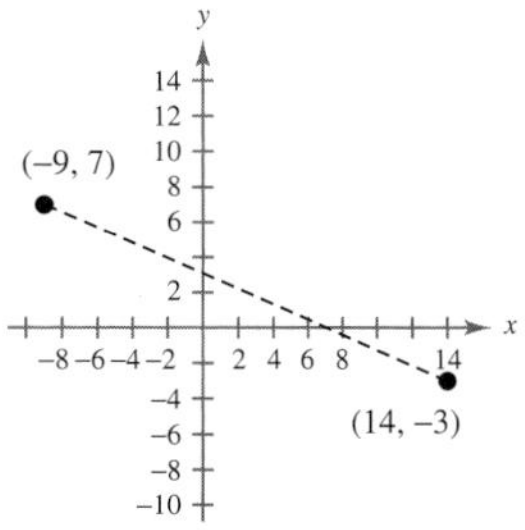

(b) $\sqrt{629}$

162. (a)

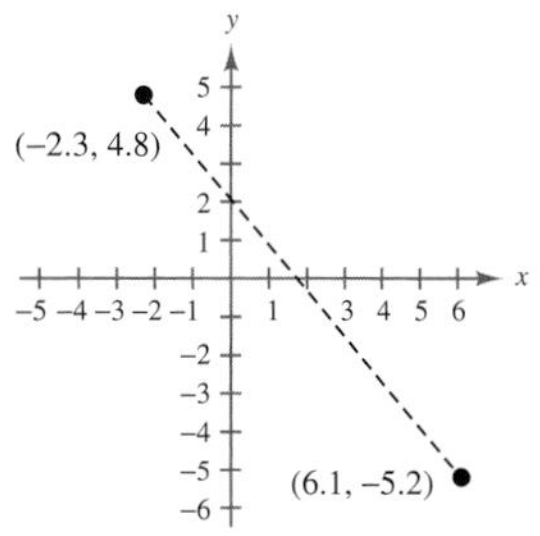

(b) $\sqrt{170.56} \approx 13.06$

164. (a)

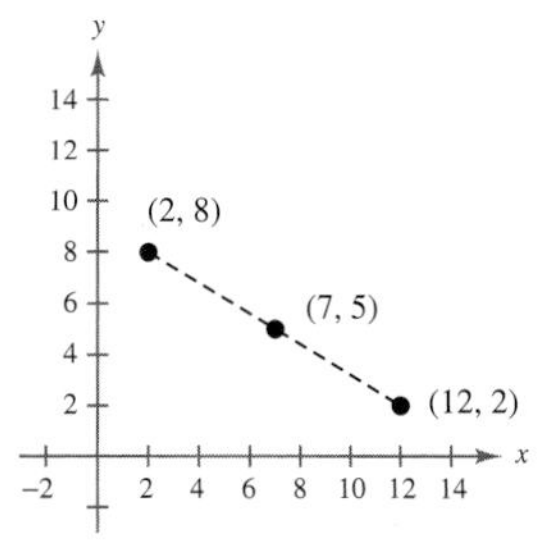

(b) $(7, 5)$

166. (a)

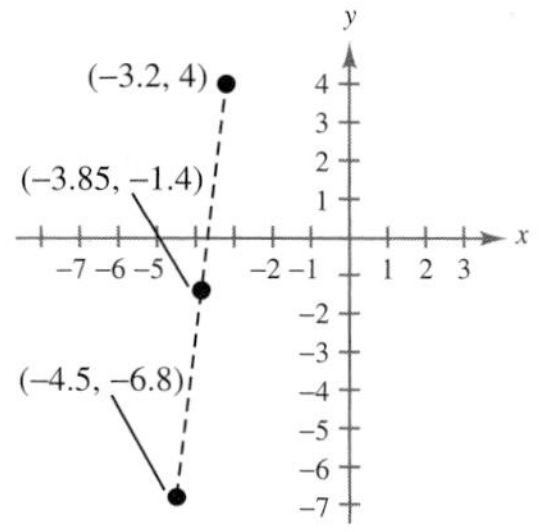

(b) $(-3.85, -1.4)$

168. False. $\frac{x^3 - 1}{x - 1} = x^2 + x + 1,\ x \neq 1$

170. False. $x^n - y^n = (x - y)(x^{n-1} + x^{n-2}y + \cdot \cdot \cdot + y^{n-1})$, for odd values of n.

172. True. When evaluating the Quadratic Formula, you obtain a negative number under the radical sign.

174. (a)

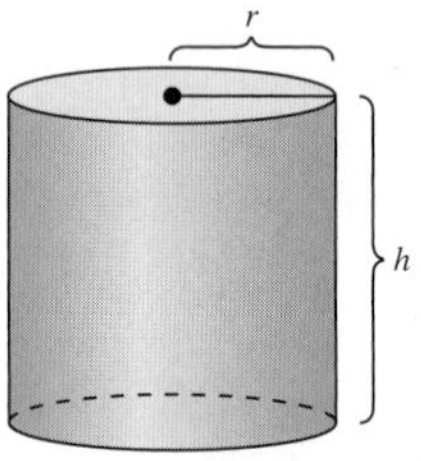

(b) $168\pi \approx 527.79$ square inches

176. Some solutions to certain types of equations may be extraneous solutions, which do not satisfy the original equation. So, checking is crucial.

178. There is no solution. The absolute value of an expression is always nonnegative.

Chapter 1

Section 1.1 *(page 107)*

2. (a) Yes (b) No **4.** (a) Yes (b) No

6.

x	-2	0	1	$\frac{4}{3}$	2
y	$-\frac{5}{2}$	-1	$-\frac{1}{4}$	0	$\frac{1}{2}$

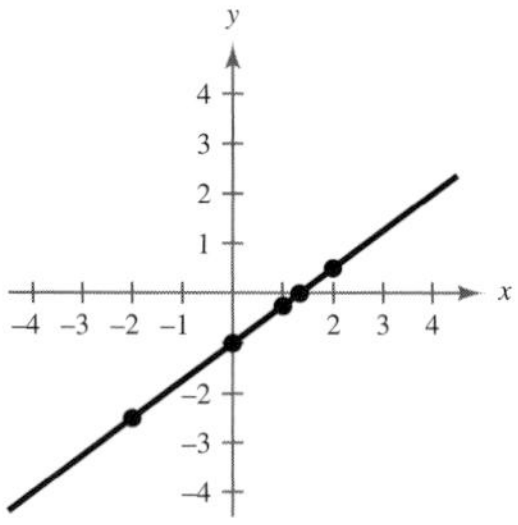

8.

x	-2	-1	0	1	2
y	1	4	5	4	1

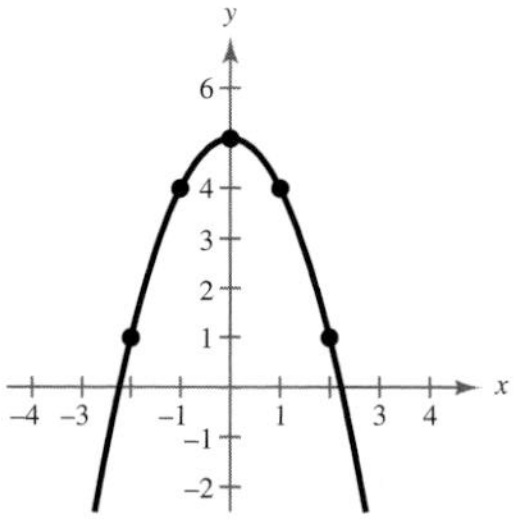

10. x-axis symmetry **12.** y-axis symmetry

14. y-axis symmetry **16.** Origin symmetry

18.

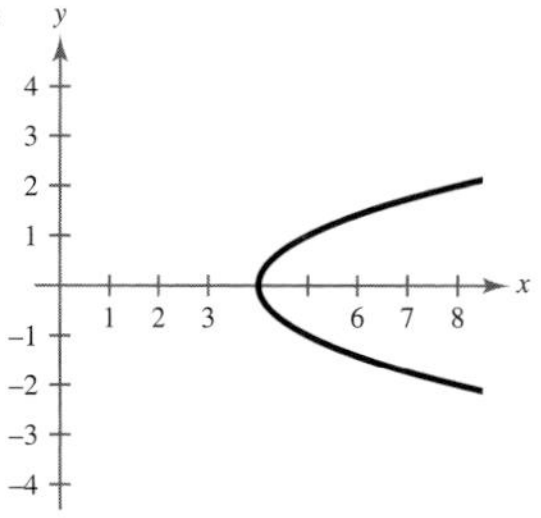

20.

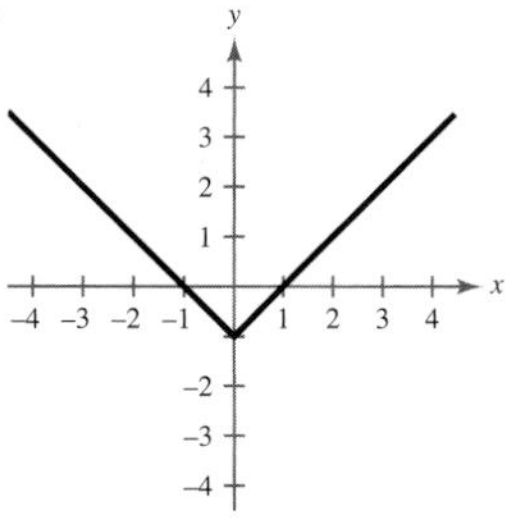

22. a **24.** d

26. x-intercept: $(-2, 0)$

y-intercept: $(0, 4)$

28. x-intercept: $(-1, 0)$

y-intercepts: $(0, \pm 1)$

30.

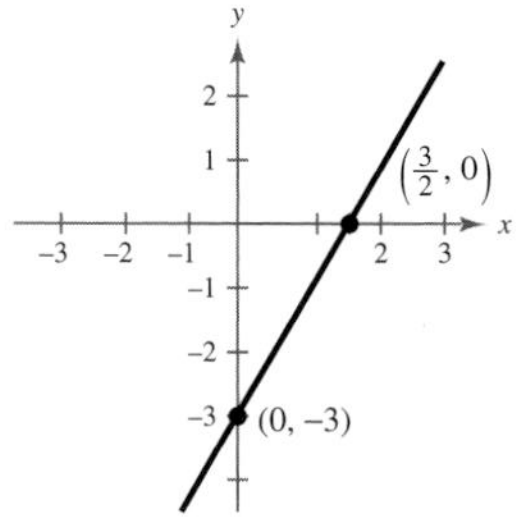

32.

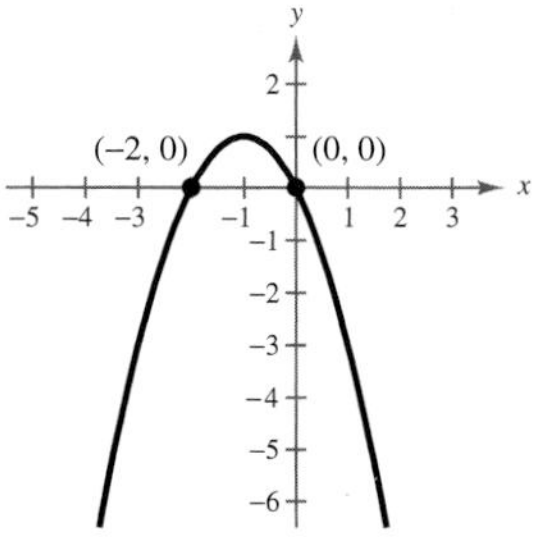

34.

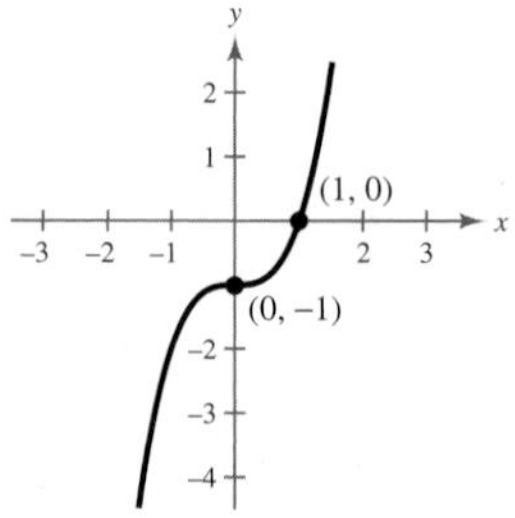

36.

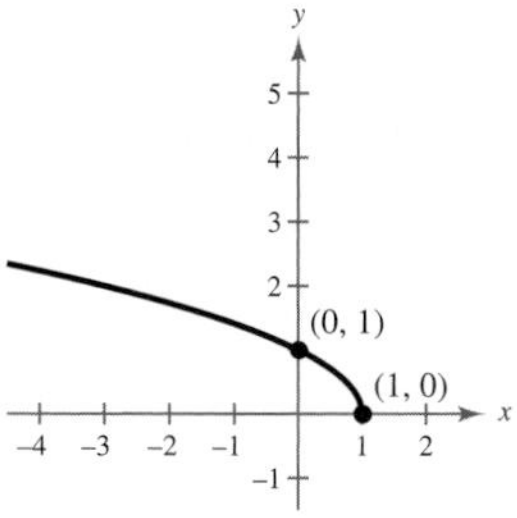

38.

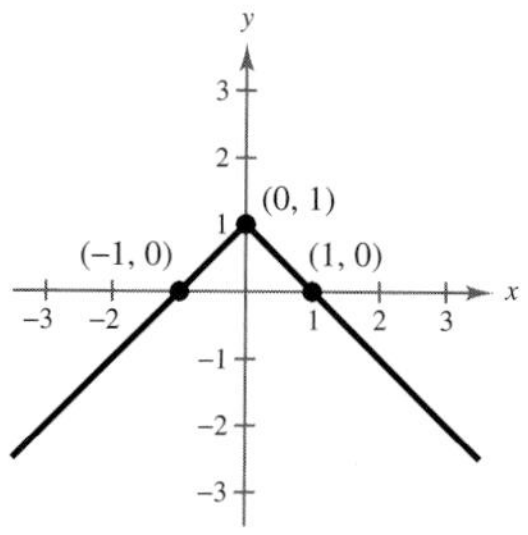

40.

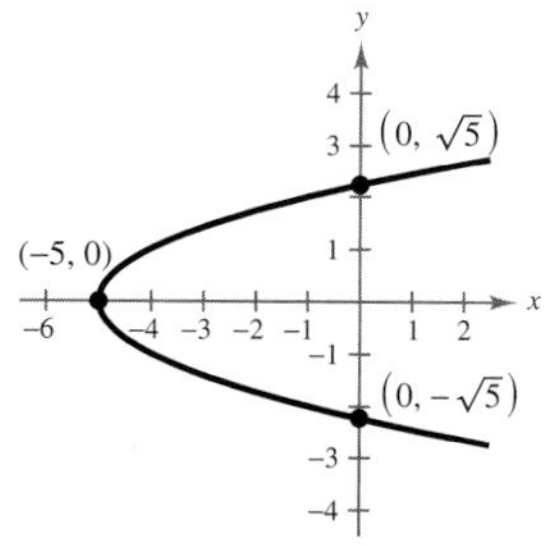

42.

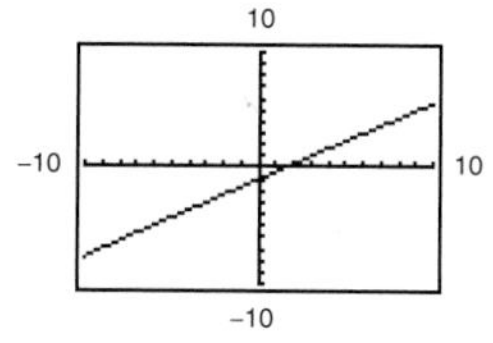

Intercepts: $\left(\frac{3}{2}, 0\right)$, $(0, -1)$

44.

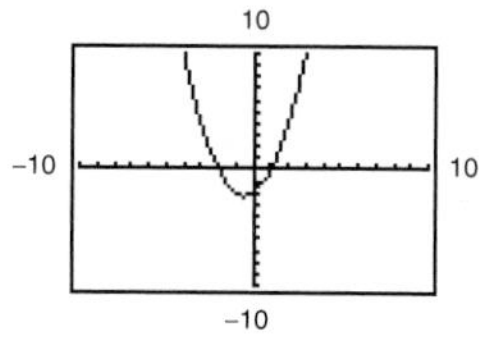

Intercepts: $(-2, 0)$, $(1, 0)$, $(0, -2)$

46.

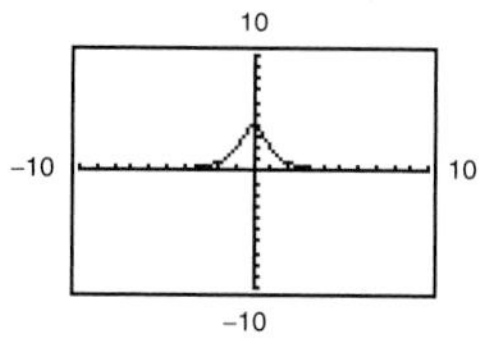

Intercept: $(0, 4)$

48.

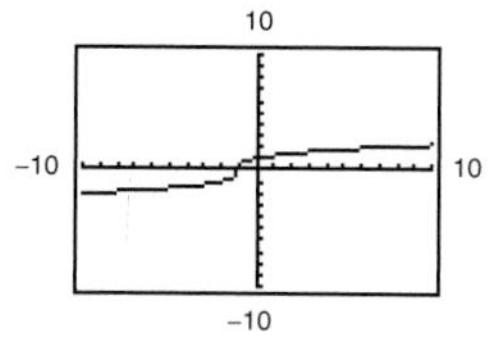

Intercepts: $(-1, 0)$, $(0, 1)$

50.

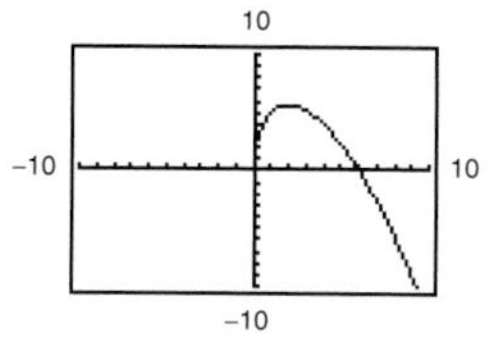

Intercepts: $(0, 0)$, $(6, 0)$

52.

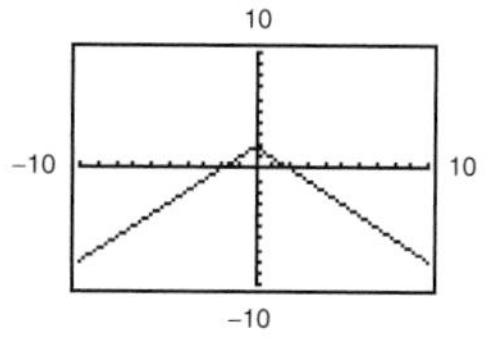

Intercepts: $(\pm 2, 0)$, $(0, 2)$

54. $x^2 + y^2 = 25$ **56.** $(x + 7)^2 + (y + 4)^2 = 64$

58. $(x - 3)^2 + (y + 2)^2 = 25$ **60.** $x^2 + y^2 = 17$

62. Center: $(0, 0)$; Radius: 4

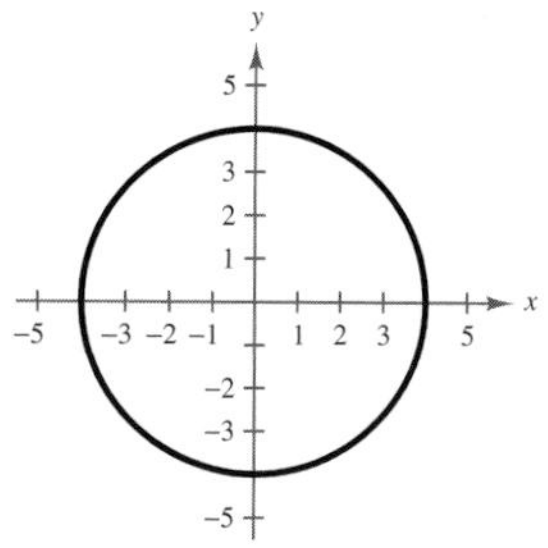

64. Center: $(0, 1)$; Radius: 1

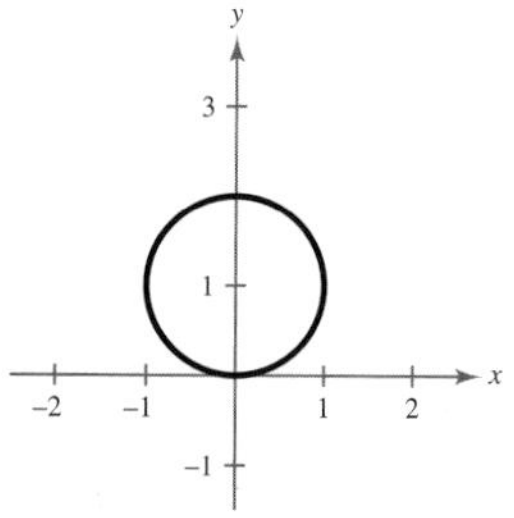

66. Center: $(2, -1)$; Radius: $\sqrt{3}$

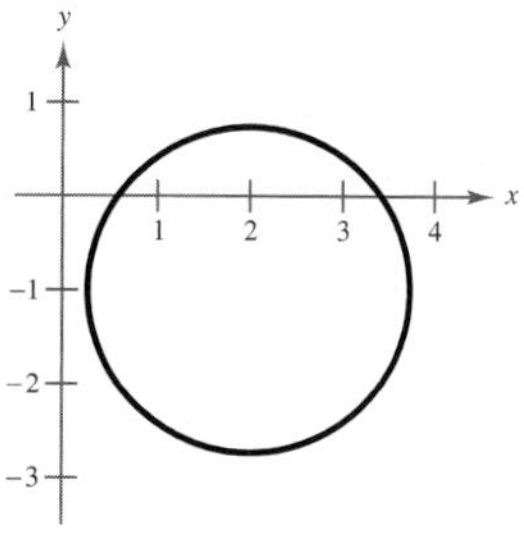

68.

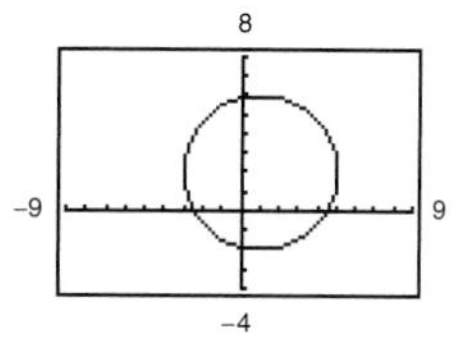

Circle

CHAPTER 1

70.

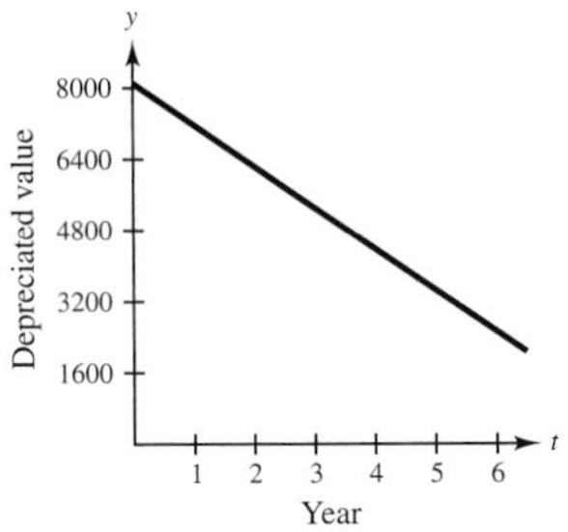

72. (a)

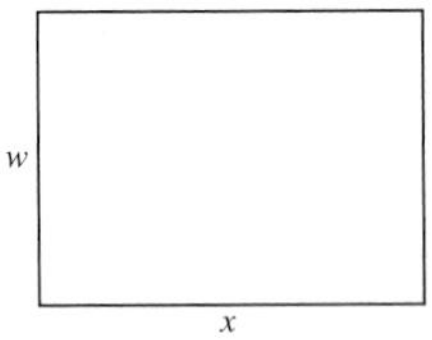

(b) Answers will vary.

(c)

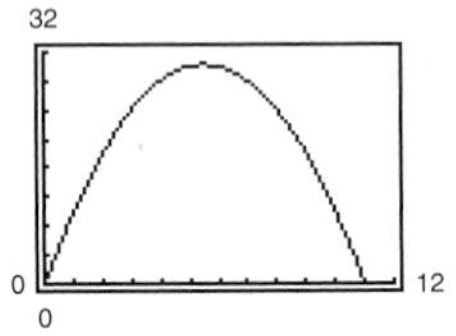

(d) $x = 5.5$, $w = 5.5$

74. (a) and (b)

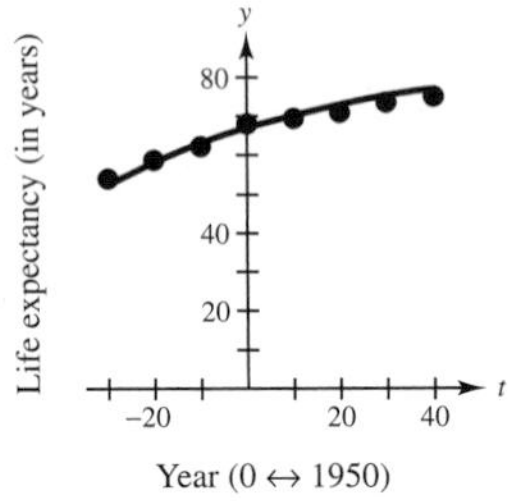

(c) The curve seems to be a good fit for the data.

(d) 2002: 78.2 years; 2004: 78.5 years

76. False. To find y-intercepts, let $x = 0$ and solve the equation for y.

78. The viewing window is incorrect. Change the viewing window. Examples will vary.

80. Assuming that the graph does not go beyond the vertical limits of the display, you will see the graph for larger values of x.

82. False. $(3 + 4)^2 = 3^2 + 2 \cdot 3 \cdot 4 + 4^2$

84. -7^4 **86.** $|x|\sqrt[4]{x}$ **88.** $5(2\sqrt{5} + 3)$ **90.** $\sqrt[6]{y}$

Section 1.2 *(page 119)*

2. (a) L_2 (b) L_1 (c) L_3

4.

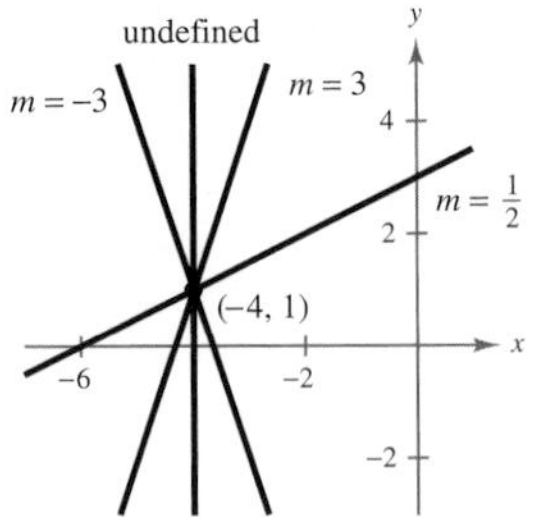

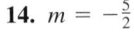

6. $\frac{8}{3}$ **8.** -1 **10.** $\frac{2}{3}$

12. $m = -4$

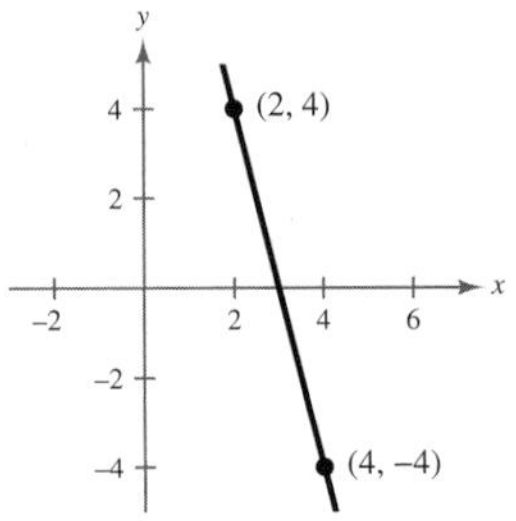

14. $m = -\frac{5}{2}$

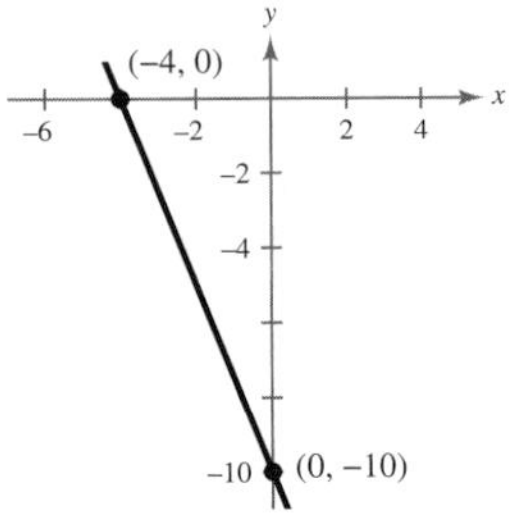

16. $m = -\frac{8}{3}$

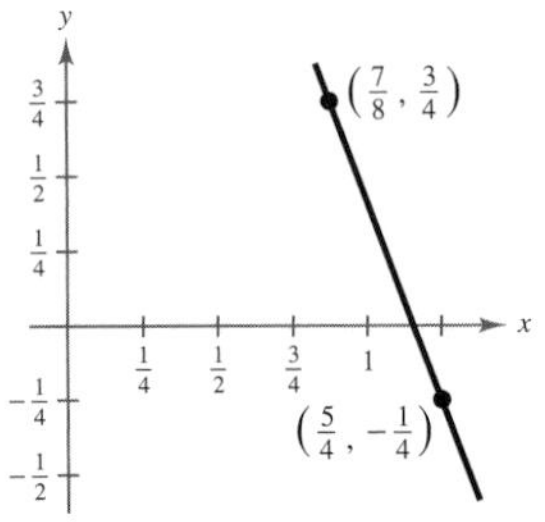

18. $m = 1.425$

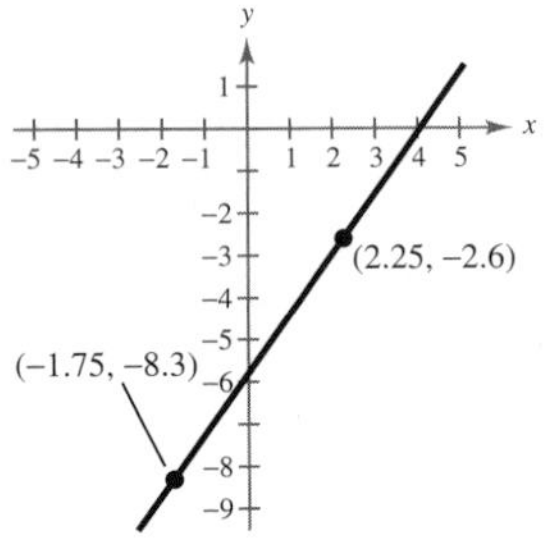

20. $(-4, 0), (-4, 3), (-4, 5)$ **22.** $(0, 4), (9, -5), (11, -7)$

24. $(-4, -1), (-2, -1), (0, -1)$

26. $(-2, -5), (1, -11), (3, -15)$

28. $(-3, -5), (1, -7), (5, -9)$

30. Neither parallel nor perpendicular **32.** Perpendicular

34. (a) Revenues increasing \$400 per day.

(b) Revenues increasing \$100 per day.

(c) No change in revenues.

36. (a) Greatest increase: 1996 to 1997
Smallest increase: 1988 to 1989

(b) $m = 0.073$

(c) Each year, the dividends per share increase by \$0.073.

38. 12 feet

40. $m = 5$; Intercept: $(0, 3)$

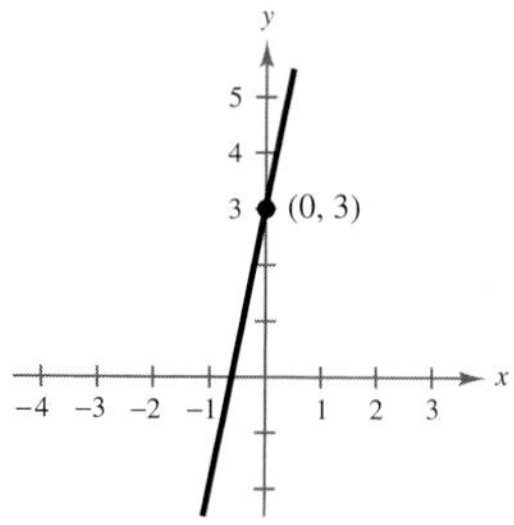

42. m is undefined. There is no y-intercept.

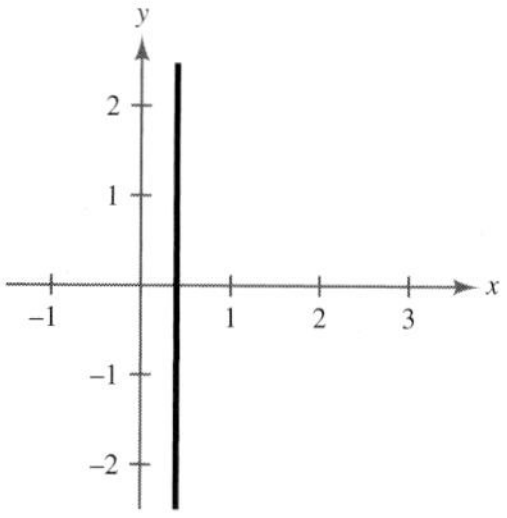

44. $m = -\frac{7}{6}$; Intercept: $(0, 5)$

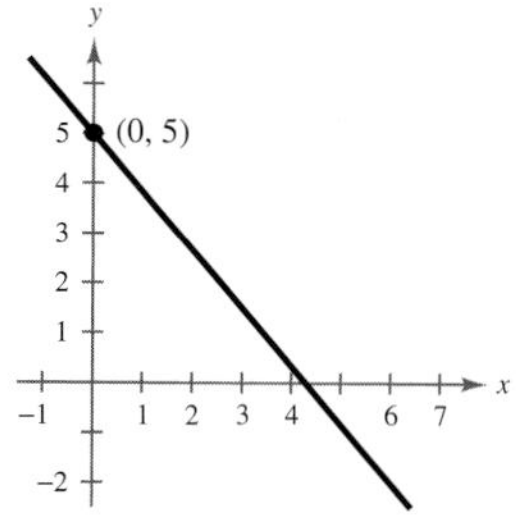

46. $y = 3x - 2$

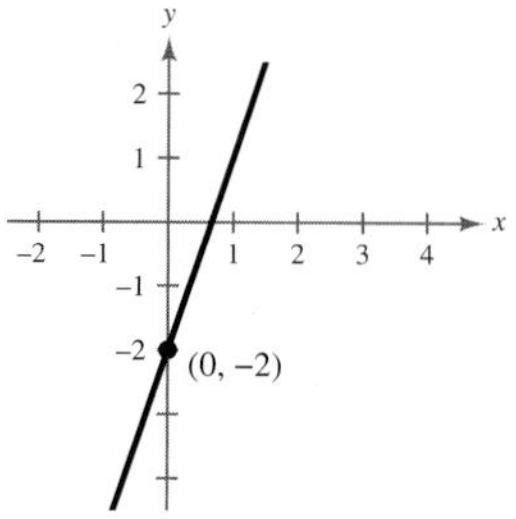

48. $y = -2x$

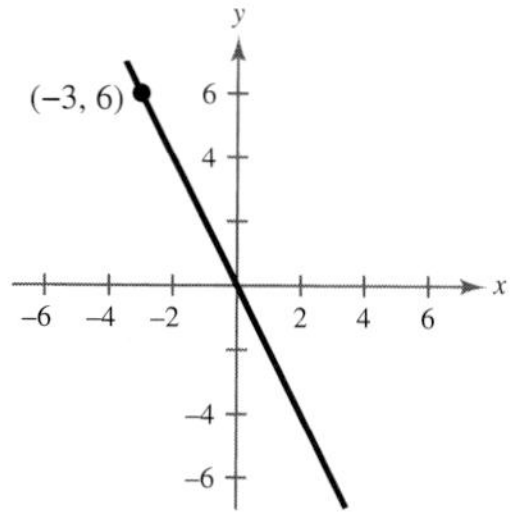

50. $y = -\frac{1}{3}x + \frac{4}{3}$

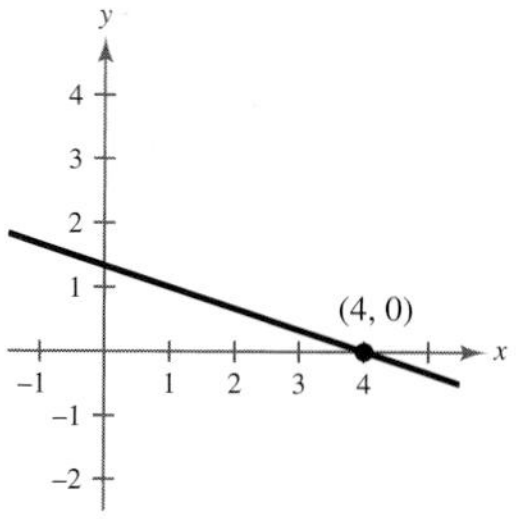

52. $x = 6$

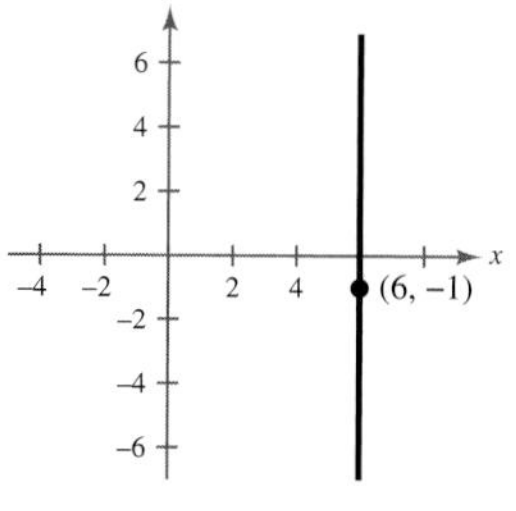

54. $y = \frac{4}{3}x - \frac{17}{6}$

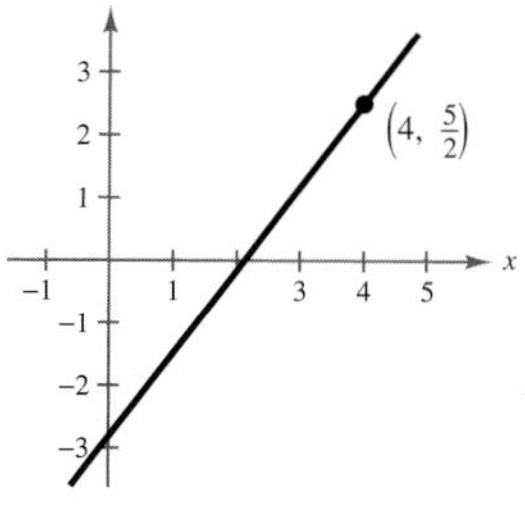

56. $y = 5x + 27.3$

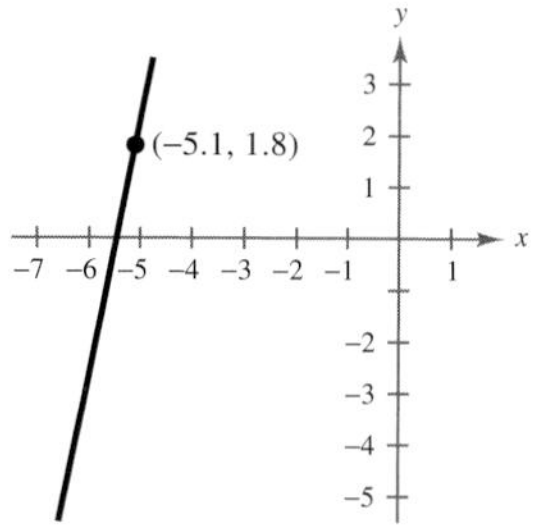

58. $y = -\frac{3}{5}x + 2$

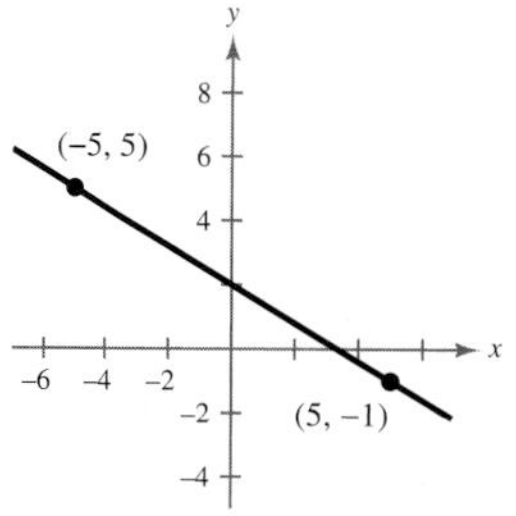

60. $x = -8$

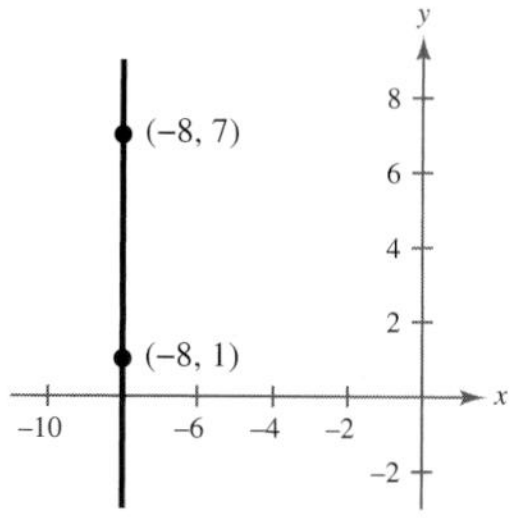

62. $y = -\frac{1}{2}x + \frac{3}{2}$

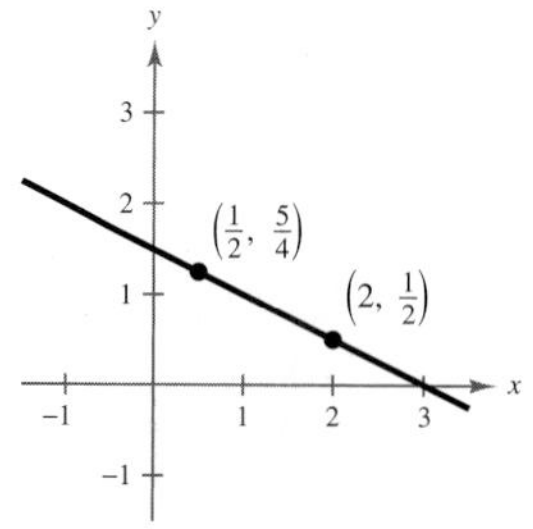

64. $y = -\frac{6}{5}x - \frac{18}{25}$

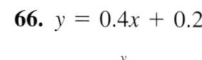

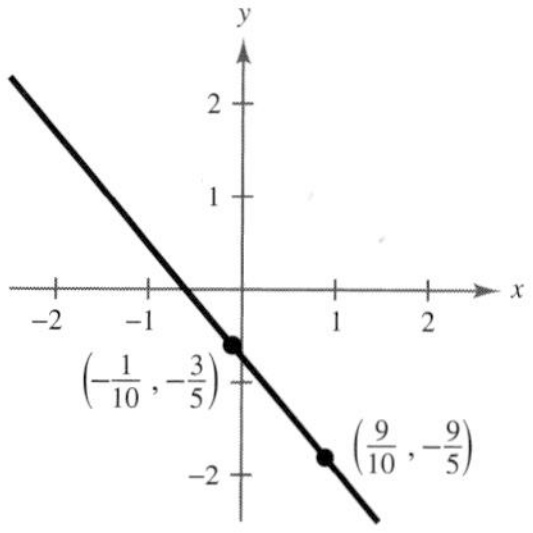

66. $y = 0.4x + 0.2$

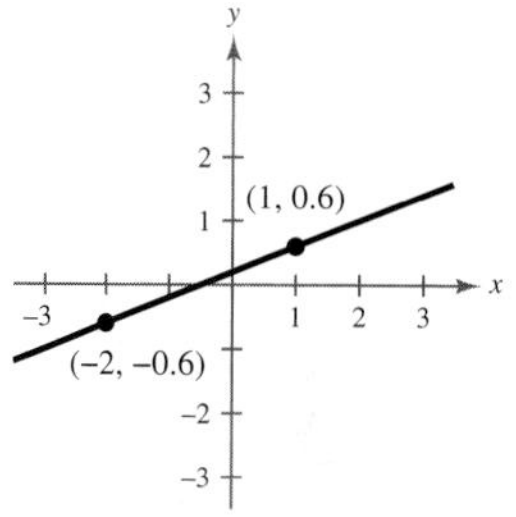

68. $3x + 2y - 6 = 0$ **70.** $12x + 3y + 2 = 0$

72. $x + y - 3 = 0$

74. (a) $y = 2x - 3$ (b) $y = -\frac{1}{2}x + 2$

76. (a) $y = -\frac{3}{4}x + \frac{3}{8}$ (b) $y = \frac{4}{3}x + \frac{127}{72}$

78. (a) $y = 0$ (b) $x = -1$

80. (a) $y = x + 4.3$ (b) $y = -x + 9.3$

82.

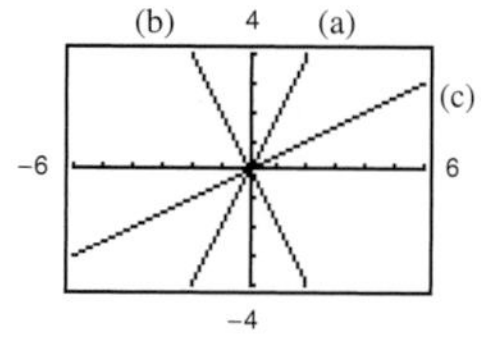

(b) is perpendicular to (c).

84. 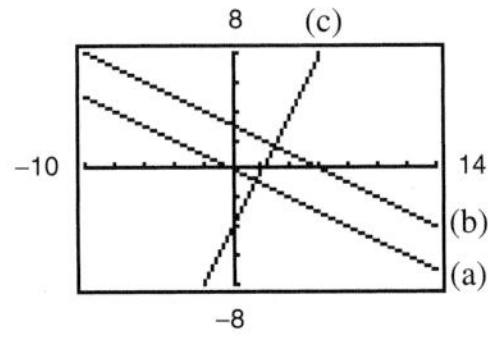

(a) is parallel to (b).

(c) is perpendicular to (a) and (b).

86. $V = 125t + 2415$

88. b; The slope is -20, which represents the decrease in the amount of the loan each week.

90. a; The slope is 0.32, which represents the increase in travel cost for each mile driven.

92. $3x - 2y - 1 = 0$ **94.** $80x + 12y + 139 = 0$

96. $y = 0.44t + 0.26$; 2000: \$1.14, 2001: \$1.58

98. $F = \frac{9}{5}C + 32$ **100.** \$39,500

102. $V = -175t + 875$ **104.** $S = 0.85L$

106. (a) $C = 16.75t + 36{,}500$ (b) $R = 27t$

(c) $P = 10.25t - 36{,}500$ (d) $t \approx 3561$ hours

108. (a) 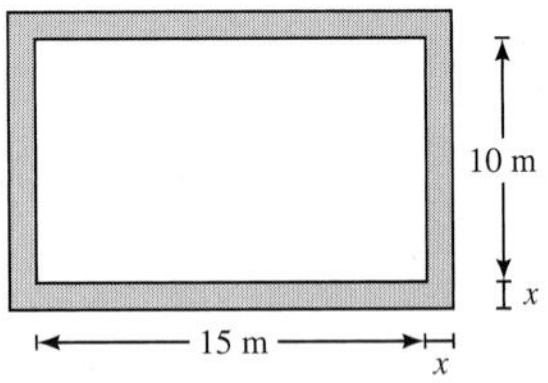

(b) $y = 8x + 50$

(c) 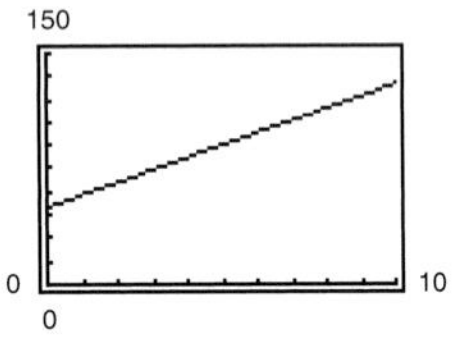

(d) $m = 8$, 8 meters

110. $C = 0.31x + 120$ **112.** $y = 95t + 475$

114. False. The slope with the greatest magnitude corresponds to the steepest line.

116. By finding the distance between each point and using the Pythagorean Theorem

118. No. The slope cannot be determined without knowing the scale on the y-axis. The slopes could be the same.

120. V-intercept: initial cost; Slope: annual depreciation

122. No. The slopes of two perpendicular lines have opposite signs (assume that neither line is vertical or horizontal).

124. c **126.** b **128.** $\frac{5}{2}$ **130.** $4 \pm \sqrt{13}$

132. $\frac{1}{9}$, 25

Section 1.3 *(page 133)*

2. No **4.** Yes

6. No, the input values of 0 and 1 each have two different output values.

8. Yes, it does not matter that each input value has the same output value.

10. (a) Not a function, because the element c in A corresponds to two elements, 2 and 3, in B.

(b) Function

(c) Not a function from A to B. (It is instead a function from B to A.)

(d) Function

12. 16 million **14.** Not a function **16.** Not a function

18. Not a function **20.** Function **22.** Not a function

24. (a) 2, 2 (b) $-3, -3$

(c) $(t + 1), (t + 1)$ (d) $(x + c), (x + c)$

26. (a) 7 (b) 0 (c) $1 - 3s$

28. (a) 0 (b) -0.75 (c) $x^2 + 2x$

30. (a) 2 (b) 5 (c) $\sqrt{x} + 2$

32. (a) $\dfrac{11}{4}$ (b) Undefined (c) $\dfrac{2x^2 + 3}{x^2}$

34. (a) 6 (b) 6 (c) $x^2 + 4$

36. (a) 6 (b) 3 (c) 10

38.

x	3	4	5	6	7
$g(x)$	0	1	$\sqrt{2}$	$\sqrt{3}$	2

40.

s	0	1	$\frac{3}{2}$	$\frac{5}{2}$	4
$f(s)$	-1	-1	-1	1	1

42.

x	1	2	3	4	5
$h(x)$	8	5	0	1	2

44. $-\frac{1}{5}$ **46.** $\pm 2\sqrt{3}$ **48.** 3, 5

50. $1, \pm 2$ **52.** $-1, 2$ **54.** $0, \pm 2$

56. All real numbers x **58.** All real numbers $y \neq -5$

60. All real numbers t **62.** $x \leq -3, x \geq 0$

64. All real numbers $x \neq 0, 2$ **66.** $x > -6$

68. All real numbers $x \neq \pm 3$

70. $\left\{\left(-2, -\frac{4}{5}\right), (-1, -1), (0, 0), (1, 1), \left(2, \frac{4}{5}\right)\right\}$

72. $\{(-2, 1), (-1, 0), (0, 1), (1, 2), (2, 3)\}$

74. $f(x) = \frac{1}{4}x$ **76.** $h(x) = 3\sqrt{|x|}$

78. $-(5 + h), \; h \neq 0$ **80.** 2

82. $-\frac{1}{t},\ t \neq 1$ **84.** $\frac{x^{2/3} - 4}{x - 8}$

86. $A = \frac{C^2}{4\pi}$ **88.** $A = \frac{\sqrt{3}}{4}s^2$

90. (a)

Units x	Price p	Profit P
110	$90 - 10(0.15)$	$110[90 - 10(0.15)] - 110(60) = 3135$
120	$90 - 20(0.15)$	$120[90 - 20(0.15)] - 120(60) = 3240$
130	$90 - 30(0.15)$	$130[90 - 30(0.15)] - 130(60) = 3315$
140	$90 - 40(0.15)$	$140[90 - 40(0.15)] - 140(60) = 3360$
150	$90 - 50(0.15)$	$150[90 - 50(0.15)] - 150(60) = 3375$
160	$90 - 60(0.15)$	$160[90 - 60(0.15)] - 160(60) = 3360$

The maximum profit is \$3375.

(b)

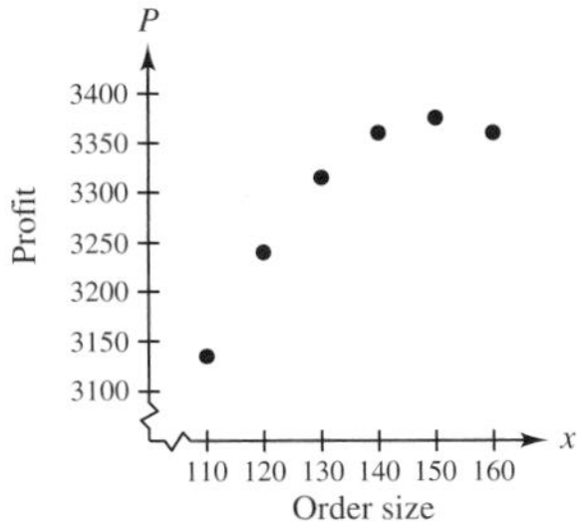

P is a function of x.

(c) $P = 45x - 0.15x^2,\ x > 100$

92. $A = 2xy = 2x\sqrt{36 - x^2},\quad 0 < x < 6$

94. $V = x^2y = x^2(108 - 4x) = 108x^2 - 4x^3,\ 0 < x < 27$

96. (a) $C = 6000 + 0.95x$ (b) $\overline{C} = \frac{C}{x} = \frac{6000}{x} + 0.95$

98. (a)

y	5	10	20
$F(y)$	26,474.08	149,760.00	847,170.49

y	30	40
$F(y)$	2,334,527.36	4,792,320.00

The deeper the water, the greater the force.

(b) 21 feet. A better estimate could be found by using smaller intervals of depth.

100. (a) -17 average decrease per year in the lynx population

(b)

t	1988	1989	1990	1991
N	9.0	19.8	43.9	53.6

t	1992	1993	1994	1995
N	30.9	16.7	10.2	6.9

The N-values are similar to the actual data.

102. False. The range is $[-1, \infty)$.

104. No. The element 3 in the domain corresponds to two elements in the range.

106. It gives a name to the relationship so it can be easily referenced. When evaluating a function, you see both the input and output values.

108. 8 **110.** $\frac{2}{3}$ **112.** $x + y - 10 = 0$

114. $10x + 18y - 49 = 0$

Section 1.4 *(page 147)*

2. Domain: all real numbers
Range: all real numbers

4. Domain: $[1, \infty)$
Range: $[0, \infty)$

6. Domain: $(-\infty, \infty)$
Range: $[0, \infty)$

8. Domain: $(-\infty, 1), (1, \infty)$
Range: $-1, 1$

10. Function **12.** Not a function **14.** Not a function

16. $-8, \frac{2}{3}$ **18.** 2, 7 **20.** $\pm 3, 4$ **22.** $0, \pm\frac{5}{3}$

24. 0, 7

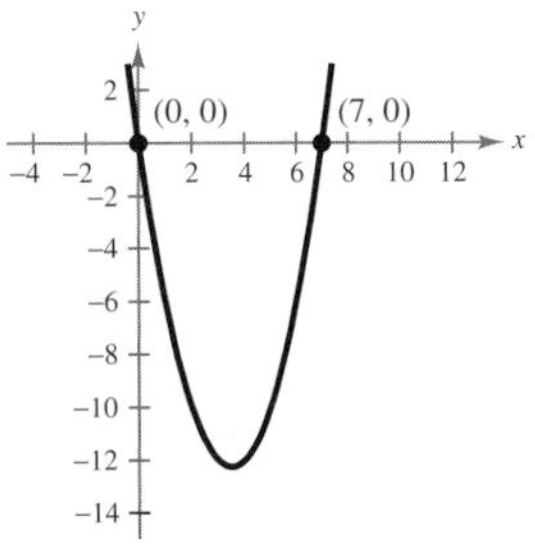

26. 26

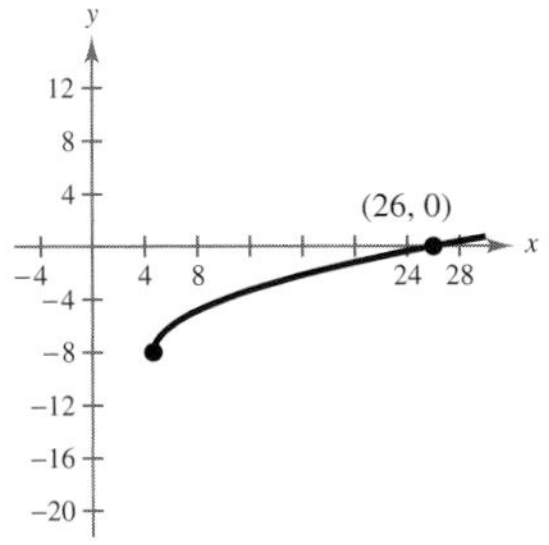

28. $\pm\frac{3\sqrt{2}}{2}$

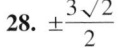

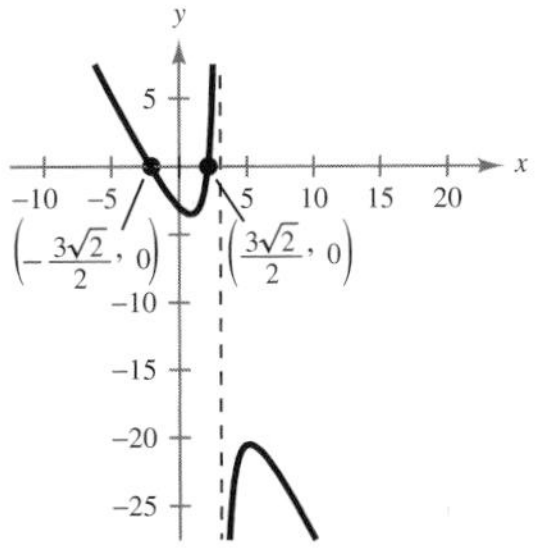

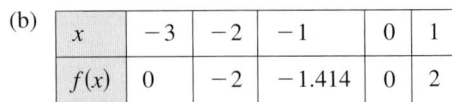

30. (a) Decreasing on $(-\infty, 2)$; Increasing on $(2, \infty)$

(b) Neither even nor odd

32. (a) Decreasing on $(-\infty, -1)$; Increasing on $(1, \infty)$

(b) Even

34. (a)

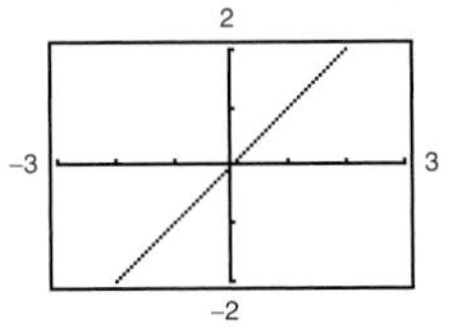

Increasing on $(-\infty, \infty)$

(b)

x	-2	-1	0	1	2
$g(x)$	-2	-1	0	1	2

36. (a)

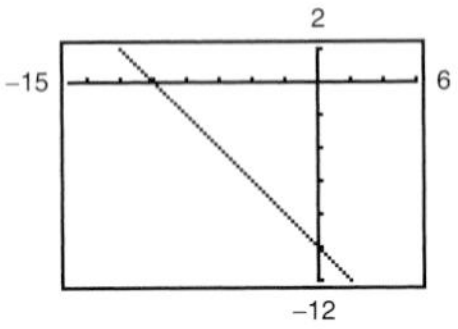

Decreasing on $(-\infty, \infty)$

(b)

x	-12	-8	-4	0	4
$f(x)$	2	-2	-6	-10	-14

38. (a)

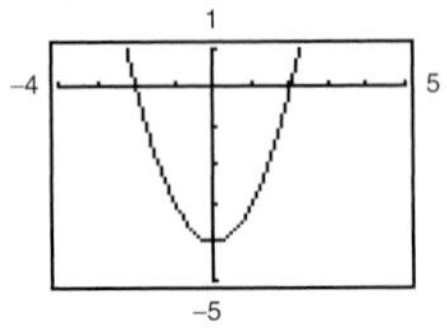

Decreasing on $(-\infty, 0)$; Increasing on $(0, \infty)$

(b)

x	-2	-1	0	1	2
$h(x)$	0	-3	-4	-3	0

40. (a)

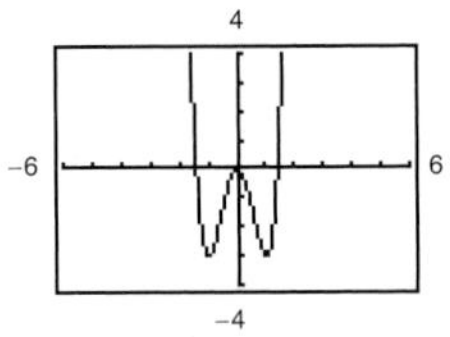

Increasing on $(-1, 0)$, $(1, \infty)$

Decreasing on $(-\infty, -1)$, $(0, 1)$

(b)

x	-2	-1	0	1	2
$f(x)$	24	-3	0	-3	24

42. (a)

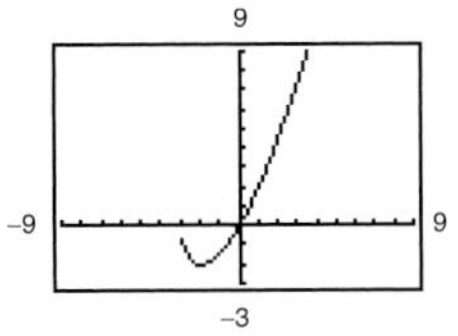

Increasing on $(-2, \infty)$

Decreasing on $(-3, -2)$

(b)

x	-3	-2	-1	0	1
$f(x)$	0	-2	-1.414	0	2

44. (a)

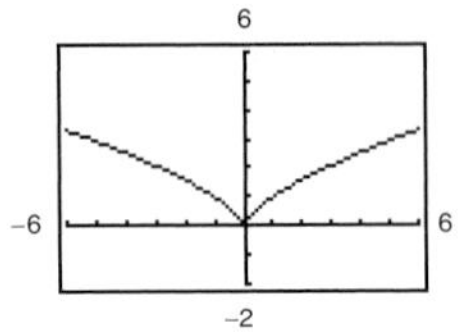

Decreasing on $(-\infty, 0)$; Increasing on $(0, \infty)$

(b)

x	-2	-1	0	1	2
$f(x)$	1.59	1	0	1	1.59

46. (a)

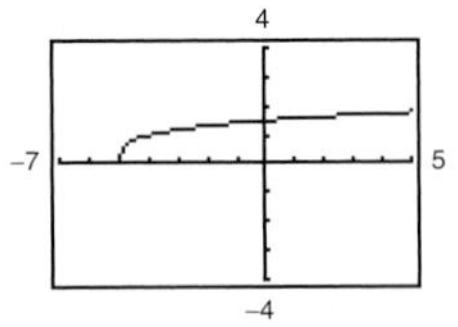

Increasing on $(-5, \infty)$

(b)

x	-5	-3	-1	1	3
$f(x)$	0	1.19	1.41	1.57	1.68

48. (a)

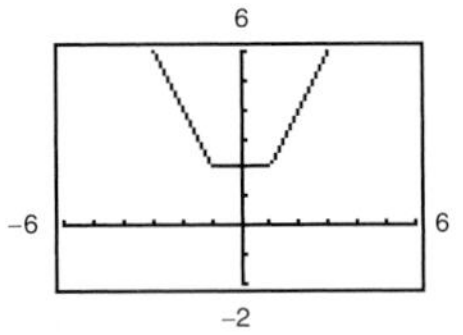

Decreasing on $(-\infty, -1)$

Constant on $(-1, 1)$

Increasing on $(1, \infty)$

(b)

x	-3	-2	-1	0	1	2	3
$f(x)$	6	4	2	2	2	4	6

50. (a)

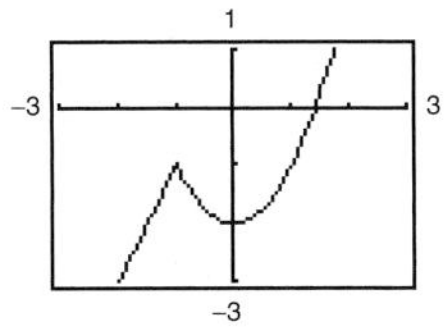

Increasing on $(-\infty, -1)$ and $(0, \infty)$

Decreasing on $(-1, 0)$

(b)

x	-2	-1	$-\frac{1}{2}$	0	1	2
$f(x)$	-3	-1	-1.75	-2	-1	2

52.

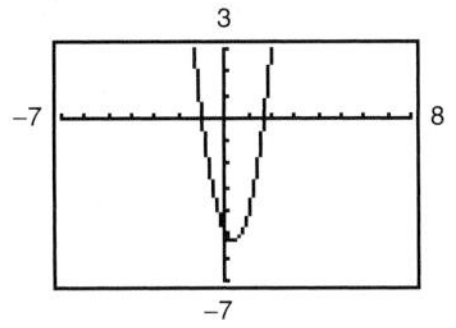

Relative minimum: $\left(\frac{1}{3}, -\frac{16}{3}\right)$

54.

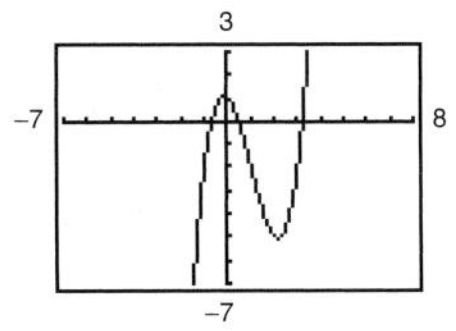

Relative maximum: $(-0.15, 1.08)$
Relative minimum: $(2.15, -5.08)$

56.

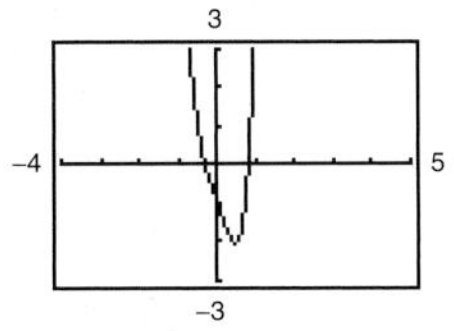

Relative minimum: $(0.45, -2.02)$

58.

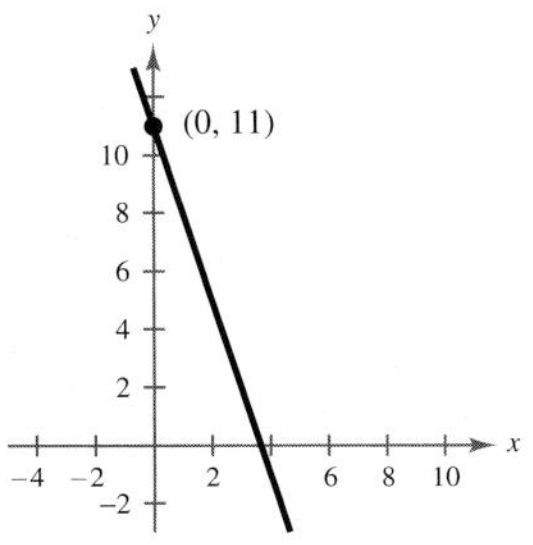

60.

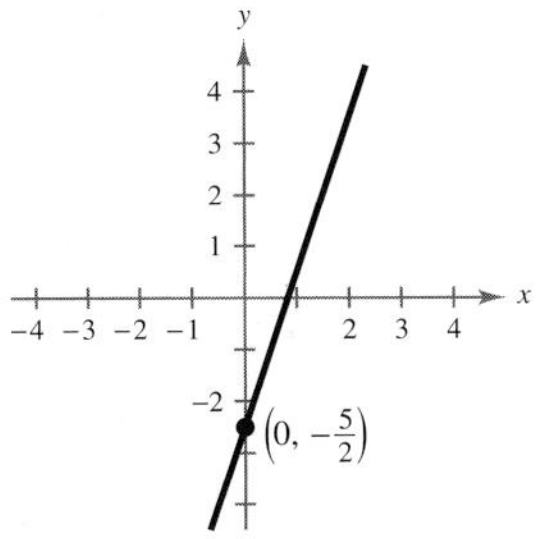

62.

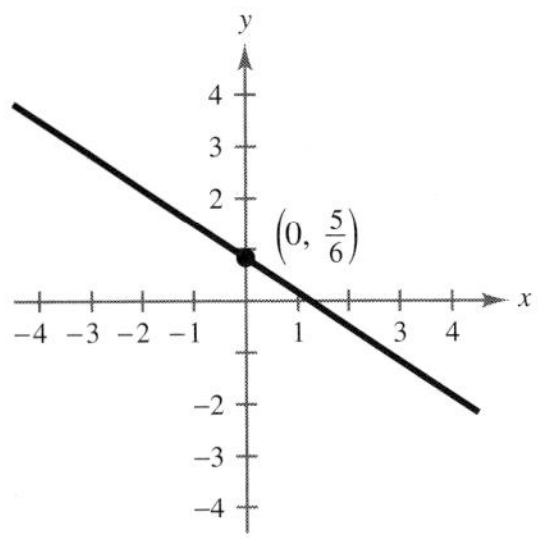

64.

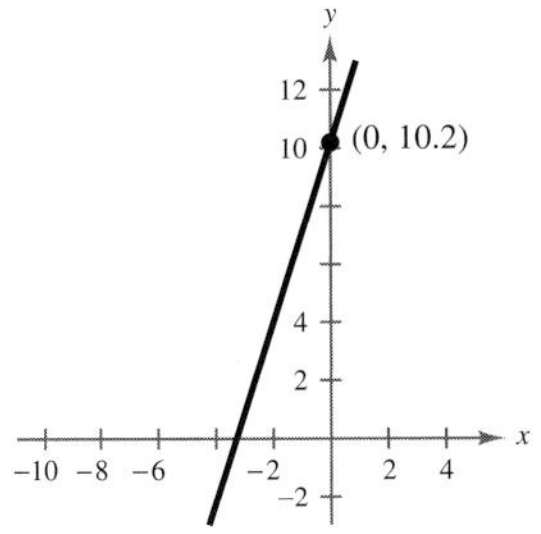

66. $f(x) = \frac{5}{2}x - \frac{1}{2}$

68. $f(x) = 5x - 6$

70. $f(x) = -\frac{1}{2}x + 7$

72. $f(x) = \frac{3}{4}x - 8$

74.

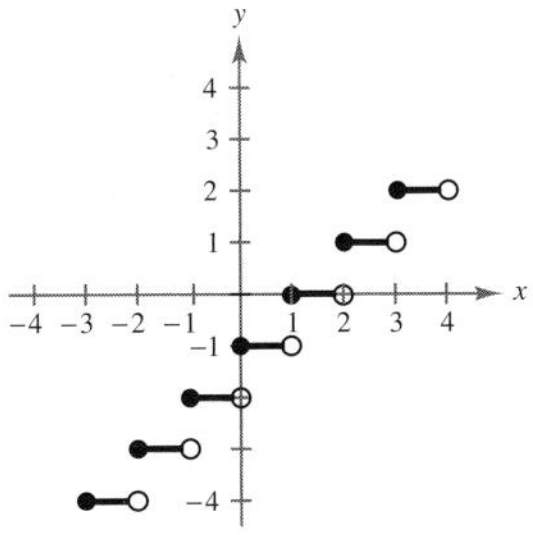

Shift 1 unit right

76.

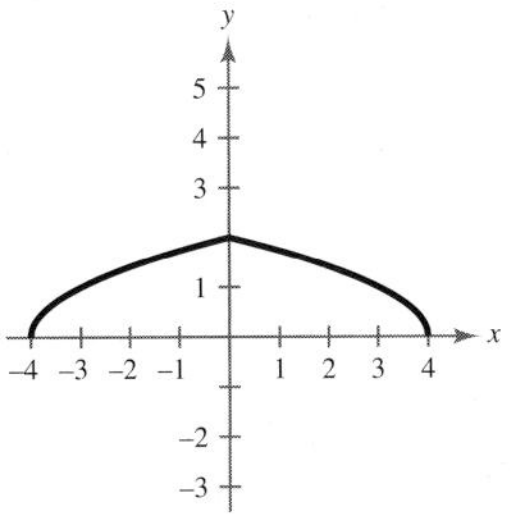

78.

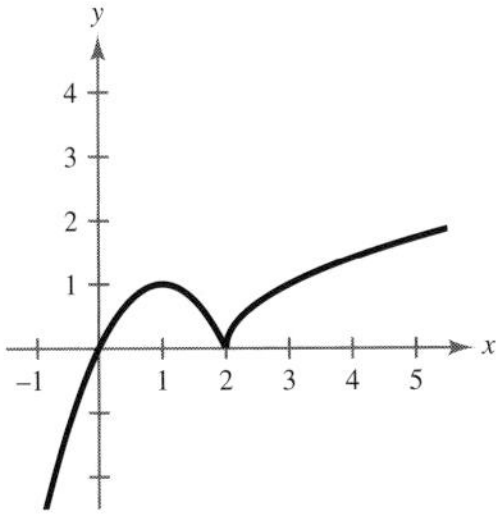

80. $\left[-\frac{1}{2}, \infty\right)$

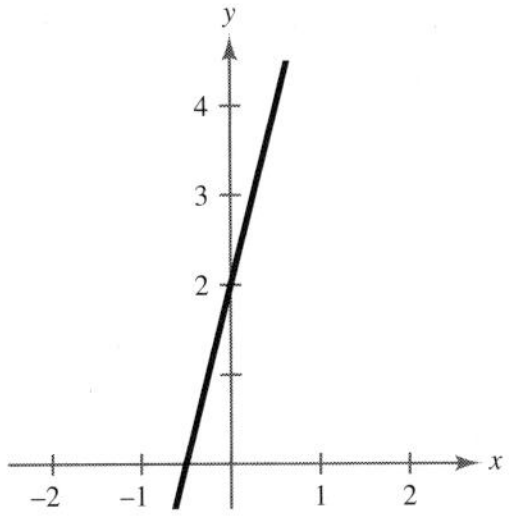

82. $(-\infty, 0]$, $[4, \infty)$

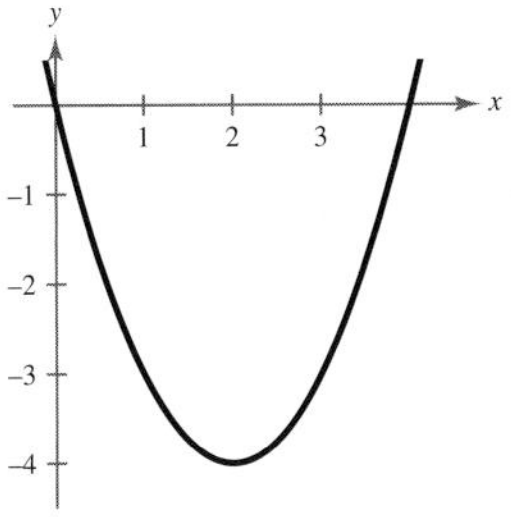

84. $[-2, \infty)$

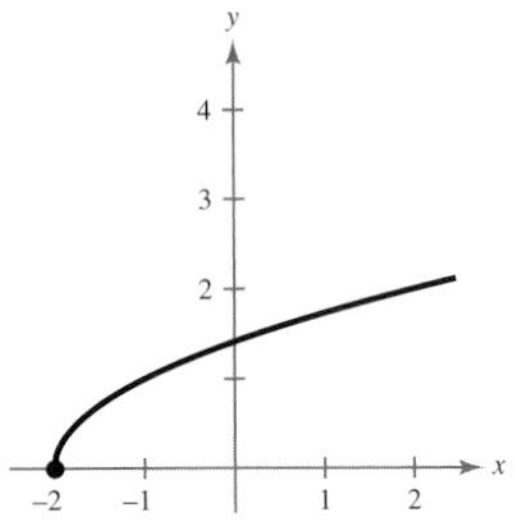

86. $f(x) < 0$ for all x

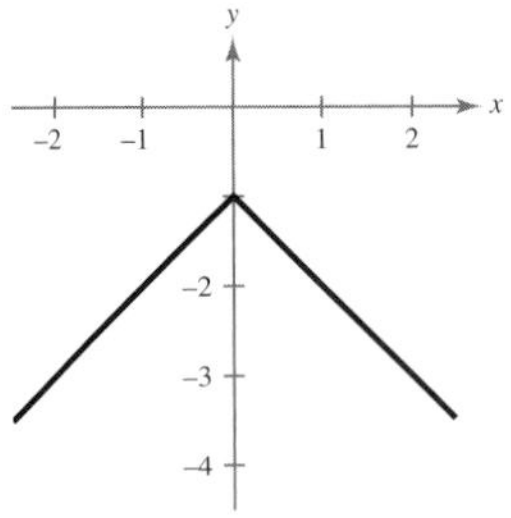

88. $(-\infty, \infty)$

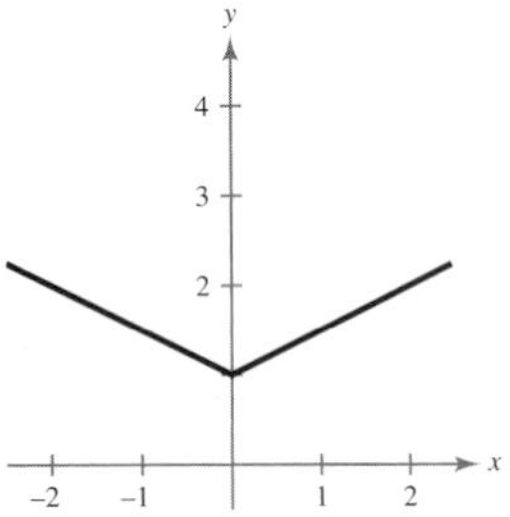

90. $\left(-\infty, \dfrac{-1-\sqrt{5}}{2}\right], \left[\dfrac{-1+\sqrt{5}}{2}, \infty\right)$

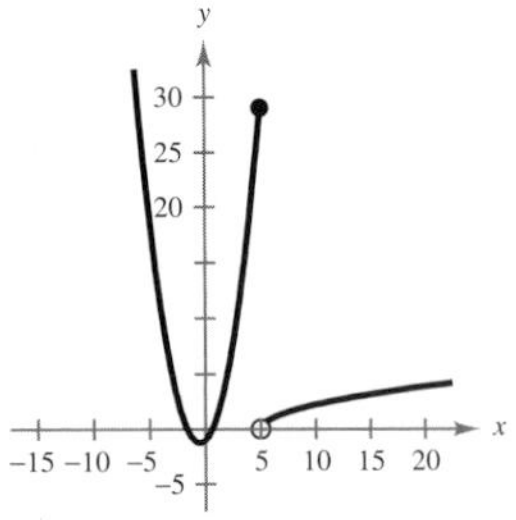

92.

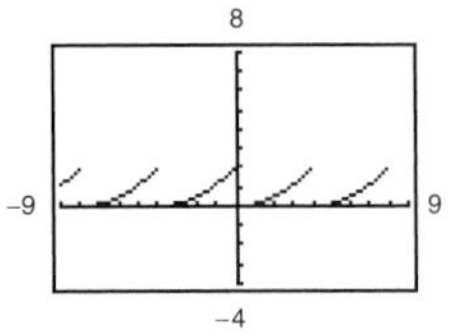

Domain: $(-\infty, \infty)$

Range: $[0, 2)$

Sawtooth pattern

94. Neither even nor odd **96.** Odd **98.** Even

100. (a) $\left(\frac{5}{3}, -7\right)$ (b) $\left(\frac{5}{3}, 7\right)$

102. (a) $(-5, -1)$ (b) $(-5, 1)$

104. $C = 9.80 + 2.50[\![x]\!], \quad x > 0$

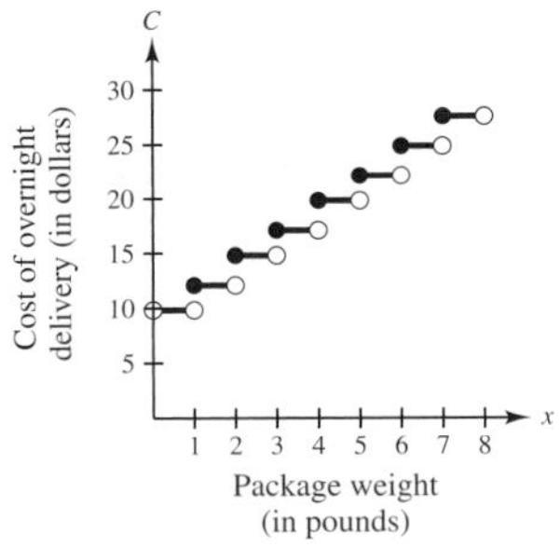

106. $h = -x^2 + 4x - 3$ **108.** $h = 2x - x^2$

110. $L = \frac{1}{2}y^2$ **112.** $L = 4 - y^2$

114. (a) $1.47x^3 - 16.41x^2 + 31.24x - 95.20$

(b) $0 \le x \le 7$

(c)

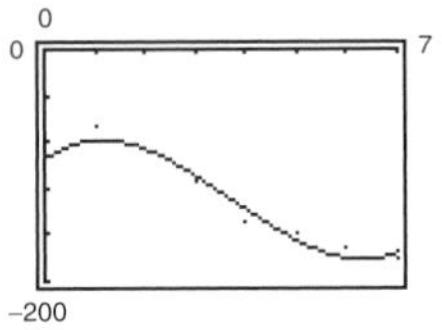

(d) 1992; 1991

116. (a) \$10,000 (b) 50,000,000 (c) 1%

118. False. The function $f(x) = \sqrt{x^2 + 1}$ has a domain of all real numbers.

120. Answers will vary.

122. (a) Even. The graph is a reflection in the x-axis.

(b) Even. The graph is a reflection in the y-axis.

(c) Even. The graph is a vertical translation of f.

(d) Neither. The graph is a horizontal translation of f.

124. No. For some values of y there correspond more than one value of x.

126. Both graphs will pass through the origin. $y = x^7$ will be symmetric with respect to the origin, and $y = x^8$ will be symmetric with respect to the y-axis.

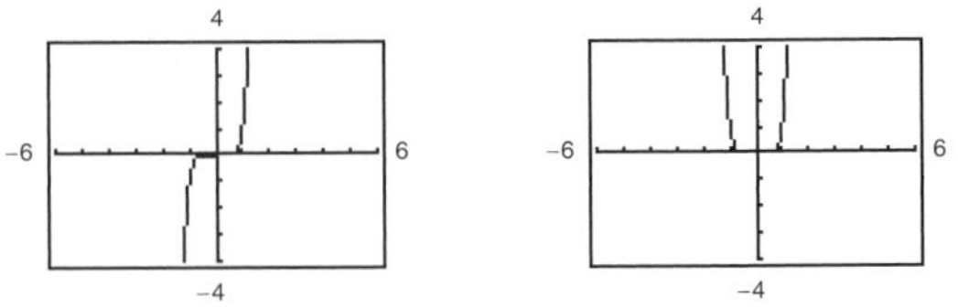

128. $-5, 15$ **130.** $\frac{5}{4}$

132. (a) -24 (b) 144 (c) $x^2 - 18x + 56$

134. (a) -3 (b) $-\frac{87}{16}$ (c) $139 - 2\sqrt{3}$

136. $-6 - h, h \neq 0$

Section 1.5 *(page 158)*

2. (a)

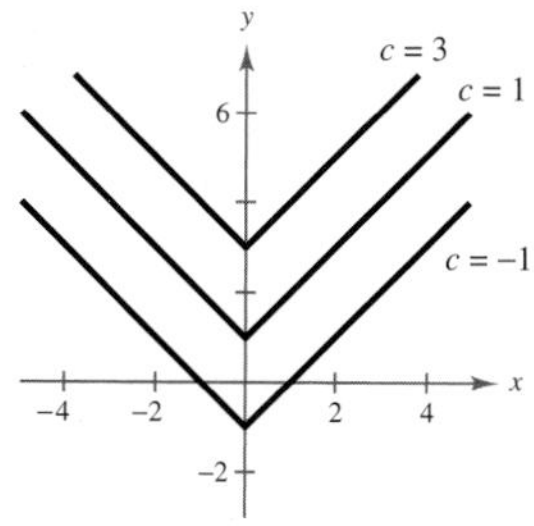

(b)

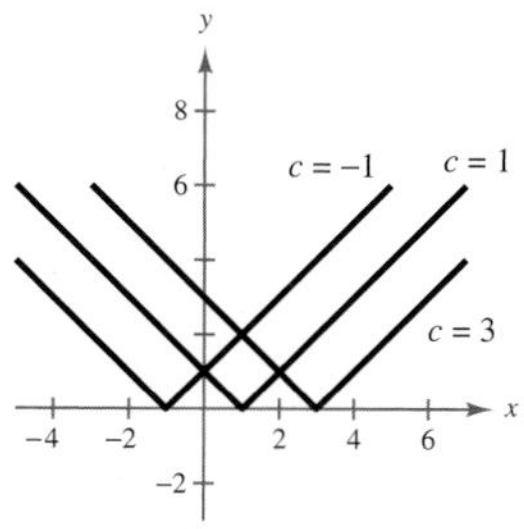

(c)

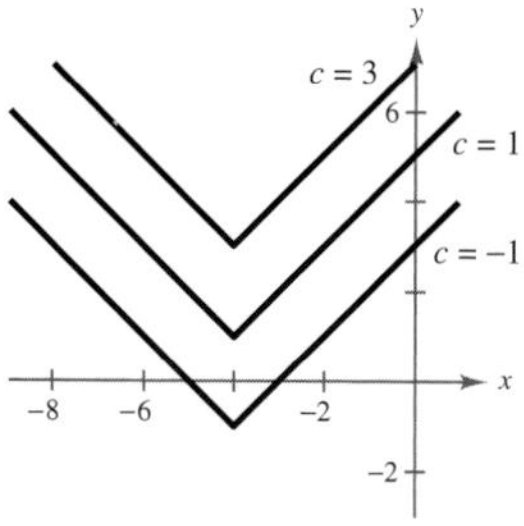

4. (a)

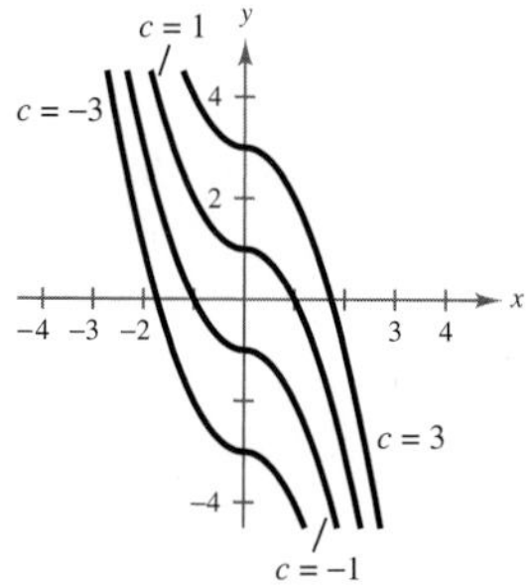

(b)

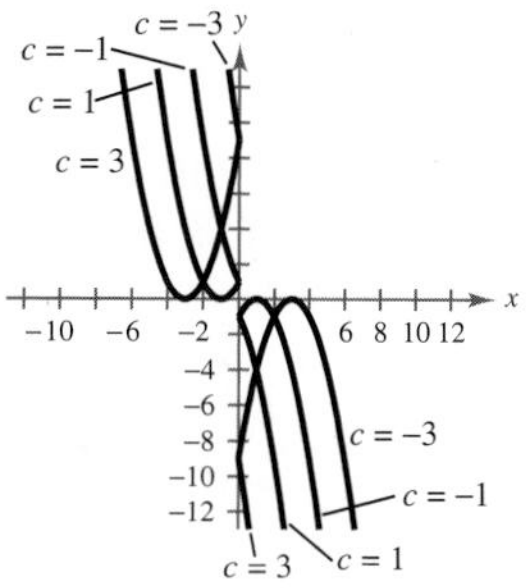

6. (a)

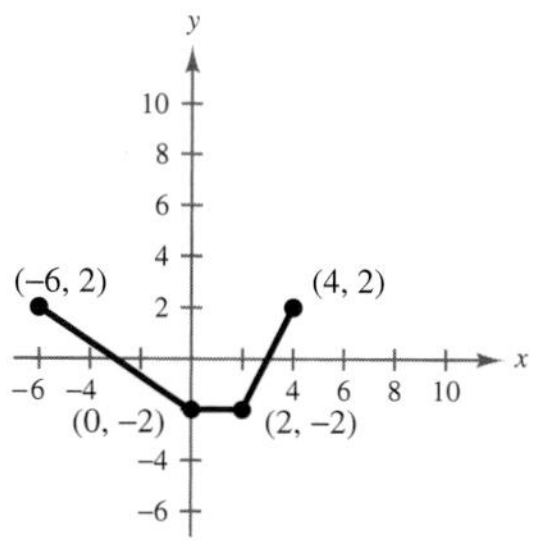

(b)

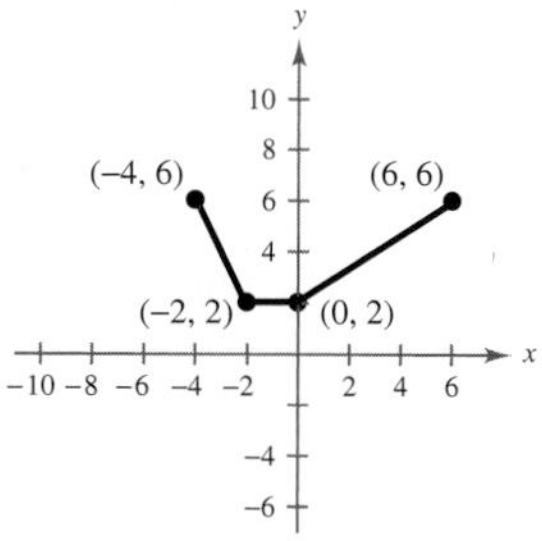

(c)

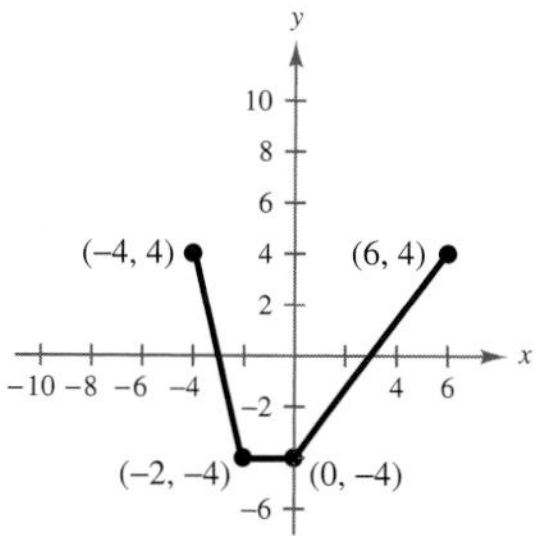

(d)

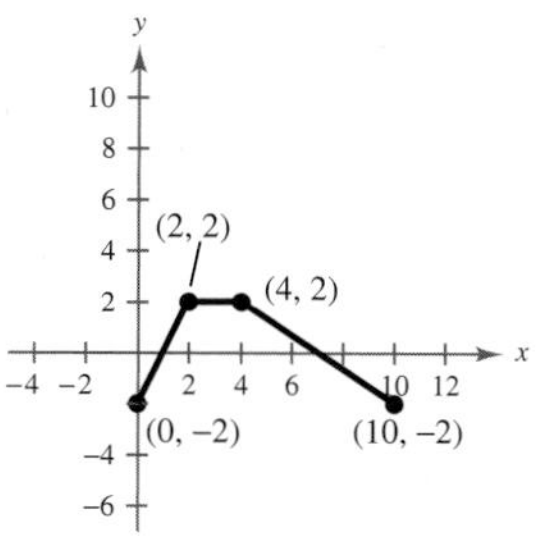

(e)

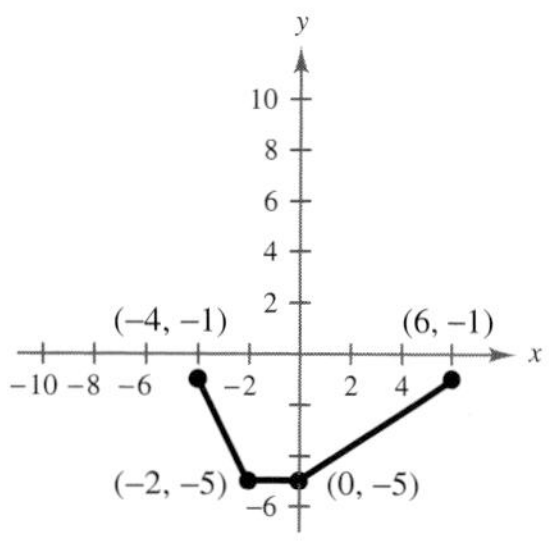

(f)

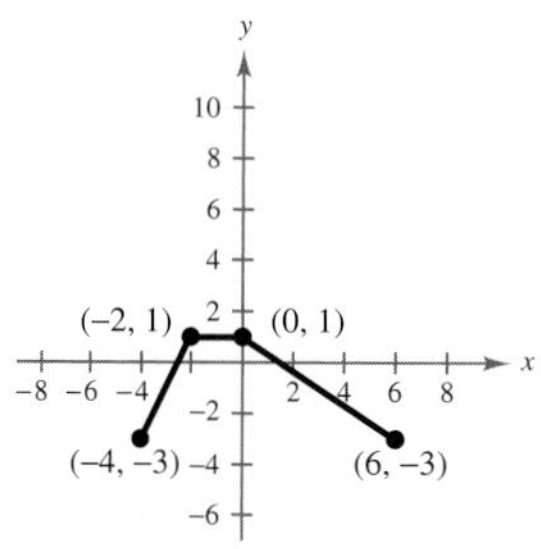

8. (a)

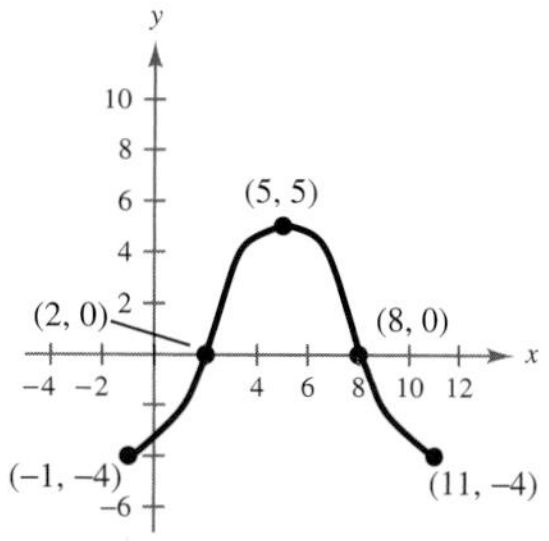

(b)

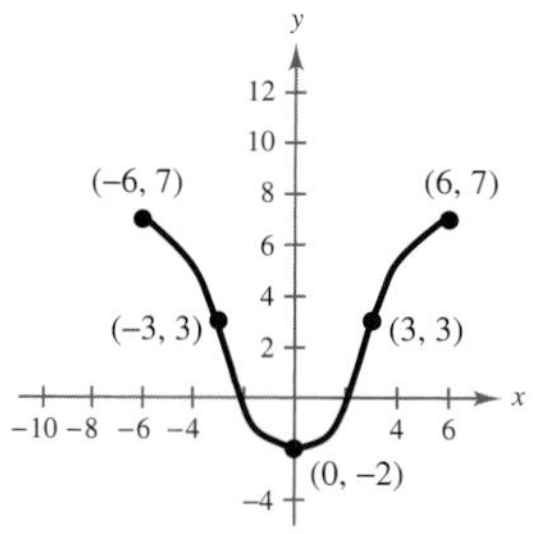

(c)

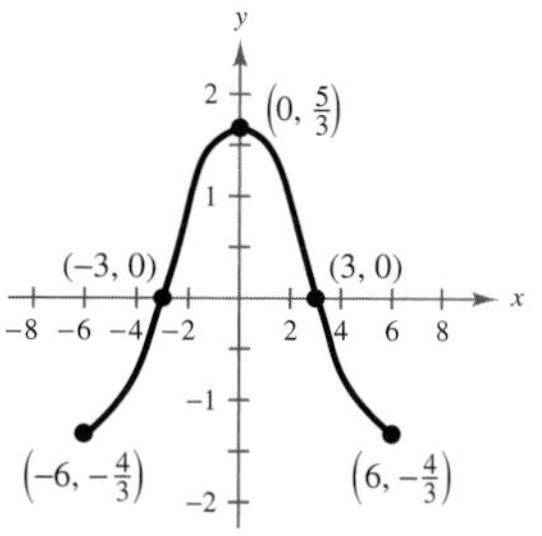

(d)

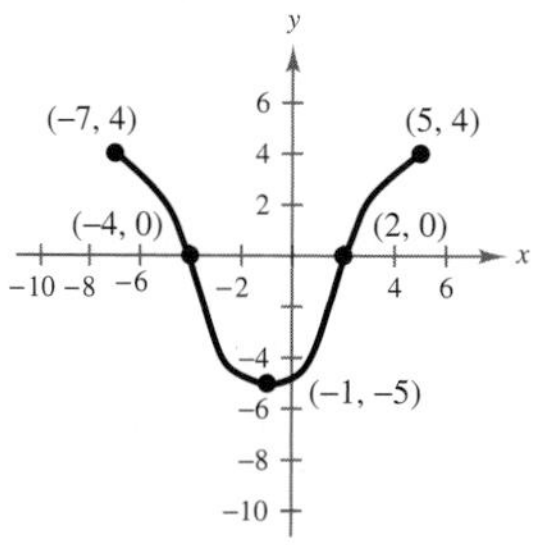

(e)

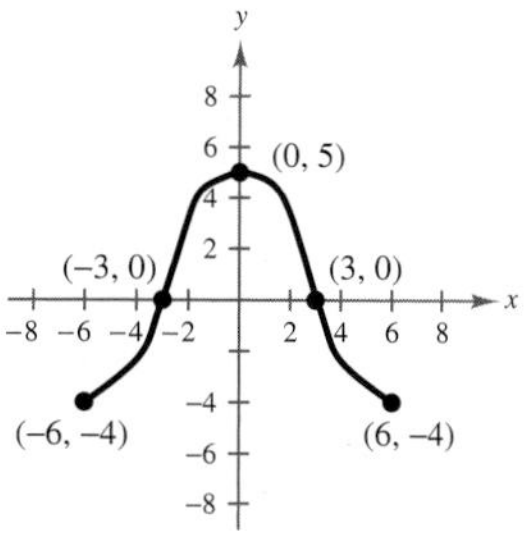

(f)

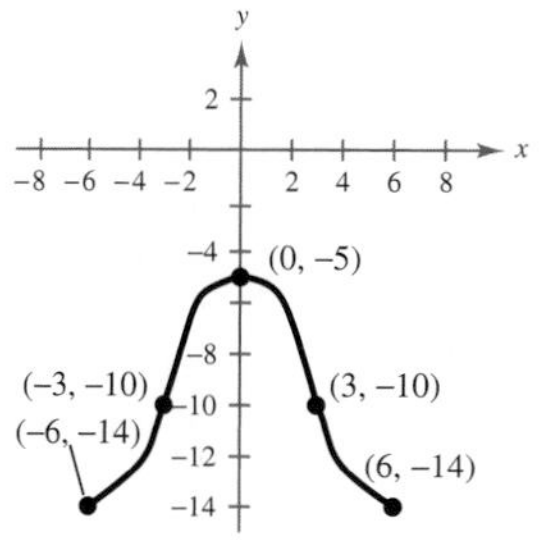

10. (a) $y = 1 - x^3$ (b) $y = (x - 1)^3 + 1$
(c) $y = -(x + 3)^3 - 1$ (d) $y = (x - 10)^3 - 4$

12. (a) $y = \sqrt{x} - 3$ (b) $y = \sqrt{x + 1} - 7$
(c) $y = -\sqrt{x - 5} + 5$ (d) $y = -\sqrt{-x + 3} - 4$

14. Vertical shrink of $y = x$; $y = \frac{1}{2}x$

16. Constant function; $y = 7$

18. Horizontal shift of $y = |x|$;

$y = |x + 2|$

20. Horizontal shift of $f(x) = x^2$ 8 units to the right

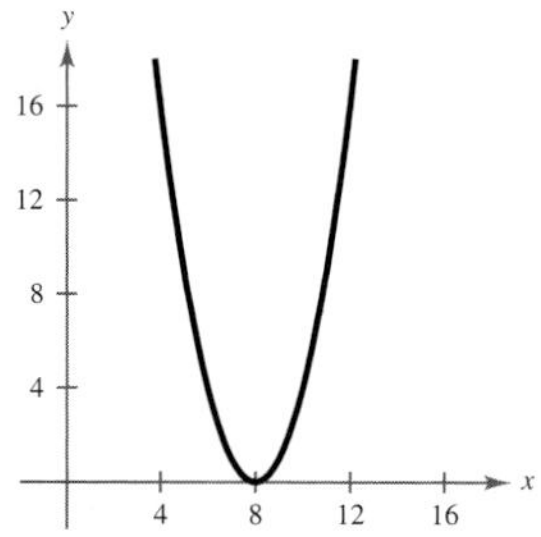

22. Reflection of $f(x) = x^3$ in x-axis and vertical shift of 1 unit downward

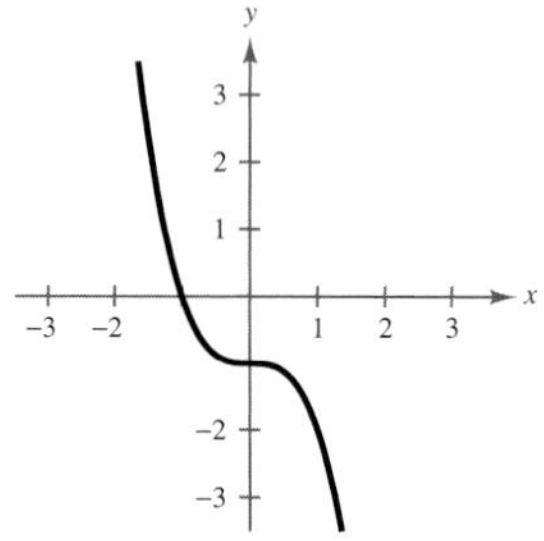

24. Reflection of $f(x) = x^2$ in x-axis, horizontal shift of 10 units to the left, and vertical shift of 5 units upward

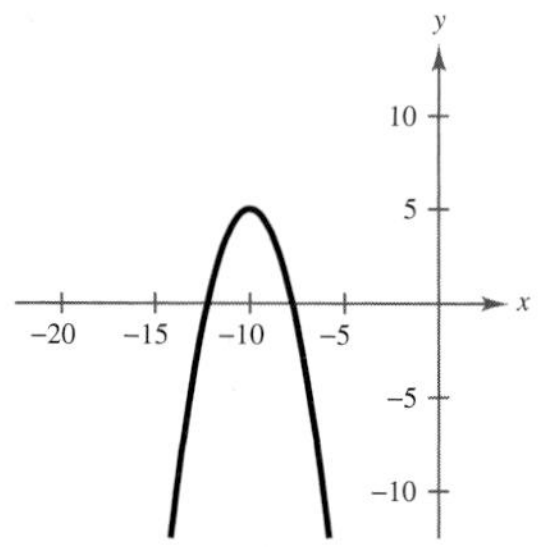

26. Reflection of $f(x) = x^3$ in x-axis, horizontal shift of 3 units to the left, and vertical shift of 10 units downward

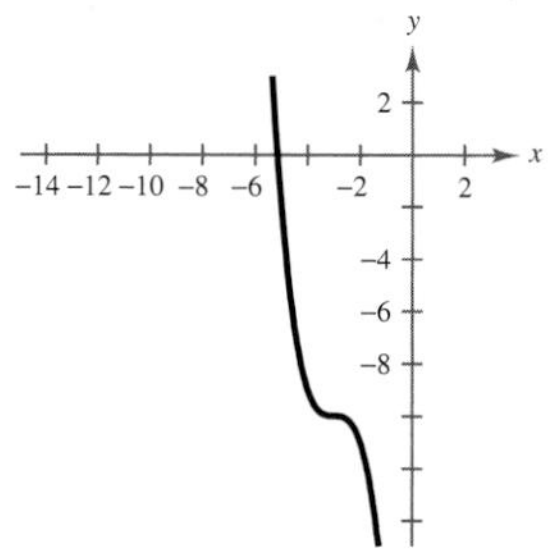

28. Reflection of $f(x) = |x|$ in x-axis, horizontal shift of 5 units to the left, and vertical shift of 6 units upward

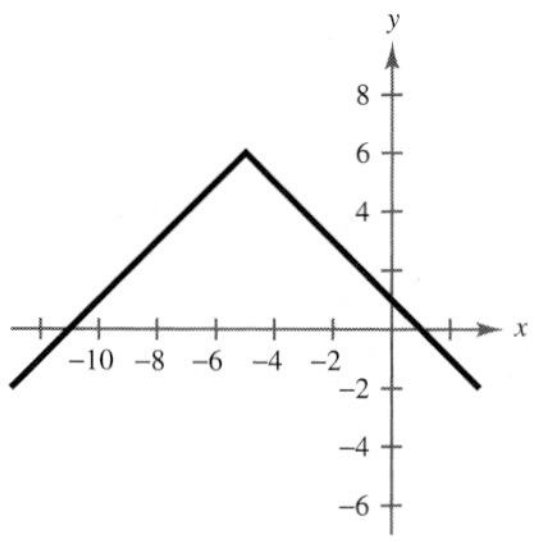

30. Reflection of $f(x) = |x|$ in y-axis, horizontal shift of 3 units to the left, and vertical shift of 9 units upward

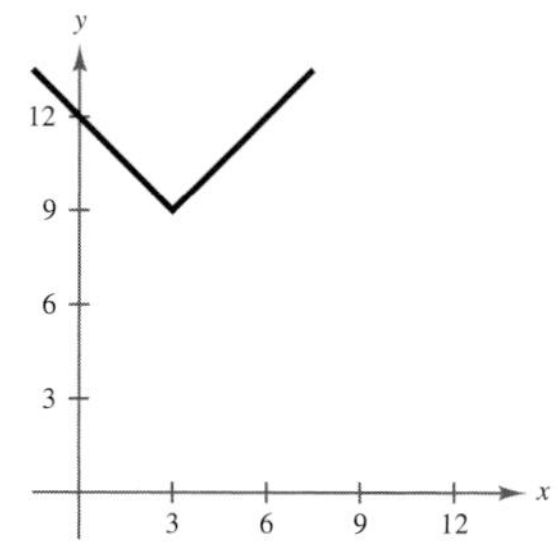

32. Horizontal shift of $f(x) = \sqrt{x}$ four units to the left and vertical shift of 8 units upward

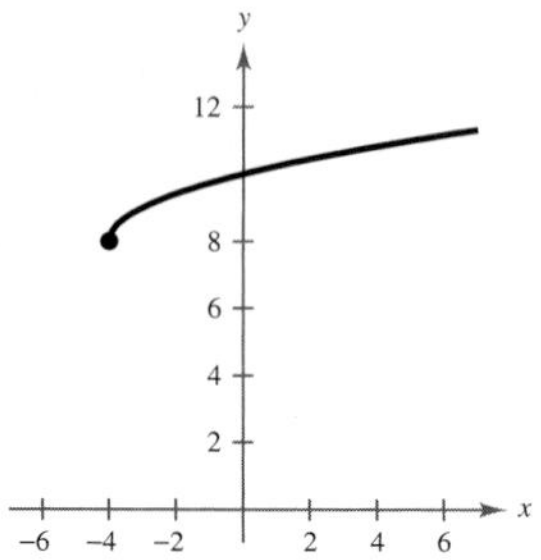

34. Reflection of $f(x) = \sqrt{x}$ in x-axis, horizontal shift of 1 unit to the left, and vertical shift of 6 units downward

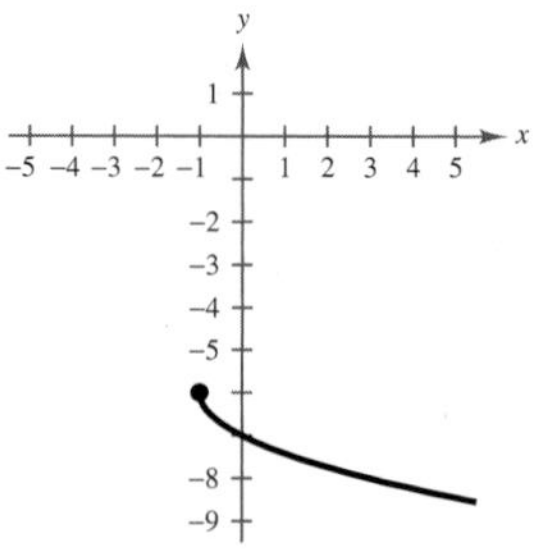

36. $f(x) = -(x + 3)^2 + 7$ **38.** $f(x) = (-x - 6)^3 - 6$

40. $f(x) = |x + 1| - 7$ **42.** $f(x) = -\sqrt{-x} - 9$

44. (a) $y = \frac{1}{4}x^3$ (b) $y = -2x^3$

46. (a) $y = 8\sqrt{x}$ (b) $y = -\frac{1}{4}\sqrt{x}$

48. Vertical stretch of $y = |x|$; $y = 6|x|$

50. Vertical stretch of $y = \sqrt{x}$; $y = 3\sqrt{x}$

52. Reflection in x-axis, vertical shift of 2 units downward, and vertical stretch of $y = |x|$;

$y = -2|x| - 2$

54. $y = |x + 4| - 2$ **56.** $y = (x - 2)^2 + 4$

58. (a)

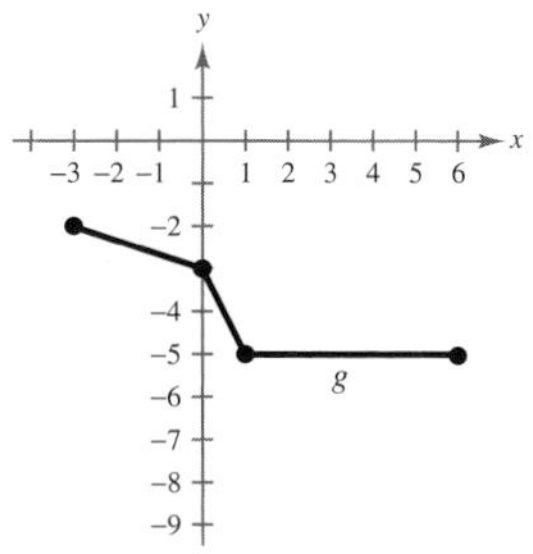

(b)

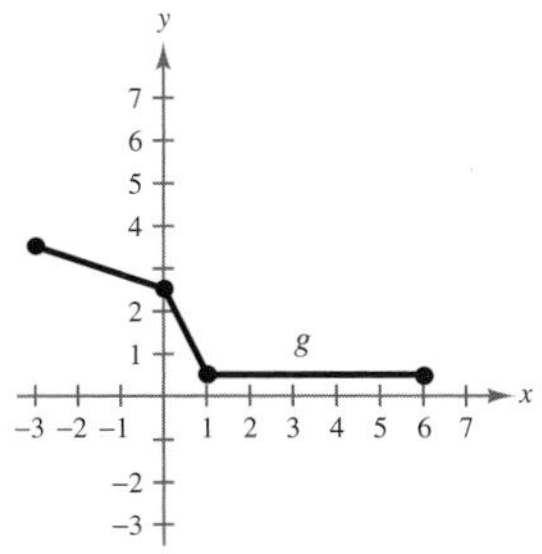

(c)

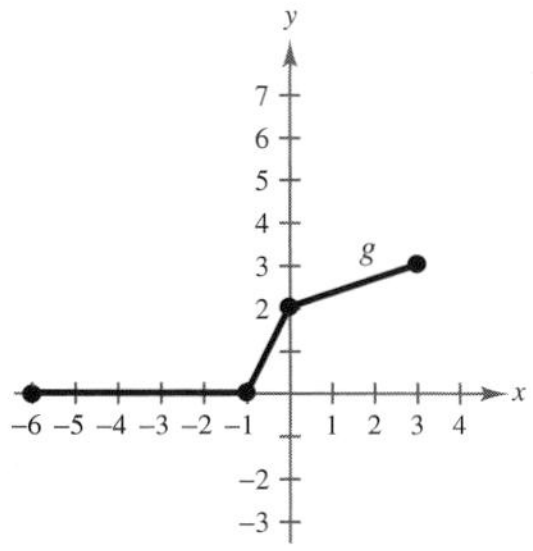

(d)

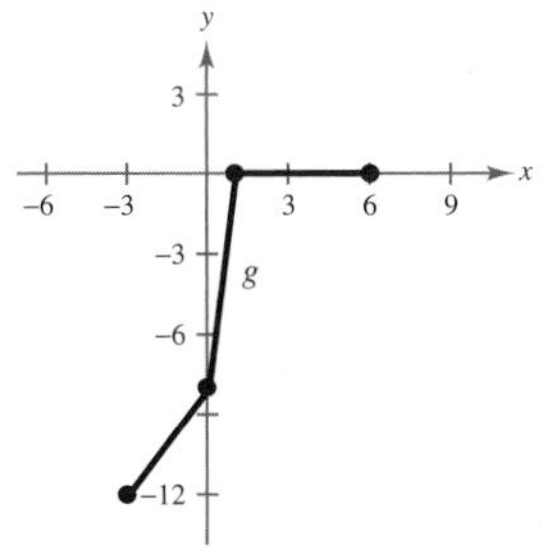

60. (a) Vertical stretch of 1.5 and vertical shift of 1.25 units downward

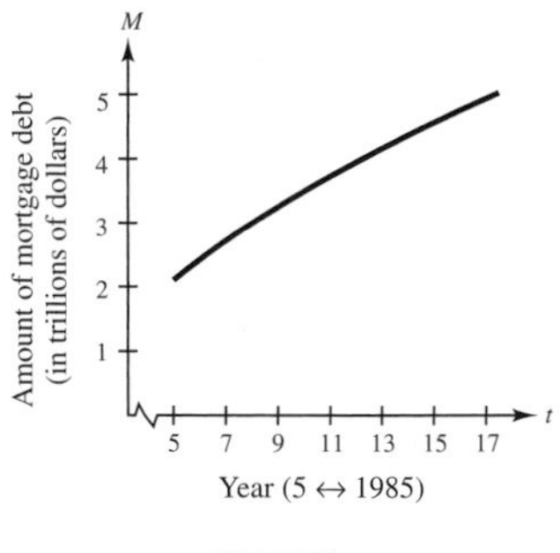

(b) $f(t) = 1.5\sqrt{t + 10} - 1.25$. By shifting the graph 10 units to the left, you obtain $t = 5$ represents 1995.

62. False. The point $(-2, -61)$ lies on the transformation.

64. If you consider the x-axis to be a mirror, the graph of $y = -f(x)$ is the mirror image of the graph of $y = f(x)$.

66. Answers will vary.

68. $\dfrac{-20}{(x + 5)(x - 5)}$ **70.** $\dfrac{3x - 5}{2(x - 5)}$

72. $\dfrac{x + 1}{x(x + 2)}, x \neq 2$ **74.** $\dfrac{x + 1}{(x - 7)(x + 3)}, x \neq 0, -1, -4$

76. (a) -3 (b) 3 (c) $\sqrt{x} - 3$ **78.** $x \geq 3, x \neq 8$

80. All real numbers

Section 1.6 *(page 168)*

2.

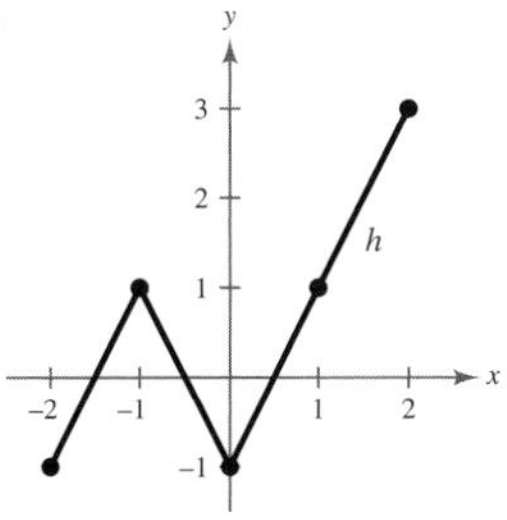

4.

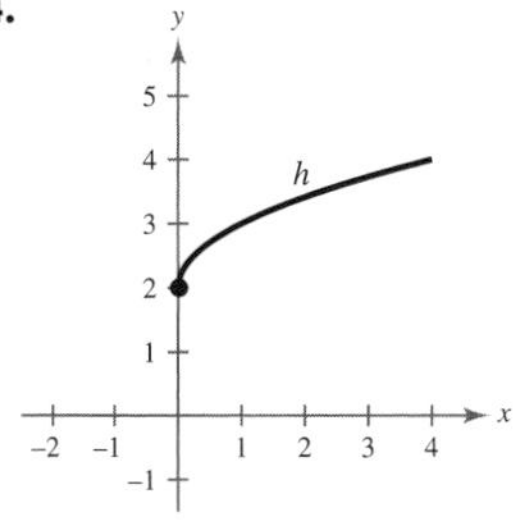

6. (a) $x - 3$ (b) $3x - 7$

(c) $-2x^2 + 9x - 10$ (d) $\dfrac{2x - 5}{2 - x}; x \neq 2$

8. (a) $2x - 1$ (b) $2x - 9$

(c) $8x - 20$ (d) $\frac{1}{2}x - \frac{5}{4}; -\infty < x < \infty$

10. (a) $\sqrt{x^2 - 4} + \dfrac{x^2}{x^2 + 1}$ (b) $\sqrt{x^2 - 4} - \dfrac{x^2}{x^2 + 1}$

(c) $\dfrac{x^2\sqrt{x^2 - 4}}{x^2 + 1}$ (d) $\dfrac{(x^2 + 1)\sqrt{x^2 - 4}}{x^2}; |x| \geq 2$

12. (a) $\dfrac{x^4 + x^3 + x}{x + 1}$ (b) $\dfrac{-x^4 - x^3 + x}{x + 1}$

(c) $\dfrac{x^4}{x + 1}$ (d) $\dfrac{1}{x^2(x + 1)}; x \neq 0, -1$

14. 7 **16.** -1 **18.** $t^2 - 3t - 1$ **20.** -370

22. $-\frac{1}{4}$ **24.** 52

26.

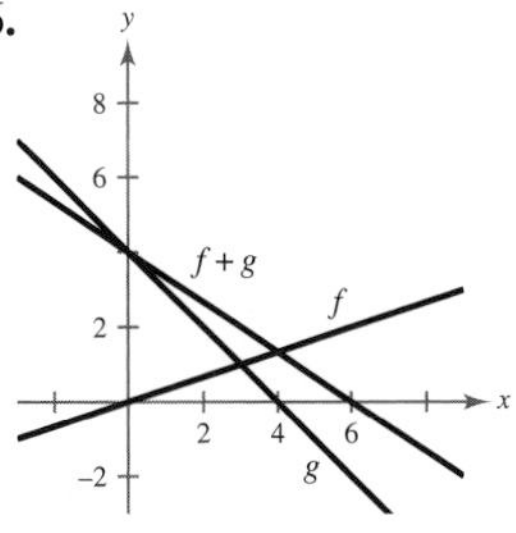

28.

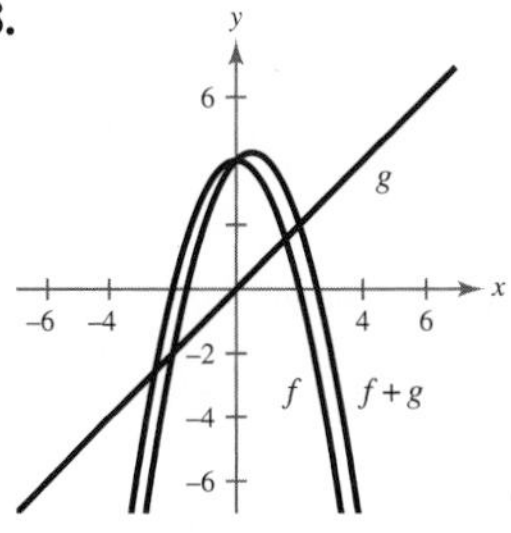

30.

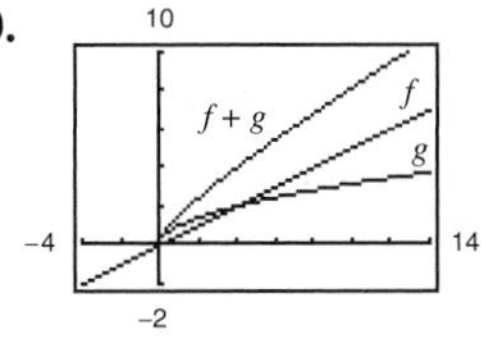

$g(x), f(x)$

32. $R_T = 734 - 7.22t - 0.8t^2$

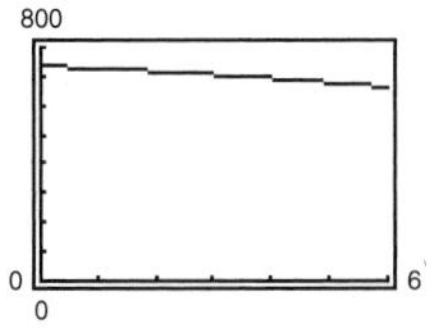

34. 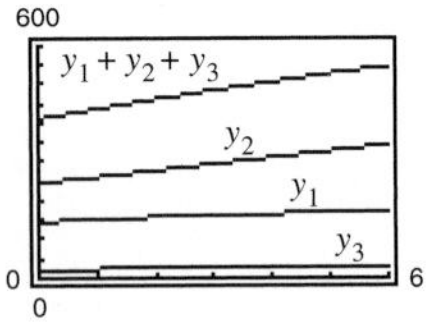

\$613.74 billion

36. $T(t) = \begin{cases} 60, & 0 \le t \le 6 \\ 24t - 84, & 6 < t < 6.5 \\ 72, & 6.5 \le t < 20.5 \\ -24t + 564, & 20.5 < t < 21 \\ 60, & 21 \le t \le 24 \end{cases}$

38. (a) x (b) x (c) $\sqrt[3]{\sqrt[3]{x-1} - 1}$

40. (a) $\dfrac{1}{x^3}$ (b) $\dfrac{1}{x^3}$ (c) x^9

42. (a) $\sqrt[3]{x^3 - 4}$ (b) $x - 4$

Domain of f, g, $f \circ g$, and $g \circ f$: all real numbers

44. (a) $x + 1$ (b) $\sqrt{x^2 + 1}$

Domain of f and $g \circ f$: all real numbers

Domain of g and $f \circ g$: $x \ge 0$

46. (a) $\sqrt{2x - 3}$ (b) $2\sqrt{x} - 3$

Domain of f and $g \circ f$: $x \ge 0$

Domain of g: all real numbers

Domain of $f \circ g$: $x \ge \frac{3}{2}$

48. (a) x^4 (b) x^4

Domain of f, g, $f \circ g$, and $g \circ f$: all real numbers

50. (a) $\dfrac{3}{x^2 + 2x}$ (b) $\dfrac{x^2 + 2}{x^2 - 1}$

Domain of f and $g \circ f$: all real numbers $x \ne \pm 1$

Domain of g: all real numbers

Domain of $f \circ g$: all real numbers $x \ne 0, -2, \pm 1$

52. (a) -1 (b) 0 **54.** (a) 2 (b) 2

56. $f(x) = x^3$, $g(x) = 1 - x$

58. $f(x) = \sqrt{x}$, $g(x) = 9 - x$

60. $f(x) = \dfrac{4}{x^2}$, $g(x) = 5x + 2$

62. $f(x) = \dfrac{27x + 6\sqrt[3]{x}}{10 - 27x}$, $g(x) = x^3$

64. $(A \circ r)(t) = 0.36\pi t^2$; $A \circ r$ represents the area of the circle at time t.

66. False. $(f \circ g)(x) = 6x + 1$ and $(g \circ f)(x) = 6x + 6$

68. $g(f(x))$ represents 3 percent of an amount over \$500,000.

70. Odd **72.** $-2x - h$

74.

$$\frac{\sqrt{2(x+h)+1} - \sqrt{2x+1}}{h} = \frac{2}{\sqrt{2(x+h)+1} + \sqrt{2x+1}}$$

76. $x + y + 3 = 0$ **78.** $5x - 7y - 35 = 0$

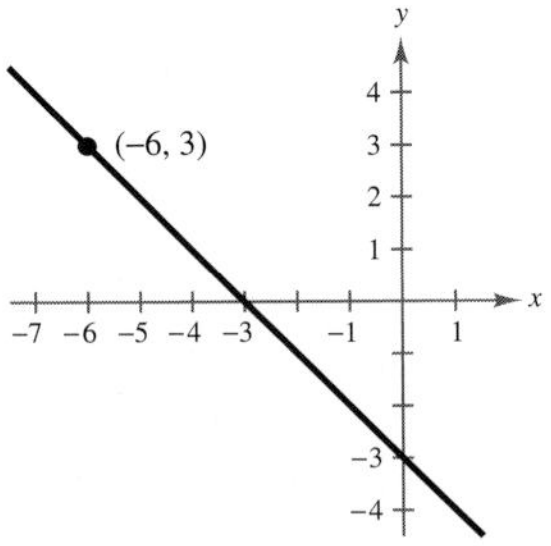

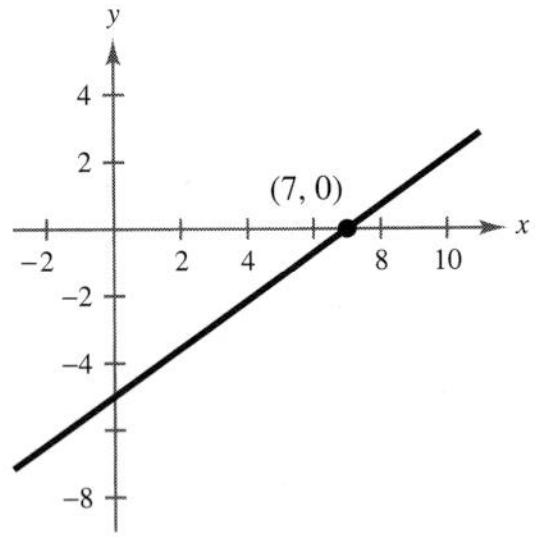

Section 1.7 *(page 177)*

2. b **4.** d **6.** $f^{-1}(x) = 3x$ **8.** $f^{-1}(x) = x + 4$

10. $f^{-1}(x) = 5x + 1$ **12.** $f^{-1}(x) = \sqrt[5]{x}$

14. (a) $f(g(x)) = f(x + 5) = (x + 5) - 5 = x$

$g(f(x)) = g(x - 5) = (x - 5) + 5 = x$

(b) 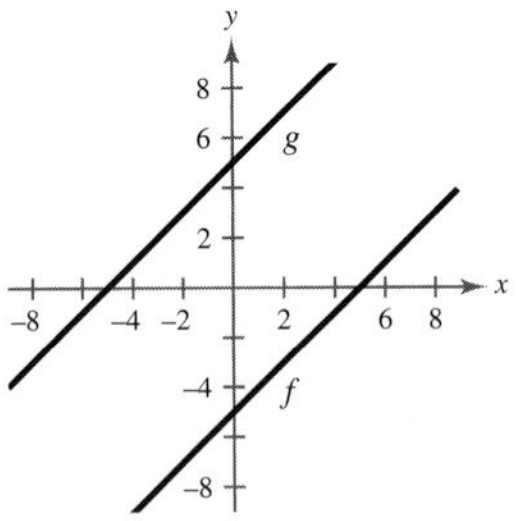

16. (a) $f(g(x)) = f\left(\dfrac{3 - x}{4}\right) = 3 - 4\left(\dfrac{3 - x}{4}\right) = x$

$g(f(x)) = g(3 - 4x) = \dfrac{3 - (3 - 4x)}{4} = x$

(b) 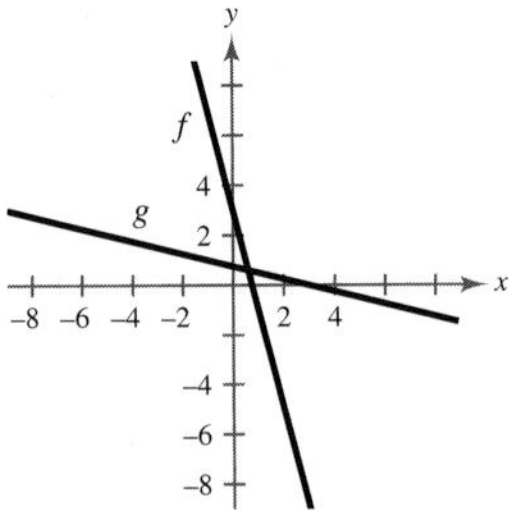

18. (a) $f(g(x)) = f\left(\dfrac{1}{x}\right) = \dfrac{1}{1/x} = x$

$g(f(x)) = g\left(\dfrac{1}{x}\right) = \dfrac{1}{1/x} = x$

(b)

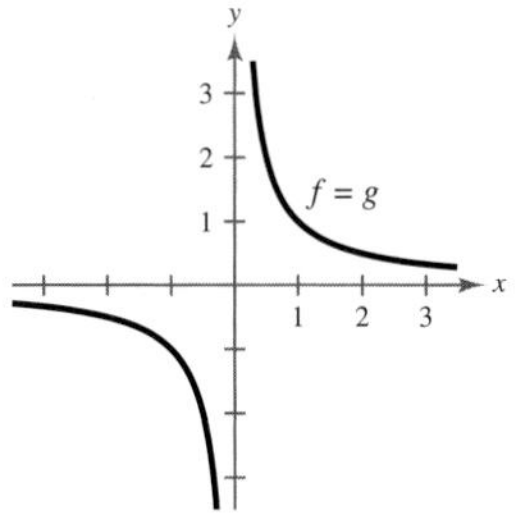

20. (a) $f(g(x)) = f\left(\sqrt[3]{1-x}\right)$

$$= 1 - \left(\sqrt[3]{1-x}\right)^3 = x$$

$$g(f(x)) = g(1 - x^3)$$

$$= \sqrt[3]{1 - (1 - x^3)} = x$$

(b)

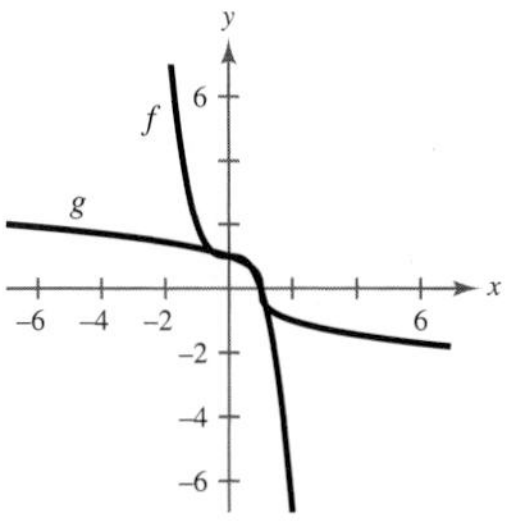

22. (a) $f(g(x)) = f\left(\dfrac{1-x}{x}\right), \quad 0 < x \le 1$

$$= \frac{1}{1 + (1-x)/x} = \frac{x}{x + 1 - x} = x$$

$$g(f(x)) = g\left(\frac{1}{1+x}\right), \quad x \ge 0$$

$$= \frac{1 - 1/(1+x)}{1/(1+x)} = \frac{1 + x - 1}{1} = x$$

(b)

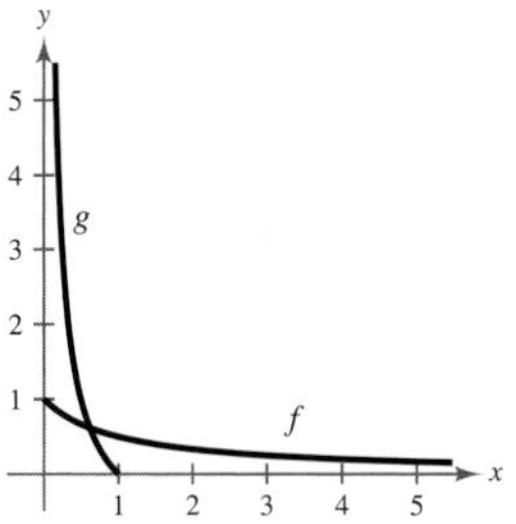

24. (a) $f(g(x)) = f\left(\dfrac{2x+3}{x-1}\right) = \dfrac{\left(\dfrac{2x+3}{x-1}\right) + 3}{\left(\dfrac{2x+3}{x-1}\right) - 2}$

$$= \frac{2x + 3 + 3x - 3}{2x + 3 - 2x + 2} = x$$

$$g(f(x)) = g\left(\frac{x+3}{x-2}\right) = \frac{2\left(\dfrac{x+3}{x-2}\right) + 3}{\left(\dfrac{x+3}{x-2}\right) - 1}$$

$$= \frac{2x + 6 + 3x - 6}{x + 3 - x + 2} = x$$

(b)

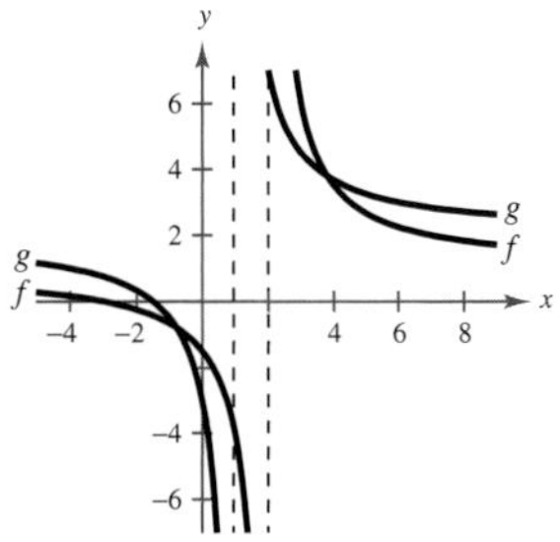

26. Yes

28.

x	-10	-7	-4	-1	2	5
$f^{-1}(x)$	-3	-2	-1	0	1	2

30. No **32.** Yes

34.

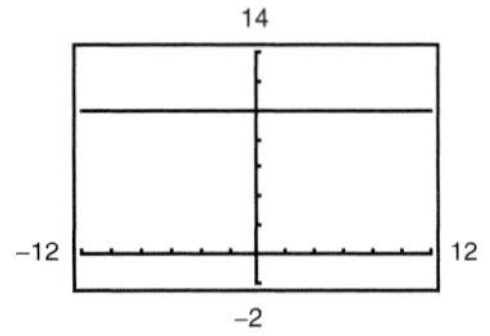

The function does not have an inverse.

36.

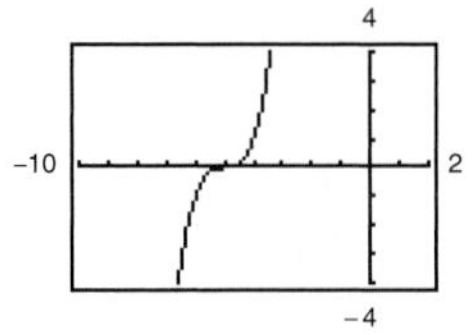

The function has an inverse.

38.

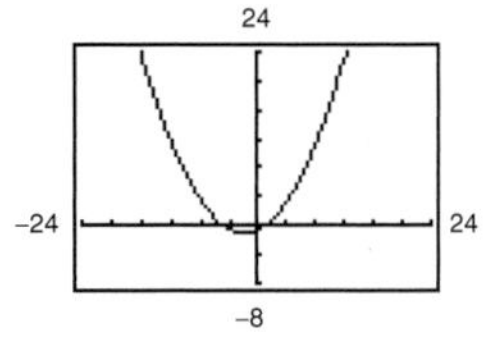

The function does not have an inverse.

40. $f^{-1}(x) = \dfrac{x-1}{3}$

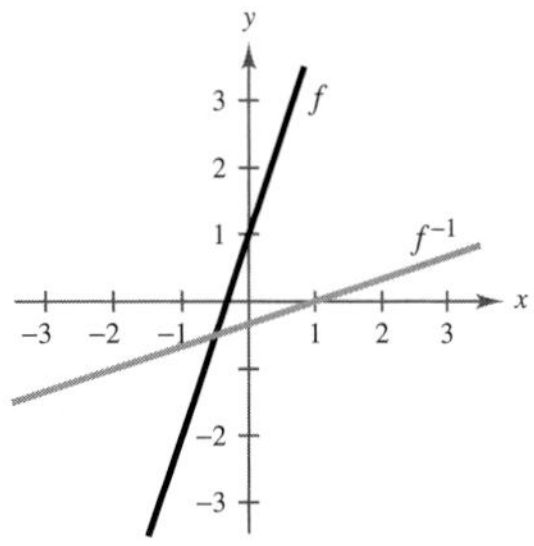

42. $f^{-1}(x) = \sqrt[3]{x - 1}$

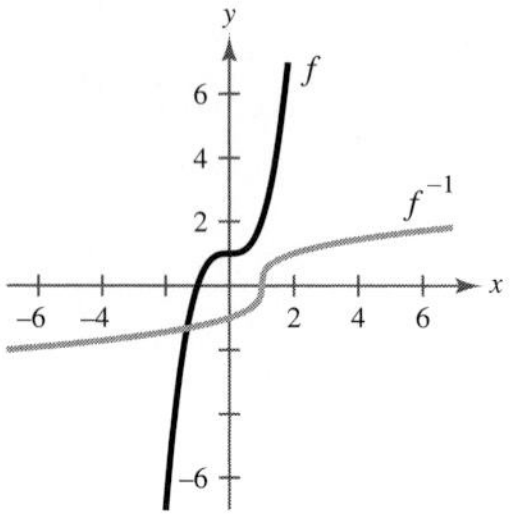

44. $f^{-1}(x) = \sqrt{x}$

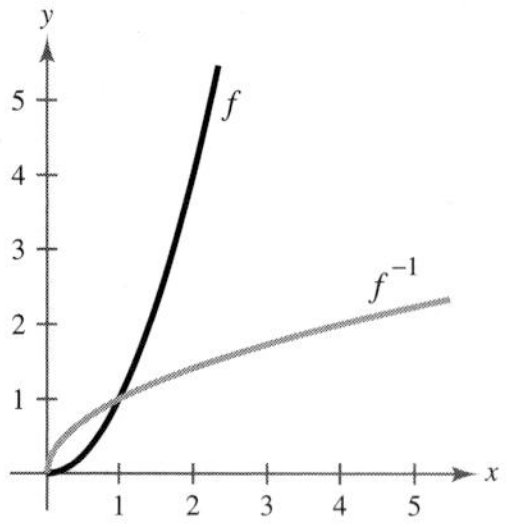

46. $f^{-1}(x) = -\sqrt{x + 2}$

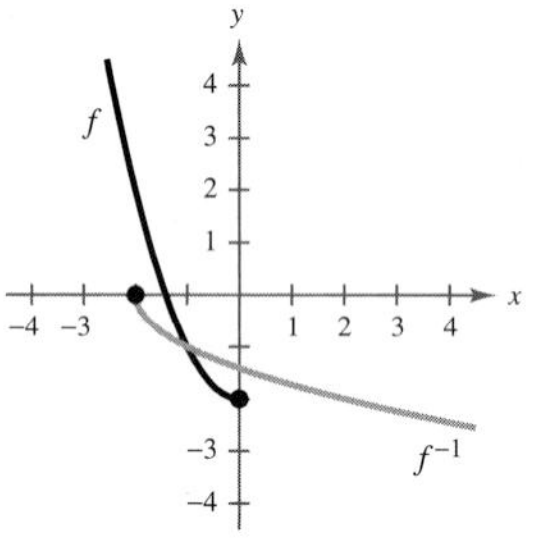

48. $f^{-1}(x) = -\dfrac{2}{x}$

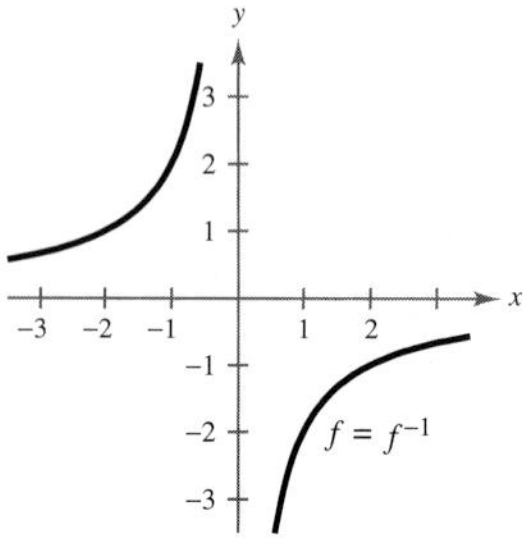

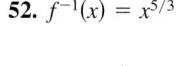

50. $f^{-1}(x) = \dfrac{-2x - 3}{x - 1}$

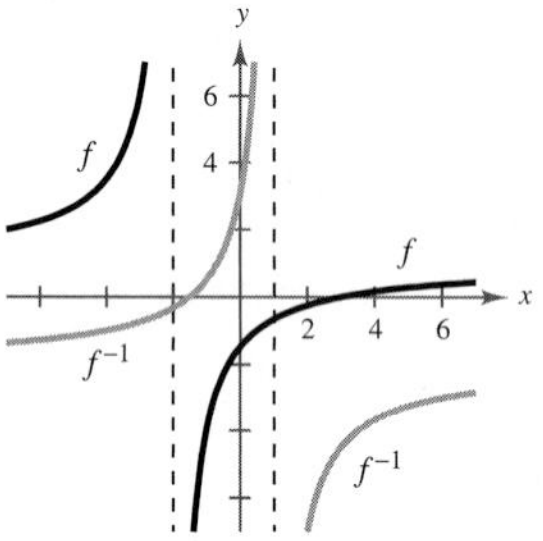

52. $f^{-1}(x) = x^{5/3}$

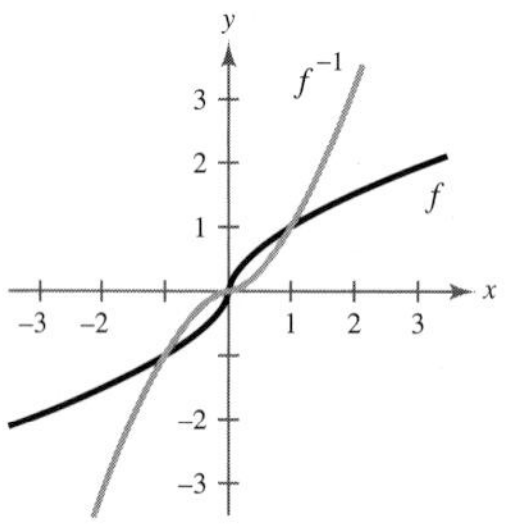

54. $f^{-1}(x) = \dfrac{-6x - 4}{2x - 8}$

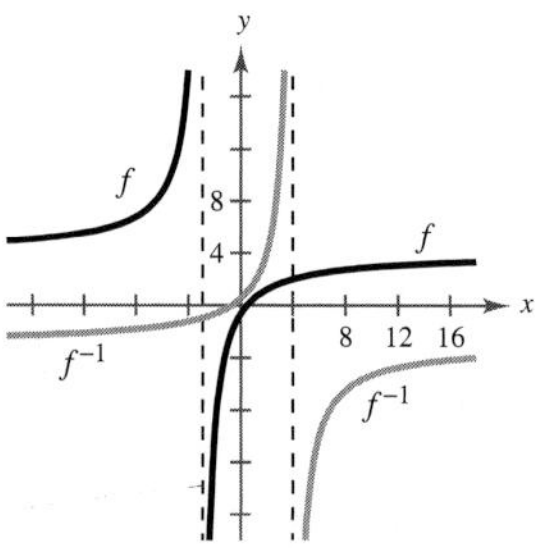

56. No inverse **58.** $f^{-1}(x) = \dfrac{x - 5}{3}$

60. $f^{-1}(x) = \dfrac{5x - 4}{3}$ **62.** No inverse

64. No inverse **66.** $f^{-1}(x) = 2 - x, \quad x \geq 0$

68. $f^{-1}(x) = x^2 + 2, \quad x \geq 0$ **70.** 0 **72.** $-\sqrt[9]{4}$

74. $2\sqrt[3]{x + 3}$ **76.** $\dfrac{x - 3}{2}$ **78.** $\dfrac{x - 3}{2}$

80. (a) Answers will vary.

(b) $y = \dfrac{80 - x}{0.35}$

$x =$ total cost; $y =$ number of pounds of the less expensive commodity

(c) $62.5 \leq x \leq 80$ (d) 20

82. (a) Yes

(b) f^{-1} yields the year for a given total value of new car sales.

(c) 5

84. (a) 5

(b) f^{-1} yields the year for a given average local bill for cellular phones.

(c) $f(t) = -5.36t + 78.95$

(d) $f^{-1}(t) = \dfrac{t - 78.95}{-5.36}$

(e) 12.677

86. False. $f(x) = x^2$ has no inverse.

88. True. If $f(x) = x^3$, then $f^{-1}(x) = \sqrt[3]{x}$.

90.

x	1	3	4	6
y	1	2	6	7

x	1	2	6	7
$f^{-1}(x)$	1	3	4	6

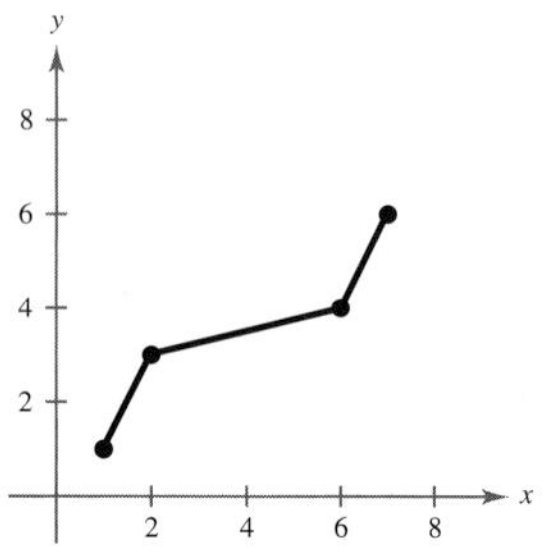

92.

x	-2	-1	3	4
y	6	0	-2	-3

x	-3	-2	0	6
$f^{-1}(x)$	4	3	-1	-2

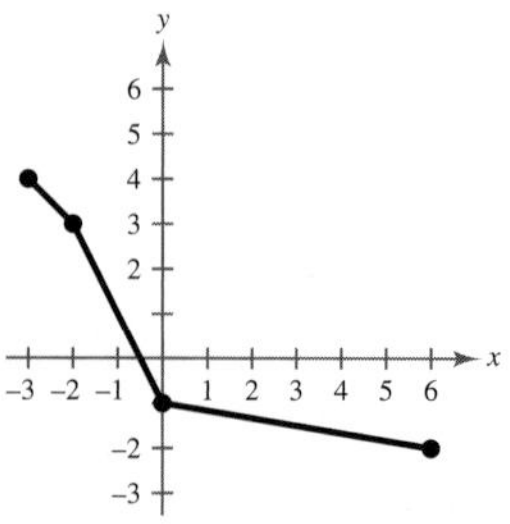

94. $k = \frac{1}{4}$ **96.** $5 \pm 2\sqrt{2}$ **98.** $-1, -\frac{1}{3}$

100. $-1, 3$ **102.** $\frac{11}{8}$ **104.** $x \geq -6$

106. All real numbers $x \neq -\frac{7}{5}$ **108.** 31.7 feet, 16 trips

110. $b = \sqrt{10}$ feet, $h = 2\sqrt{10}$ feet

Section 1.8 *(page 187)*

2.

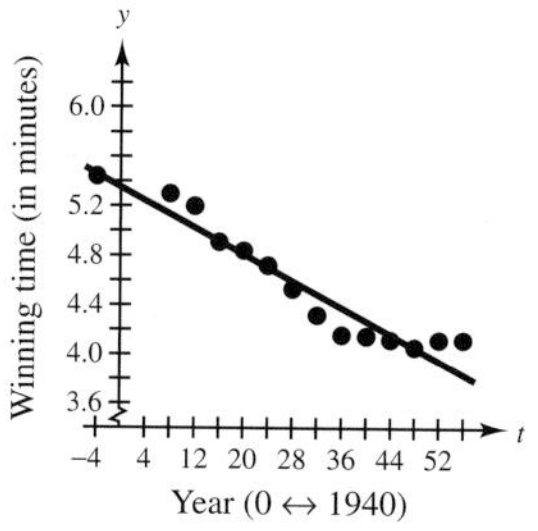

The model is a "good fit" for the actual data.

4. Directly

6.

x	2	4	6	8	10
$y = kx^2$	8	32	72	128	200

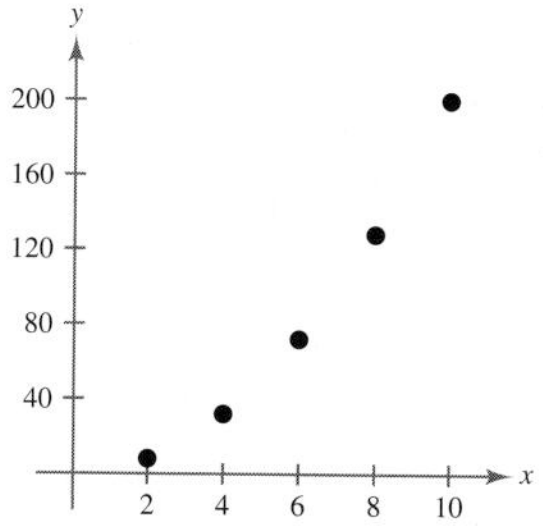

8.

x	2	4	6	8	10
$y = kx^2$	1	4	9	16	25

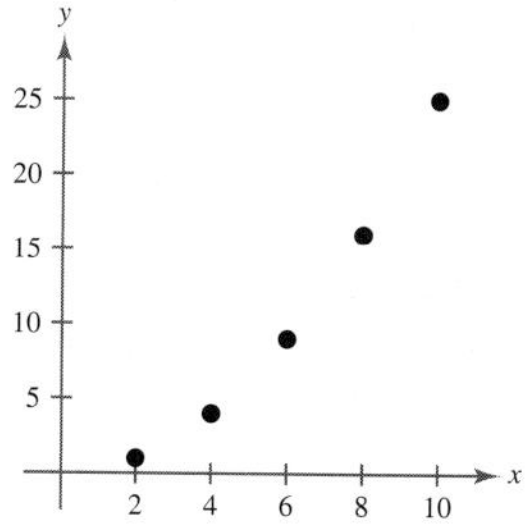

10.

x	2	4	6	8	10
$y = k/x^2$	$\frac{5}{4}$	$\frac{5}{16}$	$\frac{5}{36}$	$\frac{5}{64}$	$\frac{1}{20}$

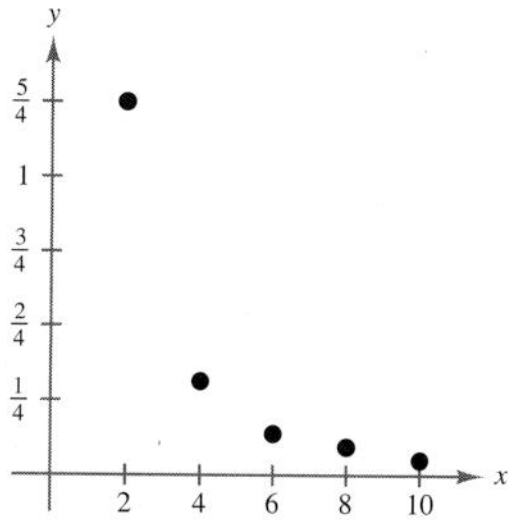

12.

x	2	4	6	8	10
$y = k/x^2$	5	$\frac{5}{4}$	$\frac{5}{9}$	$\frac{5}{16}$	$\frac{1}{5}$

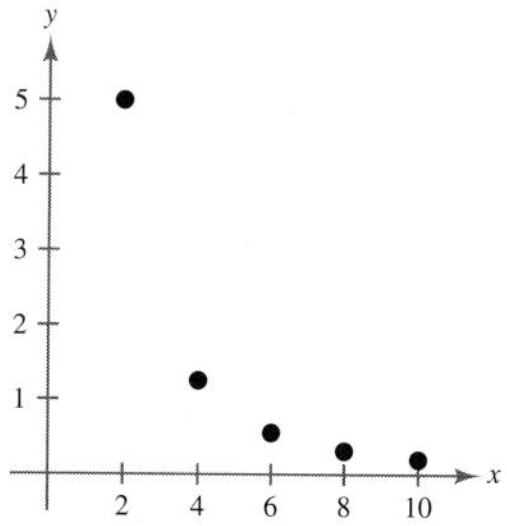

14. $y = \frac{2}{5}x$ **16.** $y = \dfrac{120}{x}$ **18.** $y = 7x$

20. $y = \frac{290}{3}x$ **22.** $I = 0.0375P$

24. Model: $y = \frac{53}{14}x$

Gallons	5	10	20	25	30
Liters	18.9	37.9	75.7	94.6	113.6

26. $y = 0.07x$; \$37.84

28. $293\frac{1}{3}$ newtons

30. Combined lifting force $= 2F = 120$ pounds

32. $V = ke^3$ **34.** $h = \dfrac{k}{\sqrt{s}}$ **36.** $z = kx^2y^3$

38. $R = k(T - T_e)$ **40.** $R = kS(L - S)$

42. The surface area of a sphere varies directly as the square of its radius.

44. The volume of a right circular cylinder is jointly proportional to its height and the square of its radius.

46. ω varies directly as the square root of g and inversely as the square root of W.

48. $y = \dfrac{75}{x}$ **50.** $z = 2xy$ **52.** $P = \dfrac{18x}{y^2}$

54. $v = \dfrac{24pq}{287s^2}$ **56.** $\dfrac{k(2v)^2}{kv^2} = 4$

58. Diameter $= 2r = 0.054$ inch

60. $208\frac{1}{3}$ feet **62.** ≈ 441 units

64. (a) The safe load is unchanged.

(b) The safe load is eight times as great.

(c) The safe load is four times as great.

(d) The safe load is one-fourth as great.

66. (a)

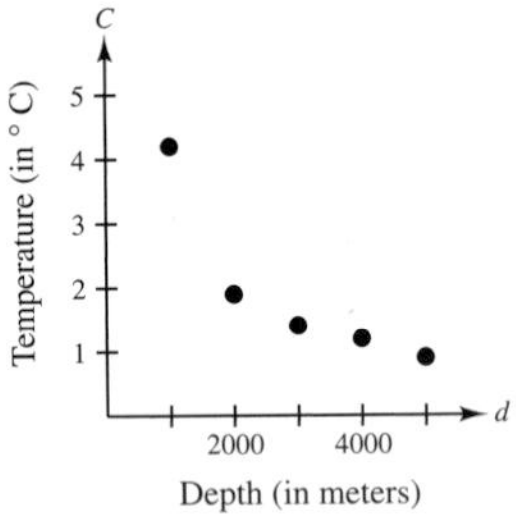

(b) Yes. $k = 4200$

(c)

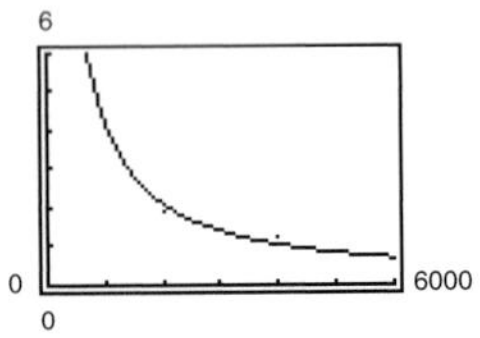

(d) ≈ 1400 meters

68. The illumination is reduced to one-fourth of the original. Explanations will vary.

70. (a) $y = 1.10t + 126.9$

(b)

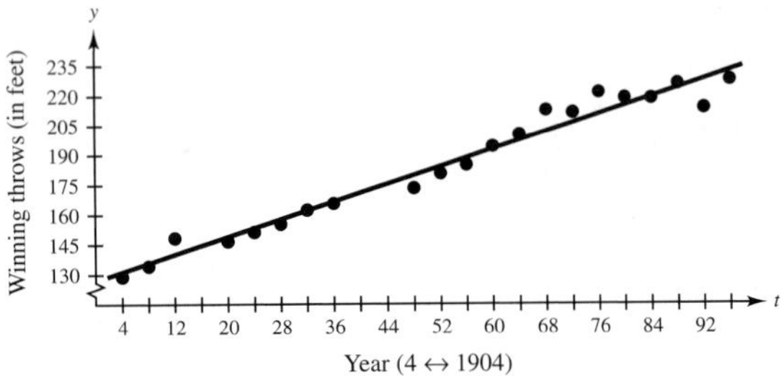

(c) 236.9 feet (d) Answers will vary.

72. (a) $y = -3.24x + 12{,}368$

(b)

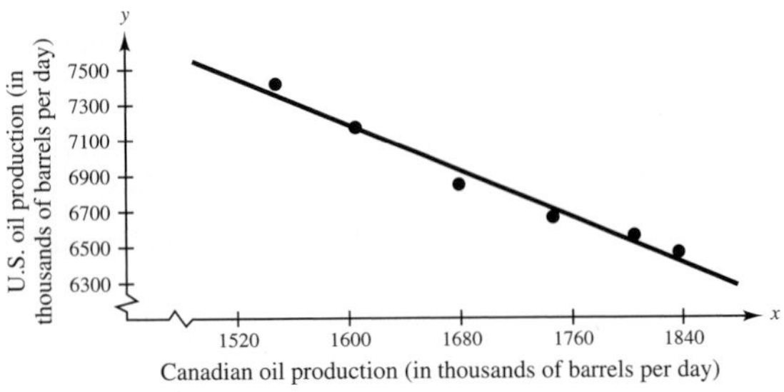

(c) 5888 thousand barrels per day

(d) For each 1000-barrel increase in Canadian oil production, the U.S. oil production decreases by 3.24 thousand barrels.

74. False. y will increase if k is positive and y will decrease if k is negative.

76. Good approximation **78.** Poor approximation

80.

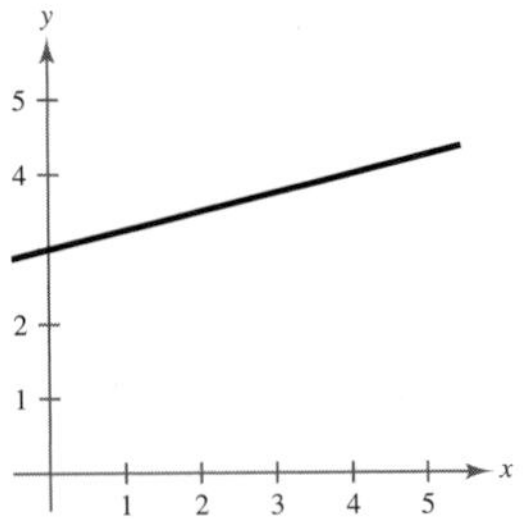

$y = \frac{1}{4}x + 3$

82.

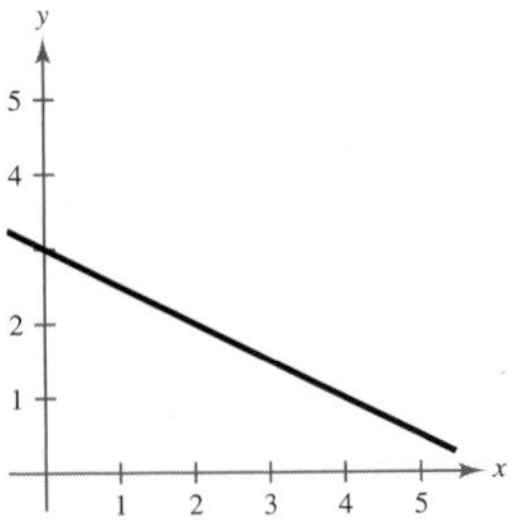

$y = -\frac{1}{2}x + 3$

84. The accuracy is questionable when based on limited data.

86. $-1 < x < 3$

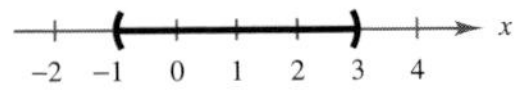

88. $x = 0,\ x \geq 8$

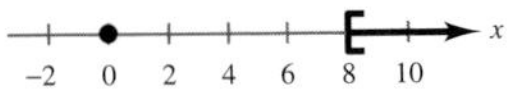

90. (a) 6 (b) 9 (c) 383

92. All real numbers **94.** $x \geq 3,\ x \neq 6$

Review Exercises *(page 194)*

2.

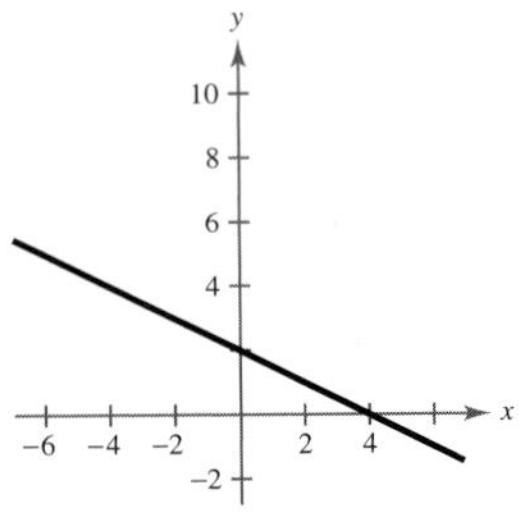

4.

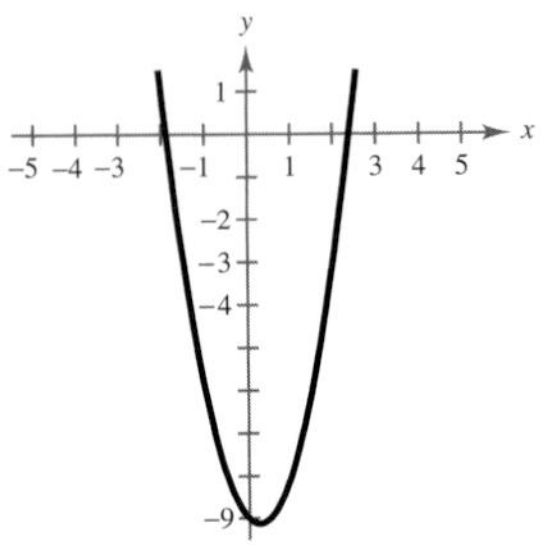

6. x-intercepts: $(0, 0)$, $(8, 0)$

y-intercept: $(0, 0)$

8. x-intercepts: $(0, 0)$, $(\pm 3, 0)$

y-intercept: $(0, 0)$

10.

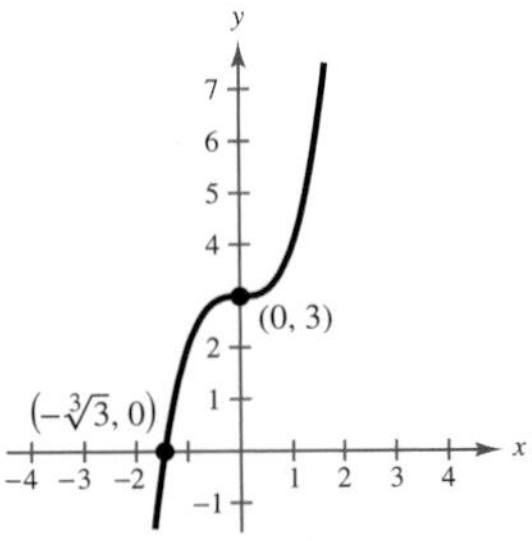

12.

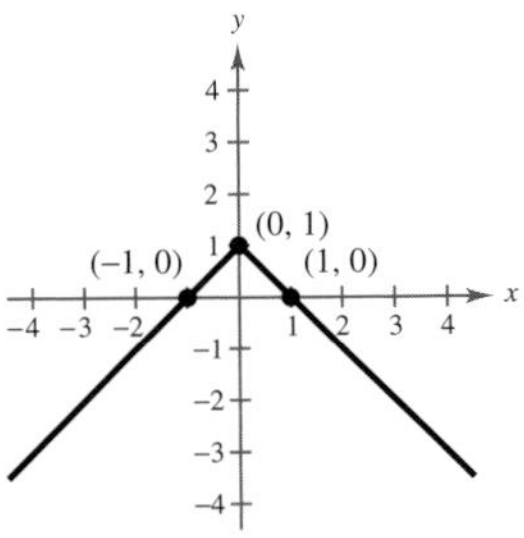

14. Center: $(0, 0)$

Radius: 2

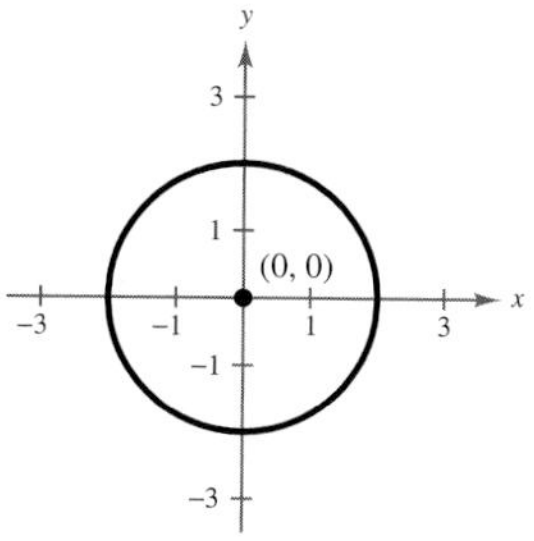

16. Center: $(0, 8)$

Radius: 9

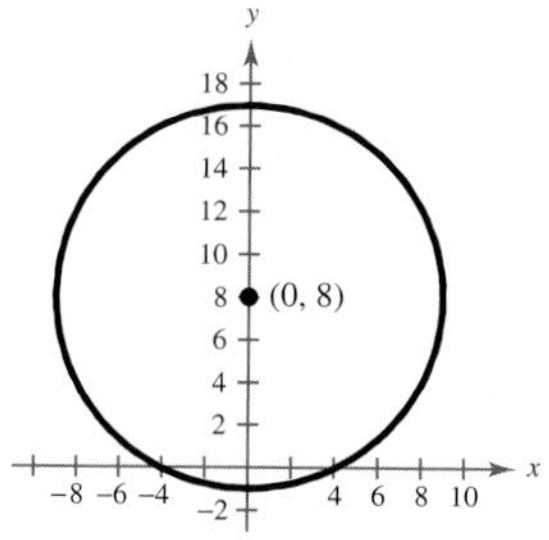

18. $(x - 1)^2 + \left(y + \frac{13}{2}\right)^2 = \frac{85}{4}$

20.

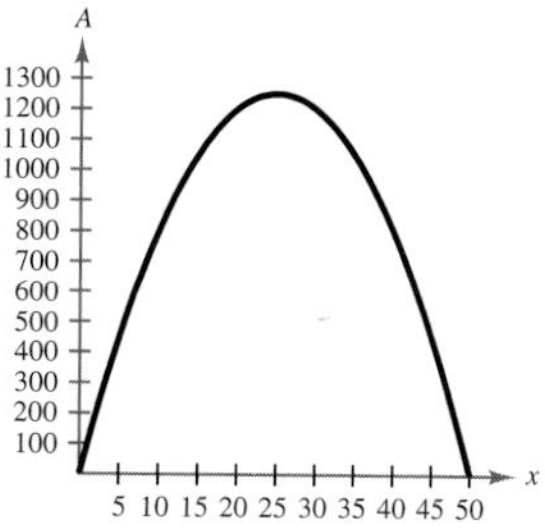

Maximum area: 1250 square feet

Length: 50 feet; Width: 25 feet

22. Slope: Undefined
y-intercept: None

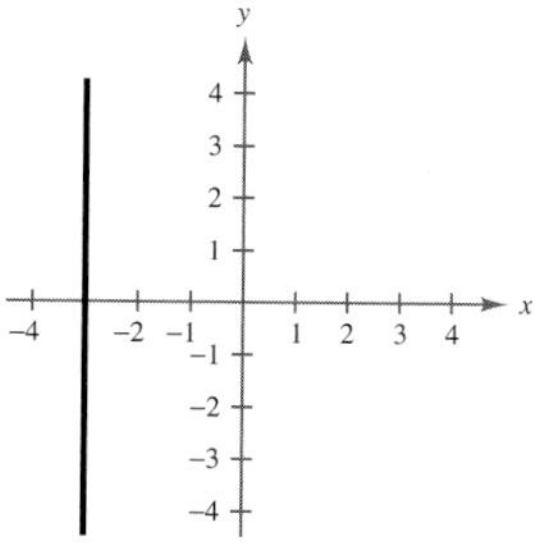

24. Slope: -10
y-intercept: $(0, 9)$

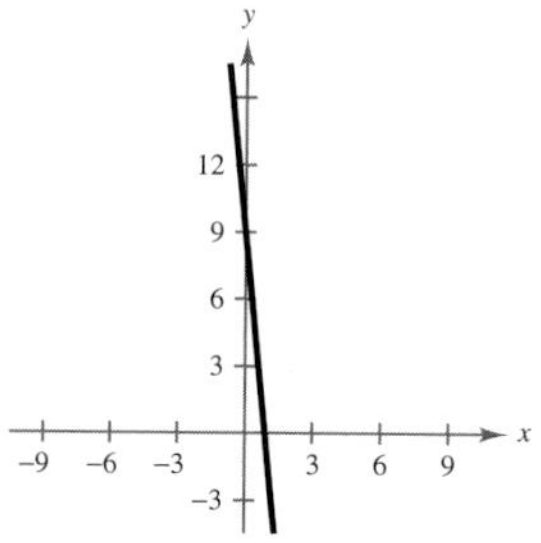

26.

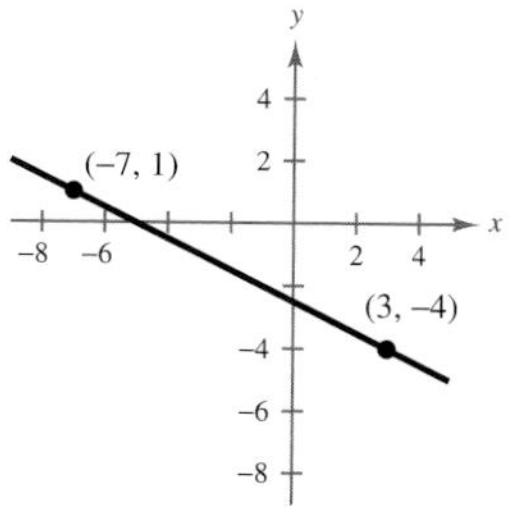

$m = -\frac{1}{2}$

28.

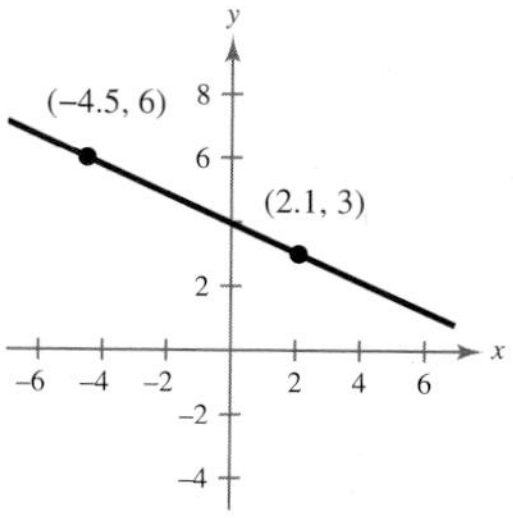

$m = -\frac{5}{11}$

30. $x = 0$ **32.** $4x + 3y - 8 = 0$

34. $3x - 2y - 10 = 0$

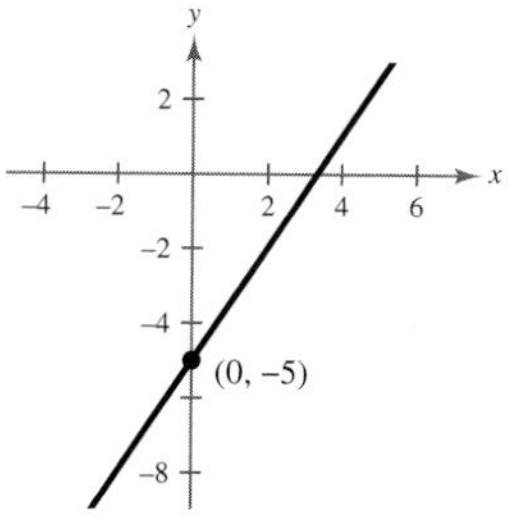

36. $x + 2y - 4 = 0$

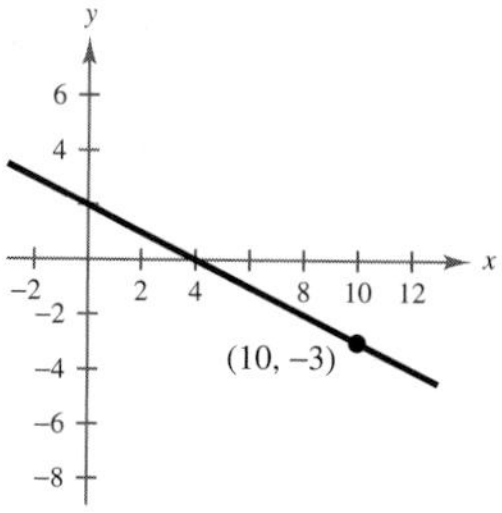

38. (a) $5x - 4y - 23 = 0$ (b) $4x + 5y - 2 = 0$

40. $V = 850t + 12{,}500$ **42.** No **44.** Yes

46. (a) 5 (b) 17 (c) $t^4 + 1$ (d) $-x^2 - 1$

48. $-5 \le x \le 5$

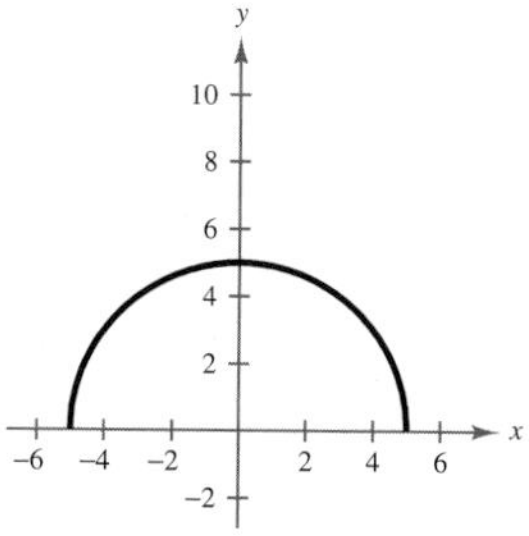

50. All real numbers $x \neq 3, -2$

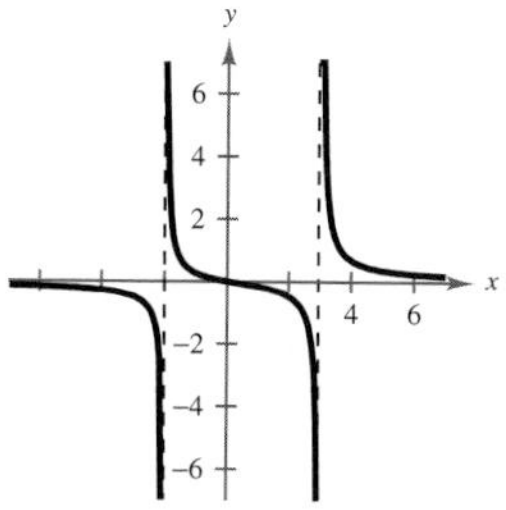

52. (a) 16 feet per second (b) 1.5 seconds

(c) -16 feet per second

54. Function **56.** Not a function **58.** $\frac{7}{3}, 3$ **60.** $-\frac{3}{8}$

62.

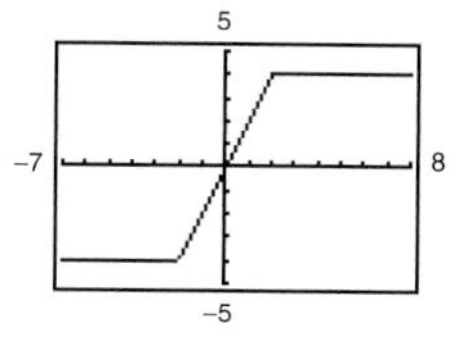

Increasing on $(-2, 2)$

Constant on $(-\infty, -2)$ and $(2, \infty)$

CHAPTER 1

64.

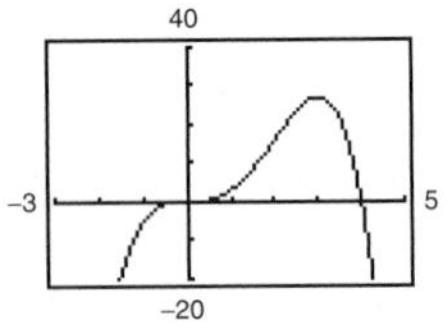

Increasing on $(-\infty, 3)$

Decreasing on $(3, \infty)$

66. $f(x) = -3x$

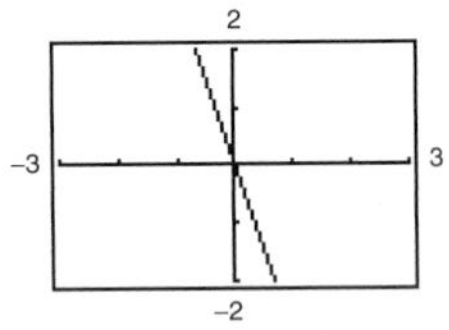

68. $f(x) = \frac{5}{3}x + \frac{10}{3}$

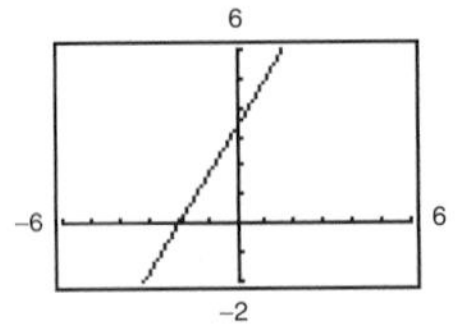

70.

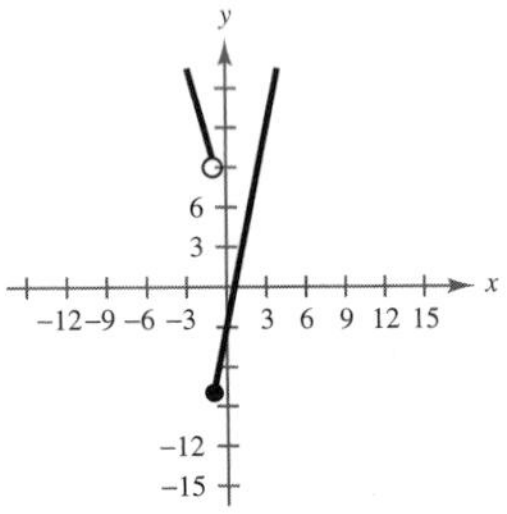

72. Neither **74.** Odd

76. Vertical shift of 4 units upward and horizontal shift of 4 units to the left of $y = x^3$

78. Vertical shift of 9 units downward

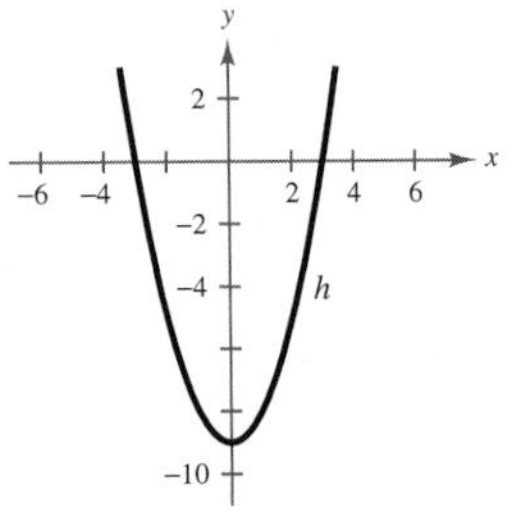

80. Horizontal shift of 3 units to the left and vertical shift of 5 units downward

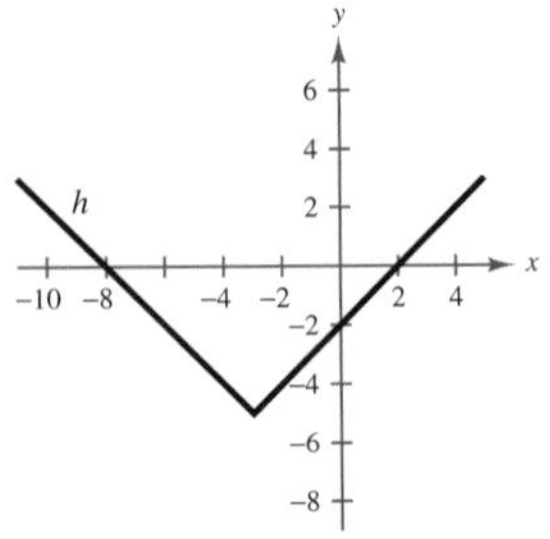

82. Reflection in x-axis, horizontal shift of 1 unit to the left, and vertical shift of 9 units upward

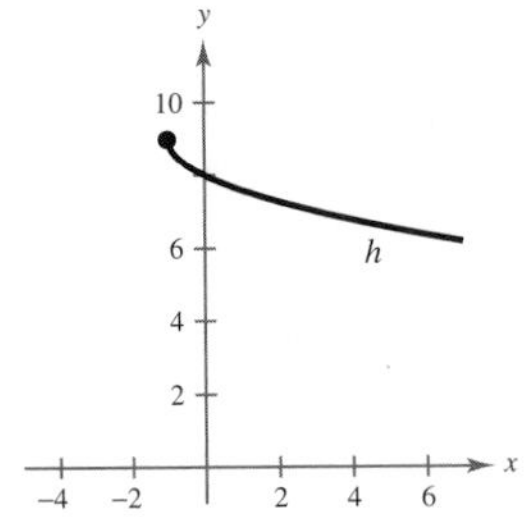

84. -7 **86.** 23

88. $y_1 = 0.80t^2 + 3.34t + 24.23$

$y_2 = -0.43t^2 + 18.14t + 62.89$

90. $f^{-1}(x) = \dfrac{x}{6}$

$f(f^{-1}(x)) = 6\left(\dfrac{x}{6}\right) = x$

$f^{-1}(f(x)) = \dfrac{6x}{6} = x$

92. $f^{-1}(x) = x + 7$

$f(f^{-1}(x)) = x + 7 - 7 = x$

$f^{-1}(f(x)) = x - 7 + 7 = x$

94.

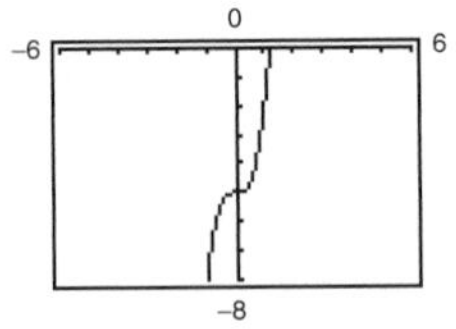

Yes

96.

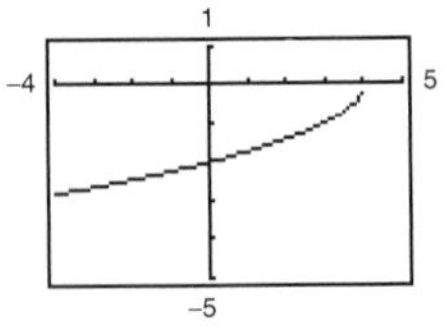

Yes

98. (a) $f^{-1}(x) = 2x + 6$

(b)

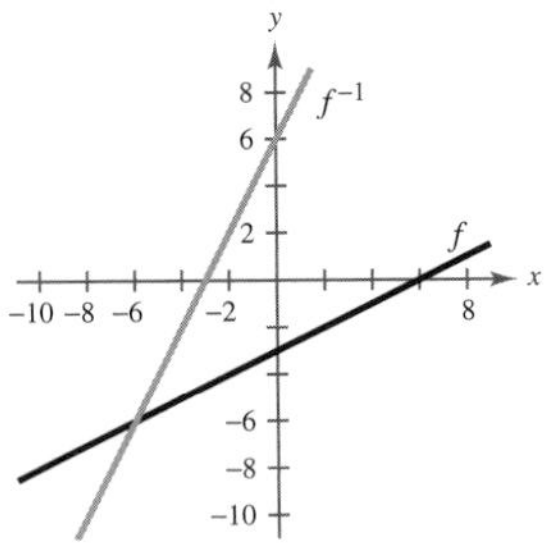

(c) $f^{-1}(f(x)) = f^{-1}\left(\frac{1}{2}x - 3\right)$

$= 2\left(\frac{1}{2}x - 3\right) + 6$

$= x - 6 + 6$

$= x$

$f(f^{-1}(x)) = f(2x + 6)$

$= \frac{1}{2}(2x + 6) - 3$

$= x + 3 - 3$

$= x$

100. (a) $f^{-1}(x) = x^2 - 1, \quad x \geq 0$

(b)

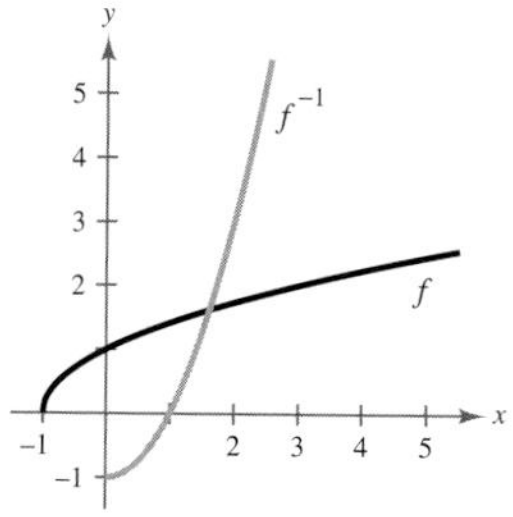

(c) $f^{-1}(f(x)) = f^{-1}(\sqrt{x+1})$

$= (x + 1) - 1$ (b)

$= x$

$f(f^{-1}(x)) = f(x^2 - 1), \quad x \geq 0$

$= \sqrt{x^2 - 1 + 1}$

$= x$

102. $x \geq 4$; $f^{-1}(x) = \sqrt{\dfrac{x}{2}} + 4$

104. (a)

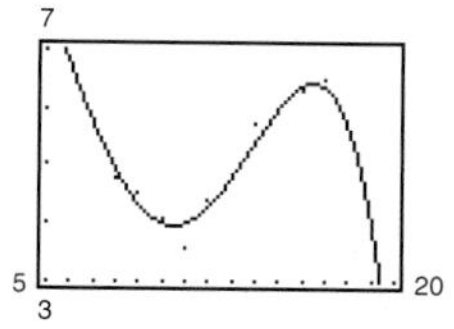

The model is a good fit for the data.

(b) Recession; yes (c) 0.30; more

(d) $-\$3.47$ billion. The model is not accurate in predicting future sales because the sales become negative.

106. 2438.7 kilowatts **108.** $y = \dfrac{49.5}{x}$

110. (a) $y_1 = 0.414t + 13.213$

$y_2 = 0.437t + 12.904$

(b)

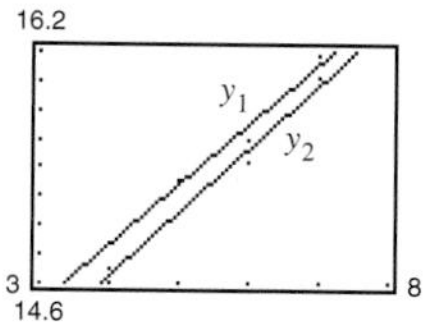

(c) The slope represents the average increase of hourly wages per year.

(d) Mining: \$18.18; construction: \$18.15

112. True. If $f(x) = x^3$ and $g(x) = \sqrt[3]{x}$, then the domain of g is all real numbers, which is equal to the range of f, and vice versa.

114. The Vertical Line Test is used to determine if a graph of y is a function of x. The Horizontal Line Test is used to determine if a function has an inverse function.

Chapter 2

Section 2.1 *(page 208)*

2. c **4.** h **6.** a **8.** d

10. (a)

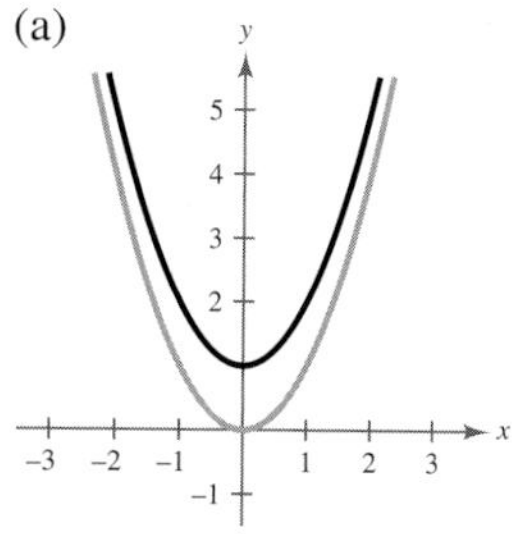

Vertical translation

(b)

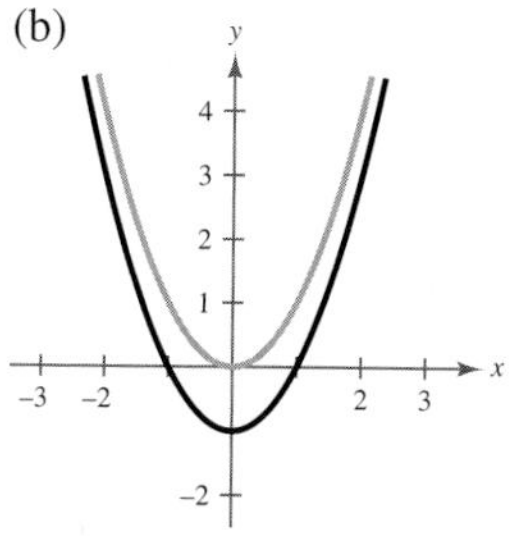

Vertical translation

(c)

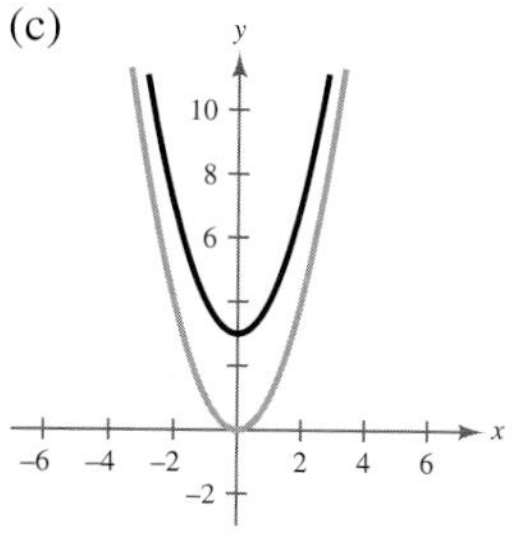

Vertical translation

(d)

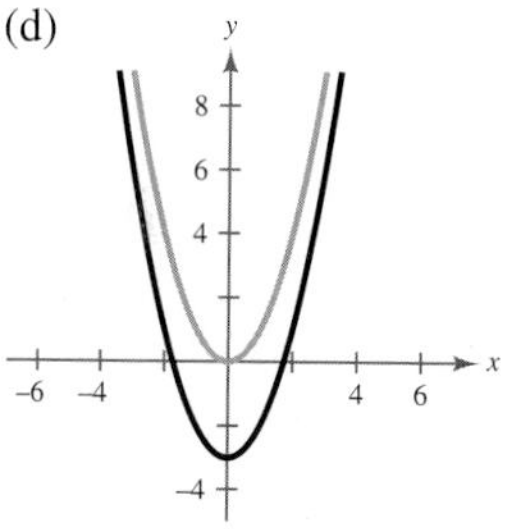

Vertical translation

12. (a)

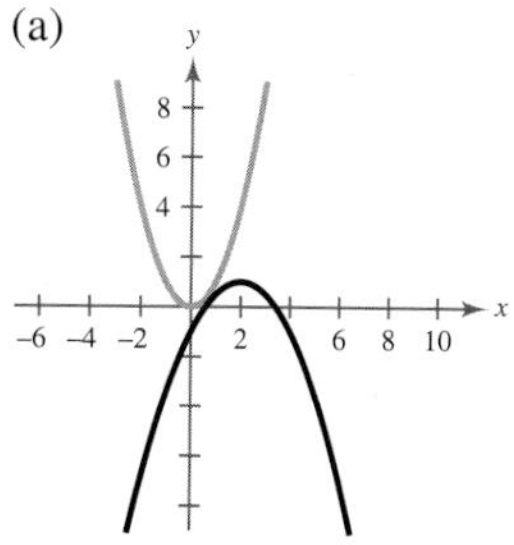

Horizontal translation, vertical shrink, reflection in the x-axis, and vertical translation

(b)

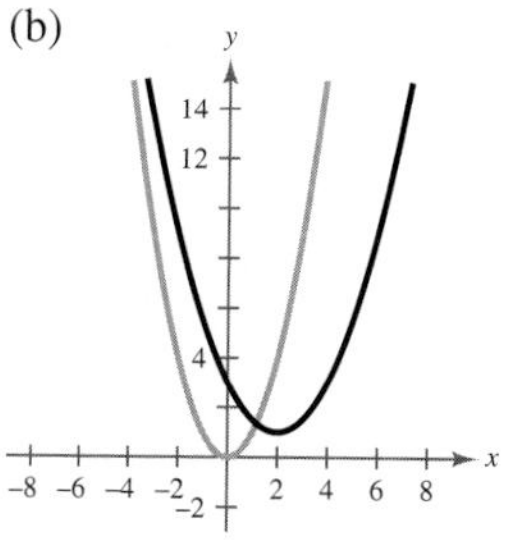

Vertical shrink, horizontal translation, and vertical translation

(c)

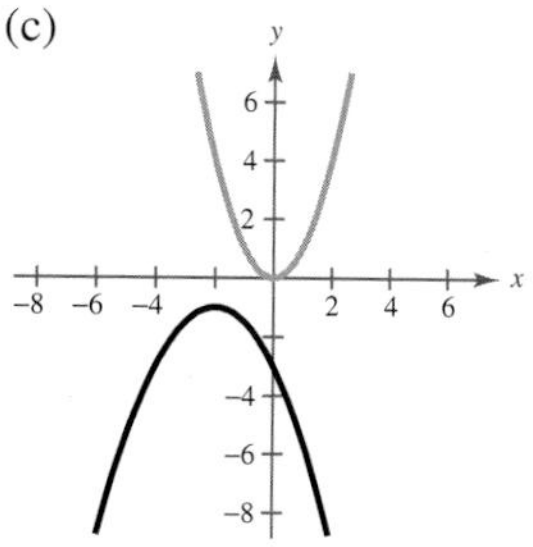

Reflection in the x-axis, vertical shrink, horizontal translation, and vertical translation

(d)

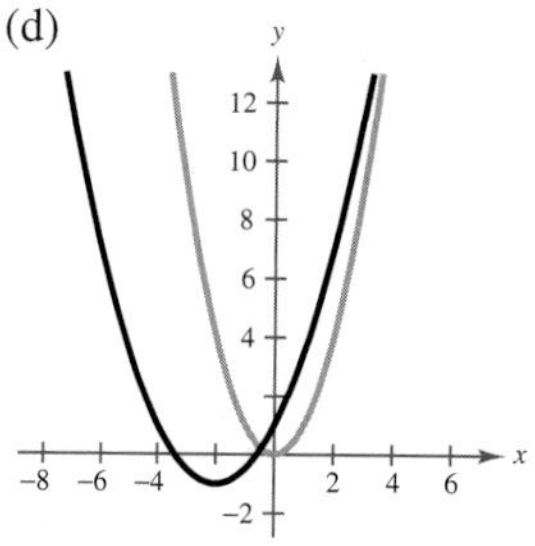

Vertical shrink, horizontal translation, and vertical translation

14.

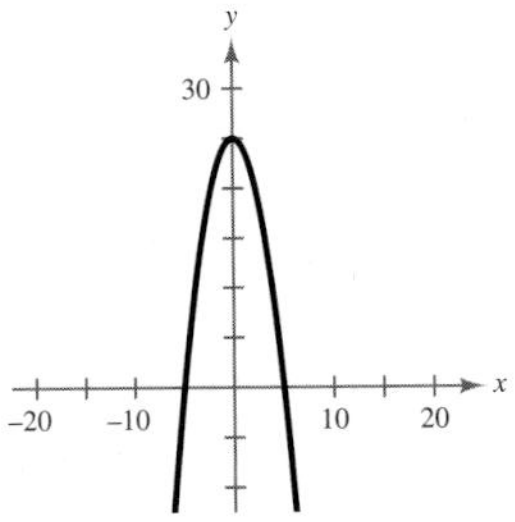

Vertex: (0, 25)

x-intercepts: $(\pm 5, 0)$

16.

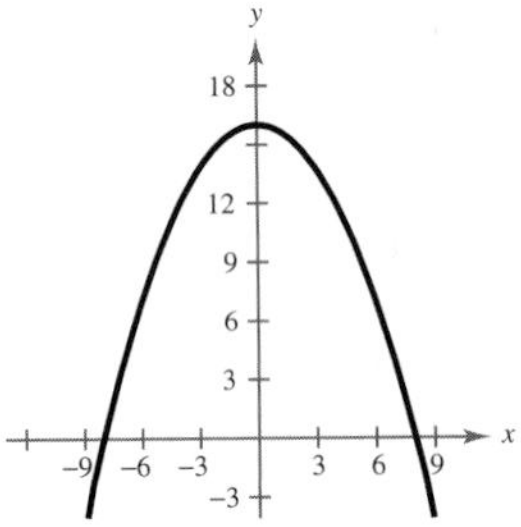

Vertex: (0, 16)

x-intercepts: $(\pm 8, 0)$

18.

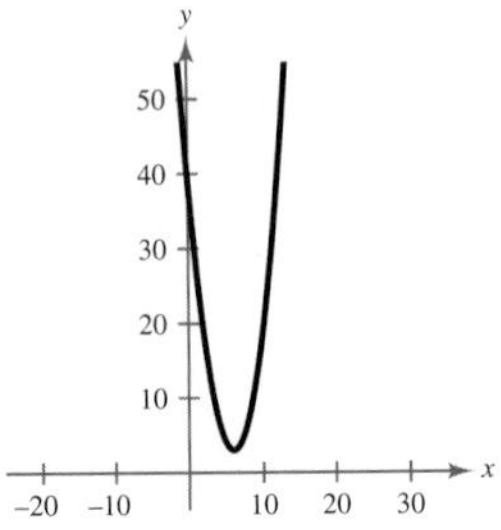

Vertex: (6, 3)

No x-intercept

20.

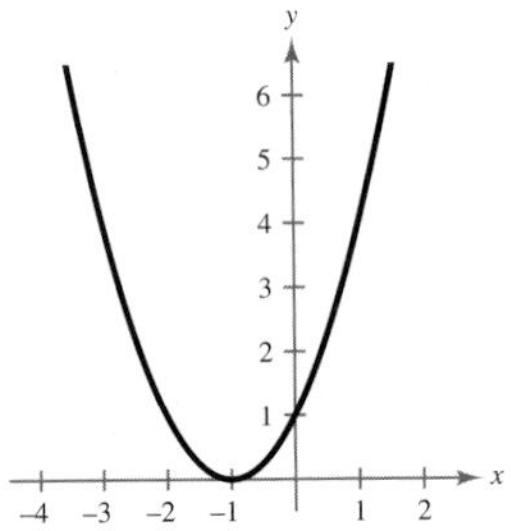

Vertex: $(-1, 0)$

x-intercept: $(-1, 0)$

22.

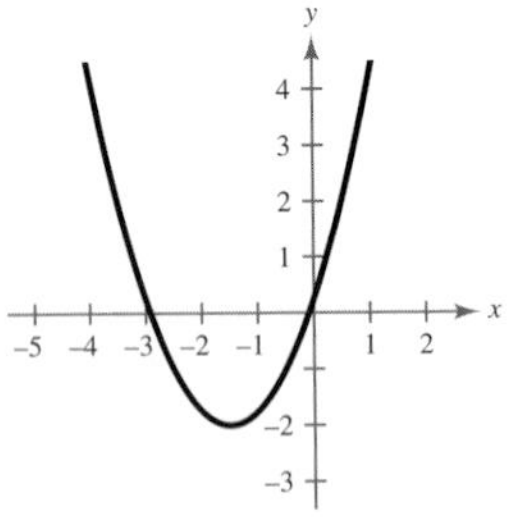

Vertex: $\left(-\frac{3}{2}, -2\right)$

x-intercepts: $\left(-\frac{3}{2} \pm \sqrt{2}, 0\right)$

24.

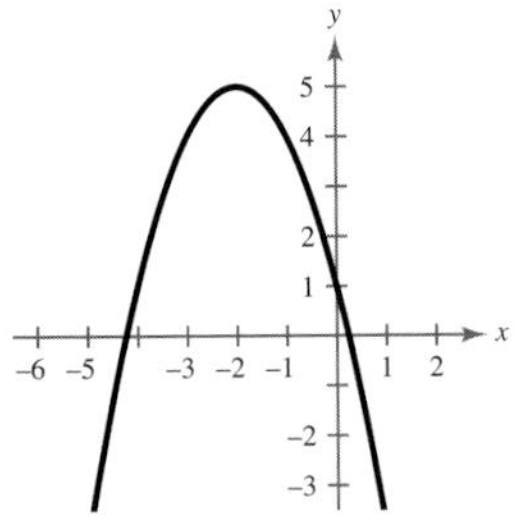

Vertex: $(-2, 5)$

x-intercepts: $\left(-2 \pm \sqrt{5}, 0\right)$

26.

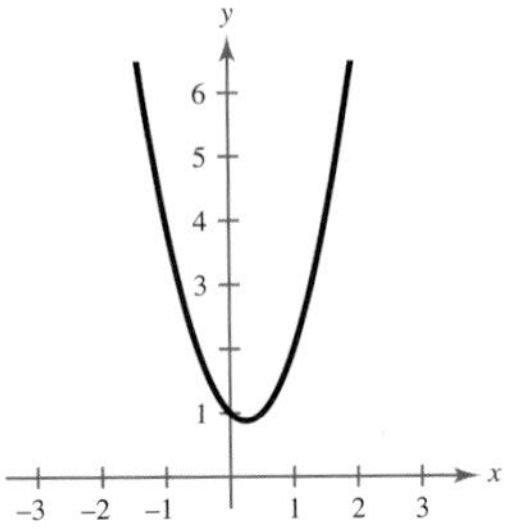

Vertex: $\left(\frac{1}{4}, \frac{7}{8}\right)$

No x-intercept

28.

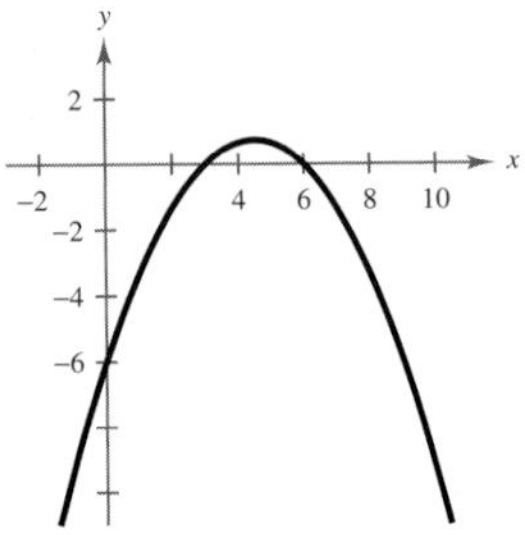

Vertex: $\left(\frac{9}{2}, \frac{3}{4}\right)$

x-intercepts: (3, 0), (6, 0)

30.

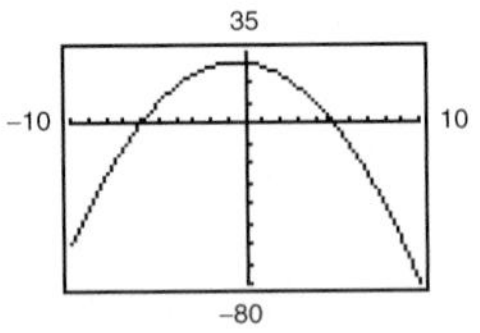

Vertex: $\left(-\frac{1}{2}, \frac{121}{4}\right)$

x-intercepts: $(-6, 0)$, $(5, 0)$

32.

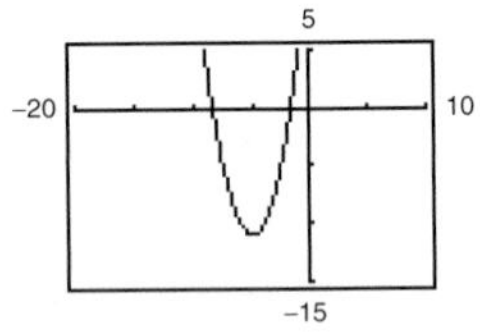

Vertex: $(-5, -11)$

x-intercepts: $\left(-5 \pm \sqrt{11}, 0\right)$

34.

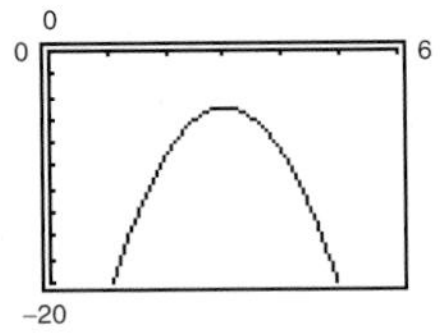

Vertex: $(3, -5)$

No x-intercept

36.

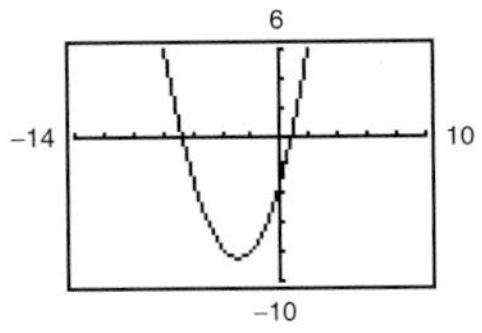

Vertex: $\left(-3, -\frac{42}{5}\right)$

x-intercepts: $\left(-3 \pm \sqrt{14}, 0\right)$

38. $y = -x^2 + 1$ **40.** $y = (x + 2)^2 - 1$

42. $y = 2(x - 2)^2$ **44.** $f(x) = (x - 4)^2 - 1$

46. $f(x) = -\frac{1}{4}(x - 2)^2 + 3$ **48.** $f(x) = 2(x + 2)^2 - 2$

50. $f(x) = \frac{19}{81}\left(x - \frac{5}{2}\right)^2 - \frac{3}{4}$

52. $f(x) = -450(x - 6)^2 + 6$

54. (3, 0). The x-intercept and solution of the equation are the same.

56. $(-3, 0), \left(\frac{1}{2}, 0\right)$. The x-intercepts and solutions of the equation are the same.

58.

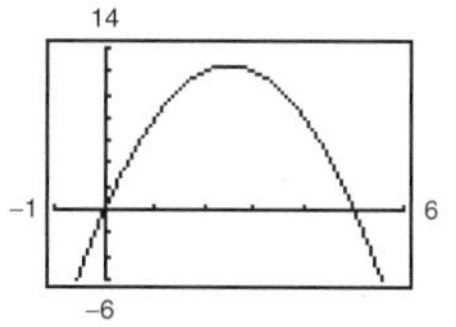

$(0, 0), (5, 0)$

60.

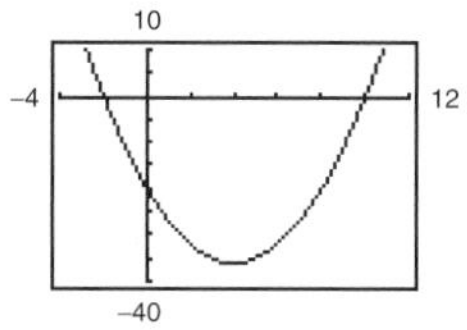

$(-2, 0), (10, 0)$

62.

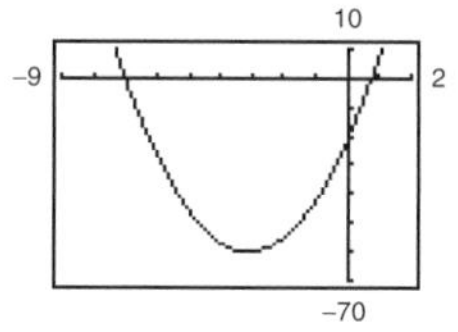

$(-7, 0), \left(\frac{3}{4}, 0\right)$

64.

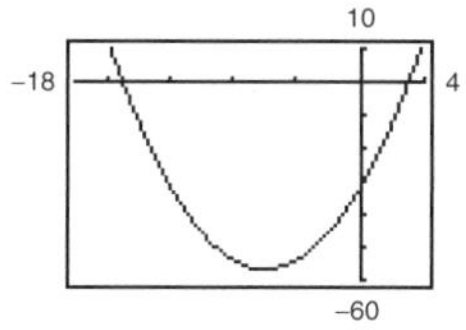

$(-15, 0), (3, 0)$

66. $f(x) = x^2 - 25$
$g(x) = -x^2 + 25$

68. $f(x) = x^2 - 12x + 32$
$g(x) = -x^2 + 12x - 32$

70. $f(x) = 2x^2 + x - 10$
$g(x) = -2x^2 - x + 10$

72. $\frac{S}{2}, \frac{S}{2}$ **74.** 21, 7

76. (a) $A = x(18 - x), \quad 0 < x < 18$

(b)

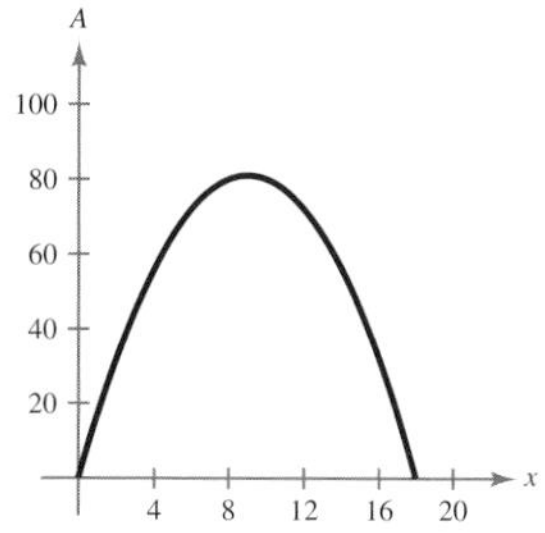

(c) 9 meters × 9 meters

78. (a) $r = \frac{1}{2}y; d = y\pi$ (b) $y = \frac{200 - 2x}{\pi}$

(c) $A = x\left(\frac{200 - 2x}{\pi}\right); \ x = 50, y = \frac{100}{\pi}$

80. 250,000 units **82.** 1222 units **84.** \$2000

86. (a)

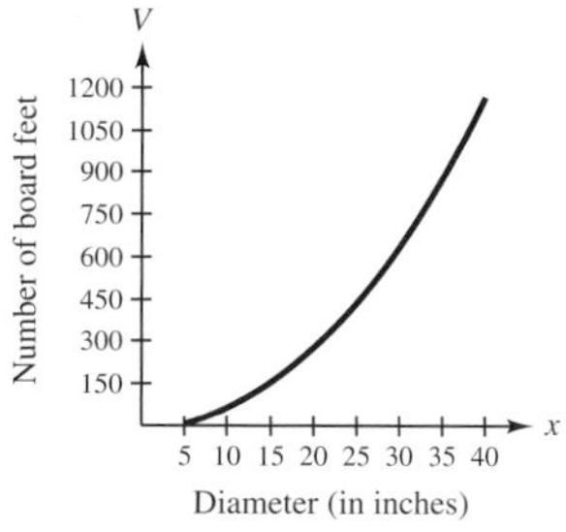

(b) 166.69 board feet (c) 26.6 inches

88. (a)

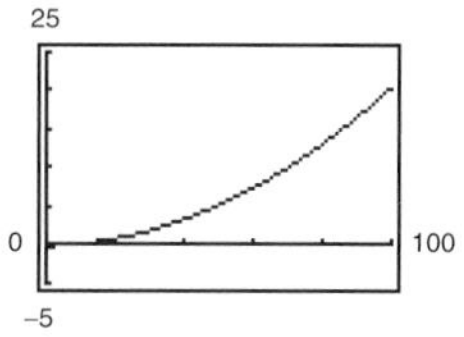

(b) 69.6 miles per hour

90. (a)

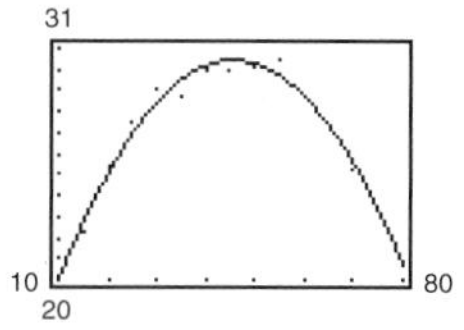

$y = -0.0082x^2 + 0.746x + 13.47$

(b) 45.5 miles per hour

92. True. The equation has no real solution, so the graph has no x-intercepts.

94. $f(x) = a\left(x + \frac{b}{2a}\right)^2 + \frac{4ac - b^2}{4a}$

96. Yes. A graph of a quadratic equation whose vertex is (0, 0) has only one x-intercept.

98. (a)

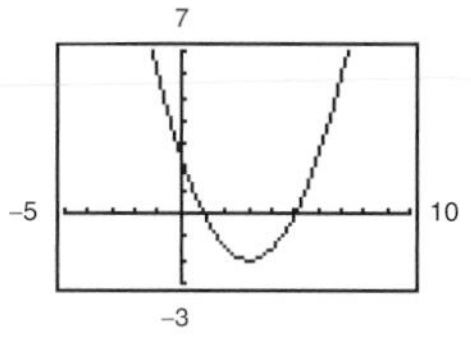

(b)

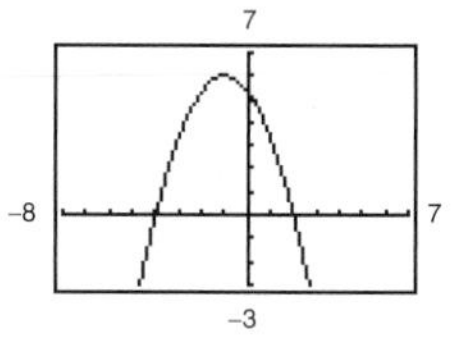

100. $y = \frac{3}{2}x - \frac{13}{4}$ **102.** $y = -3x - 20$ **104.** 7

106. $-\frac{4}{3}$ **108.** 72

Section 2.2 *(page 222)*

2. g **4.** f **6.** e **8.** b

10. (a)

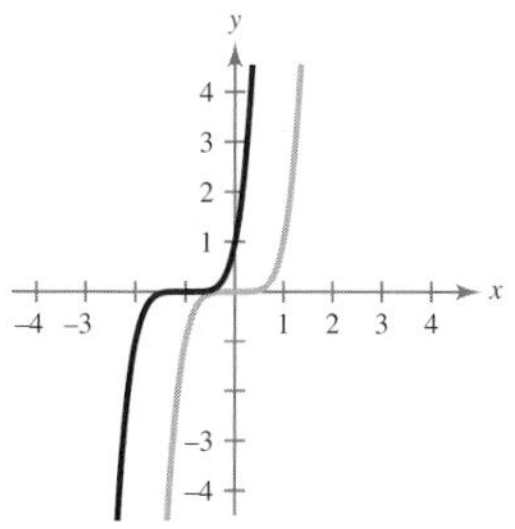

(b)

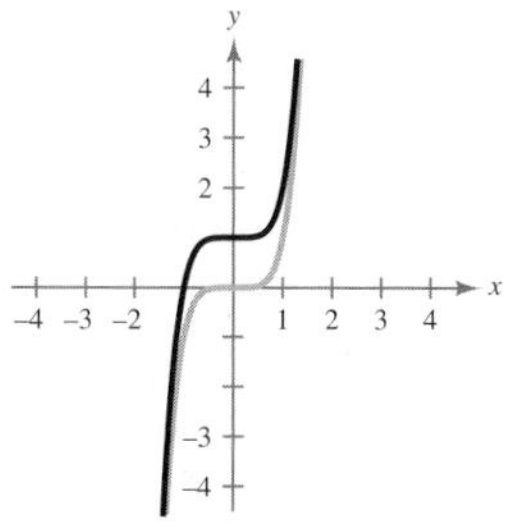

(c)

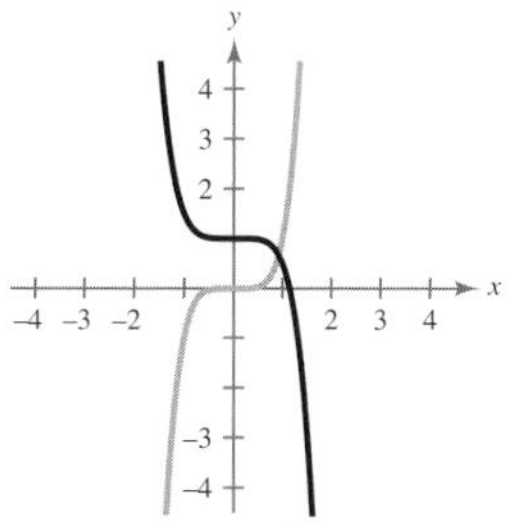

(d)

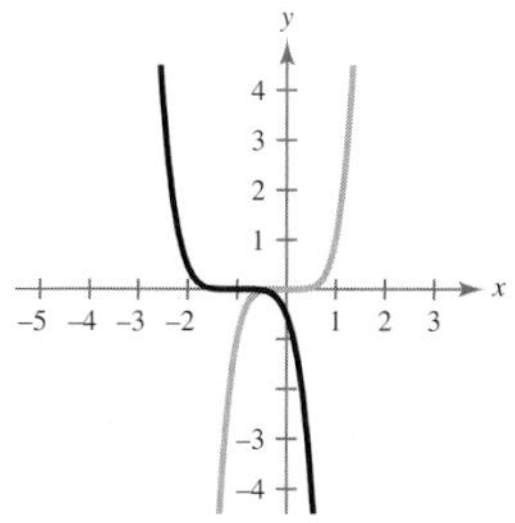

12. (a)

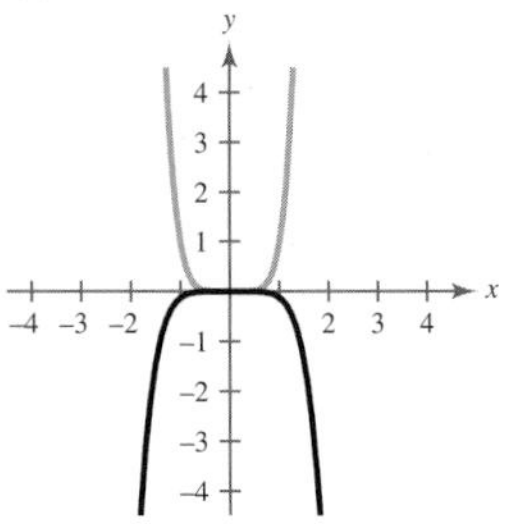

(b)

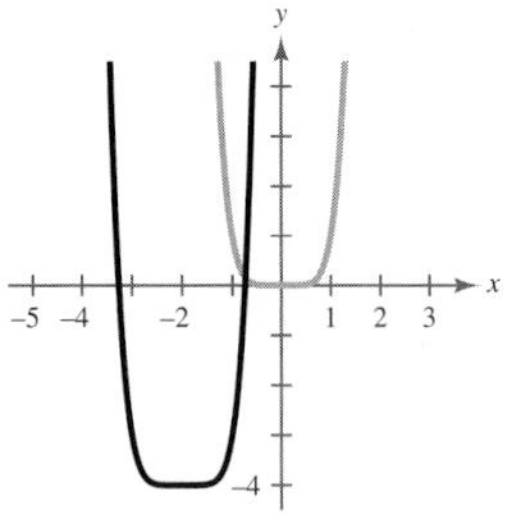

(c)

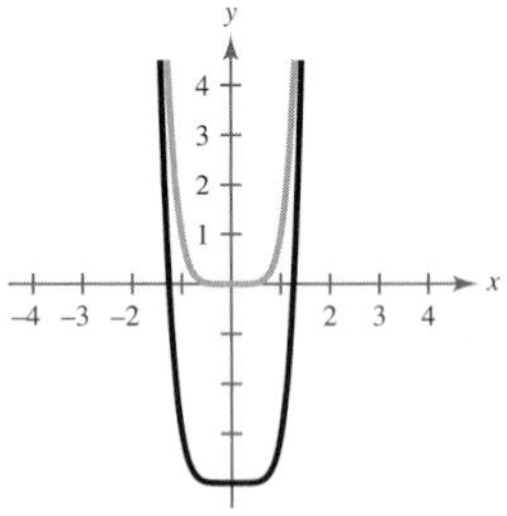

(d)

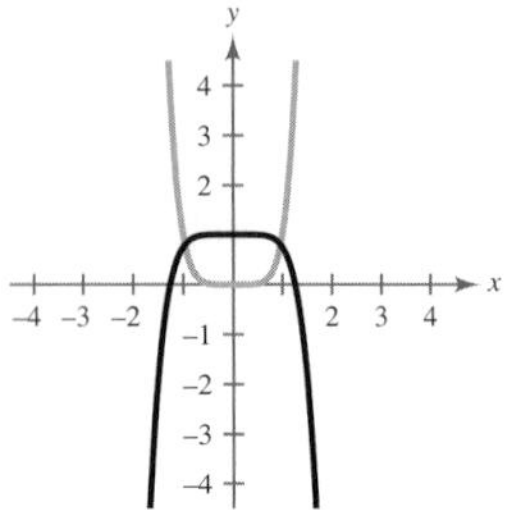

14. Rises to the left. Rises to the right.

16. Falls to the left. Falls to the right.

18. Falls to the left. Rises to the right.

20. Rises to the left. Rises to the right.

22. Rises to the left. Falls to the right.

24.

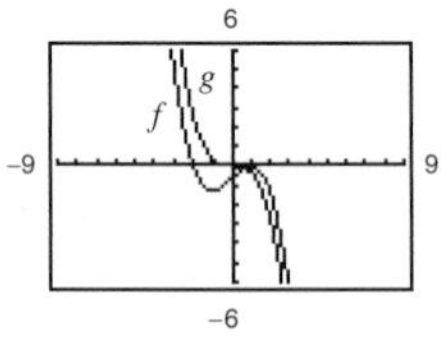

26.

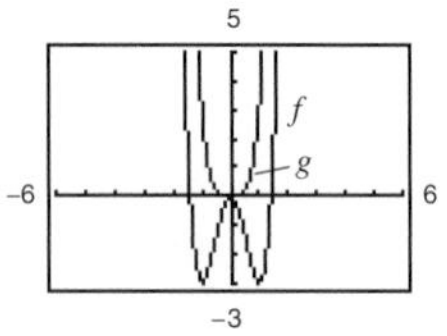

28. ± 7 **30.** -5 **32.** $\dfrac{-5 \pm \sqrt{37}}{2}$ **34.** $1 \pm \sqrt{2}$

36. $-4, 0, 5$ **38.** $0, \pm\sqrt{2}$ **40.** $\pm\sqrt{5}$ **42.** $4, \pm 5$

44.

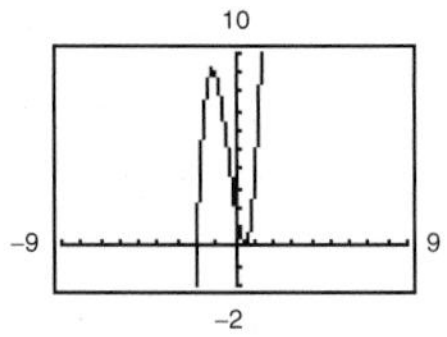

$(-2, 0), \left(\frac{1}{2}, 0\right)$

46.

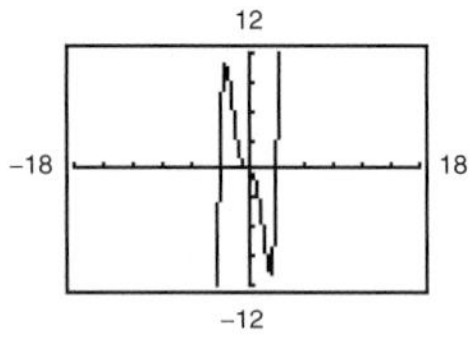

$(0, 0), (\pm 3, 0)$

48. $f(x) = x^2 + 3x$ **50.** $f(x) = x^2 - x - 20$

52. $f(x) = x^3 - 7x^2 + 10x$ **54.** $f(x) = x^5 - 5x^3 + 4x$

56. $f(x) = x^3 - 10x^2 + 27x - 22$

58. $f(x) = x^2 + 12x + 32$

60. $f(x) = x^3 - 9x^2 + 6x + 56$

62. $f(x) = x^3 - 27x^2 + 243x - 729$

64. $f(x) = x^4 - 4x^3 - 23x^2 + 54x + 72$

66. $f(x) = x^5 - 10x^4 + 14x^3 + 88x^2 - 183x + 90$

68.

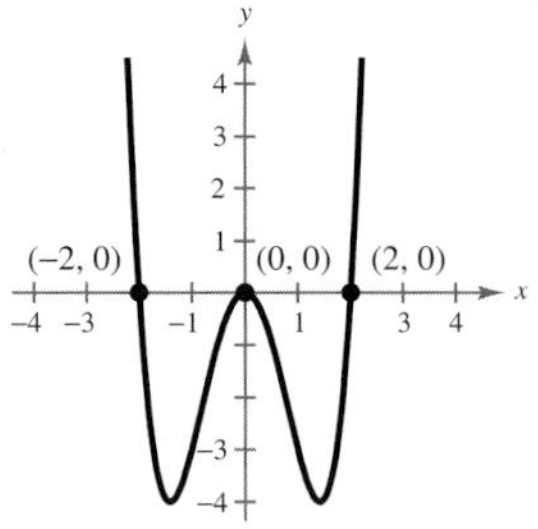

70.

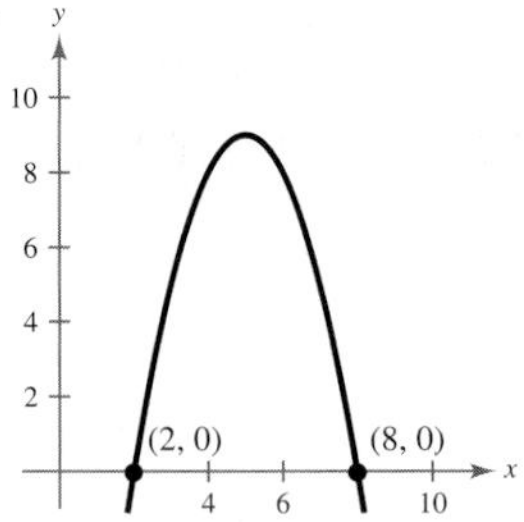

72.

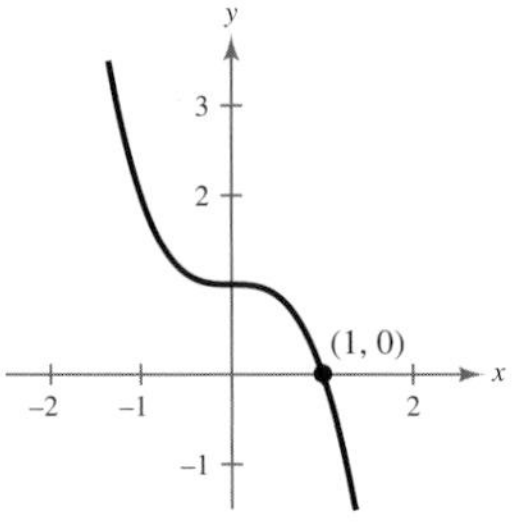

74.

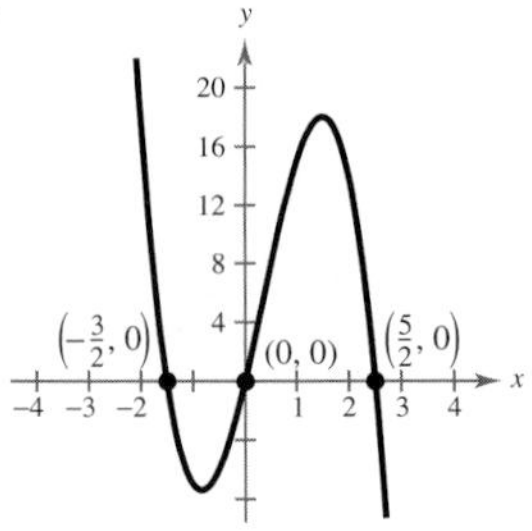

76.

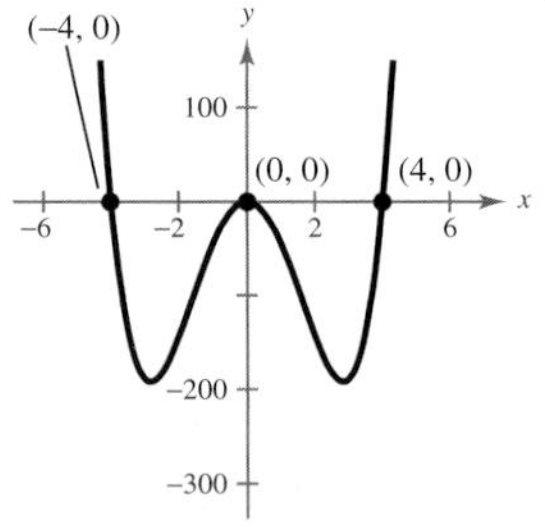

78.

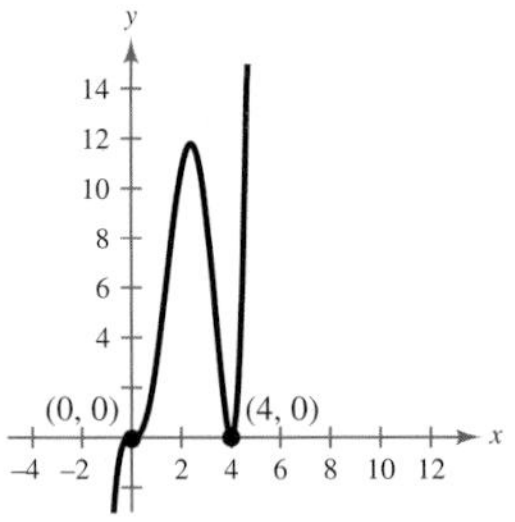

80.

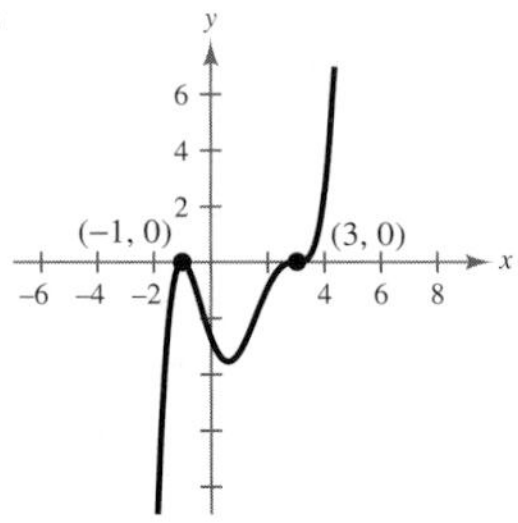

82.

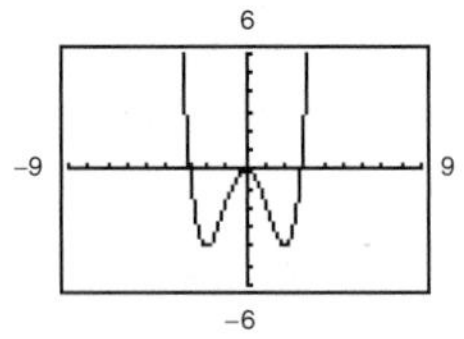

Zeros: $\pm 2\sqrt{2}$, odd multiplicity; 0, even multiplicity

84.

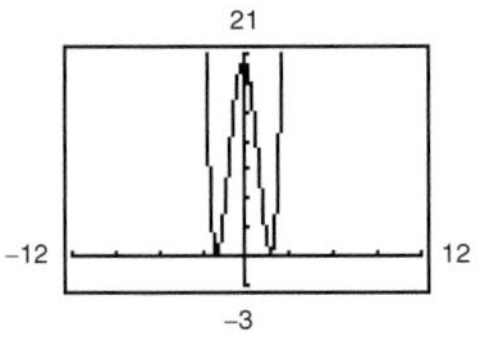

Zeros: $-2, \frac{5}{3}$, even multiplicity

86. (0, 1), (6, 7), (11, 12)

88. (0, 1), $(-1, 0)$, (3, 4), $(-4, -3)$

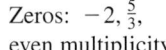

90. (a) Answers will vary. (b) $0 < x < 6$

(c)

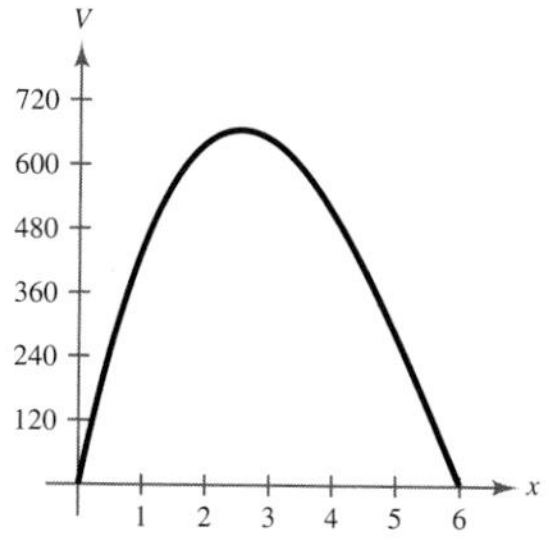

$x \approx 2.54$

92.

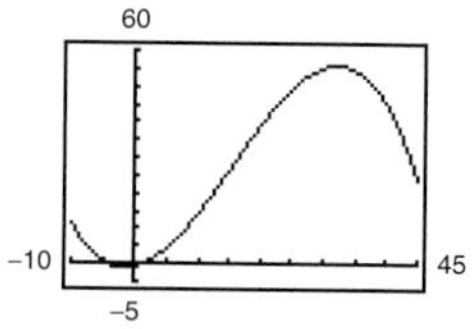

$t \approx 15$

94. True. $f(x) = (x - 1)^6$ has one repeated solution.

96.

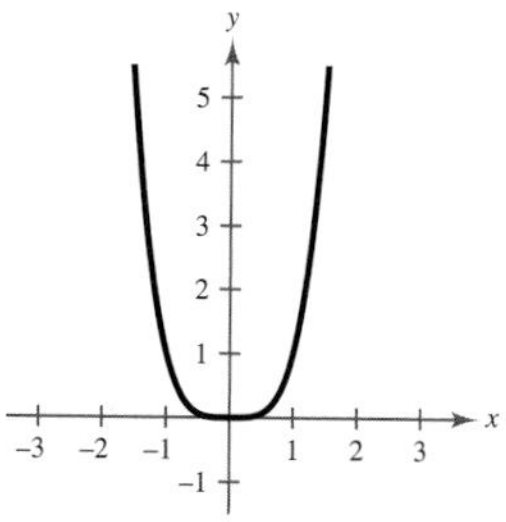

(a) Vertical shift of 2 units; Even

(b) Horizontal shift of 2 units; Neither even nor odd

(c) Reflection in the y-axis; Even

(d) Reflection in the x-axis; Even

(e) Horizontal stretch; Even

(f) Vertical shrink; Even

(g) $g(x) = x^3$; Odd

(h) $g(x) = x^{16}$; Even

98. $-\frac{7}{2}, 4$ **100.** $-\frac{5}{4}, \frac{1}{3}$ **102.** $1 \pm \sqrt{22}$

104. $\dfrac{-5 \pm \sqrt{185}}{4}$ **106.** $(5x - 8)(x + 3)$

108. $x^2(4x + 5)(x - 3)$

110. (a) The graph does not pass the Horizontal Line Test.

(b) 3

Section 2.3 *(page 233)*

2. Answers will vary. **4.** Answers will vary.

6.

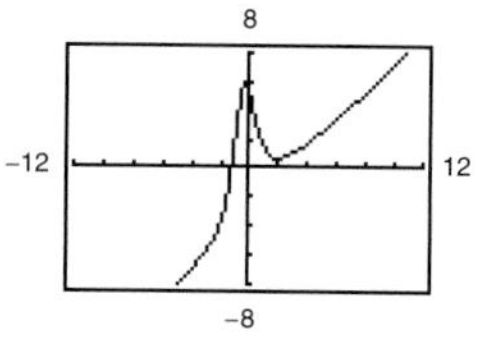

8. $5x + 3$ **10.** $2x^2 - 4x + 3$

12. $x^2 + 7x + 18 + \dfrac{42}{x - 3}$ **14.** $4 - \dfrac{9}{2x + 1}$

16. $x - \dfrac{x + 9}{x^2 + 1}$ **18.** $x^2 + \dfrac{x^2 + 7}{x^3 - 1}$

20. $2x - \dfrac{17x - 5}{x^2 - 2x + 1}$ **22.** $5x^2 + 3x - 2$

24. $9x^2 - 16$ **26.** $3x^2 + 2x + 12$

28. $5x^2 - 10x + 26 - \dfrac{44}{x + 2}$

30. $x^4 - 16x^3 + 48x^2 - 144x + 312 - \dfrac{856}{x + 3}$

32. $x^2 + 9x + 81$

34. $-3x^3 + 6x^2 - 12x + 24 - \dfrac{48}{x + 2}$

36. $-x^2 + 3x - 6 + \dfrac{11}{x + 1}$

38. $3x^2 + \dfrac{1}{2}x + \dfrac{3}{4} + \dfrac{49}{8x - 12}$

40. $f(x) = (x + 2)(x^2 - 7x + 3) + 2, f(-2) = 2$

42. $f(x) = \left(x - \frac{1}{5}\right)(10x^2 - 20x - 7) + \frac{13}{5}, f\left(\frac{1}{5}\right) = \frac{13}{5}$

44. $f(x) = \left(x + \sqrt{5}\right)\left[x^2 + \left(2 - \sqrt{5}\right)x - 2\sqrt{5}\right] + 6,$
$f\left(-\sqrt{5}\right) = 6$

46. $f(x) = \left(x - 2 - \sqrt{2}\right)\left[-3x^2 + \left(2 - 3\sqrt{2}\right)x + 8 - 4\sqrt{2}\right],$
$f\left(2 + \sqrt{2}\right) = 0$

48. (a) 14 (b) 3122 (c) 434 (d) 2

50. (a) -2.5 (b) 20 (c) 65.5 (d) 5668

52. $(x + 4)(x + 2)(x - 6)$; Zeros: $-4, -2, 6$

54. $(3x - 2)(4x - 3)(4x - 1)$; Zeros: $\frac{2}{3}, \frac{3}{4}, \frac{1}{4}$

56. $\left(x - \sqrt{2}\right)\left(x + \sqrt{2}\right)(x + 2)$; Zeros: $\sqrt{2}, -\sqrt{2}, -2$

58. $\left(x - 2 + \sqrt{5}\right)\left(x - 2 - \sqrt{5}\right)(x + 3)$;
Zeros: $2 - \sqrt{5}, 2 + \sqrt{5}, -3$

60. (a) Answers will vary. (b) $3x - 1$
(c) $f(x) = (3x - 1)(x + 3)(x - 2)$ (d) $\frac{1}{3}, -3, 2$
(e)

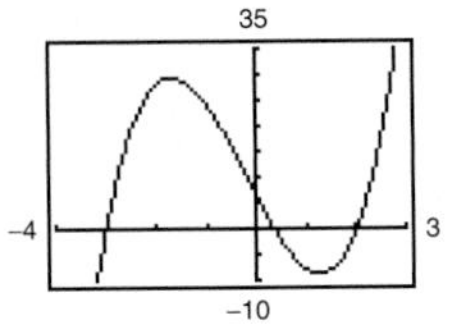

62. (a) Answers will vary. (b) $(4x + 3), (2x - 1)$
(c) $f(x) = (4x + 3)(2x - 1)(x + 2)(x - 4)$
(d) $-\frac{3}{4}, \frac{1}{2}, -2, 4$
(e)

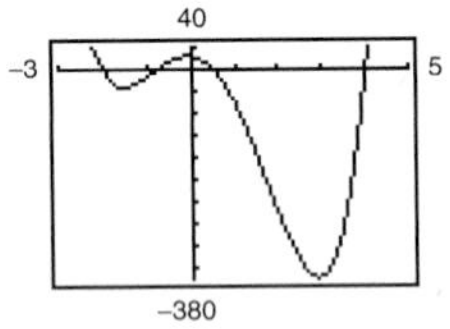

64. (a) Answers will vary. (b) $x - 3$
(c) $f(x) = (x - 3)(2x + 5)(5x - 3)$ (d) $3, -\frac{5}{2}, \frac{3}{5}$
(e)

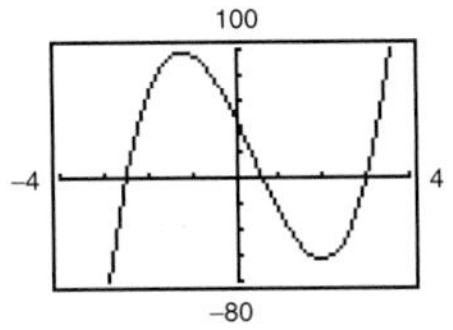

66. (a) Answers will vary. (b) $x - 4\sqrt{3}$
(c) $f(x) = (x - 4\sqrt{3})(x + 4\sqrt{3})(x + 3)$
(d) $\pm 4\sqrt{3}, -3$
(e)

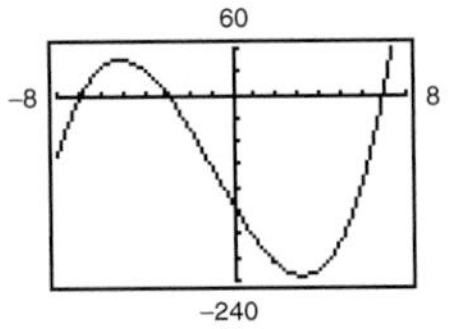

68. (a) Zeros are 4 and $\approx \pm 1.414$.
(b) $g(x) = (x - 4)(x - \sqrt{2})(x + \sqrt{2})$

70. (a) Zeros are 6, ≈ 0.764, and ≈ 5.236.
(b) $f(s) = (s - 6)[s - (3 + \sqrt{5})][s - (3 - \sqrt{5})]$

72. $x^2 - 7x - 8, \ x \neq -8$ **74.** $2x^2 + 5x + 2, \ x \neq 1$

76. $x^2 + 9x - 1, \ x \neq \pm 2$

78. (a) and (b)

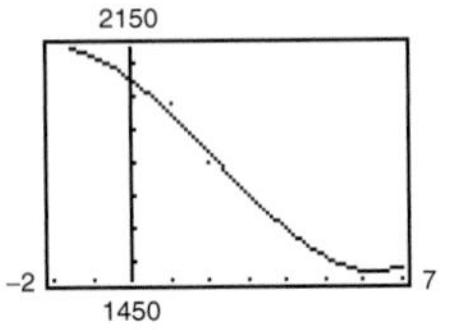

$M = 2.412t^3 - 15.557t^2 - 90.488t + 2056.617$

(c)

t	-1	0	1	2
M	2129	2057	1953	1833

t	3	4	5	6
M	1710	1600	1517	1475

(d) 2389 thousand. No, because the model will approach infinity quickly.

80. True.
$6x^6 + x^5 - 92x^4 + 45x^3 + 184x^2 + 4x - 48$
$= (2x - 1)(x + 1)(x - 2)(x - 3)(3x + 2)(x + 4)$

82. $x^{2n} - x^n + 3$

84. Multiply the divisor and the quotient to obtain the dividend.

86. $c = 42$

88. Because $f(x)$ is in factored form, it is easier to evaluate directly.

90. $\pm\dfrac{5}{3}$ **92.** $-\dfrac{7}{5}, 2$ **94.** $\dfrac{-3 \pm \sqrt{3}}{2}$

96. $f(x) = x^3 - 7x^2 + 12x$ (Answer is not unique.) **98.** $f(x) = x^3 + x^2 - 7x - 3$ (Answer is not unique.)

Section 2.4 *(page 241)*

2. $a = 13, \ b = 4$ **4.** $a = 0, \ b = -\frac{5}{2}$ **6.** $3 + 4i$
8. $1 + 2\sqrt{2}i$ **10.** $2i$ **12.** 45 **14.** $4 + 2i$
16. $0.02i$ **18.** $8 + 4i$ **20.** $-3 - 11i$ **22.** 4
24. $17 + 18i$ **26.** $-4.2 + 7.5i$ **28.** $-5\sqrt{2}$
30. -75 **32.** $6 - 22i$ **34.** $32 - 72i$
36. $(21 + 5\sqrt{2}) + (7\sqrt{5} - 3\sqrt{10})i$
38. $-5 - 12i$ **40.** $-8i$ **42.** $7 + 12i, \ 193$
44. $-3 - \sqrt{2}i, \ 11$ **46.** $-\sqrt{15}i, \ 15$
48. $1 - \sqrt{8}, \ -7$ **50.** $7i$ **52.** $\frac{5}{2} + \frac{5}{2}i$ **54.** $4 + i$
56. $8 - 4i$ **58.** $\frac{60}{169} - \frac{25}{169}i$ **60.** $\frac{12}{5} + \frac{9}{5}i$
62. $\frac{5}{17} - \frac{20}{17}i$ **64.** $-3 \pm i$ **66.** $\frac{1}{3} \pm 2i$
68. $\dfrac{1}{8} \pm \dfrac{\sqrt{11}}{8}i$ **70.** $\dfrac{3}{7} \pm \dfrac{\sqrt{34}}{14}i$ **72.** $\dfrac{1}{3} \pm \dfrac{\sqrt{23}}{3}i$
74. $-1 + 6i$ **76.** $-5i$ **78.** $-375\sqrt{3}i$
80. i **82.** (a) 8 (b) 8 (c) 8
84. (a) $\frac{53}{17} - \frac{33}{34}i$ (b) $\frac{11{,}240}{877} + \frac{4630}{877}i$
86. True. $x^4 - x^2 + 14 = 56$
$(-i\sqrt{6})^4 - (-i\sqrt{6})^2 + 14 \stackrel{?}{=} 56$
$36 + 6 + 14 \stackrel{?}{=} 56$
$56 = 56$
88. $\sqrt{-6}\sqrt{-6} = \sqrt{6}i\sqrt{6}i = 6i^2 = -6$
90. Answers will vary. **92.** $x^3 + x^2 + 2x - 6$

94. $4x^2 - 20x + 25$ **96.** 14 **98.** $-\frac{4}{3}$

100. $r = \dfrac{\sqrt{\alpha m_1 m_2 F}}{F}$ **102.** 88.9 kilometers per hour

Section 2.5 *(page 253)*

2. $0, -3, \pm 1$ **4.** $-5, 8$ **6.** $3, 2, \pm 3i$

8. $\pm 1, \pm 2, \pm 4, \pm 8, \pm 16$ **10.** $\pm 1, \pm 2, \pm \frac{1}{2}, \pm \frac{1}{4}$

12. $-2, -1, 3$ **14.** $1, 2, 6$ **16.** 3 **18.** $\frac{1}{3}, 3$

20. $\pm 1, 5, \frac{5}{2}$ **22.** $0, -1, -3, 4$ **24.** $-2, 0, 1$

26. (a) $\pm 1, \pm 2, \pm 4, \pm 8, \pm 16, \pm \frac{1}{3}, \pm \frac{2}{3}, \pm \frac{4}{3}, \pm \frac{8}{3}, \pm \frac{16}{3}$

(b)

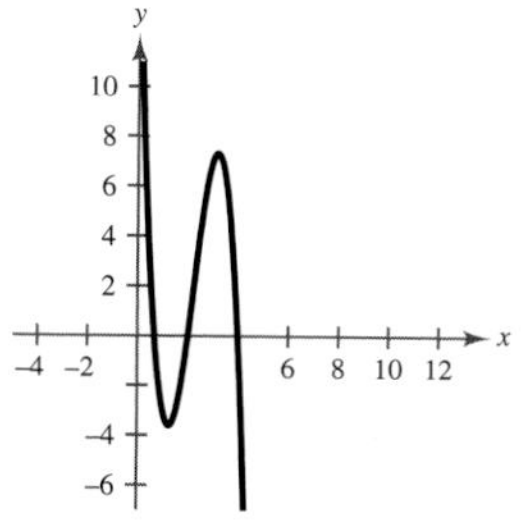

(c) $\frac{2}{3}, 2, 4$

28. (a) $\pm 1, \pm 3, \pm 5, \pm 15, \pm \frac{1}{2}, \pm \frac{3}{2}, \pm \frac{5}{2}, \pm \frac{15}{2}, \pm \frac{1}{4}, \pm \frac{3}{4}, \pm \frac{5}{4}, \pm \frac{15}{4}$

(b)

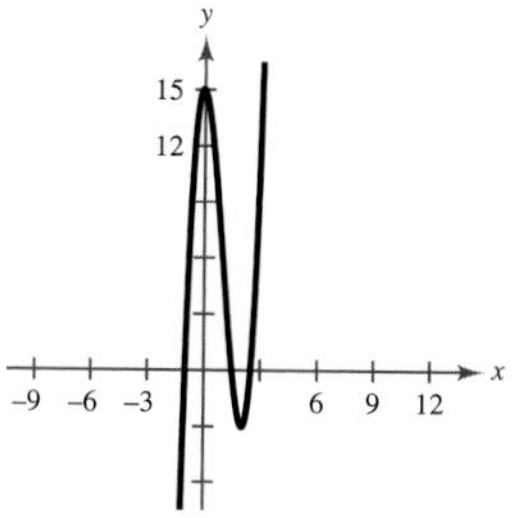

(c) $-1, \frac{3}{2}, \frac{5}{2}$

30. (a) $\pm 1, \pm 2, \pm 4, \pm \frac{1}{2}, \pm \frac{1}{4}$

(b)

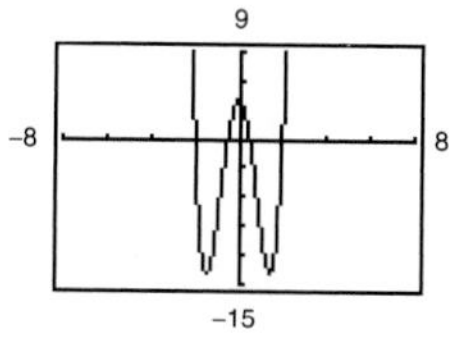

(c) $\pm 2, \pm \frac{1}{2}$

32. (a) $\pm 1, \pm 2, \pm 3, \pm 6, \pm 9, \pm 18, \pm \frac{1}{2}, \pm \frac{3}{2}, \pm \frac{9}{2}, \pm \frac{1}{4}, \pm \frac{3}{4}, \pm \frac{9}{4}$

(b)

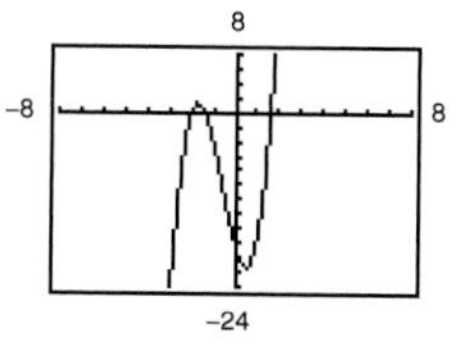

(c) $-2, \dfrac{1}{8} \pm \dfrac{\sqrt{145}}{8}$

34. (a) $\pm 2, \approx \pm 1.732$

(b) $P(t) = (t + 2)(t - 2)(t + \sqrt{3})(t - \sqrt{3})$

36. (a) $\pm 3, 1.5, 0.333$

(b) $g(x) = (x + 3)(x - 3)(2x - 3)(3x - 1)$

38. $x^3 - 4x^2 + 9x - 36$ **40.** $x^3 - 10x^2 + 33x - 34$

42. $x^4 + 8x^3 + 9x^2 - 10x + 100$

44. (a) $(x^2 - 6)(x^2 - 2x + 3)$

(b) $(x + \sqrt{6})(x - \sqrt{6})(x^2 - 2x + 3)$

(c) $(x + \sqrt{6})(x - \sqrt{6})(x - 1 - \sqrt{2}i)(x - 1 + \sqrt{2}i)$

46. (a) $(x^2 + 4)(x^2 - 3x - 5)$

(b) $(x^2 + 4)\left(x - \dfrac{3 + \sqrt{29}}{2}\right)\left(x - \dfrac{3 - \sqrt{29}}{2}\right)$

(c) $(x + 2i)(x - 2i)\left(x - \dfrac{3 + \sqrt{29}}{2}\right)\left(x - \dfrac{3 - \sqrt{29}}{2}\right)$

48. $-1, \pm 3i$ **50.** $-3, 5 \pm 2i$ **52.** $-\frac{2}{3}, 1 \pm \sqrt{3}i$

54. $-2, -1 \pm 3i$

56. $\dfrac{1 \pm \sqrt{223}i}{2}$; $\left(x - \dfrac{1 - \sqrt{223}i}{2}\right)\left(x - \dfrac{1 + \sqrt{223}i}{2}\right)$

58. $-5 \pm \sqrt{2}$; $(x + 5 + \sqrt{2})(x + 5 - \sqrt{2})$

60. $\pm 5, \pm 5i$; $(y + 5)(y - 5)(y + 5i)(y - 5i)$

62. $1 \pm i, 1$; $(x - 1)(x - 1 + i)(x - 1 - i)$

64. $3 \pm 2i, -4$; $(x + 4)(x - 3 + 2i)(x - 3 - 2i)$

66. $-2 \pm \sqrt{3}i, -5$; $(x + 5)(x + 2 + \sqrt{3}i)(x + 2 - \sqrt{3}i)$

68. $1 \pm \sqrt{3}i, -\frac{2}{3}$; $(3x + 2)(x - 1 + \sqrt{3}i)(x - 1 - \sqrt{3}i)$

70. $-3, \pm i$; $(x + 3)^2(x + i)(x - i)$

72. $\pm 2i, \pm 5i$; $(x + 2i)(x - 2i)(x + 5i)(x - 5i)$

74. $1 \pm 2i, \frac{1}{2}$ **76.** $\frac{1}{3} \pm \frac{2}{3}i, 1$ **78.** $1 \pm \sqrt{3}i, 2$

80. Two or no positive zeros

82. Two or no positive zeros

84. One or three positive zeros

86. One or three negative zeros

88. Answers will vary. **90.** Answers will vary.

92. $-\frac{3}{2}, \frac{1}{3}, \frac{3}{2}$ **94.** $\frac{2}{3}$ **96.** $-3, \frac{1}{2}, 4$

98. $-2, -\frac{1}{3}, \frac{1}{2}$ **100.** a **102.** c

104.

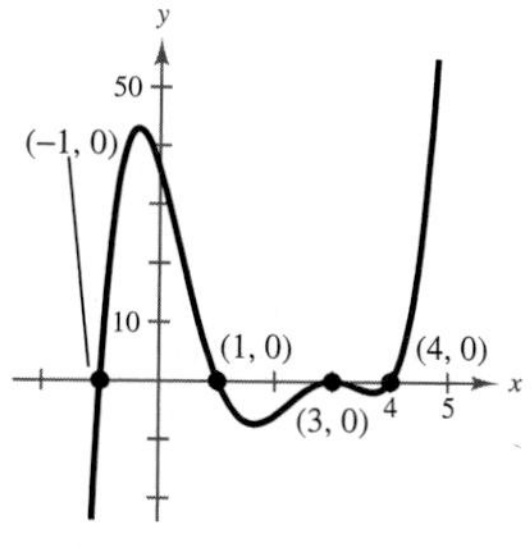

(Answers will vary.)

CHAPTER 2

106. (a) $-2, 0, 2$

(b) The graph touches the x-axis at $x = 0$.

(c) The least possible degree of the function is 4 because there are at least four real zeros (0 is repeated) and a function can have at most the number of real zeros equal to the degree of the function. The degree cannot be odd by the definition of multiplicity.

(d) Negative. From the information in the table, you can conclude that the graph will eventually fall to the left and fall to the right.

(e) $f(x) = -x^4 + 4x^2$ (Answer is not unique.)

(f)

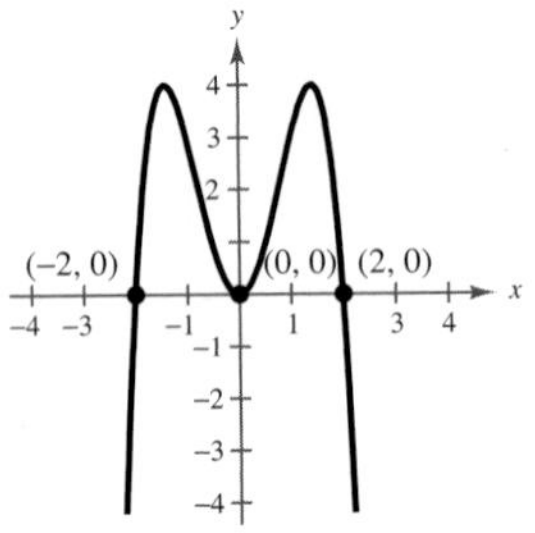

108. (a) Answers will vary.

(b)

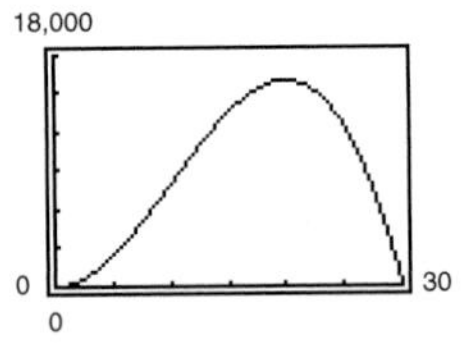

20 in. × 20 in. × 40 in.

(c) $15, \dfrac{15 \pm 15\sqrt{5}}{2}$; $\dfrac{15 - 15\sqrt{5}}{2}$ is negative.

110. $x \approx 31.5$, or \$315,000

112. (a)

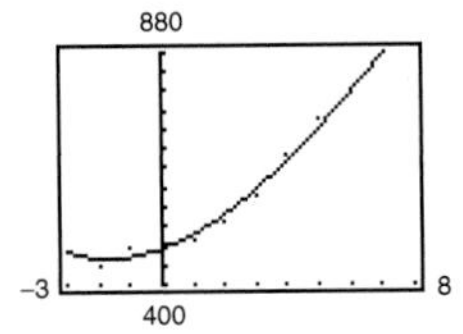

(b) 1995

(c) No. The right hand behavior will eventually decrease.

114. No. Setting $P = 9{,}000{,}000$ and solving the resulting equation yields imaginary roots.

116. False. f does not have real coefficients.

118. r_1, r_2, r_3 **120.** $\dfrac{r_1}{2}, \dfrac{r_2}{2}, \dfrac{r_3}{2}$

122. $-r_1, -r_2, -r_3$ **124.** (a) No (b) No

126. (a) Not correct because f has $(0, 0)$ as an intercept.

(b) Not correct because the function must be at least a fourth-degree polynomial.

(c) Correct function

(d) Not correct because k has $(-1, 0)$ as an intercept.

128. $12 + 11i$ **130.** 106 **132.** $18 + 26i$

134.

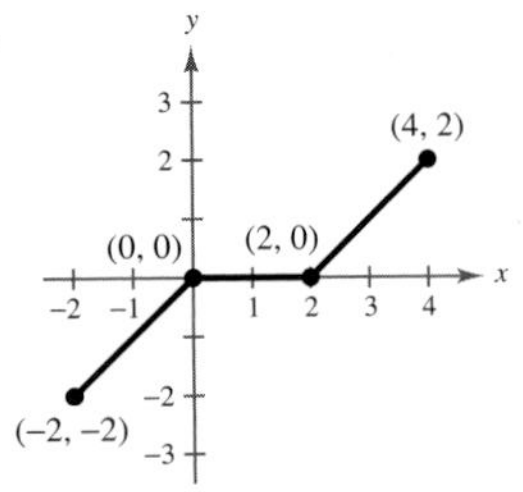

136.

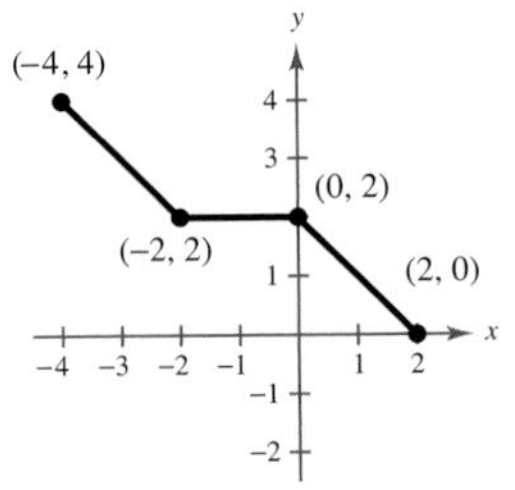

138.

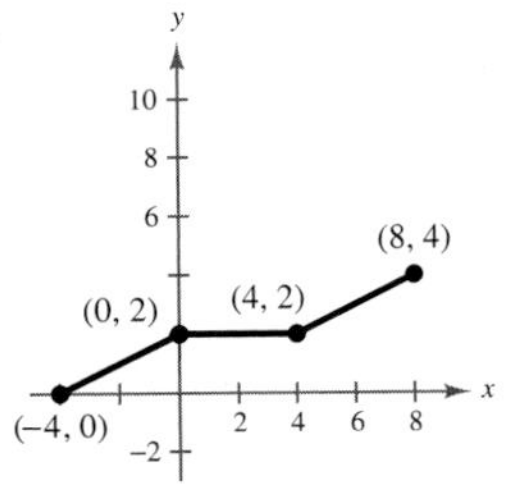

Section 2.6 *(page 267)*

2. (a)

x	$f(x)$	x	$f(x)$	x	$f(x)$
0.5	-5	1.5	15	5	6.25
0.9	-45	1.1	55	10	$5.\overline{55}$
0.99	-495	1.01	505	100	$5.\overline{05}$
0.999	-4995	1.001	5005	1000	$5.\overline{005}$

(b) Vertical asymptote: $x = 1$

Horizontal asymptote: $y = 5$

(c) Domain: all real numbers $x \neq 1$

4. (a)

x	$f(x)$	x	$f(x)$	x	$f(x)$
0.5	4	1.5	4	5	0.50
0.9	20	1.1	20	10	$0.\overline{22}$
0.99	200	1.01	200	100	$0.\overline{02}$
0.999	2000	1.001	2000	1000	$0.\overline{002}$

(b) Vertical asymptote: $x = 1$

Horizontal asymptote: $y = 0$

(c) Domain: all real numbers $x \neq 1$

6. (a)

x	$f(x)$	x	$f(x)$	x	$f(x)$
0.5	$-2.\overline{66}$	1.5	4.8	5	$0.8\overline{33}$
0.9	-18.95	1.1	20.95	10	$0.\overline{40}$
0.99	-199	1.01	201	100	0.04
0.999	-1999	1.001	2001	1000	0.004

(b) Vertical asymptotes: $x = \pm 1$

Horizontal asymptote: $y = 0$

(c) Domain: all real numbers $x \neq \pm 1$

8. Domain: all real numbers $x \neq 2$

Vertical asymptote: $x = 2$

Horizontal asymptote: $y = 0$

10. Domain: all real numbers $x \neq -\frac{1}{2}$

Vertical asymptote: $x = -\frac{1}{2}$

Horizontal asymptote: $y = -\frac{5}{2}$

12. Domain: all real numbers $x \neq -1$

Vertical asymptote: $x = -1$

14. Domain: all real numbers

Horizontal asymptote: $y = 3$

16. a **18.** c **20.** b **22.** None **24.** 2

26. (a) Intercept: $\left(0, -\frac{1}{3}\right)$

(b) Vertical asymptote: $x = 3$
Horizontal asymptote: $y = 0$

(c) No symmetry

(d) and (e)

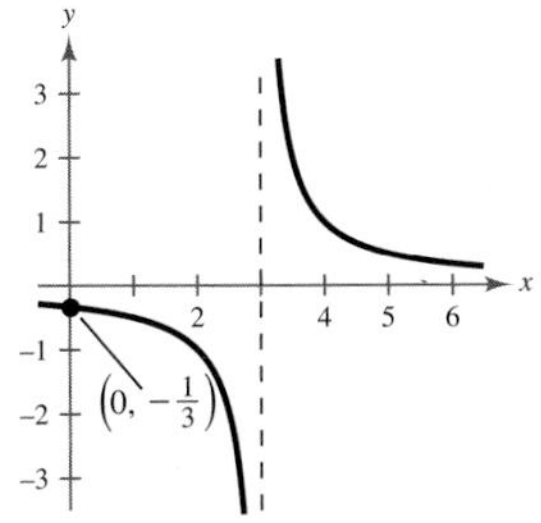

28. (a) Intercept: $\left(0, \frac{1}{3}\right)$

(b) Vertical asymptote: $x = 3$
Horizontal asymptote: $y = 0$

(c) No symmetry

(d) and (e)

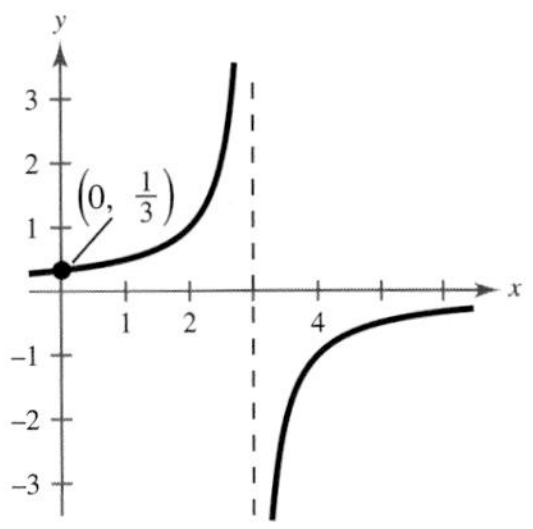

30. (a) Intercepts: $(0, 1)$, $\left(\frac{1}{3}, 0\right)$

(b) Vertical asymptote: $x = 1$
Horizontal asymptote: $y = 3$

(c) No symmetry

(d) and (e)

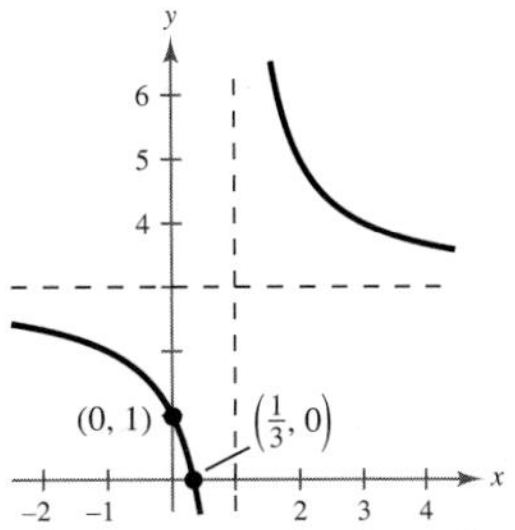

32. (a) Intercepts: $\left(-\frac{\sqrt{6}}{2}, 0\right)$, $\left(\frac{\sqrt{6}}{2}, 0\right)$

(b) Vertical asymptote: $x = 0$
Horizontal asymptote: $y = 2$

(c) y-axis

(d) and (e)

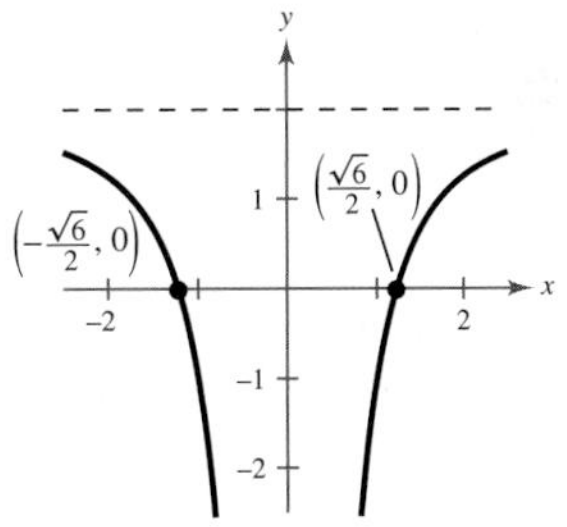

34. (a) Intercept: $\left(\frac{1}{2}, 0\right)$

(b) Vertical asymptote: $x = 0$
Horizontal asymptote: $y = -2$

(c) No symmetry

(d) and (e)

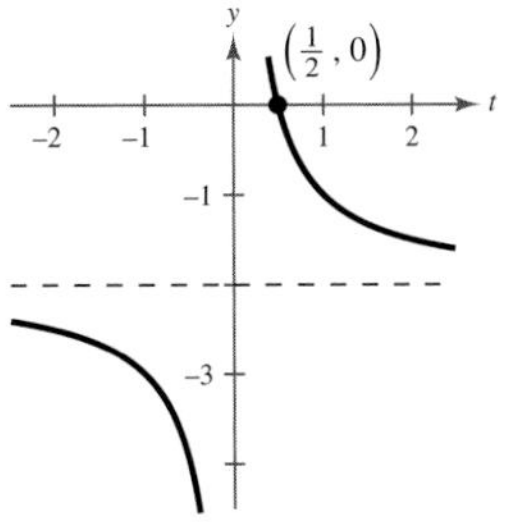

36. (a) Intercept: $(0, 0)$

(b) Vertical asymptotes: $x = \pm 3$
Horizontal asymptote: $y = 0$

(c) Origin

(d) and (e)

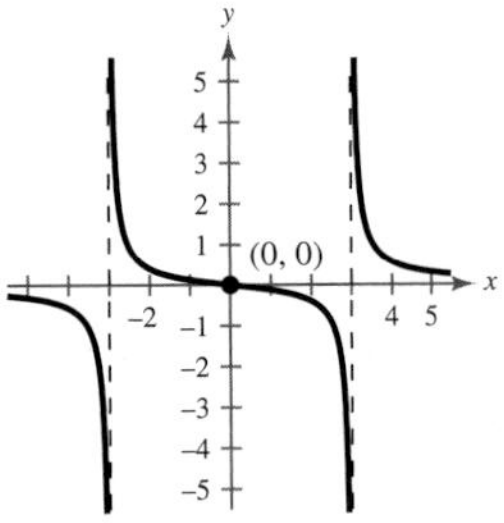

38. (a) Intercept: $\left(0, -\frac{1}{4}\right)$

(b) Vertical asymptote: $x = 2$
Horizontal asymptote: $y = 0$

(c) No symmetry

(d) and (e)

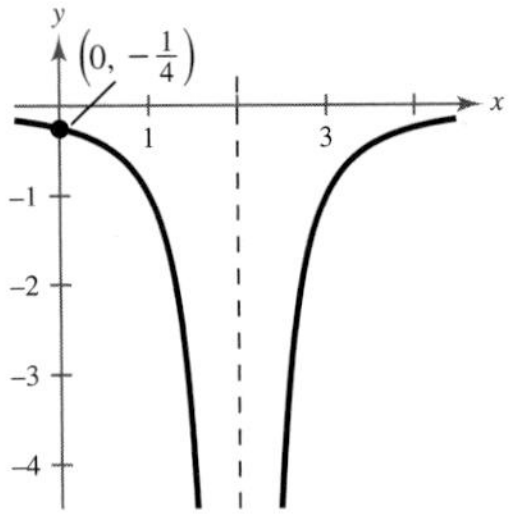

40. (a) No intercepts

(b) Vertical asymptotes: $x = 0$, $x = 2$
Horizontal asymptote: $y = 0$

(c) No symmetry

(d) and (e)

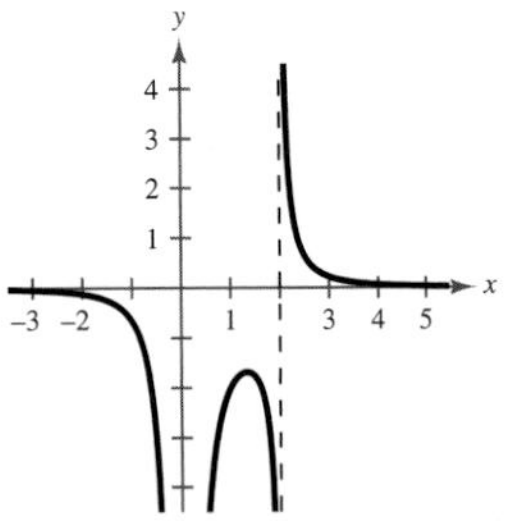

42. (a) Intercept: $(0, 0)$

(b) Vertical asymptotes: $x = -2$, $x = 1$
Horizontal asymptote: $y = 0$

(c) No symmetry

(d) and (e)

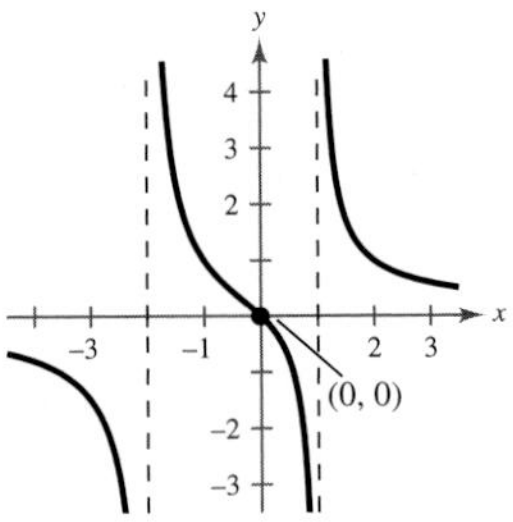

44. (a) Intercept: $\left(0, -\frac{1}{5}\right)$

(b) Vertical asymptotes: $x = -5$, $x = 3$
Horizontal asymptote: $y = 3$

(c) No symmetry

(d) and (e)

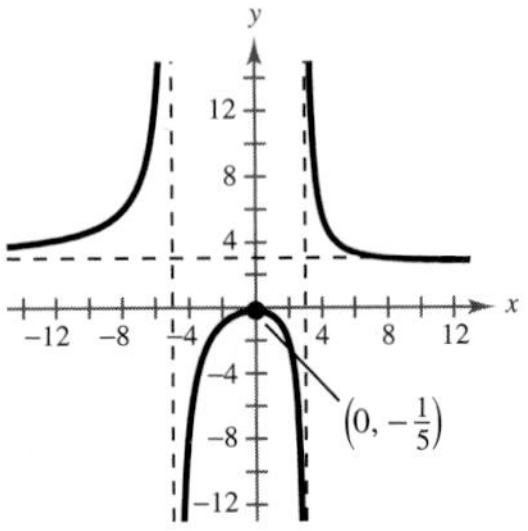

46. (a) Domain of f: all real numbers $x \neq 0, 2$

Domain of g: all real numbers

(b) Vertical asymptote: none

(c)

x	-1	0	1	1.5	2	2.5	3
$f(x)$	-1	Undef.	1	1.5	Undef.	2.5	3
$g(x)$	-1	0	1	1.5	2	2.5	3

(d)

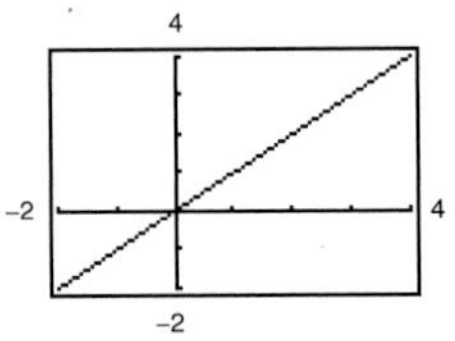

(e) Because there are only a finite number of pixels, the utility may not attempt to evaluate the function where it does not exist.

48. (a) Domain of f: all real numbers $x \neq 3, 4$

Domain of g: all real numbers $x \neq 4$

(b) Vertical asymptote: $x = 4$

(c)

x	0	1	2	3	4	5	6
$f(x)$	$-\frac{1}{2}$	$-\frac{2}{3}$	-1	Undef.	Undef.	2	1
$g(x)$	$-\frac{1}{2}$	$-\frac{2}{3}$	-1	-2	Undef.	2	1

(d)

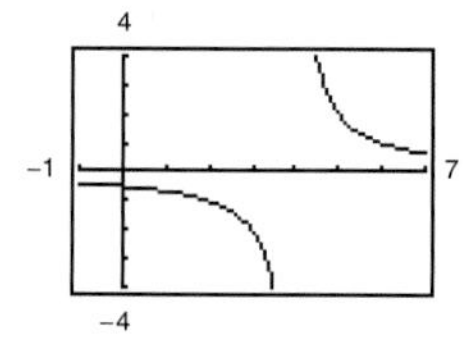

(e) Because there are only a finite number of pixels, the utility may not attempt to evaluate the function where it does not exist.

50.

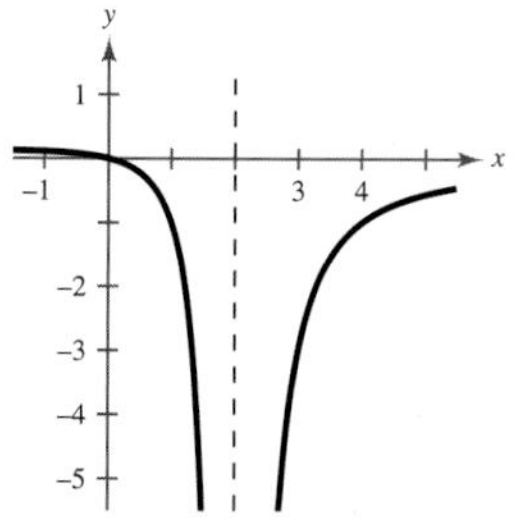

Domain: all real numbers $x \neq 2$

Vertical asymptote: $x = 2$

Horizontal asymptote: $y = 0$

52.

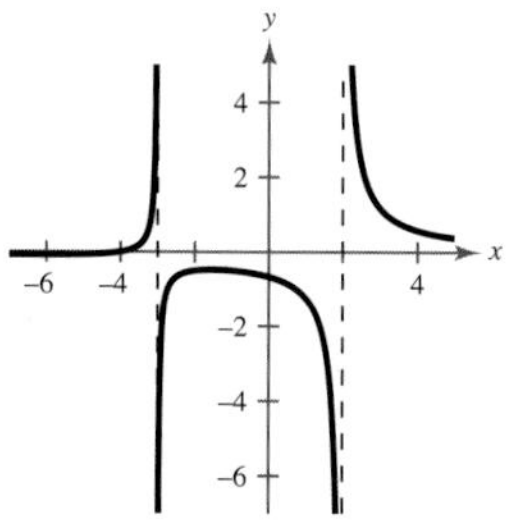

Domain: all real numbers $x \neq -3, 2$

Vertical asymptotes: $x = -3, x = 2$

Horizontal asymptote: $y = 0$

54.

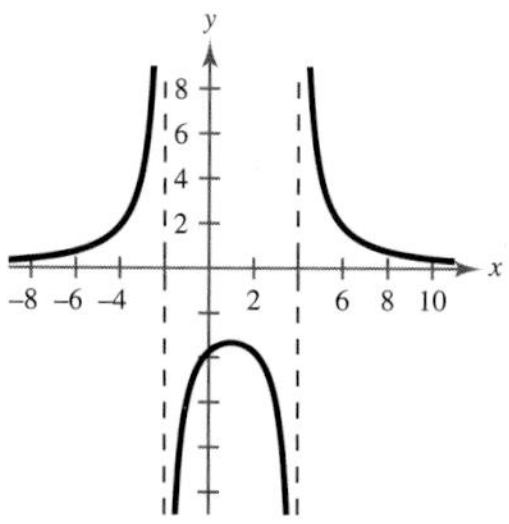

Domain: all real numbers $x \neq -2, 4$

Vertical asymptotes: $x = -2, x = 4$

Horizontal asymptote: $y = 0$

56. (a) Intercepts: $(-1, 0), (1, 0)$

(b) Vertical asymptote: $x = 0$
Slant asymptote: $y = -x$

(c) Origin

(d) and (e)

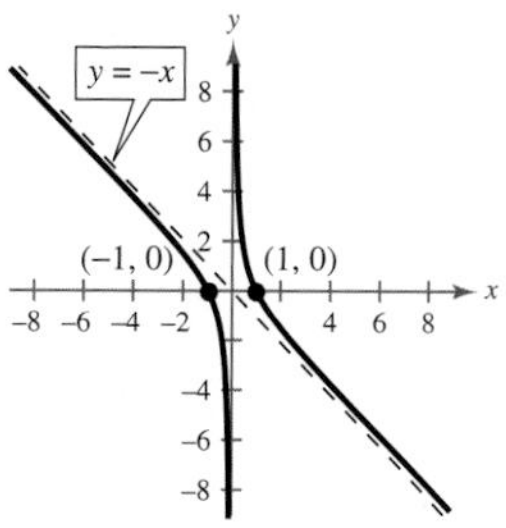

58. (a) Intercept: $(0, 0)$

(b) Vertical asymptote: $x = 1$
Slant asymptote: $y = x + 1$

(c) No symmetry

(d) and (e)

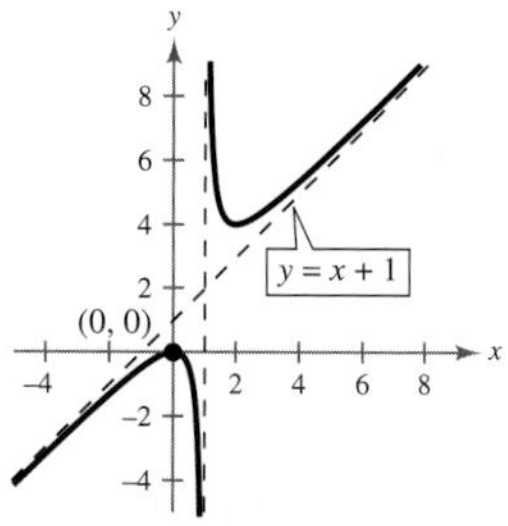

60. (a) Intercept: $(0, 0)$

(b) Vertical asymptotes: $x = \pm 2$
Slant asymptote: $y = \frac{1}{2}x$

(c) Origin

(d) and (e)

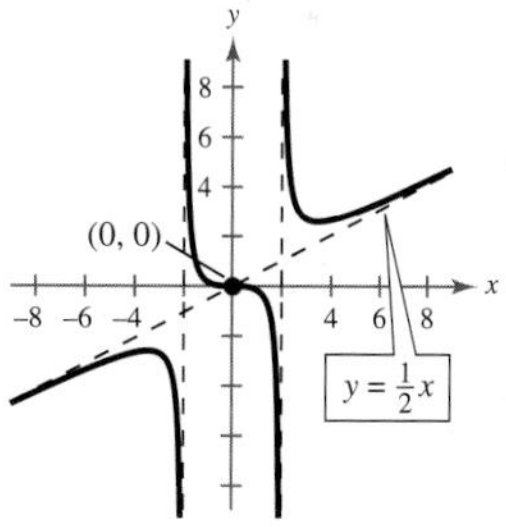

62. (a) Intercept: $\left(0, -\frac{5}{2}\right)$

(b) Vertical asymptote: $x = 2$
Slant asymptote: $y = 2x - 1$

(c) No symmetry

(d) and (e)

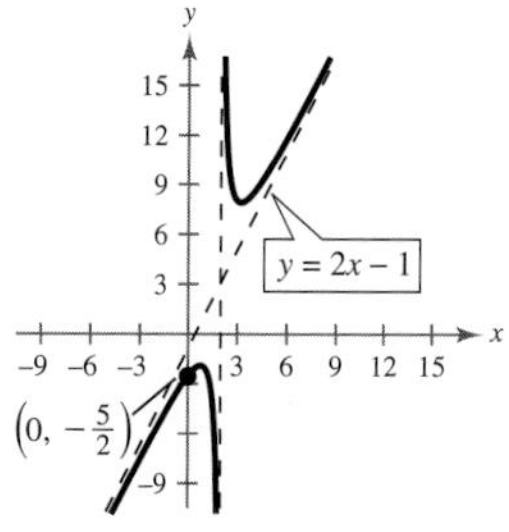

64.

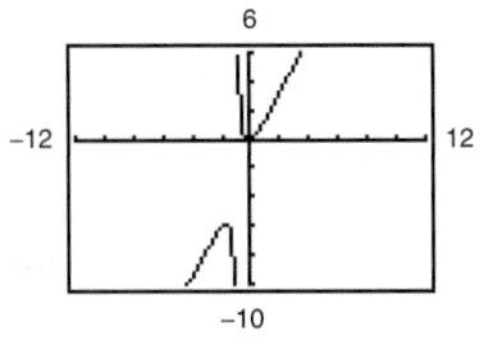

Domain: all real numbers $x \neq -1$

Vertical asymptote: $x = -1$

Slant asymptote: $y = 2x - 1$

$y = 2x - 1$

66.

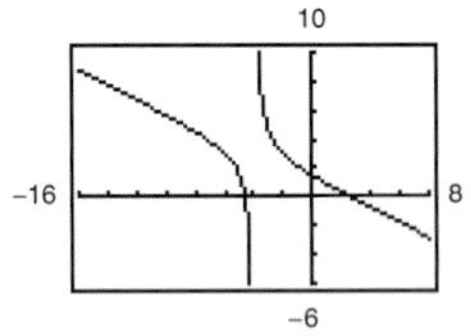

Domain: all real numbers $x \neq -4$

Vertical asymptote: $x = -4$

Slant asymptote: $y = -\frac{1}{2}x + 1$

$y = -\frac{1}{2}x + 1$

68. (a) and (b) $(0, 0)$ **70.** (a) and (b) $(1, 0), (2, 0)$

72. (a) \$4411.76 (b) \$25,000 (c) \$225,000

(d) No. The function is undefined at $p = 100$.

74. (a) Answers will vary. (b) $[0, 950]$

(c)

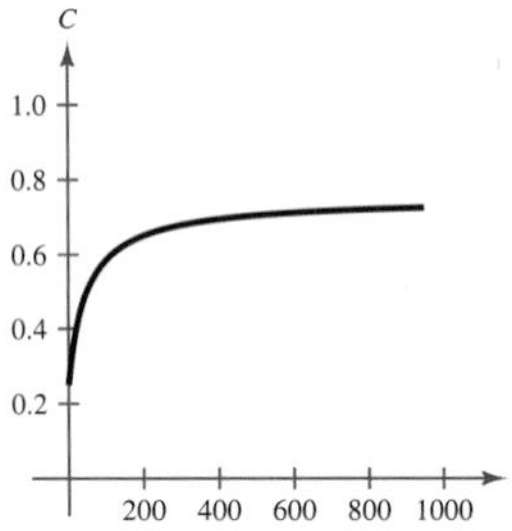

Increases more slowly; 0.75

76. 12.8×8.5 inches

78. (a) Answers will vary.

(b) Vertical asymptote: $x = 25$

Horizontal asymptote: $y = 25$

80. False. The graph of $f(x) = \dfrac{x}{x^2 + 1}$ crosses $y = 0$, which is a horizontal asymptote.

82. $f(x) = \dfrac{1}{1 + x^2}$ **84.** $f(x) = \dfrac{-6x^2 + 15x + 1}{x(2x - 5)}$

86. An asymptote of a graph is a line to which the graph becomes arbitrarily close as $|x|$ or $|y|$ increases without bound.

88.

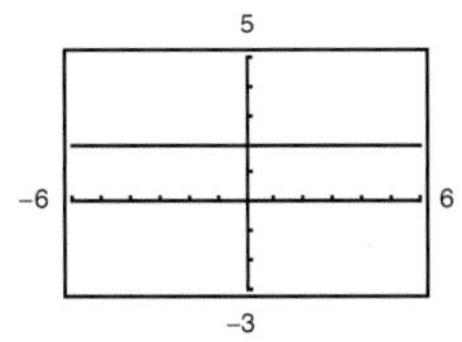

The fraction is not reduced.

90. $(x - 7)(x - 8)$ **92.** $(x - 5)(x^2 + 4)$

94. $x \geq \frac{10}{3}$

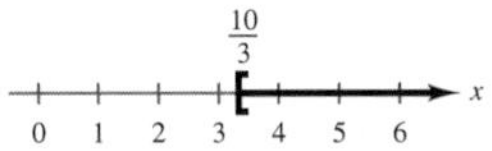

96. $-3 < x < 7$

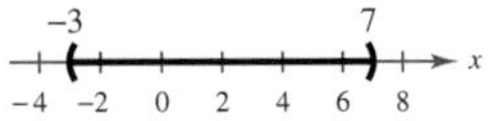

Section 2.7 *(page 277)*

2. c **4.** a **6.** $\dfrac{A}{x + 3} + \dfrac{B}{x + 1}$ **8.** $\dfrac{A}{x} + \dfrac{B}{x^2} + \dfrac{C}{4x + 11}$

10. $\dfrac{A}{x + 2} + \dfrac{B}{(x + 2)^2} + \dfrac{C}{(x + 2)^3} + \dfrac{D}{(x + 2)^4}$

12. $\dfrac{A}{2x} + \dfrac{B}{x^2 + 4}$ **14.** $\dfrac{A}{x} + \dfrac{B}{x^2} + \dfrac{C}{3x - 1} + \dfrac{D}{(3x - 1)^2}$

16. $\dfrac{1}{6}\left(\dfrac{1}{2x - 3} - \dfrac{1}{2x + 3}\right)$ **18.** $\dfrac{1}{x - 3} - \dfrac{1}{x}$

20. $\dfrac{1}{x - 2} - \dfrac{1}{x + 3}$ **22.** $\dfrac{1}{x + 3}, x \neq -1$

24. $\dfrac{1}{2}\left(\dfrac{3}{x - 4} - \dfrac{1}{x}\right)$ **26.** $\dfrac{2}{x - 1} - \dfrac{1}{(x - 1)^2}$

28. $\dfrac{2}{x} + \dfrac{1}{x^2} - \dfrac{2}{x - 1} + \dfrac{7}{(x - 1)^2}$

30. $\dfrac{1}{3}\left(\dfrac{1}{x - 1} - \dfrac{x - 1}{x^2 + x + 1}\right)$

32. $\dfrac{1}{5}\left(\dfrac{9}{x - 3} + \dfrac{1}{x + 2} - \dfrac{10}{x - 2}\right)$

34. $\dfrac{2}{x^2 + 4} + \dfrac{x}{(x^2 + 4)^2}$ **36.** $\dfrac{1}{x} - \dfrac{x - 1}{x^2 + 1}$

38. $\dfrac{2}{x + 1} - \dfrac{x - 1}{x^2 - 2x + 3}$ **40.** $1 - \dfrac{5x + 6}{x^2 + x + 6}$

42. $x - 1 + \dfrac{1}{5}\left(\dfrac{27}{x + 4} + \dfrac{3}{x - 1}\right)$

44. $2x + 3 + \dfrac{6}{2x - 1} + \dfrac{4}{(2x - 1)^2} + \dfrac{1}{(2x - 1)^3}$

46. $\dfrac{2}{x} + \dfrac{4}{x + 1} - \dfrac{3}{x - 1}$ **48.** $\dfrac{1}{2}\left[-\dfrac{1}{x} + \dfrac{5}{x + 1} - \dfrac{3}{(x + 1)^2}\right]$

50. $\dfrac{1}{2}\left[\dfrac{1}{x + 2} - \dfrac{1}{(x + 2)^2} + \dfrac{1}{x - 2} + \dfrac{1}{(x - 2)^2}\right]$

52. $x - 1 + \dfrac{1}{x + 2} + \dfrac{1}{x - 1}$

54. $\frac{2}{x} + \frac{4}{x^2 + 1}$

$y = \frac{2(x + 1)^2}{x(x^2 + 1)}$ $\qquad y = \frac{2}{x}, y = \frac{4}{x^2 + 1}$

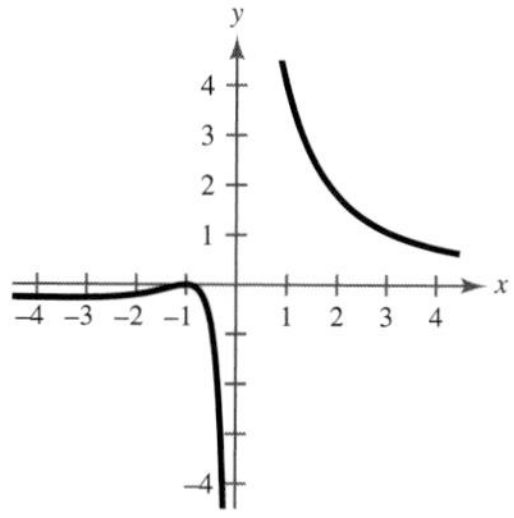
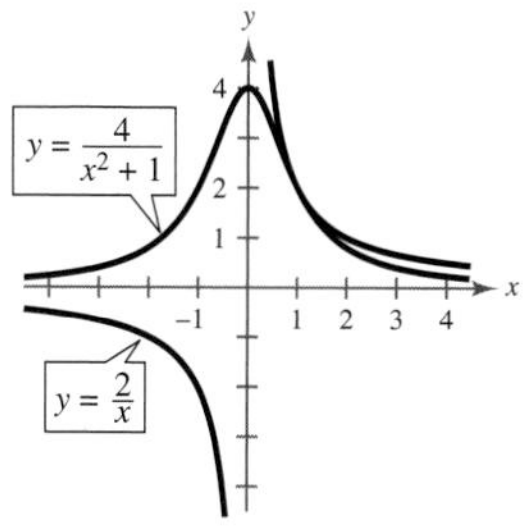

The vertical asymptotes are the same.

56. $\frac{3}{x^2} + \frac{5}{x^2 - 10x + 26}$

$y = \frac{2(4x^2 - 15x + 39)}{x^2(x^2 - 10x + 26)}$ $\qquad y = \frac{3}{x^2}, y = \frac{5}{x^2 - 10x + 26}$

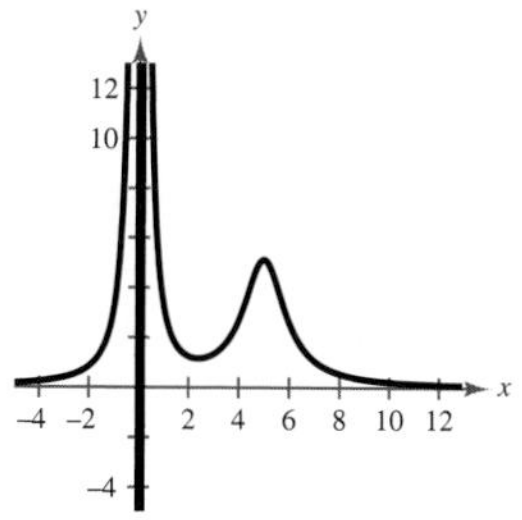
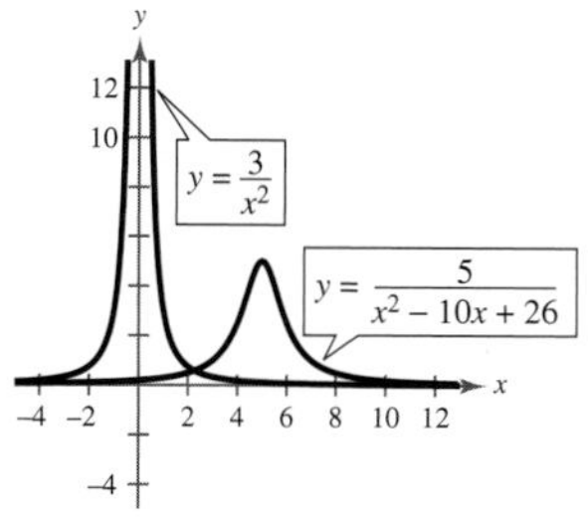

The vertical asymptotes are the same.

58. False. The partial fraction decomposition is

$\frac{A}{x + 10} + \frac{B}{x - 10} + \frac{C}{(x - 10)^2}.$

60. $\frac{1}{2a}\left(\frac{1}{a + x} + \frac{1}{a - x}\right)$ **62.** $\frac{1}{a}\left(\frac{1}{y} + \frac{1}{a - y}\right)$

64.

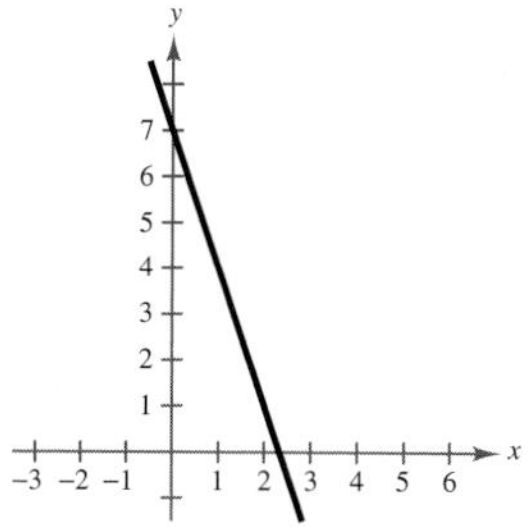

66.

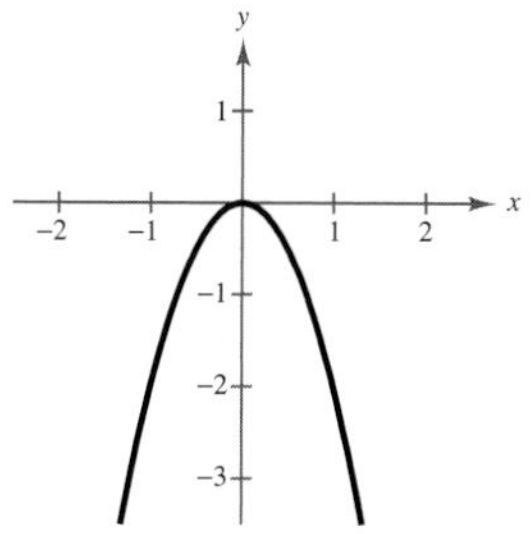

68.

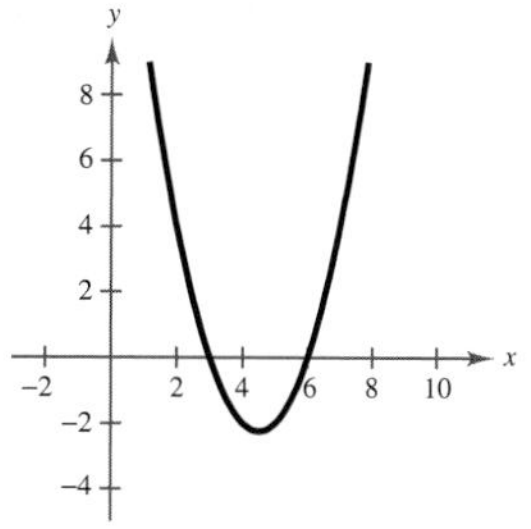

70.

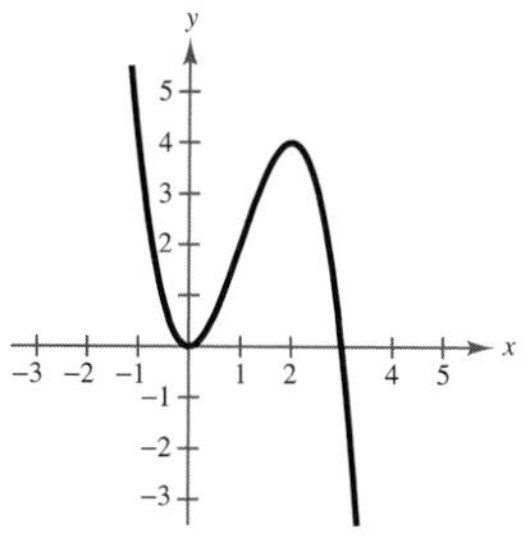

72.

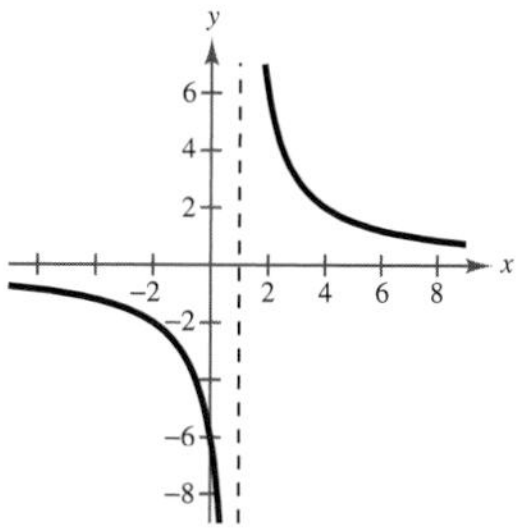

74.

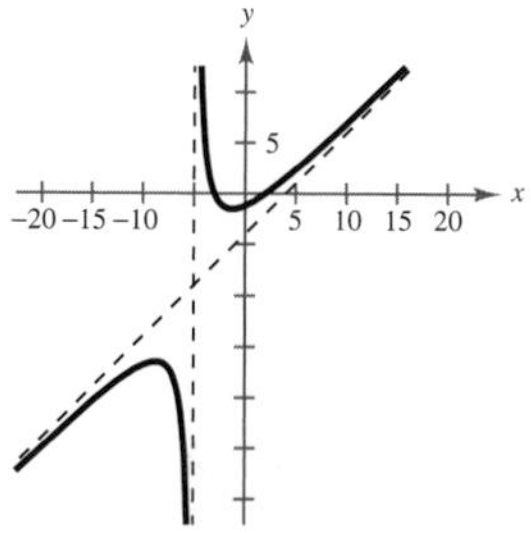

76.

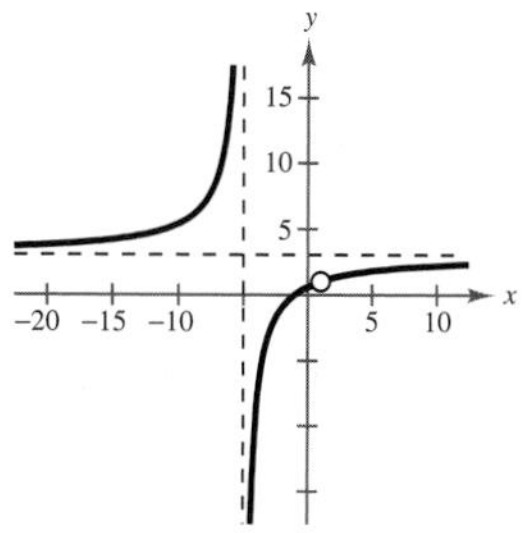

Review Exercises *(page 280)*

2. (a)

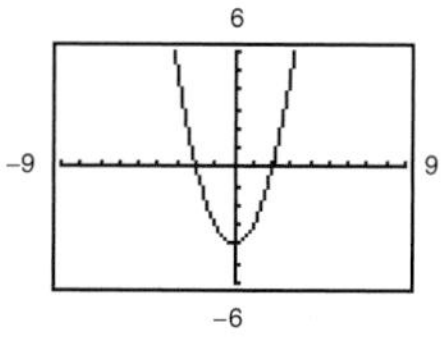

Vertical translation

(b)

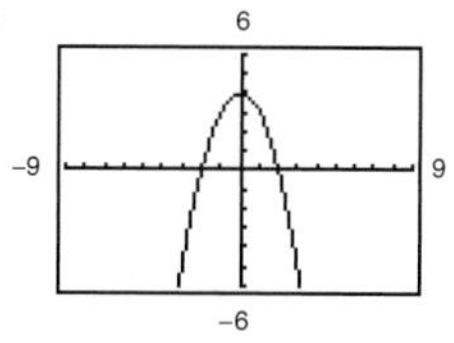

Reflection in the x-axis and vertical translation

(c)

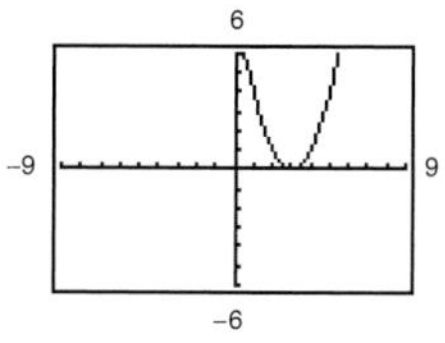

Horizontal translation

(d)

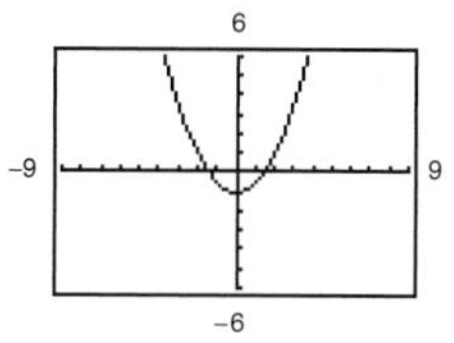

Vertical shrink and vertical translation

4. $f(x) = \frac{1}{4}(x - 2)^2 + 2$ **6.** $f(x) = \frac{1}{3}(x - 2)^2 + 3$

8. $f(x) = -(x-3)^2 + 9$

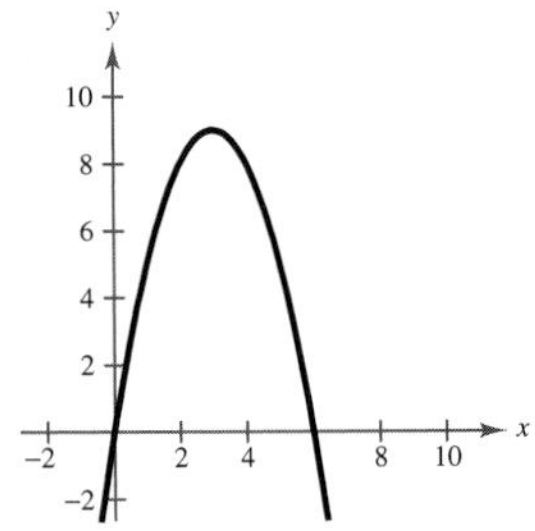

10. $h(x) = -(x-2)^2 + 7$

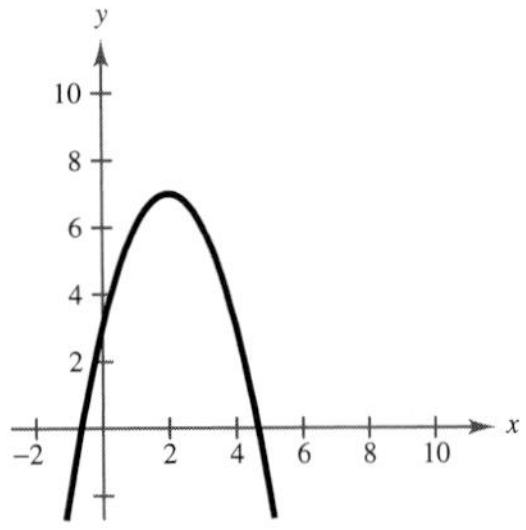

12. $f(x) = (x-4)^2 - 4$

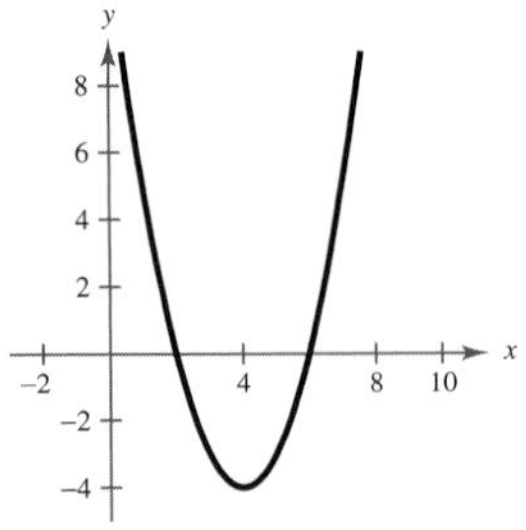

14. $f(x) = (x-3)^2 - 8$

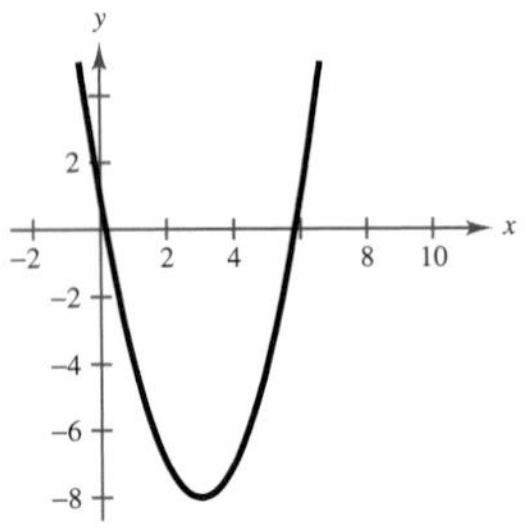

16. $f(x) = 4\left(x + \frac{1}{2}\right)^2 + 4$

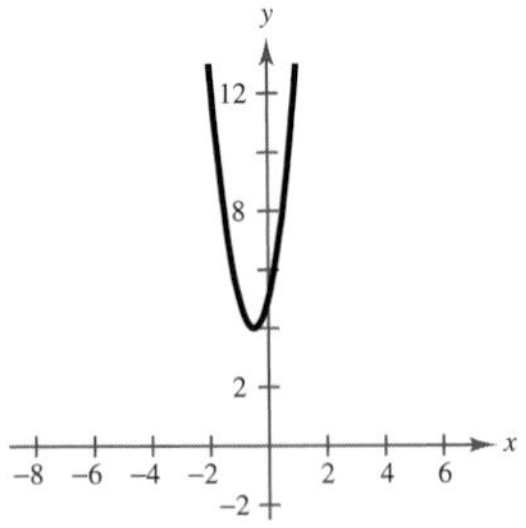

18. $f(x) = 3(x-2)^2 - 1$

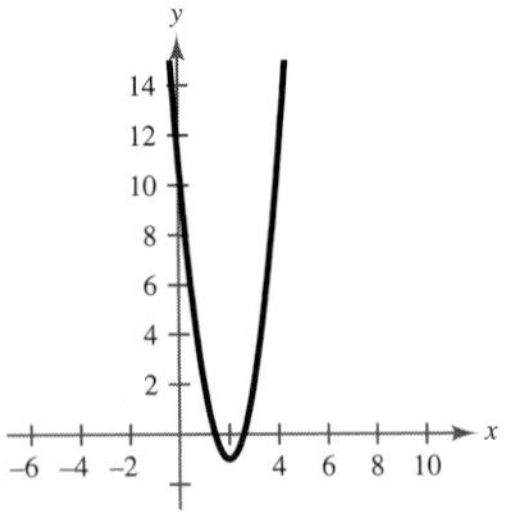

20. \$1020 **22.** 25

24.

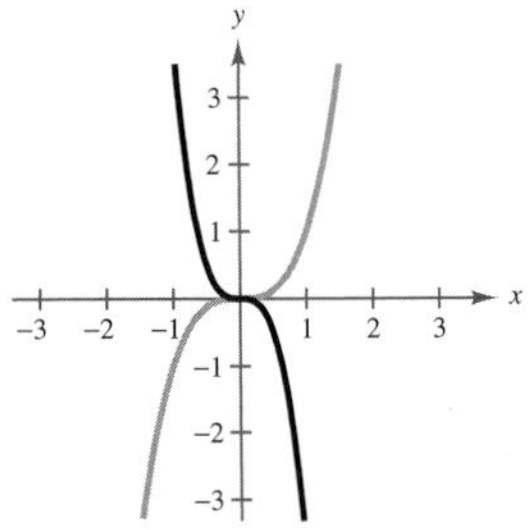

26.

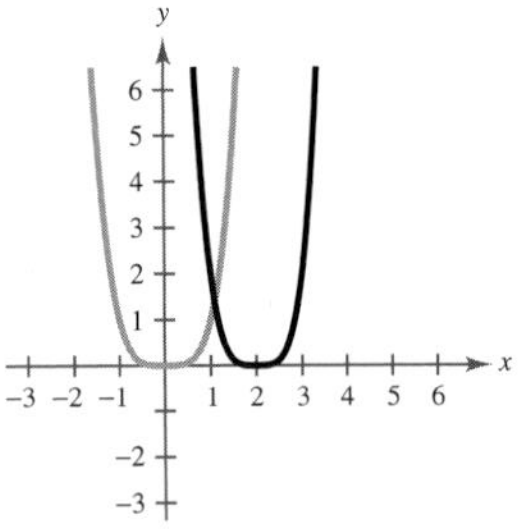

28.

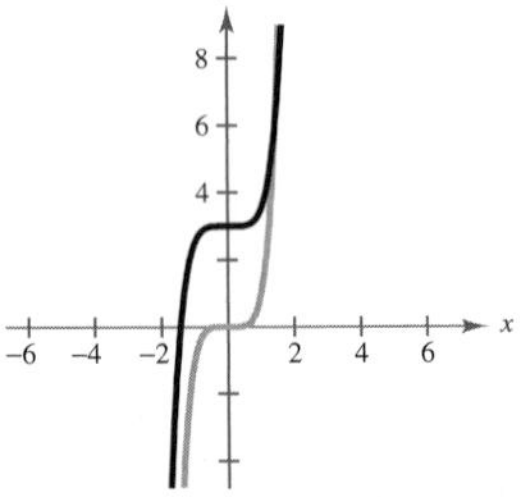

30. Falls to the left.
Rises to the right.

32. Rises to the left.
Falls to the right.

34. $0, -3$

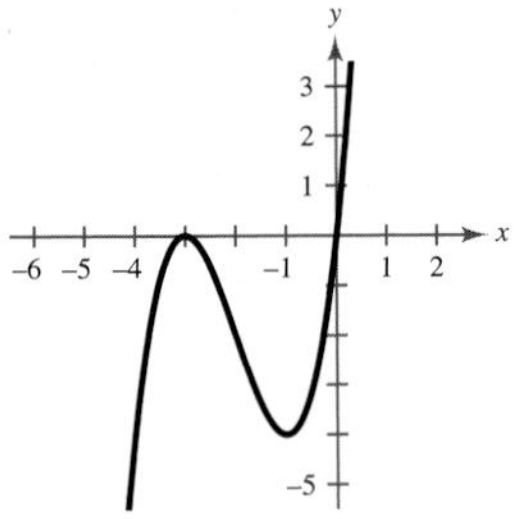

36. $0, 8$

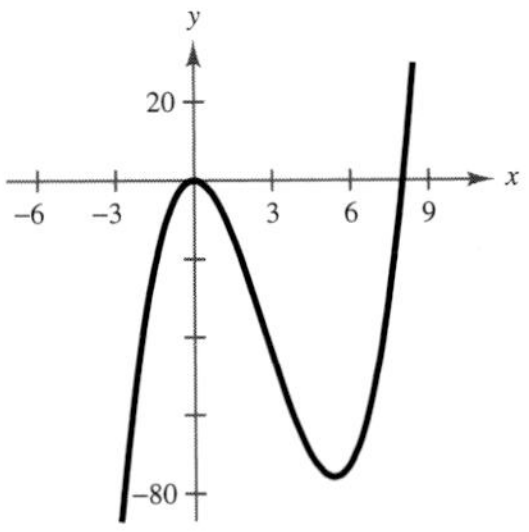

38. $0, 2, -1$

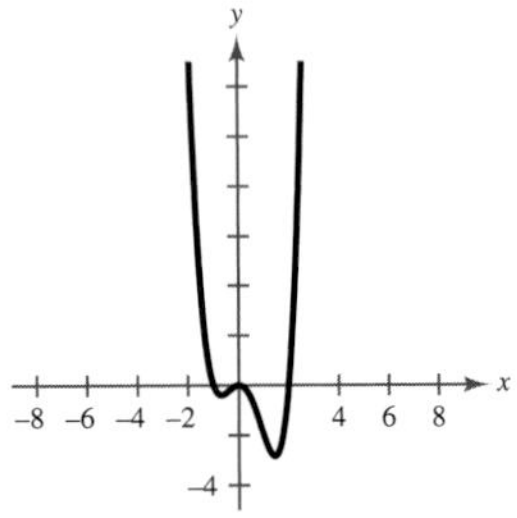

40. $(-5, -4)$ **42.** $(-2, -1), (-1, 0)$

44. $\dfrac{4}{3} + \dfrac{29}{3(3x-2)}$ **46.** $3x^2 + 3 + \dfrac{3}{x^2 - 1}$

48. $3x^2 + 5x + 8 + \dfrac{10}{2x^2 - 1}$

50. $0.1x^2 + 0.8x + 4 + \dfrac{19.5}{x-5}$ **52.** $3x^2 + 11x - 4$

54. (a) Yes (b) No (c) Yes (d) No

56. (a) -3276 (b) 0

58. (a) Answers will vary. (b) $(2x+5), (x-3)$
(c) $f(x) = (2x+5)(x-3)(x+6)$ (d) $-\frac{5}{2}, 3, -6$
(e)

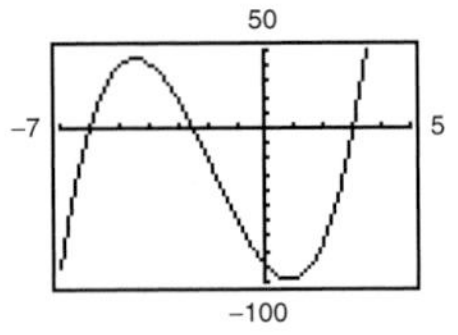

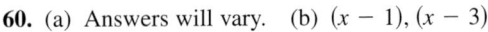

60. (a) Answers will vary. (b) $(x-1), (x-3)$
(c) $f(x) = (x-1)(x-3)(x-2)(x-5)$ (d) 1, 2, 3, 5
(e)

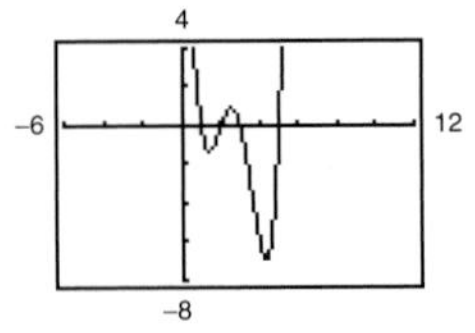

62. $6 + 2i$ **64.** $-1 + 3i$ **66.** $3 + 7i$

68. $40 + 65i$ **70.** $-4 - 46i$ **72.** $\dfrac{23 + 10i}{17}$

74. $\dfrac{21 - i}{13}$ **76.** $\pm\dfrac{\sqrt{3}}{3}i$ **78.** $1 \pm 3i$ **80.** 0, 2, 2

82. 8, 1 **84.** $-4, 6, \pm 2i$

86. $\pm 1, \pm 3, \pm 5, \pm 15, \pm\frac{1}{2}, \pm\frac{3}{2}, \pm\frac{5}{2}, \pm\frac{15}{2}, \pm\frac{1}{4}, \pm\frac{3}{4}, \pm\frac{5}{4}, \pm\frac{15}{4}$

88. $-1, -3, 6$ **90.** 1, 8 **92.** $-4, 3$

94. $3x^4 - 14x^3 + 17x^2 - 42x + 24$

96. (a) $(x + 1)(x - 1)(x^2 - 2x + 5)$
(b) $(x + 1)(x - 1)(x^2 - 2x + 5)$
(c) $(x + 1)(x - 1)(x - 1 - 2i)(x - 1 + 2i)$

98. $12, \pm i$

100. Two or no positive real zeros, one negative real zero

102. Domain: all real numbers $x \neq -12$

104. Domain: all real numbers $x \neq 6, 4$

106. Vertical asymptote: $x = -3$
Horizontal asymptote: $y = 0$

108. Vertical asymptotes: $x = -2, x = 2$
Horizontal asymptote: $y = 1$

110. No intercepts
y-axis
Vertical asymptote: $x = 0$
Horizontal asymptote: $y = 0$

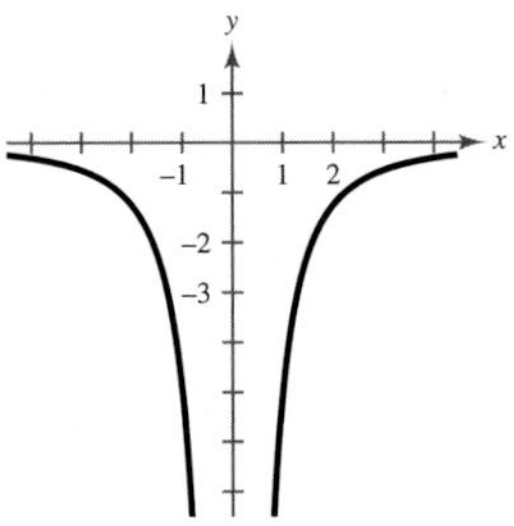

112. Intercepts: $(-2, 0), (0, 2)$
No symmetry
Vertical asymptote: $x = 1$
Horizontal asymptote: $y = -1$

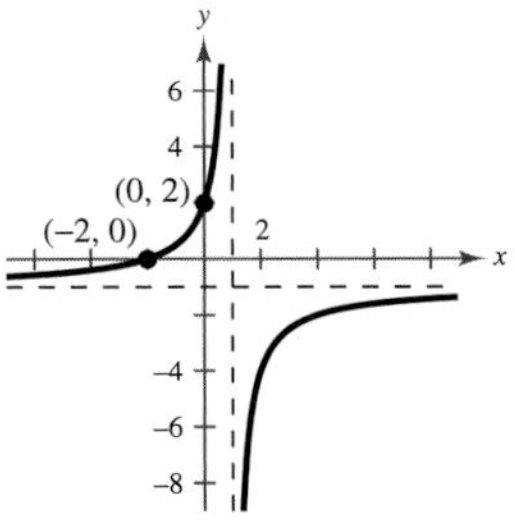

114. Intercept: (0, 0)
y-axis
Horizontal asymptote: $y = 1$

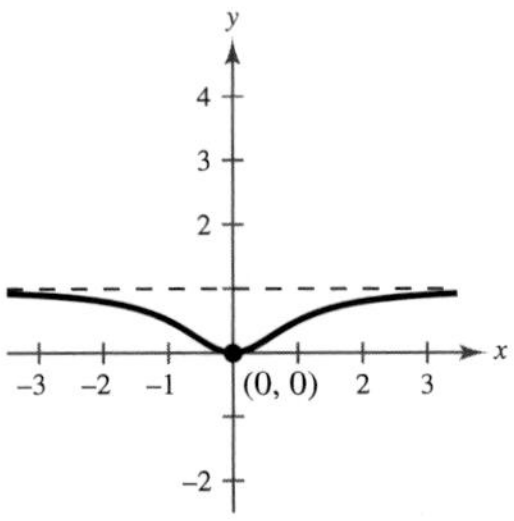

116. Intercept: (0, 0)
Origin
Horizontal asymptote: $y = 0$

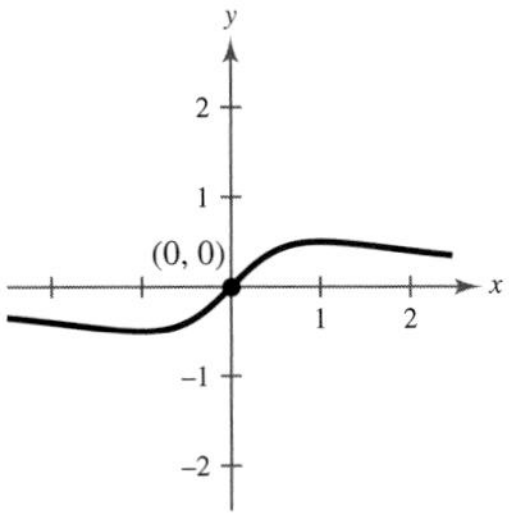

118. Intercept: (0, 0)
y-axis
Horizontal asymptote: $y = -6$

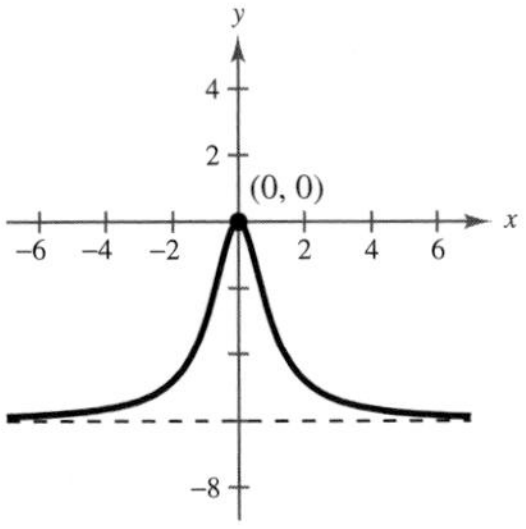

120. Intercept: (0, 0)
Origin
Vertical asymptotes: $x = \pm 1$
Horizontal asymptote: $y = 0$

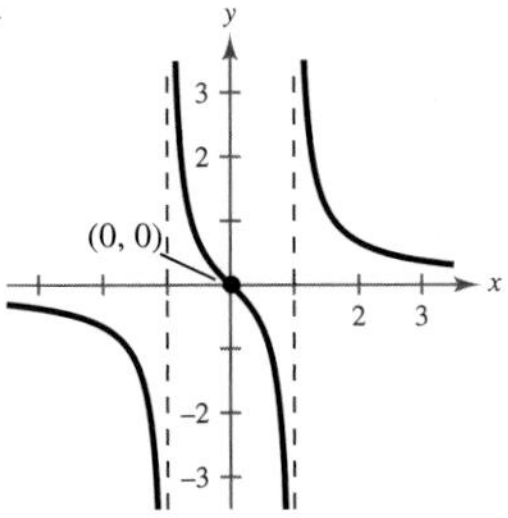

122. Slant asymptote: $y = 2x$

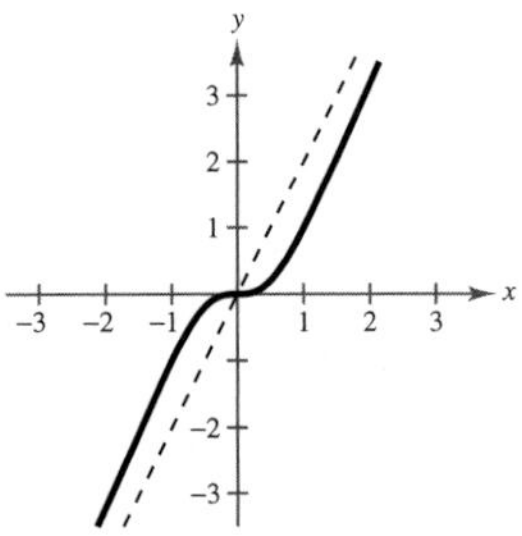

124. Slant asymptote: $y = x + 1$

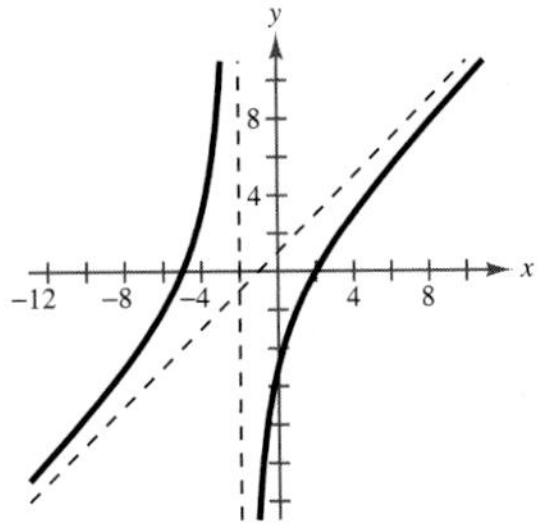

126. As x increases, the cost approaches the horizontal asymptote, $\overline{C} = 0.5$.

128. $x \approx 0.346$ inch **130.** $\dfrac{A}{x} + \dfrac{B}{x + 20}$

132. $\dfrac{A}{x} + \dfrac{B}{x^2} + \dfrac{C}{x - 5}$ **134.** $\dfrac{3}{x + 2} - \dfrac{4}{x + 4}$

136. $1 - \dfrac{25}{8(x + 5)} + \dfrac{9}{8(x - 3)}$

138. $\dfrac{1}{2}\left(\dfrac{3}{x - 1} - \dfrac{x - 3}{x^2 + 1}\right)$ **140.** $\dfrac{3x}{x^2 + 1} + \dfrac{x}{(x^2 + 1)^2}$

142. False. A fourth-degree polynomial can have at most four zeros, and complex zeros occur in conjugate pairs.

144. Answers will vary. **146.** 4

Chapter 3

Section 3.1 *(page 296)*

2. 1767.767 **4.** 472,369.379 **6.** 16.380

8. 1.649 **10.** 24.533 **12.** c **14.** b

16. Shift the graph of f 1 unit upward.

18. Reflect the graph of f in the y-axis and shift f 3 units to the right.

20. Reflect the graph of f in the x-axis and y-axis and shift f 6 units to the right.

22. Reflect the graph of f in the x-axis and y-axis and shift f 8 units upward.

24.

x	-2	-1	0	1	2
$f(x)$	0.028	0.167	1	6	36

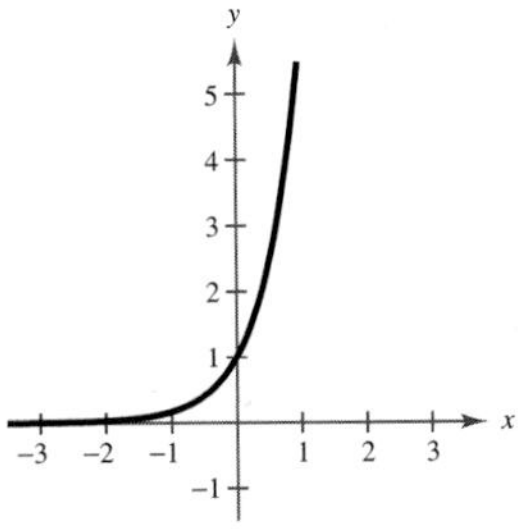

26.

x	-2	-1	0	1	2
$f(x)$	36	6	1	0.167	0.028

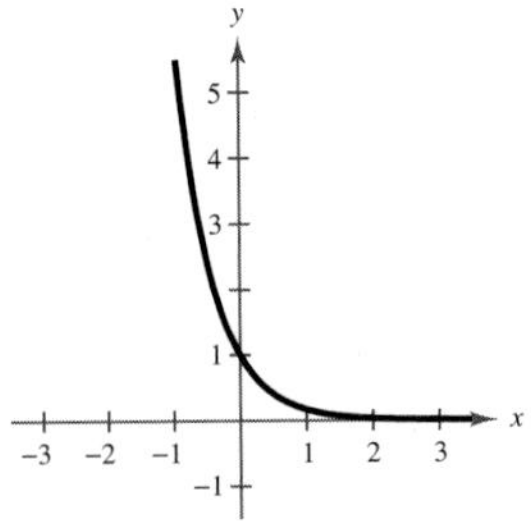

28.

x	-4	-3	-2	-1	0
$f(x)$	0.111	0.333	1	3	9

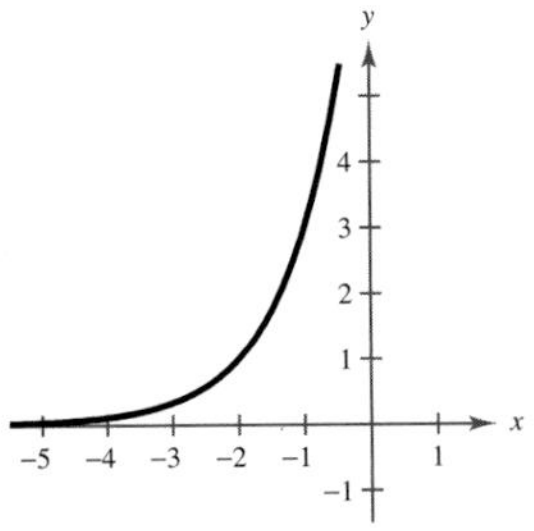

30.

x	-2	-1	0	1	2
$f(x)$	7.389	2.718	1	0.368	0.135

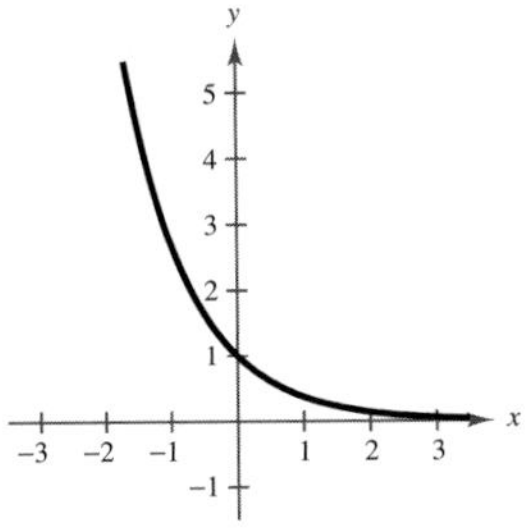

32.

x	-2	-1	0	1	2
$f(x)$	5.437	3.297	2	1.213	0.736

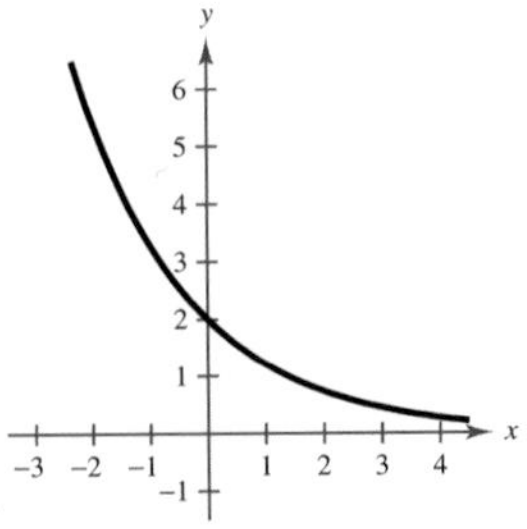

34.

x	0	2	4	5	6
$f(x)$	2.007	2.050	2.368	3	4.718

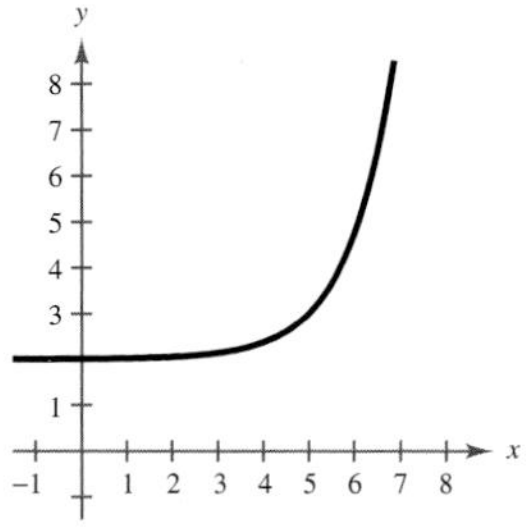

36.

x	-1	0	1	3	4
$f(x)$	-3.004	-3.016	-3.063	-4	-7

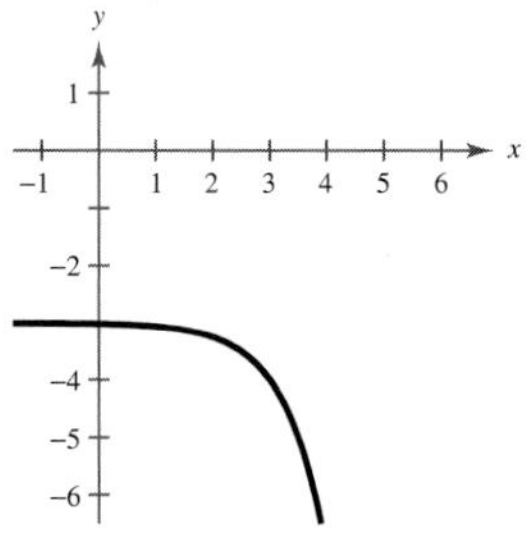

38.

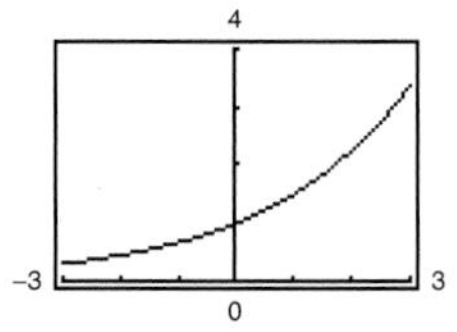

40.

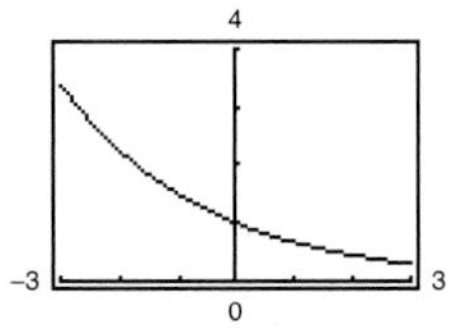

42.

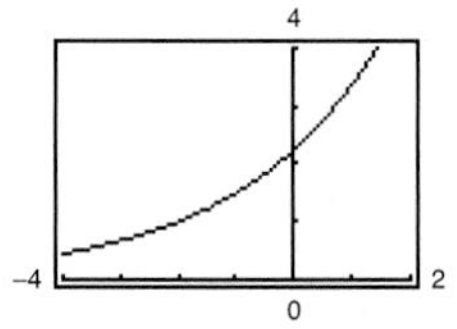

44.

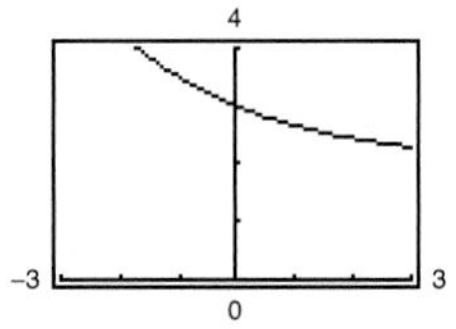

46.

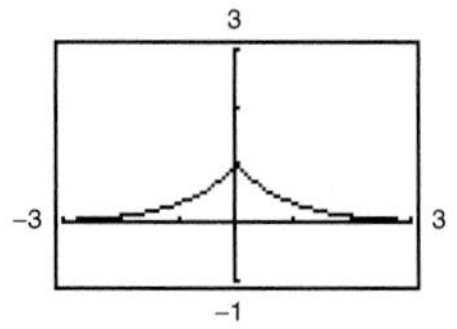

48.

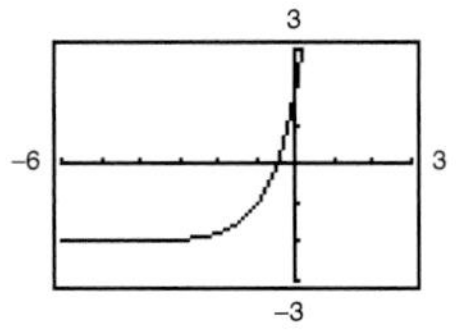

50.

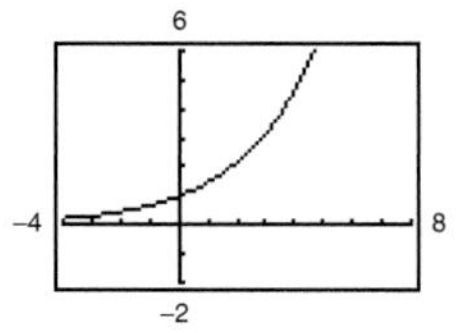

52.

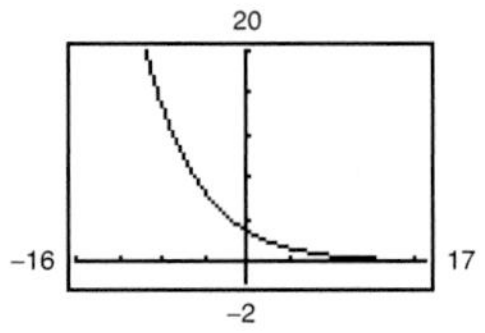

54.

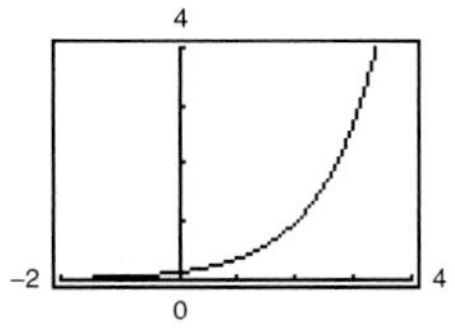

56.

n	1	2	4	12
A	\$1790.85	\$1806.11	\$1814.02	\$1819.40

n	365	Continuous
A	\$1822.03	\$1822.12

58.

n	1	2	4
A	\$10,285.72	\$10,640.89	\$10,828.46

n	12	365	Continuous
A	\$10,957.45	\$11,021.00	\$11,023.18

60.

t	1	10	20
A	\$12,742.04	\$21,865.43	\$39,841.40

t	30	40	50
A	\$72,595.77	\$132,278.12	\$241,026.44

62.

t	1	10	20
A	\$12,934.61	\$25,404.00	\$53,780.27

t	30	40	50
A	\$113,852.83	\$241,026.44	\$510,252.98

64. \$212,605.41

66.

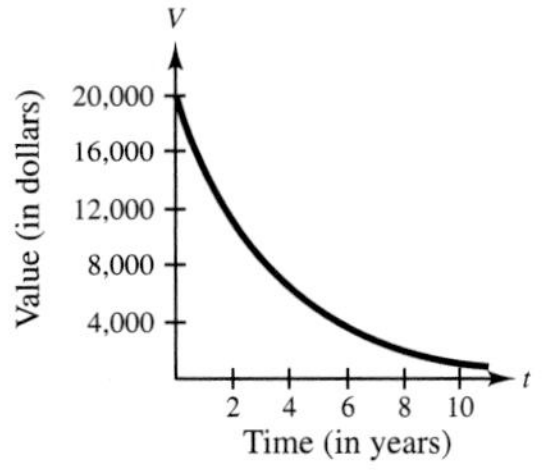

\$11,250

68. (a)

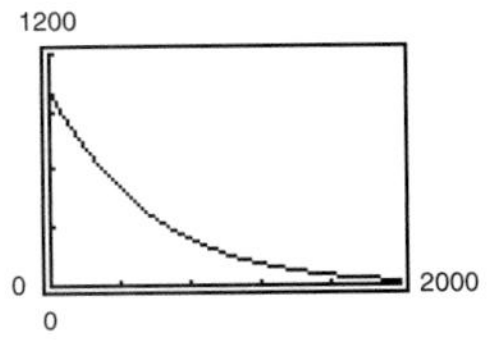

(b) \$421.12

(c) \$350

70. (a) 3351 (b) 4492

72. (a) 10 units (b) 7.85 units

(c)

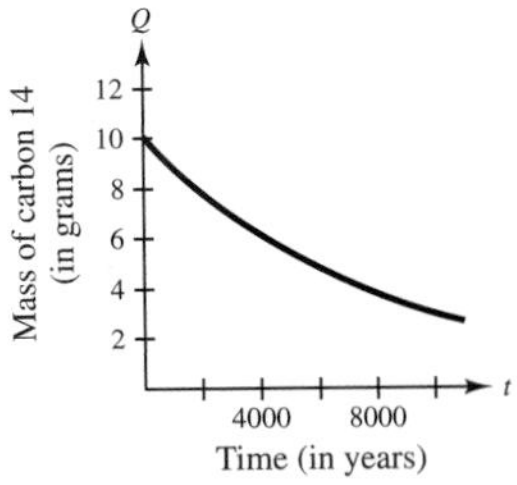

74. (a)

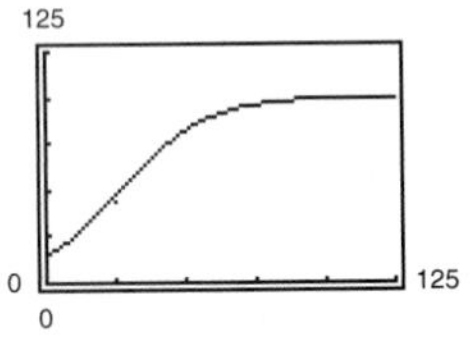

(b)

x	0	25	50	75	100
y	15	47	82	96	99

(c) 64.7% (d) 36.9

76. False. e is an irrational number.

78. $g(x) = h(x)$ **80.** None are equal.

82. (a) $x > 0$ (b) $x < 0$

84. (a)

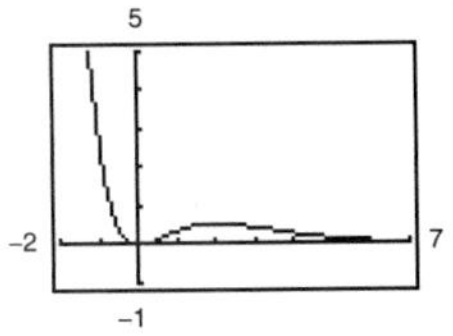

Decreasing: $(-\infty, 0), (2, \infty)$

Increasing: $(0, 2)$

Relative maximum: $(2, 4e^{-2})$

Relative minimum: $(0, 0)$

(b)

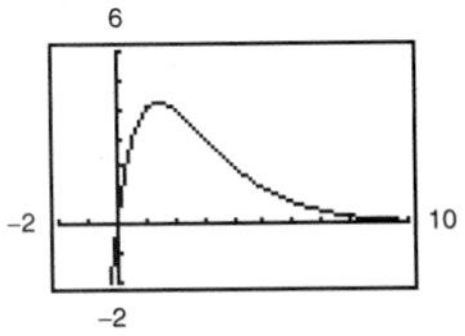

Decreasing: $(1.44, \infty)$

Increasing: $(-\infty, 1.44)$

Relative maximum: $(1.44, 4.25)$

86. The exponential function increases at a faster rate.

88.

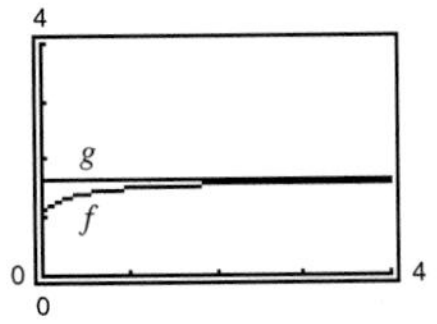

As $x \to \infty, f(x) \to g(x)$.

90. c, d

92.

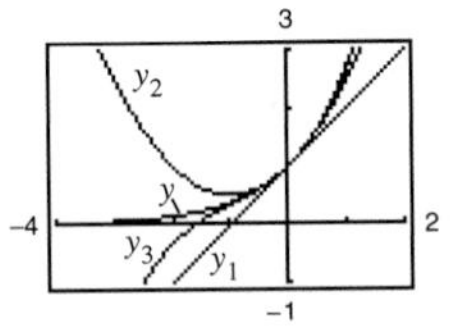

94. (a) $f(u + v) = a^{u+v} = a^u \cdot a^v = f(u) \cdot f(v)$

(b) $f(2x) = a^{2x} = (a^x)^2 = [f(x)]^2$

96. $y = \frac{1}{3}(4 - x^2)$

98. $y = \begin{cases} x - 2, & x \ge 2 \\ -(x - 2), & x < 2 \end{cases}$

100.

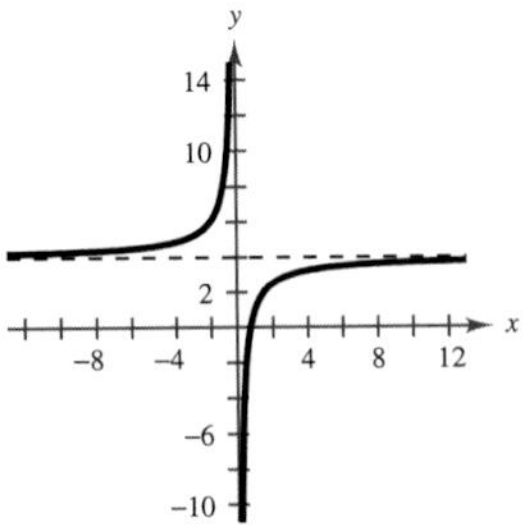

102.

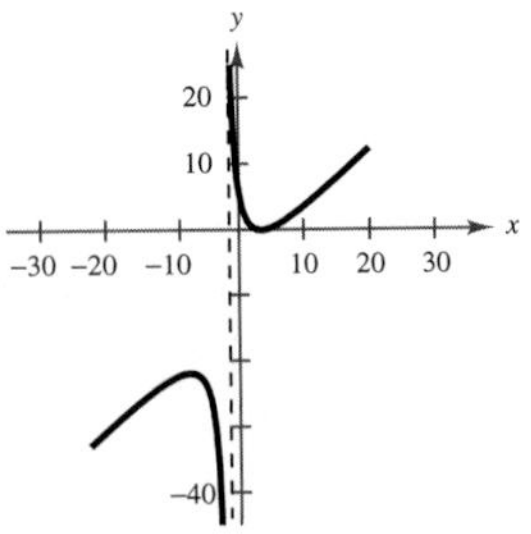

Section 3.2 *(page 307)*

2. $3^4 = 81$ **4.** $10^{-3} = \frac{1}{1000}$ **6.** $16^{3/4} = 8$

8. $e^{1.386\ldots} = 4$ **10.** $\log_8 64 = 2$ **12.** $\log_9 27 = \frac{3}{2}$

14. $\log_{10} 0.001 = -3$ **16.** $\ln 4 = x$ **18.** $\log_u w = v$

20. -3 **22.** $\frac{2}{3}$ **24.** 3 **26.** 1 **28.** $\frac{5}{2}$

30. -2 **32.** -3 **34.** 2.161 **36.** 1.097

38. 1.869 **40.** -0.575 **42.** -1.444 **44.** -0.693

46. f **48.** e **50.** a

52. Domain: $(0, \infty)$

Intercept: $(1, 0)$

Vertical asymptote: $x = 0$

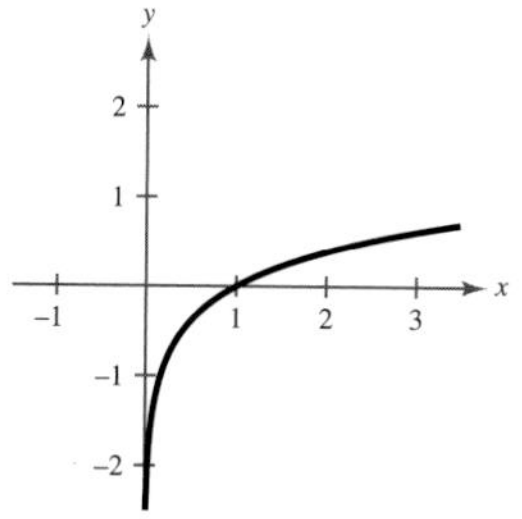

54. Domain: $(3, \infty)$

Intercept: $(4, 0)$

Vertical asymptote: $x = 3$

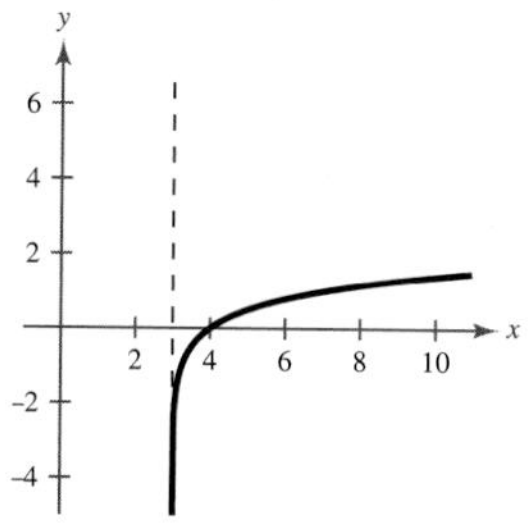

56. Domain: $(1, \infty)$

Intercept: $\left(\frac{626}{625}, 0\right)$

Vertical asymptote: $x = 1$

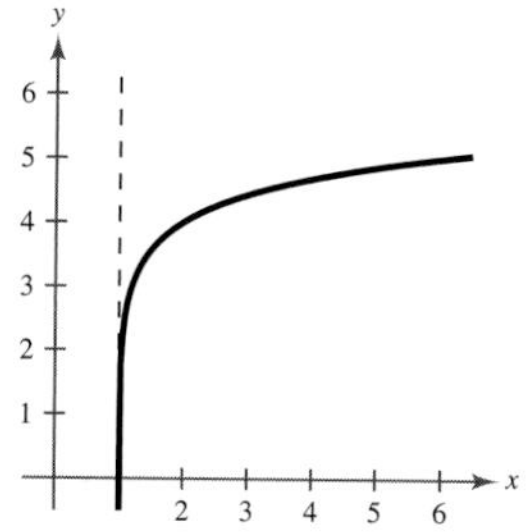

58. Domain: $(-\infty, 0)$

Intercept: $(-1, 0)$

Vertical asymptote: $x = 0$

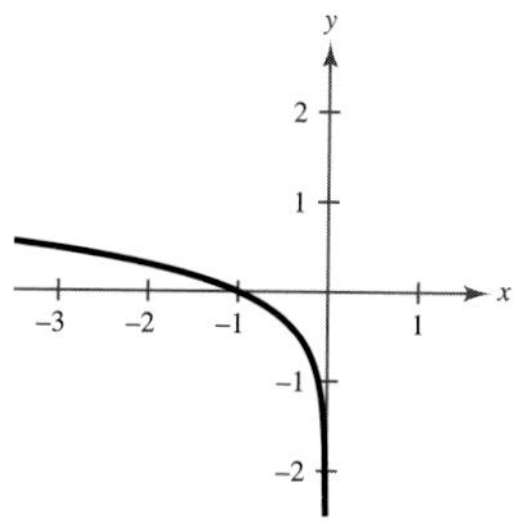

60. Domain: $(-1, \infty)$

Intercept: $(0, 0)$

Vertical asymptote: $x = -1$

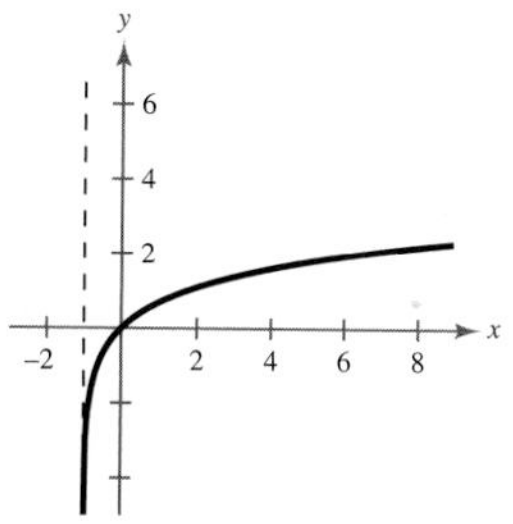

62. Domain: $(-\infty, 3)$

Intercept: $(2, 0)$

Vertical asymptote: $x = 3$

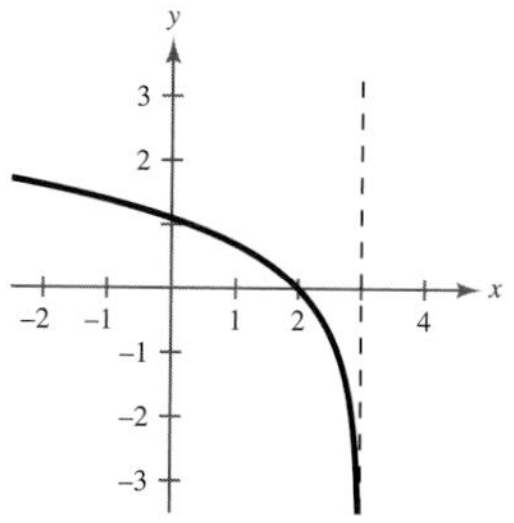

64.

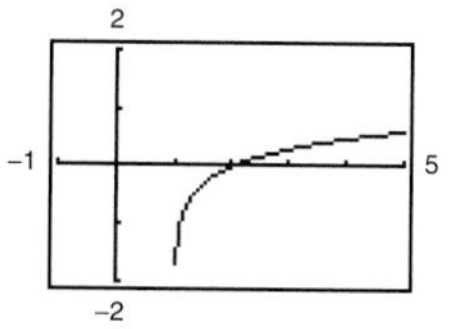

66.

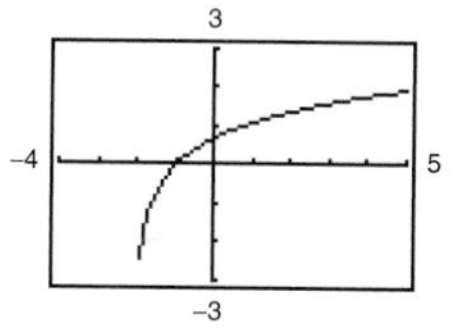

68.

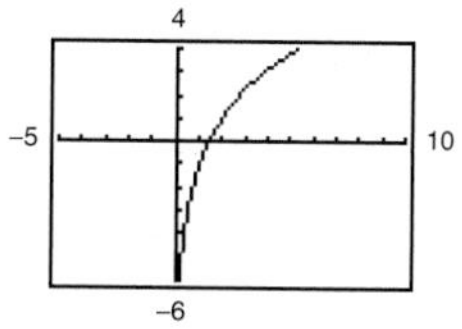

70. 23.68 years

72. (a)

K	1	2	4	6	8	10	12
t	0	7.3	14.6	18.9	21.9	24.2	26.2

The number of years required to multiply the original investment by K increases with K. However, the larger the value of K, the fewer the years required to increase the value of the investment by an additional multiple of the original investment.

(b)

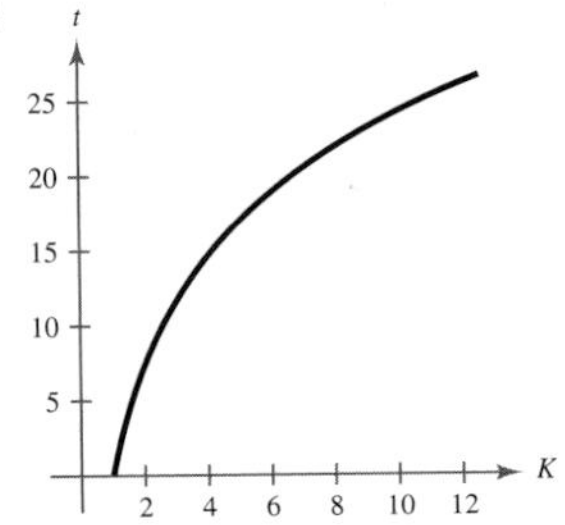

74. (a) 15 cubic feet per minute

(b) 380 cubic feet

(c) 380 square feet

76. (a) 120 decibels

(b) 100 decibels

(c) No, the difference results from the logarithmic relationship between intensity and number of decibels.

78. 20 years **80.** Total amount: \$301,123.20

Interest: \$151,123.20

82. False. Reflecting $g(x)$ about the line $y = x$ will determine the graph of $f(x)$.

84.

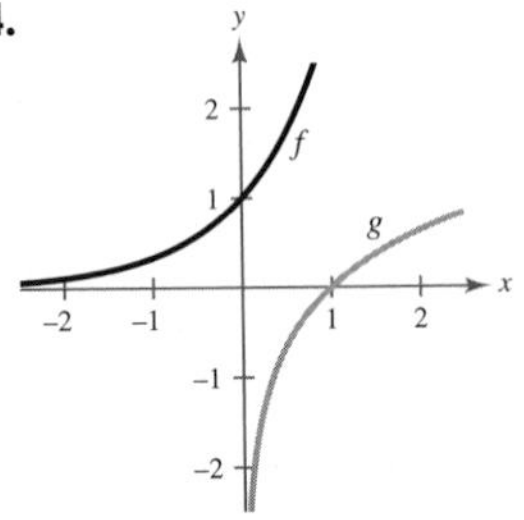

$g = f^{-1}$

86.

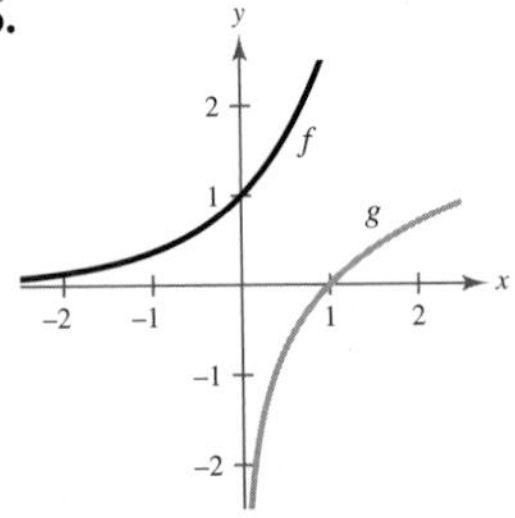

$g = f^{-1}$

88. (a)

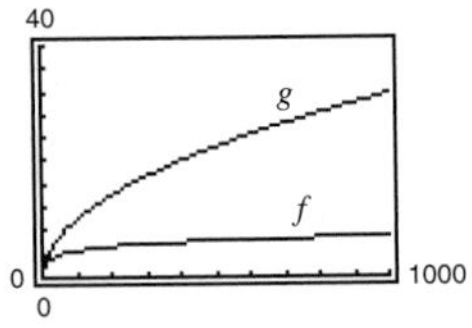

$g(x)$; The natural log function grows at a slower rate than the square root function.

(b)

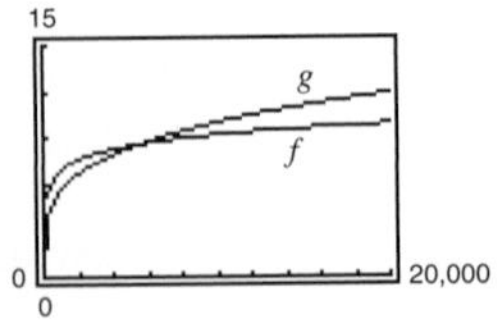

$g(x)$; The natural log function grows at a slower rate than the fourth root function.

90.

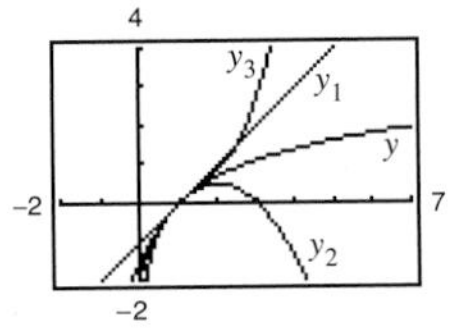

92. (a) $(0, \infty)$ (b) $f^{-1}(x) = 10^x$ (c) $3 < x < 4$

(d) $0 < x < 1$ (e) 10 (f) $10^{2n} : 1$

94. (a)

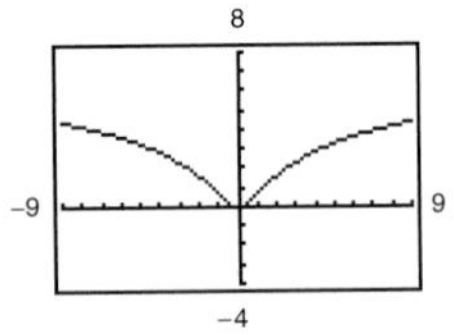

(b) Increasing: $(0, \infty)$

Decreasing: $(-\infty, 0)$

(c) Relative minimum: $(0, 0)$

96. (a)

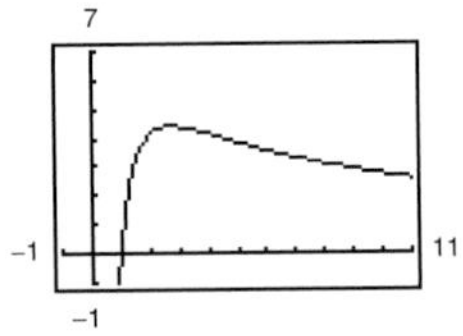

(b) Increasing: $(0, 2.72)$

Decreasing: $(2.72, \infty)$

(c) Relative maximum: $(2.72, 4.41)$

98. $9.25 + 0.75q$ **100.** $w^2 + 10w$

102. Vertical asymptote: $x = 0$

Slant asymptote: $y = 2x$

104. Vertical asymptote: $x = 7$

106. 4.482 **108.** 0.010

Section 3.3 *(page 315)*

2. 0.712 **4.** -1.161 **6.** -0.694 **8.** -3.823

10. (a) $\dfrac{\log_{10} x}{\log_{10} 3}$ (b) $\dfrac{\ln x}{\ln 3}$ **12.** (a) $\dfrac{\log_{10} x}{\log_{10} \frac{1}{3}}$ (b) $\dfrac{\ln x}{\ln \frac{1}{3}}$

14. (a) $\dfrac{\log_{10} \frac{3}{4}}{\log_{10} x}$ (b) $\dfrac{\ln \frac{3}{4}}{\ln x}$ **16.** (a) $\dfrac{\log_{10} x}{\log_{10} 7.1}$ (b) $\dfrac{\ln x}{\ln 7.1}$

18. $f(x) = \dfrac{\log_{10} x}{\log_{10} 4} = \dfrac{\ln x}{\ln 4}$

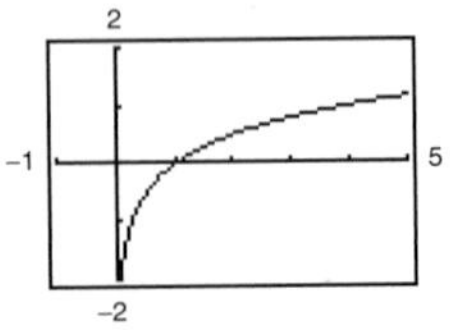

20. $f(x) = \dfrac{\log_{10} x}{\log_{10} \frac{1}{4}} = \dfrac{\ln x}{\ln \frac{1}{4}}$

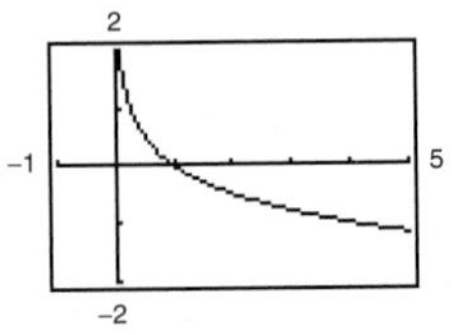

22. $f(x) = \dfrac{\log_{10} x}{\log_{10} 12.4} = \dfrac{\ln x}{\ln 12.4}$

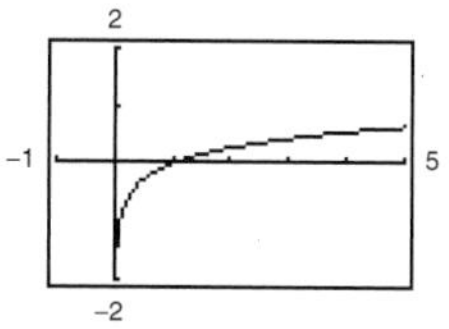

24. $\log_{10}10 + \log_{10}z$

26. $\log_{10}y - \log_{10}2$ **28.** $-3\log_6 z$ **30.** $\frac{1}{3}\ln t$

32. $\ln x + \ln y - \ln z$

34. $\ln(x+1) + \ln(x-1) - 3\ln x$

36. $\ln x - \frac{1}{2}\ln(x^2+1)$ **38.** $\frac{1}{2}(2\ln x - 3\ln y)$

40. $\ln x + \frac{1}{2}\ln(x+2)$ **42.** $\frac{1}{2}\log_b x + 4\log_b y - 4\log_b z$

44. $\ln ty$ **46.** $\log_5 \dfrac{8}{t}$ **48.** $\log_6 \dfrac{1}{16x^4}$

50. $\log_7(z-2)^{2/3}$ **52.** $\ln 64z^5$

54. $\ln \dfrac{x^3y^4}{z^4}$ **56.** $\ln \dfrac{z^4(z+5)^4}{(z-5)^2}$ **58.** $\ln\left(\dfrac{x}{x^2-1}\right)^2$

60. $\ln\left[x^6(x-1)\sqrt{x+1}\right]$ **62.** $\ln 5\sqrt{5}t^6$

64. $\log_7\sqrt{70} = \frac{1}{2}(\log_7 7 + \log_7 10)$
$= \frac{1}{2} + \log_7\sqrt{10}$; Properties 1 and 3

66. $\frac{1}{3}$ **68.** -3

70. -16 is not in the domain of $\log_2 x$. **72.** 3 **74.** 12

76. 0 **78.** $\frac{3}{4}$ **80.** $4 + 4\log_2 3$ **82.** $\log_{10}3 - 2$

84. $\ln 6 - 2$ **86.** $\beta = 10(\log_{10}I + 12)$; 60 decibels

88. True; Property 1 **90.** False. $f\left(\sqrt{x}\right) = \frac{1}{2}f(x)$

92. True **94.** Answers will vary.

96.

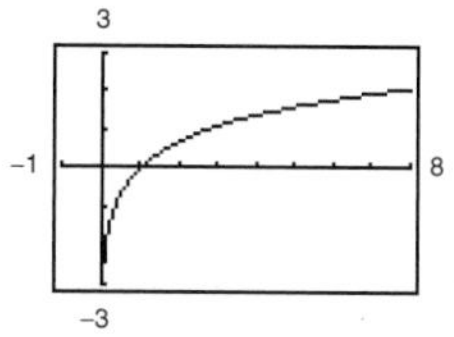

98. $\ln 2 \approx 0.6931$ $\ln 10 \approx 2.3025$
$\ln 3 \approx 1.0986$ $\ln 12 \approx 2.4848$
$\ln 4 \approx 1.3862$ $\ln 15 \approx 2.7080$
$\ln 5 \approx 1.6094$ $\ln 16 \approx 2.7724$
$\ln 6 \approx 1.7917$ $\ln 18 \approx 2.8903$
$\ln 8 \approx 2.0793$ $\ln 20 \approx 2.9956$
$\ln 9 \approx 2.1972$

100. $\dfrac{27y^3}{8x^6}$ **102.** $\dfrac{(xy)^2}{x+y}$ **104.** 0.129 **106.** 23.836

108. 2712.655 **110.** -0.204 **112.** 1.294

Section 3.4 *(page 324)*

2. (a) No (b) No

4. (a) Yes (b) Yes, approximate (c) Yes

6. (a) Yes (b) Yes, approximate (c) No

8. 5 **10.** 6 **12.** $\frac{2}{3}$ **14.** -3

16. 2 **18.** 8 **20.** 5 **22.** $\ln 4 \approx 1.386$

24. $e^{-7} \approx 0.000912$ **26.** 5 **28.** 0.001 **30.** 1

32. $\left(\frac{2}{3}, 9\right)$ **34.** $(5, 0)$ **36.** $2x - 1$

38. $x^3, x > 0$ **40.** x^4 **42.** $x^2, x \neq 0$

44. $x^3 - 8, x > 0$ **46.** $\ln\frac{91}{4} \approx 3.125$

48. $\ln\dfrac{25}{3} \approx 2.120$ **50.** $\dfrac{\ln 50}{2} \approx 1.956$

52. $-\frac{1}{4}\ln\frac{3}{40} \approx 0.648$ **54.** $\ln 2 \approx 0.693$; $\ln 3 \approx 1.099$

56. $\ln 7 \approx 1.946$ **58.** $\log_{10}570 \approx 2.756$

60. $\dfrac{\ln 3000}{5\ln 6} \approx 0.894$ **62.** $-\dfrac{\ln(0.10)}{3\ln 4} \approx 0.554$

64. $\dfrac{-\ln 64 - \ln 431}{\ln 8} \approx -4.917$

66.

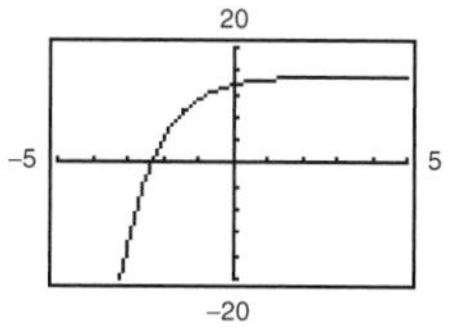

-2.322

68.

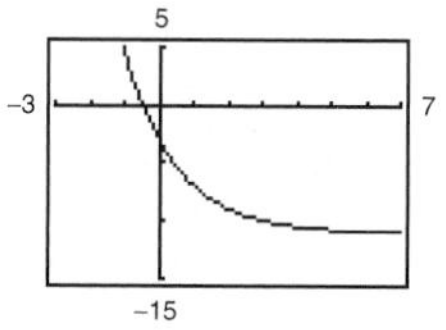

-0.478

70.

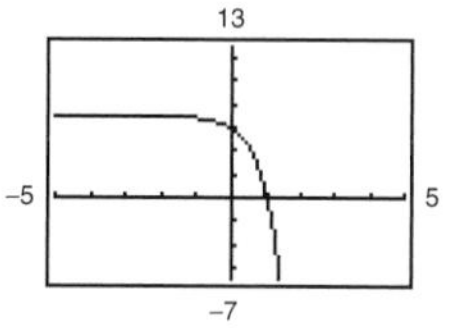

1.081

72.

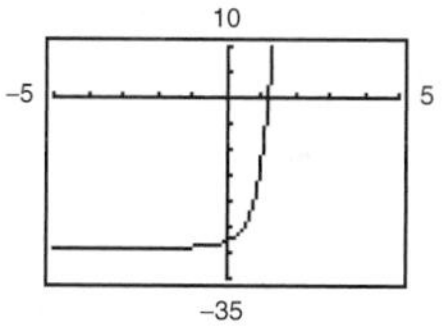

1.236

74. $6 + \log_{10}\frac{7}{5} \approx 6.146$ **76.** $6 - \dfrac{\ln 5}{\ln 3} \approx 4.535$

78. $\dfrac{\ln 21}{9\ln 3.938225} \approx 0.247$ **80.** $\dfrac{\ln 30}{3\ln\left(16 - \frac{0.878}{26}\right)} \approx 0.409$

82. $\dfrac{\ln 31}{6} \approx 0.572$ **84.** $e^2 \approx 7.389$ **86.** $\dfrac{e}{4} \approx 0.680$

88. $e^{7/2} \approx 33.115$ **90.** $e^{10} + 8 = 22{,}034.466$

92. $\dfrac{-1 + \sqrt{1+4e}}{2} \approx 1.223$ **94.** $\dfrac{-3 + \sqrt{9+4e}}{2} \approx 0.729$

96. 2.547 **98.** ± 1000 **100.** $10^{11/5} + 2 \approx 160.489$

102. 2 **104.** 9 **106.** $\dfrac{1225 + 125\sqrt{73}}{2} \approx 1146.500$

108.

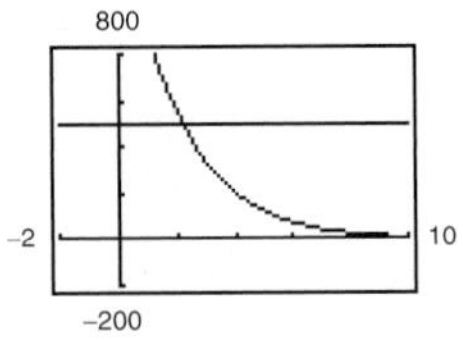

$(2.197, 500)$

110.

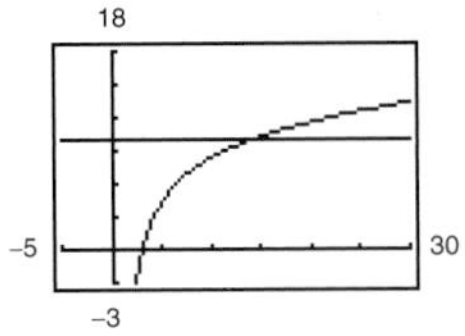

$(14.182, 10)$

112. 5.8 years **114.** 9.2 years

116. (a) 303 units (b) 528 units

118. 12.76 inches

120. (a)

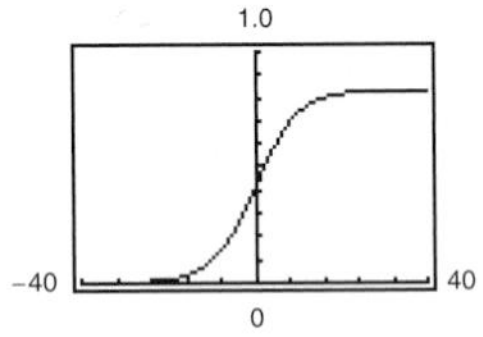

(b) Horizontal asymptotes: $P = 0$, $P = 0.83$

The proportion of correct responses will approach 0.83 as the number of trials increases.

(c) ≈ 5 trials

122. (a)

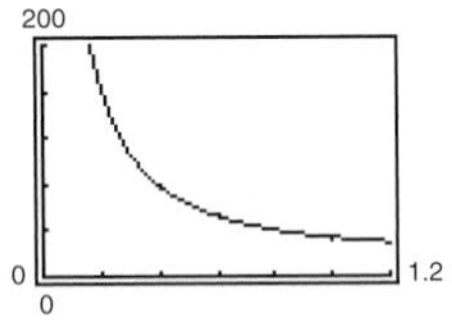

The model appears to fit the data well.

(b) 1.2 meters (c) No

124. $\log_b(u + v) = (\log_b u)(\log_b v)$

False.

$2.04 \approx \log_{10}(10 + 100) \neq (\log_{10} 10)(\log_{10} 100) = 2$

126. $\log_b \dfrac{u}{v} = \log_b u - \log_b v$

True by Property 2 in Section 3.3.

128. Yes. Time to double: $t = \dfrac{\ln 2}{r}$;

Time to quadruple: $t = \dfrac{\ln 4}{r} = 2\left(\dfrac{\ln 2}{r}\right)$

130. $4|x|y^2\sqrt{3y}$ **132.** $5\sqrt[3]{3}$ **134.** $M = kp^3$

136. $d = kab$ **138.** $c = \dfrac{k}{\sqrt{w}}$ **140.** 1.262

142. 1.486

Section 3.5 *(page 335)*

2. e **4.** a **6.** f

	Initial Investment	*Annual % Rate*	*Time to Double*	*Amount after 10 Years*
8.	\$20,000	10.5%	6.60 yr	\$57,153.02
10.	\$10,000	5.7762%	12 yr	\$17,817.97
12.	\$600	34.66%	2 yr	\$19,205.00
14.	\$8986.58	8%	8.66 yr	\$20,000.00

16. \$4214.16

18. (a) 6.94 years (b) 6.63 years

(c) 6.602 years (d) 6.601 years

20.

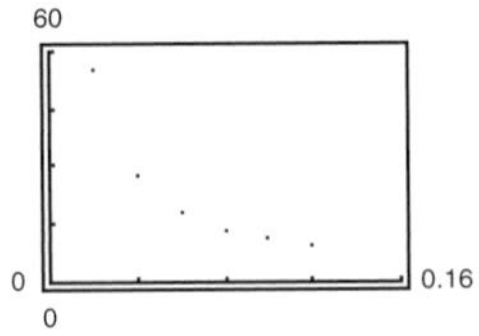

Use PwrReg: $t = 1.099r^{-1}$

22.

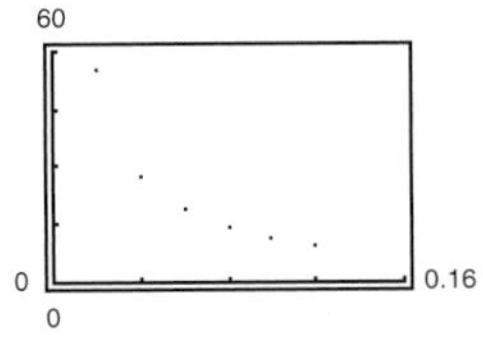

Use PwrReg: $t = 1.222r^{-1}$

24.

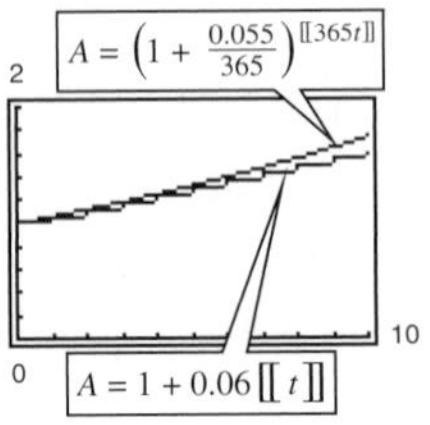

Daily compounding

	Isotope	*Half-life (Years)*	*Initial Quantity*	*Amount after 1000 Years*
26.	^{226}Ra	1620	2.30 g	1.5 g
28.	^{14}C	5730	3 g	2.66 g
30.	^{239}Pu	24,360	0.41 g	0.4 g

32. $y = \frac{1}{2}e^{0.5756x}$ **34.** $y = e^{-0.4621x}$ **36.** 2011

38. $k = 0.0084$; 166,203 **40.** 3.15 hours **42.** 95.8%

44. (a) $V = -4500t + 22{,}000$ (b) $V = 22{,}000e^{-0.263t}$

(c)

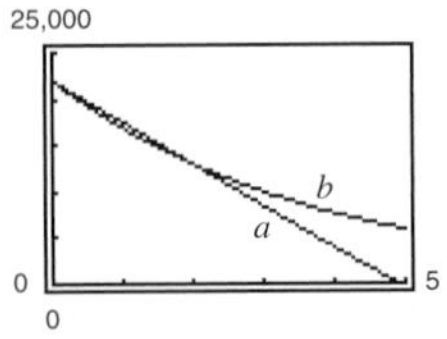

Exponential

(d) 1 year: Straight-line, \$17,500;
Exponential, \$16,912
3 years: Straight-line, \$8500;
Exponential, \$9995

(e) Decreases \$4500 per year

46. (a) $S(t) = 100(1 - e^{-0.1625t})$

(b)

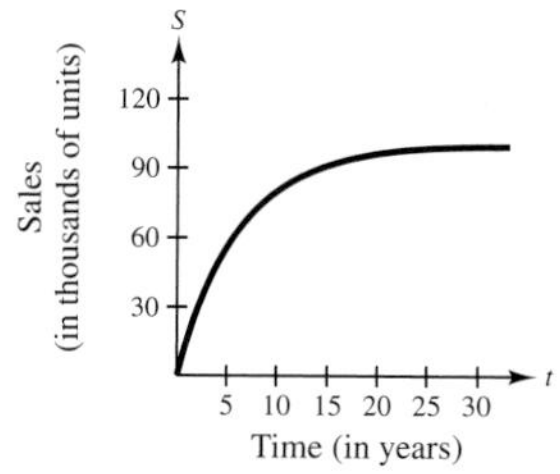

(c) 55,625

48. (a) $S = 10(1 - e^{-0.0575x})$ (b) 3314 units

50. (a) $N = 30(1 - e^{-0.050t})$ (b) 36 days

(c) No. It is not a linear function.

52. (a) 7.91 (b) 7.68 (c) 5.40

54. (a) 20 decibels (b) 70 decibels

(c) 95 decibels (d) 120 decibels

56. 95% **58.** 4.64 **60.** 1.58×10^{-6} moles per liter

62. 10^7

64. (a)

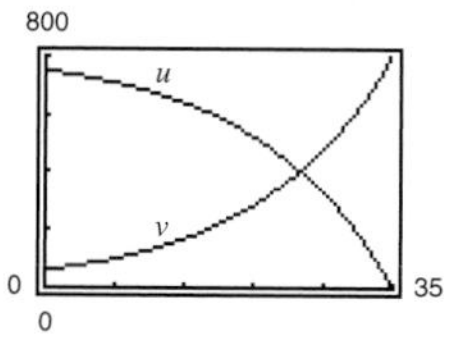

(b) Interest; $t \approx 26$ years

(c)

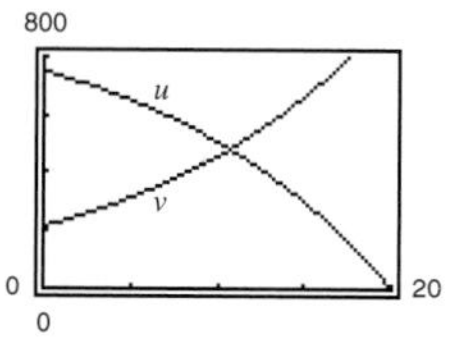

Interest; $t \approx 11$ years; The interest is still the majority of the monthly payment in the early years, but now the principal and interest are nearly equal when $t \approx 11$ years.

66. (a) $t_3 = 0.2729s - 6.0143$

$t_4 = 1.5385e^{0.02913s}$ or $t_4 = 1.5385(1.0296)^s$

(b)

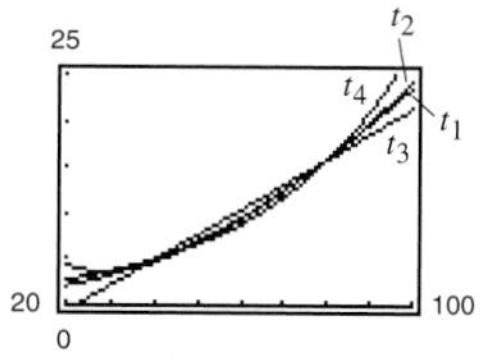

(c)

s	30	40	50	60	70	80	90
t_1	3.6	4.6	6.7	9.4	12.5	15.9	19.6
t_2	3.3	4.9	7.0	9.5	12.5	15.9	19.9
t_3	2.2	4.9	7.6	10.4	13.1	15.8	18.5
t_4	3.7	4.9	6.6	8.8	11.8	15.8	21.2

(d) Model t_1: Sum = 2.0
Model t_2: Sum = 1.1
Model t_3: Sum = 5.6
Model t_4: Sum = 2.7
Quadratic model fits best.

68. False. The domain can be the set of real numbers for a logistics growth function.

70. False. The graph of $f(x)$ is the graph of $g(x)$ shifted upward 5 units.

72. Answers will vary. **74.** $4x^2 - 12x + 9$

76. $2x^2 + 3 + \dfrac{3}{x - 4}$

78.

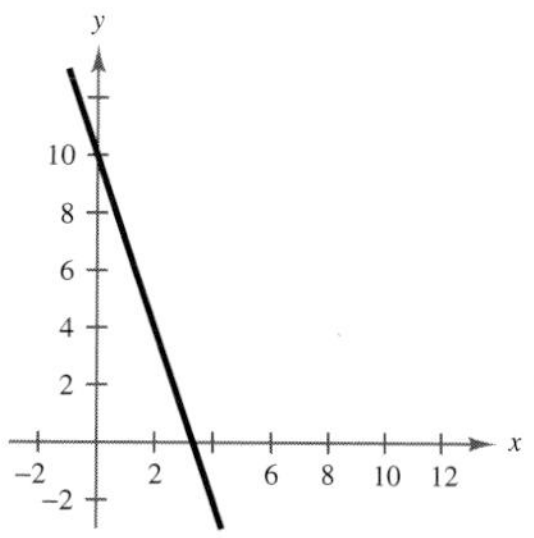

80.

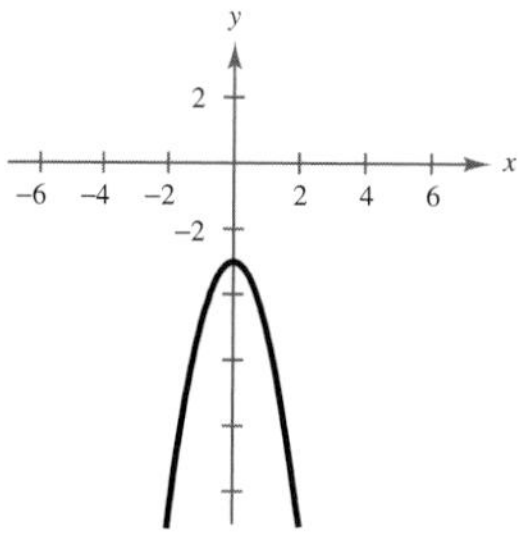

82.

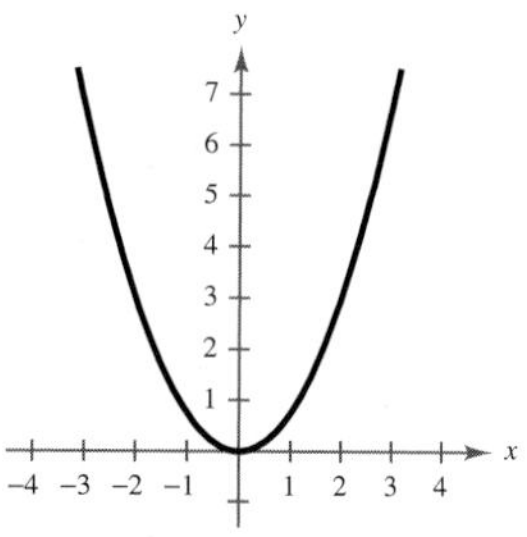

84.

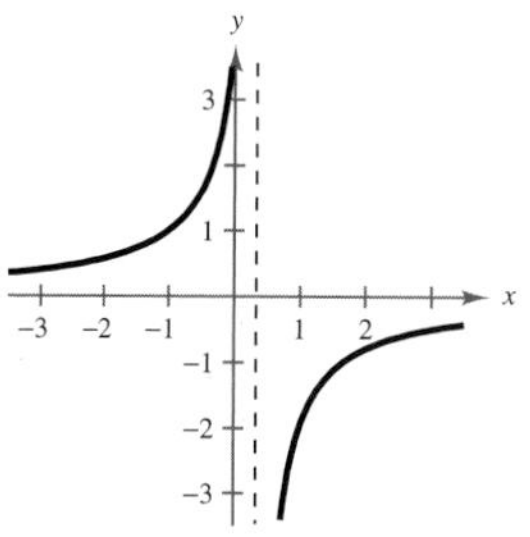

86.

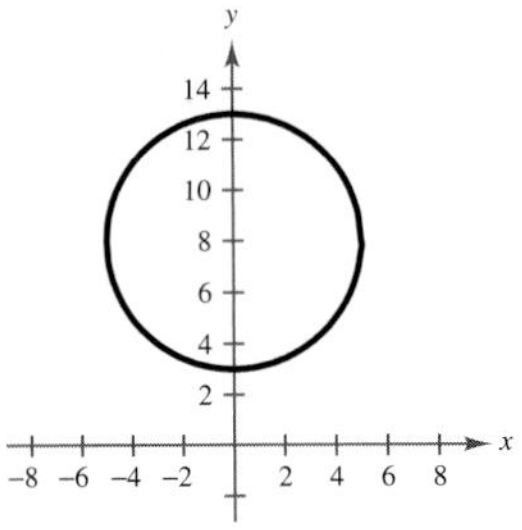

88.

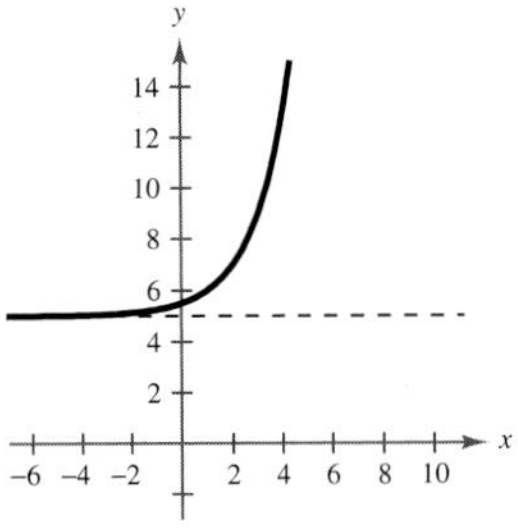

90.

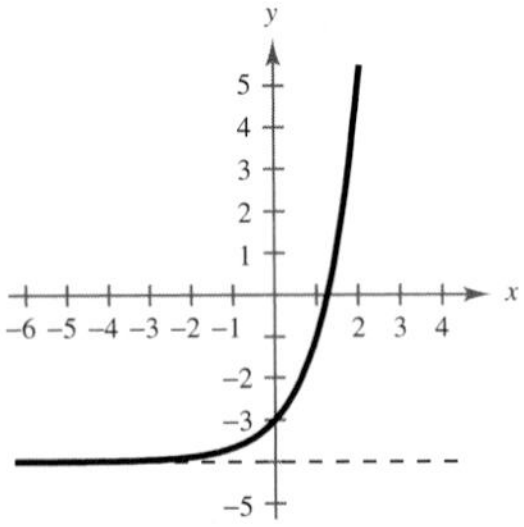

Review Exercises *(page 342)*

2. -3.863 **4.** 4.181 **6.** 0.002 **8.** d **10.** b

12.

x	-2	-1	0	1	2
$f(x)$	-3.063	-3.25	-4	-7	-19

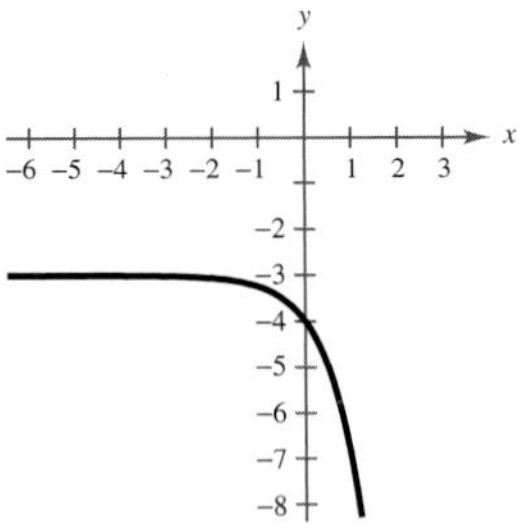

14.

x	-3	-1	0	1	3
$f(x)$	0.020	0.142	0.377	1	7.023

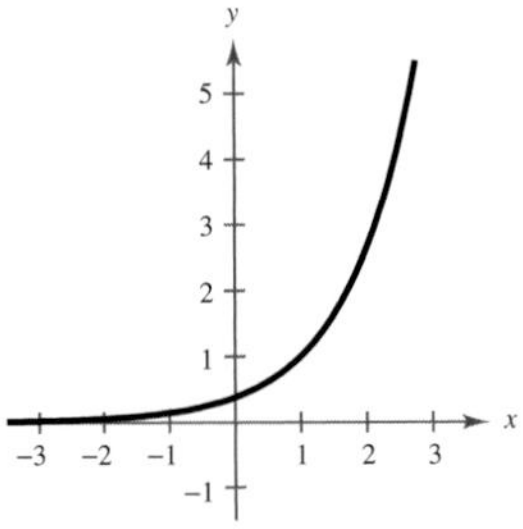

16.

x	0	5	6	7	8	9
$f(x)$	-4.984	-4.5	-4	-3	-1	3

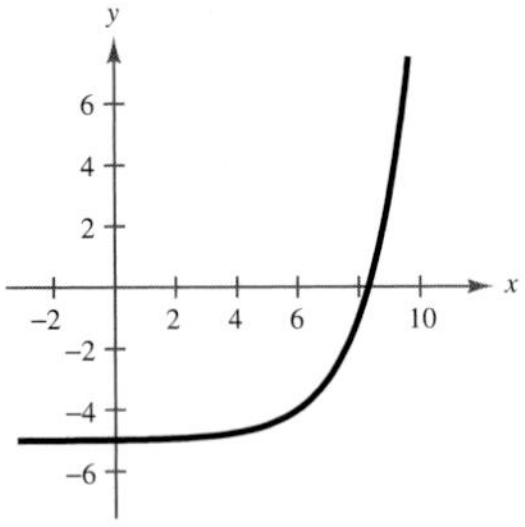

18.

x	-3	-2	-1	0	2
$f(x)$	3	-4	-4.875	-4.984	-5

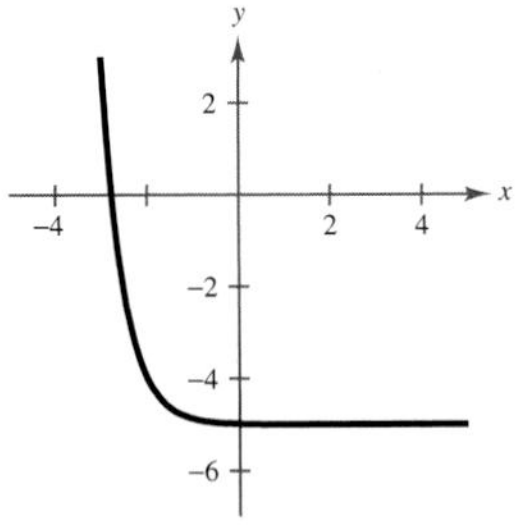

20. 1.868 **22.** 1.320

24.

x	-2	-1	0	1	2
$h(x)$	-0.72	0.35	1	1.39	1.63

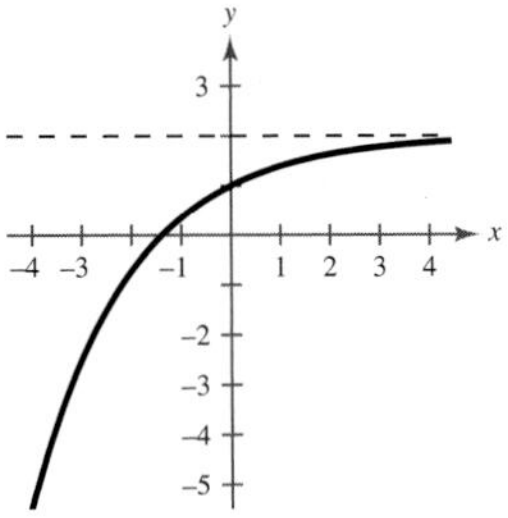

26.

t	$\frac{1}{2}$	1	2	3	4
$s(t)$	0.07	0.54	1.47	2.05	2.43

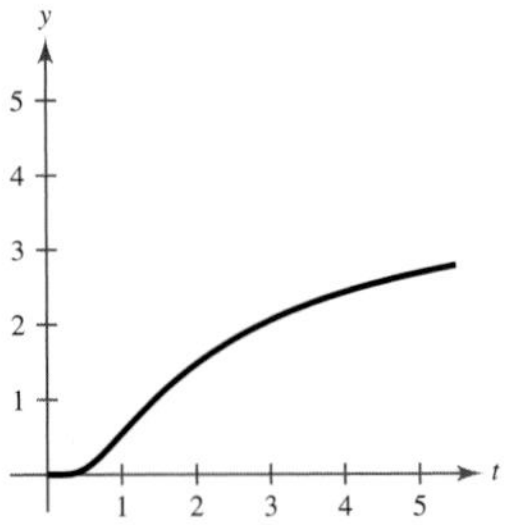

28.

n	1	2	4	12
A	\$8643.88	\$8799.58	\$8880.43	\$8935.49

n	365	Continuous
A	\$8962.46	\$8963.38

30.

t	1	10	20
P	\$188,381.07	\$109,926.55	\$60,419.23

t	30	40	50
P	\$33,208.39	\$18,252.42	\$10,032.13

32. (a)

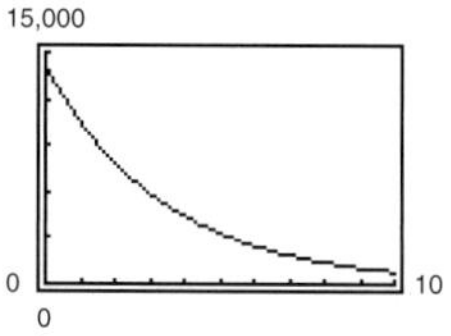

(b) \$7875

(c) At the beginning. Yes

34.

s	50	55	60	65	70
y	28	26.4	24.8	23.4	22.0

36. $\log_{25} 125 = \frac{3}{2}$ **38.** $\frac{1}{2}$ **40.** -1

42.

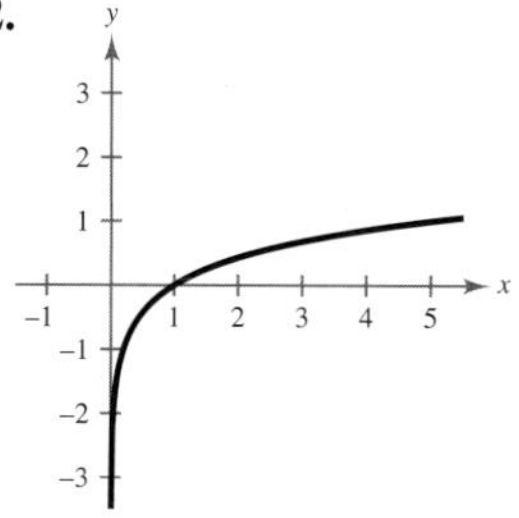

Vertical asymptote: $x = 0$

44.

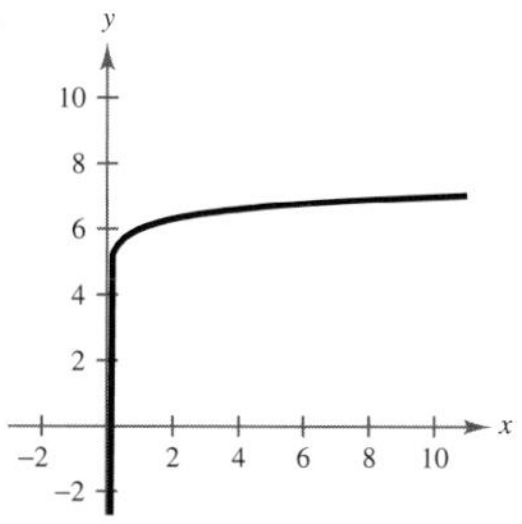

Vertical asymptote: $x = 0$

46.

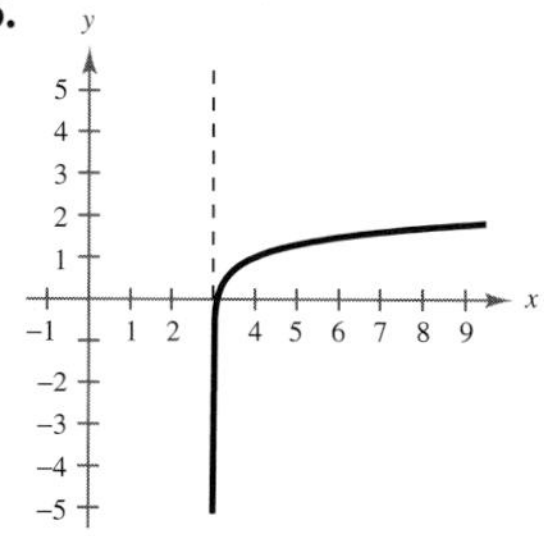

Vertical asymptote: $x = 3$

48. -0.020 **50.** 7 **52.** -1.530

54.

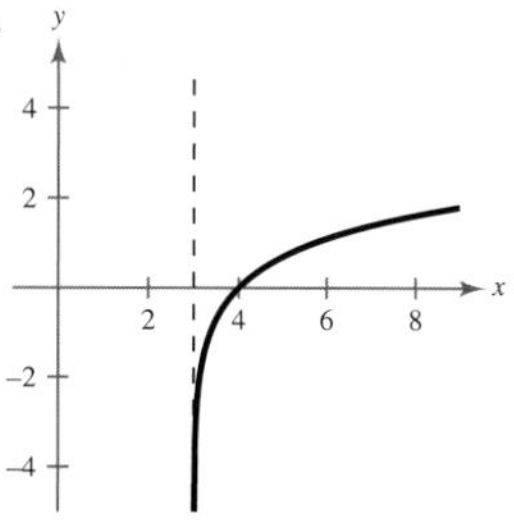

Vertical asymptote: $x = 3$

56.

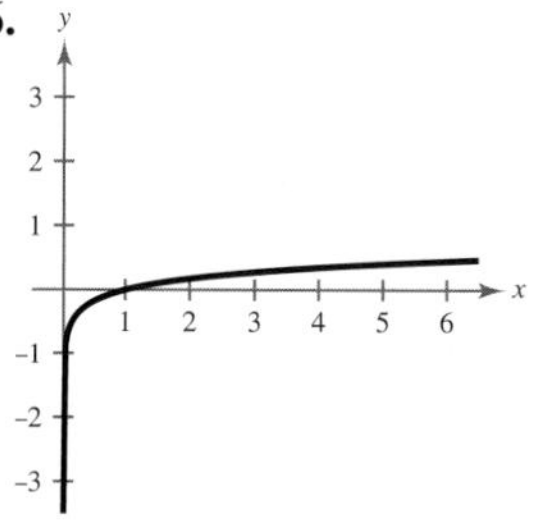

Vertical asymptote: $x = 0$

58. 1.585 **60.** -2.322

62.–66. Answers will vary.

68. $\frac{1}{2}\log_7 x - \log_7 4$ **70.** $\ln|x-1| - \ln|x+1|$

72. $\log_6 \frac{y}{z^2}$ **74.** $\ln\left|\frac{(x-2)^5}{(x+2)x^3}\right|$

76. 3 **78.** -3 **80.** 2401 **82.** $\ln 12 \approx 2.485$

84. $-\frac{\ln 44}{5} \approx -0.757$ **86.** $\ln 22 \approx 3.091$

88. $\frac{\ln 17}{\ln 5} \approx 1.760$ **90.** $\ln 2 \approx 0.693$, $\ln 5 \approx 1.609$

92. 0.39, 7.48 **94.** 2.45 **96.** $\frac{1}{3}e^{8.2} \approx 1213.650$

98. $\frac{1}{4}e^{7.5} \approx 452.011$ **100.** $3e^2 \approx 22.167$

102. $e^4 - 1 \approx 53.598$ **104.** No solution **106.** 0.900

108. 1.64 **110.** No solution **112.** 15.2 years

114. (a) 1151 units (b) 1325 units

116. b **118.** d **120.** c **122.** 98.6%

124. $y = 2e^{0.1014x}$

126. (a)

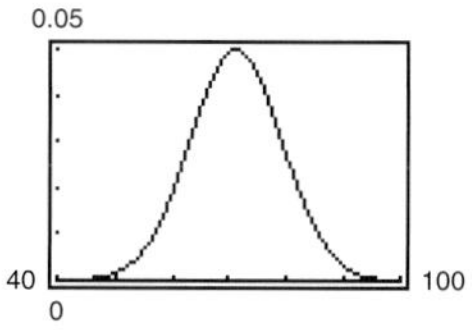

(b) 71

128. $10^{-3.5}$ watt per square centimeter

130. True by the inverse properties

132. False. $\ln x + \ln y = \ln(xy) \neq \ln(x + y)$

134. True. $\log_{10}\left(\frac{10}{x}\right) = \log_{10} 10 - \log_{10} x = 1 - \log_{10} x$

136. $b < d < a < c$

b and d are negative.

a and c are positive.

CHAPTER 3

Chapter 4

Section 4.1 *(page 359)*

2. 5.5 **4.** -4 **6.** 6

8. (a) Quadrant III (b) Quadrant III

10. (a) Quadrant IV (b) Quadrant III

12. (a) Quadrant IV (b) Quadrant II

14. (a) (b)

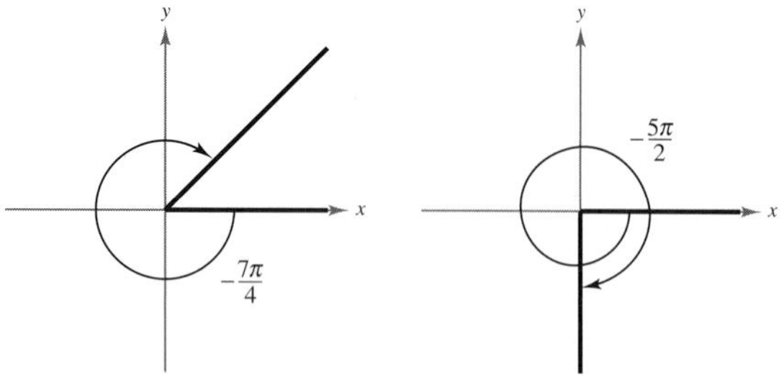

16. (a) (b)

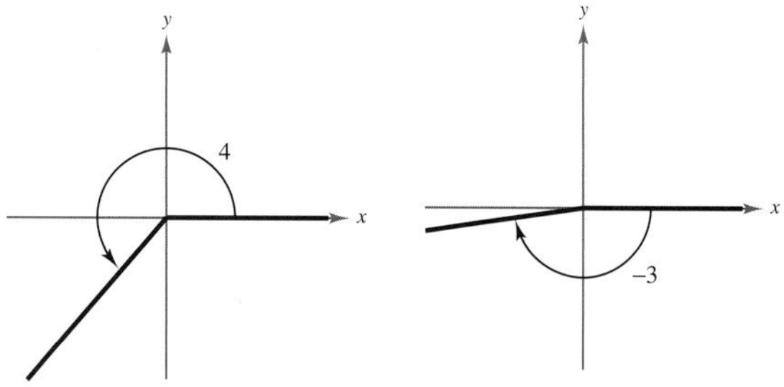

18. (a) $\frac{19\pi}{6}, -\frac{5\pi}{6}$ (b) $\frac{\pi}{6}, -\frac{23\pi}{6}$

20. (a) $\frac{7\pi}{4}, -\frac{\pi}{4}$ (b) $\frac{28\pi}{15}, -\frac{32\pi}{15}$

22. (a) Complement: $\frac{5\pi}{12}$; Supplement: $\frac{11\pi}{12}$

(b) Complement: none; Supplement: $\frac{\pi}{12}$

24. (a) Complement: none; Supplement: $\pi - 3 \approx 0.14$

(b) Complement: $\frac{\pi}{2} - 1.5 \approx 0.07$

Supplement: $\pi - 1.5 \approx 1.64$

26. (a) $\frac{7\pi}{4}$ (b) $\frac{2\pi}{3}$ **28.** (a) $-\frac{3\pi}{2}$ (b) $\frac{4\pi}{5}$

30. 1.525 **32.** -0.842 **34.** 6.021 **36.** 0.009

38. (a) $-105°$ (b) $20°$ **40.** (a) $330°$ (b) $408°$

42. $81.818°$ **44.** $1170°$ **46.** $864°$

48. $-32.659°$ **50.** $120°$ **52.** $-330°$ **54.** $10°$

56. (a) Quadrant I (b) Quadrant III

58. (a) Quadrant II (b) Quadrant IV

60. (a) (b)

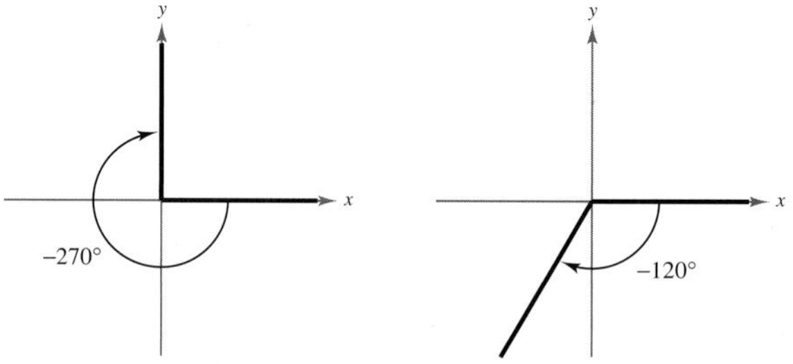

62. (a) (b)

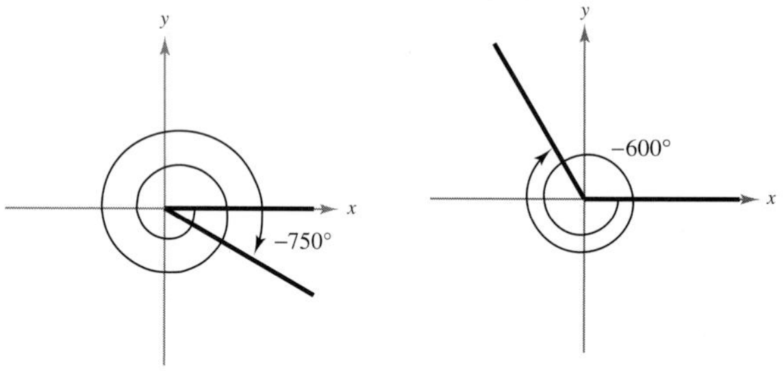

64. (a) $480°, -240°$ (b) $300°, -60°$

66. (a) $300°, -60°$ (b) $590°, -130°$

68. (a) Complement: $87°$; Supplement: $177°$

(b) Complement: $26°$; Supplement: $116°$

70. (a) Complement: none; Supplement: $50°$

(b) Complement: none; Supplement: $10°$

72. (a) $245.167°$ (b) $2.2°$

74. (a) $-135.01°$ (b) $-408.272°$

76. (a) $-345°\,7'\,12''$ (b) $0°\,27'$

78. (a) $-0°\,21'\,18''$ (b) $0°\,47'\,11.4''$

80. $\frac{29}{10}$ radians **82.** $-\frac{4}{5}$ radian

84. $\frac{4}{7}$ radian **86.** 2 radians **88.** $3\pi \approx 9.42$ feet

90. $5\pi \approx 15.71$ centimeters **92.** 686.36 miles

94. 4045.67 miles **96.** 0.078 radian $\approx 4.49°$ **98.** $275°$

100. (a) 3400π radians per minute; 1700π radians per minute

(b) 850 revolutions per minute

102. (a) $\frac{14\pi}{3}$ feet per second (b) ≈ 10 miles per hour

104. True. Let α and β equal coterminal angles, and let n be an integer.

$\alpha = \beta + n(360°)$

$\alpha - \beta = n(360°)$

106. Increases. The linear velocity is proportional to the radius.

108. The arc length increases. If θ is constant, the length of the arc is proportional to the radius $(s = r\theta)$.

110.

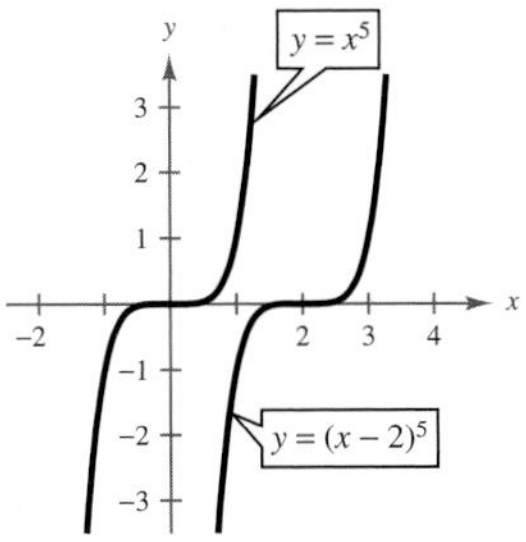

112.

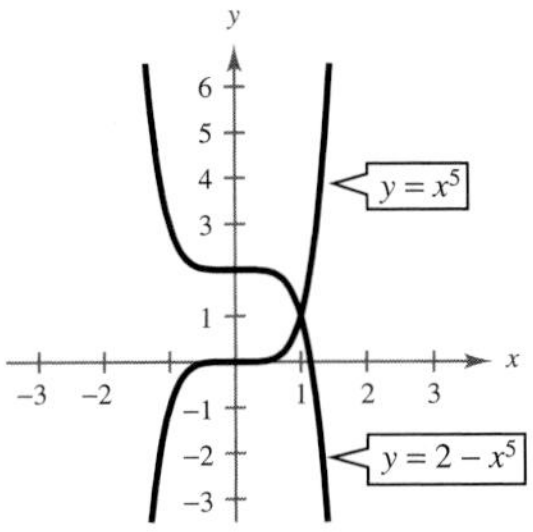

114.

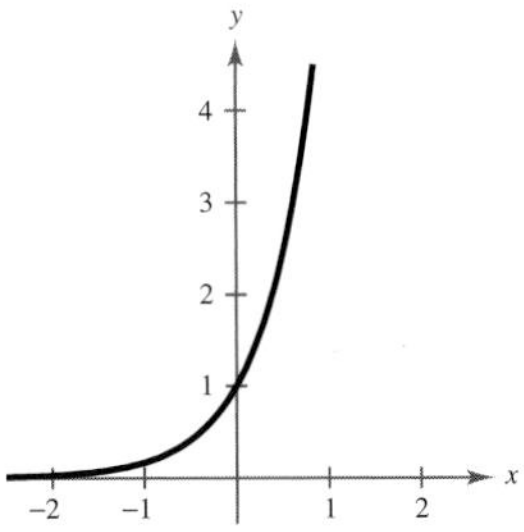

116.

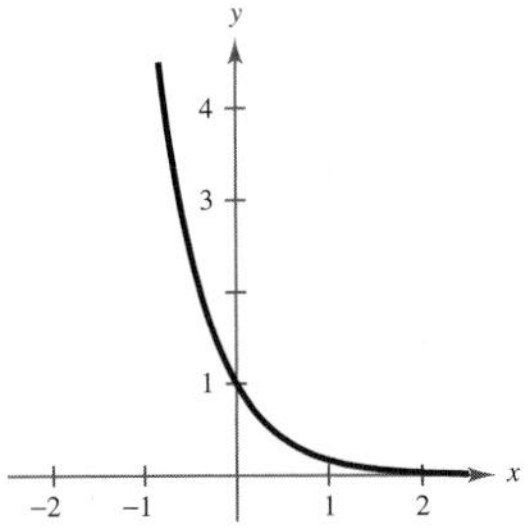

118.

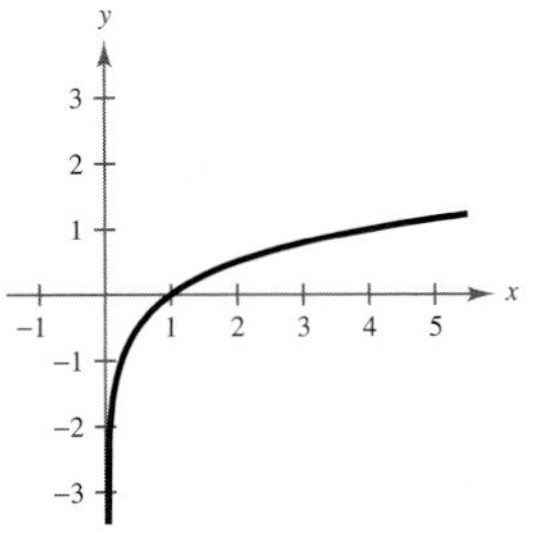

120.

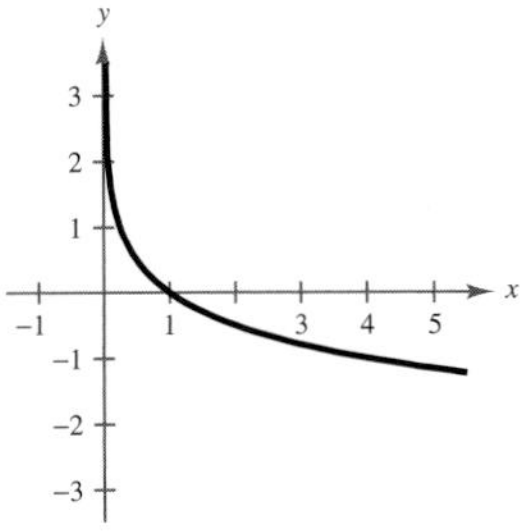

122.

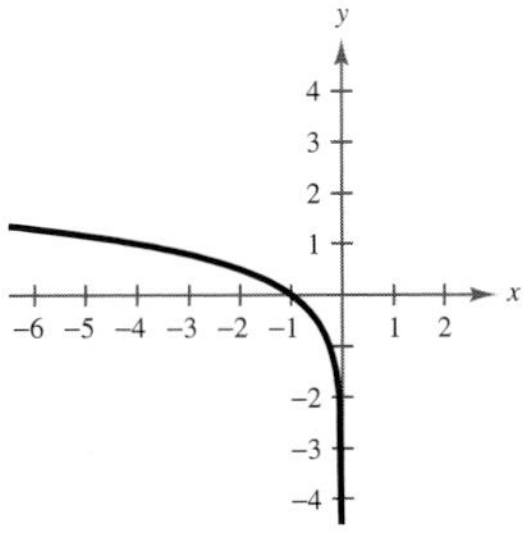

124. $\dfrac{\sqrt{2}}{2}$ **126.** $\sqrt{2}$ **128.** $2\sqrt{10}$ **130.** $12\sqrt{2}$

Section 4.2 *(page 368)*

2. $\sin\theta = \frac{5}{13}$ $\csc\theta = \frac{13}{5}$
$\cos\theta = \frac{12}{13}$ $\sec\theta = \frac{13}{12}$
$\tan\theta = \frac{5}{12}$ $\cot\theta = \frac{12}{5}$

4. $\sin\theta = -\frac{3}{5}$ $\csc\theta = -\frac{5}{3}$
$\cos\theta = -\frac{4}{5}$ $\sec\theta = -\frac{5}{4}$
$\tan\theta = \frac{3}{4}$ $\cot\theta = \frac{4}{3}$

6. $\left(\frac{1}{2}, \frac{\sqrt{3}}{2}\right)$ **8.** $\left(-\frac{\sqrt{2}}{2}, -\frac{\sqrt{2}}{2}\right)$ **10.** $\left(\frac{1}{2}, -\frac{\sqrt{3}}{2}\right)$

12. $(-1, 0)$

14. $\sin\frac{\pi}{3} = \frac{\sqrt{3}}{2}$
$\cos\frac{\pi}{3} = \frac{1}{2}$
$\tan\frac{\pi}{3} = \sqrt{3}$

16. $\sin\left(-\frac{\pi}{4}\right) = -\frac{\sqrt{2}}{2}$
$\cos\left(-\frac{\pi}{4}\right) = \frac{\sqrt{2}}{2}$
$\tan\left(-\frac{\pi}{4}\right) = -1$

18. $\sin\left(-\frac{4\pi}{3}\right) = \frac{\sqrt{3}}{2}$
$\cos\left(-\frac{4\pi}{3}\right) = -\frac{1}{2}$
$\tan\left(-\frac{4\pi}{3}\right) = -\sqrt{3}$

20. $\sin\frac{5\pi}{3} = -\frac{\sqrt{3}}{2}$
$\cos\frac{5\pi}{3} = \frac{1}{2}$
$\tan\frac{5\pi}{3} = -\sqrt{3}$

22. $\sin(-2\pi) = 0$
$\cos(-2\pi) = 1$
$\tan(-2\pi) = 0$

24. $\sin\frac{5\pi}{6} = \frac{1}{2}$
$\cos\frac{5\pi}{6} = -\frac{\sqrt{3}}{2}$
$\tan\frac{5\pi}{6} = -\frac{\sqrt{3}}{3}$
$\csc\frac{5\pi}{6} = 2$
$\sec\frac{5\pi}{6} = -\frac{2\sqrt{3}}{3}$
$\cot\frac{5\pi}{6} = -\sqrt{3}$

26. $\sin\frac{3\pi}{2} = -1$
$\cos\frac{3\pi}{2} = 0$
$\tan\frac{3\pi}{2}$ is undefined.
$\csc\frac{3\pi}{2} = -1$
$\sec\frac{3\pi}{2}$ is undefined.
$\cot\frac{3\pi}{2} = 0$

28. $\sin\left(-\frac{3\pi}{2}\right) = 1$ $\csc\left(-\frac{3\pi}{2}\right) = 1$
$\cos\left(-\frac{3\pi}{2}\right) = 0$ $\sec\left(-\frac{3\pi}{2}\right)$ is undefined.
$\tan\left(-\frac{3\pi}{2}\right)$ is undefined. $\cot\left(-\frac{3\pi}{2}\right) = 0$

30. $\cos 5\pi = \cos\pi = -1$ **32.** $\sin\frac{9\pi}{4} = \sin\frac{\pi}{4} = \frac{\sqrt{2}}{2}$

34. $\sin(-3\pi) = \sin(-\pi) = 0$

36. $\cos\left(-\frac{8\pi}{3}\right) = \cos\frac{4\pi}{3} = -\frac{1}{2}$

38. (a) $-\frac{3}{8}$ (b) $-\frac{8}{3}$ **40.** (a) $-\frac{3}{4}$ (b) $-\frac{4}{3}$

42. (a) $-\frac{4}{5}$ (b) $-\frac{4}{5}$ **44.** 1.7321 **46.** 0.6421

48. -0.8011 **50.** -4.4014 **52.** -0.7833

54. (a) 0.7 (b) -0.8 **56.** (a) 4.0, 5.4 (b) 0.7, 5.6

58. $0.8415 = \sin(0.25 + 0.75)$

$\neq \sin 0.25 + \sin 0.75 = 0.9290$

60. (a) 0.2500 foot (b) 0.0138 foot (c) -0.1501 foot

62. True. The tangent function has a period of π.

64. (a) Origin (b) $\sin(t_1 + \pi) = -\sin t_1$

(c) $\cos(t_1 + \pi) = -\cos t_1$

66. Odd **68.** $f^{-1}(x) = \frac{2}{3}(x + 1)$

70. $f^{-1}(x) = \sqrt{x^2 + 4},\ x \geq 0$

Section 4.3 *(page 377)*

2. $\sin\theta = \frac{5}{13}$ $\csc\theta = \frac{13}{5}$

$\cos\theta = \frac{12}{13}$ $\sec\theta = \frac{13}{12}$

$\tan\theta = \frac{5}{12}$ $\cot\theta = \frac{12}{5}$

4. $\sin\theta = \dfrac{\sqrt{2}}{2}$ $\csc\theta = \sqrt{2}$

$\cos\theta = \dfrac{\sqrt{2}}{2}$ $\sec\theta = \sqrt{2}$

$\tan\theta = 1$ $\cot\theta = 1$

6. $\sin\theta = \frac{8}{17}$ $\csc\theta = \frac{17}{8}$

$\cos\theta = \frac{15}{17}$ $\sec\theta = \frac{17}{15}$

$\tan\theta = \frac{8}{15}$ $\cot\theta = \frac{15}{8}$

The triangles are similar, and corresponding sides are proportional.

8. $\sin\theta = \dfrac{\sqrt{5}}{5}$ $\csc\theta = \sqrt{5}$

$\cos\theta = \dfrac{2\sqrt{5}}{5}$ $\sec\theta = \dfrac{\sqrt{5}}{2}$

$\tan\theta = \dfrac{1}{2}$ $\cot\theta = 2$

The triangles are similar, and corresponding sides are proportional.

10.

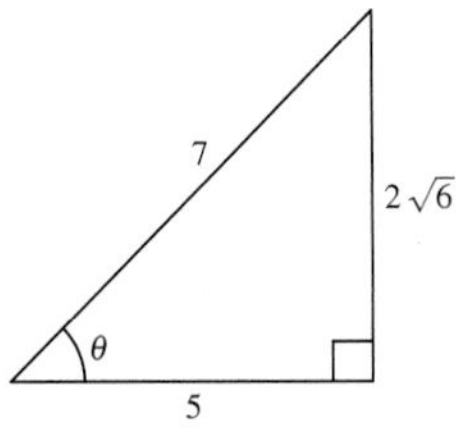

$\sin\theta = \dfrac{2\sqrt{6}}{7}$

$\tan\theta = \dfrac{2\sqrt{6}}{5}$

$\csc\theta = \dfrac{7\sqrt{6}}{12}$

$\sec\theta = \dfrac{7}{5}$

$\cot\theta = \dfrac{5\sqrt{6}}{12}$

12.

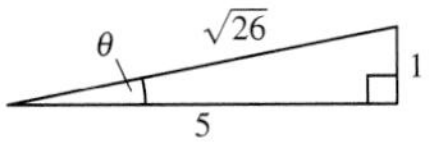

$\sin\theta = \dfrac{\sqrt{26}}{26}$ $\csc\theta = \sqrt{26}$

$\cos\theta = \dfrac{5\sqrt{26}}{26}$ $\sec\theta = \dfrac{\sqrt{26}}{5}$

$\tan\theta = \dfrac{1}{5}$

14.

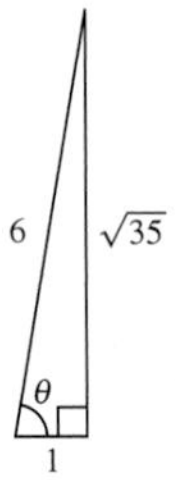

$\sin\theta = \dfrac{\sqrt{35}}{6}$

$\cos\theta = \dfrac{1}{6}$

$\tan\theta = \sqrt{35}$

$\csc\theta = \dfrac{6\sqrt{35}}{35}$

$\cot\theta = \dfrac{\sqrt{35}}{35}$

16.

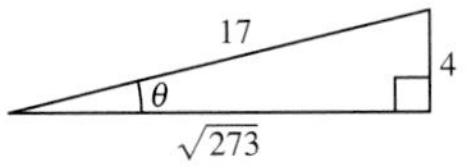

$\sin\theta = \dfrac{4}{17}$

$\cos\theta = \dfrac{\sqrt{273}}{17}$ $\sec\theta = \dfrac{17\sqrt{273}}{273}$

$\tan\theta = \dfrac{4\sqrt{273}}{273}$ $\cot\theta = \dfrac{\sqrt{273}}{4}$

18. (a) 2 (b) $\dfrac{\sqrt{3}}{3}$ (c) $\dfrac{\sqrt{3}}{2}$ (d) $\sqrt{3}$

20. (a) $\dfrac{1}{5}$ (b) $\dfrac{\sqrt{6}}{12}$ (c) $2\sqrt{6}$ (d) $\dfrac{2\sqrt{6}}{5}$

22. (a) $\dfrac{1}{5}$ (b) $\dfrac{\sqrt{26}}{26}$ (c) $\dfrac{1}{5}$ (d) $\dfrac{\sqrt{26}}{5}$

24. (a) 1 (b) $\dfrac{\sqrt{2}}{2}$ (c) $\sqrt{2}$

26. (a) $\dfrac{\sqrt{3}}{2}$ (b) 1 (c) $\dfrac{2\sqrt{3}}{3}$

28. (a) 0.4348 (b) 0.4348

30. (a) 0.9598 (b) 0.9609

32. (a) 0.9964 (b) 1.0036

34. (a) 1.7946 (b) 0.5572

36. (a) 2.6695 (b) 0.0699

38. (a) $45° = \dfrac{\pi}{4}$ (b) $45° = \dfrac{\pi}{4}$

40. (a) $60° = \dfrac{\pi}{3}$ (b) $60° = \dfrac{\pi}{3}$

42. (a) $60° = \dfrac{\pi}{3}$ (b) $45° = \dfrac{\pi}{4}$

44. (a) $10° \approx 0.175$ (b) $29° \approx 0.506$

46. (a) $22° \approx 0.384$ (b) $68° \approx 1.187$

48. $9\sqrt{3}$ **50.** $20\sqrt{2}$ **52.–60.** Answers will vary.

62. 18 feet **64.** 137.6 feet

66. (a)

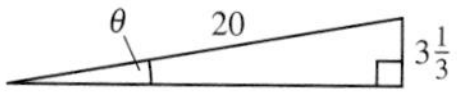

(b) $\sin\theta = \dfrac{3\frac{1}{3}}{20}$ (c) 9.59°

68. 6.57 centimeters

70. $\sin 75° \approx 0.97$ $\csc 75° \approx 1.04$

$\cos 75° \approx 0.26$ $\sec 75° \approx 3.86$

$\tan 75° \approx 3.73$ $\cot 75° \approx 0.27$

72. True, $\sec x = \csc(90 - x)$ **74.** True, -1 for all θ

76. False, $0.4663 = \tan 25° \neq (\tan 5°)(\tan 5°) = 0.0077$.

78. Yes. Tan θ is equal to opp/adj. You can find the value of the hypotenuse by the Pythagorean Theorem, then you can find sec θ, which is equal to hyp/adj.

80. (a)

θ	0°	18°	36°	54°	72°	90°
$\sin\theta$	0	0.3090	0.5878	0.8090	0.9511	1
$\cos\theta$	1	0.9511	0.8090	0.5878	0.3090	0

(b) Increasing function

(c) Decreasing function

(d) As the angle increases, the length of the side opposite the angle increases relative to the length of the hypotenuse and the length of the side adjacent to the angle decreases relative to the length of the hypotenuse. Thus the sine increases and the cosine decreases.

82. $\dfrac{2t+3}{4-t}$, $t \neq \pm\dfrac{3}{2}, -4$ **84.** $\dfrac{1}{4}$, $x \neq 0, 12$

86. $x = 2$ **88.** $x = \frac{2}{3}$

Section 4.4 *(page 387)*

2. (a) $\sin\theta = -\frac{5}{13}$ $\csc\theta = -\frac{13}{5}$

$\cos\theta = -\frac{12}{13}$ $\sec\theta = -\frac{13}{12}$

$\tan\theta = \frac{5}{12}$ $\cot\theta = \frac{12}{5}$

(b) $\sin\theta = \dfrac{\sqrt{2}}{2}$ $\csc\theta = \sqrt{2}$

$\cos\theta = -\dfrac{\sqrt{2}}{2}$ $\sec\theta = -\sqrt{2}$

$\tan\theta = -1$ $\cot\theta = -1$

4. (a) $\sin\theta = \dfrac{\sqrt{10}}{10}$ $\csc\theta = \sqrt{10}$

$\cos\theta = \dfrac{3\sqrt{10}}{10}$ $\sec\theta = \dfrac{\sqrt{10}}{3}$

$\tan\theta = \dfrac{1}{3}$ $\cot\theta = 3$

(b) $\sin\theta = -\dfrac{\sqrt{2}}{2}$ $\csc\theta = -\sqrt{2}$

$\cos\theta = \dfrac{\sqrt{2}}{2}$ $\sec\theta = \sqrt{2}$

$\tan\theta = -1$ $\cot\theta = -1$

6. $\sin\theta = \frac{15}{17}$ $\csc\theta = \frac{17}{15}$

$\cos\theta = \frac{8}{17}$ $\sec\theta = \frac{17}{8}$

$\tan\theta = \frac{15}{8}$ $\cot\theta = \frac{8}{15}$

8. $\sin\theta = -\dfrac{2\sqrt{29}}{29}$ $\csc\theta = -\dfrac{\sqrt{29}}{2}$

$\cos\theta = -\dfrac{5\sqrt{29}}{29}$ $\sec\theta = -\dfrac{\sqrt{29}}{5}$

$\tan\theta = \dfrac{2}{5}$ $\cot\theta = \dfrac{5}{2}$

10. $\sin\theta = -\dfrac{31\sqrt{1157}}{1157}$ $\csc\theta = -\dfrac{\sqrt{1157}}{31}$

$\cos\theta = \dfrac{14\sqrt{1157}}{1157}$ $\sec\theta = \dfrac{\sqrt{1157}}{14}$

$\tan\theta = -\dfrac{31}{14}$ $\cot\theta = -\dfrac{14}{31}$

12. Quadrant I **14.** Quadrant IV

16. $\sin\theta = -\frac{3}{5}$ $\csc\theta = -\frac{5}{3}$

$\cos\theta = -\frac{4}{5}$ $\sec\theta = -\frac{5}{4}$

$\tan\theta = \frac{3}{4}$ $\cot\theta = \frac{4}{3}$

18. $\sin\theta = -\frac{15}{17}$ $\csc\theta = -\frac{17}{15}$

$\cos\theta = \frac{8}{17}$ $\sec\theta = \frac{17}{8}$

$\tan\theta = -\frac{15}{8}$ $\cot\theta = -\frac{8}{15}$

20. $\sin\theta = \dfrac{1}{4}$ $\csc\theta = 4$

$\cos\theta = -\dfrac{\sqrt{15}}{4}$ $\sec\theta = -\dfrac{4\sqrt{15}}{15}$

$\tan\theta = -\dfrac{\sqrt{15}}{15}$ $\cot\theta = -\sqrt{15}$

22. $\sin\theta = 0$ $\csc\theta$ is undefined.

$\cos\theta = -1$ $\sec\theta = -1$

$\tan\theta = 0$ $\cot\theta$ is undefined.

24. $\sin\theta = -1$ $\quad\csc\theta = -1$

$\cos\theta = 0$ $\quad\sec\theta$ is undefined.

$\tan\theta$ is undefined. $\quad\cot\theta = 0$

26. $\sin\theta = -\dfrac{\sqrt{10}}{10}$ $\quad\csc\theta = -\sqrt{10}$

$\cos\theta = -\dfrac{3\sqrt{10}}{10}$ $\quad\sec\theta = -\dfrac{\sqrt{10}}{3}$

$\tan\theta = \dfrac{1}{3}$ $\quad\cot\theta = 3$

28. $\sin\theta = -\frac{4}{5}$ $\quad\csc\theta = -\frac{5}{4}$

$\cos\theta = \frac{3}{5}$ $\quad\sec\theta = \frac{5}{3}$

$\tan\theta = -\frac{4}{3}$ $\quad\cot\theta = -\frac{3}{4}$

30. 0 **32.** Undefined **34.** 0 **36.** Undefined

38. $\theta' = 51°$

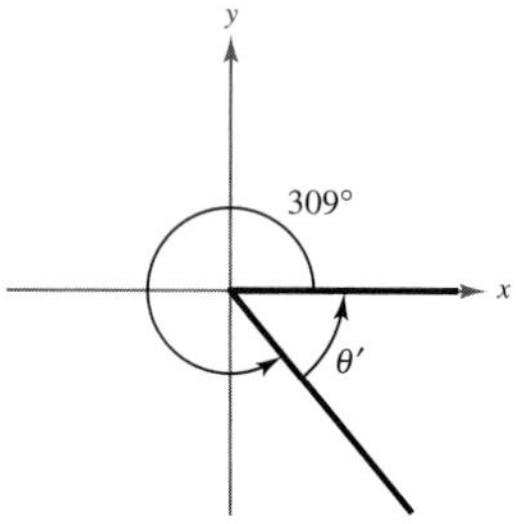

40. $\theta' = 35°$

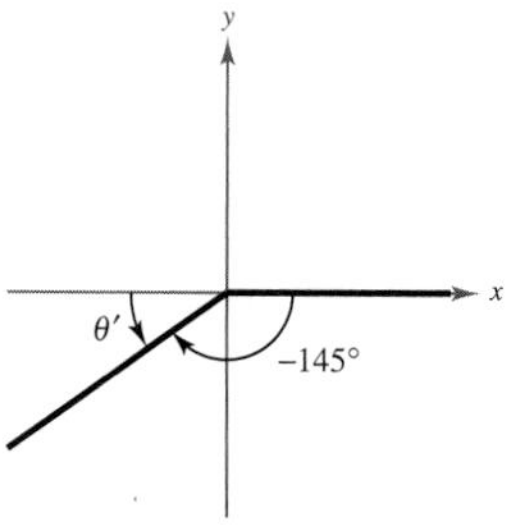

42. $\theta' = \dfrac{\pi}{4}$

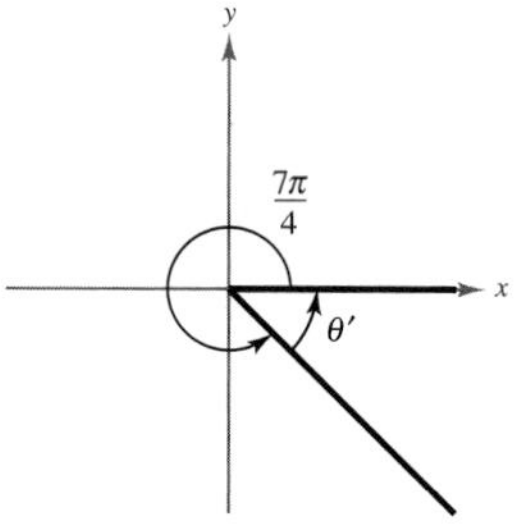

44. $\theta' = \dfrac{\pi}{3}$

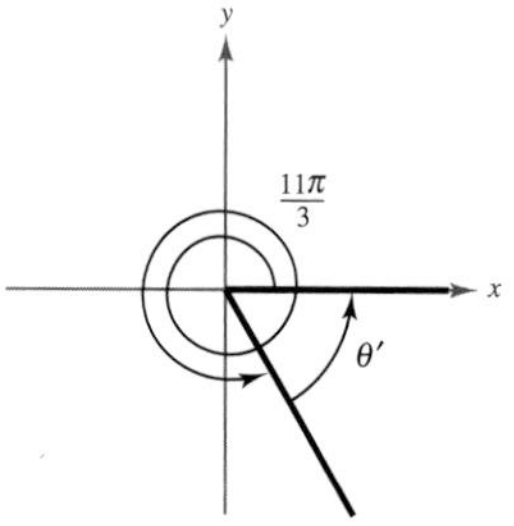

46. $\sin 300° = -\dfrac{\sqrt{3}}{2}$

$\cos 300° = \dfrac{1}{2}$

$\tan 300° = -\sqrt{3}$

48. $\sin(-405°) = -\dfrac{\sqrt{2}}{2}$

$\cos(-405°) = \dfrac{\sqrt{2}}{2}$

$\tan(-405°) = -1$

50. $\sin(-840°) = -\dfrac{\sqrt{3}}{2}$

$\cos(-840°) = -\dfrac{1}{2}$

$\tan(-840°) = \sqrt{3}$

52. $\sin\dfrac{\pi}{4} = \dfrac{\sqrt{2}}{2}$

$\cos\dfrac{\pi}{4} = \dfrac{\sqrt{2}}{2}$

$\tan\dfrac{\pi}{4} = 1$

54. $\sin\left(-\dfrac{\pi}{2}\right) = -1$

$\cos\left(-\dfrac{\pi}{2}\right) = 0$

$\tan\left(-\dfrac{\pi}{2}\right)$ is undefined.

56. $\sin\dfrac{10\pi}{3} = -\dfrac{\sqrt{3}}{2}$

$\cos\dfrac{10\pi}{3} = -\dfrac{1}{2}$

$\tan\dfrac{10\pi}{3} = \sqrt{3}$

58. $\sin\left(-\dfrac{25\pi}{4}\right) = -\dfrac{\sqrt{2}}{2}$

$\cos\left(-\dfrac{25\pi}{4}\right) = \dfrac{\sqrt{2}}{2}$

$\tan\left(-\dfrac{25\pi}{4}\right) = -1$

60. -1.4142 **62.** 2.0000 **64.** 0.2245

66. -0.3640 **68.** 0.6052

70. (a) $45° = \dfrac{\pi}{4}, 315° = \dfrac{7\pi}{4}$ (b) $135° = \dfrac{3\pi}{4}, 225° = \dfrac{5\pi}{4}$

72. (a) $60° = \dfrac{\pi}{3}, 300° = \dfrac{5\pi}{3}$ (b) $120° = \dfrac{2\pi}{3}, 240° = \dfrac{4\pi}{3}$

74. (a) $60° = \dfrac{\pi}{3}, 120° = \dfrac{2\pi}{3}$ (b) $240° = \dfrac{4\pi}{3}, 300° = \dfrac{5\pi}{3}$

76. 29.00°, 331.00° **78.** 220.65°, 319.35°

80. 0.018, 3.124 **82.** 0.175, 3.317 **84.** 1.898, 4.385

86. $\dfrac{\sqrt{10}}{10}$ **88.** $-\sqrt{3}$ **90.** $\dfrac{\sqrt{65}}{4}$

92. (a) 27,700 units (b) 33,000 units (c) 22,800 units
(d) 28,100 units

94. (a) 2 centimeters (b) 0.11 centimeter
(c) -1.2 centimeters

96. (a) 12 miles (b) 6 miles (c) 6.9 miles

98. False. Let $n = 1$ and $\theta = 225°$. $0 \le 135 \le 360$, but $360n - \theta = 135$ is not the reference angle. The reference angle would be 45°.

100. Determine the trigonometric function of the reference angle and prefix the appropriate sign.

102.

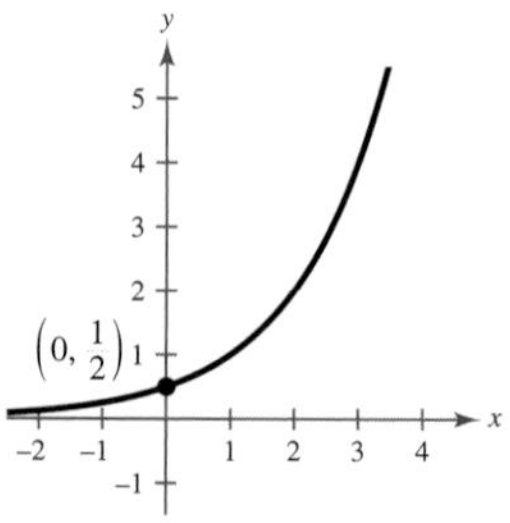

Intercept: $\left(0, \frac{1}{2}\right)$
Asymptote: $y = 0$
Domain: all real numbers
Range: $y > 0$

104.

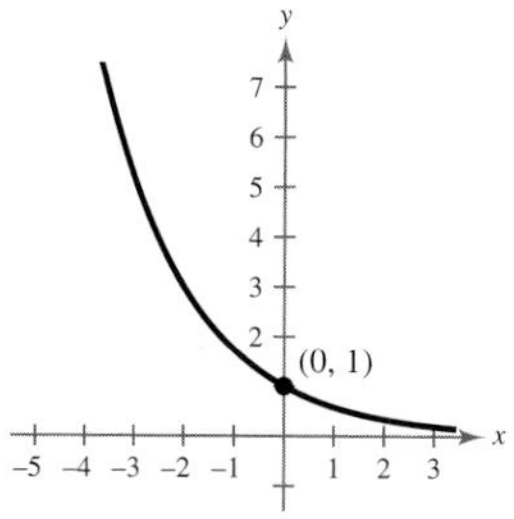

Intercept: (0, 1)
Asymptote: $y = 0$
Domain: all real numbers
Range: $y > 0$

106.

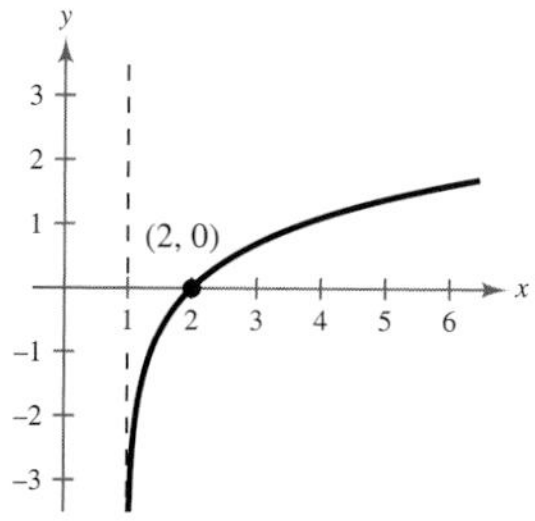

Intercept: (2, 0)
Asymptote: $x = 1$
Domain: $x > 1$
Range: all real numbers

108.

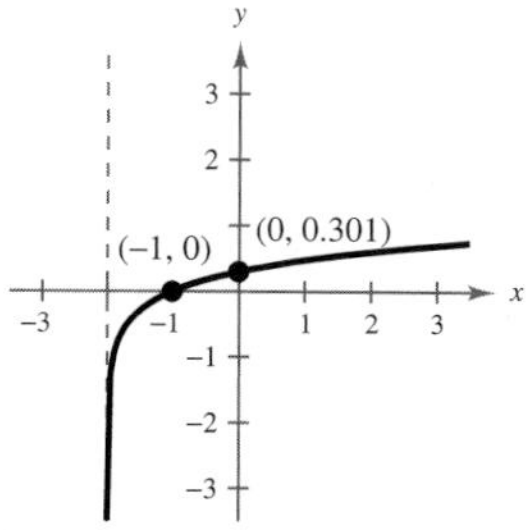

Intercepts:
$(-1, 0)$, $(0, 0.301)$
Asymptote: $x = -2$
Domain: $x > -2$
Range: all real numbers

Section 4.5 *(page 397)*

2. Period: $\frac{2\pi}{3}$
Amplitude: 2

4. Period: 6π
Amplitude: 3

6. Period: 4
Amplitude: $\frac{3}{2}$

8. Period: 3π
Amplitude: 1

10. Period: $\frac{\pi}{4}$
Amplitude: $\frac{1}{3}$

12. Period: 8π
Amplitude: $\frac{5}{2}$

14. Period: 20
Amplitude: $\frac{2}{3}$

16. g is a shift of f π units to the left.

18. g is a reflection of f in the y-axis.

20. The period of g is one-third that of f.

22. g is a shift of f 2 units downward.

24. The period of g is $\frac{1}{3}$ that of f.

26. g is a shift of f 2 units upward.

28.

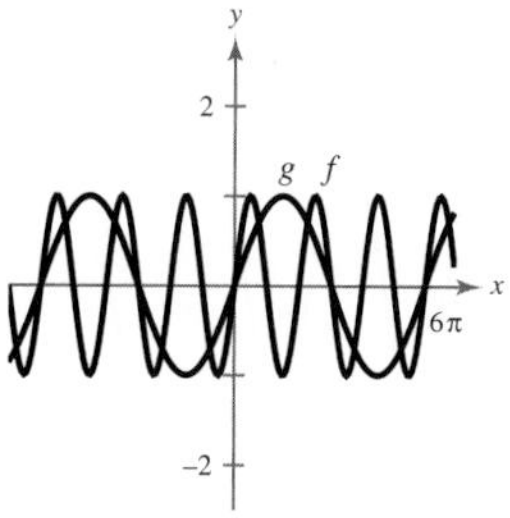

30.

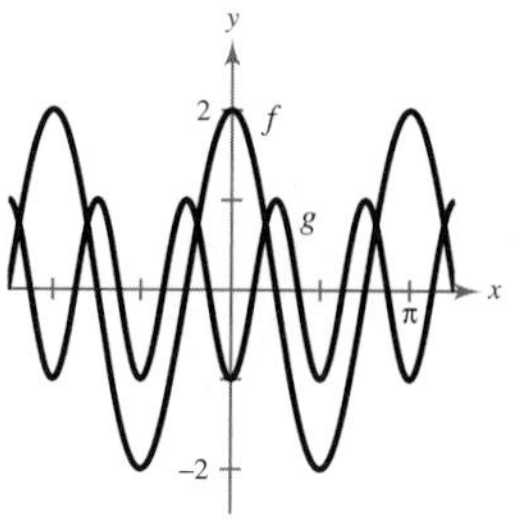

32.

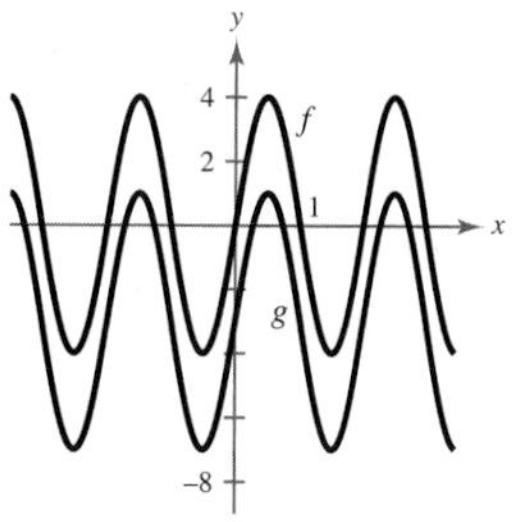

34.

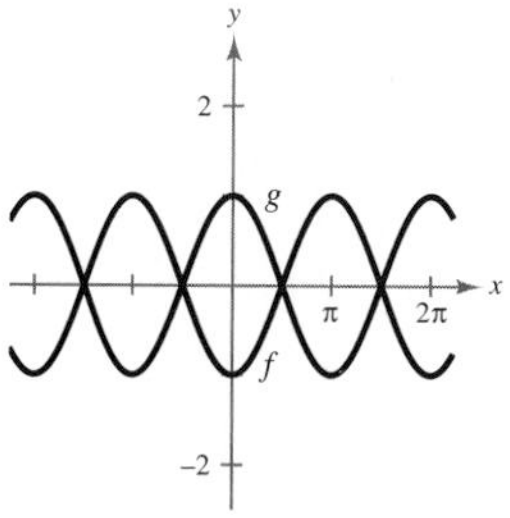

36.

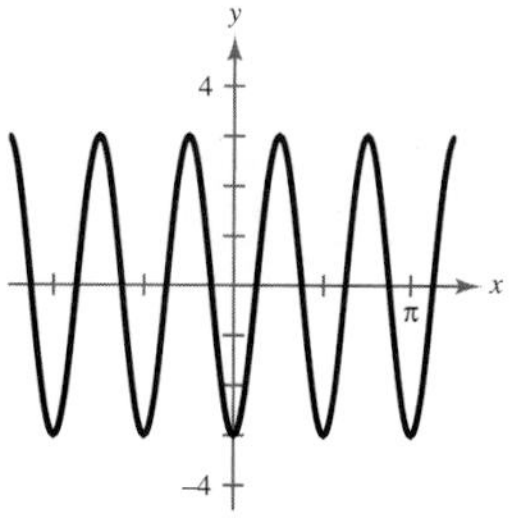

38.

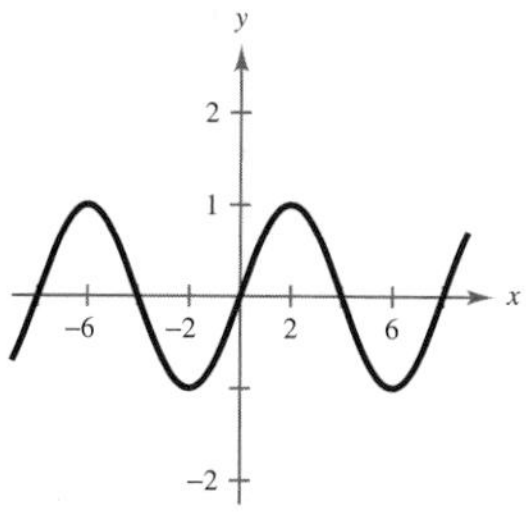

40.

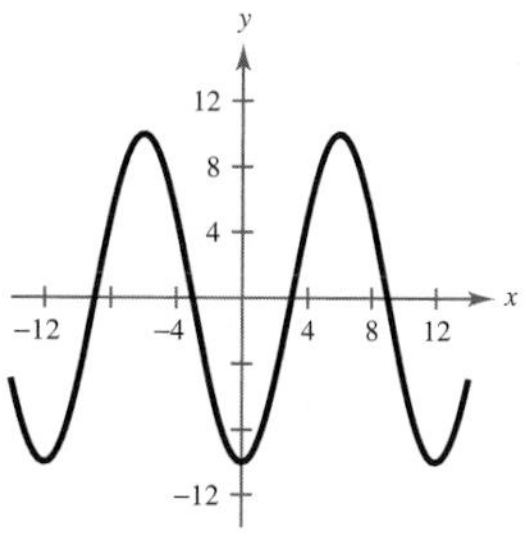

42.

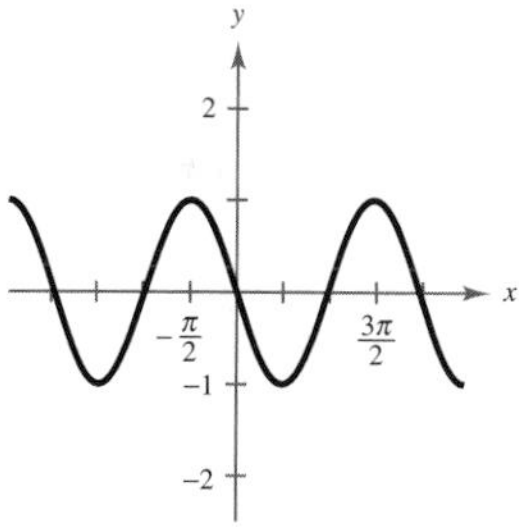

44.

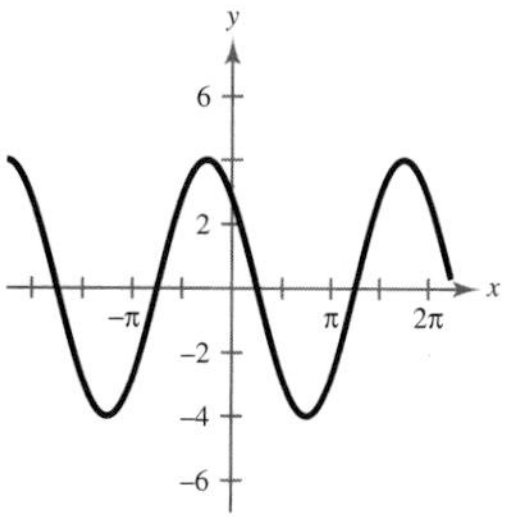

46.

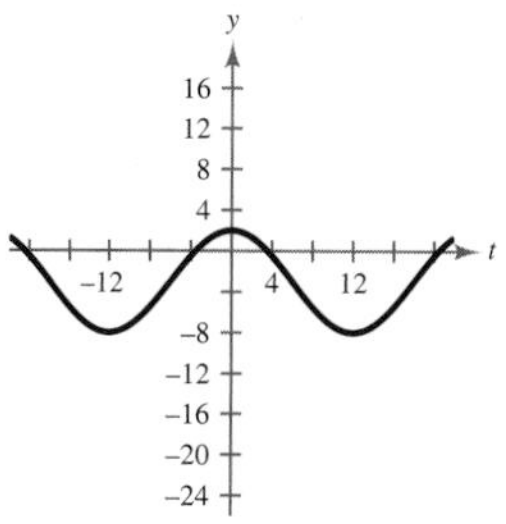

48.

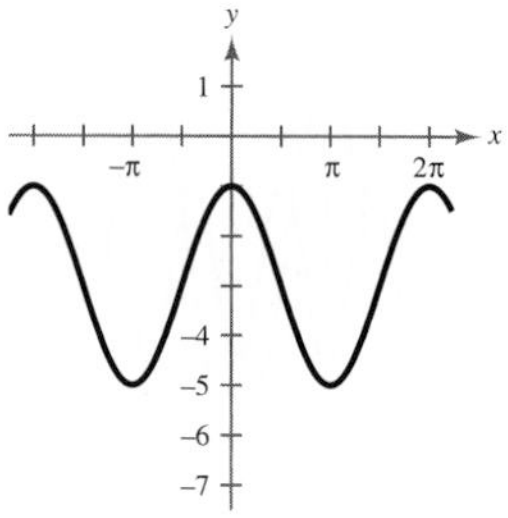

50.

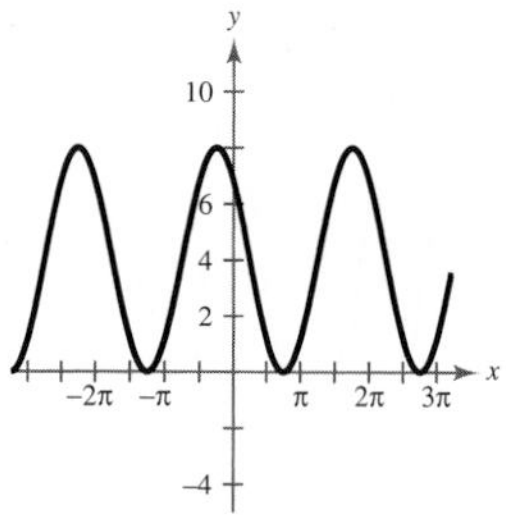

52.

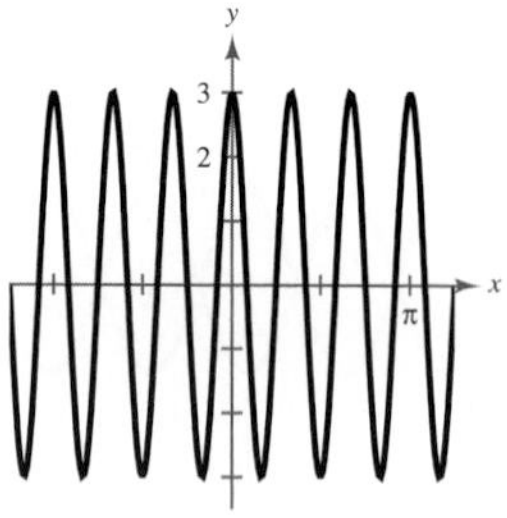

54.

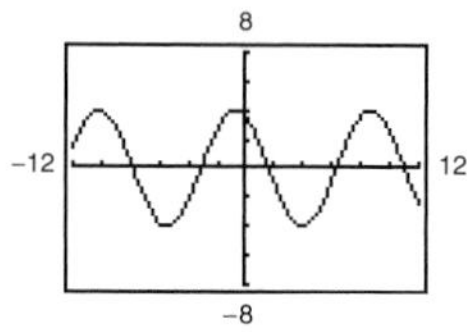

56.

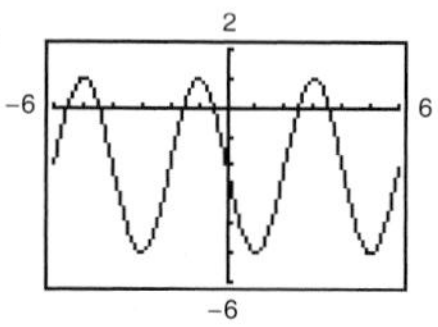

58.

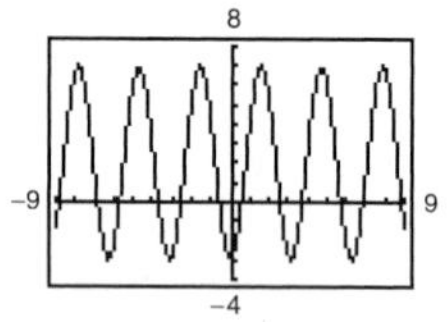

60.

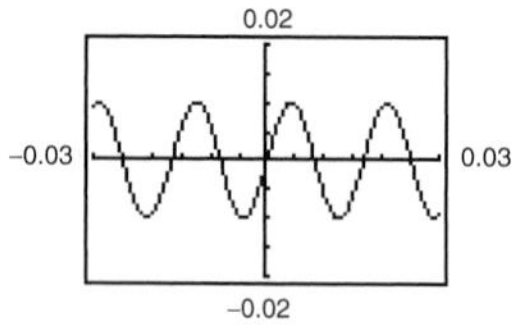

62. $a = 2, d = -1$ **64.** $a = -1, d = -3$

66. $a = 2, b = \frac{1}{2}, c = 0$ **68.** $a = 2, b = \frac{\pi}{2}, c = -\frac{\pi}{2}$

70.

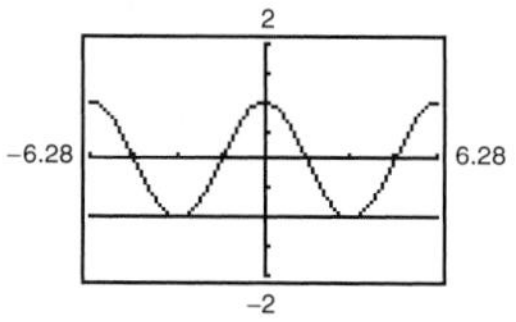

$x = \pi, -\pi$

72.

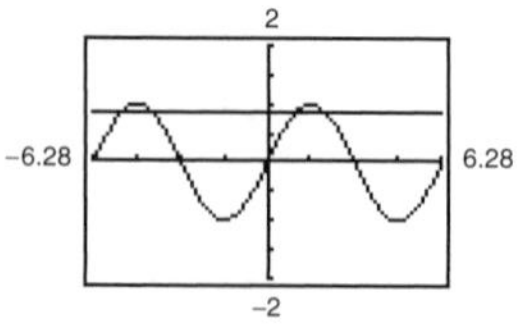

$x = \frac{\pi}{3}, \frac{2\pi}{3}, -\frac{4\pi}{3}, -\frac{5\pi}{3}$

74. (a) 4 seconds (b) 15 cycles per minute

(c)

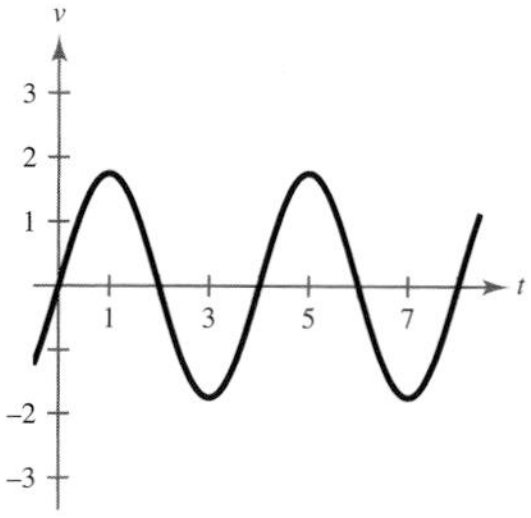

76. (a) $\frac{6}{5}$ seconds (b) 50 heartbeats per minute

78.

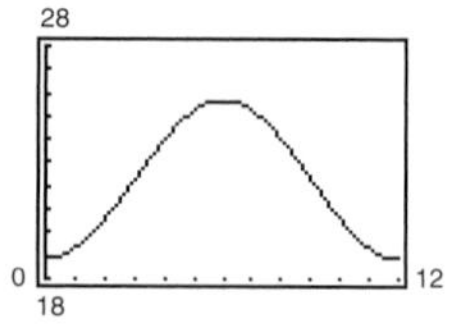

80. (a) 365. Yes. One year is 365 days.

(b) The constant 30.3

(c)

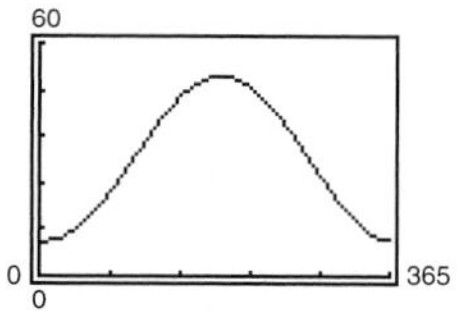

$124 < t < 252$

82. False. The graph of $f(x) = \sin(x + 2\pi)$ translates the graph of $f(x) = \sin x$ exactly one period to the left so that the two graphs look identical.

84.

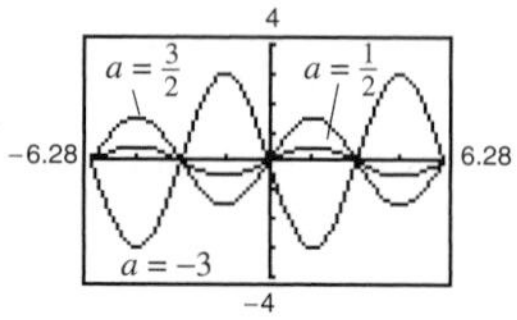

Amplitude changes

86.

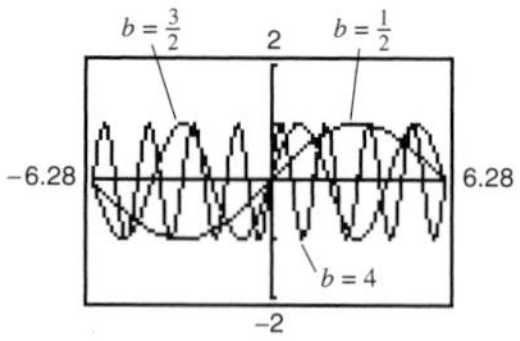

Period changes

88.

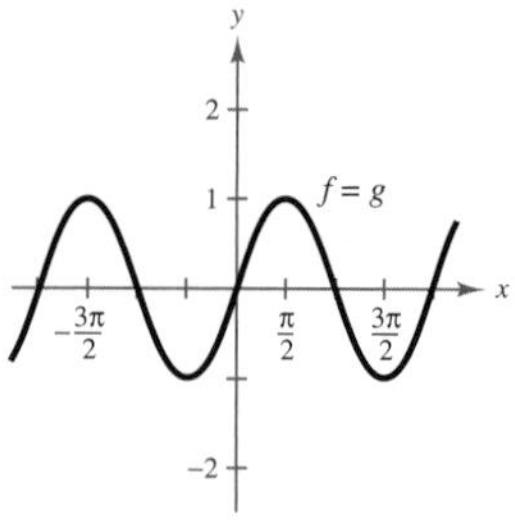

Conjecture:

$\sin x = \cos\left(x - \frac{\pi}{2}\right)$

90.

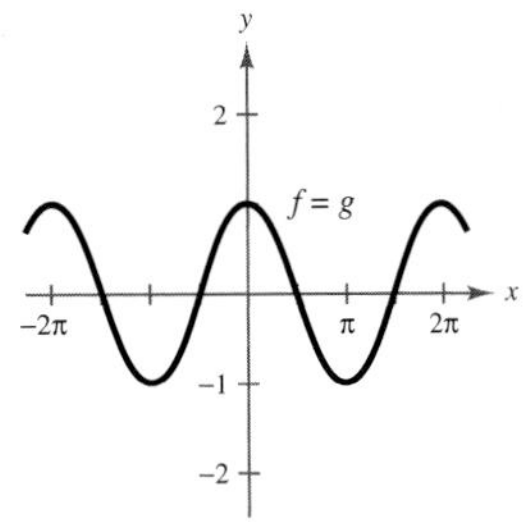

Conjecture:

$$\cos x = -\sin\left(x - \frac{\pi}{2}\right)$$

92. (a)

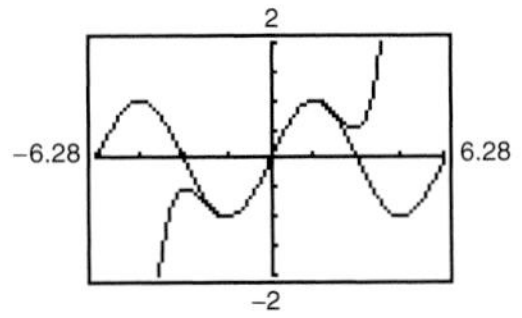

The graphs appear to coincide from $-\frac{\pi}{2}$ to $\frac{\pi}{2}$.

(b)

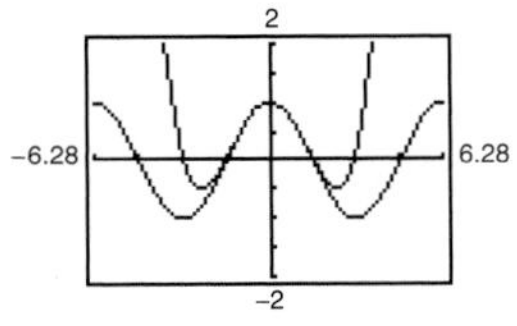

The graphs appear to coincide from $-\frac{\pi}{2}$ to $\frac{\pi}{2}$.

(c) $-\frac{x^7}{7!}, -\frac{x^6}{6!}$

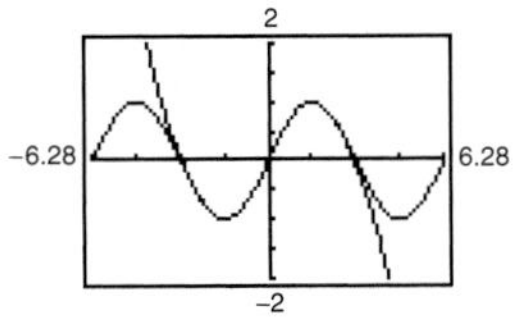

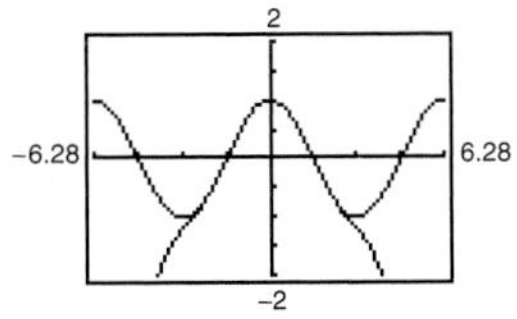

The accuracy increased.

94. (a) Even (b) Even **96.** $\frac{1}{2}\log_{10}(x - 2)$

98. $3 \ln t - \ln(t - 1)$ **100.** $\log_{10} \sqrt{xy}$

102. $\ln \frac{3x}{y^4}$

Section 4.6 *(page 408)*

2. c, 2π **4.** f, 4 **6.** b, 4

8.

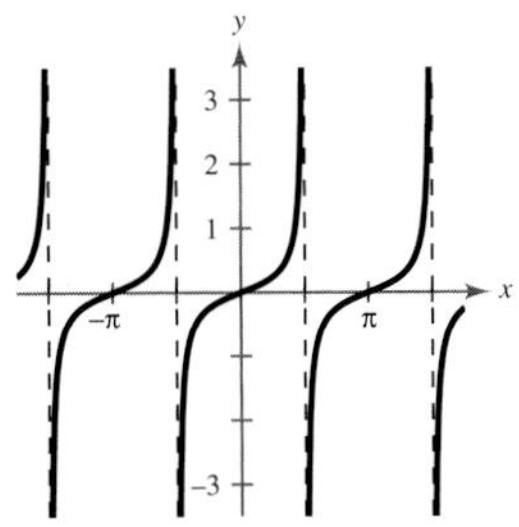

10.

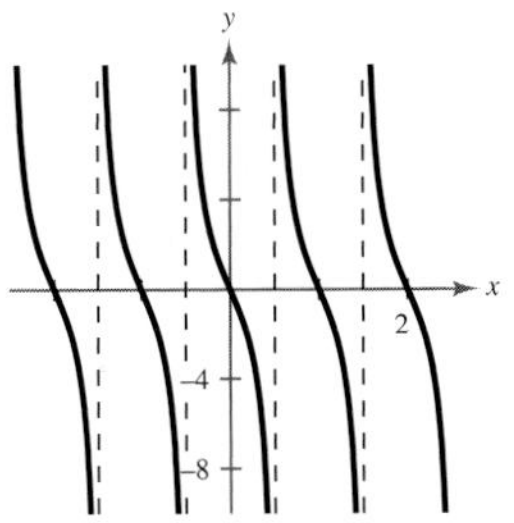

12.

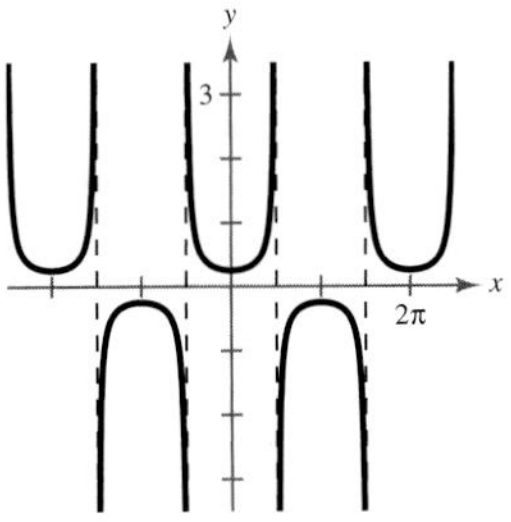

14.

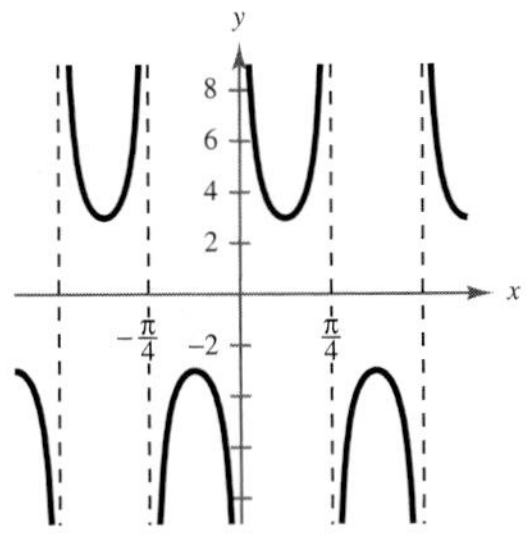

16.

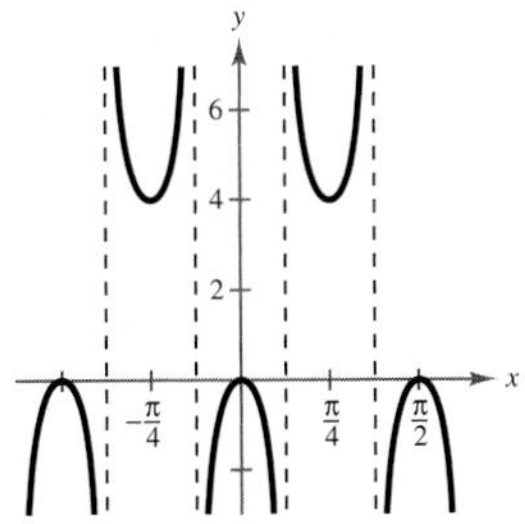

18.

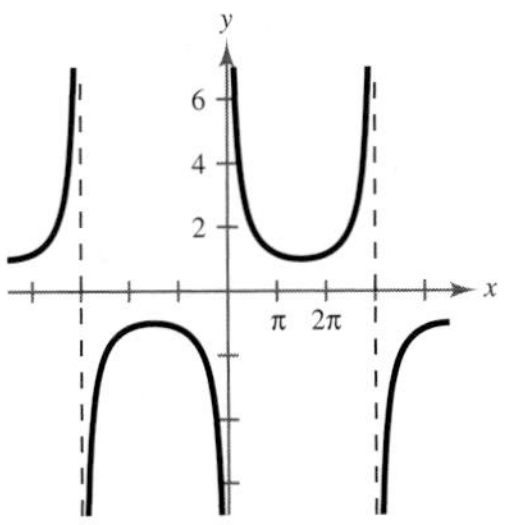

20.

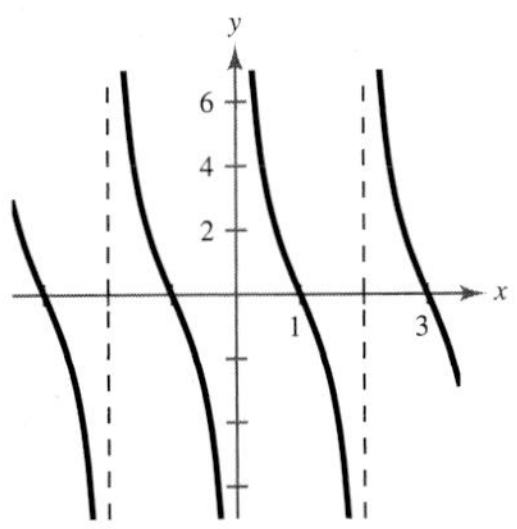

22.

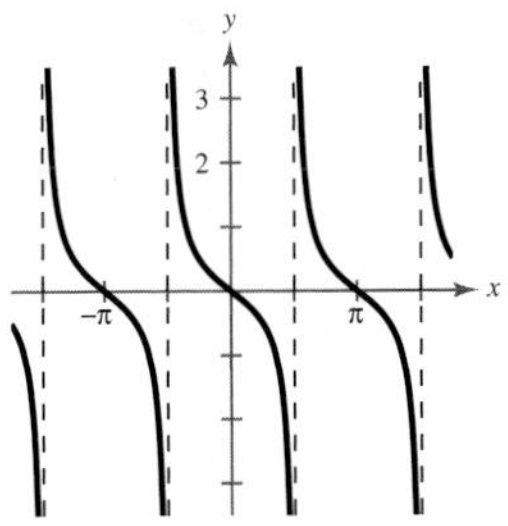

24.

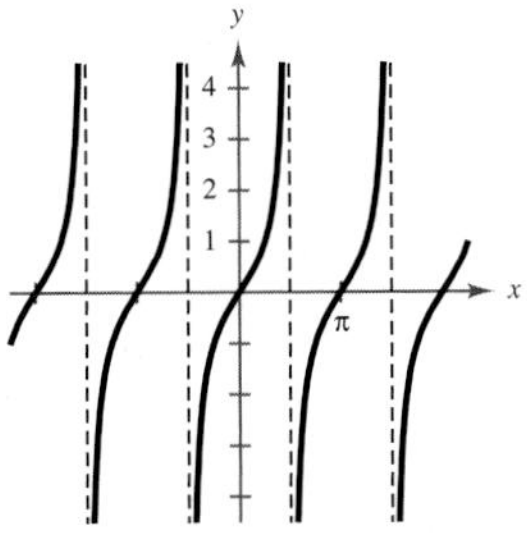

26.

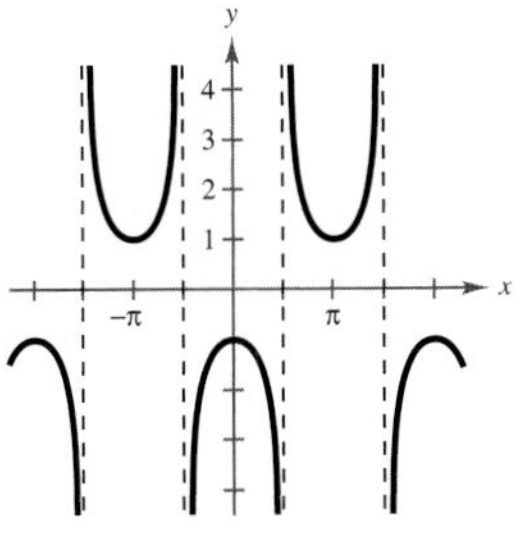

28.

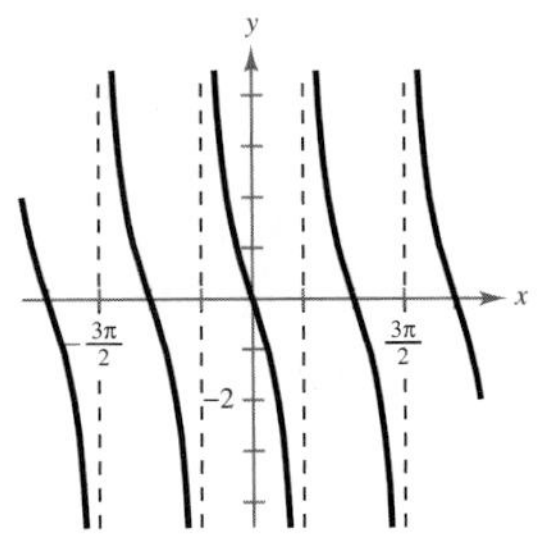

30.

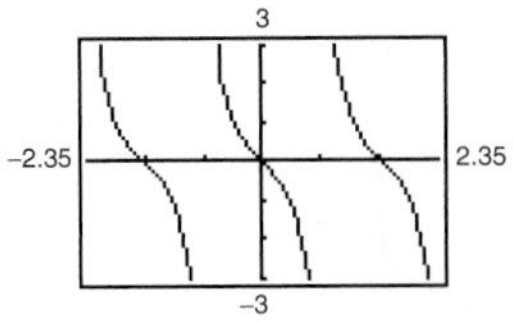

32.

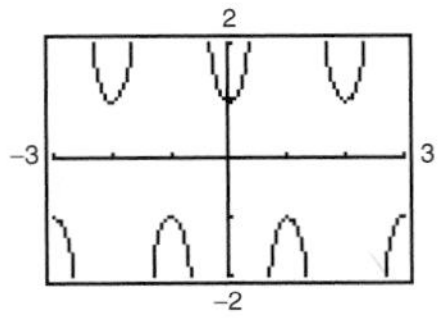

34.

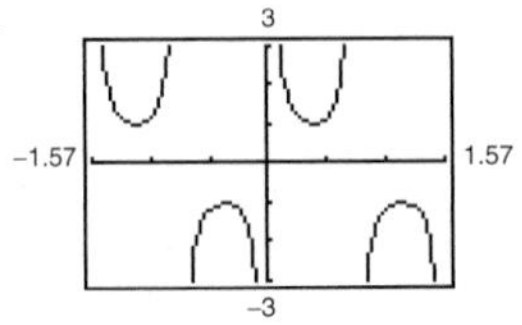

36.

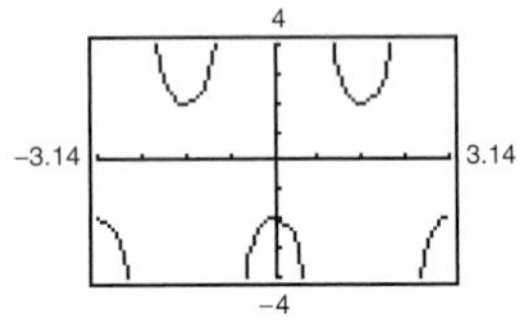

38.

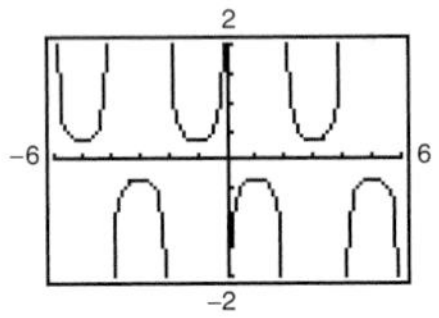

40. $-\frac{5\pi}{3}, -\frac{2\pi}{3}, \frac{\pi}{3}, \frac{4\pi}{3}$

42. $-\frac{7\pi}{4}, -\frac{3\pi}{4}, \frac{\pi}{4}, \frac{5\pi}{4}$

44. $-\frac{5\pi}{3}, -\frac{\pi}{3}, \frac{\pi}{3}, \frac{5\pi}{3}$

46. $-\frac{2\pi}{3}, -\frac{\pi}{3}, \frac{4\pi}{3}, \frac{5\pi}{3}$

48. Odd

50. (a)

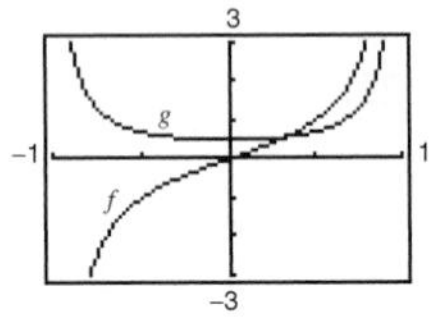

(b) $\left(-1, \frac{1}{3}\right)$

(c) Same interval

52.

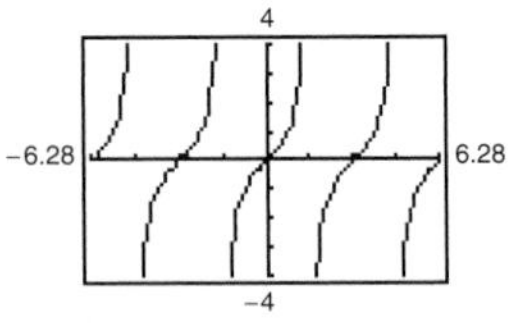

The expressions are equivalent.

54.

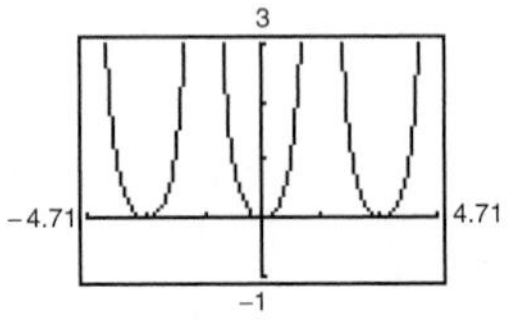

The expressions are equivalent.

56. a, $f \to 0$ as $x \to 0$.

58. c, $g \to 0$ as $x \to 0$.

60.

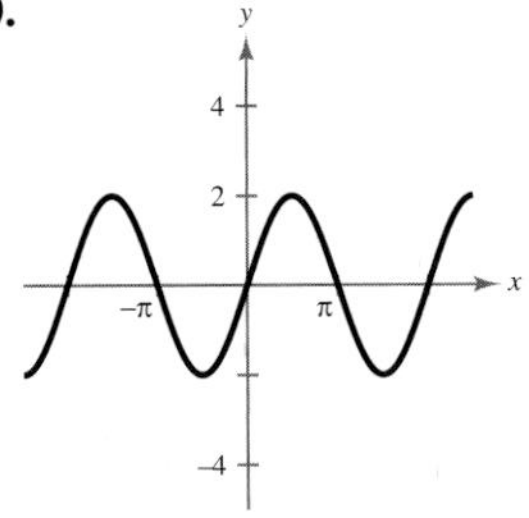

The functions are equal.

62.

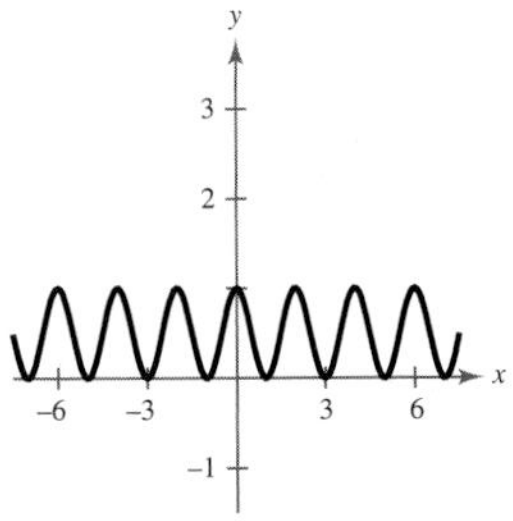

The functions are equal.

64.

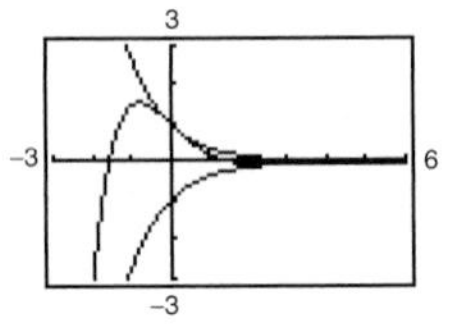

As $x \to \infty, f(x) \to 0$.

66.

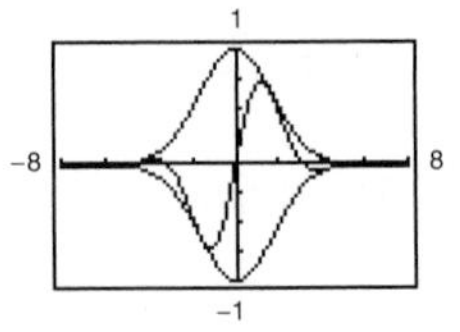

As $x \to \infty$, $h(x) \to 0$.

68.

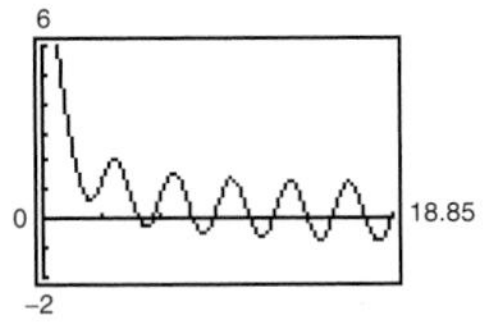

As $x \to 0$, $y \to \infty$.

70.

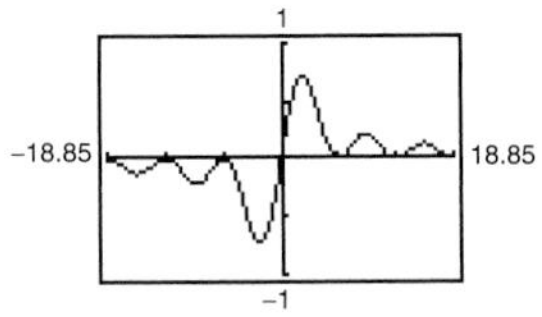

As $x \to 0, f(x) \to 0$.

72.

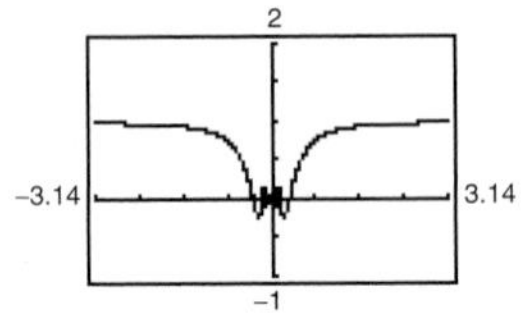

As $x \to 0$, $h(x)$ oscillates.

74. $d = 27 \sec x$

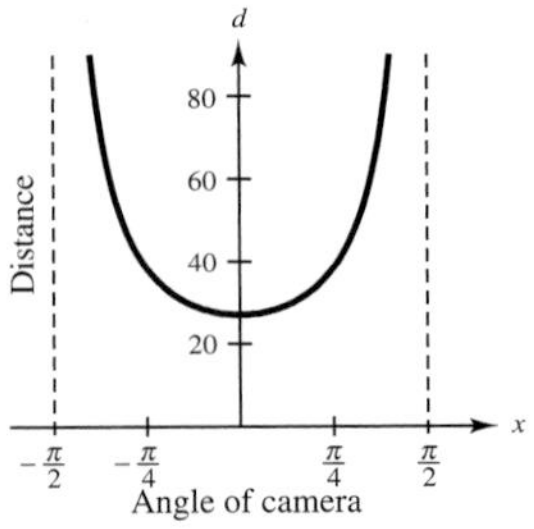

76. As the predator population increases, the number of prey decreases. When the number of prey is small, the number of predators decreases.

78. (a)

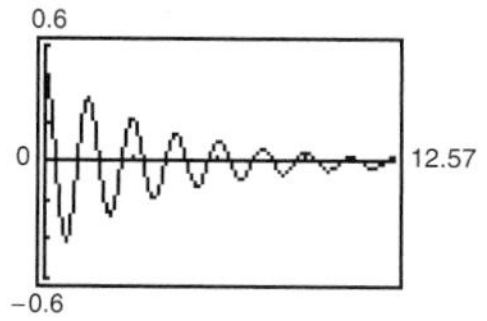

(b) Damped sine wave; goes to 0 as t increases.

80. True. $y = \sec x$ is equal to $y = 1/\cos x$, and if the reciprocal of $y = \sin x$ is translated $\pi/2$ units to the left, then

$$\frac{1}{\sin\left(x + \frac{\pi}{2}\right)} = \frac{1}{\cos x} = \sec x.$$

82. As x approaches π from the left, f approaches ∞. As x approaches π from the right, f approaches $-\infty$.

84.

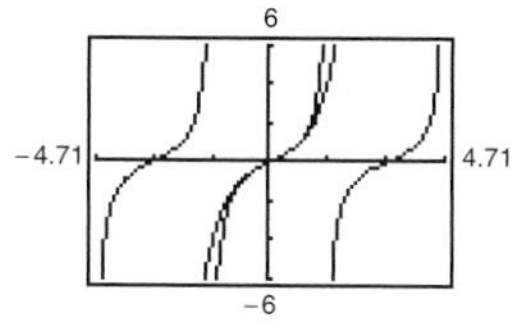

The graphs appear to coincide on the interval $-1.1 \le x \le 1.1$.

86. (a)

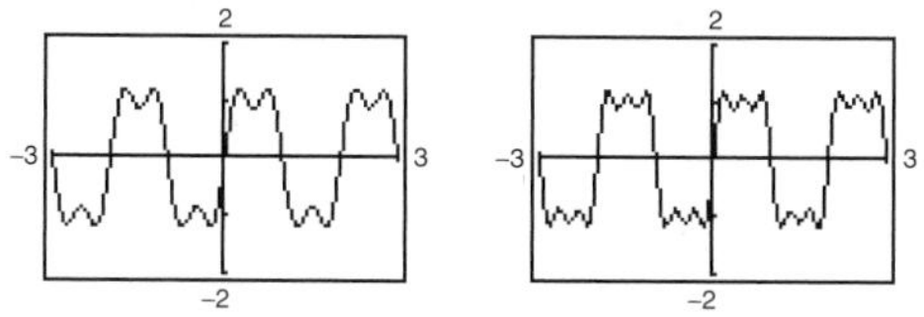

(b) $y_3 = \dfrac{4}{\pi}\left[\sin(\pi x) + \dfrac{1}{3}\sin(3\pi x) + \dfrac{1}{5}\sin(5\pi x) + \dfrac{1}{7}\sin(7\pi x)\right]$

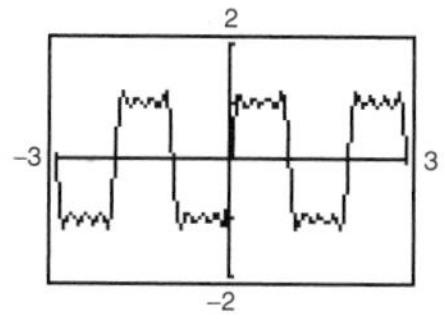

(c) $y_4 = \dfrac{4}{\pi}\left[\sin(\pi x) + \dfrac{1}{3}\sin(3\pi x) + \dfrac{1}{5}\sin(5\pi x) + \dfrac{1}{7}\sin(7\pi x) + \dfrac{1}{9}\sin(9\pi x)\right]$

88. $\dfrac{1}{3}\left(\dfrac{\log_{10} 98}{\log_{10} 8}\right) \approx 0.735$

90. $\dfrac{1}{365}\left(\dfrac{\log_{10} 5}{\log_{10} 1.00041096}\right) \approx 10.732$

92. $\dfrac{14 - e^{68}}{2} \approx -1.702 \times 10^{29}$

94. $e^{10} - 4 \approx 22{,}022.466$ **96.** $\sqrt{65} \approx 8.062$

Section 4.7 *(page 418)*

2. 0 **4.** $\dfrac{\pi}{2}$ **6.** $-\dfrac{\pi}{4}$ **8.** $-\dfrac{\pi}{4}$ **10.** $\dfrac{\pi}{3}$ **12.** $\dfrac{\pi}{4}$

14. $-\dfrac{\pi}{6}$ **16.** 0 **18.** 0.47 **20.** 2.35 **22.** 1.50

24. 1.31 **26.** -0.13 **28.** 1.23 **30.** 1.91

32. -1.50 **34.** $\pi, \dfrac{2\pi}{3}, \dfrac{\sqrt{3}}{2}$

36.

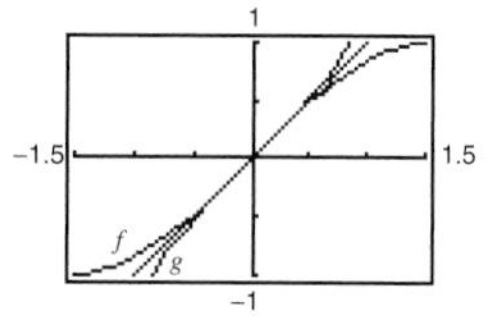

38. $\theta = \arccos \dfrac{4}{x}$

40. $\theta = \arctan \dfrac{x+1}{10}$ **42.** $\theta = \arctan \dfrac{1}{x+1},\ x \ne 1$

44. 25 **46.** -0.2 **48.** $\dfrac{\pi}{2}$ **50.** $\dfrac{5}{3}$ **52.** $\dfrac{2\sqrt{5}}{5}$

54. $-\dfrac{13}{5}$ **56.** $-\dfrac{3\sqrt{7}}{7}$ **58.** $\dfrac{8}{5}$ **60.** $\dfrac{x}{\sqrt{x^2+1}}$

62. $\sqrt{9x^2+1}$ **64.** $\dfrac{1}{\sqrt{2x - x^2}}$ **66.** x

68. $\dfrac{\sqrt{r^2 - (x-h)^2}}{r}$

70.

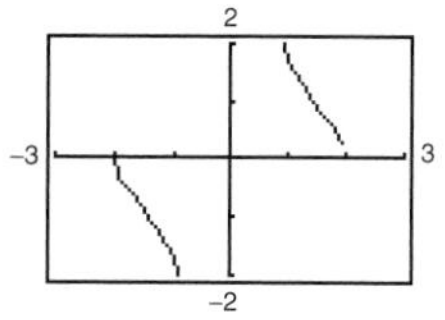

Asymptote: $x = 0$

72. $\dfrac{x}{6}$ **74.** $\dfrac{\sqrt{4x - x^2}}{x - 2}$

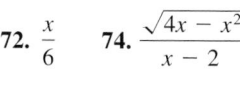

76.

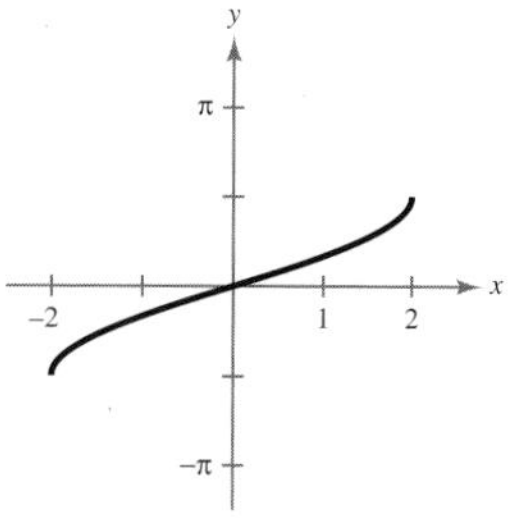

78.

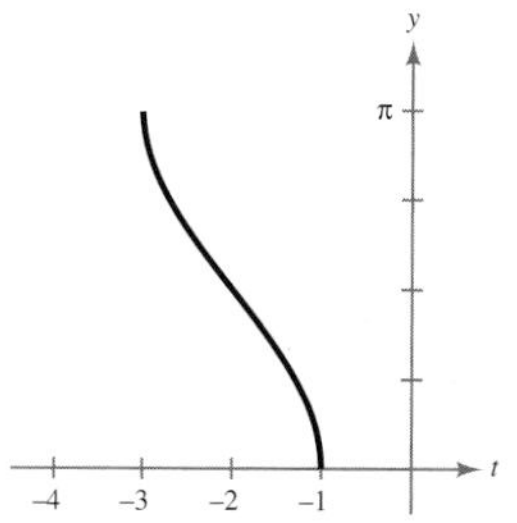

80.

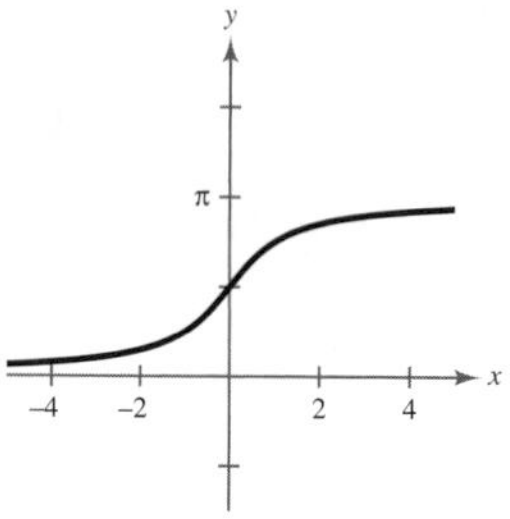

82.

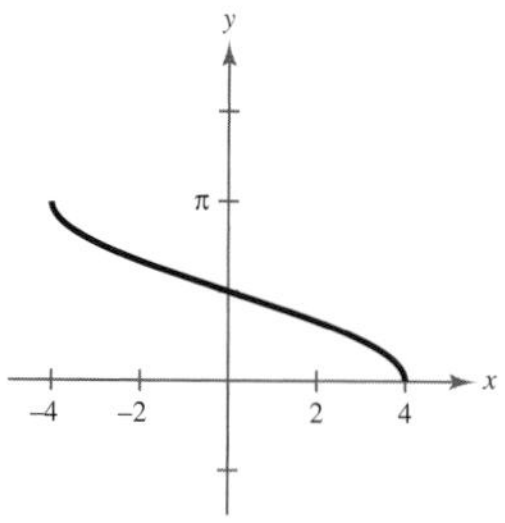

84.

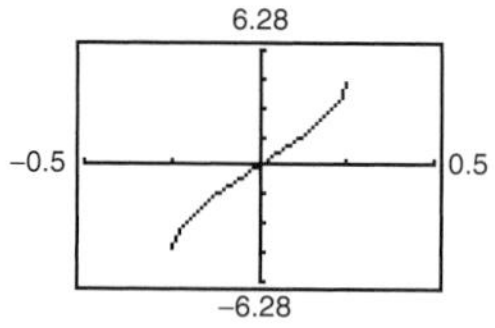

86.

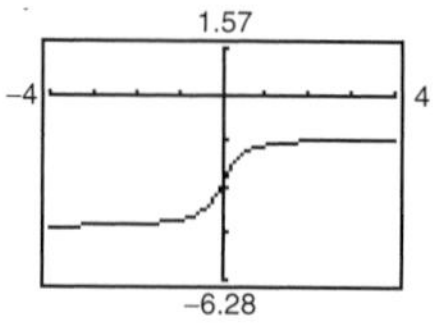

88.

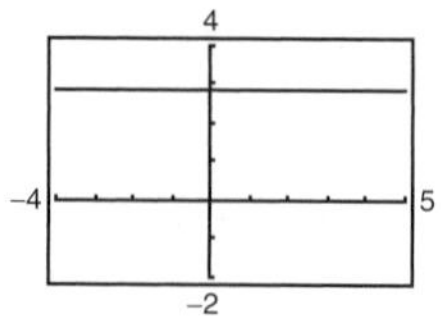

90. $5 \sin\left(\pi t + \arctan \frac{4}{3}\right)$

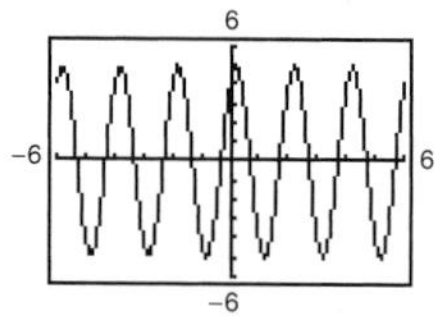

The graph implies that the identity is true.

92. (a) $\theta = \arctan \dfrac{s}{750}$ (b) 0.38, 1.01

94. (a) $\theta = \arctan \dfrac{6}{x}$ (b) 0.71, 1.41

96. False. $\dfrac{5\pi}{6}$ is not in the range of the arcsine.

98. (a) $\dfrac{\pi}{4}$ (b) $\dfrac{\pi}{2}$ (c) 1.25 (d) 2.03

100. Domain: $(-\infty, -1] \cup [1, \infty)$

Range: $[0, \pi/2) \cup (\pi/2, \pi]$

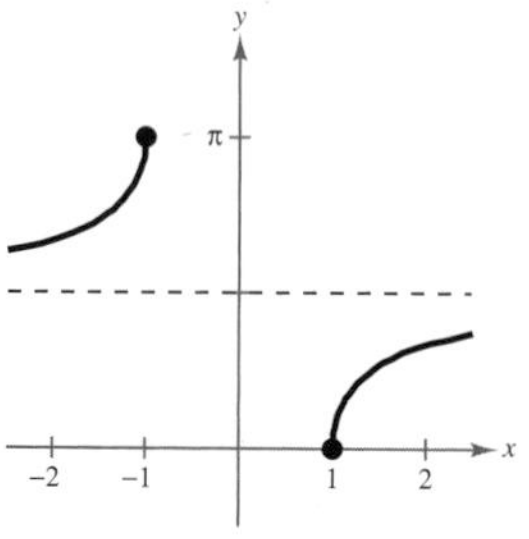

102. (a) $\dfrac{\pi}{4}$ (b) 0 (c) $\dfrac{5\pi}{6}$ (d) $\dfrac{\pi}{6}$

104. (a) $f \circ f^{-1}$

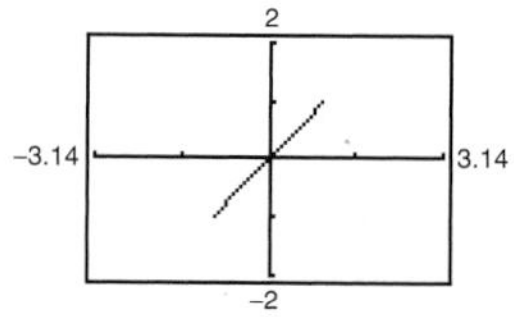

$f^{-1} \circ f$

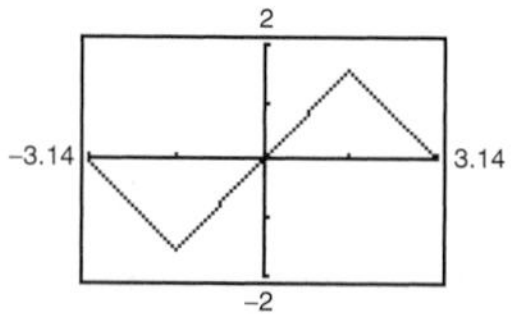

(b) The domains and ranges of the functions are restricted. The graphs of $f \circ f^{-1}$ and $f^{-1} \circ f$ differ because of the domains and ranges of f and f^{-1}.

106.–110. Answers will vary.

112.

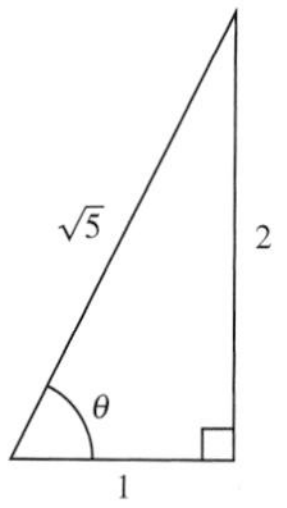

114.

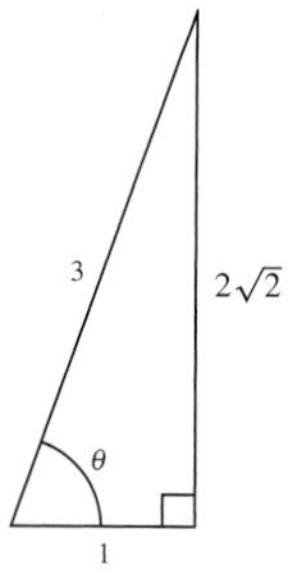

116. 0.051 **118.** 2.718×10^{-8} **120.** 3 miles per hour

Section 4.8 *(page 428)*

2. $a \approx 8.82$
$b \approx 12.14$
$A = 36°$

4. $b \approx 274.27$
$c \approx 277.24$
$B = 81.6°$

6. $b \approx 24.49$
$A \approx 45.58°$
$B = 44.42°$

8. $a \approx 9.36$
$A \approx 81.97°$
$B = 8.03°$

10. $b \approx 30.73$
$c \approx 33.85$
$A = 24°48'$

12. 1.62 meters **14.** 2.80 feet **16.** 1648.5 feet

18. 81.2 feet **20.** 123.5 feet **22.** 65 feet

24. (a)

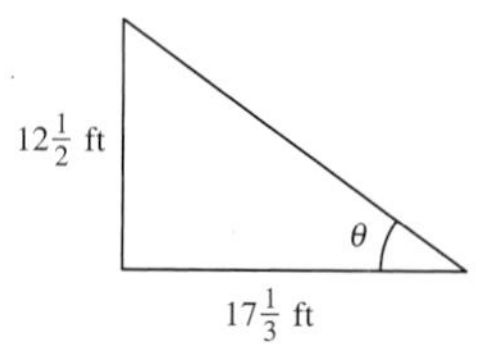

(b) $\tan \theta = \dfrac{12\frac{1}{2}}{17\frac{1}{3}}$

(c) 35.8°

26. 2.06° **28.** 117.7 seconds **30.** 6.8°; 2516.3 feet

32. 107 miles south; 54 miles west **34.** 5.46 kilometers

36. S 35.3° W **38.** 11.8 kilometers

40. ≈ 1.02 miles or ≈ 5410 feet **42.** 52.1° **44.** 54.7°

46. 25 inches **48.** 9.06 centimeters

50. $a \approx 21.6$ feet, $b \approx 7.2$ feet, $c \approx 13$ feet

52. (a) $\frac{1}{2}$ (b) 10 (c) $\frac{1}{40}$

54. (a) $\dfrac{1}{64}$ (b) 396 (c) $\dfrac{1}{792}$ **56.** $d = 3 \sin\left(\dfrac{\pi t}{3}\right)$

58. $d = 2 \cos\left(\dfrac{\pi t}{5}\right)$ **60.** $d = \dfrac{7}{4} \cos \dfrac{\pi t}{5}$

62. False. The tower is leaning, so it is not perfectly vertical and does not form a right angle with the ground.

64. (a)

θ	L_1	L_2	$L_1 + L_2$
0.1	$\dfrac{2}{\sin 0.1}$	$\dfrac{3}{\cos 0.1}$	23.0
0.2	$\dfrac{2}{\sin 0.2}$	$\dfrac{3}{\cos 0.2}$	13.1
0.3	$\dfrac{2}{\sin 0.3}$	$\dfrac{3}{\cos 0.3}$	9.9
0.4	$\dfrac{2}{\sin 0.4}$	$\dfrac{3}{\cos 0.4}$	8.4

(b)

θ	L_1	L_2	$L_1 + L_2$
0.5	$\dfrac{2}{\sin 0.5}$	$\dfrac{3}{\cos 0.5}$	7.6
0.6	$\dfrac{2}{\sin 0.6}$	$\dfrac{3}{\cos 0.6}$	7.2
0.7	$\dfrac{2}{\sin 0.7}$	$\dfrac{3}{\cos 0.7}$	7.0
0.8	$\dfrac{2}{\sin 0.8}$	$\dfrac{3}{\cos 0.8}$	7.1

7.0 (minimum length)

(c) $L = L_1 + L_2 = \dfrac{2}{\sin \theta} + \dfrac{3}{\cos \theta}$

(d)

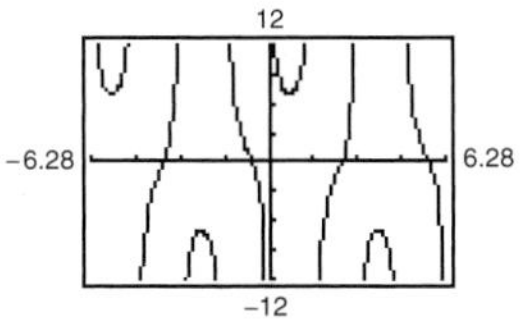

7.0 (minimum length)

66. (a)

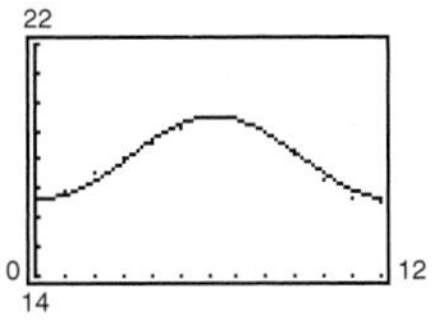

(b) 12. Yes. One period is 1 year.

(c) 1.41. 1.41 represents the maximum change in time from the average time ($d = 18.09$) of sunset.

68.

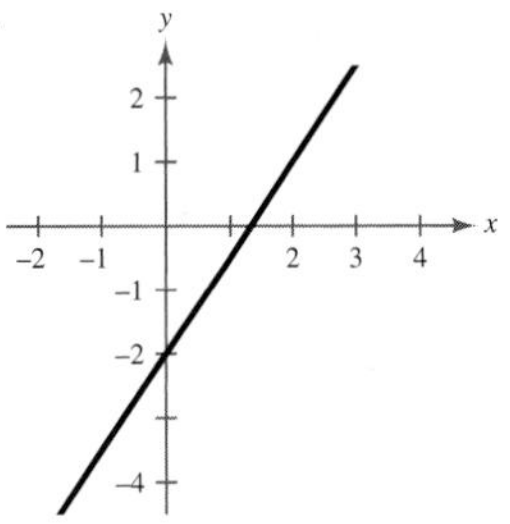

70.

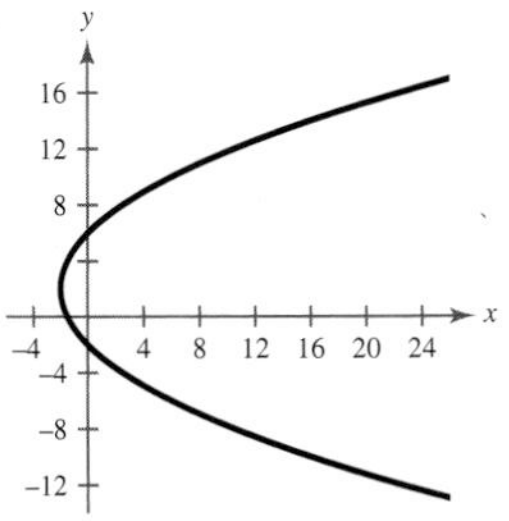

72.

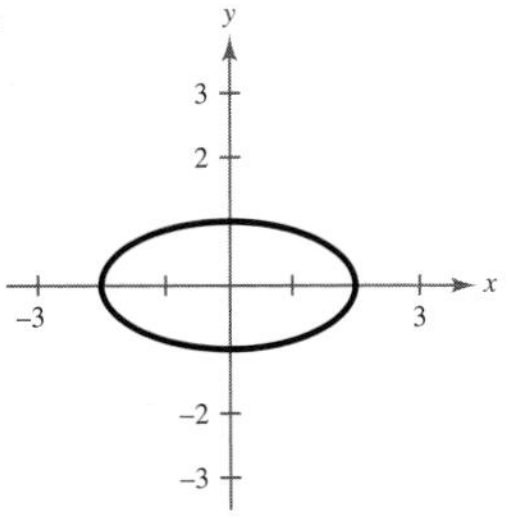

74.

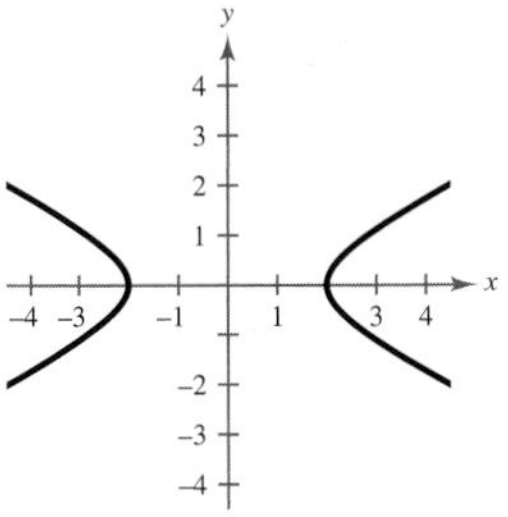

76.

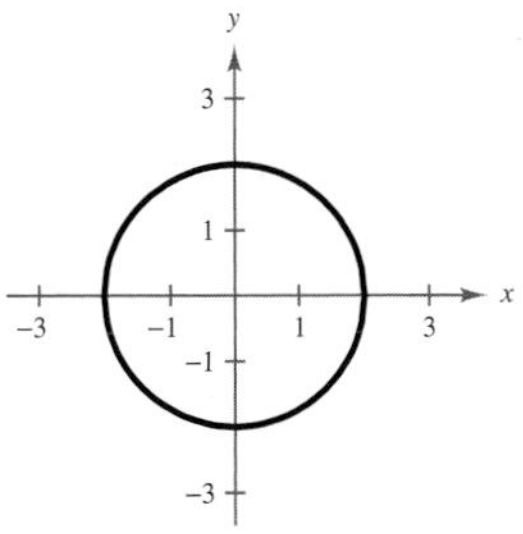

Review Exercises *(page 434)*

2. 2 **4.** −3.5

6.

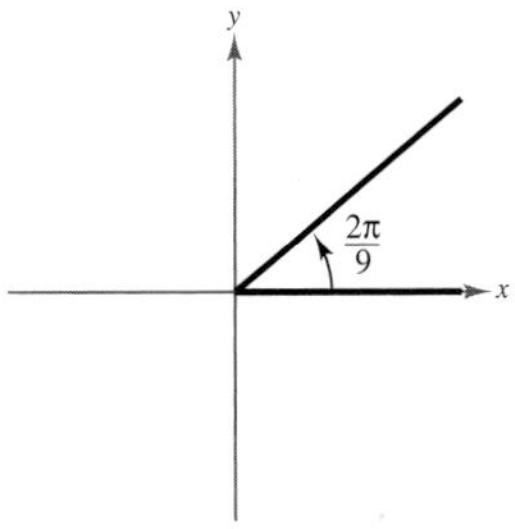

$\dfrac{20\pi}{9}, -\dfrac{16\pi}{9}$

8.

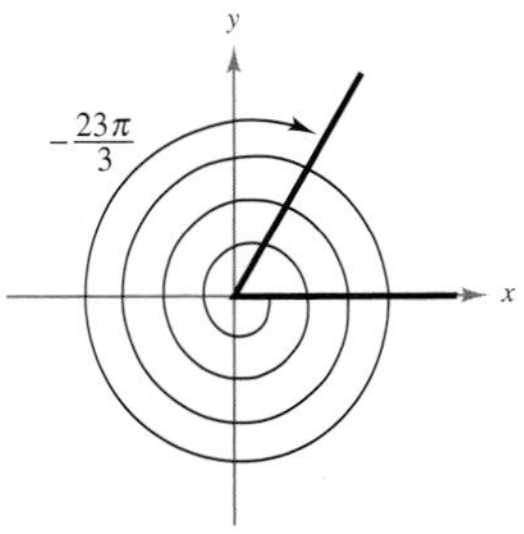

$\dfrac{\pi}{3}, -\dfrac{17\pi}{3}$

10.

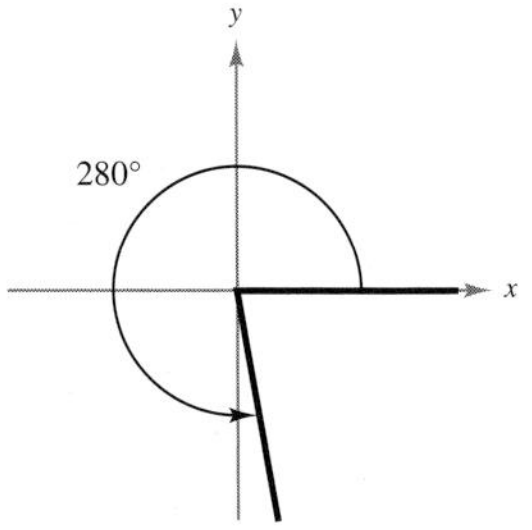

640°, $-80°$

12.

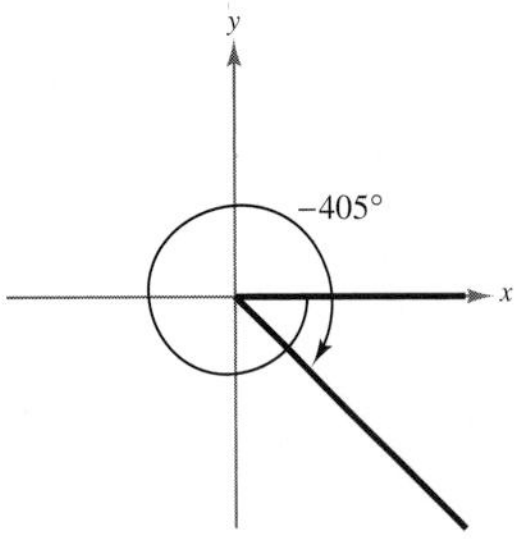

315°, $-45°$

14. $-330°$ **16.** 326.59° **18.** -2.2253 **20.** 3.4432

22. 212.1 inches per second $\approx$ 12.06 miles per hour

24. $\left(-\frac{\sqrt{2}}{2}, \frac{\sqrt{2}}{2}\right)$ **26.** $\left(-\frac{1}{2}, \frac{\sqrt{3}}{2}\right)$

28. $\sin\frac{\pi}{4} = \frac{\sqrt{2}}{2}$ $\csc\frac{\pi}{4} = \sqrt{2}$
$\cos\frac{\pi}{4} = \frac{\sqrt{2}}{2}$ $\sec\frac{\pi}{4} = \sqrt{2}$
$\tan\frac{\pi}{4} = 1$ $\cot\frac{\pi}{4} = 1$

30. $\sin 2\pi = 0$ $\csc 2\pi$ is undefined.
$\cos 2\pi = 1$ $\sec 2\pi = 1$
$\tan 2\pi = 0$ $\cot 2\pi$ is undefined.

32. $\cos 4\pi = \cos 2\pi = 1$

34. $\cos\left(-\frac{13\pi}{3}\right) = \cos\frac{5\pi}{3} = \frac{1}{2}$

36. -1.14 **38.** -0.34

40. $\sin\theta = \frac{\sqrt{2}}{2}$ $\csc\theta = \sqrt{2}$
$\cos\theta = \frac{\sqrt{2}}{2}$ $\sec\theta = \sqrt{2}$
$\tan\theta = 1$ $\cot\theta = 1$

42. $\sin\theta = \frac{5}{9}$ $\csc\theta = \frac{9}{5}$
$\cos\theta = \frac{2\sqrt{14}}{9}$ $\sec\theta = \frac{9\sqrt{14}}{28}$
$\tan\theta = \frac{5\sqrt{14}}{28}$ $\cot\theta = \frac{2\sqrt{14}}{5}$

44. (a) $\frac{1}{4}$ (b) $\sqrt{17}$ (c) $\frac{\sqrt{17}}{17}$ (d) $\frac{\sqrt{17}}{4}$

46. (a) $\frac{1}{5}$ (b) $2\sqrt{6}$ (c) $\frac{\sqrt{6}}{12}$ (d) 5 **48.** 5.24

50. 5.39 **52.** 0.20 **54.** 19.5 feet

56. $\sin\theta = -\frac{4}{5}$ $\csc\theta = -\frac{5}{4}$
$\cos\theta = \frac{3}{5}$ $\sec\theta = \frac{5}{3}$
$\tan\theta = -\frac{4}{3}$ $\cot\theta = -\frac{3}{4}$

58. $\sin\theta = -\frac{\sqrt{26}}{26}$ $\csc\theta = -\sqrt{26}$
$\cos\theta = -\frac{5\sqrt{26}}{26}$ $\sec\theta = -\frac{\sqrt{26}}{5}$
$\tan\theta = \frac{1}{5}$ $\cot\theta = 5$

60. $\sin\theta = 0.8$ $\csc\theta = 1.25$
$\cos\theta = 0.6$ $\sec\theta \approx 1.67$
$\tan\theta \approx 1.33$ $\cot\theta = 0.75$

62. $\sin\theta = -\frac{3\sqrt{13}}{13}$ $\csc\theta = -\frac{\sqrt{13}}{3}$
$\cos\theta = -\frac{2\sqrt{13}}{13}$ $\sec\theta = -\frac{\sqrt{13}}{2}$
$\tan\theta = \frac{3}{2}$ $\cot\theta = \frac{2}{3}$

64. $\sin\theta = \frac{2}{3}$
$\cos\theta = -\frac{\sqrt{5}}{3}$
$\tan\theta = -\frac{2\sqrt{5}}{5}$
$\sec\theta = -\frac{3\sqrt{5}}{5}$
$\cot\theta = -\frac{\sqrt{5}}{2}$

66. $\sin\theta = -\frac{5\sqrt{41}}{41}$
$\cos\theta = -\frac{4\sqrt{41}}{41}$
$\csc\theta = -\frac{\sqrt{41}}{5}$
$\sec\theta = -\frac{\sqrt{41}}{4}$
$\cot\theta = \frac{4}{5}$

68. $\cos\theta = \frac{\sqrt{3}}{2}$
$\tan\theta = -\frac{\sqrt{3}}{3}$
$\csc\theta = -2$
$\sec\theta = \frac{2\sqrt{3}}{3}$
$\cot\theta = -\sqrt{3}$

70. $\sqrt{2}$ **72.** -1

74. $-\frac{1}{2}$ **76.** -0.14

78. 0.09 **80.** 0 **82.** 4.38

84.

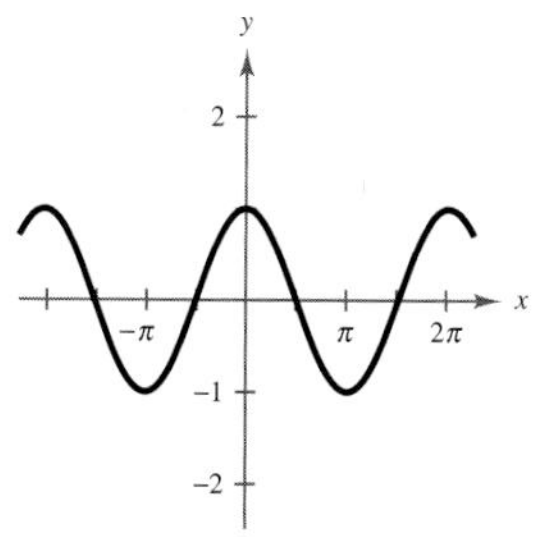

86.

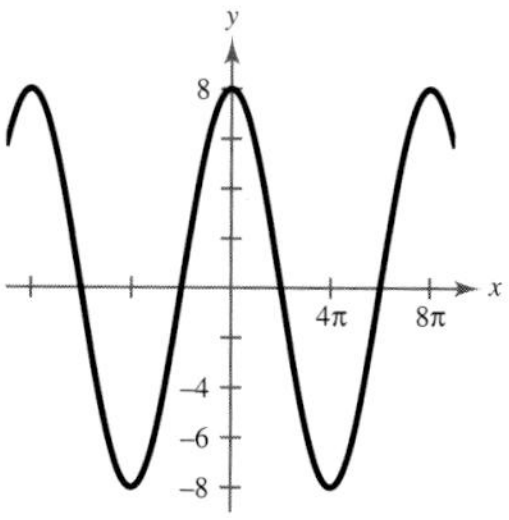

88.

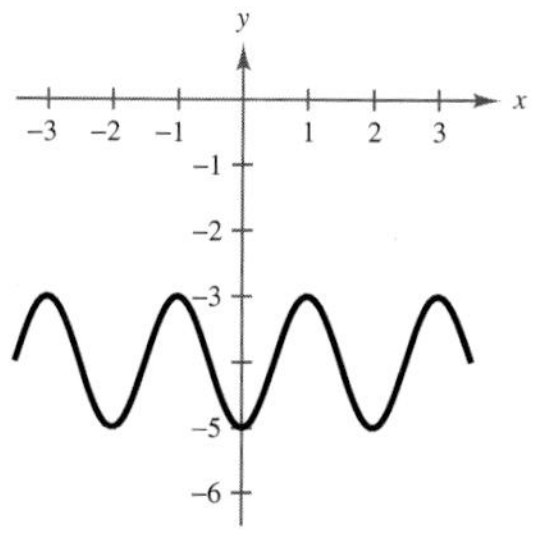

90.

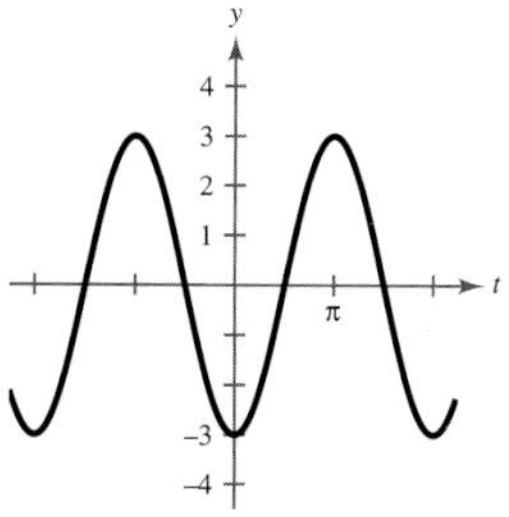

92. $y = 2\cos 528\pi\left(x - \dfrac{\pi}{2}\right)$

94.

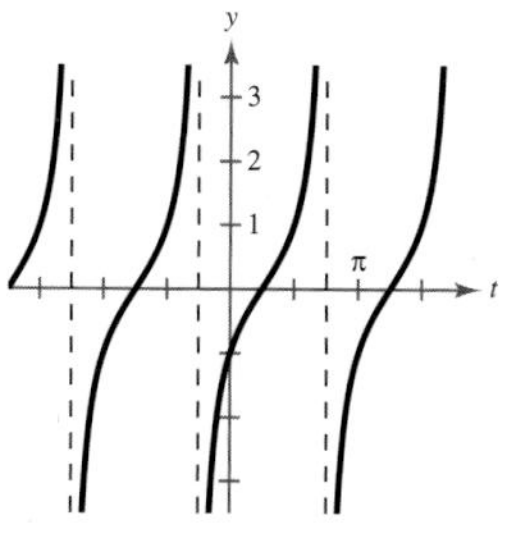

96.

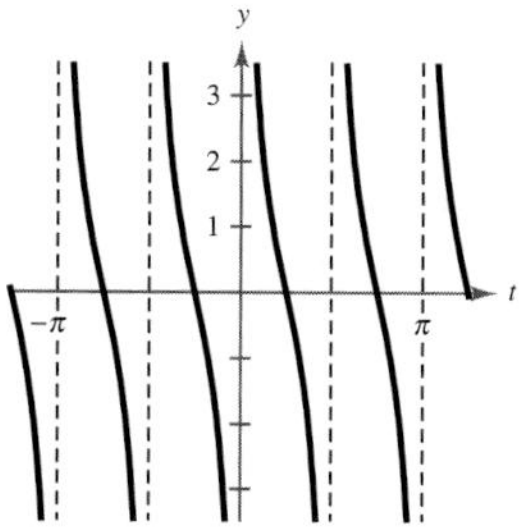

98.

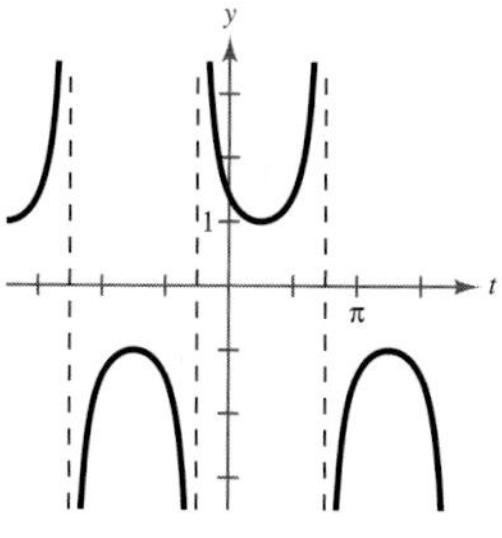

100.

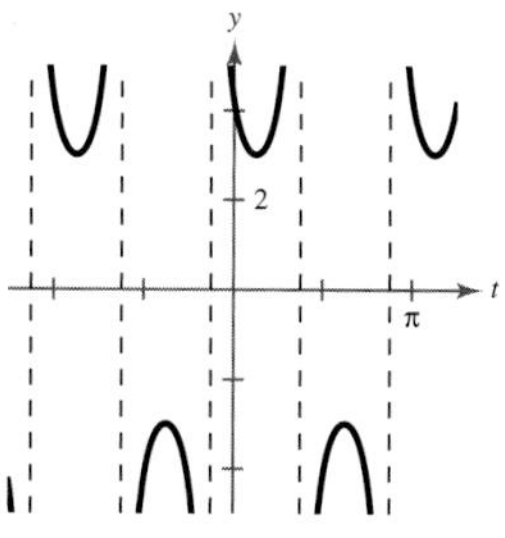

102.

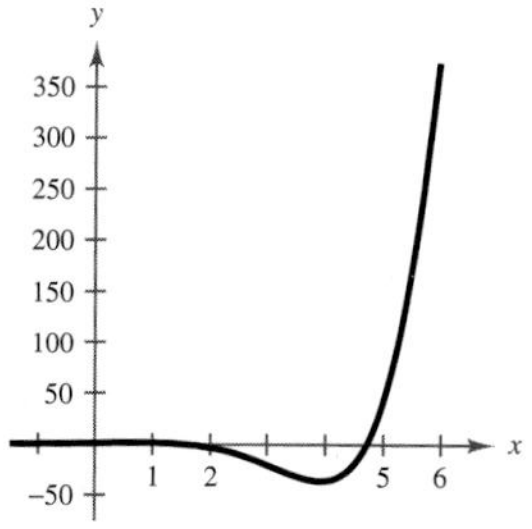

104. $-\dfrac{\pi}{2}$

106. 0.21 **108.** 1.10 **110.** $\dfrac{\pi}{4}$ **112.** $\dfrac{\pi}{6}$

114. 2.66 **116.** 1.17 **118.** 1.56 **120.** 1.45

122. 0.25 **124.** π **126.** $\frac{4}{3}$ **128.** $-\frac{5}{12}$

130. 9.6 feet **132.** $d = 0.75\cos\left(\dfrac{2\pi t}{3}\right)$

134. False. The inverse sine, $y = \arcsin x$, could be defined as a function over an interval that is greater than the interval defined as $-\pi/2 \le y \le \pi/2$, such as $\pi/2 \le y \le 3\pi/2$.

136. False. $3\pi/4$ is not in the range of the arctangent function.

138. a; the period is 2π and, because $a < 0$, the graph is reflected in the x-axis.

140. c; the period is 4π and the amplitude is 2.

142. (a) Equal; two-period shift

(b) Not equal; $f\left(t + \frac{1}{2}c\right)$ is a horizontal translation and $f\left(\frac{1}{2}t\right)$ is a period change.

(c) Not equal; for example, $\sin\left[\frac{1}{2}(\pi + 2\pi)\right] \neq \sin\left(\frac{1}{2}\pi\right)$.

144. (a) The displacement is increased.

(b) The friction damps the oscillations more quickly.

(c) The frequency of the oscillations increases.

146. (a)

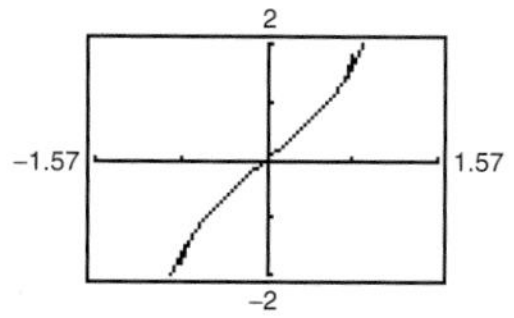

The approximation is accurate over the interval $-1 \le x \le 1$.

(b)

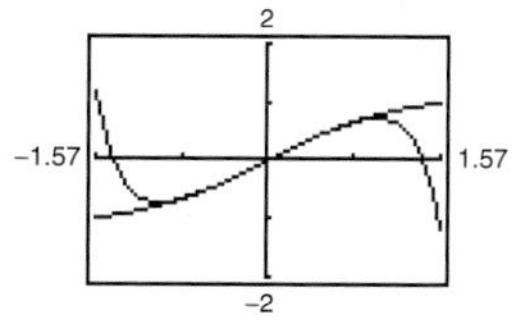

The approximation is accurate over the interval $-1 \le x \le 1$.

(c) $\dfrac{x^9}{9}$

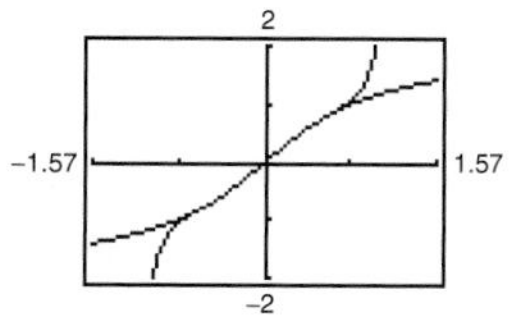

The accuracy improved.

Chapter 5

Section 5.1 *(page 447)*

2. $\sin x = -\frac{1}{2}$

$\csc x = -2$

$\sec x = -\frac{2\sqrt{3}}{3}$

$\cot x = \sqrt{3}$

4. $\sin \theta = \frac{3}{5}$

$\cos \theta = \frac{4}{5}$

$\sec \theta = \frac{5}{4}$

$\cot \theta = \frac{4}{3}$

6. $\cos \phi = -\frac{3\sqrt{10}}{10}$

$\tan \phi = -\frac{1}{3}$

$\csc \phi = \sqrt{10}$

$\sec \phi = -\frac{\sqrt{10}}{3}$

8. $\sin x = \frac{3}{5}$

$\tan x = \frac{3}{4}$

$\csc x = \frac{5}{3}$

$\sec x = \frac{5}{4}$

$\cot x = \frac{4}{3}$

10. $\sin x = \frac{\sqrt{15}}{4}$

$\cos x = \frac{1}{4}$

$\tan x = \sqrt{15}$

$\csc x = \frac{4\sqrt{15}}{15}$

$\cot x = \frac{\sqrt{15}}{15}$

12. $\sin \theta = -\frac{1}{5}$

$\cos \theta = -\frac{2\sqrt{6}}{5}$

$\tan \theta = \frac{\sqrt{6}}{12}$

$\sec \theta = -\frac{5\sqrt{6}}{12}$

$\cot \theta = 2\sqrt{6}$

14. $\sin \theta = 1$

$\cos \theta = 0$

$\csc \theta = 1$

$\sec \theta$ is undefined.

$\cot \theta = 0$

16. a **18.** f **20.** c **22.** c **24.** a **26.** d

28. $\sin \beta$ **30.** 1 **32.** $\cot \theta$ **34.** $\cos^2 x$

36. $\sin^2 \theta$ **38.** $\sin x$ **40.** $\sec t$ **42.** $2 \sec \phi$

44. $\csc \theta \sec \theta$ **46.** $\cos^2 x$ **48.** 1 **50.** $\cos x + 2$

52. $\sin^4 x$ **54.** $\sec^2 x + \tan^2 x$ **56.** $\tan^2 x(\sec x - 1)$

58. -1 **60.** $9 \cos^2 x$ **62.** $-2 \cot^2 x$ **64.** $-\cot x$

66. $5(\sec x - \tan x)$ **68.** $\tan^4 x(\csc x - 1)$

70.

x	0.2	0.4	0.6	0.8	1.0
y_1	0.0428	0.2107	0.6871	2.1841	8.3087
y_2	0.0428	0.2107	0.6871	2.1841	8.3087

x	1.2	1.4
y_1	50.3869	1163.61
y_2	50.3869	1163.61

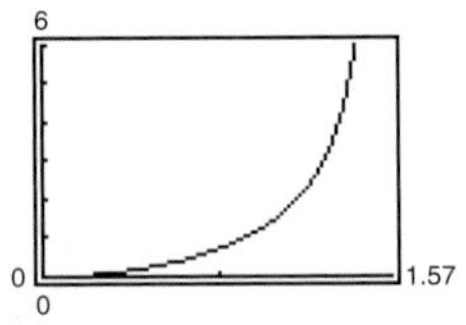

$y_1 = y_2$

72.

x	0.2	0.4	0.6	0.8	1.0
y_1	0.0403	0.1646	0.3863	0.7386	1.3105
y_2	0.0403	0.1646	0.3863	0.7386	1.3105

x	1.2	1.4
y_1	2.3973	5.7135
y_2	2.3973	5.7135

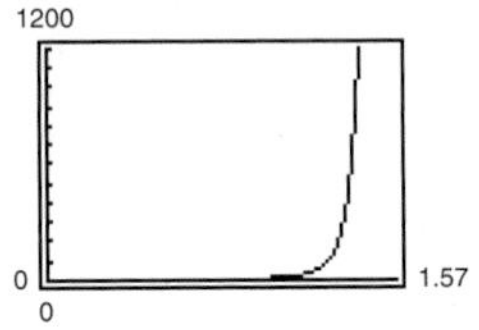

$y_1 = y_2$

74. $\cot x$ **76.** $\sec \theta$ **78.** $8 \sin \theta$ **80.** $2 \tan \theta$

82. $10 \sec \theta$ **84.** $6 \cos \theta = 3$; $\sin \theta = \pm\frac{\sqrt{3}}{2}$; $\cos \theta = \frac{1}{2}$

86. $10 \sin \theta = -5\sqrt{3}$; $\sin \theta = -\frac{\sqrt{3}}{2}$; $\cos \theta = \frac{1}{2}$

88. $\frac{\pi}{2} \le \theta \le \frac{3\pi}{2}$ **90.** $0 < \theta < \pi$ **92.** $\ln|\tan x|$

94. 0 **96.** (a) $(\tan 346°)^2 + 1 = (\sec 346°)^2 \approx 1.0622$

(b) $(\tan 3.1)^2 + 1 = (\sec 3.1)^2 \approx 1.0017$

98. (a) $\sin(-250°) = -\sin 250° \approx 0.9397$

(b) $\sin\left(-\frac{1}{2}\right) = -\sin\left(\frac{1}{2}\right) \approx -0.4794$

100. False. The sign of a value of a trigonometric function depends in which quadrant the angle lies.

102. 1, 1 **104.** ∞, 0

106. Not an identity because $\cos\theta = \pm\sqrt{1 - \sin^2\theta}$

108. Not an identity because $\dfrac{\sin k\theta}{\cos k\theta} = \tan k\theta$

110. Identity because $\sin\theta \cdot \dfrac{1}{\sin\theta} = 1$

112. $\cos\theta = \pm\sqrt{1 - \sin^2\theta}$

$\tan\theta = \pm\dfrac{\sin\theta}{\sqrt{1 - \sin^2\theta}}$

$\csc\theta = \dfrac{1}{\sin\theta}$

$\sec\theta = \pm\dfrac{1}{\sqrt{1 - \sin^2\theta}}$

$\cot\theta = \pm\dfrac{\sqrt{1 - \sin^2\theta}}{\sin\theta}$

114. $x - 25$ **116.** $4z + 12\sqrt{z} + 9$

118.

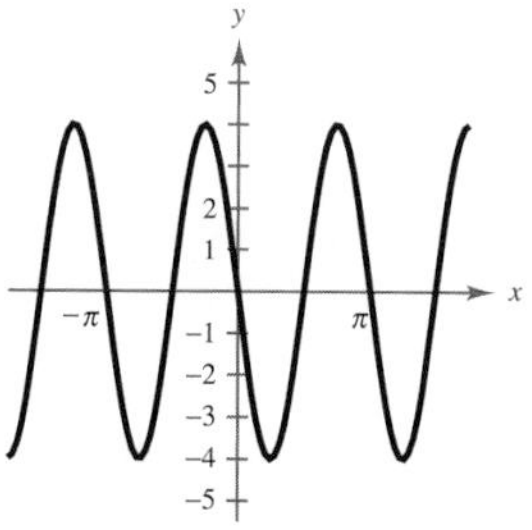

120.

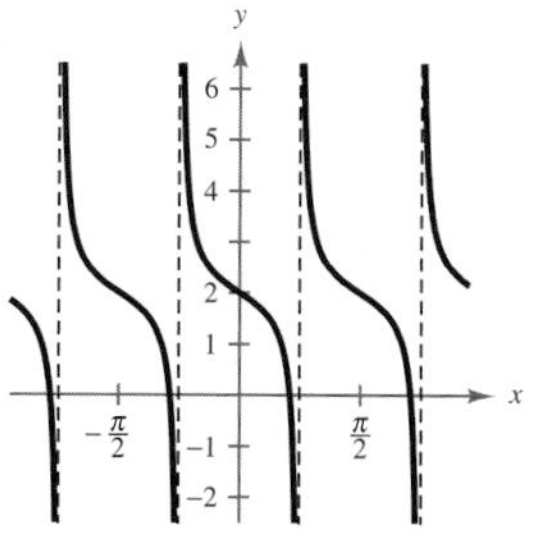

Section 5.2 *(page 455)*

2.–60. Answers will vary. **62.** 1 **64.** 2

66. False. An identity is an equation that is true for all real values of θ.

68. Not an identity because $\sin\theta = \pm\sqrt{1 - \cos^2\theta}$

Possible answer: $\dfrac{7\pi}{4}$

70. Not an identity because $\sqrt{\tan^2 x} = |\tan x|$

Possible answer: $\dfrac{3\pi}{4}$

72. Answers will vary. **74.** $2 + (3 - \sqrt{26})i$

76. $-8 + 4i$ **78.** $3 \pm \sqrt{3}i$ **80.** $-1 \pm \sqrt{3}i$

82. $\dfrac{3 \pm \sqrt{183}i}{8}$ **84.** $\dfrac{1 \pm \sqrt{967}i}{22}$

Section 5.3 *(page 464)*

2.–6. Answers will vary. **8.** $\dfrac{7\pi}{6} + 2n\pi, \dfrac{11\pi}{6} + 2n\pi$

10. $\dfrac{2\pi}{3} + n\pi$ **12.** $\dfrac{\pi}{3} + n\pi, \dfrac{2\pi}{3} + n\pi$

14. $\dfrac{\pi}{6} + n\pi, \dfrac{5\pi}{6} + n\pi, \dfrac{\pi}{3} + n\pi, \dfrac{2\pi}{3} + n\pi$

16. $\dfrac{\pi}{3} + n\pi, \dfrac{2\pi}{3} + n\pi$ **18.** $\dfrac{\pi}{9} + \dfrac{n\pi}{3}, \dfrac{2\pi}{9} + \dfrac{n\pi}{3}$

20. $\dfrac{\pi}{4} + \dfrac{n\pi}{2}, \dfrac{2\pi}{3} + 2n\pi, \dfrac{4\pi}{3} + 2n\pi$

22. $0, \pi$ **24.** $\dfrac{\pi}{2}, \dfrac{3\pi}{2}, \dfrac{2\pi}{3}, \dfrac{4\pi}{3}$ **26.** $\dfrac{\pi}{3}, \dfrac{5\pi}{3}$ **28.** 0

30. $\dfrac{7\pi}{6}, \dfrac{3\pi}{2}, \dfrac{11\pi}{6}$ **32.** $\dfrac{\pi}{3}, \dfrac{5\pi}{3}$ **34.** $\dfrac{2\pi}{3} + n\pi, \dfrac{5\pi}{6} + n\pi$

36. $\dfrac{\pi}{12} + \dfrac{n\pi}{2}, \dfrac{5\pi}{12} + \dfrac{n\pi}{2}$ **38.** $\dfrac{8\pi}{3} + 4n\pi, \dfrac{10\pi}{3} + 4n\pi$

40. $-\frac{1}{4}, \frac{3}{4}, \frac{7}{4}, \frac{11}{4}$ **42.** ± 2 **44.** $-5, 4$; No solution

46. 0.7854, 2.3562, 3.6652, 3.9270, 5.4978, 5.7596

48. 0.5236, 2.6180 **50.** 4.9172

52. 0.5153, 2.7259, 3.6569, 5.8675

54. 0.5236, 0.7297, 2.4119, 2.6180

56. 0.5880, 2.0344, 3.7296, 5.1760 **58.** 1.7794, 4.5038

60. $\arctan(-2) + \pi, \arctan(-2) + 2\pi, \dfrac{\pi}{4}, \dfrac{5\pi}{4}$

62. $\dfrac{\pi}{6}, \dfrac{5\pi}{6}$

64. (a)

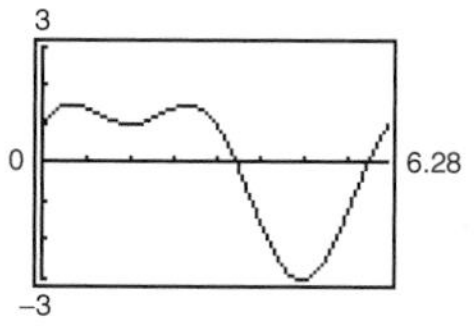

Maximum: $\left(\dfrac{\pi}{6}, \dfrac{3}{2}\right)$

Minimum: $\left(\dfrac{\pi}{2}, 1\right)$

Maximum: $\left(\dfrac{5\pi}{6}, \dfrac{3}{2}\right)$

Minimum: $\left(\dfrac{3\pi}{2}, -3\right)$

(b) $\dfrac{\pi}{6}, \dfrac{\pi}{2}, \dfrac{5\pi}{6}, \dfrac{3\pi}{2}$

66. 0.739

68. (a) All real numbers except $x = 0$

(b) y-axis symmetry

(c) y approaches 1.

(d) Four solutions: $\pm\pi, \pm 2\pi$

70. 1.96 seconds **72.** 36.9°, 53.1°

74. (a)

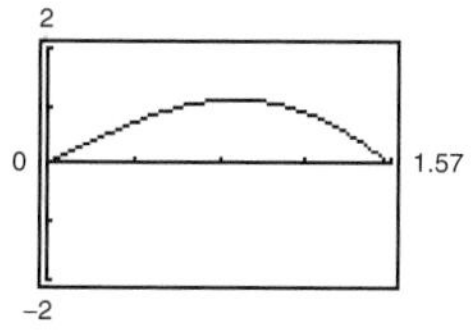

$A \approx 1.12$

(b) $0.6 < x < 1.1$

76. (a)

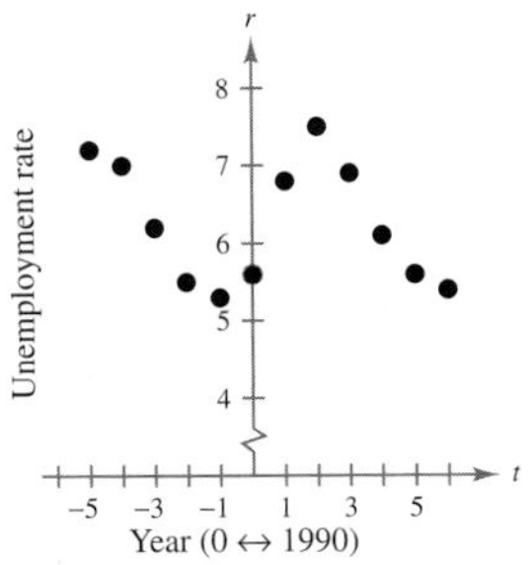

(b) (3)

(c) The constant, 6.26

(d) ≈7 years

(e) 1997

78. False. There is no value of x for which $\sin x = 3.4$.

80. 3

82. $B = 60°$, $a \approx 24.25$, $c \approx 12.12$

84. $C = 19°$, $a \approx 15.4$, $c \approx 5.0$

86. $\sin\theta = -\frac{1}{2}$, $\cos\theta = -\frac{\sqrt{3}}{2}$, $\tan\theta = \frac{\sqrt{3}}{3}$

88. $\sin\theta = \frac{\sqrt{2}}{2}$, $\cos\theta = -\frac{\sqrt{2}}{2}$, $\tan\theta = -1$

90. $\sin\theta = \frac{1}{2}$, $\cos\theta = \frac{\sqrt{3}}{2}$, $\tan\theta = \frac{\sqrt{3}}{3}$

92. 1.36°

Section 5.4 *(page 472)*

2. (a) $-\frac{\sqrt{6}+\sqrt{2}}{4}$ (b) $\frac{\sqrt{2}+1}{2}$

4. (a) 0 (b) $\frac{\sqrt{3}-1}{2}$

6. (a) $\frac{\sqrt{6}+\sqrt{2}}{4}$ (b) $\frac{\sqrt{2}-\sqrt{3}}{2}$

8. $\sin 165° = \frac{\sqrt{2}}{4}(\sqrt{3}-1)$

$\cos 165° = -\frac{\sqrt{2}}{4}(\sqrt{3}+1)$

$\tan 165° = -2+\sqrt{3}$

10. $\sin 255° = -\frac{\sqrt{2}}{4}(\sqrt{3}+1)$

$\cos 255° = \frac{\sqrt{2}}{4}(1-\sqrt{3})$

$\tan 255° = 2+\sqrt{3}$

12. $\sin\frac{7\pi}{12} = \frac{\sqrt{2}}{4}(\sqrt{3}+1)$

$\cos\frac{7\pi}{12} = \frac{\sqrt{2}}{4}(1-\sqrt{3})$

$\tan\frac{7\pi}{12} = -2-\sqrt{3}$

14. $\sin\left(-\frac{\pi}{12}\right) = \frac{\sqrt{2}}{4}(1-\sqrt{3})$

$\cos\left(-\frac{\pi}{12}\right) = \frac{\sqrt{2}}{4}(\sqrt{3}+1)$

$\tan\left(-\frac{\pi}{12}\right) = -2+\sqrt{3}$

16. $\sin(-105°) = -\frac{\sqrt{2}}{4}(\sqrt{3}+1)$

$\cos(-105°) = \frac{\sqrt{2}}{4}(1-\sqrt{3})$

$\tan(-105°) = 2+\sqrt{3}$

18. $\sin 15° = \frac{\sqrt{2}}{4}(\sqrt{3}-1)$

$\cos 15° = \frac{\sqrt{2}}{4}(1+\sqrt{3})$

$\tan 15° = 2-\sqrt{3}$

20. $\sin\left(-\frac{7\pi}{12}\right) = -\frac{\sqrt{2}}{4}(\sqrt{3}+1)$

$\cos\left(-\frac{7\pi}{12}\right) = \frac{\sqrt{2}}{4}(1-\sqrt{3})$

$\tan\left(-\frac{7\pi}{12}\right) = 2+\sqrt{3}$

22. $\sin\frac{5\pi}{12} = \frac{\sqrt{2}}{4}(1+\sqrt{3})$

$\cos\frac{5\pi}{12} = \frac{\sqrt{2}}{4}(\sqrt{3}-1)$

$\tan\frac{5\pi}{12} = \sqrt{3}+2$

24. $\sin 190°$ **26.** $\tan 80°$ **28.** $\cos\frac{12\pi}{35}$

30. $\cos(3x-2y)$ **32.** $\frac{\sqrt{2}}{2}$ **34.** $\frac{\sqrt{2}}{2}$ **36.** $\frac{\sqrt{3}}{3}$

38. $\frac{56}{65}$ **40.** $-\frac{33}{65}$ **42.** $\frac{65}{33}$ **44.** $-\frac{16}{63}$ **46.** $\frac{4}{5}$

48. $\frac{117}{44}$ **50.** $-\frac{125}{44}$ **52.** $\frac{2x^2-\sqrt{1-x^2}}{\sqrt{4x^2+1}}$

54. $\frac{x+x\sqrt{1-x^2}}{\sqrt{x^2+1}}$ **56.–64.** Answers will vary.

66. $-\cos x$ **68.** $\tan\theta$ **70.** $\frac{\pi}{3}, \frac{5\pi}{3}$

72. $0, \frac{\pi}{3}, \pi, \frac{5\pi}{3}$ **74.** $0, \pi$ **76.** Answers will vary.

78. False. $\cos(u + v) = \cos u \cos v \mp \sin u \sin v$

80. True.

$$\sin\left(x - \frac{\pi}{2}\right) = \sin x \cos \frac{\pi}{2} - \cos x \sin \frac{\pi}{2} = -\cos x$$

82. and 84. Answers will vary.

86. (a) $5 \sin(2\theta + 0.9273)$ (b) $5 \cos(2\theta - 0.6435)$

88. (a) $\sqrt{2} \sin\left(2\theta - \frac{\pi}{4}\right)$ (b) $-\sqrt{2} \cos\left(2\theta + \frac{\pi}{4}\right)$

90. $-\frac{5\sqrt{2}}{2} \sin \theta - \frac{5\sqrt{2}}{2} \cos \theta$ **92.** $15°$ **94.** $u + v = w$

96. Answers will vary. **98.** $f^{-1}(x) = -8x + 7$

100. $f^{-1}(x) = x^2 + 16$ **102.** $3x^2$ **104.** $x^2 + 10x$

Section 5.5 *(page 482)*

2. $\frac{1}{4}$ **4.** $\frac{8}{17}$ **6.** $\frac{17}{15}$ **8.** $\frac{15}{8}$

10. $\frac{\pi}{2}, \frac{7\pi}{6}, \frac{3\pi}{2}, \frac{11\pi}{6}$ **12.** $\frac{\pi}{2}, \frac{3\pi}{2}, \frac{\pi}{4}, \frac{3\pi}{4}, \frac{5\pi}{4}, \frac{7\pi}{4}$

14. $\frac{\pi}{2}, \frac{7\pi}{6}, \frac{11\pi}{6}$ **16.** $\frac{\pi}{6}, \frac{\pi}{2}, \frac{5\pi}{6}, \frac{3\pi}{2}$

18. $0, \frac{\pi}{4}, \frac{\pi}{2}, \frac{3\pi}{4}, \pi, \frac{5\pi}{4}, \frac{3\pi}{2}, \frac{7\pi}{4}$

20. $3 \cos 2x$ **22.** $\cos 2x$

24. $\sin 2u = -\frac{4\sqrt{5}}{9}$
$\cos 2u = -\frac{1}{9}$
$\tan 2u = 4\sqrt{5}$

26. $\sin 2u = -\frac{8}{17}$
$\cos 2u = \frac{15}{17}$
$\tan 2u = -\frac{8}{15}$

28. $\sin 2u = -\frac{4\sqrt{2}}{9}$
$\cos 2u = \frac{7}{9}$
$\tan 2u = -\frac{4\sqrt{2}}{7}$

30. $\frac{1}{128}(35 - 48 \cos 2x + 28 \cos 4x - 16 \cos 2x \cos 4x + \cos 8x)$

32. $\frac{1}{128}(3 - 4 \cos 4x + \cos 8x)$

34. $\frac{1}{16}(1 - \cos 2x - \cos 4x + \cos 2x \cos 4x)$

36. $\frac{\sqrt{17}}{17}$ **38.** $\frac{\sqrt{17}}{4}$ **40.** 4

42. $\sin 165° = \frac{1}{2}\sqrt{2 - \sqrt{3}}$
$\cos 165° = -\frac{1}{2}\sqrt{2 + \sqrt{3}}$
$\tan 165° = \sqrt{3} - 2$

44. $\sin 67° 30' = \frac{1}{2}\sqrt{2 + \sqrt{2}}$
$\cos 67° 30' = \frac{1}{2}\sqrt{2 - \sqrt{2}}$
$\tan 67° 30' = 1 + \sqrt{2}$

46. $\sin \frac{\pi}{12} = \frac{1}{2}\sqrt{2 - \sqrt{3}}$
$\cos \frac{\pi}{12} = \frac{1}{2}\sqrt{2 + \sqrt{3}}$
$\tan \frac{\pi}{12} = 2 - \sqrt{3}$

48. $\sin \frac{7\pi}{12} = \frac{1}{2}\sqrt{2 + \sqrt{3}}$
$\cos \frac{7\pi}{12} = -\frac{1}{2}\sqrt{2 - \sqrt{3}}$
$\tan \frac{7\pi}{12} = -2 - \sqrt{3}$

50. $\sin \frac{u}{2} = \frac{\sqrt{5}}{5}$
$\cos \frac{u}{2} = \frac{2\sqrt{5}}{5}$
$\tan \frac{u}{2} = \frac{1}{2}$

52. $\sin \frac{u}{2} = \frac{1}{2}\sqrt{\frac{10 + 3\sqrt{10}}{5}}$
$\cos \frac{u}{2} = -\frac{1}{2}\sqrt{\frac{10 - 3\sqrt{10}}{5}}$
$\tan \frac{u}{2} = -3 - \sqrt{10}$

54. $\sin \frac{u}{2} = \frac{3\sqrt{14}}{14}$
$\cos \frac{u}{2} = \frac{\sqrt{70}}{14}$
$\tan \frac{u}{2} = \frac{3\sqrt{5}}{5}$

56. $|\cos 2x|$ **58.** $-\left|\sin\left(\frac{x - 1}{2}\right)\right|$ **60.** $0, \frac{\pi}{3}, \frac{5\pi}{3}$

62. $0, \frac{\pi}{2}, \frac{3\pi}{2}$ **64.** $2\left[\sin \frac{7\pi}{6} - \sin\left(-\frac{\pi}{2}\right)\right]$

66. $\frac{3}{2}[\cos(-\alpha) - \cos 5\alpha]$ **68.** $\frac{1}{2}[\cos(-2\theta) + \cos 6\theta]$

70. $\frac{1}{2}(\sin 2x + \sin 2y)$ **72.** $\frac{1}{2}(\cos 2\pi - \cos 2\theta)$

74. $3(\sin 60° + \sin 30°)$ **76.** $2 \cos 75° \cos 45°$

78. $2 \cos \pi \sin \frac{\pi}{4}$ **80.** $2 \sin 2\theta \cos \theta$

82. $2 \sin 3x \cos 2x$ **84.** $2 \cos(\phi + \pi) \cos \pi$

86. $2 \sin x \cos \frac{\pi}{2}$ **88.** $0, \frac{\pi}{4}, \frac{\pi}{2}, \frac{3\pi}{4}, \pi, \frac{5\pi}{4}, \frac{3\pi}{2}, \frac{7\pi}{4}$

90. $0, \frac{\pi}{2}, \pi, \frac{3\pi}{2}, \frac{\pi}{4}, \frac{3\pi}{4}, \frac{5\pi}{4}, \frac{7\pi}{4}$ **92.** $\frac{144}{169}$ **94.** $\frac{36}{65}$

96.–110. Answers will vary.

112.

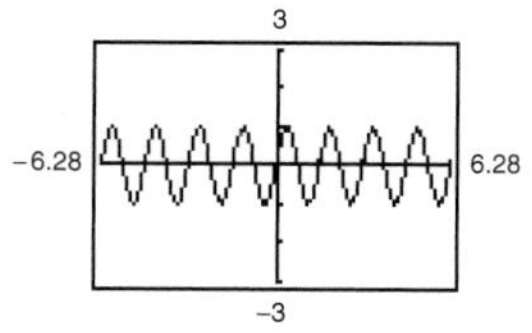

114.

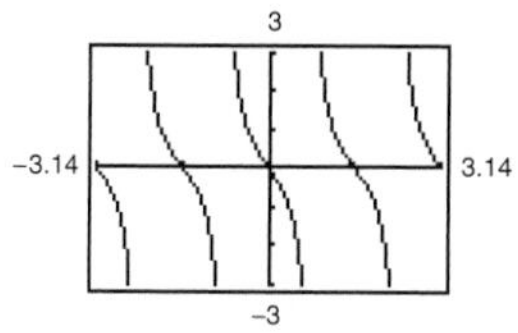

116.

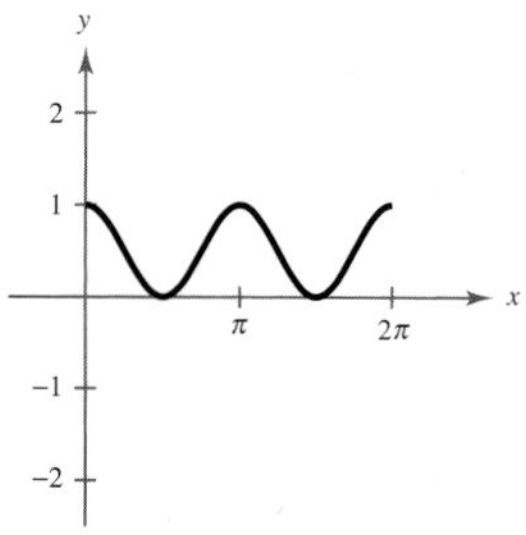

118. $2x^2 - 1$

120. $r = \frac{1}{16} v_0^2 \sin\theta\cos\theta$ **122.** $x = 2r(1 - \cos\theta)$

124. False. $\sin\frac{u}{2} = \sqrt{\frac{1-\cos u}{2}}$ when $\frac{u}{2}$ is in the second quadrant.

126. (a)

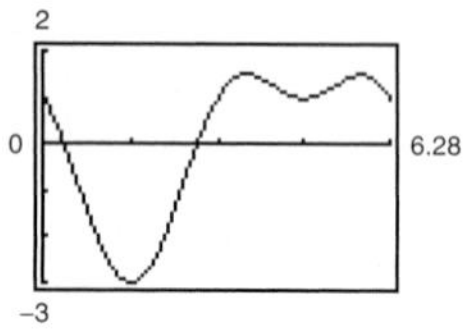

Minima: $\left(\frac{\pi}{2}, -3\right), \left(\frac{3\pi}{2}, 1\right)$

Maxima: $\left(\frac{7\pi}{6}, \frac{3}{2}\right), \left(\frac{11\pi}{6}, \frac{3}{2}\right)$

(b) $\frac{\pi}{2}, \frac{7\pi}{6}, \frac{3\pi}{2}, \frac{11\pi}{6}$

128. (a)

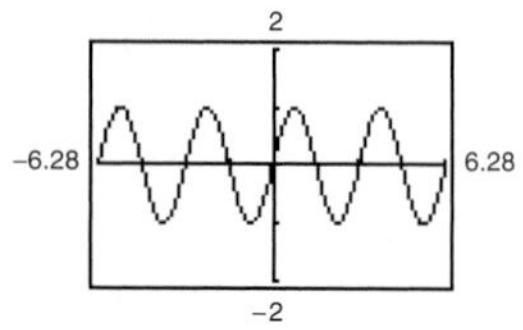

(b) $g(x) = \sin 2x$

(c) Answers will vary.

130. 1.5 hours **132.** ≈ 127 feet

Review Exercises *(page 487)*

2. $\csc x$ **4.** $\cot x$ **6.** $|\sec x|$

8. $\sin\theta = \frac{2\sqrt{13}}{13}$

$\cos\theta = \frac{3\sqrt{13}}{13}$

$\csc\theta = \frac{\sqrt{13}}{2}$

$\cot\theta = \frac{3}{2}$

10. $\cos\theta = \frac{1}{9}$

$\tan\theta = 4\sqrt{5}$

$\csc\theta = \frac{9\sqrt{5}}{20}$

$\sec\theta = 9$

$\cot\theta = \frac{\sqrt{5}}{20}$

12. $\sec\theta\csc\theta$ **14.** $\cos^2 x$ **16.** $\tan u \sec u$ **18.** 1

20. $2\tan^2 x - 2\sec x\tan x + 1$ **22.** $1 + \sin x$

24.–30. Answers will vary.

32. $\frac{\pi}{3} + 2n\pi, \frac{2\pi}{3} + 2n\pi$ **34.** $\frac{\pi}{6} + n\pi$

36. $\frac{\pi}{3} + 2n\pi, \frac{2\pi}{3} + 2n\pi, \frac{4\pi}{3} + 2n\pi, \frac{5\pi}{3} + 2n\pi$

38. $0, \frac{2\pi}{3}, \frac{4\pi}{3}$ **40.** $0, \frac{\pi}{2}, \pi$ **42.** $\frac{\pi}{8}, \frac{3\pi}{8}, \frac{9\pi}{8}, \frac{11\pi}{8}$

44. $0, \frac{\pi}{8}, \frac{3\pi}{8}, \frac{5\pi}{8}, \frac{7\pi}{8}, \frac{9\pi}{8}, \frac{11\pi}{8}, \frac{13\pi}{8}, \frac{15\pi}{8}$ **46.** $0, \pi$

48. $\arctan(-4) + \pi, \arctan(-4) + 2\pi, \arctan 3, \pi + \arctan 3$

50. $\sin 285° = -\frac{\sqrt{2}}{4}\left(\sqrt{3} + 1\right)$

$\cos 285° = \frac{\sqrt{2}}{4}\left(\sqrt{3} - 1\right)$

$\tan 285° = -2 - \sqrt{3}$

52. $\sin\frac{25\pi}{12} = \frac{\sqrt{2}}{4}\left(\sqrt{3} - 1\right)$

$\cos\frac{25\pi}{12} = \frac{\sqrt{2}}{4}\left(\sqrt{3} + 1\right)$

$\tan\frac{25\pi}{12} = 2 - \sqrt{3}$

54. $\sin 15°$ **56.** $\tan 35°$ **58.** $-\frac{3}{52}\left(5 + 4\sqrt{7}\right)$

60. $\frac{1}{52}\left(5\sqrt{7} + 36\right)$ **62.** $\frac{1}{52}\left(5\sqrt{7} - 36\right)$

64. $\frac{\pi}{4}, \frac{7\pi}{4}$ **66.** $\frac{\pi}{6}, \frac{11\pi}{6}$

68.

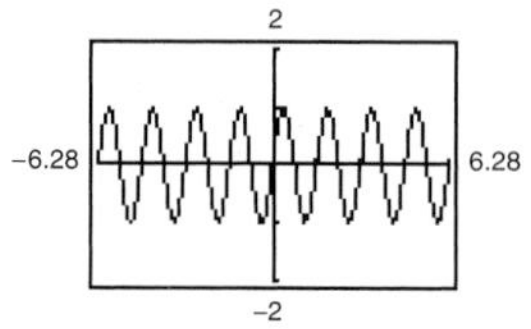

70. $\sin 2u = \frac{24}{25}$

$\cos 2u = -\frac{7}{25}$

$\tan 2u = -\frac{24}{7}$

72. $\theta = 15°$ or $\frac{\pi}{12}$ **74.** $\frac{1 + \cos 6x}{2}$ **76.** $\frac{1 - \cos 2x}{2}$

78. $\sin 15° = \frac{1}{2}\sqrt{2 - \sqrt{3}}$

$\cos 15° = \frac{1}{2}\sqrt{2 + \sqrt{3}}$

$\tan 15° = 2 - \sqrt{3}$

80. $\sin\left(-\frac{17\pi}{12}\right) = \frac{1}{2}\sqrt{2 + \sqrt{3}}$

$\cos\left(-\frac{17\pi}{12}\right) = -\frac{1}{2}\sqrt{2 - \sqrt{3}}$

$\tan\left(-\frac{17\pi}{12}\right) = -2 - \sqrt{3}$

82. $|\tan 3x|$ **84.** $\frac{1}{2}\sin\frac{\pi}{3}$ **86.** $\frac{1}{2}(\cos 2\theta + \cos 8\theta)$

88. $2 \sin 75° \cos 15°$ **90.** $-2 \sin x \sin\frac{\pi}{6}$

92. (a) $y = \frac{1}{2}\sqrt{10}\sin\left(8t - \arctan\frac{1}{3}\right)$

(b) $\frac{1}{2}\sqrt{10}$ feet (c) $\frac{4}{\pi}$ cycles per second

94. False. Using the sum and difference formula, $\sin(x + y) = \sin x \cos y + \cos x \sin y$.

96. True by the product-to-sum formula

98. $\sin x$, $\csc x$, $\tan x$, $\cot x$

100. $-1 \le \sin x \le 1$ for all x **102.** $y_1 = 1 - y_2$

104. $-3.1395, -2.0000, -0.4378, 2.0000$

Chapter 6

Section 6.1 *(page 500)*

2. $A = 35°, a \approx 11.88, b \approx 13.31$

4. $A = 35°, a \approx 36.50, b \approx 11.05$

6. Two solutions:

$B \approx 45.79°, C \approx 74.21°, b \approx 7.45$

$B \approx 14.21°, C \approx 105.79°, b \approx 2.55$

8. $B = 101.1°, a \approx 1.35, b \approx 3.23$

10. $C = 166°5', a \approx 3.30, c \approx 8.05$

12. $A \approx 174°41', C \approx 2°34', a \approx 11.98$

14. $B = 75.48°, C \approx 4.52°, b \approx 122.87$

16. $A \approx 44°14', B \approx 50°26', b \approx 38.67$

18. $A = 48°, b \approx 2.29, c \approx 4.73$

20. Two solutions:

$B \approx 72.2°, C = 49.8°, c \approx 10.27$

$B \approx 107.8°, C = 14.2°, c \approx 3.30$

22. $B \approx 36.82°, C \approx 67.18°, c \approx 32.30$

24. $B \approx 48.74°, C \approx 21.26°, c \approx 48.23$

26. $B = 22°, C = 136°, c \approx 1.32$

28. (a) $b \le 10, b = \frac{10}{\sin 60°}$ (b) $10 < b < \frac{10}{\sin 60°}$

(c) $b > \frac{10}{\sin 60°}$

30. (a) $b \le 315.6, b = \frac{315.6}{\sin 88°}$ (b) $315.6 < b < \frac{315.6}{\sin 88°}$

(c) $b > \frac{315.6}{\sin 88°}$

32. 474.9 **34.** 4.5 **36.** 159.3

38. (a)

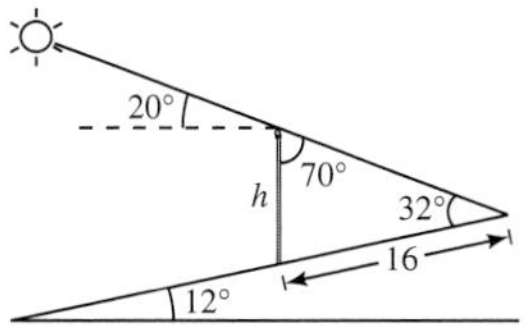

(b) $\frac{16}{\sin 70°} = \frac{h}{\sin 32°}$

(c) 9 meters

40. S 60° W

42. (a)

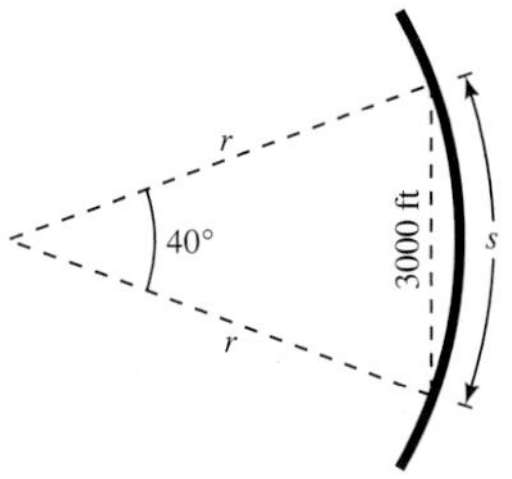

(b) 4385.71 feet

(c) 3061. 80 feet

44. From Pine Knob: 42.4 kilometers

From Colt Station: 15.5 kilometers

46. 4.55 miles

48. True. The longest side of a triangle is always opposite the largest angle.

50. (a) $\alpha = \arcsin(0.5 \sin \beta)$

(b)

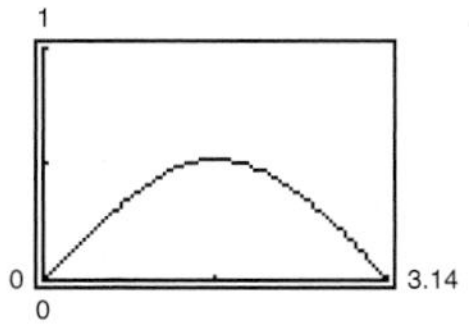

Domain: $0 < \beta < \pi$

Range: $0 < \alpha \le \frac{\pi}{6}$

(c) $c = \frac{18 \sin[\pi - \beta - \arcsin(0.5 \sin \beta)]}{\sin \beta}$

(d)

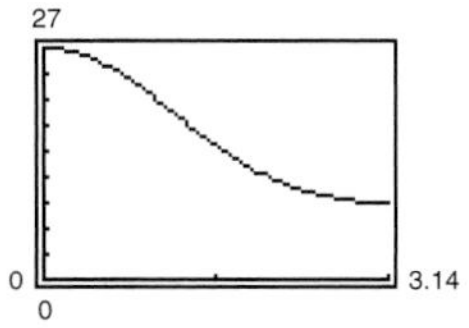

Domain: $0 < \beta < \pi$

Range: $9 < c < 27$

(e)

β	0.4	0.8	1.2	1.6
α	0.1960	0.3669	0.4848	0.5234
c	25.95	23.07	19.19	15.33

β	2.0	2.4	2.8
α	0.4720	0.3445	0.1683
c	12.29	10.31	9.27

As β increases from 0 to π, α increases then decreases, and c decreases from 27 to 9.

52. $\cos x = -\frac{\sqrt{21}}{5}$

$\tan x = -\frac{2\sqrt{21}}{21}$

$\csc x = \frac{5}{2}$

$\sec x = -\frac{5\sqrt{21}}{21}$

$\cot x = -\frac{\sqrt{21}}{2}$

54. $\sin x = -\frac{\sqrt{15}}{4}$

$\cos x = -\frac{1}{4}$

$\tan x = \sqrt{15}$

$\csc x = -\frac{4\sqrt{15}}{15}$

$\cot x = \frac{\sqrt{15}}{15}$

56. $\cos x$ **58.** $\sin^2 x$

Section 6.2 *(page 507)*

2. $A \approx 61.2°, B \approx 19.2°, C \approx 99.6°$

4. $A \approx 53.7°, B \approx 21.3°, c \approx 12.0$

6. $A \approx 39.35°, B \approx 16.74°, C \approx 123.91°$

8. $A \approx 86.7°, B \approx 31.8°, C \approx 61.5°$

10. $B \approx 16.5°, C \approx 108.5°, a \approx 8.64$

12. $A \approx 37°6', C \approx 67°34', b \approx 9.94$

14. $A \approx 157°2', B \approx 7°43', c \approx 4.21$

16. $A \approx 23.6°, B \approx 53.4°, c \approx 0.91$

	a	b	c	d	θ	ϕ
18.	25	35	52.20	31.22	60°	120°
20.	40	60	63.25	80	104.5°	75.5°
22.	35.18	25	50	35	68.7°	111.3°

24. 54 **26.** 1350.22 **28.** 0.61

30.

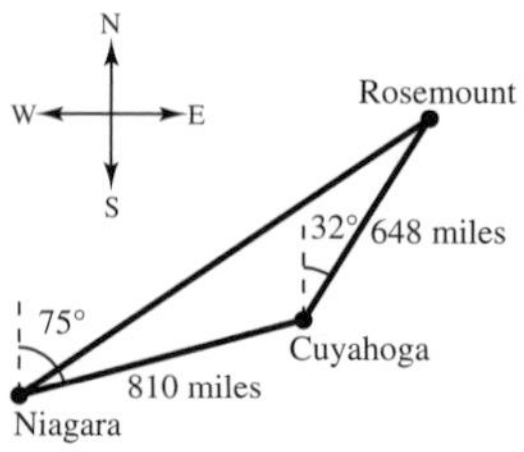

1357.8 miles, N 56° E

32. 41.2°, 52.9° **34.** 127.2° **36.** 131.1 feet, 118.6 feet

38. (a) N 59.7° E (b) N 72.8° E **40.** 103.9 feet

42. 3.8 miles

44. (a) $49 = 2.25 + x^2 - 3x\cos\theta$

(b) $x = \frac{1}{2}\left(3\cos\theta + \sqrt{9\cos^2\theta + 187}\right)$

(c)

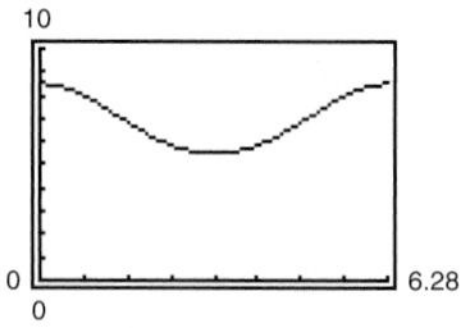

(d) 6 inches

46. 2.76 feet **48.** 6577.8 square meters

50. False. To solve an SSA triangle, the Law of Sines is needed.

52. (a) 570.60 (b) 5910.68 (c) 177.09

54. Answers will vary. **56.** $-\frac{\pi}{2}$ **58.** $\frac{\pi}{3}$

60. $-\frac{\pi}{3}$ **62.** $\frac{1}{\sqrt{1 - 4x^2}}$ **64.** $\frac{1}{x - 2}$

66. $\cos\theta = 1$

$\sec\theta = 1$

$\csc\theta$ is undefined.

68. $\tan\theta = -\frac{\sqrt{3}}{3}$

$\sec\theta = \frac{2\sqrt{3}}{3}$

$\csc\theta = -2$

Section 6.3 *(page 520)*

2. $\mathbf{v} = \langle -4, -2\rangle$; $\|\mathbf{v}\| = 2\sqrt{5}$ **4.** $\mathbf{v} = \langle 4, 6\rangle$; $\|\mathbf{v}\| = 2\sqrt{13}$

6. $\mathbf{v} = \langle 7, 0\rangle$; $\|\mathbf{v}\| = 7$ **8.** $\mathbf{v} = \langle 8, -8\rangle$; $\|\mathbf{v}\| = 8\sqrt{2}$

10. $\mathbf{v} = \langle 12, 29\rangle$; $\|\mathbf{v}\| = \sqrt{985}$

12. $\mathbf{v} = \langle 7, -24\rangle$; $\|\mathbf{v}\| = 25$

14.

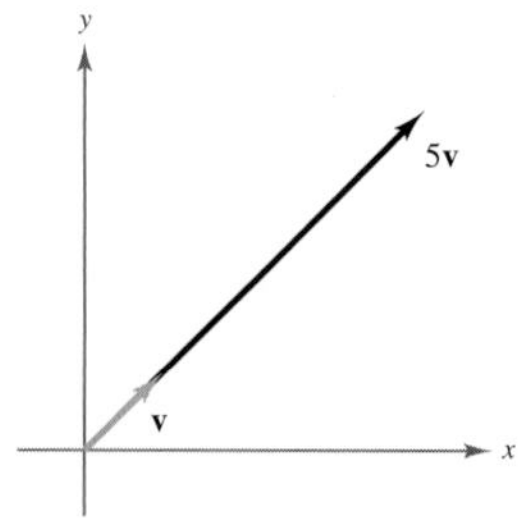

16.

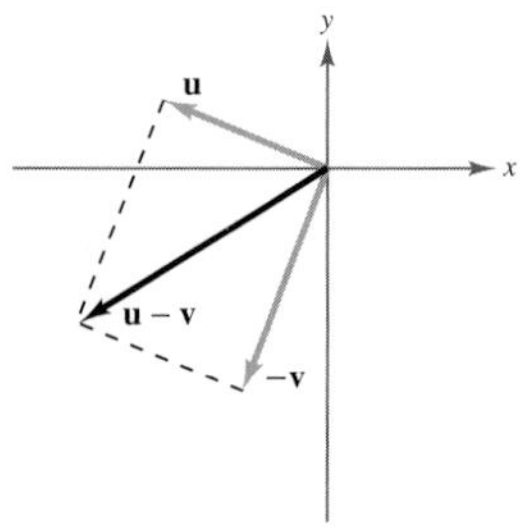

18.

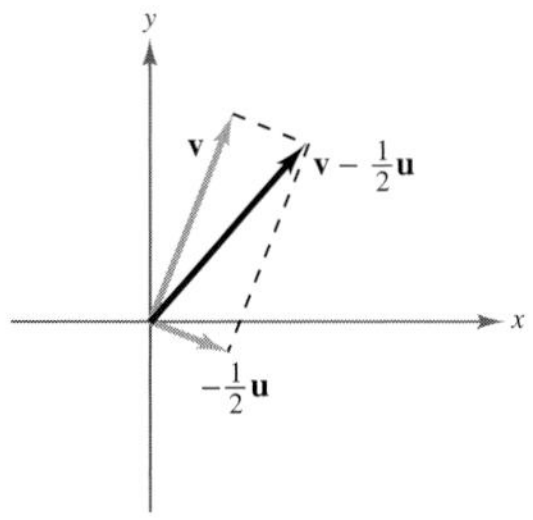

20. (a) $\langle 6, 3 \rangle$

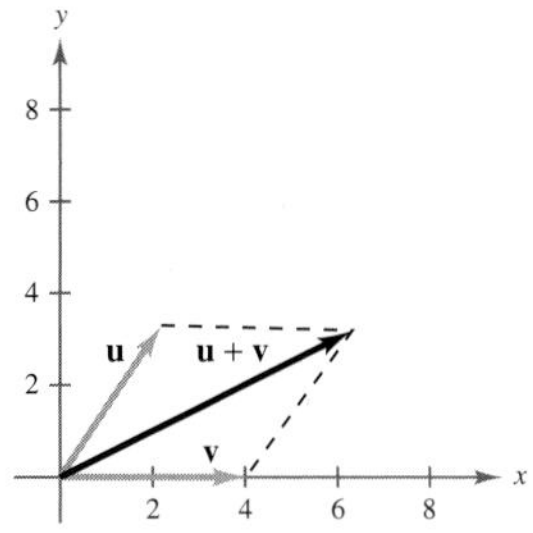

(b) $\langle -2, 3 \rangle$

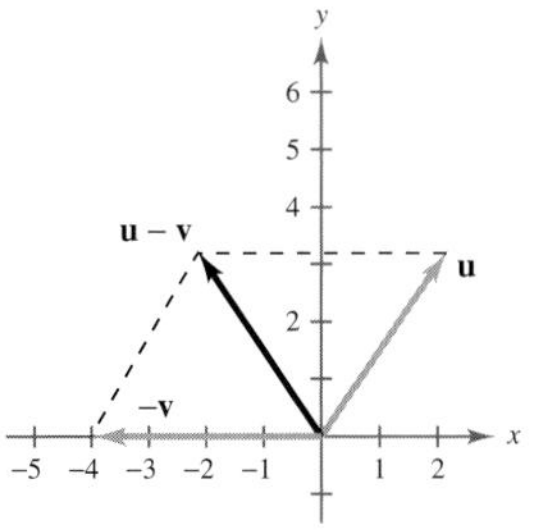

(c) $\langle -8, 6 \rangle$

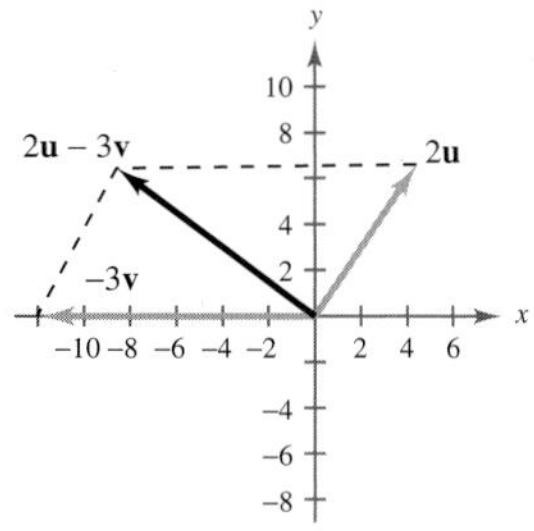

22. (a) $\langle 2, 1 \rangle$

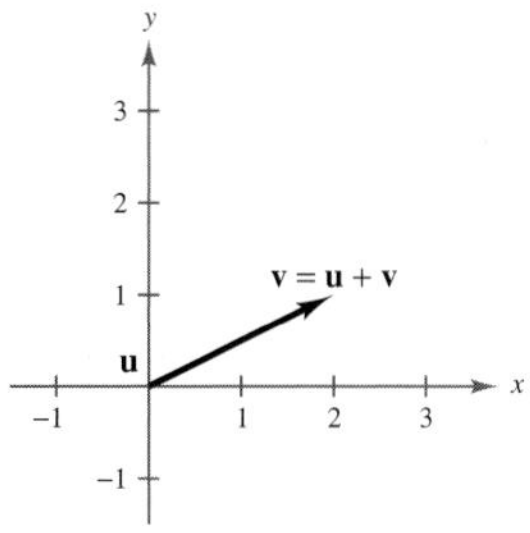

(b) $\langle -2, -1 \rangle$

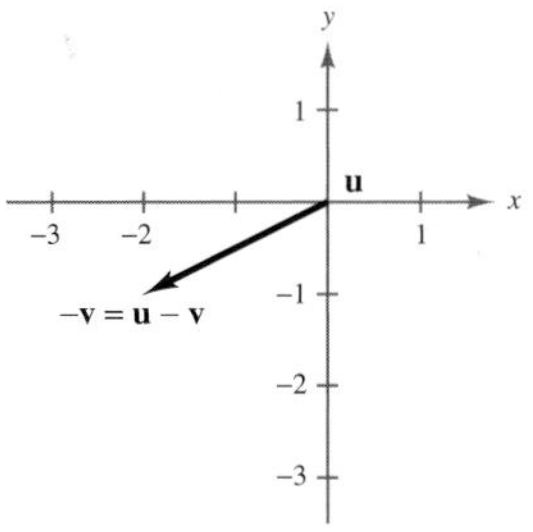

(c) $\langle -6, -3 \rangle$

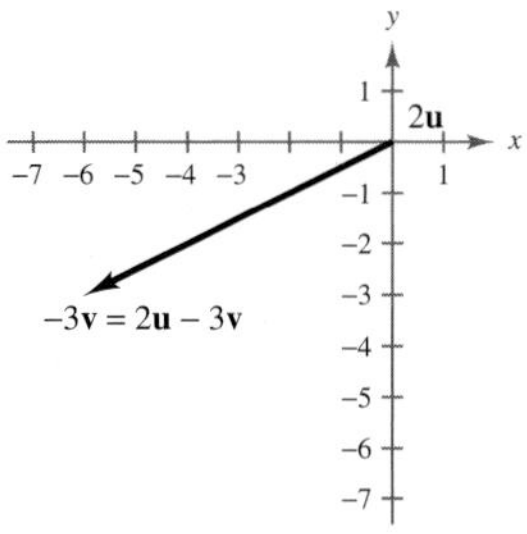

24. (a) $-3\mathbf{i} + 3\mathbf{j}$

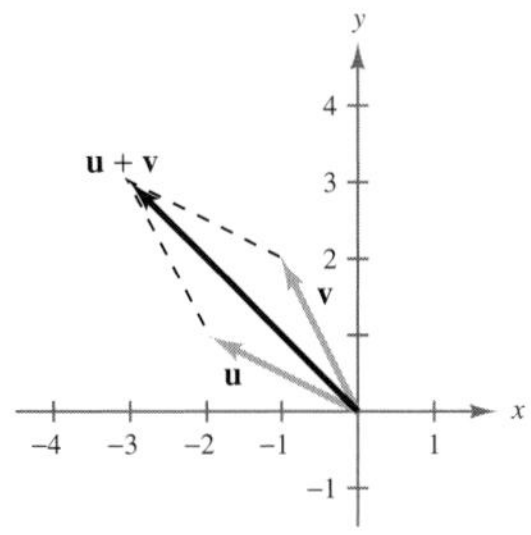

(b) $-\mathbf{i} - \mathbf{j}$

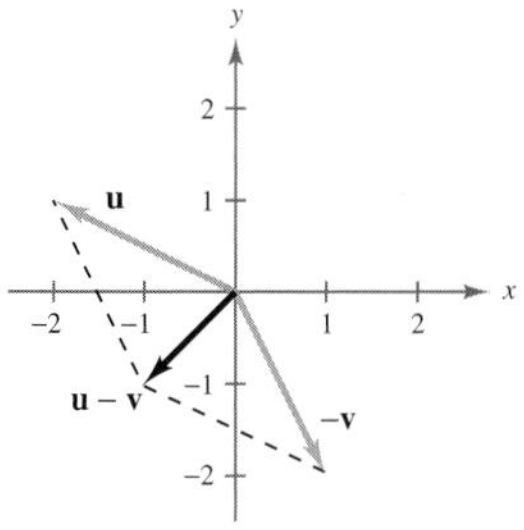

(c) $-\mathbf{i} - 4\mathbf{j}$

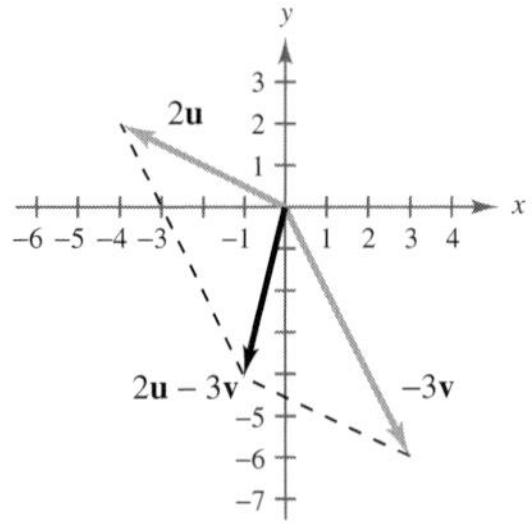

26. (a) $2\mathbf{i} + 3\mathbf{j}$

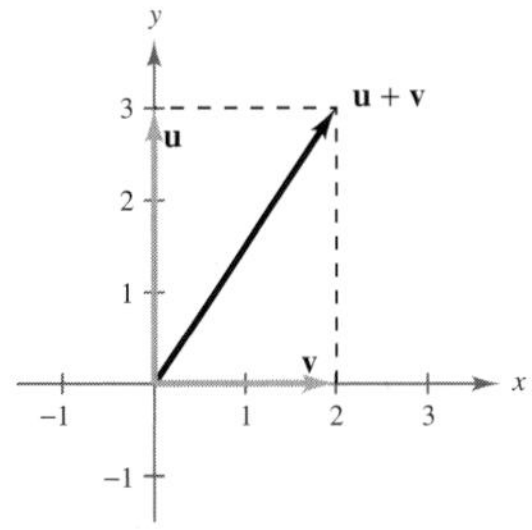

(b) $-2\mathbf{i} + 3\mathbf{j}$

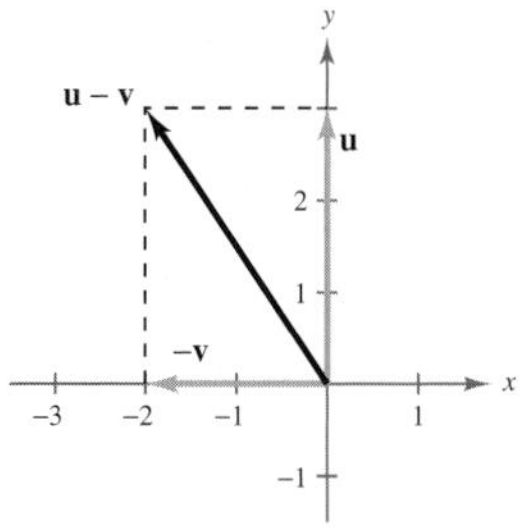

(c) $-6\mathbf{i} + 6\mathbf{j}$

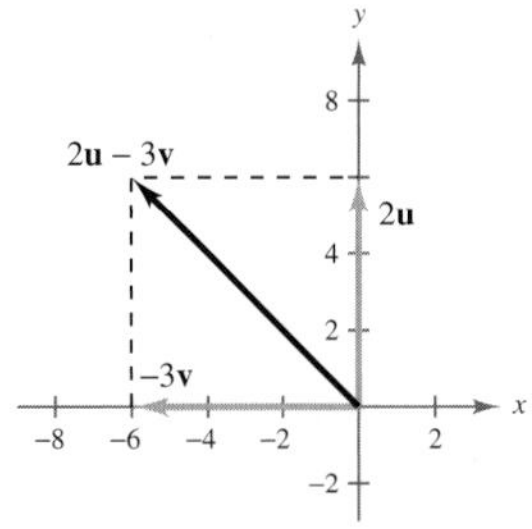

28. $\langle 0, -1 \rangle$ **30.** $\left\langle \frac{5}{13}, -\frac{12}{13} \right\rangle$ **32.** $\frac{1}{\sqrt{2}}\mathbf{i} + \frac{1}{\sqrt{2}}\mathbf{j}$

34. $-\mathbf{i}$ **36.** $-\frac{3}{\sqrt{58}}\mathbf{i} + \frac{7}{\sqrt{58}}\mathbf{j}$ **38.** $\left\langle -\frac{6}{\sqrt{2}}, \frac{6}{\sqrt{2}} \right\rangle$

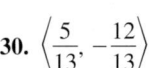

40. $\langle -10, 0 \rangle$

42. $\mathbf{v} = \langle \frac{3}{4}, \frac{3}{2} \rangle$

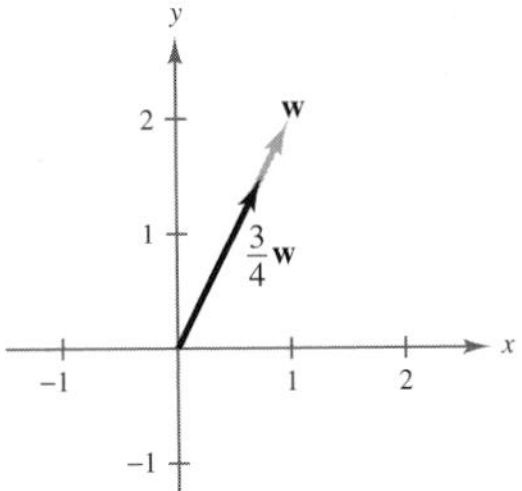

44. $\mathbf{v} = \langle -1, 3 \rangle$

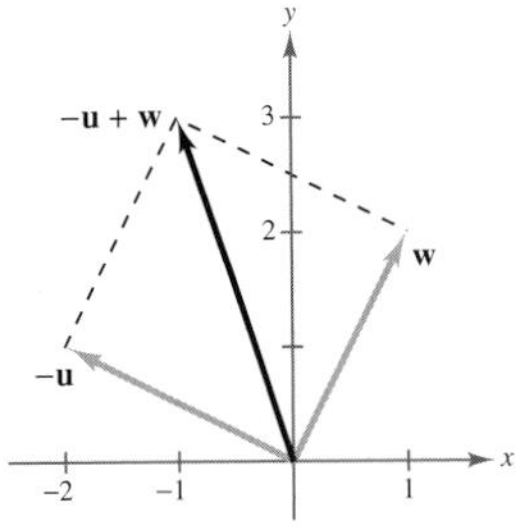

46. $\mathbf{v} = \langle 0, -5 \rangle$

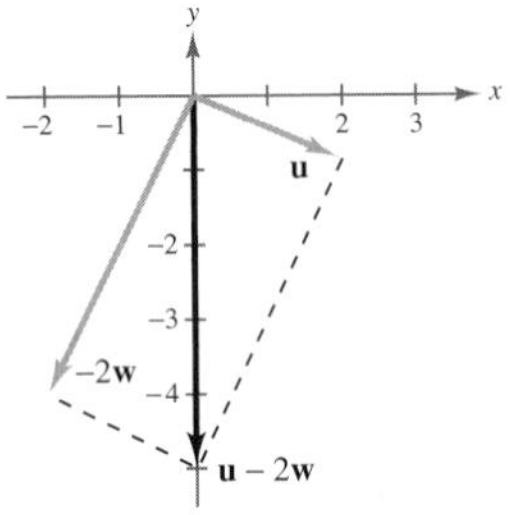

48. $\|\mathbf{v}\| = 8$; $\theta = 135°$ **50.** $\|\mathbf{v}\| = \sqrt{41}$; $\theta = 141.3°$

52. $\mathbf{v} = \left\langle \frac{\sqrt{2}}{2}, \frac{\sqrt{2}}{2} \right\rangle$

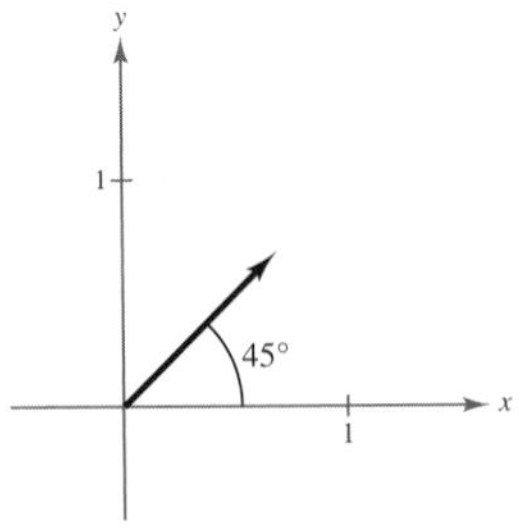

54. $\mathbf{v} = \left\langle \frac{5\sqrt{2}}{4}, \frac{5\sqrt{2}}{4} \right\rangle$

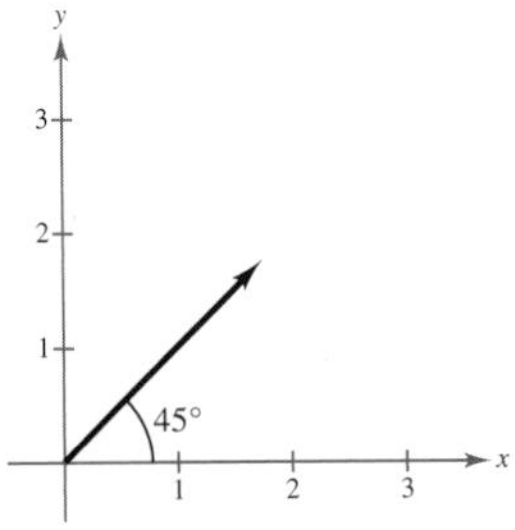

56. $\mathbf{v} = \langle 0, 4\sqrt{3} \rangle$

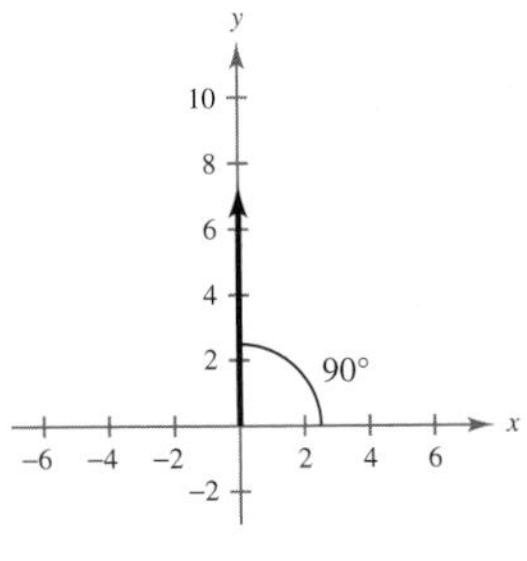

58. $\mathbf{v} = \langle \frac{9}{5}, \frac{12}{5} \rangle$

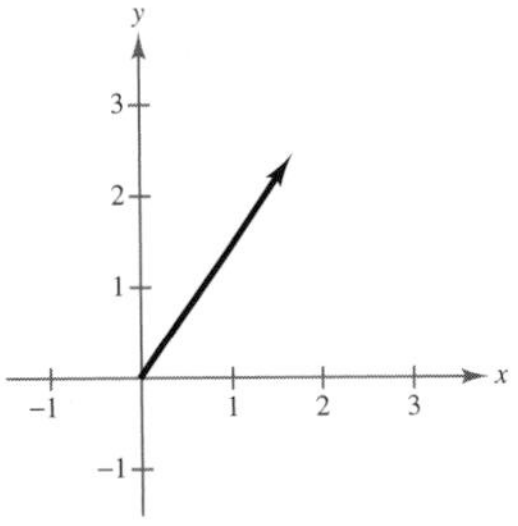

60. $\langle 2, 4 + 2\sqrt{3} \rangle$ **62.** $\langle 33.04, 53.19 \rangle$ **64.** 78.7°

66. 90° **68.** 47.4° **70.** 8.7°; 2396.19 newtons

72. 37.5°; 58.6 pounds

74. Vertical component: $1200 \sin 6° \approx 125.4$ feet per second

Horizontal component:

$1200 \cos 6° \approx 1193.4$ feet per second

76. $T_{AC} \approx 3611.1$ pounds

$T_{BC} \approx 2169.5$ pounds

78.

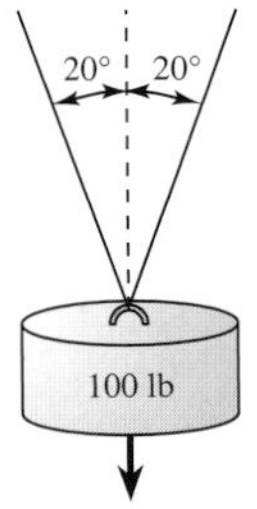

53.2 pounds

80.

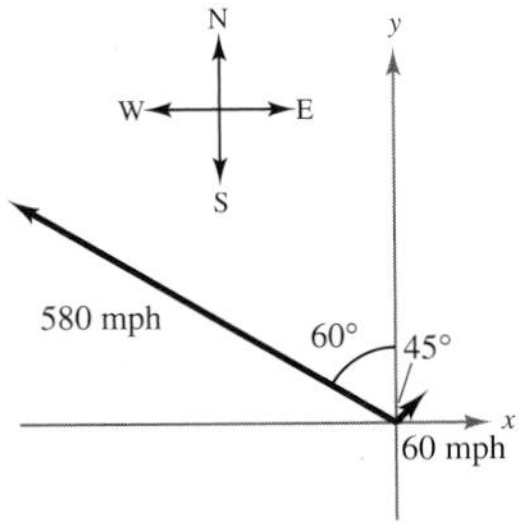

N 54.1° W; 567.4 miles per hour

82. $\sqrt{2}$ pounds, 1 pound

84. True, by the definition of a unit vector.

86. True. If $a^2 + b^2 = 1$, then $\|\mathbf{u}\| = \sqrt{a^2 + b^2} = 1$.

88. (a) $5\sqrt{5 + 4\cos\theta}$

(b)

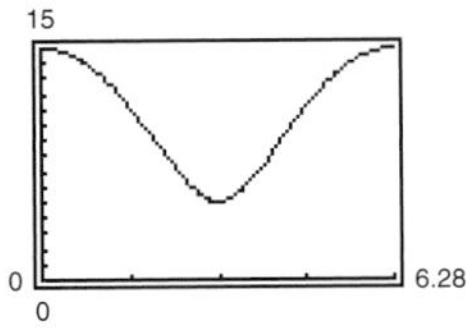

(c) Range: $[5, 15]$

Maximum is 15 when $\theta = 0$.

Minimum is 5 when $\theta = \pi$.

(d) The magnitudes of $\mathbf{F}_1$ and $\mathbf{F}_2$ are not the same.

90. Answers will vary. **92.** $\langle 10, 50 \rangle$ or $\langle -10, -50 \rangle$

94. $8 \cos \theta$ **96.** $125 \tan^3 \theta$

98. $n\pi, \frac{5\pi}{4} + 2n\pi, \frac{7\pi}{4} + 2n\pi$

100. $\frac{\pi}{2} + n\pi, \frac{5\pi}{4} + 2n\pi, \frac{7\pi}{4} + 2n\pi$

102. $n\pi$

Section 6.4 *(page 530)*

2. 9 **4.** 13 **6.** 8, scalar **8.** 6, scalar

10. $2\sqrt{5}$ **12.** 20 **14.** 21 **16.** 33.69°

18. 7.13° **20.** 53.13° **22.** 93.18° **24.** $\frac{\pi}{4}$

26. 41.4°, 64.2°, 74.4° **28.** 21.8°, 41.6°, 116.6°

30. $12{,}500\sqrt{3}$ **32.** 0 **34.** Neither **36.** Neither

38. Orthogonal **40.** $\langle 0, 0 \rangle$, $\langle 4, 2 \rangle$

42. $\frac{14}{17}\langle -4, -1 \rangle$, $\frac{5}{17}\langle 1, -4 \rangle$ **44.** $\langle 3, 8 \rangle$, $\langle -3, -8 \rangle$

46. $3\mathbf{i} - \frac{5}{2}\mathbf{j}$, $-3\mathbf{i} + \frac{5}{2}\mathbf{j}$ **48.** 14

50. $61,700.63; vector operation: $1.05\mathbf{v}$

52. (a) 4175.2 pounds (b) 29,708.0 pounds

54. 12,000 foot-pounds **56.** 10,282.651.78 newton-meters

58. True. $\cos 90° = 0$

60. (a) **u** and **v** are parallel. (b) **u** and **v** are orthogonal.

62. and 64. Answers will vary. **66.** $\frac{\pi}{2}, \frac{5\pi}{4}, \frac{3\pi}{2}, \frac{7\pi}{4}$

68. $\frac{7\pi}{6}, \frac{3\pi}{2}, \frac{11\pi}{6}$ **70.** $-\frac{323}{325}$ **72.** $-\frac{253}{204}$

Section 6.5 *(page 540)*

2.

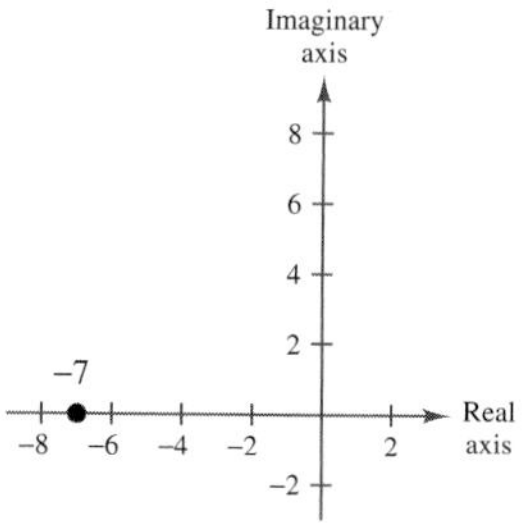

7

4.

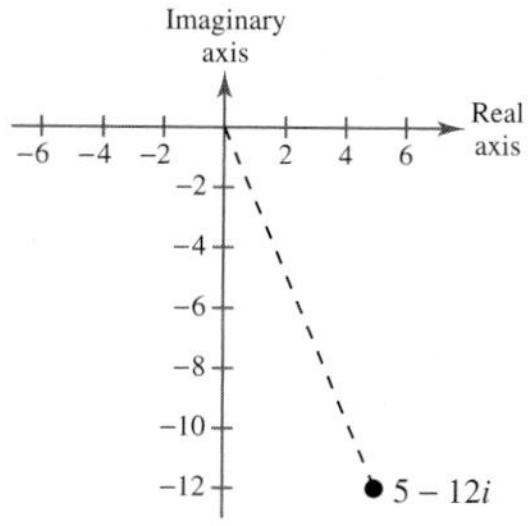

13

6.

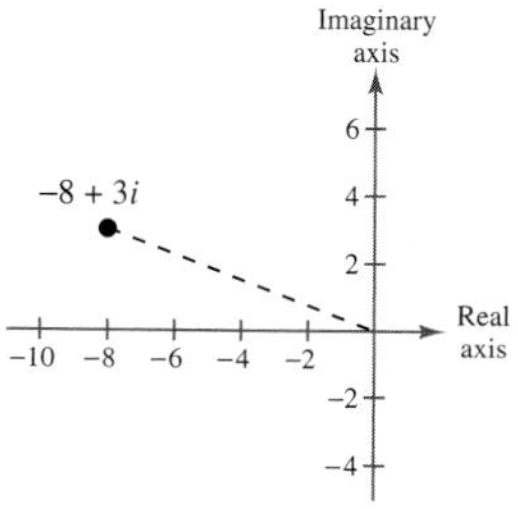

$\sqrt{73}$

8. $2(\cos \pi + i \sin \pi)$ **10.** $2\left(\cos \frac{2\pi}{3} + i \sin \frac{2\pi}{3}\right)$

12.

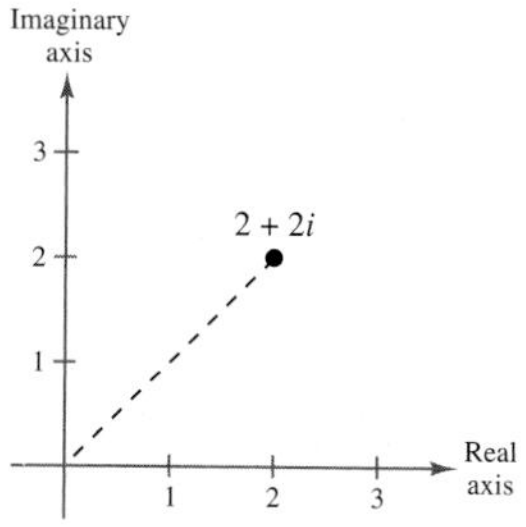

$2\sqrt{2}\left(\cos \frac{\pi}{4} + \text{i} \sin \frac{\pi}{4}\right)$

14.

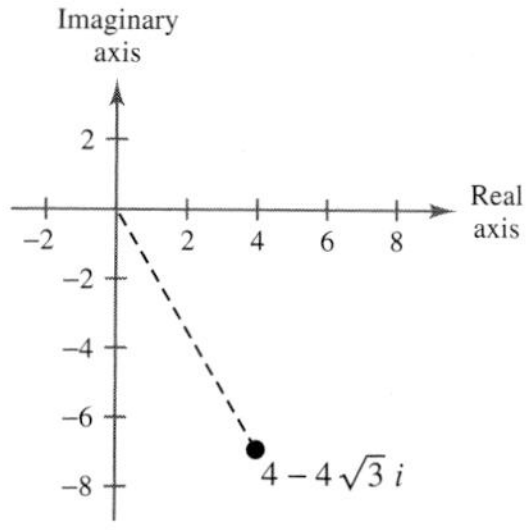

$8\left(\cos \frac{5\pi}{3} + i \sin \frac{5\pi}{3}\right)$

16.

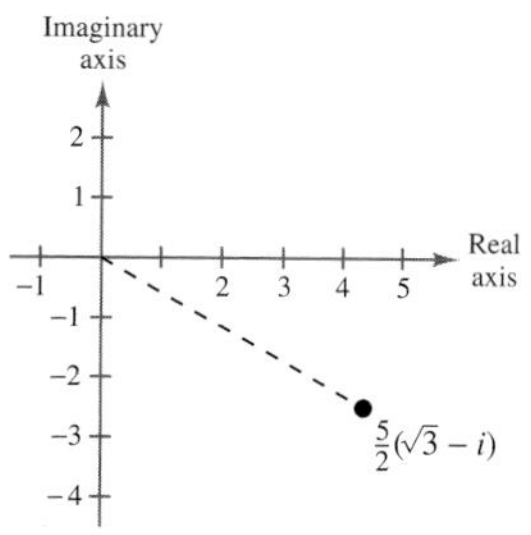

$5\left(\cos \frac{11\pi}{6} + i \sin \frac{11\pi}{6}\right)$

18.

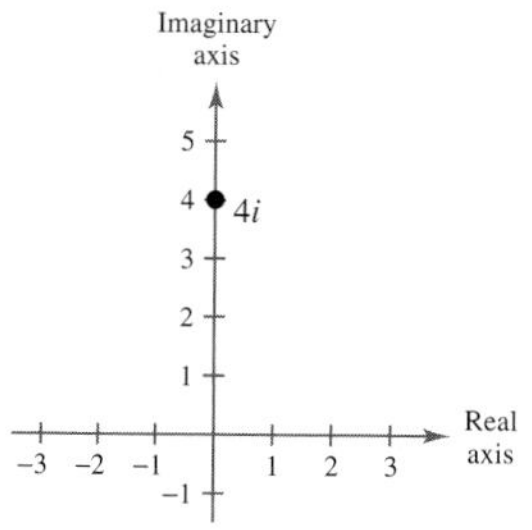

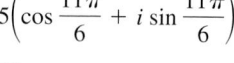

$4\left(\cos \frac{\pi}{2} + i \sin \frac{\pi}{2}\right)$

20.

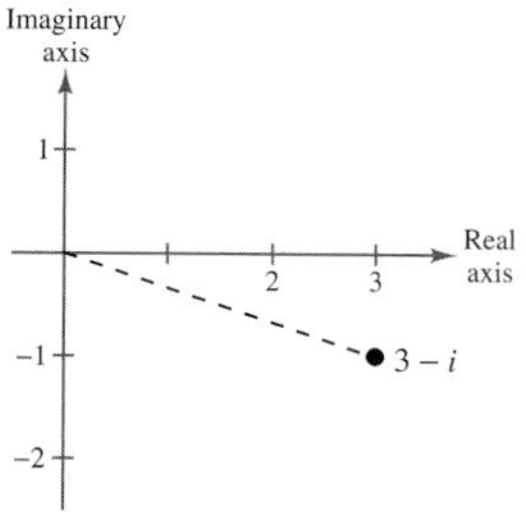

$\sqrt{10}(\cos 5.96 + i \sin 5.96)$

22.

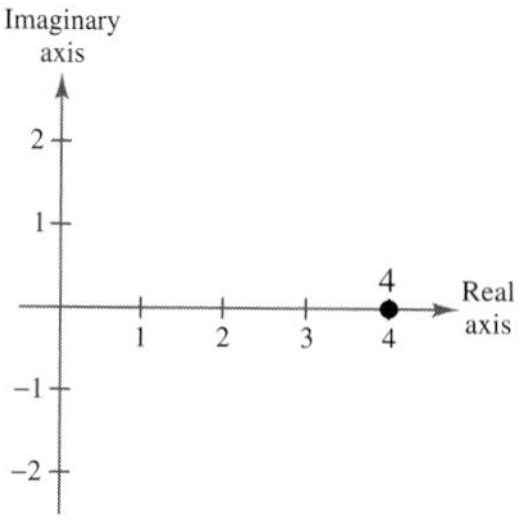

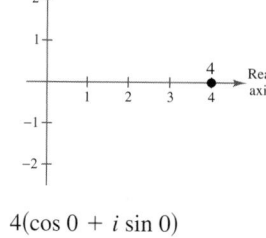

$4(\cos 0 + i \sin 0)$

24.

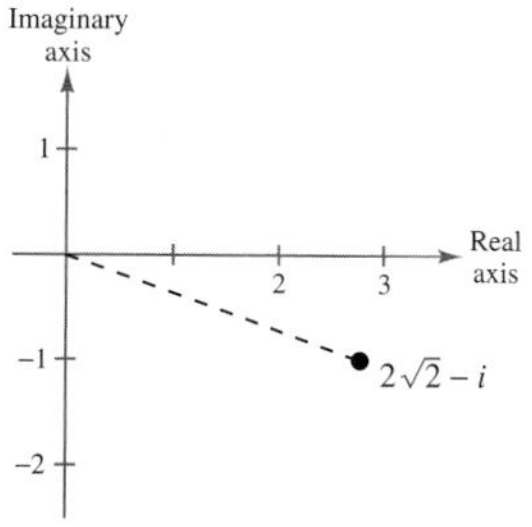

$3(\cos 5.94 + i \sin 5.94)$

26.

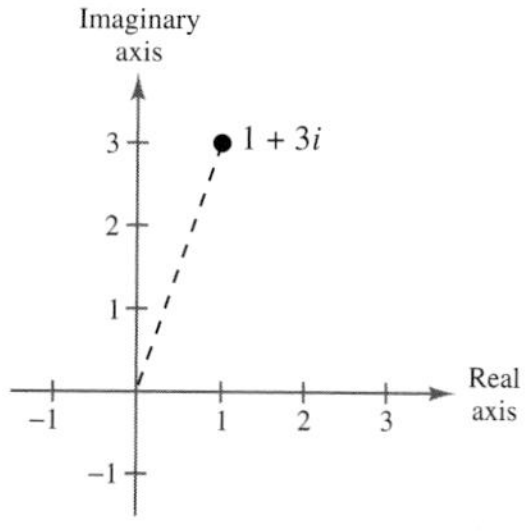

$\sqrt{10}(\cos 1.25 + i \sin 1.25)$

28. $8.54(\cos 0.36 + i \sin 0.36)$

30. $5.10(\cos 3.34 + i \sin 3.34)$

32. $9.80(\cos 5.86 + i \sin 5.86)$

34. $11(\cos 3.75 + i \sin 3.75)$

36.

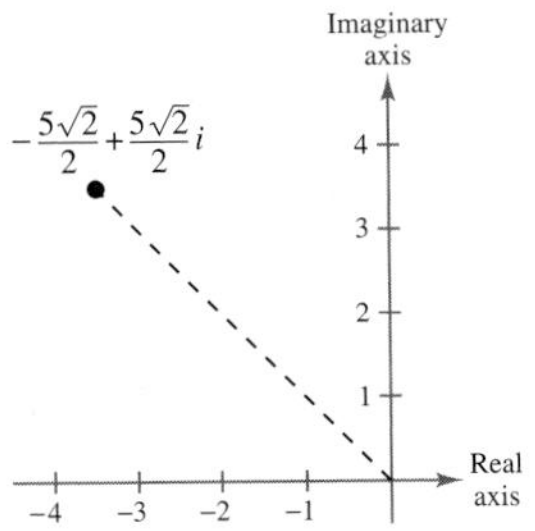

$-\dfrac{5\sqrt{2}}{2} + \dfrac{5\sqrt{2}}{2}i$

38.

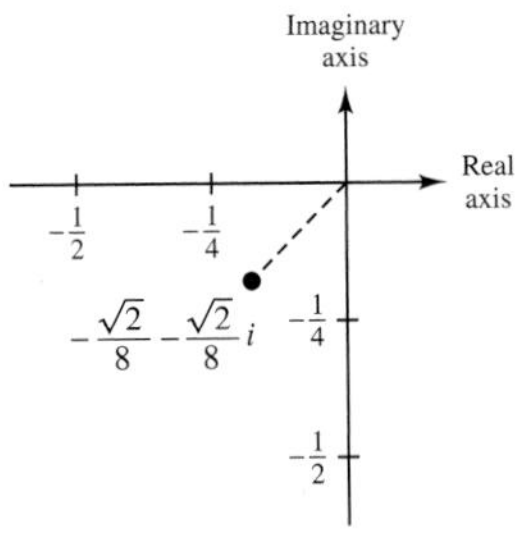

$-\dfrac{\sqrt{2}}{8} - \dfrac{\sqrt{2}}{8}i$

40.

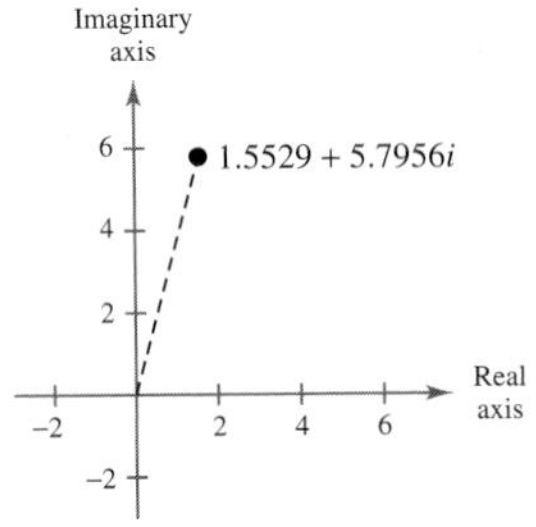

$1.5529 + 5.7956i$

42.

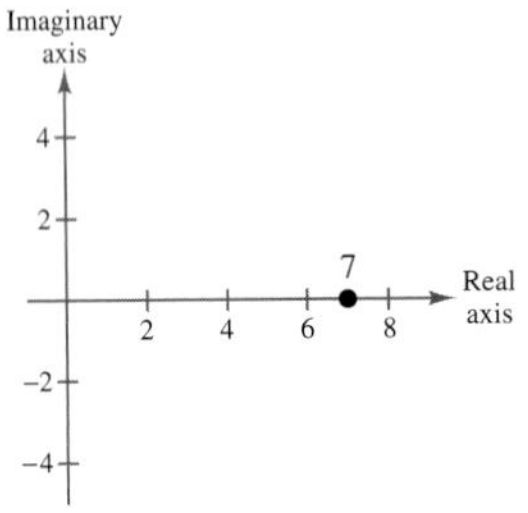

7

44.

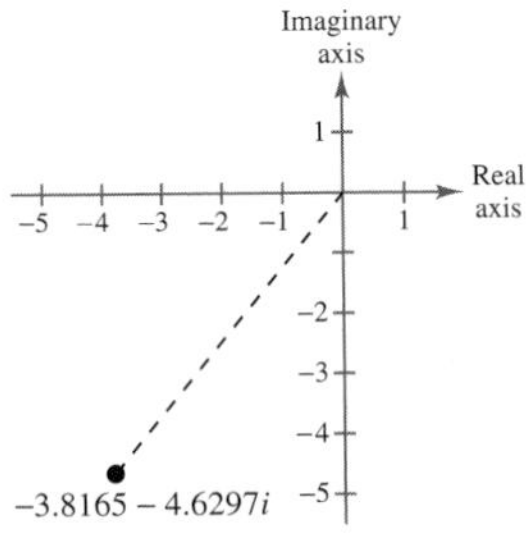

$-3.8165 - 4.6297i$

46. $3.09 + 9.51i$

48. $4.77 + 7.63i$ **50.** $3\left(\cos\dfrac{13\pi}{12} + i\sin\dfrac{13\pi}{12}\right)$

52. $0.4(\cos 40° + i\sin 40°)$ **54.** $\cos 25° + i\sin 25°$

56. $\frac{1}{2}(\cos 80° + i\sin 80°)$ **58.** $\frac{5}{4}[\cos(2.2) + i\sin(2.2)]$

60. $\frac{6}{7}(\cos 300° + i\sin 300°)$

62. (a) $\left[2\left(\cos\dfrac{\pi}{6} + i\sin\dfrac{\pi}{6}\right)\right]\left[\sqrt{2}\left(\cos\dfrac{\pi}{4} + i\sin\dfrac{\pi}{4}\right)\right]$

(b) $2\sqrt{2}\left(\cos\dfrac{5\pi}{12} + i\sin\dfrac{5\pi}{12}\right) \approx 0.732 + 2.732i$

(c) $(\sqrt{3} - 1) + (\sqrt{3} + 1)i \approx 0.732 + 2.732i$

64. (a) $[4(\cos 0 + i\sin 0)]\left\{2\left[\cos\left(\dfrac{5\pi}{3}\right) + i\sin\left(\dfrac{5\pi}{3}\right)\right]\right\}$

(b) $8\left[\cos\left(\dfrac{5\pi}{3}\right) + i\sin\left(\dfrac{5\pi}{3}\right)\right] = 4 - 4\sqrt{3}i$

(c) $4 - 4\sqrt{3}i$

66. (a) $\left[2\left(\cos\dfrac{\pi}{3} + i\sin\dfrac{\pi}{3}\right)\right] \div$
$\{3\sqrt{5}[\cos(-0.46) + i\sin(-0.46)]\}$

(b) $\dfrac{2\sqrt{5}}{15}(\cos 1.51 + i\sin 1.51) \approx 0.018 + 0.298i$

(c) $\approx 0.018 + 0.298i$

68. (a) $\left[4\left(\cos\dfrac{\pi}{2} + i\sin\dfrac{\pi}{2}\right)\right] \div [2\sqrt{5}(\cos 2.68 + i\sin 2.68)]$

(b) $\dfrac{2}{\sqrt{5}}[\cos(-1.11) + i\sin(-1.11)] \approx 0.400 - 0.800i$

(c) $\dfrac{2}{5} - \dfrac{4}{5}i = 0.400 - 0.800i$

70.

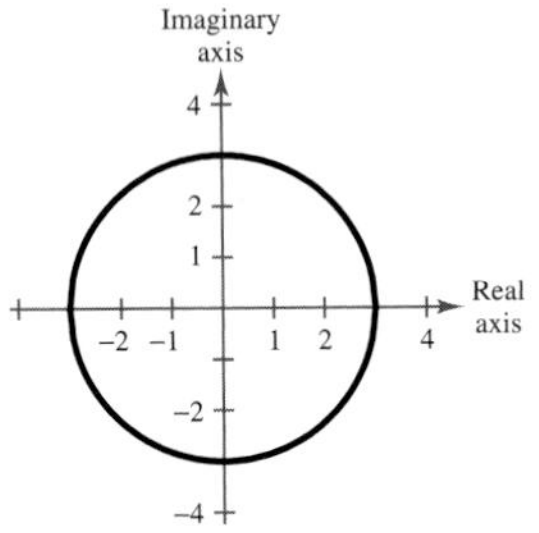

72.

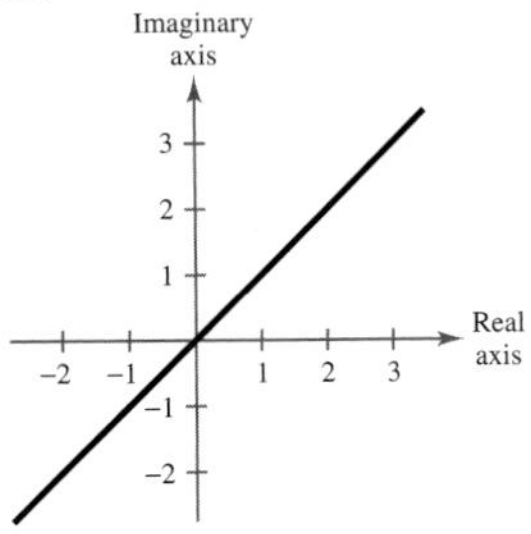

74. $-512i$ **76.** $-239 + 28{,}560i$ **78.** -32

80. $-\dfrac{81}{2} - \dfrac{81\sqrt{3}}{2}i$ **82.** 256 **84.** 1

86. $-43\sqrt{5} + 4i$ **88.** $44.4539 + 252.1108i$

90. $-32\sqrt{2} + 32\sqrt{2}i$

92. (a) $4(\cos 30° + i\sin 30°)$
$4(\cos 210° + i\sin 210°)$

(b)

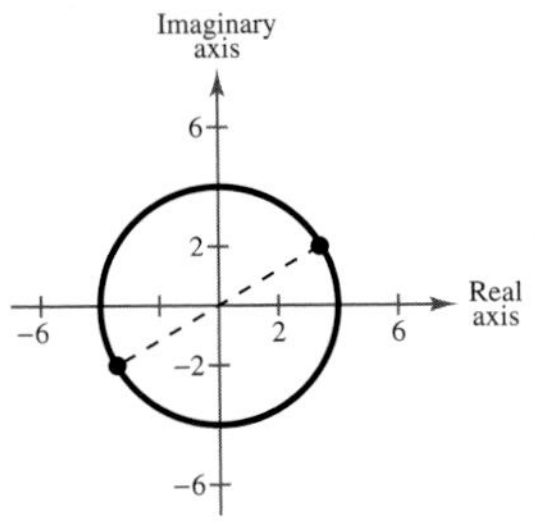

(c) $4\left(\dfrac{\sqrt{3}}{2} + \dfrac{1}{2}i\right) = 2\sqrt{3} + 2i$

$4\left(-\dfrac{\sqrt{3}}{2} - \dfrac{1}{2}i\right) = -2\sqrt{3} - 2i$

94. (a) $2\left(\cos \frac{\pi}{6} + i \sin \frac{\pi}{6}\right)$

$2\left(\cos \frac{17\pi}{30} + i \sin \frac{17\pi}{30}\right)$

$2\left(\cos \frac{29\pi}{30} + i \sin \frac{29\pi}{30}\right)$

$2\left(\cos \frac{41\pi}{30} + i \sin \frac{41\pi}{30}\right)$

$2\left(\cos \frac{53\pi}{30} + i \sin \frac{53\pi}{30}\right)$

(b)

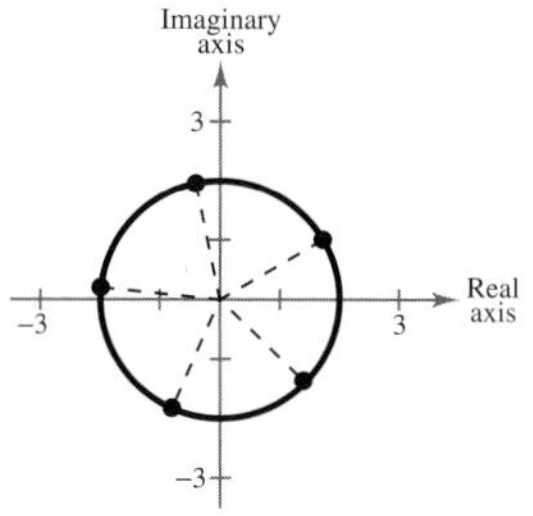

(c) $\sqrt{3} + i, -0.4158 + 1.9563i, -1.9890 + 0.2091i,$
$-0.8135 - 1.8271i, 1.4863 - 1.3383i$

96. (a) $5\left(\cos \frac{\pi}{8} + i \sin \frac{\pi}{8}\right)$

$5\left(\cos \frac{5\pi}{8} + i \sin \frac{5\pi}{8}\right)$

$5\left(\cos \frac{9\pi}{8} + i \sin \frac{9\pi}{8}\right)$

$5\left(\cos \frac{13\pi}{8} + i \sin \frac{13\pi}{8}\right)$

(b)

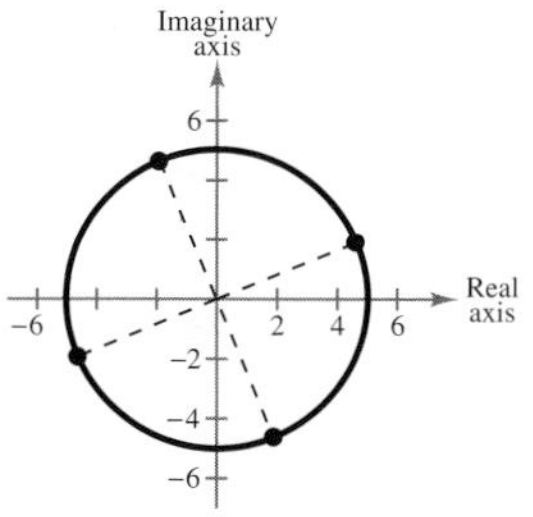

(c) $4.6194 + 1.9134i, \ -1.9134 + 4.6194i,$
$-4.6194 - 1.9134i, 1.9134 - 4.6194i$

98. (a) $2\left(\cos \frac{\pi}{4} + i \sin \frac{\pi}{4}\right)$

$2\left(\cos \frac{11\pi}{12} + i \sin \frac{11\pi}{12}\right)$

$2\left(\cos \frac{19\pi}{12} + i \sin \frac{19\pi}{12}\right)$

(b)

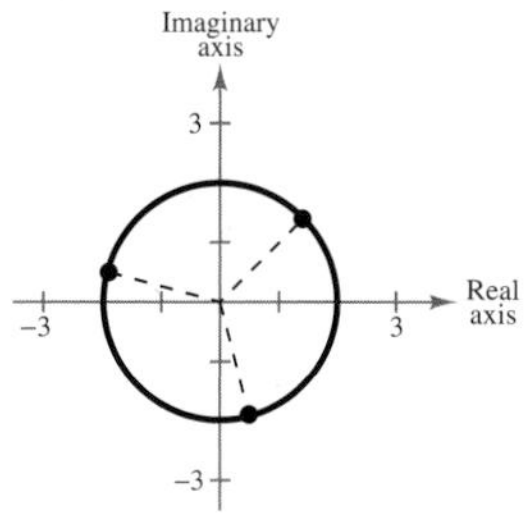

(c) $\sqrt{2} + \sqrt{2}i, -1.9319 + 0.5176i, 0.5176 - 1.9319i$

100. (a) $\cos \frac{\pi}{8} + i \sin \frac{\pi}{8}$

$\cos \frac{5\pi}{8} + i \sin \frac{5\pi}{8}$

$\cos \frac{9\pi}{8} + i \sin \frac{9\pi}{8}$

$\cos \frac{13\pi}{8} + i \sin \frac{13\pi}{8}$

(b)

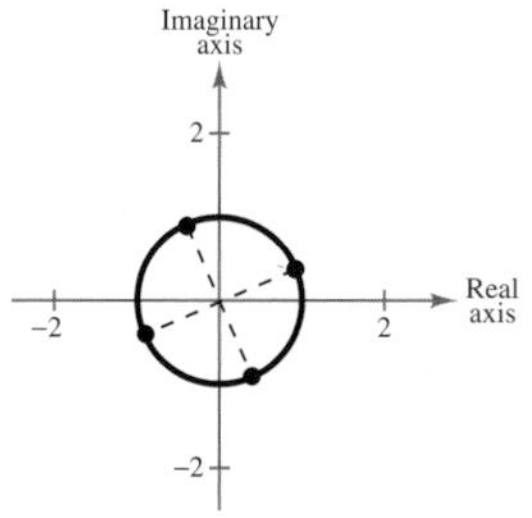

(c) $0.9239 + 0.3827i, \ -0.3827 + 0.9239i,$
$-0.9239 - 0.3827i, \ 0.3827 - 0.9239i$

102. (a) $10(\cos 0 + i \sin 0)$

$10\left(\cos \frac{2\pi}{3} + i \sin \frac{2\pi}{3}\right)$

$10\left(\cos \frac{4\pi}{3} + i \sin \frac{4\pi}{3}\right)$

(b)

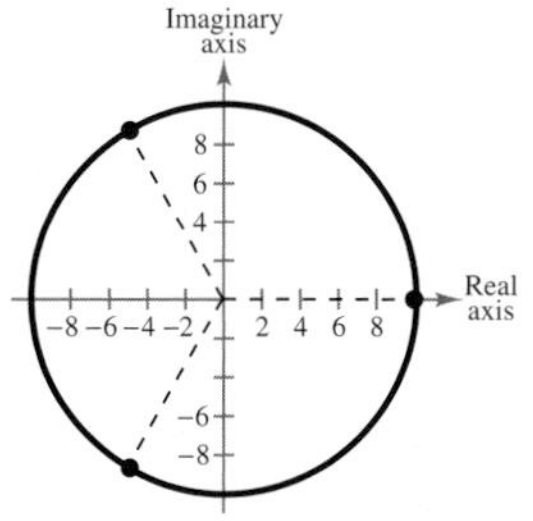

(c) $10, -5 + 5\sqrt{3}i, -5 - 5\sqrt{3}i$

104. (a) $\sqrt{2}\left(\cos\frac{\pi}{4} + i\sin\frac{\pi}{4}\right)$

$\sqrt{2}\left(\cos\frac{3\pi}{4} + i\sin\frac{3\pi}{4}\right)$

$\sqrt{2}\left(\cos\frac{5\pi}{4} + i\sin\frac{5\pi}{4}\right)$

$\sqrt{2}\left(\cos\frac{7\pi}{4} + i\sin\frac{7\pi}{4}\right)$

(b)

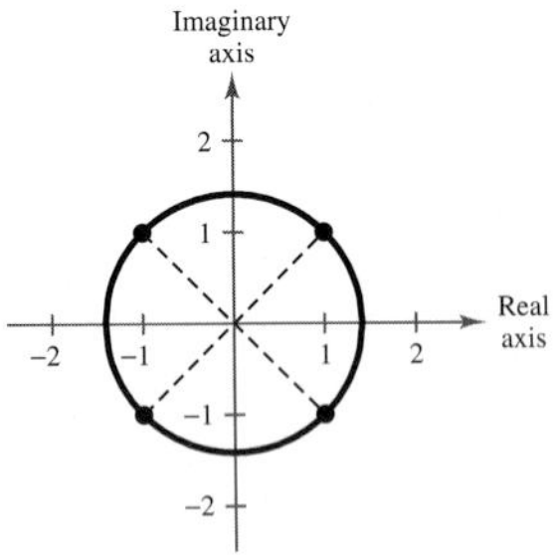

(c) $1 + i,\ -1 + i,\ -1 - i,\ 1 - i$

106. (a) $2\left(\cos\frac{\pi}{12} + i\sin\frac{\pi}{12}\right)$

$2\left(\cos\frac{5\pi}{12} + i\sin\frac{5\pi}{12}\right)$

$2\left(\cos\frac{3\pi}{4} + i\sin\frac{3\pi}{4}\right)$

$2\left(\cos\frac{13\pi}{12} + i\sin\frac{13\pi}{12}\right)$

$2\left(\cos\frac{17\pi}{12} + i\sin\frac{17\pi}{12}\right)$

$2\left(\cos\frac{7\pi}{4} + i\sin\frac{7\pi}{4}\right)$

(b)

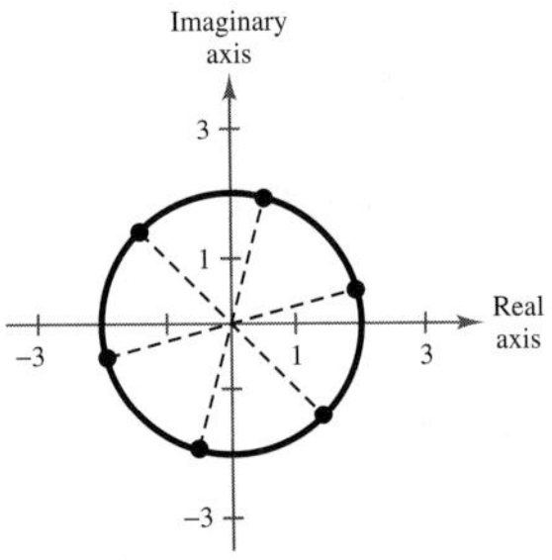

(c) $1.9319 + 0.5176i, 0.5176 + 1.9319i,$
$-\sqrt{2} + \sqrt{2}i, -1.9319 - 0.5176i,$
$-0.5176 - 1.9319i, \sqrt{2} - \sqrt{2}i$

108. $\cos\frac{\pi}{3} + i\sin\frac{\pi}{3}$

$\cos\pi + i\sin\pi$

$\cos\frac{5\pi}{3} + i\sin\frac{5\pi}{3}$

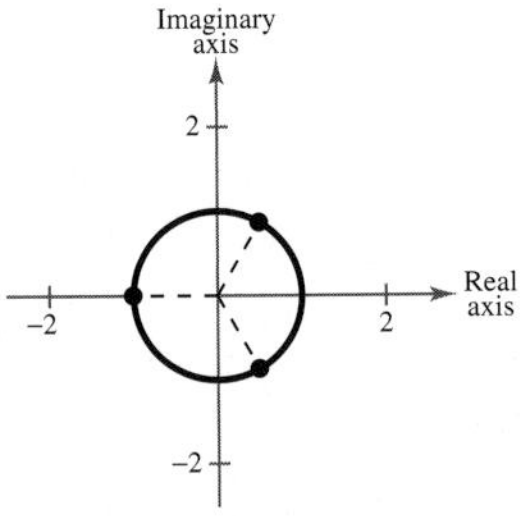

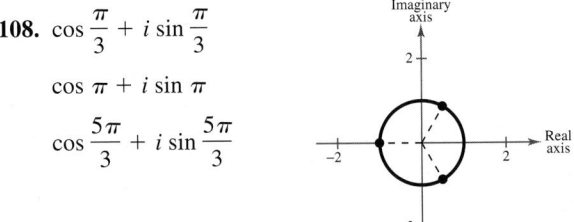

110. $3(\cos 0 + i\sin 0)$

$3\left(\cos\frac{2\pi}{3} + i\sin\frac{2\pi}{3}\right)$

$3\left(\cos\frac{4\pi}{3} + i\sin\frac{4\pi}{3}\right)$

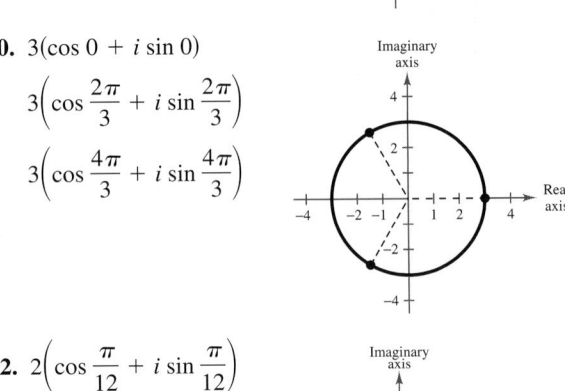

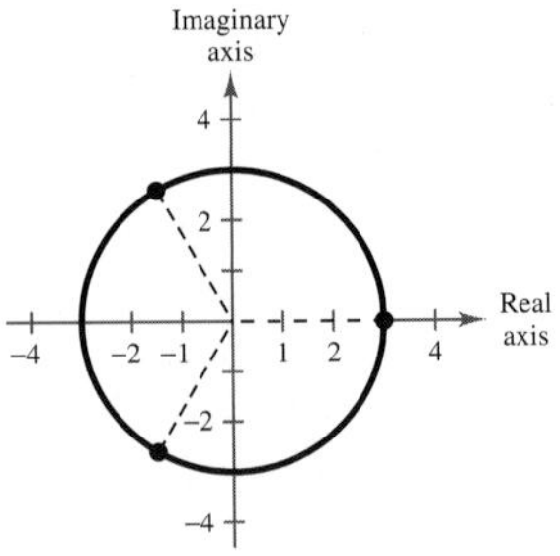

112. $2\left(\cos\frac{\pi}{12} + i\sin\frac{\pi}{12}\right)$

$2\left(\cos\frac{5\pi}{12} + i\sin\frac{5\pi}{12}\right)$

$2\left(\cos\frac{3\pi}{4} + i\sin\frac{3\pi}{4}\right)$

$2\left(\cos\frac{13\pi}{12} + i\sin\frac{13\pi}{12}\right)$

$2\left(\cos\frac{17\pi}{12} + i\sin\frac{17\pi}{12}\right)$

$2\left(\cos\frac{7\pi}{4} + i\sin\frac{7\pi}{4}\right)$

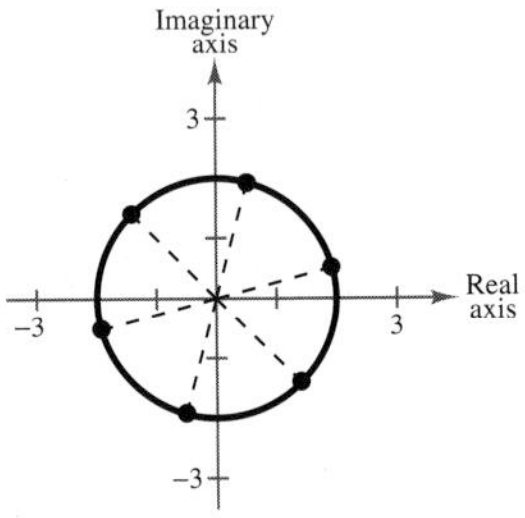

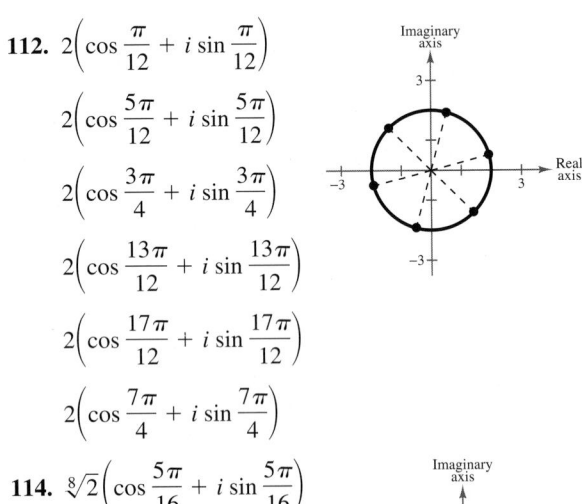

114. $\sqrt[8]{2}\left(\cos\frac{5\pi}{16} + i\sin\frac{5\pi}{16}\right)$

$\sqrt[8]{2}\left(\cos\frac{13\pi}{16} + i\sin\frac{13\pi}{16}\right)$

$\sqrt[8]{2}\left(\cos\frac{21\pi}{16} + i\sin\frac{21\pi}{16}\right)$

$\sqrt[8]{2}\left(\cos\frac{29\pi}{16} + i\sin\frac{29\pi}{16}\right)$

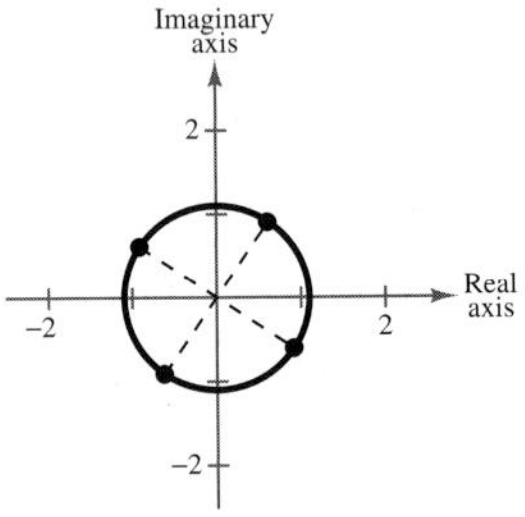

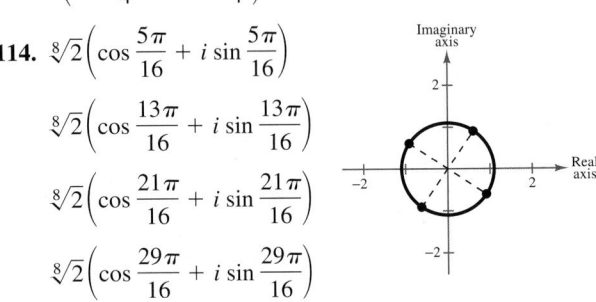

116. False. They are equally spaced along the circle centered at the origin with radius $\sqrt[n]{r}$.

118. False. The complex number needs to be converted to trigonometric form before using DeMoivre's Theorem.
$(4 + \sqrt{6}i)^8 = \left[\sqrt{22}(\cos 0.55 + i\sin 0.55)\right]^8$

120.–124. Answers will vary.

126. (a) $3(\cos 45° + i \sin 45°)$

$3(\cos 135° + i \sin 135°)$

$3(\cos 225° + i \sin 225°)$

$3(\cos 315° + i \sin 315°)$

(b) -81

128. $A = 24°, b \approx 75.24, c \approx 82.36$

130. $A = 84°, a \approx 2009.43, c \approx 2020.50$

132. $A = 8°30', a \approx 1.01, b \approx 6.73$

134. $\frac{1}{8}; \frac{1}{24}$ **136.** $\frac{1}{12}; 0$

Review Exercises *(page 544)*

2. $C = 37°, b \approx 38.90, c \approx 27.31$

4. $A = 150°, a \approx 48.24, b \approx 16.75$

6. $C = 40°, a \approx 162.42, b \approx 115.29$

8. $A = 80°, b \approx 334.95, c \approx 219.04$ **10.** No solution

12. Two solutions:

$A \approx 40.92°, C \approx 114.08°, c \approx 8.64$

$A \approx 139.08°, C \approx 15.92°, c \approx 2.60$

14. 15.757 **16.** 44.08 **18.** 4.8

20. $A \approx 29.69°, B \approx 52.41°, C \approx 97.90°$

22. $A \approx 29.93°, B \approx 86.18°, C \approx 63.90°$

24. $A \approx 35°, C \approx 35°, b \approx 6.55$

26. $A \approx 45.76°, B \approx 91.24°, c \approx 21.42$

28. 615.1 meters **30.** 9.798 **32.** 8.36

34.

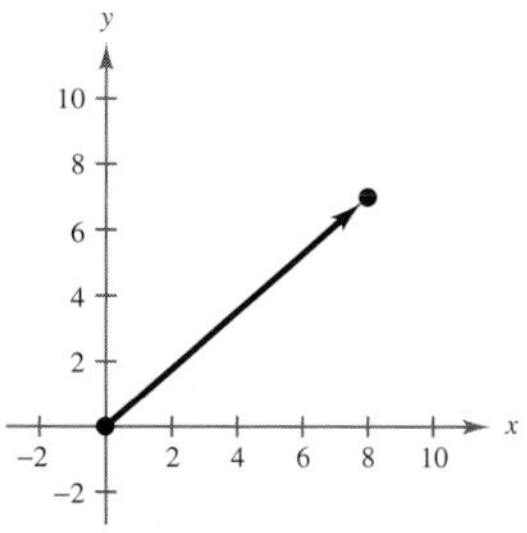

36.

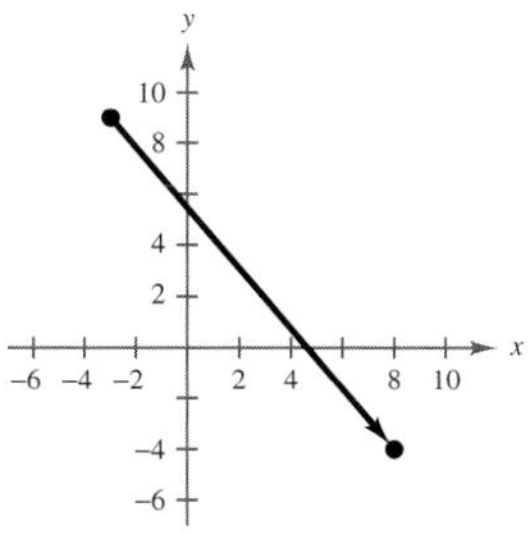

38. $\langle 7, -5 \rangle$ **40.** $\langle 7, -7 \rangle$ **42.** $\langle -4, 4\sqrt{3} \rangle$

44. $\langle 22, -7 \rangle$

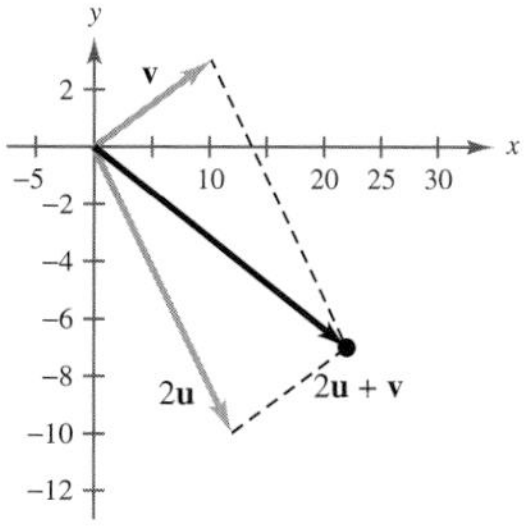

46. $\langle 30, 9 \rangle$

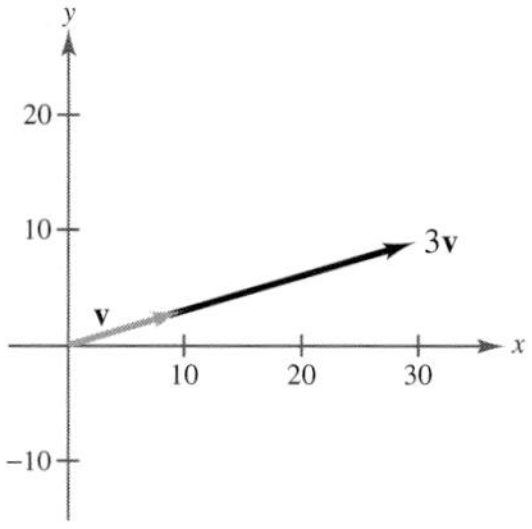

48. $-3\mathbf{i} + 4\mathbf{j}$ **50.** $6\mathbf{i} + 4\mathbf{j}$

52. $10\sqrt{2}(\mathbf{i} \cos 135° + \mathbf{j} \sin 135°)$ **54.** $\|\mathbf{v}\| = 7; \theta = 60°$

56. $\|\mathbf{v}\| = \sqrt{41}; \theta = 38.7°$ **58.** $\|\mathbf{v}\| = 3\sqrt{2}; \theta = 225°$

60. The resultant force is 133.92 pounds and 5.5° from the 85-pound force.

62. 740.5 kilometers per hour; N 32.1° E **64.** -140

66. -136 **68.** 5; scalar **70.** -6; scalar

72. 105° **74.** 22.4° **76.** Parallel **78.** Orthogonal

80. $\langle 5, 0 \rangle, \langle 0, 6 \rangle$ **82.** $\frac{25}{29}\langle -5, 2 \rangle, \frac{19}{29}\langle 2, 5 \rangle$ **84.** -132

86.

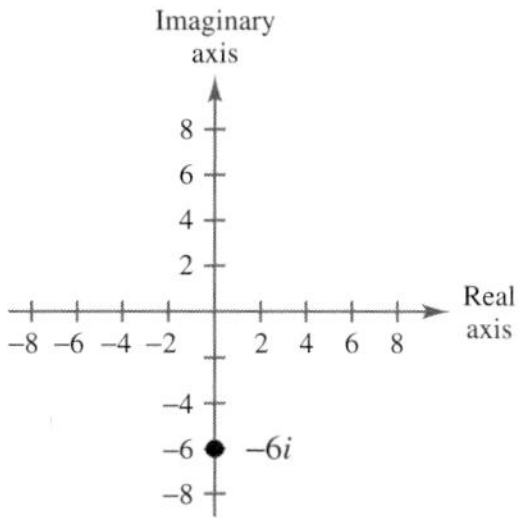

6

88.

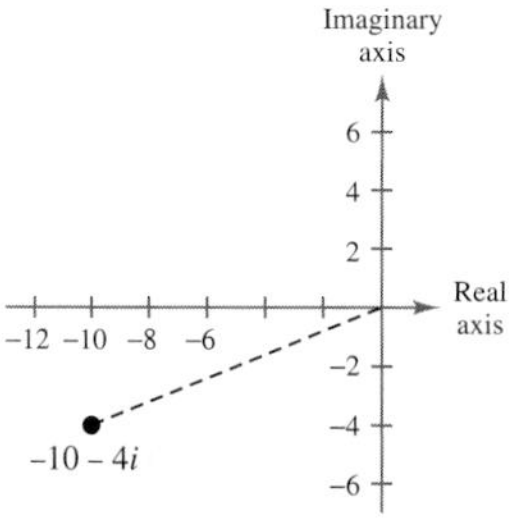

$2\sqrt{29}$

90. $13(\cos 1.176 + i \sin 1.176)$ **92.** $7(\cos \pi + i \sin \pi)$

94. (a) $z_1 = 3\sqrt{2}\left(\cos \frac{5\pi}{4} + i \sin \frac{5\pi}{4}\right)$

$z_2 = 4\left(\cos \frac{\pi}{6} + i \sin \frac{\pi}{6}\right)$

(b) $z_1 z_2 = 12\sqrt{2}\left(\cos \frac{17\pi}{12} + i \sin \frac{17\pi}{12}\right)$

$\frac{z_1}{z_2} = \frac{3\sqrt{2}}{4}\left(\cos \frac{13\pi}{12} + i \sin \frac{13\pi}{12}\right)$

96. $-16 - 16\sqrt{3}i$ **98.** 16

100. (a) $4(\cos 60° + i \sin 60°)$

$4(\cos 150° + i \sin 150°)$

$4(\cos 240° + i \sin 240°)$

$4(\cos 330° + i \sin 330°)$

(b) $-128 - 128\sqrt{3}i$

102. $4(\cos 0 + i \sin 0) = 4$

$4\left(\cos \frac{\pi}{2} + i \sin \frac{\pi}{2}\right) = 4i$

$4(\cos \pi + i \sin \pi) = -4$

$4\left(\cos \frac{3\pi}{2} + i \sin \frac{3\pi}{2}\right) = -4i$

104. $2(\cos 0 + i \sin 0) = 2$

$2\left(\cos \frac{2\pi}{5} + i \sin \frac{2\pi}{5}\right) = 0.6180 + 1.9021i$

$2\left(\cos \frac{4\pi}{5} + i \sin \frac{4\pi}{5}\right) = -1.6180 + 1.1756i$

$2\left(\cos \frac{6\pi}{5} + i \sin \frac{6\pi}{5}\right) = -1.6180 - 1.1756i$

$2\left(\cos \frac{8\pi}{5} + i \sin \frac{8\pi}{5}\right) = 0.6180 - 1.9021i$

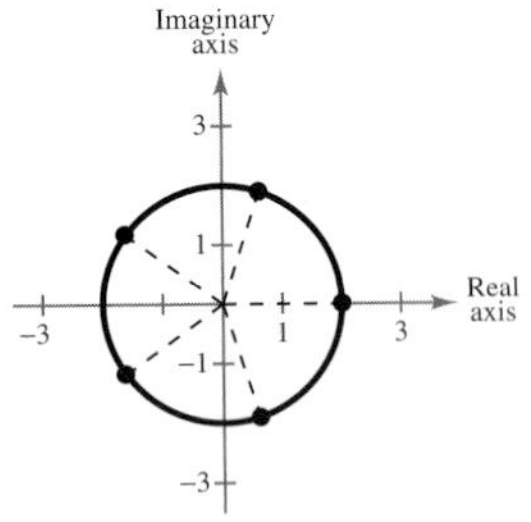

106. $\cos 0 + i \sin 0 = 1$

$\cos \frac{\pi}{2} + i \sin \frac{\pi}{2} = i$

$\cos \frac{2\pi}{3} + i \sin \frac{2\pi}{3} = -\frac{1}{2} + \frac{\sqrt{3}}{2}i$

$\cos \frac{4\pi}{3} + i \sin \frac{4\pi}{3} = -\frac{1}{2} - \frac{\sqrt{3}}{2}i$

$\cos \frac{3\pi}{2} + i \sin \frac{3\pi}{2} = -i$

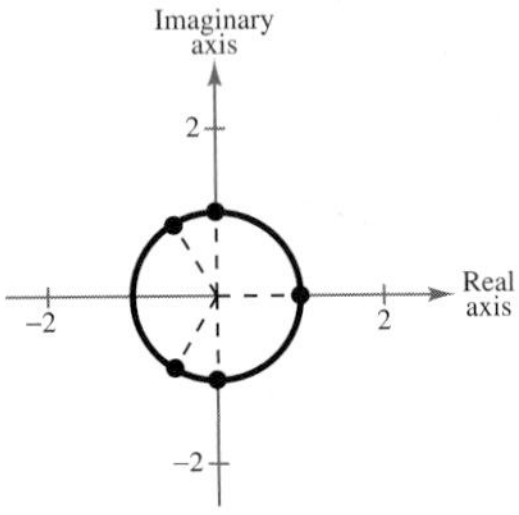

108. False. There may be no solution, one solution, or two solutions.

110. $a^2 = b^2 + c^2 - 2bc \cos A$,

$b^2 = a^2 + c^2 - 2ac \cos B$,

$c^2 = a^2 + b^2 - 2ab \cos C$

112. Direction and magnitude

114. a; The angle between the vectors is acute.

116. The diagonal of the parallelogram with **u** and **v** as its adjacent sides

118. (a) 3

(b) On the circle 120°, 210°, and 300° from the positive x- axis

Chapter 7

Section 7.1 *(page 561)*

2. a and d **4.** b and c **6.** $(-1, 3)$

8. $(-\sqrt{3}, 2 + 3\sqrt{3}), (0, 2), (\sqrt{3}, 2 - 3\sqrt{3})$

10. $(0, 0), (2, -2), (-2, 2)$ **12.** $(0, 2), (1, 0), (-1, 0)$

14. $(0, 4), (1, 2), (2, 0)$ **16.** $(-3, 2)$ **18.** $\left(\frac{4}{3}, \frac{4}{3}\right)$

20. $\left(2, \frac{5}{2}\right)$ **22.** $\left(\frac{208}{17}, \frac{88}{17}\right)$ **24.** No solution

26. $(0, 0), (12, 6)$ **28.** $(0, 0)$ **30.** $(2, -2)$

32. $(5, 3)$ **34.** $(-3, 0), (3, 6)$ **36.** $(3, 1), (15, 7)$

38. $(2, 2)$ **40.** No solution

42. $(3, 4), (3, -4)$

44.

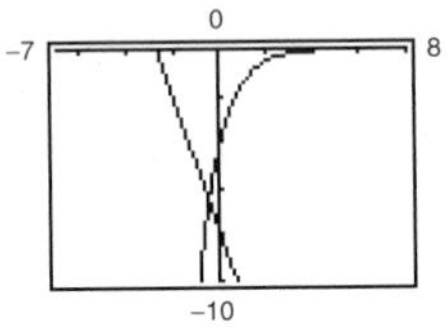

$(-0.49, -6.53)$

46.

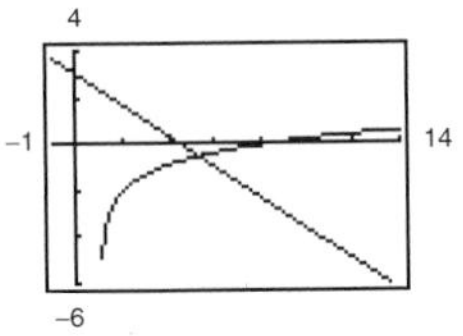

$(5.31, -0.54)$

48.

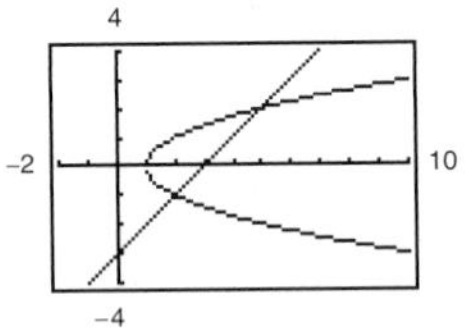

$(5, 2), (2, -1)$

50.

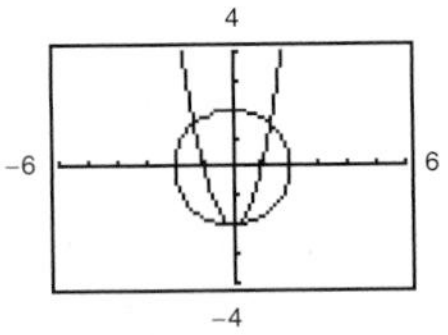

$(0, -2), \left(\pm\frac{1}{2}\sqrt{7}, \frac{3}{2}\right)$

52. No solution **54.** $(3, 4), (5, 0)$

56. No solution **58.** $(-1.96, 0.14), (1.06, 2.88)$

60. $(0, -1), (2, 1), (-1, -5)$ **62.** $(1, 0), (5, 2)$

64. 233,333 units **66.** 1464 units

68. (a) 400 units (b) 1080 units

70. (a) $\begin{cases} x + y = 20{,}000 \\ 0.065x + 0.085y = 1600 \end{cases}$

(b)

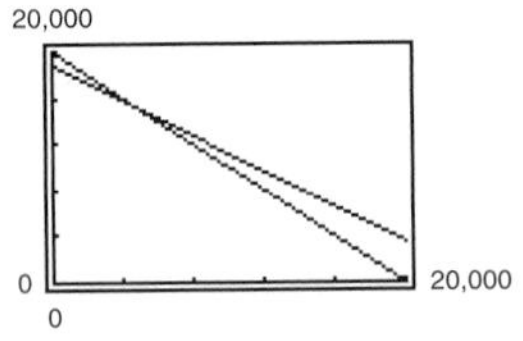

Decreases; interest is fixed at \$1600.

(c) \$5000

72. \$500,000

74.

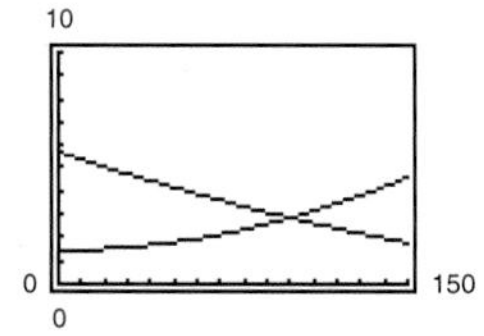

(99.99, 2.85)

76. 60×80 centimeters **78.** 42×63 feet

80. $\sqrt{2} \times \sqrt{2} \times 2$ inches

82. False. To solve a system of equations by substitution, you can solve for either variable in one of the two equations and then back-substitute.

84. An ordered pair that satisfies each equation in the system

86. Graphical solutions may be approximate.

88. (a) $b = 1$

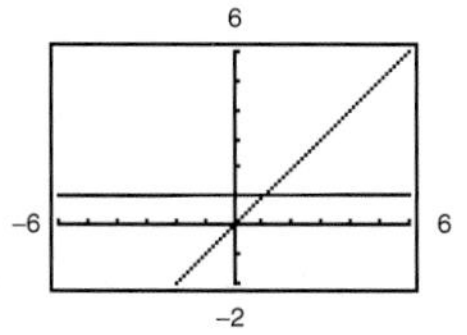

$b = 2$

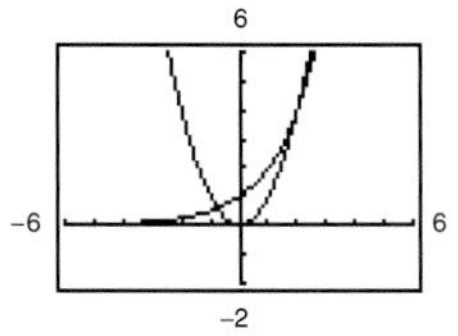

$b = 3$

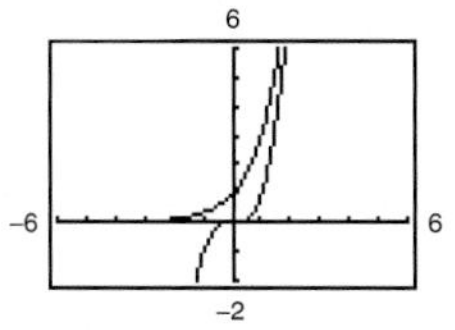

$b = 4$

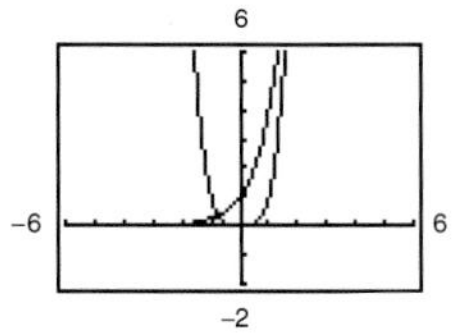

(b) Three points of intersection

90. $2x - 6.5y + 19 = 0$ **92.** $x - 4 = 0$

94. $45x + 29y - 127 = 0$

96. Domain: All real numbers $x \neq -\frac{2}{3}$

Horizontal asymptote: $y = \frac{2}{3}$

Vertical asymptote: $x = -\frac{2}{3}$

98. Domain: All real numbers $x \neq 0$

Horizontal asymptote: $y = 3$

Vertical asymptote: $x = 0$

100.

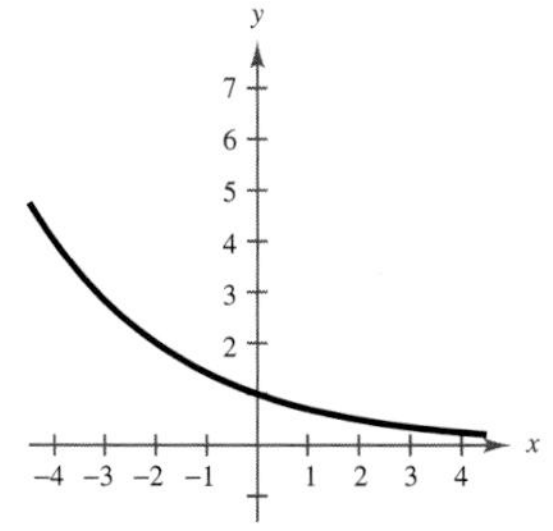

102.

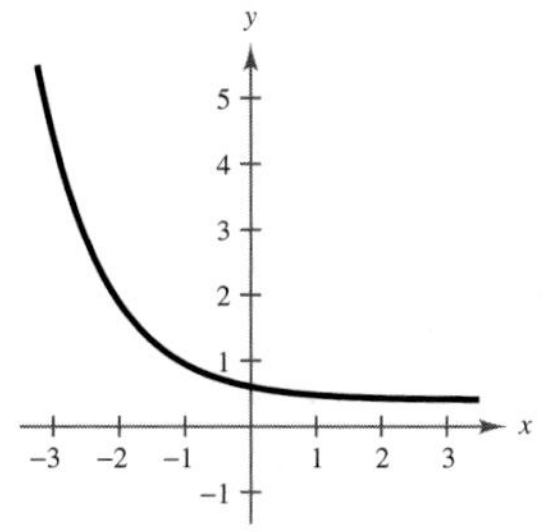

Section 7.2 *(page 573)*

2. $(-2, 1)$ **4.** $(3, 3)$ **6.** No solution

8. $(a, 3a + 5)$ **10.** $\left(-\frac{35}{12}, -\frac{41}{36}\right)$ **12.** $\left(3, \frac{7}{5}\right)$

14. $(5, 1)$ **16.** $\left(\frac{5}{6}, \frac{5}{6}\right)$ **18.** $(-17, 5)$

20. No solution **22.** $(a, 4 - 4a)$

24. $(a, -0.875a + 0.75)$ **26.** $(101, 96)$

28. $\left(-\frac{14}{9}, \frac{20}{9}\right)$ **30.** $(7, 1)$

32.

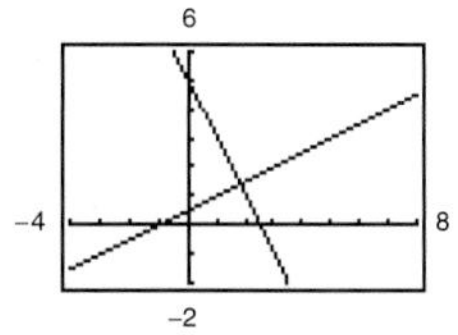

Consistent, one solution

34.

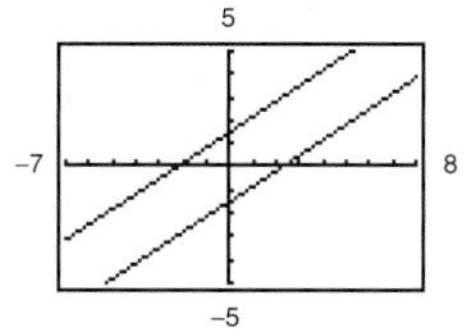

Inconsistent

36.

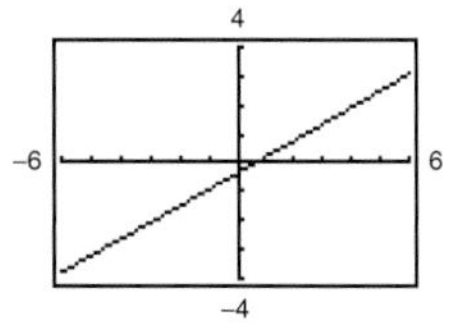

Consistent, infinitely many solutions

38.

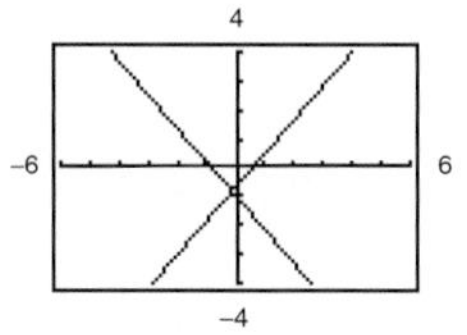

Consistent, one solution

40.

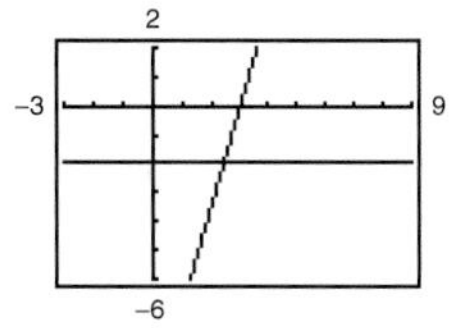

$(3, -2)$

42.

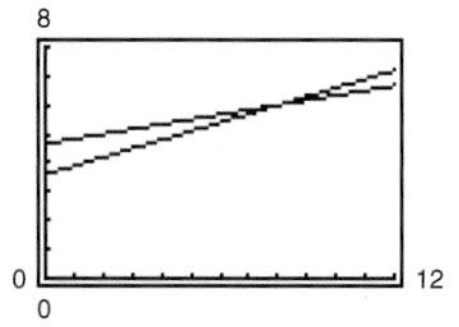

$(8, 6)$

44.

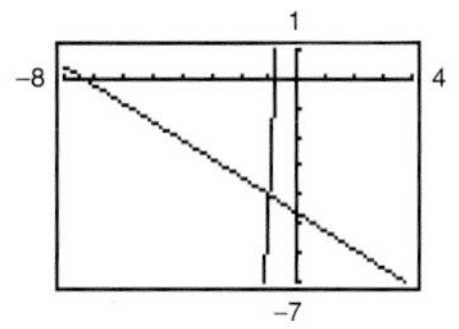

$(-1, -4)$

46.

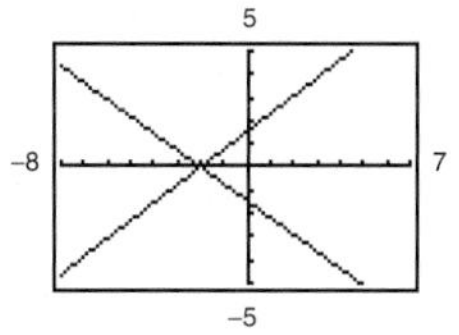

$(-2, 0)$

48. $(-2, 5)$ **50.** $(1, 3)$ **52.** $(-23, 61)$ **54.** $(3, 2)$

56. $(500, 75)$ **58.** $(250{,}000, 350)$

60. First plane: 880 kilometers per hour

Second plane: 960 kilometers per hour

62. (a) $\begin{cases} x + y = 500 \\ 87x + 92y = 44{,}500 \end{cases}$

(b)

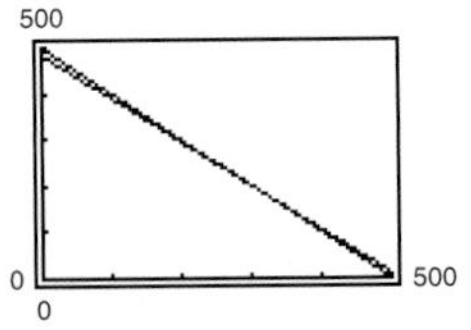

Decreases

(c) 87 octane: 300 gallons; 92 octane: 200 gallons

64. \$20,000

66. Before noon: 81 jackets; After noon: 133 jackets

68. Child: 48 balloons; Father: 32 balloons

70. $y = 0.22x + 1.9$ **72.** $y = -0.583x + 5.390$

74. $y = \frac{1}{2}x - \frac{3}{4}$

76. (a) and (b) $y = 14x + 19$

(c)

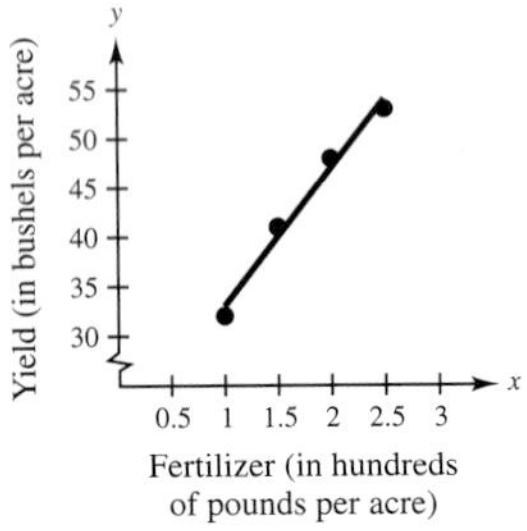

(d) 41.4 bushels per acre

78. False. Solving a system of equations algebraically will always give an exact solution.

80. $\begin{cases} x + y = 9 \\ 3x - 2y = 12 \end{cases}$ **82.** $\begin{cases} 2x + 2y = 11 \\ x - 4y = -7 \end{cases}$

84. (39,600, 398). It is necessary to change the scale on the axes to see the point of intersection.

86. No. Two lines will intersect only once or will coincide, and if they coincide the system will have infinitely many solutions.

88. $k = -4$

90. $x \le -\frac{22}{3}$

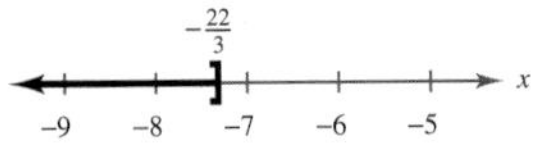

92. $x \le \frac{19}{16}$

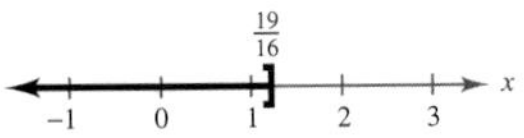

94. $-2 < x < 18$

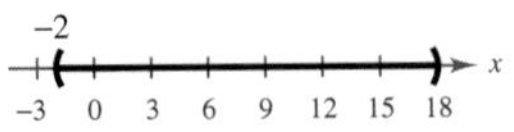

96. $-5 < x < \frac{7}{2}$

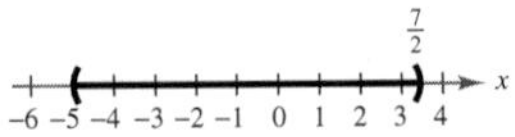

98. $-\dfrac{6}{x+5} + \dfrac{7}{x+6}$ **100.** $\ln 6x$ **102.** $\log_9 \dfrac{12}{x}$

104. No solution

Section 7.3 *(page 585)*

2. b **4.** a and d

6. $(2, -3, -2)$ **8.** $(8, -2, 6)$ **10.** $\left(-2, -\frac{10}{3}, -4\right)$

12. $\begin{cases} x - 2y + 3z = 5 \\ -x + 3y - 5z = 4 \\ 4y - 9z = -10 \end{cases}$

Required step in putting the system in row-echelon form

14. $\left(\frac{5}{3}, \frac{1}{3}, 1\right)$ **16.** $(5, -3, 3)$ **18.** $\left(1, \frac{1}{2}, -3\right)$

20. No solution **22.** $\left(\frac{3}{10}, \frac{2}{5}, 0\right)$

24. $\left(-\frac{1}{2}a + \frac{5}{2}, 4a - 1, a\right)$ **26.** $\left(-4a + 13, -\frac{15}{2}a + \frac{45}{2}, a\right)$

28. $(-5a + 3, -a - 5, a)$ **30.** $\left(-\frac{3}{8}a - \frac{1}{4}, -\frac{3}{4}a + \frac{5}{2}, a\right)$

32. $(1, 0, 3, 2)$ **34.** $\left(3, \frac{7}{2}, \frac{1}{2}\right)$ **36.** $(0, 0, 0)$

38. $\left(\frac{1}{5}a + \frac{1}{5}, -\frac{3}{5}a + \frac{2}{5}, a\right)$

40. $y = -x^2 + 2x + 3$

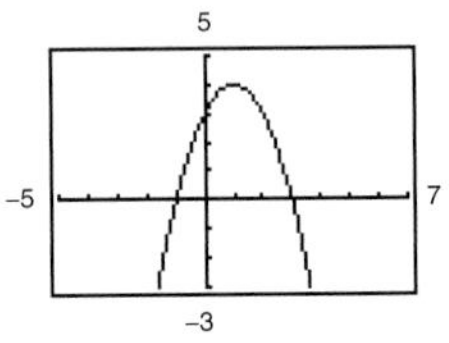

42. $y = -2x^2 + 5x$

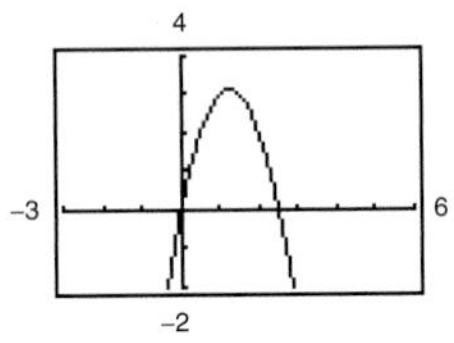

44. $x^2 + y^2 - 6y = 0$

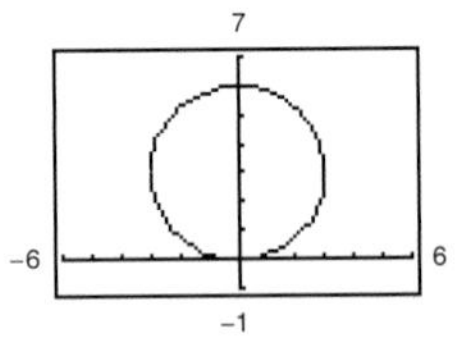

46. $x^2 + y^2 - 3x + 2y = 0$

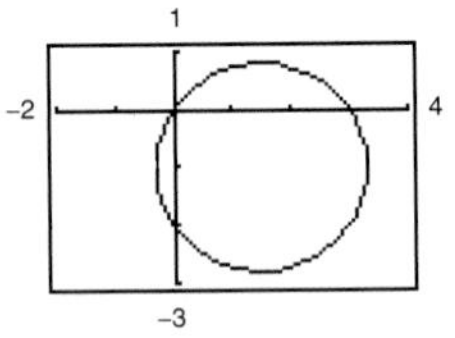

48. $s = -16t^2 + 64t$ **50.** $s = -16t^2 + 16t + 132$

52. 3-point baskets: 8
2-point baskets: 32
Free-throws: 16

54. \$625,000 at 8%
\$50,000 at 9%
\$125,000 at 10%

56. $406{,}250 - \frac{1}{2}s$ in certificates of deposit,
$-31{,}250 + \frac{1}{2}s$ in municipal bonds,
$125{,}000 - s$ in blue-chip stocks,
s in growth stocks

58. (a) No 10% solution, $8\frac{1}{3}$ liters of 20% solution, $1\frac{2}{3}$ liters of 50% solution

(b) No 20% solution, $6\frac{1}{4}$ liters of 10% solution, $3\frac{3}{4}$ liters of 50% solution

(c) 7 liters of 20% solution, 1 liter of 10% solution

60. $I_1 = 1, I_2 = 2, I_3 = 1$

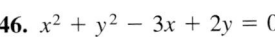

62. $t_1 = 128$ pounds
$t_2 = 64$ pounds
$a = 0$ feet per second squared

64. $\dfrac{1}{x-1} - \dfrac{1}{x+2}$ **66.** $-\dfrac{2}{x} + \dfrac{6}{5(x-2)} + \dfrac{4}{5(x+3)}$

68. $y = \frac{3}{7}x^2 + \frac{6}{5}x + \frac{26}{35}$ **70.** $y = -\frac{5}{4}x^2 + \frac{9}{20}x + \frac{199}{20}$

72. (a) $-0.0075x^2 + 1.3x + 20$

(b)

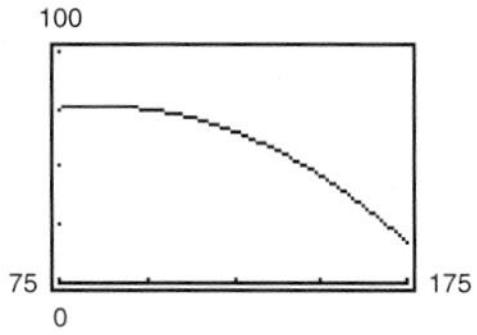

(c) 24.25%

74. $x = 2$
$y = 2$
$\lambda = -4$

76. $x = 25$
$y = 50$
$\lambda = -51$

78. True. If a system of three linear equations is inconsistent, then it has no points common to all three equations.

80. The first system is inconsistent because 4 times Equation 1 added to Equation 2 yields $0 = 20$.

82. $\begin{cases} 3x + y - z = 9 \\ x + 2y - z = 0 \\ -x + y + 3z = 1 \end{cases}$ **84.** $\begin{cases} x + 2y - 4z = -5 \\ -x - 4y + 8z = 13 \\ x + 6y + 4z = 7 \end{cases}$

86. 6.375 **88.** 80,000

90. (a) $-4, 0, 3$

(b)

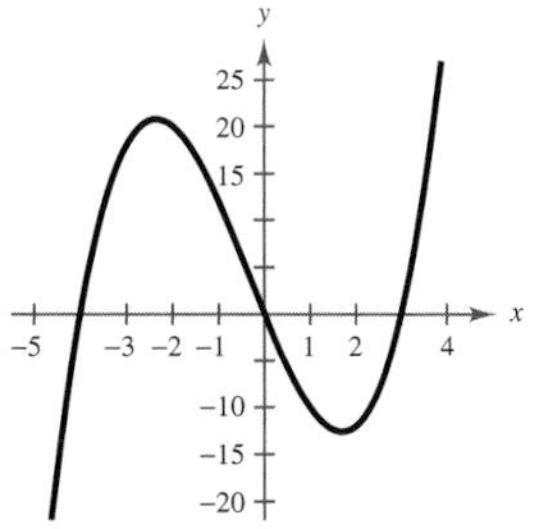

92. (a) $-4, -\frac{3}{2}, 3$

(b)

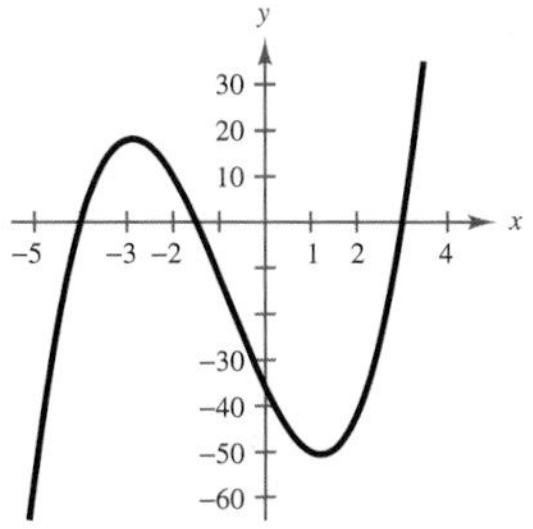

94.

x	-2	-1	0	1	2
y	5.793	4.671	4	3.598	3.358

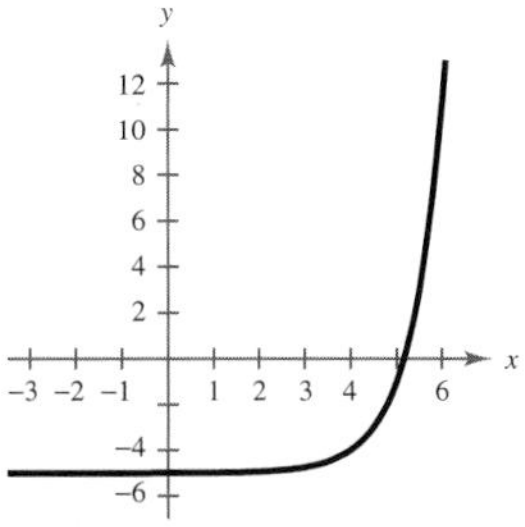

96.

x	-2	0	2	4	5
y	-5	-4.996	-4.938	-4	-1

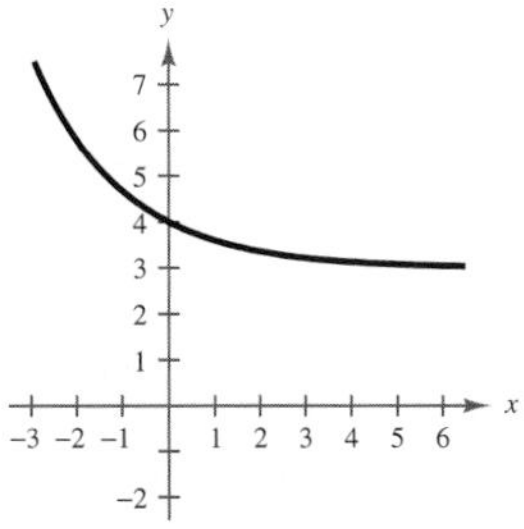

98. (40, 40)

Section 7.4 *(page 597)*

2.

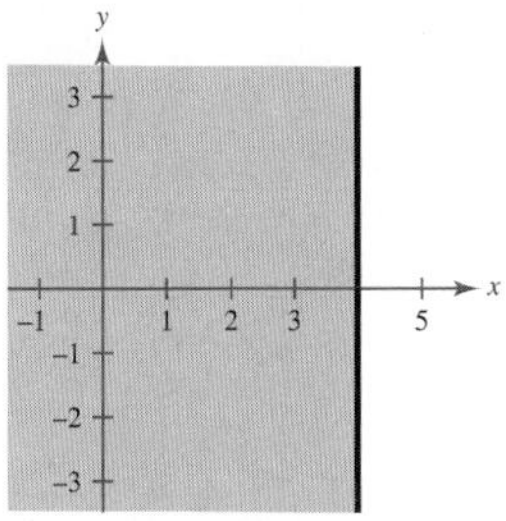

4.

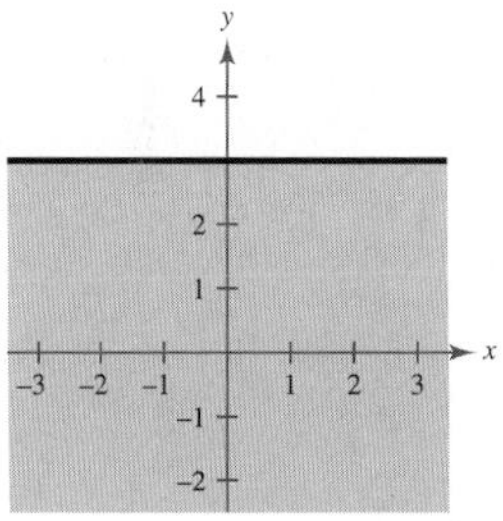

6.

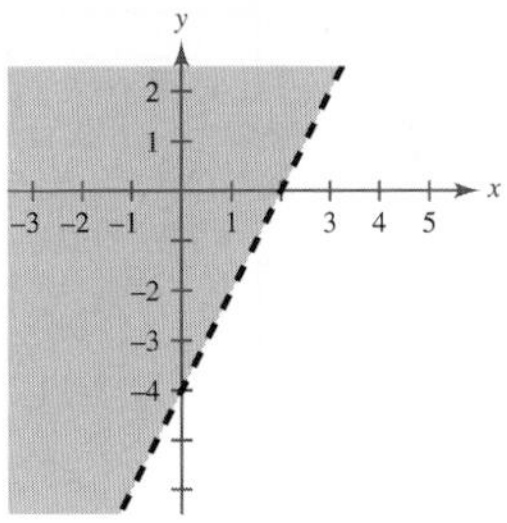

8.

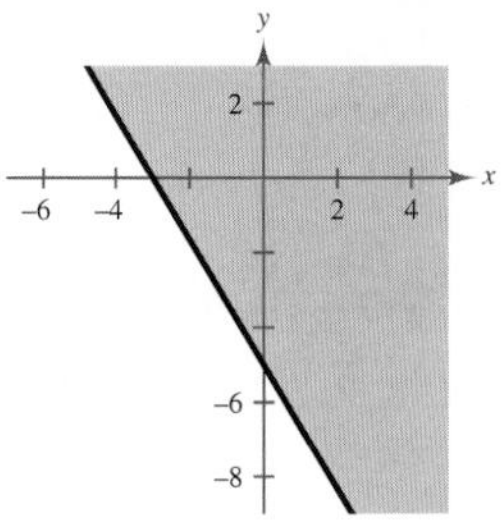

10.

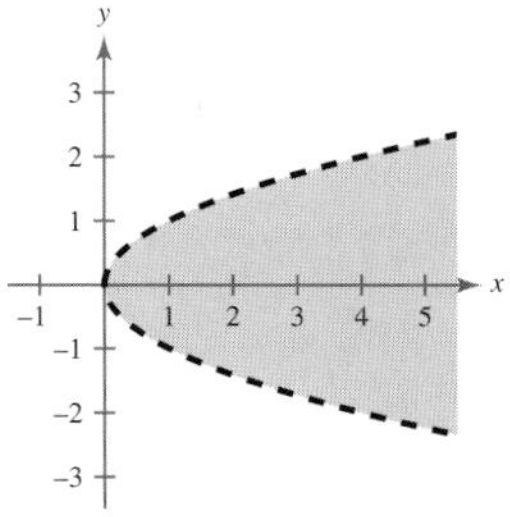

12.

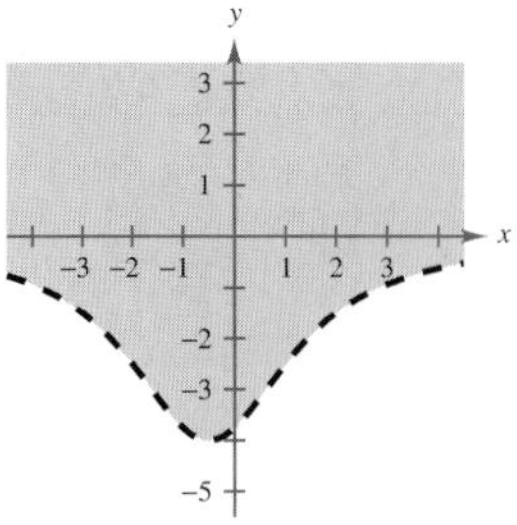

14.

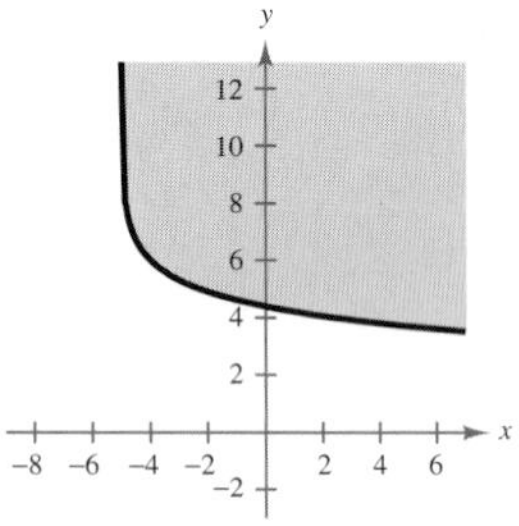

16.

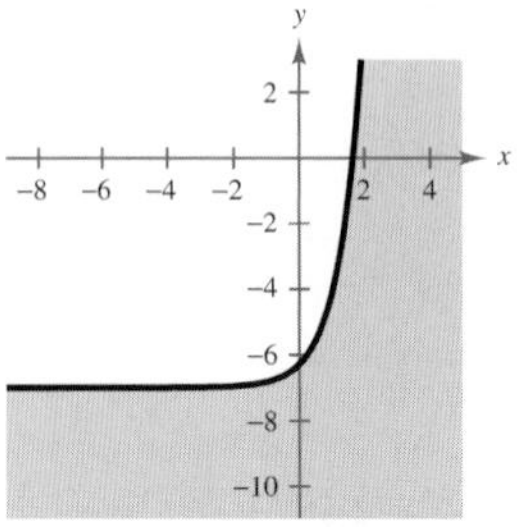

18.

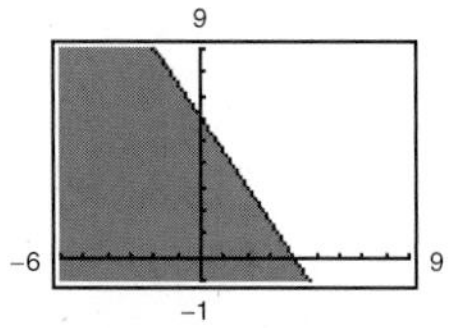

20.

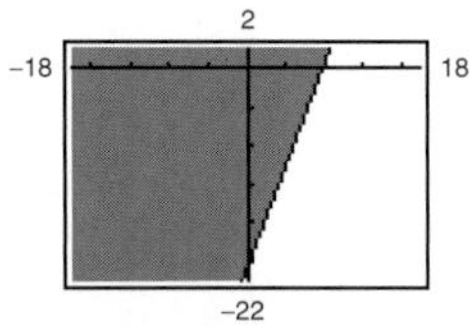

22.

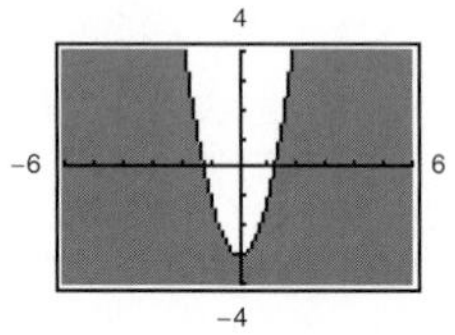

24.

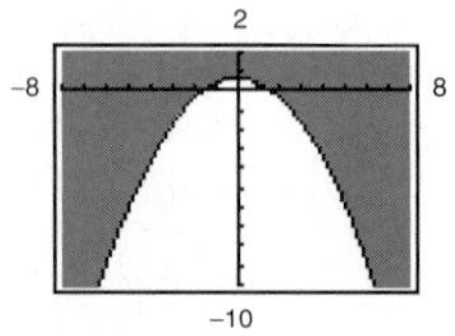

26. $y \ge x^2 - 4$ **28.** $x^2 + y^2 \le 9$ **30.** a **32.** d

34.

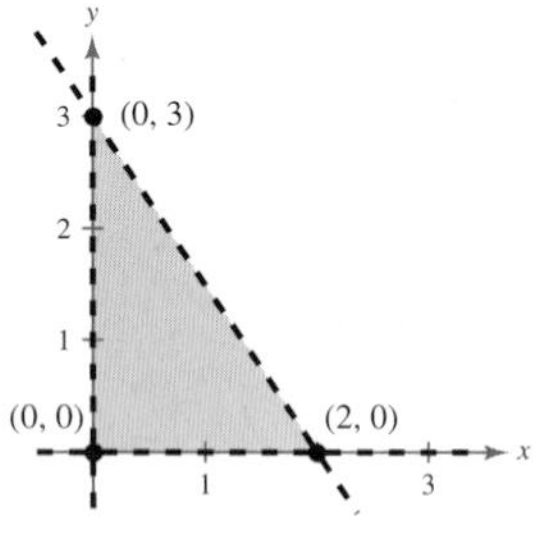

36.

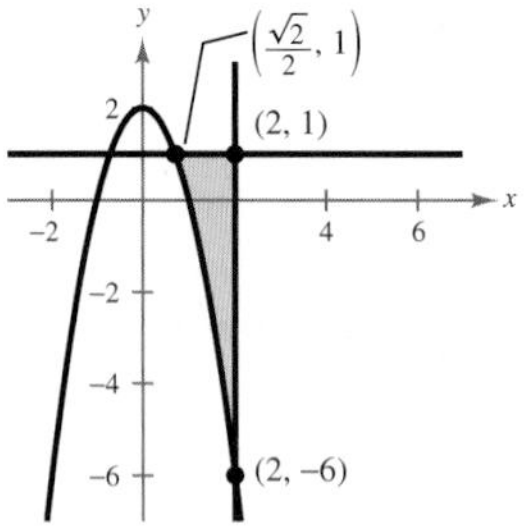

38.

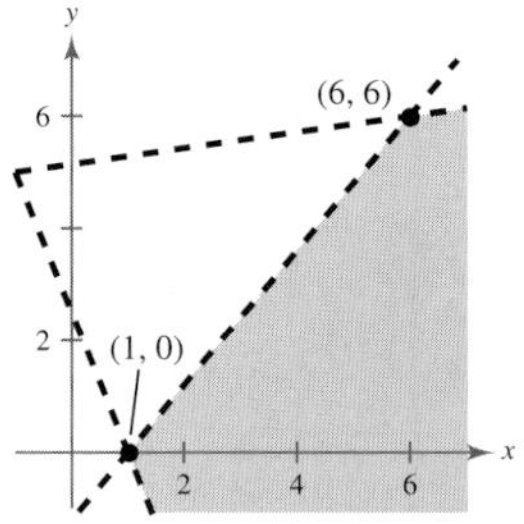

40.

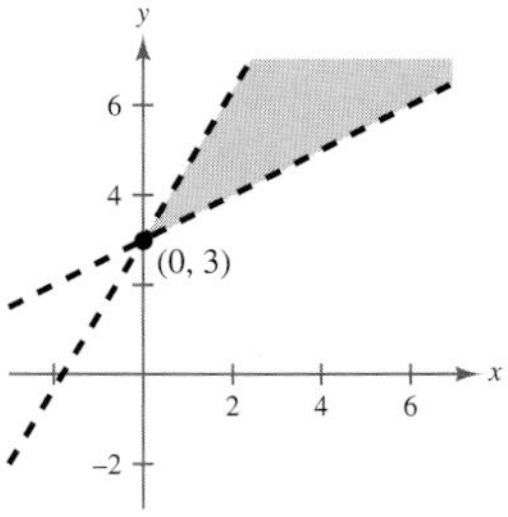

42.

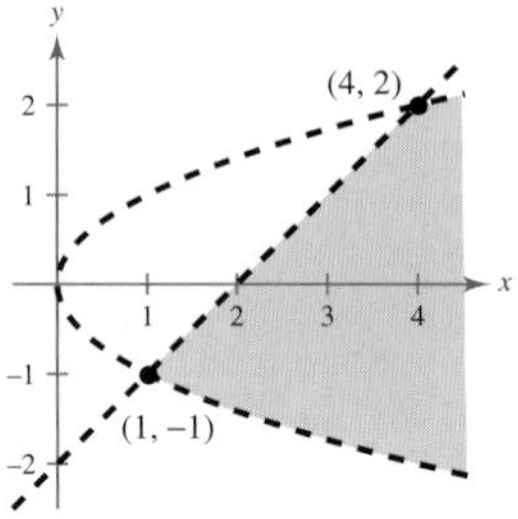

44.

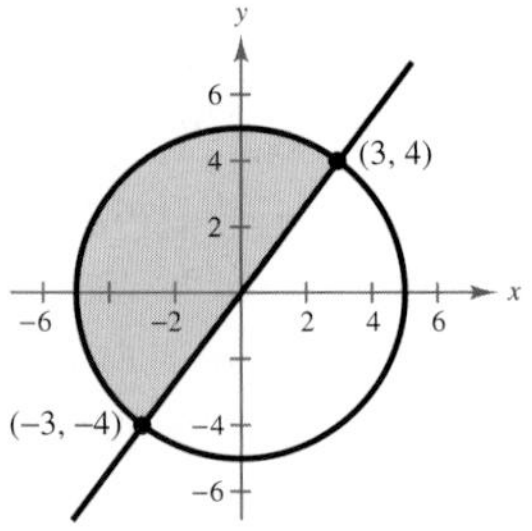

46.

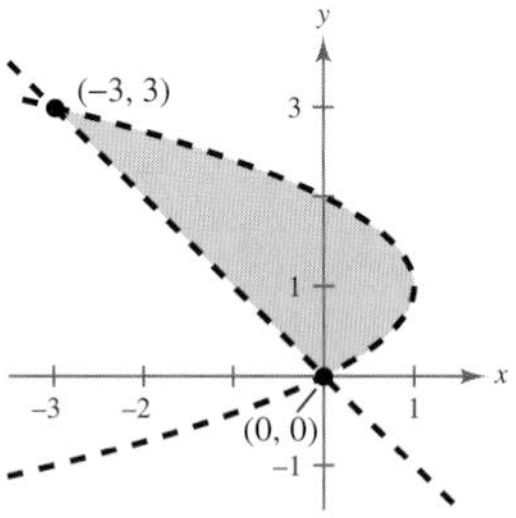

48.

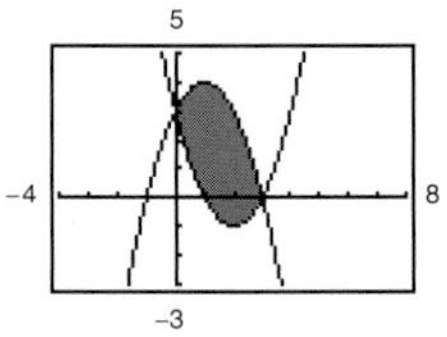

50.

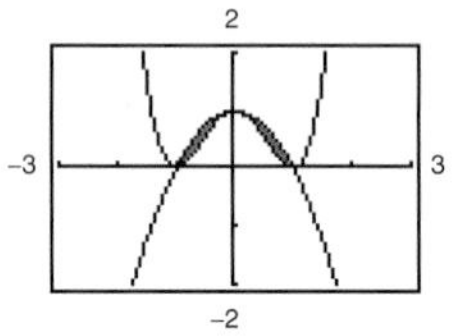

52.

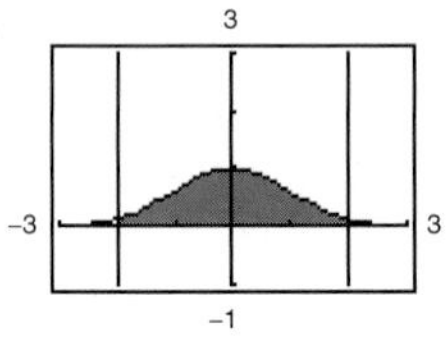

54. $\begin{cases} y < 6 - 2x \\ y \ge x - 3 \\ x \ge 1 \end{cases}$ **56.** $x^2 + y^2 > 4$

58. $\begin{cases} x^2 + y^2 \le 16 \\ x \le y \\ x \ge 0 \end{cases}$ **60.** $\begin{cases} 4x - y \ge 0 \\ 4x - y \le 16 \\ 0 \le y \le 4 \end{cases}$

62. $\begin{cases} y \le x + 1 \\ y \le -x + 1 \\ y \ge 0 \end{cases}$

64. Consumer surplus: 6250

Producer surplus: 12,500

66. Consumer surplus: 6,250,000

Producer surplus: 15,625,000

68. $\begin{cases} x \geq 2y \\ 8x + 12y \leq 200 \\ x \geq 4 \\ y \geq 2 \end{cases}$

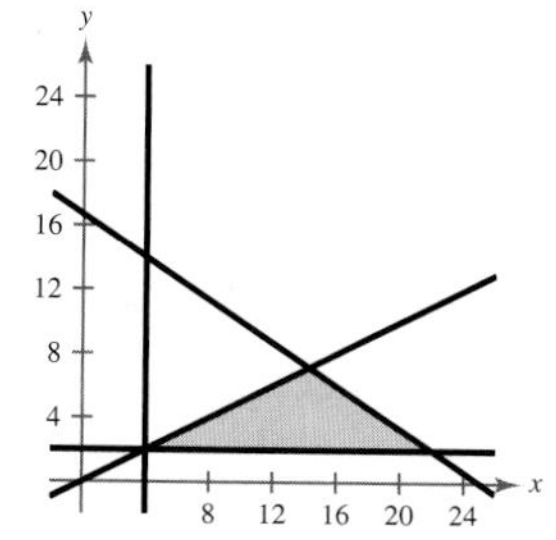

70. $\begin{cases} x + y \leq 3000 \\ 30x + 20y \geq 75{,}000 \\ x \leq 2000 \\ x \geq 0 \\ y \geq 0 \end{cases}$

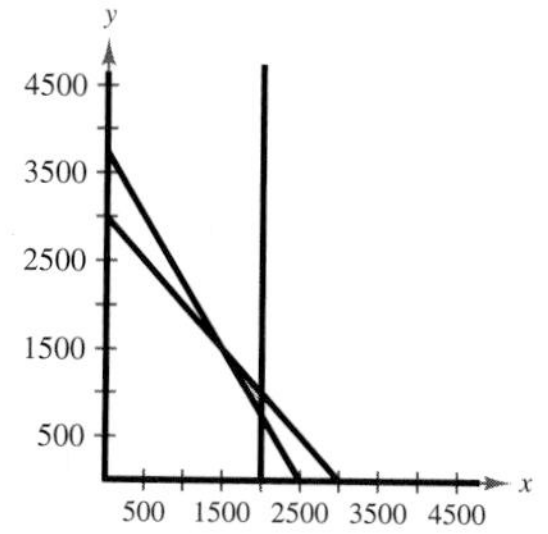

72. $\begin{cases} 20x + 10y \geq 300 \\ 15x + 10y \geq 150 \\ 10x + 20y \geq 200 \\ x \geq 0 \\ y \geq 0 \end{cases}$

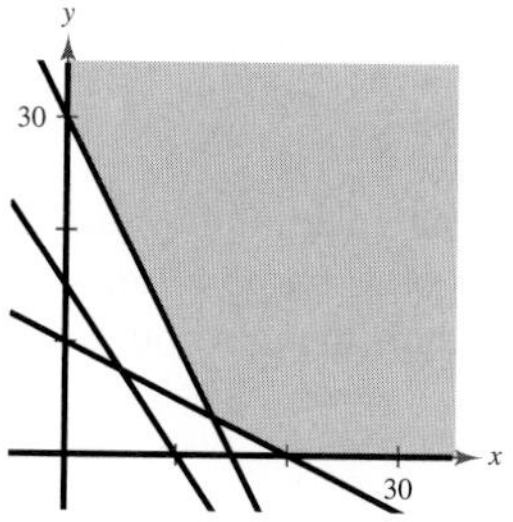

74. True. The figure is a rectangle with a length of 9 units and a width of 11 units.

76. d **78.** c

80. (a) The boundary would be included in the solution.

(b) The solution would be the half-plane on the opposite side of the boundary.

82. Test a point on either side.

84. $5x + 3y - 8 = 0$ **86.** $28x + 17y + 13 = 0$

88. $x + y + 1.8 = 0$

90. $M = 1.686t + 40.081$;

$M = -0.243t^2 + 3.871t + 35.871$

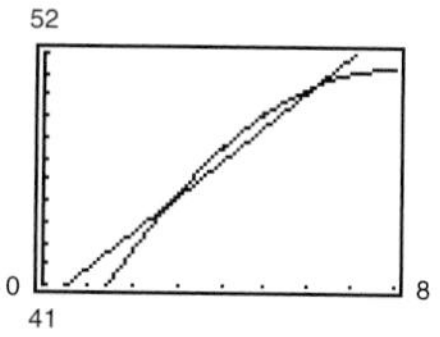

The quadratic model represents the data more closely than the linear model.

92. 4.081 **94.** 3082.955 **96.** 0.027

98. $(-1, 2, -3)$

Section 7.5 *(page 607)*

2. Minimum at $(0, 0)$: 0

Maximum at $(0, 4)$: 32

4. Minimum at $(0, 0)$: 0

Maximum at $(2, 0)$: 14

6. Minimum at $(0, 2)$: 10

Maximum at $(4, 3)$: 31

8. Minimum at $(0, 2)$: 2

Maximum at $(4, 3)$: 11

10. Minimum at $(450, 0)$: 11,250

Maximum at $(0, 800)$: 28,000

12. Minimum at $(450, 0)$: 6750

Maximum at $(0, 800)$: 16,000

14.

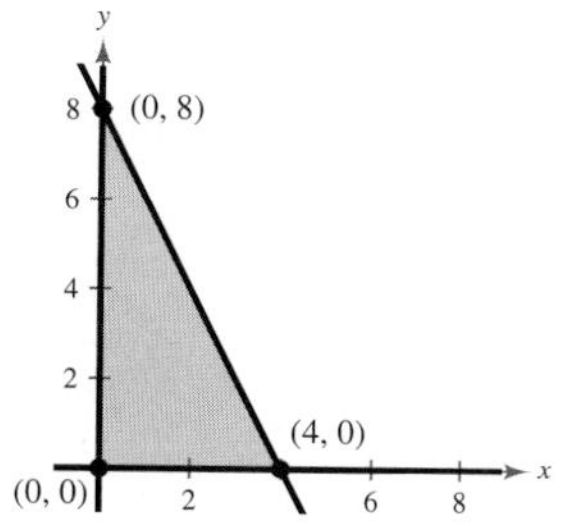

Minimum at $(0, 0)$: 0

Maximum at $(0, 8)$: 64

16.

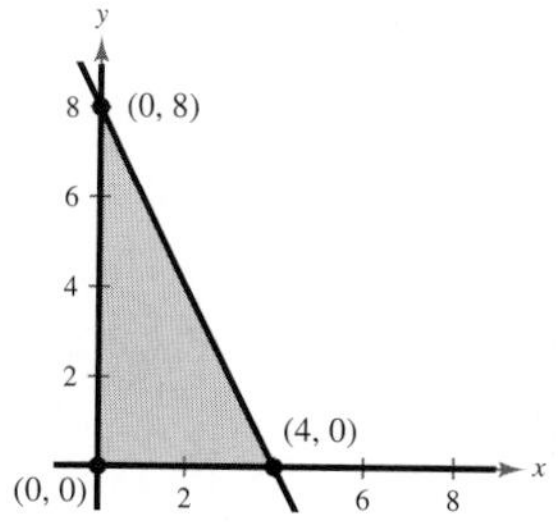

Minimum at $(0, 0)$: 0

Maximum at $(4, 0)$: 28

18. 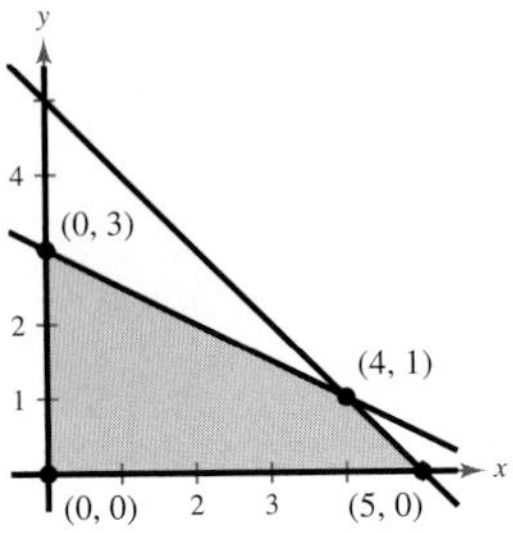

Minimum at $(0, 0)$: 0
Maximum at $(4, 1)$: 21

20. 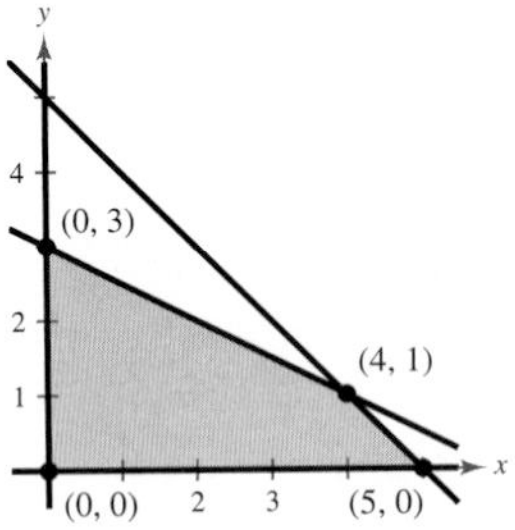

Minimum at $(0, 3)$: -3
Maximum at $(5, 0)$: 10

22. 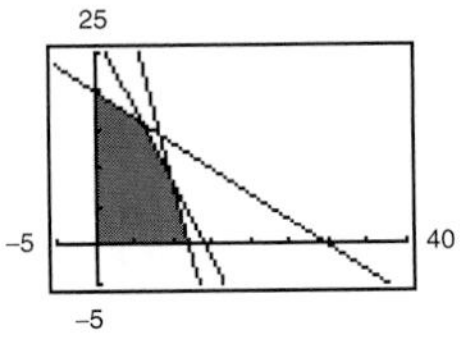

Minimum at any point on the line segment connecting $(0, 0)$ and $(0, 20)$: 0

Maximum at $(12, 0)$: 12

24. 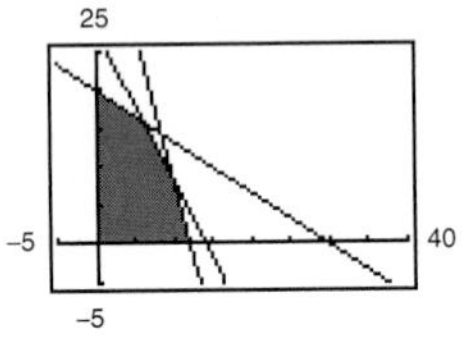

Minimum at any point on the line segment connecting $(0, 0)$ and $(12, 0)$: 0

Maximum at $(0, 20)$: 20

26. Maximum at $(5, 0)$: 25

28. Maximum at any point on the line segment connecting $(3, 6)$ and $(5, 0)$: 15

30. Maximum at $\left(\frac{22}{3}, \frac{19}{6}\right)$: $\frac{82}{3}$

32. Maximum at $\left(\frac{21}{2}, 0\right)$: 42

34. 1000 units of model A

500 units of model B

Maximum profit: \$76,000

36. 60 acres of crop A

90 acres of crop B

Maximum profit: \$29,550

38. A: $\frac{1}{6}$ gallon

B: $\frac{5}{6}$ gallon

Minimum cost: \$1.255 per gallon

40. No audits

48 tax returns

Maximum revenue: \$14,400

42. 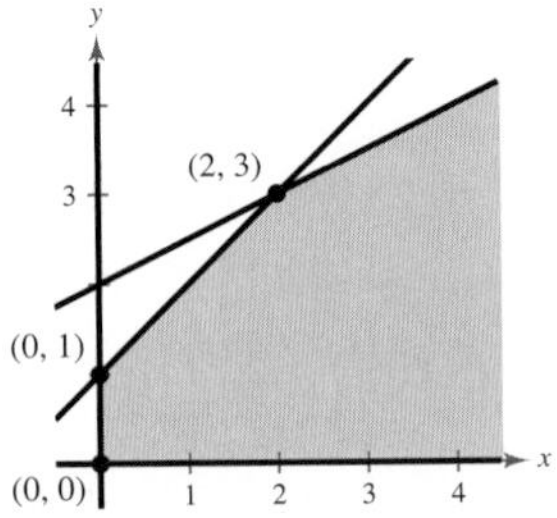

The constraints do not form a closed set of points. Therefore, $z = x + y$ is unbounded.

44. 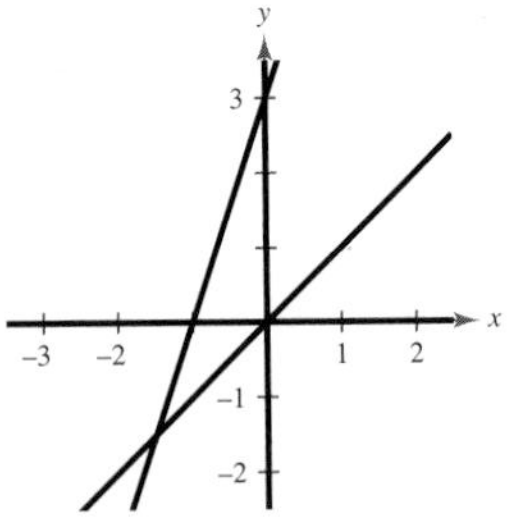

The feasible set is empty.

46. 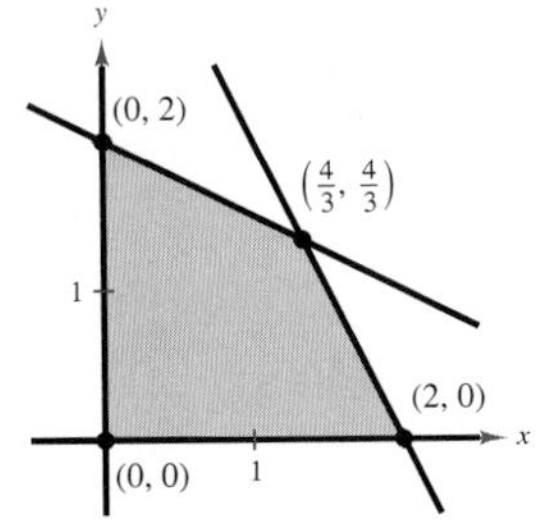

The maximum, 4, occurs at any point on the line segment connecting $(0, 2)$ and $\left(\frac{4}{3}, \frac{4}{3}\right)$.

48. True. If an objective function has a maximum value at more than one vertex, then any point on the line segment connecting the points will produce the maximum value.

50. (a) $t \le 6$ (b) $t \ge 6$

52. $z = x + y$

54. $z = -10x + y$

56. $\dfrac{1}{x - 2}$, $x \ne 0, -2$

58. $\dfrac{(x + 1)(x + 3)}{3}$

60. $\ln 4 \approx 1.386$, $\ln 6 \approx 1.792$

62. $-\ln 6 \approx -1.792$

64. $e - 9 \approx -6.282$

Review Exercises *(page 612)*

2. (13, 0), (5, 12) **4.** (5, 2), (2, −1)

6. (0, 0), (−3, 3) **8.** (4, 4) **10.** (9.68, −0.84)

12. Sales greater than $500,000

14. $\left(\frac{5}{2}, 3\right)$ **16.** (−0.5, 0.8) **18.** (0, 0)

20. $\left(\frac{8}{5}a + \frac{14}{5}, a\right)$

22.

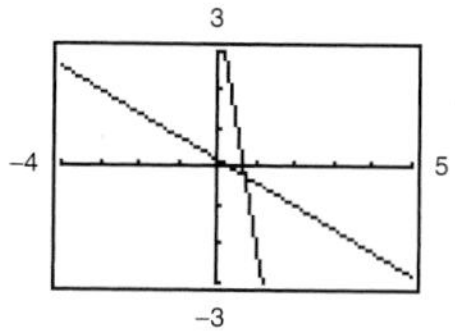

Consistent, one solution

24.

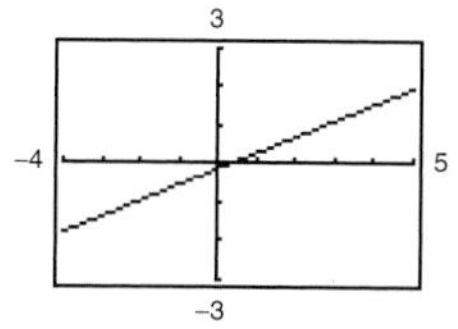

Consistent, infinitely many solutions

26. 90% solution: 125 liters
50% solution: 75 liters

28. 193.75 miles per hour;
218.75 miles per hour

30. (250,000, 95) **32.** (5, −8, 3) **34.** $\left(\frac{38}{17}, \frac{40}{17}, -\frac{63}{17}\right)$

36. $\left(-\frac{3}{4}, 0, -\frac{5}{4}\right)$ **38.** $y = 3x^2 + 11x - 14$

40. $x^2 + y^2 - 2x + 2y - 23 = 0$

42. (a) $y = 1.146x - 1.607$

(b)

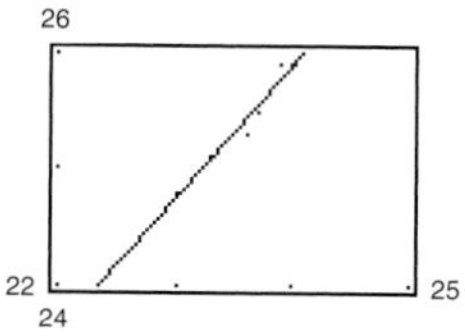

(c) The line seems to be a good model for the data.

(d) A 1-year change in x results in a 1.146-year change in y.

44. $(-3a + 2, 5a + 6, a)$

46. $16,000 at 7%
$13,000 at 9%
$11,000 at 11%

48.

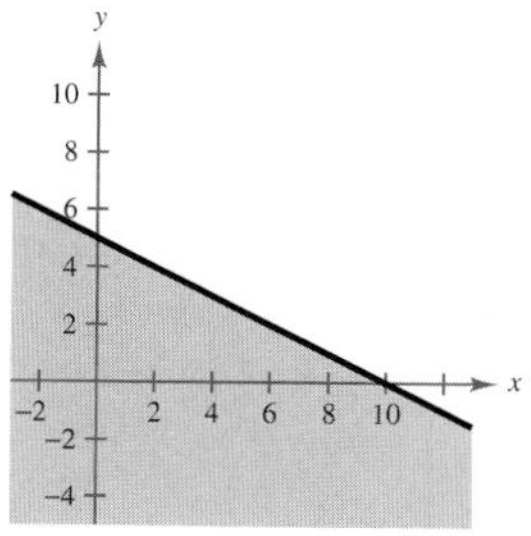

50.

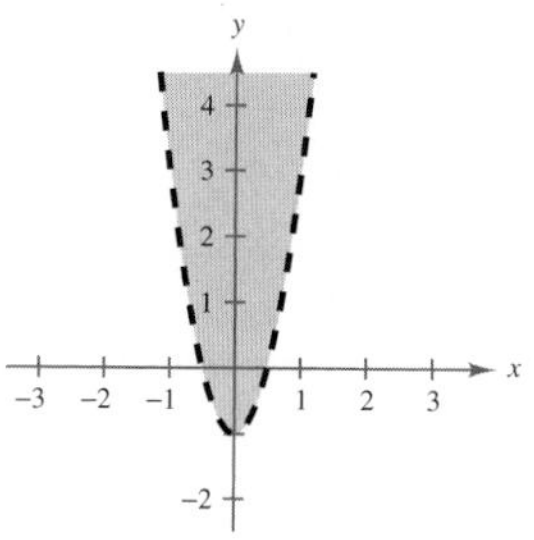

52.

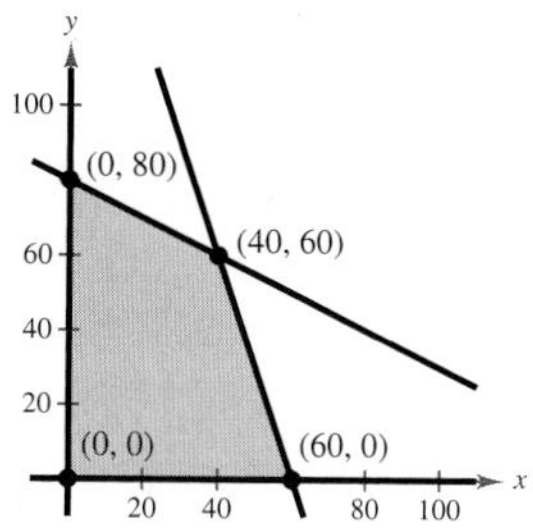

54.

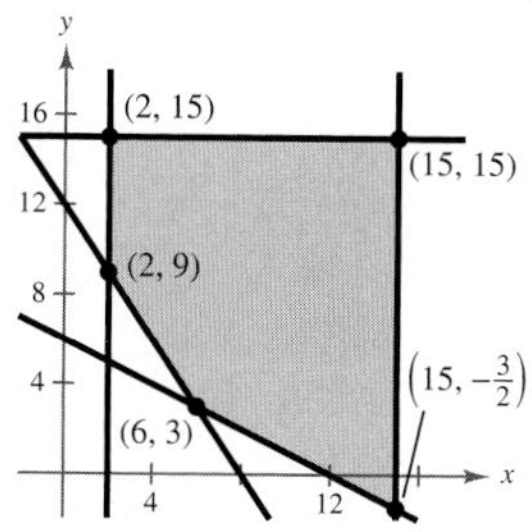

56.

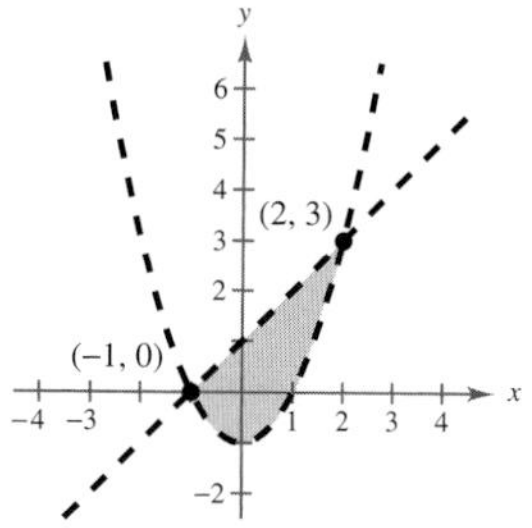

58.

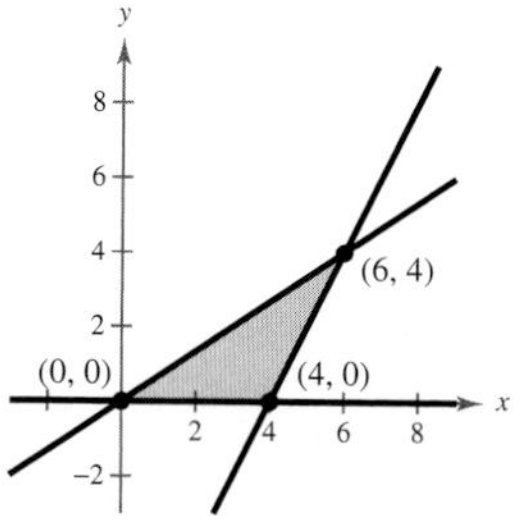

60. $\begin{cases} x + y \le 1500 \\ x \ge 400 \\ y \ge 600 \end{cases}$

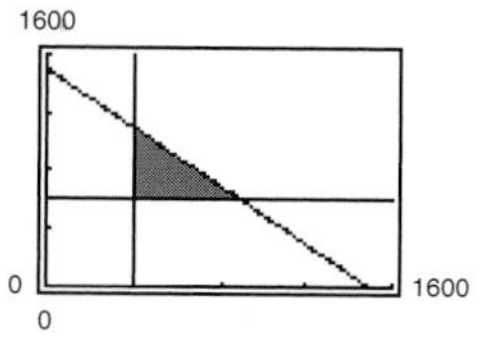

62. Consumer surplus: $4,500,000
Producer surplus: $9,000,000

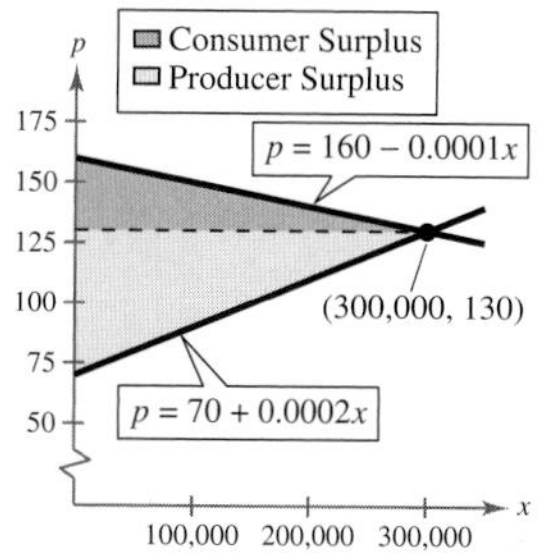

64. Maximum at (5, 8): 47 **66.** Minimum at (15, 0): 26.25

68. 72 haircuts, 0 permanents; Maximum revenue: $1800

70. Three bags of brand X
Two bags of brand Y
Minimum cost: $105

72. False. To represent a region covered by an isosceles trapezoid the last two equality signs should be ≤.

CHAPTER 7

74. $\begin{cases} x + y = 2 \\ x - y = -14 \end{cases}$ **76.** $\begin{cases} 3x + y = 7 \\ -6x + 3y = 1 \end{cases}$

78. $\begin{cases} x + y + z = 6 \\ x + y - z = 0 \\ x - y - z = 2 \end{cases}$ **80.** $\begin{cases} 2x + 2y - 3z = 7 \\ x - 2y + z = 4 \\ -x + 4y - z = -1 \end{cases}$

82. $k_1 = -\frac{10}{3}, k_2 = \frac{16}{3}$

84. There will be a contradictory equation of the form $0 = N$, where N is a nonzero real number.

86. The lines are distinct and parallel.

$\begin{cases} x + 2y = 3 \\ 2x + 4y = 9 \end{cases}$

Chapter 8

Section 8.1 *(page 630)*

2. 1×4 **4.** 3×4 **6.** 2×3

8. $\begin{bmatrix} 7 & 4 & \vdots & 22 \\ 5 & -9 & \vdots & 15 \end{bmatrix}$

10. $\begin{bmatrix} -1 & -8 & 5 & \vdots & 8 \\ -7 & 0 & -15 & \vdots & -38 \\ 3 & -1 & 8 & \vdots & 20 \end{bmatrix}$

12. $\begin{bmatrix} 9 & 2 & -3 & \vdots & 20 \\ 0 & -25 & 11 & \vdots & -5 \end{bmatrix}$

14. $\begin{cases} 7x - 5y = 0 \\ 8x + 3y = -2 \end{cases}$ **16.** $\begin{cases} 4x - 5y - z = 18 \\ -11x + 6z = 25 \\ 3x + 8y = -29 \end{cases}$

18. $\begin{cases} 6x + 2y - z - 5w = -25 \\ -x + 7z + 3w = 7 \\ 4x - y - 10z + 6w = 23 \\ 8y + z - 11w = -21 \end{cases}$

20. Reduced row-echelon form

22. Row-echelon form

24. $\begin{bmatrix} 1 & 2 & \frac{8}{3} \\ 4 & -3 & 6 \end{bmatrix}$

26. $\begin{bmatrix} 1 & 2 & 4 & \frac{3}{2} \\ 1 & -1 & -3 & 2 \\ 2 & 6 & 4 & 9 \end{bmatrix}$ $\begin{bmatrix} 1 & 2 & 4 & \frac{3}{2} \\ 0 & -3 & -7 & \frac{1}{2} \\ 0 & 2 & -4 & 6 \end{bmatrix}$

28. Add 3 times Row 1 to Row 2.

30. Add 2 times Row 1 to Row 2.

Add 5 times Row 1 to Row 3.

32. (a) $\begin{bmatrix} 7 & 1 \\ 0 & 2 \\ -3 & 4 \\ 1 & 5 \end{bmatrix}$ (b) $\begin{bmatrix} 1 & 5 \\ 0 & 2 \\ -3 & 4 \\ 7 & 1 \end{bmatrix}$ (c) $\begin{bmatrix} 1 & 5 \\ 0 & 2 \\ 0 & 19 \\ 7 & 1 \end{bmatrix}$

(d) $\begin{bmatrix} 1 & 5 \\ 0 & 2 \\ 0 & 19 \\ 0 & -34 \end{bmatrix}$ (e) $\begin{bmatrix} 1 & 5 \\ 0 & 1 \\ 0 & 19 \\ 0 & -34 \end{bmatrix}$ (f) $\begin{bmatrix} 1 & 0 \\ 0 & 1 \\ 0 & 0 \\ 0 & 0 \end{bmatrix}$

The matrix is in reduced row-echelon form.

34. $\begin{bmatrix} 1 & 2 & -1 & 3 \\ 0 & 1 & -2 & 5 \\ 0 & 0 & 1 & -1 \end{bmatrix}$ **36.** $\begin{bmatrix} 1 & -3 & 0 & -7 \\ 0 & 1 & 1 & 2 \\ 0 & 0 & 0 & 0 \end{bmatrix}$

38. $\begin{bmatrix} 1 & 3 & 0 \\ 0 & 0 & 1 \\ 0 & 0 & 0 \end{bmatrix}$ **40.** $\begin{bmatrix} 1 & 0 & 0 & 0 \\ 0 & 1 & 0 & 0 \\ 0 & 0 & 1 & 0 \\ 0 & 0 & 0 & 1 \end{bmatrix}$

42. $\begin{bmatrix} 1 & 0 & 0 & 2 \\ 0 & 1 & 2 & -6 \end{bmatrix}$

44. $\begin{cases} x + 5y = 0 \\ y = -1 \end{cases}$ **46.** $\begin{cases} x + 2y - 2z = -1 \\ y + z = 9 \\ z = -3 \end{cases}$

$(5, -1)$ $(-31, 12, -3)$

48. $(-6, 10)$ **50.** $(5, -3, 0)$ **52.** $(-1, 3)$

54. $(9, 13)$ **56.** $(-2, -1)$ **58.** $(3a + 5, a)$

60. $(8, 10, 6)$ **62.** $(-6, 8, 2)$ **64.** $(5, -1, -2)$

66. $\left(-\frac{3}{2}a + \frac{3}{2}, \frac{1}{3}a + \frac{1}{3}, a\right)$

68. $(-3b + 96a + 100, b, 52a + 54, a)$

70. Inconsistent **72.** $(-5a, a, 3)$

74. $(-1, 1, 3, 1)$ **76.** $(-2a, -a, a, a)$

78. No **80.** No

82. $I_1 = 2, I_2 = 3, I_3 = 1$

84. \$150,000 at 7%

\$750,000 at 8%

\$600,000 at 10%

86. $y = x^2 + 2x + 5$

88. (a) $y = -0.004x^2 + 0.367x + 5$

(b)

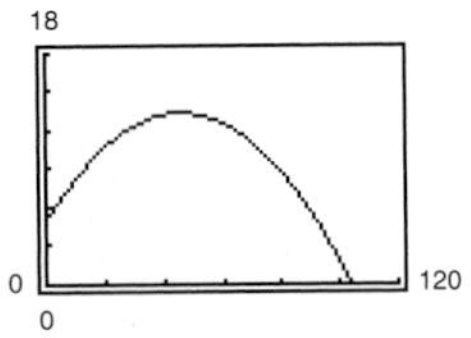

13 feet, 104 feet

(c) 13.403 feet, 103.719 feet

90. (a) $x_1 = s, x_2 = t, x_3 = 600 - s, x_4 = s - t,$
$x_5 = 500 - t, x_6 = s, x_7 = t$

(b) $x_1 = 0, x_2 = 0, x_3 = 600, x_4 = 0, x_5 = 500,$
$x_6 = 0, x_7 = 0$

(c) $x_1 = 0, x_2 = -500, x_3 = 600, x_4 = 500,$
$x_5 = 1000, x_6 = 0, x_7 = -500$

92. False. It is a 2×4 matrix.

94. False. Gaussian elimination reduces a matrix until a row-echelon form is obtained and Gauss-Jordan elimination reduces a matrix until a reduced row-echelon form is obtained.

96. (a) There exists a row with all zeros except for the entry in the last column.

(b) There are fewer rows with nonzero entries than there are variables.

98. They are the same.

100.

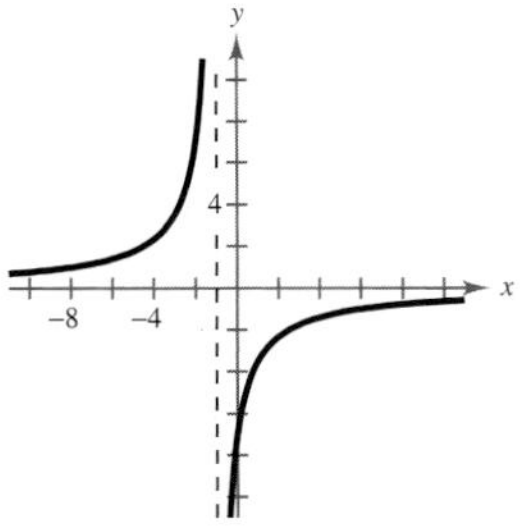

Vertical asymptote: $x = -1$

Horizontal asymptote: $y = 0$

102.

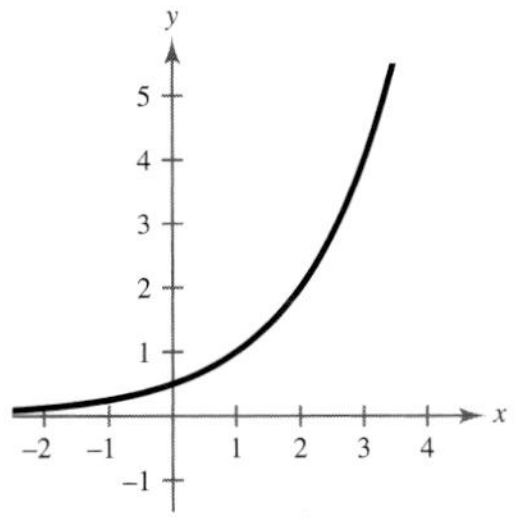

104.

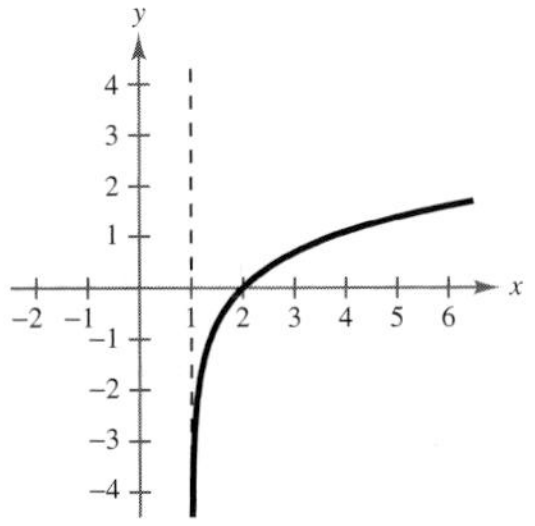

Section 8.2 *(page 645)*

2. $x = 13,\ y = 12$ **4.** $x = -4,\ y = 9$

6. (a) $\begin{bmatrix} -2 & 0 \\ 6 & 3 \end{bmatrix}$ (b) $\begin{bmatrix} 4 & 4 \\ -2 & -1 \end{bmatrix}$

(c) $\begin{bmatrix} 3 & 6 \\ 6 & 3 \end{bmatrix}$ (d) $\begin{bmatrix} 9 & 10 \\ -2 & -1 \end{bmatrix}$

8. (a) $\begin{bmatrix} 4 & -2 & 5 \\ -4 & 0 & 2 \end{bmatrix}$ (b) $\begin{bmatrix} 0 & 4 & -3 \\ 2 & -2 & 6 \end{bmatrix}$

(c) $\begin{bmatrix} 6 & 3 & 3 \\ -3 & -3 & 12 \end{bmatrix}$ (d) $\begin{bmatrix} 2 & 9 & -5 \\ 3 & -5 & 16 \end{bmatrix}$

10. (a) $\begin{bmatrix} -4 & 9 & 1 \\ 5 & -6 & -5 \\ 15 & -5 & -2 \\ 3 & 10 & -10 \\ -4 & 0 & -2 \end{bmatrix}$ (b) $\begin{bmatrix} 2 & -1 & -1 \\ 1 & 2 & 9 \\ -5 & 13 & 0 \\ -3 & 6 & -2 \\ -4 & -2 & 2 \end{bmatrix}$

(c) $\begin{bmatrix} -3 & 12 & 0 \\ 9 & -6 & 6 \\ 15 & 12 & -3 \\ 0 & 24 & -18 \\ -12 & -3 & 0 \end{bmatrix}$ (d) $\begin{bmatrix} 3 & 2 & -2 \\ 5 & 2 & 20 \\ -5 & 30 & -1 \\ -6 & 20 & -10 \\ -12 & -5 & 4 \end{bmatrix}$

12. (a), (b), and (d) not possible

(c) $\begin{bmatrix} 9 \\ 6 \\ -3 \end{bmatrix}$

14. $\begin{bmatrix} -5 & 6 \\ -2 & -2 \end{bmatrix}$ **16.** $\begin{bmatrix} \frac{19}{2} & 2 & -7 & \frac{9}{2} \end{bmatrix}$

18. $\begin{bmatrix} -\frac{11}{3} & -\frac{31}{3} \\ 1 & \frac{3}{2} \\ -8 & -1 \end{bmatrix}$ **20.** $\begin{bmatrix} -440 & 495 \\ -495 & 1375 \end{bmatrix}$

22. $\begin{bmatrix} 132 & 168 \\ -108 & 60 \\ -348 & 60 \end{bmatrix}$ **24.** $\begin{bmatrix} -2 & -\frac{5}{2} \\ 0 & 0 \\ 5 & -\frac{7}{2} \end{bmatrix}$ **26.** $\begin{bmatrix} 2 & -5 \\ -5 & 0 \\ 5 & 6 \end{bmatrix}$

28. (a) $\begin{bmatrix} -3 & 3 \\ 12 & -12 \end{bmatrix}$ (b) $\begin{bmatrix} 0 & 0 \\ 3 & -15 \end{bmatrix}$ (c) $\begin{bmatrix} 3 & -6 \\ 6 & 15 \end{bmatrix}$

30. (a) $\begin{bmatrix} 4 & 2 \\ -2 & 4 \end{bmatrix}$ (b) $\begin{bmatrix} 4 & 2 \\ -2 & 4 \end{bmatrix}$ (c) $\begin{bmatrix} 0 & -2 \\ 2 & 0 \end{bmatrix}$

32. (a) $[12]$ (b) $\begin{bmatrix} 6 & 4 & 2 \\ 9 & 6 & 3 \\ 0 & 0 & 0 \end{bmatrix}$ (c) Not possible

34. Not possible

36. $\begin{bmatrix} -1 & 19 \\ 4 & -27 \\ 0 & 14 \end{bmatrix}$ **38.** $\begin{bmatrix} 1 & 0 & 0 \\ 0 & 1 & 0 \\ 0 & 0 & \frac{7}{2} \end{bmatrix}$

40. $\begin{bmatrix} 60 & -20 & 10 & 60 \\ 72 & -24 & 12 & 72 \end{bmatrix}$

42. $\begin{bmatrix} 252 & 30 \\ 298 & 452 \\ 217 & 180 \end{bmatrix}$ **44.** Not possible

46. $\begin{bmatrix} -249 & 42 & -417 \\ 124 & 104 & 284 \\ 232 & 176 & 520 \end{bmatrix}$ **48.** $\begin{bmatrix} 27 & -6 \\ -6 & -27 \end{bmatrix}$

50. $\begin{bmatrix} 12 & 6 \\ -4 & -2 \\ 20 & 10 \\ 28 & 14 \end{bmatrix}$

52. (a) $\begin{bmatrix} 2 & 3 \\ 1 & 4 \end{bmatrix}\begin{bmatrix} x_1 \\ x_2 \end{bmatrix} = \begin{bmatrix} 5 \\ 10 \end{bmatrix}$ (b) $\begin{bmatrix} -2 \\ 3 \end{bmatrix}$

54. (a) $\begin{bmatrix} -4 & 9 \\ 1 & -3 \end{bmatrix}\begin{bmatrix} x_1 \\ x_2 \end{bmatrix} = \begin{bmatrix} -13 \\ 12 \end{bmatrix}$ (b) $\begin{bmatrix} -23 \\ -\frac{35}{3} \end{bmatrix}$

56. (a) $\begin{bmatrix} 1 & 1 & -3 \\ -1 & 2 & 0 \\ 1 & -1 & 1 \end{bmatrix}\begin{bmatrix} x_1 \\ x_2 \\ x_3 \end{bmatrix} = \begin{bmatrix} 9 \\ 6 \\ -5 \end{bmatrix}$ (b) $\begin{bmatrix} 0 \\ 3 \\ -2 \end{bmatrix}$

58. (a) $\begin{bmatrix} 1 & -1 & 4 \\ 1 & 3 & 0 \\ 0 & -6 & 5 \end{bmatrix}\begin{bmatrix} x_1 \\ x_2 \\ x_3 \end{bmatrix} = \begin{bmatrix} 17 \\ -11 \\ 40 \end{bmatrix}$ (b) $\begin{bmatrix} 4 \\ -5 \\ 2 \end{bmatrix}$

60. Not possible **62.** 2×2 **64.** Not possible

66. 2×2 **68.** 2×3

70. $\begin{bmatrix} 110 & 99 & 77 & 33 \\ 44 & 22 & 66 & 66 \end{bmatrix}$

72. [\$497,500 \$494,500]

The entries represent the costs of the three models of the product at the two warehouses.

74. $\begin{bmatrix} 0.40 & 0.15 & 0.15 \\ 0.28 & 0.53 & 0.17 \\ 0.32 & 0.32 & 0.68 \end{bmatrix}$

P^2 gives the proportion of the voting population that changed parties or remained loyal to their party from the first election to the third.

76. $P^3 = \begin{bmatrix} 0.300 & 0.175 & 0.175 \\ 0.308 & 0.433 & 0.217 \\ 0.392 & 0.392 & 0.608 \end{bmatrix}$

$P^4 = \begin{bmatrix} 0.250 & 0.188 & 0.188 \\ 0.315 & 0.377 & 0.248 \\ 0.435 & 0.435 & 0.565 \end{bmatrix}$

$P^5 = \begin{bmatrix} 0.225 & 0.194 & 0.194 \\ 0.314 & 0.345 & 0.267 \\ 0.461 & 0.461 & 0.539 \end{bmatrix}$

$P^6 = \begin{bmatrix} 0.213 & 0.197 & 0.197 \\ 0.311 & 0.326 & 0.280 \\ 0.477 & 0.477 & 0.523 \end{bmatrix}$

$P^7 = \begin{bmatrix} 0.206 & 0.198 & 0.198 \\ 0.308 & 0.316 & 0.288 \\ 0.486 & 0.486 & 0.514 \end{bmatrix}$

$P^8 = \begin{bmatrix} 0.203 & 0.199 & 0.199 \\ 0.305 & 0.309 & 0.292 \\ 0.492 & 0.492 & 0.508 \end{bmatrix}$

Approaches the matrix

$\begin{bmatrix} 0.2 & 0.2 & 0.2 \\ 0.3 & 0.3 & 0.3 \\ 0.5 & 0.5 & 0.5 \end{bmatrix}$

78. False. For most matrices, $AB \neq BA$.

80. $AC = BC = \begin{bmatrix} 2 & 3 \\ 2 & 3 \end{bmatrix}$

82. $A^2 = \begin{bmatrix} -1 & 0 \\ 0 & -1 \end{bmatrix}$ and $i^2 = -1$.

$A^3 = \begin{bmatrix} -i & 0 \\ 0 & -i \end{bmatrix}$ and $i^3 = -i$.

$A^4 = \begin{bmatrix} 1 & 0 \\ 0 & 1 \end{bmatrix}$ and $i^4 = 1$.

84. Diagonal matrix whose entries are the products of the corresponding entries of A and B

86. $-\frac{1}{4}, \frac{3}{2}$ **88.** $0, -9, \frac{5}{3}$ **90.** $\frac{5}{2}, \pm\sqrt{6}$

92. $(-1, 3)$ **94.** $(4, 1)$

Section 8.3 *(page 655)*

2.–10. $AB = I$ and $BA = I$

12. $\begin{bmatrix} 7 & -2 \\ -3 & 1 \end{bmatrix}$ **14.** $\begin{bmatrix} -19 & -33 \\ -4 & -7 \end{bmatrix}$

16. $\begin{bmatrix} 0 & -1 \\ 1 & 11 \end{bmatrix}$ **18.** $\begin{bmatrix} \frac{4}{5} & -\frac{3}{5} \\ -\frac{1}{5} & \frac{2}{5} \end{bmatrix}$

20. Does not exist

22. $\begin{bmatrix} -13 & 6 & 4 \\ 12 & -5 & -3 \\ -5 & 2 & 1 \end{bmatrix}$

24. Does not exist

26. $\begin{bmatrix} 1 & -\frac{3}{2} & -4 & \frac{13}{5} \\ 0 & \frac{1}{2} & 1 & -\frac{4}{5} \\ 0 & 0 & -\frac{1}{2} & \frac{1}{10} \\ 0 & 0 & 0 & \frac{1}{5} \end{bmatrix}$

28. $\begin{bmatrix} -10 & -4 & 27 \\ 2 & 1 & -5 \\ -13 & -5 & 35 \end{bmatrix}$ **30.** $\begin{bmatrix} 1 & -1 & 0 \\ 7 & -8.5 & 1 \\ -8 & 10 & 1 \end{bmatrix}$

32. Does not exist

34. $\begin{bmatrix} 3.75 & 0 & -1.25 \\ 3.458\overline{3} & -1 & -1.375 \\ 4.1\overline{6} & 0 & -2.5 \end{bmatrix}$

36. $\begin{bmatrix} 27 & -10 & 4 & -29 \\ -16 & 5 & -2 & 18 \\ -17 & 4 & -2 & 20 \\ -7 & 2 & -1 & 8 \end{bmatrix}$

38. $\begin{bmatrix} -24 & 7 & 1 & -2 \\ -10 & 3 & 0 & -1 \\ -29 & 7 & 3 & -2 \\ 12 & -3 & -1 & 1 \end{bmatrix}$ **40.** $\begin{bmatrix} -\frac{5}{61} & -\frac{12}{61} \\ \frac{8}{61} & \frac{7}{61} \end{bmatrix}$

42. $\begin{bmatrix} -\frac{2}{9} & -\frac{1}{3} \\ -\frac{5}{9} & -\frac{4}{3} \end{bmatrix}$ **44.** $\begin{bmatrix} -\frac{32}{143} & \frac{81}{143} \\ \frac{60}{143} & \frac{9}{143} \end{bmatrix}$

46. $(6, 3)$ **48.** $(-7, -4)$ **50.** $(1, 7, -9)$

52. $(-32, -13, -37, 15)$ **54.** $\left(\frac{1}{2}, \frac{1}{3}\right)$

56. $(6, -2)$ **58.** $\left(-\frac{1}{13}, \frac{8}{13}\right)$ **60.** $(-12, 10)$

62. $(5, 8, -2)$

64. $(-7, 3, -2)$ **66.** $(10, -3, 5)$

68. $(6.21, -0.77, -2.67, 2.40)$

70. \$4000 in AAA-rated bonds
\$2000 in A-rated bonds
\$4000 in B-rated bonds

72. \$200,000 in AAA-rated bonds
\$100,000 in A-rated bonds
\$200,000 in B-rated bonds

74. $I_1 = 2$ amperes
$I_2 = 3$ amperes
$I_3 = 5$ amperes

76. True. If A and B are both square matrices and $AB = I_n$, it can be shown that $BA = I_n$.

78. Answers will vary.

80. (a) Answers will vary.

(b) $A^{-1} = \begin{bmatrix} \frac{1}{a_{11}} & 0 & 0 & \cdots & 0 \\ 0 & \frac{1}{a_{22}} & 0 & \cdots & 0 \\ 0 & 0 & \frac{1}{a_{33}} & \cdots & 0 \\ \vdots & \vdots & \vdots & \cdots & \vdots \\ 0 & 0 & 0 & \cdots & \frac{1}{a_{nn}} \end{bmatrix}$

82. $x = -5 \ln \frac{1}{5} \approx 8.05$ **84.** $x = \dfrac{1 + \sqrt{5}}{2} \approx 1.62$

86. $\begin{bmatrix} -\frac{3}{4} & \frac{1}{4} & 0 \\ -\frac{1}{2} & -3 & \frac{9}{4} \end{bmatrix}$ **88.** $\begin{bmatrix} 28 & -7 \\ 1 & 9 \end{bmatrix}$

Section 8.4 *(page 663)*

2. -8 **4.** -11 **6.** 14 **8.** 0 **10.** 0

12. 34 **14.** 18 **16.** $\frac{10}{9}$ **18.** -0.022

20. -11.217 **22.** -20

24. (a) $M_{11} = 2, M_{12} = -3, M_{21} = 0, M_{22} = 11$
(b) $C_{11} = 2, C_{12} = 3, C_{21} = 0, C_{22} = 11$

26. (a) $M_{11} = -2, M_{12} = 7, M_{21} = 5, M_{22} = -6$
(b) $C_{11} = -2, C_{12} = -7, C_{21} = -5, C_{22} = -6$

28. (a) $M_{11} = 38, M_{12} = -8, M_{13} = -26, M_{21} = -4,$
$M_{22} = 4, M_{23} = -2, M_{31} = -5, M_{32} = 5, M_{33} = 5$
(b) $C_{11} = 38, C_{12} = 8, C_{13} = -26, C_{21} = 4,$
$C_{22} = 4, C_{23} = 2, C_{31} = -5, C_{32} = -5, C_{33} = 5$

30. (a) $M_{11} = 36, M_{12} = -42, M_{13} = 85, M_{21} = -82,$
$M_{22} = -12, M_{23} = -68, M_{31} = 24,$
$M_{32} = -28, M_{33} = -51$
(b) $C_{11} = 36, C_{12} = 42, C_{13} = 85, C_{21} = 82,$
$C_{22} = -12, C_{23} = 68, C_{31} = 24, C_{32} = 28,$
$C_{33} = -51$

32. (a) 151 (b) 151 **34.** (a) 650 (b) 650

36. (a) -1167 (b) -1167 **38.** 3

40. -2 **42.** -5 **44.** 2 **46.** -66

48. -108 **50.** 0 **52.** -100 **54.** 223

56. 105 **58.** 7441 **60.** -48

62. (a) 0
(b) -1
(c) $\begin{bmatrix} -2 & -5 \\ 4 & 10 \end{bmatrix}$
(d) 0

64. (a) -17
(b) -6
(c) $\begin{bmatrix} 4 & 22 \\ -1 & 20 \end{bmatrix}$
(d) 102

66. (a) -23
(b) 1
(c) $\begin{bmatrix} -9 & 4 & 1 \\ -5 & -10 & 6 \\ 4 & -1 & -1 \end{bmatrix}$
(d) -23

68. (a) 0
(b) -7
(c) $\begin{bmatrix} 7 & -4 & 9 \\ 8 & -6 & 3 \\ 6 & -2 & 15 \end{bmatrix}$
(d) 0

70.–74. Answers will vary. **76.** $-1, 3$ **78.** $-2, 3$

80. $3x^2 + 3y^2$ **82.** e^{-2x} **84.** x

86. True. If a square matrix has two columns that are equal, then elementary column operations can be used to create a column with all zeros.

88. (a) For an $n \times n$ matrix $(n > 2)$ with consecutive integer entries, the determinant appears to be 0.
(b) Answers will vary.

90. Yes. $|2A| = 8|A| = 8(5) = 40$

92.

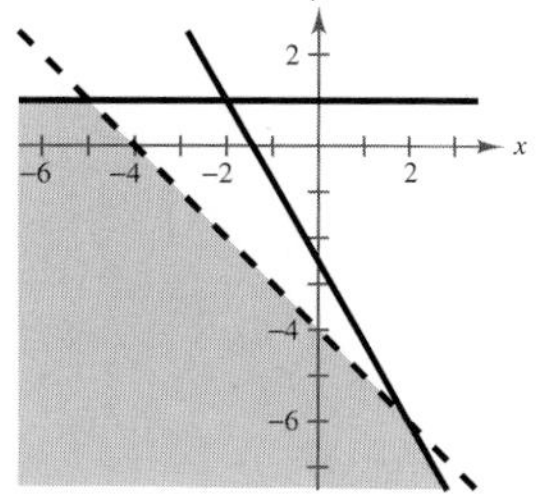

94. $\begin{bmatrix} -1 & -\frac{4}{3} \\ \frac{1}{2} & \frac{5}{6} \end{bmatrix}$ **96.** $\begin{bmatrix} -\frac{1}{4} & \frac{1}{6} & \frac{1}{3} \\ -\frac{1}{4} & \frac{1}{2} & 1 \\ -\frac{1}{2} & \frac{1}{3} & \frac{5}{3} \end{bmatrix}$

Section 8.5 *(page 675)*

2. $(-3, -5)$ **4.** $\left(\frac{8}{5}, -\frac{83}{10}\right)$ **6.** $(5, 8, -2)$

8. $(0, 3, -2)$ **10.** $(-3, -1, 2)$

12. Cramer's Rule does not apply.

14. $\frac{33}{2}$ **16.** $\frac{31}{2}$ **18.** 55 **20.** $\frac{25}{2}$ **22.** $\frac{23}{2}$

24. 19 or 3 **26.** 6 or 0 **28.** 3100 square feet

30. Not collinear **32.** Collinear **34.** Not collinear

36. $x = 3$ **38.** $x + y = 0$ **40.** $7x - 6y - 28 = 0$

42. $3x - 2y + 6 = 0$

44. Uncoded: [16 12 5], [1 19 5], [0 19 5], [14 4 0], [13 15 14], [5, 25, 0]

Encoded: 43 6 9 −38 −45 −13 −42 −47 −14 44 16 10 49 9 12 −55 −65 −20

46. 13 19 10 −1 −33 −77 3 −2 −14 4 1 −9 −5 −25 −47 4 1 −9

48. 58 122 139 1 −37 −95 40 67 55 23 17 −19 47 88 88 14 21 11

50. BRONCOS WIN SUPER BOWL

52. HAVE A GREAT WEEKEND

54. RETURN AT DAWN

56. CANCEL ORDERS SUE

58. False. If the determinant of the coefficient matrix is zero, the system has either no solution or infinitely many solutions.

60. $(-6, 4)$ **62.** $(-1, 0, -3)$

64.

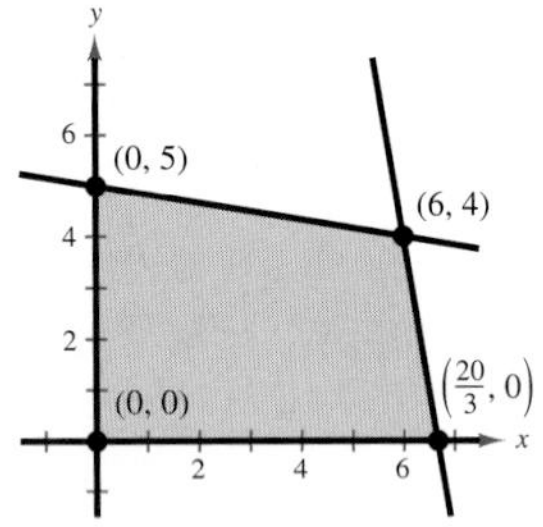

Minimum at (0, 0): 0

Maximum at (6, 4): 52

66. 37 **68.** 59

Review Exercises *(page 679)*

2. 2×4 **4.** 1×5

6. $\begin{bmatrix} 8 & -7 & 4 & \vdots & 12 \\ 3 & -5 & 2 & \vdots & 20 \\ 5 & 3 & -3 & \vdots & 26 \end{bmatrix}$

8. $\begin{cases} 13x + 16y + 7z + 3w = 2 \\ x + 21y + 8z + 5w = 12 \\ 4x + 10y - 4z + 3w = -1 \end{cases}$

10. $\begin{bmatrix} 1 & 0 & 0 & 0 \\ 0 & 1 & 0 & 0 \\ 0 & 0 & 1 & 0 \\ 0 & 0 & 0 & 1 \end{bmatrix}$

12. Inconsistent **14.** Consistent, infinitely many solutions

16. $(-9, -4)$ **18.** $(0.6, 0.5)$

20. $\left(\frac{1}{2}, -\frac{1}{3}, 1\right)$ **22.** Inconsistent

24. Inconsistent **26.** $\dfrac{10}{x-4} - \dfrac{10}{x-3} - \dfrac{7}{(x-3)^2}$

28. $\left(\frac{31}{42}, \frac{5}{14}, \frac{13}{84}\right)$ **30.** $(6, -2, 0)$

32. Inconsistent **34.** $x = 8, y = 0$

36. $x = 12, y = -2$ **38.** Yes

40. Not possible because matrices are not of the same order.

42. Not possible because matrices are not of the same order.

44. $\begin{bmatrix} 2 & 1 & 12 \\ -13 & 1 & -17 \\ -30 & -54 & 40 \end{bmatrix}$ **46.** $\begin{bmatrix} 6 & -8 \\ -11 & 54 \\ -44 & 2 \end{bmatrix}$

48. $\begin{bmatrix} -\frac{13}{6} & 1 \\ -\frac{1}{3} & -\frac{17}{6} \\ 0 & \frac{10}{3} \end{bmatrix}$ **50.** $\begin{bmatrix} -\frac{13}{3} & -\frac{10}{3} \\ 4 & -5 \\ -\frac{26}{3} & -\frac{16}{3} \end{bmatrix}$

52. Not possible because the number of columns of A does not equal the number of rows of B.

54. Yes

56. Not possible because the number of columns of the first matrix does not equal the number of rows of the second matrix.

58. $\begin{bmatrix} 4 & 6 & 3 \\ 0 & 6 & -10 \\ 0 & 0 & 6 \end{bmatrix}$ **60.** $[4 \quad 10]$ **62.** $\begin{bmatrix} -12 & 9 \\ -246 & 144 \end{bmatrix}$

64. $\begin{bmatrix} 13 & 24 \\ 20 & 4 \end{bmatrix}$ **66.** $\begin{bmatrix} 2 & 3 & 1 \\ 2 & -3 & -3 \\ 4 & -2 & 3 \end{bmatrix}\begin{bmatrix} x \\ y \\ z \end{bmatrix} = \begin{bmatrix} 10 \\ 22 \\ -2 \end{bmatrix}$

68. and 70. $AB = I$ and $BA = I$

72. $\begin{bmatrix} 4 & -5 \\ 5 & -6 \end{bmatrix}$

74. $\begin{bmatrix} 13 & 6 & -4 \\ -12 & -5 & 3 \\ 5 & 2 & -1 \end{bmatrix}$ **76.** $\begin{bmatrix} \frac{1}{5} & \frac{1}{5} \\ \frac{1}{10} & -\frac{1}{15} \end{bmatrix}$

78. $\begin{bmatrix} \frac{1}{2} & -1 & -\frac{1}{2} \\ \frac{1}{2} & -\frac{2}{3} & -\frac{5}{6} \\ 0 & \frac{2}{3} & \frac{1}{3} \end{bmatrix}$ **80.** $\begin{bmatrix} 1 & -1 \\ 4 & -\frac{7}{2} \end{bmatrix}$ **82.** $\begin{bmatrix} 2 & \frac{20}{3} \\ \frac{1}{10} & \frac{1}{6} \end{bmatrix}$

84. $(36, 11)$ **86.** $(-6, -1)$ **88.** $(2, -1, -2)$

90. $(6, 1, -1)$ **92.** $(-3, 1)$ **94.** $(1, 1, -2)$

96. -42 **98.** 550

100. (a) $M_{11} = 4, M_{12} = 7, M_{21} = -1, M_{22} = 2$
(b) $C_{11} = 4, C_{12} = -7, C_{21} = 1, C_{22} = 2$

102. (a) $M_{11} = 30, M_{12} = -12, M_{13} = -21,$
$M_{21} = 20, M_{22} = 19, M_{23} = 22, M_{31} = 5,$
$M_{32} = -2, M_{33} = 19$
(b) $C_{11} = 30, C_{12} = 12, C_{13} = -21,$
$C_{21} = -20, C_{22} = 19, C_{23} = -22,$
$C_{31} = 5, C_{32} = 2, C_{33} = 19$

104. 130 **106.** 279 **108.** $(4, 7)$ **110.** $(-1, 4, 5)$

112. 6 carnations; 6 roses **114.** $y = x^2 + 2x + 3$

116. 16 **118.** $\frac{7}{4}$ **120.** $x - 2y + 4 = 0$

122. $2x + 6y - 13 = 0$

124. Uncoded: $[12 \;\; 15 \;\; 15], [11 \;\; 0 \;\; 15], [21 \;\; 20 \;\; 0],$
$[2 \;\; 5 \;\; 12], [15 \;\; 23 \;\; 0]$
Encoded: $-21 \;\; 6 \;\; 0 \;\; -68 \;\; 8 \;\; 45 \;\; 102 \;\; -42 \;\; -60 \;\; -53$
$20 \;\; 21 \;\; 99 \;\; -30 \;\; -69$

126. SEE YOU FRIDAY

128. False. The matrix must be square.

130. The matrix must be square and its determinant nonzero.

132. No. The first two matrices describe a system of equations with one solution. The third matrix describes a system with infinitely many solutions.

134. $\lambda = \pm 2\sqrt{10} - 3$

Chapter 9

Section 9.1 *(page 695)*

2. 2, 7, 12, 17, 22 **4.** $\frac{1}{2}, \frac{1}{4}, \frac{1}{8}, \frac{1}{16}, \frac{1}{32}$

6. $-\frac{1}{2}, \frac{1}{4}, -\frac{1}{8}, \frac{1}{16}, -\frac{1}{32}$ **8.** $\frac{1}{3}, \frac{1}{2}, \frac{3}{5}, \frac{2}{3}, \frac{5}{7}$ **10.** $2, \frac{14}{9}, \frac{28}{19}, \frac{16}{11}, \frac{74}{51}$

12. 0, 2, 0, 2, 0 **14.** $\frac{2}{3}, \frac{4}{9}, \frac{8}{27}, \frac{16}{81}, \frac{32}{243}$

16. $10, \frac{10}{\sqrt[3]{4}}, \frac{10}{\sqrt[3]{9}}, \frac{10}{\sqrt[3]{16}}, \frac{10}{\sqrt[3]{25}}$ **18.** 1, 1, 2, 6, 24

20. $-\frac{1}{2}, \frac{2}{3}, -\frac{3}{4}, \frac{4}{5}, -\frac{5}{6}$ **22.** 0.3, 0.3, 0.3, 0.3, 0.3

24. $-5, -4, 9, 40, 95$ **26.** -240 **28.** 2520 **30.** $\frac{37}{130}$

32.

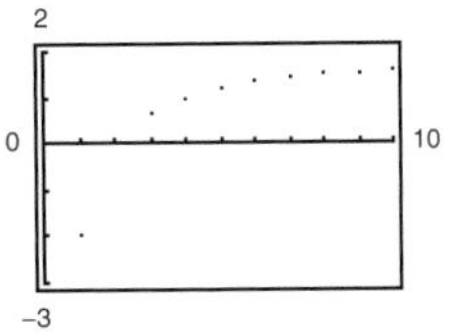

34.

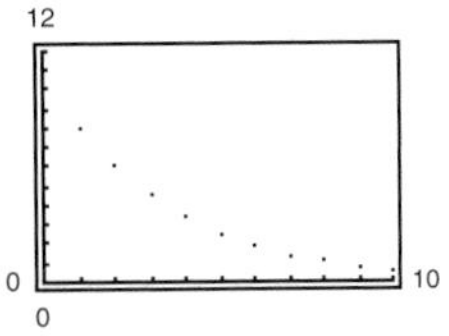

36.

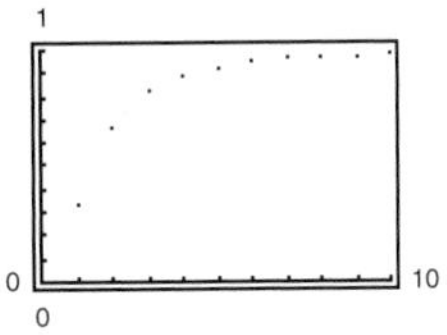

38. b **40.** a

42. $a_n = 4n - 1$ **44.** $a_n = (-1)^{n+1}(2n)$

46. $a_n = \frac{(-1)^{n+1}}{2^n}$ **48.** $a_n = \frac{2^{n-1}}{3^n}$ **50.** $a_n = \frac{1}{n!}$

52. $a_n = \frac{2^{n-1}}{(n-1)!}$ **54.** $a_n = 1 + \frac{2^n - 1}{2^n}$

56. 15, 18, 21, 24, 27 **58.** 32, 16, 8, 4, 2

60. 25, 20, 15, 10, 5 **62.** $14, -28, 56, -112, 224$
$a_n = 30 - 5n$ $a_n = 14(-2)^{n-1}$

64. $\frac{1}{336}$ **66.** 600 **68.** $(n + 2)(n + 1)$ **70.** $3n + 1$

72. 57 **74.** 25 **76.** 110 **78.** $\frac{124}{429}$ **80.** 238

82. 11 **84.** 6.06 **86.** $\frac{3}{8}$ **88.** $\sum_{i=1}^{15} \frac{5}{1 + i}$

90. $\sum_{k=1}^{6} \left[1 - \left(\frac{k}{6}\right)^2\right]$ **92.** $\sum_{n=0}^{7} \left(-\frac{1}{2}\right)^n$ **94.** $\sum_{k=1}^{10} \frac{1}{k(k+2)}$

96. $\sum_{k=1}^{6} \frac{k!}{2^k}$ **98.** $\frac{242}{243}$ **100.** $-\frac{51}{32}$ **102.** $\frac{1}{9}$ **104.** $\frac{2}{9}$

106. (a) $A_1 = \$101.00,\ A_2 = \$203.01,\ A_3 = \$306.04,$
$A_4 = \$410.10,\ A_5 = \$515.20,\ A_6 = \$621.35$
(b) $A_{60} = \$8248.64$
(c) $A_{240} = \$99{,}914.79$

108. $a_{-1} = \$2.85,\ a_0 = \$3.30,\ a_1 = \$3.70,\ a_2 = \$4.06,$
$a_3 = \$4.39,\ a_4 = \$4.70,\ a_5 = \$4.99,\ a_6 = \$5.26$

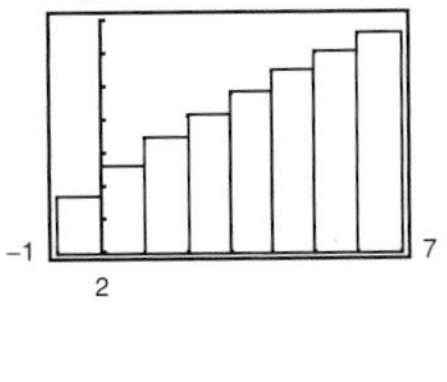

CHAPTER 9

110. \$6.25 **112.** Answers will vary. **114.** \$1.301

116. Answers will vary.

118. True, because

$2^1 + 2^2 + 2^3 + 2^4 = 2^{3-2} + 2^{4-2} + 2^{5-2} + 2^{6-2}$.

120. (a) $\begin{bmatrix} 10 & 19 \\ -12 & -5 \end{bmatrix}$ (b) $\begin{bmatrix} -30 & -69 \\ 44 & 26 \end{bmatrix}$

(c) $\begin{bmatrix} 56 & -43 \\ 48 & 114 \end{bmatrix}$ (d) $\begin{bmatrix} 48 & -72 \\ 36 & 122 \end{bmatrix}$

122. (a) $\begin{bmatrix} -1 & 0 & 0 \\ 2 & 0 & 4 \\ 1 & -1 & 1 \end{bmatrix}$ (b) $\begin{bmatrix} 3 & 4 & 0 \\ -3 & 1 & -14 \\ -4 & 3 & -1 \end{bmatrix}$

(c) $\begin{bmatrix} 12 & 0 & -8 \\ 1 & 21 & 2 \\ -6 & -1 & 8 \end{bmatrix}$ (d) $\begin{bmatrix} 20 & 4 & 8 \\ 2 & 15 & -4 \\ 1 & -6 & 6 \end{bmatrix}$

124. -126 **126.** $-11{,}758$

Section 9.2 *(page 705)*

2. Arithmetic sequence, $d = 3$

4. Not an arithmetic sequence

6. Arithmetic sequence, $d = -\frac{1}{2}$

8. Arithmetic sequence, $d = 0.4$

10. Not an arithmetic sequence

12. 97, 94, 91, 88, 85
Arithmetic sequence, $d = -3$

14. 1, 5, 9, 13, 17
Arithmetic sequence, $d = 4$

16. 1, 2, 4, 8, 16
Not an arithmetic sequence

18. 2, 8, 24, 64, 160
Not an arithmetic sequence

20. 6, 11, 16, 21, 26; $d = 5$; $a_n = 5n + 1$

22. 72, 66, 60, 54, 48; $d = -6$; $a_n = -6n + 78$

24. 0.375, 0.625, 0.875, 1.125, 1.375; $d = 0.25$;
$a_n = 0.25n + 0.125$

26. $5, \frac{17}{4}, \frac{7}{2}, \frac{11}{4}, 2$ **28.** 16.5, 16.75, 17, 17.25, 17.5

30. 1, 6, 11, 16, 21 **32.** 22.45, 20.725, 19, 17.275, 15.55

34. $a_n = 4n + 11$ **36.** $a_n = -\frac{2}{3}n + \frac{2}{3}$

38. $a_n = 5yn - 6y$ **40.** $a_n = -5n + 15$

42. $a_n = 5n - 9$ **44.** $a_n = -15n + 265$

46. d **48.** a

50.

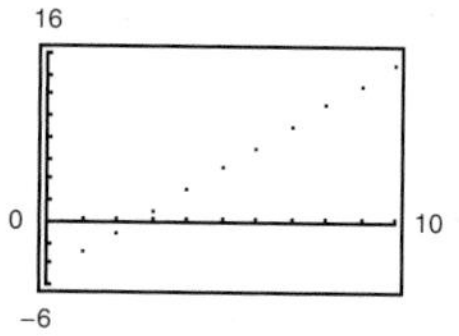

52.

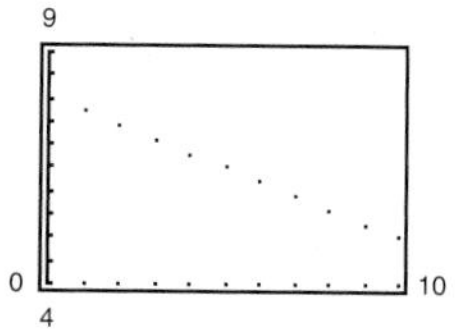

54. 1850 **56.** 23 **58.** 375 **60.** 16,100

62. 10,100 **64.** 26,425 **66.** 2500 **68.** 218,625

70. 44,625 **72.** -896.375 **74.** 1402.5

76. (a) \$45,550 (b) \$247,050

78. 2430 seats **80.** 203 bricks **82.** 161.7 meters

84. True by the formula for the sum of a finite arithmetic sequence,

$$S_n = \frac{n}{2}(a_1 + a_n)$$

86. First term plus $(n - 1)$ times the common difference

88. 1220

90. (a) Yes. Common difference: C

(b) Yes. $C \times$ (common difference)

(c) No

92. $S_n + 5n$ **94.** (a, a, a) **96.** $(0, 0, 0)$

98. $(-3, 7)$ **100.** $(0, 0, 0)$

Section 9.3 *(page 714)*

2. Geometric sequence, $r = 4$

4. Not a geometric sequence

6. Geometric sequence, $r = 0.2$

8. Geometric sequence, $r = -\frac{2}{3}$

10. Not a geometric sequence **12.** 6, 12, 24, 48, 96

14. $1, \frac{1}{3}, \frac{1}{9}, \frac{1}{27}, \frac{1}{81}$ **16.** $6, -\frac{3}{2}, \frac{3}{8}, -\frac{3}{32}, \frac{3}{128}$

18. $3, 3\sqrt{5}, 15, 15\sqrt{5}, 75$ **20.** $5, 10x, 20x^2, 40x^3, 80x^4$

22. 81, 27, 9, 3, 1; $r = \frac{1}{3}$; $a_n = 243\left(\frac{1}{3}\right)^n$

24. 5, -10, 20, -40, 80; $r = -2$; $a_n = -\frac{5}{2}(-2)^n$

26. 48, -24, 12, -6, 3; $r = -\frac{1}{2}$; $a_n = -96\left(-\frac{1}{2}\right)^n$

28. $\dfrac{10{,}935}{128}$ **30.** $-\dfrac{1}{4096}$ **32.** $27\sqrt{3}$

34. 1342.139 **36.** 12 **38.** $\frac{256}{243}$ **40.** c **42.** d

44.

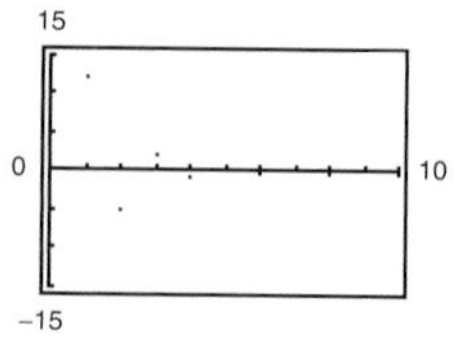

46.

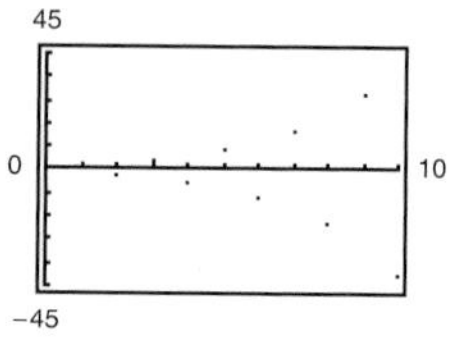

48. 171 **50.** $\frac{1365}{32}$ **52.** 592.647 **54.** 3949.147

56. 3.750 **58.** $\sum_{n=1}^{8} 7(2)^{n-1}$ **60.** $\sum_{n=1}^{6} 15\left(-\frac{1}{5}\right)^{n-1}$

62. $\sum_{n=1}^{5} 32\left(\frac{3}{4}\right)^{n-1}$ **64.** 6 **66.** $\frac{6}{5}$ **68.** $\frac{10}{9}$ **70.** 5

72. -12.5 **74.** 27 **76.** Undefined **78.** $\frac{11}{37}$ **80.** $\frac{25}{18}$

82.

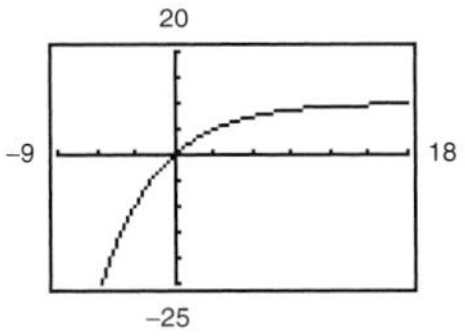

Horizontal asymptote: $y = 10$

Corresponds to the sum of the series

84. (a) \$11,652.39 (b) \$12,002.55
(c) \$12,188.60 (d) \$12,317.01
(e) \$12,380.41

86. 368,318 **88.** \$3698.34 **90.** Answers will vary.

92. (a) \$84,714.78 **94.** (a) \$76,122.54
(b) \$85,196.05 (b) \$76,533.16

96. \$222,289.91 **98.** \$39.7 billion

100. \$3,623,993.23

102. False. A sequence is geometric if the ratios of consecutive terms are the same.

104. The value of a real number between -1 and 1, raised to a power, approaches zero.

106. $x^2 + 2x$ **108.** $3x^2 + 6x + 1$

110. $x(3x + 8)(3x - 8)$ **112.** $(3x + 1)(2x - 5)$

114. $\frac{3x}{x - 3}, \ x \neq -3$ **116.** $\frac{2x + 1}{3}, \ x \neq 0$

118. $\frac{5x^2 + 9x - 30}{(x + 2)(x - 2)}$

Section 9.4 *(page 726)*

2. $\frac{1}{2(k + 3)}$ **4.** $\frac{k + 1}{3}(2k + 3)$

6.–18. Answers will vary. **20.** 465 **22.** 3025

24. 61,776 **26.** 43,890 **28.** 195

30. $S_n = \frac{n}{2}(-3n + 53)$ **32.** $S_n = \frac{6}{5}\left[1 - \left(-\frac{3}{2}\right)^n\right]$

34. $S_n = \frac{n}{2(n + 2)}$ **36.–48.** Answers will vary.

50. 1, 3, 5, 7, 9 **52.** 4, 2, -2, -4, -2

54. 0, 3, 6, 9, 12
First differences: 3, 3, 3, 3
Second differences: 0, 0, 0
Linear

56. 3, 1, -2, -6, -11
First differences: -2, -3, -4, -5
Second differences: -1, -1, -1
Quadratic

58. 0, 1, 3, 6, 10
First differences: 1, 2, 3, 4
Second differences: 1, 1, 1
Quadratic

60. 2, 4, 6, 8, 10
First differences: 2, 2, 2, 2
Second differences: 0, 0, 0
Linear

62. 1, 2, 6, 15, 31
First differences: 1, 4, 9, 16
Second differences: 3, 5, 7
Neither

64. $a_n = n^2 - n + 3$ **66.** $a_n = \frac{1}{2}n^2 + n - 3$

68. True. P_7 may be false.

70. True. If the second differences are all zero, then the first differences were all the same and the sequence is arithmetic.

72. See page 719. **74.** $(2, 4)$ and $\left(-\frac{1}{2}, \frac{1}{4}\right)$

76. $(1, 2, 1)$ **78.** $4x^4 - 4x^2 + 1$

80. $-64x^3 + 240x^2 - 300x + 125$

Section 9.5 *(page 733)*

2. 28 **4.** 1 **6.** 792 **8.** 210 **10.** 4950

12. 8 **14.** 20

16. $x^6 + 6x^5 + 15x^4 + 20x^3 + 15x^2 + 6x + 1$

18. $a^5 + 25a^4 + 250a^3 + 1250a^2 + 3125a + 3125$

20. $y^5 - 10y^4 + 40y^3 - 80y^2 + 80y - 32$

22. $c^3 + 3c^2d + 3cd^2 + d^3$

24. $x^4 + 8x^3y + 24x^2y^2 + 32xy^3 + 16y^4$

26. $32x^5 - 80x^4y + 80x^3y^2 - 40x^2y^3 + 10xy^4 - y^5$

28. $125 - 225y + 135y^2 - 27y^3$

30. $x^{12} + 6x^{10}y^2 + 15x^8y^4 + 20x^6y^6 + 15x^4y^8 + 6x^2y^{10} + y^{12}$

32. $\frac{1}{x^6} + \frac{12y}{x^5} + \frac{60y^2}{x^4} + \frac{160y^3}{x^3} + \frac{240y^4}{x^2} + \frac{192y^5}{x} + 64y^6$

34. $3x^5 + 15x^4 + 26x^3 + 18x^2 + 3x - 1$

36. $81 - 216z + 216z^2 - 96z^3 + 16z^4$

38. $64v^6 + 576v^5 + 2160v^4 + 4320v^3 + 4860v^2 + 2916v + 729$

40. 3,247,695 **42.** 720 **44.** 16,128 **46.** 45

48. $8t^{3/2} - 12t + 6t^{1/2} - 1$

50. $u^3 + 10u^{12/5} + 40u^{9/5} + 80u^{6/5} + 80u^{3/5} + 32$

52. $4x^3 + 6x^2h + 4xh^2 + h^3$ **54.** $-\dfrac{1}{x(x + h)}$

56. $-38 - 41i$ **58.** $-10 + 198i$ **60.** $184 - 440\sqrt{3}i$

62. 1049.890 **64.** 467.721

66.

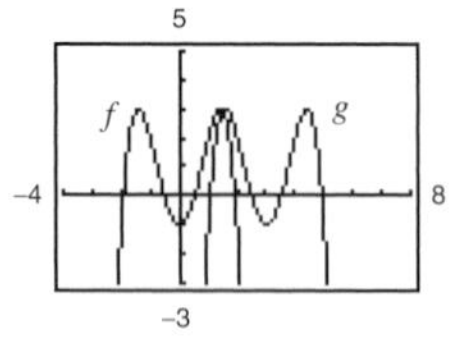

g is shifted 3 units to the right of f.

$g(x) = -x^4 + 12x^3 - 50x^2 + 84x - 46$

68. 0.273 **70.** 0.171

72. (a) $g(t) = 0.0348t^2 + 5.8043t + 95.588$

(b)

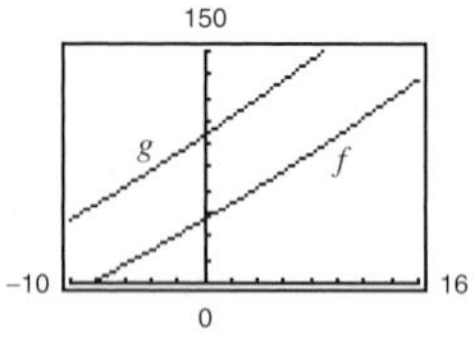

74. True. The coefficients from the Binomial Theorem can be used to find the numbers in Pascal's Triangle.

76. The first and last numbers in each row are 1. Every other number in each row is formed by adding the two numbers immediately above the number.

78. $n + 1$ terms **80. and 82.** Answers will vary.

84. $g(x)$ is shifted 8 units up from $f(x)$.

86. $g(x)$ is the reflection of $f(x)$ in the y-axis.

88.

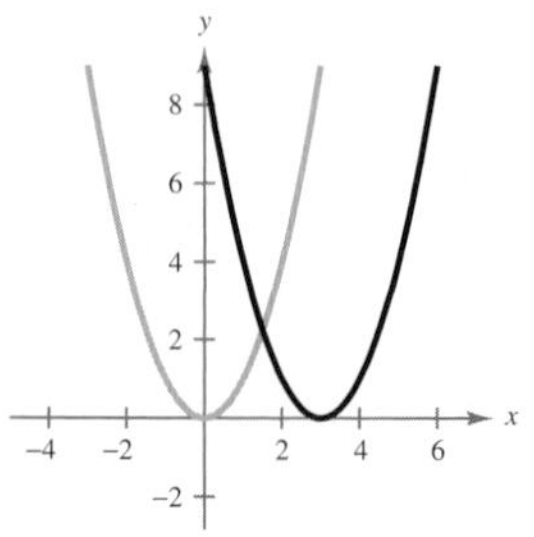

$g(x) = (x - 3)^2$

90.

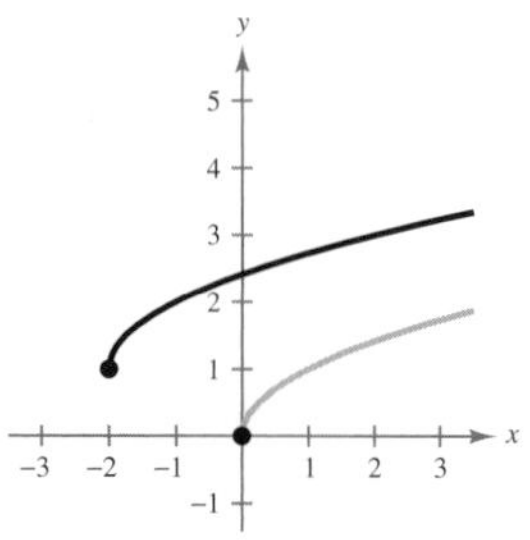

$g(x) = \sqrt{x + 2} + 1$

Section 9.6 *(page 743)*

2. 6 **4.** 3 **6.** 4 **8.** 6 **10.** 60 **12.** 30

14. 4096 **16.** 1440 **18.** 5,760,000

20. (a) 9000 (b) 4536 (c) 4000 (d) 4500

22. 125,000

24. (a) 40,320 (b) 576

26. 120 **28.** 380 **30.** 840 **32.** $n = 9$ or $n = 10$

34. 9,034,502,400 **36.** 1,814,400 **38.** 120 **40.** 56

42. 34,650 **44.** ABCD, DBCA, ACBD, DCBA

46. ABC, ABD, ABE, ABF, ACD, ACE, ACF, ADE, ADF, AEF, BCD, BCE, BCF, BDE, BDF, BEF, CDE, CDF, CEF, DEF

48. 720 **50.** 24 **52.** 66 **54.** 15,890,700

56. 24,040,016

58. (a) 35 (b) 63 (c) 203 **60.** 3744

62. (a) 3 (b) 28 (c) 66 (d) 190

64. 9 **66.** 35 **68.** True

70. ${}_{10}P_6 > {}_{10}C_6$. Changing the order of any of the six elements selected results in a different permutation but the same combination.

72. and 74. Answers will vary.

76. The symbol ${}_nP_r$ denotes the number of ways to choose and order r elements out of a collection of n elements.

78. (a) 2 (b) 4 (c) $\sqrt{x} + 2$

80. (a) 29 (b) -3 (c) 445

82. $y^6 - 12y^5 + 60y^4 - 160y^3 + 240y^2 - 192y + 64$

84. $x^8 - 4x^6y^2 + 6x^4y^4 - 4x^2y^6 + y^8$

Section 9.7 *(page 754)*

2. $\{2, 3, . . . , 12\}$

4. {(red, red), (red, blue), (red, black), (blue, blue), (blue, black)}

6. {SSS, SSF, SFS, SFF, FSS, FSF, FFS, FFF}

8. $\frac{1}{2}$ **10.** $\frac{1}{2}$ **12.** $\frac{10}{13}$ **14.** $\frac{6}{13}$ **16.** $\frac{7}{12}$

18. $\frac{1}{9}$ **20.** $\frac{19}{36}$ **22.** $\frac{1}{15}$ **24.** $\frac{11}{15}$ **26.** 0.64

28. $\frac{1}{6}$ **30.** 0.16 **32.** $\frac{13}{100}$

34. (a) 36,630,000 (b) 27% (c) 56%

36. (a) 34% (b) 45% (c) 23%

38. (a) $\frac{5}{12}$ (b) $\frac{7}{12}$ (c) $\frac{5}{36}$

40. 19%

42. (a) $\frac{3}{28}$ (b) $\frac{15}{28}$ (c) $\frac{9}{14}$

44. (a) $\frac{3}{8}$ (b) $\frac{19}{30}$ **46.** (a) $\frac{1}{24}$ (b) $\frac{1}{6}$ **48.** $\frac{6}{4165}$

50. (a) $\frac{91}{323}$ (b) $\frac{455}{969}$ (c) $\frac{232}{323}$

52. (a) $\frac{1}{4}$ (b) $\frac{1}{2}$ (c) $\frac{841}{1600}$ (d) $\frac{1}{40}$

54. (a) 0.81 (b) 0.01 (c) 0.99

56. (a) $\frac{1}{16}$ (b) $\frac{1}{8}$ (c) $\frac{15}{16}$ **58.** 0.1024

60. (a) $\frac{\pi}{4}$ (b) Approximations will vary.

62. False. The complement of the event is to roll a number greater than or equal to 3 and its probability is $\frac{2}{3}$.

64. Meteorological records indicate that over an extended period of time with similar weather conditions it will rain 40% of the time.

66. $\frac{-3 \pm \sqrt{57}}{4}$ **68.** $0, \pm 1$ **70.** ± 4 **72.** -1

74. 3 **76.** $\ln 28 \approx 3.332$

78. $\ln 3 \approx 1.099, \ln 4 \approx 1.386$ **80.** $\ln 3.2 \approx 1.163$

82. $e^{-3/4} \approx 0.472$ **84.** $\frac{e^3}{2} \approx 10.043$

Review Exercises *(page 760)*

2. $5, \frac{10}{3}, 3, \frac{20}{7}, \frac{25}{9}$ **4.** 0, 2, 6, 12, 20 **6.** 12 **8.** $\frac{1}{8}$

10. 56 **12.** 6.17 **14.** 35 **16.** $\sum_{k=1}^{9} \frac{k}{k+1}$

18. $\frac{1}{3}$ **20.** $\frac{1}{10}$ **22.** 676

24. Not an arithmetic sequence

26. Arithmetic sequence, $d = -\frac{1}{9}$ **28.** $a_n = -3n + 28$

30. $a_n = nx - 3x$ **32.** 52 **34.** 250 **36.** 3050

38. 73.5 miles **40.** 2, 4, 8, 16, 32

42. $2, 2\sqrt{6}, 12, 12\sqrt{6}, 72$ or $2, -2\sqrt{6}, 12, -12\sqrt{6}, 72$

44. $a_n = 216\left(\frac{1}{6}\right)^{n-1}$, 259.2 **46.** $a_n = 5(0.2)^{n-1}$, 6.25

48. 121 **50.** $\frac{364}{243}$ **52.** 720 **54.** 25 **56.** 1493.50

58. $\frac{3}{2}$ **60.** 2 **62.** $\frac{13}{9}$ **64.** 4,371,379.65

66. and 68. Answers will vary. **70.** 385 **72.** 12,201

74. $S_n = 4n(18 - n)$ **76.** $S_n = \frac{144}{13}\left[1 - \left(-\frac{1}{12}\right)^n\right]$

78. $-3, -7, -13, -21, -31$

First differences: $-4, -6, -8, -10$

Second differences: $-2, -2, -2$

Quadratic

80. 0, 1, 1, 2, 2

First differences: 1, 0, 1, 0

Second differences: $-1, 1, -1$

Neither

82. 120 **84.** 220 **86.** 126 **88.** 10

90. $\frac{64}{x^6} - \frac{576}{x^4} + \frac{2160}{x^2} - 4320 + 4860x^2 - 2916x^4 + 729x^6$

92. $2187x^7 + 5103x^6y^2 + 5103x^5y^4 + 2835x^4y^6 + 945x^3y^8 + 189x^2y^{10} + 21xy^{12} + y^{14}$

94. $-236 - 115i$ **96.** 11 **98.** 50,000 **100.** 720

102. 1140 **104.** $\frac{1}{120}$

106. (a) 41.6% (b) 80% (c) 7.4%

108. $\frac{5}{324}$ **110.** $\frac{31}{32}$ **112.** True by Properties of Sums

114. True because $2^1 + 2^2 + 2^3 + 2^4 + 2^5 + 2^6 = 2^{3-2} + 2^{4-2} + 2^{5-2} + 2^{6-2} + 2^{7-2} + 2^{8-2}$

116. (a) Odd-numbered terms are negative.

(b) Even-numbered terms are negative.

118. (a) Arithmetic. There is a constant difference between consecutive terms.

(b) Geometric. Each term is a constant multiple of the previous term. In this case the common ratio is greater than 1.

120. Increased powers of real numbers between 0 and 1 approach zero.

122. a **124.** c **126.** $0 \le p \le 1$, closed interval

128. Meteorological records indicate that over an extended period of time with similar weather conditions it will rain 60% of the time.

Chapter 10

Section 10.1 *(page 774)*

2. 1 **4.** $-\sqrt{3}$ **6.** $-\frac{\sqrt{3}}{3}$ **8.** -0.2677

10. 2.0344 radians, 116.6° **12.** 1.1071 radians, 63.4°

14. 1.9513 radians, 111.8° **16.** 0.6023 radian, 34.5°

18. 2.0344 radians, 116.6° **20.** 2.4669 radians, 141.3°

22. $\frac{\pi}{4}$ radian, 45° **24.** $\frac{\pi}{4}$ radian, 45°

26. 1.1071 radians, 63.4° **28.** 1.4109 radians, 80.8°

30. 1.0808 radians, 61.9° **32.** 1.0240 radians, 58.7°

34. $(-3, 2)$, 35.8°; $(1, 3)$, 94.4°; $(2, 0)$, 49.8°

36. $(-3, 4)$, 32.5°; $(2, 1)$, 16.9°; $(-2, 2)$, 130.6°

38. $\frac{4\sqrt{5}}{5} \approx 1.7889$ **40.** $\frac{5\sqrt{2}}{2} \approx 3.5355$ **42.** 4

44. $9\sqrt{2} \approx 12.7279$ **46.** (a) $\frac{33\sqrt{29}}{29}$ (b) $\frac{33}{2}$

48. (a) $\frac{31\sqrt{389}}{389}$ (b) $\frac{31}{2}$ **50.** $\frac{9}{5}$

52. 0.2027, 1049 feet **54.** 31.0°

56. (a) 1.2016 radians (b) 5971 feet, 2420 feet

58. False. Substitute $\tan\theta_1$ and $\tan\theta_2$ for m_1 and m_2 in the formula for the angle between two lines.

60. (a) $d = \dfrac{3|m+1|}{\sqrt{m^2+1}}$

(b)

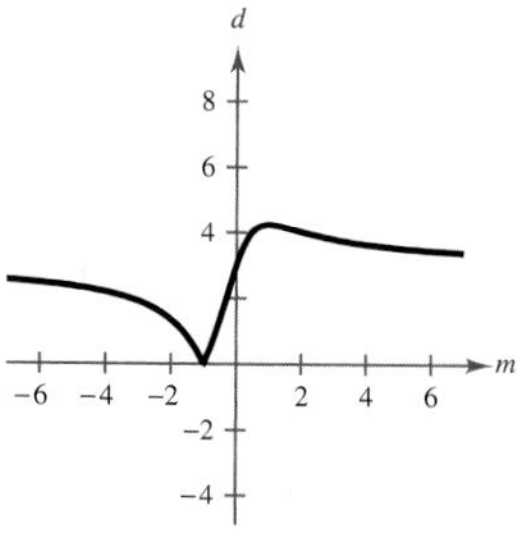

(c) $m = 1$

(d) Yes. $m = -1$

(e) $d = 3$. As the line approaches the vertical, the distance approaches 3.

62. x-intercept: $(-9, 0)$

y-intercept: $(0, 81)$

64. x-intercept: none

y-intercept: $(0, 133)$

66. x-intercepts: $(-11, 0), (2, 0)$

y-intercept: $(0, -22)$

68. $f(x) = 2\left(x - \frac{1}{4}\right)^2 - \frac{169}{8}$

Vertex: $\left(\frac{1}{4}, -\frac{169}{8}\right)$

70. $f(x) = -(x+4)^2 + 1$

Vertex: $(-4, 1)$

72. $f(x) = -8\left(x + \frac{17}{8}\right)^2 + \frac{121}{8}$

Vertex: $\left(-\frac{17}{8}, \frac{121}{8}\right)$

Section 10.2 *(page 782)*

2. An ellipse is formed when a plane intersects only the top or bottom half of a double-napped cone but is not parallel or perpendicular to the axis of the cone, is not parallel to the side of the cone, and does not intersect the vertex.

4. A hyperbola is formed when a plane intersects both halves of a double-napped cone, is parallel to the axis of the cone, and does not intersect the vertex.

6. b **8.** f **10.** c

12. Vertex: $(0, 0)$

Focus: $\left(0, -\frac{1}{8}\right)$

Directrix: $y = \frac{1}{8}$

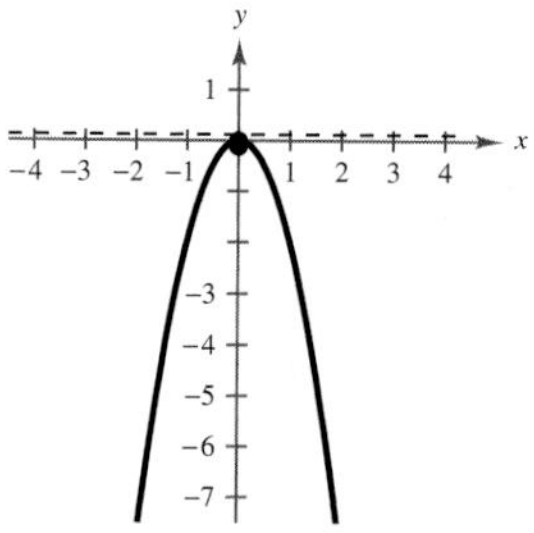

14. Vertex: $(0, 0)$

Focus: $\left(\frac{3}{4}, 0\right)$

Directrix: $x = -\frac{3}{4}$

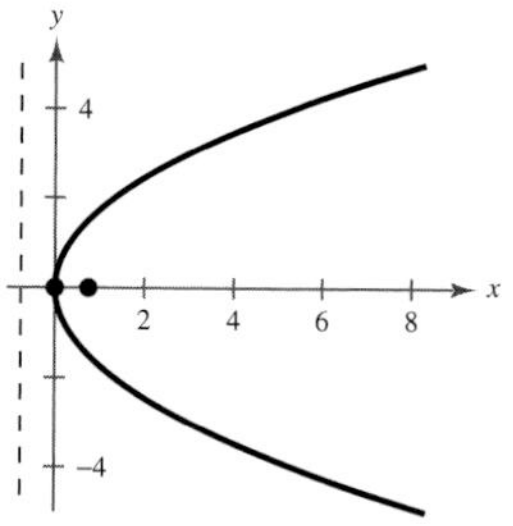

16. Vertex: $(0, 0)$

Focus: $\left(-\frac{1}{4}, 0\right)$

Directrix: $x = \frac{1}{4}$

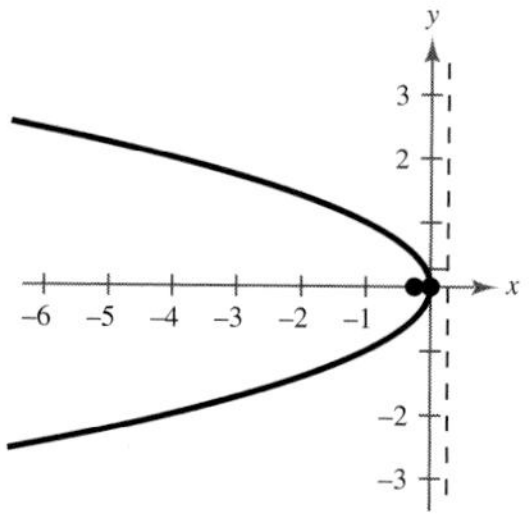

18. Vertex: $(-5, 1)$

Focus: $\left(-\frac{21}{4}, 1\right)$

Directrix: $x = -\frac{19}{4}$

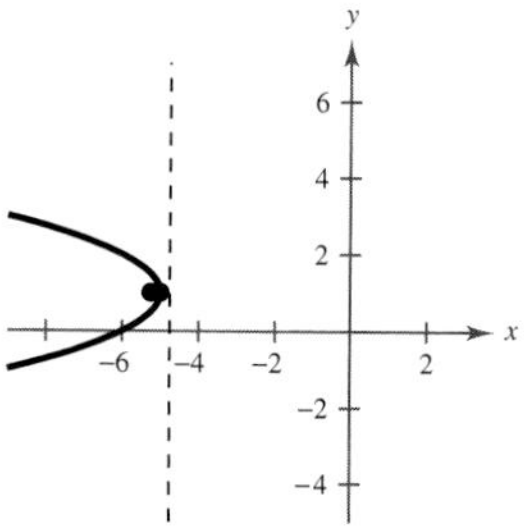

20. Vertex: $\left(-\frac{1}{2}, 1\right)$

Focus: $\left(-\frac{1}{2}, 2\right)$

Directrix: $y = 0$

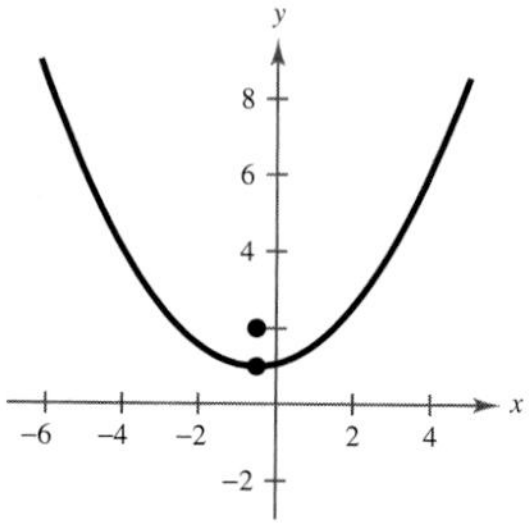

22. Vertex: $(8, -1)$

Focus: $(9, -1)$

Directrix: $x = 7$

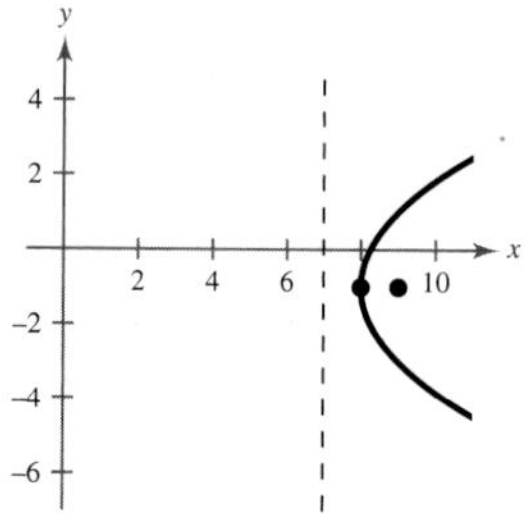

24. Vertex: $(-1, 2)$

Focus: $(0, 2)$

Directrix: $x = -2$

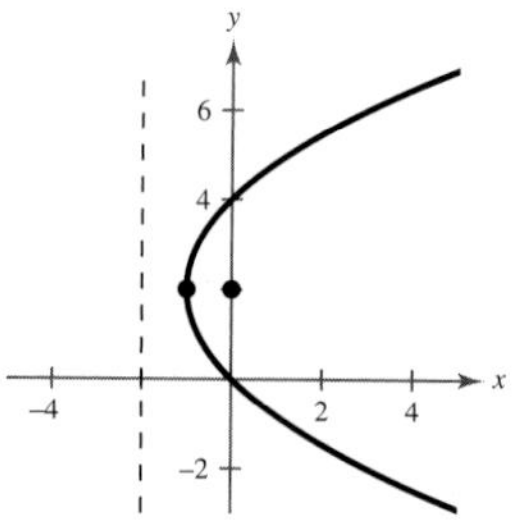

26. Vertex: $(1, -1)$

Focus: $(1, -3)$

Directrix: $y = 1$

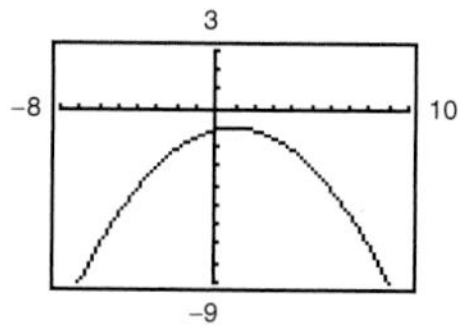

28. Vertex: $(-1, 0)$

Focus: $(0, 0)$

Directrix: $x = -2$

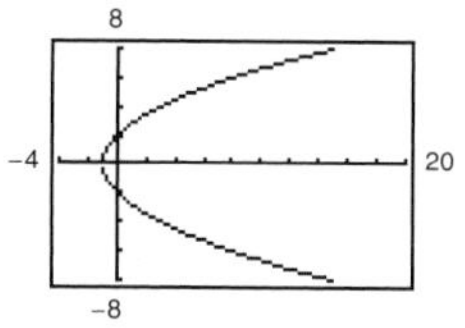

30.

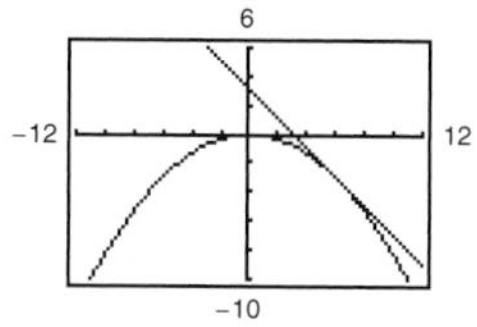

$(6, -3)$

32. $y^2 = -18x$ **34.** $y^2 = 8x$ **36.** $x^2 = -8y$

38. $x^2 = -12y$ **40.** $y^2 = 12x$ **42.** $x^2 = -3y$

44. $(y - 3)^2 = -2(x - 5)$ **46.** $(x - 3)^2 = 3(y + 3)$

48. $(x + 1)^2 = -8(y - 2)$ **50.** $(y - 1)^2 = -12(x + 2)$

52. $x^2 = -16(y - 4)$ **54.** $y = -\sqrt{2(x - 4)} - 1$

56. $6x + 2y + 9 = 0; \left(-\frac{3}{2}, 0\right)$

58. $8x + y - 8 = 0; (1, 0)$

60.

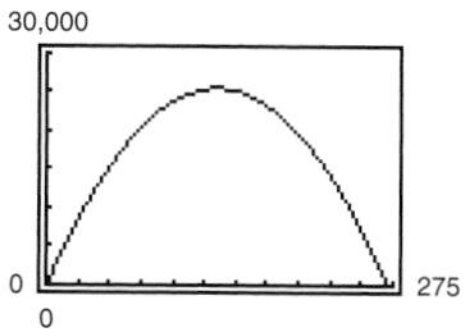

$x = 135$

62. (a)

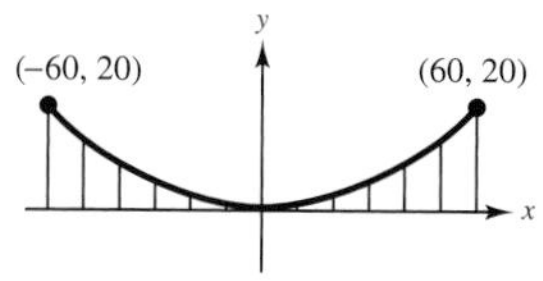

(b) $y = \dfrac{x^2}{180}$

(c)

x	0	20	40	60
y	0	$2\frac{2}{9}$	$8\frac{8}{9}$	20

64. $y^2 = 640x$

66. (a)

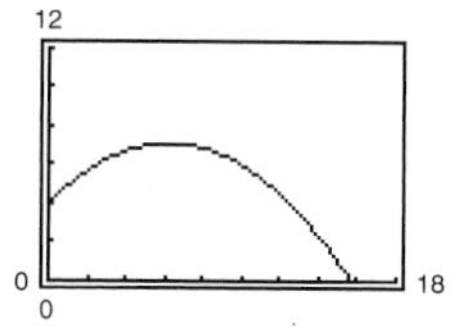

(b) Highest point: $(6.25, 7.125)$

Range: 15.69 feet

68. False. If the graph crossed the directrix there would exist points nearer the directrix than the focus.

70. (a)

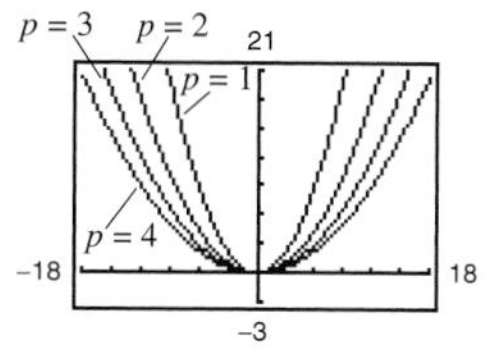

As p increases, the graph becomes wider.

(b) $(0, 1)$, $(0, 2)$, $(0, 3)$, $(0, 4)$

(c) 4, 8, 12, 16; $4p$

(d) Easy way to determine two additional points on the graph

72. $m = \dfrac{x_1}{2p}$

74. One or three positive real zeros, one negative real zero

76. Four, two, or zero positive real zeros, one negative real zero

78. $\pm\frac{1}{2}, \pm 1, \pm 2, \pm\frac{5}{2}, \pm 5, \pm 10$

80. $\pm\frac{1}{3}, \pm\frac{2}{3}, \pm 1, \pm 2, \pm\frac{11}{3}, \pm\frac{22}{3}, \pm 11, \pm 22$ **82.** $\frac{3}{2}, \pm 5i$

84.

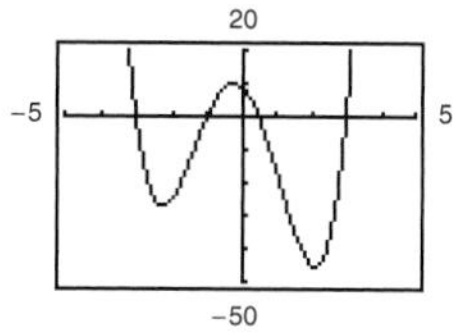

$\pm 3, -1, \frac{1}{2}$

Section 10.3 *(page 792)*

2. c **4.** f **6.** e

8. Center: $(0, 0)$

Vertices: $(0, \pm 12)$

Foci: $\left(0, \pm 3\sqrt{7}\right)$

Eccentricity: $\dfrac{\sqrt{7}}{4}$

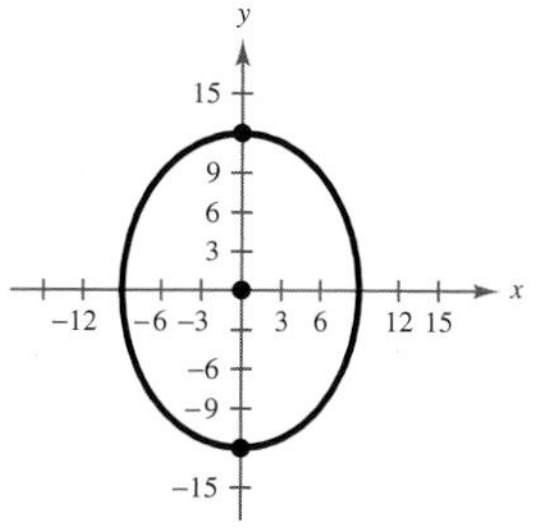

10. Center: $(0, 0)$

Vertices: $(\pm 8, 0)$

Foci: $(\pm 6, 0)$

Eccentricity: $\dfrac{3}{4}$

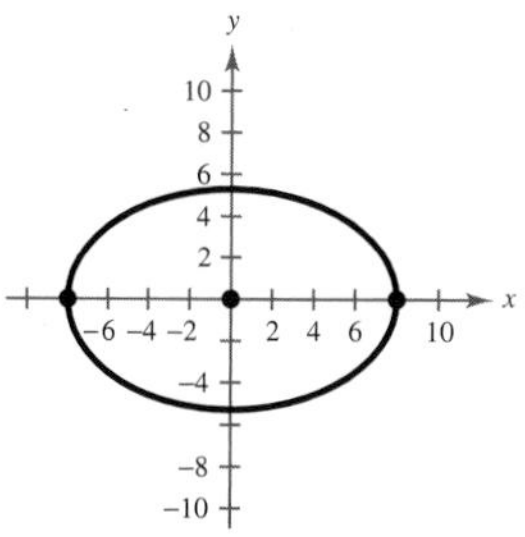

12. Center: $(4, -3)$

Vertices: $(4, 1), (4, -7)$

Foci: $(4, -1), (4, -5)$

Eccentricity: $\frac{1}{2}$

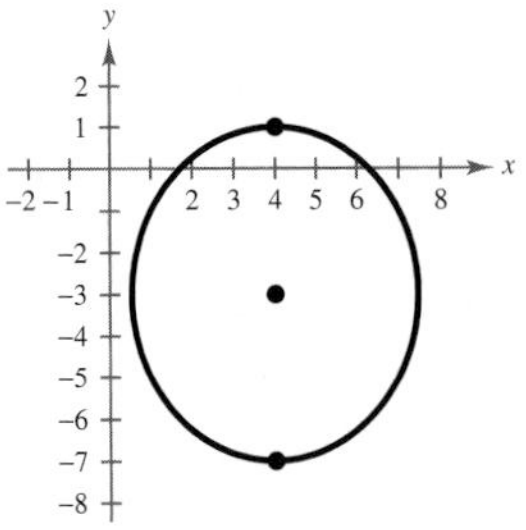

14. Center: $(-2, -4)$

Vertices: $(-3, -4), (-1, -4)$

Foci: $\left(\dfrac{-4 \pm \sqrt{3}}{2}, -4\right)$

Eccentricity: $\dfrac{\sqrt{3}}{2}$

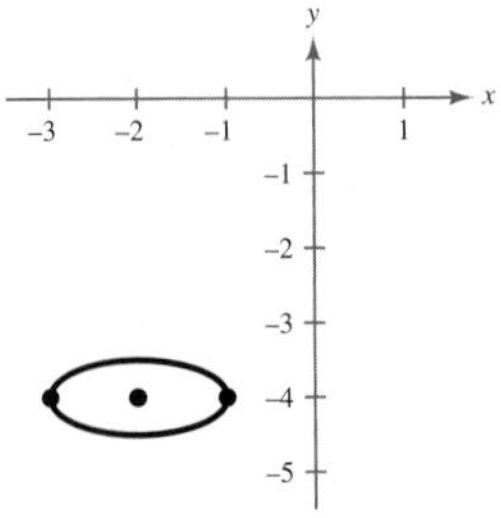

16. Center: $(3, -5)$

Vertices: $(3, 1), (3, -11)$

Foci: $(3, -5 \pm 2\sqrt{5})$

Eccentricity: $\dfrac{\sqrt{5}}{3}$

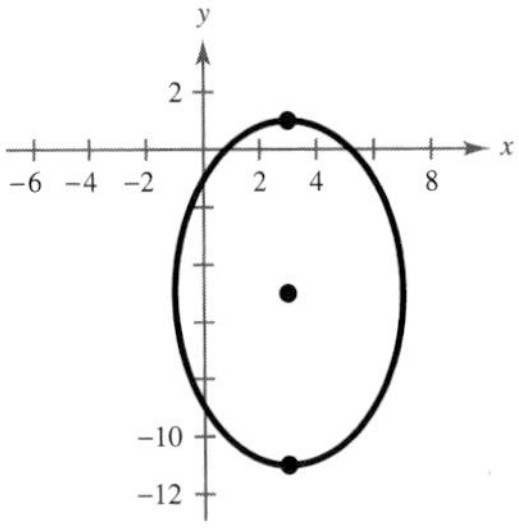

18. Center: $(-3, 1)$

Vertices: $(-3, 7), (-3, -5)$

Foci: $(-3, 1 \pm 2\sqrt{6})$

Eccentricity: $\dfrac{\sqrt{6}}{3}$

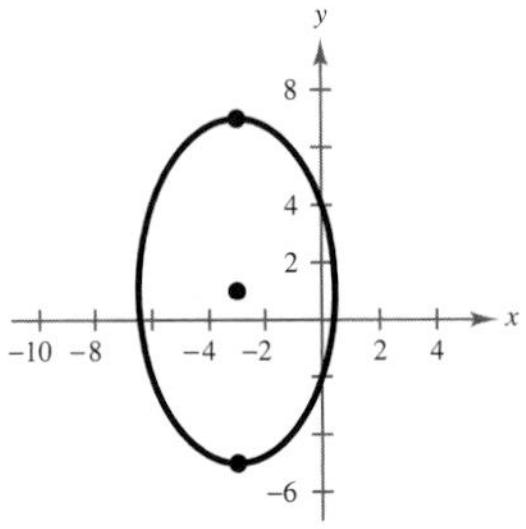

20. Center: $\left(3, -\dfrac{5}{2}\right)$

Vertices: $\left(9, -\dfrac{5}{2}\right), \left(-3, -\dfrac{5}{2}\right)$

Foci: $\left(3 \pm 3\sqrt{3}, -\dfrac{5}{2}\right)$

Eccentricity: $\dfrac{\sqrt{3}}{2}$

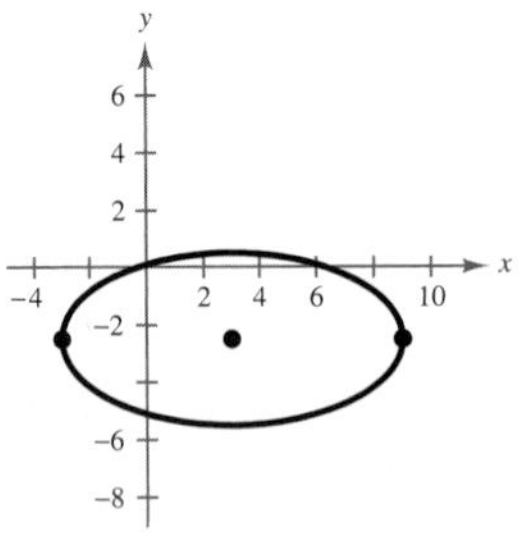

22. Center: $(2, 1)$

Vertices: $\left(\frac{7}{3}, 1\right), \left(\frac{5}{3}, 1\right)$

Foci: $\left(\frac{34}{15}, 1\right), \left(\frac{26}{15}, 1\right)$

Eccentricity: $\frac{4}{5}$

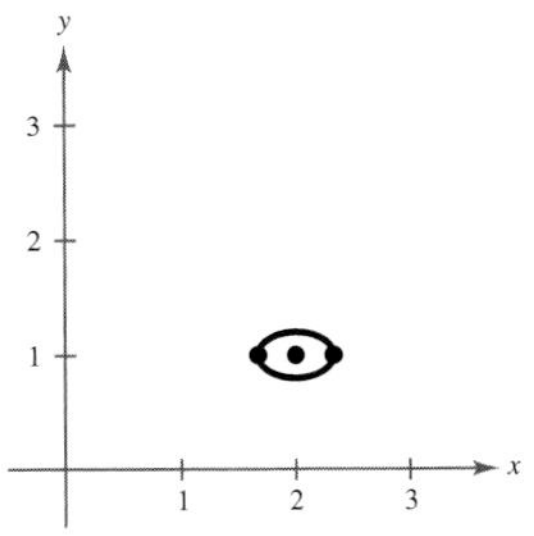

24.

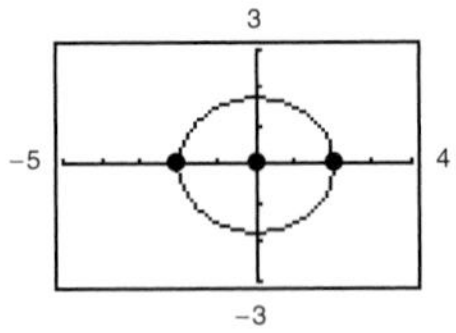

Center: $(0, 0)$

Vertices: $(\pm 2, 0)$

Foci: $(\pm 1, 0)$

26.

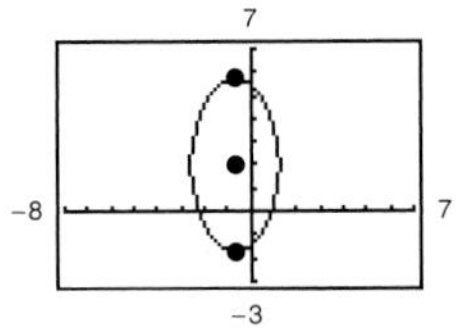

Center: $\left(-\dfrac{2}{3}, 2\right)$

Vertices: $\left(-\dfrac{2}{3}, 2 \pm \dfrac{2\sqrt{31}}{3}\right)$

Foci: $\left(-\dfrac{2}{3}, 2 \pm \dfrac{\sqrt{93}}{3}\right)$

28. $\dfrac{x^2}{4} + \dfrac{4y^2}{9} = 1$ **30.** $\dfrac{x^2}{48} + \dfrac{y^2}{64} = 1$

32. $\dfrac{x^2}{16} + \dfrac{y^2}{12} = 1$ **34.** $\dfrac{x^2}{4} + \dfrac{y^2}{16} = 1$

36. $\dfrac{(x-4)^2}{9} + \dfrac{y^2}{16} = 1$

38. $\dfrac{(x-2)^2}{4} + \dfrac{(y+1)^2}{1} = 1$

40. $\dfrac{(x-2)^2}{16} + \dfrac{y^2}{12} = 1$

42. $(x-2)^2 + \dfrac{4(y+1)^2}{9} = 1$

44. $\dfrac{(x-3)^2}{36} + \dfrac{(y-2)^2}{32} = 1$

46. $\dfrac{(x-5)^2}{16} + \dfrac{(y-6)^2}{36} = 1$ **48.** $\dfrac{x^2}{48} + \dfrac{y^2}{64} = 1$

50. (a)

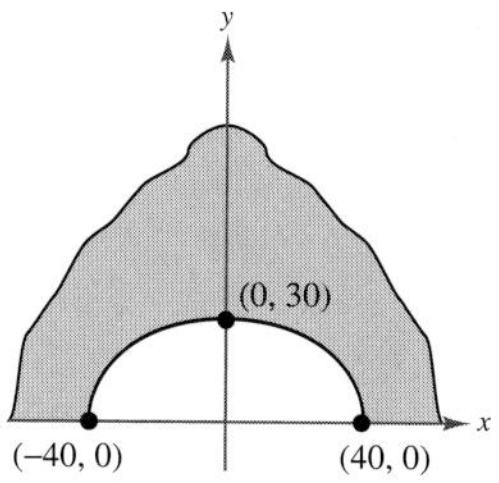

(b) $\frac{x^2}{1600} + \frac{y^2}{900} = 1$ (c) 14.5 feet

52. $\frac{x^2}{327.25} + \frac{y^2}{19.34} = 1$ **54.** $e \approx 0.052$

56.

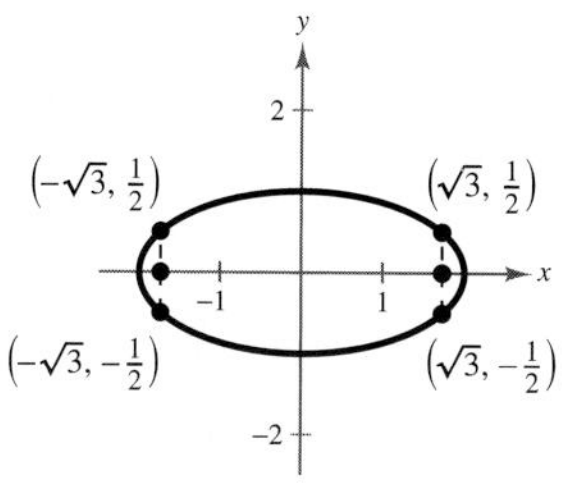

58.

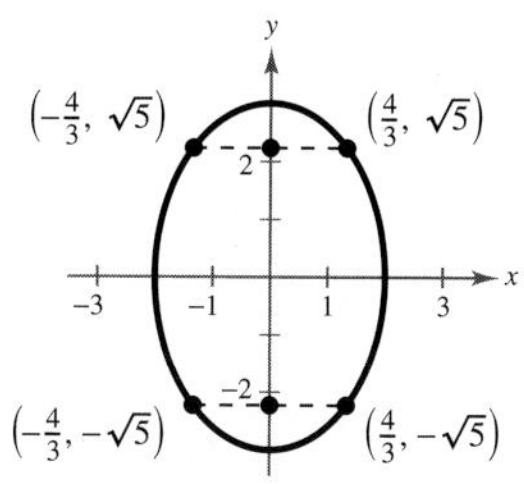

60. False. The equation of an ellipse is second degree in x and y.

62. True. The area of the circle is 16π. The area of the ellipse is $4\pi b$, where $b < 4$ because $2a = 8$ is the major axis length.

64. (a) $2a$

(b) The sum of the distances from the two fixed points is constant.

66. Arithmetic **68.** Geometric **70.** $a_n = 3n + 10$

72. $a_n = 3.5 + 1.5n$ **74.** 1093 **76.** 15.10

Section 10.4 *(page 802)*

2. c **4.** d

6. Center: $(0, 0)$

Vertices: $(\pm 3, 0)$

Foci: $(\pm\sqrt{34}, 0)$

Asymptotes: $y = \pm\frac{5}{3}x$

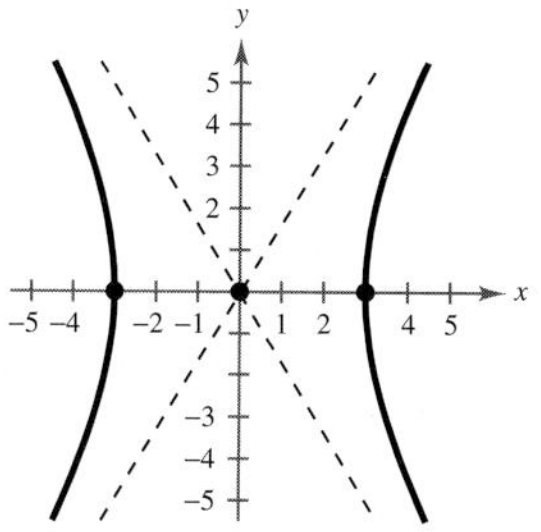

8. Center: $(0, 0)$

Vertices: $(\pm 6, 0)$

Foci: $(\pm 2\sqrt{10}, 0)$

Asymptotes: $y = \pm\frac{1}{3}x$

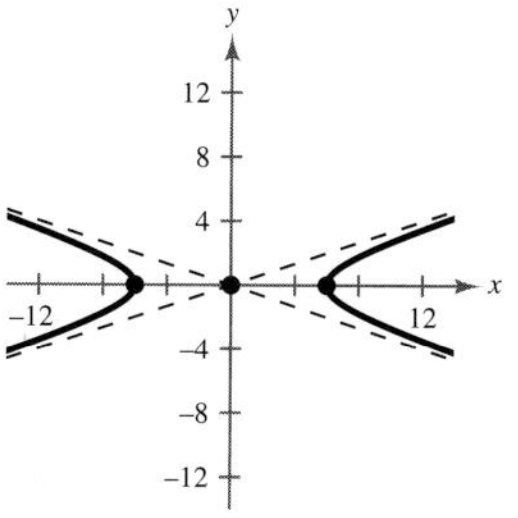

10. Center: $(-3, 2)$

Vertices: $(9, 2), (-15, 2)$

Foci: $(10, 2), (-16, 2)$

Asymptotes:

$y = 2 \pm \frac{5}{12}(x + 3)$

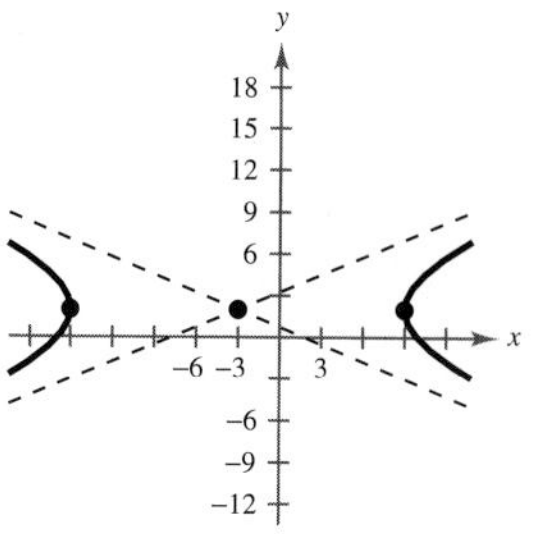

12. Center: $(-3, 1)$

Vertices: $\left(-3, \frac{3}{2}\right), \left(-3, \frac{1}{2}\right)$

Foci: $\left(-3, 1 \pm \frac{\sqrt{5}}{4}\right)$

Asymptotes:

$y = 1 \pm 2(x + 3)$

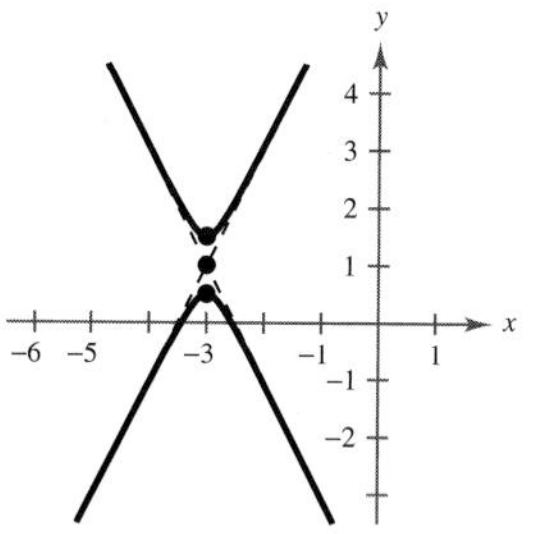

14. Center: $(0, 2)$

Vertices: $(-6, 2), (6, 2)$

Foci: $(\pm 2\sqrt{10}, 2)$

Asymptotes: $y = 2 \pm \frac{1}{3}x$

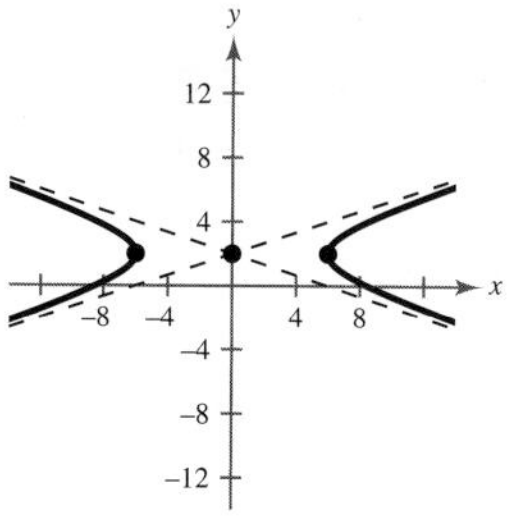

16. The graph of this equation is two lines intersecting at $(1, -2)$.

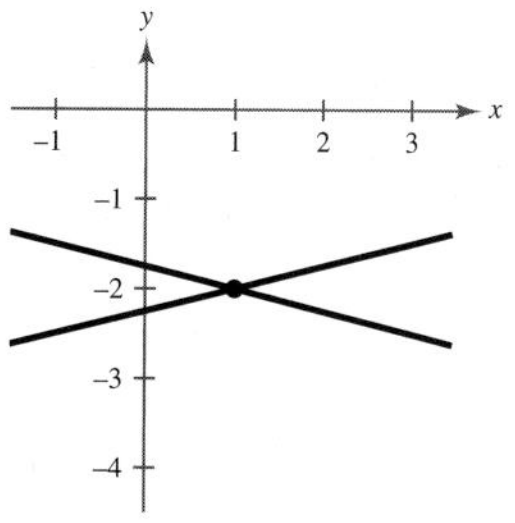

18. Center: $(0, 0)$

Vertices: $(0, \pm\sqrt{3})$

Foci: $(0, \pm 3)$

Asymptotes: $y = \pm\frac{\sqrt{2}}{2}x$

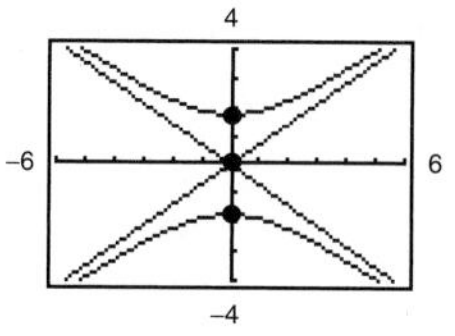

20. Center: $(-3, 5)$

Vertices: $\left(-\frac{10}{3}, 5\right), \left(-\frac{8}{3}, 5\right)$

Foci: $\left(-3 \pm \frac{\sqrt{10}}{3}, 5\right)$

Asymptotes: $y = 5 \pm 3(x + 3)$

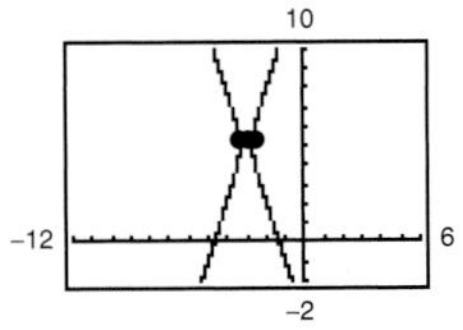

22. $\frac{x^2}{16} - \frac{y^2}{20} = 1$ **24.** $\frac{y^2}{9} - x^2 = 1$

26. $\frac{x^2}{64} - \frac{y^2}{36} = 1$ **28.** $\frac{y^2}{9} - \frac{(x-2)^2}{27} = 1$

30. $\frac{x^2}{4} - \frac{(y-1)^2}{5} = 1$ **32.** $\frac{x^2}{4} - \frac{7(y-1)^2}{12} = 1$

34. $\frac{y^2}{4} - \frac{(x-1)^2}{4} = 1$ **36.** $\frac{(y-3)^2}{9} - \frac{(x-3)^2}{9} = 1$

38. $\frac{(y-2)^2}{4} - \frac{(x-3)^2}{9} = 1$ **40.** $x \approx 110.3$ miles

42. Circle **44.** Hyperbola **46.** Ellipse

48. Parabola

50. True. For a hyperbola, $c^2 = a^2 + b^2$. The larger the ratio of b to a, the larger the eccentricity of the hyperbola, $e = c/a$.

52. Answers will vary. **54.** $x^3 + x^2 + 2x - 6$

56. $x^2 - 2x + 1 + \frac{2}{x+2}$ **58.** $x(x + 4)(x - 4)$

60. $2x(x - 6)^2$ **62.** $2(2x + 3)(4x^2 - 6x + 9)$

Section 10.5 *(page 810)*

2. $(3\sqrt{2}, 0)$ **4.** $\left(\frac{3+\sqrt{3}}{2}, \frac{1-3\sqrt{3}}{2}\right)$

6. $(\sqrt{3} + 2, 2\sqrt{3} - 1)$

8. $\frac{(x')^2}{4} - \frac{(y')^2}{4} = 1$

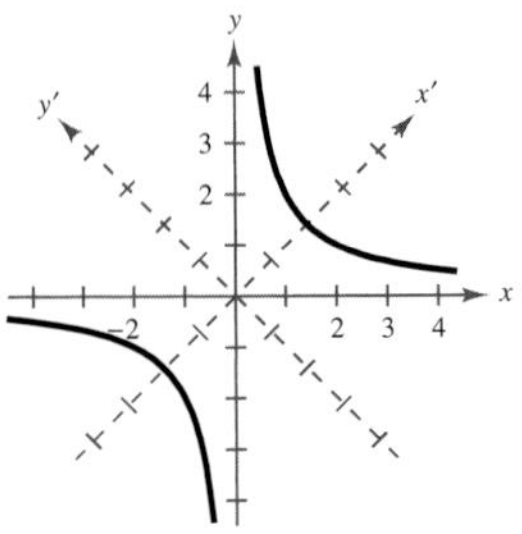

10. $\frac{[y' + (3\sqrt{2}/2)]^2}{10} - \frac{[x' - (\sqrt{2}/2)]^2}{10} = 1$

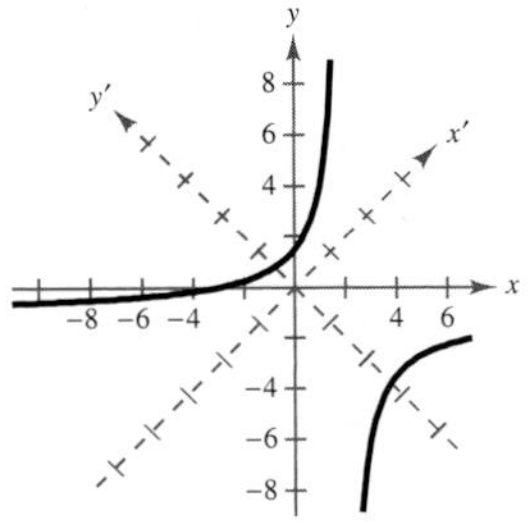

12. $\frac{(x')^2}{4} - \frac{(y')^2}{4} = 1$

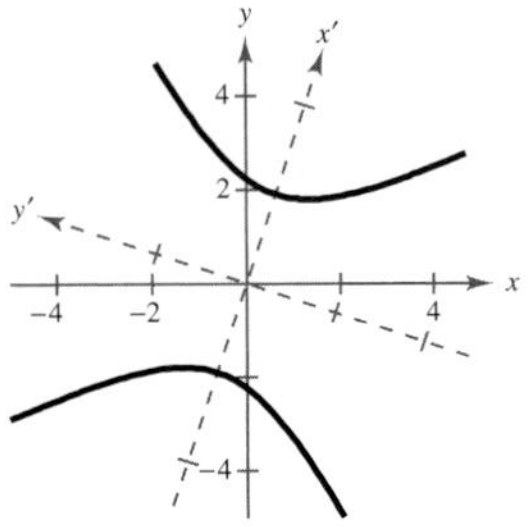

14. $(x')^2 + \frac{(y')^2}{4} = 1$

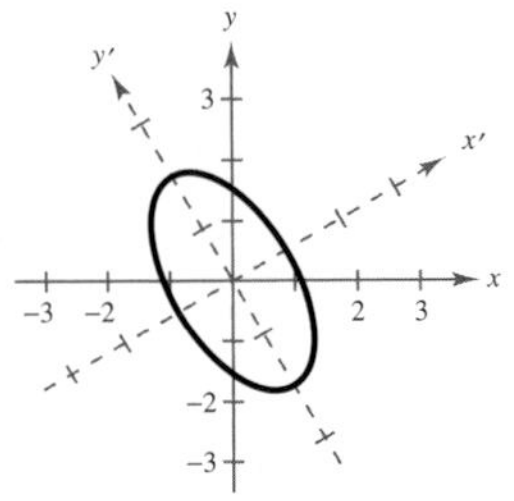

16. $(y')^2 = 4(x' - 1)$

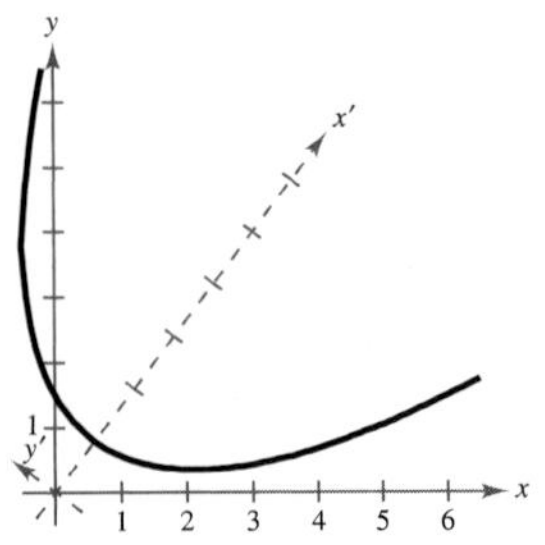

18. $y' = \frac{1}{4}(x')^2$

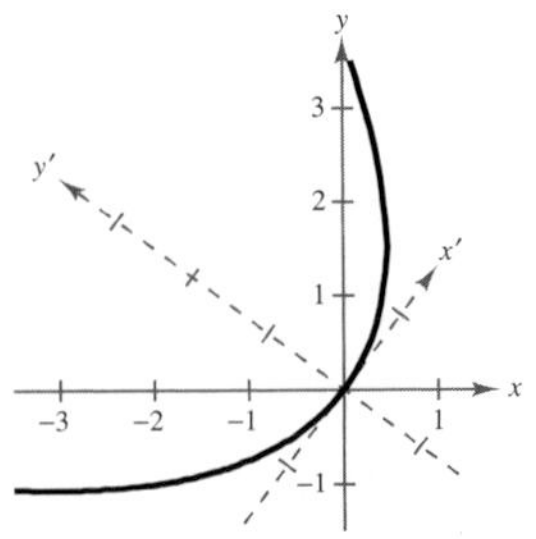

20.

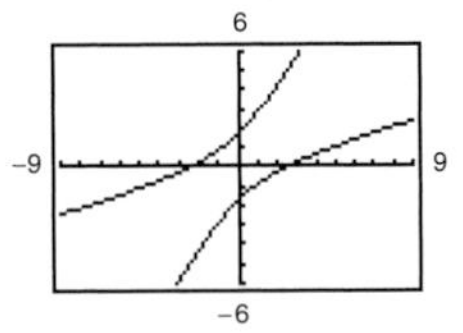

$\theta = 37.98°$

22.

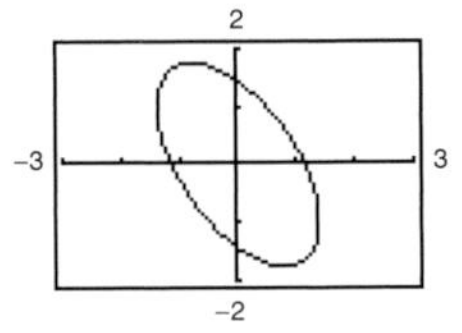

$\theta = 33.69°$

24.

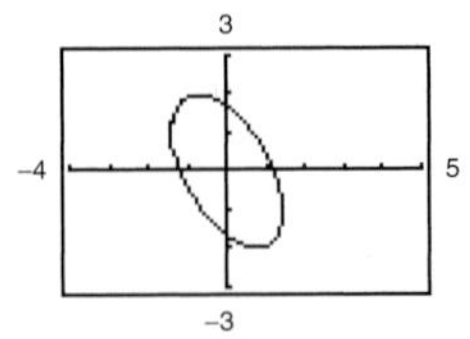

$\theta = 28.15°$

26.

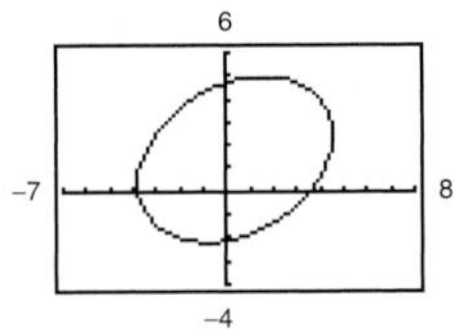

$\theta = 31.72°$

28. f **30.** a **32.** c

34. Hyperbola

$$y = \frac{4x \pm \sqrt{16x^2 + 8(x^2 - 6)}}{-4}$$

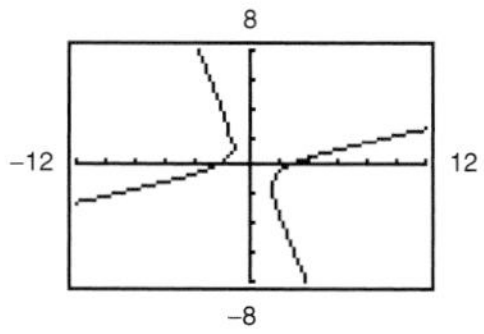

36. Ellipse

$$y = \frac{-(4x - 4) \pm \sqrt{(4x - 4)^2 - 20(2x^2 + 3x - 20)}}{10}$$

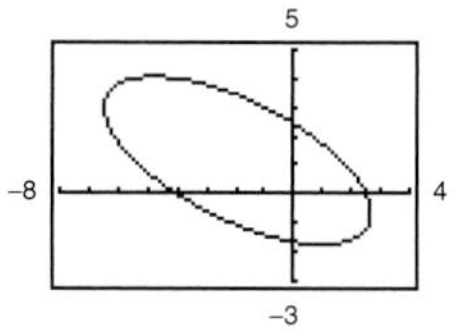

38. Parabola

$$y = \frac{60x - 9 \pm \sqrt{(60x - 9)^2 - 3600x^2}}{50}$$

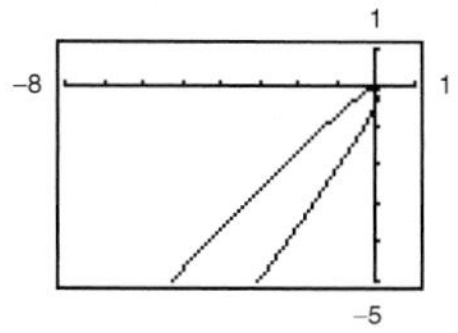

40. Ellipse

$$y = \frac{-(x + 1) \pm \sqrt{(x + 1)^2 - 16(x^2 + x - 4)}}{8}$$

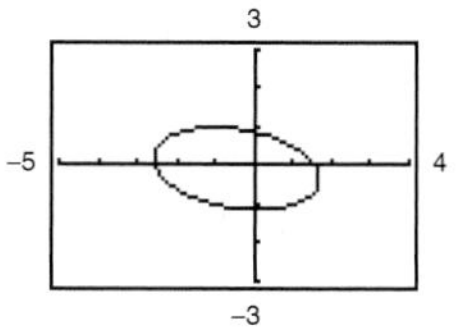

42.

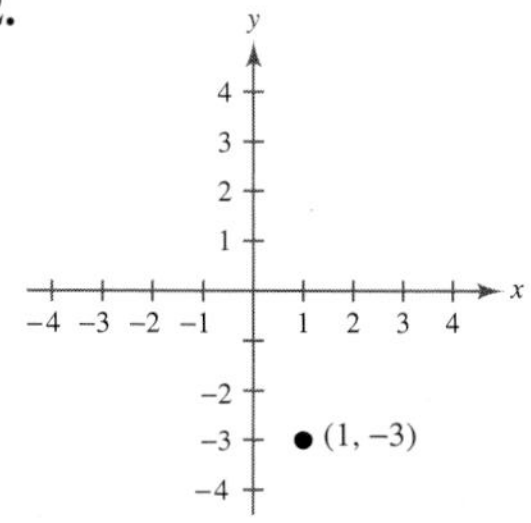

44.

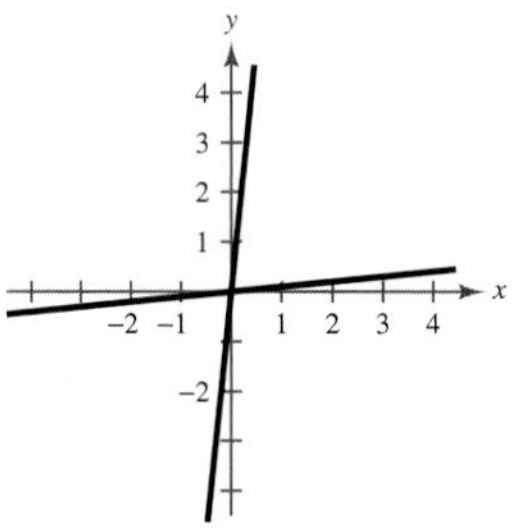

46. $(-7, 0), (-1, 0)$ **48.** $(14, -8), (6, -8)$

50. $(1, 0)$ **52.** $(-2, 8), \left(5, 8 \pm 4\sqrt{21}\right)$

54. $(\pm 3, 2)$ **56.** $(1, 1), (3, 1)$

58. $\left(\frac{1}{6}(3 + \sqrt{30}), \frac{1}{6}(3 - \sqrt{30})\right), \left(\frac{1}{6}(3 - \sqrt{30}), \frac{1}{6}(3 + \sqrt{30})\right)$

60. False. The coefficients of the new equation after it has been rotated are obtained by making the substitutions $x = x'\cos\theta - y'\sin\theta$ and $y = x'\sin\theta + y'\cos\theta$.

62. Major axis: 4; Minor axis: 2

64. No zeros **66.** 3 **68.** ± 1

70.

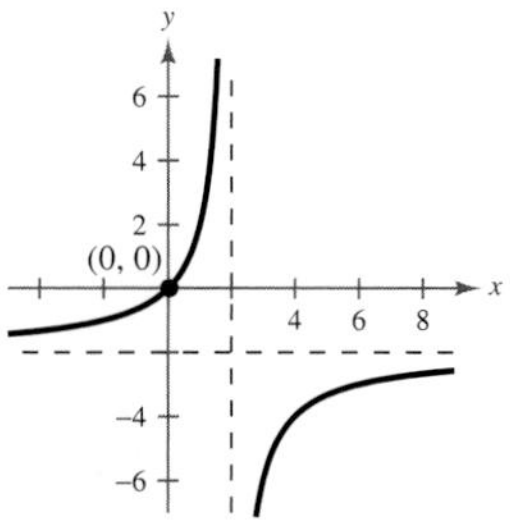

Intercept: $(0, 0)$

Asymptotes: $x = 2, y = -2$

72.

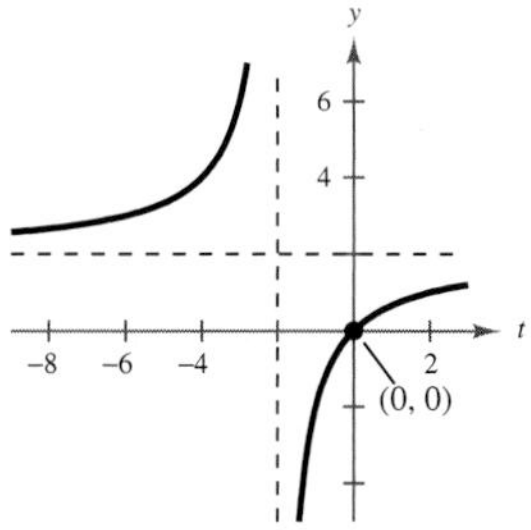

Intercept: $(0, 0)$

Asymptotes: $t = -2, y = 2$

74.

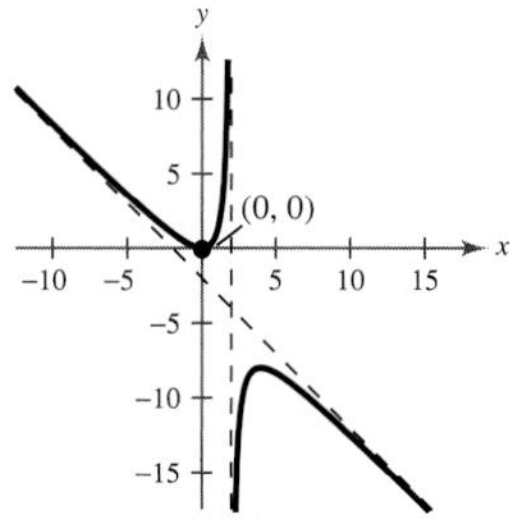

Intercept: $(0, 0)$

Asymptotes: $x = 2, y = -x - 2$

76.

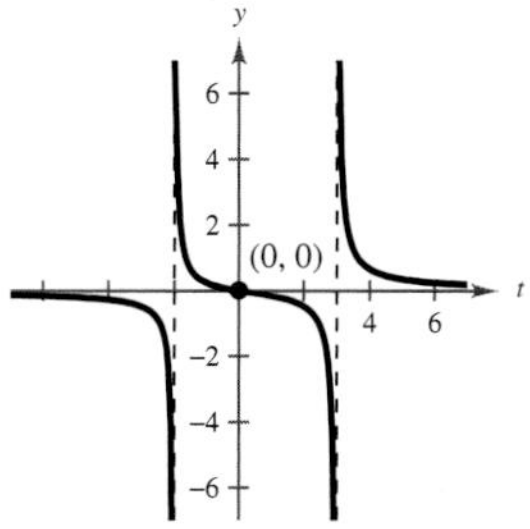

Intercept: $(0, 0)$

Asymptotes: $t = -2, t = 3, y = 0$

Section 10.6 *(page 817)*

2. (a)

t	$-\frac{\pi}{2}$	$-\frac{\pi}{4}$	0	$\frac{\pi}{4}$	$\frac{\pi}{2}$
x	0	2	4	2	0
y	-2	$-\sqrt{2}$	0	$\sqrt{2}$	2

(b)

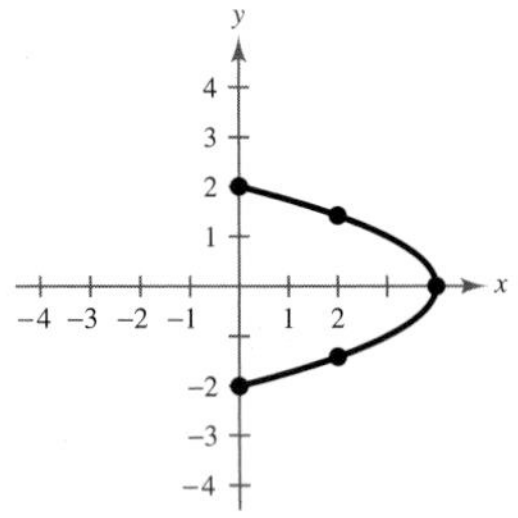

(c) $x = -y^2 + 4$

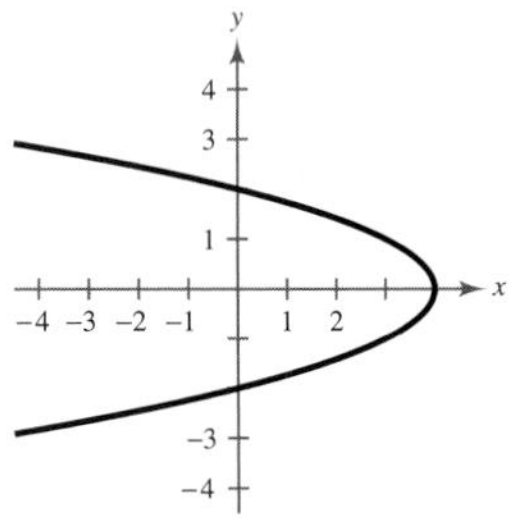

The graph of the rectangular equation continues the graph into the second and third quadrants.

4.

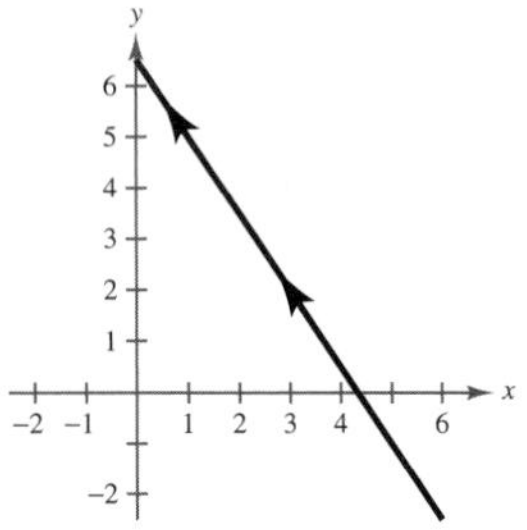

6.

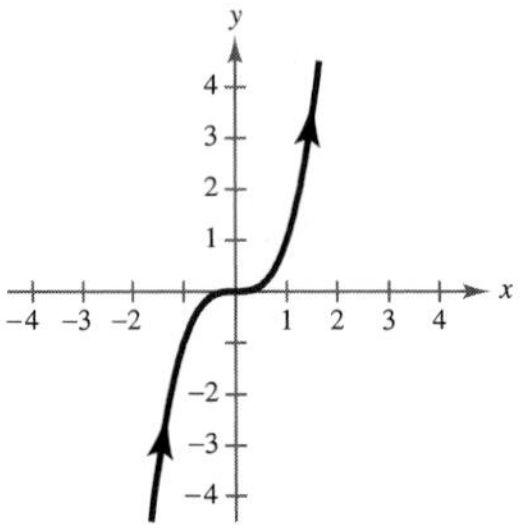

8.

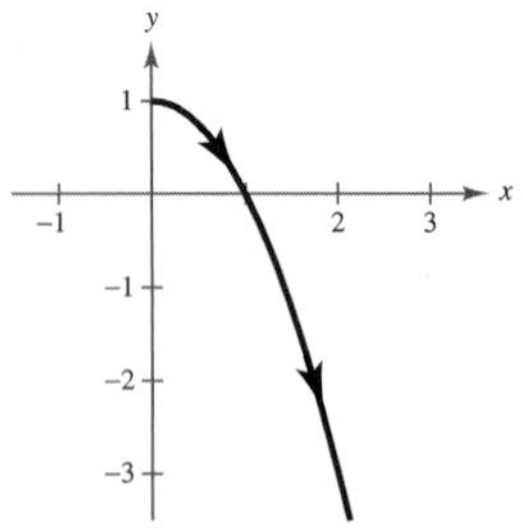

10.

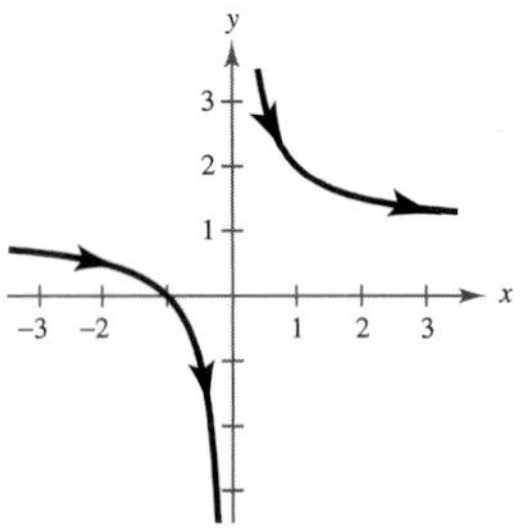

12.

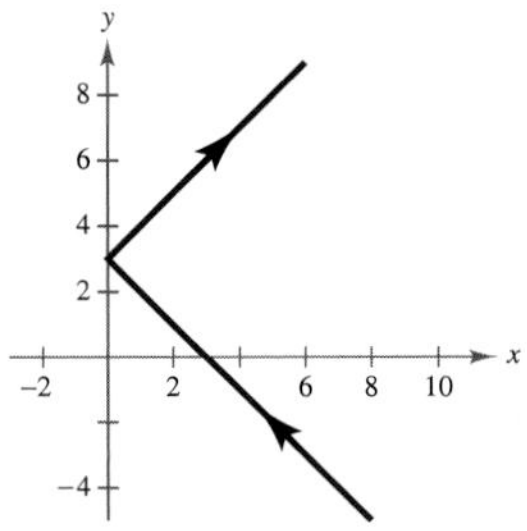

14.

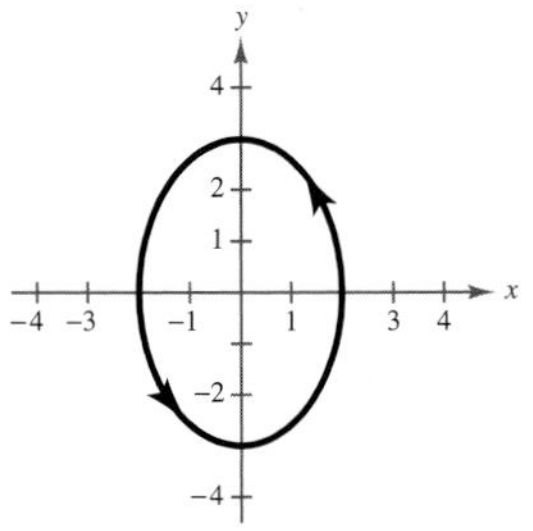

16.

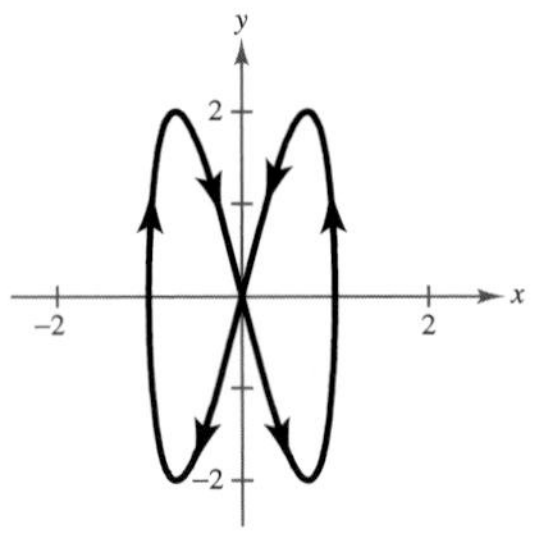

18.

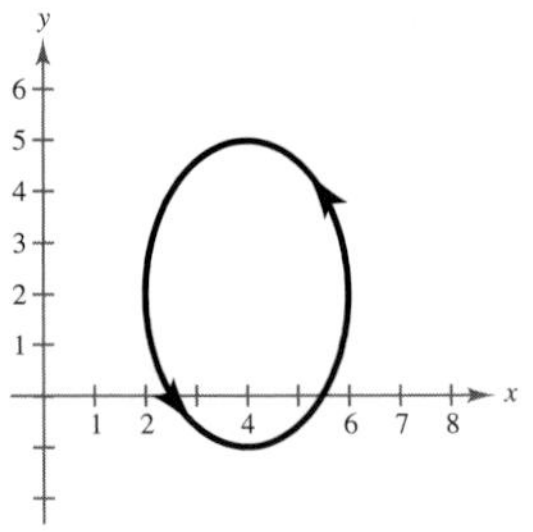

20.

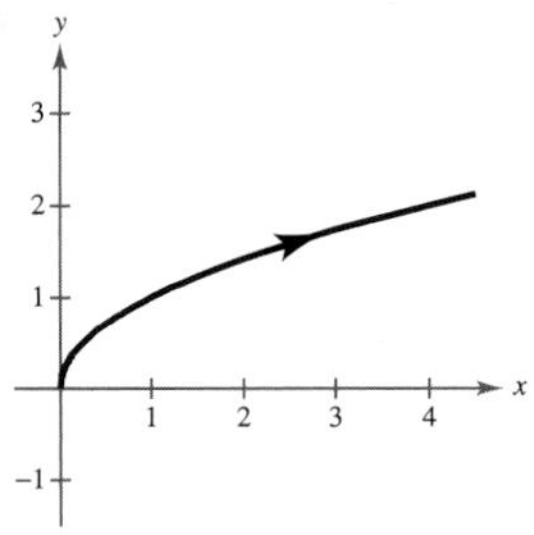

22.

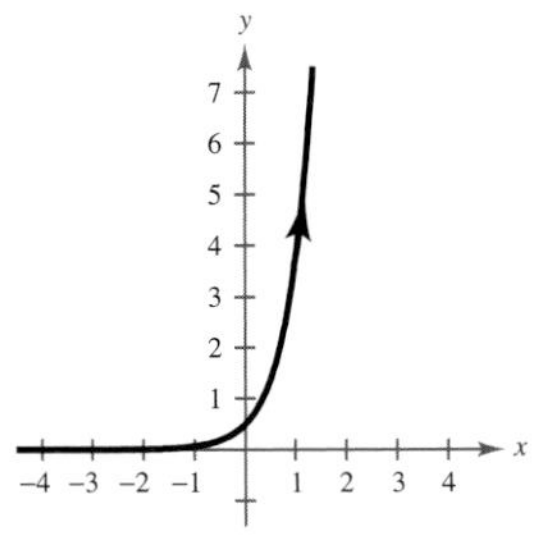

24. Each curve represents a portion of the parabola $y = x^2 - 1$.

	Domain	*Orientation*
(a)	$(-\infty, \infty)$	Left to right
(b)	$[0, \infty)$	Depends on t
(c)	$[-1, 1]$	Depends on t
(d)	$(0, \infty)$	Left to right

26. $(x - h)^2 + (y - k)^2 = r^2$

28. $\dfrac{(x - h)^2}{a^2} - \dfrac{(y - k)^2}{b^2} = 1$

30. $x = 2 + 4t$, $y = 3 - 6t$

32. $x = -3 + 5\cos\theta$, $y = 2 + 5\sin\theta$

34. $x = 4 + 5\cos\theta$, $y = 2 + 4\sin\theta$

36. $x = 2\sqrt{3}\tan\theta$, $y = 2\sec\theta$

38. (a) $x = t,\ y = \frac{1}{3}(t + 2)$

(b) $x = -t + 2,\ y = -\frac{1}{3}(t - 4)$

40. (a) $x = t,\ y = t^3$

(b) $x = -t + 2,\ y = (-t + 2)^3$

42. (a) $x = t,\ y = 2 - t$

(b) $x = -t + 2,\ y = t$

44. (a) $x = t, y = \dfrac{1}{2t}$

(b) $x = -t + 2, y = \dfrac{1}{-2t + 4}$

46.

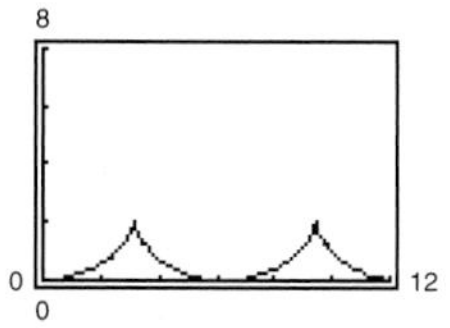

48.

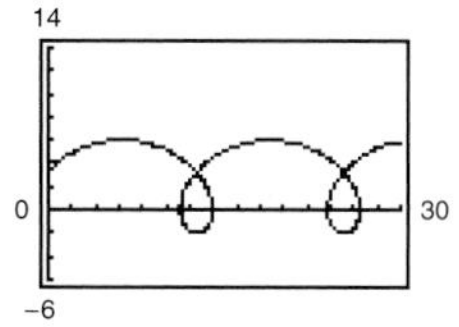

50.

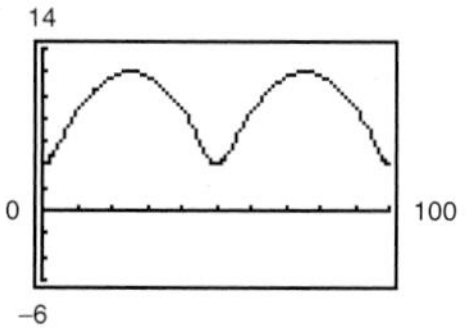

52.

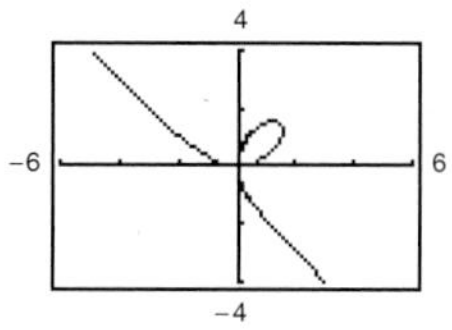

54. c

Domain: $[-4, 4]$

Range: $[-6, 6]$

56. a

Domain: $(-\infty, \infty)$

Range: $[-2, 2]$

58. (a)

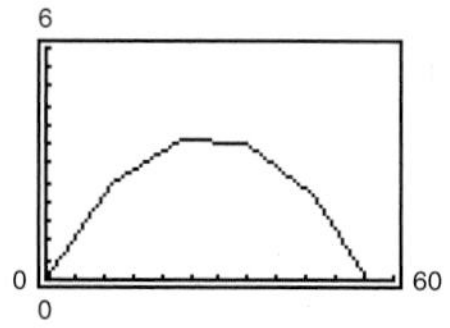

Maximum height: 3.8 feet

Range: 56.3 feet

(b)

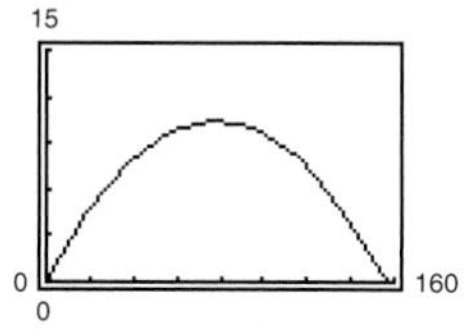

Maximum height: 10.5 feet

Range: 156.3 feet

(c)

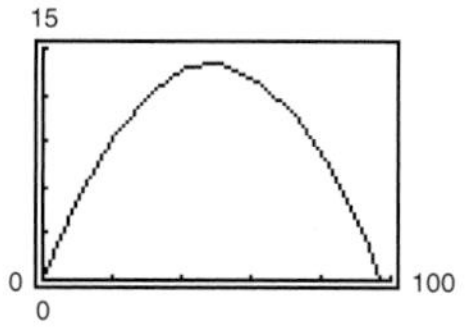

Maximum height: 14.1 feet

Range: 97.4 feet

(d)

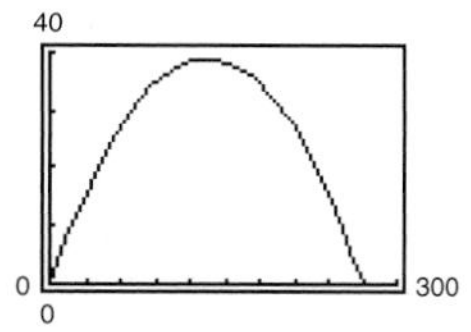

Maximum height: 39.1 feet

Range: 270.6 feet

60. (a) $x = (240 \cos 10°)t$

$y = 5 + (240 \sin 10°)t - 16t^2$

(b) 643 feet

(c)

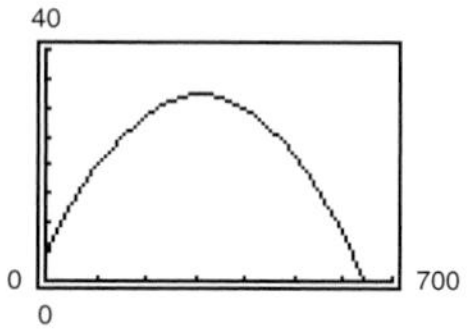

32.1 feet

(d) 2.72 seconds

62. (a) $h = 7, v_0 = 40, \theta = 45°$

$x = (40 \cos 45°)t$

$y = 7 + (40 \sin 45°)t - 16t^2$

(b)

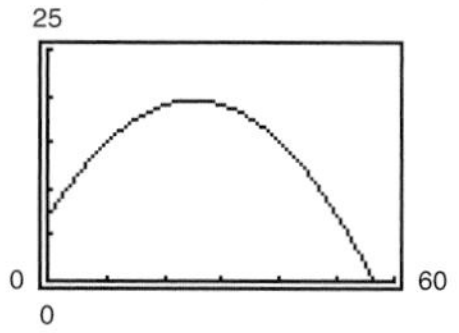

(c) Maximum height: 19.5 feet

Range: 56.2 feet

64. $x = 3 \cos \theta - \cos 3\theta$

$y = 3 \sin \theta - \sin 3\theta$

66. False. $y = x$ for $x \ge 0$

68. $(-2, 3)$ **70.** $(3, 1, -2)$

72. $y = -x^2 + 4x - 6$ **74.** $y = 2x^2 + 28x + 90$

Section 10.7 *(page 824)*

2.

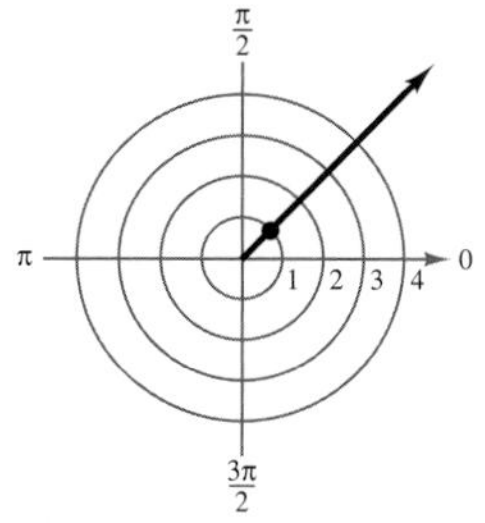

$\left(-1, \dfrac{5\pi}{4}\right), \left(1, \dfrac{\pi}{4}\right)$

4.

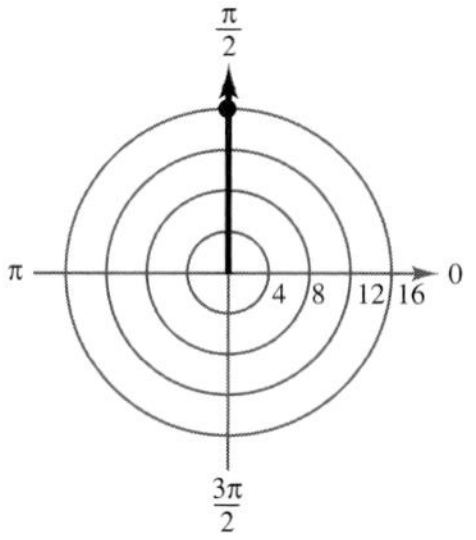

$\left(16, \dfrac{\pi}{2}\right), \left(-16, \dfrac{3\pi}{2}\right)$

6.

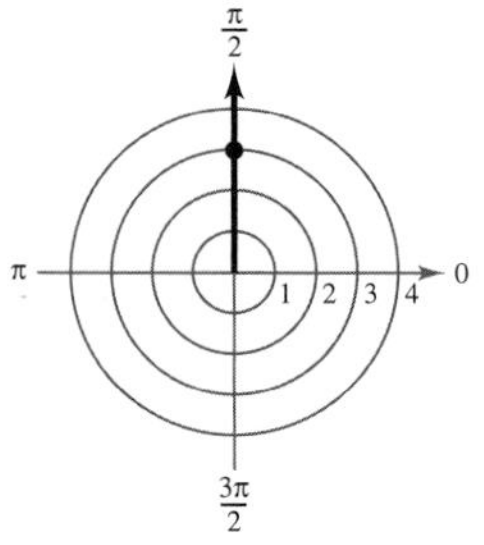

$(-3, 4.7132), (3, 1.5716)$

8.

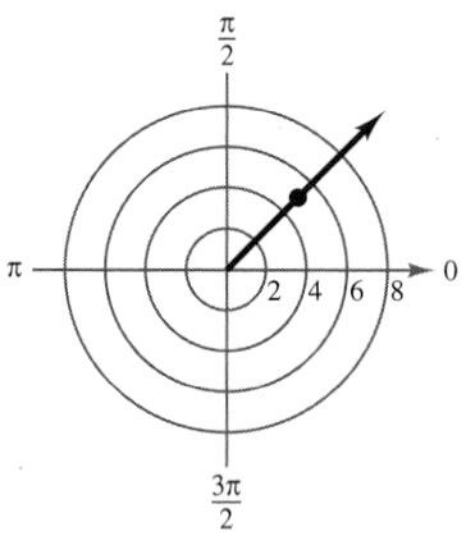

$(-5, 3.9232), (5, 0.7816)$

10. $(0, -3)$ **12.** $(0, 0)$ **14.** $(\sqrt{3}, 1)$

16. $(-7.7258, -2.8940)$ **18.** $\left(3\sqrt{2}, \dfrac{5\pi}{4}\right)$

20. $\left(5, \dfrac{3\pi}{2}\right)$ **22.** $(\sqrt{10}, 5.9614)$ **24.** $\left(2, \dfrac{11\pi}{6}\right)$

26. $(13, 1.1760)$ **28.** $(\sqrt{29}, 2.7611)$ **30.** $\left(6, \dfrac{\pi}{4}\right)$

32. $(2.3049, 0.7086)$ **34.** $r = 4$ **36.** $r = 10 \sec \theta$

38. $r = \dfrac{2}{3 \cos \theta + 5 \sin \theta}$ **40.** $\theta = \dfrac{\pi}{4}$

42. $r^2 = 9 \cos 2\theta$ **44.** $r = 3a$ **46.** $r = 4a \sec \theta$

48. $r = 2a \sin \theta$ **50.** $x^2 + y^2 - 3x = 0$

52. $x^2 + y^2 = 16$ **54.** $(x^2 + y^2)^2 = 2xy$

56. $y^2 = 2x + 1$ **58.** $2x - 3y = 6$

60. $x^2 + y^2 = 64$

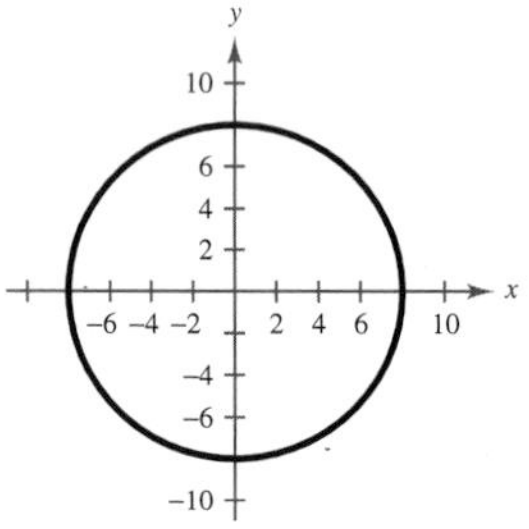

62. $x + y = 0$

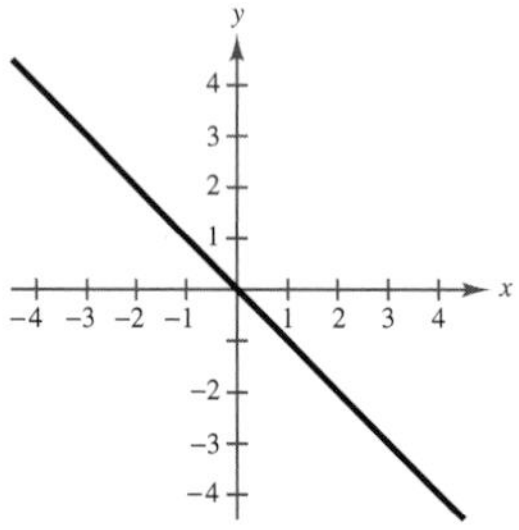

64. $y - 2 = 0$

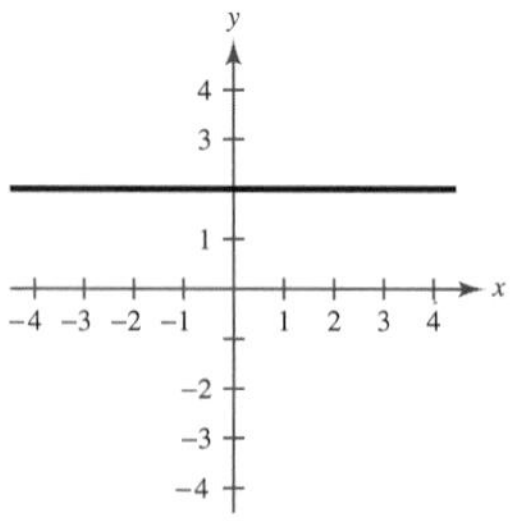

66. False. If $r_1 = -r_2$, then (r_1, θ) and (r_2, θ) are different points.

68. $\left(x - \frac{1}{2}\right)^2 + \left(y - \frac{3}{2}\right)^2 = \frac{5}{2}$; circle

70. (a) Horizontal: x-coordinate changes
Vertical: y-coordinate changes
(b) Horizontal: r and θ both change
Vertical: r and θ both change
(c) Unlike r and θ, x and y measure horizontal and vertical changes, respectively.

72. $\left(-\frac{5}{26}, \frac{55}{26}\right)$ **74.** $\left(\frac{295}{89}, \frac{844}{89}, -\frac{672}{89}\right)$ **76.** $\left(-2, \frac{3}{2}, 1, 4\right)$

78. Collinear **80.** Not collinear

Section 10.8 *(page 832)*

2. Cardioid **4.** Lemniscate **6.** Circle

8. Polar axis **10.** Polar axis **12.** Pole

14. Maximum: $|r| = 18$ when $\theta = 0$

Zero: $r = 0$ when $\theta = \dfrac{2\pi}{3}, \dfrac{4\pi}{3}$

16. Maximum: $|r| = 3$ when $\theta = \dfrac{\pi}{4}, \dfrac{3\pi}{4}, \dfrac{5\pi}{4}, \dfrac{7\pi}{4}$

Zero: $r = 0$ when $\theta = 0, \dfrac{\pi}{2}, \pi, \dfrac{3\pi}{2}$

18.

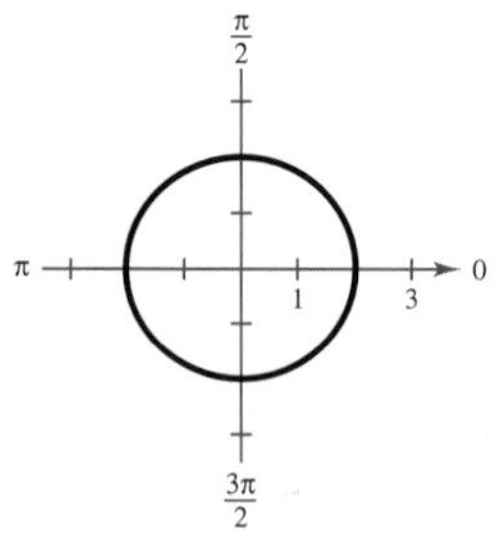

20.

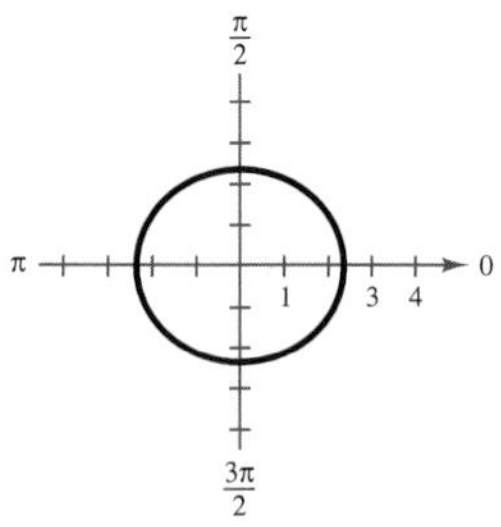

22.

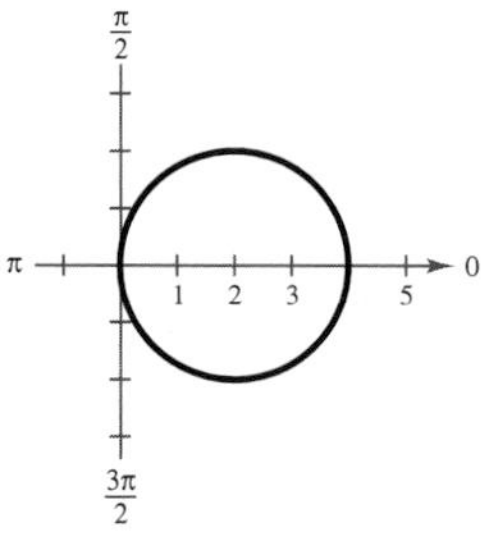

24.

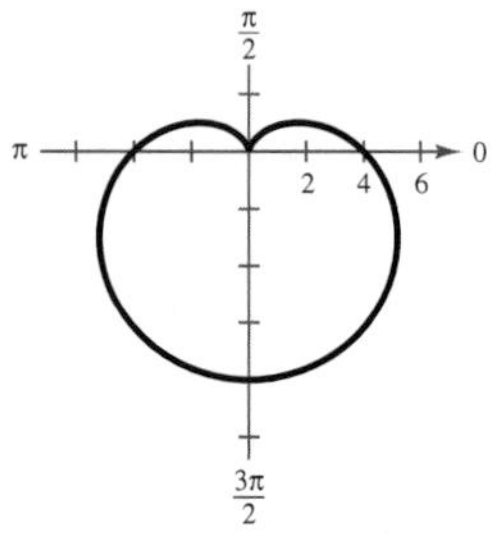

26.

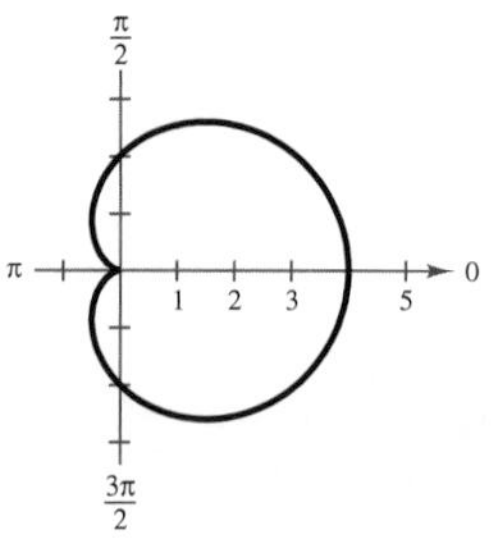

28.

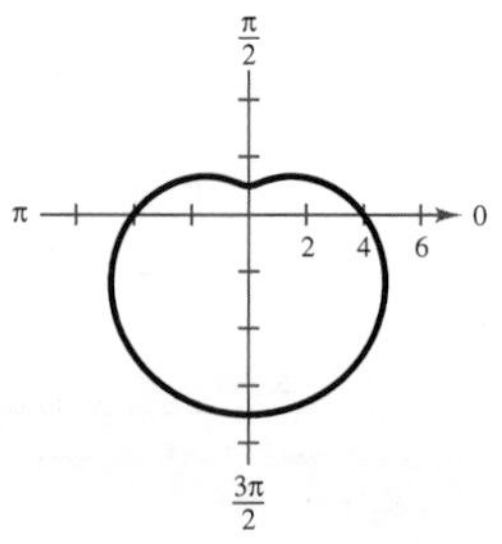

30.

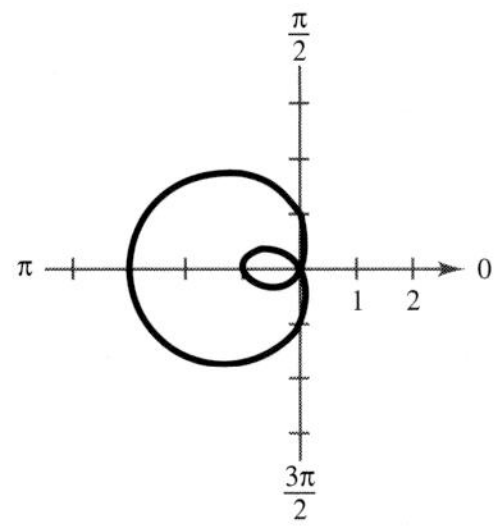

32.

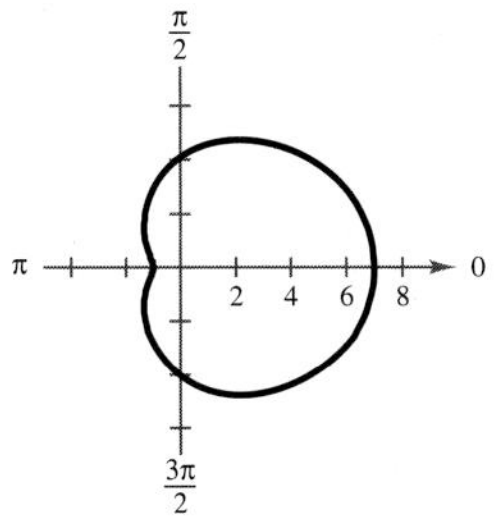

34.

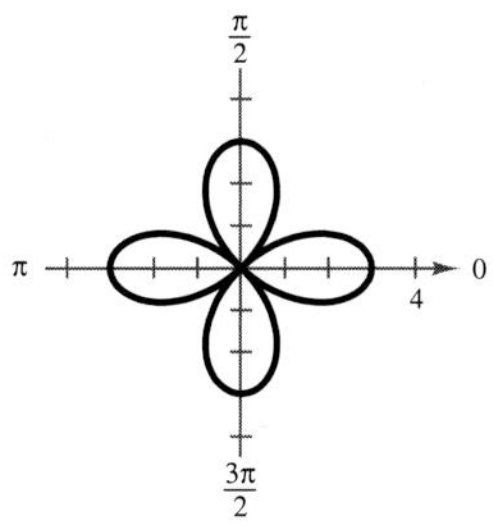

36.

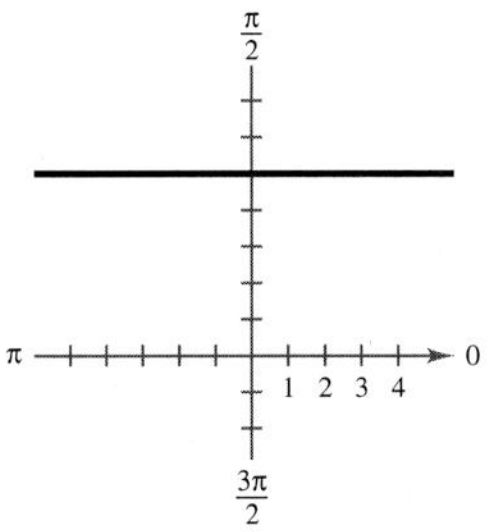

38.

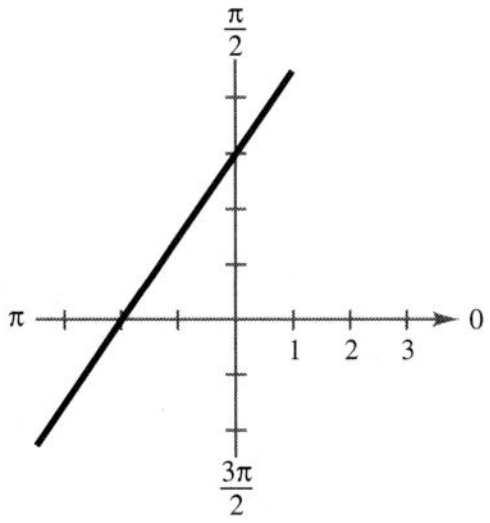

40.

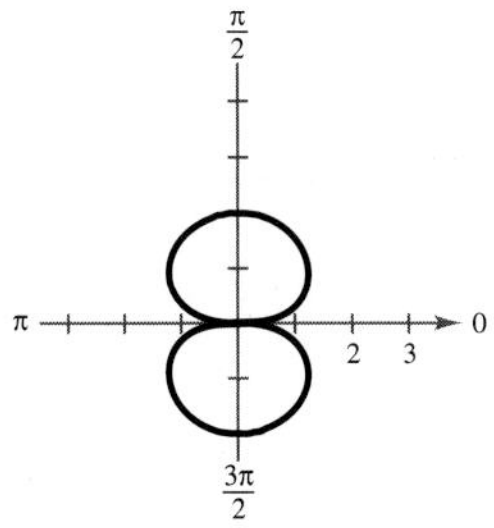

42.

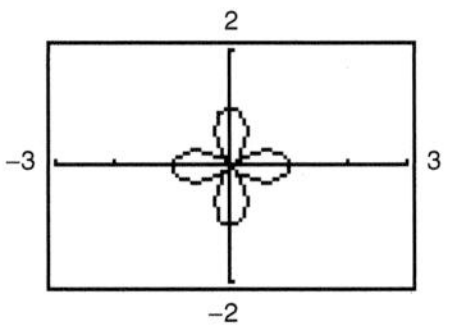

44.

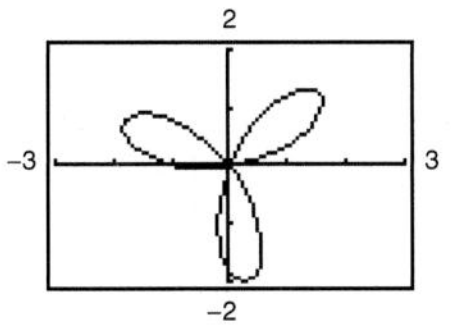

46.

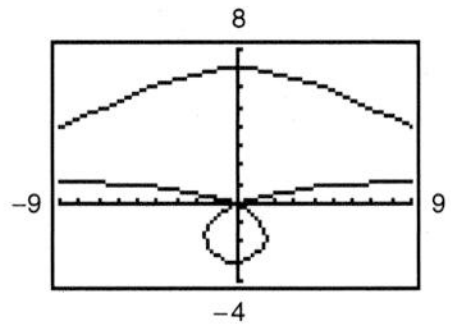

48.

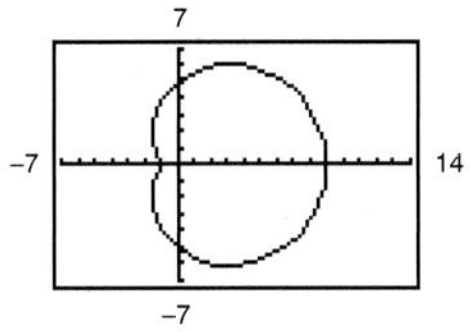

$0 \le \theta < 2\pi$

50.

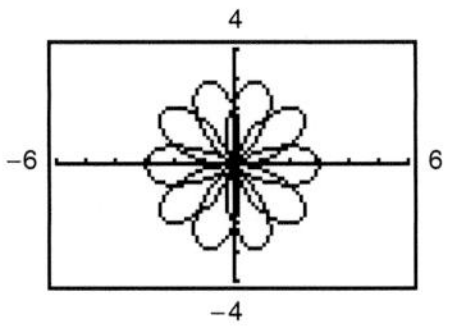

$0 \le \theta < 4\pi$

52.

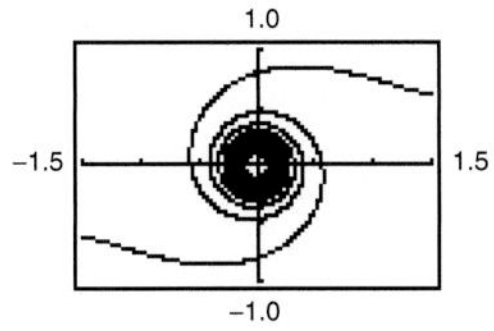

$0 \le \theta < \infty$

54.

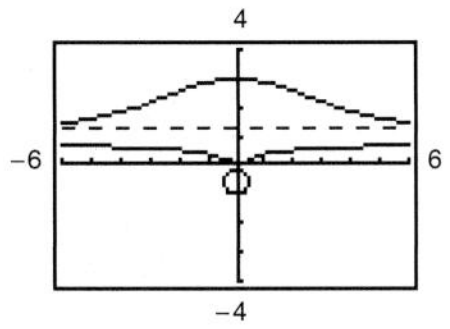

56.

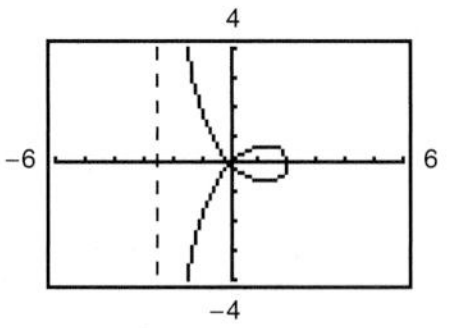

58. False. For a graph symmetric with respect to the pole, one portion of the graph coincides with the other portion when rotated π radians about the pole.

60. (a)

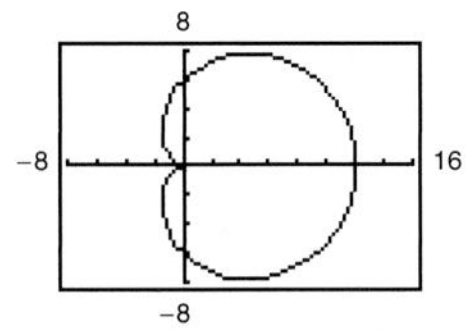

(b)

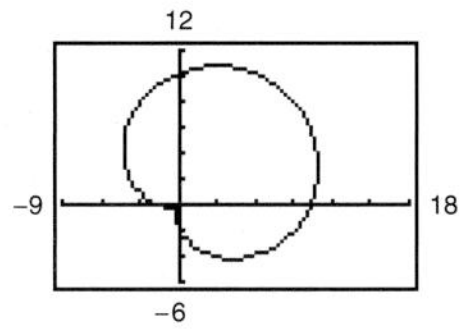

(c)

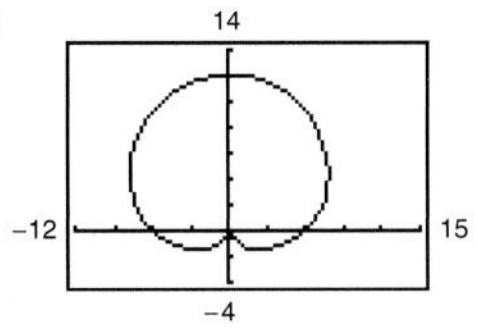

The angle ϕ controls rotation of the axis of symmetry.

$r = 6(1 + \sin\theta)$

62. Answers will vary.

64. (a) $r = 4\sin\left(\theta - \frac{\pi}{6}\right)\cos\left(\theta - \frac{\pi}{6}\right)$

(b) $r = -4\sin\theta\cos\theta$

(c) $r = 4\sin\left(\theta - \frac{2\pi}{3}\right)\cos\left(\theta - \frac{2\pi}{3}\right)$

(d) $r = 4\sin\theta\cos\theta$

66. (a)

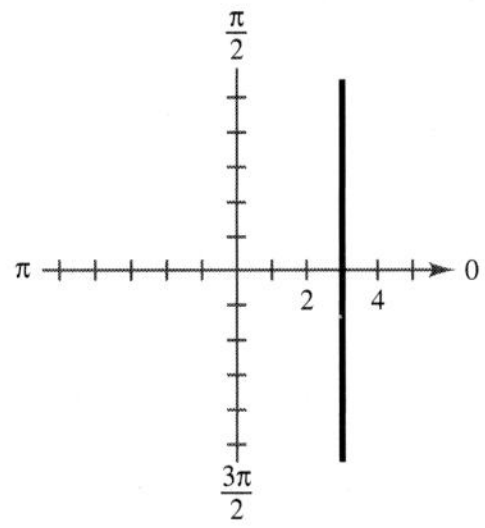

(b)

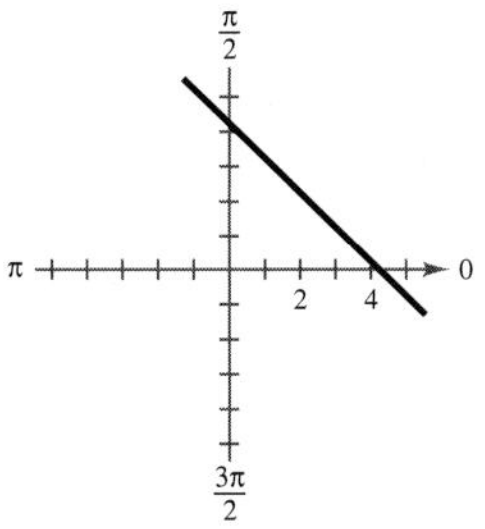

(c)

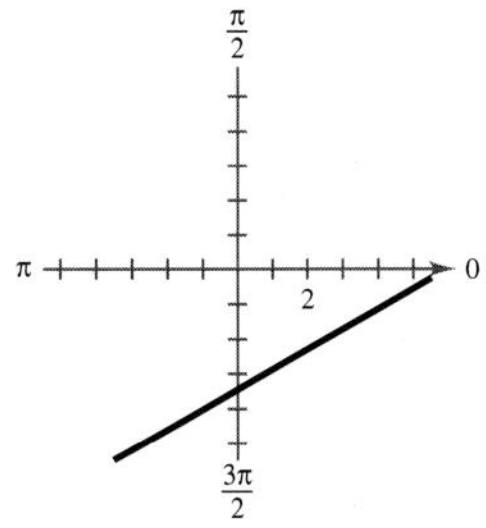

(d)

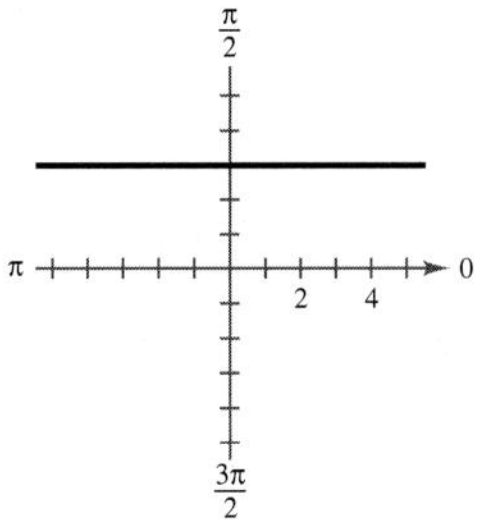

68. (a)

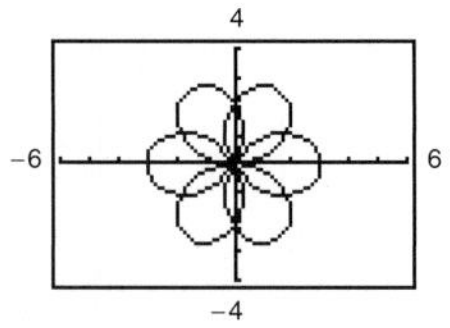

$0 \leq \theta < 4\pi$

(b)

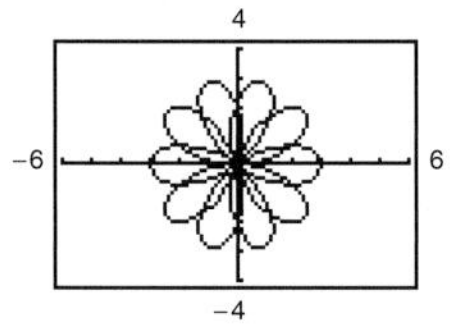

$0 \leq \theta < 4\pi$

(c) Yes

70. $\ln \frac{47}{6} \approx 2.058$ **72.** $\frac{1}{2}(\log_5 60) \approx 1.272$

74. $\frac{1}{4}e^{9/2} \approx 22.504$ **76.** No zeros **78.** 3

Section 10.9 *(page 838)*

2.

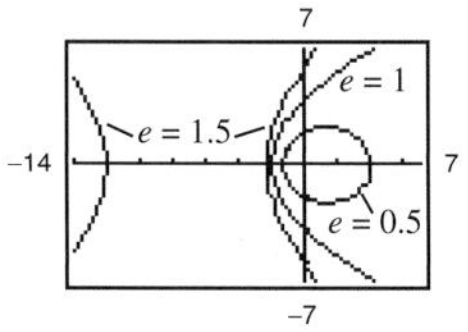

4.

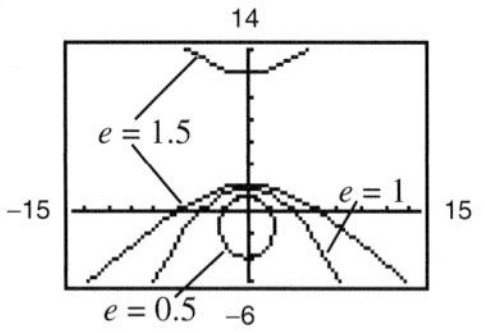

6. c **8.** e **10.** b

12. Parabola

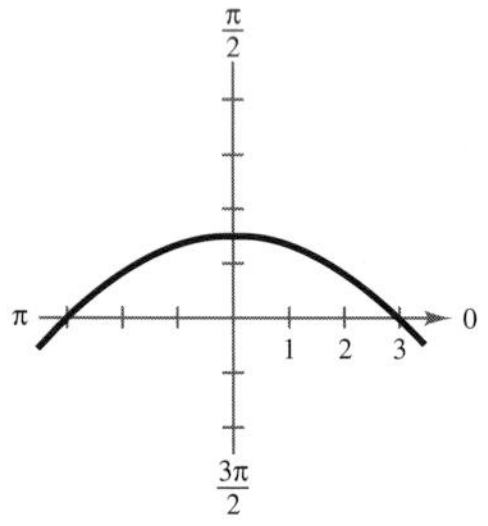

14. Parabola

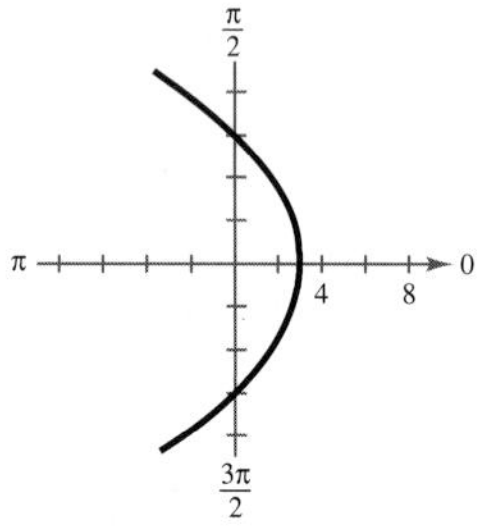

16. Ellipse

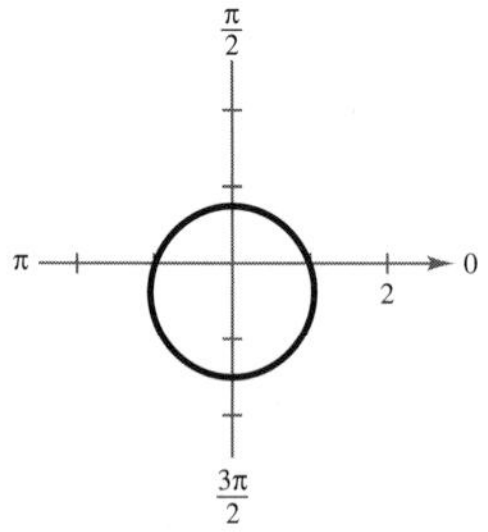

18. Ellipse

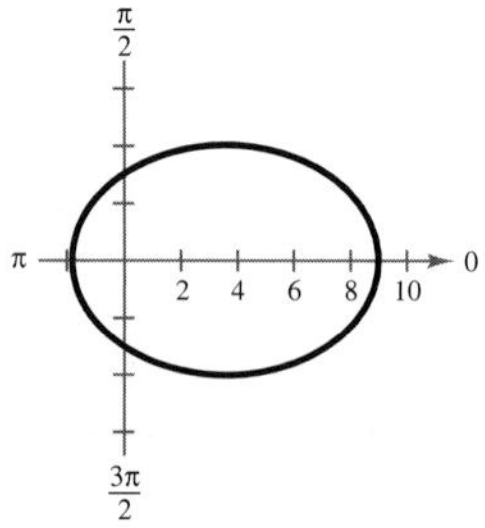

20. Hyperbola

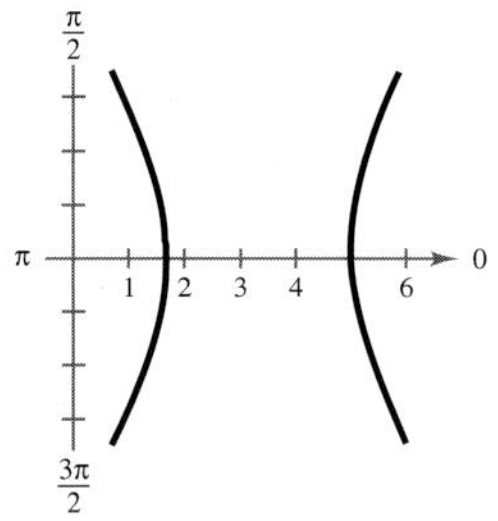

22. Hyperbola

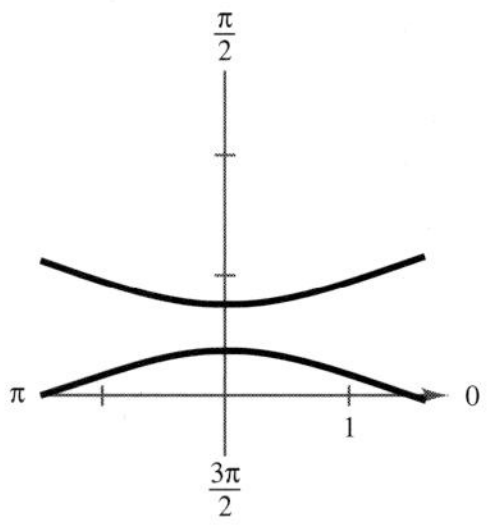

24. Hyperbola

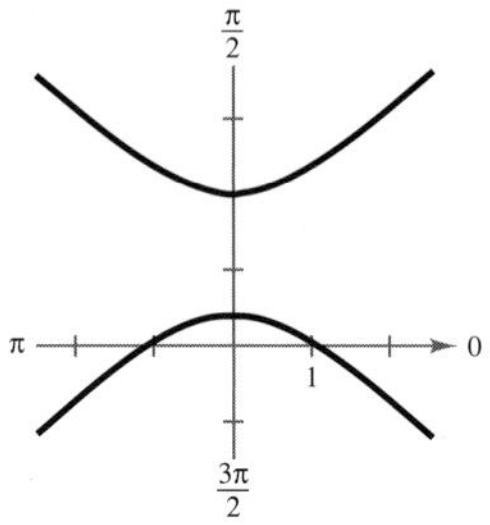

26.

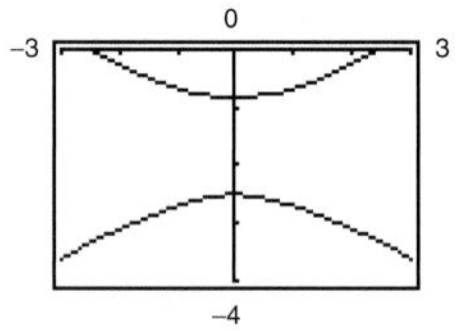

Hyperbola

28.

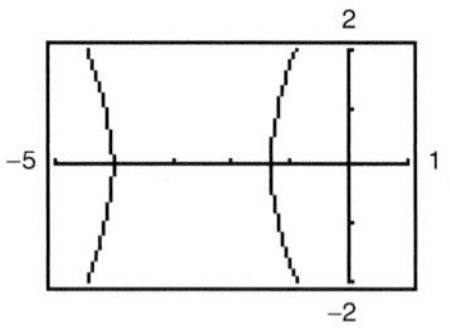

Hyperbola

30.

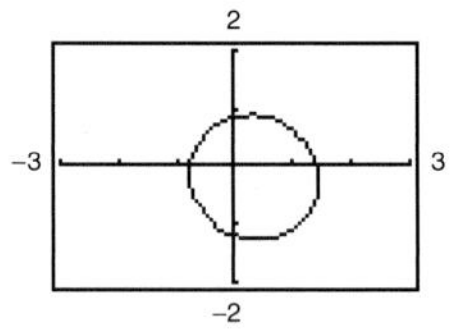

32.

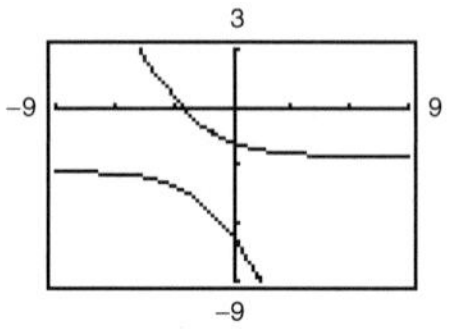

34. $r = \dfrac{2}{1 - \sin\theta}$ **36.** $r = \dfrac{9}{4 - 3\sin\theta}$

38. $r = \dfrac{3}{2 - 3\cos\theta}$ **40.** $r = \dfrac{12}{1 + \cos\theta}$

42. $r = \dfrac{20}{1 + \sin\theta}$ **44.** $r = \dfrac{8}{3 + \sin\theta}$

46. $r = \dfrac{16}{3 + 5\cos\theta}$ **48.** $r = \dfrac{8}{3 + 5\sin\theta}$

50. Answers will vary.

52. $r = \dfrac{1.4248 \times 10^9}{1 - 0.0543\cos\theta}$

Perihelion: 1.3514×10^9 km

Aphelion: 1.5066×10^9 km

54. $r = \dfrac{3.4459 \times 10^7}{1 - 0.2056 \cos \theta}$

Perihelion: 2.8583×10^7 miles

Aphelion: 4.3377×10^7 miles

56. $r = \dfrac{7.7658 \times 10^8}{1 - 0.0484 \cos \theta}$

Perihelion: 7.4073×10^8 km

Aphelion: 8.1607×10^8 km

58. 394.4 miles

60. True. The graphs represent the same hyperbola.

62. Answers will vary. **64.** $r^2 = \dfrac{400}{25 - 9 \cos^2 \theta}$

66. $r^2 = \dfrac{36}{10 \cos^2 \theta - 9}$ **68.** $r^2 = \dfrac{225}{25 - 16 \cos^2 \theta}$

70. $\dfrac{\pi}{3} + 2n\pi, \dfrac{5\pi}{3} + 2n\pi$ **72.** $\dfrac{\pi}{3} + n\pi, \dfrac{2\pi}{3} + n\pi$

74. $\dfrac{\pi}{3} + 2n\pi, \dfrac{5\pi}{3} + 2n\pi$ **76.** $-\dfrac{7\sqrt{2}}{10}$ **78.** $\dfrac{\sqrt{2}}{10}$

80. $\sin 2u = -\dfrac{\sqrt{3}}{2}$

$\cos 2u = -\dfrac{1}{2}$

$\tan 2u = \sqrt{3}$

Review Exercises *(page 842)*

2. 149° **4.** 40.6° **6.** 25.35° **8.** 81.98°

10. $\dfrac{6\sqrt{5}}{5}$ **12.** Parabola **14.** $y^2 = -8(x - 2)$

16. $(x - 2)^2 = 8(y - 2)$ **18.** $y = 4x + 8, (-2, 0)$

20. $y^2 = 6x$ **22.** $\dfrac{(x - 2)^2}{3} + \dfrac{(y - 2)^2}{4} = 1$

24. $\dfrac{(x - 2)^2}{4} + (y - 1)^2 = 1$

26. Longest distance: 36 feet

Shortest distance: 28 feet

Distance between foci: $16\sqrt{2}$ feet

28. Center: $(-2, 3)$

Vertices: $(3, 3), (-7, 3)$

Foci: $\left(-2 \pm \sqrt{21}, 3\right)$

Eccentricity: $\dfrac{\sqrt{21}}{5}$

30. Center: $(5, -3)$

Vertices: $(5, 3), (5, -9)$

Foci: $\left(5, -3 \pm \sqrt{35}\right)$

Eccentricity: $\dfrac{\sqrt{35}}{6}$

32. $\dfrac{x^2}{4} - \dfrac{(y - 2)^2}{12} = 1$ **34.** $\dfrac{5y^2}{16} - \dfrac{5(x - 3)^2}{4} = 1$

36. Center: $(-1, -3)$

Vertices: $(-1, -1), (-1, -5)$

Foci: $\left(-1, -3 \pm \sqrt{29}\right)$

Asymptotes: $y = -3 \pm \frac{2}{5}(x + 1)$

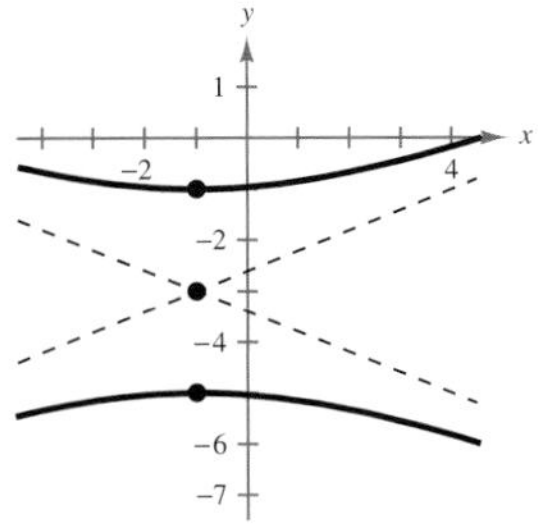

38. Center: $(0, 1)$

Vertices: $(0, 3), (0, -1)$

Foci: $\left(0, 1 \pm \sqrt{5}\right)$

Asymptotes: $y = 1 \pm 2x$

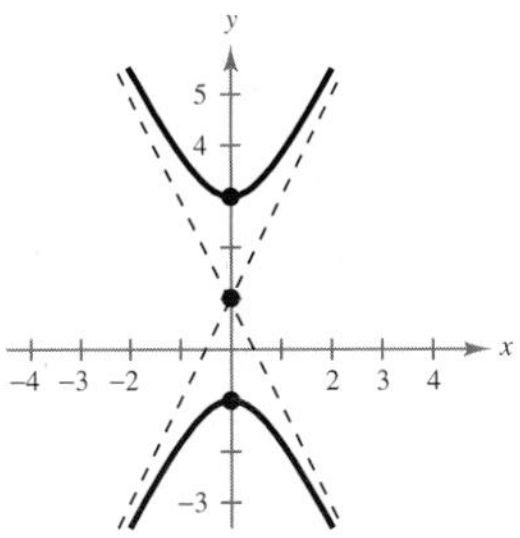

40. $\dfrac{576x^2}{25} - \dfrac{576y^2}{2279} = 1, \dfrac{64(x - 1)^2}{25} - \dfrac{64y^2}{39} = 1$

42. Parabola

44. $\dfrac{(x')^2}{1/4} - \dfrac{(y')^2}{1/6} = 1$ **46.** $y' = -4(x')^2 - 8x'$

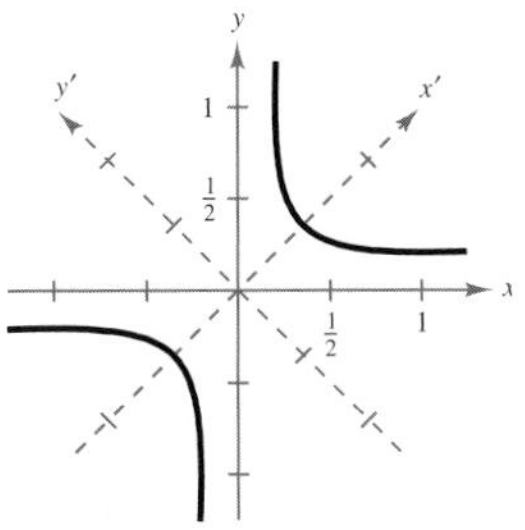

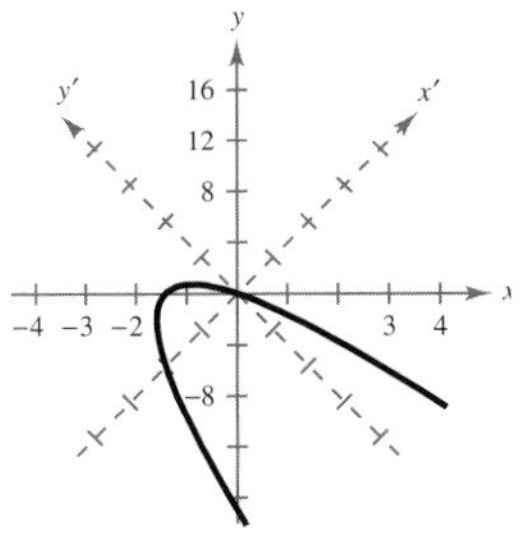

48. Ellipse

$y = \dfrac{8x \pm \sqrt{64x^2 - 28(13x^2 - 45)}}{14}$

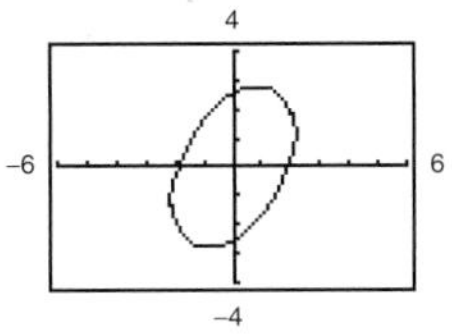

50. Hyperbola

$$y = \frac{10x \pm \sqrt{100x^2 - 4(x^2 + 1)}}{2}$$

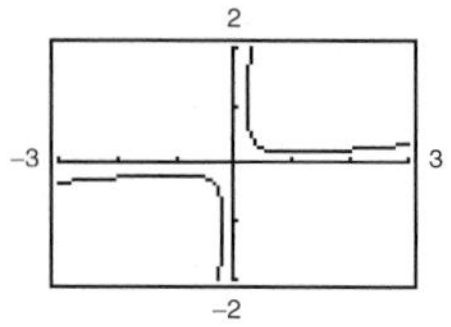

52. $x = \frac{3}{2}, y = \frac{3}{2}$ **54.** $x = \frac{3\sqrt{2}}{2}, y = 1$

56.

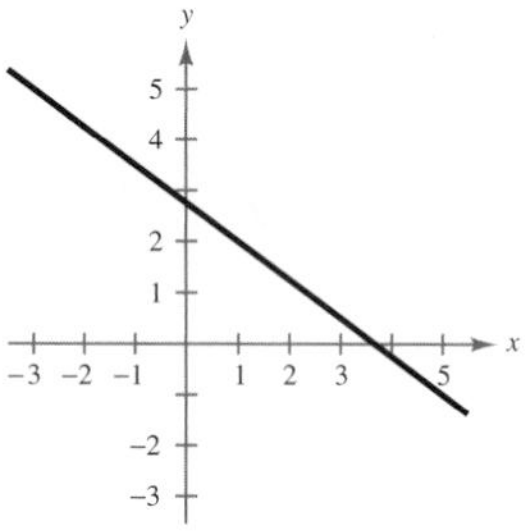

$y = -\frac{3}{4}x + \frac{11}{4}$

58.

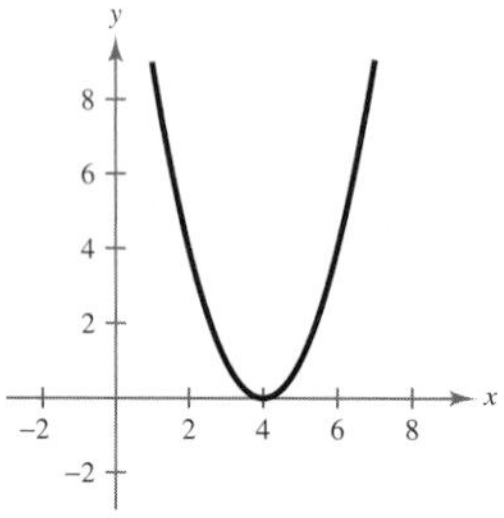

$y = (x - 4)^2$

60.

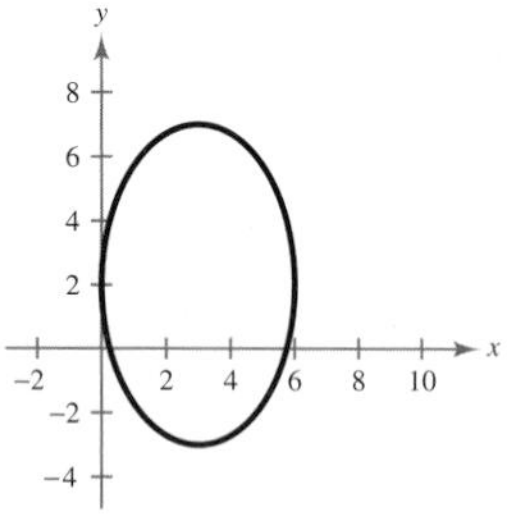

$$\frac{(x - 3)^2}{9} + \frac{(y - 2)^2}{25} = 1$$

62. $x = 3 \tan \theta;\ y = 4 \sec \theta$ **64.** Answers will vary.

66.

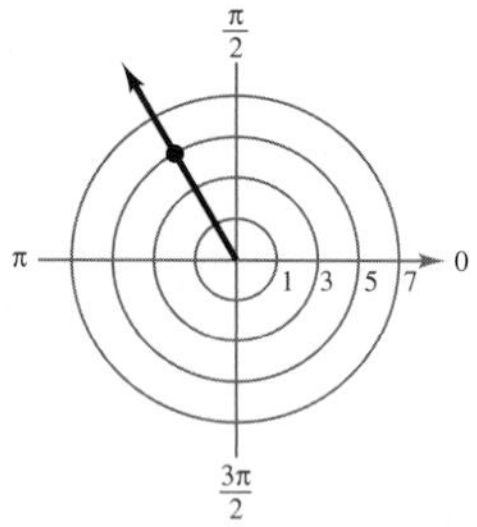

$\left(-\frac{5}{2}, \frac{5\sqrt{3}}{2}\right)$

68.

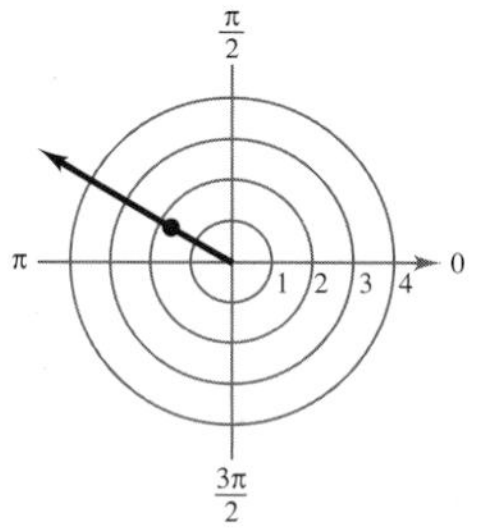

$(-1.5017, 0.8630)$

70. $\left(\sqrt{10}, \frac{3\pi}{4}\right), \left(-\sqrt{10}, \frac{7\pi}{4}\right)$

72. $(5, 5.356), (-5, 2.214)$ **74.** $x^2 + y^2 = 100$

76. $(x^2 + y^2)^2 - x^2 + y^2 = 0$ **78.** $r = 4 \cos \theta$

80. Spiral

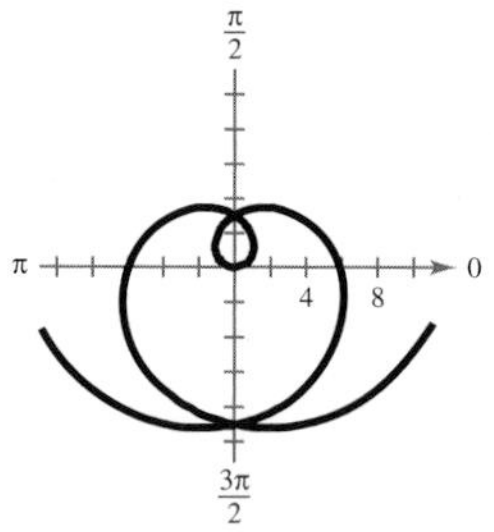

82. Rose curve

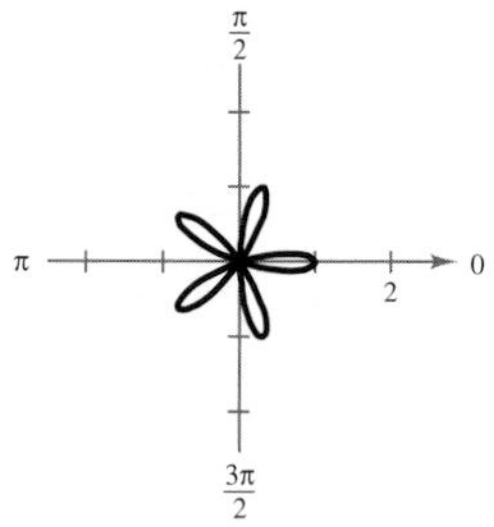

84. Limaçon

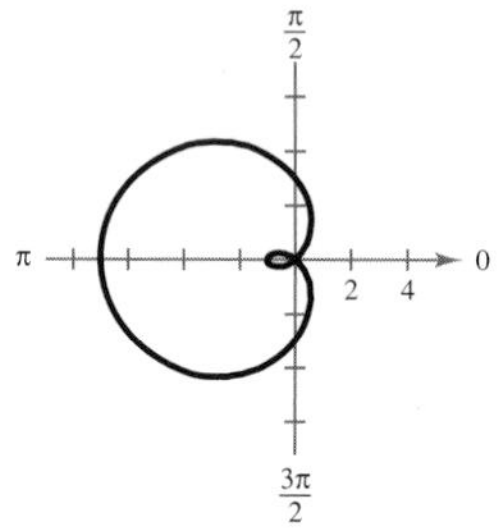

86. Symmetry: polar axis

Maximum value of $|r|$: $|r| = 10$ when $\theta = \pi$

Zeros of r: $r = 0$ when $\theta = 0, 2\pi$

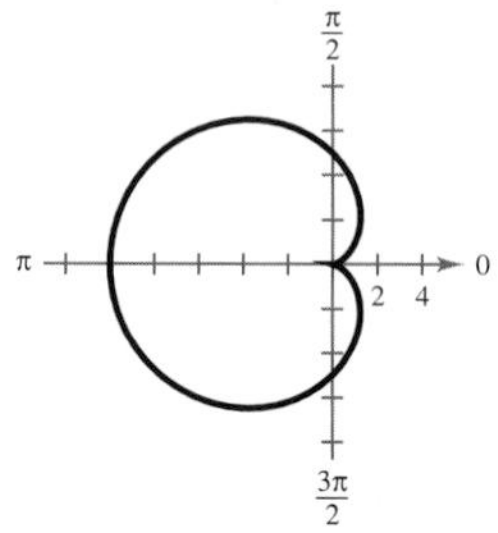

88. Symmetry: $\theta = \frac{\pi}{2}$, polar axis, pole

Maximum value of $|r|$: $|r| = 1$ when $\theta = 0, \pi$

Zeros of r: $r = 0$ when $\theta = \frac{\pi}{4}, \frac{3\pi}{4}$

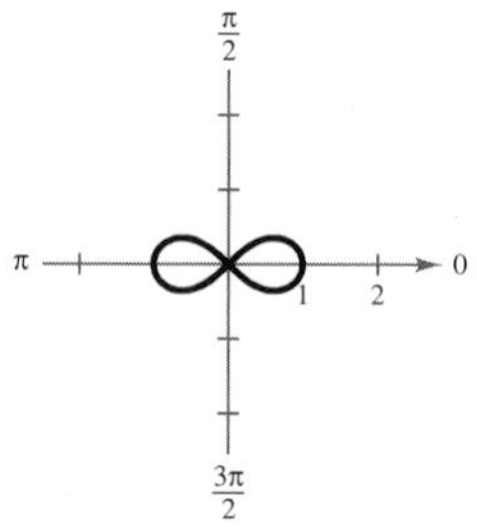

90. Limaçon **92.** Lemniscate

94. Eccentricity: 1; parabola

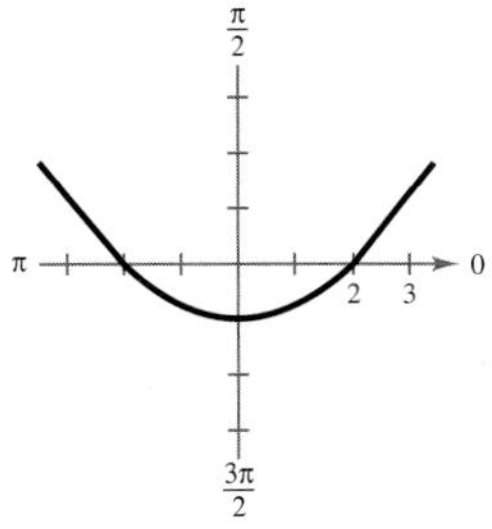

96. Eccentricity: $\frac{5}{4}$; hyperbola

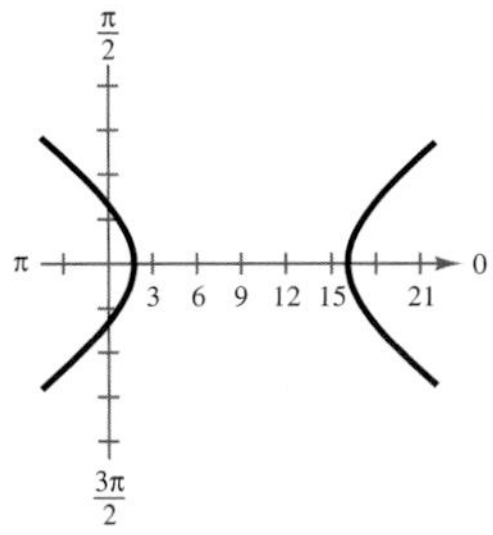

98. $r = \dfrac{4}{1 + \sin \theta}$ **100.** $r = \dfrac{7}{3 + 4 \cos \theta}$

102. $r = \dfrac{12{,}000{,}000}{1 + \sin \theta}$; 89,600,000 miles

104. False. The equation of a hyperbola is a second-degree equation.

106. False. $(2, \pi/4)$, $(-2, 5\pi/4)$, and $(2, 9\pi/4)$ all represent the same point.

108. Yes. The orientation would be reversed.

110. (a) Symmetric to the pole

(b) Symmetric to the polar axis

(c) Symmetric to $\pi/2$

Explorations

Chapter P

(page 5)

a. 6 **b.** 1 **c.** 3 **d.** 3

Absolute value expressions are never negative.

(page 30)

$$u^6 - v^6 = (u^3 + v^3)(u^3 - v^3)$$
$$= (u + v)(u - v)(u^2 - uv + v^2)(u^2 + uv + v^2),$$
$$x^6 - 1 = (x + 1)(x - 1)(x^2 - x + 1)(x^2 + x + 1),$$
$$x^6 - 64 = (x + 2)(x - 2)(x^2 - 2x + 4)(x^2 + 2x + 4)$$

Chapter 1

(page 110)

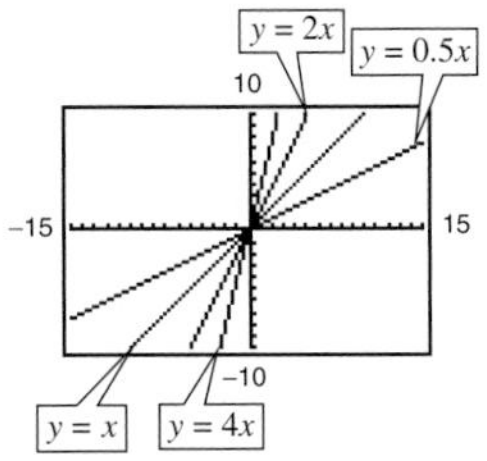

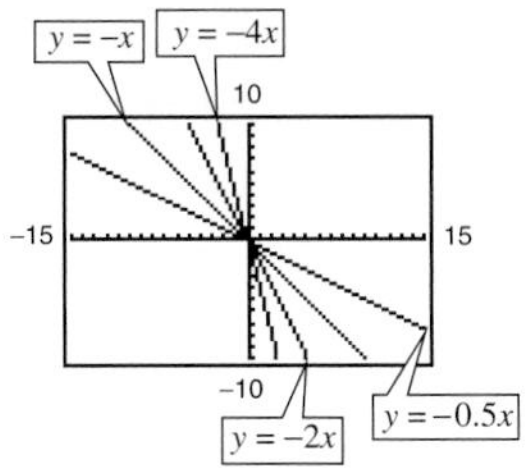

The line $y = 4x$ The line $y = -4x$

As $|m|$ increases, the line rises or falls faster.

(page 146)

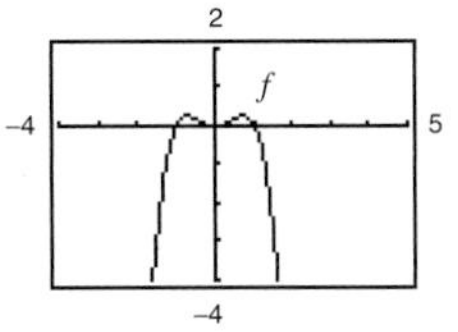

Even

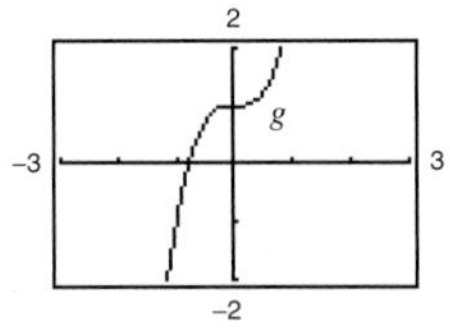

Neither

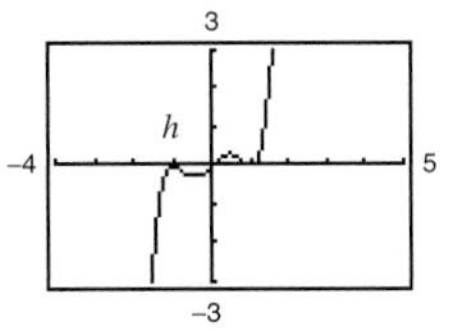

Odd

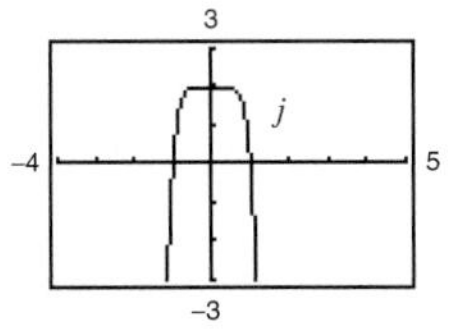

Even

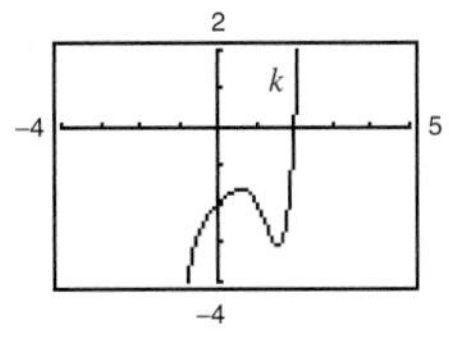

Neither

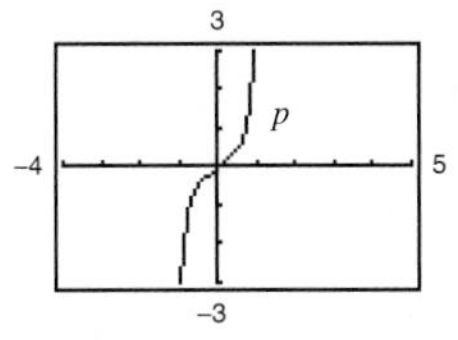

Odd

Equations of odd functions contain only odd powers of x. Equations of even functions contain only even powers of x. Odd functions have all variables raised to odd powers, and even functions have all variables raised to even powers. A function that has variables raised to even and odd powers is neither odd nor even.

(page 154)

a.

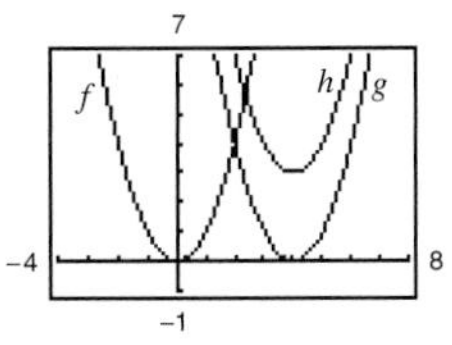

g is a right shift of 4 units. h is a right shift of 4 units and an upward shift of 3 units.

EXPLORATIONS

b.

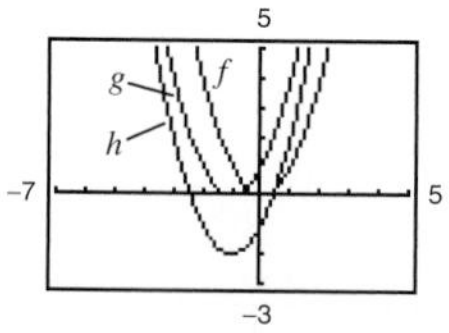

g is a left shift of 1 unit. h is a left shift of 1 unit and a downward shift of 2 units.

c.

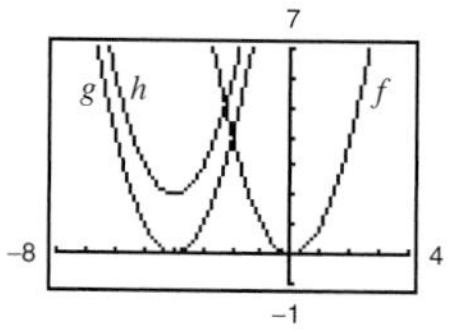

g is a left shift of 4 units. h is a left shift of 4 units and an upward shift of 2 units.

(page 157)

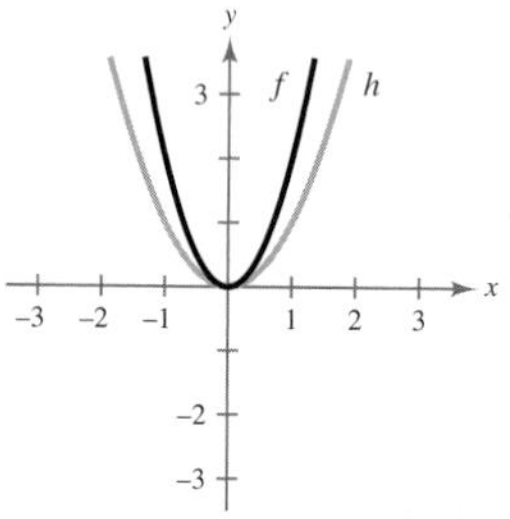

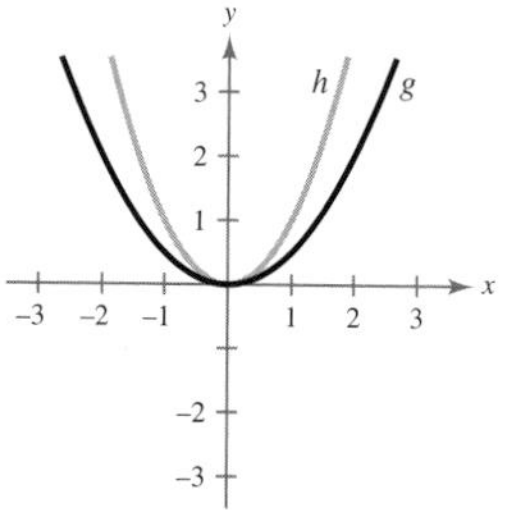

The graph becomes narrower. The graph becomes wider.

The larger the coefficient of the x^2-term, the narrower the graph. The smaller the coefficient of the x^2-term, the wider the graph.

(page 167)

Option (b), because option (a) takes a 10% discount on a smaller amount.

Option (b) represents $f(g(x))$; option (a) represents $g(f(x))$.

(page 172)

x	-10	0	7	45
$f(f^{-1}(x))$	-10	0	7	45
$f^{-1}(f(x))$	-10	0	7	45

The functions are inverses of each other.

Chapter 2

(page 204)

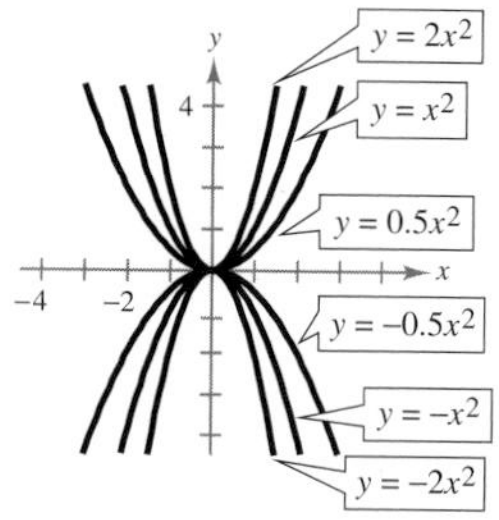

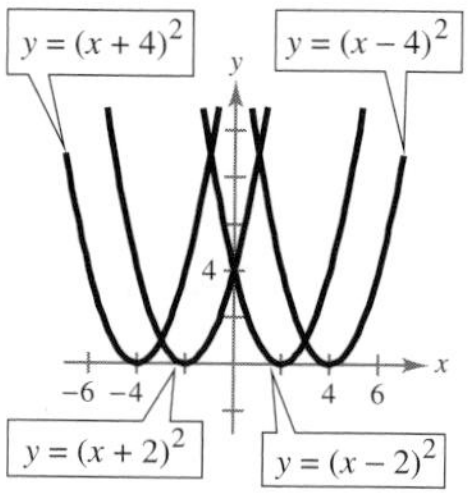

As $|a|$ gets larger, the parabola becomes narrower.

As h gets larger, the parabola shifts to the right.

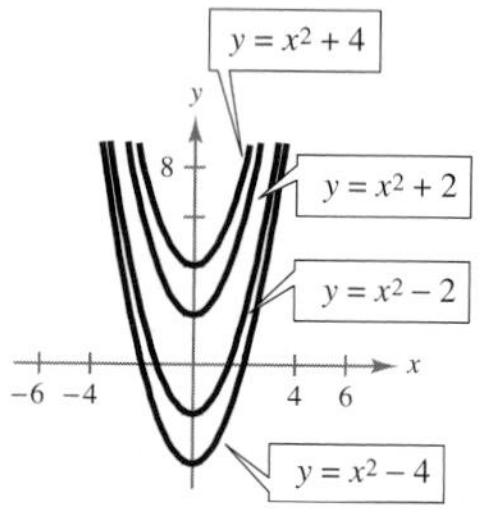

As k gets larger, the parabola shifts upward.

(page 215)

a. Third degree, odd; 1, greater than 0

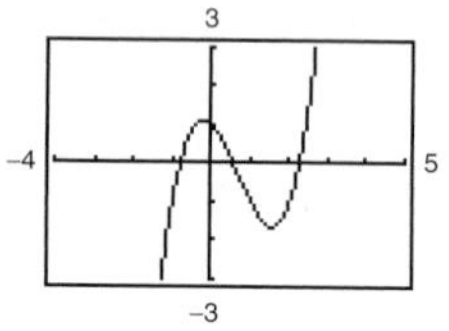

b. Fifth degree, odd; 2, greater than 0

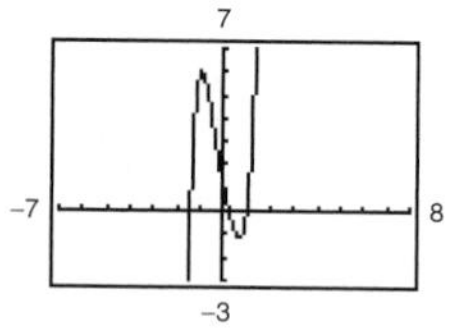

c. Fifth degree, odd; -2, less than 0

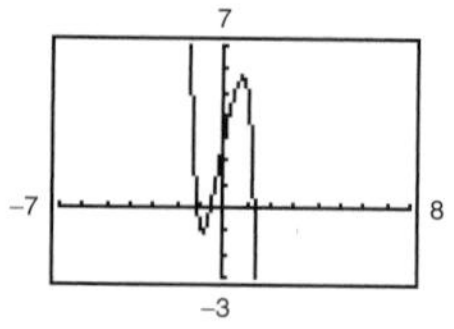

d. Third degree, odd; -1, less than 0

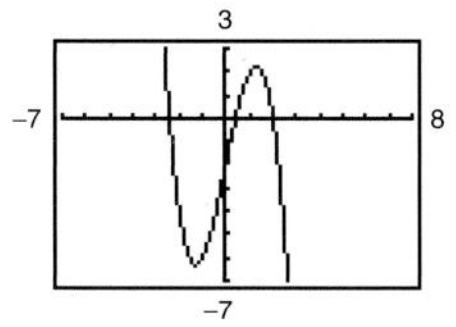

e. Second degree, even; 2, greater than 0

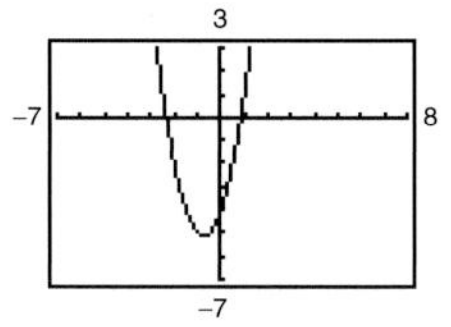

f. Fourth degree, even; 1, greater than 0

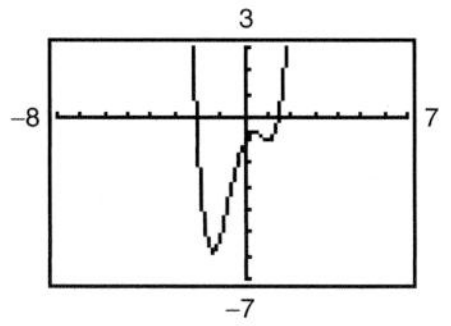

g. Second degree, even; 1, greater than 0

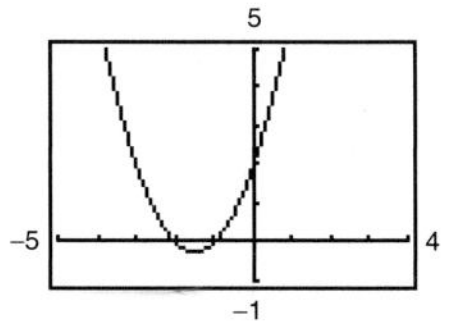

h. Sixth degree, even; -1, less than 0

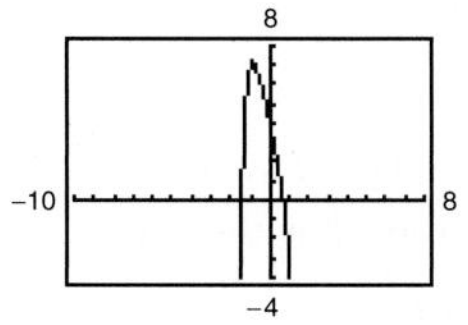

When the degree of the function is odd and the leading coefficient is positive, the graph falls to the left and rises to the right, but if the leading coefficient is negative, the graph falls to the right and rises to the left. When the degree of the function is even and the leading coefficient is positive, the graph rises to the left and right, but if the leading coefficient is negative, the graph falls to the left and right.

(page 238)

$i, -1, -i, 1, i, -1, -i, 1;$

The pattern repeats the first four results. Divide the exponent by 4.

If the remainder is 1, the result is i.

If the remainder is 2, the result is -1.

If the remainder is 3, the result is $-i$.

If the remainder is 0, the result is 1.

Chapter 3

(page 290)

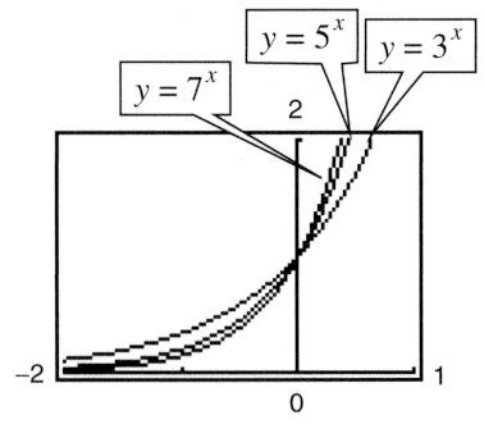

The graphs rise to the right and intersect at $(0, 1)$. $y = 3^x$. $y = 7^x$. $y = 7^x$. $y = 3^x$.

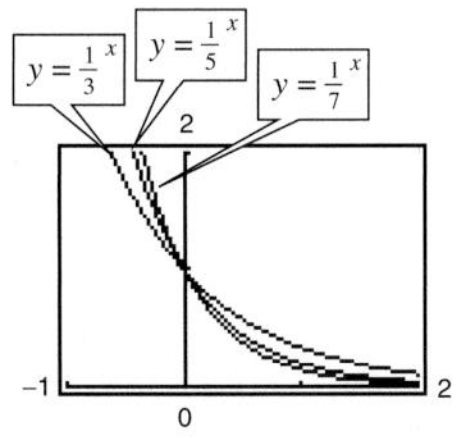

The graphs fall to the right and intersect at $(0, 1)$. The smaller the value of b, the more sharply the graph falls.

(page 301)

x	-2	-1	0	1	2
$f(x) = 10^x$	$\frac{1}{100}$	$\frac{1}{10}$	1	10	100

x	$\frac{1}{100}$	$\frac{1}{10}$	1	10	100
$f(x) = \log_{10} x$	-2	-1	0	1	2

The domain of $f(x) = 10^x$ is equal to the range of $f(x) = \log_{10} x$, and vice versa. $f(x) = 10^x$ and $f(x) = \log_{10} x$ are inverses of each other.

(page 313)

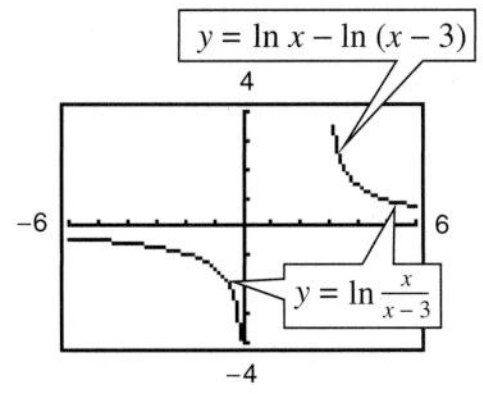

No; the domain of the first graph is $(3, \infty)$ and the domain of the second graph is $(-\infty, 0) \cup (3, \infty)$.

(page 322)

a. 7% **b.** 7.251% **c.** 7.186% **d.** 7.45%

Savings plan (d) will have the greatest effective yield. Savings plan (d) will have the highest balance after 5 years.

Chapter 4

(page 366)

1.

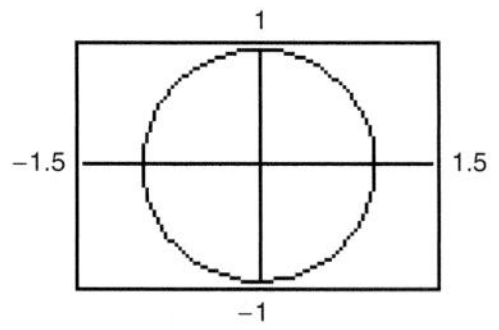

The graph is a circle.

2. The t-values represent the central angle in radians. The x- and y-values represent the location in the coordinate plane.

3. $-1 \le x \le 1, -1 \le y \le 1$

(page 393)

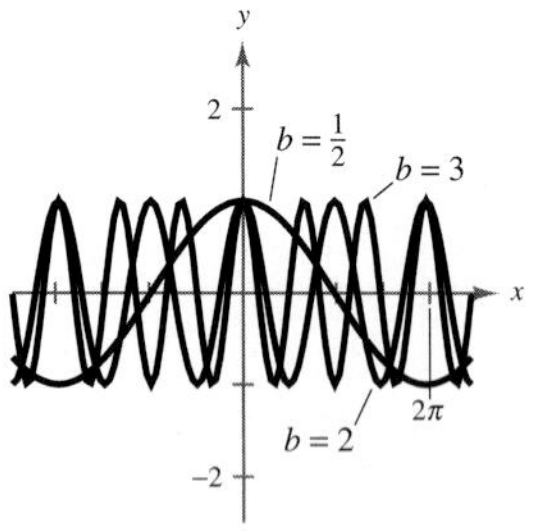

b affects the period of the graph.

$b = \frac{1}{2}, \frac{1}{2}$ cycle;

$b = 2$, 2 cycles;

$b = 3$, 3 cycles

(page 394)

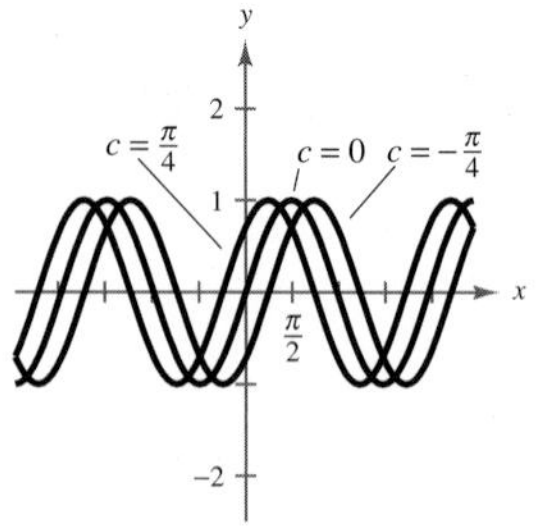

c shifts the graph horizontally.

Chapter 5

(page 461)

Yes. Preferences will vary.

(page 468)

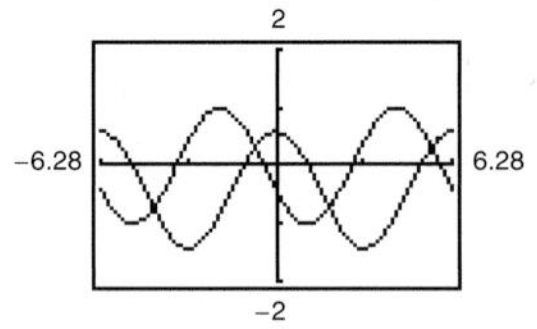

The graphs are different. No, it is not true.

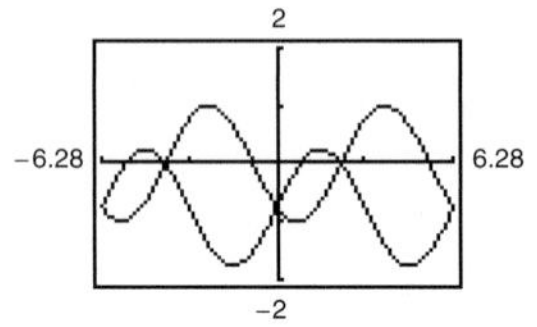

The graphs are different. No, it is not true.

Chapter 6

(page 504)

Pythagorean Theorem.

The Pythagorean Theorem is just a special case of the more general Law of Cosines.

(page 537)

The given equation can be written as

$$x^4 = -16 = 16(\cos \pi + i \sin \pi)$$

which means that you can solve the equation by finding the four fourth roots of -16. Each of these roots has the form

$$\sqrt[4]{16}\left(\cos\frac{\pi + 2\pi k}{4} + i \sin\frac{\pi + 2\pi k}{4}\right).$$

Finally, using $k = 0, 1, 2,$ and 3, you obtain the roots

$$2\left(\cos\frac{\pi}{4} + i\sin\frac{\pi}{4}\right) = 2\left(\frac{\sqrt{2}}{2} + \frac{\sqrt{2}}{2}i\right) = \sqrt{2} + \sqrt{2}i$$

$$2\left(\cos\frac{3\pi}{4} + i\sin\frac{3\pi}{4}\right) = 2\left(-\frac{\sqrt{2}}{2} + \frac{\sqrt{2}}{2}i\right) = -\sqrt{2} + \sqrt{2}i$$

$$2\left(\cos\frac{5\pi}{4} + i\sin\frac{5\pi}{4}\right) = 2\left(-\frac{\sqrt{2}}{2} - \frac{\sqrt{2}}{2}i\right) = -\sqrt{2} - \sqrt{2}i$$

$$2\left(\cos\frac{7\pi}{4} + i\sin\frac{7\pi}{4}\right) = 2\left(\frac{\sqrt{2}}{2} - \frac{\sqrt{2}}{2}i\right) = \sqrt{2} - \sqrt{2}i.$$

Chapter 7

(page 555)

No, the coordinates are approximations.

(page 557)

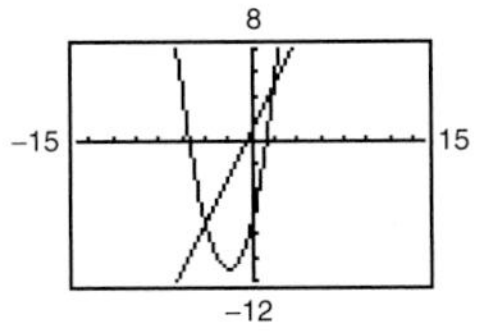

Two solutions

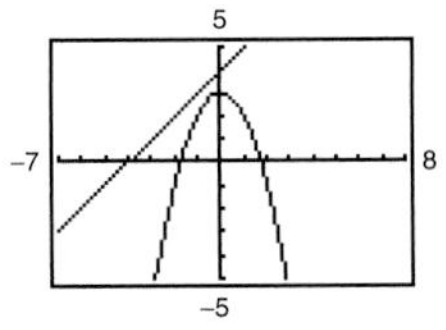

No solution

The number of points of intersection of the graph of a system of equations is the number of solutions the system has.

(page 567)

a.

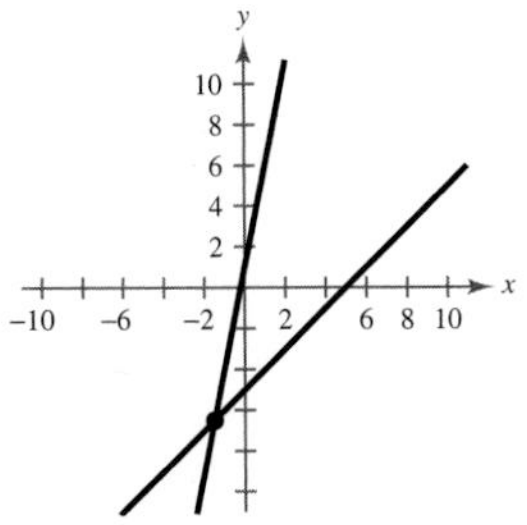

One solution. Graphs have one point of intersection.

b.

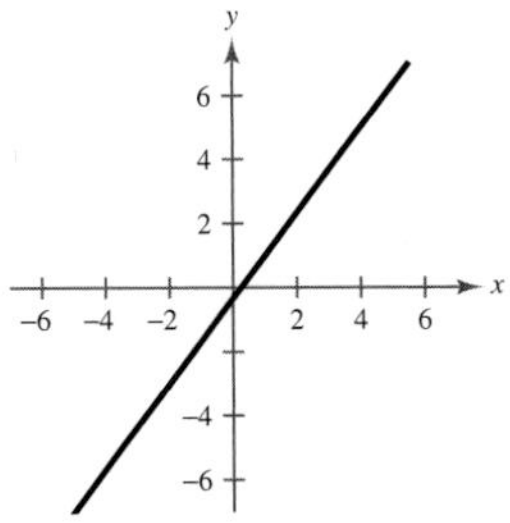

Infinitely many solutions. Graphs coincide.

c.

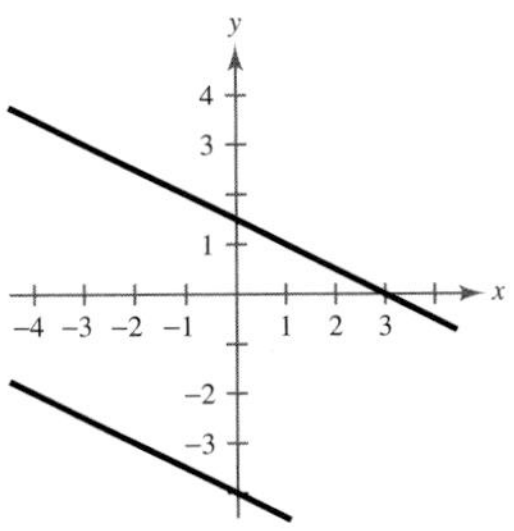

No solution. Graphs do not intersect.

Chapter 8

(page 641)

No, $AB = \begin{bmatrix} 4 & 7 \\ 8 & 15 \end{bmatrix}$, $BA = \begin{bmatrix} 3 & 4 \\ 11 & 16 \end{bmatrix}$.

Matrix multiplication is not commutative.

(page 653)

An error message appears because

$1(6) - (-2)(-3) = 0.$

(page 659)

An error message appears because the matrix is not square.

Chapter 9

(page 689)

$1, -\frac{1}{3}, \frac{1}{5}, -\frac{1}{7}, \frac{1}{9}$; No; the terms have opposite signs.

(page 712)

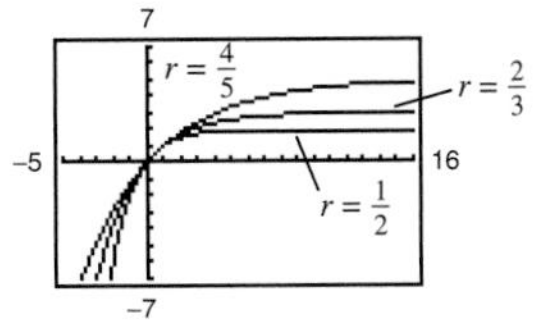

As $x \to \infty$, $y \to 2$ for $r = \frac{1}{2}$, $y \to 3$ for $r = \frac{2}{3}$, and $y \to 5$ for $r = \frac{4}{5}$.

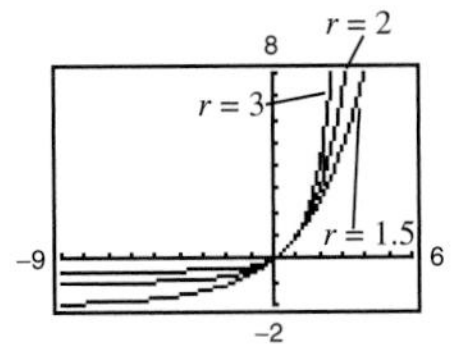

As $x \to \infty$, $y \to \infty$ for $r = 1.5, 2,$ and 3.

(page 730)

n	r	${}_nC_r$	${}_nC_{n-r}$
9	5	126	126
7	1	7	7
12	4	495	495
6	0	1	1
10	7	120	120

This illustrates the symmetry of Pascal's Triangle.

(page 753)

Assuming 365 different birthdays in a year, the probability that at least two out of 23 people will have a birthday on the same day of the year is ≈ 0.51.

The probability that at least two out of 50 people will have a birthday on the same day of the year is ≈ 0.97.

Chapter 10

(page 796)

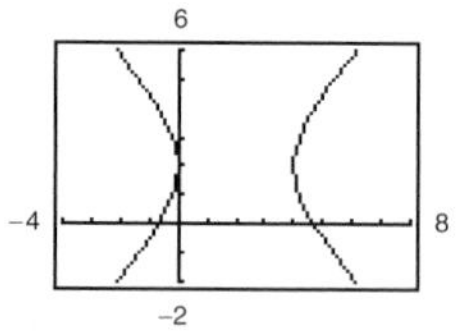

The graphs are the same.

(page 814)

t should be greater than -1. The upper bound of t varies, but one possibility is $t = 20$.

(page 821)

a. Yes. $\theta \approx 3.927, x \approx -2.121, y \approx -2.121$

b. Yes. Answers and explanations will vary.

Technology

Chapter 1

(page 117)

The lines appear perpendicular with the square setting.

(page 129)

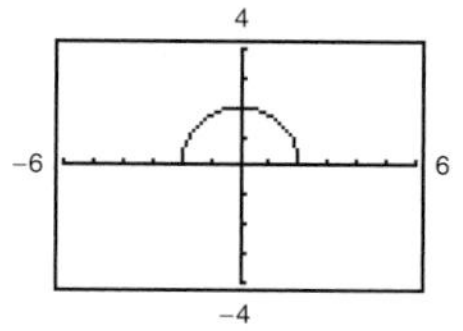

Domain: $[-2, 2]$

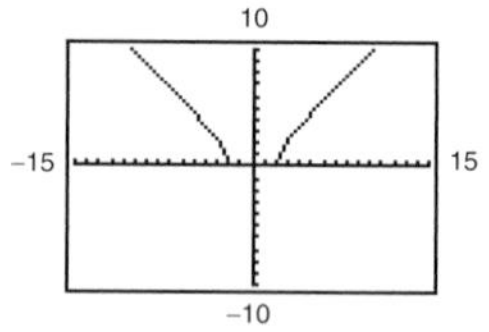

Domain: $(-\infty, -2] \cup [2, \infty)$

Yes, for -2 and 2.

Chapter 2

(page 218)

(a)

Chapter 3

(page 293)

$A = 5466.09, A = 5466.35, A = 5466.36, A = 5466.38.$

Arguments will vary.

(page 329)

$P = 5303(1.014)^t$. The models are almost the same.

Chapter 4

(page 391)

No graph is visible. Try $-\pi \le x \le \pi$ and $-0.5 \le y \le 0.5$ as a viewing window.

Chapter 7

(page 556)

The point of intersection (7000, 5000) agrees with the solution.

(page 558)

(a) (2, 0) (b) $(0, -1), (2, 1)$ (c) None

Chapter 8

(page 638)

$$A + B = \begin{bmatrix} 1 & 1 \\ 1 & -5 \end{bmatrix}$$

Chapter 10

(page 816)

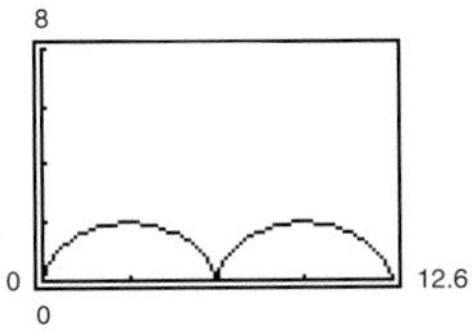

Writing About Mathematics

Chapter P

Section P.4 (page 44)

x	-3	-2	-1	0	1	2	3
$\frac{x^2 - 3x + 2}{x - 2}$	-4	-3	-2	-1	0	Undefined	2
$x - 1$	-4	-3	-2	-1	0	1	2

The expressions are not equivalent because they do not have the same domain.

Section P.6 (page 69)

(a) Let a, b, c, and d be real numbers.

1. Transitive Property: If a is less than b and b is less than c, then a is less than c.
2. Addition of Inequalities: If a is less than b and c is less than d, the sum of a and c is less than the sum of b and d.
3. Addition of a Constant: If a is less than b, the sum of a and c is less than the sum of b and c.
4. Multiplication by a Constant:

   (i) For $c > 0$, if a is less than b, the product of a and c is less than the product of b and c.

   (ii) For $c < 0$, if a is less than b, the product of a and c is greater than the product of b and c.

(b) Examples will vary. Possibilities include:

1. Transitive Property: $5 < 7$ and $7 < 13$, so $5 < 13$.
2. Addition of Inequalities: $3 < 8$ and $0 < 10$, so $3 + 0 < 8 + 10$ because $3 < 18$.
3. Addition of a Constant: $5 < 12$, so $5 + 2 < 12 + 2$ because $7 < 14$.
4. Multiplication by a Constant: $1 < 6$, so $1 \cdot 2 < 6 \cdot 2$ because $2 < 12$, and $1 < 6$, so $1 \cdot -2 > 6 \cdot -2$ because $-2 > -12$.

(c) One example is:

Transitive Property:

$5 < 7$

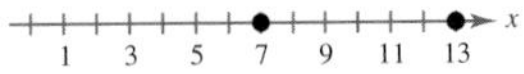

$7 < 13$

thus,

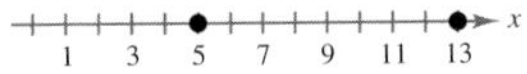

$5 < 13$

Section P.8 (page 86)

$(-x, y)$ reflects the point in the y-axis.

$(x, -y)$ reflects the point in the x-axis.

$(-x, -y)$ reflects the point through the origin.

Chapter 1

Section 1.2 (page 118)

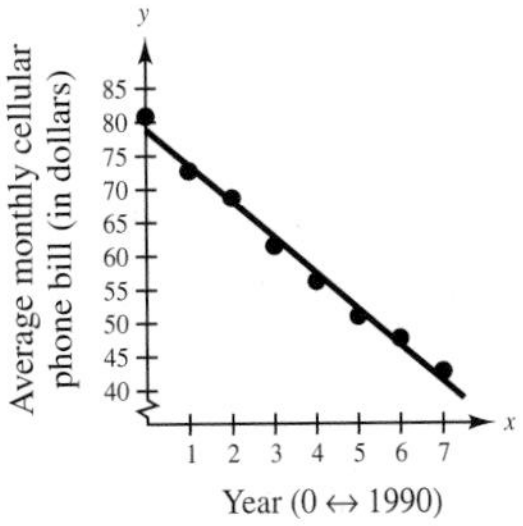

Best-fitting line from least squares regression: $y = -5.361x + 78.95$. Students' linear equations should be close to the least squares regression line, but they will not be exact. Models will vary depending on what each student judged to be the "best-fitting line."

The slope of -5.361 is the average rate at which the average monthly cellular phone bill decreased each year. The intercept of 78.95 represents the average monthly cellular phone bill in 1990. The actual values and the values obtained through the model are very close. An estimate of the average monthly bill in 2000 is \$25.35.

Section 1.3 (page 132)

$$f(x) = \begin{cases} 0.5x^2 - 1.47x + 6.3, & 1 \le x \le 6 \\ -1.97x + 26.33, & 6 < x \le 12 \end{cases}$$

Students can determine the domain of each part by inspection of a graph of the data with the two models.

$f(5) = 11.45$, $f(11) = 4.66$; These values represent the revenue for the months of May and November, respectively.

These values are quite close to the actual data values.

Section 1.6 (page 167)

a.

x	1	2	3	4	5	6
$g(x)$	0	1	2	0	1	-1

b.

x	1	2	3	4	5	6
$h(x)$	2	1	-3	-1	0	5

Section 1.7 (page 176)

a. The function does not have an inverse, because the retail price cannot be determined exactly because the sales tax is rounded.

b. The function does have an inverse $\left(y = \frac{5}{9}(x - 32)\right)$ because $f(f^{-1}(x)) = x$ and $f^{-1}(f(x)) = x$.

Chapter 2

Section 2.2 (page 221)

To obtain a polynomial whose zeros are j and k, you can write $f(x) = (x - j)(x - k)$. This method extends to any number of zeros.

a. $2x^2 + 9x - 5$ and $x^2 + \frac{9}{2}x - \frac{5}{2}$

b. $6x^2 - 41x - 7$

c. $x^3 - 3x^2 - 10x + 24$ and $-x^3 + 3x^2 + 10x - 24$

d. $x^3 - \frac{37}{12}x^2 - \frac{1}{4}x + \frac{3}{2}$

e. $x^4 + 12x^3 + 54x^2 + 108x + 81$

f. $x^4 + 2x^3 - 13x^2 - 14x + 24$

Section 2.6 (page 266)

No

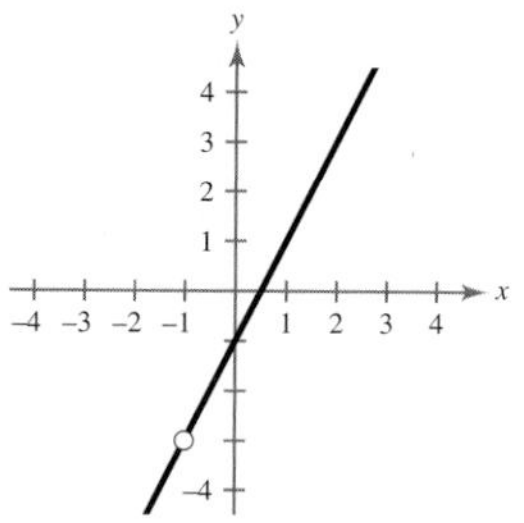

a. No

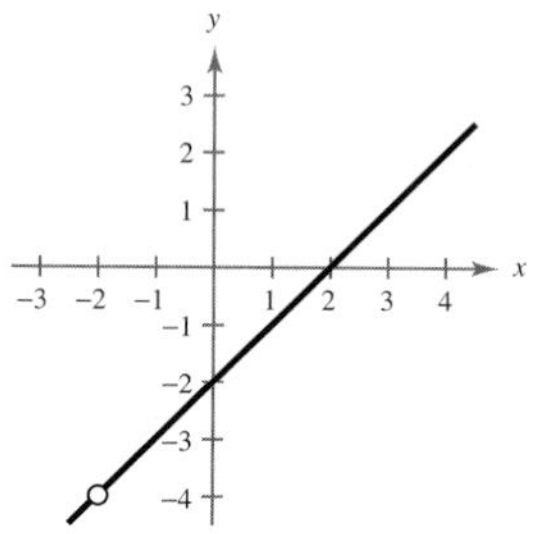

b. Yes

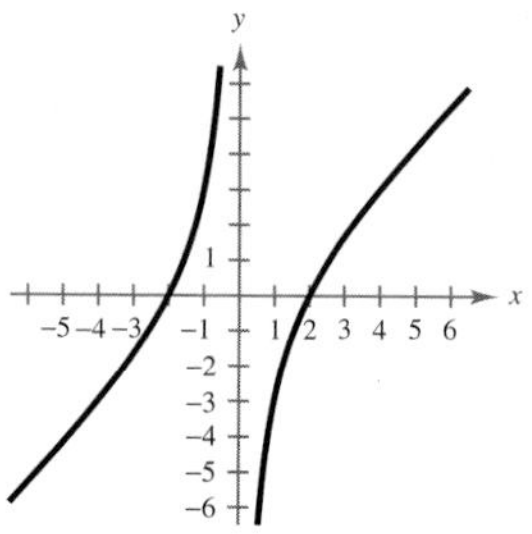

c. No

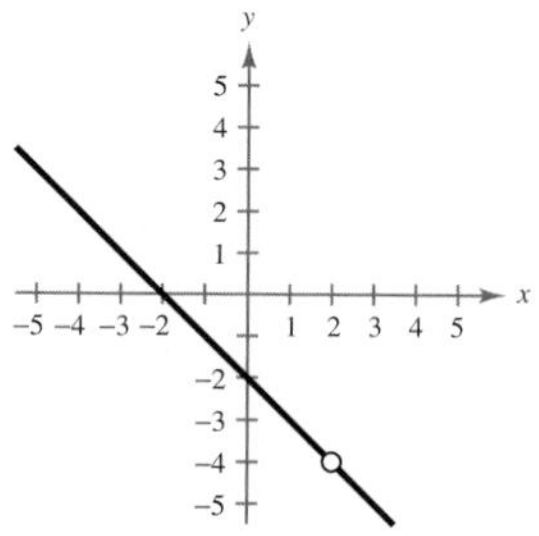

A rational function will not have a vertical asymptote when the function can be factored and reduced so that a variable factor no longer appears in the denominator.

Section 2.7 (page 276)

Because $\frac{x^2 + 1}{x(x - 1)}$ is an improper fraction, the student should begin by dividing to obtain $\frac{x^2 + 1}{x(x - 1)} = 1 + \frac{x + 1}{x(x - 1)}$, and decompose $\frac{x + 1}{x(x - 1)}$ into its partial fractions by the usual methods.

Chapter 3

Section 3.1 (page 295)

$g(x)$ can be generated by f_5.

$h(x)$ can be generated by f_2 or f_3.

All six functions have $(0, 8)$ as their y-intercept.

The functions f_2 and f_3 are the same because

$$f_2(x) = 8\left(\tfrac{1}{2}\right)^x = 2^3(2)^{-x} = 2^{(3-x)} = \left(\tfrac{1}{2}\right)^{-(3-x)}$$

$$= \left(\tfrac{1}{2}\right)^{(x-3)} = f_3(x).$$

The functions f_1 and f_6 are the same because

$$f_1(x) = 2^{(x+3)} = 2^x 2^3 = (8)2^x = f_6(x).$$

Examples of exponential functions with y-intercepts of $(0, -3)$ will vary. Possibilities include $f(x) = -3(2^x)$ and $f(x) = \left(\tfrac{1}{3}\right)^x - 4$.

Section 3.2 (page 306)

Approximately 11 months. Explanations will vary.

Section 3.4 (page 323)

(a)

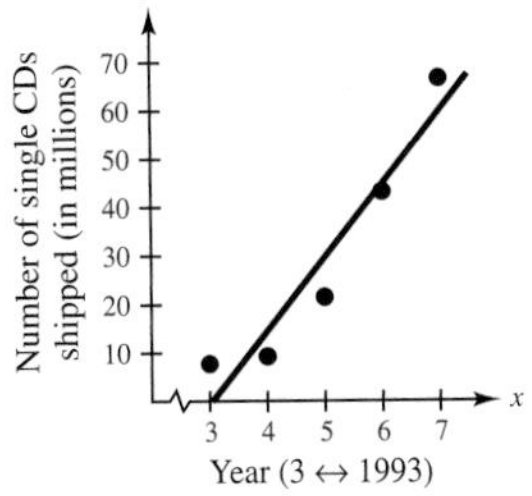

Linear model: $y = 15.17x - 46.15$

According to the model, the annual shipment of single CDs reached 100 million in 1999.

(b)

$\ln x$	1.0986	1.3863	1.6094	1.7916	1.9459
$\ln y$	2.0541	2.2300	3.0681	3.7658	4.2002

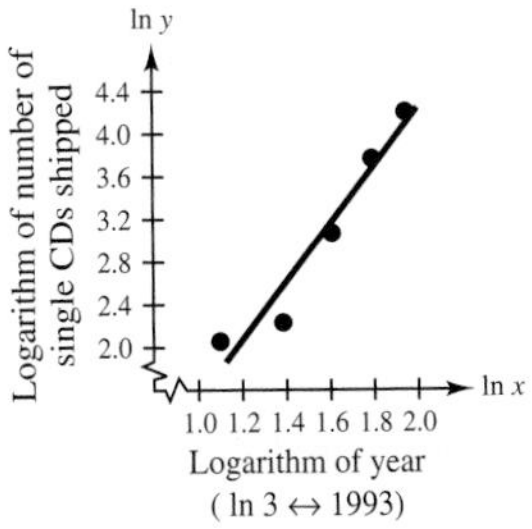

Logarithmic model: $\ln y = 2.70612 \ln x - 1.1752$

According to the model, the annual shipment of single CDs reached 100 million (i.e., $\ln y$ will reach 4.60517) when $\ln x = 2.1361$ (or $x \approx 8.5$) in 1999.

(c)

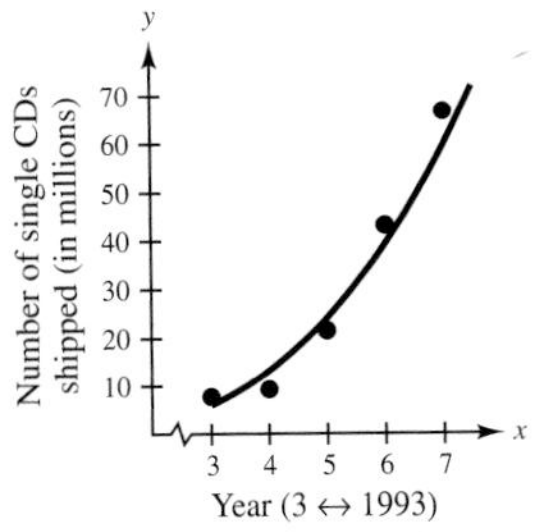

$y = 0.3088x^{2.7061}$ seems to fit the data better and will probably better predict the number of single CDs shipped.

Section 3.5 (page 334)

The quadratic model is better.

Chapter 4

Section 4.6 (page 407)

$h(x) = x + \sin x$

(a) $f(x) = x$ and $g(x) = \sin x$

(b) Tables will vary. One example is:

x	-3	-2	-1
$f(x) = x$	-3	-2	-1
$g(x) = \sin x$	-0.14112	-0.90930	-0.84147
$h(x) = f(x) + g(x)$	-3.14112	-2.90930	-1.84147

x	0	1	2
$f(x) = x$	0	1	2
$g(x) = \sin x$	0	0.84147	0.90930
$h(x) = f(x) + g(x)$	0	1.84147	2.90930

x	3
$f(x) = x$	3
$g(x) = \sin x$	0.14112
$h(x) = f(x) + g(x)$	3.14112

(c) A graph of h can be formed by adding the graphs of f and g point by point.

$h(x) = \cos x - \sin 3x$

(a) $f(x) = \cos x$ and $g(x) = \sin 3x$

(b) Tables will vary. One example is:

x	-3	-2	-1
$f(x) = \cos x$	-0.98999	-0.41615	0.54030
$g(x) = \sin 3x$	-0.41212	0.27942	-0.14112
$h(x) = f(x) - g(x)$	-0.57787	-0.69557	0.68142

x	0	1	2
$f(x) = \cos x$	1	0.54030	-0.41615
$g(x) = \sin 3x$	0	0.14112	-0.27942
$h(x) = f(x) - g(x)$	1	0.39918	-0.13673

x	3
$f(x) = \cos x$	-0.98999
$g(x) = \sin 3x$	0.41212
$h(x) = f(x) - g(x)$	-1.40211

(c) A graph of h can be formed by subtracting the graph of g from the graph of f point by point.

Examples will vary. One example is $f(x) = \sin x$ and

$g(x) = \cos\left(x + \frac{\pi}{2}\right)$.

Section 4.8 (page 427)

(a) Amplitude modulation

(b) Frequency modulation

Explanations will vary.

Chapter 5

Section 5.1 (page 446)

$1 + \tan^2 u = \sec^2 u$. Answers will vary.

Section 5.2 (page 454)

This particular set of range settings actually shows the statement's left side and right side as coinciding, when an algebraic verification shows that they are not equal (using a different set of range settings will confirm this). A trigonometric identity should always be verified algebraically first, and then graphically. Showing that the graphs of the left and right sides of a statement appear to coincide is not strong enough proof to conclude that the statement is an identity.

Section 5.3 (page 463)

Equations a and b. Either $-2 \le \sqrt{b^2 - 4c} + b \le 2$ or $-2 \le \sqrt{b^2 - 4c} - b \le 2$ must be true.

Section 5.5 (page 481)

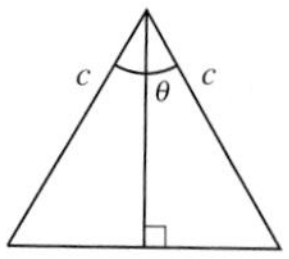

Let c be the length of the two equal sides of an isosceles triangle, and let θ be the angle between the two equal sides.

$$\begin{aligned}\text{Area} &= \left(\frac{1}{2}\text{ base}\right)(\text{height})\\ &= \left(c \sin\frac{\theta}{2}\right)\left(c \cos\frac{\theta}{2}\right)\\ &= c^2\sqrt{\frac{1 - \cos\theta}{2}}\sqrt{\frac{1 + \cos\theta}{2}}\\ &= c^2\sqrt{\frac{1 - \cos^2\theta}{4}}\\ &= \frac{1}{2}c^2\sqrt{\sin^2\theta}\\ &= \frac{1}{2}c^2\sin\theta\end{aligned}$$

Examples will vary.

Chapter 6

Section 6.1 (page 499)

Yes

(a) $A = 40°,\ a \approx 12.86,\ b \approx 15.32$

(b) $A = 40°,\ b \approx 11.92,\ c \approx 15.56$

It is probably easier to use the right triangle definitions of sine, cosine, and tangent to solve the triangle.

Section 6.2 (page 506)

(a) Area $= \frac{1}{2}(2)(4)\sin 50° \approx 3.064$ square feet

(b) Area $= \sqrt{\frac{9}{2}\left(\frac{9}{2} - 2\right)\left(\frac{9}{2} - 3\right)\left(\frac{9}{2} - 4\right)} \approx 2.905$ square feet

(c) Area $= \frac{1}{2}(2)(4) = 4$ square feet

(d) Area $= \sqrt{6(6 - 3)(6 - 4)(6 - 5)} = 6$ square feet

Section 6.5 (page 539)

$e^{a+bi} = e^a(\cos b + i \sin b)$

Let $a = 0$ and $b = \pi$.

$$\begin{aligned}e^{0+\pi i} &= e^0(\cos\pi + i\sin\pi)\\ e^{\pi i} &= -1\end{aligned}$$

$e^{\pi i} + 1 = 0$

Chapter 7

Section 7.1 (page 560)

a. Agency B cost equation: $y = 100 + 0.25x$

b.

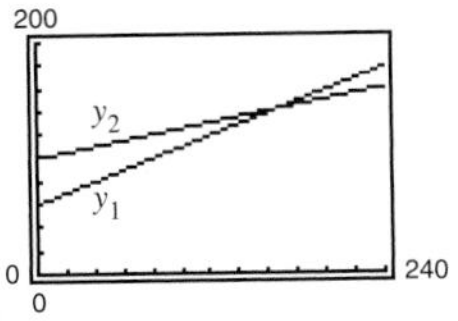

Point of intersection: (167.08, 141.77). This gives the point at which the plans from both agencies cost the same. (\$141.77 for 167.08 miles)

c. Agency A. If you plan to drive the truck less than 167 miles, it is less expensive to choose Agency A.

d. Agency B. If you plan to drive more than 167 miles, choose Agency B.

Section 7.4 (page 596)

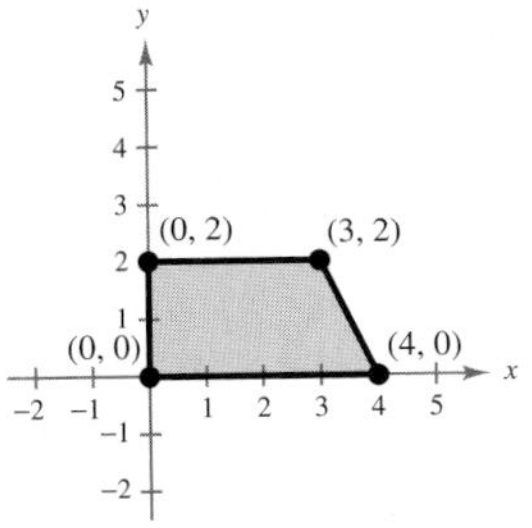

$$\begin{cases} 0 \le y \le 2 \\ x \ge 0 \\ y \le -2x + 8 \end{cases}$$

Section 7.5 (page 606)

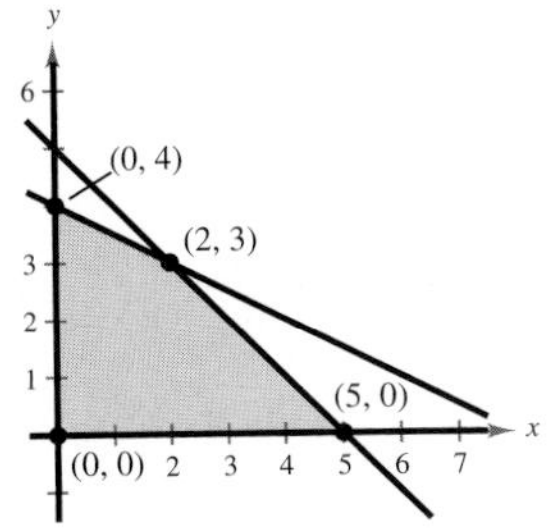

Answers are not unique.

a. $z = -x + y$ **b.** $z = 2x + 3y$

c. $z = x - y$ **d.** $z = -x - y$

Chapter 8

Section 8.2 (page 644)

Problems will vary.

Section 8.5 (page 674)

Answers will vary.

Chapter 9

Section 9.2 (page 704)

a. $-7, -4, -1, 2, 5, 8, 11;\ a_{n+1} = a_n + 3$

b. $17, 23, 29, 35, 41, 47, 53, 59, 65, 71;\ a_{n+1} = a_n + 6$

c. Not possible

d. $4, 7.5, 11, 14.5, 18, 21.5, 25, 28.5, 32, 35.5, 39;$

$a_{n+1} = a_n + 3.5$

e. Not possible

Section 9.3 (page 713)

Sequences will vary. One possibility (in inches, when starting with an 8-foot piece of string) is: 96, 48, 24, 12, 6, 3, $1\frac{1}{2}$, $\frac{3}{4}$, $\frac{3}{8}$, $\frac{3}{16}$, $\frac{3}{32}$, $\frac{3}{64}$ (and can go no further).

Formula: $a_n = 96\left(\frac{1}{2}\right)^{n-1}$.

You could theoretically make an infinite number of cuts, but realistically, the string would become so small that it would be impossible to cut.

Section 9.5 (page 732)

a. Student used the wrong sign and raised $3y$ to the incorrect power in the expansion. Correct solution:

$$-5(2x)^4(3y)^1 = -240x^4y$$

b. Student gave the fifth term instead of the fourth term. Correct solution:

$$_6C_3\left(\frac{1}{2}x\right)^3(7y)^3 = 20\left(\frac{1}{2}x\right)^3(7y)^3$$
$$= 857.5x^3y^3$$

Chapter 10

Section 10.1 (page 773)

The angle of inclination is the positive angle made with the x-axis measured counterclockwise. Therefore, there are two cases: (1) a line with positive slope, implying $\theta < 90°$, and (2) a line with negative slope, implying $90° < \theta < 180°$. The angle between two lines is, by definition, the smaller of the two angles formed by the intersecting lines. The two angles formed are supplementary; therefore, one angle must be acute—i.e., less than 90°—or, if the lines are perpendicular, both angles equal 90°.

False. The inclination is the *positive angle measured counterclockwise* from the x-axis, not necessarily the angle between the line and the x-axis.

Section 10.2 *(page 781)*

The resulting surface has the property that all incoming rays parallel to the axis are reflected through the focus of the parabola.

Section 10.3 *(page 791)*

a. Answers will vary.

b.

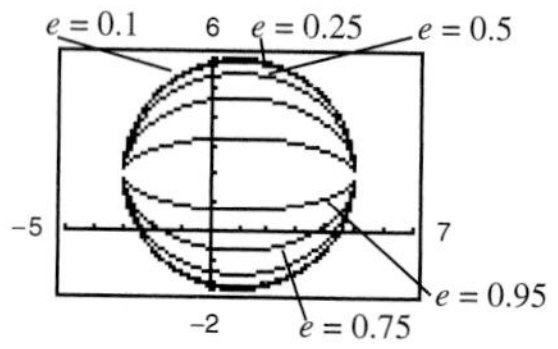

As e approaches 0, the shape of the ellipse approaches the shape of a circle.

c. When $e = 0$, the graph of the equation is circular. The equation of the ellipse is

$$\frac{(x-h)^2}{a^2} + \frac{(y-k)^2}{a^2} = 1,$$

otherwise known as the equation of a circle.

Section 10.4 *(page 801)*

a.

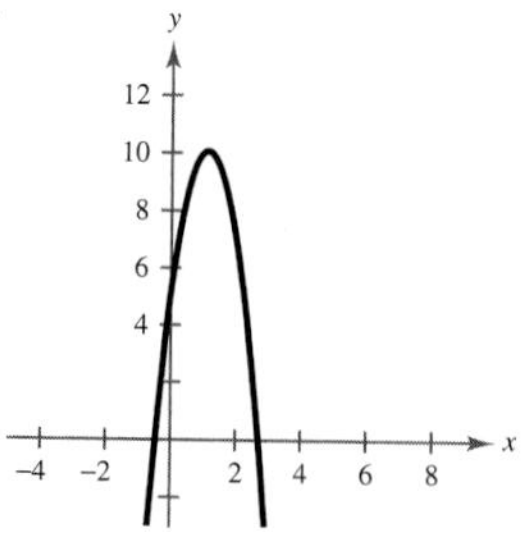

b.

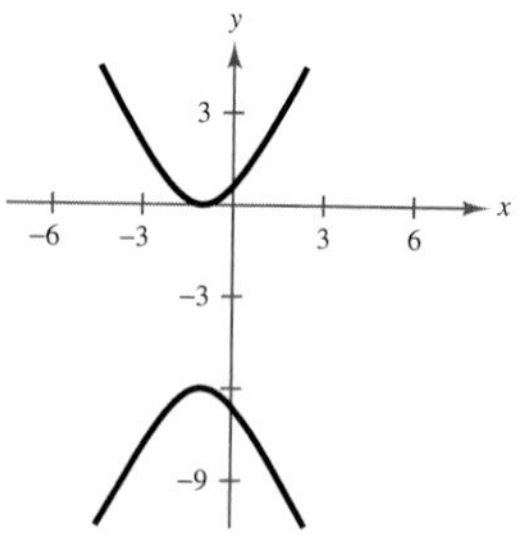

c.

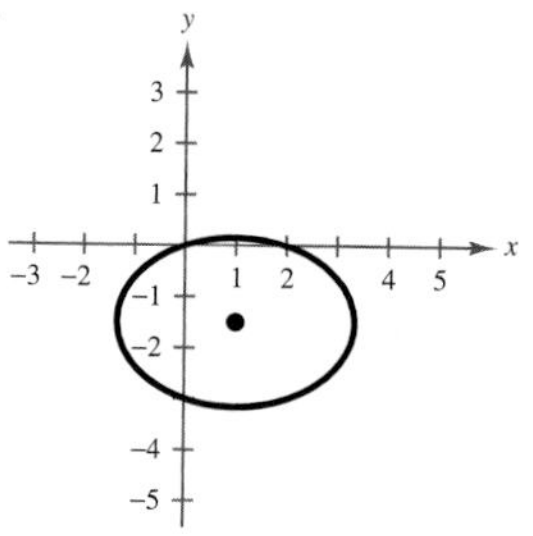

d.

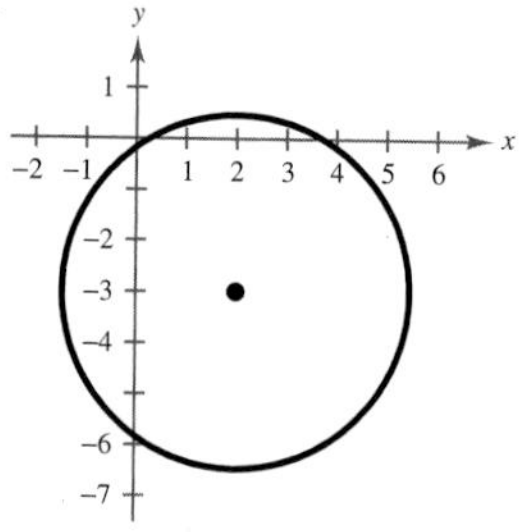

Discussions will vary.

Section 10.5 *(page 809)*

$f(x) = 1/x \Rightarrow y = 1/x \Rightarrow xy = 1 \Rightarrow xy - 1 = 0 \Rightarrow B = 1$ and $F = -1$. So, $B^2 - 4AC = (1)^2 - 4(0)(0) = 1 > 0$.

Therefore, the graph of $f(x) = 1/x$ is a hyperbola.

Chapter Projects

Chapter P *(page 96)*

1. As x gets closer and closer to 0, the height increases without bound. No.

2. As x gets closer and closer to $\sqrt{108}$, the height gets closer and closer to (but does not become) 0. No.

3.

Base, x	5.9	5.99	5.999
Volume, V	215.9105	215.9991005	215.999991

Base, x	6.001	6.01	6.1
Volume, V	215.999991	215.9990995	215.9095

As x gets closer and closer to 6, the volume gets closer and closer to 216.

4. $x \approx 4.24$, $2x \approx 8.49$, $h \approx 5.66$

Chapter 1 *(page 198)*

1. (a) (0, 40): Because $(x, p) = (0, 40)$, no units are sold at a price of \$40.

(b) $(\sqrt{8}, 0)$: Because $(x, p) = (\sqrt{8}, 0)$, no more than $\sqrt{8}$ units are "given away" at a price of \$0.

2. Maximum revenue: \$43.546484 million at $x = 1.6329933$

$p = 40 - 5(1.6329933)^2 \approx \$26.67$

Yes, yes.

3.

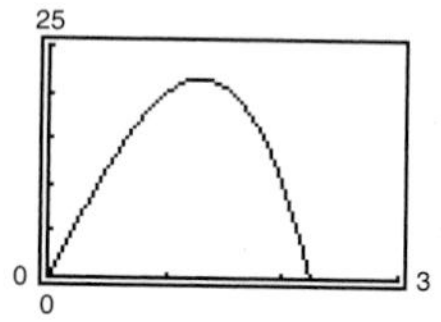

1.29 million units for a profit of \$31.68 per unit

Chapter 2 *(page 284)*

1. 1.893

2.

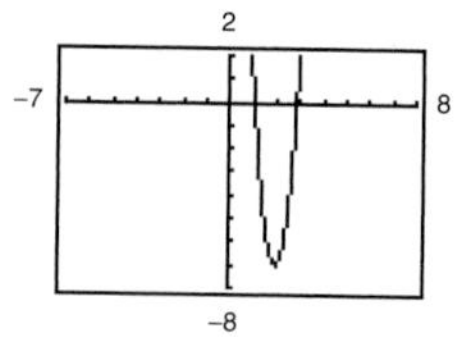

(a) Answers will vary.

$1.05477 \le x \le 1.05484$

$-3.81470 \le y \le 3.81470$

(b) Answers will vary.

$2.8409925 \le x \le 2.8410688$

$-3.81470 \le y \le 3.81470$

3. $(-0.034, 2.102), (3.236, 2.765)$

4.

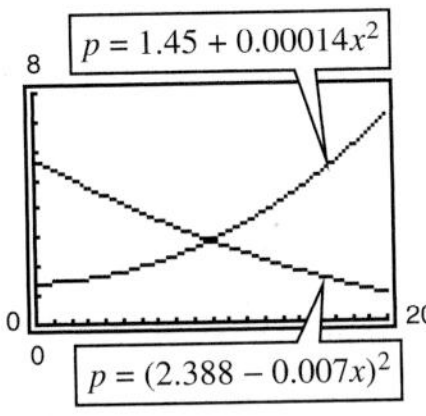

100 bushels per day at $2.85 per bushel

Chapter 3 *(page 346)*

1. An annual percentage rate of 8.05% compounded monthly would produce a larger balance, because this option has a slightly higher effective yield ($\approx 8.35\%$ compared with $\approx 8.33\%$ for the other plan).

2. 32 years, 5 months; 104 years, 1 month

3. (b), if $t > 16.8$ years

Chapter 4 *(page 438)*

1.

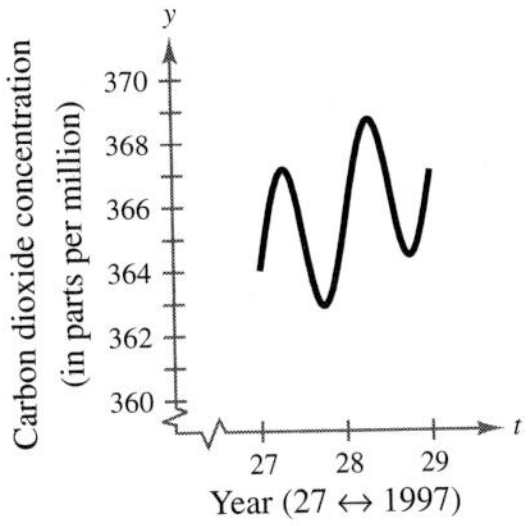

Maximum $y \approx 368.69$ when $t \approx 28.27$

Minimum $y \approx 362.88$ when $t \approx 27.73$

2.

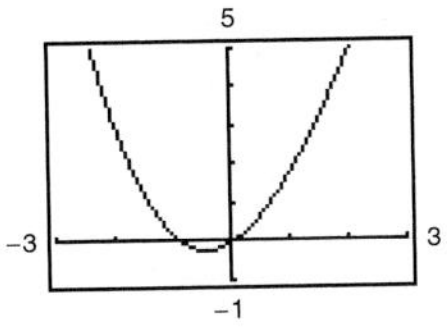

3.

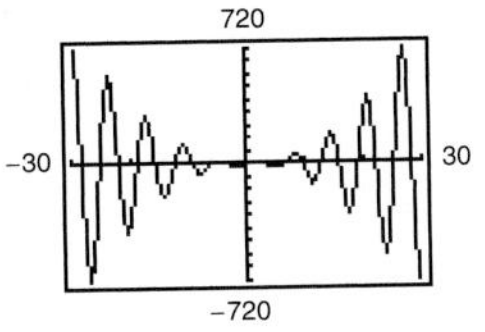

4.

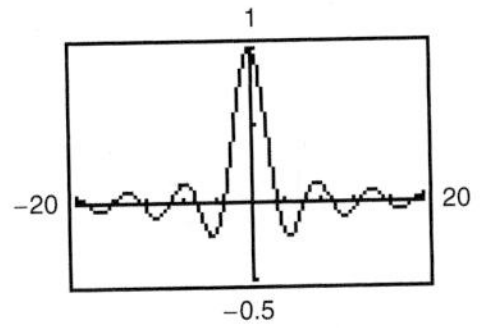

Chapter 5 *(page 490)*

(a) 14.15 feet (b) 0.95 second

(c) $t \approx 1.88$ seconds (d) ≈ 155.46 feet

1. $t \approx 1.88$ seconds, $x(t) \approx 155.46$ feet

2. Maximum height ≈ 68.06 feet

Range ≈ 471.55 feet

Tmin = 0
Tmax = 5
Tstep = .05
Xmin = -50
Xmax = 500
Xscl = 50
Ymin = -20
Ymax = 80
Yscl = 20

3. 72.31 feet, 97.43 feet, 110.79 feet, 110.79 feet, 97.43 feet
$\theta = 45°$ should be used to produce a maximum range.

4. It takes twice as long for the projectile to return to the ground as it does for the projectile to reach its maximum height.

5. Answers will vary.

$\theta = 45°$

Chapter 6 *(page 548)*

(a)

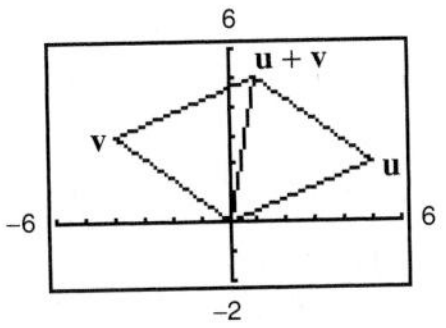

(b) 394 miles per hour, N 49.2° W

1.

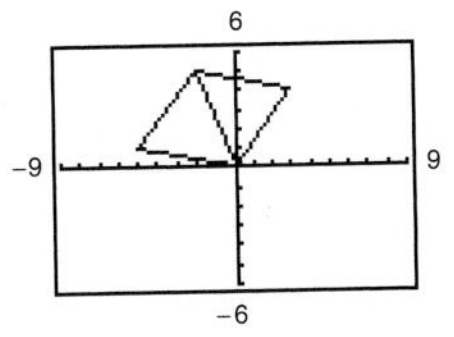

$\mathbf{u} + \mathbf{v} = -2\mathbf{i} + 5\mathbf{j}$

2.

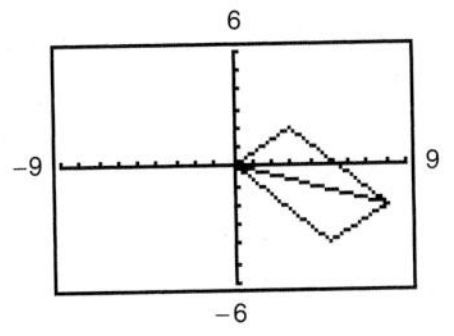

$\mathbf{u} + \mathbf{v} = 8\mathbf{i} - 2\mathbf{j}$

3.

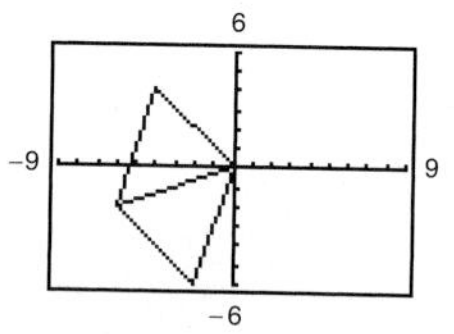

$\mathbf{u} + \mathbf{v} = -6\mathbf{i} - 2\mathbf{j}$

4.

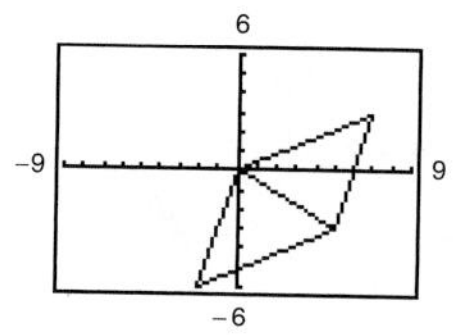

$\mathbf{u} + \mathbf{v} = 5\mathbf{i} - 3\mathbf{j}$

5. A lower speed. The plane is now traveling at about 394 miles per hour because of wind resistance.

6. Try substituting different values for the wind's velocity into the program and approximating the components of the resultant vector. Use the approximations to find the velocity. To obtain the answer analytically, you could solve the following equation for w, the wind's velocity.

$$\tan 140^\circ = \frac{400 \sin 150^\circ + w \sin 50^\circ}{400 \cos 150^\circ + w \cos 50^\circ}$$

$w = 69.5$ mph

7. Try substituting different values for the wind's direction into the program and approximating the components of the resultant vector. Use the approximations to find the direction. To obtain the answer analytically, solve the following equation for θ, the wind's direction.

$$\tan 140^\circ = \frac{400 \sin 150^\circ + 75 \sin \theta}{400 \cos 150^\circ + 75 \cos \theta}$$

θ is about N 62.2° E.

Chapter 7 *(page 616)*

1. 1998

2. $y = 41.1 + 0.61t$; 49 million

3. The average circulation per morning newspaper decreased from about 73,882 papers sold per day in 1990 to about 64,397 papers sold per day in 1997.

Chapter 8 *(page 684)*

6600 grams of alloy X, 7500 grams of alloy Y, 8800 grams of alloy Z

Chapter 9 *(page 764)*

1. (a) 3

(b) 4, 10; No. The function is not a straight line, which means that the average rates of change will be different for different intervals.

(c) A speedometer shows how fast you are traveling at a specific time. The average speed is the total distance traveled divided by the total time for the trip.

2. (a) $\frac{1}{150}, \frac{3}{100}, \frac{1}{150}, \frac{1}{60}, \frac{1}{200}, \frac{1}{100}, \frac{1}{75}, \frac{3}{400}, \frac{1}{400}$

(b) The average rate of change was largest between 1974 and 1975 and smallest between 1995 and 1999.

(c) No. When the average rate of change is zero, it means that the cost of postage did not change during that interval.

Chapter 10 *(page 846)*

(a) $\dfrac{(x-6)^2}{81} + \dfrac{y^2}{45} = 1$

(b) Center: $(6, 0)$

Foci: $(12, 0)$, $(0, 0)$

Eccentricity: $\frac{2}{3}$

The results are the same.

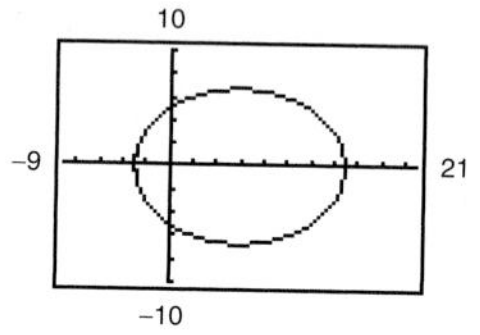

(c) $x = 6 + 9 \cos \theta$

$y = \sqrt{45} \sin \theta$

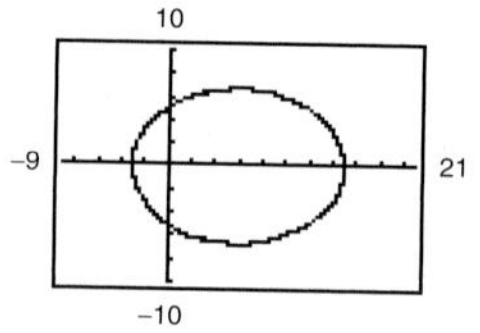

1.

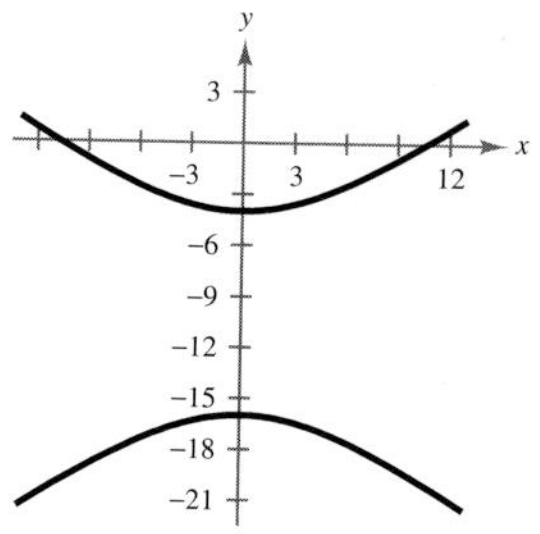

Hyperbola

2. $\dfrac{(y+10)^2}{36} - \dfrac{x^2}{64} = 1$

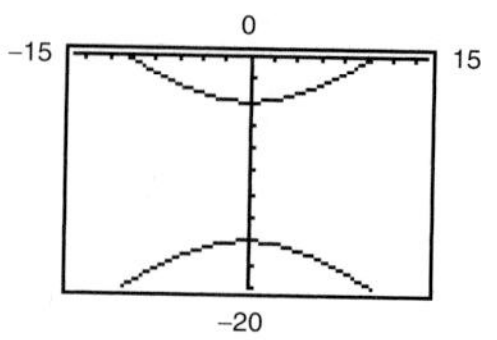

Yes. Explanations will vary.

3. $x = 8 \tan \theta$

$y = 6 \sec \theta - 10$

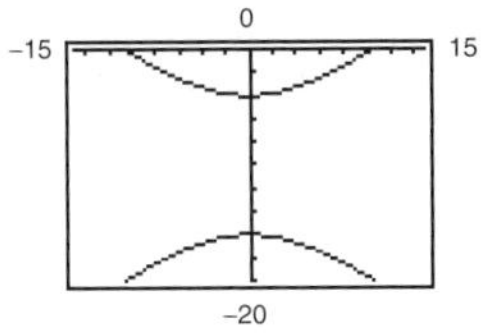

Yes

4. Yes. The parametric equations represent an ellipse. According to Kepler's Laws and Newton, all heavenly bodies, including comets, move in elliptical orbits with the sun at one focus.

Appendix B

Section B.1 *(page A19)*

2. (a) 900 pounds (b) $[600, 1300]$

4. Exam 1:

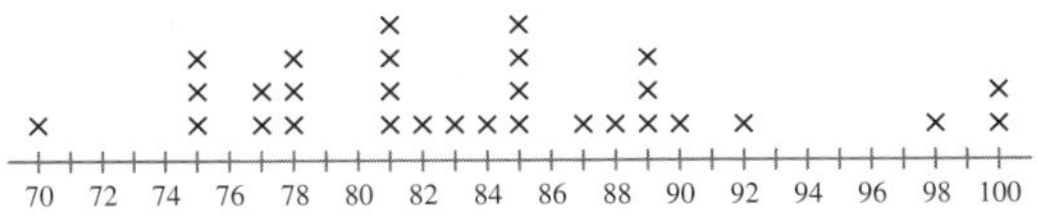

81 and 85

Exam 2:

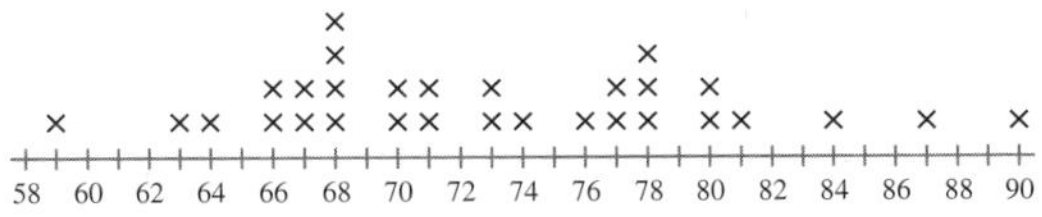

68

6.

Leaves (Exam 2)	*Stems*	*Leaves (Exam 1)*
9	5	
8 8 8 8 7 7 6 6 4 3	6	
8 8 8 7 7 6 4 3 3 1 1 0 0	7	0 5 5 5 7 7 8 8 8
7 4 1 0 0	8	1 1 1 1 2 3 4 5 5 5 5 7 8 9 9 9
0	9	0 2 8
	10	0 0

Exam 1

8. Organize the data by using a stem-and-leaf plot.

Stem	*Leaves*
1	3.0 3.0 5.7 8.7 8.9 9.1
2	1.1 1.3 1.5 1.6 1.8 1.8 3.3 3.8 5.8 6.2 9.0 9.2
3	0.7 1.0 2.1 2.3 3.6 4.4 8.0 9.8
4	2.1 4.6 7.5 9.0

Section B.2 *(page A26)*

2. Mean: 33.86; median: 33; mode: 32

4. Mean: 32.43; median: 33; mode: 32

6. Mean: 34; median: 33; mode: 32

8. (a) Mean: 14.86; median: 14; mode: 13

Each is increased by 6.

(b) Each will increase by k.

10. Mean: 320; median: 320; mode: 320

12. (a) Average number of hits = 1

(b) Batting average = 0.250

14. One possibility: $\{4, 4, 6, 7.5, 8.5\}$

16. Median and mode give most representative descriptions.

18. $\bar{x} = 7, v = 22, \sigma \approx 4.69$ **20.** $\bar{x} = 2, v = 0, \sigma = 0$

22. $\bar{x} = 3, v = 4, \sigma = 2$

24. $\bar{x} = 1.1, v = 0.38, \sigma \approx 0.616$ **26.** 13.53

28. 2.19 **30.** 1.92

32. (a) $\bar{x} = 15;\ \sigma = 3.19$ (b) $\bar{x} = 15;\ \sigma = 2.83$

(c) $\bar{x} = 25;\ \sigma = 2.83$ (d) $\bar{x} = 5;\ \sigma = 3.19$

34. All numbers must be equal.

36. $\bar{x} = 381.9,\ v = 5947.29,\ \sigma = 77.119$
90% lie within two standard deviations.

38.

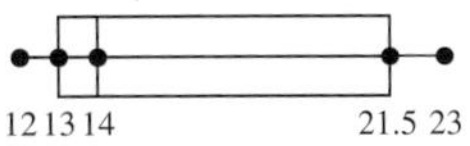

40.

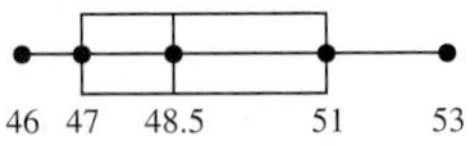

42.

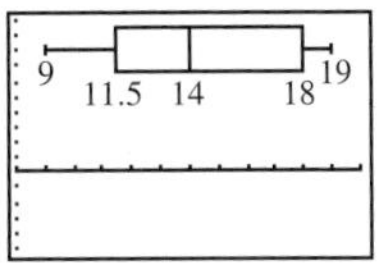

44.

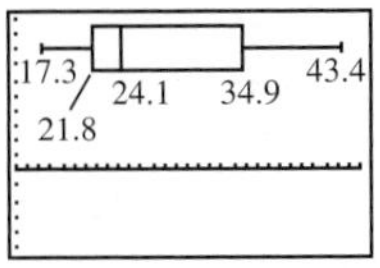

46.

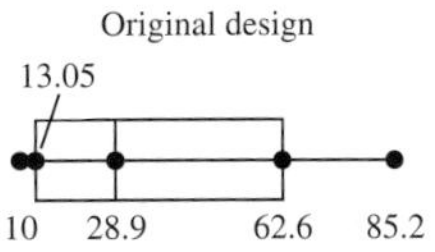

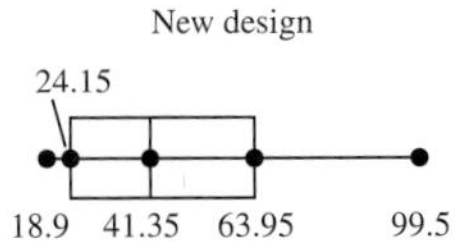

From the plots, you can see that the lifetimes of the units in the sample made by the new design are greater than the lifetimes of the units in the sample made by the original design. (The median increased by more than 12 months.)

APPENDIX

Index of Applications

Biology and Life Sciences

Business

Chemistry and Physics

Construction

Consumer

Geometry

Interest Rate

Miscellaneous

Time and Distance

U.S. Demographics

Index

D

E

F

Q

R

S

T

FORMULAS FROM GEOMETRY

Triangle:

$h = a \sin \theta$

$\text{Area} = \frac{1}{2}bh$

(Laws of Cosines)

$c^2 = a^2 + b^2 - 2ab \cos \theta$

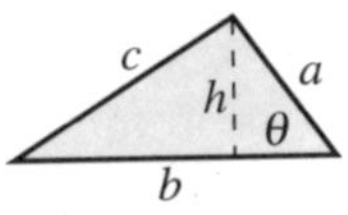

Right Triangle:

(Pythagorean Theorem)

$c^2 = a^2 + b^2$

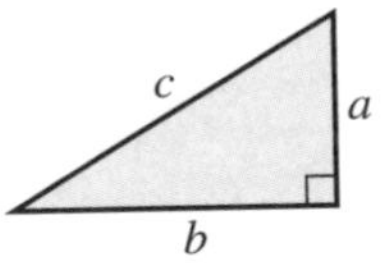

Equilateral Triangle:

$h = \frac{\sqrt{3}s}{2}$

$\text{Area} = \frac{\sqrt{3}s^2}{4}$

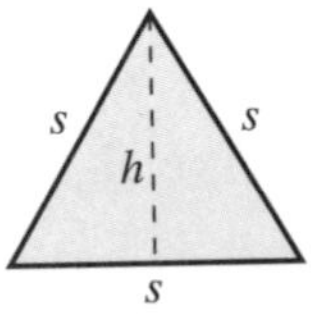

Parallelogram:

$\text{Area} = bh$

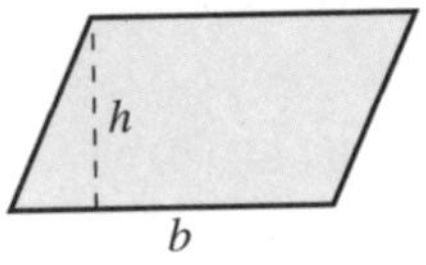

Trapezoid:

$\text{Area} = \frac{h}{2}(a + b)$

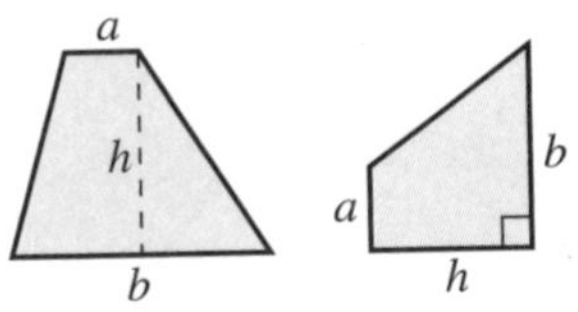

Circle:

$\text{Area} = \pi r^2$

$\text{Circumference} = 2\pi r$

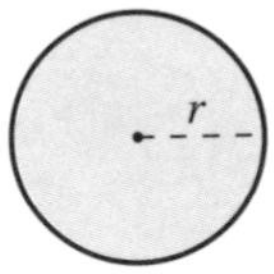

Sector of Circle:

(θ in radians)

$\text{Area} = \frac{\theta r^2}{2}$

$s = r\theta$

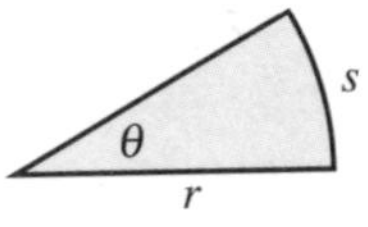

Circular Ring:

(p = average radius,

w = width of ring)

$\text{Area} = \pi(R^2 - r^2)$

$= 2\pi pw$

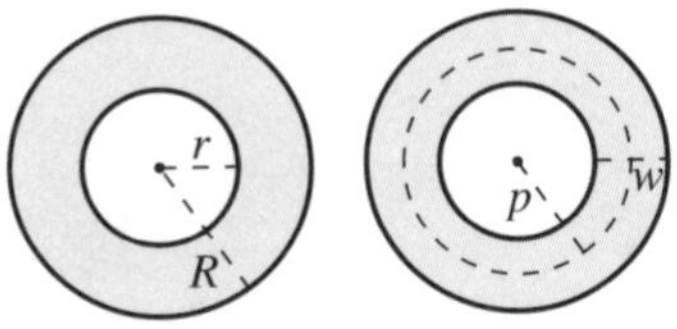

Sector of Circular Ring:

(p = average radius,

w = width of ring,

θ in radians)

$\text{Area} = \theta pw$

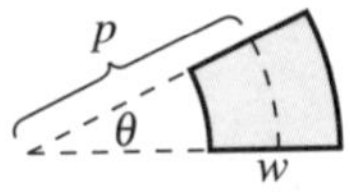

Ellipse:

$\text{Area} = \pi ab$

$\text{Circumference} \approx 2\pi\sqrt{\frac{a^2 + b^2}{2}}$

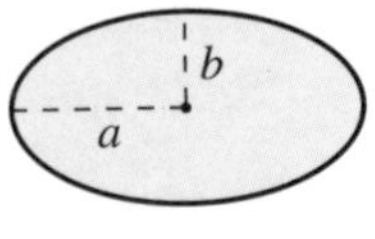

Cone:

(A = area of base)

$\text{Volume} = \frac{Ah}{3}$

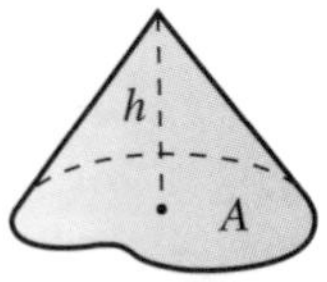

Right Circular Cone:

$\text{Volume} = \frac{\pi r^2 h}{3}$

$\text{Lateral Surface Area} = \pi r\sqrt{r^2 + h^2}$

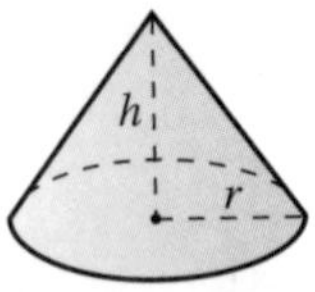

Frustum of Right Circular Cone:

$\text{Volume} = \frac{\pi(r^2 + rR + R^2)h}{3}$

$\text{Lateral Surface Area} = \pi s(R + r)$

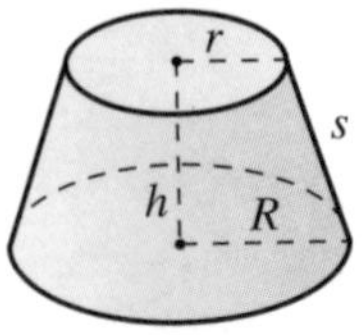

Right Circular Cylinder:

$\text{Volume} = \pi r^2 h$

$\text{Lateral Surface Area} = 2\pi rh$

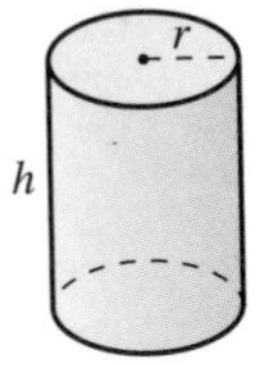

Sphere:

$\text{Volume} = \frac{4}{3}\pi r^3$

$\text{Surface Area} = 4\pi r^2$

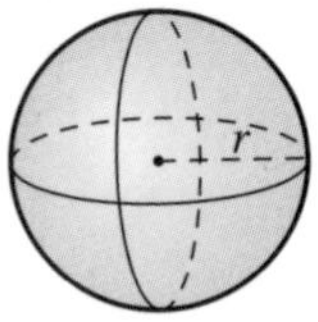

Wedge:

(A = area of upper face,

B = area of base)

$A = B \sec \theta$

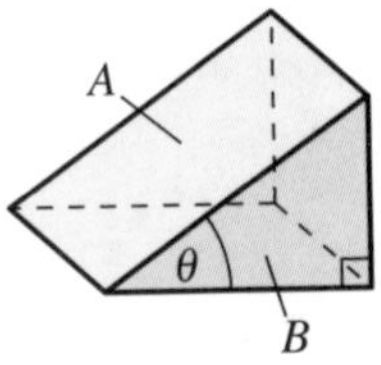

Definition of the Six Trigonometric Functions

Right triangle definitions, where $0 < \theta < \pi/2$.

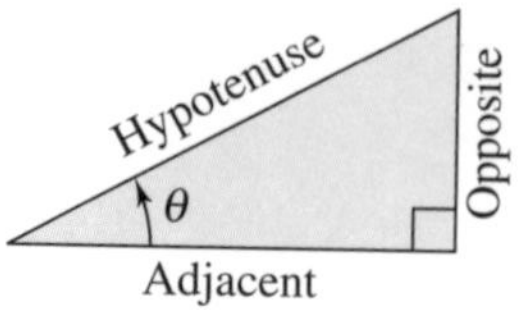

$$\sin\theta = \frac{\text{opp.}}{\text{hyp.}} \qquad \csc\theta = \frac{\text{hyp.}}{\text{opp.}}$$

$$\cos\theta = \frac{\text{adj.}}{\text{hyp.}} \qquad \sec\theta = \frac{\text{hyp.}}{\text{adj.}}$$

$$\tan\theta = \frac{\text{opp.}}{\text{adj.}} \qquad \cot\theta = \frac{\text{adj.}}{\text{opp.}}$$

Circular function definitions, where θ *is any angle.*

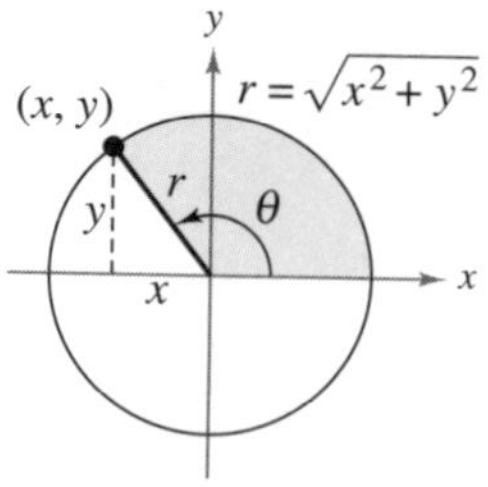

$$\sin\theta = \frac{y}{r} \qquad \csc\theta = \frac{r}{y}$$

$$\cos\theta = \frac{x}{r} \qquad \sec\theta = \frac{r}{x}$$

$$\tan\theta = \frac{y}{x} \qquad \cot\theta = \frac{x}{y}$$

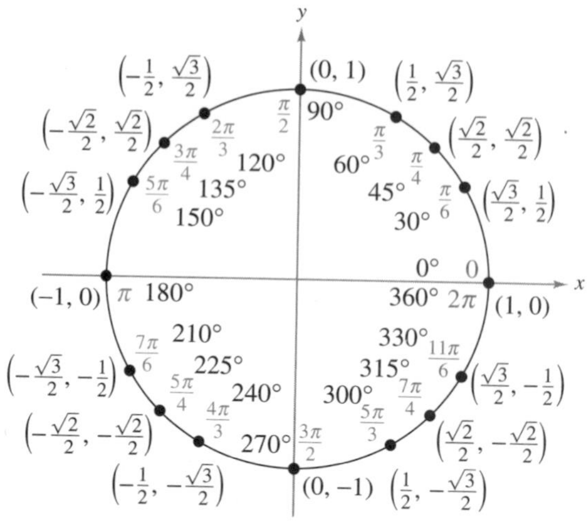

Reciprocal Identities

$$\sin u = \frac{1}{\csc u} \qquad \cos u = \frac{1}{\sec u} \qquad \tan u = \frac{1}{\cot u}$$

$$\csc u = \frac{1}{\sin u} \qquad \sec u = \frac{1}{\cos u} \qquad \cot u = \frac{1}{\tan u}$$

Quotient Identities

$$\tan u = \frac{\sin u}{\cos u} \qquad \cot u = \frac{\cos u}{\sin u}$$

Pythagorean Identities

$$\sin^2 u + \cos^2 u = 1$$

$$1 + \tan^2 u = \sec^2 u \qquad 1 + \cot^2 u = \csc^2 u$$

Cofunction Identities

$$\sin\left(\frac{\pi}{2} - u\right) = \cos u \qquad \cot\left(\frac{\pi}{2} - u\right) = \tan u$$

$$\cos\left(\frac{\pi}{2} - u\right) = \sin u \qquad \sec\left(\frac{\pi}{2} - u\right) = \csc u$$

$$\tan\left(\frac{\pi}{2} - u\right) = \cot u \qquad \csc\left(\frac{\pi}{2} - u\right) = \sec u$$

Even/Odd Identities

$$\sin(-u) = -\sin u \qquad \cot(-u) = -\cot u$$

$$\cos(-u) = \cos u \qquad \sec(-u) = \sec u$$

$$\tan(-u) = -\tan u \qquad \csc(-u) = -\csc u$$

Sum and Difference Formulas

$$\sin(u \pm v) = \sin u \cos v \pm \cos u \sin v$$

$$\cos(u \pm v) = \cos u \cos v \mp \sin u \sin v$$

$$\tan(u \pm v) = \frac{\tan u \pm \tan v}{1 \mp \tan u \tan v}$$

Double-Angle Formulas

$$\sin 2u = 2 \sin u \cos u$$

$$\cos 2u = \cos^2 u - \sin^2 u = 2\cos^2 u - 1 = 1 - 2\sin^2 u$$

$$\tan 2u = \frac{2 \tan u}{1 - \tan^2 u}$$

Power-Reducing Formulas

$$\sin^2 u = \frac{1 - \cos 2u}{2}$$

$$\cos^2 u = \frac{1 + \cos 2u}{2}$$

$$\tan^2 u = \frac{1 - \cos 2u}{1 + \cos 2u}$$

Sum-to-Product Formulas

$$\sin u + \sin v = 2 \sin\left(\frac{u + v}{2}\right) \cos\left(\frac{u - v}{2}\right)$$

$$\sin u - \sin v = 2 \cos\left(\frac{u + v}{2}\right) \sin\left(\frac{u - v}{2}\right)$$

$$\cos u + \cos v = 2 \cos\left(\frac{u + v}{2}\right) \cos\left(\frac{u - v}{2}\right)$$

$$\cos u - \cos v = -2 \sin\left(\frac{u + v}{2}\right) \sin\left(\frac{u - v}{2}\right)$$

Product-to-Sum Formulas

$$\sin u \sin v = \frac{1}{2}[\cos(u - v) - \cos(u + v)]$$

$$\cos u \cos v = \frac{1}{2}[\cos(u - v) + \cos(u + v)]$$

$$\sin u \cos v = \frac{1}{2}[\sin(u + v) + \sin(u - v)]$$

$$\cos u \sin v = \frac{1}{2}[\sin(u + v) - \sin(u - v)]$$